WATER RESOURCES ENGINEERING (Vol. II)

# IRRIGATION ENGINEERING AND HYDRAULIC STRUCTURES

## Other Publications by the Same Author

(1) "*Water Resources Engineering (Vol. I)—Hydrology, Flood Control and Ground Water Engineering*" (27th Revised 2021 Edition)

(2) "*Environmental Engineering (Vol. I)—Water Supply Engineering*" (35th Revised 2021 Edition)

(3) "*Environmental Engineering (Vol. II)—Sewage Disposal and Air Pollution Engineering*" (41st Revised 2021 Edition)

(4) "*Physical and Engineering Geology*" (5th Reprint of 7th Edition)

(5) "*Geotech Engineering—Soil Mechanics and Foundation Engineering*" (13th Revised 2021 Edition)

(6) "*Ecology and Environmental Studies*" (4th Revised 2015 Edition)

(7) "*Environmental Studies and Green Technology*" (2nd Reprint of 1st Edition)

(8) "*Environmental Studies (as per UGC Syllabus)*" (1st Edition 2020)

(9) "*Elementary Irrigation Engineering*" (Newly written 2021 edition)

**Note :** Orders for procuring all the above listed books may be sent to M/s Khanna Publishers at B-35/9, G.T. Karnal Road, Industrial Area, (Near Telephone Exchange), Delhi-110033 *Phones* : 011-27224179 · *Mob.* 09811541460

# IRRIGATION ENGINEERING AND HYDRAULIC STRUCTURES

For
[Civil, Agriculture, and Environmental Engineering Degree Students; U.P.S.C. and Other State Service Competitions including GATE Exams. ; AMIE (Section B) Exams and For Professionals]

*(Containing updated Questions of Civil Services and Engineering Services Competitions)*

*By*
**SANTOSH KUMAR GARG**
*B.Sc. Engg. (Civil)*
*Topper (1969 Batch), Delhi University*
*Life Member—Indian Water Resources Society*

***FORMERLY***
*Superintending Engineer*
*Flood Control and Irrigation Department,*
*Govt. of NCT of Delhi*

**(38th Revised 2023 Edition)**

**KHANNA PUBLISHERS®**
*Investing in Learning®*
*Operational Office :*
4575/15, Onkar House, Opp. Happy School,
Ground Floor, Daryaganj, New Delhi 110 002
*Phones* : 011-45033819 • *Mob*. 09811541460
*email* : contactus@khannapublishers.in

*Published by :*
Romesh Chander Khanna & Vineet Khanna
*for* **KHANNA PUBLISHERS**
2-B, Nath Market, Nai Sarak
Delhi- 110 006 (India)

**Website : www.khannapublishers.in**



*Hologram & Description*

**To all readers of our books:** In order to prevent yourself from being defrauded by pirated books, please make sure that there exists an Hologram on the top cover of the book with the specifications indicated below. If you find any book without Hologram of this Description, please do not buy the pirated copy, and inform us through e-mail at **contactus@khannapublishers.in**

Thanking you
(Khanna Publishers)

**ISBN No. 978-81-7409-047-8**

***First Edition : August, 1976***
***38th Revised Edition : July, 2023***
***First Reprint : August, 2025***

***Price :*** **₹ 829.00**

# Dedication

*This book is dedicated to the memory of those unknown workers who have lost their lives in the execution of world's gigantic water resources projects.*

# FOREWORD

I feel both pleasure and pride in being called upon to write the Foreword to this book. This is Shri Garg's second book and he has fully justified the promise shown by his first one—"Water Resources and Hydrology". This is not surprising as he has had a brilliant academic career, culminating in **First Class First** Degree in Civil Engineering from Delhi University.

Shri Garg has fully covered the subject of irrigation engineering and the design of hydraulic structures connected with irrigation. The book is expected to be specially useful for students of engineering colleges, and those preparing for the A.M.I.E. Examinations in their spare time. It will also be useful for practising engineers concerned with irrigation engineering. The language has been kept simple, making the book extremely readable.

I wish the author all success in this as well as in future efforts in this direction.

**Pritam Singh**
Chief Engineer, Irrigation & Flood
Control Deptt., Govt. of Delhi,
on Deputation from Central Water
Commission, Govt. of India,
New Delhi.

New Delhi
May, 1976

# PREFACE TO THE 37th REVISED EDITION

The first edition of this book was published in August 1976, and soon thereafter it gained enormous popularity due to its simple language and coverage of the entire subject of irrigation engineering with detailed designs of various irrigation & hydraulic structures. The book was liked and appreciated by the learned teachers and professors of the various engineering colleges, including the prestigious IIT's and NITs. So much, so that several solved numerical examples from this book were directly set in the Civil Engineering papers of the Engineering Services Competitions, such as for the years 1978, 1980, 1983, 1986, etc. Even thereafter, solved questions from this book have continuously been found to have appeared in the Civil Services as well as the Engineering Services competitions, year after year. ***Even after 45 years of its first edition, the book continues to possess its highly acclaimed character, and is still considered to be the prized possession by the students, teachers, as well as professionals alike.***

The growing popularity of the book continued to push its sales, encouraging the author to go on revising and enlarging the text from edition to edition. So far, 36 editions and their reprints have been sold out with more than 3 Lakh copies, and the present edition is its 37th one.

The M.K.S. units initially adopted in the book have subsequently been changed over the years to **S.I. Units,** as these modern international units have gained enough popularity in our country. Various other changes have also been affected, edition after edition, to keep pace with the changing times and to make explanations easier and more definite.

*Modern equations developed a few years ago for the design of irrigation channels in India to avoid the use of Kennedy's Charts, have also been given in this book, which will help in obtaining more definite and quick solutions for the optimal designs of alluvial canals, besides developing computer programming for their quick designs. These concepts are not at all available in any other book.*

From its 24th edition, the subject of hydrology was removed from this book as to contain its size, and this book on irrigation was titled as Vol. II of Water Resources Engineering, keeping its old sub title as "*Irrigation Engineering & Hydraulic Structures*" intact. The already existing sister publication on Hydrology and Water Resources, had also been printed in a similar size and retitled as "Water Resources Engineering Vol. I" with its sub title as "***Hydrology, Flood Control and Ground Water Engineering***".

In its 36th edition, several changes were again made in the text including rewriting the "Introduction". Moreover, in order to further reduce the number of pages of this voluminous book, the chapter on "Ground Water and Tubewells" was removed from this volume, since it already stood included in Vol. I. This has helped to reduce the thickness of this popular book by about 200 pages, as to make it more easily handable.

In this revised 37th edition, several changes have further been made in the text, including correcting certain typing errors of the previous edition. Further, the Question Banks of the Engineering Services Exams as well as of the Civil Services Exams. have been updated up to the year 2020, besides updating the text at several places.

*Acknowledgments.* In going through the thirty five editions of this book, the author has accumulated an enormous amount of intellectual debts from students and teachers over the years, who had written to him about their satisfaction with the book, and extended suggestions for the improvement of its various editions. If the book today, is so strikingly popular, they all deserve much of the credit. The author certainly wants them to know of his gratitudes and openness to further suggestions, and is hence enlisting their names below :

*Sh. R. Palnivelu* of Government College Coimbatore ; *Sh. M.P. Raghavan* of Government College Salem ; *Sh R. Chandrasekharan* and *Dr. A. Achutyanan* of Regional Engineering College Calicut ; *Sh. P. Sen,* the then Chief Engineer in Central Water Commission ; *Sh. Punya Prasad Khanal,* an old student of Jaipur Engineering College ; *Sh. V.K. Raju,* Asstt engineer in I & P.D. Deptt. Govt of Andhra Pradesh ; *Mr K. Chandrasekharan Nair,* Asstt Professor of Regional Engineering College Calicut ; *Sh. Mohammad S. Mian,* Head of Water Engg. Natural Resources Development College Lusaka ; *Sh. Biswaroop Ghosh,* an Ex. Student of Bengal Engineering College Calcutta ; *Sh. Soumenda Ghosh,* a student of Burla Engineering College, Sambalpur (Orissa) ; *Sh. K.R. Ganesh Hegde* & Sh. *L.V.N. Malleswara Rao,* both Lecturers in the College of Engineering Visakhapatnam (Andhra Pradesh); *Mr A.L. Lalani,* Lecturer in L.E. College of Engineering Morbi (Gujarat) ; *Mr. Sasank Sekhar Hota* — Lecturer in Barapada School of Engineering and Technology Balasore (Orissa) ; *Mr. STATES KABASIA* from Nairobi (Kenya) ; *Mr. Shri Ram,* Lecturer in Madan Mohan Malviya Engineering College, Gorakhpur (UP) ; *Mr. Girma G. Hawariat from Harar Ethopia ; Mr. K.R. Thamhane, Mr. Liladhar Roratkar* and *Mr. R.S. Mirje* all students from Govt. College of Engineering Shivaji Nagar (Pune) ; *Professor Arun Kumar* from Delhi College of Engineering ; *Mr. Bishoke Sil* of Irrigation and Waterways Directorate of Govt of West Bengal ; *Sh. Vijay Singh,* a final year student of H.B.T.I., Kanpur ; *Sh. Avadh Pratap Singh,* a final year student of MITS Gwalior ; Sh. *Udo Daniel Poly-Unwana* of Nigeria ; *Sh. Sanjoy Nandi* of Bankura, West Bengal, etc.

The author further wants to record his highest degree of gratitudes to **Prof. M.D. Pol** of Aurangabad for sending & suggesting modifications at several places in the text, based on which, large scale changes were made in its 19th edition. The author also expresses his gratitudes to **Prof. Dr. Kapil Gupta** of IIT Bombay for sending his suggestions, based on which, several changes were affected in the 24$^{th}$ edition in Chapter 5.

The author also wishes to record his heartiest thanks and gratitudes to the **Late Mrs. Rajeshwari Garg**, his beloved wife, without whose untiring efforts and tolerance, he would never had been able to write and publish this book at the young age of 30 years.

The author also takes this opportunity to express his reverred gratitudes towards **Prof. Dr. R. Narayanan**, the then Head of the Civil Engineering Department at Delhi College of Engg, who encouraged him to take up writing work, when he was highly depressed in life, at its prime stage due to medical problems, resulting in the removal of his left lung, and the loss of Class I job through the Engineering Services Exam.

New Delhi
November 01, 2021

**SANTOSH KUMAR GARG**
*Res.* : 3A/2 WEA, Sat Nagar
KAROL BAGH,
New Delhi-110005
*Mobile for sms only* : **(+91) 9811125410**

# Contents

CHAPTERS | PAGES

CHAPTERS PAGES

CHAPTERS PAGES

CHAPTERS | PAGES

| CHAPTERS | | | PAGES |
|---|---|---|---|

CHAPTERS PAGES

## List of Plates, and Chart Figs Pasted in the book



## List of Colour Figs/Photos pasted in the book.

**Also available by the same author "Water Resources Engineering (Vol. I) — Hydrology, Flood Control and Ground Water Engineering"**

***Chapterwise contents of this related Vol-I:***

1. Introduction to the Subject with Indian Potentials
2. Requirements of Water for Various Uses
3. Precipitation-yielding Fresh Water as Surface Flow
4. Measuring Precipitation and Hydrological Analysis of Precipitation Data.
5. Generation of Runoff (Streamflow) from Rainfall
6. Abstraction from Rainfall for the Balance to Appear as Surface Runoff
7. Computing Runoff from a Given Rainfall.
8. Streamflow Measurement and Gauge Discharge Curve
9. Food Control Measures
10. Flood Preparedness and Emergency Evacuation
11. Flood Forecasting Techniques.
12. Ground Water Extractions for Irrigation and Other Uses.
13. Plannaing and Development of Multipurpose Water Resources Projects.
14. Engineering Economy in Water Resources Planning
15. Computer Applications in the Design of Water Resources Systems
16. Chapterwise Multi-choice Objective Questions
17. Test Paper on Objective Questions
18. Syllabus and Conventional Questions of the Engineering Services Exam.s
19. Syllabus and Conventional Questions of Civil Services Exams.
20. Certain Additional Typical Important Numerical Examples from the AMIE Exams.
21. Objective Questions of the Engineering Services Exams.
22. Objective Questions of the AMIE Exams.

# *Introduction to S.I. Units*

Different countries have been following different systems of units, like f.p.s. system, c.g.s. system, m.k.s. system, etc. However, in order to avoid consequent confusion in international trade and to introduce uniformity over the entire world, it was decided during the 11th General Conference on Weights and Measures held in Paris in the year 1960, that a new system called **S.I. (System International),** be adopted all over the globe. Although, however, our country still largely continues to follow the m.k.s. system, and S.I. units still remain in its infancy.

Be that as it may, the S.I. unit system is somewhat similar to the m.k.s. system, because of the fact that the basic units of length, mass and time are the same in both the systems, *i.e.*, the length is measured by metre (m), the mass is measured by kilogram (kg), and time by seconds (s). This system is, rather a rationalisation of m.k.s. system, as it removes confusion between kg and kg wt (also usually expressed as kg), as adopted in m.k.s. system.

**S.I. Unit of Force.** The basic difference between m.k.s. units and S.I. units is in the unit of force. Whereas, in m.k.s. system, the unit of force is kg wt or kgf (kg force) usually written as kg ; in S.I. units, the unit of force is *Newton* (*N*).

***One Newton (N)*** *is defined as the force required to move a body of mass equal to 1 kg, as to generate in it, an acceleration equal to 1 m/sec*$^2$. Since

Force = mass × acceleration, we have by definition of Newton

$$1 \text{ Newton} = 1 \text{ kg} \times 1 \text{ m/sec}^2$$

or
$$1\ N = 1 \text{ kg m/sec}^2$$

*Hence, Newton (N) will have the dimensions of kg m/sec*$^2$.

Moreover, the relation between two force units *i.e.* kgf and N would be worked out as :

$$1 \text{ kgf} = 1 \text{ kg wt}$$
$$= 1 \text{ kg} \cdot g$$

where $g$ is acceleration due to gravity $= 9.807 \text{ m/sec}^2$

$$= 1 \text{ kg} \times 9.807 \text{ m/sec}^2$$
$$= 9.807 \text{ kg m/sec}^2$$
$$= 9.807 \text{ N}$$

or
$$\boxed{1 \text{ kgf} = 9.807 \text{ N}}$$

Newton is, thus a very small unit of force (approximately 1/10th of kg wt) and hence, it is customary to use a bigger unit, such as, **kilo-Newton (kN)** or **mega Newton (MN)** for expressing force ; kilo newton being $10^3$ N, and mega newton being $10^6$ N, as usual.

**S.I. Unit of Stress or Pressure.** Since the force per unit area is *stress* or *pressure*, we would evidently have S.I. unit of pressure as Newton/m$^2$ (N/m$^2$), **and this is known as Pascal Pa**).

The m.k.s. unit for pressure being kgf/m$^2$, usually written as kg/m$^2$ or its bigger unit of kg/cm$^2$. The relation between these two units can be derived as follows :

$$\text{kgf/m}^2 = 9.807 \text{ N/m}^2$$

or
$$9.807 \text{ N/m}^2 = 1 \text{ kg/m}^2$$

(kgf/m$^2$ being usually expressed as kg/m$^2$)

or $$9.807 \text{ N/m}^2 = \frac{1}{10^4} \text{ kg/cm}^2$$

or $$1 \text{ kg/cm}^2 = 9.807 \times 10^4 \text{ N/m}^2$$

or $$\boxed{1 \text{ kg/cm}^2 = 98.07 \text{ kN/m}^2}$$

**Kilo Pascal** (kPa), the bigger S.I. unit of pressure would then be

$$\text{kPa} = \text{kN/m}^2 = \frac{1}{98.07} \text{ kg/cm}^2$$

or $$100 \text{ kPa} = \frac{100}{98.07} \text{ kg/cm}^2$$
$$= 1.02 \text{ kg/cm}^2$$
$$= 1.02 \times 10 \text{ m head of water}$$
$$= 10.2 \text{ m of water head.}$$

$\therefore$ $$\boxed{100 \text{ kPa} = 1.02 \text{ kg/cm}^2 = 10.2 \text{ m of water head}}$$

**S.I. Unit of Energy.** When the applied force of 1 newton displaces a body by 1 metre in the direction of application of the force, then work is said to be done equal to 1 newton metre (1 N-m), called 1 Joule (J). The S.I. Unit of work or energy is, thus, **Joule** (J), and has the dimensions of N.m *i.e.* kg $\frac{\text{m}}{\text{sec}^2}$ m

or $$1 \text{ Joule} = 1 \text{ Nm} = 1 \text{ kg m}^2/\text{sec}^2$$

The m.k.s. unit for energy is kgf.m (commonly written as kg.m).

$\therefore$ $$1 \text{ kg.m} = 9.807 \text{ N.m} = 9.807 \text{ J}$$

$\therefore$ $$\boxed{1 \text{ kg.m of energy} = 9.807 \text{ N.m} = 9.807 \text{ J}}$$

*kilo joule* (kJ) will also be used as unit of energy and would be equal to 1000 J = 1 kN.m.

**S.I. Unit of Power.** Power, as we know is the rate of doing work, *i.e.* energy per second. Its m.k.s. unit being kg-m /sec (strictly speaking kgfm/sec). The S.I. unit for power would evidently be Joule per second (j/s), or N.m/s, and this is called watt.

$$1 \text{ watt} = 1 \text{ j/s} = 1 \text{ Nm/s}$$

It may also be pointed out here, that the term *horse power,* being used as the unit of power in m.k.s. and fps systems, is not at all used in S.I. units. The relation between j/s (*i.e.*, watt) and kgf.m/sec can be worked out as below :

$$1 \text{ kgf m/sec} = 9.807 \text{ N.m/sec}$$
$$= 9.807 \text{ j/s} = 9.807 \text{ W}$$

or $$1 \text{ kgm/sec} = 9.807 \text{ W}$$

$$1 \text{ H.P. (metric)} = 75 \text{ kgm/sec}$$
$$= 75 \times 9.807 \text{ W} = \textbf{735 watts}$$
$$1 \text{ H.P. (fps)} = 550 \text{ ft lb/sec}$$
$$= 550 \times 0.138 \times 9.807 \text{ watts} = \textbf{746 watts.}$$

**S.I. Unit of Temperature.** The unit of temperature in S.I. units is **kelvin** (K), and is equal to the temperature interval which is equal to temperature interval corresponding to 1°C. The relation between K and °C is given by :

$$°\text{C} + 273° = \text{K}$$

## Interconversion of MKS and S.I. Units

| *Item* | *MKS unit* | *S.I. unit* | *Conversion of MKS units to S.I. units* | *Conversion of S.I units to M.K.S. units* |
|---|---|---|---|---|
| *(1)* | *(2)* | *(3)* | *(4)* | *(5)* |
| Length | m | m | — | — |
| Area | $m^2$ | $m^2$ | — | — |
| Volume | $m^3$ | $m^3$ | — | — |
| Discharge | $m^3$/sec (cumec) | $m^3$/s (cumec) | — | — |
| Force | kg* | Newton (N) | 1 kg* = 9.807 N (1 N = $10^5$ dyne) | 1 N = 0.102 kg* |
| Pressure (Force/area) | kg*/$m^2$ | N/$m^2$ or Pascal (Pa) | 1 kg*/$m^2$ = 9.807 Pa | 1 Pa = 0.102 kg*/$m^2$ |
| Density ($\rho$) (mass/volume) | kg/$m^3$ | kg/$m^3$ | — | — |
| Unit wt. ($w$ or $\gamma = \rho.g$) wt./volume | kg*/$m^3$ | N/$m^3$ | 1.kg*/$m^3$ = 9.807 N/$m^3$ | 1 N/$m^3$ = 0.102 kg*/$m^3$ |
| Energy (*Force × Distance*) | kg*.m | N.m or Joule (J) (1 J = 1 N.m) | 1 kg* m = 9.807 J | 1 J = 0.102 kg*m |
| Power (*energy/sec*) | kg*/m/sec or H.P. (1 H.P. = 75 kg*m/sec) | J/s or watt (W) (1 W = 1 J/s) | 1 kg*.m/sec = 9.807 W 1 H.P. = 735 W. | 1 W = 0.102 kg*m/sec |
| Dynamic viscosity ($\mu$ or $\eta$) | kg*sec/$m^2$ | N-s/$m^2$ | 1 k* sec/$m^2$ = 9.807 N-s/$m^2$ | 1 N-s/$m^2$ = 0.102 kg*-sec/$m^2$ = 10 **Poise** (dyne sec/$m^2$) |
| Kinematic viscosity $\left(\nu = \frac{\mu}{\rho}\right)$ | $m^2$/sec 1 $m^2$/sec = $10^4$ **Stoke** ($cm^2$/sec ) | $m^2$/s | — | — |

*kg* means kg wt or kgf.

# *Introduction to the Subject*

Water is the greatest resource of humanity. It not only helps in survival but also helps in making life comfortable and luxurious. Besides various other uses of water, the largest use of water in the world is made for irrigating lands. ***Irrigation, infact, is nothing but "a continuous and a reliable water supply to the different crops in accordance with their different needs".*** When sufficient and timely water does not become available to the crops, the crops fade away, resulting in lesser crop yield, consequently creating famines and disasters. Irrigation can, thus, save us from such disasters.

The fact that the provision of irrigation facilities can enhance the yield of our crops by a large extent, can be found from the fact that in Madhya Pradesh State, the crops yield is only 40% more than that in Punjab State, while the cropped area in Madhya Pradesh is about 3 times that in Punjab. The reason is that the gross area under irrigation in Madhya Pradesh is only about 40% as compared to about 98% in Punjab. On an average in India, the yield from irrigated land is about 2.5 tonnes/hectare ; while that from the unirrigated land is about 0.5 tonne/hectare.

It can, therefore, be concluded that if full irrigation facilities are not developed, the production of food grains shall be reduced, as the yield of different crops will be reduced. And, if sufficient food grains are not available, the people will remain hungry, leading to all round chaoes, looting, and economical destruction of the country, hampering its progress & prosperity. In the light of these facts, it can be easily emphasised that 'irrigation' is inevitable, at least in every tropical* or subtropical* country like India.

The adoption of irrigation practice in our country is not a new thing, as it appears to be, because sufficient proofs are available in Indian history, which confirm that irrigation was being practised not only during the periods of Mughals and Aryans, but even during the periods of Pandavas (about 3150 B.C.). Besides the various ancient books which confirm the above facts, there are ruins of various ancient irrigation works, a few of which are till today existing. For example, the most famous old irrigation work, which is functioning even today, is the **'Grand Anicut'** which was built by Chola rulers in the first century A.D. on the Cauvery river. At the site of the Anicut, the river divides into two branches ; the right branch, which is a lower one, is called the *Coleroon*. The Grand Anicut must have been constructed to prevent the flow of water into this low levelled branch (Coleroon) to ensure supplies into Cauvery river, so as to irrigate the fertile Tanjavoor delta land. The importance of irrigation was well recognised by the Mughal rulers, and as such, the *Western Yamuna Canal* which was built by Ferozshah Tughlaq in the year 1355, was got renovated by Emperor Akbar for irrigating lands in the Hissar district (now in Haryana

* **Tropical climate** is hot all the year with much rain. **Sub-tropical climate** is the one in which the heat is under control with less hot summers and less cool winters. **Temperate climate** is represented by warm summers, cool winters, with rain falling evenly throughout the year, or there may be rainy and dry seasons

State) in the year 1568. The *Eastern Yamuna Canal*, was also built by Emperor Mohammad Shah Abdali (1712—1748)

During the British regime, starting from the year 1818, considerable attention was paid towards utilising the surface waters for irrigation, so as to overcome the frequent famines that used to occur in our country. The first public work of the Britishers was the re-renovation of the historic Western Yamuna Canal, which used to provide water to Delhi and the neighbouring areas, but had fallen into disuse during the long period of turmoil. This activity was of course undertaken from considerations of demonstrating the new rule of the *East India Company,* as the emerging rulers of India. It was very much appreciated by the people, and more than that, it was found to be financially very remunerative, yielding a return of 14% in 1847. The Eastern Yamuna Canal built by emperor Mohammad Shah Abdali, was also soon repaired and renovated in the year 1830.

The 1040 km long *Upper Ganga Canal System*, the longest in the world, was also planned, executed and completed in 1854 for carrying a discharge of 191 cumecs to irrigate 0.62 Mha of land, at a cost of ₹ 2.15 million, *about half of the cost of construction of Taj Mahal*. It is still one of the longest canal systems in the world. Modern engineering in India, infact, started from that date, when the first diploma engineering college was also started at Roorkee in 1846, to provide supporting junior staff for the construction of upper Ganga Canal system. Considerable development of irrigation, subsequently took place in the Indus basin* also.

Besides the construction of several irrigation canal projects, such as the *Upper Ganga Canal, the Sirhind Canal, the Lower Chenab Canal* in the Ganga and Indus basins, a lot of work was also done in South India on various rivers, such as Cauvery, Godavari, and Krishna. Construction was taken up and completed on *Khadakvasla dam and its link canals, the Periyar dam and its link canals,* etc. It may, however, be stated that the primary emphasis *during the British regime was on developing extensive irrigation through the construction of diversion headworks and surface water canals.* The storage irrigation was not executed to any appreciable extent. The diversion arrangements, when first developed, were also temporary, which used to wash away during the next monsoon floods, and had to be constructed again every year. The **extensive irrigation** (covering a large command area though irrigating only a part of it) was a preferred strategy to overcome the frequent famines over large commands, besides preventing water-logging, since drainage was not developed to keep the costs down. *Thus, only about 30% of the canal command was irrigated.* The process ended with the delivery of 1 cusec (0.0283 $m^3/s$) of water at the farm head, and further use and management of water was left to the farmers. There was little agriculture extension, and only three waterings were scheduled. *As spring season approached, river flows dwindled, and since there was no upstream storage, the supplies became inadequate. The irrigation for summer (Kharif) period agriculture was not developed. Very few storage and virtually no ground water development was undertaken. There was little hydro-electric development except for some run-of-canal works on Ganga Canal,*

* The Indus river having five famous tributaries—the *Satluj*, the Beas, the *Ravi*, the *Chenab*, and the *Jhelum* do constitute the Indus basin.

*and those too, developed just prior to the independence. There was no concern for development of drinking water and sanitation facilities, much less flood mitigation. Environmental concerns were totally neglected, as almost all the lean season flows were diverted for irrigation, and the rivers, downstream of the diversion head works, were left almost bone dry. In the downstream reaches, even the regenerated ground water flows were not allowed to flow freely, and were again diverted for irrigation, again leaving the rivers dry*. But then, it was perhaps the need of the hour, as it was necessary to stabilise agriculture and mitigate famines. The large scale efforts made by the British Rulers resulted in providing irrigation to 28.2 Mha of area, out of total net cultivated area of 116.8 Mha, thus bringing about 24% of the net cultivated area under irrigation. In order to develop irrigation in the country, the first irrigation commission* was even appointed in 1901, which submitted its report in 1903, based on which, large scale canal development works were undertaken, besides preparing the blue prints of various important storage projects, including that of the Bhakra dam. *Although in undivided India, about 24% of the country's cropped area came under irrigation, yet with the unfortunate partition of the country, 31% of the country's irrigated area stood transferred to Pakistan, as against the transfer of only 18% of its population.*

***The independent India in the year 1947 was thus left high and dry with only 19% of its cropped area under irrigation ; whereas Pakistan at the time of its creation found 44% of its cropped area under irrigation.***

The Indian Government of the independent India, at its very outset, was thus faced to force acute famines and food grain shortages, because most of the country's agricultural area was dependent upon natural rainfall, which proved highly unreliable and erratic. Large scale import of wheat from USA as soft term assistance under the famous PL-480 programme, did help the nation to overcome hunger, deaths, and destruction at that time.

Large scale efforts were then made by the Indian Government under the prime ministership of our beloved leader Pt. Jawahar Lal Nehru, to develop and harness our vast water resources, so as to ensure collection of water during monsoon, and its subsequent use for irrigation during non-monsoon period. **Several dams/reservoirs**** were, therefore, planned and constructed across various rivers to store water during rainy season to reduce the fury of floods, and long **canal networks** constructed to move down the stored water during dry weather to the fields, to ensure irrigation supplies to the crops. Hydro-power development was also undertaken as a side product.

---

* The second irrigation commission was appointed in 1969, and it submitted its report in March 1972.

** From less than 300 large dams existing at the beginning of planned development, the number of large dams constructed so far (excluding about 500 dams under construction) has gone upto about **4000,** as to create **live storage capacity of about 226 B.cum** (billion cubic metre) on different rivers by the completed projects, and an additional 60 Bcum shall be created by the ongoing projects. Projects to further create 108 Bcum of live storage are also under consideration of the Govt. of India. Out of such a large number of dam reservoirs, CWC-GoI regularly monitors the storage position of 71 major reservoirs, which together account for a design storage of 113 Bcum.

and dedicating the **Bhakra Dam** to the Nation on 22.10.1963, called it a ***'New temple of resurgent India'*** and *'Symbol of India's progress'*. The great Pt. Nehru gave a great impetus to the construction of such new temples, and did all what he could, for the execution of several multi-purpose projects, by investing huge funds at a time when money was badly required for other important sectors, like Education, Housing, Roads, Railways, Hospitals, etc.

With the construction of multipurpose dams, hydropower also became available, which helped to develop tubewell irrigation in the country. This was an important change that took place from early 1960's, when rapid development of tube-wells took place. So much so that, by the end of 11th plan (year 2011–2012), we were using ground water for irrigating about 61% of our gross irrigated area, as against irrigating 33% of the area in 1965–66. Whereas, although development of minor irrigation through the use of ground water has helped in controlling water-logging in areas where canal irrigation did exist, yet it has resulted in abnormal fall in the ground water tables in several areas, where no attention has been paid to the ground water recharge. The lowered water-tables over vast areas of the country has resulted in making the available ground waters saline and contaminated with uorides and arsenic, making them unfit for drinking and other domestic requirements, for which they have largely been used, since olden days*.

**Table I. Planwise Expenditure Incurred on Irrigation in India upto year 2011-12**

| S. No. | Plan | Period | *Expenditure Incurred (₹ Crores) in* | | |
|---|---|---|---|---|---|
| | | | *Major** and Medium** Irr. Projects* | *Minor Irr. i/c CAD Projects*** | *Total* |
| 1. | First Plan | 1951–56 | 376 | 66 | 442 |
| 2. | Second Plan | 1956–61 | 380 | 162 | 542 |
| 3. | Third Plan | 1961–66 | 576 | 443 | 1,019 |
| 4. | Annual Plans | 1966–69 | 430 | 561 | 991 |
| 5. | Fourth Plan | 1969–74 | 1,242 | 1,174 | 2,416 |
| 6. | Fifth Plan | 1974–78 | 2,516 | 1,410 | 3,926 |
| 7. | Annual Plans | 1978–80 | 2,079 | 1,345 | 3,424 |
| 8. | Sixth Plan | 1980–85 | 7,369 | 4,160 | 11,529 |
| 9. | Seventh Plan | 1985–90 | 11,107 | 7,627 | 18,734 |
| 10. | Annual Plan | 1990–92 | 5,459 | 3,650 | 9,109 |
| 11. | Eighth Plan | 1992–97 | 21,072 | 13,885 | 34,957 |
| 12. | Ninth Plan | 1997–2002 | 49,289 | 13,760 | 63,049 |
| 13. | Tenth Plan | 2002–2007 | 83,647 | 16,459 | 1,00,106 |
| 14. | Eleventh Plan | 2007–2012 | 1,65,350 | 46350 | 2,11,700 |
| Grand Total | | | **3,50,892** | **1,11,052** | **4,61,944** |

Inspite of all such shortcomings and drawbacks, impressive achievements in developing and harnessing our water resources have been made after

* Ground water still provides about 80% of the domestic water supply in rural areas, and fulfils about 50% of urban and industrial water requirements (Planning Commission, 1999).

** Projects having CCA upto 2000 ha are classified as ***minor irrigation projects***. Those having CCA of more than 10,000 ha are classified as ***major projects*** ; and others between 2000 to 10,000 hectares are ***medium projects***.

Inspite of all such shortcomings and drawbacks, impressive achievements in developing and harnessing our water resources have been made after independence to create an irrigation potential of about 113.53 Mha (by the end of 11th plan ; *i.e.*, March 2012), by incurring an expenditure of about 4.62 Lakh crores of rupees, including an expenditure of about ₹ 1.11 lakh crores on minor irrigation and Command Area Development (CAD) projects, as shown in Table II and I.

The plan wise investments made under irrigation sector during 1951-2012 are re ected in table I.

The Planwise irrigation potential developed and utilised up to March 2012 (end of $11^{th}$ plan) in India are shown in Table II. It can be seen from this table that the created irrigation potential has gone up from **22.60 Mha** in 1951, to **113.53 Mha** by the end of 11th plan (March, 2012), against the ultimate irrigation potential of **140 Mha** from conventional in– basin storage techniques (Table III). Out of the total created potential of **113.53 Mha**, we have, however, utilised only **87.86 Mha** during the year 2011–12. This has resulted

**Table II. Planwise Position of Irrigation Potential Created and Utilised (Cummulative Figs. in Mha)**

| Plan | *Potential Created* | | | | | *Potential Utilised* | | | | |
|---|---|---|---|---|---|---|---|---|---|---|
| | *Major & Medium* | *Minor* | | | *Total Major Medium & Minor* | *Major & Medium* | *Minor* | | | *Total Major Medium & Minor* |
| | | *Surface Water* | *Ground Water* | *Total* | | | *Surface* | *Ground* | *Total* | |
| Pre-plan upto 1951 | 9.70 | 6.40 | 6.50 | 12.90 | 22.60 | 9.70 | 6.40 | 6.50 | 12.90 | 22.60 |
| Ist Plan (51-56) | 12.20 | 6.43 | 7.63 | 14.06 | 26.26 | 10.98 | 6.43 | 7.63 | 14.06 | 25.04 |
| II Plan (56-61) | 14.33 | 6.45 | 8.30 | 14.75 | 29.08 | 13.05 | 6.45 | 8.30 | 14.75 | 27.80 |
| III Plan (61-66) | 16.57 | 6.48 | 10.52 | 17.00 | 33.57 | 15.17 | 6.48 | 10.52 | 17.00 | 32.17 |
| Annual Plans (66-69) | 18.10 | 6.50 | 12.50 | 19.00 | 37.10 | 16.75 | 6.50 | 12.50 | 19.00 | 35.75 |
| IV Plan (69-74) | 20.70 | 7.00 | 16.50 | 23.50 | 44.20 | 18.39 | 7.00 | 16.50, | 23.50 | 41.89 |
| V Plan (74-78) | 24.72 | 7.50 | 19.80 | 27.30 | 52.02 | 21.16 | 7.50 | 19.80 | 27.30 | 48.46 |
| Annual Plans (78-80) | 26.61 | 8.00 | 22.00 | 30.00 | 56.61 | 22.64 | 8.00 | 22.00 | 30.00 | 52.64 |
| VI Plan (80-85) (Reapprised) | 27.70 | 9.70 | 27.82 | 37.52 | 65.22 | 23.57 | 9.01 | 26.24 | 35.25 | 58.82 |
| VII Plan (85-90) | 29.92 | 10.90 | 35.62 | 46.52 | 76.44 | 25.47 | 9.97 | 33.15 | 43.12 | 68.59 |
| Annual Plans (90-92) | 30.74 | 11.46 | 38.89 | 50.35 | 81.09 | 26.31 | 10.29 | 36.25 | 46.54 | 72.85 |
| VIII Plan (92-97) | 32.95 | 12.51 | 40.80 | 53.31 | 86.26 | 28.44 | 11.07 | 37.70 | 48.77 | 77.21 |
| IX Plan (1997-2002) | 37.05 | 13.60 | 43.30 | 56.90 | 93.95 | 30.95 | 11.44 | 38.55 | 49.99 | 80.94 |
| Xth Plan (2002-2007) | 41.64 | 14.00 | 46.10 | 60.10 | 101.74 | 33.68 | 12.00 | 39.48 | 51.48 | 85.16 |
| XIth Plan (2007-2012) | 47.97 | 15.0* | 50.56* | 65.56 | **113.53** | 34.95* | 12.50* | 40.41 | 52.91 | **87.86** |

*Provisional figures.

## Table III : Statewise Ultimate Irrigation Potential

| S. No. | Name of State | From Major and medium Surface Water Schemes | From Minor Irrigation | | | Total (Major, Medium and minor) |
|---|---|---|---|---|---|---|
| | | | Surface Water | Ground Water | Total Minor | |
| 1. | Andhra Pradesh* (including Telangana) | 5.000 | 2.300 | 3.960 | 6.260 | 11.260 |
| 2. | Arunachal Pradesh | 0.00 | 0.150 | 0.018 | 0.168 | 0.168 |
| 3. | Assam | 0.970 | 1.000 | 0.900 | 1.900 | 2.870 |
| 4. | Bihar | 5.224 | 1.547 | 4.117 | 5.664 | 10.888 |
| 5. | Chhatisgarh | 1.147 | 0.081 | 0.490 | 0.571 | 1.718 |
| 6. | Goa | 0.062 | 0.025 | 0.029 | 0.054 | 0.116 |
| 7. | Gujarat | 3.000 | 0.347 | 2.756 | 3.103 | 6.103 |
| 8. | Haryana | 3.000 | 0.050 | 1.462 | 1.512 | 4.512 |
| 9. | Himachal Pradesh | 0.050 | 0.235 | 0.068 | 0.303 | 0.353 |
| 10. | Jammu & Kashmir | 0.250 | 0.400 | 0.708 | 1.108 | 1.358 |
| 11. | Jharkhand | 1.277 | 0.353 | 0.830 | 1.183 | 2.460 |
| 12. | Karnataka | 2.500 | 0.900 | 2.574 | 3.474 | 5.974 |
| 13. | Kerala | 1.000 | 0.800 | 0.879 | 1.679 | 2.679 |
| 14. | Madhya Pradesh | 4.853 | 2.119 | 9.242 | 11.361 | 16.214 |
| 15. | Maharashtra | 4,100 | 1.200 | 3.652 | 4.852 | 8.952 |
| 16. | Manipur | 0.135 | 0.100 | 0.369 | 0.469 | 0.604 |
| 17. | Meghalaya | 0.020 | 0.085 | 0.063 | 0.148 | 0.168 |
| 18. | Mizoram | 0.000 | 0.065 | 0.005 | 0.070 | 0.070 |
| 19. | Nagaland | 0.010 | 0.070 | 0.005 | 0.075 | 0.085 |
| 20. | Orissa | 3.600 | 1.000 | 4.203 | 5.203 | 8.803 |
| 21. | Punjab | 3.000 | 0.050 | 2.917 | 2.967 | 5.967 |
| 22. | Rajasthan | 2.750 | 0.600 | 1.778 | 2.378 | 5.128 |
| 23. | Sikkim | 0.020 | 0.050 | 0.000 | 0.050 | 0.070 |
| 24. | Tamil Nadu | 1.500 | 1.200 | 1.832 | 4.032 | 5.532 |
| 25. | Tripura | 0.100 | 0.100 | 0.081 | 0.181 | 0.281 |
| 26. | Uttar Pradesh | 12.154 | 1.186 | 16.295 | 17.481 | 29.635 |
| 27. | Uttarakhand | 0.346 | 0.014 | 0.504 | 0.518 | 0.864 |
| 28. | West Bengal | 2.300 | 1.300 | 3.318 | 4.618 | 6.918 |
| 29. | Total UTs | 0.098 | — | — | 0.046 | 0.144 |
| | Grand Total (All India) | **58.465** ; Say **58.46** Mha | **17.372** | **64.171** | **81.438** ; Say **81.43** Mha | **139.903** ; Say **140** Mha |

in irrigating about 46% of our total gross cropped area of 189 Mha, which has helped the country to achieve self sufficiency in food grains, the production of which has touched about 250 million tonnes (2012) as against our consumption of about 182 Mt. by 1210 Million (year 2012) population at an average annual per capita consumption of about 150 kg/year. The actual 250 Mt production of food grains is found to be almost in consonance with the theoretical production at our observed production rates of 2.5 t/ha for irrigated land, and

* Now bifurcated into two States of Andhra Pradesh and Telangana since June 2014.

1.0 t/ha for unirrigated (rain-fed) land, for 67% of the cropped area growing food grain crops, as : 87.86 Mha × 0.67 (use factor for food grain crops) × 2.5 t/ha + (189 – 87.86) 0.67 × 1.0 t/ha = 214 Mt.

### Table IV : Statewise Extent of Cultivation and Irrigation done in India (2011-12)

| S. No. | Name of State | Geographical area (Mha) | Total cultivable area i.e., CCA in (Mha) | Net cropped Area(Mha) | Gross croped Area (Mha) | Cross irrigated Area (Mha) | %age of gross area under irrigation $= \frac{G.Irr}{G\ Cropped} \times 100 = \frac{(7)}{(6)} \times 100$ | Annual intensity of irrigation $= \frac{Cross\ Irr}{CCA} = \frac{(7)}{(4)} \times 100$ |
|---|---|---|---|---|---|---|---|---|
| (1) | (2) | (3) | (4) | (5) | (6) | (7) | (8) | (9) |
| 1. | Andhra Pradesh (*i.e.* Telangana) | 27.51 | 15.93 | 11.161 | 13.759 | 6.785 | 49.31 | 42.59 |
| 2. | Arunachal Pradesh | 8.37 | 0.27 | 0.215 | 0.281 | 0.057 | 20.28 | 21.11 |
| 3. | Assam | 7.84 | 3.22 | 2.811 | 4.174 | 0.163 | 3.91 | 5.06 |
| 4. | Bihar | 9.42 | 6.94 | 5.396 | 7.647 | 5.158 | 67.75 | 74.32 |
| 5. | Chhatisgarh | 13.52 | 5.60 | 4.677 | 5.664 | 1.648 | 29.10 | 29.43 |
| 6. | Goa | 0.38 | 0.22 | 0.132 | 0.164 | 0.041 | 25.00 | 18.64 |
| 7. | Gujarat | 19.60 | 12.31 | 10.302 | 13.086 | 6.305 | 48.18 | 51.22 |
| 8. | Haryana | 4.42 | 3.77 | 3.513 | 6.489 | 5.680 | 87.53 | 150.66 |
| 9. | Himachal Pradesh | 5.57 | 0.82 | 0.543 | 0.940 | 0.187 | 19.89 | 22.80 |
| 10. | Jammu & Kashmir | 22.22 | 1.05 | 0.746 | 1.161 | 0.480 | 41.34 | 45.71 |
| 11. | Jharkhand | 7.97 | 4.19 | 1.250 | 1.430 | 0.152 | 10.63 | 3.62 |
| 12. | Karnataka | 19.18 | 12.89 | 9.941 | 12.059 | 4.137 | 34.31 | 32.09 |
| 13. | Kerala | 3.89 | 2.45 | 2.040 | 2.662 | 0.546 | 20.51 | 22.29 |
| 14. | Madhya Pradesh | 30.92 | 17.23 | 15.237 | 22.516 | 8.228 | 36.54 | 47.75 |
| 15. | Maharashtra | 30.77 | 21.00 | 17.386 | 22.011 | 4.089 | 18.58 | 19.47 |
| 16. | Manipur | 2.23 | 0.16 | 0.348 | 0.348 | 0.069 | 19.82 | 43.31 |
| 17. | Meghalaya | 2.24 | 1.10 | 0.285 | 0.339 | 0.080 | 23.60 | 7.27 |
| 18. | Mizoram | 2.11 | 0.58 | 0.097 | 0.097 | 0.013 | 16.46 | 2.24 |
| 19. | Nagaland | 1.66 | 0.64 | 0.379 | 0.474 | 0.092 | 19.41 | 14.38 |
| 20. | Orissa | 15.57 | 8.09 | 4.394 | 4.964 | 1.437 | 28.95 | 17.76 |
| 21. | Punjab | 5.04 | 4.37 | 4.134 | 7.905 | 7.771 | **98.30** | 177.83 |
| 22. | Rajasthan | 34.22 | 25.71 | 18.034 | 24.505 | 8.903 | 36.33 | 34.63 |
| 23. | Sikkim | 0.71 | 0.11 | 0.077 | 0.137 | 0.019 | 13.87 | 17.27 |
| 24. | Tamil Nadu | 13.01 | 8.44 | 4.986 | 5.890 | 3.519 | 59.75 | 41.69 |
| 25. | Tripura | 1.25 | 0.31 | 0.256 | 0.371 | 0.129 | 34.77 | 41.61 |
| 26. | Uttar Pradesh | 24.09 | 19.43 | 16.623 | 25.939 | 19.792 | 76.30 | 101.86 |
| 27. | Uttarakhand | 5.35 | 1.42 | 0.714 | 1.132 | 0.555 | 49.03 | 39.08 |
| 28. | West Bengal | 8.88 | 5.93 | 5.198 | 9.353 | 5.437 | 58.13 | 91.69 |
| 29. | Total UTs | 1.09 | 0.21 | 0.081 | 0.120 | 0.060 | 50.00 | 28.57 |
| Grand Total (All India) | | **328.13** | **184.35** | **140.974** | **195.632** | **91.530** | **46.79** | **49.65** |

The impressive development of an irrigation potential of about 113 Mha by the end or 11th plan has been achieved by the execution of about 1300 major, medium and ERM** irrigation projects, besides a huge number of minor irrigation projects.

The net and gross cropped areas as well as net and gross irrigated areas in various States of India are shown in table IV, along with computing annual intensities of irrigation in Col. (9) and plotted in Fig. I. The data used here is of year 2011-12. The extent of the use of the different sources of water for irrigation is also given in table V.

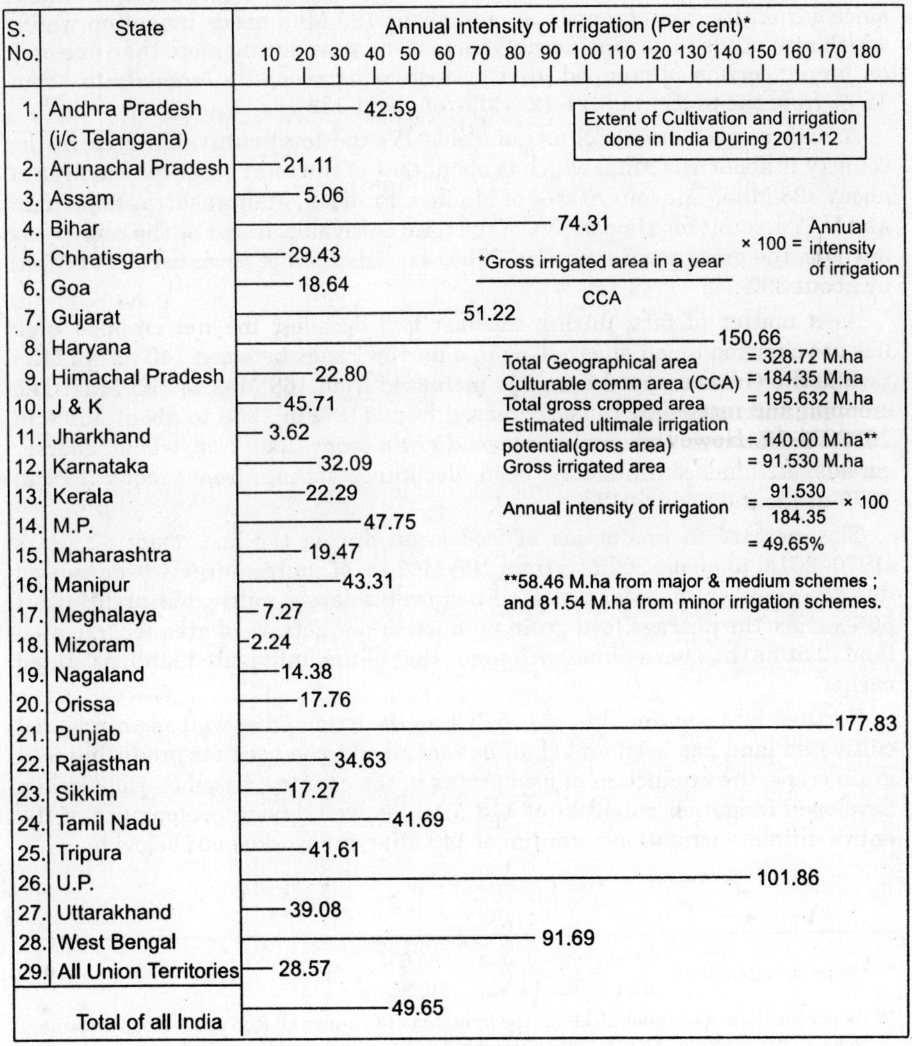


**Fig. 1**

* Extension, Renovation and Modernisation (EMR) projects.

**Table V. Extent of Irrigation being done from various sources in India (Year 2011-12)**

| S.No. | Surface | % age of Irrigation (app.) |
|---|---|---|
| 1. | Canals | 25% surface Irr. (39%) |
| 2. | Tanks and other minor surface sources | 14% |
| 3. | Tubewells and other wells | 61% Ground water Irr. |
| | Total | 100% |

Inspite of such large scale efforts made by India, in having generated an irrigation potential of about 113 Mha, we are still, much short of our requirement, since our entire gross cropped area of about 189 Mha needs irrigation water, which also needs to be considerably increased by producing more than one crop on larger portion of cropped area by increasing ***cropping intensity*** to about 150% from the present (2011-12) value of about 139%*.

As can be seen from Col. (4) of Table IV, the total cultivable area in the country is about 184 Mha, which is about 60% of the total geographical area of about 328 Mha. The four States of Madhya Pradesh, Maharashtra, Rajasthan and U.P. account for almost 50% of the total cultivable area. For the country as a whole, the gross sown area (195 Mha) exceeds the net sown area (141 Mha) by about 39%.

As a matter of fact, during the last two decades, the net cropped area has not changed much, fluctuating in a narrow range between 140 to 143 Mha ; while the Gross cropped area has increased from 166 Mha to 195 Mha. The cropping intensity has, thus, increased from 118% in 1980 to about 139% in 2011-12. However, the area under *food grain crops* (like rice, wheat, coarser cereals, etc.) has continuously been declining, falling from about 75% in 1970 to about 67% in 2012.

The increase in production of food grain during the last about 42 years (1970–2012) to about 252 Mt. from 108 Mt has, of course, largely been caused due to extension of irrigation facilities over a larger cultivated area (about 50%), since the average food grain production per hectare of area for irrigated land (2.5t/ha) has been almost 2.5 times that of the unirrigated land, as stated earlier.

Further, by assuming that only 67%** of the irrigated as well as unirrigated cultivated land has been and shall be continued to be used for producing food grain crops, the production of food grains in the country, based on the already developed irrigation potential of 113 Mha, as well as on development of the entire ultimate irrigation potential of 140 Mha, are worked out below :

---

* Cropping intensity $= \dfrac{\text{Cross cropped area}}{\text{Net cropped area}} = \dfrac{195.632}{140.974} \approx 139\%$

** As per the data of the year 2011-12, the irrigated area under all food grain crops amounted to 61.47 Mha, while the total gross irrigated area was 91.53 Mha, which means that the percentage of total irrigated area under food grain crops was $= \dfrac{61.47}{91.53} \times 100 = 67\%$.

## PROPOSED INTERBASIN WATER TRANSFEF

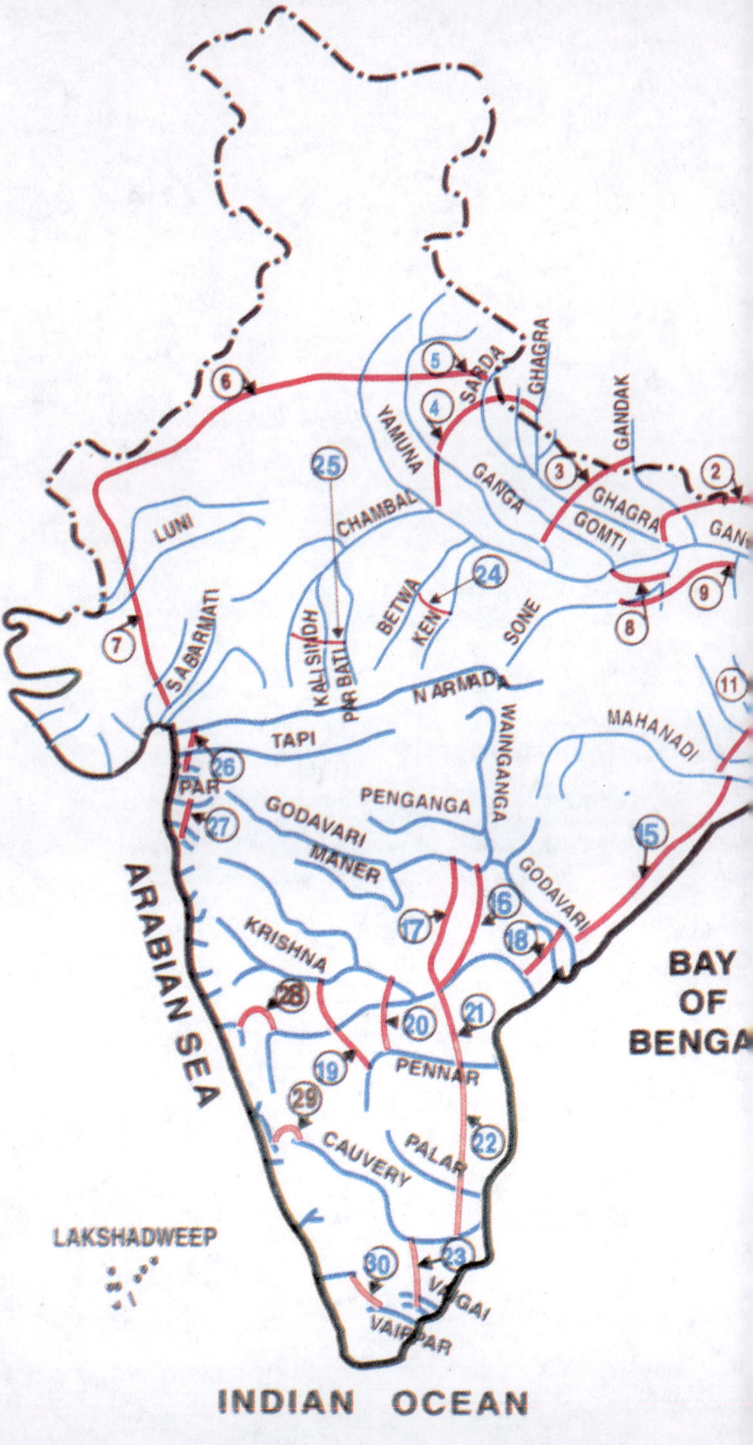

## LINKS UNDER STUDY

### HIMALAYAN COMPONENT
### NAME OF THE LINKS

1. Manas-Sankosh-Tista-Ganga
2. Kosi-Ghagra
3. Gandak-Ganga
4. Ghagra-Yamuna
5. Sarda-Yamuna
6. Yamuna-Rajasthan
7. Rajasthan-Sabarmati
8. Chunar-Sone Barrage
9. Sone Dam-Southern Tributaries of Ganga
10. Ganga-Damodar-Subernarekha
11. Subernarekha-Mahanadi
12. Kosi-Mechi
13. Farakka-Sunderbans
14. Jogigopa-Tista-Farakka (Alternative to Sl. No. 1)

### PENINSULAR COMPONENT
### NAME OF THE LINKS

15. Mahanadi (Manibhadra)-Godavari (Dowlaiswram)
16. Godawari (Inchampalli low dam) Krishna (Nagarjunasagar Tail Pond/Pulichintala)
17. Godavari (Inchampalli)-Krishna (Nagarjunasagar)
18. Godavari (Polavaram)-Krishna (Vijayawada)
19. Krishna (Almatti)-Pennar
20. Krishna (Srisailam)-Pennar
21. Krishna (Nagarjunasagar)-Pennar (Somasila)
22. Pennar (Somasila)-Cauvery (Grand Anicut)
23. Cauvery (Kattalai)-Vaigai-Gundar
24. Ken-Betwa
25. Parbati-Kalisindh-Chambal
26. Par-Tapi-Narmada
27. Damanganga-Pinjal
28. Bedti-Varda
29. Netravati-Hemavati
30. Pamba-Achanikovil-Vaippar

**LEGEND**

International Boundary

River

Links

(*i*) Likely food production with utilisation of entire developed irrigation potential of about 113 Mha

= 113 Mha × 0.67 (use factor for food grain crops) × 2.5 t/ha + (195 – 113) Mha × 0.67 × 1.0 t/ha

= 189 + 55 = **244 Mt**

(*ii*) Likely food production with utilisation of entire ultimate irrigation potential of 140 Mha

= 140 Mha × 0.67 × 2.5 t/ha + (195 – 140) Mha × 0.67 × 1.0 t/ha

= 235 + 37 = **272 Mt**

It can, thus, be seen that even if we are able to create and utilise the entire ultimate irrigation potential of 140 Mha, we can produce only about 272 M.t of food grains, which may just suffice for our future population of the year 2025, which is projected to be about 1400 million (M) at a moderate average food grain per capita consumption level of about 180 kg/year (*i.e.* 0.5 kg/day).

It can, thus, be concluded that even if we develop the entire ultimate irrigation potential of 140 Mha, which by better management of minor irrigation projects, has already been increased by 27 Mha from its earlier estimated figure of 113 Mha, we can grow food grains, which would be just sufficient to meet our future requirements upto the year 2025, or so.

*Efforts will, therefore, have to be made to plan something more for the future, rather than simply targeting to exploit the ultimate irrigation potential of our conventional inbasin development methods.* One of the proposed method to achieve this aim is to increase the cropping intensity and to promote ***dry land farming***. These measures, however, are not going to create any prominent impact on increasing food grain production. *The only viable method left to meet our future requirement is the* ***interbasin transfer of water from surplus regions to the deficit regions.*** *It may therefore, become imperative to transfer waters from surplus river basins like Brahmaputra, Barak, Mahanadi, Godavari, etc. to the water deficit areas.* ***Interbasin transfer of water***, which is very much required for the country, is not a new concept, since such transfers had been practised even in the 18th and 19th centuries, when canals like *Western Yamuna Canal, Agra Canal, Kurnool-Cuddapah Canal, Periyar-Vaigai Canal,* etc. were constructed to carry waters of river basins of surplus waters over long distances to the regions of deficit waters. *Indira Gandhi Canal (Rajasthan Canal), Ravi-Beas link,* and *Beas-Sutlej link,* are the modern examples of inter-basin transfer projects of 20th century. **Several additional links** are still needed to be planned, cleared and got executed to ensure optimum development of country's water resources, NWDA (National Water Development Agency) has infact, suggested linking of several *Himalayan as well as Peninsular rivers,* through the 19 links under *Himalayan river development plan* and 17 links under the *Peninsular river development plan.* These links are shown in Colour Fig. 1. All such links are likely to help in creating an additional irrigation potential of about 35 Mha, besides providing drinking water, generating 34 M kW of hydropower and providing large scale protection against oods.

Although GOI is making large scale efforts to expedite the execution of atleast some of these river linking projects, but the progress is very slow, primarily because the involved rivers are inter-state rivers, and the different states do not easily agree to sharing of the benefits or to loss some of their areas due to inundation likely to be caused by the construction of dams, during the execution of such links.

Be that as it may, the following three links are under advance stage of approvals.

(*i*) ***Ken-Betwa Link*** to link Ken and Betwa rivers flowing through UP and MP. This link is likely to provide irrigation water to water deficient regions of Central Madya Pradesh to help irrigate additional 4.46 lakh hectares of land. Work is likely to start soon.

(*ii*) ***Damanganga-Pinjal Link*** and ***Par-Tapi-Narmada Link*** in Gujarat and Maharashtra are also in advance stages, as their DPR's (Detailed Project Reports) have already been completed.

In addition to developing and harnessing greater water supplies for use by developing inter-basin transfers of water, there also exists an urgent need to **conserve our water supplies,** and to reduce our water demands by developing modern technologies.

While industrial water demand can be reduced by developing recycling of waste water and by using superior modern methods to reduce consumption of water, the irrigation water demand can be reduced by using farming techniques like **sprinkler irrigation***, and **drip irrigation**** (called ***micro irrigation***), on a large scale, which utilize the valuable water in an economical manner, by avoiding its wastage through infiltration and evaporation. Realizing its necessity, the Govt. of India has recently started paying greater attention to promote the use of *micro-irrigation*, by ensuring supply of equipment and know-how at subsidized rates to the farmers. A task force to suggest institutional mechanism needed for promoting such technologies and to ensure reaching of the intended benefits to the target group, has also been constituted by the Union Ministry of Finance, Govt. of India.

The present status of use of these modern technologies in the country is, however, limited, since only about 5.37 Mha* is under micro irrigation (2.55 Mha under sprinkler and 2.82 Mha under drip irrigation). Even this figure has been achieved due to extra ordinary efforts of the new central government since 2014, in the form of providing benefits under the 'Per Drop More Benefit' component of the *Pradhan Mantri Krishi Sinchayee Yojna (PMKSY-PMDC).* Rajasthan and Haryana States lead in the usage of sprinkler irrigation, whereas, Sikkim, Karnataka, Andhra Pradesh, Gujarat and Maharashtra are presently leading in drip irrigation.

It would also be pertinent to mention here that the adoption of intensive irrigation in certain States like Punjab, Orissa, Andhra Pradesh, etc. in the past years, had caused alarming rise in the ground water-tables, which resulted in causing water-logging and reduced crop yields. The non-provision of surface and sub-surface drainage and improper management of irrigation command area had compounded the problem to such an extent that a lot of land has become saline and even alkaline, making it unfit for cultivation. The land affected due to these reasons is estimated to be about 6 Mha (2.7 Mha affected by water-logging + 3.06 Mha affected by salinity + 0.24 Mha affected by alkalinity), **Conjunctive**

---

*Explained in articles 1.6(6) and 1.6(7), respectively.

*This figure was only 1.15 Mha in 2012.

**use** of surface and ground water collectively together is an accepted solution for removing water-logging problem, to enhance our crop yields. There is a large scope of making such improvements in alluvial plains, particularly in the *Gangetic plain, the coastal areas of Orissa and Anhdra Pradesh, the Brahmaputra Valley, the Cauvery delta,* and in parts of the Narmada basin.

Although irrigation facilities are so very important and essential for the development of any nation, they are complex and intricate. *Proper development of water resources involves the most economical exploitation of the entire available water of a country to meet the needs of its various regions, not only for irrigation but also for drinking, domestic, and industrial requirements, which are in no way, less important. Such optimum developments of water resources must also give due consideration to* **ecological aspects** by proper allocation of low river flows for **environmental conservation,** *rather than diverting them entirely for irrigation or water supply demands.* Aiming for such an **optimum and integrated utilisation** of the country's entire water-resources, makes this field very complex, intricate, and interesting too. It is here, that the real job of an Irrigation engineer or more precisely, a water-resources expert, comes into play.

*How to plan our available water sources to fulfil requirements of various sectors, and to ensure maximum benefits, keeping in view the socialistic distribution of the achieved benefits within the available funds, is the real tough job. The design of various irrigation works is another job, which must be accomplished with a fair degree of economy and accuracy.* Here is a book, which provides the basic fundamental principles, broad guidelines, and details of this intricate and interesting field in a simple language. It is hoped that this book will prove useful to every student, to every teacher, as well as to every design and field engineer.

Although irrigation facilities are so very important and essential for the development of any nation, they are complex and intricate. *Proper development of water resources involves the most economical exploitation of the entire available water of a country to meet the needs of its various regions, not only for irrigation but also for drinking, domestic, and industrial requirements, which are in no way, less important. Such optimum developments of water resources must also give due consideration to* **ecological aspects** by proper allocation of low river flows for **environmental conservation**, *rather than diverting them entirely for irrigation or water supply demands.* Aiming for such an **optimum and integrated utilisation** of the country's entire water-resources, makes this field very complex, intricate, and interesting too. It is here, that the real job of an Irrigation engineer or more precisely, a water-resources expert, comes into play.

*How to plan our available water sources to fulfil requirements of various sectors, and to ensure maximum benefits, keeping in view the socialistic distribution of the achieved benefits within the available funds, is the real tough job. The design of various irrigation works is another job, which must be accomplished with a fair degree of economy and accuracy.* Here is a book, which provides the basic fundamental principles, broad guidelines, and details of this intricate and interesting field in a simple language. It is hoped that this book will prove useful to every student, to every teacher, as well as to every design and field engineer.

# 1

# Irrigation Technqiues and Quality of Irrigation Water

## 1.1. Definition of Irrigation

Plants are living beings and do require water and air for their survival, as do human beings require. Their requirement of water varies with their type. Different types of plants require different quantities of water, and at different times, till they grow up completely. Water is normally supplied to these plants by nature through direct rain or through the flood waters of rivers which inundate large land areas during floods. The flood water may saturate the land before the flood subsides. The water absorbed by the land during floods, supplements the water requirement of the crop during dry season. These natural processes, whereby, the water is supplied to the crops for their growth, are dependent upon 'Nature' or 'God', whatever we may call it. Sometimes, there may be very heavy rains creating serious floods and damaging the crops, and sometimes, there may not be any rains at all, creating scarcity of water for the crops. Thus, famine and scarcity conditions are created. In his bid to control the nature, man discovered various methods by which the water can be stored during the periods of excess rainfall, and to use that stored water during periods of 'less rainfall' or 'no rainfall'. The art or the science by which it is accomplished, is generally, termed, as irrigation. ***Irrigation may, therefore, be defined as the science of artificial application of water to the land, in accordance with the 'crop requirements' throughout the 'crop period' for full-fledged nourishment of the crops.***

## 1.2. Necessity of Irrigation in India

India is a tropical country with a vast diversity of climate, topography and vegetation. Rainfall in India, varies considerably in its place of occurrence, as well as in its amount. Even at a particular place, the rainfall is highly erratic and irregular, as it occurs only during a few particular months of the year. Crops cannot, therefore, be raised successfully, over the entire land, without providing artificial irrigation of fields.

More than seventy percent of our population, directly depends on agriculture, and the remaining depends indirectly on agriculture. Out of a total geographical area of about 328 million hectares, about 184 million hectares is the cultivable area. In order to save this area from the complete wishes and vagaries of nature, and to ensure full growth of crops, it is necessary to provide adequate artificial irrigation facilities. In order to achieve this, the Indian Government is trying hard and spending enormously to provide irrigation facilities for the entire cultivable land. So far developed irrigation facilities in India have been shown under "Introduction" in the previous pages.

## 1.3. Advantages of Irrigation

Every irrigation project is designed, keeping in view of its economics, *i.e.* the expenditure likely to be incurred and the benefit likely to occur. There is a capital investment on the project and the future recurring charges for maintenance and operation. The project estimate is generally sanctioned when the benefit gives at least about 8% interest on the capital outlay. Sometimes, unproductive projects are also sanctioned in view of their *general public benefits.*

There is hardly any point in emphasizing the importance and advantages of irrigation during the times of acute food shortages and growing population of our country.

Even then, some of the advantages of irrigation are summarised below:

(1) **Increase in Food Production.** Irrigation helps in increasing crop yields, and hence, to attain self-sufficiency in food.

(2) **Optimum Benefits.** Optimum utilisation of water is made possible by irrigation. By optimum utilisation, we generally mean, obtaining maximum crop yield with required amount of water. In other words, yield will be smaller for any quantity lesser than or in excess of this optimum quantity.

(3) **Elimination of Mixed Cropping.** In the areas, where irrigation is not assured, generally mixed cropping is adopted. *By mixed cropping, we mean, sowing together of two or more crops in the same field.* If the weather conditions are not favourable to one of the crops, they may be better suitable for the other; and thus, the farmer may get at least some yield. Mixed cropping, is thus, found necessary and also economical when irrigation facilities are lacking, and especially during periods of *Crash programmes* in under-developed countries. But if irrigation is assured, mixed cropping can be eliminated.

Mixed cropping is generally not acceptable, because different crops require different types of field preparations and different types of waterings, manurings, etc. If two crops are mixed together, the field preparations, waterings, manurings, etc. cannot be made to suit the special needs of either. Moreover, during the time of harvesting, the crops get intermixed with each other, reducing the purity of each other. But, when regular and permanent water supply is assured, a single superior crop can be sown, depending upon the conditions of the soil and the needs of the country.

(4) **General Prosperity.** Revenue returns with well developed irrigation, are sometimes, quite high, and helps in all round development of the country and prosperity of the entire nation and community.

(5) **Generation of Hydro-electric Power.** Cheaper power generation can be obtained from water development projects, primarily designed for irrigation alone. Canal outlets from dams and Canal falls on irrigation canals can be used for power generation. For example, Ganga and Sarda Canals, constructed for irrigation, are now generating hydro-electric power as a side product, up to about 80,000 kilo-watts.

(6) **Domestic Water Supply.** Development of irrigation facilities in an area helps in augmenting the water supply in nearby villages and towns, where other sources of water are not available or are scarcely available. It also helps in providing drinking water for animals, and water for swimming, bathing, etc.

(7) **Facilities of Communications.** Irrigation channels are generally provided with embankments and inspection roads. These inspection paths provide good roadways to the villagers for walking, cycling or sometimes even for motoring.

(8) **Inland Navigation.** Sometimes, larger irrigation canals can be used and developed for navigation purposes.

(9) **Afforestation.** Trees are generally grown along the banks of the channels, which increase the timber wealth of the country and also help in reducing soil erosion and air polluition.

## 1.4. Disadvantages and Ill-Effects of Irrigation

(1) Irrigation may contribute in various ways to the problem of water pollution. One of these is the seepage into the ground water of the nitrates, that have been applied to the soil as fertilizer. Sometimes, up to 50% of nitrates applied to the soil, sinks into the underground reservoir. The underground water may thus get polluted, and if consumed by people through wells, etc., it is likely to cause diseases such as anemia. Whether it will ultimately affect the fishing on way to the sea, or as the tides carry the polluted water ahead into the ocean, is yet a matter of research.

(2) Irrigation may result in colder and damper climate, resulting in marshy lands and breeding of mosquitoes, causing outbreak of diseases like malaria and dengu.

(3) Over-irrigation may lead to water-logging* and may reduce crop yields.

(4) Procuring and supplying irrigation water is complex and expensive in itself. Sometimes, subsidised cheaper water has to be provided at the cost of the government, which raduces revenue returns.

## 1.5. Types of Irrigation

Irrigation may broadly be classified into :

1. *Surface irrigation* ; and 2. *Sub-surface irrigation*

(1) **Surface irrigation** can be further classified into :

(*a*) Flow irrigation ; and (*b*) Lift irrigation.

When the water is available at a higher level, and it is supplied to lower level, by the mere action of gravity, then it is called *Flow Irrigation.* But, if the water is lifted up by some mechanical or manual means, such as by pumps, etc. and then supplied for irrigation, then it is called *Lift Irrigation.* Use of wells and tubewells for supplying irrigation water fall under this category of irrigation.

*Flow irrigation* can be further sub-divided into :

(*i*) *Perennial irrigation* ; and (*ii*) *Inundation* or *Flood irrigation.*

(*i*) ***Perennial Irrigation.*** In perennial system of irrigation, constant and continuous water supply is assured to the crops in accordance with the requirements of the crop, throughout the 'crop period'. In this system of irrigation, water is supplied through canal distribution system taking-off from above a weir or a reservoir.

When irrigation is done by diverting the river runoff into the main canal by constructing a diversion weir or a barrage across the river, then it is called as **direct irrigation.** Ganga Canal System is an example of this type of irrigation. But, if a dam is constructed across a river to store water during monsoons, so as to supply water in the off-taking channel during periods of low flow, then it is termed as **storage irrigation.** Ram-Ganga dam project in U.P. is an example of storage type of irrigation system. In the regions of peninsula India, where rivers are generally seasonal, storage irrigation is an absolute necessity, whereas, in Indo-

*For detailed description of water-logging, please refer Chapter 6.

gangetic region, direct irrigation is feasible, since the rivers are perennial, getting their supplies from the melting of snow. Direct irrigation is always simple, easy and economical. The perennial system of irrigation is most important and is mostly practised in India. This system of irrigation is also called *controlled irrigation*.

(*ii*) ***Inundation Irrigation.*** This type of irrigation is also called ***Flood Irrigation***. In this method of irrigation, soil is kept submerged and thoroughly flooded with water, so as to cause thorough saturation of the land. The moisture soaked by the soil, when occasionally supplemented by natural rainfall or minor waterings, brings the crop to maturity. It is usually practised on tracts near rivers and in alluvial and delataic regions where the river water level during hot weather season is sufficiently high to supply water to the land by flow, or partly by flow and partly by lift. This system of irrigation is also called *uncontrolled irrigation*.

Comparison between Perennial Irrigation and Inundation Irrigation is shown in Table 1.1 below :

**Table 1.1. Comparison between Perennial Irrigation and Inundation Irrigation.**

| *S. No.* | *Perennial irrigation* | *Inundation irrigation* |
|---|---|---|
| 1. | The irrigation water is available throughout the year | 1. The irrigation water is available in rainy season only |
| 2. | Hydraulic structures are necessary, (such as diversion head work, cross drainage works, etc.) | 2. No hydraulic structure is necessary |
| 3. | The canal water contain practically no silt and hence chemical manure is essential. | 3. The canal water contains plenty of silt which makes the land fertile. |
| 4. | Large area can be included under this system | 4. Large area cannot be included under this system |
| 5. | Negligible silting takes place in the canal bed. | 5. The silting of the canal bed is a major problem. |
| 6. | Water tax can be imposed. | 6. Water tax cannot be imposed. |
| 7. | Initial cost is high. | 7. Initial cost is low. |
| 8. | Technical persons are always required for the operation of the irrigation system. | 8. No technical persons are required for the operation of the irrigation system. |
| 9. | The main canal is provided with head regulator, and hence there is no possibility of over irrigation. | 9. The main canal is not provided with regulator and hence there is a possibility of over irrigation. |

(2) **Sub-surface Irrigation.** It is termed as sub-surface irrigation, because in this type of irrigation, water does not wet the soil surface. The underground water nourishes the plant roots by capillarity. It may be divided into the following two types :

(*a*) Natural sub-irrigation ; and (*b*) Artificial sub-irrigation.

(*a*) ***Natural sub-irrigation.*** Leakage water from channels, etc., goes underground, and during passage through the sub-soil, it may irrigate crops, sown on lower lands, by capillarity. Sometimes, leakage causes the water-table to rise up, which helps in irrigation of crops by capillarity. When underground irrigation is achieved, simply by natural processes, without any additional extra efforts, it is called natural sub-irrigation.

(*b*) ***Artificial sub-irrigation.*** When a system of open jointed drains is artificially laid below the soil, so as to supply water to the crops by capillarity, then it is known as artificial sub-irrigation. It is a very costly process and hence, adopted in India on a very small scale. It may be recommended only in some special cases with favourable soil conditions and for cash crops of very high return. Sometimes, irrigation water may be intentionally collected in some ditches near the fields, the percolation water may then come up to the roots through capillarity.

## 1.6. Techniques of Water Distribution in the Farms

There are various ways in which the irrigation water can be applied to the fields. Their main classification is as follows :

(1) Free flooding (2) Border flooding
(3) Check flooding (4) Basin flooding
(5) Furrow irrigation method (6) Sprinkler irrigation method
(7) Drip irrigation method.

These methods are briefly discussed below :

(1) **Free flooding or Ordinary flooding.** In this method, ditches are excavated in the field, and they may be either on the contour or up and down the slope. Water from these ditches, flows across the field. After the water leaves the ditches, no attempt is made to control the flow by means of levees, etc. Since the movement of water is not restricted, it is sometimes called **wild flooding.** Although the initial cost of land preparation is low, labour requirements are usually high and water application efficiency is also low. Wild flooding, is most suitable for close growing crops, pastures, etc., particularly where the land is steep. Contour ditches called laterals or subsidiary ditches, are generally spaced at about 20 to 50 metres apart, depending upon the slope, texture of soil, crops to be grown, etc. This method may be used on rolling land (topography irregular) where borders, checks, basins and furrows are not feasible.

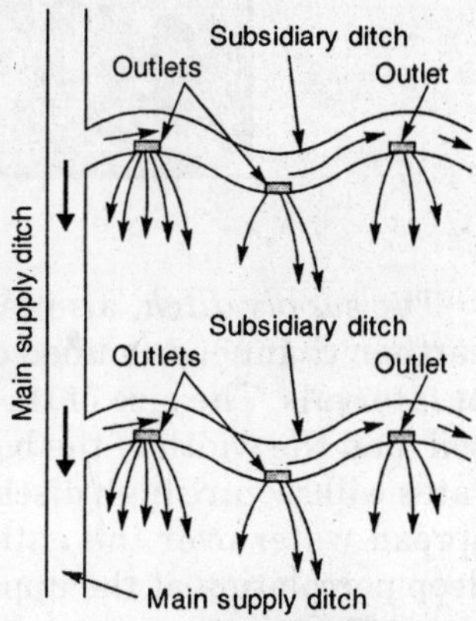

Fig. 1.1. Free flooding (plan view).

(2) **Border flooding.** In this method, the land is divided into a number of strips, separated by low levees called *borders*. The land areas confined in each strip is of the order of 10 to 20 metres in width, and 100 to 400 metres in length, as shown in Fig. 1.2. Ridges between borders should be sufficiently high to prevent overtopping during irrigation.

To prevent water from concentrating on either side of the border, the land should be levelled perpendicular to the flow. Water is made to flow from the supply ditch into each strip. The water flows slowly towards the lower end, and infiltrates into the soil as it advances. When the advancing water reaches the lower end of the strip, the supply of water to the strip is turned off.

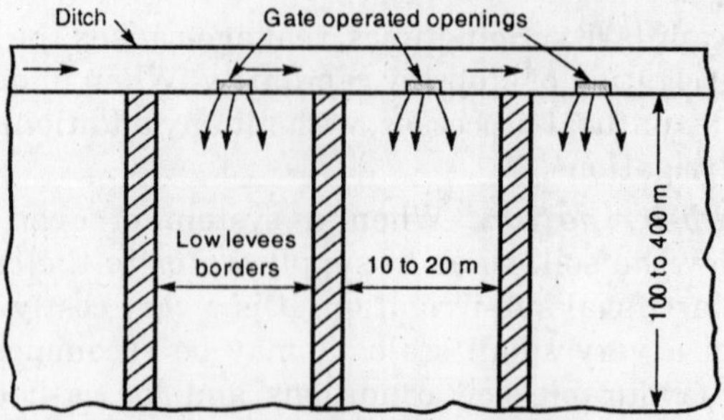


(*a*) Plan view.

(*b*) Photographic view.

Fig. 1.2. Border flooding.

The *supply ditch,* also called *irrigation stream,* may either be in the form of an earthen channel or a lined channel or an underground concrete pipe having risers at intervals. The size of the supply ditch depends upon the infiltration rate of the soil, and the width of the border strip. Coarse textured soils with high infiltration rates will require high discharge rate and therefore larger supply ditch, in order to spread water over the entire strip *rapidly,* and to avoid excessive losses due to deep percolation at the upper reaches. On the other hand, fine textured soils with low infiltration rates, require smaller ditches to avoid excessive losses due to run off at the lower reaches.

A relationship between the discharge through the supply ditch ($Q$), the average depth of water flowing over the strip ($y$), the rate of infiltration of the soil ($f$), the area of the land irrigated ($A$), and the approximate time required to cover the given area with water ($t$), is given by the equation :

$$t = 2.3 \cdot \frac{y}{f} \cdot \log_{10}\left(\frac{Q}{Q - fA}\right) \qquad \ldots(1.1)$$

where $Q$ = Discharge through the supply ditch

$y$ = Depth of water flowing over the border strip

$f$ = Rate of infiltration of soil

$A$ = Area of land strip to be irrigated

$t$ = Time required to cover the given area $A$.

The above equation can be obtained by considering small area ($dA$) of the border strip of area $A$, as shown in Fig. 1.3. Let us also assume that in time $dt$, water advances over this area $dA$. Now, the volume of water that flows to cover this area would be $ydA$, because $y$ is the water depth over this area. Also, during the same time $dt$, the volume of water that percolates into the soil over the area $A$ would be $f . A . dt$. The total quantity of water supplied to the strip during time $dt$ would be $Q . dt$, and also equal to $y . dA + f . A . dt$.

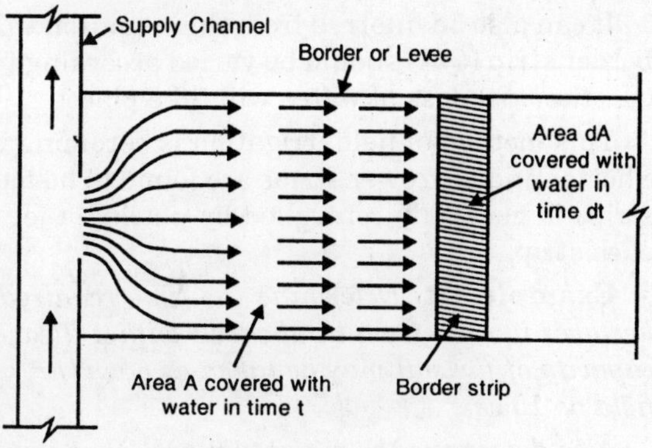


Fig. 1.3.

$$\therefore \quad Q . dt = y . dA + f . A . dt \qquad ...(i)$$

$$\text{or} \quad dt = \left(\frac{y . dA}{Q - f . A}\right) \qquad ...(ii)$$

Considering $y$, $f$, and $Q$ as constants, and integrating the above equation, we get

$$t = \frac{y}{f} \log_e \left(\frac{Q}{Q - f . A}\right) + \text{a constant of integration } (K)$$

But at $t = 0$, $A = 0$

$$\therefore \quad K = \frac{y}{f} \log_e 1 = 0 \qquad (\because \log_e 1 = 0)$$

$$\therefore \quad t = \frac{y}{f} \log_e \left(\frac{Q}{Q - fA}\right)$$

$$\text{or} \quad t = 2.3 . \frac{y}{f} \log_{10} \left(\frac{Q}{Q - fA}\right),$$

which is the above given equation (1.1).

This equation can be further written as

$$\frac{t . f}{2.3y} = \log_{10} \left(\frac{Q}{Q - fA}\right). \quad \text{Now, let } \frac{t . f}{2.3y} = x$$

$$\text{Then,} \quad x = \log_{10} \left(\frac{Q}{Q - fA}\right) \quad \text{or} \quad 10^x = \left(\frac{Q}{Q - fA}\right)$$

$$\text{or} \quad Q.10^x - f.A.10^x = Q \quad \text{or} \quad Q(10^x - 1) = f . A . 10^x$$

$$\text{or} \quad A = \frac{Q(10^x - 1)}{f . 10^x}$$

Further, considering the maximum value of $\dfrac{10^x - 1}{10^x} \approx 1$, we get $$A_{max} = \frac{Q}{f} \qquad ...(1.2)$$

This equation enables us to determine the maximum area that can be irrigated with a supply ditch of discharge $Q$ and soil having infiltration capacity, $f$.

It can also be inferred from this equation that the discharge per unit area of the border strip ($Q/A$) should be varied according to the infiltration capacity of the soil ($f$), otherwise loss of water will take place.

This method of field irrigation is becoming very popular among the cultivators. Shorter and narrower strips are found to be more efficient. Entry of water into the strips is generally controlled by placing a jet in the supply ditch, at the head of each strip.

**Example 1.1.** *Determine the time required to irrigate a strip of land of 0.04 hectares in area from a tube-well with a discharge of 0.02 cumec. The infiltration capacity of the soil may be taken as 5 cm/hr, and the average depth of flow on the field as 10 cm.*

*Also determine the maximum area that can be irrigated from this tube-well.*

**Solution.** $A = 0.04$ hectares $= 0.04 \times 10^4$ m$^2$ $= 400$ m$^2$.

$$Q = 0.02 \text{ cumecs} = 0.02 \text{ m}^3/\text{sec} = 0.02 \times 60 \times 60 \text{ m}^3/\text{hr}$$
$$= 72 \text{ m}^3/\text{hr}.$$

$$f = 5 \text{ cm/hr} = \frac{5}{100} \text{ m/hr} = 0.05 \text{ m/hr.} ; y = 10 \text{ cm} = 0.1 \text{ m.}$$

Using eqn. (1.1), we have

$$t = 2.3 \frac{y}{f} \log_{10}\left(\frac{Q}{Q - fA}\right)$$

or $$t = 2.3 \frac{0.10}{0.05} \log_{10}\left(\frac{72}{72 - 0.05 \times 400}\right) \text{hr}$$

$$= 2.3 \times 2 \log_{10} (72/52) \text{ hr} = 0.65 \text{ hr} = 39 \text{ minutes.}$$ **Ans.**

Maximum area that can be irrigated is given by eqn. (1.2) as

$$A_{max} = \frac{Q}{f} = \frac{72}{0.05} \text{ m}^2 = 1440 \text{ m}^2.$$

$$= \frac{1440}{10^4} \text{ hectares} = 0.144 \text{ hectares.}$$ **Ans.**

**Note.** After irrigating this much of area, surface flow will stop, and deep percolation will start.

(3) **Check flooding.** Check flooding is similar to ordinary flooding except that the water is controlled by surrounding the *check* area with low and flat levees. Levees are generally constructed along the contours, having vertical interval of about 5 to 10 cm. These levees are connected with cross-levees at convenient places as shown in Fig. 1.4. The confined plot area varies from 0.2 to 0.8 hectare.

In check flooding, the *check* is filled with water at a fairly high rate and allowed to stand until the water infiltrates.

This method is suitable for more permeable soils as well as for less permeable soils. The water can be quickly spread in case of high permeable soils, thus reducing the percolation losses. The water can also be held on the surface for a longer time in case of less permeable soils, for assuring adequate penetration.

These checks, are sometimes used to absorb water, where the stream-flow is diverted during periods of high run off.

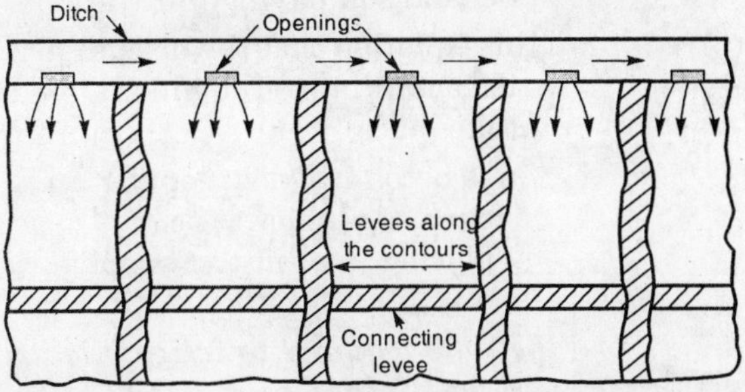


(*a*) Plan view.

(*b*) Photographic view.

Fig. 1.4. Check flooding.

(4) **Basin flooding.** This method is a special type of check flooding and is adopted specially for orchard trees. One or more trees are generally placed in the basin, and the surface is flooded as in check method, by ditch water, as shown in Fig. 1.5.

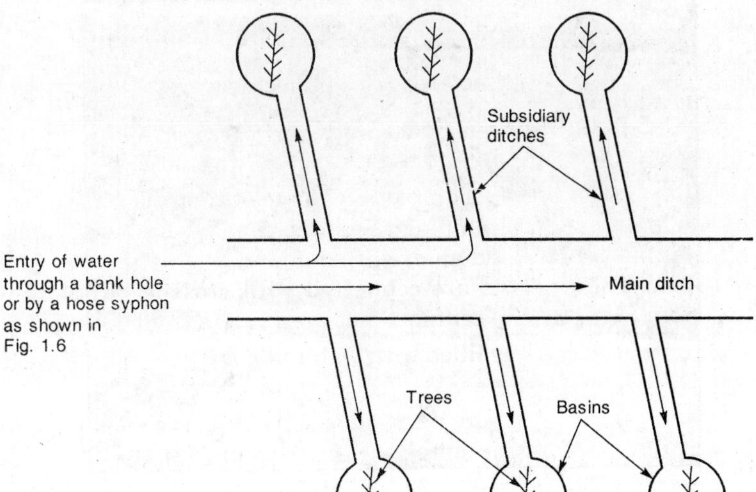


(*a*) Plan view

(*b*) Photographic view.
Fig. 1.5. Basin flooding.

(5) **Furrow irrigation method.** In flooding methods, described above, water covers the entire surface; while in furrow irrigation method (Fig. 1.6), only one-

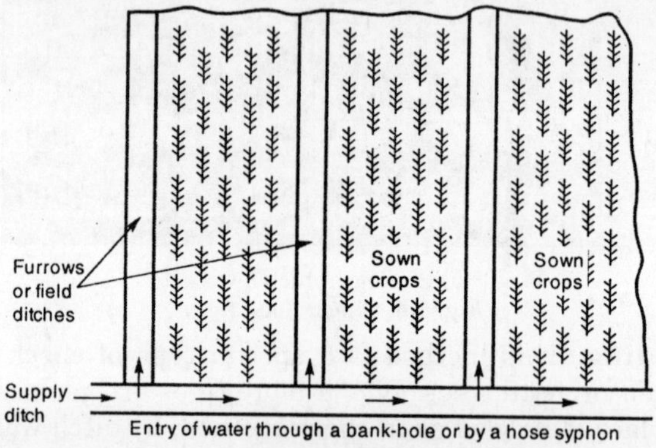


(*a*) Plan view.

(*b*) Photographic view (using plastic syphons).
Fig. 1.6. Furrow Irrigation.

fifth to one-half of the land surface is wetted by water. It therefore, results in less evaporation, less pudding of soil, and permits cultivation sooner after irrigation.

Furrows are narrow field ditches, excavated between rows of plants and carry irrigation water through them. Spacing of furrows is determined by the proper spacing of the plants. Furrows vary from 8 to 30 cm deep, and may be as much as 400 metres long.

Excessive long furrows may result in too much percolation near the upper end, and too little water near the down-slope end. Deep furrows are widely used for row crops. Small shallow furrows called *corrugations*, are particularly suitable for relatively irregular topography and close growing crops, such as meadows and small grains.

Water may be diverted into the furrows by an opening in the bank of the supply ditch or preferably by using a rubber hose tubing, which can be primed by immersion in the ditch. The use of hose, prevents the necessity of breaking the ditch bank, and provides a uniform flow into the furrow. Fig. 1.7 shows such different devices which are used for distributing water into the fields from the supply ditch.

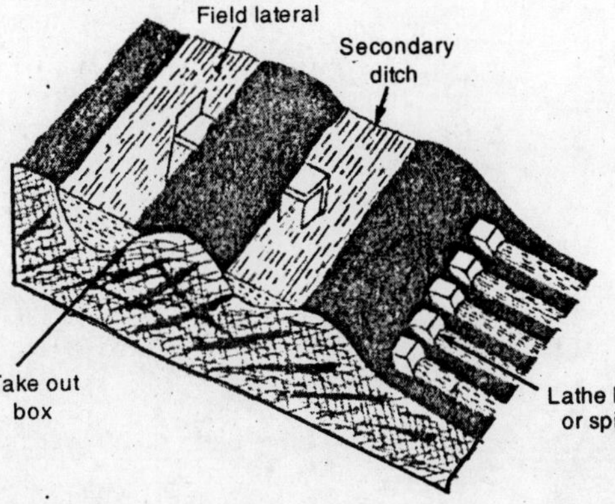


(*a*) Spiles or Lathe Box

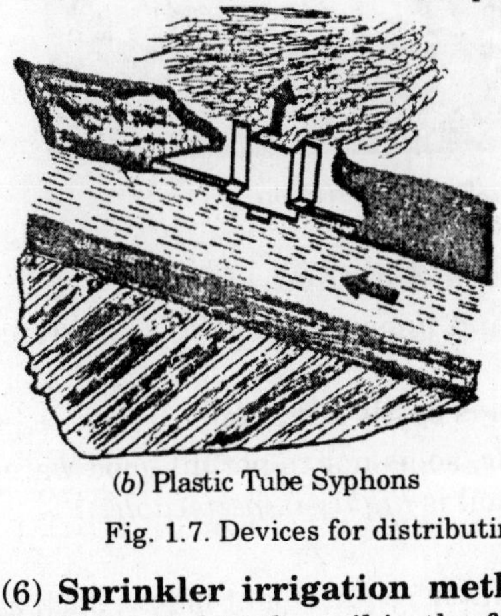

(*b*) Plastic Tube Syphons

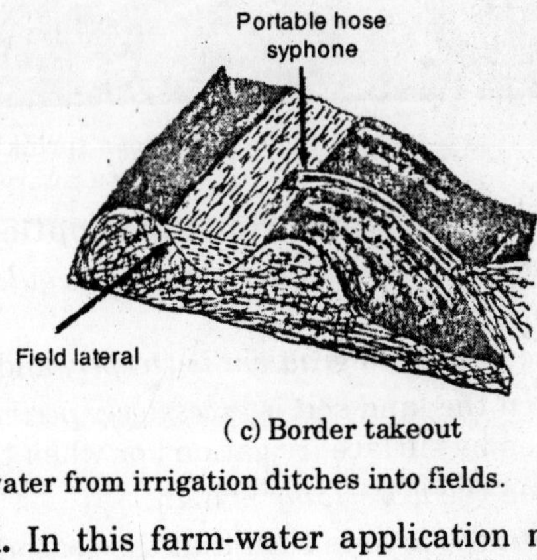


( c) Border takeout

Fig. 1.7. Devices for distributing water from irrigation ditches into fields.

(6) **Sprinkler irrigation method.** In this farm-water application method, water is applied to the soil in the form of a spray through a network of pipes and pumps. It is a kind of an artificial rain and, therefore, gives very good results. It is a costly process and widely used in U.S.A. It can be used for all types of soils and

for widely different topographies and slopes. It can advantageously be used for many crops, because it fulfils the normal requirement of uniform distribution of water. This method possesses great potentialities for irrigating areas in Rajasthan in India, where other types of surface or sub-surface irrigation are very difficult.

Inspite of the numerous advantages which this method possesses over the other methods, it has not become popular in India for the simple reason that ours is a poor and a developing nation. This method is not only costly but requires a lot of technicalities. The correct design and efficient operation are very important for the success of this method. Special steps have to be taken for preventing entry of silt and debris, which are very harmful for the sprinkler equipment. Debris-choke nozzles, interfere with the application of water on the land : while the abrasive action of silt causes excessive wear on pump impellers, sprinkler nozzles, and bearings. The system is to be designed in such a way that the entire sprayed water seeps into the soil, and there is no run off from the irrigated area.

Fig. 1.8. Photoview of a sprinkler system in operation (using a rotary head).

The **conditions favouring the adoption of this method,** are :

(*i*) When the land *topography is irregular,* and hence unsuitable for surface irrigation.

(*ii*) When the land *gradient is steeper,* and soil is easily *erodible.*

(*iii*) When the land soil is *excessively permeable,* so as not to permit good water distribution by surface irrigation ; or when the soil is *highly impermeable.*

(*iv*) When the *watertable* is high.

(*v*) When the area is such that the seasonal water requirement is low, such as *near the coasts.*

(*vi*) When the *crops to be grown are such :*

(*a*) as to require humidity control, as in tobacco ;

(*b*) crops having shallow roots ; or

(*c*) crops requiring high and frequent irrigation.

(*vii*) When the water is available with difficulty and is scarce.

***Types of sprinkler systems.*** A sprinkler system can be classified under three heads, as :

1. Permanent system ; 2. Semi-permanent system ; and
3. Portable system.

Earlier, the *fixed over-head perforated pipe installations* were being used for sprinkler network; but with the advent of light weight steel pipes and quick couplers, portable systems have been designed.

In *permanent system,* pipes are permanently buried in such a way that they do not interfere with the farming operations. In the *semipermanent system,* the main lines are buried in the ground, while the laterals are portable. In the *portable* system, the mains as well as laterals are portable. These portable networks can be moved from farm to farm.

In the sprinkler irrigation network, we have the mains and the sub-mains, through which water under pressure is made to flow. *Revolving sprinkler heads* are then usually mounted on *rising pipes* attached to the laterals. The water jet comes out through the revolving sprinkler heads, with force. When sprinkler heads are not provided, perforations are made in the pipes, and they are provided with nozzles, through which water jets out and falls on the ground. Generally, such a perforated pipe system operates at low heads ; whereas, the revolving head sprinklers operate on high as well as low heads, depending upon the type of rotary head used.

The **advantages** of sprinkler irrigation are enumerated below :

(*i*) Seepage losses, which occur in earthen channels of surface irrigation methods, are completely eliminated. Moreover, only *optimum quantity of water* is used in this method.

(*ii*) Land levelling is not required, and thus avoiding removal of top fertile soil, as happens in other surface irrigation methods.

(*iii*) No cultivation area is lost for making ditches, as happens in surface irrigation methods. It, thus, results in increasing about 16% of the cropped area.

(*iv*) In sprinkler system, the water is to be applied at a rate lesser than the infiltration capacity of the soil, and thus avoiding surface run off, and its bad effects, such as loss of water, washing of top soil, etc.

(*v*) Fertilizers can be uniformly applied, because they are mixed with irrigation water itself.

(*vi*) This method leaches down salts and prevents water-logging or salinity.

(*vii*) It is less labour oriented, and hence useful where labour is costly and scarce.

(*viii*) Upto 80% efficiency can be achieved, *i.e.* upto 80% of applied water can be stored in the root zone of plants.

The **limitations** of sprinkler irrigation are also enumerated below :

(*i*) High winds may distort sprinkler pattern, causing non-uniform spreading of water on the crops.

(*ii*) In areas of high temperature and high wind velocity, considerable evaporation losses of water may take place.

(*iii*) They are not suited to crops requiring frequent and larger depths of irrigation, such as paddy.

(*iv*) Initial cost of the system is very high, and the system requires a high technical skill.

(*v*) Only sand and silt free water can be used, as otherwise pump impellers lifting such waters will get damaged.

(*vi*) It requires larger electrical power.

(*vii*) Heavy soil with poor intake cannot be irrigated efficiently.

(*viii*) A constant water supply is needed for commercial use of equipment.

**Note.** The widely known Indian Commercial Company which specialises in installing sprinkler irrigation systems, is "Premier Irrigation Equipments (Pvt. Ltd.), Calcutta," with its Delhi office also. This firm and its publications can be referred to, for specialised knowledge in this branch of field irrigation.

(7) **Drip Irrigation Method.** Drip irrigation, also called **trickle irrigation,** is the latest field irrigation technique, and is meant for adoption at places where there exists acute scarcity of irrigation water and other salt problems. In this method, water is *slowly* and directly applied to the root zone of the plants, thereby minimising the losses by evaporation and percolation.

This system involves laying of a system of *head, mains, sub-mains, laterals,* and *drop nozzles.* Water oozes out of these small drip nozzles uniformly and at a very small rate, directly into the plant roots area.

The head consists of a pump to lift water, so as to produce the desired pressure of about 2.5 atmosphere, for ensuring proper flow of water through the system. The lifted irrigation water is passed through a fertilizer tank, so as to mix the fertilizer directly in the irrigation water, and then through a filter, so as to remove the suspended particles from the water, to avoid clogging of drip nozzles.

The *mains and sub-mains* are the specially designed small sized pipes, made of flexible material like black PVC. These are generally buried or laid on the ground, as shown in Fig. 1.9. Their sizes should be sufficient to carry the design discharge of the system.

The *laterals* are very small sized (usually 1 to 1.25 cm dia.), specially designed, black PVC pipes, taking off from the mains or sub-mains, Laterals can usually be

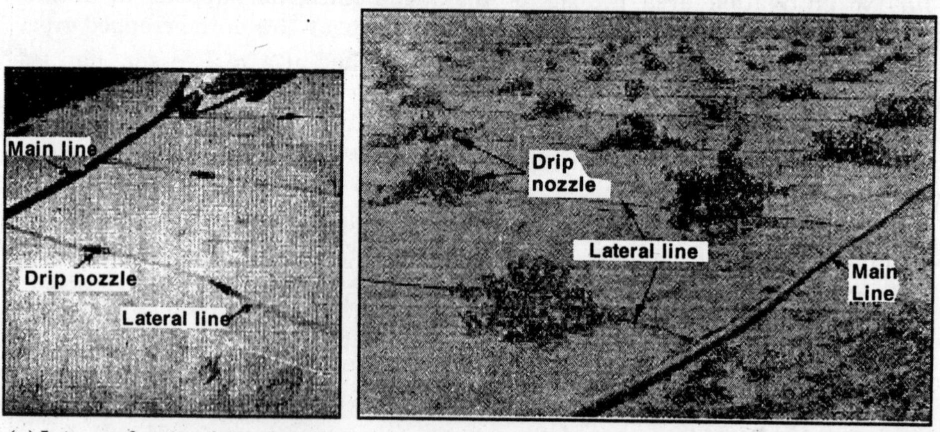


(*a*) Layout of mains, laterals and drip nozzles.

(*b*) Field irrigated by drip system laid on the ground.

Fig. 1.9. Two photoviews of Drip irrigation system.

up to 50 m long, and one lateral line is laid for each row of crop. *Hardie Biwall* is a patented name of a special dual chambered micro tubing, manufactured from a linear low density polyethylene, and is being used these days for laterals.

The *drip nozzles,* also called *emitters,* or *valves,* are fixed on laterals, at regular intervals of about 0.5 to 1 m or so, discharging water at very small rates of the order of 2 to 10 litres per hour.

Like the sprinkler system, this method also involves specialised knowledge, and is not being adopted by our ordinary farmers. This method, is however, being used for small nurseries, orchards, or gardens. The widely known Commercial Indian Company, which specialises in this field irrigation method, is known as "Jain Irrigation Systems Ltd., Jalgaon" (Post Box No. 20), Pin Code : 425001. This firm can be contacted in special needs for layout of such an irrigation system.

## 1.7. Quality of Irrigation Water

Just as every water is not suitable for human beings, in the same way, every water is not suitable for plant life. Water containing impurities, which is injurious to plant growth and is not satisfactory for irrigation is called the unsatisfactory water.

The quality of suitable irrigation water is very much influenced by the constituents of the soil which is to be irrigated. A particular water may be harmful for irrigation on a particular soil, but the same water may be tolerable or even useful for irrigation on some other soil. The various types of impurities, which make the water unfit for irrigation, are classified as :

(*i*) Sediment concentration in water

(*ii*) Total concentration of soluble salts in water

(*iii*) Proportion of sodium ions to other cations

(*iv*) Concentration of potentially toxic elements present in water

(*v*) Bicarbonate concentration as related to the concentration of calcium plus magnesium

(*vi*) Bacterial contamination.

The effects of these impurities are discussed below :

(1) **Sediment.** The effect of sediment present in the irrigation water depends upon the type of irrigated land. When fine sediment from water is deposited on sandy soils, the fertility is improved. On the other hand, if the sediment has been derived from the eroded areas, it may reduce the fertility or decrease the soil permeability. Sedimented water creates troubles in irrigation canals, as it increases their siltation and maintenance costs. In general, ground water or surface water from reservoirs, etc. does not have sufficient sediment to cause any serious problems in irrigation.

(2) **Total concentration of soluble salts.** Salts of calcium, magnesium sodium and potassium, present in the irrigation water may prove injurious to plants. When present in excessive quantities, they reduce the osmotic activities of the plants, and may prevent adequate aeration, causing injuries to plant growth. The injurious effects of salts on the plant growth depend upon the concentration of salts left in the soil.

The concentration of salts in water, may not appear to be harmful to the plants, but the concentration of salts which remain in the soil after the saline water is used up by the plants is much more than the first, and may prove to be harmful. In other words, at the beginning of irrigation with undesirable water, no harm may be visible, but with the passage of time, the salt concentration in the soil may increase to a harmful level, as the soil solution gets concentrated by evaporation.

Hence, the effects of salts on plant growth depend largely upon the total amount of salts present in the soil solution. The salinity concentration of the soil solution ($C_s$) after the consumptive water ($C_u$) has been extracted from the soil, is given by

$$C_s = \frac{C \cdot Q}{[Q - (C_u - R_e)]} \quad \text{...(1.3)}$$

where $Q$ = The quantity of water applied

$C_u$ = Consumptive use of water, *i.e.* the total amount of water used by the plant for its growth

$R_e$ = Useful or Effective rainfall

$C_u - R_e$ = Used up irrigation water

$C$ = Concentration of salt in irrigation water

$CQ$ = Total salt applied to soil with $Q$ amount of irrigation water.

The salt concentration is generally expressed by ppm (parts per million) or by mg/l (milligram per litre), both units being equal. The critical salt concentration in the irrigation water depends upon many factors, yet however, ***amounts in excess of 700 ppm are harmful to some plants, and more than 2000 ppm are injurious to all crops.***

The salt concentration is generally measured by determining the electrical conductivity of water. They are directly proportional to each other. Electrical conductivity is expressed in micro mhos per centimetre. When its value is up to 250 micro mhos/cm at 25°C, it is called low conductivity water (C1) ; when its value is between 250 to 750, it is called *Medium Conductivity water* (C2) ; when its

**Table 1.2**

| *S. No.* | *Type of water* | *Use in irrigation* |
|---|---|---|
| 1. | Low salinity water (C1). Conductivity between 100 to 250 micro mhos/cm at 25°. | Can be used for irrigation for almost all crops and for almost all kinds of soils. Very little salinity may develop, which may require slight leaching; but it is permissible under normal irrigation practices except in soils of extremely low permeabilities. |
| 2. | Medium salinity water (C2). Conductivity between 250 to 750 micro mhos/cm at 25°C. | Can be used, if a moderate amount of leaching occurs. Normal salt-tolerant plants can be grown without much salinity control. |
| 3. | High salinity water (C3). Conductivity between 750 to 2250 micro mhos/cm at 25°C. | Cannot be used on soils with restricted drainage. Special precautions and measures are undertaken for salinity control and only high-salt tolerant plants can be grown. |
| 4. | Very high salinity water (C4). Conductivity more than 2250 micro mhos/cm at 25°C. | Generally not suitable for irrigation. |

value is between 750 to 2250, it is called *High Conductivity water* (C3) ; and the values above 2250, are classified as very *High Conductivity water* (C4). The suitabilities of these four types of waters for irrigation supplies are discussed in Table 1.2. This classification which is shown in Table 1.2 was given by U.S.D.A. Handbook No. 60 (1954).

(3) **Proportion of sodium ions to other cations.** Most of the soils contain calcium and magnesium ions and small quantities of sodium ions. *The percentage of the sodium ions is generally less than 5% of the total exchangeable cations.* If this percentage increases to about 10% or more, the aggregation of soil grains breaks down. The soil becomes less permeable and of poorer tilth. It starts crusting when dry and its pH increases towards that of an alkaline soil. High sodium soils are, therefore, plastic, sticky when wet, and are prone to form clods, and they crust on drying.

The proportion of sodium ions present in the soils, is generally measured by a factor called ***Sodium-Absorption Ratio (SAR)*** and represents the sodium hazards of water. SAR is defined as :

$$SAR = \frac{Na^{+}}{\sqrt{\dfrac{Ca^{++} + Mg^{++}}{2}}} \quad ...(1.4)$$

where, the concentration of the ions is expressed in equivalent per million (epm) ; epm is obtained by dividing the concentration of salt in mg/1 or ppm by its combining weight (*i.e.* Atomic wt. ÷ Valence.)

When the value of SAR lies between 0 to 10, it is called *Low Sodium Water* (S1); when between 10 to 18, it is called *Medium Sodium Water* (S2) ; between 18 to 26;

**Table 1.3**

| *S. No.* | *Type of water* | *Use in irrigation* |
|---|---|---|
| 1. | Low sodium water (S1). SAR value lying between 0 to 10. | Can be used for irrigation on almost all soils and for almost all crops except those which are highly sensitive to sodium, such as stone-fruit trees and adocados, etc. |
| 2. | Medium sodium water (S2). SAR value lying between 10 to 18. | Appreciably hazardous in fine textured soils, which may require gypsum, etc. ; but may be used on course-textured or organic soils with good permeability. |
| 3. | High sodium water (S3). SAR value lying between 18 to 26. | May prove harmful on almost all the soils,, and do require good drainage, high leaching, gypsum addition etc. for proper irrigation. |
| 4. | Very high sodium water (S4). SAR value above 26. | Generally, not suitable for irrigation. |

it is called *High Sodium Water* (S3) ; and when value of SAR is more than 26,it is called *Very High Sodium Water* (S4). The suitabilities of these four kinds of water for irrigation are discussed in Table 1.3. The SAR value can be reduced by adding gypsum ($CaSO_4$) to the water or to the soil.

Depending upon the *Electrical conductivity EC* (representing salt content) of water, the *exchangeable sodium percentage ESP* (representing percentage of sodium w.r.to total exchangeable cations), and the pH value of the soil, the soils are classified as *saline, alkaline,* or *saline-alkali,* as shown in table 1.4 below :

**Table 1.4. The Classification of Saline and Alkaline Soils**

| *Sl. No.* | *Classification* | *Electrical Conductivity (EC) in micro-mho / cm* | *Exchangeable Sodium Percentage (ESP)* | *pH value* |
|---|---|---|---|---|
| 1. | **Saline soil** or *white alkali* | > 4000 | < 15 | ≤ 8.5 |
| 2. | **Alkaline soil** or *Non-saline alkali* or *Sodic soil* or *Black alkali,* | <4000 | >15 | 8.5 to 10.0 |
| 3. | *Saline-alkali soil,* | > 4000 | > 15 | < 8.5 |

(4) **Concentration of potentially toxic elements.** A large number of elements such as boron, selenium, etc. may be toxic to plants. Traces of *Boron* are essential to plant growth, but its concentrations above 0.3 ppm may prove toxic to certain plants. The concentration above 0.5 ppm is dangerous to nuts, citrus fruits and *deciduous* fruits. Cotton, Cereals and certain truck crops are moderately tolerant to boron, while Dates, Beets, Asparagus etc. are quite tolerant. Even for the most tolerant crops, the boron concentration should not exceed 4 ppm. Boron is generally present in various soaps. The waste water containing soap, etc. should, therefore, be used with great care in irrigation.

Selenium, even in low concentration, is toxic, and must be avoided.

(5) **Bicarbonate concentration as related to concentration of calcium plus magneisum.** High concentration of bi-carbonate ions may result in precipitation of calcium and magnesium bicarbonates from the soil-solution, increasing the relative proportion of sodium ions and causing sodium hazards.

(6) **Bacterial contamination.** Bacterial contamination of irrigation water is not a serious problem, unless the crops irrigated with highly contaminated water are directly eaten, without being cooked. Cash crops like cotton, nursery stock, etc. which are processed after harvesting, can, therefore, use contaminated waste waters, without any trouble.

**Example 1.2.** *(a) What is the classification of irrigation water having the following characteristics : Concentration of Na, Ca and Mg are 22, 3 and 1.5 milli-equivalents per litre respectively, and the electrical conductivity is 200 micro mhos per cm at 25°C ? (b) What problems might arise in using this water on fine textured soils ? (c) What remedies do you suggest to overcome this trouble ?*

**Solution.** (*a*)

$$\text{SAR} = \frac{\text{Na}^+}{\sqrt{\dfrac{\text{Ca}^{++} + \text{Mg}^{++}}{2}}} = \frac{22}{\sqrt{\dfrac{3 + 1.5}{2}}}$$

$$= \frac{22}{\sqrt{2.25}} = \frac{22}{1.5} = 14.67.$$

If SAR is between 10 to 18, then it is classified as medium Sodium water and is represented by S2 (See Table 1.4).

Electrical conductivity is 200 micro-mhos per cm at 25°C.

According to Table 1.2, the water is called of Low conductivity (C1) if the value of electrical conductivity is between 100 to 250 micro mhos per cm at 25°C. It is, therefore, C1 water.

Hence, the given water is classified as C1-S2 *water*. **Ans.**

(*b*) In fine-textured soils, the medium sodium (S2) water may create the following problems :

(*i*) Soil becomes less permeable.

(*ii*) It starts crusting when dry.

(*iii*) It becomes plastic and sticky when wet.

(*iv*) Its pH increases towards that of alkaline soil.

(*c*) Gypsum ($CaSO_4$) addition, either to soil or to water is suggested to overcome sodium hazards posed by the given water.

## PROBLEMS

1. Define irrigation and explain its necessity in a tropical country like India. What are the advantages and ill-effects of assured irrigation ?
2. What is meant by surface and sub-surface irrigation ; and what are their types ? Discuss briefly the various techniques used for distributing water in the farms.
3. What is meant by 'Furrow Irrigation" and 'Sprinkler Irrigation' ? Which one is preferred in India and why ?
4. 'The sprinkler system of irrigation is an excellent method but not used in India'. Discuss critically and briefly.
5. What is meant by 'Border flooding', and how does it differ from 'Check flooding' and 'Free flooding'?
6. Describe briefly the necessity and importance of irrigation works in our country? What are different types of irrigation ? Write brief notes on each of them.
7. (*a*) What are the benefits that can be accrued from Irrigation projects ?

   (*b*) What is 'flood irrigation' ? Where is it practised ?

   (*c*) How is the Flow irrigation different from the Lift irrigation ?
8. Discuss critically the quality standards required for irrigation water.

9. 'All the waters are not fit for irrigating crops'. Discuss briefly and critically the above statement.

10. What is meant by C2—S2 water ? Discuss its usefulness for irrigating fine textured soils.

11. Write short notes on :

(i) Lift irrigation.

(ii) Mixed cropping.

(iii) Ill-effects of irrigation.

(iv) Border strip and Sprinkler methods of irrigating fields.

(v) Sodium-Absorption-Ratio (SAR).

(vi) Salt concentration of irrigation waters and their utility in irrigation.

(vii) Sodium hazards of irrigation waters.

(viii) Boron concentration in irrigation waters.

(ix) Drip irrigation method.

# 2

# Water Requirements of Crops

## 2.1. General

Every crop requires a certain quantity of water after a certain fixed interval, throughout its period of growth. If the natural rain is sufficient and timely so as to satisfy both these requirements, no irrigation water is required for raising that crop. In England, for example, the natural rain falling regularly throughout the year, satisfies both these requirements for practically all the crops, and, therefore, irrigation is not significantly needed in England. But in a tropical country like India, the natural rainfall is either insufficient, or the water does not fall regularly, as required by the crops. Since the magnitude as well as the frequency of the rainfall varies throughout a tropical country, certain crop may require irrigation in certain part of the country, and the same crop may not require any irrigation in some other part of the country. The area where irrigation is a must for agriculture is called the **arid region,** while the area in which inferior crops can be grown without irrigation is called a **semi-arid region.**

The term 'Water requirements of a crop' means *the total quantity and the way in which a crop requires water, from the time it is sown to the time it is harvested.* It is very clear from the above discussion, that the water requirement, will vary with the crop as well as with the place. In other words, *different crops will have different water requirements, and the same crop may have different water requirements at different places of the same country ;* depending upon the variations in climates, type of soils, methods of cultivation, and useful rainfalls, etc.

## 2.2. Crop Period or Base Period

The time period that elapses from the instant of its sowing to the instant of its harvesting is called the *crop-period.* The time between the first watering of a crop at the time of its sowing to its last watering before harvesting is called the *Base period* or the *Base of the crop.* Crop period is slightly more than the base period, but for all practical purposes, they are taken as one and the same thing, and generally expressed in days. Hence, in future, the terms like *growth period, crop period, base period,* etc., will be used as synonyms, each representing crop period, and will be represented by $B$ (in days).

## 2.3. Duty and Delta of a Crop

**2.3.1. Delta.** Each crop requires a certain amount of water after a certain fixed interval of time, throughout its period of growth. The depth of water required every time, generally varies from 5 to 10 cm depending upon the type of the crop, climate and soil. The time interval between two such consecutive waterings is called the **frequency of irrigation,** or **rotation period**. The rotation period may vary between 6-15 days for different crops. The summation of the total water depth supplied during the base period of a crop, for its full growth, will evidently represent the total quantity of water required by the crop for its full-fledged nourishment. *This total quantity of water required by the crop for its full growth* (maturity) may be expressed in hectare-metre (Acre-ft) or in million cubic metres (million cubic-ft) or simply as depth to which water would stand on the irrigated area, if the total quantity supplied were to stand above the surface without percolation or evaporation. *This total depth of water (in cm) required by a crop to come to maturity is called its delta* (Δ).

**Example 2.1.** *If rice requires about 10 cm depth of water at an average interval of about 10 days, and the crop period for rice is 120 days, find out the delta for rice.*

**Solution.** Water is required at an interval of 10 days for a period of 120 days. It evidently means that 12 no. of waterings are required, and each time, 10 cm depth of water is required. Therefore, total depth of water required

$$\Delta = 12 \times 10 \text{ cm} = 120 \text{ cm}.$$

Hence Δ for rice = 120 cm. **Ans.**

**Example 2.2.** *If wheat requires about 7.5 cm of water after every 28 days, and the base period for wheat is 140 days, find out the value of delta for wheat.*

**Solution.** Assuming the base period to be representing the crop period, as per usual practice, we can easily infer that the water is required at an average interval of 28 days up to a total period of 140 days. This means that $\frac{140}{28} = 5$ no. of waterings are required.

The depth of water required each time = 7.5 cm.

∴ Total depth of water reqd. in 140 days = 5 × 7.5 cm = 37.5 cm

Hence, Δ for wheat = 37.5 cm. **Ans.**

**2.3.2. Delta for certain crops.** The average values of deltas for certain crops are shown in Table 2.1. These values represent the total water requirement of the crops. The actual requirement of irrigation water may be less, depending upon the useful rainfall. Moreover, these values represent the values on field, *i.e.* 'delta on field' which includes the evaporation and percolation losses.

**Table 2.1. Average Approximate Values of Δ for Certain Important Crops in India**

| *S. No.* | *Crop* | *Delta on field* |
|---|---|---|
| *(1)* | *(2)* | *(3)* |
| 1. | Sugarcane | 120 cm (48″) |
| 2. | Rice | 120 cm (48″) |
| 3. | Tobacco | 75 cm (30″) |
| 4. | Garden fruits | 60 cm (24″) |
| 5. | Cotton | 50 cm (22″) |
| 6. | Vegetables | 45 cm (18″) |
| 7. | Wheat | 40 cm (16″) |
| 8. | Barley | 30 cm (12″) |
| 9. | Maize | 25 cm (10″) |
| 10. | Fodder | 22.5 cm (9″) |
| 11. | Peas | 15 cm (6″) |

**2.3.3. Duty of Water.** The 'duty' of water is the relationship between the volume of water and the area of the crop it matures. It may be defined as the number of hectares of land irrigated for full growth of a given crop by supply of 1 $m^3$/sec of water continuously during the entire base period (B) of that crop. Thus, if water flowing at a rate of one cubic metre per second, runs continuously for $B$ days, and matures 200 hectares, then the duty of water for that particular crop will be defined as 200 hectares per cumec to the base of $B$ days. The duty is generally represented by the letter $D$.

**2.3.4. Relation between duty and delta.** Let there be a crop of base period $B$ days. Let one cumec of water be applied to this crop on the field for $B$ days.

Now, the volume of water applied to this crop during $B$ days

$$= V = (1 \times 60 \times 60 \times 24 \times B)\ \text{m}^3$$
$$= \mathbf{86\,400}\ \boldsymbol{B}\ \text{(cubic metre)}$$

By definition of duty ($D$), one cubic metre supplied for $B$ days matures $D$ hectares of land.

$\therefore$ This quantity of water ($V$) matures $D$ hectares of land or $10^4\ D$ sq. m of area.

Total depth of water applied on this land

$$= \frac{\text{Volume}}{\text{Area}} = \frac{86{,}400\ B}{10^4\ D} = \frac{8.64\ B}{D}\ \text{metres}$$

By definition, this total depth of water is called delta ($\Delta$).

$$\therefore \qquad \Delta = \frac{8.64\ B}{D}\ \text{metres} \qquad \text{...(2.1)}$$

$$\text{or} \qquad \Delta = \frac{864\ B}{D}\ \text{cm.} \qquad \text{...(2.2)}$$

where, $\Delta$ is in cm, $B$ is in days ; and $D$ is duty in hectares/cumec.

**Example 2.3.** *Find the delta for a crop when its duty is 864 hectares/cumec on the field, the base period of this crop is 120 days.*

**Solution.** $\Delta\ (\text{cm}) = \frac{864\ B}{D}$ where $B$ is in days and $D$ is in hectares/cumec

In this question, $B = 120$ days and $D = 864$ hectares/cumec

$$\therefore \qquad \Delta = \frac{864 \times 120}{864} = \mathbf{120\ cm.} \qquad \textbf{Ans.}$$

**2.3.5. Duty at various places.** In a large canal irrigation system, the water from its source, first of all, flows into the main canal ; from the main canal, it flows into the branch canal ; from the branch canal, it flows into the distributary ; from the distributary, it flows into the minor ; and then into the field channels (water-courses) ; and finally into the fields. A systematic layout of a canal system is shown in Fig. 2.1.

During the passage of water from these irrigation channels, water is lost due to evaporation and percolation. These losses are called Transit losses or Transmission or Conveyance losses in channels.

Duty of water for a crop, is the number of hectares of land, which the unit flow of water for $B$ days can irrigate. Therefore, if the water requirement of a particular crop at a particular location is more, then lesser number of hectares of land it will irrigate. Hence, *if water consumed by a crop of a given base period is more, its duty will be less. It, therefore, becomes clear that the duty of water at the head of the water-course will be less than the duty of water 'on the field' ; because when water flows from the head of the water-course and reaches the field, some water is lost enroute as transit losses.* Applying the same reasoning, it can be established *that duty of water at the head of a minor will be less than that at the head of the water-course ; duty at the head of a distributary will be less than that at the head of a minor, duty at the head of a branch canal will be less than that at the head of a distributary, and duty at the head of a main canal will be less than the duty at the head of a branch canal.* Duty of water, therefore, varies from one place to another, and *increases* as one moves *downstream* from the head of the main canal towards the head of the branches or water-courses. *The duty at the head of water-course (i.e.* at the outlet point of the minor), is quite important, and is called the **outlet discharge factor.** This outlet point is generally the end point of Irrigation Department. The control of Irrigation Department finishes at the outlet point, and the water is carried into the fields through water-courses by the beneficiary cultivators themselves.

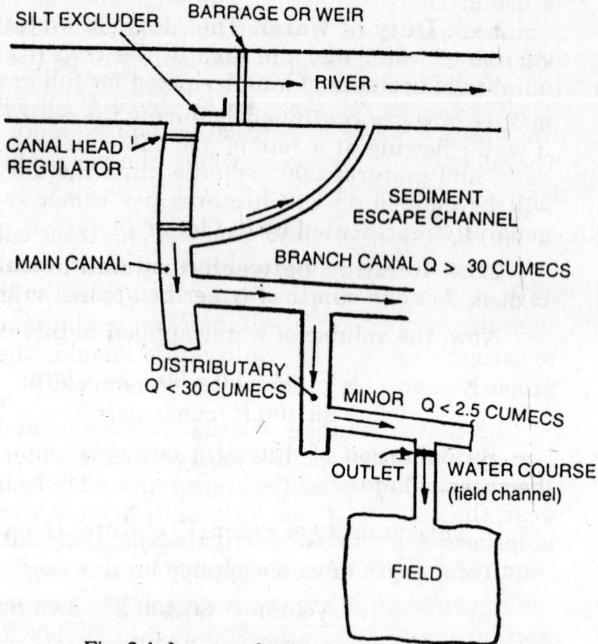


Fig. 2.1. Layout of a canal system.

**2.3.6. Flow duty and Quantity duty.** In direct irrigation, duty is always expressed in hectares/cumec. It is then called as *flow-duty* or duty.

In storage irrigation, duty may, sometimes be expressed in hectares/million cubic metre of water available in the reservoir. It eventually means that every million cubic metre of water available in the reservoir will mature so many hectares of a particular crop. Hence, the irrigation capacity of the reservoir is directly known. When duty is expressed in this manner, it is called *Quantity duty* or *Storage duty.*

**2.3.7. Factors on which duty depends.** Duty of irrigation water depends upon the following factors :

(*i*) *Type of crop.* Different crops require different amount of water, and hence, the duties for them are different. A crop requiring more water will have less flourishing acreage for the same supply of water as compared to that requiring less water. Hence, *duty will be less for a crop* requiring *more* water and *vice versa.*

(*ii*) *Climate and season.* As stated earlier, duty includes the water lost in evaporation and percolation. These losses will vary with the season. Hence, duty varies from season to season, and also from time to time in the same season. The figures for duties which we generally express are their average values considered over the entire crop period.

(*iii*) *Useful rainfall.* If some of the rain, falling directly over the irrigated land, is useful for the growth of the crop, then so much less irrigation water will be required to mature the crop. More the useful rainfall, less will be the requirement of irrigation water, and hence, more will be the duty of irrigation water.

(*iv*) *Type of soil.* If the permeability of the soil under the irrigated crop is high, the water lost due to percolation will be more and hence, the duty will be less. Therefore, for sandy soils, where the permeability is more, the duty of water is less.

(*v*) *Efficiency of cultivation method.* If the cultivation method (including tillage and irrigation) is faulty and less efficient, resulting in the wastage of water, the duty of water will naturally be less. If the irrigation water is used economically, then the duty of water will improve, as the same quantity of water would be able to irrigate more area. Cultivators should, therefore, be trained and educated properly to use irrigation water economically.

**2.3.8. Importance of duty.** It helps us in designing an efficient canal irrigation system. Knowing the total available water at the head of a main canal, and the overall duty for all the crops required to be irrigated in different seasons of the year, the area which can be irrigated can be worked out. Inversely, if we know the crops area required to be irrigated and their duties, we can work out the discharge required for designing the channel.

**2.3.9. Duty for certain crops.** The average values of duties for certain important Indian crops are tabulated in Table 2.2.

**Table 2.2. Average Approximate Values of Duty for Certain Important Crops in India**

| *Crop* | *Duty in hectares / cumec* |
|---|---|
| Sugarcane | 730 |
| Rice | 775 |
| Other Kharif | 1500 |
| Rabi | 1800 |
| Perennials | 1100 |
| Hot fodder | 2000 |

**2.3.10. Measures for Improving duty of water.** The duty of canal water can certainly be improved by effecting economy in the use of water by resorting to the following precautions and practices.

(1) *Precautions in field preparation and sowing :*

(*i*) Land to be used for cultivation should, as far as possible, be levelled.

(*ii*) The fields should be properly ploughed to the required depth.

(*iii*) Improved modern cultivation methods may preferably be adopted.

(*iv*) Porous soils should be treated before sowing crops to reduce seepage of water.

(*v*) Alkaline soils should be properly leached before sowing.

(*vi*) Manure fertilisers should be added to increase water holding capacity of the soil.

(*vii*) *Rotation of crops** should be preferred, as this will ensure increased crop yields with minimum use of water.

* explained in article 2.5.5.

(2) *Precautions in handling irrigation supplies :*

(*i*) The source of irrigation water should be situated within the prescribed limits, and should be capable of delivering sufficient quantity of satisfactary quality of irrigation water.

(*ii*) Canals carrying irrigation supplies should be lined to reduce seepage and evaporation*, thereby reducing on field requirement of water and consequently improving the duty of water.

(*iii*) Water courses may preferably be lined ; or R.C.C. pipes may be used for the same to reduce on field requirement of water, thereby improving duty.

(*iv*) Free flooding of fields should be avoided and furrow irrigation method may preferably be adopted, if surface irrigation is resorted to.

(*v*) Sub surface irrigation and Drip irrigation may be preferred to ordinary surface irrigation.

(*vi*) If canals are not lined, then two canals running side by side may be preferred to a single canal, as this will reduce the FSL, thereby reducing percolation losses.

(*vii*) Irrigation supplies should be economically used by proper control on its distribution, volumetric assessment, and by imparting proper education to the farmers.

## 2.4. Crop Seasons and Indian Agriculture

More than 70% of the Indian population is directly or indirectly connected with agriculture. The chief crops of India are rice, wheat, sugarcane, tea, cotton, groundnut, jute, coffee, rubber, garden crops (like coconuts, orange, etc), etc. Different types of soils are needed for raising different types of crops. For example, heavy retentive soil (40% clay) is favourable for raising crops like sugarcane, rice, etc., requiring more water. Light sandy soil (2 to 8% clay) is suitable for crops like gram, fodder, etc. requiring less water. Medium or normal soil (having about 10—20% of clay) is suitable for crops like wheat, cotton, maize, vegetables, oil seeds, etc. requiring normal amount of water.

From the agricultural point of view, the year can be divided into two principal cropping seasons, *i.e. Rabi* and *Kharif.* Normally, **Rabi** *starts from 1st October and ends on 31st March ; while* **Kharif** *starts from 1st April and ends on 30th September.* These dates are not rigid dead lines. The time may vary up to 1—3 months on either side. Sugarcane, which is an important cash crop, extends over both seasons.

*The* **Kharif crops** *are rice, bajra, jowar, maize, cotton, tobacco, groundnut, etc. The* **Rabi crops** *are wheat, barley, gram, linseed, mustard, potatoes, etc.* Kharif crops are also called 'Summer crops' and Rabi crops as 'Winter crops'. Kharif crops require about two to three times the quantity of water required by the Rabi crops.

The above distinction of seasons is well applicable to North India, but in South India, there is no such marked distinction between the different seasons. In fact, in South India, there is no clear cut winter, spring, summer and autumn seasons, as they are in North India. Except Bombay-Deccan, where there are five crop seasons, there are only three crop seasons in the remaining parts of the country. These three classifications of seasons are :

(*i*) Hot weather or Kharif season. (*ii*) Monsoon season. (*iii*) Winter or Rabi season.

---

* Water moving with higher velocity, will ensure reduced evaporation.

When a crop requires water for its crop season and also for some time in the beginning of the next crop season, allowance has to be made for this overlap. This allowance is known as **overlap allowance,** Sugarcane is an example of this kind of crop. Some important Indian crops, their periods of growth, water requirements, seed requirements, yields, etc. are shown in Table 2.3.

**Table 2.3. Irrigation Requirements of Certain Important Indian Crops**

| *S. No.* | *Crop* | *Period of growth* | *Av. water depth reqd. (in cm)* | *Irrigation requirements and remarks* | *Average quantity of seed required in kg / hectare* | *Average quantity of yield obtained in kg / hectare* |
|---|---|---|---|---|---|---|
| *(1)* | *(2)* | *(3)* | *(4)* | *(5)* | *(6)* | *(7)* |
| | **Kharif Crops** | | | | | |
| *(i)* | Maize, high yielding | June to Sept.-Oct. | 45 | Four or five waterings. Sensitive to drought and floods. Responsible to fertilizers. | 15 | 3,000 |
| *(ii)* | Bajra (spiked millets, or Pearl millets) high yielding | July to Nov. | 30 | Water should not stand. Irrigation as required. Resistant to drought and flooding. | 3.75 | 2,000 |
| *(iii)* | Juar (Great millets), high yielding | Sown in July as fodder and cut green more than once. | 30 | Same as above. | 12.5 | 3,000 |
| *(iv)* | Ground-nut | May to Nov-Dec. | 45 | 'Paleo' reqd. before sowing. | — | 1,600 |
| *(v)* | Cotton | May–June to Nov.–Jan. | 25—40 | Three or four irrigations are required. Damage up to the extent of 50% may be caused by flooding, rains, etc. | — | 500 |
| *(vi)* | Pulses like Arhar, etc. | July-Aug. to Nov.-Dec. | 30 | Irrigated when leaves get dried. | 12.5 | 700 |
| *(vii)* | Transplanted Rice (Paddy), high yielding | July to Nov. | 125—150 | Standing water of 5 to 8 cm gives best results. | 30 to 35 kg of seed is sufficient to raise nursery to transplant one hectare | 4500 |
| *(viii)* | Til | July-Aug. to Oct.-Nov. | — | Generally not irrigated but better to irrigate once. | 1.25 | 350 |

| (1) | (2) | (3) | (4) | (5) | (6) | (7) |
|---|---|---|---|---|---|---|
| | **Rabi Crops** | | | | | |
| *(i)* | Wheat (ordinary) | Oct. to March—April | 37.5 | Three-four waterings of 7—10 cm depth. | 80—100 | 1500 |
| *(ii)* | High yielding Wheat | Oct. to March—April | 45 | Five-six waterings of 7—10 cm depth. | 100—125 | 4000 |
| *(iii)* | Gram (high yielding) | Sept.-Oct. to March | 30 | Irrigated when leaves get dry. | 12.5 | 3500 |
| *(iv)* | Barley | Oct. to Mar.-April | 30 | Two waterings; one at jointing and another at booting stage. | 120 | 1300 |
| *(v)* | Potatoes | Sept.-Oct. to Feb. | 60—90 | Usually irrigated; sown in high hills up to early April. Second crop in plains is sometimes, taken in Feb.-April. | 15,000 | 35,000 |
| *(vi)* | Tobacco | Oct.-Feb. to Feb.-May | 60 | Four to five waterings. | | 4,500 |
| *(vii)* | Linseed *i.e.* Alsi | Oct.-Nov. to March | 45—50 | Irrigated at intervals of 15 days. Resistant to drought but damaged by frost and flooding. | | 700 |
| *(viii)* | Mustard | Oct. to Feb.-Mar. | 45 | Watered at intervals of 7—10 days. | 33 | 1000 to 1600 |
| | **Overlapping crop but generally classified under Rabi crop** | | | | | |
| *(i)* | Sugarcane | Feb.-March to Dec.-March | 90 | 5 or 6 waterings of 10 cm or more. | 500 | 25,000—30,000 |

## 2.5. Certain Important Definitions

**2.5.1. Crop ratio.** The area to be irrigated for Rabi crop is generally more than that for the Kharif crop. This ratio of proposed areas, to be irrigated in Rabi season to that in the Kharif season is called, Rabi-Kharif ratio or Crop ratio. This ratio is generally 2 : 1, *i.e.* Rabi area is twice the Kharif area.

**2.5.2. Paleo irrigation.** Sometimes, in the initial stages before the crop is sown, the land is very dry. This particularly happens at the time of sowing of Kharif crops because of hot September, when the soil may be too dry to be sown easily. In such a case, the soil is moistened with water, so as to help in sowing of the crops. This is known as Paleo irrigation.

**2.5.3. Kor-watering.** The first watering which is given to a crop, when the crop is a few centimetres high, is called kor-watering. It is usually the maximum single watering followed by other waterings at usual intervals, as required by drying of leaves.

The optimum depth of kor-watering for different crops are different. *For example, the optimum depth of kor-watering for Rice is 19 cm., for Wheat (in U.P.) is about 13.5 cm, and for Sugarcane is 16.5 cm.*

The kor-watering must be applied within a fixed limited period, called *Kor-period.* If the plants fail to receive this water in time or in sufficient quantity, then they do suffer a significant loss. The kor-period depends upon the climate. It is less for humid climates and more for dry climates. The *kor-period for rice varies from 2 to 4 weeks, and that for wheat varies from 3 to 8 weeks.*

**2.5.4. Cash crops.** A cash crop may be defined as a crop which has to be encashed in the market for processing, etc. as it cannot be consumed directly by the cultivators. All non-food crops, are thus, included in cash crops. Crops like jute, tea, cotton, tobacco, sugarcane, etc. are, therefore, called cash crops. The food crops like wheat, rice, barley, maize, etc. are excluded from the list of cash crops.

**2.5.5. Crops rotation.** When the same crop is grown again and again in the same field, the fertility of land gets reduced as the soil becomes deficient in plant foods favourable to that particular crop. In order to enhance the fertility of the land and to make the soil regain its original structure, it is often found necessary and helpful to give some rest to the land. This can be achieved either by allowing the land to lie fallow without any cultivation for some time, or to grow crops which do not mainly require those salts or foods which were mainly required by the earlier grown crop. This method of growing different crops in rotation, one after the other, in the same field, is called *Rotation of Crops.* A cash crop may be followed by a fodder crop, which, in turn, may be followed by a *soil-renovating crop* like gram, which being a **liguminous crop**, helps in giving nitrogen to the fields, thereby renovating the soil. The cultivators who are fond of sowing cash crops always, should be educated and made to understand the advantages of sowing crops in rotation.

The rotation of crops will help in extracting different foods from the soil, and thus avoiding the general deficiency of any particular type(s) of element(s). Moreover, if only one type of crop is grown in the same field, numerous insects and pests (of similar nature) will get developed. The crop rotation will also help in checking such growths. *Crop rotation will thus help in increasing the fertility of soil, and reducing the diseases and wastage due to insects, and hence increasing the overall crop yield.*

In general, the following rotations of crops may be adopted depending upon the soil conditions :

(*i*) Wheat—Juar—Gram
(*ii*) Rice—Gram*
(*iii*) Cotton—Wheat—Gram*— Fallow (up to July)
(*iv*) Cotton—Juar—Gram.
(*v*) Sugarcane (18 months) — Thadwa — Wheat or gram — Fallow (upto July).

## 2.6. Optimum Utilisation of Irrigation Water

If a crop is sown and produced under absolutely identical conditions, using different amounts of water depths, the yield is found to vary. The yield increases with water, reaches a certain maximum value and then falls down, as shown in Fig. 2.2. *The quantity of water at which the yield is maximum, is called the optimum water depth.*

---

* After harvesting a heavy water consuming crop like Rice, a less water consuming short term crop like gram may be taken, which may come up on the remnant moisture and manure.

Therefore, optimum utilisation of irrigation, generally means, getting maximum yield with any amount of water. The supplies of water to the various crops should be adjusted in such a fashion, as to get optimum benefit ratio, not only for the efficient use of available water and maximum yield, but also to prevent water-logging of the land in question. To achieve economy in the use of water, it is necessary that the farmers be made acquainted with the fact that only a certain fixed amount of water gives best results. More than that quantity, as well as, less than that quantity, reduces the yield. Many cultivators, till today feel, that they can increase the crop yield by using more and more water. Hence, they try to supply more water to their fields by undue tapping at the outlets. This must be checked. Moreover, farmers should be encouraged to line their water courses, thereby saving at least 20% of the costly irrigation water, which can be used to irrigate extra additional fields.

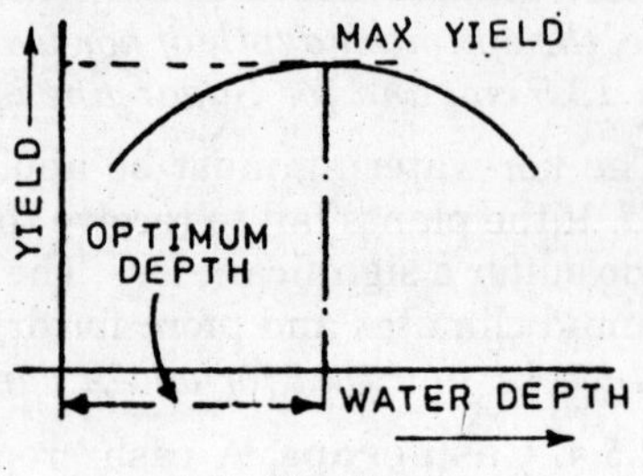

Fig. 2.2.

## 2.7. Irrigation Efficiencies

Efficiency is the ratio of the water output to the water input, and is usually expressed as percentage. Input minus output is nothing but losses, and hence, if losses are more, output is less and, therefore, efficiency is less. Hence, efficiency is inversely proportional to the losses. Water is lost in irrigation during various processes and, therefore, there are different kinds of irrigation efficiencies, as given below :

(*i*) **Efficiency of water-conveyance.** It is the ratio of the water delivered into the fields from the outlet point of the channel, to the water entering into the channel at its starting point. It may be represented by $\eta_c$. It takes the conveyance or transit losses into consideration.

(*ii*) **Efficiency of water-application.** It is the ratio of the quantity of water stored into the root zone of the crops to the quantity of water actually delivered into the field. It may be represented by $\eta_a$. It may also be called **on farm efficiency,** as it takes into consideration the water lost in the farm.

(*iii*) **Efficiency of water-storage.** It is the ratio of the water stored in the root zone during irrigation to the water needed in the root zone prior to irrigation (*i.e.* field capacity – existing moisture content). It may be represented by $\eta_s$.

(*iv*) **Efficiency of water use.** It is the ratio of the water beneficially used, including leaching water, to the quantity of water delivered. It may be represented by $\eta_u$.

**Example 2.4.** *One cumec of water is pumped into a farm distribution system. 0.8 cumec is delivered to a turn-out, 0.9 kilometre from the well. Compute the conveyance efficiency.*

**Solution.** By definition, $\eta_c = \frac{\text{Output}}{\text{Input}} \times 100 = \frac{0.8}{1.0} \times 100 = 80\%$ **Ans.**

**Example 2.5.** *10 cumecs of water is delivered to a 32 hectare field, for 4 hours. Soil probing after the irrigation indicates that 0.3 metre of water has been stored in the root zone. Compute the water application efficiency.*

**Solution.** Volume of water supplied by 10 cumecs of water applied for 4 hours

$$= (10 \times 4 \times 60 \times 60)\ \text{m}^3 = 1,44,000\ \text{m}^3$$
$$= 14.4 \times 10^4\ \text{m}^3 = 14.4\ \text{m} \times 10^4\ \text{m}^2 = 14.4\ \text{ha.m.}$$

$(\because \quad 10^4\ \text{m}^2 = 1\ \text{hectare})$

$\therefore$ Input = *14.4 ha.m* ...(*i*)

*Output* = 32 hectares land is storing water upto 0.3 m depth.

$\therefore$ Output = 32 × 0.3 ha.m = *9.6 ha.m* ...(*ii*)

$$\text{Water application efficiency, } (\eta_a) = \frac{\text{Output}}{\text{Input}} \times 100 = \frac{9.6}{14.4} \times 100 = \mathbf{66.67\%} \quad \textbf{Ans.}$$

(*v*) **Uniformity coefficient or Water distribution efficiency**. The effectiveness of irrigation may also be measured by its water distribution efficiency ($\eta_d$), which is defined below :

$$\eta_d = \left(1 - \frac{d}{D}\right) \qquad \text{...(2.3)}$$

where $\eta_d$ = Water distribution efficiency.

$D$ = Mean depth of water stored during irrigation.

$d$ = Average of the absolute values of deviations from the mean.

The water distribution efficiency represents the extent to which the water has penetrated to a uniform depth, throughout the field. When the water has penetrated uniformly throughout the field, the deviation from the mean depth is zero and water distribution efficiency is 1.0.

**Example 2.6.** *The depths of penetrations along the length of a boarder strip at points 30 metres apart were probed. Their observed values are 2.0, 1.9, 1.8, 1.6 and 1.5 metres. Compute the water distribution efficiency.*

**Solution.** The observed depths at five stations are 2.0, 1.9, 1.8, 1.6 and 1.5 metres, respectively.

$$\text{Mean depth} = D = \frac{2.0 + 1.9 + 1.8 + 1.6 + 1.5}{5} = \frac{8.8}{5} = 1.76 \text{ metres}$$

Values of deviations from the mean are (2.0 – 1.76), (1.9 – 1.76), (1.8 – 1.76), (1.6 – 1.76), (1.5 – 1.76) *i.e.*

0.24, 0.14, 0.04, – 0.16 and – 0.26.

The absolute values of these deviations from the mean, are 0.24, 0.14, 0.04, 0.16 and 0.26.

The average of these absolute values of deviations from the

$$\text{mean} = d = \frac{0.24 + 0.14 + 0.04 + 0.16 + 0.26}{5} = \frac{0.84}{5} = 0.168 \text{ metre}$$

The water distribution efficiency

$$= \left(1 - \frac{d}{D}\right) = \left[1 - \frac{0.168}{1.76}\right] = 1 - 0.095 = 0.905$$

Hence, the water distribution efficiency = **0.905**. **Ans.**

**Example 2.7.** *A stream of 130 litres per second was diverted from a canal and 100 litres per second were delivered to the field. An area of 1.6 hectares was irrigated in 8 hours. The effective depth of root zone was 1.7 m. The runoff loss in the field was 420 cu. m. The depth of water penetration varied linearly from 1.7 m at the head end of the field to 1.1 m at the tail end. Available moisture holding capacity of the soil is 20 cm per metre depth of soil. It is required to determine the water conveyance efficiency, water application efficiency, water storage efficiency, and water distribution efficiency. Irrigation was started at a moisture extraction level of 50% of the available moisture.*

**Solution.**

(*i*) Water conveyance efficiency ($\eta_c$)

$$= \frac{\text{Water delivered to the fields}}{\text{Water supplied into the canal at the head}} \times 100$$

$$= \frac{100}{130} \times 100 = \mathbf{77\%} \quad \textbf{(say)} \qquad \textbf{Ans.}$$

(*ii*) Water application efficiency ($\eta_a$)

$$= \frac{\text{Water stored in the root zone during irrigation}}{\text{Water delivered to the field}} \times 100$$

Water supplied to field during 8 hours @ 100 litres per second

$= 100 \times 8 \times 60 \times 60$ litres $=$ 2880 cu. m

Runoff loss in the field = 420 cu. m.

∴ the water stored in the root zone

$= 2880 - 420 = 2460$ cu. m

∴ Water application efficiency ($\eta_a$)

$$= \frac{2460}{2880} \times 100 = \mathbf{85.4\%} \qquad \textbf{Ans.}$$

(*iii*) Water storage efficiency

$$= \frac{\text{Water stored in the root zone during irrigation}}{\text{Water needed in root zone prior to irrigation}} \times 100$$

Moisture holding capacity of soil

= 20 cm per m depth × 1.7 m depth of root zone = 34 cm.

Moisture already available in root zone at the time of start of irrigation

$$= \frac{50}{100} \times 34 = 17 \text{ cm.}$$

Additional water required in root zone

$= 34 - 17 = 17$ cm.

$= \frac{17}{100} \times (1.6 \times 10^4)$ cu. m = 2720 cu. m (*i.e.* Depth × Plot area)

But actual water stored in root zone = 2460 cu. m

∴ Water storage efficiency ($\eta_s$) $= \frac{2460}{2720} \times 100 = \mathbf{90\%}$ **(say)** **Ans.**

(*iv*) Water distribution efficiency, $\eta_d = \left(1 - \frac{d}{D}\right)$

where $D$ = mean depth of water stored in the root zone

$$= \frac{1.7 + 1.1}{2} = 1.4 \text{ m.}$$

$d$ is computed as below :

Deviation from the mean at upper end (absolute value)

$$= |1.7 - 1.4| = 0.3$$

Deviation from the mean at lower end $= |1.1 - 1.4| = 0.3$.

$d$ = Average of the absolute values of deviations from mean $= \frac{0.3 + 0.3}{2} = 0.30$.

Using Eq. (2.3), we have, $\eta_d = \left(1 - \frac{0.30}{1.4}\right) =$ **0.786 or 78.6%** **Ans.**

## 2.8. Consumptive Use or Evapotranspiration ($C_u$)

Consumptive use for a particular crop may be defined as the total amount of water used by the plants in transpiration (building of plant tissues, etc.) and evaporation from adjacent soils, in any specified time. The values of consumptive use ($C_u$) may be different for different crops, and may be different for the same crop at different times and places.

In fact, the consumptive use for a given crop at a given place may vary throughout the day, throughout the month and throughout the crop period. Values of daily consumptive use or monthly consumptive use, are generally determined for a given crop and at a given place. Values of monthly consumptive use over the entire crop period, are then used to determine the irrigation requirement of the crop.

**2.8.1. Transpiration.** The Consumptive use* or Evapotranspiration* ($C_u$), as stated above, infact, *is the combined measure of transpiration loss (transpiration)* and the evaporation lost from the soil body** surrounding the plant.

The **transpiration** is defined as the process by which the water leaves a living plant during photosynthis, through its leaves, to enter the atmosphere as water vapour. Transpiration, infact, occurs, when the plant manufactures carbohydrates for its growth, by the process of ***photosynthesis***. What actually happens is, that the water extracted by the plant-root system from the soil, is transferred to the plant leaves; and on the surface of each plant leaf, photosynthesis occurs, where the air enters the *stomata* openings in the leaf surface, and the *chloroplasts* within the leaf utilise carbon dioxide from the air along with a very very small portion of the available moisture, to manufacture carbohydrates. *As air enters the leaf, the water escapes through the stomata,* and is evaporated (lost) to the atmosphere. It should be understood that transpiration occurs during photosynthesis, which usually occurs during the sun-light; and hence, most of the transpiration will occur during the day light hours. So much so that about 95% of the total transpiration occurs during the day time alone.

On the other hand, the evaporation continues throughout the day and night, although its rate may be different. Moreover, the transpiration loss, not only depends upon the available sun light, but also upon the available moisture, since

* Strictly speaking, the consumptive use may be defined as the amount of water used in the evapotranspiration from an area under vegetation, plus the water used by the plants in their metabolic process for building of plant tissues. Since the quantity of water used in metabolic process is insignificant compared to evapotranspiration, *the terms 'consumptive use ($C_u$)' and 'evapotranspiration ($E_t$)' are considered as one and the same thing.*

** This also includes the water body, when plants grow within a water body.

plants transpire less moisture when its availability becomes scarce. It will also be influenced by the type and characteristics of the plants. Different plants will evidently transpire different amount of water; and, hence, their water consuming characteristics are compared by a term, called *transpiration ratio.*

**2.8.1.1. Transpiration ratio** (T.R.) is defined as the ratio of the mass of water transpired by the plant in its full growth period, to the mass of the dry matter produced.

Thus,

$$\text{T.R.} = \frac{\text{Total mass of water transpired by the plant during its full growth}}{\text{Mass of dry matter produced}}$$

Evidently, the TR has no Units. ...(2.4)

"Mass of dry matter produced," is generally taken as the entire mass of the plant including its roots; but sometimes, only the marketed or the useful crop, such as grain of wheat, gram, etc. is weighed.

For a given kind of plant, the T.R. varies widely, depending upon the region and its climate. T.R. of a crop will increase if the water requirement of the crop is more. That is why the T.R. of wheat is of the order of 300–600 (average = 450) and that of the rice is of the order of 600–800 (average = 700).

**2.8.1.2. Transpiration measurement.** In order to measure the transpiration, various methods are used by botanists. Phytometer method, which is generally adopted by engineers, is, however, discussed here.

**Phytometer method.** This is the most important and widely used method. The instrument consists of a closed and water-tight tank, containing sufficient earth to nourish the plant. It is sealed, so as to prevent any automatic inflow or outflow of water. Water is applied, only artificially, till the plant growth is completed. The instrument is weighed in the beginning as well as at the end of the experiment. Water required during the growth is measured, and the loss due to transpiration can be calculated by the equation :

$$T = (M_1 + M) - M_2 \qquad ...(2.5)$$

where $M_1$ = Initial mass of the instrument

$M_2$ = Final mass of the instrument

$M$ = Total mass of water added during full growth of the plant.

$T$ = Loss due to transpiration.

But one must not forget that the transpirational loss calculated above, is not the actual value of the field, but is only of the lab. Some influence is produced by the protected growth and due to the difference left in exactly reproducing the field soil conditions in the lab. So the values obtained by the above test must be multiplied by a suitable factor to obtain the possible field results. The effect of the shape of phytometer upon the water requirement of the plant is still not known. The growth in screened shelters has, however, been found to reduce the water requirement by about 10% or so.

**2.8.1.3. Potential and Actual Evapotranspiration.** As defined above, the transpiration is the sum of the water lost to the atmosphere by the plants through transpiration, and the water evaporated from the soil or water body, surrounding the plant. This loss will evidently depend upon the available moisture for both the processes of transpiration and evaporation. When sufficient moisture is freely available to completely meet the needs of the vegetation fully covering an area, the resulting evapo-transpiration is called the **potential evapo-transpiration** (PET) or consumptive use ($C_u$) for that vegetated area. The real evapo-transpiration, occurring in a specific situation in the field, however, is called the *actual evapo transpiration* (AET).

It has further been established that *whereas, PET critically depends upon the climatological factors, rather than on the characteristics of plants and soils; the AET is largely affected by the characteristics of soil and vegetation.*

The Actual evapo-transpiration (AET) in the field can be measured by an instrument called a *lysimeter*. A **lysimeter**, like a phytometer, is a special water-tight tank, containing a block of soil, and is installed in a field of growing plants. Same plants are grown in the lysimeter, as in the surrounding field. The amount of water required to maintain constant moisture condition within the tank are measured either volumetrically or gravimetrically through an arrangement made in the lysimeter. This value of water required, when averaged over the crop period, in mm/day, will give us the value of the *actual evapo-transpiration* (AET) in mm/day for that crop/vegetation, under these specific field conditions.

A lysimeter should be so designed, as to accurately reproduce the field conditions of soil, moisture content, and type and size of the vegetation of the surrounding area. The instrument should also be buried in the field, so that the soil is at the same level inside as well as outside the container. Such field studies are time consuming and costlier, and are generally adopted only for research studies on crops.

## 2.9. Effective Rainfall or Useful Rainfall ($R_e$)

A portion of the rain falling over a cropped area during the growing period of the crop is utilised by the crop to fulfil its evapotranspiration needs, thereby reducing the irrigation requirement of the crop. This portion or ratio of rainfall is known as the effective or useful rainfall, and is denoted as $R_e$ or $R_u$. This value can be worked out by subtracting from the total rainfall, the following abstractions:

(*i*) Evaporation, (*ii*) rainfall lost through deep percolation below the root zone; and (*iii*) water lost as surface runoff.

Since these abstractions cannot be easily evaluated in their mm values, a ratio factor is used on empirical recommendations, to estimate the useful rainfall value as :

Useful rainfall (mm) = Total rain fall (mm) × average ratio factor ($r$), such as 0.6 or 0.7 or so, for the given rain depending on its areal extent.

This ratio factor ($r$) will be less for a wide spread rain falling over the 100% area than the scattered rain falling over lesser area.

A rain falling over, say 60% of the area, is designated as the rain having ***chance percent*** of ***occurence*** equal to 60%. Such a rain will eventually fall over 60% of the area with 100% chance.

Various average ratio values ($r$) are given/suggested for rains having different values of their percent chance of occurrence, as shown in Table 2.4. Different ratio values are suggested for places having different average annual rainfall values. Thus, by knowing the average annual rainfall value of the place of cropping, we can find the ratio ($r$) for the given rain. The chance percentage of the occurrence of the rain can be assumed depending upon the areal extent of its occurrence. Thus, a scattered rain having say 40% chance of occurrence will be the one which will fall/cover only the 40% of the cropped

**area. Thus, a wide spread rain will have higher chance of occurrence, while a scattered rain will have lesser value of its chance of occurrence. The $R_e$ values can thus be computed for rains occurring on daily basis or weekly or monthly basis, while analysing irrigation demand.**

**Table 2.4. Average Ratio Values (*r*) Applicable to Effective Rainfall**

| *Average annual rainfall in cm.* | *Assumed r values for per cent chance of occurrence* | | | | |
|---|---|---|---|---|---|
| | *50* | *60* | *70* | *80* | *90* |
| 10 | 0.84 | 0.72 | 0.61 | 0.50 | 0.38 |
| 20 | 0.90 | 0.81 | 0.71 | 0.62 | 0.51 |
| 30 | 0.93 | 0.85 | 0.78 | 0.69 | 0.58 |
| 40 | 0.95 | 0.88 | 0.81 | 0.73 | 0.63 |
| 50 | 0.96 | 0.90 | 0.83 | 0.75 | 0.67 |
| 60 | 0.97 | 0.91 | 0.84 | 0.78 | 0.70 |
| 70 | 0.97 | 0.92 | 0.86 | 0.80 | 0.72 |
| 80 | 0.98 | 0.93 | 0.87 | 0.81 | 0.74 |
| 90 | 0.98 | 0.93 | 0.88 | 0.82 | 0.75 |
| 100 | 0.98 | 0.94 | 0.89 | 0.83 | 0.76 |
| 120 | 0.98 | 0.94 | 0.90 | 0.85 | 0.78 |
| 140 | 0.99 | 0.95 | 0.91 | 0.86 | 0.80 |
| 160 | 0.99 | 0.95 | 0.91 | 0.87 | 0.82 |
| 180 | 0.99 | 0.95 | 0.92 | 0.88 | 0.84 |
| 200 | 0.99 | 0.95 | 0.92 | 0.89 | 0.85 |

**In addition to this *ratio method, Food and Agriculutre Organisation (FAO)* of United Nations has suggested two simple empircal equations to estimate monthly effective rainfall value ($R_e$) in mm from the total monthly rainfall value ($R$) in mm. These equations are given below :**

**(*i*) $R_e = (0.8\,R - 25 \text{ mm})$ if $R > 75$ mm/month ...(2.6)**

**and (*ii*) $R_e = (0.6\,R - 10 \text{ mm})$ if $R < 75$ mm/month ...(2.7)**

**where $R$ = total rainfall in mm/month**

**Example 2.8.** *Calculate the effective rainfalls for the following monthly rainfall figures : R = 30, 70, and 100 mm*

**Solution.**

(*i*) *When R = 30 mm*

Since $R < 75$ mm/month, using Eq. (2.7), we get

$$R_e = (0.6\,R - 10 \text{ mm}) = (0.6 \times 30 \text{ mm} - 10 \text{ mm})$$
$$= 18 - 10 = 8 \text{ mm} \quad \textbf{Ans.}$$

(*ii*) For $R = 70$ mm,

Since $R < 75$ mm/month, use Eq. (2.7) as :

$$R_e = (0.6\,\text{R} - 10 \text{ mm}) = (0.6 \times 70 \text{ mm} - 10 \text{ mm})$$
$$= 42 - 10 = 32 \text{ mm} \quad \textbf{Ans.}$$

(*iii*) For $R = 100$ mm.

Since $R > 75$ mm, use Eq. (2.6) as :

$$R_e = (0.8\,\text{R} - 25 \text{ mm}) = (0.8 \times 100 \text{ mm} - 25 \text{ mm})$$
$$= 80 - 25 = 55 \text{ mm.} \quad \textbf{Ans.}$$

## 2.10. Consumptive Irrigation Requirement (CIR)

It is the amount of *Irrigation water* required in order to meet the evapotranspiration needs of the crop during its full growth. It is, therefore, nothing but the consumptive use itself, but exclusive of effective precipitation, stored soil moisture, or ground water. When the last two are ignored, then we can write

$$\text{C.I.R.} = C_u - R_e \qquad \text{...(2.8)}$$

## 2.11. Net Irrigation Requirement (NIR)

It is the amount of irrigation water required in order to meet the evapotranspiration need of the crop as well as other needs such as leaching. Therefore, N.I.R. = $C_u - R_e$ + Water lost as percolation in satisfying other needs such as leaching.

## 2.12. Factors Affecting Consumptive Use

Consumptive use or evapotranspiration depends upon all those factors on which evaporation and transpiration depend* ; such as temperature, sunlight, humidity, wind movement, etc.

**Example 2.9.** *The following table gives the values of consumptive uses and effective rainfalls for the periods shown against them, for a Jowar crop sown at Bellary in Karnataka State. The period of growth is from 16th October to 2nd Feb., i.e. (110 days). Determine the net irrigation requirement of this crop, assuming that water is not required for any other purpose except that of fulfilling the evapotranspiration needs of the crop.*

**Table 2.5 (*a*)**

| *Dates* | | $C_u$ *in mm.* | $R_e$ *in mm.* |
|---|---|---|---|
| (1) | | (2) | (3) |
| October | 16—31 | 37.0 | 30.8 |
| November | 1—30 | 84.2 | 20.4 |
| December | 1—31 | 154.9 | 6.7 |
| January | 1—31 | 188.1 | 2.4 |
| February | 1—2 | 13.3 | 1.0 |

**Solution.** The given table is extended as shown in Table 2.5 (*b*). Values in col. (4) are obtained by subtracting values of col. (3) from values of col. (2).

**Table 2.5 (*b*)**

| *Dates* | | $C_u$ | $R_e$ | $NIR = C_u - R_e$ |
|---|---|---|---|---|
| (1) | | (2) | (3) | (4) |
| October | 16—31 | 37.0 | 30.8 | 6.2 |
| November | 1—30 | 84.2 | 20.4 | 63.8 |
| December | 1—31 | 154.9 | 6.7 | 148.2 |
| January | 1—31 | 188.1 | 2.4 | 185.7 |
| February | 1—2 | 13.3 | 1.0 | 12.3 |
| | | | | Σ = 416.2 mm<br>**= 41.62 cm** |

Hence, the net irrigation requirement = **41.62 cm.** **Ans.**

* Explained in article 6.2.2.3 in Water Resources Engg. Vol. I– "Hydrology, Flood Control and Ground Water Engineering" by the same author.

## 2.13. Estimation of Consumptive Use

Although various methods have been developed in order to estimate evapotranspiration (consumptive use) values of different crop in an area, or for areas vegetated with the same cropping pattern, but the most simple and commonly used methods are :

(1) Blaney-Criddle Equation, and

(2) Hargreaves class *A* pan evaporation method.

(3) Penman's equation.

They are described below :

**2.13.1. Blaney-Criddle Formula.** It states that the monthly consumptive use is given by

$$C_u = \frac{k.p}{40}\,[1.8t + 32] \qquad \text{...(2.9)}$$

where, $C_u$ = Monthly consumptive use in cm.

$k$ = **Crop factor,** determined by experiments for each crop, under the environmental conditions of the particular area.

$t$ = Mean monthly temperature in °C.

$p$ = Monthly per cent of annual day light hours that occur during the period.

If $\frac{p}{40}\,[1.8\,t + 32]$ is represented by $f$, we get

$$C_u = k \cdot f \qquad \text{...(2.10)}$$

This formula has been extensively used throughout the world for estimating seasonal water requirements. However, it was found that the values of $k$ based on seasonal determinations were too low for the short periods between irrigations. This led to further developments and finally the formula was expressed as

$$C_u = k\,\Sigma f \qquad \text{...(2.11)}$$

where $C_u$ = Seasonal consumptive use, *i.e.* consumptive use during the period of growth for a given crop in a given area.

The above formula involves the use of crop factor, the value of which is to be determined for each crop and for different places. At present, this information is not available in India. Moreover, this formula does not take into consideration the factors such as humidity, wind velocity, elevation, etc. on which consumptive use depends. Hargreaves class *A* pan evaporation method is, therefore, generally used in India.

**Example 2.10.** *Wheat is to be grown at a certain place, the useful climatological conditions of which are tabulated below in Table 2.6 (a). Determine the evapotranspiration and consumptive irrigation requirement of wheat crop. Also determine the field irrigation requirement if the water application efficiency is 80%. Make use of Blaney-Criddle equation and a crop factor equal to 0.8.*

**Table 2.6 (*a*)**

| *Month* | *Monthly temp. in °C averaged over the last 5 years* | *Monthly per cent of day time hr. of the year computed from the Sun-shine Tables* | *Useful rainfall in cm, averaged over the last 5 years* |
|---|---|---|---|
| (1) | (2) | (3) | (4) |
| November | 18.0 | 7.20 | 1.7 |
| December | 15.0 | 7.15 | 1.42 |
| January | 13.5 | 7.30 | 3.01 |
| February | 14.5 | 7.10 | 2.25 |

**Solution.** Blaney-Criddle equation is given by Eq. (2.11) as :

$$C_u = k \cdot \Sigma f \qquad \text{where, } f = \frac{p}{40}(1.8t + 32)\text{ ; and } k = 0.8 \text{ (given)}$$

Values of *f* are worked out in col. (5) of Table 2.6(*b*)

**Table 2.6 (*b*)**

| *Month* | *t* | *p* | $R_e$ | $f = \frac{p}{40}(1.8t + 32)$ |
|---|---|---|---|---|
| (1) | (2) | (3) | (4) | (5) |
| November | 18.00 | 7.20 | 1.70 | 11.6 |
| December | 15.00 | 7.15 | 1.42 | 10.5 |
| January | 13.50 | 7.30 | 3.01 | 10.3 |
| February | 14.50 | 7.10 | 2.25 | 10.3 |
| | | | Σ = 8.38 cm | Σ = 42.7 cm |

$$C_u = k \cdot \Sigma f = 0.8 \times 42.7 = 34.16 \text{ cm.}$$

Hence, Consumptive use $C_u$ = 34.16 cm. **Ans.**

Consumptive irrigation requirement

$$= C_u - R_e = 32.72 - 8.38 = 24.34 \text{ cm. } \textbf{Ans.}$$

Field irrigation requirement = C.I.R./$\eta_a$ = 24.34/0.8 = 30.43 cm. **Ans.**

**Example 2.11.** *Determine the volume of water required to be diverted from the head works to irrigate area of 5000 ha using the data given in the table below. Assume 80% as the effective precipitation to take care of the consumptive use of the crop. Also assume 50% efficiency of water application in the field and 75% as the conveyance efficiency of canal.*

**Table 2.7 (*a*)**

| *Mouth* | *Temp. F* | *Precentage hrs of sunshine* | *Rainfall mm* | *Consumptive coefficient or Crop factor (k)* |
|---|---|---|---|---|
| (1) | (2) | (3) | (4) | (5) |
| June | 70.8 | 9.90 | 75 | 0.80 |
| July | 74.4 | 10.20 | 108 | 0.85 |
| August | 72.8 | 9.60 | 130 | 0.85 |
| September | 71.6 | 8.40 | 115 | 0.85 |
| October | 69.3 | 7.86 | 105 | 0.65 |
| November | 55.2 | 7.25 | 25 | 0.65 |
| December | 47.1 | 6.42 | 0 | 0.60 |
| January | 48.8 | 8.62 | 0 | 0.60 |
| February | 53.9 | 9.95 | 0 | 0.65 |
| March | 60.0 | 8.84 | 0 | 0.70 |
| April | 62.5 | 8.86 | 0 | 0.70 |
| May | 67.4 | 9.84 | 0 | 0.75 |

**Solution.** The given Table 2.7 (*a*) is extended as shown in Table 2.7 (*b*) to compute monthly values of consumptive use ($C_u$) by the Eqn. $C_u = k \cdot f$, where $f = \frac{p}{40}$ [1.8*t* + 32], where *t* is in °C. When *t* is in °F as given in col (2) of Table 2.7 (*a*), then the eqn. becomes

$$C_u = k\left(\frac{p}{40}\, t\right) \text{ where } t \text{ is °F.} \qquad \text{...(2.9a)}$$

Monthly values of *f* are hence worked out in col. (6) of Table 2.7 (*b*).

**Table 2.7 (*b*)**

| *Month* | *Temp. °F* | *Percetage hrs of sun shine p* | *Rainfall cm* | *Consumptive coefficient or Crop factor k* | $C_u = kf = \frac{k \cdot p}{40} \cdot t$ $\frac{col(5) \times col(3) \times col(2)}{40}$ *cm.* |
|---|---|---|---|---|---|
| (1) | (2) | (3) | (4) | (5) | (6) |
| June | 70.8 | 9.90 | 7.5 | 0.80 | 14.02 |
| July | 74.4 | 10.20 | 10.8 | 0.85 | 16.13 |
| August | 72.8 | 9.60 | 13.0 | 0.85 | 14.85 |
| September | 71.6 | 8.40 | 11.5 | 0.85 | 12.78 |
| October | 69.3 | 7.86 | 10.5 | 0.65 | 8.85 |
| November | 55.2 | 7.25 | 2.5 | 0.65 | 6.50 |
| December | 47.1 | 6.42 | 0 | 0.60 | 4.54 |
| January | 48.8 | 8.62 | 0 | 0.60 | 6.31 |
| February | 53.9 | 9.95 | 0 | 0.65 | 8.71 |
| March | 60.0 | 8.84 | 0 | 0.70 | 9.28 |
| April | 62.5 | 8.85 | 0 | 0.70 | 9.68 |
| May | 67.4 | 9.84 | 0 | 0.75 | 12.44 |
| Σ | | | 55.8 | | 124.09 cm |

Total consumptive use = Σ col. (6) = 124.09 cm.

Useful rainfall = 80% of total precipitation (given) = 80% × 55.8 cm. = 44.64 cm.

∴ Net irrigation requirement $NIR = C_u - R_e$ = 124.09 – 44.64 = 79.45 cm.

$$\text{Field irrigation requirement} = FIR = \frac{N \cdot I \cdot R}{\eta_a} = \frac{79.45}{50\%} = \frac{79.45}{0.5} = 158.9 \text{ cm}$$

$$\text{Gross irrigation requirement at headworks} = \frac{FIR}{\eta_c}$$

where $\eta_c$ = conveyance efficiency = 75% = 0.75

$$\therefore \quad GIR = \frac{158.9}{0.75} = 211.87 \text{ cm.}$$

Volume of water required for 5000 ha area

$$= \frac{211.87}{100} \text{ m} \times (5000 \times 10^4 \text{ m}^2) = \mathbf{105.93 \text{ M.m}^3}. \qquad \textbf{Ans.}$$

**Example 2.12.** *The monthly consumptive use values for Paddy are tabulated in Table 2.8. Determine the total consumptive use. What is the average monthly consumptive use and peak monthly consumptive use?*

**Table 2.8**

| Dates | | Rice (Loam Soil) $C_u$ in cm |
|---|---|---|
| June | 1—30 Nursery | 29.69 |
| July | 1—12 Nursery | 8.76 |
| July | 13—31 | 14.38 |
| August | 1—31 | 22.73 |
| September | 1—30 | 21.29 |
| October | 1—31 | 25.50 |
| November | 1—24 | 15.06 |

**Solution.** The summation of consumptive uses

$= 29.69 + 8.76 + 14.38 + 22.73 + 21.29 + 25.50 + 15.06 = 137.41$ cm

Hence, total consumptive use for paddy = 137.41 cm. **Ans.**

Average daily consumptive use

$$= \frac{137.41}{\text{Period of growth in days}} = \frac{137.41}{30 + 31 + 31 + 30 + 31 + 24}$$

$$= \frac{137.41}{177} = 0.77 \text{ cm} = 7.7 \text{ mm} \quad \textbf{Ans.}$$

Average monthly consumptive use = 0.77 × 30 = 23.1 mm **Ans.**

Peak monthly consumptive use

= **26.69 cm.** (Highest value given) **Ans.**

**2.13.2. Hargreaves Class A pan evaporation method.** In this method, evapotranspiration (consumptive use) is related to pan evaporation by a constant *K*, called consumptive use coefficient. The formula can be written as

$$\frac{\text{Evapotranspiration } (E_t \text{ or } C_u)}{\text{Pan evaporation } (E_p)} = K$$

or $$\boxed{E_t \quad \text{or} \quad C_u = K \cdot E_p} \qquad \text{...(2.12)}$$

Consumptive use coefficient (*K*) is different for different crops and is different for the same crop at different places. It also varies with the crop growth, and is different at different crop stages for the same crop. The above relationship is now available for various crops from many countries such as Israel, Philippines, U.S.A. and India. Research stations constantly go on reporting more and more data. Where specific data are not available, average values can be used as recommended by Hargreaves, and given in Table 2.9. The crops have been divided into 8 groups and the coefficients have been suggested for average conditions of soil, etc.

(*i*) *Group A.* The important crops include :

Sugar, Beats, Maize, Cotton, Jowar, Bean, Peas, Potatoes, etc.

(*ii*) *Group B.* This group consists of deciduous fruits and some field crops. Important crops are :

Tomatoes, Hybrid Walnuts, Plumes, Olives, and some group *A* crops that fail to produce maximum vegetative cover and maximum growth ratios.

(*iii*) *Group C.* $\frac{E_t}{E_p}$ ratios are of the order of 0.6. It includes crops like Melons, Onions, Carrots, Hops, Grapes, etc.

(iv) *Group D*. The maximum $\frac{E_t}{E_p}$ ratio is about 0.90 and usually occurs at about 75 to 80% completion of crop vegetative cycle. The important crops are : Wheat, Barley, Celery, Flax and other small grains, etc.

(v) *Group E*. Ratios of $\frac{E_t}{E_p}$ vary from 0.7 to 1.10. The model value being 0.90. The important crops are :
Pastures, Orchard with cover crop, Plantain, etc.

(vi) *Group F*. It includes citrus crops such as Oranges, Grape fruit, etc. The $\frac{E_t}{E_p}$ ratios are fairly constant throughout the year and average to a value of about 0.60.

(vii) *Group G*. $\frac{E_t}{E_p}$ values generally increase with crop and vary from 0.66 to 1.00. It includes *Sugarcane and Alfalfa*.

(viii) *Paddy or Rice*. $\frac{E_t}{E_p}$ increases from 0.80 to 1.30, with crop growth and then falls down, reaching its maximum value somewhere near 50% growth, as shown in Table 2.9.

**Table 2.9. Hargreave's Average Value of Consumptive Use Coefficient K ($E_t = KE_p$)**

| *Per cent of crop growing season* | *Consumptive use coefficient (K) to be multiplied by class A Pan Evaporation ($E_p$), i.e. $E_t = K \cdot E_p$* | | | | | | | |
|---|---|---|---|---|---|---|---|---|
| | *Group A* | *Group B* | *Group C* | *Group D* | *Group E* | *Group F* | *Group G* | *Rice* |
| (1) | (2) | (3) | (4) | (5) | (6) | (7) | (8) | (9) |
| 0 | 0.20 | 0.15 | 0.12 | 0.08 | 0.90 | 0.60 | 0.50 | 0.80 |
| 5 | 0.20 | 0.15 | 0.12 | 0.08 | 0.90 | 0.60 | 0.55 | 0.90 |
| 10 | 0.36 | 0.27 | 0.22 | 0.15 | 0.90 | 0.60 | 0.60 | 0.95 |
| 15 | 0.50 | 0.38 | 0.30 | 0.19 | 0.90 | 0.60 | 0.65 | 1.00 |
| 20 | 0.64 | 0.48 | 0.38 | 0.27 | 0.90 | 0.60 | 0.70 | 1.05 |
| 25 | 0.75 | 0.56 | 0.45 | 0.33 | 0.90 | 0.60 | 0.75 | 1.10 |
| 30 | 0.84 | 0.63 | 0.50 | 0.40 | 0.90 | 0.60 | 0.80 | 1.14 |
| 35 | 0.92 | 0.69 | 0.55 | 0.46 | 0.90 | 0.60 | 0.85 | 1.17 |
| 40 | 0.97 | 0.73 | 0.58 | 0.52 | 0.90 | 0.60 | 0.90 | 1.21 |
| 45 | 0.99 | 0.74 | 0.60 | 0.58 | 0.90 | 0.60 | 0.95 | 1.25 |
| 50 | 1.00 | 0.75 | 0.60 | 0.65 | 0.90 | 0.60 | 1.00 | 1.30 |
| 55 | 1.00 | 0.75 | 0.60 | 0.71 | 0.90 | 0.60 | 1.00 | 1.30 |
| 60 | 0.99 | 0.74 | 0.60 | 0.77 | 0.90 | 0.60 | 1.00 | 1.30 |
| 65 | 0.96 | 0.72 | 0.58 | 0.82 | 0.90 | 0.60 | 0.95 | 1.25 |
| 70 | 0.91 | 0.68 | 0.55 | 0.88 | 0.90 | 0.60 | 0.90 | 1.20 |
| 75 | 0.85 | 0.64 | 0.51 | 0.90 | 0.90 | 0.60 | 0.85 | 1.15 |
| 80 | 0.75 | 0.56 | 0.45 | 0.90 | 0.90 | 0.60 | 0.80 | 1.10 |
| 85 | 0.60 | 0.45 | 0.36 | 0.80 | 0.90 | 0.60 | 0.75 | 1.00 |
| 90 | 0.46 | 0.35 | 0.28 | 0.70 | 0.90 | 0.60 | 0.70 | 0.90 |
| 95 | 0.28 | 0.21 | 0.17 | 0.60 | 0.90 | 0.60 | 0.55 | 0.80 |
| 100 | 0.20 | 0.20 | 0.17 | 0.60 | 0.90 | 0.60 | 0.50 | 0.20 |

The coefficients shown in Table 2.9 are only average values and care must be taken while using them. Local values when available should only be used. The factors which increase or decrease the evapotranspiration may be taken into consideration. For example, taller and more uneven vegetation tends to result in greater turbulence and more efficient utilisation of radiation in the production of water use. *Dark green vegetation produces higher rate of absorption of solar energy, and hence, its evapotranspiration rate will be higher than that for light green vegetation.* For this reason, the plant diseases, causing yellowing of the leaves of the plants, greatly reduce evapotranspiration.

Values of $K$ for certain crops reported from India and U.S.A. are given in Tables 2.10 and 2.11 respectively.

**Table 2.10. Values of $K$ for Certain Indian Crops ; ($E_t = K \cdot E_p$)**

| *% of crop growing season* | *Wheat Ludhiana (India)* | *Wheat Poona (India)* | *Cotton Poona (India)* | *Maize Ludhiana (India)* |
|---|---|---|---|---|
| (1) | (2) | (3) | (4) | (5) |
| 0 | 0.14 | 0.30 | 0.22 | 0.40 |
| 5 | 0.17 | 0.40 | 0.22 | 0.42 |
| 10 | 0.23 | 0.51 | 0.23 | 0.47 |
| 15 | 0.33 | 0.62 | 0.24 | 0.54 |
| 20 | 0.45 | 0.73 | 0.26 | 0.63 |
| 25 | 0.60 | 0.84 | 0.35 | 0.75 |
| 30 | 0.72 | 0.92 | 0.58 | 0.85 |
| 35 | 0.81 | 0.96 | 0.80 | 0.96 |
| 40 | 0.88 | 1.10 | 0.95 | 1.04 |
| 45 | 0.90 | 1.10 | 1.03 | 1.07 |
| 50 | 0.91 | 1.00 | 1.08 | 1.09 |
| 55 | 0.90 | 0.91 | 1.08 | 1.10 |
| 60 | 0.89 | 0.80 | 1.07 | 1.11 |
| 65 | 0.86 | 0.65 | 1.05 | 1.10 |
| 70 | 0.83 | 0.51 | 1.00 | 1.07 |
| 75 | 0.80 | 0.40 | 0.93 | 1.04 |
| 80 | 0.76 | 0.30 | 0.85 | 1.00 |
| 85 | 0.71 | 0.20 | 0.73 | 0.97 |
| 90 | 0.65 | 0.12 | 0.62 | 0.89 |
| 95 | 0.58 | 0.10 | 0.50 | 0.81 |
| 100 | 0.51 | 0.10 | 0.40 | 0.70 |
| *Seasonal value K* | 0.61 | 0.61 | 0.68 | 0.86 |

**Class A pan evaporation ($E_p$) measurements.** $E_p$ can be experimentally determined by directly measuring the quantity of water evaporated from the standard class A pan. This pan is 1.2 m in diameter, 25 cm deep, and bottom is raised 15 cm above the ground surface. The depth of water is to be kept in a fixed range such that the water surface is at least 5 cm, and never more than 7.5 cm, below the top of the pan.

The pan evaporation ($E_p$) can also be determined by using the **Christiansen formula,** which states

$$E_p = 0.459\, R \cdot C_t \cdot C_w \cdot C_h \cdot C_s\, C_e \qquad \text{...(2.13)}$$

**Table 2.11. Values of K for Certain U.S.A. Crops ; ($E_t = K \cdot E_p$)**

| *% of crop growing season* | *Sugarcane* Hawaii (USA)* | *Rice Los Banos (USA)* | *Maize Albama (USA)* | *Jowar Albama (USA)* | *Sugarcan* Hargreaves* |
|---|---|---|---|---|---|
| (1) | (2) | (3) | (4) | (5) | (6) |
| 0 | 0.34 | 1.00 | 0.40 | 0.42 | 0.48 |
| 5 | 0.37 | 1.02 | 0.40 | 0.44 | 0.50 |
| 10 | 0.40 | 1.03 | 0.43 | 0.46 | 0.53 |
| 15 | 0.44 | 1.05 | 0.46 | 0.48 | 0.55 |
| 20 | 0.50 | 1.07 | 0.52 | 0.50 | 0.60 |
| 25 | 0.60 | 1.09 | 0.59 | 0.52 | 0.67 |
| 30 | 0.72 | 1.11 | 0.67 | 0.56 | 0.75 |
| 35 | 0.86 | 1.13 | 0.76 | 0.59 | 0.80 |
| 40 | 0.93 | 1.16 | 0.85 | 0.64 | 0.85 |
| 45 | 0.98 | 1.18 | 0.93 | 0.71 | 0.87 |
| 50 | 1.02 | 1.20 | 1.02 | 0.79 | 0.89 |
| 55 | 1.05 | 1.21 | 1.09 | 0.92 | 0.90 |
| 60 | 1.07 | 1.22 | 1.14 | 1.01 | 0.90 |
| 65 | 1.10 | 1.22 | 1.19 | 1.07 | 0.85 |
| 70 | 1.13 | 1.21 | 1.21 | 1.09 | 0.75 |
| 75 | 1.16 | 1.19 | 1.23 | 1.09 | 0.71 |
| 80 | 1.19 | 1.16 | 1.22 | 1.05 | 0.68 |
| 85 | 1.20 | 1.10 | 1.16 | 0.99 | 0.59 |
| 90 | 1.20 | 1.03 | 1.06 | 0.91 | 0.51 |
| 95 | 1.19 | 0.96 | 0.95 | 0.82 | 0.50 |
| 100 | 1.19 | 0.86 | 0.75 | 0.70 | 0.50 |
| *Seasonal K* | 0.89 | 1.10 | 0.86 | 0.75 | 0.69 |

where $R$ = Extra-terrestrial radiation in the same units as $E_p$ in cm or mm (Table 2.12),

$C_t$ = Coefficient for temperature, and is given by

$$C_t = 0.393 + 0.02796\, T_c + 0.0001189\, T_c^{\,2} \quad \text{...(2.14)}$$

where, $T_c$ is the mean temperature in °C.

$C_w$ = Coefficient for wind velocity, and is given as

$$= 0.708 + 0.0034\, W - 0.0000038\, W^2 \quad \text{...(2.15)}$$

where, $W$ is the mean wind velocity at 0.5 metre above the ground in km/day.

$C_h$ = Coefficient for relative humidity, and is given by

$$C_h = 1.250 - 0.0087H + 0.75 \times 10^4\, H^2 - 0.85 \times 10^{-8}\, H^4 \quad \text{...(2.16)}$$

where $H$ is the mean percentage relative humidity at noon or average relative humidity for 11 and 18 hours.

---

*Value of K for sugarcane reported from Hawaii and found by Hargreaves are campared in col. (2) and col. (6).

**Table 2.12. Mean Monthly Values of Extra-terrestrial Radiation R in cm.**

| *Latitudes in degrees* North | *Jan.* | *Feb.* | *March* | *April* | *May* | *June* | *July* | *Aug.* | *Sept.* | *Oct.* | *Nov.* | *Dec.* |
|---|---|---|---|---|---|---|---|---|---|---|---|---|
| 45 | 15.621 | 19.990 | 31.953 | 41.072 | 50.317 | 52.146 | 52.426 | 46.101 | 35.204 | 25.730 | 16.891 | 13.843 |
| 40 | 19.609 | 23.393 | 35.027 | 42.926 | 50.003 | 52.146 | 52.730 | 47.498 | 37.770 | 29.312 | 20.701 | 17.831 |
| 35 | 23.546 | 26.670 | 37.871 | 44.450 | 51.384 | 51.918 | 52.730 | 48.565 | 40.056 | 32.664 | 24.460 | 21.819 |
| 30 | 27.407 | 29.794 | 40.411 | 45.669 | 51.460 | 51.384 | 52.451 | 49.276 | 42.037 | 35.814 | 28.092 | 25.781 |
| 25 | 31.140 | 32.690 | 42.647 | 46.558 | 51.206 | 50.571 | 51.841 | 49.682 | 43.688 | 38.710 | 31.572 | 29.642 |
| 20 | 34.722 | 35.382 | 44.552 | 47.117 | 50.597 | 49.428 | 50.902 | 49.708 | 45.009 | 41.300 | 34.900 | 33.376 |
| 15 | 38.100 | 37.821 | 46.126 | 47.320 | 49.657 | 47.980 | 49.657 | 49.403 | 45.989 | 43.612 | 37.973 | 36.906 |
| 10 | 41.250 | 40.005 | 47.346 | 47.168 | 48.388 | 46.253 | 48.082 | 48.717 | 46.609 | 45.593 | 40.818 | 40.259 |
| 5 | 44.120 | 41.885 | 48.209 | 46.660 | 46.787 | 44.221 | 46.178 | 47.701 | 46.888 | 47.244 | 43.409 | 43.358 |
| 0 | 46.736 | 43.485 | 48.692 | 45.822 | 44.882 | 41.910 | 43.993 | 46.335 | 46.812 | 48.539 | 45.695 | 46.177 |

$= 0.542 + 0.008 - 0.78 \times 10^{-4}\ S^2$

$= 0.542 + 0.008\ S - 0.78 \times 10–4\ S^2$

$+ 0.62 \times 10^{-6}\ S^3$ ...(2.17)

where $S$ is the mean sunshine percentage.

$C_e$ = Coefficient of elevation

$= 0.97 + 0.00984\ E$ ...(2.18)

where $E$ is the elevation in 100 metres.

Values of $R$ for different latitudes are tabulated in Table 2.12.

**Example 2.13.** (*a*) *Determine the pan evaporation from the following data for the month of April, using Christiansen method.*

*Latitude 15° 19′ N Elevation + 449 metres*

*Month : April.*

*Mean Temperature 31.8°C.*

*Mean wind velocity at 0.5 m above the ground = 183 kilometres per day.*

*Mean relative humidity = 40%*

*Mean sunshine per cent = 89%*

*Use tables for extra-terrestrial radiation [(Table 2.12)(b)] What is the consumptive use for April in this country for a crop having a consumptive use coefficient equal to 0.80.*

**Solution.** Find the value of $R$ from Table 2.12 for the month of April and for a latitude of 15° 19′ $N$. It comes out to be about **47.3 cm**. Now, using eqs. (2.14) to (2.18), we get

$$C_t = 0.393 + 0.02796 \times 31.8 + 0.0001189\ (31.8)^2 = 1.403$$

$$C_w = 0.708 + 0.0034 \times 183 - 0.0000038\ (183)^2 = 1.200.$$

$$C_h = 1.250 - 0.0087 \times 40 + 0.75 \times 10^4\ (40)^2 - 0.85 \times 10^{-8}\ (40)^4 = 1.000$$

$$C_s = 0.542 + 0.008 \times 89 - 0.78 \times 10^{-4}\ (89)^2 + 0.62 \times 10^{-6}\ (89)^3 = 1.073.$$

$$C_e = 0.97 + 0.00984 \times 4.49 = 1.014.$$

(*a*) Pan evaporation $E_p$ is given by eq. (2.12) as :

$$\mathbf{E_p = 0.459\ R \cdot C_t \cdot C_w \cdot C_h \cdot C_s \cdot C_e}$$

$$= 0.459\ (47.3)\ (1.403)\ (1.200)\ (1.000)\ (1.073)\ (1.014)$$

$$= 39.8 \text{ cm. } \textbf{Ans.}$$

(*b*) $\mathbf{E_t} = K \cdot E_p = 0.8 \times 39.8 = 31.84$ cm

Hence, the required value of consumptive use = 31.84 cm. **Ans.**

**Example 2.14.** *Determine the consumptive use and net irrigation requirement for Jowar sown at Bellary (Karnataka) from the following data:*

**Table 2.13 (*a*)**

| *Dates and period of growth* | *Pan evaporation $E_p$ cm* | *Consumptive use coefficient (K)* | *Effective precipitation in cm.* |
|---|---|---|---|
| *(1)* | *(2)* | *(3)* | *(4)* |
| Oct. 16—31 | 8.49 | 0.44 | 3.42 |
| Nov. 1—30 | 15.57 | 0.54 | 2.19 |
| Dec. 1—31 | 16.59 | 0.94 | 0.54 |
| Jan. 1—31 | 19.10 | 0.99 | 0.15 |
| Feb. 1—2 | 1.54 | 0.73 | 0.02 |

**Solution.**

Proceed as shown in Table 2.13 (*b*).

**Table 2.13 (*b*)**

| *Dates* | $E_p$ *(cm)* | *K* | $K{\cdot}E_p = E_t = C_u$ *in cm* | $R_e$ *in cm* | $C_u - R_e =$ *N.I.R.* |
|---|---|---|---|---|---|
| *(1)* | *(2)* | *(3)* | *(4)* | *(5)* | *(6)* |
| Oct. 16—31 | 8.49 | 0.44 | 3.74 | 3.42 | 0.32 |
| Nov. 1—30 | 15.57 | 0.54 | 8.41 | 2.19 | 6.22 |
| Dec. 1—31 | 16.59 | 0.94 | 15.59 | 0.54 | 15.05 |
| Jan. 1—31 | 19.10 | 0.99 | 18.91 | 0.15 | 18.76 |
| Feb. 1—2 | 1.54 | 0.73 | 1.12 | 0.02 | 1.10 |
| | | | $\Sigma = 47.77$ | | $\Sigma = 41.45$ |

First of all, calculate the values of $C_u$ as in col. (4) Table 2.13 (*b*) and then determine N.I.R. as in col. (6) of this Table. Table is otherwise self-explanatory.

Total consumptive use = **47.77 cm**
Net Irrigation requirement = **41.45 cm** **Ans.**

**Note :** Based on the above method of calculating NIR, sample calculations for computing irrigation requirements. For cotton and wheat crops are shown in Table 2.14 and 2.15, respectively.

**Example 2.15.** *For use in Hargreaves method, in respect of Group D wheat crop, the following K values are suggested :*

| *% age of crop growing season* | 0 | 5 | 10 | 15 | 20 | 25 | 30 | 35 | 40 | 45 | 50 |
|---|---|---|---|---|---|---|---|---|---|---|---|
| *K* | *0.08* | *0.08* | *0.15* | *0.19* | *0.27* | *0.33* | *0.40* | *0.46* | *0.52* | *0.58* | *0.65* |
| *% age of crop growing season* | *55* | *60* | *65* | *70* | *75* | *80* | *85* | *90* | *95* | *100* | |
| *K* | *0.71* | *0.77* | *0.82* | *0.88* | *0.90* | *0.90* | *0.80* | *0.70* | *0.60* | *0.60* | |

*Wheat is grown between 1 November to 15 March. The effective rainfall in the individual months is taken as follows :*

*Nov.–0.5 cm; Dec.–1.6 cm; Jan–3.2 cm; Feb.–2.7 cm; March–Nil. Percolation losses are nil throughout, Field Irrigation Efficiency is 70% and Gross Irrigation Efficiency is 75%. The mean class A pan evaporation through the respective months are:*

*Nov.–13 cm; Dec.–11 cm; Jan.–7 cm; Feb.–15 cm; March–10 cm. Compute the gross irrigation requirement for the whole crop period including 7 cm of net pre-sowing requirement.* *(Engineering Services, 1989)*

**Solution.** Follow the sample calculations given for wheat crop in table 2.15, and compute G.I.R. as equal to 55.02 cm, as shown in table 2.16.

**2.13.3. Penman's Equation.** While the *Blaney Criddle equation* (1975) and the *Hargreaves class A pan equation using Christiansen formula* (1968) had been in use for the last many years for computing the consumptive use, $C_u$ (*i.e.* evapotranspiration, $E_t$) values, and net irrigation requirements for different crops; the *Penman equation* (1998) has, however, more recently been introduced for determing the consumptive use of different areas or different segments of a basin, depending upon the type of vegetation covering each sub-basin. The

**Table 2.14. Sample Calculations for Determining Irrigation Requirement of Cotton Crop. (in cm)**

*Cotton (Oct. 11—Feb. 26 = 139 days)*

| Dates | No. of days up to mid-point of interval | % of growing season = col. $(2) \times \frac{100}{139}$ | Pan Evap. $= E_p$ | $K$ | $E_pK = C_u$ | Effective rainfall $R_e$ | N.I.R. in cm | F.I.R. $\frac{N.I.R.}{\eta_a}$ | GIR $= \frac{FIR}{\eta_c}$ |
|---|---|---|---|---|---|---|---|---|---|
| (1) | (2) | (3) | (4) | (5) | (6) | (7) | (8) | (9) | (10) |
| **Oct. 11—31** | 11 | 7.9 | 14.73 | 1.23 | 3.38 | — | 3.38 | 3.96 | 5.28 |
| **Nov. 1—30** | 35 | 25.2 | 18.85 | 0.36 | 6.78 | — | 6.78 | 7.98 | 10.65 |
| **Dec. 1—31** | 67 | 48.2 | 17.60 | 1.00 | 17.60 | — | 17.60 | 20.70 | 27.60 |
| **Jan. 1—31** | 98 | 70.5 | 19.63 | 0.99 | 19.43 | — | 19.43 | 22.86 | 30.48 |
| **Feb. 1—26** | 126 | 90.6 | 19.94 | 0.61 | 12.17 | — | 12.17 | 14.30 | 18.07 |
| | | | | | $\Sigma = 59.36$ | | $\Sigma = 59.36$ | $\Sigma = 69.80$ | $\Sigma = 92.80$ |

**FIR = Field Irrigation Requirement (in cm) = $\frac{\text{NIR}}{\eta_a} = \frac{\text{NIR}}{0.85}$**

**FIR includes percolation losses in the field water-courses, field channels and in field application of water.**

**GIR = Gross Irrigation Requirement (in cm) = $\frac{\text{FIR}}{\eta_c} = \frac{\text{FIR}}{0.85}$**

**GIR = FIR + Conveyance losses in distributaries up to the field.**

**Table 2.15. Sample Calculations for Determining Irrigation Requirement for Wheat Crop (in cm)**

*Wheat* : Period of growth : 1 Nov. to 15 March (135 days)

| *Dates* | *No. of days up to midpoint of intervals* | *% of growing season = col.* $(2) \times \frac{100}{135}$ | *Pan evaporation* $(E_p)$ *in cm* | *Coeff. (K)* | $E_t = K \cdot E_p = C_u$ *Consumptive use in cm* | $R_e$ *in cm* | *N.I.R.* $= C_u - Re$ *cm* | *F.I.R.* $= \frac{NIR}{\eta_a} = \frac{NIR}{0.7}$ |
|---|---|---|---|---|---|---|---|---|
| *(1)* | *(2)* | *(3)* | *(4)* | *(5)* | *(6)* | *(7)* | *(8)* | *(9)* |
| Nov. 1—30 | 15 | 11 | 15 | 0.16 | 2.40 | 0.4 | 2.00 | 2.86 |
| Dec. 1—31 | 46 | 34 | 12.8 | 0.46 | 5.89 | 1.6 | 4.29 | 6.13 |
| Jan. 1—31 | 77 | 57 | 9.4 | 0.73 | 6.86 | 3.2 | 3.66 | 5.23 |
| Feb. 1—28 | 106 | 78 | 15.4 | 0.90 | 13.86 | 2.2 | 11.66 | 16.66 |
| March 1—15 | 128 | 95 | 10.1 | 0.60 | 6.06 | Nil | 6.06 | 8.66 |
| | | | | | $\Sigma = 35.07$ | | $\Sigma = 27.67$ | $\Sigma = 39.54$ |

*Note* :

1. The peak requirement during a month will be more than the average ; allowance should therefore, be made for this, while designing channel capacity.
2. Effective rainfall $(R_e)$ should be taken from the previous records, and the mean value should not be taken. On the other hand, the values which are available in 75% to 80% of the years should be taken.

**Table 2.16. Solution Table for Example 2.15. Growth period = 1 Nov. to 15 March (135 days)**

| *Dates* | *No. of days upto mid point of interval from start* | *%age of growing season col. (2)* × $\frac{100}{135}$ | *Pan evaporation* $(E_p)$ *in cm given value in Ex. 2.15* | *Coefficient (K) interpolated values for %age values of col. (3) from given values* | $E_t = KE_p = C_u$ *i.e. col. (4) × col. (5); consumptive use in cm* | $R_e$ *= Effective rainfall in cm (given)* | *Net irr. requirement NIR* $= C_u - R_e$ *= col. (6) – col. (7)* |
|---|---|---|---|---|---|---|---|
| (1) | (2) | (3) | (4) | (5) | (6) | (7) | (8) |
| Nov. 1—30 | 15 | 11 | 13 | 0.16* | 2.08 | 0.50 | 1.58 |
| Dec. 1—31 | 46 | 34 | 11 | 0.46 | 5.06 | 1.60 | 3.46 |
| Jan. 1—31 | 77 | 57 | 9 | 0.73 | 6.57 | 3.20 | 3.37 |
| Feb. 1—28 | 106 | 78 | 15 | 0.90 | 13.50 | 2.70 | 10.80 |
| March 1—15 | 128 | 95 | 10 | 0.60 | 6.00 | Nil | 6.0 |
| | | | | | Σ = 33.21 cm | 8.00 cm | 25.21 cm |

*Given value of $K$ for 10% of crop growing season = 0.15
value of $K$ for 15% of crop growing season = 0.19
∴ value of $K$ for 11% of crop growing season

$$= 0.15 + \frac{0.19 - 0.15}{4\%} \times 1\% = 0.16$$

N.I.R. = 25.21 cm

$$\text{F.I.R.} = \frac{N.I.R.}{\eta_f} = \frac{25.21}{0.70} = 36.01 \text{ cm}$$

$$\text{G.I.R.} = \frac{F.I.R.}{\eta_c} = \frac{36.01}{0.75} = 48.02 \text{ cm}$$

Net pre-sowing requirement before 1 Nov. = 7 cm.

∴ Total gross irrigation requirement = 48.02 + 7 = 55.02 cm. **Ans.**

advantage with this equation lies in the fact that the different specified values of *coefficient of reflection (albedo)*, a factor used in this equation, are available for different types of areas, which can be used in Penman's equation to compute consumptive use (*i.e.* Potential evapotranspiration, PAT) values for different segments of command area.

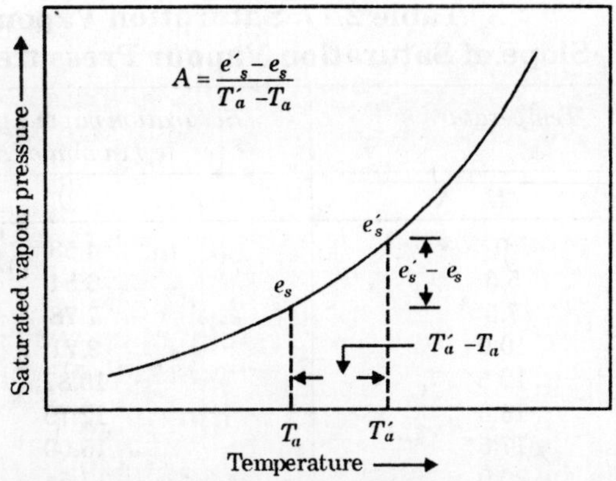


**Fig. 2.3.** Saturation vapour pressure *vs.* Temp. curve.

Penman's equation for computation of *PET* or $C_u$ for an area, has a sound theoretical reasoning, and it is not a simple empirical equation. This equation has, infact, been derived by intelligently combining the *energy balance* and *mass transfer approaches* of the computations of transpiration and evaporation, respectively. Hence, although slightly complicated mathematical conceptual work is involved here, yet its use is becoming more and more popular, in today's modern computer age.

Penman's equation, incorporating some of the modifications suggested by other investigators, is given as:

$$E_t = \frac{A \cdot H_n + E_a \cdot \gamma}{A + \gamma} \quad \text{...(2.19)}$$

where $E_t$ = Daily potential evapotranspiration

$A$ = Slope of the saturation vapour pressure *Vs* Temperature curve at the mean air temperature, as shown in Fig. 2.3, and values given in Table 2.17.

$H_n$ = Net incoming solar radiation or energy, expressed *in mm of evaporable water per day**

$E_a$ = A parameter including wind velocity and saturation deficit, as given by Eqn. (2.22) in mm/day

$\gamma$ = psychromatic constant = 0.49 mm of Hg/°C

The net radiation ($H_n$) in the above equation is estimated by the equation

$$H_n = H_c\,(1 - r)\left(a + b \cdot \frac{n}{N}\right) - \sigma \cdot T_a^{\,4}\,(0.56 - 0.092\sqrt{e_a} \times \left(0.10 + 0.90\,\frac{n}{N}\right) \quad \text{...(2.20)}$$

where $H_c$ = mean incident solar radiation at the top of the atmosphere on a horizontal surface, expressed in mm of evaporable water per day. This value is a function of latitude ($\phi$) of

---

*Since the energy required by water in evaporation equals 585 calories/gm = 585 calories/cc (1 gm = 1 cc), we have, Energy as 1 mm of evaporable water from an area of A hectares.

$= \left(\frac{1}{10}\text{ cm}\right)(A \times 10^8\text{ cm}^2) \times 585\text{ cal/cm}^3$ = **585 × 10⁷ A calories.**

**Table 2.17. Saturation Vapour Pressure ($e_s$), and Slope of Saturation Vapour Pressure *Vs* Temperature Curve (A)**

| *Temperature* | *Saturation vapour pressure ($e_s$) in mm of Hg* | *Slope A in mm / °C* |
|---|---|---|
| *(1)* | *(2)* | *(3)* |
| 0 | 4.58 | 0.30 |
| 5.5 | 6.54 | 0.45 |
| 7.5 | 7.78 | 0.54 |
| 10.0 | 9.21 | 0.60 |
| 12.5 | 10.87 | 0.71 |
| 15.0 | 12.79 | 0.80 |
| 17.5 | 15.00 | 0.95 |
| 20.0 | 17.54 | 1.05 |
| 22.5 | 20.44 | 1.24 |
| 25.0 | 23.76 | 1.40 |
| 27.5 | 27.54 | 1.61 |
| 30.0 | 31.82 | 1.85 |
| 32.5 | 36.68 | 2.07 |
| 35.0 | 42.81 | 2.35 |
| 37.5 | 48.36 | 2.62 |
| 40.0 | 55.32 | 2.95 |
| 45.0 | 71.20 | 3.66 |

the place and the period of the year, as per the mean monthly values given in table 2.19.

$r$ = **reflection coefficient** (*albedo*) of the given area. Usual values of this coefficient for different types of areas are given in Table 2.18.

$a$ = a constant depending upon the latitude ($\phi$) and is given as

$a = 0.29 \cos \phi$ ...(2.21)

$b$ = a constant having average value = 0.52

$n$ = actual duration of bright sunshine in hours

$N$ = maximum possible hours of bright sunshine (mean value). This value is a function of latitude ($\phi$), and its values are given in table 2.20 for each month of the year

$\sigma$ = *Stefan-Bolzman constant*
$= 2.01 \times 10^{-9}$ mm/day

$T_a$ = mean air temperature in °K = 273 + °C

$e_a$ = actual mean vapour pressure in the air in mm of Hg.

The parameter $E_a$ of Penman's equation (2.19) is estimated as :

$$E_a = 0.35 \left(1 + \frac{V_2}{160}\right) (e_s - e_a) \text{ mm/day} \quad \text{...(2.22)}$$

where $V_2$ = mean wind speed at 2 m above the ground in km/day

$e_s$ = saturation vapour pressure at mean air temperature in mm of Hg (Table 2.17)

$e_a$ = actual mean vapour pressure of air in mm of Hg.

With the help of the above equation, and using the values of $A$, $e$, $r$, $H_c$ and $N$ from Tables 2.17 to 2.20, $E_t$ or $C_u$ can be determined for the given area. This equation can also be used to compute the evaporation from a water surface (lake, etc.) by using $r = 0.05$. Due to its general applicability, this equation is widely used these days in India, the U.K., the Australia, and in some parts of U.S.A.

**Table 2.18. Values of Reflection Coefficient r (albedo)**

| *Surface* | *Range of r values* |
|---|---|
| Close grained crops | 0.15 – 0.25 |
| Bare lands | 0.05 – 0.45 |
| Water surface | 0.05 |
| Snow | 0.45 – 0.90 |

**Table 2.19. Mean Monthly Solar Radiation at Top of Atmosphere $H_c$ in mm of evaporable water/day**

| *North Latitude* | *Jan.* | *Feb.* | *Mar.* | *Apr.* | *May* | *Jun.* | *Jul.* | *Aug.* | *Sep.* | *Oct.* | *Nov.* | *Dec.* |
|---|---|---|---|---|---|---|---|---|---|---|---|---|
| 0° | 14.5 | 15.0 | 15.2 | 14.7 | 13.9 | 13.4 | 13.5 | 14.2 | 14.9 | 15.0 | 14.6 | 14.3 |
| 10° | 12.8 | 13.9 | 14.8 | 15.2 | 15.0 | 14.8 | 14.8 | 15.0 | 14.9 | 14.1 | 13.1 | 12.4 |
| 20° | 10.8 | 12.3 | 13.9 | 15.2 | 15.7 | 15.8 | 15.7 | 15.3 | 14.4 | 12.9 | 11.2 | 10.3 |
| 30° | 8.5 | 10.5 | 12.7 | 14.8 | 16.0 | 16.5 | 16.2 | 15.3 | 13.5 | 11.3 | 9.1 | 7.9 |
| 40° | 6.0 | 8.3 | 11.0 | 13.9 | 15.9 | 16.7 | 16.3 | 14.8 | 12.2 | 9.3 | 6.7 | 5.4 |
| 50° | 3.6 | 5.9 | 9.1 | 12.7 | 15.4 | 16.7 | 16.1 | 13.9 | 10.5 | 7.1 | 4.3 | 3.0 |

**Table 2.20. Mean Monthly Values of Possible Sunshine Hours (N)**

| *North Latitude* | *Jan.* | *Feb.* | *Mar.* | *Apr.* | *May* | *Jun.* | *Jul.* | *Aug.* | *Sep.* | *Oct.* | *Nov.* | *Dec.* |
|---|---|---|---|---|---|---|---|---|---|---|---|---|
| 0° | 12.1 | 12.1 | 12.1 | 12.1 | 12.1 | 12.1 | 12.1 | 12.1 | 12.1 | 12.1 | 12.1 | 12.1 |
| 10° | 11.6 | 11.8 | 12.1 | 12.4 | 12.6 | 12.7 | 12.6 | 12.4 | 12.9 | 11.9 | 11.7 | 11.5 |
| 20° | 11.1 | 11.5 | 12.0 | 12.6 | 13.1 | 13.3 | 13.2 | 12.8 | 12.3 | 11.7 | 11.2 | 10.9 |
| 30° | 10.4 | 11.1 | 12.0 | 12.9 | 13.7 | 14.1 | 13.9 | 13.2 | 12.4 | 11.5 | 10.6 | 10.2 |
| 40° | 9.6 | 10.7 | 11.9 | 13.2 | 14.4 | 15.0 | 14.7 | 13.8 | 12.5 | 11.2 | 10.0 | 9.4 |
| 50° | 8.6 | 10.1 | 11.8 | 13.8 | 15.4 | 16.4 | 16.0 | 14.5 | 12.7 | 10.8 | 9.1 | 8.1 |

**Example 2.16.** *Compute the total consumptive use (Cu) from a drainage basin located near Gurgaon (Haryana) during the month of April by Penman's formula. The following data is given:*

*Latitude of place* = *28° N*

*Elevation* = *RL 220 m.*

**Meteorologically observed data during April**

*Mean monthly temperature* = *40°C*

*Mean relative humidity* = *35%*

*Mean observed sunshine hours per day* = *13h*

*Mean wind velocity at 2 m height* = *72 km/day/(3 km/h)*

**Data available and obtained from standard charts**

*(i) Slope of the $e_s$ Vs T° chart at 40°C* = *2.95 mm of Hg/°C*

*(ii) Saturation pressure at 40°C* = *55.32 mm of Hg*

*Mean monthly solar radiation at top of atmosphere during April for 28°N latitude* = *14.9 mm of evaporable water / day*

*Mean monthly value of possible sunshine hours for April for 28°N latitude*
*= 12.9 h*

*Albedo for the area* *= 0.25.*

**Solution.** Penman's equation (2.19) is given as :

$$E_t = C_u = \frac{A\,H_n + E_a\,.\,\gamma}{A + \gamma}$$

where, $A$ = 2.95 mm of Hg/°C

$H_n$ = to be computed by eq. (2.20)

$E_a$ = given by eq. (2.22) as :

$$= 0.35\left(1 + \frac{V_2}{160}\right)(e_s - e_a) \text{ mm/day}$$

where $V_2$ = 72 km/d (given)

$e_a$ = (Relative humidity × $e_s$

= 35% × 35.52 = 19.36 mm Hg

$$\therefore \quad E_a = 0.35\left[1 + \frac{72}{160}\right](55.32 - 19.36)$$

= 18.07 mm/day

$\gamma$ = constant = 0.49 mm of Hg/°C

$H_n$ is given by eqn. (2.20) as :

$$H_n = H_c\,(1 - r)\left(a + b \cdot \frac{n}{N}\right) - \sigma\, T_a^{\,4}\,(0.56 - 0.092\,\sqrt{e_a}) \times \left(0.10 + 0.90\,\frac{n}{N}\right)$$

where $H_c$ = mean monthly incident solar radiation at top ol atmosphere = 14.9 mm of evaporable water/day

$r$ = 0.25

$a$ = 0.29 cos $\phi$ = 0.29 cos 28° = 0.256

$b$ = 0.52

$n$ = 13 h

$N$ = 12.9 h

$\sigma$ = 2.01 × $10^{-9}$ mm/day

$T_a$ = (40°C + 273) = 313°K

$e_a$ = (R.H.) $e_s$ = 35% × 55.32 = 19.36 mm of Hg

Substituting values, we get

$$H_n = 14.9\,(1 - 0.25)\left[0.256 + 0.52 \times \frac{13}{12.9}\right]$$

$$- \left[\{2.01 \times 10^{-9} \times (313)^4\} \times \{0.56 - 0.092\sqrt{19.36}\} \times \left\{0.10 + 0.90 \times \frac{13}{12.9}\right\}\right]$$

= 14.9 (0.75) (0.78) – (19.292) (0.155) (1.007)

= 8.716 – 3.011 = **5.705** mm of evaporative water/day

Now,
$$C_u = \frac{AH_n + E_a\gamma}{A + \gamma} = \frac{2.95 \times 5.705 + 18.07 \times 0.49}{2.95 + 0.49}$$

$$= \frac{49.648 + 8.854}{3.44} = \mathbf{17.01 \text{ mm/day.}} \quad \textbf{Ans.}$$

**2.13.4. Comparison of Blaney-Criddle Equation, Hargreaves-Christiansen Equation and Penman's Equation.** These three empirical equations have been developed by the various researchers over the last 40 years to estimate evapotranspiration ($E_t$) values for different crops, or area segments vegetated with the same cropping pattern, under different climatic variables. Since the suggested empirical equations are often subjected to rigorous local calibrations, they can not have a global validity. It however, becomes difficult to suggest as to which equation should be used in a particular case. *A study made in Chandigarh region has, however, shown that the annual evapotranspiration values obtained from Penman's equation are quite close to the values obtained from the actual field observations made in pan evaporation method ; while the values obtained by Blaney-Criddle equation were on much higher side (about 30% higher), and the values obtained by Hargreaves-Christiansen equation were on lower side (about 15-20% lower).* No definite conclusion can, hence, be drawn regarding the decision on the global use of a particular equation.

Although, use of Penman's equation is being largely advocated these days, yet since the equation needs elaborate data, it may not be always feasible to use this equation. Moreover, this equation can be used for generalised vegetated areas, and not for individual crops, since the value range of reflection coefficient *i.e. albedo* (*r*), as used in this equation, is given for areas having close grained crops, as to vary between 0.15-0.25 (Pl. see Table 2.18)

### 2.14. Soil-Moisture-Irrigation Relationship

The water below the watertable is known as ground water and above the watertable as soil-moisture.

Extending down from the ground surface, is the soil zone or the root zone, which is defined as being the depth of overburden that is penetrated by the roots of vegetation, as shown in Fig. 2.4. This zone is the most important from irrigation point of view, because it is this zone, from which the plants do take their water supplies. When water falls over the ground, a part of it gets absorbed in this root zone, and the rest flows downward under the action of gravity and is called *gravity water*.

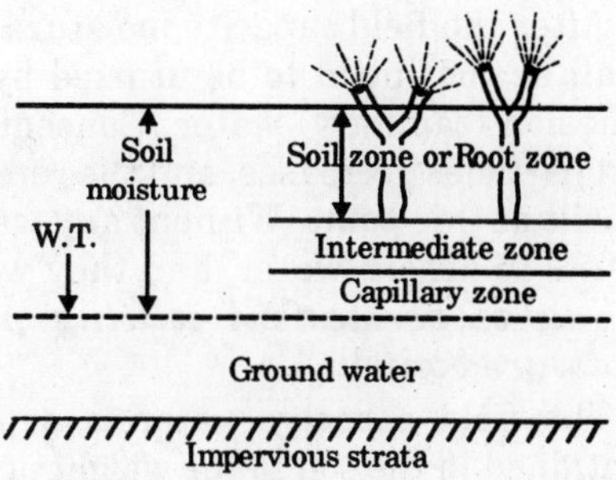

Fig. 2.4.

**2.14.1. Field Capacity.** Consider what happens to water that enters soil after heavy rain or after the application of irrigation water. At first, the pores* in the soil mass are filled with water. This is called saturation. For well draining soils like sandy and silty loams, this water does not normally remain in the soil pores for more than 2 to 5 days. It moves downward through the soil profile by gravity, to join the water table in well drained soils, or is removed by tile drains in heavy retentive cohesive soils.**

When all the gravity water has drained down to the water-table or to tile drains a certain amount of water is retained on the surfaces of soil grains by molecular attraction and by loose chemical bonds.

There infact, exists two attractive forces. one is the *cohesive attraction* between the water molecules, which have positively charged and negatively charged ends. Second, the water molecules are attracted to soil grains by *adhesive force,* which is

---

*Any volume of soil contains solid materials (about 50%) and pore spaces (space between soil particles). This pore space is about 50% of any volume of soil, and known as void space, which is filled with air and water.

**In heavy retentive cohesive soils, this extra gravity water can be removed with tile drains, as not to damage the crops by blocking the air space and causing exygen shortage to plants.

much stronger than the cohesive force.

The water which remains attached to soil grains by these two forces cannot be easily drained under the action of gravity, and is called *field capacity water* or *field capacity.*

The **field capacity water**, thus, consists of two parts. (*i*) One part is that which is attached to soil molecules by cohesive attraction between water molecules. The force of surfaces tension* helps this water to stick against the gravitational force, and this water can be easily extracted by plants by ***capillarity***. This water is called ***capillary water***. (*ii*) The other part is that which *adheres* to soil particles by loose chemical bonds between water and soil grains (adhesion). This water cannot be easily removed by capillarity, as it can be removed only by heating. Hence, this water is not available to the plants, and is called the ***hygroscopic water***.

When drainage of gravity water stops, the soil moisture level is at field capacity. At this point, air fills the larger pores (voids) and thick film of water (cohesive water) surrounds such soil particles. This is ideal moisture and plants grow best at this moisture level.

After the field capacity moisture is reached on stoppage of drainage, the water shall be continued to be utilised by plants by capallirity to start depleating the cohesion (capillary) water. Consequently the size of the water films around the soil particles decreases, and the remaining water is held more tightly. Plants start to wilt at this point. Without any water input, it becomes increasingly difficult for plants to absorb water, and they wilt permanently. The moisture at his point is known as ***permanent wilting point moisture content*** **or** ***wilting piont moisture content.***

The field capacity water is expressed as the ratio of the weight of water contained in the soil to the weight of the dry soil retaining that water : *i.e.*

$$\text{Field Capacity} = \frac{\text{Wt. of water retained in a certain vol. of soil}}{\text{Wt. of the same volume of dry soil}} \times 100 \quad ...(2.23)$$

If we consider 1 $m^2$ area of soil and $d$ metre depth of root zone, then the volume of soil is $d \times 1 = d$ cubic metres. If the dry unit wt. of soil is $\gamma_d$ kN/$m^3$**, then the Wt. of $d$ cubic metres of soil is $\gamma_d \cdot d$ kN. If $F$ is the field capacity, then

$$F = \frac{\text{Wt. of water retained in unit area of soil}}{\gamma_d . d}$$

or Wt. of water retained in unit area of soil $= \gamma_d \, . \, d \, . \, F$ kN/$m^2$

∴ $$\text{Vol. of water stored in unit area of soil} = \frac{\gamma_d \, . \, d \, . \, F \text{ kN/m}^2}{\gamma_w \text{ kN/m}^3}$$

or Total water storage capacity of soil in (m depth of water)

$$= \frac{\gamma_d \, . \, d \, . \, F}{\gamma_w} \text{ m}$$

where $F$ = the field capacity m.c.

$d$ = depth of root zone in m

$\gamma_w$ = the unit wt. of water

---

****Surface tension*** is defined as the property of the surface of a liquid that allows it to resist an external force, due to coehsive nature of the water.

**It is the unit wt. of the dried soil sample and not of the soil solids. It may sometimes hence be called as **apparent unit wt.**

$\gamma_d$ = the dry unit wt. of soil.

Hence, the depth of water stored in the root zone in filling the soil upto field capacity

$$= \frac{\gamma_d \cdot d \cdot F}{\gamma_w} \text{ metres} \qquad ...(2.24)$$

The knowledge of field capacity is very important, because it is the field capacity water which can supply water for plant nourishment. The larger part of applied water drains down and joins the watertable and is thus a waste from irrigation point of view.

As expressed earlier, the total field capacity water cannot be utilised by the plants. The plants can extract water from the soil till the permanent wilting point is reached. The **permanent wilting point** is that water content at which plant can no longer extract sufficient water for its growth, and wilts up. It is the point at which permanent wilting of plants take place. *It, therefore, becomes evident that the water which is available to the plants, is the difference of field capacity water and permanent wilting point water. This is known as available moisture or maximum storage capacity of soil. Hence, the available water or* **available moisture content** may be defined as the difference in water content of the soil between field capacity and permanent wilting point. The water left in the soil after the permanent wilting point is reached, cannot be removed, and is known as, unavailable moisture or ***Hygroscopic water*** (See Fig. 2.5).

**2.14.2. Readily available moisture.** It is that portion of the available moisture which is most easily extracted by the plants, and is approximately 75 to 80% of the available moisture.

**2.14.3. Soil-moisture deficiency.** The water required to bring the soil moisture content of a given soil to its field capacity is called the *field moisture deficiency* or *soil-moisture deficiency.*

**2.14.4. Equivalent moisture.** Just as the field capacity is the water retained by a saturated soil after being acted upon by gravity; similarly, *equivalent moisture* is the water retained by a saturated soil after being centrifuged for 30 minutes by a centrifugal force of 1000 times that of gravity. Therefore, it is slightly less, or at the most equal to the field capacity.

## 2.15. Estimating Depth and Frequency of Irrigation on the Basis of Soil Moisture Regime Concept

Water or soil moisture is consumed by plants through their roots. It, therefore, becomes necessary that sufficient moisture remains available in the soil from the surface to the root zone depth. As explained earlier, the soil moisture in the root zone can vary between field capacity (upper limit) and wilting point moisture content (lower limit) as shown in Fig. 2.5.

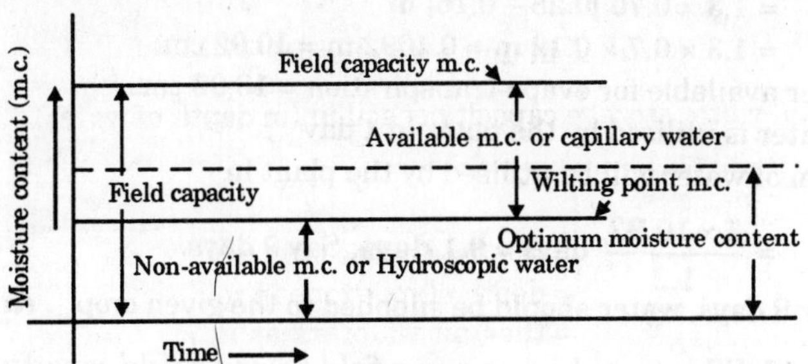


Fig. 2.5.

It is also evident from the previous discussion that the soil moisture is not allowed to be depleted up to the wilting point, as it would result in considerable fall in crop yield. The optimum level up to which the soil moisture may be allowed to be depleted in the root zone without fall in crop yield, has to be worked out for every crop and soil, by experimentation. *The irrigation water should be supplied as soon as the moisture falls up to this optimum level (fixing irrigation frequency) and its quantity should be just sufficient to bring the moisture content up to its field capacity, making allowance for application losses (thus fixing water depth).*

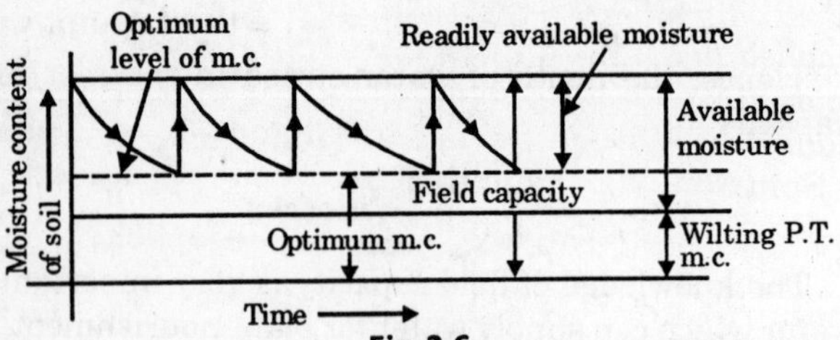


**Fig. 2.6.**

Water will be utilised by the plants after the fresh irrigation dose is given, and soil moisture will start falling. It will again be recouped by a fresh dose of irrigation, as soon as the soil moisture reaches the optimum level, as shown in Fig. 2.6.

**Example 2.17.** *After how many days will you supply water to soil in order to ensur sufficient irrigation of the given crop, if*

*(i) Field capacity of the soil* = *28%*
*(ii) Permanent wilting point* = *13%*
*(iii) Dry density of soil* = *1.3 gm/c.c.*
*(iv) Effective depth of root zone* = *70 cm*
*(v) Daily consumptive use of water for the given crop = 12 mm.*
*Assume any other data, not given.* *(Engineering Services, 1974*

**Solution.** We know, by definition of available moisture, that the availabl moisture = Field capacity – Permanent wilting = 28 – 13 = 15%.

Let us assume that the readily available moisture or the optimum soil moistur level is 80% of available moisture.

*i.e.*, Readily available moisture = 0.80 × 15% = 12%

∴ Optimum moisture = 28 – 12 = 16%

It means that the moisture will be filled by irrigation between 16% and 28%.

Depth of water stored in root zone between these two limits

$$= \frac{\gamma_d \cdot d}{\gamma_w} \text{ [Field capacity m.c. – Optimum m.c.]}$$

$$\text{where } \frac{\gamma_d}{\gamma_w} = \frac{\rho_d \cdot g}{\rho_w \cdot g} = \frac{\rho_d}{\rho_w} = \frac{1.3 \text{ gm/cc}}{1.0 \text{ gm/cc}} = 1.3$$

$$d = 0.7 \text{ m (given)}$$

$$= 1.3 \times 0.70\ [0.28 - 0.16] \text{ m}$$

$$= 1.3 \times 0.7 \times 0.12 \text{ m} = 0.1092 \text{ m} = 10.92 \text{ cm.}$$

Hence, water available for evapo-transpiration = 10.92 cm.

1.2 cm of water is utilised by the plant in 1 day

∴ 10.92 cm of water will be utilised by the plant in

$$= \frac{1 \times 10.92}{1.2} \text{ days} = \mathbf{9.1 \text{ days}}\text{; Say 9 days.}$$

Hence, after 9 days, water should be supplied to the given crop. **Ans.**

**Example 2.18.** *Wheat is to be grown in a field having a field capacity equal to 27% and the permanent wilting point is 13%. Find the storage capacity in 80 cm depth of the soil, if the dry unit weight of the soil is 14.72 kN/m$^3$. If irrigation water is to be supplied when the average soil moisture falls to 18%, find the water depth required to be* *supplied to the field if the field application efficiency is 80%. What is the amount of water needed at the canal outlet if the water lost in the water-courses and the field channels is 15% of the outlet discharge ?*

**Solution.** Maximum storage capacity or Available moisture.

$$= \frac{\gamma_d \cdot d}{\gamma_w}\left[\frac{\text{Field capacity m.c.}}{100} - \frac{\text{Wilting pt. m.c.}}{100}\right]$$

where $\gamma_d$ = 14.72 kN/m$^3$

$d$ = depth of root zone = 0.8 m

∴ Max. storage capacity or max. Available moisture

$$= \frac{14.72}{9.81} \times 0.8\,[0.27 - 0.13] \qquad [\because \gamma_w = 9.81 \text{ kN/m}^3]$$

$$= 1.2\,[0.14] = 0.168 \text{ metres} = 16.8 \text{ cm} \quad \textbf{Ans.}$$

Since the moisture is allowed to vary between 27% and 18%, the deficiency created in this fall

$$= \frac{14.72}{9.81} \times 0.8\,[0.27 - 0.18]$$

$$= 1.2 \times 0.09 = 0.108 \text{ metres} = \textbf{10.8 cm.}$$

Hence, 10.8 cm depth of water is the net irrigation requirement.

Quantity of water required to be supplied to the field (F.I.R.) $= \dfrac{\text{N.I.R.}}{\eta_a}$

or $$\text{F.I.R.} = \frac{\text{N.I.R.}}{0.80} = \frac{10.8}{0.80} = 13.5 \text{ cm.} \quad \textbf{Ans.}$$

Quantity of water needed at the canal outlet

$$= \frac{\text{F.I.R.}}{\eta_c} = \frac{13.5}{0.85} = 15.88 \text{ cm.} \quad \textbf{Ans.}$$

**Example 2.19.** *800 m$^3$ of water is applied to a farmer's rice field of 0.6 hectares. When the moisture content in the soil falls to 40% of the available water between the field capacity (36%) of soil and permanent wilting point (15%) of the soil crop combination, determine the field application efficiency. The root zone depth of rice is 60 cm. Assume porosity = 0.4.* **(Civil Services, 1994)**

**Solution.** We have defined Field Capacity m.c. (F) as :

$$F = \frac{\text{Wt. of water contained in a certain vol. of soil}}{\text{Wt. of the same volume of dry soil}} \quad \text{...(Eq. 2.23)}$$

(*i.e.* wt. of dry soil retaining that water)

If a saturated soil contains volume equal to $V$, and the volume of its voids is $V_v$, then the weight of water contained in this soil = $\gamma_w \cdot V_v$ ; where $\gamma_w$ is the unit wt. of water. The wt. of this soil of $V$ m$^3$ after it is oven dried to remove water and to fill the voids with air, is given by $\gamma_d \cdot V$ ; where $\gamma_d$ is the *dry unit wt. of the soil.*

∴ $$F = \frac{\gamma_w \cdot V_v}{\gamma_d \cdot V} \qquad \text{But } \frac{V_v}{V} = n \text{ (porosity)}$$

∴ $$F = \frac{\gamma_w}{\gamma_d} \cdot n$$

∴ $$\frac{\gamma_d}{\gamma_w} = \frac{n}{F} = \frac{0.4}{0.36} = 1.11 \qquad \because n = \text{Porosity} = 0.4 \text{ (given)}$$

$F = \text{F.C.} = 0.36$

Max. quantity of water stored between field capacity (F.C.) and permanent wilting point (P.W.)

$= \left(\dfrac{\gamma_d}{\gamma_w}\right) d \cdot (w_F - w_P)$; where $d$ = root zone depth = 0.6 m (given)

$= 1.11 \times 0.60\,[0.36 - 0.15] = 0.14$ m.

Deficiency of water created when irrigation is done

$= 60\% \times 0.14$ m

[∵ irrigation water is applied when m.c. falls to 40% of m.c. available between F.C. and W.P.]

= **0.084 m**

Hence, irrigation water is supplied to fill up 0.084 m depth of water.

∴ Vol. of irrigation water required to fill up the created deficiency

= 0.084 m × (0.6 hect.)

$= 0.084 \text{ m} \times (0.6 \times 10{,}000) \text{ m}^2 = \mathbf{504\ m^3}$ ...(i)

Actual irrigation water supplied = 800 $m^3$ ...(ii)

∴ Efficiency of field application $= \dfrac{504}{800} = \mathbf{63\%}$. **Ans.**

**Example 2.20.** *Work out the irrigation schedule based on the soil moisture concept, given the following information. Also extract the data on the total depth of irrigation water required and the respective dates of irrigation water supply :*

(a) *The crop is grown in an appropriate soil with no restrictive layers within the top 1.5 m depth of soil.*

(b) *Normal root zone depth of the crop is 1.2 m.*

(c) *Bulk density of soil is 1.35.*

(d) *Field capacity is 18% and permanent wilting point is 7%.*

(e) *Moisture level in the soil is to be maintained at not less than one-third of available retention. Irrigation will then be done over a duration of 2 days at a uniform rate of supply and at a uniform rate of advance to fully and just compensate for the depletion.*

(f) *No extra water is ever required for leaching.*

(g) *Sowing is done on 1 November when the soil moisture is left just at field capacity in the entire root zone.*

(h) *For the crop, at the location, the average evapotranspiration rates are :*

| | | |
|---|---|---|
| *1 Nov. — 30 Nov.* | : | *1.1 mm/day* |
| *1 Dec. — 31 Dec.* | : | *1.7 mm/day* |
| *1 Jan. — 31 Jan.* | : | *2.4 mm/day* |
| *1 Feb. — 28 Feb.* | : | *1.5 mm/day* |
| *1 March — 25 March* | : | *3.5 mm/day* |

(i) *Harvesting is done on or after 26 March.*

(j) *There is expected an effective rainfall of 24 mm during 4 January to 19 January, both days inclusive, with uniform intensity.*

(k) *By the end of the crop growth season, only the minimum water needed to be left unused in the root zone.* (*Engineering Services, 1990*)

**Solution.**

Max. moisture retained by soil = Field capacity = 18%

Permanent wilting *i.e.* below which soil cannot extract water for plant's growth = 7%

∴ Max. moisture available for plant's growth *i.e. available moisture retention*

= 18 − 7 = 11%

$\frac{1}{3}$ available moisture = $\frac{11}{3}$ % = 3.67%

∴ Moisture level at which irrigation must start = Minimum m.c. at which plants start wilting + $\frac{1}{3}$ of available moisture (given) = 7% + 3.67% = 10.67%.

This means that we will start irrigation as soon as m.c. falls to 10.67%, and will, thus, fill the soil with moisture till it rises to 18% (field capacity).

Irrigation water required to increase m.c. of soil in root zone from 10.67% to 18% is obtained by equation :

$$= \frac{\gamma_d \cdot d}{\gamma_w} \left[ \begin{array}{c} \text{Upper limit m.c.} \quad \text{Lower limit m.c.} \\ \text{as fraction} - \text{as fraction} \end{array} \right]$$

where $\gamma$ = Unit wt. (Apparent ) of soil

$w$ = Unit wt. of water

∴ $\frac{\gamma}{w}$ = Density (apparent) of soil = 1.35 (given)

$d$ = root zone depth = 1.2 m (given)

$$= 1.35 \times 1.2 \left[ \frac{18}{100} - \frac{10.67}{100} \right] = 1.35 \times 1.2[0.18 - 0.1067] = \mathbf{11.87\ cm.}$$

In other words, as and when this 11.87 cm depth of stored moisture gets consumed by evapotranspiration, irrigation water will be supplied.

From the given consumptive uses, we find that

Water consumed from 1st Nov. to 3 Jan. (when rains start)

= 1.1 mm/day × 30 days (*i.e.*, between 1 Nov. – 30 Nov.)

+ 1.7 mm/day × 31 days (*i.e.*, between 1 Dec. – 31 Dec.)

+ 2.4 mm/day × 3 days (*i.e.*, between 1 Jan. – 3 Jan.)

= 92.9 mm = **9.29 cm** ...(*i*)

Hence, water withdrawn from soil during Ist Nov. to 3 Jan. = **9.29 cm** (< permissible 11.87 cm).

No irrigation is, thus, required till then.

During rains (between 4 Jan. to 19 Jan.), effective rain water received by soil per day

$$= \frac{2.4 \text{ cm}}{16 \text{ days}} = 0.15 \text{ cm/day}$$

Consumptive use of 0.24 cm/day during this period, means that an amount of 0.24 – 0.15 = 0.09 cm/day of moisture is only consumed from soil, *i.e.*

Additional water consumed from soil during 4 Jan. – 19 Jan.

= 0.09 cm/day × 16

= **1.44 cm** ...(*ii*)

Hence, water withdrawn from soil during 1st Nov.—19 Jan.

= (*i*) + (*ii*) = 9.29 + 1.44 = 10.73 cm.

Balance water left in soil to be withdrawn before irrigation

= 11.87 – 10.73 = 1.14 cm.

This is consumed @ 2.4 mm/day in $x$ days,

$$\text{where} \quad x = \frac{1.14}{0.24} = 4.75 \text{ days, say 4 days.}$$

*Hence, Ist irrigation will be needed after 4 days from 20th Jan., i.e., on 24th Jan. This irrigation is to be done over 2 days (given), i.e., on 24 Jan. and 25 Jan.*

*First irrigation water required on 24 Jan. and 25 Jan.*

$$= (10.73 + 4 \times 0.24) + \frac{2.4}{10} \text{ cm/day} \times 2 \text{ days}$$

(to compensate for depletion in 2 days)

$$= 11.69 + 0.48 = \mathbf{12.17 \text{ cm.}} \quad \textbf{Ans.}$$

With effect from 26 Jan., water is again consumed as below:

Between 26 Jan.–31 Jan. = 0.24 cm/day × 6 days = 1.44 cm
Between 1 Feb.–28 Feb. = 0.15 cm/day × 28 days = 4.20 cm
Between 1 March–25 March = 0.35 cm/day × 25 days = 8.75 cm
Total = 14.39 cm
> 11.87 cm

Hence, another water is required after $x$ days of March, where

$$x = \frac{11.87 - (1.44 + 4.5)}{0.35} = 17.8 \text{ days } i.e., \text{ 17 days}$$

Hence, 2nd irrigation should start on **18th March** and water depth now required is only

$$= 14.39 - 11.87 = \mathbf{2.52 \text{ cm.}}$$

Thus, only 2.52 cm irrigation water is required at 2nd time.

Hence, *the required irrigation schedule is*

(*i*) 1st watering on 29th and 30 Jan. = 12.17 cm of water depth
(*ii*) 2nd watering on 18th March = 2.52 cm of water depth **Ans.**

**Example 2.21.** *A sandy loam soil holds water at 140 mm/m depth between field capacity and permanent wilting point. The root depth of the crop is 30 cm and the allowable depletion of water is 35%. The daily water use by the crop is 5 mm/day. The area to be irrigated is 60 ha and water can be diverted at 28 l.p.s. The surface irrigation application efficiency is 40%. There are no rainfall and ground water contribution.*

**Determine :**

(*i*) *allowable depletion depth between irrigations.* (*ii*) *frequency of irrigation* (*iii*) *net application depth of water* (*iv*) *volume of water required* (*v*) *time to irrigate 4 ha plot.* (*Engineering Services 1999*)

**Solution.** Moisture holding capacity of soil = 140 mm/m depth

Depth of root zone = 30 cm = 0.3 m

$\therefore$ Moisture holding capacity of root zone = $140 \dfrac{\text{mm}}{\text{m}} \times 0.3 \text{ m} = 42 \text{ mm} = \mathbf{4.2 \text{ cm}}$

Allowable depletion = 35%

(*i*) $\therefore$ Available moisture depth or Allowable depletion depth between irrigations = 35% × 4.2 cm = **1.47 cm. Ans.**

Daily use of water = consumptive use = 5 mm/day

(*ii*) $\therefore$ Frequency of irrigation

$$= \frac{\text{Available moisture}}{\text{Moisture consumed per day}} = \frac{1.47 \text{ cm}}{0.5 \text{ cm/day}}$$

= 2.94 days., say **3 days. Ans.**

Net water depth to be applied while irrigating each time after 3 days

= 3 × 0.5 = **1.5 cm Ans.** (in place of 1.47 cm)

Field irrigation requirement

$$= \frac{\text{Net irrigation requirement}}{\text{Efficiency of irrigation}} = \frac{1.5}{0.4} = 3.75 \text{ cm}$$

(*iv*) ∴ Quantity. of water required in the fields

$= 3.75$ cm of water depth $= 3.75$ cm × Area of field

$$= 3.75 \text{ cm} \times 60 \text{ ha} = \frac{3.75 \text{ m}}{100} \times (60 \times 10^4) \text{ m}^2 = 22{,}500 \text{ m}^3$$

Hence, volume of water required to irrigate 60 ha area, each time at 3 days interval = **22,500 m³** **Ans.**

(*v*) *Time to irrigate 4 ha when irrigation water is supplied @ 28 lps :*

Volume of water required to irrigate 4 ha plot

$$= 3.75 \text{ cm} \times 4 \text{ ha} = \frac{3.75}{100} \times (4 \times 10^4) \text{ m}^3 = 1500 \text{ m}^3$$

Time during which 1500 m³ of water can be supplied @ 28 Lps.

$$= \frac{1500 \times 10^3 \ . \ L}{28 \text{ Lps}} = \frac{1500 \times 10^3}{28} \text{ s}$$

$$= \frac{1500 \times 10^3}{28} \times \frac{1}{60 \times 60} \text{ hr} = \mathbf{14.88 \text{ hr.}} \quad \textbf{Ans.}$$

**Example 2.22.** *Determine the field capacity of a soil for the following data :*
*(i) Depth of root zone = 1.8 m; (ii) Existing moisture = 8% ; (iii) Dry density of soil 1450 kg/m³ ; (iv) Quantity of water applied to soil = 650 m³ ; (v) Water lost du to deep percolation and evaporation = 10% ; (vi) Area to be irrigated = 1000 m*

**Solution.** Volume of total water applied = 650 m³.

Water wasted = 10% of 650 m³ = 65 m³.

Water used in raising m.c. up to field capacity = 650 – 65 = 585 m³.

Depth of water used in raising m.c. up to to field capacity from the existing 8%

$$= \frac{585 \text{ m}^3}{\text{Area} = 1000 \text{ m}^2} = 0.585 \text{ m}$$

But water depth required to increase m.c. in root zone from one level to another higher level

$$= \frac{\gamma_d \cdot d}{\gamma_w} \text{ [upper limit m.c. – lower limit m.c., as fractions]}$$

$$\therefore \quad 0.585 = \frac{1.45 \text{ t/m}^3}{1 \text{ t/m}^3} \times 1.8 \text{ m [F.C.} - 0.08]$$

or $(\text{F.C.} - 0.08) = 0.224$

or $\text{F.C.} = 0.224 + 0.08 = 0.304$

Hence, Field capacity = 30.4%. **Ans.**

**Example 2.23.** *The following data pertains to healthy growth of a crop :*

*(i) Field capacity of soil = 30% ;*

*(ii) Permanent wilting percentage = 11%;*

*(iii) Density of soil = 1300 kg/m³ ;*

*(iv) Effective depth of root zone = 700 mm ;*

*(v) Daily consumptive use of water for the given crop = 12 mm*

*For healthy growth, moisture content must not fall below 25% of the water holding capacity between the field capacity and the permanent wilting point. Determine the watering interval in days.* (*Engg. Services, 1992*)

**Solution.** Max. water holding capacity of soil

= Field capacity m.c. – Permanent wilting point m.c.

= 30% – 11% = 19%

25% of Max. holding capacity = 0.25 × 19% = 4.75%

The left out available m.c. must at no stage be allowed to fall below this value and watering should be applied as soon as the left out m.c. falls to such a low level. Hence lower limit to which m.c. can be allowed to fall = 4.75% above wilting point m.c. (at which plants fade away) = 4.75% + 11% = 15.75%.

In other words, the m.c. will be allowed to vary between 30% and 15.75%. Water depth stored between these two limits

$$= \frac{\gamma_d \cdot d}{\gamma_w} [30\% - 15.75\%]$$

$$= \frac{1300}{1000} \times 0.7\,[0.30 - 0.1575] = 0.1297 = 12.97 \text{ cm}$$

Since consumptive use is 12 mm/day ;

12 mm of water is consumed in = 1 day

$\therefore$ 12.97 cm of water will be consumed in $= \frac{1}{12} \times (12.97 \times 10) = 10.8$ days.

The watering should, therefore, be applied after every 10 days.

**Note.** 10.8 days cannot be rounded to 11 days, as otherwise the left out m.c. will fall down below the minimum required. **Ans.**

**Example 2.24.** *A certain crop is grown in an area of 3000 hectares, which is fed by a canal system. The data pertaining to irrigation are as follows :*

| | |
|---|---|
| *Field capacity of soil* | *= 26%* |
| *Optimum moisture* | *= 12%* |
| *Permanent wilting point* | *= 10%* |
| *Effective depth of root zone* | *= 80 cm* |
| *Apparent relative density of soil* | *= 1.4* |

*If the frequency of irrigation is 10 days and the overall irrigation efficiency is 22%, find :*

(*i*) *the daily consumptive use ; and*

(*ii*) *the water discharge in* $m^3/s$ *required in the canal feeding the area.*

(*Civil Services, 1990*)

**Solution.** Follow Example 2.14 and 2.15, and determine :

Depth of water used by plants for growth, which is supplemented by irrigation after every 10 days

$$= \frac{\gamma_d \cdot d}{\gamma_w} [\text{Field capacity m.c.} - \text{Optimum m.c.}]$$

$$= \frac{1.4 \times 0.8 \text{ m}}{1} [0.26 - 0.12] = 0.1568 \text{ m} = 15.68 \text{ cm.}$$

(*i*) Daily water consumed by plants ; *i.e.*, daily consumptive use

$$= \frac{15.68 \text{ cm}}{10 \text{ days}} = 1.568 \text{ cm./day} \quad \textbf{Ans.}$$

[$\because$ Frequency of irrigation = 10 days]

(*ii*) Total irrigation water required i/c losses in field and conveyance

$$= \frac{\text{N.I.R.}}{\eta_{irrigation}} = \frac{1.568}{0.22} \text{ cm/day} = 7.127 \text{ cm/day}$$

Field area $= 3000 \text{ ha} = 3000 \times 10^4 \text{ m}^2$

$\therefore$ Discharge required in irrigation canal

$$= \frac{7.127}{100} \times \frac{3000 \times 10^4}{24 \times 60 \times 60} \text{ m}^3/\text{s} = \mathbf{24.75 \text{ m}^3/\text{s}}. \quad \textbf{Ans.}$$

**Example 2.25.** *During a particular stage of the growth of a crop, consumptive use of water is 2.5 mm/day. Determine the interval in days between irrigations, and the depth of water to be applied when the amount of water available in the soil is 50% of the maximum depth of available water in the root zone, which is 80 mm. Assume irrigation efficiency to be 60%.* (*Civil Services, 1993*)

**Solution.** *Follow Example 2.14, and compute :*

Readily available moisture in root zone = 50% × 80 mm = 4 cm.

Consumptive use = 2.5 mm/day = 0.25 cm/day

$\therefore$ 0.25 cm of water is consumed by crop in = 1 day

4 cm of water will be consumed by crop in $= \frac{1}{0.25} \times 4 = \mathbf{16}$ **days.**

Also, the depth of water to be recouped through irrigation = 4 cm.

Actual water required to be applied to the field

$$= \frac{\text{Water required to be recouped or absorbed}}{\text{Efficiency of irrigation application}}$$

$$= \frac{4 \text{ cm}}{60\%} = \frac{4 \text{ cm}}{0.6} \text{ cm} = \mathbf{6.67 \text{ cm}}. \quad \textbf{Ans.}$$

Hence, 6.67 cm depth of irrigation water shall have to be applied to the fields at an interval of 16 days. **Ans.**

## PROBLEMS

**1.** (*a*) **What is meant by 'Duty' and 'Delta' of canal water ? Derive a relationship between duty and delta for a given base period.**

(*b*) **Find the delta for sugarcane when its duty is 730 hectares/cumec on the field and the base period of the crop being 110 days.** [**Ans.** 130 cm]

(*c*) **Define and explain the following terms as used in relation to water requirements of crops :**

(*i*) Base period. (*ii*) Intensity of irrigation.

(*iii*) Cash crops.

**2.** (*a*) **What do you understand by 'Duty' of canal water and what is its importance ? Explain how does duty differs from that at the head of a water-course and that at the head of a canal bringing water to the watercourse.**

(*b*) **Mention the approximate values of Duty and Delta for rice, wheat and sugarcane in your region.**

**3.** (*a*) **Define 'Duty' and 'Delta', and derive their relationship.**

(*b*) **What are the factors on which duty depends ?**

(*c*) **How can the duty be improved and what will be the gain ?**

(*d*) **What is meant by 'Flow duty' and 'Quantity duty' ?**

**4.** **What is meant by 'duty' ? Enumerate the different terms by which duty can be improved. What are the factors affecting duty ? The base period of paddy is 120 days. If the duty for this crop is 900 hectares per cumec, find the value of delta.** [**Ans.** 115 cm]

**5.** **Describe briefly the factors affecting duty.**
**Water is released at the rate of 5 cumecs at the head sluice. If the duty at the field is 100 hectares/cumec and the loss of water in transit is 30%, find the area of the land that can be irrigated.** [**Ans.** 350 hectares]

**6.** **What is meant by "Duty of water" ? Explain the influence of several factors which affect duty. What are the different ways in which duty can be expressed ?**
**A reservoir with a live storage capacity of 300 million cubic metres is able to irrigate an ayacut of 40,000 hectares with 2 fillings each year. The crop season is 120 days. What is the duty ?** [**Ans.** 691 hectares/cumec]

**7.** **Name the principal kharif crops of your region, and detail the agricultural and climatic requirements for sowing, growth and harvesting of one of the principal ones. Give the normal requirement of seed per hectare and the average yield per hectare of the crop. Suggest ways to increase the "duty" in an irrigation system.**

**8.** **(*a*) Explain as how the following factors affect the 'duty' of a crop :**
(*i*) Soil and sub-soil condition. (*ii*) Stage of growth.
(*iii*) Temperature (*iv*) Rainfall.
**(*b*) Compute the depth and frequency of irrigation required for a certain crop with data given below :**
**Root zone depth = 100 cm.** **Field ccapacity = 22%**
**Wilting point = 12%** **Apparent specific gravity of soil* = 1.50**
**Consumptive use = 25 mm/day** **Efficiency of irrigation = 50%**
**Assume 50% depletion on available moisture before application of irrigation water at field capacity.**
[**Hint.** Follow example 2.14, and work out :
Readily available moisture = 5%, and finally work out :
Depth of water stored in root zone = 7.5 cm
Frequency of irrigation = 3 days **Ans.**

**9.** **Explain with neat sketch the layout of a modern canal system, carrying water from a barrage. Discuss as to how the duty of water increases as we move downstream from the head of the main canal towards the head of the watercourse.**

**10.** Write short notes on :
(*i*) Optimum utilisation of irrigation water. (*ii*) Crop rotation.
(*iii*) Consumptive use and its estimation.
(*iv*) Water distribution efficiency.
(*v*) Net irrigation requirement (NIR).
(*vi*) Outlet factor.
(*vii*) Estimating depth and frequency of irrigation on the basis of soil moisture regime concept.
(*viii*) Crop seasons in India and their principal crops.

**11.** Define and explain the following terms :
(*i*) Cash crops. (*ii*) Field capacity.
(*iii*) Available moisture. (*iv*) Soil moisture deficiency.
(*v*) Crop ratio. (*vi*) Overlap allowance.
(*vii*) Paleo irrigation. (*viii*) Kor water depth.

**12.** **How will you proceed for determining the field irrigation requirement (FIR) for an important crop like wheat ? Explain with reference to a sample table, with assumed monthly values of pan evaporations.** [**Hint.** Please see Table 2.15]

**13.** **Name any two methods used for estimating consumptive use of water for a particular crop at a particular place. Explain in details the one which is most widely used in your region, and the reasons for preferring that particular method.**

---

* **Apparent sp. gr. of soil** = $\frac{\gamma_d}{\gamma_w}$, **where $\gamma_d$ is the dry unit wt. of soil (*i.e.* the soil containing air filled voids.**

3

# Canal Irrigation System

## 3.1. General

A direct irrigation scheme which makes use of a weir or a barrage, as well as a storage irrigation scheme which makes use of a storage dam or a storage reservoir, necessitates the construction of a network of canals, as explained earlier. The entire system of main canals, branch canals, distributaries and minors is to be designed properly for a certain realistic value of peak discharge that must pass through them, so as to provide sufficient irrigation water to the commanded* areas. These canals have to be aligned and excavated either in alluvial soils or non-alluvial soils ; depending upon which they are called *alluvial canals or non-alluvial canals,* as explained below.

## 3.2. Alluvial and Non-alluvial Canals

(*i*) **Alluvial Soils and Alluvial Canals.** The soil which is formed by transportation and deposition of silt through the agency of water, over a course of time, is called the alluvial soil. Say for example, in the deltaic region** a river carries heavy charge of silt, which gets deposited on the adjoining land, as and when the river overtops its banks during flood season. The process of silt deposition may continue over long periods of time, resulting in the formation of a soil called *Alluvial Soil. The soil which is so formed by the continuous deposition of silt-from the water flowing through a given area, is hence, called the alluvial soil.* The area of alluvial soil is even, and is having a flat surface slope. Hard foundations, are generally not available in this kind of soil. In prehistoric periods, the entire Indo-Gangetic Plain was, perhaps a depression, and was filled up with constant silt deposition dropped from the water flowing through this area, resulting in the formation of an alluvial-soil region. The rivers flowing through such alluvial areas, have a tendency to shift their courses. The river bed consists of sand of considerable thickness, and is, therefore, permeable. Whenever, an irrigation structure is to be constructed on such a river, special precautions and design methods are to be adopted. Most of our North Indian rivers, which pass through alluvial soils, do pose these problems. The canals when excavated through such soils, are called *Alluvial Canals.* Canal irrigation (Direct irrigation using a weir or a barrage) is generally preferred in such areas, as compared to the storage irrigation (*i.e.* by using a dam). Alluvial soil is very fertile, as it can absorb a fair percentage of rainfall and retain it in the substratum, making it highly productive, as water remains available within the root zone of crops.

* Gross and Net command or commanded area is defined under article 3.6.

** A river before joining the sea gets divided into a number of streams, forming the shape of a delta (Δ), and this region is called the deltaic region.

(*ii*) **Non-alluvial Soils.** Mountaineous regions may go on disintegrating over a period of time, resulting in the formation of a rocky plain area, called non-alluvial area. It has an uneven topography, and hard foundations are generally available. The rivers, passing through such areas, have no tendency to shift their courses, and they do not pose much problems for designing irrigation structures on them. Canals, passing through such areas are called *Non-alluvial Canals.* Major portion of Maharashtra State is non-alluvial. Storage irrigation is preferred to canal irrigation in this type of soil. Non-alluvial soils may be permeable or impermeable, but generally, they are non-permeable.

## 3.3. Alignment of Canals

Irrigation canals can be aligned in any of the following three ways :

(*i*) as *watershed canal* or *ridge canal.*

(*ii*) as *contour canal* ; and

(*iii*) as *side-slope canal.*

These three types of canals are discussed below :

(*i*) **Watershad Canal or Ridge Canal.** The dividing ridge line between the catchment areas* of two streams (drains) is called the *water-shed*, or the **ridge**. Thus, between two major streams, there is the main watershed (ridge line), which divides the drainage area of the two streams, as shown in Fig. 3.1. Similarly, between a main stream and any of its tributary, there are *subsidiary watersheds* (ridge lines), dividing the drainage between the two streams on either side.

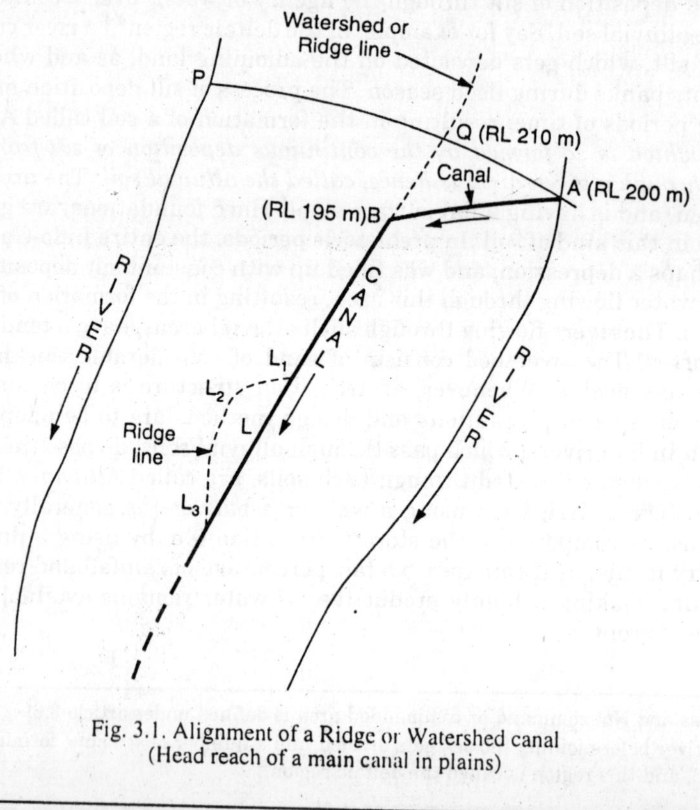


Fig. 3.1. Alignment of a Ridge or Watershed canal
(Head reach of a main canal in plains)

* The area from which rain water flows into a drain or a stream, is known as its *catchment area.*

For a canal system in plain areas, where land slopes are relatively flat and uniform, it is often necessary and advantageous to align canals on the watersheds (ridge lines) of the areas to be irrigated. *The canal which is aligned along any natural watershed (ridge line) is called a watershed canal, or a ridge canal.* The natural limits of the command area of such irrigation channels would be the drainage area on either side of the channel. Aligning a canal (main canal or branch canal or distributary) on the ridge, ensures gravity irrigation on both sides of the canal. Moreover, since the drainage flows away from the ridge, no drainage can cross a canal aligned on the ridge. *Thus, a canal aligned on the watershed saves the cost of construction of cross-drainage works.*

However, the main canal has to be taken off from a river, which is the lowest point in the cross-section, and this canal must mount the watershed (ridge) in as short a distance as possible. Since the available ground slope in the head reaches of a canal is usually much higher than the required canal bed slope, the canal generally needs only a short distance to reach the ridge line. This is illustrated in Fig. 3.1, in which the main canal takes off from a river at point $A$, and mounts the watershed at point $B$. Let the canal bed level at $A$ be 200 m and the elevation of the highest point $Q$ along the section $PQA$ be 210 m. Assuming that the ground slope is 1 m per km, the distance of the point $B$ ($RL$ 195 m) from $Q$ ($RL$ 210 m) on the watershed would be 15 km. If the designed canal bed slope is 1 in 4000 (*i.e.* 0.25 m per km), then the length $AB$ of the canal would be 20 km. In this length $AB$, the canal would cross small streams, and hence, construction of cross-drainage structures would be necessary for this length. As a matter of fact, the alignment $AB$ is influenced considerably by the need of providing suitable locations for the cross-drainage structures. The exact location of $B$ would be determined by trial, so that the alignment $AB$ results in an economic as well as an efficient canal system. It can also be seen that on the watershed side of the canal $AB$, the ground (*i.e.* area $AQB$) is higher than the ground area on valley side (*i.e.* river side). Therefore, this canal portion $AB$ can irrigate only on one side (*i.e.* on left side) of the canal.

When once the canal has reached the watershed (ridge line), it is generally kept on the watershed, except where : (*i*) localities are settled on the watershed ; or (*ii*) where the watershed is looping and not running straight, as shown by $L_1L_2L_3$ in Fig. 3.1. In a situation of a *looping ridge line*, the canal alignment may be taken straight along $L_1L_3$ by leaving the ridge line. The area between the canal and the watershed in the region $L$ (Fig. 3.1) can be irrigated by a distributary which takes off at $L_1$ and follows the watershed alignment along $L_1L_2L_3$. In the region $L$, the main canal may also have to cross some small streams, and hence some cross-drainage structures may have to be constructed.

If the watershed is passing through villages or towns, the canal may have to leave the watershed (ridge line) for some distance.

*The depressions in the ridge line may also necessitate construction of viaducts or syphons to maintain the canal FSL.*

(*ii*) **Contour Canals.** The above arrangement of providing the canal along the ridge line are, however, not found economical in hill areas, since the conditions in hills are vastly different compared to those of plains. In hills, the river flows in the valley well below the watershed. Infact, the ridge line (watershed) may be hundred of metres above the river. It therefore becomes virtually impossible to take the canal on top of such a higher ridge line. In such conditions, *contour canals* (Fig. 3.2) are usually constructed. *Contour channels follow a contour, except for giving the required longitudinal slope to the canal.*

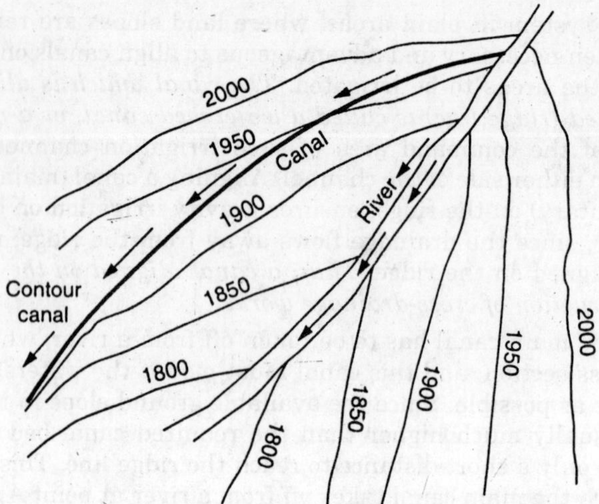


Fig. 3.2. Alignment of a Contour canal.
(Head reach of main canal in hills)

Since the river slope is much steepter than the canal bed slope, the canal encompasses more and more area between itself and the river. It may be noted that more fertile areas in the hills are located at lower levels only.

*A contour canal irrigates only on one side because the area on the other side is higher, as can be seen in Fig. 3.2.*

As the drainage flow is always at right angles to the ground contours, such a channel would definitely have to cross natural drains and streams, necessitating construction of cross-drainage structures.

(*iii*) **Side slope canal.** A side slope canal is that which is aligned at right angles to the contours ; *i.e.* along the side slopes, as shown in Fig. 3.3. Since such a canal runs parallel to the natural drainage flow, it usually does not intercept drainage channels, thus, avoiding the construction of cross-drainage structures.

In order to finalise the canal networks for an irrigation project, trial alignment of canals are initially marked on the map prepared during the detailed survey. A large-scale map is required to work out the details of individual canals. However, a small-scale map showing the entire command area of the irrigation project is also desirable. The alignents of canals marked on the map are transferred on the field, and adjustments and changes are made, wherever found necessary. These adjustments are transferred on the map as well. The alignment on the field is marked by small masonry pillars erected at every 200 metres distance or so. The centre line on top of these pillars coincides with the exact alignment of the given canal. In between the adjacent pillars, a small trench may be excavated in the ground, to mark the canal alignment.

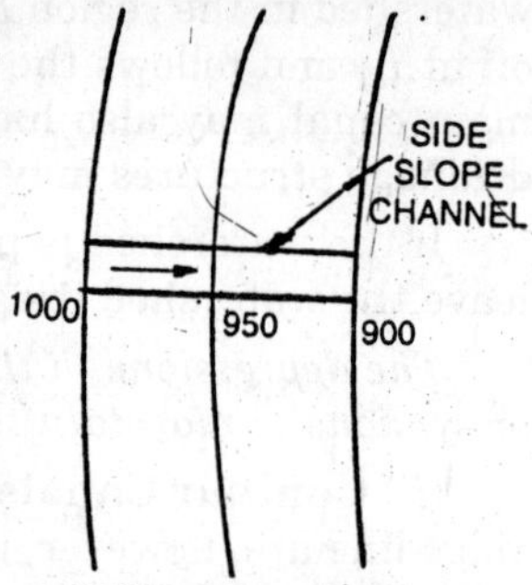


Fig. 3.3. Alignment of a Side Slope canal.

## 3.4. Distribution System for Canal Irrigation

It has been emphasized earlier that the direct irrigation scheme using a weir or a barrage, as well as the storage irrigation scheme using a dam or a reservoir, require a network of irrigation canals of different sizes and capacities. The entire network of irrigation channels (Fig. 3.4) is called the *Canal System.* The canal system, as explained earlier in Chapter 2, consists of :

(*i*) Main canal ;
(*ii*) Branch canals ;
(*iii*) Distributaries, also called *major distributaries* ;
(*iv*) Minors, also called *minor distributaries,*
(*v*) Watercourses.

In case of direct irrigation scheme, a weir or a barrage is constructed across the river, and water is headed up on the upstream side. The arrangement is known as *Head Works* or *Diversion Head Works.* It will be explained in details in Chapter 9. Water is diverted into the main canal by means of a diversion weir. A head regulator is provided at the head of the main canal, so as to regulate the flow of water into the main canal.

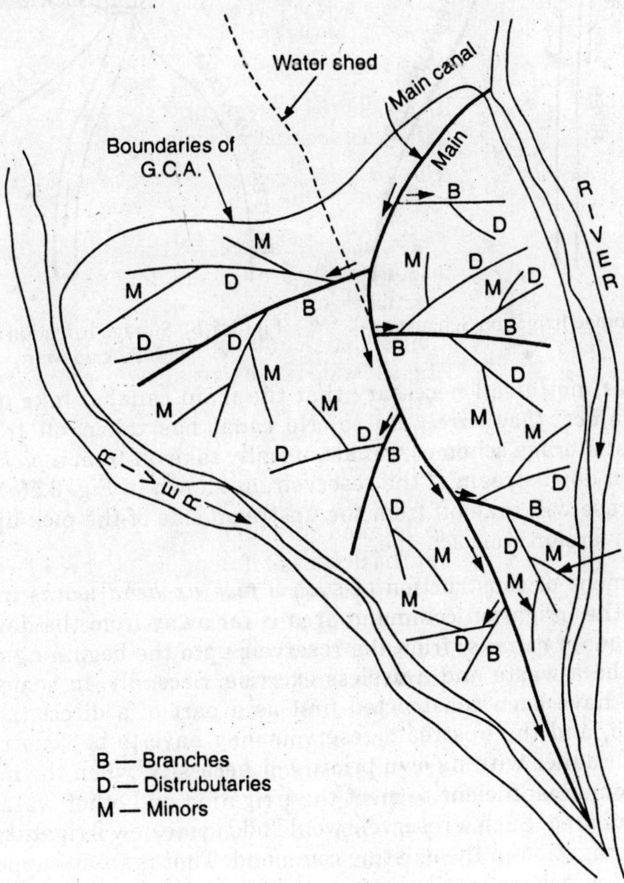


Fig. 3.4. Layout of an Irrigation canal network.

In storage irrigation scheme, a dam is constructed across the river, thus forming a reservoir on the upstream side of the river. The water from this reservoir is taken into the main canal through the outlet sluices. There are generally two main canals, which off-take from the reservoir from the left side or the right side, and are hence called the *Left Bank Canal* and the *Right Bank Canal*, respectively, as shown in Fig. 3.5($a$).

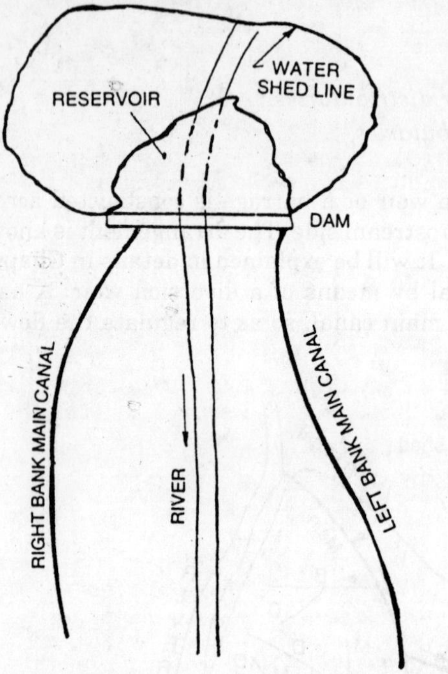


Fig. 3.5($a$). Storage Irrigation Scheme.

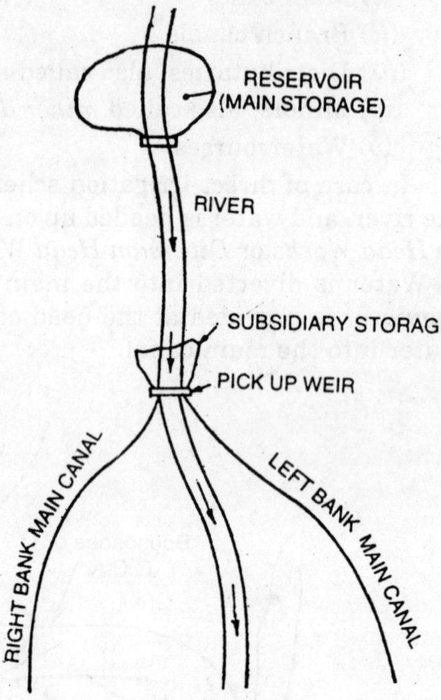


Fig. 3.5($b$). Storage Irrigation Scheme with Pick up weir.

It is, however, not at all necessary that the main canal(s) take off from the reservoir, *and infact, they rarely do so*. No canal has taken off from Bhakra reservoir. In most storage schemes, canals, usually take off from a *pick up weir or a barrage*, located downstream of the reservoir, as shown in Fig. 3.5($b$). The main canal in such a case will take off from the upstream side of the pick up weir, just as in a normal diversion weir scheme.

The requirement of construction of such a *pick up head works* may become necessary when the irrigation command area is far away from the dam site, and constructing the main canal(s) from the reservoir upto the beginning of the command area may be a waste and a useless exercise. Secondly, in some cases, the headwork might have been constructed first as a part of a direct (non-storage) irrigation scheme, and the upstream reservoir may have to be constructed on a later date, in accordance with its own priority or necessity, when the natural lean season flows become insufficient to meet the irrigation and other water demands of the downstream area. Such a reservoir would add some new irrigation command and firm up the irrigation in the existing command. That is what happened in the case of Bhakra dam, where canal system was already in existence as a part of a non storage scheme, purely based on perennial river supplies. Similar constructions are expected in case of Yamuna river, where *Western Yamuna Canal* (*WYC*) and *Eastern Yamuna Canal* (*EYC*) taking off from the **Tajewala headworks*** have been in existence for the last more than 100 years, but without the backing of any upstream reservoir. Such a direct irrigation scheme proved satisfactory in olden times, when population was less and the river used to get sufficient snowmelt throughout the lean season. However, with the increasing population and consequently increasing water demand along with dwindling snowmelt due to extension of habitat in upper hills and green house effects of urbanisation, the available natural flows in lean season have become insufficient, necessitating construction of reservoir(s) upstream. That is why, there are plans to construct three reservoirs at *Renuka, Kishau* and *Lakhwar-Vyasi* in the Yamuna catchment upstream. When these dame are constructed, there will be very little addition to irrigation command area, but they will mostly firm up the irrigation in the already existing *WYC* and *EYC* command, and make available water for other needs. The (Hathnikund) headworks constructed as a part of direct irrigation scheme will then, become a part of a storage irrigation scheme.

*It can be easily understood that a reservoir redistribute the water in time (storing water in rainy season and releasing it in lean season) ; while the barrage and the canal system will redistribute it in space, (taking it upto the fields.*

In both these irrigation schemes, when once the water reaches into the main canal, the problem left is to distribute this water upto the fields. The purpose is achieved through a network of channels, as described below :

(*a*) **Main Canal (Head reach).** The canal headworks are generally situated on the river flowing in a valley, and the canal should reach the ridge line in the shortest possible distance. The canal, in this reach, must, therefore, be aligned very carefully, and has to be generally excavated in deep cuttings below N.S.L. (natural surface level). Sometimes, it has to cross various drainage lines. Many a times, straight alignment has to be sacrificed and detours need to be accepted, in order to achieve a good site for cross drainage works.

(*b*) **Main Canal (Portion below head reach.)** Attempts are made to align the canal along the ridge and somewhat central to the command area. Sometimes, ridge line has to be sacrificed, to bypass towns and villages, etc. Main canal is not required to do any irrigation.

(*ii*) **Branch Canals.** *Branch canals* are taken off from the main canal on eather side to take irrigation water to the whole tract required to be irrigated. Very little irrigation is infact, done from the branch canals themselves, as they serve to supply water primarily the distributaries. Attempts are made to align them along subsidiary ridges. Discharge in a branch channel, is generally, more than 30 cumec.

(*iii*) **Distributaries.** Smaller channels which take off from the branch canals and distribute their supply through outlets into minors or water courses, are called disributaries. They are aligned either as ridge canals or as contour canals. Discharge in a distributary is generally less than 30 cumec.

(*iv*) **Minors.** Sometimes, the country is such that the distance between the distributary outlet and the farmer's field is very long ; say more than 3 km or so. In such a case, small channels called minors, are taken off from the distributaries, so as to supply water to the cultivators at the point nearer to their fields. Discharge in a minor, is generally, less than 2.5 cumec.

---

* The age old Tajewala headworks has recently been replaced by newly constructed Hathnikund barrage headworks.

(*v*) **Watercourses.** These are not the government channels and belong to the cultivators. They are small channels, which are excavated and maintained by the cultivators at their own costs, to take water from the government-owned outlet points, provided in the distributary or the minor.

### 3.5. Curves in Channels

Attempts are made to align the channels straight as far as possible. But many a times, the curves become inescapable. Whenever, a curve is proposed, while aligning unlined channels, it should be as gentle as possible. A curve causes disturbance of flow and results in silting on the inside (*i.e.* convex side) and scouring on the outside (*i.e.* concave side).

Pitching is, therefore, sometimes proposed on the concave side, so as to avoid scouring. If the discharge is more, the curve should be more gentle and should, therefore, have more radius. The generally adopted minimum values of radii for different discharges are tabulated below in Table 3.1.

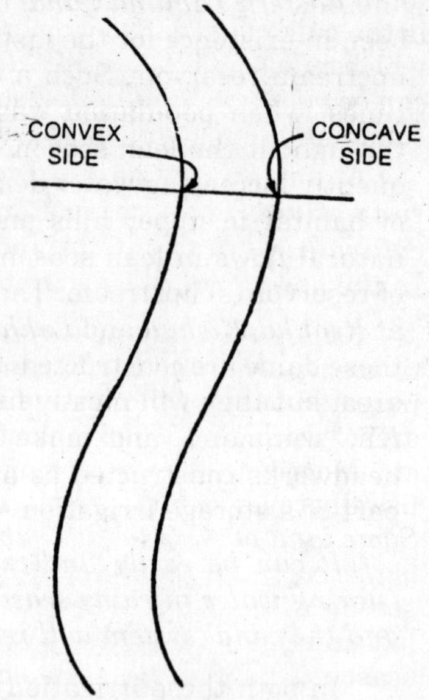


Fig. 3.6.

**Table 3.1**

| *Discharging Capacity of Channels (in cumecs)* | *Minimum recommended curve Radius (in metres)* |
|---|---|
| Over 100 | 1500 |
| 30—100 | 900 |
| 15—30 | 600 |
| 3—15 | 300 |
| 0.5—3 | 150 |
| Less than 0.5 | 100 |

### 3.6. Certain Important Definitions

Before we discuss the techniques, which are employed to determine the 'design discharge' for a canal, let us first of all, define certain important terms, which often come across in the design of irrigation canal systems.

**3.6.1. Gross Command Area (G.C.A.).** It is the total area, bounded within the irrigation boundary of a project, which can be economically irrigated without considering the limitation of the quantity of available water. *It includes the cultivable as well as the un-cultivable area. For example, ponds, residential areas, roads, reserved forests, etc. are the uncultivable areas of the gross command area.*

A given irrigation canal system lies in a *doab* (*i.e. the area between two drainages*), and can economically irrigate the *doab*. It is, obviously, uneconomical to use the irrigation system to irrigate across the two drainages. Thus, the boundaries of gross command of a canal system is, fixed by the drainages on both sides of the main irrigation canal, as shown in Fig. 3.4. The gross command area, evidently represents the geographical area of the *doab*.

**3.6.2. Culturable or Cultivable Command Area (CCA).** Culturable area is the cultivable part of the gross command area, and includes all land of GCA on which cultivation is possible. *It will, thus, include pastures and fallow lands, which can be made cultivable.* Obviously, it does not include uncultivable part of the gross command, like populated areas, ponds, roads, reserved forests, ushar land, etc.

At any given time, however, all the cultivable land may not be actually under cultivation. Therefore, sometimes, the CCA is divided into two categories ; *i.e.* (*i*) *cultivated portion of CCA* ; and (*ii*) *Cultivable but not cultivated portion of CCA*. In the absence of detailed data, CCA may be assumed to be equal to 80% of GCA.

**3.6.3. Intensity of Irrigation (Seasonal and Annual).** The entire cultivated portion of the culturable command area (CCA) is not proposed to be irrigated at one time (in one season) to avoid **intensive irrigation** of a particular area, which may cause harmful effects like water logging*, salinity and malaria, etc. Moreover, due to shortage of irrigation water, larger area of the command is usually covered with partial coverage of the fields of different sub-areas or pockets. *Some land of a particular sub area is thus, either allowed to take rest, or is shown with crops which do not require irrigation water.* Fields left unirrigated in one season will be supplied water in the next season, when some other fields will be left unirrigated. The fields are thus, supplied water in rotation over different crop seasons, thereby irrigating only about 40 to 60% of the fields of various sub areas over a season. The irrigation water is, thus, usually supplied extenstively, covering a larger area. Such as **extensive irrigation**, covering a larger portion of the cultivable command, is preferred to the *intensive irrigation* of the smaller portion of the command, to avoid general famine conditions over some area (which would be totally left unirrigated, if the entire water is used in intensive irrigation of some other portion). Extensive irrigation also helps in avoiding harmful effects of over-irrigation (like salinity and water-logging), which is caused by intensive irrigation.

Due to these reasons, only a small percentage of CCA is brought under irrigation over a given season. *This percentage of CCA proposed to be irrigated in a given season is called the intensity of irrigation of that season, or seasonal intensity of irrigation.* Say for example, the sanctioned intensity of irrigation under Bhakra canal system is only 27.6% for Kharif season, and 34.4% for Rabi season.

Sometimes, the intensity of irrigation is worked out over the entire year (inclusive of two or more crop seasons), when it is called the *annual irrigation intensity* or *annual intensity of irrigation* (AII), which may be defined as the percentage of CCA which may be irrigated annually. AII is, thus, obtained by dividing the *gross irrigated area* (*i.e.* total area irrigated once in the year + the area irrigated more than once in that year) by the CCA. The annual intensity of irrigation is the sum total of intensities of irrigation of all the seasons of the year. The annual intensity of irrigation for Bhakra canal system is thus, computed by adding the intensity of irrigation for kharif season and that for the Rabi season, giving combined annual irrigation intensity of 27.6 + 34.4 = 62%.

The annual irrigation intensity is usually found to be in the range of 40 to 60%, but needs to be raised to the range of 100 to 180% by cultivating larger parts of CCA with more than one crop in a year, and through improved management and economical utilisation of the available irrigation water. From the perusal of col. (7) of table IV under the first chapter on "Introduction to the subject", it can be seen that in India, U.P. and Punjab, States have already exceeded this figure of AII beyond 100%, to 139.6% and 133.4%, respectively, with Haryana (90.2%) and West Bengal (73.3%) following these two best irrigated States. Other States

* **Explained in details in chapter 6.**

are far behind to such an extent that the overall annual intensity of irrigation in the country is only 45.9%.

**3.6.4. Net and Gross Sown Areas.** Sometimes, two crops in two seasons are grown during a particular year on the same area. Hence, such an area will be sown more than once during a given year. If this area is added to the area which is sown only once (and called the *net shown area*), then we get what is known as the *gross sown area*, or the *gross cropped area*. Hence,

Gross cropped or Gross sown area (during a year)

= Net cropped area, *i.e.* area sown once in a year

+ Area shown more than once during the same year ...(3.1)

**3.6.5. Net and Gross Irrigated Areas.** Based on the above analogy, the area which is irrigated once during a year is called the *net irrigated area*, and when to this is added the area irrigated more than once, we obtain the *gross irrigated area*.

∴ Gross irrigated area (in a given year)

= Net irrigated area (*i.e.* area irrigated once in a year)

+ Area irrigated more than once during the same year. ...(3.2)

**3.6.6. Area to be Irrigated.** The area proposed to be irrigated in any one crop season or over any given year, is called the *area to be irrigated in that season or the year*, respectively. It is obtained by multiplying CCA by the seasonal or annual intensity of irrigation, as the case may be. The areas to be irrigated are usually worked out separately for each crop season, because the water requirement of the crops of two seasons are quite different.

**3.6.7. Time Factor.** To check the dangers of over irrigation, leading to water-logging and salinity, no distributary is allowed to operate on all the days during any crop season. *The ratio of the actual operating period of a distributary to the crop period is called the time factor of the distributary*. For the Bhakra canal system, for example, the time factors for Kharif and Rabi seasons are fixed at 0.80 and 0.72, respectively, which means that each distributary would receive its full supply for a period of $0.80 \times 180 = 144$ days, and $0.72 \times 180 = 129$ days, respectively, in each crop season of 180 days.

For computing the design capacity of a distributary, therefore, the computed water requirement (for the crops proposed to be grown and irrigated by that distributary) should be divided by the time factor, since this factor is less than 1.

**3.6.8. Capacity Factor.** The capacity factor for a canal is the ratio of the mean supply discharge in a canal during a period to its designed full capacity. Since canals have to run almost to their full designed capacities during Kharif (summer) season, this value usually varies from 0.9 to 0.95 for Kharif season. However, since water requirement during Rabi (winter) season reduces to about $\frac{2}{3}$rd times* the full supply, the capacity factor usually varies from 0.60 to 0.70 for Rabi season. To improve this factor during Rabi season, the cropped area in Rabi season is, hence, usually increased.

**3.6.9. Full supply coefficient.** Full supply coefficient is the design duty at the head of the canal. In other words, the number of hectares irrigable per cumec

* The winter (Rabi) crops usually mature on about 52% of the supply required by crops in summer ; but due to lower winter discharges in canals, the percentage of losses enroute becomes higher in winter season. The actual water supply at the head of canal is, therefore, found to be about 66% of full supply.

of the canal capacity at its head, is known as the full supply coefficient of the canal. It can hence, be represented by the equation :

Full supply coefficient

$$= \frac{\text{Area estimated to be irrigated during base period}}{\text{Design full supply discharge at the head of the canal}} \quad ...(3.3)$$

This factor is also called the **Duty on capacity.**

**3.6.10. Nominal Duty.** It is the ratio of the area actually irrigated by the cultivators to the mean supply discharge let out from the outlet of the distributary over the crop period.

For example, let $x$ cumec of water is released daily from the outlet of a distributary for 100 days (says) in a total crop period of 125 days (say). Then, the mean supply discharge over the crop period will be $\frac{x \times 100}{125} = 0.8x$ (cumec).

If the area of crop irrigated by this discharge is A hectares, then the nominal duty will be given as :

$$\text{Nominal duty} = \frac{A(ha)}{0.8x\ (\text{cumec})} = 1.25\frac{A}{x}\ \text{ha/cumec.}$$

**3.6.11. Crop Ratio.** Crop ratio is defined as the ratio of the land area of a crop irrigated under Kharif season to the land area of that crop irrigated under Ravi season.

For example, if $x$ is the crop ratio and $A$ is the total area of the a crop irrigated/ grown over both the Kharif and Rabi seasons, then the area grown in Kharif season $(A_K) = x$ [Area grown under Rabi season $(A_R)$]

$\therefore$ $$A_K = x \cdot A_R$$

But $$A_K + A_R = A$$

$\therefore$ $$(x+1)\,A_R = A$$

and $$A_R = \frac{A}{x+1} \quad ...(3.4)$$

Also $$A_K = \frac{x}{x+1}(A) \quad ...(3.5)$$

**3.6.12. Intensity of Cropping.** Intensity of cropping or *cropping intensity* is defined as the ratio of the 'Gross Cropped Area' to the 'Net Cultivated Area' or 'Net Cropped Area'.

$$\text{Thus, Intensity of cropping} = \frac{\text{Gross cropped area}}{\text{Net cultivated area or Net cropped area}} \quad ...(3.6)$$

Since CCA (Culturable Command Area) is the area which can be irrigated from a scheme and is fit for cultivation, net cultivated area can be equated with CCA.

**Example :** As per 2013-14 statistical data of India, the gross cropped area is 200.9 Mha, while the net cultivated (or shown or cropped) area is 141.4 Mha.

$$\text{Thus, the cropping intensity} = \frac{200.9}{141.4} \times 100 = 142\%.$$

## 3.7. Computing the Design Capacity of an Irrigation Canal

Whenever one plans for supplying irrigation water, one has to think of the likely crops that would be sown in any one season. The peak rate of water requirement of all the crops in each season of a year is also needed to be worked out. The capacity of the canal should be sufficient to fulfil the maximum of the peak demand of all the crops that are required to be irrigated at any one time amongst all the seasons. It is explained below in details :

The most important Rabi crop is *wheat,* which requires water from *December to March,* during the Rabi season. Similarly, *Paddy* (Rice) is the most important Kharif crop, requiring water from *June to November.* So it can be presumed that when Rabi crops require water, Kharif crops do not, and *vice-versa.* Sugarcane and garden crops are **perennial crops**, requiring water throughout the year. Hence, the canal may be designed for a capacity equal to the greater of the water requirement of Rabi and Sugar cane (plus garden crops if any) ; or Kharif and Sugarcane (and garden crops, if any). This is a very simple method for fixing the capacity of the channel. The entire Rabi area is supposed to be sown with wheat and entire Kharif area with Paddy ; Sugarcane area (including garden crop area, if any) is excluded from both. Their water requirements in cumecs are worked out for both seasons, separately. The water requirement of sugarcane and garden crops are separately worked out each, in cumecs. The sum of cumecs required by the Rabi plus Sugarcane (including garden crop, if any) and Kharif plus Sugarcane (including garden crop if any), are separately worked out. The canal capacity may then be fixed for the maximum of the two values.

The most important point which must be kept in mind while fixing the channel capacity is that, we must take into account the **keenest demand** of the crop and not the **average demand**. For example, let Rice require 120 cm of water during 120 days, thus giving an average outlet factor of 864 hectares/cumec

$$\left( i.e.\ D = \frac{864B}{\Delta} = \frac{864 \times 120}{120} = 864 \right)$$

But a canal designed on this average outlet factor will prove to be very inadequate, as it will fail to supply the required water to the crop at its **peak demand**, *i.e.* at the time of **kor-watering**, as explained below :

The **kor depth** for Rice is about 19 cm and the **kor period** is about 2 weeks (*i.e.* 14 days). It means that 19 cm of water depth must be supplied in about 14 days. The outlet factor for this, works out to be 637 hectares/cumec as

$$D = \frac{864B}{\Delta} = \frac{864 \times 14}{19} = 637$$

Now, discharge required to mature $A$ hectares of land for an outlet factor of 864 is $\frac{A}{864}$ cumecs ; while that for an outlet factor of 637 is $\frac{A}{637}$ cumecs. Out of these two values, the second value $\left( i.e.\ \frac{A}{637} \right)$ is more, and hence, the discharge required for fulfilling the kor demand of the crop is more than that for the average demand.

Similarly, wheat requires about 40 cm of water in a total base period of about 160 days; thus giving an average outlet factor of 3464 hectares/cumec. But the kor water depth required by wheat is about 14 cm in about four weeks, giving an outlet factor of 1728 hectares/cumec. Applying the previous reasoning, we can say that the discharge required to fulfil kor-demand is much more than that for the average-demand (almost double of average). Hence, it follows that the peak demand, *i.e.* kor-demand of the crop should be taken into account while fixing the capacity of a canal. Moreover, the provision for canal losses should also be made, while deciding the final capacity of the canal.

The method, described above, for determining the canal capacity, is a simplified approximate process. To be more precise, we can find out the monthly or fortnightly water requirements of various crops (as was explained in Chapter 2). The water depth required in this interval is multiplied by the crop area, so as to give the volume of water required in this interval. Dividing the volume by interval, we can find out the discharge required in each interval, by various crops. The summation of which for all the crops will give us the discharge required by all the crops in each interval. The canal may then be designed for maximum of these values.

It is evident that in order to be more precise, the interval should be as small as possible. Generally, monthly water requirement studies are conducted, and the canal capacity is increased by 20 to 25% to cater for the peak demand in the month.

**Example 3.1.** *The gross commanded area for a distributary is 6000 hectares, 80% of which is culturable irrigable. The intensity of irrigation for Rabi season is 50% and that for Kharif season is 25%. If the average duty at the head of the distributary is 2000 hectares/cumec for Rabi season and 900 hectares/cumec for Kharif season, find out the discharge required at the head of the distributary from average demand considerations.*

**Solution.** G.C.A. = 6000 hectares

$$\text{C.C.A.} = 6000 \times \frac{80}{100} = 4800 \text{ hectares.}$$

Area to be irrigated in Rabi season

$$= \text{C.C.A.} \times \text{Intensity of Irrigation}$$

$$= 4800 \times \frac{50}{100} = 2400 \text{ hectares.}$$

Area to be irrigated in Kharif season

$$= 4800 \times \frac{25}{100} = 1200 \text{ hectares.}$$

Water required at the head of the distributary to irrigate **Rabi** area

$$= \frac{2400}{2000} \text{ cumecs} = \mathbf{1.20 \text{ cumecs.}} \qquad ...(i)$$

Water required at the head of the distributary to irrigate **Kharif** area

$$= \frac{1200}{900} \text{ cumecs} = \mathbf{1.33 \text{ cumecs.}} \qquad ...(ii)$$

Thus, the requirement in Kharif season is 1.33 cumecs and that in Rabi season is 1.20 cumecs. The required discharge is maximum of the two, *i.e.* **1.33 cumecs. Ans.**

Hence, the distributary should be designed for 1.33 cumecs discharge at its head, from average demand considerations. The head regulator should be sufficient to carry 1.33 cumec ; and in Rabi season, only 1.20 cumecs will be released.

**Example 3.2.** *Determine the discharge required at the head of the distributary in Example 3.1 given above, for fulfilling maximum crop requirement. Assume suitable values of kor depth and kor period.*

**Solution.** Let us assume a kor period of 4 weeks for Rabi crop (wheat) and 2.5 weeks for Kharif crop (rice). Also assume, Kor depth of 13.5 cm for Rabi (wheat) and 19 cm for Kharif (rice) crop.

Now, outlet factor for rabi $= \dfrac{864 \times B}{\Delta} = \dfrac{864 \times (4 \times 7)}{13.5} = 1792$ hectare/cumec.

Outlet factor for Kharif $= \dfrac{864 \times (2.5 \times 7)}{19} = 796$ hectares/cumec.

Area to be irrigated in Rabi season (worked out in previous example)
= 2400 hectares.

Area to be irrigated in Kharif season (worked out in previous example)
= 1200 hectares.

Water reqd. at the head of the distributary to irrigate Rabi area

$$= \frac{2400}{1792} = 1.34 \text{ cumecs.} \qquad ...(i)$$

Water required at the head of the distributary to irrigate Kharif area

$$= \frac{1200}{796} = 1.51 \text{ cumecs.} \qquad ...(ii)$$

The required discharge is maximum of the two, *i.e.* **1.51 cumecs. Ans.**

*Note :* The required discharge from kor demand considerations have gone up to 1.51 cumecs from 1.33 cumecs (worked out in the previous Example), *i.e.* an increase of about 14%.

**Example 3.3.** *The culturable commanded area of a watercourse is 1200 hectares. Intensities of sugarcane and wheat crops are 20% and 40% respectively. The duties for the crops at the head of the watercourse are 730 hectares/cumec and 1800 hectares/cumec, respectively. Find (a) the discharge required at the head of the water-course (b) determine the design discharge at the outlet, assuming a time factor equal to 0.8.*

**Solution.** C.C.A. = 1200 hectares

Intensity of irrigation for sugarcane = 20%

∴ Area to be irrigated under sugarcane $= 1200 \times \dfrac{20}{100} = 240$ hectares

Intensity of irrigation for wheat = 40%

∴ Area to be irrigated under wheat $= 1200 \times \dfrac{40}{100} = 480$ hectares

Duty for sugarcane = 730 hectares/cumec

Duty for wheat = 1800 hectares/cumec.

∴ Discharge required for sugarcane $= \dfrac{240}{730}$ cumec = 0.329 cumec

and Discharge required for wheat $= \dfrac{480}{1800}$ cumec = 0.271 cumec.

Now, sugarcane requires water for all the 12 months and wheat requires water for only rabi season. Hence, the water requirement at the head of the watercourse at any time of the year will be the summation of the two, *i.e.* equal to 0.329 + 0.271 = 0.6 cumec.

(*a*) Hence, the discharge required at the head of the water-course is **0.6 cumec. Ans.**

*Note :* The discharge during rabi season will be 0.6 cumec and for the rest of the year, it will be 0.329 cumec.

(*b*) Time factor = 0.8 ; since the channel runs for fewer days than the crop days, therefore, the actual design discharge at the outlet

$$= \frac{0.6}{0.8} = 0.75 \text{ cumec.} \quad \textbf{Ans.}$$

**Example 3.4.** *The culturable commanded area for a distributary is 15,000 hectares. The intensity of irrigation (I.I.) for Rabi (wheat) is 40% and for Kharif (rice) is 15%. If the total water requirement of the two crops are 37.5 cm and 120 cm and their periods of growth are 160 days and 140 days respectively ; (a) Determine the outlet discharge from average demand considerations ; (b) Also determine the peak demand discharge, assuming that the kor water depth for two crops are 13.5 cm and 19 cm and their kor periods are 4 weeks and 2 weeks respectively.*

**Solution.** C.C.A. = 15,000 hectares.

I.I. For wheat (Rabi) = 40%

I.I. For rice (Kharif) = 15%

Wheat (Rabi) area = 15,000 × 0.40 = 6000 hectares

Rice (Kharif) area = 15,000 × 0.15 = 2250 hectares

$\Delta$ for wheat = 37.5 cm.

$\Delta$ for rice = 120 cm.

$B$ for wheat = 160 days

$B$ for rice = 140 days

Now $$D = \frac{864B}{\Delta}$$

$$\text{Average duty } (D) \text{ for wheat} = \frac{864 \times 160}{37.5} = 3686 \text{ hectares/cumec}$$

$$\text{Average duty } (D) \text{ for rice} = \frac{864 \times 140}{120} = 1008 \text{ hectares/cumec}$$

$$\text{Outlet discharge required for wheat} = \frac{\text{Area}}{\text{Duty}} = \frac{6000}{3686} = 1.63 \text{ cumecs}$$

$$\text{Outlet discharge required for rice} = \frac{2250}{1008} = 2.23 \text{ cumecs.}$$

The required design discharge at outlet (from average demand considerations) is maximum of the two values, *i.e.* **2.23 cumecs. Ans.**

(*b*) Kor water depth for wheat = 13.5 cm.

Kor period for wheat = 4 weeks = 4 × 7 = 28 days

Kor water depth for rice = 19 cm.

Kor period for rice = 2 weeks = 2 × 7 = 14 days.

$$\text{Duty for wheat (for kor demand)} = \frac{864 \times B}{\Delta}$$

$$= \frac{864 \times \text{Kor period for wheat}}{\text{Kor water depth for wheat}} = \frac{864 \times 28}{13.5} = 1792 \text{ ha/cumec}$$

Duty for rice (for Kor demand)

$$= \frac{864 \times \text{Kor period for rice}}{\text{Kor water depth for rice}} = \frac{864 \times 14}{19} = 637 \text{ ha./cumec}$$

Outlet discharge required for wheat (for kor demand)

$$= \frac{\text{Area}}{\text{Duty}} = \frac{6000}{1792} = 3.35 \text{ cumec.} \qquad ...(i)$$

Outlet discharge required for rice (for kor demand)

$$= \frac{2250}{637} = 3.53 \text{ cumecs} \qquad ...(ii)$$

The required design discharge at the outlet, from peak demand considerations, is maximum of these two values, *i.e.* = **3.54 cumecs.** **Ans.**

**Example 3.5.** *At a certain place in India, the transplantation of rice takes 16 days, and the total depth of water required by the crop is 60 cm on the field. During this transplantation period of 16 days, rain starts falling and about 10 cm of rain is being utilised to fulfil the rice demand. Find the duty of irrigation water required for rice during transplanting period. (a) Assuming 25% losses of water in water courses, find the duty of water at the head of the watercourse. (b) Find the duty of water at the head of distributary, assuming 15% losses from the distributary head to the watercourse head.*

**Solution.** Total depth of water required for transplanting rice = 60 cm.

Useful rainfall = 10 cm.

Extra water depth required to be supplemented by irrigation = 60 – 10 = 50 cm.

Period in which this water is required = 16 days.

$$\text{Duty of irrigation water} = \frac{864B}{\Delta} = \frac{864 \times 16}{50} = \mathbf{276.5 \text{ hectares/cumec}}$$

(*a*) Assuming 25% losses in the watercourse, we have

Duty at the head of watercourse = 276.5 × 0.75 = 207.4 hectares/cumec. **Ans.**

(*b*) Similarly, duty at the head of the distributary (assuming 15% losses)

= 207.4 × 0.85 = 176.3 hectares/cumec. **Ans.**

**Example 3.6.** *A main canal, which offtakes from a storage reservoir, has to irrigate crops in a certain country having three seasons in a year. Data for the irrigated crops, is given in table 3.2.*

**Table 3.2**

| *S. No.* | *Name of crop* | *Crop period in days* | *Area to be irrigated in hectares* | *Duty at the head of the main canal in hectares/cumec* |
|---|---|---|---|---|
| 1. (*a*) | Sugarcane | 280 | 315 | 630 |
| (*b*) | Overlap for Sugarcane in hot weather | 100 | 70 | 630 |
| 2. | Jowar (Rabi) | 120 | 4800 | 1600 |
| 3. | Bajra (Monsoon) | 120 | 5600 | 2800 |
| 4. | Vegetables (Hot season) | 120 | 350 | 700 |

(*a*) *Find the discharge required at the head of the main canal, taking time factor for the main canal as 0.7.*

*(b) What should be the gross storage capacity of the reservoir ? Assume suitable factors, wherever needed.*

**Solution.** Water required by various crops, is as follows (figures taken from Table 3.2).

(*i*) Water required for sugarcane (whole year) $= \frac{315}{630} = 0.5$ cumec.

(*ii*) Water required for overlap sugarcane (Hot season) $= \frac{70}{630} = 0.11$ cumec.

(*iii*) Water required for jowar (Rabi season) $= \frac{4800}{1600} = 3.0$ cumecs

(*iv*) Water required for bajra (Monsoon season) $= \frac{5600}{2800} = 2.0$ cumecs.

(*v*) Water required for vegetables (Hot season) $= \frac{350}{700} = 0.5$ cumec.

Total water required in each season is as follows :

(*a*) in hot season $= 0.5 + 0.11 + 0.5 = 1.11$ cumecs ;

(*b*) in monsoon season $= 0.5 + 2.0 = 2.5$ cumecs ;

(*c*) in Rabi season $= 0.5 + 3.0 = 3.5$ cumecs.

It is evident that the maximum water is required in Rabi season, *i.e.* equal so **3.5 cumecs.**

Now, Time factor = 0.7 ;

Therefore, full supply discharge (based on average demand or duty)

$= \frac{3.5}{0.7} = 0.5$ cumecs.

Assuming the peak demand discharge to be 25% more than the average, the full supply discharge on peak demand $5 \times 1.25$ cumec = 6.25 cumecs.

(*a*) Hence the F.S.Q. at the head of the main canal (assuming negligible seepage losses from the head of the main canal to the fields)

= 6.25 cumecs **Ans.**

(*b*) To find out the gross storage capacity of the reservoir, let us work out the volume of water required by various crops as follows :

1. Water required by sugarcane $= 0.5 \times 280 \times 24 \times 60 \times 60 \text{ m}^3$

(*i.e.* discharge × days × secs in one day)

Water required by overlap sugarcane $= 0.11 \times 100 \times 24 \times 60 \times 60 \text{ m}^3$

Water required by jowar $= 3.0 \times 120 \times 24 \times 60 \times 60 \text{ m}^3$

Water required by bajri $= 2.0 \times 120 \times 24 \times 60 \times 60 \text{ m}^3$

Water required by vegetables $= 0.5 \times 120 \times 24 \times 60 \times 60 \text{ m}^3$

Total water required

$= \Sigma = 24 \times 60 \times 60[0.5 \times 280 + 0.11 \times 100 + 3.0 \times 120 + 2.0 \times 120 + 0.5 \times 120]$

$= 24 \times 60 \times 60[140 + 11 + 360 + 240 + 60]$ cubic metres

$= (24 \times 60 \times 60 \times 811)$ cubic metres $= 70.07 \times 10^6$ cubic metres

= 70.07 million cubic metres = **70.07 M.m³**

Assuming conveyance loss in canal network = 20%, we get

Total water required from the reservoir $= \frac{70.07}{0.8} = 87.59 \text{ M.m}^3$

Also, assume reservoir losses due to evaporation and absorption to be 10%, we get,

$$\text{Reservoir storage required} = \frac{87.59}{0.9} = 97.32 \text{ M.m}^3$$

Further, to allow for late and irregular monsoon, let us assume a carry over storage of 5% of 97.32 M.m$^3$ = 5% × 97.32 = 4.87 M.m$^3$

∴ Required live storage = 97.32. + 4.87 = 102.19 M.m$^3$

Also, gross storage = live storage + dead storage.

Assuming dead storage at 10% of gross storage, we get

$$\text{Gross storage (G)} = \frac{102.19}{0.9} = 113.54 \text{ M.m}^3$$

Hence, the Gross storage of the reservoir is **113.54 M.m$^3$. Ans.**

**Example 3.7.** *A pump is installed on a well to lift water and to irrigate rice crop, sown over 3 hectares of land. If duty for rice is 864 hectares/cumec on the field and pump efficiency is 48% ; determine the minimum required input (H.P.) of the pump, if the lowest well water level is 8 metres below the highest portion of the field. Assume negligible field channel losses.*

**Solution.** Area of rice to be irrigated = 3 hectares.

Duty of water for rice = 864 hectares/cumec.

Discharge required for rice for fulfilling its duty demand

$$= \frac{3}{864} \text{ cumec.} = \frac{1}{288} \text{ cumec.}$$

∴ Volume of water lifted per second $= \frac{1}{288}$ m$^3$

∴ Weight of water lifted per second $= \frac{1}{288}$ m$^3$ × 9.81 kN/m$^3$ = 0.0341 kN/sec.

(∵ unit wt. of water = 9.81 kN/m$^3$)

Minimum static lift of pump = 8 metres

∴ Work done by the pump in lifting this water

= 0.0341 kN/sec. × 8 m = 0.273 kN.m/sec. = 0.273 kWatt.

$$\text{The output of the pump (H.P.)} = \frac{0.273}{0.735} = 0.37$$

(∵ 1 metric H.P. = 0.735 k.watt)

∴ Input H.P. of the pump $= \frac{\text{Output}}{\eta} = \frac{0.37}{0.48} = 0.77$ ; **say 0.8 H.P. Ans.**

**Example 3.8.** *Monthly water requirement studies as shown in Table 3.3 were conducted on various crops that are required to be grown.*

*A reservoir is proposed to be constructed to command an area equal to 1,20,000 hectares. The various crops are : Paddy, Groundnut, Maize, Cotton, Sugarcane and Chillies. The areas under irrigation of these crops are going to be : 20%, 5%, 5%, 10%, 10%, 3% of command respectively. Determine the annual storage required for the reservoir, assuming canal losses as 25% of head discharge, and reservoir evaporation and dead storage losses as 20% of gross capacity.*

**Table 3.3**

| *Month* | *Field irrigation requirement in cm (FIR)* | | | | | |
|---|---|---|---|---|---|---|
| | *Paddy* | *Groundnut* | *Maize* | *Cotton* | *Sugarcane* | *Chillies* |
| (1) | (2) | (3) | (4) | (5) | (6) | (7) |
| June 1—30 | 19.3 | | | | 25.9 | |
| July 1—31 | 6.0 | | | | 7.6 | |
| Aug. 1—31 | 7.4 | | | | 6.8 | |
| Sept. 1—30 | 7.9 | | | | 6.0 | 1.7 |
| Oct. 1—31 | 29.9 | | 3.4 | 4.0 | 34.7 | 23.5 |
| Nov. 1—30 | 20.7 | 6.3 | 15.1 | 8.0 | 42.3 | 22.4 |
| Dec. 1—31 | | 16.2 | 23.8 | 20.6 | 18.0 | 16.6 |
| Jan. 1—31 | | 21.6 | 20.5 | 22.8 | 22.0 | 10.8 |
| Feb. 1—28 | | 13.4 | | 14.3 | 25.0 | |
| March 1—31 | | | | | 36.5 | |
| April 1—30 | | | | | 40.8 | |
| May 1—31 | | | | | 50.0 | |

**Solution.** The areas to be irrigated for Paddy, groundnut, Maize, Cotton, Sugarcane and Chillies respectively are : 24,000, 6,000, 6,000, 12,000, 12000, and 3,600 hectares. The water required at field *i.e.* at outlet point is worked out for each crop (monthwise) as shown in Table 3.4. The table is self explanatory.

**Table 3.4**

| *Month* | *Water required in hectare-metre (i.e. Area × Depth in metres given in question)* | | | | | |
|---|---|---|---|---|---|---|
| | *Paddy* 24000 hectares | *Groundnut* 6000 hectares | *Maize* 6000 hectares | *Cotton* 12000 hectares | *Sugarcane* 12000 hectares | *Chillies* 3600 hectares |
| (1) | (2) | (3) | (4) | (5) | (6) | (7) |
| June 1—30 | 4632 | | | | 3081 | |
| July 1—31 | 1440 | | | | 912 | |
| Aug. 1—31 | 1776 | | | | 816 | |
| Sept. 1—30 | 1896 | | | | 720 | 61.2 |
| Oct. 1—31 | 7176 | | 204 | 480 | 4164 | 846.0 |
| Nov. 1—30 | 4968 | 378 | 906 | 960 | 5076 | 806.4 |
| Dec. 1—31 | | 972 | 1428 | 2472 | 2160 | 597.6 |
| Jan. 1—31 | | 1296 | 1230 | 2736 | 2640 | 388.8 |
| Feb. 1—28 | | 804 | | 1716 | 3000 | |
| Mar. 1—31 | | | | | 4380 | |
| April 1—30 | | | | | 4896 | |
| May 1—31 | | | | | 6000 | |
| Σ | 21888 | 3450 | 3768 | 8364 | 37845 | 2700 |

Grand total = 78,015 hectare metres

Water required at outlet = 78,015 ha.m

$$\text{Water required at canal head} = \frac{78015}{0.75} = 1{,}04{,}020 \text{ ha.m}$$

$$\text{Gross storage required} = \frac{1{,}04{,}020}{0.80} = 1{,}30{,}025 \text{ ha.m}$$

$$= 1{,}30{,}025 \times 10^4 \text{ m}^3 \approx \mathbf{1300\ M.m^3}$$

**Example 3.9.** *Fix the channel capacity for the crop pattern of example 3.8. Assume any suitable data you need.*

**Solution.** Table 3.4 is extended in Table 3.5, so as to show the monthly water requirement of all the crops in col. (8). Discharge required in each interval is woked out in col. (9). The table is otherwise self-explanatory.

**Table 3.5**

| *Month* | *Paddy* | *Groundnut* | *Maize* | *Cotton* | *Sugercane* | *Chillies* | *Total vol. of water reqd. by all crops (V) in ha m = summation of col. (2) to (7)* | *Discharge reqd. in ha. m/day = col. 8 / interval in days* |
|---|---|---|---|---|---|---|---|---|
| *(1)* | *(2)* | *(3)* | *(4)* | *(5)* | *(6)* | *(7)* | *(8)* | *(9)* |
| June 1—30 | 4632 | | | | 3081 | | 7713 | 257.1 |
| July 1—31 | 1440 | | | | 912 | | 2352 | 75.8 |
| August 1—31 | 1776 | | | | 816 | | 2592 | 83.5 |
| September 1—30 | 1896 | | | | 720 | 61.2 | 2677.2 | 89.2 |
| October 1—31 | 7176 | | 204 | 480 | 4164 | 846.0 | 12870 | 415 |
| November 1—30 | 4968 | 378 | 906 | 960 | 5076 | 806.4 | 13094.4 | **436.48*** |
| December 1—31 | | 972 | 1428 | 2472 | 2160 | 597.6 | 7629.6 | 246 |
| January 1—31 | | 1296 | 1230 | 2736 | 2640 | 388.8 | 8290.8 | 267 |
| February 1—28 | | 804 | | 1716 | 3000 | | 5520 | 197 |
| March 1—31 | | | | | 4380 | | 4380 | 142 |
| April 1—30 | | | | | 4896 | | 4896 | 163 |
| May 1—31 | | | | | 6000 | | 6000 | 193 |

*Maximum value.

The maximum value, out of all the discharges in col. (9) of table 3.5 is 436.48 ha-m/day

$$= \frac{436.48 \times 10^4}{24 \times 60 \times 60} \text{ cubic metre per second}$$

$$= \frac{436.48 \times 10^4}{86400} = \frac{436.48}{8.64} = \mathbf{50.6\ cumecs.}$$

Let us assume that the peak demand of the month is 1.25 times the average demand.

Then, the peak demand of the most demanding month

$= 1.25 \times 50.6 = 63.25$ cumecs, say 64 cumecs

Hence, the channel capacity at the outlet point. = **64 cumecs. Ans.**

*Note.* The above channel capacity is at the outlet point, as the FIRs were taken into consideration.

**Example 3.10.** *The base period, intensity of irrigation, and duty of water for various crops under the canal system are given. Determine the reservoir capacity if the culturable commanded area is 40,000 hectares, canal losses are 25%, and reservoir losses are 15%.*

| Crop | Base period days | Duty of the water at the field hec/cumec | Intensity of irrigation % |
|---|---|---|---|
| Wheat, | 120 | 1800 | 20 |
| Sugarcane | 360 | 1700 | 20 |
| Cotton | 180 | 1400 | 10 |
| Rice | 120 | 800 | 15 |
| Vegetables | 120 | 700 | 15 |

(*Engg. Services, 1994*)

**Solution.** Follow Example 3.6, and compute Areas of different crops under irrigation by multiplying *CCA* (40,000 hectares) with intensity of irrigation, as given in column (5) of table 3.6. Water required by each crop in cumecs is then determined in column (6) by dividing crop area in hec by duty in hec/cumec. The volume of water required by each crop in *B* days is finally computed in column (7) by multiplying base period in secs with water required in m³/sec *i.e.* column (2) × 24 × 60 × 60 × column (6)

**Table 3.6**

| Crop | Base period (B) in days | Duty at the field in hec/cumec | I.I. % | Area of each crop under irrigation CCA × I.I. = 40,000 × Col. (4) | Water required by each crop in m³/sec $= \frac{\text{Col. (5)}}{\text{Col. (3)}}$ | Volume of water required in m³ in (B) days = Col. (2) × 24 × 60 × 60 × Col. (6) |
|---|---|---|---|---|---|---|
| (1) | (2) | (3) | (4) | (5) | (6) | (7) |
| Wheat | 120 | 1800 | 20 | 8,000 | 4.444 | 46.08 M-m³ |
| Sugarcane | 360 | 1700 | 20 | 8,000 | 4.706 | 146.37 M-m³ |
| Cotton | 180 | 1400 | 10 | 4,000 | 2.857 | 44.43 M-m³ |
| Rice | 120 | 800 | 15 | 6,000 | 7.50 | 77.76 M-m³ |
| Vegetables | 120 | 700 | 15 | 6,000 | 8.571 | 88.87 M-m³ |
| Σ | | | 80% | 32,000 | | 403.51 M-m³ |

Hence, total water required at the field = 403.51 Mm³.

Canal losses = 25%

$\therefore$ Water required at Canal Head $= \frac{403.51}{0.75}$ Mm³ = 538.013 Mm³

Reservoir losses = 15%

$\therefore$ Gross Reservoir Capacity $= \frac{538.013}{0.85}$ Mm³ = 632.96 Mm³

say **633 Mm³ Ans.**

**Example 3.11.** *A reservoir is proposed to be constructed to command an area of 1,20,000 hectares. It is anticipated that ultimately sugar and paddy would both be irrigated equal to 20% of the command and Rabi equal to 50% of the command, making a total annual irrigation equal to 90% of the commanded area.*

*(a) Work out the storage required for the reservoir, assuming water requirement as prevailing in any area you are familiar with. Assume canal losses at 25% of the head discharge, and reservoir evaporation and dead storage losses at 20% of gross capacity.*

*(b) For the above crop pattern, fix the channel capacity in the head reaches.*

**Solution.** Total commanded area = 1,20,000 hectares
Area under Sugar = 0.20 × 1,20,000 = 24,000 hectares
Area under Paddy = 0.20 × 1,20,000 = 24,000 hectares
Area under Rabi = 0.50 × 1,20,000 = 60,000 hectares

Let us assume water depth requirements, average duties, on capacity duties, etc. for the three crops, as shown in Table 3.7.

**Table 3.7**

| *Crop* | *Total water depth reqd. in cm* | *Av. Duty in hectares/cumec* | *Duty on capacity, i.e. max demand duty in hectares/cumec* |
|---|---|---|---|
| Sugar | 90 | 2600 | 600 |
| Paddy | 120 | 864 | 637 |
| Rabi | 40 | 3464 | 1728 |

The quantity of water required for all crops is worked out as shown in Table 3.8.

**Table 3.8**

| *Crop* | *Δ in cm (assumed)* | *Hectares under cultivation* | *Water reqd. in hectare-metres = col. (2) × col. (3)/100* |
|---|---|---|---|
| Sugar | 90 | 24000 | 21600 |
| Paddy | 120 | 24000 | 28800 |
| Rabi | 40 | 60000 | 24000 |
| | | | Σ = 74400 |

Total volume of water required at outlet

= 74400 hectare metres = 74400 × $10^4$ cubic metres

= 744 million cubic metres (M.$m^3$)

*Canal losses = 25% of head discharge (given)*

∴ Total volume of water required at the head of the canal system

$$= \frac{744}{0.75} \text{ M.m}^3 = 992 \text{ M.m}^3$$

or Net Reservoir Storage = **992 M.m³**

Reservoir evaporation and dead storage loss = 20% of Gross storage *(given)*

(*a*) Gross Storage of the required reservoir = $\frac{992}{0.80}$ = **1240 M.m³** **Ans.**

(*b*) *Channel capacity.* Channel capacity can be fixed on the basis of 'on capacity outlet factors' assumed in col. (4) Table 3.6. Let us work out the capacity discharge required for each crop as shown in col (4) of Table 3.9.

**Table 3.9**

| *Crop* | *'Capacity Duty' hectares/cumec (assumed)* | *Hectares under cultivation (given)* | *Capacity in cumecs* |
|---|---|---|---|
| Sugar | 600 | 24000 | 40 |
| Paddy | 637 | 24000 | 37.7 |
| Rabi | 1728 | 60000 | 34.7 |

Cumecs reqd. for Sugar and Paddy = 40 + 37.7 = 77.7 cumec

Cumecs reqd. for Sugar and Rabi = 40 + 34.7 = 74.7 cumec.

Assuming that the Paddy does not require water when Rabi requires, and Sugar may require its peak demand at a time when either Rabi or Paddy is in full demand ; we can fix the capacity of the channel as maximum of these two, *i.e.* **77.7 cumec** (It is very hypothetical). In practice, however, demand of sugar may arise only when Rabi or Paddy are in very small demand. In such a hypothetical case, assuming that the peak demands of Sugar and Rabi or Paddy (one of them) may arise at the same time, we may fix channel capacity as :

Channel capacity at the outlet = 77.7 cumec

$$\therefore \quad \text{Channel capacity in the head reaches} = \frac{77.7}{0.75} \text{ cumec} = \mathbf{103.6 \text{ cumec.}} \quad \textbf{Ans.}$$

### 3.8. Losses of Water in Canals

During the passage of water from the main canal to the outlet at the head of the watercourse, water may be lost either by evaporation from the surface or by seepage through the peripheries of the channels. These losses are sometimes very high, of the order of 25 to 50% of the water diverted into the main canal. In determining the designed channel capacity, a provision for these water losses must be made. The provision for the water lost in the watercourses and in the fields is however, already made in the outlet discharge factor, and hence, no extra provision is made on that account. Evaporation and seepage losses of channels are discussed below :

(1) **Evaporation.** The water lost by evaporation is generally very small, as compared to the water lost by seepage in certain channels. Evaporation losses are generally of the order of 2 to 3 per cent of the total losses. They depend upon all those factors*, on which the evaporation depends, such as temperature, wind velocity, humidity, etc. In summer season, these losses may be more but seldom exceed about 7% of the total water diverted into the main canal.

(2) **Seepage.** There may be two different conditions of seepage, *i.e.* (*i*) Percolation, (*ii*) Absorption.

(*i*) ***Percolation.*** In percolation, there exists a zone of continuous saturation from the canal to the water-table and a direct flow is established. Almost all the water lost from the canal, joins the ground water reservoir.

The loss of water depends upon the difference of the top water surface level of the channel and the level of the water-table. (*i.e.* *H*) as shown in Fig. 3.7.

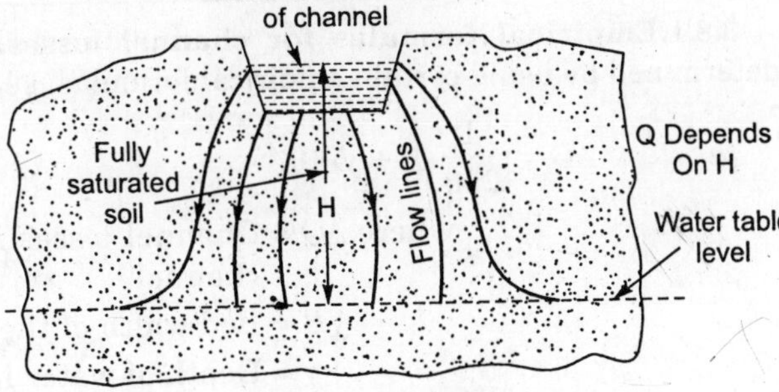


Fig. 3.7. Percolation

(*ii*) ***Absorption.*** In absorption, a small saturated soil zone exists round the canal section, and is surrounded by zone of decreasing saturation. A certain zone just above the water-table is saturated by capillarity. Thus, there exists an unsaturated soil zone between the two saturated zones, as shown in Fig. 3.8.

*Discussed in "*Water Resources Engg. (Vol. I)–Hydrology and Water Resources*".

In this case, the rate of loss is independent of seepage head ($H$) but depends only upon the water head $h$ (*i.e.* distance between water surface level of canal and the bottom of the saturated zone) plus the capillary head $h_c$, as shown in Fig. 3.8.

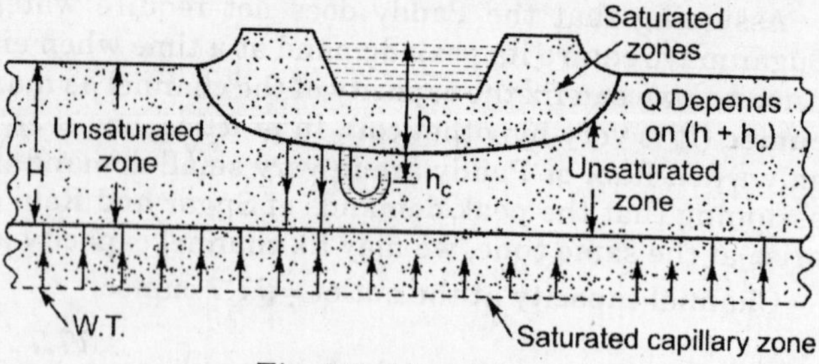


Fig. 3.8. Absorption.

**The seepage losses depend upon the following factors :**

(*i*) Type of seepage, *i.e.* whether 'percolation' or 'absorption'.

(*ii*) Soil permeability.

(*iii*) The condition of the canal ; the seepage through a silted canal is less than that from a new canal.

(*iv*) Amount of silt carried by the canal ; the more the silt, lesser are the losses.

(*v*) Velocity of canal water ; the more the velocity, the lesser will be the losses.

(*vi*) Cross-section of the canal and its wetted perimeter.

For designs, a combined figure for seepage losses as well as for evaporation losses, expressed as cumec per million sq. m of wetted perimeter, may be taken, as tabulated in Table 3.10.

**Table 3.10. Channel Losses**

| *S. No.* | *Type of soil through which channel is excavated* | *Total loss in cumec/million sq. m of wetted area* |
|---|---|---|
| 1. | Rocks | 0.9 |
| 2. | Black cotton soil | 1.6 |
| 3. | Alluvial red soil | 2.5 |
| 4. | Decayed rock, gravel, etc. | 3.0 |
| 5. | Loose sandy soil | 5.5 |

**3.8.1. Empirical formulas for channel losses.** The channel losses can be determined by using certain empirical formulas, such as

(*a*) $$\Delta Q = \frac{1}{200}(B + y)^{2/3} \quad ...(3.7)$$

where $\Delta Q$ = Channel losses in cumecs per km length of channel,

$B$ = Bed width of the channel in metres.

$y$ = Depth of water in the channel in metres.

This formula is generally used in U.P.

(*b*) $$\Delta Q = 1.9 \cdot Q^{1/6} \quad ...(3.8)$$

where $\Delta Q$ = Losses in cumecs per million sq. m. of wetted perimeter.

$Q$ = Discharge in cumecs.

The above formula is commonly used in Punjab.

## 3.9. Canal Regulation

The amount of water which can be diverted from the river into the main canal depends upon :

(*i*) The water available in the river ;

(*ii*) The capacity of the main canal ; and

(*iii*) the share of other canals taking off from the river.

The flow from the main canal is further diverted to the branch canals ; and from the branch canals to the distributaries. The distribution of water (flow) depends upon the water demand of various canals. *The method of distribution of available supplies into various canals of a canal system is called the canal regulation.*

Obviously, when there exists a significant demand of water anywhere in the command area of a canal, the said canal has to be kept flowing. The canal can, however, be closed, if the water demand falls below a specified quantity. The canal shall be reopened again when the water demand exceeds the specified minimum quantity. Normally, there always exists a demand in some part of the command area of any major canal. Such major canals can, hence, be closed only for a very small period (say hardly for three to four weeks in a year). These canals, thus run almost continuously. They may, however, carry discharges much less than their full capacity, either when there is less demand, or when the available supplies are insufficient.

When the demand is less, only the distributaries which need water are kept running, while others (including those which have very little demand) are closed. In case of higher demand but insufficient supplies, either all the smaller channels are made to run simultaneously and continuously with reduced supplies ; or some channels are closed turn by turn and the remaining ones are made to run to their full or near-full capacities. *The first alternative of simultaneously running all the distributaries with reduced supplies is generally not preferred, as it causes channel siltation, weed growth, increased seepage, water-logging and low heads on the outlets. The second alternative of running the distributaries turn by turn with full supplies is, hence, generally preferred, since it does not have these disadvantages, and allows sufficient time for inspections and repairs of the channels.*

A roster is, therefore, usually prepared for indicating the allotted supplies to different channels, showing the schedule of their closure and operation.

A ***flexible regulation*** is always preferred, so that the supplies are allocated in accordance with the anticipated demand. The allocation of supplies should hence be varied on the basis of the information provided by the canal revenue staff, who is supposed to keep a close watch on the crop condition and irrigation water demand of their respective distributaries and outlets.

The discharge in a canal is usually regulated at its *head regulator**, which is generally designed as a meter, as to indicate the discharge being let out from it. However, when a head regulator cannot be used as a discharge meter, a depth guage is provided at about 200 m distance downstream of the head regulator. The gauge-discharge curve of the canal is kept ready, as to indicate the discharge for the observed gauge reading. By manipulating the opening through the regulator gates, the desired discharge can be obtained in the given canal.

---

***Discussed in details in chapter 13.**

## 3.10. Distribution of Water into the Fields Through Water Courses

The water from a distributary or a minor is allowed to flow into the water course, through an opening, called *module* or *outlet**. When once the water reaches the water course, the problem of its equitable distribution among the various farmers or fields arises. The release of water into the water course and its sharing by farmers with different field holdings, largely depends upon the available supply. Since the water supply is usually limited, the following two possible alternatives for its distribution become available :

(1) The canal irrigation is restricted to a limited area, which can be fully supported with the lowest available supply. This does not lead to optimum utilisation of available water and may cause *intensive irrigation* and its ill effects.

Agricultural production and protection against famine would also not be optimum. The production may be maximum per unit of land covered (if the farmers donot over-irrigate), though it would certainly not be per unit of water available. This approach would, however, not require a precise or sophisticated method for distribution of irrigation water amongst the field owners. The delivery system for this alternative can be either continuous or demand-based, depending upon the availability of water. A continuous delivery system can be effectively used for large farms and continuous terraced rice fields. Though ideal, a demand-based delivery system is not practical on large irrigation systems.

(2) In the second and usually adopted alternative, irrigation is extended to a much larger area than what could be supported by the lowest available supply. This *extensive irrigation often creates perpetual scarcity of irrigation water but ensures that a comparatively much less quantity of water remains unutilised. Agricultural production and protection again famine will be at optimum levels. The crop production would be maximum per unit of available water, though it may not be optimum per unit of land covered.* Since this method of delivering water is in the interest of a larger section of farmers, it is usually adopted in our country, inspite of the fact that this method of distribution requires elaborate control and monitoring on the release of water from the different outlets into the different water courses, and to further ensure that the farmers in possession of different land holdings (land areas) do share water as per their decided shares and utilise it on the decided days, rather than permitting the upstream farmers to utilise larger amounts of irrigation water and thereby leaving very little or no water for farmers at the tail end of the conveyance system. Such a system of distributing irrigation water among the farmers is popularly known as the Warawandi, and is explained below :

**3.10.1. Warabandi.** *Warabandi* (Fig. 3.9) is an integral management system for conveying and distributing irrigation water from the source (river or reservoir) down to the farm gates (*i.e. nakkas*) of various land holders, so as to ensure supplies upto the tail end of the command area. The entire water conveyance system is divided into the following three categories :

(*i*) *Primary distribution system* ;

(*ii*) *Secondary distribution system* ; *and*

(*iii*) *Tertiary distribution system.*

Conveyance of water from the source into the *main canal*, feeding two or more *branch canals* (which may operate by rotation) and may not carry the total required supply, constitute what is known as the **primary distribution system.** This primary distribution system runs throughout the irrigation season with varying supply.

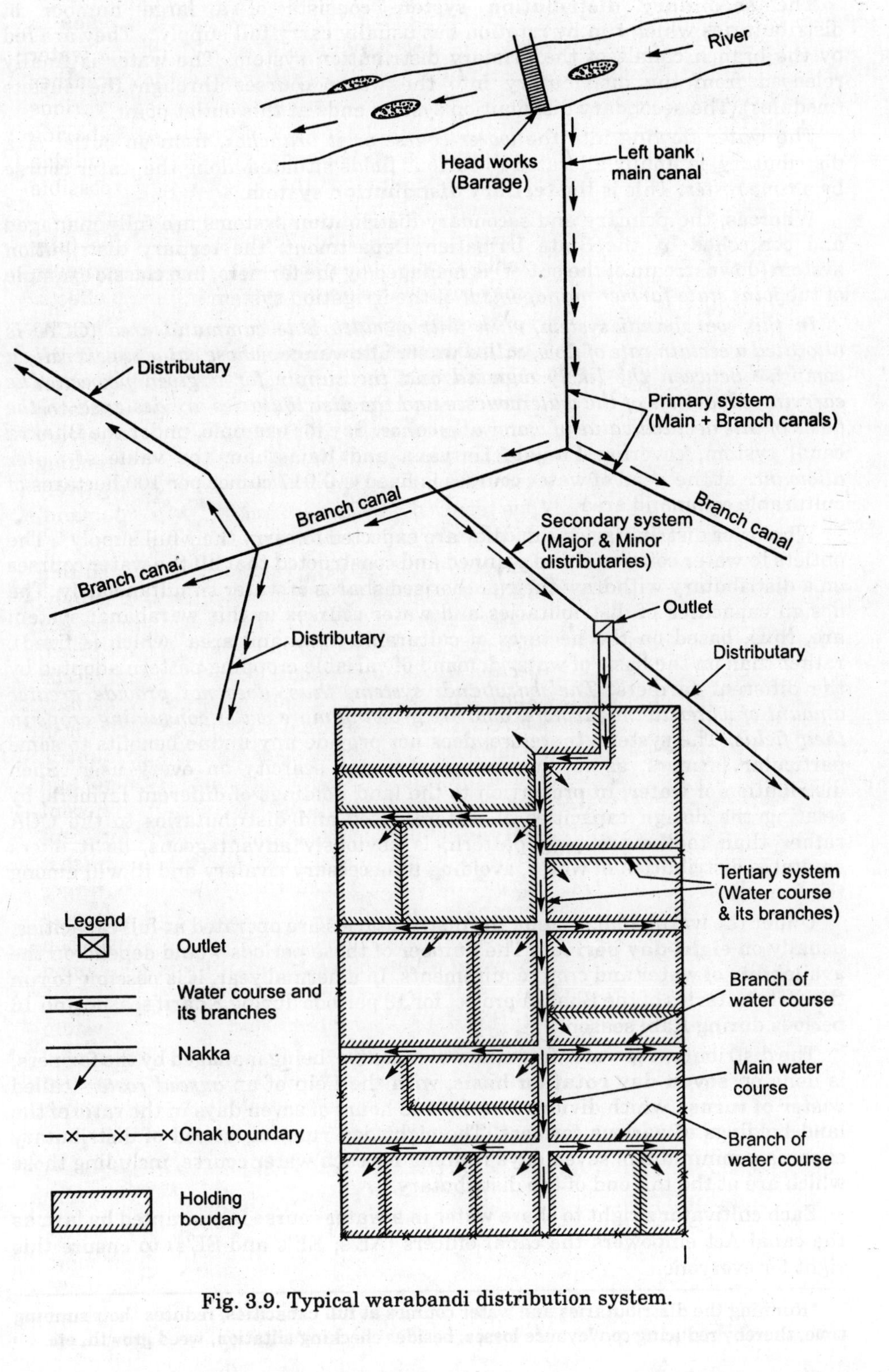


**Fig. 3.9. Typical warabandi distribution system.**

The **secondary distribution system** consists of a large number of distributaries which run by rotation but usually carry full supply*. They are fed by the branch canals of the primary distribution system. The water is finally released from the distributary into the water courses through the outlets (modules). The secondary distribution system ends at this outlet point.

The water flowing into the *water course or it branches*, from an outlet of a distributary, is finally allocated to various fields situated along the water course by a *time roster*. This is the **tertiary distribution system**.

Whereas, the primary and secondary distribution systems are fully managed and controlled by the State Irrigation Department, the tertiary distribution system (downstream of the outlet) is managed by the farmers, in a classic example of the *joint state-farmer management* of the irrigation system.

*In this warabandi system, each unit of culturable cammand area (CCA) is allocated a certain rate of flow, called* **water allowance**, *whose value is generally a comprise between the likely demand and the supply for a given project. The carrying capacities of the watercourses and the distributaries are designed on the basis of this decided value of water allowance*. Say for example, under the Bhakra canal system, covering Punjab, Haryana and Rajasthan, the value of *water allowance* at the head of water courses is fixed @ 0.017 cumec per 100 hectares of culturable command area.

Whenever distributaries run, they are expected to carry their full supply*. The outlets to water courses are so planned and constructed that all the water courses on a distributary withdraw their authorised shares of water simultaneously. The design capacities of distributaries and water courses in this warabandi system are, thus, based on the hectares of culturable command area (which is fixed), rather than on the basis of water demand of variable cropping pattern adopted by the different farmers. *The warabandi system, thus, does not provide greater amount of water to the farmers, who are growing more water consuming crops in their fields.* The system, therefore, does not provide any undue benefits to some particular farmers, as it imposes equal water scarcity on every user. Such distribution of water, in proportion to the land holdings of different farmers, by relating the design capacities of watercourses and distributaries to the CCA rather than to the cropping pattern, is obviously advantageous, as it offers equitable distribution of water, avoiding unnecessary rivalary and ill will among the farmers.

Under the warabandi system, the distributaries are operated at full capacities, usually on **eight-day periods**. The number of these periods would depend on the availability of water and crop requirements. In a normal year, it is possible to run the distributaries of the Bhakra project for 18 periods during Kharif season and 16 periods during Rabi season.

The distribution of water from the water course, being managed by the farmers, is done on **seven-day rotation basis**, with the help of an *agreed roster* (called **roster of turns**), which divides 24 × 7 = 168 hours of seven days in the ratio of the land holdings of various farmers. The eight day running period of distributary ensures a minimum of seven days running for each water course, including those which are at the tail end of the distributary.

Each cultivator's right to share water in a water course is guaranted by law, as the canal Act empowers the canal officers (AE's, EE's and SE's) to ensure this right for everyone.

---

***Running the distributaries and water courses at full capacities, reduces their running time, thereby reducing conveyance losses, besides checking siltation, weed growth, etc.**

Whenever a distributary is running, a water course receives its share of water at a constant rate round the clock and water distribution proceeds from head to tail. Each cultivator on the watercourse is entitled to receiving the entire water in a water course only on specific week day and at a specific time (including night time). There is no provision in the system to compensate a defaulting farmer, who has failed to utilise his turn for any reason.

**Roster of Turn.** The cycle of turns on a *water course or its branch* starts from its head, proceeds downstream and ends at the tail. Before a farmer receives his share of water, some time is spent in filling the water-course length between the point of his taking over and the beginning of his holding. This time is called ***bharai*** (*i.e.* the filling time), which is debited to a common pool time of 168 hours and credited to the account of the concerned farmer. The supply in the water course has to be stopped when the tail end farmer is having his turn, since the water filled in the water-course during the entire ***bharai period*** can be discharged only into the field of the tail-end farmer ; and hence, normally, the total time spent on filling the entire length of the water course should be recovered from him in lieu of this ; but then, he does not receive this water (flowing after closure of outlet) at a constant rate. Since such a supply, beyond a limit, is not efficient for field applications, the tail end farmer is compensated for it, and is allowed a certain discount on the recovery of the *bharai* time. This value of *bharai* is called *Jharai* (*i.e.* emptying time). Evidently, precise determination of ***Jharai*** time is not feasible, and its present values are favourable to the tail-end farmers. After allowing for *bharai* and *jharai* timings, the flow time (*FT*) for a unit area of land holding of an individual farmer is given as :

Flow time (*FT*) for unit area in hours ($h$)

$$= \frac{168h - [\text{Total bharai time in } h - \text{Total jharai time in } h]}{\text{Total area to be irrigated by the given outlet or watercourse}} \quad ...(3.9)$$

Flow time in hour for an individual farmer

$$= [FT \text{ for unit area} \times \text{area of farmer's fields} + (\text{his } bharai - \text{his } jharai)] \quad ...(3.7)$$

Obviously, *bharai* is usually zero in the case of last farmer (at tail) and *jharai* is zero for all except the last farmer. It may also be noted that the losses in the watercourse are not reflected anywhere in the above equations.

## PROBLEMS

**1.** (*a*) Differentiate between "Alluvial" and "Non-alluvial" canals.

(*b*) What are the different ways in which the irrigation canals can be aligned ?

(*c*) Name the various types of canals which are required to be constructed while planning a canal irrigation system.

**2.** Give an account of the investigations and surveys required while planning an irrigation canal project in a given tract of land. Discuss the factors governing the selection of alignment of the main canal and its branches.

**3.** (*a*) "The construction and design of canals in Tamil Nadu is easier compared to that in U.P." Discuss critically the above statement with reference to the type of soils available in these two States of India.

(*b*) State how will you fix up the following in an irrigation canal :

(*i*) Alignment ; (*ii*) Full supply level ; (*iii*) Full supply discharge.

**4.** (*a*) Discuss the factors that are considered in fixing the alignment of a main canal (*b*) The left bank canal of a storage irrigation scheme carries a discharge of 10 cumec and has a culturable commanded area of 8000 hectares. The intensity of Rabi crop is 70% and base period is 110 days. The right bank canal of the scheme on the other hand, carries a discharge of 24 cumec and commands a culturable area of 15000 hectares. The intensity of rabi crops is 80% and the base period is 110 days. Compare the efficiencies of the two canal systems.

**5.** (*a*) Define and explain the following terms :

(*i*) Culturable commanded area. (*ii*) Intensity of irrigation
(*iii*) Nominal duty (*iv*) Full supply coefficient
(*v*) Time factor (*vi*) Capacity factor.

(*b*) The discharge available from a tubewell is 136 cubic metres per hour. Assuming 3000 hours of working for this tubewell in a year, estimate the culturable area that this tubewell can command. The intensity of irrigation is 50% and the average water depth required for the rabi and kharif crops is 51 cm

$$\left[\textbf{Hint. } \text{CCA} = \frac{136 \times 3000}{0.51} \cdot \frac{1}{0.5} \cdot \frac{1}{10^4} \text{ hectares} = 160 \text{ hectares} \quad \textbf{Ans.}\right]$$

**6.** Write short notes on :

(*i*) Alluvial and Non-alluvial soils. (*ii*) Water-shed canals and contour canals
(*iii*) Distribution system for canal irrigation (*iv*) Water losses in irrigation canals.
(*v*) Curves in irrigation channels.

**7.** (*a*) Explain in detail, as to how will you proceed for determining the approximate value of design discharge of an irrigation canal. What procedure are required for its precise computations ?

(*b*) Discuss the precautions which you would adopt while designing irrigation channels aligned on curvilinear paths.

**8.** (*a*) How is the capacity of an irrigation channel decided for Indian irrigation conditions? Also discuss, as to why the kor demand of course should be considered in place of average demand while computing this capacity.

(*b*) A field channel has a culturable commanded area of 2000 hectares. The intensity of irrigation for gram is 30% and for wheat is 50%. Gram has a crop period of 18 days and kor depth of 12 cm, while wheat has a kor period of 15 days and a kor depth of 15 cm. Calculate the discharge of the field channel.

[**Hint.** Duty for gram = 864 × 18/12 = 1296 ha/cumec
Duty for wheat = 864 × 15/15 = 864 ha/cumec
Area of gram irrigated = 2000 × 0.30 = 600 ha.
Discharge for gram = 600/1296 cumec = 0.46 cumec
Area of wheat irrigated = 50% × 2000 = 1000 ha
Discharge for wheat = 1000/864 = 1.16 cumec
Total discharge required = 0.46 + 1.16 = 1.62 cumec. **Ans.**]

**9.** A watercourse commands an irrigation area of 800 hectares. The intensity of irrigation of rice in this area is 50%. The transplantation of rice crop takes 15 days and total depth of water required by the crop is 60 cm on the field during the transplantation period, given that the rain falling on the field during this period is 15 cm. (*i*) Find the duty of irrigation water for the crop on the field during transplantation ; (*ii*) at the head of the distributary, assuming losses of water to be 20% in the water courses (*iii*) Calculate the discharge required in the watercourse.

[**Ans.** (*i*) 216 hectare/cumec (*ii*) 172.8 hectare/cumec, (*iii*) 2.32 cumec.]

**10.** How is the water lost while flowing through irrigation canals ? What are the factors affecting these losses, and how will you estimate these losses ? How do these losses affect the duty of water and what precautions will you undertake in order to minimise them ?

**11.** What are the possible causes of water losses in a canal ? What are the methods adopted for reducing such losses ?

4

# Sediment Transport and Design of Irrigation Channels

Whenever water flows in a channel (natural or artificial), it tries to scour its surface. Silt or gravel or even larger boulders are detached from the bed or sides of the channel. These detached particles are swept downstream by the moving water. This phenomenon is known as ***Sediment Transport.***

## 4.1. Importance of Sediment Transport

(*i*) The phenomenon of sediment transport causes large scale scouring and siltation of irrigation canals, thereby increasing their maintenance. Many poorly designed artificial channels get silted up so badly, that they soon become inoperable, causing huge economic loss to the public exchequer. The artificial channels should, therefore, be properly designed, and should not fail to carry the sediment load admitted at the canal headworks.

(*ii*) The design and execution of a flood control scheme is chiefly governed by the peak flood levels, which, in turn, depend upon the scour and deposition of sediment. Firstly, the bed levels may change by direct scouring or deposition of sediment, and thereby changing the flood levels. Secondly, the scouring and silting on the river banks may create sharp and irregular curves, which increase the flow resistance of the channel, and thereby, raising the flood levels for the same discharge.

(*iii*) Silting of reservoirs* and rivers is another important aspect of sediment transport. The storage capacity of the reservoir is reduced by its silting, thereby, reducing its use and life. Natural rivers used for navigation are frequently silted up, reducing the clear depth (draft) required for navigation. Sediment deposited in these rivers and harbours may often require costly dredgings.

Sediment transport, thus, poses numerous problems, and is a subject of great importance, and possesses enough potential for further research and development.

## 4.2. Sediment Load

The sediment in a canal is a burden to be borne by the flowing water, and is, therefore, designated as sediment load.

**Bed Load and Suspended Load.** The sediment may move in water either as bed load or as suspended load. Bed load is that in which the sediment moves along the bed with occasional jumps into the channel. While, the suspended load is the one in which the material is maintained in suspension due to the turbulence of the flowing water.

*For details of reservoir silting, please see Chapter 18.

## 4.3. Bed Formation (Practical Aspect)

The channel bed may be distorted into various shapes by the moving water, depending upon the discharge or the velocity of the water.

At low velocities, the bed does not move at all, but it goes on assuming different shapes as the velocity increases. *Let us see what happens to a channal bed made of fine sand (lesser than 2mm dia) when the velocity is gradually increased in steps.*

When the velocity is gradually increased, then first of all, a stage is reached, when the sediment load comes just at the point of motion. This stage is known as **threshold stage of motion.** On further increase of velocity, the bed develops *ripples* of the saw-tooth type, as shown in Fig. 4.1 (*a*). Such ripples can also be seen in sand on any beach. As the velocity is increased further, larger periodic irregularities appear, and are called *Dunes.* When they first appear, ripples are superimposed on them [Fig. 4.1 (*b*)]. But at still higher velocities, the ripples disappear and only the dunes are left [Fig. 4.1 (*c*)]. Dunes may form in any grain size of sediment, but ripples do not occur if the

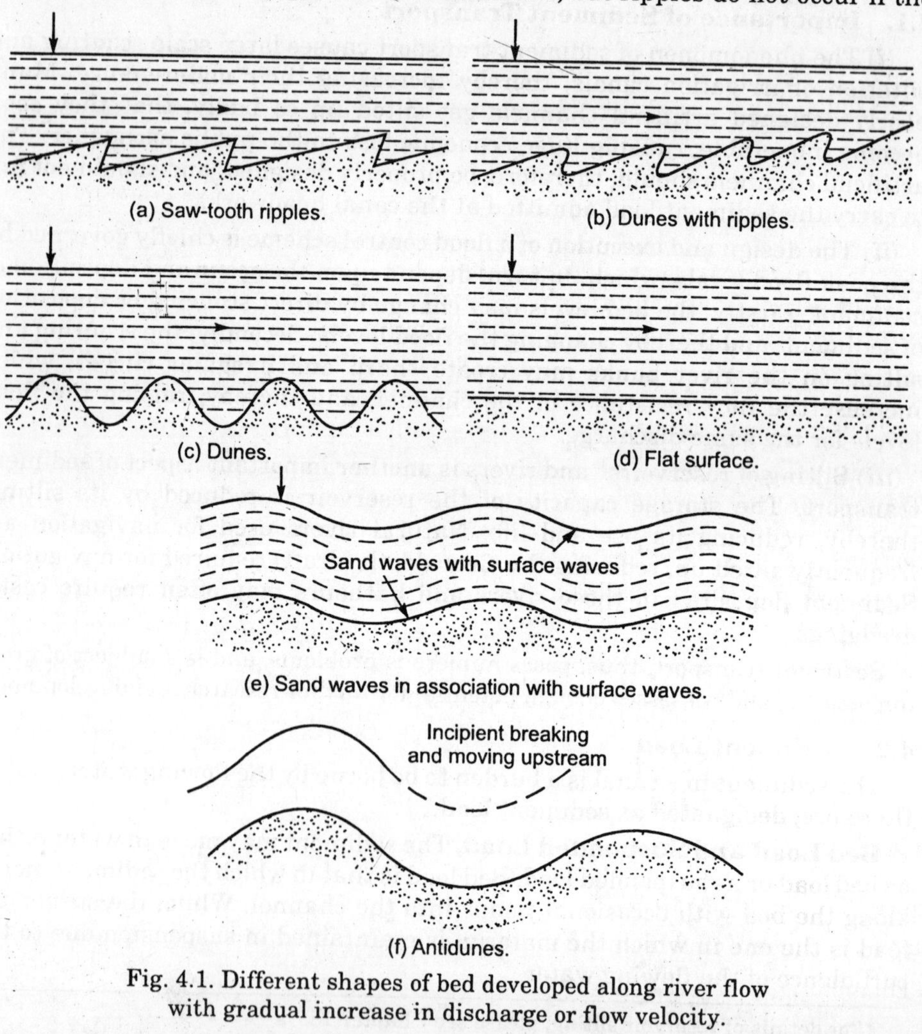


Fig. 4.1. Different shapes of bed developed along river flow with gradual increase in discharge or flow velocity.

size of the bed particles is coarser than 0.6 mm. Dunes are much larger (in length and height) and more rounded than ripples. So much so that ripples seldom exceed 40 cm in length (between two adjacent troughs) and 4 cm in height (trough to crest), while the dunes in laboratory flumes may have 3m length and height up to about 40 cm, while in large rivers, they may be as high as 15 metres, with several hundred metres length. Crests of both, do not extend across the entire width of the stream, *i.e.*, both formations tend to occur in the form of "short crested waves". The flow conditions remain sub-critical in both these regimes. While most of the sediment particles move along the bed, some finer particles of sediment may go in suspension.

When the velocity is increased beyond formation of dunes, the dunes are erased by the flow, leaving very small undulations or virtually a flat surface with sediment particles in motion [Fig. 4.1 (*d*)]. Further increase in velocity, results in the formation of sand waves in association with surface waves [Fig. 4.1 (*e*)]. As the velocity is further increased, so as to make the Froude number (*i.e.* $V/\sqrt{gy}$ ) exceeding unity, the flow becomes super-critical, and the surface waves become so steep that they break intermittently and move upstream, although the sediment particles keep on moving downstream only [Fig. 4.1 (*f*)]. Sand Waves are then called **anti-dunes,** since the direction of movement of bed forms in this regime is opposite to that of the dunes. The sediment transport rate in this regime is obviously very high. The resistance to flow is, however, small compared to that of the ripple and dune regime. In case of canals and natural streams, anti-dunes rarely occur.

## 4.4. Mechanics of Sediment Transport

In the study of mechanics of sediment transport, we will throughout assume that the soil is *incoherent.* By incoherent soil, we mean that there are no cohesive forces between the particles, or in other words $c = 0$, such as in sands or gravels.

Most of our river beds are made up of sands and gravels, and hence, we confine ourselves to the mechanism of movement of such a soil only. Though cohesive clays, etc. are also sometimes met with, but no systematic study upon such soils has been undertaken and only very little work has been done in that direction.

By assuming the soil to be incoherent, each soil grain can be studied individually. The basic mechanism behind the phenomenon of sediment transport is the drag force exerted by water (or fluid) in the direction of flow, on the channel bed. This force, which is nothing but a pull of water on the wetted area, is known as *Tractive Force* or *Shear Force* or *Drag Force.*

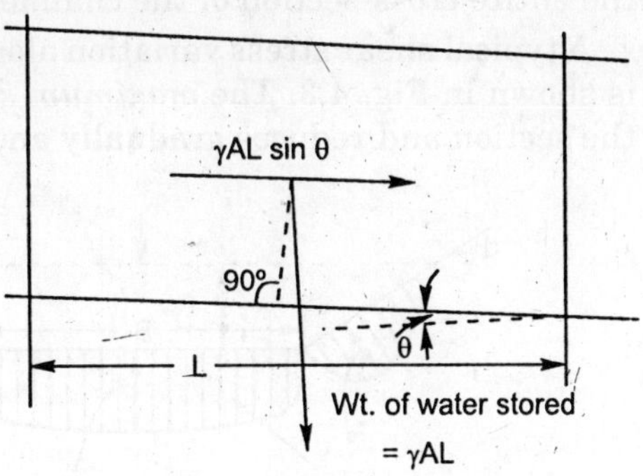

Fig. 4.2

Let us consider a channel of length $L$ and cross-sectional area $A$.

The volume of water stored in this channel reach $AL$

Wt. of water stored = $\gamma_w AL$

where $\gamma_w$ = unit wt. of water = $\rho_w\, g$, where $\rho_w$ is the density of water.

Horizontal component of this wt. = $\gamma_w AL \sin\theta = \gamma_w ALS$

where $S$ = channel bed slope.

This horizontal force exerted by water is nothing but Tractive force.

Average Tractive force per unit of wetted area

$$= \text{Unit Tractive Force } (\tau_0) = \frac{\gamma_w\, ALS}{\text{Wetted area}}$$

$$= \frac{\gamma_w\, ALS}{\text{Wetted perimeter} \times \text{Length}} = \frac{\gamma_w\, ALS}{P.L} = \gamma_w\left(\frac{A}{P}\right)S. = \gamma_w R\,.\,S \quad \left(\because \ \frac{A}{P} = R\right)$$

where $R$ = is the hydraulic mean depth of channel.
$S$ = channel bed slope
$\gamma_w$ = unit wt. of water
$P$ = wetted perimeter

Hence, **Average Unit Tractive force,** also called Shear stress

$$= \tau_0 = \gamma_w\, RS \qquad \text{...(4.1)}$$

We can hence conclude that the Applied Tractive Stress (average) exerted by the flowing water under uniform flow conditions is given by Eq. (4.1) as equal to $\gamma_w$. $RS$.

However, the maximum stress induced by any flow occurs at the point of greatest depth or at the centre of any channel with horizontal bed, and is given by:

$$\tau_{max} = \gamma_w\,.\,y\,S \qquad \text{...(4.1a)}$$

where $\gamma_w$ = unit wt of water
$y$ = water depth in channel
$S$ = channel bed slope.

It has further been established by Lane by using membrane technology *that the tractive or shear stress in channels, except for wide open channels, is not uniformly distributed along the wetted perimeter*. Variations do occur across the entire cross-section of the channel.

A typical shear stress variation along the boundary of a trapezoidal channel is shown in Fig. 4.3. The *maximum shear at bed* ($\tau_{bm}$) occurs at the centre of the section and reduces gradually and then abruptly to zero at both corners.

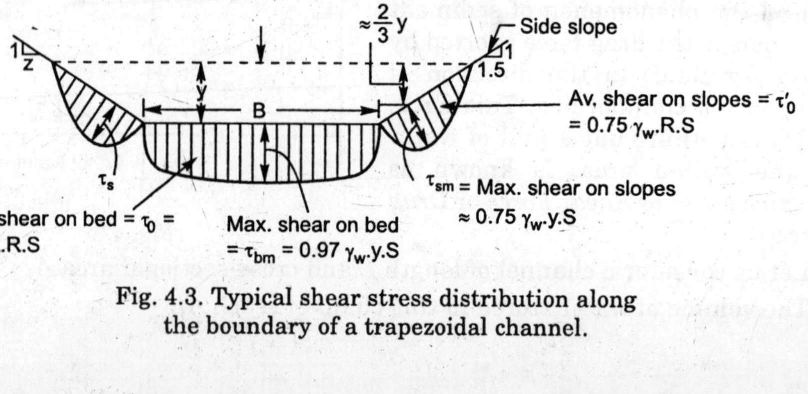


Fig. 4.3. Typical shear stress distribution along the boundary of a trapezoidal channel.

The value of $\tau_{bm}$ is approximately given as :

$$\tau_{bm} = 0.97^* \ \tau_{max} = 0.97^* . \gamma_w . y.S \quad ...(4.1b)$$

Along the side slopes, the maximum shear ($\tau_{sm}$) occurs at approximately two-thirds of the depth of flow with magnitude given as :

$$\tau_{sm} = 0.75 \ \gamma_w .y. S \quad ...(4.1c)$$

Before entering into the mathematical aspect of sediment transport, we will again visualise the "threshold movement of the sediment," and its applications for design of non-scouring channels.

**4.4.1. Threshold Motion of the Sediment.** When the velocity of flow through a channel is very small, the channel bed does not move at all, and the channel behaves as a **rigid boundary channel.** As the flow velocity increases steadily, a stage is reached when the shear force exerted by the flowing water on the bed particles will just exceed the force opposing their movement. At this stage, a few particles on the bed will just start moving intermittently. This condition is called the **incipient motion condition** or simply the **critical condition** or the **threshold point.**

A knowledge of the velocity at which such a critical condition occurs is quite helpful in designing stable non-scouring channels admitting clear waters, since this velocity will help us in fixing the hydraulic mean depth ($R$)** and bed slope ($S_0$) of the channel. The knowledge of this incipient motion condition is also required in some of the methods adopted for computation of sediment load.

The experimental data on incipient motion condition was analysed using *critical tractive force approach,* for the first time by Shield, so as to help in designing stable channels in alluviums. This design basically assumes the entry of clean and clear water in the channel, which is designed to develop non-scouring highest possible flow velocity at the peak flow.

## 4.5. Shield's Entrainment Method for Design of Non-Scouring Stable Channels having Protected Side Slopes in Alluviums

Shield was the first investigator, who provided a semi-theoretical analysis of the problem of incipient condition of bed motion, and used it for designing non-scouring channels. He defined the **critical tractive stress** ($\tau_c$) as that average shear stress ($\tau_0$) acting on the bed of the channel, at which the sediment particle just begins to move. According to him, the bed particle begins to move when the drag force ($F_1$) exerted by the fluid on the particle, just equals or exceeds the resistance ($F_2$) offered by the particle to its movement.

(*i*) The drag force ($F_1$) exerted by the flow is given by :

$$F_1 = K_1 \left[ C_D \, . \, d^2 \, . \frac{1}{2} . \rho_w \, . \, V_0^2 \right] \quad ...(4.2)$$

where $K_1$ = a factor depending on the shape of the particle
$C_D$ = coefficient of drag

---

*This value is usually found to vary between 0.9 to 1.0 for $B = y$ to $B \geq 6y$ and for trapezoidal channels having 1.5H : IV side slopes. For all general uses, the factor 0.97 is accepted.

**This will help in fixing depth ($y$) and bed width ($b$).

$d$ = The dia of the particle

$\rho_w$ = The density of the flowing fluid *i.e.* water

$V_0$ = The velocity of flow at the top of the particle *i.e.* at the bottom of the channel

Using *Karman-Prandtl equation* for the velocity distribution along a channel *x*-section, the velocity of flow at the bottom of the channel ($V_0$) can be expressed as :

$$\frac{V_0}{V^*} = f_1\left(\frac{Vd}{\nu}\right) = f_1 \cdot R_e^* \quad \text{...(4.3)}$$

or

$$V_0 = V^* f_1 \cdot R_e^* \quad \text{...(4.4)}$$

where $V^*$ = **Shear friction velocity** $= \sqrt{\frac{\tau_0}{\rho_w}}$

where $\tau_0$ is the shear stress acting on the boundary of the channel.

$\nu$ = Kinematic viscosity of the flowing fluid *i.e.* water.

$R_e^*$ = **Particle Reynold Number** = $V^* d/\nu$

Also, the coefficient of drag $C_D$ is given by :

$$C_D = f' \cdot \left(\frac{V_0 \cdot d}{\nu}\right) = f'\left(\frac{V^* f_1 R_e^*}{\nu} \cdot d\right)$$

or

$$C_D = f_2 \cdot \left(\frac{V^* d}{\nu}\right) R_e^* = f_2 \cdot R_e^{*2} \quad \text{...(4.5)}$$

Substituting values of $V_0$ and $C_D$ from Eqns. (4.4) and (4.5) in Eq. (4.2), we get

$$F_1 = K_1\left[\left(f_2 \cdot R_e^{*2}\right) \cdot d^2 \cdot \frac{1}{2}\rho_w \cdot (V^* f_1 \cdot R_e^*)^2\right]$$

or

$$F_1 = \left[K_1 \cdot f_2 \cdot f_1^2 \cdot \frac{1}{2}d^2 \cdot \rho_w \cdot V^{*2} R_e^{*4}\right] \quad \text{...(4.6)}$$

(*ii*) The particle resistance ($F_2$) is further given by :

$$F_2 = K_2\left[d^3 \cdot (\rho_s - \rho_w)g\right] \quad \text{...(4.7)}$$

where $\rho_s$ = density of particle

$\rho_w$ = density of fluid or water

$S_s$ = Specific gravity of particle $= \frac{\rho_s}{\rho_w}$

$\gamma_w$ = unit wt. of fluid or water.

$K_2$ = a factor dependent on the shape of the particle and internal friction of soil.

$$\therefore \quad F_2 = K_2\left[d^3 \cdot \left(\frac{\rho_s}{\rho_w} - 1\right)\rho_w \cdot g\right] = K_2\left[d^3 \cdot (S_s - 1)\gamma_w\right]$$

$$= K_2 \cdot \gamma_w \, d^3 \, (S_s - 1) \quad \text{...(4.8)}$$

At critical condition, equating Eqs. (4.6) and (4.8), and introducing subscript (*c*) to denote critical conditions, we get

$$K_1 f_2 \cdot f_1^{\,2} \frac{1}{2} \cdot d^2 \cdot \rho_w \cdot V_{(c)}^{*2}\ R_{e(c)}^{*4} = K_2 \cdot \left[\gamma_w \cdot d^3 \cdot (S_s - 1)\right] \quad ...(4.9)$$

or
$$\frac{\rho_w \cdot V_{(c)}^{*2}}{\gamma_w\ d\,(S_s - 1)} = \left(\frac{2K_2}{K_1 \cdot f_2 \cdot f_1^{\,2}}\right) \cdot R_{e(c)}^{*-4}\ ;\ \text{But } \rho_w \cdot V_{(c)}^{*2} = \tau_c$$

$$\therefore \quad \frac{\tau_c}{\gamma_w \cdot d \cdot (S_s - 1)} = F \cdot R^*_{e\,(c)} \quad ...(4.14)$$

(some function of $R_{e(c)}$)

The left hand side term is a dimensionless number and is called the **Shield's Entrainment function,** and is usually denoted by $F_s$

$$\therefore \quad F_s = F \cdot (R^*_{e(c)}) \quad ...(4.10a)$$

$R^*_{e(c)}$ is $R^*_e$ at critical stage of bed movement in a channel in alluviums.

The above mathematical work shows that $F_s$ *i.e.* $\dfrac{\tau_c}{\gamma_w \cdot d\,(S_s - 1)}$ is a function of $R^*_e$ at critical stage of bed movement ; and based on the experimental work done by Shield, graphs have been plotted between $F_s$ and $R^*_e$, as shown in Fig. 4.4. The obtained curve, forms a suitable basis for the design of channels, where it is required to prevent bed movement or to keep it to the minimum.

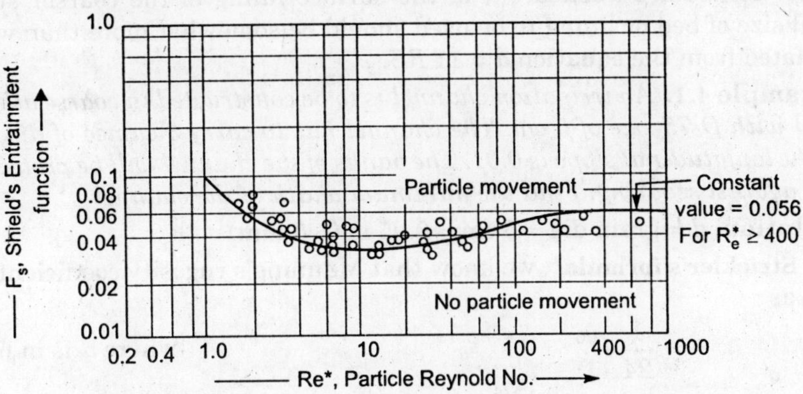


Fig. 4.4. Shield's curve for incipient motion condition.

The application of this curve becomes more simple when Particle Reynold number is more than 400, and as such $F_s$ becomes constant and equal to 0.056. Particle Reynold number, representing roughness, has been found to be more than 400, when the particle size exceeds 6 mm, such as for 'coarse alluvium soils'.

---

*Mittal and Swamee has worked out a general relation between $\tau_c$ and $d$ which gives results within + 5% of the values given by Shield's curve, for all values of *d*. The relation for water and soil of $S_s$ = 2.65, is given by equation

$$\tau_c(\text{N/m}^2) = 0.155 + \frac{0.409 d^2_{mm}}{\sqrt{1 + 0.177 d^2_{mm}}} \quad ...(4.12)$$

Hence, for **designing non-scouring channels in coarse alluviums**

$$\frac{\tau_c}{\gamma_w d(S_s - 1)} = 0.056 \text{ (for } d > 6 \text{ mm)}^* \qquad ...(4.11)$$

where $\gamma_w$ = unit wt. of water = 9.81 kN/m³ or 1 t/m³ or 1000 kgf/m³.

The average shear stress caused on the bed of a channel by the flowing water is given by Eq. (4.1) as :

$$\tau_0 = \gamma_w RS$$

where, $R$ = Hydraulic mean radius of the channel, *i.e.* $A/P$.

$S$ = Bed slope.

Moreover, $\tau_0$ should be $\leq \tau_c$ ; (for non-scouring channels)

or $\tau_0 \leq \gamma_w d\,(S_s - 1)\,(0.056)$ or $\gamma_w RS \leq \gamma_w d(S_s - 1)\,(0.056)$

or $RS \leq d\,(S_s - 1)\,(0.056)$ or $RS \leq d\,(2.65 - 1)\,(0.056)$

or $RS \leq \dfrac{d}{11}$

or $d \geq 11\,RS$ ...(4.13)

Equation (4.13) gives the minimum size of the bed material or lining stone that will remain at rest in a channel of given $R$ and $S$.

Since with the passage of time, the channel bed becomes Armored (*i.e.* the smaller stones are flushed out of the surface lining of the coarser stones), actual size of bed or lining to be used should be somewhat more than what is calculated from the equation $d = 11\,RS$.

**Example 4.1.** *An irrigation channel is to be constructed in coarse alluvium gravel with D-75 size of 5 cm. The channel has to carry 3 cumec of discharge and the longitudinal slope is 0.01. The banks of the channel will be protected by grass against scouring. Find the minimum width of the channel.*

**Solution.** $d$ = grain dia = 5 cm = 0.05 m (> 6 mm)

By **Strickler's formula***, we know that Manning's rugosity coefficient ($n$) is given as :

$$n = \frac{1}{24} d^{1/6} \qquad \text{where } d \text{ is in metres}$$

or $$n = \frac{1}{24} \times \left(\frac{5}{100}\right)^{1/6} = \frac{1}{24}(0.05)^{0.16} = \frac{1}{24} \times 0.619 = 0.0258.$$

Now, $$d \geq 11\,RS, \text{ or } R \leq \frac{d}{11\,S} = \left(\frac{5}{100} \times \frac{1}{11 \times 0.01}\right) = 0.455 \text{ m},$$

or $R_{max} = 0.455$ m

---

* Strictly speaking, the **Strickler's formula** is applicable to the *rigid boundary channels* only and not to the *moveable boundary channels*, since it gives the roughness coefficient ($n'$) due to grain roughness alone, and does not account for form roughness, which is caused by the undulations in the bed of a moveable boundary channel. The true value of $n$ in Manning's equation will, on the other hand, represent roughness of bed consisting of grain roughness as well as the form roughness, together. This formula is therefore valid for rivers with beds of coarse materials, practically free from ripples.

Now using $V = \frac{1}{n} R^{2/3} S^{1/2}$ (Manning's formula), we have

$$V_{max} = \frac{1}{0.0258} (0.455)^{2/3} (0.01)^{1/2}$$

$$= \frac{1}{0.0258} (0.592)(0.1) = 2.29 \text{ m/sec.}$$

Assuming $R = y$ (1st app.), $Q = AV = by \times V = b.RV$

or $Q = b \,.\, RV$, if $R$ and $V$ are taken maximum $b$ will be minimum.

$\therefore \quad 3 = b_{min} \times 0.455 \times 2.29$

or $b_{min} = \frac{3}{0.455 \times 2.29} = 2.88$ m.

Use a conservative value of base width as 3 m. **Ans.**

**Example 4.2.** *Water flows at a depth of 0.6 m in a wide stream having a bed slope of 1 in 2500. The median diameter of the sand bed is 1.0 mm. Determine whether the soil grains are stationary or moving, and comment as to whether the stream bed is scouring or non-scouring.*

**Solution.** Since the given size of bed particles is 1.0 mm, which is less than 6 mm, we can not use Shield's Eq. (4.11), since $R_e^*$ in this case will be less than 400. We will, therefore, use the general Eq. (4.12), which is valid for all sizes of $d$.

$$\tau_c \,(\text{N/m}^2) = 0.155 + \frac{0.409\, d^2_{mm}}{\sqrt{1 + 0.177\, d^2_{mm}}} \qquad \text{...(Eq. 4.12)}$$

$$= 0.155 + \frac{0.409 \times 1}{\sqrt{1 + 0.177 \times 1}} = \mathbf{0.53\ N/m^2}.$$

Also using Eq. (4.1), we have

$$\tau_0 = \gamma_w \times R \,.\, S = 9.81 \times 0.6 \times \frac{1}{2500} \text{ kN/m}^2$$

$$\left( \because \begin{array}{l} \gamma_w = 9.81 \text{ kN/m}^3 \text{ for water} \\ R \approx y \text{ for wide streams} = 0.6 \text{ m} \end{array} \right)$$

$$= 2.35 \times 10^{-3} \text{ kN/m}^2$$

$$= 2.35 \text{ N/m}^2$$

$$= 2.35 \text{ N/m}^2; \text{ which is more than } 0.53 \text{ N/m}^2.$$

Since $\tau_0 > \tau_c$, the soil grains will not be stationary, and the scouring and sediment transport will occur. **Ans.**

**Example 4.3.** *A wide unlined channel carrying silt free water has a depth of 2.0 m. The maximum tractive stress permissible on the bed to prevent scour is 0.20 kg/m². What is the maximum slope that can be given to the channel.*

**Solution.** Use : $\tau_0 = \gamma RS$,

$\therefore \quad \tau_{0\,(max)} = \gamma R.S._{max} = 1000 \times 2.0 \times S_{max} = 0.20$ kg/cm²

$\therefore \quad S_{max} = 1$ in 10,000. **Ans.**

## 4.6. Stability of Channel Slopes (Design of Non-Scouring Channels with Unprotected Side Slopes)

Upto now, we have considered the stability of horizontal beds, where the shear stress $\tau_0$ (given by $\tau_0 = \gamma_w RS$) was the only disturbing force. But on side slopes of channels, one more disturbing force, *i.e.*, the component of the weight of the particle, also comes into picture.

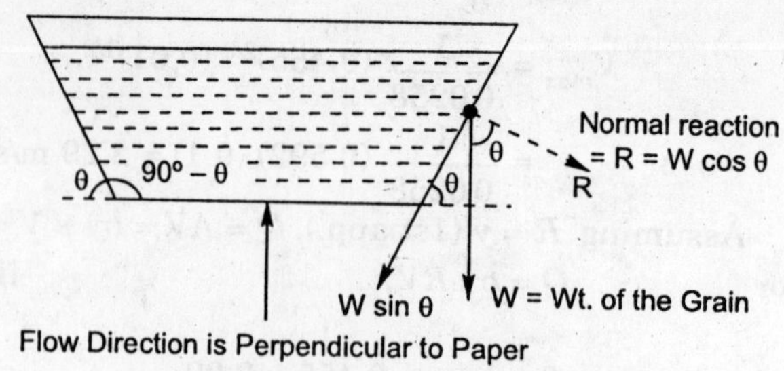


Fig. 4.5 (*a*) Forces acting on a grain on the side slope of a channel.

We will now consider a grain on the side slope of a channel. Various forces acting on this grain are shown in Fig. 4.5 (*a*).

Now, let $\tau_c'$ be the shear stress required to move the grain of weight $W$ on the side slope. The free-body diagram of various forces acting on the grain is shown in Fig. 4.5 (*c*).

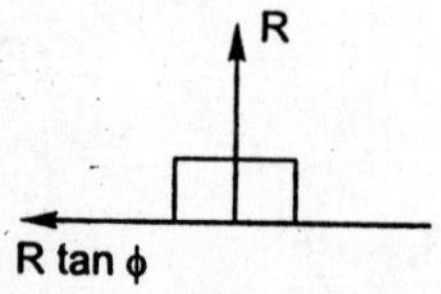


(*b*) where $\phi$ = Angle of repose of soil

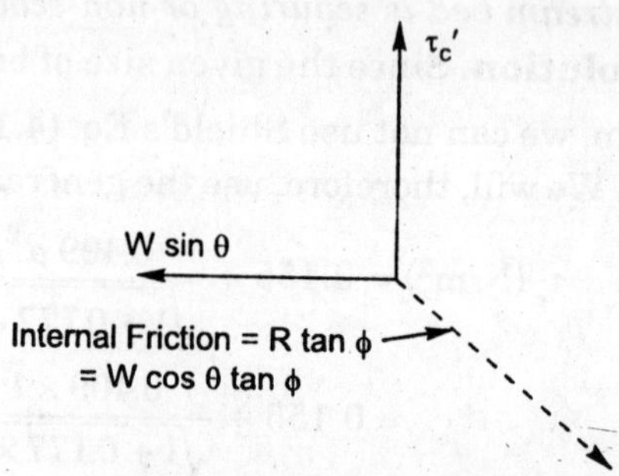


(*c*) Free body diagram of forces

Fig. 4.5

Let $\tau_c$ represents the critical shear stress or the shear stress required to move a similar grain on a horizontal bed, as shown in Fig. 4.6.

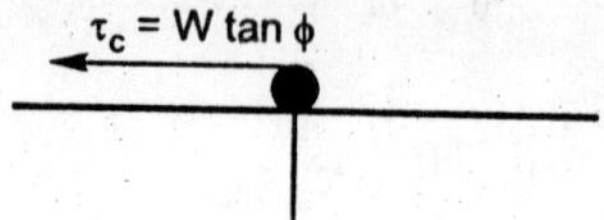


Fig. 4.6

Now, $\tau_c = W \tan \phi$

From Fig. 4.5 (*c*),

$$(\tau_c')^2 + (W \sin \theta)^2 = [(W \cos \theta) \tan \phi]^2$$

or
$$\tau_c'^2 + \left[\frac{\tau_c}{\tan \phi} \sin \theta\right]^2 = \left[\frac{\tau_c}{\tan \phi} \cos \theta \tan \phi\right]^2 \qquad \left(\because\ W = \frac{\tau_c}{\tan \phi}\right)$$

or
$$\tau_c'^2 + \frac{\tau_c^2}{\tan^2 \phi} \sin^2 \theta = \tau_c^2 \cos^2 \theta$$

or
$$\tau_c'^2 = \tau_c^2 \left[\cos^2 \theta - \frac{\sin^2 \theta}{\tan^2 \phi}\right] \qquad \text{...(4.14)}$$

or
$$\frac{\tau_c'^2}{\tau_c^2} = \cos^2 \theta - \frac{\sin^2 \theta}{\tan^2 \phi} = \cos^2 \theta \left[1 - \frac{\tan^2 \theta}{\tan^2 \phi}\right]$$

or $$\frac{\tau_c'}{\tau_c} = \cos\theta\sqrt{1-\frac{\tan^2\theta}{\tan^2\phi}} \qquad ..(4.15)$$

Equation (4.14) can also be written as

$$\left(\frac{\tau_c'}{\tau_c}\right)^2 = \left[\cos^2\theta + (\sin^2\theta - \sin^2\theta) - \frac{\sin^2\theta}{\tan^2\phi}\right] = \left[1-\sin^2\theta - \frac{\sin^2\theta}{\tan^2\phi}\right]$$

$$= \left[1-\sin^2\theta\left(1+\frac{1}{\tan^2\phi}\right)\right] = \left[1-\sin^2\theta \times \frac{1+\tan^2\phi}{\tan^2\phi}\right]$$

$$= \left[1-\sin^2\theta \cdot \frac{\sec^2\phi}{\tan^2\phi}\right] = \left[1-\frac{\sin^2\theta}{\sin^2\phi}\right]$$

or $$\boxed{\frac{\tau_c'}{\tau_c} = \sqrt{1-\frac{\sin^2\theta}{\sin^2\phi}}} \qquad ...(4.16)$$

The above equation shows that $\tau_c' < \tau_c$ : which means *that the shear stress required to move a grain on the side slopes is less than the shear-stress required to move the grain on canal bed.*

Moreover, on the channel bed, the average value of actual shear stress generated by the flowing water in a channel of given $R$ and $S$, is given by equation (4.1), as $\tau_0 = \gamma_w RS$

while on slopes, this value is given by $(\tau_0') = 0.75\gamma_w RS$ ...(4.17)

The above distribution of shear-stresses was shown earlier in Fig. 4.3.

**Example 4.4.** *A canal is to be designed to carry a discharge of 56 cumec. The slope of the canal is 1 in 1000. The soil is coarse alluvium having a grain size of 5 cm. Assuming the canal to be unlimited and of a trapezoidal section, determine a suitable section for the canal, $\phi$ may be taken as 37°.*

**Solution.** First of all, let us choose suitable side slopes, such that $\theta < \phi$. Let $\theta$ be 30°.

Now $$\frac{\tau_c'}{\tau_c} = \sqrt{1-\frac{\sin^2\theta}{\sin^2\phi}} = \sqrt{1-\frac{\sin^2 30°}{\sin^2 37°}} = \sqrt{1-\left(\frac{0.50}{0.62}\right)^2} = 0.557$$

$$\therefore \quad \frac{\tau_c'}{\tau_c} = 0.557$$

Therefore, minimum shear stress required to dislodge the grain on side slope is given by

$$\tau_c' = 0.557\,\tau_c.$$

Hence, for stability, the shear stress actually going to be generated on the slopes of a channel of given $R$ and $S$ must be less than or equal to $0.557\,\tau_c$.

*i.e.*, $\tau_0' \le 0.557\,\tau_c$ (for stability)

But the shear stress (average) actually going to be generated on the side slopes of a channel of given $R$ and $S$, from Eq. (4.17), is

$$= \tau_0' = 0.75\,\gamma_w RS$$

$$0.75\,\gamma_w RS \le 0.557\,\tau_c$$

But $\tau_c = \gamma_w RS = \dfrac{\gamma_w \cdot d}{11}$ $\left[\because \tau_c = 0.056 \times \gamma_w . d(S_s - 1) \text{ Eq.(4.11)} = 0.056 \times \gamma_w \times d(1.65) = \dfrac{d}{11}\gamma_w\right]$

$\therefore$ $0.75\, \gamma_w RS \le 0.557 \cdot \dfrac{\gamma_w d}{11}$

or $RS \le \dfrac{0.557}{0.75 \times 11} d$

or $RS \le 0.0676\, d$

or $RS \le (0.0676 \times 0.05)$ m.

or $R \times \dfrac{1}{1000} \le (0.0676 \times 0.05)$ when $R$ is in metres

or $R \le 3.38$ m

$\therefore$ $y \le 3.38$ m. $(\because\ y \approx R)$

With 20% factor of safety, use $y = 2.8$ m.

**Hence, choose the depth as 2.8 m.** Let us now choose the base width $b$ in such a way as to have a discharge of 56 cumec.

Let us use hit and trial method to determine $b$.

$$A = y\,(b + x) = 2.8\,(b + 4.85)$$

and $P = b + 2\sqrt{x^2 + y^2}$ (w.r. to Fig. 4.7)

$$= b + 2\sqrt{23.6 + 7.84} = b + 11.22$$

$$n = \frac{1}{24} d^{1/6} = \frac{1}{24} \times (0.05)^{1/6} = 0.0258$$

$$V = \frac{1}{n} R^{2/3} \cdot S^{1/2}$$

$$= \frac{1}{0.0258} R^{2/3} \sqrt{\frac{1}{1000}} = 1.223\, R^{2/3}.$$

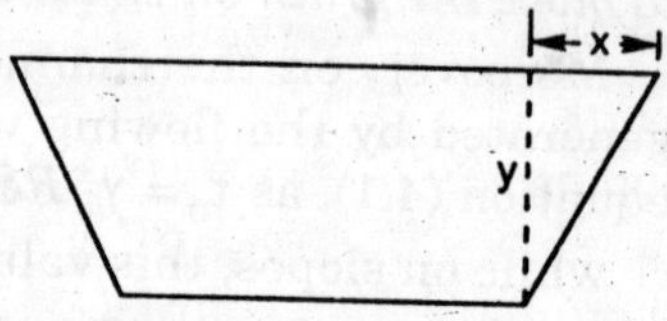


$\dfrac{y}{x} = \tan 30° = \dfrac{1}{\sqrt{3}}$

or $x = 2.8\sqrt{3} = 4.85$ m

Fig. 4.7

Choose a number of trial values of $b$ and assuming a depth of 2.8 m, proceed by the given table 4.1, till a discharge of 56 cumec is reached at $b = 6.3$ m.

**Table 4.1**

| *b in metres* | *A = 2.8 × (b + 4.85) m²* | *P in m = (b + 11.22)* | *R = A/P m* | *R^{2/3}* | *1.223 R^{2/3} = V m/sec* | *Q = A × V cumecs* |
|---|---|---|---|---|---|---|
| 3 | 22 | 14.22 | 1.544 | 1.337 | 1.635 | 36 |
| 4 | 24.8 | 15.22 | 1.625 | 1.382 | 1.695 | 42 |
| 5 | 27.6 | 16.22 | 1.695 | 1.422 | 1.745 | 48.3 |
| 6 | 30.4 | 17.22 | 1.764 | 1.470 | 1.800 | 54.8 |
| 6.3 | 31.22 | 17.52 | 1.782 | 1.470 | 1.798 | **56.1 ; say 56** |

Hence, use 6.3 m base width and 2.8 m depth **Ans.**

**Example 4.5.** *A trapezoidal channel with side slopes 1.5 horizontal : 1 vertical is required to carry 15 m³/s of flow with a bed slope of 1 in 4000. If the channel is lined, the Manning's coefficient n will be 0.014, and it will be 0.028 if the channel is unlined. Calculate the average boundary shear stress if a* *hydraulically efficient lined channel is adopted. What percentage of earthwork is saved in a lined section, relative to an unlined section, when hydraulically efficient section is used in both the cases ? The free board can be assumed to be 0.75 m in both the cases and the lining can be assumed to be up to the top of the section.* *(Civil services, 1991)*

**Solution.** $Q = 15$ m³/s (given)

$N = 0.014$ for lined section, and

0.028 for unlined section (given)

$$S = \frac{1}{4000} \text{ (given)}$$

**A hydraulically and economically efficient channel** is the one *which possesses minimum perimeter for a given area ;*

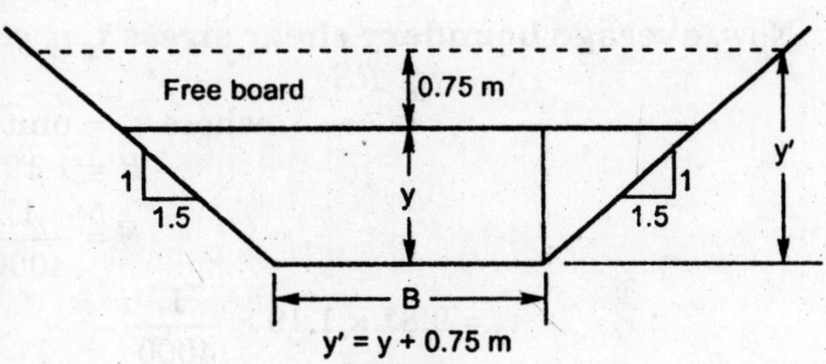


Fig. 4.8

For a trapezoidal channel having 1.5 $H$ : 1V side slopes, bed width $B$, and depth $y$, we have

$P = B +$ 2 [sloping side] $= B + 2\,[1^2 + 1.5^2]\,y = B + 2 \times 1.8y$

or $$P = B + 3.6y \qquad ...(i)$$

Also $$A = [B + y \times 1.5]\,y = (B + 1.5y)\,y$$

or $$\frac{A}{y} = B + 1.5y$$

or $$B = \frac{A}{y} - 1.5y \qquad ...(ii)$$

Substituting in (*i*), we get

$$P = \left[\frac{A}{y} - 1.5y\right] + 3.6y$$

or $$P = \frac{A}{y} + 2.1y \qquad ...(iii)$$

For minimum perimeter (with constant $A$), $\frac{dP}{dy} = 0$

$$\therefore \quad \frac{dP}{dy} = A(-1y^{-2}) + 2.1 \times 1 = -\frac{A}{y^2} + 2.1 = 0$$

or $$\frac{A}{y^2} = 2.1 \quad \text{or} \quad \mathbf{A = 2.1\,y^2}$$

Substituting in (*iii*), we get

$$P = \frac{2.1y^2}{y} + 2.1y = 4.2y$$

$$\therefore \quad R = \frac{A}{P} = \frac{2.1y^2}{4.2y} = \frac{\mathbf{y}}{\mathbf{2}}$$

**For a channel having 1.5 $H$ : 1 $V$ slopes, therefore, the most efficient section will have $A = 2.1y^2$, and $R = \frac{y}{2}$.**

**Now, using $Q = \frac{1}{n} A \cdot R^{2/3} S^{1/2}$, we have**

$$15 = \frac{1}{0.014} \times (2.1\, y_1^2)\left(\frac{y_1}{2}\right)^{2/3} \frac{1}{\sqrt{4000}}$$

or $\quad y_1 = (10.039)^{3/8} \quad$ or $\quad y_1 = \mathbf{2.37\ m}$

$\therefore \quad B_1 = 0.6 \times 2.37 = \mathbf{1.42\ m} \quad \therefore \quad R_1 = \frac{y_1}{2} = 1.19$ m

**Now, average boundary shear stress $\tau_0$ *is given by eqn. (4.1) as :***

$$\tau_0 = \gamma_w \cdot RS$$

where $\gamma_w$ = unit wt. of water = 9.81 kN/m³

$R = 1.19$ m

$S = \frac{1}{4000}$

$$\therefore \quad \tau_0 = 9.81 \times 1.19 \times \frac{1}{4000}$$

$= 2.918 \times 10^{-3}$ kN/m² = **2.918 N/m². Ans.**

**Now, we compute $X$-sectional areas of excavation in both cases :**

**Using free-board of 0.75 m**

***(for lined section)*** $= A_1 = (B_1 + 1.5\, y'_1)\, y'_1$

where $y_1' = y_1 + 0.75 = 2.37 + 0.75 = 3.12$ m

or $\quad A_1 = (1.42 + 1.5 \times 3.12)\ 3.12\ \text{m}^2 = \mathbf{19.04\ m^2}$ ...(*i*)

***For unlined section of bed width $B_2$ and depth $y_2$,*** **we have, $n = 0.028$**

$$\therefore \quad 15 = \frac{1}{0.028}(2.1\, y_2^2)\left(\frac{y_2}{2}\right)^{2/3} \cdot \frac{1}{\sqrt{4000}}$$

or $\quad y_2 = (2 \times 10.039)^{3/8} = 3.08$ m

$\therefore \quad B_2 = 0.6 y_2 = 0.6 \times 3.08 = 1.85$ m

$A_2 = (B_2 + 1.5\, y_2')\, y_2' \quad$ where $y_2' = 3.08 + 0.75 = 3.83$ m

$= (1.85 + 1.5 \times 3.83)\ 3.83\ \text{m}^2 = \mathbf{29.09\ m^2}$ ...(*ii*)

**Percentage saving earth work due to lining**

$$= \frac{A_2 - A_1}{A_2} \times 100 = \frac{29.09 - 19.04}{29.09} \times 100 = \mathbf{34.55\%} \quad \textbf{Ans.}$$

**Example 4.6.** ***A most efficient trapezoidal section is required to give a maximum discharge of 21.5 m³/s of water. The slope of the channel bottom is 1 in 2500, Taking C = 70 m^1/2/s in Chezy's equation, determine the dimensions of the channel. Also determine the value of Manning's 'n', taking the value of velocity of flow as obtained for the channel by Chezy's equation.*** **(*Engg. Services 1997*)**

**Solution.** We shall first derive the equations to be used in the question as follows.

The conveyance (*i.e.* discharge carrying capacity) of a channel section of a given area increases with a decrease in its perimeter. Hence, a channel section having the minimum perimeter for a given area of flow provides the maximum value of conveyance of discharge. With the slope, roughness coefficient, and area of flow being fixed, a minimum perimeter section will represent the

**hydraulically efficient section**, as it conveys the maximum discharge for a given area. ***Evidently, such a section will always have the minimum excavation.*** Such an efficient channel section is also called the ***most economical section*** or the **best section.**

For the most efficient channel, therefore, the wetted perimeter must be minimum for a given area. Thus, for a trapezoidal channel of bed width $B$, depth $y$, and side slopes $m : 1$ $(H : V)$, we have

$$A = (B + m \cdot y)\, y \qquad ...(i)$$

$$P = 2 \cdot \sqrt{(1+m^2)}\, y + B \qquad ...(ii)$$

or $$P = 2\sqrt{(1+m^2)} \cdot y + \left(\frac{A}{y} - m \cdot y\right)$$

or $$P = \frac{A}{y} - m \cdot y + 2\,(1+m^2)^{1/2} \cdot y$$

or $$\frac{dP}{dy} = -A \cdot y^{-2} - m + 2\sqrt{1+m^2} = 0$$

or $$A/y^2 = 2\sqrt{1+m^2} - m$$

or $$\boxed{A = \left[2\sqrt{1+m^2} - m\right] y^2} \qquad ...(iii)$$

But from (i) $$B = \frac{A}{y} - my = \left[2\sqrt{1+m^2} - m\right] y - my$$

or $$B = 2\left[\sqrt{1+m^2} - m\right] y \qquad ...(iv)$$

**Note : The second derivative of $P$ with respect to $y$ is worked out to be $2A/y^3$, which is + ve, and hence the condition obtained above is for minimum $P$.**

Using the above worked out relation, we can write

$$P = 2\sqrt{1+m^2} \cdot y + B = 2\sqrt{1+m^2} \cdot y + 2\left[\sqrt{1+m^2} - m\right] y$$

or $$P = 2 \cdot y\left[2\sqrt{1+m^2} - m\right]$$

$$\therefore \quad R = \frac{A}{P} = \frac{\left[2\sqrt{1+m^2} - m\right] y^2}{\left[2\sqrt{1+m^2} - m\right] 2y} = \frac{y}{2}$$

or $$\boxed{R = y/2} \qquad ...(1)$$

In the expression for $P$, there are two variables $y$ and $m$. A second condition for min. $P$ can be obtained by equating $\frac{dP}{dm} = 0$, holding $y$ as constant.

$\therefore$ Use $P = 2y\left[2\sqrt{1+m^2} - m\right]$ from (iv) above

or $$P = 4 \cdot y\sqrt{1+m^2} - 2y \cdot m$$

or $$\frac{dP}{dm} = 4 \cdot y\,\frac{1}{2}\,(1+m^2)^{-1/2} \cdot 2m - 2y \times 1 \qquad \text{(with constant } y)$$

$$= \frac{4y \cdot m}{\sqrt{1+m^2}} - 2y = 0$$

or $\dfrac{4ym}{\sqrt{1+m^2}} = 2y$ or $\dfrac{2m}{\sqrt{1+m^2}} = 1$

or $\sqrt{1+m^2} = 2m$ or $(1 + m^2) = 4m^2$ or $3m^2 = 1$

or $$m = \frac{1}{\sqrt{3}} \qquad ...(2)$$

**Hence, for the most efficient trapezoidal channel, side slopes should be $\frac{1}{\sqrt{3}}$ : 1 (H : V)**

**Using the above two conditions, for the most efficient channel ; *i.e.* $m = \frac{1}{\sqrt{3}}$ and $R = y/2$, we can solve the given question as follows :**

**Using $Q = C \cdot \sqrt{R \cdot S} \times A$, we have**

$$Q = 21.5 \text{ m}^3\text{/s}, C = 70 \sqrt{\text{m}} \text{ / sec and } S = 1/2500$$

$$\therefore \quad 21.5 = 70 \times \sqrt{R} \cdot \frac{1}{\sqrt{2500}} \times A$$

or $$21.5 = \frac{70}{50}\sqrt{R} \cdot A$$

For the most efficient trepezoidal channel, we have side slopes $m : 1$ $(H : V)$ where $m = \frac{1}{\sqrt{3}}$ ; and $R = \frac{y}{2}$.

Also, $$A = \left[2\sqrt{1+m^2} - m\right] y^2 \quad \text{... from Eq. } (iii)$$

$$= \left[2\sqrt{1+\frac{1}{3}} - \frac{1}{\sqrt{3}}\right] y^2 = (2.3094 - 0.5773)y^2 = 1.732y^2$$

$$\therefore \quad 21.5 = \frac{70}{50} \cdot \sqrt{\frac{y}{2}} \times 1.732y^2$$

or $$\frac{21.5 \times 50 \times \sqrt{2}}{70 \times 1.732} = y^{2.5} \quad \text{or} \quad \mathbf{y = 2.75 \text{ m} \quad Ans.}$$

Also $A = 1.732 \cdot (2.75)^2 = 13.10$

But $A = (B + m \cdot y)\, y = \left(B + \frac{1}{\sqrt{3}} \times 2.75\right) 2.75$

$\therefore$ $13.10 = (B + 0.577 \times 2.75)\, 2.75$ or $13.10 = (B + 1.588)\, 2.75$

or **B = 3.18 m**

The channel dimensions are thus worked out as :

$B = 3.18$ m and $y = 2.75$ m

Side slopes = $\frac{1}{\sqrt{3}}$ : 1 (*i.e.* sides inclined at 60° to horizontal)

Velocity as per Chezy's Equation determined above is

$$V = C \cdot \sqrt{RS} = 70 \cdot \sqrt{R} \cdot \frac{1}{50} = 70 \cdot \sqrt{\frac{y}{2} \cdot \frac{1}{50}}$$

$$= 70 \times \sqrt{\frac{2.75}{2} \cdot \frac{1}{50}} = 1.64 \text{ m/s}$$

With Manning's equation

$$V = \frac{1}{n} \cdot R^{2/3} \cdot \sqrt{S} = \frac{1}{n} \cdot \left(\frac{y}{2}\right)^{2/3} \cdot \sqrt{S} = \frac{1}{n} \times \left(\frac{2.75}{2}\right)^{2/3} \cdot \frac{1}{50}$$

or $$1.64 = \frac{1}{n} \cdot (1.375)^{2/3} \cdot \frac{1}{50}$$

or $$n = \frac{(1.375)^{2/3}}{50 \times 1.64} \quad \text{or} \quad n = 0.015 \textbf{ Ans.}$$

### 4.7. Design of Stable Channels in India

So long as the average shear stress ($\tau_0$) acting on the boundary of an alluvial channel is less than the critical shear stress ($\tau_c$), the channel shape remains unchanged, and hence the channel can be considered of rigid boundary. The resistance equations, such as those given by Chezy's formula and Manning's formula, remain well applicable to such channels. However, as soon as the sediment movement starts, undulations develop on the bed, which increases the boundary resistance of the channel. Besides this, some energy is required to move the grains. The suspended load, carried due to turbulence in the flow, further affects the resistance of the alluvial streams. All these factors render the evaluation of resistance of alluvial streams to be a very complex problem, and the complexity further increases if one includes the effects of the channel shape, non-uniformity of sediment size, discharge variation, and other such factors. None of the resistance equations developed so far, takes all these factors into account. The direct accurate mathematical solution to the design of channels in alluvial soils is, therefore, not an easy job ; and hence in India, alluvial channels are designed on the basis of hypothetical theories given by Kennedy and Lacey. These theories are based on experiments and experience gained on the existing channels over the past many years. The semi-theoretical approach discussed in the previous article, is however, used to design channels in countries like America and Europe.

**4.7.1. Problem in India.** In prehistoric periods, the area bounded by Indo-Gangetic plain, *i.e.* the area starting from Himalayas to Vindhya mountains, used to be in the form of depressions with water flowing over it. But with the passage of time, it was filled up with loosely filled fine silt particles, thereby forming nothing, but what is known as **alluvial soil.** Almost all the north Indian rivers flow through such soils and, therefore, do carry a certain amount of sediment. Artificial channels have to carry their water supply from such rivers, and thus carrying sediment.

We also know that water moving with a given velocity and a certain depth can carry in suspension, only a certain amount of silt of a certain nature. If water of a given velocity and depth is not fully charged with silt (that it can carry in suspension) it will scour the bed and sides of the channel, till it is fully charged with silt. *Hence, if the velocity of flow in the channel is more, the bed and the banks are likely to be eroded, and similarly, if the velocity is less, the silt which was formerly carried in suspension is likely to be dropped.*

Silting and scouring in channels is not very uncommon and is understandable, but must be avoided by proper designs. Scouring lowers the full supply level and causes loss of command. It may also cause breaching of canal banks and failure of foundations of irrigation structures. Silting interferes with the proper working of a channel, as the channel section gets reduced by siltation, thereby reducing the discharging capacity of the channel.

*Therefore, while thinking to design a properly functioning channel, one must think to design such a channel in which neither silting nor scouring takes place. Such channels are known as stable channels or regime channels.*

**4.7.2. Regime Channels.** A channel is said to be in a state of 'Regime', if the flow is such that 'silting and scouring' need no special attention. Such a state is not easily possible in rivers, but in artificial channels, such a state can be obtained by properly designing the channel.

*The basis for designing such an ideal, non-silting, non-scouring channel is that, whatever silt has entered the channel at its head is kept in suspension, so that it does not settle down and deposit at any point of the channel. Moreover, the velocity of the water should be such that it does not produce local silt by erosion of channel bed and slopes.*

**4.7.3. Kennedy's Theory (1895).** R.G. Kennedy, an Executive Engineer of Punjab P.W.D, carried out extensive investigations on some of the canal reaches in the upper Bari Doab Canal System. He selected some straight reaches of the canal section, which had not posed any silting and scouring problems during the previous 30 years or so.

From the observations, he concluded that the silt supporting power in a channel cross-section was mainly dependent upon the generation of the eddies, rising to the surface. These eddies are generated due to the friction of the flowing water with the channel surface. The vertical component of these eddies try to move the sediment up, while the weight of the sediment tries to bring it down, thus keeping the sediment in suspension. So if the velocity is sufficient to generate these eddies, so as to keep the sediment just in suspension, silting will be avoided. Based upon this concept, he defined the **critical velocity** ($V_0$) in a channel as the *mean velocity* (across the section) which will just keep the channel free from silting or scouring, and related it to the depth of flow by the equation

$$V_0 = c_1 \cdot y^{c_2}$$

where $c_1$ and $c_2$ are constants depending upon silt charge.

$c_1$ and $c_2$ were found to be 0.55 and 0.64 (in M.K.S. or S.I. units), respectively.

Therefore, $V_0 = 0.55\, y^{0.64}$ ...(4.18)

Since this formula was worked out especially for the upper Bari Doab canal system, it could not have been applicable in toto to other canals or canal systems due to variation in the type of soil (or silt) at various canal sites. Realising this lacuna, Kennedy later introduced a factor (m) in this equation, to account for the type of soil through which the canal was to pass. This factor, which was dependent upon the silt grade, was named as **critical velocity ratio (C.V.R.)** and denoted by *m*.

The equation for critical velocity was, thus, modified as :

$$\boxed{V_0 = 0.55\, m y^{0.64}} \quad ...(4.19)$$

where $V_0$ = Critical velocity in the channel in m/s.

$y$ = water depth in channel in m

$m = C.V.R$

For sands coarser than the standard, the values of $m$ were given from 1.0 to 1.2 ; and for sands finer than the standard, $m$ was valued between 1.0 to 0.7, as shown in Table 4.2.

**Table 4.2. Recommended Values of C.V.R. (m)**

| *S.No.* | *Type of silt* | *Value of m* |
|---|---|---|
| 1. | Silt of River Indus (Pakistan) | 0.7 |
| 2. | Light sandy silt in North Indian Rivers | 1.0 |
| 3. | Light sandy silt, a little coarser | 1.1 |
| 4. | Sandy, loamy silt | 1.2 |
| 5. | Debris of hard soil | 1.3 |

**Design procedure.** Determine the critical velocity $V_0$ by the above Eq. (4.19) by assuming a trial depth, and then determine area by dividing discharge by velocity. Then determine channel dimensions. Finally, compute the actual mean velocity ($V$) that will prevail in the channel of this cross-section, by using Kutter's formula, Manning's formula, etc. If the two velocities $V_0$ and $V$ work out to be the same, then the assumed depth is all right, otherwise change it and repeat the procedure, till $V$ and $V_0$ become equal.

**Kutter's Formula**

$$V = \left[\frac{\frac{1}{n} + \left(23 + \frac{0.00155}{S}\right)}{1 + \left(23 + \frac{0.00155}{S}\right)\frac{n}{\sqrt{R}}}\right]\sqrt{RS} \qquad \ldots(4.20)$$

**Manning's Formula**

$$V = \frac{1}{n} R^{2/3} \cdot S^{1/2} \qquad \ldots(4.21)$$

where $V$ = Velocity of flow in metres/sec.
$R$ = Hydraulic mean depth in metres.
$S$ = Bed slope of the channel.
$n$ = Rugosity coefficient.

The values of $n$ in both these equations depend upon channel condition and also upon discharge. The values of $n$ may be taken as given in Table 4.3.

**Table 4.3. Recommended Values of Manning's Coefficient *n* for Unlined Channels**

| *Condition of channel* | *Value of n* |
|---|---|
| Very good | 0.0225 |
| Good | 0.025 |
| Indifferent | 0.0275 |
| Poor | 0.030 |

The Central Board of Irrigation and Power (India) has recommended the following values of $n$ for different discharges.

**Table 4.4. Values of Manning's *n* for Different Discharges**

| *Discharge in cumec* | *Value of n for Unlined Channels* |
|---|---|
| 14 to 140 | 0.025 |
| 140 to 280 | 0.0225 |
| 280 and above | 0.020 |

**Chezy's Formula**

$$V = C\sqrt{RS} \quad \text{...(4.22)}$$

where $C$ = a constant depending upon the shape and surface of the channel.

$R$ and $S$ have the same meaning as in eq. (4.21).

The actual mean velocity($V$) generated in the channel can be computed by any of these three resistance equations, but generally Kutter's equation is used with Kennedy's theory.

**Example 4.7.** *Design an irrigation channel to carry 50 cumecs of discharge. The channel is to be laid at a slope of 1 in 4000. The critical velocity ratio for the soil is 1.1. Use Kutter's rugosity coefficient as 0.023.*

**Solution.** $Q = 50$ cumecs, $S = \frac{1}{4000}$, $m = 1.1$, $n = 0.023$

Use equation (4.19), as, $V_0 = 0.55m \cdot y^{0.64}$

Assume a depth equal to 2 m

$$V_0 = 0.55 \times 1.1 \times (2)^{0.64} = 0.605 \times 1.558 = \mathbf{0.942} \text{ m/sec}$$

$$A = \frac{Q}{V_0} = \frac{50}{0.942} = 53.1 \text{ m}^2.$$

Assume side slopes as $\frac{1}{2} : 1 \left(\frac{1}{2} H : 1V\right)$

Now, $A = y\left(b + y \cdot \frac{1}{2}\right)$

$\therefore \quad 53.1 = 2\,(b + 1)$

or $\quad 26.55 = b + 1$

or $\quad b = 25.55$ m

and $\quad P = b + 2\sqrt{\left(1 + \frac{1}{4}\right)} \times y$

or $\quad P = b + 2\,\frac{\sqrt{5}}{2}\,y = 25.55 + \sqrt{5} \times 2 = 30.03$

$$R = \frac{A}{P} = \frac{53.1}{30.03} = 1.77 \text{ m.}$$

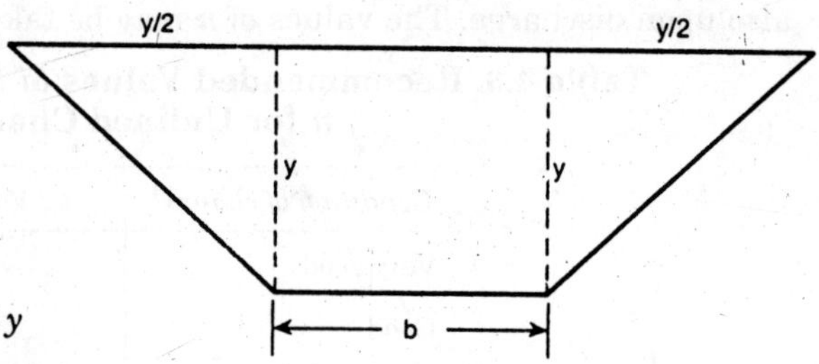


Fig. 4.9.

But, from eqn. (4.20),

$$V = \left[ \frac{\frac{1}{n} + \left(23 + \frac{0.00155}{S}\right)}{1 + \left(23 + \frac{0.00155}{S}\right)\frac{n}{\sqrt{R}}} \right] \sqrt{RS}$$

$$\therefore \quad V = \left[ \frac{\frac{1}{0.023} + 23 + \frac{0.00155}{1/4000}}{1 + \left(23 + \frac{0.00155}{1/4000}\right)\frac{0.023}{\sqrt{1.77}}} \right] \sqrt{1.77 \times \frac{1}{4000}}$$

$$= \left[ \frac{43.5 + (23 + 6.2)}{1 + \frac{29.2 \times 0.023}{1.33}} \right] \left[ 1.33 \times \frac{1}{63.3} \right]$$

$$= \frac{72.7}{1 + 0.505} \times 1.33 \times \frac{1}{63.3} = \frac{72.7}{1.505} \times 1.33 \times \frac{1}{63.3}$$

$$= 1.016 \text{ m/sec} > 0.942 \text{ ; or } V > V_0.$$

In order to *increase* the critical velocity ($V_o$),, we have to *increase* the depth.

So increase the depth.

**Use 3 m depth :**

$$V_0 = 0.605 \times (3)^{0.64} = 0.605 \times 2.02 = 1.22 \text{ m/sec.}$$

$$A = \frac{50}{1.22} = 40.8 \text{ m}^2.$$

$$40.8 = 3\left(b + \tfrac{1}{2} \cdot 3\right) \qquad \text{or} \qquad 13.6 - 1.5 = b = 12.1 \text{ m.}$$

$$P = 12.1 + 2 \times \frac{\sqrt{5}}{2}\, 3 = 12.1 + 6.72 = 18.82$$

$$R = \frac{A}{P} = \frac{40.8}{18.82} = 2.17 \text{ ; therefore } \sqrt{R} = 1.47.$$

$$V = \frac{43.5 + 29.2}{1 + \frac{29.2 \times 0.023}{1.47}} + \left[ 1.47 \times \frac{1}{63.3} \right] = \frac{72.7}{1.45} \times 1.47 \times \frac{1}{63.3}$$

$$= 1.16 \text{ m/sec.} < 1.22 \text{ ; or } V < V_0, \text{ So reduce the depth.}$$

**Use 2.5 m depth**

$$V_0 = 0.605 \times (2.5)^{0.64} = 0.605 \times 1.797 = 1.087 \text{ m/sec.}$$

$$A = \frac{50}{1.087} = 46$$

$$46 = 2.5\left(b + \tfrac{1}{2} 2.5\right)$$

$$18.4 - 1.25 = b = 17.15 \text{ m}$$

$$P = 17.15 + \sqrt{5} \times 2.5 = 17.15 + 5.58 = 22.73$$

$$R = \frac{A}{P} = \frac{4}{22.73} = 2.02 \text{ ; therefore } \sqrt{R} = 1.42$$

$$V = \frac{72.7}{1 + \frac{29.2 \times 0.023}{1.42}} (1.42) \left( \frac{1}{63.3} \right) = \frac{72.7}{1.472} \times \frac{1.42}{63.3}$$

$$= 1.1 \text{ m/sec} > 1.087 \text{ ; } V > V_0$$

So increase the depth.

**Use 2.7 m depth**

$$V_0 = 0.605 \times 1.189 = 1.147$$

$$A = \frac{50}{1.147} = 43.5$$

$$43.5 = 2.8\,(b + \tfrac{1}{2} 2.8)$$

$$15.54 - 1.4 = b = 14.14 \text{ m}$$

$$P = 14.14 + \sqrt{5} \times 2.8 = 14.14 + 6.26 = 20.40$$

$$R = \frac{43.5}{20.4} = 2.13, \text{ therefore, } \sqrt{R} = 1.46$$

$$\therefore \quad V = \left[\frac{72.7}{1 + \dfrac{29.2 \times 0.023}{1.46}}\right]\left[\frac{1.46}{63.3}\right] = \left[\frac{72.6}{1.46}\right]\left[\frac{1.46}{68.3}\right]$$

$$= 1.148 \text{ m/sec} \approx 1.147 \text{ or } V \approx V_0.$$

Actual velocity $V$ tallies with $V_0$.

Hence, use the depth equal to 2.7 m and base width 14.14 m. (say 14.2 m) with slopes $\frac{1}{2}$ : 1 of trapezoidal section. **Ans.**

**Example 4.8.** *Design an irrigation channel to carry 40 cumecs of discharge, with B/D, i.e. base width to depth ratio as 2.5. The critical velocity ratio is 1.0. Assume a suitable value of Kutter's rugosity coefficient and use Kennedy's method.*

**Solution.** $V_0 = 0.55\,(y)^{0.64}$ $(\because m = 1)$

Here $y = D$

$$\therefore \quad V_0 = 0.55 \cdot D^{0.64}$$

$$Q = AV$$

Using $\frac{1}{2}$ : 1 slopes, area ($A$) of trapezoidal section is given as :

$$A = BD + 2 \cdot \frac{1}{2} \cdot D\frac{D}{2} = D\left[B + \frac{D}{2}\right]$$

$$\therefore \quad 40 = D\left[B + \frac{D}{2}\right]V_0 \qquad \left(\because \frac{B}{D} = 2.5\right)$$

But $B/D = 2.5$ ; or $B = 2.5D$

$$\therefore \quad 40 = D\,[2.5\,D + 0.5D]\,V_0 = D\,[3D]\,V_0 = 3D^2 \cdot V_0$$

But $V_0 = 0.55 \cdot D^{0.64}$ $\quad \therefore \quad 40 = 3D^2\,(0.55 \cdot D^{0.64})$

or $$D^{2.64} = \frac{40}{3 \times 0.55} = 24.2$$

or $$D = (24.2)^{\frac{1}{2.64}} = (24.2)^{0.379} = \mathbf{3.34\ m}$$

Now $B = 2.5\,D = 2.5 \times 3.34 = \mathbf{8.35\ m}$

Now determine the slope $S$

$$A = 3D^2 = 3 \times (3.34)^2 = 33.5 \text{ m}^2.$$

$$P = \left[B + 2 \cdot \frac{\sqrt{5}}{2} D\right] = (8.35 + \sqrt{5} \times 3.34) = (8.35 + 7.46) = 15.81 \text{ m}$$

$$R = \frac{33.5}{15.81} = 2.12, \text{ or } \sqrt{R} = 1.456$$

$$V_0 = 0.55\,(3.34)^{0.64} = 0.55 \times 2.163 = \mathbf{1.19}$$

Assume $n = 0.023$.

Using Eq. (4.20), we get

$$V = \left[\frac{\frac{1}{0.023} + \left(23 + \frac{0.00155}{S}\right)}{1 + \left(23 + \frac{0.00155}{S}\right)\frac{0.023}{1.456}}\right] 1.456\sqrt{S} \qquad \text{...}(i)$$

Assume $S = 1/4000$.

Putting this value of $S$ and computing the value of $V$, we get

$$V = \frac{43.5 + (23 + 6.2)}{1 + 29.2 \times \frac{0.023}{1.456}} \times \frac{1.456}{63.3} = 1.14$$

$$1.14 < 1.19 \quad \text{or} \quad V < V_0$$

Therefore, to increase the value of $V$, we must increase/steepen the slope ; hence, use a slope = 1 in 3700 (say)

Putting $S = \frac{1}{3700}$ in $(i)$ above, we get

$$V = 1.189 \approx 1.19$$

or $V = V_0$ for value of $S = \frac{1}{3700}$. So use $S = \frac{1}{3700}$

Hence, use a trapezoidal channel section as follows :

Depth = 3.34 m
Base width = 8.35 m
Side slopes = $\frac{1}{2}H : 1V$
Bed slope = 1 in 3700 **Ans.**

**4.7.3.1. Use of Garret's Diagrams for Applying Kennedy's Theory.** A lot of mathematical calculations are required in designing irrigation channels by the use of Kennedy's method. To save mathematical calculations, graphical solution of Kennedy's and Kutter's equations, was evolved by Garret. The original diagrams given by him were in F.P.S. system, but here they have been changed into M.K.S./S.I. system. The diagrams are shown in **Plates 4.1 (*a*), (*b*) and (*c*)**. The procedure adopted for design of irrigation channels using Garret's diagrams is explained below :

(*i*) The discharge, bed slope, rugosity coefficient, value of C.V.R. are given for the channel to be designed.

(*ii*) Find out the point of intersection of the given slope line and discharge curve. At this point of intersection, draw a vertical line intersecting the various bed width curves.

(*iii*) For different bed widths ($B$), the corresponding values of water depth ($y$) and critical velocity ($V_0$) can be read on the right hand ordinate. Each such pair of bed width ($B$) and depth ($y$) will satisfy Kutter's equation, and is capable of carrying the required discharge at the given slope and rugosity coefficient. Choose one such pair and determine the actual velocity of flow ($V$).

(*iv*) Determine the critical velocity ratio ($V/V_0$) taking $V$ as calculated and $V_0$ as read.

(*v*) If the value of C.V.R. is not the same as given in question, repeat the procedure with other pairs of $B$ and $y$.

The diagrams have been drawn for a trapezoidal channel with side slopes as $\frac{1}{2}$ $H:1V$ ($\frac{1}{2}$ : 1) on the assumption that irrigation channels adopt approximately this shape, even though they were constructed on different side slopes.

Another important point which should be noted in these diagrams is that from the *Nomogram* provided at the top, the same curves can be used for different values of rugosity coefficient $n$. In the nomogram, a vertical arrow has been shown. It represents the value of $n$ for which the curves have been drawn. When the same curves are used for some other value of $n$ (marked on right and left sides of central value), the point of intersection of discharge and slope curves, has to be shifted to the extent given in the nomogram and also in the same direction, for drawing the vertical line.

**Example 4.9.** *Design an irrigation channel to carry 30 cumec of discharge. The channel is to be laid at a slope of 1 in 5000. The critical velocity ratio for the soil is 1.06. Use Kutter's rugosity coefficient as 0.0225.*

**Solution.** $Q = 30$ cumec, $S = \dfrac{1}{5000} = 0.2 \times 10^{-3} = 0.2$ m/km

$m = 1.1, \quad n = 0.0225$

$$\frac{V}{V_0} = m = 1.1, \quad \therefore \quad V = 1.1\, V_0$$

Using Plate 4.1 (*b*), find out the point of intersection of the given slope line ($S = 0.2$) and given discharge curve ($Q = 30$ cumecs). Draw a vertical line through this point. Choose a point of say bed width = 12 m, as first approximation. Calculate $m$ as shown in the last column of Table 4.5.

**Table 4.5**

| *S.No.* | *B metres* | *y metres* | $V_0$ *m/sec* | $A = (2B + y)\dfrac{y}{2}\,m^2$ | $V = \dfrac{Q}{A} = \dfrac{30}{A}$ *m/sec.* | $\dfrac{V}{V_o} = m$ |
|---|---|---|---|---|---|---|
| 1 | 12.0 | 2.4 | 0.96 | $(24 + 2.4)\dfrac{2.4}{2} = 31.68$ | 0.95 | 0.99 |
| 2 | 13.0 | 2.30 | 0.93 | 30.25 | 0.99 | 1.06 |

The first approximation of using $B = 12.0$ m gives a value of $m$ equal to 0.99 ; while as per given data its value should be 1.1. So we have to decrease $V_0$ in order to increase $m$. Therefore, in order to decrease the critical velocity, we have to reduce depth and thus to increase $B$. Hence, in the 2nd approximation, we increase $B$ from 12.0 to 13 m, and find that $m$ comes out to be 1.06, which tellies with its given value. Hence, choose the final values of $B = 13.0$ m, $y = 2.3$ m for the channel of bed slope 1 in 5000 and side slopes $\frac{1}{2}H : 1V$ **Ans.**

**4.7.4. Lacey's Theory (1939).** Lacey, an eminent civil engineer of U.P. Irrigation Department, carried out extensive investigations on the design of stable channels in alluviums. On the basis of his research work, he found many drawbacks in Kennedy's Theory (1895) and he put forward his new theory. The essential points which he argued, and the design procedure which he suggested, is briefly described here.

**4.7.4.1. Lacey's regime channels.** It was stated by Kennedy that a channel is said to be in a state of 'regime' if there is neither silting nor scouring in the channel. But Lacey came out with the statement that even a channel showing no

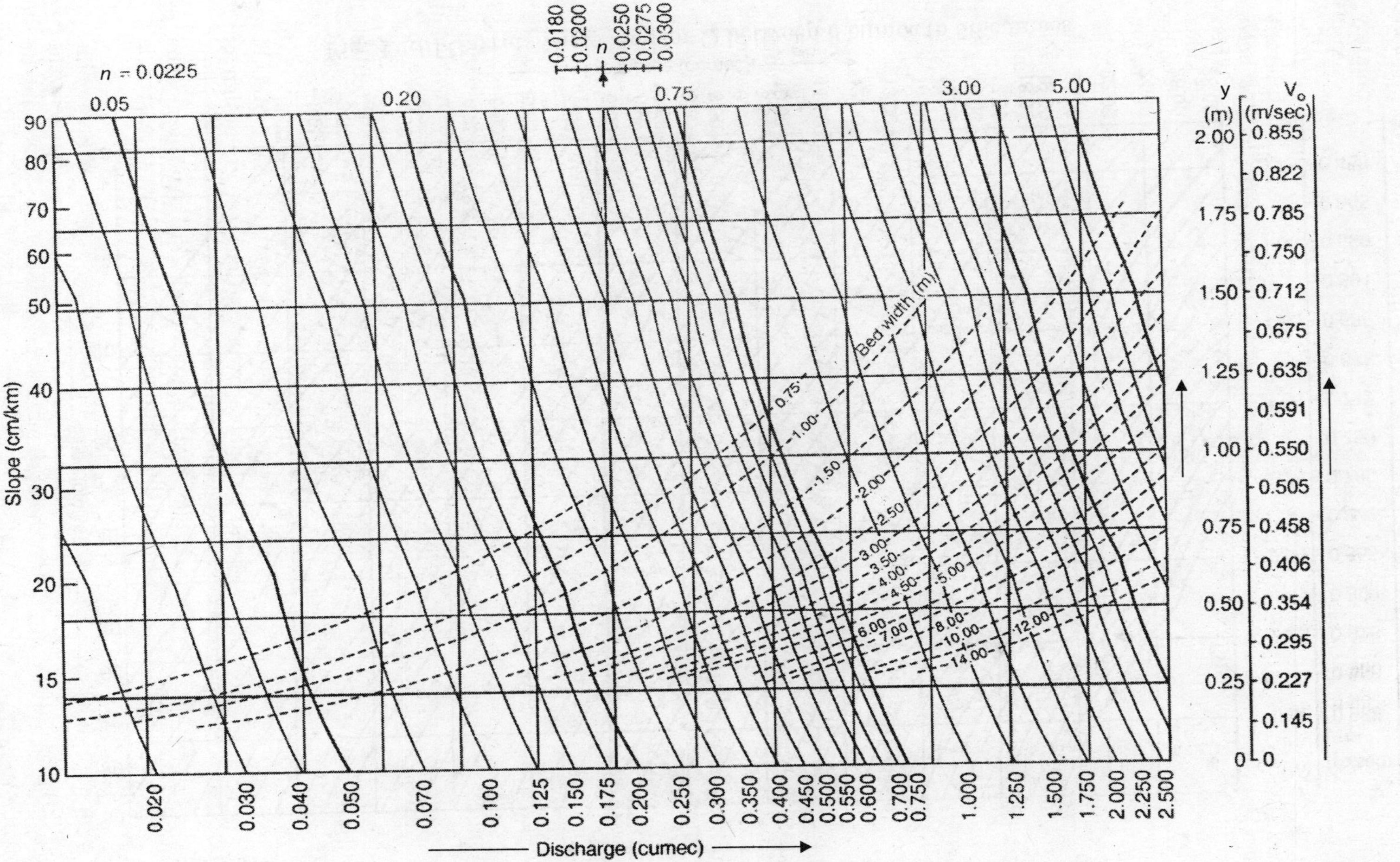


Fig. 4.1(*a*) Garrets Diagrams for Q between 0.02 cumec to 5.0 cumecs

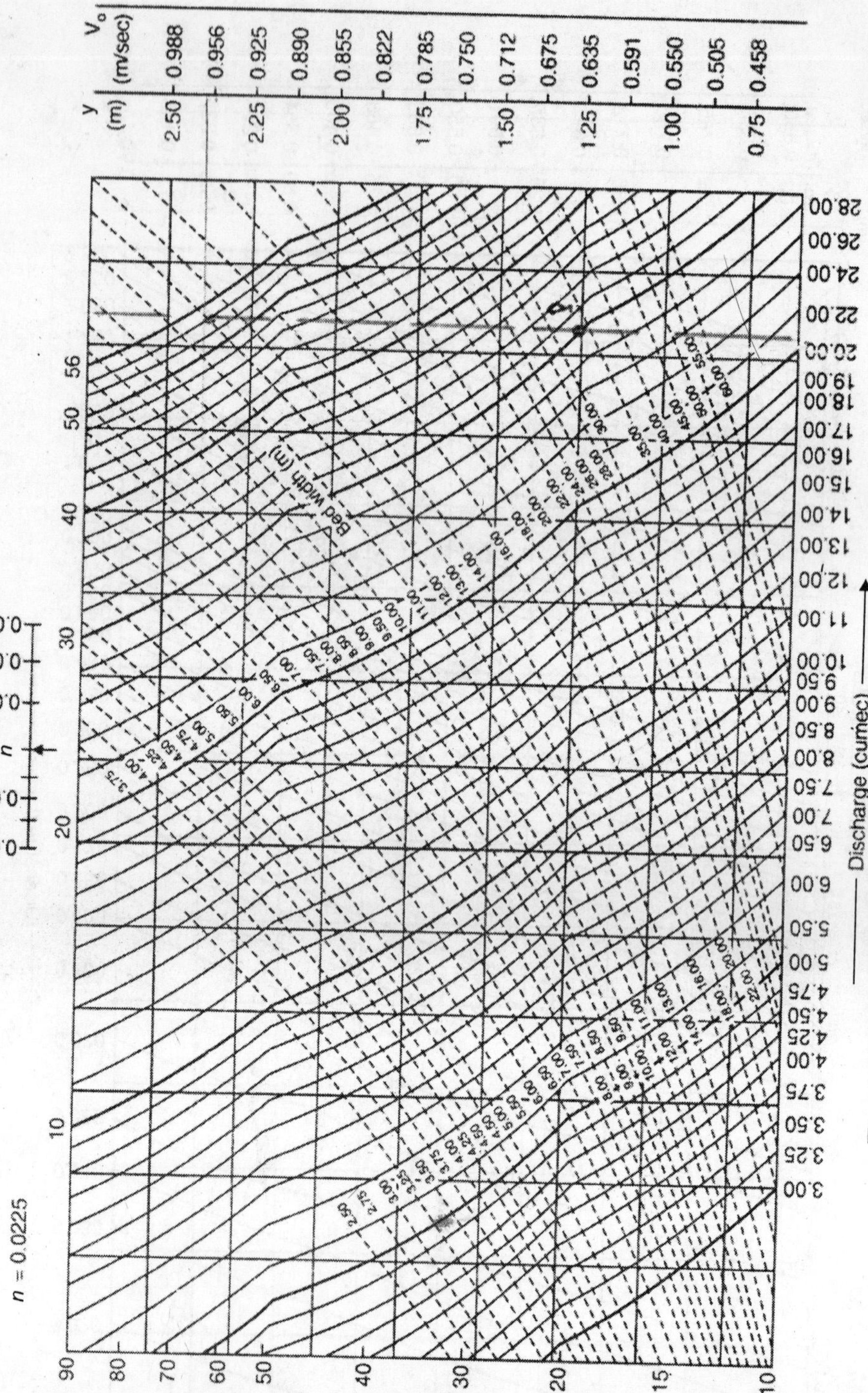


Fig. 4.1(b) Garrets Diagrams for Q between 3 cumec to 56 cumecs

**Plate No. 4.1(c)**

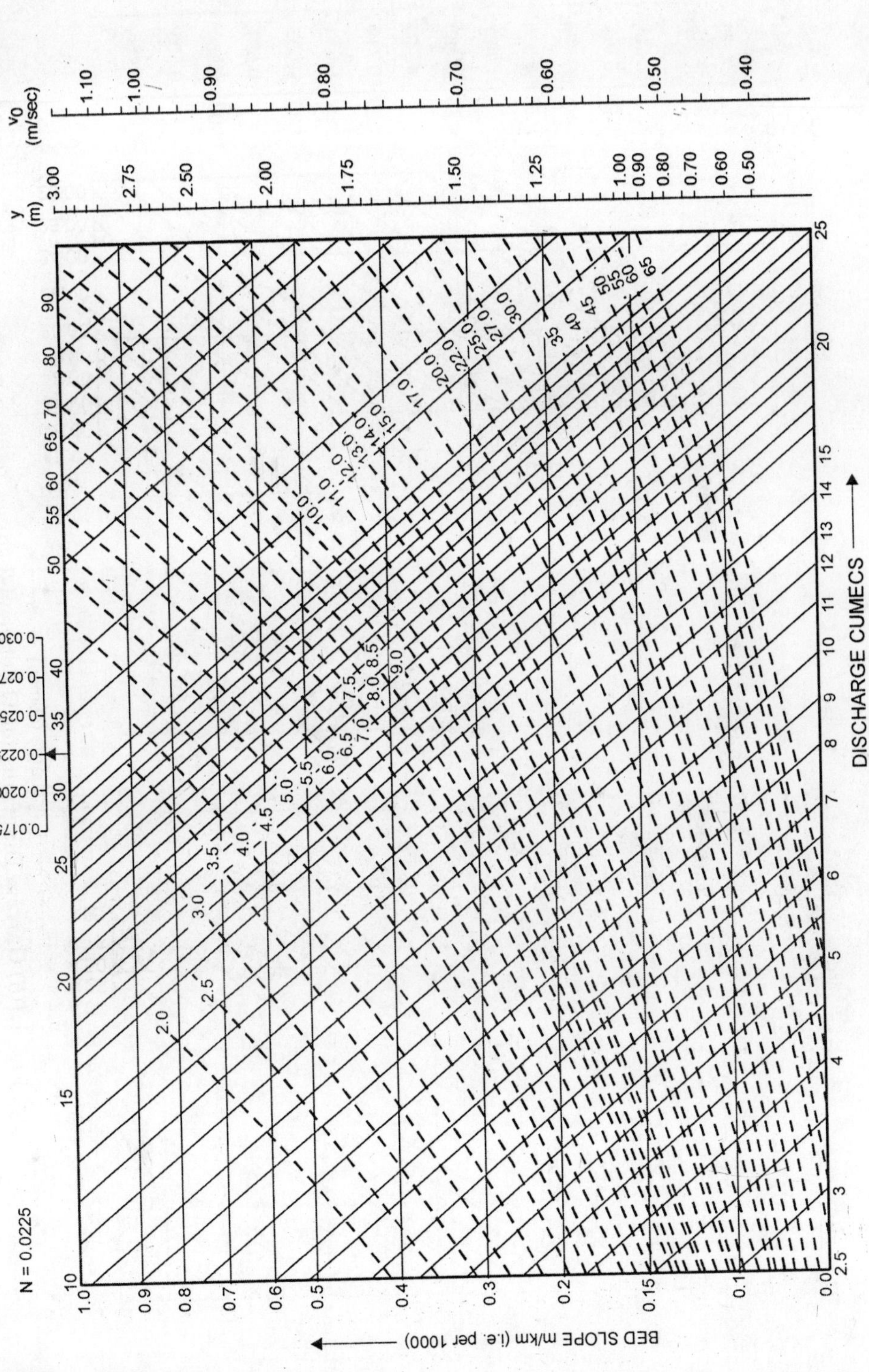
N = 0.0225
Rugosity Coefficient
0.0175
0.0200
0.0225
0.0250
0.0275
0.0300
y (m)
v0 (m/sec)
DISCHARGE CUMECS
BED SLOPE m/km (i.e. per 1000)

silting no scouring may actually not be in regime. He, therefore, differentiated between three regime conditions : (*i*) True regime ; (*ii*) *Initial regime ;* and (*iii*) *Final regime.*

According to him, a channel which is under 'initial' regime, is not a channel in regime (though outwardly it appears to be in regime, as there is no silting or scouring) and hence, regime theory is not applicable to such channels. His theory is therefore applicable only to those channels, which are either in *true regime* or *in final regime.*

**4.7.4.2. True regime.** A channel shall be in regime, if there is neither silting nor scouring. For this condition to be satisfied, the silt load entering the channel must be carried through, by the channel section. Moreover, there can be only one channel section and one bed slope at which a channel carrying a given discharge and a particular quantum and type of silt, would be in regime. Hence, an artificially constructed channel having a certain fixed section and a certain fixed slope can behave in regime only if the following conditions are satisfied :

(*i*) *Discharge is constant ;*

(*ii*) *Flow is uniform ;*

(*iii*) *Silt charge is constant* ; *i.e.* the amount of silt is constant ;

(*iv*) *Silt grade is constant* ; *i.e.*, the type and size of silt is always the same, and

(*v*) *Channel is flowing through a material which can be scoured as easily as it can be deposited (such soil is known as* **incoherent alluvium***), *and is of the same grade as is transported.*

Hence, a designed channel shall be in 'true regime' if the above conditions are satisfied. But in practice, all these conditions can never be satisfied. *And, therefore, artificial channels can never be in 'true regime'; they can either be in initial regime or final regime,* as explained below :

**4.7.4.3. Initial regime and Final regime.** When only the bed slope of a channel varies due to dropping of silt, and its cross-section or wetted perimeter remains unaffected, even then the channel can exhibit 'no silting no scouring' properties, called *Initial regime.* Thus, when water flows through an excavated channel with somewhat narrower dimensions and defective slopes, the silt carried by the water may get dropped in the upper reaches, thereby increasing the channel bed slope. Consequently, the velocity is increased, and a non-silting equilibrium is established, called 'Initial regime'. Sides of such channels are subjected to a lateral restraint and could have scoured if the bank soil would have been a true alluvium. But in practice, they may either be grassed or be of clayey soil, and therefore, they may not get eroded at all. Hence, such channels will exhibit 'non-silting, non-scouring' properties, and they will appear to be in regime ; but in fact, they are not. *They have achieved only a working stability due to the rigidity of their banks.* Their slopes and velocities are higher and cross-sections narrower than what would have been if the sides were not rigid. *Such channels are termed as channels in initial regime, and regime theory is not applicable to them, as they are infact, not the channels in alluvium.*

---

* **incoherent** soil, as defined in article 4.4, is the soil, which does not possess any cohesion between its soil grains. The alluvial soil is defined in article 3.2, as the soil formed by continuous deposition of silt by the moving water. Such sandy soil deposits which donot possess any cohesion ($c = 0$) and are formed by the agency of water are called incoherent alluvium.

But, if there is no resistance from the sides, and all the variables such as perimeter, depth, slope, etc. are equally free to vary and finally get adjusted according to discharge and silt grade, then the channel is said to have achieved permanent stability, called ***Final Regime***. Regime theory is applicable to such channels only, and not to all regime channels, as was envisaged by Kennedy.

*Such a channel in which all variables are equally free to vary, has a tendency to assume a semi-elliptical section.* The coarser the silt, the flatter is the semi-ellipse, *i.e.* greater is the width of the water-surface. The finer the silt, the more nearly the section attains a semi-circle (Fig. 4.10).

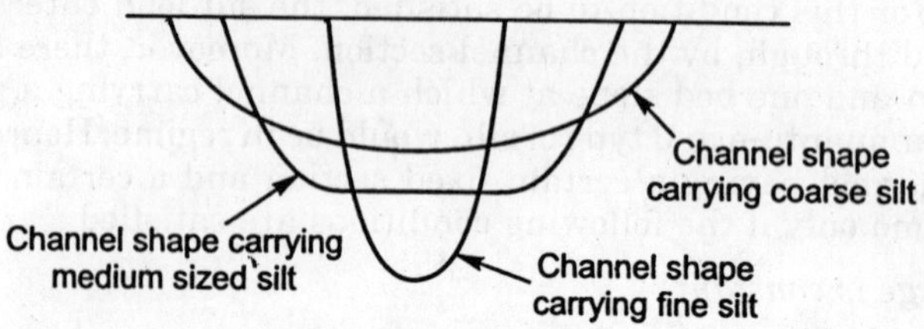


**Fig. 4.10**

The second point which Lacey argued was that the *sediment is kept in suspension not only by the vertical component of the eddies which are generated on the channel bed, but also by the eddies generated on the sides of the channel.* Kennedy had neglected the eddies that are generated on the sides of the channel, by presuming that such eddies has horizontal movement for greater part, and, therefore, do not have sediment supporting power. Lacey thus, argued that the silt supporting power of a channel is proportional to the wetted perimeter of the channel and not to its width, as was presumed by Kennedy. *Lacey, therefore, assumed hydraulic mean depth (R) as variable, unlike Kennedy, who assumed depth (y) as variable.*

Thirdly, Lacey argued that the grain size of the material forming the channel is an important factor, and should need much more rational attention than what was given to it by Kennedy (different values of critical velocity ratio (m) for different types of soils). He, therefore, introduced a term called **silt factor (f)** in his equations, and connected it to the average particle size (as per equation (4.32)).

**4.7.4.4. *Lacey's Regime Equations.*** On the basis of arguments mentioned above, Lacey plotted a large collection of data to obtain a relationship between (*i*) *V vs R* and (*ii*) *A vs V*. Lacey also introduced a factor *f*, known as **silt factor** in the regime relationship. The *first two basic equations* originally suggested by Lacey in F.P.S. units, converted in M.K.S. units, are given below:

(*i*) $V = \sqrt{\frac{2}{5}} \cdot \sqrt{fR}$ ...(4.23)

(*ii*) $Af^2 = 140\ V^5$ ...(4.24)

where $A$ = Area of channel section

$V$ = Velocity of flow in the channel

$R$ = Hydraulic mean depth of channel $= \frac{A}{P}$.

$f$ = Silt fector.

In addition to the above two **fundamental** equations, Lacey obtained a third relationship expressing $V$ in terms of $R$ and $S$ by plotting a large collections of data from a number of sources, as :

$$V = 10.8 \cdot R^{2/3} \cdot S^{1/3} \qquad \text{...(4.25)}$$

where $S$ = Slope of water surface (which equals to bed slope for ordinary flows)

This equation is very useful for finding velocities during high floods.

The above three basic equations are further used to derive certain relations for being used in the design of Laceys regime channels, as discussed below :

(1) **Perimeter-Discharge Relation (i.e. P-Q Relation).** By combining equations (4.23) and (4.24), a relation between $P$ and $Q$ can be obtained, as below :

From Eq. (4.23), we have

$$V = \sqrt{\frac{2}{5}} \cdot \sqrt{fR} \qquad \text{...(Eq. 4.23)}$$

Raising both sides to fourth power, we get

$$V^4 = \frac{4}{25} \cdot f^2 R^2 \qquad \text{...}(i)$$

Also from Eq. (4.24), we have

$$Af^2 = 140V^5 \qquad \text{...(Eq. 4.24)}$$

or
$$f^2 = \frac{140V^5}{A} \qquad \text{...}(ii)$$

Eliminating $f^2$ between $(i)$ and $(ii)$, we get

$$V^4 = \frac{4}{25} \cdot \left(\frac{140V^5}{A}\right) \cdot R^2$$

or
$$V^4 = \frac{4}{25}\left(\frac{140V^5}{A}\right)\left(\frac{A}{P}\right)^2$$

or
$$V^4 = \frac{4}{25} \cdot 140V^5 \frac{A}{P^2}.$$

or
$$\frac{25}{4} = 140 \cdot \frac{V \cdot A}{P^2}$$

or
$$\frac{25}{4} = 140 \cdot \frac{Q}{P^2} \qquad [\because \quad V \cdot A = Q]$$

or
$$P^2 = 140 \,.\, Q \times \frac{4}{25} = 22.4\, Q.$$

or
$$\boxed{P = 4.75\sqrt{Q}} \qquad \text{...(4.26)}$$

Equation (4.26) is a very important equation in as much as it is universally true. Power of $Q$ does not vary, though coefficient 4.75 varies between certain range over different regions.

(2) ***V-Q-f* Relation.** A relation expressing $V$ in terms of $Q$ and $f$ can be obtained from Eq. (4.24) as follows :

From Eq. (4.24), we have

$$Af^2 = 140V^5$$

Multiplying both sides by $V$, we get

$$A \cdot V \cdot f^2 = 140 \cdot V^6$$

But $A \cdot V = Q$ (Basic equation of discharge)

$$\therefore \quad Qf^2 = 140 \cdot V^6$$

or
$$V^6 = \frac{Qf^2}{140}$$

or
$$\boxed{V = \left[\frac{Qf^2}{140}\right]^{1/6}} \quad ...(4.27)$$

(3) **Regime Slope Equation:** A relation expressing regime slope ($S$) in terms of $Q$ and $f$ can be derived as follows :

From Eq. (4.25), we have

$$V = 10.8 \cdot R^{2/3} \cdot S^{1/3} \quad ...(\text{Eq. } 4.25)$$

Cubing both sides, we get

$$V^3 = 1260\, R^2 \cdot S \quad ...(i)$$

or
$$S = \frac{V^3}{1260R^2}$$

or
$$S = \left(\frac{V^2}{R}\right)^{5/3} \frac{1}{1260 \cdot (VR)^{1/3}}$$

or
$$S = \left[\frac{V^2}{R}\right]^{5/3} \cdot \frac{1}{1260 \cdot \left(V \cdot \dfrac{A}{P}\right)^{1/3}} \quad \left[\because \; R = \frac{A}{P}\right]$$

or
$$S = \left(\frac{V^2}{R}\right)^{5/3} \cdot \frac{P^{1/3}}{1260 \cdot Q^{1/3}} \quad [\because \; V \cdot A = Q]$$

But from Eq. (4.23), $V^2 = \dfrac{2}{5} \cdot fR$;

or
$$\frac{V^2}{R} = \frac{2}{5} f.$$

$$\therefore \quad S = \left(\frac{2}{5} f\right)^{5/3} \cdot \frac{P^{1/3}}{1260\, Q^{1/3}}$$

Also
$$P = 4.75 \sqrt{Q} \quad [\text{From Eq. } 4.26]$$

$$\therefore \qquad S = \left(\frac{2}{5}f\right)^{5/3} \cdot \frac{[4.75\,Q^{1/2}]^{1/3}}{1260 \cdot Q^{1/3}}$$

or
$$S = \frac{\left(\frac{2}{5}\right)^{5/3} \cdot f^{5/3} \cdot (4.75)^{1/3} \cdot Q^{1/6}}{1260\,Q^{1/3}}$$

or
$$\boxed{S = \left[\frac{f^{5/3}}{3340 \cdot Q^{1/6}}\right]} \qquad \text{...(4.28)}$$

(4) **Regime Scour Depth Relation for Rivers**

From Eq. (4.23) :

$$V^2 = \frac{2}{5} \cdot fR$$

or
$$\boxed{R = \frac{5}{2}\frac{V^2}{f}} \qquad \text{...(4.23}a\text{)}$$

But from Eq. (4.27), we have

$$V^2 = \left[\frac{Qf^2}{140}\right]^{2/6} = \left[\frac{Qf^2}{140}\right]^{1/3}$$

Substituting this value of $V^2$ in Eq. (4.23*a*), we get

$$\therefore \qquad R = \frac{5}{2}\left[\frac{Qf^2}{140}\right]^{1/3} \cdot \frac{1}{f}$$

or
$$\boxed{R = 0.47\left[\frac{Q}{f}\right]^{1/3}} \qquad \text{...(4.29)}$$

This $R$ is known as **Lacey's Regime Scour Depth** or **Lacey's Normal Regime Scour Depth,** and may be represented by $R_r$. The above scour depth equation will be applicable only when the river width equals the regime width of $4.75\sqrt{Q}$.

When the actual river width does not equal the regime width of $4.75\sqrt{Q}$, the scour depth can be computed in terms $q$ (rather than in terms of $Q$) where $q$ is the discharge per unit width. In such a case,

$$q = \frac{Q}{\text{Width}} = \frac{Q}{B} = \frac{A.V}{B}$$

But
$$R = \frac{A}{P} \approx \frac{A}{B} \quad \text{or} \quad B = \frac{A}{R}$$

$$\therefore \qquad q = \frac{AV}{B} = \frac{AV}{A/R} = R.V.$$

or
$$q = R \cdot V \qquad \text{...(4.30)}$$

Substituting for $R$ from Eq. (4.23*a*)

$$q = \left[\frac{5}{2}\frac{V^2}{f}\right] V = \frac{5}{2}\frac{V^3}{f}$$

Substituting for $V$ from Eq. (4.27), we get

$$q = \frac{5}{2f} \cdot \left(\frac{Qf^2}{140}\right)^{3/6}$$

or $$q = = 0.21 \cdot Q^{1/2}$$

or $$Q = \left[\frac{q}{0.21}\right]^2 = 22.67q^2.$$

Substituting this value of $Q$ in Eq. (4.29), we get

$$R = 0.47 \left[\frac{22.67q^2}{f}\right]^{1/3} \quad \text{or} \quad \boxed{R = 1.35 \left[\frac{q^2}{f}\right]^{1/3}} \qquad ...(4.31)$$

The above scour depth value is usually represented by $R'$.

(5) **Silt Factor-Grain Size Relation.** The value of silt factor ($f$) considered in various equations given above, has been related to mean particle size of silt (*i.e.* alluvial soil through which channel is constructed) by equation :

$$\boxed{f = 1.76 \cdot \sqrt{d_{mm}}} \qquad ...(4.32)$$

where $d_{mm}$ = Average particle size of alluvium in mm. Some typical values of $d_{mm}$ for different types of alluvium soils are given in Table 4.6.

**Table 4.6. Values of Particle size ($d_{mm}$) for Various types of Alluvial materials for use in Eq. (4.32)**

| *S. No.* (1) | *Type of material (soil)* (2) | *Av. grain size in mm* ($d_{mm}$) (3) |
|---|---|---|
| 1 | *Silt* | |
| | Very fine | 0.05 to 0.08 |
| | Fine | 0.12 |
| | Medium | 0.16 |
| | Standard | 0.32 ($f$ = 1.0) |
| 2 | *Sand* | |
| | Medium | 0.51 |
| | Coarse | 0.73 |
| 3 | *Bajri and Sand,* | |
| | Fine | 0.89 |
| | Medium | 1.29 |
| | Coarse | 2.42 |
| 4 | *Gravel* | |
| | Medium | 7.28 |
| | Heavy | 26.10 |
| 5 | *Boulders* | |
| | Small | 50.10 |
| | Medium | 72.50 |
| | Large | 188.80 |

**4.7.4.5. Design procedure for channel design by Lacey's theory**

(1) Calculate the velocity from equation

$$V = \left[\frac{Qf^2}{140}\right]^{1/6} \text{ m/sec.} \qquad \text{...(Eq. 4.27)}$$

where $Q$ is in cumecs ;

$V$ is in m/s ; and

$f$ is the **silt factor**, given by

$$f = 1.76 \cdot \sqrt{d_{mm}} \qquad \text{...(Eq. 4.32)}$$

where $d_{mm}$ = Average particle size in mm, as given in Table 4.6.

(2) Work out the hydraulic mean depth ($R$) from the equation

$$R = \frac{5}{2}\left(\frac{V^2}{f}\right) \qquad \text{...(Eq. 4.23a)}$$

where $V$ is in m/sec ; $R$ is in m.

(3) Compute area of channel section $A = \dfrac{Q}{V}$

(4) Compute wetted perimeter, $P = 4.75\sqrt{Q}$ ...(Eq. 4.26)

where $P$ is in m ; $Q$ is in m³/sec.

(5) Knowing these values, the channel section is known ; and finally the bed slope $S$ is determined by the equation

$$S = \left[\frac{f^{5/3}}{3340\, Q^{1/6}}\right] \qquad \text{...(Eq. 4.28)}$$

where $f$ is the silt factor, given by Eq. (4.32)

$Q$ is the discharge in cumec.

***Lacey's Regime Width and Scour Depth for Alluvial Rivers :***

For wide streams or rivers, as we know, wetted perimeter $P$, approximately equals the river width. Therefore, according to Lacey, for alluvial rivers :

*Lacey's regime width* = W = $4.75\sqrt{Q}$ ...(4.26*a*)

For such streams, Lacey has also defined the regime scour depth, as

*Lacey's Regime Scour Depth** = $\mathbf{R'}_r = 0.473\,(Q/f)^{1/3}$ ...(Eq. 4.29)

The above scour depth equation will be applicable only when the river width equals the regime width of $4.75\sqrt{Q}$. For any other value of active river width, the normal scour depth is given by the equation :

*Lacey's Normal Scour Depth** (**R′**)

$$= \mathbf{R'} = 1.35\,(q^2/f)^{1/3} \qquad \text{...(Eq. 4.31)}$$

where $q$ is the discharge intensity per unit width of stream = $Q/L$, where $L$ is the actual river width at the given site.

---

*Symbol R is frequently used for representing Lacey's scour depth, but to avoid confusion with the symbol R used for the hydraulic Radius (i.e., hydraulic mean depth = $A/P$), we are using the symbol **R′** here for scour depth. When river width equals the regime width, then the regime scour depth is being represented by $\mathbf{R'}_r$.

**Example 4.10.** *Design a regime channel for a discharge of 50 cumecs and silt factor 1.1, using Lacey's Theory.*

**Solution.** $Q$ = 50 cumecs, $f$ = 1.1

$$V = \left[\frac{Qf^2}{140}\right]^{1/6} = \left[\frac{50 \times (1.1)^2}{140}\right]^{1/6} = 0.869 \text{ m/s}$$

$$A = \frac{Q}{V} = \frac{50}{0.869} = 56.3 \text{ m}^2$$

$$R = \frac{5}{2} \cdot \frac{V^2}{f} = \frac{5}{2} \cdot \frac{(0.869)^2}{1.1} = 1.675 \text{ m.}$$

$$P = 4.75\sqrt{Q} = 4.75 \cdot \sqrt{50} = 33.56 \text{ m}$$

For a trapezoidal channel with $\frac{1}{2}$ $H$ : 1$V$ slopes

$$P = b + \sqrt{5}\, y \quad \text{and} \quad A = \left(b + \frac{y}{2}\right)y$$

$$\therefore \quad 33.56 = b + \sqrt{5} \cdot y \qquad \text{...}(i)$$

$$\text{and} \quad 56.3 = by + \frac{y^2}{2} \qquad \text{...}(ii)$$

From Eq. ($i$), we get, $b = 33.56 - 2.24y$

Putting this value of $b$ in Eq. ($ii$), we get

$$56.3 = [33.56 - 2.24y]\, y + \frac{y^2}{2}$$

$$= 33.56y - 2.24y^2 + 0.5y^2 = 33.56y - 1.74y^2$$

$$\text{or} \quad 1.74y^2 - 33.56y + 56.3 = 0$$

$$\text{or} \quad y^2 - 19.3y + 32.4 = 0$$

$$\therefore \quad y = \frac{19.3 \pm \sqrt{372 - 129.6}}{2} = \frac{19.3 \pm \sqrt{242.4}}{2} = \frac{19.3 \pm 15.6}{2}$$

Neglecting unfeasible + ve sign, we get

$$y = \frac{19.3 - 15.6}{2} = 1.65 \text{ m}$$

$\therefore$ $y$ = 1.65 m. **Ans.**

$b$ = 33.56 – 2.24 × 1.65 = **29.77 m**

or $b$ = 29.77 m. **Ans.**

$$S = \frac{f^{5/3}}{3340\, Q^{1/6}} = \frac{(1.1)^{5/3}}{3340 \cdot (50)^{1/6}} = \frac{1}{5420}$$

Use a bed slope of 1 in 5420. **Ans.**

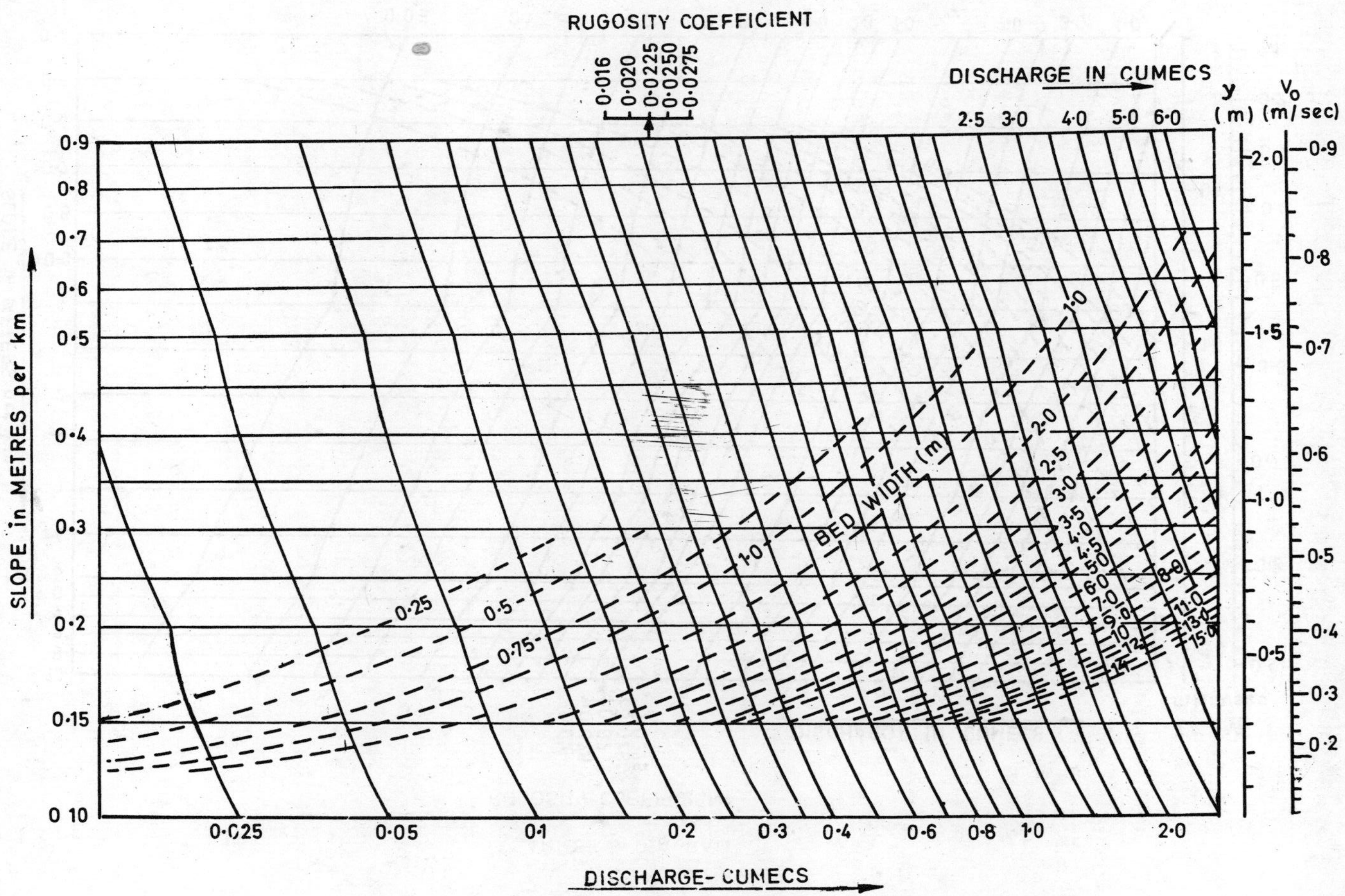

RUGOSITY COEFFICIENT
0·016
0·020
0·0225
0·0250
0·0275
DISCHARGE IN CUMECS
2·5
3·0
4·0
5·0
6·0
y (m)
$V_0$ (m/sec)
SLOPE in METRES per km
0·9
0·8
0·7
0·6
0·5
0·4
0·3
0·2
0·15
0 10
0·025
0·05
0·1
0·2
0·3
0·4
0·6
0·8
1·0
2·0
DISCHARGE- CUMECS
BED WIDTH (m)
0·25
0·5
0·75
1·0
2·0
2·5
3·0
3·5
4·0
4·5
5·0
6·0
7·0
8·0
9·0
10
11·0
12
13·0
14
15·0
2·0
1·5
1·0
0·5
0·9
0·8
0·7
0·6
0·5
0·4
0·3
0·2

GARRET'S DIAGRAM

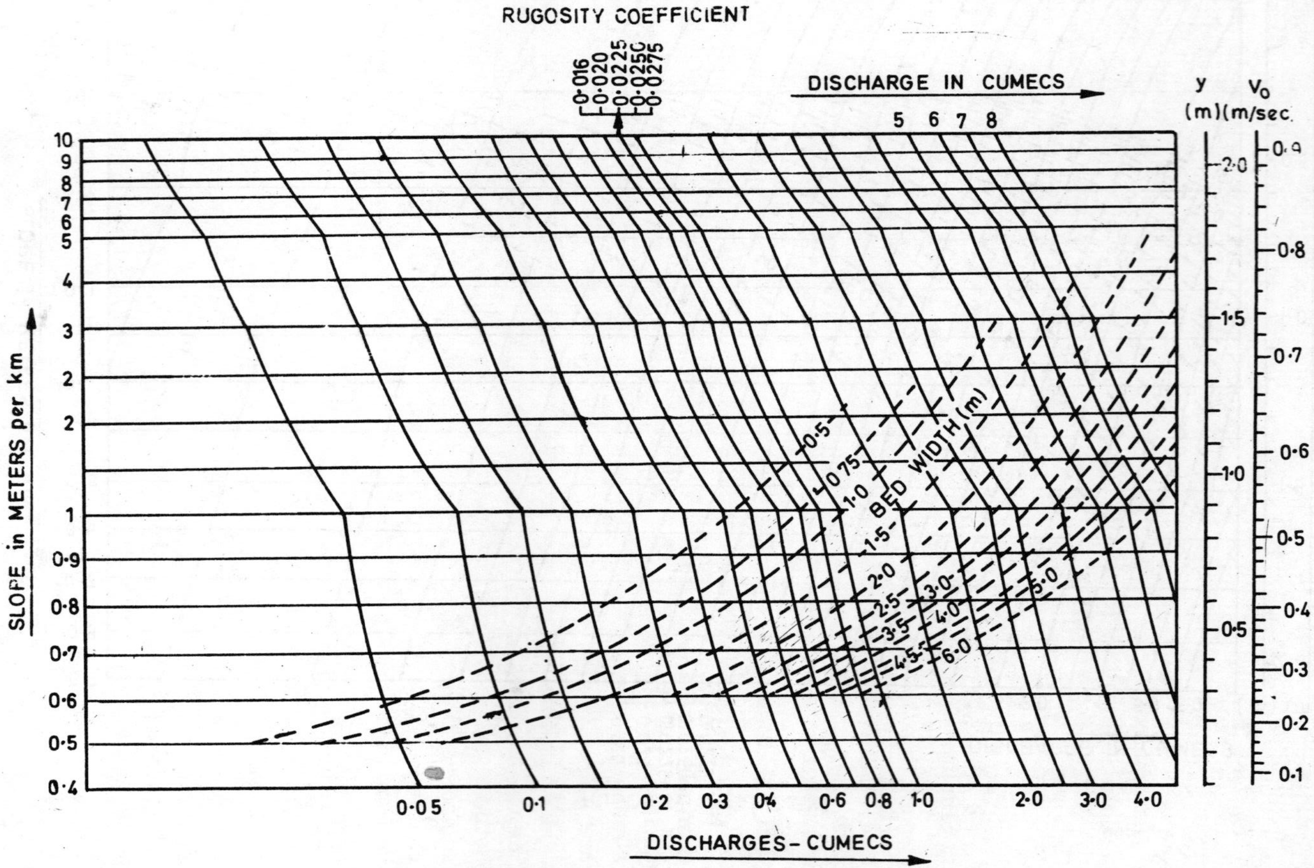

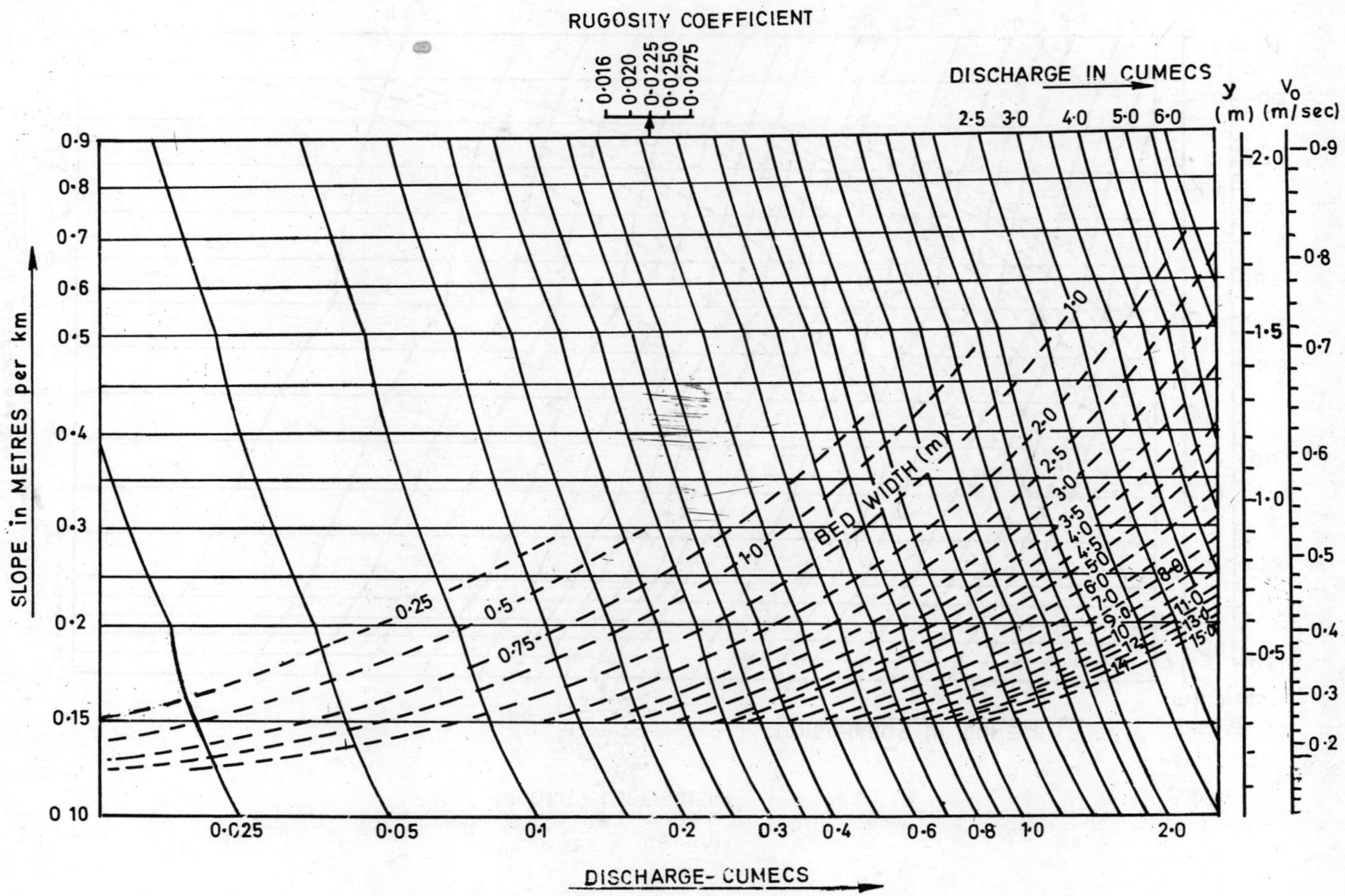

RUGOSITY COEFFICIENT
0·016
0·020
0·0225
0·0250
0·0275
DISCHARGE IN CUMECS
2·5 3·0 4·0 5·0 6·0
y (m)
$V_0$ (m/sec)
SLOPE in METRES per km
0·9 0·8 0·7 0·6 0·5 0·4 0·3 0·2 0·15 0 10
BED WIDTH (m)
0·25 0·5 0·75 1·0 2·0 2·5 3·0 3·5 4·0 4·5 5·0 6·0 7·0 8·0 9·0 10 11·0 12 13·0 14 15·0
0·025 0·05 0·1 0·2 0·3 0·4 0·6 0·8 1·0 2·0
DISCHARGE- CUMECS

GARRET'S DIAGRAM

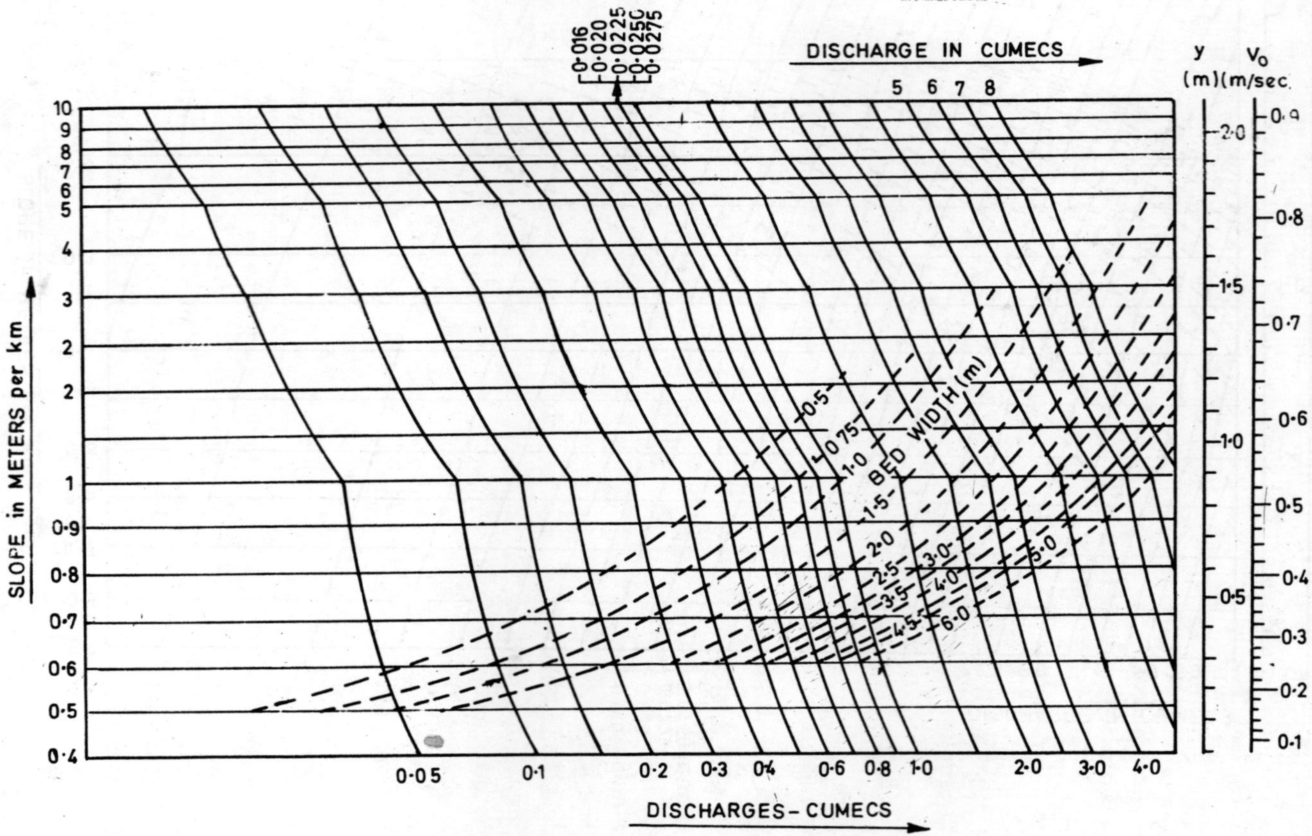

**Example 4.11.** *Design a Lacey's regime channel, flowing through material of average diameter 0.328 mm, for the following data :*

*Culturable commanded area = 1,00,000 hectare*

*Intensity of irrigation :*

| | |
|---|---|
| *Rabi* | = 40% |
| *Kharif* | = 30% |

*Outlet discharge factor :*

| | |
|---|---|
| *Rabi* | = 1,800 hec/cumec |
| *Kharif* | = 800 hec/cumec |

*Assume conveyance losses to be 10%.* (*Civil Services, 1994*)

**Solution.** Area irrigated in Rabi = 40,000 hec

$$\therefore \quad Q \text{ for Rabi} = \frac{40{,}000 \text{ hec}}{1800 \text{ hec/cumec}} = 22.2 \text{ m}^3/\text{s}$$

$$\text{Similarly } Q \text{ for Kharif} = \frac{30{,}000 \text{ hec}}{800 \text{ hec/cumec}} = 37.5 \text{ m}^3/\text{s}$$

Hence, design the channel on Lacey's theory for $Q = 37.5$ m³/s and silt factor $f = 1.76\sqrt{d_{mm}} = 1.76\sqrt{0.328} = 1.0$, by following Example 4.10, and compute $b = 25.06$ m, $y = 1.8$ m and $S = 1$ in 6110 for a channel having side slopes of $\frac{1}{2}$H : 1*V*. **Ans.**

**4.7.4.5. Use of Lacey's diagrams.** Lacey's equations for design of irrigation channels (explained earlier) have been converted into graphical solutions. Pl. see enclosed **Plate Figs 4.2 (*a*), (*b*) and (*c*)**. From these diagrams, the slopes and dimensions of any channel can be easily determined, if discharge and silt factor are known.

**4.7.5. Comparison of Kennedy's and Lacey's Theories and Improvements over Lacey's Theory.** (1) The concept of silt transportation is the same in both the theories. Both the theories agree that the silt is carried by the vertical component of the eddies generated by the friction of the flowing water against the channel surface. The difference is that Kennedy considered a trapezoidal channel section and, therefore, he neglected the eddies generated from the sides, on the presumption that these eddies has horizontal movement for greater part and, therefore, did not have silt supporting power. For this reason, Kennedy's critical velocity formula was derived only in terms of depth of flow ($y$). On the other hand, Lacey considered that an irrigation channel achieves a cup-shaped section (semi-ellipse) and that the entire wetted perimeter ($P$) of the channel contributes to the generation of silt supporting eddies. He, therefore, used hydraulic mean radius ($R = A/P$) as a variable in his regime velocity formula instead of depth ($y$).

(2) Kennedy stated all the channels to be in a state of regime provided they did not silt or scour. But, Lacey differentiated between the two regime conditions, *i.e.* Initial regime and final regime.

(3) According to Lacey, the grain size of the material forming the channel is an important factor and should need much more rational attention than what was given to it by Kennedy. Kennedy has simply stated that Critical velocity ratio ($V/V_0 = m$) varies according to the silt conditions (*i.e.* silt grade and silt charge).

**Plate No. 4.2(*a*)**

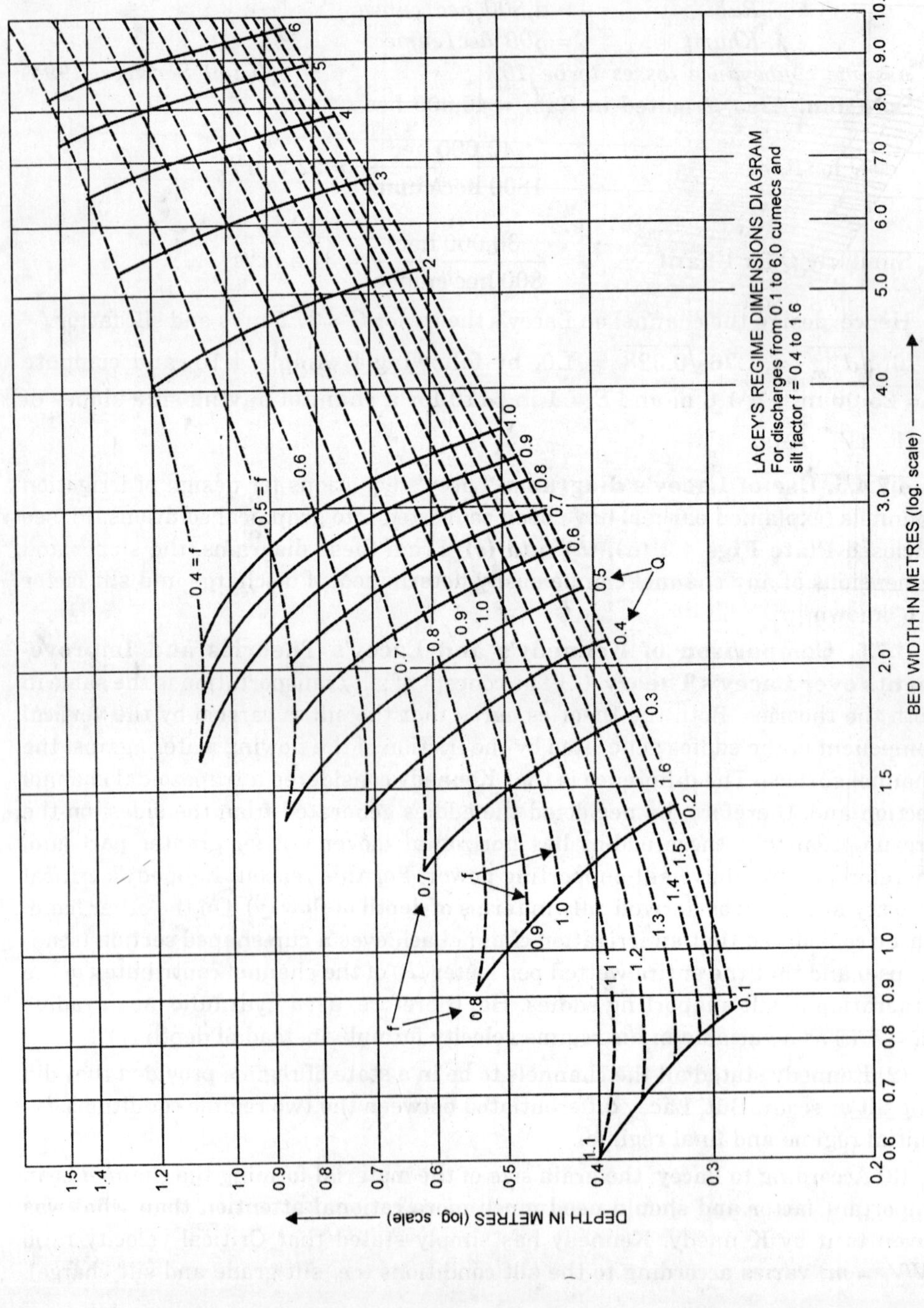


Plate No. 4.2(a) : Lacey's diagram for Q = 0.1 cumec to 6 cumecs and f = 0.4 to 1.6

**Plate No. 4.2(*b*)**

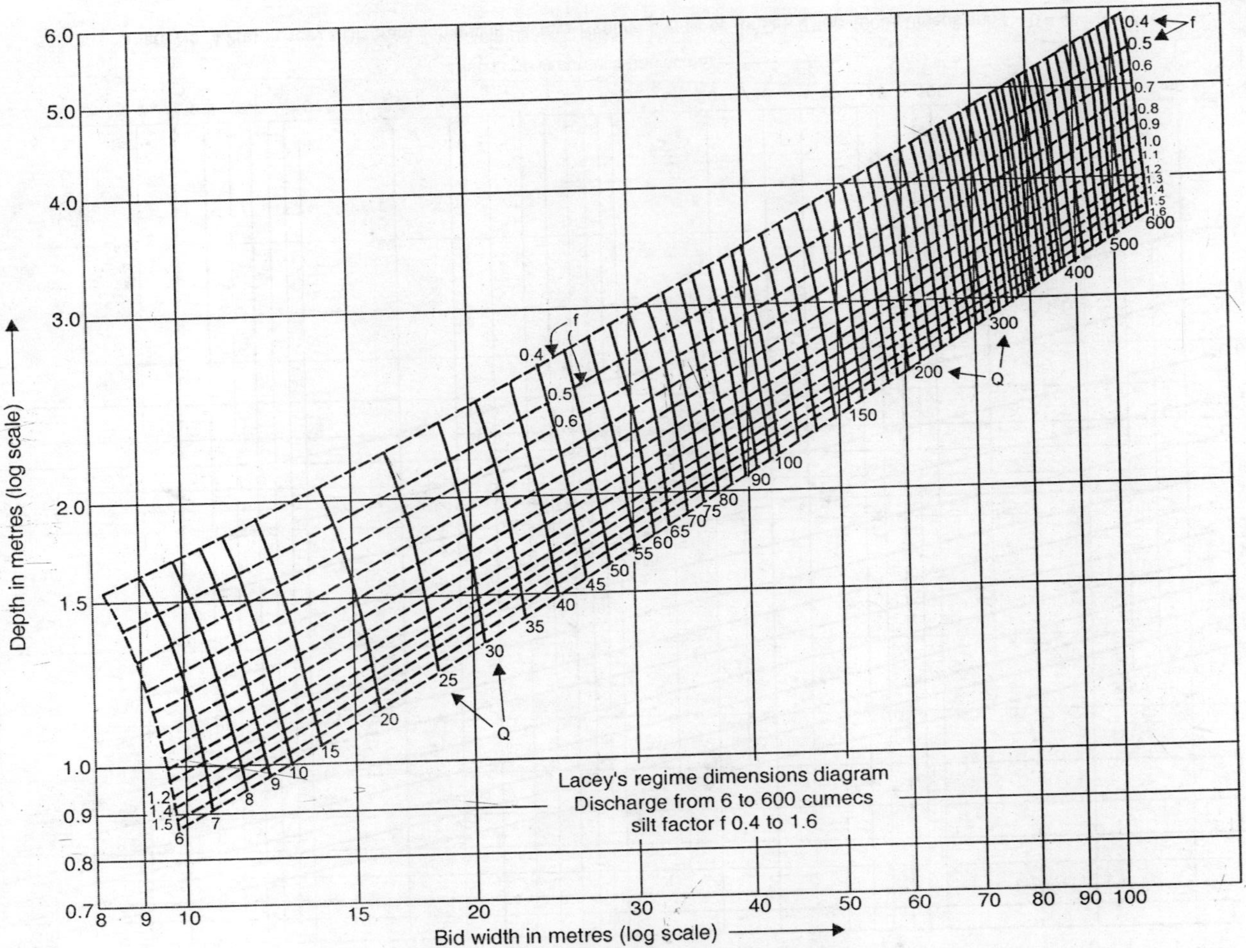


**Plate No.4.2(*b*)**: Lacey's Regime Diagram for Q = 6 to 600 cumecs and f = 0.4 to 1.6.

## Plate No. 4.2(c)

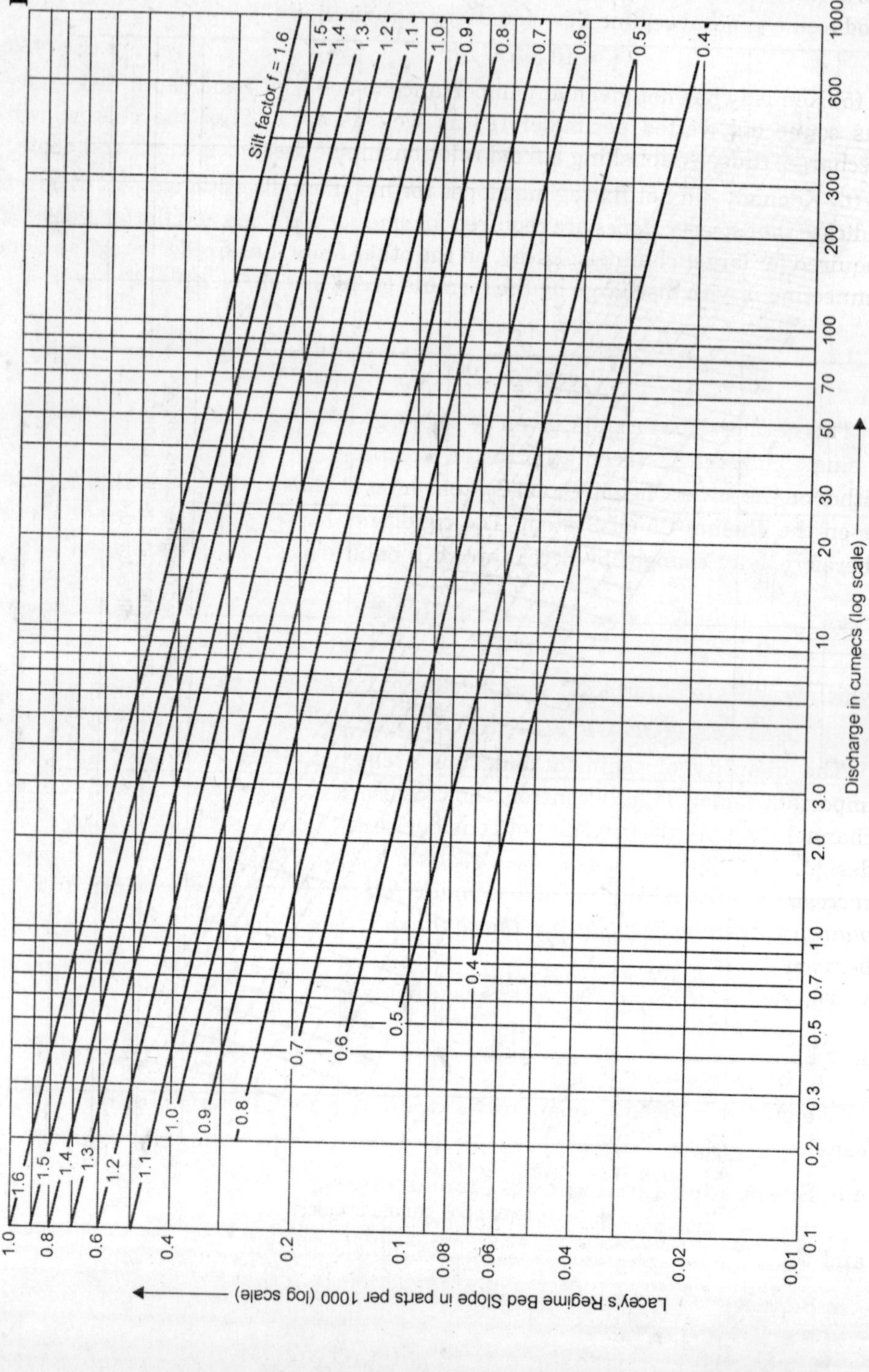


**Plate No. 4.2(c)** : Lacey's Diagram computing Lacey;s Regime bed slope for Q = 0.1 to 1000 cumecs and f = 0.4 to 1.6

Lacey, however, connected the grain size ($d$) with his silt factor ($f$) by the equation $f = 1.76 \cdot \sqrt{d_{mm}}$ . The silt factor ($f$) occurs in all those Lacey's equations, which are used to determine channel dimensions.

(4) Kennedy has used Kutter's formula for determining the actual generated channel velocity. The value of Kutter's rugosity coefficient ($n$) is again a guess work. Lacey, on the other hand, after analysing huge data on regime channels, has produced a general regime flow equation, stating that

$$V = 10.8\, R^{2/3}\, S^{1/3} \quad \text{...(Eq. 4.25)}$$

(5) Kennedy has not given any importance to bed width and depth ratio. Lacey has connected wetted perimeter ($P$) as well as area ($A$) of the channel with discharge, thus, establishing a fixed relationship between bed width and depth.

(6) Kennedy did not fix regime slopes for his channels, although his diagrams indicate that *steeper slopes* are required for *smaller channels* and flatter slopes are required for larger channels. Lacey, on the other hand, has fixed the regime slope, connecting it with discharge by the formula given by eqn. (4.28) as

$$S = \frac{f^{5/3}}{3340 \cdot Q^{1/6}}, \quad \text{...(Eq. 4.28)}$$

This regime slope formula, given by Lacey, gives excessive slope values. Not even a single channel has been constructed according to this regime slope equation, either on the lower Chenab Canal System or on the Lower Bari Doab Canal System or on the Jhelum Canal System. The rigidity of this regime slope equation was, therefore, later changed by Lacey (in C.B.I. publication No. 20) to the form

$$S \propto \frac{f^{5/3}}{q^{1/3}} \quad \text{...(4.28 a)}$$

which was his final formula, showing no rigidity of the constants.

## ESTIMATION OF TRANSPORTED SEDIMENT IN A CANAL

The quantity of sediment entering a channel from the head-works is an important factor. It fully controls the cross-section and shape of the true regime channel. Regime theories do not consider this vital important factor in channel design. *However, it has now been realised that the channel design shall not be successful unless full provision is made for the effects produced by the actual quantity of sediment moving in the channel. Hence, a detailed study of the subject becomes important.*

The sediment moving in water has already been classified into

(*i*) *Bed Load*, and

(*ii*) *Suspended Load*

We shall now derive equations for evaluating the amounts of these loads in a canal.

### 4.8. Suspended Load and its Measurement

After the threshold of motion has been passed due to increase in flow velocity, and the sediment movement is well established, some of the sediment will be





carried in suspension. *The material is kept in suspension by the turbulence, or in other words, by the generation of the eddies that rise from the regions of higher sediment concentration to the regions of lower sediment concentrations.*

In a laminar flow, the shear stress caused at the base is due to one factor, and that is due to the difference of velocities at the top and at the base, as shown in Fig. 4.11. Therefore,

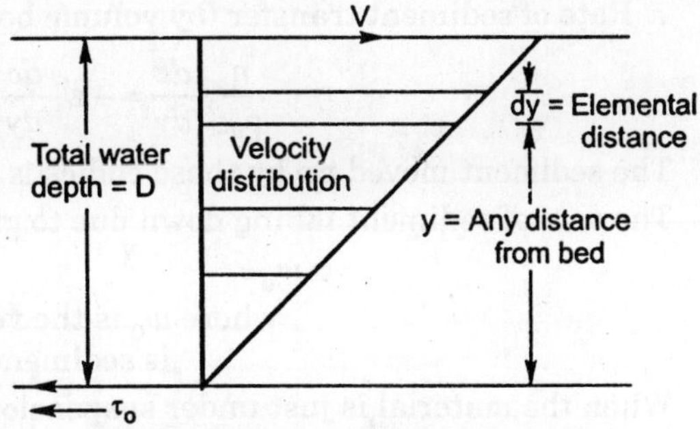


$\tau_o$ = Shear stress at bed = Rate of change of velocity with distance = $\mu \frac{\partial V}{\partial y}$

**Fig. 4.11**

$$\tau_0 = \mu \frac{\partial V}{\partial y} \text{ for a laminar flow,}$$

where μ is the **dynamic viscosity** of the water (fluid.) But in a turbulent flow, another factor that comes into play, is the jumping of particles from higher velocity region to lower velocity region. This is known as momentum transfer or mass exchange. Due to this transference of mass or momentum between the two adjacent fluid layers, an effective shear stress is caused at the interface between the layers, as shown in Fig. 4.12.

The shear stress produced by this action is given by $\eta \cdot \frac{dV}{dy}$, where η is the **eddy viscosity,** and is defined as the *rate of mass exchange per unit area* between the adjacent layers.

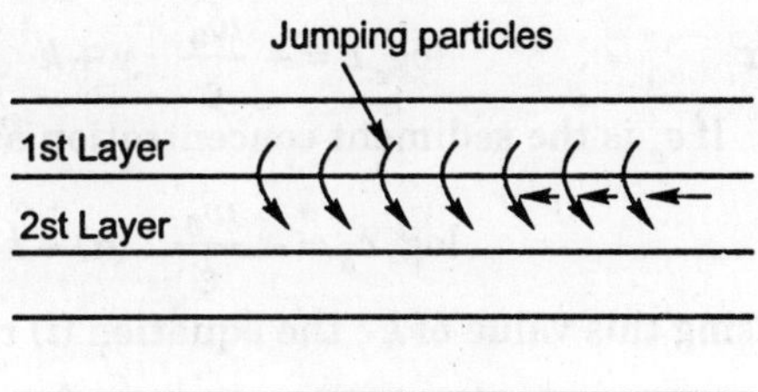


**Fig. 4.12**

Just as we define *kinematic viscosity* *of water* $\nu = \left(\frac{\mu}{\rho_w}\right)$, we here define $\varepsilon = \frac{\eta}{\rho_w}$

where ε is defined as the *rate of volume exchange per unit area,* between the *adjacent layers,* and is called **eddy kinematic viscosity.**

This momentum transfer caused by the generation of eddies brings about the transfer of sediment from the regions of higher concentration to the regions of lower concentration.

If $\frac{dc}{dy}$ represents the gradient of sediment concentration between the adjacent layers (Fig. 4.13), then the rate of sediment transfer (in terms of mass between the adjacent layers)

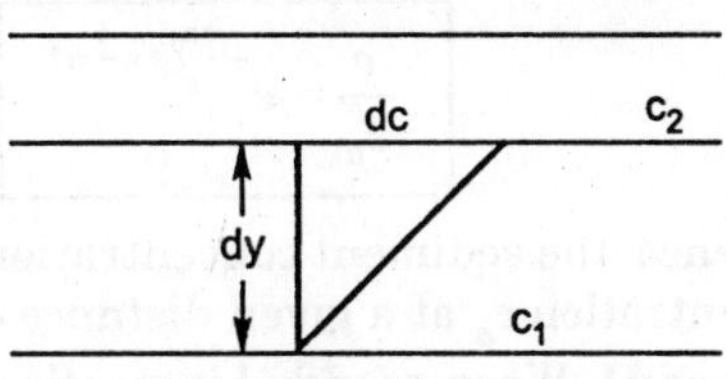


**Fig. 4.13**

$$= -\eta \frac{dc}{dy}$$

(– ve sign indicates that concentration is decreasing in the positive direction of *y*.)

∴ Rate of sediment transfer (by volume between the adjacent layers)

$$= -\frac{\eta}{\rho_w} \cdot \frac{dc}{dy} = -\varepsilon \cdot \frac{dc}{dy}$$

The sediment moved up by these eddies is brought down by its self-weight. The rate of sediment falling down due to gravity

$$= w_0 c$$

where $w_0$ is the velocity of free fall of the grain, and $c$ is sediment concentration.

When the material is just under suspension, the two forces must be equal.

Therefore, $$-\varepsilon \frac{dc}{dy} = w_0 c$$

or $$w_0 c + \varepsilon \frac{dc}{dy} = 0 \qquad ...(4.33)$$

is the governing differential equation.

Integrating both sides after separating the variables, we get

$$\int \frac{dc}{c} = -\int \frac{w_0}{\varepsilon} dy$$

Two cases now arise.

**Case I.** ***When ε (eddy kinematic viscosity) is constant***

$$\int \frac{dc}{c} = -\frac{w_0}{\varepsilon} \int dy$$

or $$\log_e c = -\frac{w_0}{\varepsilon} \cdot y + k \qquad ...(i)$$

If $c_a$ is the sediment concentration at a distance $a$ from the bed, then

$$\log_e c_a = -\frac{w_0}{\varepsilon} \cdot (a) + k \quad \text{or} \quad k = \log_e c_a + \frac{w_0}{\varepsilon} (a)$$

using this value of $k$ ; the equation $(i)$ reduces to

$$\log_e c = -\frac{w_0}{\varepsilon} \cdot y + \left( \log_e c_a + \frac{w_0}{\varepsilon} \cdot a \right)$$

or $$\log_e c - \log_e c_a = -\frac{w_0}{\varepsilon} \cdot y + \frac{w_0}{\varepsilon} \cdot a$$

or $$\log_e \frac{c}{c_a} = -\frac{w_0}{\varepsilon} \cdot (y - a)$$

or $$\boxed{\frac{c}{c_a} = e^{-\frac{w_0}{\varepsilon}(y-a)}} \qquad ...(4.34)$$

Hence, the sediment concentration $c$ at any distance $y$ can be determined, if the concentration $c_a$ at a given distance $a$ is known.

**Case II.** ***When ε (eddy kinematic viscosity) is not constant.***

Generally, ε is not constant, and it has been shown by Prandtl that $\varepsilon = l^2 \cdot \frac{dV}{dy}$

where, $l$ is '**mixing length**' and is analogous to 'mean free path' of kinetic theory of gases. The mixing length is a measure of turbulence and is an indicator of the size of the generated eddies.

Now, we know that $c = (\mu + \eta) \cdot dV/dy$.

But $\mu$ is very small compared to $\eta$, and can be neglected.

Therefore, $c = \eta\,(dV/dy)$

In the close neighbourhood of a solid boundary, it has been found from dimensional considerations that the 'Mixing length' ($l$) is directly proportional to the distance ($y$) from the boundary

or $l \propto y$ or $l = Ky$

where $K$ is known as **Von-Karman Universal constant** and is equal to 0.4 for all homogeneous fluids.

Now, $$\tau = \eta \cdot \frac{dV}{dy} = \rho_w \cdot \varepsilon \cdot \frac{dV}{dy} = \rho_w \cdot \left(l^2 \cdot \frac{dV}{dy}\right)\frac{dV}{dy}$$

$$= \rho_w \cdot (Ky)^2 \cdot \left(\frac{dV}{dy}\right)^2 = \rho_w K^2 \cdot y^2 \cdot \left(\frac{dV}{dy}\right)^2.$$

Again in the same immediate neighbourhood, we can assume, $\tau = \tau_0$.

$\therefore$ $$\tau_0 = \rho_w \cdot K^2 \cdot y^2 \cdot \left(\frac{dV}{dy}\right)^2 \quad \text{or} \quad \sqrt{\frac{\tau_0}{\rho_w}} = K \cdot y \cdot \frac{dV}{dy}$$

But $$\sqrt{\frac{\tau_0}{\rho_w}} = V^* \qquad \therefore \quad V^* = K \cdot y \cdot \frac{dV}{dy}$$

or $$\boxed{\frac{dV}{dy} = \frac{V^*}{K \cdot y}} \qquad \text{...(4.35)}$$

The equation (4.35) gives the well known law of logarithmic velocity distribution ; given by

$$V = \frac{V^*}{K} \log_e y + \text{constant} \qquad \text{...(4.36)}$$

But, in fact, the shear stress at any distance $y$ from the boundary is given by

$$\tau = \tau_0 \left(1 - \frac{y}{D}\right) \qquad \text{...(4.37)}$$

where $D$ is the total depth.

Also assume that at sufficient distance ($y$) from the boundary, the mixing length is given by

$$l = K \cdot y \cdot \sqrt{1 - \frac{y}{D}}$$

Now, $$\varepsilon = l^2 \cdot \frac{dV}{dy} = K^2 \cdot y^2 \cdot \left(1 - \frac{y}{D}\right) \cdot \frac{dV}{dy}$$

But $$\frac{dV}{dy} = \frac{V^*}{K \cdot y} \text{ from equation (4.35)}$$

$\therefore$ $$\varepsilon = K^2 \cdot y^2 \cdot \left[1 - \frac{y}{D}\right] \cdot \frac{V^*}{Ky}$$

or $$\varepsilon = V^* \, Ky \left[1 - \frac{y}{D}\right] \qquad \text{...(4.38)}$$

The differential equation (4.33) was

$$w_0 c + \varepsilon \cdot \frac{dc}{dy} = 0$$

Putting the value of ε from Eqn. (4.38) in the above equation, we get :

$$w_0 c + V^* K{\cdot}y\left[1 - \frac{y}{D}\right] \cdot \frac{dc}{dy} = 0 \quad \text{or} \quad w_0 c + V^* K{\cdot}y\left[\frac{D-y}{D}\right] \cdot \frac{dc}{dy} = 0$$

or $$D \cdot w_0 c + V^* Ky\,(D-y) \cdot \frac{dc}{dy} = 0 \quad \text{or} \quad D \cdot w_0 c = -\,V^* Ky\,(D-y) \cdot \frac{dc}{dy}$$

or $$\frac{D{\cdot}dy}{y\,(D-y)} = -\frac{V^* K}{w_0} \cdot \frac{dc}{c}$$

Integrating, we get

$$D \cdot \int \frac{dy}{y\,(D-y)} = -\frac{V^*{\cdot}K}{w_0} \int \frac{dc}{c} \qquad ...(4.39)$$

Let us evaluate the L.H.S. integral first, *i.e.*

$$D \cdot \int \frac{dy}{y\,(D-y)}.$$

Let $$\frac{1}{y\,(D-y)} = \frac{A}{y} + \frac{B}{D-y} \quad \text{or} \quad A\,(D-y) + B{\cdot}y = 1$$

Equating coefficients on both sides, we get

$$AD = 1, \quad \text{or} \quad A = \frac{1}{D}$$

and $$-A + B = 0 \quad \text{or} \quad B = A = \frac{1}{D}$$

∴ $$\frac{1}{y\,(D-y)} = \frac{1}{D{\cdot}y} + \frac{1}{D\,(D-y)} = \frac{1}{D}\left[\frac{1}{y} + \frac{1}{D-y}\right]$$

or $$D\int \frac{dy}{y\,(D-y)} = D \cdot \frac{1}{D} \cdot \int \left[\frac{1}{y} + \frac{1}{D-y}\right] dy = \int \frac{1}{y}\,dy + \int \frac{1}{D-y}\,dy$$

$$= \log_e y - \log_e \overline{D-y} = \log_e \left(\frac{y}{D-y}\right)$$

Putting this value in Eqn. (4.39), we get

$$\log_e \left(\frac{y}{D-y}\right) = -\frac{V^* K}{w_0} \int \frac{dc}{c}$$

or $$\log_e \left(\frac{y}{D-y}\right) = -\frac{V^* K}{w_0} \log_e c + \text{constant of integration}$$

or $$\log_e \left(\frac{y}{D-y}\right) = -\frac{V^* K}{w_0} \log_e c + K'$$

or $$-\log_e c = \frac{w_0}{V^* K} \log_e \left(\frac{y}{D-y}\right) + K'' \qquad ...(i)$$

If $y = a$, $c = c_a$

∴ $$-\log_e c_a = \frac{w_0}{KV^*} \log_e \left(\frac{a}{D-a}\right) + K''$$

or $$K'' = -\log_e c_a - \frac{w_0}{KV^*}\log_e\left(\frac{a}{D-a}\right) \quad ...(ii)$$

Putting this value of $K''$ in $(i)$, we get

$$-\log_e c = \frac{w_0}{KV^*}\log_e\left(\frac{y}{D-y}\right) - \log_e c_a - \frac{w_0}{KV^*}\log_e\left(\frac{a}{D-a}\right)$$

or $$\log_e c - \log_e c_a = \frac{w_0}{KV^*}\log_e\left(\frac{a}{D-a}\right) - \frac{w_0}{KV^*}\log_e\frac{y}{D-y}$$

or $$\log_e\frac{c}{c_a} = \frac{w_0}{KV^*}\left[\log_e\frac{a}{D-a}\times\frac{D-y}{y}\right] = \log_e\left[\frac{a(D-y)}{y\,(D-a)}\right]^{w_0/KV^*}$$

or $$\boxed{\frac{c}{c_a} = \left[\frac{a\,(D-y)}{y\,(D-a)}\right]^{w_0/KV^*}} \quad ...(4.40)$$

where, $c$ is the sediment concentration at a distance $y$, and this can be determined by the above **Rouse equation,** by knowing the sediment concentration ($c_a$) at a known distance '$a$' apart ; $D$ being the total water depth.

$w_0$ is the fall velocity of a grain in still water, and $V^*$ is the **shear friction velocity,** given by

$$V^* = \sqrt{\frac{\tau_0}{\rho_w}} \quad ...(4.41)$$

where $\tau_0$ is the shear stress at the bottom, and $\rho_w$ is the density of water.

$K$ is the Von-Karman Universal Constant = 0.4.

$D$ is the total water depth.

The equation (4.40) is known as the *suspended load concentration equation.* The limitation of this equation is that it cannot be used directly to predict the sediment concentration at any point, unless the sediment concentration at some known distance '$a$' above the channel bed is preknown. Moreover, this equation has been derived on the assumptions of : (*i*) *two dimensional steady flow;* (*ii*) Constant fall velocity ($w_0$), and (*iii*) Constant Value of Karman's Constant (K) ; although however, it is a known fact today, that both $w_0$ as well as $K$ vary with sediment concentration and turbulence.

Inspite of its above limitations, this equation (4.40) remains an important equation for determining sediment concentration in a channel at any depth $y$ above the bed of the channel. The computation of sediment concentrations at different depths above the bed of the channel, can further be used to evaluate the total suspended load transported by the channel, as discussed below :

**Rate of suspended load transport.** As stated above, equation (4.40) will help us to compute the sediment concentrations at different depths, and thus plot a curve between sediment concentration ($c$) and depth ($y$) as shown in Fig. 4.14 ($a$). This curve is known as **sediment concentration curve.** When once this variation of sediment concentration with depth is known, it can be integrated over the depth to obtain total suspended load. This integration also requires plotting of **velocity distribution curve,** which is shown in Fig. 4.14 ($b$).

With reference to Fig. 4.14, let us consider a flow strip of unit width and of thickness $dy$ at an elevation $y$.

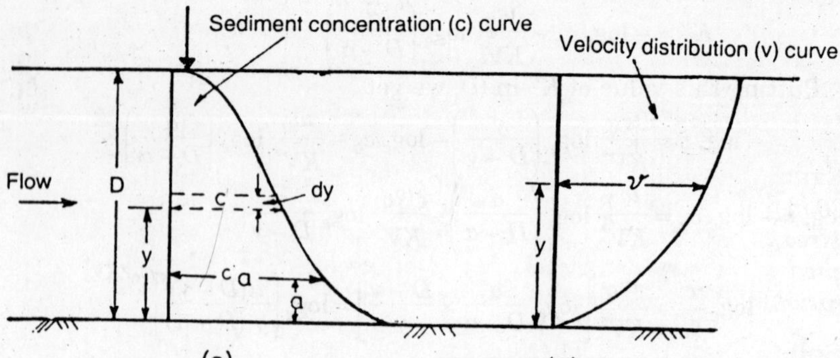


Fig. 4.14. Sediment concentration and Velocity distribution curves for a channel.

*The volume of suspended sediment transported past this strip in a unit time*

$$= \left(\frac{c}{100}\right) v \cdot dy \quad \text{...(4.41)}$$

where $c$ = The sediment concentration (volume expressed as percentage at any elevation $y$, where the flow velocity is $v$.

or
$$q_s = \int \frac{c}{100} \cdot v \cdot dy \quad \text{...[4.42 (a)]}$$

Since the suspended sediment moves only on top of the bed layers, the lower limit of integration can be considered as equal to the thickness of the bed layer, which is approximated to be equal to $2d$* ($d$ being grain size), and the upper limit will certainly be $D$ (total water depth). Hence,

$$q_s = \int_{y=2d}^{y=D} \frac{c}{100} \cdot v \cdot dy \quad \text{...(4.42)}$$

where $q_s$ = Rate of suspended load transported in $m^3$/sec.

The reference concentration $c_a$ required to compute various values of $c$ by Eq. (4.40), as pointed out earlier, needs to be predetermined for the above computations. When this depth '$a$' is considered equal to '$2d$', the sediment concentration $c_a$ or '$c_{2d}$' can be considered to be equal to the concentration of bed load, since bed load is assumed to be existing up to '$2d$' above the bed*. On the basis of computations carried out for bed load rate ($q_b$) in $m^3$/sec, discussed in the next article, Einstein has further computed this sediment concentration $c_a$ at depth $a \neq 2d$ above bed, as equal to:

$$c_a = c_{2d} = \frac{q_b}{23.2 V^{*\prime} d} \times 100 \quad \text{...(4.43)}$$

where $V^{*\prime}$ = shear friction velocity

$$= \sqrt{\frac{\tau_0}{\gamma_w}} = \sqrt{\frac{\gamma_w \cdot R' \cdot S}{\rho_w}}$$

$$= \sqrt{\frac{\rho_w g \cdot R' S}{\rho_w}} = \sqrt{g R' S}. \quad \text{...(4.44)}$$

---

* Sediment confined along and above the bed, up to a depth $2.d$ ($d$ being grain size), is treated as bed load ($q_b$).

where $R'$ is the modified value of $R$ due to ripples, as explained in the next article, and equals $R' = R \cdot \left(\frac{n'}{n}\right)^{3/2}$ as per equation (4.47).

**Example 4.12.** *In a wide stream, a suspended load sample taken at a height of 0.30 m from the bed indicated a concentration of 1000 ppm of sediment by weight. The stream is 5.0 m deep and has a bed slope of 1/4000. The bed material can be assumed to be of uniform size with a fall velocity of 2.0 cm/s. Estimate the concentration of the suspended load at mid-depth.* (*Civil Services, 1991*)

**Solution.** Using Eq. (4.40), we have

$$\frac{c}{c_a} = \left[\frac{a\,(D-y)}{y\,(D-a)}\right]^{\frac{w_0}{KV^*}} \qquad \text{(Eq. 4.40)}$$

where $D$ = 5 m

$S = \frac{1}{4000}$

$y$ = up distance from bed of channel, where sediment concentration $c$ is desired *i.e.* at mid depth $= \frac{5}{2} = 2.5$ m

$c_a$ = known sediment concentration = 1000 ppm (of sediment by wt.)

$a$ = known distance where sediment concentration is $c_a$ = 0.3 m (given)

$w_0$ = fall velocity of sediment grain in still water = 2.0 cm/s (given) = 0.02 m/s

$K$ = Von-karman constant = 0.4 (assumed)

$V^*$ = shear friction velocity

$$= \sqrt{\frac{\tau_0}{\rho_w}} = \sqrt{\frac{\gamma_w RS}{\rho_w}} = \sqrt{\frac{\rho_w\, gRS}{\rho_w}} = \sqrt{g \cdot RS}$$

where $R = D$ for wide channels = 5 m.

or $$V^* = \sqrt{9.81 \times 5 \times \frac{1}{4000}} = 0.111 \text{ m/s}$$

$c$ = ?

Using the above equation (4.40), we have

$$\frac{c}{1000 \text{ ppm}} = \left[\frac{0.3\,(5-2.5)}{2.5\,(5-0.3)}\right]^{\frac{0.02}{0.4 \times 0.111}} = \left(\frac{0.75}{11.75}\right)^{0.45} = 0.29$$

or $c$ = **290 ppm. Ans.**

### 4.9. Bed Load and Its Measurement

Bed load, as explained earlier, is the sediment material that remains in the bottom layers of the flow, and its movement takes place by *rolling, sliding,* and *hopping* (*i.e.* **saltation**) depending upon the velocity of flow. The bed load can be measured, though quite unsatisfactorily, by various **samplers,** such as *box type sampler, slot type sampler,* etc. Bed load is also sometimes estimated by assuming it to be between 3 to 25% of the total suspended load, depending upon the nature of response of the load material to the forces from physical, chemical and biological sectors. A figure of 10% is more commonly adopted.

**Mathematical equations for bed load.** The suspended load is the sediment which remains suspended in the water flowing in the channel, and this is caused by the forces of turbulence generated by the flow. The bed load, on the other hand, is the sediment which moves along the bed of the channel, and this movement is caused by the shear stress ($\tau_0$) developed by the flowing water along the channel bed. This shear force, called the *drag force* or the *tractive force* is thus fully responsible for the bed load movement. Evidently, the amount of bed load transported in a channel of given $R$ and $S$ will depend on the shear stress ($\tau_0$) developed on the bed of that channel. This bed shear or unit tractive force ($\tau_0$) is given by the already indicated equation (4.1) as :

$$\tau_0 = \gamma_w RS$$

where $R$ = Hydraulic mean depth of the channel $\approx y$ (or $D$) for wider channels.

$S$ = Bed slope

$\gamma_w$ = Unit wt. of water = 9.81 kN/m$^3$.

= $9.81 \times 10^3$ N/m$^3$

We further know that a certain minimum value of shear stress is required to move the grain, depending upon the internal friction of soil. It is called critical shear stress and is represented by $\tau_c$. For usual turbulent flow and for quartz grains, the value of $\tau_c$ has been approximated as

$$\tau_c = 0.687\, d_a \qquad ...(4.45)$$

where $d_a$ = arithmatic average diameter of sediment in mm (generally varying between $d_{50}$ and $d_{60}$).

$\tau_c$ = critical shear stress in N/m$^2$.

when the unit tractive force caused by the flowing water ($\tau_0$) exceeds the critical unit tractive force ($\tau_c$) ; naturally, sediment starts moving. The rate of bed-load transported, must, therefore, be a function of ($\tau_0 - \tau_c$). But the problem becomes complicated, when we take into account the fact that as soon as the grain starts moving, the channel bed develops ripples, and a large part of shearing force is absorbed by the form resistance caused by these ripples. A part of the tractive force is, therefore, lost in overcoming ripples, and it does not play any role in transporting bed material. The quantity of shear stress lost in this process is unknown, and no perfect mathematical solution has been but forward to work out this quantity.

It has been widely suggested that the tractive force is reduced by ripples, in the following ratio :

$$\tau_0' = \tau_0 \left[\frac{n'}{n}\right]^{3/2} \qquad ...(4.46)$$

where $\tau_0'$ = Resultant unit tractive force, left after the ripple resistance is overcome (*i.e.*, $\tau_0$ − tractive force lost in overcoming ripples).

$\tau_c$ = Original tractive force exerted by the water flowing in the channel of given $R$ and $S$, and is equal to $\gamma_w RS$.

$n'$ = Rugosity coefficient that should come into play, theoretically, in an unrippled channel of given $R$ and $S$. Its value may be obtained by Strickler's formula i.e.

$n' = \frac{1}{24} d^{1/6}$, where $d$ is the effective grain diameter *i.e.* **median size** ($d_{50}$) of the bed sediment in metres.

$n =$ Rugosity coefficient, actually observed by experiments on the rippled bed of the channel and its value is generally taken as 0.020 for discharges over 11 cumecs, and 0.0225 for smaller canals.

From Eqn. (4.46), and using $\tau_0 = \gamma_w RS$ from Eq. (4.1), we get

$$\tau_0' = \gamma_w RS \cdot [n'/n]^{3/2}$$

or $$\tau_0' = \gamma_w \left[R\,(n'/n)^{3/2}\right] S$$

or $$\tau_0' = \gamma_w R'S.$$

$$\text{where } R' = R \cdot (n'/n)^{3/2} \quad \text{...(4.47)}$$

where $R'$ is the corresponding hydraulic mean depth that would exist in the channel, if the bed was unrippled. In other words, if we use the value of $R'$ in all our calculations instead of $R$, we can forget about bed ripples, *i.e.* the effect of ripples is only to reduce the hydraulic mean depth to a value $R'$ from $R$.

Certain empirical formulas, used these days for determining bed load transport, are given below. The Einstein's semi-theoretical formula based on various assumptions is the most important and widely used these days.

**4.9.1. Empirical Formula by DU-Bois.** The first equation on the rate of bed load transport was proposed by DU-Bois, who assumed that the rate of bed load transportation was proportional to the excess of prevailing tractive force over the critical value required to initiate movement. Thus,

$$q_b = K_b \cdot \tau_0 (\tau_0 - \tau_c) \quad \text{...(4.48)}$$

where $q_b =$ Bed load (volume) transported in m$^3$ per second per unit width of channel.

$\tau_0 =$ Average shear stress on the channel boundary (N/m$^2$).

$\tau_c =$ Minimum shear stress required to move the grain, called critical shear stress (N/m$^2$).

$K_b =$ a constant depending upon the grain size, and given as :

$K_b = \frac{1.798 \times 10^{-3}}{(d)^{3/4}}$, where $d$ is the effective grain diameter *i.e.* median size ($d_{50}$) of the bed sediment, in mm ...(4.48*a*)

**4.9.2. Shield's Formula (1936).** A dimensionally homogeneous equation for sediment of uniform size, taking into account the effect of specific gravity of sediment ($S_s$), was proposed by shield as :

$$\frac{qS_s}{q_b} = 10\left[\frac{\tau_0 - \tau_c}{\gamma_w \cdot d\,(S_s - 1)}\right] \quad ...(4.49)$$

where $q_b$ = Bed load transported in m³/sec per m width of channel

$S_s$ = Specific gravity of the bed grain

$q$ = Discharge per unit width in m²/sec

$\gamma_w$ = Unit weight of fluid in kN/m³

$d$ = Dia. of bed grain in m

$\tau_o$ and $\tau_c$ are stresses in kN/m².

**4.9.3. Meyer-Peter's Formula (1948).** On the lines already discussed, Meyer and Peter has suggested that the unit tractive force causing bed load to move, is reduced by ripples, in the ratio of

$$\tau_0' = \tau_0 \cdot \left(\frac{n'}{n}\right)^{3/2} \quad ...(4.50)$$

The effective unit tractive force going to cause bed load transportation, is then given by

$$\tau_{eff} = \left[\tau_0\left(\frac{n'}{n}\right)^{3/2} - \tau_c\right]$$

On these concepts, Meyer and Peter (1948) has suggested the following formula for calculating the quantity of bed load transport

$$\boxed{g_b = 0.417\left[\tau_0\left(\frac{n'}{n}\right)^{3/2} - \tau_c\right]^{3/2}} \quad ...(4.51)$$

where $g_b$ is the rate of bed load transport (by wt.) in N per m width of channel per second. ($g_b = q_b \cdot \gamma_w \cdot S_s$)

where $q_b$ = Vol. of sediment transport per metre width of channel per second.

$\gamma_w$ = Unit wt. of water.

$S_s$ = Sp. gravity of grain.

$n'$ = Manning's coefficient pertaining to grain size on an unrippled bed and given by **Strickler's formula**

*i.e.* $n' = \frac{1}{24} \cdot d^{1/6}$ where $d$ is the median size ($d_{50}$) of the bed sediment in metres.

$n$ = the actual observed value of rugosity coefficient on rippled channels. Its value is generally taken as 0.020 for discharges of more than 11 cumecs, and 0.0225 for lower discharges.

$\tau_c$ = Critical shear stress required to move the grain in N/m², and given by equation (4.45) as

$\tau_c = 0.687\, d_a$ ;

where $\tau_c$ is in N/m² and $d_a$ is the mean or average size of the sediment in mm. This arithmetic average size is usually found to vary between $d_{50}$ and $d_{60}$.

$\tau_0$ = Unit tractive force produced by the flowing water *i.e.* $\gamma_w RS$. Truly speaking, its value should be taken as the unit tractive force produced by the flowing water on bed = 0.97 $\gamma_w\, RS$.

**Example 4.13.** *Design a channel which has to carry 25 cumecs with a bed load concentration of 40 p.p.m. by wt. The median grain diameter of the bed material may be taken as 0.3 mm. Use, Lacey's Regime perimeter and Meyer-peter's formulas.*

**Solution.** Quantity of bed load transported by wt. is 40 parts per million of flowing water

$$= \frac{40}{10^6} \text{ [wt. of flowing water]}$$

Quantity (wt.) of bed load transported per second

$$= \frac{40}{10^6}\left[25\,\frac{\text{m}^3}{\text{sec}} \times 9.81 \times 10^3\,\frac{\text{N}}{\text{m}^3}\right] = 9.81 \text{ N/sec.}$$

Lacey's Regime perimeter of the channel ($P$)

$$= 4.75 \cdot \sqrt{Q} = 4.75 \cdot \sqrt{25} = 23.75 \text{ m.}$$

Let the channel bed width ($B$) be kept as 21 m.

Hence, the rate of bed load transport per unit width (by wt.)

$$g_b = \frac{9.81}{21} \text{ N/m/sec.} = \mathbf{0.467\ N/m/sec.}$$

Meyer Peter's equation (4.51) is

$$g_b = 0.417 \cdot \left[\tau_0 \cdot \left(\frac{n'}{n}\right)^{3/2} - \tau_c\right]^{3/2}$$

where $n' = (1/24) \cdot d^{1/6}$

where $d$ is median size of sediment in metres $= \frac{0.3}{1000}$ m (given)

$$\therefore \quad n' = \frac{1}{24} \cdot \left(\frac{0.3}{1000}\right)^{1/6} = 0.0108$$

The value of $n$ may be taken as 0.020, because the discharge is more than 11 cumecs or so, and the channel may be taken in good shape and smooth soil.

$$\therefore \quad \frac{n'}{n} = \frac{0.0108}{0.0200} = \mathbf{0.54}$$

From equation (4.45),

$$\tau_c = 0.687 \cdot d_a$$

where $d_a$ is the average particle size in mm ; and we can assume it to be either 0.3 mm or slightly more than that, since mean size is usually found to vary between $d_{50}$ and $d_{60}$. Let us assume it to be equal to $d_{50}$ *i.e.* 0.3 mm.

$$\therefore \quad \tau_c = 0.687 \times 0.3 = \mathbf{0.206\ N/m^2}$$

Putting these values in Meyer-Peter's formula, we get

$$g_b = 0.417\left[\tau_0 \cdot \left(\frac{n'}{n}\right)^{3/2} - \tau_c\right]^{3/2} \text{ N/m/sec.} \qquad \text{where } \tau_0 = \gamma_w RS$$

$$\therefore \quad 0.467 = 0.417\left[\gamma_w RS \cdot \left(\frac{n'}{n}\right)^{3/2} - \tau_c\right]^{3/2}$$

or $$0.467 = 0.417\left[(9.81 \times 10^3) \times RS\,(0.54)^{3/2} - 0.206\right]^{3/2}$$

$$0.467 = 0.417 \times (9.81)^{3/2}\left[1000RS\,(0.54)^{3/2} - 0.021\right]^{3/2}$$

or $$\left(\frac{0.467}{0.417 \times (9.81)^{3/2}}\right)^{2/3} = \left[1000\,RS\,(0.54)^{3/2} - 0.021\right]$$

or $$(0.0365)^{0.67} = 1000 \times 0.397\,RS - 0.021$$

or $$0.11 = 397\,RS - 0.021$$

or $$0.131 = 397\,RS$$

or $$\mathbf{RS = 0.00033} \qquad \text{...}(i)$$

Using Manning's equation

$$Q = \frac{1}{n} \cdot AR^{2/3} \cdot S^{1/2}$$

$$\text{where } \frac{A}{P} = R\,;\ \text{or } A = P \cdot R = 23.75 \cdot R$$

$$Q = 25 = \frac{1}{0.020} \times (23.75\,R)\,R^{2/3} \cdot S^{1/2}.$$

or $$\frac{25 \times 0.02}{23.75} = R^{5/3} \cdot S^{1/2}$$

or $$0.021 = R^{5/3} \cdot S^{1/2} \qquad \text{...}(ii)$$

Solving (*i*) and (*ii*), we get

$$RS = 0.00033\,;\quad S = \frac{0.00033}{R}$$

$$\therefore \quad 0.021 = R^{5/3} \times \left(\frac{0.00033}{R}\right)^{1/2} \quad \text{or} \quad \frac{0.021}{\sqrt{0.00033}} = R^{7/6}$$

$$R = \mathbf{1.13\ m}$$

$$\therefore \quad S = \frac{0.00033}{1.13} \quad \text{or} \quad S = \frac{1}{3424}\,;\ \text{say } \mathbf{\frac{1}{3400}}.$$

**Let $y$ be the depth of water in the trapezoidal channel of $\frac{1}{2}$ : 1 slopes**

$$\therefore \quad P = 21 + \sqrt{5} \cdot y$$

$$A = 21y + \frac{y^2}{2}$$

$$R = \frac{A}{P} = \frac{21y + \frac{y^2}{2}}{21 + 2.24 \cdot y} = 1.13 \text{ (worked out earlier)}$$

or $$42y + y^2 = 2.26\,(21 + 2.24\,y)$$

or $$y^2 + 42y = 47.46 + 5.06\,y$$

or $$y^2 + 36.94y - 47.46 = 0$$

or $$y = \frac{-36.94 \pm \sqrt{(36.94)^2 + 4 \times 47.46}}{2} = \frac{-36.94 \pm 39.43}{2}$$

$$= \frac{-36.94 + 39.43}{2} = \mathbf{1.25\ m} \qquad \text{(using only the feasible + ve sign)}$$

**Use Depth = 1.25 m.**

Hence, use the following channel dimensions.

$$\left.\begin{aligned} B &= 21 \text{ m} \\ y &= 1.25 \text{ m} \\ S &= \frac{1}{3400} \\ \text{Side slopes} &= \tfrac{1}{2}\,H : 1V. \end{aligned}\right\} \textbf{Ans.}$$

**4.9.4. Einstein's Formula (1950).** Einstein has put forward a semi-theoretical approach to the problem of bed load transport. In his formula, it is assumed, that every particle after it is dislodged from the bed, travels a certain minimum distance $L$ (which is proportional to the grain diameter) before coming to rest.

Now, let us take a bed area of length $L$ and of unit width. This is clear from the above assumption, that each and every grain that is dislodged from this area will cross the outer boundary ($AB$) of this area (because each particle has to travel a minimum distance ($L$). See Fig. 4.15.

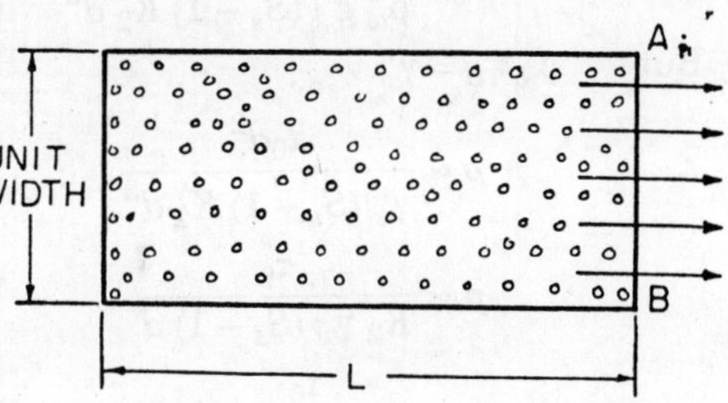


Fig. 4.15.

Now, if $K_1 d^2$ is the bed area covered by each grain (where $K_1$ is a constant depending upon the shape of the grain), then the number of grains dislodged from this area is given by

$$\frac{L \times 1}{K_1 d^2}$$

If $p$ is the probability of the grain being dislodged in the given second ($\because$ in one particular second, the particles will not be dislodged from the entire area), then the number of particles dislodged per second is given by

$$\frac{L \times 1}{K_1 d^2} \cdot p$$

If $K_2 d^3$ represents the volume of each grain, then the volume of sediment transported per second per unit width is given by

$$q_b = \frac{L \times 1}{K_1 d^2} \cdot p \cdot K_2 \cdot d^3 = \frac{K_2}{K_1} L \cdot p \cdot d.$$

But $L \propto d$ (by assumption made in the beginning)

or $$L = K_3 d$$

$$\therefore \quad q_b = \frac{K_2}{K_1} K_3\, p (d)^2 \qquad \text{...(4.52)}$$

Einstein further postulated that :

$$\text{the probability, } p \propto \frac{\text{Lift force which the flow can exert on the grain}}{\text{Submerged weight of the grain}}$$

But, Lift force $\propto \rho_w V^2 d^2 = \tau_0 d^2$

and Submerged weight of the grain $\propto$ [Submerged density× volume) $\cdot g$

$$\left[ \because \begin{array}{r} \text{Mass} = \text{Volume} \times \text{Density} \\ \text{Weight} = \text{Mass} \times g. \end{array} \right]$$

$\therefore$ Submerged weight of the grain $\propto [\rho_w (S_s - 1) g (K_2 d^3)]$

Using these two values, we get

$$p \propto \frac{\tau_0 d^2}{\rho_w (S_s - 1) g K_2 d^3}$$

or $$p \propto \frac{\tau_0 d^2}{\rho_w g \cdot (S_s - 1) K_2 d^3}$$

But $\rho_w \cdot g = \gamma_w.$

$$\therefore \quad p \propto \frac{\tau_0 d^2}{\gamma_w (S_s - 1) K_2 \cdot d^3}$$

or $$p \propto \frac{\tau_0}{K_2 \gamma_w (S_s - 1) d}$$

or $$p \propto \frac{\tau_0}{\gamma_w \cdot d (S_s - 1)} \qquad \text{...(4.53)}$$

But $$\frac{\tau_0}{\gamma_w d (S_s - 1)} = F_s = \text{Shield's entrainment function}$$

therefore, $p \propto F_s.$

But $F_s$ is dimensionless and $p$ has the units of time$^{-1}$, therefore, introducing time, we get

$$p = \frac{1}{t} f(F_s), \qquad f(F_s) \text{ means, a function of } F_s.$$

Further, it was argued by Einstein that

$$t = \frac{d}{w_0} \text{ i.e., time } (t) = \frac{\text{Distance travelled equal to diameter}}{\text{Fall velocity}}$$

Therefore, $$p = \frac{w_0}{d} f(F_s)$$

or $$p = f(F_s)\left[\frac{w_0}{d}\right] \quad \text{...(4.54)}$$

Now Eq. (4.52) reduces to

$$q_b = \frac{K_2}{K_1} K_3 \cdot f(F_s)\left[\frac{w_0}{d}\right] \cdot d^2$$

or $$\frac{q_b}{w_0 d} = \frac{K_2}{K_1} K_3 \cdot f(F_s)$$

or $$\frac{q_b}{w_0 d} = \text{F}(F_s) \text{ (some other function of } (F_s)) \quad \text{...(4.55)}$$

Now, $\frac{q_b}{w_0 d}$ is generally represented by $\phi$, and $F_s$ is represented by $\frac{1}{\psi}$; where $\phi$ is known as **Einstein's bed load function.**

Therefore, we get,

$$\phi = F\frac{1}{\psi} \quad \text{...(4.56)}$$

It was also given by Einstein, that

$$w_0 = G\sqrt{gd(S_s - 1)} \quad \text{...(4.57)}$$

where $$G = \sqrt{\frac{2}{3} + \frac{36\nu^2}{gd^3(S_s - 1)}} - \sqrt{\frac{36\nu^2}{gd^3(S_s - 1)}} \quad \text{...(4.58)}$$

where $\nu$ is the kinematic viscosity of water in $m^2$/sec.
when $d \geq 1.6$ mm, $G \approx \sqrt{(2/3)} \approx 0.81$

The above $\phi$–$\psi$ relationship is represented by the curve, shown in Fig. 4.16, and can be used to compute bed load transport rate of a given channel; or conversely to design stable channels for a given bed load transport.

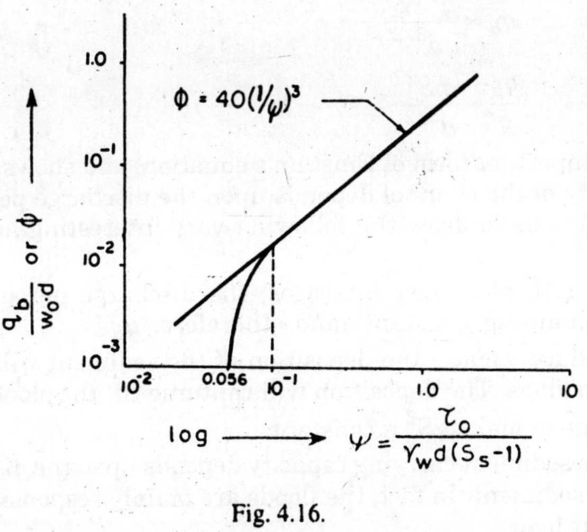


Fig. 4.16.

The straight line portion of this curve, follows the relation

$$\phi = 40 \cdot \left(\frac{1}{\psi}\right)^3 \qquad ...(4.59)$$

or

$$\boxed{\frac{q_b}{w_0 d} = 40 \cdot \left[\frac{\tau_0}{\gamma_w d\,(S_s - 1)}\right]^3} \qquad ...(4.60)$$

Now, $w_0 = G \cdot \sqrt{gd\,(S_s - 1)}$

$$\therefore \quad \frac{q_b}{G \cdot d\,\sqrt{gd\,(S_s - 1)}} = 40\left[\frac{\tau_0}{\gamma_w d\,(S_s - 1)}\right]^3 \qquad ...(4.61)$$

But $\tau_0 = \gamma_w RS$.

$$\therefore \quad \frac{q_b}{G \cdot d\,\sqrt{gd\,(S_s - 1)}} = 40\left[\frac{\gamma_w\,RS}{\gamma_w \cdot d\,(S_s - 1)}\right]^3$$

or

$$\frac{q_b}{G \cdot d\,\sqrt{gd\,(S_s - 1)}} = 40\left[\frac{R^3 \cdot S^3}{d^3\,(S_s - 1)^3}\right]$$

or

$$q_b \propto \frac{R^3 \cdot S^3 \cdot d\sqrt{d}}{d^3}$$

or

$$q_b \propto \frac{R^3 S^3}{d^{3/2}} \qquad ...(4.62)$$

Now, if it is assumed that the channel is wide and Chezy's $C$ is constant, then

$$Q = C \cdot A\,\sqrt{RS}$$

$\therefore \quad q = C \cdot y \cdot \sqrt{RS}$ But $R \approx y$ for wide channels or rivers.

$$\therefore \quad q = C \cdot R \cdot \sqrt{RS} = C \cdot R^{3/2}\,S^{1/2}$$

or

$$q^2 \propto R^3 \cdot S \qquad ...(4.63)$$

From Eq. (4.62), $q_b \propto \dfrac{R^3 \cdot S \cdot S^2}{d^{3/2}}$

or

$$q_b \propto \frac{q^2 \cdot S^2}{d^{3/2}} \qquad ...(4.64)$$

or

$$\frac{q_b}{q} \propto \frac{q \cdot S^2}{d^{3/2}} \qquad ...(4.65)$$

This is an important form of Einstein's equation, and shows that the sediment carrying capacity of the channel depends upon the discharge per unit width. This relationship helps us to draw the following very interesting and important conclusions:

(1) If the width of a river increases, the discharge per unit width ($q$) will decrease ($Q$ remaining constant) and, therefore, $q_b/q$ *i.e.*, sediment carrying capacity will reduce. Hence, the deposition of the sediment will start, which will increase the bed slope. The deposition will continue till the slope $S$ is increased to a value sufficient to make $qS^2$ a constant.

(2) Since the sediment carrying capacity depends upon the discharge, the floods will carry more sediment. In fact, the floods are mainly responsible for most of the annual sediment load.

(3) If a branch channel starts from a main channel, then an interesting phenomenon may happen, as described below :

Let a branch channel off-takes from a main channel, as shown in Fig. 4.17.

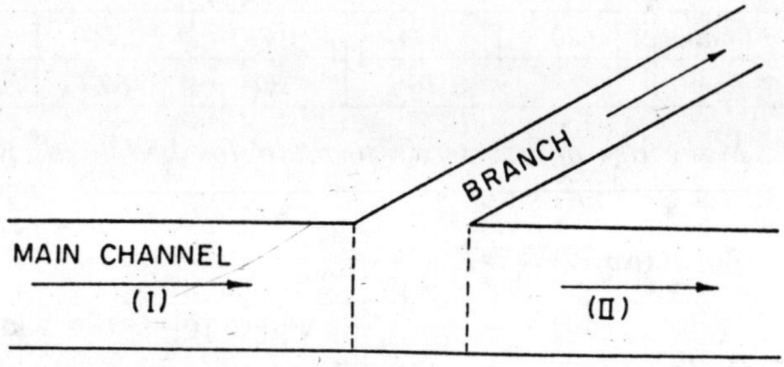

Fig. 4.17.

Now, after the branch channel starts extracting some discharge, the discharge in the main channel reduces, thereby reducing its sediment carrying capacity. Hence, deposition of sediment will start, which further diverts the flow towards the off-taking channel, causing more reduction in discharge in the main channel and more silt deposition ; causing more inflow towards the branch channel. The process may continue till the entire water gets diverted towards the branch channel.

The prime cause for this phenomenon is the existence of a favourable gradient in the branch channel, but the mechanism which keeps the process going, is necessarily, what is explained above by the Einstein's equation.

A typical example exhibiting the importance of this phenomenon is quoted below:

Mississippi river is connected to another old river Atchafalya by a branch channel, which off-takes from the Mississippi river at about 480 kilometres from its mouth, as shown in Fig. 4.18.

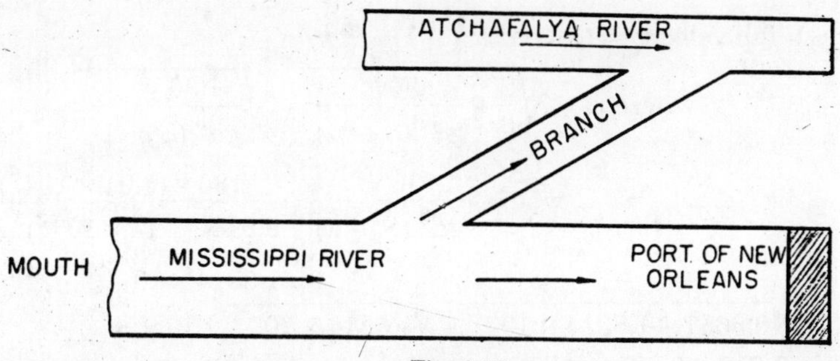

Fig. 4.18.

The branch channel has been capturing flow from the main channel for the last many decades. In fact, engineers have tried to promote this flow as a means of diverting flood waters. About 26% of discharge was flowing towards this channel at one stage, with tendency of further increase. But, if the proportion is allowed to become rather higher and reaches to a value of the order of 40% or so, the process would become almost irreversible, and the whole flow may be rapidly diverted towards the Atchafalya river through this branch channel. If such a state arises, the port of New Orleans (situated at the join of Mississippi river with sea) could become redundant.

**Example 4.14.** *Determine the rate of bed load transport in a wide alluvial stream for the following data :*

*Depth of flow* = *4.5 m*

*Velocity of flow* = *1.3 m/s*

*Bed slope* = $2.0 \times 10^{-4}$

*Size distribution of the sediment :*

| d (mm) | 0.20 | 0.44 | 0.78 | 1.14 | 1.65 | 3.6 | 5.2 |
|---|---|---|---|---|---|---|---|
| % finer | 2 | 10 | 30 | 50 | 70 | 80 | 100 |

*Make use of Einstein's formula for bed load. Kinematic viscosity of water = $1.01 \times 10^{-6}\ m^2/s$.*

**Solution.** $V = \frac{1}{n} \cdot R^{2/3} \cdot \sqrt{S}$

where $R \approx D$ (for wide streams) = 4.5 m (given)

$V = 1.3$ m/s (given)

$S = 2 \times 10^{-4}$ (given)

$\therefore \quad 1.3 = \frac{1}{n} \cdot (4.5)^{2/3} \cdot \sqrt{2 \times 10^{-4}}$

or $\quad n = \mathbf{0.02965}$

Also, $\quad n' = \frac{1}{24} \cdot (d_{50})^{1/6}$

where $d_{50} = 1.14$ mm from the given table

$= 1.14 \times 10^{-3}$ m

$= (1/24) \cdot (1.14 \times 10^{-3})^{1/6} = \mathbf{0.0135}$

$\frac{n'}{n} = \frac{0.0135}{0.02965}\ \mathbf{0.454}$

$R' = R \cdot (n'/n)^{3/2} = 4.5\ (0.454)^{3/2} = \mathbf{1.377\ m}$

Using equation (4.57), we have

$w_0 = \text{fall velocity} = G \cdot \sqrt{gd\ (S_s - 1)}$

where $G = \sqrt{\frac{2}{3} + \frac{36 \nu^2}{gd^3\ (S_s - 1)}} - \sqrt{\frac{36\ \nu^2}{gd^3\ (S_s - 1)}}$

where, $\frac{36\ \nu^2}{gd^3\ (S_s - 1)} = \frac{36 \times (1.01 \times 10^{-6})^2}{9.81 \times (1.14 \times 10^{-3})^3 \cdot 1.65}$

$= 1.5314 \times 10^{-3}$

$G = \sqrt{0.6667 + 1.5314 \times 10^{-3}} - \sqrt{1.5314 \times 10^{-3}}$

$= 0.817 - 0.039 = 0.778$

$\therefore \quad w_0 = 0.778 \sqrt{9.81 \times (1.14 \times 10^{-3})\ (1.65)} = \mathbf{0.105\ m/s}$

Using equation (4.60), we have

$$\frac{q_b}{w_0 \cdot d} = 40 \left[\frac{\gamma_w R'S}{\gamma_w d\ (S_s - 1)}\right] = 40 \left[\frac{R'S}{d\ (S_s - 1)}\right]^3 \qquad \because \quad \tau_0 = \gamma_w R'S$$

To appreciate bed ripples, $R'$ is used in place of $R$

or $$\frac{q_b}{0.105 \times 1.14 \times 10^{-3}} = 40 \left[\frac{1.377 \times 2.0 \times 10^{-4}}{1.14 \times 10^{-3}\ (1.65)}\right]^3$$

$\therefore$ Bed load by vol.

$= q_b = 1.5027 \times 10^{-5}$ m$^3$/sec/m width of channel.

*Bed load by weight* ($g_b$)

$$g_b = q_b \cdot \gamma_w \cdot S_s$$

$$= 1.5027 \times 10^{-5} \times (9.81 \times 10^3) \times 2.65 \text{ N/sec/m} = \mathbf{0.39 \ N/s/m} \quad \textbf{Ans.}$$

**Example 4.15.** *A wide irrigation channel is designed to have hydraulic mean depth of 3 m and bed slope equal to $1.6 \times 10^{-4}$. The bed sediment has an average median size of 0.3 mm. If the specific gravity of the bed soil is taken as 2.65 and the observed Manning's n to be 0.020, compute the rate of bed load transported by the channel in N/s/m width of channel.*

*Also compute the suspended load concentrations at different depths to plot the sediment concentration curve for this channel. Make use of Einstein's formulas. Kinematic viscosity of water at given temperature may be taken to be $1.01 \times 10^{-6}$ m²/s.*

**Solution.** Bed load transport rate as volume ($q_b$) is given by Einstein's equation (4.60) as :

$$\frac{q_b}{w_0 \cdot d} = 40 \left[ \frac{\gamma_w R'S}{\gamma_w \cdot d\,(S_s - 1)} \right]^3 = 40 \cdot \left[ \frac{R'S}{d\,(S_s - 1)} \right]^3$$

To appreciate bed ripples, $R'$ is used in place of $R$.

$$\text{where } w_0 = G \cdot \sqrt{gd\,(S_s - 1)} \quad \text{...(4.57)}$$

$$\text{where } G = \sqrt{\frac{2}{3} + \frac{36 \cdot \nu^2}{gd^3\,(S_s - 1)}} - \sqrt{\frac{36\,\nu^2}{gd^3\,(S_s - 1)}}$$

where $\nu$ = kinematic viscosity of water at 20° C = $1.01 \times 10^{-6}$.

$$\frac{36\,\nu^2}{gd^3\,(S_s - 1)} = \frac{36 \times (1.01 \times 10^{-6})^2}{9.81 \times (0.3 \times 10^{-3})^3\,(1.65)} = \frac{36 \times 1.01 \times 1.01 \times 10^{-12}}{9.81 \times (0.3)^3\,(1.65) \times 10^{-9}} = 0.084$$

$$\therefore \quad G = \sqrt{0.667 + 0.084} - \sqrt{0.084} = 0.867 - 0.290 = \mathbf{0.577}$$

$$w_0 = 0.577\,\sqrt{9.81 \times (0.3 \times 10^{-3})\,(1.65)} = \mathbf{0.04 \ m/sec.}$$

Using equation (4.60) :

$$\frac{q_b}{0.04 \times (0.3 \times 10^{-3})} = 40 \left[ \frac{R'S}{d\,(S_s - 1)} \right]^3$$

$$\text{where } R' = R \left( \frac{n'}{n} \right)^{3/2}$$

$$\text{where } n' = \frac{1}{24} \cdot d^{1/6} = \frac{1}{24}\,(0.3 \times 10^{-3})^{1/6} = 0.01078$$

$$n = 0.020 \text{ (given)}$$

$$\frac{n'}{n} = \frac{0.01078}{0.020} = 0.539$$

$$R = 3 \text{ m (given)}$$

$$R' = 3 \times (0.539)^{3/2} = \mathbf{1.187 \ m}$$

$$\therefore \quad \frac{q_b}{0.040 \times (0.3 \times 10^{-3})} = 40 \left[ \frac{1.187 \times (1.6 \times 10^{-4})}{(0.3 \times 10^{-3})\,(1.65)} \right]^3 = 40\,(0.3837)^3 = 2.259$$

or

$$q_b = (2.259 \times 0.040 \times 0.3 \times 10^{-3})$$

$$= \mathbf{2.711 \times 10^{-5}\ m^3/sec/m} \text{ width of channel}$$

$$g_b = q_b \cdot \gamma_w \cdot S_s$$

$$= 2.711 \times 10^{-5} \frac{m^3}{sec.m} \times \left(9.81 \times 10^3 \frac{N}{m^3}\right) \times 2.65$$

$$= 0.7 \text{ N/s/m width of channel.}$$

Hence, bed load transported per m width of channel

$$= \mathbf{0.7\ N/s/m\ width.\ Ans.}$$

For computing suspended load, we compute from Eq. (4.43),

$$c_a = c_{2d} = \frac{q_b}{23.2 \times V^{*\prime} \times d} \times 100\%$$

or
$$c_a = \frac{2.711 \times 10^{-5}}{23.2 \times V^{*\prime} \times (0.3 \times 10^{-3})} \times 100\%$$

$$\text{where } V^{*\prime} = \sqrt{\frac{\tau_0'}{\rho_w}} = \sqrt{\frac{\rho_w g R' S}{\rho_w}} = \sqrt{gR'S}$$

$$= \sqrt{9.81 \times 1.187 \times 1.6 \times 10^{-4}} = \mathbf{0.043\ m/sec}$$

$$\therefore \quad c_a = \frac{2.711 \times 10^{-5}}{23.2 \times 0.043 \times (0.3 \times 10^{-3})} \times 100\% = \mathbf{9.058\% = 90{,}580\ ppm.}$$

Now, using Eq. (4.40), we have

$$\frac{c}{c_a} = \left[\frac{a\,(D-y)}{y\,(D-a)}\right]^{\frac{w_0}{KV^*}}$$

$$\text{where } V^* = \sqrt{gRS}$$

$$= \sqrt{9.81 \times 3 \times 1.6 \times 10^{-4}} = 0.069 \text{ m/sec}$$

$$K = 0.4 \text{ (constant)}$$

$$a = 2d = 2 \times 0.3 \times 10^{-3} \text{ m}$$

$$\therefore \quad \frac{w_0}{KV^*} = \frac{0.04}{0.4 \times 0.069} = 1.457$$

$$\therefore \quad \frac{c}{9.058\%} = \left[\frac{(2 \times 0.3 \times 10^{-3})(3-y)}{y\,(3 - 2 \times 0.3 \times 10^{-3})}\right]^{1.457} = 2 \times 10^{-4}\left[\frac{3-y}{y}\right]^{1.457}$$

or
$$c = 1.812 \times 10^{-3}\left(\frac{3-y}{y}\right)^{1.457}$$

The different values of $c$ at different values of $y$, say 0.05 m, 0.1 m, 0.15 m, 0.2 m, 0.5 m, 1.0 m, 2.0 m, 3.0 m are now computed by using the above equation, as :

$$c_{0.05m} = 1.812 \times 10^{-3}\left(\frac{3-0.05}{0.05}\right)^{1.457} = 0.6891\% = 6891 \text{ ppm}$$

$$c_{0.1\,m} = 1.812 \times 10^{-3}\left(\frac{3-0.1}{0.1}\right)^{1.457} = 0.2447\% = 2447 \text{ ppm}$$

$$c_{0.15\,m} = 1.812 \times 10^{-3}\left(\frac{3-0.15}{0.15}\right)^{1.457} = 0.1264\% = 1264 \text{ ppm}$$

$$c_{0.2\,m} = 1.812 \times 10^{-3}\left(\frac{3-0.2}{0.2}\right)^{1.457} = 0.0847\% = 847 \text{ ppm}$$

$$c_{0.5\,m} = 1.812 \times 10^{-3}\left(\frac{3-0.5}{0.5}\right)^{1.457} = 0.0188\% = 188 \text{ ppm}$$

$$c_{1.0\,m} = 1.812 \times 10^{-3}\left(\frac{3-1.0}{1.0}\right)^{1.457} = 0.0050\% = 50 \text{ ppm}$$

$$c_{2.0\,m} = 1.812 \times 10^{-3}\left(\frac{3-2.0}{2.0}\right)^{1.457} = 0.0005\% = 5 \text{ ppm}$$

$$c_{2.5\,m} = 1.812 \times 10^{-3}\left(\frac{3-2.5}{2.5}\right)^{1.457} = 0.00025\% = 2.5 \text{ ppm}$$

$$c_{3\,m} = 1.812 \times 10^{-3}\left(\frac{3-3}{3}\right)^{1.457} = \text{zero.}$$

## DESIGN PROCEDURE FOR IRRIGATION CHANNELS

After having discussed the various theories for designing irrigation canals, we shall now discuss, as to how in actual practice, an irrigation canal is designed and constructed.

### 4.10. Cross-section of an Irrigation Canal

A typical and most desired section of a canal is shown in Fig. 4.19. This section is 'partly in cutting and partly in filling', and aims at balancing the quantity of earth work in 'excavation' with that in 'filling'.

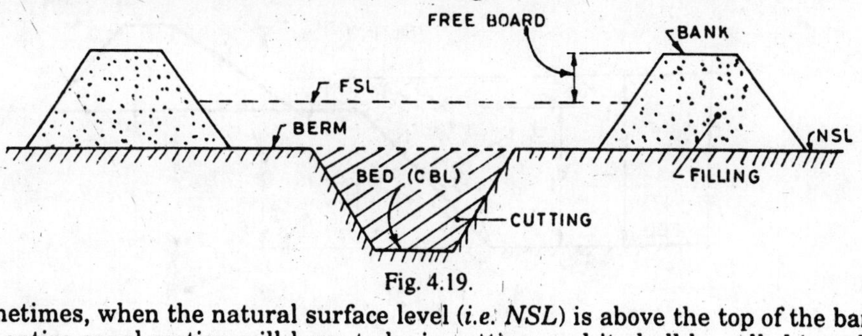


Fig. 4.19.

Sometimes, when the natural surface level (*i.e. NSL*) is above the top of the bank, the entire canal section will have to be in cutting, and it shall be called 'canal in cutting'. Similarly, when the *NSL* is lower than the Bed level of the canal, the entire canal section will have to be built in filling, and it is called 'canal in filling', or 'canal in banking'.

**Side Slopes.** The side slopes should be such that they are stable, depending upon the type of the soil. A comparatively steeper slope can be provided in cutting rather than in filling, as the soil in the former case shall be more stable.

**$1\,H : 1\,V$ (1 : 1) to $1\frac{1}{2}\,H : 1\,V$ ($1\frac{1}{2}$ : 1) slope in cutting, and $1\frac{1}{2}\,H : 1V$ to $2\,H : 1\,V$ in filling, are generally adopted.**

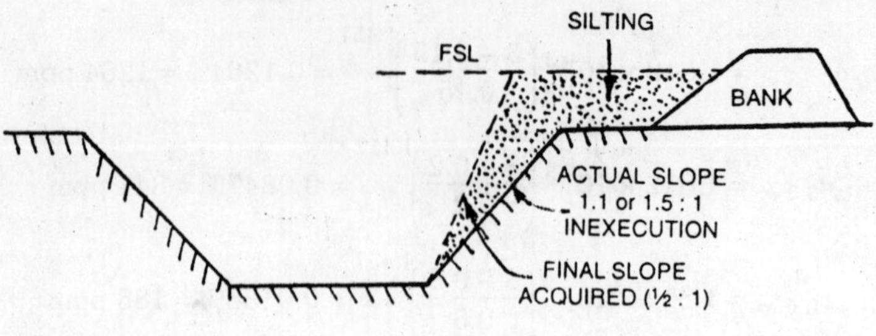


Fig. 4.20.

In case of channels with silt laiden water, the actual capacity of the channel is worked out with $\frac{1}{2}$ : 1 side slopes, even though flatter slopes such as 1 : 1 to $1\frac{1}{2}$ : 1 may be constructed at the time of execution. This is because of the fact that the sides of such a channel gets silted up to a slope $\frac{1}{2}$ : 1 with the passage of time, as shown in Fig. 4.20.

**Berms.** Berm is the horizontal distance left at ground level between the toe of the bank and the top edge of cutting.

The berm is provided in such a way that the bed line and the bank line remains parallel. If $s_1 : 1$ is the slope in cutting and $s_2 : 1$ in filling, then the initial berm width = $(s_2 - s_1)\, d_1$. Since ground level (NSL) fluctuates considerably, while canal bed level (CBL) varies very slightly, $d_1$ shall vary ; and, therefore, the berm width shall vary.

After the water flows in the channel for some time, the silt gets deposited on the sides giving them a slope of $\frac{1}{2}$ : 1. The position of the berm, therefore, changes from ground level to FSL, as shown in Fig. 4.21 and its width becomes equal to $(s_2 - \frac{1}{2}) \cdot y$. If $s_2 = 1\frac{1}{2}$, then the final berm width = $y$, *i.e.* equal to the depth of the canal.

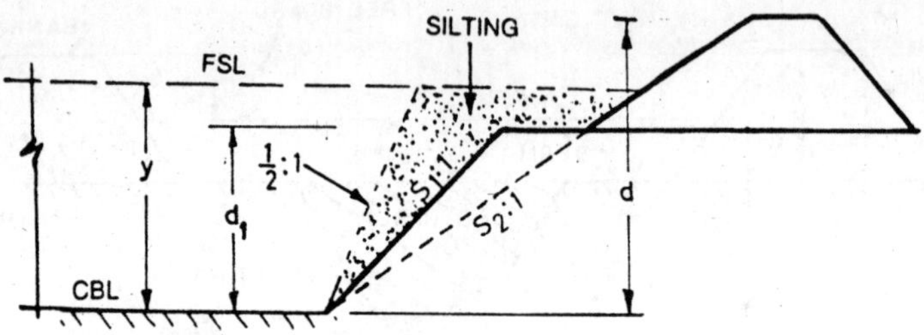


Fig. 4.21.

**The berms when fully formed, serve the following purposes :**

(*i*) The silt deposited on the sides is very fine and impervious. It, therefore, serves as a good lining for reducing losses, leakage and consequent breaches, etc.

(*ii*) They help the channel to attain regime conditions, as they help in providing a wider waterway, if required. Even fluctuations of discharge do not produce much fluctuations in depths because of wider waterway.

(*iii*) They give additional strength to the banks and provide protection against erosion and breaches.

(*iv*) The possibility of breaches gets reduced because the saturation line comes more in the body of the embankment.

(*v*) They protect the banks from erosion due to wave action.

(*vi*) They provide a scope for future widening of the canal.

(*vii*) Berms can be used as borrow pits for excavating soil to be used for filling.

**Freeboard.** The margin between FSL and bank level is known as freeboard. The amount of freeboard depends upon the size of the channel. The generally provided values of freeboard are given in Table 4.7.

**Table 4.7. Values of Free-boards in Canals**

| *Discharge in cumecs* | *Extent of freeboard in metres* |
|---|---|
| 1 to 5 | 0.50 |
| 5 to 10 | 0.60 |
| 10 to 30 | 0.75 |
| 30 to 150 | 0.90 |

**Banks.** The primary purpose of banks is to retain water. They can be used as means of communication and as inspection paths. They should be wide enough, so that a minimum cover of 0.5 metre is available above the saturation line, as shown in Fig. 4.22. High banks will have to be designed as earth dams.

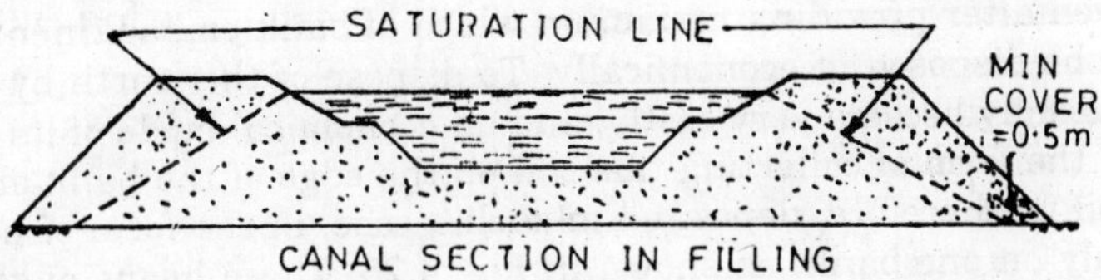


Fig. 4.22.

**Service Roads.** Service roads are provided on canals for inspection purposes, and may simultaneously serve as the means of communication in remote areas. They are provided 0.4 m to 1.0 m above FSL, depending upon the size of the channel.

**Dowlas.** As a measure of safety in driving, dowlas 0.3 m high and 0.3 to 0.6 m wide at top, with side slopes of $1\frac{1}{2} : 1$ to $2 : 1$, are provided along the banks, as shown in Fig. 4.23. They also help in preventing slope erosion due to rains, etc.

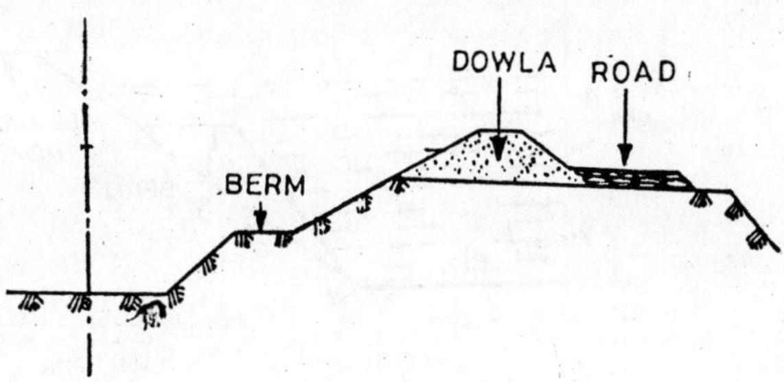


Fig. 4.23.

**Back Berm or Counter Berms.** Even after providing sufficient section for bank embankment, the saturation gradient line may cut the downstream end of the bank. In such a case, the saturation line can be kept covered at least by 0.5 metre with the help of counter berms, as shown in Fig. 4.24.

The straight saturation gradient line may be drawn with the following slopes.

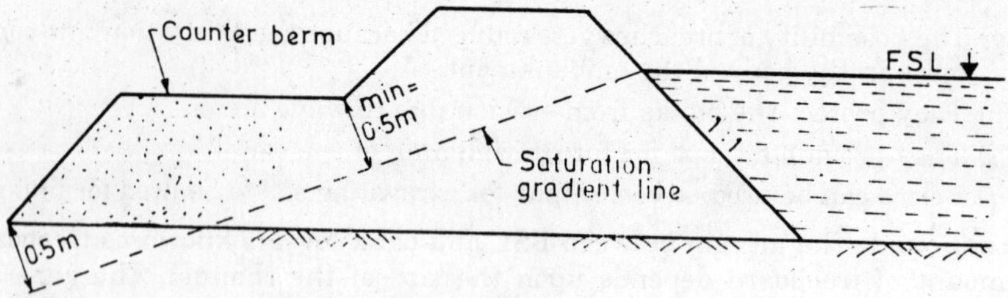


Fig. 4.24.

**Table 4.8. Assumed Values of Saturation Gradients in Different Soils**

| *Type of soil* | *Slope (H : V)* |
|---|---|
| Clay | 1 in 4 |
| Clayey Loam | 1 in 6 |
| Loam | 1 in 8 |
| Loamy sand | 1 in 10 |
| Sand | 1 in 15 |

**Spoil Banks.** When the earthwork in excavation exceeds earthwork in filling, even after providing maximum width of bank embankments, the extra earth has to be disposed of economically. To dispose of this earth by mechanical transport, etc. may become very costly, and an economical mode of its disposal may be found in the form of collecting this soil on the edge of the bank embankment itself. The soil is, therefore, deposited in such a case, in the form of heaps on both banks or only on one bank, as shown in Fig. 4.25. These heaps of soil are discontinued at suitable intervals and longitudinal drains running by their sides are excavated for the disposal of rain water. Cross drains through the spoil banks may also be excavated, if needed.

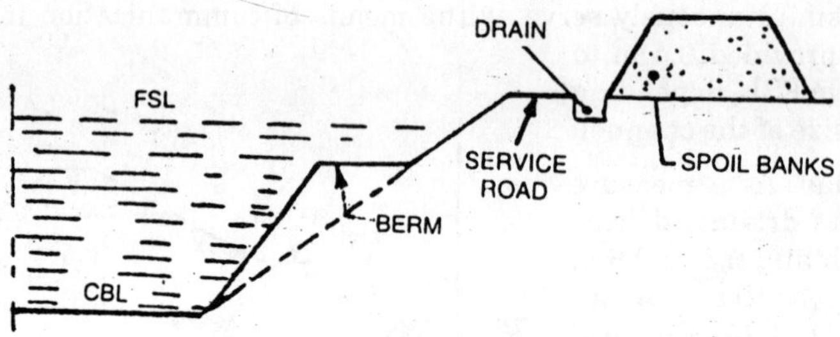


Fig. 4.25. Spoil bank.

**Borrow Pits.** When earthwork in filling exceeds the earthwork in excavation, the earth has to be brought from somewhere. The pits, which are dug for bringing earth, are known as borrow pits. If such pits are excavated outside the channel, they are known as *external borrow pits,* and if they are excavated somewhere within the channel, they are known as *internal borrow pits.* It is a very costly affair to bring soil from distances. Even in the nearby areas, these pits may cause mosquito nuisance due to collection of rain water in these pits, and hence, external borrow pits are not preferred.

Internal borrow pits are, therefore, excavated on the bed of the canal, as shown in Fig. 4.26, when needed.

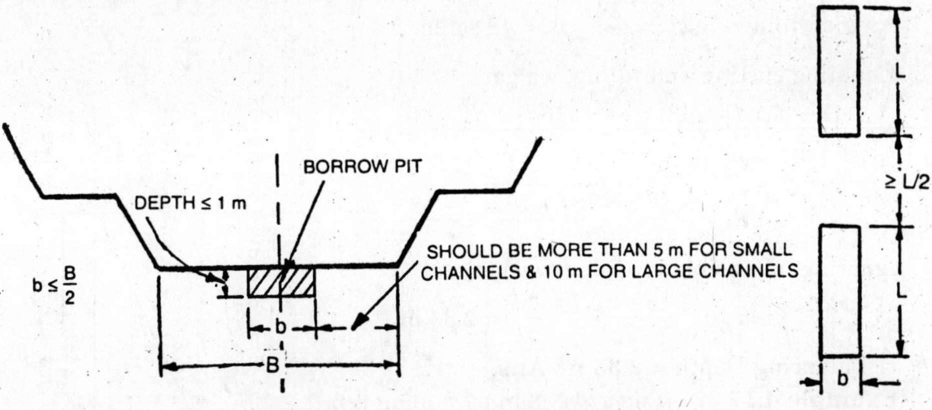


(a) Elevation. (b) Plan.

Fig. 4.26. Internal borrow pits.

The borrow pits should start from a point at a distance more than 5 m from the toe for small channels, and 10 m for large channels. The width of these pits $b$, should be less than half the width of the canal $B$, and should be dug in the centre. The depth of these pits should be equal to or less than 1 m.

Longitudinally, these pits should not run continuous, but a minimum space of $L/2$ should be left between two consecutive pits, (where $L$ is the length of one pit) as shown in Fig. 4.26 (*b*).

### 4.11. Balancing Depth for Excavating Canals

It was pointed out earlier that the maximum economy can be achieved in canal construction, if the earthwork in excavation equals the earthwork in filling. Such a thing is possible when a canal is constructed partly in filling and partly in cutting, which mostly happens in practical life.

If this balance between cutting and filling can occur, then the need for spoil banks or borrow pits is entirely eliminated, and moreover, earthwork has to be paid only once in a single item.

*For a given cross-section of a channel, there can be only one depth, for which such a balance between cutting and filling will occur. This depth is known as the Balancing depth.* This depth can be worked out easily by equating the areas of cutting and filling, as illustrated in the following example.

**Example 4.16.** *Calculate the balancing depth for a channel section having a bed width equal to 18 m and side slopes of 1 : 1 in cutting and 2 : 1 in filling. The bank embankments are kept 3.0 m higher than the ground level (berm level) and crest width of banks is kept as 2.0 m.*

**Solution.** The channel section is shown in Fig. 4.27. Let $d_1$ be the balancing depth, *i.e.* the depth for which excavation and filling becomes equal.

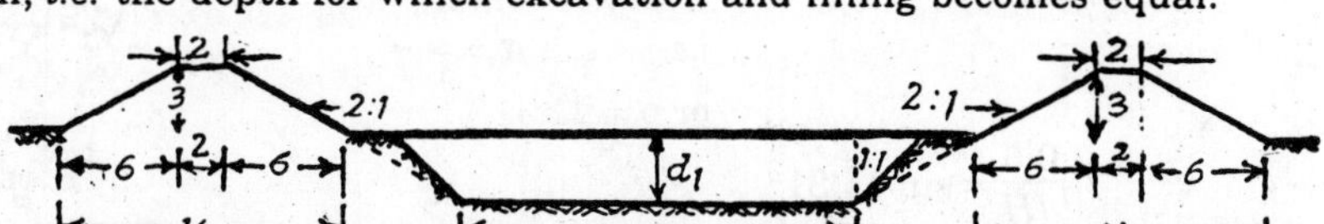


Fig. 4.27.

Area of cutting = $(18 + d_1)\, d_1$ sq. m.

Area of filling $= 2\left[\frac{2+14}{2} \times 3\right] = 48$ sq. m.

Equating cutting and filling, we get

$$(18 + d_1)\, d_1 = 48 \quad \text{or} \quad 18d_1 + d_1^2 = 48$$

or $\quad d_1^2 + 18d_1 - 48 = 0$

or $\quad d_1 = \frac{-18 \pm \sqrt{324 + 192}}{2} = \frac{-18 \pm 22.7}{2}$

Ignoring unfeasible – ve sign, we get

$$d_1 = \frac{-18 + 22.7}{2} = \frac{4.7}{2} = 2.35 \text{ m}$$

∴ Balancing Depth = **2.35 m. Ans.**

**Example 4.17.** *An irrigation channel having a full supply level of 4 m above the existing ground level is provided with banks 3 m wide at top. The side slopes are 2H : 1 V ; and the slope of the hydraulic gradient line through the bank soil is 5 : 1. Assuming a free board of 1 m, calculate the minimum width and height of counter-berm needed to ensure that the seepage of water does not pose any problem for the safety of the canal banks.*

**Solution.** The given values are used in sketching the *X*-section of the Canal, as shown Fig. 4.28. The H.G. line is kept covered by 0.5 m (see $EF = HI = 0.5$ m).

The counter berm is started from the point $E$ on the $d/s$ slope, at a height $h$ metre above the bed level. The width of the counter berm is kept as equal to $EG$ ; say $B$ metre. *Now, From H.G. line slope,* we have

$$NJ = 4 \times 5 = 20 \text{ m}$$

(1 : 5 slope of H.G. line starting at ht. = 4 m)

But $NJ = NM + ML + LK + KJ$

or $\quad 20 = 2 + 3 + 10 + KJ$

∴ $\quad KJ = 5$ m.

But $\quad B = EG = KH$ (In ∥ gm $EKHG$)

or $\quad B = KJ + JH$

or $\quad B = 5 + JH \qquad ...(i)$

where $JH$ is determined from $\Delta HIJ$ ; as :

$$\frac{0.5}{JH} = \sin \alpha$$

where α is the angle which H.G. line makes with horizontal ; *i.e.*, $\tan \alpha = \frac{1}{5}$ ; or $\alpha = 11.31°$

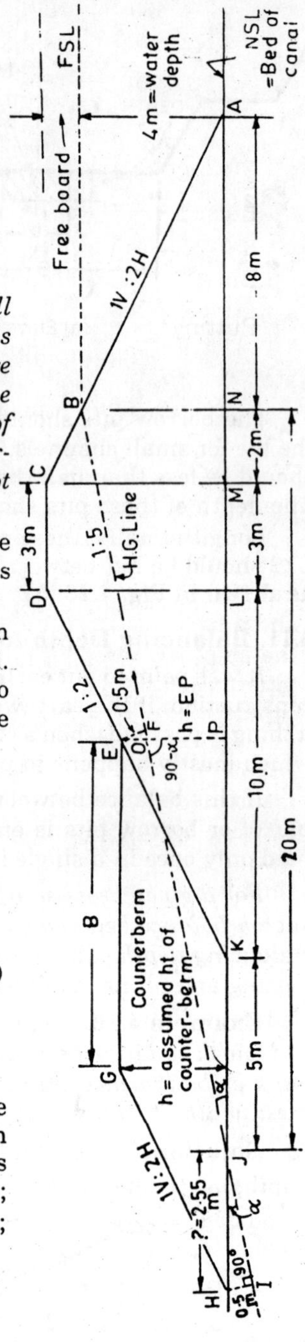


Fig. 4.28

∴ $\quad \frac{0.5}{JH} = \sin 11.31°$

or $\quad JH = \frac{0.5}{\sin 11.31°} = \mathbf{2.55\ m} \qquad ...(ii)$

∴ $B = 5 + 2.55 = \mathbf{7.55\ m}$ **Ans.**

*To determine height of counter berm (h),* we calculate as below :

$$h = EP = EO + OP$$

where $EO$ is calculated from $\Delta EOF$ (Fig. 4·28 *a*) as

$$\frac{EF}{EO} = \frac{0.5}{EO} = \cos \alpha$$

or $$EO = \frac{0.5}{\cos \alpha} = \frac{0.5}{\cos 11.35°} = 0.51 \text{ m}$$

∴ $$OP = h - 0.51 \quad ...(iii)$$

But $OP/PJ = 1/5$ (slope of H.G. line)

or $(h - 0.51)/PJ = 1/5$ ...(*iv*)

But $PJ = PK + KJ = 2h + 5$ ...(*v*)

Putting in (*iv*) $$\frac{h - 0.51}{2h + 5} = \frac{1}{5}$$

$$5h - 2.55 = 2h + 5$$

$$3h = 7.55$$

$$\mathbf{h = 2.52\ m.\ Ans.}$$

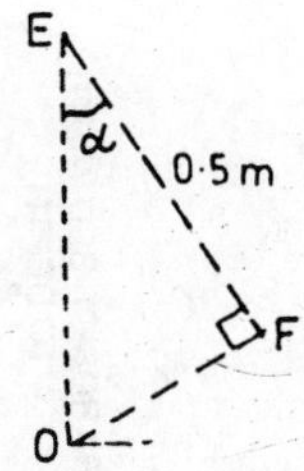

Fig. 4.28 (*a*)

Hence, width of counter berm = 7.55 m
ht. of counterberm = 2.52 m **Ans.**

### 4.12. Fixing the L-Section of the Canal and Other Design Considerations

If a channel is designed according to Lacey's theory, it shall have a fixed slope and a fixed section for a given discharge and silt factor. But on the other hand, if the channel is designed on Kennedy's theory, it can have different sections for different slopes. In practice, it has been found that Lacey's slope equation gives excessive slopes. Lacey himself had changed his fixed slope equation afterwards, as pointed out earlier, making it flexible. *The slope of the channel is, however, fixed on available country slope consistent with economy.* A steeper slope governed by maximum permissible velocity, will be most economical, but it will lower the FSL, causing less irrigation. Hence, the maximum possible irrigation would indicate flatter slopes governed by minimum permissible velocity. A via media between these two limits must be adopted for selecting a suitable bed slope for the channel.

If the chosen designed slope is found to be flatter than the natural available slope, the difference can be adjusted by providing suitably designed falls (explained in chapter 12). But if the designed slope is steeper than that available, then adjustments are made to change the design slope, so as to make it as near to the available slope as possible.

Since a change in depth causes non-uniform flow, it is desirable to change the depth as less as possible. For this reason, the channels in the upper reaches are generally designed with large bed width to depth ratio.

Moreover, in Kennedy's theory, there can be various combinations of width and depth for a given slope. The width and depth ratio can be controlled using the following empirical formulas ;

1. (*a*) *For channels upto 15 cumecs*

$$y = 0.5\sqrt{B} \quad ...(4.62)$$

where $y$ = Depth of water in channel
$B$ = Base width of channel.

1. (*b*) For channels of 15 cumecs and above, depths of the following order may be provided :

**Table 4.9. Assumed Values of Water Depths for Canals**

| *Discharge Q (cumecs)* | *Depth y in metres* |
|---|---|
| 15 | 1.7 |
| 30 | 1.8 |
| 75 | 2.3 |
| 150 | 2.6 |
| 300 | 3.0 |

2. C.W.C. has recommended a graphical relation giving $B/y$ ratios for various discharges (0 to 300 cumecs) as shown in Fig. 4.29.

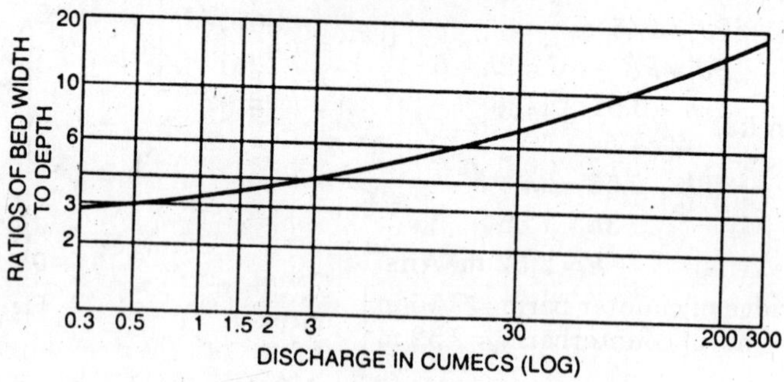


Fig. 4.29.

The limiting maximum permissible velocities for unlined channels in different soils are given in Table 4.10.

**Table 4.10. Max. Permissible Velocities in Unlined Channels**

| *Type of soil* | *Maximum permissible velocity in m/sec* |
|---|---|
| Rock and gravel | 1.5 |
| Murum, hard soil, etc. | 1.0 to 1.1 |
| Sandy loam, black cotton soil etc. | 0.6 to 0.9 |
| Very light loose sand to average sandy soils | 0.3 to 0.6 |
| Ordinary soils | 0.6 to 0.9 |

After marking the ground line, a trial bed line, and water depths, the FSL line on the *L*-section of the channel is checked, as it is governed by the following considerations :

(*i*) The FSL of the channel should be about 10 to 20 cm above the ground line for most of its length.

(*ii*) The water depth ($y$) or FSL governs the depth of excavation ($d_1$) and hence, attempts should be made to equal $d_1$ with balancing depth, so that the channel becomes in balanced earth work.

(*iii*) For branch canal, etc., where some distributaries have to take-off from it, the FSL of the channel should be kept 15 cm higher than the FSL of the off-taking channel, so as to allow for losses in head regulator.

Suitable adjustments in bed slope, depth, etc. can be made for fixing FSL on these considerations.

After fixing *L*-section of the channel, the cross-section can be fixed on the basis of various canal standards, given in Table 4.11.

## Table 4.11. Canal Standards

| S. No. | Item | *Size of canal* | | | | | |
|---|---|---|---|---|---|---|---|
| | | *Minor Distributaries* | | *Major Distributaries* | | *Main and Branch Canals* | |
| | | *below 0.3 cumec* | *0.3 to 1 cumec* | *1 to 5 cumecs* | *5 to 10 cumecs* | *10 to 30 cumecs* | *30 to 150 cumecs* |
| 1. | Maximum width of the bank crest | 1.0 m | 1.5 m | 2 m | 2.25 m | 2.5 m | 3 m |
| 2. | Width of roadway | Nil | 3.5 m | 3.5 m | 5 m | 5 m | 6 m |
| 3. | Free board | 0.3 m | 0.4 m | 0.5 m | 0.6 m | 0.75 m | 0.9 m |
| 4. | Depth of earth cover over saturation gradient | 0.5 m | 0.5 m | 0.5 m | 0.5 m | 0.8 m | 1 m |
| 5. | Width of berms | Maximum berm width = 0.6 m+ $\frac{1}{4}$th of the width of combined side slopes of cutting and embankment. Maximum berm width = 0.6 m+ $\frac{1}{2}$ the width of combined slopes | | | | | |
| 6. | Width of land to be acquired (clear of banks) when canal cutting is deeper than the balancing depth. | Half the height of the bank above ground, subject to minimum of 1.5 m. | | | | | |
| 7. | Width of land to be acquired (clear of banks) when canal cutting is lesser than the balancing depth. | Full height of bank above ground level plus 1.5. | | | | Full height of bank+ 5 m. | Full height of bank+ 5 m |

**Design Procedure**

1. First of all, the longitudinal section of the existing ground along the proposed canal alignment, is plotted on a suitable scale (say 1 cm = 100 metres for horizontal scale, and 2 cm = 1 m for vertical scale).

2. A suitable channel slope is assumed from Lacey's diagrams for an approximate discharge and silt factor. It is made consistent with the existing slope.

3. A trial slope line is marked for drawing FSL line, keeping in view the guidelines already given. The depth and position of falls, etc. is decided tentatively.

4. The channel is then designed from its tail reach towards its head reach, kilometre to kilometre.

5. The discharge required in the channel in the given reach, for required irrigation potential, is worked out, and losses are added, so as to calculate the required discharge. The cumulative losses at a particular km of channel is the sum total of the losses occurring between the particular km and the next below it (*i.e.* for example at km 3, the total losses would be equal to losses between km 3 and km 4, plus, total losses below km 4).

Since the section at km 4 will be worked out earlier, the losses between km 3 and km 4 can be worked out on the wetted perimeter calculated on the basis of the channel section at km 4. The required discharge is increased by 10%, so as to obtain the design discharge.

6. The channel is now designed for this discharge and assumed bed slope, by Kennedy's theory, using Garret's diagrams. The bed width and depth ratios are kept within specified limits. Sometimes, even the slope may have to be changed for keeping them so.

7. The bed slope, FSL, falls, etc. are all adjusted using intelligence, judgement and knowledge. The bed levels, water depths, etc. are drawn on *L*-section. The *X*-sections at every km are then drawn, using canal standards.

The calculations of discharge and channel dimensions, etc. can be carried out in a tabular form called "Schedule of Area Statistics and Channel Dimensions'. A specimen form of this table is given in table 4.12.

**Tabel 4.12. Specimen Table for Schedule of Area Statistics and Channel Dimensions**

| *Below km* | *Gross commanded area* | *Culturable commanded area* | *Area to be irrigated* | | |
|---|---|---|---|---|---|
| | | | *Rabi* | *Sugar* | *Rice* |
| *(1)* | *(2)* | *(3)* | *(4)* | *(5)* | *(6)* |
| *Outlet discharge factor* | *Outlet discharge* | *Losses in reach* | *Total losses* | *Total discharge* | *Bed slope* |
| *(7)* | *(8)* | *(9)* | *(10)* | *(11)* | *(12)* |
| *Bed width* | *Water depth* | *Height of banks* | *Width of banks* | *Velocity* | *CVR* $V/V_0$ |
| *(13)* | *(4)* | *(15)* | *(16)* | *(17)* | *(18)* |

**Example 4.18.** *A distributary canal takes off from a branch canal having CBL at 204.0 m and FSL at 205.8 m. The gross commanded area at the head of the distributary is 30,000 hectares, and after each km it is reduced by 5,000 hectares. Out of this command, the culturable area is only 75%. The intensity of irrigation for the Rabi and Kharif seasons is 32% and 15% respectively. Design suitable channel sections for the first 3 km of this distributary, assuming the following data :*

(i) *Total losses below km 3 = 0.44 cumec.*

(ii) *Channel losses occur @ 2 cumecs/million square metres of wetted perimeter.*

(iii) *Kor period for Rabi (wheat) = 4 weeks*

(iv) *Kor depth for Rabi = 14 cm*

(v) *Kor period for Kharif (Rice) = 2.5 weeks*

(vi) *Kor depth for Kharif = 20 cm.*

(vii) *Manning's n = 0.0225.*

(viii) *Critical velocity ratio = 0.95.*

*The ground levels at every 200 metres, along the line of the proposed alignment, have been obtained and are tabulated in table 4.13.*

**Table 4.13**

| *Distance from head in metres* | *Reduced level (G.T.S) in metres* |
|---|---|
| 0 | 205.20 |
| 200 | 205.30 |
| 400 | 205.25 |
| 600 | 205.00 |
| 800 | 204.90 |
| 1,000 | 204.30 |
| 1,200 | 204.30 |
| 1,400 | 204.20 |
| 1,600 | 204.20 |
| 1,800 | 204.10 |
| 2,000 | 204.05 |
| 2,200 | 204.00 |
| 2,400 | 203.95 |
| 2,600 | 203.95 |
| 2,800 | 203.90 |
| 3,000 | 203.80 |

**Solution.** The channel is to be designed from its tail (where the losses are known) towards its head, km by km. The gross commanded areas and culturable commanded areas at various km are, first of all, worked out in Table 4.14.

**Table 4.14**

| *Below km*<br>*(1)* | *Gross commanded area in hectares*<br>*(2)* | *Gross culturable area in hectares*<br>*(3)* |
|---|---|---|
| 0 (*i.e.* head) | 30,000 | 22,500 |
| 1 | 25,000 | 18,750 |
| 2 | 20,000 | 15,000 |
| 3 | 15,000 | 11,250 |

Outlet discharges for the two crop seasons are determined as given below :

(*i*) *For Rabi,* $D = \frac{8.64\,B}{\Delta}$

where, $B = 4$ weeks = 28 days

$\Delta = 14$ cm = 0.14 m.

$\therefore \quad D = \frac{8.64 \times 28}{0.14} = 1728$ hectares/cumec.

(*ii*) *For Kharif,* $D = \frac{8.64\,B}{\Delta}$

where, $B = 2.5$ weeks = $2.5 \times 7$ days = 17.5 days

$\Delta = 20$ cm = 0.20 m.

$\therefore \quad D = \frac{8.64 \times 17.5}{0.20} = 756$ hectares/cumec

Intensity of irrigation for Rabi = 32%

and Intensity of irrigation for Kharif = 15%.

If $G$ is the gross culturable area at any point, then 0.32 $G$ is the Rabi area and 0.15$G$ is the Kharif area.

Discharge reqd. for this Rabi area $= \frac{0.32\,G}{\text{Outlet factor for Rabi}} = \frac{0.32\,G}{1728} = \frac{G}{5400}$

Similarly, discharge reqd. for Kharif area $= \frac{0.15\,G}{756} = \frac{G}{\mathbf{5040}}$

Since the discharge required for Kharif crop is more than that required for Rabi crop, the outlet factor of Kharif crop becomes the controlling factor. Discharges needed at various kilometres for the given command are worked out in Table 4.15.

**Table 4.15**

| *Below km* | *Gross culturable area from Table 4.14 col. (3)* | *Discharge required for Kharif crop in cumecs* = $\frac{\text{Col. (2)}}{5040}$ |
|---|---|---|
| *(1)* | *(2)* | *(3)* |
| 0 | 22,500 | 4.46 |
| 1 | 18,750 | 3.72 |
| 2 | 15,000 | 2.98 |
| 3 | 11,250 | 2.23 |

(*i*) **Design at km 3**

Losses below km 3 = 0.44 cumec (given)

Discharge reqd. for crop at this point (Table 4.15) = 2.23 cumecs.

Total discharge required = 2.23 + 0.44 = 2.67 cumecs

Design discharge = 10% more than required.

$= 1.1 \times 2.67 = 2.937$ cumecs, say **2.94 cumecs**

$Q = 2.94$ cumecs

$\frac{V}{V_0} = \text{C.V.R.} = 1.0$

$n = 0.0225$

Lacey's regime slope for this discharge and silt factor = 1, is approximately 22 cm per km. Let us kep the slope at 22.5 cm/km*:

Assume bed slope = 22.5 cm per km.

or $$S = \frac{0.225}{1000} = \frac{1}{4444}.$$

From Garret's diagrams [Plate 4.1 (*b*)], assuming $\frac{1}{2}$ : 1 side slopes, the channel section is designed as shown in Table 4.16.

**Table 4.16**

| *Discharge* | *S* | *B* *m* | *y* *m* | $A = \left(B + \frac{y}{2}\right) y$ $m^2$ | $V = \frac{Q}{A}$ *m/sec* | $V_0$ *m/sec* | $\frac{V}{V_0}$ | *Remarks* |
|---|---|---|---|---|---|---|---|---|
| 2.94 cumecs | $\frac{1}{4444}$ | 5.0 | 0.98 | 5.38 | 0.55 | 0.53 | 1.04 | much larger than 0.95 |
| | | 4.5 | 1.05 | 5.28 | 0.56 | 0.58 | 0.96 | O.K. |

Hence, adopt $B = 4.5$ m

$y = 1.05$ m

$S = \frac{1}{4444}$ (*i.e.* 22.5 cm per km.)

These dimensions, quite nearly satisfy the bedwidth-depth relationship, given by $y = 0.5 \sqrt{B}$ ; and hence, the assumed slope is all right and can be adopted.

(*ii*) **Design at km 2**

Outlet discharge required below km 2, from Table 4.15, col. (3) = 2.98 cumecs.

Losses below km 3 = 0.44 cumec.

**Losses in channel between km 3 to km 2 :** For the calculation of these, the perimeter of the section at km 3 shall be taken, as the section at km 2 is not known so far.

Wetted perimeter $= B + \sqrt{5} \cdot y = 4.5 + \sqrt{5} \times 1.08 =$ **6.92 m**

Loss @ 2 cumecs/million sq. m. $= 2 \times \left[\frac{6.92 \times 1000}{10^6}\right] =$ **0.014 cumec.**

**Total Losses below km 2**

= Losses below km 3 + Losses between km 3 and km 2

= 0.44 + 0.014 = 0.454 cumec

Total discharge required at km 2 = 2.98 + 0.454 = 3.434 cumecs

Design discharge = 1.1 × 3.434 = 3.7874 ; say **3.79 cumecs**

Use the same slope of 22.5 cm in 1 km, *i.e.* $\frac{1}{4444}$.

Using Garret's diagrams [Plate 4.1 (*b*)], we design the channel section as shown in Table 4.17.

---

* The bed slope is generally kept in multiples of 2.5 cm/km, to obtain bed levels in fraction of 0.005 m, *i.e.* the fractions up to which levels can be read on levelling staff.

**Table 4.17**

| $Q$ | $S$ | $B$ $m$ | $y$ $m$ | $A = \left(B + \frac{y}{2}\right) y\ m^2$ | $\frac{Q}{A} = V\ m/sec$ | $V_0\ m/sec$ | $\frac{V}{V_0}$ | *Remarks* |
|---|---|---|---|---|---|---|---|---|
| 3.79 cumecs | $\frac{1}{4444}$ | 6.0 | 1.05 | 6.85 | 0.55 | 0.58 | 0.95 | O.K. |

Hence, adopt $B = 6.0$ m

$y = 1.05$ m

$$S = \frac{1}{4444}$$

(*iii*) **Design at km 1**

Outlet discharge required below km 1, from Table 4.15, col. (3) = 3.72 cumecs.

Losses below km 2 worked out earlier = 0.454 cumec.

**Losses between km 2 and km 1 :** To work out these losses, the perimeter of the section at km 2 shall be taken, as the section at km 1 is not known so far.

$\therefore$ Wetted perimeter $= B + \sqrt{5}y$

$= 6.0 + \sqrt{5} \times 1.08 = 6.0 + 2.42 = \mathbf{8.42\ m.}$

Losses @ 2 cumecs/million sq. m. (*i.e.* in length of 1 km, *i.e.* 1000 m)

$$= 2 \times \left(\frac{8.42 \times 1000}{10^6}\right) = 0.017 \text{ cumec.}$$

**Total Losses below km 1** $= 0.454 + 0.017 = 0.471$ cumec

Total discharge required at km 1

$= 3.72 + 0.471 = 4.191$ cumecs

Design discharge $= 1.1 \times 4.191 = \mathbf{4.61}$ **cumecs.**

Let us adopt a slope of 20 cm in 1 km, *i.e.* $S = \frac{1}{5000}$

Using Garret's diagrams [Plate 4.1 (*c*)], the required channel section is designed as shown in Table 4.18.

**Table 4.18**

| $Q$ | $S$ | $B$ $m$ | $y$ $m$ | $A = \left(B + \frac{y}{2}\right) y,\ m^2$ | $\frac{Q}{A} = V$ $m/sec$ | $V_0$ $m/sec$ | $\frac{V}{V_0}$ | *Remarks* |
|---|---|---|---|---|---|---|---|---|
| 4.61 cumecs | $\frac{1}{5000}$ | 6.0 | 1.2 | 7.92 | 0.582 | 0.615 | 0.947 | O.K. |

Hence adopt $B = 6.0$ m

$y = 1.2$ m

$$S = \frac{1}{5000} \text{ (20 cm in 1 km.)}$$

(*iv*) **Design at 0 km.**

Outlet discharge required at 0 km, from Table 4.15, col. (3) = 4.46 cumecs

Losses below km 1, as worked out earlier = 0.471 cumec.

**Losses between km 0 to 1 :** To work out these losses, the perimeter of the section at km 1 shall be taken, as the section at km 0 is not known so far.

Now, wetted perimeter $= B + \sqrt{5} \cdot y$

$$= 6.0 + \sqrt{5} \times 1.2 = 6 + 2.68 = 8.68 \text{ m}$$

Losses @ 2 cumecs/million sq. m. in a length of 1 km

$$= 2\left[\frac{8.68 \times 1000}{10^6}\right] = 0.01736 = \text{say } 0.017 \text{ cumec.}$$

**Total losses below km 0** $= 0.471 + 0.017 = 0.488$ cumec

Total discharge required at 0 km

$$= 4.46 + 0.488 = 4.948 \text{ cumecs.}$$

Design discharge $= 1.1 \times 4.948 =$ **5.44 cumecs.**

Let us adopt a slope of 20 cm in 1 km, *i.e.* $S = \frac{1}{5000}$

Using Garret's diagrams [Plate 4.1 (*c*)], the required channel section is worked out as shown in Table 4.19.

**Table 4.19**

| $Q$ | $S$ | $B$ $m$ | $y$ $m$ | $A = \left(B + \frac{y}{2}\right) y$ $m^2$ | $V = \frac{Q}{A}$ $m/sec$ | $V_0$ $m/sec$ | $\frac{V}{V_0}$ | *Remarks* |
|---|---|---|---|---|---|---|---|---|
| 5.44 cumecs | $\frac{1}{5000}$ | 7.2 | 1.20 | 9.36 | 0.58 | 0.615 | 0.944 | O.K. |

Hence, adopt $B = 7.2$ m

$y = 1.2$ m

$S = \frac{1}{5000}$ (20 cm in 1 km)

All the data worked out above, has been entered at their proper places in the 'schedule of area statistics and channel dimensions' (Table 4.20). The table has been completed with the help of Canal Standards already given in Table 4.11. The L-section of the distributary is drawn, as shown in Fig. 4.30, starting from the head (*i.e.* 0 km) by keeping its FSL at head at 0.2 m below the FSL of the branch channel. The cross-sections at various km are drawn in Figs. 4.31 (*a*), (*b*), (*c*) and (*d*), by assuming $1\frac{1}{2}$ : 1 slopes in filling and 1 : 1 slopes in cutting. The dowel level is kept 0.15 m higher than the bank level, and road level at 0.15 m below the bank level.

**Note.** The slopes of 1 : 1 shall afterwards become $\frac{1}{2}$ : 1, due to silting, and that is why in design calculations, $\frac{1}{2}$ H : 1 V slopes are taken.

**Table 4.20**
**Schedule of Area Statistics and Channel Dimensions**

| *Below km* | *Gross commanded area (hectares)* | *Culturable commanded area (hectares)* | *Controlling area to be irrigated (15%) in hectares* | *Controlling outlet discharge factor* | *Losses in reach (cumecs)* | *Total losses (cumecs)* | *Design discharge (cumecs)* | *Bed slope (cm/km)* | *Bed width (m)* | *Water depth (metres)* | *Freeboard from Table 14.11 in metres* | *Ht. of bank above ground decided after drawing L-section* | *Width of bank from Table 14.11* | *Velocity in m/sec* | $\frac{V}{V_0} = m$ |
|---|---|---|---|---|---|---|---|---|---|---|---|---|---|---|---|
| (1) | (2) | (3) | (4) | (5) | (6) | (7) | (8) | (9) | (10) | (11) | (12) | (13) | (14) | (15) | (16) |
| 0 | 30,000 | 22,500 | 3,775 | 756 | 0.017 | 0.488 | 5.44 | 20 | 7.2 | 1.2 | 0.6 | 1.00 | 2.25 | 0.58 | 0.944 |
| 1 | 25,000 | 18,750 | 2,812.5 | 756 | 0.010 | 0.471 | 4.61 | 20 | 6.0 | 1.2 | 0.5 | 1.00 | 2.0 | 0.582 | 0.947 |
| 2 | 20,000 | 15,000 | 2,250.0 | 756 | 0.014 | 0.454 | 3.79 | 22.5 | 6.0 | 1.05 | 0.5 | 0.90 | 2.0 | 0.55 | 0.95 |
| 3 | 15,000 | 11,250 | 1,687.5 | 756 | — | 0.440 | 2.94 | 22.5 | 4.2 | 1.05 | 0.5 | 0.925 | 2.0 | 0.56 | 0.96 |

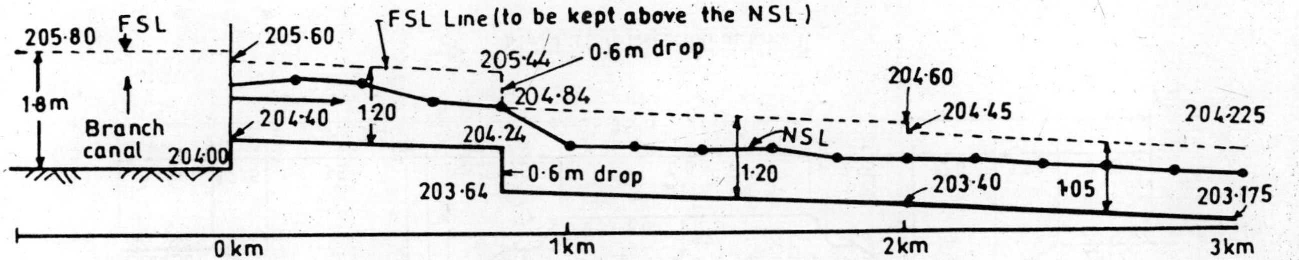


| Chainage in metres | 0 | 200 | 400 | 600 | 800 | 1000 | 1200 | 1400 | 1600 | 1800 | 2000 | 2200 | 2400 | 2600 | 2800 | 3000 |
|---|---|---|---|---|---|---|---|---|---|---|---|---|---|---|---|---|
| Cutting depth | 0.80 | 0.94 | 0.93 | 0.72 0.66 | 1.26 | 0.70 | 0.74 | 0.68 | 0.72 | 0.76 | 0.65 | 0.644 | 0.638 | 0.782 | 0.678 | 0.62 |
| Designed discharge | 5.44 $m^3/s$ | | | | | 4.61 $m^3/s$ | | | | | 3.79 $m^3/s$ | | | | | 2.94 $m^3/s$ |
| Bed slope | 20 cm/km | | | | | 20 cm/km | | | | | 22.5 cm/km | | | | | 22.5 cm/km |
| Depth | 1.20 m | | | | | 1.20 m | | | | | 1.05 m | | | | | 1.05 m |
| Bed width | 7.2 m | | | | | 6.0 m | | | | | 6.0 m | | | | | 4.2 m |
| Designed FSL (m) | 205.60 | | | 205.44 | 204.84 | 204.80 | | | | 204.60 | 204.45 | | | | | 204.255 |
| Ground level available (m) | 205.20 | 205.30 | 205.25 | 205.0 | 204.90 | 204.30 | 204.30 | 204.20 | 204.20 | 204.10 | 204.05 | 204.00 | 203.95 | 203.95 | 203.90 | 203.80 |
| Designed bed level (m) | 204.40 | 204.36 | 204.32 | 204.28 204.24 | 203.64 | 203.60 | 203.56 | 203.52 | 203.48 | 203.44 | 203.40 | 203.355 | 203.310 | 203.265 | 203.220 | 203.175 |
| | 0 km (3 km) | | | | | | (1 km) | | | | | (2 km) | | | | |

Fig. 4.30. L-Section of the Distributary.

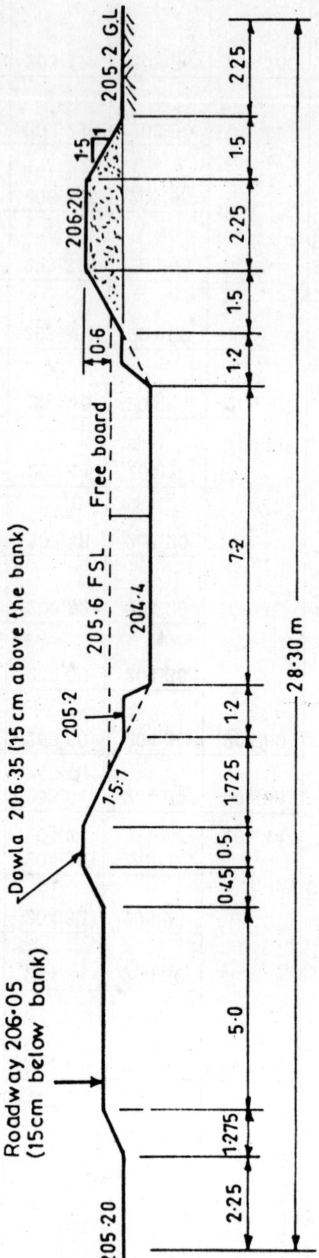


Fig. 4.31. (*a*) Section at km–0.

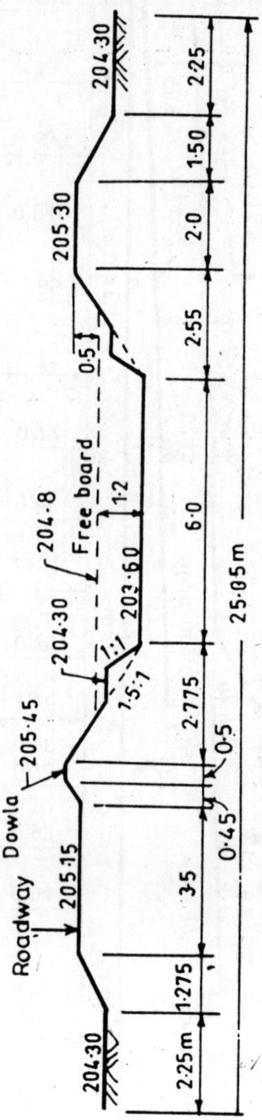


Fig. 4.31. (*b*) Section at km–1.

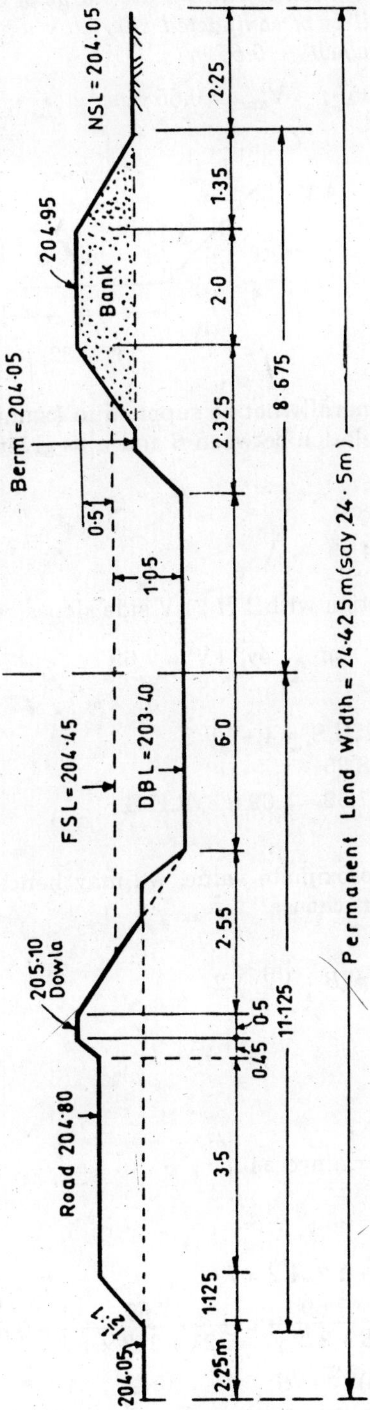


Fig. 4.31. (*c*) Section at km–2.

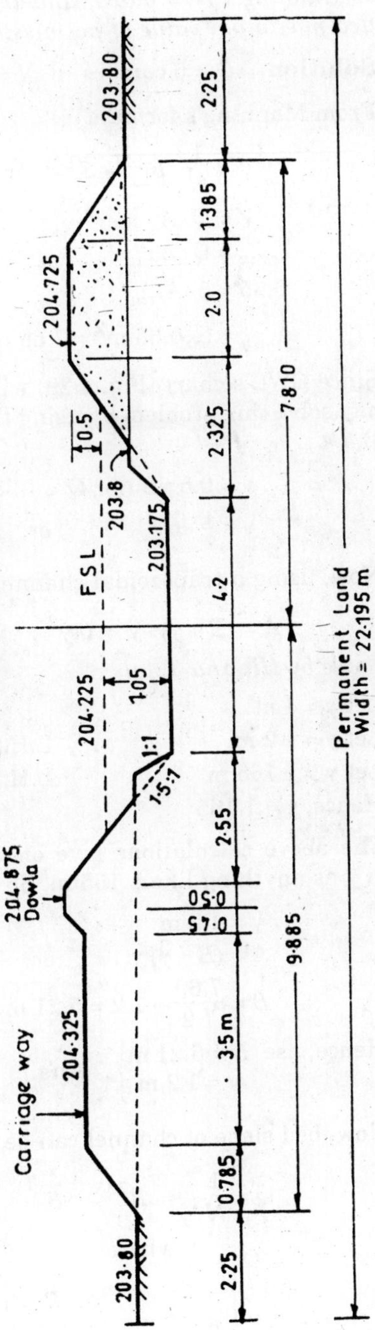


Fig. 4.31. (*d*) Section at km–3.

**Example 4.19.** *Design a reservoir fed canal to carry a discharge of 5 cumecs. Given Manning's N = 0.025, and the soil is of compacted silty clay with recommended maximum value of permissible velocity = 0.65 m/s.*

**Solution.** $Q = 5$ cumecs ; $N = 0.025$ ; $V_{max} = 0.65$ m/s.

From Manning's formula

$$V = \frac{1}{N} \cdot R^{2/3} \cdot S^{1/2}. \text{ and } Q = A \cdot V$$

or $V = Q/A$

$\therefore$ $V_{max} = Q/A_{min}$

or $0.65 = 5/A_{min}$

$A_{min} = 5/0.65 \text{ m}^2 = 7.69 \text{ m}^2$

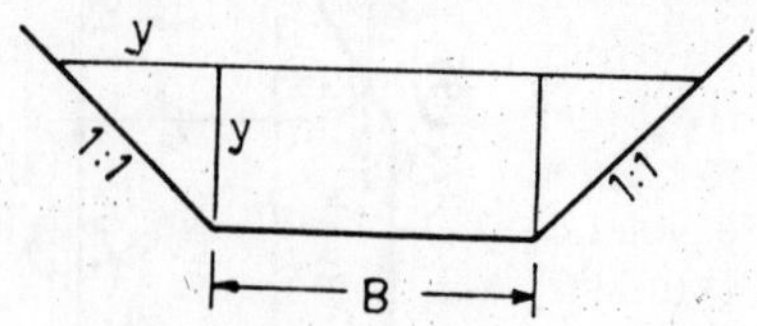

Fig. 4.32.

Since CWC's chart (Fig. 4.29) will generally not be supplied in examinations, we may solve this problem by using the relation between $B$ and $y$, as given by eqn. (4.53), *i.e.*,

$$y = 0.5\sqrt{B} \text{ (for } Q < 15 \text{ cumecs)}$$

or $y = \frac{1}{2}\sqrt{B}$. or $B = 4y^2$.

Now, using a trapezoidal channel section with $1H : 1V$ side slopes, we have

$$A = (B + y) \cdot y. = (4y^2 + y)\, y \quad \text{or} \quad 4y^3 + y^2 = 7.69$$

*Solve by Hit and Trial*

Let $y = 1$ m L.H.S. = 4 + 1 = 5

Let $y = 1.2$ m L.H.S. = 8.35

Let $y = 1.165$ m L.H.S. = 7.68 ≈ 7.69 (*i.e.* R.H.S.)

Hence, $y \approx 1.165$

The above calculations give only approximate value, we may hence choose depth $y$ as anything like 1.165 m. So let us choose

$y = 1.2$ m

$A = (B + y)\, y$ or $7.69 = (B + 1.2)\, 1.2.$

or $B = \frac{7.69}{1.2} - 1.2 = \mathbf{5.21\ m.}$

Hence, use $B = 5.21$ m, $y = 1.2$ m } **Ans.**

Now, bed slope of channel can be determined as :

$$Q = \frac{1}{N} \cdot A \cdot R^{2/3} \cdot \sqrt{S}$$

where $A = (5.21 + 1.2)\, 1.2 = 7.69$

$$R = \frac{A}{P} = \frac{7.69}{(B + 2\sqrt{2} \cdot y)} = \frac{7.69}{5.21 + 2\sqrt{2} \times 1.2} = 0.894 \text{ m}$$

$\therefore$ $5 = (1/0.025) \times 7.69 \times (0.894)^{2/3} \cdot \sqrt{S}$

or $5 = 285.45\sqrt{S}$ or $\sqrt{S} = 1/57.08$

or $S = \frac{1}{3260}$ **Ans.**

### 4.13. Maintenance of Irrigation Canals

Irrigation canals are nothing but earthwork constructions, and as such, very much susceptible to damage. They, therefore, require a lot of maintenance, upkeep, watch and ward, etc. as to ensure their continuous efficient functioning. Various problems which are posed by the irrigation canals during their use, and as such need constant attention, are : (*i*) silting of canals ; (*ii*) weed and plant growth, (*iii*) failure of weaker banks ; (*iv*) hollows created by burrowing animals, crabs, etc. in the filled up sections, causing seepage and piping ; (*v*) canal breaches due to piping, overflowing of canal, etc. Some of these factors, such as the (*i*) and (*ii*) reduce the efficiency of the canals, whereas the others may result in complete stoppage of canal supplies to the irrigated fields. All these causes of troubles and their remedial measures are discussed below :

(*i*) **Silting of Canals.** Irrigation canals normally get silted during their course of flow. Whenever the flow velocity in the channel reduces, the silt carried by the water in suspension gets deposited on the bed and sides of the canal. The silt so deposited reduces the effective canal cross-section and the carrying capacity of the channel.

In order to prevent too much of silt deposition, irrigation channels, (particularly those which are constructed in alluvial soils) must be properly designed, so as to ensure a velocity which neither causes any silting nor scouring in the channel. These design principles have already been discussed. However, even in such channels, the discharge may sometimes be less than the designed; and in that eventuality, silting may take place in the channels.

Moreover, attempts should be made to remove the silt from the water entering the canal itself. From the headworks, for this purpose, almost silt-free water is to be admitted in the canal. However, inspite of using all these measures, certain amount of silt will definitely enter the canal, and certain amount may be added to it, by some scouring from within the channel. This silt often gets deposited on the bed and sides of the canal.

The deposited silt normally deforms the shape of a channel, and reduces its carrying capacity. It must, therefore, be removed periodically by *desilting operations.*

The desilting can be carried out by actually excavating the silt either manually (when the canal is dry) or by dredgers (when the canal is either dry or running). However, if the canal functions properly and is in régime and taking its fully supply, it is not necessary to clear the silt to the theoretical cross-section. But if the canal is not functioning properly, desilting is required, but only to the extent of clearing a portion of it as to get the canal back into efficient working order.

The extent of silting that has taken place in a canal can be found from measuring the silt depth over bed bars.*

The excavated silt from the canals should be disposed of suitably depending upon its quality. Moreover, the excavated silt should not be deposited above the canal banks, as in that case, the same silt may find its way back into the desilted channel.

---

*** Bed bars are the masonry walls partly extending into the canal bed and flush with it, and partly flush with the sides of the channel as shown in Fig. 13.23. For larger canals, a bed bar may consist of concrete block with its upper face flush with theoretical bed level of the canal as shown in Fig. 13.24. They serve as permanent marks of reference to indicate the correct alignment and theoretical bed levels of the channel.**

(*ii*) **Weed and Plant growth.** Irrigation canals normally get infested with aquatic weed growth like bushes, hyacinth, etc. Various types of aquatic weeds and plants get grown in canals. The silted canals provide excellent base for weed growth. The sun rays also help in promoting weed growth. The weed growth hinders the flow of water, reduces the carrying capacities of channels, and thereby impairing their efficiencies. The weed growth may also sometimes impart harmful qualities to irrigation water.

Weed growths may be controlled by maintaining higher velocities in canals. Moreover, many a times, the weed growths will have to be physically removed from the canals in order to improve their efficiencies.

In case of newly constructed canals, regular inspections should be undertaken to locate any spots where the weed growth has set in. Weeds from such spots should be removed complete, so that the infestation does not spread. In the case of old canals, where aquatic weed growth is profuse, suitable mechanical or chemical methods may be employed, at as early stage as possible.

(*iii*) **Failure of Weaker Banks.** The banks of the canal should be made in full designed width (as per I.S. provisions), and should be maintained as such. Due to constant use by men and animals, the canal banks get eroded at various places. They must be repaired, and all cuts and breaches filled up with suitable soil and proper tamping.

Moreover, grass or turfing should not be scrapped while smoothening the bank surfaces, as it helps in stabilising the soil and thus preventing its erosion. However, long grasses will have to be cut as far as necessary to smoothen the surface of the bank and to avoid holes being hidden under high grasses.

Sometimes, when a canal-reach runs in filling, the banks are subjected to water pressure. In such a case, there exists every possibility of damage occurring to the banks. In such cases, the banks must be given extra strength by increasing their sections. In order to keep the cost of construction low, additional soil for increasing the bank section may be obtained by the natural silting process. For this purpose, favourable conditions are created to cause silting internally. In this system, sufficiently wider section is provided, and a part of it later gets silted, forming the additional bank as shown in Fig. 4.33. Such a method is possible only for new

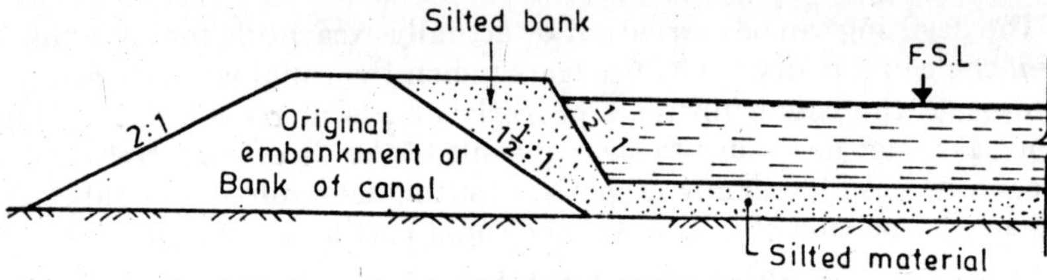


Fig. 4.33. Internal silting method for widening canal banks in canals constructed in filling.

channels. To accelerate silting, low submergible spurs projecting from the banks into the channel may be constructed.

(*iv*) **Canal Breaches.** Canal breach is, infact, an opening or a gap developed in the canal bank due to erosion of some portion of the bank. When once water starts flowing out of such a breach, it starts becoming deeper & wider, unless remedial measures are taken immediately to plug the breach. The breaches in canal banks may be caused due to various reasons, such as :

(*a*) breach due to faulty design or construction of the canal banks ;

(b) breach due to overflow of the canal ;

(c) breach due to seepage or piping ;

(d) breach due to intentional cuts made by cultivators.

These reasons are briefly summarised below :

(a) **Breach due to faulty design or construction of the banks.** If the width of a canal bank is not kept sufficient, or if its height is not as per the correct design requirements, the bank may fail to hold the water properly. Similarly, if the soil material used at site is poor and not up to the specifications or if the rolling and watering, etc. has not been properly done at the time of construction, the banks may not be so strong as were expected to be, and hence fail to hold the canal water without being damaged. This continuous recurring damage to the bank may ultimately lead to a full-fledged cut and breach within the bank length.

(b) **Breach due to overflow of canal.** Sometimes, the discharge in the canal may exceed its design capacity, and water may overtop the canal bank(s). The overflow will definitely damage the bank(s) considerably.

(c) **Breach due to piping.** Sometimes, the banks may not be overtopped, but the water depth may go above the full supply depth due to increased discharge or reduced silted canal section. In such a case, the hydraulic gradient line within the 'filled' canal bank section will also rise (as shown in Fig. 4.34). And if the provided

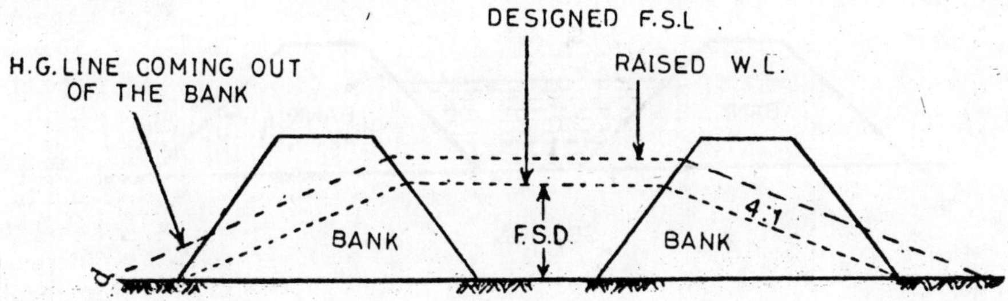


Fig. 4.34.

cover over the hydraulic gradient line is not sufficient, it may pass out of the banks, thus dislodging the soil particles from the outer slopes of the banks. It will consequently weaken and erode the banks, which may ultimately lead to a complete breach within the bank, and water may rush out of the canal.

Sometimes, the water may start leaking through the holes created within the body of the bank, by insects or burrowing animals. The leaking water will go on removing the sand particles alongwith ; and if went unchecked, the holes may go on increasing in size, and ultimately resulting in a rush of water through the canal bank, leading to sinking of the bank soil, and formation of a full fledged breach. Normally, however, these holes are smaller in size and do not pose a serious problem, if timely inspection and remedial plugging is done. As a preventive measure, sometimes, a sand core may be provided within the bank embankment section, which will settle and fill the holes. In such a case, of course, the banks will settle down, but at least breaches will be avoided.

(d) **Breach due to intentional cuts made by cultivators.** Sometimes, when the area gets flooded due to excessive rain or poor drainage, the cultivators cut the canal bank in order to pass the drainage water to the other side. This sometimes proves rather detrimental to their expectations, as the canal supplies may also enter the low lying area, thus further aggravating the water congestion in the area.

Sometimes, similarly, the cultivators cut the banks to obtain illegal water supplies from the canal. These small cuts may widen and take the shape of big breaches in a short time, consequently damaging the land and crops.

**Closure of breaches.** In case of *small minors and distributaries,* a breach may be closed by dumping huge quantities of earth instantaneously from both sides of the gap. As the water of the breach spreads on the adjoining land, there is usually no outside nearby place to borrow earth for closing the breach. The earth has, therefore, to be collected either by cutting the outer slope of the existing bank, or from the spoil banks (if existing), or berms of the canal. It is essential to store huge quantities of earth on both sides of the gap before closing the breach. Moreover, the closing process should be started from both sides of the breach, at the same time, as pointed out earlier.

In case of *bigger canals,* such as a major distributary or a branch canal, the above procedure cannot be adopted, as the huge discharge from such a canal may completely wash off the dumped soil. In such a case, it is, therefore, necessary to, first of all, reduce the flow through the breach. This should be done by driving a double line of stakes or wooden piles (ballies) in the opening of the breach, as shown in Fig. 4.35. The space between the pile lines is filled with planks or bushes, etc. The filler material can be secured by placing sand bags on the top. If the breach is very wide, another line of defence may be provided.

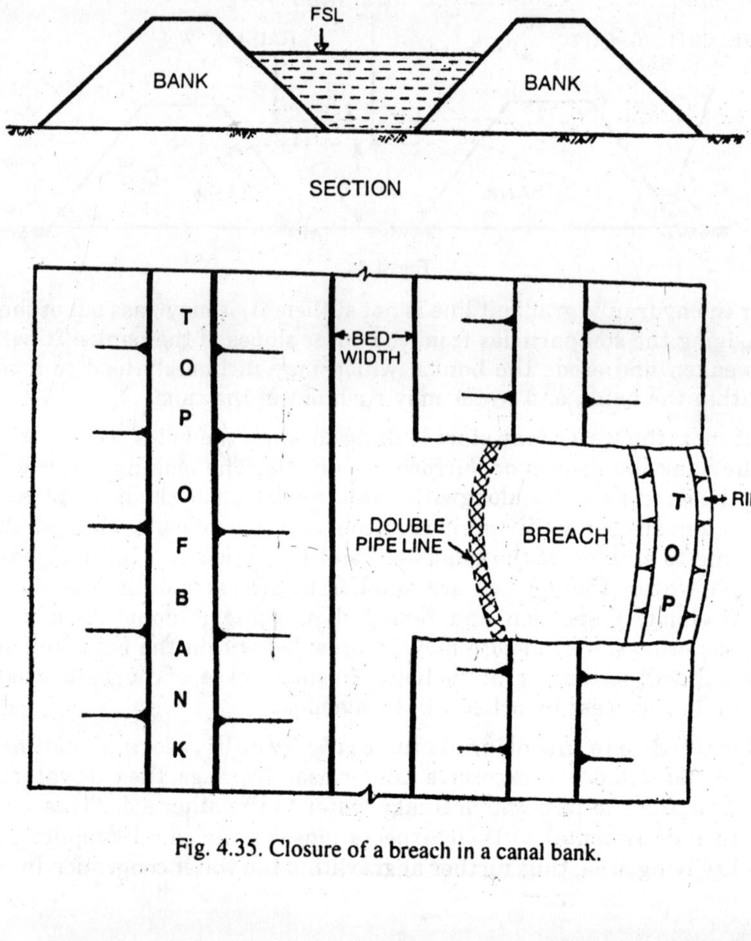


Fig. 4.35. Closure of a breach in a canal bank.

This will reduce the outflow from the breach. No dumping of earth in the breach should progress before the flow through the breach has been arrested to some extent in this fashion. Meanwhile earth should be piled up on the bank on both sides of the breach.

This deposited earth should then be dumped instantaneously from both sides to form a *ring bund* on the outer side of the breach as shown. The opening is then properly filled with suitable earth in layers, each layer being properly compacted. All jungle from the ring bund site should be removed before earth work progresses. Moreover, the earth should be freed from grasses or bushes, etc., before dumping.

### 4.14. Modern Simplified Equations for Optimal Designs of Alluvial Canals

While discussing the design procedure by Kennedy's theory, we have illustrated that the alluvial canals are normally designed to satisfy known values of discharge ($Q$), rugosity coefficient ($n$), and silt grade ($m$). The value of longitudinal slope ($S$) is also usually predecided on the basis of ground considerations. Thus, the design is usually carried out for known values of $Q$, $n$, $m$ & $S$ to satisfy Kutter's equation as well as Kennedy's equation, both. Such a design will give us dimensions of a channel which will be hydraulically adequate, (indicated by Kutter's Eqn.) and will also be non-silting-non scouring (indicated by Kennedy's eqn.). *But infact, such a channel may not behave satisfactorily from considerations of silt transportation.* This is verified by experience in the field. Evidently, some guidance for fixing a suitable bed width - depth ratio (*$b/y$ ratio* ) is required. These ratios were fixed in various areas on the basis of experience, or by empirical equations and charts like the ones given in eqn. (4.62) and Fig. 4.29. The resulting solution obtained for channel dimensions (with given values of $Q$, $n$, $m$ *and* $S$) is finally compared with the recommended values of $b/y$ from Fig. 4.29, or Eq. (4.62). If the two values differ significantly, suitable modification in the bed slope ($S$) would become necessary. Since the slope is decided on the basis of ground profile, it will limit the range of slope. Within this range of slope, one can obtain different combinations of $b$ and $y$ satisfying Kennedy's and Kutter's equations and the curve in Fig. 4.29. All such computations are cumbersome and time consuming. To avoid time consumption, Garret's diagrams are used, which provide graphical solution to Kennedy's and Kutter's equations. However, this method involves interpolation errors. Considering these drawbacks of the age old design method, a need was being felt to develop explicit type of equations, which will satisfy both Kennedy's and Kutter's equations, without involving trial and interpolation errors.

Some modern equations have, therefore been developed to overcome the above shortcomings of the old age method, and are given below.

Considering a trapezoidal channel section with $\frac{1}{2}H : 1V$ side slope, we have from Kennedy's Eqn.

$$V = 0.55\, m \cdot y^{0.64} \quad ...(i)$$

Also from continuity eqn., we have

$$Q = A \cdot V = y\left(b + \frac{1}{2} \cdot y\right) \cdot V \quad ...(ii)$$

Eliminating $V$ between ($i$) and ($ii$), we get

$$Q = y\,(b + 0.5y)\, 0.55 \cdot m \cdot y^{0.64}$$
$$= 0.55m \cdot y^{1.64}\,(b + 0.5y)$$

$$= 0.55 \cdot m \cdot y^{1.64} \cdot y \left( \frac{b}{y} + 0.5 \right)$$

or $$Q = 0.55\, m \left( \frac{b}{y} + 0.5 \right) y^{2.64}$$

Substituting $b/y = r$, we get

$$Q = 0.55\, m \cdot (r + 0.5)\, y^{2.64}$$

or $$y = \left[ \frac{Q}{0.55 \cdot m\,(r + 0.5)} \right]^{\frac{1}{2.64}}$$

or $$y = \left[ \frac{1.818\, Q}{(r + 0.5)\, m} \right]^{0.3788} \quad ...(4.63)$$

where $b/y = r$ ...(4.64)

One can compute both $b$ and $y$ from Eq. (4.63) & (4.64), if a suitable explicit equation written symbolically as :

$$r = f(Q, n, m, S) \quad ...(iii)$$

is available, which satisfies both Kennedy's and Kutter's equations. A large number of canal designs can be obtained by varying the parameters $Q$, $n$, $m$ and $S$ in their practical ranges and a suitable correlation can be established between $r$, $Q$, $n$, $m$ and $S$.

By varying the parameters $r$, $Q$, $n$, $m$ and $S$, in their practical ranges, 1296 number of canal designs were obtained by satisfying both the Kennedy's and Kutter's equations. Analysing all the resulting solutions obtained earlier, it is found that a unique relation could be established between $r$, $Q$, $n$, $m$ and $S$. By a method of curve fitting, the following empirical formula has been obtained :

$$\boxed{r = \left[ \frac{1.607\, S^{1.63} \cdot Q^{0.033}}{n^{3.26} \cdot m^{3.293}} - 0.258 \right]^{-0.915}} \quad ...(4.65)$$

For known values of $Q$, $n$, $m$ and $S$, value of $r$ can be computed from the above eqn. With this value of $r$, the value of $y$ can be calculated from Eq. (4.63), and then the value of $b$ can be calculated by Eqn. (4.64). The resulting solution may finally be compared with the curve of Fig. 4.29. If the matching differs widely, a modification in $S$ value would be necessary and may be done in the following manner.

The curve of Fig. 4.29 can be represented by an equation :

$$r = \left[ 15 + 6.44\, Q \right]^{0.382} \quad ...(4.66)$$

Eq. (4.66) can be considered as the optimality condition for the value of $r$. On combining Eqns. (4.65) and (4.66), we get

$$S = \frac{n^2 \cdot m^{2.02}}{1.338 \cdot Q^{0.02}} \left[ 0.258 + (15 + 6.44\, Q)^{-0.417} \right]^{0.6135} \quad ...(4.67)$$

Eq. (4.67) will give optimal value of $S$, which will satisfy the curve of Fig. 4.29. However, if a particular value of $S$ is to be maintained, the design may be carried out by relaxing the requirement of $r$.

**Design Steps**

For given values of $Q$, $n$ and $m$ the proposed design steps are :

(*i*) Compute the optimal value of $S$-from Eqn. (4.67).

(*ii*) Find the required value of $r$ from Eq. (4.66).

(*iii*) Find $y$ from Eq. (4.63).

(*iv*) Find $b$ from Eq. (4.64).

*If the slope is to be maintained at a given value, then the computation steps will be as follows :*

(*i*) Find $r$-from Eq. (4.65)

(*ii*) Find $y$-from Eq. (4.63)

(*iii*) Find $b$-from Eq. (4.64).

**Example 4.20.** *Design an irrigation channel to carry 50 cumecs of Discharge. The channel is to be laid at a slope of 1 in 4000. The critical velocity ratio (m) for the soil is 1.1. use Kutter's rugosity coefficient (n) as 0.023. Solve by using the modern equations.*

**Note.** *This problem is the same as was there in Example 4.6, which was solved by old methods.*

**Solution.** $Q = 50 \text{ m}^3/\text{s} \quad ; \quad S = \dfrac{1}{4000}$

$m = 1.1; \qquad n = 0.023$

(*i*) *Find r by using Eq. (4.65) as :*

$$r = \left[\frac{1.607 \cdot S^{1.63} \cdot Q^{0.033}}{n^{3.26} \cdot m^{3.293}} - 0.258\right]^{-0.915}$$

Substituting the given values of $S$, $Q$, $n$ and $m$, we get

$$r = \left[\frac{1.607 \cdot \left(\dfrac{1}{4000}\right)^{1.63} \cdot (50)^{0.033}}{(0.023)^{3.26} \cdot (1.1)^{3.293}} - 0.258\right]^{-0.915}$$

$$= \left[\frac{1.607 \times 1.3447 \times 10^{-6} \times 1.1378}{4.5628 \times 10^{-6} \times 1.36869} - 0.258\right]^{-0.915}$$

$$= (0.3937 - 0.258)^{-0.915} = (0.1357)^{-0.915}$$

$$= \frac{1}{(0.1357)^{0.915}} = 6.219$$

(*ii*) *Now find y, by using Eq. (4.63) as :*

$$y = \left[\frac{1.818\, Q}{(r + 0.5)\, m}\right]^{0.3788} = \left[\frac{1.818 \times 50}{(6.219 + 0.5)\, 1.1}\right]^{0.3788} = 2.587 \text{ m ; say } 2.59 \text{ m.}$$

(*iii*) *Find b from Eq. (4.64) as :*

$$\frac{b}{y} = r \qquad \therefore \quad b = yr = 2.59 \times 6.219 = 16.1 \text{ m}$$

The dimensions of the designed channel will hence be :

$$\left.\begin{aligned} B &= \text{bed width } (b) = 16.1 \text{ m} \\ y &= \text{water depth} = 2.59 \text{ m} \end{aligned}\right] \textbf{Ans.}$$

**Note 1.** Compare this design with the design done in Example 4.6, where we obtained $B = 14.14$ m and $y = 2.7$ m

**Note 2.** These modern equations can be easily used to design canals, which would not only satisfy Kutter's & Kennedy's Eqns. but also provide desired $B/y$ ratios.

**Note 3.** Such equations can be easily used in digital computer programmings for design of irrigation canals.

**Example 4.21.** *Design an irrigation channel to carry 40 cumecs of discharge with an optimum value of Bedwidth/Depth ratio. The value of rugosity coefficient n = 0.023, and critical velocity ratio, m = 1.*

**Solution.** (*i*) Use Eq. (4.67) to determine optimal value of bed slope $S$, as

$$S = \frac{n^2\, m^{2.02}}{1.338\, Q^{0.02}} \left[0.258 + (15 + 6.44\, Q)^{-0.417}\right]^{0.6135}$$

$$\text{where } n = 0.023$$
$$m = 1$$

Substituting values, we get

$$S = \frac{(0.023)^2\,(1)^{2.02}}{1.338\,(40)^{0.02}} \left[0.258 + (15 + 6.44 \times 40)^{-0.417}\right]^{0.6135}$$

$$= 3.6725 \times 10^{-4} \left[0.3544\right]^{0.6135} = \mathbf{1.944 \times 10^{-4}}$$

(*ii*) Find $r$ from Eq. (4.66)

$$r = \left[15 + 6.44\, Q\right]^{0.382} = \left[15 + 6.44 \times 40\right]^{0.382} = \mathbf{8.519}$$

(*iii*) Find $y$ from Eq. (4.63) as :

$$y = \left[\frac{1.818\, Q}{(r + 0.5)\, m}\right]^{0.3788}$$

$$= \left[\frac{1.818 \times 40}{(8.519 + 0.5)\, 1}\right]^{0.3788} = 2.20 \text{ m}$$

(*iv*) Find $b$ from Eq. (4.64) as :

$$\frac{b}{y} = r$$

or
$$b = yr = 2.20 \times 8{,}519 = 18.78 \text{ m}$$

The designed dimensions of the channel are :

$$\left.\begin{aligned} &\text{bed width, } b = 18.78 \text{ m} \\ &\text{water depth, } y = 2.20 \text{ m} \\ &\text{Bed slope, } S = 1.944 \times 10^{-4} = \frac{1}{5144} \end{aligned}\right] \textbf{Ans.}$$

Please compare the results which were obtained for this data in Example 4.7 as

$$b = 8.35 \text{ m}, \; y = 3.34 \text{ m}, \; S = 1 \text{ in } 3700].$$

**Example 4.22.** *Make an alternative design for the canal at km 3 in Example 4.16 by using Modern Equations in place of using Garret's Diagrams as were used in Table 4.16.*

**Solution.** The data relating to canal design at km 3, was :

$$Q = 2.94 \text{ m}^3/\text{s} ; \qquad n = 0.0225$$

$$S = \frac{1}{4444} = 0.225 \times 10^{-3}$$

Side slopes $= \frac{1}{2} : 1$ ; CVR $= m = 1$

Now, instead of designing this canal section by Garret's diagrams, we can design it by using modern equations of optimal canal design, as follows :

(*i*) Find *r* from Eq. (4.65) as :

$$r = \left[ \frac{1.607\,(0.225 \times 10^{-3})^{1.63} \cdot (2.94)^{0.033}}{(0.0225)^{3.26} \cdot (1)^{3.293}} - 0.258 \right]^{-0.915} = 4.659$$

(*ii*) Find *y* from Eq. (4.63) as :

$$y = \left[ \frac{1.818 \times 2.94}{(4.659 + 0.5)\, 1} \right]^{0.3788} = \mathbf{1.01}$$

(*iii*) Find *b* from Eq. 4.64

$$b = y \cdot r = 1.013 \times 4.659 = \mathbf{4.72\ m}$$

Hence, adopt $\left.\begin{matrix} B = 4.72 \text{ m} \\ y = 1.01 \text{ m} \end{matrix}\right]$ Compare these figures with the values worked out in Table 4.16 by Garret's diagrams as :

$$\left.\begin{matrix} B = 4.5 \text{ m} \\ y = 1.05 \text{ m} \end{matrix}\right]$$

**Note.** In this manner, the use of Garret's diagrams can be completely avoided.

## PROBLEMS

1. What is sediment ? How is sediment transported in streams ? What are dunes and antinodes ? Write an equation giving the relation between total sediment transport and streamflow. Explain how would you estimate the life of a reservoir.
2. What is the difference between suspended load and bed load ?
   State the better empirical bed load formulae. Give briefly the theory of distribution and transportation of suspended matter. What is Einstein's approach ?
   How is suspended sediment measured and expressed ?
3. (*a*) Explain the following terms connected with sediment transport phenomenon:
   (*i*) Suspended load (*ii*) Bed load
   (*iii*) Regime channel (*iv*) Lacey's silt factor
   (*v*) Threshold motion of the sediment.

   (*b*) In the design of a trapezoidal canal, the following dimensions are obtained :

   side slopes = 2 vertical to 1 horizontal
   bed width = 21 m
   depth = 1.5 m
   bed slope = $2.25 \times 10^{-3}$

   Manning's friction factor for the canal was $n = 0.022$. Using Kennedy's equation on silt theory, given by $V_0 = 0.55, D^{0.64}$ with usual notations, find whether the chosen canal section is satisfactory.

4. (*a*) Discuss the mechanics involved in Sediment transport.

   (*b*) Explain in details, the Shield's method for design of channels with pitched slopes.

   (*c*) An irrigation channel is to be constructed in coarse alluvium gravel with 4 cm size. The channel has to carry 5 cumecs of discharge and the longitudinal slope is 0.04. The banks of the channel will be pitched against scouring. Find the minimum width of the channel. **[Hint : Follow example 4.1.]**

5. (*a*) Derive an expression for the average tractive force per unit of wetted area that is generated in a trapezoidal channel section of given $R$ and $S$. How does this shear stress distribution at banks differs from that at bed ?

   (*b*) Also derive an expression to prove that the shear stress required to move a grain on the bank of a channel is lesser than the shear stress required to move the grain on the channel bed.

6. (*a*) Discuss critically the statement "the banks of an unlined canal are more susceptible to erosion than its bed, and hence the stability of the banks and not of its bed is the governing factor in unlined canal designs".

   (*b*) A canal is to be designed to carry a discharge of 600 cumecs. The bed slope is kept as 1 in 1600. The soil is coarse alluvium having a grain size of 5 cm. Assuming the canal to be unlined with unprotected banks and of a trapezoidal section, determine a suitable section for the canal ; $\phi$ for the soil may be taken as 37°. **[Hint : Follow example 4.3]**

7. What is meant by "regimen of a river" ?

   Compare briefly the silt theories of Kennedy and Lacey.

   Design a regime channel to carry a discharge of 50 cumecs. Assume silt factor as 1.0.

8. When do you classify the channel as having attained regime condition ?

   Describe briefly the observations of Lacey on the regime of river.

   Design a channel using Lacey's theory to carry a discharge of 100 cumecs. Assume silt factor as 1.

9. Describe briefly the two recognised silt theories. Explain how one theory is an improvement over the other.

   What are defects in both these theories.

10. What is meant by 'regime' ? Differentiate between regime in natural rivers and in artificial channels.

    Design an irrigation channel section for the following data :

    Discharge = 40 cumecs

    Silt factor = 1.0

    Side slopes = $\frac{1}{2}$ : 1 ;

    Determine the longitudinal slope also.

11. Discuss the salient features of Kennedy's theory for the design of earth channels based on the critical velocity concept, and mention its limitations.

    Also, design an earth canal section to carry 50 cumecs discharge at a slope of 0.25 m/km, having been given that $N = 0.0225$, and $m = 1.00$, where the symbols have their usual meaning.

12. Explain what is meant by 'unlimited incoherent alluvium' in the context of Lacey's theory and discuss the concept of regime embodied in this theory for the design of earth channels in alluvium.

    Design an irrigation channel to carry 40 cumecs at a slope of 1 in 5000 with $N = 0.0225$ (Manning's) and Kennedy's $m = 0.9$.

13. (*a*) Differentiate between 'Initial regime' and 'Final regime'.

    (*b*) Design an irrigation channel section for the following data :

    Discharge = 30 cumecs ; Silt factor = 1.0 ;Side slopes = $\frac{1}{2}$ : 1.

    Draw the complete channel cross-section assuming it to be in part cutting and part filling.

**14.** (*a*) Discuss briefly the problems that arise and the methods which are adopted while designing the sections for irrigation canals in India.

(*b*) An area of 40,000 ha. has to be irrigated by a canal for growing wheat ; water requirement for which is 10 cm per month. Design and draw a suitable canal section with the data given below :

Mean slope of the ground = 1 in 3400,

Manning's roughness coefficient = $N = 0.025$,

Side slopes = 1.1.

Use Kennedy's formula. Try a depth equal to 2 m.

$$\left(\textbf{Hint:}\quad Q = \frac{40.000 \times 10^4 \times 0.1}{30 \times 24 \times 60 \times 60}\ \text{cumecs}\right)$$

**15.** Write detailed notes on any two of the following :

(*i*) Importance of sediment transport in designing earthen irrigation canals.

(*ii*) Kennedy's and Lacey's silt theories for designing irrigation canals in India.

(*iii*) Comparison of Kennedy's and Lacey's silt theories, and further improvements over Lacey's theory.

(*iv*) Use of Garret's Diagrams for designing irrigation canals.

(*v*) Suspended load and its measurement.

(*vi*) Bed load and its measurement.

(*vii*) Popular Methods of designing irrigation canals in western countries.

**16.** (*a*) What is meant by 'Suspended load' in an irrigation canal ?

(*b*) Derive an expression for measuring the suspended load concentration at any distance *y* from the bed of the channel, provided the concentration at a known distance is already known. **[Hint :** Derive equation 4.34 and 4.40]

**17.** (*a*) Define bed load, and discuss Meyer Peter's formulae for determining bed load in irrigation canal.

(*b*) Design a channel which has to carry 30 cumecs of discharge with a bed load concentration of 40 p.p.m. by weight. The average grain diameter of the bed material may be taken as 0.35 mm. Use Lacey's regime perimeter and Meyer-Peter's formulas.

**[Hint :** Follow example 4.10]

**18.** Derive Einestein's formula for bed load transport in a form which can help in deducing that the 'silt carrying capacity of a river channel increases during floods'. What other interesting conclusions can be drawn from this formula ? **[Hint :** Derive Equation 4.65]

**19.** (*a*) Draw a typical canal cross-section which is partly constructed in cutting and partly in filling. Discuss briefly its various components, such as : side slopes, berms, banks, service road, dowla, spoil banks, etc.

(*b*) What is meant by 'Balancing depth and how is it determined' ?

(*c*) The following data refer to an irrigation canal

Bed width = 1.0 m

Side slopes = 2 : 1 (in filling) and 1 : 1 (in cutting)

Top width of embankment on either side of canal = 3 m

Full supply depth = 5 m.

Free board = 1 m.

Determine the balancing depth. Draw a neat cross-section of the canal illustrating the various dimensions and the levels.

(Take ground level as 100.00 m)

**20.** Write short notes on :

(*i*) Free board in canals. (*ii*) Berms and counter berms.

(*iii*) Requirements of permanent land widths for constructing canals.

(*iv*) Spoil Banks. (*v*) Borrow pits.

(*vi*) Balancing depth. (*vii*) Fixing bed width ratio for an irrigation canal.
(*viii*) Canal standards. (*ix*) Schedule of area statistics.

**21.** **Describe briefly as to what data you will collect and how you will proceed for fixing the L-section of an irrigation canal. Also discuss step by step, the procedure that you will adopt for designing the entire length of the channel section, if the culturable commanded area and the details of the crops along the canal are given.**

# 5

# *Lining of Irrigation Canals and Economics of Lining*

## 5.1. General

Next to China, India has the largest area under irrigation. The irrigated area is constantly and continuously being increased, so as to ensure assured irrigation and to avoid crop failures due to famines and non-availability of water. To secure the benefits of irrigated land, a tremendous amount of capital has been invested in this country in the irrigation projects. So much so, that an expenditure of about Rs. 1,15,000 crores has been incurred in our country on major and medium irrigation projects since independence and up to the end of ninth five-year plan (1997–2002). This data confirms the fact that irrigation water is a costly commodity, and as such, there should be no wastage during its carriage from the reservoirs to the fields. Most of the canals, constructed in India to carry irrigation water, are unlined, and hence, a large part of the costly irrigation water is lost in percolation and absorption as seepage loss. No doubt, there are regions where the soil is such that seepage losses are very small, and there is no justification for lining them, but at the same time, it is also true that there are areas where 25 to 50% of the water is lost in seepage. This is a very serious loss and proportionately reduces the irrigation potential of the supplied water. Such seepage loss of the costly irrigation water must, therefore, be minimised. The seepage can be avoided or minimised by *lining the irrigation canals.*

By lining the canal, we mean that the earthen surface of the channel is lined with a stable (inerodible) lining surface, such as concrete, tiles, asphalt, etc. Depending upon the type of lining adopted, the seepage losses can be reduced to 2%–5% of their original values by lining the canals. Apart from reduction in seepage losses, there are various other advantages of lining the canals. All these advantages are described below in details :

## 5.2. Advantages of Lining

(1) **Seepage Control.** It has been emphasised above, that the seepage losses are considerably reduced if the channels are lined. A lined canal costs about 2 to $2\frac{1}{2}$ times as much as an unlined canal, but where seepage is heavy, the saving of costly irrigation water may itself be sufficient to fully justify the capital expenditure on lining. It should also be kept in mind that heavy seepage losses in canals would necessitate the construction of larger reservoirs and bigger dams. Prevention of seepage by lining would, therefore, reduce their impounding capacity, and hence, lower the construction costs of these works. The idea of the extent to which seepage takes place in different kinds of channels can be obtained from Table 5.1.

**Table 5.1. Seepage Losses in Channels**

| *S. No.* | *Type of lining* | *Initial rate of seepage in cumecs per million sq. m. of wetted area before lining* | *Stabilised rate of seepage in cumecs per million sq. m. of wetted area after lining* |
|---|---|---|---|
| 1. | Unlined channels. | 7.4 | 3.4 |
| 2. | 30 cm × 15 cm × 5 cm tiles using cement mortar 1 : 3 (1 cement ; 3 sand in between two layers. | 0.17 | 0.009 |
| 3. | 10 cm thick cement concrete 1 : 3 ; 6 (1 cement : 3 sand : 6 brick ballast. | 0.13 | 0.007 |
| 4. | 10 cm thick cement lime concrete 1 : 5 : 12 : 24 (1 cement : 5 lime : 12 surkhi : 24 brick ballast). | 0.40 | 0.13 |

(2) **Prevention of Water-logging.** Uncontrolled seepage through unlined canals, often raises the water-table in the surrounding fields up to or near to the ground level, as to bring the crop (plant) roots within the capillary fringe. This, in turn, brings up the alkali salts near to the ground surface, which soon renders the land unfit for cultivation. Such a land is usually called *thur*, and the phenomenon of rise of water table is known as water-logging.* It is well-known that large tracts of lands in Punjab (India and Pakistan) have been rendered uncultivable in this manner. Thus, the very canal, built to increase food production, have within a few years, caused the ruin of a large portion of agriculturable land. Lining of canals prevents seepage and, thus, protects cultivable land. Combined with land drainage schemes, lining helps to reclaim water-logged areas.

(3) **Increase in Channel Capacity.** The capacity of a chosen canal section can be considerably increased by lining it. The reason is very simple. The lining presents a smooth surface and, therefore, causes less resistance to the flow of water. The water, therefore, flows faster, and hence, more of it is carried per second than that in an unlined canal. Flow of water in unlined canals, is often, further retarded by the vegetation on the sides and bottom. And since capacity is a function of velocity, the higher the velocity, the greater is the capacity of the channel.

We know that Kutter's formula for channel velocity is

$$V = \left[ \frac{23 + \dfrac{0.00155}{S} + \dfrac{1}{n}}{\left(23 + \dfrac{0.00155}{S}\right)\dfrac{n}{\sqrt{R}} + 1} \right] \sqrt{RS}$$

where $n$ is the coefficient of roughness. The velocity, therefore, varies inversely with $n$. For unlined earth channels in good condition, the value of $n$ is about 0.0225 against 0.012 for cement mortar, 0.015 for concrete, and 0.018 for brick lined channels. Table 5.2 gives commonly used $n$ values for different types of linings.

* For details, please see Chapter 6.

**Table 5.2. Values of Mannings (*n*) for Lined Channels**

| *Type of Lining* | *Values of n for straight alignment** |
|---|---|
| Cast insitu concrete trowel finish | 0.015—0.018** |
| Cement plastered masonry | 0.012—0.015 |
| P.C.C. slabs or tiles | 0.018—0.020 |
| Brick lining | 0.018—0.020 |
| Split Boulder lining | 0.020—0.025 |
| Round Boulder lining | 0.025—0.035 |

Lining, therefore, increases channel capacity and consequently reduces the required channel section. Hence, for a new designed project, a lined channel will require lesser dimensions and hence, lesser earth work. The consequent saving in earth work handling (*i.e.* excavation and filling) and acquisition of land, thus, become possible by canal lining.

(4) **Increase in Commanded Area.** A lined canal can be designed not only smaller in cross-section but also shorter in length. The steeper gradients can be provided because higher velocities are permissible (as the material is less erodible) and a shorter alignment can, therefore, be selected. On the other hand, flatter slopes can be provided without silting on a lined channel compared to these on an unlined channel. It can, therefore, help to bring high areas under command.

(5) **Reduction in Maintenance Costs.** The up-keep of unlined canals involve huge recurring expenditure, generally charged under the head of account : *A.R.* and *M.O.* of the canal system (*i.e.* Annual repair and maintenance of the canal system). This expenditure may be required on :

(*i*) periodical removal of silt deposited on the bed and sides of the canal section ;

(*ii*) minor repairs like plugging of cracks, cuts and uneven settlements of banks ; and

(*iii*) removal of weeds and water plants.

The provision of lining reduces these charges considerably, as the cost of upkeep of a lined canal is comparatively negligible.

The above three heads are discussed below in details :

(*i*) *Removal of silt.* A lined canal is not susceptible to erosion. It is usually designed to carry the sediment load likely to enter at the canal headworks. Moreover, on account of the high velocities in lined channels, the sand blown into it during sand storms, which may occur during summer in areas like Rajasthan deserts, etc. is readily carried away. This eliminates or considerably reduces the annual expenditure required on unlined channels for desilting.

(*ii*) *Minor repairs.* Periodical plugging of holes burrowed by rats, insects, etc., is constantly required in unlined channels, failing which, breaches of channel embankments may occur. The provision of adequate lining, reduces the danger of these breaches, and lesser vigilance is required.

* For canals with non-straight alignments, loss of head by resistance to flow shall increase, and hence the given *n* values may be slightly increased.

** Lower values may be attained in canals with relatively higher discharges and in absolutely straight reaches.

(*iii*) *Removal of weeds.* Huge money is spent in removing weeds and water plants like hyaocinth, etc., from the canals. Lining eliminates or reduces this expenditure considerably, as the plants flow down the canal due to higher flow velocites in lined canals.

(6) **Elimination of Flood Dangers.** An unlined canal founded on weaker foundations is always in danger, and a breach may occur at any time. Instances have occurred where small breaches in unlined canals resulted in washing away of considerable length of embankment—leading to flooding of certain areas and causing scarcity of irrigation water in others, as the canal was out of service at a critical time for crops. A strong concrete lining removes all such dangers.

## FINANCIAL JUSTIFICATION AND ECONOMICS OF CANAL LINING

In certain cases, the lining of a channel may be required from purely technical considerations. For example, a canal constructed on a high fill or a canal founded partly on rock and partly on permeable strata, may be unsafe, unless it is lined. Sometimes a hard lined surface may be required to withstand the high flow velocities, as in power channels. Apart from these special circumstances, the engineer is required to produce a good economic justification for the capital outlay that is likely to be invested on lining. In considering the economy of canal lining, it is necessary to evaluate the tangible (which can be measured in terms of money) and additional benefits, and then to compare these with the cost of lining. *Benefit cost ratio* can, therefore, be worked out, so as to justify the necessity of lining.

Mathematically speaking, expenditure on a project is justified if the resultant annual benefits exceed the annual costs (including interest on the capital expenditure) *i.e. Benefit cost ratio is more than one.* The justification for lining the existing channels is different from that of constructing new lined channels in a new project. It is because of the fact that a large number of additional advantages ; such as lesser earth-work-handling, lesser land acquisition, lesser impounding reservoir capacities, etc., are obtained in a new project, by adopting lining for new canals.

### 5.3. Justification for Lining the Existing Canals

(*i*) **Annual Benefits.** Irrigation water is sold to the cultivators at a certain rate. Let this rate be rupees $R_1$ per cumec. If $m$ cumecs of water is saved by lining the canal, annually, then the money saved by lining $= mR_1$ rupees.

Lining will also reduce maintenance cost. The average cost of annual upkeep of unlined channel can be worked out from previous records. Let it be Rs. $R_2$. If $p$ is the percentage fraction of the saving achieved in maintenance cost by lining the canal, then the amount saved $= p \cdot R_2$ rupees.

$\therefore$ *The total annual benefits* $= mR_1 + p \cdot R_2$ ...(5.1)

(The value of $p$ is generally taken as 0.4)

(*ii*) **Annual Costs.** If the capital expenditure required on lining is $C$ rupees, and the lining has a life of say $Y$ years, then the annual depreciation charges will be $\frac{C}{Y}$ rupees. If $r$ percent is the rate of annual simple interest, then a locked up capital of $C$ rupees would earn, annually $C\left(\frac{r}{100}\right)$ rupees as interest charges, and since the capital value of the asset decreases from $C$ to zero in Y years, the average annual interest cost may be taken as $\frac{C}{2}\left(\frac{r}{100}\right)$ rupees.

∴ The total *annual costs of lining*

$$= \frac{C}{Y} + \frac{C}{2} \times \frac{r}{100} \qquad \text{...(5.2)}$$

$$\textit{Benefit cost ratio} = \frac{\text{Annual Benefits}}{\text{Annual Costs}}$$

$$= \left[\frac{m{\cdot}R_1 + p{\cdot}R_2}{\frac{C}{Y} + \frac{C}{2} \times \frac{r}{100}}\right] \qquad \text{...(5.3)}$$

If $p$ is taken as 0.4, then

$$\text{Benefit cost ratio} = \left\{\frac{m{\cdot}R_1 + 0.4\,R_2}{\frac{C}{Y} + \frac{C}{2} \times \frac{r}{100}}\right] \qquad \text{...(5.4)}$$

For project justification, benefit cost ratio must be greater than unity.

In addition to the benefits grouped above *i.e.* (water saving and reduction in maintenance cost) there may be benefits, like prevention of water–logging, reduced cost of drainage for adjoining lands, reduced risk of breaching, reduced incidence of malaria and other diseases in damp areas. Actual evaluation of these benefits is very difficult and may be approximated, based on experience, and may be taken into account for evaluating the annual benefit cost ratio.

**Example 5.1.** *An unlined canal giving a seepage loss of 3.3 cumecs per million sq. metres of wetted area is proposed to be lined with 10 cm thick cement concrete lining, which costs Rs. 180.00 per 10 sq. m. Given the following data, work out the economics of lining and benefit cost ratio.*

*Annual revenue per cumec of water from all crops= Rs. 3.5 lakhs*

*Discharge in the channel= 83.5 cumecs.*

*Area of the channel= 40.8 sq. m.*

*Wetted perimeter of the channel= 18.8 metres*

*Wetted perimeter of the lining= 18.5 metres*

*Annual maintenance cost of unlined channel per 10 sq. m.= Re. 1.00*

*Assume additional suitable data, if required.*

**Solution.**

Let us consider 1 km reach of canal.

Therefore, the wetted surface per km. = 18.8 × 1,000 = 18,800 sq. m.

(*i*) **Annual Benefits**

(*a*) *Seepage.* Seepage loss in unlined canal @ 3.3 cumecs per million sq m

$$= \frac{3.3}{10^6} \times 18{,}800 \text{ cumecs/km.} = \left(\frac{62{,}040}{10^6}\right) \text{cumecs/km}$$

Assume seepage loss in lined channel at 0.01 cumec per million sq m of wetted perimeter.

∴ Seepage loss in lined canal

$$= \frac{0.01}{10^6} \times 18{,}800 = \frac{188}{10^6} \text{ cumecs/km}$$

$$\text{Net saving } = \left(\frac{62{,}040}{10^6} - \frac{188}{10^6}\right) = \frac{61{,}852}{10^6} = 0.06185 \text{ cumec/km}$$

Annual revenue saved per km of channel

$$= \text{Rs. } \frac{61852}{10^6} \times 3.5 \text{ lakhs} = \text{Rs. } 0.21648 \text{ lakhs} = \textbf{Rs. 21,648}$$

(*b*) *Saving in Maintenance*

Annual maintenance cost of unlined channel for 10 sq. m. = Re. 1.00

Total wetted perimeter per 1 km length = 18,800 sq. m.

∴ Annual maintenance charge for unlined channel per km = Rs. 1,880.00

Assume that 40% of this is saved in lined channel

Annual saving in maintenance charges = Rs. 0.4 × 1,880.00 = Rs. 752.00

∴ Total annual benefits per km = Rs. 21,648.00 + Rs. 752.00 = **Rs. 22,400**

(*ii*) **Annual Costs.** Area of lining per km of channel

$$= 18.5 \times 1{,}000 = 18{,}500 \text{ sq. m.}$$

Cost of lining per km of channel @ Rs. 180.00 per 10 sq. m.

$$= \text{Rs. } \frac{18{,}500 \times 180.00}{10} = \text{Rs. } 3{,}33{,}000$$

Assume life of lining as 40 years

$$\text{Depreciation cost per year} = \text{Rs. } \frac{3{,}33{,}000}{40} = \text{Rs. } 8325.00$$

Assume 5% rate of interest.

$$\text{Average annual interest} = \frac{1}{2} C \cdot \frac{r}{100} = \frac{1}{2} \text{ Rs. } 3{,}33{,}000 \times \frac{5}{100} = \text{Rs. } 8325$$

$$\text{Total annual cost} = \text{Rs. } 8{,}325 + 8{,}325 = \textbf{Rs. 16,650}$$

$$\text{Benefit cost ratio} = \frac{\text{Annual benefits}}{\text{Annual costs}} = \frac{22{,}400}{16{,}650} = \textbf{1.35.}$$

B.C. Ratio is more than unity, and hence, the lining is justified.

**Example 5.2.** *An existing unlined channel is having the following dimensions :*

| | |
|---|---|
| *Width of the bottom* | *= 1.52 m.* |
| *Side slopes* | $= 1\frac{1}{2} : 1$ |
| *Depth of flow* | *= 0.91 metre* |
| *Bed-slope* | *= 0.0006.* |

*It is proposed to line this channel for the same discharging capacity. Find out the dimensions of the lined channel and work out the economics of the concrete lining, if following data are given :*

| | |
|---|---|
| *Length of irrigation season* | *= 150 days* |
| *Saving in seepage loss by lining the canal* | *= 1.5 per cent per km* |
| *Cost of water* | *= Rs. 150.00 per hectare metre* |
| *Cost of concrete lining* | *= Rs. 16.00 per sq. m.* |
| *Cost of reshaping and trimming canal* | *= Rs. 4.00 per sq. m.* |
| *Life of lining* | *= 40 years.* |
| *Interest rate* | *= 7 per cent.* |

*Annual maintenance and operational cost (per km per year)*
*for unlined canals in earth* = *Rs. 1000.00*
*and for concrete lined canals* = *Rs. 200.00*
*Other additional benefits* = *Rs. 350.00*
*Assume any other necessary data, if not given.*

**Solution.** Let us first of all work out the lined channel section which is required to pass the same discharge as is being passed by the given unlined channel.

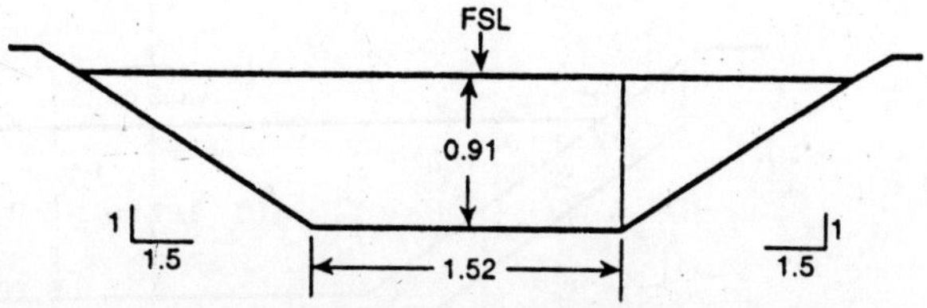


Fig. 5.1. Existing unlined canal section.

**Discharge capacity of the existing unlined channel**

If $y$ is the water depth in channel, then

$$\text{Sloping side} = y \cdot \sqrt{(3/2)^2 + (1)^2} = \frac{\sqrt{13}}{2} y = 1.8y = 1.8 \times 0.91 = 1.64 \text{ m}$$

$$\text{Wetted perimeter} = 1.52 + 2 \times 1.8y = 1.52 + 3.6y = 1.52 + 3.6 \times 0.91 = \mathbf{4.8 \text{ m}}$$

$$\text{Area} \quad (A) = \left[\frac{1.52 + (3y + 1.52)}{2}\right] y = (1.52 + 1.5y)\, y = (1.52 + 1.5 \times 0.91)\, 0.91$$

$$= 0.91\,(1.52 + 1.36) = 0.91 \times 2.88 = \mathbf{2.62 \text{ sq. m.}}$$

$$R = \frac{A}{P} = \frac{2.62}{4.8} \text{ m} = 0.546 \text{ m.}$$

**By Manning's Formula**

$$V = \frac{1}{n} \cdot R^{2/3} \cdot S^{1/2}$$

Assume $n$ for unlined channel = 0.025

Velocity through unlined channel is given as :

$$V = \frac{1}{0.025} (0.546)^{2/3} \times (0.0006)^{1/2} = \mathbf{0.654 \text{ m/sec.}}$$

$$Q = A.V. = 2.62 \times 0.654 = \mathbf{1.715 \text{ cumecs.}}$$

In a lined channel, the water will flow more rapidly and the depth of flow will be less for the same discharge. Assume $n$ for lined channel = 0.014.

**Discharge through lined channel** is given by :

$$Q = \frac{1}{n} \cdot A \cdot R^{2/3} \cdot S^{1/2} = \frac{1}{n} \cdot A \left(\frac{A}{P}\right)^{2/3} \cdot S^{1/2}$$

or
$$1.715 = \frac{1}{0.014} \, \frac{(1.52y + 1.5y^2)^{5/3}}{(1.52 + 2 \times 1.8y)^{2/3}} \times (0.0006)^{1/2}$$

or
$$1.715 = \frac{1}{0.014} \times \frac{2.45}{100} \left[\frac{(1.52y + 1.5y^2)^{5/3}}{(1.52 + 3.6y)^{2/3}}\right]$$

or
$$\frac{1.715 \times 1.4}{2.45} = \frac{(1.52y + 1.5y^2)^{5/3}}{(1.52 + 3.6y)^{2/3}} \quad \text{or} \quad 0.98 = \frac{(1.52y + 1.5y^2)^{5/3}}{(1.52 + 3.6y)^{2/3}}.$$

Solving the above equation by hit and trial, we get

$$y = 0.67 \text{ m.}$$

If we use 0.5 m as free-board, (Table 5.4) then perimeter of lining is

$$P = 1.52 + 2 \times \sqrt{2.25 + 1} \times 1.17 = \mathbf{5.74\ m}$$

Perimeter per km of lined channel = 5.74 × 1000 = **5740 sq. m.**

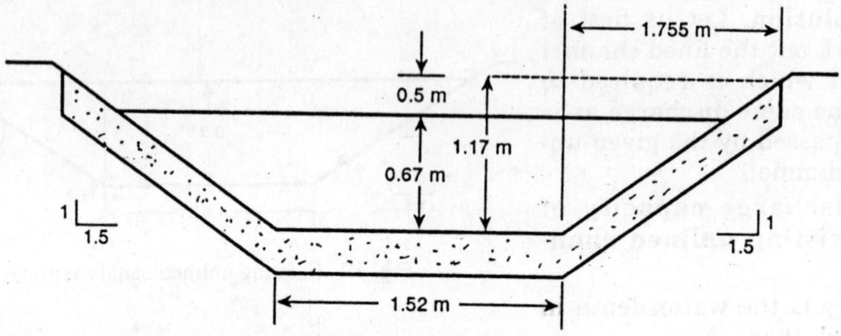


Fig. 5.2. Designed lined canal section.

(*i*) **Annual Benefits**

*Seasonal flow.* Discharge in the channel = 1.715 cumecs

1 cumec flowing for 150 days

$$= (1 \times 60 \times 60 \times 24 \times 150)\ \text{m}^3$$

1.715 cumecs flowing for 150 days

$$= (1.715 \times 3600 \times 24 \times 150)\ \text{m}^3$$

$$= \frac{1.715 \times 3600 \times 3600}{10^4} \text{ hectare–metre} = 2220 \text{ ha. m.}$$

∴ Seasonal flow = **2,220 hectare metres.**

Saving in seasonal seepage loss per km = 1.5% of flow

$$= \frac{1.5 \times 2220}{100} = \mathbf{33.3\ hectare\ metres.}$$

Annual Money saved in seepage loss per km

= 33.3 × Rs. 150 = Rs. 4995

Annual Saving in maintenance charges (per km)

= Rs. 1000 – Rs. 200 = Rs. 800

Other additional annual benefits = Rs. 350 (given)

*Total annual benefits per km* = 4995 + 800 + 350 = **Rs. 6145**

(*ii*) **Annual Costs**

Perimeter of lining = 5740 sq. m.

The cost of lining per km of channel @ Rs. 16 per sq. m.

= Rs. 16 × 5740 = Rs. 91,840

Life of lining = 40 years

Annual depreciation charges = Rs. $\frac{91,840}{40}$ = Rs. 2296

Annual interest charges

$$= \frac{1}{2}\left(C \cdot \frac{r}{100}\right) = \frac{1}{2} \times 91840 \times \frac{7}{100} = \text{Rs. } 3214$$

*Total annual costs per km* = 2296 + 3214 = **Rs. 5510** ...(*ii*)

Benefit-cost ratio $= \dfrac{\text{Annual benefits}}{\text{Annual costs}} = \dfrac{\text{Rs. } 6145}{\text{Rs. } 5510} = \mathbf{1.11}$

Since the Benefit-cost ratio is greater than unity, the lining is justified.

## 5.4. Justification for Lining Canals on New Projects

It has been stated earlier that the benefits obtained by lining canals in a new project are many more than what are obtained by lining the existing canals. The smaller canal sections, smaller earth work, smaller land acquisition, smaller and possibly fewer canal structures, reduced storage and diversion capacity, etc. are the additional benefits that are obtained in a new project.

These benefits may also be added in finding out the annual benefits for a new project.

## DESIGN OF LINED IRRIGATION CHANNELS

Irrigation canals should be aligned and laid out, so that the velocity of flow is uniform under all conditions, and so that the water reaches the irrigated area at an elevation sufficient to ensure even and economical distribution. High velocities of flow can be permitted by taking the advantage of hard wearing surface, so as to ensure a hydraulically efficient channel.

Very high flow velocities, even if not damaging to lining, do entail extra expenditure at turnouts, and require higher walls to take care of pulsations or wave action. While aligning the channel, sharp curves should also be avoided, as they not only reduce the velocity of flow, but also require higher walls on the outside to retain the water as it rounds the curve.

## 5.5. Channel Cross-sections

Generally, two types of channel sections are adopted, *i.e.*

(*i*) Triangular channel section for smaller discharges, (*ii*) Trapezoidal channel section for larger discharges. In order to increase $A/P$ ratio, the corners are rounded and attempts are made to use deeper sections by limiting depth, etc. The sections and their properties, most commonly used, are shown in Figs. 5.3 and 5.4.

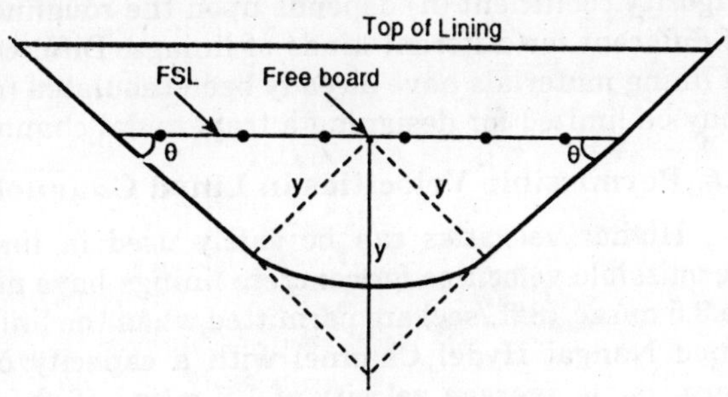


Fig. 5.3. Triangular section.

(*i*) **Triangular Section (Fig. 5.3)**

Let central depth = radius of circle = $y$

$$\text{Area} = \pi y^2 \cdot \frac{\theta}{\pi} + 2 \times \tfrac{1}{2}\, y \cdot y \cot\theta$$

$$= \frac{\pi \cdot y^2 \cdot \theta}{\pi} + y^2 \cot\theta$$

$$= y^2\, [\theta + \cot\theta]$$

$$\text{Perimeter} = 2\pi \cdot y \times \theta/\pi + 2 \cdot y \cot\theta$$

$$= 2y \cdot \theta + 2y \cdot \cot\theta = 2y\,(\theta + \cot\theta)$$

Hence, Area for Fig. 5.3 $= y^2\,(\theta + \cot\theta)$ ...(5.5)

Perimeter for Fig. 5.3 $= 2y\,(\theta + \cot\theta)$ ...(5.6)

Hydraulic mean depth for Fig. 5.3

$$= \frac{y^2\,(\theta + \cot\theta)}{2y\,(\theta + \cot\theta)} = \frac{y}{2} \quad \text{...(5.7)}$$

(*ii*) **Trapezoidal Section (Fig. 5.4)**

$$\text{Area} = B \cdot y + 2 \cdot \left(\pi \cdot y^2 \frac{\theta}{2\pi}\right) + 2\left(\tfrac{1}{2} \cdot y \cdot y \cot \theta\right)$$

or

$$A = B \cdot y + y^2\,\theta + y^2 \cot \theta$$

Hence, area for Fig. 5.4

$$= \boxed{y\,(B + y\,\theta + y \cot \theta)} \qquad \text{...(5.8)}$$

$$\text{Perimeter } (P) = B + 2\left(2\pi \cdot y \cdot \frac{\theta}{2\pi}\right) + 2\,(y \cot \theta)$$

Hence, perimeter for Fig. 5.4 = $\boxed{B + 2y\theta + 2y \cot \theta}$ ...(5.9)

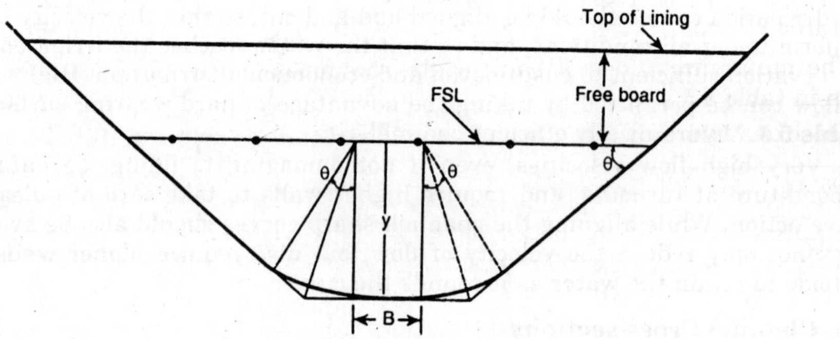


Fig. 5.4. Trapezoidal section.

The channel is, then designed, according to Manning's formula. The value of rugosity coefficient ($n$) depends upon the roughness of the channel boundary, and is different for different kinds of linings. Different values of $n$ for different types of lining materials have already been tabulated in Table 5.2. The velocity or depth may be limited for designing a trapezoidal channel section.

## 5.6. Permissible Velocities in Lined Channels

Higher velocities can be safely used in lined channels. Though maximum permissible velocities for concrete linings have not been established, velocities up to 2.5 m/sec. (8 ft./sec) are permitted when the lining is not reinforced. The concrete lined Nangal Hydel Channel with a capacity of 355 cumecs is designed for a permissible average velocity of 1.8 m/sec. (6 ft./sec.). If still higher velocities are desired, the lining can be reinforced accordingly. Asphaltic concrete, which has less resistance to abrasion, can withstand maximum velocity up to 1.5 m/sec. (5 ft./sec.). Values of limiting velocities in different kinds of linings, as per Indian Standards, are tabulated in Table 5.3.

**Table 5.3. Max. Permissible Velocities in different types of Linings**

| *Type of Lining* | *Permissible Velocity* |
|---|---|
| Cement concrete lining (Unreinforced) | 2.0 to 2.5 m/sec. |
| Burnt clay tile lining | 1.8 m/sec. |
| Boulder lining | 1.5 m/sec. |

## 5.7. Free-Board in Lined Canals.

The distance measured above the full supply level and up to the top of lining is known as the free-board for lined canals. The minimum values of free-boaxd for lined canals are specified by BIS code IS : 10430-1982, as shown in table 5.4.

**Table 5.4. Values of Free-board for Lined Canals as for IS : 10300-1982**

| *S. No.* | *Type of canal* | *Discharge (Q) in cumecs* | *Value of free-board in m* |
|---|---|---|---|
| 1. | Main and Branch canals | Q > 10 cumecs | 0.75 |
| 2. | Branch canals and Distributaries | (*i*) Q < 10 cumecs but > 5 cumecs | 0.60 |
| | | (*ii*) Q = 5.0 to 1.0 cumecs | 0.50 |
| 3. | Minors | Q < 1.0 cumecs | 0.30 |
| 4. | Water courses | Q < 0.06 | 0.1 to 0.15 |

## 5.8. Bank Widths

The minimum values of bank width recommended by IS : 10300-1982 are shown in table 5.5.

**Table 5.5. Values of Bank Widths for Lined Canals as per IS : 10300-1982**

| *S. No.* | *Discharge in cumecs* | *Minimum Top Width of Bank* | |
|---|---|---|---|
| | | *Inspection Bank (m)* | *Non inspection Bank (m)* |
| 1. | 0.15 – 7.5 | 5.0 | 1.5 |
| 2. | 7.5 – 10.0 | 5.0 | 2.5 |
| 3. | 10.0 – 15.0 | 6.0 | 2.5 |
| 4. | 15.0 – 30.0 | 7.0 | 3.5 |
| 5. | 30.0 and above | 8.0 | 5.0 |

**Example 5.3.** *Design a lined channel to carry a discharge of 15 cumecs. The available and accepted country slope is 1 in 9000. Assume suitable values of side slopes and good brick work in lining.*

**Solution.** Let us first of all assume that the side slopes of the channel be $1\frac{1}{4}$ : 1 (*i.e.* $1\frac{1}{4}$ *H* : 1 *V*) and the value of Manning's rugosity coefficient be 0.015 for good brick work. The channel section may be designed as triangular (as given in Fig. 5.3) because the discharge is small.

Considering Fig. 5.3, we have

$$\tan\theta = \frac{1}{1\frac{1}{4}} = \frac{1}{1.25} \qquad \text{or} \quad \cot\theta = 1.25$$

$$\therefore \qquad \theta = 0.675 \text{ radians}$$

Using $$A = y^2\,(\theta + \cot\theta) \qquad ...(5.5)$$

and $$P = 2y\,(\theta + \cot\theta) \qquad ...(5.6)$$

We get $$A = y^2\,(0.675 + 1.25) = 1.925\,y^2$$

$$P = 2y\,(0.675 + 1.25) = 3.85\,y$$

$$R = y/2 = 0.5\,y \qquad ...(5.7)$$

Now, using Manning's formula

$$Q = \frac{1}{n}\cdot A{\cdot}R^{2/3}{\cdot}S^{1/2}$$

$Q$ is given to be 15 cumecs.

$$\therefore \quad 15 = \frac{1}{0.015} \times (1.925y^2)\cdot(0.5y)^{2/3}\,\frac{1}{\sqrt{9000}}$$

$$\text{or} \quad 15 = \frac{1.925}{0.015 \times 94.8}\,y^2\,(0.63y^{2/3}) = 0.852\,y^{8/3}$$

$$\therefore \quad y^{8/3} = 17.6$$

$$\text{or} \quad y = (17.6)^{3/8} = 2.93 \text{ metres ; say } \mathbf{2.95\ m.}$$

With 0.75 m free-board (Table 5.4), the total depth = **13.7 m.**

Hence, use the section shown in Fig. 5.3 with 3.7 m depth, with $1\frac{1}{4}$ : 1 side slopes.

**Example 5.4.** *Design a concrete lined channel to carry a discharge of 350 cumecs at a slope of 1 in 5,000. The side slopes of the channel may be taken as $1\frac{1}{2}$ : 1. The value of n for lining is 0.014. Assume limiting velocity in the channel as 2m / sec.*

**Solution.** $$V = \frac{1}{n}\cdot R^{2/3}\cdot S^{1/2}$$

$$\therefore \quad 2 = \frac{1}{0.014} R^{2/3}\cdot\frac{1}{\sqrt{5000}}$$

$$\text{or} \quad (2 \times 0.014 \times 70.8) = R^{2/3}$$

$$\text{or} \quad R = (1.98)^{3/2} = \mathbf{2.79\ m.} \qquad ...(i)$$

The channel section is assumed to be trapezoidal as shown in Fig. 5.4. For $1\frac{1}{2}$ : 1 slopes ; we have, $\cot\theta = 1.5$ or $\theta = 34.1° = 0.59$ radians.

Using $$A = y\,(B + y\theta + y\cot\theta) \qquad ...(5.8)$$

$$P = B + 2y\theta + 2y\cot\theta \qquad ...(5.9)$$

we get $$A = y\,(B + 0.59\,y + 1.5\,y) = y\,(B + 2.09\,y)$$

$$P = B + 1.18\,y + 3y \quad = B + 4.18\,y$$

But $$A = \frac{Q}{V} = \frac{350}{2} = 175 \text{ sq.m.}$$

$$\therefore \quad 175 = y\,(B + 2.09\,y)$$

$$\text{or} \quad 175/y = B + 2.09\,y$$

$$\text{or} \quad \left(B = \frac{175}{y} - 2.09\,y\right) \qquad ...(ii)$$

From (*i*), we get

$$2.79 = R = \frac{A}{P} = \frac{175}{B + 4.18\,y} \qquad \text{or } 2.79 = \frac{175}{\left(\frac{175}{y} - 2.09y\right) + 4.18\,y}$$

$$\text{or} \quad \left(\frac{175}{y} - 2.09\,y\right) + 4.18\,y = \frac{175}{2.79} = 62.7$$

$$\text{or} \quad 175 - 2.09y^2 + 4.18y^2 = 62.7y$$

$$\text{or} \quad 2.09\,y^2 - 62.7\,y + 175 = 0 \quad \text{or} \quad y^2 - 30\,y + 83.7 = 0$$

$$\text{or} \quad y = \frac{30 \pm \sqrt{900 - 334.8}}{2} = \frac{30 \pm 23.8}{2} = \frac{30 - 23.8}{2} = \mathbf{3.2\ metres.}$$

(ignoring unfeasible + ve sign)

But $$B = \frac{175}{y} - 2.09\, y = \frac{175}{3.2} - 2.09 \times 3.2 = 54.7 - 6.7 = \mathbf{48\ m.}$$

Using free-board of 0.75 m (Table 5.4), we have depth of lined canal = 3.2 + 0.75 = 3.95 m ; Bed width = 48 m

Hence $B$ = Bed width = **48 m**
$D$ = Total Depth = **3.95 m** ] **Ans.**

**Example 5.5.** *Design a concrete lined channel to carry a discharge of 350 cumecs at a slope of 1 in 6400. The side slopes of the channel may be taken as $1\frac{1}{2}$ : 1. The value of n for lining material may be taken as 0.013. Assume limiting water depth of the channel as 4.0 m.*

**Solution.** The channel section is assumed to be trapezoidal as shown in Fig. 5.4. Now, we have

$$A = y\,(B + y\theta + y \cot \theta) \qquad ...(5.8)$$
$$P = B + 2y\cdot\theta - 2y\cdot\cot \theta \qquad ...(5.9)$$

For $1\frac{1}{2}$ : 1 slope, cot θ = 1.5 and θ = 0.59 radian.

$$\therefore \quad A = y\,(B + 0.59\,y + 1.5\,y) = y\,(B + 2.09\,y) \qquad ...(i)$$
$$P = B + 1.18\,y + 3y \qquad = B + 4.18\,y \qquad ...(ii)$$
$$y = \mathbf{4.0\ m\ (given)}$$

Therefore, from (*i*)

$$A = 4\,(B + 2.09 \times 4) = 4\,(B + 8.36) = \mathbf{4B + 33.44} \qquad ...(iii)$$

From (*ii*) $$P = B + 4.18 \times 4 = B + 16.72 \qquad ...(iv)$$

$$R = \frac{A}{P} = \frac{4B + 33.44}{B + 16.72} \qquad ...(v)$$

$$Q = \frac{1}{n} \cdot A{\cdot}R^{2/3} \cdot S^{1/2}.$$

$$\therefore \quad 350 = \frac{1}{0.013} \times (4B + 33.44) \left(\frac{4B + 33.44}{B + 16.72}\right)^{2/3} \frac{1}{\sqrt{6400}}$$

or $$350 \times 0.013 \times 80 = \frac{(4B + 33.44)^{5/3}}{(B + 16.72)^{2/3}}$$

or $$364 = \frac{(4B + 33.44)^{5/3}}{(B + 16.72)^{2/3}} \qquad ...(A)$$

**Solving Eq. (A) by Hit and Trial**

(*i*) Use $B$ = 30 m.

$$\text{L.H.S.} = 364\ ;\ \text{R.H.S.} = \frac{(120 + 33.44)^{5/3}}{(30 + 16.72)^{2/3}} = \frac{(153.44)^{5/3}}{(46.72)^{2/3}} = \mathbf{339}$$

(*ii*) Use $B$ = 32 m.

$$\text{R.H.S.} = \frac{(161.44)^{5/3}}{(48.72)^{2/3}} = \mathbf{359}$$

(*iii*) Use $B$ = 32.5 m.

$$\text{R.H.S.} = \frac{(163.44)^{5/3}}{(49.22)^{2/3}} = 364;\ \therefore\ OK$$

Hence, use $\boldsymbol{B}$ = **32.5 m** and $\boldsymbol{y}$ = **4.0 m**

Using free-board of 0.75 m, we get total depth of lined channel = 4.0 m + 0.75 = 4.75 m.

Hence, use $\left.\begin{aligned} \boldsymbol{B} &= \mathbf{32.5\ m} \\ \boldsymbol{D} &= \mathbf{4.75\ m} \end{aligned}\right]$ **Ans.**

## TYPES OF LININGS, THEIR CONSTRUCTION, AND USES

Various types of canal linings, which are commonly adopted are enumerated below:

(*A*) **Hard Surface Linings**

(1) *Cast insitu Cement Concrete lining*
(2) *Shotcrete or Plaster lining*
(3) *Cement Concrete tile lining or Brick lining*
(4) *Asphaltic Concrete lining*
(5) *Boulder lining*

(*B*) **Earth Type Linings**

(1) *Compacted earth lining*
(2) *Soil cement lining*

All the above types of linings are discussed below :

### 5.9. Hard Surface or Rigid Linings

**5.9.1. Cast Insitu Cement Concrete Lining.** Cement concrete lining made from M15 cement concrete mix (1:2:4) is considered a *good quality type of lining.* Such linings usually give very satisfactory service, and are widely used owing to their durability, impermeability, hydraulic efficiency, and for providing weed free surface. Despite the fact that the initial cost of C.C. lining is high, yet its long life and minimum maintenance cost usually makes it an economical type of lining over the life span. For this reason, concrete linings have been extensively used in America and other developed nations. India being a developing nation, however, finds it difficult to afford their high initial costs, and as such, their use is restricted only to important projects. Concrete linings have been used in India in the canals of various important projects, such as Bhakra-Nangal Project (Fig. 5.5), Tungabadhra Project, Amaravati Project, Sarda Canal in U.P., etc. About 570 km length of channels have been lined in concrete in Bhakra-Nangal Project.

**5.9.1.1. Sub-grade preparations.** Cement concrete linings are best suited for main canals which have to carry huge flows and at higher velocities. However, being rigid, its success depends to a great extent upon a stable foundation. Particular care should, therefore, be taken on thorough compaction of side slopes of the canals, on which the lining is to be laid. A firm foundation eliminates the possibility of cracking due to settlement of the subgrade. Natural earth in cutting, is usually, satisfactory ; but embankments in filling must be compacted by some suitable means.

Ordinarily, any rock or earth or predominantly sandy soil, is suitable as a subgrade material for cement concrete canal linings. However, when unusual subgrade conditions are encountered, such as expansive clays, susceptible to volume changes, then proper moisture and density control of the soil should be ensured during placing the lining. If the subgrade becomes too dry, it should be thoroughly wetted several hours before the lining is placed, so that the subgrade will be moist but not muddy at that time. Highly expensive soils like *black cotton soils,* need special treatment in the form of laying a layer of CNS (Cohesive non-swelling soil) material like *murum,* etc. in depth of about 300-1000 mm over the given soil surface before laying the lining.

Fig. 5.5. Photographic view of one of the most unique artificial channels in the world, the Nangal power canal–constituting a part of the Bhakra-Nangal project–is lined with concrete throughout its length of 64 km.

Occasionally, it may be necessary to construct a concrete lined canal in areas where the ground water is likely to rise above the invert level of lining. In such cases, it may be necessary to lay drains underneath or alongside the canal, so as to relieve the hydrostatic pressure, which might cause uplift pressure on the lining during periods of low flow or no flow.

**5.9.1.2. Thickness.** A thickness of about 5 to 15 cm of cement concrete is generally used for larger canals, depending upon the canal capacity*, the nature of canal*, and the special requirements of imperviousness and structural strength to resist cracking, on slight movement of subgrade. *Minimum thickness used in*

---

* *Hydel canals* often require greater thickness of lining than the *irrigation canals*, because of the drawdown effects and non-possibility of closure of hydel canals for repairs. Similarly, deeper canals will have greater thickness than shallow depth canals. Minimum thickness of canal lining, based on canal capacity, is shown in table 5.6 below :

**Table 5.6. Suggested Thickness of C.C. Lining for Irrigation Canals**

| *S. No.* | *Capacity of canal in cumecs* | *Depth of water in m* | *Thickness of C.C. lining in cm* |
|---|---|---|---|
| 1 | 0—5 | 0—1 | 5—6 |
| 2 | 5—50 | 1—2.5 | 6—7.5 |
| 3 | 50—200 | 2.5—4.5 | 7.5—10 |
| 4 | 200—300 | 4.5—6.5 | 9—10 |
| 5 | 300—700 | 6.5—9.0 | 12—15 |

*India is 7.5 cm.* Stable side slopes of the order of $1\frac{1}{2} : 1$ or $1\frac{1}{4} : 1$ should be adopted. Steeper slopes entail extra earth pressure and consequently require more thickness, and hence are uneconomical.

**5.9.1.3. Laying of Concrete.** The levelled and dressed sub-grade soil surface shall be moistened thoroughly before laying cement concrete, so that the moisture is not withdrawn by the sub grade from the cement concrete. In case an **inverted filter** is laid over the sub grade to take care of the differential hydrostatic pressure and draw-down in canals, it shall be properly designed keeping the coarsest filter material (gravel) layer in immediate contact with C.C. lining. *To make such filter blanket effective and to prevent ingress of concrete into it, a tar paper may be placed over the filter blanket before placing cement concrete.*

**Hand placing** of cement concrete is usually adopted in India, particularly for smaller canal sections. The cement concrete shall be dumped and spread on the bed and side slopes of the canal, usually in panels of not more than 3 m side, with suitable joints in between. The *construction joints* so formed will also serve the purpose of *contraction joints*, which are basically required for shrinkage of concrete and to take care of temperature stresses caused by the changes of temperature.

In order to prepare uniform panels, *screed guides* shall be laid on the sub grade, and the cement concrete shall be screeded up to the grade to proper thickness. Before laying the cement concrete for lining, precast cement concrete sleepers on side slopes and cast in situ C.C. sleepers on bed shall be placed under the joints, to serve as templates for accurate dressing of subgrade and to reduce the seepage through the joints. The sleepers shall be 20 cm wide and 10 cm deep for canals with capacity more than 15 cumecs ; and 15 cm wide and 7.5 cm deep for canals with capacity less than 15 cumecs. The sleepers shall be placed centrally below the joints. The C.C. used for sleepers shall be of the same grade as for lining. Mechanical Vibration of cement concrete is always preferred, for which, screed vibrators should be used.

**Mechanical placing** of cement concrete for laying C.C. lining is usually adopted on large sized projects in developed countries, by using *slip-form machines* supported on rails placed along both berms of the canal. Concrete for slip form should be *air-entrained* as to provide a more workable and slippable mix. The percentage of air recommended in shown in table 5.7.

**Table 5.7. Percentage of air for Air-entrained cement concrete for C.C. lining to be placed by slip form machines**

| *S. No.* | *Maximum Aggregate size in mm* | *Air percent by volume* |
|---|---|---|
| 1 | 10 | 8.0 |
| 2 | 12.5 | 7.0 |
| 3 | 20 | 6.0 |
| 4 | 25 | 5.0 |
| 5 | 40 | 4.5 |

For lining small to moderate sized canals, *sub grade guided slip forms* are used. The slip form is supported directly on the sub grade and operated longitudinally along it. Concrete is screeded on the bed along the canal length and on sides from bottom to top.

For larger canals of considerable length, *rail guided slip-forms* are used. Slip-forms supported on rails are placed along both berms of the canal and are operated longitudinally and on the side slopes from bottom to top.

For hand placing with the light machines, where concrete in screeded from bottom to the top of the side slope, the consistency of concrete should be such that it will barely stay on the slope. A slump of 60 to 70 mm is generally allowed. For heavier longitudinally operating slip-form machines, a slump of 50 mm at the laying point is permitted. There should be a close control on the consistency and workability of concrete, and the slumps should not vary more than 20 mm, otherwise there will be interference with the progress and quality of the work.

**5.9.1.4. Finishing.** The surface of concrete finished against forms shall be smooth and free from projections, honey-coming and other objectionable defects. Immediately on removal of forms, all unsightly ridges or lips shall be removed, and undesireable local bulging on exposed surfaces shall be remedied by tooling and rubbing. Repairs to concrete surfaces and additions where required shall be made by cutting regular openings into the concrete and placing fresh concrete to the required lines. The chipped openings shall be sharp and shall not be less than 70 mm in depth. The fresh concrete shall be reinforced and chipped and trowelled to the surface of the openings. The mortar shall be placed in layers not more than 20 mm in thickness after being compacted, and each layer shall be compacted thoroughly. All exposed concrete surfaces shall be cleaned of impurities, lumps of mortar or grout, and unsightly strains.

**5.9.1.5. Curing.** Subsequent to laying of concrete lining and after a period of 24-36 hours, the lining shall be cured for at least 28 days.

On bed, this may be done by constructing 150 mm deep earthen bunds across the bed, so that a small depth of water will regularly stand on the bed.

The curing of side slopes may be done by constructing masonry drains with weep holes or perforated pipes on the coping at the top of lining, or by sprinklers.

**5.9.1.6. Surface drainage.** Concurrently with the curing operation, surface drainage arrangement of the bank such as construction of dowels, bank surface slope away from the lining, and construction of longitudinal drain on the outer wedge shall be completed. This is necessary to prevent surface and subgrade erosion and consequent damage to the lining.

**5.9.1.7. Joints in cement concrete lining.** Cracks in C.C. lining usually occur due to :

(*i*) *Warping stresses* caused due to the difference in temperature between the atmosphere and the concrete lining, or due to moisture potential between the two faces of the C.C. slab ; and

(*ii*) *Tensile stresses* caused by the differential temperature variation between the upper and lower faces of the C.C. slab.

Cracking caused due to above causes is controlled either by providing reinforcement (as discussed in artile 5.9.1.8), or by providing joints in concrete lining, as discussed below :

The various types of joints which may be provided in C.C. lining are :

(*i*) Expansion joints ; (*ii*) Construction joints ; and

(*iii*) Contraction joints.

**Expansion joints** are usually not provided in C.C. lining, except where structures intersect the canal. At the intersecting structure, an expansion joint of 25 mm width, filled with sealing compound, is provided.

**Construction joints** which are left during casting of cement concrete lining as pointed out earlier, do serve the purpose of **contraction joints**, which are specifically required to take care of shrinkage and temperature stresses. Each *construction joint* will oppose contraction stresses, and hence will be a *contraction* *joint.* Therefore, practically, there is no difference between the two types of joints. *Theoretically, however, the joints left due to difficulty in laying the cement concrete at a stretch, are called construction joints ; whereas, the joints left intentionally for making provision for shrinkage and temperature stresses, are called contraction joints.*

The most commonly adopted type of construction joint (serving as contraction joint also) provided while placing cement concrete lining in panels on top of C.C. sleepers, is shown in Fig. 5.6 (read in conjunction with Fig. 5.7 and table 5.8). Other types of construction joints, which are in use, are shown in Fig. 5.8.

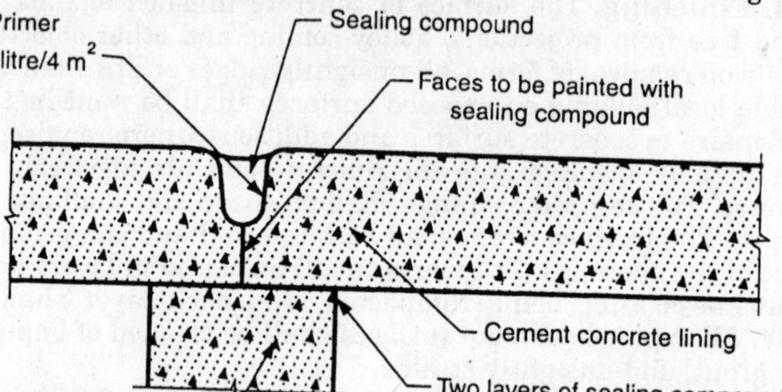


Fig. 5.6. Commonly adopted construction joints in cement concrete lining (IS : 3873-1978).

**Table 5.8. Details of Groove to be provided in Expansion joint of Fig. 5.6 and lone contraction joints provided when lining is laid at a single stretch without leaving construction joints. (IS 3873-1978)**

| *Values of distances b and c for various values of thickness (t) of lining (Fig. 5.7)* | | | *App. groove spacing (c/c)* |
|---|---|---|---|
| *t (mm)* | *b (mm)* | *c (mm)* | *m* |
| 50 | 8 | 17 | 3 |
| 65 | 8 | 20 | 3 |
| 75 and 80 | 11 | 27 | 4 to 5 |
| 90 | 11 | 30 | 4 to 5 |
| 100 and more | 11 | 33 | 4 to 5 |

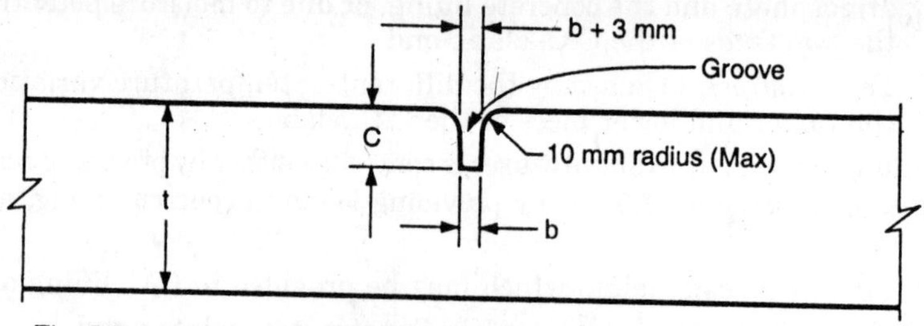


Fig. 5.7. Details of groove to form part of expansion joint of Fig. 5.6, or as an independent contraction joint.

**Note 1.** Allowable tolerance on *b* and *c* shall be ± 1.5 mm.

**Note 2.** The groove shall be cleaned and made free from foreign substances, and then filled with hot applied sealing compound, after curing period of cement concrete is over. During the curing period, course sand is filled in the groove which can be easily blown out while filling sealing compound.

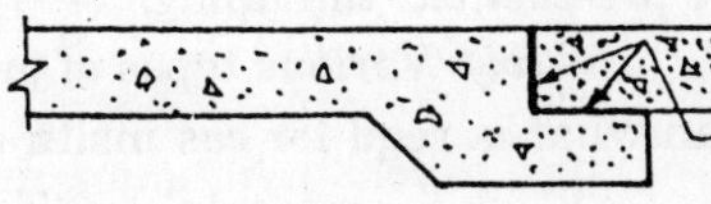


(*a*) Lap joint.

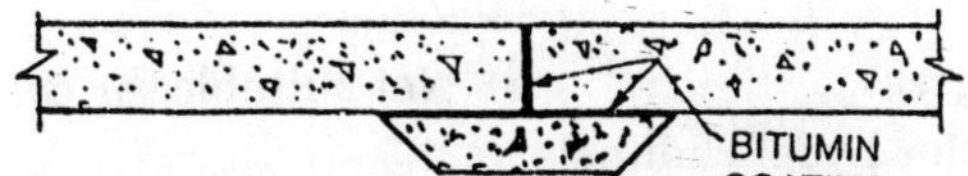


(*b*) Lap joint with separate fillet.

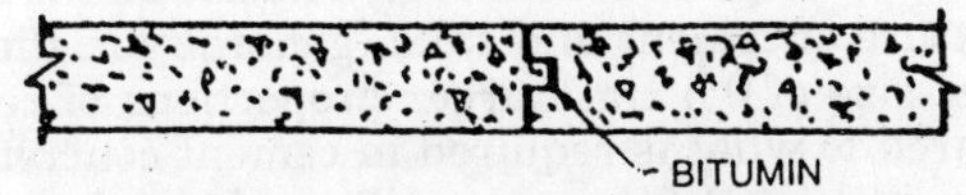


(*c*) Tongue and groove joint.

Fig. 5.8. Various types of construction joints in C.C. Linings.

**5.9.1.8.** *Reinforcement.* It has now been widely accepted that normal steel reinforcement (0.25 to 0.3% of concrete) adds practically nothing to the structural strength of uncracked lining. But it has been found that reinforcement reduces the width of shrinkage cracks, thereby reducing seepage and prevents possible faulting of the cracked slabs where unstable subgrades are encountered. In long slabs of more than 15 m or so, the intervals between transverse cracks was found to be more in unreinforced slabs as compared to those in reinforced slabs. If transverse joints can be provided at intervals sufficient to control intermediate cracking, the use of reinforcement is of no material advantage and is not justified. However, transverse expansion and contraction joints can be avoided by providing longitudinal steel reinforcement of the order of 0.5 per cent, and transverse steel reinforcement of the order of 0.25 per cent of the cross-sectional area of concrete. Such a lining will involve only the construction joints to be left after days work and longitudinal steel to be taken continuously through these joints.

**Advantages of cast in situ cement concrete linings.**

(*i*) Longer life than that of any other type.

(*ii*) Least permeable of all types.

(*iii*) Most resistant to erosion.

(*iv*) Permits fast construction by mechanical means.

(*v*) Low recurring maintenance charges.

**Disadvantages of cast-in-situ cement concrete linings.**

(*i*) Higher initial cost.

(*ii*) Greater possibility of temperature cracking.

(*iii*) Less flexible and easily affected by adverse subgrade conditions.

(*iv*) Skilled supervision and construction necessary.

**Advantages of pre-cast cement concrete tile linings (discussed in article 5.9.3)**

(*i*) Higher strength for equivalent thickness.

(*ii*) Avoids plaster finish. (*iii*) No lead of raw materials.

(*iv*) Lesser skilled labour, as campared to that required for cast in situ c.c. lining, is required for pre-cast c.c. tile lining.

(*v*) Easy to repair.

(*vi*) Various types of joints possible.

(*vii*) Lesser formwork than what is reqd for cas insitu c.c. lining, is required here.

**Disadvantages of pre-cast cement concrete tile linings (discussed in article 5.9.3.)**

(*i*) Slow progress.

(*ii*) Not suitable for curves.

(*iii*) Too light for hydel channels.

**5.9.2. Shotcrete Lining.** Shotcrete is a technical term used to designate cement mortar applied under pressure through a nozzle on the surface of the channel. It consists of a mixture of cement and sand (generally in the ratio of 1:4). Sand is having a maximum size of 0.5 cm. Larger proportions of cement are required in shotcrete as compared to what is required in cement concrete. Wire mesh reinforcement is generally, although not necessarily, used in shotcrete canal linings.

Shotcrete is of immense use for smaller jobs, because of the lighter equipment and small crew required to carry out the work. It is also useful for repair works and in rehabilitation of old canals. Shotcrete linings can be placed in an irregular canal section, thus eliminating the necessity of trimming the section, as is required for concrete canal linings when placed with a slip-form. Shotcrete lining is generally laid in a thickness of about 3.5 cm.

Excavation, compaction, curing etc. for a shotcrete lining are the same as those required for a cement concrete lining.

**5.9.3. Cement Concrete Tile Lining or Brick Lining.** Such types of linings are very popular in India, because of certain inherent advantages in their use. These *advantages* are :

(*i*) Bricks or concrete tiles can be laid by ordinary masons, and specially skilled labour, as reqd for cast insitu c.c. lining, is not required.

(*ii*) Rigid quality control is not required.

(*iii*) No expansion joints are required.

(*iv*) Rounded sections can be easily laid without using any formwork.

(*v*) Larger number of labour is required, thus providing greater employment potential.

(*vi*) Isolated damaged portions can be repaired easily.

(*vii*) Bricks can be plastered to increase the carrying capacity of canal with the same section, and also to help increase the life span of lining.

The specifications may provide either a single or a double layer of c.c. tiles or bricks laid in mortar. Sometimes, a layer of tiles is laid over a layer of brick masonry. The top layer of tiles is generally laid in 1 : 3 cement mortar over 15 mm thick layer of plaster in 1 : 3 cement plaster. The size of tiles is generally restricted to 30 × 150 × 53 mm. Typical sections are shown in Fig. 5.9 (*a*) and (*b*).

This type of lining gives a very satisfactory service. Even if there is a settlement of subgrade, the mortar joints between the bricks or tiles provide for numerous cracks so fine that seepage would be insignificant. In case of abnormal settlement, the local area can be repaired without any problem.

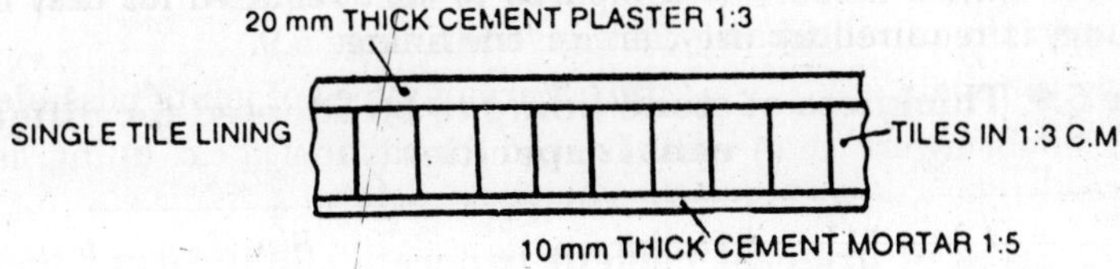


Fig. 5.9. (*a*) Single Tile Lining.

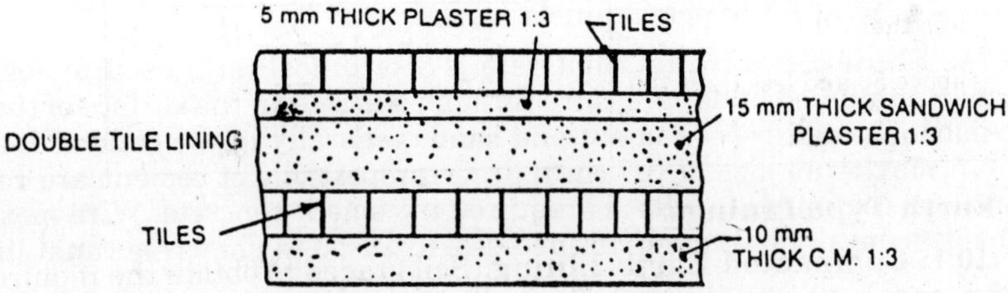


Fig. 5.9. (*b*) Double Tile Lining.

**5.9.4. Asphaltic Concrete Lining.** Asphalt has been used as a lining material at a very few places. It is still under the stage of evolution. Asphaltic concrete is a carefully controlled mixture of asphalt and graded stone aggregate, mixed and placed under elevated temperature. It provides a fairly cheap lining, especially where the asphalt is available within the country. It is flexible and readily confirms to the subgrade. The disadvantages or the limitations of this type of lining are :

(*i*) It does not decrease the rugosity coefficient of the channel.

(*ii*) It permits certain type of weed growth.

**5.9.5. Boulder Lining.** Boulder lining, also called **dry stone lining** or **stone pitching**, consists of lining the side slopes* of an earthen canal by proper placement and packing of stones, either after laying a filter layer over the soil surface, or without any such filter, depending upon the site requirement. Such a lining does not prevent seapage of canal water, though helps in retaining the shape of canal section, thereby reducing maintenance cost. The stones to be used for lining are rounded or sub-angular river cobbles, or blasted rock pieces with sufficient base area, so as to remain stable in their position.

Stones are generally placed on the levelled subgrade, and hand packed after dividing side slope length into compartments by the construction of *Dhamalis (ribs)* of stone masonry constructed at suitable intervals. The dhamalis do rest on a c.c. or R.R. masoury toe wall constructed on the drain bed along the Junction of drain bed and side slope. Usually, dhamalis in R.R. masonry @15 m centre to centre, and 0.6 m in width, having depth equal to the depth of pitching (22.5 cm or so) are usually provided.

The biggest advantage of such a lining however, is that it is a **pervious lining** allowing free flow of water from the submerged or saturated subgrade into the canal. Such a lining, therefore, does not need any drainage arrangements, in the form of pressure relief valves, etc. as may be required in concrete or brick linings. Such a lining may therefore be preferred when watertable is very high in the area, higher than DBL or even FSL of the canal.

---

* Such dry stone lining is generally adopted only on the side slopes of canals or drains, and the canal beds are rarely lined with such a lining.

The thickness of the lining and the size of the stones, which may be adopted depending upon the canal capacity, are given in table 5.9.

**Table 5.9. Thickness of stone lining to be adopted for different canal capacities**

| *S. No.* | *Canal / Drain capacity in cumecs* | *Thickness of stone lining recommended in cm.* | *Size of stones to be used* | |
|---|---|---|---|---|
| | | | *Av. dia along longest axis in cm* | *Min. dia. at any section in cm.* |
| 1. | less than 50 | 15 | 15 | 7.5 |
| 2. | above 50 and less than 100 | 22.5 | 22.5 | 11.0 |
| 3. | 100 and above | 30 | 30 | 15.0 |

## 5.10. Earth Type Linings

**5.10.1. Compacted Earth Lining.** Soil graded to obtain the required characteristics and containing enough fines, so as to make it impervious, is thoroughly compacted at optimum moisture content*, and is used to provide a lining of 30 to 90 cm thickness. The use of this type of lining is restricted to the availability of suitable soils in the area, through which the canal is being constructed. It would be uneconomical to transport the selected soils from outside.

**5.10.2. Soil-Cement Lining.** Portland cement up to the extent of 2 to 8% is added to the soil having a high percentage of fines. The mixture is first mixed dry. Water is then added so as to bring the soil to its optimum moisture content, and again mixed thoroughly. Material is then placed at site and compacted. Curing is then required for atleast seven days by covering with wet sand, etc.

## 5.11. Requirements of Good Lining

The canal linings should generally possess the following **essential requirements.**

(1) *economy ;* (2) *structural stability ;*
(3) *durability ; and* (4) *repairability,* as discussed below :

(1) **Economy.** The selection of a suitable type of lining for any canal project is mainly a question of economics and availability of *material, skilled and unskilled labour, construction machinery and equipment, and time available for completing the work.*

*The type of lining selected should not only be economical in initial cost, but also in repair and maintenance cost.*

Economic analysis of different types of linings, which can technically be used on a given project, is therefore of vital importance to us.

(2) **Structural stability.** The lining should be able to withstand the *differential sub-soil water pressure*** from behind the lining due to sub-grade*** getting saturated through seepage or rain or due to sudden drawdown of the canal. The

---

* Please refer chapter 8 in "Soil Mechanics and Foundation Engineering" by the same author to know about compaction of earth under OMC.

** Special drainage arrangements are provided for this, as discussed in article 5.13.

*** Subgrade here means the soil surface of the ground against which the lining is placed on bed as well as sides of the canal.

lining should also be sufficiently heavy and strong to withstand the effect due to local cavity formation, if any, behind the lining.

(3) **Durability.** The canal lining should be able to withstand the natural wear and tear, such as the effect of *velocity of water, rain, sunshine, frost, thawing (applicable only in cold countries), thermal and moisture changes, and chemical action of salts,* etc. It should also be able to withstand the damaging effects caused by *cattle traffic, weed and rodent growth,* etc.

(4) **Repairability.** Since the lining will get damaged with its use over a period of time, it should be such that it can be repaired easily and economically. *Brick tile* or *concrete tile* or *stone boulder linings,* or *precast slab lining* can be easily repaired, as compared to *cast in situ concrete lining.*

In addition to the above *essential requirements,* the following may be the **additional requirements**, *depending on the need of a particular project:*

(*i*) *impermeability ;*

(*ii*) *hydraulic efficiency ; and*

(*iii*) *resistance to erosion,* as discussed below :

(*i*) **Impermeability.** The permeability of lining may decide the quantum of seepage loss from a canal, which also is governed by the depth of water in the canal, and the type of subgrade soil. The permissible values of seepage losses from a canal for a particular area will depend upon the local conditions, such as the *values of land and water, population intensity,* etc. The lining chosen for a particular project should confirm to the allowable water losses.

(*ii*) **Hydraulic Efficiency.** The hydraulic efficiency of a canal, generally reduces with time, since the surface of lining gets eroded, increasing the friction factor (*n*), and thereby reducing its carrying capacity. The lining chosen for a given project must therefore, be such that its surface can be easily restored to original smoothness, if no reduction in its carrying capacity can be permitted. Cement plastered brick lining can be preferable in such a case.

(*iii*) **Resistance to erosion.** Sometimes, a canal may have to transport a considerable amount of sediment load, which may damage the lining by abrasion. Hence, in such a canal, the lining chosen should be able to withstand such abrasion. *Cement concrete and stone boulder linings may provide better abrasion resistance, as compared to brick tile lining.*

## 5.12. Factors Responsible for Selection of a Particular Type of Lining

Depending upon the project requirements as discussed above, the following factors may guide the selection of a particular lining for a given canal project :

(1) **Size and Importance of the canal.** For smaller canals, which may be used only intermittently, one may choose a lining, the construction of which may require little equipment and machinery. Larger canals on the other hand, may permit the use of cast in situ operations, with more elaborate T & P articles brought to site. Moreover, larger & important canals may require continuous operations, and, hence may need stronger linings, such as concrete lining.

(2) **Canal Slopes and Alignments.** These factors also need consideration since frequent changes in alignment and *steeper slopes* may encounter higher flow velocities, leading to selection of stronger linings. The limiting safe velocities, in generally used types of linings, are given in table 5.10, and may serve as a guide in this regard.

**Table 5.10. Limiting Velocities in Different Types of Linings**

| *S. No.* | *Type of lining* | *Safe limiting velocity* |
|---|---|---|
| 1. | Boulder lining | 1.5 m/s |
| 2. | Burnt clay tile lining | 1.8 m/s |
| 3. | Cement concrete lining | 2.7 m/s* |

(3) **Climate of the Area.** Higher quality linings should be used in areas which are susceptible to severe frosts and temperature changes such as in western countries.

(4) **Availability of Materials.** Every type of lining will require certain materials and ingredients. Some of these materials may be easily available at a certain place, and others may not be easily available. The type of lining should be such that the required materials are most easily available locally or in the vicinity of the area from where they can be carted to site with least cost.

(5) **Initial Expenditure.** Mathematically speaking, the most economical type of lining is the one which shows maximum annual benefit-cost ratio. This lining may have higher initial cost but longer life, than some other kind of lining having lesser value of annual benefit cost ratio. From long term planning point of view, the first type of lining should be chosen. But sometimes, the initial cost may be too high to be borne by the State, and hence, the lining with lesser initial cost may have to be adopted, even though its benefit-cost ratio may be less.

Keeping in mind the above factors, the suitable type of linings for different sizes of canals are given for *general guidance* in table 5.11.

**Table 5.11. General Guidelines for Choice of Linings for different sized Canals**

| *S. No.* | *Size of canal* | *Choice of Lining* |
|---|---|---|
| 1. | Canals with bed width upto 3 m | (*i*) *Single burnt clay tile lining* ; or *brick lining* where seepage considerations are important.<br>(*ii*) *P.C.C. slab lining* ; and<br>(*iii*) *flexible membrane lining*, with adequate earth/tile cover. |
| 2. | Canals with bed width between 3 to 8 m | (*i*) *single burnt clay tile lining* ;<br>(*ii*) *P.C.C. slab lining* ; and<br>(*iii*) *flexible membrane lining in the bed and rigid lining on the sides.* This may be adopted where channels have become stable and no danger of scour is expected. |
| 3. | Canals with bed width greater than 8 m | (*i*) *insitu cement concrete lining in bed as well as on sides* ;<br>(*ii*) *insitu cement concrete lining in bed and PCC slab lining on sides* ; and<br>(*iii*) *burnt clay tile lining* (*single layer on bed and double layer on sides*) to be adopted at places where aggregates for manufacture of concrete are not available economically. |

## 5.13. Under Drainage of Lined Canals (*i.e.* Drainage Behind Linings)

Many of the linings fail due to build up of water pressure behind the lining. Such a water pressure may essentially develop, when *the watertable rises above the canal bed* and there may be low flow or no flow in the canal. Such a situation will always cause an uplift on the lining, equal to the difference in the water heads, as shown in Fig. 5.10.

* Other factors, like loss of head due to friction, would, however, limit the velocity in concrete lining to 2 m/s, or so.

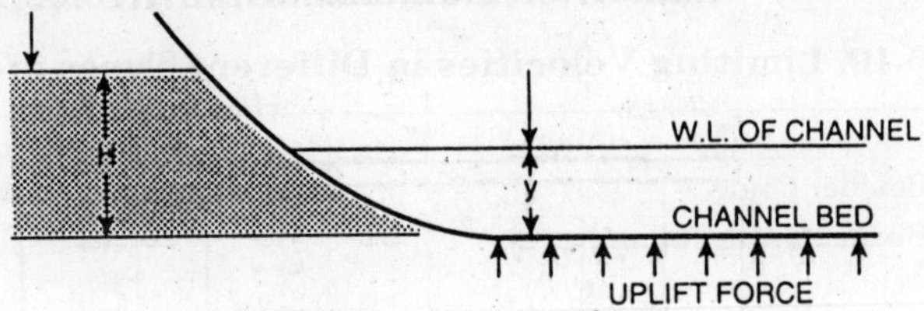


Uplift force = $H$ metres of water
If $H > y$, a net uplift equal to $(H-y)$ metres of water will be exerted on channel bed. $(H-y)$ is called *Differential Head.*
Fig. 5.10.

Such hydrostatic pressures may also be caused on the lining by the seeping rain water in the backfill, even *when the watertable remains below the canal bed, if the backfill is of low drainage* (*i.e.* of low permeability). In such a case, the backfill may get saturated over a period of time due to seepage of water through the joints and cracks; and in case of drawdown in the canal, the water in the backfill may not be drained out as quickly as the occuring drawdown, leading to building up of pressure behind the lining.

In all such cases, properly designed drainage arrangements must be provided, to help in reducing such hydrostatic pressures to safe limits. Such drainage arrangements may be provided in the form of *pressure release valves* for releasing the external water from the subgrade into the canal, through *drainage pockets* or through well connected *pipe drains* laid on the bed and sides of the canal, below the lining.

However, no such drainage arrangements may be necessary if the subgrade is made of clear gravel or sand of good permeability (more than $10^{-3}$ cm/sec or so) and the watertable is also not likely to go above the canal bed, because in such a case, there will be no appreciable time lag in the dissipation of drawdown pore pressures in the subgrade.

Drainage arrangement to be adopted will therefore depend upon : (*i*) *the position of the watertable* ; (*ii*) *upon the type of soil below the lining* ; and (*iii*) *the type of lining to be adopted.* Soil investigation is thus, of vital importance in designing such drainage arrangements.

**5.13.1. Drainage Arrangements.** Drainage arrangements, as stated above, may be accomplished by the following methods :

(1) **By providing drainage relief pockets**. Drainage relief pockets filled with *graded filter,* containing *gravel, coarse sand,* and *fine sand,* and provided with *pressure relief valves*, as shown in Fig. 5.11, may be used. The graded filter must, however, be properly designed on the considerations indicated below :

**Design of graded filter.** *The graded filter should be designed in such a way that there is no loss of bed soil. The gradation curve of the bed soil should be obtained from the sieve analysis. The $D_{15}$ size of the adjoining layer A of the filter material should be atleast 4 times as large as $D_{15}$ size of the bed soil, and less than 4 times its $D_{85}$ size. Design of the other layers should be made in the similar way, till the requirement of the filter opening is met.*

Such *drainage relief pockets* may be provided at **isolated locations** in the bed as well as sides of the lined canal, below the lining at suitable spacings. They may be cubical in shape with 60 to 90 cm size, and may be provided @ 15 to 20 m intervals, along rows in such a way that atleast one row is provided on the bed for

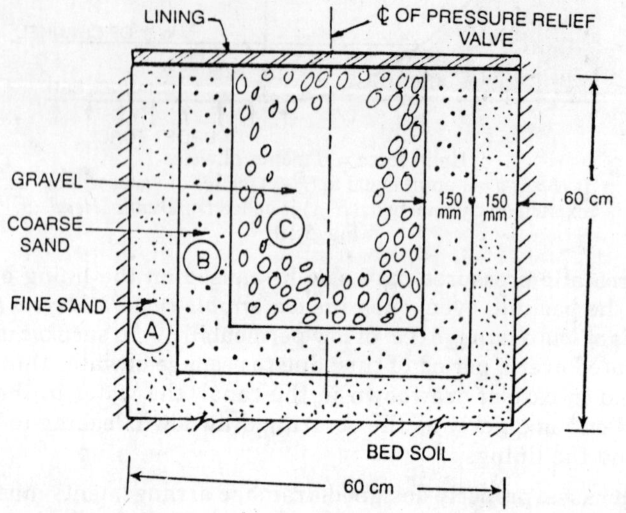


Fig. 5.11. Details of graded inverted filter to be used in drainage relief pockets.

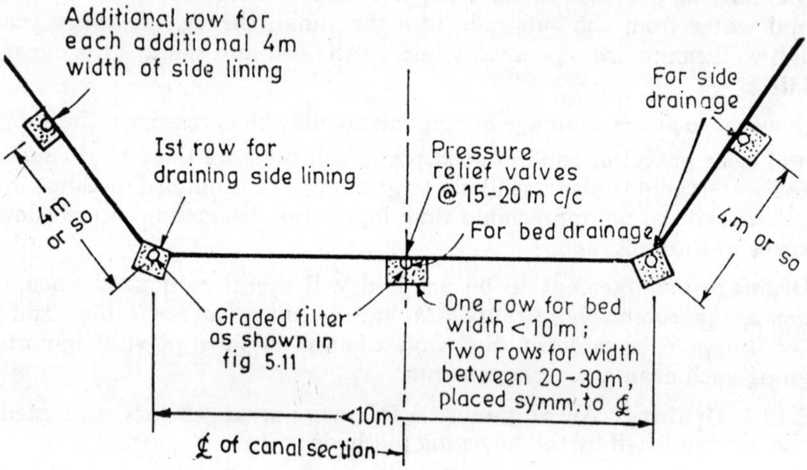


Fig. 5.12. Showing the locations of isolated drainage relief pockets.

every 10 m bed width, and one row on sides for every 4 m side width, as shown in Fig. 5.12.

Better drainage may sometimes be provided by providing **continuous drainage pockets** throughout the perimeter of the lined section of the canal, in 1 to 2 m width @ 15 to 20 m spacings, as shown in Fig. 5.13.

(2) By providing **open jointed pipe drains** in 50 cm × 50 cm sized trench, surrounded by graded gravel ( 5 to 20 mm in size) and discharging the drained water into the canal through pressure relief valves located at suitable intervals.

Such pipe drains will run logitudinally in a trench excavated below the lining on the canal bed along the length of canal, and transverse to the length of canal on the side slopes.

Water from the drains is collected in suitable masonry or precast concrete boxes, having pressure release valves placed on the top of the boxes, as shown in Fig. 5.14.

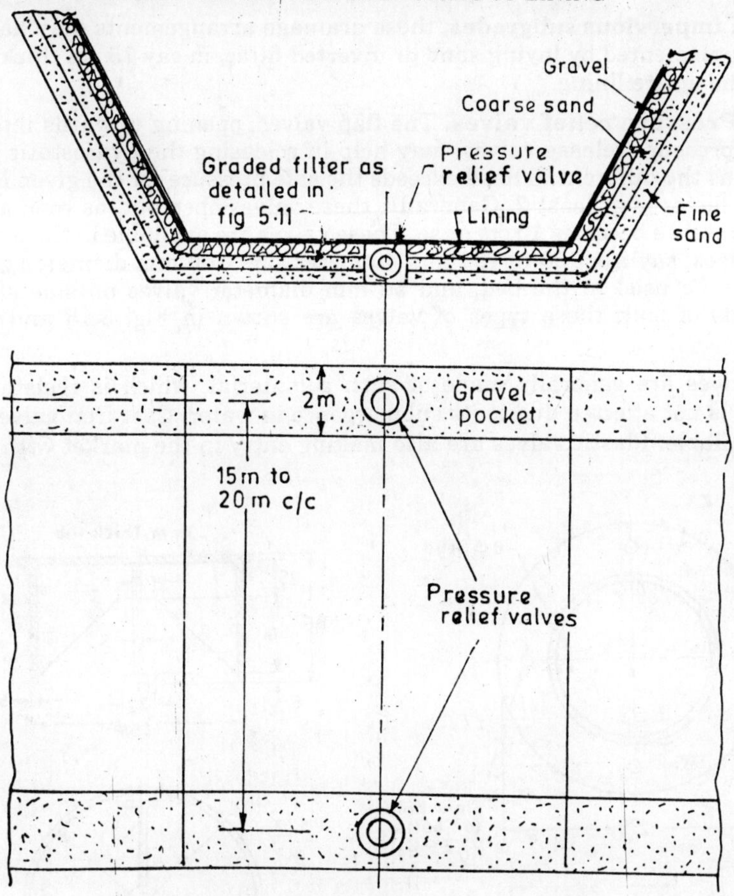


Fig. 5.13. Continuous drainage relief pockets running across the entire section in 1 to 2 m width at regular intervals along the lining length.

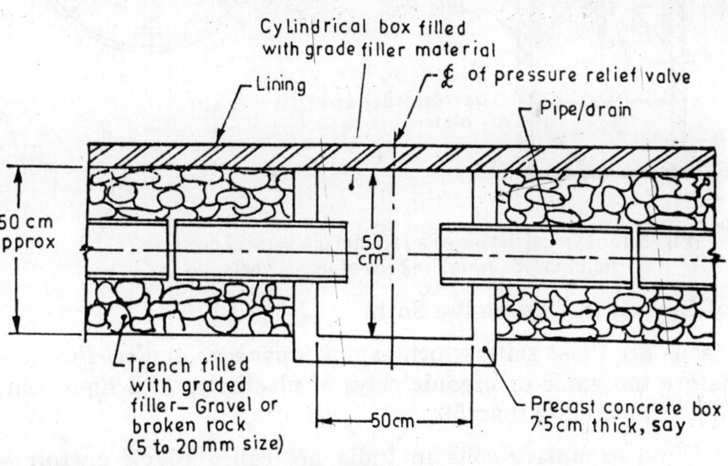


Fig. 5.14. Outlets for open jointed pipe drains.

This water is finally released into the canal through opening of the flap shutter of a pressure release valve, as and when the differential head exceeds 10 cm or so.

In case of **impervious subgrades,** these drainage arrangements may have to be further supplemented by laying sand or inverted filter, in say 75 cm thickness, underneath the entire lining.

**5.13.1.1. Pressure relief valves.** The flap valves, opening upwards into the canal, called pressure release valves, may help in releasing the hydrostatic pressure, as soon as the differential head exceeds the safe pressure for the given lining with a safety factor of at least 2. Generally, these valves open out, as soon as the differential pressure becomes 10 cm or so. These valves are available in the market in different sizes, say from 50 to 150 mm diameter size. 150 mm diameter valves may generally be used in the bed, and 50 mm diameter valves on side slopes. Typical details of both these types of valves are shown in Fig. 5.15 and 5.16, respectively.

These valves are generally made of such a material, which is resistant to abrasion and is not affected due to its submergence in water. Cast iron valves are generally available. Plastic valves are also making entry in the market with good potential.

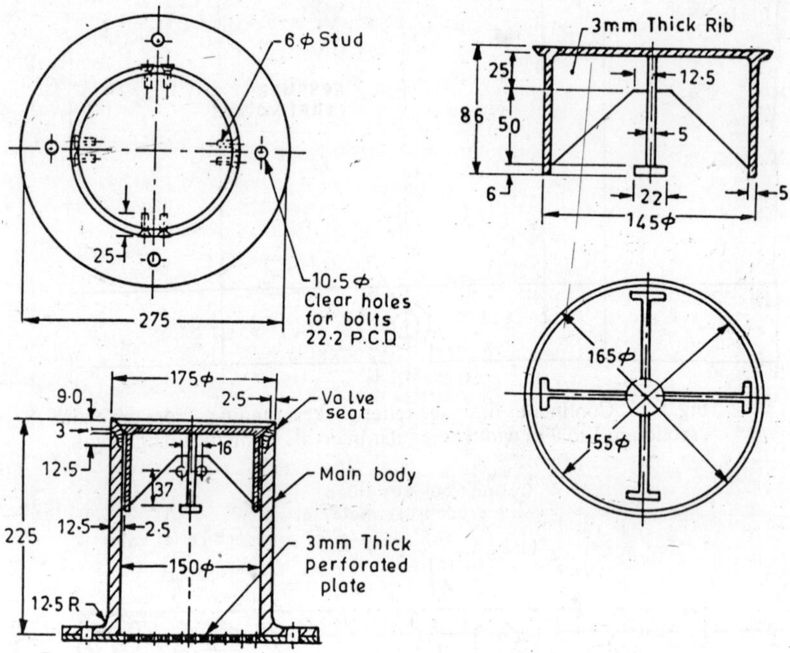


Fig. 5.15. Typical details of a 150 mm dia vertical pressure relief valve, useful on beds of lined canals.

## 5.14. Lining of Canals in Expansive Soils

Expansive soils are those soils, which exhibit *shrinkage* and *swelling* properties. Such soils are inorganic or organic clays of plasticity, with high compressibility, and liquid limits more than 50.

Commonly found expansive soils in India are called **black cotton soils.** Different varieties of black cotton soils, varying from *black* to *yellowish* to grey are found in our country. The chief property of such a soil is its *swelling pressure.* The larger is the swelling pressure, the more problemsome is the soil.

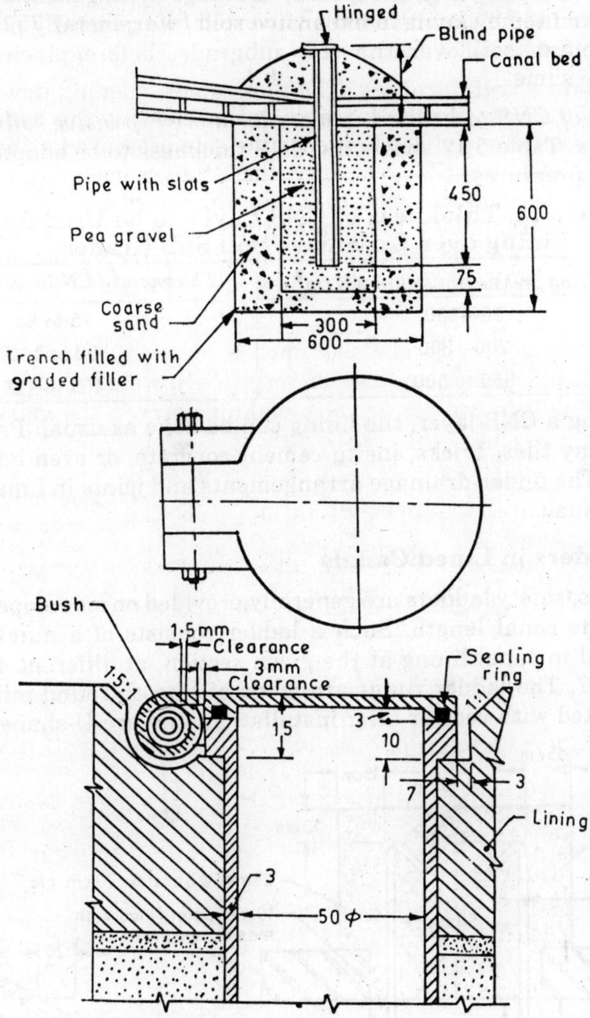


Fig. 5.16. Typical details of a 50 mm dia vertical pressure relief valve, useful on sides of lined canals.

For canals excavated in such soils, lining is found to be very advantageous, because earthen canals in such soils pose serious problems of instability of slopes ; and it is thus, very difficult to maintain their regular sections.

But when lining materials are directly placed in contact with the expansive soils, they undergo deformations by *heaving*, disturbing the lining and throwing the canal out of commission. This deformation is dependent upon the swelling pressure developed by the given expansive soil, when it imbibes water in its intra-layers.

In order to counteract such swelling pressure, it is found advantageous to place a layer of some *cohesive non-swelling soil* (CNS)* like *muram* having some cohesion, over the given expansive soil surface. The larger the thickness of CNS introduced over the expansive soil, the lesser would be the resulting swelling and deformations.

* CNS material shall be non-swelling with a min. of 10 $KN/m^2$ (1020 $kg/m^2$) of cohesion, and swelling pressure not more than 15 $kN/m^2$ at optimum moisture content. Most **murums** with some cohesive property perform satisfactorily as CNS material.

Hence, in order to line canals in expansive soils, we generally place a layer of cohesive non-swelling soil over the soil subgrade, before placing the lining materials over the same.

*The thickness of CNS to be used is more for more expansive soils and less for less expansive soils.* Table 5.12 shows the CNS thickness to be adopted for soils of different swelling pressures.

**Table 5.12. Thickness of CNS Layer to be Used for Lining over Expansive Soil Sub-Grades**

| *S. No.* | *Swelling pressure of soil in* $KN/m^2$ | *Thickness of CNS material in cm** |
|---|---|---|
| 1. | 50—150 | 75 to 85 |
| 2. | 200—300 | 90 to 100 |
| 3. | 350 to 500 | 105 to 125 |

After providing a CNS layer, the lining can be done as usual. Precast cement concrete burnt clay tiles, bricks, insitu cement concrete, or even boulder linings may be adopted. The under-drainage arrangements and joints in lining would also be provided, as usual.

## 5.15. Safety Ladders in Lined Canals

In large canals, safety ladders are generally provided on side slopes, at suitable intervals along the canal length. Such a ladder consists of a number of *ladder rungs,* constructed in canal lining at the given section, at different elevations, as shown in Fig. 5.17. The ladder rungs are made of smooth round mild steel bars, galvanised or coated with coaltar after installation, and are U-shaped as shown.

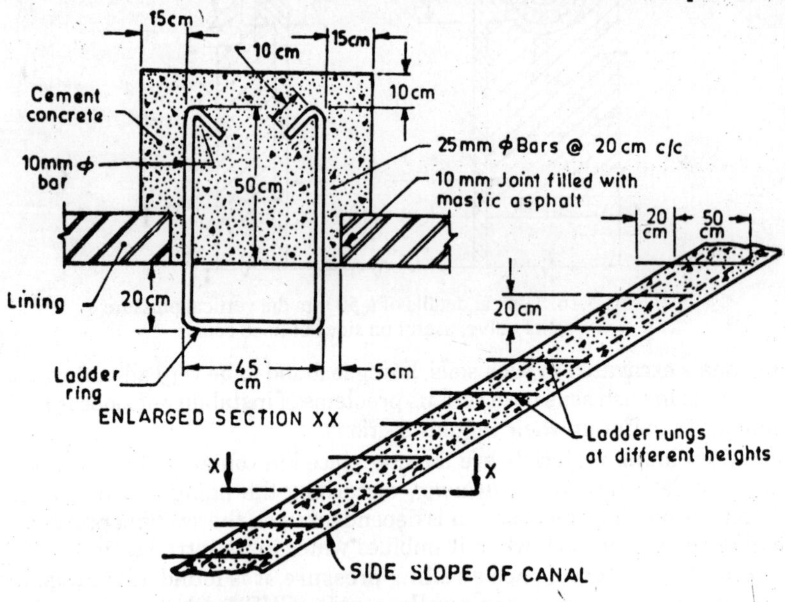


Fig. 5.17. Details of safety ladder.

The ladders are provided on both banks, alternatively at about 300 metres-staggered distance, in straight reaches. Such ladders shall also always be constructed on both banks at about 30 m upstream of the point, where the canal enters some underground structure.

* Optimum thickness can be determined only be actual experiments in the field or the lab.

These ladders can be used by persons or cattle, that may be swept away with the flowing water in the canal. Such a person, who may otherwise get drowned, may catch hold of one such ladder rung at the level of the flowing water, and can climb the higher rungs to get out of the canal, easily.

## PROBLEMS

**1.** What is meant by 'canal lining', and what are its advantages ?

Enumerate the different types of canal linings, and discuss the design and construction features of concrete linings.

**2.** (*a*) Enumerate the different types of lining materials, and discuss the factors which are responsible for selecting a particular material in a particular project.

(*b*) Design a triangular concrete lined channel to carry a discharge of 20 cumecs at a slope of 10 cm/km. The side slopes of the channel are $1\frac{1}{4}$ : 1. The value of $n$ may be taken as 0.015. **[Hint** : Follow example 5.3].

**3.** (*a*) Explain the necessity of lining of canals. Enumerate the various types of linings practised in Tamil Nadu State.

(*b*) Design a trapezoidal concrete lined channel to carry a discharge of 350 cumecs at a slope of 1 in 6400. The side slopes of the channel may be taken as $1\frac{1}{2}$ : 1. The value of $n$ for the lining material may be taken as 0.013. Assume the limiting $B/D$ ratio to be 5. **[Ans.** Use $B = 20$ m, $D = 4$ m]

**4.** What are the main types of channel linings ?

Design a concrete lined trapezoidal channel to carry a discharge of 200 cumecs at a slope of 1 in 5000. The side slopes of the channel are 1 : 1 and Manning's coefficient of rugosity may be taken as 0.014. Assume the limiting velocity in the channel as 2 m per second.

**[Hint** : Follow example 5.4]

**5.** (*a*) How will you justify economically the necessity of lining an existing canal ? What added benefits you will expect if the canal to be lined is new and yet to be constructed ? **[Hint** : Please see articles 5.3 and 5.4]

(*b*) What is meant by 'under-drainage of lined canals', and how is it provided ?

**6.** Write short notes on the following :

(*i*) Lining of irrigation canals.

(*ii*) Under-drainage of lined canals.

(*iii*) Use of cement concrete for lining canals.

(*iv*) Economics of canal lining.

(*v*) Various types of canal linings.

(*vi*) Financial justification for lining new canals.

(*vii*) Advantages of lining irrigation canals.

(*viii*) Safety ladders for large canals.

# 6

# *Reclaimation of Water-Logged and Saline Soils for Agricultural Purposes*

## 6.1. Definition of Salinity and Water-logging

*An agricultural land is said to be water-logged, when its productivity gets affected by the high watertable.* The productivity of land infact, gets affected when the root zone of the plants gets flooded with water, and thus become ill-aerated. Ill aeration reduces crop yield, as explained below :

The life of a plant, infact, depends upon the nutrients like nitrates, and the form in which the nitrates are consumed by the plants is produced by the bacteria, under a process called *nitrification.* These bacteria need oxygen for their survival. The supply of oxygen gets cutoff when the land becomes ill aerated, resulting in the death of these bacteria, and fall in the production of plant's food (*i.e.* nitrates) and consequent reduction in the plant growth, which reduces the crop yield. Apart from ill-aeration of the plants, many other problems are created by water-logging, as discussed below :

(*i*) The normal cultivation operations, such as tilling, ploughing, etc. cannot be easily carried out in wet soils. In extreme cases, the free water may rise above the surface of the land, making the cultivation operations impossible. In ordinary language, such a land is called a swampy land.

(*ii*) Certain water loving plants like grasses, weeds, etc. grow profusely and luxuriantly in water-logged lands, thus affecting and interfering with the growth of the crops.

(*iii*) Water-logging also leads to salinity, as explained below :

If the watertable has risen up, or if the plant roots happen to come within the capillary fringe, water is continuously evaporated by capillarity. Thus, a continuous upward flow of water from the watertable to the land-surface, gets established. With this upward flow, the salts which are present in the water, also rise towards the surface, resulting in the deposition of salts in the root zone of the crops. The concentration of these alkali salts present in the root zone of the crops has a corroding effect on the roots, which reduces the osmotic activity of the plants and checks the plant growth, and the plant ultimately fades away. Such soils are called **saline soils**. From the above discussion, it becomes evident that the water-logging ultimately leads to salinity, the result of which is, the reduced crop yield. For this reason, salinity and water-logging are treated as a twin problem; under the head 'salinity and water-logging'. *Whenever there is water-logging, salinity is a must.*

## 6.2. Causes of Water-logging

Water-logging is the rise of watertable, which may occur due to the following factors :

**(i) Over and Intensive Irrigation.** When a policy of intensive irrigation is adopted, then, the maximum irrigable area of a small region is irrigated. This leads to, too much of irrigation, in that region, resulting in heavy percolation and subsequent rise of watertable. For this reason, to avoid water-logging, a policy of *extensive irrigation* (*i.e.* irrigation spread over wider regions) should supersede the policy of *intensive* irrigation.

**(ii) Seepage of Water from the Adjoining High Lands.** Water from the adjoining high lands may seep into the sub-soil of the affected land and may raise the watertable.

**(iii) Seepage of Water through the Canals.** Water may seep through the beds and sides of the adjoining canals, reservoirs, etc., situated at a higher level than the affected land; resulting in high watertable. This seepage is excessive, when soil at the site of canals, reservoirs, etc. is very pervious.

**(iv) Impervious Obstruction.** Water seeping below the soil moves horizontally (*i.e.* laterally) but may find an impervious obstruction, causing the rise of watertable on the upstream side of the obstruction. Similarly, an impervious stratum may occur below the top layers of pervious soils. In such cases, water seeping through the pervious soils will not be able to go deep, and hence, quickly resulting in high watertable.

**(v) Inadequate Natural Drainage.** Soils having less permeable sub-stratum (such as clay) below the top layers of pervious soils, will not be able to drain the water deep into the ground, and hence, resulting in high water level in the affected soil.

**(vi) Inadequate Surface Drainage.** Storm water falling over the land and the excess irrigation water should be removed and should not be allowed to percolate below. If proper drainage is not provided, the water will constantly percolate and will raise the level of the underground reservoir.

**(vii) Excessive Rains.** Excessive rainfall may create temporary water-logging, and in the absence of good drainage, it may lead to continued water-logging.

**(viii) Submergence due to Floods.** If a land continuously remains submerged by floods, water loving plants like grasses, weeds, etc. may grow, which obstruct the natural surface drainage of the soil, and thus, increasing the chances of water-logging.

**(ix) Irregular or Flat Topography.** In steep terrain, the water is drained out quickly. On flat or irregular terrain having depressions, etc., the drainage is very poor. All these factors lead to greater detention of water on the land, causing more percolation and raised watertable.

## 6.3. Water-logging Control

It is evident that water-logging can be controlled only if the quantity of water into the soil below is checked and reduced. To achieve this, the inflow of water into the underground reservoir should be reduced and the outflow from this reservoir should be increased, as to keep the highest position of water-table at least about 3m below the ground surface. The various measures adopted for controlling water-logging are enumerated below :

**(1) Lining of Canals and Water Courses.** Attempts should be made to reduce the seepage of water from the canals and water courses. This can be achieved by lining them. It is a very effective method to control water-logging.

(2) **Reducing the Intensity of Irrigation.** In areas where there is a possibility of water-logging, intensity of irrigation should be reduced. Only a small portion of irrigable land should receive canal water in one particular season. The remaining areas can receive water in the next season, by rotation.

(3) **By Introducing Crop-rotation.** Certain crops require more water and others require less water. If a field is always sown with a crop requiring more water, the chances of water-logging are more. In order to avoid this, a high water requiring crop should be followed by one requiring less water, and then by one requiring almost no water. Rice may be followed by wheat, and wheat may be followed by a dry crop such as cotton.

(4) **By Optimum Use of Water.** It is a known fact that only a certain fixed amount of irrigation water gives best productivity. Less than that and more than that, reduces the yield. But most of our cultivators are unaware of these technicalities, and they feel that by using more water they can increase crop yield. Therefore, they try to use more and more water. This can be checked by educating the cultivators by proper propaganda. Moreover, the revenue should not be charged on the basis of irrigated area but should be charged on the basis of the quantity of water utilised. A strict watch should also be kept at the outlet, in order to stop undue tapping.

(5) **By Providing Intercepting Drains.** Intercepting drains along the canals should be constructed, wherever necessary. These drains can intercept and prevent the seeping canal water from reaching the area likely to be water-logged.

(6) **By Provision of an Efficient Drainage System.** An efficient drainage system should be provided in order to drain away the storm water and the excess irrigation water. A good drainage system consists of surface drains as well as sub-surface drains (described in details a little later).

(7) **By Improving the Natural Drainage of the Area.** To reduce the percolation, the water should not be allowed to stand for a longer period. Some relief in this direction can be obtained by removing the obstructions from the path of natural flow. This can be achieved by removing bushes, jungles, forests, etc. and improving the slopes of the natural drainage lines.

(8) **By Adopting Consumptive Use of Surface and Subsurface Water.** The introduction of lift irrigation to utilize ground water helps in lowering the water-table in a canal irrigated area, where water-table tends to go up. *Hence, the ground water should also be used in conjunction with canal water for irrigation,* as the continuous use of ground water will not allow any appreciable rise in the level of water-table, due to continuous seepage of canal water.

*This combined use of subsurface water (ground water) and the surface water (canal water) in a judicious manner, as to derive maximum benefits,* called **conjunctive use**, should hence be adopted to control water-logging.

## 6.4. Reclamation of Saline and Alkaline Lands

Land reclamation is a process by which an unculturable land is made fit for cultivation. Saline and water logged lands give very less crop yields, and are, therefore, almost unfit for cultivation, unless they are reclaimed. Before summarising the remedies for reclaiming such lands, we shall first review the process, whereby, a land becomes, 'saline' or in extreme cases 'alkaline'.

Every agricultural soil contains certain mineral salts in it. Some of these salts are beneficial for plants as they provide the plant foods, while certain others prove injurious to plant growth. These injurious salts are called *alkali salts* and their common examples are $Na_2CO_3$, $Na_2SO_4$, and NaCl. **$Na_2CO_3$ or black alkali* is the most harmful; and NaCl is the least harmful**. These salts are soluble in water.

* It is known as black alkali because it dissolves some organic constituents of soil, which when in solution with it, appear black. The ground, therefore, gets spotted with patches of black stain.

If the watertable rises up, or if the plant's roots happen to come within the capillary fringe, water from the watertable starts flowing upward. The soluble alkali salts also move up with water and get deposited in the soil within the plant roots as well as on the surface of the land. *This phenomenon of salts coming up in solution and forming a thin (5 to 7.5 cm) crust on the surface, after the evaporation of water, is called effeorescence.* Land affected by effeorescence is called *saline soil*. The salty water surrounding the roots of the plants reduces the osmotic activity of the plants, as explained below :

Since the plant roots act as semi-permeable membranes, so we have almost pure water on one side of the membrane (*i.e.* the water already extracted by the roots) and highly concentrated salt solution on the other side. Now, from the knowledge of physical chemistry, we can conclude that pure water from within the roots will start flowing out of the roots by **'osmosis'** towards the salt solution, until the pressure on pure water side becomes equal to the osmotic pressure of the salt solution. The plant will, hence, die due to lack of water, as shown in Fig. 6.1.

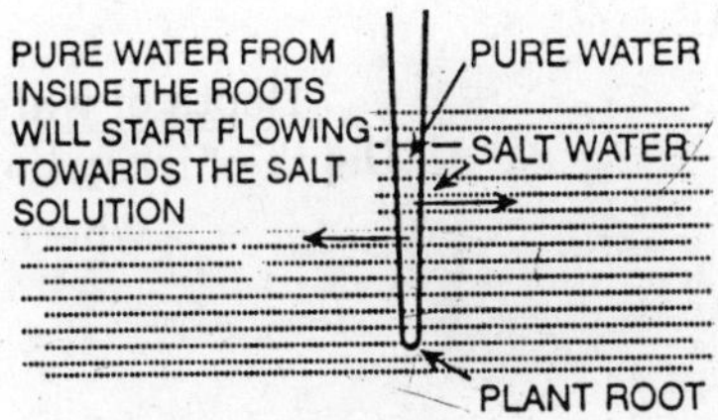

Fig. 6.1

Such a salt affected soil is unproductive and is known as "saline soil". If the salt effeorescence continues for a longer period, a base exchange reaction sets up, particularly if the soil is clayey, thus *sodiumising the clay*, making it impermeable and, therefore, ill-aerated and highly unproductive. Such soils are called *alkaline soils*. The reclamation of alkaline lands is more difficult.

*Reclamation of salt affected lands.* It is evident from the above discussion that effeorescence can be avoided if the watertable is maintained sufficiently (about 3m) below the roots, so that the capillary water is not able to reach the root zone of the plant. *Hence, all those measures which were suggested for preventing water-logging hold good for preventing salinity of lands also.* An efficient drainage system consisting of surface drains as well as sub-surface drains (explained in the next article) must be provided in order to control and lower the watertable in saline lands. After the high watertable has been lowered by suitable drainage, the soil is freed from the existing salts by a process, called *Leaching*.

**6.4.1. Leaching.** In this process, the land is flooded with adequate depth of water. The alkali salts present in the soil, get dissolved in this water, which percolate down to join the watertable or drained away by surface and sub-surface drains. The process is repeated till the salts in the top layer of the land are reduced to such as extent that some salt resistant crop can be grown. This process is known as leaching. High salt resistant crops like fodder, berseem, bajra etc., are now grown on this leached land for one or two seasons or till the salinity is reduced to such an extent that an ordinary crop like wheat, cotton, citrus garden crops, etc. can be grown. The land is then said to have been reclaimed.

When sodium carbonate ($Na_2CO_3$) is present in the saline soil, gypsum ($CaSO_4$) is generally added to the soil before leaching and thoroughly mixed with water. $Na_2CO_3$ reacts with $CaSO_4$ forming $Na_2SO_4$, which can be leached out as explained earlier.

**6.4.1.1. Leaching requirement (LR) of a soil.** In order to maintain status quo on the salinity of a given soil, and to avoid any further increase in its salinity, it is necessary to apply water to the soil in excess of the consumptive use (*i.e.* the requirement to meet evapotranspiration needs). This excess water will flow down beyond the root zone of the crop to the underground drainage system or to the underground reservoir, washing down the excess salts, which otherwise would have been deposited in the soil to further increase the salinity of the soil.

This excess water, which is required to meet the leaching needs, is generally expressed as the percentage of the total irrigation water applied to the soil (field) to meet the consumptive use as well as the leaching needs. This percentage quantity of water required for maintaining equilibrium in the salt content of the soil, has been computed to be expressed by the following equation :

$$LR \text{ (Leaching Requirement)} = \frac{D_d}{D_i}$$

$$= \frac{\text{Depth of water drained out per unit area}}{\text{Depth of irrigation water applied per unit of area}} \quad ...(6.1)$$

where $D_i$ = Total irrigation water depth applied.

$= C_u + D_d$

$C_u \downarrow$ Consumptive use + $D_d \downarrow$ Drained outwater depth

$$\therefore \quad L.R. = \frac{D_d}{D_i} = \frac{D_i - C_u}{D_i} \quad ...(6.2)$$

For salt equilibrium, the ratio $D_d/D_i$ is found to be equal to $C_i/C_d$; where $C_i$ is the salt content of irrigation water, and $C_d$ is the salt content of drainage or leached water. Since the salt content is directly proportional to the Electrical conductivity* (*EC*), $\frac{C_i}{C_d}$ will be equal to $\frac{EC_{(i)}}{EC_{(d)}}$ where $EC_{(i)}$ is the *electrical conductivity of irrigation water*; and $EC_{(d)}$ is the *electrical conductivity of drained water* (leached water or leaching water). Hence, equation (6.1) can be written as :

$$L.R. = \frac{D_d}{D_i} = \frac{EC_{(i)}}{EC_{(d)}} \quad ...(6.3)$$

The EC of drainage water or leaching water, *i.e.* $EC_{(d)}$, may be assumed on the basis of permissible salt tolerance limit of the grown crop, but is generally assumed to be twice the EC value of the *saturation soil extract*** *i.e.*, $EC_{(e)}$. Hence, eqn. (6.3) can also be written as :

$$L.R. = \frac{D_d}{D_i} = \frac{EC_{(i)}}{EC_{(d)}} = \frac{EC_{(i)}}{2EC_{(e)}} \quad ...(6.4)$$

**Example 6.1.** *Estimate the leaching requirement when electrical conductivity (EC) value of a saturated extract of soil is 10 m mho/cm at 25% reduction in the yield of a crop. The EC of irrigation water is 1.2 m.mho/cm. What will be the required depth of water to be applied to the field if the consumptive use requirement of the crop is 80 mm? EC value of the leaching water may be suitably assumed.*

---

* **Electrical conductivity** is a measure of salt content in a given water sample, and is dealt in more details in article 1.7(2).

** The water solution extracted from a soil at its saturation percentage.

**Solution.** The given values are :

$EC_{(e)}$ = E.C. value of saturated soil extract = 10 milli mho/cm

$EC_{(i)}$ = E.C. value of irrigation water = 1.2 milli mho/cm

$C_u$ = Consumptive use = 80 mm

The *Leaching Requirement* (LR) is given by the Eqn. (6.3) as :

$$LR = \frac{D_d}{D_i} = \frac{EC_{(i)}}{EC_{(d)}}$$

where $EC_{(d)}$ is the E.C. value of leaching water, which may be assumed to be equal to $2 \cdot EC_{(e)} = 2 \times 10$ m.mho/cm = 20 milli mho/cm

Substituting the above values, we get

$$L.R. = \frac{EC_{(i)}}{EC_{(d)}} = \frac{1.2 \text{ milli mho/cm}}{20 \text{ milli mho/cm}} = \frac{1.2}{20} \times 100\% = 6\%$$

Hence, the Leaching Requirement is 6%. **Ans.** ...(1)

Now using eqn. (6.2), we have

$$LR = \frac{D_d}{D_i} = \frac{D_i - C_u}{D_i} = \frac{D_i - 80 \text{ mm}}{D_i} \times 100\% \quad ...(2)$$

Equating (1) and (2), we have

$$6 = \left(\frac{D_i - 80 \text{ mm}}{D_i}\right) \times 100$$

or $\quad 6\,D_i = 100\,D_i - 8000$ mm $\quad$ or $\quad 94\,D_i = 8000$ mm

or $\quad D_i = \frac{8000}{94}$ mm = **85.1 mm**

Hence, the required water depth for irrigation = **85.1 mm Ans.**

## LAND DRAINAGE

Surface irrigation is a blessing only if it is practised with great care. Only optimum amount of water should be supplied to the crop, in accordance with the requirement of that crop, and the properties of the soil must be given full consideration. Excess water, which the root zone of the soil fails to absorb, may percolate and help in raising the watertable. Sometimes, this gravity water may encounter an impervious stratum and may not be drained up to the watertable. As explained earlier, this excess water is not only a waste but may be harmful to crop yield also. If such conditions are likely to occur, it becomes necessary that the excess water is removed and drained out from below the soil and discharged back either into a river, a canal, etc. or somewhere else. *Hence, while designing a canal irrigation network, it is sometimes desirable to provide a suitable drainage system, for removing the excess irrigation water.* This may be necessary in areas of high watertable and in river deltas, when irrigation facilities are extended to such areas. Drainage system is also required for draining out the storm water, and thus to prevent its percolation and to ensure easy disposal. Two types of drainage can be provided, *i.e.*,

(1) Surface drainage, (2) Sub-surface drainage, called Tile-drainage or Underground drainage. These are explained below :

## 6.5. Surface Drainage or Open Drainage

Surface drainage is the removal of excess rain water falling on the fields or the excess irrigation water applied to the fields, by constructing open ditches, field drains, and other related structures. The land is sloped towards these ditches or drains, as to make the excess water flow in to these drains.

When irrigation is extended to arid regions, drainage ditches become necessary to remove water required for leaching undesirable salts from the soil, and to dispose off the excess rainfall.

The open drains, which are constructed to remove the excess irrigation water collected in the depressions on the fields, as well as the storm (rain) water, are broad and shallow, and are called *shallow surface drains*. These drains carry the runoff to the *outlet drains*, which are large enough to carry the flood water of the catchment area from the shallow surface drains, and are of sufficient depths to provide outlets even for the underground tile drains, if provided. These outlet drains may be called *deep surface drains*.

Surface drains constructed for removing excess irrigation water applied to the farms and the storm water, cannot and should not be deep enough, as to interfere with the agricultural operations. They are, therefore, designed as shallow surface drains.

*Land grading,* which results in a continuous land slope towards the field drains, is an important part of a surface drainage system. Land grading or land leveling is also necessary for surface irrigation.

The **shallow surface drains** are trapezoidal in cross-section. Strictly speaking, they should be designed to carry the normal storm water from the fields, plus the excess irrigation water. Many a times, the excess irrigation water is neglected and these drains are designed only for the runoff resulting from the average storms. It is neither economical nor desirable to design these drains for exceptional storms. Kutter's or Manning's equations may be used to design these drains, keeping the velocity within the limits of the critical velocity, and thereby avoiding silting or scouring. Manning's equation is, however, generally used for the design of shallow as well as deep surface drains.

**Deep surface drains** or **outlet drains** carry the storm water discharge from the shallow surface drains, and the seepage water coming from the underground tile drains. They are, therefore, designed for the combined discharge of the shallow surface drains as well as that of the tile drains. Generally, a cunnette of about 0.6 m depth is provided in the centre of the drain-bed, so as to carry the seepage water of the underground tile drains. A steeper slope is given to the cunnette and it is lined, so as to withstand higher flow velocities, and thus, to inhibit weed growth. The full section would be operative only during the rainy season, as otherwise, the flow will be confined only within the cunnette.

**6.5.1. Surface Inlet.** The surface water from the pot holes, depressions, road ditches, farm steads, etc. may be removed either by connecting them with the shallow surface drains, sometimes called *random field drains* (shown in Fig. 6.2), or by constructing an intake structure called an *open inlet* or *surface inlet* (Fig. 6.3). A *surface inlet* is a structure constructed to carry the pit water into the sub-surface or tile drain. A cast iron pipe or a manhole constructed of brick or monolithic concrete, is sufficient and satisfactory. Manholes with sediment basins are sometimes used as surface inlets.

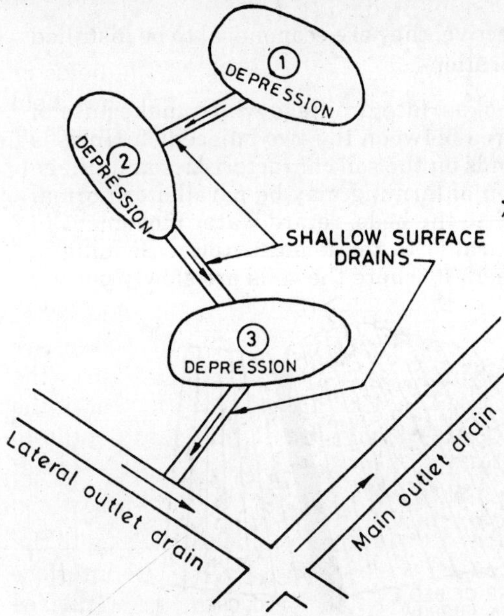


Fig. 6.2. Random field-drain (shallow surface drain) system for surface drainage.

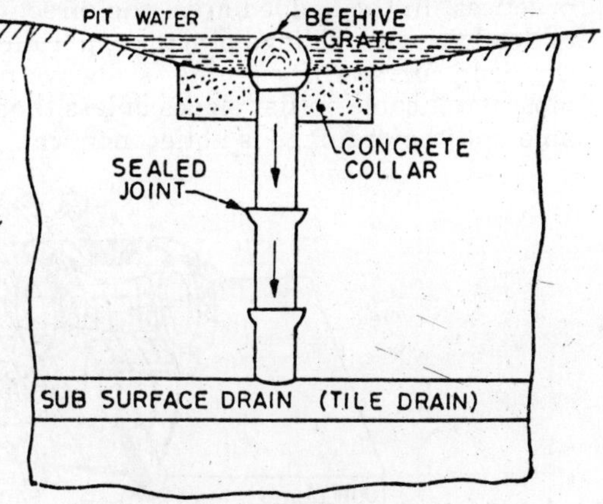


Fig. 6.3. surface inlet draining the surface water into a tile drain.

At the surface of the ground, a concrete collar extending around the intake is constructed on the riser to prevent growth of vegetation and to hold it in place. On the top of the riser, beehive grate or some other suitable grate is provided, so as to prevent trash from entering the tile. When the inlet is constructed in a cultivated field, the area immediately around the intake should be kept in grass.

When the surface inlet is connected to a main tile drain, it is a good practice to offset the surface inlet from the main. Such constructions may eliminate failure of the system, if the surface inlet structure should become damaged.

**6.5.2. French Drain.** When the quantity of water to be removed from the pits or depressions is small, a **blind inlet** may be installed over the tile drain. The blind inlet is also called *french drain.* These are constructed by back filling the trench of the tile drain with graded materials, such as gravel and coarse sand, or with corn cobs, straw and similar substances, as shown in Fig. 6.4.

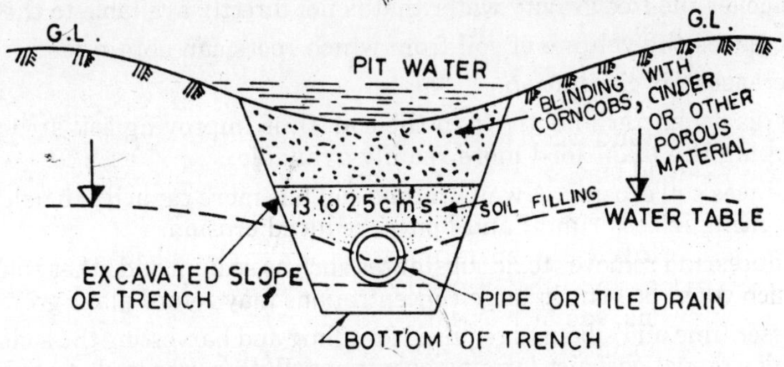


Fig. 6.4. Blind inlet or French drain.

Such inlets are not permanently effective. The voids in the backfill of the blind inlet become filled up with the passage of time, thereby reducing its effectiveness.

Even though they are not permanently effective, they are economical to be installed and do not interfere with the farming operations.

**6.5.3. Bedding.** Bedding is a method of surface drainage which makes use of dead furrows, as shown in Fig. 6.5. The area between the two adjacent furrows is known as a bed. The depth of the bed depends on the soil characteristics and tillage practices. In the bedded area, the direction of farming may be parallel or normal to dead furrows. Tillage practices, parallel to the beds, retard water movement to the dead furrows. Ploughing is always parallel to the dead furrows. Bedding is most practicable on flat slopes of less than 15%, where the soils are slowly permeable and the drainage is not economical.

Fig. 6.5. Cross-section of bed showing method of construction.

## 6.6. Sub-surface Drainage or Tile Drainage

Plants need air as well as moisture in their root zones for their survival. Excess irrigation farm water is free to move into the underground tile drains, if provided*. This water, if not removed, retards the plant growth, because it fills the soil voids and restricts proper aeration. *Surface drains are, therefore, needed for removing the excess farm water, for most of the cultivated crops on flat or undulating topography. Sub-surface drains, on the other hand, are required for soils with poor internal drainage and a high watertable. If no impervious layer occurs below the farm land and the watertable is low (lower than about 3m from the ground), internal soil drainage may be sufficient and no tile drains needed. For maximum productivity of most of the crops, both surface as well as sub-surface drains may sometimes,however, become, essential*, particularly in areas of higher water-tables.

**Advantages of Tile Drains.** Tile drainage helps in increasing crop yields by draining the water or by lowering the watertable in the following manner :

(*i*) Removes the free gravity water that is not directly available to the plants.

(*ii*) Increases the volume of soil from which roots can obtain food.

(*iii*) Increases air circulation.

(*iv*) Increases bacterial activity in the soil, thus improving soil structure and making the plant food more readily available.

(*v*) Reduces soil erosion. A well drained soil has more capacity to hold rainfall, resulting in less runoff and hence, reduced erosion.

(*vi*) Reduces and removes toxic substances such as sodium and other soluble salts, which when present in large concentrations may retard plant growth.

(*vii*) Lesser time and labour is required for tilling and harvesting the soils, as these drains do not obstruct farming operations. With a crop such as corn, a delay in planting may decrease the yields. Planting in wet soils is also likely to decrease yields. All such troubles are removed in tile drained soils.

* Underground tile drains are not very common in India, where water-tables are generally falling and availability of water is becoming scarce.

(*viii*) Tile drains permit deep roots development by lowering the watertable, especially during spring months, as shown in Fig. 6.6.

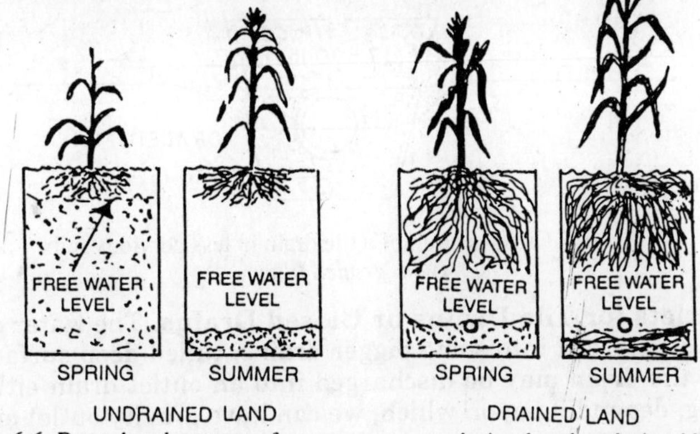


Fig. 6.6. Root development of crops grown on drained and undrained land.

**Note :** A plant having deep roots can extract water from greater depths and hence, can withstand droughts better than the one having shallow roots. Moreover, a deep rooted plant is larger and, therefore, capable of more transpiration and hence, giving increased yields.

Providing underground tile drains, however, is a costly proposition and may be required only in areas of high water-table, and where the ground soil has a poor internal drainage capacity.

**6.6.1. Envelope Filters.** Tile drains, are usually, pipe drains made up of porous earthenware and are circular in section. The diameters may vary from 10 to 30 cm or so. These drains are laid below the ground level, butting each other with open joints. The trenches in which they are laid, are back filled with sand and excavated material, as shown in Fig. 6.7. As far as possible, the tile drains should not be placed below less pervious strata. Because in that case, they may remain dry even though the land above the impervious strata may be water-logged, as the water will not be able to reach the drain.

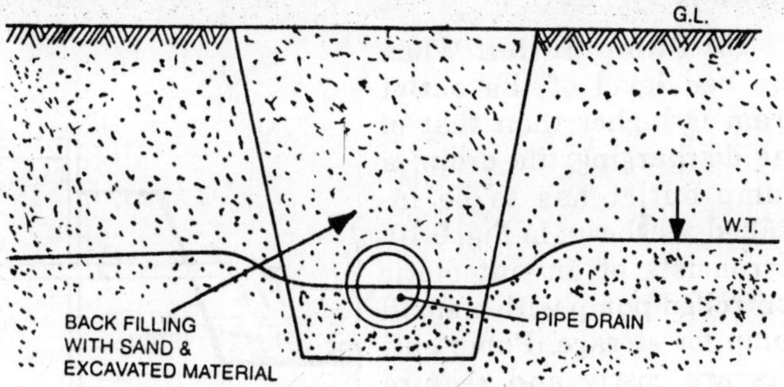


Fig. 6.7. Cross-section of a tile drain in pervious soils (without any filter).

When tile drains are placed in less pervious soils, they are generally surrounded by graded gravel filters, called envelope filters (Fig. 6.8). The envelope filter serves two functions : (*i*) it prevents the inflow of the soil into the drain, and (*ii*) it increases the effective tile diameter, and thus increases the inflow rate. The filter consists of different gradations, such as gravel, coarse sand, bajri, etc., The coarsest material is placed immediately over the tile, and the size is gradually reduced towards the surface. The minimum thickness of the filter is about 7.5 cm. The graded filter may sometimes be substituted by a single gradation, depending upon the availability and cost considerations.

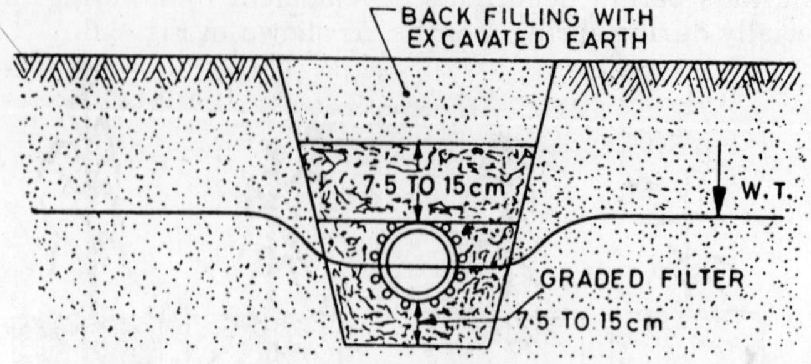


Fig. 6.8. Cross-section of a tile drain in less pervious soils (*with graded filter*).

**6.6.2. Outlets for Tile Drains or Closed Drains.** The water drained by the tile drains is discharged into some bigger drains, called deep surface drains. The water from a tile drain may be discharged into an outlet drain either by gravity or by pumping, depending upon which, we can have gravity outlet or pump outlet, as described below :

(*a*) **Gravity outlet.** If the bed level and the full supply level (FSL) of the outlet drain is lower than the invert level of the tile drain, then the water can be discharged easily into the outlet drain by the mere action of gravity. Corrugated metal pipe with a flap shutter to prevent entry of rodents and back flow from the outlet drain into the tile drain, is generally provided at the outfall point, as shown in Fig. 6.9.

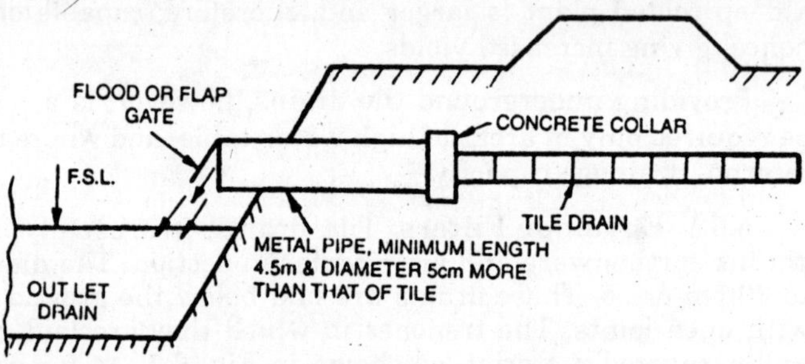


Fig. 6.9. Gravity outlet (outfall) for tile drain.

(*b*) **Pump outlet.** When the bed level of the outlet drain is higher than that of the discharging tile drain; a pump outlet has to be installed, as shown in Fig. 6.10. It consists of an automatic controlled pump with a small sump for storage. Pump outlets are costly and require technicality. Possibility of deepening the outlet drain should, therefore, be investigated. The cost of installing and maintaining a pump outlet should be compared with that of excavating and maintaining a deeper outlet drain, before making a final selection.

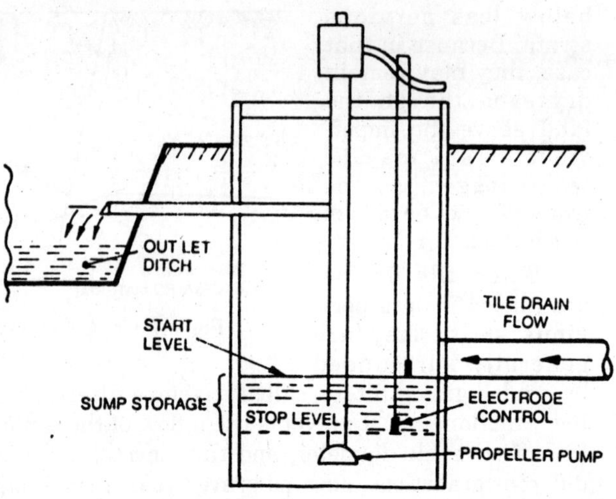


Fig. 6.10. Pump outlet (outfall) for tile drain.

**6.6.3. Drawdown Curve or Movement of Water into the Tile Drains.** In a fully saturated soil, water flows into the tile drain along the path shown in Fig. 6.11 (*a*). Since the quantity of water moving between any two flow lines is the same, the drawdown will be more near the tile than at the points farther away. After the saturated soil has drained for a day or so, the resulting watertable will be, as shown in Fig. 6.11 (*a*). With series of tile drains, the sub soil water level directly over the drains, is lower than the level midway between them as shown in Fig. 6.11 (*b*).

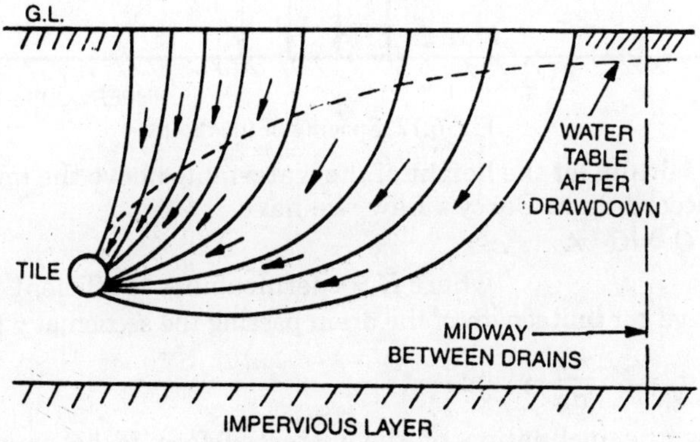


Fig. 6.11. (*a*) Drawdown curve with a single tile drain.

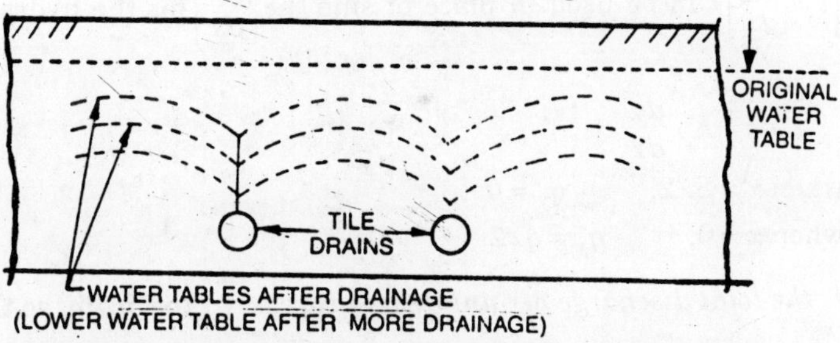


Fig. 6.11. (*b*) Drawdown curve with a series of tile drains.

When a filter is provided around the tile drains to surround the drains with more pervious soil, then the overall drawdown will be more. The rate of drop of watertable mainly depends upon the soil permeability and spacing of the drains. In this case, the water has to travel more distance horizontally than vertically before it reaches the drain, the horizontal permeability of the soil is more important. The permeabilities of most of the soils decrease with depth. This change in permeability affects the shape of the flow lines and the rate of the fall of watertable.

**6.6.4. Depth and Spacing of the Tile Drains.** The closed drains are generally spaced at such a distance as to be capable of lowering the watertable sufficiently below the root zone of the plants. For most of the plants, the top point of the watertable must be at least 1.0 to 1.5 metres below the ground level; although this distance may vary from 0.7 to 2.5 m., depending upon the soil and the crop.

The tile drains may be placed at about 0.3 metre below the desired highest level of the watertable. A fair idea of the spacing between the tile drains can be obtained based on the above theory, as follows :

Let $S$ be spacing between the drains, and $a$ be the depth of impervious stratum from the centre of the drains, as shown in Fig. 6.12. Let the maximum height of the drained watertable above the impervious layer be $b$. At any distance $x$ from

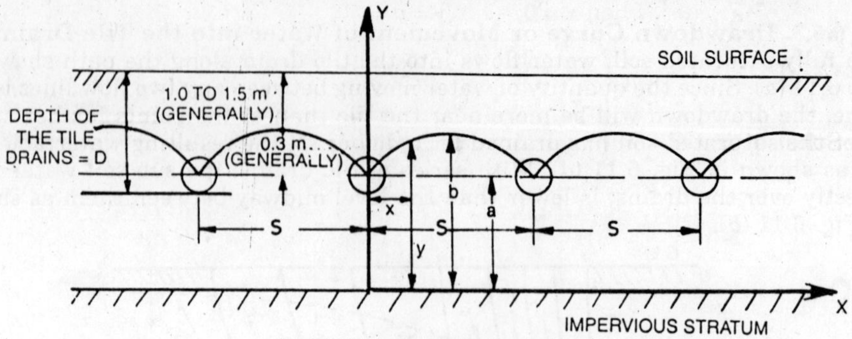


Fig. 6.12. Spacing of tile drains.

the centre of a drain, let the height of the watertable above the impervious stratum be $y$. Then, according to Darcy's Law, we have

$$Q = KI \cdot A$$

where $K =$ permeability coefficient in m/sec.

$\therefore$ Discharge per unit length of the drain passing the section at $y$ ($q_y$) is given as :

$$q_y = K \cdot \frac{dy}{ds} \cdot y$$

Assuming the inclination of the water surface to be small, such that the tangent $\left(i.e.\ \frac{dy}{dx}\right)$ can be used in place of sine $\left(i.e.\ \frac{dy}{ds}\right)$ for the hydraulic gradient, we get

$$q_y = K \cdot \frac{dy}{dx} \cdot y \qquad \text{...(6.5)}$$

But when $x = S/2$, $\quad q_y = 0$.

and, when $x = 0$, $\quad q_y = q/2$

where *q is the total discharge per unit length carried by the drain*, so that $\frac{1}{2}q$ enters the drain from either side.

Also assuming that $q$ is inversely proportional to the distance from the drain, we can write

$$q_y = \frac{1}{2}q - \frac{1}{2}q\,\frac{x}{S/2} = \frac{1}{2}q\left(1 - \frac{x}{S/2}\right)$$

or $$= \frac{q}{2S}\left[S - 2x\right] \qquad \text{...(6.6)}$$

Equating Eqs. (6.6) and (6.5), we get

$$\frac{q}{2S}(S - 2x) = K\,y \cdot \frac{dy}{dx}$$

Rearranging and integrating, we get

$$\int \frac{q}{2SK}(S - 2x)\,dx = \int y\,dy.$$

Assuming the soil permeability to be constant, we get

$$\frac{q}{2K.S}\left[Sx - \frac{2x^2}{2}\right] = \frac{y^2}{2} + C \qquad \text{...(6.7)}$$

when $x = 0, \quad y = a$

$$\therefore \quad \frac{q}{2KS}[0] = \frac{a^2}{2} + C \quad \text{or} \quad C = -\frac{a^2}{2}$$

Substituting $C = -\frac{a^2}{2}$, equation (6.7) becomes

$$\frac{q}{2KS}\left[Sx - x^2\right] = \frac{y^2}{2} - \frac{a^2}{2}$$

or
$$\frac{q}{2KS}x(S - x) = \frac{y^2 - a^2}{2}$$

or
$$q = \frac{KS(y^2 - a^2)}{(S - x)x} \quad \text{...(6.8)}$$

Also, when $x = S/2, y = b$, equation (6.8) then becomes

$$q = \frac{KS\cdot(b^2 - a^2)}{\left(S - \frac{S}{2}\right)\frac{S}{2}} \quad \text{or} \quad q = \frac{KS(b^2 - a^2)}{\frac{S^2}{4}}$$

or
$$\mathbf{q} = \frac{\mathbf{4K}}{\mathbf{S}}(\mathbf{b^2 - a^2}) \quad \text{...(6.9)}$$

or
$$\boxed{S = \frac{4K}{q}(b^2 - a^2)} \quad \text{...(6.10)}$$

Equation (6.10) can be used to predict the spacing ($S$) between the drains, if $q$ is known. $q$ will depend on the infiltration discharge into the ground, which should be removed by the drains. Different values have been suggested. Generally, a value equal to 1% of the average annual rainfall of a place is considered to be drained by the tile drains in 24 hours. If the average annual rainfall of the place is $P_{AA}$ (metres), then

$$q = \left(\frac{0.01 \times P_{AA}}{24 \times 3600}\right)(S \times 1) \text{ cumecs/m length of drains}$$

$$= \frac{0.01 \times P_{AA} \cdot S}{24 \times 3600} = \frac{P_{AA.} \cdot S}{86{,}40{,}000} = \frac{P_{AA} \cdot S}{8.64 \times 10^6} \quad \text{...(6.11)}$$

Equating with equation (6.9), we get

$$q = \frac{4K\cdot(b^2 - a^2)}{S} = \frac{P_{AA}\cdot S}{8.64 \times 10^6}$$

or
$$S = \sqrt{\frac{(8.64 \times 10^6)\,4K\cdot(b^2 - a^2)}{P_{AA}}} \quad \text{...(6.12)}$$

Hence, spacing ($S$) can be determined easily by using eq. (6.12).

**Example 6.2.** *In a tile drainage system, the drains are laid with their centres 1.5 m below the ground level. The impervious layer is 9.0 m below the ground level and the average annual rainfall in the area is 80 cm. If 1% of the annual rainfall is to be drained in 24 hours to keep the highest position of the watertable to 1 metre below ground level, determine the spacing of the drain pipes. Coefficient of permeability may be taken as 0.001 cm/sec.*

**Solution.** Although eqn. (6.12) can be directly used in this question, since that eqn. has been derived for designing the drains to take 1% of the average annual rainfall in 24 hours, which tallies with the given data, yet it would be prudent to use the basic eqn. (6.10) for determining the spacing of tile drains, and separately compute $q$, as:

$$q = \frac{\frac{1}{100} \times \left(\frac{80}{100}\right) \times (S \times 1)}{(24 \times 60 \times 60)} \text{ m}^3/\text{s}$$

$$= \frac{0.8S}{8.64 \times 10^6} \text{ m}^3/\text{s/m length of tile drain}$$

Using eqn. (6.10), we have

$$S = \frac{4K}{q}(b^2 - a^2)$$

where $b$ = ht. of W.T. above the impervious layer = 9 m − 1 m = 8 m

$a$ = depth of impervious stratum below the centre of the drains = 9 − 1.5 = 7.5 m

$K$ = 0.001 cm/s = $\frac{0.001}{100}$ m/s

Substituting values, we get

$$S \cdot q = 4 \times \frac{0.001}{100}(8^2 - 7.5^2)$$

or $$S \times \left(\frac{0.8S}{8.64 \times 10^6}\right) = \frac{4 \times 0.001}{100}(8^2 - 7.5^2)$$

or $$0.8S^2 = \frac{4 \times 0.001}{100} \times (8.64 \times 10^6)(8^2 - 7.5^2) = 2678.4$$

or $$S = \sqrt{\frac{2678.4}{0.8}} = \sqrt{3348} = \mathbf{57.86\ m} \quad \textbf{Ans.}$$

**6.6.5. Drainage Coefficient (D.C.).** The rate at which the water is removed by a drain is called the *drainage coefficient*. It is expressed as the depth of water in cm or metres, to be removed in 24 hours from the drainage area. The drainage coefficient largely depends upon the rainfall but varies with the type of soil, type of crop, and degree of surface drainage, etc. Its recommended value is 1% of the average annual rainfall to be removed per day.

In irrigated areas, the discharge through the tiles may vary between 10 to 50% of the total water applied. Since the entire area is not irrigated at the same time, the drainage area to be used to calculate tile flow is not the same as the entire tiled area, but is estimated from the area irrigated. A suitable value of drainage coefficient (DC) may be taken for the calculations, depending upon the local recommendations. Values of 1 to 2.5 cm/day for mineral soils and 1.25 to 10 cm/day for organic soils for different crops, have been suggested for humid regions, by U.S. Soil Conservation Service.

**Example 6.3.** *A tile drainage system draining 12 hectares, flows at a design capacity for two days, following a storm. If the system is designed using a D.C. of 1.25 cm, how many cubic metres of water will be removed during this period ?*

**Solution.** D.C. of 1.25 cm means that 1.25 cm of water depth from the drainage area shall be removed by the drain in 24 hours.

∴ Volume of water entering the drain per day

$$= \left(\frac{1.25}{100} \text{ m/day}\right) \times (12 \times 10^4 \text{ m}^2) = 1{,}500 \text{ m}^3/\text{day}$$

Volume of water passing the drain within 2 days of flow

$$= 2 \times 1500 = \mathbf{3000\ m^3} \quad \textbf{Ans.}$$

**6.6.6. Drainage Area.** The area actually drained by the tile drain system is called its *drainage area*. Sometimes, the surface water is also to be removed by the tiles. In that case, the watershed area will be the drainage area, even though it may not be entirely tiled.

**6.6.7. Size of the Tile Drains.** The tile drains are designed according the Manning's formula to carry a certain discharge decided by D.C. and drainage area. The drains are laid on a certain longitudinal slope varying from 0.05 to 3%. A desirable minimum working grade is 0.2%. Where sufficient slope is not available, the grade may be reduced to 0.1%. Depending upon the available slope of the soil surface and the depth of the outlet, suitable value of longitudinal slope can be given to the tiles. Their sizes can be easily evaluated from Manning's formula. 10 to 15 cm tiles are minimum recommended sizes. The minimum size for perforated tubing or pipes can be reduced, as in that case, the misalignment at joints or cracks is not a problem.

**Example 6.4.** *Determine the size of a tile at the outlet of a 6 hectare drainage system, if the D.C. is 1 cm and the tile grade is 0.3%. Assume the rugosity coefficient for the tile drain material as 0.011.*

**Solution.** 1 cm D.C. means that 1 cm of water from an area of 6 hectares is entering the tiles per day.

$\therefore$ Volume of water passing the drain in 1 day $= \left(\frac{1}{100} \times 6 \times 10^4\right) = 600\ m^3/day$

$\therefore$ Volume of water passing the drain in 1 second $= \left(\frac{600}{24 \times 3600}\right) = \frac{1}{144}\ m^3/s$

$$\therefore \quad Q = \frac{1}{144}\ m^3/s \qquad \text{Now,} \quad Q = \frac{1}{n} \cdot A.R^{2/3} \cdot S^{1/2}$$

For a circular drain of diameter $D$, we have

$$A = \frac{\pi D^2}{4},\ P = \pi D,\ R = \frac{D}{4}$$

or $$\frac{1}{144} = \frac{1}{0.011} \times \left(\frac{\pi D^2}{4}\right) \cdot \left(\frac{D}{4}\right)^{2/3} \cdot \left(\frac{0.3}{100}\right)^{1/2}$$

or $$\frac{1}{144} \times \frac{0.011 \times 4}{\pi} = \frac{D^2 \cdot D^{2/3}}{(4)^{2/3}} \times \frac{1}{\sqrt{333.3}}$$

or $$\frac{0.011 \times 4 \times 2.52 \times 18.26}{144 \times \pi} = D^{8/3}$$

or $$D = (0.00447)^{3/8} = 0.132 \text{ metre} = 13.2 \text{ cm}$$

**Use 15 cm dia. pipe. Ans.**

**Example 6.5.** *Sugarcane (root zone depth 1.8 m) is grown in a particular area where the ground water table is 2.0 m below ground. If the size of the soil pores is 0.08 mm in diameter and surface tension = 0.054 N/m, is the field water-logged ? If so, determine the vertical location of closed drains below ground, spaced at 15 m. Take drainage coefficient as 0.116 cumecs/km$^2$. Coefficient of permeability as $10^{-6}$ m/s, and the impervious stratum to occur at 7.0 m, below ground.*

**Solution.** We have stated in article 6.4 that if the water table rises up or if the plant roots happen to come within the capillary fringe, the salinity and water-logging of the land occurs. Here we have to see as to whether the plant roots (extending 1.8 m below the ground level) do come within the water-table or extended water table (due to capillary fringe). The actual water table is 2 m below ground and hence is not directly affecting the roots. But the capillary fringe is to be seen.

The height to which water rises by capillary action in a soil is given by eqn. (6.3) in "Soil Mechanics and Foundation Engineering" by the same author, as

$$H_e = \frac{4T \cdot \cos \alpha}{d \cdot \gamma_w}$$

where $\alpha$ = 0 for max. capillary height

$d$ = Dia of pores = 0.08 mm = $0.08 \times 10^{-3}$ m

$\gamma_w$ = Unit wt. of water = $9.81 \times 10^3$ N/m$^3$

$T$ = Surface tension = 0.054 N/m

$$\therefore \quad H_e = \frac{4 \times 0.054 \times 1}{0.08 \times 10^{-3} \times (9.81 \times 10^3)} = \frac{4 \times 0.054}{0.08 \times 9.81} = 0.28 \text{ m}$$

The water-table, thus, effectively stands at 2 – 0.28 = 1.72 m below the ground, thereby causing slight water logging, since the roots extend up to 1.8 m below the ground. **Ans.**

In order to design the tile drains, we represent the given conditions in Fig. 6.13 as follows. Let the tile drains be laid at a height '$a$' above the top of the impervious layer.

From Eqn. (6.9), we have

$$q = \frac{4K}{S}(b^2 - a^2) \qquad ...(i)$$

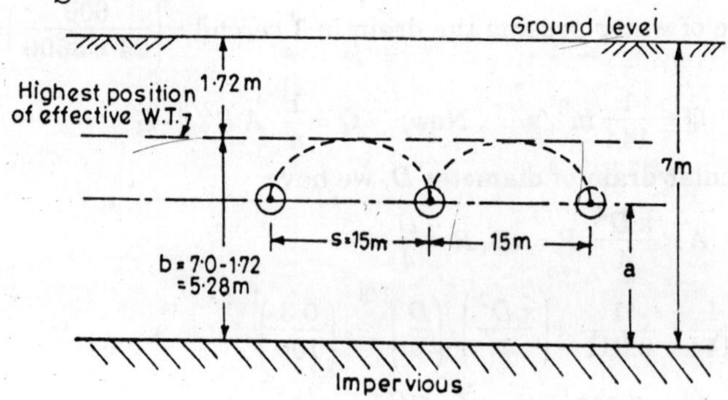


Fig. 6.13.

where $S$ = 15 m (given)

$K$ = $10^{-6}$ m/s (given)

$b$ = 7 – 1.72 = 5.28 m

$a$ = ?

$q$ = discharge carried by each tile drain in cumecs/m run

The tiles have a drainage coefficient of 0.116 cumecs /km$^2$. Each tile carries discharge from 15 m width and 1 m length (considering 1 m length ⊥ to diagram).

∴ The discharge carried by each tile drain of 1 m length, $q$ is given as :

$$q = 0.116 \frac{\text{Cumecs}}{\text{km}^2} \times \text{Area in km}^2$$

$$= 0.116 \times \left(\frac{15}{1000} \times \frac{1}{1000}\right) \text{cumces} = 1.74 \times 10^{-6} \text{ cumces/m run}$$

Substituting values, in (i) we get

$$1.74 \times 10^{-6} = \frac{4 \times 10^{-6}}{15} (5.28^2 - a^2)$$

or $$a = \mathbf{4.62\ m}$$

The tile drains should therefore be laid at 7 – 4.62 = **2.38 m** below the ground. **Ans.**

**6.6.8. Materials of Tile Drains.** Pipes used for tile drains are generally made from materials, such as clay or concrete, in short lengths. Sometimes, they may be made of bituminous fibre or steel. Corrugated plastic perforated pipes are gaining popularity, because of its lightness and reduced labour in handling it.

**Comparison between clay and concrete pipes.** Good quality concrete pipes are very resistant to freezing and thawing, but may deteriorate in alkaline and acidic soils. Clay pipes, on the other hand, are not affected by acid or alkaline soils. When subjected to continuous freezing or thawing conditions, concrete pipes are found to be safer than the clay pipes, although clay pipes are resistant to frost damage. Both the types of pipes should have sufficient strength, so as to withstand static and impact loads transmitted from the soil above.

Good clay or concrete pipes should have the following characteristics:

(*i*) Resistant to weathering and deterioration in the soil.

(*ii*) Low water absorption, *i.e.* high density.

(*iii*) Uniformity in shape and wall thickness, etc.

(*iv*) Freedom from defects, such as cracks, etc.

(*v*) Resistant to freezing, thawing and frost damage.

(*vi*) Sufficient strength to withstand static and impact loads for which designed.

**6.6.9. Layout of Tile Drains.** The tile drains may be aligned in different fashions, depending upon the topography of the area. Generally, laterals (branch drains) run through most of the drainage area and join the mains, which in turn, outfall into some deep open drain. The depth of the deepest tile drain shall be kept within the range of 3 metres from the surface, since the root zones of crops get affected by water within this depth range. A simple network of the drainage arrangement is shown in Fig. 6.14.

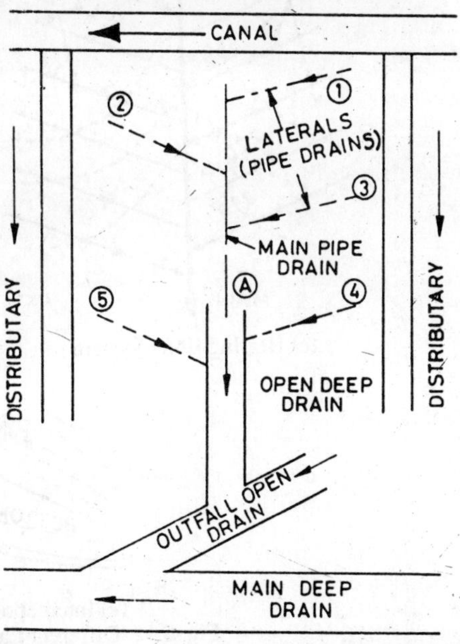


Fig. 6.14. General layout of a tile drain network.

Various possible **alternative layouts for the tile drainage system** are shown in Fig. 6.15, and are discussed below:

(1) **Natural system.** The natural system is generally adopted *in rolling topography, where drainage of isolated areas is required.* The mains and the connected laterals are provided in natural course, as shown in Fig 6.15 (*a*). This system is suitable when the land is not to be completely drained. The system is quite flexible and permits location of drains where they are most needed.

(2) **Grid iron system**. This drainage system, consisting of laterals and mains (or submains), is shown in Fig 6.15 (*b*). In this

system, the laterals are provided only on one side of the main, as shown. *This system is adopted when the land is practically level; or where the land surface slopes away from the sub main on one side, and when the entire area has to be drained.*

(3) **Herring bone system**. In this layout pattern, laterals join the mains (or submains) from each side, alternatively, as shown in Fig 6.15(*c*). This layout is adopted, when the main (or sub-main) is laid in depression. *The land along the main is double drained, but since it exists in depression, it probably requires more drainage than the land on the adjacent slopes.*

(4) **Double main system.** This system has two mains with separate laterals for each main, as shown in Fig 6.15(*d*). *This layout is adopted when the bottom of depression is wide.* This arrangement reduces the length of the laterals and eliminates the break in slope of the lateral at the edge of the depression.

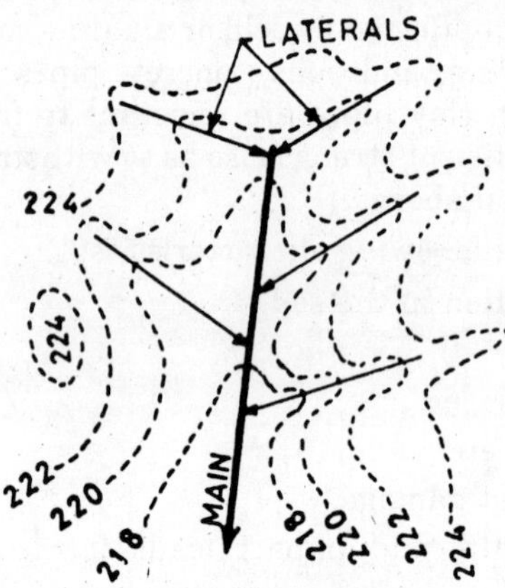


(*a*) Natural System

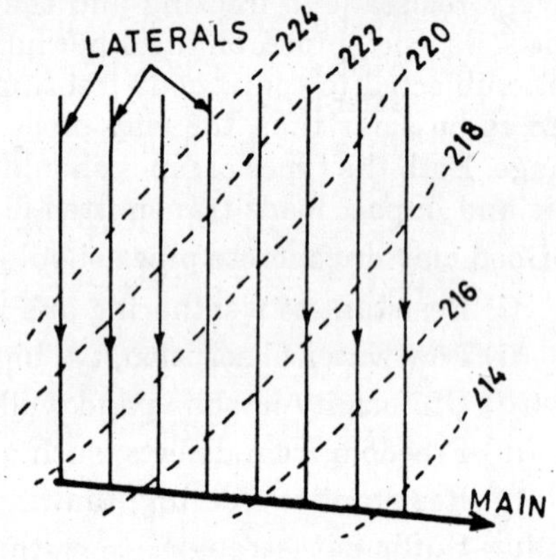


(*b*) Grid Iron System.

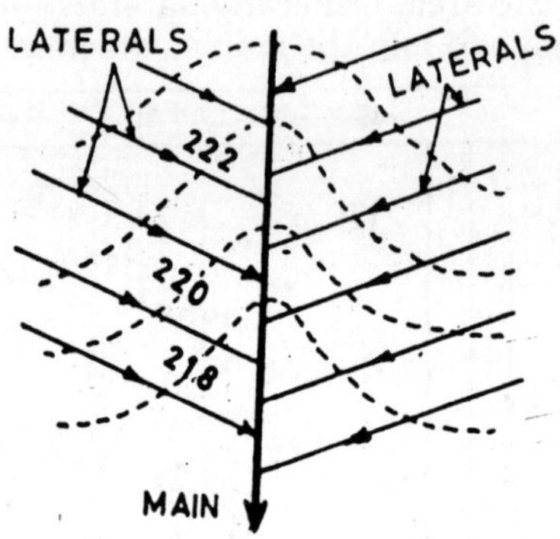


(*c*) Herring Bone System

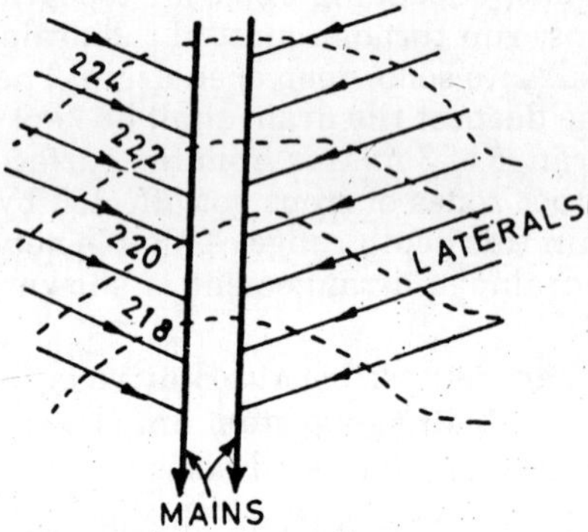


(*d*) Double Main System.

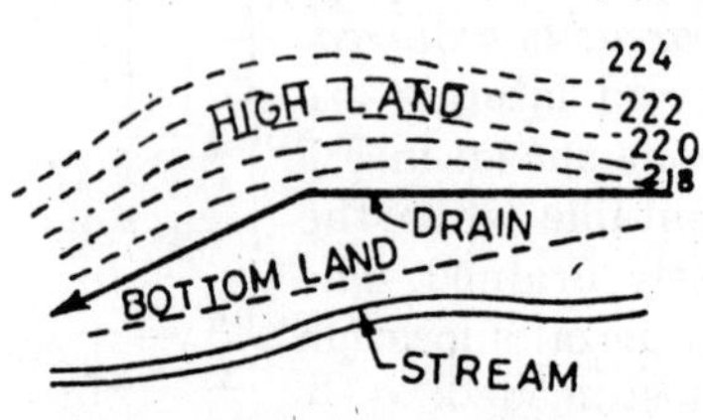


(*e*) Intercepting Tile Drains System.

Fig. 6.15. Different Layouts of Tile Drainage Systems.

(5) **Intercepting tile drains system**. In this system, there is no lateral drain. A main (or submain) is provided at the toe of the slope, as shown in Fig 6.15(*e*). *This arrangement is preferred when the main source of drainage is from a hilly land.*

In general, all mains and submains should be kept away from trees, as the roots of trees can easily enter the open joints of the tile drains, and thereby blocking them.

## PROBLEMS

**1.** (*a*) What is meant by "water-logging"?

(*b*) What are the principal causes and effects of water-logging in a canal irrigated farm?

(*c*) What precautions and measures will you adopt to prevent water logging of irrigated lands?

**2.** (*a*) What are the principal causes of water-logging in a canal irrigated tract?

(*b*) What precautions will you observe in constructing a new canal system to avoid water-logging?

[**Hint**. Provide lined canals equipped with under-drainage system]

(*c*) What steps will you take to improve an already water-logged tract?

**3.** What is meant by water-logging? What are its ill effects? Describe some anti-water-logging measures with suitable sketches.

**4.** Discuss different types of drainage systems provided in irrigated tracts as precautions against water-logging. Give salient features of an open drain system.

**5.** (*a*) What is meant by "saline" and "alkaline" soils?

(*b*) What precautions will you adopt to prevent salinity of irrigated land?

(*c*) How will you proceed to reclaim saline land?

**6.** (*a*) Discuss critically the statement : "intensive irrigation leads to reduced crop yields".

(*b*) What are "tile drains" and how do they help in preventing water logging? How will you decide the depth and spacing of tile drains?

**7.** (*a*) What is meant by "Land Drainage", and where is it needed?

(*b*) What are the two principal types of drainage systems necessary for draining irrigated tracts to avoid their water-logging?

**8.** (*a*) What is meant by "open drainage" and "closed drainage" as applied to draining irrigated tracts?

(*b*) A tile drainage system draining 15 hectares, flows at a design capacity for three days, following a storm. If the system is designed using a drainage coefficient of 2.25 cm, how much water will be removed during this period?

[**Hint.** Follow example 6.3] [**Ans.** 10125 $m^3$]

**9.** (*a*) What is meant by "Drainage coefficient" and how does it help in determining the sizes of tile drains to be laid for preventing water-logging of irrigated farms?

(*b*) What are the different materials which are commonly used for the drains, and what are their comparative merits and demerits?

(*c*) Determine the size of a circular tile drain, draining 6 hectares of a drainage area, if the drainage coefficient (D.C.) is 1.5 cm and the tile grade is 0.4%. Assume the rugosity coefficient for the tile material as 0.013. [**Hint.** Follow example 6.4]

**10.** What is meant by "tile drains" and what are the different methods of aligning them? How will you proceed for fixing their depth, spacing and sizes? What are their advantages?

**11.** Write notes on any two of the following:

(*i*) Open drainage (*ii*) Tile drainage

(*iii*) Surface inlet (*iv*) Blind inlet or French drain

(*v*) Envelope filters (*vi*) Gravity outlets

(*vii*) Pump outlets (*viii*) Movement of water into tile drains

(*ix*) Random and Herring bone types of tile drainage systems.

**12.** Fill in the blanks with appropriate words:

(*i*) If the soil is more permeable, the chances of water-logging are ...... (less/more)

(*ii*) Intensive irrigation should be ......... in areas susceptible to water-logging. (adopted/avoided)

(*iii*) Extensive irrigation should be ......... in areas susceptible to water-logging. (adopted/avoided)

(*iv*) ........ irrigation helps in checking water-logging. (Flow/Lift)

(*v*) The remedial measures for water-logging are ......

(*vi*) Drainage schemes are implemented for ..... water-logged areas.

[**Ans.** (*i*) more; (*ii*) avoided; (*iii*) adopted (*iv*) lift; (*v*) installing lift irrigation schemes, and implementing drainage schemes; (*iv*) reclaiming.]

# 7

# Irrigation Revenue Rates

## 7.1. Methods of Pricing Irrigation Water

When water is supplied to the cultivators for irrigating their crops, then naturally, it has to be charged and priced at some specified rate. This rate may be based either : (*i*) on flat rate basis ; or (*ii*) on the basis of the actual quantity of water consumed by each cultivator, as discussed below :

**7.1.1. Flat Rate Pricing.** When pricing of irrigation water is done on flat rate or crop rate basis, then the rates are specified per hectare of each crop sown and cultivated by the cultivators. Different rates for different crops are generally specified, depending upon the water requirement of the crop.

In other words, a crop requiring the highest quantity of irrigation water, such as sugarcane will be priced at the highest rate ; whereas the crop requiring the lowest quantity of water, such as Groundnut, shall be rated at the lowest rate. Different crops will, thus, generally be rated at reducing rates, with their reducing irrigation requirements, as shown in Table 7.1 (*a*).

**Table 7.1. (*a*) Water Requirements, and Specified Irrigation Water Rates for Different Crops in Maharashtra State of India (1990-data)**

| *S.No.* | *Name of the crop sown* | *Water required as depth of irrigation (Δ) in cm* | *Vol. of water required per ha of crop in $m^3$* | *Specified Irrigation Revenue Rate in Rs/ha of each crop.* |
|---|---|---|---|---|
| (1) | (2) | (3) | (4) | (5) |
| 1. | Sugar-cane (*perennial*) | 270 | 27,000 | 750/– |
| 2. | L.S. Cotton (*Kharif-Ravi*) | 90 | 9,000 | 400/– |
| 3. | Vegetables (*Ravi*) | 75 | 7,500 | — |
| 4. | Wheat (*Ravi*) | 45 | 4,500 | 75/– |
| 5. | H.Y. Maize (*Kharif*) | 45 | 4,500 | — |
| 6. | H.Y. Jowar (*Kharif*) | 22 | 2,200 | — |
| 7. | Groundnut (*Kharif*) | 15 | 1,500 | 300/– |

**Note.** It can also be seen from Col (5) of the above table that although the specified revenue rates are generally in accordance with the general pricing formula of specifying higher rates for higher water requirement, yet there is a deviation in the case of ground-nut crop, which is priced at four times the rate of wheat, although its water consumption is one third of that of wheat. One of the reasons for pricing groundnut at higher rate is the *higher earning capacity* of this crop, as compared to that of wheat, as shown in Col. (7) of table 7.1 (*b*). The second reason for higher revenue rate for groundnut compared to wheat, is the fact that groundnut is a cash crop, while wheat is a food crop.

**Table 7.1. (*b*) Earnings provided by different crops in Maharashtra State (1990-Data)**

| *Sl. No.* | *Name of crop* | *Yield per ha in Qtl (100 kg) Qtl/ha* | *Sale price in Rs./Qtl* | *Gross income per ha in Rs./ha (3)×(4)* | *Estimated cost of cultivation (without accounting for irrigation charges in) Rs/ha* | *Net income in Rs/ha (5)—(6)* |
|---|---|---|---|---|---|---|
| (1) | (2) | (3) | (4) | (5) | (6) | (7) |
| 1. | Sugar-cane | 865 | 30 | 25,950 | 12,450 | 13,500 (Max.) |
| 2. | L.S. cotton | 19 | 498 | 9462 | 5,002 | 4,460 |
| 3. | Vegetables | 198 | 110 | 21,780 | 10,325 | 11,455 |
| 4. | Wheat | 25 | 290 | 7,250 | 3,640 | 3,610 |
| 5. | H.Y. Maize | 30 | 185 | 5,550 | 2,500 | 3,050 |
| 6. | H.Y. Jowar | 27 | 200 | 5,400 | 2,530 | 2,870 |
| 7. | Ground nut | 17 | 575 | 9,775 | 4,854 | 4921 |

**Note.** A comparison between Col. (5) of table 7.1 (*a*) and Col (7) of table 7.1 (*b*) will reflect that the specified irrigation revenue rates for different crops are varying between 2 to 9% of their earning capacities. It also reflects that the rates are highly beneficial to sugarcane, which is a highly earning crop.

It can, therefore, be concluded that the specified rates in a *crop rate system of pricing,* mainly depends upon two factors ; viz.

(*i*) *water requirement of each crop* ; and

(*ii*) *earning capacity of that crop.*

**7.1.1.1. Demerits of the crop rate system.** This method does not offer any incentive to the cultivators for economising in the use of water, since they will be charged at flat rate for the area of the crops grown. The cultivators will, therefore, never bother to avoid wastage of water, and will also try to use more and more quantity of water, under the wrong notion that more water will flourish their crops more.

This method, therefore, may lead to over-utilisation of water, particularly in command areas falling in the head-reaches of canals. *This may ultimately lead to water-logging and salinity of the irrigated lands, besides leading to the wastage of the precious and scarce irrigation water, which will ultimately cause scarcity of water at the tail reaches of the canal command, thus leading to unequitable distribution of water in the command area of the irrigation system.*

**7.1.1.2. Field Assessment of grown crops in crop rate revenue system.** The field assessment of the area of each crop grown by each cultivator is generally carried out and recorded by the area *Patwari* at the time of sowing, and also at the time of maturity and harvesting. At the end of the crop period, a *demand statement,* called **Khatauni,** for each cultivator is prepared.

The recorded measurements of crops carried out by Patwaris, are generally test checked by officers above Patwaris, such as *Tehsildars (Ziledars)*, SDM, *ADM*

*(Revenue)*, etc, working under the Revenue Department of the Government. In States, where the engineers are also designated as canal officers, such powers are also vested in the hands of sub-divisional and divisional engineers. Checking of irrigated areas, thus, constitute an important part of the duties of engineers. In such cases, divisional engineers (*i.e.* executive engineers) will also have to raise the revenue bills against the cultivators by signing the Khatuni's on 31st June and 31st Dec. every year, for Ravi and Kharif crops, respectively.

*In case of failure of crops or reduced production, due to reasons beyond the control of cultivators, full or part remissions can be granted by canal officers. Additional fines and levies may also be imposed by canal officers on cultivators, in case of unauthorised water use, or for damaging the outlets, or cutting the canal banks, or even for wasting water.* The decision of the divisional canal officer is generally final and binding in all such cases, because his decisions can be appealed against only at a very high level, *i.e.* to the commissioner (Revenue division), who is normally a senior IAS officer above the Deputy Commissioner of the District.

**7.1.2. Volumetric Method of Pricing.** In this method, pricing of irrigation water is done on the basis of actual consumption of water by each cultivator, or by a chunk of cultivators, such as by a cooperative society of the cultivators.

This can be done only when measuring devices are installed at the outlets of irrigation minors or water-courses. The method, therefore, requires costly installations, like those of *water-meters* or *venturi-meters,* or *modular outlets.*

Due to the high installation and maintenance cost of such measuring devices as required in this system, this method of revenue assessment has generally not been adopted in our country. This method, therefore, although has its inherent advantages leading to economic use of water, does not find much practical use in India.

The method can be made use in assessing canal irrigation revenues, only if, Irrigation Departments supply water to cooperatives or Panchayats rather than dealing with the individual cultivators. If this can be achieved, then naturally, very less number of meters would be required and this system of charging can be implemented.

This method of pricing irrigation water, can of course, be used much easily in *tubewell irrigation,* where the billing can be done on the basis of number of units of electrical energy consumed in running a particular tubewell-supplying water to a particular chunk of cultivators. This method of charging by the number of electrical energy units consumed or by the number of running hours of the tubewell, is largely used where private tubewells supply water to the cultivators, although its use on Govt. irrigation supplies is still not being adopted on a large scale. The States have simply specified slightly different rates for lift irrigation supplies, and tubewell supplies are covered under those rates.

**7.1.2.1. Merits and demerits of volumetric method.** This pricing method offers its best merit of economical use of water, thus, avoiding the wastage of this preciously scarce commodity, and also avoiding water-logging and salinity of irrigated lands.

The *demerits* of this method are that :

(*i*) it requires costly measuring devices, atleast at each outlet head, and round-the-year-staff to maintain and look after these devices ;

(*ii*) it promotes unauthorised drawals of irrigation supplies, and also manipulation and mishandling of the measuring devices at the outlet heads ;

(*iii*) it forces the chunk of cultivators to pay equally in accordance with the water supplied at the outlet head, although the cultivators at the tail end of the water-course gets lesser supplies due to wastage of water enroute the length of the water course ; and

(*iv*) silt debris may retard the flow through the metered outlets.

## 7.2. Economic Water Rates Vs Prevailing Revenue Rates in India

We have stated earlier that in our country, the Govt revenue rates for irrigation waters are generally fixed by considering ; (*i*) *The water requirement of the crop* ; and (*ii*) *the earning capacity of the crop*. However, the most important factor, *i.e. the real cost of the irrigation water,* is generally being ignored.

This real cost of the water can be worked out on *Economic considerations,* by computing the cost of the irrigation project, and recovering the same during its effective life span, along with costs of interest and annual operating and maintenance charges of the irrigation system. Such a rate, when worked out in terms of Rs. per unit quantity of water obtained, is known as the *economic water rate.*

Such a rate will evidently depend upon *the cost of the irrigation project, its effective life span, the rate of interest considered on the investment, and actual O and M charges on the system.* The *annual recovery factor,* computed on Economic considerations as explained in chapter 12 of "Hydrology and Water Resources Engg"., can, thus, give us the cost of irrigation water per year from a particular project, and it can finally be used to work out the unit cost of such water.

Such economic water rates have been computed to be generally much higher than the rates being changed by the government bodies in the country.

For example, economic water rate on a typical tubewell minor irrigation project, in Gujarat State has been computed to be Rs. 0.30P per kilolitre (on 1990 rates) as against the rate being charged by the State Tubewell Corporation @Rs. 0.18P per kilo litre. Similarly, for a typical interstate minor lift irrigation scheme, executed in Gujarat and Orissa States, the economic rate has been worked out to be Rs. 1.53P per kilolitre (on 1990 rates), as against the rate of Rs. 0.18P per kilolitre being charged by the Gujarat State Corporation, and Rs. 0.02P per kilolitre being charged by the Orissa State.

It has thus, generally been seen that the rates, being charged by the Govt. Corporations or State Irrigation Departments, are *highly subsidised* and *too low* incomparison to the actual cost of water. *So much so, that the rates being charged are not even capable to repay the O* and *M charges alone, what to talk of recovering the capital investment on the project.*

Considering the urgent necessity of improving the rates, so as to reduce the heavy losses and subsidies, being incurred by the Government in the operations of irrigation systems, the newly framed *National Water Policy (1987)*, has specified that :

"The irrigation water rates should be such as to convey the scarcity value of resources to the users, and to fasten the motivation for economy in water use. *The revenue rates should, therefore, be adequate to recover the annual operation and maintenance charges and a part of the fixed capital cost.* Efforts should be made to reach this ideal, over a period of time, while ensuring the assured and timely supplies of irrigation water. The water rates for surface water and ground water should be rationalised with due regard to the interests of small and marginal farmers".

From the above, it can be seen that the built in requirement of the National policy is that the water rates should recover annual *R* and *M* charges and part of the capital cost, if not the full cost. Even if for the time being, we neglect recovering of the capital cost, we must rationalise our irrigation revenue rates in such a fashion as to fully recover our annual running and maintenance expenditure on the irrigation systems.

*It has, however, been computed that the average annual irrigation revenue realisation in the country is only about 12% of the total O and M expenditure alone, with Statewise variation ranging between 32% in Delhi to 0.11% in Nagaland with Punjab not collecting any revenue at all, since it has totally abolished water charges for irrigation water supplies wef Feb. 1997.*

There is, therefore, an urgent need to reconsider and revise the water rates, even at the cost of inflation and possible rise in the rates of crops.

At present, different revenue rates are prevailing in different States of the country even for the same crop ; and even in the same State, different rates are prevailing for different irrigation systems. There is no uniform method of computing the revenue rates ; and the rates are generally being decided and guided by the whims and fencies of the political powers. A few examples of the non-uniformity and irrationality prevailing in our present revenue rates, are highlighted below :

In Gujarat State, the revenue rates for lift irrigation on canals are $\frac{1}{2}$ times, and on rivers and reservoirs $\frac{1}{3}$rd times the flow irrigation rates, provided the water is lifted by the cultivators at their own cost. In Maharashtra, the water rates for all types of lift irrigations to $\frac{1}{10}$th times (when cost of lifting is met by cultivators) are hardly $\frac{1}{5}$th to $\frac{1}{10}$ times the rates prescribed for flow irrigation. In Haryana State, the rates are reduced only to $\frac{1}{2}$ times of the flow irrigation rates, when water is privately lifted from any State irrigation work. In Jammu and Kashmir State, the rates prescribed for lift irrigation (cost of lifting is met by Govt.) are about 15 times the rates prescribed for flow irrigation ; whereas, in Kerala, the prescribed rates for such Govt. lift irrigation schemes are only about $1\frac{1}{2}$ times the rates for flow irrigation. States of Andhra Pradesh, Assam, Bihar, Chhatisgarh, Jharkhand, Karnataka, Madhy Pradesh, Tamil Nadu & West Bengal have not prescribed any separate rates for lift irrigation, and flow irrigation rates remain applicable to lift irrigation rates also. The State of Punjab has completely abolished the water rates wef 14.2.1997. While Himachal Pradesh has levied extremely low rates — lowest amongst all the States. Besides Punjab, some other States like Arunachal Pradesh, Sikkim, Meghalya, Mizoram, and UT's like Andaman and Nicobar Islands, Chandigarh and Lakshadweep have also not levied any irrigation revenue rates, so far, although the States of Sikkim and of Arunachal Pradesh are in the process of introducing such rates.

While the rich-farmers State like Punjab has completely abolished the irrigation revenue rates, the progressive States of Maharashtra and Gujarat have not only revised their water rates in the year 2001, but have also made an inbuilt provision of automatic annual increase in their notified rates. Say for example, the Maharashtra State has specified a uniform 15% annual increase over the rates notified on 1.9.2001. Similarly, the Gujarat State has provided for 25% annual increase in notified water rates of Paddy, and 15% annual increase in rates notified for all other crops wef 16.2.2001.

It can also be seen from Table 7.2 that certain States and UT's like Andhra Pradesh, Delhi, Goa, Himachal Pradesh, J & K, Karnataka, Kerala, Manipur, West Bengal, etc. have not revised their irrigation revenue rates for the last more than 10 to 50 years. NCT of Delhi is in worst condition in this respect, as it has not revised its rates since the year 1951. There is an urgent need for revision of the water rates in all such States.

**Table 7.2. Table Showing the Dates on Which Irrigation Revenue Rates were Last Notified in Various Indian States**

| *State* | *Date on which irrigation revenue rates were last notified* | *State* | *Last date on which irrigation revenue rates were notified* |
|---|---|---|---|
| Andhra Pradesh | 7.1.1196 | Maharashtra | 1.9.2001 |
| Assam | 30.3.2000 | Manipur | Year 1977-78 |
| Bihar | Nov. 2001 | Mizoram | Not specified |
| Chhattisgarh | 15.6.1999 | Orissa | 5.4.2002 for flow irrigation<br>July 1997 for lift irrigation |
| Delhi | Year 1951 | Punjab | Abolished wef 14.2.1997 |
| Goa | 11.2.1988 | Rajasthan | 24.5.1999 |
| Gujarat | 16.2.2001 | Tamil Nadu | 6.11.1987 |
| Haryana | 27.7.2000 | Tripura | Yet to start |
| Himachal Pradesh | Year 1977 | U.P. | 18.9.1995 |
| J & K | 1.4.1976 | Uttaranchal | 18.9.1995 |
| Jharkhand | 14.11.1995 | West Bengal | 6.4.1997 |
| Karnataka | 1.7.1985 | Andaman and Nicobar Island | 1.11.1992 |
| Kerala | 18.9.1974 | Dadar and Nagar Haveli | 29.1.1996 |
| Madhya Pradesh | 15.6.1999 | Daman and diu | Year 1980 |

The prevailing irrigation water rates in different States of the country, for flow irrigation are compared in Table 7.3; while those for lift irrigation are compared in table 7.4.

With a view to rationalise the irrigation water rates in different States, the Planning Commission, Govt. of India, appointed a *"Committee on Pricing of Irrigation Water"*, vide notification dated 23-10-1991, under the chairmanship of Dr. A. Vaidyanathan, Professor of Madras Institute of Development studies and former Member of Planning Commission, with 16 members from Central Water Commission (CWC) and Other State Governments. This committee has submitted its report in September, 1992. But the same appears to have not yet been implemented. Some of the important recommendations of this Committee are briefly pointed out in article 7.3.

**Table 7.3. Comparison of Prevailing Irrigation Revenue Rates (Maximum) in Various States of India for Flow Irrigation**

| S.No. | Name of State | Water rates for flow irrigation in Rs. per hectare of each crop grown | | | | | | | | | | | | |
|---|---|---|---|---|---|---|---|---|---|---|---|---|---|---|
| | | *Rice or paddy (Kharif)* | *Sugarcane* | *Wheat* | *Cotton* | *Oil seeds* | *Pulses* | *Vegetables* | *Groundnut* | *Jute* | *Toba-cco* | *Gram* | *Barley* | *Maize* |
| (1) | (2) | (3) | (4) | (5) | (6) | (7) | (8) | (9) | (10) | (11) | (12) | (13) | (14) | (15) |
| 1. | Andhra Pradesh | 494 | 494 | — | 494 | 494 | 494 | — | — | — | — | — | — | — |
| 2. | Assam | 281 | 222 | 563 | — | 99 | — | — | — | 150 | — | — | — | — |
| 3. | Bihar | 217 | 371 | 185 | — | 247 | 99 | 148 | — | 99 | — | 99 | 148 | — |
| 4. | Chhatisgarh | 494 | 741 | 200 | 371 | 247 | 247 | 494 | 247 | 247 | 494 | 247 | 494 | — |
| 5. | Delhi (NCT) | 99 | — | 45 | — | 30 | — | 89 | — | — | — | — | 44 | 30 |
| 6. | Goa | 150 | 300 | — | — | 100 | — | 100 | 100 | — | — | — | — | 60 |
| 7. | Gujarat* | 300 | 2750 | 240 | 275 | 440 | 130 | 220 | 440 | — | 330 | 90 | — | 90 |
| 8. | Haryana | 148 | 197 | 124 | 124 | 124 | 99 | 148 | 86 | — | 148 | 99 | 124 | 86 |
| 9. | Himachal Pradesh | 24 | 41 | 7 | 17 | 16 | 12 | 8 | — | — | 21 | 11 | 16 | 14 |
| 10. | Jammu and Kashmir | 49 | 49 | 25 | — | 25 | 20 | 37 | — | — | — | — | — | 25 |
| 11. | Jharkhand | 109 | 371 | 185 | — | 99 | 99 | 185 | — | 99 | — | 99 | 185 | — |
| 12. | Karnataka | 247 | 988 | 148 | 148 | — | — | — | 148 | — | 87 | — | — | 87 |
| 13. | Kerala | 99 | — | — | — | — | — | — | — | — | — | — | — | — |
| 14. | Madhya Pradesh | 200 | 741 | 200 | 371 | — | 247 | 494 | 247 | — | 494 | 247 | 494 | — |

Contd....

* With annual increase of 25% for Paddy and 15% for other crops since 16.2.2001.

| (1) | (2) | (3) | (4) | (5) | (6) | (7) | (8) | (9) | (10) | (11) | (12) | (13) | (14) | (15) |
|---|---|---|---|---|---|---|---|---|---|---|---|---|---|---|
| 15. | Maharashtra* | 180 | 4763 | 360 | 280 | 548 | — | 270 | 548 | — | — | — | — | — |
| 16. | Manipur | 38 | — | 38 | — | 23 | 23 | 23 | — | — | — | — | — | — |
| 17. | Orissa | — | 500 | 170 | — | 170 | 170 | 280 | 170 | 84 | 420 | — | — | 140 |
| 18. | Punjab | Charges abolished wef 14.2.1997 | | | | | | | | | | | | |
| 19. | Rajasthan | 114 | 247 | 148 | 178 | 114 | 79 | 109 | — | — | 178 | 182 | 79 | 67 |
| 20. | Tamil Nadu | 37 | 49 | — | 62 | — | — | 50 | — | — | 49 | — | — | — |
| 21. | Tripura (Proposed rates) | — | — | 313 | — | — | 313 | 313 | 313 | — | — | — | — | — |
| 22. | Uttar Pradesh | 287 | 474 | 287 | 114 | — | — | 287 | — | — | 306 | — | 287 | — |
| 23. | Uttaranchal | 143 | — | 143 | — | — | — | — | — | — | 153 | — | — | — |
| 24. | West Bengal | 37 | — | 49 | — | — | — | — | — | — | — | — | — | — |
| 25. | U. Ts | | | | | | | | | | | | | |
| | (*a*) Dadra and Nagar Haveli | 140 | 830 | — | — | — | — | 110 | — | — | — | — | — | — |
| | (*b*) Daman and Diu | 200 | — | — | — | — | — | — | — | — | — | — | — | — |

* With annual increase of 15% wef 1.9.2001

**Note.** States which have not levied any irrigation revenue charges so far, have not been included in this table.

**Table 7.4. Comparison of Prevailing Irrigation Water Rates (Maximum) in Various States of India for Lift Irrigation**

| S.No. | Name of State | *Water rates for Lift irrigation in Rs per hectare of each cnp grown* | | | | | | | | | | | | |
|---|---|---|---|---|---|---|---|---|---|---|---|---|---|---|
| | | *Rice or paddy (Kharif)* | *Sugarcane* | *Wheat* | *Cotton* | *Oil seeds* | *Pulses* | *Vegetables* | *Groundnut* | *Jute* | *Tobacco* | *Gram* | *Barley* | *Maize* |
| (1) | (2) | (3) | (4) | (5) | (6) | (7) | (8) | (9) | (10) | (11) | (12) | (13) | (14) | (15) |
| 1. | Andhra Pradesh* | 494 | 494 | — | 494 | 494 | 494 | — | — | — | — | — | — | — |
| 2. | Assam* | 281 | 222 | 563 | — | 99 | — | — | — | 150 | — | — | — | — |
| 3. | Bihar* | 217 | 371 | 185 | — | 247 | 99 | 148 | — | 99 | — | 99 | 148 | — |
| 4. | Chhatisgarh* | 494 | 741 | 200 | 371 | 247 | 247 | 494 | 247 | 247 | 494 | 247 | 494 | — |
| 5. | Delhi (NCT) | 148 | — | 67 | — | 45 | — | 133 | — | — | — | — | 67 | 44 |
| 6. | Goa | 300 | 600 | — | — | — | 120 | 200 | 200 | — | — | — | — | 120 |
| 7. | Gujarat** | 170 | 420 | 100 | 90 | 180 | 55 | 90 | 90 | — | — | 45 | 113 | 35 |
| 8. | Haryana | 74 | 99 | 62 | 62 | — | 49 | 74 | 43 | — | 74 | 49 | 62 | 43 |
| 9. | Himachal Pradesh | 49 | 82 | 29 | 32 | 32 | 24 | 32 | — | — | 41 | 29 | 32 | 28 |
| 10. | Jammu & Kashmir | 247 | 716 | 124 | — | — | — | — | — | — | — | — | — | — |
| 11. | Jharkhand* | 109 | 371 | 185 | — | 99 | 99 | 185 | — | 99 | — | 99 | 185 | — |
| 12. | Karnataka* | 247 | 988 | 148 | 148 | — | — | — | 148 | — | 87 | — | — | 87 |
| 13. | Kerala | 149 | — | — | — | — | — | — | — | — | — | — | — | — |
| 14. | Madhya Pradesh* | 200 | 741 | 200 | 371 | — | 247 | 494 | 247 | — | 494 | 247 | 494 | — |

*Contd....

| (1) | (2) | (3) | (4) | (5) | (6) | (7) | (8) | (9) | (10) | (11) | (12) | (13) | (14) | (15) |
|---|---|---|---|---|---|---|---|---|---|---|---|---|---|---|
| 15. | Maharashtra | — | 495 | — | 100 | — | — | — | 100 | — | — | — | — | — |
| 16. | Manipur | 38 | — | 38 | — | — | 23 | 23 | — | — | — | — | — | — |
| 18. | Mizoram | No water rates are in Vogue | | | | | | | | | | | | |
| 17. | Orissa | 516 | 1398 | 646 | 861 | 345 | 215 | 1292 | 684 | 430 | 1034 | — | — | — |
| 18. | Punjab | Charges abolished wef 14.2.1997 | | | | | | | | | | | | |
| 19. | Rajasthan | 395 | 573 | 296 | 356 | 227 | 158 | 217 | — | — | 356 | 207 | 158 | 133 |
| 20. | Tamil Nadu* | 37 | 49 | — | 62 | — | — | 50 | — | — | 49 | — | — | — |
| 21. | Tripura (Proposed rates) | — | — | — | — | — | — | — | 313 | 313 | — | — | — | — |
| 22. | Uttar Pradesh | 143 | 237 | 143 | 57 | — | — | 143 | — | — | 153 | — | 143 | — |
| 23. | Uttaranchal | 287 | — | 287 | — | — | — | — | — | — | 306 | — | — | — |
| 24. | West Bengal* | 37 | — | 49 | — | — | — | — | — | — | — | — | — | — |
| 26. | U. Ts | | | | | | | | | | | | | |
| | (a) Dadra and Nagar Haveli | 75 | 100 | 100 | — | — | 100 | 275 | — | — | — | 100 | — | — |
| | (b) Daman and Diu* | 200 | — | — | — | — | — | — | — | — | — | — | — | — |

* Separate rates for Lift irrigation are not given, and hence rates shown are the same as for flow irrigation.

** Specified rates here are for lift irrigation done by cultivators at their own cost from canals & irrigation projects.

### 7.3. Important Recommendations of the Committee on Pricing of Irrigation Water (1992)

(1) **Paras 3.11 and 3.12.** A revision in the level and structure of water rates is, thus, necessary in the interest of both efficiency and equity. The revision should be such as to achieve full cost of recovery in due course and in the process promote saving, create disincentives for waste and thereby enable the service area to be expanded and more reliable service assured.

(2) **Paras 3.13 and 3.16.** Revision of water rates should go hand in hand with measures to improve the quality of service and to keep a check on costs.

(3) **Paras 3.54 and 3.56.** The objective of the *first phase* should be to rationalise and simplify the existing system of assessment (based on cropwise irrigated area on an individual basis) to a system of season-specific area rates. It is possible to estimate the relative water consumption per hectare irrigated in different seasons. The variable part of the tarrif in the case of major and medium projects and such of these minor works as are still under State management should be fixed on this basis. *We should urge that all minor systems be turned over to users immediately after completion.* Both categories of projects will pay a flat basic rate per ha. *The level of cost-recovery to be aimed at the first phase should at least cover the O and M costs and 1% interest on capital employed.*

(4) **Para 3.57.** Irrigated area under a crop which spreads over two seasons will be charged at the rates applicable to both seasons, and perennials for all three seasons ; but crops like paddy which take a lot of water for non-consumptive uses need specific treatment. Where paddy is a significant but not a dominant crop, some differentiation may have to be made. In each season, therefore, we need to distinguish at best three categories, viz. paddy, sugarcane, and, perennials and other crops.

(5) **Para 3.58.** In the *second phase* to be implemented in the course of the next decade, the aim would be to shift to a fully volumetric system. Additional investments to modify the distribution system for effective regulation of volume delivered at outlets (estimated at approximately Rs. 5,000 crores) will be needed. As system efficiency and productivity improve, the targets of cost recovery can be progressively increased.

(6) **Paras 3.59 and 3.61.** There are many important matters of detail to be decided in shifting to the volumetric system of charging. These are best decided in consultation with user's representatives. *The most crucial and also the most difficult task in this phase will be to promote the creation of sufficiently large farmer's groups.*

(7) **Para 3.62. Phase III,** which will spread over a much longer period should seek to extend and consolidate the system of farmer group management, and implement, with the involvement and participation of such groups, a programme for upgrading the system to a higher level of efficiency in water use and therefore of productivity. Besides substantial investments in conjunctive use and distribution networks, the techniques of water management will have to become tighter and more sophisticated.

(8) **Para 4.5.** There is a case for earmarking the whole or a substantial part of the receipts from each irrigation system towards the operation and maintenance of that system. *In the long run, there is a case for moving towards the conversion of each irrigation system into an independent self-financing system, whether through the formation of corporations or otherwise.*

(9) **Paras 4.9 and 4.10.** We recommend that States set up special expert groups to work out appropriate norms and a procedure for periodic monitoring and updating for different agro-climatic regions and broad categories of projects.

(10) **Para 4.26.** The States should form a high-powered autonomous Board which may be called "Irrigation and Water-Pricing Board" to review the policy, establish the norms regarding maintenance costs for various components and staff costs, assess the actual expenditure in relation to these norms, and determine the parameters and criteria for revising water rates. There should be a mandatory review of all these matters every five years with an opportunity for users to present their views.

(11) **Paras 5.1 and 5.2.** Among the various problems faced in the matter of assessment, unauthorised irrigation and the incorrect reporting of crops & irrigated areas are the major ones. *There are also delays in raising demands. In spite of low and subsidised water rates, actual revenue recoveries are substantially below the demands. Large arrears have been allowed to accumulate and these tend to be eventually written off.* The existing mechanisms for preventing unauthorised, excessive and wasteful use of water as well as for the recovery of outstanding dues have not proved very effective. Lack of coordination among different agencies involved in assessment and collection also aggravates the problem.

(12) **Paras 5.9. and 5.15.** There is considerable diversity in the mechanism for the assessment and collection of irrigation revenues. The limited data that we have seen suggest that the ratio of accumulated arrears to annual demand is generally much higher in States where the Irrigation Department is responsible for both the assessment and collection than in States where both functions are vested in the Revenue Department or where they are divided between the Irrigation and Revenue Departments. *Having considered the matter, the committee is of the view that the assessment function is best entrusted to the Irrigation Department. As for collection, States may choose one of two options — (1) entrusting both assessment and collection to the Irrigation Department, and (2) making the Irrigation Department responsible for assessment and the Revenue Department for collections — in the light of their specific circumstances and experience. Where alternative (1) is preferred, it would be necessary to empower the Irrigation Department officials to recover arrears of irrigation dues under the Revenue Recovery Laws.*

(13) **Paras 5.16 and 5.18.** The Committee would like to emphasise the need for purposive and strong measures to ensure the accurate assessment of irrigation charges and their prompt and full collection. We suggest that a regular system of independent verification of actual irrigation on a sample basis be introduced on all major and medium project commands. At the same time, we strongly recommend that a serious effort be made by the Irrigation Departments to use remote sensing as an independent source of information on irrigated area, which can be used along with sample verification to test the veracity of records maintained by the field staff. Such independent checks linked to a system of penalties for inaccurate (and rewards for accurate) recording would minimise the loss from under-assessment.

(14) **Para 5.19.** We are of the view that the practice of waiving or suspending collections of irrigation charges on account of drought is not justified in respect of areas actually irrigated.

(15) **Paras 5.20 and 5.21.** The reluctance of the governments to support the agencies concerned in enforcing the regulations has led to a situation in which these agencies have practically given up even raising demands for betterment levies. Very little is done to take cognizance of the widespread violations of rules and even less to enforce what little penalties are levied. We need hardly emphasise

that such laxity has serious consequences not just in terms of revenue but for the efficient management of the system.

(16) **Para 5.22.** With a view to improving collections, the States should consider switching from the existing system of supplying water on credit to one of supply against advance payment. The collection performance relative to demand should be an important consideration for deciding the allocation of *O* and *M* funds to individual systems. We also recommend that proceeds from the collection of accumulated arrears from a system be used for making up the cumulative effects of past neglect in the maintenance of that system.

(17) **Paras 5.23 to 5.25.** Until a system of group assessment on a volumetric basis is introduced, the State agencies will need to verify and record the area irrigated by plots in order to determine the dues from individual farmers. The proposed system of season-hectare assessment (*i.e.*, assessment on the basis of area irrigated in each season) will substantially simplify the task. In the case of minor surface works, since assessment will be at a flat rate per hectare of command, there is no need for recording cropwise area irrigated for the assessment of water rates.

(18) **Para 5.26.** The aim should be to increase user participation in management initially at the level of distributaries and minors, and in due course at the level of the system as a whole. Each system should become an autonomous entity which manages its own finances both for operation and eventually for the expansion/improvement of facilities.

(19) **Para 7.1 to 7.6.** For the purposes of illustrating the application of the suggested approach to the revision of water rates, if we use the norm suggested by the Jakhade Committee, namely, Rs. 180 per ha. gross irrigated area, with adjustments for inflation since 1987 and for departmental overheads at 25 per cent of the norm so adjusted, and add interest @ 1% on capital, *the total recovery in Phase I should average around Rs. 340 per ha.* As against this, the estimated gross receipts from major and medium projects in 1989-90 was Rs. 68 ; the actual irrigation revenue works out in 1989-90 to Rs. 50 per ha. Assuming conservatively the additional revenue on account of an increase in the rates for industrial use at Rs. 10 per ha, **the recovery from irrigation charges in Phase-I has to be Rs. 310 per ha compared to the present realisation of Rs. 50 per ha.**

(20) **Para 7.9.** A basic levy at the rate of Rs. 50 per ha is recommended for all lands in the cultivable commands of major and medium as well as minor works. This is intended as a fee for the right to get water from the system (a sort of "demand charge").

(21) **Para 7.10 to 7.12.** The rate per unit of water needs to be equalised across crops. The additional revenue through such rationalisation will be sizeable, the increase ranging from 18 per cent to 140 per cent of the revenues at current rates if the per ha. cm. rates for all crops are made equal to the irrigation rate now charged for ID crops, and from 50 to 325 per cent if they are made equal to the highest irrigation rate per ha.cm.

(22) **Para 7.13.** The level of rates will also have to be raised. The extent of increase required, depending as it does on the potential for rationalisation, cannot be quantified. It is also likely to vary from States to States. Nevertheless, on the average, the required revenue by way of irrigation charges (Rs. 310 per ha) will still be barely 6% of the gross produce per hectare of the irrigated area, and that without taking any account of likely improvements in productivity.

(23) **Para 7.15.** As a measure of inducement for farmer's groups to take over greater responsibility, we suggest that when the proposed revisions are implemented, the rates for group delivery be fixed at substantially lower levels than for individual delivery, while keeping the basic fee of Rs. 50 per ha. common.

(24) **Para 7.22.** The implementation of the approach suggested here will require expeditious action on the part of each State to set up task forces, with adequate expert staff and authority for collecting the necessary data, to determine *O* and *M* norms by region and category of projects ; undertake sample studies in the field to determine the extent of under-assessment and under-collection at existing rates ; determine the per hectare rates applicable to paddy and other seasonal crops by season and for perennials in terms of volume of irrigation required and costs connected with carry-over between seasons ; and work out the existing and projected use by non-agricultural users and determine the rates to be charged to such users, the appropriate contractual arrangements, and other relevant details.

The committee report, however, appears to be lying in the waste paper baskets, of the States, and the States are simply not inclined to take politically inconvenient decisions of increasing the water rates. Let us appeal to the politicians of this country to forget their political benefits infront of the overall benefits of the country.

## PROBLEMS

1. Discuss the two important methods, which are used to price the irrigation water supplies for collecting irrigation revenues. Compare the merits and demerits of both these methods. Also discuss as to which of these two methods is generally being used in our country and why.
2. (*a*) Explain the various points which need to be considered while fixing irrigation water rates. *(Pune Univ. May 1995)*

   (*b*) Explain the crop rate basis of assessment of water charges. State the other method. *(Pune Univ. Nov. '95)*
3. (*a*) Enumerate the merits and demerits of volumetric assessment of water charges. *(Pune Univ. Nov. '96)*

   (*b*) How have the irrigation revenue rates been fixed in India ?

   Compare the irrigation revenue rates prevailing in different States of India, explaining the urgent necessity of revision of these rates. What has the GOI done to effect this revision and/or to switch to volumetric assessment system.
4. (*a*) "The revenue rates for canal irrigation water should differ from the revenue rates for lift irrigation supplies". Critically discuss this statement in special reference to Indian States.

   (*b*) Point out the major recommendations of the Expert Committee, constituted by GOI in 1992, for revision of irrigation revenue rates.

# 8

# *Rivers, Their Behaviour, Control and Training*

## 8.1. Importance of Rivers and Necessity of Controlling Them

Water is the most important natural resource not only of a state or of a country, but of the entire humanity. The prosperity of a nation depends primarily upon the exploitation of this resource. Thus, it can be stated that the primary wealth of a nation is water, which flows in rivers and streams. This itself establishes the importance of rivers, and no other explanation is required to stress their importance.

Importance of rivers has been recognised since time immemorial, and rivers have occupied a very prominent place in every stage of human development. With the incoming of modernisation and mechanisation of human life, water and hence, rivers are becoming more and more important. They provide us water for industrial uses and are a source of cheap energy. Rivers have always been satisfying our domestic, municipal, irrigation, and other demands, and that is why most of our cities were established in the vicinity of the natural rivers. In the primitive times, there was absolutely no control on these natural rivers, and hence, they used to cause tremendous devastations and troubles to human beings. But with the development of science and technology, and in his bid to control nature, man has devised and is devising means and ways to control these mighty rivers. In the above context, we have to study the rivers, their behaviour, and to develop means and ways, so as to ensure an effective control upon them.

Rivers take off from mountains, flow from the mountainous plain terrains, and finally join the oceans. They are formed along more or less defined channels, drain away the land water obtained by precipitation and snow melting in high altitudes, and discharge the unutilised waters back into the sea, thus completing the hydrological cycle. The rivers not only carry this huge amount of water but also carry a tremendous amount of silt and sediment which is washed down from the catchment area or gets eroded from their beds and banks. Before we discuss the techniques that are often adopted for controlling and improving the rivers, let us, first of all, discuss the various types of rivers and their characteristics.

## 8.2. Types of Rivers and Their Characteristics

**8.2.1. Classification of Rivers on the Basis of the Topography of the River Basin.** Depending upon the topography of the basin, the river reaches can be classified into two main classes, *i.e.*

(*i*) *Rivers in hills* (Upper reaches) ;

(*ii*) *Rivers in alluvial plains*, known as rivers in flood plains (Lower reaches) ; and

(*iii*) *Tidal rivers*

All these three types of river reaches are described below :

(*i*) **Rivers in Hills (Upper Reaches)**. The rivers generally take off from the mountains and flow through the hilly regions before traversing the plains. These upper reaches of the rivers may be termed as *Rivers in Hills*. They can be further sub-divided into :

(*a*) Incised or Rocky River stage ; and

(*b*) Boulder River stage.

(*a*) *Rocky Stage or Incised River Stage*. In this type, the flow channel is generally formed by the process of degradation (erosion). The sediment transported in this reach is often different from the river bed material, since most of it comes from the catchment due to denundation and soil erosion. These river-reaches are highly steep with swift flow, and forming rapids along their courses. The beds and banks of such rivers are less susceptible to erosion. The bed-load carried by such river-reaches cannot be determined on the basis of usual bed load transportation formulas, derived on the basis of bed characteristics.

(*b*) *Boulder River Stage*. The river bed in these reaches consists of a mixture of boulders, gravels, shingles and alluvial sand-deposits created by itself. Still these river reaches differ considerably from those carrying sand and silt and flowing through plains. In the latter stage, the river flows through deep well defined beds and wider flood plains and develops zig-zag courses. On the other hand, in the boulder stage, *the river flows through wide shallow beds and interlaced channels, and develops a straighter course*. During a flood, the boulders, shingles and gravels are transported downstream, but as the flood subsides, the material gets deposited in heaps. The water, then unable to shift these heaps, go round them, and the channel often wanders in new directions, often attacking the banks and consequently widening the bed.

(*ii*) **Rivers in Alluvial Flood Plains (Lower Reaches)**. The chief characteristics of these river reaches is the zig-zag fashion in which they flow, called **meandering.** They meander freely from one bank to another and carry sediment which is similar to bed material. Material gets eroded constantly from the concave bank (outer edge) of the bend and gets deposited either on the convex side (inner edge) of the successive bends or between two successive bends to form a bar, as shown in Fig. 8.1.

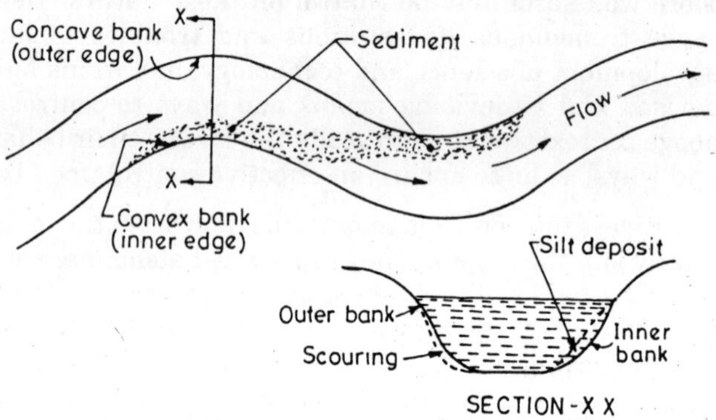


Fig. 8.1. Silting and scouring in a meandering river.

When once a straight moving river, just slightly deviates from its axis, the unbalance created goes on multiplying *with constant erosion from the concave side and deposition on the convex side*. If unchecked, the process continues, resulting in the formation of large meanders.

**Rivers in flood plains can be further classified as :**

(*a*) *Aggrading* ; (*b*) *Degrading* ; (*c*) *Stable* ; (*d*) *Braided* ; and (*e*) *Deltaic*.

If the river is collecting sediment and is building up its bed, it is called an *aggrading* or of an *accreting type*. If the bed is getting scoured year to year, it is called a *degrading type*. If there is no silting or scouring, it is called a *stable river*. It is not necessary that a river reach should be of one type in its entire alluvial length, rather it is generally of more than one type of reach, in its length. In other words, the same river reach may behave as of aggrading or of degrading or of stable type. How and under what circumstances, the river may change its type, will become very clear, if we study these types in details, as given below :

(*a*) *Aggrading or Accreting Type*. An **aggrading river** is a silting river, as shown in Fig. 8.2. Such a river increases its bed slope, which is called *building up of slope*. The silting may be due to various reasons, such as heavy sediment load; construction of an obstruction across the river, such as a dam or a weir; sudden intrusion of sediment from a tributary; etc. This type of river, usually, has straight and wide reaches with shoals in the middle, which shift with floods, dividing the flow into a number of braided channels.

(*b*) *Degrading Type*. If the river bed is constanly getting scoured (eroded) to reduce and dissipate available excess land slope, as shown in Fig. 8.3, then the river is known as a **degrading river.** It may be found either above a *cutoff** or below a dam or a weir or a barrage, etc. For example, the Colorado River (U.S.A.) became a degrading type on the downstream, after the construction of the Boulder dam.

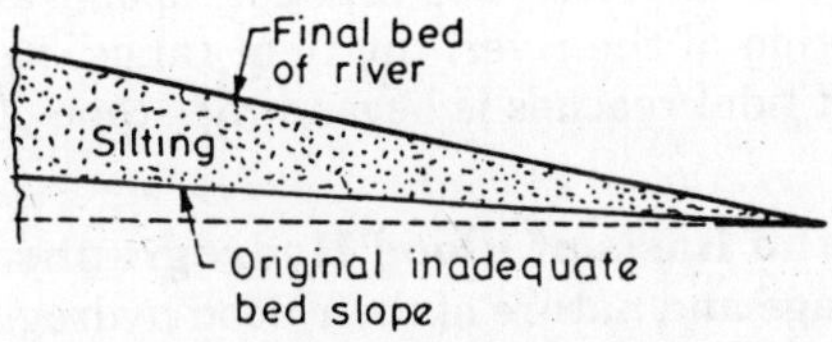


Fig. 8.2. Aggrading River

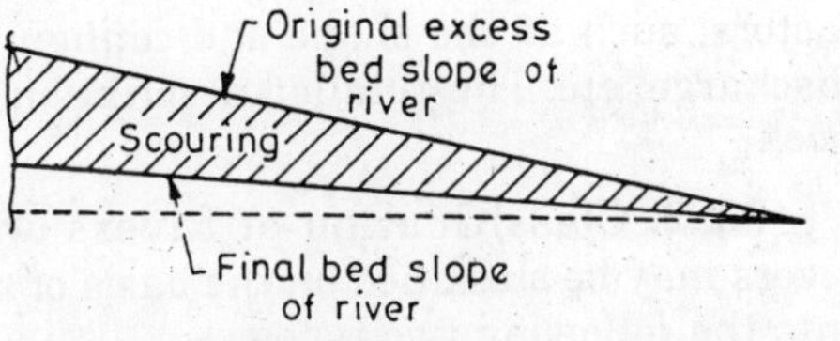


Fig. 8.3. Degrading River.

(*c*) *Stable Type.* A river which does not change its alignment, slope and its regime significantly, is called a *stable river.* Changes such as silting or scouring or advancement of delta into the sea may take place, but they are negligible and may fail to produce any change in the regime of the channel, except, perhaps, that the river may shift within its *Khadirs***. Most of the sediment load carried by them is brought to the sea.

The behaviour of a particular reach (whether to be aggrading, degrading or stable) depends mainly upon the variations of silt charge (size as well as quantity) and flow discharge with time.

(*d*) *Braided Rivers.* When a river flowe in two or more channels around alluvial islands (as shown in Fig. 8.4), it is known as a **braided river.**

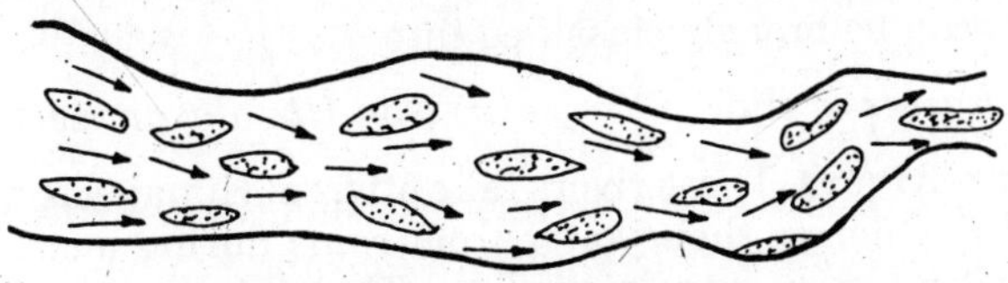

Fig. 8.4. A typical braided reach of a river.

* Explained later.

** The extreme limits within which a river is known to have ever wandered.

The braided pattern in a river develops after local deposition of coarser material, which cannot be transported under prevailing conditions of flow, and which subsequently grows into an island consisting of coarse as well as fine material.

(*e*) *Deltaic Rivers.* A river before it joins the sea, gets divided into branches, thus forming a Δ shaped delta, as shown in Fig. 8.5.

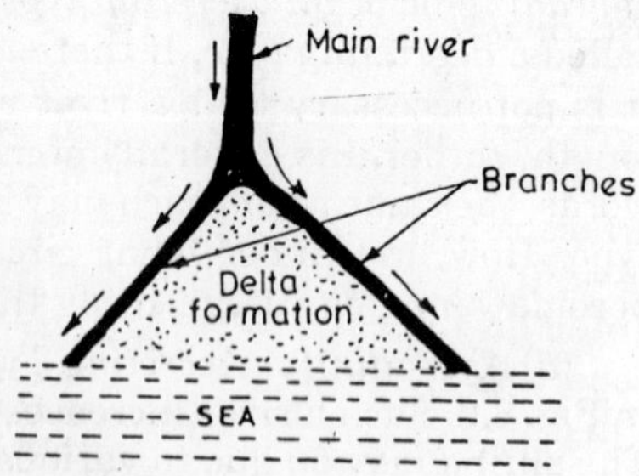

Fig. 8.5. Delta formation.

As the river approaches the sea, its velocity is reduced, and consequently the channel gets silted and water level rises resulting in spills and eventual formations of new channels. These branches multiply in their number as the river approaches the sea. Causes of delta formation and factors responsible for its shape and growth etc., are beyond the scope of this book. The **delta river** indicates a stage, rather than a type of a river.

(*iii*) **Tidal Rivers**. The tail reaches of the rivers adjoining the oceans are affected by the tides in the ocean. The ocean water enters the river during the flood tide and goes out into the ocean during the ebb tide. The river, therefore, undergoes periodical rise and fall in its water level, depending upon the nature of the tide. The distance upto which the tidal effect is experienced, depends upon various factors, such as the shape and configuration of the river, the tidal range, freshet discharge, etc. The detailed description of tidal reaches is beyond the scope of this book.

**8.2.2. Classification of Rivers on the Basis of Flood Hydrographs.** The rivers may be classified on the basis of stage and nature of their flood hydrograph, into the following two types :

(*i*) *Flashy rivers* ; and (*ii*) *Virgin rivers.*

(*i*) **Flashy Rivers**. If the flood rise and flood fall in a river is sudden, then it is called a *flashy river*. In flashy rivers, the flood flows, therefore, occur suddenly, and rise and fall of water level is very quick. The flood hydrographs are very steep, indicating floods, all of a sudden.

(*ii*) **Virgin Rivers.** In arid zones (deserts), a river water may completely dry before it joins another river or the ocean. Such a river is called a *virgin river*. After flowing for a certain distance from its source, the water of such a river disappears due to high percolation or due to excessive evaporation. There are several virgin rivers in the States of Kutch and Rajasthan in India.

## 8.3. Indian Rivers and Their Classification

Indian rivers can be broadly classified into :

(*i*) *Himalayan rivers* ; and (*ii*) *Non-Himalayan rivers*.

(*i*) **Himalayan Rivers**. These rivers take off from Himalayas and flow through alluvial plains. They derive their water from rains during monsoon and winter, and from the melting snow during summer. These rivers are, therefore, almost perennial and can give dependable yields throughout the year. These rivers carry huge sediment, because of two reasons : (*i*) The Himalayan rocks are soft and friable (*ii*) The Himalayan zone, particularly the north-eastern part, is susceptible to earthquake disturbances, causing landslides and increased rock sediment. The important Himalayan rivers are :

*Indus, Jhelum, Chinab, Ravi, Beas, Sutlej, Ganga, Gandak, Ghaghar, Gomti, Kosi, Brahmaputra, etc.*

Due to the heavy rainfall in the month of July and August (causing about $\frac{3}{4}$th of average annual rainfall), these North Indian rivers rise in high floods. The surface runoffs so caused are usually much more than the normal average flow of the entire year. The flood discharge may be as great as 50 to 100 times the normal winter flow. The sections required for passing flood discharges are vastly out of proportion to the sections required for normal winter flows. These large variations in discharge and sediment load make the hydraulics of these rivers very complex and cause them to meander.

(*ii*) **Non-Himalayan Rivers**. They are non-perennial rivers. They receive their major water supply only in rainy seasons ; and for the rest of the year, they may draw water from ground water as base flow. They flow in Central and South India and takeoff from Aravali, Vindhya and Satpura mountain ranges. *Chambal, Mahanadi, Cauvery, Godavari, Tapti, Narmada, etc.*, are the important examples of these rivers. These rivers are much more stable than the Himalayan rivers, and pose lesser problems, as they flow through non-alluvial soils. The dividing line between Himalayan & Non-Himalayan rivers is the boundary hetched by the left bank of river *Sutlej*, Right bank of river *Yamuna* and *Ganga* (after Allahabad) and left bank of river *Brahmaputra*, as can be seen in the map of Indian rivers, shown in the enclosed coloured photo Fig. 8.1.

**Note** : For regional classifications of Indian rivers, please refer article 1.5 in "Hydrology and Water Resources Engineering" by the same Author.

## 8.4. Behaviour of Rivers

The chief factor which is responsible for moulding the behaviour of rivers is the silt and sediment that flows in the river. The sediment carried by the river poses numerous problems, such as : increasing of flood levels, silting of reservoirs, silting of irrigation and navigation channels, meandering of rivers, splitting up of a river into a number of interlaced channels, etc. The *meandering* causes the rivers to leave their original courses, forces them to flow along new courses, and thus devastating vast areas of land and affecting important and valuable nearby structures, such as bridges, railway lines, roads, etc. We shall now discuss the behaviour of the rivers in special situations, such as in bends, meanders and cutoffs, in details.

**8.4.1. Straight Reaches.** In a straight reach of a river, the river cross-section is in the shape of a trough, with high velocity flow in the middle of the section.

Since the velocity is higher in the middle, the water surface level will be lower in the middle and higher at the edges, as shown in Fig. 8.6. Due to the existence of this transverse gradient from sides towards the centre, transverse rotary currents get developed, as shown in Fig. 8.6. However, straight reaches are very few in alluvial rivers.

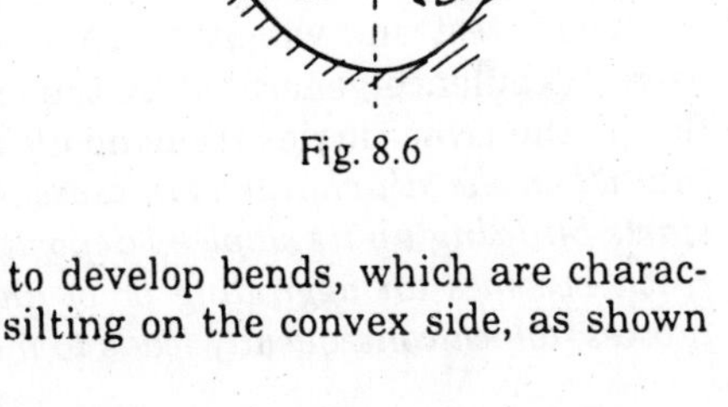


Fig. 8.6

**8.4.2. Bends.** Every alluvial river tends to develop bends, which are characterised by scouring on the concave side and silting on the convex side, as shown in Fig. 8.7. The silting and scouring in a bend may continue due to the action of the centrifugal force.

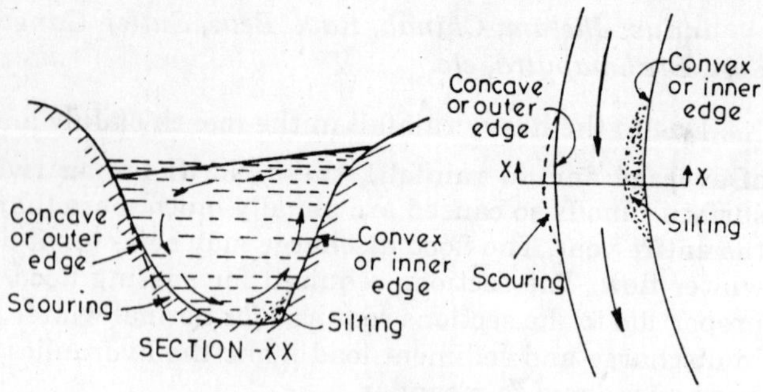

Fig. 8.7

When the flow moves round a bend, a centrifugal force is exerted upon the water, which results in the formation of transverse slope of water surface from the convex edge to the concave edge, creating greater pressure near the convex edge (see Fig. 8.7). To keep its own level, water tends to move from the convex-side towards the concave side. However, the topmost water surface movement is prevented by the centrifugal force. Moreover, towards the bottom, the velocities are much less than towards the top, and enough centrifugal force is not available to counteract the tendency of water at the top to move inwards. Hence, the water dives in, from the top at the concave end, and moves at the bottom towards the convex end. These rotary currents cause the erosion of concave edge and deposition on convex edge, forming shoals on the convex sides. When once a bend is formed, the flow tends to make the curvature larger and larger.

**8.4.3. Meanders.** When once a river deviates from its axial path and a curvature is developed (either due to its own characteristics or due to the impressed external forces), the process moves downstream by building up shoals on the convex side by means of secondary currents. The formation of shoals on the convex side, results in further shifting of the outer bank by erosion on the concave side.

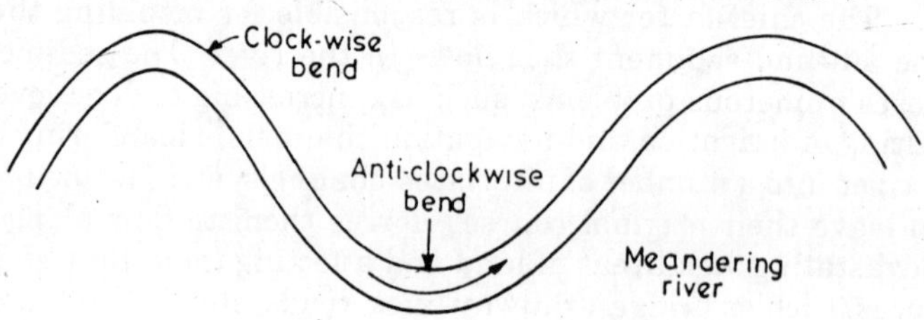

Fig. 8.8. Meandering River.

*Formation of successive bends of reverse order may lead to the formation of a complete S curve called meander. When consecutive curves of reverse order connected with short straight reaches called (crossings) are developed in a river reach,* the river is stated to be a meandering river (Fig. 8.8). In order to study the behaviour of a meandering river, the river may be supposed to follow a *sine curve.*

**Causes of meandering.** There was a belief that meandering is caused due to the presence of an excessive bed slope in the river that the excess energy is dissipated by increasing the channel length by meandering. The above theory has been proved to be incorrect.

The latest and widely accepted theory behind meandering is based upon the extra turbulence generated by the excess of river sediment during floods. During floods, the river carries tremendous amount of silt charge. *It has been established that when the silt charge is in excess of the quantity required for stability, the river starts building up its slope by depositing the silt on its bed. In other words, the river reach becomes an aggrading or of an accreting type. This accretion, is the primary process, which consequently leads to meandering, as explained below :*

The increase in river slope tends to increase the width of the channel, if the banks are not resistant. The banks are, thus, attacked by water, and in the process, one bank is likely to be attacked slightly more than the other, causing a slight deviation of flow. This slight deviation from uniform axial flow, *helps in moving more and more flow towards one bank than towards the other.* The process continues with more and more vigour, causing more and more flow towards the former bank and forming shoals along the latter, thus, accentuating the curvature of flow, and finally, producing meanders in its wake.

*The concave bank goes on eroding and the convex bank goes on silting.* Further, it has been established that collection of sediment on the convex side is independent of the concave side happenings, and deposition and formation of convex shoals will continue, irrespective of what happens on the concave side. When the concave side is rivetted, or pitched, the river gets deepened as the convex bar continues to progress towards the pitching. Due to the shift of the convex bar, the width between the banks reduces, which increases the velocity, and thus provides a current strong enough to prevent further, extension of convex-bar. In that case, the extension of shoal towards concave side is stopped, but the shoal continues to gain in height.

There are four variables, which govern the meandering process. They are : (*i*) Valley slope, (*ii*) Silt grade and silt charge, (*iii*) Discharge, (*iv*) Bed and side materials and their susceptibility to erosion. All these factors considerably affect the meandering patterns, and all of them are interdependent. On the basis of this theory and experiments, various empirical formulas have been developed for connecting various meander parameters ; *but much remains to be learnt about the meander process.*

**Meander parameters.** The various meander parameters are shown in Fig. 8.9, and are defined below :

*Meander Length* ($M_L$). It is the axial length of one meander, *i.e.* the tangential distance between the corresponding points of a meander.

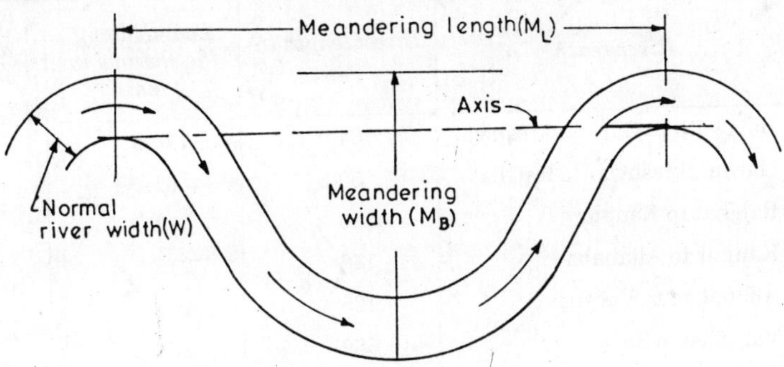


Fig. 8.9. Meander Parameters.

*Meander Belt* ($M_B$). It is the distance between the outer edges of clockwise and anti-clockwise loops of the meander.

*Meander Ratio.* It is the ratio of meander belt to meander length, *i.e.* $M_B/M_L$

*Tortuosity.* It is the ratio of the length along the channel (*i.e.* arcual length) to the direct axial length of the river reach.

*Crossings or Cross-overs.* The short straight reaches of the river, connecting two consecutive clockwise and anti-clockwise loops, are called crossings or cross-overs.

Experiments have been and are being conducted, so as to establish some reliable relationship between these parameters.

A large data on Indian rivers was collected by Jefferson and analysed by Inglis, on the basis of which, he was able to give the following tentative relationship-between different parameters :

**Table 8.1** *

| *Rivers in flood plains (where W is the river width)* | *Incised Rivers (Rocky Rivers in hills*)* |
|---|---|
| $M_L = 6.06\ W$ | $M_L = 11.45\ W$ |
| $M_B = 17.40\ W$ | $M_B = 27.3\ W$ |
| $M_B = 2.86\ M_L$ | $M_B = 2.2\ M_L$ |
| $M_L = 53.6\ Q^{1/2}$ | $M_L = 46\ Q^{1/2}$ |
| $M_B = 153.5\ Q^{1/2}$ | $M_B = 102.3\ Q^{1/2}$ |
| $W = 8.85\ Q^{1/2}$ | $W = 4.53\ Q^{1/2}$ |

It is generally observed that the total length of a meandering river along the channel remains more or less the same from year to year. If the length is shortened by the development of a cut-off, the original length is regained by rapid erosion and increase in meander belt downstream of cut-off.

It has also been found that the steep rivers meander to a lesser extent compared to those with flatter gradients.

The extent of turtuosity was found to vary throughout the course of the river. How the tortuosity varies, from the point a river becomes alluvial down to its mouth, can be estimated from the meandering of Ganga from Balawalli to the sea, as shown in Table 8.2.

**Table 8.2**

| *S.No.* | *River reach* | *Direct length in km* | *Length along the channel in km* | *Percentage of tortuosity or percentage of meander* |
|---|---|---|---|---|
| 1. | Balawalli to Garhmukteshwar | 104.0 | 116.8 | 12 |
| 2. | Garhmukteshwar to Rajghat | 59.2 | 67.2 | 14 |
| 3. | Rajghat to Kanpur | 280 | 312 | 11 |
| 4. | Kanpur to Allahabad | 176 | 216 | 23 |
| 5. | Allahabad to Varanasi | 136 | 208 | 51 |
| 6. | Varanasi to Sara | 680 | 864 | 27 |
| 7. | Sara to Sea | 296 | 320 | 8 |

For natural rivers, the discharge which determines the meander length and meander belt, may be called the *dominant discharge*. **Dominant discharge** is different from the maximum discharge, because at maximum discharge, the main flow swings away from the bend and erosion at the bank is reduced. Thus, the meander pattern is developed due to the combined effect of the discharge cycle during a flood season, and not by the maximum discharge. In most rivers, the

* **Incised river**, as defined in article 8.2.1(*a*) represents rocky stage of a river in hills.

dominant discharge varies between $\frac{1}{2}$ to $\frac{2}{3}$ of the maximum discharge. Meander length for rivers in flood plains may, therefore, be given by

$$M_L = 65.8\sqrt{Q_{dominant}} \qquad \text{...(8.1)}$$

where $Q_{dominant} = \frac{1}{2}$ to $\frac{2}{3}$ (generally 9/16) of maximum discharge.

**Cut off.** In an excessively meandering river, a particular bend may sometimes be abandoned by the formation of a straighter and a shorter channel (*AB*), as shown in Fig. 8.10. The process, whereby, this chord channel is developed or the chord

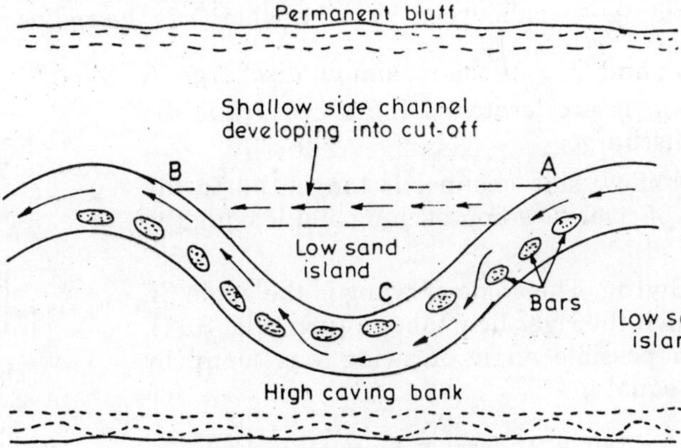


Fig. 8.10. Cut-off development.

channel itself, is termed as cut-off. It is evident that a meander increases the river length but a cut-off reduces the river length. Hence, a cut-off is a natural phenomenon for counterbalancing the otherwise ever increasing length of the river course due to the development of meanders.

**Development of a cut-off.** A meandering river flowing along the curved path has a shallow side channel besides its main curvilinear path, as shown in Fig. 8.9. This might be the remains of its old course or may be created by floods spilling over the river bank. During high floods, excessive deepening of the pools occur, and is supplemented by the growth of bars at the inflections. Both these factors tend the water to flow more and more towards the side channel. When the flow starts reducing in the main channel, more and more silting occurs in the main channel, which further increases the flow in the side channel. The process continues and finally, a time may come when the entire water starts flowing from the full developed chord channel and the curvilinear path gets silted up.

The period, over which the opening of a chord channel matures into a full-fledged cut-off may vary considerably, depending upon the local conditions. It may take six months, a year, or even five or ten years.

Whenever, a cut-off starts developing, panic and chaos is created over a certain period for miles above and below the newly formed short-cut, as the affected settlements like Jhuggis or agricultural operations, will have to be adjusted. Banks start caving, new channels are formed, old channels may get silted up, until atlast, a temporary stability is created during low floods. The period of adjustment and agitation starts again with the next flood. The process continues, until, full equilibrium conditions are established.

**Cut-off ratio.** The ratio of the length of the bend to that of the chord, *i.e.* *ACB*/*AB* (Fig. 8.10) is called the cut-off ratio. This ratio varies depending upon the characteristics of the river at site, such as the discharge, the river flood stage, surface fall, bed material and its suitability for the growth of protective grass and weeds, etc. Hence, only a certain minimum value of cut-off ratio is not sufficient for the development of a cut-off. In other words, a cut-off ratio of 1.7 to 3.0 on a certain reach may result in the formation of a cut-off, while even a cut off ratio equal to 8.0 to 10.0 may not result in the formation of a cut off on some other reach. Normally, cutoff may be developed, if the following conditions are satisfied :

(*i*) The cut-off ratio varies from 1.7 to 3.0 or more.

(*ii*) The ratio $\frac{r}{\sqrt{Q_{max}}}$ is between 13 to 24, where $r$ is the radius of curvature of the loop (bend) ; and $Q_{max}$ is the maximum discharge. A cut-off formation is accelerated if the curvature is too sharp for the discharge.

(*iii*) The shallow side channel is tangential to the main direction of river flow approaching and leaving the cut.

**Angle of Swing.** The angle of swing is the angle (in degrees) by which the river flow takes a turn (Fig. 8.11). The maximum possible angle of swing was found by Chately, to be equal to

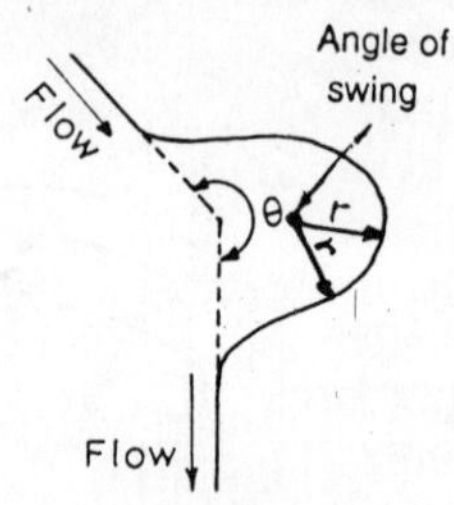


Fig. 8.11. Angle of swing.

$$\theta = 180° + 2\left[\text{Vers}^{-1}\left(\frac{\text{Chord}}{2 \times \text{Radius}}\right)\right] \quad ...(8.2)$$

This was found on the consideration that when arc/chord ratio exceeds 1.571, *i.e.* when the angle of swing exceeds 180°, conditions favourable for a cut-off are obtained.

## 8.5. Control and Training of Rivers

River training, in its wider aspects, covers all those engineering works which are constructed on a river, so as to guide and confine the flow to the river channel, and to control and regulate the river bed configuration, thus ensuring safe and effective disposal of floods and sediment loads. Stabilising and training the river along a certain alignment with a suitable waterway is, therefore, the first and formost aim of river training.

**8.5.1. Objectives of River Training.** The river training works may serve the following objectives or advantages :

(*i*) To prevent the river from changing its course and to avoid outflanking of structures like bridges, weirs, aqueducts, etc.

(*ii*) To prevent flooding of the surrounding countries by providing a safe passage for the flood waters without overtopping the banks.

(*iii*) To protect the river banks by deflecting the river away from the attacked banks.

(*iv*) To ensure effective disposal of sediment load.

(*v*) To provide minimum water depth required for navigation.

**8.5.2. Classification of River Training.** Depending upon the purpose for which a river training programme is undertaken, the river training works may be classified into the following three categories :

(1) High water training or Training for discharge.

(2) Low water training or Training for depth.

(3) Mean water training or Training for sediment.

(1) **High water training or Training for discharge.** High water training is undertaken with the primary purpose of flood control. It, therefore, aims at providing sufficient river cross-section for the safe passage of maximum flood, and is concerned with making the adjoining area flood-proof, by construction of dykes or levees, etc.

(2) **Low water training or Training for depth.** Low water training is undertaken with the primary purpose of providing sufficient water depth in navigable channels during low water periods. It may be accomplished by concentrating and enhancing the flow in the desired channel by closing other channels, by the process of *bandalling* by contracting the width of the channel with the help of 'groynes', etc.

(3) **Mean water training or Training for sediment.** Mean water training aims at efficient disposal of suspended load and bed load, and thus, to preserve the channel in good shape. The maximum accretion capacity of a river occurs in the vicinity of mean water or dominant discharge. Therefore, the changes in the river bed are attempted in accordance with that stage of flood flow. The mean water training is the most important type and forms the basis on which the former two are planned.

**8.5.3. Methods of River Training.** The chief aim of river training is to achieve *ultimate stability* of the river with the aid of river-training measures. The *stability* of a river does not mean that changes like scouring and silting of bed, advancement of delta into the sea, etc. will not take place. It only means, that the river attains an equilibrium stage, and no significant change occurs in its alignment, slope, regime, etc. The regime may change within a year but shows little variation from year to year, except that, the river may meander within its *khadirs* (*i.e.* the extreme lines within which the river is ever known to wander).

It was discussed earlier, that a river adjusts its alignment, perimeter, area, slope, etc.; with respect to the discharge and sediment load, either by aggrading, meandering or by degrading. Aggrading and meandering is one and the same thing except that aggrading is the initial stage and meandering is the final stage.

*It may be concluded that meandering type is the full and final development of an alluvial river. The other two types are the interim phases and are maintained so long as the factors causing them remain operative. Aggrading rivers are, therefore, not equally amenable to river training on account of their instability.* River training works undertaken on aggrading or degrading rivers may, therefore, fail to impose any stability on such rivers. For example, bank protection works undertaken on an aggrading river may either be destroyed by severe erosion or get buried under sediment deposition. *Soil conservation measures in the upper reaches and construction of check dams on tributaries are the most effective measures to be undertaken for controlling aggrading rivers,* before taking up any river training works on such rivers. Similarly, the training works undertaken on degrading rivers may fail due to scour and undermining of foundations by bed scour.

The scouring tendency of degrading rivers must be controlled *by building cross bars, weirs, etc., before attempting any other* river training works on such rivers.

The following are the generally adopted methods for training rivers, including bank protection :

(1) Marginal embankments or Levees.

(2) Guide banks.

(3) Groynes or Spurs.

(4) Artificial cut-offs.

(5) Pitching of banks and provision of launching aprons.

(6) Pitched islands.

(7) Miscellaneous methods, such as Sills, Bandalling, etc.

(1) **Marginal Embankments or Levees**. Marginal embankments are generally earthen embankments, running parallel to the river, at some suitable distance from it. They may be constructed on both sides of the river or only one one side, for some suitable river length, where the river is passing through towns or cities or any other places of importance. These embankment-walls, retain the flood water and thus, preventing it from spreading into the nearby lands and towns. *A levee or a dyke is mainly used for flood protection by controlling the river and not by training the river.*

The alignment of levees should follow the normal meandering pattern of the river. The retirement of the levees has to be governed by technical as well as economical and political considerations, because the land falling within the levees, is either to be acquired by the government or remains susceptible to floods. The levees, are many a times, pitched on the upstream side (*i.e.* water side). Launching apron may also be provided, if the bank or levee is close to the main river channel.

**Design of levee section.** Levees are just like earthen dams with the difference that they are very long, come in operation discontinuously and for a short time, and have limited possibilities for selection of their alignment along favourable geological strata. Their sections should be designed in such a way as to keep the seepage gradient inside the body of the embankment by at least one metre from below the top surface of the embankment. The normal value of usually adopted seepage gradient varies between 4 : 1 to 6 : 1 (*i.e.* H : V) depending upon the character of the soil which may-necessitate river side slopes varying between 2 : 1 to 5 : 1, land side slopes between 2 : 1 to 7 : 1, and top width between 2.5 to 10 m. The top level of the levee should be decided by leaving a sufficient freeboard varying between 0.3 to 1.5 m above the high flood level.

The levee sections which are usually adopted for different heights are shown in Fig. 8.12.

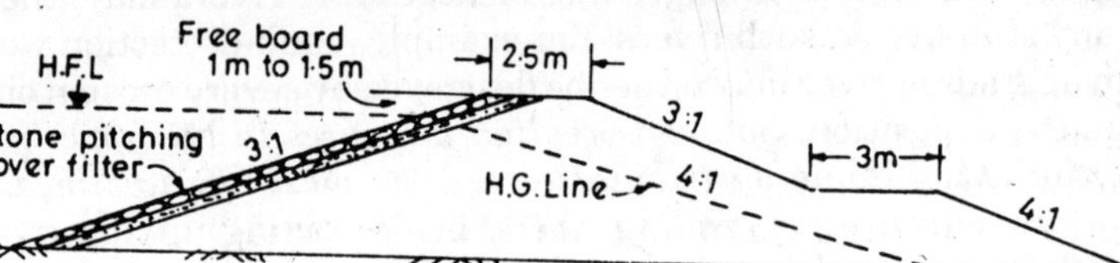


(*a*) For heights below 2.5 m.

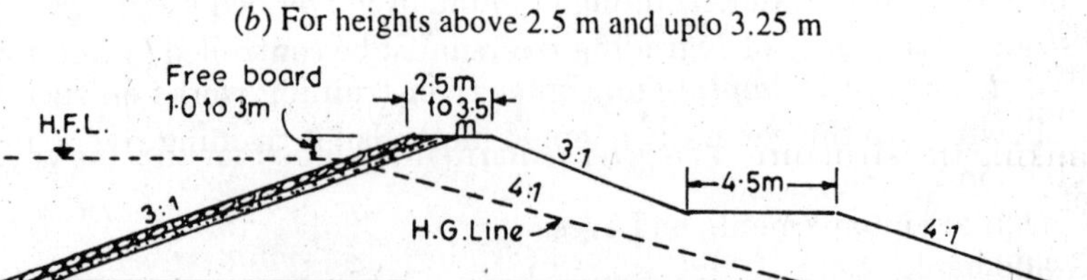


(*b*) For heights above 2.5 m and upto 3.25 m

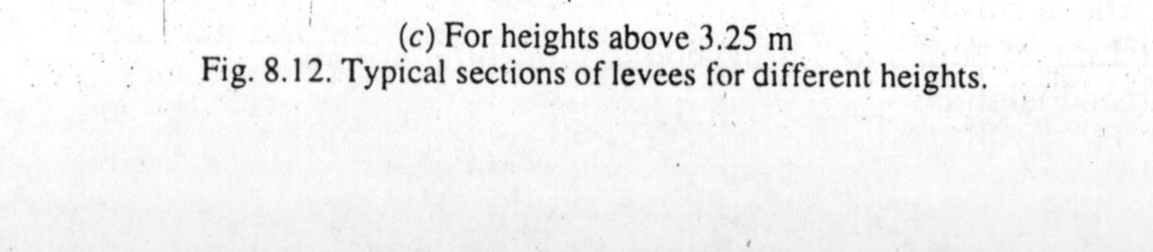


(*c*) For heights above 3.25 m

Fig. 8.12. Typical sections of levees for different heights.

**Effects of levees on river bed**. In order to assess the effects of dikes or levees on rivers, we shall, first of all, discuss the hydraulics of an undiked river during floods, and then discuss the hydraulics of a diked river.

During floods, water spills over the natural banks of a river. The velocity of this water will be reduced because it spreads over the adjoining land. Due to this reduction in velocity, soil gets deposited as the silt deposition varies inversely with the velocity. In an undiked alluvial river, the flood water spreads over a vast area, so that the velocity becomes so small that almost the entire silt gets deposited on the flood area, and clear water flows back into the main river as the flood recedes.

When the river is enclosed by dikes, the water spills over the natural banks of the river and will spread in the area confined between the river and the dikes. Due to this, the velocity of flow reduces and silt gets deposited. But infact, this reduction in velocity in a diked river is much less than that in an undiked river, because the spread area is less in a diked river. Hence, it can be concluded that the silt deposition will be less in a diked river compared to that in an undiked river.

At the same time, it can also be concluded that in a diked river, the silt will be deposited only in the confined fixed area (*i.e.* area enclosed between dikes and river), and therefore, the bed level of the river as well as that of this enclosed area will increase. Hence, the land enclosed between the dikes will appear to be higher than the adjoining land, and this probably leads to false common man belief that "the dikes cause ill-effect of raising the river-bed".

On the other hand, in an undiked river, the level of the entire land will increase and there will not appear any appreciable raising of the bed. Since the clear-cut raising of the bed is not generally visible on undiked rivers, people get misguided, and they feel that there was no bed rise in undiked rivers, and they become against diking stating "dikes cause the ill-effects of raising the river-bed", although it is not a correct statement.

(2) **Guide Banks**. If an engineering structure, such as, a weir or a barrage or a bridge, etc. is constructed across a river, the river width is reduced and trained in such a fashion, as to ensure not only a safe and expeditious disposal of flood water but also to ensure a permanent reasonable width of the waterway for the river flow. It has already been discussed that the alluvial rivers do shift their courses. Now, if today a structure such as a bridge, is constructed across the existing river width, the other day, the river may shift and there may not be any river below the existing bridge at all, and the river may be found to be flowing away from it, necessitating the construction of another structure. This may lead to the extension of the structure for the entire river length between its *khadirs*.

But, it is unwise and uneconomical to span the entire width of the river and to expose the structure to vagaries of attack and deep scour. Hence, a structure such as a weir, or a barrage, or a bridge, etc. is extended in a smaller width of the river, and river water is trained to flow almost axially through this trough without out-flanking the structure. The river is normally trained for this purpose with the help of a pair of guide banks.

The guide banks are generally provided in pairs, symmetrical in plan and may either be kept parallel or may diverge slightly up-stream of the works. Symmetrical and parallel guide banks are usually adopted, unless, the local conditions warrant otherwise.

Before the water enters into the trough formed between these two guide banks, the flow may have to be partially controlled and directed with the help of marginal bunds or by groynes or both. The guide banks, usually, consist of two heavily built embankments in the river in the shape of a *bell mouth* (named after the name of its inventer – Mr. Bell). The portion of the river between the normal river banks and the guide banks is closed by ordinary embankments. Sometimes, one of the guide bank may become unnecessary and may be dispensed with, for economical reasons. This may happen either when the khadir bank is very near the works-site, thus serving the purpose of guide bank itself, or when the marginal bund is highly resistant to scouring, general layout of guide banks, plan, section, etc. are shown in Fig. 8.13.

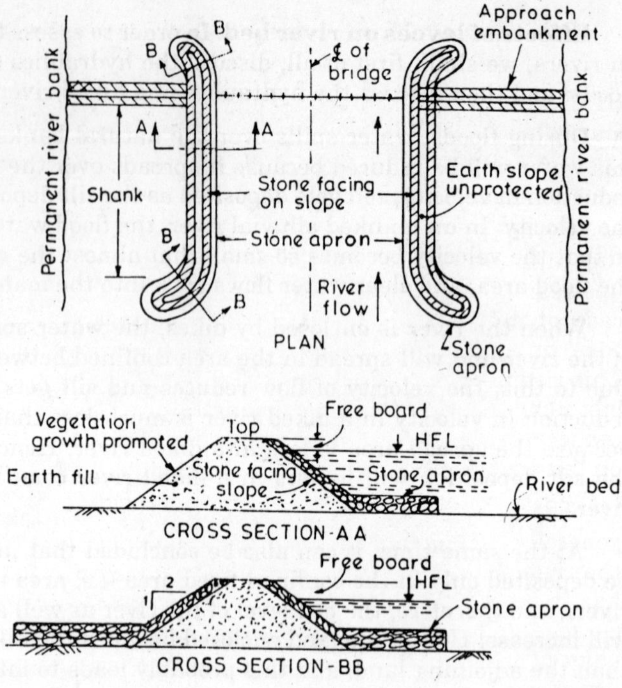


Fig. 8.13. Guide Bank details.

Before we discuss the principles applied for designing the guide banks, it shall be worthwhile to speak something about the selection of the works-site.

**Selection of works-site.** It is evident that an engineering structure, such as a bridge, or a weir should be spanned in that portion of the river, where the distance between the khadir banks is minimum. This reduces the extent of possible embayment at the back of the guide banks (discussed at little later) and permits shorter guide banks. Further, on a meandering river, the river section at the bend is always wide and non-uniform ; deep on the concave bank and shallow on the convex. While the transition reaches connecting two adjacent bends are narrower and uniform in depth. *Bridges should therefore, preferably be built in these transition reaches, rather than on the bends.*

Further, in case of bridges, the river bed at the proposed bridge site should consist of deep strata of erodible land, so that after constriction, the river may be able to deepen the bed to gain an adequate waterway. If the bed consists of stiff clay etc., the constriction can be done only after due allowance is made for afflux, which may make it constlier.

In case of weirs and barrages, the usual practice is to construct the weir or the barrage, outside the main river channel in a minor creek which is dry in winter, and then to divert the main river channel through it. At the weir site, the river width is constricted. The meandering river upstream has, therefore, to be trained to flow between the two abutments of the weir, without causing any damage.

*Principles and Factors Governing the Design of Guide Banks*

(*i*) **Top level of guide banks.** The top level of guide banks is governed by HFL, afflux, velocity head, and freeboard. It can be obtained by adding all these four values.

*Afflux*

By **afflux**, we mean the rise in the high flood level of the river, upstream of the weir (or the bridge in case of non-erodible soils), or barrage, as a result of its construction. This rise in water level is maximum just near the site of constriction and reduces as we go away from it, upstream. The afflux, extends for a long distance on the upstream side, as shown in Fig. 8.14.

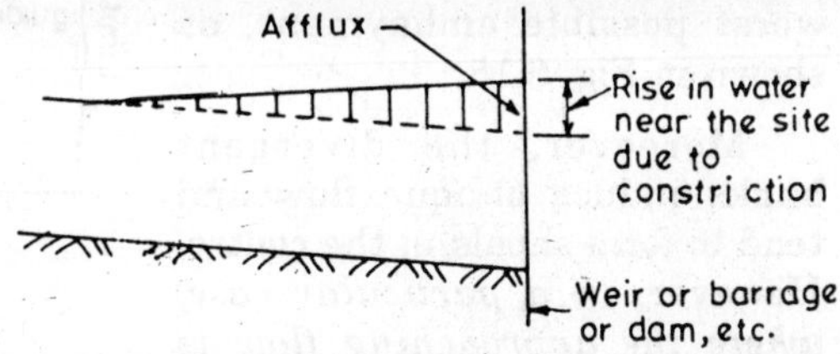


Fig. 8.14. Afflux formation at constriction.

The amount of afflux, in case of weirs or barrages founded on alluvial soils, is generally limited to 1 to 1.2 m, or more commonly as 1.0 m. In steep reaches of the river with boulder or rocky bed and in flashy rivers, a higher value of afflux has to be taken. The bed erosion and afflux are interlinked. *In case, a bridge is founded on a river, in which full bed scour develops before high floods, a negligible value of afflux may be taken.* While in boulder beds and flashy rivers, the time for bed scour may not be adequate, causing very high afflux. The amount of afflux governs the top levels of the guide banks as well as that of the marginal bunds*. The distance to which the afflux appreciably extends on the upstream side, governs the length and sections of the marginal bunds.

*The Waterway and the Discharge per metre run*

The waterway is the actual width from which the water has to flow after the river is constricted. Since this width is reduced at the works-site, the depth of water, and hence, the HFL will go up. The rise in this HFL is nothing but afflux. The limit placed on afflux automatically places a limit on the constriction or on the waterway, since the two are interdependent. In addition to high afflux, the insufficient waterway will cause excessive velocities, causing dangerously deep scour along the guide banks at the bridge piers, etc. On the other hand, too long a bridge or weir may cause slack velocities, causing formation of shoals with consequent non-uniform flow distribution and an oblique flow.

A likely figure to be adopted for waterway may be given by Lecey's Regime perimeter ($P$), given by $P = 4.75 \sqrt{Q}$ ; (because in larger rivers, width is approximately equal to its perimeter). In case of a bridge, obstruction caused by piers should be accounted for, and the above equation should be taken to represent the clear effective waterway. It should be remembered that the regime conditions are disturbed after the construction of the weir, and the above formula is not strictly applicable. Most of the existing weirs and bridges have been provided with a clear waterway from 10% to 50% more than that given by Lacey's Regime perimeter.

(*ii*) **Shape of guide banks in plan.** As stated earlier, the guide banks are generally provided in pairs, symmetrical in plan and may either be kept parallel or may diverge slightly upstream of the works. The diverging guide banks may be favoured on the ground that they cover larger width of the khadirs and exert an attracting influence

* Please see chapter 19 on "Diversion Head Works".

on the flow. But they are not recommended mainly because for equal bank lengths, they provide relatively less protection to the approach embankment, under the worst possible embayment, as shown in Fig. 8.15.

Moreover, the divergent banks induce oblique flow and tend to form shoals in the centre. However, *in a particular case, where the approaching flow is oblique, it may become necessary to provide divergent banks in order to keep the flow active in adjacent spans.* For example in Yamuna River Bridge in Delhi (opposite old fort), a splay of 10° was given in the left guide bank so as to improve the intensity of flow in the left spans.

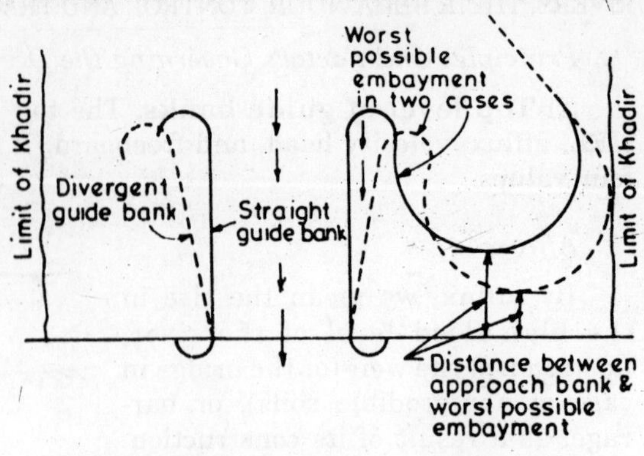


Fig. 8.15. Extent of protection provided by the straight and divergent guide banks.

**Note.** Divergent banks exercise attracting influence on the flow and hence, suitable for oblique flows but bringing the embayment nearer the approach bank and hence, not suitable in ordinary cases.

In a river which is likely to meander on both sides, it is absolutely essential to make the banks symmetrical in order to straighten the current under all possible conditions, and thus, to ensure uniform distribution of discharge and scour. Symmetrical and parallel guide banks are, hence, usually adopted, unless the local conditions warrant otherwise.

(*iii*) **Length of the guide banks.** Spring and Gales have corelated the length of the guide bank with the length of structure between the abutments ($L$). According to Spring, the length of the upstream part of the guide bank should be equal to 1.1 L or even longer if required to obviate the possibility of the river curving at the back and cutting into the approach bank. The length of the guide bank on the downstream side should be between 0.1 $L$ to 0.2 $L$.

According to Gales, the upstream length of the guide bank is 1.25 $L$ for flood discharges up to 20,000 cumecs, and 1.5 $L$ for flood discharges more than 20,000 cumecs. The length of the guide bank on the downstream side should be taken equal to 0.25 $L$. A convergence of 1 in 20 in the former case and 1 in 40 in the letter case has also been suggested by him. His recommendations are based on the assumption that the structure is constructed well within the *khadirs* and is provided with two training bunds. Based on Gales recommendations, the general layout dimensions of guide banks, are shown in Fig. 8.16.

However, in certain cases, where the *khadir* is wide, the Gale's guide bunds may fail to provide enough protection to the approaching embankments. In such cases, the lengths of the guide banks and their layout should be such as to ensure the safety of the approach embankments against the worst possible embayment, in case the river embays considerably behind the training works. Only one loop is likely to be formed if the khadir bank is nearer, as shown in Fig. 8.17 ; and if the distance of the khadir edge is large, a double loop, as shown in Fig. 8.18 ; may be formed. Fig. 8.19 shows the anticipated double loop (dotted) and the actual developed loop on Gomti river u/s of Amghat road bridge.

*Radius of the worst embayment.* To determine the radius of the sharpest loop (*i.e.* minimum radius = $r_{min}$) the river course for the past several years is plotted and average values of $M_L$, $M_B$, $B$ and $C$ (where $B$ and $C$ are the arc and chord

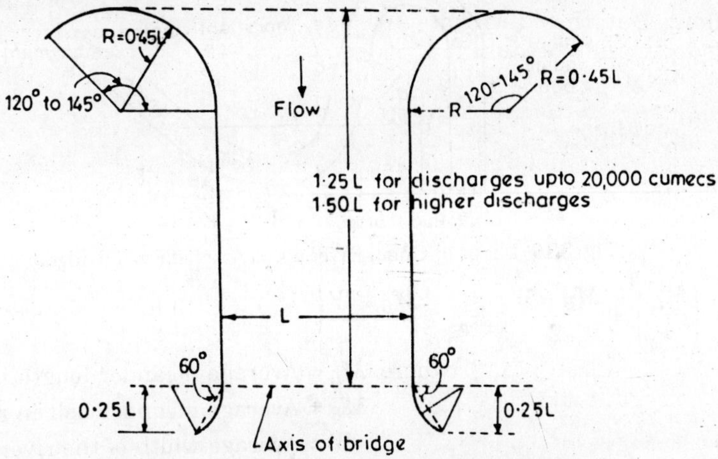


Fig. 8.16. Gale's recommendations for guide banks.

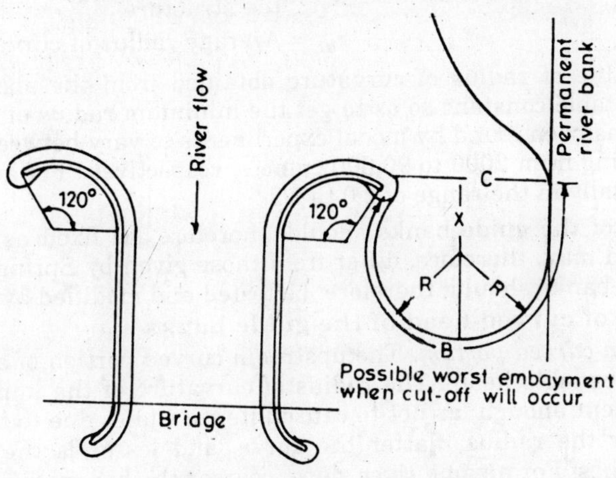


Fig. 8.17. Single loop formation.

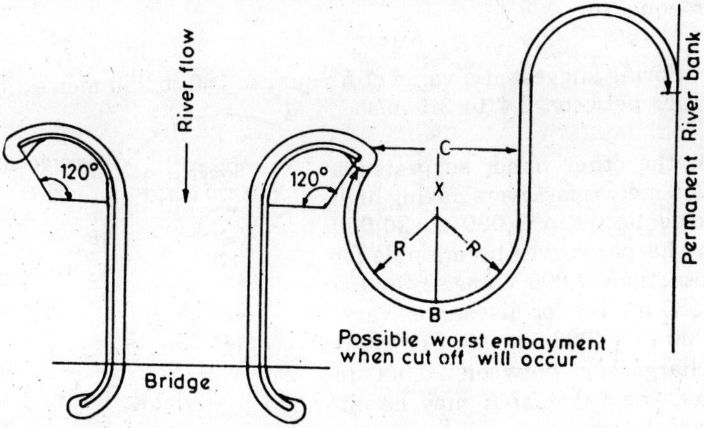


Fig. 8.18. Double loop formation.

**length) of the river are taken. The average value of the radius of the loop $r_{av}$ may then be determined from the formula**

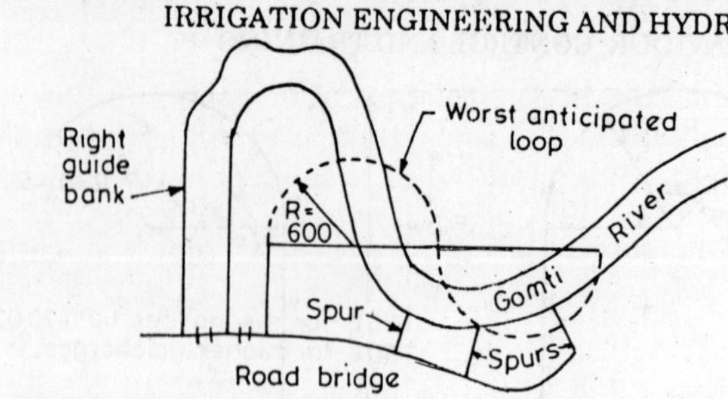


Fig. 8.19. Loops of Gomti River u/s of Amghat Road Bridge.

$$\left(\frac{M_L}{4}\right)^2 = \frac{M_B - W}{2}\left[2r_{av} - \frac{M_B - W}{2}\right] \quad ...(8.3)$$

where, $M_L$ = Average meander length in metres

$M_B$ = Average meander belt in metres

$W$ = Average width of the river in the main channel during floods in the vicinity of the structure

$r_{av}$ = Average radius of curvature.

The average value of radius of curvature obtained from the above equation is divided by a suitable constant so as to get the minimum radius of the worst loop. This constant has been found by model experiments to vary between 2.5 to 1.7 for discharges varying from 2000 to 90,000 cumecs, respectively. For acute meanders, $B/C$ ratio is usually in the range of 2.0 to 2.2.

The length of the guide banks should, therefore, be fixed as per the above requirement and may, therefore, differ from those given by Spring or Gales. The lengths of guide banks should, therefore, be tested and modified by model studies.

(*iv*) **Radius of curved head of the guide banks**

(*a*) *Upstream curved portion.* The upstream curved portion of a guide bank is called 'the impregnable head'. The radius of curvature of the impregnable head should be sufficient enough, as not to cause intense eddies due to the curved flow near it. Greater the radius, flatter the curve, and lesser is the probability of formation of eddies. For a same river slope, coarser the bed material, shorter can be the radius, depending on the expected velocity. A safe value for the radius ($R$) may be taken equal to

$$R = 0.45\ L.$$

However, Spring suggested a value of $R$ equal to 180 to 250 metres for rivers having velocities between 2.4 to 3.1 m/sec., respectively.

Gales, on the other hand, suggested a value of $R$ = 250 metres for rivers having high flood discharge between 7,000 to 20,000 cumecs (with sharper curves permissible for discharges less than 7,000 cumecs); and a value of $R$ = 580 metres for discharges varying from 40,000 to 70,000 cumecs. For intermediate discharges *i.e.* between 20,000 to 40,000 cumecs, the value of $R$ may be obtained by interpolation.

The upstream curve is extended to subtend an angle of 120° to 140° at its centre, as shown in Fig. 8.20.

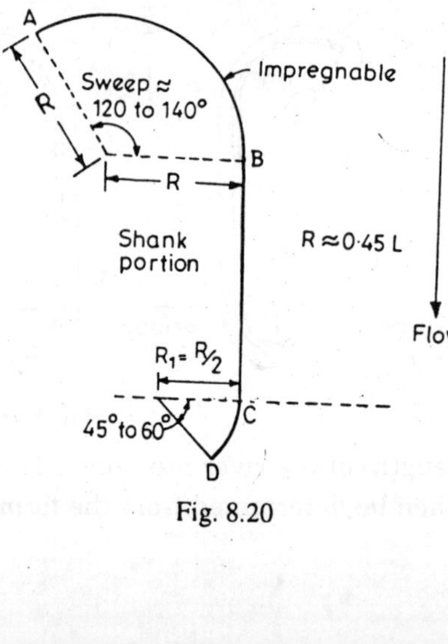


Fig. 8.20

(*b*) *Downstream curved portion.* On the downstream, the river fans out so as to attain its normal width. The downstream portion of the guide bank ensures the safety of approach embankments and prevents the river from attacking them. This purpose can be well served *by providing short guide bund with sharp curved head.* A radius equal to half the radius at the upstream side, may be adopted here and sweep angle of 45° to 60° may be provided, as shown in Fig. 8.20.

(*v*) **The shank portion and section of bund.** The straight portion of the guide bank *BC* (Fig. 8.19) is called the shank portion. The guide banks should have a minimum top width of 4 metres so as to provide sufficient carriage-way. Extra width may, however, be provided for storing pitching materials, etc. Side slopes may vary from 1.5 *H* : 1 *V* to 2.5 *H* : 1 *V*, (but generally kept as 2 *H* : 1 *V*), depending on the construction materials and the height of the bund. A freeboard of 1.2 to 1.5 m is generally provided.

(*vi*) **Slope pitching.** The sloping water side of the entire guide bund as well as the sloping rear side of the curved portions are pitched with *one man stone* (*i.e.* a stone which can be lifted by one person—weighing 40 to 50 kg) or concrete blocks. The pitching should extend up to 1 m higher than HFL. The rear side of the shank portion is not pitched, but is generally coated with 0.3 to 0.6 m earth for encouraging vegetation growth, so as to make it resistant against rain, wind, etc. [Refer sections *AA* and *BB* in Figs. 8.22 (*a*) & (*b*)]

The thickness of the pitching on the river side may be calculated by the formula

$$t = 0.06\, Q^{1/3} \qquad ...(8.4)$$

where, $t$ is the thickness of stone pitching in metres

$Q$ is the discharge in cumecs.

The thickness of pitching should be 25% more at the impregnable head than for the rest of the bund.

(*vii*) **Launching apron**. Whenever, a sloping face is protected by stone pitching against scour, the pitching is extended beyond the toe on the bed in the form of packed stones, as shown in Fig. 8.21. This stone dumping is known as *Launching apron.*

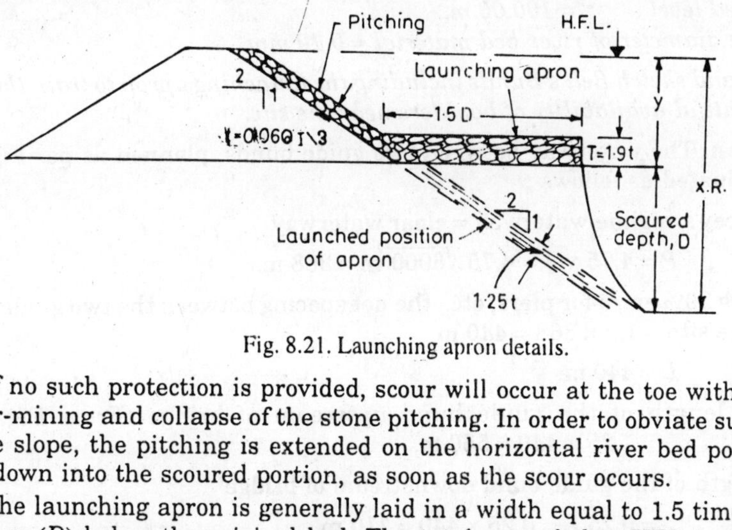


Fig. 8.21. Launching apron details.

If no such protection is provided, scour will occur at the toe with consequent under-mining and collapse of the stone pitching. In order to obviate such a danger to the slope, the pitching is extended on the horizontal river bed portion, which falls down into the scoured portion, as soon as the scour occurs.

The launching apron is generally laid in a width equal to 1.5 times the depth of scour (*D*) below the original bed. The total scour below HFL is taken as *xR*, where *R* is the Lacey's normal scoured depth given by equation

$$R = 0.47 \left(\frac{Q}{f}\right)^{1/3} \quad \text{...(8.5)}$$

where $Q$ is the discharge and $f$ is the silt factor ; and values of $x$ are tabulated for different places, in Table 8.3.

**Table 8.3**

| *S. No.* | *Location* | *Mean value of x* | *D = xR – water depth above bed* |
|---|---|---|---|
| 1 | Noses of guide banks | 2.25 | $2.25R - y$ |
| 2 | Transitions from noses to straight portions | 1.50 | $1.5R - y$ |
| 3 | Straight reaches of guide bunds | 1.25 | $1.25R - y$ |

Generally, a scour slope of 2 : 1 is assumed and the thickness of the launched apron should be 1.25 $t$, where $t$ is the thickness of stone pitching. Then the volume of stone required in the launched apron per unit length perpendicular to paper

$$= \sqrt{2^2 + 1^2} \cdot D \cdot (1.25\, t) = 1.25\, t \times \sqrt{5}\, D = \mathbf{2.8\, t\,.\, D.}$$

If the width of unlaunched apron is 1.5 $D$, then the thickness of the unlaunched ($T$) apron is given as :

$$T = \frac{2.8\, t..D}{1.5\, D} = 1.87\, t \text{ ; say } 1.9\, t.$$

Hence $\boxed{\mathbf{T = 1.9\, t}}$ ...(8.6)

where $t$ is given by Eq. (8.4).

The width of the apron (1.5 $D$) will be different in different portions of guide bank, depending upon the values of $D$ as per Table 8.3.

**Example 8.1.** *The following hydraulic data pertains to a bridge site of a river.*

*Maximum discharge = 6,000 cumecs.*
*Highest flood level = 104.00 m.*
*River bed level = 100.00 m.*
*Average diameter of river bed material = 0.10 mm.*

*Design and sketch Bell's Bunds including the launching apron to train the river. Assume plentiful availability of boulders near the site.*

**Solution.** The waterway between two guide bunds, planned as per Fig. 8.13 may be evaluated as follows :

The Lacey's regime waterway = clear waterway

$$P = 4.75\sqrt{Q} = 4.75\sqrt{6000} \text{ m} = 368 \text{ m}.$$

Allowing 20% extra for piers, etc., the net spacing between the two guide bunds at the bridge site = 1.2 × 368 = 440 m

Hence $L = 440$ m.

∴ The length of the guide bund upstream of bridge (From Fig. 8.16)
= 1.25 × 440 = 550 m.

The length of the guide bund downstream of bridge
= 0.25 $L$ = 0.25 × 440 = 110 m.

The radius of the curved head (upstream portion) may be kept
= 0.45 $L$ = 0.45 × 440 = say 194 m.

The u/s end of guide bund may therefore, be curved by 130° (between 120 to 145°) with a radius of 194 m.

The d/s end of the guide bund may, therefore, be curved in such a way as to make an angle of 60°, as shown in Fig. 8.22 (*a*).

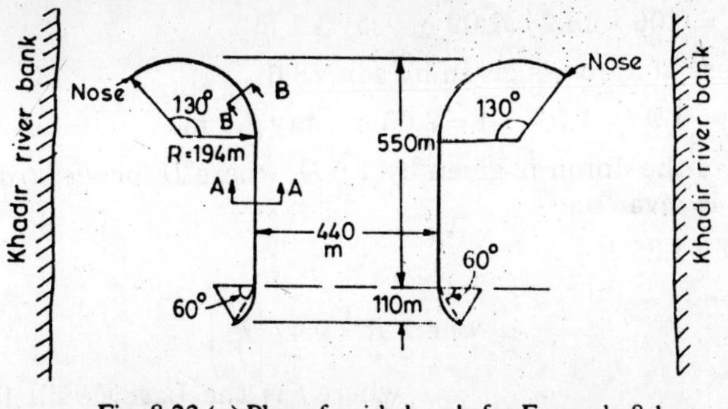


Fig. 8.22 (*a*) Plan of guide bunds for Example 8.1.

*Cross-sections of the guide bunds*

The given HFL at bridge site = 104.00 m.

Assuming a free-board of 1.5 m, and nil value of afflux, and ignoring velocity head*, we have the

Top level of guide bund = 104.00 + 1.5 = 105.5.

To be more safe and making an allowance for future settlement etc. ; let us adopt the top level of bund as 106.00 m.

Now, height of the bund above river bed level

= 106.00 – 100.00 = 6.00 m.

Assuming the top width of the bund as 5 m, and side slopes as 2 : 1, we have the required sections of the bund as shown in Figs. 8.22 (*b* & *c*).

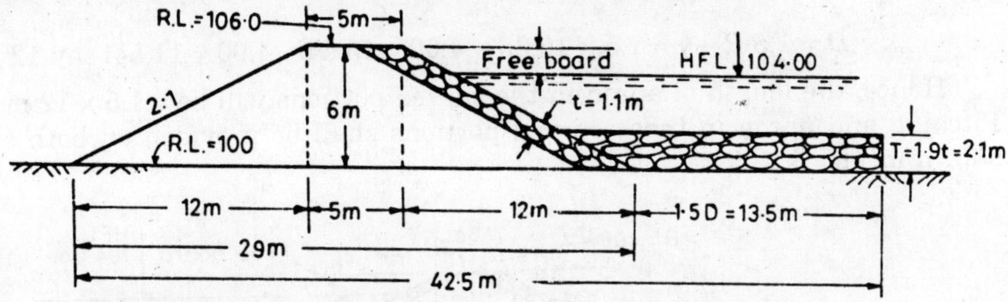


Fig. 8.22 (*b*) Cross-section (*AA*) of the straight reach of Bell's Bund.

The stone pitching and an apron must be provided on the slope on water side for the entire length of the bund ; however, the rear side will also be pitched for curved portions of the bund.

---

*Vel. head $= \dfrac{V^2}{2g}$

where, $V = \dfrac{6000}{PR} = \dfrac{6000}{368 \times 10.36} = 1.57$ m/.sec.

(*R* is calculated afterwards)

$\therefore$ Vel. head $= \dfrac{(1.57)^2}{2 \times 9.81} = 0.13$ m.

*Design of stone pitching and apron*

The thickness of stone pitching on the side ($t$) is given by Eq. (8.4), as :

$$t = 0.06\, Q^{1/3} = 0.06 \times (6000)^{1/3}$$
$$= 0.06 \times 18.2 = 1.09 \text{ m} \text{ ; say } 1.1 \text{ m}$$

The thickness of apron is given by eqn. (8.6)

$$= 1.9\, t = 1.9 \times 1.1 = 2.09 \text{ m ; } \textbf{say 2.1 m.}$$

The length of the apron is given by 1.5 $D$, where $D$ for *the straight reaches* of the guide bund is given as :

$$D = 1.25\, R - y$$

$$\text{where } R = 0.47 \left(\frac{Q}{f}\right)^{1/3} \quad ...(8.6)$$

where $f$ is the Lacey/s silt factor given by Eqn. (4.24) as, $f = 1.76 \sqrt{d_{mm}}$

where, $d_{mm}$ is the av. particle dia in mm = 0.1 mm (given)

$$f = 1.76 \sqrt{0.10} = 1.76 \times 0.322 = \mathbf{0.556.}$$

$$\therefore\ R = 0.47 \left(\frac{6000}{0.556}\right)^{1/3} = 0.47\ (10780)^{1/3}$$
$$= 0.47 \times 22 = \mathbf{10.36 \text{ m.}}$$

$y$ = Depth of water above river bed level

$$= 104.0 - 100.00 = \mathbf{4.00 \text{ m.}}$$

$$\therefore \quad D = 1.25 \times 10.36 - 4.00 = 13 - 4.0 = \mathbf{9.0 \text{ m.}}$$

Length of apron = 1.5 $D$ = 1.5 × 9.0 = 13.5 m.

For the curvilinear transition portions of the guide bund, however, $D$ is given by

$$D = 1.5\, R - y = 1.5 \times 10.36 - 4.00 = 15.54 - 4.00 = 11.54\text{; say } 12 \text{ m.}$$

Hence, the length of apron in the curved portions will be = 1.5 × 12 m = 18 m. Pitching and apron in these curved portions shall be provided on both sides, as shown in Fig. 8.22 ($c$).

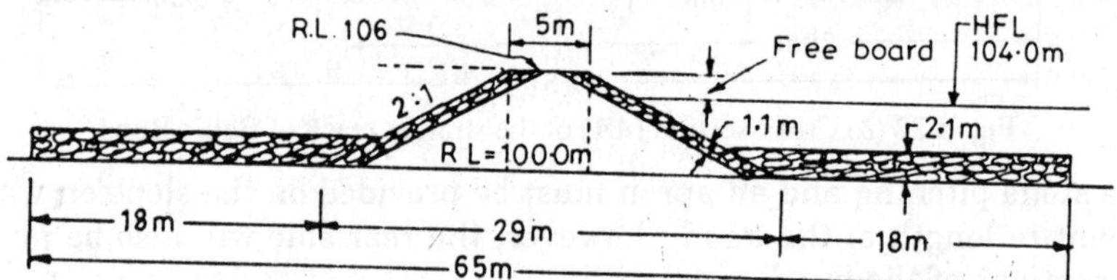


Fig. 8.22 ($c$) Cross-section ($BB$) of the curved portions of the Bell's bund.

The length of apron at noses should be increased to 1.5 $D$ (where $D = 2.24 \times 10.36 - 4.00 = 19.36$) *i.e.* say 30 m.

(3) **Spurs or Groynes.** Groynes are the embankment type structures, constructed transverse to the river flow, extending from the bank into the river. That is why, they may also be called 'Transverse Dykes'. They are constructed, in order to protect the bank from which they are extended, by deflecting the current away from the bank. As the water is unable to take a sharp embayment, the bank gets

protected for certain distance upstream and downstream of the groyne. However, the nose of the groyne is subjected to tremendous action of water, and has to be heavily protected by pitching, etc. The action of eddies reduces from the head towards the bank, and, therefore, the thickness of slope pitching and apron can be reduced accordingly.

**Types of alignment.** The groynes may be built either perpendicular to the bank line or they may be inclined upstream or downstream, as shown in Fig. 8.23.

A groyne aligned perpendicular to the bank line [Fig. 8.23(*a*)] is known as an **ordinary groyne** or a **normal groyne.**

A groyne pointing upstream [Fig. 8.23 (*b*)] has the property of *repelling* the flow away from it, and scour holes caused by the formation of vertical eddies are developed away from the bank, and near the head of the groyne. Such groynes, are called **repelling groynes.**

On the other hand, a groyne pointing downstream [Fig. 8.23 (*c*)], has the property of attracting the flow towards it, and is called an **attracting groyne.**

In an attracting groyne, scour holes are developed nearer the bank, as compared to those in a repelling groyne. Since such attracting groynes bring the water current as well as scour holes nearer the bank and make it more susceptible to damage, they are generally not used.

The groynes are, therefore, generally aligned either perpendicular to the bank or *pointing upstream*. The 'perpendicular alignment' is generally used on convex banks, and the 'upstream pointing alignment' is generally used on concave banks.

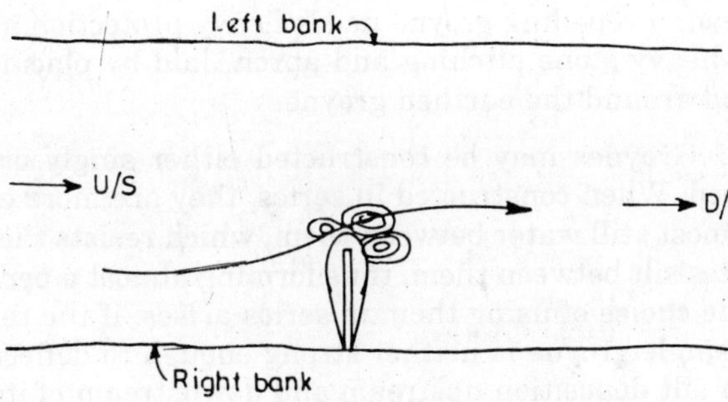


(*a*) Normal groyne

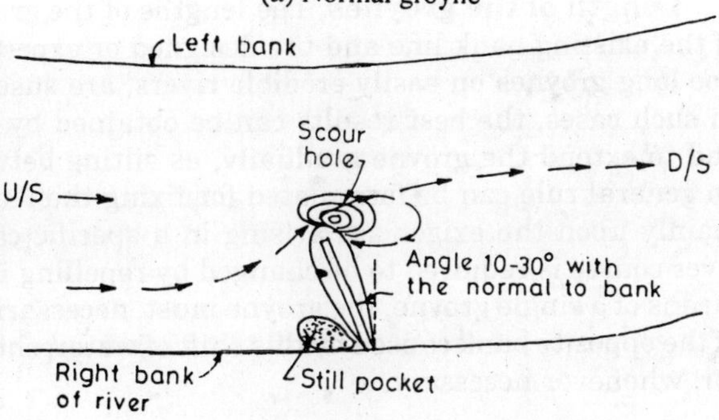


(*b*) Repelling groyne

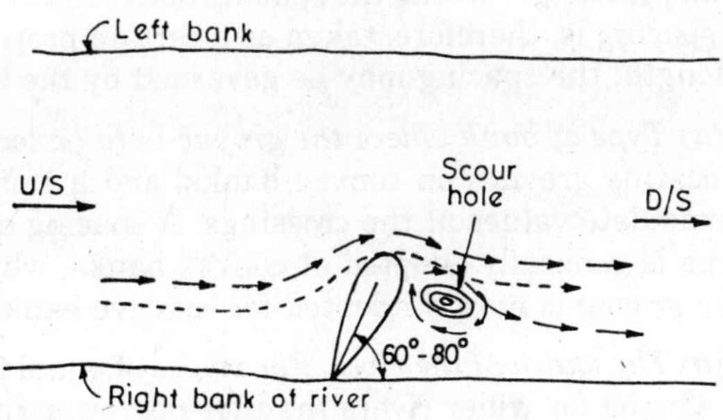


(*c*) Attracting groyne

Fig. 8.23

When the length of an upstream pointing groyne is small, such that it changes only the direction of flow, without repelling it, it is called a **deflecting groyne**, instead of calling it a *repelling groyne.*

The repelling groynes are generally found to serve the desired results, provided they are properly located with due regard to their positioning in relation to the meander length. *It is desirable to test their performance in hydraulic models before constructing them in actual field.* These model studies in India are undertaken by Central Water and Power Research Station (CWPRS), Poona. *The angle of deflection upstream is generally kept at about 10 to 30° with a line normal to the bank.*

On the upstream side of a repelling groyne, a still water pocket is formed, where the suspended sediment carried by the river, gets deposited. The head of such a groyne, however, causes severe disturbance in the flow at its nose, where heavy scour takes place. Due to this heavy attack of swirling river current at the head, a repelling groyne needs heavy protection at and near its head, in the form of heavy stone pitching and apron, laid by placing stones filled in wire crates in and around the earthen groyne.

Groynes may be constructed either singly or in series, depending upon the need. When constructed in series, they are more effective, as they create a pool of almost still water between them, which resists the current and gradually accumulates silt between them, thus forming almost a permanent bank after certain time. The choice of using them in series arises, if the reach to be protected is long, or if a single groyne is neither strong enough to deflect the current nor quite effective for silt deposition upstream and downstream of itself.

**Length of the groynes.** The lengths of the groynes depend upon the position of the existing bank line and the designed or expected bank line for trained river. Too long groynes on easily erodible rivers, are susceptible to damage and failure. In such cases, the best results can be obtained by starting with a shorter length and to extend the groyne gradually, as silting between them proceeds. However, no general rule can be formulated for fixing the length of the groynes. It depends mainly upon the exigencies arising in a specific case. For example, if the entire river course is required to be changed by repelling it towards the opposite bank by means of a single groyne, the groyne must, necessarily be, sufficiently long. Erosion of the opposite bank caused by this shift of water, should be anticipated and allowed for, whenever necessary.

**Spacing of the groynes.** As each groyne can protect a certain length, the primary factor governing the spacing between two adjacent groynes, is their length. The spacing is, therefore, taken as a certain proportion of their length. Apart from the length, the spacing may be governed by the following factors :

(*a*) *Type of bank where the groyne is to be located.* Larger spacing is required for locating groynes on convex banks, and a smaller one for concave banks with intermediate values at the crossings. A spacing of 2 to 2.5 times the length of the groyne is generally adopted at convex banks, while a spacing equal to the length of the groyne is mostly adopted for concave banks.

(*b*) *The width of the river.* For rivers of equal flood discharges, a larger spacing is preferred for wider rivers than for narrower rivers.

(*c*) *Type of groyne.* A higher value of spacing may be used for permeable groynes as compared to that required for impermeable groynes.

**Types of groynes.** Based upon their material of construction, the groynes may be divided into two types, namely :

(*i*) *Impermeable groynes* : and (*ii*) *Permeable groynes.*

(*i*) **Impermeable groynes.** Impermeable groynes are also called as solid groynes or embankment groynes. These groynes may be rockfill embankments or earthern embankments, armoured with stone pitching, concrete blocks, etc. The head of the groyne needs special protection and is generally provided with a launching apron in addition to the increased pitching, as shown in Fig. 8.24 (*b*). These groynes are called 'impermeable groynes' because they do not allow any significant flow through them.

Their design is the same as that for the guide bund with an apron. Commonly adopted dimensions of the groyne embankment are :

Freeboard = 1 m
Top width = 3 m
Side slopes = 2 : 1
Head = square and having a slope of 5 : 1
Apron = as per standard design requirements.

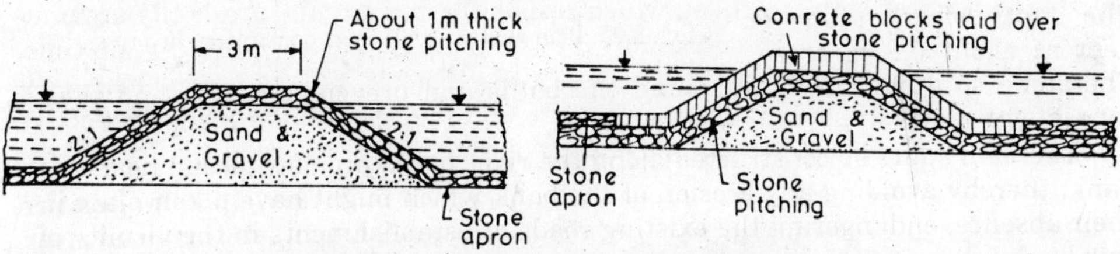


(*a*) Normal section (*b*) Section at the nose (*i.e.* the head)
Fig. 8.24. Impermeable groyne-section.

(*ii*) **Permeable groynes.** Permeable groynes do permit restricted flow through them. The groynes are more or less temporary structures and are susceptible to damage by floating debris, etc. The common materials used as permeable groynes, are :

(*a*) Trees used as groynes, called *Tree Spurs.*

(*b*) Timber stakes or wooden piles called *Balli Spurs.*

(*c*) Stone filled in balli crates.

(*d*) Stone filled in wire crates.

Permeable groynes, simply obstruct the flow, reducing its velocity and causing silt deposition. *They are, therefore, best suited for rivers carrying huge sediment load in suspension.* In comparatively clear rivers, they reduce the erosive strength of the current, and thus, prevent local bank erosion. Permeable spurs do not change the flow abruptly as is done by impermeable spurs, and hence, intense and serious eddies and scour holes are not developed. They are cheaper and perhaps the best for silt laiden rivers. When the groyne is to be submerged, then permeable groynes give much better results, because they do not generate so strong turbulence as is generated by submerged—impermeable groynes, making them susceptible to be washed away due to over-topping.

*Balli spurs* are being successfully used these days on many rivers in plains, and are becoming very popular.

In such spurs, sal ballis are driven vertically into the river bed, projecting from the bank to be protected, towards the interior of the river, at suitable spacings of about 1 m or so in two rows in zig-zag fashion. The top level of the ballis is kept at about 15 cm above the dry weather flow level of the river. The ballis are connected at top by cross ballis and longitudinally by longitudinal ballis, as shown in Fig. 8.25.

The length of the spur which projects into the river is kept at about 15 to 20 m or so, and the spacing between the spurs is kept at about 50 m or so. The length and spacings are interdependent on each other. Stone is dumped into the bottom round the spur as shown. The depth of the stone is kept at about 0.8 m or so. The stone so dumped will surround the ballis and prevent scouring and their consequent removal.

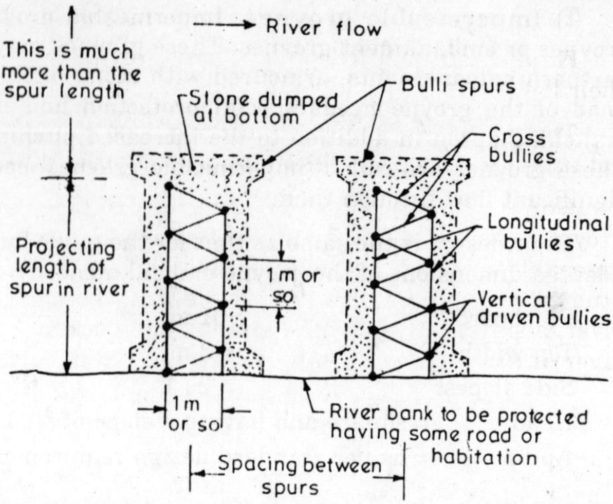


Fig. 8.25. Plan view of Balli spurs constructed in a river.

The balli spurs so constructed along the river bank, promote silting near this bank; thereby avoiding any erosion of the bank which might have taken place in their absence, endangering the existing roads or establishments in the vicinity of this bank.

Balli spurs constructed on Yamuna river in Delhi at Barapula near Ring road and at Khijrabad village, under the supervision of this author, had proved very successful, leading to enough silting and protecting the bank reach from future erosion. However, their effect was felt downstream, where erosion did take place at Jogabai village.

Although such permeable spurs are excellent tools for training heavily silted rivers, yet their biggest drawback is their short life. They are not strong enough, as to resist shocks and pressure from debris, floating ice and logs of wood, etc. These are, therefore, not suitable for upper reaches of rivers. Moreover during floods, submerged groynes are a danger to navigation.

**T-shaped groynes.** Denehey's T-shaped groyne is a special type of groyne developed in India. It is an ordinary groyne provided with an extra cross groyne at the head giving it a T-shape. The cross groyne protects the main groyne on the same principles as the main groyne saves the bank. The longer arm (*ab*) of the T is provided on the upstream, and the shorter one (*bc*), on the downstream, as shown in Fig. 8.26.

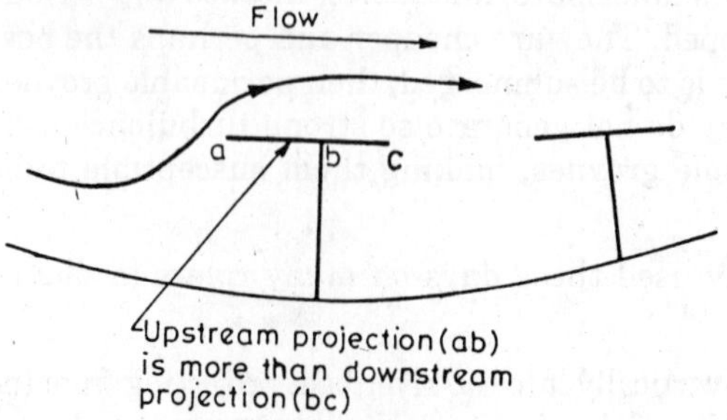


Fig. 8.26. T-shaped groyne

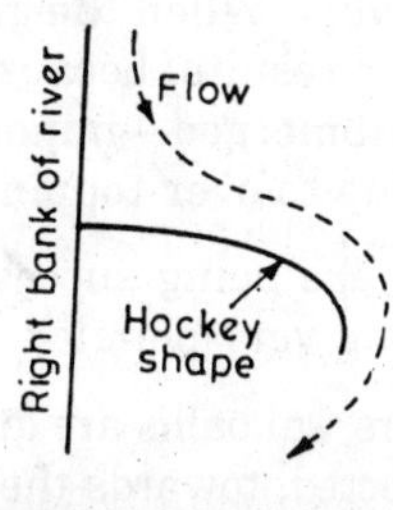


Fig. 8.27. Hockey shaped groyne.

These groynes are usually spaced at about 800 metres apart.

**Hockey-shaped groynes.** These groynes are shaped like a hockey stick at their lower end, as shown in Fig. 8.27.

These groynes exert an 'attracting' type of influence on the flow and hence are not useful for bank protection for repelling the current away from it.

**Concluding remarks on groynes.** In the end, it may be repeated that the groynes should be constructed after model studies. Their design is not much amenable to theoretical investigation and has to be checked and tested with model studies. Whereas, a series of groynes may be useful for general deflection of the river, a single groyne placed suitably may best serve the purpose of controlling a river at a certain works-site, such as shown in Fig. 8.28. (*a*) and (*b*).

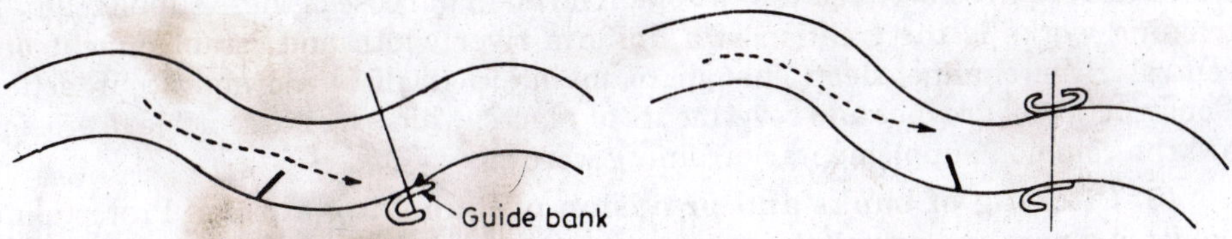


Water attracted towards guide bank, protecting the free bank
(*a*) attracting groyne

Water deflected away from concave side reducing meander tendencies
(*b*) deflecting groyne

Fig. 8.28

(4) **Artificial cut-offs.** It was discussed earlier, that a considerably meandering river may attain its original straight flow by means of a cut-off. A cut-off channel may develop by itself or may be induced artificially. When a meander goes on increasing and may endanger some valuable land or property, then the river course may be straightened by inducing an artificial cut-off. The newly developed course will be far away from the establishments which otherwise would have been affected. For inducing an artificial cut-off, only a pilot channel is required to be excavated in case of rivers having easily erodible beds. The flood water will gradually enlarge **the pilot cut** to the required cross-section and will abandon the old curved channel, provided, the circumstances for the development of this cut-off are favourable.

*It has been established that the side pilot channel will be self scouring, provided* $\left(\frac{R}{L^2}\right)$ *is greater for the cut than for the original course* (where $R$ is the hydraulic mean depth $\left(R = \frac{A}{P}\right)$ and $L$ is the length along the cut or along the river course as the case may be. Hence, a deeper and more hydraulically efficient channel $\left(\text{such that } \frac{R}{L^2}\right.$ is greater for the cut than for the main course) is required to be excavated for inducing a cut-off. *The pilot cut should be made as deep as possible.* As tractive force is directly proportional to depth, a deeper cut would be helpful for rapid development. The width of the cut is relatively less important and is governed by the minimum working space required for the excavating machines, etc.

Moreover, the alignment of the cut should be tangential to the main direction of flow approaching and leaving *the cut*.

It was discussed earlier that a cut-off brings violent changes in the river regime. To keep this violence within permissible limits, a full scale diversion of the river through an artificial cut-off is seldom accomplished at any one time. The complete formation of a cut-off is generally brought about slowly and slowly. The common practice is to make pilot cuts which carry 8 to 10 per cent of discharge in the beginning, and are developed subsequently to carry about 40 to 50 per cent of the total river discharge.

Since a cut-off helps in reducing flood heights and flood periods, it helps in shortening the navigation course and also helps in protecting endangered establishments. It is, thus, a useful river training measure. However, cut-offs as river training measure are not enough by themselves. Although they correct the instability and inefficiency at sharp bends, they do little to correct conditions in the reaches between these bends. For this reason, extensive supplementary training works are required between the cut-offs. The main purpose of these supplementary training works is the creation of a uniform river width and establishment of a central river channel deep enough to maintain itself by normal scour action. Construction of groynes and revetments at places, which are susceptible to erosion, are the common supplementary training works.

(5) **Pitching of banks and provision of launching aprons**. Protection of banks is a part and parcel of river training works. Sometimes 'Bank protection' is separated from the 'River Training' under the argument that the former has a limited function of protecting the bank, while the latter aims at training the river to a desired stable course. However, since bank protection itself helps in training the river, the two are generally treated as one and the same thing.

Banks of a river are directly protected by stone pitching, or by concrete blocks, or by brick lining or by growing vegetative cover, etc. Concrete blocks are very costly, and stone pitching is mostly adopted, if available without much difficulty. The banks of the river are made stable by giving them a stable slope varying from 1 : 1 to 2 : 1 depending upon the material of the bank. They are then pitched, so as to make them strong enough to resist erosion.

Sometimes, as discussed earlier, a launching apron is projected from the toe of the bank into the river, so as to prevent scour at the toe and the consequent fall of slope pitching. The standard design of an apron has already been discussed and is based on the principle that the stone in the apron has to launch into the deepest scour ($D$) possible at the location, to a slope of 2 : 1, at an average thickness of 1.25 times the thickness of pitching provided at the bank. The deepest scour may be taken equal to $x$ times the Lacey's Regime scoured depth ; $x$ generally varying from 1 to 2.5 or 3 for different places.

(6) **Pitched islands.** A 'pitched island' is an artificially constructed island in the river bed and is protected by stone pitching on all sides. Because of the turbulence generated by the island in its vicinity, the river channel around the island gets deepened and thus, attracting the river towards itself and holding it permanently. Pitched islands may therefore, help in attracting the current towards themselves and thus, reduce undue concentration on the opposite banks.

The device of pitched island as a river training measure is of a recent origin and is still in the experimental stages. Nevertheless, *experience* has proved that pitched island is a very effective means of controlling and training rivers, especially in the vicinity of control points, such as a bridge or a weir or a barrage, etc. However, model studies must be conducted before adopting pitched island, as a river training measure.

(7) **Miscellaneous methods.** (*i*) *Submerged Dikes* (called Sills) are used as a river training measure in particular situations. Sometimes, a river may create deep channels in the vicinity of certain pucca structures and are required to be corrected. In such situations, **sills** are placed across the scoured portion of the bed, with their top levels at or slightly below the designed bed level aspired to be achieved after correcting the deep scours. They are spaced closely, so as to ensure their proper functioning.

(*ii*) *Closing Dykes* are sometimes used to close a particular flow, so that the river flow may be directed in some other desired direction.

**Conclusions and remarks.** River training works depend mainly upon the type of the river, its regime and upon the characteristics of its flow. *There are no two countries with the same hydrology ; in the same country, there are no two rivers with the same regime ; and even in the same river, there are no two floods of the same characteristics. River training works, therefore, vary from place to place and from time to time and also depend upon the finances available at a particular time.* A particular method of river training, which has been successful in a particular reach on a particular river, may not be the best solution for the other rivers or for other reaches of the same river.

Hence, no particular method of river control can be stated to be the best. It all depends upon the exigencies arising out of a particular problem. However, in general, it can be stated that no river training method is of immense use in isolation, unless it is supplemented by other measures. A comprehensive planning and model testing is necessary before any measures are undertaken to control and train the river. If roughly planned, costly structures may either be washed away in a few days or may create worse or irreversible consequences causing tremendous losses and devastations.

## PROBLEMS

1. Comment on the statement—"Rivers have played an important part in the development of civilisation right from the early ages."
2. (*a*) What is meant by "meandering of rivers" and what are its causes ? How does the increasing river length due to meandering is counteracted in nature ?

   (*b*) Write short notes on the following river training works :

   (*i*) Levees (*ii*) Guide banks (*iii*) Repelling groynes (*iv*) Cut-off
3. Give a typical plan and section of Bell's bund, and explain its design procedure step by step.

   Differentiate between 'attracting' and 'repelling' spurs explaining their action in river training.
4. What are 'cut-offs' ? How are they used as a method of river training ?

   Draw a neat plan and section of a Bells Bund and give salient steps of its design procedure.
5. (*a*) What are the different types of rivers and their characteristics ?

   (*b*) Name the important Himalayan Indian rivers, and compare their behaviours with the other Non-Himalyan Indian rivers.

   (*c*) Name the important types of river training methods indicating the purpose for which each type is adopted.
6. Explain how do the following assist in river control :

   (*i*) Spurs (*ii*) Revetment (*iii*) Guide bunds.

   Draw a neat plan and section of an earthen Tee-spur showing protective apron.

   State the good and evil effects of embankment construction for flood control.

**7.** What is the difference between permeable and impermeable spurs ? Describe briefly the type of spurs commonly used in river control.

Data given below refer to a Bell's bund :

| | |
|---|---|
| High flood level | 206.00 m. |
| Low water bed level | 202.00 m. |
| Maximum scour depth (Lacey's formula) R (with factor or safety) | 9.5 m. |
| Free-board | 1.5 m. |
| Side slope of embankment | 2 : 1 |
| Thickness of stone apron when launched | 40 cm. |

Determine volume of stone required per metre length of shank. (Assume 25% loss). Draw a neat sectional view of the bund with necessary dimensions.

**8.** (*a*) What is meant by 'river training' and what are the different objectives served by it.

(*b*) Enumerate the different methods which are used for controlling and training rivers, and describe any one of these methods in details. Also discuss the circumstances under which a particular method of river training is to be adopted.

**9.** (*a*) Explain with neat sketches the different types of 'spurs' which are commonly used for controlling and training Indian rivers ?

(*b*) Discuss briefly the usefulness of the 'Balli spurs' on North Indian rivers.

(*c*) What are 'cut-offs' and how are they artificially induced ? What are their advantages and disadvantages ?

**10.** Write short notes on any three of the following :

(*i*) Importance of rivers and necessity of controlling them.

(*ii*) Aggrading and degrading rivers.

(*iii*) Meandering of rivers and meander parameters.

(*iv*) Cut-offs and cut-offs ratio.

(*v*) High water, low water, and mean water training.

(*vi*) Pitching of Bank's and Pitched island's as the means of training rivers.

(*vii*) Groynes, their types and uses.

(*viii*) Use of Levees for protecting cities from floods.

9

# Diversion Head Works

It was stated earlier that the main permanent canal, forming the primary part of a direct irrigation scheme, takes off from a diversion weir or a barrage. In fact, these permanent canals take off from rivers and the arrangements are so well made at their heads, that a constant and a continuous water supply is ensured into the canal, even during the periods of low flow. *The works, which are constructed at the head of the canal, in order to divert the river water towards the canal, so as to ensure a regulated continuous supply of silt-free water with a certain minimum head into the canal, are known as Diversion Head Works.*

## 9.1. Weir and Barrage

In general, the above purpose can be accomplished by constructing a barrier across the river, so as to raise the water level on the upstream side of the obstruction, and thus, to feed the main canals taking off from its upstream side at one or both of its flanks. The ponding of water can be achieved either only by a permanent pucca raised crest across the river or by a raised crest supplemented by falling counter-balanced gates or shutters, working over the crest. *If the major part or the entire ponding of water is achieved by a raised crest and a smaller part or nil part of it is achieved by the shutters, then this barrier is known as a weir [Fig. 9.1 (a) and (b)]. On the other hand, if most of the ponding is done by gates and a smaller or nil part of it is done by the raised crest, then the barrier is known as a Barrage or a River Regulator [See Fig. 9.2 (a) and (b)].*

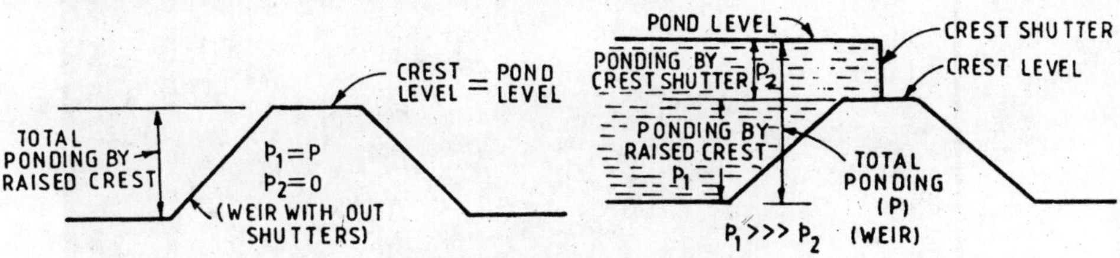


(*a*) Weir without shutters (*b*) Weir with shutters
Fig. 9.1

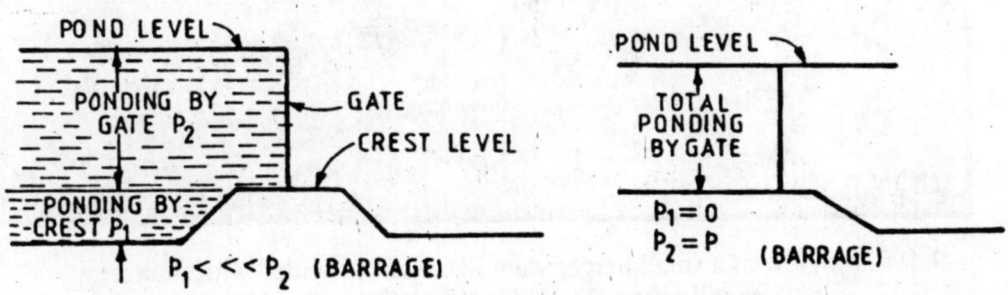


(*a*) Barrage with a small raised crest (*b*) Barrage without any raised crest
Fig. 9.2

If most of the ponding or the entire ponding is done by a permanent raised crest, as in a weir, then the afflux caused during high floods is quite high. On the other hand, if most of the ponding is done by gates, as in a barrage, then the gates can be opened during high floods and the afflux (*i.e.* rise in HFL near the site) will be nil or minimum. Hence, the latter device *i.e.* a barrage, gives less afflux and a

Fig. 9.3. Photoview of a Barrage (Girija Barrage on Ghagra river in U.P.)

Fig. 9.4. Photoview of a small height. dam like weir, provided with shutters at top of its solid crest (Devri Dam near Jhansi in U.P.)

better control upon the river flow, because the inflow and outflow can be controlled to a much greater extent by suitable manipulations of its gates.

Moreover, a gate controlled weir (*i.e.* a barrage) can be provided with a roadway across the river at a small additional cost. The choice between a weir with shutters and one with counter-balanced gates (*i.e.* a barrage) is largely a matter of cost and convenience in working. A shuttered weir will be relatively cheaper but will lack the effective control possible in the case of a barrage. Moreover, a barrage type construction can be easily supplemented with a roadway across the river at a small additional cost. *Hence, barrages are almost invariably constructed these days on all important rivers.*

## 9.2. Gravity and Non-Gravity Weirs

When the weight of the weir (*i.e.* its body and floor) balances the uplift pressure caused by the head of the water seeping below the weir, it is called a *Gravity weir.* On the other hand, if the weir floor is designed continuous with the divide piers as reinforced structure, such that the weight of concrete slab together with the weight of divide piers, keep the structure safe against the uplift ; then the structure may be called as a *Non-gravity Weir.* In the latter case, RCC has to be used in place of brick piers, but in particular cases, considerable savings may be obtained, as the weight of the floor can be much less than what is required in a gravity weir.

## 9.3. Layout of a Diversion Head Works and its Components

A typical layout of a canal head-works is shown in Fig. 9.5. Such a head-works consists of :

(1) *Weir proper.*

(2) *Under-sluices.*

(3) *Divide wall,* dividing the river width into two portions ; one is called the weir portion, and the other portion from which the canal takes off, is having openings* and called the 'under-sluice-pocket' or *'under-sluices'* or *'weir scouring sluices'.* If there are two canals, taking off from each flank, then there will be two divide walls and two under-sluices.

(4) *River training works,* such as marginal bunds, guide banks, groynes, etc.

(5) *Fish Ladder.*

(6) *Canal Head Regulator.*

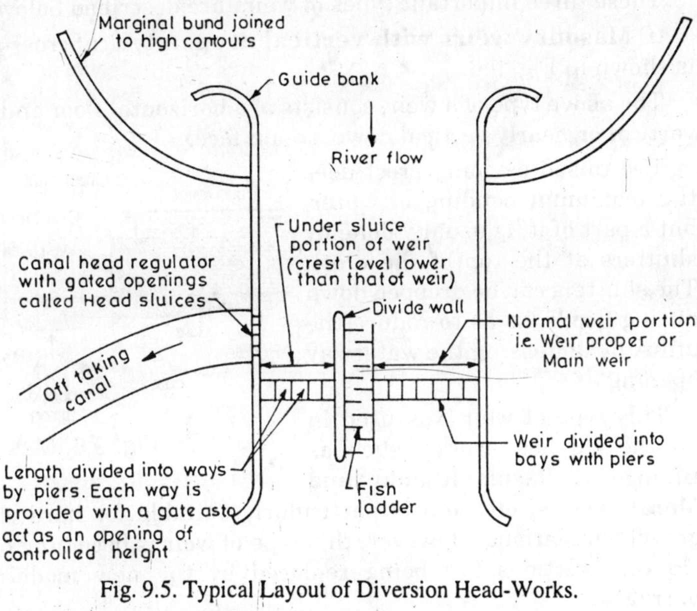


Fig. 9.5. Typical Layout of Diversion Head-Works.

---

* These openings are made by partial lifting of gates of the under-sluice ways. During high-floods the gates of the under-sluice ways shall be fully lifted depending on excess discharge in the river.

(7) *Weir's ancillary works,* such as shutters, gates, etc.

(8) *Silt Regulation Works.*

The various parts of a diversion head-works scheme and their design considerations are explained in details in the following pages.

**9.3.1. The Diversion Weir and its Types.** *Alignment.* As stated earlier, ***a diversion weir*** or ***an anicut or an intake weir*** is a raised pucca structure with or without shutters and laid across the river width. It is, essentially, of a height say up to 9 metres or so. The height of shutters over the weir crest seldom exceeds 1.2 metres or so. The entire length of the weir is divided into a number of bays by means of divide piers so as to avoid cross-flow in floods. As far as possible, the weirs should be aligned at right angle to the direction of the main river current. This ensures lesser length of the weir, better discharging capacity and lesser cost. This right-angled alignment is better and, therefore, common, especially, when the river bed is silty or sandy. Sometimes, the weir may be aligned at an oblique angle to the direction of the river current, and thereby, obtaining more safe and better foundations. In such a case, the weir will be of greater length, will have less discharging power and will be costlier. Moreover, due to non-axial flow, cross-currents may be developed, which may undermine the weir foundation. An oblique alignment may sometimes become necessary, when the river bed consists of gravel and shingle, which could otherwise enter the head regulator of the main canal and get deposited into the head reach of the main canal.

**Types of Weirs**. The weirs may be divided into the following three classes :

(*i*) *Masonry weirs with vertical drop ;*

(*ii*) *Rock-fill weirs with sloping aprons* ; and

(*iii*) *Concrete weirs with sloping glacis.*

These three important types of weirs are described below :

(*i*) **Masonry weirs with vertical drop**. A typical cross-section of such a weir is shown in Fig. 9.6.

The above type of a weir, consists of a horizontal floor and a masonry crest with vertical or nearly vertical downstream face.

The raised masonry crest does the maximum ponding of water, but a part of it, is usually done by shutters at the top of the crest. The shutters can be dropped down during floods, so as to reduce the afflux by increasing the waterway opening.

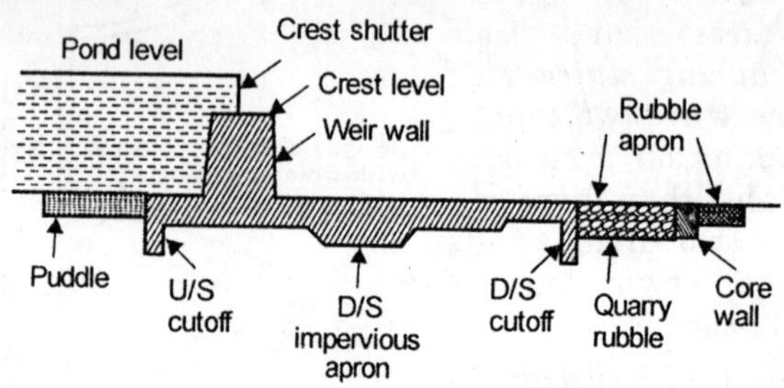


Fig. 9.6. Masonry Weir

This type of weir was used in all the old head-works, such as Bhimgoda, Rasul, Khanki and Marala works, etc. and is particularly suitable for hard clay and consolidated gravel foundations. However, this type of weir is becoming obsolete and even the old constructions are being replaced by the new modern concrete weirs or barrages.

(*ii*) **Rock-fill weirs with sloping aprons.** Such a weir is also called — '*Dry Stone Slope Weir*'. A typical cross-section of such a weir is shown in Fig. 9.7. It is the simplest type of construction, and is suitable for fine sandy foundations like those in alluvial areas in North India.

The old Okhla weir* across the Yamuna river in Delhi provides an example of such a weir. Such a weir requires huge quantities of stone and is economical only when stone is easily available. The stability of such a weir is not amenable to theoretical treatment.

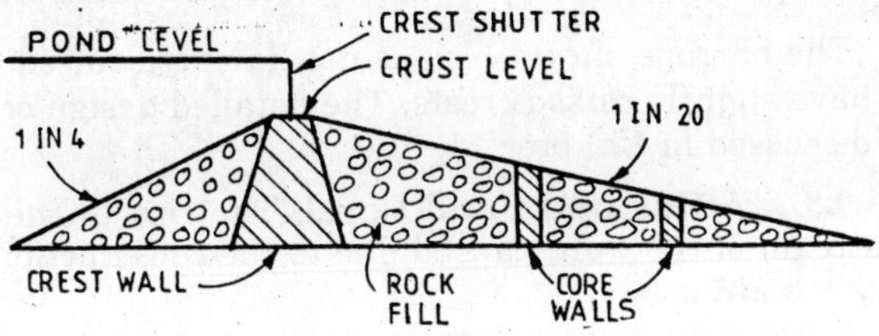


Fig. 9.7. Rock-fill Weir.

However, with the development of concrete glacis weirs, the above type is also becoming obsolete.

(*iii*) **Modern concrete weirs with sloping downstream glacis.** Weirs of this type are of recent origin and their design is based on modern concepts of sub-surface flow (*i.e.* Khosla's Theory). A typical cross-section of such a weir is shown in Fig. 9.8. Sheet piles of sufficient depths are driven at the ends of upstream and downstream floor. Sometimes, an intermediate pile line is also provided. *The hydraulic jump is formed on the downstream sloping glacis, so as to dissipate the energy of the flowing water.*

This type of weirs are now exclusively used, especially, on permeable foundations, and are generally provided with a low crest with counter-balanced gates, thus, making it a barrage. A typical cross-section of a barrage is shown in Fig. 9.9.

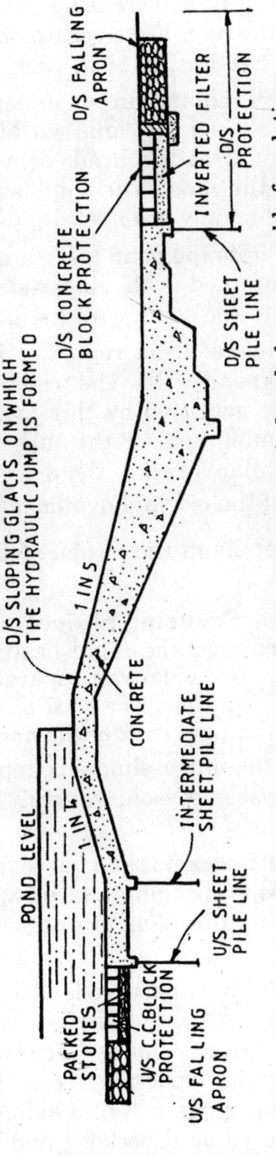


Fig. 9.8. A typical cross-section of a modern concrete weir founded on permeable foundations.

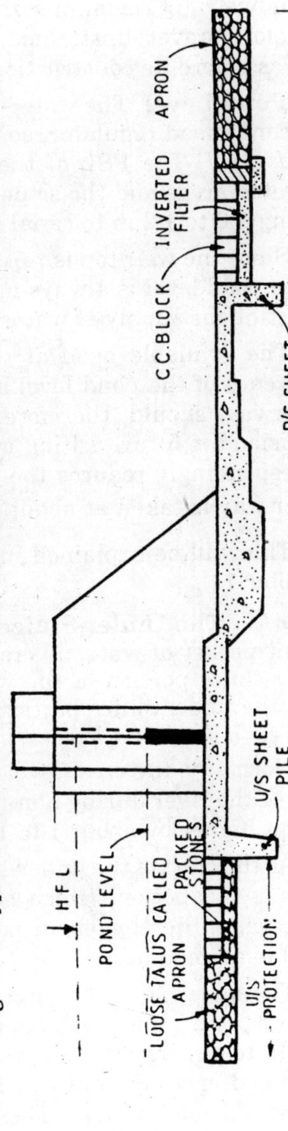


Fig. 9.9. A typical cross-section of a Barrage founded on pervious foundations.

* Now replaced by a new barrage, about 2 km downstream. The old weir has however, not been dismantled.

The barrage shown here is not having a raised crest at all, but barrages can do have slightly raised crests. The detailed design of such a weir or a barrage will be discussed in Chapter 11.

**9.3.2. Afflux and Pond Level.** The rise in the maximum flood level (HFL) upstream of the weir, caused due to the construction of the weir across the river, is called **afflux.**

In the initial stages, the water level is raised on account of afflux up to small distance equal to the length of the back water curve. But with the passage of time, the river bed rises due to silting caused by the reduction in flow-velocity upstream of the weir. ($Q$ remaining same, $y$ or $A$ increasing and $V$ decreasing). The effect, therefore, travels upstream till the river bed slope upstream of the weir is the same as was before its construction. (See Fig. 11.18 in Chapter 11).

**Pond Level**. The water-level required in the under-sluice pocket upstream of the canal head regulator, so as to feed the canal with its full supply, is known as *Pond Level*. The FSL of the canal at the head, depends upon the level of the irrigated areas and the slope of the canal. The pond level is generally obtained by adding 1.0 to 1.2 m to canal full supply level.

Since the weir top is raised upto the pond level, a minimum water level equal to the pond level is always maintained in the under-sluice pocket, so as to ensure a continuous supply of water into the canal with its full supply level.

The available head at the canal head regulator is, therefore, equal to the difference of the pond level and canal FSL. The width of regulator openings (*i.e.* waterway) should, therefore, be governed by this available head. But to make provision for future silting of canal, and for the subsequent rise in FSL, (which correspondingly reduces the available head), the designed available head for its water-way is taken at about $\frac{1}{2}$ of the original available value.

This will be explained in details in the design of Canal Head Regulators (in Chapter 11).

**9.3.3. The Under-Sluices or Scouring Sluices.** A comparatively less turbulent pocket of water is created near the canal head regulator by constructing under-sluice portion of the weir. A divide wall separates the main weir portion from the under-sluice portion of the weir. *The crest of the under-sluice portion of the weir is kept at a lower level than the crest of the normal portion of the weir.*

Normally, the crest level of the under-sluices is kept equal to the deepest bed level of the river during non-monsoon season; whereas, the crest level of the 'weir' is kept higher by about 1 to 1.5 m.

As the crest of the under-sluice pocket is at a low level, a deep channel develops towards this pocket, which helps in bringing low dry weather discharge towards this pocket, thereby, ensuring easy diversion of water into the canal through the canal head regulator.

The under-sluiced length of weir or barrage is divided into a number of ways by piers, and separate gates are installed on these ways. Each way can thus, be opened to any desired height by lifting its gate. Each way can thus, act as a gate controlled opening, and will help in bypassing the excess supplies to the downstream side of the river. These openings will also help in scouring and removing the deposited silt from the under-sluiced pocket ; and hence are also called the *scouring sluices.*

The crest level of the head regulator is also kept higher than the crest level of the under-sluices, so that only silt free water is admitted into the canal through the *head sluices*. Silt gets deposited over the under-sluice floor, and may be periodically removed over the crest of the under-sluices and towards the downstream side of the river by opening these gate-controlled openings (undersluices). Since the under-sluices help in removing the silt from near the head regulator, they are also called ***scouring sluices.***

The spans of under-sluices (*i.e.* span between two adjoining piers) should be wide enough, to be efficient in scouring action. Stoney gates which can be easily operated from platform at top, are widely used for controlling the flow through the under-sluices. In closed position, the top of these gates should be above the pond level in the pocket, so as to ensure a continuous and a constant supply of water into the off-taking canal through the canal head regulator.

Apart from these two primary objectives served by the under-sluices in all types of weirs and barrages, they may serve a third function in the case of weirs provided with shutters. That is : they help in passing the dry weather flow and low floods without dropping the weir shutters, the raising of which entails a good deal of labour and time. Whenever, water is surplus to canal requirement, it has to be passed across the weir. The surplus water is passed through the undersluices, so long as it does not exceed their discharging capacity, and the weir shutters are kept raised up. The weir shutters are dropped only to pass down the high floods.

*Design considerations.* Sill of the under-sluice pocket is kept at or slightly above the deepest river bed and about 0.9 to 1.8 metres below the sill of the canal head regulator. The length of the undersluice pocket between the divide wall and the head regulator may be taken as 1.5 times the upstream length of the divide wall, as shown in Fig. 9.10.

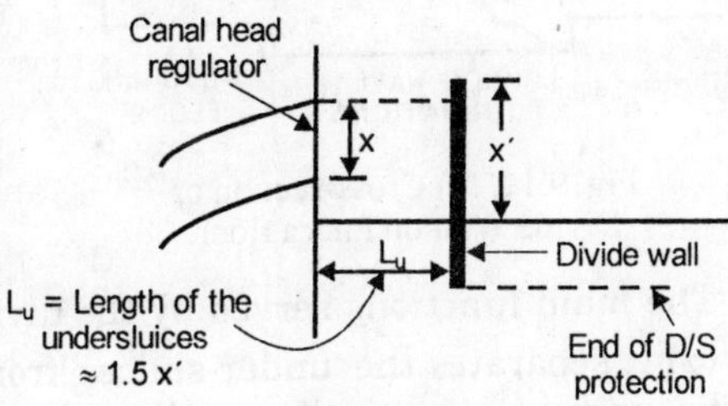

**Fig. 9.10.**

However, this length is governed by the discharging capacity of the undersluices, which should be sufficient to enable them to serve their main functions, described earlier. The discharging capacity of under-sluices may be selected as follows :

(*i*) They should be able to ensure sufficient scouring capacity, for which the discharging capacity should be atleast twice the full supply discharge of the main canal at its head.

(*ii*) They should be able to pass the dry weather-flow and low floods during the months excluding the rainy season, without the necessity of dropping the weir shutters.

(*iii*) They should be able to dispose of 10 to 15% of the high flood discharge during severe floods.

The spans of the under-sluices should be wide enough (usually 10 to 20 m) in order to be efficient in scouring action. The section of the under-sluices will be similar to that of a weir and will be designed to ensure–proper hydraulic jump formation (from surface flow considerations) and for uplift pressure for pervious foundations (by Khosla's Theory) as per procedure adopted for a normal weir design. Discharge intensity being greater in the under-sluices, its floor and apron will be stronger as compared to those of weir proper. The floor downstream of the crest, should not be kept level with the crest level, because the formation of hydraulic jump as destroyer of energy is very uncertain on a level floor.

**9.3.4. The Divide Wall.** The 'divide wall' is a masonry or a concrete wall constructed at right angle to the axis of the weir, and separates the 'weir proper' from the 'under-sluices'. The divide wall extends on the upstream side beyond the beginning of the canal head regulator; and on the downstream side, it extends up to the end of loose protection of the under-sluices (*see* Fig. 9.10). The top width of divide wall is about 1.5 to 2.5 metres. These walls are founded on wells closely spaced beyond the pucca floor upto the end. The wells are taken well below the deepest possible scour. Typical cross-section of the divide wall on pucca floor and beyond the pucca floor are shown in Fig. 9.11 (*a*) and (*b*).

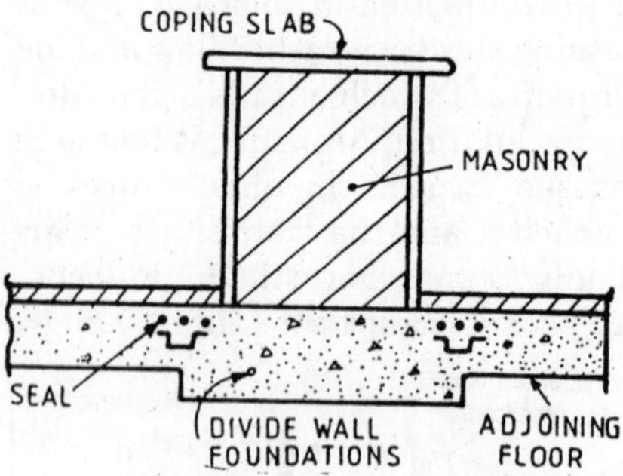


Fig. 9.11. (*a*) Cross-section of Divide Wall on Pucca floor.

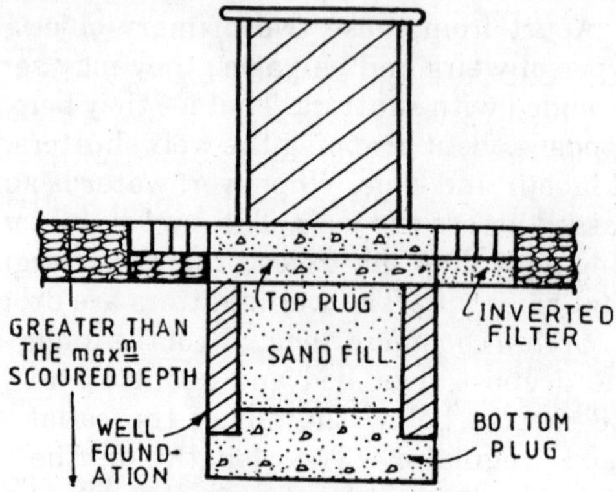


Fig. 9.11. (*b*) Cross-section of Divide Wall beyond Pucca floor.

The main functions served by the divide wall are :

(*a*) It separates the 'under-sluices' from the 'weir proper'. Since the crest level of the under-sluices is lower than that of the weir proper, the two must be separated, and this is being done by the Divide wall.

(*b*) It helps in providing a comparatively less turbulent pocket near the canal head regulator, resulting in deposition of silt in this pocket and, thus, to help in the entry of silt-free water into the canal.

(*c*) Divide wall may keep the cross-currents, if at all they are formed, away from the weir. Cross-currents lead to vortices and deep scours, and therefore, prove hazardous to weirs. For this purpose, additional divide walls projecting at right angle to the weir may sometimes be constructed at equal or desired spacings. They may, then, act as groynes, protecting, in turn, the weir from cross-currents, as shown in Fig. 9.12.

These cross-currents may sometimes develop, when there is a natural tendency for the main river current to attack the bank opposite to the canal head regulator. In such a case, the water will be forced by the weir to flow towards the head regulator, and cross-current will develop. This function of the divide walls, is not fully established, and the most modern practice is only to construct the divide wall for separating the 'under-sluices' from 'weir proper'.

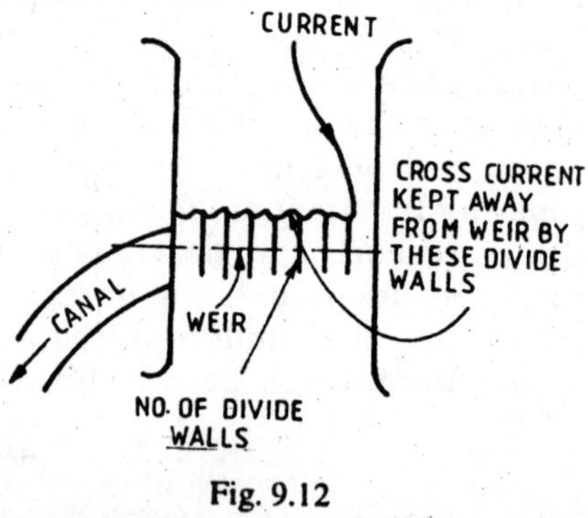


Fig. 9.12

*Design considerations.* The divide walls can be designed as cantilever retaining walls subjected to silt pressure and water pressure from the under-sluice side. For the worst case, the design should be checked for full silt pressure on the pocket side with equal water on two sides during low floods, and also for maximum differential water-head, when full discharge is passing through the under-sluices and no discharge is passing the weir. The value of differential pressure, may be taken, arbitrarily, as 1.0 m for waterhead and about 2.0 m for silt pressure.

**9.3.5. River Training Works.** River training works are required near the weir site in order to ensure a smooth and an axial flow of water, and thus, to prevent the river from outflanking the works due to a change in its course. The river training works required on a canal headworks, are :

(*i*) *Guide banks* ; (*ii*) *Marginal bunds* ; and

(*iii*) *Spurs or groynes*.

The guide banks force the river into a restricted channel, and thus, ensuring a smooth and an almost axial flow near the weir site. The design and other details of the guide banks have already been discussed in the previous chapter.

Marginal bunds are provided on the upstream side of the works in order to protect the area from submergence due to rise in HFL, caused by the afflux. These bunds are, therefore, continued till they join contours higher than the new HFL. Construction of these bunds would be justified only, when the value of land saved is more than the cost of the marginal bunds. In certain cases, when the ponding is high and the watershed is low, the construction of marginal bunds may almost become obligatory. The layout of the marginal bunds is very important in economising the overall cost of the training works and subsequent cost of maintenance. Apart from their primary purpose of protecting the adjoining area from river spills caused by afflux, these bunds do help in partially controlling and guiding the river flow between the guide banks. For this reason, the length of the guide banks do depend on the layout of the marginal bunds.

Marginal bunds are nothing but earthen embankments, protected by groynes, wherever needed. They are designed on the principles of design of earthen dams or dikes. The design aspects of groynes have already been discussed in the previous chapter.

**9.3.6. Fish Ladder.** Large rivers are generally inhabited by several types of fish, many of which are *migratory*. Such migratory type of fish, called *anadromous fish*, move from one part of the river to another part, according to the season. In India, only one such migratory fish is found, and this specie is known as **Hilsa**. *Salman, Steel head trout,* etc. are the other species of such anadromous fish, found in other countries.

*Generally speaking for India, these anadromous fish have been found to be moving from upstream (hills) to downstream (plains) in the beginning of the winter season in search of warmer waters, and return to their spawning grounds upstream, slightly before monsoons, in the month of May and June.* If no arrangement is made in the weir or the dam, to enable the fish to pass upstream, then such migratory fish have been found to be striking and striking against the water current in their efforts to move up, till death. Non-provision of an arrangement for fish to pass upstream, may thus, lead to large scale destruction of fish life.

It has been established that most kinds of fish can travel upstream, against a flow velocity of about 3 to 3.5 m/sec. Usually, there is a head difference of 5 to 6 metres between the upstream of the weir and the downstream water level in the river. If simply an open gap is left in the weir for this migration, the velocity of flow through such an opening will be very high. Therefore, even the strongest fish will not be able to travel upstream ; resulting in large scale destruction of fish near the downstream end of the fish gap. This had actually happened on the river Ganges at old Bhimgoda weir in Hardwar (U.P.), where a suitable arrangement was later on made for this migration.

A structure which enables the fish to pass upstream is called a *'fish ladder'*. It is a device by which the flow energy can be dissipated in such a manner as to provide smooth flow at sufficiently low velocity, not exceeding 3 to 3.5 m/sec. This object is generally accomplished by providing a narrow opening adjacent to the divide wall and provide suitable baffles or staggering devices in it, so as to control the flow velocity.

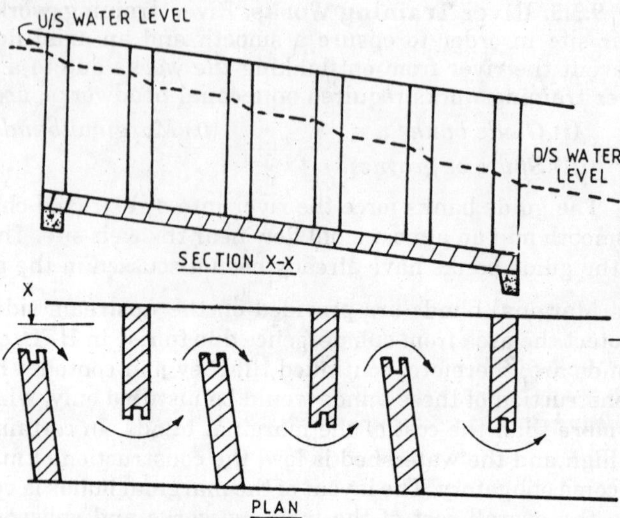

Fig. 9.13. (*a*) Section and plan of a typical fish ladder.

*Pool type and Steep Channel type* are commonly used types of fish ladders in weirs and barrages. Typical sketches (section, plan and photographic view) of ordinarily used **steep channel type** of a fish ladder, are shown in Fig. 9.13 (*a*) and (*b*).

In this type of ladder, the space between the weir and the divide wall, is provided with oblique walls, and holes are also staggered, so that the fish can take rest after passing one hole before they move on to the other. Sometimes, a fish ladder is provided within the divide wall itself, as shown in Fig. 9.14.

*Fish ladders are generally not found to be economical on dams higher than 100 m or so. For such cases, fish are usually transported upstream by mechanical means, using tramways or*

Fig. 9.13. (*b*) Photographic view of a fish ladder.

Fig. 9.14. Photographic view of a fish ladder provided in the divide wall.

*tank trucks. In dams, the fish also need to be protected from the turbines and spillways. If the head is less than 5 m or so, the fish can pass through the turbines and spillways with very low mortality.*

*Various types of screens and louvers are also used to divert the fish into a safe by-pass, which may be in the form of a short length of a pressure pipe, through which the fish pass downstream, or the fish may be diverted into a pond, from which they can be extracted for downstream transport by tank trucks.*

**9.3.7. The Canal Head Regulator or Head Sluices.** A canal head regulator (C.H.R.) is provided at the head of the off-taking canal, and serves the following functions :

(*i*) *It regulates the supply of water entering the canal.*

(*ii*) *It controls the entry of silt in the canal.*

(*iii*) *It prevents the river floods from entering the canal.*

A typical cross-section of a head regulator is shown in Fig. 9.15. The regulator is generally aligned at right angle to the weir, but slightly larger angles (between

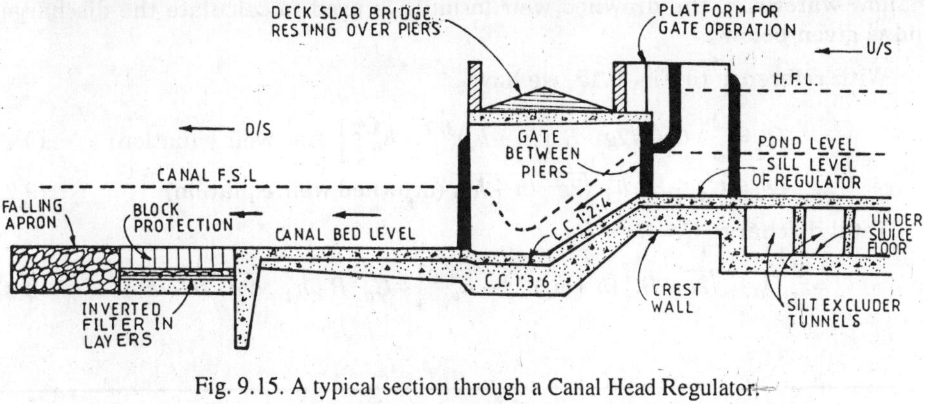


Fig. 9.15. A typical section through a Canal Head Regulator.

90° to 110°) are now considered preferable for providing smooth entry of water into the regulator, as shown in Fig. 9.16.

The regulator is provided with gates. Smaller spans were used in the past, but with the improvements in gate constructions, larger spans, say between 6 to 10 metres, are generally used these days, if found economical.

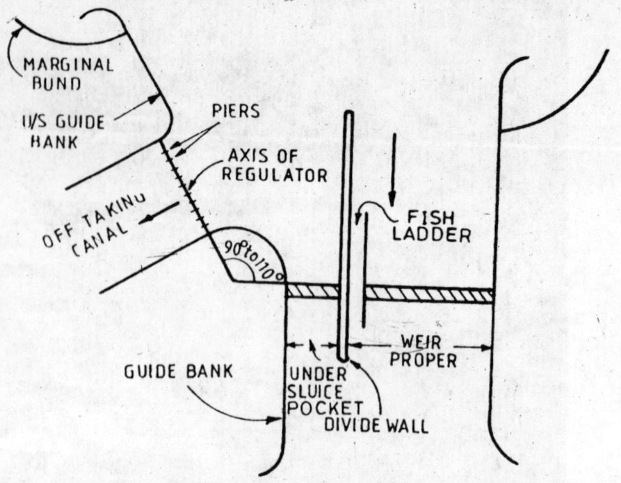


Fig. 9.16. Alignment of a Canal Head Regulator.

The water from the under-sluice pocket is made to enter the regulator bays, so as to pass the full supply discharge into the canal. The maximum height of these gated openings, called **head sluices** will be equal to the difference of Pond Level and Crest Level of the regulator. The entry of silt into the canal is controlled by keeping the crest of the head regulator by about 1.2 to 1.5 metres higher than the crest of the under-sluices. If a Silt-Excluder* is provided, the regulator crest is further raised by about 0.6 to 0.7 metre. Silt gets deposited in the pocket, and only the clear water enters the regulator bays. The deposited silt can be easily scoured out periodically, and removed through the under-sluice openings.

The crest level of the regulator, generally called the **sill level,** is not only governed by silt considerations, but is also governed by the discharge considerations. The full supply discharge has to pass through the regulator openings, the height of which will be equal to the difference of pond level and sill level. *The smaller the height of the openings, the larger will be the width of the openings. In other words, if the sill level is raised too much, the clear waterway of the regulator will increase enormously, and, thereby making it more costly.*

The waterway of the regulator is fixed so that the full supply discharge of the canal could easily pass over the crest of the regulator with designed pond level, with ample factor of safety to allow for the future silting up of the canal and the subsequent rise in its FSL. Since a regulator is provided with a very wide and a shallow waterway, the drowned weir formula is used to calculate the discharge, and is given below :

With reference to Fig. 9.17, we have

$$Q_1 = \frac{2}{3} \cdot C_{d_1} \sqrt{2g} \cdot B.\left[(h + h_v)^{3/2} - h_v^{3/2}\right] \text{(free weir equation)} \quad ...(9.1)$$

$$Q_2 = C_{d_2} \cdot B \cdot h_1 \sqrt{2g \cdot (h + h_v)} \text{ (drowned weir equation)} \quad ...(9.2)$$

Total discharge $Q = Q_1 + Q_2$

or $$Q = \frac{2}{3} C_{d_1} \cdot \sqrt{2g} \cdot B.\left[(h + h_v)^{3/2} - h_v^{3/2}\right] + C_{d_2} B \cdot h_1 . \sqrt{2g \cdot (h + h_v)} \quad ...(9.3)$$

* Explained in the next article.

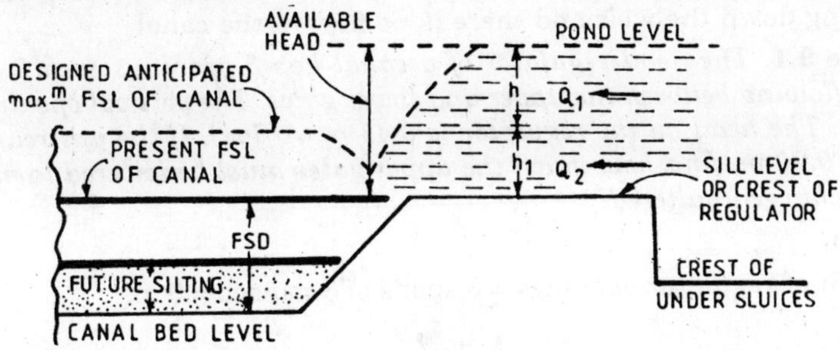


Fig. 9.17

The values of $C_{d_1}$ and $C_{d_2}$ may be taken as 0.577 and 0.80, respectively.

where $h$ = Difference of the upstream and downstream water level, *i.e.* (Pond level — Maximum anticipated FSL of canal after making due allowance for future silting up of canal).

*h is called the utilized head or working head or designed head for the regulator, and is generally taken as half of the available head, where available head is equal to the difference of Pond Level and actual FSL of canal.*

$h_v$ = Head due to velocity of approach

$B$ = Clear width of waterway

$h_1$ = Depth of downstream water level above the crest *i.e.* (maximum anticipated FSL of canal minus sill level)

When all other variables are fixed and known, value of clear waterway width ($B$) can be calculated.

The width of the waterway ($B$) calculated above, generally works out to be more than the normal width of the canal down-stream. In such a case, the sill level may be lowered, so as to increase the head and to decrease waterway to make it equal to the width of the canal. But the sill level is also governed by silt exclusion considerations, and therefore, many a times, it may not be possible to lower the sill level. In such a case, the calculated value of waterway is provided and the normal required width of the canal is obtained by contracting the wings.

As stated earlier, in a canal head regulator, the gate controlled openings are provided from the sill level to the pond level. During high floods, the water level in the river pocket will, however, be much higher than the pond level. To avoid spilling of this water over the canal regulator gates, a R.C.C. wall is provided from the pond level up to river HFL. This wall spans for the entire length of the regulator and will rest over the piers of the regulator bays. This wall is known as **breast wall,** and will be subjected to vertical self-weight and horizontal water pressure acting against it from the upstream side. Its design may be done as cantilever retaining wall for the horizontal water pressure, and as a girder spanning between the supports for its self-weight.

The length and thickness of horizontal floor, glacis, protection aprons, etc. is worked out on the same principles as are applicable to weir design, explained later in chapter 11. Usually, the worst condition for uplift pressure occurs, when high flood is passing down the weir and there is no flow in the canal.

**Example 9.1.** *The head regulator of a canal has 3 openings each 3 m wide. The water is flowing between the upper and lower gates. The vertical opening of the gate is 1.0 m. The head on the regulator is 0.45 m (Afflux). If the upstream water level rises by 0.20 m, find how much the upper gates must be lowered to maintain the canal discharge unaltered.*

**Solution.**

The width of regulator openings = 3 spans of 3 m each = 9 m.

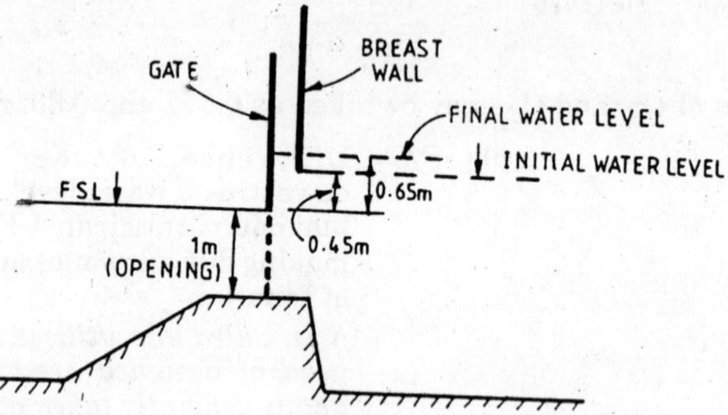


Fig. 9.18

When the gate opening is 1 m. The discharge can be calculated by submerged orifice formula, *i.e.*

$$Q = C_d \cdot A \cdot \sqrt{2gh} = 0.62\,(1 \times 9)\,\sqrt{2 \times 9.81 \times 0.45} \qquad ...(i)$$

In the 2nd case, when upstream water level rises by 0.2 m, let the gate opening be $x$ metre to keep the discharge unaltered.

$$\therefore \qquad Q = 0.62\,(x \times 9)\,\sqrt{2 \times 9.81 \times 0.65} \qquad ...(ii)$$

Equating (*i*) and (*ii*), we have

$$0.62\,(1 \times 9)\,\sqrt{2 \times 9.81 \times 0.45} = 0.62\,(x \times 9)\,\sqrt{2 \times 9.81 \times 0.65}$$

or $\quad 1\sqrt{0.45} = x\sqrt{0.65} \quad$ or $\quad 0.45 = x^2\,(0.65)$

or $\quad x = 0.83$ m

Hence, the gates must be lowered by an amount, 1 – 0.83 = 0.17 m **Ans.**

## BARRAGE REGULATION AND SILT CONTROL AT HEAD WORKS

The supplies entering a canal which takes off from the upstream side of a weir or a barrage, can be regulated into the following two ways :

(*i*) **Still Pond Regulation Method.** In this method of operation, all the gates of the scouring sluices are kept closed when the canal is running. Only as much discharge is drawn into the under-sluice pocket as much is required for the canal ; the surplus being escaped over some other section of the weir or the barrage. The velocity of water in the under-sluice pocket, therefore, gets reduced, as the smaller discharge enters through the same waterway. The low velocity causes the sediment to settle down and relatively clearer water enters the canal.

The silt is, thus allowed to accumulate in the pocket till it reaches to within 0.5 m below the crest of the regulator. The canal is then closed and the scouring sluices opened, till the entire silt deposit gets washed away on the downstream of river. The scouring operation takes about 24 hours, and for this much of time, the canal supply has to be stopped. After the silt deposit is washed out, the scouring sluices are closed and the canal supply is restored again. This method of barrage regulation is called still pond regulation and is very useful in controlling the amount of silt entering the canal. However, this method of regulation is practicable only where the sill of the canal head regulator is sufficiently higher than the upstream floor of the under-sluices. Moreover, it leads to closure of the off-taking canal atleast once a month, resulting in the wastage of discharge and loss of irrigation to that much extent.

(*ii*) **Semi-Open flow Operation**. In this method, water in excess of the canal requirement, is allowed to enter into the under-sluice pocket. The gates of the scouring sluices are kept partially opened. The total water entering the pocket, thus, gets divided into two parts, in front of the head regulator. The top water (above the sill level of head regulator) enters the canal through the head regulator, and the bottom water escapes downstream through the under-sluices. Due to this, certain velocity is maintained in the pocket and silt remains in suspension. The turbulence created may sometimes cause even the coarser silt to rise up and to enter into the canal. The silt control, is thus reduced in this method. However, the advantage of this method is that the silt is constantly and continuously scoured out, and the canal has not to be closed, as is required to be done in still pond regulation method.

Even in spite of using still pond regulation techniques, certain amount of silt may enter the canal. And also, silt will definitely enter the canal in semi-open flow operation. In order to avoid this silt, special silt control works are constructed, which remove the silt from the canal water either before the water enters the canal or after the water has entered and travelled a certain distance in the canal. These devices are explained below in details :

**9.3.8. Silt Control Devices.** The entry of silt into a canal, which takes off from a Head-Works, can be reduced by constructing certain special works, called silt control works. These works may be classified into the following two types :

(*a*) **Silt Excluders**. Silt excluders are those works which are *constructed on the bed of the river*, upstream of the head regulator. The clearer water enters the head regulator and the silted water enters the silt excluder. In this type of works, the silt is, therefore, removed from the water before it enters the canal.

(*b*) **Silt Ejectors.** Silt ejectors, also called **silt extractors,** are those devices which extract the silt from the canal-water after the silted water has travelled a certain distance in the off-take canal. These works are, therefore, constructed on the bed of the canal, and a little distance downstream from the head regulator.

*Principles of Silt Control*

Before we describe the mechanism and functioning of these silt control devices, we shall explain the basic principle on which the silt is removed from the water.

The fundamental principle behind silt control is : that most of the silt tries to settle down in water, thus, confining itself mostly in the bottom layers of water. We also know that the silt is kept in suspension by the force of the vertical eddies generated by the friction of the flowing water against the bed. In other words, if this bed friction is more, the upward force of eddies shall be more, and hence, lesser chances of silt settlement will exist. Hence, if the friction can be reduced by constructing a smooth approach channel, more settlement of silt and its consequent removal, is possible. Further, the silt from the bottom layers can be removed, by separating the top layers and the bottom layers, without causing any disturbance to the flow. The chances of less disturbance and that of providing a smooth approach channel can be better attained in a canal rather than in the river bed. Hence, the works *which are constructed in the canal* (*i.e.* the silt extractors or silt ejectors) will *definitely be superior and more effective than the works* which are constructed in the river bed (*i.e.* the silt excluders). *Silt extractor is therefore, better than a silt excluder.* However, the silt extractor shall be costlier because surplus water has to be taken into the canal from the head, and an escape channel which will feed the highly silted water back into the river, shall have to be constructed.

**9.3.8.1. Description and design of a silt excluder.** A silt excluder consists of a number of rectangular tunnels running parallel to the axis of the head regulator and terminating near the under-sluiced weir. The tunnel nearest to the crest of the head regulator has to be at least of the same length as the head regulator. Other tunnels may be shorter in length. The roof slab of the excluder tunnels is kept at the same level as that of the regulator crest, as shown in Figs. 9.15 and 9.19. The bottom layer of water which is highly charged with silt and sediment will pass down the tunnels and escape over the floor of the under-sluice way(s), since the gates of the under sluice way(s) shall be kept open upto the top of the tunnels. The clearer water over the top of the roof of the excluder tunnels, will thus enter the canal through the head regulator. Usually, two or three bays of the under-sluices of the weir or the barrage are covered by the excluder. *However, excluder covering only one bay has been designed and is usually adopted for sandy rivers.*

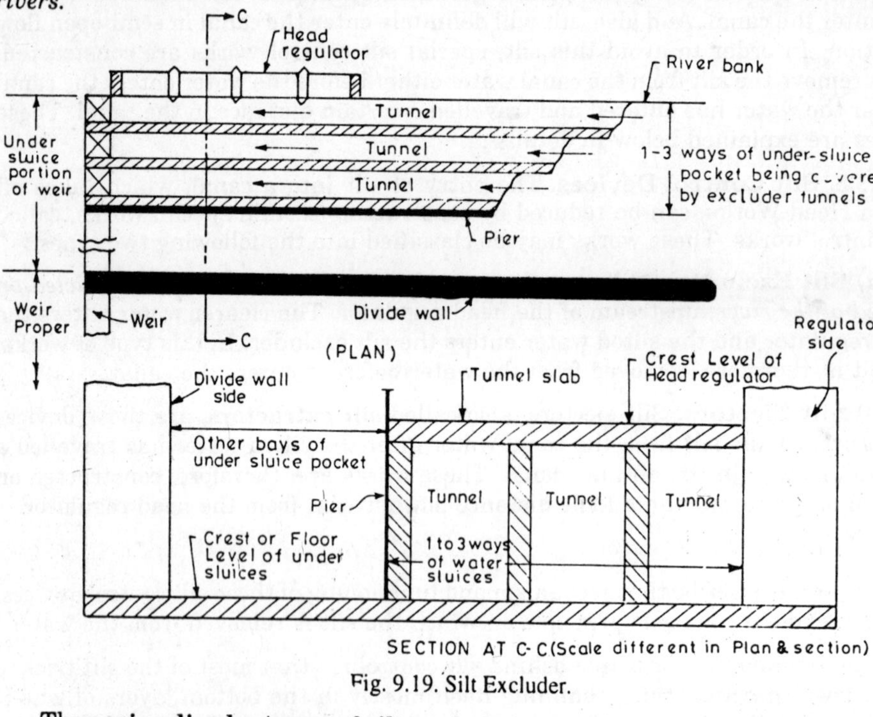


Fig. 9.19. Silt Excluder.

The **major disadvatages of silt excluders** are :

(*i*) Simultaneous use of under sluices-masked by the silt excluder tunnels, for silt exclusion and their assigned purpose of passing high river discharge, cannot be obtained.

(*ii*) The tunnels constructed on the river bed will have to be of robust construction, as to withstand river action.

(*iii*) In the presence of tunnels, on the river bed in the under-sluice pocket, securing of good approach conditions become difficult. If these tunnels were not there, a curved off-take or a curved approach could be provided at a small cost.

The **major advantages of the pocket excluders** are :

(*i*) As the tunnels are large, there are unlikely to be choked by rolling or suspended debris.

(*ii*) Since the same gates of the scouring sluices can be utilised for the excluder tunnels, some economy in cost is secured.

(*iii*) At the weir, sufficient head for operating the excluder is generally available.

**Design of tunnels.** A theoretical design of a silt excluder is confined mainly to find out the area of the tunnel openings required to pass the designed discharge and to determine its structural requirements. On the basis of past experience and model studies, the design of an excluder is finalised. The Punjab Irrigation Research Institute has carried out extensive research in the design of excluders. Based on their research findings, it has been found that if the excluder discharge is restricted to 15 to 20% of the canal discharge, satisfactory silt exclusion can be obtained. A minimum velocity of 2 to 4.5 m/sec must be maintained through the tunnels in order to keep them free from sediment. A value of 3 m/sec is usually adopted for ordinary straight reaches. A higher value of 4 to 4.5 m/sec may be adopted for boulder-stage rivers. A lower value of about 2 m/sec may be adopted for sandy rivers. After fixing the discharge and velocity, the cross-sectional area of the excluder tunnel-openings, can be determined. Knowing the height, the required width can be found and divided into a suitable number of bays.

By model studies, it has been found that the tunnels should be located at selected positions, rather than distributed uniformly over the entire length of the regulator. The position of tunnels being more important than their number. River curvature must be taken into consideration as far as possible. It was found by experience at Khanki that three tunnels were more efficient than six. Also, a smaller number of pocket bays covered by the excluder, give better results as do the openings of the tunnels confined to the mouth. Side openings into the tunnels, have been found to decrease the efficiency. Usually, 4 to 6 tunnels are generally provided.

**Approach and exit.** At the entrance, the tunnels are generally given a bell mouthed shape so as to increase the zone of suction. The radius of bell mouthing varies from 2 to 6 times the tunnel width ; the radii increasing for tunnels away from canal head regulator and for shorter tunnels. At the exit end, the tunnels are throttled for restricting the discharge to the desired value and to increase the velocity to prevent deposition of silt.

**Height of tunnels**. The height of tunnels generally varies from 0.5 to 0.6 m for sandy rivers, and 0.8 to 1.2 m for boulder stage rivers.

**Body and roof of tunnels.** The roof of tunnels must be strong enough to withstand the full water load coming from its top, with no water inside. Such a condition may arise when the tunnels are closed from upstream for repairs or for some other purpose. Moreover, the tunnel walls and roofs must be strong enough, as not to be, damaged by debris, boulders, shingles, etc., during floods. This is all the more important, since damage to excluder is not discovered as they are always under water and difficult to repair. Moreover, even the slight damage to the tunnel roof has found to have made the excluder less efficient.

**Outfall channel**. It should also be ensured that a channel to take the silt laden water into the river is maintained downstream of the excluder. No separate outfall channel is generally required for sandy rivers as a continuous flow through the alluvial bed, itself makes a channel there. However, if gravel and boulders are to be excluded and the main channel is away, a lined channel may be necessary, or a channel may have to be created by suitable training measures. This was done in the case of Tajewala Head Works on Yamuna river, where two short spurs induced the river to flow on the side of the under-sluices, and thus eliminated the construction of a separate outfall channel.

**Losses in tunnels.** The following head losses may be taken into account for calculating the tunnel sections :

(i) *Frictional Loss.*

$$h_f = \frac{n^2 \cdot V^2 \cdot L}{R^{4/3}} \text{ (Manning's formula).} \quad ...(9.4)$$

where $h_f$ = Head loss in metres.
$L$ = length of tunnel in metres.
$R$ = Hydraulic mean depth in metres.
$n$ = Manning's Rugosity coefficient.

(ii) *Loss due to Bend.*

$$h_b = F\left(\frac{V^2}{2g}\right)\left(\frac{\theta}{180°}\right) \quad ...(9.5)$$

where $F = 0.124 + 3.134\left(\frac{S}{2r}\right)^{1/2}$ ...(9.6)
$\theta$ = Angle of deviation in degrees
$S$ = Width of tunnel in metres.
$r$ = Radius of bend along centre line of tunnel in metres.
$h_b$ = Head loss in metres.

(iii) *Transitional loss due to change of velocity in contraction.*

$$h_c = 0.1\left[\frac{{V_2}^2}{2g} - \frac{{V_1}^2}{2g}\right] \quad ...(9.7)$$

where $V_1$ = Velocity in canal upstream in m/sec.
$V_2$ = Velocity in tunnel in m/sec
$h_c$ = Head loss in metres.

(iv) *Transitional loss due to change of velocity in expansion.*

$$h_e = 0.1\left[\frac{{V_2}^2}{2g} - \frac{{V_3}^2}{2g}\right] \quad ...(9.8)$$

where $V_2$ = Velocity in tunnel.
$V_3$ = Velocity downstream of tunnel.

**9.3.8.2. Description and Design of a Silt Extractor or a Silt Ejector.** The typical layout of a silt ejector is shown in Fig. 9.20 (*a* and *b*). It essentially consists of a horizontal diaphragm slab, a little distance above the canal bed, which separates out the bottom layers. Under the diaphragm, which is normally spanning the entire width of the canal, there are tunnels or compartments which extract the

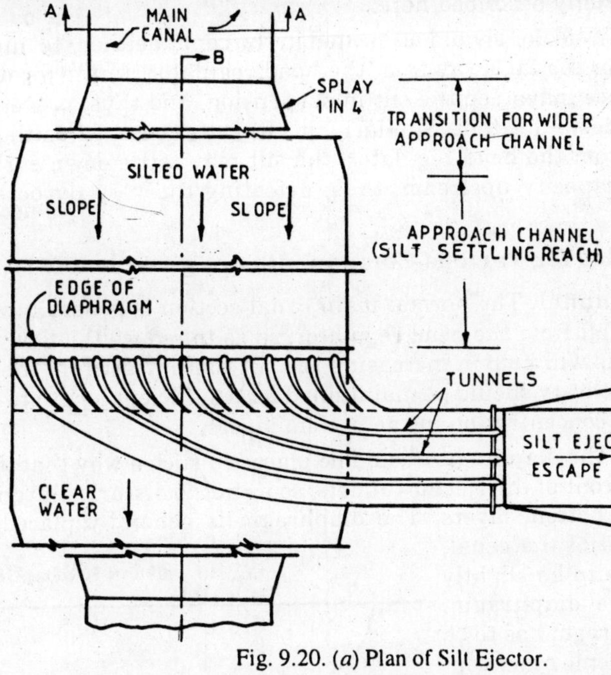


Fig. 9.20. (*a*) Plan of Silt Ejector.

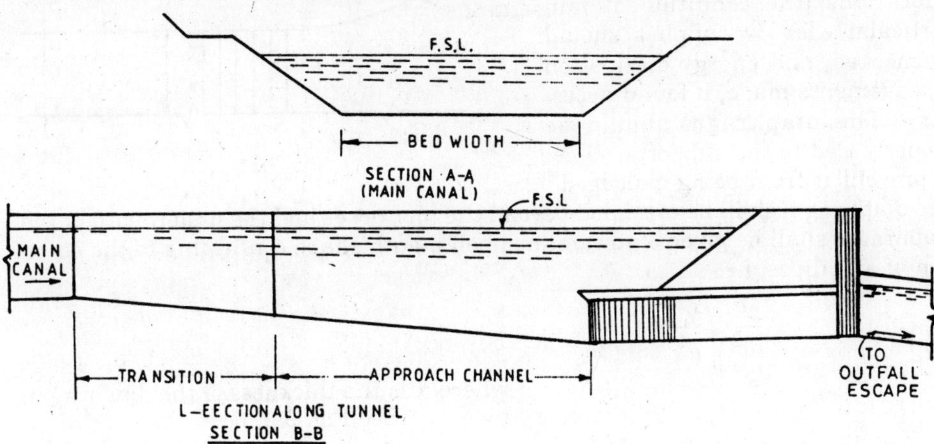


Fig. 9.20. (*b*) Sections of a Silt Ejector.

highly silted bottom water into an escape channel. The tunnel entrances should be designed in such a way that there is no disturbance of flow at entry, and the escape flow is quickly accelerated under the diaphragm, so as to prevent the clogging of tunnels. These objectives are achieved by dividing the entire span width into a number of compartments or tunnels by means of *streamlined vanes*, which are gradually converging (thus gradually decreasing the sectional area and correspondingly increasing the flow velocity). These main compartments are sub-divided into smaller compartments or sub-tunnels by vanes of radii varying from 3 to 4 times the width of the sub-tunnels to avoid cross flow in the transition section. The tunnel discharge is regulated by means of gates at the outlet end. The escape channel should have sufficient slope to carry the silt laden water without getting silted up, and should lead the water by the most economical path back to the river. The principles governing the location and design of various components of a silt ejector are briefly described here.

**Location**. The silt ejector should be located in the canal neither too much near the head regulator nor too far away from the head regulator. In the former case, the residual turbulence may keep the silt in suspension, and thus, preventing its extraction up to the desired extent. Similarly, in the latter case, when the ejector is located far away from the head regulator, the silt may settle down earlier and reduce the channel capacity upstream, thus, defeating the very purpose of the ejector.

**Design Principles for Various Components are given below :**

(*i*) **Approach channel.** The normal main canal section is generally widened after the canal takes off from the head regulator, so as to reduce the flow velocity to a desired level. This will help in increasing the silt concentration in the bottom layers. The reduced velocity should be maintained for a sufficient length to achieve the desired sediment concentration in the bottom layers.

(*ii*) **Diaphragm**. The diaphragm should be placed in such a way that it causes least disturbance in front of the ejector tunnels, so as not to disturb silt concentration attained in the bottom layers. The diaphragm is generally placed at the downstream bed level of the canal, *i.e.* the canal bed has to be slightly depressed under the diaphragm. However, if the diaphragm has to be placed higher due to some other considerations, the condition of fall, particularly for low supplies, should be checked, and energy dissipation arrangements made, if found necessary. The diaphragm should be properly tied to the supports, so as to prevent it from being dislodged.

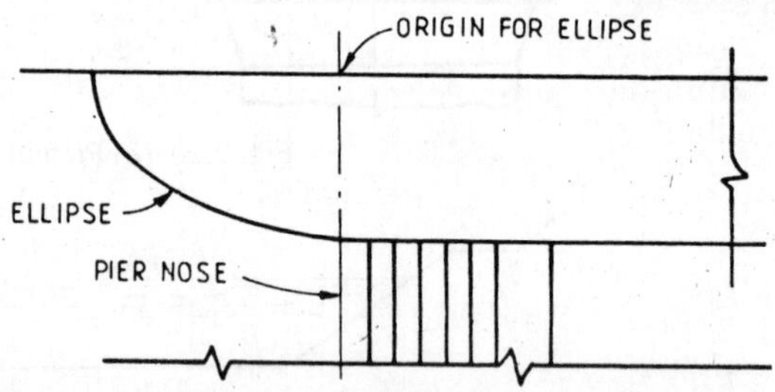


Fig. 9.21

The diaphragm shall be extended beyond the pier noses and the underneath of the diaphragm shall be given a streamlined bell mouth shape confirming to the equation of an ellipse, *i.e.*

$$\frac{x^2}{4a^2} + \frac{y^2}{a^2} = 1$$

where $a$ is the thickness of the diaphragm. (See Fig. 9.21).

(*iii*) **Tunnels**. The tunnels or compartments shall be constructed by gradually converging vanes, as explained earlier. The upstream noses of the vanes and piers may have cut-water shapes, while the downstream end of vanes may be fish-tailed. The tunnel dimensions at the entry and exit may be so fixed, as to ensure velocities that would carry the size of sediment to be removed. The section of the sub-tunnels at the entry may be so chosen that the velocity of flow at the intake is slightly higher than the velocity of bottom filaments of water upstream of ejector. The section of sub-tunnels up to their exit, where they end into the main tunnels, may be reduced gradually in such a way that there is an overall increase of 10 to 15 per cent in the velocity of emerging flow.

At the exit of sub-tunnels, the section of the main tunnel may be designed such that the flow velocities of the combined discharge are not less than the velocities emerging from the sub-tunnel. The section at the exit of main tunnels may be so designed, as to attain a velocity 2.5 to 6 m/sec depending on the grade of sediment to be ejected ; but in all cases, the exit velocity has to be more than the critical velocity, so as to ensure super-critical flow. The height of the tunnels is generally kept from 0.5 to 0.75 metres.

The width of the tunnels may be so adjusted as to have equal losses in each tunnel.

**Note**. As the lengths of tunnels are different, the head loss in them would be different, if all of them were to take the same discharge. As the difference between the water level in the canal and that in the escape channel is the same, the head available for all the tunnels is the same. Hence, the shorter tunnel would have a higher velocity of flow through it, than in the others, and this would lead to disturbance at entry. To counteract this possibility, either the shorter tunnel may serve a larger width of the canal than a longer one, or the tunnel cross-sections may be altered, each serving an equal width. The adjustment of width is recommended by BIS.

(*iv*) **Gate control regulation.** The discharge from ejector is controlled by gated regulation at the downstream end of the tunnels. The amount of discharge passing down the ejector and the frequency of its operation would vary in different parts of the year depending on the silt load carried by the canal. Proper gate regulation is, therefore, required, as per the intelligence and initiative of the maintenance engineers. The gate must be operated occasionally, so as to ensure it to be in working order.

(*v*) **Escape channel and escape discharge.** The outflow from the ejector is taken to a natural drainage through an escape channel. The escape channel should be designed to have a self-cleansing velocity, so that the ejected material is transported without deposition. Adequate fall between the F.S.L. of the escape channel at its tail end and the normal H.F.L. of the natural drain, is desirable for efficient functioning of the channel.

The escape discharge is generally governed by : (*i*) amount of discharge required to remove the desired sediment size and load ; and (*ii*) minimum discharge required for flushing individual tunnels. Normally, an escape discharge varying from 10 to 20% of the full supply discharge of the canal, downstream of the ejector, is sufficient for this purpose.

### Efficiency of Excluders and Ejectors

The efficiency of excluders and ejectors, can be expressed by

$$\eta = \frac{I_u - I_d}{I_u} \times 100 \text{ (per cent)} \quad ...(9.9)$$

where $\eta$ = Per cent efficiency.

$I_u$ = Silt intensity in canal upstream of the ejector.

$I_d$ = Silt intensity in canal downstream of the ejector.

## 9.4. Emerging Trends in River Water Management through Construction of Rubber Dams

***Rubber dams*** are *weir systems* in which an elongated rubber body is firmly connected to the foundation of the weir system. Such a structure is made of fabric reinforced rubber bag, which can be inflated or deflated by air or water or by both.

It acquires a balloon shape when inflated compared to a pucca conventional water retaining structure like a gated barrage on a weir. Rubber dams thus offer a different type of a hydraulic structure, which could also serve the purpose of releasing the surplus water over the body of the dam by emptying the filled air or water from the dam body.

Rubber dams have widely been constructed in China over the past 40 years or so, as ***cheaper water conservation structures.*** Such dams have also been installed in USA, France and Japan, etc., although the concept is relatively new to India. Such a structure is constructed in specific cases, where it may prove to be a good alternative to a costlier permanent structure.

Combine with other water management methods of urban areas, such a dam could prove to be an effective tool of river water management, especially for cities near the banks of rivers, particularly for *small and medium rivers.*

Rubber dams have wide global prospects, since they can be used for irrigation, hydropower generation, environmental improvement, and recreation purposes. Besides using them for storing water, they can also be used to release the surplus flood water over the body of the dam by emptying the filled water or air from the dam body.

Rubber dams have certain definite advantages within the applicability range. Rubber is inflatable and deflectable material and may be quite useful in cold areas where the temperature may be as low as –40°C. The water filled ***rubber dams can be constructed for water heads from 0.3 m to more than 3 m or so.***

As stated above, inflatable rubber dams (weirs) can be filled with water, or air, or both. They are low pressure structures, typically in the range of 27 to 70 kPa [*i.e.*, 4 to 10 pounds per sq. inch (psi)].

The ***operating principle*** **of an inflatable weir** is shown in Fig. 9.22.

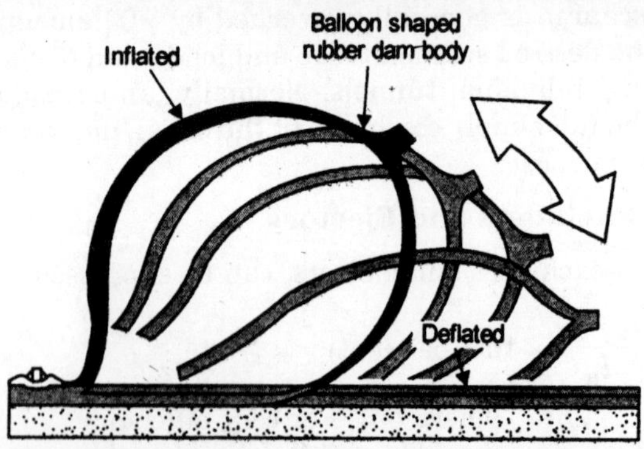


**Fig. 9.22.** Operating principle of an inflatable weir (rubber dam).

**9.4.1. Historical Development of Rubber Dams.** The first inflatable rubber dam was constructed in USA in the year 1959, as a part of a water supply project for the city of Los Angeles in California. The American engineer who

Fig. 9.23($a$) Photoview of Gayaji Rubber Dam

Fig. 9.23(*b*) Photoview of Gayaji Rubber Dam

installed the first and subsequent rubber dam was Mr. Imberston. In 1960, the first electricity board ordered its first rubber dam in *Aubas town* on the *Vezere River* in *France*. In the following decade, ten inflatable rubber dams were installed in France, of which four were for the French Electricity Board. Half of these dams are still operating and working efficiently. Since then, about 2000 rubber dams were installed in USA and 1500 rubber dams were installed in Japan.

**India's Position.** The concept of construction of rubber dam or *inflated weir,* as it is sometimes called, was first put into practice in India by constructing a minor weir of length 60 m and height 3.3 m in the year 2006 across *Jhanjhavathi* river* at Rajyalakshmipuram village in a northern coastal district of Andhra Pradesh Vizianagaram.

Thereafter, two more *rubber weirs* were installed on *Musi river* in Hyderabad in the year 2008, at one km apart, under the ambitious Musi beautification programme. One of these small weirs is located near the Andhra Pradesh High Court (length 80 m and height 1.3 m), and the other is located near Salarjung museum (length 73 m and height 1.3 m).

The *longest rubber dam of India* has recently been constructed in the year 2022 on ***Falgu river*** near Gaya town in Bihar State of India. The foundation stone of this ambitious project was laid in September 2020, when the Covid pandemic was at its peak. The dam was completed and inaugurated in September 2022. The dam is located on Falgu river, which is a vast stretch of sand dunes. This dam is expected to change this sandy landscape, to a greener one, besides providing water supply to Gaya town and nearby areas. This dam has been named as ***Gayaji dam***. It has been constructed at a cost of Rs. 312 crores. This dam is made of 17 mm thick rubber, and is 411 m long, 95.5 m wide, and 3 m high. The rubber membrane, *i.e.*, balloon work was done by an Australian company, named *Rubina Proprietary (Pty.) Ltd.*

The water will be stored in Falgu river up to about two and a half kilometre upstream. The stored water in Falgu river will be supplemented by pumping water from tubewells, which have been installed for timely replacement of water in the river. Water harvesting shall be done by stoppage of water in the Falgu river. To stop the flow of surface water, sheet pile of 1031 m length and diaphagram wall of 300 m has been provided. Thus, the water will be available round the year in atleast 0.6 m depth for devotees to take bath before performing *Tarpan* and *Pind daan* for their ancestors, at the holy city of Gaya. Infact, devotees do come in large numbers from around the world to Gaya to perform Pind Daan and Tarpan especially during *pitra paksha* (*shradh* period) for their ancestors as well as for themselves in advance.

The Gayaji dam is bullet proof and experts claim that it will not deteriorate for 100 years.

Two photo viewes of Gayaji dam are placed at Figs. 9.23(*a*) and (*b*).

**9.4.2. Material Used for Rubber Dams.** The main material used for rubber dams is a "*rubber bag*", which controls the flow of water. Some characteristics of rubber bag are as follows :

(*i*) It is multi-layer fabric of synthetic fiber (usually nylon). The fabric is quite flexible and exhibits good wear-resistance characteristics.

---

*Jhanjhavathi is a main tributary of Nagaveli river. Both of these rivers originate in the adjoining State of Odisha.

(ii) The fabric bag should be water resistance, water tolerant, corrosion resistance, and durable in atmosphere.

(iii) The layers of rubber coated fabric are jointed in the longitudinal direction. The actual number of layers of rubber coated fabric for each rubber body depends on the height of the dam.

**9.4.3. Classification of Rubber Dams.** They are of the following two types:

**(1) Air Filled Rubber Dams.** In these types of rubber dams, the inflation medium used is air. This type of rubber dam is particularly useful in cold countries where freezing of water is the main problem with water filled rubber dams. The inflating air pressure given by any manufacturer is 30 kPa. At this pressure, dam is fully inflated and it gains circular balloon shape, as shown in Fig. 9.24.

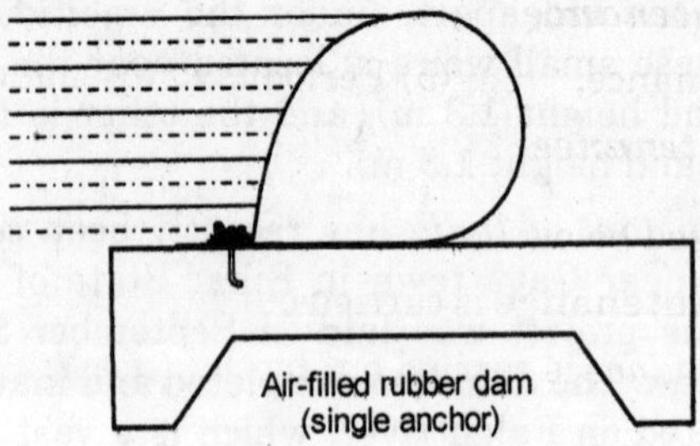


**Fig. 9.24.**

**(2) Water Filled Rubber Dam.** In these types of rubber dams, the inflating medium used is water, as shown in Fig. 9.25. This has special advantage over the air filled rubber dams that this has easy deflating process. But in case of cold regions, water may get freezed and to avoid it, continuous circulation is required, thereby increasing the maintenance cost.

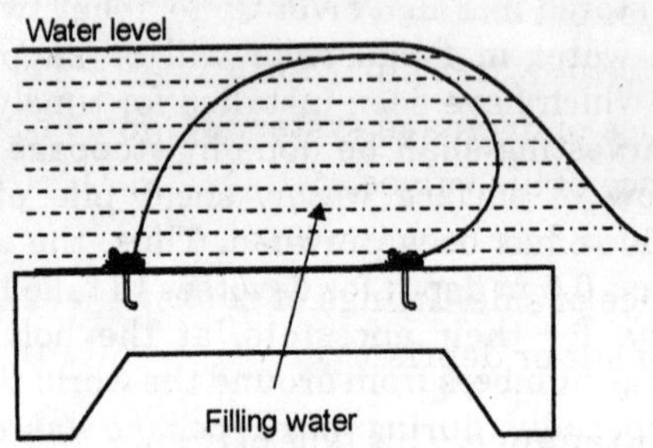


**Fig. 9.25.**

**(3) Combined Air and Water Filled Rubber Dam.** This type of dam uses both the medium *i.e.*, air and water for inflating. At first, water is filled in the tube and then air is filled in the tube. This is a recent development. This type of rubber dam is easy to deflate and it is assumed that this dam deflates completely.

**9.4.4. Advantages of Inflatable Weirs (Rubber Dams)**

1. Easy construction process.
2. Rubber dams have a simple inflation and deflation mechanism.
3. A light upper structure and uniform load of rubber dam body minimizes uneven foundation settlement.

4. The flexible structure of rubber dams virtually eliminate the influence of downstream dirt and sediment.
5. The rubber dams can be installed on virtually any side slop.
6. Rubber dams require only light foundation, leading to considerable saving in time and cost.
7. Rubber dam have low maintenance cost.

**9.4.5. Maintenance of Rubber Dams.** The safety of the dam relates to responsible operation, maintenance and inspection of its structural elements, so that there is an adequate margin of safety to the collapse or failure of the dam.

The various maintenance jobs can be divided into two categories.

**1. *Routine Maintenance***

(*a*) Daily maintenance, (*b*) Periodic maintenance

**2. *Special Maintenance***

These are discussed below :

**(1) Routine Maintenance** is carried out either daily or periodically, as below:

(*a*) *Daily Maintenance* is carried out by the departmental labour through out the year.

Daily maintenance of dam involves following jobs:

(*i*) Daily inspection of dam section.
(*ii*) Tightening of loose bolts.

(*b*) *Periodic maintenance* includes:

(*i*) Maintenance of reinforced rubber material.
(*ii*) Maintenance of water pipes for water-filled rubber dams.
(*iii*) Maintenance of drainage system
(*iv*) Maintenance of air valves.
(*v*) Maintenance of air or water mechanism systems.
(*vi*) Maintenance of side-fittings of rubber tube.
(*vii*) Removal of silt or debris.

**(2) Special Maintenance.** This type of maintenance is carried out whenever necessary in case of any accidents related to the rubber tube.

## PROBLEMS

**1.** Draw a neat sketch of a River-regulator and explain its salient components. Write short notes on :
(*i*) Silt ejector. (*ii*) Scouring sluices. (*iii*) Dividing groyne.
(**Hint** : Dividing groyne means divide wall.)

**2.** (*a*) Differentiate between a weir and a barrage.
(*b*) Draw a neat layout of Diversion head-works and indicate the various components of the system. Briefly indicate the function of each component.

**3.** (*a*) What is the purpose served by "scouring sluices" at weirs ? What are the important points to be borne in mind in designing these sluices ?

(*b*) How does "scouring-sluices" differ from those of "head-sluices" ?

(*c*) Why is it necessary to provide a "fish ladder" on large rivers, and how does it help in achieving the required objective ?

**4.** Draw a neat layout of a river regulator, and indicate suitable locations for the following :

(*i*) A navigational lock (*ii*) A fish ladder (*iii*) Scouring sluices (*iv*) Head sluices

Draw a typical cross-section of a head regulator, and indicate the various components of the same.

**5.** What are the functions of a canal head regulator ?

Design a head sluice for a canal, intended to irrigate 24,000 hectares at a duty of 800 hectares per cumec. It is in a main canal taken from a river and has the following data :

| | |
|---|---|
| Head available to permit full supply in canal | = 0.30 m. |
| Bed width of canal | = 20 m. |
| FSD in canal | = 1.5 m |
| Depth in the river at max. water level | = 4.2 m |

**6.** Enumerate and explain briefly the different methods for control of entry of silt into canals. Give a neat sketch of any one silt exclusion work to show its structural details and principle of functioning.

Show the component parts of a diversion head-works on a neat sketch. Mention the functions and important design considerations pertinent to the divide wall.

**7.** (*a*) Illustrate with neat sketch the following parts of a barrage : (*i*) Sheets piles, (*ii*) Fish ladder, (*iii*) Silt excluder, (*iv*) Divide wall.

(*b*) Briefly describe a system of regulation that is practised to minimise sediment load entering a canal.

**8.** (*a*) Differentiate between the following : (*i*) A barrage and a dam. (*ii*) Scouring sluices and head sluices. (*iii*) Surplus weir and storage weir. (*iv*) Gravity and non-gravity weirs. (*v*) Silt excluders and silt ejectors. (*vi*) River regulator and a canal regulator.

(*b*) What is meant by "afflux", and how does it effect the design of weirs and barrages ?

**9.** (*a*) What is the difference between a weir and a barrage ? Why does a barrage preferred to a weir in modern days ?

(*b*) How does a diversion weir aligned ? Draw a neat sketch showing the different components of a diversion weir scheme.

(*c*) What are the different construction materials which may be used for weirs, and how are the weirs classified on this score ?

**10.** (*a*) What is meant by an "Intake weir" and how does it differ from a "surplus weir" ?

(*b*) What are "under-sluices", and what are their functions in a river regulator ?

(*c*) Draw a neat sketch of a typical fish ladder indicating the purposes served by it.

**11.** (*a*) What are "Divide walls", and how do they help in a diversion weir scheme ? Draw neat sketches showing the cross-section of divide walls on pucca floor as well as beyond the pucca floor. Also discuss the design considerations that are involved in designing divide walls.

(*b*) Discuss the use of 'guide banks' and 'marginal bunds' in a river regulator scheme.

**12.** (*a*) What are the two principal methods of regulating the canal supplies in a diversion head works scheme ? What are their comparative merits and demerits ?

(*b*) Differentiate between a 'Silt extractor' and a 'Silt excluder'. Draw a neat sketch and discuss the principles involved in designing the different components of a silt extractor.

13. (*a*) Differentiate between the following, indicating the preference, which you will give to one, in modern days, with proper reasoning : (*i*) Weir and barrage, (*ii*) Silt excluder and silt-ejector, (*iii*) Rock-fill weirs and sloping glacis concrete weirs, (*iv*) Still pond regulation and semi open flow operation at a barrage.

(*b*) Draw a neat sketch of a silt excluder and indicate briefly the principles involved in designing its different components. How will you express its efficiency ?

14. Describe briefly some of the effects of silting in rivers. Describe briefly a method of removal of silt accumulation behind a river regulator.

# Hydraulic Jump and Its Usefulness in the Design of Irrigation Structures

—10

## 10.1. General

Hydraulic jump is the jump of water that takes place when a super-critical flow changes into a sub-critical flow. When a stream of water moving with a high velocity and low depth (*i.e.* supercritical flow) strikes another stream of water moving with a low velocity and high depth (*i.e.* sub-critical flow), a sudden rise in the surface of the former takes place. This phenomenon is called ***Hydraulic jump***, and is generally accompanied by a large scale turbulence, dissipating most of the kinetic energy of super-critical flow. Such a phenomenon may occur in a canal below a regulating sluice, at the bottom of a spillway, or at a place where a steep channel slope suddenly turns flat.

When water falls over a spillway or a vertical or a glacis fall, it acquires a lot of momentum and velocity. This high velocity, if allowed to persist, shall cause large scale erosion and scouring of the downstream soil of the work, and hence, it must be checked or controlled. In such situations, the phenomenon of hydraulic jump can be used with great advantage for dissipating the kinetic energy of the water.

If the jump is low, *i.e.* the change in the depth is small, the water shall not rise abruptly but will pass through a number of undulations. Such a low jump is called *an undular jump*. In this case, the energy dissipation shall be low and incomplete. But on the other hand, if the jump is high, *i.e.* when the change in depth is large, the jump shall occur abruptly and is called a *direct jump*. Such a jump involves large scale dissipation of energy and is of considerable importance in the design of hydraulic-cum-irrigation structures.

It may be noted that the depth before the jump is always less than the depth after the jump. The depth before the jump is called the **initial depth** ($y_1$) and the depth after the jump is called the **sequent depth** ($y_2$). These depths are shown on specific energy curve (Fig. 10.1) and must be differentiated from *alternate depths* $y_1$ and $y_2'$. $y_2'$ is the depth that shall occur in a sub-critical flow if there was no loss of energy in the jump formation ; while $y_2$ is the actual depth that occurs after the jump, involving the energy loss $H_L$.

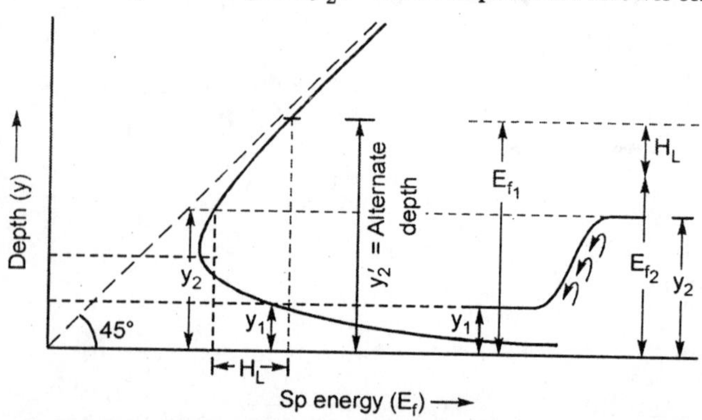


Fig. 10.1. Hydraulic jump interpreted by specific energy curve.

## 10.2. Types of Jump

Depending upon the incoming Froude No. $F_1$, the jump on a horizontal floor can be classified, as follows :

(*i*) For $F_1 = 1$, the flow is critical and hence, no jump can form.

(*ii*) For $F_1 = 1$ to 1.7, the water surface shows undulations and the jump is called *undular jump*.

(*iii*) For $F_1 = 1.7$ to 2.5, a series of small rollers develop on the surface of the jump, but the down-stream water surface remains smooth. The velocity throughout is fairly uniform and the energy loss is low. This jump is called a *weak jump*.

(*iv*) For $F_1 = 2.5$ to 4.5, there is an oscillating jet entering the jump bottom to surface and back again with no periodicity. Each oscillation produces a large wave of irregular period, which, very commonly in canals, can travel for miles, doing unlimited damage. This jump is called an *oscillating jump*. The *value of incoming Froude number for barrages and canal regulators generally lies in this zone* or in the previous zone *i.e.* zone of weak jump. The energy dissipators and stilling basins for such conditions need careful design. Different States and Organisations have recommended standard design criteria for the design of stilling basins. A description of these standard basins is available in I.S : 4997—1968. A little description of these stilling basins is given in chapter 20.

(*v*) For $F_1 = 4.5$ to 9.0, the jump is well balanced and its performance is at its best. The energy dissipation ranges from 45 to 70%. The jump is called a *steady jump*.

(*vi*) For $F_1 = 9.0$ and larger, it is a *strong jump* and the energy dissipation may reach up to 85%.

The ranges of Froude number given above, are not definite and may overlap to a certain extent depending on local conditions.

In the subsequent pages, we shall explain the mathematical treatment of hydraulic jump phenomenon on a level surface as well as on a sloping glacis. The methods of plotting the hydraulic jump profiles are also given.

## 10.3. Momentum Formula

In Fig. 10.2, a stream of water passing through a hydraulic jump has its velocity, depth and energy of flow as $V_1$, $y_1$ and $E_{f1}$ respectively before the jump, and $V_2$, $y_2$ and $E_{f2}$ after the jump. Let $H_L$ be the loss of energy in the jump.

Then $H_L = E_{f1} - E_{f2}$.

By equation of continuity, we can say that the discharge passing per unit width of the water-way ($q$) is given as :

$$q = V_1 \cdot y_1 = V_2 \cdot y_2$$

or
$$V_1 = \frac{Q}{y_1} \qquad \text{...(10.1)}$$

and
$$V_2 = \frac{Q}{y_2} \qquad \text{...(10.2)}$$

According to Newton's second law of motion, the rate of change of momentum must be equal to the applied force. This force is applied due to difference in hydrostatic pressure (forces) from upstream (section 1-1) to downstream (section 2-2). The hydrostatic pressure distribution diagram is shown in Fig. 10.2.

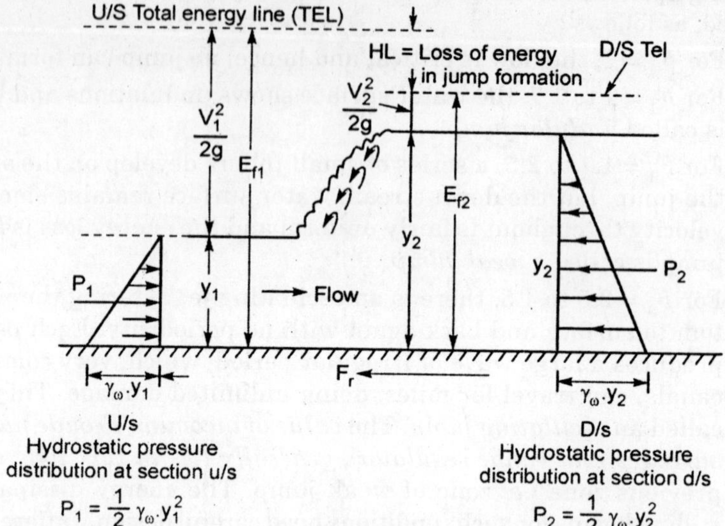


**Fig. 10.2.**

**Considering per metre width.** The force $P_1$ from U/s to D/s = $\frac{1}{2}\gamma_w \cdot y_1{}^2$

and, the force $P_2$ from D/s to U/s = $\frac{1}{2}\gamma_w \cdot y_2{}^2$.

If Fr is the frictional force acting along the channel bed, then the net force acting in the direction of flow = $P_1 - P_2 - F_r$

where $P_1 = \frac{1}{2} \cdot \gamma_w \cdot y_1^2$ ; and $P_2 = \frac{1}{2}\gamma_w \cdot y_2^2$

$$\therefore \text{Net force} = \frac{1}{2}\gamma_w \cdot y_1^2 - \frac{1}{2}\gamma_w \cdot y_2^2 - F_r \qquad ...(10.3a)$$

Moreover, this force is equal to the change of momentum of the control volume of water under consideration. Since momentum = $m.V$ (*i.e.*, mass × velocity), we have change of momentum per second.

$$= m\,(v_2 - v_1) = \rho \cdot q\,(v_2 - v_1)$$

($\because$ $m$ = mass of water = density of water × discharge per unit width)

or
$$m\,(v_2 - v_1) = \left(\frac{\gamma_w}{g}\right) \cdot q\,(V_2 - V_1)$$

Thus, charge in momentum $= \frac{\gamma_w}{g} \cdot q\,(V_2 - V_1)$ ...(10.3)

Equating (10.3) and (10.3*a*), we get

$$\frac{1}{2} \cdot \gamma_w \cdot (y_1^2 - y_2^2) - F_r = \frac{\gamma_w}{g}(V_2 - V_1)$$

or
$$\frac{1}{2} \cdot \gamma_w \cdot (y_2^2 - y_1^2) + F_r = (V_1 - V_2)$$

Now, substituting $V_1 = \frac{q}{y_1}$, and $V_2 = \frac{q}{y_2}$, we get

$$\frac{1}{2}\gamma_w\left(y_2^2 - y_1^2\right) + F_r = \frac{\gamma_w \cdot q}{g}\left[\frac{q}{y_1} - \frac{q}{y_2}\right]$$

$$\frac{1}{2}\gamma_w\left(y_2^2 - y_1^2\right) + F_r = \frac{\gamma_w \cdot q^2}{g}\left[\frac{y_2 - y_1}{y_1 y_2}\right] \qquad ...(10.4)$$

If friction is neglected, then $F_r = 0$; the equation (10.4) then reduces to

$$\frac{1}{2}\gamma_w\left(y_2^2 - y_1^2\right) = \frac{\gamma_w \cdot q^2}{g}\left[\frac{y_2 - y_1}{y_1 y_2}\right] q \qquad ...(10.4a)$$

or $$\left(y_2^2 - y_1^2\right) = \frac{2q^2}{g}\left[\frac{y_2 - y_1}{y_1 \cdot y_2}\right]$$

or $$(y_2 + y_1)(y_2 - y_1) = \frac{2q^2}{g}\left[\frac{y_2 - y_1}{y_1 y_2}\right]$$

or $$y_1 y_2 (y_2 + y_1) = \frac{2q^2}{g} \qquad ...(10.5)$$

or $$y_1 y_2^2 + y_1^2 y_2 = \frac{2q^2}{g}$$ or $$y_2^2 + y_1 \cdot y_2 = \left(\frac{2q^2}{g}\right)\frac{1}{y_1} = \frac{2q^2}{q y_1}$$

or $$y_2^2 + y_1 y_2 - \frac{2q^2}{g y_1} = 0$$ or $$y_2 = \frac{-y_1 \pm \sqrt{y_1^2 + \dfrac{8q^2}{g y_1}}}{2}$$

Neglecting unfeasible – ve sign, we get

$$y_2 = \frac{-y_1}{2} + \sqrt{\frac{y_1^2}{4} + \frac{2q^2}{g y_1}} \qquad ...(10.6)$$

$$y_2 = \frac{y_1}{2}\left[-1 + \sqrt{1 + \frac{8q^2}{g y_1^3}}\right] = \frac{y_1}{2}\left[-1 + \sqrt{1 + \frac{8V_1^2 y_1^2}{g y_1^3}}\right] = \frac{y_1}{2}\left[-1 + \sqrt{1 + 8\frac{V_1^2}{g y_1}}\right]$$

or $$y_2 = \frac{y_1}{2}\left[-1 + \sqrt{1 + 8F_1^2}\right] \qquad ...(10.7)$$

where $F_1$ = Incoming Froude number $\dfrac{V_1}{\sqrt{g y_1}}$

Similarly, we can show that

$$y_1 = \frac{y_2}{2}\left[-1 + \sqrt{1 + 8F_2^2}\right] \qquad ...(10.8)$$

where $F_2$ = Outging Froude number $\dfrac{V_2}{\sqrt{g y_2}}$.

The assumptions made in the derivation of the above formula are :

(i) *The jump takes place abruptly.*

(ii) *Flow remains streamlined throughout.*

(iii) *Friction is negligible.*

(iv) *Bed is horizontal.*

This Eq. (10.4) can be used to determine any one of the four parameters ($y_1$, $y_2$, $q$ and $Fr$), if other three are known.

Secondly, in the case of a sloping channel, the bed does no longer remain horizontal, and the horizontal component of the weight of flowing water will also be involved besides $P_1$, $P_2$, $F_r$ in evaluating the applied force, as shown in Fig. 10.3.

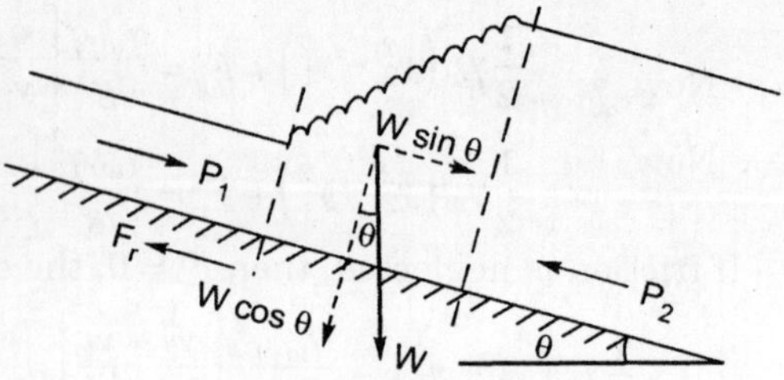

Fig. 10.3.

In this case, applied force will be equal to : $= [P_1 - P_2 - F_r + W \sin \theta]$

where $W$ = Wt. of water in controlled volume per unit width $= \gamma_w . V_0 = \gamma_w .(q . L_o)$

where $L_0$ = Length of channel between 1-1 and 2-2

Hence, the equations (10.7) & (10.8) derived above for $y_2$ and $y_1$ shall not hold good precisely. But if the slope of the glacis is not too pronounced, the gravity component $W \sin \theta$ can be neglected on the grounds that the error so introduced is practically cancelled by the errors involved in assuming the channel to be frictionless, since both $F_r$ and $W \sin \alpha$ acts in opposite directions. Hence, the equations derived for a horizontal bed may be applied to a sloping channel also, within the limits of accuracy.

**Example 10.1.** *A hydraulic jump occurs in a horizontal rectangular channel, which is 1 m wide. The pre-jump depth is 5 cm and the post-jump depth is 20 cm. Assuming the channel to be frictionless, estimate the discharge of water. If the friction is not neglected, and the friction force is estimated to be 20 N over the jump-length, what will be the estimated discharge?* *(Engineering Services, 2018)*

**Solution.** Width of channel = 1 m

$$y_1 = 5 \text{ cm}$$
$$y_2 = 20 \text{ cm}$$

**Case I** : When $F_r = 0$, use the basic equation (10.4) connecting $y_1$ and $y_2$ as

$$\frac{1}{2} . \gamma_w . \left( y_2^2 - y_1^2 \right) + F_r = \frac{\gamma_w . q^2}{g} \left[ \frac{y_2 - y_1}{y_1 y_2} \right] \quad ...(10.4)$$

Here given values are : $F_r = 0$

$y_1 = 5 \text{ cm} = 0.05 \text{ m}$

$y_2 = 20 \text{ cm} = 0.20 \text{ m}$

$\gamma_w = 9.81 \text{ kN/m}^3$ (assumed)

$q = ?$

Substituting given values, we get

$$\frac{1}{2} .(9.81).\left(0.2^2 - 0.05^2\right) + 0 = \frac{9.81 . q^2}{9.81} \left[ \frac{0.20 - 0.05}{0.20 \times 0.05} \right]$$

or
$$\frac{1}{2} \times 0.375 = \frac{q^2}{9.81} [15]$$

or
$$q = 0.111 \text{ m}^3\text{/sec} \quad \textbf{Ans.}$$

**Case II :** When Fr = 20 N

Now, we have $F_r = 20 \text{ N} = \frac{20}{1000} \text{ kN} = 0.020 \text{ kN}$

Now, Eq. (10.4) reduces to

$$\frac{1}{2}(9.81)(0.2^2 - 0.05^2) + 0.020 = \frac{9.81q^2}{9.81}\left[\frac{0.20 - 0.05}{0.20 \times 0.05}\right]$$

$$0.2039 = 15\, q^2$$

or $$q = 0.117 \text{ m3/sec} \qquad \textbf{Ans.}$$

**Example 10.2.** *Water discharging from an overfall spillway flows to a jump type basin. The discharge intensity is 30 m³/s/m and the water depth is 1.5 m. Find the conjugate depth for the jump to form. Also calculate the length of the jump basin.*

**Solution.** Conjugate depth is given by equation (10.6) as :

$$= y_2 = -\frac{y_1}{2} + \sqrt{\frac{y_1^2}{4} + \frac{2q^2}{gy_1}} \qquad ...(i.e. \text{ Eq. } 10.6)$$

$$= -\frac{1.5}{2} + \sqrt{\frac{1.5^2}{4} + \frac{2 \times 30^2}{9.81 \times 1.5}}$$

$$= -0.75 + \sqrt{0.5625 + 122.324} = -0.75 + 11.08 = 10.33 \text{ m.} \quad \textbf{Ans.}$$

Length of jump basin $= 5(y_2 - y_1) = 5[10.33 - 1.5] = 44.15$ m. **Ans.**

### 10.3.1. Loss of Energy ($H_L$) in the Standing Wave

$$H_L = E_{f_1} - E_{f_2}$$

$$\text{where } E_{f_1} = y_1 + \frac{V_1^2}{2g} \qquad ...(10.9)$$

$$E_{f_2} = y_2 + \frac{V_2^2}{2g} \qquad ...(10.10)$$

$$\therefore \quad H_L = \left(y_1 + \frac{V_1^2}{2g}\right) - \left(y_2 + \frac{V_2^2}{2g}\right)$$

$$= (y_1 - y_2) + \frac{1}{2g}(V_1^2 - V_2^2) = (y_1 - y_2) + \frac{1}{2g}\left[\frac{q^2}{y_1^2} - \frac{q^2}{y_2^2}\right]$$

or $$H_L = (y_1 - y_2) + \frac{q^2}{2g}\left[\frac{y_2^2 - y_1^2}{y_1^2 y_2^2}\right] \qquad ...(10.11)$$

From equation (10.5), we get

$$y_1 y_2 (y_2 + y_1) = \frac{2q^2}{g} \quad \text{or} \quad \frac{q^2}{2g} = \frac{y_1 y_2}{4}(y_2 - y_1)$$

Substituting this value in equation (10.11), we get

$$H_L = (y_1 - y_2) + \frac{y_1 y_2}{4}(y_2 - y_1)\left(\frac{y_2^2 - y_1^2}{y_1^2 y_2^2}\right) = (y_2 - y_1)\left[\frac{(y_2 + y_1)^2}{4y_1 y_2} - 1\right]$$

$$= (y_2 - y_1)\left[\frac{(y_2 + y_1)^2 - 4y_1 y_2}{4y_1 y_2}\right] = (y_2 - y_1)\left[\frac{(y_2 + y_1)^2}{4y_1 y_2}\right]$$

or $$H_L = \frac{[y_2 - y_1]^3}{4y_1 y_2} \qquad ...(10.12)$$

Usually, in any hydraulic jump, the following eight variables are involved: $E_{f_1}$, $V_1$, $y_1$, $E_{f_2}$, $V_2$, $y_2$, $q$ and $H_L$. These variables are related by six independent equations, as given below :

$$y_1 y_2 (y_1 + y_2) = \frac{2q^2}{g} \qquad ...(10.5)$$

$$H_L = \frac{(y_2 - y_1)^3}{4 y_1 y_2} \qquad ...(10.12)$$

$$E_{f_1} = y_1 + \frac{V_1^2}{2g} \qquad ...(10.9)$$

$$E_{f_2} = y_2 + \frac{V_2^2}{2g} \qquad ...(10.10)$$

$$V_1 = \frac{q}{y_1} \qquad ...(10.1)$$

$$V_2 = \frac{q}{y_2} \qquad ...(10.2)$$

Hence, if any two variables are known, the remaining six can be worked out by using these six equations, mathematically. The mathematical solution is complicate and to avoid large scale calculations, *Blench* has given some curves by taking $q$ and $H_L$ as known variables (as in actual problems, the discharge intensity *q and the drop in the total energy level* $H_L$ *are generally known).*

Blench had given curves, relating $H_L$ and $E_f$ for different values of $q$ (Plate 10.1). These curves are known as ***Blench Curves***, and are very useful in determining the location of the jump on a sloping glacis, as explained in article 10.4.

**When Blench curves are not available,** the following mathematical equations can be used to evaluate $y_1$ and $y_2$ by known values of $q$ and $H_L$ :

$$\text{Compute } y_c = \text{critical depth} = \sqrt[3]{\frac{q^2}{g}} \qquad ...(10.11)$$

Now express

$$\frac{y_1}{y_c} = X \text{ (say)} \qquad ...(10.12)$$

$$\frac{y_2}{y_c} = Y \text{ (say)} \qquad ...(10.13)$$

$$\frac{H_L}{y_c} = Z \text{ (say)} \qquad ...(10.14)$$

where $H_L$ and $y_c$ are known, hence $Z$ is known.

Relations between $Z$ and $X$ ; and $Z$ and $Y$ are worked out as :

$$Z = \frac{-X^6 + 20X^3 + 8 - (X^4 + 8X)^{3/2}}{16X^2} \qquad ...(10.15)$$

and

$$Z = \frac{-Y^6 - 20Y^3 - 8 - (Y^4 + 8Y)^{3/2}}{16Y^2} \qquad ...(10.16)$$

For different values of $X$ and $Y$, values of $Z$ can be tabulated, and curves $X - Z$ and $Y - Z$ plotted, to read value of $X$ and $Y$ for known value of $Z$.

To obtain more direct solution, Swamee C.P. has obtained an approximate solution to $Y - Z$ curves, and related $Y$ with $Z$ by the eqn. :

$$Y = 1 + 0.93556\, Z^{0.368} \text{ for } Z < 1 \quad \text{...(10.17)}$$

and
$$Y = 1 + 0.93556\, Z^{0.240} \text{ for } Z > 1 \quad \text{...(10.18)}$$

From the known value of $Z$, value of $Y$ *i.e.* $y_2/y_c$ can be calculated to finally compute the value of $y_2$.

With known values of $Z$ and $X$, value of $X$ *i.e.* $y_1/y_c$ can also be computed by using the eqn. :

$$Z = \frac{(Y - X)^3}{4XY} \quad \text{...(10.19)}$$

This computed value of $X$ *i.e.* $y_1/y_c$ can be used to compute $y_1$.

Values of $E_{f_1}$ and $E_{f_2}$ can also be computed by using the values of $X$ and $Y$, as follows :

$$E_{f_1} = \varepsilon = X + \frac{1}{2X^2} \quad \text{...(10.20)}$$

$$\frac{E_{f_2}}{y_c} = \eta = Y + \frac{1}{2Y^2} \quad \text{...(10.21)}$$

**Example 10.3.** *Given that the unit discharge in a rectangular channel is 18 m³/sec and that the head loss across a hydraulic jump that forms in this channel is 1.1 m, estimate the pre-jump and post jump depths.* *(Engg. Services, 1990)*

**Solution.** From eqns. (10.5), we have

$$y_1 y_2 (y_1 + y_2) = \frac{2q^2}{g} = \frac{2 \times 18^2}{9.81} = 66.06 \quad \text{...}(i)$$

[$q$ = unit discharge *i.e.*, discharge per unit width]

Also, from Eq. (10.12), we have

and
$$H_L = \frac{(y_2 - y_1)^3}{4 y_1 y_2} \qquad \text{where } H_L = 1.1 \text{ m (given)}$$

$$\therefore \quad (y_2 - y_1)^3 = 4.4\, y_1\, y_2 \quad \text{...}(ii)$$

Putting $y_2 = m y_1$ in (*ii*), we get

$$(m - 1)^3 \,.\, y_1^3 = 4.4 \,.\, m y_1^2$$

$$\therefore \quad y_1 = \frac{4.4m}{(m-1)^3} \quad \text{...}(iii)$$

Putting $y_2 = m y_1$ and this value of $y_1$ in (*i*), we get

$$y_1 \,.\, m \,.\, y_1 (y_1 + m \,.\, y_1) = 66.06$$

or
$$m(m + 1) \,.\, y_1^3 = 66.06$$

or
$$m(m + 1)\left[\frac{4.4m}{(m-1)^3}\right]^3 = 66.06$$

or
$$\frac{m(m+1)(m)^3}{(m-1)^9} = \frac{66.06}{4.4^3} \quad \text{or} \quad \frac{m^4(m+1)}{(m-1)^9} = 0.775$$

Solving by hit and trial, we get, $m = 2.93$

Substituting it in (*iii*), we get

$$y_1 = \frac{4.4 \times 2.93}{(1.93)^3} = \mathbf{1.79\ m}$$

$$y_2 = 2.93 \times 1.79 = 5.25 \text{ m}$$

**Ans.**

**Example 10.4.** *A sluice across a channel 6 metre wide discharges into a stream 1 metre deep. What is the flow rate when the depth upstream of the sluice is 7 metre? On the downstream side, concrete blocks have been placed to create conditions for hydraulic jump to occur. Determine the force on the blocks if the downstream depth is 3 metre.* *(Engg. Services, 1992)*

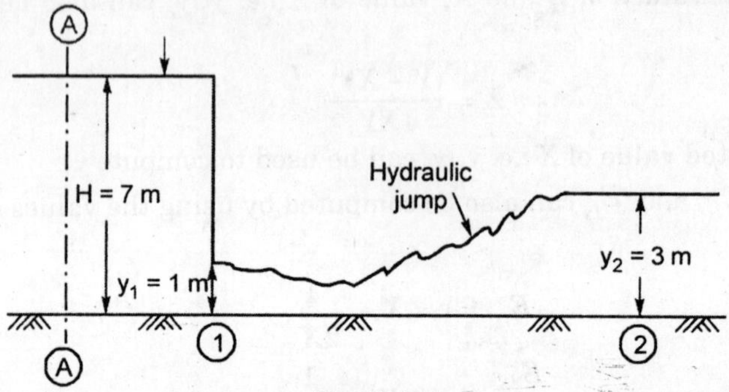


Fig. 10.4.

**Solution.** Discharge through the sluice gate

$Q = C_d\,(6 \times 1)\,\sqrt{2gH}$ as width of channel = 6 m

area of flow = 6 m × 1 m

where $C_d = C_c \cdot C_v$

Assuming $C_v = 0.95$

$C_c = 0.62$

$C_d = C_v \cdot C_c = 0.59$

$Q = 0.59 \times 6 \times 1 \times \sqrt{2 \times 9.81 \times 7} = 41.49$ m³/s **Ans.**

(*b*) Force acting on concrete blocks is given by the rate of change of momentum per unit channel width.

$$= \left(\frac{\gamma_w}{g}\right) q\,(V_1 - V_2)\text{ kg}$$

$$\text{where } q = \frac{Q}{6} = \frac{41.49}{6} = 6.91\text{ m}^2/\text{s}$$

$$V_1 = \frac{q}{C_c\, y_1} = \frac{6.91}{0.95 \times 1}\text{ m/s} = 7.28\text{ m/s}$$

$$V_2 = \frac{q}{y_2} = \frac{6.91}{3} = 2.3\text{ m/s}$$

$$= \frac{9.81 \times 6.91}{9.81}(7.28 - 2.3)\text{ kN} = 34.41\text{ kN.} \quad \textbf{Ans.}$$

## 10.4. Location and Profile of the Jump on a Sloping Glacis

**10.4.1. Position of the Jump.** Let water be passing over the crest of a work (such as a weir, etc.) with a certain head, as shown in Fig. 10.5. Then discharge per unit width = $C_d H^{3/2}$, where $H$ is the head measured from the total energy line to the crest and $C_d$ is the coefficient of discharge. Knowing $H$, the level of u/s TEL is known. For the given discharge, the depth of water on the d/s is known from gauge discharge curves of the channel. This fixes TEL on the d/s. The difference in the levels of u/s TEL and d/s TEL gives $H_L$.

Knowing $q$ and $H_L$, $E_{f_2}$ can be obtained from Blench Curves (Plate 10.1). Subtracting $E_{f_2}$ from d/s TEL, the level at which the jump will form can be easily obtained and hence the position of point $P$ is fixed. This is how the position of the jump on the sloping glacis is obtained.

Knowing $E_{f_2}$, $E_{f_1}$ can also be calculated by using, $E_{f_1} - E_{f_2} = H_L$. The corresponding values of $y_1$ and $y_2$ for the known values of $E_{f_1}$ and $E_{f_2}$ can be directly read from the 'Energy of Flow Curves' given by Montague **(Plate 10.2)**.

**10.4.2. Profile Before the Jump.** The water surface profile before the point of jump formation ($P$) can be easily plotted with the help of Montague's curves (Plate 10.2) as follows : For different points on the glacis from $B$ to $P$, (Fig. 10.5), $E_{f_1}$ will vary, being equal to u/s TEL minus glacis level at the considered point. Since glacis level is different at different points in the length $BP$, the $E_{f_1}$ will be different at different points and shall go on increasing. For these different values of $E_{f_1}$, different corresponding values of $y_1$ can be tabulated from Montague's Curves (Plate 10.2). These values of $y_1$ shall go on reducing till the point $P$ is reached. These values can be plotted over the glacis, and hence, water surface profile before the jump can be plotted easily as shown in Fig. 10.6.

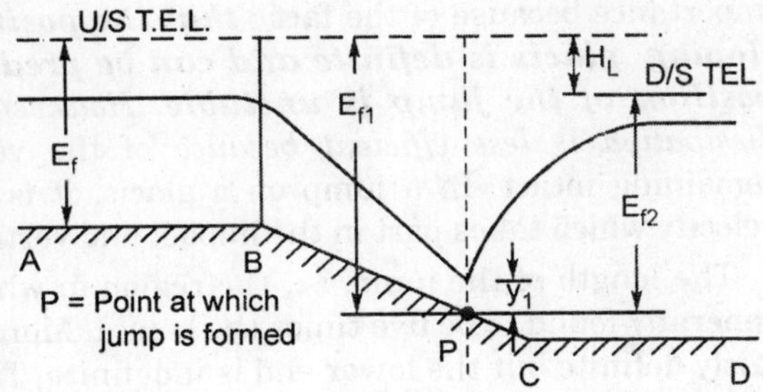

Fig. 10.5.

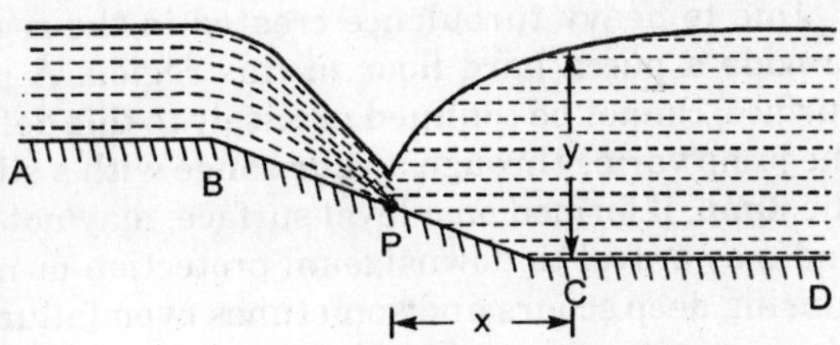

Fig. 10.6.

**10.4.3. Profile after the Jump.** To plot the water surface profile after the jump point ($P$), it is necessary to know the Incoming Froude No. $F_1$.

$$F_1 = \frac{V_1}{\sqrt{gy_1}} = \frac{q}{\sqrt{gy_1^3}}$$

Knowing $q$ and $y_1$, $F_1$ can be determined. Graphs are available between $\left(\frac{x}{y_1}\right) Vs \left(\frac{y}{y_1}\right)$ for different values of Froude No. $F_1$, as shown in **Plate 10.3($a$)**. Taking different points beyond $P$ along the profile, different values of $x$ and hence that of $x/y_1$, can be tabulated. Corresponding to these values of $x/y_1$ for a fixed $F_1$, different values of $y/y_1$, can be read out from Plate 10.3($a$). Hence, different values of $x$ and $y$ are known, where $(x, y)$ is any pt. in the direction of flow w.r.t. the point $P$ (*i.e.* glacis level at the point of jump formation) as origin, as shown in Fig. 10.7. Hence, the water surface profile after the jump point can be easily plotted.

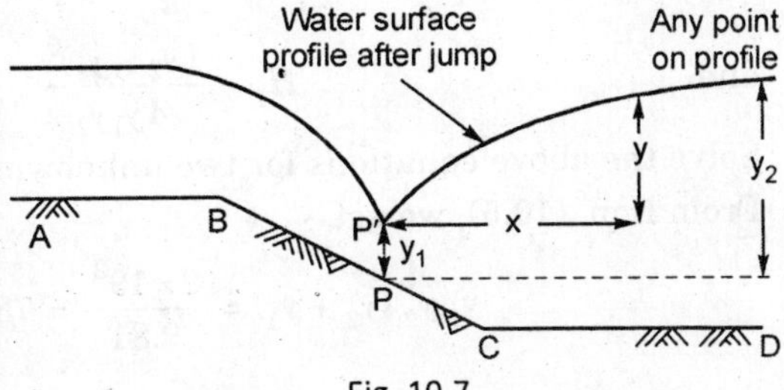

Fig. 10.7.

The water surface profile after the jump can also be plotted with the help of curves shown in **Plate 10.3(*b*)**. Values of $y/_{y_2 - y_1}$ for known values of $x/_{y_2 - y_1}$ can be read out for a given value of $F_1$ ; where $(x, y)$ in this case are the ordinates of any point on the profile w.r.t. the point of jump formation ($P'$) as origin, as shown in Fig. 10.7. Hence, for any assumed values of $x$, corresponding values of $y$ can be worked out and profile plotted easily.

## 10.5. Hydraulic Jump on a Sloping Glacis as Energy Dissipator

Much has already been said about the use of hydraulic jump phenomenon as energy dissipation device in the design of hydraulic and irrigation structures. The use of 'sloping glacis' for bringing out hydraulic jump to occur is of utmost importance because of the fact : ***that the position of the hydraulic jump on a sloping glacis is definite and can be predicted, while on a level floor the position of the jump is unstable.*** *However, on a 'sloping glacis', the energy dissipation is less efficient because* of the vertical component of the velocity remaining intact. In a jump on a glacis, it is only the horizontal component of velocity which takes part in the impact and vertical component remains uneffected.

The length of the jump, *i.e.* the region in which heavy turbulence is created is generally found to be five times the height. Moreover, the start point of the jump is fairly definite but the lower end is indefinite. This point is, however, taken as the place where water surface becomes sensibly level.

Due to heavy turbulence created in the region of the jump. It is necessary to provide a pucca hard floor in this region. A jump, if allowed to form on a level surface, cannot be confined precisely to this definite region, because the position of the jump varies through a wide range with a slight change in the discharge. Hence the jump, if formed on a level surface, may not confine itself to the pucca platform and may travel to downstream protection or natural erodible bed of the channel causing deep scours and sometimes even failure of the structure. *Hence, a sloping glacis is always preferred to a horizontal bed for affecting hydraulic jump phenomenon, because the position of the jump on a 'glacis' is always definite although less efficient.* Hence, a sloping glacis having a slope of 2 : 1 to 5 : 1 is generally provided, and hydraulic jump is made to occur on the glacis itself, and in no case, at a point lower than the toe of the glacis.

**Example 10.5.** *At an energy dissipater structure below a low spillway, the discharge is 19 m²/s, and the energy loss is 1 m at the hydraulic jump forming therein. Determine the depth of flow at both ends of the jump.(Engg. Services, 2002)*

**Solution.** $q$ = 19 m²/s ; $H_L$ = 1 m ; $y_1$ = ? ; $y_2$ = ?

Use equations as :

$$y_1 y_2 (y_2 + y_1) = \frac{2q^2}{g} \quad \text{...(Eqn. 10.5)}$$

and

$$H_L = \frac{(y_2 - y_1)^3}{4 y_1 y_2} \quad \text{...(Eqn. 10.12)}$$

Solve the above equations for two unknowns, $y_1$ and $y_2$, as follows :

From Eqn. (10.5), we get

$$y_1 y_2 (y_2 + y_1) = \frac{2 \times 19^2}{9.81} = 73.6 \quad \text{...(}i\text{)}$$

From Eqn. (10.12), we get $1 = \dfrac{(y_2 - y_1)^3}{4.y_1y_2}$ ...(*ii*)

Let $\dfrac{y_2}{y_1} = x$

$\therefore \quad y_2 = y_1 . x$

Substituting in (*i*), we get

$$y_1 \cdot y_1 x (y_1 x + y_1) = 73.6$$

or $y_1^2 . x . y_1 (x + 1) = 73.6$

or $x . y_1^3 (x + 1) = 73.6$ ...(*iii*)

Also from (*ii*), $1 = \dfrac{(y_1x - y_1)^3}{4.y_1.x.y_1} = \dfrac{y_1^3(x-1)^3}{4.x.y_1^2} = \dfrac{y_1}{4x}(x-1)^3$

or $4x = y_1 (x - 1)^3$

or $y_1 = \dfrac{4x}{(x-1)^3}$ ...(*iv*)

Substituting in (*iii*), we get

$$x . \left[\frac{4x}{(x-1)^3}\right]^3 [x + 1] = 73.6$$

Solve by hit and trial, for *x*, as :

$$x = \mathbf{2.548}$$

$$\therefore \quad y_1 = \frac{4 \times 2.548}{(2.548 - 1)^3} = \mathbf{2.748\ m}$$

$$y_2 = y_1 x = 2.748 \times 2.548 = \mathbf{7.0\ m}$$

Hence, the depth at both ends, of jump, $y_1$ and $y_2$ will be 2.748 m and 7.0 m, respectively. **Ans.**

**Example 10.6.** *Water discharges at the rate of 298 m³/sec over a spillway 14 m wide into a stilling basin of the same width. The water level behind the spillway is 61 m over datum. The river water level in the downstream is 29.2 m over datum. The river bed is paved flush with the bed level to just adequate length to form the jump basin. Assuming that no energy is dissipated in the flow down the spillway, find the basin invert level required for a hydraulic jump to be formed within the basin.* (*Engg. Services, 2003*)

[**Solution.** Here $H_L$ = (61.0 – 29.2) m = 31.8 m

$$q = \frac{Q}{B} = \frac{298}{14} = 21.29\ \text{m}^3/\text{s/m}$$

This is a case where $H_L$ and $q$ are known, and we have to find the value of $y_2$, since the invert level of jump basin will be equal to ($RL$ 29.2 – $y^2$). To determine $y_2$, proceed as given in article 10.3.1—when Blench curves are not available.

From Eq. (10.4), determine, $y_c = \sqrt[3]{\dfrac{q^2}{g}} = \sqrt[3]{\dfrac{(21.29)^2}{9.81}} = 3.59$ m

Now use Eq. (10.14) as : $Z = \dfrac{H_L}{y_c} = \dfrac{31.8}{3.59} = 8.86$

Now use Eq. (10.18) as : $Y = 1 + 0.93556 . Z^{0.24}$ for Z > 1

For $Z$ = 8.86 (> 1) $Y = 1 + 0.93556 \times (8.86)^{0.24} = 2.58$

Now, use Eq. (10.13) as : $\frac{y_2}{y_c} = Y$

or $$\frac{y_2}{3.59} = 2.58 \quad \text{or} \quad y_2 = 9.26 \text{ m}$$

The basin invert level = 29.2 – 9.26 = **19.94 m** above datum. **Ans.**]

## PROBLEMS

1. (*a*) What is meant by 'Hydraulic jump' and how does it helps in designing irrigation structures?
   (*b*) Differentiate between 'Sequent depth' and 'Alternate depth'.
   (*c*) Derive an expression for expressing energy dissipation obtained by a jump, in terms of initial depth ($y_1$) and sequent depth ($y_2$).
2. (*a*) What is a hydraulic jump ? How does it help in dissipating the energy of the water falling over a weir or a dam. What would happen if this energy is not properly dissipated ?
   (*b*) What is the importance of "Incoming Froude number", and how does it help in indicating the success of a jump formation ?
   (*c*) How would you fix the jump portion when water is flowing over a sloping glacis. Also explain as to how the pre-jump as well as post jump profile can be plotted and what useful purpose will be served by such plottings ?

Plate No. 10.1

BLENCH CURVES

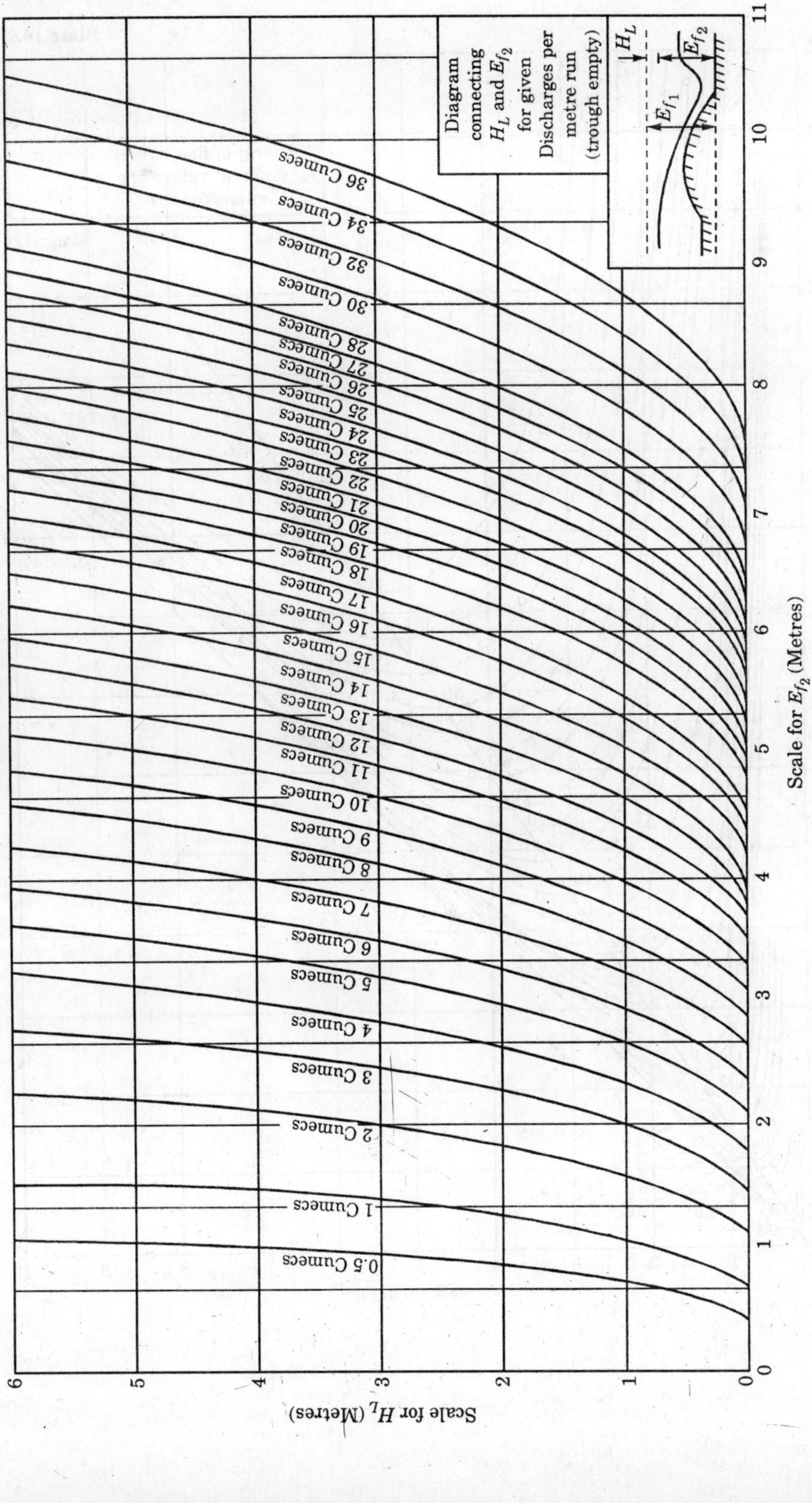

Plate 10.2.

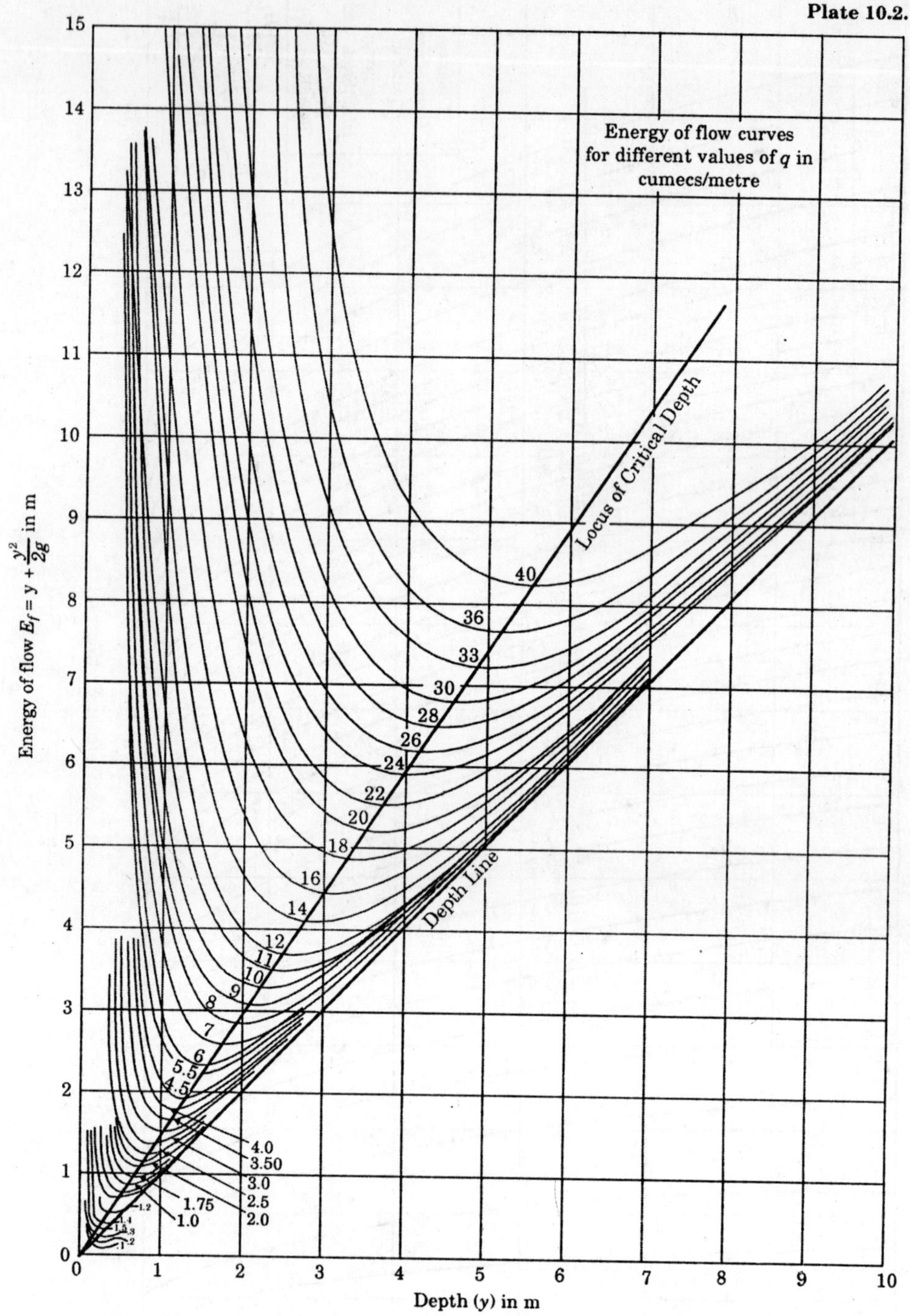
Energy of flow curves
for different values of $q$ in
cumecs/metre
Locus of Critical Depth
Depth Line
40
36
33
30
28
26
24
22
20
18
16
14
12
11
10
9
8
7
6
5.5
4.5
4.0
3.50
3.0
2.5
2.0
1.75
1.0
1.2
Energy of flow $E_f = y + \frac{y^2}{2g}$ in m
Depth ($y$) in m
0
1
2
3
4
5
6
7
8
9
10
11
12
13
14
15

Plate 10.3(*a*)

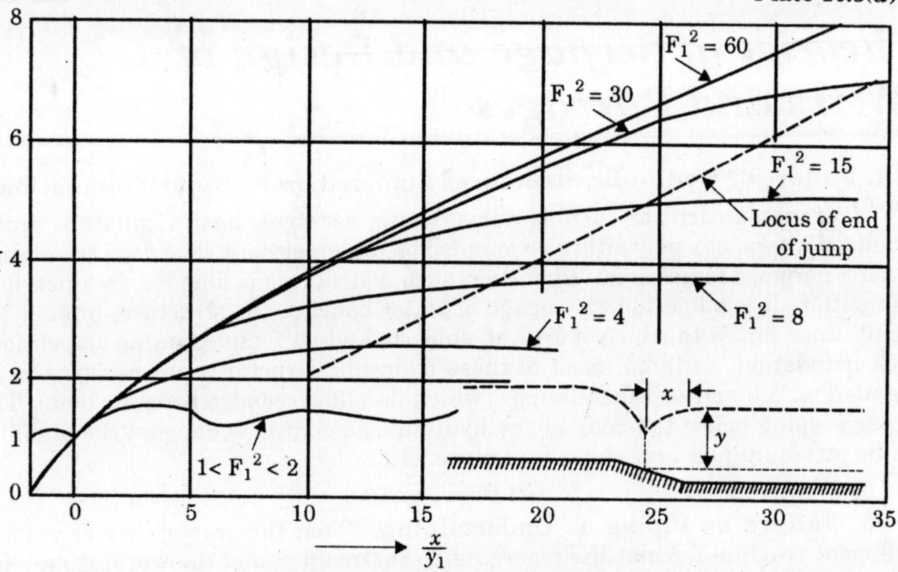


**Curve for plotting post jump profile**

Plate 10.3(*b*)

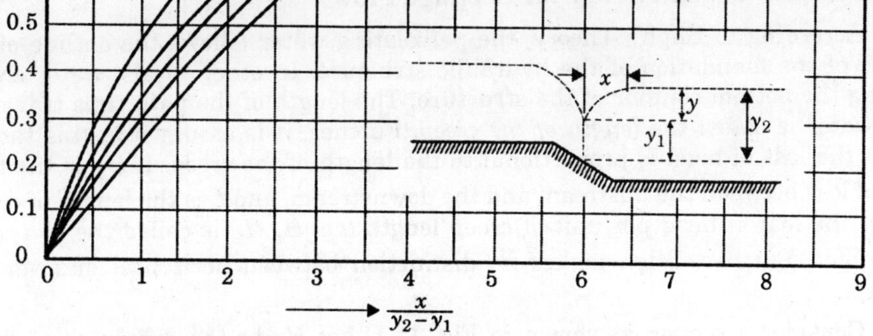


**Alternate curve for plotting post jump profile.**

# 11

# *Theories of Seepage and Design of Weirs and Barrages*

## 11.1. Failure of Hydraulic Structures Founded on Pervious Foundations

Hydraulic structures such as dams, weirs, barrages, head regulators, cross-drainage works, etc. may either be founded on an impervious solid rock foundation or on a pervious foundation. Whenever, such a structure is founded on a pervious foundation, it is subjected to seepage of water beneath the structure, in addition to all other forces to which it will be subjected when founded on an impervious rock foundation. In India, most of these hydraulic structures are required to be founded on alluvial soil foundations, which do allow seepage beneath them. The water seeping below the body of the hydraulic structure, endangers the stability of the structure and may cause its failure, either by :

(*i*) *Piping* ; or by (*ii*) *Direct uplift*.

(*i*) **Failure by Piping or Undermining**. When the seepage water retains sufficient residual force at the emerging downstream end of the work, it may lift up the soil particles. This leads to increased porosity of the soil by progressive removal of soil from beneath the foundation. The structure may ultimately subside into the hollow so formed, resulting in the failure of the structure.

(*ii*) **Failure by Direct Uplift.** The water seeping below the structure, exerts an uplift pressure on the floor of the structure. If this pressure is not counter-balanced by the weight of the concrete or masonry floor, the structure will fail by a rupture of a part of the floor.

The above concepts of the failure of hydraulic structures due to sub-surface flow were introduced by Bligh, on the basis of experiments and the research work conducted after the failure of Khanki weir, which was designed on experience and intuition without any rational theory.

## 11.2. Bligh's Creep Theory for Seepage Flow

According to Bligh's Theory, the percolating water follows the outline of the base of the foundation of the hydraulic structure. In other words, water creeps along the bottom contour of the structure. The length of the path thus traversed by water is called the *length of the creep*. Further, it is assumed in this theory, that the loss of head is proportional to the length of the creep. If $H_L$ is the total head loss between the upstream and the downstream, and $L$ is the length of creep, then the loss of head per unit of creep length (*i.e.* $H_L/L$) is called the *hydraulic gradient*. Further, Bligh makes no distinction between horizontal and vertical creep.

Consider a section as shown in Fig. 11.1. Let $H_L$ be the difference of water levels between upstream and downstream ends. (No water is shown on d/s side in Fig. 11.1). Water will seep along the bottom contour as shown by arrows. It starts percolating at $A$ and emerges at $B$. The total length of creep is given by

$$L = d_1 + d_1 + L_1 + d_2 + d_2 + L_2 + d_3 + d_3$$
$$= (L_1 + L_2) + 2\,[d_1 + d_2 + d_3]$$
$$= b + 2\,(d_1 + d_2 + d_3)$$

Head loss per unit length, or **hydraulic gradient**

$$= \left[\frac{H_L}{b + 2.\,(d_1 + d_2 + d_3)}\right] = \frac{H_L}{L}$$

Head losses equal to $\left(\frac{H_L}{L} \times 2d_1\right)$, $\left(\frac{H_L}{L} \times 2d_2\right)$ and $\left(\frac{H_L}{L} \times 2d_3\right)$; will occur respectively, in the planes of three vertical cut offs. The hydraulic gradient line (H.G. Line) can then be drawn as shown in Fig. 11.1.

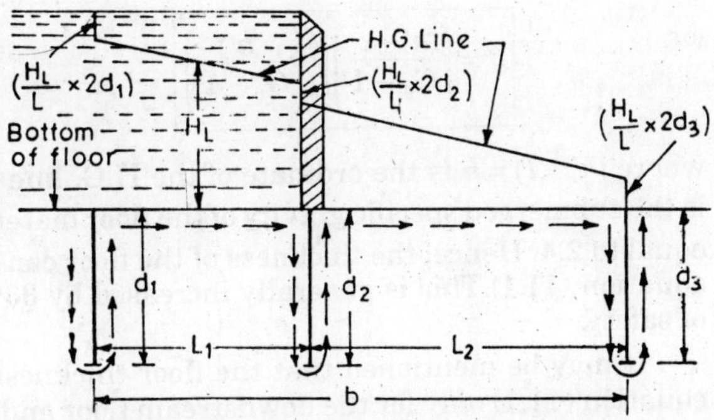


Fig. 11.1. Bligh's Creep.

(*i*) **Safety Against Piping or Undermining**. According to Bligh, the safety against piping can be ensured by providing sufficient creep length, given by $L = C.H_L$, where $C$ is Bligh's coefficient for the soil. Different values of $C$ for different types of soils are tabulated in Table 11.1.

**Table 11.1. Values of Bligh's Safe Hydraulic Gradient for Different Types of Soils**

| *S.No.* | *Type of soil* | *Value of C* | *Safe Hydraulic gradient should be less than* |
|---|---|---|---|
| 1. | Fine micaceous sand (as in North Indian Rivers) | 15 | 1/15 |
| 2. | Coarse grained sand (as in Central and South Indian Rivers) | 12 | 1/12 |
| 3. | Sand mixed with boulder and gravel, and for loam soil | 5 to 9 | 1/5 to 1/9 |
| 4. | Light sand and mud | 8 | 1/8 |

*Note* : The hydraulic gradient, *i.e.* $H_L/L$ is then equal to $1/C$. Hence, it may be stated that the hydraulic gradient must be kept under a safe limit in order to ensure safety against piping.

(*ii*) **Safety against uplift pressure.** The ordinates of the H.G. line above the bottom of the floor represent the residual uplift water head at each point. Say for example, if at any point, the ordinate of H.G. line above the bottom of the floor is 1 m, then 1 m head of water will act as uplift at that point. If $h'$metres is this ordinate, then water pressure equal to $h'$ metres will act at this point, and has to be counterbalanced by the weight of the floor of thickness say $t$.

∴ Uplift pressure caused by the seeping water $= \gamma_w\, h'$

where $\gamma_w$ is the unit wt. of water

Downward pressure due to weight of the concrete floor of thickness $t$

$$= (\gamma_w \cdot G_c) \, t$$

where $G_c$ is the specific gravity of the floor material (i.e. concrete).

For equilibrium

$$\gamma_w \cdot h' = \gamma_w \cdot G_c \cdot t$$

$$h' = G_c \, t$$

Subtracting $t$ on both sides, we get

$$(h' - t) = (G_c . t - t) = t \, (G_c - 1)$$

or

$$\boxed{t = \left(\frac{h' - t}{G_c - 1}\right) = \left(\frac{h}{G_c - 1}\right)} \quad ...(11.1)$$

where $(h' - t) = h$ is the ordinate of the H.G. line above the top of the floor. $(G_c - 1)$ is the submerged specific gravity of the floor material. For concrete, $G_c$ may be taken equal to 2.4. Hence, the thickness of the floor can be easily determined by using the equation (11.1).This is generally increased by 33%, so as to allow a suitable factor of safety.

It may be mentioned that the floor thickness has to be designed according to equation (11.1) only for the downstream floor and for the worst conditions *i.e.* when maximum ordinates of H.G. line occur. The water standing on the upstream floor, more than counterbalances the uplift caused by the same water, and hence, *only a nominal floor thickness is required on the upstream side*, so as to resist wear, impact of flowing water, etc.

Hence, while designing aprons of hydraulic structures on Bligh's theory for sub-surface flow, the floor thickness is designed in accordance with the above rules, and sufficient length of pucca floor given by Equation $L = C.H_L$ is provided, so as to ensure a safe value of hydraulic gradient. This will be discussed in details in article 11.5.

## 11.3. Lane's Weighted Creep Theory

Bligh, in his theory, had calculated the length of the creep, by simply adding the horizontal creep length and the vertical creep length, thereby making no distinction between the two creeps. However, Lane, on the basis of his analysis carried out on about 200 dams all over the world, stipulated that the horizontal creep is less effective in reducing uplift (or in causing loss of head) than the vertical creep. He, therefore, suggested a weightage factor of $\frac{1}{3}$ for the horizontal creep, as against 1.0 for the vertical creep.

Thus, in Fig. 11.1, the total **Lane's creep length** $(L_l)$ is given as :

$$L_l = (d_1 + d_1) + \frac{1}{3} L_1 + (d_2 + d_2) + \frac{1}{3} L_2 + (d_3 + d_3)$$

$$= \frac{1}{3} \cdot (L_1 + L_2) + 2\,(d_1 + d_2 + d_3) = \frac{1}{3} \cdot b + 2\,(d_1 + d_2 + d_3)$$

To ensure safety against piping, according to this theory, the creep length $L_l$ must not be less than $C_1 H_L$, where $H_L$ is the head causing flow, and $C_1$ is Lane's creep coefficient given in table 11.2.

**Table 11.2. Values of Lane's Safe Hydraulic Gradient for Different Types of Soils**

| S.No. | Type of soil | Value of Lane's Coefficient, $C_1$ | Safe Lane's Hydraulic gradient should be less than |
|---|---|---|---|
| 1. | Very find sand or silt | 8.5 | 1/8.5 |
| 2. | Fine sand | 7.0 | 1/7 |
| 3. | Coarse sand | 5.0 | 1/5 |
| 4. | Gravel and sand | 3.5 to 3.0 | 1/3.5 to 1/3 |
| 5. | Boulders, gravels and sand | 2.5 to 3.0 | 1/2.5 to 1/3 |
| 6. | Clayey soils | 3.0 to 1.6 | 1/3 to 1/1.6 |

Lane's theory was an improvement over Bligh's theory, but however, was purely empirical without any rational basis, and hence, is generally not adopted in any designs. Bligh's theory, though is still used (even after the invention of modern Khosla's theory), but Lane's theory is practically nowhere used, and is having only a theoretical importance.

## 11.4. Khosla's Theory and Concept of Flow Nets

Many of the important hydraulic structures, such as weirs and barrages, were designed on the basis of Bligh's theory between the period 1910 to 1925. In 1926-27, the upper Chenab canal syphons, designed on Bligh's theory, started posing undermining troubles. Investigations started, which ultimately lead to Khosla's theory.

A detailed description of this theory is available in C.B.I. publication No. 12, which is available at Publication's Division at Civil Lines, Delhi. The main principles of this theory are summarised below :

(1) The seeping water does not creep along the bottom contour of pucca floor as stated by Bligh, but on the other hand, this water moves along a set of streamlines as shown in Fig. 11.3. This steady seepage in a vertical plane for a homogeneous soil can be expressed by Laplacian equation.

$$\frac{d^2\phi}{dx^2} + \frac{d^2\phi}{dz^2} = 0$$

where $\phi$ = Flow potential = $Kh$ ; where $K$ is the Coefficient of permeability of soil as defined by Darcy's law, and $h$ is the residual head at any point within the soil.

The above equation represents two sets of curves intersecting each other orthogonally (Fig. 11.2). One set of lines is called the *Streamlines*, and the other set is called the *Equipotential lines.* The resultant flow diagram showing both the sets of curves is called a *Flow Net.*

**11.4.1. Stream Lines.** The streamlines represent the paths along which the water flows through the sub-soil. Every particle entering the soil at a given point upstream of the structure, will trace out its own path and will represent a streamline. The first streamline follows the bottom contour of the structure, and is the same as Bligh's path of creep. The remaining streamlines follow smooth curves, transiting slowly from the outline of the foundation to a semi-ellipse, as shown in Fig. 11.2.

**11.4.2. Equipotential Lines.** (1) Treating the downstream bed as datum and assuming no water on the downstream side, it can be easily stated that every streamline possesses a head equal to $H_1$ while entering the soil (Fig. 11.2); and when it emerges at the down-stream end into the atmosphere, its head is zero. Thus, the head $H_1$ is entirely lost during the passage of water along the streamline.

Further, at every intermediate point in its path, there is certain residual head ($h$) still to be dissipated in the remaining length to be traversed to the downstream end. This fact is applicable to every streamline, and hence, there will be points on different streamlines having the same value of residual head $h$. If such points are joined together, the curve obtained is called an *equipotential line.*

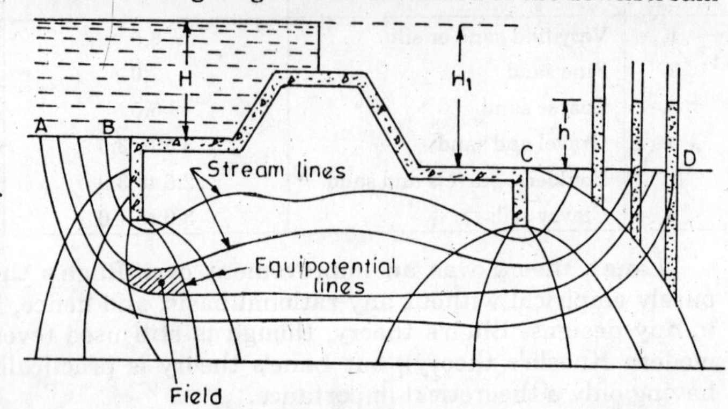


Fig. 11.2. Khosla's Flow Net.

Every water particle on line $AB$ is having a residual head $h = H_1$ and on $CD$ is having a residual head $h = 0$, and hence, $AB$ and $CD$ are equipotential lines.

Since an equipotential line represents the joining of points of equal residual head, hence if piezometers were installed on an equipotential line, the water will rise in all of them up to the same level, as shown in Fig. 11.2.

(2) The seepage water exerts a force at each point in the direction of flow and tangential to the streamlines, as shown in Fig. 11.3. This force ($F$) has an upward component from the point where the streamline turns upward. For soil grains to remain stable, the upward component of this force should be counterbalanced by the submerged weight of the soil grain. This force has the maximum disturbing tendency at the exit end, because the direction of this force at the exit point is vertically upward, and hence full force acts as its upward component. For the soil grain to remain stable, the submerged weight of soil grain should be more than this upward disturbing force. The disturbing force at any point is proportional to the gradient of pressure of water at that point (*i.e.* $dp/dl$). This gradient of pressure of water at the exit end, is called the **exit gradient.** In order that the soil particles at exit remain stable, the upward pressure at exit should be safe. In other words, the exit gradient should be safe.

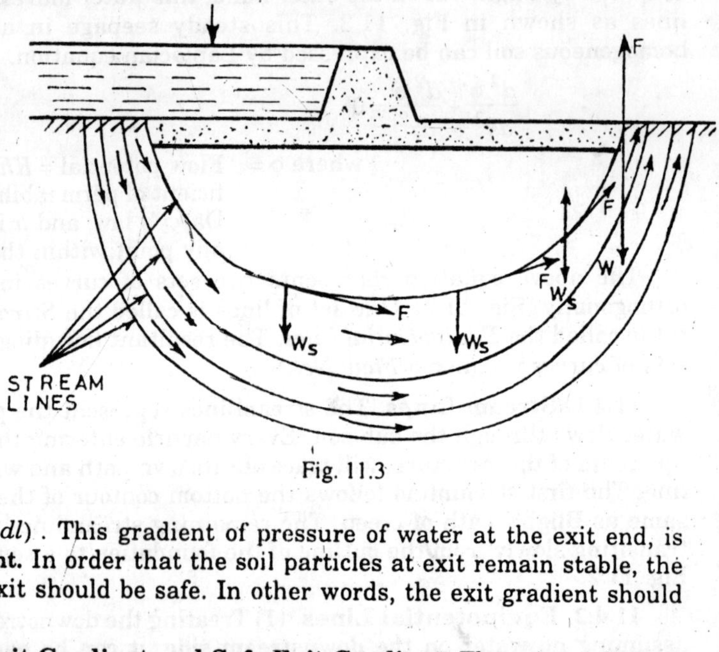


Fig. 11.3

**11.4.3. Critical Exit Gradient and Safe Exit Gradient.** The exit gradient is said to be *critical*, when the upward disturbing force on the grain is just equal to the submerged weight of the grain at the exit. When a factor of safety equal to 4 or 5 is used, the exit gradient can then be taken as the *safe exit gradient*. In other words, an exit gradient equal to $\frac{1}{4}$ to $\frac{1}{5}$ of the critical exit gradient is ensured, so as to keep the structure safe against piping.

Now, the submerged weight ($W_s$) of a unit volume of soil is given as :

$$W_s = \gamma_w (1 - n)(G - 1)$$

where $\gamma_w$ = unit weight of water.
$G$ = sp. gravity of soil particles
$n$ = porosity of the soil material.

For critical conditions to occur at the exit point,

$$F = W_s.$$

where $F$ is the upward disturbing force on the grain

$$\text{Force } F = \text{pressure gradient at that point} = \frac{dp}{dl} = \gamma_w \cdot \frac{dh}{dl},$$

where $\frac{dh}{dl}$ represents the rate of loss of head or the gradient at the exit end.

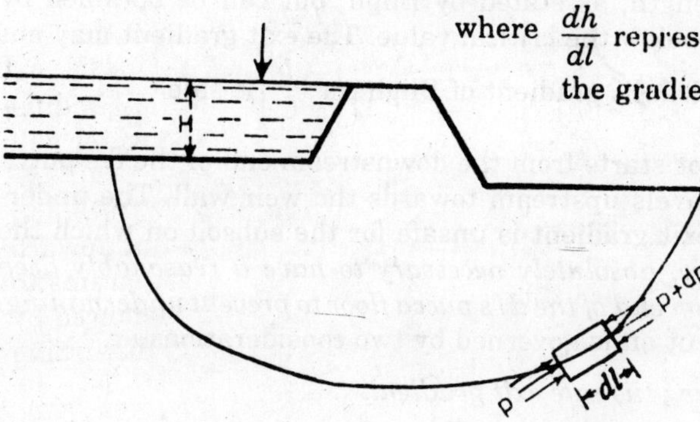


$p = \gamma_w \cdot h$

$p + dp = \gamma_w (h + dh)$

where, $h$ = Residual head still to be dissipated, called the hydrostatic excess head, at the considrad point.

Fig. 11.4

$$\therefore \quad \gamma_w \cdot \frac{dh}{dl} = \gamma_w (1 - n)(G - 1)$$

$$\text{or} \quad \frac{dh}{dl} = (1 - n)(G - 1) \quad \text{...(11.2)}$$

Under critical conditions, the *critical exit gradient* is equal to $(1 - n)(G - 1)$. For most of the river sands, $G \cong 2.65$ and $n \cong 0.4$, then the value of critical exit gradient

$$= (1 - 0.4)(2.65 - 1) = 0.6 \times 1.65 = 0.99 \cong 1.0$$

*Hence, an exit gradient equal to $\frac{1}{4}$ to $\frac{1}{5}$ of the critical gradient means that an exit gradient equal to $\frac{1}{4}$ to $\frac{1}{5}$ has to be provided for keeping the structure safe against piping.*

Values of safe exit gradient for some of the subsoils are given in Table 11.3.

**Table 11.3. Values of Khosla's Safe Exit Gradient for Different Types of Soils**

| *Type of soil* | *Khosla's Safe Exit Gradient* |
|---|---|
| Shingle | $\frac{1}{4}$ to $\frac{1}{5}$ (0.25 to 0.20) |
| Course sand | $\frac{1}{5}$ to $\frac{1}{6}$ (0.20 to 0.17) |
| Fine sand | $\frac{1}{6}$ to $\frac{1}{7}$ (0.17 to 0.14) |

Khosla's theory of flow nets made it very clear that the loss of head does not take place uniformly, in direct proportion to the creep length, as stated by Bligh. In fact, it depends upon the whole geometry of the figure, *i.e.* the shape of foundation, depth of impervious boundary and levels of u/s and d/s beds. When the *equipotential lines are traced to be closer, the rate of loss of head will definitely be quicker and vice versa.*

It can, hence be concluded that the safety against piping can not be obtained by providing sufficient floor length, as stated by Bligh, but can be obtained by keeping the exit gradient well below the critical value. The exit gradient may not be safe even if the average hydraulic gradient of Bligh $\left(i.e.\ \frac{l}{C}\right)$ is safe.

(3) Undermining of the floor starts from the downstream end of the d/s pucca floor, and if not checked, it travels upstream towards the weir wall. The undermining starts only when the exit gradient is unsafe for the subsoil on which the weir is founded. *It is, therefore, absolutely necessary to have a reasonably deep vertical cut-off at the downstream end of the d / s pucca floor to prevent undermining.* The depth of this d/s vertical cut off is governed by two considerations *i.e.*

(*i*) *maximum depth of scour* ; (*ii*) *safe exit gradient.*

While designing a weir, downstream cutoff from the maximum scoured depth considerations is, first of all, provided, and checked for exit gradient. If a safe value of exit gradient is not obtained, then the depth of cutoff is increased. The depth of cutoff is also governed and limited by practical considerations, as the construction of a very deep cutoff may be difficult or unpracticable at site.

A weir or a barrage may fail not only due to seepage (*i.e.* sub-surface flow) as stated by Bligh, but may also fail due to the surface flow. The surface flow (*i.e.* when flood water flows over the weir crest) may cause scour, dynamic action ; and in addition, will cause uplift pressures in the jump trough* (if the hydraulic jump forms on the downstream). These uplift pressures must be investigated for various flow conditions. The maximum uplift due to this dynamic action (*i.e.* for surface flow) should then be compared with the maximum uplift under steady seepage (*i.e.* for sub-surface flow) ; and the maximum of the two chosen for designing the aprons and the floors of the weirs. All these modern aspects and other details about designing weirs on permeable foundations, as per the Khosla's theory, have been discussed in article 11.6.

Khosla's theory differs from Bligh's theory in all the above respects, but owing to the simplicity, Bligh's theory is still used for design of small works. A minimum practical thickness for the floor and a deep vertical cutoff at the downstream end

* For detailed description, please refer to Article 11.6.1.1.

is, however, always provided, in addition to the requirements of Bligh's theory. However, on major works, Bligh's theory should never be used, as it may lead to expensive and unsafe erroneous designs.

**11.4.4. Khosla's Method of Independent Variables for Determination of Pressures and Exit Gradient for Seepage Below a Weir or a Barrage.** In order to know as to how the seepage below the foundation of a hydraulic structure is taking place, it is necessary to plot the flownet. In other words, we must solve the Laplacian equations. This can be accomplished either by mathematical solution of the Laplacian equations, or by Electrical analogy method, or by Graphical sketching by adjusting the streamlines and equipotential lines w.r.t. the boundary conditions. These are complicated methods and are time consuming. Therefore, for designing hydraulic structures such as weirs or barrages on pervious foundations, Khosla has evolved a simple, quick and an accurate approach, called the *Method of Independent Variables.*

In this **method of independent variables,** a complex profile like that of a weir is broken into a number of simple profiles, each of which can be solved mathematically. Mathematical solutions of flownets for these simple standard profiles have been presented in the form of equations given in Fig. 11.5, and curves given in **Plate 11.1,** which can be used for determining the percentage pressures at the various key points. The simple profiles which are most useful are :

(*i*) A straight horizontal floor of negligible thickness with a sheet pile line on the u/s end or at the d/s end [Fig. 11.5 (*a*) and (*b*)].

(*ii*) A straight horizontal floor depressed below the bed but without any vertical cut-offs [Fig. 11.5 (*c*)].

(*iii*) A straight horizontal floor of negligible thickness with a sheet pile line at some intermediate point [Fig. 11.5 (*d*)].

The key points are the junctions of the floor and the pile lines on either side, and the bottom point of the pile line, and the bottom corners in the case of a depressed floor. The percentage pressures at these key points for the simple forms into which the complex profile has been broken is valid for the complex profile itself, if corrected for

(*a*) *correction for the mutual interference of piles ;*

(*b*) *correction for thickness of floor ;*

(*c*) *correction for the slope of the floor.*

These corrections are described below :

(*a*) **Correction for the mutual interference of piles.** The correction $C$ to be applied as percentage of head due to this effect, is given by

$$C = 19\sqrt{\frac{D}{b'}}\left[\frac{d+D}{b}\right] \qquad ...(11.3)$$

where $b'$ = The distance between two pile lines.

$D$ = The depth of the pile line, the influence of which has to be determined on the neighbouring pile of depth $d$. $D$ is to be measured *below the level at which interference is desired.*

$d$ = The depth of the pile on which the effect is considered.

$b$ = Total floor length.

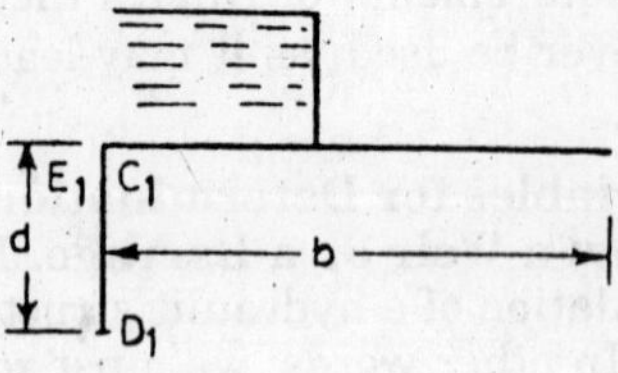


$$\phi_{C_1} = 100 - \phi_E \quad ...(11.4)$$

$$\phi_{D_1} = 100 - \phi_D \quad ...(11.5)$$

(a)

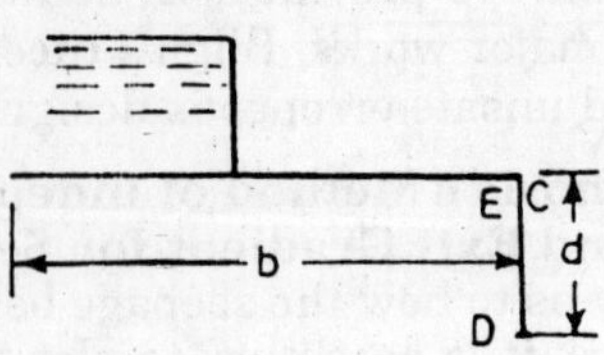


$$\phi_E = \frac{1}{\pi} \cos^{-1}\left(\frac{\lambda - 2}{\lambda}\right) \quad ...(11.6)$$

$$\phi_D = \frac{1}{\pi} \cos^{-1}\left(\frac{\lambda - 1}{\lambda}\right) \quad ...(11.7)$$

$$\text{where } \lambda = \frac{1 + \sqrt{1 + \alpha^2}}{2} \quad ...(11.8)$$

$$\alpha = b/d$$

(for respective cases) ...(11.9)

(b)

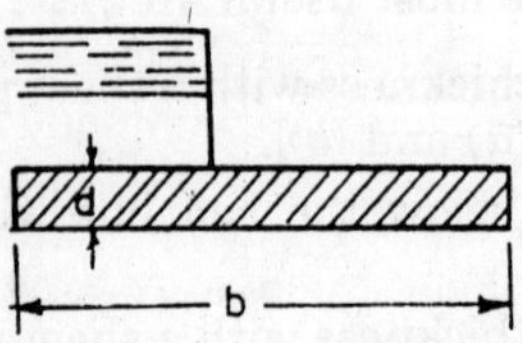


$$\phi'_D = \frac{2}{3}(\phi_E - \phi_D) + \frac{3}{\alpha^2} \quad ...(11.10)$$

$$\phi'_{D_1} = 100 - \phi'_D \quad ...(11.11)$$

(c)

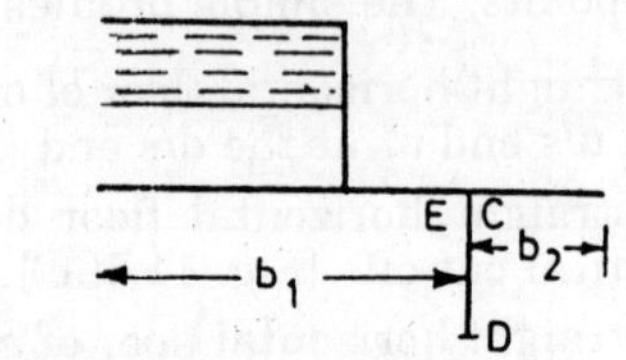


$$\phi_E = \frac{1}{\pi} \cos^{-1}\left(\frac{\lambda_1 - 1}{\lambda}\right) \quad ...(11.12)$$

$$\phi_D = \frac{1}{\pi} \cos^{-1}\left(\frac{\lambda_1}{\lambda}\right) \quad ...(11.13)$$

$$\phi_C = \frac{1}{\pi} \cos^{-1}\left(\frac{\lambda_1 + 1}{\lambda}\right) \quad ...(11.14)$$

$$\text{where } \lambda = \frac{\sqrt{1 + \alpha_1^2} + \sqrt{1 + \alpha_2^2}}{2} \quad ...(11.15)$$

$$\lambda_1 = \frac{\sqrt{1 + \alpha_1^2} - \sqrt{1 + \alpha_2^2}}{2} \quad ...(11.16)$$

$$\alpha_1 = b_1/d \quad ...(11.17)$$

$$\text{and } \alpha_2 = b_2/d \quad ...(11.18)$$

(d)

Fig. 11.5. Khosla's simple profiles for a weir of complex profile.

This correction *is positive for the points in the rear or back water; and substractive for the points forward in the direction of flow. This equation does not apply to the effect of an outer pile on an intermediate pile, if the intermediate pile is equal to or smaller than the outer pile and is at a distance less than twice the length of the outer pile.*

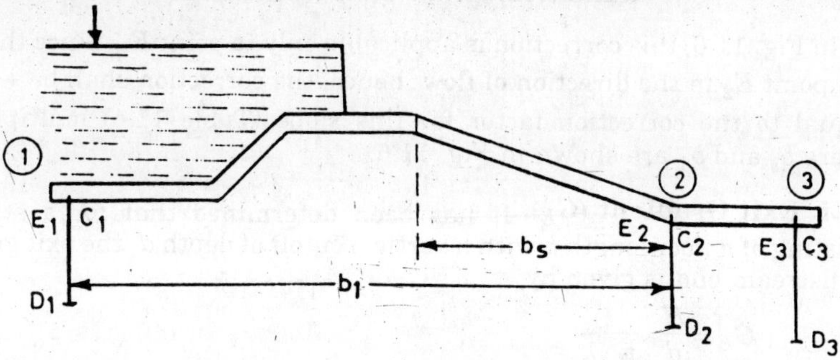


Fig. 11.6

Suppose in the above Fig. 11.6, we are considering the influence of the pile No. (2) on pile No. (1) for correcting the pressure at $C_1$. Since the point $C_1$ is in the rear, this correction shall be + ve. While the correction to be applied to $E_2$ due to pile No. (1) shall be negative, since the point $E_2$ is in the forward direction of flow. Similarly, the correction at $C_2$ due to pile No. (3) is positive, and the correction at $E_2$ due to pile No. (2) is negative.

(*b*) **Correction for the thickness of floor.** In the standard form profiles, the floor is assumed to have negligible thickness. Hence, the percentage pressures calculated by Khosla's equations or graphs shall pertain to the top levels of the floor. While the actual junction points $E$ and $C$ are at the bottom of the floor. Hence, the pressure at the actual points are calculated by assuming a straight line pressure variation. Since the corrected pressure at $E_1$ should be less than the calculated pressure at $E_1'$, the correction to be applied for the point $E_1$ shall be – ve. Similarly, the pressure calculated $C_1'$ is less than the corrected pressure at $C_1$, and hence, the correction to be applied at point $C_1$ is + ve.

Fig. 11.7

(*c*) **Correction for the slope of the floor**. A correction is applied for a sloping floor, and is taken as + ve for the down, and – ve for the *up* slopes following the direction of flow. Values of correction of standard slopes such as, 1 : 1, 2 : 1, 3 : 1, etc. are tabulated in Table 11.4.

**Table 11.4**

| *Slope Horizontal : Vertical* | *Correction factor* |
|---|---|
| 1 : 1 | 11.2 |
| 2 : 1 | 6.5 |
| 3 : 1 | 4.5 |
| 4 : 1 | 3.3 |
| 5 : 1 | 2.8 |
| 6 : 1 | 2.5 |
| 7 : 1 | 2.3 |
| 8 : 1 | 2.0 |

*The correction factor given above is to be multiplied by the horizontal length of the slope and divided by the distance between the two pile lines between which the sloping floor is located. This correction is applicable only to the key points of the pile line fixed at the start or the end of the slope.*

Thus, in Fig. 11.6, this correction is applicable only to point $E_2$. Since the slope is down at point $E_2$ in the direction of flow, hence, the correction shall be + ve and will be equal to the correction factor for this slope (Table 11.4) multiplied by $b_s/b_1$, where $b_s$ and $b_1$ are shown in Fig. 11.6.

**11.4.4.1. Exit Gradient ($G_E$).** It has been determined that for a standard form consisting of a floor length $b$ with a vertical cutoff of depth $d$, the exit gradient at its downstream end is given by

$$G_E = \frac{H}{d} \cdot \frac{1}{\pi\sqrt{\lambda}} \quad \text{...(11.19)}$$

$$\text{where } \lambda = \frac{1 + \sqrt{1 + \alpha^2}}{2} \text{ ; and } \alpha = \frac{b}{d}.$$

From the curve of **Plate 11.2** ; for any value of α, *i.e.* $\frac{b}{d}$, the corresponding value of $\frac{1}{\pi\sqrt{\lambda}}$ can be read. Knowing $H$ and $d$, the value of $G_E$ can be easily calculated. The exit gradient so calculated must lie within the safe limits, shown in Table 11.3.

The uplift pressures must be kept as low as possible consistent with the safety at the exit, so as to keep the floor thickness to the minimum.

It is obvious from equation (11.19), that if $d = 0$ ; $G_E$ is infinite. *Hence, it becomes absolutely essential to provide a vertical cutoff at the downstream end.*

**Example 11.1.** *Determine the percentage pressures at various key points in Fig. 11.8. Also determine the exit gradient and plot the hydraulic gradient line for pond level on u/s and no flow on d/s.*

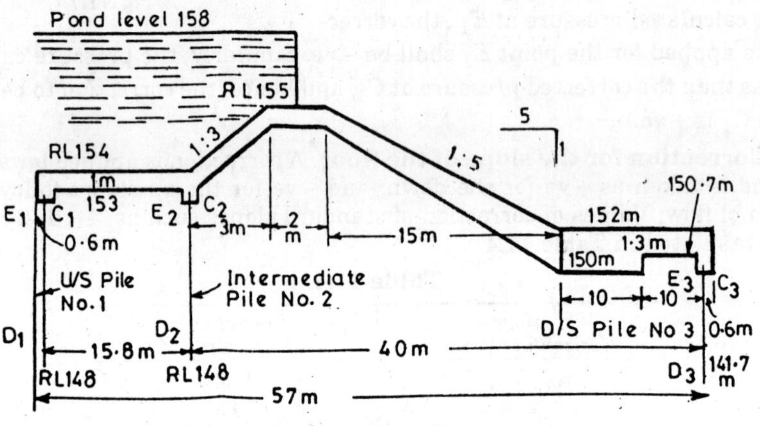


Fig. 11.8

**Solution.**

**(1) For Upstream Pile Line No. (1)**

**Total length of the floor = $b$ = 57.0 m.**

**Depth of u/s pile line = $d$ = 154.00 – 148.00 = 6.0 m**

$$\alpha = \frac{b}{d} = \frac{57.0}{6.0} = 9.5$$

$$\frac{1}{\alpha} = \frac{1}{9.5} = 0.105$$

From curve Plate 11.1 (*a*)

$$\phi_{C_1} = 100 - 29 = 71\%$$

$$\phi_{D_1} = 100 - 20 = 80\%$$

These values of $\phi_{C_1}$ and $\phi_{D_1}$ must be corrected for three corrections as below :

**Corrections for $\phi_{C_1}$**

(*a*) *Correction at $C_1$ for Mutual Interference of Piles.* $\phi_{C_1}$ is affected by intermediate pile No. 2.

$$\text{Correction} = 19\sqrt{\frac{D}{b'}}\left(\frac{d+D}{b}\right) \quad \text{... Eq. (11.3)}$$

where $D$ = Depth of pile No. 2.
= 153.00 – 148.00 = 5.0 m
$d$ = Depth of pile No. 1
= 153.0 – 148.00 = 5.0 m.
$b'$ = Distance between two piles = 15.8 m.
$b$ = Total floor length = 57.0 m.

$$\text{Correction} = 19\sqrt{\frac{5}{15.7}}\left[\frac{5+5}{57}\right] = 1.88\%$$

Since the point $C_1$ is in the rear in the direction of flow from pile No. (2), the correction is + ve.

∴ Correction due to pile interference on

$C_1$ = 1.88% (+ ve) ...(*i*)

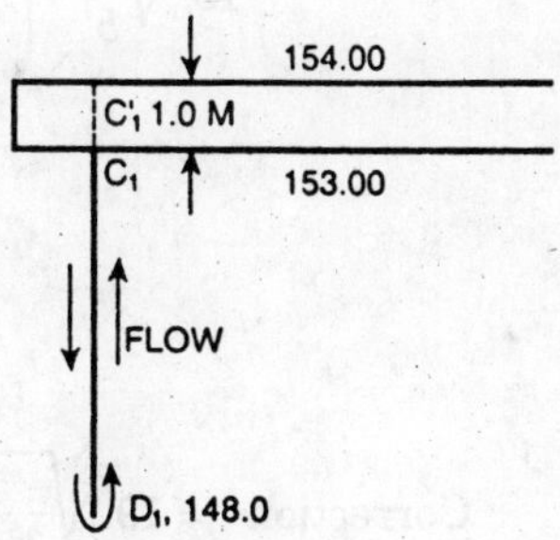

Fig. 11.9

(*b*) *Correction at $C_1$ due to thickness of floor.*

Pressure calculated from curve is at $C_1'$, (Fig. 11.9) but we want the pressure at $C_1$. Pressure at $C_1$ shall be more than at $C_1'$ as the direction of flow is from $C_1$ to $C_1'$ as shown; and hence, the correction will be + ve and

$$= \left[\frac{80\% - 71\%}{154.0 - 148.0}\right] \times (154.00 - 153.00) = \frac{9}{6} \times 1 = 1.5\% \text{ (+ ve)} \quad ...(ii)$$

(*c*) *Correction due to slope at $C_1$ is nil, as this point is neither situated* at the start nor at the end of a slope.

∴ Corrected $\phi_{C_1} = 71\% + 1.88\% + 1.5\% = 74.38\%$

**Hence corrected** $\phi_{C_1}$ **= 74.38% Ans.**

and $\phi_{D_1} = 80\%$

**(2) For Intermediate Pile Line No. (2)**

$d$ = 154.00 – 148.00 = 6.0 m

$b$ = 57.0 m

$$\alpha = \frac{b}{d} = \frac{57.0}{6.0} = 9.5$$

Using curves of Plate 11.1 (*b*), we have $b_1$ in this case

$$= 0.6 + 15.8 = 16.4 \text{ m}$$

and $b = 57.0$ m

$$\therefore \quad \frac{b_1}{b} = \frac{16.4}{57.0} = 0.298.$$

$$1 - \frac{b_1}{b} = 1 - 0.298 = 0.702.$$

$$\phi_{E_2} = 100 - 30\% = 70\%$$

(where 30% is $\phi_C$ for a base ratio of 0.702 and $\alpha = 9.5$)

$$\phi_{C_2} = 56\% \text{ (for a base ratio of 0.298 and } \alpha = 9.5)$$

$$\phi_{D_2} = 100 - 37 = 63\%$$

(where 37% is $\phi_D$ for a base ratio of 0.702 and $\alpha = 9.5$)

**Corrections for $\phi_{E_2}$**

(*a*) *Correction at $E_2$ for sheet pile lines.* Pile No. (1) will affect the pressure at $E_2$, and since $E_2$ is in the forward direction of flow from pile No. (1), this correction shall be – ve. The amount of this correction is given as :

$$19\sqrt{\frac{D}{b'}} \cdot \left[\frac{d+D}{b}\right] \qquad \text{i.e. Eq. (11.3)}$$

where $D$ = Depth of pile No. 1, the effect of which is considered = 153.0 – 148.0 = 5.0 m

$d$ = depth of pile No. 2, the effect on which is considered = 153.0 – 148.0 = 5.0 m.

$b'$ = Distance between the two piles = 15.8 m.

$b$ = Total floor length = 57.0 m.

$$\text{Correction} = 19\sqrt{\frac{5}{15.8}}\left(\frac{5+5}{57}\right) = 1.88\% \text{ (– ve)}$$

(*b*) *Correction at $E_2$ due to floor thickness*

$$= \frac{\text{Obs. } \phi_{E_2} - \text{Obs. } \phi_{D_2}}{\text{Distance between } E_2D_2} \times \text{Thickness of floor}$$

$$= \left(\frac{70\% - 63\%}{154.0 - 148.0}\right) \times 1.0 = \frac{7}{6} \times 1.0 = 1.17\%.$$

Since the pressure observed is at $E_2'$ and not at $E_2$, (Fig. 11.10) and by looking at the direction of flow, it can be stated easily that the pressure at $E_2$ shall be less than that at $E_2'$, hence, this correction is *negative*.

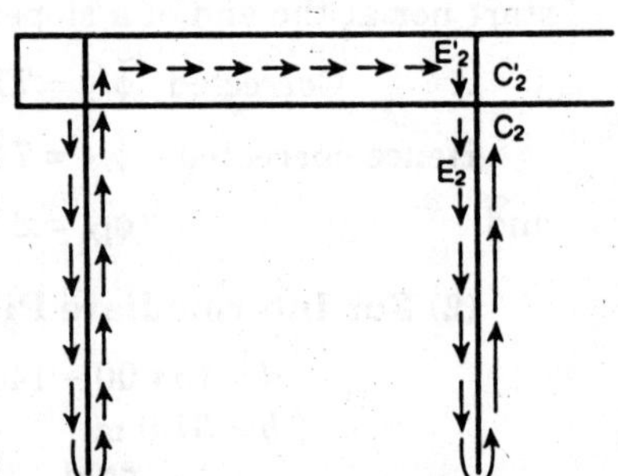

Fig. 11.10

∴ Correction at $E_2$ due to floor thickness

$$= 1.17\% \text{ (– ve)}.$$

(*c*) *Correction at $E_2$ due to slope is nil*, as the point $E_2$ is neither situated at the start of a slope nor at the end of a slope.

Hence, the corrected percentage pressure at $E_2$

$$= \textbf{corrected } \phi_{E_2} = 70\% - 1.88\% - 1.17\% = \textbf{66.95\% Ans.}$$

**Corrections for $\phi_{C_2}$**

(*a*) *Correction at $C_2$ due to pile interference.* Pressure at $C_2$ is affected by pile No. (3) and since the point $C_2$ is in the rear in the direction of flow from pile No. (3), this correction is + ve. The amount of this correction is given as :

$$= 19\sqrt{\frac{D}{b'}} \times \left(\frac{D+d}{b}\right) \qquad \textit{i.e. } \text{Eq. (11.3)}$$

where $D$ = Depth of pile No. (3), the effect of which is considered below the level at which interference is desired
= 153.0 – 141.7 = 11.3 m

$d$ = Depth of pile No. 2, the effect on which is considered
= 153.0 – 148.0 = 5.0 m.

$b'$ = Distance between pile 2 and pile 3 = 40.0 m

$b$ = Total floor length = 57.0 m.

$$\text{Correction} = 19\sqrt{\frac{11.0}{40.0}}\left(\frac{11.3+5.0}{57.0}\right) = \textbf{2.89\%} \text{ (+ ve)}$$

(*b*) *Correction at $C_2$ due to floor thickness.* From Fig. 11.10, it can be easily stated that the pressure at $C_2$ shall be more than that at $C_2'$, and since the observed pressure is at $C_2'$, this correction shall be + ve and its amount is the same as was calculated for the point $E_2$ = 1.17%.

Hence, correction at $C_2$ due to floor thickness = 1.17% (+ ve).

(*c*) *Correction at $C_2$ due to slope.* Since the point $C_2$ is situated at the start of a slope of 3 : 1, *i.e.* an up slope in the direction of flow ; the correction is *negative.*

Correction factor for 3 : 1 slope from Table 11.4 = 4.5

Horizontal length of the slope = 3 m.

Distance between two pile lines between which the sloping floor is located = 40.0 m.

∴ Actual correction = 4.5 × (3/40) = 0.34% (– ve)

Hence, **corrected $\phi_{C_2}$**

$$= 56\% + 2.89\% + 1.17\% - 0.34\% = \textbf{59.72\%}$$

(3) **For Downstream Pile Line No. (3)**

$$d = 152.0 - 141.7 = 9.3 \text{ m}$$

$$b = 57.0 \text{ m.}$$

$$\frac{1}{\alpha} = \frac{d}{b} = \frac{10.3}{57.0} = 0.181.$$

From curves of Plate 11.1 (*a*), we get

$$\phi_{D_3} = 32\% \text{ and } \phi_{E_3} = 38\%$$

**Corrections for $\phi_{E_3}$**

(a) *Correction due to piles.* The pressure at point $E_3$ is affected by pile No. 2, and since $E_3$ is in the forward direction of flow from pile No. 2, this correction is *negative*, and its amount is given as

$$= 19\sqrt{\frac{D}{b'}}\left(\frac{d+D}{b}\right) \quad \text{... Eq. (11.3)}$$

where $D$ = Depth of pile No. 2, the effect of which is considered
= 150.7 – 148.0 = 2.7 m.
$d$ = Depth of pile No. 3, the effect on which is considered
= 150.7 – 141.7 = 9.0 m.
$b'$ = Distance between piles = 40.0 m.
$b$ = Total floor length = 57.0 m.

$$\text{The correction} = 19\sqrt{\frac{2.7}{40}} \times \left(\frac{9+2.7}{57}\right) = 1.02\% \text{ (– ve)}$$

(b) *Correction due to floor thickness*

From Fig. 11.11, it can be stated easily that the pressure at $E_3$ shall be less than at $E_3'$, and since the pressure observed from curves is at $E_3'$ ; this correction shall be – ve and its amount

$$= \frac{38\% - 32\%}{152.0 - 141.7} \times 1.3 = \frac{16}{10.3} \times 1.3 = 0.76\% \text{ (– ve)}$$

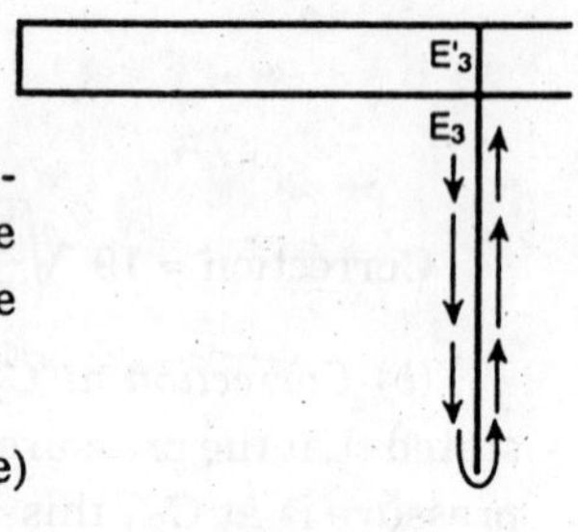

Fig. 11.11

(c) *Correction due to slope at $E_3$* is nil, as the point $E_3$ is neither situated at the start nor at the end of any slope.

Hence, **corrected $\phi_{E_3}$ = 38% – 1.02 – 0.76% = 36.22% Ans.**

The corrected pressures at various key points are tabulated below in Table 11.5.

**Table 11.5**

| *Upstream Pile No. 1* | *Intermediate Pile No. 2* | *Downstream Pile No. 3* |
|---|---|---|
| $\phi_{E_1}$ = 100% | $\phi_{E_2}$ = 66.95% | $\phi_{E_3}$ = 36.22% |
| $\phi_{D_1}$ = 80.0% | $\phi_{D_2}$ = 63.0% | $\phi_{D_3}$ = 32.0% |
| $\phi_{C_1}$ = 74.38% | $\phi_{C_2}$ = 59.72% | $\phi_{C_3}$ = 0% |

**Exit Gradient**

Let the water be headed up to pond level, *i.e.* on *RL* 158.0 m on the upstream side with no flow downstream.

The maximum seepage head = $H$ = 158.0 – 152.0 = 6.0 m
The depth of d/s cut-off = $d$ = 152.0 – 141.7 = 10.3 m
Total floor length = $b$ = 57.0 m.

$$\alpha = \frac{b}{d} = \frac{57.0}{10.3} = 5.53$$

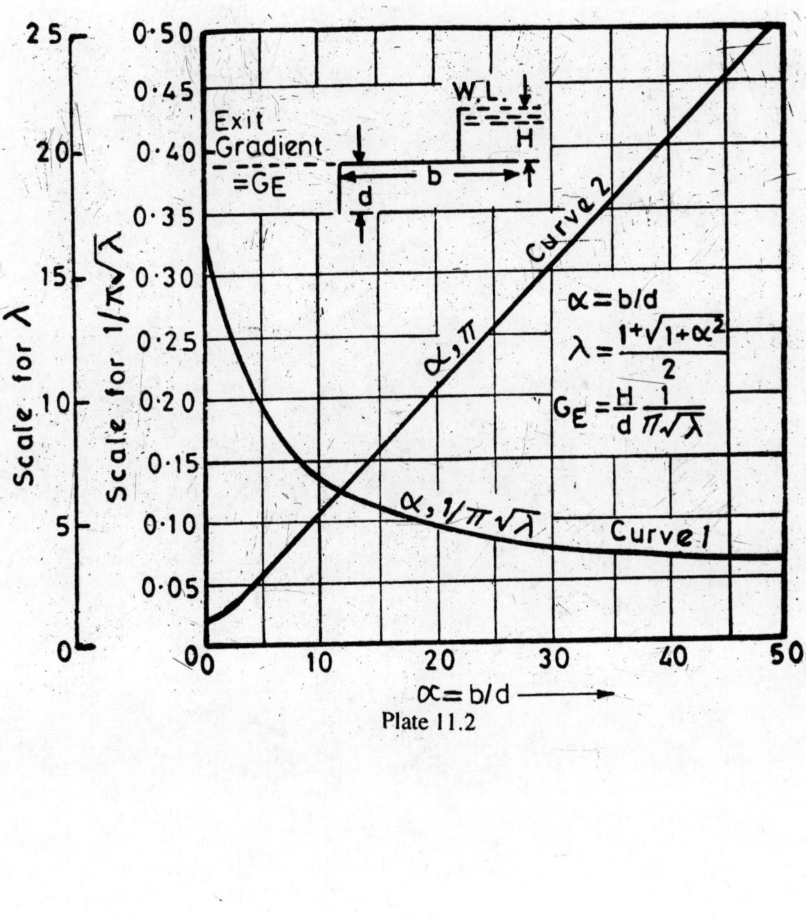

Scale for λ
Scale for 1/π√λ
Exit Gradient =GE
W.L.
H
b
d
Curve 2
α, π
α=b/d
λ = (1+√(1+α²))/2
GE = H/d 1/(π√λ)
α, 1/π√λ
Curve 1
α=b/d


Plate 11.2

For a value of $\alpha = 5.53$, $\dfrac{1}{\pi\sqrt{\lambda}}$ from curves of Plate 11.2 is equal to 0.18.

$$\text{Hence,} \quad G_E = \frac{H}{d} \cdot \frac{1}{\pi\sqrt{\lambda}} = \frac{6.0}{10.3} \times 0.18 = \frac{1}{9.53} = 0.105$$

Hence, the exit gradient shall be equal to 0.105, *i.e.* **1 in 9.53**, which is very much safe, as can be seen from the safe values given in table 11.3.

**Plotting the Hydraulic Gradient Line**

The percentage pressures, computed and tabulated in Table 11.5, can be used to work out the elevation of H.G. line above the datum, as given in Table 11.6.

**Table 11.6**

| *Flow condition* | *Upstream water level in metres* | *Downstream water level in metres* | *Head in metres* | *Height/Elevation of Sub-soil H.G. Line above Datum* | | | | | | | | |
|---|---|---|---|---|---|---|---|---|---|---|---|---|
| | | | | *Upstream Pile Line* | | | *Intermediate Pile Line* | | | *Downstream Pile Line* | | |
| | | | | $\phi_{E_1}$ 100% | $\phi_{D_1}$ 80% | $\phi_{C_1}$ 74.38% | $\phi_{E_2}$ 66.95% | $\phi_{D_2}$ 63.0% | $\phi_{C_2}$ 59.72% | $\phi_{E_3}$ 36.22% | $\phi_{D_3}$ 32.0% | $\phi_{C_3}$ 0% |
| Pond level u/s with no flow d/s | 158.00 | 152.00 | 6.0 | 6.0 | 4.8 | 4.46 | 4.02 | 3.78 | 3.58 | 2.68 | 1.92 | 0.0 |
| | | | | 158.0 | 156.8 | 156.46 | 156.02 | 155.78 | 155.58 | 154.17 | 153.92 | 152.0 |

The subsoil H.G. Line is then plotted in Fig. 11.12.

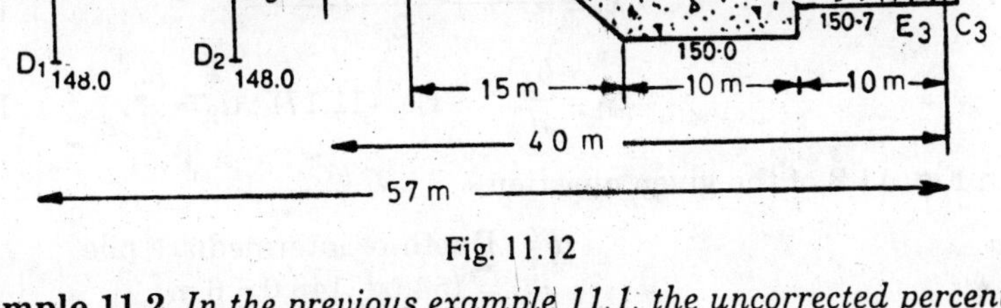


Fig. 11.12

**Example 11.2.** *In the previous example 11.1, the uncorrected percent residual pressures at* $C_1, D_1, E_2, D_2, C_2, E_3$ and $D_3$, *were all computed with the use of Khosla's charts. It is now desired to compare these pressures analytically by using the respective formulas, if the charts are not available.*

**Solution.**

(1) **For Upstream Pile Line No. 1.**

$$b = 57 \text{ m}$$

$$d = 154.00 - 148.00 = 6.0 \text{ m}$$

$$\alpha = \frac{b}{d} = \frac{57}{6} = 9.5.$$

Now, $\phi_E$ for such a case [Refer Fig. 11.5 (*a*)] is given by

$$\phi_E = \frac{1}{\pi}\cos^{-1}\left(\frac{\lambda - 2}{\lambda}\right) \qquad \text{... Eq. (11.6)}$$

$$\text{where } \lambda = \frac{1 + \sqrt{1 + \alpha^2}}{2} \qquad \text{... Eq. (11.8)}$$

$$\therefore \quad = \frac{1 + \sqrt{9.5^2}}{2} = 5.28$$

$$\therefore \quad \phi_E = \frac{1}{\pi}\cos^{-1}\left(\frac{5.28 - 2}{5.28}\right) = \frac{1}{\pi} \times 51.65° \times \frac{\pi}{180°} = 0.287 \text{ ; } i.e. \text{ } 28.7\%.$$

$$\therefore \quad \phi_{C_1} = 100 - \phi_E = 100 - 28.7 = 71.3\% \text{ **Ans.**}$$

(as against 71% read out earlier)

Similarly, from Fig. 11.5 (*a*)

$$\phi_{E_1} = 100 - \phi_D \quad \text{where } \phi_D = \frac{1}{\pi}\cos^{-1}\left(\frac{\lambda - 1}{\lambda}\right)$$

$$= \frac{1}{\pi}\cos^{-1}\left(\frac{5.28 - 1}{5.28}\right) = 0.199 \text{ ; say } 19.9\%$$

$$\therefore \quad \phi_{D_1} = 100 - 19.9 = 80.1\% \text{ **Ans.**}$$

(as against 80% readout earlier)

(2) **Intermediate Pile Line No. 2.** As given in Fig. 11.5 (*d*),

$$\phi_{E_2} = \phi_E = \frac{1}{\pi}\cos^{-1}\left(\frac{\lambda_1 - 1}{\lambda}\right) \qquad \text{... Eq. (11.12)}$$

$$\phi_{C_2} = \phi_C = \frac{1}{\pi}\cos^{-1}\left(\frac{\lambda_1 + 1}{\lambda}\right) \qquad \text{... Eq. (11.14)}$$

$$\phi_{D_2} = \phi_D = \frac{1}{\pi}\cos^{-1}\left(\frac{\lambda_1}{\lambda}\right) \qquad \text{... Eq. (11.13)}$$

$$\text{where } \lambda = \frac{\sqrt{1 + {\alpha_1}^2} + \sqrt{1 + {\alpha_2}^2}}{2} \qquad \text{... Eq. (11.15)}$$

$$\lambda_1 = \frac{\sqrt{1 + {\alpha_1}^2} - \sqrt{1 + {\alpha_2}^2}}{2} \qquad \text{... Eq. (11.16)}$$

$$\alpha_1 = \frac{b_1}{d} \quad \text{... Eq. (11.17) ; } \alpha_2 = \frac{b_2}{d}. \qquad \text{... Eq. (11.18)}$$

From Fig. 11.8 of the given question,

$d$ = Depth of intermediate pile
= 154.0 – 148.0 = 6 m

$\alpha_1$ = Floor length *U/S* of intermediate pile
= 16.4 m.

$\alpha_2$ = Floor length *D/S* of intermediate pile
= 40.6 m.

$$\therefore \quad \alpha_1 = \frac{16.4}{6} = 2.73$$

$$\alpha_2 = \frac{40.6}{6} = 6.77$$

$$\therefore \quad \lambda = \frac{\sqrt{1+2.73^2} + \sqrt{1+6.77^2}}{2} = \frac{2.907 + 6.843}{2} = 4.875$$

$$\lambda_1 = \frac{2.907 - 6.843}{2} = -1.968$$

Using, $\phi_{E_2} = \frac{1}{\pi} \cos^{-1}\left(\frac{\lambda_1 - 1}{\lambda}\right)$

$$\phi_{E_2} = \frac{1}{\pi} \cos^{-1}\left(\frac{-1.968 - 1}{4.875}\right) = 0.708 \text{ ; } i.e. \text{ } 70.8\%. \textbf{ Ans.}$$

(as against 70% read out earlier).

Now, using Eq. (11.13), we have

$$\phi_{D_2} = \phi_D = \frac{1}{\pi} \cos^{-1}\left(\frac{\lambda_1}{\lambda}\right)$$

$$= \frac{1}{\pi} \cos^{-1}\left(\frac{-1.968}{4.875}\right) = 0.632 = 63.2\% \textbf{ Ans.}$$

(as against 63% read out earlier)

Now, using Eq. (11.14), we have

$$\phi_{C_2} = \phi_C = \frac{1}{\pi} \cos^{-1}\left(\frac{\lambda_1 + 1}{\lambda}\right)$$

$$= \frac{1}{\pi} \cos^{-1}\left(\frac{-1.968 + 1}{4.875}\right) = 0.564 = 56.4\%.$$

(as against 56% read out earlier)

(3) **For Downstream Pile line No. 3.** W.R. to Fig. 11.5 (*b*)

$b = 57.0$ m

$d = 152.0 - 141.7 = 10.3$ m

$$\alpha = \frac{b}{d} = \frac{57}{10.3} = 5.53$$

$$\lambda = \frac{1 + \sqrt{1+\alpha^2}}{2} = \frac{1 + \sqrt{1 + 5.53^2}}{2} = 3.31$$

$$\therefore \quad \phi_{E_2} = \phi_E = \frac{1}{\pi} \cos^{-1}\left(\frac{\lambda - 2}{\lambda}\right) \qquad \text{... Eq. (11.6)}$$

$$= \frac{1}{\pi} \cos^{-1}\left(\frac{3.31 - 2}{3.31}\right) = \frac{1}{\pi} 66.65° \times \frac{\pi}{180°} = 0.37 \text{ ; } i.e. \text{ } 37\%. \textbf{ Ans.}$$

(as against 38% read out earlier)

$$\phi_{D_2} = \phi_D = \frac{1}{\pi} \cos^{-1}\left(\frac{\lambda - 1}{\lambda}\right) \qquad \text{... Eq. (11.7)}$$

$$= \frac{1}{\pi} \cos^{-1}\left(\frac{3.31 - 1}{3.31}\right) = 0.254 \text{ ; } i.e., \text{ } 25.4\%. \textbf{ Ans.}$$

(as against 26% read out earlier)

**Example 11.3**. *An impervious floor of a weir on permeable soil is 16 m long and has sheet piles at both the ends. The upstream pile is 4 m deep and the downstream pile is 5 m deep. The weir creates a net head of 2.5 m. Neglecting the thickness of the* *weir floor, calculate the uplift pressures at the junction of the inner faces of the pile with the weir floor, by using Khosla's theory.* (*Civil Services, 1990*)

**Solution.** In the given question, since Khosla's curves are not supplied, *it becomes necessary to remember and use* the formulas on which those Khosla's curves are based. These formulas are already mentioned in Fig. 11.5.

For the given question, pressures are required at inner junctions of both piles. Refer Fig. 11.5 (*a*) and (*b*). Pressures are thus required at $C_1$ for *U/S* pile, and at *E* at *D/S* pile line.

(*a*) **U/S Pile Line** ($\phi_{C_1} = ?$)

$$d = 4 \text{ m}$$
$$b = 16 \text{ m}$$
$$\alpha = \frac{b}{d} = \frac{16}{4} = 4$$
$$\phi_{C_1} = 100 - \phi_E \qquad \text{... Eq. (11.4)}$$

From Eq. (11.5) & (11.7), we have

$$\phi_E = \frac{1}{\pi}\cos^{-1}\left(\frac{\lambda - 2}{\lambda}\right) \quad \text{where } \lambda = \frac{1 + \sqrt{1 + \alpha^2}}{2} = \frac{1 + \sqrt{1 + 4^2}}{2} = 2.5616$$

$$\therefore \quad \phi_E = \frac{1}{\pi}\cos^{-1}\left(\frac{2.5616 - 2}{2.5616}\right) = \frac{1}{\pi} \times (77.34°)$$

$$= \frac{1}{\pi} \times (77.34°) \times \frac{\pi}{180°} = 0.4296 \text{ ; say } 0.43 \text{ } i.e. \text{ } 43\%.$$

$$\therefore \quad \phi_{C_1} = 100 - 43 = 57\%.$$

*Correction due to D/S pile*

$$C = +19 \cdot \left(\frac{D}{b'}\right)\left(\frac{d + D}{b}\right) \qquad \text{... Eq. (11.3)}$$

where $D$ = Depth of D/S influencing pile = 5 m
$d$ = Depth of U/S pile being influenced = 4 m
$b$ = $b'$ = 16 m

$$C = +19 \cdot \left(\frac{5}{16}\right)\left(\frac{4 + 5}{16}\right) = +6\%.$$

$$\therefore \phi_{C1 \text{ (corrected)}} = 57\% + 6\% = 63\%$$

$$\therefore \quad P_{C_1} = 63\% \times 2.5 \text{ m} = \mathbf{1.575 \text{ m}}. \textbf{ Ans.}$$

(*b*) **For D/S Pile Line** ($\phi_E = ?$)

$$d = 5 \text{ m}$$
$$b = 16 \text{ m}$$
$$\alpha = \frac{b}{d} = \frac{16}{5} = 3.2$$

From Eq. (11.8), we get

$$\lambda = \frac{1 + \sqrt{1 + \alpha^2}}{2} = \frac{1 + \sqrt{1 + 3.2^2}}{2} = 2.176$$

$$\phi_E = \frac{1}{\pi}\cos^{-1}\left(\frac{\lambda - 2}{\lambda}\right) \qquad \text{...Eq. (11.6)}$$

or

$$\phi_E = \frac{1}{\pi}\cos^{-1}\left(\frac{2.176 - 2}{2.176}\right)$$

$$= \frac{1}{\pi} \times 85.35° \times \frac{\pi}{180°} = 0.474 = 47.4\%.$$

*Correction due to U/S pile.*

$$C = (-)\ 19 \cdot \sqrt{\frac{4}{16}} \times \left(\frac{5+4}{16}\right) = -5.3\%$$

$$\phi_{E\ (corrected)} = 47.4 - 5.3 = 42.1\%$$

$$P_E = 42.1\% \times 2.5 \text{ m} = 1.05 \text{ m.} \quad \textbf{Ans.}$$

**Example 11.4.** *The concrete floor of a head regulator is level with the channel bed (except for the short crest hump) and is 13 m long. The floor is provided with cut off walls at its upstream and downstream ends. The depth of upstream cutoff is 1.5 m (below the floor level) and that of the downstream wall is 2.0 m. Using Khosla's theory (see Fig. 11.13 for definition, sketch and formula), determine the thickness of the floor at its mid length and also at its junction with the upstream and downstream cutoff walls. The floor thickness may not be less than 30 cm anywhere. The upstream FSL is 1.5 m above the floor level. If the permissible exit gradient is 0.18, is the floor safe against failure by piping ?* *(Civil Services, 1982)*

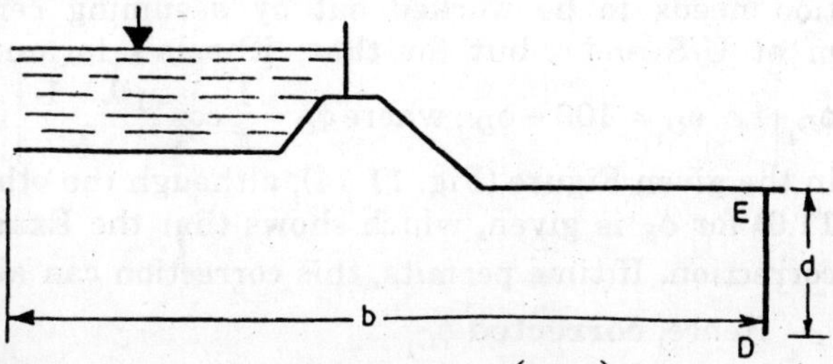


$$\phi_E = \frac{1}{\pi} \cos^{-1}\left(\frac{\lambda - 2}{\lambda}\right)$$

$$\lambda = \frac{1 + \sqrt{1 + \alpha^2}}{2}$$

$$\alpha = \frac{b}{d}$$

Fig. 11.13. Definition sketch for Khosla's theory for uplift pressures.

**Solution.**

**(1) For U/S cut off wall.**

W.r. to Fig. 11.5 (*a*) and (*b*), we have

$b = 13$ m

$d$ = depth of pile line from top of floor level = 1.5 m.

$$\alpha = \frac{b}{d} = \frac{13}{1.5} = 8.67$$

$$\lambda = \frac{1 + \sqrt{1 + \alpha^2}}{2} = \frac{1 + \sqrt{1 + 8.67^2}}{2} = 4.86$$

Now, $\phi_{C_1} = 100 - \phi_E$ where $\phi_E = \frac{1}{\pi} \cos^{-1}\left(\frac{\lambda - 2}{\lambda}\right)$

$$= \frac{1}{\pi} \cos^{-1}\left(\frac{4.86 - 2}{4.86}\right) = \frac{1}{\pi} \times 53.9° \times \frac{\pi}{180°} = 0.30 \text{ ; i.e., } 30\%$$

$$\therefore \quad \boxed{\phi_{C_1} = 100 - 30 = 70\%.,}$$

(*i*) *Correction for D/S pile line.*

$$\text{Correction} = 19 \cdot \sqrt{\frac{D}{b'}} \left(\frac{d + D}{b}\right) \qquad \text{... Eq. (11.3)}$$

where $D$ = Depth of influencing D/S pile = 2 m

$d$ = Depth of U/S pile getting influenced = 1.5 m

$b'$ = Distance between two piles = 13

$$= (+)\ 19 \cdot \sqrt{\frac{2}{13}} \left(\frac{1.5 + 2}{13}\right) = 2.0$$

(Since point $C_1$ is in the rear in the direction of flow from the d/s pile, the correction is positive)

*(ii) Correction for floor thickness.* Strictly speaking, this positive correction needs to be worked out by assuming certain floor thickness of say 1 m at U/S end ; but for that, Khosla's formula (*i.e.* Eq. 11.5 & 11.7) for $\phi_{D_1}\left[i.e.\ \phi_{D_1} = 100 - \phi_D \text{ ; where } \phi_D = \frac{1}{\pi}\cos^{-1}\frac{\lambda - 1}{\lambda}\right]$ is required, which is not shown in the given Figure (Fig. 11.14), although the other applicable formula (*i.e.* Eq. 11.6) for $\phi_E$ is given, which shows that the Examiner wants us to ignore this correction. If time permits, this correction can also be worked out.

Hence, **corrected** $\phi_{C_1}$

$$= 70 + 2 = 72\% \qquad \text{(Ignoring + ve correction due to floor thickness)}$$

*(a)* $P_{C_1}$ = Residual pressure causing uplift at start point (*i.e.* inner edge of U/S cutoff)

$$= \text{Corrected } \phi_{C_1} \times 1.5 \text{ m} = 72\% \times 1.5 \text{ m} = 1.08 \text{ m}$$

∴ **Depth of floor required at start point $C_1$**

$$= \frac{1.08}{G - 1} = \frac{1.08}{1.24} = 0.87 \text{ m ; say 1 m}$$

Use 1.0 m depth (to be conservative for not accounting + ve correction for floor thickness). **Ans.**

(2) **For D/S cutoff wall**

$$b = 13 \text{ m}$$

$$d = 2 \text{ m}$$

$$\alpha = \frac{13}{2} = 6.5$$

$$\lambda = \frac{1 + \sqrt{1 + 6.5^2}}{2} = 3.79$$

$$\phi_E = \frac{1}{\pi}\cos^{-1}\left(\frac{\lambda - 2}{\lambda}\right) = \frac{1}{\pi}\cos^{-1}\left(\frac{3.79 - 2}{3.79}\right) = 0.3434 \text{ ; say } 34.34\%.$$

*Correction for the effect of U/S pile line.* It is – ve, as the point is forward in the direction of flow

$$= (-)\ 19 \cdot \sqrt{\frac{D}{b'}} \cdot \left(\frac{d + D}{b}\right) = (-)\ 19 \cdot \sqrt{\frac{1.5}{13}} \left(\frac{2 + 1.5}{13}\right) = (-)\ 1.74\%.$$

∴ **corrected** $\phi_E = 34.34 - 1.74 = 32.6\%$.

*(b)* $P_E$ = Residual pressure causing uplift at end point (*i.e.* inner edge of D/S cutoff)

$$= \text{Corrected } \phi_E \times 1.5 \text{ m} = 32.6\% \times 1.5 \text{ m} = 0.489 \text{ m}$$

**Thickness of floor required at end point (E)**

$$= \frac{0.489}{1.24} = 0.394, \text{ say 0.4 m.}$$ **Ans.** ; which eventually is more than the min. specified thickness of 0.3 m ; and hence OK

(*c*) **Thickness of floor at mid length** can be the average of the two thicknesses, because uplift is varying from start to end, from 72% to 32.6% ; and its value at mid length is just average of the two. Hence, use floor thickness at mid point as

$$= \frac{1.0 + 0.4}{2} = \mathbf{0.7\ m.\ Ans.}$$

Hence, floor thickness at start = 1.0 m. **Ans.**
floor thickness at end = 0.4 m. **Ans.**
floor thickness at mid length = 0.7 m. **Ans.**

(3) **Exit Gradient.** Exit gradient is given by Eqn. (11.19) as

$$G_E = \frac{H}{d} \cdot \frac{1}{\pi \cdot \sqrt{\lambda}}$$

where $H$ = Total head = 1.5 m (given)

$$\lambda = \frac{1 + \sqrt{1 + \alpha^2}}{2}$$

$$\alpha = \frac{b}{d}$$

$d$ = Depth of D/S cutoff = 2 m

$$\alpha = \frac{b}{d} = \frac{13}{2} = 6.5$$

$$\therefore \quad \lambda = \frac{1 + \sqrt{1 + 6.5^2}}{2} = 3.79,$$

$$\therefore \quad G_E = \frac{1.5}{2} \frac{1}{\pi \cdot \sqrt{3.79}} = \mathbf{0.123 < 0.18} \text{ (Permissible)}$$

Hence, the floor is safe against piping. **Ans.**

## 11.5. Design of a Vertical Drop Weir on Bligh's Theory

Many of the vertical drop weirs, such as shown in Fig. 9.6, have been designed on Bligh's theory ; and even though this theory has now been replaced by modern Khosla's theory, yet it is still used at certain places and especially for minor works, owing to its simplicity. The design of a vertical drop weir, on the basis of this theory is explained below :

**Design of Pucca-floor and Aprons.** As discussed earlier the total length of the pucca floor of the weir (including twice the length of cut-off, if provided) is designed in accordance with the equation $L = CH_L$, and the thickness of the floor is designed by using the equation $t = 1.33 \left(\frac{h}{G-1}\right)$. Bligh has further given certain empirical formulas for determining the length of the downstream pucca floor ($L_2$). The cut-offs may be provided as per the provision of Khosla's theory (explained in article 11.4). The balance floor length [*i.e.* (the total length) – (the downstream length + twice the cut-off length)] is then provided under the crest and on the upstream side as shown in Fig. 11.14. The lengths of the upstream and downstream loose stone talus* (or aprons) which are provided in order to prevent the scour from

* In continuation to the upstream pucca floor, a length $L_4$ (as shown in Fig. 11.14) of loose stone talus is provided to keep the erosion and scour away from the upstream pucca floor. Similarly, a length $L_3$ of loose stone talus is provided on the downstream of pucca floor, so as to dissipate the residual energy and to keep the scour away from the pucca floor. As soon as the scour occurs, the talus falls into the scour, and thus preventing the scour to travel and reach upto the

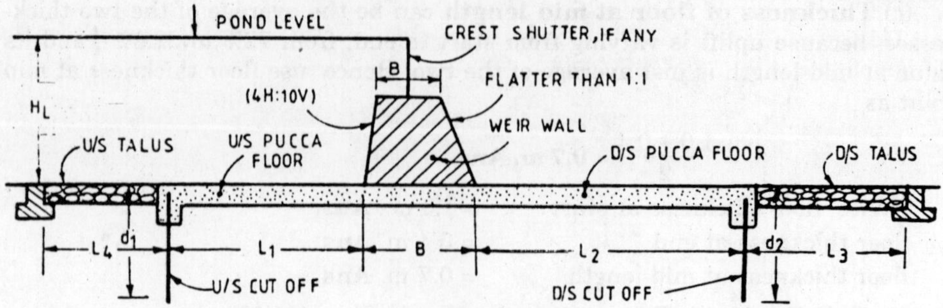


Fig. 11.14. Dimensions of Vertical Drop Weir based on Bligh's Theory.

reaching the pucca floor are also worked out by the empirical formulas put forward by Bligh.

The various empirical formulas put forward by Bligh (w.r.t. Fig. 11.14) are given below :

(a) $L_2 = 2.21\, C \cdot \sqrt{\frac{H_L}{13}}$ for weirs having crest shutters. ..(11.20)

(b) $L_2 = 2.21\, C \cdot \sqrt{\frac{H_L}{10}}$ for weirs having no crest shutters . ..(11.21)

where $H_L$ = the total head loss.

$L_2$ = the length of d/s pucca floor.

(c) $L_2 + L_3 = 18\, C \cdot \sqrt{\frac{H_L}{13} \cdot \frac{q}{75}}$ for weirs having crest shutters ...(11.22)

(d) $L_2 + L_3 = 18\, C \cdot \sqrt{\frac{H_L}{10} \cdot \frac{q}{75}}$ for weirs having no crest shutters. ...(11.23)

where $q$ = the discharge intensity in cumecs/metre.

$L_3$ = the length of d/s loose stone talus.

The length of upstream talus ($L_4$) may be kept equal to half the length of d/s talus. Thus

$$L_4 = \frac{L_3}{2} \qquad ...(11.24)$$

The above formulas are applicable for designing the proper weir portion ; whereas for designing the 'undersluice' portion of the weir, the following modified formulas are used.

(i) $L_2 = 3.87\, C \cdot \frac{H_L}{13}$ for 'undersluices' having crest shutters . ..(11.25)

(ii) $L_2 = 3.87\, C \cdot \frac{H_L}{10}$ for 'undersluices' having no crest shutters . ...(11.26)

(iii) $L_2 + L_3 = 27\, C \cdot \sqrt{\frac{H_L}{13} \cdot \frac{q}{75}}$ for 'undersluices' having crest shutters ...(11.27)

(iv) $L_2 + L_3 = 27\, C \cdot \sqrt{\frac{H_L}{10} \cdot \frac{q}{75}}$ for undersluices having no crest shutters ...(11.28)

---

pucca floor. The rational design of the talus has been discussed a little later, under "Design of Protection Works", in article 11.6.1.

**Design of Weir Wall.** Bligh has further given certain empirical formulas for the design of weir wall. According to him, the top width of weir wall ($B'$) is given as :

$$B' = \frac{H}{\sqrt{G-1}} \quad ...(11.29)$$

where $B'$ = Top width of weir wall and is generally 1.5 to 1.8 m.

$H$ = Head of water over the weir wall at the time of max. flood.

$G$ = Specific gravity of floor material.

Further, the crest width should also be greater by 0.6 m than the height of the crest shutters ; if any.

The bottom width ($B$) of the weir wall may be obtained by providing suitable side slopes. The u/s batter may by kept as $4H : 10V$ and the d/s batter should not be flatter than 1 : 1. The bottom width ($B$) of the weir wall should not be less than

$$B = \frac{H + \text{Height of weir}}{\sqrt{G-1}} \quad ...(11.30)$$

The crest level of the weir wall and the height of solid masonry weir is determined from the considerations of afflux. The afflux produced should not exceed the allowable value, generally kept less than 1.5 m or so. If the crest level, works out to be practically equal to the pond level, then a solid masonry weir can be provided ; and if it is much less than the pond level, then the balance may be provided by crest shutters.

**Example 11.5.** *A weir with a vertical drop has the following particulars :*

*Nature of bed : Coarse sand with the value of Bligh's C = 12*
*Flood discharge = 300 cumecs*
*Length of weir = 40 m*
*Height of weir above low water = 2 m*
*Height of falling shutter = 0.6 m*
*Top width of weir = 2.0 m*
*Bottom width of weir = 3.5 m*

*Design the length and thickness of aprons and draw the cross-section of the weir.* *(Madras University, 1975)*

**Solution.** Total max. head loss = $H_L = 2 + 0.6 = 2.6$ m

Total length of creep required including creep along cut-off

$= L = C.H_L = 12 \times 2.6 = 31.2$ m.

The length of downstream floor is given by Eq. (11.20) as

$$L_2 = 2.21\, C \cdot \sqrt{\frac{H_L}{13}}$$

or
$$L_2 = 2.21 \times 12 \cdot \sqrt{\frac{2.6}{13}} = 2.21 \times 12 \times 0.447 = 11.8 \text{ m; say } 12 \text{ m}$$

The bottom width of weir = $B$ = 3.5 m.

*Provision of cut-offs*

Let us first calculate as to what will be the head over the weir when high flood discharge is passing.

Use $q = 1.7\, H^{3/2}$

where $q = \frac{Q}{L} = \frac{300}{40} = 7.5$

$\therefore$ $7.5 = 1.7\, H^{3/2}$

$$\therefore \quad H^{3/2} = \frac{7.5}{1.7} = 4.41$$

or $$H = (4.41)^{2/3} = 2.68 \text{ m}$$

Head over the weir crest = **2.68 m.**

$\therefore$ U/s HFL (assuming bed level as 100.0 m) and crest level as 102 m)

$$= 102.0 + 2.68 = \mathbf{104.68\ m}$$

Now $R$ = Lacey's regime scoured depth

$$= 1.35\left(\frac{q^2}{f}\right)^{1/3}$$

assuming $f = 1$

$$R = 1.35\left(\frac{7.5^2}{1}\right)^{1/3} = 1.35 \times 3.84 = \mathbf{5.18\ m}$$

Depth of u/s sheet pile from below u/s HFL

$$= 1.5\,R = 1.5 \times 5.18 = 7.8 \text{ m}$$

$\therefore$ Level of bottom of U/s sheet pile

$$= 104.68 - 7.8 = 96.88 \text{ m}$$

Provide a depth of 100 – 96.88 = 3.02, say 3 m for u/s cut-off.

Since afflux is not known, the d/s HFL is not known; and the exact calculation for the depth of d/s sheet pile cannot be done. Use a similar d/s cut off of 3 m depth below the weir floor.

Now, total creep length provided except u/s floor

$$= 2 \times 3 \text{ m} + 3.5 \text{ m} + 12 \text{ m} + 2 \times 3 \text{ m} = 27.5 \text{ m}$$

The balance length, *i.e.* 31.2 – 27.5 = 3.7 m; say 5 m; is therefore, provided as u/s floor as shown in Fig. 11.15. Hence, total creep length provided = 27.5 + 5 = 32.2 m.

Now using Eq. (11.22), we have

$$L_2 + L_3 = 18\,C \cdot \sqrt{\frac{H_L}{13} \cdot \frac{q}{75}}$$

where $L_3$ is the length of downstream loose talus

$$q = \text{Discharge intensity} = \frac{Q}{L} = \frac{300}{40} = 7.5$$

$$L_2 + L_3 = 18 \times 12\sqrt{\frac{2.6}{13} \cdot \frac{7.5}{75}}$$

$$= 18 \times 12 \times 0.141 = 31.6 \text{ m}$$

But $L_2 = 11.8$ m

$$\therefore \quad L_3 = 31.6 - 11.8 = \mathbf{19.8\ m}\text{ ; say 20 m.}$$

Hence provide (say 1 m thick) d/s loose talus of 20 m in length.

(**Note** : This length of 20 m can be partly provided as blocks over inverted filter and partly

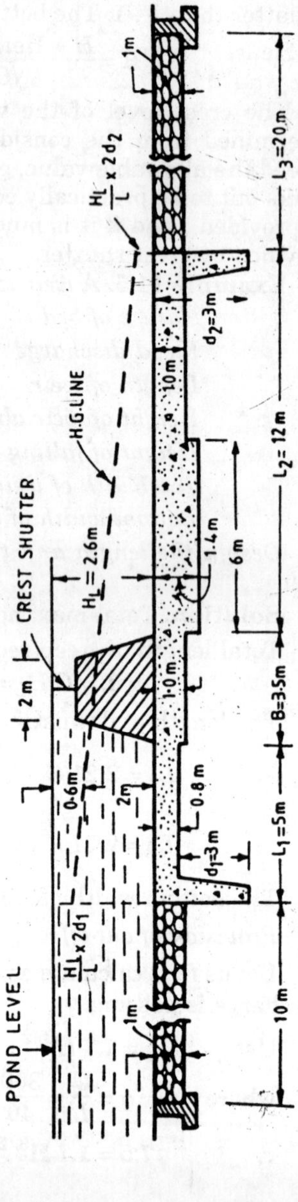


Fig. 11.15

as launching apron. The length of u/s talus $L_4$ may be kept as equal to 1/2 the length of d/s talus, *i.e.* 10 m.

*D/s Floor Thickness*

The H.G. line is now plotted as shown in Fig. 11.15. The max. ordinate of the H.G. line above the bottom of the floor for the downstream portion at the junction of weir wall

$$= h = \frac{2.6}{32.3} \times 18 = 1.45 \text{ m}$$

The thickness of D/s floor at this point is then obtained by Eq.

$$t = 1.33\left(\frac{h}{G-1}\right) = 1.33 \cdot \left(\frac{1.45}{2.4-1}\right)$$

$$= \frac{1.33 \times 1.45}{1.4} = 1.35; \text{ say } \mathbf{1.4\ m}.$$

Hence, provide 1.4 m thickness for D/s floor from just near its junction with weir wall. The thickness can be curtailed as below.

The thickness required at half way of D/s floor length

$$= \frac{1.33\, h}{1.4}$$

$$\text{where } h = \frac{2.6}{32.3} \times 12 = 0.97 \text{ m}$$

$$= \frac{1.33 \times 0.97}{1.4} = 0.92 \text{ m; say 1 m.}$$

Further, provide a nominal thickness of 0.8 m below the u/s floor and 1 m below the weir wall. The complete details are shown in Fig. 11.15.

## 11.6. Design of Modern Weirs and Barrages Founded on Permeable Foundations on the Basis of Khosla's Theory

The complete design of a modern glacis-weir or a barrage can be divided into two main aspects, *i.e.*

(1) *Hydraulic Design* ; and

(2) *Structural Design.*

**11.6.1. Hydraulic Design for the Weir.** The 'hydraulic design' involves determining the section of the weir and the details of its u/s cutoff, crest, glacis, floor, d/s cutoff, protection works u/s and d/s, etc. The hydraulic design of weirs on permeable foundations may be classified into :

(*i*) *Design for Sub-surface Flow;* and

(*ii*) *Design for Surface Flow.*

The effects of sub-surface flow or seepage flow on the stability of a hydraulic structure founded on permeable foundations have been thoroughly described earlier and the same holds good for weirs or barrages. *Khosla's method of independent variables is invariably used for determining the uplift pressures exerted by the seeping water on the floor of the weir. The safety of the structure against piping has to be checked by keeping the exit gradient within safe limits.*

The maximum uplift pressure shall occur when the pond is full and there is no water flowing down the weir. But when flood water passes over the weir, entirely new conditions are superimposed. The formation of hydraulic jump causes uplift or unbalanced head in the jump trough, which may be larger than that under steady seepage, as explained below.

**11.6.1.1. Uplift pressure in the jump trough.** The maximum difference of head and hence the maximum uplift pressures are normally imposed on the structure when water is ponded upto the highest level on the upstream side without any discharge passing down the weir. The hydraulic gradient line under such a situation shall be as shown in Fig. 11.16 (*a*).

The pressure distribution along the length of the weir section is also shown in Fig. 11.16 (*a*).

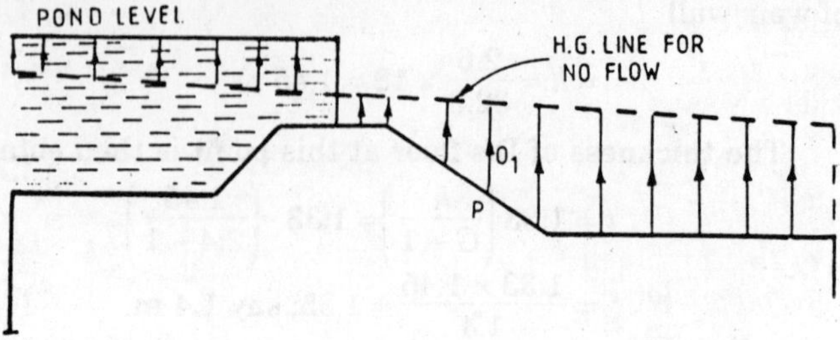


Fig. 11.16 (*a*). Uplift pressures in jump trough for no flow.

When a certain discharge is passing over the weir and hydraulic jump is forming, the seepage head is the difference of the water level of upstream and downstream, which is generally much smaller than the seepage head in case there is no flow. The hydraulic gradient line for this case has also been plotted along with water surface profile, as shown in Fig. 11.16 (*b*).

The ordinate of the uplift pressure is to be measured from the H.G. line to the water surface, as the rest of the uplift is exactly counterbalanced by the weight of the water standing on the floor. The pressure distribution is also shown in Fig. 11.16 (*b*). It is quite evident from these figures that the uplift pressure on the upstream side is more than counterbalanced by the water standing on the floor. Thus, it becomes very clear that there is absolutely no uplift acting on u/s floor, and hence, only a nominal floor thickness is to be provided on the upstream side, and no design is required for u/s floor thickness.

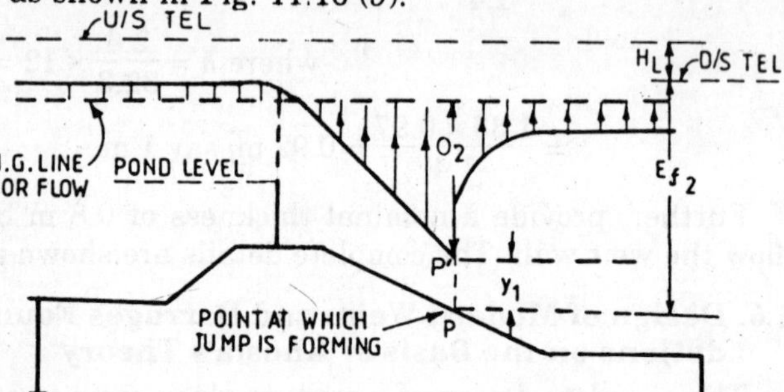


Fig. 11.16 (*b*). Uplift pressures in jump trough with a certain flow.

Further, if Fig. 11.16 (*a*) and (*b*) are imposed upon each other, it is found that the uplift ordinate $O_2$ in the second case is much larger than the uplift ordinate $O_1$ in the first case at the corresponding point $P$ ; where the jump is forming and also at some other points in its vicinity. Since the point of jump formation $P$ is likely to shift with the variation in discharge passing over the weir, the entire glacis may have to be designed for the second condition; while most of the remaining floor may have to be designed for the first condition. In other words, the requirement of floor thickness should be obtained by taking the larger of the two ordinates, and dividing it by the submerged density of the floor material (concrete), *i.e.* $(G_c - 1) = (2.24 - 1) = 1.24$ for concrete.

The water surface profile after the jump and before the jump can be plotted as explained in Chapter 10, and then, H.G. line can be plotted and the net ordinate determined easily. The floor thickness shall have to be designed to withstand the larger pressure.

The uplift pressure due to dynamic action (*i.e.* hydraulic jump) are further reduced due to the following reasons :

(*i*) The backward rolling flow of water in the trough

(*ii*) The uplift pressure due to jump is maximum at the point of the jump but reduces rapidly on either side. As the floor has beam action, it may be designed for average uplift ordinate rather than for maximum uplift ordinate.

(*iii*) The vertical component of the momentum remains unaffected in the jump, which exerts a downward pressure in the vertical direction.

*Due to these reasons, the uplift pressure due to dynamic action is reduced to* $\frac{2}{3}$*rd of the theoretical value, and the design is done for maximum of either the* $\frac{2}{3}$*rd of the largest ordinate due to dynamic action or the largest ordinate due to steady seepage.*

*Uplift pressures at the point of jump formation may also approximately be taken as* = $[50\% \; (y_2 - y_1) + \phi \cdot H_L]$, *where* $\phi$ *is the percentage of pressure at the jump location. (Generally used in the design of falls, etc.)*

The other aspects of surface flow are scour considerations. The sheet piles at the ends must go below the deepest anticipated scour level.

For a discharge intensity $q$, the normal depth of scour ($R$) is given by Lacey's equation (4.20) *i.e.*,

$$R = 1.35 \left[ \frac{q^2}{f} \right]^{1/3} \qquad \text{...(Eq. 4.20)}$$

The value of $q$ will be different for the weir and for the undersluice section, and should, therefore, be taken separately for each. For the design of sheet piles, it is just enough to take them down to the level obtained by measuring the normal depth of scour $R$, below the H.F.L. Though sometimes, even 1.5 $R$ on the upstream side and 2.0 $R$ on the d/s side is taken in conservative designs. A value of 1.25 $R$ on u/s side and 1.5 $R$ on d/s is widely accepted as a *via media.*

**11.6.1.2. Length of pucca concrete floor.** The total length of the pucca floor is mainly governed by the exit gradient considerations. For a safe exit gradient and a depth of downstream cut-off suitable from scour considerations, the length of the horizontal floor '$b$' can be worked out as $b = \alpha \cdot d$ ($\alpha$ is known when Exit gradient ($G_E$) is fixed).

The main turbulence of the hydraulic jump is generally confined to a length equal to five times the jump height. Hence, a pucca floor equal to or more than 5 $(y_2 - y_1)$ in length is provided after the lowest point of jump formation, *i.e.* the end point of glacis. *The glacis should be sloped down at a slope of 3 : 1 to 5 : 1 for maximum dissipation of energy coupled with economy and stability of the jump.* The top width of the crest is generally kept as 2.0 to 3.0 metres from practical considerations, and the upstream slope to the crest is kept as 1 : 1 to 3 : 1. The length of the upstream floor may be adjusted so as to provide the necessary total floor length $b$, calculated above, as $b = \alpha.d$.

By providing a deeper d/s cut-off, it is possible to reduce the floor length '$b$' and *vice-versa,* and hence $b$ and $d$ can be mutually adjusted to provide the most economical and suitable combination and to keep a safe exit gradient.

**11.6.1.3. Design of protection works.** Protection works are required on the upstream as well as on the downstream in order to obviate the possibility of scour hole travelling close to the pucca floor of the weir and to relieve any residual uplift pressure through the filter. The arrangement consists of: (*i*) Inverted filter, and (*ii*) Launching apron, as explained below.

(*A*) *Downstream Loose Protections*

(*i*) **Inverted filter**. Just after the end of the concrete floor, an '*Inverted filter*', 1.5 to 2 *D* long is generally provided, where *D* is the depth of scour below the original river bed. The total scour below HFL is taken as $xR$, where *R* is Lacey's normal scour depth and values of *x* for different classes of scour are tabulated below in Table 11.7. *Value of x is generally taken as 1.5 for design of d/s protection works and 1.25 for design of u/s protection works.*

**Table 11.7**

| *Class of scour* | *Reach* | *Mean value of x* | *D = xR – Water depth above bed* |
|---|---|---|---|
| *A* | Straight | 1.25 | $1.25\,R - y$ |
| *B* | Moderate Bend | 1.50 | $1.50\,R - y$ |
| *C* | Severe Bend | 1.75 | $1.75\,R - y$ |
| *D* | Right-angled Bend | 2.00 | $2\,R - y$ |

The depth of inverted filter is kept equal to the depth of d/s *launching apron.* It generally consists of 1.0 to 1.2 m deep concrete blocks with open joints laid over 0.6 m thick graded filter material. The openings between the blocks are filled with *clean bajri.*

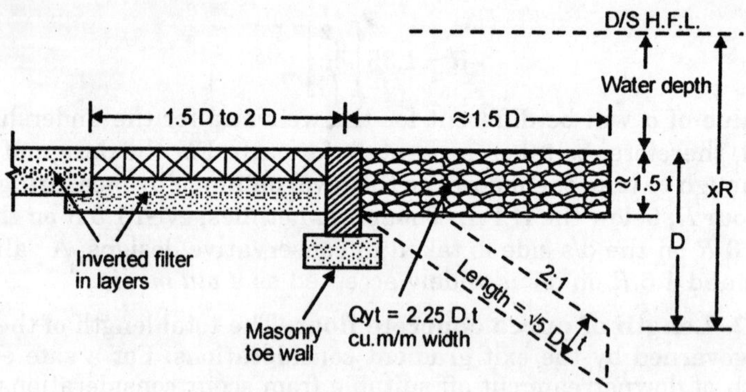


Fig. 11.17. Details of d/s loose protection.

An 'inverted filter' invariably reduces the possibility of piping, as it allows free flow of seepage water through itself without allowing the foundation soil to be lifted upward. *The filter, therefore, consists of layers of materials of increasing permeability from bottom to top.* The gradation should be such that while it allows free flow of seepage water, the foundation material does not penetrate to clog the filter. The design criteria to satisfy these conditions are discussed in chapter 20 on "Earth Dams". To prevent filter from dislocation under surface flow, concrete or masonry blocks are laid over the filter material.

(*ii*) **Launching apron**. After the inverted filter, the loose apron called 'launching apron' is provided for a length, generally equal to 1.5 *D*, where *D* has the same meaning as given above in inverted filter calculations.

The apron generally launches to a slope of 2 : 1, and if *t* is the thickness of the apron in the launched position, length being $\sqrt{5}\ D$; the volume of stone per metre width will then be

$$= \sqrt{5} \cdot D \cdot t = 2.24\ D.t; \quad \text{oor say } \mathbf{2.25\ D.t.}$$

Hence, Volume of stone per metre is given by 2.25 $D.t.$cu. m/metre.

Since the volume of stone should be the same in launched and unlaunched apron, and if the unlaunched apron is laid in a length equal to 1.5 $D$, the thickness of the unlaunched apron is given as :

$$= \frac{2.25\, D.t}{1.5\, D} = 1.5\, t$$

If $t$ is taken as 1 m ; then an apron of 1.5 m thickness and 1.5 $D$ metres long has to be laid as d/s launching apron. Different values of $t$ have been recommended by different investigators.

Blench has recommended $t$ equal to 1.5 to 2 times the size of the stone ($d_s$) required by equation (11.15) as :

$$V_{av} = 4.915\, d_s^{1/2} \qquad \text{...(11.31)}$$

where $V_{av}$ = average velocity of flow in m/sec.

$d_s$ = mean diameter of stone in metres.

Spring has recommended $t$ at about 0.9 m. Gales recommended $t$ varying from 1.35 to 1.9 m, for discharge varying from 7,000 to 70,000 cumecs. Blench's recommendations are quite adoptive.

*(B) Upstream Loose Protections*

Just upstream of the concrete floor of the weir, block protection is provided. It generally consists of concrete blocks laid over packed stone, for a length equal to $D$. ($D = xR - y$, where $x$ = 1.0 to 1.5, generally taken as 1.25).

Upstream of the blocks, a launching apron is provided in the same way as described for the downstream portion, except that the proper value of $x$ should be chosen.

Toe walls are always generally constructed in between the 'filter' and the 'apron' as shown in Fig. 11.17.

**11.6.2. Structural Design for the Weir Floor.** The concrete floor is usually designed for the uplift pressures as a pure gravity section at each point. Sometimes, when $H_L$ is enormous, the uplift pressures are too high, and as such, the thickness of the required concrete gravity section becomes enormous and, therefore, becomes very costly. In such cases, the floor may be designed as a reinforced concrete raft structure held down by the weight of the raft and piers. A raft, in such cases, may be cheaper and more desirable as the thin section of raft reduces deep excavations and dewatering problems. Recently, Farraka Barrage, Narora Barrage, and Durgapur Barrage have been designed and constructed as reinforced raft structures. However, the treatment of this type of design is a simple structural problem. The total downward weight of raft and piers is counterbalanced by a uniform uplift, and design is done by working out bending moments, etc.

**11.6.3. Effects Produced by Weir on River Regime and Retrogression of Downstream Levels.** Before we start with the actual design of a weir, let us first review the effects that are produced by the weir construction, on the regime of the river. The first effect produced by the construction of a weir across a river, is that : *the downstream bed of the river goes on eroding, consequently causing progressive lowering of the downstream levels. This progressive lowering of the downstream levels is known as Retrogression of downstream levels or* **retrogression.**

The basic cause for retrogression is the variation in the silt carrying capacity of the channel. As soon as a weir is constructed, the water starts ponding on its upstream side, causing the water surface slope to flatten for some distance behind the weir. This reduces the silt carrying capacity of the river in this reach, and consequently the silt deposition starts, *i.e.* the river starts dropping sediment, and this leads to formation of shoals and islands on the upstream side. This clearer water passes over the weir and picks up sediment from its downstream bed, so as to fulfil the increased demand of the silt carrying capacity of the channel in the downstream, *i.e.* the sediment deficit caused on the upstream is made up by eroding extra sediment from the downstream. This causes the progressive lowering of the downstream bed levels. The process continues for a number of years till the river starts to regain its original slope in the upstream portion by extending the afflux more and more upstream. The stage is gradually reached, when the upstream pond absorbs no more silt. As the off-taking channel takes comparatively silt free water, the sediment will go downstream, while the discharge going down will be below normal, and hence, the sediment taken shall be more than the 'carrying capacity of the river' ; consequently resulting in sediment deposition on the downstream river bed and long range recovery of d/s bed levels.

A provision must, therefore, be made for retrogression of d/s bed, while designing a weir, as it shall lower the d/s TEL and increase $H_L$. Hence, if retrogression is not taken into account, it may lead to undermining of floor. Observations on various weirs in Punjab have exhibited a retrogression of 1.2 to 2.2 metres. The retrogression of such a high magnitude has been observed at low water levels, but at high flood levels, the maximum retrogression is between 0.3 to 0.5 metres.

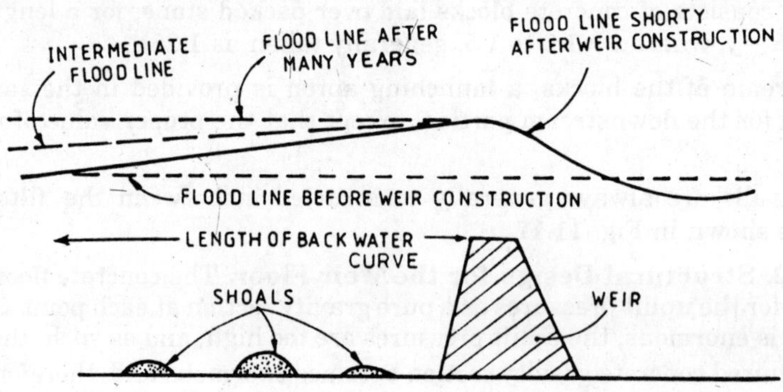


Fig. 11.18. Figure showing the effects of weir construction on flood levels.

*Hence, a retrogression of 0.5 m at high flood stage, and a higher retrogression varying linearly upto 2.0 m at lower discharges, is generally considered in the design of weirs or barrages.*

The recovery of downstream bed levels, sometimes continues even beyond the original bed levels. This may lead to reduced control on silt regulation. Hence, sufficient margin must be provided between the canal full supply level and the pond level, so as to allow raising of the crest of the canal Head Regulator, if found necessary, in future.

***Since the afflux extends on the upstream of the weir with the passage of time, it increases the high flood levels on the upstream, even beyond the original backwater curve. Hence, the marginal bunds will have to be extended upstream as soon as the above effects come into picture. Since it happens after many years, it is economical to construct marginal afflux bunds only for the backwater length in the beginning, and to extend them upstream afterwards.***

**11.6.4. Factors Governing the Design of a Weir or a Barrage** The following data must be collected before a weir or a barrage can be designed :

(*i*) High flood levels for the river at the weir site.

(*ii*) High flood or maximum flood discharge for the river at the weir site.

(*iii*) River cross-section at the weir site.

(*iv*) The stage discharge curve for the river at the weir site.

All the above informations can be obtained from Topographical maps of the area and by consulting the Hydrological and Meteorological Departments of the area.

**Factors to be Decided.** Besides Retrogression, certain other factors which have to be decided while designing a weir or a barrage are :

(*i*) Crest Levels.

(*ii*) Afflux.

(*iii*) Waterway and the discharge per metre.

(*iv*) Pond Level.

They are described below :

(*i*) **Crest Levels.** It has been stated earlier that the weir consists of two parts :

(*a*) The main weir section, called Weir Bay Section.

(*b*) Undersluice Section.

The undersluice section is kept at a lower level, so as to provide deeper silent river pocket near the canal head regulator. The undersluice crest is, therefore, kept slightly lower than the barrage bay crest so as to attract a deep current infront of the canal head regulator, so that dry weather flow may remain rear the regulator.

*The undersluice crest is generally kept as near the bed level in the existing deepest channel, as is practically possible.*

*The crest level of other barrage bays is generally kept 1.0 to 1.5 metres higher than the crest level of the undersluices. It is also guided by the general bed level of the river in the barrage bay portion.*

It can be easily seen that the afflux and discharge per metre are dependent upon the crest levels. If crest level is low, afflux shall be less, and since the depth of water over the crest will be more, it shall lead to higher discharge per metre. *A low set barrage, with increased depth of water over the crest may, therefore, result in the increase in height of gates, thickness of floor, and cost of superstructure above floor level.*

(*ii*) **Afflux**. It was defined earlier that the rise in the maximum flood level of the river upstream of the weir after construction is known as afflux. This afflux is confined only to a short reach (equal to the length of the back water curve) in the beginning, but extends gradually very far up, as explained earlier.

The amount of afflux will determine the top levels of guide banks and marginal bunds. The lengths and sections of marginal bunds are also dependent upon it. It will govern the dynamic action downstream of the work as well as the depth and location of the hydraulic jump. *By providing a higher afflux, the waterway and, therefore, the length of the weir can be reduced, but it will increase the cost of training works and the risk of failure by outflanking*. At the same time, the discharge intensity and the consequent scour shall go up, and hence, the *sections of loose protections upstream and downstream as well as the depths of pile lines at either ends shall have to be increased,* thereby making it costly. *It is, therefore, always desirable to limit the afflux to a safe value of 1.0 to 1.2 metres, more commonly 1.0 metre.* However, in steep reaches with rocky bed, a higher value of afflux may be permitted.

(*iii*) **Waterway and discharge per metre.** The waterway and afflux are correlated. If afflux is increased, waterway is reduced and *vice-versa*. Hence, a limit placed on maximum afflux shall limit the minimum waterway. It shall be seen that the cost of works as a whole is minimum for a certain waterway and afflux. Attempts should, therefore, be made to attain the most economical combination of these two factors. This can be made by trial and error, generally limiting the maximum value of afflux.

A likely figure for the waterway is obtained by Lacey's wetted perimeter formula, given by eq. (4.18) as

$$P = 4.75\sqrt{Q} \quad \text{... Eq. (4.18)}$$

A waterway equal to 1.2 to 1.4 $P$ is generally assumed in rivers in plains. Some engineers preferred to keep a shorter waterway inspite of costlier works, as it was thought that a shorter waterway reduces shoaling, but as a matter of fact, it is not so.

(*iv*) **Pond levels.** The pond level is the minimum water level required in the undersluice pocket upstream of the canal head regulator, so as to feed the canal with its full supply. *The pond level is generally obtained by adding 1.0 to 1.2 metres to the canal FSL.*

Water in the undersluice pocket has to be maintained at Pond Level, even during dry weather flow. This can be accomplished either by a raised crest or by shutters or by a combination of both. A permanently raised crest will lead to higher afflux during floods and is likely to result in loss of control on the river.

In modern design of a barrage, the entire ponding is done by gates which are opened during floods and the crest level of the undersluices is generally taken as the available river bed level in the deepest channel. *No raised crest is thus generally provided for the undersluices.* A raised crest is provided where possible, as it improves the coefficient of discharge.

The **discharge formulas** to be used in the design of a weir or a barrage are

(*a*) **For a broad crested weir :**

$$Q = 1.7\,(L - K.n.H)\,H^{3/2} \quad \text{...(11.32)}$$

where $Q$ = Discharge in cumecs

$H$ = Total head in metres including velocity head

$n$ = No. of end contractions (Twice the number of gated bays)

$L$ = Clear waterway length in metres.

$K$ = coefficient of end contraction varying from 0.1 for thick blunt peir noses to 0.04 for thin pointed noses : generally taken as 0.1 in ordinary calculations.

(*b*) **For a sharp crested weir :**

$$Q = 1.84 \cdot (L - KnH)\,H^{3/2}. \quad \text{...(11.33)}$$

$Q$, $L$, $n$, $K$ and $H$ have the same meanings as given above.

If the head over the weir crest is more than 1.5 times the width of the weir, the weir behaves as a *sharp crested weir.*

**Concentration factor.** While calculating the cistern levels and the depths of cutoffs, the possibility of non-uniform flow is taken into account by providing a suitable concentration factor. This factor is chosen arbitrarily. Generally, a 20% concentration is taken at any particular section. Hence, the calculated maximum discharge intensity is increased by 20% in designs. No concentration of flow is taken while designing protection works.

### Example on the design of a Barrage

**Example 11.6.** *A barrage is to be constructed on a river having a high flood discharge of about 8,100 cumecs, with the given data as follows :*

| | |
|---|---|
| *Average bed level of the river* | = *257.0 m* |
| *High Flood Level (before construction of barrage)* | = *262.2 m* |
| *Permissible afflux* | = *1.0 m* |
| *Pond Level* | = *260.6 m.* |

*Stage Discharge curve for the river at the barrage site is given in Fig. 11.19. Prepare a complete hydraulic design for the under-sluice section as well as for other barrage bay section, on the basis of Hydraulic jump theory and Khosla's theory. A safe exit gradient of 1/6 may be assumed. 0.5 metres retrogression and 20% discharge concentration may be assumed where non-uniform flow is likely to occur. Assume any other data if not given.*

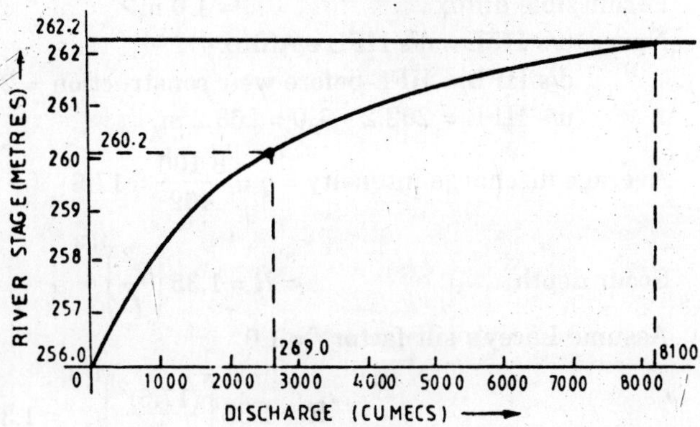


Fig. 11.19

**Solution. Fixing the Crest Levels and Waterway**

**Crest Levels.** The average bed level of the river is given to be 257.0 m, which may be taken as the crest level of the undersluices. The upstream floor of the undersluices may be kept the same, and thus there will be no raised crest in the undersluices. The crest level of other barrage bays may be kept 1.0 to 1.5 m higher than the crest level of undersluices. Let us keep it 1.3 m higher, *i.e.* say 257.0 + 1.3 = 258.3 m.

**Waterway** : The waterway, as per Lacey's wetted perimeter equation, is given by

$$P = 4.75\sqrt{Q} = 4.75\sqrt{8100} = 4.75 \times 90 = \mathbf{427.5\ m.}$$

Now let us provide a waterway approximately equal to 1.2 $P$ by trial, in such a way that approximately 15 to 20% of the discharge is passed through the undersluices and the total provided waterway should be able to pass the entire discharge.

**Assume the waterway as below**

(*a*) **Undersluice portion :**

| | |
|---|---|
| 5 bays of 15 m each | 75 m |
| 4 piers of 2.5 m each | 10 m |
| Overall waterway | 85 m |

(*b*) **Other barrage bays portion**

| | |
|---|---|
| 27 bays of 12 m each | 324.00 m |
| 25 piers of 2 m each | 50.00 m |
| Overall waterway | 374.00 m |

Assume a divide wall of 3.0 m thickness. Hence, total overall waterway provided between abutments = 85 + 374 + 3 = **462 m.**

Now, let us check whether the maximum flood can pass through this waterway with the maximum permissible afflux of 1.0 metre or not.

HFL before barrage construction = 262.2 m.

Permissible afflux = 1.0 m.

Now u/s HFL = d/s HFL + Afflux

d/s HFL = HFL before weir construction = 262.2 m

∴ u/s HFL = 262.2 + 1.0 = 263.2 m.

Average discharge intensity $= q = \dfrac{8{,}100}{462} = 17.6$

Scour depth $= R = 1.35 \left(\dfrac{q^2}{f}\right)^{1/3}$

Assume Lacey's silt factor $f = 1.0$

$$R = 1.35 \left[\frac{(17.6)^2}{1}\right]^{1/3} = 1.35\,(17.6)^{0.667} = 9.20 \text{ m.}$$

Velocity of approach $= V = \dfrac{q}{R} = \dfrac{17.6}{9.20} = 1.90$ m/sec.

Velocity Head $= \dfrac{V^2}{2g} = \dfrac{(1.90)^2}{2 \times 9.81} = 0.19$ m.

u/s TEL = u/s HFL + Velocity Head = 263.2 + 0.19 = 263.39 m.

Head (*i/c* velocity head) over the undersluice crest

= u/s TEL – Undersluice crest level = 263.39 – 257.0 = 6.39 m.

Head (*i/c* velocity head) over the crest of other barrage bays

= u/s TEL – Crest level of barrage bays = 263.39 – 258.3 = **5.09 m.**

Discharge passing through undersluices is given by the discharge formula for a broad crested weir; since the crest and u/s floor are at the same level, and the width of the crest is sufficient, it will behave as a broad crested weir.

$$\therefore \quad Q_1 = 1.7\,(L - K\,n\,H)\,H^{3/2} \qquad \text{...(Eq. 11.33)}$$

where $L$ is the clear waterway

$Q_1$ = Discharge through undersluices

$= 1.7\,[75 - 0.1 \times 10 \times 6.39]\,(6.39)^{3/2}$ (assuming $K = 0.1$)

$= 1.7\,[75 - 6.39]\,(6.39)^{3/2} = 1.7 \times 68.61 \times 16.1 =$ **1,880 cumecs.**

Let us keep the width of the crest of other barrage bays portion as 2.0 m. Since the head over the other barrage bays crest is 5.09 m., which is more than 1.5 times the width of the crest, it shall behave like a sharp crested weir. The discharge is then given by

$$Q = 1.84\,(L - K\,nH)\,H^{3/2} \qquad \text{... Eq. (11.33)}$$

$\therefore$ $Q_2$ = Discharge through other barrage bays

$$= 1.84\,(324.0 - 0.1 \times 52 \times 5.09)\,(5.09)^{3/2}$$

(assuming $K = 0.1$)

$$= 1.84\,(324.0 - 26.4)\,(5.09)^{3/2}$$

$$= 1.84 \times 297.6 \times 11.4 = \mathbf{6,\,260\ cumecs}$$

Total discharge that can pass through the barrage

$$= Q_1 + Q_2 = 1{,}880 + 6{,}260 = 8{,}140 \text{ cumecs}$$

$$= 8{,}140 \text{ cumecs} > 8{,}100 \text{ cumecs.}$$

Hence, the assumed waterway and crest levels are in order.

Actual overall waterway provided = 462 m against Lacey's wetted perimeter of 427.5 m.

$$\therefore \text{ Looseness factor} = \frac{462}{427.5} = \mathbf{1.08}$$

The design of undersluice section and that of other barrage bays section shall be carried out separately.

**Design of Undersluice Portion**

There are two major flow conditions : (*i*) When high flood is passing ; and (*ii*) When flow is at Pond Level (with all gates open). Let us calculate $q$ and $H_L$ for these two conditions.

**(1) High flood condition**

(*a*) *Assuming no concentration and retrogression* [Fig. 11.20 (*a*)]

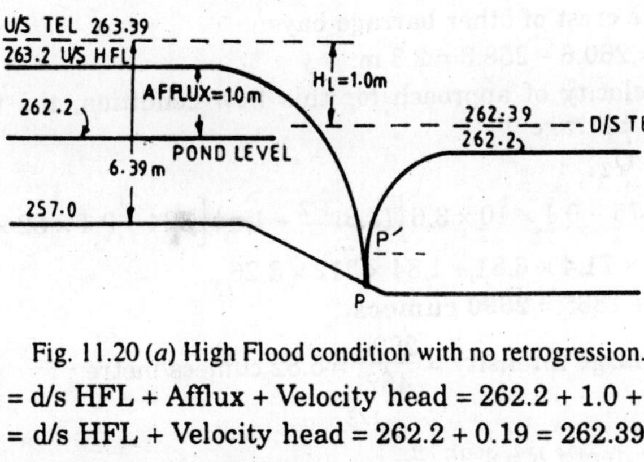


Fig. 11.20 (*a*) High Flood condition with no retrogression.

u/s TEL = d/s HFL + Afflux + Velocity head = 262.2 + 1.0 + 0.19 = 263.39 m.

d/s TEL = d/s HFL + Velocity head = 262.2 + 0.19 = 262.39 m.

Head Loss = $H_L$ = 263.39 − 262.39 = **1.0 m**

Discharge intensity between piers = $q$

$$= CH^{3/2} = 1.7\,(6.39)^{3/2} = 1.7 \times 16.1 = 27.4 \text{ cumecs/metre.}$$

*(b) High flood condition with 20% concentration and 0.5 m retrogression [Fig. 11.20 (b)]*

The discharge intensity is increased by 20%. Therefore, new discharge intensity

$= 1.2 \times 27.4 = 32.9$ cumecs/metre.

New head (*i/c* velocity head) required for this discharge intensity to pass

$$= \left(\frac{32.9}{1.7}\right)^{2/3} = (19.5)^{0.667} = \mathbf{7.2\ m.}$$

The conditions are shown in Fig. 11.20 (*b*).

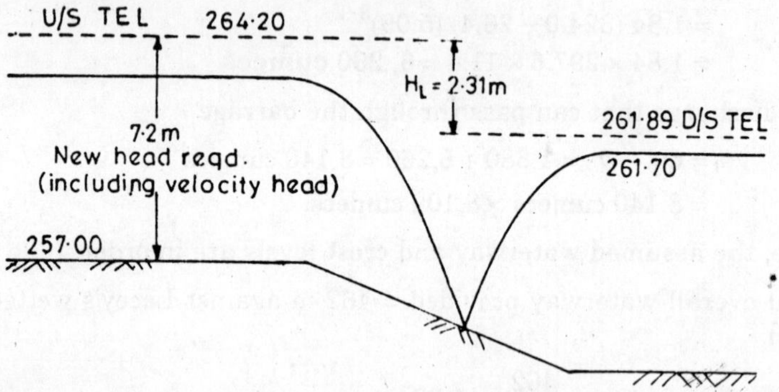


Fig. 11.20 (*b*) High flood condition with 20% concentration and 0.5 m retrogression.

u/s TEL = 257.0 + 7.2 = 264.2 m.

The d/s HFL is depressed by 0.5 m due to retrogression, *i.e.* it becomes 261.7 m

$\therefore$ d/s TEL = 261.7 + 0.19 = 261.89 m

$\therefore$ $H_L = 264.20 - 261.89 = \mathbf{2.31\ m.}$

**(2) Pond Level Flow Condition**

*(a) With no concentration and retrogression*

Pond Level (given) = 260.6 m.

Head over the crest of undersluices under this condition

$= 260.6 - 257.0 = 3.6$ m

Head over the crest of other barrage bays

$= 260.6 - 258.3 = 2.3$ m.

Neglecting velocity of approach for this flow condition, the total discharge passing down the barrage

$$Q = Q_1 + Q_2$$

$$= 1.7\left[75 - 0.1 \times 10 \times 3.6\right](3.6)^{3/2} + 1.84\left[324 - 0.1 \times 52 \times 2.3\right](2.3)^{3/2}$$

$$= 1.70 \times 71.4 \times 6.81 + 1.84 \times 312 \times 3.26$$

$$= 825 + 1865 = \mathbf{2690\ cumecs.}$$

Average discharge intensity $= \frac{2690}{462} = 5.82$ cumecs/metre

Normal scour depth $R = 1.35\left(\frac{q^2}{f}\right)^{1/3}$

$$= 1.35 \times \left[\frac{(5.82)^2}{1}\right]^{1/3} = 1.35 \times (5.82)^{0.667} = 4.45\ \text{m}$$

Velocity of approach $= V = \dfrac{q}{R} = \dfrac{5.82}{4.45} = 1.31$ m/sec.

Velocity head $= \dfrac{V^2}{2g} = \dfrac{(1.31)^2}{2 \times 9.81} = 0.088 =$ **say 0.09 m.**

These conditions are shown in Fig. 11.21 (*a*).

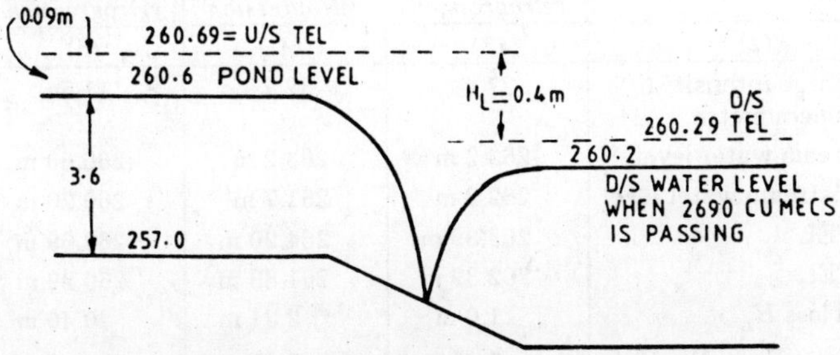


Fig. 11.21 (*a*) Pond level flow with no concentration and retrogression.

$\therefore$ U/s TEL = 260.6 + 0.09 = 260.69 m.

The d/s water level when a discharge of 2,690 cumecs is passing can be found from the Stage-Discharge curve of the river, given in Fig. 11.19. It is found to be 260.2 m.

$\therefore$ D/s TEL = 260.2 + 0.09 = 260.29 m.

$H_L$ = U/s TEL – D/s TEL

$H_L = 260.69 - 260.29 =$ **0.40 m.**

Discharge intensity between piers $= 1.7 \times (3.6)^{3/2}$

$= 1.7 \times 6.81 = 11.54$ cumecs/metre.

(*b*) *Pond Level flow with 20% concentration and 0.5 m retrogression.*

New discharge intensity $= 1.2 \times 11.54 =$ **13.86** cumecs/metre

New head reqd. (*i/c* velocity head) over the crest $= \left(\dfrac{13.86}{1.7}\right)^{2/3} = 4.06$ m.

U/s TEL = 257.0 + 4.06 = 261.06 m.

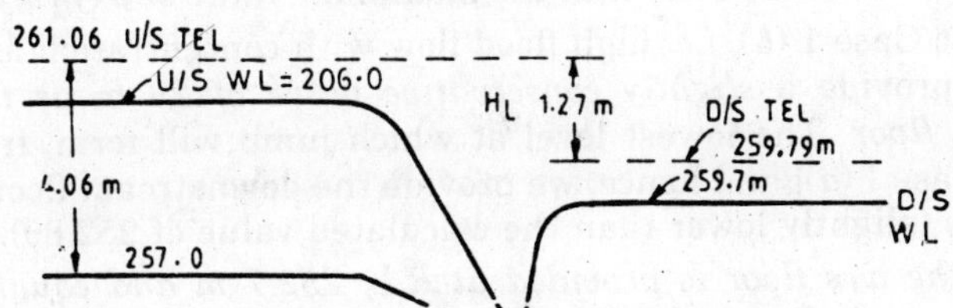


Fig. 11.21 (*b*) Pond level flow with 20% concentration and 0.5 m retrogression.

d/s HFL which was 260.2 m is depressed by 0.5 m. The new d/s HFL shall be 259.7 m.

$\therefore$ D/s TEL = 259.7 + 0.09 = 259.79 m.

$H_L$ = U/s TEL – D/s TEL

$\therefore$ $H_L = 261.06 - 259.79 =$ **1.27 m.**

The values of $q$, $H_L$, the water levels and energy levels for all these four cases are tabulated in Table 11.8.

**Table 11.8. Undersluice Portion of Barrage**

| S. No. | Item | High Flood Flow: condition 1(a) without concentration and retrogression | High Flood Flow: condition 1(b) with concentration and retrogression | Pond Level Flow: condition 2(a) without concentration and retrogression | Pond Level Flow: condition 2(b) with concentration and retrogression |
|---|---|---|---|---|---|
| (1) | (2) | (3) | (4) | (5) | (6) |
| 1. | Discharge intensity ($q$) in cumecs/metre | 27.4 | 32.7 | 11.54 | 13.86 |
| 2. | Upstream water level | 263.2 m | 263.2 m | 260.60 m | 260.6 m |
| 3. | Downstream water level | 262.2 m | 261.7 m | 260.20 m | 259.7 m |
| 4. | U/s TEL | 263.39 m | 264.20 m | 260.69 m | 261.06 m |
| 5. | D/s TEL | 262.39 m | 261.89 m | 260.29 m | 259.79 m |
| 6. | Head loss $H_L$ | 1.0 m | 2.31 m | 0.40 m | 1.27 m |
| 7. | $E_{f2}$ (from Plate No. 10.1) | 7.40 m | 9.00 m | 4.00 m | 5.00 m |
| 8. | Level at which jump will form, *i.e.* (D/s TEL $- E_{f2}$) | 254.99 m | 252.89 m | 256.29 m | 254.79 m |
| 9. | $E_{f1} = E_{f2} + H_L$ | 8.40 m | 11.31 m | 4.40 m | 6.27 m |
| 10. | $y_1$ corresponding to $E_{f1}$ (Plate No. 10.2) | 2.4 m | 2.5 m | 1.65 m | 1.30 m |
| 11. | $y_2$ corresponding to $E_{f2}$ (Plate No. 10.2) | 6.40 m | 8.0 | 3.30 m | 4.5 m |
| 12. | Length of concrete floor required = 5 $(y_2 - y_1)$ | 20.0 m | 27.5 m | 8.25 m | 16.00 m |
| 13. | Froude No. $F_1 = \frac{q}{\sqrt{g y_1^3}}$ | 2.36 m | 2.66 m | 1.74 m | 2.97 m |

All these four cases of jump formation are separately analysed. Using Blench curves, the values of $E_{f2}$, $E_{f1}$, position of jump point $P$, $y_2$, $y_1$, etc. are all calculated as explained earlier and shown in Table 11.8. The table is self-explanatory. It can be seen from this Table 11.8, that the maximum value of 5 $(y_2 - y_1)$ is 27.5 metres for the worst Case 1 ($b$), *i.e.* high flood flow with concentration and retrogression. Hence, we provide a *slightly conservative value of 29 m as the length of the downstream floor*. The lowest level at which jump will form, from this table, is 252.89 m [Case I ($a$)] and hence, we provide the downstream floor at a level of say *252.7 metres* (slightly lower than the calculated value of 252.89).

*Hence, the d/s floor is provided at R.L. 252.7 m and equal to* **29** *metres in length,* as shown in Fig. 11.22. The glacis is provided in 3 : 1 slope with a horizontal length of **12.9** metres, as shown in Fig. 11.22. Let us now calculate the total floor length required from Khosla's Safe Exit Gradient Theory, so that the balance length of floor can be provided on the u/s side. Before this is calculated, we shall have to decide the depth of d/s cut off provided from scour considerations.

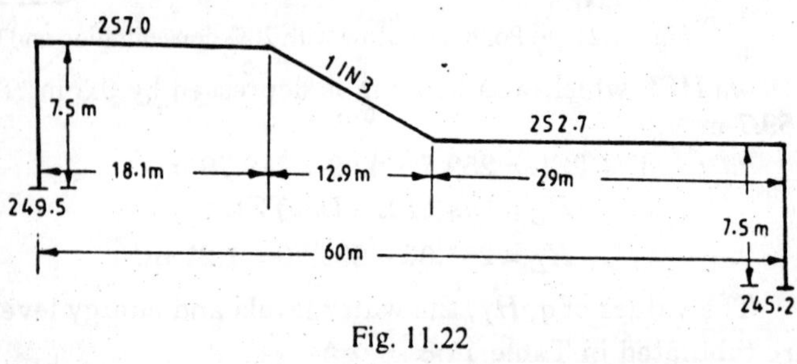


Fig. 11.22

**Depth of Sheet Pile Lines from Scour Considerations**

Total discharge passing through undersluices = 1,880 cumecs (calculated earlier)
Overall waterway of undersliuces = 85 m.
Average discharge intensity = 1880/85 = 22.3 cumecs/metre.
Depth of scour

$$R = 1.35\left(\frac{q^2}{f}\right)^{1/3}$$

$$= 1.35\left(\frac{22.3^2}{1}\right)^{1/3} = 1.35(22.3)^{2/3} = 1.35 \times 8 = 10.8 \text{ m, say 11 m.}$$

Let us provide a downstream cut-off at 1.5 $R$ below the d/s water level (which is 261.7 m with retrogression). Hence, the R.L. of bottom of d/s cut-off

$$= 261.7 - 1.5 \times 11.0 = 261.7 - 16.5 = 245.2 \text{ m.}$$

*Let us provide the d/s cut-off upto a bottom level of 245.2 m, i.e.* a depth of 7.5 metres, as shown in Fig. 11.22.

*U/s cut-off.* Let us provide u/s cut-off at depth of 1.25 $R$ (*i.e.* 1.25 × 11.0 = 13.75 m) from top of u/s water level.

Level of bottom of u/s cut-off = 263.20 – 13.75 = 249.45 m.

Let us, therefore, provide the u/s cut-off up to a bottom level of 249.5 m, *i.e.* 7.5 m deep, as shown in Fig. 11.22.

**Total Floor length and Exit Gradient**

Safe-exit gradient ($G_E$) = 1/6

Maximum Static Head ($H$) = 260.6 – 252.7 = **7.9 m**

Depth of d/s cut-off ($d$) = 252.7 – 245.2 = **7.5 m**

$$G_E = \frac{H}{d} \cdot \frac{1}{\pi\sqrt{\lambda}}$$

$$\therefore \quad \frac{1}{6} = \frac{7.9}{7.5} \cdot \frac{1}{\pi\sqrt{\lambda}}$$

$$\frac{1}{\pi\sqrt{\lambda}} = \frac{7.5}{7.9} \times \frac{1}{6} = 0.158$$

From Plate 11.2, value of α for a value of $\frac{1}{\pi\sqrt{\lambda}}$ as 0.158, comes out to be approximately 8.0.

$$\therefore \quad b = \alpha \cdot d = 8 \times 7.5 = \mathbf{60\ m.}$$

Hence balance of 18.1 m *i.e.* [60 m – 12.9 m (glacis) – 29 m (d/s floor) = 18.1 m] can be provided as u/s floor length (shown in Fig. 11.22) so as to make the total floor length equal to 60.0 metres

**Uplift Pressures**

To calculate uplift pressures from Khosla's theory, let us first assume floor thickness at u/s cut-off as 1.0 m and at d/s cut-off as 1.5 m. Fig. 11.22 then becomes as shown in Fig. 11.23. No intermediate cut-off is necessary and hence not provided.

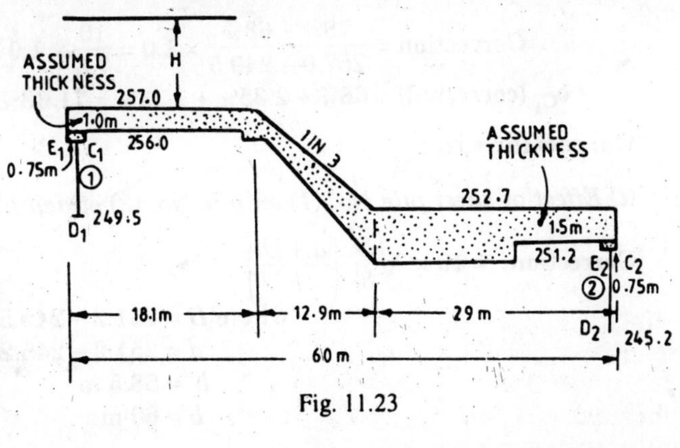


Fig. 11.23

**Upstream Pile No. (1)**

$$b = 60 \text{ m}$$
$$d = 7.5 \text{ m}$$
$$\frac{1}{\alpha} = \frac{d}{b} = \frac{7.5}{60} = 0.125$$

*From Plate 11.1 (a)*

$$\phi_{E_1} = 100\%$$
$$\phi_{C_1} = 100 - \phi_E = 100 - 32 = 68\%$$
$$\phi_{D_1} = 100 - \phi_D = 100 - 22 = 78\%$$

**D/s Pile Line**

$$d = 7.5 \text{ m}$$
$$b = 60 \text{ m}$$
$$\frac{1}{\alpha} = \frac{d}{b} = \frac{7.5}{60} = 0.125$$

*From Plate 11.1 (a)*

$$\phi_{C_2} = 0\%$$
$$\phi_{E_2} = \phi_E = 32\%$$
$$\phi_{D_2} = \phi_D = 22\%$$

**Let us correct these pressures**

$$\phi_{C_1} = 68\%$$
$$\phi_{E_2} = 32\%$$

**Corrections to $\phi_{C_1}$ :**

*(i) Effect of sheet pile No. (2) on pile No. (1) of depth d*

$$\text{Correction} = 19 \times \sqrt{\frac{D}{b'}}\left(\frac{d+D}{b}\right) \quad \text{... Eq. (11.3)}$$

where $D$ = depth of pile (2) below the point $C_1$, *i.e.* the point at which interference is desired

$= 256.0 - 245.2 = \mathbf{10.8\ m}$

$d = 256.0 - 249.5 = 6.5$ m

$b' = 58.5$ m ; $b = 60$ m

$$\text{Correction} = 19 \times \sqrt{\frac{10.8}{58.5}}\left[\frac{6.5 + 10.8}{60}\right] = 19 \times \frac{1}{2.33} \times \frac{17.3}{60} = 2.35\% \text{ (+ ve)}$$

*(ii) Correction for depth ;*

$$\text{Correction} = \frac{78\% - 68\%}{257.0 - 249.5} \times 1.0 = \frac{10}{7.5} \times 1.0 = 1.33\% \text{ (+ ve)}$$

$$\phi_{C_1} \text{ (corrected)} = 68\% + 2.35\% + 1.33\% = \mathbf{71.68\%}.$$

**Corrections to $\phi_{E_2}$**

*(i) Effect of sheet pile No. (1) on pile No. (2) of depth d*

$$\text{Correction} = 19 \times \sqrt{\frac{D}{b'}}\left(\frac{d+D}{b}\right) \quad \text{... Eq. (11.3)}$$

where $D = 251.2 - 249.5 = 1.7$ m

$d = 251.2 - 245.2 = 6.0$ m

$b' = 58.5$ m

$b = 60$ m

$$\text{Correction} = 19 \times \sqrt{\frac{1.7}{58.5}}\left(\frac{6.0+1.7}{60.0}\right) = 19 \times \frac{1}{5.86} \times \frac{7.7}{60} = 0.42\% \ (-\text{ ve})$$

(*ii*) Correction due to thickness of floor :

$$\text{Correction} = \left(\frac{32\% - 22\%}{252.7 - 245.2}\right) 1.5 = \frac{10}{7.5} \times 1.5 = 2.0\% \ (-\text{ ve})$$

$\phi_{E_2}$ (Corrected) = 32% – 0.42% – 2.0% = **29.58%**

The levels of H.G. line at key points for various flow condition are tabulated in Table 11.9 as explained earlier.

**Table 11.9**

| *Condition of flow* | *u/s water level in metres* | *d/s water level in metres* | *Head in metres (H)* | *Height/Elevation of Subsoil H.G. Line above Datum* | | | | | |
|---|---|---|---|---|---|---|---|---|---|
| | | | | *Upstream Pile Line No. (1)* | | | *Downstream Pile Line No. 2* | | |
| | | | | $\phi_{E_1}$ *100%* | $\phi_{D_1}$ *78%* | $\phi_{C_1}$ *= 71.68%* | $\phi_{E_2}$ *= 29.58%* | $\phi_{D_2}$ *22%* | $\phi_{C_2}$ *0%* |
| | | | | 7.9 | 6.17 | 5.66 | 2.33 | 1.74 | 0 |
| No flow, maximum static Head | 260.6 | 252.7 (No water d/s) | 7.9 | 260.6 | 258.87 | 258.36 | 255.03 | 254.44 | 252.70 |
| | | | | 1.5 | 1.17 | 1.07 | 0.44 | 0.33 | 0 |
| (High Flood with concentration and retrogression) | 263.2 | 261.7 | 1.5 | 263.2 | 262.87 | 262.77 | 262.14 | 262.03 | 261.7 |
| | | | | 0.9 | 0.70 | 0.64 | 0.27 | 0.20 | 0 |
| Flow at pond level (with concentration and retrogression) | 260.6 | 259.7 | 0.9 | 260.6 | 260.40 | 260.34 | 259.97 | 259.9 | 259.7 |

We shall now determine the *hydraulic jump profiles* for the two flow conditions, *i.e.* High flood flow with concentration and retrogression, and pond level flow with concentration and retrogression.

(*a*) **Pre-jump profile** : It is worked out in Table 11.10 as explained earlier.

**Table 11.10. Pre-jump Profile Calculations**

| *Distance from the d/s end of the crest, i.e. the start of glacis, in metres* | *Glacis level in metres* | *High Flood Flow q = 32.9 cumecs/metre* | | *Pond Level Flow q = 13.86 cumecs/metre* | |
|---|---|---|---|---|---|
| | | $E_{f_1}$ *u/s TEL – Glacis Level, i.e. 264.20 – col. (2)* | $y_1$ *from Plate 10.2* | $E_{f_1}$ *u/s TEL – Glacis Level, i.e. 261.06 – Col. (2)* | $y_1$ *from Plate 10.2* |
| *(1)* | *(2)* | *(3)* | *(4)* | *(5)* | *(6)* |
| 0 | 257.0 | 7.2 | — | 4.06 | — |
| 3 | 256.0 | 8.2 | 3.5 | 5.06 | 1.8 |
| 6 | 255.0 | 9.2 | 3.0 | 6.06 | 1.4 |
| 6.63 | 254.79 Point at which jump is formed at Pond Level | 9.41 | 2.9 | 6.27 | 1.3 |
| 9.00 | 254.00 | 10.20 | 2.7 | | |
| 12.33 | 252.89 Point at which jump is formed at high flood | 11.31 | 2.5 | | |

(*b*) **Post jump Profile**

From Table 11.8

Froude No. for high flood condition, F = 2.66, or $F^2$ = 7.1

Depth $y_1$ for high flood condition = 2.5 m.

Froude No. for pond level flow condition = 2.97, or $F^2$ = 8.8

Depth $y_1$ for pond level flow condition = 1.3 m.

From Plate 10.3 (*a*), the Table 11.11 is worked out :

**Table 11.11. Post Jump Profile Calculations.**

| $\frac{x}{y_1}$ on Plate 10.3 (a) | High Flood Flow $F^2 = 7.1, y_1 = 2.5$ m | | | Pond Level Flow $F^2 = 8.8, y_1 = 1.3$ m | | |
|---|---|---|---|---|---|---|
| | $\frac{y}{y_1}$ from Plate 10.3 (a) | y | x = Col. (1) × 2.5 | $\frac{y}{y_1}$ from Plate 10.3 (a) | y | x = Col. (1) × 1.3 |
| (1) | (2) | (3) | (4) | (5) | (6) | (7) |
| 1.2 | 1.3 | 2.85 | 2.5 | 1.3 | 1.69 | 1.3 |
| 2.5 | 1.9 | 4.75 | 6.25 | 1.9 | 2.47 | 3.25 |
| 5.0 | 2.5 | 6.25 | 12.5 | 2.6 | 3.38 | 6.5 |
| 10.0 | 3.0 | 7.5 | 25.0 | 3.2 | 4.16 | 13.0 |

The Hydraulic jump profiles for these two flow conditions are plotted along with the H.G. lines for their respective conditions, as shown in Figs. 11.24 (*a*) and (*b*). From

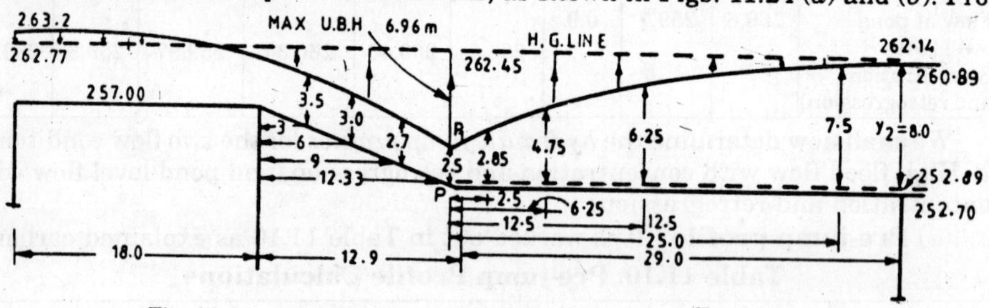


Fig. 11.24. (*a*). Unbalanced head in jump trough at high flood flow.

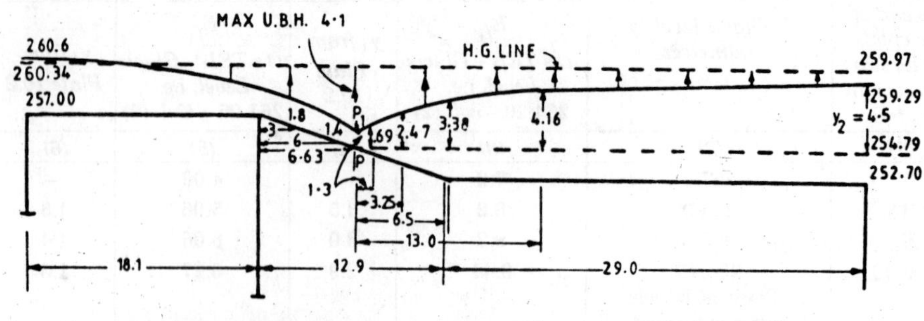


Fig. 11.24. (*b*). Unbalanced head in jump trough at pond level flow.

these figures, it is found that the maximum unbalanced head (U.B.H.) in the jump trough is equal to 6.96 m. The thickness for glacis shall, therefore, be designed for 2/3rd of this head, *i.e.* equal to 4.64 *metres*, or for the static condition head, whichever is greater. The H.G. line for the static head at pond level is also plotted in Fig. 11.24 (*c*).

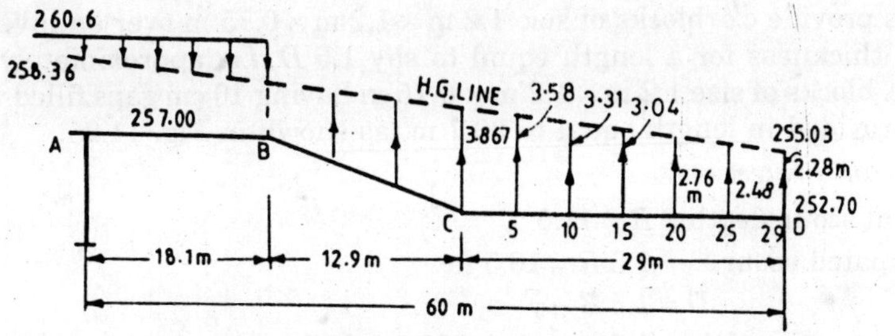


Fig. 11.24. (*c*). Unbalanced head in jump trough at static head at pond level.

It can be seen from the study of all these three figures, *i.e.* Fig. 11.24 (*a*), (*b*) and (*c*), that the dynamic action (*i.e.* flow) governs the floor thickness upto a length of about 5 m beyond the toe of the glacis, and the static head governs the rest of the downstream floor thickness.

Thickness required at point of jump

$$= \frac{4.64}{1.24} = 3.74 \text{ m}$$

Thickness required at 5 m beyond toe

$$= \frac{3.58}{1.24} = 2.88 \text{ m}$$

Thickness required at 15 m beyond toe

$$= \frac{3.04}{1.24} = 2.45 \text{ m}$$

Thickness required at 20 m beyond toe

$$= \frac{2.76}{1.24} = 2.23 \text{ m}$$

Thickness required at 25 m beyond toe

$$= \frac{2.48}{1.24} = 2.0 \text{ m}$$

Thickness required at 29 m beyond toe

$$= \frac{2.28}{1.24} = 1.84 \text{ m}$$

Provide floor thicknesses as shown in Fig. 11.25.

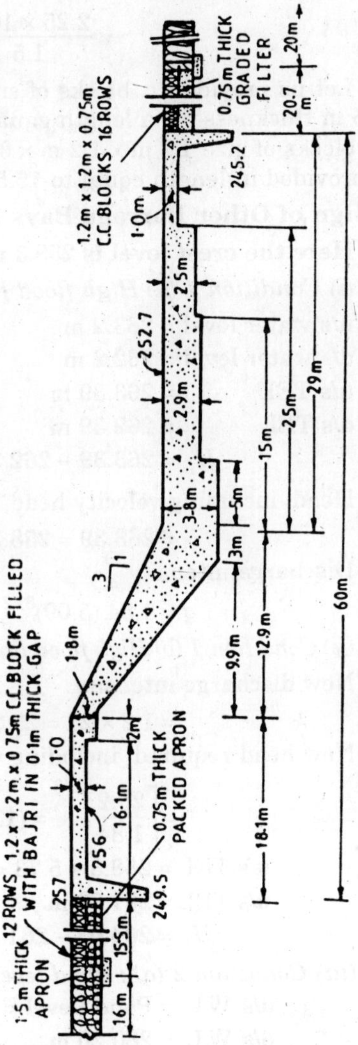


Fig. 11.25. Section of undersluice portion of Barrage.

## Protection Works

*Downstream Protection*

Normal scour depth $= R = 11.0$ m

$$D = 2R - y$$

$$= 2 \times 11.0 - (261.7 - 252.7)$$

$$= 22 - 9 = 13 \text{ m}.$$

Provide a launching apron equal 1.5*D*, *i.e.* say 20 m in length and of such a thickness (*t*) that its quantity is approximately 2.25 *D* cu. m/m.

$\therefore \quad t = \frac{2.25 \times 13.0}{20} = 1.46$ m ; **Say 1.5 m thick.**

Let us provide c.c. blocks of size 1.2 m × 1.2 m × 0.75 m over an inverted filter of 0.75 m thickness for a length equal to say 1.5 *D*, *i.e.* approximately 20 m. 16 rows of c.c. blocks of size 1.2 m × 1.2 m × 0.75 m having 10 cm gaps filled with bajri, shall be provided in length equal to 20.7 m, as shown in Fig. 11.25.

*Upstream Protection*

Normal scour depth = $R$ = 11.0

Anticipated scour = 1.5 $R$ = 16.5 m

$$D = 1.5\,R - y$$
$$= 1.5 \times 11.0 - (263.2 - 257.0) = 16.5 - 6.2 = 10.3 \text{ m}$$

Provide a launching apron of 1.5 m thickness in a length

$$= \frac{2.25 \times 10.3}{1.5} = 15.5 \text{ ; \textbf{say 16 m in length.}}$$

Let us provide c.c. blocks of size 1.2 m × 1.2 m × 0.75 m over packed apron of 0.75 m thickness for a length equal to say 1.5 *D*, *i.e.* 15.5 m in length. 12 rows of c.c. blocks of size 1.2 m × 1.2 m × 0.75 m having 10 cm gaps filled with bajri shall be provided in length equal to 15.5 m, as shown in Fig. 11.25.

## Design of Other Barrage Bays Portion

**Here the crest level is 258.3 m.** The four cases are analysed as given below :

(*i*) *Condition 1 (a) High flood flow with no concentration and retrogression.*

u/s water level = 263.2 m

d/s water level = 262.2 m

u/s TEL = 263.39 m

d/s TEL = 262.39 m

$$H_L = 263.39 - 262.39 = 1.0 \text{ m.}$$

Head, including velocity head, over the crest

$$= 263.39 = 258.3 = 5.09 \text{ m}$$

Discharge intensity

$$q = 1.84\,(5.09)^{3/2} = 1.84 \times 11.4 = 21 \text{ cumecs/metre.}$$

(*ii*) *Condition 1 (b) High flood flow with 20% concentration and 0.5 m retrogression*

New discharge intensity

$$= 1.2 \times 21 = 25.2 \text{ cumecs/metre}$$

New head required, including velocity head, for this discharge to pass.

$$= \left[\frac{25.2}{1.84}\right]^{2/3} = (13.7)^{0.667} = 5.73 \text{ m.}$$

u/s TEL = 258.3 + 5.73 = 264.03 m

u/s TEL = 261.89 m

$$H_L = 264.03 - 261.82 = 2.14 \text{ metres.}$$

(*iii*) *Condition 2 (a) : Pond level flow with no concentration and retrogression.*

u/s W.L. = Pond Level = 260.6 m

d/s W.L. = 260.20 m

u/s TEL = 260.69 m

d/s TEL = 260.29 m

$$H_L = 260.69 - 260.29 = 0.4 \text{ m}$$

Head, including velocity head, over the crest = 260.69 – 258.3 = 2.39 m

Discharge intensity = 1.84 $(2.39)^{3/2}$ = 1.84 × 3.69 = 6.78 cumecs/metre.

*(iv) Condition 2 (b) : Pond level flow with 20% concentration and 0.5 m retrogression.*

New discharge intensity = 1.2 × 6.78 = 8.13 cumecs/metre.

New head required, including velocity head, for this discharge to pass

$$= \left[\frac{8.13}{1.84}\right]^{3/2} = (4.42)^{0.667} = 2.69 \text{ m}$$

u/s TEL = 258.3 + 2.69 = 260.99 m

d/s TEL = 260.29 – 0.5 = 259.79 m

$H_L$ = 260.99 – 259.79 = **1.2 m.**

These values are tabulated in Table 11.12 and the table is completed as shown. The lowest level at which jump will form is found from this table as 254.39 m.

**Table 11.12. Other Barrage Bays Portion**

| S. No. | Item | *High Flood Flow* | | *Pond level Flow* | |
|---|---|---|---|---|---|
| | | *Condition 1 (a) without concentration and retrogression* | *Condition 1 (b) with 20% concentration and 0.5 m retrogression* | *Condition 2 (a) without concentration and retrogression* | *Condition 2 (b) with 20% concentration and 0.5 m retrogression* |
| *(1)* | *(2)* | *(3)* | *(4)* | *(5)* | *(6)* |
| 1. | Discharge intensity $q$ in cumec/metre | 21.0 | 25.2 | 6.78 | 8.13 |
| 2. | Upstream water level | 263.2 m | 263.2 m | 260.6 m | 260.6 m |
| 3. | Downstream water level | 262.2 m | 262.2 m | 260.2 m | 259.7 m |
| 4. | u/s TEL | 263.39 m | 264.03 m | 260.69 m | 260.99 m |
| 5. | d/s TEL | 262.39 m | 261.89 m | 260.29 m | 259.79 m |
| 6. | Head loss $H_L$ | 1.0 m | 2.14 m | 0.4 m | 1.2 m |
| 7. | $E_{f_1}$ from plate 10.1 | 6.3 m | 7.5 m | 3.0 m | 3.7 m |
| 8. | Level at which jump will form, *i.e.* d/s TEL – $E_{f_2}$ | 256.09 m | **254.39 m*** | 257.29 m | 256.09 m |
| 9. | $E_{f_1} = E_{f_2} + H_L$ | 7.3 m | 9.64 m | 3.4 m | 4.9 m |
| 10. | $y_1$ corresponding to $E_{f_1}$ (plate 10.2) | 2.1 m | 1.9 m | 1.1 m | 0.9 m |
| 11. | $y_2$ corresponding to $E_{f_2}$ (plate 10.2) | 5.7 m | 6.8 m | 2.7 m | 3.3 m |
| 12. | Length of concrete floor required, *i.e.* 5 $(y_2 - y_1)$ | 5 (3.6) = 18.0 m | 5 (4.9) = 24.5 m | 5 (1.6) = 8.0 m. | 5 (2.4) = 12.0 m |
| 13. | Froude No. $F_1 = \frac{q}{\sqrt{g y_1^3}}$ | 2.52 | 3.09 | 1.91 | 3.04 m |

*Lowest point at which jump will form.

Hence, provide the downstream floor at level, say 254.2 m. (slightly lower than the calculated value of 254.39 m), and equal to *26* metres in length. The glacis is provided in 3 : 1 slope with a horizontal length of 3 (258.3 – 254.2) = 3 × 4.1 = 12.3 metres. Crest width provided is 2 metres and upstream glacis in a slope of 1 : 1 with a horizontal length equal to 258.3 – 257.0 = 1.3 metres, as shown in Fig. 11.26.

**Depth of Sheet Piles from Scour**

Discharge passing = 6,260 cumecs

Overall waterway = 374 metres

Average discharge intensity = $\frac{6,260}{374}$ = 16.8 cumecs/metre

$$R = 1.35\left[\frac{(16.8)^2}{1}\right]^{1/3} = 1.35 \times 6.6 = 9.07 \text{ m.}$$

Let us provide a downstream cutoff upto depth 1.5 $R$ below the d/s water level which is 261.7 m with retrogression. Hence, the R.L. of bottom of d/s cut-off

$$= 261.7 - 1.5 \times 9.07 = 261.7 - 13.6 = 248.1 \text{ m.}$$

Let us provide d/s cut-off upto a bottom level of 248.1 m, *i.e.* depth of 6.1 m, as shown in Fig. 11.26.

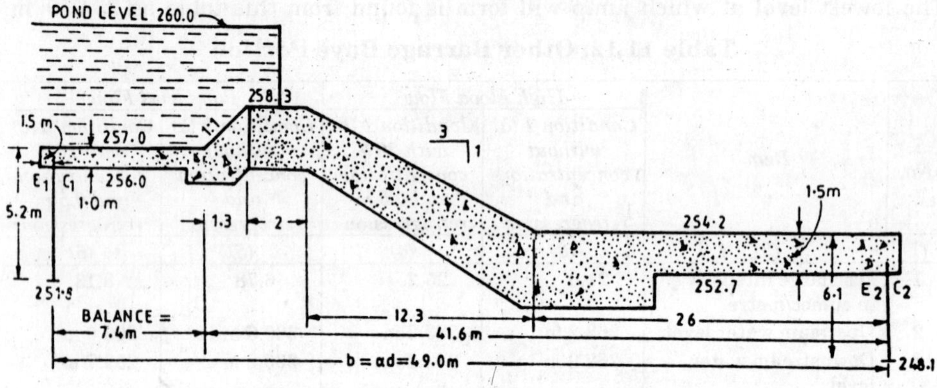


Fig. 11.26

*U/s Cutoff.* Let us provide u/s cutoff up to 1.25 $R$ level ; *i.e.* up to

$$= 1.25 \times 9.07 = 11.4 \text{ m below the W.L.}$$

∴ The level of bottom of u/s cutoff = 263.2 – 11.4 = 251.8 m.

Let us provide u/s cutoff upto a bottom level of 251.8 m, *i.e.* for a depth of 5.2 m, as shown in Fig. 11.26.

**Total Floor Length and Exit Gradient**

$$G_E = \frac{1}{6}.$$

Maximum Static Head = $H$ = 260.6 – 254.2 = 6.4 m

Depth of d/s cutoff = $d$ = 254.2 – 248.1 = 6.1 m

$$G_E = \frac{H}{d} \cdot \frac{1}{\pi \cdot \sqrt{\lambda}}$$

$$\therefore \quad \frac{1}{6} = \frac{6.4}{6.1} \cdot \frac{1}{\pi \cdot \sqrt{\lambda}}$$

or $$\frac{1}{\pi \cdot \sqrt{\lambda}} = \frac{1}{6} \times \frac{6.1}{6.4} = 0.159.$$

From Plate 11.2, value of $\alpha$ for a value of $\frac{1}{\pi\sqrt{\lambda}}$ as 0.159 comes out to be approximately 8.0.

∴ $b = \alpha d = 8 \times 6.1 = 48.8$ m ; say provide $b = 49$ m.

∴ Balance length of 49 – 1.3 – 2.0 – 12.3 – 26.0 = 49 – 41.6 = 7.4 m is provided as upstream floor, as shown in Fig. 11.26.

**Uplift Pressures**

Let us assume floor thickness as 1.0 m at u/s cutoff end and 1.5 m at d/s cutoff end, as shown in Fig. 11.26.

**Upstream Pile No. (1)**

$$b = 49 \text{ m}$$
$$d = 5.2 \text{ m}$$
$$\frac{1}{\alpha} = \frac{d}{b} = \frac{5.2}{49.0} = 0.106.$$

From Plate 11.1 (*a*)

$$\phi_{E_1} = 100\%$$
$$\phi_{C_1} = 100 - \phi_E = 100 - 29 = 71\%$$
$$\phi_{D_1} = 100 - \phi_D = 100 - 20 = 80\%$$

**Downstream Pile No. (2)**

$$d = 6.1 \text{ m}$$
$$b = 49.0 \text{ m}$$
$$\frac{1}{\alpha} = \frac{d}{b} = \frac{6.1}{49.0} = 0.125$$

From Plate 11.1 (*a*)

$$\phi_{C_2} = 0\%$$
$$\phi_{E_2} = \phi_E = 32\%$$
$$\phi_{D_2} = \phi_D = 22\%$$

**Let us correct these pressures**

$$\phi_{C_1} = 71\%$$
$$\phi_{E_2} = 32\%$$

**Corrections to $\phi_{C_1}$**

*(i) Effect of sheet pile No. (2) on pile No. (1) of depth d*

$$\text{Correction} = 19 \times \sqrt{\frac{D}{b'}}\left(\frac{d + D}{b}\right) \qquad \text{... Eq. (11.18)}$$

where $d = 256.0 - 251.8 = 4.2$ m

$D = 256.0 - 248.1 = 7.9$ m

$b' = 47.5$ m

$b = 49.0$ m

$$\text{Correction} = 19\sqrt{\frac{7.9}{47.5}}\left(\frac{4.2 + 7.9}{49}\right)$$
$$= 19 \times \frac{1}{2.45} \times \frac{12.1}{49} = 1.49\% \text{ (+ ve).}$$

*(ii)* Correction for depth of floor

$$\text{Correction} = \frac{80\% - 71\%}{257.0 - 251.8} \times 1.0 = \frac{9}{5.2} \times 1.0 = 1.73\% \text{ (+ ve)}$$

$\phi_{C_1}$ *(corrected)* = 71% + 1.49 + 1.73% = **74.22%**

**Corrections to $\phi_{E_2}$**

*(i) Effect of sheet pile No. (1) on sheet pile No. (2) of depth d*

$$\text{Correction} = 19 \times \sqrt{\frac{D}{b'}}\left(\frac{d+D}{b}\right) \qquad \text{... Eq. (11.18)}$$

where $d = 252.7 - 248.1 = 4.6$ m

$D = 252.7 - 251.8 = 0.9$ m

$b' = 47.5$ m

$b = 49.0$ m

$$\text{Correction} = 19 \times \sqrt{\frac{0.9}{47.5}}\left(\frac{4.6+0.9}{49}\right) = 0.29\% \ (-\text{ve})$$

*(ii) Correction due to thickness of floor*

$$\text{Correction} = \frac{32\% - 22\%}{254.2 - 248.1} \times 1.5 = 2.46\% \ (-\text{ve})$$

$\phi_{E_2}$ *(corrected)* = 32 – 0.29 – 2.46 = **29.25%**

The levels of H.G. line at key points for various flow conditions are tabulated in Table 11.13.

**Table 11.13**

| *Condition of flow* | *U/s water level in metres* | *D/s water level in metres* | *Head in metres H* | *Height/Elevation of sub soil H.G. line above datum* | | | | | |
|---|---|---|---|---|---|---|---|---|---|
| | | | | *Upstream pile line No. (1)* | | | *Downstream pile line No. (2)* | | |
| | | | | $\phi_{E_1}$ *100%* | $\phi_{D_1}$ *80%* | $\phi_{C_1}$ *74.22%* | $\phi_{E_2}$ *29.25%* | $\phi_{D_2}$ *22%* | $\phi_{C_2}$ *0%* |
| No flow, maximum static head | 260.6 | 254.2 (No water) | 6.4 | 6.4 | 5.12 | 4.75 | 1.87 | 1.41 | 0 |
| | | | | 260.6 | 259.32 | 258.95 | 256.07 | 255.61 | 254.2 |
| High flood flow with concentration and retrogression | 263.2 | 261.7 | 1.5 | 1.5 | 1.2 | 1.11 | 0.44 | 0.33 | 0 |
| | | | | 263.2 | 262.90 | 262.81 | 262.14 | 262.03 | 261.7 |
| Flow at pond level with concentration and retrogression | 260.6 | 259.7 | 0.9 | 0.9 | 0.72 | 0.67 | 0.26 | 0.20 | 0 |
| | | | | 260.60 | 260.42 | 260.37 | 259.96 | 259.90 | 259.7 |

We shall now determine the hydraulic jump profiles for the two flow conditions.

**Pre-jump Profile** is worked out in Table 11.14.

**Table 11.14. Pre-jump Profile Calculations**

| *Distance from the start of 3 : 1 glacis* | *Glacis level in metres* | *High Flood Flow with concentration and retrogression $q = 25.2$ cumecs / metre* | | *Pond Level Flow with concentration and restogression $q = 8.13$ cumecs / metre* | |
|---|---|---|---|---|---|
| | | $E_{f_1}$ = *u/s TEL – glacis level i.e. 264.03 —col. (2)* | $y_1$ *From Plate 10.2* | $E_{f_1}$ = *u/s TEL – glacis level i.e. 260.99 – col. (2)* | $y_1$ *From Plate 10.2* |
| (1) | (2) | (3) | (4) | (5) | (6) |
| 0 | 258.3 | 5.73 | — | 2.69 | — |
| 3 | 257.3 | 6.73 | 3.0 | 3.69 | 1.15 |
| 6 | 256.3 | 7.73 | 2.25 | 4.69 | 1.00 |
| 6.63 | 256.09 | 7.94 | 2.2 | 4.90 | 0.90 |
| | Point at which jump forms for Pond Level flow | | | | |
| 9.0 | 255.3 | 8.73 | 2.1 | — | — |
| 11.73 | 254.39 | 9.64 | 1.9 | — | — |
| | Point at which jump forms for High Flood flow | | | | |

**Post Jump Profile.** If $(x, y)$ are the ordinates of any point w.r. to the point of jump $P'$ as the origin, the profile can be worked out using plate 10.3 (*b*) as shown in Table 11.15. This profile can also be worked out by using plate 10.3 (*a*) as was done in the design of undersluice portion (Table 11.11). We have used both these plates just to explain their respective use.

**Table 11.15. Post Jump Profile Calculations**

| *Values of x where x is the horizontal distance from the point of jump* | *High Flood Flow* $y_2 - y_1 = 4.9, F_1 = 3.09$ | | | *Pond Level Flow* $y_2 - y_1 = 2.4, F_1 = 3.04$ | | |
|---|---|---|---|---|---|---|
| | $\frac{x}{y_2 - y_1}$ | $\frac{y}{y_2 - y_1}$ *From Plate 10.3 (b)* | $y$ | $\frac{x}{y_2 - y_1}$ | $\frac{y}{y_2 - y_1}$ *From Plate 10.3 (b)* | $y$ |
| 2 m | 0.41 | 0.16 | 0.79 | 0.83 | 0.33 | 0.79 |
| 4 m | 0.81 | 0.32 | 1.57 | 1.67 | 0.51 | 1.22 |
| 6 m | 1.20 | 0.38 | 1.86 | 2.50 | 0.68 | 1.63 |
| 8 m | 1.61 | 0.50 | 2.45 | 3.33 | 0.77 | 1.85 |
| 10 m | 2.01 | 0.58 | 2.84 | 4.17 | 0.88 | 2.11 |
| 15 m | 3.06 | 0.76 | 3.73 | 6.25 | 0.98 | 2.35 |
| 20 m | 4.08 | 0.86 | 4.21 | 8.25 | 1.00 | 2.40 |
| 25 m | 5.10 | 0.95 | 4.67 | 10.4 | — | — |

Hydraulic jump profile for two flow conditions, their H.G. lines and the uplift pressure diagrams are plotted in Fig. 11.27 (*a*) and (*b*). The H.G. line and the uplift pressure diagram for maximum static head is also plotted in Fig. 11.27 (*c*). By studying these diagrams, it is found that the maximum design head for dynamic

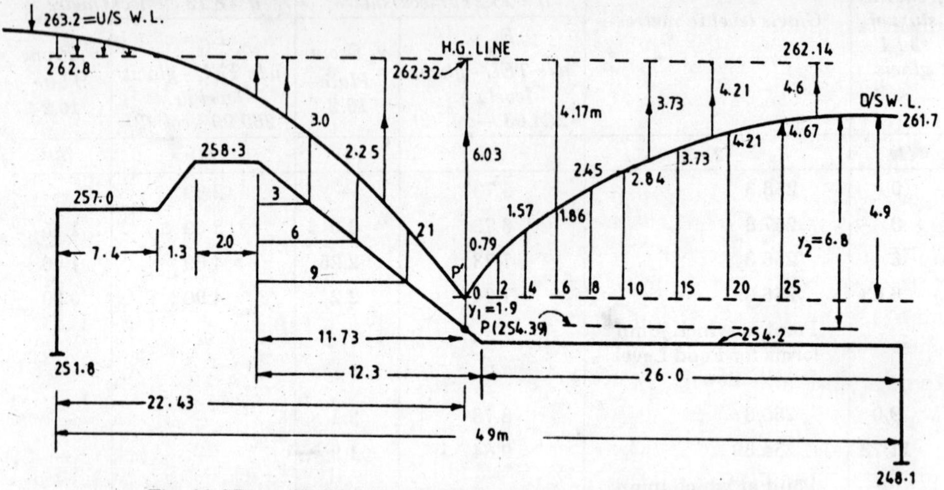


Fig. 11.27. (*a*). Uplift pressures in the jump trough for high flood flow.

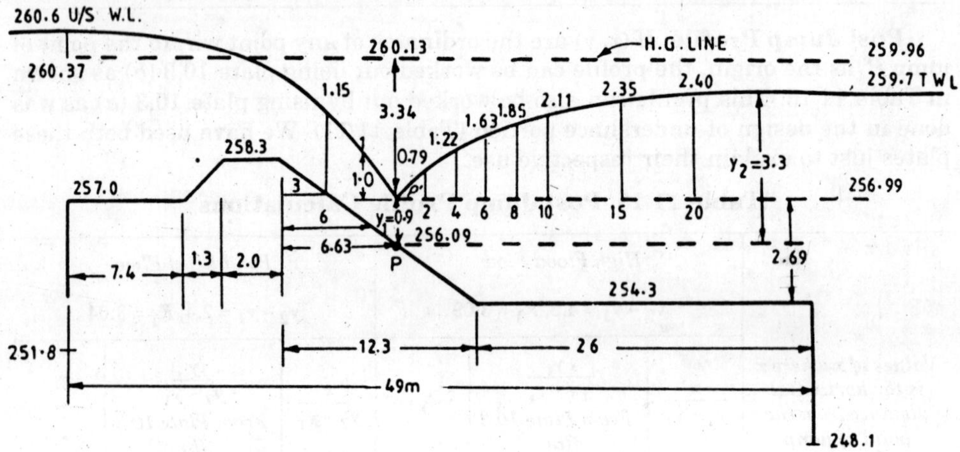


Fig. 11.27. (*b*). Uplift pressures in the jump trough for pond level flow.

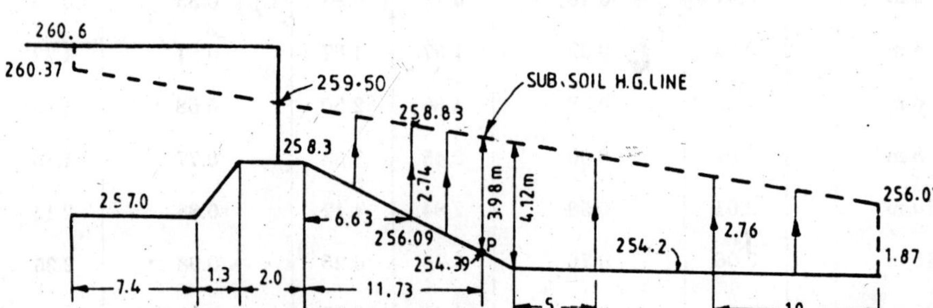


Fig. 11.27. (*c*). Uplift pressures in the jump trough for max. static head at pond level.

action is $\frac{2}{3} \times 6.03 = 4.02$ m [Fig. 11.27 (*a*)]. The maximum ordinate for uplift in static head condition = 4.12 m [Fig. 11.27 (*c*)]. Hence, static head condition becomes the governing factor. So minimum thickness required.

$$= \frac{4.12}{1.24} = 3.32 \text{ m.}$$

So we *provide 3.5 m.*

For the rest of the downstream floor beyond 5 m of the toe of the glacis, the thickness is decided by dividing the uplift ordinates of Fig. 11.27 (*c*) by 1.24; and by making suitable adjustments, the depths are provided as shown in Fig. 11.28.

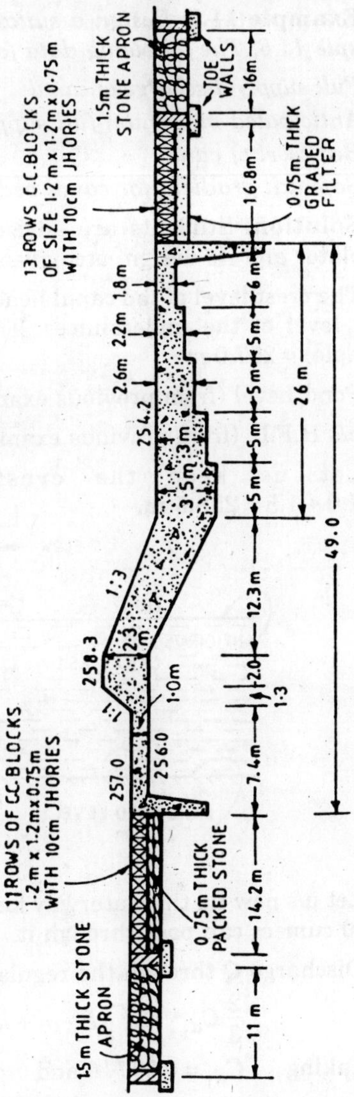


Fig. 11.28. Section of other Barrage Bays Portion

**Protection Works**

(*i*) *Downstream Protection*

Normal scour depth

$R = 9.07$ m

$D = 2R - y$

$= 2 \times 9.07 - (261.7 - 254.2)$

$= 18.14 - 7.5 = 10.64$ m

Provide a launching apron equal to 1.5 D, *i.e.* say 16 m in length and of thickness say 1.5 m.

Let us provide C.C. blocks of size 1.2 m × 1.2 m × 0.75 m over a graded filter of 0.75 m thickness for a length equal to 1.5 D, *i.e.* approximately 16 m. 13 rows of C.C. blocks of size 1.2 m × 1.2 m × 0.75 m having 10 cm gaps filled with bajri shall hence, be provided in length equal to 16.8 metres.

(*ii*) *Upstream Protection*

Normal scour depth

$= R = 9.07$ m

$D = 1.5\, R - y$

$= 1.5 \times 9.07 - (263.2 - 257.0)$

$= 13.6 - 6.2 = 7.4$ m.

Provide a launching apron of thickness 1.5 m in a length

$$= \frac{2.25 \times 7.4}{1.5} = 11.1 \text{ m.} = \textbf{Say 11 m.}$$

Let us provide C.C. blocks of size 1.2 m × 1.2 m × 0.75 m over packed stone of 0.75 m thickness for a length equal to say 1.5 D *i.e.* 13.6 m. Hence, provide 11 rows of C.C. blocks of size 1.2 m × 1.2 m × 0.75 m having 10 cm jhories (*i.e.* gaps filled with bajri), in a length equal to 14.2 metres, as shown in Fig. 11.28.

## Canal Head Regulator Design

**Example 11.7.** *Design a suitable head regulator for the barrage designed in example 11.6. The following data for the off-taking canal are also given :*

*Full supply discharge of canal* = *180 cumecs*
*Anticipated maximum full supply level of canal* = *260.2 m*
*Bed level of canal* = *257.2 m*
*Safe exit gradient for canal bed material* = *1/5.*

**Solution.** Students are advised to go through the theory for Canal Head Regulator and its design, etc. given in Chapter 9, before solving this problem.

The crest level of the canal head regulator is kept 1.2 to 1.5 m higher than the crest level of the undersluices. The crest level of undersluices (from previous example) = 257.0 m

Pond Level (from previous example = 260.6 m

u/s H.F.L. (from previous example) = 263.2 m

Let us keep the **crest level** of regulator (*i.e.* sill level) = 257.0 + 1.5 = **258.5 m.**

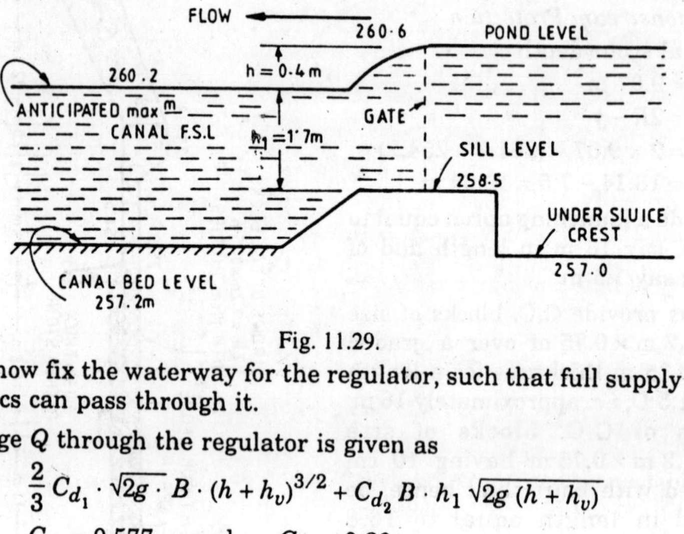


Fig. 11.29.

Let u/s now fix the waterway for the regulator, such that full supply discharge of 180 cumecs can pass through it.

Discharge $Q$ through the regulator is given as

$$\frac{2}{3} C_{d_1} \cdot \sqrt{2g} \cdot B \cdot (h + h_v)^{3/2} + C_{d_2} B \cdot h_1 \sqrt{2g\,(h + h_v)}$$

Taking $C_{d_1} = 0.577$ and $C_{d_2} = 0.80$

and neglecting velocity head $h_v$, we get

$$Q = \frac{2}{3} \times 0.577 \cdot \sqrt{2g} \cdot B \,.\, h^{3/2} + 0.80 \,.\, B \,.\, h_1 \sqrt{2gh.}$$

Here in this example, we have

$$h = 0.4 \text{ m} \;;\; h_1 = 1.7 \text{ m} \;; \text{and } Q = 180 \text{ cumecs}$$

$$\therefore \quad 180 = \frac{2}{3} 0.577 \sqrt{19.62.} \quad B\,(0.4)^{3/2} + 0.80\,B \,.\, (1.7) \sqrt{19.62 \times 0.4}$$

$$= B\,(1.69 \times 0.253 + 1.36 \times 2.8) = 4.237 \text{ B}$$

or
$$\text{B} = \frac{180}{4.237} = 42.5 \text{ m}$$

**Provide 6 bays of 7.5 m each giving a clear waterway of 45 m.** Provide 5 piers of 1.5 m each ; Overall waterway of the regulator = 45 + 5 × 1.5 = 52.5 m.

**Hydraulic Calculations for Various Flow Conditions**

(*i*) *Full supply discharge passing down the regulator during high flood.* When u/s water level is 263.2 m (*i.e.* High flood), water shall pass over the regulator and the gated opening provided between the sill level and pond level shall have to be adjusted by partially opening this gate, as shown in Fig. 11.30.

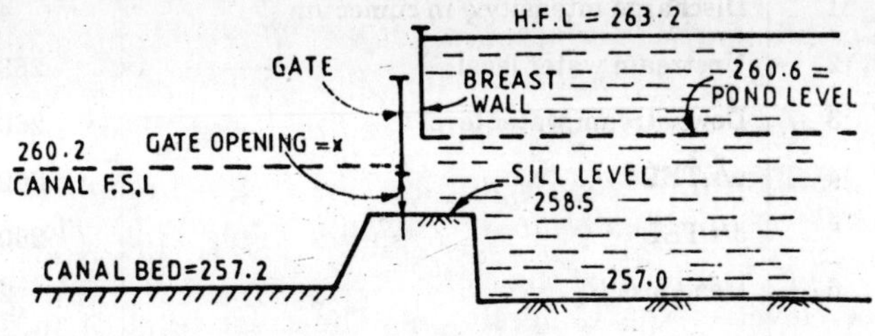


Fig. 11.30

Let the gate opening be $x$ metres. The discharge can then be calculated by submerged orifice formula. *i.e.*

$$Q = C_d \cdot A \cdot \sqrt{2gh}$$

where $C_d = 0.62$

$A = B \cdot x = 45 \cdot x \text{ m}^2$

$Q = 180$ cumecs.

$h$ = head causing flow = 263.2 – 260.2 = 3.0 m.

$\because \quad 180 = 0.62\,(45x)\,\sqrt{2 \times 9.81 \times 3}$

or $\quad 180 = 0.62 \times 45 \times 7.67\, x$

or $\quad x = \dfrac{180}{0.62 \times 45 \times 7.67} = \mathbf{0.84\ m.}$

Velocity of flow through opening

$$= \frac{180}{45 \times 0.84} = 4.76 \text{ m/sec}$$

Loss of head at entry $= 0.5\dfrac{V^2}{2g} = \dfrac{0.5 \times (4.76)^2}{2 \times 9.81} = 0.58$ metres

TEL just u/s of gate = 263.2 + 0.19 = 263.39 m

[$\because$ Vel. head is calculated in previous example = 0.19 m]

TEL just downstream of gate = 263.39 – 0.58 = 262.81 m.

Downstream water level = 260.2 m.

Head loss $= H_L = 262.81 - 260.2 = \mathbf{2.61\ m.}$

Discharge intensity $q = \dfrac{180}{45} = 4.4$ cumecs/metre.

(*ii*) *Full supply discharge passing down the regulator at pond level.*

$$H_L = 260.6 - 260.2 = 0.4 \text{ m}$$

$$q = 4.0 \text{ cumecs/metre.}$$

Hydraulic jump calculations for the two flow conditions are tabulated in Table 11.16.

**Table 11.16**

| *S.No.* | *Item* | *High flood flow condition* | *Pond level flow condition* |
|---|---|---|---|
| 1 | Discharge intensity $q$ in cumecs/m | 4.0 | 4.0 |
| 2 | Upstream water level | 263.2 m | 260.6 m |
| 3 | Downstream water level | 260.2 m | 260.2 m |
| 4 | u/s TEL | 262.81 m | 260.6 m |
| 5 | d/s TEL | 260.2 m | 260.2 m |
| 6 | Head loss $H_L$ | 2.61 m | 0.4 m |
| 7 | $E_{f_2}$ (from Plate 10.1) | 2.75 m | 2.05 m |
| 8 | Level at which jump will form *i.e.* d/s TEL – $E_{f_2}$ | 257.45 m | 258.15 m |
| 9 | $E_{f_1} = E_{f_2} + H_L$ | 5.36 m | 2.54 m |
| 10 | $y_1$ corresponding to $E_{f_1}$ (Plate 10.2) | 0.45 m | 0.68 m |
| 11 | $y_2$ corresponding to $E_{f_2}$ (Plate 10.2) | 2.67 m | 1.90 m |
| 12 | Length of concrete floor required = $5\,(y_2 - y_1)$ | 5 × 2.22 = 11.1 | 5 × 1.22 = 6.1 m |
| 13 | Froude No. $F_1 = \dfrac{q}{\sqrt{g y_1^{\,3}}}$ | | |

*Provide downstream floor at R.L. 257.2 metres with a horizontal length of 13 metres.*

**Depth of Sheet Piles from Scour Considerations**

Discharge intensity $q$ = 4.0 cumecs/metre

*D/s Sheet Pile.* Depth of scour '$R$' $= 1.35 \cdot \left(\dfrac{q^2}{f}\right)^{1/3}$

$$= 1.35 \cdot \left(\frac{4^2}{1}\right)^{1/3} = 1.35 \times 2.52 = 3.4$$

Let us provide a d/s cutoff upto 1.5 $R$ *i.e.* 5.1 m below the d/s water level which is 260.2 m.

∴ R.L. of bottom of d/s cutoff = 260.2 – 5.1 = 255.1 m.

This value gives only 257.2 – 255.1 = 2.1 m deep cutoff, which is small. *So let us provide d/s cutoff up to a bottom level of 253.0 m i.e. 4.2 m deep below the d/s floor level.*

*U/s Sheet Pile.* Provide upstream sheet pile line down to the elevation 249.5 m (*i.e.* the same as that of the undersluices).

**Total Floor Length and Exit Gradient**

The worst condition of flow occurs when the maximum flood is passing in the river (u/s water level 263.2 m) and there is no water in the canal (*i.e.* the canal is completely closed). The bed level of d/s floor = 257.2 m.

Maximum static head under this condition = $H = 263.2 - 257.2 = 6.0$ m

Depth of d/s cutoff = $(d) = 4.2$ m.

$$G_E = 1/5 \text{ (Given)}$$

$$G_E = \frac{H}{d} \cdot \frac{1}{\pi\sqrt{\lambda}}$$

$$\therefore \quad \frac{1}{5} = \frac{6.0}{4.2} \cdot \frac{1}{\pi\sqrt{\lambda}} \qquad \text{or} \qquad \frac{1}{\pi\sqrt{\lambda}} = \frac{1}{5} \times \frac{4.2}{6.0} = \mathbf{0.140}$$

From Plate 11.2

$$\alpha = 9$$

$$\therefore \quad b = \alpha \cdot d = 9 \times 4.2 = 37.8 \text{ m}$$

**Adopt total floor length = 38 metres.**

The floor length shall be provided as below and as shown in Fig. 11.31.

| | |
|---|---|
| D/s horizontal length | = 13.0 m |
| D/s glacis length with 3 : 1 slope | = 4.2 m. |
| Crest width | = 2.0 m |
| Balance provided as upstream floor | = 18.8 m |
| | Total = 38.0 m |

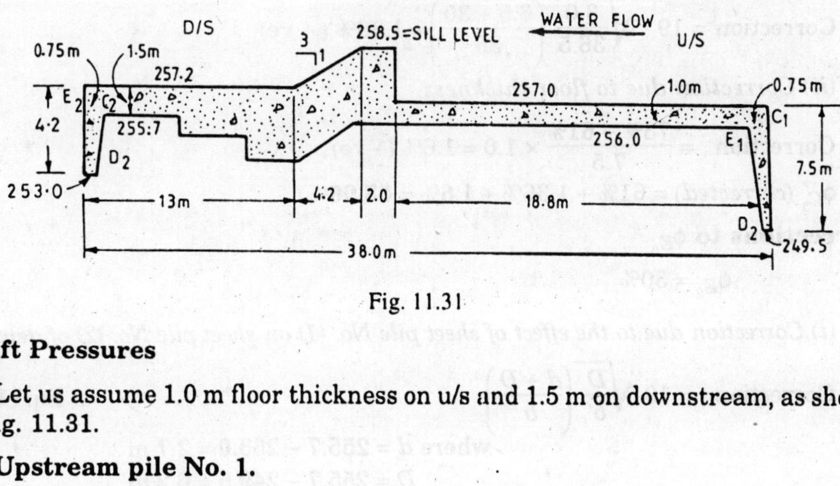


Fig. 11.31

**Uplift Pressures**

Let us assume 1.0 m floor thickness on u/s and 1.5 m on downstream, as shown in Fig. 11.31.

**Upstream pile No. 1.**

$$b = 38 \text{ m}$$

$$d = 7.5 \text{ m}$$

$$\frac{1}{\alpha} = \frac{d}{b} = \frac{7.5}{38} = 0.197.$$

**From Plate 11.1,**

$$\phi_{E_1} = 100\%$$

$$\phi_{D_1} = 100 - \phi_D = 100 - 27\% = 73\%$$

$$\phi_{C_1} = 100 - \phi_E = 100 - 39\% = 61\%$$

**Downstream pile No. (2)**

$$b = 38 \text{ m}$$
$$d = 4.2$$
$$\frac{1}{\alpha} = \frac{d}{b} = \frac{42}{38} = 0.11.$$

From Plate 11.1

$$\phi_{E_2} = \phi_E = 30\%$$
$$\phi_{D_2} = \phi_D = 20\%$$
$$\phi_{C_2} = 0\%$$

These pressures are to be corrected as follows :

**Corrections to $\phi_{C_1}$**

*(i) Correction due to the effect of sheet pile No. (2) on pile No. (1) of depth d.*

$$\text{Correction} = 19 \cdot \sqrt{\frac{D}{b'}} \cdot \left[\frac{d+D}{b}\right] \quad \text{... Eq. (11.18)}$$

where $d = 256.0 - 249.5 = 6.5$ m
$D = 256.0 - 253.0 = 3.0$ m
$b' = 36.5$ m
$b = 38$ m.

$$\text{Correction} = 19 \cdot \sqrt{\frac{3.0}{36.5}} \left(\frac{6.5 + 30}{38}\right) = 1.36\% \text{ ( + ve)}$$

*(ii) Correction due to floor thickness*

$$\text{Correction} = \frac{73\% - 61\%}{7.5} \times 1.0 = 1.6\% \text{ (+ ve)}$$

$\phi_{C_1}$ *(corrected)* = 61% + 1.36% + 1.6% = **63.96%**

**Corrections to $\phi_{E_2}$**

$$\phi_{E_2} = 30\%$$

*(i) Correction due to the effect of sheet pile No. (1) on sheet pile No. (2) of depth d.*

$$\text{Correction} = 19 \sqrt{\frac{D}{b'}} \left(\frac{d+D}{b}\right) \quad \text{... Eq. (11.18)}$$

where $d = 255.7 - 253.0 = 2.7$ m
$D = 255.7 - 249.5 = 6.2$ m
$b' = 36.5$ m
$b = 38$ m.

$$\text{Correction} = 19 \cdot \sqrt{\frac{6.2}{36.5}} \left(\frac{2.7 + 6.2}{38}\right) = 1.84\% \text{ (– ve).}$$

*(ii) Correction for floor thickness*

$$\text{Correction} = \frac{30\% - 20\%}{257.2 - 253.0} \times 1.5 = 3.57\% \text{ (– ve)}$$

$\phi_{E_2}$ *(corrected)* = 30 – 1.84 – 3.57 = **24.59%**

The levels of the H.G. lines at key points for different flow conditions are tabulated in Table 11.17.

**Table 11.17**

| Condition of flow | U/s water level in metres | D/s water level in metres | Head in metres (H) | *Height/elevation of sub-soil H.G. line above datum* | | | | | |
|---|---|---|---|---|---|---|---|---|---|
| | | | | *Upstream pile line No. (1)* | | | *Downstream pile line No. (2)* | | |
| | | | | $\phi_{E_1}$ 100% | $\phi_{D_1}$ 73% | $\phi_{C_1}$ 63.96% | $\phi_{E_2}$ 24.59% | $\phi_{D_2}$ 20.0% | $\phi_{C_2}$ 0% |
| | | | | 6.0 | 4.38 | 3.84 | 1.48 | 1.20 | 0 |
| No flow, maximum static head | 263.2 | 257.2 No water in canal | 6.0 | 263.2 | 261.58 | 261.04 | 258.68 | 258.40 | 257.2 |
| | | | | 3.0 | 2.19 | 1.92 | 0.74 | 0.60 | 0 |
| High flood on barrage and full supply discharge through canal head regulator | 263.2 | 260.2 (Canal FSL) | 3.0 | 263.2 | 262.39 | 262.12 | 260.94 | 260.8 | 260.2 |
| | | | | 0.4 | 0.29 | 0.26 | 0.10 | 0.08 | 0 |
| Flow at pond level | 260.6 | 260.2 (Canal FSL) | 0.4 | 260.6 | 260.49 | 260.46 | 260.30 | 260.28 | 260.2 |

**Floor Thicknesses**

The maximum static head on the floor occurs for the worst condition *i.e.* when high flood is passing down the barrage but the canal is empty and regulator gates are closed.

∴ The maximum static head = 263.2 – 257.2 = 6.0 metres

The subsoil H.G. line shall be drawn for the maximum static head only (Fig. 11.32) since the floor thicknesses are governed by this critical condition.

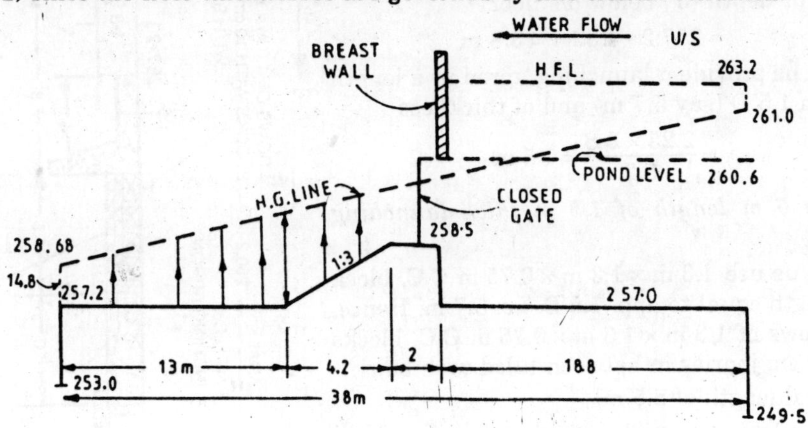


Fig. 11.32. Max. Static Head condition.

$$\text{Thickness reqd. at d/s end of floor} = \frac{1.48}{1.24} = 1.19 \text{ m.}$$

$$\text{At 4 m from end, d/s unbalanced head} = 1.48 + \frac{2.36}{38} \times 4 = 1.73 \text{ m.}$$

$$\therefore \text{Thickness reqd.} = \frac{1.73}{1.24} = 1.4 \text{ m.}$$

*Provide 1.5 metre thickness in this 4 m reach*

At 8 m from d/s end, unbalanced head

$$= 1.48 + \frac{2.36}{38} \times 8 = 1.98 \text{ m}$$

$$\text{Thickness reqd.} = \frac{1.98}{1.24} = 1.61 \text{ m.}$$

*Provide 1.8 m thickness in this portion*

At 13 m from d/s end, unbalanced head

$$= 1.48 + \frac{2.36}{38} \times 13 = 2.29 \text{ m}$$

$$\text{Thickness required} = \frac{2.29}{1.24} = 1.85 \text{ m}$$

*Provide 2 m thickness in this portion.*

The provided floor thicknesses are shown in Fig. 11.33.

**Protection works**

(*i*) *U/s Protection.* It shall be equal to what was provided in upstream of under-sluices

(*ii*) *D/s Protection*

$$R = 3.4 \text{ m (calculated earlier)}$$

Anticipated scour = $2R = 2 \times 3.4 = 6.8$ m.

The level of d/s scour hole

$$= \text{d/s water level} - 2R$$
$$= 260.2 - 6.8 = 253.4 \text{ m}$$

Scour depth ($D$) below d/s floor

$$= 257.2 - 253.4 = \mathbf{3.8 \text{ m}}$$

Let us provide a launching apron in a length equal to 1.5 D (say 5.7 m) and of thickness

$$= \frac{2.25 \times 3.8}{5.7} = 1.5 \text{ m.}$$

*Use 6 m length of 1.5 m thick launching apron.*

Let us use 1.3 m × 1.3 m × 0.75 m C.C. block in a length equal to app. 1.5 D, *i.e.* 5.7 m. Hence, use 4 rows of 1.3 m × 1.3 m × 0.75 m C.C. blocks with 10 cm jhories in between filled with bajri, in a total length of 5.5 m.

The protection works are shown in Fig. 11.33.

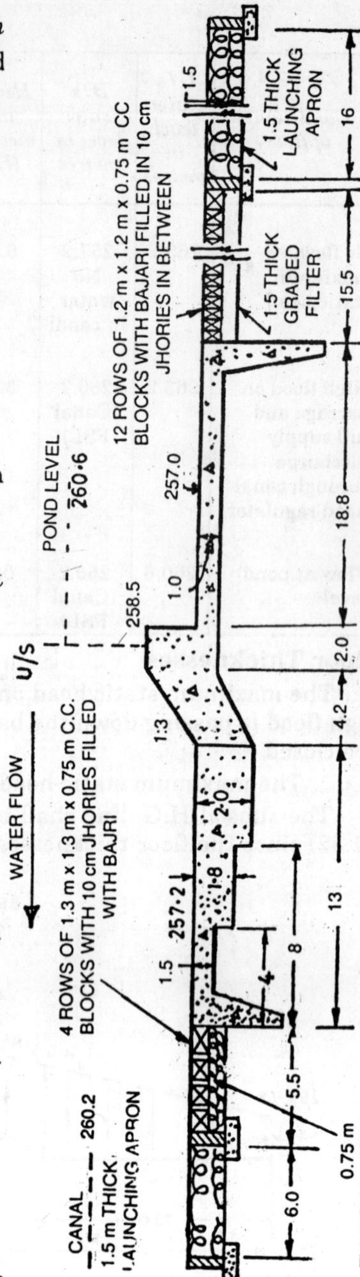


Fig. 11.33. Section of Head Regulator.

## CERTAIN IMPORTANT INDIAN BARRAGES

### 11.7. Data Pertaining to Certain Important Barrages of India

The data pertaining to the location, purpose, flood discharge, and other characteristics of certain important barrages/weirs constructed on Indian rivers, are given below :

**11.7.1. Durgapur Barrage.** Durgapur barrage (Fig. 11.34) is situated at ***Durgapur in Burdwan Distt., West Bengal State*** (on Damodar river). It helps in serving *irrigation, industrial, water supply,* and *navigation* needs. The work on

Fig. 11.34. (*a*). Photoview of Durgapur Barrage.

this barrage was started in the year 1952, and completed in 1955. The *design flood discharge* for the barrage site is 15,576 cumecs (5.5 lakh cusecs). Other details of the barrage and its canal head regulator are given below.

**Hydraulic Particulars of Durgapur Barrage**

| | |
|---|---|
| Pond Level | ...211.5 m |
| Width of river | ...2,000 m |
| Length of barrage/anicut | ...692.2 m |
| No. of under-sluice bays | ...10 |
| No. of barrage bays | ...24 |
| Width of under-sluice bays | ...18 m each |
| Width of barrage-bays | ...18 m each |
| Thickness of intermediate piers | ...2.13 m |
| **Gates.** (*a*) *Number* | ...24 in weir bays |
| | ...10 in undersluice (barrage) bays. |
| (*b*) *Size* Barrage | ...4.87 m× 18 m |
| Left undersluice | ...5.38 m × 18 m |
| Right undersluice | ...5.79 m × 18 m. |
| Sediment excluding devices | ...Silt excluder and under-sluices. |
| Length of upstream divide wall | ...80.47 m. |

**Canal Head Regulator Details**

| | *Left* | *Right* |
|---|---|---|
| **Width of head regulator** | **59.44 m** | **13.71 m** |
| **No. of bays and their width** | **8 of 6.09 m each** | **2 of 6.09 m each** |
| **Thickness of piers** | **2.44 m** | **1.52 m** |
| **Orientation with respect to barrage axis** | **Perpendicular to the barrage axis** | |
| **Location of head regulator from barrage axis** | **16 m from G.L. of barrage axis** | |
| **Energy dissipating devices below head regulator** | **c.c. blocks and boulder pitching apron** | |
| **Max. discharge of canal** | **260.18 cumecs** | **64.31 cumecs** |

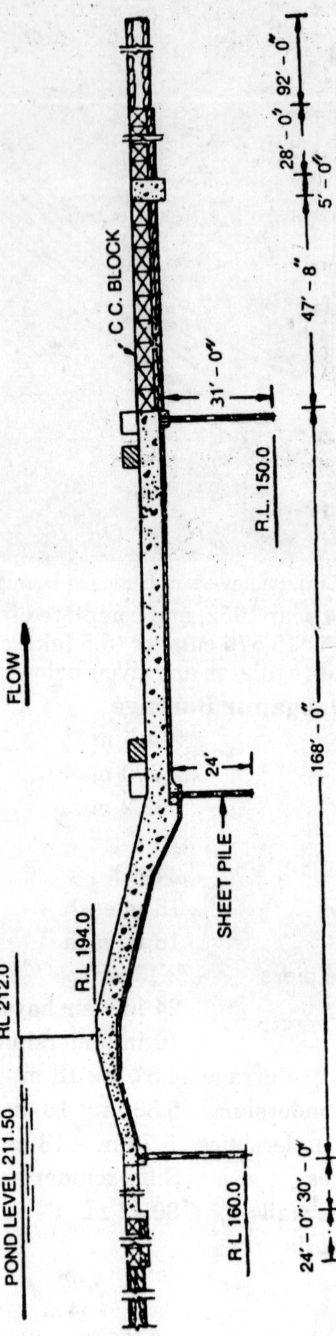


Fig. 11.34. (*b*). Section through under-sluices of Durgapur Barrage.

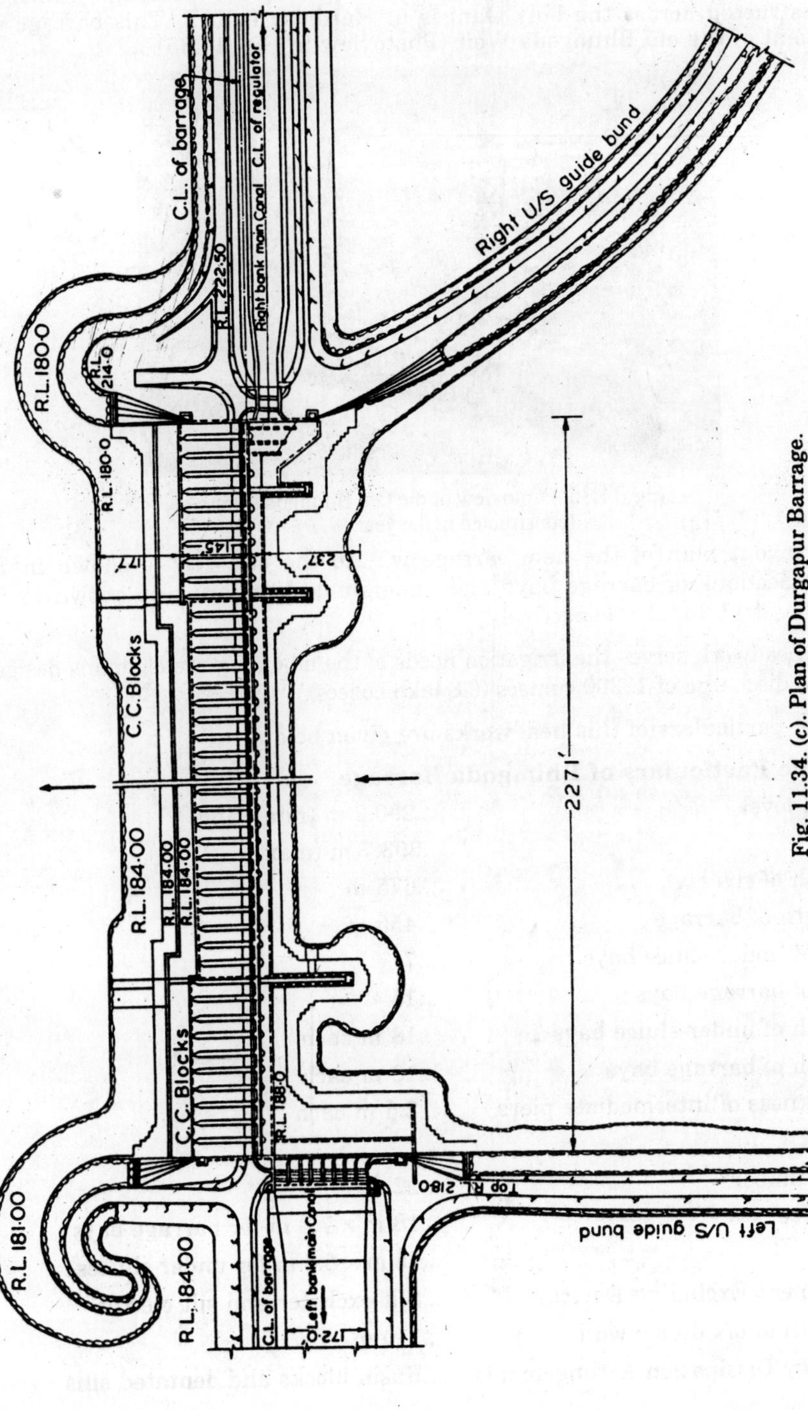


Fig. 11.34. (*c*). Plan of Durgapur Barrage.

**11.7.2. Bhimgoda Barrage.** Bhimgoda barrage is a new barrage, which has been constructed across the holy Ganges at Hardwar in U.P. This barrage is a replacement of the old Bhimgoda Weir (Photoview in Fig. 11.35).

Fig. 11.35. Photoview of the Old Bhimgoda Weir (constructed in the year 1920).

The layout plan of the new barrage w.r. to the old weir is shown in Fig. 11.36(*a*). Sections of barrage bays and undersluice bays are also shown in Fig. 11.36 (*b*) and 11.36 (*c*) ; respectively.

This headwork serves the irrigation needs of the adjoining areas, and is designed for a flood discharge of 19300 cumecs (6.8 lakh cusecs).

Other particulars of this headworks are given below.

**Hydraulic Particulars of Bhimgoda Barrage**

| | |
|---|---|
| Pond level | ...290.2 m (min.) and 293.7 m (max.) |
| Width of river | ...675 m |
| Length of barrage | ...455 m |
| No. of under-sluice bays | ...7 |
| No. of barrage bays | ...15 |
| Width of under-sluice bays | ...18 m each |
| Width of barrage bays | ...18 m each |
| Thickness of intermediate piers | ...2.5 m each |
| *Gates* | |
| (*a*) Number | ...22 |
| (*b*) Size | ...18 m × 7.8 m for barrage bays<br>...18 m × 8.4 m for under-sluices |
| Sediment Excluding Devices | ...Silt excluder and silt ejector |
| Length of u/s divide wall | ...110 m |
| Energy Dissipation Arrangements | ...Basin blocks and dentated sills |

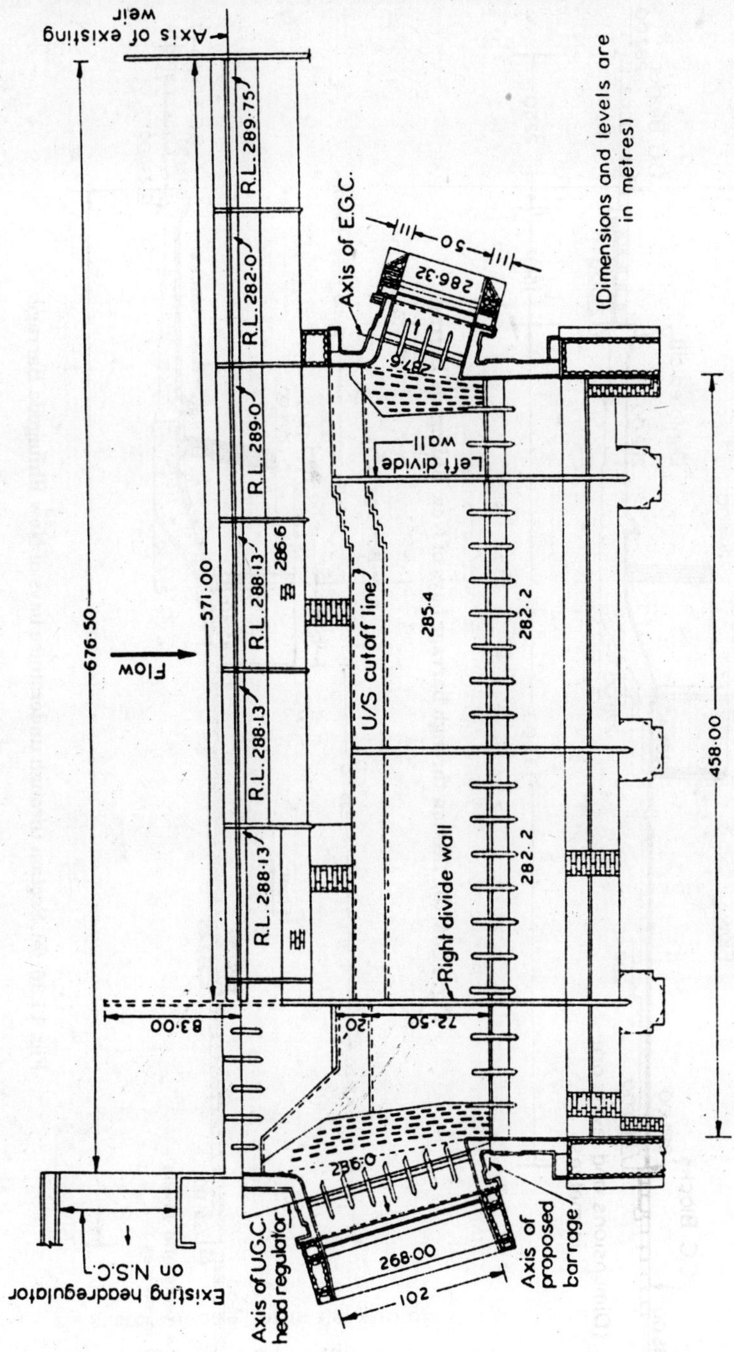


Fig. 11.36 (*a*). Layout Plan of New Bhimgoda Headworks (Barrage at Ganges river in U.P.

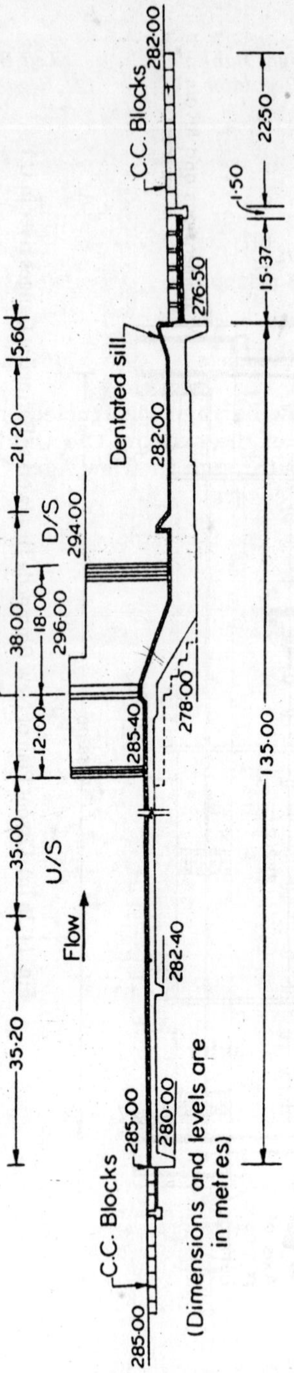


Fig. 11.36. (*b*). Section through barrage bays of New Bhimgoda Barrage

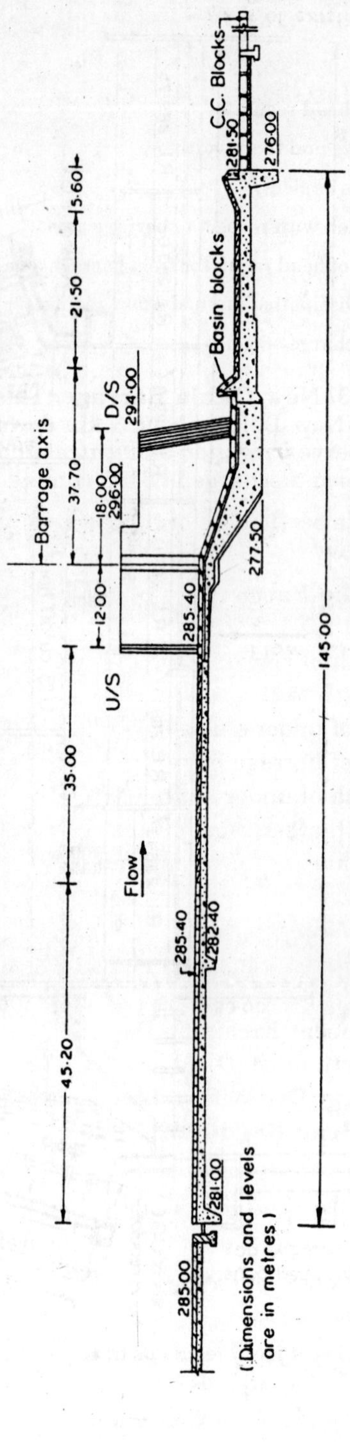


Fig. 11.36. (*c*). Section through undersluice bays of New Bhimgoda Barrage.

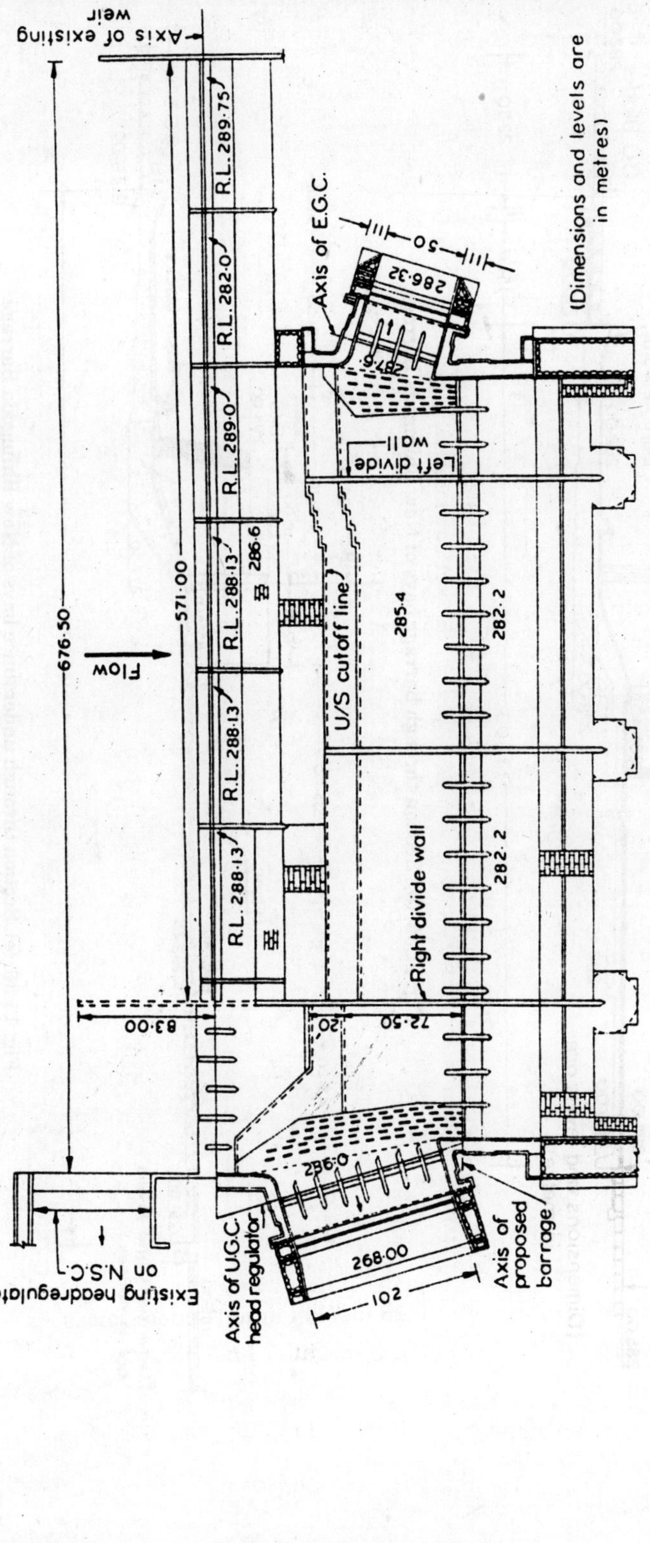


Fig. 11.36 (*a*). Layout Plan of New Bhimgoda Headworks (Barrage at Ganges river in U.P.

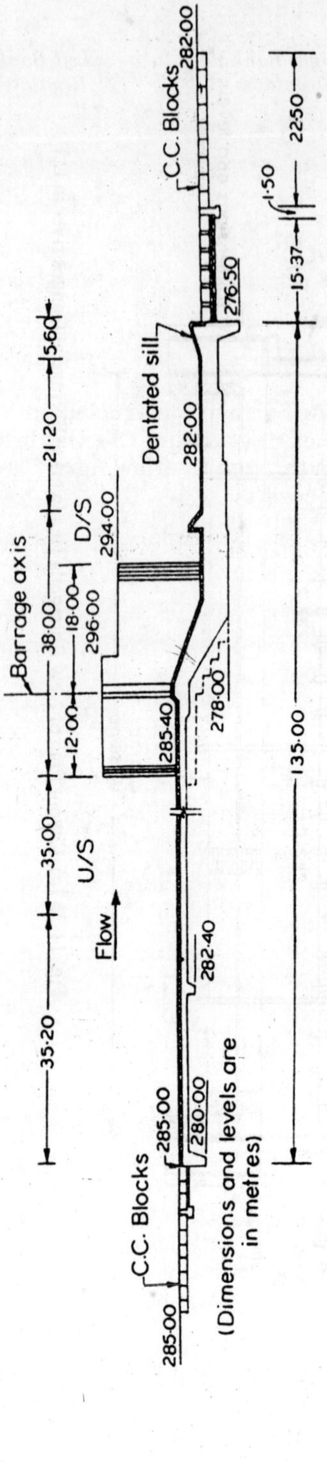


Fig. 11.36. (*b*). Section through barrage bays of New Bhimgoda Barrage

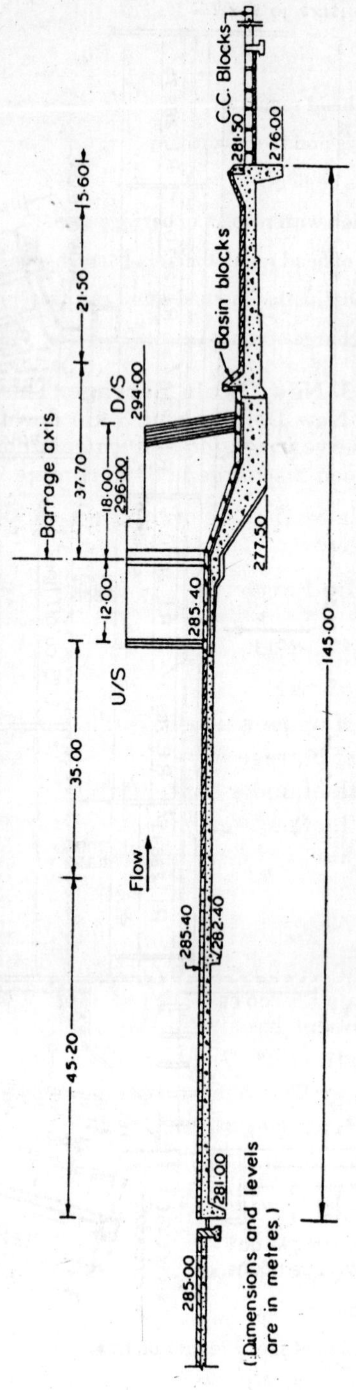


Fig. 11.36. (*c*). Section through undersluice bays of New Bhimgoda Barrage.

**Canal Head Regulator Details**

| | *Right Bank Regulator* | *Left Bank Regulator* |
|---|---|---|
| Width of head regulator | 102 m | 50 m |
| No. of bays and their width | 8 of 11 m each | 4 of 11 m each |
| Thickness of piers | 2 m | |
| Orientation with respect to barrage axis | 107° | |
| Location of head regulator from barrage axis | 66 m (app.) | 40 m (app.) |
| Energy Dissipation method below regulator | Basin blocks | |
| Max. discharge of canal | 410 cumecs | 164 cumecs |

**11.7.3. New Okhla Barrage.** This is a new barrage constructed on Yamuna river, at New Delhi, about 3 km downstream of the existing Old Okhla weir. It aims to serve *irrigation* and *water supply* needs through the New Agra Canal. The design flood discharge is 8495 cumecs. (3 lakh cusecs).

Other particulars of this barrage (Fig. 11.37) and its canal head regulator are given below :

**Hydraulic Particulars of New Okhla Barrage**

Pond level ...201.35 m
Width of river ...445.73 m
Length of barrage ...552 m
No. of under-sluice bays ...5
No. of barrage bays ...22
Width of under-sluice bays ...18.3 m each
Width of barrage bays ...18.3 m each
Thickness of intermediate pier ...2.13 m
*Gates*
(*a*) *Number* ...27
(*b*) *Size* For under-sluices : 5 of 18.3 m × 6 m each
For others : 22 of 18.3 m × 5.1 m each
Sediment Excluding Devices ...Silt excluder
Length of u/s Divide wall ...85.42 m
Energy Dissipation Arrangements ...Baffle blocks and Dentated sills

**Canal Head Regulator Details**

Width of head regulator ...80.85 m
No. of bays and their width ...9 of 7.65 m each
Thickness of piers ...1.5 m each
Orientation with respect to barrage axis ...100°
Location of Head regulator from barrage axis ...22.25 m u/s on right bank,
Energy dissipation arrangements below regulator ...Baffle blocks and end sills
Max. discharge of canal ...242.4 cumecs.

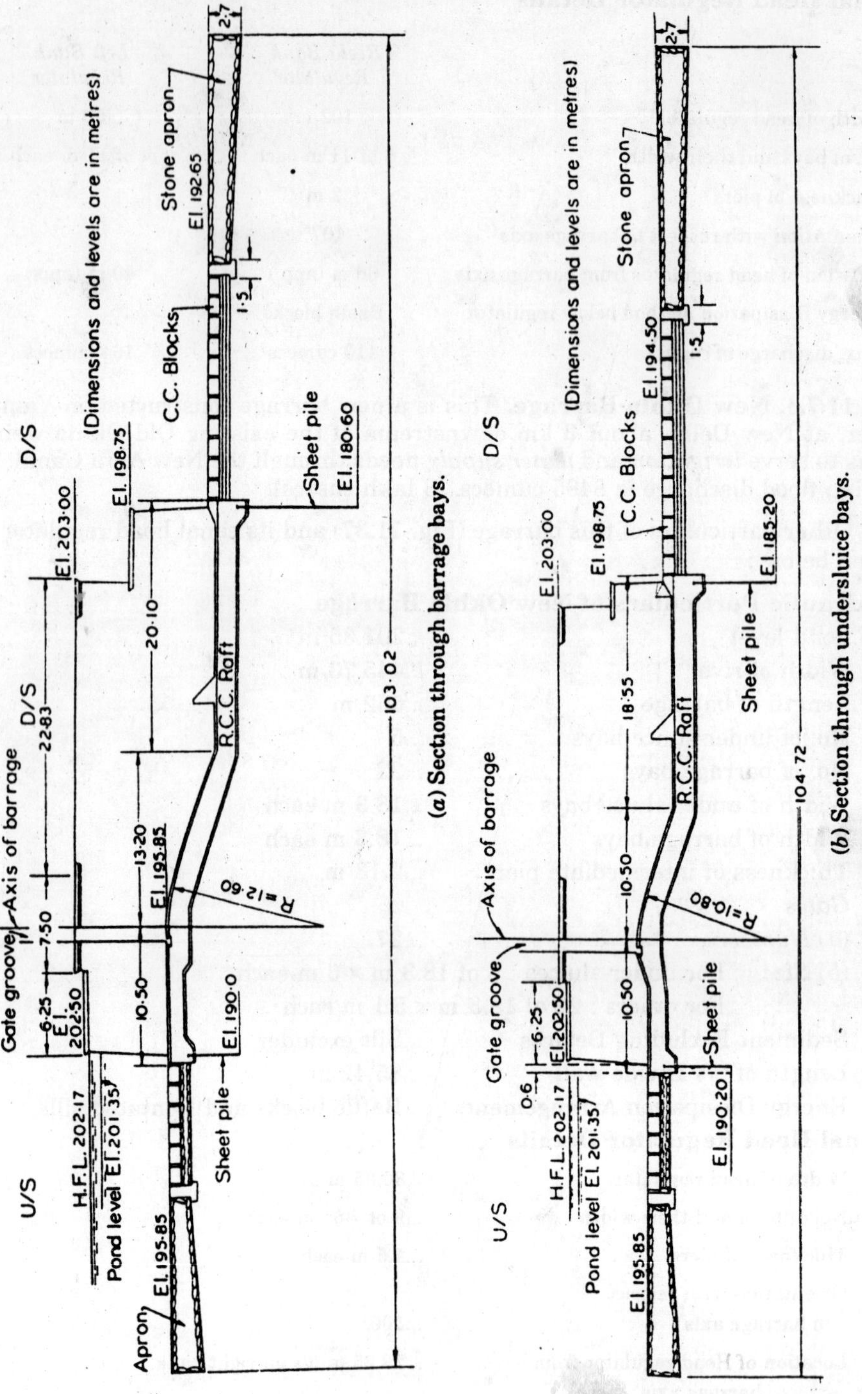


(a) Section through barrage bays.

(b) Section through undersluice bays.

Fig. 11.37. New Okhla Barrage at New Delhi on Yamuna river.

**Coloured Fig. 11.1[illegible].** Two views of Mettur dam.

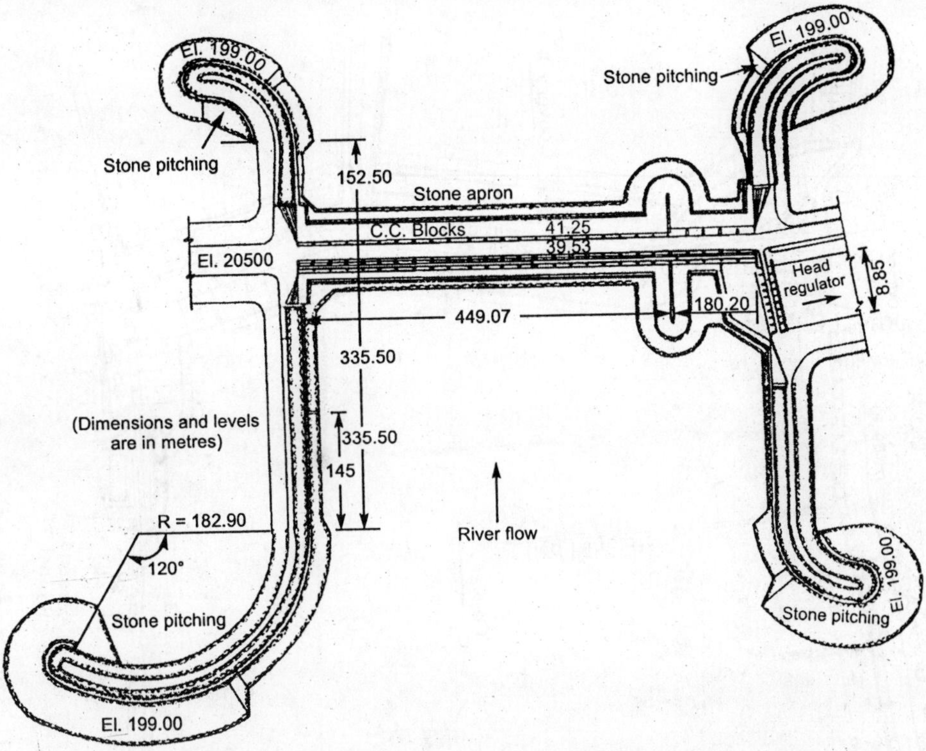


**Fig. 11.37.** (*c*) Layout Plan of New Okhla Barrage at New Delhi on Yamuna river.

**11.7.4. Grand Anicut.** The famous old *Grand Anicut* (Fig. 11.38 *a* & *b*) is located at 16 km from Tiruchirapalli (Tamil Nadu State) on Cauvery river, and is situated 193 km downstream of the famous Mettur dam (Colour Photo Fig. 11.1) near Salem.

**Fig. 11.38.** (*a*) Photoview (close) of Grand Anicut.

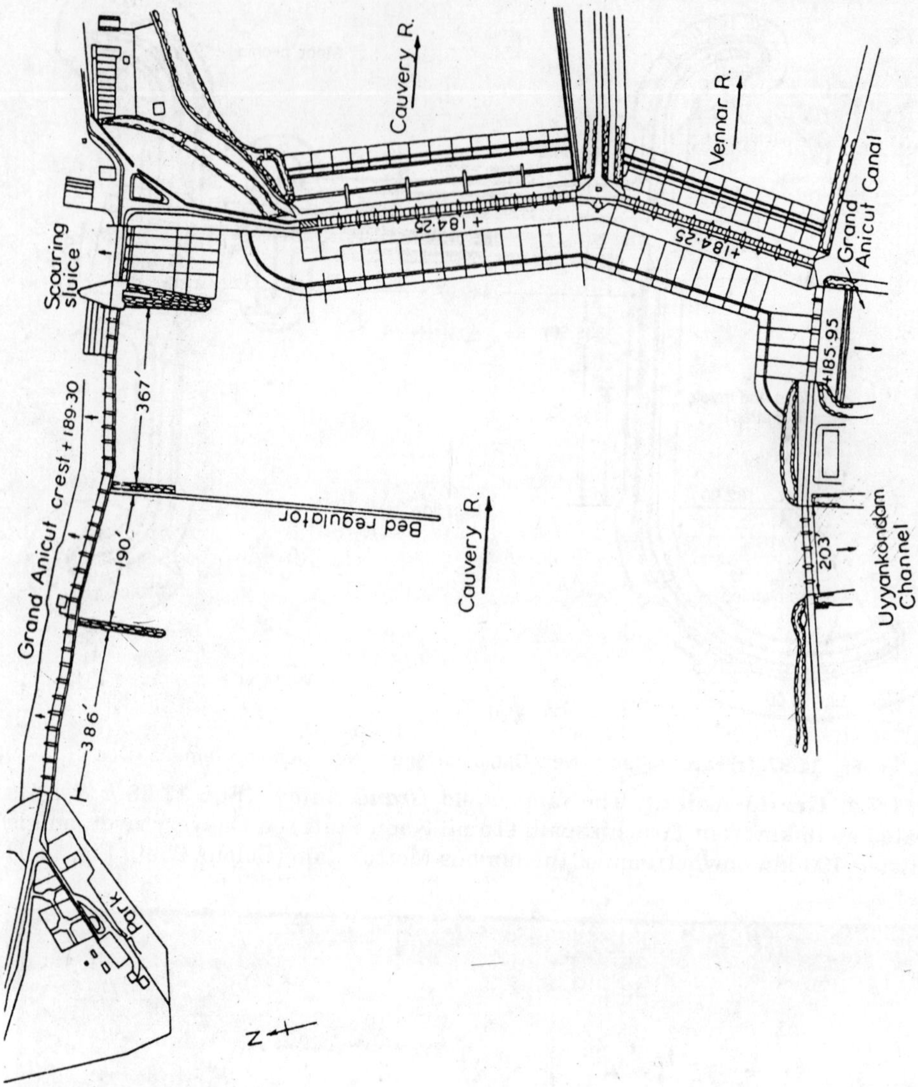


Fig. 11.38 (*b*) General Plan of Anicut Complex (Grand Anicut).

It serves the *irrigation* needs of the adjoining areas. The design flood discharge is 5094 cumecs (1.8 lakh cusecs). It was constructed in the year 200 A.D. Other particulars of this project are given below :

**Hydraulic Particulars of the Grand Anicut**

| | |
|---|---|
| Pond Level | ...61.567 m |
| Width of river | ...419 m |
| Length of anicut | ...329 m |
| No. of Under-sluice bays | ...5 |
| No. of Weir bays | ...30 |
| Width of Under-sluice bays | ...6.1 m each |
| Width of weir bays | ...9.75 m each |

| | |
|---|---|
| Thickness of intermediate pier | ...1.22 m (weir) and 1.83 m (under-sluices) |
| *Gates* | |
| (*a*) Number | ...30 |
| (*b*) Size | ...9.75 m × 1.52 m |
| Sediment Excluding Devices | ...Under-sluices |
| Energy Dissipation Arrangements | ...Stepped apron |

**11.7.5. Prakasm Barrage.** The famous Prakasm barrage (Fig. 11.40) is located at Vijaywada (in Andhra Pradesh State) on Krishna river. It basically serves the *irrigation* needs of the adjoining areas. The Design flood discharge is as high as 33,984 cumecs (12 lakh cusecs). Its construction started in 1954 and got completed in 1957.

Fig. 11.40. (*a*) Photoview of Prakasm Barrage.

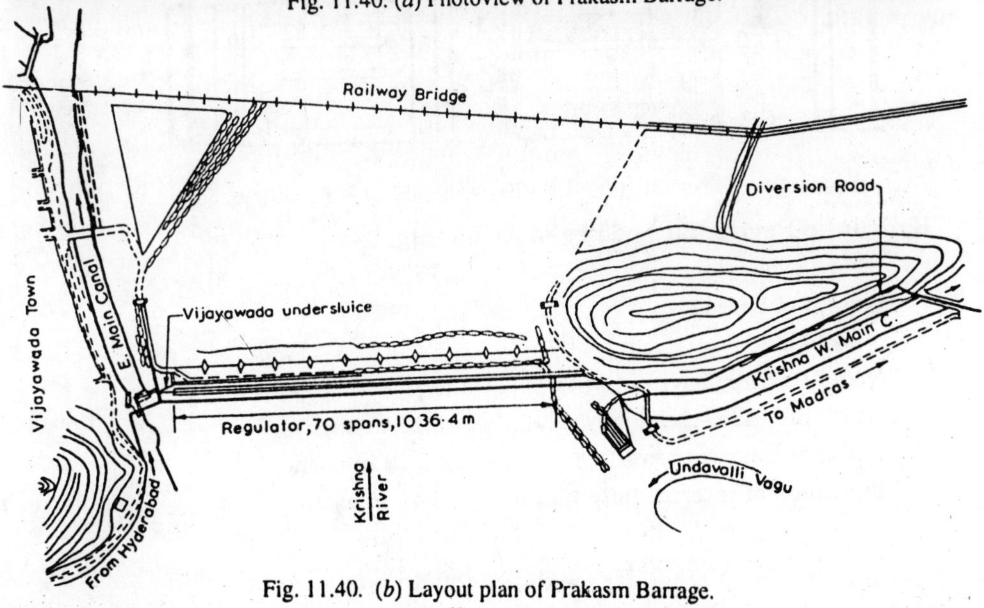


Fig. 11.40. (*b*) Layout plan of Prakasm Barrage.

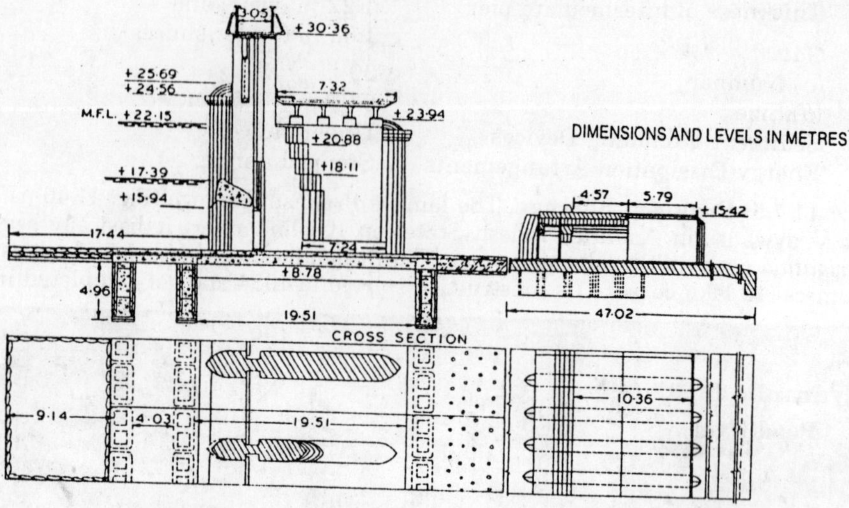


Fig. 11.40. (*c*) Plan (Sitanagram side) of Prakasm Barrage.

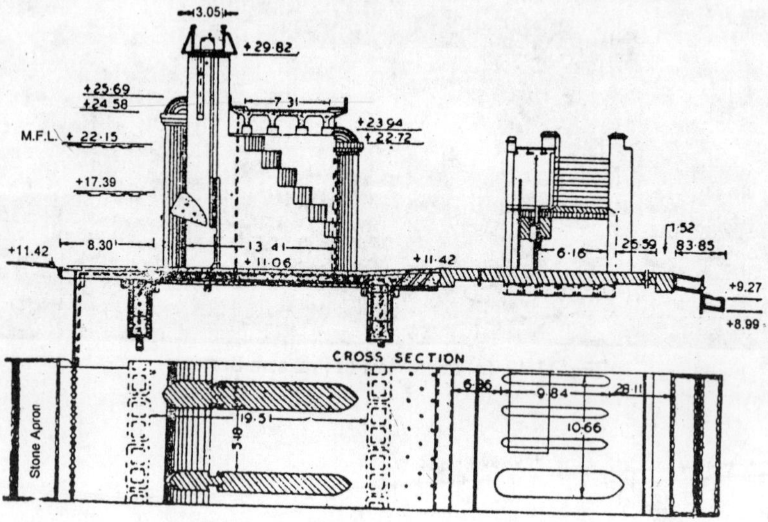


Fig. 11.40. (*d*) Plan (Vijaywada side) of Prakasm Barrage.

## Hydraulic Particulars of Prakasm Barrage

| | |
|---|---|
| Pond Level | ...17.38 m |
| Length of barrage | ...1138.73 m |
| No. of Under-sluice bays | ...14 |
| No. of Barrage bays | ...70 |
| Width of Under-sluice bays | ...5.18 m each |
| Width of Barrage bays | ...12.19 m each |
| Thickness of intermediate piers | ...2.44 m each |

**Gates**

| | |
|---|---|
| (*a*) *Number* | 70 |
| (*b*) *Size* | 12.19 m × 3.66 m |
| Energy Dissipation Arrangements | Cistern |

**11.7.6. Shah Nehar Barrage.** Shar Nehar barrage (Fig. 11.41) is located 5 km downstream of *Beas dam,* in Talwara township (Punjab State) on the Beas river. It caters to the *irrigation* needs of the adjoining areas. The design flood discharge at the barrage site is 11,043 $m^3$/sec (3.9 lakh cusecs).

Other particulars of this diversion headworks are given below :

**Hydraulic Particulars of the Shah Nehar Barrage**

| | |
|---|---|
| Pond level | ...330.7 m |
| Width of the river | ...609.6 m |
| Length of barrage | ...561.87 m |
| No. of Under-sluice bays | ...4 |
| No. of Barrage bays | ...46 |
| Width of Under-sluice bays | ...9.15 m each |
| Width of Barrage bays | ...9.15 m each |
| Thickness of intermediate pier | ...2.13 m (at top) |
| **Gates :** (*a*) *Number* | ...50 |
| (*b*) *Height* | ...6.7 m |
| Length of Upstream Divide Wall | ...41.775 m |
| Energy Dissipation Arrangements | ...Hydraulic jump on cistern |

**Canal Head Regulator Details**

| | |
|---|---|
| Width of Head regulator | ...67.1 m |
| No. of bays and their width | ...8 bays of 7.32 m clear span |
| Thickness of piers | ...1.22 m (at top) |
| Orientation with respect to barrage axis | at 100° |
| Energy Dissipation Devices below regulator | ...Hydraulic jump in cistern |
| Max. Discharge of Canal | ...325.64 cumecs |

Fig. 11.41. (*a*) Photoview of Shah Nehar Barrage.

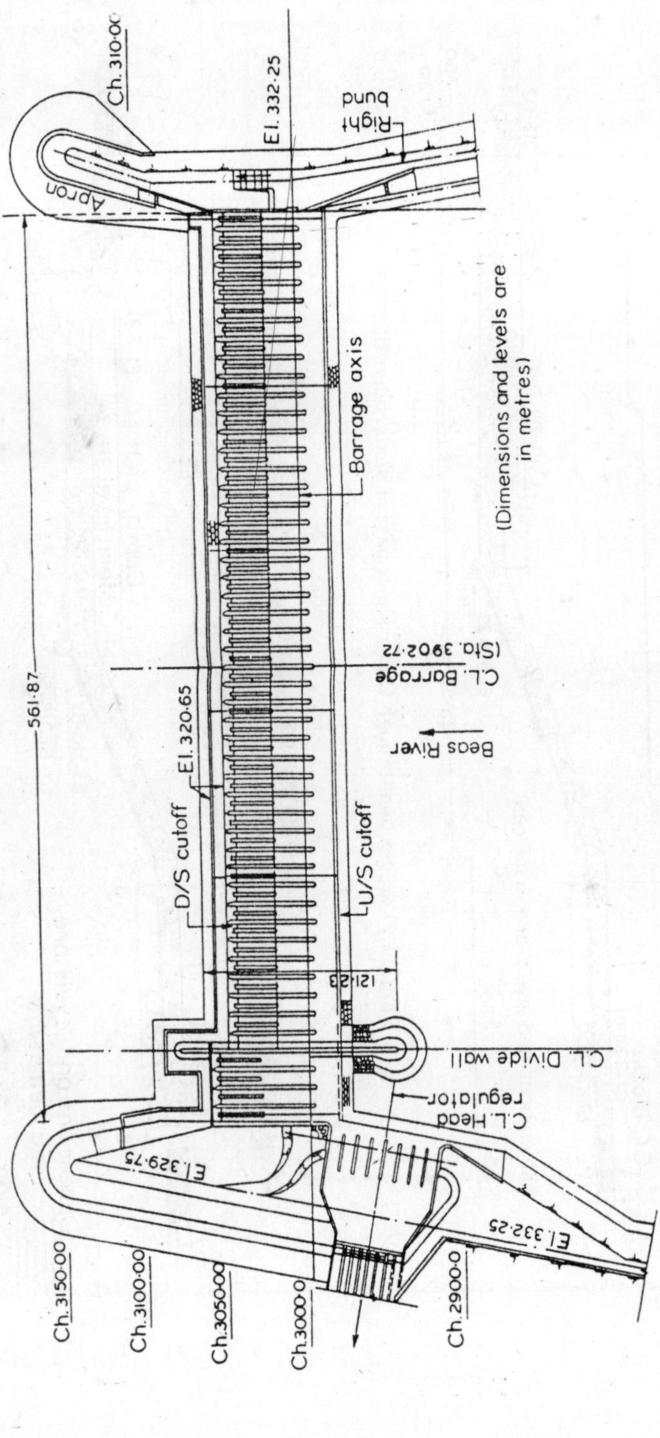


Fig. 11.41 (b) Layout plan of Shah Nehar Barrage on Beas river in Punjab.

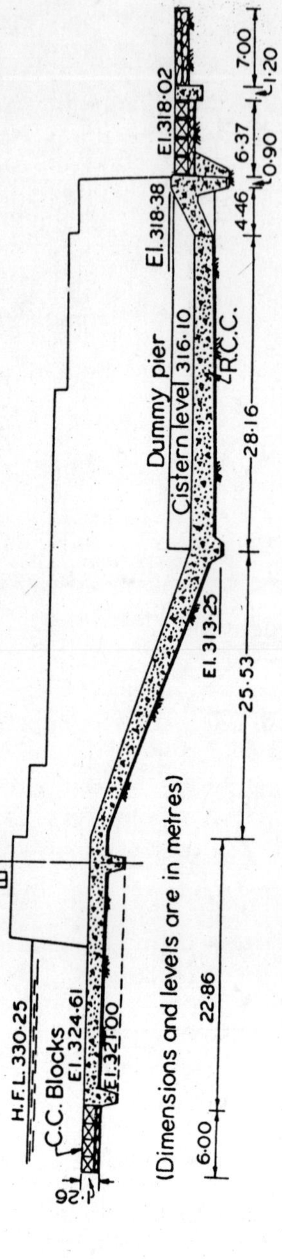


Fig. 11.41. (*c*) Section through Under-sluice bays of Shah Nehar Barrage

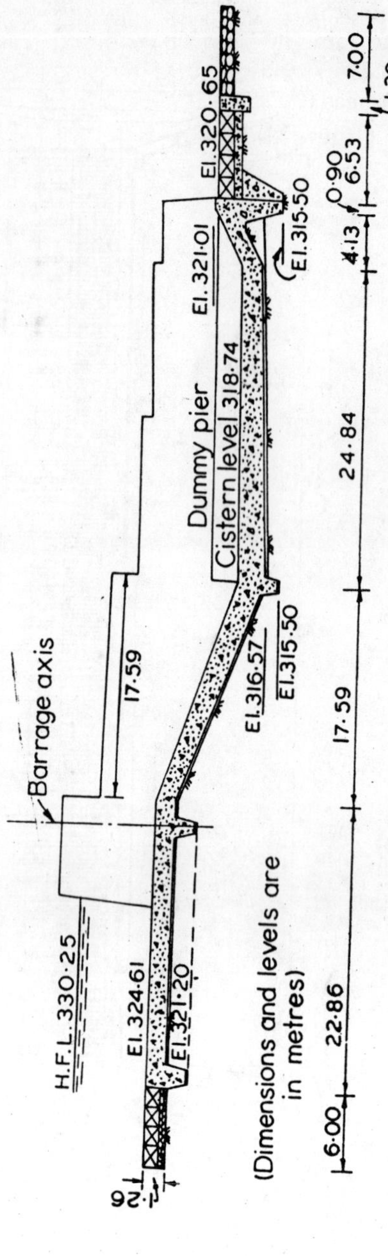


Fig. 11.41. (*d*) Section through Spillway bays of Shah Nehar Barrage.

**11.7.7. Tajewala Head Works.** Tajewala headworks (Fig. 11.42) was constructed as long back as the year 1873, across Yamuna river. This barrage/anicut is located 37 km from Jagadhri in Yamuna Nagar district in Haryana, and the famous Western Yamuna Canal (W.J.C.) and the Eastern Yamuna Canal (E.J.C.) take off from this headworks. The design capacity of this barrage is 10,024 cumecs (3.5 lakh cusecs).

Other particulars of this barrage are given below :

Fig. 11.42. (*a*) Photoview of Tajewala Headworks.

Fig. 11.42. (*b*) Closeview of Tajewala weir bays portion.

**Hydraulic Particulars of Tajewala Head works**

| | |
|---|---|
| Length of the barrage/anicut | ...753 m |
| Width of the under sluice bays | ...7 spans — 6 m<br>10 spans—8 m<br>8 spans—7 m |
| Width of barrage bays | ...564 m |
| Pond Level | ...324 m |
| Sediment Excluding Devices | ...Shingle Excluder |
| Length of upstream Divide wall | ...No. Divide wall |
| Energy Dissipation devices | ...Friction Blocks |

**Western Yamuna Canal Head Regulator Details**

| | |
|---|---|
| Width of head regulator | ...150 m |
| No. of bays and their width | ...13.7 m each and one of 5.4 m |
| Thickness of piers | ...1.37 m |

Orientation w.r.t. to Barrage axis ...90°
Energy Dissipation devices ...Friction Blocks
Max. Discharge of canal ...453 cumecs.

Since the year 2002, the very old Tajewala headworks has been replaced by the newly constructed barrage 3 km upstream at *Hathnikund.* The ***Hathnikund barrage*** was actually constructed 3 km upstream of Tajewala headworks, between October 1996 and June 1999, to replace the old Tajewala headworks, which is now out of service.

The plans to replace the old Tajewala headworks had infact been prepared in the early 1970s, but an agreement between the governments of Haryana and Himachal Pradesh (which share water it diverts) was not made until July 1994. Although this new barrage was completed in 1999, it was not operational until March 2002 because of delay in completion of ancillory works.

This new barrage now diverts water into the western as well East[illegible]na Canals. The small reservoir/pond created by the barrage also serves as a wetland for 31 species of water bird.

This new barrage is 360 m long and is composed of ten main flood gates along with five undersluices on its right side and three on its left side. The maximum discharge capacity of the barrage is 28200 $m^3/s$ (9,95,874 cuft/sec), which caters to 1 in 500 year flood frequency.

**11.7.8. Pick-Up Weir at Ottakkal.** This pick-up weir (Fig. 11.43) is located at Ottakkal, Quilon district in Kerala State on Kallada river. The design flood discharge is 2830 cumecs (1 lakh cusecs).

Other particulars of this diversion headworks are given below :

**Hydraulic Particulars of the Ottakkal weir**

Width of the river ...133 m
Length of the weir ...131.04 m
No. of undersluice bays ...1
No. of weir bays ...7, and R.B.C. head regulator bay
Width of undersluice bays ...1.83 m each
Width of weir bays ...15.24 m each
Energy Dissipation Arrangement ...Nil

**Canal Head Regulator Details**

Width of Head regulator ...11.59 m
No. of bays and their width ...3 of 3.05 m each
Thickness of piers ...1.22 m each
Orientation with respect to weir axis ...90°
Location of Head regulator from weir axis ...R.B.C.* head regulator at the right, and L.B.C.** head regulator at 120 m upstream of the weir axis and 90 m from left edge of the river.
Max. Canal Discharge ...39.08 cumecs (R.B. Canal)
...220 cumecs (L.B. Canal)

*Right Bank Canal
**Left Bank Canal

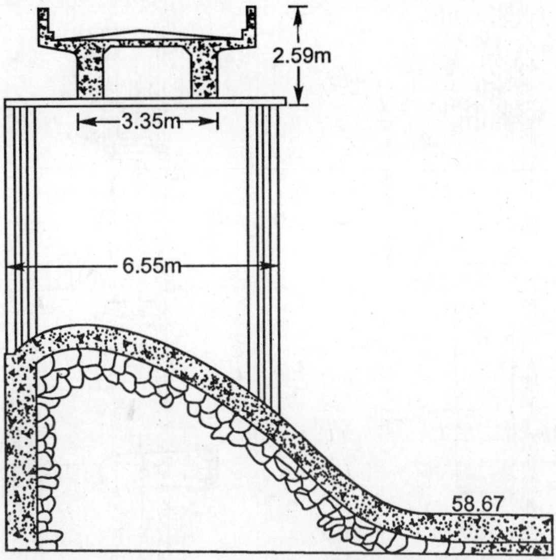


Fig. 11.43. (*a*) Sectional elevation of Ottakkal weir.

Fig. 11.43. (*b*) Photoview of Ottakkal weir.

## PROBLEMS

1. (*a*) What are the Main causes of failures of weirs on permeable foundations, and what remedies would you suggest to prevent them ?

   (*b*) Explain the surface flow considerations involved in the design of thickness of the sloping glacis and the downstream floor of a weir for different flow considerations.

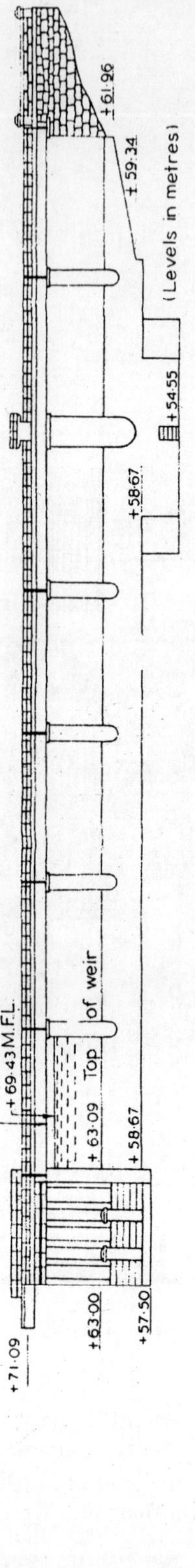


(*c*) Downstream elevation of Ottakal weir.

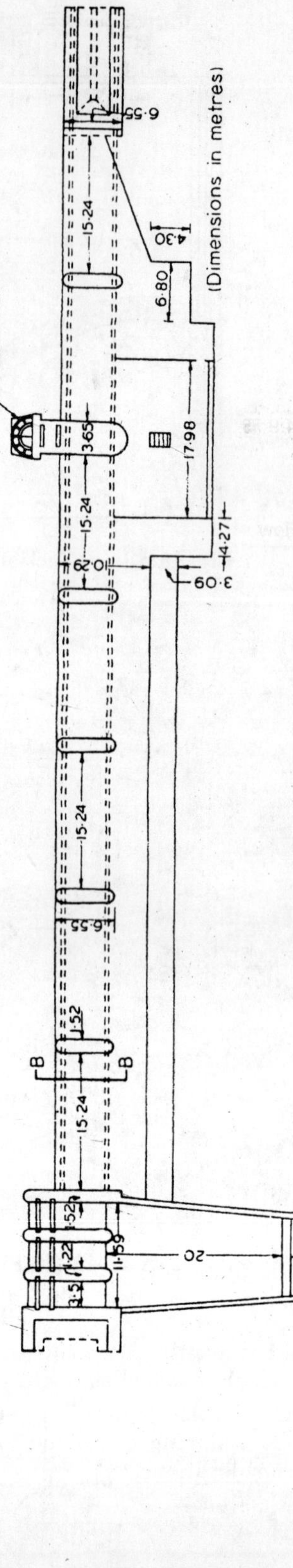


Fig. 11.43. (*d*) Plan at bottom of Ottakkal weir.

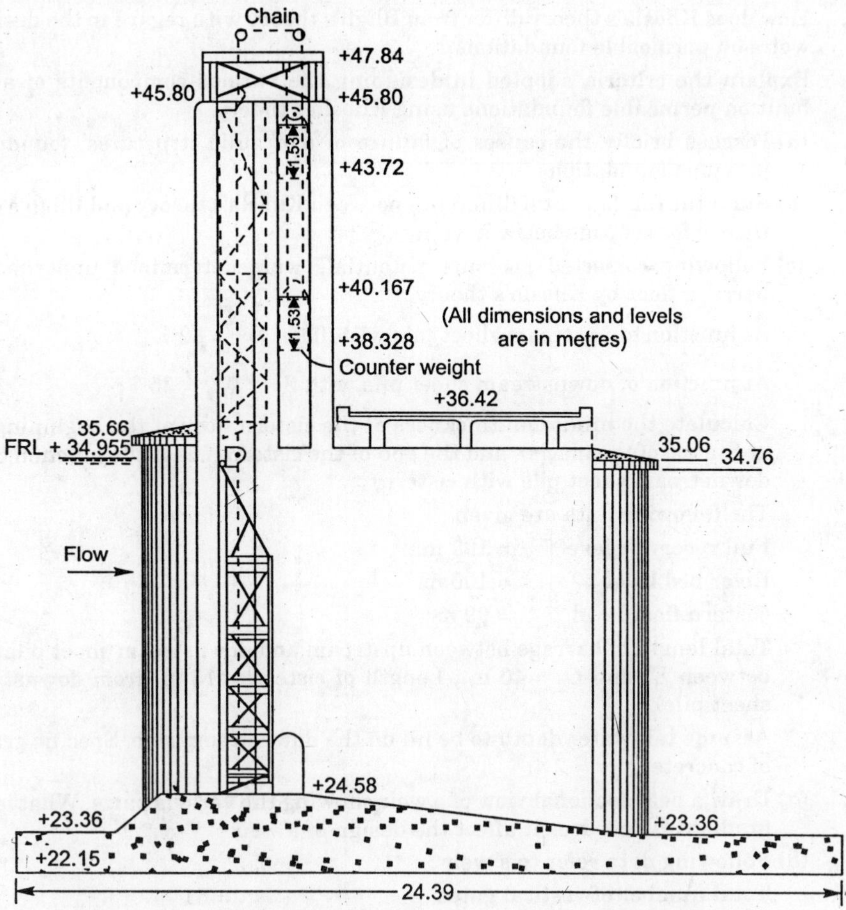


Fig. 11.43. (*e*) Cross-section of Spillway Gate at Ottakkal weir.

2. Discuss the main causes of failure of weirs founded on pervious foundation. Also discuss the important theories which have been put forward for designing such weirs to avoid there failure due to the above causes.
3. What are the different types of weirs ? Explain with neat sketches circumstances under which each type is adopted.

   What is meant by "piping" in a hydraulic structure ? What are ill-effects of piping ? What are the precautionary methods to avoid the ill effect of piping ?
4. Briefly explain the salient features of Khosla's theory and how it is used in the design of permeable foundations ?

   How does Lane's theory differ from Bligh's creep theory ?
5. (*a*) What is meant by "piping" on foundation of a weir. Explain Bligh's method of safe guarding the foundation against the ill effects of piping.

   (*b*) Briefly outline Khosla's theory on the design of weirs on permeable foundation. Enumerate the various corrections that are needed in the application of this theory.
6. Explain briefly Khosla's exit gradient concept.

   When a weir is constructed across a river ; it disturbs the existing regime. Explain the changes that occur till the river attains regime conditions again.

**7.** How does Khosla's theory differ from Bligh's theory with regard to the design of weirs on permeable foundations ?

Explain the criteria adopted in designing the various components of a weir built on permeable foundations using Khosla's theory.

**8.** (*a*) Discuss briefly the causes of failure of hydraulic structures, founded on pervious foundations.

(*b*) State the fundamental difference between Khosla's theory and Bligh's creep theory for seepage below a weir.

(*c*) Following corrected pressure potentials were determined underneath a barrage floor by Khosla's theory :

At junction of upstream sheet pile with floor $\phi_{E_1} = 82\%$

At junction of downstream sheet pile with floor $\phi_{C_2} = 35\%$

Calculate the minimum thickness of the cistern floor at the beginning (*i.e.* at the toe of the glacis), and the end of the cistern (*i.e.* at the junction of the downstream sheet pile with cistern)

The following data are given :

Full reservoir level = 105 m.

River bed level = 100 m.

Cistern floor level = 99 m.

Total length of barrage between upstream and downstream sheet piles (*i.e.* between $E_1$ and $C_2$ = 40 m ; Length of cistern = 15 m (from downstream sheet pile).

Assume tail water depth to be nil on the downstream side. Specific gravity of concrete floor = 2.4.

**9.** (*a*) Draw a neat sectional view of a weir showing the various parts. What is exit gradient ? How does it affect the design of a weir.

(*b*) Following data refer to a weir :

Total number of vertical gates = 51

Span of each gate = 10 m.

Full reservoir level (u/s) = 110 m.

Crest level = 106 m.

Coefficient of end contraction for piers = 0.02

Coefficient of discharge (in Francis formula) $C_d$ = 1.70 $m^{1/2}$/sec.

Compute the max. flood discharge which can safely pass over the weir without exceeding the full reservoir level. Neglect Velocity of approach.

[**Solution** Use $Q = 170\ L_e \cdot H^{3/2}$, where $H = 110 - 106 = 4$ m

$L_e = L - [0.1 \times$ No. of end contractions for abutments $+ 0.02 \times$ No. of end contractions for piers$]\ H$

$L_e = 51 \times 10 - [0.1 \times 2 + 0.02 \times 50]\ H = 501.2$ m

($\because$ each pier gives 2 contractions, and No. of piers = 50)

$\therefore$ $Q = 1.7 \times 501.2 \times (4)^{3/2}$ cumecs. = **6820 cumecs. Ans.**]

**10.** Use Khosla's curves to calculate the percentage uplift pressures at the three cut-offs for a barrage foundation profile, shown in Fig. 11.44 applying corrections as applicable.

(Given slope correction for 1 in 4 slope is 3.3%)

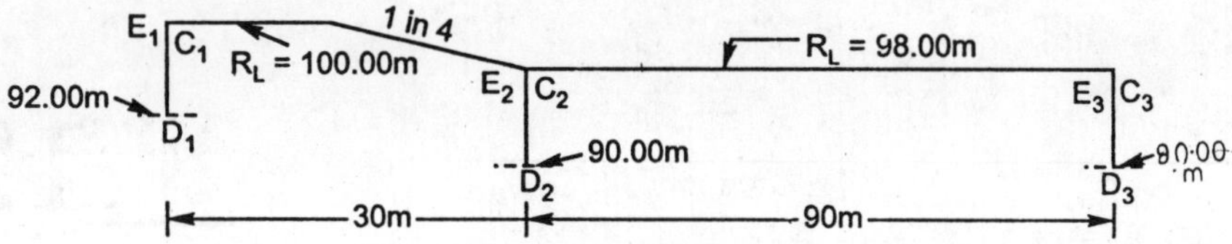


Fig. 11.44

Having determined the percentage uplift pressures, explain how the foundation floor thicknesses are determined corresponding to a known specific gravity of the material of the floor.

**11.** Write short notes on any three of the following :

(*i*) Stream-lines and Equipotential lines.

(*ii*) Bligh's creep theory for seepage flow.

(*iii*) Khosla's theory and concept of flow net.

(*iv*) Exit gradient and its importance.

(*v*) Khosla's method of independent variables for determining pressures and exit gradient for seepage below a weir.

(*vi*) Design of inverted filters and launching aprons for weirs (u/s as well as d/s).

(*vii*) Retrogression of levels due to weir construction.

(*viii*) Factors governing the design of weirs.

(*ix*) Uplift pressures in jump trough and design of weir floor thickness.

**12.** Design a barrage across a river at a site which is situated at 25 km downstream of a 120 m high dam, from the following data :

High flood discharge = 6100 cumecs

High flood level (before the construction of barrage) = 229.9 m

River bed level (winter) = 224.3 m Pond level = 228.15 m

Permissible afflux = 1.0 m Canal discharge = 151.5 cumecs.

Canal bed width = 50 m Canal full supply depth = 3 m.

Canal bed level = 224 m Angle of off-take = 110°

Silt factor in the river = 1.5

The gauge discharge curve of the river is given below in Fig. 11.45.

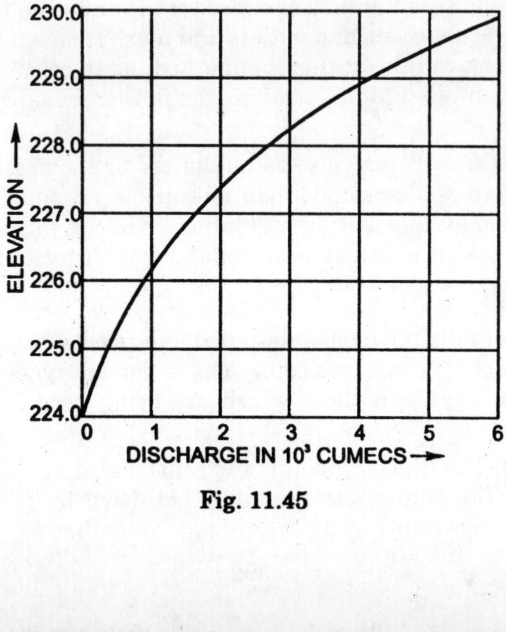


Fig. 11.45

# 12

# *Canal Falls*

## 12.1. Definition and Location of Canal Falls

**12.1.1. Definition.** Whenever the available natural ground slope is steeper than the designed bed slope of the channel, the difference is adjusted by constructing vertical 'falls' or 'drops' in the canal bed at suitable intervals, as shown in Fig. 12.1.

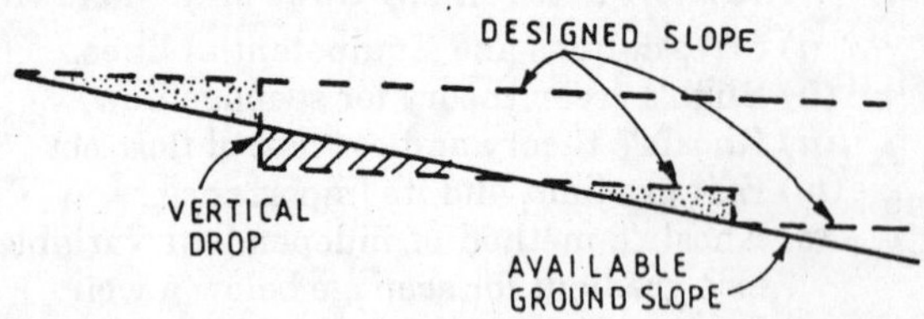


Fig. 12.1

Such a drop in a natural canal bed will not be stable and, therefore, in order to retain this drop, a masonry structure is constructed. Such a pucca structure is called a *canal fall* or a *canal drop.*

**12.1.2. Proper location.** The location of a fall in a canal depends upon the topography of the country through which the canal is passing. In case of the main canal, which does not directly irrigate any area, the site of a fall is determined by considerations of economy in 'cost of excavation and filling' *versus* 'cost of fall'. The excavation and filling on two sides of a fall should be tried to be balanced, because the unbalanced earthwork is quite costly. By providing a larger drop in one step, the quantity of unbalanced earth work increases, but at the same time, the number of fall reduces. An economy between these two factors has to be worked out before deciding the locations and extent of falls.

In case of branch canals and distributary channels, the falls are located with consideration to commanded area. The procedure is to fix the FSL required at the head of the off-taking channels and outlets and mark them on the L-section of the canal. The FSL of the canal can then be marked, as to cover all the commanded points, thereby deciding suitable locations for falls in canal FSL, and hence, in canal beds.

The location of the falls may also be influenced by the possibility of combining it with a bridge, regulator, or some other masonry work, since such combinations often result in economy and better regulation. When a fall is combined with a regulator and a bridge, it is called a fall-regulator with road bridge.

## 12.2. Types of Falls

Various types of falls have been designed and tried since the inception of the idea of 'falls construction' came into being. The important types of such falls, which were used in olden days and those which are being used in modern days, are described below:

**(1) Ogee Falls.** The 'Ogee type fall' was constructed in olden days on projects like Ganga canal. The water was gradually led down by providing convex and concave curves, as shown in Fig. 12.2.

The performance of such an ogee fall was found to have the following major defects :

(*i*) There was heavy draw-down on the upstream side, resulting in lower depths, higher velocities and consequent bed erosion. Draw-down would also affect the supply in a distributary, situated just upstream of fall.

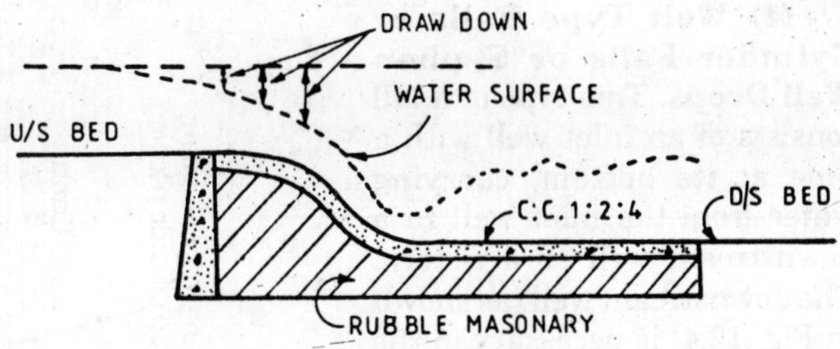

Fig. 12.2. Ogee fall.

(*ii*) Due to smooth transition, the kinetic energy of the flow was not at all dissipated, causing erosion of downstream bed and banks.

A 'raised crest' was soon added to restrict the draw-down and a long protection was provided on the downstream side. Later, it was converted into a much better type of fall, called a 'Vertical Impact type'.

(2) **Rapids**. In Western Yamuna canal, long rapids at slopes of 1 : 15 to 1 : 20 (*i.e.*, gently sloping glacis) with boulder facings, were provided. They worked quite satisfactorily, but were very expensive, and hence became obsolete.

(3) **Trapezoidal Notch Falls**. The trapezoidal notch fall was designed by Ried in 1894. It consists of a number of trapezoidal notches constructed in a high crested wall across the channel with a smooth entrance and a flat circular lip projecting downstream from each notch to spread out the falling jet (See Fig. 12.3).

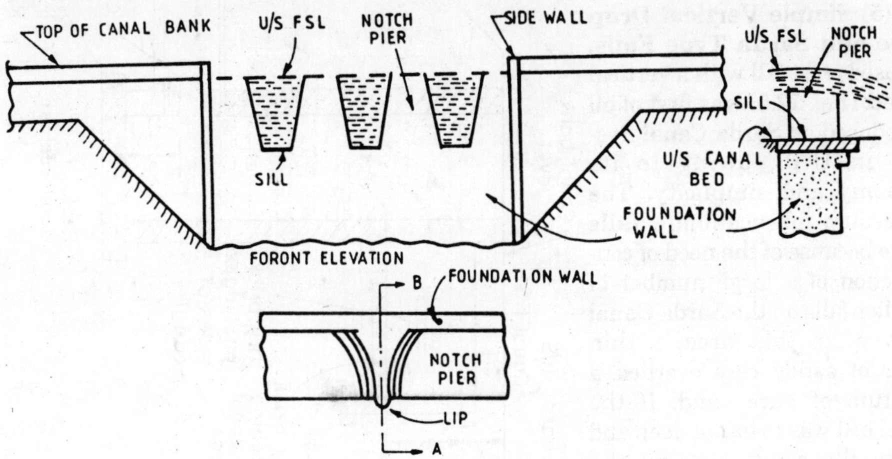

Fig. 12.3. Trapezoidal Notch fall.

The notches could be designed to maintain the normal water depth in the upstream channel at any two discharges, as the variation at intermediate values is small. Hence, the depth discharge relationship of the channel remains practically unaffected by the introduction of the fall. In other words, there would neither be drawdown nor heading up of water, as the channel approaches the fall. These falls remained quite popular, till simpler, economical, and better modern falls were developed.

**(4) Well Type Falls or Cylinder Falls or Syphon Well Drops.** This type of a fall consists of an inlet well with a pipe at its bottom, carrying water from the inlet well to a downstream well or a cistern. The downstream well (as shown in Fig. 12.4) is necessary in the case of falls greater than 1.8 m and for discharges greater than 0.29 cumecs. The water falls into the inlet well, through a trapezoidal notch constructed in the steining of the well, from where it emerges near the bottom, dissipating its energy in turbulence inside the well.

This type of falls are very useful for affecting larger drops for smaller discharges. They are commonly used as tail escapes for small canals, or where high levelled smaller drains do outfall into a low levelled bigger drain.

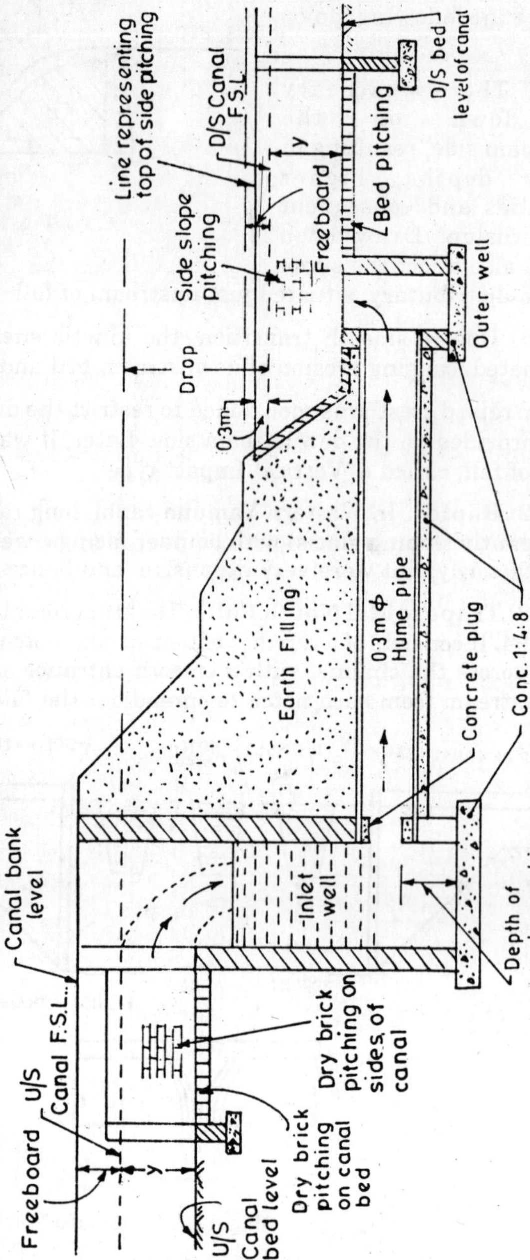

Fig. 12.4. Siphon Well drop.

**(5) Simple Vertical Drop Type and Sarda Type Falls.** A raised crest fall with a vertical impact (Fig. 12.5) was first of all introduced on Sarda Canal System in U.P., owing to its economy and simplicity. The necessity for economic falls arose because of the need of construction of a large number of smaller falls on the Sarda Canal System. In that area, a thin layer of sandy clay overlied a stratum of pure sand. If the canal bed was to be cut deep and up to the sand stratum, the seepage losses would have been tremendous. Hence, the depth of cutting had to be kept low, necessitating the construction of a large number of smaller falls.

In this type of a high crested fall, the nappe impinges into the water cushion below. *There is no clear hydraulic jump and the energy dissipation is brought about by the turbulent diffusion, as the high velocity jet enters the deep pool of water downstream.*

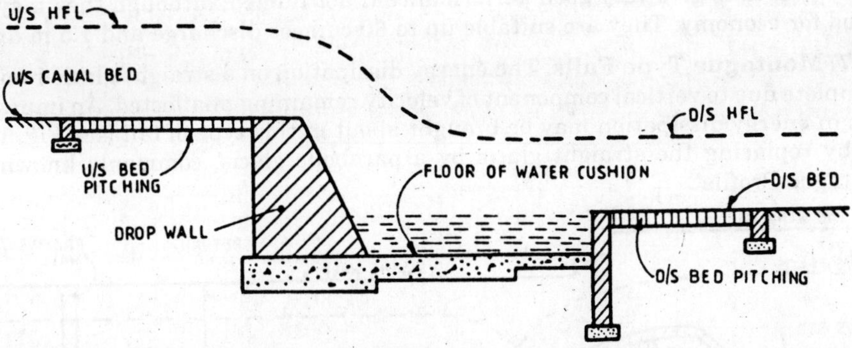


Fig. 12.5. Simple Vertical Drop fall.

Two types of crests which are used in Sarda type falls are shown a little later in Figs. 12.10 (*a*) and (*b*).

Sarda type fall is a high crested fall, and if the discharge in the canal varies (say between 50 to 100%), the water will head up on the upstream side at low discharges. The reach upstream of the crest will silt up as the clearer water will pass downstream of the crest. Due to reduction in silt in the d/s discharge, there may be a tendency of scouring on d/s, so as to make up the silt loss. Hence, this type of fall is not quite suitable for canals in which discharge varies within a wide range. A trapezoidal notch fall, although costlier than Sarda type or glacis type fall, is free from such troubles and, therefore, preferred for canals where the discharge is very small and also varies over a wide range.

**(6) Straight Glacis Falls**. In this type of a modern fall, a 'straight glacis' (generally sloping 2 : 1) is provided after a 'raised crest' (see Fig. 12.6). The hydraulic jump is made to occur on the glacis, causing sufficient energy dissipation.

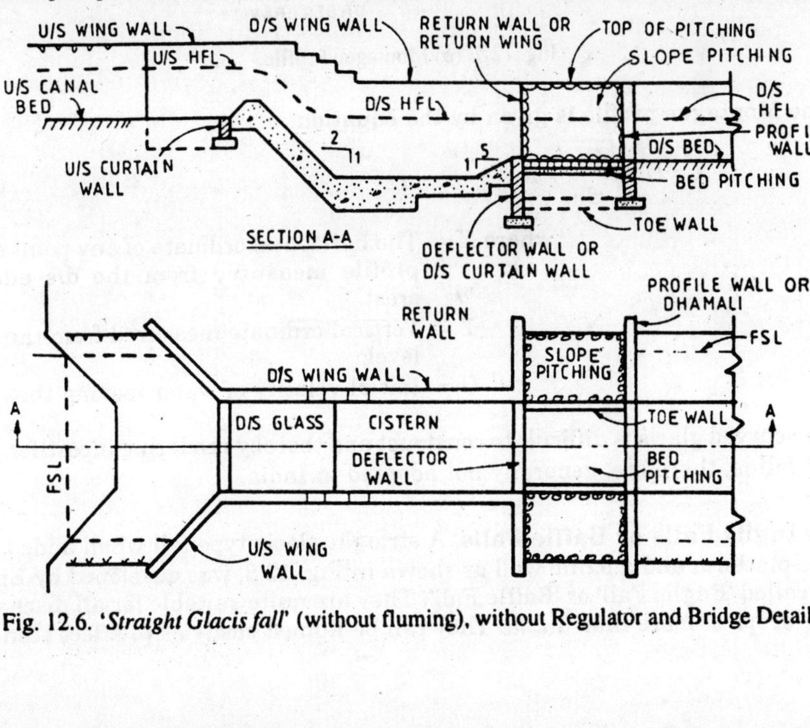


Fig. 12.6. '*Straight Glacis fall*' (without fluming), without Regulator and Bridge Details.

This type of falls give very good performance if not flumed, although they may be flumed for economy. They are suitable up to 60 cumecs discharge and 1.5 m drop.

(7) **Montague Type Falls.** The energy dissipation on a straight glacis remain incomplete due to vertical component of velocity remaining unaffected. An improvement in energy dissipation may be brought about in this type of fall [see Fig. 12.7 (*a*)], by replacing the straight glacis by a parabolic glacis, commonly known as 'Montague Profile'.

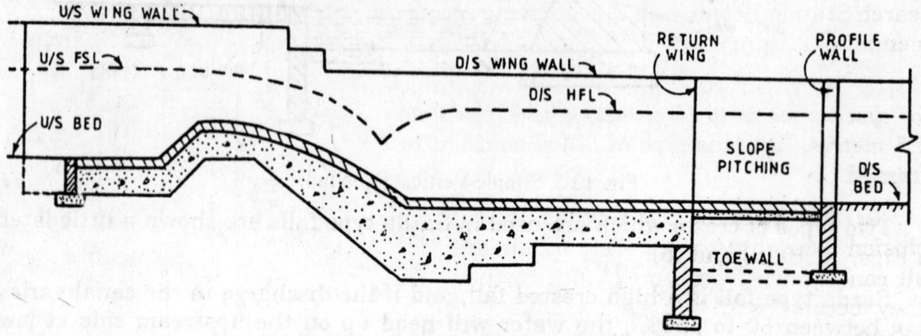


Fig. 12.7. (*a*) Montague Type fall.

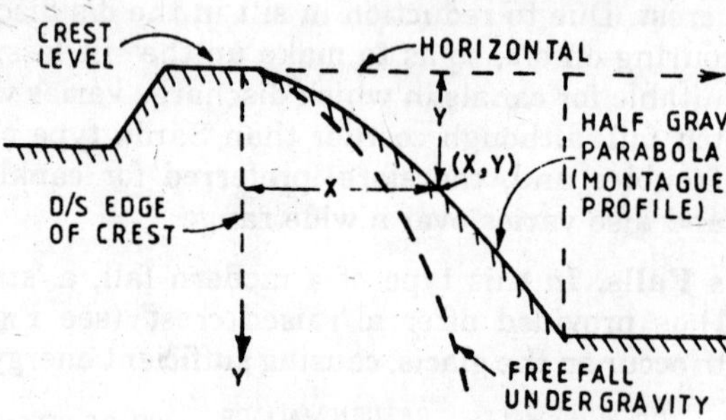


Fig. 12.7. (*b*) Montague Profile.

The Montague profile is given by the equation.

$$X = U\sqrt{\frac{4Y}{g}} + Y \quad \text{...(12.1)}$$

where $X$ = The horizontal ordinate of any point of the profile measured from the d/s edge of crest.

$Y$ = Vertical ordinate measured from the crest level.

$U$ = Initial velocity of water leaving the crest.

The curved glacis is difficult to construct and thereby rendering it costlier. This type of fall is, therefore, generally not adopted in India.

(8) **Inglis Falls or Baffle Falls.** A straight glacis type fall when added with a baffle platform and a baffle wall as shown in Fig. 12.8, was developed by Englis, and is called 'Englis Fall' or 'Baffle Fall'. They are quite suitable for all discharges and for drops of more than 1.5 m. They can be flumed easily as to affect economy.

The baffle wall is provided at a calculated height and a calculated distance from the toe of the glacis, so as to ensure the formation of the jump on the baffle platform, as shown in Fig. 12.8.

**Comparison of Different Types of Falls.** Extensive testing of various modern types of falls was carried out in 1952 at Research Station Poona, and the following recommendations were made :

(*i*) 'Vertical drop' falls are quite suitable *for discharges upto 15 cumecs and drops upto 1.5 metres.* But this type of fall should not be flumed.

(*ii*) For 'Straight glacis' type falls, the conclusion was, that they work satisfactorily for all conditions, if unflumed ; but in that case they become costly. *Even then, they can be adopted suitably for discharges upto 60 cumecs and drops upto 1.5 metres, and can even be flumed.*

(*iii*) 'Baffle fall' or 'English fall' may be used for all discharges when drop is more than 1.5 metres. *This type of fall functions very satisfactorily, either flumed or unflumed, so long as it is undrowned.*

(*iv*) Well type falls are suitable and economical for high drops and very low discharges. They can hence be easily used, as tail escapes of small channels.

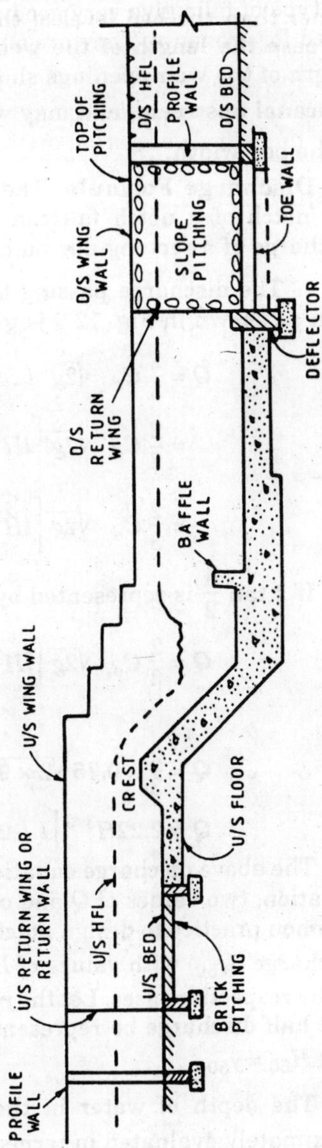


Fig. 12.8. 'Baffle fall' or 'English-fall'.

**Meter and Non-Meter Falls**. Meter falls are those which can be used to measure the discharge of the canal. If the discharge cannot be measured accurately at the site of the fall, then it is called a non-meter fall. Vertical drop fall is not suitable as a meter due to the formation of partial vacuum under the nappe. Glacis type fall is quite suitable as a metering device. *Since a sharp crest does not give a constant coefficient of discharge with varying heads, while a broad crest does so reasonably ; a fall to be used as meter must be provided with a broad crest.* Generally, a flumed glacis fall or a flumed baffle fall, is used as meter, while an unflumed glacis fall is used as a non-meter fall.

## DESIGN PRINCIPLES OF VARIOUS TYPES OF FALLS

### 12.3. Design of a Trapezoidal Notch Fall

As pointed out earlier, a notch fall provides a proportionate fall, in the sense that there is no heading up or drawdown of water level in the canal near the fall. The whole width of the channel is divided into a number of notches. The crest (*i.e.*

the sill level or the level of the bottom of notch) may be kept higher than the bed level of the canal, which will tend to increase the length of the weir, but in no case, the total length of the weir openings should exceed the bed width of the canal upstream, and may well be reduced to about $\frac{7}{8}$th of the bed width.

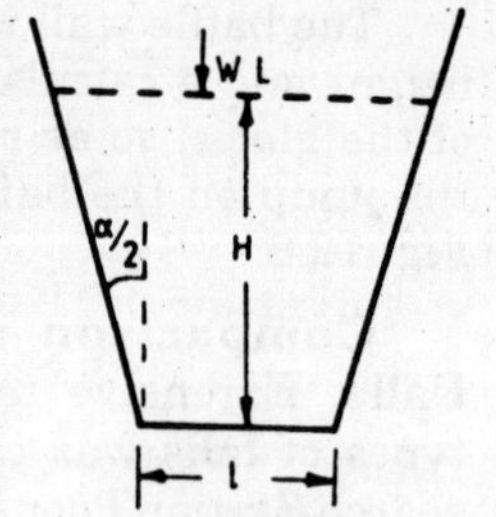


Fig. 12.9
Trapezoidal notch

**Discharge Formula.** The discharge passing through one notch of a notch fall can be obtained by adding the discharge of a rectangular notch and a V-notch.

∴ The discharge passing through a trapezoidal notch such as shown in Fig. 12.9 is given by

$$Q = \frac{2}{3} C_d \cdot \sqrt{2g}\ l \,.\, H^{3/2} + \frac{8}{15} \cdot C_d \,.\, \sqrt{2g} \tan\frac{\alpha}{2} H^{5/2}$$

$$= \frac{2}{3} C_d \cdot \sqrt{2g}\left[lH^{3/2} + \frac{4}{5}\tan\frac{\alpha}{2} H^{5/2}\right]$$

$$= \frac{2}{3} C_d \cdot \sqrt{2g}\left[lH^{3/2} + \frac{2}{5}\left(2\tan\frac{\alpha}{2}\right) H^{5/2}\right] \quad ...(12.2)$$

If $2 \tan\frac{\alpha}{2}$ is represented by $n$, then

$$Q = \frac{2}{3} C_d \cdot \sqrt{2g}\left[lH^{3/2} + 0.4 \,.\, nH^{5/2}\right] \quad ...(12.3)$$

where, $C_d$ = Coefficient of discharge ≈ 0.75

$$\therefore \quad Q = \frac{2}{3} \times 0.75\sqrt{2 \times 9.81}\left[lH^{3/2} + 0.4\, nH^{5/2}\right]$$

or
$$Q = 2.22H^{3/2}\left[l + 0.4nH\right] \quad ...(12.4)$$

The above discharge equation contains two unknowns $l$ and $n$. For solving this equation, two values of $Q$ and corresponding values of $H$ must be assumed. It is a common practice to design notches for full supply discharge ($Q_{100}$) and half supply discharge ($Q_{50}$) with values of $H$ equal to the normal water depths in the channel in the respective cases. Let the normal water depths in the channel at full discharge and half discharge be represented by $y_{100}$ and $y_{50}$ respectively. Then $H_{100} = y_{100}$, and $H_{50} = y_{50}$.

The depth of water in the channel at 50% discharge (*i.e.* $y_{50}$) can be approximately evaluated in terms of full supply depth ($y_{100}$) as follows :

Let $V = C \cdot y^{0.64}$ ...(Kennedy's Eq. for Vel. in channels)

Now $Q = A.V.$

∴ $Q = B \,.\, y \,.\, C \,.\, y^{0.64}$ Using $A \approx B.y$ (neglecting $sy^2$)

or $Q = C \,.\, B \,.\, y^{1.64}$

∴ $Q_{100} = C \,.\, B \,.\, y_{100}^{1.64}$

and $Q_{50} = C \,.\, B \,.\, y_{50}^{1.64}$

or $$\frac{Q_{50}}{Q_{100}} = \left(\frac{y_{50}}{y_{100}}\right)^{1.64}$$

or $$\frac{y_{50}}{y_{100}} = \left(\frac{Q_{50}}{Q_{100}}\right)^{1/1.64} = (0.5)^{1/1.64} = 0.66$$

$\therefore$ $$y_{50} = 0.66 \cdot y_{100} \qquad ...(12.5)$$

**Number of Notches.** The number of notches should be so adjusted by hit and trial method that the top width of the notch lies between $\frac{3}{4}$ to full water depth above the cill of the notch. This hit and trial procedure would become clear when we solve a numerical example.

**Notch Piers.** The thickness of notch piers should not be less than half the water depth and may be kept more if they have to carry a heavy super structure. The top length of piers should not be less than their thickness.

In plan, the notch profile is set back by 0.5 m from the downstream face of the notch fall for larger canals, and by 0.25 m for distributaries. All curves are circular arcs, and all centres lie in the plane of the profile. The splay upstream from the notch section is 45°, and the downstream splay is kept 22.5°. The lip is circular and is corbelled out by 0.8 m on larger canals, and by 0.6 m on distributaries.

**Example 12.1.** *Design the size and number of notches required for a canal drop with the following particulars :*

| | |
|---|---|
| *Full supply discharge* | *= 4 cumecs* |
| *Bed width* | *= 6.0 m* |
| *F.S. depth* | *= 1.5 m* |
| *Half supply depth* | *= 1.0 m* |

*Assume any other data if required.*

**Solution.** The bed width of the canal is 6 m. Each notch at top should be roughly equal to F.S. depth *i.e.* 1.5 m. So let us, in the first trial, provide 3 notches.

$\therefore$ Full supply discharge through each notch $= \frac{4}{3} = 1.33$ cumecs

From Eq. (12.4) we have

$$Q = 2.22 H^{3/2} [l + 0.4\, nH]$$

Using $Q_{100} = 2.22\, (y_{100})^{3/2}\, [l + 0.4n\, y_{100}]$

where $Q_{100} = 1.33$ cumecs

$y_{100} = 1.5$ m

$\therefore$ We have $1.33 = 2.22 \,.\, (1.5)^{3/2}\, [l + 0.4n \times 1.5]$

or $$1.33 = 2.22 \times 1.84\, [l + 0.6n]$$

or $$l + 0.6n = 0.326 \qquad ...(i)$$

Now, using

$$Q_{50} = 2.22 \,.\, (y_{50})^{3/2}\, [l + 0.4n \,.\, y_{50}]$$

where $Q_{50} = \frac{1.33}{2} = 0.67$ cumecs

$y_{50} = 1.0$ m

$\therefore$ $$0.67 = 2.22 \,.\, (1.0)^{3/2}\, [l + 0.4n \times 1]$$

or $$l + 0.4n = 0.3 \qquad ...(ii)$$

Subtracting (*ii*) from (*i*) we get

$$0.2n = 0.026$$

or $$n = 0.13.$$

Putting the value of $n$ in (*ii*) we get

$$l + 0.4 \times 0.13 = 0.3$$

or $$l = 0.248; \text{ say. } \mathbf{0.25\ m.}$$

By this trial, we find the top width

$$= 0.25 + 2 \tan \alpha \cdot H = 0.25 + n.\, H$$

$$= 0.25 + 0.13 \times 1.5 = 0.25 + 0.195$$

$$= 0.445; \text{ say } 0.45 \text{ m, which is much less than the full depth of } 1.5 \text{ m.}$$

In order to increase the top width, and to make it near 1 to 3/4th *FSD*, it is necessary to increase $l$ and $n$ : which can be done by reducing the number of notches.

The values of $l$ and $n$ obtained for 3 notches will increase in direct proportion, when number of notches are reduced.

In other words, the values $l$ of and $n$ will become 3 times, when number of notches are reduced 3 times. Thus, when we provide only one notch instead of 3 notches, the values of $n$ and $l$ will triple. Similarly, when we use 2 notches against 3, *i.e.*, $\frac{1}{1.5}$ times the values $n$ and $l$ will become 1.5 times of those obtained for 3 notches.

*Hence when we use 2 notches*, values will be

$$n = 1.5 \times 0.13 = 0.20$$

$$l = 1.5 \times 0.25 = 0.3 \text{ m}$$

and Top width = $1.5 \times 0.45 = \mathbf{0.68\ m.}$

Since the width is still quite low, we may use only one notch.

*When we use only one notch, the values will be*

$$n = 3 \times 0.13 = 0.39$$

$$l = 3 \times 0.25 = 0.75 \text{ m}$$

Top width = $3 \times 0.45 = 1.35 \approx$ FSD (O.K.)

$$\frac{\alpha}{2} = \tan^{-1} \frac{n}{2} = \tan^{-1} \frac{0.39}{2} = 11°$$

Since this condition gives us top width = 1.35 m, which is O.K., we may provide one notch, centrally placed in the given channel of 6 m width. The section of the notch to be adopted is also shown in Fig. 12.10.

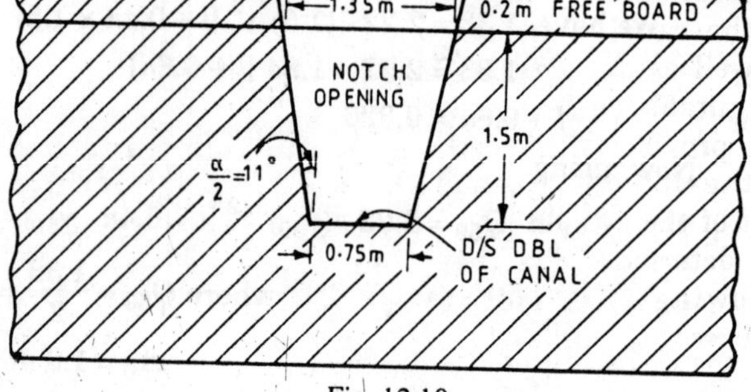


Fig. 12.10

**Check for raised crest if possible.** It has also been noticed that when lesser number of notches are provided, with their bottoms kept at U/S DBL of canal, the concentration of flow gets increased considerably. To avoid such an eventuality, it is preferable to increase the number of notches, and this may sometimes be achieved by providing the notches in the raised crest. In other words, the bottom of notch opening will be kept higher than U/S DBL of canal. This raising may be between 10% to 30% of full depth. The design calculations are hence to be repeated to compute $n$ and $l$ with a raised crest, whenever a detailed designing is being done, and number of notches determined are low.

*These calculations for the above question 12.1 will be as follows :*

Let us assume a raised crest equal to 20% of FSD

$$= 20\% \times 1.5 \text{ m} = 0.3 \text{ m}.$$

Using Eqn. (12.4) as

$$Q = 2.22\, H^{3/2}\, [l + 0.4nH], \text{ we have}$$

$$Q_{100} = 2.22\,(1.5 - 0.3)^{3/2}\, [l + 0.4n\,(1.5 - 0.3)]$$

or $$Q_{100} = 2.92\,(l + 0.48n) \qquad \text{...}(iii)$$

Also $$Q_{50} = 2.22\,(1.0 - 0.3)^{3/2}\, [l + 0.4n\,(1.0 - 0.3)]$$

$$[\because \text{FSD at } \tfrac{1}{2} \text{ discharge} = 1.0 \text{ m (given)}]$$

or $$Q_{50} = 1.3\,(l + 0.28\,n) \qquad \text{...}(iv)$$

But $$Q_{100} = 2Q_{50}$$

$\therefore$ $$2.92\,(l + 0.48\,n) = 2 \times 1.3\,(l + 0.28\,n)$$

This gives a negative value of $n$, which is not feasible, and hence such a raised crest may not be feasible in this particular case. Hence, the design made earlier, and shown in Fig. 12.10, holds good.

## 12.4. Design of a Syphon Well Drop

A syphon well drop, such as shown in Fig. 12.4, is generally adopted for smaller discharges and larger drops. The main features of the design involves determining the size of the inlet well and that of the hume pipe. A suitable size for the outer well, a proper provision of water cushion at the bottom of the inlet well, the bed and side slope pitchings in the canal upstream as well as downstream for suitable lengths, are also provided. The size of the inlet well and that of the syphon pipe are determined on the following considerations w.r. to Fig. 12.11.

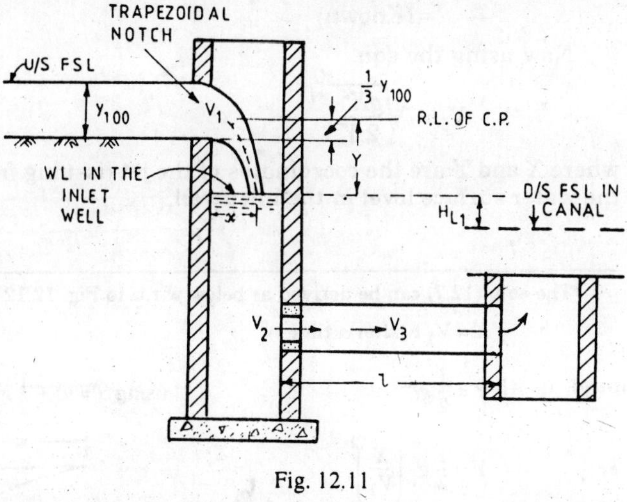


Fig. 12.11

First of all, the size of the trapezoidal notch is determined so as to pass the designed discharge by using eqn. (12.4) in the same way as is done for a trapezoidal notch. Then let $V_1$ be the velocity over the notch, $V_2$ be the velocity of entry in the pipe, and $V_3$ be the velocity through the pipe. All these values of velocities can be determined easily as below :

$$V_1 = \frac{\text{Full supply discharge}}{\text{Area of flow over the notch}}$$

$$V_2 = \frac{\text{Full supply discharge}}{\text{Area of opening at entry (for assumed dia of opening)}}$$

$$V_3 = \frac{\text{Full supply discharge}}{\text{Area of pipe (for assumed dia )}}$$

The head loss between the inlet well and the d/s FSL is then given by $H_{L_1}$ as

$$H_{L_1} = 0.5\frac{V_2^2}{2g} \text{ (i.e. loss due to entry)} + \frac{(V_2 - V_3)^2}{2g} \text{ (i.e. the loss due to sudden enlargement)} + \frac{f' L V_3^2}{2gd} \text{ (i.e. the loss in the assumed pipe length } L\text{)} + \frac{V_3^2}{2g} \text{ (loss due to exit).} \quad ...(12.6)$$

Knowing all the above values, $H_{L_1}$ can be determined, and thus the R.L. of water surface inlet well (*i.e.* d/s FSL + $H_{L_1}$) can be determined .

Now, approximate *R.L.* of centre of pressure (C.P.) of the trapezoidal waterway through the notch

$$= \text{u/s canal bed level} + \frac{1}{3}\text{ FSD.}$$

= (which can be calculated)

Then, the height ($Y$) of the centre of pressure above the water level in the inlet well

= R.L. of C.P. – R.L. of water level in inlet well

= (Known)

Now using the eqn.

$$V_1 = \sqrt{\frac{gX^2}{2.Y}}\text{*} \quad ...(12.7)$$

where $X$ and $Y$ are the coordinates of the jet (issuing from centre of pressure) w.r.t. the water surface level in the inlet well.

---

*The eqn. (12.7) can be derived as below w.r.t. to Fig. 12.12 :

$X = V_1 t$ (after a time $t$)

$Y = \frac{1}{2}gt^2$ $\left[\because \text{ using } S = ut + \frac{1}{2}gt^2 \text{ and } u = 0, \text{ we have } S = \frac{1}{2}gt^2\right]$

or $Y = \frac{1}{2}g \cdot \left(\frac{X}{V_1}\right)^2$

or $V_1 = \sqrt{\frac{gX^2}{2y}}$

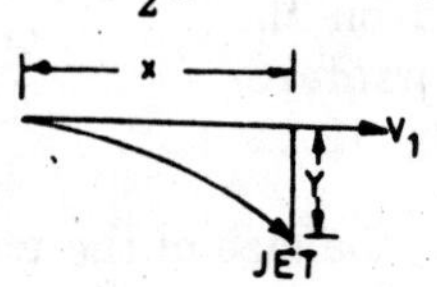


Fig. 12.12

The value of $X$ can be determined. Finally the dia of the inlet well may be kept at about 1.5 times the value of $X$. The entire procedure will become more clear when we solve the following numerical example.

**Example 12.2.** *Design the salient dimensions of a syphon well drop for the following particulars :*

| | |
|---|---|
| *Fall* | = *3.8 m* |
| *General ground level* | = *+ 163.36 m* |
| *Full supply depth* | = *75 cm* |
| *Bed level upstream* | = *+ 162.83* |
| *Discharge* | = *1 cumec* |
| *Bed width upstream and downstream* | = *2.4 m* |

**Solution.** For a trapezoidal notch, we have the discharge eqn. (12.4) as

$$Q = 2.22 \cdot H^{3/2}\left[l + 0.4\,n\,H\right]$$

At full supply discharge, we have

$$Q_{100} = 2.22\,(y_{100})^{3/2}\left[l + 0.4\,n\,y_{100}\right]$$

where $y_{100} = F.S.D. = 0.75$ m

$Q_{100} = F.S.Q = 1$ cumec

$$\therefore \quad 1 = 2.22\,(0.75)^{3/2}\left[l + 0.4n\,(0.75)\right]$$

$$\text{or} \quad 0.71 = l + 0.3\,n \qquad \ldots(i)$$

At 50% full discharge, we have

$$Q_{50} = 2.22\,y_{50}^{3/2}\left[l + 0.4n\,y_{50}\right]$$

where $y_{50} \approx 0.66\,y_{100}$ (eqn. 12.5)

$= 0.66 \times 0.75 = 0.5$ m.

$Q_{50} = 0.5$ cumec

$$\therefore \quad 0.5 = 2.22\,(0.5)^{3/2}\left[[l + 0.4n\,(0.5)\right]$$

$$\text{or} \quad 0.64 = l + 0.2\,n \qquad \ldots(ii)$$

Subtracting (*ii*) from (*i*) we get

$$0.07 = 0.1\,n$$

$$\text{or} \quad n = 0.7$$

$$\therefore \quad 2\tan\frac{\alpha}{2} = 0.7, \text{ or } \frac{\alpha}{2} = 19.3°$$

Substituting this value of $n$ in (*ii*), we get

$$l = 0.64 - 0.14 = 0.50$$

Hence, provide a trapezoidal notch in the steining of the inlet well, with 0.5 m bottom width and each side inclined to an angle of 19.3° with the vertical.

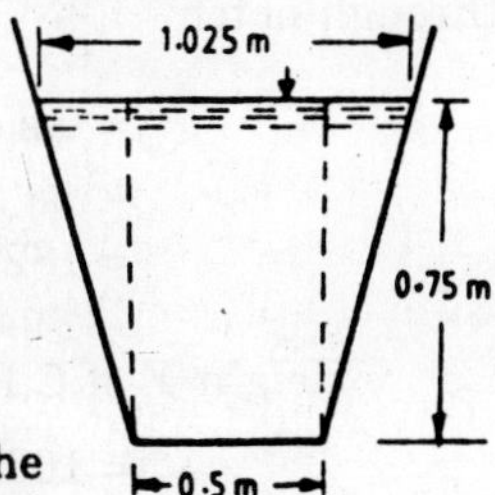

Fig. 12.13

Now, the width of water (at FSL) flowing over notch

$= 0.5 + 0.7 \times (0.75) = 0.5 + 0.525 = 1.025$ m.

Velocity ($V_1$) over the notch

$$= \frac{\text{F.S.Q.}}{\text{Area of flow over the notch}}$$

$$= \frac{1}{\frac{0.5 + 1.025}{2} \times 0.75}\ \text{m/sec} = \frac{1}{0.76 \times 0.75}\ \text{m/sec} = 1.75\ \text{m/sec}$$

Let us now assume that the diameter of the pipe used be 1 m

∴ Velocity $V_3$ through the pipe

$$= \frac{1}{\frac{\pi}{4}(1)^2} \text{ m/sec.} = 1.27 \text{ m/sec.}$$

Let us assume that the diameter of the opening at the inlet of pipe be 0.5 m

∴ The velocity of entry into the pipe ($V_2$)

$$= \frac{1}{\frac{\pi}{4}(0.5)^2} \text{ m/s} = 5.1 \text{ m/sec.}$$

Now

Loss of head between the inlet well and the d/s FSL is given by Eqn. (12.6).

$$= 0.5 \cdot \frac{V_2^2}{2g} + \frac{(V_2 - V_3)^2}{2g} + \frac{f' L V_3^2}{2gd} + \frac{V_3^2}{2g}$$

Let us assume that the length of the pipe is kept as 12 m and $f'$ = Darcey's coefficient of friction be taken as equal to 0.012, we then have

$$H_{L_1} = 0.5 \times \frac{(5.1)^2}{2 \times 9.81} + \frac{(5.10 - 1.27)^2}{2 \times 9.81} + \frac{0.012 \times 12 \times (1.27)^2}{2 \times 9.81 \times 1.0} + \frac{(1.27)^2}{2 \times 9.81}$$

$$= 0.66 + 0.77 + 0.01 + 0.08 = \mathbf{1.52 \text{ m.}}$$

∴ R.L. of water surface in the inlet well

$$= \text{d/s FSL} + 1.52 \quad \left[\begin{aligned} &\text{d/s FSL} = \text{u/s FSL} - \text{fall} \\ &= (162.83 + 0.75) - 3.8 = 159.78) \end{aligned}\right]$$

$$= 159.78 + 1.52 = \mathbf{161.30}\,.$$

Approximate R.L. of the centre of pressure (C.P.) of the trapezoidal waterway through notch

$$= \text{u/s canal bed level} + \frac{1}{3} \text{ FSD}$$

$$= 162.83 + \frac{0.75}{3} = 162.83 + 0.25 = \mathbf{163.08}$$

∴ Height $Y$ of C.P. above water level in the inlet well

$$= 163.08 - 161.30 = 1.78 \text{ m.}$$

Now using Eqn. (12.7), we have

$$V_1 = \sqrt{\frac{g \cdot X^2}{2Y}}$$

or

$$X = \sqrt{\frac{V_1^2\, 2Y}{g}} = \sqrt{\frac{(1.75)^2 \times 2 \times 1.78}{9.81}}$$

$$= 1.75 \times 0.6 = 1.05 \text{ m.}$$

Now, the dia. of the inlet well may be kept at about 1.5 *X*, *i.e.* $1.5 \times 1.05 = 1.575$ m, say 1.6 m. Keep the dia of the d/s outlet well, as say 1.2 m. Also provide a water cushion at the bottom of the inlet well. Bed and sides of channel for suitable lengths on the u/s as well as d/s side are protected by dry brick pitching. The complete details are shown in Fig. 12.14.

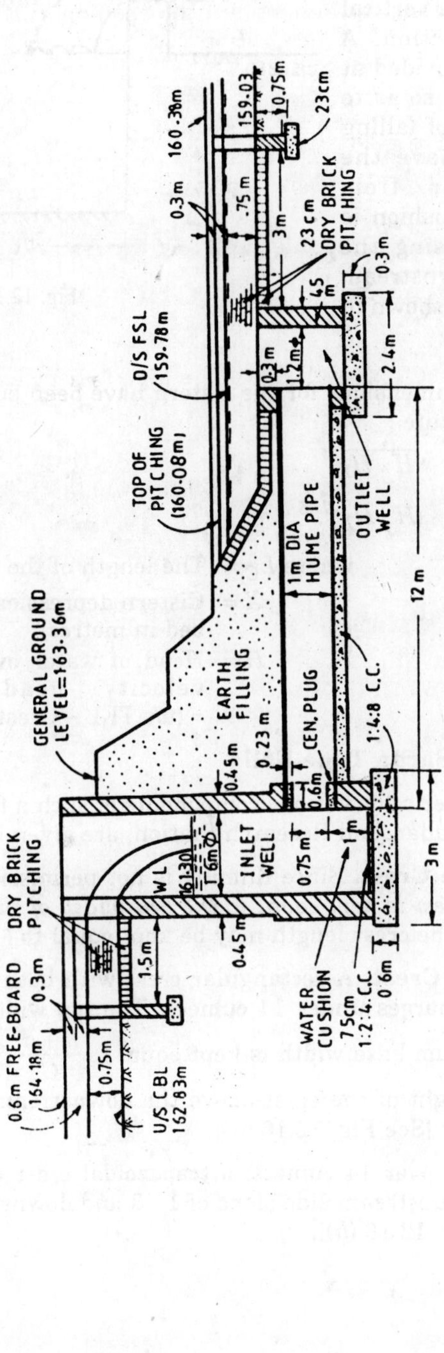


Fig. 12.14

## 12.5. Design of Simple Vertical drop Fall

In a vertical drop fall, the energy of the flowing water is dissipated by means of impact and by sudden deflection of velocity from vertical to horizontal direction. A water cushion is provided at the toe of the drop, so as to reduce the impact of falling jet and thus to save the downstream floor from scour. The water cushion is formed by depressing the floor below the downstream bed of the canal, as shown in Fig. 12.15.

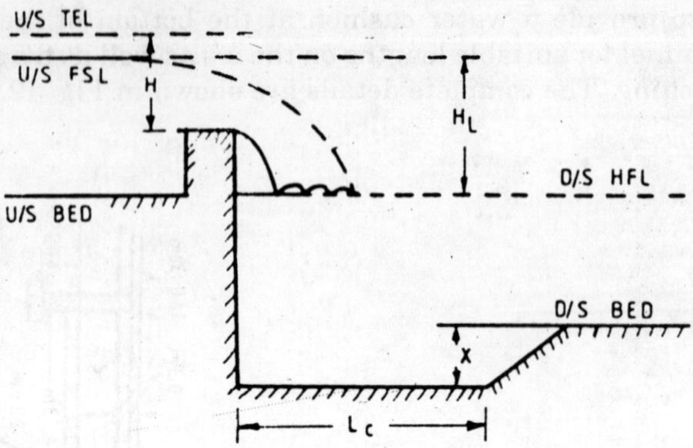


Fig. 12.15

The following dimensions for the cistern have been suggested by U.P. Irrigation Research Institute :

$$L_C = 5 \cdot \sqrt{H \cdot H_L} \quad \text{...(12.8)}$$

$$X = \frac{1}{4} \cdot (H \cdot H_L)^{2/3} \quad \text{...(12.9)}$$

where $L_C$ = The length of the cistern in metres.

$X$ = Cistern depression below the downstream bed in metres.

$H$ = Head of water over the crest, including velocity head, in metres, *i.e.* = (u/s TEL – Crest level).

## 12.6. Design of a Sarda Type Fall

The design criteria for various components of such a fall, based on the recommendations of Bahadarabad Research Station, are given below :

**Length of the Crest.** Since fluming is not permissible in this type of falls, *the length of the crest is kept equal to the bed width of the canal.* Sometimes, for future expansion, the crest length may be kept equal to ( bed width + depth).

**Shape of the Crest.** A rectangular crest with both faces vertical has been suggested for discharges under 14 cumecs. The top width is kept equal to 0.55 $\sqrt{d}$ and the minimum base width is kept equal to $\frac{h+d}{G}$ (Take $G = 2$ for masonry) where $d$ is the height of the crest above the downstream bed level and $h$ is the head over the crest [See Fig. 12.16 (*a*)].

For discharges over 14 cumecs, a trapezoidal crest with top width equal to $0.55 \cdot \sqrt{H + d}$ with upstream side slope of 1 : 3 and downstream side slope of 1 : 8 is adopted [See Fig. 12.16 (*b*)].

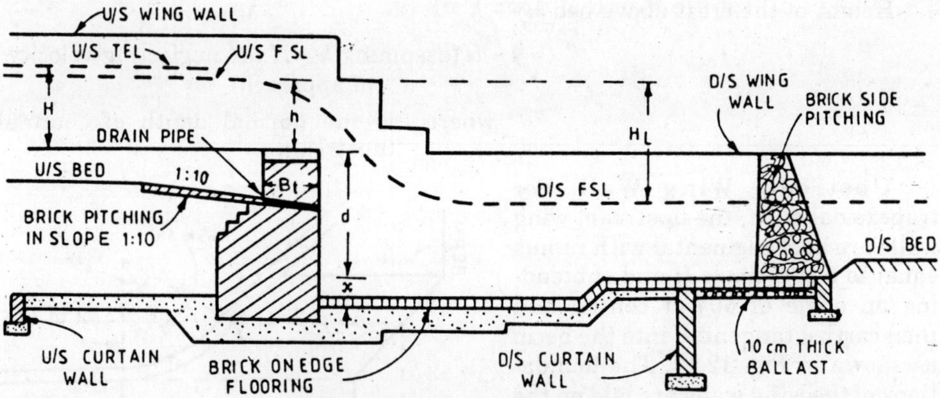


$Q$ = upto a maximum of 14 cumecs

$B_t$ = Top width of crest = $0.55\sqrt{d}$

Base width = $\dfrac{h+d}{2}$

$$Q = 1.84 \cdot LH^{3/2}\left(\frac{H}{B_t}\right)^{1/6}$$

Fig. 12.16. (*a*) Rectangular Crest for Sarda Type fall.

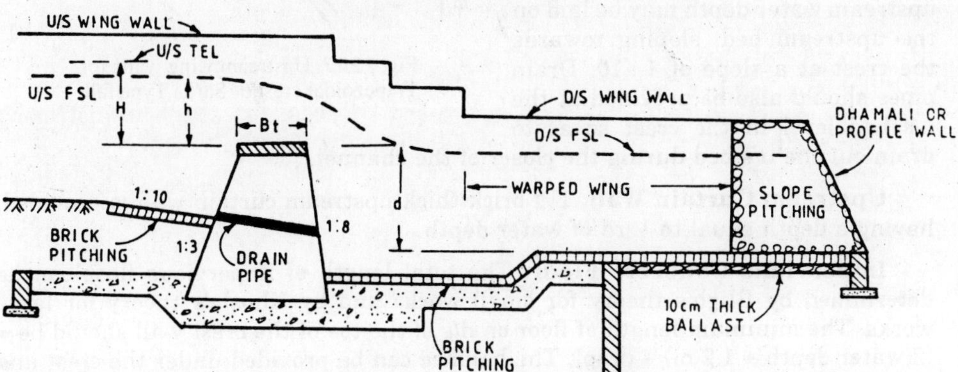


$Q$ = for 14 cumecs and over

$B_t$ = Top width of crest = $0.55\sqrt{H+d}$

Base width = as determined by the batters

$$Q = 1.99 \cdot LH^{3/2}\left(\frac{H}{B_t}\right)^{1/6}$$

Fig. 12.16. (*b*) Trapezoidal crest for a Sarda Type fall.

**Crest level.** The following discharge formula is used to determine the height of the crest.

$$Q = C_d \cdot \sqrt{2g} \cdot L \cdot H^{3/2}\left(\frac{H}{B_t}\right)^{1/6} \quad \ldots(12.10)$$

where $C_d$ = 0.415 for rectangular crest

= 0.45 for trapezoidal crest

$L$ = Length of the crest

$B_t$ = Top width of crest.

Height of the crest above bed $= y - h$

$\approx y - H$ (assuming $h \approx H$ *i.e.* neglecting velocity of approach)

where $y$ is the normal depth of channel (upstream).

**Upstream Wing Wall.** For trapezoidal crest, the upstream wing walls are kept segmental with radius equal to 5 to 6 times $H$ and subtending an angle of 60° at centre, and then carried tangential into the berm as shown in Fig. 12.17. The foundations of the wing walls are laid on the impervious concrete floor itself.

For rectangular crest (*i.e.* discharge less than 14 cumecs), the approach wings may be splayed straight at an angle of 45°.

**Upstream Protection.** Brick pitching in a length equal to upstream water depth may be laid on the upstream bed, sloping towards the crest at a slope of 1 : 10. Drain pipes should also be provided at the u/s bed level in the crest so as to drain out the u/s bed during the closer of the channel.

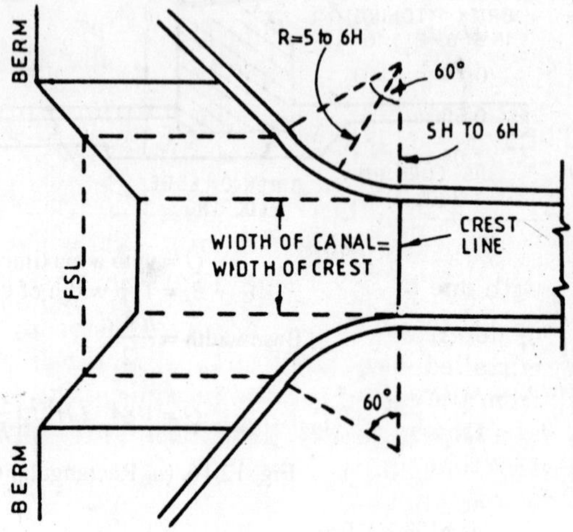


Fig. 12.17. Upstream wing walls for Trapezoidal crest of Sarda Type fall.

**Upstream Curtain Wall.** 1½ brick thick upstream curtain wall is provided, having a depth equal to ⅓rd of water depth.

**Impervious Concrete Floor.** The total length of impervious floor can be determined by Bligh's theory for small works and by Khosla's theory for large works. The minimum length of floor on d/s of the toe of the crest wall should be = [2(water depth + 1.2 m) + drop]. The balance can be provided under the crest and on upstream.

The floor thickness required on the downstream side can be worked out for uplift pressures (using minimum thickness of 0.4 m to 0.6 metre) and only a nominal thickness of 0.3 metre is provided on the upstream side. The maximum seepage head will occur when water is stored upto top of crest on u/s side and there is no flow on the downstream side.

**Cistern.** The length and depth of cistern can be worked out from equations (12.8) and (12.9).

**Downstream Protection.** The d/s bed may be protected with dry brick pitching, about 20 cm thick resting on 10 cm thick ballast. The length of the d/s pitching is given by the values of Table 12.1; or 3 times the depth of downstream water, whichever is more. The pitching may be provided between two or three curtain walls. The curtain walls may be 1½ brick thick and of depth equal to ½ the downstream depth; or as given in Table 12.1 (minimum = 0.5 m).

**Table 12.1**

| *Head over the crest H (metres)* | *Total length of d/s pitching (metres)* | *Remarks* | *Curtain walls* | |
|---|---|---|---|---|
| | | | *No.* | *Depth in metres* |
| Upto 0.3 m | 3.0 | All sloping down at 1 in 10 | 1 | 0.30 |
| 0.3 to 0.45 | 3.0 + Twice $H_L$ | Horizontal up to end of masonry wings and then sloping down at 1 : 10 | 1 | 0.30 |
| 0.45 to 0.60 | 4.5 + Twice $H_L$ | | 1 | 0.45 |
| 0.60 to 0.75 | 6.0 + Twice $H_L$ | | 1 | 0.60 |
| 0.75 to 0.90 | 9.0 + Twice $H_L$ | " | 1 | 0.75 |
| 0.90 to 1.05 | 13.5 + Twice $H_L$ | " | 2 | 0.90 |
| 1.05 to 1.20 | 18.0 + Twice $H_L$ | " | 2 | 1.05 |
| 1.20 to 1.50 | 22.5 + Twice $H_L$ | " | 3 | 1.35 |

**Slope Pitching.** After the return wing, the sides of the channel are pitched with one brick on edge. The pitching should rest on a toe wall $1\frac{1}{2}$ brick thick and of depth equal to half the downstream water depth. The side pitching may be curtailed at an angle of 45° from the end of the bed pitching, or extended straight from the end of the bed pitching.

**Downstream Wings.** Downstream wings are kept straight for a length of 5 to 8 time $\sqrt{H \cdot H_L}$ and may then be gradually wrapped. They should be taken upto the end of the pucca floor.

All wing walls must be designed as retaining walls, subjected to full pressure of submerged soil at their back when the channel is closed. Such a wall generally has a base width equal to $\frac{1}{3}$rd its height.

**Example 12.3.** *Design a 1.5 metres Sarda type fall for a canal having a discharge of 12 cumecs, with the following data :*

*Bed level upstream* = *103.0 m*
*Side slopes of channel* = *1 : 1 m*
*Bed level downstream* = *101.5 m*
*Full supply level upstream* = *104.5 m*
*Bed width u/s and d/s* = *1.0 m*
*Soil* = *Good loam*
*Assume Bligh's Coefficient* = *6*

**Solution.**

*Length of crest.* Same as d/s bed width = 10 m

*Crest level.* A rectangular crest is provided, since the discharge is less than 14 cumecs. The discharge formula is given by

$$Q = 1.84 \, . \, L \, . \, H^{3/2} \left[\frac{H}{B_t}\right]^{1/6}$$

Assume top width of the crest as 0.8 m.

$$\therefore \quad 12 = 1.84 \times 10 \times H^{3/2} \times \frac{H^{1/6}}{(0.8)^{1/6}}$$

or $$H^{5/3} = \frac{12 \times 0.964}{1.84 \times 10} = 0.628$$

or $$H = (0.628)^{3/5} = 0.755 \text{ m} \; ; \text{ Say } \mathbf{H = 0.76 \text{ m}}.$$

Velocity of approach

$$= V_a = \frac{\text{Discharge}}{\text{Area}} = \frac{12}{(10 + 1.5)\ 1.5} \quad (\because \text{ Depth of water } = 1.5 \text{ m})$$

$$= \frac{12}{11.5 \times 1.5} = 0.696 \text{ m/sec.}$$

$$\text{Velocity head } = \frac{V_a^2}{2g} = 0.025 \text{ m.}$$

$$\text{u/s TEL} = \text{u/s FSL} + \text{Velocity Head} = 104.5 + 0.025 = 104.525 \text{ m}$$

R.L. of the crest

$$= (\text{u/s TEL} - H) = 104.525 - 0.755 = \mathbf{103.77 \text{ m.}}$$

**Use crest level of 103.77 metres**

Height of the crest above d/s floor = 103.77 – 103.0 = **0.77 m.**

*Shape of the crest.*

Width of the crest $(B_t) = 0.55 \,.\, \sqrt{d}$

where $d$ = Height of the crest above d/s bed

$$= 103.77 - 101.5 = 2.27 \text{ m}$$

$$\therefore \quad B_t = 0.55 \cdot \sqrt{d} = 0.55 \cdot \sqrt{2.27} = \mathbf{0.825 \text{ m.}}$$

**Keep 0.85 m width of the crest**

$$\text{Thickness at base} = \frac{h + d}{2} = \frac{(0.755 - 0.025) + 2.27}{2} = \frac{0.73 + 2.27}{2} = 1.5 \text{ m.}$$

The top shall be capped with 20 cm thick C.C. 1 : 2 : 4

*Upstream wing wall.* It shall be splayed straight at an angle of 45° from the u/s edge of the crest and shall be embedded by 1.0 m into the berm. On the d/s side, wing walls are kept straight and parallel up to the end of the floor and joined to return walls, as shown in Fig. 12.19.

*Upstream protection.* 1.5 m long brick pitching (equal to u/s water depth) is laid on the u/s bed, sloping down towards the crest at 1 : 10, and *three drain pipes of 15 cm diameter at the u/s bed level should be provided in the crest so as to drain out the u/s bed during the closure of the canal.*

*Upstream curtain wall.* Maximum depth of u/s curtain wall

$$= \frac{y_u}{3} = \frac{1.5}{3} = 0.5 \text{ m.}$$

Provide 0.4 m × 0.8 m deep curtain wall on the u/s.

*Cistern.* Depth of cistern, is given by Eq. 12.9 as

$$X = \frac{1}{4}\,[H \cdot H_L)^{2/3} = \frac{1}{4}\Big[0.76 \times 1.5\Big]^{2/3} = \frac{1}{4} \times (1.14)^{0.667} \qquad \text{...(12.9)}$$

$$= \frac{1}{4} \times 1.091 = 0.273 \text{ m} \,;\ \textbf{Say 0.3 m deep.}$$

∴ R.L. of cistern = 101.5 – 0.3 = **101.2 m.**

Length of cistern $= 5\sqrt{H \cdot H_L}$

$$= 5 \times \sqrt{0.76 \times 1.5} = 5 \times \sqrt{1.14} = 5.34 \text{ m} \,;\ \textbf{say 5.5 m.}$$

**Provide 5.5 m long cistern at R.L. 101.2 m.**

*Impervious floor.*

Maximum Static Head

= (Crest level – d/s bed level)

= 103.77 – 101.5 = 2.27 m.

Total floor length required

= *C.H.*; where *C* is Bligh's coefficient & *H* here is max. static head.

= 6 × 2.27 = 13.62 m. ; say **13.7 m.**

Minimum d/s floor length required

= [2 (Water depth + 1.2) + $H_L$)

= 2 (1.5 + 1.2) + 1.5 = 2 (2.7) + 1.5 = 5.4 + 1.5 = 6.9 m ; say **7 m.**

Provide 7 m d/s floor and the balance 6.7 m under and upstream of the crest, as shown in Fig. 12.18.

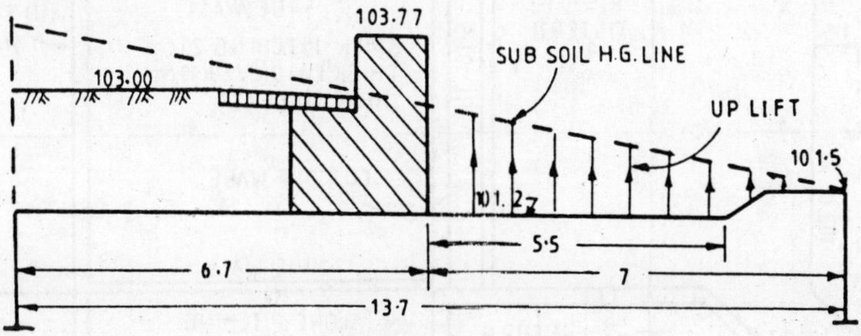


Fig. 12.18

**Floor Thicknesses**. H.G. line for the maximum static head is shown in Fig. 12.18.

*Maximum unbalanced uplift at the d/s toe of the crest*

$$= 0.3 + \frac{(103.77 - 101.3)}{13.7} \times 7 = 0.3 + 1.16 = 1.46 \text{ m}$$

Thickness required $\frac{1.46}{1.24} = 1.29$ m ; say 1.3 m.

Provide 1.1 m thick concrete overlain with 0.2 m thick brick pitching.

*Unbalanced head at 3 m from the toe of the crest*

$$= 0.3 + \frac{2.27}{13.7} \times 4 = 0.3 + 0.67 = 0.97$$

Thickness required $= \frac{0.97}{1.24} = 0.78$ m ; say 0.8 m.

Use 0.6 m thick concrete with 0.2 m brick layer.

*Unbalanced head at 5 m from the toe*

$$= 0.3 + \frac{2.27}{13.7} \times 2 = 0.3 + 0.33 = 0.63 \text{ m.}$$

Thickness required $= \frac{0.63}{1.24} = 0.51$ m ; Say 0.55 m.

Use 0.35 m thick concrete with 20 cm thick brick layer, as shown in Fig. 12.19.

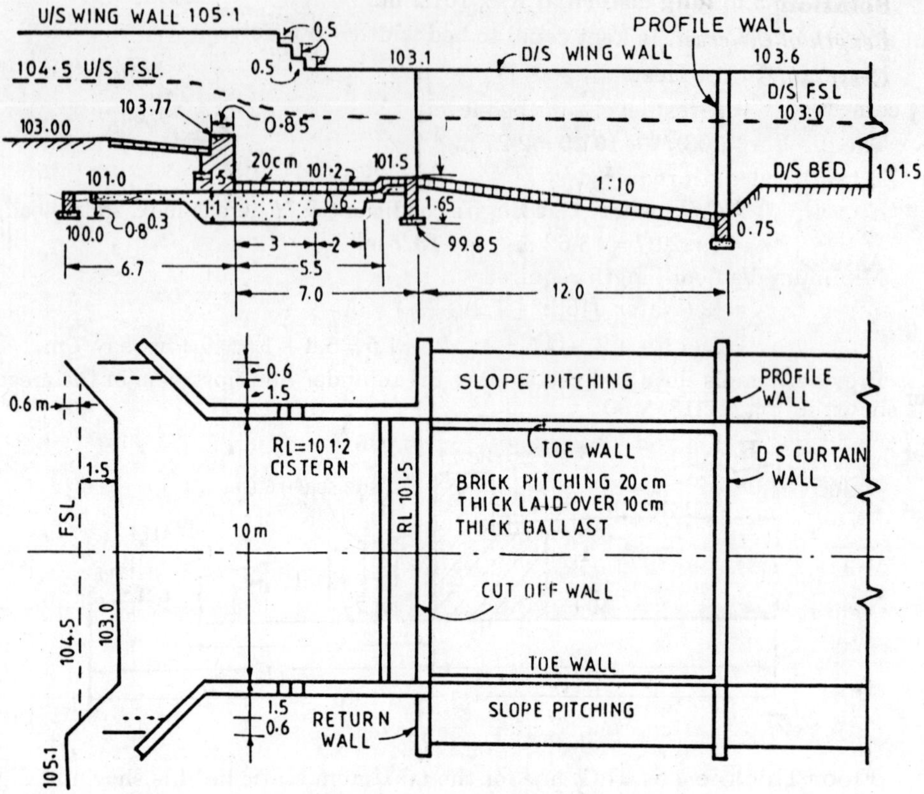


Fig. 12.19. Details of the Sarda Type fall (rectangular crest) of example 12.3.

*D/s Curtain Wall.* The curtain wall at the d/s end of the floor should be 0.75 m deep (for $H = 0.76$ m in Table 12.1)

Provide 0.4 m × 1.65 m deep curtain wall at d/s end of floor, *i.e.* upto a level of 101.5 – 1.65 = 99.85 metres, *i.e.* the deepest foundation level.

*Downstream pitching.* From Table 12.1,

Total length of d/s pitching

$$= 9 + 2 \times 1.5 = 12 \text{ metres.}$$

Pitching is kept sloped at 1 : 10. A curtain wall of 0.4 m × 0.75 m shall be provided at the end of the pitching, as shown in Fig. 12.19.

**Example 12.4.** *Design a 1.5 metres Sarda Type fall for a canal carrying a discharge of 40 cumecs with the following data :*

*Bed level upstream* = *105.0 m.*

*Bed level downstream* = *103.5 m.*

*Side slopes of channel* = *1 : 1*

*Full supply level upstream* = *106.8 m.*

*Full supply level downstream* = *105.3 m.*

*Berm level u/s* = *107.4 m*

*Bed width u/s and d/s* = *30 m.*

*Safe exit gradient for Khosla's Theory* = *1/5.*

**Solution.**

*Length of the crest.* Is kept equal to bed width = 30 metres.

*Crest level.* A trapezoidal crest is provided, since the discharge is more than 14 cumecs.

The discharge formula is given by

$$Q = 1.99\, LH^{3/2} \left(\frac{H}{B_t}\right)^{1/6}$$

Assume $B_t = 1.0$ m.

$$\therefore \qquad 40 = 1.99 \times 30 \frac{H^{3/2} \cdot H^{1/6}}{(1)^{1/6}}$$

$$\text{or} \qquad H^{5/3} = \frac{40}{1.99 \times 30} = 0.796$$

$$\text{or} \qquad H = (0.796)^{3/5} = 0.862 \text{ m; say } \mathbf{0.865 \text{ m.}}$$

Velocity of approach

$$\therefore \qquad = V_a = \frac{40}{(30 + 1.8)\, 1.8} = 0.7 \text{ m/sec} \qquad (\because \text{ Full supply depth} = 1.8 \text{ m})$$

$$\text{Velocity head} = \frac{(0.7)^2}{2g} = 0.025 \text{ m.}$$

u/s TEL = u/s FSL + Velocity head

= 106.8 + 0.025 = **106.825 m.**

∴ R.L. of the crest

= u/s TEL − $H$ = 106.825 − 0.862 = **105.963 m.**

**Adopt a crest level of 105.97 m.**

*Shape of the crest.* Adopt crest width at top

$$B_t = 0.55 \sqrt{H + d}$$

where $H = 0.865$ m

$d$ = Height of the crest above d/s bed

= 105.97 − 103.5 = 2.47 m.

$$\therefore \qquad B_t = 0.55 \sqrt{0.865 + 2.47} = 0.55 \sqrt{3.335} = \mathbf{1.0 \text{ m.}}$$

Adopt a trapezoidal crest **with top width of 1.0 m** and u/s slope 1 : 3 and d/s slope 1 : 8.

*Upstream Wing Walls.* Radius of the wings should be 5 to 6 times the head over the crest = 5 × 0.865 = 4.325, to 6 × 0.865 = 5.19 m. Use 5.0 m radius for the wings. U/s wing walls shall be kept segmental with 5 m radius subtending an angle of 60° at centre and then carried tangentially into the berm.

*Downstream Wing Walls.* The downstream wings shall be kept straight up to a distance of say $6 \cdot \sqrt{H \cdot H_L}$, *i.e.* $6 \cdot \sqrt{0.865 \times 1.5} = 6.8$ m; say 7 m, and then warped in a slope of 1 : 1 and shall be taken upto the end of pucca floor.

*Upstream Protection.* Brick pitching equal to u/s water depth *i.e.* 1.8 m is laid on the u/s towards the crest at 1 : 10 slope. Provide 20 cm drain holes in the entire length at 3 m c/c to drain out the u/s bed during the closure of the canal.

*Upstream Curtain Wall.* The minimum depth of curtain wall = $\frac{1}{3}$rd water depth, *i.e.* $\frac{1}{3} \times 1.8 = 0.6$ m. Provide 0.7 m deep masonry wall over 0.3 m thick concrete.

Thus, provide a curtain wall 0.4 m × 1.0 m deep on the u/s.

*Downstream Curtain Wall.* Minimum thickness

$$= \frac{\text{Depth}}{2} = \frac{1.8}{2} = 0.9 \text{ m.}$$

or from Table 12.1, it is equal to 0.75 m.

Provide a d/s curtain wall 0.4 m × 1.4 m over 0.3 m cement concrete. Thus, total depth of d/s curtain wall shall be 1.7 m with its bottom level at 101.8 m.

*Cistern.* Depth of cistern

$$= X = \frac{1}{4}(\text{H} \cdot H_L)^{2/3} \qquad \text{...(12.9)}$$

$$\therefore \quad X = \frac{1}{4}\left[(0.865 \times 1.5)^{2/3}\right] = \frac{1}{4} \times 1.19 = 0.3 \text{ m (say)}$$

**R.L. of Cistern** = *103.5 – 0.3 = 103.2 m.*

Length of cistern = $5\sqrt{H \cdot H_L} = 5 \cdot \sqrt{0.865 \times 1.5} = 5 \times 1.14 = 5.7$ m.

**Provide 5.7 m long Cistern**

**Total Floor Length and Exit Gradient**

$$G_E = \frac{H}{d} \cdot \frac{1}{\pi\sqrt{\lambda}} \text{ where } H \text{ here is max. static head}$$

Maximum static head ($H$) is caused when water is stored upto the crest level and there is no water d/s.

$$H' = 105.97 - 103.5 = 2.47 \text{ m.}$$

$$d = 1.7 \text{ m (\textit{i.e.} Depth of d/s curtain wall)}$$

$$G_E = 1/5 \text{ (given)}$$

$$\therefore \quad \frac{1}{5} = \frac{2.47}{1.7} \cdot \frac{1}{\pi\sqrt{\lambda}}$$

or $$\frac{1}{\pi\sqrt{\lambda}} = \frac{1.7}{2.47} \times \frac{1}{5} = 0.137$$

From plate 11.2

$$\alpha = 10 \quad \text{or} \quad \frac{b}{d} = 10$$

or $b = 10 \times 1.7 = 17$ m. ; **Use 18 m.**

Minimum floor length required on the d/s

$$= 2 \text{ (Water depth } + 1.2) + H_L$$

$$= 2\,(1.8 + 1.2) + 0.865 = 6.865 \text{ m ; \textbf{say 7 m.}}$$

Provide the balance length of 11 m under and upstream of the crest, as shown in Fig. 12.20.

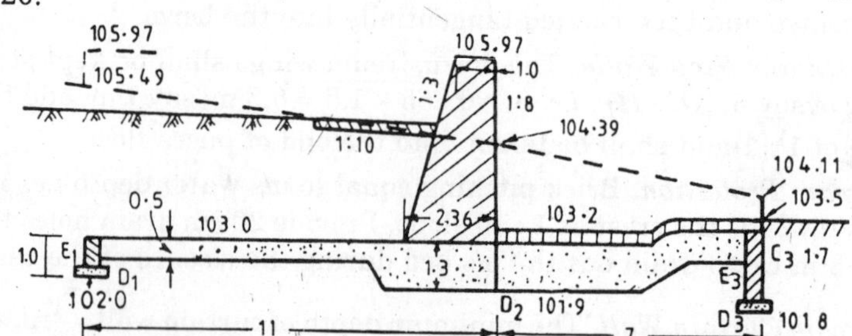


Fig. 12.20

**Uplift Pressure Calculations**

Assume u/s floor thickness as 0.5 m, and d/s floor thickness as 0.8 m, and floor thickness at toe of the crest as 1.3 m.

(1) *Upstream Wall.*

$$b = 18 \text{ m} \quad \text{and} \quad d = 1.0 \text{ m}.$$

$$\frac{1}{\alpha} = \frac{d}{b} = \frac{1.0}{18.0} = 0.056$$

From Plate 11.1 (*a*)

$$\phi_{E_1} = 0\%$$

$$\phi_{D_1} = 100 - \phi_D = 100 - 16 = 84\%$$

$$\phi_{C_1} = 100 - \phi_E = 100 - 23 = 77\%.$$

*Correction to* $\phi_{C_1}$ *for depth of floor*

$$= \frac{84\% - 77\%}{103.0 - 102.0} \times 0.5 = \frac{7}{1} \times 0.5 = 3.5\% \text{ (+ ve)}$$

*Correction due to influence of other wall is very small and neglected.*

$$\therefore \quad \phi_{C_1} \textbf{ (corrected)} = 77\% + 3.5\% = \mathbf{80.5\%}$$

(2) *Toe of Crest.*

$$b_1 = 11 \text{ m} \quad \text{and} \quad b = 18 \text{ m}$$

$$\frac{b_1}{b} = \frac{11}{18} = 0.61$$

$$d = 103.2 - 101.9 = 1.3 \text{ m}$$

$$\alpha = \frac{b}{d} = \frac{18}{1.3} = 13.9$$

From Plate 11.1(*b*) $\phi_{D_2} = 36\%$

(3) *Downstream Curtain Wall.*

$$d = 1.7 \text{ m} \quad \text{and} \quad b = 18 \text{ m}$$

$$\frac{d}{b} = \frac{1.7}{18} = 0.094$$

From Plate 11.1(*a*)

$$\phi_{E_3} = \phi_E = 29\%$$

$$\phi_{D_3} = \phi_D = 20\%$$

$$\phi_{C_3} = 0\%$$

*Correction for depth to* $\phi_{E_2} = \dfrac{29\% - 20\%}{1.7} \times 0.8 = \dfrac{9}{1.7} \times 0.8 = \mathbf{4.2\%}$ (– ve)

$\phi_{E_3}$ **(corrected)** $= 29 - 4.2 = \mathbf{24.8\%}$

The levels of H.G. line for **maximum static head** are worked out in Table 12.2 and plotted in Fig. 12.20.

**Table 12.2**

| *Condition of flow* | *u/s* W.L. *in metres* | *d/s* W.L. *in metres* | *Head H in metres* | *Height/Elevation of H.G. line above datum* | | | | | | |
|---|---|---|---|---|---|---|---|---|---|---|
| | | | | $\phi_{E_1}$ *100%* | $\phi_{D_1}$ *84%* | $\phi_{C_1}$ *80.5%* | $\phi_{D_2}$ *36%* | $\phi_{E_3}$ *24.8%* | $\phi_{D_3}$ *20%* | $\phi_{C_3}$ *0%* |
| Maximum static head, *i.e.* Water up to crest level on u/s and no water downstream | 105.97 | 103.50 | 2.47 | 2.47 | 2.08 | 1.99 | 0.89 | 0.61 | 0.50 | 0.90 |
| | | | | 105.97 | 105.58 | 105.49 | 104.39 | 104.11 | 104.00 | 103.5 |

**Floor Thicknesses**

Provide a nominal thickness of 0.4 m under u/s floor.

*Unbalanced head at d/s toe of glacis* = 104.39 – 103.2 = 1.19 m

Thickness required = $\dfrac{1.19}{1.24}$ = 0.97 m ; Use 1.2 m.

Provide 1.0 m thick C.C. overlain by 0.2 m thick brick pitching.

*Unbalanced head at d/s end of floor* = 104.11 – 103.5 = 0.61 m.

Thickness required = $\dfrac{0.61}{1.24}$ = 0.5 m ; Use 0.7 m.

Use 0.5 m thick C.C. overlain by 0.2 m thick brick pitching.

*Unbalanced head at 3 m from d/s toe of crest*

$$= 0.91 + \frac{0.28}{7} \times 3 = 0.91 + 0.12 = 1.03 \text{ m}$$

Thickness required = 1.03/1.24 = 0.83 m.

Provide 0.8 m thick C.C. laid over by 0.2 m thick bricks.

Thicknesses provided are shown in Fig. 12.21.

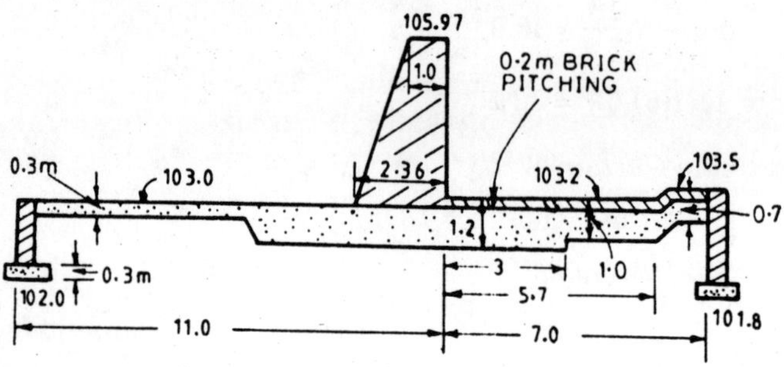


Fig. 12.21

**Downstream Pitching.** From Table 12.1, the length of d/s pitching required

$$= 9 + 2H_L = 9 + 2 \times 1.5 = \mathbf{12\ m.}$$

The pitching is kept sloping at 1 : 10 and a curtain wall of 0.4 m × 0.75 m is provided at the end of this pitching. The bed pitching and side slope pitching are separated by a toe wall 0.4 m × 0.75 m. Slope pitching is curtailed at an angle of 45° from the end of the bed pitching.

*Note* : Two rows of friction blocks of size 1.2 m × 0.6 m × 0.6 m staggered at a distance of 1.0 m from toe of the crest on the d/s cistern and two rows of cube blocks of size 0.25 m × 0.25 m × 0.25 m staggered at the d/s end of the floor, may be provided in conservative designs as an additional source of energy dissipation, as explained a little later under the heading of 'Roughening Devices'.

## 12.7. Design of a Straight Glacis Fall

**Cistern Element.** In a Glacis type fall, the kinetic energy of water is dissipated by the formation of a hydraulic jump on a sloping glacis. It was discussed earlier in Chapter 10, that for a given drop in energy line $H_L$ and the discharge intensity $q$, there will be a definite value of downstream specific energy $E_{f_2}$ required for jump formation. Knowing $q$ and $H_L$, $E_{f_2}$ can be read from Plate 10.1. $E_{f_2}$ measured below d/s TEL, determines the level of the point of jump formation. Theoretically, the cistern should be provided at this point. However, the depth of cistern is increased by 25% of $E_{f_2}$ in order to ensure the formation of jump on the sloping glacis itself under varying conditions of head.

If the d/s bed level is lower than the cistern level determined by measuring 1.25 $E_{f_2}$ below d/s TEL, the cistern should be provided at d/s bed level. But if d/s bed level is higher than the cistern level, then the cistern is joined at a slope of 5 : 1 to the bed, as shown in Fig. 12.22.

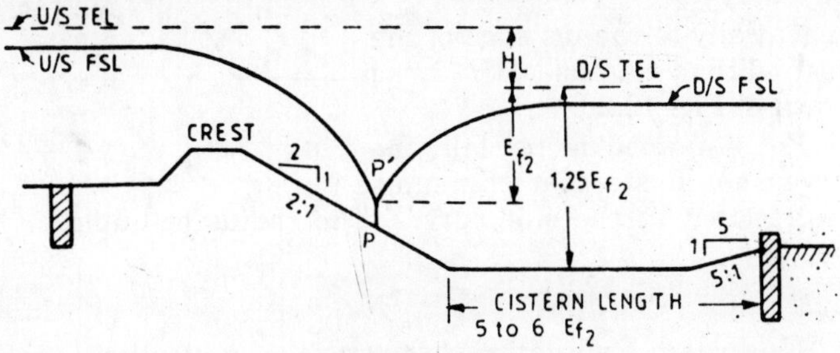


Fig. 12.22

Sufficient length of the cistern should be provided, so that the turbulence is confined within this pucca floor. To ensure this, the length of the pucca floor is generally kept 5 times $E_{f_2}$ for normal soils and 6 times $E_{f_2}$ for sandy soils.

**Crest Length.** Glacis fall may be unflumed ; and in that case, the crest length shall be equal to the bed width of the canal. They can be flumed also ; and in that case, they are generally provided with a regulator and a bridge. Such a fall is used as a meter and is called a *meter fall*. The fluming may be limited to the following percentage of bed width. This is called *fluming ratio*.

**Table 12.3. Fluming ratio**

| *Drop $H_L$* | *Clear length of crest* |
|---|---|
| 1.25 m | 65% |
| 1.25 to 3 m | 75% |
| Above 3 m | 85% |

**Crest Width.** A narrow crest, *i.e.* a sharp crest, provides a higher coefficient of discharge and thus a greater discharging capacity than that for a broad crest. A narrow crest is, therefore, preferred, because it is economical, as it reduces the cost of the crest longitudinally and transversely.

However, the disadvantage of a narrow crest is that it does not ensure a constant coefficient of discharge for varying heads. *Hence, if the fall is to be used as a meter, it must not be provided with a narrow crest, and a broad crest of minimum thickness 2.5 H is adopted.* But for an unflumed ordinary fall, a narrow crest of width $\frac{2}{3}H$ is generally adopted.

**Discharge Formula.** Discharge is given by the formula,

$$Q = C \cdot (L - 0.1n \cdot H)\, H^{3/2}$$

where $n$ = No. of end contractions
= Twice the number of bays.

The value of $C$ may be taken as 1.7 for a broad crest, and 1.84 for a sharp crest.

**Crest Level.** Crest level is given by

Crest level = u/s TEL $- H$

where u/s TEL = u/s FSL + Velocity head.

If the height of the crest works out to be more they 0.4 times the upstream water depth, the fall may be flumed or fluming ratio increased, so as to increase $H$ and to lower the crest.

**Hump.** For an unflumed non-meter ordinary fall, the bed approach may have a slope of $\frac{1}{2} : 1$ ($\frac{1}{2} H : 1V$), joined tangentially to the u/s edge of the crest with a radius of $H/2$, as shown in Fig. 12.23.

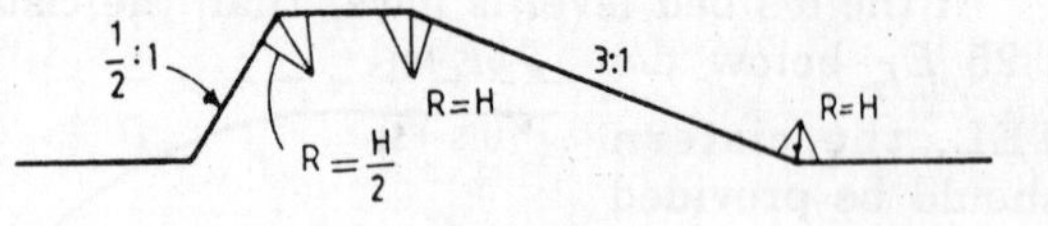


Fig. 12.23

For a flumed meter fall, the curve should start at the same cross-section as the side curves. The radius of hump or bed curve is equal to $\dfrac{L_a^2 + h^2}{2h}$ (see Fig. 12.24).

**Upstream Protection.** No upstream protection is required for an ordinary unflumed fall. However, for a flumed fall, dry brick pitching on edge may be laid both in bed and sides for a length equal to u/s water depth (in a slope of 1 : 10 in the bed).

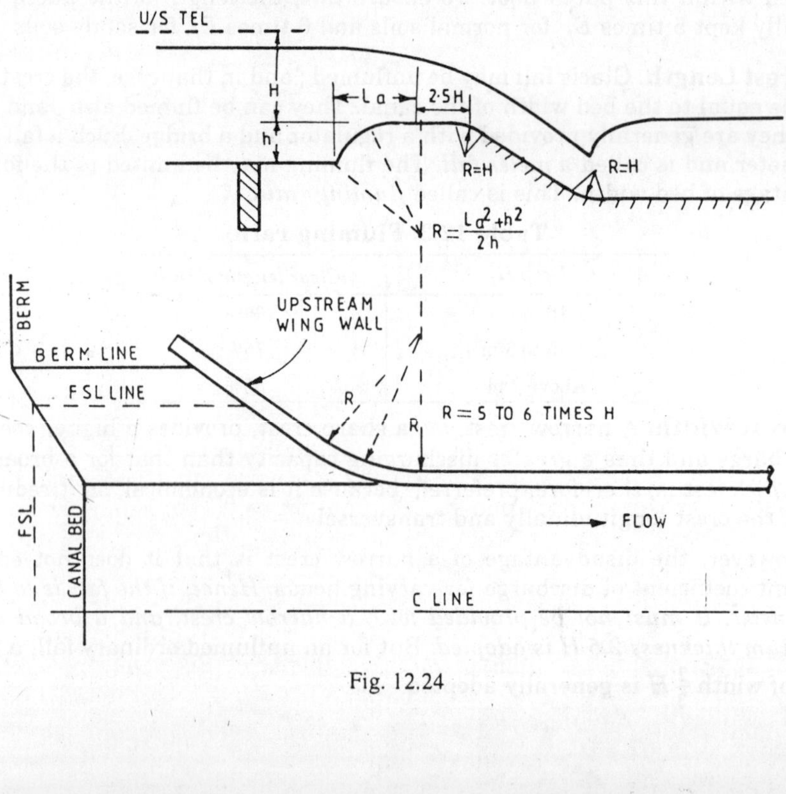


Fig. 12.24

**Downstream Glacis.** A straight glacis with 2 : 1 slope has been found to be quite suitable for this type of falls. The glacis may be joined to the crest at the u/s end, and to the floor at the d/s end, with a radius = $H$, as shown in Fig. 12.23.

**Length and Thickness of Floor.** The floor is designed as per Khosla's theory and the total length of the floor is determined from exit gradient considerations, as was done for Trapezoidal Sarda type fall.

**Downstream Protection.** With the provision of a deflector wall, no d/s bed pitching is provided, and only side slope pitching is provided for a length equal to 3 times the d/s water depth and should rest on a toe wall.

**Upstream Wing Walls.** For an unflumed non-meter fall, the side walls may be splayed at 45° from u/s edge of crest and carried into the berm for about one metre [as was done in rectangular crest—Sarda type fall (see Fig. 12.19)].

For a flumed meter fall, curved wings with a radius of 5 to 6 times $H$ subtending an angle of 60° at centre, and therefore, carried tangentially into the berm [as was done in Trapezoidal crest – Sarda type fall (see Fig. 12.17)] may be provided.

**Downstream Wing Walls.** For an unflumed fall, the walls are taken straight up to the d/s end of the floor and then joined with return walls (as was done in a rectangular crest — Sarda type fall— Fig. 12.19.

For a flumed fall, the d/s wings shall have to be expanded up to the normal width of the river. The expansion can be achieved straight with 1 in 3 slope for smaller works, and a hyperbolic 'expansion (Fig. 12.25), given by equation (12.11), may be used for large works. The hyperbolic expansion is given by equation :

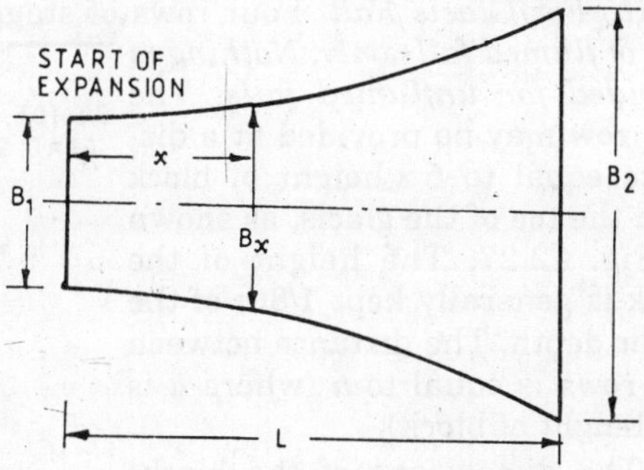


Fig. 12.25. Hyperbolic Expansion.

$$B_x = \frac{B_1 \cdot B_2 \cdot L}{L \cdot B_2 - (B_2 - B_1) \cdot x} \quad ...(12.11)$$

where $B_x$ is the width of expanding flume at any distance $x$ from the beginning of expansion.

The wings shall be embedded into the berms by at least 1 metre.

## Roughening Devices or Energy Dissipators

In Sarda type fall, where energy is dissipated by impact, roughening devices can serve as an additional source of energy dissipation. Similarly, in Glacis fall, roughening devices may be employed to arrest the energy left after jump formation. They may help in jump formation also. Various types of roughening devices used in falls are given below :

(1) **Friction Blocks.** These are the most simple and useful of all such devices. They consist of rectangular concrete blocks securely anchored into the floor. Their height is approximately $\frac{1}{4}$th water depth. The spacing between the blocks is about

twice the height of the blocks. The specific recommendations for their use in the modern falls are as given below :

(a) *For Vertical Drop Fall or Sarda Type Fall with Trapezoidal crest.* Two rows of friction blocks staggered in plan may be provided, as shown in Fig. 12.26.

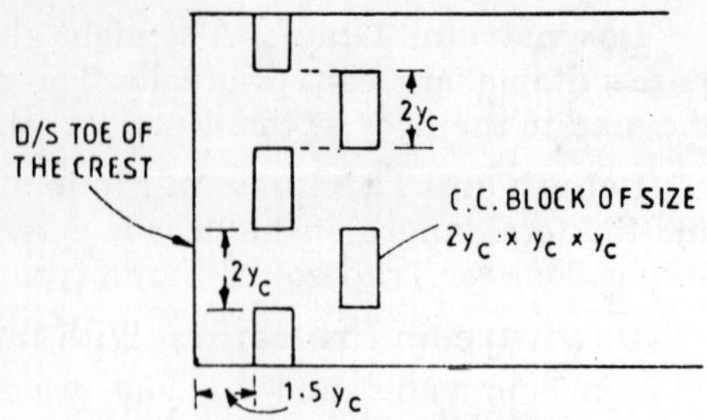

Fig. 12.26. Provision of friction blocks in Sarda type fall.

The first row may start from a distance 1.5 $y_c$ (where $y_c$ = critical depth) from the d/s toe of the crest. Spacing between the rows is kept equal to $y_c$.

Spacing between the blocks $= 2 \cdot y_c$.

The dimensions of the blocks $= 2y_c \times y_c \times y_c$

In addition, cube blocks of size $\frac{1}{8}$th to $\frac{1}{16}$th of water depth may be provided at the end of the impervious floor.

(b) *For Glacis Fall.* Four rows of staggered blocks are generally provided *in case of flumed falls only. Nothing is provided for unflumed falls.* The first row may be provided at a distance equal to 5 × height of block from the toe of the glacis, as shown in Fig. 12.27. The height of the block is generally kept 1/8th of the water depth. The distance between two rows is equal to $h$ (where $h$ is the height of block).

The dimensions of the blocks are $3h \times h \times h$.

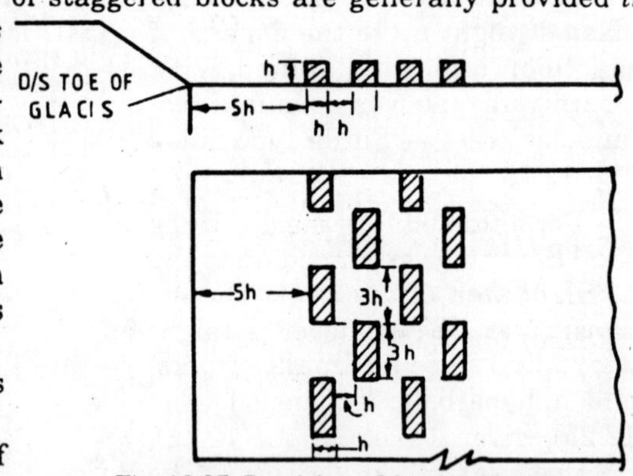

Fig. 12.27. Provision of friction blocks in flumed glacis fall.

(c) *For Baffle Fall.* Two rows of friction blocks staggered in plan may be provided only when the drop is more than 2 metres. The suitable dimensions are

Height $(h) = 0.26\, y_2$

where $y_2$ is the sub-critical depth in the canal d/s, required for formation of jump for parallel downstream sides, in metres.

Length $= h$.

Top width $= \frac{2}{3}h$.

Distance between rows $= h$.

The first row may be placed at a distance $\frac{2}{3}$rd of the cistern length from the upstream end of the cistern.

(2) **Glacis Blocks.** A single row of blocks called Glacis blocks and of the same dimensions as friction blocks may be provided just at the d/s toe of the glacis, in case of flumed falls with drops more than 2 metres, as shown in Fig. 12.28. It helps in reducing turbulence in flow, which in turn, reduces wave wash, thus ensuring uniform flow.

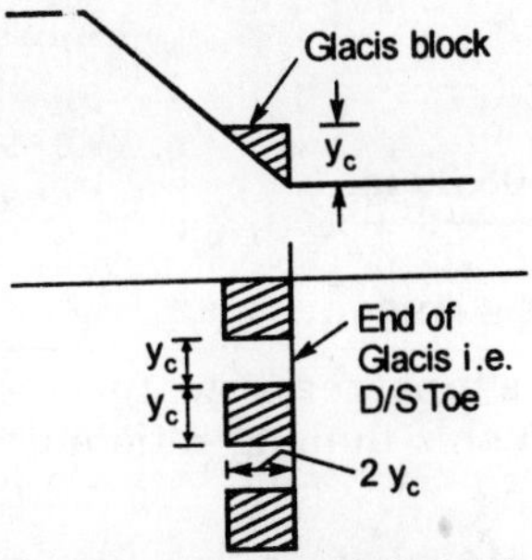


Fig. 12.28. Provision of glacis blocks in flumed falls with drops more than 2 metres.

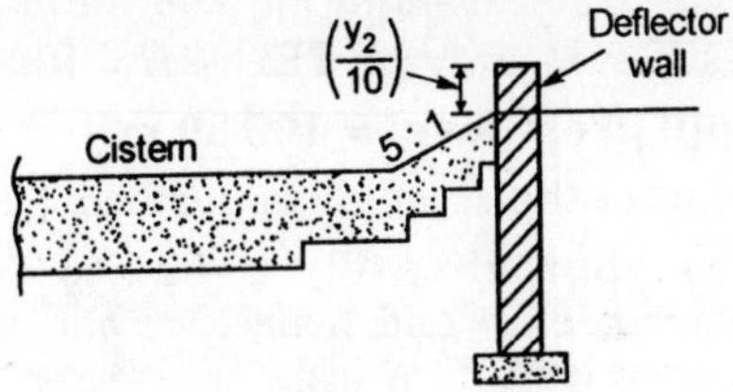


Fig. 12.29. Provision of a deflector wall.

**(3) Biff Wall or Deflector Wall.** A deflector wall of height approximately 1/10th of water depth may be provided at the downstream end of concrete floor (in cast with d/s cutoff) as shown in Fig. 12.29.

**Example 12.3.** *Design a straight flumed meter glacis fall with the following data :*

| | |
|---|---|
| *Full supply discharge of the canal* | *= 120 cumecs.* |
| *Bed level of the canal upstream* | *= 107.5 m* |
| *Bed level of the canal downstream* | *= 106.0 m* |
| *Drop (HL)* | *= 1.5 m.* |
| *FSL of the canal upstream* | *= 109.7 m* |
| *FSL of the canal downstream* | *= 108.2 m* |
| *Bed width of the canal u/s and d/s* | *= 60 m* |
| *Safe Exit Gradient for canal material* | *= 1/5.5.* |

**Solution.**

*Length of Crest.* From Table 12.3, for $H_L = 1.5$,

fluming ratio = 75%

∴ Length of the crest= 75% of Bed width of canal

$$= 75\% \times 60 \text{ m} = \mathbf{45\ m.}$$

**Adopt crest length** = 45 m.

*Crest Level.* Since the fall is to be used as a meter, a broad crest shall be provided.

∴ Discharge $Q$ is given by

$$Q = 1.70\, LH^{3/2}.$$

∴ $$120 = 1.70 \times 45\, H^{3/2}.$$

or $$H^{3/2} = \frac{120}{1.7 \times 45} = 1.57$$

or $$H = (1.57)^{2/3} = \mathbf{1.35\ m}$$

Velocity of approach $$= V_a = \frac{120}{(60 + 2.2)\, 2.2} = 0.876$$

[assuming 1 : 1 side slopes of canal and taking depth = 2.2 as given]

Velocity head $= \frac{(0.876)^2}{2 \times 9.81} = 0.039$ m; **say 0.04 m.**

u/s TEL = u/s FSL + Velocity head

= 109.7 + 0.04 = 109.74 m.

Crest Level = u/s TEL − $H$ = 109.74 − 1.35 = 108.39 m.

**Adopt crest level = 108.39 m.**

Height of the crest above u/s bed = 108.39 − 107.5 = 0.89 m.

which is approximately equal to 0.4 times the upstream water depth (2.2 × 0.4 = 0.88 m) and hence, we can go ahead with this fluming ratio and the present crest level.

*Crest Width.* Adopt crest width for broad crest

= 2.5$H$ = 2.5 × 1.35 = 3.375 m.

Provide crest width = **3.38 m.**

*Hump.* The crest shall be joined with upstream bed with a curve of radius

$$R = \frac{(L_a^2 + h^2)}{2h}$$

where $L_a$ = 2 m (assume).

$h$ = 0.89 (ht. of crest above u/s bed).

$$\therefore \text{ Radius of the curve } R = \frac{2^2 + (0.89)^2}{2 \times 0.89} = \frac{4.79}{1.78} = \mathbf{2.69\ m}$$

*Cistern.* The d/s glacis is provided in a slope of 2 : 1. The cistern level is worked out as below :

$Q$ = 120 cumecs.

Width provided = 45 m.

$q$ = Discharge intensity $= \frac{120}{45} = 2.27$ cumecs/metre

$H_L$ = 1.5 m.

From plate 10.1, $E_{f_2}$ = 1.85 m.

Level at which jump will form

= d/s TEL − $E_{f_2}$ = 108.24 − 1.85 = **106.39 m.**

1.25 $E_{f_2}$ = 2.31 m

R.L. of Cistern = d/s TEL − 1.25 $E_{f_2}$

= (108.2 + 0.04) − 2.31 = 108.24 − 2.31 = **105.93 m.**

which is lower than the d/s bed level, and hence, **adopt cistern level = 105.93 m.**

*Length of Cistern.* 5 to 6 times $E_{f_2}$, *i.e.* 5 × 1.85 = 9.25 m; or 6 × 1.85 = 11.10 m. Hence, **use 10 m length of cistern** joined to d/s bed in a slope of 5 : 1 in a length of 5 (106 − 105.93) = 0.35 m.

*Downstream Curtain Walls.* Depth of d/s curtain wall below bed $= \frac{\text{(water depth)}}{2} = \frac{2.2}{2} = 1.1$ m; or from Table 12.1, it is equal to 1.35 m (for H = 1.35 m), therefore, adopt a curtain wall of total depth equal to 1.4 m, *i.e.* 1.1 m masonry laid over 0.3 m concrete.

**Total Floor Length and Exit Gradient.** Maximum static head ($H$) is caused when water is stored up to crest level on the u/s and there is no water d/s.

Maximum static head ($H$) = 108.39 – 106.0 = 2.39 m

Depth of d/s curtain wall ($d$) = 1.4 m.

Safe Exit gradient $G_E = \dfrac{1}{5.5}$

But $$G_E = \frac{H}{d} \cdot \frac{1}{\pi\sqrt{\lambda}}$$

$$\therefore \quad \frac{1}{5.5} = \frac{2.39}{1.4} \cdot \frac{1}{\pi\sqrt{\lambda}}$$

or $$\frac{1}{\pi\sqrt{\lambda}} = \frac{1}{5.5} \times \frac{1.4}{2.39} = \mathbf{0.107}$$

From Plate 11.2, $\alpha = 17$

Total floor length required = $\alpha \cdot d$ = 17 × 1.4 = 23.8 m. ; **say 24 m.**

The distribution of floor length is shown in Fig. 12.29. 20.65 m is already provided and the balance of 3.35 m is provided as u/s floor length.

*Upstream Curtain Wall.* A curtain wall of minimum depth $\dfrac{y_u}{3} = \dfrac{2.2}{3} = 0.73$ m must be provided at the end of pucca floor. Let us provide an u/s curtain wall (cutoff) of 0.4 m × 0.8 m deep masonry wall laid over 0.3 m thick cement concrete, thus giving a total depth of cutoff as 1.1 m.

**Uplift Pressures**. Refer Fig. 12.30. Assume floor thickness on the u/s = 0.3 m and thickness on d/s = 0.8 m

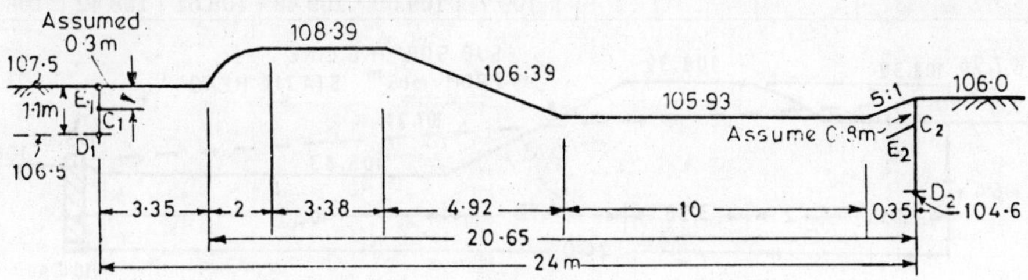


Fig. 12.30

*U/s Curtain Wall*

$$b = 24 \text{ m} \quad \text{and} \quad d = 1.1 \text{ m}$$

$$\frac{1}{\alpha} = \frac{d}{b} = \frac{1.1}{24} = 0.046$$

From Plate 11.1 ($a$)

$$\phi_{E_1} = \mathbf{100\%}$$

$$\phi_{C_1} = 100 - \phi_C = 100 - 20 = 80\%$$

$$\phi_{D_1} = 100 - \phi_D = 100 - 13 = \mathbf{87\%}$$

*Correction for depth to* $\phi_{C_1} = \dfrac{87\% - 80\%}{1.1} \times 0.3 = 1.91\%$ (+ ve)

$\phi_{C_1}$ **(corrected)** = 80 + 1.91 = **81.91%**

*D/s Curtain Wall*

$$b = 24 \text{ m} \quad \text{and} \quad d = 1.4 \text{ m}$$

$$\frac{1}{\alpha} = \frac{d}{b} = \frac{1.4}{24} = 0.0583$$

From Plate 11.1 (*a*)

$$\phi_{E_2} = \phi_E = 23\%$$
$$\phi_{D_2} = \phi_D = 15\%$$
$$\phi_{C_2} = 0\%$$

*Correction for depth to* $\phi_{E2} = \dfrac{23\% - 15\%}{1.4} \times 0.8 = 4.6\%$ (– ve)

$\phi_{E_2}$ **(corrected)** = 23% + 4.6% = **27.6%**

The levels of H.G. lines for maximum static head and flow at FSL are worked out in Table 12.4 and plotted in Figs. 12.31 (*a*) and (*b*), respectively.

**Table 12.4**

| *Condition of flow* | *u/s* W.L. *in metres* | *d/s* W.L. *in metres* | *Head in metres* | *Height / elevation of H.G. line above datum* | | | | | |
|---|---|---|---|---|---|---|---|---|---|
| | | | | $\phi_{E_1}$ *100%* | $\phi_{D_1}$ *87%* | $\phi_{C_1}$ *81.91%* | $\phi_{E_2}$ *27.6%* | $\phi_{D_2}$ *15%* | $\phi_{C_2}$ *0%* |
| Maximum Static Head, *i.e.* Water upto Crest level on u/s and no water on d/s | 108.39 | 106.0 | 2.39 | 2.39 | 2.08 | 1.96 | 0.61 | 0.36 | 0 |
| | | | | 108.39 | 108.08 | 107.96 | 106.61 | 106.36 | 106.0 |
| Flow at FSL | 109.7 | 108.2 | 1.5 | 1.5 | 1.30 | 1.23 | 0.41 | 0.22 | 0 |
| | | | | 109.7 | 109.50 | 109.43 | 108.61 | 108.42 | 108.2 |

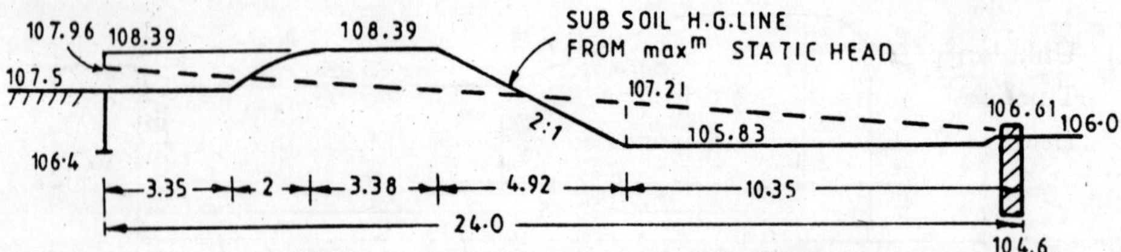


Fig. 12.31. (*a*) Maximum static head condition.

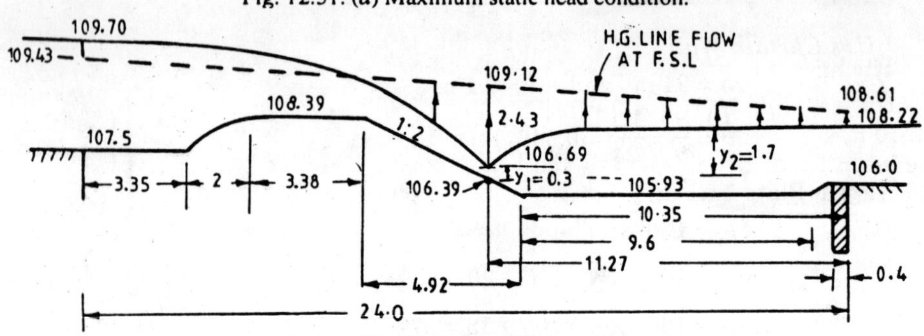


Fig. 12.31. (*b*) Flow at FSL condition.

**Floor Thicknesses.** Provide 0.3 m thickness under upstream floor.

*At Toe of Glacis*. Level of H.G. line (*for static head*) at toe of glacis.

$$= 106.61 + \frac{107.96 - 106.61}{24} \times 10.35$$

$$= 106.61 + \frac{1.35}{24} \times 10.35$$

$$= 106.61 + 0.59 = 107.20 \text{ m.}$$

Unbalanced head at the toe due to static head

$= 107.20 - 105.93 = 1.27$ m

*Jump formation*

$q = 2.27$ cumecs/metre.

$H_L = 1.5$ m.

$E_{f_2} = 1.85$ m.

$E_{f_1} = E_{f_2} + H_L = 1.85 + 1.50 = 3.35$ m.

From Plate 10.2

$y_1$ (for $E_{f_1}$) = 0.3 m.

$y_2$ (for $E_{f_2}$) = 1.7 m.

The maximum ordinate at point of jump, from Fig. 12.31 (*b*) = 2.43 m.

$\frac{2}{3}$rd of this ordinate = $\frac{2}{3} \times 2.43 = 1.62$ m, which is more than that for static head.

Hence, the floor thickness at point of jump must not be less than

$$= \frac{1.62}{1.24} = 1.31 \text{ m.}$$

**Provide 1.5 m at this point and extend upto 2 m beyond toe**

Level of H.G. line at 2 m beyond toe (due to static head)

$$= 106.61 + \frac{1.35}{24} \times 8.35 = 106.61 + 0.47 = \mathbf{107.08\ m.}$$

Unbalanced head = 107.08 – 105.93 = **1.15 m.**

Thickness required = 1.15/1.24 = 0.925 m., **Provide 1.0 m.**

Level of H.G. line at 5 m beyond toe (due to static head)

$$= 106.61 + \frac{1.35}{24} \times 5.35 = 106.61 + 0.30 = 106.91$$

Unbalanced head due to static head at this point

$= 106.91 - 105.93 = 0.98$ m.

Thickness required $= 0.98/1.24 = 0.79$ m., **Use 0.8 m.**

*Upstream Wing Walls.* Curved wing walls with radius equal to $6H = 6 \times 1.35 = 8.10$ m subtending an angle of 60° at centre and then carried tangentially into the berm for a suitable length shall be provided as shown in the attached chart Fig. 12.32.

*Downstream Expansion.* The width of 45 m shall be expanded to 60 m in a splay of 3 : 1.

Minimum length of expansion required = $\frac{3\,(60-45)}{2}$ = **22.5 m.**

The wings can be splayed straight from the d/s toe of the glacis in a length equal to 22.5 m. The bed shall have to be pitched upto the end point of wings. Slope pitching shall be provided in a length $3 \times 2.2 = 6.6$ m beyond this point.

Upstream pitching in a minimum length of 2.2 m (equal to u/s water depth) is required. It shall, however, be provided as shown in Fig. 12.32.

*Energy Dissipators.* Four rows of staggered friction blocks shall be provided. The height of the block is $\frac{1}{8}$th of water depth, *i.e.* $\frac{1}{8} \times 2.2 = 0.275$ m. The first row shall be at a distance 5 × 0.275 = 1.375 m from the d/s toe of glacis. The size of the blocks shall be 0.825 m × 0.275 m × 0.275 m. Distance between rows = 0.275 m.

A deflector wall at end point of floor is provided and is to have a height equal to $\frac{y_d}{10}$, *i.e.* $\frac{2.2}{10}$ = **0.22 m.**

The detailed arrangement is shown in attached chart Fig. 12.32.

## 12.8. Design of a Baffle Fall or Inglis Fall

Certain flumed type ordinary straight Glacis falls constructed in Punjab were later found to give some serious troubles, which gave rise to the conclusion that considerable surplus energy might remain in water even after the jump formation. One major cause of these troubles was found to be, too rapid expansion after fluming, which may generate cork screw eddies causing deep scours. Research was carried out to eliminate these defects and Baffle fall was evolved.

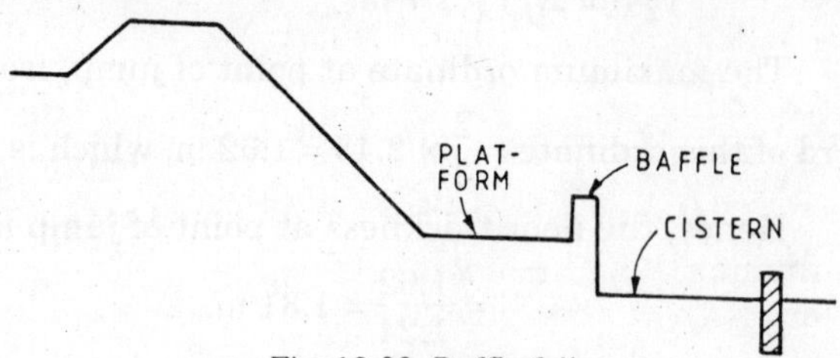

Fig. 12.33. Baffle fall.

A baffle fall makes use of the principle of horizontal impact for energy dissipation. The jump is held stable on a horizontal platform by means of a baffle wall (called baffle).

**Baffle Platform**. The horizontal platform is provided at the level at which the jump would normally form. This can be determined by Blench curves in case there is no expansion of the wings in the region of supercritical jet. If the supercritical jet is splayed, the optimum level at which the baffle platform should be provided, can be determined by designer's curves (given in C.B.I. publication No. 10). In the absence of curves, the values can be determined by using the formulas given below.

Subcritical depth ($y_2$) required for jump formation in ordinary cases without fluming is very nearly given by

$$y_2 = 0.98\, q^{0.52}\, H_L^{0.21} \quad \text{...(12.10)}$$

where $q$ = the discharge intensity in cumecs/metre
$H_L$ = Drop in metres.
$y_2$ = the subcritical depth in metres.

The subcritical depth [$y_2$ (flumed)] required for jump formation, in case there is a fluming, is given by

$$y_2\ (\text{flumed}) = y_2 + (H_X - H_L) \quad \text{...(12.11)}$$

where $H_X$ is the calculated drop in metres given by

$$H_X = \frac{H_L}{K^{0.152}} \quad \text{...(12.12)}$$

where $K$ is the fluming ratio (more than 1), *i.e.*

$$= \frac{\text{Actual width of canal before fluming}}{\text{Flumed width of canal}}$$

The R.L. of the baffle platform will then be given by

(*i*) For unflumed fall = d/s FSL $- y_2$.

(*ii*) For flumed fall = d/s FSL $- y_2$ (flumed)

**Baffle Wall.** Height of the baffle wall $= h_b = y_c - y_1$ ...(12.13)

where $y_c$ is the critical depth, given by

$$y_c = \left[\frac{q^2}{g}\right]^{1/3} \quad ...(12.14)$$

and $y_1$ is prejump super critical depth, given by

$$y_1 = 0.183\, q^{0.89} \cdot H_X^{-0.35} \quad ...(12.15)$$

$H_X = H_L$, when there is no fluming

*Thickness of the baffle wall* $= \frac{2}{3} h_b$ ...(12.16)

*Length of baffle platform* $= 5.25\, h_b$. ...(12.17)

**Cistern.** A cistern of length equal to $5\, y_2$ (flumed) (equal to $5 \cdot y_2$, for without fluming) may be provided after the baffle wall. The depth of the cistern below the downstream bed should be 10% of the downstream water depth ($y_d$), subject to a minimum of 15 cm for distributaries and minors and 30 cm for main canals and branches.

**Upstream Wings.** The upstream wings can be curved with a radius equal to $3.6\, H^{3/2}$ (where $H$ is the total head over the crest) when $H$ is more than 0.3 m, and equal to $2H$ when $H$ is less than 0.3 m. The circular wings are continued till they subtend an angle of 60° at the centre, and afterwards they can be extended tangentially for the required length into the berm or bank.

**Downstream Wings.** The downstream divergence for flumed meter falls can be provided at a slope of 1 in 3 to 1 in 4.

A milder divergence is preferable for straight Glacis fall as well as for Baffle fall (1 in 4 to 1 in 10 depending upon the bed width depth ratio of the downstream channel); but that will make the structure costlier. Hence, a divergence of 1 in 3 is generally used for meter falls of both types. The d/s wings of the unflumed baffle fall are kept as are kept for unflumed straight Glacis fall.

**Downstream Glacis.** The d/s glacis for unflumed baffle falls is kept at 2/3 : 1; but for flumed meter falls, it is kept as 2 : 1. The glacis is joined to crest at the u/s end and to the floor at the d/s end with a radius equal to $H$, as was done in the case of a glacis fall.

All other details of pitching, etc. remain the same as for straight glacis fall.

The friction blocks are either not provided (upto 2 m fall); or provided as explained earlier (for drops of more than 2 m.)

**Example 12.6.** *Design an unflumed non-meter baffle fall for the canal having the following data :*

| | |
|---|---|
| *Full supply discharge* | *= 30 cumecs* |
| *Bed level u/s* | *= 203.0 m* |
| *Bed level d/s* | *= 201.2 m* |
| *FSL u/s* | *= 204.3 m* |
| *FSL d/s* | *= 202.5 m* |
| *Bed width* | *= 28 m* |
| *Drop ($H_L$)* | *= 1.8 m* |
| *Side slopes of channel* | *= 1 : 1.* |

**Solution.**

*Crest Length.* Equal to bed width : Provide 28 m. crest length

*Crest level.* A sharp narrow crest is provided, for which

$$Q = 1.84\,LH^{3/2}$$

$$\therefore \quad 30 = 1.84 \times 28\,H^{3/2}$$

or $$H^{3/2} = \frac{30}{1.84 \times 28} = 0.582$$

or $$H = (0.582)^{2/3} = 0.697 \text{ m; } \textbf{say 0.7 m.}$$

Velocity of approach $= V_a = \dfrac{30}{(28 + 1.3)\,1.3} = 0.787$ m/sec.

Velocity Head $= \dfrac{V_a^2}{2g} = \dfrac{(0.787)^2}{2 \times 9.81} = 0.0315$ m; **say 0.03 m.**

Now, u/s TEL = u/s FSL + Velocity head = 204.3 + 0.03 = 204.33 m.
Crest Level = u/s TEL $- H$ = 204.33 – 0.7 = 203.63 m.

**Adopt crest level = 203.6 m.**

*Width of the crest* is kept equal to $\frac{2}{3}H = \frac{2}{3} \times 0.7 = 0.47$ m

Provide **crest width = 0.47 m.**

*U/s Glacis.* Glacis of $\frac{1}{2}$ : 1 joined tangentially to the crest with a radius equal to $H/2 = 0.35$ m shall be provided.

*D/s Glacis.* Glacis of $\frac{2}{3}$ : 1 joined tangentially to the baffle platform with a radius equal to $H = 0.70$ m shall be provided.

*Upstream Wings.* The u/s wing walls shall be splayed at an angle of 45° from the u/s end of the floor and shall be embedded into the bank by 1.0 m beyond FSL line.

*Downstream Wings.* Parallel vertical sides up to the end of pucca floor shall be provided, which shall be connected with the return walls at 90°.

*Upstream Protection.* No pitching is required in bed and on sides. Depth of u/s curtain wall required is

$$= \frac{y_u}{3} = \frac{1.3}{3} = 0.43 \text{ m.}$$

Provide 0.4 m × 0.6 m deep curtain wall over 0.3 m foundation concrete, thus making its overall depth as 0.9 m.

**Baffle Platform and Baffle Wall**

*Baffle Platform.* Since there is no fluming,

$$H_L = H_x = 1.8 \text{ m}$$

Now $$y_2 = 0.98 \cdot q^{0.52} \cdot H_L^{0.21} \qquad (i.e. \text{ Eq. } 12.10)$$

where $q = \dfrac{30}{28} = 1.07$ cumecs/metre

$$\therefore \quad y_2 = 0.98 \cdot (1.07)^{0.52} \cdot (1.8)^{0.21}$$

$$= 0.98 \times 1.0358 \times 1.133 = 1.15 \text{ m.}$$

R.L. of baffle platform

$$= \text{d/s FSL} - y_2 = 202.5 - 1.15 = 201.35 \text{ m.}$$

**Provide Baffle platform at R.L. = 201.35 m**

*Baffle Wall.* Height of the baffle wall

$$= h_b = y_c - y_1$$

where $$y_c = \left[\frac{q^2}{g}\right]^{1/3} = \left[\frac{(1.07)^2}{9.81}\right]^{1/3} = (0.117)^{0.33} = 0.49 \text{ m}$$

$$y_1 = 0.183 \cdot q^{0.89} \cdot (H_x)^{-0.35} \qquad (i.e. \text{ Eq. 12.15})$$

$$= 0.183\,(1.07)^{0.89}\,(1.8)^{-0.35} = \mathbf{0.16\ m}$$

∴ **Height of the baffle wall**

$$h_b = y_c - y_1 = 0.49 - 0.16 = \mathbf{0.33\ m.}$$

**Thickness of baffle wall** $= \frac{2}{3} h_b =$ **0.22 m.**

**Length of baffle platform**

$= 5.25\, h_b = 5.25 \times 0.33 = 1.73$ m. ; **say 1.8 m.**

*Cistern.* Depth of cistern below d/s bed

$= 0.1\, y_d = 0.1 \times 1.3 = 0.13$ m.

**R.L. of cistern** $= 201.2 - 0.13 =$ **201.07 m.**

**Length of cistern** $= 5 \cdot y_2 = 5 \times 1.15 = 5.75$ m; **say 5.8 m.**

*D/s Curtain Wall.* Depth of the downstream curtain wall required

$$= \frac{y_d}{2} = \frac{1.3}{2} = 0.65 \text{ m};$$

or from Table 12.1, the depth for the curtain wall is equal to 0.6 m.

Provide 0.4 m × 1.0 m deep curtain wall, over 0.3 m thick foundation concrete, thus making a total depth of curtain wall = 1.3 m

Height of deflector above d/s bed

$$= \frac{y_d}{10} = \frac{1.3}{10} = 0.13 \text{ m.}$$

Hence, the d/s curtain wall shall be raised by 0.13 m above d/s bed.

*D/s Bed Pitching.* No pitching is required.

*D/s Side Slope Pitching.* Is required in a length equal to

$$3 \cdot y_d = 3 \times 1.3 = 3.9 \text{ m.}$$

Provide 0.2 m thick dry brick pitching over 0.1 m thick brick ballast in a length equal to 3.9 m. The slope pitching shall rest on a toe wall 0.4 m thick and 0.8 m deep (overall) constructed in the bed at the junction of bed and sides. A solid profile wall called 'Dhamali' shall be constructed at the end of pitching. It shall be 0.4 m thick and plastered in cement mortar.

*Friction Blocks.* Not required.

**Total Floor Length from Exit Gradient Considerations**

Safe exit gradient = 1/5

Maximum static head ($H$) is exerted when water is stored upto crest level on u/s and there is no water on d/s.

$$\therefore \qquad H = 203.6 - 201.2 = 2.4 \text{ m.}$$

Depth of d/s curtain wall $= d = 1.3$ m

Now, $G_E = \frac{H}{d} \cdot \frac{1}{\pi \sqrt{\lambda}}$

$$\frac{1}{5} = \frac{2.4}{1.3} \cdot \frac{1}{\pi \sqrt{\lambda}}$$

or $$\frac{1}{\pi \sqrt{\lambda}} = \frac{1}{5} \times \frac{1.3}{2.4} = 0.108$$

From Plate 11.2, $\alpha = 16.5$

$\therefore$ Total floor length required $= \alpha d = 16.5 \times 1.3 = 21.45$ m;

Provide **22 m** overall length.

The floor length already provided (12.365 m) is shown in Fig. 12.34 ; the balance, *i.e.* 9.635 m is now provided on the u/s, as shown in Fig. 12.34.

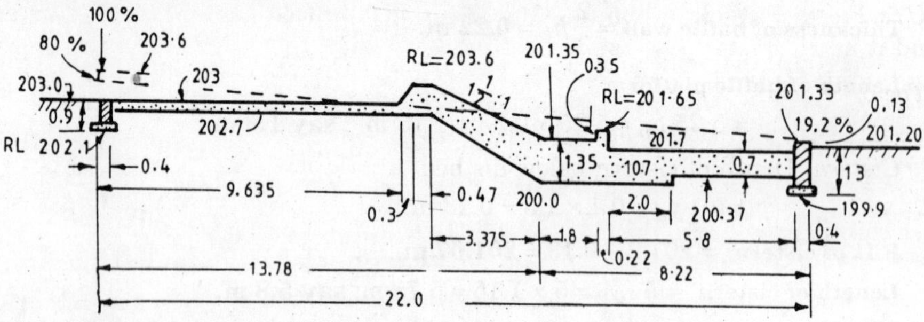


Fig. 12.34. Dimensions of the baffle fall of example 12.6.

## Calculations for Uplift Pressures

(*i*) *Upstream curtain wall*

$b = 22$ m and $d = 0.9$ m

$$\frac{1}{\alpha} = \frac{d}{b} = \frac{0.9}{22} = 0.041$$

From Plate 11.1 (*a*),

$$\phi_{E_1} = \mathbf{100\%}$$

$$\phi_{D_1} = 100 - \phi_D = 100\% - 14\% = \mathbf{86\%}$$

$$\phi_{C_1} = 100 - \phi_E = 100\% - 20\% = \mathbf{80\%}$$

Assume upstream floor thickness = 0.3 m.

*Correction to* $\phi_{C_1}$ *for depth*

$$= \frac{86\% - 80\%}{0.9} \times 0.3 = 2.0\% \text{ (+ ve)}$$

$\phi_{C_1}$ **(Corrected)** = 80% + 2% = **82%**

(*ii*) *Downstream curtain wall*

$b = 22$ m and $d = 1.3$ m

$$\frac{1}{\alpha} = \frac{d}{b} = \frac{1.3}{22} = 0.059$$

From Plate 11.1 (*a*)

$$\phi_{E_2} = \phi_E = 23\%$$

$$\phi_{D_2} = \phi_D = \mathbf{16\%}$$

$$\phi_{C_2} = \mathbf{0\%}$$

Assume d/s floor thickness near d/s curtain wall = 0.7 m

*Correction to $\phi_{E_2}$ for floor thickness*

$$= \frac{23\% - 16\%}{1.3} \times 0.7 = 3.8\% \text{ (– ve)}$$

$\phi_{E_2}$ (**Corrected**) = 23% – 3.8% = **19.2%**.

Uplift pressure at the toe of the glacis

$$= 19.2\% + \frac{80\% - 19.2\%}{22} \times 8.22 = 19.2\% + 22.7\% = 41.9\%$$

**Floor Thicknesses**

*U/s Floor.* Provide a nominal thickness of 0.3 m on the upstream side and extend it up to d/s end of crest. Its bottom level shall be at R.L. 202.7 m.

*Toe of Glacis*

Level of H.G. line at toe of glacis

$$= 201.2 + 41.9\% \times 2.4 = 201.2 + 1.0 = 202.2 \text{ m}$$

∴ Unbalanced head due to this maximum static head of 2.4 m

$$= 202.2 - 201.35 = \mathbf{0.85\ m}$$

Unbalanced head due to dynamic condition may be taken as

$$= 50\% \ (y_2 - y_1) + \% \text{ pressure} \times H_L \qquad \text{(Refer page 558 Chapter 11)}$$

$$= 50\% \ (1.15 - 0.16) + 41.9\% \times 1.8$$

$$= 0.495 + 0.754 = 1.249 \text{ m ; } \mathbf{say\ 1.25\ m.}$$

Thus, at toe of glacis, the head due to dynamic condition is more than that due to static condition. Hence, minimum thickness required at toe $= \frac{1.25}{1.24} = 1.01$ m. Provide 1.35 m thickness in the entire length of baffle platform, thus keeping its bottom at R.L. 201.0 m.

*Thickness at the start of cistern*

Percentage pressure at 2.02 m from toe of glacis (*i.e.* start point of cistern)

$$= 19.2\% + \frac{80\% - 19.2\%}{22} \times 6.2 = 19.2\% + 17.2 = 36.4\%$$

Level of H.G. line at this point

$$= 201.2 + 36.4\% \ (2.4) = 201.2 + 0.87 = 202.07 \text{ m.}$$

Maximum unbalanced head at this point

$$= 202.07 - 201.07 = 1.0 \text{ m}$$

Floor thickness required at this point

$$= \frac{1.0}{1.24} = 0.81 \text{ m}$$

Provided thickness = 201.07 – 200.0 = 1.07 m.

Thickness required at 3 m beyond the baffle wall

$$= 19.2\% + \frac{80\% - 19.2\%}{22} \times 3.2 = 19.2\% + 8.85\% = 28.05\%$$

Level of H.G. line at this point

$$= 201.2 + 28.05\% \ (2.4) = 201.2 + 0.67 = 201.87 \text{ m}$$

Maximum unbalanced head at this point

$$= 201.87 - 201.07 = 0.80 \text{ m}$$

Floor thickness required at this point = 0.80/1.24 = 0.645 m

Hence, provide 0.7 m thickness in the remaining portion, as shown in Fig. 12.34.

Full details of the fall are shown in the attached chart Fig. 12.35.

## PROBLEMS

1. (*a*) What is meant by 'Canal drops' ? Why are canal drops constructed in a canal system ?

   (*b*) Enumerate the various types of canal drops which have been used since olden days. Explain in details the design principles governing any one of the modern types.

2. Why are "drops" constructed in an irrigation canal ?

   Draw a neat sketch of a syphon well drop and explain briefly its components.

3. (*a*) What is meant by "falls" and where are they located ?

   (*b*) Discuss briefly the components of various types of falls with neat sketches. Also discuss the suitability of each type.

4. Sketch a syphon well drop to carry 0.3 cumec of water from the following data :

   | | |
   |---|---|
   | Ground level | = 30.00 m. |
   | B.L. of channel above drop | = 28.00 m. |
   | F.S.L. of channel above drop | = 30.00 m. |
   | F.S.L. of channel below drop | = 28.50 m. |
   | B.L. Of channel below drop | = 26.50 m. |

   Provide a 4 m cart track over the well drop. The earthwork connections should be clearly shown. Assume any other data that you may require.

5. (*a*) Explain why trapezoidal notches are preferred to rectangular notches in the design of canal drops.

   (*b*) Design the size and number of notches required for a canal drop with the following particulars:

   | | |
   |---|---|
   | Full supply discharge | = 20 cumecs |
   | Bed width | = 14 m. |
   | F.S. depth | = 1.9 m. |

   Assume any other data if required.

   **Ans.** [6 Notches as with $l = 0.83$ m and $n = 0.37$, provided in a crest raised above the bed level by 0.6 m.]

6. (*a*) Discuss the comparative merits and demerits of Notch falls and Sarda type falls.

   (*b*) Design 1 m Sarda type fall on a channel carrying 20 cumecs discharge whose bed width and water depth are 14 m and 1.9 m respectively.

7. (*a*) What are canal falls and why are they constructed ?

   (*b*) Design an unflumed non-meter hydraulic jump type fall (*i.e.* glacis fall) on a distributary carrying 10 cumecs discharge. The drop to be affected is 2 m, and the depth of flow and the bed width of channel are 1.4 m and 8.0 m respectively.

8. (*a*) Discuss the principle advantage offered by a baffle fall in comparison to a straight glacis fall.

(*b*) Design a flumed baffle fall for a canal having the following data :

Full supply discharge of the canal = 120 cumecs.

Bed level of canal upstream = 107.5 m.

Bed level of canal downstream = 105.5 m.

Drop ($H_L$) = 2 m.

F.S.L. of canal upstream = 109.7 m

F.S.L. of canal downstream = 107.7 m.

Bed width of canal u/s and d/s = 60 m.

Bligh's safe hydraulic gradient for the soil = $C$ = 12.

**9.** (*a*) What are "canal falls" and where are they located ?

(*b*) What is meant by "flumed falls" and what are their advantages ?

(*c*) Discuss how fluming is done for flumed falls. Write down the equations etc. that you will adopt for designing the upstream as well as downstream wing walls of flumed falls.

(*d*) What are "roughening devices" ? Discuss their use in 'falls construction' ? What roughening device would you recommend for a straight glacis fall ? (*a*) when flumed (*b*) when not flumed.

**10.** (*a*) State briefly how you will decide the location of a canal drop. Explain how a trapezoidal notch type fall helps to maintain depth discharge relation in the canal.

(*b*) Following data were observed in a canal fall :

Full supply level = 115 m.

Bed level of canal = 112 m.

Bed width of canal = 15 m.

Full supply discharge = 30 cumecs

Side slope of canal = 2 ($H$) : 1 ($V$)

Length of crest of the fall = 10 m (crest section is rectangular)

Coefficient of discharge over crest = 1.70 $m^{1/2}$/sec.

Calculate the crest level.

**[Solution.** $Q = 30$ cumecs

$y = 3$ m.

$B = 15$ m.

$A = (B + 2y)\, y$ (with $2H : 1\,V$ slopes) = 63 sq. m.

$V_a$ = velocity in channel

= velocity of approach = $Q/A = 30/63 = 0.48$

$V_a^2/2g = 0.0117$; **say 0.012 m**

$$\therefore \text{U/s T.E.L.} = \text{u/s F.S.L.} + \frac{V_a^2}{2g} = 115 + 0.012 = 115.012 \text{ m.}$$

Now,
$$Q = 1.7\, L \cdot H^{3/2}$$

$$30 = 1.7 \times 10\, H^{3/2}$$

$$H = 1.462 \text{ m.}$$

$\therefore$ Crest level = u/s T.E.L. – $H$

= 115.012 – 1.462 = 113.65 m. **Ans.]**

**11.** Write short notes on any three of the following :

(*i*) Syphon well drop.

(*ii*) Roughening Devices.

(*iii*) Wing wall for flumed falls.

(*iv*) Trapezoidal notch fall.

(*v*) Simple vertical drop type fall.

(*vi*) Sarda type fall.

(*vii*) Straight glacis fall

(*viii*) Baffle or Inglis fall.

(*ix*) Different types of falls and their suitability for a particular project.

# —13

# *Regulators Modules, and Miscellaneous Canal Structures*

## 13.1. Canal Regulation

The water which enters into the main canal from the river has to be divided into different Branches and Distributaries, in accordance with the relative urgency of demand on different channels. This process of distribution is called 'Regulation'. To distribute water effectively, the discharge has to be adjusted to any desired value. This aim is achieved by means of regulators.

## 13.2. Canal Regulation Works

The works which are constructed in order to control and regulate discharges, depths, velocities etc. in canals, are known as canal regulation works. These structures ensure the efficient functioning of a canal irrigation system, by giving full control upon the canals. The important of these structures are :

(*i*) *Canal Falls.*

(*ii*) *Canal Regulators (Head Regulator and Cross Regulator).*

(*iii*) *Canal Escapes.*

(*iv*) *Metering Flumes, etc.*

(*v*) *Canal Outlets and Modules.*

Canal falls have already been described in the previous chapter. Now we shall deal with the remaining structures.

## CANAL REGULATORS

## 13.3. Alignment of the off-taking channel

When a branch channel takes off from the main channel (called Parent Channel), the off-take alignment must be carefully designed.

The best ideal alignment is : when the off-taking channel makes zero angle with the parent channel initially and then separates out in a transition, as shown in Fig. 13.1.

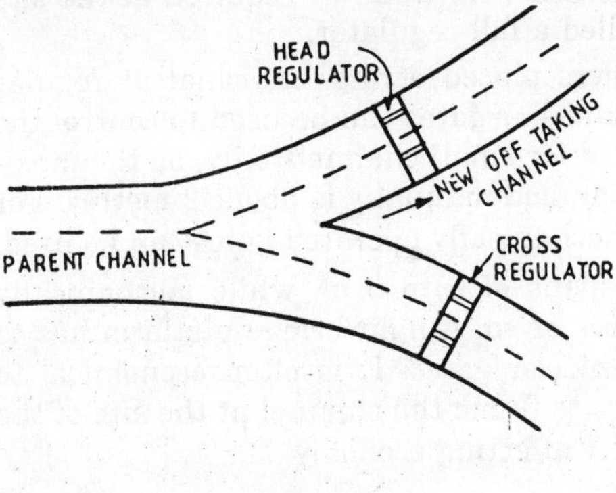


Fig. 13.1

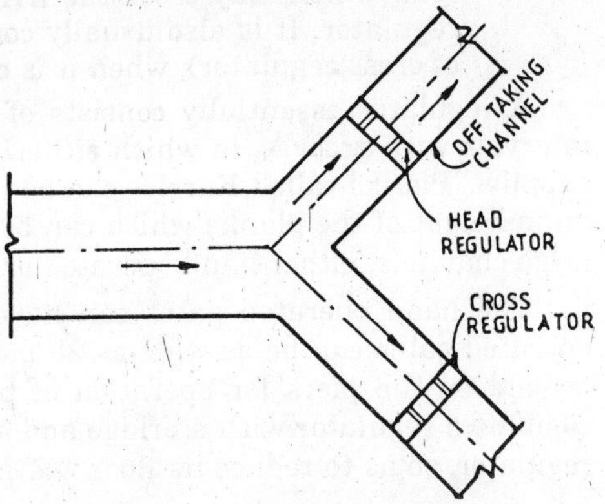


Fig. 13.2

The transitions will have to be properly designed, so as to avoid accumulation of silt jetty. As an alternative to the transitions, both the channels should make an angle with the parent channel upstream of the off-take, as shown in Fig. 13.2.

When the parent channel has to be carried straight, the edge of the canal rather than the centre line should be considered in deciding the angle of the off-take (Fig. 13.3).

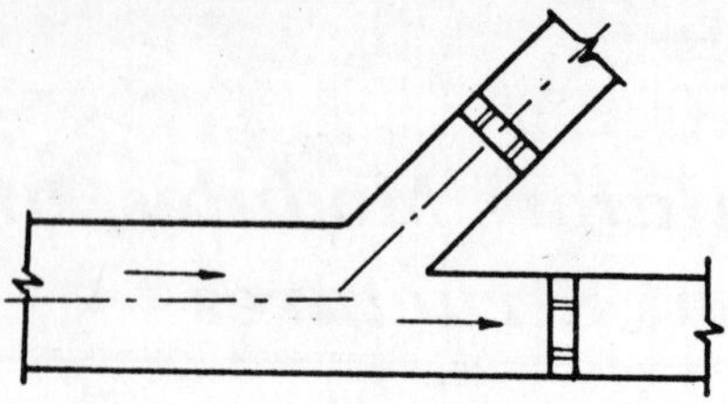

Fig. 13.3

### 13.4. Distributary Head Regulator and Cross Regulator

A distributary head regulator controls the supply of the off-taking channel ; while a cross-regulator controls the supply of the parent channel.

A head regulator provided at the head of the off-taking channel, controls the flow of water entering this new channel. While a cross-regulator may be required in the main parent channel down-stream of the off-taking channel, and is operated when necessary so as to head up water on its upstream side, thus to ensure the required supply in the off-taking channel even during the periods of low flow in the main channel.

*The main functions of a head regulator are :*

(*i*) To regulate or control the supplies entering the off-take channel.

(*ii*) To control silt entry into the off-take channel.

(*iii*) To serve as a meter for measuring discharge.

*The main functions of a cross-regulator are :*

(*i*) To effectively control the entire Canal Irrigation System.

(*ii*) When the water level in the main channel is low, it helps in heading up water on the u/s and to feed the off-take channels to their full demand in rotation.

(*iii*) They help in absorbing fluctuations in various sections of the canal system, and in preventing the possibilities of breaches in the tail reaches.

(*iv*) Cross regulator is often combined with a road bridge, so as to carry the road which may cross the irrigation channel near the site of the cross regulator. It is also usually combined with a fall (if required at the site of cross regulator); when it is called a fall-regulator.

A regulator essentially consists of piers placed across the canal at regular intervals with grooves, in which either planks or gates can be used to control the supplies. Planks called Karries can be used for small channels only, as the maximum height of the planks which can be handled manually is about 2 metres. For large channels, either hand operated or mechanically operated gates can be used.

The hand operated gates can have spans of 6 to 8 m, while mechanically operated gates can be as wide as 20 metres or so. A light bridge platform has to be laid on the piers for operation of planks or gates. It is often economical to combine a regulator with a bridge and also to flume the channel at the site of the regulator, so as to reduce its floor width for affecting economy.

## 13.5. Design of Cross Regulator and Distributary Head Regulator

**Crest Levels.** The crest of a cross regulator is generally kept at the upstream bed level of the channel. While, the crest level of the distributary head regulator is generally kept 0.3 to 1.0 m higher than the crest level of the cross regulator.

**Water-way.** The water-way can be worked out by using the drowned weir formula, given as :

$$Q = \frac{2}{3} \cdot C_{d_1} \sqrt{2g} \cdot B \cdot \left[(h + h_v)^{3/2} - h_v^{3/2}\right] + C_{d_2} \cdot B \cdot h_1 \sqrt{2g\,(h + h_v)} \quad \text{...(13.1)}$$

where $C_{d_1}$ = 0.577.

$C_{d_2}$ = 0.80.

$B$ = Clear water-way required.

$h$ = Difference of water level u/s and d/s of the crest, as shown in Fig. 13.4.

$h_1$ = Depth of the d/s water level in the channel above the crest.

$h_v$ = Head due to velocity of approach, which is very small and is generally ignored. The discharge formula then becomes.

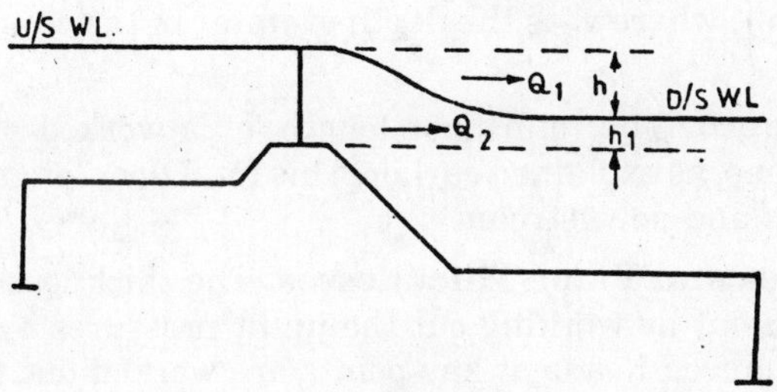


Fig. 13.4

$$Q = \frac{2}{3} \times 0.577 \sqrt{2 \times 9.81}\, B \cdot h^{3/2} + 0.80\, B \cdot h_1 \sqrt{2 \times 9.81}\, \sqrt{h}$$

$$= 1.69\, B \cdot h^{3/2} + 3.54\, B \cdot h_1 \sqrt{h}$$

or $$Q = B \cdot \sqrt{h}\left[1.69\, h + 3.54\, h_1\right] \quad \text{...(13.2)}$$

**Conditions of flow for design.** The design is done for the worst of the following two conditions :

(*a*) Full supply discharge is passing down both the channels with all gates of cross regulator and head regulator fully open.

(*b*) The discharge in the parent channel is low but the off-take channel is running full, and its FSL is maintained by the partial opening of the gates of cross-regulator.

In the first condition, $q$ and $H_L$ are fixed ; while in the second case $q$ reduces and $H_L$ increases, depending upon the low flow of the channel. The first condition of flow generally governs the design, but in certain cases, the second flow condition may become more critical.

If the low flow record of the parent channel is not given, the design can be done for the first flow condition.

**Downstream Floor Level or Cistern Level.** For the above two flow conditions, $q$ and $H_L$ are worked out. Then $E_{f_2}$ is found from Plate 10.1. The level at which jump would form, *i.e.* the level of d/s floor, is then given by d/s TEL – $E_{f_2}$.

Neglecting velocity head, d/s TEL = d/s FSL

∴ Level of d/s floor = d/s FSL – $E_{f_2}$.

If the d/s floor for the worst condition works out to be higher than the d/s bed level of the channel, the floor is provided at the bed level itself.

**Length of d/s Floor.** It is worked out by calculating 5 $(y_2 - y_1)$. If by chance, this length comes out to be small in comparison to 2/3 of total floor length (worked out by exit gradient considerations, *i.e.* $b = \alpha \cdot d$ .), then, the length of the d/s floor is kept equal to 2/3 of the total floor length.

**Cut-offs are provided as given below :**

*Upstream Cut-off.* The minimum depth of u/s cut off below u/s floor level is kept as $\left(\frac{y_u}{3} + 0.6 \text{ m}\right)$, where $y_u$ is the depth of water in the channel upstream.

*Downstream Cut-off.* The minimum depth of d/s cut-off below the d/s floor level is kept as $\left(\frac{y_d}{2} + 0.6 \text{ m}\right)$, where $y_d$ is the depth of water in the channel downstream.

**Total floor Length.** The total floor length '*b*' is worked out from safe exit gradient considerations, as explained earlier. This total floor length is then suitably distributed upstream and downstream.

**Uplift Pressures and Floor Thicknesses.** The thicknesses of the d/s floor required, are worked out by working out the uplift pressures by Khosla's theory. The maximum unbalanced heads at key points are worked out for the maximum static head. The pressure at toe of glacis is also worked out for dynamic condition. The floor thicknesses are designed for the worst case, as explained earlier in the design of weirs and falls. A nominal thickness of 0.3 to 0.5 m is provided on u/s side.

**Protection Works.** The protection works are designed for a scour depth $(D)$ equal to $\left(\frac{y_u}{3} + 0.6 \text{ m}\right)$ on the upstream, and equal to $\left(\frac{y_d}{2} + 0.6 \text{ m}\right)$ on the downstream. The 'C.C. blocks' and 'inverted filter' are provided in a length approximately equal to 1.5 D. The quantity of stone in launching apron is kept as 2.25 D cu. m/metre.

A typical design of such a regulator is illustrated in the example given below :

**Example 13.1.** *Design a cross regulator and a head regulator for a channel which takes off from the parent channel with the following data :*

| | |
|---|---|
| *Discharge of parent channel* | *= 140 cumecs* |
| *Discharge of distributary* | *= 15 cumecs* |
| *FSL of the parent channel, u/s* | *= 210.0 m* |
| *FSL of the parent channel, d/s* | *= 209.8 m* |
| *Bed width of parent channel, u/s* | *= 52 m* |
| *Bed width of parent channel, d/s* | *= 46 m* |
| *Depth of water in the parent channel d/s and u/s* | *= 2.5 m* |

*FSL of distributary* = 209.1 m

*Silt factor* = 0.8 m

*Assume safe exit-gradient* = 1/5

**Solution.**

**(A) Design of Cross Regulator**

*Crest Levels.* Crest level of cross regulator is kept the same as u/s bed level of parent channel = 210.0 – 2.5 = 207.5 m.

**Provide Crest at R.L. 207.5 m.**

$$\textit{Waterway } Q = \frac{2}{3} C_{d_1} \sqrt{2g} \cdot B \left[(h + h_v)^{3/2} - h_v^{3/2}\right] + C_{d_2} \cdot B \cdot h_1 \sqrt{2g\,(h + h_v)} \quad \text{...(13.1)}$$

*Ignoring velocity head*

$$Q = B \cdot \sqrt{h} \left[1.69\, h + 3.54\, h_1\right]. \quad \text{...(13.2)}$$

where $B$ is the clear waterway required.

In this case, $h$ = u/s FSL – d/s FSL = 210.0 – 209.8 = 0.2 m.

$h_1$ = d/s FSL – Crest Level = 209.8 – 207.5 = 2.3 m.

$$\therefore \quad 140 = B\sqrt{0.20}\ [1.69 \times 0.2 + 3.54 \times 2.3]$$

or
$$B = \frac{140}{0.447}\left[\frac{1}{8.488}\right] = 36.9 \text{ m.}$$

**Provide 5 bays of 8.0 m each** with a clear waterway = 5 × 8.0 = 40.0 m.

**Provide 4 piers of 1.5 m** width each = 6.0 m.

Thus the overall waterway = 40 + 6 = 46 m.

*Downstream Floor Level or Cistern Level*

$q$ = 140/40 = 3.5 cumecs/metre.

$H_L$ = 0.2 m.

From Plate 10.1, $E_{f_2}$ = 1.85 m.

d/s floor level = d/s FSL – $E_{f_2}$ = 209.8 – 1.85 = 207.95 m.

d/s bed level = 209.8 – 2.5 = 207.3 m.

which is lower than the calculated d/s floor level. **Hence, provide the cistern or d/s floor at R.L. 207.3 m.**

*Length of d/s floor.* Length of d/s floor required = 5 $(y_2 - y_1)$

From Plate 10.2,

For $E_{f_2}$ = 1.85 m ; $y_2$ = 1.8 m.

For $E_{f_1} = E_{f_2} + H_L$ = 1.85 + 0.2 = 2.05 m ; $y_1$ = 0.8 m.

$\therefore$ 5 $(y_2 - y_1)$ = 5 (1.8 – 0.8) = **5.0 m.**

(subject to the limitation of $\frac{2}{3}$rd of total floor length $b$).

**Vertical cut-offs**

Provide *upstream cut-off* for a depth

$$= \frac{y_u}{3} + 0.6 = \frac{2.5}{3} + 0.6 = 1.43 \text{ m}$$

below the floor, *i.e.* the level of its bottom

= 207.5 – 1.43 = **206.07 m.**

*Downstream cutoff.* Depth of d/s cutoff below floor level

$$= \frac{y_d}{2} + 0.6 = \frac{2.5}{2} + 0.6 = 1.85 \text{ m.}$$

Hence, the bottom level of d/s cutoff = 207.30 – 1.85 = **205.45 m.**

**Total Floor Length from Exit Gradient Considerations**

$$G_E = \frac{H}{d} \cdot \frac{1}{\pi \cdot \sqrt{\lambda}}$$

where $H$ = Maximum static head, which is caused when there is full water on u/s and there is no water on d/s.

$H$ = u/s FSL – d/s bed level

= 210.0 – 207.3 = 2.7 m.

$d$ = Depth of d/s cutoff = 1.85 m

$G_E = \frac{1}{5}$ (given)

$$\therefore \qquad \frac{1}{5} = \frac{2.7}{1.85} \cdot \frac{1}{\pi \cdot \sqrt{\lambda}}$$

$$\text{or} \qquad \frac{1}{\pi\sqrt{\lambda}} = \frac{1}{5} \times \frac{1.85}{2.7} = 0.137.$$

From Plate 11.2,

For $\frac{1}{\pi\sqrt{\lambda}} = 0.137$, $\alpha = 9$

$\therefore$ $b = \alpha d = 9 \times 1.85 = 16.65$ m ; **say 17 m.**

Minimum d/s floor length required

$= \frac{2}{3} \cdot b = \frac{2}{3} \times 17 = 11.3$ m; which is greater than 5.0 m, *i.e.* $5\,(y_2 - y_1)$.

Hence, **provide 11.3 m as the d/s floor length.**

Glacis length = 2 × 0.2 = 0.4 m

The balance, *i.e.* 17 – 11.3 – 0.4 = 5.3 m is provided as u/s floor length, as shown in Fig. 13.5.

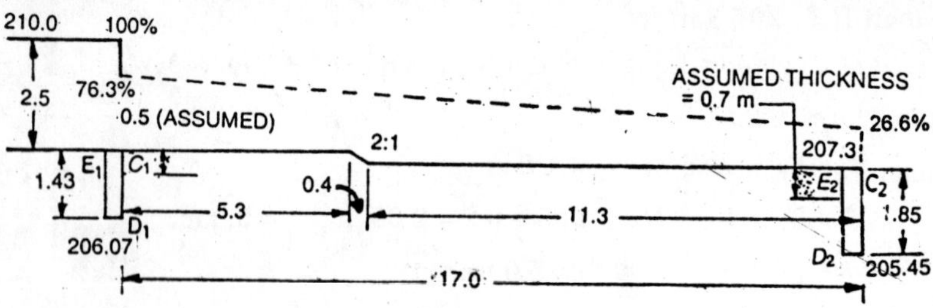


Fig. 13.5

**Calculations for Uplift Pressures**

*Upstream Cut-off*

$d = 1.43$ m.

$b = 17.0$ m.

$$\frac{1}{\alpha} = \frac{d}{b} = \frac{1.43}{17.0} = 0.085$$

From Plate 11.1 (*a*),

$$\phi_{E_1} = 100\%$$

$$\phi_{D_1} = 100\% - \phi_D = 100\% - 18.5\% = 81.5\%$$

$$\phi_{C_1} = 100\% - \phi_E = 100\% - 26.5\% = 73.5\%$$

(Assume u/s floor thickness = 0.5 m)

*Correction to $\phi_{C_1}$ for floor thickness*

$$= \frac{81.5 - 73.5}{1.43} \times 0.5 = 2.8\% \text{ (+ ve).}$$

$\therefore$ $\phi_{C_1}$ **(corrected)** = 73.5 + 2.8 = 76.3%

(*Note.* Correction due to pile interference is small and may be neglected).

*Downstream Cut-off*

$$d = 1.85 \text{ m}$$

$$b = 17.0 \text{ m.}$$

$$\therefore \quad \frac{1}{\alpha} = \frac{d}{b} = \frac{1.85}{17.0} = 0.109.$$

From Plate 11.1 (*a*),

$$\phi_{E_2} = \phi_E = 30\%$$

$$\phi_{D_2} = \phi_D = 21\%$$

$$\phi_{C_2} = 0\%$$

(Assume d/s floor thickness near d/s cut off = 0.7 m)

*Correction to $\phi_{E_2}$ due to floor thickness*

$$= \frac{30 - 21}{1.85} \times 0.7 = 3.4\% \text{ (– ve)}$$

$\therefore$ $\phi_{E_2}$ **(corrected)** = 30% – 3.4% = 26.6%

(*Note.* Correction due to interference of u/s cut off is neglected).

**Floor Thicknesses : D/s Floor**

*At Toe of Glacis*

% pressure at toe of glacis

$$= 26.6\% + \frac{76.3\% - 26.6\%}{17} \times 11.3 = 26.6\% + 33.1\% = 59.7\%$$

Maximum unbalanced head at toe of glacis due to maximum static head (2.7 m)

$$= 59.7\% \times 2.7 = 1.61 \text{ m.}$$

Head due to dynamic action can be taken as

$$50\% \, (y_2 - y_1) + \phi \, . \, H_L$$

$$= 50\% \, (1.8 - 0.8) + 59.7\% \times 0.2 = 0.5 + 0.12 = 0.62 \text{ m} < 1.61 \text{ m.}$$

Hence, static head governs the thicknesses.

$\therefore$ Thickness required at toe of glacis $= \frac{1.61}{1.24} = 1.3$ m; **Provide 1.4 m.**

*At 3 m beyond toe of glacis*

$$\% \text{ pressure} = 26.6\% + \frac{76.3\% - 26.6\%}{17} \times 8.3 = 26.6\% + 24.3\% = 50.9\%$$

Maximum unbalanced head

$= 50.9\% \times 2.7 = 1.32$ m.

Thickness required

$= \dfrac{1.32}{1.24} = 1.07$ m; **Provide 1.1 m.**

*At 8 m beyond toe of glacis*

% pressure

$= 26.6\% + \dfrac{76.3\% - 26.6\%}{17} \times 3.3$

$= 26.6\% + 9.65\% = 36.25\%$

Maximum unbalanced head

$= 36.25\% \times 2.7 = 0.98$ m.

Thickness required

$= \dfrac{0.98}{1.24} = 0.79$ m; **Provide 0.8 m**

from this point to the end, as shown in Fig. 13.6.

**U/s Floor**. Theoretically no floor thickness is required under the upstream floor, since the uplift is more than counterbalanced by the weight of the water standing over it. But a nominal thickness of 0.5 m is provided. The floor is thickened to 1.0 m under the crest in a length equal to 2.0 m as shown in Fig. 13.6.

**Upstream Protection** shall be provided for a scour depth

$= \dfrac{y_u}{3} + 0.6 \text{ m} = \dfrac{2.5}{3} + 0.6 = 1.43$ m.

*Launching apron.* Provide launching apron of thickness 1.2 m

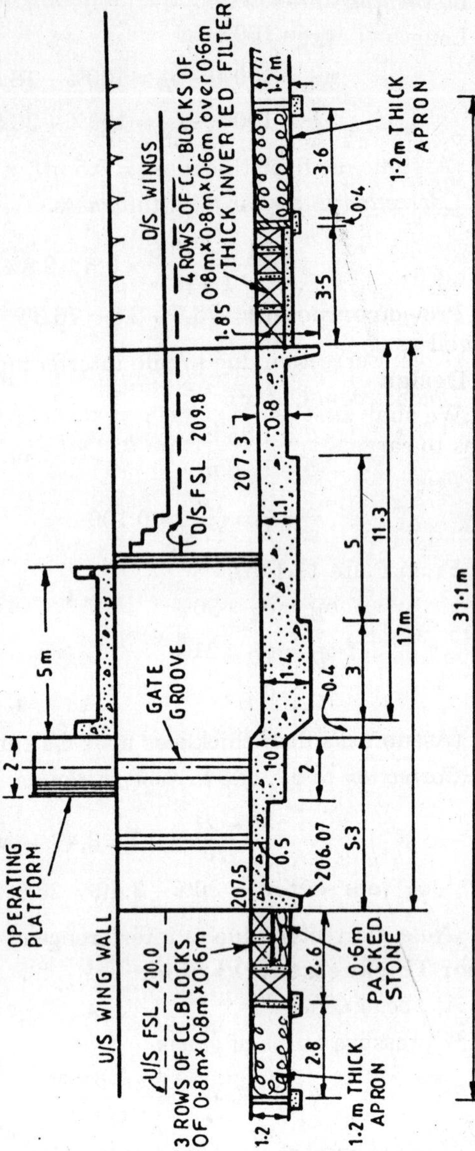


Fig. 13.6. Details of Cross Regulator.

in a length $= \dfrac{2.25 \times 1.43}{1.2} = 2.68$ m( ∵ Volume required = 2.25 $D$ cu. m/metre)

Use launching apron = 1.2 m thick laid in a length of 2.8 m.

*C.C. Blocks.* Let us provide C.C. blocks of size 0.8 m × 0.8 m × 0.6 m for a length $D = 1.5 \times 1.43 = 2.2$ m. Hence, 3 rows of C.C. blocks of size 0.8 m × 0.8 m × 0.6 m size with 10 cm. jhories filled with bajri, laid over packed stone apron of 0.6 m thickness, in a total length of 2.6 m, shall be provided as shown in Fig. 13.6.

**Downstream Protection** shall be provided for a scour depth

$$D = y_d/2 + 0.6 = 2.5/2 + 0.6 = 1.85 \text{ m.}$$

*Launching apron.* Provide a launching apron of thickness equal to 1.2 m.

Length of apron required

$$= \frac{2.25\,D}{t} = \frac{2.25 \times 1.85}{1.2} = 3.48 \text{ m; } \textbf{say 3.6 m.}$$

Provide a launching apron of 1.2 m thickness laid in a length of 3.6 m.

*Inverted filter.* Length of filter required = 1.5 × 1.85 = 2.77 m.

Provide 4 rows of C.C. blocks of size 0.8 m × 0.8 m × 0.6 m with 10 cm gaps in between, laid over an inverted filter of 0.6 m thickness, in a total length of 3.5 m, as shown in Fig. 13.6.

Provide 0.4 m thick and 1.2 m deep toe walls between blocks and apron on u/s as well as on d/s.

**(B) Design of Distributary Head Regulator**

We shall first of all determine a suitable regime section for the distributary, so as to carry the given discharge of 16 cumecs with the given silt factor = 0.8.

*From Lacey's Regime Diagrams*

Bed width of such a channel = 15.0 m.

Depth of water in this channel = 1.5 m.

**Bed Level of Distributary** = FSL – Depth = 209.1 – 1.5 = **207.6 m.**

*Crest Level.* The crest level of the Distributary Head is generally kept 0.3 to 1.0 m higher than the bed level of the parent channel. Let us keep it 0.6 m higher.

∴ Crest Level = 207.5 + 0.6 = 208.1 m.

**Hence keep the crest of Head Regulator at RL 208.1 m.**

*Waterway.* The discharge is given by

$$Q = B \cdot \sqrt{h}\,[1.69\,h + 3.54\,h_1] \quad \textit{i.e. } \text{Eq. (13.2)}$$

where $h$ = FSL of parent channel – FSL of Distributary

= 210.0 – 209.1 = 0.9 m.

$h_1$ = FSL of Distributary – Crest Level

= 209.1 – 208.1 m = 1.0 m.

$Q$ = 15 cumecs.

$B$ = Clear waterway required.

$$\therefore \quad 15 = B \cdot \sqrt{0.90}\left[1.69 \times 0.9 + 3.54 \times \sqrt{1.0}\right]$$

or $$15 = B \times 0.949\,[1.521 + 3.54]$$

or $$B = \frac{15}{0.949 \times 5.061} = \mathbf{3.12\ m.}$$

Which is very small as compared to bed width of distributary, *i.e.* 15.4 m.

Hence, provide 2 bays of 3.0 m each with 1.0 m thick pier in between.

Overall waterway provided = 6 + 1 = 7.0 m

Clear waterway provided = 6.0 m.

The wing walls shall be expanded with proper divergence, so as to provide the normal width of the channel.

*Cistern or d/s floor.* The discharge intensity $q$

$$= \frac{15}{6} = 2.5 \text{ cumecs/metre}$$

$$H_L = \text{u/s FSL} - \text{d/s FSL}$$
$$= \text{FSL of parent channel} - \text{FSL of Distributary}$$
$$= 210.0 - 209.1 = 0.9 \text{ m}$$

From plate 10.1,

$$E_{f_2} = 1.8 \text{ m}$$
$$\therefore \quad E_{f_1} = E_{f_2} + H_L = 1.8 + 0.9 = 2.7 \text{ m.}$$

From plate 10.2,

For $\quad E_{f_1} = 2.7 \text{ m} \,;\, y_1 = 0.4 \text{ m}$

For $\quad E_{f_2} = 1.8 \text{ m} \,;\, y_2 = 1.7 \text{ m.}$

**R.L. of cistern** (*i.e.* d/s floor)

$$= \text{d/s FSL} - E_{f_2} = 209.1 - 1.8 = \mathbf{207.3\ m.}$$

Length of cistern required

$$= 5\,(y_2 - y_1) = 5\,(1.7 - 0.4)$$
$$= 8.0 \text{ m (subjected to a minimum of } \tfrac{2}{3}\, b)$$

*Vertical Cutoff.* Depth of u/s cutoff below floor

$$= \frac{y_u}{3} + 0.6 = \frac{2.5}{3} + 0.6 = 1.43 \text{ m.}$$

**The bottom level of u/s cutoff**

$$= 207.5 - 1.43 = \mathbf{206.07\ m.}$$

The depth of d/s cutoff below floor

$$= \frac{y_d}{2} + 0.6 = \frac{1.5}{2} + 0.6 = 1.35 \text{ m.}$$

Provide d/s cutoff depth = 1.6 m

**The bottom of d/s cut-off**

$$= 207.3 - 1.6 = \mathbf{205.7\ m.}$$

**Total Floor Length from Exit Gradient Considerations**

Maximum static head ($H$) is caused when there is full water on u/s and no water on d/s

$$\therefore \quad H = 210.0 - 207.3 = 2.7 \text{ m.}$$
$$d = \text{Depth of d/s cutoff} = 1.6 \text{ m.}$$
$$G_E = \frac{H}{d}\,\frac{1}{\pi \cdot \sqrt{\lambda}}$$
$$\therefore \quad \frac{1}{5} = \frac{2.7}{1.6} \cdot \frac{1}{\pi\,\sqrt{\lambda}}$$

or

$$\frac{1}{\pi\,\sqrt{\lambda}} = \frac{1}{5} \times \frac{1.6}{2.7} = 0.119$$

From Plate 11.2,

For $\quad \dfrac{1}{\pi\,\sqrt{\lambda}} = 0.119 \,;\, \alpha = 13$

$$\therefore \quad b = \alpha \cdot d = 13 \times 1.6 = 20.8 \text{ m; } \mathbf{say\ 21\ m.}$$

Minimum length of d/s floor required

$$= \frac{2}{3} \times 21 = \mathbf{14\ m.} \qquad \text{(which is greater than 8 m.)}$$

**Hence, Provide d/s floor length = 14 m.**

Length of d/s Glacis = 1.6 m.

Length of Crest = 1.0 m.

Length of u/s Glacis = 0.6

Total = 17.2 m.

The remaining 3.8 m is provided on the u/s side, as shown in Fig. 13.7.

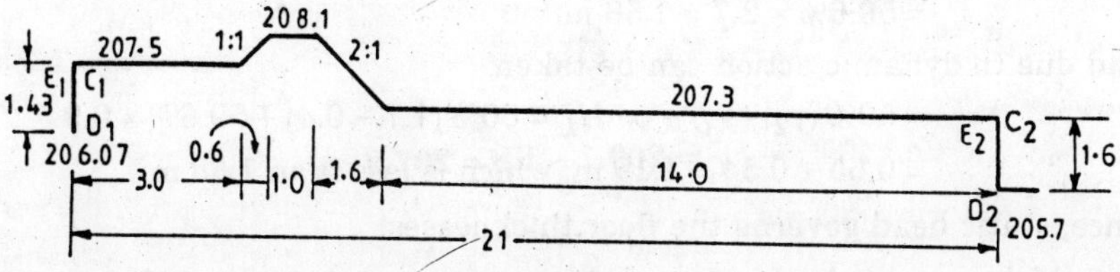


Fig. 13.7

**Calculation for Uplift Pressures**

*(i) Upstream cutoff*

$$b = 21 \text{ m.}$$

$$d = 1.43 \text{ m.}$$

$$\frac{1}{\alpha} = \frac{d}{b} = \frac{1.43}{21} = 0.068$$

From Plate 11.1 (*a*),

$$\phi_{E_1} = 100\%$$

$$\phi_{D_1} = 100\% - \phi_D = 100\% - 16.5\% = 83.5\%$$

$$\phi_{C_1} = 100\% - \phi_E = 100\% - 24\% = 76\%$$

Assume u/s floor thickness near cutoff = 0.5 m.

*Correction to $\phi_{C_1}$ due to floor thickness*

$$= \frac{83.5\% - 76\%}{1.43} \times 0.5 = 2.62\% \text{ (+ ve)}$$

*Correction due to d/s pile is neglected.*

$\phi_{C_1}$ **(corrected)** = 76% + 2.62% = 78.62%; **say 78.6%**

*(ii) Downstream cutoff*

$$b = 21.0 \text{ m.}$$

$$d = 1.6 \text{ m.}$$

$$\frac{1}{\alpha} = \frac{d}{b} = \frac{1.6}{21.0} = 0.076.$$

From Plate 11.1 (*a*),

$$\phi_{E_2} = \phi_E = 25.5\%$$

$$\phi_{D_2} = \phi_D = 17.5\%$$

$$\phi_{C_2} = 0\%.$$

Assume d/s floor thickness near cutoff = 0.8 m.

*Correction to $\phi_{E_2}$ due to floor thickness*

$$= \frac{25.5\% - 17.5\%}{1.6} \times 0.8 = 4.0\% \text{ (– ve).}$$

*Correction due to d/s pile is neglected*

$\phi_{E_2}$ **(corrected)** = 25.5% – 4.0% = **21.5%**

**Floor Thickness**

*At toe of d/s Glacis*

$$\%\text{age pressure} = 21.5\% + \frac{78.6\% - 21.5\%}{21} \times 14 = 21.5\% + 38.1\% = 59.6\%$$

Maximum unbalanced head due to static head

$$= 59.6\% \times 2.7 = 1.58 \text{ m.}$$

Head due to dynamic action can be taken

$$= 50\% \, (y_2 - y_1) + \phi \cdot H_L = 50\% \, (1.7 - 0.4) + 59.6\% \times 0.9$$

$$= 0.65 + 0.54 = 1.19 \text{ m; which is less than } 1.58 \text{ m.}$$

Hence, static head governs the floor thicknesses.

Floor thickness required at toe of glacis

$$= \frac{1.58}{1.24} = 1.28 \text{ m; } \textbf{Provide 1.4 m.}$$

*At 3 m beyond toe of d/s glacis*

$$\%\text{age pressure} = 21.5\% + \frac{78.6\% - 21.5\%}{21} \times 11$$

$$= 21.5\% + 29.9\% = 51.4\%$$

Unbalanced head = 51.4% × 2.7 = 1.39 m.

$$\text{Thickness reqd.} = \frac{1.39}{1.24} = 1.12 \text{ m; } \textbf{Provide 1.2 m.}$$

*At 6 m beyond toe of d/s Glacis*

$$\%\text{age pressure} = 21.5\% + \frac{78.6\% - 21.5\%}{21} \times 8 = 21.5\% + 21.8\% = 43.3\%$$

Unbalanced head = 43.3% × 2.7 = 1.17 m

Thickness reqd. = 1.17/1.24 = 0.945 m. **Provide 1 m.**

*At 10 m beyond tow of d/s Glacis*

$$\%\text{age pressure} = 21.5\% + \frac{78.6\% - 21.5\%}{21} \times 4 = 21.5\% + 10.9\% = 32.4\%$$

Unbalanced head = 32.4% × 2.7 = 0.875 m.

Thickness reqd. = 0.875/1.24 = 0.705 m; **Provide 0.8 m** upto the end, as shown in Fig. 13.8.

**U/s Floor.** Provide a nominal thickness of 0.5 m under the u/s floor and extend it under the crest and then join it to the bottom of d/s glacis, as shown in Fig. 13.8.

**U/s Protection**. Same as provided in the upstream of Cross Regulator.

**D/s Protection**. The d/s protection shall be provided for a scour depth ($D$)

$$= \frac{y_d}{2} + 0.6 = \frac{1.5}{2} + 0.6 = \mathbf{1.35 \text{ m.}}$$

*Launching apron*. Provide 1.2 m thick launching stone in a

$$\text{length} = \frac{2.25 \times 1.35}{1.2} = 2.54 \text{ m; } \textbf{say 2.6 m.}$$

*Inverted filter*. The length of filter reqd.

$$= 1.5D = 1.5 \times 1.35 = 2.03 \text{ m.}$$

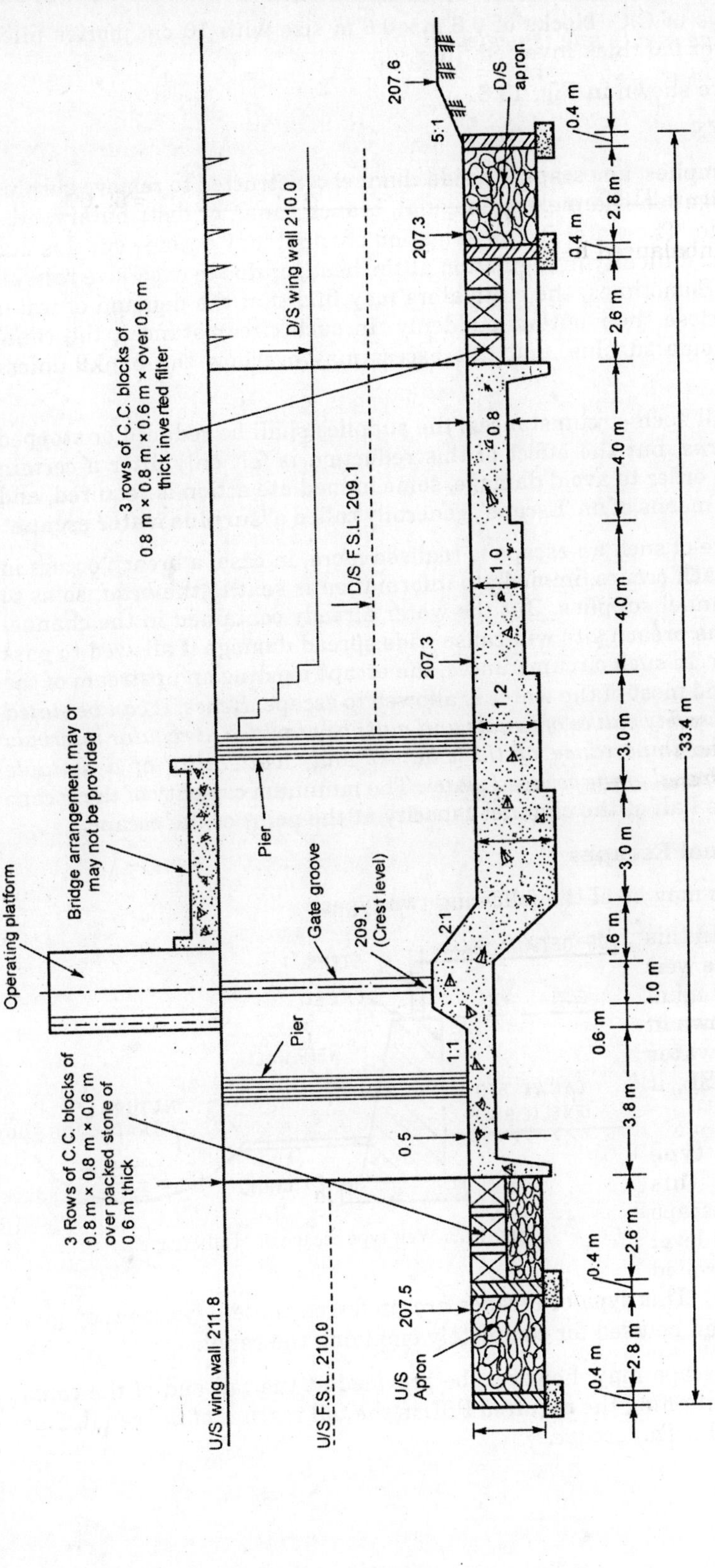


Fig. 13.8 Details of Distributary Head Regulator

Provide 3 rows of C.C. blocks of 0.8 m × 0.6 m size with 10 cm jhories filled with bajri, laid over 0.6 thick inverted filter.

The details are shown in Fig. 13.8.

## CANAL ESCAPES

As the name implies, an escape is a side channel constructed to remove surplus water from an irrigation channel (main canal, branch canal, or distributary, etc.) into a natural drain. The water in the irrigation channel may become surplus due to some mistake; or difficulty in regulation at the head; or due to excessive rainfall in upper reaches. Sometimes, the cultivators may find that the demand of water is over and may close their outlets suddenly. In such circumstances, the canal supplies shall become surplus, and this excess may overflow the banks unless escaped.

No doubt, in all such circumstances, the supplies shall be reduced or stopped from the *head works,* but the effect of this reduction is felt only after a certain time. Therefore, in order to avoid damage, some immediate action is required, and this is achieved by means of an 'Escape', generally called a **'surplus water escape'**.

The importance of such an escape is realised more, in case, a breach occurs in a channel. If a breach occurs, immediate information is sent to the *head*, so as to close down the channel supplies. But the water already contained in the channel from the head to the breach site will cause wide spread damage if allowed to pass through the breach. In such circumstances, the escape existing on upstream of the breach is opened and most of the water is allowed to escape. *Hence, it can be stated that escapes are the safety valves of canals and must be provided at regular intervals depending upon the importance of the channel and availability of a suitable drainage for the disposal of the escaped water.* The minimum capacity of the escape is generally kept as half of the channel capacity at the point of the escape.

### 13.6. Types of Canal Escapes

Such an escape may be of the following two types :

(*i*) **Weir type.** In this type, the crest of the weir wall is kept at R.L. equal to canal FSL, as shown in Fig. 13.9. When water level rises above FSL, it gets escaped.

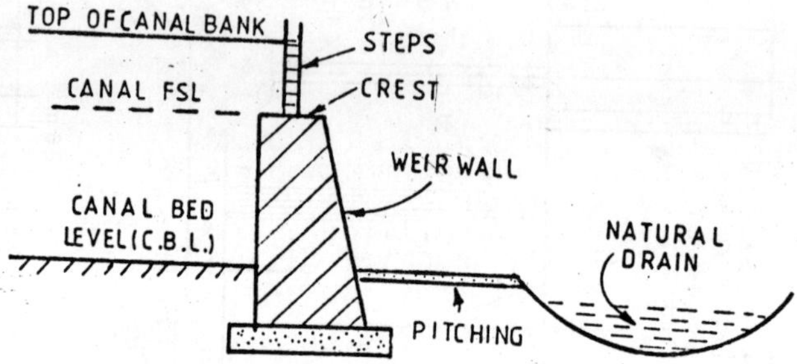

Fig. 13.9. Weir type escape or 'Tail escape'.

(*ii*) **Regulator type (Sluice type).** In this type, the sill of the escape is kept at canal bed level and the flow is controlled by gates (Fig. 13.10). This type of escapes are preferred these days, as they give better control and can be used for completely emptying the canal.

The weir type escape may, however, be provided at the tail end of the canal and is useful in maintaining the required FSL in the tail reaches of the canal. This escape is then called a *Tail escape*.

The 'regulator type' may, however, be constructed for the purpose of scouring off excess bed silt deposited in the head reaches, from time to time. In that case, it is known as a *Scouring escape*.

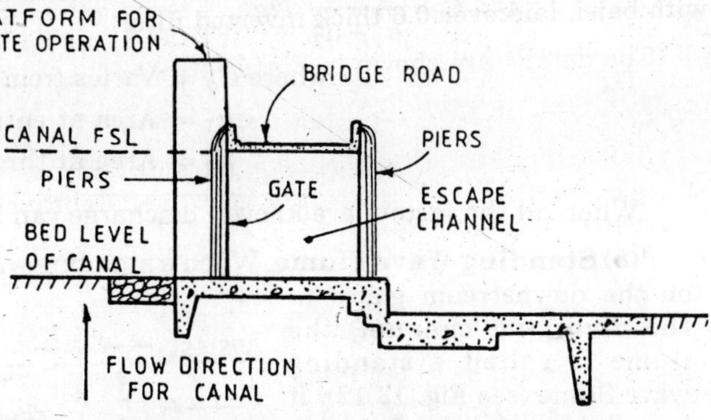


Fig. 13.10. Regulator type escape or 'Surplus water escape'.

In such a case, the sill of the regulator is generally kept at about 0.3 m below the canal bed level at the escape site. When the deposited silt is to be scoured out ; a discharge higher than the full supply discharge is let off into the canal from its *head*. The gates of the escape are raised so as to produce scouring velocities, which scour out the deposited silt. For better efficiency, a cross regulator may be provided just downstream of scouring escape. Such escapes have become obsolete, because *silt ejectors* are now-a-days provided for better efficiency.

The regulator type escape can be designed like a head regulator without any raised crest.

## METERING FLUMES

A meter is a structure constructed in a canal for measuring its discharge accurately. A metering flume is an artificially flumed (narrowed) section of the channel, which can be utilized for calculating the discharge in the channel.

The normal upstream section of the channel is narrowed by masonry walls with a splay of 1 : 1 to 2 : 1 to a rectangular section called *Throat*. From where, the channel is slowly diverged so as to attain its normal section by means of masonry wings with a splay of 2 : 1 to 10 : 1. More gradual the convergence and divergence, less will be the loss of head in the flume.

### 13.7. Types of Metering Flumes

There are two types of metering flumes which are generally used, and work on the principle of Venturimeter. They are :

(*a*) Non-modular venturi flume or drowned venturi-flume, generally called venturi flume.

(*b*) Standing wave flume; or modular venturi flume; or free flow venturi flume.

(*a*) **Venturi Flume**. A venturi flume consists of a gradually contracting channel leading to 'throat', and a gradually expanding channel leading away from it. Stilling wells are provided for measuring head at the entrance and at the throat, as shown in Fig. 13.11. If $h$ is the difference of head between the two wells, then the discharge is given by

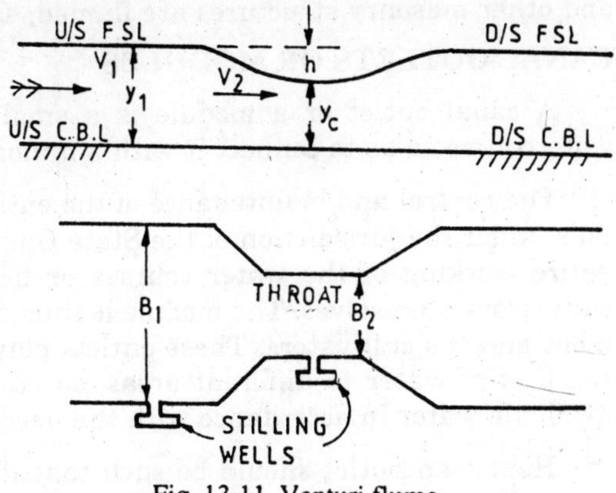


Fig. 13.11. Venturi flume.

$$Q = C_d \frac{a_1 \cdot a_2}{\sqrt{a_1^2 - a_2^2}} \sqrt{2gh} \quad \text{...(13.3)}$$

where $C_d$ = Varies from 0.95 to 1.0

$a_1$ = Area at entrance = $B_1 \cdot y_1$

$a_2$ = Area at throat = $B_2 \cdot y_2$

When all the factors are known, discharge can be calculated.

(*b*) **Standing Wave Flume.** When a standing wave, (*i.e.* hydraulic jump) forms on the downstream glacis in the diverging channel, the flume is called a standing wave flume (see Fig. 13.12). It is superior to venturi flume because its discharge depends only upon the upstream head over the crest of the throat, and also for the same upstream head, its discharging capacity is more than that of a venturi flume. The length of its throat is atleast 2 to 3 times the head over the crest. Discharge formula for such a flume is given by

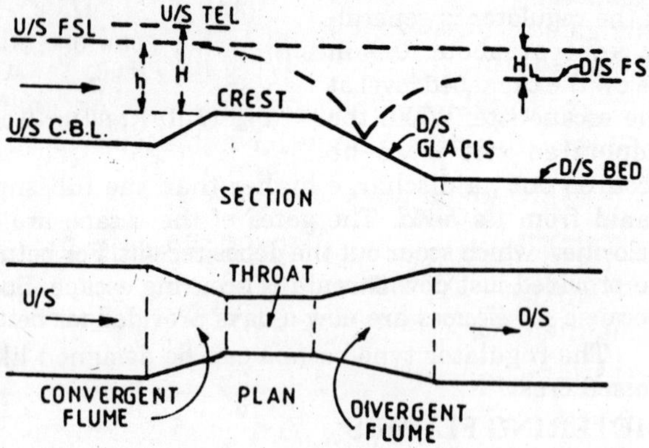


Fig. 13.12. Standing wave flume.

$$Q = 1.7\, C_d \cdot B \cdot H^{3/2} \quad \text{...(13.4)}$$

where $B$ is the width of the throat.

The head of water over the crest can be measured in a stilling well and the discharge can be easily calculated.

The only drawback of such a flume is that it requires greater loss of head ($H_L$), and where this loss of head is not available, this flume will act like a venturi flume. Due to this reason, canal fall sites (giving sufficient loss of head, $H_L$) are generally used as standing wave flumes. For this purpose, flumed meter falls are constructed as explained earlier.

Such a flume is, therefore, seldom constructed separately, and falls, regulators and other masonry structures are flumed, so as to act like a metering flume.

## CANAL OUTLETS OR MODULES

A canal outlet or a module is a small structure built at the head of the watercourse so as to connect it with a minor or a distributary channel.

The control and maintenance of the entire network of canals upto the module falls under the jurisdiction of the State Government, and beyond the module, the entire working of the water courses or field channels is taken care of by the cultivators themselves. The module is thus a connecting link between the Government and the cultivators. These outlets play a very prominent role in controlling the flow of water to different areas, so as to effect an equitable distribution of available water in accordance with the needs of the whole area.

Hence, an outlet should be such that it fits well to the decided principles of water distribution.

Moreover, it should be simple, cheaper, efficient and easy to handle. The design of the outlet is, therefore, of prime importance in efficient and effective management of canal irrigation system. The requirements of a good module are listed below :

### 13.8. Requirements of a Good Module

(1) The module should fit well to the decided principles of water distribution. For example, if the supply is to be fixed in accordance with the cultivable area commanded by the outlet, the outlet must be able to pass a constant and a fixed discharge. Similarly, if the supply is to be regulated in accordance with the area irrigated in the past year, the capacity of the outlet should be capable of being changed from year to year.

(2) The module should be simple, so that it can be easily constructed or fabricated by local masons or technicians.

(3) It should work efficiently with a small working head.

(4) The outlet should be cheaper, since they are required in large numbers.

(5) It should be sufficiently strong with no moving parts, thus avoiding periodic maintenance.

(6) The outlet should be such as to avoid interference by cultivators, thus preventing undue tapping of water by cultivators. In other words, any interference by the local people should be made difficult and easily detectable.

(7) It should draw its fair share of silt.

### 13.9. Types of Outlets (Modules)

The various available types of outlets can be classified into three classes :

(*i*) **Non-modular outlets** are those through which the discharge depends upon the difference of head between the distributary and the water-course. The discharge through such a module, therefore, varies widely with either a change in the water level of the distributary or that of the water-course. The common examples of this type of outlets are : (*i*) *open sluice,* and (*ii*) *drowned pipe outlet.*

(*ii*) **Semi-modules or Flexible modules** are those through which the discharge is independent of the water level of the water-course but depends only upon the water level of the distributary so long as a minimum working head is available. The discharge through such an outlet will, therefore, increase with a rise in the distributary water surface level and *vice versa*. The common examples of this type of modules are : *pipe outlet, venturi flume, open flume and orifice semi-module.*

(*iii*) **Rigid modules or Modular outlets** are those through which the discharge is constant and fixed within limits, irrespective of the fluctuations of the water levels of either the distributary or of the water course or of both. *Gibb's module* is a common example of such a module.

### 13.10. Criteria for Judging the Performance of Modules

The behaviour and functioning of a module can be judged by the following important terms and definitions :

**Flexibility.** Flexibility is defined as the ratio of the rate of change of discharge of the outlet to the rate of change of discharge of the distributary channel.

Thus $$F = \left[\frac{dq/q}{dQ/Q}\right] \qquad ...(13.5)$$

where $F$ = Flexibility of the outlet
$q$ = Discharge passing through the outlet
$Q$ = Discharge in the distributary channel.

Now, if $H$ is the head acting on the outlet, the discharge through the outlet ($q$) may be expressed as

$$q = C \cdot H^m \qquad ...(13.6)$$

where $C$ and $m$ are constants depending upon the type of outlet.

Similarly, the discharge passing down the distributary channel may be expressed as

$$Q = K \cdot y^n \qquad ...(13.7)$$

where $K$ and $n$ are constants and $y$ is the depth of water in the distributary.

Differentiating equations (13.6) and (13.7), we get

$$dq = C \cdot m \cdot H^{m-1}\, dH \qquad ...(13.8)$$

$$dQ = K \cdot n \cdot y^{n-1}\, dy \qquad ...(13.9)$$

Dividing equation (13.8) with (13.6), we get

$$\frac{dq}{q} = \frac{C \cdot m \cdot H^{m-1}\, dH}{C \cdot H^m} = \frac{m}{H}\, dH \qquad ...(13.10)$$

Similarly, $$\frac{dQ}{Q} = \frac{K \cdot n \cdot y^{n-1}\, dy}{K \cdot y^n} = \frac{n}{y}\, dy \qquad ...(13.11)$$

Hence, the flexibility, $$F = \frac{\frac{dq}{q}}{\frac{dQ}{Q}} = \frac{\frac{m}{H} dH}{\frac{n}{y} dy} = \frac{m}{H} \cdot \frac{y}{n} \cdot \frac{dH}{dy} = \frac{m}{n} \cdot \frac{y}{H} \frac{dH}{dy}$$

or $$F = \frac{m}{n} \cdot \frac{y}{H} \frac{dH}{dy} \qquad ...(13.12)$$

But since a change in the water depth of the distributary ($dy$) would result in an equal change in the head working on the outlet ($dH$), we have

$$dy = dH.$$

Substituting in equation (13.12), we get

$$\boxed{\text{Flexibility } = F = \frac{m}{n} \cdot \frac{y}{H}} \qquad ...(13.13)$$

**Proportionality.** The outlet is said to be proportional when the rate of change of outlet discharge equals the rate of change of channel discharge. In other words, the outlet is 'proportional' when 'flexibility' equals unity. Hence for a proportional outlet, we get

$$F = \frac{m}{n} \cdot \frac{y}{H} = 1$$

or $$\frac{H}{y} = \frac{m}{n} = \frac{\text{outlet index}}{\text{channel index}} \qquad ...(13.14)$$

**Setting.** The ratio $H/y$, *i.e.* the ratio of the depth of the sill level of the outlet below the FSL of the distributary, to the full supply depth of the distributary, is known as *setting*. Thus, for a proportional outlet

$H/y$, *i.e.* 'outlet setting' should be made.

$$= \frac{\text{outlet index}}{\text{channel index}}.$$

For a wide trapezoidal channel, the discharge is proportional to $y^{5/3}$. Therefore, for such channels

$$n = 5/3$$

Hence, the channel index is generally 5/3. The discharge through an orifice type outlet is proportional to $\sqrt{H}$; and thus for such an outlet, index $m = 1/2$.

$$\therefore \text{ Setting } = \frac{H}{y} = \frac{m}{n} = \frac{1/2}{5/3} = 0.3$$

Hence, for such a combination of an orifice type outlet and a trapezoidal channel, the setting must be equal to 0.3. In other words, a pipe or orifice type outlet shall be proportional, if the outlet is fixed or set at 0.3 times the depth below the water surface.

Similarly, for a weir type outlet, $m = 3/2$

($\because$ the discharge is proportional to $H^{3/2}$).

Hence, the setting for a combination of a weir type outlet and a trapezoidal channel

$$= \frac{m}{n} = \frac{3/2}{5/3} = 0.9$$

Hence, for the weir type outlet to be proportional, the outlet should be set at 0.9 times the depth below the water surface.

*Hyper-proportional outlet.* An outlet is known to be **hyper-proportional** if its flexibility is greater than unity.

Thus $F = \frac{m}{n} \cdot \frac{y}{H} > 1$ or $\frac{m}{n} > \frac{H}{y}$

or $\frac{H}{y} < \frac{m}{n}$ or $\text{Setting} < \frac{m}{n}$

Hence, the outlet will be hyper-proportional, if the numerical value of its setting $(H/y)$ is less than $m/n$. In other words, the outlet is hyper-proportional if set higher (*i.e.* head acting on the outlet, $H$, is less than what is required for proportionality).

*Sub-proportional outlet.* An outlet is known to be **sub-proportional** if its flexibility is less than unity. Thus

$$F = \frac{m}{n} \frac{y}{H} < 1 \quad \text{or} \quad \frac{m}{n} < \frac{H}{y} \quad \text{or} \quad \frac{H}{y} > \frac{m}{n}.$$

Hence, the outlet will be sub-proportional, if the numerical value of its setting $(H/y)$ is more than $m/n$. In other words, the outlet is sub-proportional if set lower (*i.e.* $H$ is more than what is required for proportionality).

**Sensitivity**. Sensitivity is defined as the ratio of the rate of change of discharge through the outlet to the rate of change of water level of the distributary, referred to the normal depth of the channel. For rigid modules, the discharge is fixed and hence the sensitivity is zero.

For flexible modules, where the discharge through the outlet is independent of the water level of the watercourse and depends only upon the water level of the distributary, a gauge can be fixed and calibrated so as to indicate its reading $G = 0$ when $q = 0$. Thus

$$\text{Sensitivity } = S = \left[\frac{dq/q}{dG/y}\right] \quad ..(13.15)$$

**Relation between Sensitivity and Flexibility.** We know that

$$\text{Flexibility } = F = \left[\frac{dq/q}{dQ/Q}\right] \quad \textit{i.e. } \text{Eq. (13.5)}$$

$$\text{But} \quad \frac{dQ}{Q} = \frac{n}{y}\,dy \quad \textit{i.e. } \text{Eq. (13.11)}$$

Equation (13.5) then becomes

$$F = \left[\frac{dq/q}{(n/y)\,dy}\right] = \frac{1}{n}\left[\frac{dq/q}{dy/y}\right]$$

But since, $dG = dy$ we get $F = \frac{1}{n} \cdot S$.

or

$$\boxed{S = n \cdot F} \quad ...(13.16)$$

If $n = \frac{5}{3}$, for wide trapezoidal channels, then, $S = (5/3)\,F$.

It is also evident that :

(*i*) Greater is the variation in the discharge of an outlet for a given rise or fall in water surface level of the distributary, the larger is the sensitivity of the outlet.

(*ii*) The sensitivity of a non-flexible module is zero.

## 13.11. Certain other Important Definitions Connected with Modules

**Minimum modular head.** The minimum difference between the upstream and downstream water levels, which is required to be maintained so as to enable the module to pass the designed discharge is known as *minimum modular head* or *minimum modular loss.*

**Efficiency of an outlet.** It may be defined as the ratio of the head recovered to the head put in. Lesser is the head required for functioning of the outlet ; more efficient the outlet will be. Efficiency is a measure of the conservation of head by the outlet.

**Drowning Ratio.** It is the ratio of the depth of water level over crest on the downstream of the module to the depth of water level over crest on the upstream of module. In case of a weir type outlet, the efficiency is the same as the drowning ratio.

**Modular limit and modular range.** The modular limits are the extreme values of any one or more variables, beyond which an outlet becomes incapable of acting as a module or semi-module. The range between the lowest and the highest limiting values of various such factors is known as *modular range.*

## 13.12. Types of non-Modular Outlets

A non-modular outlet may be in the form of a rectangular opening or *open sluice,* or a *simple submerged pipe*. Pipe outlet (when submerged) is a very simple type of a non-modular outlet and is extensively used in South India.

**13.12.1. Open Sluice.** An open sluice (Fig. 13.13), like a bridge opening, is a rectangular pucca opening created across the bank of the distributary, by raising two abutments at 2.5 to 3 m apart and with a horizontal pucca floor. The width $B$ of the opening and height $H$ of the opening are computed to pass the given discharge, by using the appropriate discharge formula.

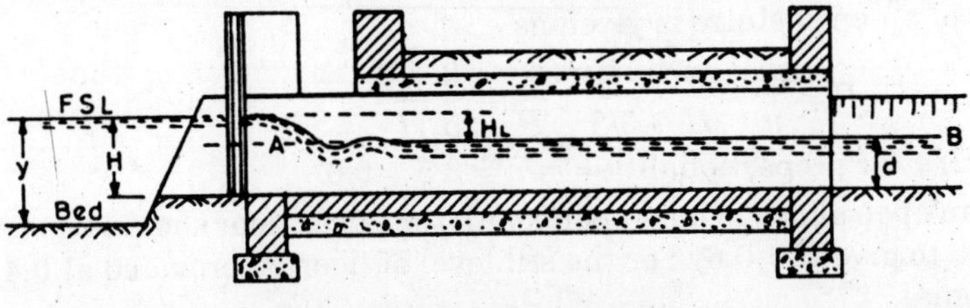


Fig. 13.13. Open Masonry Sluice.

The sill level of the outlet (*i.e.* the sill level of the pucca floor) is kept somewhat above the DBL of the distributary, as shown in Fig. 13.13. For maximum silt draw, however, the height of water level in distributary above sill level, *i.e*, $H$, has to be kept equal to FSD of distributary, $y$.

There was a time when the sectional area of the opening (B.H) was calculated by assuming a flow velocity of 0.3 m/sec through the opening, to pass the given discharge, $q$ [ *i.e.* Area $(BH) = q/0.3$ ] through the outlet. The old simple calculation method is even being used in present days for sluices off-taking from the river. But as owners of the water-courses could draw extra discharge in a simple open cut by lowering & clearing their water courses, such open cuts are nowadays designed on the basis of drowned weir formula by computing the average head loss ($H_L$). $H_L$ is given as = [the water level of distributary-water level of water course]. $H_L$ in design is, thus, taken as equal to the average of head loss values observed once a fortnight throughout the season.

The discharge through the opening is, thus, given by the *supressed weir* (*i.e.* weir without end contractions)* *formula, for submerged flow*, where discharge consists of :

(*i*) flow over upper part of section (lying above the line $AB$) of depth $H_L$ (*i.e.* $H - d$) may be considered as free discharge into the air as a weir flow, given by eqn.

$$q_1 = (2/3)\, C_{d_1} \cdot \sqrt{2g} \cdot B \cdot H_L^{3/2} \qquad \text{...(i)}$$

(*ii*) flow through the remaining lower part (lying below the line $AB$), which is computed as discharge through a submerged orifice, given as :

$$q_2 = C_{d_2} \cdot B \cdot d \sqrt{2g\, H_L}$$
$$= C_{d_2} \cdot B \cdot (H - H_L) \sqrt{2g \cdot H_L} \qquad \text{...(ii)}$$

If we assume $C_{d_1} = C_{d_2} = C_d$, then the total discharge through the outlet is given as :

$$q = q_1 + q_2 = (i) + (ii)$$
$$= (2/3)\, C_d \cdot \sqrt{2g} \cdot B \cdot H_L^{3/2} + C_d \cdot B\,(H - H_L) \cdot \sqrt{2g \cdot H_L}$$
$$= C_d \cdot \sqrt{2g} \cdot B \sqrt{H_L} \cdot \left[ \frac{2}{3} \cdot H_L + (H - H_L) \right]$$

* Weir having length equal to width of canal, across which the weir is constructed.

$$= Cd \,.\, \sqrt{2g \,.\, H_L} \,.\, B\left[H - \frac{1}{3}H_L\right] \quad \text{...(13.17)}$$

Taking $C = 0.64$, we get

$$q = 2.84\ \sqrt{H_L}\ B\left[H - \frac{1}{3}H_L\right] \quad \text{...(13.18)}$$

Flexibility of an open sluice is given as

$$= F = \frac{m}{n} \times \frac{y}{H} \times \frac{1}{5/3} \times \frac{y}{H} = \frac{3}{5}\frac{y}{H} \quad \text{...(13.19)}$$

and setting $H/y$ for proportionality $= m/n = 3/5 = 0.6$.

Hence for an open sluice to be proportional, the pucca floor should be provided in such a way as to give $H = 0.6y$ ; or the sill level of floor be provided at $0.4\,y$ above the DBL of canal.

**13.12.2. Submerged Pipe Outlet.** Two types of pipe outlets are shown in Figs. 13.14 (*a*) and (*b*). The pipe diameter varies from 10 to 30 cm. Pipes are generally embedded in concrete and are generally fixed horizontally at right angles to the direction of flow. They may also be laid sloping upwards by depressing the upstream end of the pipe [as shown in Fig. 13.14 (*b*)] so as to increase silt conductivity. In U.P., the pipes are generally laid horizontally at about 21 cm below the water surface level of the distributary channel.

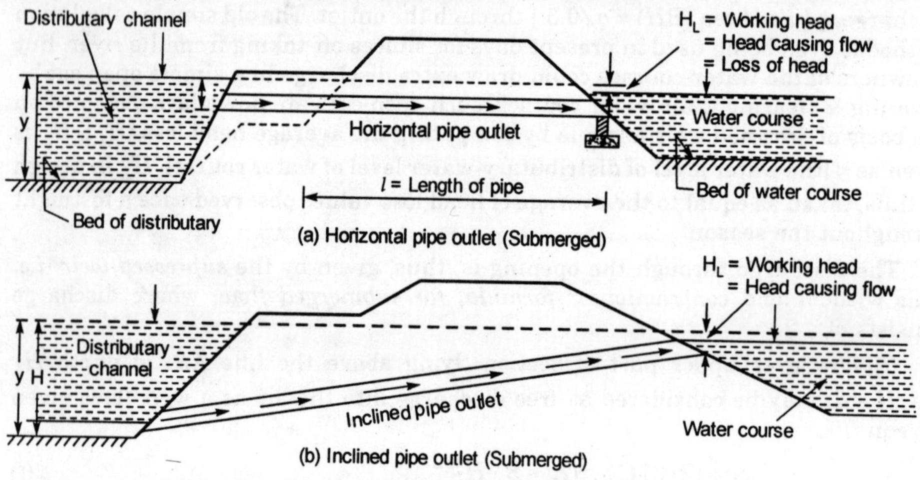


Fig. 13.14. Non-modular Outlets

The velocity through the pipe can be precisely computed by using the relation :

$H_L$ = Total loss of head = Entry loss + Frictional loss + Velocity head at exit

or $$H_L = 0.5\,\frac{V^2}{2g} + \frac{f'lV^2}{2gd} + \frac{V^2}{2g} \quad \text{...(13.20)}$$

or $$H_L = \frac{V^2}{2g}\left[1.5 + \frac{f'\,l}{d}\right] \quad \text{...(13.21)}$$

where $H_L$ is the difference in the water level of the distributary and the water course, $l$ is the length of the pipe, $d$ is the diameter of the pipe, and $f'$ is the coefficient of friction of the material of the pipe.

After computing $V$ from Eq. (13.21), the discharge $q$ can be computed as $q = V \times$ area of outlet pipe.

The discharge, however, for all practical purposes, may be easily computed by using the simple relation :

$$q = C_d \cdot A \cdot \sqrt{2gH_L} \quad ...(13.22)$$

where $q$ = Discharge through the outlet

$C_d$ = Coefficient of discharge for the pipe outlet. Its average value, determined at CWPRS poona is 0.73, although it depends upon the length & size of the pipe, besides on the pipe material.

$A$ = Area of the pipe

$H_L$ = Difference of head between the FSL of distributary and FSL of water course, usually called working head of irrigation outlet.

It becomes evident from equation (13.22), that the discharge through the outlet can be increased by increasing $H_L$ or by lowering the bed of the watercourse. Hence, for a certain water level in the distributary, the water entering high fields shall be less and that entering the lower fields shall be more ; consequently equitable distribution of water becomes difficult, as the owners of high fields are put to a disadvantage in drawal of their share.

*Although, such outlets can work even with very small available heads, and silt conduction is also quite good due to disturbance at the entrance, they may not be preferred as they do not ensure equitable distribution of water.*

Pipe outlets are more in vogue in South India, where glazed earthenware pipe lengths, 0.6m long, are used to build up the barrel. Against the usual convention of sockets facing against the flow, the pipes here are usually laid in the reverse direction, otherwise the head (front) socket will become a bellmouth at entry of discharge from the distributary, making the measurement of diameter difficult and unreliable. The spigot of the pipes are hence kept at the upstream ends.

Fig. 13.14(*c*). Pipe outlet controlled by Shutter on u/s end.

A non-modular outlet may be controlled by providing a shutter on its upstream end (see photographic view, Fig. 13.14(c)). Arrangement is provided to lock the shutter in any required position, so as to have a given discharge pass through the outlet (*i.e.* area of opening $A$ can be changed and the discharge passing is controlled). *Since the loss of head in a non-modular outlet is less than that in a modular outlet, the former is useful where much loss of head is not available.*

*Flexible modules (semi modules) and rigid modules are preferred to non-modular outlets, so as to ensure more equitable distribution of water in the water-courses, irrespective of their being at high or low levels.* However, such outlets can work as modular outlets, only within certain limits of water level in the distributary and the water course. The range over which each module works as modular outlet, is called its *working range;* or *range of modularity*. Also certain minimum difference of water levels on two sides of each module should always be there to ensure its modularity (*i.e.* its working as modular outlet). This minimum difference of levels is called *minimum modular head* or minimum *loss* of head of the modular outlet.

**Example 13.1.** *Design an irrigation outlet for the following data :*

*FSQ of outlet* = *50 lit/sec.*
*FSL in distibutary on u/s side of outlet* = *200.00 m.*
*FSL in water course on d/s side of outlet* = *199.92 m.*
*FSD in distributary on u/s side of outlet* = *1.05 m.*

**Solution.**

Available head across the outlet

= FSL of Distributary – FSL of water course
= 200.00 – 199.92 = 0.08 m.

Since the available head is very small, a non-modular outlet (such as a submerged pipe outlet) will have to be provided [ ∵ The very small head of 0.08 m clearly indicates that the d/s water level will definitely be above the opening on that side.]

The discharge in such a submerged pipe outlet is given by Eq. (13.22) as :

$$Q = C_d \cdot A \cdot \sqrt{2g\,H_L}$$

where $Q$ = 50 lit/sec = 0.05 cumec.
$A$ = Area of pipe
$H_L$ = working head or Loss of head between u/s and d/s.
= 0.08 m
$C_d$ = 0.73 for submerged flow.

Putting these values in the above Eqn. we get

$$0.05 = 0.73\,A \cdot \sqrt{2 \times 9.81 \times 0.08} \quad \text{or} \quad A = 0.055 \text{ m}^2$$

If a pipe of dia $d$ is used, then

$$\frac{\pi}{4} d^2 = 0.055 \text{ m}^2 \quad \text{or} \quad d = 0.264 \text{ m.}$$

Hence, use a pipe of dia, say 30 cm.

The R.L. of the bed of the distributary

= 200.00 – 1.05 = 198.95 m.

The pipe top can be fixed at about 22 cm below the FSL of the distributary. In other words, the pine can be laid horizontally with its invert level (or sill level)

= 200.00 – 0.22 – 0.30 = 199.48 m

*i.e.* at 199.48 – 198.95 = 0.53 m above the bed of the distributary.

### 13.13. Types of Semi-Modules or Flexible Outlets

The common types of semi-modules are :

(i) *Pipe outlet discharging freely into the air.*

(ii) *Venturi-flume outlet or Kennedy's Gauge outlet.*

(iii) *Open flume outlet.*

(iv) *Adjustable orifice semi-module.*

These outlets are discussed below in details :

**13.13.1. Free Pipe Outlet.** Pipe outlet discharging freely into the atmosphere is the simplest and the oldest type of a flexible outlet. The discharge through such an outlet will depend only upon the water level of the distributary, and will be independent of the water level of the water-course so long as the pipe is discharging freely. Silt conduction for such an outlet is quite good and efficiency is high. But a freely falling jet outlet can be provided only at a few places where sufficient level difference between the distributary and water-course is available. The discharge can be easily computed by using the equation.

$$Q = C_d \cdot A \cdot \sqrt{2g\,H_0} \qquad ...(13.23)$$

where $C_d$ is coefficient of discharge = 0.62 for average condition of free over fall.

$H_0$ = Head on u/s side measured from FSL of distributary up to the centre of pipe outlet.

$A$ = Area of cross-section of pipe

**Example 13.2.** *Design a pipe outlet for the following data :*

*Full supply discharge at the head of water course* = *90 lit/sec*

*FSL in distributary* = *205.00 m.*

*FSL in water course* = *204.00 m.*

**Solution**

Available head across the outlet

= FSL of distributary − FSL of water course

= 205.00 − 204.00 = 1 m.

This available head of 1 m is sufficient enough to make the pipe outlet discharge freely into the water course, as the d/s end of the pipe can be fixed below the water level of the water course, thus making it a semi-module.

The discharge through such an outlet is given as :

$$Q = C_d \cdot A\,\sqrt{2g\,H_0}$$

where $C_d$ = 0.62 for free pipe outlet.

$H_0$ = Head on upstream side above the centre line of pipe.

$Q$ = 9.0 lit/see = 0.09 cumec.

Assuming the dia of the pipe as 25 cm, we have

$$0.09 = 0.62\left[\frac{\pi}{4} \times (0.25)^2\right]\sqrt{2 \times 9.81\,H_o} = 0.62 \times 0.049 \times 4.43\,\sqrt{H}$$

$$\sqrt{H_o} = 0.657 \text{ m.}$$

$$\therefore \quad H_o = 0.44 \text{ m}$$

∴ R.L. of the centre of outlet pipe

= 205.00 − 0.44 = 204.56 m.

R.L. of invert of outlet pipe (*i.e.* sill level)

$$= 204.56 - \frac{0.25}{2} = 204.43 \text{ m} > \text{FSL of water course } i.e. \ 204.00 \text{ m}.$$

Hence, a pipe of 25 cm dia can be laid horizontally with its bottom or sill level at RL 204.43 m, and it will be discharging freely as a semi-module. **Ans.**

**13.13.2. Venturi Flume Outlet or Kennedy's Gauge Outlet.** Kennedy's Gauge Outlet is of a Venturi flume type and is shown in Fig. 13.15. It is made of cast iron and consists of three main parts :

(*a*) an orifice with a bell mouth entry ;

(*b*) a long expanding delivery pipe ;

(*c*) an air-vent connecting the throat of the delivery pipe to the atmosphere.

Water from the distributary enters this module through the bell-mouthed entry. The diameter is narrowest at the throat and the water jet shoots across the air in the base of the vertical air bent pipe. The base of the vent pipe is kept conical and air holes are provided at the top, so as to allow free circulation of air around the jet. In other words, the jet is equivalent to be discharging freely into the atmosphere, thereby making the discharge independent of the water level in the water-course. The jet, then passes through the cast iron expanding pipe (about 3 metres long) and then through its concrete extension and finally discharges into the watercourse, as shown in Fig. 13.15.

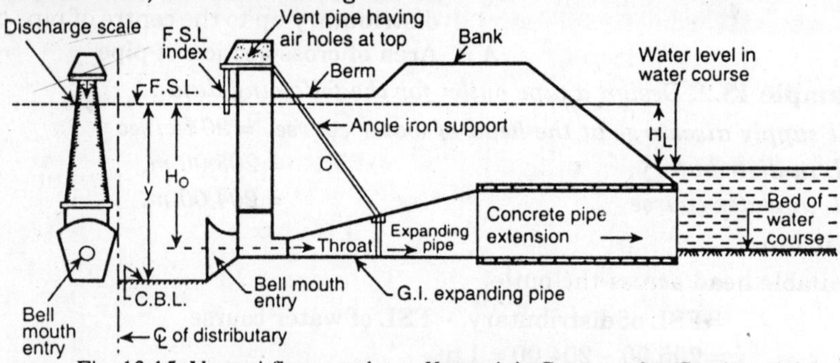


Fig. 13.15. Venturi flume outlet or Kennedy's Gauge outlet (Semi-module).

The discharge through this type of outlet is given as :

$$q = C_d \cdot A \sqrt{2g H_0} \qquad \text{...(13.24)}$$

where $q$ = Discharge through the outlet

$A$ = Area of cross-section of the pipe at the throat.

$H_0$ = Head measured from FSL of distributary to the centre of the pipe at throat.

$C_d$ = Coefficient of discharge which may be as high as 0.97.

Kennedy's gauge outlet is an old one and was designed in 1906, by R.G. Kennedy, Chief Engineer, Punjab Irrigation Department. This type of a module, though having the properties of semi-modularity, was very easy to tamper with. The farmers used to block the air holes so as to increase the outlet discharge. When the air bent is closed, the jet sucks the surrounding air, and pressure falls below the atmosphere, which increases the discharge. Hence, due to its tampering defect and due to its high cost of construction, *this type of module has almost become obsolete these days.*

**13.13.3. Open Flume Outlet.** It is a weir type outlet with a constricted throat and an expanding flume on the downstream, as shown in Fig. 13.16. Due to the constriction, a super-critical velocity is ensured in the throat and thereby allowing the formation of a jump in the expanding flume. The formation of hydraulic jump makes the outlet discharge independent of the water level in the water-course ; thus making it a semi-module. There are many types of such an outlet.

The most commonly used type of such an outlet is called the Punjab Open Flume Outlet, as the same was first used in Punjab. A typical plan and section of such a module are shown in Fig. 13.16. A photographic view of a distributary installing such modules is also shown in Fig. 13.17.

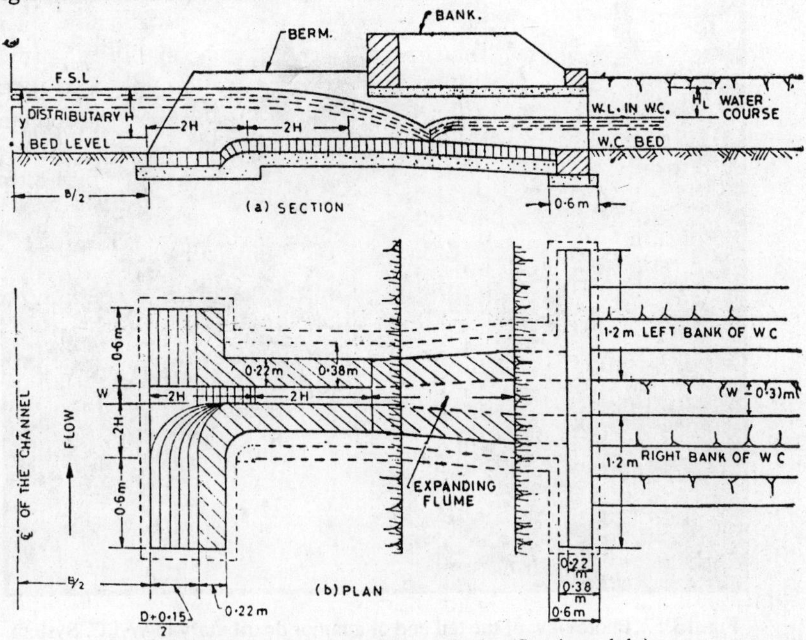


Fig. 13.16. Punjab Open Flume Outlet (Semi-module).

In such a Punjab open flume outlet, the upstream approach is made curved and the downstream approach is made straight, as shown in Fig. 13.16 (*b*). The length of the throat is kept 2*H*. The width of the flume at the downstream end is kept 0.3 m wider than the throat width (*W*), as shown. The slope of the d/s glacis is kept to the water course.

The discharge through such an outlet is given as :

$$q = C_d \cdot WH^{3/2} \qquad ...(13.25)$$

where $q$ = Discharge through the outlet

$H$ = Head over the crest measured from FSL of distributary

$W$ = Width of the throat $\approx \frac{q}{Q} A$, where

$Q$ = discharge of distributary and

$A$ = Area of waterway of distributary of bed width $B$ and depth $y$

$C_d$ = Coefficient of discharge, having a theoretical value equal to 1.71 in M.K.S. or S.I. units.

Due to the various losses, the actual value of $C_d$ is less than 1.71, and depends upon the width of the throat, as shown in Table 13.1.

**Table 13.1**

| *Throat width in metres (W)* | *Value of $C_d$* |
|---|---|
| 0.06 to 0.089 | 1.60 |
| 0.09 to 0.119 | 1.64 |
| Over 0.12 | 1.66 |

Fig. 13.17. Photoview of the tail end of a minor distributary in W.J.C. System in Haryana. Three water courses are taking off from this minor, and the three outlets provided are of open flume type.

For such a module to be proportional, setting is given by

$$\frac{H}{y} = \frac{m}{n} = \frac{3/2}{5/3} = 0.9 \qquad \text{...(13.26)}$$

Its efficiency varies between 80 to 90%. The minimum modular loss involved in such an outlet is about 0.2 *H*.

**13.13.4. Adjustable Orifice Semi-Modules.** Various types of orifice semi-modules have been designed since olden days. The one which found a lot of popularity is called **Crump's adjustable proportionate module (APM)** (see Figs. 13.18 and 13.19). Further improvements in approaches etc. have since been carried out in crump's APM, and the latest model, which is now used in Punjab and Haryana, is called an Adjustable orifice semi-module (A.O.S.M.). Typical dimensions of such an outlet are shown in the attached chart Fig. 13.20.

This type of an adjustable module is considered to be the best of all the modules and is mostly adopted. An adjustable orifice module consists of an orifice provided with a gradually expanding flume on the downstream side of the orifice. The flow through the orifice is super-critical, resulting in the formation of a hydraulic jump

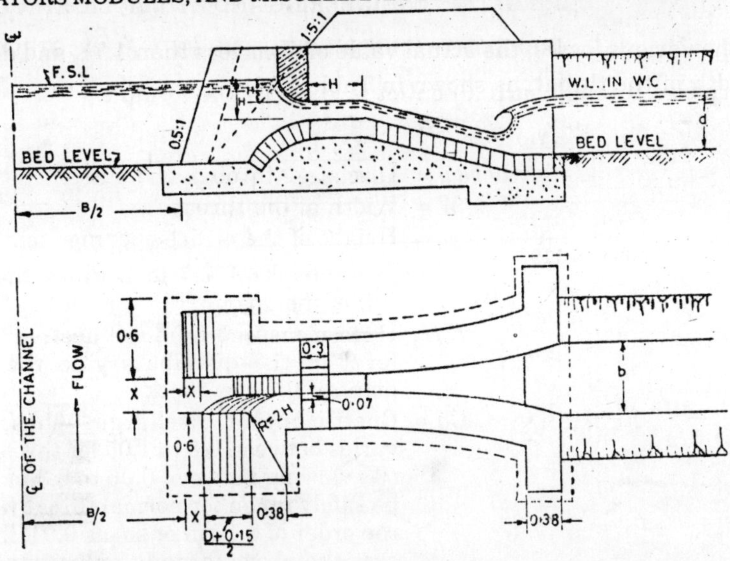


Fig. 13.18. Crump's A.P.M. outlet (Orifice-Semi-module)

(a) (b)

Fig. 13.19. Photoview of two A.P.M. Outlets. The common type of outlet used in Northern India. The distributary in Fig. 13.19 (b) is flowing full, and that in Fig. 13.19 (a) is dry. The size of the outlet is fixed on the basis of CCA. There is no control over the discharge. The water flows through the outlet so long as the canal water surface is above the bottom of the outlet opening.

in the expanding flume position. The formation of jump makes the discharge independent of the water level in the water-course.

The principal features of an adjustable orifice module are similar to those of a flumed regulator with horizontal crest and curved water approach on the upstream, and downstream wings expanding to the width of the water course, $b$. But unlike gates, it is provided with cast iron roof block, around which masonry is done. The opening height, $y_0$, can be changed by suitably adjusting the roof block, which can be easily done after dismantling the masonry around it. Since the roof block cannot be readjusted without breaking the masonry around it, the opening, $y_0$, and hence the outlet discharge, cannot be easily tempered with by the cultivators. The module is thus perfectly rigid, and at the same time adjustable in dimensions at a slight cost of re-doing the masonry.

The discharge through such an outlet is given by the formula :

$$q = C_d \cdot (W \cdot y_0) \cdot \sqrt{2g\, H_s} \quad \text{...(13.27)}$$

where $q$ = Discharge through the outlet

$W$ = Width of the throat

$y_0$ = Height of the orifice opening (smallest)

= generally kept 1.5 to 2 times the width ($W$) of the opening.

$H_s$ = Head measured from the upstream water level in the distributary to the lowest point of the roof block.

$C_d$ = Coefficient of discharge, whose value varies between 0.8 to 1.05 for throat width ($W$) varying between 0.06 to 0.3 m. It may be safely taken for normal throat width (of the order of 0.12 m or so) as 0.91, in which case the above formula reduces to

$$q = 4.04 \cdot W \cdot y_0 \sqrt{H_s} \quad \text{...(13.28)}$$

This type of adjustable modules are provided in eight different standard widths = 0.06, 0.075, 0.10, 0.12, 0.15, 0.19, 0.24 and 0.3 metres.

The *minimum modular head* loss involved in such a module is given by the formula

$$H_L = 0.82\, H_s - \frac{W}{2} \quad \text{...(13.29)}$$

Originally, when this module had a setting (*i.e.* $H/y$) of 6/10, it aimed at exact proportionality and, therefore, used to be called A.P.M. (Adjustable Proportional Module). But experience showed that the channels using such modules silted very badly, and hence proportionality was sacrificed to enable the outlet to carry higher silt charge by increasing the setting to 8/10. Since then, the outlet is known at A.O.S.M. (Adjustable Orifice Semi-module).

A few rules which are of use in fixing the dimensions of the latest model of AOSM have been indicated in the Fig. 13.20 itself.

*Advantages of this type of outlet are :*

(*i*) The adjustment can be made by raising or lowering the roof blocks at low costs by dismantling the masonry in which the roof bolts are fixed.

(*ii*) Any undue tampering of roof blocks by the cultivators can be easily detected, as it requires the breaking of the visible masonry key.

(*iii*) It is simple and cheaper.

(*iv*) Silt conduction is good at various settings, such as shown below

| *Setting* | *Silt Conduction* |
|---|---|
| $\frac{6}{10}$ th Setting | 99.5% |
| $\frac{8}{10}$ th Setting | 109.7% |
| $\frac{10}{10}$ th Setting | 121.9% |

## 13.14. Types of Rigid Modules

There are a few types of rigid modules which have no moving parts, such as :

(i) *Gibb's module ;*
(ii) *Khanna's rigid module* ; and
(iii) *Foote module.*

Out of all these modules, Gibb's module is the most important and widely used. It is described below :

**13.14.1. Gibb's Rigid Module.** The plan and section of Gibb's module are shown in Fig. 13.21. Gibb's module was designed by A.S. Gibb, formely Executive Engineer in Punjab Irrigation Deptt. The discharge range of Gibb's module varies between 0.03 to 0.45 cumec (1 to 16 cusecs). For lower values of discharge, *i.e.* between 1 to 3 cusecs, it is often economical to construct it in R.C.C., but for higher discharges (greater than 3 cusecs) it may be constructed in brick masonry.

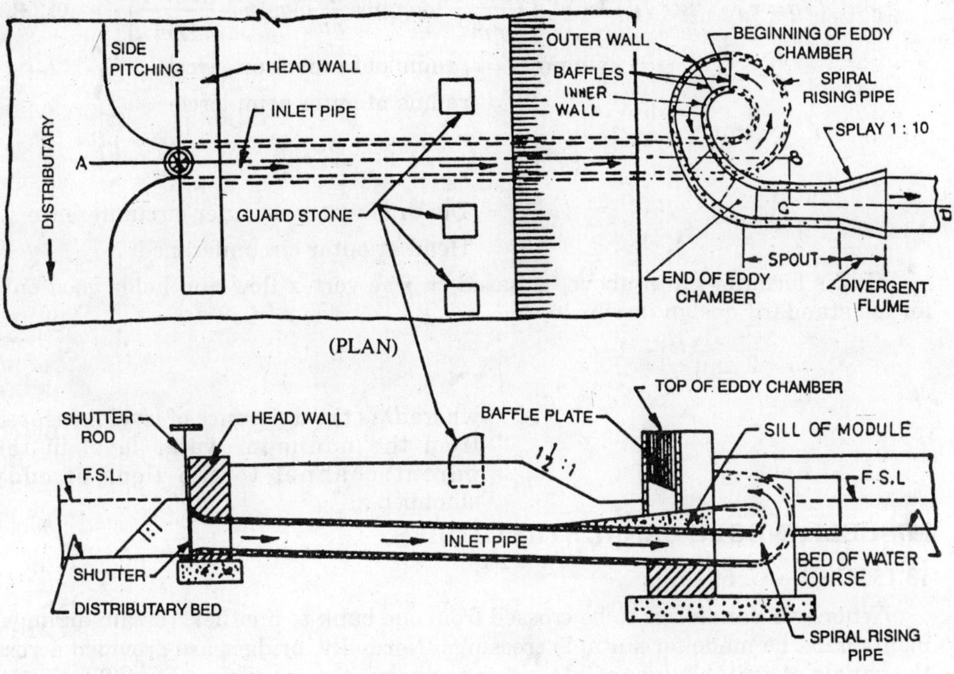


(Section A-B-C-D)
Fig. 13.21. Gibb's Module.

Gibb's module consists of an inlet pipe having a bell mouth. Water entering this inlet pipe is directed to the eddy chamber through a 180° rising spiral pipe in which free vortex flow is developed. The upstream end of the inlet pipe is controlled by a shutter. The rectangular eddy chamber having a horizontal floor is generally semi-circular in plan (although it may be a circle or 1½ circle depending upon discharge and range of working required for the module). The eddy chamber is fitted with a number of baffle plates (minimum six numbers) placed equidistant along its circular path. The lower edge of each baffle, slopes up from the inner wall to the outer wall of the eddy chamber.

The free vortex flow generated in the eddy chamber ensures, that, velocity × radius = constant, for all filaments. Thus, the water at the outer wall of the chamber having greater radius will have lesser velocity, resulting in a rise of water level there. Hence, the water surface level in the eddy chamber will slope down from the outer wall towards the inner wall. The baffle plates inserted in the eddy chamber, sloping up towards the outer wall, at required heights above the sill of the module, will help in preventing any excess amount of discharge likely to pass through the module. If the head causing flow increases, water banks up at the outer wall and impinges against the baffles, imparting an upward rotational flow to the water, which spins round in the compartment between two successive baffles, and finally drops on the oncoming stream of water ; thus dissipating excess energy. This helps in keeping the discharge constant for a wide range of variations in the head. The number of baffles coming into the action depends upon the variation in the head, as more and more baffles come into action with more and more discharge. The device thus maintains a constant discharge.

The discharge through the Gibb's module is given by Gibb's formula as :

$$q = r_2 \cdot \sqrt{2g}\left(y_1 + h_2\right)^{1.5}\left[\frac{m^2 - 1}{m^3}\log_e m + \frac{1}{m}\log_e m - \frac{m^2 - 1}{2m^3}\right] \quad \text{...(13.30)}$$

where $r_1$ = radius of inner semi-circle

$r_2$ = radius of outer semi-circle

$m = \frac{r_1}{r_2}$

$y_1$ = Depth of water at inner circumference

$h_2$ = Head at outer circumference.

Gibb's formula given above, is based on free vortex flow and holds good only for his standard design in which

$$m = 2 \text{ and } \frac{h_2}{D} = \frac{1}{7}$$

where $D$ is the difference of level measured from the minimum water level in the parent channel to the floor of eddy chamber.

## MISCELLANEOUS CANAL STRUCTURES

### 13.15. Cattle Crossings

Whenever canals are to be crossed from one bank to another, certain arrangements must be made for suitable crossings. Normally, bridges are provided across the canals at suitable key points, so as to permit the movement of traffic, cattle, and human beings, etc. across the canals. However, in remote village areas where no major movement is involved, certain arrangements are made so that the cattle, bullock carts, etc. can swim across the canals. Such crossings are called *cattle crossings;* or *cart crossings.*

For canals, carrying smaller water depths, of the order of 0.75 m or so, a ramp is provided for the entry, and another ramp opposite to the entry ramp is provided for the exit. The cattle or cart moves from one bank on the pitched ramp, comes on the bottom of canal within the water, and goes across by walking over the exit ramp.

But for larger canals, having water depth greater than 1.2 m or so, the cattle will have to swim as they cannot straightaway walk down from one bank to another (*via* water of course). For such canals, therefore, the cattle will swim over the deeper portion of the channel, and be carried downstream by some distance depending upon the width of the channel. A crossing with two ramps on one side and one ramp on the other side located exactly mid way, as shown in Fig. 13.22, will then be necessary. The cattle entering the upper ramp on the side having two ramps will get out at the ramp on the other side : while that entering from there will find exit at the lower ramp on the opposite side, as shown.

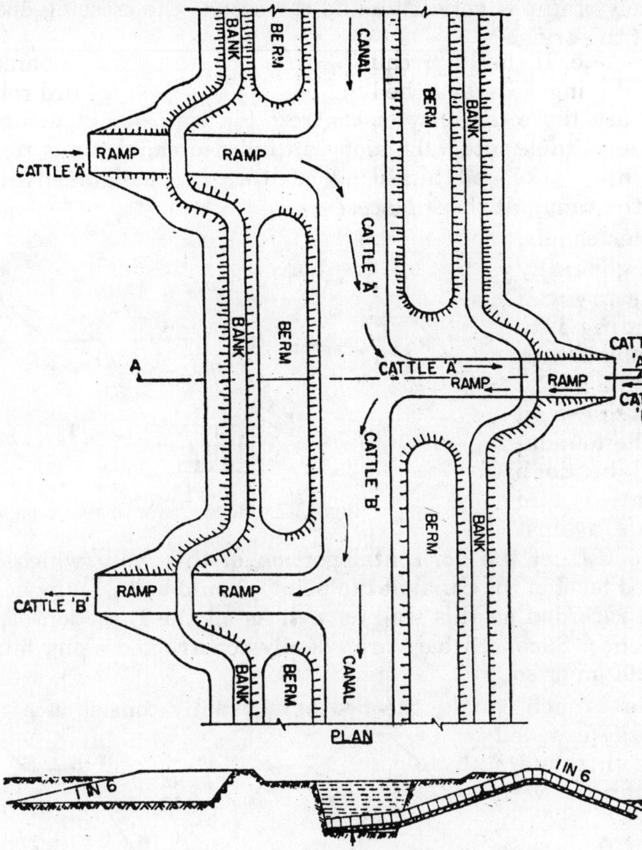


Fig. 13.22. Cattle crossing.

The exact distance to be kept between the two ramps on the same side should theoretically be

$$\frac{2BV}{U} \quad \text{...(13.31)}$$

where $B$ = width of the channel
$V$ = Av. flow velocity of water in the channel
$U$ = the speed with which cattle swim.

It is better to keep the distance a little longer than shorter, as even if the cattle reach the opposite side much too nearer, they will get at the ramp, but going upstream will be rather difficult.

The cattle crossings though cheaper are generally not preferred, compared to bridges due to the following reasons :

(*i*) they cannot be of much use on canals carrying considerable water depths.

(*ii*) they cannot be used by children or by those human beings who do not know swimming and when channel depth is more than that permitting straight walk over.

(*iii*) the cattle, after crossing one bank and reaching the other, do take rest and ramble about on the bank, causing a lot of damage to the bank.

Cattle crossings, therefore, require regular attention and repairs. They are, therefore, not preferred these days. They may however, be adopted in lone areas, where the likely traffic is very little and the cost of the crossing does not exceed 66% of that of the bridge.

## 13.16. Bed Bars

Bed bars are those masonry or concrete toe wall like structures which are constructed, at suitable intervals, along an unlined canal, so as to serve as permanent reference marks, and thus to indicate the correct alignment and theoretical bed levels of the canal, at these places.

For smaller canals, the bed bar generally consist of masonry wall partly extending into the bed and flush with it, and partly flush with the bank, as shown in Fig. 13.23. The foundations should be deep and substantial, and should be safe against scouring. The R.L. of the horizontal portion of this wall (which indicates the theoretical bed level of the canal at the point) should be taken by accurate double levelling and recorded on this wall as well as on the L section of the canal for further reference. Such bed bars are usually constructed along minor canals at intervals of 200 m or so.

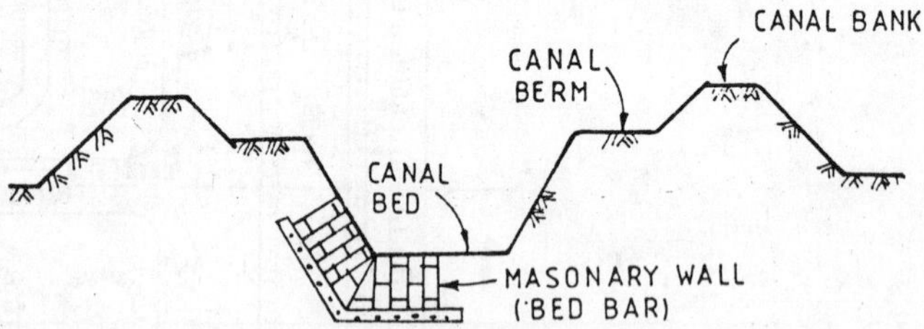


Fig. 13.23. Bed bars for smaller canals.

For larger branch canals, the bed bars usually consist of a masonry or a concrete block provided at suitable intervals along the canal, with their centres at the centre line of the canal, and to flush with the theoretical bed level of the canal at that point, as shown in Fig. 13.24. The depth of the block and its cross-section should be substantial enough as not to be disturbed by the action of the flowing water. It should also be safe against scouring. A typical size is 1.0 m × 1.2 m × 1.5 m deep block.

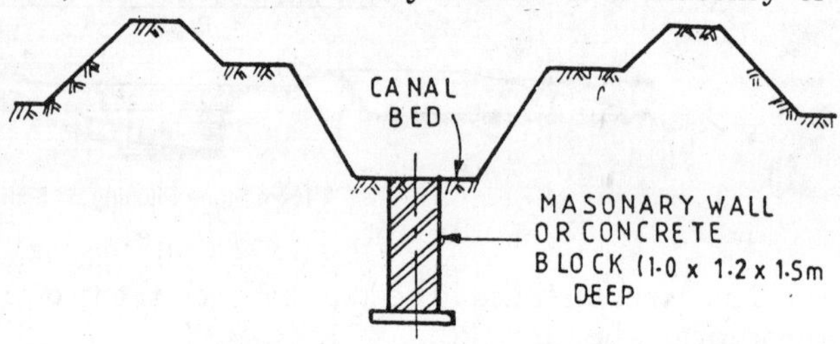


Fig. 13.24. Bed bars for larger branch canals.

The bed bars help in keeping a continuous watch on the behaviour of the canal w.r.t. its silting or scouring tendencies. They also help in regrading or desilting the canal section during its maintenance.

## PROBLEMS

1. What is meant by canal regulation ? Enumerate the different engineering structures which may be required to be constructed in a canal project in order to regulate effectively the functioning of the different canals of a canal system ? Discuss briefly the necessity, design and other details of any two such regulation works.
2. (*a*) What is meant by "canal regulation" and what are the different "canal regulation works" ?

(*b*) Describe the necessity and functioning of a "Distributary head regulator" and a "Cross-regulator" in a canal project. Also discuss the procedure that you will adopt for designing these regulation works.

**3.** Design a cross regulator and a head regulator for a distributary channel taking off from the parent channel, for the following data :

| | |
|---|---|
| Discharge of the parent channel | = 100 cumecs. |
| Discharge of the distributary | = 15 cumecs. |
| FSL of the parent channel | = (u/s)/(d/s) = 208.10/207.90 |
| Bed width of parent channel | = (u/s)/(d/s) = 42 m/38 m |
| Full supply water depth in the parent channel | = (u/s )/( d/s) = 2.5 m/2.5 m |
| FSL of distributary | = 207.10 m |
| Bed width of distributary | = 15 m |
| Full depth of water in the distributary | = 1.5 m |
| Permissible Khosla's safe exit gradient | = 1/6. |

**4.** (*a*) What is meant by "Canal Escapes" ? How do they help in protecting the adjoining areas against flooding due to some breach in the canal embankment ?

(*b*) Describe with neat sketches the different types of canal escapes, that may be constructed on modern canal projects.

**5.** (*a*) What is a "Metering flume" and how does it help in measuring the discharge of a canal ? Describe with neat sketches, the different types of such flumes and their functioning.

(*b*) What are "modules" and what is their importance in an irrigation canal system ?

**6.** What are "modules" ? What are the requirements of a good module ? Describe briefly with neat sketches the important types of modules and their suitabilities for a particular project.

**7.** (*a*) What are 'outlets' ? Enumerate the different types of outlets which are in common use on canal projects. Describe briefly with a neat sketch the functioning, use and design, of an 'Adjustable proportionate module' (APM).

(*b*) What is meant by the term 'Setting of the outlet' ? A semi-modular outlet has a 'Setting' such that its 'flexibility' is equal to 1. Indicate the variations in the discharge of the outlet when the discharge in the parent channel increases or decreases by a certain percentage.

[**Ans.** The discharge of the outlet will increase or decrease by the same percentage by which the discharge in the parent channel increases or decreases.]

**8.** (*a*) What is meant by the terms 'flexibility', 'proportionality', 'setting' and 'sensitivity' as applied to modules. Derive equations for them and discuss relation between these terms.

(*b*) Write note on any two of the following :

(*i*) Flexible module or Semi-modular type canal outlet
(*ii*) Standing wave flume
(*iii*) Canal escapes
(*iv*) Gibb's rigid module
(*v*) Proportional, sub-proportional, and hyper-proportional outlets
(*vi*) APM type outlet
(*vii*) Open flume outlet
(*viii*) Pipe outlets
(*ix*) Kennedy's gauge outlet
(*x*) Cattle crossings
(*xi*) Bed bars .

# 14

# *Cross Drainage Works*

## 14.1. Introduction

A cross drainage work is a structure which is constructed at the crossing of a canal and a natural drain, so as dispose of drainage water without interrupting the continuous canal supplies. In whatever way the canal is aligned, such cross drainage works generally become unavoidable. In order to reduce the cross drainage works, the artificial canals are generally aligned along the ridge line called watershed. When once the canal reaches the watershed line, cross drainage works are generally not required, unless the canal alignment is deviated from the watershed line. However, before the watershed is reached, the canal which takes off from the river has to cross a number of drains, which move from the watershed towards the river, as shown in Fig. 14.1. At all such crossings $c_1$, $c_2$, $c_3$, $c_4$, etc. cross drainage works are required.

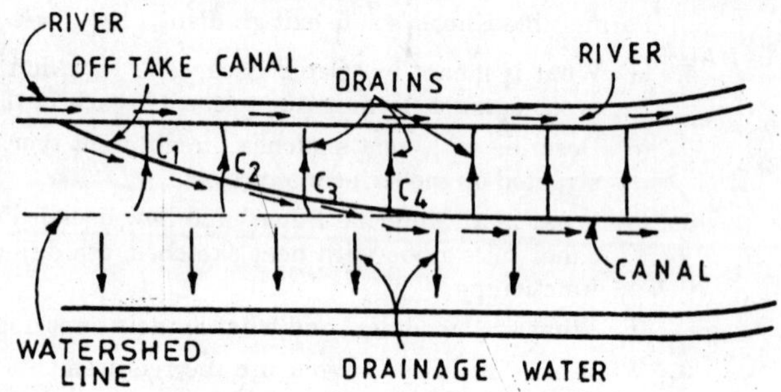


Fig. 14.1

A cross drainage work is generally a costly construction and must be avoided as far as possible. Since a watershed canal* crosses minimum number of drains, such an alignment is preferred to a contour canal which crosses maximum number of drains. The number of cross drainage works may also be reduced by diverting one drain into another and by changing the alignment of the canal, so that it crosses below the junction of two drains.

## 14.2. Types of Cross-drainage works

The drainage water intercepting the canal can be disposed of in either of the following ways :

(1) By passing the canal over the drainage. This may be accomplished either through (*i*) an *aqueduct;* or through a (*ii*) *syphon-aqueduct.*

(2) By passing the canal below the drainage. This may be accomplished either through (*i*) a *super-passage;* or through a (*ii*) *canal syphon* generally called a *syphon.*

(3) By passing the drain through the canal, so that the canal water and drainage water are allowed to intermingle with each other. This may be accomplished through (*i*) a *level crossing*; or through (*ii*) *inlets and outlets.*

All these different types of cross drainage works are described below in details.

---

* Also called a ridge canal.

**14.2.1. Aqueduct and Syphon Aqueduct.** In these works, the canal is taken over the natural drain, such that the drainage water runs below the canal (see Fig. 14.2) either freely or under syphoning pressure. *When the HFL of the drain is sufficiently below the bottom of the canal, so that the drainage water flows freely under gravity, the structure is known as an Aqueduct (Fig. 14.3). However, if the HFL of the drain is higher than the canal bed and the water passes through the aqueduct barrels under syphonic action, the structure is known as Syphon Aqueduct (See Fig. 14.4).*

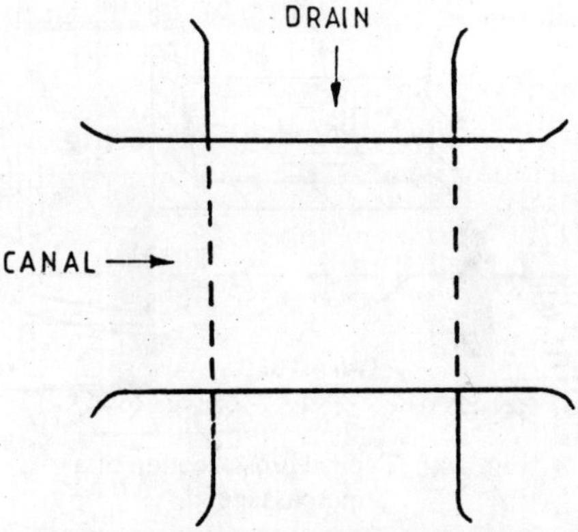


Fig. 14.2. Canal taken over the drain in an aqueduct or a syphon aqueduct (Line Plan of Crossing).

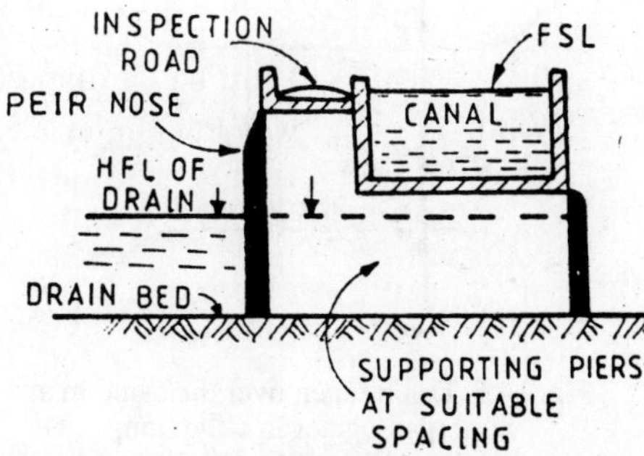


Fig. 14.3. Typical cross-section of an aqueduct.

In this type of works, the canal water is taken across the drainage in a trough supported on piers. An inspection road is generally provided alongwith the trough, as shown. An aqueduct is just like a bridge except that instead of carrying a road or a railway, it carries a canal on its top. An aqueduct is provided when sufficient level difference is available between the canal and the natural drainage, and canal bed level is sufficiently higher than the torrent level. In Sirsa, a city near Roper in Punjab, an excellent aqueduct having 20 spans of about 13 m each has been constructed to carry a canal having bed width of 28 metres and a discharge of about 360 cumecs, with a torrent discharge of about 4300 cumecs. A difference of 3.3 metres was available between the bed level of canal and that of torrent in this case.

In the case of a syphon aqueduct, the drain bed is generally depressed and provided with pucca floor, as shown in Fig. 14.4. On the upstream side, the drainage bed may be joined to the pucca floor either by a vertical drop (when drop is of the order of 1 m) or by a glacis of 3 : 1 (when drop is more). The downstream rising slope should not be steeper than 5 : 1.

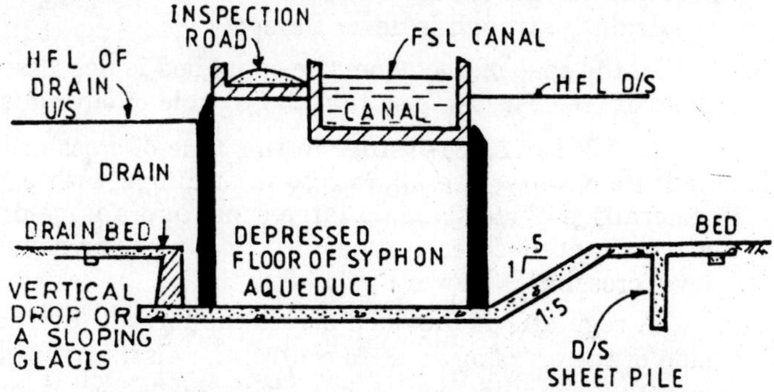


Fig. 14.4. Typical cross-section of a Syphon Aqueduct.

In this type of cross-drainage works (*i.e.* when the canal is taken over the drainage), the canal remains open to inspection throughout, and the damage caused by floods are rare. However, during heavy floods, the foundations of the work may be susceptible to scour; or waterway of the drain may get choked with debris, trees, etc.

**14.2.2. Super-passage and syphon.** In these works, the drain is taken over the canal such that the canal water runs below the drain (Fig. 14.5) either freely or under syphoning pressure. *When the FSL of the canal is sufficiently below the bottom of the drain trough, so that the canal water flows freely under gravity, the*

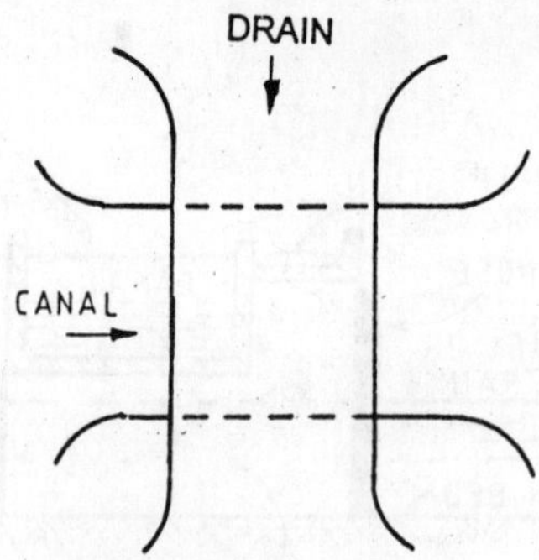


Fig. 14.5. Drain taken over the canal in a Superpassage or in a Syphon. (Line Plan of Crossing)

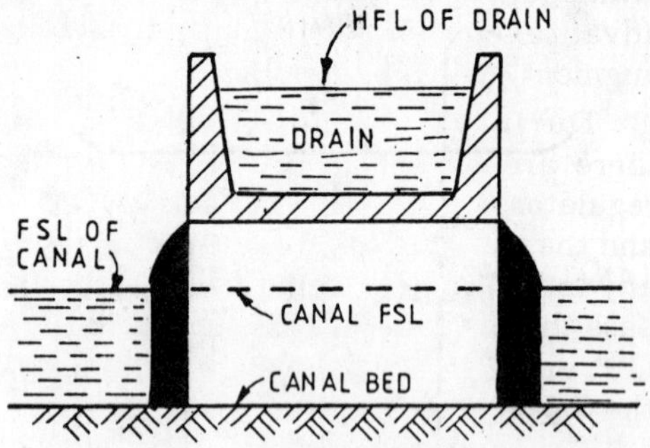


Fig. 14.6. Typical cross-section of a Superpassage.

*structure is known as a Superpassage (Fig. 14.6). However, if the FSL of the canal is sufficiently above th: bed level of the drainage trough, so that the canal flows under syphonic action under the trough, the structure is known as a canal syphon or a Syphon (Fig. 14.7).*

A superpassage is thus the reverse of an aqueduct, and similarly, a syphon is a reverse of an aqueduct syphon. However, in this type of cross-drainage works, the inspection road cannot be provided along the canal and a separate bridge is required for the road-way. For affecting economy, the canal may be flumed, but the drainage trough is never flumed.

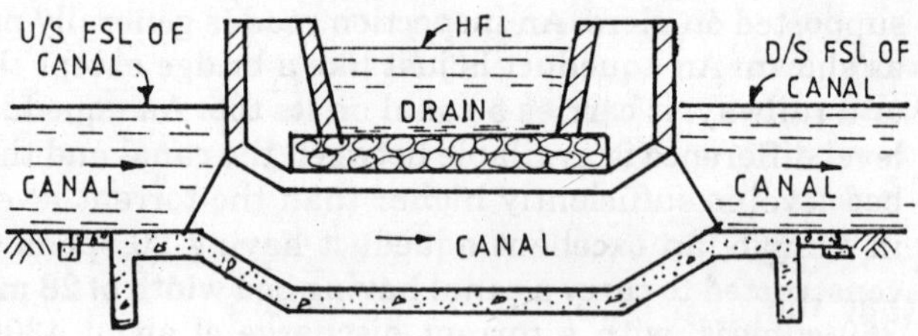


Fig. 14.7. Typical cross-section of a Canal Syphon (generally called a Syphon).

In the case of a syphon, the canal bed is depressed and a ramp is provided at the exit (see Fig. 14.7) so that the trouble of silting is minimised.

**14.2.3. Level Crossing.** In this type of cross-drainage work, the canal water and drain water are allowed to intermingle with each other. A level crossing is generally provided when a large canal and a huge drainage (such as a stream or a river) approach each other practically at the same level. A typical layout of a level crossing is shown Fig. 14.8.

A regulator is provided across the torrent (drainage) just on the downstream side of the crossing so as to control the discharge passing into the torrent. At the outgoing canal, a regulator is also provided so as control the discharge into the canal. A regulator at the end of the incoming canal is also sometimes required.

The arrangement is practically the same as is provided on a canal headworks. This arrangement is generally provided when a huge sized canal crosses a large torrent carrying a very high but short lived* flood discharge. In this arrangement, the perennial drainage discharge is sometimes advantageously used, so as to augment the canal supplies.

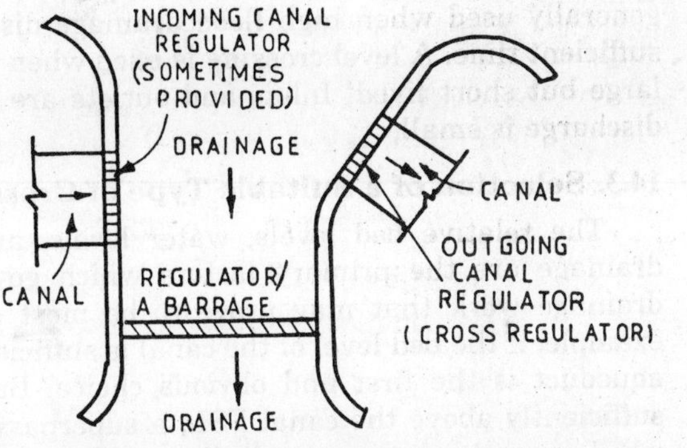


Fig. 14.8. Typical layout of a Level Crossing.

During dry season, when there are no floods, the torrent regulator is generally kept closed and the outgoing canal regulator is kept fully open, so that the canal flows without any interruption. During floods, however, the torrent regulator is opened so as to pass the flood discharge. A beautiful level crossing has been provided in Eastern U.P. under Sarda Sahayak Pariyojna ; where a canal carrying 370 cumecs crosses the Sarda river carrying a high flood discharge of the order of 10,000 cumecs.

**14.2.4. Inlets and Outlets.** An inlet is a structure constructed in order to allow the drainage water to enter the canal and get mixed with the canal water and thus to help in augmenting canal supplies. Such a structure is generally adopted when the drainage discharge is small and the drain crosses the canal with its bed level equal to or slightly higher than the canal F.S.L. Moreover, for the canal to remain in regime, the drain water must not admit heavy load of silt into the canal. Thus, in an inlet, the drainage water is simply added to the canal.

But, when the drainage discharge is high or if the canal is small, so that the canal section cannot take the entire drainage water, an outlet may sometimes be constructed to escape out the additional discharge at a suitable site, a little downstream along the canal. It is not necessary that the escaped discharge should be equal to the admitted discharge.

Similarly, it is also not necessary, that the number of inlets and outlets should be the same. There may be one outlet for two or three inlets. The outlet is generally combined with some other work where arrangement for escaping is in any case to be provided or may be added at a small extra cost.

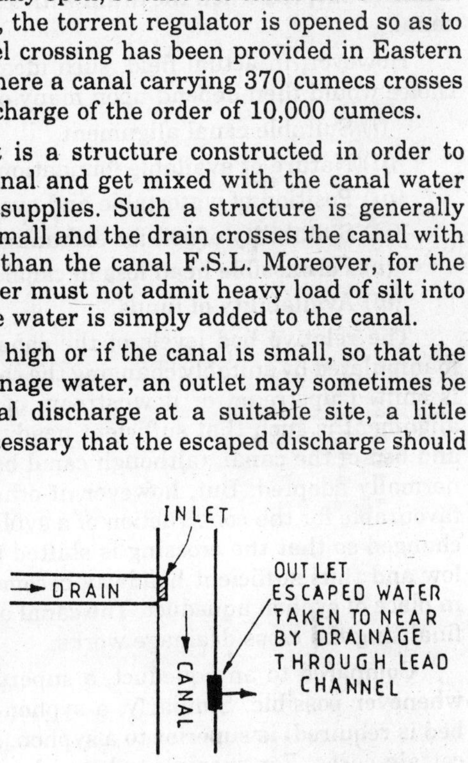


Fig. 14.9. Inlet and outlet (Plan).

An *inlet* essentially consists of an open cut in a canal bank, suitably protected by pitching, to admit the upland drainage water into the canal. The bed and sides of the canal are also pitched for a certain distance upstream and downstream of the inlet. Similarly the *outlet* is another open cut in the canal bank with bed and sides of the cut properly pitched. The escaping water from the outlet is taken away by a lead channel to some nearby drain, on the downstream side of the surface outlet.

This type of cross-drainage work (*i.e.* those requiring intermingly of canal water with drainage water) are inferior to aqueduct or superpassage type of works,

* When floods are intermittent and not continuous.

but they are cheaper. Hence, the aqueduct or superpassage type of works are generally used when high flood drainage discharge is large and continues for a sufficient time. A level crossing is used when the high flood drainage discharge is large but short lived. Inlets and outlets are used when the high flood drainage discharge is small.

## 14.3. Selection of a Suitable Type of Cross-Drainage Work

The relative bed levels, water levels, and discharge of the canal and the drainage are the primary factors which govern and dictate the type of cross drainage work that may prove to be most suitable at a particular place. For example, if the bed level of the canal is sufficiently above the HFL of the drain, an aqueduct is the first and obvious choice. But, if the bed level of the drain is sufficiently above the canal FSL, a superpassage may be constructed. Similarly, when a canal carries a small discharge compared to the drain, the canal may be taken below the drain by constructing a syphon, as against a syphon aqueduct which is adopted when the drain with smaller discharge can be taken below a large canal.

However, in actual field, such ideal conditions may not be available and the choice would then depend upon many other factors, such as :

(*i*) Suitable canal alignment.

(*ii*) Nature of available foundation.

(*iii*) Position of watertable and availability of dewatering equipment.

(*iv*) Suitability of soil for embankment.

(*v*) Permissible head loss in canal.

(*vi*) Availability of funds.

The relative bed levels of the canal and the drainage may be changed and manipulated by suitably changing the canal alignment, so that the point of crossing is shifted upstream or downstream of the drainage. For example, if the canal alignment is such that sufficient headway is not available between HFL of drain and bed of the canal, (although canal bed is higher) a syphon aqueduct has to be normally adopted. But, however, if other conditions (enumerated above) are not favourable for the construction of a syphon-aqueduct, the canal alignment may be changed so that the crossing is shifted to the downstream where drainage bed is low and thus sufficient headway becomes available for constructing an aqueduct in place of syphon aqueduct. The canal alignment is, therefore, finalised only after finalising the cross drainage works.

Compared to an aqueduct, a superpassage is inferior and should be avoided whenever possible. Similarly, a syphon-aqueduct (unless large drop in drainage bed is required) is superior to a syphon. A level crossing may become inevitable in certain cases. For example, when a large canal crosses a large torrent at almost equal bed levels, a level crossing may remain to be the only answer. An inlet may be adopted when a small drain crosses the canal with its bed level equal to canal FSL or slightly higher than it (*i.e.* bed conditions similar to those favouring the choice of a syphon or a superpassage). Inlets, though cheaper, are not preferred these days because their performance has not been very satisfactory.

## 14.4. Various Types of Aqueducts and Syphon-Aqueducts

They may be classified into three types depending on the sides of the aqueduct :

**Type I**. In this type, the sides of the aqueduct are earthen bank with complete earthen slopes. The length of the culvert through which the drainage water has to pass under the canal should not only be sufficient to accommodate the water section of the canal but also the earthen banks of the canal with adequate slopes (Fig. 14.10).

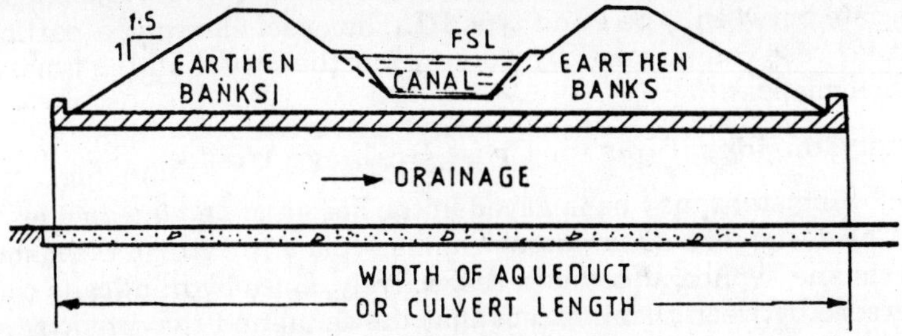


Fig. 14.10. Aqueduct (Type I)

**Type II.** In this type, the canal continues in its earthen section over the drainage, but the outer slopes of canal banks are replaced by retaining walls, thereby, reducing the length of the drainage culvert by that much extent (Fig. 14.11).

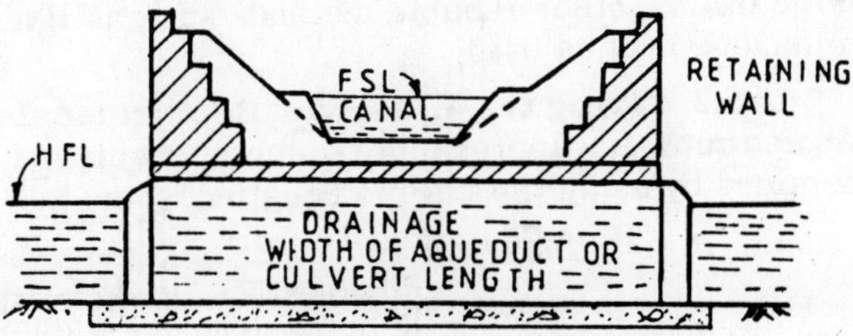


Fig. 14.11. Aqueduct (Type II)

**Type III.** In this type, earthen section of the canal is discontinued and the canal water is carried in a masonry or a concrete trough. The canal is generally flumed in this case, so as to effect economy in construction.

The culvert length or width of aqueduct is maximum in Type I and minimum in Type III. An intermediate value exists in Type II.

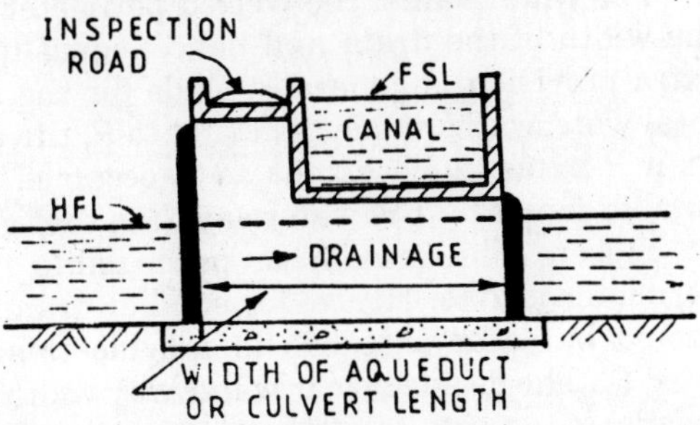


Fig. 14.12. Aqueduct (Type III)

**Selection of the Suitable Type.** *The selection of a particular type out of three types of aqueducts or syphon-aqueducts lies on the considerations of economy. The cheapest of the three types at a particular place shall be the obvious choice.*

In fact, in all cases, the cost of abutments and wing walls, is independent of the length of the culvert along the canal. In type I, no canal wings are required since the canal section is not at all changed. However, in this type, the width of the aqueduct is the largest. Type I will, therefore, prove economical only where the length of the aqueduct is small and where the cost of bank connections would be large in comparison to the savings obtained from the reduction in the width of the aqueduct.

In type III, the width of the aqueduct is minimum but the cost of bank connections is maximum. This type is, therefore, suitable where the length of the aqueduct is very large and where the cost of bank connections would be small in comparison to the savings obtained from the reduction in the width of the aqueduct.

*On the basis of above discussion, it can be concluded that the choice of a particular type depends mainly upon the length of the aqueduct (i.e. the width of the drainage) in relation to the size of the canal. On a very small drain, type III is most economical; while on a very wide drainage, type I is most economical.* Type II is intermediate between type I and type III. The exact choice of a particular type in a particular case can be made by working out the cost of all the types and then choosing the cheapest.

## 14.5. Design Considerations for Cross Drainage Works

The following steps may be involved in the design of an aqueduct or a syphon-aqueduct. The design of a superpassage and a syphon is done on the same lines as for aqueducts and syphon aqueducts, respectively, since hydraulically there is not much difference between them, except that the canal and the drainage are interchanged by each other.

**14.5.1. Determination of Maximum Flood Discharge.** The high flood discharge for smaller drains may be worked out by using empirical formulas ; and for large drains, other reliable methods such as Hydrograph analysis, Rational formula, etc. may be used.

**14.5.2. Fixing the Waterway Requirements for Aqueducts and Syphon-Aqueducts.** An approximate value of required waterway for the drain may be obtained by using the Lacey's equation, given by

$$P = 4.75 \cdot \sqrt{Q}$$

where $P$ = is the watted perimeter in metres
$Q$ = Total discharge in cumecs.

For wide drains, the wetted perimeter may be approximately taken equal to the width of the drain and hence, equal to the waterway required. However, no extra provision is generally made for the space occupied by piers. Hence, if the total waterway provided is equal to $P$, the effective or clear waterway will be less than $P$ by as much extent as is occupied by pier widths. For smaller drains, a smaller figure for the waterway than that given by Lacey's regime perimeter, may be chosen. The maximum permissible reduction in waterway from Lacey's perimeter is 20%. Hence, for smaller drains, the width of the waterway provided should be so adjusted as to provide this required perimeter (minimum value = $0.8\,P$). The decided clear water way width is provided in suitable number of bays (spans).

*Size of the Barrels.* After having fixed the waterway width & number of compartments (bays), the height of the drain barrels has to be fixed. In case of an aqueduct, the canal trough is carried clear above the drain HFL, and drain bed is not to be depressed. Hence, the height of bay openings is automatically fixed in aqueducts, as equal to the difference between HFL and DBL of drain.

However, in syphon-aqueducts, the required area of the drainage waterway can be obtained by dividing the drainage discharge by the permissible velocity through the barrels. This velocity through the barrels is generally limited to 2 to 3 m/sec. The waterway area is then divided by the decided waterway width of the drain openings, to compute the height of the openings, and the extent of depressed floor.

Due to the reduction in the width of the drainage, afflux is produced near the work site. The afflux will increase more and more, if the waterway is reduced more and more. The value of afflux is limited, so that there is no flooding of the country-side. The afflux may be calculated by using Unwin's formula as explained below in the following article.

**14.5.3. Afflux and Head Loss through Syphon Barrels.** It was stated earlier that the velocity through syphon barrels is limited to a scouring value of about 2 to 3 m/sec. A higher velocity may cause quick abrasion of the barrel surfaces by rolling grit, etc. and shall definitely result in higher amount of afflux on the upstream side of the syphon or syphon-aqueduct, and thus, requiring higher and longer marginal banks.

The head loss ($h$) through syphon barrels and the velocity ($V$) through them are generally related by Unwin's formula*, given as :

$$h = \left[1 + f_1 + f_2 \frac{L}{R}\right] \frac{V^2}{2g} - \frac{V_a^2}{2g} \quad \text{...(14.1)}$$

where $L$ = Length of the barrel.
$R$ = Hydraulic mean radius of the barrel.
$V$ = Velocity of flow through the barrel.
$V_a$ = Velocity of approach and is often neglected.
$f_1$ = Coefficient of head loss at entry
= 0.505 for unshaped mouth
= 0.08 for bell mouth.
$f_2$ = is a coefficient such that the loss of head through the barrel due to surface friction is given by

$$f_2 \cdot \frac{L}{R} \cdot \frac{V^2}{2g}; \text{ where } f_2 \text{ is given as :}$$

$$f_2 = a\left(1 + \frac{b}{R}\right) \quad \text{...(14.2)}$$

where the values of $a$ and $b$ for different materials may be taken as given in Table 14.1.

**Table 14.1**

| *Material of the surface of barrel* | *a* | *b* |
|---|---|---|
| Smooth iron pipe | 0.00497 | 0.025 |
| Encrusted pipe | 0.00996 | 0.025 |
| Smooth cement plaster | 0.00316 | 0.030 |
| Ashlar or brick work | 0.00401 | 0.070 |
| Rubble masonry or stone pitching | 0.00507 | 0.250 |

After having fixed the velocity ($V$) through the barrels, the head ($h$) required to generate that much velocity can be found by using the equation (14.1).

*The total head loss consists of three losses, *i.e.*

Entry loss $= f_1 \frac{V^2}{2g}$, *(ii)* Friction loss $= \frac{f_2 L V^2}{2gR}$, *(iii)* Exit loss $= \frac{V^2}{2g}$.

The d/s HFL of the drain remains unchanged by the construction of works, and thus the u/s HFL can be obtained by adding $h$ to the d/s HFL. The u/s HFL, therefore, gets headed up by an amount equal to $h$ and is known as afflux. The amount of afflux is limited because the top of guide banks and marginal bunds, etc. are governed by this raised HFL. So a limit placed on afflux will limit the velocity through the barrels and *vice versa*. Hence, by premitting a higher afflux and, therefore, a higher velocity through the barrels, the cross-sectional area of syphon barrels can be reduced, but there is a corresponding increase in the cost of guide banks and marginal bunds and also the length of d/s protection is increased. Hence, an economic balance should be worked out and a compromise obtained between the barrel area and afflux. Moreover, in order to reduce the afflux for the same velocity, the entry is made smooth by providing bell mouthed piers and surface friction is reduced by keeping the inside surface of the barrels as smooth as possible.

**14.5.4. Fluming of the Canal.** The contraction in the waterway of the canal (*i.e.* fluming of the canal) will reduce the length of barrels or the width of the aqueduct. This is likely to produce economy in many cases. The fluming of the canal is generally not done when the canal section is in earthen banks. Hence, the canal is generally not flumed in works of Type I and Type II. However, fluming is generally done in all the works of Type III.

The maximum fluming is generally governed by the extent that the velocity in the trough should remain subcritical (of the order of 3 m/sec). Because, if supercritical velocities are generated, then the transition back to the normal section on the downstream side of the work may involve the possibility of the formation of a hydraulic jump. This hydraulic jump, where not specifically required and designed for, would lead to undue loss of head and large stresses on the work. The extent of fluming is further governed by the economy and permissible loss of head. The greater is the fluming, the greater is the length of transition wings upstream as well as downstream. This extra cost of transition wings is balanced by the saving obtained due to the reduction in the width of the aqueduct. Hence, an economic balance has to be worked out for any proposed design.

After deciding the normal canal section and the flumed canal section, the transition has to be designed so as to provide a smooth change from one stage to the other, so as to avoid sudden transition and the formation of eddies, etc. For this reason, the u/s or approach wings should not be steeper than $26\frac{1}{2}^\circ$ (*i.e.* 2 : 1 splay) and the d/s or departure wings should not be steeper than $18\frac{1}{2}^\circ$ (*i.e.* 3 : 1 splay). Generally, the normal earthen canal section is trapezoidal, while the flumed pucca canal section is rectangular. It is also not necessary to keep the same depth in the normal and flumed sections. Rather, it may sometimes be economical to increase the depth and still further reduce the channel width in cases where a channel encounters a reach of rocky terrain and has to be flumed to curtail rock excavation. But an increase in the water depth in the canal trough will certainly increase the uplift pressures on the roof as well as on the floor of the culvert, thus requiring larger roof and floor sections and lower foundations. Due to these reasons, no appreciable economy may be obtained by increasing the depth.

The following methods may be used for designing the channel transitions :

(*i*) Mitra's method of design of transitions (when water depth remains constant).

(*ii*) Chaturvedi's method of design of transitions (when water depth remains constant).

(*iii*) Hind's method of design of transitions (when water depth may or may not vary).

(*i*) **Mitra's Hyperbolic Transition when water depth remains constant.** Shri A.C. Mitra, Chief Engineer, U.P. Irrigation Deptt. (Retd.), has proposed a hyperbolic transition for the design of channel transitions. According to him, the channel width at any section *X–X*, at a distance *x* from the flumed section (Fig. 14.13) is given by

$$B_x = \frac{B_n \cdot B_f \cdot L_f}{L_f B_n - (B_n - B_f)\, x} \qquad ...(14.3)$$

where $B_n$ = Bed width of the normal channel section.
$B_f$ = Bed width of the flumed channel section.
$B_x$ = Bed width at any distance $x$ from the flumed section.
$L_f$ = Length of transition.

*Derivation of equation (14.3) is given below :*

The above transition equation (*i.e.* equation 14.3) was derived on the basis that the rate of change of velocity per unit length of the transition remains constant

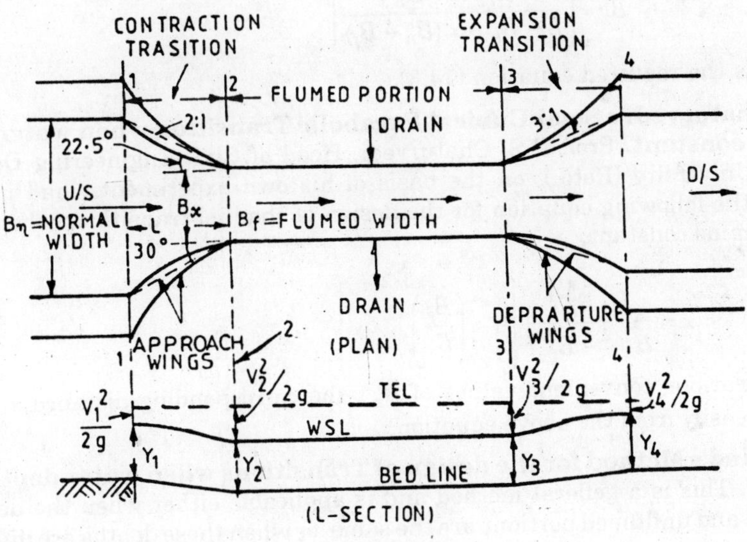


Fig. 14.13

throughout the transition length. Thus, if $V_n$, $V_f$ and $V_x$ represent velocities at the corresponding sections, we have

$$\frac{V_f - V_x}{x} = \frac{V_f - V_n}{L_f} \qquad ...(i)$$

Now, since depth $y$ is assumed to be constant and the total discharge $Q$ is also constant, we have

Velocity × Area = Discharge

$$\therefore\ V_f \cdot B_f \cdot y = V_x \cdot B_x \cdot y = V_n \cdot B_n \cdot y = Q$$

(assuming rectangular section throughout).

$$\therefore \quad V_f \cdot B_f = V_x \cdot B_x = V_n \cdot B_n$$
$$= Q/y = \text{constant} = K \text{ (say)}$$

or
$$V_f = K/B_f$$
$$V_x = K/B_x$$
$$V_n = K/B_n$$

Substituting these values in equation (*i*) we get

$$\left[\frac{\frac{K}{B_f} - \frac{K}{B_x}}{x}\right] = \left[\frac{\frac{K}{B_f} - \frac{K}{B_n}}{L_f}\right]$$

or
$$\frac{B_x - B_f}{B_f \cdot B_x \cdot x} = \frac{B_n - B_f}{L_f \cdot B_f \cdot B_n}$$

or
$$B_x \cdot L_f B_f \cdot B_n - L_f \cdot B_f^2 \cdot B_n = B_n B_f B_x \cdot x - B_f^2 B_x \cdot x$$

or
$$B_x \cdot L_f \cdot B_f \cdot B_n - B_n \cdot B_f \cdot B_n \cdot x + B_f^2 \cdot B_x \cdot x = L_f \cdot B_f^2 \cdot B_n$$

or
$$B_x \cdot B_f [L_f \cdot B_n - B_n \cdot x + B_f \cdot x] = L_f \cdot B_f^2 \cdot B_n$$

or
$$B_x [L_f \cdot B_n - x (B_n - B_f)] = L_f \cdot B_f \cdot B_n$$

or
$$B_x = \left[\frac{L_f \cdot B_f \cdot B_n}{L_f \cdot B_n - x (B_n - B_f)}\right] \quad \text{...(14.3)}$$

This is the required equation (14.3).

(*ii*) **Chaturvedi's Semi-Cubical Parabolic Transition when water depth remains constant.** Prof. R.S. Chaturvedi, Head of Civil Engineering Deptt. in Roorkee University (Retd.), on the basis of his own experiments, had in 1963, proposed the following equation for the design of channel transitions when water depth remains constant.

$$x = \frac{L \cdot B_n^{3/2}}{B_n^{3/2} - B_f^{3/2}} \left[1 - \left(\frac{B_f}{B_x}\right)^{3/2}\right] \quad \text{...(14.4)}$$

Choosing various convenient values of $B_x$ ; the corresponding distance $x$ can be computed easily from the above equation.

(*iii*) **Hind's Method for the design of Transitions when water depth may also vary.** This is a general method and is applicable either when the depth in the flumed and unflumed portions are the same, or when these depths are different.

In Fig. 14.13, the contraction transition (*i.e.* the approach transition) starts at section 1-1 and finishes at section 2-2. The flumed section continues from section 2-2 to section 3-3. The expansion transition starts at section 3-3 and finishes at section 4-4. From section 4-4 onwards, the channel flows in its normal cross-section and the conditions at this section are completely known. Let $V$ and $y$ with appropriate subscripts refer to velocities and depths at different sections.

The FSL at section 4–4 = Bed level at section 4–4 + $y_4$ = (known)

$\therefore$ TEL at section 4–4 = FSL at section 4–4 + $\frac{V_4^2}{2g}$ = (known)

Between section 3-3 and 4-4, there is an energy loss in the expansion, which is generally taken as equal to $0.3\left(\frac{V_3^2 - V_4^2}{2g}\right)$

$$\therefore \text{ TEL at section 3–3} = \text{TEL at section 4–4 (known)} + 0.3\left(\frac{V_3^2 - V_4^2}{2g}\right)$$

As the trough dimensions at section 3-3 are known, $V_3$ is also known, and hence, TEL at section 3-3 can be computed. Knowing TEL at 3-3 ; FSL at 3-3 can be calculated by subtracting $\frac{V_3^2}{2g}$ from TEL. Similarly, bed level at 3-3 can also be computed by substracting $y_3$ from FSL at 3-3.

Between sections 2-2 and 3-3, the channel flows in a trough of constant cross-section. The only loss in the trough ($H_L$) is the friction loss which can be computed with Manning's formula, *i.e.*,

$$\left(Q = \frac{1}{n} A \cdot R^{2/3} \cdot S^{1/2}\right)$$

or
$$Q = \frac{1}{n} A \cdot R^{2/3} \cdot \sqrt{\frac{H_L}{L}}$$

or
$$H_L = \frac{Q^2 \cdot n^2 \cdot L}{A^2 \cdot R^{4/3}}.$$

Adding this head loss $H_L$ to TEL of section 3-3, the TEL at section 2-2 is obtained. The FSL at section 2-2 can then be obtained by subtracting $\frac{V_2^2}{2g}$ from TEL of 2-2. Similarly, the bed level at section 2-2 can be easily obtained by further substracting $y_2$ from FSL at 2-2. Since the depth and velocity are constant in the trough, the TEL, FSL and bed lines are all parallel to each other from section 2-2 to 3-3.

Between section 1-1 and 2-2, there is a loss of energy due to contraction. This loss is generally taken as equal to $0.2\left[\frac{V_2^2 - V_1^2}{2g}\right]$.

Thus the TEL at section 1-1

$$= \text{TEL at section 2–2} + 0.2\left(\frac{V_2^2 - V_1^2}{2g}\right) = \text{(known)}.$$

Knowing TEL at section 1-1, FSL at 1-1 can be obtained by subtracting $V_1^2/2g$ from TEL at 1-1. Similarly, bed level at 1-1 can be obtained by subtracting $y_1$ from FSL at 1-1.

The bed level, FSL and TEL having been determined at all the four sections, the total energy line may be drawn by assuming it to be a straight line between adjacent sections. The bed line may also be drawn straight between adjacent sections, provided, the rise or fall in bed is small. The corners should, however, be rounded off in this case. However, if the change in bed level is considerable, the bed line in the transition section should be drawn as a smooth reverse curve, tangential to the bed lines at ends.

*Water surface in Transition* : In the contraction transition between section 1-1 to 2-2, there will be a drop in water surface due to the drop in energy line and also due to the increased velocity head at 2-2. This drop in water surface has to be negotiated by a smooth curve tangential at both ends. This can be easily accomplished by using two parabolic curves meeting tangentially at the centre of the transition, as shown in Fig. 14.14.

Let $2X_1 = L$ = The length in which fluming has been done.

$2Y_1$ = Total difference in water levels between section 1-1 and 2-2.

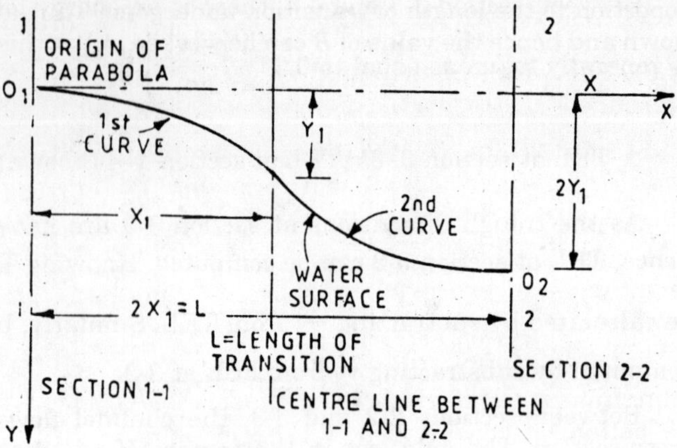


Fig. 14.14. Water Surface profile for Transition Contraction.

The distance of the middle point of transition will be $X_1$ and drop in water surface will be $Y_1$. The equation of the first parabolic curve, with origin at water surface of section 1-1 (*i.e.* $O_1$) is given by

$$Y = C \cdot X^2$$

when $\quad Y = Y_1$, and $X = X_1$

$$C = \frac{Y_1}{X_1^2}$$

Therefore, the equation of parabola becomes

$$Y = \left[\frac{Y_1}{X_1^2}\right] X^2 \qquad ...(14.5)$$

Using the above equation, the first parabolic curve can be easily plotted. Similarly, the second parabolic curve can be plotted by taking the origin at $O_2$ on section 2-2.

The water surface in the expansion transition between sections 3-3 and 4-4 can also be plotted in a similar fashion, where there will be a rise in the water surface from section 3-3 to 4-4, as shown in Fig. 14.13.

After having plotted the water surface profile over the entire length, the velocity head say ($h_v$) can be found by measuring the vertical distance between TEL and water surface line at any point. The velocity head can then be converted into equivalent velocity by using $V = \sqrt{2g \cdot h_v}$. Hence, the velocity at each point can be known. The cross-sectional area required to pass the given discharge at each point can be found by dividing discharge by velocity at that point (*i.e.* $A = Q/V$).

In trapezoidal channel of water depth $y$, the bed width $B$, and side slopes $s : 1$ ; area is given by

$$A = BD + s \cdot y^2. \qquad ...(14.6)$$

In flared wings, the side slopes are generally brought to vertical from an initial slope of $s$ : 1 and, therefore, the side slopes at any point can be interpolated in proportion to the length of transition undergone. Thus at any point, $A$, $y$ and $s$ are known and hence the value of $B$ can be worked out at this pt. by using the equation (14.6). The width of the canal at various points in the transition can thus be determined. Hence, all the dimensions of the transition are fully found out.

**14.5.5. Design of Pucca Canal Trough.** The canal trough is designed as follows :

*For an Aqueduct.* In case of an aqueduct, the bottom of the canal *i.e.* the roof of the culvert is subjected to the dead weight and the vertical load of water from the top, as shown in Fig. 14.15.

Since in an aqueduct, there is no uplift from the underside acting on the bottom of canal bed, the canal bottom has to be designed for taking the dead weight and full water load (when canal is running full) by either providing a thickness sufficient to take this much of load merely by gravity or by providing a reinforced concrete slab with reinforcement at its bottom.

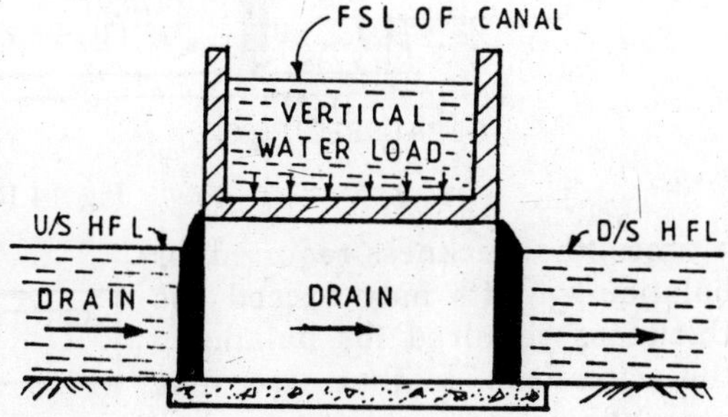

Fig. 14.15

The side walls of the canal are to be designed as retaining walls. They may be made of masonry or R.C.C. It is preferable to have an entire R.C.C. section. The retaining walls will be designed to carry the entire horizontal force exerted by the canal water and shall, therefore, carry reinforcement on the water face.

*For an Aqueduct Syphon.* In case of an aqueduct syphon, besides the vertical load of canal water, one more force comes into action *i.e.* the uplift pressure exerted by the drain water. The roof of the culvert *i.e.* the bottom-slab of canal should now be designed to withstand these two forces independently. Although these two forces act in a opposition to each other but still, under the worst circumstances, there may be times when only one of them may be acting. For example, when drainage is flowing at its maximum high flood level, canal may be empty. Similarly, there may not be any drainage water touching the slab when the canal may be running full. Hence, the slab should be designed for :

(*i*) full water load and dead weight, with no uplift
(*ii*) full-uplift with no water load.

The amount of the uplift pressures exerted by the drain water on the roof of the culvert can be evaluated by drawing the hydraulic gradient line (H.G. line), as shown Fig. 14.16.

The uplift pressure at any point under the roof of the culvert will be equal to the vertical ordinate between the hydraulic gradient line and the underside of the canal trough at that point, as shown in Fig. 14.16. From this uplift diagram of Fig. 14.16, it is very evident that the maximum uplift occurs at the upstream end point near the entry. The uplift pressure on the underside of the trough at the upstream end will be =[u/s water level – Entry loss – the level of the underside of the trough]. The slab thickness should be designed to withstand this maximum uplift.

When the slab is designed to counter-balance the maximum uplift, merely be gravity, it is sometimes found that the slab thickness required is less than what is required for the first condition (*i.e.* when designed for full water load). But many

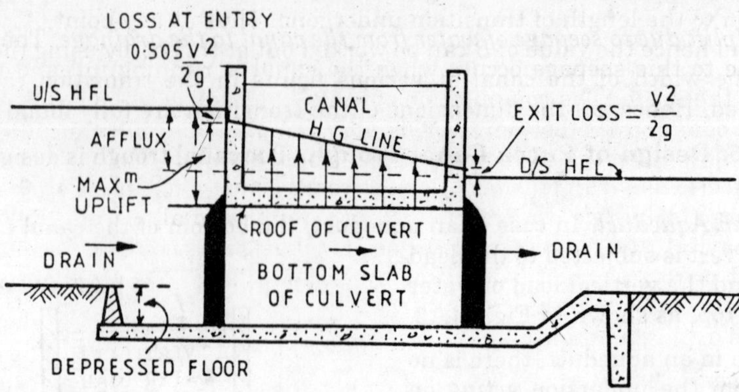


Fig. 14.16

a times, the thickness required for balancing uplift may exceed the thickness required for balancing the water load. In that case, it is generally not advisable to increase the thickness because any increase in thickness will result in lowering the levels of both the roof of the culvert as well as the bottom slab of culvert. This, in turn, will increase the uplift on the roof slab as well as on the bottom slab of culvert. Hence, in such a case, the thickness of the roof slab is generally provided from the considerations of water load and the remaining uplift is resisted by bending by providing top reinforcement in the roof slab. In such a case, the roof will have to be anchored to the bottom slab through piers by steel bolts, etc. so as to provide necessary end reactions for upward bending. A typical anchoring arrangement is shown in Fig. 14.17. *A better alternative may be to provide full fledged R.C.C. box culvert.*

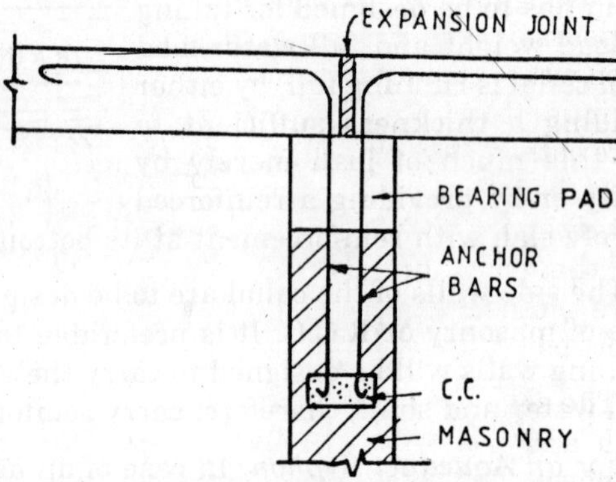


Fig. 14.17. Typical details of anchoring.

**14.5.6. Design of Bottom Floor of Aqueduct and Syphon Aqueduct.** The floor of the aqueduct or syphon-aqueduct is subjected to uplift due to two causes :

(*a*) *Uplift due to water-table.* This force acts where the bottom floor is depressed below the drainage bed, especially in syphon aqueducts.

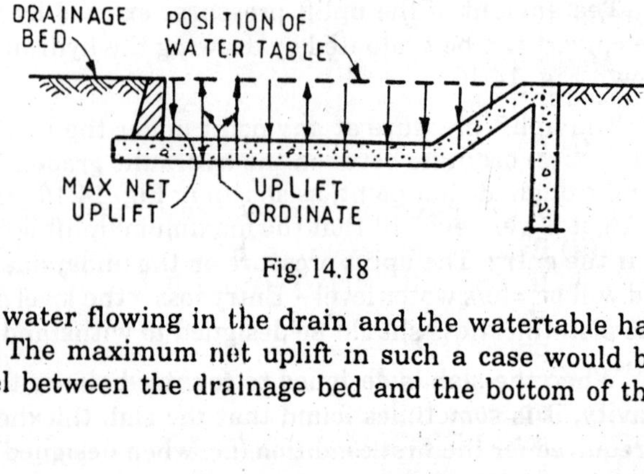


Fig. 14.18

The maximum uplift under the worst condition would occur when there is no water flowing in the drain and the watertable has risen up to the drainage bed. The maximum net uplift in such a case would be equal to the difference in level between the drainage bed and the bottom of the floor, as shown in Fig. 14.18.

(*b*) *Uplift due to seepage of water from the canal to the drainage.* The maximum uplift due to this seepage occurs when the canal is running full and there is no water in the drain. The computations of this uplift, exerted by the water seeping from the canal on the bottom floor, is very complex and difficult, due to the fact that the flow takes place in three dimensional flownet. The flow cannot be approximated to a two dimensional flow, as there is no typical place across which the flow is practically two dimensional. Hence, for the smaller works, Bligh's Creep theory may be used for assessing the seepage pressures. But, for the larger works, the uplift pressures must be checked by model studies. The seepage pressure can be evaluated by Bligh's theory as explained below :

The seepage flow occurs from the beginning of pucca canal trough (point $a$) and reappears in the drainage bed on either side of the impervious floor along the centre of the floor of the first culvert bay (say point $c$ or point $d$) in Fig. 14.19. Point $b$ is the point under the centre of the floor of the first culvert bay.

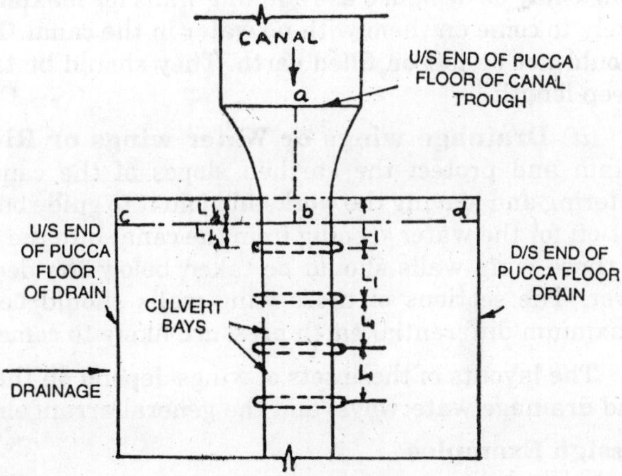

Fig. 14.19

The seepage path from $a$ to $b$ and from $b$ to $c$ can be known. The total creep length will then be equal to $= ab + bc$. If $H$ is the total seepage head (*i.e.* $H$ = FSL of canal − d/s bed level of drain), the residual head at the point $b$ (*i.e.*, $H_b$) is then given by Bligh's theory as equal to

$$H_b = H - \left[\frac{H}{ab + bc} \times ab\right]$$

or $$H_b = \left[\frac{H}{ab + bc} \times bc\right] \quad \ldots(14.7)$$

The floor of the syphon-aqueduct must be designed for the total uplift which is equal to the sum of the uplift due to seepage plus the uplift due to static head. The total uplift may be partly resisted by the wt. of the floor and partly by bending in reinforcement.

*Methods of reducing uplift on the floor.* The uplift on the bottom floor may be reduced in two ways :—

(*i*) By extending the impervious canal trough on either side of the drainage so as to increase the creep length $ab$. A puddle apron may be used in place of concrete floor, if clay is easily available.

(*ii*) By providing drainage holes in the culvert floor so as to release the uplift. If such relief holes are provided in the bottom floor ; an inverted filter, should also be provided below the floor. The inverted filter would help in preventing the soil particles from getting out of the holes. The performance of such holes may not prove very successful in actual field as it appears to be on paper. Because, if these get choked or if there occurs some defect in filter system, there may be a danger of failure of work by excessive uplift or by undermining.

**14.5.7. Design of Bank Connections.** Two set of wings are required in aqueducts and syphon-aqueducts. These are :

(*i*) *Canal Wings* or Land Wings.

(*ii*) *Drainage Wings* or Water Wings.

(*i*) **Canal wings or Land wings.** These wings provide a strong connection between the masonry or concrete sides of a canal trough and earthen canal banks. These wings are generally warped in plan so as to change the canal section from trapezoidal to rectangular. They should be extended upto the end of splay. These wings may be designed as retaining walls for maximum differential earth pressure likely to come on them with no water in the canal. The foundations of these wings should not be left on filled earth. They should be taken deep enough to give safe creep length.

(*ii*) **Drainage wings or Water wings or River wings**. These wing walls retain and protect the earthen slopes of the canal, guide the drainage water entering and leaving the work, and join it to guide banks and also provide a vertical cut-off for the water seeping from the canal into the drainage bed. The foundations of these wing walls should be taken below the deepest anticipated scour in the river. The sections of these wing walls should be capable of withstanding the maximum differential earth pressure likely to come on them.

The layouts of these sets of wings depend on the extent of contraction of canal and drainage waterways, and the general arrangement of the work.

**Design Examples**

**Example 14.1.** *Design a suitable cross-drainage work, given the following data at the crossing of a canal and a drainage.*

**Canal**

*Full supply discharge* = *32 cumecs*

*Full supply level* = *R.L. 213.5*

*Canal bed level* = *R.L. 212.0 m.*

*Canal bed width* = *20.*

*Trapezoidal canal section with* $1\frac{1}{2}H : 1V$ *slopes.*

*Canal water depth* = *1.5 m.*

**Drainage**

*High flood discharge* = *300 cumecs.*

*High flood level* = *210.0 m.*

*High flood depth* = *2.5 m.*

*General ground level* = *212.5 m.*

**Solution.** Since the drainage is of a large size, work of type III will be adopted. Further, because the canal bed level (212.0 m) is much above the H.F.L. of drainage (*i.e.* 210.0 m) an **aqueduct** will be constructed. The earthen banks of the canal will be discontinued and the canal water taken in a concrete trough. For effecting economy, the canal shall be flumed.

**Step 1. Design of Drainage Waterway**

Lacey's regime perimeter $= P = 4.75\sqrt{Q}$

where $Q$ = High flood discharge of drain = 300 cumecs (given)

$$P = 4.75 \cdot \sqrt{300} = 82.3 \text{ m.}$$

Let the clear span between piers be 9 m and the pier thickness be 1.5 m.

Using 8 bays of 9 m each, clear waterway = 8 × 9 = 72 m.

Using 7 piers of 1.5 each, length occupied by piers = 7 × 1.5 = 10.5 m.

Total length of waterway = 72 + 10.5 = **82.5 m**

**Step 2. Design of Canal Waterway**

Bed width of canal = 20.0 m.

Let the width be flumed to 10.0 m.

Providing a splay of 2 : 1 in contraction, the length of contraction transition

$$= \frac{20-10}{2} \times 2 = \mathbf{10.0\ m}$$

Providing a splay of 3 : 1 in expansion, the length of expansion transition

$$= \frac{20-10}{2} \times 3 = \mathbf{15\ m}$$

Length of the flumed rectangular portion of the canal between abutments = 82.5 m (provided).

In transitions, the side slopes of the canal section will be warped in plan from the original slope of $1\frac{1}{2}$ : 1 to vertical.

**Step 3. Head loss and bed levels at different sections.** (Fig. 14.20).

**At Section 4-4**

At section 4-4, where the canal returns to its normal section, we have

Area of trapezoidal canal section

$$= (B + 1.5y)\,y$$

$$= (20 + 1.5 \times 1.5)\,1.5 = 22.5 \times 1.5 = 33.75\ m^2$$

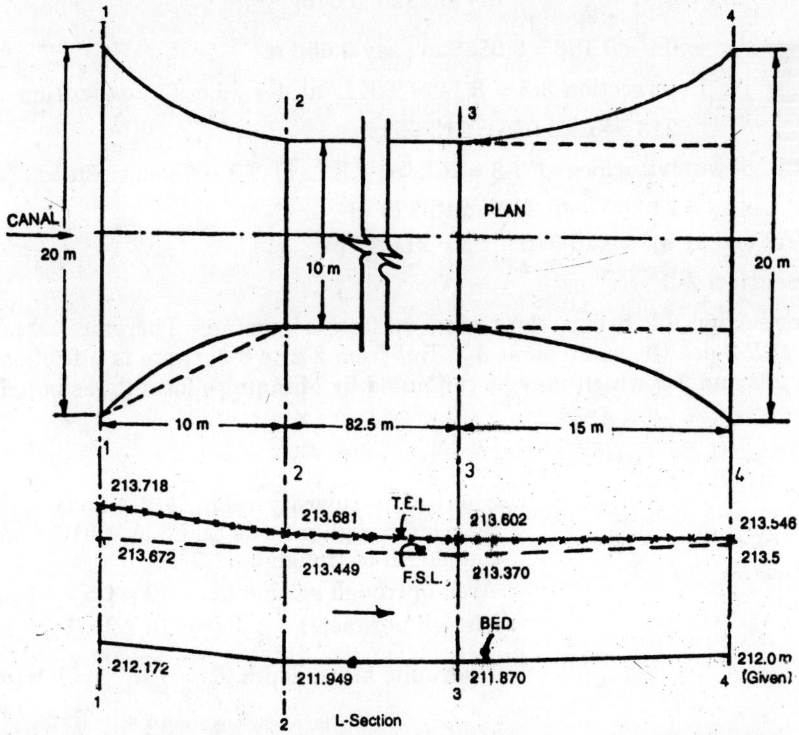


Fig. 14.20. Plan and Section of Canal Trough in Example 14.1.

Velocity $= V_4 = \left(\frac{Q}{A}\right) = \frac{32}{33.75} = \mathbf{0.947\ m/sec}$

Velocity head $= \frac{V_4^2}{2g} = \frac{(0.947)^2}{2 \times 9.81} = 0.046$ m

R.L. of bed at 4-4 = **212.0 m** (given)

R.L. of water surface at 4-4 = 212.0 + 1.5 = **213.5 m**

R.L. of T.E.L. at 4.4 = 213.5 + 0.046 = 213.546 m

The known condition of 4-4 shall now be utilised for finding the bed levels etc. at 3.3.

**At Section 3-3**

Keeping the same depth of 1.5 m throughout the channel, we have at section 3.3 :

Bed width = 10 m

Area of channel = 10 × 1.5 = 15 sq m

$$\text{Velocity} = V_3 = \frac{32}{15} = 2.13 \text{ m/sec}$$

Velocity head $= \frac{V_3^2}{2g} = \frac{(2.13)^2}{2 \times 9.81} = 0.232$ m

Assuming that the loss of head in expansion from section 3-3 to section 4-4 is taken as

$$= 0.3 \left[\frac{V_3^2 - V_4^2}{2g}\right] = 0.3\ [0.232 - 0.046]$$

$$= 0.3 \times 0.186 = 0.0558 \text{ m ; say } \mathbf{0.056\ m}$$

R.L. of T.E.L. at section 3-3 = R.L. of T.E.L. at 4-4 + Loss in expansion

= 213.546 + 0.056 = 213.602 m

∴ R.L. of water surface at 3-3 = R.L. of T.E.L. at 3-3 – Velocity Head

= 213.602 – 0.232 = **213.370 m**

R.L. of bed at 3.3 = 213.370 – 1.5 = **211.87 m**

**At Section 2-2**

From section 2-2 to 3-3, the trough section is constant. Therefore, area and velocity at 2-2 are the same as at 3-3. But from 2-2 to 3-3, there is a friction loss between 2-2 and 3-3 which may be computed by Manning's formula as equal to

$$H_L = \frac{n^2 \cdot V^2 \cdot L}{R^{4/3}}.$$

where $n$ is rugosity coefficient whose value in concrete trough may be taken as 0.016; and $L$ is the length of trough = 82.5 m.

Area of trough section ($A$) = 10 × 1.5 = 15 sq m

Wetted perimeter ($P$) = 10 + 2 × 1.5 = 13 m

Hydraulic mean depth ($R$) $= \frac{A}{P} = \frac{15}{13} = 1.16$ m

Velocity in trough $= \frac{Q}{A} = \frac{32}{15} = 2.13$ m/sec

$$\therefore \quad H_L = \frac{(0.016)^2 \times (2.13)^2 \times 82.5}{(1.16)^{4/3}} = 0.0787 \text{ m; say } \mathbf{0.079 \text{ m}}$$

R.L. of T.E.L. at 2-2 = R.L. of T.E.L. at 3-3 + Friction loss in trough

= 213.602 + 0.079 = 213.681 m

R.L. of water surface at 2-2

= 213.681 – 0.232 = **213.449 m**

R.L. of bed at 2-2

= 213.449 – 1.5 = **211.949 m**

**At Section 1-1**

Loss of head in contraction transition from 1-1 to 2-2

$$= 0.2\left(\frac{V_2^2 - V_1^2}{2g}\right)$$

$$= 0.2\left[\frac{(2.13)^2 - (0.947)^2}{2 \times 9.81}\right] = 0.2\,[0.232 - 0.046] = \mathbf{0.037 \text{ m}}$$

R.L. of T.E.L. at 1-1 = R.L. of T.E.L. at 2-2 + Loss in contraction

= 213.681 + 0.037 = **213.718 m**

R.L. of water surface at 1-1

= 213.718 – 0.046 = **213.672 m**

R.L. of bed at 1-1

= 213.672 – 1.5 = **212.172 m**

All the bed levels, F.S.L. and T.E.L. are plotted in Fig. 14.20.

**Step 4. Design of Transitions**

(*a*) *Contraction Transition.* Since the depth is kept constant, the transition can be designed on the basis of Mitra's Hyperbolic transition equation (14.2) given as :

$$B_x = \frac{B_n \cdot B_f L_f}{L_f B_n - x\,(B_n - B_f)}$$

where $B_f = 10$ m

$B_n = 20$ m

$L_f = 10$ m

Substituting we get

$$B_x = \frac{20 \times 10 \times 10}{10 \times 20 - x\,(20 - 10)} = \frac{2{,}000}{200 - 10x}$$

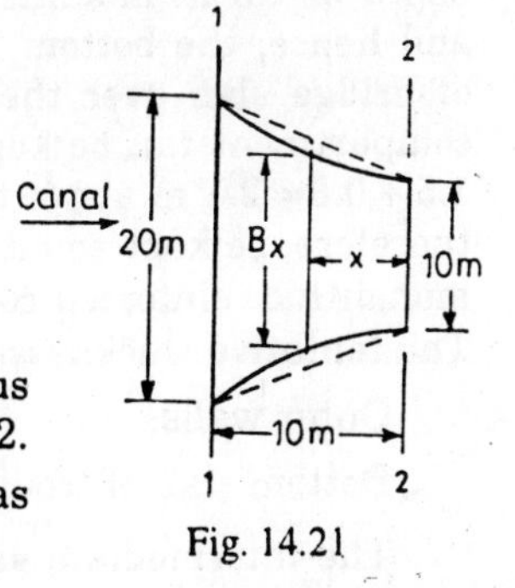

Fig. 14.21

For various values of $x$ lying between 0 to 10 m, various values of $B_x$ are worked out, as shown below in Table 14.2. The distance $x$ is measured from flumed section *i.e.* 2-2, as shown in Fig. 14.21.

**Table 14.2**

| *x in metres* | 0 | 2 | 4 | 6 | 8 | 10 |
|---|---|---|---|---|---|---|
| $B_x = \frac{2,000}{200 - 10x}$ *in metres* | 10.0 | 11.11 | 12.5 | 14.29 | 16.67 | 20.0 |

The contraction transition can be plotted with these values.

*Expansion Transition.* In this case $B_n = 20$ m, $B_f = 10$ m, and $L_f = 15$ m.

Using Eqn. (14.2), we get

$$B_x = \frac{B_n \cdot B_f \cdot L_f}{L_f \cdot B_n - x\,(B_n - B_f)}$$

$$= \frac{20 \times 10 \times 15}{15 \times 20 - x\,(20 - 10)} = \frac{3,000}{300 - 10x}.$$

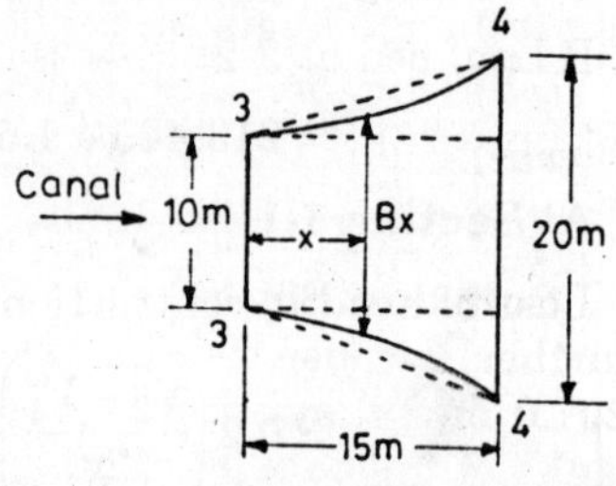


Fig. 14.22

For various values of $x$ lying between 0 to 15 m, various values of $B_x$ are worked out by using the above equation, as shown in Table 14.3.

**Table 14.3**

| *x in metres* | 0 | 2 | 4 | 6 | 8 | 10 | 12 | 14 | 15 |
|---|---|---|---|---|---|---|---|---|---|
| $B_x = \frac{3,000}{300 - 10x}$ *in metres* | 10.0 | 10.71 | 11.54 | 12.5 | 13.64 | 15.0 | 16.67 | 18.75 | 20.0 |

The expansion transition can be easily plotted with these values.

**Step 5. Design of Trough**

The trough shall be divided into two equal compartments of 5 m each and separated by an intermediate wall of 0.3 m thickness. The inspection road shall be carried on the top of left compartment as shown in Fig. 14.23.

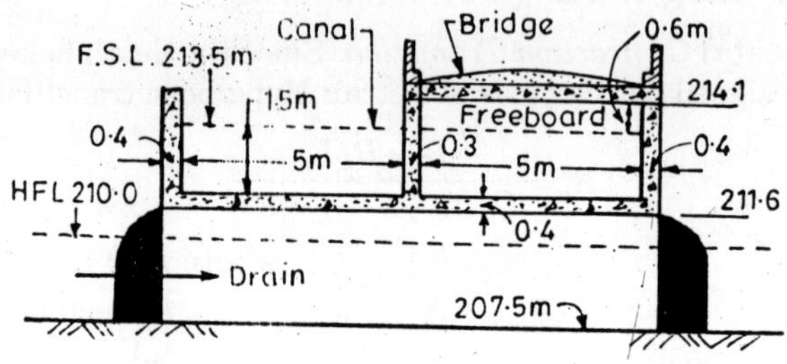


Fig. 14.23

A freeboard of 0.6 m above the normal water depth of 1.5 m is sufficient, and hence, the bottom level of bridge slab over the left compartment can be kept at 1.5 + 0.6 = 2.1 m above the bed level of the trough. The height of the trough will, therefore, be kept equal to 2.1 m. The entire trough section will be constructed in monolithic reinforced concrete and can be designed by usual structural methods. The tentative thicknesses may be used as follows :

Outer walls = 0.4 m thick

Bottom slab of trough = 0.4 m thick

The intermediate partition wall is to be extended in the transitions so as to provide the necessary clear width of 10 m. The detailed drawing of the aqueduct is illustrated in attached chart Fig. 14.24.

**Example 14.2.** *Design a syphon aqueduct if the following data at the crossing of a canal and a drainage are given :*

*(i) Discharge of canal* = *40 cumecs.*

*(ii) Bed width of canal* = *30 m.*

*(iii) Full supply depth of canal* = *1.6 m.*

*(iv) Bed level of canal* = *206.4 m.*

*(v) Side slopes of canal* = $1\frac{1}{2}$ *H : 1 V.*

*(vi) High flood discharge of drainage* = *450 cumecs*

*(vii) High flood level of drainage* = *207.0 m.*

*(viii) Bed level of drainage* = *204.5 m.*

*(ix) General ground level* = *206.5 m.*

**Solution.** Since the drainage is of a large size, work of type III will be adopted. Further, because the canal bed level (206.4 m) is slightly below the drainage HFL (207.0 m) ; **a syphon aqueduct** is required and is also asked for. The earthen banks of the canal will be discontinued and the canal water taken in a concrete trough. For affecting economy, the canal shall be flumed.

**Step 1. Design of Drainage Waterway**

Lacey's regime perimeter $= P = 4.75\sqrt{Q}$

$$= 4.75\sqrt{450} = 100.8 \text{ m}$$

Provide 11 clear spans of 8 m each and let the width of each pier be 1.5 m.

The length occupied by 11 bays of 8 m each $= 11 \times 8 = 88$ m

The length occupied by 10 piers of 1.5 each $= 10 \times 1.5 = 15$ m

Total length of waterway $= 88 + 15 =$ **103 m.**

Let us, now, limit the velocity through syphon-barrels, to a value, say 2 m/sec.

Height of barrels required

$$= \frac{\text{Discharge}}{\text{Velocity} \times \text{clear width of waterway}} = \frac{450}{2 \times 88} \text{ m} = 2.56 \text{ m.}$$

Hence, provide 11 rectangular barrels, each 8 m wide and 2.5 m high.

$$\text{Actual velocity through barrels} = \frac{450}{11 \times 8 \times 2.5} = 2.05 \text{ m/sec.}$$

**Step 2. Design of Canal Waterway**

Normal bed width of canal = 30 m

Let the width be reduced to 15 m.

Providing a splay of 2 : 1 in contraction, the length of contraction transition

$$= \frac{30 - 15}{2} \times 2 = 15 \text{ m.}$$

Providing a splay of 3 : 1 in expansion, the length of expansion transition

$$= \frac{30 - 15}{2} \times 3 = 22.5 \text{ m.}$$

Length of flumed rectangular portion of the canal between abutments = 103 m (provided). In transitions, the side slopes of the canal section shall be warped in plan from the original slope of $1\frac{1}{2} H : 1 V$ to vertical.

**Step 3. Design of Bed Levels at Different Sections** (Fig. 14.25)

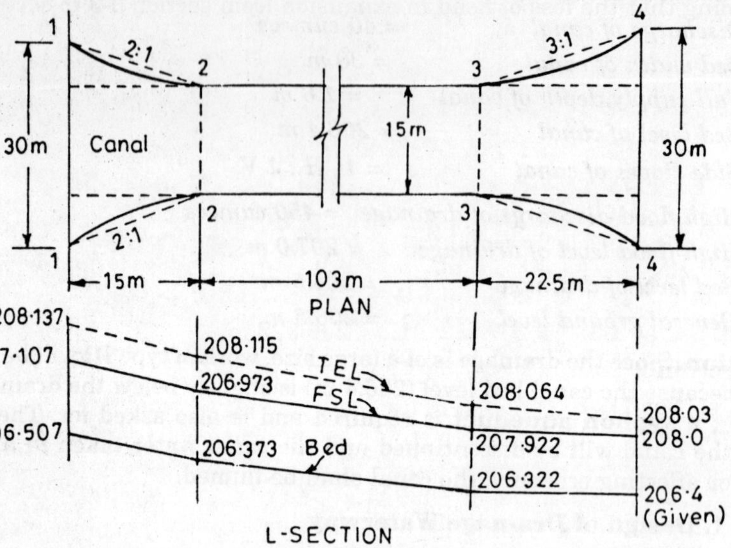


Fig. 14.25. Plan and Section of Canal Trough in Example 14.25.

**At Section 4-4**

When the canal returns to its normal section, we have the known conditions as follows :

Area of trapezoidal canal section

$$= (B + 1.5y)\, y$$

where $B$ = Bed width = 30 m

$y$ = Depth = 1.6 m

$$= [30 + 1.5 \times 1.6]\, 1.6 = 32.4 \times 1.6 = 51.84 \text{ sq. m.}$$

$$\text{Velocity of flow} = V_4 = \frac{Q}{A} = \frac{40}{51.84} = 0.77 \text{ m/sec.}$$

$$\text{Velocity head} = \frac{V_4^2}{2g} = \frac{(0.77)^2}{2 \times 9.81} = 0.030 \text{ m}$$

R.L. of canal bed at 4-4 = **206.4 m** (given)

Water depth = 1.6 m (given)

R.L. of water surface at 4-4 = 206.4 + 1.6 = **208.0 m**

R.L. of T.E.L. at 4-4 = 208.0 + 0.03 = 208.03 m.

**At Section 3-3**

Assuming a constant depth of 1.6 m throughout the channel, we have at section 3-3, a rectangular channel, as follows :

Bed width = 15 m

Depth = 1.6 m (assumed constant)

Area = 15 × 1.6 = 24 sq. m.

Velocity = $V_3$ = 40/24 = 1.67 m/sec.

$$\text{Velocity head} = \frac{V_3^2}{2g} = \frac{(1.67)^2}{2 \times 9.81} = 0.142 \text{ m}$$

Assuming that the loss of head in expansion from section 3-3 to section 4-4 is taken as

$$= 0.3\left[\frac{V_3^2 - V_4^2}{2g}\right]$$

$$= 0.3\ [0.142 - 0.030] = 0.3 \times 0.112 = 0.0336 \text{ m; } \textbf{say 0.034 m.}$$

R.L. of T.E.L. at 3-3 = R.L. of T.E.L. at 4-4 + Loss in expansion

= 208.030 + 0.034 = 208.064 m.

R.L. of water surface at 3-3

= 208.064 − 0.142 = **207.922 m.**

R.L. of bed at 3-3 = 207.922 − 1.6 = **206.322 m.**

**At Section 2-2**

From section 2-2 to section 3-3, the trough section is constant. Therefore, the area and velocity at 2-2 are the same as at 3-3. There is a friction loss between 2-2 and 3-3, which may be computed by Manning's formula, as equal to

$$H_L = \frac{n^2 V^2 L}{R^{4/3}}$$

where $n$ is rugosity coefficient, whose value in a concrete trough may be taken as 0.016 and $L$ is the length of channel = 103 m.

Area of trough section ($A$) = 15 × 1.6 = 24 sq. m

Wetted perimeter = 15 + 2 × 1.6 = 18.2 m

$$\text{Hydraulic mean depth} = R = \frac{A}{P} = \frac{24}{18.2} = 1.32 \text{ m}$$

$$\text{Velocity in trough} = \frac{Q}{A} = \frac{40}{24} = 1.67 \text{ m/sec.}$$

$$\text{Head loss, } H_L = \frac{n^2 \cdot V^2 \cdot L}{R^{4/3}} = \frac{(0.016)^2 \times (1.67)^2 \times 103}{(1.32)^{4/3}} = \mathbf{0.051 \ m.}$$

R.L. of T.E.L. at 2-2 = R.L. of T.E.L. at 3-3 + Head loss in trough

= 208.064 + 0.051 = 208.115 m.

R.L. of water level at 2-2 = 208.115 − 0.142 = **207.973 m.**

R.L. of bed at 2-2 = 207.973 − 1.6 = **206.373 m.**

**At Section 1.1**

Loss of head in contraction transition from section 1-1 to section 2-2 may be taken as

$$= 0.2\left[\frac{V_2^2 - V_1^2}{2g}\right] = 0.2\left[\frac{(1.67)^2 - (0.77)^2}{2 \times 9.81}\right] = 0.0224 \text{ m ; } \textbf{say 0.022 m.}$$

R.L. of T.E.L. at 1-1 = R.L. of T.E.L. at 2-2 + Loss in contraction

= 208.115 + 0.022 = 208.137 m

R.L. of water surface at 1-1

= 208.137 − 0.030 = **208.107 m.**

R.L. of bed at 1-1 required to maintain constant depth

= 208.107 − 1.6 = **206.507 m.**

All these levels are plotted and shown in Fig. 14.25.

**Step 4. Design of Transitions**

(*a*) *Contraction Transition.* Since depth is kept constant, the transition shall be designed on the basis of Mitra's Hyperbolic transition equation, (14.2) given by

$$B_x = \frac{B_n \cdot B_f \cdot L_f}{B_n \cdot L_f - x\,(B_n - B_f)}$$

where $B_f = 15$ m ; $B_n = 30$ m ; $L_f = 15$ m.

Substituting, we get

$$B_x = \frac{30 \times 15 \times 15}{30 \times 15 - x\,(30 - 15)} = \frac{6750}{450 - 15x} = \frac{450}{30 - x}$$

For various values of $x$ lying between 0 to 15 m, various values of $B_x$ are worked out, as shown in Table 14.4. The distance $x$ is is measured from the flumed section 2-2.

**Table 14.4**

| $x$ *in metres* | 0 | 2 | 4 | 6 | 8 | 10 | 12 | 14 | 15 |
|---|---|---|---|---|---|---|---|---|---|
| $B_x = \frac{450}{30 - x}$ *in metres* | 15.0 | 16.04 | 17.27 | 18.72 | 20.42 | 22.5 | 25.0 | 28.1 | 30.0 |

*Expansion Transition.* In this case, we have

$B_n = 30$ m

$B_f = 15$ m

$L_f = 22.5$ m

$$B_x = \frac{B_n \cdot B_n \cdot L_f}{B_n \cdot L_f - x\,(B_n - B_f)} \quad \textit{i.e. } \text{Eq. (14.2)}$$

$$= \frac{30 \times 15 \times 22.5}{30 \times 22.5 - x\,(30 - 15)} = \frac{675}{45 - x}$$

For various values of $x$ lying between 0 to 22.5 m, corresponding values of $B_x$ are worked out, as shown in Table 14.5. The distance $x$ is measured from the flumed section 3-3.

**Table 14.5**

| $x$ *in metres* | 0 | 2 | 4 | 6 | 8 | 10 | 12 | 14 | 16 | 18 | 20 | 22.5 |
|---|---|---|---|---|---|---|---|---|---|---|---|---|
| $B_x = \frac{675}{45 - x}$ *in metres* | 15.0 | 15.7 | 16.46 | 17.3 | 18.25 | 19.3 | 20.4 | 21.75 | 23.3 | 25.0 | 27.0 | 30.0 |

**Step 5. Design of Trough**

The trough shall be divided into three equal compartments, each 5 m wide, separated by 0.3 m thick partition walls (2 Nos.). The inspection road (5 m wide) shall be carried on the extreme left compartment, as shown in Fig. 14.28. A free-board of 0.6 m above the normal water depth of 1.6 m is sufficient, and hence, the bottom level of bridge slab may be kept at 1.6 + 0.6 = 2.2 m above the bed level of the trough. The height of the trough will also be kept as equal to 2.2 m. The entire trough section can be designed as monolithic reinforced concrete structure by the usual structural methods. The tentative thicknesses may be used as follows :

Outer walls = 0.4 m thick

Bottom slab of trough = 0.4 m thick.

The intermediate walls shall be extended into transitions, so as to provide the necessary clear width of 15 m.

Now, the overall outer width of trough (including walls)

$$= 15 + 2 \times 0.3 + 2 \times 0.4 = 15 + 0.6 + 0.8 = \mathbf{16.4\ m.}$$

Hence, the length of syphon barrel = **16.4 m**

**Step 6. Head Loss Through the Syphon Barrels**

The head loss through the syphon barrels is given by Unwin's formula as equal to (neglecting vel. of approach)

$$h = \left[1 + f_1 + f_2 \cdot \frac{L}{R}\right]\frac{V^2}{2g} \quad \textit{i.e.}\ \text{Eq. (14.1)}$$

where $V$ = velocity through barrels = 2.05 m/sec

$f_1$ = coefficient of loss of head at entry,

= 0.505 for unshaped mouth.

$f_2 = a\left(1 + \frac{b}{R}\right)$ where the values of $a$ and $b$ are taken from table 14.1 for cement plastered barrels as

$a$ = 0.00316

$b$ = 0.030

$R$ = Hydraulic mean depth for barrel.

$$= \frac{A}{P} = \frac{8 \times 2.5}{2\,(8 + 2.5)} = \frac{20}{21} = 0.953$$

$L$ = Length of barrel = 16.4 m.

Substituting these values, we get

$$f_2 = 0.00316\left[1 + \frac{0.030}{0.953}\right] = \mathbf{0.00326}$$

$$\therefore \quad h = \left[1 + 0.505 + 0.00326\left(\frac{16.4}{0.953}\right)\right]\frac{(2.05)^2}{2 \times 9.81} = \mathbf{0.333\ m}$$

High flood level of Drainage is given = 207.0 m

∴ d/s H.F.L. = 207.0 m

Afflux ($h$) = 0.333 m

u/s H.F.L. = d/s H.F.L. + Afflux (or loss of head)

= 207.0 + 0.333 = **207.333 m.**

**Step 7. Uplift Pressure on Roof of barrels**

R.L. of bottom of trough = R.L. of canal bed – Slab thickness

= 206.4 – 0.4 = 206.0 m

$$\text{Loss of head at entry of barrel} = 0.505\,\frac{V^2}{2g} = 0.505 \times \frac{(2.05)^2}{2 \times 9.81} = 0.108\ \text{m.}$$

Uplift on the roof

= u/s H.F.L. – Loss at entry – Level of underside of roof slab

= 207.333 – 0.108 – 206.0

= 1.225 m of water = 12.25 kN/m $^2$ (1.225 t/m$^2$)

(Assuming unit wt. of water = 10 kN/m$^3$ or 1 t/m$^2$)*

The concrete trough slab is 0.4 m thick and will thus exert a downward load of

$0.4 \times 24 = 0.96$ kN/m$^2$ (assuming unit wt. of concrete = 24 kN/m$^2$)

The balance of the uplift pressure *i.e.* 12.25 – 9.6 = 2.65 kN/m$^2$ has to be resisted by the reinforcement to be provided at the top in the roof slab. The roof slab has to be designed for full canal water load (1.6 m of water) plus self weight, when the drainage water is low and not exerting any uplift. Suitable reinforcement at bottom of the slab may be provided for this downward force, as worked out below :

**Step 8. Design of roof of barrel**

Uplift to be balanced by top reinforcement = 2.65 kN/m$^2$

Downward water load acting when there is no uplift = 1.6 m of water = 16 kN/m $^2$

Load due to self wt. of slab = $0.4 \times 24 = 9.6$ kN/m$^2$

Total downward load = 16 + 9.6 = 25.6 kN/m$^2$

An intermediate wall of 0.3 m thickness has been provided in the trough, having clear span of 5 m, the effective span is, therefore, = 5 + 0.3 = 5.3 m.

Let us consider 1 m wide strip of slab.

Maximum sagging bending moment in slab due to downward loads

$$= \frac{25.6 \times (5.3)^2}{10} \text{ kN–m} = 71.7 \text{ kN–m} = 71.7 \times 10^5 \text{ N.cm}$$

Maximum Hogging moment due to residual uplift acting from below

$$= \frac{2.65 \times (5.3)^3}{10} \text{ kN–m} = 7.42 \text{ kN–m} = 7.42 \times 10^5 \text{ N.cm}$$

$$\text{Max. shear force} = \frac{wl}{2} = \frac{25.6 \times 5.0}{2} \text{ kN} = 64 \text{ kN}$$

Using 1 : 2 : 4 cement concrete, we have

Effective depth ($d$) of slab required

$$d = \sqrt{\frac{M}{Qb}} = \sqrt{\frac{71.7 \times 10^5}{87 \times 100}} = \mathbf{28.8\ cm}$$

Provided overall thickness is 40 cm and thus provided effective depth

$d = 37.5$ cm.

Steel required at the bottom of the slab

$$= \frac{71.7 \times 10^5}{12000 \times 0.87 \times 37.5} \text{ cm}^2\text{/m length} = 18.3 \text{ cm}^2$$

(using reduced stress in steel as 12000 N/cm$^2$)

* Strictly speaking, unit wt. of water = 9.81 kN/m$^3$ ; but to ease in calculations we have taken unit wt. of water = 10 kN/m$^3$ (1 t/m$^3$) and of concrete = 24 kN/m$^3$.

Provide 16 mm $\phi$ bars @ 10 cm c/c in the bottom of the slab.

Steel required at top

$$= \frac{7.42 \times 10^5}{12000 \times 0.87 \times 37.5}$$

$= 1.9\ cm^2$ (Nominal)

Provide 10 mm $\phi$ bars @ 15 cm/cc.

Also provide 10 mm $\phi$ bars @ 20 cm c/c as distribution reinforcement both at top as well as at bottom, as shown in Fig. 14.26.

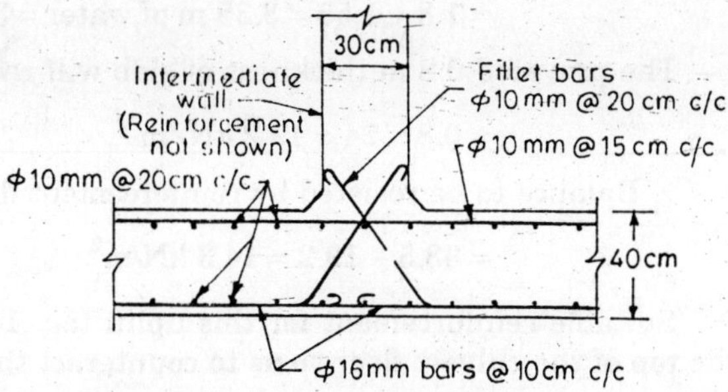


Fig. 14.26. Reinforcement in roof slab.

**Step 9. Uplift on the bottom floor of barrel**

(a) *Static Head*

R.L. of barrel floor = R.L. of trough bottom − Height of barrel

= 206.0 − 2.5 = 203.5 m.

Let us assume that a thickness of 0.8 m is provided.

∴ R.L. of bottom of floor

= 203.5 − 0.8 = 202.7 m

Bed level of drain = 204.5 m.

Assuming that the water-table has gone upto bed level of drain, the static uplift on the floor (refer Fig. 14.18)

= 204.5 − 202.7 = 1.8 m of water

(b) *Seepage Head.* The seepage head will be maximum when the canal is running full and the drain is dry.

Thus, the total seepage head = F.S.L. of canal − bed level of drain

= 208.0 − 204.5 = 3.5 m.

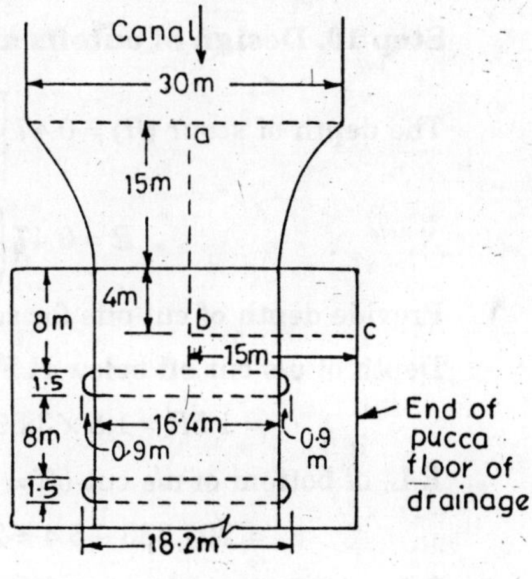


Fig. 14.27

The residual seepage head at a point '*a*' in the centre of the first barrel (Fig. 14.27) has been calculated by Bligh's theory, as follows :

Assuming that the total length of drainage floor = 30 m.

The seepage line *abc* will traverse creep lengths as follows :

*ab* = Length of *u/s* transition + Half the barrel span = 15 + 4 = 19 m.
*bc* = 15 m (Half of the total length of 30 m = assumed)

Total creep length = 19 + 15 = 34 m.

Residual seepage head at $b = 3.5\left[1 - \frac{19}{34}\right] = 3.5 \times \frac{15}{34} = 1.55$ m.

Total uplift = Static head + Seepage head

$$= 1.8 + 1.55 = 3.35 \text{ m of water} = 33.5 \text{ kN/m}^2$$

The provided 0.8 m thickness of slab will resist due to its own wt., an uplift

$$= 0.8 \times 24 = 19.2 \text{ kN/m}^2.$$

∴ Balance to be resisted by reinforcement due to bending action

$$= 33.5 - 19.2 = 14.3 \text{ kN/m}^2.$$

Suitable reinforcement for this uplift (*i.e.* 14.3 kN/m$^2$) has to be provided at the top of the culvert floor so as to counteract the bending action.

**Note.** The length of the floor has been provided equal to 32 m, as shown in Fig. 14.28, on the following considerations.

| | | |
|---|---|---|
| (*i*) | Length of floor required under barrel | = 16.4 m |
| (*ii*) | Extra floor length required to accommodate pier noses on both sides = 2 × 1.0 | = 2.0 m |
| (*iii*) | Horizontal length of d/s ramp joining to bed level at a slope of 5 : 1 = 5 (204.5 – 203.5) | = 5.0 m |
| (*iv*) | Width of d/s cut-off beyond ramp | = 0.6 m |
| (*v*) | Length of extra floor provided on u/s side | = 0.6 m |
| | | Total length = 30.0 m |

**Step 10. Design of cutoffs and protection works for the drainage floor**

The depth of scour ($R$) $= 0.47\left[\dfrac{Q}{f}\right]^{1/3}$ ; assuming $f = 1.0$

$$\therefore \qquad R = 0.47\left(\frac{450}{1}\right)^{1/3} = 0.47 \times 7.65 = 3.59 \text{ m}$$

Provide depth of cut-offs for scour hole of 1.5 $R$ on both sides

Depth of u/s cut off below H.F.L.

$$= 1.5R = 1.5 \times 3.59 = 5.4 \text{ m}$$

R.L. of bottom of u/s cut-off = u/s H.F.L. – 5.4 m

= 207.333 – 5.4 = 201.933 m ; say **201.93 m.**

R.L. of bottom of d/s cut-off = d/s H.F.L. – 5.4

= 207.0 – 5.4 = **201.6 m**

Length of u/s protection (*i.e.* 40 cm thick brick pitching)

= 2 [R.L. of u/s bed – R.L. of bottom of u/s cut-off]

= 2 [203.50 – 201.93]

= 2 × 1.57 = 3.14 m ; say **3.2 m**

Similarly, length of d/s brick pitching

= 2 [R.L. of d/s bed – R.L. of bottom of d/s cutoff]

= 2 [204.5 – 201.6] = 2 × 2.9 = **5.8 m**

The pitchings may be supported by 0.4 m wide and 1 m deep toe walls, as shown in Fig. 14.28.

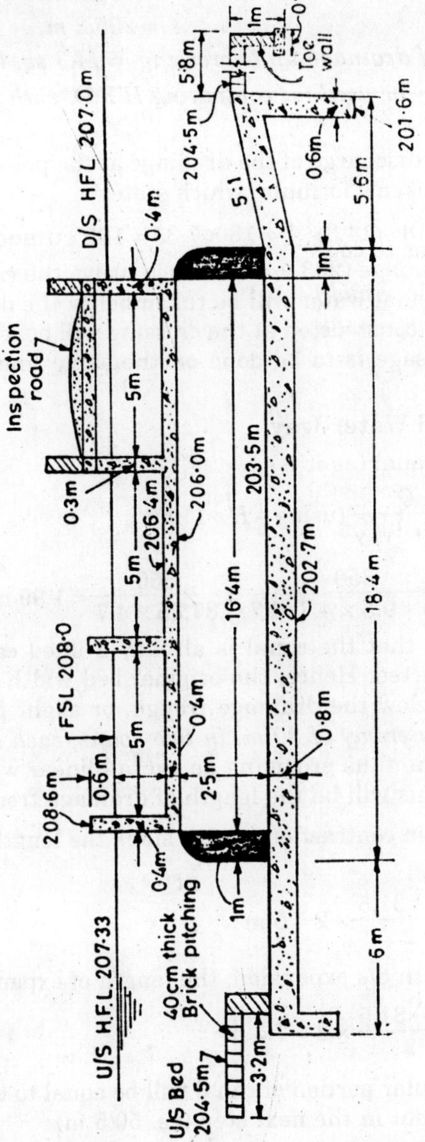


Fig. 14.28. Section of Syphon Aqueduct in Example 14.2

**Example 14.3.** *Design a suitable cross drainage work, given the following data at the site of the crossing of two streams of water :*

*(a) Irrigation channel :*

| | |
|---|---|
| *Full supply discharge* | *= 350 cumecs* |
| *Full supply level* | *= 202.5 m.* |
| *Canal bed level* | *= 197.8 m.* |
| *Canal bed width* | *= 35 m.* |
| *Full supply depth* | *= 4.7 m.* |
| *Side slopes* | *= $\frac{1}{2}H : IV$* |

*(b) Natural Drainage :*

*Drainage bed level* = *203.9 m.*

*High flood level* = *205.2 m.*

*Catchment area of drainage up to crossing* = *14.3 sq. km.*

*The Dicken's formula may be used for computing H.F.Q. with its coefficient as 18 (SI or MKS units).*

**Solution.** The high flood discharge of the drainage at the point of the crossing may be obtained by using Dicken's formula, which states

$$Q = C \cdot A^{3/4} = 18 \cdot (14.3)^{3/4} = 18 \times 7.33 = \mathbf{132\ cumecs.}$$

Since the bed level of drainage (203.9 m) is much above the canal *FSL* (202.5 m), *i.e.* by about 1.4 m ; the canal water will be taken below the drainage. Hence, the cross-drainage work to be constructed at the crossing will be a **super-passage.** The design of the super passage is to be done on the same lines as that of an aqueduct, and is given below :

**Step 1. Design of Canal Water-way**

The flow velocity in the canal (as it is)

$$\frac{\text{Discharge}}{\text{Area}} = \frac{Q}{(B + \frac{1}{2}y)\,y} \text{ (using } \tfrac{1}{2} H : 1V)$$

$$= \frac{350}{(35 + 0.5 \times 4.7)\,4.7} = \frac{350}{37.35 \times 4.7} = 1.99 \text{ m/sec.}$$

This high velocity shows that the canal is already a lined canal, and much more fluming can not be affected. Hence, the original bed width of 35 m can be continued as canal barrels below the drainage trough, or slight fluming may be done. *Let us adopt a clear waterway of 30 m, in two spans, each of 15 m,* with a central pier of width say 1.5 m, thus providing an overall linear waterway of 31.5 m between abutments, and this will be the length of drainage troughs.

Providing a splay of 2 : 1 in contraction (on u/s side), the length of contraction transition

$$= \frac{35 - 31.5}{2} \times 2 = 5 \text{ m}$$

Providing a splay of 3 : 1 in d/s expansion, the length of expansion transition

$$= \frac{35 - 31.5}{2} \times 3 = 7.5 \text{ m}$$

Length of flumed rectangular portion of canal will be equal to the width of the drainage troughs, as worked out in the next step (*i.e.* 50.5 m).

The piers, abutments, wing walls and return walls of the canal will be designed as those of a bridge, taking the load of drainage trough (including the load of water and inspection road, etc.), instead of a bridge deck slab.

**Step II. Design of Drainage Waterway**

Lacey's regime perimeter $P = 4.75\sqrt{Q} = 4.75\sqrt{132} = 54.5$ m.

The total length of the waterway provided is generally chosen equal to *P*, although it can be slightly reduced to affect economy, but too much contraction of the drainage poses problems, and hence too much fluming is never done.

Let us provide 6 R.C.C. compartments, each of clear width equal to 8.0 m, thus giving

Clear waterway = 8.0 × 6 = 48 m.

Using 5 partition walls of 0.3 m thick each,

length occupied by walls = 5 × 0.3 = 1.5 m

∴ Total length of waterway provided = 48 + 1.5 = **49.5 m**; against P of 54.5 m (which is not too less as to cause troubles)

The two side walls of the R.C.C. drainage trough may be kept 0.4 m thick each, with 49.5 m as aggregate waterway between them. Thus, end to end length of drainage trough = 49.5 + 0.8 = 50.5 m.

Thus, the length of the rectangular portion of canal will also be equal to 50.5 m.

Since the drainage has also been slightly flumed and kept lesser than *P*, so let us design its contraction and expansion lengths.

Assuming 2 : 1 convergence, we have the length of contraction transition

$$= \frac{54.5 - 49.5}{2} \times 2 = 5 \text{ m}$$

Assuming 3 : 1 splay in expansion, we have the length of expansion transition

$$= \frac{54.5 - 49.5}{2} \times 3 = 7.5 \text{ m}$$

The wing walls will be constructed to reduce the drainage waterway width from 54.5 m to 49.5 m on upstream, and return walls will be constructed to expand the drainage waterway from 49.5 m to 54.5 m on downstream. These wings will be extended so as to enter the berms of the drain.

The length of the drainage pucca rectangular trough will be equal to 31.5 m (*i.e.* equal to the rectangular waterway of canal). A line plan of the crossing is shown in Fig. 14.29.

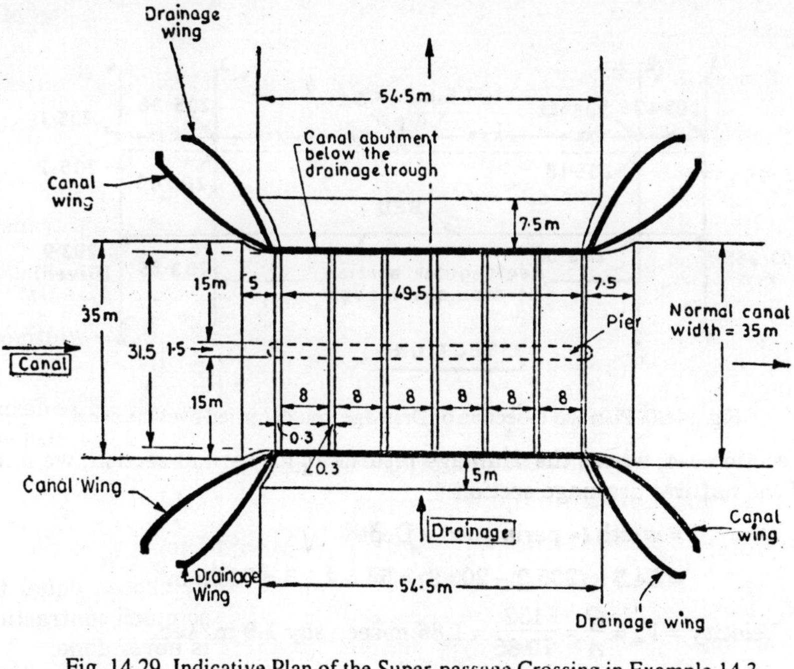


Fig. 14.29. Indicative Plan of the Super-passage Crossing in Example 14.3.

**Step 3. Head Loss and Bed Levels at different Sections along the length of drainage trough.**

**At section 4-4** (Refer Fig. 14.30)

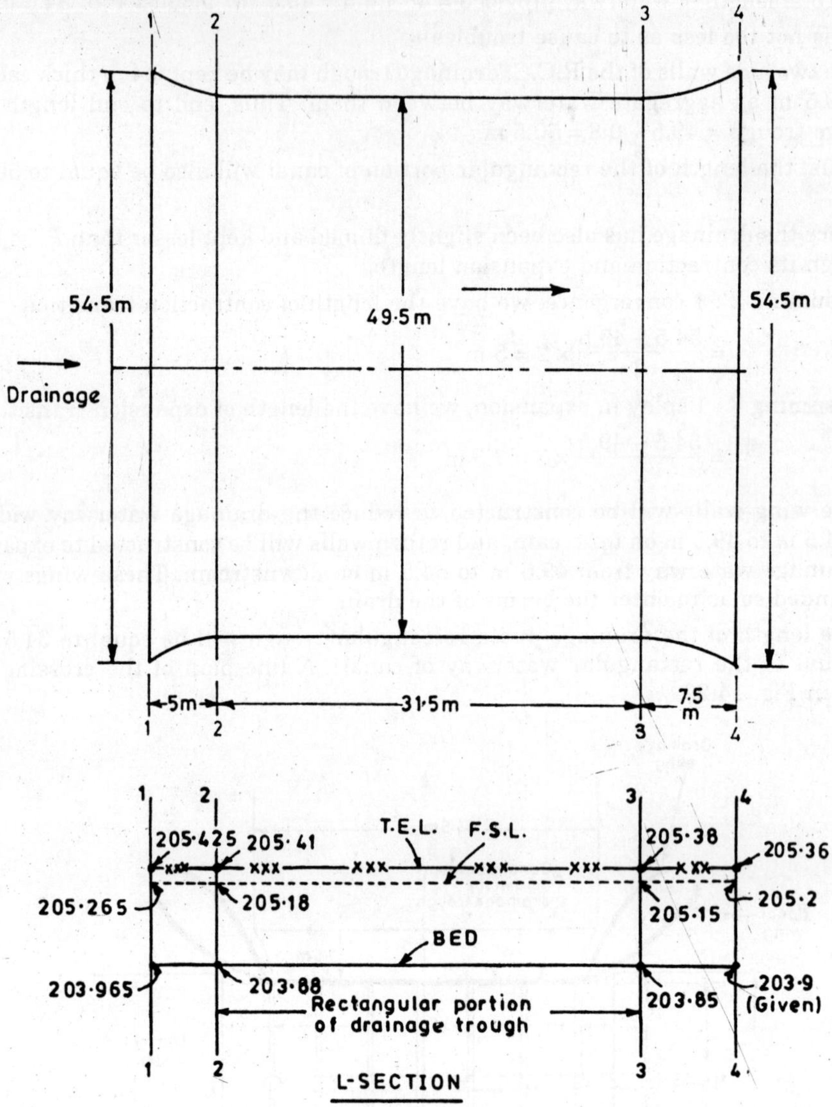


Fig. 14.30. Plan and L-section of Drainage trough carried over the canal.

**At section 4-4, where the drainage returns to its normal section, we have the area of the natural drainage section**

$$= \text{width } (\approx \text{perimeter}) \times \text{Depth}$$

$$= 54.5 \times (205.2 - 203.9) = 54.5 \times 1.3 = 70.85 \text{ m}^2$$

$$\text{Velocity} = V_4 = \frac{Q}{A} = \frac{132}{70.85} = 1.86 \text{ m/sec.; say } \mathbf{1.9 \text{ m/sec.}}$$

Velocity Head $= \frac{V_4^2}{2g} = \frac{(1.9)^2}{2 \times 9.81} = 0.16$ m

R.L. of bed at 4-4 = **203.9 m** (given)

R.L. of water surface at 4-4

= 203.9 + 1.3 (as the normal depth in drainage = 1.3 m)
= 205.2 m

R.L. of *TEL* at 4-4 = 205.2 + 0.16 = 205.36 m

**At Section 3-3** (Refer Fig. 14.30)

Keeping the same depth of 1.3 m throughout the drainage, we have at section 3-3.

Clear waterway = 8 × 6 = 48 m

Area of flow = 48 × 1.3 = 62.4 m$^2$

Velocity $= \frac{132}{62.4} = 2.12$ m/sec.

Velocity Head $= \frac{(2.12)^2}{2 \times 9.81} = 0.23$ m.

Assuming that the loss of head in expansion from section 3-3 to section 4-4 is taken as $= 0.3\left(\frac{V_3^2 - V_4^2}{2g}\right)$

= 0.3 (0.23 – 0.16) = 0.3 × 0.07 = 0.021 m; say **0.02 m**

∴ R.L. of TEL at 3-3

= R.L. of TEL at section 4-4 + Loss in expansion
= 205.36 + 0.02 = 205.38 m

∴ R.L. of water surface at 3-3

= R.L. of TEL at 3-3 – Velocity head
= 205.38 – 0.23 = **205.15 m**

∴ R.L. of bed at 3-3 = 205.15 – 1.3 = **203.85 m**

**At Section 2-2**

From section 2-2 to 3-3, the trough section is constant. Therefore, area and velocity at 2-2 are the same as at 3-3. But from 2-2 to 3-3, there is a friction loss between 2-2 and 3-3, which may be computed by Mannings formula, as :

$$H_L = \frac{n^2 \cdot V^2 \cdot L}{R^{4/3}}$$

where $n$ is the rugosity coefficient whose value in concrete trough may be taken as 0.016;

and $L$ is the length of trough = 31.5 m.

The area of trough section ($A$) = 48 × 1.3 = 62.4 sq m

Wetted perimeter ($P$) = 48 + 2 × 1.3 = 50.6 m

Hydraulic mean depth ($R$) $= \frac{A}{P} = \frac{62.4}{50.6} = 1.23$ m

Velocity in trough $= \frac{Q}{A} = \frac{132}{62.4} = 2.12$ m/sec

$$\therefore \qquad H_L = \frac{(0.016)^2 \times (2.12)^2 \times 31.5}{(1.23)^{4/3}} = 0.0275 \text{ m} \text{ ; } \textbf{say 0.03 m}$$

R.L. of TEL at 2-2 = R.L. of TEL at 3-3 + Friction loss ($H_L$) in trough

$$= 205.38 + 0.03 = 205.41 \text{ m}$$

R.L. of water surface at 2-2

$$= 205.41 - 0.23 = 205.18 \text{ m}$$

∴ R.L. of bed at 2-2 = 205.18 m − 1.3 = **203.88 m**

**At Section 1-1**

Loss of head in contraction transition from 1-1 to 2-2

$$= 0.2\left(\frac{V_2^2 - V_1^2}{2g}\right) = 0.2\ [0.23 - 0.16] = \textbf{0.014 m; say 0.015 m}$$

R.L. of TEL at 1-1 = R.L. of TEL at 2-2 + Loss in contraction

$$= 205.41 + 0.015 = 205.425 \text{ m}$$

R.L. of water surface at 1-1 = 205.425 − 0.16 = 205.265 m

R.L. of bed at 1-1 = 205.265 − 1.3 = **203.965 m**

All these bed levels, FSL and TEL are plotted in Fig. 14.30.

**Step 4. Design of Transitions for the Drainage**

(*a*) *Contraction Transition.* Since the depth is kept constant, the transitions can be designed on the basis of Mitra's hyperbolic transition equation given by *eqn.* (14.2) as:

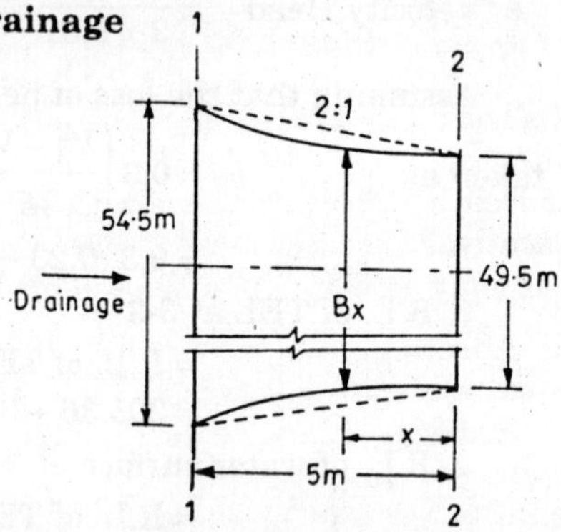


Fig. 14.31

$$B_x = \frac{B_n \cdot B_f \cdot L_f}{L_f B_n - x\,(B_n - B_f)}$$

where $B_f = 49.5$ m

$B_n = 54.5$ m

$L_f = 5$ m.

Substituting, we get

$B_x$ = width at any distance $x$ from the flumed section 2–2

$$= \frac{54.5 \times 49.5 \times 5}{5 \times 54.5 - x\,(54.5 - 49.5)} = \frac{2700}{54.5 - x}$$

For various values of $x$ lying between 0 to 5 m, values of $B$ are worked out as shown in Table 14.6.

**Table 14.6**

| $x$ *in metres* | 0 | 1 | 2 | 3 | 4 | 5 |
|---|---|---|---|---|---|---|
| $B_x = \dfrac{2700}{54.5 - x}$ *in metres* | 49.5 | 50.5 | 51.5 | 52.5 | 53.5 | 54.5 |

The contraction transition can be plotted with these values.

*Expansion Transition.* In this case $B_x$ = 54.5 m, $B_f$ = 49.5 m, $L_f$ = 7.5 m.

Using Eq. (14.2), we get

$$B_x = \frac{B_n \cdot B_f \cdot L_f}{L_f \cdot B_n - x\,(B_n - B_f)}$$

$$= \frac{154.5 \times 49.5 \times 7.5}{7.5 \times 54.5 - x\,(54.5 - 49.5)}$$

$$= \frac{4050}{81.7 - x}$$

For various values of $x$ lying between 0 to 7.5 m, various values of $B_x$ are worked out by using the above equation, as shown in Table 14.7.

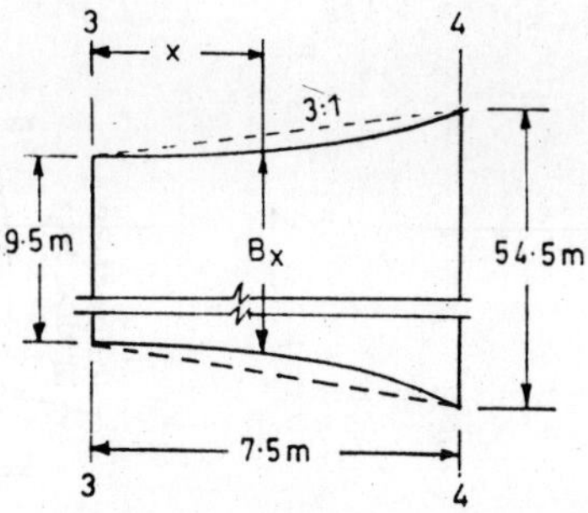


Fig. 14.32

**Table 14.7**

| *x in metres* | 0 | 2 | 4 | 6 | 7.5 |
|---|---|---|---|---|---|
| $B_x = \frac{4050}{81.7 - x}$ *in metres* | 49.5 | 50.7 | 52.0 | 53.3 | 54.5 |

The expansion transition can be plotted with these values.

**Step 5. Design of Drainage Trough**

The RCC drainage trough, as pointed out earlier, has been divided into six compartments of 8 m clear width each, and separated by intermediate walls (5 No.) each 0.3 m thick (tentative thickness). The end walls of the trough have tentatively been kept as 0.4 m wide each, as shown in Fig. 14.33. An inspection road may be carried on the top of the end compartment as shown.

A free-board of 0.6 m above the normal water depth of 1.3 m is sufficient, and hence, the bottom-level of bridge slab over the end compartment may be kept 1.3 + 0.6 = 1.9 m above the bed level of trough (Different bed levels along the length of trough are shown in Fig. 14.30.). The height of the trough walls will also be kept equal to 1.9 m. The entire trough section will be constructed in monolithic reinforced concrete, and can be designed by usual structural methods. The tentative thickness for the walls have already been assumed and should be verified. A detailed drawing for the crossing can be prepared on the same lines on which the drawing 14.24 was prepared.

**Example 14.4.** *Design a suitable cross drainage work, given the following data :*

(*a*) *Irrigation channel*

| | |
|---|---|
| *F.S.Q.* | *= 354 cumecs.* |
| *Bed width* | *= 24 m.* |
| *FSL* | *= 207.60 m.* |
| *C.B.L.* | *= 201.4 m.* |
| *Side slopes* | $= \frac{1}{2}H : IV.$ |

(*b*) *Natural drainage*

| | |
|---|---|
| *H.F.Q.* | *= 600 cumecs.* |
| *Drainage bed level* | *= 203.6 m.* |
| *H.F.L.* | *= 206.3 m.* |

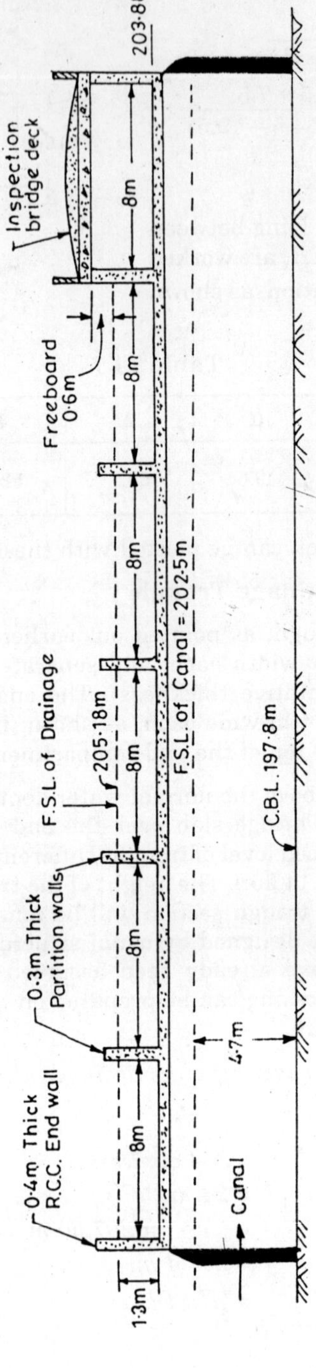


**Fig 14.33. Section of Drainage Trough passing over the Canal (Levels of Trough bed and FSL at the start of rectangular portion of Trough are shown).**

**Solution.** Since the bed of the natural drainage (203.6 m) is higher than the canal bed (201.4 m) by a sufficient margin of 2.2 m, we may think of taking the canal below the drainage. But since the FSL of canal (207.6 m), is much above the drainage bed (203.6 m), the simple super passage cannot be provided. The canal water will, therefore, have to be syphoned below the drainage. Since the discharge of the canal is low compared to that of the drainage, such a crossing will not prove uneconomical also. Hence, a **canal syphon** will be designed as below :

**Step I. Design of Canal Waterway**

Let the velocity through the canal barrels be limited to 3 m/sec.

$$\therefore \text{ Area of waterway of barrels required} = \frac{354}{3} = 118 \text{ m}^2$$

Let us provide 4 spans each having a clear waterway of 5.4 m, thus giving a clear waterway of $4 \times 5.4 = 21.6$ m. Let R.C.C. piers (three No.) each of width 0.6 m are provided ; thus occupying a width $3 \times 0.6 = 1.8$ m.

Total overall waterway provided between abutments

$= 21.6 + 1.8$

$= 23.4$ m ; which is O.K., as it is less than 24 m (*i.e.* the width of canal)

Assuming 0.3 m as the width of each bearing on abutments, the total length of drainage trough $= 23.4 + 0.6 =$ **24 m.**

Thus, the canal on the u/s and d/s sides is joined to the barrels aggregate width of 23.4 m by suitable transitions and approaches.

The height of the barrel

$$= \frac{\text{Discharge}}{\text{Velocity } \times \text{ Waterway width}} = \frac{354}{3 \times 21.6} = 5.47 \text{ m.}$$

So let us keep 5.5 m height of the barrels, thus maintaining an actual velocity through them $= \dfrac{354}{5.5 \times 21.6} = 2.98$ m/sec.

**Step 2. Design of Drainage Waterway**

Lacey's regime perimeter $= P = 4.75\sqrt{Q} = 4.75 \times \sqrt{600} = 116$ m.

Let us assume that the rectangular R.C.C. drainage trough is provided in such a width that its perimeter becomes equal to that required by Lacey's equation, thus giving width of the trough $= 116 - 2\ (2.7) = 110.6$ m ; say 111 m.

Let us divide the drainage trough into 9 (compartments each of 12 m width (clear width) with 0.4 m thick partition walls (8 No.). Thus the overall width of the trough provided will be equal to $9 \times 12 + 8 \times 0.4 = 108.0 + 3.2 = 111.2$ m. Two end walls each of 0.4 m thickness can be provided, thus giving the total width of drainage trough end to end = 112 m ; and this will be equal to the length of the canal barrels.

**Step 3. Design of Bed Levels along the Drainage Trough**

Since the drainage waterway has not been flumed, the only loss of head in TEL when the water flows in the trough is due to frictional loss. Assuming that the drainage levels (given) are on the d/s of the crossing.

The bed level of drainage trough (d/s) = 203.6 m.

$$\text{Velocity in trough, } V = \frac{600}{7\ (15.5 \times 2.7)} = 2.05 \text{ m/sec.}$$

Head loss due to friction $(H_L) = \frac{n^2 \cdot V^2 L}{R^{4/3}}$

$$\text{where } R = \frac{A}{P} = \frac{7\,(15.5 \times 2.7)}{7\,(15.5 + 5.4)} = 2.0$$

$$\therefore \quad H_L = \frac{(0.016)^2 \times (2.05)^2 \times 24}{(2)^{4/3}} = 0.0102 \text{ m.}$$

Hence, the trough should be given a slope $(S_0)$ [as to pass the full drainage discharge] given by

0.0102 m loss in 24 m length

1 m loss will occur in a length $= \frac{24}{0.0102}$ m = 2340 m.

Hence, the slope required to be given to trough is 1 in 2340 ; and bed level at the d/s of crossing = 203.6 m; and bed level at u/s of crossing = 203.61 m.

**Step 4. Design of Transitions**

As no fluming either of canal or of drainage has been done, the question of designing transitions does not arise. The wing walls and return walls shall, however, be provided, as shown in the attached chart Fig. 14.36, after properly ensuring their structural safety.

**Step 5. Design of Trough**

The trough shall be made of reinforced cement concrete and shall be designed by the usual structural methods. A bridge for inspection can be provided on one end compartment, as shown in Fig. 14.36. The trough is divided into 9 compartments each of 12 m clear width as already decided. The tentative thicknesses of partition walls (0.4 m) and end walls (0.4 m) have also been made.

**Step 6. Head Loss Trough Syphon Barrels**

The head loss trough syphon barrels is given by Eq. (14.1) as

$$h = \left(1 + f_1 + f_2 \cdot \frac{L}{R}\right)\frac{V^2}{2g} \quad \text{(neglecting velocity of approach)}$$

where $V$ is the velocity through the syphon barrels = 2.98 m/sec.

$f_1$ = coeff. of loss of head at entry = 0.505 for unshaped mouth

$f_2 = a\left[1 + \frac{b}{R}\right]$ where the values of $a$ and $b$ are taken from Table 14.1 for cement plastered barrels as

$a$ = 0.00316

$b$ = 0.030

$R$ = Hydraulic mean depth for barrel

$$= \frac{A}{P} = \frac{5.4 \times 5.5}{2\,(5.4 + 5.5)} = \frac{5.4 \times 5.5}{21.8} = 1.36 \text{ m}$$

$L$ = Length of barrels = 112 m.

Substituting these values, we get

$$f_2 = 0.00316\left[1 + \frac{0.030}{1.36}\right] = 0.00323$$

$$\therefore \quad h = \left[1 + 0.505 + 0.00323\left(\frac{112}{1.36}\right)\right]\frac{(2.98)^2}{2 \times 9.81} = 0.855; \text{ say } 0.86 \text{ m.}$$

FSL of canal (given) = 207.6 m

∴ d/s FSL = 207.6 m

Afflux ($h$) = 0.86 m.

∴ u/s FSL = d/s FSL + Afflux = 207.6 + 0.86 = **208.46 m**

**Step 7. Uplift Pressure on Roof of Barrels**

R.L. of bottom of trough

= R.L. of drainage bed – Slab thickness

= 203.6 – 0.8 (assuming 0.8 m thick roof slab)

= 202.8 m

Loss of head at entry of barrel $= 0.505 \dfrac{V^2}{2g} = \dfrac{0.505 \times (2.98)^2}{2 \times 9.81} = 0.23$ m

Uplift on the roof = u/s HFL – Loss at entry – Level of under-side of roof slab

= 208.46 – 0.23 – 202.8

= 5.43 m of water head = 54.3 kN/m$^2$.

The concrete trough slab is 0.8 m thick and will thus exert a downward load of $0.8 \times 24 = 19.2$ kN/m $^2$.

The balance of the uplift pressure, *i.e.* $54.3 - 19.2 = 35.1$ kN/m$^2$ has to be resisted by reinforcement to be provided at the top in the roof slab. The roof slab has to be designed for full drainage water load (2.7 m of water) plus self-weight, when the canal water is low and not exerting any uplift. Suitable reinforcement, may be provided for this downward force, as worked out below :

**Step 8. Design of roof barrel**

Uplift to be balanced by top reinforcement = 35.1 kN/m$^2$

Downward water load acting when there is no uplift

= 2.7 m of water = 27 kN/m$^2$

Load due to self-weight of slab $= 0.8 \times 24 = 19.2$ kN/m$^2$

Total downward load (when there is no uplift) $= 27 + 19.2 = 46.2$ kN/m$^2$

Walls of 0.4 m thickness have been provided in the trough having clear span of 12 m ; the effective span between any two walls is, therefore = 12 + 0.4 = 12.4 m

Let us consider 1 m wide strip of slab.

Maximum sagging bending moment in slab due to downward loads (when there is no uplift)

$$= \frac{46.2\,(12.4)^2}{10} \text{ kN.m. } \left(i.e.\ \frac{wl^2}{10}\right) = 708 \text{ kN–m} = 708 \times 10^5 \text{ N.cm}$$

Maximum hogging bending moment due to residual uplift acting from below

$$= \frac{35.1\,(12.4)^2}{10} \text{ kN} \cdot \text{m} = 538 \text{ kN} \cdot \text{m} = 538 \times 10^5 \text{ N} \cdot \text{cm}$$

Max. shear force $= \dfrac{wl}{2} = \dfrac{46.2 \times 12.4}{2}$ kN = 287 kN

Using 1 : 2 : 4 cement concrete, we have

Economical Effective depth ($d$) of slab required

$$d = \sqrt{\frac{M}{Qb}} = \sqrt{\frac{708 \times 10^5}{87 \times 100}} \text{ cm} = 90 \text{ cm.}$$

Provided overall thickness is 80 cm and may be kept the same. Thus, effective depth provided, $d = 77.5$ cm.

Steel required at the bottom of the slab

$$= \frac{708 \times 10^5}{12000 \times 0.87 \times 77.5} \text{ cm}^2/\text{m length}$$

(using reduced stress in steel as 12000 kN/cm²)

$$= 87.8 \text{ cm}^2/\text{m length of slab.}$$

Provide 32 mm dia bars @ 9 cm c/c in the bottom of the slab.

Steel required at top

$$= \frac{538 \times 10^5}{1200 \times 0.87 \times 77.5} = 66.7 \text{ cm}^2/\text{m length of slab.}$$

Provide 32 mm dia bars @ 12 cm c/c.

Also provide 0.15% of gross cross-sectional area of concrete as distribution steel.

∴ Area of Dist. steel required

$$= \frac{0.15 \times 80 \times 100}{100} \text{ cm}^2 = 12 \text{ cm}^2$$

∴ Spacing of 12 mm ϕ bars required

$$= \frac{\text{Area of one bar} \times 100}{\text{Area required}} = \frac{1.13 \times 100}{12} = 9.4 \text{ cm; say 10 cm c/c}$$

Hence, use 12 mm ϕ bars @ 10 cm c/c as distribution steel, both at top as well as bottom, as shown in Fig. 14.34.

The partition walls will be designed as beams carrying the bending load from the roof slab.

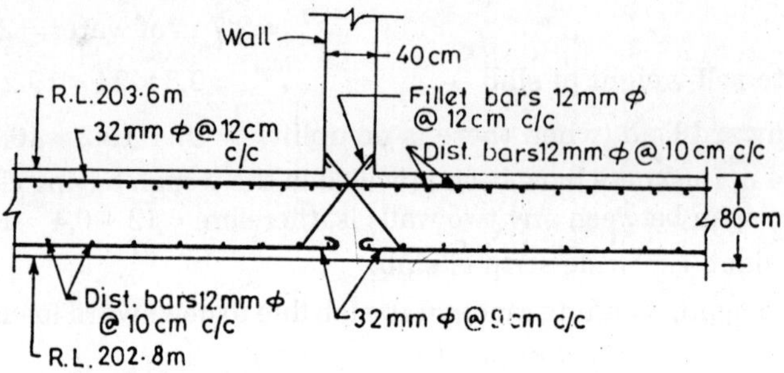


Fig. 14.34. Reinforcement in roof slab.

**Step 9. Uplift on the bottom floor of syphon barrels**

(*a*) *Static Head*

R.L. of barrel floor = R.L. of trough bottom – Height of barrel

$$= 202.8 - 5.5 = 197.3 \text{ m}$$

Let us assume that a thickness of 0.8 m is provided.

∴ R.L. of bottom of floor = 197.3 – 0.8 = 196.5 m

Bed level of canal = 201.4 m

Assuming that the water-table has gone upto bed level of the canal, the static uplift on the floor (refer Fig. 14.18)

= 201.4 – 196.5 = **4.9 m of water**

(*b*) *Seepage Head*. The seepage head will be maximum when the drain trough is running full and the canal barrels are dry.

Thus, the total seepage head = H.F.L. of drain – Bed level of canal

$$= 206.3 - 201.4 = 4.9 \text{ m}.$$

The residual seepage head at a point '*a*' in the centre of the first barrel (Fig. 14.35) has been calculated by Bligh's theory as follows :

Assuming that the total length of drainage floor = 142.6 m

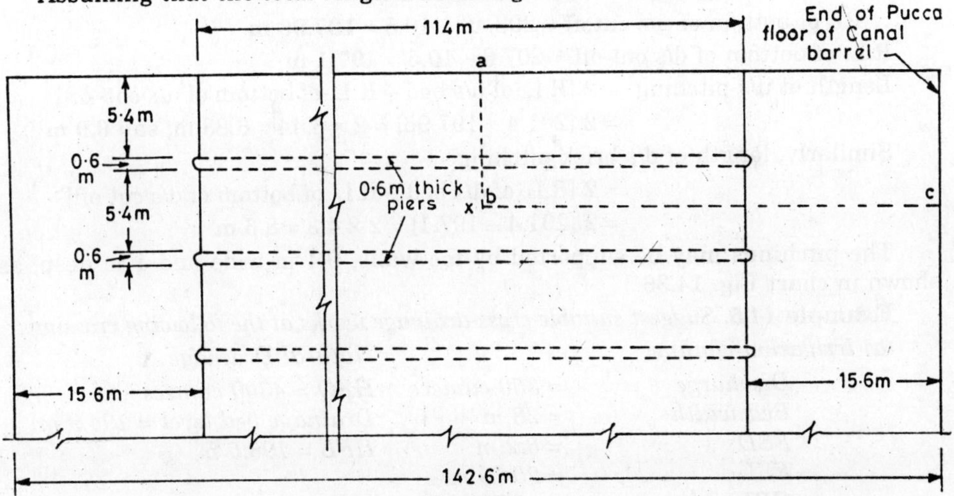


Fig. 14.35

The seepage line of drain water *abc* will traverse creep lengths as follows :

$$ab = 5.4 + 0.6 + 2.7 = 8.7 \text{ m} \; ; \; bc = \frac{142.6}{2} = 71.3 \text{ m}$$

Total creep length = 8.7 + 71.3 = 80 m

$$\text{Residual seepage head at } b = 4.9\left[1 - \frac{8.7}{80}\right] = 4.9 \times \frac{71.3}{80} = 4.37 \text{ m}$$

Total uplift

$$= \text{Static head } + \text{Seepage head} = 4.9 + 4.37 = 9.27 \text{ m} = 92.7 \text{ kN/m}^2$$

The provided 0.8 m thickness will resist due to its own weight, an uplift

$$= 0.8 \times 24 = 19.2 \text{ kN/m}^2$$

∴ Balance to be resisted by reinforcement due to bending action

$$= 92.7 - 19.2 = 73.5 \text{ kN/m}^2$$

Max. hogging B.M. in bottom slab

$$= \frac{73.5 \times (5.4 + 0.6)^2}{10} = \frac{73.5 \times 6^2}{10} = 264 \text{ kN} \cdot \text{m} = 264 \times 10^5 \text{ N–cm}$$

$$\text{Steel required} = \frac{264 \times 10^5}{12000 \times 8.7 \times 77.5} \text{ cm}^2\text{/m length} = 32.7 \text{ cm}^2$$

Use 32 mm ϕ bars @ 24 cm c/c at top in the barrel floor (*i.e.* bottom floor). 12 mm ϕ bars @ 10 cm c/c both ways at bottom, and one way also at top, may be provided as distribution bars.

The hydraulic dimensions and detailings are shown in attached chart Fig. 14.36.

**Step 10. Design of Cutoff's and protection works for the canal floor**

Lacey's normal depth of scour $= R = 0.47\left(\frac{Q}{f}\right)^{1/3}$ ; assuming $f = 1$

$$R = 0.47 \times \left(\frac{354}{1}\right)^{1/3} = 7 \text{ m below FSL}$$

Provide depth of cut off's for scour holes of 1.5 $R$ on both sides

or $1.5R = 1.5 \times 7 = 10.5$ m below FSL

$\therefore$ RL of bottom of u/s cutoff = 208.46 – 10.5 = **197.96 m**

R.L. of bottom of d/s cut-off = 207.6 – 10.5 = **197.1 m**

Length of u/s pitching* = 2 [R.L. of u/s bed – R.L. of bottom of u/s cut–off]
= 2 [201.4 – 197.96] = 2 × 3.44 = 6.88 m; **say 6.9 m**

Similarly, length of d/s brick pitching
= 2 [R.L. of d/s bed – R.L. of bottom of d/s cut off]
= 2 [201.4 – 197.1] = 2 × 4.3 = 8.6 m.

The pitchings may be supported by toe walls, 0.4 m wide and 1 m deep, as shown in chart Fig. 14.36.

**Example 14.5.** *Suggest suitable cross-drainage works at the following crossings :*

*(a) Irrigation channel* | *Natural Drainage*

| | | |
|---|---|---|
| *Discharge* | *= 350 cumecs* | *HFQ = 4300 cumecs* |
| *Bed width* | *= 28 m* | *Drainage bed level = 194.9 m* |
| *FSD* | *= 6.2 m* | *HFL = 198.5 m* |
| *FSL* | *= 204.3 m* | |
| *CBL* | *= 198.1 m* | |
| *Natural ground level = 179.3* | | |

*(b) Irrigation channel* | *Natural Drainage*

| | | |
|---|---|---|
| *Full supply discharge* | *= 350 cumecs* | *H.F.Q. = 390 cumecs* |
| | | *Drainage bed level = 207.4 m* |
| *Bed width* | *= 24 m* | *HFL = 209.3 m* |
| *FSD* | *= 6.2 m* | *Drainage bed slope = 1 in 500* |
| *FSL* | *= 215.6 m* | |
| *CBL* | *= 209.4 m* | *Springing level = 207.6 m* |
| *Canal bed slope = 1 in 10,000.* | | |

**Solution.** (*a*) In this case, the canal bed level (198.1 m) is higher than the drainage bed level (194.9) by 3.2 m and, therefore, we may think of taking the canal above the drainage. Further, drainage FSL (198.5 m) is above the canal bed (198.1 m), we will have to syphon the drainage water ; and thus a *syphon aqueduct* can be constructed at the site. However, since syphoning of drainage may not prove economical, and on the other hand, the canal bed may be raised slightly so as to pass drainage water freely below the canal trough (as the difference in CBL and drainage *HFL* is quite small *i.e.* 0.3 m). In that case, the minimum canal bed level to be kept at the crossing will be equal to

= Drainage HFL + thickness of canal trough slab + Min. free-board
= 198.5 + 0.6 + 0.9 = 200 m.

---

*Although the canal is evidently a pitched canal as the velocity $\left(i.e. \frac{354}{30.2 \times 6.2} = 1.89 \text{ m/sec.}\right)$ is high, still these pitching lengths which would be necessary to be properly maintained for the safety of the structure have been worked out. Moreover, due to pitched canal, the scouring tendencies etc. will be less, however, for safety and conservative design purposes, the various data have been worked out assuming that in a worst case, the pitching of canal may get broken.

Therefore, bed of canal trough will have to be raised by an amount equal to 200 – 198.1 = 1.9 m at the site of the crossing. Hence, hump is provided along the length of trough with flat u/s slope and d/s slope to join to the natural canal bed levels on both sides. The crossing will then be designed as an **aqueduct.** The design procedure as used in example 14.1 can be adopted.

(*b*) In this case, the canal bed level (209.4 m) is higher than the drainage bed level (207.4 m) by about 2 m ; so we may think of taking the canal above the drainage. Further, the drainage HFL (209.3 m) is below the canal bed level (209.4 m) by only 0.1 m ; which is less than the free-board and slab thickness, hence, the drainage will have to be syphoned below the canal. The other possibility of raising the canal bed is not being adopted here, as the drainage is not too large in comparison with canal discharge.

The work-will, therefore, be designed as a **syphon aqueduct**, and design procedure adopted as in example 14.2.

## 14.6. Provision of Joints and Water Bars in R.C.C. Ducts of Aqueducts and Super Passages

R.C.C. box troughs (or ducts) carry water in aqueducts and super-passages, as explained in the previous articles. The box trough will span across the drain or the canal, rested at suitable intervals (spans) on piers or abutments. The construction of such R.C.C. troughs will involve various types of joints, such as *construction joints, expansion joints,* etc. Water will leak through these joints, unless proper arrangements are made to prevent such leakages. *Water bars* or *water-stops* are provided to stop such leakages through the joints. The water bars may be made of metals like copper sheets or galvanised iron sheets, or of rubbers (natural as well as synthetic), or of P.V.C., etc.

**14.6.1. Types of Joints in R.C.C. Constructions.** The various types of joints that may be involved in R.C.C. constructions can be divided into the following types :

(1) *Movement joints* including the :

(*i*) *Contraction joints*, (*ii*) *Expansion joints*, and (*iii*) *Sliding joints* ; and

(2) *Construction joints.*

These joints are discussed below :

1 (*i*). **Contraction joints.** It is a movement joint with a deliberate discontinuity but no initial gap between the concrete on either side of the joint, the joint being intended to accommodate contraction of concrete. (See Fig. 14.37).

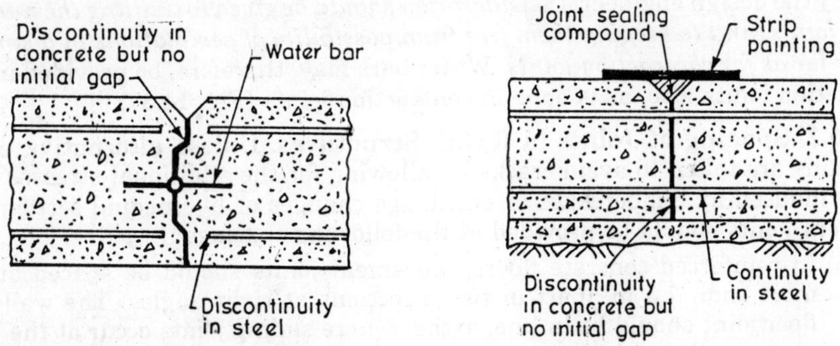


(*a*) Complete contraction joint (*b*) Partial contraction joint

Fig. 14.37. Typical contraction joints.

A distinction should also be made between a *complete contraction joint* (See Fig 14.37 *a*) in which both concrete and reinforcing steel are interrupted ; and a *partial contraction* joint (See Fig. 14.37 *b*) in which only the concrete is interrupted, the reinforcing steel running through.

1(*ii*). **Expansion joints.** A movement joint with complete discontinuity in both reinforcement and concrete, and intended to accommodate either expansion or contraction of the structure (See Fig. 14.38) is known as an expansion joint.

In general, such a joint requires the provision of an initial gap between the adjoining parts of a structure, which by closing or opening, accommodates the expansion or contraction of the structure. Design of the joint as to incorporate sliding surfaces is not, however, precluded and may sometimes be advantageous.

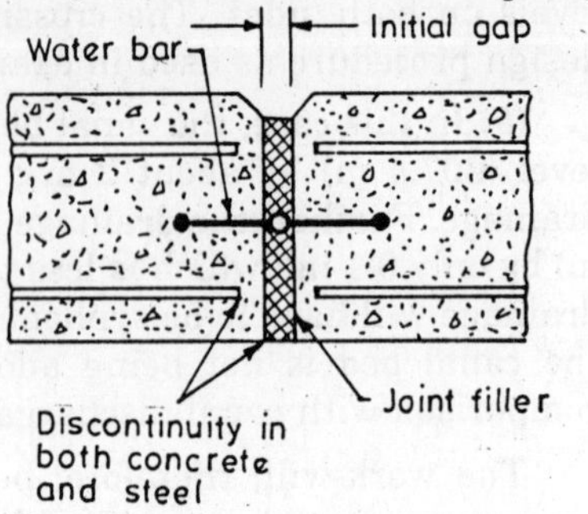

Fig. 14.38. A typical Expansion joint.

1(*iii*). **Sliding joints.** A movement joint with complete discontinuity in both reinforcement and concrete at which special provision is made to facilitate relative movement in place of the joint is known as a sliding joint. A typical application is in between a wall and a floor in same cylindrical tank designs (See Fig. 14.39)

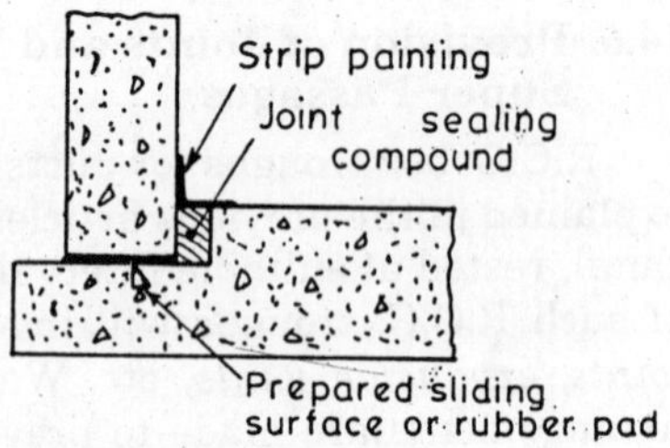

Fig. 14.39. A typical sliding joint.

(2) **Construction joints.** A joint in the concrete introduced for convenience in construction at which special measures are taken to achieve subsequent continuity without provision for further relative movement, is called a construction joint. A typical application is between successive lifts in pouring concrete in walls of a reservoir or a box trough. (See Fig. 14.40).

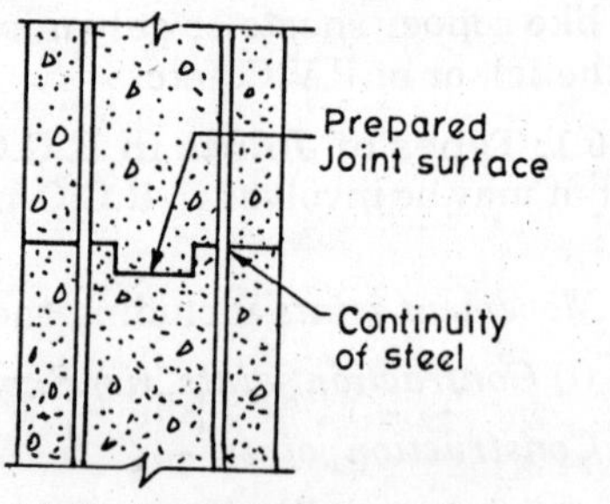

Fig. 14.40. A typical construction joint.
(Provision of water bar may also be made not shown here)

The position and arrangement of all the construction joints should be predetermined by the design engineer. *Consideration should be given to limiting the number of such joints, and to keeping them free from possibility of percolations in a similar manner for as for contraction joints.* Water bars may, therefore, be provided in the construction joints also, as is done in contraction joints (Fig. 14.37).

**14.6.2. Spacing of Joints in R.C.C. Structures.** Unless alternative effective means are taken to avoid cracks by allowing for the additional stresses that may be induced by temperature or shrinkage changes or by unequal settlement, movement joints should be provided at the following spacings :

(*i*) In reinforced concrete floors, movement joints should be spaced at not more than 7.5 m apart in two directions at right angles. The wall and floor joint should be in line, except where sliding joints occur at the base of the wall, in which case correspondence is not so important.

(*ii*) For floors with only nominal percentage of reinforcement (smaller than the minimum specified), the concrete floor should be cast in panels with sides not more than 4.5 m.

(*iii*) In concrete walls, the vertical movement joints should normally be placed at a maximum spacing of 7.5 m in reinforced walls and 6 m in unreinforced walls. The maximum length desirable between vertical movement joints will depend upon the tensile strength of the walls, and may be increased by suitable reinforcement. Thus when a sliding layer is placed at the foundation of a wall, the length of wall that can be kept free of cracks depends upon the capacity of wall section to resist the friction induced at the plane of sliding. Approximately, the wall has to stand the effect of a force at the plane of sliding equal to weight of half the length of wall multiplied by the coefficient of friction.

(*iv*) Amongst the movement joints in floors and walls as mentioned above, expansion joints should normally be provided at a spacing of not more than 30 m between successive expansion joints or between the end of the structure and the next expansion joint, all other joints being of the contraction type.

(*v*) When, however, the temperature changes to be accommodated are abnormal or occur more frequently than usual as in the case of storage of warm liquids or in uninsulated roof slabs, a smaller spacing than 30 m should be adopted, (that is a greater proportion of the movement joints should be of the expansion type). When the range of temperature is small, for example, in certain covered structures, or where restraint is small, for example, in certain elevated structures, none of the movement joints provided in small structures upto 45 m length need be of the expansion type. Where sliding joints are provided between the walls and either the floor or roof, the provision of movement joints in each element can be considered independently.

**14.6.3. Width of Gap in Expansion Joint.** An expansion joint requires the provision of an initial gap between the concrete faces on the two sides of the joint. The initial width of this gap should be sufficient to accommodate freely the maximum expansion of the structure.

In determining this initial width, consideration should be given to the requirements of the jointing materials. These will normally require the maintenance of certain minimum width of gap during maximum expansion of the structure. The joint should also be suitably treated as to maintain water-tightness during movement of the joint.

**14.6.4. Jointing Materials.** Jointing materials may be classified as follows :

(*i*) *joint fillers* ;

(*ii*) *Water bars* and *joint cover plates*; and

(*iii*) *Joint sealing compounds* (including primers where required)

They are discussed below :

(*i*) **Joint fillers.** Joint fillers are usually compressible sheet or strip materials used as spacers. They are fixed to the face of the first placed concrete and against which the second placed concrete is cast. With an initial gap of about 30 mm, the maximum expansion or contraction that the filler materials may allow may be of the order of 10 mm.

Joint fillers, as at present available, can not by themselves function as watertight expansion joints. They may be used as support for an effective joint sealing compound in floor and floor joints. But they can only be relied upon as spacers to provide the gap in an expansion joint, the gap being bridged by a water bar (See Fig. 14.42).

(*ii*) **Water bars.** Water bars or *water stops* are the preformed strips of impermeable materials which are embedded in concrete during construction, so as to span across the joint, as to provide a permanent water tight seal during the whole range of joint movement.

The water bars are usually made of *metal* sheets like copper, galvanised iron etc. ; or may be made of *rubber* (natural or synthetic) ; or of plastics like *P.V.C.* (Polyvinyl chloride). *Metal bars* can be used in water conveyance structures or dams, where the foundations are not expected to yield appreciably ; otherwise they will snap. *Annealed copper* is the most common metal used for water stops, in thickness not less than 1.5 mm; but it is likely to become brittle in course of time and may then crack. *Monel metal* is considered more durable though sometimes 20 gauge *galvanised iron* or *steel* strips are also used.

The most usual shape of metal bars is in the form of strips with central longitudinal corrugation, to obtain a V, U or M shape. The total width of the water bar may range from 150 mm to 300 mm. For box aqueducts etc., 180 to 225 mm wide water bar may suffice; while for dams, 300 mm wide bars may be preferred. *Rubber water bars* are, however, preferred to metal bars for their economy, and suitability to some what yielding foundations. Metal and rubber water bars are invariably used in ducts of aqueducts and super-passages in expansion joints; *PVC water bars*, on the other hand, are, usually adopted in contraction and construction joints.

Typical use of various types of water bars in contraction or construction joints are shown in Fig. 14.41 (*a*) to (*d*); whereas Fig. 14.42 (*a*) to (*g*) shows the use of various type of water bars in expansion joints.

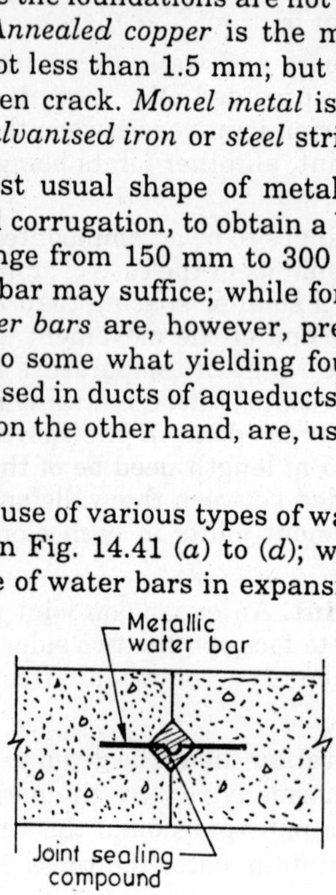

(*a*) metal water bar with central corrugation

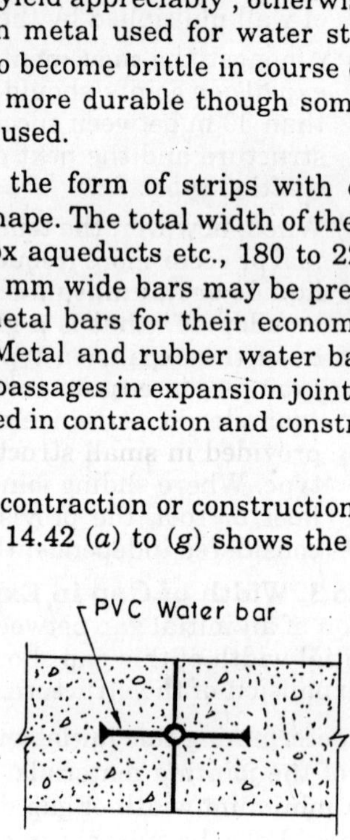

(*b*) PVC water bar

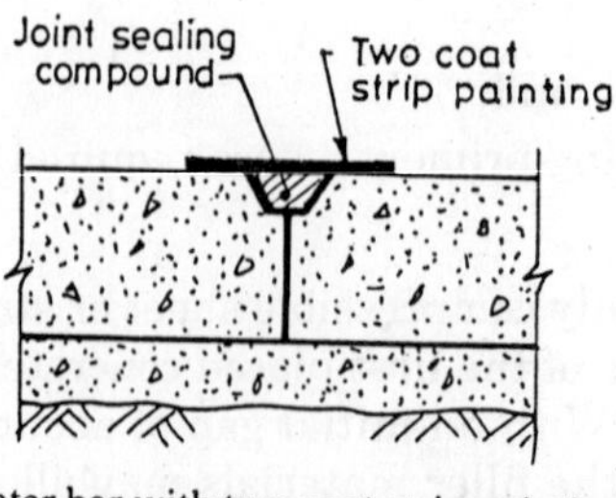

(*c*) metal water bar with two coat strip painting

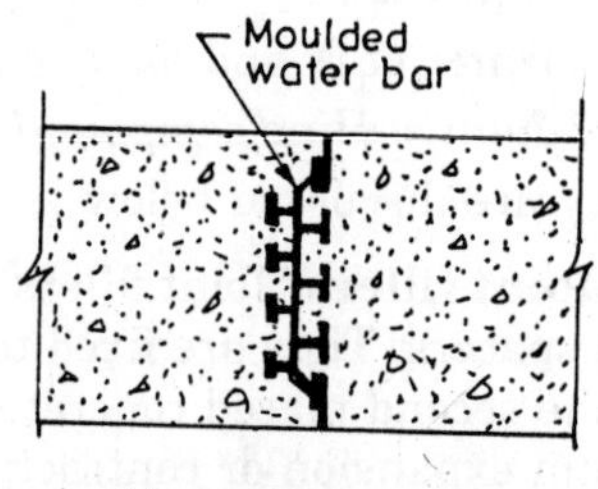

(*d*) moulded water bar

Fig. 14.41. Typical details showing use of water bars surrounded with joint sealing material in Contraction Joints.

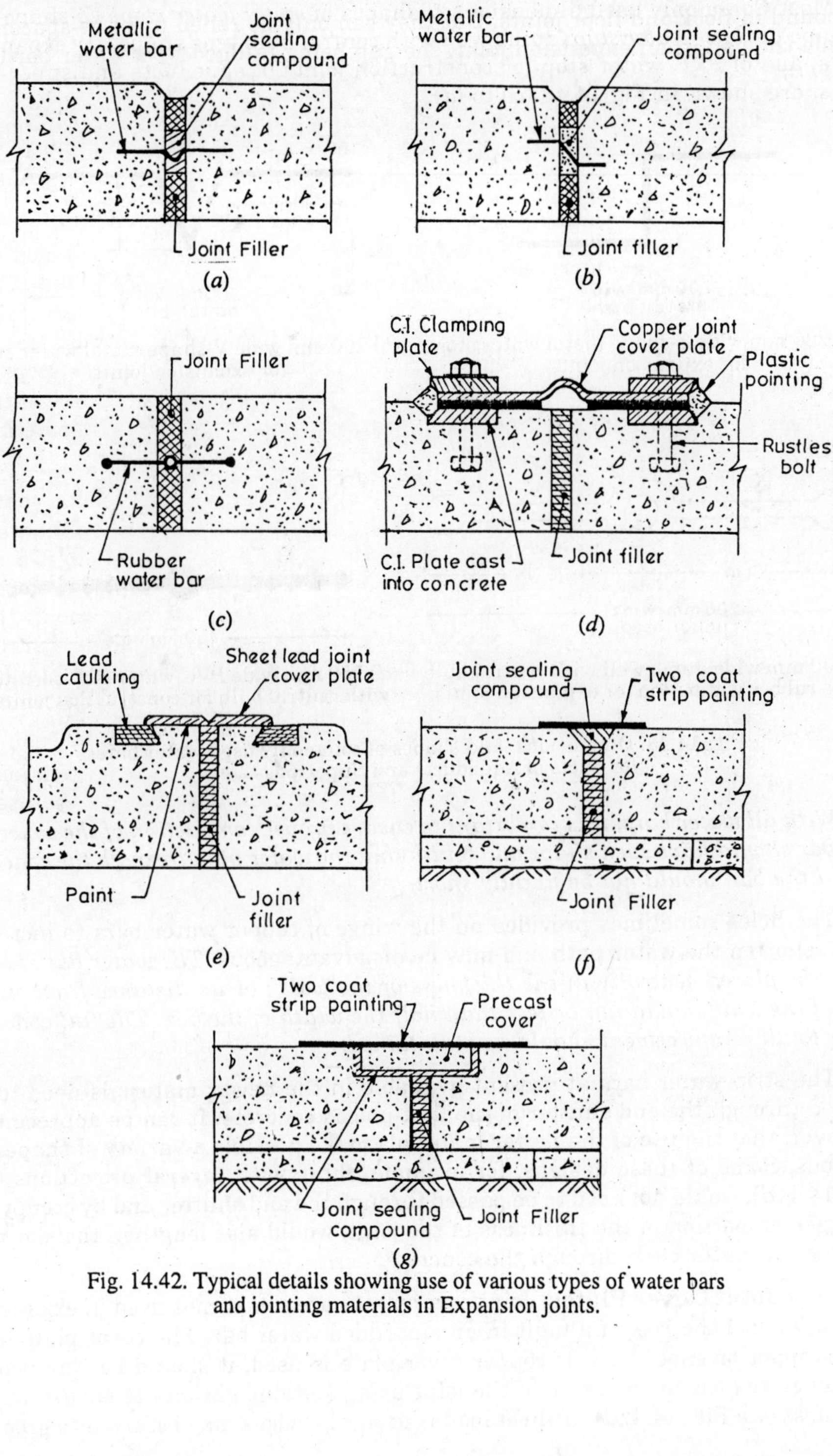


Fig. 14.42. Typical details showing use of various types of water bars and jointing materials in Expansion joints.

Most Commonly used dimensions & shapes of *metal water stops* (Z-shape anu V-shape), *rubber water stop* (dumb well with central bulb type shape) for expansion joints; and of PVC water stop for construction joints of aqueducts and super-passages, are shown in Fig 14.43 (*a*) to (*d*).

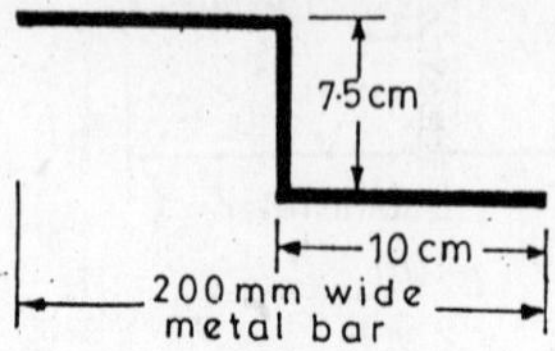


(*a*) 200 mm wide Z-shape metal water stop for expansion joints

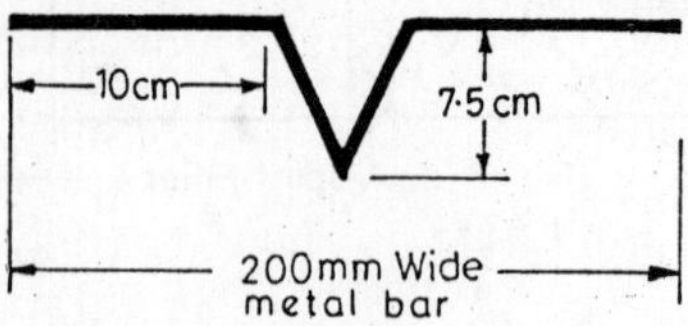


(*b*) 200 mm wide V-shape metal water stop for expansion joints.

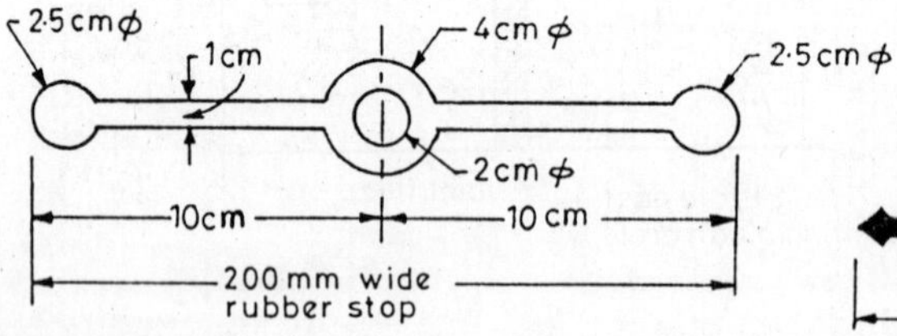


(*c*) 200 mm wide dumb well with central bulb type rubber water stop for expansion joints

(*d*) 150 mm wide *PVC* water stop serrated with central bulb for construction joints.

Fig. 14.43. Dimensions and shapes of commonly used water bars for ducts of Aqueducts and Super-passages.

*With all water bars, it is important to ensure proper compaction of the concrete. The bar should have such shape and width that the water path through the concrete round the bar should not be unduly short.*

The holes sometimes provided on the wings of copper water bars to increase bond, shorten the water path and may be disadvantageous. *The water bar should either be placed centrally in the thickness of the wall ; or its distance from either face of the wall should not be less than half the width of the bar. The full concrete cover to all reinforcement should be maintained.*

The strip water bars at present available in the newer materials need to be passed through the end shutter of the first-placed concrete. It can be appreciated, however, that the use of the newer materials make possible a variety of shapes or sections. Some of these designs, for example, those with several projections (*see* Fig. 14.41*d*), would not need to be passed through the end shutter and by occupying a bigger proportion of the thickness of the joint, would also lengthen the shortest alternative water path through the concrete.

(*ii b*) **Joint Cover Plates**. Joint cover plates are sometimes used in expansion joints to avoid the risk of a fault in an embedded water bar. The cover plate may be of copper or sheet lead. If copper cover plate is used, it should be clamped to the concrete face on each side of the joint using suitable gaskets to ensure water tightness (*see* Fig. 14.42*d*). If sheet lead is used, the edges may return into grooves formed in the concrete and be made completely watertight by lead caulking (*see* Fig. 14.42 *e*). Faces of the concrete to which sheet lead is to be fixed should be painted with bituminous or other suitable composition, and the lead sheet should be similarly coated before fixing.

(*iii*) **Joint Sealing Compounds.** Joint sealing compounds are impermeable ductile materials which are required to provide a watertight seal by adhesion to the concrete throughout the range of joint movement. The commonly used materials are based on asphalt, bitumen, or coal tar pitch with or without fillers, such as limestone or slate dust, asbestos fibre, chopped hemp, rubber or other suitable material. These are usually applied after construction or just before the reservoir is put into service by pouring in the hot or cold state, by trowelling or gunning or as performed strips ironed into position. They may also be applied during construction, such as by packing round the corrugation of a water bar. A primer is often used to assist adhesion and some local drying of the concrete surface with the help of a blow lamp is advisable. The length of the shortest water path through the concrete should be extended by suitably painting the surface of the concrete on either side of the joint.

The main difficulties experienced with this class of material are in obtaining permanent adhesion to the concrete during movement of the joint, whilst at the same time ensuring that the material does not slump or is not extruded from the joint.

In floor joints, the sealing compound is usually applied in a chase formed in the surface of the concrete along the line of the joint (*see* Fig. 14.42*c*). The actual minimum width will depend on the known characteristics of the material. In the case of an expansion joint, the lower part of the joint is occupied by a joint filler (*see* Fig. 14.42*f*). This type of joint is generally quite successful, since retention of the material is assisted by gravity and, in many cases, sealing can be delayed until just before the reservoir is put into service so that the amount of joint opening subsequently to be accommodated is quite small. The chase should not be too narrow or too deep to hinder complete filling and the length of the shortest water path through the concrete should be extended by suitably painting the surface of the concrete on either side of the joint. Here again a wider joint demands a smaller percentage distortion in the material.

An arrangement incorporating a cover slab, similar to that shown in Fig. 14.42 g, may be advantageous in reducing dependence on the adhesion of the sealing compound in direct tension.

Use of sealing compounds for vertical joints is not very successful. A stepped-joint instead of a straight through-joint with a water bar incorporated in the joint and sealing compound packed round the corrugation of the water bar would be much more successful.

**Complete typical details of a 30 mm wide expansion joint adopted in the construction of an Aqueduct** are shown in Fig. 14.44, where two water stops (225 mm wide rubber water stop and 300 mm wide V-shaped copper water stop) were used to ensure complete water-tightness.

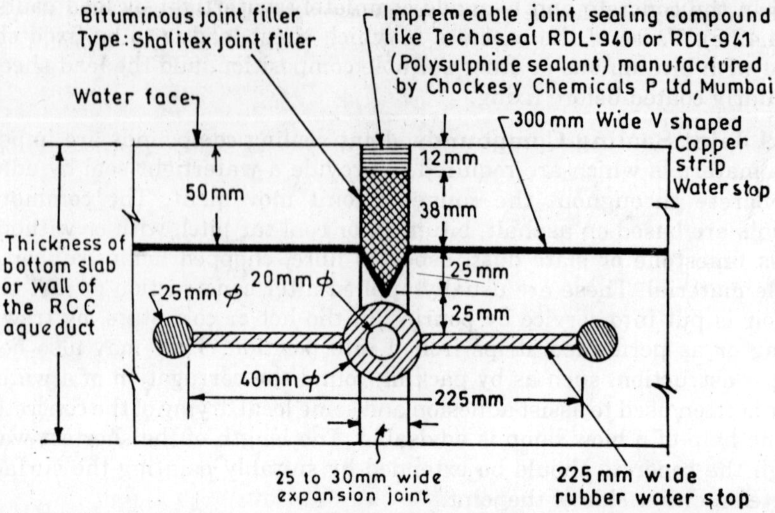


Fig. 14.44. Details of typical Expansion joint in the base slab and walls of R.C.C. duct or trough.

## PROBLEMS

1. (*a*) What is meant by a "Cross-Drainage Works"? Explain as to why such works are not met within a ridge canal system.

   (*b*) Write short notes on :

   (*i*) Syphon (*ii*) Super passage (*iii*) Syphon aqueduct.

2. Write short notes on :

   (*i*) Aqueduct (*ii*) Syphon aqueduct

   (*iii*) Canal syphon (*iv*) Level crossing.

3. Design and give a dimensional sketch of an aqueduct to carry water of an earth canal over a drainage with the following data :

| | |
|---|---|
| RL of bed of drainage | = 520.00 m |
| HFL of drainage | = 523.00 m |
| Bed width of drainage | = 50.00 m |
| Side slopes of drainage at crossing | = 1/2 : 1 |
| R.L. of ground | = 525.00 m |
| R.L. of bed of canal | = 524.50 m |
| Discharge of canal | = 30 cumecs |
| Depth of water in canal | = 1.70 m |
| Bed width of canal | = 22.0 m. |

4. (*a*) Name the different types of cross drainage works. Explain how you would avoid one type of cross drainage work and prefer to adopt another type by simply changing the alignment of the canal taking off from a head-work.

   (*b*) Under what conditions of drainage and canal crossings are syphons provided ? Draw a plan and section through a typical branch canal syphon, and suggest a method for reducing uplift on the floor of the work.

5. (*a*) What are the different types of cross drainage works that are necessary on a canal alignment? State briefly the conditions under which each one is used.

   (*b*) Discuss with neat sketches, the three different types of aqueducts which can possibly be constructed depending upon the size of the drainage to be passed below the canal. Also discuss the factors governing the choice of any of these three types of aqueducts.

6. Show by rough sketches the type of cross drainage work suitable for each of the following cases:

(a)

| | Drain | | Canal |
|---|---|---|---|
| Discharge in cumecs | 200 | | 20 |
| BL | 25.00 m | | 30.00 m |
| FSL | | 31.50 m | |
| HFL | 28.00 m | | |

[**Ans.** Aqueduct of Fig. 14.3 is recommended]

(b)

| | Drain | | Canal |
|---|---|---|---|
| Discharge in cumecs | 2 cumecs | | 400 cumecs |
| BL | 52.2 m | | 48.00 m |
| FSL | | 53.00 m | |
| HFL | 53.2 m | | |

[**Ans.** Syphon of Fig. 14.7 is recommended]

7. Discuss the various types of cross drainage works used in canal systems. What considerations govern the selection of the different types of works, mainly depending upon the levels of the canal and the drainage. Illustrate by drawing a neat sketch of each type of structure.

8. Give neat sketch of suitable designs of aqueducts for each of the following crossings :
(*a*) A major canal over a small drainage. [**Ans.** Aqueduct of type I, Fig. 14.10]
(*b*) A canal carrying low discharge over a large drainage. [**Ans.** Aqueduct of Type II, Fig. 14.11]
(*c*) A major canal over a large drainage. [**Ans.** Aqueduct of Type III, Fig. 14.12]
Explain in detail the method of design of fluming required in the crossing at (*c*) above.

9. (*a*) State under what circumstances you will recommend the use of the following cross drainage structures :
(*i*) Syphon (*ii*) Inlet (*iii*) Aqueduct
(*b*) Following data were collected from a syphon aqueduct :

| | |
|---|---|
| Diameter of the barrel (single) | = 3 m |
| Length of the barrel | = 90 m |
| Discharge through the barrel | = 25 cumecs |
| Friction factor (in Darcy-Weisbach formula) | = 0.013 |
| Coefficient of bend loss (2 bends) | = 0.10 |
| Coefficient of head loss in expansion at outlet | = 0.20 |
| Coefficient of head loss in contraction at inlet | = 0.10. |

Determine afflux. Neglect velocity head in drainage channel.

[**Solution :** Total head loss $= h = \underbrace{0.10\left(\frac{V^2}{2g}\right)}_{\text{entry loss}} + \underbrace{2 \times 0.10\left(\frac{V^2}{2g}\right)}_{\text{bend loss}} + \underbrace{0.2\left(\frac{V^2}{2g}\right)}_{\text{outlet expansion loss}} + \underbrace{\frac{f' LV^2}{2gd}}_{\text{friction loss in pipe}}$

$$= 0.5\,\frac{V^2}{2g} + \frac{0.013 \times 90}{3}\,\frac{V^2}{2g} = 0.89\,\frac{V^2}{2g}$$

Velocity in the barrel $= \dfrac{Q}{A} = \dfrac{25}{\frac{\pi}{4} \times (3)^2}$ m/sec. = 3.52 m/sec.

$\therefore \qquad h = 0.89 \times \dfrac{(3.52)^2}{2 \times 9.81} = 0.57$ m. **Ans.**]

**10.** (*a*) **What are cross-drainage works ? What is the necessity of such a work in a canal project, and how does this necessity is fulfilled by such works ?**

(*b*) **Enumerate the different methods which may be used for designing the canal transitions for a flumed canal, and the conditions under which each can be used. Describe in details a method of designing canal transitions when water depth may or may not remain constant.**

**11.** (*a*) **What is meant by cross drainage works and what is their importance in a canal project ?**

(*b*) **Describe briefly the step by step procedure that you will adopt for designing an unflumed syphon aqueduct.**

**12. Design a syphon aqueduct with the following data :**

*For Canal*

| | |
|---|---|
| Discharge | = 56 cumecs |
| Bed width | = 32 m |
| F.S. depth | = 1.98 m |
| R.L. of bed | = 267.00 m |

*For Drainage*

| | |
|---|---|
| High flood discharge | = 425 cumecs |
| HFL | = 268.20 m |
| General bed level of low water cross-section | = 265.50 m |
| General ground level | = 267.20 m |

**13. Design a suitable cross drainage work, if the crossing in problem 12 above is shifted upstream along the drainage so that the HFL is 269.55 m, the bed level of the drainage is 267.00 m, and general ground level is 269.05 m. The canal data remaining the same.**

# 15

# *Construction of Culverts and Small Road Bridges Across Drains and Canals**

## 15.1. Introduction

Whenever a new road has to cross an existing drain or a canal, or sometimes, when a drain has to cross an existing road ; a small bridge or a culvert** is constructed at the point of crossing. The number of such culverts required in any road or canal project is generally very large and account for a large percentage of the total expenditure on the project. These small bridges or culverts should therefore, be designed safely and economically, so as to avoid unnecessary over investments. In the pages below, we shall, therefore, discuss the important considerations that are involved in their designs.

## 15.2. Data Collection

After a proper site of construction has been decided on some straight reach of the stream, the highest flood level at the site should be determined. The longitudinal section of the stream extending by about 200 to 1500 metres upstream and downstream should be plotted depending on the catchment area of the stream (at the point of construction) varying from 250 to 1240 hectares or more. The actual flow velocity may be observed during times of floods, which may provide a good check on the velocity calculated theoretically.

## 15.3. High Flood Discharge Computations

After collecting the required necessary data, the maximum discharge that is likely to reach at the site may be calculated by using any of the various methods given in article 7.17 of "Water Resources Engg. Vol. I—Hydrology, Flood Control and Ground Water Engg." by same author. The various empirical formulas given for this purpose should only be used by experts, and a new designer should try to use the Rational formula only.

Road bridges and culverts are normally designed for 1 in 50 year frequency discharge, although important rail bridges may be designed for 1 in 100 years discharge.

After the design flood discharge has been finalised, the total span and number of spans of the bridge should be decided, as given in articles 15.4 and 15.8.

## 15.4. Linear Waterway of the Bridge

The linear waterway of a bridge across a purely alluvial stream is generally kept equal to its regime width ($W$) as given by Lacey. The Lacey's regime width of the stream is taken equal to its wetted perimeter ($P$), given by the equation,

$$W = P = 4.75\ Q^{1/2}$$

$$\therefore \quad L = 4.75\ Q^{1/2} \qquad \text{...(15.1)}$$

where $Q$ is the dominant discharge of the river

---

* The discussions here are based on Indian Standards laid down by Indian Road Congress in its various publications which are hereby duely acknowledged.

** A bridge with span upto 8 m is called a culvert.

Hence, the existing width of the stream is contracted at the bridge site by upstream wing walls (or training works) so as to make it equal to $W$, and the bridge may then be spanned across it. The contraction of the stream up to the regime width leads to economy, as the bridge length is reduced ; but further contraction beyond regime width are generally not economical, because any contraction beyond $W$ (Lacey's regime width) will increase the discharge intensity through the bridge spans, and thus increasing the Lacey's regime scour depth ($R$). Due to the increase in scour depth, the foundations of the bridge piers, abutments, and cut off walls will have to be taken deeper. The necessity of providing deeper foundations in streams contracted beyond Lacey's regime width, will increase the cost of foundations, although the length of the bridge superstructure will reduce. *Much of the saving in cost expected from decreasing the length of the bridge in such a case will, therefore, generally be eaten up by the corresponding increase in the depth of the foundations and the size of the training works.* Hence, in alluvial streams, it is generally futile to contract the waterway beyond the regime width, as no appreciable saving can be obtained.

However, in streams which are not alluvial and are either semi-alluvial (sides rigid and bed alluvial) or purely rigid, the waterway may be chosen equal to the actual surface width of the stream measured from edge to edge of water along the H.F.L., on the plotted cross-section. The waterway in purely rigid streams may be kept still less, if appreciable saving is obtained and afflux is not detrimental. For streams which overflow their banks and create very wide surface widths with shallow side sections, the bridge may span the active channel, but this waterway should not be so low as to cause excessive detrimental afflux. The waterway, in such a case has, therefore, to be chosen by judgement and intelligence of the designer with these guidelines.

**Note.** When the natural available width of the stream is found nowhere near the value given by Lacey's equation. (*i.e.* $W = 4.75\sqrt{Q}$), we must expect to find that the banks, even though not rocky, are not friable enough to be treated as incoherent alluvium and hence, the stream is no longer a stream in alluvium. Such streams may then be *treated as semi-alluvial streams.*

## 15.5. Scour Depth Computations

It was explained earlier in chapter 4, that, if a constant discharge passes through a straight stable reach of an alluvial stream for an infinite time, the boundary of its cross-section would ultimately become elliptical. This will happen when the stream has achieved a state of regime, and in such a case, the depth in the middle of the stream would be equal to the normal scour depth. This **normal depth of scour** (*i.e.* ***Lacey's regime scour depth***) is generally taken as a guideline for fixing the depth of the foundations.

The *Lacey's regime scour depth* ($R_r'$) is calculated as given below :

(*a*) For a purely alluvial stream having a span length ($L$) equal to $W$ (given by $W = 4.75\sqrt{Q}$).

$$\textit{Regime scour depth} = R_r' = 0.473\left(\frac{Q}{f}\right)^{1/3} \quad \text{...(15.2)}$$

where $f$ is obtained by the equation

$$f = 1.76\sqrt{d_{mm}},\ d_{mm} \text{ being average particle size in mm} \quad \text{...(15.3)}$$

(*b*) For *purely alluvial streams* which are contracted beyond $W$ at the bridge site, the value of Lacey's Scour depth ($R'$) may be computed by using the formula :

$$\text{Scour depth} = R' = 1.35\left(\frac{q^2}{f}\right)^{1/3} \quad \text{...(15.4)}$$

where $q$ is the discharge intensity per unit width of the stream at bridge site, and equals $Q/L$, where $L$ is the width of the stream at the bridge site.

$R'$ can also be worked out by using the formula

$$R' = \left(\frac{W}{L}\right)^{0.61} \times R_r' \quad \text{...(15.5)}$$

where $W$ = the Lacey's regime width for alluvial streams

$R_r'$ = the Lacey's regime scour depth given by Eq. (15.2)

**Note.** The above calculated values of $R_r'$ and $R'$ are depths below water surface elevation of the stream.

(*c*) For *quasi-alluvial streams* (*i.e.* bed is not pitched and sides are banked with stable soils or pitchings), having the bridge span length equal to the length of water line between banks, measured along HFL (called natural unobstructed width of the stream $W'$), the normal secoured depth ($D'$) may be computed by using the formula :

$$D' = \frac{1.21\, Q^{0.63}}{f^{0.33}\, W'^{0.60}} \quad \text{...(15.6)}$$

where $D'$ = the scour depth in quasi alluvial streams having $L = W'$

$W'$ = the natural unobstructed width of the stream, as shown in Fig. 15.1.

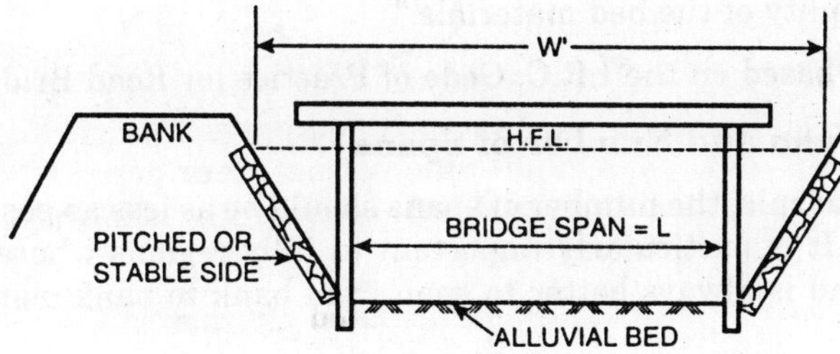


Fig. 15.1. Semi-Alluvial streams.

(*d*) For *quasi-alluvial streams which are contracted* so as to keep $L$ lesser than $W'$, the normal scour depth ($D$) is given by :

$$D = D'\left(\frac{W'}{L}\right)^{0.61} \quad \text{...(15.7)}$$

where $D'$ is given by Eq. (15.6).

(*e*) For *purely rigid streams* (such as lined pucca drains), however, from theoretical considerations, there will be no scouring, and the foundations of the bridge sub-structure will have to be taken below the bed, not by scour depth considerations, but only by nominal amount, so as to provide sufficient grip on them.

### 15.6. Maximum Scour Depth Computations

In natural streams, the scouring is not uniform along the bed width. It is not uniform even in straight reaches. Particularly at the bends and also around obstructions to the flow (*e.g.*, the piers of a bridge), there is a deeper scour than the normal scour. The maximum possible scour depth should, therefore, be fairly estimated and the foundations of the structure marginally taken below that level, so as to obtain a factor of safety on that maximum scour. The maximum scour depth is generally taken as $x$ times the normal scour depth $R$. The value of $x$ is generally taken as 1.5 for a single span structure (with no piers) on a straight reach of the stream, and is taken as 2.0 on bad sites on curves or for multispan structures.

### 15.7. Depth of Bridge Foundations ($D_f$)

The foundations of bridges on erodibe beds are taken down below the H.F.L. by an amount equal to 4/3 times the maximum scour depth, or equal to maximum scour depth plus 2 metres for arched bridges and 1.2 metres for other bridges, whichever is more.

This formula is applicable only when no bed floor is provided under the structure and the stream is free to scour as it may. The structure is then said to have deep foundations and it is always considered better to have such foundations as far as possible. But in certain cases, especially on small culverts, bed floors are generally provided. In such a case, the following in considered a safe practice on erodible beds :

"Keep the top of the floor about 0.3 metre below the bed level. Take the foundations of the abutments 1.2 metres below the top of the floor. Provide an upstream curtain wall 1 to 1.5 metres deep and down-stream curtain wall 1.5 to 2.5 metres deep from the top floor depending on the velocity of flow through the structure and erodibility of the bed materials."

These rules are based on the I.R.C. Code of Practice for Road Bridges.

### 15.8. Total Clear Span and Number of Spans

As a fundamental rule, the number of spans should be as less as possible, since piers obstruct flow. It is particularly important in hilly regions where torrential velocities prevail, and is always better to span from bank to bank using no peirs if possible.

By increasing the length of one span, the number of spans can be reduced. In such a case, the cost of super-structure increases and that of sub-structure decreases. *The most economical span length is one for which the cost of sub-structure is equal to the cost of super-structure.* This economy should, therefore, be worked out for large bridges and where difficult conditions are anticipated in sinking of foundations. Any deviations from the ideal economic case has to be justified strongly. When the regulators, barrages, etc. are provided with bridges, the span is generally guided by the economy as well as the availability of gate sizes.

**Length of one span.** The economical length of one span say ($l$) may be chosen as per the following guidelines :

(*a*) *For R.C.C. slab bridges*

$$l = 1.5\,H \qquad ...(15.8)$$

where $H$ = Total height of abutment or pier from the bottom of its foundations to its top

= $D_f$ + Vertical clearance*, *i.e.* free-board

$l$ = Clear span length.

(*b*) *For masonry arch bridges*

$$l = 2H \qquad ...(15.9)$$

where $H$ = Total height between the bottom of the foundations to the intrados of the key stone

$l$ = Clear span length.

**Number of spans.** When $L$ is more than the economical span length ($l$), the number of spans required ($N$) is tentatively found by using

$$L = N.l \qquad ...(15.10)$$

Since $N$ must be a whole number (preferably an odd number), $l$ may have to be modified suitably. Varying span lengths may sometimes be adopted, so as to keep as close as possible to the requirements of economy and to cause least obstruction to the flow.

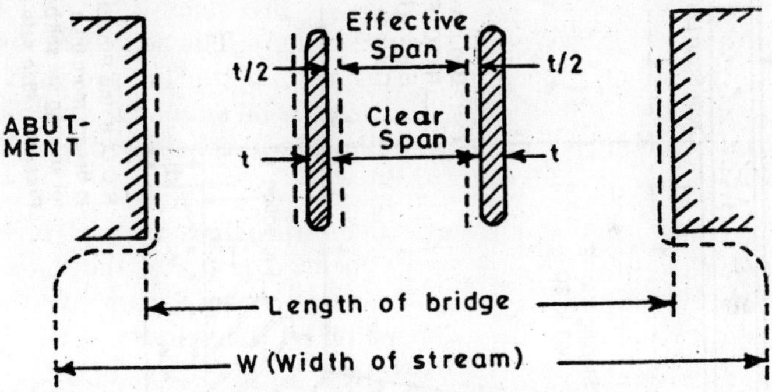


Fig. 15.2. (*a*) Plan at a Bridge site.

**Effective clear span.** It is assumed that each bridge pier causes contraction in the filaments of water equal to half its thickness on each side, and the obstruction at the ends due to abutments or pitched slope is ignored. Hence, the finally selected number and lengths of spans should be such that the sum of all the effective span lengths should be equal to the required linear waterway ($L$), and this will be equal to

$$L = \Sigma \text{ [of all the effective spans]}$$

$$= \Sigma \text{ [of all the clear span} - \text{Thickness of piers]}$$

$$L = \Sigma \text{ of all the clear spans} - \Sigma \text{ thickness of piers} \qquad ...(15.11)$$

The pier thickness above is taken as the mean value for the height of the pier from HFL to maximum scour depth.

* See Fig. 15.2 (*b*).

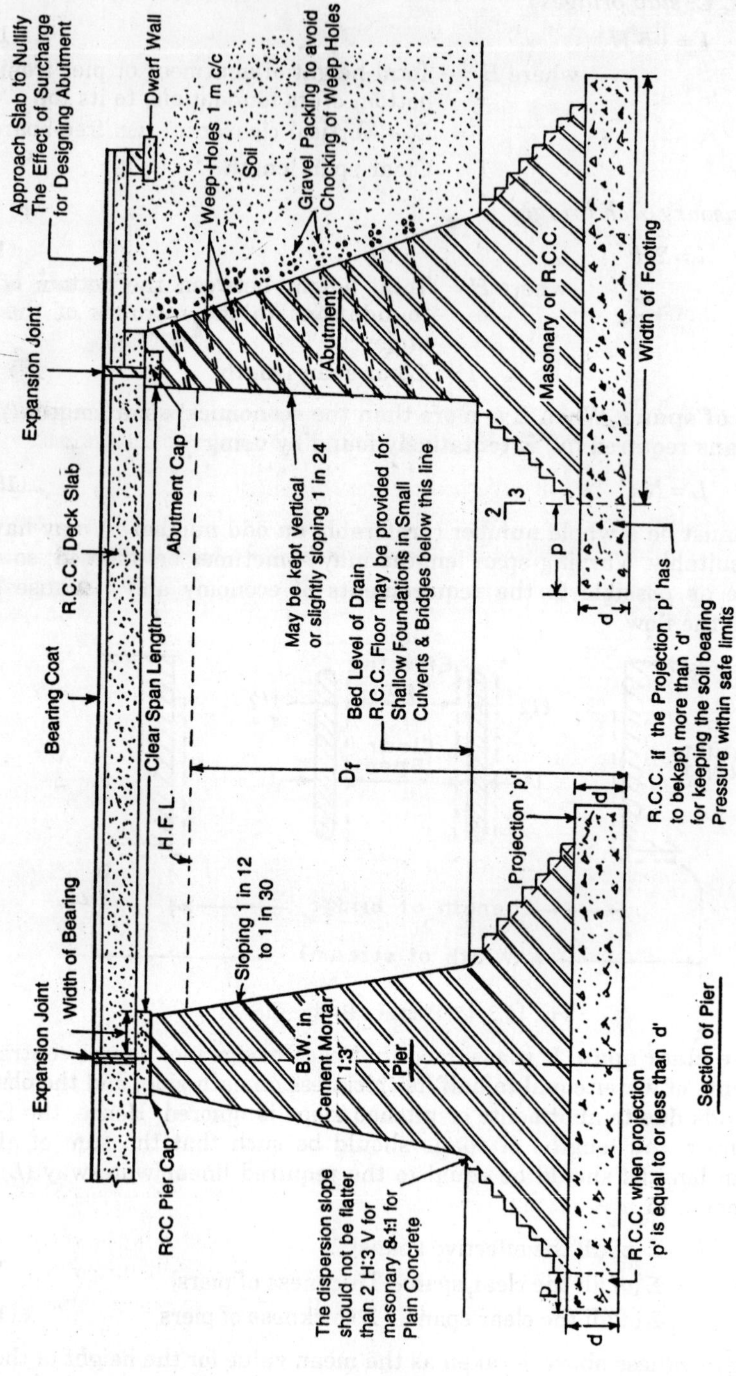


Fig. 15.2. (b) Section of abutment and piers for a slab bridge.

## 15.9. Vertical Clearance and Some Other Tentative Dimensions

They may be fixed as per I.R.C. codes for road bridges as follows :

**Table 15.1. Value of Vertical Clearance**

| *Discharge in cumecs* | *Vertical clearance or Free-board in metres* |
|---|---|
| Below 0.3 | 0.15 |
| 0.3 to 3 | 0.3 |
| 3 to 30 | 0.6 |
| 30 to 300 | 0.9 |
| 300 to 3000 | 1.2 |
| Above 3000 | 1.5 |

**Table 15.2. Minimum Top widths of PCC or RCC Piers and Abutments**

| *Clear span in metres* | *Width of abutment in metres* | *Width of pier in metres* |
|---|---|---|
| 3 | 0.38 | 0.45 |
| 6 | 0.45 | 0.55 |
| 9 | 0.55 | 0.65 |
| 12 | 0.65 | 0.75 |
| 15 | 0.75 | 0.88 |

**Table 15.3. Minimum Top Widths of Masonry Piers and Abutments**

| *Span* | *3 m* | *6 m* | *12 m* | *24 m* | *40 m* | *50 m and above* |
|---|---|---|---|---|---|---|
| Top width of piers carrying simply supported spans (in m) | 0.5 | 1.0 | 1.2 | 1.6 | 2.0 | 2.2 |
| Top width of abutment and of piers carrying continuous spans (in m) | 0.4 | 0.75 | 1.0 | 1.3 | 1.7 | 1.9 |

## 15.10. Afflux Computations

When a bridge is constructed across a contracted stream, water on the upstream will rise up. The maximum rise in water level near the bridge site is nothing but afflux. The greater is this afflux, lesser will be the available clearance for the same deck level. Since the 'clearance' is the distance between the u/s H.F.L. and the bottom of the bridge slab, it is nothing but freeboard. Hence, it is necessary to calculate afflux to examine its effects on clearance and also upon the regime of the channel u/s of the bridge, and also to determine the top levels of training walls, etc.

The afflux ($h$), the discharge ($Q$), the unobstructed stream width ($W$), and the provided linear waterway of the bridge ($L$) are all interconnected. Greater is the reduction in the linear waterway, the greater is the afflux. Since the downstream depth is not affected by the bridge, as the same is governed by the hydraulic characteristics of the d/s channel, it can be safely assumed that the u/s depth which prevailed before the bridge construction is the same as the d/s depth ($y_d$) that prevails even after the bridge construction.

Hence, $y_d$ is the depth that prevailed at the bridge site before the construction of the bridge. To estimate afflux, we must know $y_d$. This can be found from the available gauge discharge curve of the stream, or may be calculated by the hydraulic parameters of the channel.

The discharge passing through the bridge openings can be calculated in the following different ways :

(1) **By Broad Crested Weir formula.** This formula is applicable so long as the afflux, *i.e.* [upstream depth ($y_u$) — downstream depth ($y_d$)] is not less than $\frac{1}{4} \cdot y_d$. In this formula, the discharge $Q$ is dependent only upon $y_u$ and is independent of $y_d$. The fact that the downstream depth $y_d$ is having no effect on the discharge, nor on the upstream depth $y_u$, when the afflux is not less than $\frac{1}{4} y_d$, is due to the formation of the 'standing wave'.

The discharge $Q$ is then given by

$$Q = 1.71 \cdot C_d L \left[ y_u + \frac{V_a^2}{2g} \right]^{3/2} \quad \text{...(15.12)}$$

where $\frac{V_a^2}{2g}$ = head due to velocity of approach.

$y_u$ = upstream water depth

$C_d$ is the coefficient of discharge and accounts for frictional losses. The various values of $C_d$ for different types of openings are given in Table 15.4.

**Table 15.4. Values of $C_d$ for Different Types of Bridge Openings**

| *Type of bridge Opening* | *Value of $C_d$* |
|---|---|
| Narrow bridge opening with or without floors | 0.94 |
| Wide bridge opening with floors | 0.96 |
| Wide bridge opening with no bed floors | 0.98 |

(2) **By the Orifice Formula.** When the downstream depth is more than 80% of the upstream depth (*i.e.* the afflux is less than $\frac{1}{4} y_d$), the weir formula is not valid, as the performance of the bridge opening gets affected by the downstream depth ($y_d$). In such a case, the discharge can be calculated by using the Orifice formula given by

$$Q = C_0 \sqrt{2g} \cdot L \cdot y_d \cdot \left[ h + (1 + e) \frac{V_a^2}{2g} \right]^{1/2} \quad \text{...(15.13)}$$

The values of coefficients $e$ and $C_0$ may be taken from the curves given in Figs. 15.3 and 15.4.

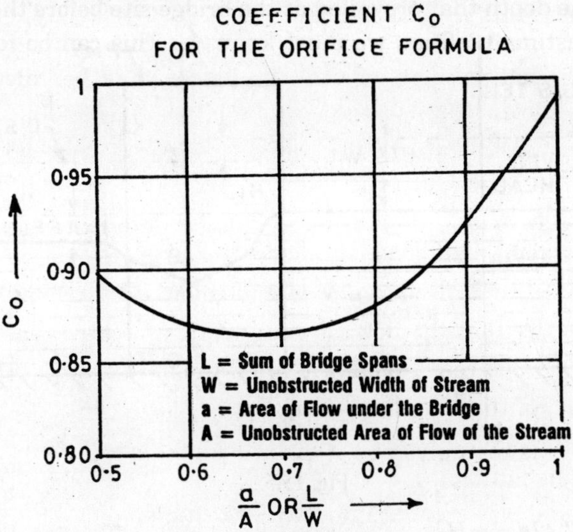


Fig 15.3

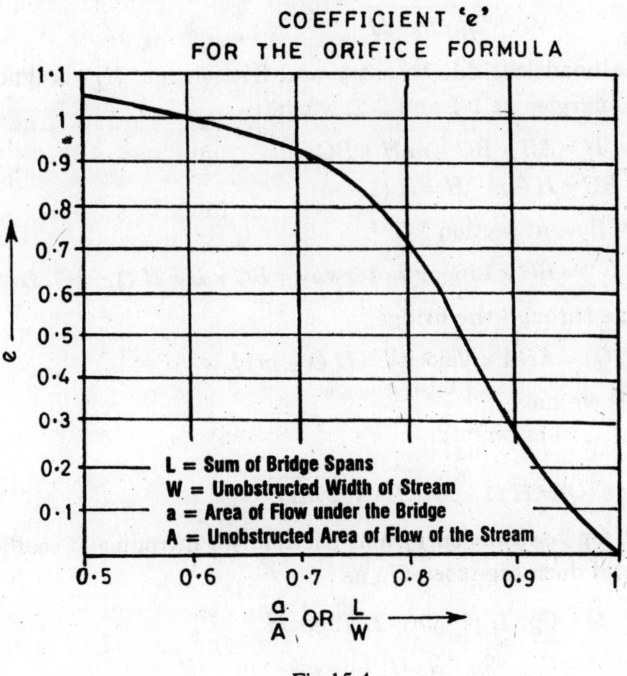


Fig 15.4

The derivations of these two formulas are given below :

**Derivation of broad crested weir formula applied to bridge openings**

With reference to Fig. 15.5, we have

The total energy ($H$) at section 1-1. $= y_u + V_a^2/2g$.

At section 2-2, let the velocity head $AB$ is a certain fraction of $H$, say let it be $m \cdot H$.

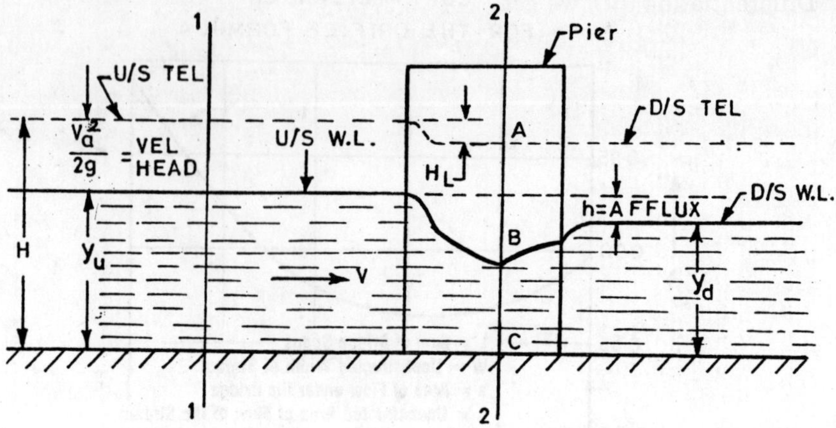


Fig. 15.5.

$$V^2/2g = mH$$

where $V$ is the velocity through the bridge opening.

$$\therefore \quad V = \sqrt{2g \cdot m \cdot H} \qquad ...(i)$$

Now, if the head loss due to entry and friction (*i.e.* $H_L$) is ignored, then by equating total energies at 1-1 and 2-2, we get

$$H = AB + BC = m \cdot H + BC$$

$$\therefore \quad BC = H - m \cdot H = H(1 - m)$$

The area of flow at section 2-2

$$= BC \times \text{Linear waterway} = BC \times L = H(1 - m) \cdot L$$

$\therefore$ Discharge through the bridge

$$(Q) = \text{Area} \times \text{Velocity} = H(1 - m)\, L \times V \qquad ...(ii)$$

But from (*i*) we have

$$V = \sqrt{2g \cdot mH}$$

$\therefore$ Discharge $(Q) = H(1 - m) \;\; L \cdot \sqrt{2gmH}$

To account for losses at entry and friction, we introduce a coefficient $C_d$ and get the final total discharge formula as

$$Q = C_d \cdot H(1 - m) \cdot L\sqrt{2gmH}$$

$$= C_d \cdot \sqrt{2g} \cdot L \cdot H \cdot (1 - m) \cdot \sqrt{m} \cdot \sqrt{H}$$

$$= C_d \sqrt{2g} \sqrt{m}\, (1 - m) \cdot \sqrt{m} \cdot L \cdot H^{3/2}$$

or
$$Q = C_d \cdot \sqrt{2g} \cdot L \cdot H^{3/2} \cdot [m^{1/2} - m^{3/2}] \qquad ...(iii)$$

The depth $BC$ adjusts itself in such a way that the discharge passing through it, is maximum. Hence,

$$\frac{dQ}{dm} = 0$$

Differentiating (*iii*), we get

$$\frac{dQ}{dm} C_d \cdot \sqrt{2g} \cdot L \cdot H^{3/2}\left[\tfrac{1}{2}m^{-1/2} - \tfrac{3}{2}m^{1/2}\right] = 0$$

$$\therefore \qquad \frac{1}{2}m^{-\frac{1}{2}} - \frac{3}{2}m^{\frac{1}{2}} = 0$$

or $$m^{-\frac{1}{2}} = 3m^{\frac{1}{2}}$$

or $$m = \tfrac{1}{3}.$$

Hence, $$Q = C_d \cdot \sqrt{2g} \cdot L \cdot H^{3/2} \cdot \left[(\tfrac{1}{3})^{1/2} - (\tfrac{1}{3})^{3/2}\right]$$

$$= C_d \cdot \sqrt{2 \times 9.81}\left[0.574 - 0.189\right] L \cdot H^{3/2}.$$

$$= 1.71\, C_d \cdot L \cdot H^{3/2}$$

or $$Q = 1.71 \cdot C_d \cdot L \cdot \left[y_u + \frac{V_a^2}{2g}\right]^{3/2} \qquad ...(15.12)$$

This is the required equation (15.12).

**Derivation of orifice formula applied to bridge openings**

With reference to Fig. 15.6.

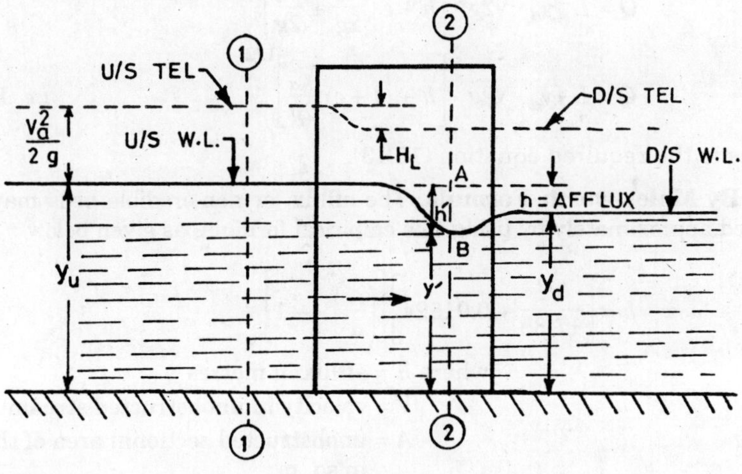


Fig 15.6

Applying Bernoulli's Equation at sections 1-1 and 2-2 and ignoring the loss ($H_L$) due to entry and friction, we get

$$y_u + \frac{V_a^2}{2g} = BC + \frac{V^2}{2g} = y' + \frac{V^2}{2g}$$

or $$\frac{V^2}{2g} = \left[y_u - y' + \frac{V_a^2}{2g}\right]$$

or $$V = \sqrt{2g\left(y_u - y' + \frac{V_a^2}{2g}\right)}$$

Putting $y_u - y' = h'$, we get

or $$V = \sqrt{2g\left(h' + \frac{V_a^2}{2g}\right)}$$

The discharge through section 2-2 is given by

$$Q = \text{Area} \times \text{Velocity}$$

$$= (L \cdot y') \times \sqrt{2g\left(h' + \frac{V_a^2}{2g}\right)}$$

When afflux is small $\left(i.e. < \frac{y_d}{4}\right)$, $y'$ is nearly equal to $y_d$, then

$$Q = L \cdot y_d \sqrt{2g} \cdot \left[h' + \frac{V_a^2}{2g}\right]^{1/2} \quad ...(ii)$$

Now $h$ is less than $h'$ because when the flow emerges from the bridge, the water surface rises due to recovery of some velocity head as depth. If $e\,(V_a^2/2g)$ is the velocity head that is converted into potential head, we have

$$h' = h + e\left(\frac{V_a^2}{2g}\right) \quad ...(iii)$$

Substituting in (*ii*), we get

$$Q = L \cdot y_d \cdot \sqrt{2g} \cdot \left[h + e \cdot \frac{V_a^2}{2g} + \frac{V_a^2}{2g}\right]^{1/2}$$

$$Q = L \cdot y_d \cdot \sqrt{2g} \cdot \left[h + (1+e)\frac{V_a^2}{2g}\right]^{1/2} \quad i.e.\ \text{Eq. (15.13)}$$

This is the required equation (15.13).

(3) **By Moles-worth Formula.** The afflux for non-erodible beds may be easily computed approximately by using this empirical formula, as given below :

$$h = \left[\frac{V^2}{17.88} + 0.01524\right]\left[\left(\frac{A}{a}\right)^2 - 1\right] \quad ...(15.14)$$

where $h$ = afflux in metres
$V$ = velocity in unobstructed stream in m/sec.
$A$ = unobstructed sectional area of the stream in sq. m.
$a$ = obstructed reduced sectional area of the stream in sq. m. at the bridge site

## 15.11. Structural Design and Other Detailing of Slab Bridges

While proceeding to design a small bridge across a drain or a canal, the design discharge, the total linear waterway, the span length, the floor level, the invert level of the bridge deck (after making a provision of clearance above H.F.L.), the normal and maximum scour depth, the depth of bridge foundations, etc. are, first of all, decided. The width of road bridge and number of lanes which it is going to accommodate are then decided depending upon the importance of the road.

Generally, road bridges are designed for two lanes of class $A$ or one lane of class 70 $R$ loading, whichever gives maximum reactions or stresses. The design procedure used for designing R.C.C. slab bridges or any other type of bridges is quite laborious and cumbersome, and the same is available in any good book of R.C.C. designs.

In actual practice, however, such detailed calculations for design of bridge decks, are usually avoided, particularly for small bridges and culverts, by making use of the standard deck designs of different spans, as published by the Ministry of Surface Transport (MOST) and Indian Road Congress (IRC).

These readymade designs not only provide the design of bridge deck, but also provide the maximum reactions that are likely to come on the piers or abutments, carrying such super structures. These reactions coupled with other forces can then be used for designing the piers as well as the abutments of such bridges.

The various forces acting on piers and abutments are taken in accordance with the provisions of I.R.C. publication No. 6–1966 titled *"Standard Specifications and Code of Practice for Road Bridges, Section II (Loads and Stresses)*.

The important forces which are taken in the design are, however, summarised below:

**Forces on Piers :**

(*i*) *Dead Load* (D.L.) from super-structure (available from readymade design tables).

(*ii*) *Live Load* (L.L.) from super-structure (available from readymade design tables).

(*iii*) *Impact Load** (I.L.) from super-structure (available from readymade design tables)

(*iv*) *Self weight of pier.*

(*v*) (*a*) *Longitudinal braking force, (P)* equal to 20% of first train load plus ten per cent of the load of the succeeding trains (in class 70 *R* loading, this is 70 tonnes). Thus, for ordinary two lane bridges, $P = 0.2 \times 70 = 14$ tonnes. (*b*) For multi-lane bridges, as in (*a*) above for first two lanes, plus five per cent of the loads in lanes in excess of two. Thus, for four lane bridges,

$$P = 0.3 \times 70\ t = 21 \text{ tonnes.}$$

(*vi*) The *wind load* and seismic forces (either of the two) should also be considered wherever predominant. Since the consideration of these forces permit to increase the permissible stresses in brick or concrete (materials used for piers) by 25% or so ; for ordinary designs in non-seismic or less seismic regions and where bridges are not at much height above the natural ground levels, these forces may often be neglected.

(*vii*) The *forces caused due to buoyancy of water* (*i.e.* reduction in the weight of pier material due to uplift exerted by the water surrounding the pier) may also be considered in design. Often this is also neglected when pier thickness, etc. are governed by the minimum permissible limits.

(*viii*) *Other minor forces* such as horizontal forces caused by water currents, etc.

**Forces on Abutments.** Besides D.L., L.L., self weight, seismic or wind forces and buoyancy if there, the important force which comes into play is the pressure exerted by the soil fill from behind the abutment. The soil may be dry or wet or saturated depending upon the site conditions. The horizontal braking force and the horizontal force caused by temperature variations will no doubt act in addition to

* The 50% impact load is considered to be a part of live load for calculating pressure at the bottom surface of the bed block ; and an impact load varying between 50% to zero is considered for calculating the pressure on the top 3 m of structure below the bed block. Thus, no impact is considered for examining pier or abutment sections below 3 m from bed block.

the soil pressure. All these forces may be considered and abutment designed. No tension should be permitted in masonry piers, and the base should be wide enough so that the soil pressure does not exceed the permissible bearing capacity of the soil.

The detailed designs of bridges are beyond the scope of this book, and the students may refer to some standard book on bridge designs.

## CAUSEWAYS AND BOX CULVERTS

### 15.12. Causeways

When a road embankment is constructed across a flat terrain, the runoff from the area having no deep and defined channels in it, may have to be disposed of by some suitable means. In such cases, the surface water will go on collecting on one side of the road embankment, unless drained out either through culvert openings inside the embankment or by allowing the water to submerge the dipped road at frequent intervals. *The second arrangement called a causeway consists in lowering or dipping the road to the ground level at frequent intervals, and thus allowing the water to flow across the road.* The provision of such a dip or causeway in the longitudinal profile of the road and letting the drainage water pass over them, may be feasible in less important towns, and may generally be impracticable because of various limitations, such as : (*i*) flow of water over numerous sections of the road makes its proper maintenance difficult and expensive; (*ii*) when the embankment is constructed high above the ground (which may be necessitated in swampy and water-logged areas), the provision of causeways will require dipping down the high road levels to the ground at frequent intervals and will produce a very undesirable road profile. Hence, under all such circumstances, it may be better to construct culverts at intervals rather than frequently dipping the road for crossing the surface water.

### 15.13. Pipe Culverts Flowing Full and Box Culverts

After having decided to construct a culvert across some such flat terrain, the runoff likely to be passed may be calculated. Now, the water coming towards the road in such cases is having very low velocities and cannot be passed below the road at those velocities, because in that case, the area of the opening will be very large. Hence, in such cases, the design is based on an increased velocity of flow through the culvert, and to create that velocity, the water is allowed to head up at the inlet end of the culvert. The greater is the heading up of water at the inlet end, the greater is the head causing flow ; and hence, the greater will be the flow velocity and lesser the opening. But this heading up of water can be allowed only to a safe predetermined level, and should be fixed, so that the road bank is not overtopped nor any property in the flood plain damaged.

After fixing the upstream level up to which the water will head up while passing maximum design discharge, the downstream water level is determined. This level will depend on the hydraulic conditions of the tail channel, if at all existing, and is generally taken as equal to the surface level of the natural unobstructed flow at the site before the road embankment is constructed.

The operating head, i.e. the difference of the upstream and downstream levels (say $H_L$) will give us the maximum head causing flow. The area of the opening should then be decided, so that it is sufficient to pass the design discharge.

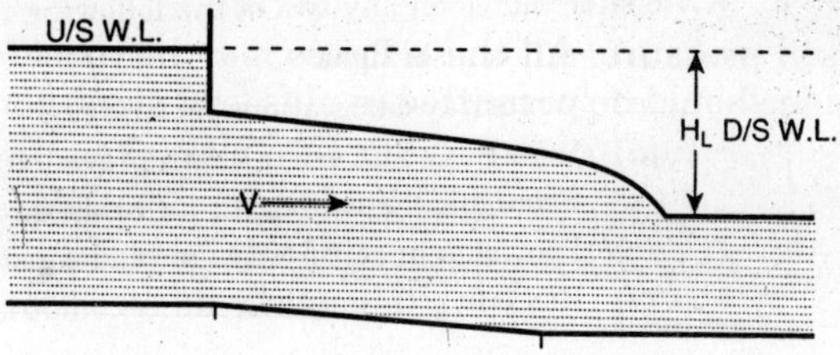


Fig. 15.7. Box or pipe culvert running full even when the outlet is not submerged.

If $V$ is the velocity through the culvert opening, which runs full in such a case, then the head loss ($H_L$) will be equal to the sum of the entrance loss, friction loss in the barrel, and the velocity head in the barrel. The entrance loss depends on the type of entrance provided, and may be taken equal to 0.505 ($V^2/2g$) for a square edged entrance, and equal to 0.05 ($V^2/2g$) for a well rounded entrance. The friction loss may be calculated by Manning's formula as $\dfrac{n^2 \cdot V^2 \cdot L}{R^{4/3}}$

Hence, $H_L$ = Entrance loss + Friction loss + Velocity head in barrel

$$= K_e \frac{V^2}{2g} + \frac{n^2 \cdot L \cdot V^2}{R^{4/3}} + \frac{V^2}{2g}$$

where $K_e = 0.505$ for square edged entrance

$= 0.05$ for bell mouthed entrance.

$$\therefore \quad H_L = K_e \cdot \frac{V^2}{2g} + \frac{V^2}{2g} + \frac{n^2 \cdot L}{R^{4/3}} \cdot 2g \left(\frac{V^2}{2g}\right)$$

$$= \frac{V^2}{2g}\left[1 + K_e + \frac{n^2 \cdot L \cdot 2g}{R^{4/3}}\right] = \frac{V^2}{2g}\left[1 + K_e + K_f\right]$$

where $K_f = \dfrac{n^2 \cdot L \cdot 2g}{R^{4/3}}$

or $$V = \sqrt{\frac{2g \cdot H_L}{1 + K_e + K_f}}$$

$$\therefore \quad A \cdot V = A \cdot \sqrt{\frac{2g\, H_L}{1 + K_e + K_f}}.$$

Hence, knowing the area, the discharging capacity of the culvert, or by knowing the discharge, the required area can be easily computed. The required sized pipe (or a rectangular barrel) may then be constructed through the road enbankment.

## PROBLEMS

1. Discuss briefly the various principles and guidelines which help us in fixing the hydraulic design of a bridge culvert.
2. (a) Distinguish between 'normal scour depth' and 'maximum scour depth' in connection with bridge design. How does it help in determining the depth of bridge foundation ?

   (b) How will you fix up the number of spans for a bridge after computing its required linear waterway ?

   (c) "Reducing the waterway of a bridge will reduce the cost of superstructure, but will increase the cost of sub-structure." Discuss critically the above statement.

**3.** Write short notes on any two of the following :

(*i*) Afflux at bridges.

(*ii*) Waterway calculations for bridges.

(*iii*) Waterway and scour depth provision in bridge design.

(*iv*) Economical span length for bridges.

(*v*) Provision of cause-ways in road construction.

(*vi*) Box culverts and their afflux computations.

# 16

# *River Navigation*

## 16.1. Introduction

There shall be no exaggeration, if we define water as something which is a necessity, a comfort and a luxury too. There cannot be two opinions about water being a necessity, as there can be no life without water. Men, animals and even plants can never survive if there is no water. Besides the basic necessities for existence, water provides us various comforts and luxuries, and makes our lives easier. Besides the various uses of water, water provides us cheap* means of transport. Before the inventions of rail-road transport, water was the only means of communication between distant places. Even today, cargo – too heavy to be packed on wagon trains or trucks, is lifted and transported through big ships. In the modern affluent living, towing through water is not only required for transporting purposes, but is also required for recreational boating. However, boating and floating of ships through the natural rivers is not always safe. Rapids and sandbars may create problems, and may require considerable time to be passed. Isolated rocks, fallen trees, debris and other obstructions may create constant hazards, and may damage or even wreck the boats, steamers, or ships being towed through such waterways. It is, therefore, absolutely essential that all the waterways through which the boats or ships are to be towed, must be made completely safe.

The chief requirement for navigating through a waterway is the availability of sufficient water depth in the waterway. A minimum water depth of about 2.7 metres is generally required for navigating safely and economically ; although a depth of about 3.7 metres is generally aspired in the final development of a navigable waterway. Availability of lesser depth in the rivers may completely eliminate the possibility of towing the ships through such rivers or may cause increased unit cost of transport.

The cost of water transport is generally expressed in ₹ per tonne per kilometre. The total cost per kilometre of travelled distance divided by the transported tonnage, gives us the cost per tonne per kilometre. The total cost consists of fuel charges, wages of crews, etc., and other operating and incidental charges. The wages and other operating charges depend only upon the time of transit ; while the fuel consumption charges depend on one more factor *i.e.* the available draft. *The consumption of fuel decreases as the depth of water between the bottom of the ship and the bottom of the river increases.* This is because, when the clear depth between the barge and the river reduces, an increased drag force is developed which increases the consumption of fuel. So much so, that the fuel consumption can be reduced by about 25%, if this clear space is increased from 0.6 m to 1.5 m or so. In order to maintain this minimum depth available, the ship will have to be loaded only to such an extent that a certain minimum clear space remains available between the bottom of the ship and the bottom of the river. Hence, the amount of loading of the barge is dependent and limited by the availability of sufficient water depth at the shallowest point in the route. From this discussion, it

*Water transport is cheaper because on water, the coefficient of friction being small, one horse power can pull 4 tonnes ; while on road and rail, it can move approximately 0.15 tonne and 0.5 tonne, respectively.

can be easily concluded that a certain minimum depth depending upon the type and extent of traffic is always required to be maintained in the river, so as to keep it fit for shipping or towing of barges. When this available depth or draft is not sufficient or if the channel width, channel alignment, locking time, current velocity, terminal facilities, etc. are not adequate, engineering measures are undertaken so as to increase the available draft and to improve other requirements. By undertaking more and more measures, and thereby spending more and more money on improving a waterway for navigation, we can make heavy and heavier transport feasible through that waterway. A compromise between the money invested on these technical works and corresponding advantages likely to be obtained by way of better and cheaper water transport, must be weighed against each other, while working out a financial justification for such works. Besides the direct tangible benefits such as increased and cheaper communications for commerce and trade, etc. ; the intangible benefits such as recreational boating and better defence facilities for carrying military equipments, during war times, should also be given due importance, while working out the economic justification for such improvement works.

## 16.2. Various Requirements of Navigable Waterways

There is no rigidity about the requirements of a good navigable waterway, since it all depends upon the extent and type of traffic likely to pass through it. However, the various general requirements are enumerated below :

(1) Sufficient water depth is available so as to pass the more heavily loaded barges cheaply and economically.

(2) The width of the waterway is sufficiently more than the width of the tow itself.

(3) The radii of the bends should not be sharp and should be high enough to allow the maximum length of the ship to pass through them.

(4) The alignment of the waterway should be as straight as possible, because a highly irregular alignment increases the circuitry or length in excess of air-line distance which the barge tow must travel. The existing channels generally have a length about 50 per cent greater than the air-line distances.

(5) The flow velocities should not be high, as they may cause substantial reduction in the true speed for tows moving upstream and thereby increasing the time of transit and the cost of transport per kilometre. The speed of most of the barge tows in still water is of the order of 2.8 m/sec. The flow velocities of the order of 1 m/sec may, therefore, cause sufficient reduction in true speed (*i.e.* 2.8 – 1.0 = 1.8 m/sec) and hence, should not exceed such a value.

(6) In order to minimise the transit time, the time required for the tow to pass through locks should be minimum. In certain cases, where the lock is not large enough to accept the entire tow, the tow is generally broken and taken through the lock in portions. This increases the time lost in locking, and thereby increasing the transit time and the cost of transport. Hence, sufficient sized locks should be ensured for economic and better transport.

(7) Efficient and adequate terminal facilities for unloading the barges for transferring the cargo effectively, must be ensured for economic and better navigation.

## 16.3. Various Measures Adopted for Achieving Navigability

There are three basic methods which are generally adopted for improving a river for navigation. They are :

(1) *Open channel methods*

(2) *Lock and dam arrangements* ; and

(3) *Canalisation.*

They are described below :

**16.3.1. Open Channel Methods.** In the open channel methods, the existing waterway is improved to such an extent as to make navigation possible. This improvement in natural waterway is possible only if the following conditions are satisfied :

(*i*) Sufficient discharge is available in the river throughout the year or at least for a reasonable portion of the year.

(*ii*) The existing river is having a satisfactory alignment without excessively sharp bends.

(*iii*) The river bed slope is reasonably flat so that the flow velocities are not excessive (*i.e.* they are within 1 m/sec or so).

(*iv*) The river width is not too small and is such that it can be improved economically for the modern barge tows.

(*v*) The material of the river bed and banks should permit satisfactory treatment by one or more of the open channel methods.

If the above requirements are approximately satisfied, the channel can be economically improved and made fit for navigation. But if the available conditions are far too short of requirements, open channel methods may prove to be highly uneconomical, and, therefore, should not be considered. However, these requirements and factors may be controlled to some extent by some suitable measures. Say for example, if the discharge in the river during lean periods is very low, while the average annual flow is adequate, reservoirs may be constructed so as to store water and augment the supplies during lean weather flows. Similarly, very sharp bends may be eliminated by cut off channels, provided the resulting channel slopes remain within limits.

The **various works and techniques** that may be involved in improving the channel by the Open channel methods are :

(*a*) *Constructing and Regulating the flow through Storage reservoirs.*

(*b*) *Excavation and Dredging.*

(*c*) *Contraction works.*

(*d*) *Bank stabilisation.*

(*e*) *Straightening the waterway by artificial cut offs.*

(*f*) *Removal of snag, debris and other obstructions.*

These techniques are generally required together as one of them may rarely provide the necessary required improvement. These techniques are described below :

(*a*) **Storage reservoirs.** The storage reservoirs generally store water during high flows and can release the required amount of water during lean-flows, so as to make downstream navigation possible even during periods of low weather flows. However, the construction and planning of storage reservoirs for navigation alone is not generally justified economically. Hence, reservoirs are mostly planned under Multipurpose projects, where navigation may be one purpose of that project. Moreover, the storage reservoirs can augment low supplies for navigation, only if the reservoir is situated at the head of a relatively short navigable reach. This is because ; *as the distance from the reservoir to the navigable river-reach increases, reservoir-releases have to be increased so as to allow for transit losses due to seepage, evaporation, etc. The releases must also be made much in advance so as* *to allow for travel time to the navigable reach and their quantity has to be sufficient even after reduction due to channel storage.*

(*b*) **Excavation and dredging.** Huge amounts of excavations are generally required for clearing sand bars and filled channel sections in order to make it fit for navigation. Besides the basic initial excavations, continuous desilting and proper maintenance is required in order to keep the waterway fit for navigation. These excavations from the bed and banks of the waterway are generally carried out by *dredging* by means of dredgers. Three types of **dredgers** are generally used. They are :

(*i*) ***Dipper dredgers.*** They are merely floating power shovels and are used on small projects.

(*ii*) ***Ladder dredgers.*** They have an endless chain of buckets for bringing the excavated material upto the surface. The cuttings carried by buckets are discharged on a *belt conveyer* which are disposed of through a *stacker conveyor* at the rear of the dredger. Since the stacker conveyors (generally called *spoil stackers*) are limited in length to about 100 metres or so, ladder dredgers cannot be used when the excavated materials (*i.e.* spoils) are to be discharged at a considerable distance from the dredge.

(*iii*) ***Suction dredgers.*** In these dredgers, the cuttings and water are collected in suction pipes, and the mixture is then discharged by pumping through a pipe supported by floats (called spoil pipe) into the desired spoil area. A line diagram of this operational process is shown in Fig. 16.1.

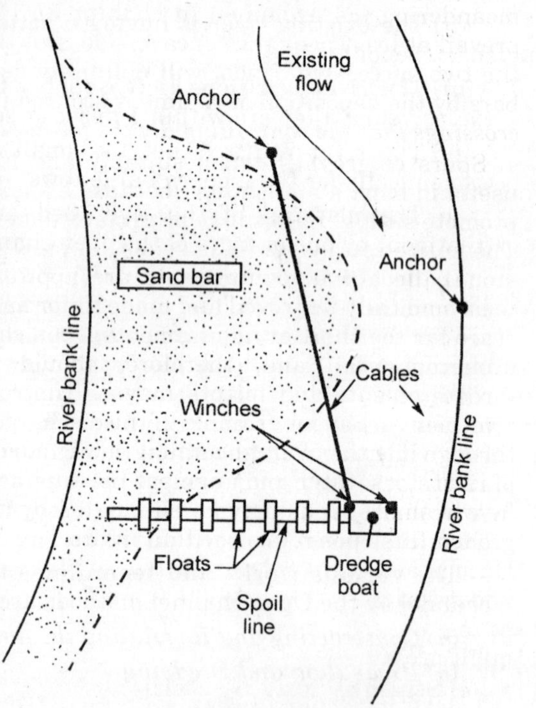


**Fig. 16.1.** Line plan of an ordinary suction dredge called Dust pan dredge.

A suction dredge cannot operate in rocky or boulder river reaches. The suction head of these dredges is provided with jets or rotating blades so as to cut or loosen the bed material, and also with suction openings through which the soil and water mixture enters into the suction pipe. These dredges can make cuts of about 10 m width through sand bars, and various such parallel cuts can be made in order to achieve a wider channel.

(*c*) **Contraction works.** Contraction works are those engineering works which are constructed in order to change a wide shallow river into a narrow deep river ; or to close off the river creeks (small branches) and thus to divert the entire water into the main river. When the material of bed and banks of a river is coarse grained with little cohesion; a shallow wide channel, or a number of coarse channels at low waters, will develop. Such situations may be corrected with the help of spurs or groynes. Under the process, rivers carrying huge sediment loads can be corrected with the help of properly placed permeable spurs called sal balli spurs made of sal ballies, driven at some suitable distance centre to centre in rows across the river current and braced at top. The function of permeable sal balli spurs is to slow the current, and thus promote silting in the dyked area. The concentration of flow in the narrower section also encourages deepening of the channel. Several years are allowed for the effect of the structures to develop.

Similarly, the rivers carrying a little sediment load can be corrected by properly placed impermeable spurs or jetties which shall divert the flow, thereby confining the entire water in a smaller width and thus deepening the same.

(*d*) **Bank stabilisation.** A good navigable channel must have stable banks. When the river banks are not stable and start caving, the river starts meandering, creating bends, which may obstruct the path of longer barge tows. Moreover, scouring at concave banks and silting at convex banks take place due to meandering (as explained in chapter 8). Hence at bends, sufficient depth will prevail at least near the concave side. But the tangents, *i.e.* the crossings joining the two successive bends, will definitely develop shallower channels with cross bars by the deposition of sediment scoured from the upstream bend. *It is in these crossings that the controlling depths for navigation occur.*

Spurs or groynes, when suitably and intelligently placed, may prove to be useful in bank stabilisation ; because a spur placed along the concave bank shall promote silting. Banks may be protected more easily by *pitching* or by *rivetments*. The entire concave bank is generally protected by pitching. The loosely dumped stone called apron or riprap is generally used, and it is extended from top of bank to beyond the toe of the underwater slope. This extension of rivetment in the bed is essential so as to avoid the failure of rivetment due to scour and consequent undermining of the underwater edge of the rivetment. The rivetment must be flexible so as to adept itself to the surface on which it is placed. Moreover, the rivetment must be relatively impervious so as to avoid, the washing of fines through it. It must also be strong enough to resist the flow currents. Various types of rivetments are used. Concrete mattresses in the form of concrete blocks placed in wire meshes may sometimes be used, when ordinary stone dumping over a graded filter is not provided due to non-availability of stone in the nearby areas. Uncompacted asphalt paving is also finding a use in developed countries, and is under serious investigations. Compacted asphalt paving and monolithic concrete paving are not generally used, as they are liable to be cracked and damaged by uplift pressures.

(*e*) **Straightening by artificial cut-off.** It was explained in chapter 8, as to how and when a cut off develops in a meandering river. Since the development of a cut-off eliminates sharp bends which are undesirable for navigation, artificial cut-offs may sometimes be used advantageously. A pilot cut is made and allowed to develop (Fig. 16.2). These cut offs have been used with success at various places, provided the banks are stabilised to avoid future caving and meandering.

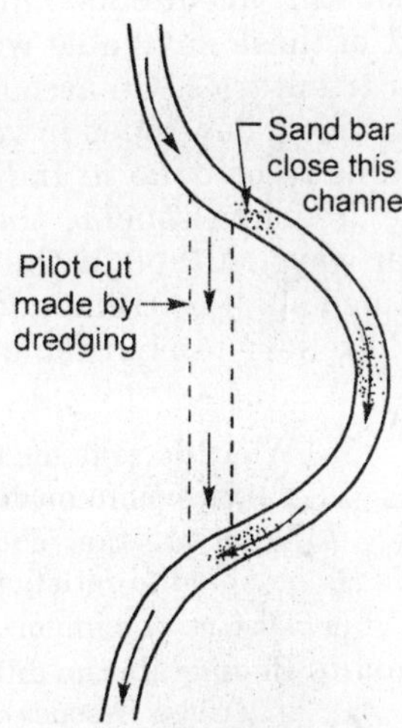


**Fig. 16.2**
Development of a cut off.

(*f*) **Removal of snag, debris and other obstructions.** Presence of debris, trees, isolated rocks, and other obstructions, not only pose a direct hazard to the barge tows, but also promote the formation of sand bars. They must, therefore, be removed effectively in order to ensure safe and economical navigation. Different methods and equipments may be used in different cases, depending upon the circumstances of each case. Tractors, winches, derrick barges, explosives, etc. may be required in the process of clearing the waterway of these obstructions.

**16.3.2. Lock and Dam Arrangements.** The arrangement consists of dams which create a series of slack water pools through which the traffic can move with locks to lift the vessels from one pool to the next. Lock and dam construction may be adopted where existing site conditions are not favourable for adopting open channel methods described earlier. *This arrangement is a second choice to open channel methods.* In this arrangement, water is required for lockages, sanitary releases, evaporation, percolation, etc. This requirement of water is much less than that required for open channel procedures. Hence, when the available water is less, these arrangements may have to be adopted.

The slack water pools behind the dams, will submerge the rapids and channel bends, and thus overcoming those problems. Further, because of their relatively large areas of cross-sections, the velocities in these pools shall be low enough as to cause lesser reduction in true speed of the barge tow moving upstream.

Lock and dam arrangements are suitable only on rivers bringing only a little sediment load. Because, highly silt laden river-waters shall fill up the pools rapidly. Moreover, suitable sites for construction of small dams must be available for providing such arrangements. Further detailed descriptions of these arrangements are beyond the scope of this book.

**16.3.3. Canalisation.** A totally new channel-cut is provided artificially around an otherwise impassable obstruction or between two navigable rivers. Such a cut is generally economical only when a short length of new channel opens a large length of existing waterways. Construction of a new channel connection between two existing waterways is also sometimes adopted, so as to ensure a continuous traffic way. However, canalisation is a costly process, as the per kilometre cost of canal, capable of passing modern barge tows, is normally very high, and are adopted when very short lengths are required.

## 16.4. India's Navigable Waterways

Unfortunately, there is no river in India running from North to South or *vice versa*. Most of the Indian rivers run from West to East or from East to West. This is the major handicap in the development of navigation in this country. Although there are about 15,000 kilometres length of navigable waterways in India, but most of them are in very bad conditions due to silting and poor maintenance. Most of these waterways are fit only for boating except for about 2,500 kilometres length, in which the steamers can ply. Out of the total navigable lengths of about 15,000 km ; about 8000 km is on rivers, and 7000 km is on canals and backwaters. Most of these navigable waterways exist in Ganga and Brahmaputra rivers and their tributaries. On account of its inherent advantages, the water transport has been widely developed in various countries (especially in U.S.A.), but a lot of work remains to be done in India. Let us hope that as soon as our Government finds some free investment, the same will be invested in improving our navigable water-ways. Although the initial investment may be high, but the advantages obtained shall be tremendous, as it will not only provide us cheap transport, but would relieve the burden upon our rail and road transport.

## PROBLEMS

1. (*a*) What do you mean by 'River navigation'? What is the importance of river navigation in modern system of transport ?
   (*b*) What are the requirements of navigable waterways, and what measures are adopted to satisfy these requirements ?
2. (*a*) Discuss the importance and necessity of improving India's waterways.
   (*b*) Enumerate the different methods which are adopted for achieving navigability in a river. Discuss one of these methods in details. Also discuss the pros and cons of different methods and their use for a particular project.
3. Write a descriptive note on 'River navigation'.

# 17

# *Dams in General and a few Dams in Particular*

## 17.1. Definition and Uses of Dams

A dam may be defined as an obstruction or a barrier built across a stream or a river. At the back of this barrier, water gets collected, forming a pool of water. The side on which water gets collected is called the upstream side, and the other side of the barrier is called the downstream side. The lake of water which is formed upstream is often called a *reservoir*, or a *dam reservoir*, or a river reservoir, or a *storage reservoir*.

The water collected in this reservoir can be supplied for irrigating farm lands through a system of canal network, or may be supplied for drinking purposes. The lake so formed can be used for recreation uses. The energy of this collected water can be used to turn a mill to grind wheat or to turn the blades of a tubrine to generate electrical power. And in times of floods, the dams can serve as protections for the towns and cities farther down the river.

Apart from these numerous advantages and uses (such as navigation, irrigation, electricity, flood control, etc.) of a dam, it sometimes helps us in planning war strategy and helps us in controlling the advancement of enemies and their forces. Dams have been frequently opened in times of war. The Dutch breached their dikes during Second World War to bedevil the invading Germans. Chinese used to destroy their dikes to flood out the enemy. Russian army retreating from the Nazi marauders, partly destroyed the famous Dneprostroi Dam in the Ukraine to keep its power plant from falling into the hands of Hitler's men.

## 17.2. History of Dam Construction.

It is very difficult to say as where and when the first man-made dam was built. Archaeological evidences help in estimating that the very first man made dam is at least 3000—5000 years old. Whenever it was built, that first dam was almost certainly an irrigation dam. Its designer might have observed beavers at work or he might have thought of it the some other way. Beavers are mammel creatures belonging to the family of rats, mice, squirrels. They live under water and are generally 1.2 m long including more than 0.3 m of tail, and weigh upto 25 to 30 kg. These rodents produce fur. These creatures create dam type barriers to create a place where their family can live. They provide themselves with comfortable ponds to live in, by building dam type barriers across the stream with the help of trees which they themselves cut.

These elegant structures, built out of logs, buttressed with twigs and branches and sealed with mud and stone help us many a times in controlling silt entry into the stream. A beaver dam accumulates silt brought down by its stream. When a beaver dam is breached, the silted water pours through, and the fertile silt is deposited over the wide area. This creates what farmers call a *beaver's meadow,* where crops grow particularly well.

Beavers are sometimes encouraged to build dams in areas where man can not reach easily to construct man-made dams.

### 17.3. Construction of Modern Dams

The first modern dam of the world was perhaps constructed on the Nile river, in Egypt at *Aswan.* It was completed in 1902 and was a major engineering project.

This famous **Aswan Dam** was designed primarily to control the flooding of the Nile river, to promote irrigation in the Nile Valley, and to further navigation along the river.

Aswan is 1200 km from the mouth of the Nile. The site was chosen because the river at Aswan is shallow and has a granite bed, on which a firm foundation could be erected.

The first step in damming any major river is to divert the flow of water so as to permit dam construction. This was accomplished at Aswan, by constructing a circular earthen barrier around the area chosen from the dam foundations. At low river, this enclosure was pumped out. All the work had to be completed before the river flooded again. The foundation was laid round the clock with a huge labour force. So much so that 3,600 tons of masonry were put in places in a single day.

This 120′ high dam, running for about $1\frac{1}{2}$ miles from shore to shore, 100 feet thick at its base and 24′ wide at top, costed about 1.5 crore dollars, when completed in 1902. A roadway ran along the upper rim. About 1 billion tons of water was its storage capacity, and 180 sluices were constructed into this barrier. Water could be collected during the rainy season and released through the sluice gates during the summer and drought.

**Roosevelt Dam.** The next famous dam of the world was completed in 1911 on the Salt River of Arizona (U.S.A.) and was called Roosevelt dam.

The Aswan dam was constructed with stone, while the Roosevelt dam was constructed with solid blocks of concrete : which was of the type known as 'Solid Masonry Gravity Dam' : which simply means that it was built with solid blocks of concrete, which hold back the flow of water by sheer weight. This type of dam is one of the most ancient.

The construction was started in 1905. Despite the numerous troubles in its construction, the dam was completed in 1911 with a reservoir capacity of $\frac{1}{2}$ billion gallons of water. The completed dam, 280 feet high, was *158 feet thick* and spanning for a length of 1,125 feet. Had the dam been built without the benefit of 19th century engineering advances in design and stress control, it would have had to be 700 to *1000 feet thick* at base ; a far more expensive and cumbersome thing to build.

The Roosvelt dam was a fascinating advancement, but was soon overshadowed by other vastly greater dams, such as *Hoover Dam*, (726′ high), *Bhakra Dam* (740′ high), etc. All of them are 'solid concrete gravity type' dams.

### 17.4. Various Types of Dams

Before we describe some of the famous dams of the world, it is worth while to classify the various types of dams.

Most engineers recognise seven general types of dams. Three of them are ancient in origin, and four have come into general use only in the last about 100 years or so.

***The three older types of dams are :***

(1) *Earth Dams*

(2) *Rook-fill dams*

(3) *Solid masonry gravity dams.*

These three types of dams are discussed below :

(1) **Earth Dams**. Earth dams are made of soil that is pounded down solidly. They are built in areas where the foundation is not strong enough to bear the weight of a concrete dam, and where earth is more easily available as a building material compared to concrete or stone or rock.

Some important earth dams of the world are :

(*i*) Green mountain dam on Colorado river in U.S.A.

(*ii*) Swift dam in Washington in U.S.A.

(*iii*) Side flanks of Nagarjun Sagar dam in India.

(*iv*) Trinity Dam in California in U.S.A.

(*v*) Maithan Dam in India (which is partly Earthen and partly Rockfill).

(2) **Rockfill Dams.** Rockfill are formed of loose rocks and boulders piled in the river bed. A slab of reinforced concrete is often laid across the upstream face of a rockfill dam to make it water-tight.

Some important rock-fill dams of the world are ;

(*i*) The Salt Springs Dam in California (345′ high) in U.S.A.

(*ii*) The San Gabriel No. 1 Dam (321′ high) in U.S.A.

(*iii*) Cougar Dam on Mc-Knezie River in Oregon (445′ high) in U.S.A.

(3) **Solid-masonry Gravity dams.** These are familiar to us by now, after we have talked about Aswan, Roosevelt, Hoover, and above all Bhakra dam.

These big dams are expensive to be built but are more durable and solid than earth and rock dams. They can be constructed on any dam site, where there is a natural foundation strong enough to bear the great weight of the dam.

These three types of dams were all found in ancient days. In recent times, *four other types of dams have come into practice.* They are :

(4) *Hollow masonry gravity dams* ;

(5) *Timber dams* ;

(6) *Steel dams* ; and

(7) *Arch dams.*

(4) **The hollow masonry gravity dams**. These are essentially designed on the same lines on which the solid masonry gravity dams are designed. But they contain less concrete or masonry ; about 35 to 40% or so. Generally, the weight of water is carried by a deck of R.C.C. or by arches that share the weight burden. They are difficult to build and are adopted only if very skilled labour is easily available, otherwise the labour cost is too high to build its complex structure.

(5) **Steel dams.** These are not used for major works. Today, steel dams are used as temporary coffer dams needed for the construction of permanent dams. Steel coffer dams are usually reinforced with timber or earthfill.

(6) **Timber dams.** These are short lived, since in a few years time, rotting sets in. Their life is not more than 30 to 40 years and must have regular maintenance during that time. However they are valuable in agricultural areas, where a cattle

raiser may need a pool for his live stock to drink from, and for meeting other such low-level needs.

(7) **Arch dams.** Arch dams are very complex and complicated. They make use of the horizontal arch action in place of weight to hold back the water. They are best suited at sites where the *dam must be extremely high and narrow.* Some examples are :

(*i* *Sautet dam* on the Drac River in France, 414′ high, but only 230′ long at top and 85′ long at bottom of the gorge, 56′ thick at bottom and 8′ thick at top.

(*ii*) *The Tignes dam* in France (592′ high).

(*iii*) *Mauvoisin dam* on the Drause River in Switzerland, (780′ high).

(*iv*) *Idduki* dam in Kerala State, across the Periyar river, which is the only arch dam in India. It is 366 m (1200′) long double curvature arch dam, made in concrete, and has a height of about 170 m (560′).

## 17.5. Problems in Dam Construction

Dams are extremely useful things. Anyone who lives in Punjab or at Asansol in West Bengal, knows how valuable dams are. The farmers of Punjab and people getting electricity from the Bhakra sing praises for it. The people of areas benefitted by various dams and other ancillary works on Damodar river are really thankful to those human beings who have miraculousy harnessed the Damodar river for them. The prosperity and welfare of millions and billions of people depend directly on these towering handsome dams with which the nation's rivers have been harnessed.

But dams can cause problems too. Dams have drawbacks and disadvantages also. Let us here discuss some of the negative features of dams and let us see what can be done to overcome them. There are four major problems, in general, which are posed by such huge constructions. They are:

(1) Fish Problem ;
(2) Submergence Problem ;
(3) Failure Problem; and
(4) Bomb Problem

They are described below:

(1) **Fish Problem.** On large rivers, in late summer season, fish move from downstream to upstream to lay their eggs. These eggs are fertilised by male fish. The old fish may get exhausted and the new born fish again move downstream. They, after two to three years, return to their ancestral spawning place and may die after getting exhausted, while the newborns move downstream. The cycle goes on for years.

The fish which move to their ancestral spawning place (upstream) are called *anadromous fish. Salmon and Hilsa* are typical examples of such a fish. These are commercially valuable fish, and important industries are dependent on them.

When a dam barrier is constructed on a river, these fish can not move upstream to lay their eggs; because it is impossible for these fish to overtop such a barrier. But surprisingly, even when they find a barrier in their path of advancement towards their ancestral spawning ground, these fish do not return to their downstream dwelling place (*i.e.* sea). However, they go on fighting against the barrier, trying furiously to overtop it, till they get exhausted and die down. This results in a serious large scale killing of fish, causing great damage to fish industry and economy of the nations.

In the beginning, much attention was not paid to this problem; but a little later, it was realised, and serious attempts were made to find out solution to the problem. Sometimes, fish were trapped on one side of the dam and passed on to the other side by giant steel and plastic nets. An external arrangement called *Fish Ladder* was also devised.

**Fish Ladder.** Just as river-going vessels can bypass a dam by using a navigation lock, so a series of 'locks' enable the fish to get over the dam. A separate channel is created, consisting of a series of little dams that form a row of pools, rising up over the big dam to reservoir level. The salmon, entering the lowest rung of the ladder at the base of the dam, could leap from pool to pool until they had crested the dam. Then, they could continue on through the reservoir to the spawning grounds. The new born fish called *finger lings* could later return to the sea (downstream) in the same fashion via the ladder. A section, plan and photographic view of a fish ladder has already been shown in the chapter on Weirs.

In the beginning, the fish ladders worked better in theory than in practice. The fish seemed to prefer to mill ground in splashing water under spillway, instead of entering the ladder. This difficulty was overcome by careful design that put the fish ladder in the place where it was most likely to attract the fish. Another problem was that the slow moving water was stranger to fish and they tended to collect in the lower pools without going onward.

Millions and billions were spent into fish-ladder research. Improvements in design made the fish ladder more attractive to fish, more like the rapids they were accustomed to.

Fish ladders are not always practicable from engineering stand point. In such cases, other steps have to be taken to protect the fish.

Meanwhile, other experiments are going forward to see if fish can be successfully induced to spawn in waters other than their own ancestral spawning grounds. In the long run, it may save millions of currency to construct *fish hatcheries* instead of *fish-ladders*. There are many possible solutions to the problem of anadromous fish, and research is being undertaken in different regions of the world to find out a better solution to the problem.

(2) **Submergence Problem.** Whenever a dam is constructed across a river to store water on the upstream side, a large area gets submerged due to the rise in the water levels. The entire area which gets submerged, forming a reservoir, has to be calculated and acquired before a dam can be constructed. The owners of the land have to be persuaded, adequately compensated, and well settled somewhere else, before, the work can be taken up in hand. Hence it is necessary to investigate the probable damage caused by this submergence.

(3) **Failure Problem.** We try our best to build dams to last as long as possible. Every person whosoever has worked on a dam hopes that the dam will live as long as the pyramids of Egypt. But many a times, the dam give way under the continued insistent pressure of the water penned up behind them. This failure of the dam may be caused either due to bad workmanship or due to faulty design or due to the occurrence of unanticipated floods.

Luckily, these disasters have been comparatively rare in this century. Dams used to give way easily in olden times, but due to engineering advancement in modern times, their failure has been considerably reduced.

These huge structures are now properly designed, keeping in view the various forces which they are going to face. Proper and rational design, good supervision and constant vigil and watch during maintenance period ensures their safety and makes us fairly confident of it. Bhakra Dam on Satluj River in India and Boulder Dam on Colorado River in U.S.A. *cannot fail* in one attempt, how furiously these rivers may try to move their foundations. We are fairly confident of this, but sometimes the confidence is *rudely and cruelly repaid with tragedies.*

Dams may sometimes fail due to excessive and unanticipated earthquakes. The *Koyna Dam* in India was at the verge of failure in 1968 earthquake. Thanks to the efforts of the Indian engineers who saved that dam by toiling hard day and night. A very confident dam called *Vega de Tera Dam* is Spain failed in January 1959. The people were tucked in the town of Rivaldelago. The disaster caused was tremendous. Rivaldelago was flattened. Telephone poles were snapped like matchsticks. Within moments, 123 villages were drowned. Several hundred luckier ones were saved, but were rendered homeless. This was a case where a dam had simply not been built strong enough to bear the full weight of its intended reservoir. Heavy rains wrecked it. *Faulty design and bad engineering must be blamed.*

Another important dam called *The Malpasset Dam,* a 200 feet high arch dam on the Reyran River, was completed in 1954. This dam gave way in December 1956, causing 421 persons to die in floods. Investigations revealed that the dam had failed because the foundation rock has shifted along a thin clay seam in the left abutment, making the dam unstable and vulnerable to any serious stress.

We learnt from our mistakes;and several other dams of the same type, then under construction in Europe, were quickly resurveyed to find the possibility of such a geological formation. This was very very small comfort to the relatives of those who died when Malpasset failed; but at least, we should learn from our mistakes and there should be no such repetitions.

(4) **The Bomb Problem.** The dams create dangers in wars, especially in modern atomic age. One single atom bomb may cause the failure of Hoover Dam (Boulder Dam) or Bhakra Dam. The resultant failure of such a dam will create catastrophes, but also, it will get contaminated by radioactivity from which there could be no escape.

This is an important point which is generally stressed by opponents of big dams. But the only answer to this argument is that it would not be advisable to deprive ourselves of the benefits of big dams simply because they are hazards in war time. After all, an atom bomb dropped in Calcutta, Delhi, or New York would also cause tremendous damage and catastrophe, but this *does not mean* that we should not *develop big cities.*

Atomic war is dangerous to every aspect of living and not only to the construction of dams. We don't refuse riding in automobiles or aeroplanes because of the fear of accidents. *Certain risk has to be accepted if there is to be progress.*

So, without denying the very great damage that could be caused by atomic explosions at our dams, we must go on building dams. We need them and we must devote our energies to the cause of continued peace, so that bombs will never be able to fall. We may also take more precautions, and anticraft guns and radars can be established at and in the vicinity of such important works. The use of atomic energy for peaceful purposes and a general feeling of brotherhood is the only possible way to reduce such threats.

## 17.6. Selection of the Type of Dam and Their Classifications

(1) *Classification.* Dams can be classified in various ways depending upon the purpose of the classification.

(1) ***Classification According to the Material used for Dam Construction :***

The dams classified according *to the material used* for construction are: Solid masonry gravity dams, Earthen dams, Rockfill dams, Hollow masonry gravity dams, Timber dams, Steel dams, and R.C.C. Arch dams. They have already been explained in a previous article.

(2) ***Classification According to Use***

(*i*) **Storage Dams.** They are constructed in order to store water during the periods of surplus water supply, to be used later during the periods of deficient supply. The stored water may be used in different seasons and for different uses. They may be further classified depending upon the specific use of this water, such as navigation, recreation, water supply, fish, electricity, etc.

(*ii*) **Diversion Dams.** These small dams are used to raise the river water level, in order to feed an off-taking canal and or some other conveyance systems. They are very useful as irrigation development works. A diversion dam is generally called a **weir** or a **barrage.**

(*iii*) **The Detention Dams.** They detain food-waters temporarily so as to retard flood runoff and thus minimise the bad effects of sudden flood.

Detention dams are sometimes constructed to trap sediment. They are often called debris dams.

(3) ***Classification According to Hydraulic Designs***

(*i*) **Overflow Dams.** They are designed to pass the surplus water over their crest. They are often called Spillways. They should be made of materials which will not be eroded by such discharges.

(*ii*) **Non-overflow Dams.** They are those which are not designed to be over-topped. This type of design gives us wider choice of materials including earthfill and rockfill dams.

Many a times, the overflow dam and the non-overflow dam are combined together to form a composite single structure.

(*iii*) **Rigid Dams and Non-rigid Dams.** Rigid dams are those which are constructed of rigid materials like masonry, concrete, steel, timber, etc.; while non-rigid dams are constructed of earth and rock-fill. They have already been explained.

### 17.7. Factors Governing the Selection of a Particular Type of Dam

Whenever we decide to construct a dam at a particular place, the first baffling problem which faces us, is to choose the kind of the dam. *Which type will be the most suitable and most economical?* Two, three kinds of dams may be technically feasible, but only one of them will be the most economical. Various designs and their estimates have to be prepared before signalling one particular type. The various factors which must be thoroughly considered before selecting one particular type are described below:

(1) **Topography.** Topography dictates the first choice of the type of dam. For example:

(*i*) A narrow U-shaped valley, *i.e.* a narrow stream flowing between high rocky walls, would suggest a concrete overflow dam.

(*ii*) A low, rolling plain country, would naturally suggest an earth fill dam with a separate spillway.

(*iii*) The availability of a 'Spillway Site' is very important while selecting a particular kind of a dam.

(*iv*) A narrow V-shaped valley indicates the choice of an arch dam. It is preferable to have the top width of the valley less than one-fourth of its height. But a separate site for the spillway must also be available.

(2) **Geology and Foundation Conditions.** The foundations have to carry weight of the dam. The dam site must be thoroughly surveyed by geologists, so as to detect the thickness of the foundation strata, presence of faults, fissured materials, and their permeability, slope, and slip, etc.

The various kinds of foundations generally encountered are discussed below:

(*i*) *Solid Rock Foundations.* Solid rock foundations such as granite, gneiss, etc. have a strong bearing power. They offer high resistance to erosion and percolation. Almost every kind of dam can be built on such foundations. Sometimes, seams and fractures are present in these rocks. They must be grouted and sealed properly.

(*ii*) *Gravel Foundations.* Coarse sands and gravels are unable to bear the weight of high concrete gravity dams and are suitable for earthen and rock-fill dams. Low concrete gravity dams up to a height of 15 m may also be suggested on such foundations.

These foundations have high permeability and, therefore, subjected to water percolation at high rates. Suitable cut-offs must be provided to avoid danger of undermining.

(*iii*) *Silt and Fine Sand Foundations.* They suggest the adoption of earth dams or very low gravity dams (upto height of 8 m). A rockfill dam on such a foundation is not suitable. Seepage through such a foundation may be excessive. Settlement may also be a problem. They must be properly designed to avoid such dangers. The protection of foundations at the downstream toe from erosion must also be ensured.

(*iv*) *Clay Foundations.* Unconsolidated and high moisture clays are likely to cause enormous settlement of the dam. They are not fit for concrete gravity dams or for rock-fill dams. They may be accepted for earthen dams, but that too, after special treatment.

(*v*) *Non-uniform Foundations.* At certain places, a uniform foundation of the types described above may not be available. In such a case, a non-uniform foundation of rock and soft material may have to be used if the dam is to be built. Such unsatisfactory conditions have to be dealt with by special designs. However, every problem is an individual problem and a solution has to be found by experienced engineers. For example— The Jawahar Sagar Dam in Rajasthan offered such a problem. A bed of clay was encountered, between the base of the dam and solid rock foundation. It was not economically feasible to remove this clay bed. The solution adopted was to anchor the base of the dam to the foundations below, by means of prestressed cables.

(3) **Availability of Materials.** In order to achieve economy in the dam, the materials required for its construction must be available locally or at short distances from the construction site.

Sometimes, good soil is easily available, which naturally calls for an earthen dam. If sand, cement and stone, etc., are easily available, one should naturally think of a concrete gravity dam. If the material has to be transported from far off distances, then a hollow concrete dam (Buttress) is a better choice.

(4) **Spillway Size and Location.** Spillway, as defined earlier, diposes of the surplus river discharge. The capacity of the spillway will depend on the magnitudes of the floods to be by- passed. The spillway will, therefore, become much more important on streams with large flood potential. On such rivers, the spillway may

become dominant structure, and the type of dam may become the secondary consideration.

The cost of constructing a separate spillway may be enormous or sometimes a suitable separate site for a spillway may not be available. In such cases, combining the spillway and the dam into one structure may be desirable, indicating the adoption of a concrete overflow dam.

At certain places, where excavated material from a separate spillway channel may be utilised in dam embankment, an earthfill dam may prove to be advantageous. Small spillway requirement often favours the selections of earth fill or rockfill dams even in narrow dam sites.

The practice of building a concrete spillway on earth and rock embankments is being discouraged these days, because of their conservative design assumptions and the vigil and watch that has to be kept during their operations.

(5) **Earthquake Zone.** If the dam is to be situated in an earthquake zone, its design must include the earthquake forces. Its safety should be ensured against the increased stress induced by an earthquake of worst intensity. The type of structures best suited to resist earthquake shocks without danger are earthen dams and concrete gravity dams.

(6) **Height of the Dam.** Earthen dams are usually not provided for heights more than 30 m or so. Hence, for greater heights, gravity dams are generally preferred.

(7) **Other Considerations.** Various other factors such as, the life of the dam, the width of the roadway to be provided over the dam, problem of skilled labour, legal and aesthetic point must also be considered before a final decision is taken. Overall cost of construction and maintenance and the funds available will finally decide the choice of a particular kind of a *dam at a particular place.*

## 17.8. Selection of Dam Site

The selection of a site for constructing a dam should be governed by the following factors:

(1) Suitable foundations (as determined in the previous article) must be available.

(2) For economy, the length of the dam should be as small as possible, and for a given height, it should store the maximum volume of water. It, therefore, follows, that the river valley at the dam site should be narrow but should open out upstream to provide a large basin for a reservoir. A general configuration of contours for a suitable site is shown in Fig. 17.1.

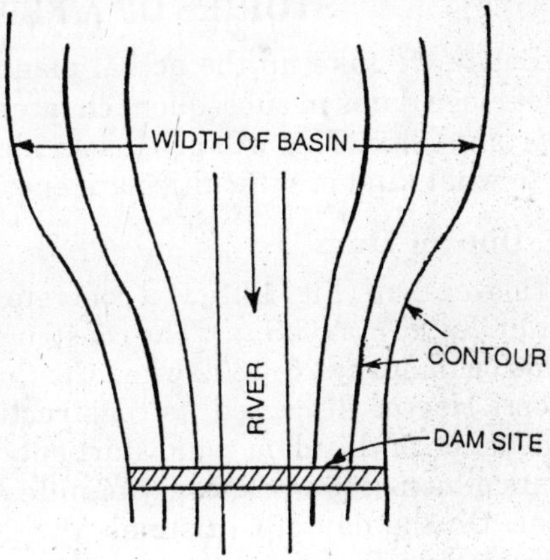


Fig. 17.1

(3) The general bed level at dam site should preferably be higher than that of the river basin. This will reduce the height of the dam and will facilitate the drainage problem.

(4) A suitable site for the spillway should be available in the near vicinity. If the spillway is to be combined with the dam, the width of the gorge should be such as to accommodate both.

The best dam site is one, in which a narrow deep gorge is separated from the flank by a hillock with its surface above the dam, as shown in Fig. 17.2.

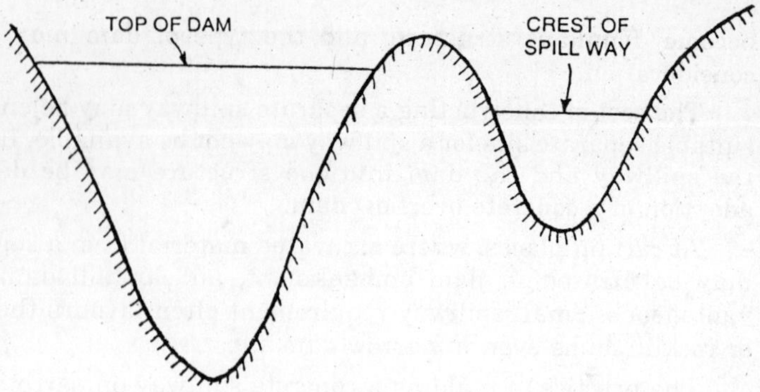


Fig. 17.2

If such a site is available, the spillway can be located separately in the flank, and the main valley spanned by an earthen or similar dam. Sometimes, the spillway and concrete masonry dam may be compositely spanned in the main gorge, while the flanks are in earth at low cost.

(5) Materials required for the construction should be easily available, either locally or in the near vicinity, so that the cost of transporting them is as low as possible.

(6) The reservoir basin should be reasonably water-tight. The stored water should not escape out through its side walls and bed.

(7) the value of land and property submerged by the proposed dam should be as low as possible.

(8) The dam site should be easily accessible, so that it can be economically connected to important towns and cities by rails, roads, etc.

(9) Site for establishing labour colonies and a healthy environment should be available in the near vicinity.

## STORIES OF A FEW IMPORTANT DAMS

Before we take up the actual planning and design of concrete gravity dams and earthen dams in subsequent chapters, let us narrate the stories of certain such important dams. This will give us an idea as to what actually happens in the field and to what kind of difficulties are encountered and how they are overcome.

### 17.9. Hoover Dam

Hoover dam (Fig. 17.3) is a concrete gravity dam, constructed on the Colorado river in California (U.S.A.). The construction of this dam was taken off the drawing boards on January 26, 1892 when the Colorado River Commission presided by Mr. Harbert Hoover discussed its construction in their first meeting. But the actual construction of this dam could start only in late 1930. The construction took about 2 years of non-stop work, every minute of the hour, 24 hours a day, and 365 days a year. On the day of Christmas, the pouring of concrete was tremendous even under searing desert sun by day and under floodlights at night.

**Name Problem.** The name of the Hoover Dam became a political issue. It was first of all named as *Boulder Dam.* Later when Herbert Hoover became the President of America, its name was changed to *Hoover Dam.* In 1933, when the Democrats replaced the Republicans, President Roosevelt changed this name again to *Boulder Dam.* It was known as *Boulder Dam* for 14 years. When again in 1946, Republicans came to power, the name was once again changed to *Hoover Dam,* which still exists today.

**Fig. 17.3.** Photo view of Hoover Dam.

**Dimensions of the Dam.** The original dimensions of the dam are given below in F.P.S. units:

| | |
|---|---|
| Height | = 726 ft |
| Span | = 1300 ft from rim to rim of the Canyon |
| Thickness at the base | = 660 ft |
| Thickness at the top | = 45 ft |
| Name of the reservoir formed at the back of the dam | = Lake Mead |
| Length of the reservoir | = 115 miles. |

**Something about the river.** This big dam straddles the Colorado river. The Colorado river rises in the State of Colorado, runs down through Utah and into Arizona, then California, emptying finally into the Gulf of California in Mexico.

The Colorado is a river with muscles. It is a very strong river-containing huge amounts of silt and mud in it. It cuts numerous deep narrow canyons while it flows. The mile deep walls of the Grand canyon proves the power of this river.

It is a young river and all such rivers have strong powers for forming deep narrow gorges (*i.e.* canyons). They are the most turbulent, and the deepest digging. They are the best sources of hydroelectric power. Irrigation or navigation is not possible on this kind of rivers, unless, they are tamed.

Attempts were made to tame this river by constructing dikes or levees, but failed.

Spring floods of 1904 brought havoc. Another catastrophe occurred in 1916. Gila River, a tributary of Colorado, flooded at a rate of about 5,560 cubic metres/sec. The town of Yuma in Arizona, where the Gila empties into the Colorado, was submerged to a depth of 1.2 metres.

It was extremely desirable to tame this river. But how? A solution was dreamt of. The idea of constructing a big dam was visualised by Arthur Davis. It was an ambitious plan.

**Planning for the Dam.** A scheme was planned. A concrete gravity dam, of about 700 to 750 feet height was thought of. A reservoir at the back of the dam could hold every drop of water, the Colorado could send in any 2 years of steady flow. It was to be the biggest man-made lake in the world. It was going to be a multi-purpose dam, generating about 6 billion kilowatt-hours of electrical energy each year for the growing cities of southern California. The reservoir would hold the flood waters, and the outlets would release the required amounts of water to the downstream farmers.

Flood control, irrigation and electricity were the three main purposes of this project. Supply of drinking water to 13 cities of California and creation of a recreational and navigable lake reservoir were the additional advantages.

**Selection of a suitable site.** A thorough search was made for the spectacular gorges of the Colorado, seeking the best possible site. The engineers studied 70 such sites before chosing 'Black canyon' on the border between Arizona and Nevada, 48 km from the city of Las Vegas (Nevada).

The preliminary survey of *Black canyon* took about 3 years. Here, the river flows through cliffs 1000-2000 feet (300 m to 600 m) high. At the water line, the rock walls were 350 feet (105 m) apart.

Engineers roamed in heat and sun, testing the rocks, drilling into it, to make sure that it could stand the burden of enormous weight of the concrete that would be laid upon it. *Their conclusion was that it could. They recommended that the giant dam be built.*

After the technical green-signal was obtained, some political issues such as to who will be benefitted and up to what extent, were settled with a great difficulty. The rift between the different States always persists in such huge projects.

**Construction of the Dam.** The real work began in late 1930. Herbert Hoover was the President of America at that time. He himself was an engineer. The work started under his vigilance, Arthur Powell Davis, 70 years old and about to retire, saw his life long dream fulfilled as he stood high above the Colorado and watched thousands of workers working hard with picks and dynamite, far below.

The dam site was very hot. It was not a congenial surroundings to work. But a *dam had to be built.*

First of all, a town was built to house the workers. A permanent city, now called Boulder-city, was settled at an expenditure of about $70 million. This was spread in an area of about 300 acres and could accommodate 5,000 workers.

Whenever a dam is constructed, the water of the river is first of all to be diverted, so that the construction could start. This is called by engineers as 'to turn the river off.' 4 tunnels, each 56 feet wide and 4000 feet long, were dug into the solid rock of the canyon walls. These ***bypass tunnels*** received the flow of the river and carried it down to the downstream, to a point beyond the construction site. About $1\frac{1}{2}$ million cubic feet of rock had to be removed for building these four tunnels.

Then, the coffer bund was built. A coffer dam is a temporary retaining embankment upstream of the site. Huge amounts of rock and earth was heaped up, forcing the river into four bypass tunnels. The bed of the river was thus laid bare, downstream of coffer bund to carry out construction at the site of the dam.

Workers then descended into this river bed to lay the foundation for the dam. Seven million tons of concrete had to be laid. It was 660 feet thick at the base.

*The dam curves upstream, so that the water load is held back in part by the walls of the canyon. The completion of the dam took about 2 years of non-stop work.*

After completion, the bypass tunnels were blocked up, and the water started coming and collecting against the dam. The lake formed on the upstream side was called ***Lake Mead***. Electrical power houses, which are of the size of 20-storey sky-scrapers were constructed. The lake, the power houses, the dam galleries etc. are open for visitors and for inspection. It was a great achievement indeed.

### 17.10. Bhakra Dam

Bhakra Dam is a concrete gravity dam. It is 740 feet **(226 m)** high, spanning the V-shaped gorge in the lower Shivalik hills, across Satluj river. The dam is 1700 ft long at the top and only 325 ft at the bottom. The thickness of the dam at foundations is 1320 ft and it tapers to 30 ft at the top where a road runs. *Bhakra dam was the highest concrete gravity dam of the world when built, thus surpassing the existing* 726 ft (221 m) high Hoover dam. But the highest concrete gravity dam of the world, at present, is ***Grand Dixence dam*** in Switzerland ***(284 m high)***. Bhakra dam is situated in Himachal Pradesh State of India near a village used to be called Bhakra. It has been constructed on Satluj river. Satluj is a river coming from Himalayas. It is a perennial river but carries enormous water during floods and rains.

Downstream and upstream views of Bhakra dam are shown in pasted **Colour Photo Fig. 17.1 (*a*) and (*b*).**

This dam has given tremendous prosperity to India and has given her a high name in the world. The various functions served by this dam are :

(*i*) Flood control

(*ii*) Irrigation

(*iii*) Electricity

(*iv*) Fish development.

Bhakra Project is not a single Bhakra Dam but consists of the following :

(*i*) 740 ft high Bhakra Dam

(*ii*) 95 ft high Nangal Dam

(*iii*) Nangal Hydel Channel

(*iv*) Ganguwal and Kotta Power houses

(*v*) Bhakra Canal System.

**Planning and Construction.** The survey works for the construction of this multi-purpose project started in 1919. From 1919 to 1930, the survey continued and various sites were considered for various purposes. From 1932 to 1946, the work was interrupted, and finally in 1946, a railway line was spread in this area, and then the actual construction started.

After selecting a suitable narrow canyon for the construction of Bhakra Dam, the dam site was dewatered after the river was 'turned off'. Two diversion tunnels, one in either abutment were constructed in order to carry the river water. Two coffer dams enclosing the foundation area, were also constructed. Both the tunnels were 50 ft in finished diameter, half a mile long and were lined with 3.6 ft thick heavily reinforced concrete. The work on these two tunnels was started in 1948 and was completed in 1953. The total expenditure incurred on them was approximately ₹ 3.6 crores.

After the construction of these tunnels, two earth (rolled) and rock coffer dams were constructed to enclose the operation area of the dam site. The construction of 215 ft high coffer dam began in 1956. The water was forced into the tunnels and taken downstream beyond the construction point. The 15 ft high downstream coffer dam did not let the water come back in the pit.

The foundations of these rolled fill earthen dams or dikes were carried down to 60 ft and 70 ft below the river bed level. After the construction of coffer dams, the dewatering of the area was expeditiously completed and the excavation of the river bed started. The rock was blasted by an explosive. Desert-shovels were employed for loading the material in special trucks having capacity of 914 cubic yards. The material was carried and dumped at the dumping site. The total excavation was of the order of 700 million cubic yards. About one million cubic yard was excavated per day.

While the excavation was going on, a large constructional plant, capable of producing high quality concrete economically and efficiently was also installed. This required a $4\frac{1}{2}'$ long belt conveyor system, aggregate processing plant, cement handling plant, cooling plant, batching and mixing plants, high steel tristles, revolving cranes and other electricity operated cranes, etc. etc. It was an expensive, completely automised concrete plant. It was capable of handling about *600 tons* of concrete per hour.

Then, the actual concreting work of the dam started in November 1956. Huge electric cranes including 6 cantilever-cranes and 2 stiff-legged-cranes were employed for concreting of dam.

A very low heat cement was used in the construction of this dam. Because, when cement sets, it produces a large amount of heat which is liable to cause cracks in the structure. So in order to avoid this cracking, steel pipes were embedded in the structure and ice-cold water was circulated through them.

The first stage of the dam (390 ft.) was completed by 1959. Work in the 2nd stage was interrupted by a flood in the diversion tunnel that drowned ten workers and damaged the power house. The tunnel and plugging of the dam was completed in 1962.

**Dimensions and other data about Bhakra Dam** are given below :

| | |
|---|---|
| Height of the dam | = 740′ |
| Length at the bottom | = 325′ |
| Length at the top | = 1200′ |
| Breadth or thickness at the bottom | = 1320′ |
| Breadth or Thickness at the top | = 30′ |
| Concrete required for the dam | = 55 lakh cubic yard |
| Electricity generated | = 12 lakh Kilowatt |
| Name of the reservoir formed | = **Govind Sagar** |
| Length of the reservoir | = 90 miles |
| Maximum depth | = 740′ |
| Minimum depth | = 300′ |
| Total irrigation | = 1 crore acre of area |
| Population of these villages | = 30,000 |
| Total storage capacity | = 8 million acre ft. |
| Catchment area | = 22,000 sq. miles |
| Live storage | = 6.35 million acre ft. |
| No. of Inspection galleries | = 46 |

**Power Houses.** There are two power houses called (*i*) The Left Power House; and (*ii*) The Right Power House. The left power house is more important and was first constructed.

The left power house is a reinforced concrete structure. It is standing on a stone consisting partly of clay stone and partly of sand stone. The foundation chosen were adverse to the geological conditions required for a good dam site. The permeability of sand stone is high, and thus there was a possibility of leakage and danger of undermining of the foundations. It necessitated a heavily reinforced raft foundation. The left power house building is seven storeyed and its construction took about two to three years. It required about

635,000 cubic yards of concrete, about one million sq. ft. of finishing work and 6,000 tons of reinforced steel.

The machinery used in this power house is of cosmopolitan in nature. The turbines were supplied by Japan, the generators and transformers by U.K., and overhead cranes by Yogoslavia.

The local manufacture of 10 ft diameter penstocks, made out of 1.75″ thick steel plate was considered as a big engineering feat of the country in such a short period of independence. The left power house had costed nearly ₹ 10 crores. The work on right power house was also completed afterwards.

**Problem of Wood Transport.** Like all other perennial rivers of India, the River Satluj has been the cheapest means of transporting wood from the Himalayan forests down to the plains. But since the dam is very high above the river water level, it has become impossible to use this river for this purpose.

The problem is overcome by bringing logs of woods from Govind Sagar to Nangal railway station by means of an aerial rope-way. This is about $5\frac{1}{2}$ miles long. A wooden log loom has been put across Govind Sagar to obstruct timber. From there, the timber is taken by the inclined carriage way to the loading station on the upstream of the right side.

**17.10.1. Nangal Dam.** Nangal Dam is 95 ft high subsidiary dam, 8 miles downstream of Bhakra on Satluj. It falls within the jurisdiction of Punjab State in India. The length of the dam is 1,000 ft. The object of this dam is to head up water of the river Satluj and then divert it into the canal off-taking from the left bank of the river. The canal is called Nangal Hydel Canal, and is a 40 miles long concrete lined canal.

This dam has 29 strong gates of span 30 ft each. An enormous tunnel called the Inspection gallery has been made in the river Satluj in the lower portion of this dam. This was the first tunnel constructed by the Government under Bhakra Project Scheme. In order to enter into this tunnel, one has to go 70 ft down. The tunnel goes across Satluj River.

Bhakra Nangal Project is something tremendous, stupendous, something which shakes up and thrills us when we see it. It marks the India's progress after her Independence. It is something which cannot be forgotten easily, if we see it once.

## 17.11. Nagarjuna Sagar Dam

The multipurpose Nagarjuna Sagar dam is located across Krishna River, near Nandikonda village in Nalgonda District (Andhra Pradesh). It is named after Buddhist Savant, Acharya Nagarjuna, who lived at the spot about 2,000 years ago to fulfil a mission.

This Dam irrigates in Guntur, Kurnool, Nellore, Nalgonda, Khamman and Krishna Districts. Its irrigation potential is about 35 lakh acres of land and electrical potential is 1 lakh kilowatt hours of firm power (guaranteed-power generation) and 4.6 lakh kilowatt hours of seasonal power.

Two photoviews of this dam are shown in **Colour Photo Fig. 17.2 (*a*) and (*b*)**.

**Salient Features of the Dam.** The river gorge is blocked by a masonry dam 409 ft (124.6 m) high above the deepest foundation level and 4756 ft long. The full reservoir level is +590.0 ft above the mean sea level (M.S.L.). The most unique feature of this dam is the adoption of stone masonry for its construction, deviating from the traditional concrete. The use of stone had resulted in a large saving and had created huge employment potential for a large labour force. Hence, Nagarjuna Sagar Project ranks first in the man-power utilisation among the modern gigantic projects of its own kind in the world. It was designed and executed entirely by Indian engineers. On either side of the masonry dam, earth dams have been constructed for a length of about two miles, the maximum height being 85 ft. The spillway crest has been installed with 26 Radial gates each of size 45′ × 44′. Other component-works of this mighty dam include 8 penstock pipes on the left side, three Power sluices and 9 irrigation sluices on the right side, two chute sluices and a diversion-cum-irrigation tunnel. Two canals-off-take on either side of the dam for irrigation. The expenditure on the project was of the order of ₹ 80 crores. The crest level of the dam is +605 ft and the crest level of spillway is +546 ft. Various details of the dam are given below :

(1) **Location.** Lat. = 16° 34′ North, Long. = 71° 19′ East, $1\frac{1}{2}$ mile downstream of Nandikonda village Miryalaguda Taluk, Nalgonda District, 90 miles from Hyderabad.

(2) **States Covered.** ANDHRA PRADESH

(3) **Hydrology**

| | |
|---|---|
| (*a*) Watershed area at dam site | ... 83,087 sq. miles |
| (*b*) Maximum flood discharge (observed) | ... 11.7 lakh cumec. |

(4) **Reservoir**

| | |
|---|---|
| (*a*) Full Reservoir level | ... + 590′.0 |
| (*b*) Maximum water level | ... + 594′.0 |
| (*c*) Gross storage capacity | ... 9.37 M.a.ft |
| (*d*) Net (live) storage capacity | ... 5.51 M.a.ft |
| (*e*) No. of villages submerged | ... 57 |
| (*f*) Population displaced | ... 4,824 families |

(5) **Masonry dam**

| | |
|---|---|
| (*a*) Total length | ... 4,756 ft |
| (*b*) Spillway length | ... 1545 ft |
| (*c*) Non-overflow length including power dam | ... 3,211 ft |
| (*d*) Height of dam (maximum) | ... 409 ft |
| (*e*) Base width (maximum) | ... 320 ft |
| (*f*) Top width | ... 28 ft |

| | |
|---|---|
| (*g*) Top level | ... + 605′.0 |
| (*h*) Top of crest in spillway | ... + 546′.0 |
| (*i*) Chute sluices 2 No. 10′ × 25′ provided in blocks 25 and 51 with sill level elevation | ... + 450′.0 |
| (*j*) 8 No. of 16′ diameter penstocks provided one in each in blocks 16 to 23 with central line in elevation | ... + 405′.0 |
| (*k*) 3 No. power sluices 15′ × 38′ provided (two in block 71 and one in block 72) with sill level at elevation | ... + 479′.0 |
| (*l*) Radial Crest Gates 26 No. 45′ × 44′, with crest elevation | ... + 546′.0 |
| (*m*) Thickness of spillway pipes | ... + 15′ |
| (*n*) Nine No. sluices 10′ × 15′ are provided for right canal head regulator with sill at elevation | ... + 489′.0 |

(6) **Quantities of Work**

| | |
|---|---|
| (*a*) Excavation for foundations | ... + 42.07 M. cft |
| (*b*) Volume of masonry and concrete | ... 198 M. cft |
| (*c*) Quantity of cement | ... About 11 lakh tons |
| (*d*) Quantity of steel | ... About 60,000 tons |

(7) **Earth Dams**

| | |
|---|---|
| (*a*) Length of Left Earth Dam | ... 8,400 ft |
| (*b*) Length of Right Earth Dam | ... 2,800 ft |
| | 11,200 ft |
| (*c*) Maximum height above foundation level | ... 85 ft |
| (*d*) Top width | ... 30 ft |
| (*e*) Top level | ... 610′ |
| (*f*) Excavation of foundation | ... 9.2 M. cft |
| (*g*) Earthwork for embankment | ... 88 M. cft |

(8) **Irrigation-cum-Diversion Tunnel (horse-shoe in Section)**

| | |
|---|---|
| (*a*) Length | ... 2,590 ft |
| (*b*) Diameter | ... 27 ft |

(9) **Power Plant**

| | |
|---|---|
| (*a*) Left side 8 No. of penstocks 16 ft. diameter to develop | ... 3 lakh kW |
| (*b*) Right side 3 power sluices of size 15′ × 38′ to develop | ... 0.6 lakh kW |

(10) **Man-Power** ... 50,000

**Organisation.** In the earlier years of construction, a Control Board was the overall incharge of the Nagarjuna Sagar Project including technical, financial and administration aspects. The project was then located in the Andhra Pradesh and the Hyderabad States. After the reorganisation of States, the project came entirely in Andhra Pradesh State. The State Government assumed its full responsibility w.e.f. 1.8.1959. The Control Board became an Advisory Body to the State Government, whose advice is accepted as a matter of convention.

The programme of construction of dam and canals was so adjusted that partial benefits started accruing even from 1967. The reservoir was able to deliver water to first crop of Krishna delta in time, for raising substantial acreage in 2nd crop. Letting out waters in the two canals was inaugurated, by Smt. Indira Gandhi, the then Prime Minister of India, on 4.8.1967.

**Rehabilitation.** The man-made lake of Nagarjuna Sagar extended in an area of about 110 sq miles (280 sq km) and stretches as far back as up to Srisailam – 120 km upstream, making a navigable channel, fully submerged lands of 52 villages and partially submerged that of 5 villages.

About 4,900 families were displaced and rehabilitated in 24 Rehabilitation centres. Good facilities, liberal compensation and other amenities were provided to those who had to be uprouted.

**Nagarjunakonda Excavations.** As the Nagarjuna Sagar reservoir was to completely submerge the famous relics of Nagarjuna Konda, which was the seat of Ikshwaku Kings of Vijaypuri (flourishing trading centre) and one of the principal centres of Mahayana System of Buddhism, the whole area was excavated by the Archeological Department of the Government of India. The more important of the relics were located in a museum constructed for the purpose, on the top of an adjoining hill. Practically, all the relics were unearthed and shifted to the museum. Expenditure to the extent of 12 lakh rupees was debited to the Project and the balance was met by the Archeological department. This museum now finds itself located on an island existing into the reservoir formed by the dam, 14 km upstream of the dam, and is the second largest island museum in the world (the largest being in Aswan dam reservoir in Egypt). There now also exists a steamer ride from the Nagarjuna Sagar to this museum these days - an excellent recreation at the dam reservoir.

## PROBLEMS

1. (*a*) What is meant by a "dam and a reservoir" ? What are the different materials that are commonly used for dam construction and what are their comparative advantages and disadvantages ?

   (*b*) Discuss the geological and tropological features which affect the selection of the type of dam.

2. (*a*) What are 'arch' and 'buttress' dams ? Illustrate with sketches and mention site conditions favourable for construction of such dams.

   (*b*) Discuss the factors which are considered in the selection of the site for a proposed dam. It is assumed that the type of the dam has already been selected for the project.

3. (*a*) What useful purpose is served by a dam ? What are the ill-effects of dam construction ?

   (*b*) How do you classify dams according to :

   (*i*) their use ;

   (*ii*) their hydraulic designs ;

   (*iii*) their materials of construction.

   (*c*) Discuss the various factors which govern the selection of a particular type of dam for a particular project.

4. "Dams are the sources of sorrow and grief". Debate the above statement giving points in favour as well as against it.

5. Narrate briefly the story of construction of any major dam of India.

Coloured Fig. 17.1 (*a*) Upstream view of Bhakra Dam.

**Colour Photo Fig.** 17.1 (*b*) Downstream view of Bhakra Dam.

**Colour Photo Fig.** 17.1. Photoviews of Bhakra Dam.

**Coloured Fig. 17.2 (*a*):** Photoview of Nagarjuna Sagar dam from Left Earthen Dam side. The Reservoir on upstream side is full during monsoon days. (Photographed by the Author on 26-09-2007)

**Coloured Fig. 17.2 (*b*):** Downstream View of Nagarjuna Sagar Dam showing water spilling over the spillway crest after the reservoir got filled up to the full reservoir level

**Colour Photo Fig.** 17.2 Photoviews of Nagarjuna Sagar Dam.

18

# *Reservoirs and Planning for Dam Reservoirs*

## 18.1. Definition and Types

When a barrier is constructed across some river in the form of a dam, water gets stored on the upstream side of the barrier, forming a pool of water, generally called a *dam reservoir* or an *impounding reservoir* or *a river reservoir,* or a *storage reservoir.*

The quality of water stored in such a reservoir is not much different from that of a natural lake. The water so stored in a given reservoir during rainy season can be easily used almost throughout the year, till the time of arrival of the next rainy season, to refill the emptying reservoir again.

Depending upon the purpose served by a given reservoir, the reservoirs may be broadly divided into the following three types.

(1) *Storage or Conservation reservoirs* ;

(2) *Flood Control reservoirs* ; and

(3) *Multipurpose reservoirs.*

The fourth type of a reservoir is a simple storage tank constructed within a city water supply system, and is called a *Distribution reservoir* ; and such a reservoir is evidently not a river reservoir, but is a simple storage tank.

**18.1.1. Storage or Conservation Reservoirs.** A city water supply, irrigation water supply, or a hydroelectric project drawing water directly from a river or a stream may fail to satisfy the consumers demands during extremely low flows; while during high flows, it may become difficult to carry out their operations due to devastating floods. *A storage or a conservation reservoir can retain such excess supplies during periods of peak flows, and can release them gradually during low flows as and when the need arises.*

Incidentally, in addition to conserving water for later use, the storage of flood waters may also reduce flood damage below the reservoir. Hence, a reservoir can be used for controlling floods either solely or in addition to other purposes. In the former case, it is known as a 'Flood Control Reservoir' or a 'Single Purpose Flood Control Reservoir'; and in the latter case, it is called a 'Multipurpose Reservoir'.

**18.1.2. Flood Control Reservoirs.** A flood control reservoir, generally called a flood-mitigation reservoir, stores a portion of the flood flows in such a way as to minimise the flood peaks at the areas to be protected downstream. To accomplish this, the entire inflow entering the reservoir is discharged till the outflow reaches the safe capacity of the channel downstream. The inflow in excess of this rate is stored in the reservoir, which is then gradually released, so as to recover the storage capacity for the next flood.

The flood peaks at the downstream of the reservoir are thus reduced by an amount *AB*, as shown in Fig. 18.1. A flood control reservoir differs from a conservation reservoir only in its need for a large sluiceway capacity to permit rapid drawdown before or after a flood.

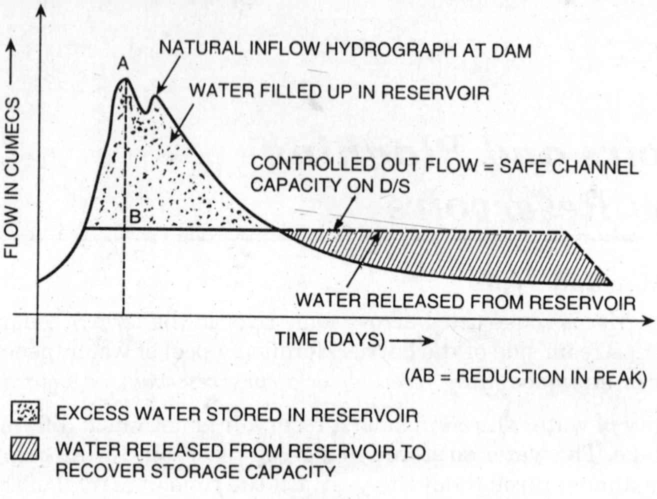


Fig. 18.1.

**Types of flood control reservoirs.** There are two basic types of flood-mitigation reservoirs; *i.e.*

(*i*) *Storage reservoirs or Detention basins* ; and

(*ii*) *Retarding basins or Retarding reservoirs.*

A reservoir having gates and valves installation at its spillway and at its sluice outlets is known as a storage reservoir; while on the other hand, a reservoir with uncontrolled and ungated outlets is known as a **retarding basin.**

*Functioning and Advantages of a Retarding Basin*

A retarding basin is usually provided with an uncontrolled spillway and an uncontrolled orifice type sluiceways. The automatic regulation of outflow, depending upon the availability of water, takes place from such a reservoir. The maximum discharging capacity of such a reservoir should be equal to the maximum safe carrying capacity of the channel downstream. As floods occur, the reservoir gets filled, and discharges through sluiceways. As the reservoir elevation increases, the outflow discharge increases. The water level goes on rising until the flood has subsided, and the inflow becomes equal to or less than the outflow. After this, the water gets automatically withdrawn from the reservoir until the stored water is completely discharged. The advantages of a retarding basin over a gate controlled detention basin are:

(*i*) Cost of the gate installation is saved.

(*ii*) There are no gates and hence, the possibility of human error and negligence in their operation is eliminated.

(*iii*) Since such a reservoir is not always filled, much of the land below the maximum reservoir level will be submerged only temporarily and occasionally, and can be successfully used for agriculture, although no permanent habitation can be allowed on this land.

*Functioning and Advantages of a Storage Reservoir*

A storage reservoir with gated spillway and gated sluiceways, provides more flexibility of operation, and thus gives us better control and increased usefulness of the reservoir. Storage reservoirs are, therefore, preferred on large rivers, which require better control; while retarding basins are preferred on small rivers. In storage reservoirs, the flood crest downstream, can be better controlled and regulated properly, so as not to cause their coincidence. This is the biggest advantage of such a reservoir and outweighs its disadvantages of being costly and involving risk of human error in installation and operation of gates.

**18.1.3. Multipurpose Reservoirs.** A reservoir planned and constructed to serve not only one purpose but various purposes together is called a multipurpose reservoir. Reservoir, designed for one purpose, incidentally serving other purposes, shall not be called a multipurpose reservoir, but will be called so, only if designed to serve those purposes also in addition to its main purpose. Hence, a reservoir designed to protect the downstream areas from floods and also to conserve water for water supply, irrigation, industrial needs, hydroelectric purposes, etc. shall be called a multipurpose reservoir. Bhakra dam and Nagarjun Sagar dam are the important multipupose projects of India.

**18.1.4. Distribution Reservoirs.** A distribution reservoir is a small storage reservoir constructed within a city water supply system. Such a reservoir can be filled by pumping water at a certain rate and can be used to supply water even at rates higher than the inflow rate during periods of maximum demands (called critical periods of demand). Such reservoirs are, therefore, helpful in permitting the pumps or the water treatment plants to work at a uniform rate, and they store water during the hours of no demand or less demand, and supply water from their 'storage' during the critical periods of maximum demand.

In this chapter, we shall however, confine ourselves to the river reservoirs only.

## 18.2. Capacity-Elevation and Area-Elevation Curves of a Reservoir Site

Whatever be the size or use of a reservoir, the main function of a reservoir is to store water and thus to stabilize the flow of water. Therefore, the most important physical characteristic of a reservoir is nothing but its *storage capacity*. The capacity of reservoirs on dam sites, is determined from the contour maps of the area. A topographic survey of the dam site is carried out, and a contour map such as shown in Fig. 18.2 is drawn. The area enclosed within each contour can be measured with a planimeter.

In fact, the general practice adopted for capacity computations is to actually survey the site contours only at vertical distances of 5 m or so. The areas of the intervening contours at smaller intervals of say 0.5 m or so, are then interpolated by taking the square root of the surveyed contours, and to assume that the square root of the interpolated ones, vary in exact proportion to their vertical distance apart. For example, suppose the area of the reservoir at 200 m contour is $A_1$ hectares, and that at 205 m contour is $A_2$ hectares; then

The area at 200.5 m contour is

$$= \left[\sqrt{A_1} + \frac{200.5 - 200}{205 - 200}(\sqrt{A_2} - \sqrt{A_1})\right]^2$$

$$= \left[\sqrt{A_1} + \frac{0.5}{5.0}(\sqrt{A_2} - \sqrt{A_1})\right]^2 \qquad \text{...[18.1 (a)]}$$

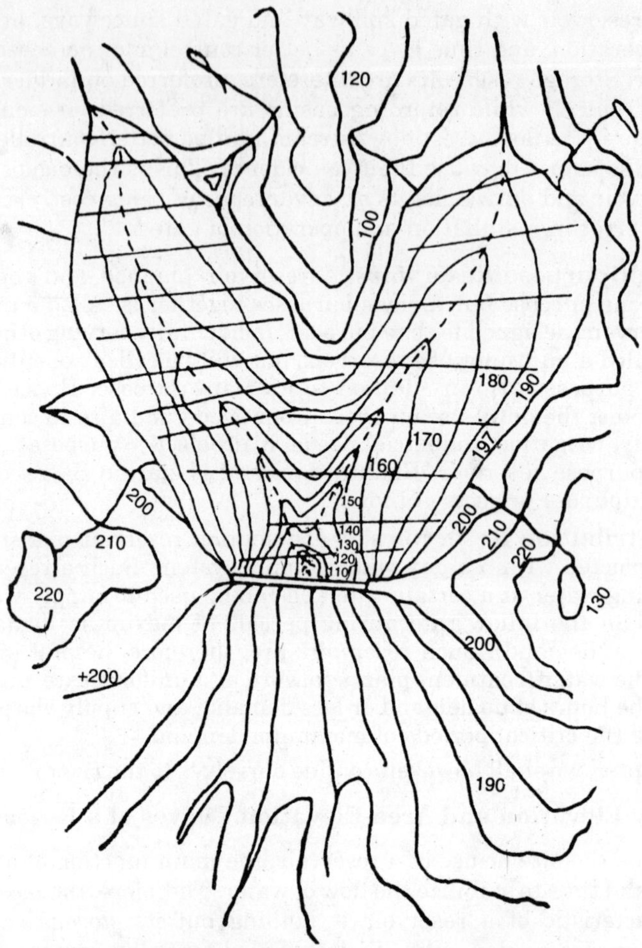


Fig. 18.2. A typical contour map of a dam site.

Similarly, the area at 201.0 m contour

$$= \left[ \sqrt{A_1} + \frac{201.0 - 200}{205 - 200} (\sqrt{A_2} - \sqrt{A_1}) \right]^2$$

$$= \left[ \sqrt{A_1} + \frac{1.0}{5.0} (\sqrt{A_2} - \sqrt{A_1}) \right]^2 \quad ...[18.1\ (b)]$$

Similarly, the area at 204.5 m contour

$$= \left[ \sqrt{A_1} + \frac{204.5 - 200}{205 - 200} (\sqrt{A_2} - \sqrt{A_1}) \right]^2$$

$$= \left[ \sqrt{A_1} + \frac{4.5}{5.0} (\sqrt{A_2} - \sqrt{A_1}) \right]^2 \quad ...[18.1\ (c)]$$

In this way, the areas can be computed at sufficiently low contour intervals (0.5 m), which can be used to determine the incremental volumes ($\Delta S$) stored between two successive contours, by using the simple average method, *i.e.* by multiplying the average of the two areas at the two elevations, by the elevation difference ($\Delta h$). The summation of these incremental volumes below any elevation, is the storage volume below that level.

Instead of using the simple average formula, *i.e.*

$$\Delta S = \frac{a_1 + a_2}{2} (\Delta h), \text{ sometimes the formula}$$

$$\Delta S = \frac{\Delta h}{3} \left[ a_1 + a_2 + \sqrt{a_1 a_2} \right] \text{ can also be used, where}$$

$a_1, a_2, \ldots$represent the areas at 0.5 m contour interval.

**Prismoidal formula** can also be used preferably where three consecutive sections at equal height are taken. According to this,

$$\Delta S = \text{Storage} = \frac{\Delta h}{6} \left[ A_1 + 4A_2 + A_3 \right] \qquad \ldots(18.2)$$

where $A_1$, $A_2$ and $A_3$ are the areas of succeeding contours, and $\Delta h$ is the vertical distance between two alternate contours.

It thus becomes evident that the areas at different elevations (contours), as well as the storage at different elevations, can be mathematically worked out, and both plotted on a graph paper, to obtain **Area-Elevation curve, and Storage-Elevation curve** (*Capacity-Elevation curve*), respectively, as shown in Fig. 18.3.

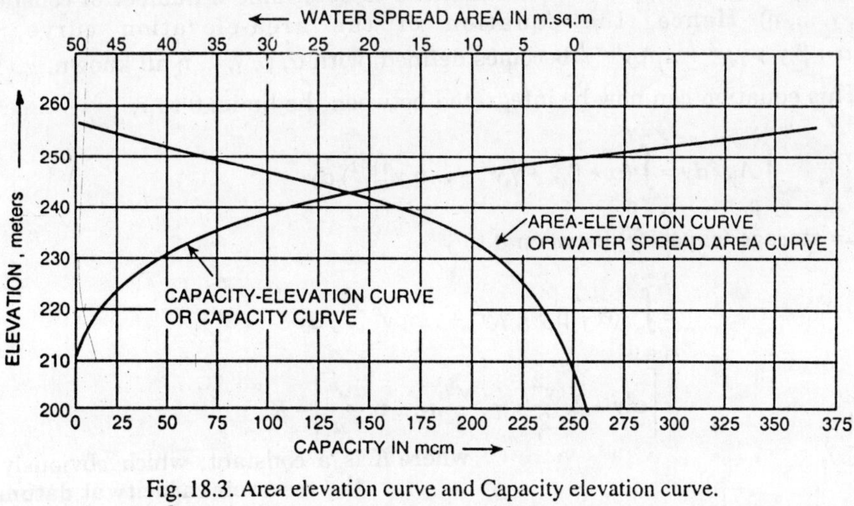


Fig. 18.3. Area elevation curve and Capacity elevation curve.

Another method for computing capacity may be the use of integration technique. Infact, the area-elevation curve, when integrated, will yield nothing but the capacity-elevation curve. Hence the surveyed areas at large-intervals, may be plotted on a simple graph paper, and a smooth curve, *i.e.* area-elevation curve, is first of all drawn. The equation of this curve is now obtained by statistical methods, which can be integrated to obtain the equation of the capacity-elevation curve, as follows:

The equation of the area-elevation curve, will generally be of the form:

$$A = \alpha + \beta \,.\, h + \gamma \,.\, h^2 + \ldots \eta \,.\, h^{n-1} \qquad \ldots(18.3)$$

where $A$ represents the area at any elevation $h$; and $\alpha, \beta, \gamma \ldots \eta$ are all constants.

This equation can be determined and then integrated to obtain the storage (capacity), as explained below:

Let $y$ be the height of the water surface in the reservoir above any assumed datum, over which the storage/capacity is to be worked out.

Let $A_y$ represents symbolically the area of the contour at this height. Then, assume that the *equation of area-elevation curve* is given by :

$$A_y = \alpha + \beta \cdot y + \gamma \cdot y^2 + \ldots \eta \cdot y^{n-1} \quad \ldots(18.3\ a)$$

where $\alpha, \beta, \gamma, \ldots, \eta$ are all constants.

From the actual survey, or from the points falling on the area-elevation curve, the area of any required number of contours ($n$) are known.

Let them be $A_0, A_1, A_2$.....corresponding to the known level values, *i.e.*, heights $0, y_1, y_2, \ldots$ above the datum.

Substituting these values in equation (18.3 $a$), we get

$$A_0 = \alpha$$

$$A_1 = \alpha + \beta \cdot y_1 + \gamma \cdot y_1^2 + \ldots \eta \cdot y_1^{n-1}$$

$$A_2 = \alpha + \beta \cdot y_2 + \gamma \cdot y_2^2 + \ldots \eta \cdot y_2^{n-1}$$

$$A_3 = \alpha + \beta \cdot y_3 + \gamma \cdot y_3^2 + \ldots \eta \cdot y_3^{n-1}$$

Thus, we get n simultaneous equations to determine n number of constants $(\alpha, \beta, \gamma, \ldots, \eta)$ Hence, the equation of the area-elevation curve, *i.e.* $A_y = \alpha + \beta.y + \gamma.y^2 + \ldots \eta.y^{n-1}$ becomes defined, with $\alpha, \beta, \gamma, \ldots \eta$ all known.

This equation can now be integrated between the limits 0 to $y$,

$$\int_{y=0}^{y=y} A_y \cdot dy = \int_{y=0}^{y=y} (\alpha + \beta.y + \gamma.y^2 + \ldots \eta.y^{n-1})\, dy$$

or $S_y$ = storage (capacity) between 0 to $y$

$$= \int_{y=0}^{y=y} (\alpha + \beta.y + \gamma.y^2 + \ldots \eta.y^{n-1})\, dy$$

$$= \left[\alpha.y + \beta.\frac{y^2}{2} + \gamma.\frac{y^3}{3} + \ldots \eta.\frac{y^n}{n}\right] + K$$

where $K$ is a constant, which obviously is the reservoir capacity at datum.

The use of this method can be best understood by solving a numerical example.

**Example 18.1.** *A contour survey of a reservoir site gives the following data:*

| *Contour value* | *Area* |
|---|---|
| *At 200 m contour* | *6.0 hectares* |
| *At 210 m contour* | *18.1 hectares* |
| *At 220 m contour* | *34.0 hectares* |

*The capacity of the reservoir upto 200 m elevation is found to be 14.1 ha. m. Determine the general equation for the area-elevation curve and capacity-elevation curve. Also determine the reservoir capacity at RL 225 m.*

**Solution.** Use equation (18.3 *a*), wherein $y$ is the height above *RL* 200 m, as

$$A_y = \alpha + \beta \, . \, y + \gamma \, . \, y^2$$

Now substituting the given values, we have

$$A_0 = 6.0 \text{ hectares}$$

$\therefore$ $6.0 = \alpha + \beta\,(0) + \gamma\,(0) = \alpha$

or $\alpha = \mathbf{6.0}$ ...(*i*)

Also $A_1 = 18.1$ hectares, at $y_1 = 10$ m (given)

$\therefore$ $18.1 = \alpha + \beta\,.\,(10) + \gamma\,.\,(10)^2$

or $18.1 = 6.0 + 10\beta + 100\gamma$

or $12.1 = 10\beta + 100\gamma$

or $\beta + 10\,\gamma = 1.21$ ...(*ii*)

Also $A_2 = 34.0$ hectares at $y_2 = 20$ m (given)

$\therefore$ $34.0 = \alpha + \beta\,(20) + \gamma\,(20)^2 = 6.0 + 20\beta + 400\gamma$

or $20\beta + 400\gamma = 28.0$

$\beta + 20\gamma = 1.4$ ...(*iii*)

Solving (*ii*) and (*iii*), we get

$\beta + 10\gamma = 1.21$

$\beta + 20\gamma = 1.40$

or $10\gamma = 0.19$ or $\gamma = \mathbf{0.019}$

$\therefore$ $\beta + 0.19 = 1.21$ or $\beta = \mathbf{1.02}$

Hence, the equation of area-elevation curve is given by

$$A_y = 6.0 + 1.02y + 0.019y^2 \quad \textbf{Ans.}$$

Integrating this equation, we get

$$S_y = 6.0y + 1.02\frac{y^2}{2} + 0.019\frac{y^3}{3} + K$$

The constant $K$ is obviously the reservoir capacity (storage) upto *RL* 200.0 m, which is given to be 14.1 ha. m.

$$\therefore \quad S_y = 6.0y + 0.51y^2 + 0.0063y^3 + 14.1$$

Hence, the capacity-elevation curve is given by

$$S_y = 0.0063y^3 + 0.51y^2 + 6.0y + 14.1 \quad \textbf{Ans}$$

To determine the capacity at *RL* 225 m, we have to substitute

$y = 225 - 200 = 25$ m in the above eqn. to obtain the requisite capacity in ha.m., *i.e.*

Required capacity at *RL* 225 m

$$= 0.0063\,(25)^3 + 0.51\,(25)^2 + 6.0 \times 25 + 14.1$$

$$= 98.96 + 318.75 + 150.0 + 14.1 = \mathbf{581.81\ ha.m.} \quad \textbf{Ans.}$$

**Conclusions.** It can thus be seen that the capacity of a reservoir by this method can be determined by surveying only a few contours. It is also found that the method does not give more than 3% error, when it is cross-checked with the capacity worked out by surveying large number of contours. This error is not considered much, in the light of the fact that the areas of contours are themselves not very precise figures.

Sometimes, storage capacity may be expressed as a single term function of $y$, such as

$$S_y = K.y^n \qquad \text{where } K \text{ and } n \text{ are constants.}$$

Infact, in practical life, no one bothers about the equations, and only curves, as shown in Fig. 18.3 are drawn. The required capacity at any elevation is read out, from such a curve.

It may also be pointed out here that the best capacity curve for a reservoir is the one in which the rise to the straight line is the quickest. It results from a cup-shaped catchment, with gentle longitudinal slope. Fig. 18.4 shows the good and bad capacity curves.

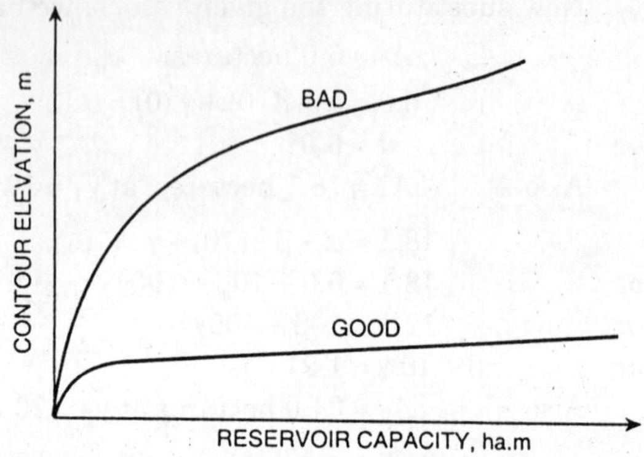


Fig. 18.4. Good and Bad capacity curves for a reservoir

## 18.3. Storage Zones of a Reservoir

These zones are defined w.r. to Fig. 18.5.

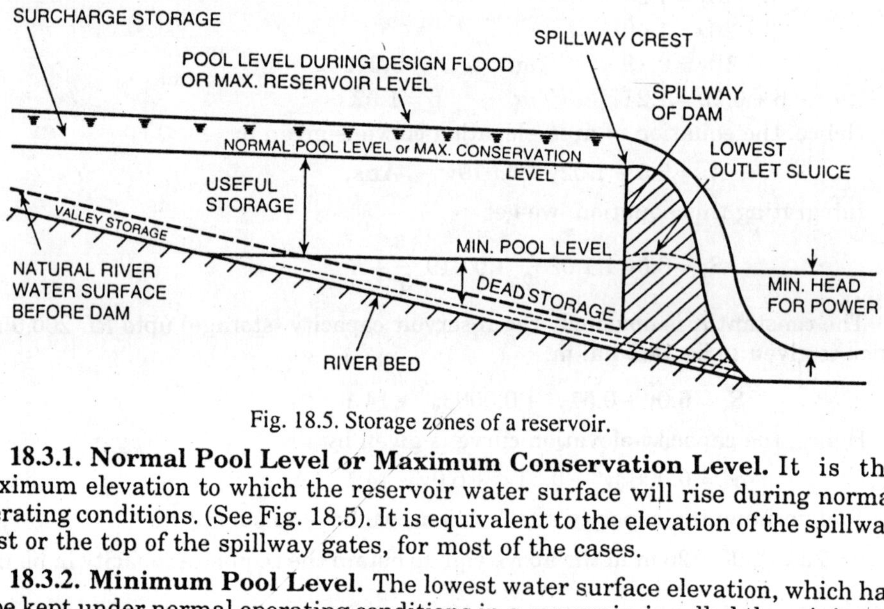


Fig. 18.5. Storage zones of a reservoir.

**18.3.1. Normal Pool Level or Maximum Conservation Level.** It is the maximum elevation to which the reservoir water surface will rise during normal operating conditions. (See Fig. 18.5). It is equivalent to the elevation of the spillway crest or the top of the spillway gates, for most of the cases.

**18.3.2. Minimum Pool Level.** The lowest water surface elevation, which has to be kept under normal operating conditions in a reservoir, is called the minimum pool level (See Fig. 18.5). This level may be fixed by the elevation of the lowest outlet in the dam or may be guided by the minimum head required for efficient functioning of turbines.

**18.3.3. Useful and Dead Storage.** *The volume of water stored in a reservoir between the minimum pool and normal pool levels is called the useful storage.* Water stored in the reservoir below the minimum pool level is known as the *Dead Storage*, and it is not of much use in the operation of the reservoirs. The useful storage may be subdivided into conservation storage and flood-mitigation storage, in a multipurpose reservoir.

**18.3.4. Maximum Pool Level or Full Reservoir Level.** During high floods, water is discharged over the spillway, but will cause the water level to rise in the reservoir above the normal pool level. The maximum level to which the water rises during the worst design flood is known as the *maximum pool level.*

**18.3.5. Surcharge Storage.** The volume of water stored between the normal pool level and the maximum pool level is called *surcharge storage.* Surcharge storage is an uncontrolled storage, in the sense that it exists only till the flood is in progress and cannot be retained for later use.

**18.3.6. Bank Storage.** When the reservoir is filled up, certain amount of water seeps into the permeable reservoir banks. This water comes out as soon as the reservoir gets depleted. This volume of water is known as the *bank storage,* and may amount to several percent of the reservoir volume depending upon the geological formations. The bank storage effectively increases the capacity of the reservoir above that indicated by the elevation capacity curve of the reservoir.

**18.3.7. Valley Storage.** Even before a dam is constructed, certain variable amount of water is stored in the stream channel, called *valley storage.* After the reservoir is formed, the storage increases, and the actual net increase in the storage is equal to the storage capacity of the reservoir minus the natural valley storage. The valley storage thus reduces the effective storage capacity of a reservoir. It is not of much importance in conservation reservoirs, but the available storage for flood mitigation is reduced, as given by the following relation:

Effective storage for flood mitigation

= Useful Storage + Surcharge Storage – Valley Storage corresponding to the rate of inflow in the reservoir.

## DESIGNING RESERVOIR CAPACITY

### 18.4. Catchment Yield and Reservoir Yield

Long range runoff from a catchment is known as the yield of the catchment. Generally, a period of one year is considered for determining the yield value. *The total yearly runoff, expressed as the volume of water entering/passing the outlet point of the catchment, is thus known as the catchment yield, and is expressed in* $M\,m^3$ *or M.ha.m.*

The annual yield of the catchment upto the site of a reservoir, located at the given point along a river, will thus indicate the quantum of water that will annually enter the reservoir, and will thus help in designing the capacity of the reservoir. This will also help to fix the outflows from the reservoir, since the outflows are dependent upon the inflows and the reservoir losses.

The amount of water that can be drawn from a reservoir, in any specified time interval, called the *reservoir yield,* naturally depends upon the inflow into the reservoir and the reservoir losses, consisting of reservoir leakage and reservoir evaporation.

The annual inflow to the reservoir, *i.e.* the *catchment yield,* is represented by the *mass curve of inflow;* whereas, the outflow from the reservoir, called the *reservoir yield,* is represented by the *mass demand line* or *the mass curve of outflow.* Both these curves decide the reservoir capacity, provided the reservoir losses are ignored or separately accounted.

The inflows to the reservoir are however, quite susceptible of variation in different years, and may therefore vary throughout the prospective life of the reservoir. *The past available data of rainfall or runoff in the catchment is therefore used to work out the optimum value of the catchment yield.* Say for example, in the past available records of say 35 years, the minimum yield from the catchment in the worst rainfall year may be as low as say 100 M.ha.m; whereas, the maximum yield in the best rainfall year may be as high as say 200 M.ha.m. The question which then arises would be as to whether the reservoir capacity should correspond to 100 M. ha.m yield or 200 M. ha.m yield. If the reservoir capacity is provided corresponding to 100 M. ha.m yield, then eventually the reservoir will be filled up every year with a dependability of 100%; but if the capacity is provided corresponding to 200 M.ha.m yield, then eventually the reservoir will be filled up only in the best rainfall year (*i.e.* once in 35 years) with a dependability of about $\frac{1}{35} \times 100 \approx 3\%$.

In order to obtain a sweet agreement, a via media is generally adopted and an intermediate *dependability percentage value* ($p$), such as 50% to 75%,* may be used to compute the **dependable yield** or **the design yield.** The yield which corresponds to the worst or the most critical year on record is however, called the **firm yield** or the **safe yield.** Water available in excess of the firm yield during years of higher inflows, is designated as the **secondary yield.** Hydropower may be developed from such secondary water, and sold to the industries *'on and when available basis'.* The power commitments to domestic consumers must, however, be based on the firm basis, and should not exceed the power which can be produced with the firm yield, unless thermal power is also available to support the hydroelectric power.

The arithmetic average of the firm yield and the secondary yield is called the **average yield.**

**18.4.1. Computing the Design or the Dependable Catchment Yield.** The dependable yield, corresponding to a given dependability percentage $p$, is determined from the past available data of the last 35 years or so. The yearly rainfall data of the reservoir catchment is generally used for this purpose, since such long runoff data is rarely available . The rainfall data of the past years is therefore used to work out the dependable rainfall value corresponding to the given dependability percentage $p$. This dependable rainfall value is then converted into the dependable runoff value by using the available empirical formulas connecting the yearly rainfall with the yearly runoff.

It is, however, an adopted practice in Irrigation Departments to plan the reservoir project by computing the dependable yield from the rainfall data, but to start river gauging as soon as the site for the reservoir is decided, and then correlate the rainfall-runoff observations to verify the correctness of the assumed empirical relation between the rainfall and the runoff. Sometimes, on the basis of such observations, the initially assumed yield value may have to be revised.

The procedure which is adopted to compute the dependable rainfall value for a given dependability percentage $p$ is explained below, and has been further used in solving example 18.2.

(*i*) The available rainfall data of the past $N$ years is first of all arranged in the descending order of magnitude.

---

* This percentage will depend upon the risk which can be absorbed for the proposed use of water. Say for example, city water supply projects can absorb lesser risk as compared to the irrigation projects, and hence should consider higher dependability percentage value.

(*ii*) The order number *m*, given by the equation

$$m = N \cdot \frac{p}{100} \qquad ...(18.4)$$

is then computed, and rainfall value corresponding to this order number in the tabulated data will represent the required dependable rainfall value.

(*iii*) If the computed value of *m* is a fraction, then the arithmetic mean of the rainfall values corresponding to whole number *m* values above and below this fraction value is taken as the **dependable rainfall** value.

This method of computing the rainfall value of the given dependability (such as 50%, 75%, etc.) will become more clear when we solve example 18.2. The rainfall value of given dependability can finally be converted into the dependable yield, by using empirical methods, discussed in article 18.4.1.1.

The dependable yield of a given dependability (*p* per cent) can also be determined from the stream gauging, by computing the annual yields, from the observed discharges, and arranging the annual yield values in descending order to determine order numbers (instead of rainfall) as exhibited in solved example 18.3.

**Example 18.2.** *The yearly rainfall data for the catchment of a proposed reservoir site for 35 years is given in Table 18.1. Compute from this data, the values of dependable rainfalls for 60% and 75% dependability percentage.*

**Table 18.1**

| *Year* | *Rainfall in cm* | *Year* | *Rainfall in cm* |
|---|---|---|---|
| 1956 | 98 | 1974 | 88 |
| 1957 | 100 | 1975 | 94 |
| 1958 | 101 | 1976 | 107 |
| 1959 | 99 | 1977 | 110 |
| 1960 | 85 | 1978 | 208 |
| 1961 | 112 | 1979 | 114 |
| 1962 | 116 | 1980 | 104 |
| 1963 | 78 | 1981 | 120 |
| 1964 | 160 | 1982 | 108 |
| 1965 | 66 | 1983 | 102 |
| 1966 | 184 | 1984 | 80 |
| 1967 | 90 | 1985 | 109 |
| 1968 | 76 | 1986 | 122 |
| 1969 | 118 | 1987 | 115 |
| 1970 | 86 | 1988 | 140 |
| 1971 | 92 | 1989 | 138 |
| 1972 | 96 | 1990 | 60 |
| 1973 | 93 | | |

**Solution.** The given data of table 18.1 is arranged in descending order, mentioning serial number (order number *m*) in front of each, as shown in Table 18.2.

**Table 18.2**

| *S.No. i.e order number (m)* | *Rainfall in desecending order in cm* | *S. No. i.e. order number (m)* | *Rainfall in descending order in cm* |
|---|---|---|---|
| 1 | 208 | 19 | 101 |
| 2 | 184 | 20 | 100 |
| 3 | 160 | 21 | 99 |
| 4 | 140 | 22 | 98 |
| 5 | 138 | 23 | 96 |
| 6 | 122 | 24 | 94 |
| 7 | 120 | 25 | 93 |
| 8 | 118 | 26 | 92 |
| 9 | 116 | 27 | 90 |
| 10 | 115 | 28 | 88 |
| 11 | 114 | 29 | 86 |
| 12 | 112 | 30 | 85 |
| 13 | 110 | 31 | 80 |
| 14 | 109 | 32 | 78 |
| 15 | 108 | 33 | 76 |
| 16 | 107 | 34 | 66 |
| 17 | 104 | 35 | 60 |
| 18 | 102 | | |

Now, using equation (18.4), we compute the order number ($m$) for the given dependability percentage $p = 60\%$, as:

$$m = N \cdot \frac{p}{100} = 35 \times \frac{60}{100} = 21$$

The rainfall value tabulated above in table 18.2 at order no.21 is 99 cm; and hence the required dependable rainfall = 99cm. ∴ $P_{60\%}$ = **99 cm** Ans.

Similarly, for dependability $p = 75\%$, we calculate the order no. ($m$)

$$= N \cdot \frac{p}{100} = 35 \times \frac{75}{100} = 26.25.$$

Since reqd. order no. is not an integer, the mean value of rainfall corresponding to order no. 26 and 27 will be taken as the value of $p$ of 75% dependability.

$$\therefore \quad P_{75\%} = \frac{92 + 90}{2} = 91 \text{ cm} \quad \textbf{Ans.}$$

**Example 18.3.** *The daily flows in a river for three consecutive years are given in the table by class interval alongwith the number of days the flow belonged to this class. What are the 50% and 75% dependable flows (annual and daily) for the river ?*

*Table 18.3*

| | | Year 1981 | Year 1982 | Year 1983 |
|---|---|---|---|---|
| S. No. | Daily mean discharge $m^3/s$ (range) | No. of days the flow belonged to the given range (class interval) | No. of days the flow equalled the class-range of col (2) | No. of days the flow equalled the range given in col (2) |
| (1) | (2) | (3) | (4) | (5) |
| 1 | 100-90.1 | 0 | 6 | 10 |
| 2 | 90-80.1 | 16 | 19 | 16 |
| 3 | 80-70.1 | 27 | 25 | 38 |
| 4 | 70-60.1 | 21 | 60 | 67 |
| 5 | 60-50.1 | 43 | 51 | 58 |
| 6 | 50-40.1 | 59 | 38 | 38 |
| 7 | 40-30.1 | 64 | 29 | 70 |
| 8 | 30-20.1 | 22 | 48 | 29 |
| 9 | 20-10.1 | 59 | 63 | 26 |
| 10 | 10-negligible | 54 | 26 | 13 |

***(Engg. Services, 2002)***

**Solution.** With the given daily flows, we will calculate the annual yields for the given 3 years of records, as computed in Table 18.4. Since daily discharge figs. are given in range (interval), mid value of range (interval) can be assumed to be the flow discharge, which when multiplied by no. of days, will give yield during those days. Summation of yield values over 365 days will give annual yield, as shown in Table 18.4, which otherwise is self explanatory.

**Table 18.4.**

| | | | Year 1981 | | Year 1982 | | Year 1983 | |
|---|---|---|---|---|---|---|---|---|
| S. No. | Daily mean discharge (rnage) (m3/s) in descending order | Mean value of daily discharge range | No. of days for which the flow equalled to that given in col. (3) | Yield ($Mm^3$) = (3) × (4) × 86400/$10^6$ [(3) × (4) × 0.0864] | No. of days for which flow equalled to that of col. (3) | Yield ($Mm^3$) = 0.0864 × col. (3) × col. (6) | No. of days for which flow equaled to that of col. (3) | Yield ($Mm^3$) = 0.0864 × col. (3) × col. (8) |
| (1) | (2) | (3) | (4) | (5) | (6) | (7) | (8) | (9) |
| 1 | 100-90.1 | 95.05 | 0 | 0 | 6 | 49.27 | 10 | 82.12 |
| 2 | 90-80.1 | 85.05 | 16 | 117.52 | 19 | 139.62 | 16 | 117.57 |
| 3 | 80-70.1 | 75.05 | 27 | 175.08 | 25 | 162.11 | 38 | 246.40 |
| 4 | 70-60.1 | 65.05 | 21 | 118.02 | 60 | 337.22 | 67 | 376.56 |
| 5 | 60-50.1 | 55.05 | 43 | 204.52 | 51 | 242.57 | 58 | 275.87 |
| 6 | 50-40.1 | 45.05 | 59 | 229.65 | 38 | 147.91 | 38 | 147.91 |
| 7 | 40-30.1 | 35.05 | 64 | 193.81 | 29 | 87.82 | 70 | 211.98 |
| 8 | 30-20.1 | 25.05 | 22 | 47.62 | 48 | 103.89 | 29 | 62.77 |
| 9 | 20-10.1 | 15.05 | 59 | 74.71 | 63 | 81.92 | 26 | 33.81 |
| 10 | 10-Negligible | 5.05 | 54 | 23.56 | 26 | 11.34 | 13 | 5.67 |
| | Σ | | 365 | 1186.49 | 365 | 1363.67 | 365 | 1566.61 |

The river yield for 3 year is thus, calculated by summation of col (5), (7) & (9). Let these yields be arranged in descending order.

| *Year* | *Annual yield* ($M\ m^3$) | *Order No.* ($m$) |
|---|---|---|
| 1983 | 1560.66 | 1 |
| 1982 | 1363.67 | 2 |
| 1981 | 1186.49 | 3 |
| | | $N = 3$ |

Order No. ($m$) for $p\%$ dependability

$$= \frac{P}{100} \times N \quad \text{...(Eqn. 18.4)}$$

(1) Order No for 75% dependability $= \frac{75}{100} \times 3 = 2.25.$

Since order No. of 2.25 is not an integer, the mean value of yield corresponding to S. No. 2 and 3 will be taken as yield for 75% dependability, given as

$$= \left[\frac{1363.67 + 1186.49}{2}\right] = 1275.08\ M\ m^3$$

Hence, Annual flow with 75% dependability = **1275.08 M m³** **Ans.**

∴ Daily flow (discharge) with 75% dependability

$$= \left[\frac{1275.08 \times 10^6}{365 \times 86400}\right] \text{m}^3\text{/s} = \mathbf{40.43\ m^3/s} \quad \textbf{Ans.}$$

(*ii*) Order No. for 50% dependability $= \frac{p}{100} \times N = \frac{50}{100} \times 3 = 1.5$

Since order No. of 1.5 is not an integer, the mean value of yield corresponding to order No. 1 and 2 shall be taken as the yield of 50% dependability, given as :

$$= \frac{1560.66 + 1363.67}{2} = 1462.17\ M\ m^3$$

Hence, Annual flow with 50% dependability = **1462.17 M m³ Ans.**

∴ Daily flow (discharge) with 50% dependability

$$= \frac{1462.17 \times 10^6}{365 \times 86400} \text{m}^3\text{/s} = \mathbf{46.37\ m^3/s} \quad \textbf{Ans.}$$

**18.4.1.1. Converting the dependable rainfall value into the dependable yield value.** Certain empirical relations are available for converting the yearly rainfall value for the given catchment into the yearly runoff value expected from that catchment. Some of these formulas are:

1. *Binnie's percentages;*
2. *Strange's tables;*
3. *Barlow's tables;*
4. *Lacey's formula;*
5. *Inglis formula; and*
6. *Khosla's formula.*

These empirical relations are briefly discussed below:

(1) **Binnie's percentages.** The first effort ever made in India to connect the long range rainfall and the runoff (yield), was from Sir Alexander Binnie. He made observations *on two rivers in the central provinces,* and worked out certain percentages to connect the monthly rainfall with the monthly yield for the entire monsoon period from June to October. These percentages have been further adjusted by Mr. Garret, and are given in table 18.5. From the values given in table 18.5, *the total monsoon rainfall Vs percentage of rainfall that becomes runoff* are plotted in Fig. 18.6(*a*).

**Table 18.5. Binnie's Monsoon Yield Percentages (as adjusted by Garret)**

| *Monsoon Rainfall (P) in cm* | *Monsoon Yield percentage* | *Monsoon Rainfall (P) in cm* | *Monsoon Yield percentage* |
|---|---|---|---|
| 25 | 7.0 | 107.5 | 39.5 |
| 30 | 9.0 | 110.0 | 40.3 |
| 35 | 11.0 | 112.5 | 41.0 |
| 40 | 13.0 | 115.0 | 41.7 |
| 45 | 15.0 | 117.5 | 42.4 |
| 50 | 17.0 | 120.0 | 43.1 |
| 55 | 19.0 | 122.5 | 43.7 |
| 60 | 21.0 | 125.0 | 44.4 |
| 65 | 23.0 | 127.5 | 45.0 |
| 70 | 25.0 | 130.0 | 45.6 |
| 75 | 27.0 | 132.5 | 46.2 |
| 80 | 29.0 | 135.0 | 46.8 |
| 85 | 31.0 | 137.5 | 47.3 |
| 90 | 33.0 | 140.0 | 47.9 |
| 95 | 35.0 | 142.5 | 48.4 |
| 100 | 37.0 | 145.0 | 49.0 |
| 102.5 | 37.9 | 147.5 | 49.5 |
| 105.0 | 38.7 | 150.0 | 50.0 |

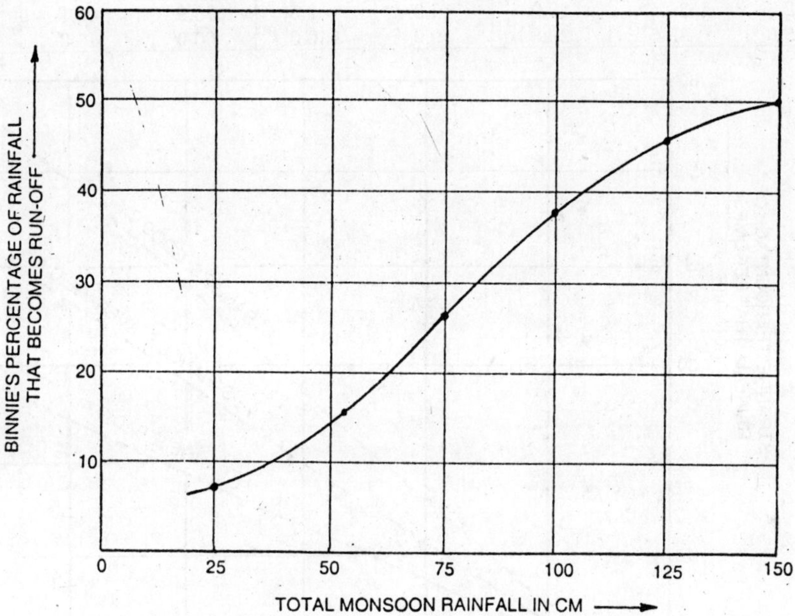


Fig. 18.6(*a*). Binnie's Monsoon Rainfall-runoff Curve.

(2) **Strange's percentages and tables.** Mr. W.L. Strange carried out investigations on catchments in Bombay Presidency, and worked out percentages for converting monsoon rainfall into monsoon yield. He even worked out such percentages for converting daily rainfalls into daily runoffs. It was an improvement over Binnie's tables, since he divided the catchments into three categories to account for the general characteristics of the catchments. The catchments prone to producing higher yields, such as those with more paved areas, etc. were categorised as

*good catchments;* and those prone to producing low yields were termed as *bad catchments*. The intermediate types were called *average catchments*. Different runoff percentages were given for different types of catchments for different values of rainfalls, as shown in table 18.6. These values have also been drawn in the shape of curves, as shown in Fig. 18.6(*b*).

**Table 18.6. Values of Strange's Run off Percentages**

| *Monsoon Rainfall in cm* | *Runoff Percentages for catchments designated as* | | | *Monsoon Rainfall in cm* | *Runoff Percentages for catchments designated as* | | |
|---|---|---|---|---|---|---|---|
| | *Good* | *Average* | *Bad* | | *Good* | *Average* | *Bad* |
| 25.0 | 4.3 | 3.2 | 2.1 | 107.5 | 40.9 | 30.6 | 20.4 |
| 30.0 | 6.2 | 4.6 | 3.1 | 110.0 | 42.0 | 31.5 | 21.0 |
| 35.0 | 8.3 | 6.2 | 4.1 | 112.5 | 43.1 | 32.3 | 21.5 |
| 40.0 | 10.5 | 7.8 | 5.2 | 115.0 | 44.3 | 33.2 | 22.1 |
| 45.0 | 12.8 | 9.6 | 6.4 | 117.5 | 45.4 | 34.0 | 23.2 |
| 50.0 | 15.0 | 11.3 | 7.5 | 120.0 | 46.5 | 34.8 | 23.8 |
| 55.0 | 17.3 | 12.9 | 8.6 | 122.5 | 47.6 | 35.7 | 23.8 |
| 60.0 | 18.5 | 14.6 | 9.7 | 125.0 | 48.8 | 36.6 | 24.4 |
| 65.0 | 21.8 | 16.3 | 10.9 | 127.5 | 49.9 | 37.4 | 24.9 |
| 70.0 | 24.0 | 18.0 | 12.0 | 130.0 | 51.0 | 38.2 | 25.5 |
| 75.0 | 26.3 | 19.7 | 13.1 | 132.5 | 52.0 | 39.0 | 26.0 |
| 80.0 | 28.5 | 21.3 | 14.2 | 135.0 | 53.3 | 39.9 | 26.6 |
| 85.0 | 30.8 | 23.1 | 15.4 | 137.5 | 54.4 | 40.8 | 27.2 |
| 90.0 | 33.0 | 24.7 | 16.5 | 140.0 | 55.5 | 41.6 | 27.7 |
| 95.0 | 35.3 | 26.4 | 17.6 | 142.5 | 56.6 | 42.4 | 28.3 |
| 100.0 | 37.5 | 28.1 | 18.7 | 145.0 | 57.8 | 43.3 | 28.9 |
| 102.5 | 38.5 | 28.9 | 19.3 | 147.5 | 58.9 | 44.1 | 29.4 |
| 105.0 | 39.5 | 29.8 | 19.9 | 150.0 | 60.0 | 45.0 | 30.0 |

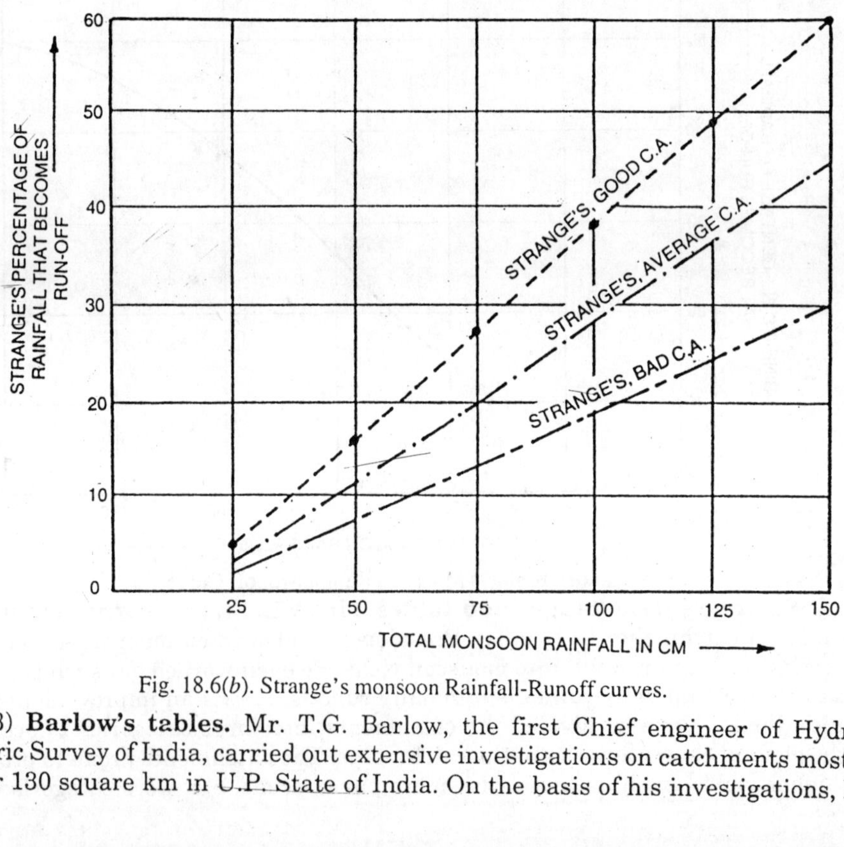


Fig. 18.6(*b*). Strange's monsoon Rainfall-Runoff curves.

(3) **Barlow's tables.** Mr. T.G. Barlow, the first Chief engineer of Hydro-Electric Survey of India, carried out extensive investigations on catchments mostly under 130 square km in U.P. State of India. On the basis of his investigations, he

divided catchments into five classes and assigned different percentages of rainfall that become runoff (over long periods) for each class of catchment, as indicated in table 18.7.

**Table 18.7. Values of Barlow's Runoff Percentages**

| *S.No.* | *Class of catchment* | *Description of the catchment* | *Runoff percentages (K)* |
|---|---|---|---|
| 1. | A | Flat, cultivated, adsorbent soils | 10 |
| 2. | B | Flat, partly cultivated stiff soils | 15 |
| 3. | C | Average catchment | 20 |
| 4. | D | Hills and plains with little cultivation | 25 |
| 5. | E | Very hilly and steep catchment with little or no cultivation | 33 |

The above values of runoff percentages are for the average type of monsoons, and are to be modified by the application of the following coefficients (Table 18.8) according to the nature of the season.

**Table 18.8. Values of Barlow's coefficient to be multiplied with K values of table 18.7 to obtain True Runoff Percentages**

| *S.No.* | *Nature of season* | *Class of catchment* | | | | |
|---|---|---|---|---|---|---|
| | | *A* | *B* | *C* | *D* | *E* |
| 1. | Light rain, no heavy down pour | 0.7 | 0.8 | 0.8 | 0.8 | 0.8 |
| 2. | Average or varying rainfall, no continuous down pour | 1.0 | 1.0 | 1.0 | 1.0 | 1.0 |
| 3. | Continuous down pour | 1.5 | 1.5 | 1.6 | 1.7 | 1.8 |

(4) **Lacey's formula.** This formula connects the monsoon *rainfall (P) with the yield (Q)* by the equation.

$$\text{Yield } (Q) = \left[\frac{P}{1 + \dfrac{304.8\ m}{p.n}}\right] \qquad ...(18.5)$$

where $m$ = a constant, called *monsoon duration factor,* the values of which are given in table 18.9.

$n$ = a constant, called *catchment factor,* the values of which are given for different classes of catchments (as defined by Barlow) in table 18.10.

**Table 18.9. Values of m to be used in Eq. (18.5)**

| *S.No.* | *Duration of Monsoon* | *Monsoon Duration factor m* |
|---|---|---|
| 1. | Bad year | 0.5 |
| 2. | Normal year | 1.2 |
| 3. | Good year | 1.5 |

**Table 18.10. Values of *n* to be used in Eq. (18.5)**

| *S.No.* | *Class of catchment* | *Description of the catchment* | *Values of n* |
|---|---|---|---|
| 1. | A | Flat cultivated, absorbent soils | 0.25 |
| 2. | B | Flat, partly cultivated stiff soils | 0.60 |
| 3. | C | Average catchments | 1.00 |
| 4. | D | Hills and plains with little cultivation | 1.70 |
| 5. | E | Very hilly and steep catchments with little or no cultivation | 3.45 |

(5) **Inglis formula.** Inglis derived his formula for catchments of West Maharashtra State of India. He divided the areas as ghat areas (*Sahyadri ranges*) where rainfall is 200 cm or more; and non-ghat areas where rainfall is less than 200 cm. His formulas are:

(*a*) *For ghat areas, with rainfall (P) equalling or exceeding 200 cm:*

$$\text{Yield} = (0.85P - 30.48)\text{ cm} \quad ...(18.6)$$

where $P$ is the rainfall in cm.

(*b*) *For non-ghat areas with rainfall P less than 200 cm.*

$$\text{Yield} = \frac{P\,(P - 17.78)}{254}\text{ cm} \quad ...(18.7)$$

where $P$ is the rainfall in cm.

(6) **Khosla's formula.** This formula is based upon the recent research work conducted in this field, and is a very simple and useful formula. It can be easily applied to the entire country, without bothering for the region of its origin. This formula states that

$$\text{Yield }(Q) = P - 0.48\,T_m \quad ...(18.8)$$

where $Q$ = the yield in cm
$P$ = the rainfall in cm
$T_m$ = mean annual temperature of the area.

**Example 18.4.** *The design annual rainfall for the catchment of a proposed reservoir has been computed to be 99 cm. The catchment area has been estimated to have the mean annual temperature of 20°C. The catchment area contributing to the proposed reservoir is 1000 sq.km. Calculate the annual design catchment yield for this reservoir. Make use of Khosla's formula.*

**Solution.** Use Khosla's formula connecting the design rainfall ($P$) with the design yield ($Q$) by Eq. (18.8) as:

$$Q = P - 0.48\,T_m$$

where $P = 99$ cm (given)
$T_m = 20°C$ (given)

$\therefore \quad Q = 99 - 0.48 \times 20 = 89.4 \text{ cm} = \mathbf{0.894\ m}$

The total yield produced in $m^3$ from the given catchment of 1000 sq.km (*i.e.* $1000 \times 10^6\ m^2$)

$$= 0.894 \times 1000 \times 10^6\ m^3 = 894\ M.m^3 \quad \textbf{Ans.}$$

### 18.4.2. Use of Flow Duration Curves for Computing Dependable Flow

As we know, the stream flow varies widely over a **water year***. This variability of stream flow can be studied by plotting *flow duration curves* for the given stream. A **flow duration curve**, also called as **discharge-frequency curve**, is a curve plotted between *stream flow (Q)* and *percent of time the flow is equalled or exceeded ($P_p$)*, and is of the type shown in Fig. 18.7 (*a*), (*b*) and (*c*).

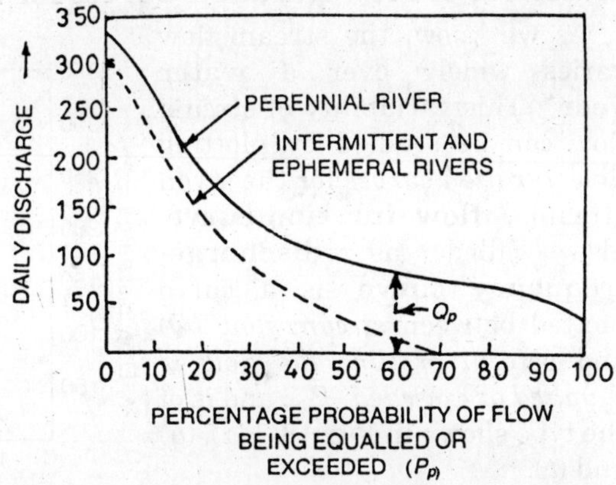


Fig. 18.7(*a*). Typical flow-duration curves on an ordinary paper.

Such a curve can be plotted by first arranging the stream flow values (*Q*) in descending order using class intervals, if the number of individual values is very large. The data to be used can be of daily values, weekly values, or monthly values. If *N* data values are used, the plotting position of any discharge (or class value) *Q* is given, as :

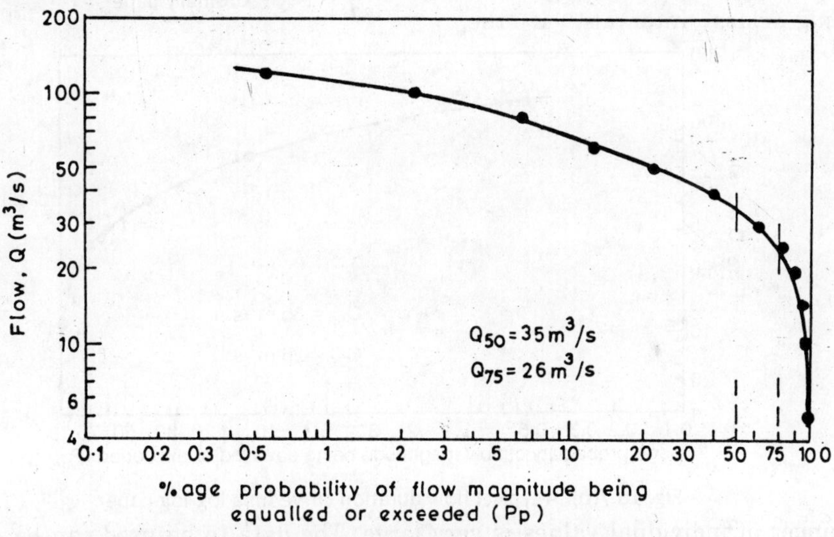


Fig. 18.7(*b*). A typical flow duration curve on a log-log paper.

$$P_p = \frac{m}{N+1} \times 100\%$$

where $m$ = is the order No. of that discharge (or class value)

$P_p$ = percentage probability of the flow magnitude being equalled or exceeded.

The ordinate *Q* at any percentage probability *p* (such as 60%), *i.e.* $Q_p$, will represent the flow magnitude of the river that will be available for 60% of the year, and is hence termed as *60% dependable flow* ($Q_{60}$). $Q_{100}$ (*i.e.* 100% dependable flow) for a perennial river can, thus, be read out easily from such a curve. $Q_{100}$ for an *ephmeral* or for an *intermittent river* shall evidently be zero.

* 1st June to 30st May.

### 18.4.2. Use of Flow Duration Curves for Computing Dependable Flow

As we know, the stream flow varies widely over a **water year***. This variability of stream flow can be studied by plotting *flow duration curves* for the given stream. A **flow duration curve**, also called as **discharge-frequency curve**, is a curve plotted between *stream flow* ($Q$) and *percent of time the flow is equalled or exceeded* ($P_p$), and is of the type shown in Fig. 18.7 (*a*), (*b*) and (*c*).

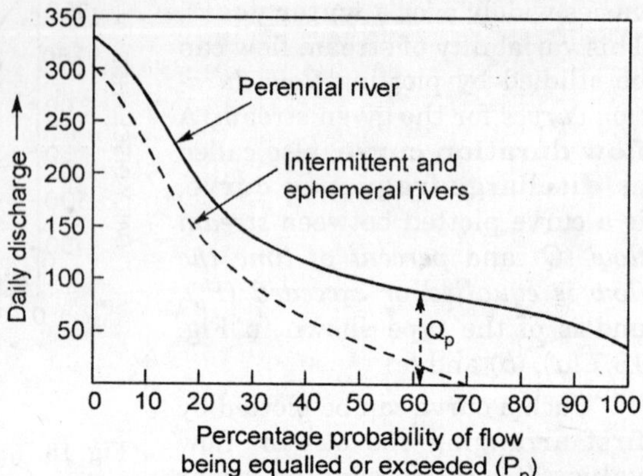


Fig. 18.7(*a*). Typical flow-duration curves on an ordinary paper.

Such a curve can be plotted by first arranging the stream flow values ($Q$) in descending order using class intervals, if the

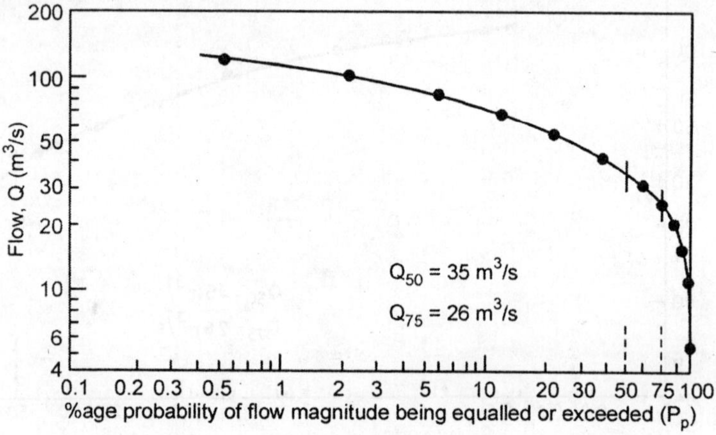


Fig. 18.7(b). A typical flow duration curve on a log-log paper.

number of individual values is very large. The data to be used can be of daily values, weekly values, or monthly values. If $N$ data values are used, the plotting position of any discharge (or class value) $Q$ is given, as :

$$P_p = \frac{m}{N+1} \times 100\% \qquad \text{...(18.9)}$$

where $m$ = is the order No. of that discharge (or class value)

$P_p$ = percentage probability of the flow magnitude being equalled or exceeded.

The ordinate $Q$ at any percentage probability $p$ (such as 60%), *i.e.* $Q_p$, will represent the flow magnitude of the river that will be available for 60% of the year, and is hence termed as *60% dependable flow* ($Q_{60}$). $Q_{100}$ (*i.e.* 100% dependable flow) for a perennial river can, thus, be read out easily from such a curve. $Q_{100}$ for an *ephmeral* or for an *intermittent river* shall evidently be zero.

*1st June to 30th May.

### 18.5. Fixing the Reservoir Capacity for the Computed Value of the Dependable Yield of the Reservoir Catchment

After deciding the dependable yield for the proposed reservoir (tank), the reservoir capacity is decided as follows:

The water demand (annual of course) is computed by estimating the *crop water requirement* (including transit losses) and any other water demand required to meet the water supply needs, or the downstream commitments of water release, if any. Reservoir losses @ about 15% of the water demand is then added to obtain the *live* or the *net storage* required to meet the given demand. *Dead storage* is now added to this live storage to obtain the *gross storage* required to meet the demand. The *reservoir capacity*, however, cannot exceed the *catchment yield* (inflow into the reservoir), and hence the reservoir capacity is fixed at a value which is lesser of the value of the assessed gross storage required to meet the demand; and ii) the assessed dependable yield for the reservoir site. The full tank level (FTL) or the full reservoir level (FRL) is finally computed from the elevation-capacity curve.

The *dead storage* level or dead storage required in the above computation is usually fixed at higher of the following values:

(*a*) dead storage = rate of silting x Life of the reservoir

(*b*) dead storage = 10% of gross storage or net water demand.

(*c*) dead storage level being equal to the full supply level of the off taking canal at the tank site

**Example 18.5.** *The lowest portion of the capacity-elevation curve of a proposed irrigation reservoir, draining 20 km² of catchment, is represented by the following data:*

| *Elevation in m* | *Capacity in ha.m* |
|---|---|
| *RL 600* | *24.2* |
| *602* | *26.2* |
| *604* | *30.3* |
| *606* | *36.8* |

*The rate of silting for the catchment has been assessed to be 300 m³/km²/year. Assuming the life of the reservoir to be 50 years, (a) compute the dead storage, and the lowest sill level (LSL), if the main canal is 6km long with a bed slope of 1 in 1000, and the canal bed level at the tail end is at RL 594.5 m. The FSD of the canal at the head is 80 cm. The crop water requirement is assessed as 250 ham.*

*(b) If the dependable yield of the catchment is estimated to be 0.29m, what will be the gross capacity of the reservoir ?*

**Solution**. The *dead storage* is first of all computed as maximum of the following three values :

(*a*) Dead storage = rate of silting × life of the reservoir

$$= 300\ \text{m}^3/\text{km}^2/\text{year} \times \underset{\text{(C.A.)}}{20\ \text{km}^2} \times \underset{\text{(Life)}}{50\ \text{year}} = 300000\ \text{m}^3$$

$$= 30\ \text{ha.m.} \quad \ldots(i)$$

(*b*) Dead storage

$$= 10\% \times \text{net water demand or crop water requirement}$$

$$= 10\% \times 250\ \text{ha.m} = 25\ \text{ha.m} \quad \ldots(ii)$$

(*c*) Dead Storage Level = FSL of canal at headworks

$$= 594.5 + (6 \times 1000)\frac{1}{1000} + 0.8 = 601.4\ \text{m}$$

Dead Storage Capacity at RL 601.4 m is interpolated

$$= 24.2 + \frac{(26.2 - 24.2)\text{ ham}}{RL\ (602 - 600)} \times RL\ (601.4 - 600) = 24.2 + \frac{2 \times 1.4}{2}$$

$$= 24.2 + 1.4 \text{ ham} = 25.6 \text{ ham} \quad ...(iii)$$

The dead storage is fixed at maximum of the three values obtained at (*i*), (*ii*) and (*iii*) above, *i.e.* 30 ham, 25 ham, and 25.6 ham. Hence, choose the dead storage at **30 ham. Ans.**

The *LSL* (*x*) corresponds to 30 ham capacity, which is computed as :

$$30 = 26.2 \text{ ham} + \frac{(30.3 - 26.2)\text{ ham}}{RL\ (604 - 602)} \times RL\ (x - 602)$$

or $$3.8 = \frac{4.1}{2}(x - 602)$$

$$(x - 602) = \frac{3.8 \times 2}{4.1} = 1.85$$

$$x = RL\ 603.85 \text{ m.}$$

The lowest sill level *i.e.* dead storage level is thus fixed at **RL 603.85 m. Ans.**

(*b*) Net water demand = Crop water requirement including transit losses = 250 ham

Reservoir losses = 15% × 250 ham = 37.5 ham

Live storage required to meet the given demand

= (250 + 37.5) ham = **287.5 ham**

Dead storage = 30 ham (computed above)

∴ Gross reservoir storage required

= Live storage + Dead storage = 287.5 + 30 = **317.5 ham**

Dependable yield = 0.29 m (depth) = 0.29 m × (20 × $10^6$ $m^2$) = **580 ham**

The gross capacity of the reservoir is fixed at the lesser of the gross storage required to meet the demand (*i.e.* 317.5 ham) and the dependable yield (*i.e.* 580 ha.m.). Hence, the reservoir capacity = **317.5 ham. Ans.**

## 18.6. Relation between Inflow, Outflow, and storage Data for a Reservoir

The *inflow* to the reservoir and the *outflow* from the reservoir are the only two factors which govern the storage capacity of a reservoir. Since the inflow to the reservoir is variable, water is stored in the reservoir to cater to the required outflow from the reservoir, particularly during the critical periods in non-monsoon season. Naturally, if more outflow is required, more capacity has to be provided.

As a matter of fact, after assessing the monthly or annual inflows into the reservoir and representing it by the mass inflow curve, the demand pattern is specified. The reservoir is then usually designed to meet this specified demand, represented by the mass outflow curve.

The reservoir capacity, the reservoir inflow, and the outflow from the reservoir are governed by the storage equation, given by:

$$\text{Inflow} - \text{Outflow} = \text{Increase in storage}$$

$$\therefore \text{Increase in Reservoir Storage} = \text{Inflow} - \text{Outflow} \quad ...(18.9)$$

## 18.7. Fixing the Reservoir Capacity from the Annual Inflow and Outflow Data

The capacity of reservoir may be determined by determining the storage needed to accommodate the given inflow minus the given outflow, as governed by the above equation. However, this study involves numerous factors, as discussed below:

Streamflow data at the reservoir site must be known. Monthly inflow rates are sufficient for large reservoirs, but daily data may be required for small reservoirs. When the inflow data at the dam site are not known, the data at a station elsewhere on the stream or on a nearby stream may be collected and adjusted to the dam site. The available data may sometimes be extended so as to include a really drought period.

Besides determining the streamflow data at the dam site, an adjustment has to be made for the water required to be passed from the reservoir to satisfy the prior water rights and to obey the agreements between various sharing States through which the river is passing.

Moreover, the construction of a reservoir increases the exposed area of the water surface above that of the natural stream, and thus, increases the evaporation losses. There is, thus, a net loss of water occurring due to reservoir construction, Sometimes, these losses may be so huge that the entire purpose of the reservoir may be defeated. Seepage from the reservoir may also add to the loss resulting from the reservoir.

All these factors make this study very very complex. An approximate easy solution for determining the reservoir capacity may be obtained graphically with the help of *mass curves* as explained in the next article. On the other hand, a tabular solution is necessary in order to account for all important factors. For more precise results and for all complex systems, computers may be used for programming the analysis (called operation study).

**Example 18.6.** *Monthly inflow rates during a low-water period at the site of a proposed dam are tabulated in Col. (2) of table 18.9. The corresponding monthly pan evaporation and precipitation at a nearby station are also tabulated in Col. (3) and Col. (4) of the same table. Prior water rights make it obligatory to release the full natural flow or 15 hectare-metres per month, whichever is minimum. If the estimated monthly demands are as given in col. (5) of table 18.9 and the net increased pool area is 400 hectares, find the required storage capacity for the reservoir. Assume pan evaporation coefficient=0.7 and also assume that only 28% of the rainfall on the land area to be flooded by reservoir has reached the stream in the past.*

**Table 18.11**

| *Month* | *Inflow at dam site in hectare metres* | *Pan evaporation in cm* | *Precipitation in cm* | *Demand in hectare-metres* |
|---|---|---|---|---|
| (1) | (2) | (3) | (4) | (5) |
| January | 1.2 | 1.8 | 1.3 | 15.8 |
| February | 0.0 | 1.8 | 1.7 | 14.3 |
| March | 0.0 | 2.6 | 0.6 | 9.6 |
| April | 0.0 | 10.2 | 0.0 | 4.8 |
| May | 0.0 | 15.4 | 0.0 | 3.5 |
| June | 0.0 | 1.6 | 1.1 | 3.4 |
| July | 240.0 | 10.8 | 16.1 | 5.0 |
| August | 480.0 | 11.7 | 16.4 | 5.0 |
| September | 1.0 | 10.8 | 2.2 | 10.0 |
| October | 0.6 | 9.6 | 0.8 | 15.6 |
| November | 0.5 | 7.8 | 0.0 | 16.8 |
| December | 0.2 | 2.0 | 0.0 | 16.8 |
| Σ | 723.5 | 101.1 | 40.2 | 120.6 |

**Solution.** Table 18.11 is extended as shown in Table 18.12. The 28% precipitation is already reaching and is included in the given inflow [Col. (2)] and hence, only 72% of the precipitation is to be included, as worked out in Col. (8). The table is otherwise self- explanatory and the monthly water drawn from the reservoir is worked out in Col. (10).

**Table 18.12. Solution Table for Example 18.6**

| *Month* | *Inflow at dam site in hectare-metres* | *Pan evaporation in cm* | *Precipitation in cm* | *Demand in hectare-metres* | *Requirement due to prior rights equal to Col. (2) or 15 ha.m. whichever is minimum* | *Evaporation in hectare-metres* $\frac{400 \times \text{Col. (3)}}{100} \times 0.7$ $= 2.8 \times \text{Col.(3)}$ | *Precipitation in hectare metres* $\frac{400 \times \text{Col. (4)}}{100} \times 0.72$ $= 2.88 \times \text{Col.(4)}$ | *Adjusted inflow. Col. (2) + Col. (8) – Col. (6) – Col. (7) in hectare metres* | *Water required from storage in ha. m. Col. (5) – Col. (9) (only +ve values)* |
|---|---|---|---|---|---|---|---|---|---|
| *(1)* | *(2)* | *(3)* | *(4)* | *(5)* | *(6)* | *(7)* | *(8)* | *(9)* | *(10)* |
| January | 1.2 | 1.8 | 1.3 | 15.8 | 1.2 | 5.03 | 3.74 | (–) 1.29 | 17.09 |
| February | 0.0 | 1.8 | 1.7 | 14.3 | 0.0 | 5.03 | 4.9 | (–) 0.13 | 14.43 |
| March | 0.0 | 2.6 | 0.6 | 9.6 | 0.0 | 7.28 | 1.73 | (–) 5.55 | 15.15 |
| April | 0.0 | 10.2 | 0.0 | 4.8 | 0.0 | 28.56 | 0.0 | (–) 28.56 | 33.36 |
| May | 0.0 | 15.4 | 0.0 | 3.5 | 0.0 | 43.1 | 0.0 | (–) 43.1 | 46.60 |
| June | 0.0 | 16.6 | 1.1 | 3.4 | 0.0 | 46.4 | 3.17 | (–) 43.23 | 46.63 |
| July | 240.0 | 10.8 | 16.1 | 5.0 | 15.0 | 30.3 | 46.4 | (+) 241.1 | Nil |
| August | 480.0 | 11.7 | 16.4 | 5.0 | 15.0 | 32.8 | 47.3 | (+) 479.5 | Nil |
| September | 1.0 | 10.8 | 2.2 | 10.0 | 1.0 | 30.3 | 6.34 | (–) 23.96 | 33.96 |
| October | 0.6 | 9.6 | 0.8 | 15.6 | 0.6 | 26.9 | 2.31 | (–) 24.39 | 39.99 |
| November | 0.5 | 7.8 | 0.0 | 16.8 | 0.5 | 21.8 | 0.0 | (–) 21.8 | 38.60 |
| December | 0.2 | 2.0 | 0.0 | 16.8 | 0.2 | 5.6 | 0.0 | (–) 5.6 | 22.40 |
| Σ | 723.5 | 101.1 | 40.2 | 120.6 | 33.5 | 283.1 | 116.49 | | 308.21 |

Finally, their summation (*i.e.* **308.21 hectare-metres**) works out to be the required storage capacity of the reservoir. **Ans.**

## 18.8. Fixation of Reservoir Capacity with the Help of Mass Curves of Inflow and Outflow

After the flow hydrographs for the stream at the dam site have been plotted for a large number of years (say 25 to 30 years), the required storage capacity for a reservoir with a given outflow pattern can be approximately calculated with the help of mass curves. A hydrograph is a plot of discharge vs. time, while a *mass curve* is a plot of accumulated flow vs. time. The area under the hydrograph between times $t = 0$ and $t = t$ will represent nothing but the accumulated flow up to the time t, and hence, the ordinate of the mass curve at time $t$.

A typical annual inflow hydrograph is shown in Fig. 18.8 (*a*), and the mass curve for this inflow hydrograph is plotted in Fig. 18.8 (*b*). The area under the first curve upto a time ($t$) is equal to the ordinate of the second curve at the same time ($t$). Adjustments for the scales and units of the two curves must be made while plotting.

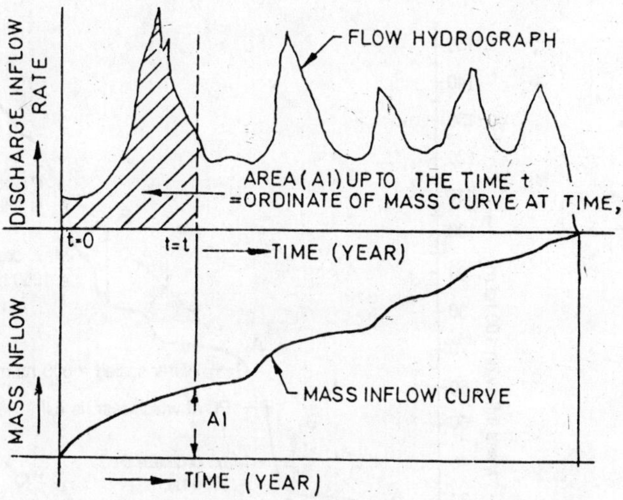

Fig. 18.8 (*a*) and (*b*).

*It is evident that a mass curve will continuously rise, as it is the plot of the accumulated inflow.* Periods of no inflow would be represented by the horizontal lines on the mass inflow curve. To differentiate such a mass curve of runoff from the mass curve of rainfall, this mass curve is usually called as the **flow mass curve** and is an integral of the flow hydrogrph.

The mass curve may also be called the *ripple diagram.* The slope of the mass curve at any time is a measure of the inflow rate at that time.

After the inflow mass curve has been plotted, the mass curve of demand may also be plotted by accumulating the required outflow. *If a constant rate of withdrawal is required from the reservoir, the mass curve of demand will be a straight line having a slope equal to the demand rate.* Demand curves or demand lines are generally straight lines (representing uniform withdrawal) although, in practice, they may be curved also [See Fig. 18.9 (*a*) and (*b*)].

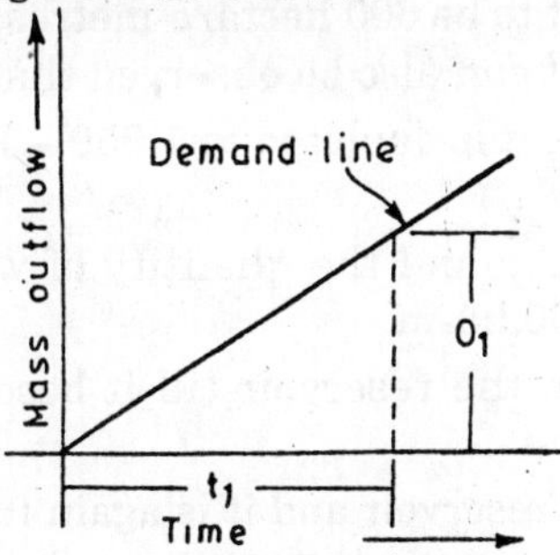

Fig. 18.9 (*a*)

$O_1/t_1$ = Slope of the line = Demand rate

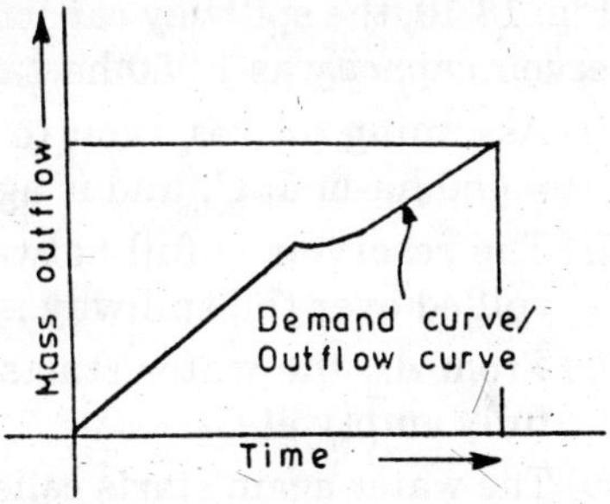

Fig. 18.9 (*b*).

### 18.8.1. Determining Reservoir Capacity for a Given Demand

The mass curve of inflow and the demand line can be used to determine the required storage capacity. In Fig. 18.10, it is evident that the demand lines drawn tangent to the high points $A_1, A_2, A_3...$ of the mass curve, represent the rate of withdrawal from the reservoir. Assuming the reservoir to be full whenever the demand line intersects the mass curve (points $F_1$, $F_2$ ...), the maximum departure ($B_1C_1$, $B_2C_2$...) between the two curves represents the reservoir capacity just required to satisfy the demand.

Fig. 18.10 has been drawn for a demand of 2,000 hectare- metres/year. The biggest departure ordinate (*i.e.* the maximum of $B_1C_1$, $B_2C_2$...) works out to be 1,950 hectare-metres/year, which represents the required storage capacity for the reservoir.

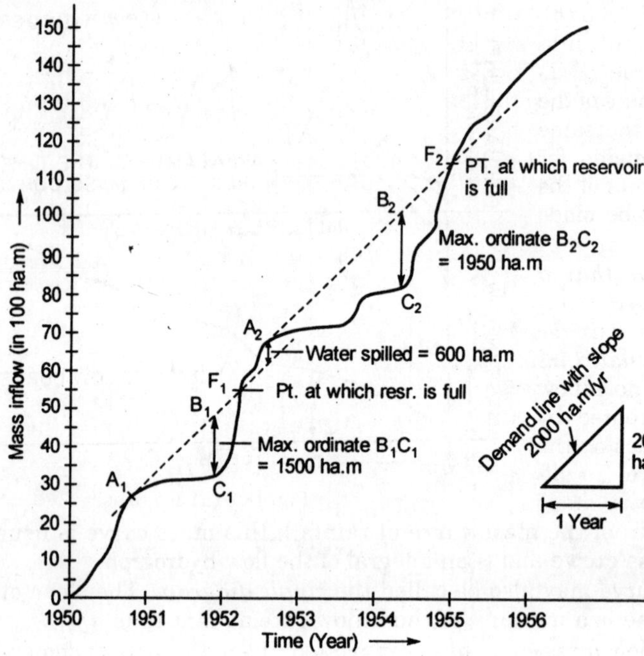


Fig. 18.10.

The vertical distance between the successive tangents $A_1B_1$, and $A_2B_2$, etc., represent the water wasted over the spillway. The spillway must have sufficient capacity to discharge this flood volume.

For Fig. 18.10, the spillway capacity works out to be 600 hectare-metres, and the reservoir capacity as 1,950 hectare-metres. It can also be observed that:

(*i*) Assuming the reservoir to be full at $A_1$, it is depleted to 1,950 – 1,500 = 450 ha-m at $C_1$ and is again full at $F_1$.

(*ii*) The reservoir is full between $F_1$ and $A_2$, and the quantity of water spilled over the spillway is equal to 600 ha-m.

(*iii*) From $A_2$, the water starts reducing in the reservoir till it becomes fully empty at $C_2$.

(*iv*) The water again starts collecting in the reservoir and it is again full at $F_2$.

**Note 1.** It may also be noted that a demand line, when extended, must intersect the mass curve. If it does not, the reservoir will not refill.

**Note 2.** When the demand curve is not a straight line, then the two mass curves are superimposed over each other in such a way that their origins and axis coincide. The larger ordinate between the two, gives the required storage capacity, as shown in Fig. 18.11.

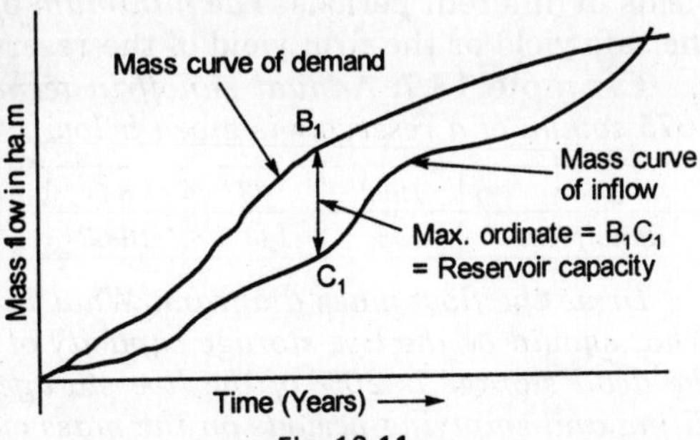


Fig. 18.11.

### 18.8.2. Fixing the Demand for a Reservoir of a Given Capacity

In the previous article, we have explained as to how the reservoir capacity can be determined for a given demand. The reverse, *i.e.* fixing the demand for a given reservoir capacity, may also be done with the help of mass curve of inflow.

In this case, the tangents are drawn to the high points ($A_1, A_2, ...$) of the mass inflow curve in such a way that the maximum departure from the mass curve is equal to the reservoir capacity. The slopes of the lines so drawn represent the demand rate which can be obtained with this capacity during different periods. The minimum value of these slopes will represent the withdrawal rate, which can certainly be obtained from the given reservoir, and will, thus, represent its *firm yield.*

**For example** in Fig. 18.12, *a mass inflow curve is given. It is further required to find out the possible yields and the safe yield for a reservoir capacity of 750 ha-m.* To determine these values, we shall proceed as follows :

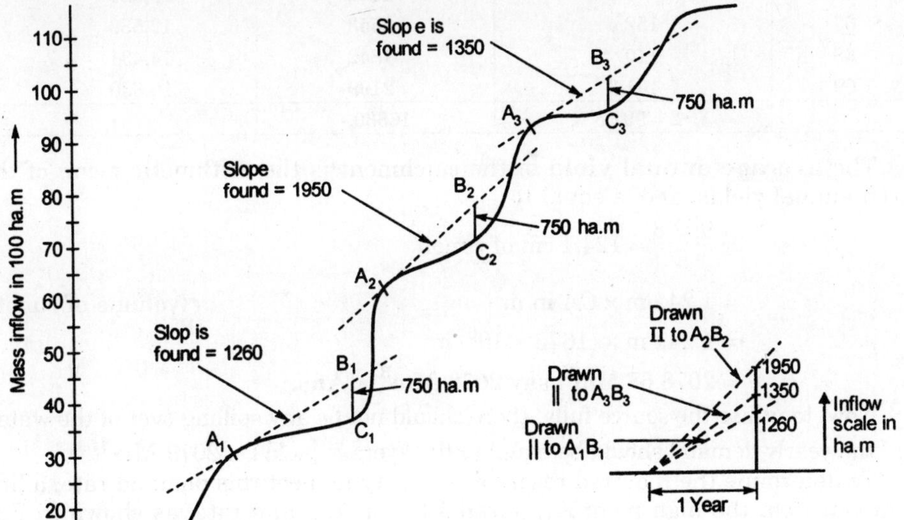


First of all, the high points $A_1$, $A_2$ and $A_3$ are marked on the mass inflow curve. The points $B_1$, $B_2$, and $B_3$ are determined in such a way that their maximum departure from the curve is equal to 750 ha-m. The tangent lines $A_1B_1$, $A_2B_2$ and

$A_3B_3$ are then drawn. The slopes of these lines are determined and are found to be 1,260, 1,950, 1,350 ha-m/year, respectively. These values represent the possible yields in different periods. *The minimum of them, i.e.* 1,260 ha-m/year represents the safe yield or the firm yield of the reservoir. **Ans.**

**Example 18.7.** *Annual runoff in terms of depth over the catchment area of 1675 sq.km. of a reservoir is given below :*

| *Year* | *1962* | *63* | *64* | *65* | *66* | *67* | *68* | *69* |
|---|---|---|---|---|---|---|---|---|
| *Runoff (cm)* | *98* | *143.5* | *168.3* | *94* | *95.3* | *152.4* | *110* | *131.3* |

*Draw the flow mass diagram. What is the average yield from the catchment? What should be the live storage capacity of the reservoir to use the source fully? If the dead storage is 20% of the live storage, what is the gross storage? Mark the filling and emptying periods on the mass curve.* (*Bhopal University, 1980*)

**Solution.** The cumulative runoff values are worked out in Col. 4 of table 18.13, and they are plotted against the values of corresponding years (col. 1) of the same table, so as to obtain the desired mass diagram [Fig. 18.13].

**Table 18.13**

| *Year* (1) | *Yearly runoff (cm)* (2) | *Runoff as volume in* $\text{M.m}^3 = \frac{Col.\ (2)}{100} \times 1675$ (3) | *Cumulative runoff as volume in* $\text{M.m}^3$ (4) |
|---|---|---|---|
| 1962 | 98 | 1642 | 1642 |
| 63 | 143.5 | 2404 | 4046 |
| 64 | 168.3 | 2819 | 6865 |
| 65 | 94 | 1575 | 8440 |
| 66 | 95.3 | 1596 | 10,036 |
| 67 | 152.4 | 2553 | 12,589 |
| 68 | 110 | 1842 | 14,431 |
| 69 | 131.3 | 2199 | 16,630 |
| | $\Sigma = 992.8$ | 16630, | |

The **average annual yield** of the catchment is the arithmetic mean of the given annual yields, and is equal to

$$= \frac{992.8}{8} = 124.1 \text{ cm of runoff}$$

$$= 1.241 \text{ m} \times CA \text{ in m}^2 \qquad \text{(volume of runoff)}$$

$$= 1.241 \text{ m} \times (1675 \times 10^6) \text{ m}^2$$

$$= 2078.68 \text{ Mm}^3; \text{ say } \mathbf{2079\ Mm^3} \quad \textbf{Ans.}$$

Now, to utilise the source fully, there should not be any spilling over of the water, and the yearly demand should be equal to the average yield *i.e.* 2079 $\text{Mm}^3$.

To determine the required reservoir capacity to meet this demand rate, a line is drawn from the high point $A_1$, parallel to this demand rate, as shown in Fig. 18.13; and the maximum departure of this line from the mass curve is read out as $B_1C_1 = 1065 \text{ Mm}^3$.

Hence, the required reservoir capacity is 1065 $\text{Mm}^3$. **Ans.**

Physically speaking, the reservoir will be empty at trough pts $C_1$ and $C_2$, and full at ridge point $A_1$.

*Alternatively,* the mass demand line can be plotted, and both curves extended back up to the origin $O$, as shown in Fig. 18.14.

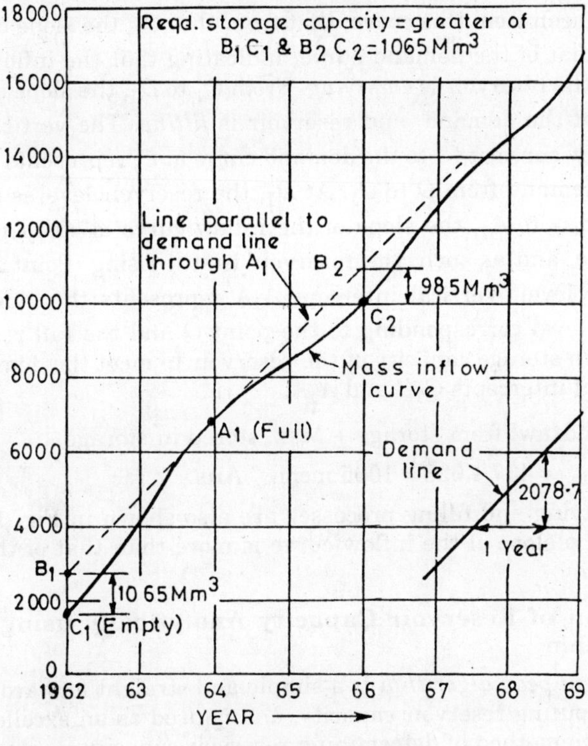


Fig. 18.13.

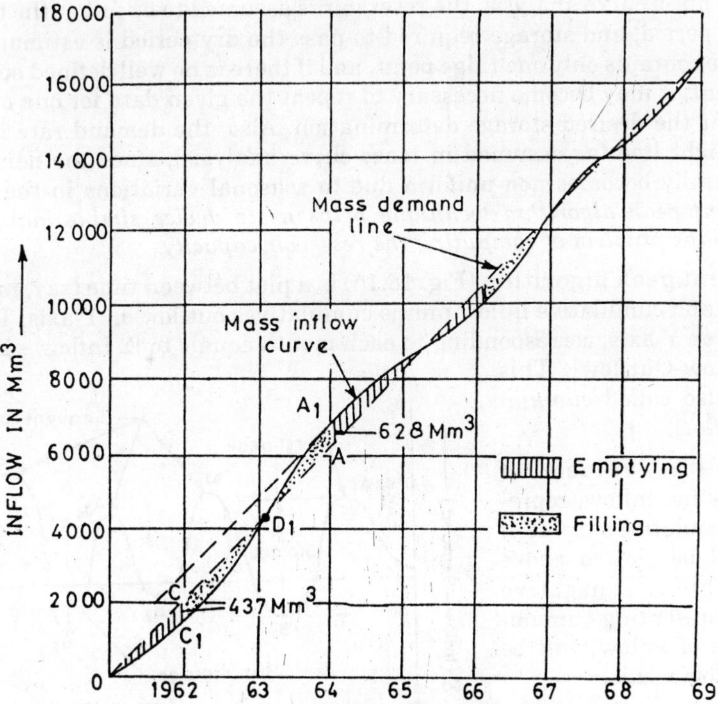


Fig. 18.14.

It can be seen from this curve, that from $O$ to $C_1$, the slope of the inflow curve is less than that of the demand curve, indicating that the inflow is less than the outflow, and the reservoir is *emptying*. From $C_1$ to $D_1$, the slope of the inflow curve is in excess of the demand, and reservoir is *filling*. The vertical intercept $C_1$ $C$, from the inflow curve at $C_1$ to the demand curve at $C$, represents the initial storage to meet the demand from $O$ to $C_1$. At $D_1$, the reservoir level is the same as at $O$. After $D_1$, and upto $A_1$, the slope of the inflow curve is steeper than that of the demand curve, and as such, the reservoir is still rising. Point $A_1$ represents the full reservoir level. Vertical intercept $A_1$A represents the storage between the initial water level corresponding to the point $O$ and the full reservoir level. The total minimum storage capacity of the reservoir to meet the demand is thus given by the vertical intercepts $C_1C$ and $A_1A$.

Max. withdrawl from storage + Max. stored in storage

$= 437 + 628 = 1065$ mcm. **Ans.**

The emptying and filling processes are also shown in Fig. 18.14 ; filling will occur when the slope of the inflow curve is more than that of the demand curve, and *vice versa*.

## 18.9. Fixation of Reservoir Capacity Analytically using Sequent Peak Algorithm

The *sequent peak algorithm* is a simple and straight forward analytical procedure, for computing reservoir capacity, and is used as an excellent alternative to the mass curve method of determining reservoir capacity.

In the mass curve analysis, the reservoir is assumed to be full at the beginning of the dry period, and storage required to pass the dry period is estimated. If the mass curve contains only one ridge point, and if there is no well defined subsequent trough point, it may become necessary to repeat the given data for one more cycle to arrive at the desired storage determination. Also, the demand rate is usually not a straight line (as assumed in mass curve analysis), since the demand (out flow) generally becomes non-uniform due to seasonal variations in the demand. *The sequent peak algorithm technique helps us to device simple mathematical solution to the problem of computing the reservoir capacity.*

**Sequent peak algorithm** (Fig. 18.15) is a plot between time (say, in months) on $X$-axis, and cumulative inflow minus cumulatives outflow on $Y$-axis. The quantity taken on $Y$-axis, corresponding to each month equals to [$\Sigma$ Inflow – $\Sigma$ outflow] = [$\Sigma$ (Inflow-Outflow]. This value is also called *cumulative net inflow*.

The $+ve$ values of cumulative net inflow, representing cumulative surplus of inflow will be plotted above $X$-axis, while its negative values, representing cumulative deficit of inflow, will be plotted below $x$-axis.

The obtained plot will consist of peaks and troughs, as shown in Fig. 18.15.

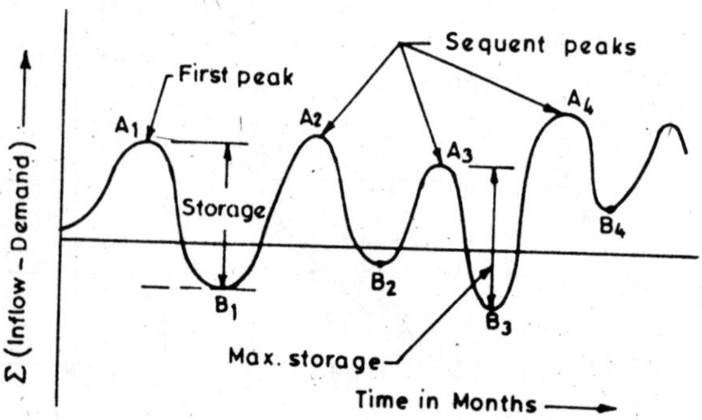


Fig. 18.15. Sequent Peak Algorithm.

The first ridge point $A_1$ in this plotting (*i.e.* first peak of cumulative net surplus) is called the *first peak* ; while subsequent ridge points $A_2, A_3, A_4$, etc., are called *sequent peaks*.

Similarly, the first trough point $B_1$ is called the first trough, while subsequent trough points $B_2, B_3$, etc. may be called the *sequent troughs*.

The difference between the first peak and first trough (height $A_1 B_1$) will, in this plot, evidently, represent the reservoir storage required under normal inflows, and is called the *normal storage* ; whereas, the maximum difference between any sequent peak and the just following trough will represent the *maximum storage* required for the reservoir.

The normal and maximum storage through sequent peak algorithm is calculated as follows :

(1) Convert the monthly inflows into the volume units for the period of the available data.
(2) Estimate the monthly volumes of all the outflows from the reservoir. This should include losses from evaporation, seepage, and other losses.
(3) Compute the cumulative values of Inflows.
(4) Compute the cumulative values of outflows.
(5) Compute the values of cumulative inflow minus cumulative outflow ; *i.e.* [ Σ Inflow – Σ Outflow].
(6) Plot a graph by taking months (time) on $X$-axis, and $\Sigma (I - 0)$ of step (5) on $Y$-axis, on an ordinary graph paper.
(7) The data will plot peaks and troughs. The second and subsequent peaks are called sequent peaks.
(8) The maximum difference between any sequent peak and the just following trough is the maximum storage required for the reservoir. The difference between the first peak and the trough following it, is the storage required under normal inflows.

The method will become more clear on solving example 18.8.

**Example 18.8.** *Monthly inflows at a proposed reservoir site for a drought period of 15 months are given along with targetted demands (found from a working table) in the table below. Compute the storage required by plotting sequent peak algorithm.*

***Table 18.14***

| *Month* | *June* | *July* | *Aug* | *Sept* | *Oct* | *Nov* | *Dec* | *Jan* | *Feb* | *March* | *April* | *May* | *June* | *July* | *Aug* |
|---|---|---|---|---|---|---|---|---|---|---|---|---|---|---|---|
| *River Inflows ($M\ m^3$)* | 250 | 350 | 400 | 200 | 150 | 150 | 100 | 50 | 150 | 300 | 400 | 450 | 150 | 200 | 450 |
| *Targetted demand $M\ m^3$* | 150 | 150 | 200 | 250 | 350 | 400 | 250 | 200 | 150 | 150 | 100 | 250 | 350 | 300 | 100 |

**Solution.** Calculations are carried out in Table 18.15 to compute $\Sigma (I - O)$ in col (6).

**Table 18.15**

| Month (1) | Inflow (I) $M\,m^3$ (2) | Outflow (O) $M\,m^3$ (3) | Cumulative Inflow $\Sigma I$ ($M\,m^3$) (4) | Cumulative Outflow $\Sigma O$ ($M\,m^3$) (5) | ($\Sigma I - \Sigma O$) (6) |
|---|---|---|---|---|---|
| June | 250 | 150 | 250 | 150 | + 100 |
| July | 350 | 150 | 600 | 300 | + 300 |
| Aug | 400 | 200 | 1000 | 500 | + 500 |
| Sept | 200 | 250 | 1200 | 750 | + 450 |
| Oct | 150 | 350 | 1350 | 1100 | + 250 |
| Nov | 150 | 400 | 1500 | 1500 | 0 |
| Dec | 100 | 250 | 1600 | 1750 | (–) 150 |
| Jan | 50 | 200 | 1650 | 1950 | (–) 300 |
| Feb | 150 | 150 | 1800 | 2100 | (–) 300 |
| March | 300 | 150 | 2100 | 2250 | (–) 150 |
| April | 400 | 100 | 2500 | 2350 | + 150 |
| May | 450 | 250 | 2950 | 2600 | + 350 |
| June | 150 | 350 | 3100 | 2950 | + 150 |
| july | 200 | 300 | 3300 | 3250 | + 50 |
| Aug | 450 | 100 | 3750 | 3350 | + 400 |

To calculate values of $\Sigma (I - O)$, values of $\Sigma I$ (cumulative inflow) are written in col (4), and values of $\Sigma O$ (cumulative outflow) are written in col (5).

Values of $(\Sigma I - \Sigma O) = \Sigma (I - O)$ are new calculated by subtracting values of col (5) from values of col (4). These values are written in col (6) with their + *ve* or – *ve* sign, representing excess & cumulative deficit, respectively. When values of $(\Sigma I - \Sigma O)$ from col (6) are plotted on *y*-axis along with corresponding values of month (col 1) taken on *x*-axis, a plot containing peaks and troughs will be obtained, as shown in Fig. 18.16, called sequent peak algorithm. The difference between first peak and the trough following it, is found to be **800 M m³**, which represents the reqd. storage capacity. **Ans.**

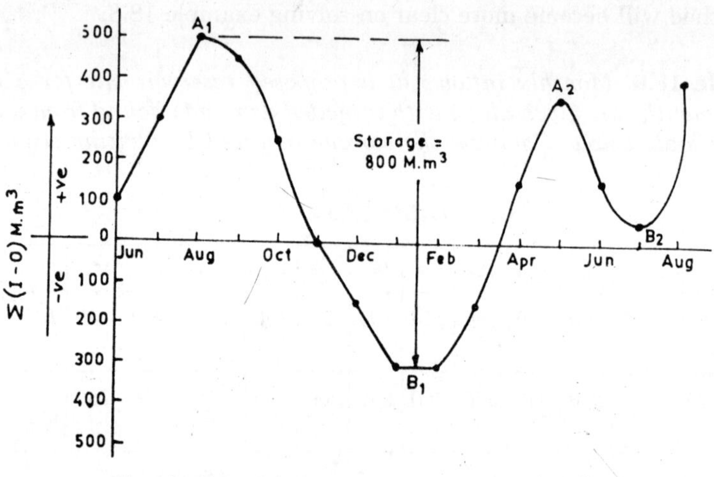


Fig. 18.16. Sequent peak algorithm for Example 18.8.

**Alternatively**, the values of $\Sigma (I - O)$ can be more easily computed by first computing + *ve* and – *ve* values of $(I - O)$, as in col (4) & (5) of Table 18.16 ; and then computing their summation values, *i.e.* values of $\Sigma (I - O)$ as in col (6) and (7) of Table 18.6. + *ve* values represent cumulative excess inflow ; while – *ve* values represent cumulative deficit. The maximum value out of all the values of Col (6) and (7) will represent the minimum storage required to accommodate the surplus (or to supply the deficit). This method thus, avoids the necessity of plotting peaks & troughs (curve) and the value of reqd storage capacity can be computed easily and mathematically, which procedure can even be programmed in a digital computer to help in estimation of required storage capacity. In present calculations, this peak value in col (6) & (7) is found to be **800 M.m$^3$. Ans.**

**Table 18.16**

| Month | Inflow I $M\,m^3$ | Outflow (O) $M\,m^3$ | (I – O) + ve values Excess ($M\,m^3$) | (I – O) – ve values Deficit ($M\,m^3$) | Σ (I – O) Cumulative excess ($M\,m^3$) | Σ (I – O) Cumulative deficit ($M\,m^3$) |
|---|---|---|---|---|---|---|
| (1) | (2) | (3) | (4) | (5) | (6) | (7) |
| June | 250 | 150 | 100 | | 100 | |
| July | 350 | 150 | 200 | | 300 | |
| Aug | 400 | 200 | 200 | | 500 | |
| Sept | 200 | 250 | | 50 | | 50 |
| Oct | 150 | 350 | | 200 | | 250 |
| Nov | 150 | 400 | | 250 | | 500 |
| Dec | 100 | 250 | | 150 | | 650 |
| Jan | 50 | 200 | | 150 | | **800*** |
| Feb | 150 | 150 | 0 | | 0 | (Peak value |
| March | 300 | 150 | 150 | | 150 | in col 6 & 7) |
| April | 400 | 100 | 300 | | 450 | |
| May | 450 | 250 | 200 | | 650 | |
| June | 150 | 350 | | 200 | | 200 |
| July | 200 | 300 | | 100 | | 300 |
| Aug | 450 | 100 | 350 | | 350 | |

**Example 18.9.** *Solve Example 18.7 analytically without using mass curve.*

**Solution.** The data given in example 18.7 is used in Table 18.17 to compute (*I* – *O*) values in col (4) and (5) [– *ve i.e.* deficit values in col (4), and + *ve i.e.* Excess values in col (5)]. The values of Σ (*I* – *O*) are finally computed in col (6) & (7). as shown :

**Table 18.17**

| Year | Inflow (I) $Mm^3$ | Outflow (O) $Mm^3$ | I – O – ve Deficit $Mm^3$ | I – O + ve Surplus $Mm^3$ | Σ (I – 0) Cumulative Deficit $Mm^3$ | Σ (I – 0) Cumulative Surplus $Mm^3$ |
|---|---|---|---|---|---|---|
| (1) | (2) | (3) | (4) | (5) | (6) | (7) |
| 1962 | 1642 | 2079 | 437 | | 437 | |
| 63 | 2404 | 2079 | | 325 | | 325 |
| 64 | 2819 | 2079 | | 740 | | → 1065* |
| 65 | 1575 | 2079 | 504 | | | |
| 66 | 1596 | 2079 | 483 | | → 987 | |
| 67 | 2553 | 2079 | | 474 | | 474 |
| 68 | 1842 | 2079 | 237 | | 237 | |
| 69 | 2199 | 2079 | | 120 | | 120 |

* Max. of all the values in col (6) & (7)

Out of all the values of col. (6) and (7), the max. value is 1065 $Mm^3$, which represents the min. storage required to accommodate this 1065 $Mm^3$ surplus water entering during the years 1963 and 1964. **Ans.**

**Gross storage reqd.** = Dead storage + Live storage

= 20% of Live storage + Live storage

= 1.2 Live storage = 1.2 × 1065 $Mm^3$ = 1278 $Mm^3$ **Ans.**

**Example 18.10.** *The yield of water in $Mm^3$ from a catchment area during each successive month is given in the table below :*

| | | | | | |
|---|---|---|---|---|---|
| *1.4* | *2.1* | *2.8* | *8.4* | *11.9* | *11.9* |
| *7.7* | *2.8* | *2.52* | *2.24* | *1.96* | *1.68* |

*Determine the minimum capacity of a reservoir required to allow the above volume of water to be drawn off at a uniform rate assuming that there is no loss of water over the spillway.*

**Solution.** The total inflow of water in 12 months

= Summation of inflow values = 57.4 $Mm^3$.

∴ Average monthly rate at which water is withdrawn to use the inflow fully = Av. demand rate = 57.4/12 = 4.78 $Mm^3$.

Now, to draw the mass curve of inflow, the cumulative inflow values are worked out in table 18.18.

**Table 18.18**

| *Month* | *Yield* ($Mm^3$) | *Cumulative yield* ($Mm^3$) |
|---|---|---|
| 1 | 1.4 | 1.4 |
| 2 | 2.1 | 3.5 |
| 3 | 2.8 | 6.3 |
| 4 | 8.4 | 14.7 |
| 5 | 11.9 | 26.6 |
| 6 | 11.9 | 38.5 |
| 7 | 7.7 | 46.2 |
| 8 | 2.8 | 49.0 |
| 9 | 2.52 | 51.52 |
| 10 | 2.24 | 53.76 |
| 11 | 1.96 | 55.72 |
| 12 | 1.68 | 57.40 |

The mass inflow curve is now plotted, as shown in Fig. 18.17. A line parallel to the demand rate line is now drawn through the high point $A_1$ on the inflow mass curve, as shown. The maximum departure between the inflow mass curve and this line, *i.e.* $B_1C_1$, gives the min. storage capacity reqd. This value is read out as 20.78 $Mm^3$. **Ans.**

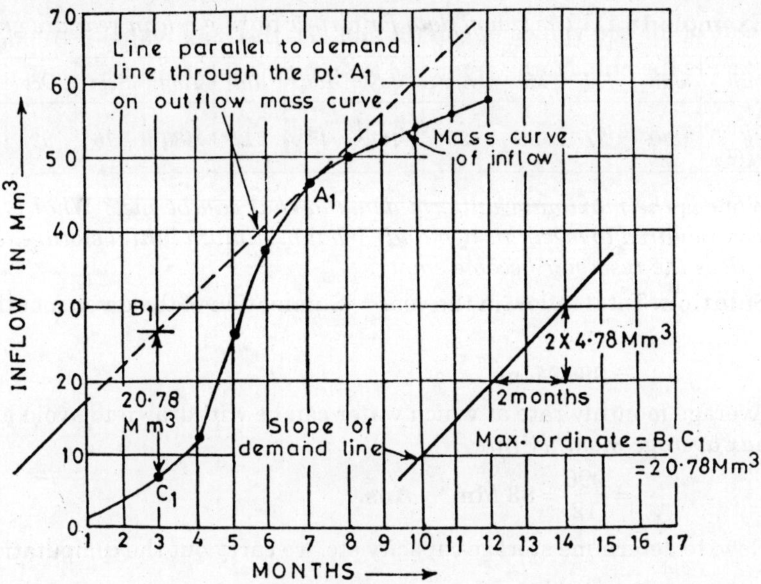


Fig. 18.17.

**Analytically**, the solution to the problem can be worked out, as shown in table 18.19.

**Table 18.19**

| *Month* | *Inflow* $Mm^3$ | *Outflow (Demand)* $Mm^3$ | *Deficit* $Mm^3$ | *Surplus* $Mm^3$ | *Cumulative Deficit* $Mm^3$ | *Cumulative Surplus* $Mm^3$ |
|---|---|---|---|---|---|---|
| *(1)* | *(2)* | *(3)* | *(4)* | *(5)* | *(6)* | *(7)* |
| 1 | 1.4 | 4.78 | 3.38 | | | |
| 2 | 2.1 | 4.78 | 2.68 | | | |
| 3 | 2.8 | 4.78 | 1.98 | | 8.04 | |
| 4 | 8.4 | 4.78 | | 3.62 | | |
| 5 | 11.9 | 4.78 | | 7.12 | | |
| 6 | 11.9 | 4.78 | | 7.12 | | |
| 7 | 7.7 | 4.78 | | 2.92 | | 20.78 |
| 8 | 2.8 | 4.78 | 1.98 | | | |
| 9 | 2.52 | 4.78 | 2.26 | | | |
| 10 | 2.24 | 4.78 | 2.54 | | | |
| 11 | 1.96 | 4.78 | 2.82 | | | |
| 12 | 1.68 | 4.78 | 3.10 | | 12.70 | |

The highest value in col. (6) and 7 (as no spilling is allowed), is 20.78 $Mm^3$, which gives the min. required reservoir storage. **Ans.**

**Example 18.11.** *The runoff data for a river during a lean year are given below :*

| *Month* | *Jan* | *Feb* | *Mar* | *Apr* | *May* | *Jun* | *Jul* | *Aug* | *Sept* | *Oct* | *Nov* | *Dec* |
|---|---|---|---|---|---|---|---|---|---|---|---|---|
| *River flow in* $10^6 m^3$ | *140* | *27* | *35* | *26* | *16* | *48* | *212* | *180* | *116* | *92* | *67* | *37* |

*What is the maximum uniform demand that can be met? What is the storage capacity required to meet this demand? What minimum initial storage is necessary? When does the reservoir become empty?*

**Solution.** Total inflow in the year = Summation of the given monthly discharges

$$= 996 \text{ M. m}^3.$$

Average monthly rate at which water can be withdrawn to avoid any wastage *i.e. max average monthly rate*

$$= \frac{996}{12} = \mathbf{83\ Mm^3} \quad \textbf{Ans.}$$

Now, to determine storage capacity etc. we carry out the computations in table 18.20.

**Table 18.20**

| *Month end* | *Monthly Inflow in reservoir in* $Mm^3$. | *Monthly outflow from reservoir in* $Mm^3$. | *Monthly deficit in reservoir (to be supplied from storage) in* $Mm^3$ | *Monthly surplus in reservoir in* $Mm^3$ | *Consecutive cumulative deficit in* $Mm^3$ | *Consecutive cumulative surplus in* $Mm^3$ | *Net water available in the reservoir in* $Mm^3$ |
|---|---|---|---|---|---|---|---|
| (1) | (2) | (3) | (4) | (5) | (6) | (7) | (8) |
| Jan | 140 | 83 | | 57 → | | → 57 | (+) 57 |
| Feb | 27 | 83 | 56 | | | | |
| March | 35 | 83 | 48 | | | | |
| April | 26 | 83 | 57 | | | | |
| May | 16 | 83 | 67 | | | | |
| June | 48 | 83 | 35 → | | → 263 | | (–) 206 |
| July | 212 | 83 | | 129 | | | |
| Aug | 180 | 83 | | 97 | | | |
| Sept | 116 | 83 | | 33 | | | |
| Oct | 92 | 83 | | 9 → | | → 268 | (+) 62 |
| Nov | 67 | 83 | 16 | | | | |
| Dec | 37 | 83 | 46 → | | → 62 | | 0 |

The highest value in col. (6) and (7) is **268 Mm³**, which represents the minimum storage capacity required to meet the demand without any spilling. **Ans.**

To compute the min. initial storage, we compute in col. (8), the net storage left in the reservoir with the above inflows and outflows. The maximum negative storage here works out to be 206 $Mm^3$. In order that the reservoir fully meets the demand with the above inflows & outflows, there should be no negative storage in it, and in the limiting case the max. negative storage should be equal to zero. Hence, the min. initial storage in the reservoir should be **206 $Mm^3$**, which will just meet the shortage created in the reservoir by *June end,* when the reservoir will become empty. **Ans.**

**Example 18.12.** *The storage capacity of a reservoir for a flood control project is to be determined. The estimated cost of damage if the emergency spillway is topped is Rs. 10 lacs for each event. The interest rate is 6% and the reservoir life is 50 years. Six reservoir designs of different storage capacities ; the probabilities of exceeding those capacities in any given year, and the estimated initial cost for construction of the reservoir are as follows :*

| *Reservior design number* | *1* | *2* | *3* | *4* | *5* | *6* |
|---|---|---|---|---|---|---|
| *Flood storage capacity in lac cubic metres* | *30* | *35* | *40* | *45* | *50* | *55* |
| *Probability of exceeding storage capacity in any given year* | *0.15* | *0.10* | *0.06* | *0.04* | *0.02* | *0.01* |
| *Estimated initial cost, in lac rupees* | *25* | *30* | *35* | *40* | *45* | *50* |

*Determine the optimal storage capacity of the reservoir.*

**Solution.** The required computations are carried out in table 18.21 below.

**Table 18.21**

| *Reservior Design No.* | *Reservior Capacity in lakh cum* | *Probability of exceeding storage capacity* | *No. of times the flood is likely to exceed the capacity in 50 yrs.* | *Damages caused due to flood exceeding capacity in Rs. Lakh = Rs. 10 lakh × Col. (4)* | *Extra benefit over proposal No. 1 of lowest storage as datum; Rs. 75 lakh – col (5)* | *Imtial cost of each proposal in Rs. Lakh* | *Extra costs involved over proposal 1 as datum, in Rs. Lakh Col. 7 – Rs. 25 lakh* | *Net additional benefit in Rs. lakh; col. (6) – col. (8)* |
|---|---|---|---|---|---|---|---|---|
| *(1)* | *(2)* | *(3)* | *(4)* | *(5)* | *(6)* | *(7)* | *(8)* | *(9)* |
| 1 | 30 | 15% | 7.5 | 75 | — | 25 | — | — |
| 2 | 35 | 10% | 5 | 50 | 25 | 30 | 5 | 20 |
| 3 | 40 | 6% | 3 | 30 | 45 | 35 | 10 | 35 |
| 4 | 45 | 4% | 2 | 20 | 55 | 40 | 15 | 40 |
| 5 | 50 | 2% | 1 | 10 | 65 | 45 | 20 | **45** |
| 6 | 55 | 1% | 0.5 | 5 | 70 | 50 | 25 | 45 |

The above table computes the damages caused by overflooding for each proposal, the max. damage of Rs. 75 lakhs being for proposal 1 - having min. reservoir capacity; and the min. damage of Rs. 5 lakh being for proposal 6 – having max. storage capacity. The additional benefits thus provided by proposals 2 to 6 over proposal 1 are computed in col (6) as 75 lakhs – col (5). Extra costs are also worked out in col (8).

These calculations reflect that in proposal 2, we spend extra Rs. 5 lakh (over proposal 1) and get a benefit of 25 lakhs (over proposal 1 of course) ; *i.e.* a net benefit of Rs. 20 lakh. Similarly, in proposal 3, we spend extra Rs. 10 lakhs & get a benefit of Rs. 45 lakhs over proposal 1; *i.e.* a net benefit of Rs. 35 lakhs. Similarly, we find that our net benefit increases up to proposal 5, and afterwards, as in proposal 6, it becomes equal to that in proposal 5. Hence, the increase in reservoir capacity up to proposal 5 gives us increasing net returns, but thereafter we don't get any extra net returns. Investing more, beyond proposal 5, would, therefore, not be optimal, and thus, proposal 5 can be considered as the most optimum design.

Hence, the optimum storage capacity of the reservoir will correspond to that of proposal 5, *i.e.* **50 lakh m$^3$. Ans.**

## 18.10. Estimation of Demands and Optimal Reservoir Operations

In the previous articles, it was explained as to how the reservoir capacity can be determined for satisfying a certain given downstream demand or *vice versa.* This demand may be constant or may vary throughout the year.

The demand pattern mainly depends upon the purpose for which the reservoir has been constructed. Hence, it has to be different for different types of reservoirs. Moreover, the demands are man made, and should be adjusted in such a way that the maximum benefits can be obtained with the minimum cost. If by reducing the demands slightly, a lot of expenditure can be saved, then we must go in for that. This is known as *optimal planning* for dam reservoirs, and involves economic considerations, discussed afterwards.

**Demand patterns for various types of reservoirs are explained below :**

*(i) Single Purpose Conservation Reservoir.* A single purpose conservation reservoir shall store water only to meet one particular demand, such as water supply or irrigation or hydroelectric power generation, etc. In all such cases, the demand for water can be predicted with a fair degree of accuracy for different parts of the year, and the storage capacity can be worked out, as explained earlier. The storage capacity so worked out should be otherwise feasible and should be justified on cost benefit considerations.

*(ii) Single Purpose Flood Control Reservoir.* In case of a single purpose flood control reservoir, there is no specific downstream demand at all, except to satisfy that the downstream release should not exceed the safe carrying capacity of the channel. Since the effect of constructing a flood control reservoir is to moderate and reduce the flood peaks by absorbing certain volume of flood, and then gradually releasing it when the flood subsides (Fig. 18.18), the capacity of such a reservoir should be sufficient to absorb this excess volume of flood.

It, therefore, becomes evident that the capacity of such a reservoir does not depend upon the pattern of demand but mainly depends upon the hydrograph of the worst flood that is likely to enter this reservoir and also upon the downstream permissible H.F.L. and the safe carrying capacity of the channel.

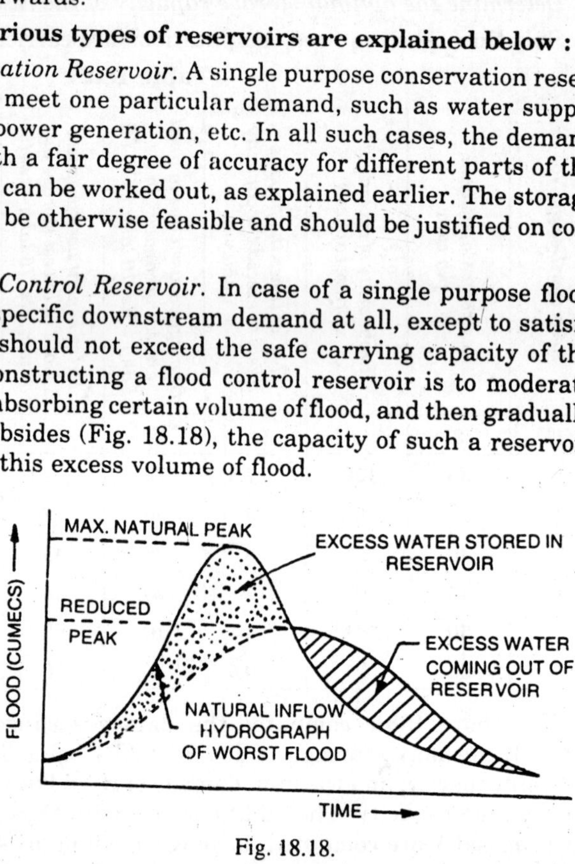

Fig. 18.18.

The hydrograph of the worst inflow flood can be

found from the hydrological investigations, and the safe peak rate of outflow can be determined from the downstream channel conditions. The capacity of the reservoir required to moderate the inflow peak to a value equal to or less than the downstream safe peak can then be found by hit and trial method with the help of *flood routing*. Flood routing is the process by which the hydrograph of the moderated flood can be determined, and is explained a little later.

(*iii*) *Multipurpose Reservoirs.* Single purpose reservoirs are seldom constructed these days. Reservoirs are therefore, generally designed to serve more than one purpose. For example, a reservoir constructed for conserving water for irrigation may be combined with its flood control purpose. The head available may simultaneously be utilised for generating hydro-electric power. This combination of irrigation, flood control, and power is generally adopted in designing multipurpose reservoirs in India.

For proper optimal planning of such a multipurpose reservoir, a schedule of operations must be finalised from a number of tentative schedules drawn on the basis of available data and past similar experiences. The schedule which gives maximum total benefits for the various design purposes, without encroaching upon the lower or upper limits of storage, is accepted for estimation and for initial operations. This schedule may be tried and further modified on the basis of the past experience and future needs.

A multipurpose reservoir constructed on a perennial snow fed river in North India (such as Bhakra reservoir) can be operated on the following lines :

Fig. 18.19 shows the typical schedule of operations for such a reservoir. The reservoir water will normally fluctuate between minimum pool level and normal pool level (*i.e.* maximum conservation level) for satisfying irrigation needs. The minimum pool level will ensure the generation of firm power, as the water level shall not normally be allowed to go below the minimum pool level. Any serious flood which may occur after the reservoir is full, shall be absorbed between the

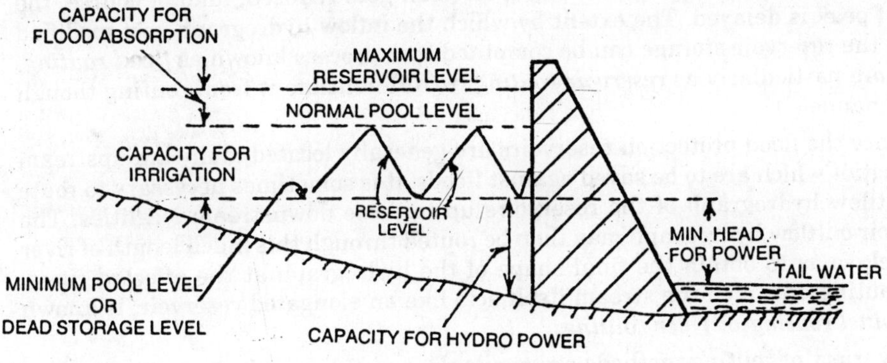


Fig. 18.19.

normal pool level and the miximum pool level, at all times. As soon as the water level reaches the normal pool level, the spillway will start functioning. But the rate of water discharged over the spillway shall be in accordance with the downstream safe carrying capacity of the channel, and hence, may be less than the rate of inflow. The water shall, then, be stored temporarily between the normal pool level and the maximum reservoir level, and discharged at a safe rate. As the flood subsides, and the excess water goes out slowly, the water level will again fall up to the normal pool level, and the reservoir will be ready to absorb another flood.

For irrigation, we generally require water from April to June for Kharif crops, and from middle of November to middle of February for Rabi crops. There are heavy rains during monsoon period of say June to middle of October. At the start of the monsoon season, the reservoir is quite depleted because of heavy outflow for Kharif crops and constant release for power generation. Thus, just before the monsoon, a large reservoir capacity is available, which not only conserves the water but also serves to control floods that may occur at this time. During the monsoon, the irrigation demand will almost be nil and water shall be released only for power generation. The reservoir level will, therefore, go on increasing steadily and will be allowed to go up to the normal pool level by the end of normal monsoon season, say middle of October. Any untimely flood occurring after the reservoir is full up to normal pool level, shall be absorbed between normal pool level and the maximum reservoir level, as explained earlier.

After the monsoon is over, the inflow is considerably reduced (almost nil) and water is constantly drawn for Rabi crops and power generation. Due to this, the reservoir level goes on falling up to about middle of February, when it is usually the lowest, somewhere near the minimum pool level (*i.e.* dead storage level). At this time, snow starts melting in the head reaches of the river, which augments the river discharge. Some winter cyclonic rains during December and January may also be helpful in augmenting the reservoir storage. From February-March onwards, there are large supplies due to melting of snow, and an equal burden is to be absorbed by the reservoir for satisfying the Kharif irrigation withdrawals. Thus, during April, May and June, the reservoir level may rise or fall slightly at slow rate; and at the start of monsoon, *i.e.* at the end of June or beginning of July, the reservoir will be quite depleted and prepared for full replenishment of its supplies.

## FLOOD ROUTING OR FLOOD ABSORPTION

The hydrograph of a flood entering a reservoir, will change in shape as it emerges out of the reservoir, because certain volume of its water is stored in the reservoir temporarily and is let off as the flood subsides. The base of the hydrograph, therefore, gets broadened, its peak gets reduced, and, of course, the time of peak is delayed. The extent by which the inflow hydrograph gets modified due to the reservoir storage can be computed by a process known as *flood routing*, and more particularly as *reservoir routing* (to differentiate it from routing though 'river channels').

Since the flood protection reservoirs are generally located many km upstream of the cities which are to be saved against floods, it is sometimes necessary to route the outflow hydrograph of the reservoirs up to these downstream localities. The reservoir outflow hydrograph may then be routed through this much length of river channel, so as to obtain the final shape of the hydrograph at the affected cities. This routing, in which the stream itself acts like an elongated reservoir, is known as *channel routing* or *river routing*.

A variety of routing methods are available, and they can be broadly classified into two categories, *viz*

(*i*) *hydrologic routing* ; *and*

(*ii*) *hydraulic routing*

*Hydrologic routing methods* employ essentially the *equation of continuity* ; *whereas the hydraulic routing methods employ the continuity equation together with the equation of motion of unsteady flow*. The basic differential equations used in hydraulic routing are popularly known as *St. Venant equations*, and afford a better description of the unsteady flow than the hydrologic methods. We shall, however, cofine ourselves in the book to the hydrologic methods only.

### 18.11. Hydrologic Reservoir Routing Methods

The passage of a flood wave through a reservoir or a river reach is an *unsteady flow* phenomenon. In hydraulics, we classify it as a *gradually varied flow*. The *equation of continuity* used in all the hydrologic routing methods, as the primary equation, states that the difference between the inflow and the outflow rate is equal to the rate of change of storage ; *i.e.*

$$I - O = \frac{dS}{dt} \quad \text{...(18.9)}$$

where $I$ = Inflow rate
$O$ = Outflow rate
$S$ = Storage

Alternatively, in a small time interval $\Delta t$, the difference between the total inflow volume and the total outflow volume is equal to the change in storage volume ; *viz.*

$$\underset{\substack{\text{(Inflow volume)} \\ I(v)}}{\bar{I}\,\Delta t} - \underset{\substack{\text{(Ouflow vol.)} \\ O(v)}}{\bar{O}\,\Delta t} = \Delta S \quad \text{...(18.10)}$$

where $\bar{I}$ = Average inflow (rate) in time $\Delta t$
$\bar{O}$ = Average outflow (rate) in time $\Delta t$
$\Delta S$ = change in storage during the time $\Delta t$

Since $\bar{I} = \frac{I_1 + I_2}{2}$ ;

$\bar{O} = \frac{O_1 + O_2}{2}$ ;

$\Delta S = S_2 - S_1$ ;

where suffixes 1 and 2 denote the beginning and the end of the time interval $\Delta t$.

Eq. (18.10) can then be written as :

$$\left(\frac{I_1 + I_2}{2}\right)\Delta t - \left(\frac{O_1 + O_2}{2}\right)\Delta t = S_2 - S_1 \quad \text{...(18.11)}$$

The time interval $\Delta t$ should be sufficiently short, so that the inflow and outflow hydrographs can be assumed to be in straight line, in that time interval. Moreover, $\Delta t$ must be shorter than the time of transit of the flood wave through the reservoir or the given river reach.

The above relationship seems to be very simple, but its evaluation is not easily possible without drastic simplifying assumptions. This is because of the fact that the relations between time and rate of inflow, elevation and storage of reservoir, and elevation and rate of outflow, cannot be expressed by simple algebraic equations. They are, respectively represented by the *inflow flood hydrograph, the elevation-storage curve, and the outflow-elevation curve.* The first two curves obviously cannot be represented by any simple equations, and the third may be represented by the spillway-discharge equation ($Q$ = 1.71 $LH^{3/2}$) only if, the discharge through the outlets is neglected. If the discharge though the outlets is also not neglected, then all the three curves will be unamenable to simple mathematical treatment, without drastic assumptions.

Several procedures have, however, been suggested by different investigators to solve the above basic equation (Eq. 18.11) by rearranging the components in different manners. Depending upon the different procedures adopted for solving the above basic equation, the following hydrologic methods may be used for reservoir routing :

(1) *Trial and Error method*

(2) *Modified Pul's method or Storage indication method ; and*

(3) *Goodrich method.*

All these methods are described below :

**18.11.1. Trial and Error Method of Reservoir Routing.** Trial and Error method is widely adopted with the assistance of computers to reduce the time taken in long calculations involved in this method. This method arranges the basic routing equation (Eq. 18.11), as follows :

$$\frac{I_1 + I_2}{2}.\Delta t = \frac{O_1 + O_2}{2}.\Delta t + (S_2 - S_1) \qquad \text{...(18.11}a\text{)}$$

The procedure involves assuming of a particular level in the reservoir at the end of the interval $\Delta t$, and computing the values on the right side of the above Eq. (18.11 *a*). The summation of $\frac{O_1 + O_2}{2}.\Delta t$ and $(S_2 - S_1)$ is then compared with the known value of $\frac{I_1 + I_2}{2}.\Delta t$. If the two values tally, then the assumed reservoir elevation at the end of the interval is supposed to be O.K. ; otherwise this is changed, and the process is repeated till the required matching is obtained.

This method gives quite reliable results, provided the chosen time interval ($\Delta t$) is sufficiently small, so that the mean of the outflow rates at the start and the end of the given interval may be taken as the average throughout the interval.

**Procedure :** The following detailed procedure may be adopted in this method, to complete the involved computations :

Data to be given : (*i*) The inflow hydrograph
(*ii*) Elevation capacity curve or Elevation area curve
(*iii*) Elevation outflow curve.

*Steps involved in computations*

(*i*) Divide the inflow flood hydrograph into a number of small intervals. The time interval should be so chosen, as not to miss the peak values.

(*ii*) Fix the normal pool level at which the spillway crest is provided, and the level at which the flood enters the reservoir ; the two are generally taken to be the same, as it is assumed that this worst design flood enters the reservoir only after the reservoir is full up to the normal pool level.

(*iii*) Work out the spillway and the outlet discharge rating curves, if not given.

(*iv*) Work out the elevation-capacity curve for the reservoir from the elevation-area curve, if the former is not given, using cone formula, *i.e.* $V = \Sigma \frac{h}{3}\left[A_1 + A_2 + \sqrt{A_1 A_2}\right]$, where $h$ is the contour interval.

....(18.12)

(*v*) Start with the first interval and compute the total inflow during the interval by multiplying the average inflow rate at the beginning and the end of the interval, with the period of the interval.

$$\therefore \qquad I_{(V)} = \frac{I_1 + I_2}{2} (\Delta t)$$

where $I_1$ = Inflow rate at the start of the interval

$I_2$ = Inflow rate at the end of the interval

$\Delta t$ = Duration of the interval

$I_{(V)}$ = Total inflow volume during the interval.

(*vi*) The reservoir level at the start of the flood (*i.e.* start of first interval) is known. Assume a trial value for the reservoir level at the end of the interval.

(*vii*) compute the total outflow during the interval

$$O_{(V)} = \frac{O_1 + O_2}{2} \cdot \Delta t$$

where $O_1$ = Outflow rate at the start of the interval, corresponding to the given reservoir level.

$O_2$ = Outflow rate at the end of the interval, corresponding to the assumed reservoir level.

$\Delta t$ = Duration of the interval

$O_{(V)}$ = Total outflow volume during the interval.

(*viii*) Using the elevation-storage curve for the reservoir, determine the storage $S_1$ and $S_2$ at the beginning and the end of the interval, corresponding to the known and the assumed reservoir levels, respectively. Their difference $S_2 - S_1 = \Delta S$, represent the amount of flood stored in the reservoir during the interval.

(*ix*) Add the volume of outflow $O_{(V)}$ obtained in step (*vii*) to the values of $\Delta S$ obtained in step (*viii*), and compare it with the inflow volume $I_{(V)}$, calculated in step (*v*). The two values must be equal (*i.e.* $I_{(V)} = O_{(V)} + \Delta S$). If this is so, the assumed reservoir level is correct, otherwise, change it and repeat the procedure till this coincidence is obtained.

(*x*) All the above steps should be repeated for other time intervals, till the entire flood is routed or still further, till the reservoir level returns to pre-flood pool level.

(*xi*) Outflow ordinates are plotted so as to obtain the outflow hydrograph. The point at which it crosses the inflow hydrograph gives the peak outflow rate. From this time, the rate of outflow begins to fall due to decrease in the inflow rate.

(*xii*) The time lag between the two peaks is evaluated as to give the time lag.

An example has been solved to make the procedure very clear.

**Example 18.13.** *The inflow flood discharges for a possible worst flood are tabulated in Table 18.22 at suitable intervals starting from 0.00 hours on august 20, 1975.*

**Table 18.22**

| *Time from start in hr* | *0* | *6* | *12* | *18* | *24* | *30* | *36* | *42* | *48* | *51* | *60* | *66* | *78* | *90* | *102* | *114* |
|---|---|---|---|---|---|---|---|---|---|---|---|---|---|---|---|---|
| *Discharge in cumecs* | *0* | *50* | *280* | *610* | *1290* | *1900* | *2130* | *1900* | *1600* | *1440* | *1060* | *780* | *500* | *370* | *220* | *130* |

*This flood approaches a reservoir with an uncontrolled spillway, the crest of which is kept at RL 140.0 m. Determine the maximum reservoir level and the hydrograph of the routed flood. Values of reservoir capacity (above spillway crest) and outflow discharge at various elevations are tabulated in Tables 18.23 and 18.24 respectively.*

**Table 18.23**

| *Elevation in metres* | *140.0* | *141.0* | *142.0* | *143.0* | *144.0* | *145.0* | *146.0* |
|---|---|---|---|---|---|---|---|
| *Reservoir storage with spillway crest as datum in million cubic metres (m.c.m.)* | *0.0* | *15.0* | *35.0* | *60.0* | *95.0* | *140.0* | *240.0* |

**Table 18.24**

| *Elevation* | *Outflow discharge in cumecs* |
|---|---|
| *140.0* | *0* |
| *141.0* | *170* |
| *142.0* | *482* |
| *143.0* | *883* |
| *144.0* | *1,360* |
| *145.0* | *1,905* |
| *146.0* | *2,500* |

**Solution.** The elevation storage curve and the elevation outflow curve are plotted with the help of Tables 18.23 and 18.24, as shown in Fig. 18.20 and 18.21 respectively. The hydrograph of the given flood is plotted in Fig. 18.22. Flood routing is carried out by hit and trial method as shown in Table 18.25 and as explained earlier. This table is otherwise self-explanatory.

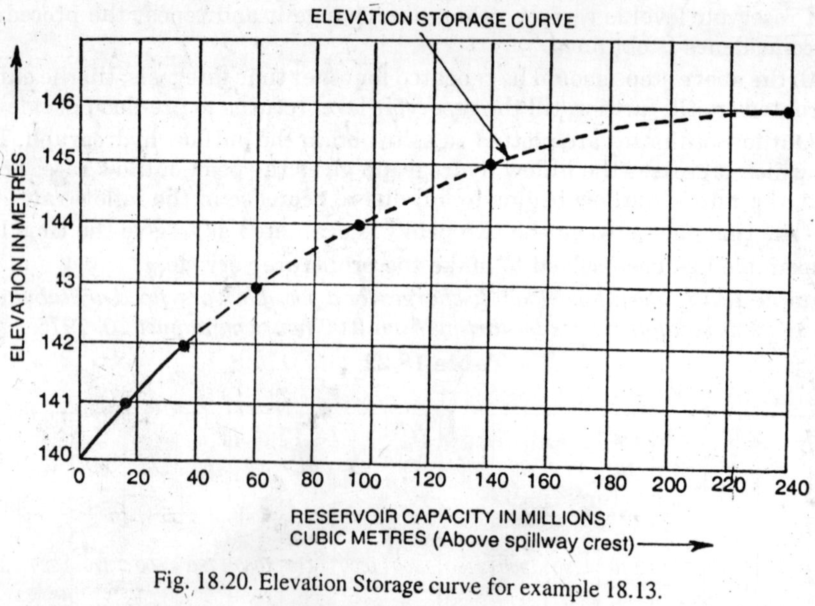


Fig. 18.20. Elevation Storage curve for example 18.13.

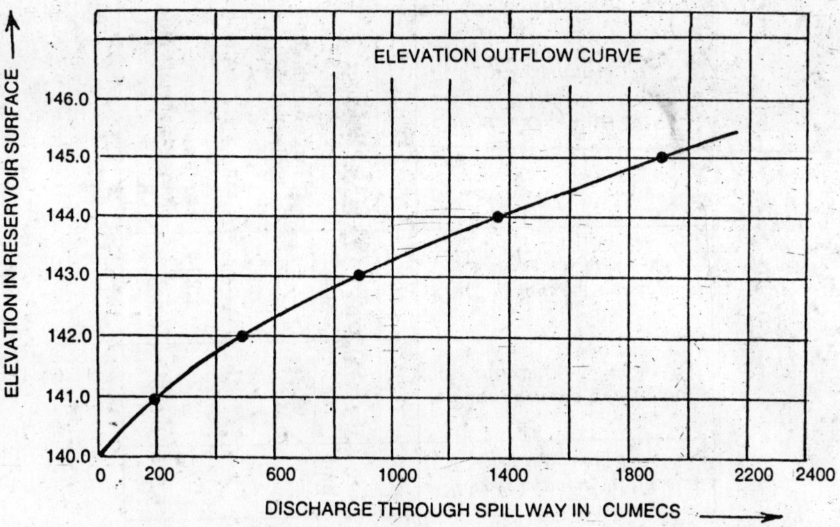


Fig. 18.21. Elevation-Outflow curve for example 18.13.

**The outflow hydrograph is plotted from col. (7) of table 18.25, as shown in Fig. 18.22 by the dotted curve. The point at which it intersects the inflow hydrograph, represents the peak of the routed flood.**

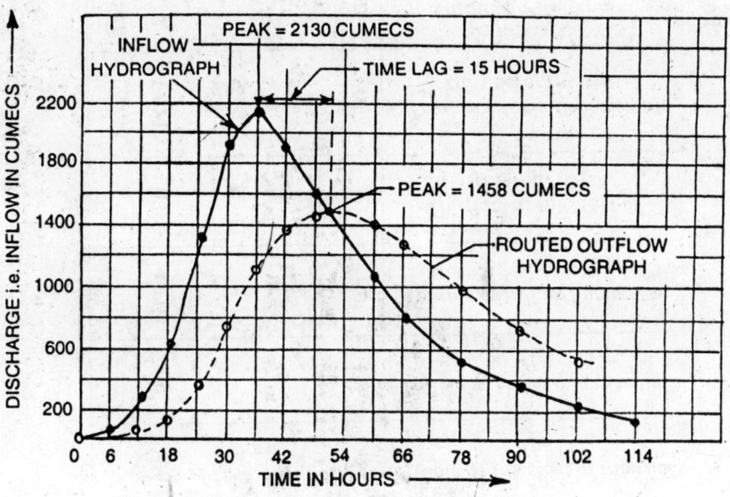


Fig. 18.22. Hydrograph of the given flood and that of the routed flood for example 18.10.

The peak of outflow flood works out to be **1458 cumecs,** while the peak of the inflow flood was 2,130 cumecs. The time lag is found to be **15 hours** (from Fig. 18.22). The maximum reservoir level is found from Table 18.25, to be **144.18 metres. Ans.**

**Table 18.25**

| *Time from start* | *Time interval (Δt)* | *Reservoir elevation at the beginning of interval* | *Inflow at the start of interval ($I_1$)* | *Inflow at the end of interval ($I_2$)* | *Volume of inflow during the interval $I_{(V)} = (I_1 + I_2)/2$* | *Outflow at the start of interval ($O_1$)* | *Trial reservoir elevation at the end of interval* | *Outflow at the end of interval ($O_2$)* | *Mean outflow volume $O_{(V)} = [(O_1 + O_2)/2]\,\Delta t$* | *Storage capacity at start of interval ($S_1$)* | *Storage capacity at the end of interval ($S_2$)* | *Change in storage during the interval $(\Delta S) = (S_2 - S_1)$* | *Outflow volume + change in storage; i.e. $O_{(V)} + \Delta S$* | *Compare $O_{(V)} + \Delta S$ with $I_{(V)}$, i.e. col. (14) with col. (6) and Remarks* |
|---|---|---|---|---|---|---|---|---|---|---|---|---|---|---|
| (1) | (2) | (3) | (4) | (5) | (6) | (7) | (8) | (9) | (10) | (11) | (12) | (13) | (14) | (15) |
| Hrs. | Hrs. | m | $I_1$ m$^3$/s | $I_2$ m$^3$/s | $I$ M.m$^3$ | $O_1$ m$^3$/s | m | $O_2$ m$^3$/s | $O_{(v)}$ M.m$^3$ | $S_1$ reckoned over spilway crest M.m$^3$ | $S_2$ M.m3 | $(S_2 - S_1)$ i.e. $(\Delta S)$ M.m$^3$ | $O_{(V)} + \Delta S$ M.m$^3$ | Whether $(O_{(V)} + \Delta S)$ is large or small as compared to $I_{(V)}$ |
| 0—6 | 6 | 140.0 | 0.0 | 50 | 0.54 | 0.0 | 140.1 | 5.67 | 0.06 | 0.0 | 1.2 | 1.2 | 1.26 | Large and hence change the trial value |
| | | | | | | | 140.03 | 0.88 | 0.01 | | 0.525 | 0.525 | 0.535 | O.K. |

...contd

**Table 18.25 (Contd.)**

| (1) | (2) | (3) | (4) | (5) | (6) | (7) | (8) | (9) | (10) | (11) | (12) | (13) | (14) | (15) |
|---|---|---|---|---|---|---|---|---|---|---|---|---|---|---|
| 6-12 | 6 | 140.03 | 50 | 280 | 3.56 | 0.88 | 140.3 | 28.7 | 0.31 | 0.525 | 4.0 | 3.475 | 3.785 | large |
| | | | | | | | 140.25 | 21.2 | 0.23 | " | 3.85 | 3.325 | 3.555 | O.K. |
| 12-18 | 6 | 140.25 | 280 | 610 | 9.62 | 21.2 | 140.5 | 60.0 | 0.88 | 3.85 | 8.0 | 4.15 | 5.03 | small |
| | | | | | | | 141.0 | 170.0 | 2.06 | " | 15.0 | 11.15 | 13.21 | large |
| | | | | | | | 140.8 | 121.6 | 1.54 | " | 13.0 | 9.15 | 10.69 | large |
| | | | | | | | 140.75 | 110.2 | 1.42 | " | 12.05 | 8.20 | 9.62 | O.K. |
| 18-24 | 6 | 140.75 | 610 | 1290 | 20.5 | 110.2 | 141.5 | 312.2 | 4.57 | 12.05 | 26.0 | 13.95 | 18.52 | small |
| | | | | | | | 141.6 | 343.0 | 4.89 | " | 27.7 | 15.65 | 20.54 | O.K. |
| 24-30 | 6 | 141.06 | 1290 | 1900 | 34.4 | 343.0 | 142.5 | 672 | 10.95 | 27.7 | 47.0 | 19.3 | 30.25 | small |
| | | | | | | | 142.6 | 713 | 11.4 | " | 50.0 | 22.3 | 33.7 | small |
| | | | | | | | 142.63 | 725 | 11.52 | " | 50.5 | 22.8 | 34.32 | O.K. |
| 30-36 | 6 | 142.63 | 1900 | 2130 | 43.6 | 725 | 143.5 | 1113 | 19.85 | 50.5 | 76.0 | 25.5 | 45.35 | large |
| | | | | | | | 143.4 | 1065 | 19.23 | " | 73.0 | 22.5 | 41.73 | small |
| | | | | | | | 143.45 | 1087 | 19.55 | " | 74.5 | 24.0 | 43.55 | O.K. |
| 36-42 | 6 | 143.45 | 2130 | 1900 | 43.6 | 1087 | 144.0 | 1360 | 26.44 | 74.5 | 95.0 | 20.5 | 46.94 | large |
| | | | | | | | 143.95 | 1335 | 26.16 | " | 92.0 | 17.5 | 43.65 | O.K. |
| 42-48 | 6 | 143.95 | 1900 | 1600 | 37.8 | 1335 | 144.5 | 1620 | 13.9 | 92.0 | 115.0 | 23.0 | 64.9 | large |
| | | | | | | | 144.2 | 1461 | 30.2 | " | 102.0 | 10.0 | 40.2 | large |
| | | | | | | | 144.15 | 1435 | 29.9 | " | 100.0 | 8.0 | 37.9 | O.K. |

...contd

**Table 18.25 (Contd.)**

| (1) | (2) | (3) | (4) | (5) | (6) | (7) | (8) | (9) | (10) | (11) | (12) | (13) | (14) | (15) |
|---|---|---|---|---|---|---|---|---|---|---|---|---|---|---|
| 48-51 | 3 | 144.15 | 1600 | 1440 | 16.4 | 1435 | 114.18 | 1458* | 15.5 | 100.0 | 101.0 | 1.0 | 16.5 | O.K. |
| 51-60 | 9 | 144.18* | 1440 | 1060 | 40.5 | 1458* | 144.10 | 1410 | 46.5 | 101.0 | 98.0 | –3.0 | 43.5 | large |
| | | | | | | | 144.05 | 1385 | 46.1 | " | 95.5 | –5.0 | 41.1 | O.K. |
| 60-66 | 6 | 144.05 | 1060 | 780 | 19.9 | 1385 | 143.8 | 1258 | 28.5 | 95.5 | 87.0 | –8.5 | 20.0 | O.K. |
| 66-78 | 12 | 143.80 | 780 | 500 | 27.8 | 1258 | 142.2 | 974 | 48.2 | 87.0 | 67.0 | –20.0 | 28.2 | O.K. |
| 78-90 | 12 | 143.20 | 500 | 370 | 18.8 | 974 | 142.60 | 713 | 35.4 | 67.0 | 50.0 | –17.0 | 18.4 | O.K. |
| 90-102 | 12 | 142.60 | 370 | 220 | 12.7 | 713 | 142.0 | 482 | 25.8 | 50.0 | 35.0 | –15.0 | 10.8 | small |
| | | | | | | | 142.08 | 510 | 26.5 | | 36.2 | –13.8 | 12.7 | O.K. |

*Results : Peak rate of outflow = 1,458 cumecs
Maximum reservoir level = 144.18 metres
Time lag = 15 hours

**Example 18.14.** *The hydrograph of inflow to a reservoir is given in the table below.*

| *Time(days)* | 0 | 2 | 4 | 6 | 8 | 10 | 12 | 14 | 16 | 18 | 20 |
|---|---|---|---|---|---|---|---|---|---|---|---|
| *Flow* ($m^3/s$) | 60 | 115 | 425 | 550 | 440 | 320 | 260 | 200 | 150 | 110 | 60 |

*The reservoir is full at the start of the flood inflow. The storage S of reservoir above the spillway crest, is given in million cubic metres by : S = 8.64 h, where h is the head in metres above the crest. The discharge over the spillway is given in cumecs by Q = 60 h. Find the head over the spillway crest at the end of 8th day of the flood.*

**Solution.** Let $h_1$ be the head over the spillway crest at the end of 2 days ; $h_2$ at the end of 4 days ; $h_3$ at the end of 6 days ; and $h_4$ at the end of 8 days. Now we will consider the position, interval by interval.

(*i*) **0—2 day interval**

*Inflow* $I_1 = 60$ cumecs.

$I_2 = 115$ cumecs.

$$\therefore \quad I_{(V)} = \left(\frac{I_1 + I_2}{2}\right) t = \left(\frac{60 + 115}{2}\right) 2 \times (24 \times 60 \times 60)\ \text{m}^3 = 15.12\ \text{M.m}^3.$$

*Outflow* $O_1 = 0$

$O_2 = 60\ h_1$

$$O_{(V)} = \frac{O_1 + O_2}{2} \cdot t = \left(\frac{60h_1}{2}\right) \times 2 \times (24 \times 60 \times 60)\ \text{m}^3 = 5.184\ h_1\ \text{M.m}^3.$$

*Storage* $S_1 = 0$

$S_2 = 8.64\ h_1\ \text{M.m}^3$

$\therefore \quad \Delta S = 8.64\ h_1$

Now using $I_{(V)} = O_{(V)} + \Delta S$, we get

$15.12 = 5.184\ h_1 + 8.64\ h_1$ or $15.12 = 13.824\ h_1$

$$\therefore \quad h_1 = \frac{15.12}{13.824} = \mathbf{1.093\ m.}$$

(*ii*) **2—4 day interval**

*Inflow* $I_1 = 115$ cumecs ; $I_2 = 425$ cumecs

$$I_{(V)} = \left(\frac{115 + 425}{2}\right) 2 \times 86400\ \text{m}^3 = 46.656\ \text{Mm}^3.$$

*Outflow* $O_1 = 60\ h_1$ ; $O_2 = 60\ h_2$

$$O_{(V)} = \left(\frac{60h_1 + 60h_2}{2}\right) 2 \times (86400)\ \text{m}^3 = (h_1 + h_2) \times 5.184\ \text{Mm}^3.$$

*Storage* $S_1 = 8.64\ h_1$ ; $S_2 = 8.64\ h_2$

$\Delta S = 8.64\ (h_2 - h_1)$

Using $I_{(V)} = O_{(V)} + \Delta S$, we get

$46.656 = (h_1 + h_2)\ 5.184 + 8.64\ (h_2 - h_1)$

Using $h_1 = 1.093$ m, we get

$46.656 = 5.666 + \mathbf{5.184\ h_2} + \mathbf{8.64\ h_2} - 9.443$

or $50.433 = 13.824\ h_2$ or $\mathbf{h_2 = 3.648\ m.}$

(*iii*) **4—6 day interval**

*Inflow* $I_1 = 425$ cumecs ; $I_2 = 550$ cumecs

$$\therefore \quad I_{(V)} = \frac{425 + 550}{2} \times 2 \times 86400 \text{ m}^3 = 84.24 \text{ M.m}^3.$$

*Outflow* $O_1 = 60h_2$ $O_2 = 60h_3$

$$O_{(V)} = \frac{60h_2 + 60h_3}{2} \times 2 \times 86400 \text{ m}^3 = (h_2 + h_3)\, 5.184 \text{ M.m}^3.$$

*Storage* $S_1 = 8.64h_2$ ; $S_2 = 8.64h_3$

$\Delta S = 8.64\,(h_3 - h_2)$

Using $I_{(V)} = O_{(V)} + \Delta S$, we get

$$84.24 = (h_2 + h_3)\, 5.184 + 8.64\,(h_3 - h_2)$$

Using $h_2 = 3.648$ m in this eqn., we get

$$84.24 = 18.911 + \mathbf{5.184\ h_3} + \mathbf{8.64\ h_3} - 31.518$$

$\therefore \quad 96.848 = 13.824\, h_3 \qquad \therefore \qquad \boldsymbol{h_3 = 7.00}$ **m.**

(*iv*) **6–8 day interval.**

*Inflow* $I_1 = 550$ cumecs ; $I_2 = 440$ cumecs

$$I_{(V)} = \left(\frac{550 + 440}{2}\right) 2\,(86400) \text{ M.m}^3 = 85.536 \text{ M.m}^3.$$

*Outflow* $O_1 = 60\, h_3$; $O_2 = 60\, h_4$

$$O_{(V)} = \frac{60h_3 + 60h_4}{2} \times 2 \times 86400 \text{ m}^3 = (h_3 + h_4)\, 5.184 \text{ M.m}^3.$$

*Storage* $S_1 = 8.64\, h_3$; $S_2 = 8.64\, h_4$

$\therefore \quad \Delta S = S_2 - S_1 = 8.64\,(h_4 - h_3)$

Using $h_3 = 7.00$ m in this eqn., we get

$$85.536 = 36.288 + \mathbf{5.184\ h_4} + \mathbf{8.64\ h_4} - 60.48$$

or $\quad 109.728 = 13.824\, h_4 \quad$ or $\quad \boldsymbol{h_4 = 7.94}$ **m.**

Hence, the head over the spillway crest at the end of the 8th day = 7.94 m. **Ans.**

**Example 18.15.** *A small reservoir has an area of 5000 hectares* ($50 \times 10^6\ m^2$) *at spillway crest level. Banks are essentially vertical above the spillway crest level. The spillway is 50 m long and has a coefficient C of 2.2 in equation* $q = CH^{3/2}$. *The inflow to the reservoir is given in the table below :*

| *Time from start (hrs.)* | *0* | *6* | *12* | *24* | *30* | *36* | *42* | *48* | *54* |
|---|---|---|---|---|---|---|---|---|---|
| *Inflow* ($m^3/s$) | *40* | *340* | *722* | *320* | *192* | *118* | *80* | *56* | *40* |

*Compute the maximum outflow discharge over the spillway; and the reservoir level to be expected if the reservoir level was at the spillway crest at the start.*

**Solution.** As done in example 18.11, let us assume that the head over the spillway crest at the end of the various intervals be $h_1, h_2, h_3, h_4$ metres. Then

**For interval 0—6 hr.**

*Inflow* $I_1 = 40$ cumecs ; $I_2 = 340$ cumecs

$$I_{(V)} = \left(\frac{40 + 340}{2}\right)(6 \times 60 \times 60)\ \text{m}^3 = 4.104\ \text{M.m}^3.$$

*Outflow* $O_1 = 0$

$$O_2 = 2.2 \times 50 \times h_1^{3/2} = 110\ h_1^{3/2} \qquad [\because Q \text{ over spillway} = \text{C.L.H.}^{3/2}]$$

$$O_{(V)} = \left(\frac{O_1 + O_2}{2}\right) \cdot t = \left(\frac{0 + 110 h_1^{3/2}}{2}\right)(6 \times 60 \times 60)\ \text{m}^3 = 1.19\ h_1^{3/2}\ \text{M.m}^3.$$

*Storage* $S_1 = 0$

$S_2 = (50 \times 10^6)\ h_1\ \text{m}^3$. [ $\because$ Storage above the spillway is rectangular = Area of the reservoir × ht. of water]

$\therefore \qquad \Delta S = 50\ h_1\ \text{M.m}^3.$

Now, using $I_{(V)} = O_{(V)} + \Delta S$, we have

$$4.104 = 1.19\ h_1^{3/2} + 50 h_1 \qquad \text{or} \qquad h_1^{3/2} + 42 h_1 - 3.45 = 0$$

Solving by hit and trial, we get

$$\mathbf{h_1 = 0.082\ m.}$$

**For interval 6–12 hr.**

$$I_{(V)} = \left(\frac{340 + 722}{2}\right)(6 \times 60 \times 60)\ \text{m}^3 = 11.47\ \text{M.m}^3$$

$$O_{(V)} = \frac{110\ h_1^{3/2} + 110\ h_2^{3/2}}{2} \cdot (6 \times 60 \times 60)\ \text{m}^3 = (h_1^{3/2} + h_2^{3/2})\ 1.19\ \text{M.m}^3$$

$$\Delta S = 50\ (h_2 - h_1)\ \text{M.m}^3$$

$$\therefore \qquad 11.47 = 1.19\ (h_1^{3/2} + h_2^{3/2}) + 50(h_2 - h_1)$$

Using $\quad h_1 = 0.082$, we have

$$9.65 = 0.02 + h_2^{3/2} + 42 h_2 - 3.44 \qquad \text{or} \qquad 13.07 = h_2^{3/2} + 42 h_2,$$

or $\quad h_2^{3/2} + 42 h_2 - 13.07 = 0 \qquad \therefore \quad \mathbf{h_2 = 0.31\ m.}$

**For interval 12—24 hr.**

*Inflow* $$I_{(V)} = \left(\frac{722 + 320}{2}\right)(12 \times 60 \times 60)\ \text{m}^3 = 22.51\ \text{M.m}^3$$

*Outflow* $O_1 = 110\ h_2^{3/2}$; $\quad O_2 = 110\ h_3^{3/2}$

$$O_{(V)} = \frac{110\ (h_2^{3/2} + h_3^{3/2})}{2} \times (12 \times 60 \times 60)\ \text{m}^3 = 2.38\ (h_2^{3/2} + h_3^{3/2})\ \text{M.m}^3$$

*Storage* $S_1 = 50\ h_2 \qquad S_2 = 50\ h_3$

$$\Delta S = 50\ (h_3 - h_2)$$

$$I_{(V)} = O_{(V)} + \Delta S.$$

$$\therefore \qquad 22.51 = 2.38\ (h_2^{3/2} + h_3^{3/2}) + 50\ (h_3 - h_2)$$

Using $\quad h_2 = 0.31$ m, we have

$$9.46 = 0.17 + h_3^{3/2} + 12.2\ h_3 - 3.78$$

$$13.07 = h_3^{3/2} + 12.2\ h_3.$$

or $h_3^{3/2} + 12.2\, h_3 - 13.07 = 0 \quad \therefore\ \mathbf{h_3 = 0.99\ m}$

**For interval 24–30 hr.**

*Inflow* $I_{(V)} = \left(\frac{320+192}{2}\right)(6 \times 60 \times 60)\ m^3 = 5.53\ M.m^3$

Outflow $O_{(V)} = \frac{110\,(h_3^{3/2} + h_4^{3/2})}{2}(6 \times 60 \times 60) = 1.19\,(h_3^{3/2} + h_4^{3/2})$

$$\Delta S = 50\,(h_4 - h_3)$$

$$I_{(V)} = O_{(V)} + \Delta S.$$

$\therefore$ $5.53 = 1.19\,(h_3^{3/2} + h_4^{3/2}) + 50\,(h_4 - h_3)$

or $4.65 = (h_3^{3/2} + h_4^{3/2}) + 9.04\,(h_4 - h_3)$

Using $h_3 = 0.99$, we have

$$4.65 = 0.99 + h_4^{3/2} + 9.04\,h_4 - 8.95$$

or $h_4^{3/2} + 9.04\,h_4 - 12.61 = 0 \quad \therefore \quad \boldsymbol{h_4 = 1.24\ m}$

**For interval 30–36 hr.**

$$I_{(V)} = \left(\frac{192+118}{2}\right)(6 \times 60 \times 60)\ m^3 = 3.35\ M.m^3$$

$$O_{(V)} = 1.19\,(h_4^{3/2} + h_5^{3/2})$$

$$\Delta S = 50\,(h_5 - h_4)$$

$\therefore$ $3.35 = 1.19\,(h_4^{3/2} + h_5^{3/2}) + 50\,(h_5 - h_4)$

or $2.81 = (h_4^{3/2} + h_5^{3/2}) + 42\,(h_5 - h_4)$

Using $h_4 = 1.24$ m, we get

$$2.81 = 1.38 + h_5^{3/2} + 42\,h_5 - 52.08$$

or $h_5^{3/2} + 42h_5 - 50.65 = 0 \quad \therefore \quad \boldsymbol{h_5 = 1.17\ m}$

Since $h_5 < h_4$, it means that the head has started reducing, reaching its max. value of **1.24 m.**

$\therefore$ max. outflow discharge over the spillway.

$= 110\,(1.24)^{3/2}$ cumecs. $= 151.8$ cumecs. **Ans.**

and, the max. reservoir level will be 1.24 m higher than the spillway crest level, which cannot be worked out as the spillway crest level is not given **Ans.**

### 18.11.2. Modified Pul's Method of Reservoir Routing

In this method of reservoir routing, the basic equation (18.11) is rearranged as:

$$\left(\frac{I_1 + I_2}{2}\right)\Delta t + \left(S_1 - \frac{O_1 \Delta t}{2}\right) = \left(S_2 + \frac{O_2 \Delta t}{2}\right) \quad \text{...(18.11 b)}$$

All the factors arranged on the left hand side of this equation are known at the beginning of a time step $\Delta t$. Hence, the value of the function $\left(S_2 + \frac{O_2 \Delta t}{2}\right)$ at the end of the interval $\Delta t$ can be calculated by evaluating the left hand side of Eq. (18.11*b*).

Since storage ($S$) at different elevations ($h$), and outflow ($O$) at different elevations ($h$) are known, the function $\left(S_2 + \frac{O \cdot \Delta t}{2}\right)$ at different elevations can be computed in the beginning. A curve between $\left(S + \frac{O \cdot \Delta t}{2}\right)$ *vs.* elevation ($h$) can thus be prepared, and is used to read out the reservoir elevation corresponding to the above computed value of $\left(S_2 + \frac{O_2 \cdot \Delta t}{2}\right)$, so as to know the value of the reservoir elevation at the end of the interval $\Delta t$.

Corresponding to this computed reservoir elevation at the end of the interval, the outflow $O_2$ at the end of the interval, can finally be read out from the outflow-elevation curve. The procedure can be repeated till the entire inflow hydrograph is analysed to compute the outflows.

*Steps involved in computations.*

(*i*) From the known *storage (S)-elevation data*, and of discharge (outflow)—elevation data, prepare curve 1, between $\left(S + \frac{O \cdot \Delta t}{2}\right)$ *vs* Elevation (Refer Fig. 18.23 of Example 18.16). $\Delta t$ may be any chosen time interval, approximately, 20 to 40% of the time of rise of the inflow hydrograph.

(*ii*) On the same plot, draw curve 2, between *outflow discharge* vs *elevation* (Refer Fig. 18.23 of Example 18.16)

(*iii*) The storage, elevation, and outflow discharge at the starting of routing (step 1) are known, to help in computing the value of $\left(\frac{I_1 + I_2}{2}\right)$ $\Delta t$ + $\left(S_2 + \frac{O_1 \cdot \Delta t}{2}\right)$, which will indicate the computed value of function $\left(S_2 + O_2 \cdot \frac{\Delta t}{2}\right)$.

(*iv*) The computed value of function $\left(S_2 + O_2 \cdot \frac{\Delta t}{2}\right)$ is used to read out the corresponding value of elevation from curve 1, and will indicate the reservoir elevation at the end of the interval.

(*v*) The above value of reservoir elevation (at the end of the interval) is used on curve 2, to read out the corresponding outflow (at the end of the interval).

(*vi*) Deducting $O_2 \cdot \Delta t$ from the computed value of $\left(S_2 + \frac{O_2 \cdot \Delta t}{2}\right)$, we can compute the value of $\left(S_2 - \frac{O_2 \Delta t}{2}\right)$ which can be used as the value of $\left(S_1 - \frac{O_2 \Delta t}{2}\right)$ for the start of the next time step.

(*vii*) The procedure is repeated till the entire inflow hydrograph is routed.

**Example 18.16.** *A reservoir has the following elevation, discharge and storage relationships :*

| *Elevation in m* | *Storage in Mm³ (mcm)* | *Outflow discharge in m³/s* |
|---|---|---|
| 300.20 | 5.5 | 0 |
| 300.70 | 6.0 | 15 |
| 301.20 | 6.6 | 40 |
| 301.70 | 7.2 | 75 |
| 302.20 | 7.9 | 115 |
| 302.70 | 8.8 | 160 |

*When the reservoir level was at 300.20 m, the following flood hydrograph entered the reservoir.*

| *Time (h)* | 0 | 3 | 6 | 9 | 12 | 15 | 18 | 21 | 24 | 27 |
|---|---|---|---|---|---|---|---|---|---|---|
| *Discharge (m³/s)* | 10 | 20 | 52 | 60 | 53 | 43 | 32 | 22 | 16 | 10 |

*Route the flood and obtain (a) the outflow hydrograph; and (b) the reservoir-elevation vs time curve during the passage of the flood wave. Adopt modified Pul's method for routing.*

**Solution.** Let us choose a time interval $\Delta t$ of 3 hr, which is the interval given in inflow discharges. This value is within 20 to 40% of the period of rise, which is 9 hours.

$$\therefore \quad \Delta t = 3 \text{ hr} = 3 \times 60 \times 60 \text{ secs.}$$
$$= 1.08 \times 10^4 \text{ secs} = \mathbf{0.0108 \times 10^6 \text{ secs}}$$

To use the modified Pul's method, we have to draw curves between outflow, (O), *vs* elvation ($h$), say curve 2; and $\left(S + \frac{O \cdot \Delta t}{2}\right) V_s h$, say curve 1. To draw curve 1, we have to compute the value of function $\left(S + \frac{O \cdot \Delta t}{2}\right)$ for different elevations ($h$), as shown in col. (5) of Table 18.26.

**Table 18.26.**

| *Elevation in m* | *Storage in m.c.m.* | *Outflow discharge in m³/s* | $\left(O.\frac{\Delta t}{2}\right) = O \times \left(\frac{0.0108 \times 10^6}{2}\right) \text{m}^3$ *i.e. col. (3) × 0.0054 in m.c.m.* | $\left(S + O\frac{\Delta t}{2}\right)$ *in m.c.m. col. (2) + col (4)* |
|---|---|---|---|---|
| (1) | (2) | (3) | (4) | (5) |
| 300.20 | 5.5 | 0 | 0.0 | 5.50 |
| 300.70 | 6.0 | 15 | 0.08 | 6.08 |
| 301.20 | 6.6 | 40 | 0.22 | 6.82 |
| 301.70 | 7.2 | 75 | 0.41 | 7.61 |
| 302.20 | 7.9 | 115 | 0.62 | 8.52 |
| 302.70 | 8.8 | 160 | 0.86 | 9.66 |

A graph is now plotted between the values of col. (1) and col. (5) of the above table as curve 1, and as shown in Fig. 18.23. Similarly, a curve (curve 2) is plotted between the values of col. (1) and col. (3), as to give outflow elevation curve.

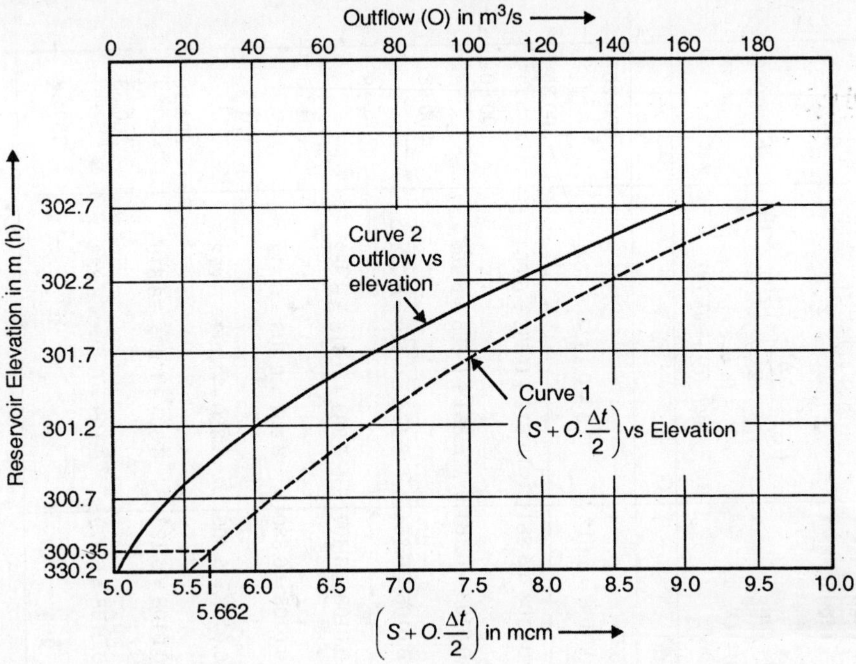


Fig. 18.23.

The requisite routing calculations are now carried out in table 18.27, as explained below :

At the start of routing, elevation is 300.2 m, $O = 0$ m³/s and $\left(S - O \cdot \frac{\Delta t}{2}\right) = (5.5 - 0 \times 0.0054) = 5.5$ mcm. Starting from this value of $\left(S - O \cdot \frac{\Delta t}{2}\right)$, Eq. (18.11 *b*) is used to get the value of $\left(S + O \cdot \frac{\Delta t}{2}\right)$ at the end of the Ist time step of 3 hr, as equal to $\left(\frac{I_1 + I_2}{2}\right)\Delta t + \left(S_1 + O_1 \cdot \frac{\Delta t}{2}\right)$; *i.e.*, 0.162 + 5.5 = 5.662 as shown in col. (6) of Table 18.27.

Corresponding to this value of 5.662 of $\left(S + O \cdot \frac{\Delta t}{2}\right)$ at the end of the interval, read out the value of Elevation from curve 1 of Fig. 18.23, as 300.25 m. Enter this value in col. (7). Corresponding to this value, also read out outflow discharge from curve 2, as 4 m³/s; and enter this Fig. in col. (8). For the next step, the initial value of $\left(S - O \cdot \frac{\Delta t}{2}\right) = \left(S + O\frac{\Delta t}{2}\right)$ of the previous step – O . $\Delta t$ = 5.662 – 4 × 0.0108 = 5.612 mcm (See col. (5), end of Ist time step or start of 2nd time step.)

**Table 18.27**

| Time (h) | Inflow in $m^3/s$ | $\bar{I} = \frac{I_1 + I_2}{2}$ in $m^3/s$ | $\left(\frac{I_1 + I_2}{2}\right)\Delta t$ in m.c.m. col. (3) × 0.0108 | $S - O \cdot \frac{\Delta t}{2}$ in m.c.m. col. (6) – O . Δt | $\left(S + O \cdot \frac{\Delta t}{2}\right)_2 = \left(S_1 - \frac{O_1 \Delta t}{2}\right) + \left(\frac{I_1 + I_2}{2}\right)\Delta t$ in m.c.m. col. (5) + col. (4) | Elevation m | O Outflow in $m^3/s$ |
|---|---|---|---|---|---|---|---|
| (1) | (2) | (3) | (4) | (5) | (6) | (7) | (8) |
| 0 | 10 | | | 5.5 | | 300.20 | 0 |
| | | 15 | 0.162 | 5.662 – 4 × 0.0108 = 5.612 | 5.5 + 0.162 = 5.662 | | |
| 3 | 20 | | | | | 300.35 | 4 |
| | | 36 | 0.389 | 6.001 – 13 × 0.0108 = 5.861 | 5.612 + 0.389 = 6.001 | | |
| 6 | 52 | | | | | 300.60 | 13 |
| | | 56 | 0.605 | 6.466 – 28 × 0.0108 = 6.164 | 5.861 + 0.605 = 6.466 | | |
| 9 | 60 | | | | | 300.98 | 28 |
| | | 56.5 | 0.610 | 6.774 – 38 × 0.0108 = 6.364 | 6.164 + 0.610 = 6.774 | | |
| 12 | 53 | | | | | 301.18 | 38 |
| | | 48 | 0.518 | 6.882 – 44 × 0.0108 = 6.407 | 6.364 + 0.518 = 6.882 | | |
| 15 | 43 | | | | | 301.25 | 44 |
| | | 37.5 | 0.405 | 6.812 – 40 × 0.0108 = 6.380 | 6.407 + 0.405 = 6.812 | | |
| 18 | 32 | | | | | 301.20 | 40 |
| | | 27 | 0.292 | 6.672 – 34 × 0.0108 = 6.305 | 6.380 + 0.292 = 6.672 | | |
| 21 | 22 | | | | | 301.10 | 34 |
| | | 19 | 0.205 | 6.510 – 28 × 0.0108 = 6.198 | 6.305 + 0.205 = 6.510 | | |
| 24 | 16 | | | | | 300.98 | 28 |
| | | 13 | 0.140 | 6.338 – 24 × 0.0108 = 6.079 | 6.198 + 0.140 = 6.338 | | |
| 27 | 10 | | | | | 300.88 | 24 |

From the computed values of outflow discharge in col. (8) of table 18.27, outflow hydrograph can be plotted over inflow hydrograph, as shown in Fig. 18.24. **Ans.**

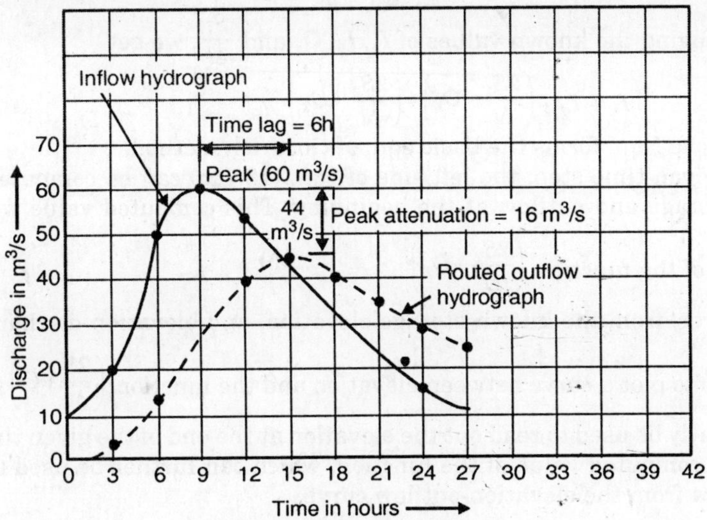


Fig. 18.24. Routed hydrograph.

A graph between the reservoir elevation *vs* time from values of col. (7) and (8) can also be plotted to obtain the reservoir elevation *vs* time curve, as desired in the questions, and as shown in Fig. 18.25. **Ans.**

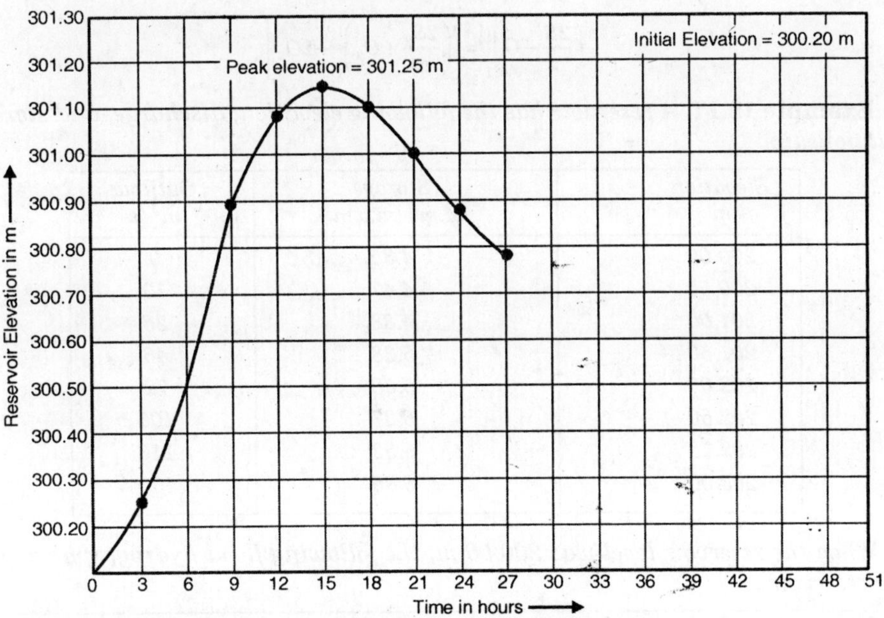


Fig. 18.25. Reservoir elevation *vs*. time curve.

### 18.11.3. Goodrich Method of Reservoir Routing

The popular Goodrich method rearranges the basic routing Eqn. (18.11); *i.e.*,

$$\left(\frac{I_1 + I_2}{2}\right)\Delta t - \left(\frac{O_1 + O_2}{2}\right)\Delta t = S_2 - S_1$$

as follows :

$$I_1 + I_2 - O_1 - O_2 = \frac{2S_2}{\Delta t} - \frac{2S_1}{\Delta t}$$

Rearranging the known values of $I_1$, $I_2$, $O_1$ and $\frac{S_1}{\Delta t}$, we get

$$I_1 + I_2 + \left(\frac{2S_1}{\Delta t} - O_1\right) = \left(\frac{2S_2}{\Delta t} + O_2\right) \quad \text{...(18.11c)}$$

The above Eqn. forms the basic eqn. of Goodrich method.

For a given time step, the left side of eqn. (18.11*c*) can be computed from the known storage and outflow at the beginning. This computed value will indicate the value of the function $\left(\frac{2S_2}{\Delta t} + O_2\right)$.

Moreover, from the known storage-elevation, and elevation-discharge data, it is possible to plot a curve between elevation and the function $\left(\frac{2S}{\Delta t} + O\right)$, which can subsequently be used to read out the elevation at the end of the given time step for the above computed value of the function, which can further be used to read out the outflow from the elevation-outflow curve.

For the next time step, $\left(\frac{2S}{\Delta t} - O\right)_1$ can be obtained by subtracting $2O_2$ from the function $\left(\frac{2S_2}{\Delta t} + O_2\right)$ computed at the end of the first step, as :

$$\left(\frac{2S}{\Delta t} - O\right)_1 = \left[\left(\frac{2S_2}{\Delta t} + O_2\right)_1 - 2O_2\right].$$

**Example 18.17.** *A reservoir has the following elevation, discharge, and storage relationship.*

| *Elevation m* | *Storage M.m³ (m.c.m.)* | *Outflow m³/s* |
|---|---|---|
| 200.00 | 4.35 | 0 |
| 200.50 | 4.47 | 10 |
| 201.00 | 4.88 | 26 |
| 201.50 | 5.38 | 46 |
| 202.00 | 5.88 | 72 |
| 202.50 | 6.37 | 100 |
| 202.75 | 6.53 | 116 |
| 203.00 | 6.86 | 130 |

*When the reservoir level was 200.50 m, the following flood hydrograph entered the reservoir :*

| *Time(h)* | 0 | 6 | 12 | 18 | 24 | 30 | 36 | 42 | 48 | 54 | 60 | 66 |
|---|---|---|---|---|---|---|---|---|---|---|---|---|
| *Inflow (m³/s)* | 10 | 30 | 86 | 140 | 126 | 96 | 76 | 60 | 46 | 36 | 26 | 20 |

*Route the flood to obtain the outflow hydrograph by Goodrich method.*

**Solution.** Since the Goodrich method uses the function $\left(\frac{2S}{\Delta t}+O\right)$, a graph has to be plotted between Elevation & this function. The different values of the function are therefore, first of all, computed for different elevations, as done in col (4) of table 18.28, by choosing $\Delta t$ = 6 hr. = 6 × 60 × 60 sec = **0.0216 × 10⁶ sec.**

**Table 18.28**

| *Elevation* <br> *m* | *Storage (S)* <br> $m^3 \times 10^6$ | *O in* <br> $m^3/s$ | $\frac{2S}{\Delta t} = \frac{2\text{ col. (2)} \times 10^6 m^3}{0.0216 \times 10^6 s}$ <br> $= \frac{2\times\text{ col. (2)}}{0.0216}$ <br> $= 92.59 \times$ *col. (2);* $m^3/s$ | $\frac{2S}{\Delta t}+O$ <br> *col. (4) + col. (3)* <br> $m^3/s$ |
|---|---|---|---|---|
| (1) | (2) | (3) | (4) | (5) |
| 200.00 | 4.35 | 0 | 403 | 403 |
| 200.50 | 4.47 | 10 | 414 | 424 |
| 201.00 | 4.88 | 26 | 452 | 478 |
| 201.50 | 5.38 | 46 | 498 | 544 |
| 202.00 | 5.88 | 72 | 544 | 616 |
| 202.50 | 6.37 | 100 | 590 | 690 |
| 202.75 | 6.53 | 116 | 605 | 721 |
| 203.00 | 6.86 | 130 | 635 | 765. |

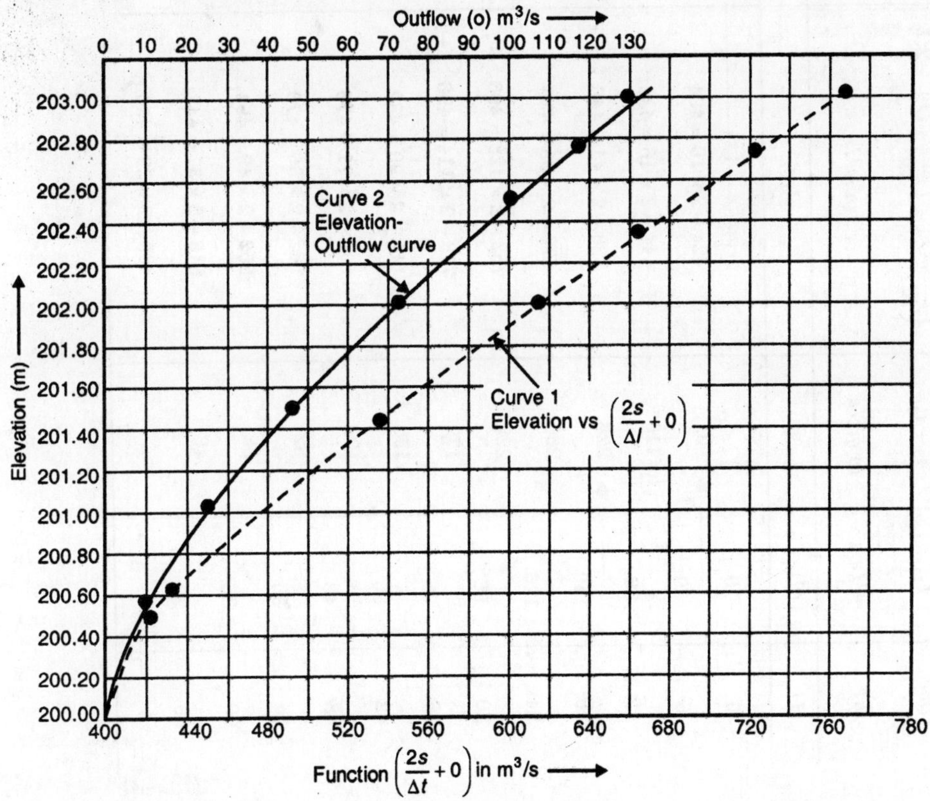


Fig. 18.26.

| Time (h) | $I_1$ $m^3/s$ | $I_1 + I_2$ in $m^3/s$ | $\left(\frac{2S}{\Delta t} - O\right)_1 = \left(\frac{2S_2}{\Delta t} + O\right)_2 - 2O_2$ in $m^3/s$ | $\left(\frac{2S}{\Delta t} + O\right)_2 = \left(\frac{2S_2}{\Delta t} - O\right)_1 + (I_1 + I_2)$ in $m^3/s$ Col. (4) + Col. (3) | Elevation m | O Outflow in $m^3/s$ from curve 2 (Fig. 18.26) for value of col. (6) |
|---|---|---|---|---|---|---|
| (1) | (2) | (3) | (4) | (5) | (6) | (7) |
| 0 | 10 | | | (424) | (200.5) | (10) |
| 6 | 30 | 40 | 424 − 2 × 10 = 404 | 404 + 40 = 444 | 200.72 | 16 |
| 12 | 86 | 116 | 444 − 2 × 16 = 412 | 412 + 116 = 528 | 201.38 | 41 |
| 18 | 140 | 226 | 528 − 2 × 41 = 446 | 446 + 226 = 672 | 202.38 | 93 |
| 24 | 126 | 266 | 672 − 2 × 93 = 486 | 486 + 266 = 752 | 202.94 | 128 |
| 30 | 96 | 222 | 752 − 2 × 128 = 496 | 496 + 222 = 718 | 202.72 | 114 |
| 36 | 76 | 172 | 718 − 2 × 114 = 490 | 490 + 172 = 662 | 202.32 | 90 |
| 42 | 60 | 136 | 662 − 2 × 90 = 482 | 482 + 136 = 618 | 201.98 | 71 |
| 48 | 46 | 106 | 618 − 2 × 71 = 476 | 476 + 106 = 582 | 201.72 | 56 |
| 54 | 36 | 82 | 582 − 2 × 56 = 470 | 470 + 82 = 552 | 201.56 | 49 |
| 60 | 26 | 82 | 552 − 2 × 49 = 454 | 454 + 62 = 516 | 201.30 | 37 |
| 66 | 20 | 46 | 516 − 2 × 37 = 442 | 442 + 46 = 488 | 201.10 | |
| | | | 488 − 2 × 29 = 430 | | | |

Curve 1 between $\left(\frac{2S}{\Delta t}+O\right)$ and Elevation is now plotted by using values of Col. (5) and of Col. (1) of table 18.28, as shown in Fig. 18.26 Outflow-Elevation curve is also plotted in the same figure, as curve 2, by using the values of col. (3) and of col (1) of table 18.28.

The routing is now carried out in table 18.29, which is self explanatory.

The outflows calculated in Col. (7) of this table (18.29) are plotted against time to draw the routed hydrograph along with the inflow hydrograph, in Fig. 18.27.

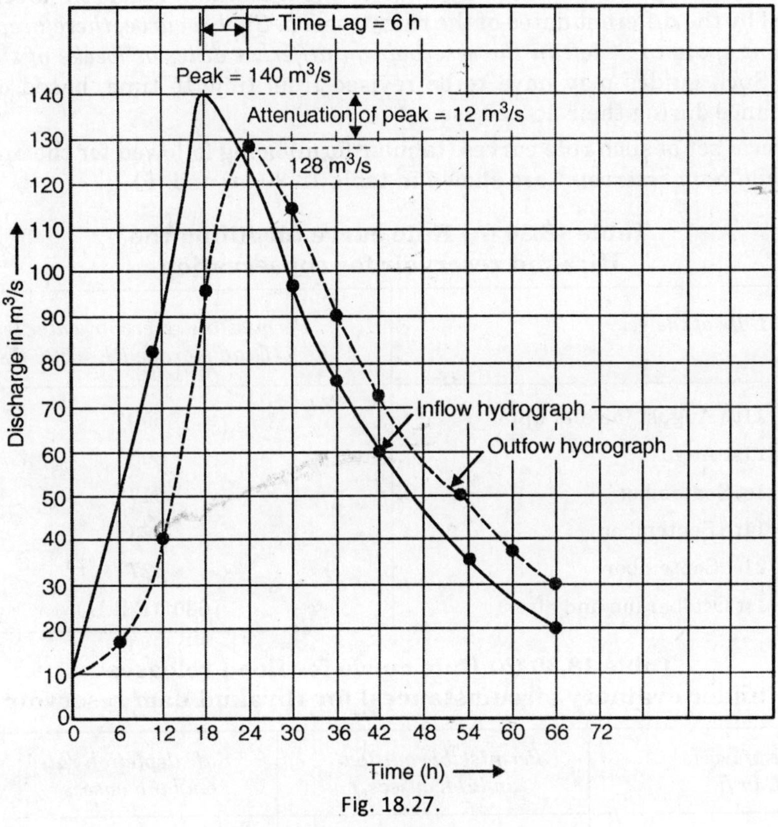


Fig. 18.27.

## RESERVOIR REGULATION

### 18.12. Rule Curves and Operating Tables for Reservoirs

Multipurpose reservoirs need to be operated and regulated efficiently with a high degree of intelligence, intuition, and experience, in order to ensure that *they are neither left partially empty by the end of the rainy season, nor they are found full at the time of arrival of a series of peak floods, leading to heavy releases, causing floods in the downstream valley.*

**Reservoir regulation committees,** consisting of experts are, therefore, generally constituted to ensure issue of proper and timely directives to the staff operating the gated openings of the reservoir, to avoid any bad and inefficient operations.

*Guiding tables and curves*, called **rule curves** or **guide curves**, are drawn in advance, and kept ready for use for the efficient regulation of the reservoir waters, with time. Such guiding curves are normally required only for the flood season, because for the rest of the year, the reservoir will only discharge water for irrigation and hydel needs. Sometimes, **reservoir regulation manuals** are also prepared and made available to the officers, dealing with the operation of the gates of the reservoir. Such manuals provide guidelines for gate operations, and for the overall maintenance and upkeep of the dam and the reservoir.

A rule curve, either in tabular or graphical form, as framed by the experts on the basis of the past data, broadly reflects the maximum reservoir levels to be achieved by the different dates of the rainy season. *Such a curve, therefore, reflects the vacant space to be left in the reservoir on different dates or weeks of the rainy season.* Such guides may have to be revised from time to time, based on their performance during their actual use.

A typical set of such rule curves (tabular form) being followed for the operation of *Hirakud dam reservoir** are shown in table 18.30 (*a*) and (*b*).

**Table 18.30 (*a*) Rule curve for filling the Hirakud reservoir for conservation**

| *Date of the year* | *Permissible reservoir water level (Dead storage RL = 590 ft)* |
|---|---|
| 11th August (on and upto) | 600′ |
| 21st August | 605′ |
| 1st September | 617′ |
| 11th September | 623′ |
| 21st September | 627′ |
| 1st October (on and after) | 630′ (F.R.L.) |

**Table 18.30 (*b*) Rule curve for flood releases (under ordinary circumstances) for Hirakud dam reservoir**

| *Range of levels RL in ft* | *Permissible outflow in lakh cusecs* | *Safe depletion rate in lakh cusecs* |
|---|---|---|
| 600-610 | 2.5 | 4.0 |
| 610-615 | 3.0 | |
| 615-620 | 4.0 | |
| 620-625 | 4.5 | |
| 625-628 | 5.5 | |
| 628-630 | 6.5 | |

*Hirakud dam is constructed on Mahanadi river is Orissa State of India. It is 3 mile long composite dam, with max. height of 200 ft. Spillway capacity at FRL of 630′ is 14.83 lakh cusecs. It intercepts a catchment of 32,200 sq. miles, and the catchment below the dam and up to the head of delta, *i.e.* Naraj, is about 18,000 sq. miles. There is no flood reserve or conservation reserve as such, and it is a general use reservoir. The river capacity at Naraj is about 9 lakh cusecs.

Prior to the above rules, Hirakud dam reservoir was following another set of rules, as given in table 18.31, which proved to be inadequate, leading to insufficient storage at the end of flood seasons for some of the years, and heavy discharges from the reservoir synchronising with floods in the lower catchment in some other years.

**Table 18.31. Old Rules, specified in the Manual for Operation of Hirakud reservoir, which have now been superceded by rules of tables 18.30 (*a*) and (*b*)**

| *S. No. of Rule* | *Rule* |
|---|---|
| **Rule 1** | There should be no full impounding till 1st of September in normal years. The entire inflow should be let out ; care being exercised that the combined discharge of reservoir outflow and the runoff from the catchment below the dam does not exceed 10 lakh cusecs at Naraj. Generally, the inflow up to 5 lakh cusecs can be passed without hesitation. When the inflow is more than 5 lakh cusecs, then the hydrometeorological conditions in the lower catchment should be carefully studied and outflow determined. |
| **Rule 2** | Safety of dam should be the prime consideration on and at every occasion when floods are absorbed in the reservoir for purposes of moderating and regulating outflow in the reservoir. A level of + 625.0' should be deemed to be the safe level normally for this purpose. |

Another typical rule curve (graphical form), as drawn and used for Maithon dam reservoir, under *D.V.C. System**, is shown in Fig. 18.28. This operation schedule shows that the reservoir, which will be quite depleted by the start of June, will be filled up fully up to *Monsoon storage level* (RL 480′), latest by 3rd week of September, and will be maintained full, till the end of November. The reservoir will be allowed to be depleted only from December onward.

Any flood, which occurs during the last week of September, October and November, will thus, find the reservoir full, and will have to be accommodated within the '*flood reserve*', *i.e.* between monsoon storage level (RL 480′) and the max. permissible operational level of gates (RL 495′).

*D.V.C. (Damodar Valley Corporation) was constituted in India in July, 1948, on the line of T.V.A. (Tennesse Valley Authority) of USA to plan and execute flood control works on river Damodar and its tributaries, which had created havoc in the year 1943, causing serious flood losses of the order of Rs. 8 crores, and cutting off the most important Calcutta city with the rest of the country for about ten weeks, due to breaches in G.T. Road and Rly link.

Consequently, 4 dams were constructed, including the Maithon and Tilaiya dams on river Barakar (a tributary of Damodar); Konar dam on river Konar, and Panchat dam on Damodar, alongwith a barrage at Durgapur (situated a few km downstream of the confluence of Damodar and Barakar. *Konar and Tilaiya* are smaller dams located in the upper reaches, whereas, Maithon and Panchat are bigger dams in lower reaches, bearing most of the flood burden.

Maithon is located about 13 km above the confluence, and 21 km northwest of Asansol and 40 km east of Dhanbad. Like Konar, it is partly earthen and partly of concrete. The primary purpose of the dam is flood control; whereas, *irrigation* and *hydropower* are secondary uses.

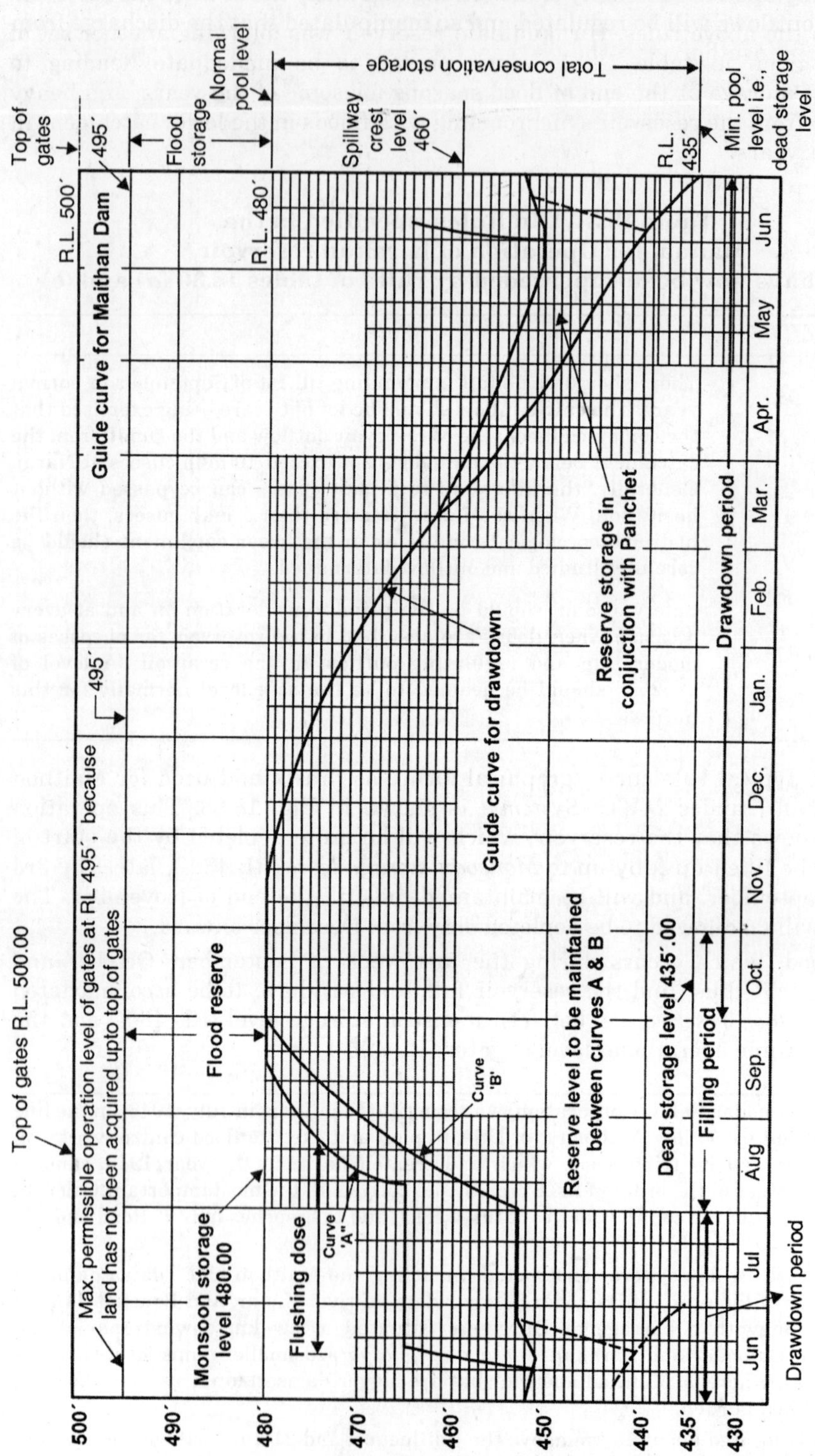


Fig. 18.28. A typical operating guide for Maithon Dam Reservior

Depending upon the intensity of the coming and likely floods after the reservoir is full, the outflows will be regulated and so manipulated that the discharge from the reservoir together with the discharge contributed by the downstream area, does not exceed the safe carrying capacity of the river farther downstream.

As and when a late flood enters the reservoir, the reservoir-gate-operators may adjust the outflow to a lower value in the beginning, but may have to increase the outflow, as the flood waves continue and the weather forecasts for continued rains are received. In this operation, a stage may reach when the outflow may become quite high, causing floods in the down valley.

While manipulating these outflows, chances of synchronisation of the discharge of the downstream uncontrolled catchment will also have to be considered properly.

Say for example, if we release 1 lakh cusecs discharge from the reservoir, and another 1 lakh cusecs gets added to it from the catchment of the downstream 1 km length of the channel, then eventually 2 lakh cusecs will be flowing in the channel farther down the 1 km length. This 2 lakh cusecs may prove quite harmful, as it may exceed the safe carrying capacity of the downstream river (below 1 km).

Another possibility for gate operators could then be to release lesser discharge, say 50,000 cusecs, which with 1 km downstream contribution of 1 lakh cusecs, will make up 1,50,000 cusecs, which may not prove dangerous to farther down. But releasing lesser quantity would cause the reservoir water level to rise rapidly, encroaching the 'flood reserve' quickly, finally leading to a stage, wherein you may have to open the entire gates, passing say 4 lakh cusecs, to avoid over-topping and failure of the dam. This 4 lakh cusecs will eventually destroy the entire downstream area, and may prove worser than what would have happened, if 1 lakh cusecs would have been allowed in the beginning itself.

You can, thus, understand and appreciate as to how important it is, to properly man the outflows from a reservoir. The intelligent and timely manipulations, accompanied by proper rain forecasts, may help us to avoid flooding of the downstream area.

## 18.13. Reservoir Sedimentation

Every river carries certain amount of sediment load. The sediment particles try to settle down to the river bottom due to the gravitational force, but may be kept in suspension due to the upward currents in the turbulent flow which may overcome the gravity force. Due to these reasons, the river carries fine sediment in suspension as suspended load, and larger solids along the river bed as bed load. When the silt laiden water reaches a reservoir in the vicinity of a dam, the velocity and the turbulence are considerably reduced. The bigger suspended particles and most of the bed load, therefore, gets deposited in the head reaches of the reservoir. Fine particles may travel some more distance and may finally deposit farther down in the reservoir, as shown in Fig. 18.29. Some very fine particles may remain in suspension for much longer period, and may finally escape from the dam along with the water discharged through the sluiceways, turbines, spillway, etc.

The deposition of sediment in the reservoir is known as 'Reservoir Silting' or 'Reservoir Sedimentation'.

The deposition of the sediment will automatically reduce the water storing capacity of the reservoir, and if this process of deposition continues longer, a stage is likely to reach when the whole reservoir may get silted up and become useless.

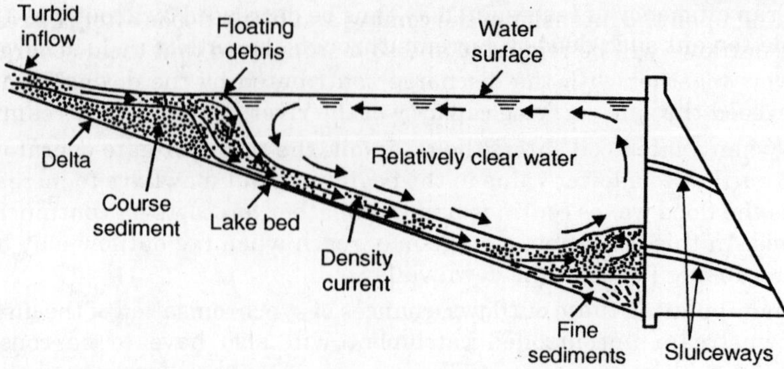


Fig. 18.29. Sediment accumulation in a typical reservoir.

Moreover, with the passage of time, the reservoir capacity will go on reducing. Thus, if today at the time of construction, a reservoir can store 10,000 cubic metres of water, tomorrow say after five years, it may be able to store only say 8,000 cubic metres of water. Therefore, in order to see that the capacity does not fall short of requirement ever during the design period, we must take this silting into account. The total volume of silt likely to be deposited during the designed life period of the dam is, therefore, estimated ; and approximately that much of volume is left unused to allow for silting, and is known as **dead storage.** The remainder is known as the ***effective storage*** or the **live storage.** The dead storage generally varies between 15 to 25% of the total capacity. For example in Bhakra dam, the gross capacity of the dam is 9,344 million cubic metres and the dead storage provided is 2,054 million cubic metres. All the outlets fetching water from the reservoir are provided above the dead storage level.

The importance of this silting can be understood by considering the following example :

Let the total capacity of a reservoir be 30 million cubic metres and the provision of dead storage be 6 million cubic metres. Let the average volume of sediment deposition be 0.15 million cubic metres per year. Then it is evident, that the dead storage will be filled up in $\frac{6}{0.15}$ = 40 years, and the total storage in about $\frac{30}{0.15}$ = 200 years.

Hence, the usefulness of this reservoir would start reducing after 40 years, and after 200 years it would be nothing but a collection of sand and sediment with no water in it, provided the siltation rate remains constant at 0.15 $M.m^3/yr$.

**18.13.1. Density Currents.** In a reservoir, the coarser sediment settles down along the bottom of the reservoir, as the muddy flow approaches the reservoir; while the finer sediment usually remains in suspension, and moves in a separate layer than the clear reservoir water, as shown in Fig. 18.29. This layer of water, containing the fine sediment, moves below the upper clearer reservoir water, as a *density current*, since its density is slightly more than the density of the main body of the reservoir water. *Because of their density difference, the water of the density current does not mix easily with the reservoir water, and maintains its identity for a considerable time.* The density current can thus be removed through the dam sluiceways, if they are located properly and at the levels of the density current. A lot of sediment load can, thus, be passed out of the reservoir, if it is possible to locate the dam outlets and sluiceways in such a fashion, as to vent out the density currents. Trap efficiency of reservoirs may thus be decreased by about 2 to 10%, if it is possible to vent such density currents through the outlets and sluiceways of the dams.

**18.13.2. Trap Efficiency.** Now, we introduce another very important term called Trap efficiency. Trap efficiency is defined as the percentage of the sediment deposited in the reservoir even inspite of taking precautions and measures to control its deposition.

Therefore, Trap Efficiency (η)

$$= \frac{\text{Total sediment deposited inthe reservoir}}{\text{Total sediment flowing in the river}} \times 100\% \quad \text{...(18.13)}$$

Most of the reservoirs trap 95 to 100% of the sediment load flowing into them. Even if various feasible silt control measures are adopted, it has not been possible to reduce this trap efficiency below 90% or so.

**18.13.3. Capacity Inflow Ratio.** The ratio of the reservoir capacity to the total inflow of water in it, is known as the capacity-inflow ratio. It is a very important factor, because the trap efficiency (η) has been found to be a function of capacity-inflow ratio *i.e.*

$$\eta = f\left(\frac{\text{Capacity}}{\text{Inflow}}\right) \quad \text{...(18.14)}$$

The graph obtained for the existing reservoirs between trap efficiency and log of $\left(\frac{\text{Capacity}}{\text{Inflow}}\right)$ has been found to be of the type shown in Fig. 18.30.

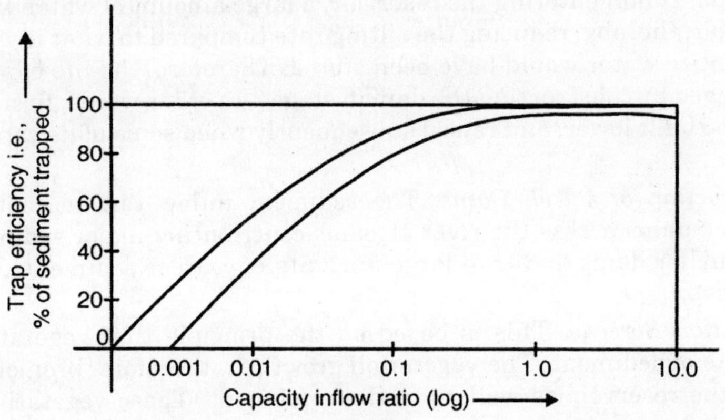


Fig. 18.30.

It is evident from the above curve, that if capacity reduces (with constant inflow), trap efficiency reduces, and hence, lesser sediment is trapped. Therefore, the silting rate in the reservoir shall be more in the beginning, and as its capacity reduces due to silting, the silting rate will reduce. Hence, the complete reservoir-silting may take longer period.

*It can also be concluded that for small reservoirs (having small capacity) on large rivers (having large inflow rates), the trap efficiency is extremely low, because the capacity inflow ratio is very small. Such reservoirs silt very little and most of their sediment is passed downstream. On the other hand, large reservoirs on smaller rivers shall silt tremendously and almost complete deposition of sediment may take place.*

**18.13.4. Silting of Power Reservoirs.** In case of reservoirs constructed solely for the purpose of power generation, the silting is comparatively less important. This is because of the fact, that for the proper and efficient functioning of a power reservoir, only a certain minimum head is necessary. This head remains available even after some silt gets deposited. So, only if sufficient water required for power generation remains available, the reservoir's efficiency remains unaffected by silting. But due to silted water, the abrasion of the blades of the turbines may occur very soon, and power production may be stopped over a considerable length of time.

**18.13.5. Silting Control in Reservoirs.** In order to increase the life of a reservoir, it is necessary to control the deposition of sediment. Various measures are undertaken in order to achieve this aim. The various methods which are adopted can be divided into two parts :

(1) *Pre-constructing measures* ; and (2) *Post-constructing measures.*

These measures are discussed below :

(1) **Pre-constructing measure.** They are those measures which are adopted before and during the execution of the project. They are innumerated below :

(*a*) *Selection of Dam Site.* The silting depends upon the amount of erosion from the catchment. If the catchment is less erodible, the silting will be less. Hence, the silting can be reduced by choosing the reservoir site in such a way as to exclude the run off from the easily erodible catchment.

(*b*) *Construction of the Dam in Stages.* The design capacity plays an important role in the silting of a reservoir. When the storage capacity is much less than the average annual runoff entering the reservoir, a large amount of water will get out of the reservoir, thereby, reducing the silting rate compared to what it would have been if the entire water would have been stored. Therefore, the life of a reservoir can be prolonged by constructing the dam in stages. In other words, first of all, the dam should be built lower, and raised subsequently when some of its capacity gets silted up.

(*c*) *Construction of Check Dams.* The sediment inflow can be controlled by building check dams across the river streams contributing major sediment load. These are smaller dams and trap large amounts of coarser sediments. They are quite expensive.

(*d*) *Vegetation Screens.* This is based on the principle that vegetations trap large amounts of sediment. The vegetation growth is, therefore, promoted at the entrance of the reservoir as well as in the catchment. These vegetative covers, through which flood waters have to pass before entering the reservoirs, are known as vegetation screens, and provide a cheap and a good method of silt control.

(*e*) *Construction of Under-sluices in the Dam.* The dam is provided with openings in its base, so as to remove the more silted water on the downstream side.

The sediment concentration will be more at some levels than at others. Therefore, sluices are located at the levels of higher sediment concentration. The method in itself, is not sufficient because the water digs out a channel behind the sluice for movement and leaves most of the sediment undisturbed. Therefore, this is simultaneously supplemented with mechanical loosening and scouring of the neighbouring sediment in order to increase its effectiveness. But to provide large sluices near the bottom of the dam, is again a structural problem. The use of this method is, therefore, limited.

(2) **Post-constructing Measures.** These measures are undertaken during the operation of the project. They are given below:

(1) *Removal of Post Flood Water.* The sediment content increases just after the floods ; therefore, attempts are generally made not to collect this water. Hence, the efforts should be made to remove the water entering the reservoir at this time.

(2) *Mechanical Stirring of the Sediment.* The deposited sediment is scoured and disturbed by mechanical means, so as to keep it in a moving state, and thus, help in pushing it towards the sluices.

(3) *Erosion Control and Soil Conservation.* This includes all those general methods which are adopted to reduce erosion of soil and to make it more and more stable. This method is the most effective method for controlling siltation, because when the soil erosion is reduced, the sedimentation problem is reduced automatically. But the methods of treating the catchment in order to minimise erosion are very costly. It has been estimated that the investment required for treating 16% of the Indian catchment area is Rs. 1,000 crores. In India, only 1.5% of the catchment area has been treated to minimise silting.

**Example 18.18.** *The following information is available regarding the relationship between trap efficiency and capacity inflow ratio.*

| *Capacity inflow ratio* | *0.1* | *0.2* | *0.3* | *0.4* | *0.5* | *0.6* | *0.7* | *0.8* | *0.9* | *1.0* |
|---|---|---|---|---|---|---|---|---|---|---|
| *Trap efficiency percent* | *87* | *93* | *95* | *95.5* | *96* | *96.5* | *97* | *97* | *97* | *97.5* |

*Find the probable life of the reservoir with an initial reservoir capacity of 30 million cubic metres, if the average annual flood inflow is 60 million cubic metres and the average annual sediment inflow is 2,00,000 tonnes. Assume a specific weight of the sediment equal to 1.2 gm per c.c. The usual life of the reservoir will terminate when 80% of its initial capacity is filled with sediment.*

**Solution.**

Average annual sediment inflow = 2,00,000 tonnes

$= 2 \times 10^5$ tonnes $= 2 \times 10^{11}$ gm

Volume of average annual sediment inflow

$$= \left(\frac{2\times10^{11}}{1.2}\right) \text{c.c.} = \frac{2\times10^{11}}{1.2\times10^{6}}\ \text{m}^3$$

$$= \frac{0.2}{1.2} \times 10^6 \text{ cubic metres} = \frac{1}{6} \text{ million cubic metre} = \frac{1}{6}\ \text{M.m}^3$$

Initial Reservoir Capacity = 30 M.m$^3$

Annual flood inflow = 60 M.m$^3$

Let us assume that 20% of the capacity, *i.e.* 6 M.m$^3$ is filled up in the first interval.

Capacity inflow ratio at the start of the interval $= \frac{30}{60} = 0.5$

Trap efficiency at the start of the interval = 0.96.

Capacity inflow ratio at the end of the interval $= \frac{24}{60} = 0.4$

Trap efficiency at the end of interval = 0.955

Average trap efficiency during the interval $= \frac{0.96 + 0.955}{2} = \mathbf{0.9575}$.

**Volume of sediment deposited annually till the 20% capacity is filled**

$$= \frac{1}{6} \times 0.9575 \text{ M.m}^3$$

∴ **No. of years during which 20% of the capacity, *i.e.* 6 M.m³ shall be filled up**

$$= \frac{6}{\frac{1}{6} \times 0.9575} \text{ years} = \frac{36}{0.9575} = \mathbf{37.6 \text{ years}}$$

**Similarly, in the 2nd interval**

Capacity inflow ratio at the start $= \frac{24}{60} = 0.4$

Capacity inflow ratio at the end $= \frac{18}{60} = 0.3$

Trap efficiency at the start = 0.955

Trap efficiency at the end = 0.95

Average trap efficiency = **0.9525.**

∴ **No. of years during which the next 20% of capacity shall be filled up**

$$= \frac{6}{\frac{1}{6} \times 0.9525} = \frac{36}{0.9525} \mathbf{37.8 \text{ years.}}$$

**Similarly, in the 3rd interval**

Capacity inflow ratio at start $= \frac{18}{60} = 0.3$

Capacity inflow ratio at the end $= \frac{12}{60} = 0.2$

Trap efficiency at the start = 0.95

Trap efficiency at the end = 0.93

Average trap efficiency during the interval = 0.94

No. of years during which the next 20% of the capacity shall be filled up

$$= \frac{6}{\frac{1}{6} \times 0.94} = \frac{36}{0.94} = \mathbf{38.3 \text{ years.}}$$

**Similarly, in the 4th interval**

Capacity inflow ratio at the start = 12/60 = 0.2

Capacity inflow ratio at the end = 6/60 = 0.1

Trap efficiency at the start = 0.93

Trap efficiency at the end = 0.87

Average trap efficiency during the interval = 0.90

No. of years during which the next 20% of the capacity shall be filled up

$$= \frac{6}{\frac{1}{6} \times 0.9} = \frac{36}{0.9} = \mathbf{40 \text{ years.}}$$

**Total probable life till 80% capacity gets filled up**

**= 37.6 + 37.8 + 38.3 + 40.0 = 153.7 years. Ans.**

**The above calculations of dividing the entire capacity into intervals (20% each in the above case) can also be carried out in a tabular form, as shown below in Table 18.32.**

**Table 18.32.**

| *Capacity* | | *Capacity/inflow* | *Trap* | *Av. Trap* | *Sediment trapped* | *Year reqd. to fill* |
|---|---|---|---|---|---|---|
| *%* | *Capacity vol. in M. cum.* | $= \frac{\text{Col. (2)}}{\text{60 Mcum}}$ | *efficiency* $\eta$ | *eff.* $\eta_{av}$ *during the interval* | *per year; col. (5) × Av. annual sediment inflow = col. 5 × 1/6 M.cum* | *up 20% capacity (6 Mcum.)* $= \frac{6}{\text{Col.(6)}}$ *in years* |
| *(1)* | *(2)* | *(3)* | *(4)* | *(5)* | *(6)* | *(7)* |
| 100 | 30 | 0.5 | 0.96 | | | |
| | | | | 0.9575 | 0.1596 | 36.6 |
| 80 | 24 | 0.4 | 0.955 | | | |
| | | | | 0.9525 | 0.1588 | 37.8 |
| 60 | 18 | 0.3 | 0.95 | | | |
| | | | | 0.94 | 0.1566 | 38.3 |
| 40 | 12 | 0.2 | 0.93 | | | |
| | | | | 0.90 | 0.15 | 40.0 |
| 20 | 6 | 0.1 | 0.87 | | | |
| | | | | | | Σ = 153.7 yrs |

**Example 18.19.** ***A proposed reservoir has a capacity of 500 ha-m. The catchment area is 125 km², and the annual streamflow averages 12 cm of runoff. If the annual sediment production is 0.03 ha.m/km², what is the probable life of the reservoir before its capacity is reduced by 10% of its initial capacity by sedimentation? The relationship between trap efficiency η(%) and capacity inflow ratio C/I, is as under :***

| *C/I* | *0.01* | *0.02* | *0.04* | *0.06* | *0.08* | *0.1* | *0.2* | *0.3* | *0.5* | *0.7* |
|---|---|---|---|---|---|---|---|---|---|---|
| η % | *43* | *60* | *74* | *80* | *84* | *87* | *93* | *95* | *96* | *97* |

***(U.P.S.C., Civil Services, 1987)***

**Solution.** **Av. annual streamflow = 12 cm of runoff**

**Area of catchment = 125 km² = 125 × 10⁶ m²**

$$\therefore \text{ Annual flood inflow} = (125 \times 10^6) \cdot \frac{12}{100} \text{ m}^3$$

$$= 15 \times 10^6 \text{ m}^3 = \mathbf{15\ M.m^3\ (Mcum).}$$

**Annual sediment inflow = 0.03 ha-m/km² of the catchment**

$$= 0.03 \times 125 \text{ ha-m} = 0.03 \times 125 \times 10^4 \text{ m}^3 = 3.75 \times 10^4 \text{ m}^3$$

$$= \frac{3.75}{100} \times 10^6 \text{ m}^3 = 0.0375 \text{ M-m}^3 \text{ (Mcum).}$$

**It means that 0.0375 Mcum of sediment flows every year into the dam/reservoir site, but the quantum of this, which is trapped in the reservoir, depends upon the**

average trap efficiency ($\eta$) during that year, and this trap efficiency, in turn, depends upon the capacity/inflow ratio.

In the question, the total capacity to be filled up by sediment is 10% of the initial reservoir capacity, *i.e.* 10% × 5 Mcum = 0.5 Mcum.

Now, we have to calculate the time during which this 0.5 Mcum of sediment will get deposited in the reservoir, as follows :

Capacity of the reservoir at the start = 5 Mcum

Capacity of the reservoir at the end

(*i.e.*, when 0.5 Mcum of sediment is filled up) = 4.5 Mcum

$$\therefore \text{ Capacity/inflow at the start} = \frac{5 \text{ Mcum}}{15 \text{ Mcum}} = 0.333$$

$\eta$ at start = 95%. Capacity/inflow at the end = $\frac{4.5}{15}$ = 0.30

$\eta$ at the end of the interval = 95%. Average $\eta$ = 95%

∴ Sediment load trapped/yr.

= 0.0375 × 95% = 0.035625

∴ No. of years during which 0.5 Mcum of sediment will get trapped

$$= \frac{0.5}{0.035625} \text{ years} = 14.04 \text{ years ; say } \mathbf{14 \text{ years.}}$$

Hence, after 14 years, 10% reservoir capacity will get filled up. **Ans.**

## 18.14. Estimating Sediment Load likely to Enter a Proposed Reservoir

The quantum of sediment flowing into a dam reservoir along with river runoff primarily depends upon the erosion characteristics of its catchment and the characteristics of the rainfalls that produce the run off entering the reservoir.

The sediment is basically produced by rains by the process of sheet erosion. The flowing water is the most active agent for erosion of soil from the land. Other agents like *wind, gravity, ice* and *human activities* do help in the erosion process. The rain drops in itself, loosen the soil perticles and break the soil lumps. The action of flowing sheet of water on the land surface helps in eroding the top soil from the ground surface and transport it down to the channels. The erosion caused by rainfall and runoff, thus, constitutes of the following two parts :

(1) **Sheet erosion.** It includes the detachment of geological material from the land surface by the impact of raindrops, and its subsequent removal by overland flow; and

(2) **Channel erosion.** It includes river bank erosion and transportation of the materials by concentrated flow.

**Human activities,** like *overgrazing of grass land, cutting of forests, forest fires, ploughing of land,* and various *mining and other excavations* etc. have magnified the problem of water erosion in river channels.

### 18.14.1. Factors Affecting the Erodibility of a Soil

The factors which effect the erodibility of a soil are given below :

(*i*) *Particle size of soil.* Larger the size of soil particles, the lesser would be its chances for erosion.

(*ii*) *Land slope.* The greater is the land slope, the greater is the action of erosive agents, the optimum being at 40° slope.

(*iii*) *Vegetation.* The thicker is the vegetation cover over a soil, the lesser will be the scope of erosion of the soil from the area.

(*iv*) *Presence of Salt and Colloidal Matter in the Soil.* The binding materials like kaolinite, montmorillonite, biotite, etc., do help in increasing the force of cohesion between the soil particles, thereby reducing the erodibility of the soil.

(*v*) *Moisture Content of Soil.* The greater is the moisture content of the soil, the lesser is the scope of its erosion.

(*vi*) *Soil compaction.* The higher is the compaction of soil, the lesser is the chance of its erosion.

(*vii*) *Soil Properties.* The characteristics of the soil, such as soil texture, structure, stratification, permeability, composition etc. do affect the soil binding, which in turn will neutralise the force of weathering agents.

(*viii*) *Human Activities.* Human activities on the land like mining, agricultural operations, construction of projects, land use, etc. do increase the erosion from the given land.

(*ix*) *Rainfall characteristics.* The intensity, duration, quantity and distribution of rainfall over space and time are some of the important factors that affect sediment yield.

### 18.14.2. Estimation of Sheet Erosion.

The sediment yield of a reservoir basically depends on sheet erosion which can be estimated by the following empirical equations :

(1) ***Musgrave Equation.*** Musgrave (1947) suggested the following equation to compute the annual gross sheet erosion from a catchment, on the basis of 19 widely scattered research stations in USA :

$$E = CR \cdot (S_0/10)^{1.35}\,(L/72)^{0.35}\,(P_{30}/1.25)^{1.75} \qquad \text{...(18.15)}$$

where $E$ = Erosion of soil lost from the catchment in inches/year

$C$ = the *soil erosion rate*, which varies from 0.43 to 0.53 inch/year depending on the soil type (which depends on texture and permeability of soil)

$R$ = the *cover factor*, which varies from 0.95 for poorly covered land to 0.10 for row crops

$S_0$ = Land slope in percentage, the default being 10%

$L$ = Length of the land slope in feet

$P_{30}$ = The max. rainfall in inches having 30 min. duration and of 2 year frequency.

It was also stipulated that the erosion value of $E$ computed above, in inch/year, can be multiplied by 150 to obtain the erosion value in ton/year/acre.

(2) ***Universal Equation.*** Agriculture Research Service of U. S. Deptt. of Agriculture developed a universal equation (1961) to predict erosion from **small catchments**. This universal equation is given as :

$$\boxed{E_a = R_f \cdot K \cdot (LS)\, C_m \cdot P} \qquad \text{...(18.16)}$$

where $E_a$ = Average soil loss in ton/acre/year.

(*i*) $R_f$ is ***Rainfall-run off factor.*** Its value takes into account the effect of rain drop impact as well as the resulting amount and rate of run off. It should also include the cumulative effects of many moderate sized storms as well as the effects of occasional severe storms.

The value of $R_f$ is generally taken as equal to *EI* (**rainfall erosion index**), which is computed as :

$$E \cdot I = E \times I_{30} \quad \text{...(18.17)}$$

where

$E$ = storm energy in 100 ft-ton/acre/inch

$I_{30}$ = Max. 30-minute intensity in inch/hr.

The value of $E$ is further computed as :

$$E = \frac{1}{100}[916 + 331 \log I] \quad \text{...(18.18)}$$

where

$E$ = Storm energy in 100 ft-ton per acre per inch.

$I$ = Storm intensity of the given storm with a limit of 3 inch/hr, since median drop size does not continue to grow beyond this limit.

The *EI* value can, thus, be computed by eqn. (18.18) for a given storm. For a specified period, the individual storm EI values can be summed up, which provide a numerical measure of the erosive potential of the rainfall within that period (one year). *In this manner, the average annual total of the storm EI values in a particular area is obtained, which is called the rainfall-erosion index for that area, which equals* $R_f$

(*ii*) *K* is **Soil erodibility factor.** Its value is determined experimentally for the given soil type. Representative values for different types of soils have been worked out and listed for different regions of USA by Soil Conservation Service of USA, as shown in Table 18.33.

**Table 18.33. Computed *K* Values for Soils on Erosion Research Stations (After Wischmeier and Smith, 1978).**

| *Soil* | *Source of Data* | *Computed K* |
|---|---|---|
| Dunkirm silt loam | Geneva, NY | 0.69[a] |
| Keene silt loam | Zanesville, OH | 0.48 |
| Shelby loam | Bethany, MO | 0.41 |
| Lodi loam | Blacksburg, VA | 0.39 |
| Fayette silt loam | LaCrosse, WI | 0.38[a] |
| Cecil sandy clay loam | Watkinsville, GA | 0.36 |
| Marshall silt loam | Clarinda, IA | 0.33 |
| Ida silt loam | Castana, IA | 0.33 |
| Mansic clay loam | Hays, KS | 0.32 |
| Hagerstown silty clay loam | State College, PA | 0.31[a] |
| Austin clay | Temple, TX | 0.29 |
| Mexico silt loam | MeCredie, MO | 0.28 |
| Honeoye silt loam | Marcellus, NY | 0.28[a] |

*Contd...*

| | | |
|---|---|---|
| Cecil sandy loam | Clemson, SC | 0.28[a] |
| Ontario loam | Geneva, NY | 0.27[a] |
| Cecil clay loam | Watkinsville, GA | 0.26 |
| Boswell fine sandy loam | Tyler, TX | 0.25 |
| Cecil sandy loam | Watkinsville, GA | 0.23 |
| Zaneis fine sandy loam | Guthrie, OK | 0.22 |
| Tifton loamy sand | Tifton, GA | 0.10 |
| Freehold loamy sand | Marlboro, NJ | 0.08 |
| Blath flaggy silt loam with surface stones > 2 inches removed | Arnot, NY | 0.05[a] |
| Albia gravely loam | Beemerville, NJ | 0.03 |

*a* Evaluated from continuous fallow. All others were computed from rowcrop data.

For soils containing less than 70% of silt and very fine sand, $K$ can be computed as:

$$K = \frac{1}{100}\,[2.1\,M^{1.14}\,(10^{-4})\,(12 - a) + 3.25\,(b - 2) + 2.5\,(c - 3)] \quad ...(18.19)$$

where $M$ = particle size parameter, defined as percent silt and very find sand (size 0.1 mm-0.002 mm) times the quantity (100-percent clay)

$a$ = percent organic matter

$b$ = soil texture code used in USDA soil classification

$c$ = profile permeabilit class

[**Note:** When the silt fraction does not exceed 70%, erodibility varies aproximately as the 1.14 power of $M$, but addition of organic matter content, soil structure, and profile-permeability class as done in Eqn. (18.19) improves the prediction accuracy.]

*(iii)* $(LS)$ = **Topographic Factor.** It is the ratio of soil less per unit area from a field slope to that from a 72.6 ft length of uniform 9% slope under otherwise identical conditions. For a specified slope and its length. $LS$ can be computed as :

$$(LS) = \left(\frac{\lambda}{72.6}\right)^{m} (65.41 \sin^2 \theta + 4.56 \sin \theta + 0.065) \quad ...(18.20)$$

where $\lambda$ = slope length in ft

$\theta$ = angle of slope

$M$ = 0.5 if the per cent sloe is 5 or more

$m$ = 0.4 on slopes of 3.5 to 4.5%

$m$ = 0.3 for slopes of 1 to 3%

$m$ = 0.2 for uniform slopes of less than 1%.

*(iv)* $C_m$ = **Soil Cover and Management Factor.** It measures the combined effect of all inter-related cover and management variable, including the type of vegetatiion, plant spacing, the stand, the quality of growth, crop sequence, tilage practices, crop residues, incorporated residues, land use residues, fertility treatmet, etc.

Values of $C_m$ for pasture, range, idle land and woodland for a combination of cover conditions are given in Table 18.34

**Table 18.34. Factor $C_m$ for permanent pasture, Range, and Idle Land (After Wischmeier and Smith 1978).**

| *Vegetative canopy* | | *Cover that contracts the soil surface* | | | | | | |
|---|---|---|---|---|---|---|---|---|
| *Type and height*[b] | *Percent* | *Percent ground cover* | | | | | | |
| | *cover*[c] | *Type*[d] | *0* | *20* | *40* | *50* | *80* | *95+* |
| No appreciable canopy | | G | 0.45 | 0.45 | 0.10 | 0.042 | 0.013 | 0.003 |
| | | W | 0.45 | 0.24 | 0.15 | 0.091 | 0.043 | 0.011 |
| Tall weeds or short brush with the average drop fall height of 20 in | 25 | G | 0.36 | 0.17 | 0.09 | 0.038 | 0.013 | 0.003 |
| | | W | 0.36 | 0.20 | 0.13 | 0.083 | 0.041 | 0.011 |
| | 50 | G | 0.26 | 0.13 | 0.07 | 0.035 | 0.120 | 0.003 |
| | | W | 0.26 | 0.16 | 0.11 | 0.076 | 0.039 | 0.001 |
| | 75 | G | 0.17 | 0.10 | 0.06 | 0.032 | 0.011 | 0.003 |
| | | W | 0.17 | 0.12 | 0.09 | 0.068 | 0.038 | 0.011 |
| Appreciable brush or bushes, with the average drop fall height of 6½ ft | 25 | G | 0.40 | 0.18 | 0.09 | 0.040 | 0.013 | 0.003 |
| | | W | 0.40 | 0.22 | 0.14 | 0.087 | 0.042 | 0.011 |
| | 50 | G | 0.34 | 0.16 | 0.08 | 0.038 | 0.012 | 0.003 |
| | | W | 0.34 | 0.19 | 0.13 | 0.082 | 0.042 | 0.011 |
| | 75 | G | 0.28 | 0.14 | 0.08 | 0.036 | 0.012 | 0.003 |
| | | W | 0.28 | 0.17 | 0.12 | 0.078 | 0.040 | 0.011 |
| Trees, but no appreciable low brush Average drop fall height of 13 ft | 25 | G | 0.42 | 0.19 | 0.10 | 0.041 | 0.013 | 0.003 |
| | | W | 0.42 | 0.23 | 0.14 | 0.089 | 0.042 | 0.011 |
| | 50 | G | 0.39 | 0.18 | 0.09 | 0.040 | 0.013 | 0.003 |
| | | W | 0.39 | 0.21 | 0.14 | 0.087 | 0.042 | 0.011 |
| | 75 | G | 0.36 | 0.17 | 0.09 | 0.039 | 0.012 | 0.003 |
| | | W | 0.36 | 0.20 | 0.13 | 0.084 | 0.041 | 0.011 |

[a] The listed C values assume that the vegetatio and mulch are randomly distributed over the entire area.

[b] Canopy height is measured as the average fall height of water drops falling from the canopy to the ground. Canopy effect is inversely proportional to the drop fall height and is negligible if fall height exceeds 33 ft.

[c] Portion of total area surface that would be hidden from view by canopy in a verticle projection (a bird's-eye view).

[d] G: cover at the surface is grass, grasslike plants, decaying compacted duff, or litter at least 2 in. deep.

W: cover at the surface mostly broadleaf herbaceous plants (as weeds with little lateral-root network near the surface) or undecayed residues or both.

(*v*) $P$ = **Support Practice Factor.** Its value depends upon crop land practice such as contour tillage, *strip cropping on the contour, and terrace systems.* Values of $P$ have been given for each of these practices by Wischmeier and Smith (19 7 8). In general, tillage and planting on the contours reduce erosion. Table 18.35 gives $P$ values for contouring.

**Table 18.35. P values (Eq. 18.16) and Slope-Length Limits for Contouring (Wischmeier and Smith)**

| *Land Slope%* | *P Value* | *Maximum Length* (feet)* |
|---|---|---|
| 1 – 2 | 0.60 | 400 |
| 3 – 5 | 0.50 | 300 |
| 6 – 8 | 0.50 | 200 |
| 9 – 12 | 0.60 | 120 |
| 13 – 16 | 0.70 | 80 |
| 17 – 20 | 0.80 | 60 |
| 21 – 25 | 0.90 | 50 |

* Limit may be increased by 25% for residue cover after crop seedlings will regularly exceed 50%.

Since it is difficult to get proper data for the use of Eq. (18.16) for a developing country like India, the ***universal equation is not popular in developing countries.***

**18.14.3. Sediment Measurement by Sample Recorrder.** The sediment produced by sheet erosion from a catchment may not always reach at the point of measurement; *i.e.*, the site of dam reservoir. Some part of sediment may be deposited en-route. *The ratio between the yield of sediment at the measuring site and the gross erosion in the catchment is called the* ***sediment delivery ratio.***

Thus, the **sediment yield** is the gross sediment yield minus the quantity of sediment deposited en route. The sediment yield, infact, is important since it is this sediment which will get deposited in the reservoir, affecting its useful life.

The *sediment yield of a reservoir* can be calculated either by using an appropriate empirical equation, out of the ones developed by various investigators; or by developing a rational appropriate relation between inflow and sediment, on the basis of actual measurements of sediment load at the site of the reservoir. The continuous measurements of suspended load and bed load at the reservoir site for a number of years will help in developing a rational relation (either a mathematical equation or a graphical curve) between sediment and the inflow into the reservoir. We will first of all discuss the method of the actual measurement of sediment and developing an appropriate relation between sediment yield ($q_s$) and the river discharge ($Q$) at the given site.

**18.14.3.1. Measuring suspended sediment load.** Sediment samplers are used to measure suspended seediment at a given site on the river to obtain the most reliable results of sediment yield. The bed load should also be calculated either by using the available empirical equations**, or on an adhoc basis as equal to 2.5-25% of the susended load, as to calculate the total sediment load (sum of suspended load and bed load).

A typical sediment sampler used in India is shown in Fig. 18.31. A depth integrating hand sampler used for small streams is also shown in Fig. 18.32.

The sampler is taken into the stream to a depth of 0.6y below the water surface, or at two depths at 0.2$y$ and 0.8$y$ depths to collect samples of sedimented water in

**Please refer article 4.9.

the sampler bottle. The collected sample of sedimented water is then analysed in the laboratory either by a *gravimetric method*, or by hydrometric *method* to

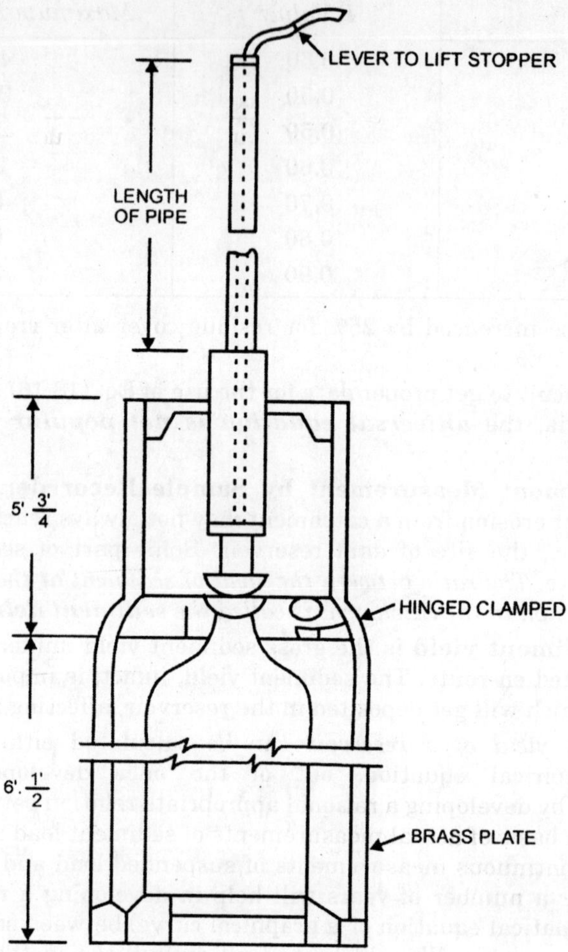


Fig. 18.31. Line diagram of a typical sediment sampler (Punjab bottle sampler).

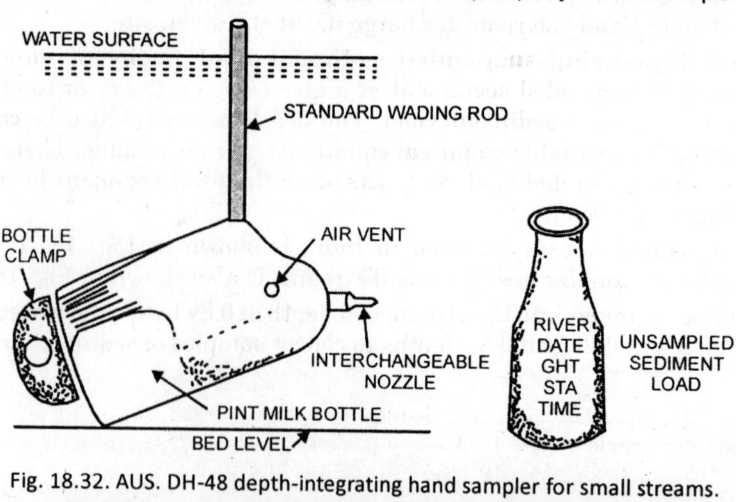


Fig. 18.32. AUS. DH-48 depth-integrating hand sampler for small streams.

determine the quantum of coarse sediment (particle size higher than 0.2 mm), *medium size sediment* (particle size between 0.075 mm to 0.2 mm) and of *fine sediment* (particle size less than 0.075 mm), separately*. Their sum will indicate the total sediment load present in the given volume of water sample. Sediment load present in the water sample is then expressed in ppm (parts for million) as:

$$\text{Sediment load in ppm} = \frac{\text{Dry mass of sediment}}{\text{Total mass of original sample including the mass of sediment and of water}} \times 10^6 \quad (18.21)$$

This value can be converted into t/day by multiplying the average unit wt. of sediment (say 1.2 t/m$^3$) with the total volume of daily inflow in m$^3$.

When a large number of such sample records become a vailable for the given site, then a curve can be plotted between the sedimet load ($q_s$) in *t*/day on *x*-axis and daily discharge in m$^3$/s on *y*-axis, as to obtain curve known as ***sediment rating*** *curve*, as shown in Fig. 18.33. Such a plotting is usually done on a log-log paper.

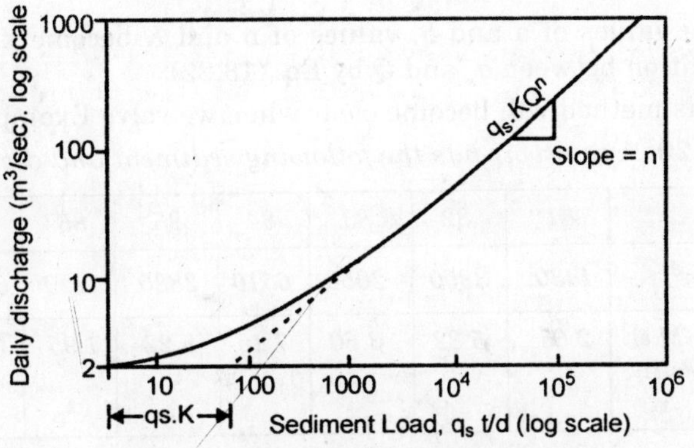


Fig. 18.33. A typical sediment rating curve.

When a standard sediment rating curve is established for a given site, the sediment yield ($q_s$) can be read out by simply knowing the discharge rate only. However, care must be taken to see that for different seasons of the year, different curves will have to be developed, since the sediment yield of a basin may vary with the season. The curve, for say March (non-monsoon period) may be entirely different from the curve for the month of say August (monsoon season).

The sediment rating curves can be used to compute daily sediment load for the given daily dscharge values, and their summation can give us the monthly or the annual sediment load.

**18.14.3.2. Mathematical equation for a sediment rating curve.** The sediment rating curve (straight line portion) giving sediment load in tonns/day ($q_s$) w.r. to daily discharge ($Q$) in m$^3$/sec can be expressed by a mathematical equation of the form:

---

* The sample collected by the sampler is first passed through a BSS-100 sieve and the coarse particles retained are taken out and over dried. Thus, the quantity of coarse sediment (higher than 0.2 mm size) is obtained. Sedimented water passing 200 m sieve is allowed to stand for 20 minutes, so that the fiiner particles settle down. The settled mass is removed by the process of decantation (pouring out water from the settled tank). The settled residue is dried and weighed as to get the mass of fine sediment (0.075 mm – 0.2 mm). Poured water contains sediment of still finer particles. To isolate this, the sample is filtered through a filter paper, and the quantity retained therein is dried and weighed. This gives the mass of fine sediment (particle size < 0.075 mm). The summation of all the three masses will give the total mass of sediment.

$$q_s = K \cdot Q^n \qquad ...(18.22)$$

Taking log on both sides,

$$\log q_s = \log K + n \cdot \log Q$$

This equation is similar to the form

$$y = a + bx,$$

where log $q_s$ is plotted on $y$-axis and log $Q$ on $x$-axis.

$$n = b \qquad ...(18.23)$$

$$\log K = a \qquad ...(18.24)$$

A mathematical solution for such an equation can be obtained by statistical **method of least squares,** where in the various known values of $x$ (*i.e.* log $Q$) and $y$ (*i.e.* log $q_s$) are analysed to estimate $a$ and $b$ values as:

$$a = \frac{\Sigma\, y . \Sigma\, x^2 - \Sigma\, x . \Sigma\, xy}{N\, \Sigma x^2 - (\Sigma x)^2} \qquad ...(18.25)$$

$$b = \frac{N\, \Sigma\, xy - \Sigma\, x . \Sigma\, y}{N . \Sigma x^2 - (\Sigma x)^2} \qquad ...(18.26)$$

By computing values of a and $b$, values of n and $K$ become known to finally compute the relation between $q_s$ and $Q$ by Eq. (18.22).

The use of this method will become clear when we solve Example 18.20.

**Example 18.20.** *A reservoir has the following sediment and discharge data:*

| *Year* | *81* | *82* | *83* | *84* | *85* | *86* | *87* | *88* |
|---|---|---|---|---|---|---|---|---|
| *Discharge (M m³)* | *1430* | *3850* | *2050* | *6510* | *2880* | *1120* | *6050* | *2220* |
| *Sediment load (M.t) as measured by silt sampler* | *2.65* | *5.82* | *3.60* | *7.15* | *5.22* | *1.95* | *6.88* | *3.94* |

*Calculate the average total sediment load / year / 100 sq. km of the catchment at the site. Develop a regression relation and predict the total and observed sediment yield for the inflow of 3450 M.m³ for the year 1978. Take the catchment area at the site as 3050 sq. km. What is the total sediment yield for 100 years? Assume bed load as 10% of suspended load.*

**Solution.** Average sediment load measured by silt sampler

$$= \frac{2.65 + 5.82 + 3.6 + 7.15 + 5.22 + 1.95 + 6.88 + 3.94}{8}$$

$$= \frac{37.21}{8} \text{ Mt/year} = 4.651 \text{ Mt/yr.}$$

Since sediment load is measured by a silt sampler, the given values of sediment are of suspended load.

Now, assume Bed load = 10% of suspended load (given)

∴ Total load = Suspended load + Bed load

∴ Total sediment load = 1.1 × 4.651 = 5.116 Mt/yr

or $(q_s)$ = 5.116 Mt/yr.

Sedimet deposited in a reservoir consolidates gradually due to the increasing silt load on it every year and the wt. of water above it. Assuming the average sediment unit weight (sediment in a reservoir consists of sand, silt and clay in water as 1.2 t/m³, we have)

∴ Total silt load/year = 5.116 Mt

$$q_s = \frac{5.116}{1.2} \text{ M. m}^3 = 4.263 \text{ M m}^3$$

Catchment area = 3050 sq. km

∴ Sediment load/100 sq. km/year = $\frac{4.263 \text{ Mm}^3}{3050} \times 100$

= **0.1397 M m³/100 sq. km/year Ans.**

Total sediment yield in 100 years

= 4.263 M m³/yr. × 100 yr. = **426.3 M m³ Ans.**

In order to develop a regression type non-linear equation between total sediment load ($q_s$) and discharge ($Q$), represented as $q_s = K \cdot Q^n$, we carry out the required calculations in table 18.36 to further evaluate the values of $a$ and $b$ by using Eqn. (18.25) and (18.26), as given below:

**Table 18.36. Computations to develop Equation $y = a + bx$ by the method of least squares.**

| *Year* | *Discharge Q (M . m³)* | *Suspended Sediment measured in M.t* | *Total Sediment volume i.e. qs* in M m³ col. (3) × 1.1**/1.2* | *log Q = x log col. (2)* | $x^2$ *(col. 5)²* | *log* $q_s$ *= y log col. (4)* | *x . y col. (5) × col (7)* |
|---|---|---|---|---|---|---|---|
| *(1)* | *(2)* | *(3)* | *(4)* | *(5)* | *(6)* | *(7)* | *(8)* |
| 81 | 1436 | 2.65 | 2.429 | 3.1553 | 9.9561 | 0.3855 | 1.2164 |
| 82 | 3850 | 5.82 | 5.335 | 3.5855 | 12.8555 | 0.7271 | 2.6070 |
| 83 | 2050 | 3.60 | 3.300 | 3.3118 | 10.9677 | 0.5185 | 1.7172 |
| 84 | 6510 | 7.10 | 6.554 | 3.8136 | 14.5433 | 0.8165 | 3.1138 |
| 85 | 2880 | 5.22 | 4.785 | 3.4594 | 11.9674 | 0.6799 | 2.3520 |
| 86 | 1120 | 1.95 | 1.788 | 3.0492 | 9.2977 | 0.2522 | 0.7690 |
| 87 | 6050 | 6.88 | 6.307 | 3.7818 | 14.3017 | 0.7998 | 3.0247 |
| 88 | 2220 | 3.94 | 3.612 | 3.3464 | 11.1981 | 0.5577 | 1.8663 |
| | | | Σ | 27.503 | 95.0875 | 4.7372 | 16.6664 |

* $q_s$ (Total sediment in Mm³) = $\frac{q_s \text{ in M.t}}{1.2 \text{ t/m}^3} \times (1.1)$ **    ** Total sediment = 1.1 (Observed sediment by soil sampler)

$= \left(\text{col. (3)} \times \frac{1.1}{1.2}\right)$; Unit wt of sediment = 1.2 t/m³

Using $$a = \frac{\Sigma\, y \,.\, \Sigma\, x^2 - \Sigma\, x \,.\, \Sigma\, xy}{N\, \Sigma x^2 - (\Sigma x)^2} \quad \text{... (i.e., Eq. 18.25)}$$

we get $$a = \frac{4.7372 \times 95.0875 - 27.503 \times 16.6664}{8 \times 95.0875 - (27.503)^2}$$

or $$a = \frac{-7.9275}{4.2850} = -1.8500$$

But $\log K = a = -1.8500$ (Eq. 18.24) $\therefore$ $K = 0.014$

Also $$b = \frac{N \,.\, \Sigma xy - \Sigma x \,.\, \Sigma y}{N \,.\, \Sigma x^2 - (\Sigma x)^2} \quad \text{...(i.e., Eq 18.26)}$$

$$= \frac{8 \times 16.6664 - 27.503 \times 4.7372}{8 \times 95.0875 - (27.503)^2} = \frac{3.0440}{4.2850} = 0.7104$$

But $b = n = 0.7104$ (From Eq. 18.23)

Regression equation is hence given as :

$$q_s = K \,.\, Q^n = 0.014\,(Q)^{0.7104} \quad \textbf{Ans.}$$

Total sediment yield for the year 1978 having inflow of 3450 M m$^3$

$= 0.014\,(3450)^{0.7104} = 4.565$ M m$^3$

$= 4.565$ M m$^3$ × 1.2 t/m$^3$ = **5.48 Mt** **Ans.**

Observed or suspended sediment load for the year 1978

$= \frac{5.48}{1.1}$ Mt = **4.98 Mt** **Ans.**

**18.14.3.3. Empirical equations for total sediment yield.** The following empirical equations have been developed by several investigators, for estimating the ***annual sediment yield of a reservoir***.

(1) *Swami's Regression Equation.* Swamy and Garde (1977) have proposed a relation correlating the cumulative volume of sediment deposited in a reservoir with the cumulative volume of water inflow, and initial bed slope of the river, as

$$V_s = C \,.\, B \,.\, (V_{ci})^{0.94}\,(S_0)^{0.84} \quad \text{...(18.27)}$$

where $V_s$ = cumulative volume of sediment deposited in the reservoir in M . m$^3$

$C$ = Regression constant with safe value of the order of 1.16. However a value less than 1.16 may be adopted depending on the reservoir

$B$ = Width of the reservoir at full reservoir level in m

$V_{ci}$ = Cumulative volume of inflow per unit width $B$ of the reservoir

$S_o$ = Bed slope of the river.

(2) *Jogelkar's Equation.* An equation proposed by Jogelkar (1960) is given as :

$$Q_s = 0.59\,(A)^{-0.24} \quad \text{...(18.28)}$$

where $Q_s$ = Annual silting rate in M . m$^3$ per 100 sq. km of catchment area

$A$ = Catchment area in sq km

(3) *Khosla's Equation*. Khosla (1953) has proposed the following empirical equation.

$$Q_s = 0.323\ (A)^{-0.28} \quad ...(18.29)$$

where $Q_s$ = Annual siltation rate in M . m$^3$/100 s.q km/year

$A$ = Catchment area in sq. km.

This equation always gives lower estimate of sediment yield at a site.

(4) *Varshneys Equations*. Varshney and Raichur have proposed the following equations for calculating sediment yield for an ungauged basin.

(*i*) *Up to 130 sq. km catchment for mountainous rivers*

$$Q_s = 0.395\ (A)^{-0.311} \quad ...(18.30)$$

where $Q_s$ = Annual sediment yield rate in M . m$^3$ per 100 sq. km of catchment

$A$ = catchment area in km$^2$.

(*ii*) *Rivers draining plain area up to 130 sq.* km

$$Q_s = 0.392\ (A)^{-0.302} \quad ...(18.31)$$

$Q_s$ and $A$ have the same meaning as in Eqn. (18.30)

(*iii*) *For area greater than 130 sq. km for North Indian catchment*

$$Q_s = 1.534\ (A)^{-0.264} \quad ...(18.32)$$

$Q_s$ and $A$ have the same meaning as in Eqn. (18.30)

(*iv*) *For South Indian Rivers up to 130 sq. km*

$$Q_s = 0.46\ (A)^{-0.468} \quad ...(18.33)$$

$Q_s$ and $A$ have the same meaning as in Eqn. (18.30).

(*v*) *For areas greater than 130 sq. km for South Indian catchments*

$$Q_s = 0.277\ (A)^{-0.194} \quad ...(18.34)$$

$Q_s$ and $A$ have the same meaning as in Eqn. (18.30).

(5) *Using Known Data of Similar Catchment.* Sediment yield of an unmeasured watershed $Q_{s2}$ can be computed from sediment yield of measured catchment $Q_{s1}$ of similar topography, land cover and land use, on area proportion basis, as

$$Q_{s2} = Q_{s1}\left(\frac{A_2}{A_1}\right) \quad ...(18.35)$$

**Example 18.21.** *Estimate the sediment load in tonne at the proposed dam site in North India with the following data using various empirical equations :*

*Catchment area* = 1830 sq. km

*Width of reservoir at FRL* = 560.0 m

*River slope at the dam site* = 0.006

*Average inflows at the site are as follows :*

| *Year,* | *1982* | *1983* | *1984* | *1985* | *1986* | *1987* | *1988* | *1989* | *1990* |
|---|---|---|---|---|---|---|---|---|---|
| *Inflow M . m*$^3$ | 2210 | 1290 | 1640 | 1780 | 2150 | 1980 | 2540 | 1285 | 1620 |

*Assume annual siltation rate per 100 sq. km from a similar catchment of 3050 sq. km to be 10.35 M . m*$^3$ */ 100 sq. km.*

**Solution.** The sedimentation rate is worked out by using various equations as described in the previous article.

(1) *By Swamy's Regression Method* : The computation is carried out in Table 18.37 by using Eqn. (18.27) as :

$$V_s = C \cdot B\,(V_{ci})^{0.94}\,(S_0)^{0.84}$$

where, $C = 1.16$

$B = 560$ m

$S_0 = 0.006$

$\therefore$ $V_s$ = cumulative volume of sediment deposited in M . m³

$$= 1.16 \times 560\,(V_{ci})^{0.94} \times (0.006)^{0.84} = 8.837\,(V_{ci})^{0.94}$$

where $V_{ci}$ is the cumulative volume of annual inflows per unit width $B$ of reservoir, over the given years.

The computations are carried out in Table 18.37, which is self explanatory.

**Table 18.37. Computation of Sediment Load by Swamy's Regression Method**

| *S.No* | *Year (2)* | *Inflow in M . m³* | *Cumulative inflow in M m³* | $V_{ci}$ *= cumulative inflow per unit width B of reservoir* $= \frac{col(4)}{560}$ | $V_s = 8.837\,(V_{ci})^{0.94}$ *= 8.837 (col 5)*$^{0.94}$ *M . m³* |
|---|---|---|---|---|---|
| *(1)* | *(2)* | *(3)* | *(4)* | *(5)* | *(6)* |
| 1 | 1982 | 2210 | 2210 | 3.946 | 32.11 |
| 2 | 1983 | 1290 | 3500 | 6.250 | 49.48 |
| 3 | 1984 | 1640 | 5140 | 9.179 | 71.01 |
| 4 | 1985 | 1780 | 6920 | 12.357 | 93.91 |
| 5 | 1986 | 2150 | 9070 | 16.196 | 121.10 |
| 6 | 1987 | 1980 | 11050 | 19.732 | 145.80 |
| 7 | 1988 | 2540 | 13590 | 24.268 | 177.11 |
| 8 | 1989 | 1285 | 14875 | 26.563 | 192.81 |
| 9 | 1990 | 1620 | 16495 | 29.455 | 212.48 |

$Q_s$ = sedimentation rate per 100 sq. km

$$= \frac{V_s\,(i.e.\text{ cumulative volume of silt})}{\text{No. of years } (i.e.\ 9)} \times \frac{100 \text{ sq. km}}{\text{Catchment area of } 1830 \text{ km}^2}$$

$$= \frac{212.48}{9} \times \frac{100}{1830}$$

$Q_s$ = **1.29 M m³/100 sq. km Ans.**

(2) *By Jogelkar's Equation*

$$Q_s = 0.59\,(A)^{-0.24}$$

$= 0.59\,(1830)^{-0.24}$ = **0.10 M m³ / 100 sq. km Ans.**

(3) *By Khosla's Equation*

$$Q_s = 0.323\,(A)^{-0.28}$$

$= 0.323\,(1830)^{-0.28}$ = **0.04 M m³ / 100 sq. km Ans.**

(4) *By Varshneys Equation for Norths Indian catchment exceeding 130 sq. km.*

$$Q_s = 1.534\,(A)^{-0.264}$$

$= 1.534\,(1830)^{-0.264}$ = **0.211 M m³ / 100 sq. km Ans.**

(5) *From similar catchment.*

$Q_{s1} = 4.56$ M m³ / 100 sq. km

$A_1 = 3050$ sq. km

$A_2 = 1830$ sq. km

$$Qs_2 = Q_{s1} \times \left(\frac{A_2}{A_1}\right) = 0.35 \times \left(\frac{1830}{3050}\right) \text{ M m}^3/100 \text{ sq. km}$$

$$= \mathbf{0.21 \text{ M m}^3/100 \text{ sq. km}} \quad \textbf{Ans.}$$

## 18.15. Reservoir Sedimentation Studies on Existing Reservoirs

Sedimentation of storage reservoirs is a natural process, since large part of the silt eroded from the catchment and transported by the river, gets deposited on the bed of the reservoir. This causes reduction in the live as well as dead storage capacities of the reservoir. Progressive loss of capacity due to sediment accumulation results in reduced benefits and may even cause operational problems. *It, therefore, becomes necessary to monitor the sedimentation rates in the existing reservoirs at regular intervals, to help in planning and executing suitable remedial measures for controlling sedimentation in order to prolong the life of the reservoir and its benefits.*

Regular monitoring and updating of elevation-capacity curve of the reservoir immensely helps in better water management. With this aim, *conventional hydrographic surveys* are conducted at regular intervals at the existing reservoirs to determine the available capacities at different elevations, and to help compute the sedimentation volume at such regular intervals. Such conventional surveys will require computation of water spread area at different water levels, and is quite a tedious and a costly process. **Remote sensing techniques** do offer a modern answer to the costly conventional surveys, as it offers a great potential for application in capacity evaluation of medium to large reservoirs. From the data provided by the remote sensing satellites, it has now been possible to compute loss of reservoir capacity due to sedimentation, and its distribution. The results obtained from this technique have been found to be quite comparable with those obtained from the costly and cumbersome conventional methods. One of the greatest advantage of this technique is that the capacity evaluation could be easily computed on yearly basis.

The methodology involved in this technique requires the use and analysis of the *satellite imageries* provided by the Remote Sensing satellites, which collect the data of the Earth surface features in different bands at regular intervals. In the case of Indian Remote Sensing Satellites IRS-1A* and IRS-1B* (both identical satellites), this periodicity was 22 days for Indian sub-continent. These two satellites together were thus capable to provide us data of all our reservoirs at 11 days interval. The third satellite of this series IRS-1C, launched on 28.01.1996, also helped in providing better pictures even of cloud bound areas. IRS-1D, launched on 29.09.1997, through our first Indian rocket launcher, also helped a lot. Subsequently launched Remote Sensing Satellites like Cartosat-1, Resourcent-1 and Resource Sat-2A, etc., are providing such services.

Due to the water withdrawals from an existing reservoir, its water spread area goes on changing throughout the year. The reservoirs are generally full just after the monsoon period in October, and get depleted to almost dead storage/minimum drawdown level just before start of monsoon season in May or June, every year.

---

*IRS-1A was launched on 19.03.1988, and IRS-1B was launched on 23.08.1991.

The satellite data of various dates during the period from October to May, provide us an array of water spread areas between maximum water level (*i.e.* around the FRL) and the minimum reservoir level (*i.e.* around the dead storage level/ minimum drawdown level). From the whole set of the satellite data, a few of them which are cloud free and of good quality and representative of the whole range of reservoir levels at close intervals, are selected for analysis.

The method of analysis depends upon the data products. The selected CCT's/ FCC's of various dates are analysed for determining the water spread areas. The corresponding water levels are obtained from the daily gauge record of the reservoir. From these, the water volumes between two consecutive water levels are computed using Prismoidal or any appropriate formula. Volume of water below the minimum water level (as recorded by the satellite) and the "new zero"* elevation, are estimated based upon the previous hydrographic surveys. In case, these informations are not available in the hydrographic survey data, then the elevation-area relationship obtained from the hydrographic surveys as well as the one obtained from the satellite data interpretation, should be extended to get the new zero elevation, and then the volume between the minimum mapped water level and this new zero level is estimated. After this, the cumulative water volume at each reservoir level is computed, and then the revised elevation-area-capacity curve is drawn. By comparing the original area-capacity curve or any other such curve, the total sediment volume and its distribution can be computed.

## 18.16. Observed Sedimentation Rates for Various Important Indian Reservoirs

The observed sedimentation rates for various important dam reservoirs in India are indicated in col. (6) and (7) of Table 18.38.

**Table 18.38. Annual Sedimentation Rates of Various Indian Reservoirs**

| *S. No.* | *Name and Location of reservoir* | *Catchment area at 100 sq. km* | *Capacity of reservoir at FRL in M . $m^3$* | *Surface area at MRL $Mm^2$* | *Annual sediment rate in ham / 100 sq.m* | *Annual volume of sediment deposit M. $m^3$* $\frac{col(6)}{100} \times col(3)$ | *Dead storage capacity provided in M . $m^3$* |
|---|---|---|---|---|---|---|---|
| *(1)* | *(2)* | *(3)* | *(4)* | *(5)* | *(6)* | *(7)* | *(8)* |
| 1 | Bhakra i.e. Govind Sagar) (HP) | 568 | 9351 | 169 | 6.00 | **34.08** | 95 |
| 2 | Gandhi Sagar (MP) | 226 | 7413 | 660 | 10.08 | 22.71 | 586 |
| 3 | Hirakud (Orissa) | 834 | 8146 | 725 | 3.89 | 32.44 | 2318 |
| 4 | Lower Bhawani (TN) | 61.5 | 929 | — | 4.10 | 2.52 | 0 |
| 5 | Maithan (DVC) | 63 | 1275 | — | 13.02 | 8.20 | 165 |
| 6 | Matatila (UP) | 207.5 | 883 | — | 3.50 | 7.26 | 50 |
| 7 | Mayurkshi (WB) | 18.6 | 617 | — | **20.09** | 3.74 | 68 |
| 8 | Nizam Sagar (AP) | 216.94 | 715 | 130 | 6.57 | 14.25 | — |
| 9 | Panchat (DVC) | 111 | 1475 | 153 | 9.92 | 11.01 | 170 |
| 10 | Ramganga (UP) | 30.76 | 2448 | — | 17.30 | 5.32 | 395 |
| 11 | Sriram Sagar or Shivaji Sagar (AP) | 8.19, | 3454 | — | 15.20 | 1.24 | 849 |
| 12 | Tawa (MP) | — | 2312 | — | 8.10 | — | 263 |

*The minimum reservoir bed level is raised due to sedimentation, which is termed as the new zero.

## 18.17. Reservoir Losses

Huge quantity of water is generally lost from an impounding reservoir due to *evaporation, absorption,* and *percolation*. Depending upon which, the following losses may occur from such a reservoir :

1. *Evaporation losses* ;
2. *Absorption losses* ; and
3. *Percolation losses* or *Reservoir leakage.*

These losses are discussed below.

**18.17.1. Evaporation Losses.** The evaporation losses from a reservoir depend upon several factors, such as : *water surface area, water depth, humidity, wind velocity, temperature, atmospheric pressure* and *quality of water* as discussed in article 6.2.2.3 of "Water Resources Engg. Vol. I". The evaporation loss from a reservoir under the given atmospheric conditions can be easily estimated by measuring the standard pan evaporation and multiplying the same by the pan coefficient. The pan coefficients and various types of pans in use are given in article 6.2.2.4 of "Water Resources Engg. Vol. I".

The evaporation losses become very significant in a hot and humid country like India ; and realistic estimation of these losses is quite important. These losses in fact vary from place to place and from season to season, and hence monthly values of these losses are usually determined. Typical average values of these losses for North and South India are given in Table 18.39.

**Table 18.39. Monthly Reservoir Evaporation Losses**

| Month | *Losses in cm* | |
|---|---|---|
| | *North India* | *South and Central India* |
| January | 7 | 10 |
| February | 9 | 10 |
| March | 13 | 18 |
| April | 16 | 23 |
| May | 27 | 25 |
| June | 24 | 18 |
| July | 18 | 15 |
| August | 14 | 15 |
| September | 14 | 15 |
| October | 13 | 13 |
| November | 9 | 10 |
| December | 8 | 10 |
| Total | 172 | 182 |

On the basis of a review conducted on 130 sample reservoirs, the Central Water Commission, in 1990, has, however, estimated the *average annual evaporation loss to be 225 cm* ; and the total water lost from all the existing reservoirs to be 27000 $Mm^3$ per annum. What a tremendous waste of precious water!

In order to control such large scale wastage of water, several methods have been devised by engineers and scientists. All these methods are based upon the efforts made to reduce the evaporation rate from the surface of the water bodies by physical or chemical means, since the basic meteorological factors affecting evaporation cannot be controlled under normal conditions. *The following methods are generally used for evaporation control* :

1. Wind breakers
2. Covering of the water surface
3. Reduction of the exposed water surface
4. Use of underground storage rather than the use of surface storage
5. Integrated operation of reservoirs
6. Use of chemicals for retarding the evaporation rate from the reservoir surface.

Out of all these methods, the last method has evoked the maximum response from all over the world, and has been considered to be the only practical solution for conservation of fresh water, inspite of its various limitations and disadvantages in high cost of application in normal conditions. The use of chemicals, called **Water Evapo-Retardants (WERs),** for controlling the evaporation rate from the surface of reservoirs is therefore, discussed here in details.

A non toxic chemical, capable of forming a thin *monomolecular film* over the water surface, is generally spread over the reservoir water surface in powder, liquid or emulsion form. *The resulting film prevents energy inputs from the atmosphere, thus reducing evaporation. Such a film, however, allows the passage of enough air through it, to avoid any harmful effects on the aquatic life due to shortage of oxygen.*

Fatty alcohols of different grades like : Cetyl alcohol ($C_{16}.H_{33}.OH$) popularly called *hexa decanol*, Stearyl alcohol ($C_{18}.H_{37}.OH$) popularly called *Octadecanol*, and Behenyl alcohol ($C_{22} \cdot H_{45}.OH$) called *docosanol*, or a mixture of these chemicals, have been generally used and found to be quite suitable. These chemicals should, however, be 99% pure for getting the desired properties of *monolayer*. National Chemical Laboratory, Pune, has developed one more compound by synthesising alkoxy ethanols.

In general, all such chemical compounds should possess the *following properties* :

(*i*) the chemical compound (WER) should be tasteless, odourless, non-toxic, non-inflammable, and should not produce any effect on the quality of water.

(*ii*) the chemical should easily spread and form an even compact cohesive and efficient monomolecular film on the water surface.

(*iii*) the thin film formed by the chemical should be pervious to oxygen and carbon dioxide, but tight enough to prevent escape of water molecules.

(*iv*) the thin film formed by the chemical should be durable, and should be able to re-seal itself, when broken due to external disturbances such as wind, waves, etc.

(*v*) the chemical and the film formed by it should not be adversely affected by the water borne bacteria, proteins and other impurities present in the water body.

The use of chemical WERs has, however not been found to be cost effective for mass scale use, and has further not been found to be suitable under the following conditions :

(*a*) when the wind velocities exceed 10 km/hr or so.

(*b*) when the temperature rises above 40°C or so.

(*c*) when the size of the water body is relatively large.

Development of cheaper **WERs** capable of withstanding higher wind speeds upto about 20 km/hr and having strong cohesive forces and properties of self spreading and re-uniting to maintain the monolayer in resilient state even at high wind velocities, is therefore of vital importance. Moreover, the life of the film formed, must be longer, so as to reduce the frequency of application to about 3 to 7 days from its present frequency of 24 hours. Development of such chemical WERs is the subject matter of present research.

Other long term evaporation control measures like *plantation of trees to act as wind breakers**, *reduction of exposed water surface by covers*, *underground storage of water*, *integrated operation of reservoirs*, etc. have been employed in some parts of the country. The effectiveness and economics of these methods are, however, yet to be established.

In India, the water conservation methods are presently being adopted only in draught prone and scarcity areas, since large scale use of such methods on all the reservoirs of the country is not found to be economical or practically unfeasible due to their large size and adverse meteorological factors.

**18.17.2. Absorption Losses.** These losses do not play any significant role in planning, since their amount, though sometimes large in the beginning, falls considerably as the pores get saturated. They certainly depend upon the type of soil forming the reservoir.

**18.17.3. Percolation Losses or Reservoir Leakage.** For most of the reservoirs, the banks are permeable but the permeability is so low that the leakage is of no importance. But in certain particular cases, when the walls of the reservoir are made of badly fractured rocks or having continuous seams of porous strata, serious leakage may occur. Sometimes, pressure grouting may have to be used to seal the fractured rocks. The cost of grouting has to be accounted in the economic studies of the project, if the leakage is large.

## 18.18. Reservoir Clearance

The removal of trees, bushes and other vegetation from the reservoir area is known as reservoir clearance. It is an expensive operation and difficult to be justified on cost-benefit considerations. Non-clearance of such vegetation may lead to the following troubles :

(*i*) decay of organic material may create undesirable odours and tastes, and hence, becomes important for water supply reservoirs.

(*ii*) trees projecting above the water surface may create undesirable appearance, when the reservoir is to be used for recreation and tourism purposes.

(*iii*) bushes, trees, etc. will float and may create debris problems at the dam.

## 18.19. Selection of a Suitable Site for a Reservoir

It is almost impossible to select a perfect ideal reservoir site. But its selection is guided by the following factors :

(*i*) A suitable dam site is available. The cost of the dam is generally a controlling factor in the selection of a reservoir site.

(*ii*) The geological formations for the reservoir banks, walls, etc. should be such as to entail minimum leakage.

(*iii*) The geology of the catchment area should be such as to entail minimum water losses through absorption and percolation.

---

*Trees having lesser evapotranspiration should only be chosen and identified.

(*iv*) The site should be such that a deep reservoir is formed. A deep reservoir is preferred to a shallow one, because of lower land cost per unit of capacity, less evaporation loss, and less possibility of weed growth.

(*v*) The reservoir site must have adequate capacity.

(*vi*) Too much silt laiden tributaries should be avoided as far as possible.

(*vii*) The reservoir basin should have a deep narrow opening in the valley, so that the length of the dam is minimum.

## 18.20. Reservoir Induced Seismicity

During the recent times, a very strange phenomenon has been observed in several dams of the world. What actually has happened is that the reservoir basins or areas which were seismologically inactive, started showing seismic activities, as the reservoir was filled up with water. The magnitude of these earthquakes is found increasing with the filling of the reservoir, giving severe shocks when the reservoir is full. In all such cases, the epicentre is always seen to be located within the reservoir or along its border.

An important Indian example of such an earthquake has been in the case of Koyna dam (Maharashtra), which is situated in the seismologically inactive zone in Peninsular India. However, when the dam got ready and water started collecting (1962) in the reservoir, the earthquakes also started occurring (1963). The frequency and intensity of these earthquakes had gone on increasing, as more and more water went on collecting in the reservoir. On 10th December, 1967, the severest shock occurred with a magnitude of about 6.5, followed by three more shocks of decreasing magnitudes. The epicentres of all these shocks were traced to be within the reservoir area.

Some other examples of such earthquake sites, noticed outside India, are at *Lake Mead* (U.S.A.) ; *Grand Val Lake* (France) ; *Vorgorno Lake* (Switzerland) ; etc.

This phenomenon has been studied by various seismologists, and it has been suggested by some that in most such cases, *there must have existed some inactive faults in the reservoir basin area, which became active again due to the extra-ordinary load of the reservoir water,* thus causing displacements along these faults and consequently resulting in earthquakes. Some other seismologists have suggested that these earthquakes are caused due to increased pore pressure in the adjoining rocks, which lowers their shearing strength, resulting in the release of tectonic strain.

Based on these explanations, the various preventive methods suggested for preventing or reducing such earthquakes, include :

(*i*) *filling of the reservoir to a limited safe level ;*

(*ii*) *reducing pore pressure by draining out water from weaker adjoining rocks; and*

(*iii*) *to actively explore the dam site for the absence of inactive faults before selecting the same.*

Such earthquakes, however, show a decreasing tendency with time. This phenomenon is indeed very complex and interesting, and still needs further research.

## 18.21. Economic Height of a Dam

The economic height of a dam is that height of the dam, corresponding to which, the cost of the dam per unit of storage is minimum. For this purpose, the estimates are prepared for construction costs, for several heights of the dam, somewhat above and below the level at which the elevation-storage curve shows a fairly high rate of increase of storage per unit rise of elevation, keeping the length of the dam moderate. The construction cost is found to increase with the dam height, as shown in Fig. 18.34.

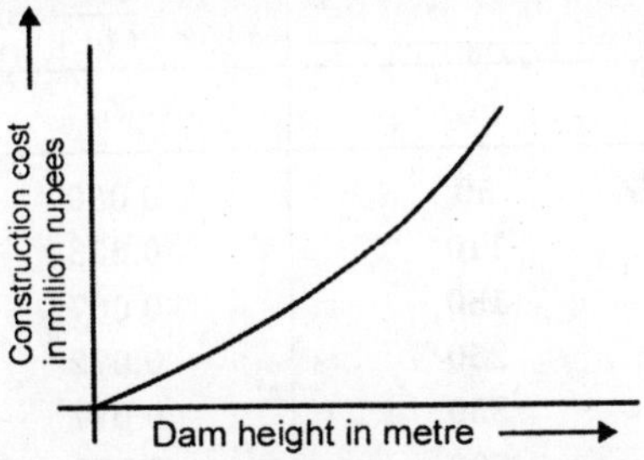


**Fig. 18.34**

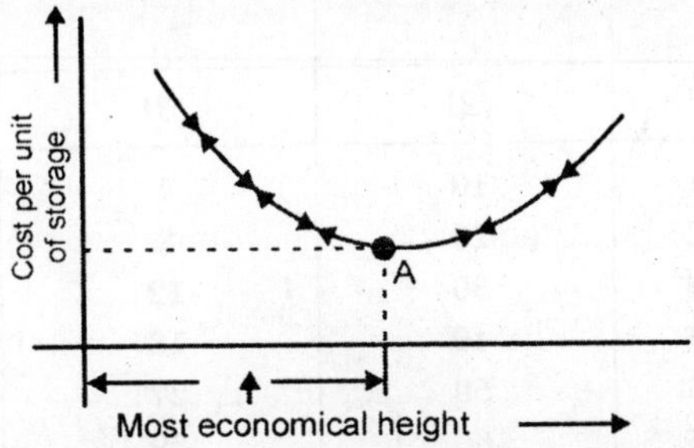


**Fig. 18.35**

For each dam height, the reservoir storage is known from the reservoir-capacity curve. The construction cost per unit of storage for all the possible dam heights can then be worked out and plotted, as shown in Fig. 18.35.

The lowest point A on this curve, gives the dam height for which the cost per unit of storage is minimum, and hence, most economical.

**Example 18.22.** *The construction costs for certain possible heights of a dam at a given site have been estimated and are tabulated in the table below. The storage capacity for all these dam heights are also given below.*

| *S.No.* | *Height of the dam in metres* | *Construction cost in million Rs.* | *Storage in million cubic metres* |
|---|---|---|---|
| *(1)* | *(2)* | *(3)* | *(4)* |
| 1 | 10 | 4 | 50 |
| 2 | 20 | 8 | 110 |
| 3 | 30 | 12 | 180 |
| 4 | 40 | 18 | 250 |
| 5 | 50 | 27 | 350 |
| 6 | 60 | 39 | 500 |
| 7 | 65 | 50 | 600 |

*Determine the most economical height of the dam from purely construction point of view.*

**Solution.** The given table is extended, so as to workout the cost per million cubic metre of storage, as shown in col. (5) of Table 18.40.

The cost per unit of storage is plotted against the height of the dam, as shown in Fig. 18.36. The most economical height is the lowest point of this curve, and it works out to be **30 metres. Ans.**

**Table 18.40.**

| S.No. | Height of the dam in metres | Construction cost in M.Rs. | Storage in $M.m^3$ | Cost per unit of storage in M.Rs. $= \frac{Col.3}{Col.4}$ |
|---|---|---|---|---|
| (1) | (2) | (3) | (4) | (5) |
| 1 | 10 | 4 | 50 | 0.080 |
| 2 | 20 | 8 | 110 | 0.073 |
| 3 | 30 | 12 | 180 | 0.067 |
| 4 | 40 | 18 | 250 | 0.072 |
| 5 | 50 | 27 | 350 | 0.077 |
| 6 | 60 | 39 | 500 | 0.078 |
| 7 | 65 | 50 | 600 | 0.083 |

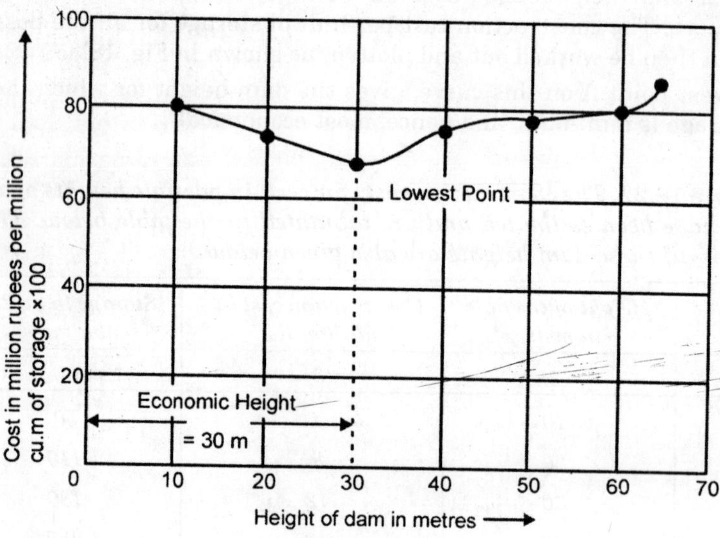


**Fig. 18.36**

## PROBLEMS

1. (*a*) What is meant by a `Reservoir'? Discuss briefly the different types of reservoirs and the purpose served by each type.
   (*b*) Describe briefly the techniques that are employed for computing the storage capacity of a reservoir for different water surface elevations.
2. (*a*) What is meant by a `Flood control reservoir' ; and what are their different types ?
   (*b*) Discuss with a neat sketch, the various storage zones of the dam reservoir.
   (*c*) What factors you will keep in mind while selecting a suitable site for a dam reservoir?
3. (*a*) Differentiate clearly between the following :
   (*i*) A flood control reservoir and a multipurpose reservoir.
   (*ii*) A retarding basin and a storage reservoir.
   (*iii*) Firm yield, design yield, and secondary yield of a reservoir.

(*b*) Briefly describe as to how you would fix the storage capacity of a reservoir and the height of the dam required for this storage.
*(Madras University,* 1976)

4. (*a*) What is the relation between 'reservoir capacity' and `reservoir yield'?
   (*b*) How would you fix the capacity of a dam reservoir at a particular river site, provided the inflow pattern and demand pattern are known. Explain the mass curve method which is used for this purpose.
5. (*a*) Explain how the storage capacity of a reservoir is fixed.
*(Madras University, 1973, 1974)*
   (*b*) Explain the mass curve method that can be used for determining :
   (*i*) Reservoir capacity for fulfilling given demand.
   (*ii*) Demand rate from a reservoir of a given capacity.
6. Discuss briefly and with necessary neat sketches, the demand patterns for the following types of reservoirs :
   (*i*) Single purpose conservation reservoir.
   (*ii*) Single purpose flood control reservoir.
   (*iii*) Multipurpose reservoir.
7. (*a*) What is meant by `flood routing through reservoirs'?
   (*b*) Describe step by step procedure that you will adopt for flood routing computations required for reservoirs under `trial and error method'.
8. Describe a method for routing flood water through a deep reservoir using the fundamental relation between inflow, outflow (discharge) and storage. Take Q = C.L. $H^{3/2}$ for the spillway. *(U.P.S.C.,Engg. Services, 1974)*
9. The initial inflow in a reservoir and outflow over the spillway was 30 m³/sec. During a storm, the following inflow rates were noted at the ends of successive half day periods : 240, 300, 240, 90, 30 and 30 m³/sec. The relationship between discharge over the spillway (*Q*), and storage rate in the reservoir $\left(\frac{S}{T}\right)$ may be expressed by the equation $Q = \frac{1}{2} \cdot \frac{S}{T}$.
   Assuming that the average inflow rate during each half day period is equal to the average of the rates occurring at the beginning and at the end of the period, find the outflow at the end of half hour period.
   Hence plot to a suitable scale the inflow and outflow hydrographs.
10. What are the factors on which the rate of silting of an impounding reservoir depends? What is trap efficiency?
    Discuss the principal measures that should be undertaken to control the inflow sediment to an impounding reservoir.
    An impounding reservoir had original storage capacity for 738 ha-m. The drainage area of the reservoir is 80 sq. km, from which, annual sediment discharges into the reservoir at the rate 0.1153 ha-m per sq. km. of the drainage area. Assuming the trap efficiency as 80 per cent, find the annual capacity loss of the reservoir in per cent per year. **[Ans. 10%]**
*(U.P.S.C., Engg. Services, 1969)*
11. Write short notes on any four of the following :
    (*i*) Reservoir losses (*ii*) Reservoir clearance
    (*iii*) Economic height of a dam
    (*iv*) Cost benefit considerations in planning dam reservoirs.
    [**Hint.** Please See Chapter 20 of "Water Resources Engineering Vol. I"]
    (*v*) Reservoir sedimentation and its control
    (*vi*) Density currents (*vii*) Trap efficiency
    (*viii*) Estimating the life of a reservoir.

# 19

# *Design and Construction of Gravity Dams*

## 19.1. Definition, etc.

A gravity dam has been defined as a structure which is designed in such a way that its own weight resists the external forces. This type of a structure is most durable and solid, and requires very little maintenance. Such a dam may be constructed of masonry or concrete. However, concrete gravity dams are preferred these days and mostly constructed. They can be constructed with ease on any dam site, where there exists a natural foundation strong enough to bear the enormous weight of the dam. Such a dam is generally straight in plan, although sometimes, it may be slightly curve. The line of the upstream face of the dam, or the line of the crown of the dam if the upstream face in sloping, is taken as the reference line for layout purposes, etc. and is known as the *Base line of the dam* or the '*Axis of the Dam*'. When suitable conditions are available, such dams car be constructed up to great heights. The highest gravity dam in the world is Grand Dixence Dam in Switzerland (284 m), followed by Bhakra dam in India (226 m); both are of concrete gravity type. The ratio of base width to height of all these structures is less than 1 : 1.

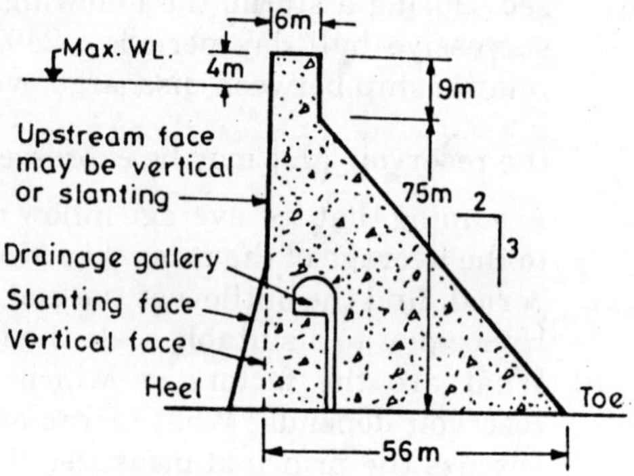

Fig. 19.1: A typical cross-section of a concrete gravity dam.

## 19.2. Typical Cross-section

A typical cross-section of a concrete gravity dam is shown in Fig. 19.1. The upstream face may by kept throughout vertical or partly slanting for some of its length, as shown. A drainage gallery is provided in order to relieve the uplift pressure exerted by the seeping water.

## 19.3. Forces Acting on Gravity Dam

The various external forces acting on a gravity dam may be :

(1) Water Pressure
(2) Uplift Pressure
(3) Pressure due to earthquake forces
(4) Silt Pressure
(5) Wave Pressure
(6) Ice Pressure
(7) The stabilising force is the weight of the dam itself.

An estimation and description of these forces is given below :

(1) **Water Pressure.** Water pressure ($P$) is the most major external force acting on such a dam. The horizontal water pressure, exerted by the weight of the water stored on the upstream side on the dam can be estimated from rule of hydrostatic pressure distribution ; which is triangular in shape, as shown in Fig. 19.2 (*a*) and (*b*). When the upstream face is vertical, the intensity is zero at the water surface and equal to $\gamma_w H$ at the base ; where $\gamma_w$ is the unit weight of water and $H$ is the depth of water : as shown in Fig. 19.2 (*a*). The resultant force due to this external water $= \frac{1}{2}\gamma_w H^2$, acting at $H/3$ from base.

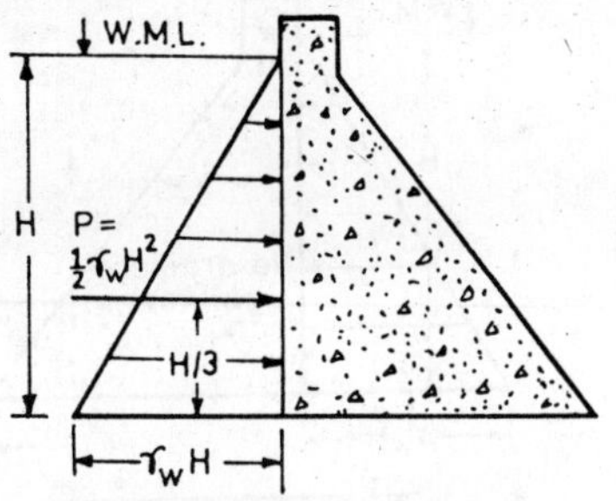


Where $\gamma_w$ = unit weight of water 9.81 kN/m$^3$ = 1000 kgf/m$^3$

Fig. 19.2. (*a*)

When the upstream face is partly vertical and partly inclined [Fig. 19.2 (*b*)], the resulting water force can be resolved into horizontal component ($P_h$) and vertical component ($P_v$). The horizontal component $P_h = \dfrac{1}{2}\gamma_w H^2$ acts at $\dfrac{H}{3}$ from the base ; and the vertical component ($P_v$) is equal to the weight of the water stored in column *ABCA* and acts at the c.g. of the area.

Similarly, if there is tail water on the downstream side, it will have horizontal and vertical components, as shown in Fig. 19.2. (*b*).

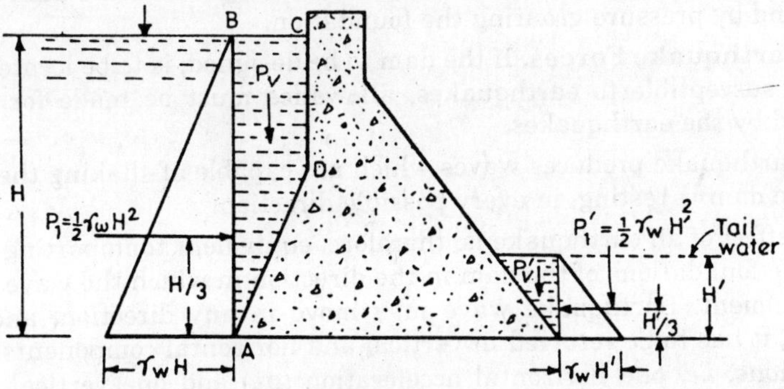


Fig. 19.2 (*b*)

(2) **Uplift Pressure.** Water seeping through the pores, cracks and fissures of the foundation material, and water seeping through dam body and then to the bottom through the joints between the body of the dam and its foundation at the base ; exert an uplift pressure on the base of the dam. It is the second major external force and must be accounted for in all calculations. Such an uplift force virtually reduces the downward weight of the body of the dam and hence, acts against the dam stability.

The amount of uplift is a matter of research and the present recommendations which are followed, are those suggested by United States Bureau of Reclamation (U.S.B.R.). According to these recommendations, the uplift pressure intensities at the heel and the toe should be taken equal to their respective hydrostatic pressures and joined by a straight line in between, as shown in Fig. 19.3 (*a*). When drainage galleries are provided to relieve the uplift, the recommended uplift at the face of

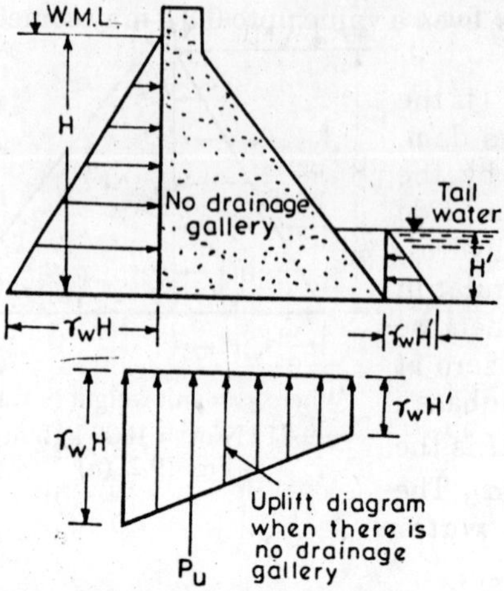


Fig. 19.3 (*a*) Uplift pressure (*U*) diagram, when no drainage gallery is provided.

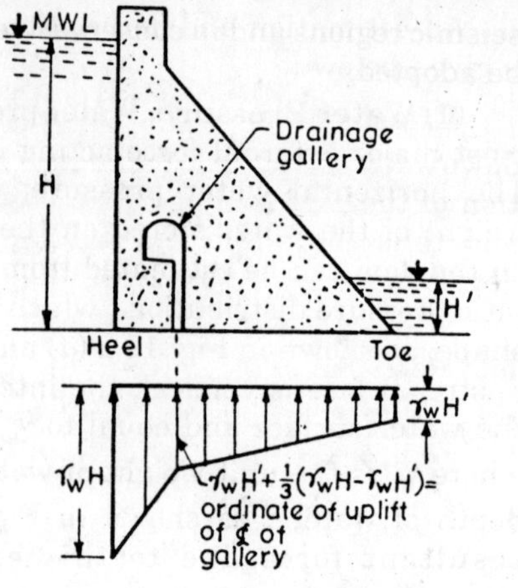


Fig. 19.3 (*b*) Uplift pressure (*U*) diagram, when drainage gallery is provided.

the gallery is equal to the hydrostatic pressure at toe ($\gamma_w \cdot H'$) plus $\frac{1}{3}$rd the difference of the hydrostatic pressures at the heel and the toe ; as shown in Fig. 19.3 (*b*) ; *i.e.* $\left[\gamma_w \cdot H' + \frac{1}{3}(\gamma_w \cdot H - \gamma_w \cdot H')\right]$. It is also assumed that the uplift pressures are not affected by the earthquake forces.

The uplift pressures can be controlled by constructing cut-off walls under the upstream face, by constructing drainage channels between the dam and its foundation, and by pressure grouting the foundation.

(3) **Earthquake Forces.** If the dam to be designed, is to be located in a region which is susceptible to earthquakes, allowance must be made for the stresses generated by the earthquakes.

An earthquake produces waves which are capable of shaking the Earth upon which the dam is resting, in every possible direction.

The effect of an earthquake is, therefore, equivalent to imparting an acceleration to the foundations of the dam in the direction in which the wave is travelling at the moment. Earthquake wave may move in any direction, and for design purposes, it has to be resolved in vertical and horizontal components. Hence, two accelerations, *i.e.* one horizontal acceleration ($\alpha_h$) and one vertical acceleration ($\alpha_v$) are induced by an earthquake. The values of these accelerations are generally expressed as percentage of the acceleration due to gravity ($g$), *i.e.* $\alpha = 0.1\,g$ or $0.2\,g$, etc.

In India, the entire country has been divided into five seismic zones depending upon the severity of the earthquakes. Zone V is the most serious zone and includes Himalayan regions of North India. A map and description of these zones is available in "Physical and Engineering Geology" by the same author, and can be referred to, in order to obtain an idea of the value of the α which should be chosen for designs. On an average, a value of α equal to 0.1 to 0.15 $g$ is generally sufficient for high dams in seismic zones. A value equal to 0.15 $g$ has been used in Bhakra dam design, and 0.2 $g$ in Ramganga dam design. However, for areas not subjected to extreme earthquakes, $\alpha_h = 0.1\,g$ and $\alpha_v = 0.05\,g$ may by used. In areas of no earthquakes or very less earthquakes, these forces may be neglected. In extremely seismic regions and in conservative designs, even a value upto 0.3 *g* may sometimes be adopted.

**Effect of vertical acceleration** ($\alpha_v$). A vertical acceleration may either act downward or upward. When it is acting in the upward direction, then the foundation of the dam will be lifted upward and becomes closer to the body of the dam, and thus the effective weight of the dam will increase and hence, the stress developed will increase.

When the vertical acceleration is acting downward, the foundation shall try to move downward away from the dam body ; thus reducing the effective weight and the stability of the dam, and hence is the worst case for designs.

Such acceleration will, therefore, exert an **inertia force** given by

$$\frac{W}{g}\alpha_v \text{ (i.e. force = Mass × Acceleration)}$$

where $W$ is the total weight of the dam.

∴ The net effective weight of the dam $= W - \frac{W}{g}\cdot\alpha_v$.

If $\alpha_v = k_v \cdot g$

[where $k_v$ is the fraction of gravity adopted for vertical acceleration, such as 0.1 or 0.2, etc.].

Then, the net effective weight of the dam

$$= W - \frac{W}{g}\cdot k_v \cdot g = W\,[1 - k_v].$$

In other words, vertical acceleration reduces the unit weight of the dam material and that of water to $(1 - k_v)$ times their original unit weights.

**Effects of horizontal acceleration** ($\alpha_h$). Horizontal acceleration may cause the following two forces :

(*i*) Hydrodynamic pressure ; and

(*ii*) Horizontal inertia force.

Both these forces are discussed below :

(*i*) *Hydrodynamic pressures.* Horizontal acceleration acting towards the reservoir causes a momentary increase in the water pressure, as the foundation and dam accelerate towards the reservoir and the water resists the movement owing to its inertia. The extra pressure exerted by this process is known as *hydrodynamic pressure.*

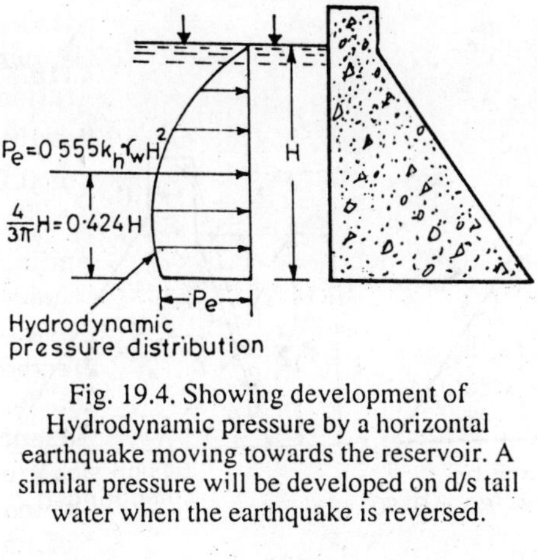


Fig. 19.4. Showing development of Hydrodynamic pressure by a horizontal earthquake moving towards the reservoir. A similar pressure will be developed on d/s tail water when the earthquake is reversed.

According to **Von-Karman**, the amount of this hydrodynamic force ($P_e$) is given by.

$$P_e = 0.555 \cdot k_h\,\gamma_w \cdot H^2 \qquad ...(19.1)$$

and it acts at the height of $\frac{4H}{3\pi}$ above the base, as shown in Fig. 19.4.

where $k_h$ is the fraction of gravity adopted for horizontal acceleration, such as 0.1, 0.2 etc.

$\gamma_w$ = unit wt. of water

Moment of this force about base

$$= M_e = P_e\left(\frac{4H}{3\pi}\right) = \mathbf{0.424\ P_e \cdot H} \qquad ...(19.2)$$

Zanger has given certain big formulas for evaluating the amount of this force and its position, etc. on the vertical as well as on an inclined faces. The results of these big formulas are quite comparable to those given by Von-Karman equation and hence, for average ordinary purposes, the Von-Karman equation (19.1) is sufficient.

**Zanger's formula for hydrodynamic force.** According to Zanger

$$P_e = 0.726\ p_e \cdot H \qquad ...(19.3)$$

where $p_e = C_m\ k_h \cdot \gamma_w \cdot H$ ...(19.4)

$$\therefore \quad P_e = 0.726\ C_m \cdot k_h \cdot \gamma_w \cdot H^2 \qquad ...(19.5)$$

where $C_m$ = Maximum value of pressure co-efficient for a given constant slope

$$= 0.735\left(\frac{\theta}{90}{}^\circ\right); \qquad ...(19.6)$$

where θ is the angle in degrees, which the u/s face of the dam makes with the horizontal.

$k_h$ = fraction of gravity adopted for horizontal acceleration ($\alpha_h$) such as $\alpha_h = k_h \cdot g$

$\gamma_w$ = unit w.t. of water

The moment of this force about the base is given as :

$$M_e = 0.299\ p_e \cdot H^2 \qquad ...(19.7)$$

$$= 0.299\ \frac{P_e}{0.726 \cdot H} \cdot H$$

or

$$M_e = 0.412\ P_e \cdot H \qquad ...(19.8)$$

It was further stated, that if the upstream face is partly inclined (Fig. 19.5 *a*), which does not extend to more than half the depth of the reservoir, it can be taken as vertical. If the slope extends to more than half the depth (Fig. 19.5 *b*), the overall slope up to the whole height may be taken as the value of θ in equation (19.6) above.

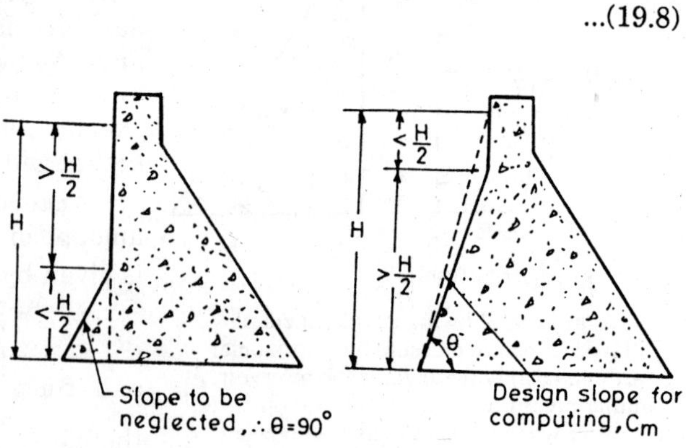


Fig. 19.5

Zanger's formulae are given in details in U.S.B.R. publication on design of small dams, which may be referred to in particular special needs.

(*ii*) *Horizontal Inertia Force.* In addition to exerting the hydrodynamic pressure, the horizontal acceleration produces an inertia force into the body of the dam. This force is generated in order to keep the body and the foundation of the dam together as one piece. The direction of the produced force will be opposite to the acceleration imparted by the earthquake.

Since an earthquake may impart either upstream or downstream acceleration, we have to choose the direction of this force in our stability analysis of dam structure, in such a way that it produces most unfavorable effects under the considered conditions.

Say for example, when the reservoir is *full*, this force would produce worst results if it additive to the hydrostatic water pressure, thus acting towards the *downstream* (*i.e.* when upstream earthquake acceleration towards the reservoir is produced). When the reservoir is *empty*, this force would produce worst results, if considered to be acting *upstream* (*i.e.* when earthquake acceleration, moving towards downstream, is produced).

Under reservoir empty conditions, earthquake forces produce effects, which may cause slight tension near the toe ; and hence stability analysis for reservoir empty case may be carried out only on the basis of wt. of the dam by ignoring earthquake forces and keeping the section free from any tension. However, for all precise designs, these forces must be fully considered, as we have done in example 19.2.

The amount of this horizontal inertial force is equal to the product of the mass of the dam and the acceleration.

∴ This horizontal Inertia force

$$= \left(\frac{W}{g}\right)\alpha_h = \frac{W}{g}\cdot k_h \cdot g = W\cdot k_h \qquad ...(19.9)$$

(where $k_h$ is the fraction of gravity adopted for horizontal acceleration, such as 0.1, or 0.2, etc.).

This force should be considered to be acting at the centre of gravity of the mass, regardless of the shape of the cross- section, and it acts horizontally downstream in worst cases, for reservoir full case.

(4) **Silt Pressure.** It has been explained under 'Reservoir Sedimentation' in chapter 18 that silt gets deposited against the upstream face of the dam. If $h$ is the height of silt deposited, then the force exerted by this silt in addition to external water pressure, can be represented by Rankine's formula as :

$$P_{silt} = \frac{1}{2}\cdot \gamma_{sub}\cdot h^2 K_a \text{ and it acts at } \frac{h}{3} \text{ from base} \qquad ...(19.10)$$

where $K_a$ is the coefficient of active earth pressure of silt $= \frac{1 - \sin\phi}{1 + \sin\phi}$ where $\phi$ is the angle of internal friction of soil, and cohesion is neglected.

$\gamma_{sub}$ = submerged unit weight of silt material.

$h$ = height of silt deposited.

If the upstream face is inclined, the vertical weight of the silt supported on the slope also acts as a vertical force.

In the absence of any reliable data for the type of silt that is going to be deposited, U.S.B.R. recommendations may be adopted. In these recommendations, deposited silt may be taken as equivalent to a fluid exerting a force with a unit wt. equal to 3.6 kN/m³ in the horizontal direction and a vertical force with a unit wt. of 9.2 kN/m³. Hence, the total horizontal force will be $3.6 \frac{h^2}{2} = 1.8\, h^2$ kN/m run, and vertical force will be $9.2 \,.\, \frac{h^2}{2} = 4.6\, h^2$ kN/m run.

In most of the gravity-dam designs, the silt pressure is neglected. The basis for neglecting this force is that :

Initially, the silt load is not present, and by the time it becomes significant, it gets consolidated to some extent and, therefore, acts less like a fluid. Moreover, silt deposited in the reservoir is somewhat impervious and, therefore, will help to minimise the uplift under the dam.

**(5) Wave Pressure.** Waves are generated on the surface of the reservoir by the blowing winds, which causes a pressure towards the downstream side. Wave pressure depends upon the wave height. ***Wave height*** may be given by the equation,

$$h_w = 0.032\sqrt{V \,.\, F} + 0.763 - 0.271\,(F)^{1/4} \text{ for F < 32 km, and} \quad \text{...(19.11)}$$

$$h_w = 0.032\sqrt{V \,.\, F} \text{ for F > 32 km} \quad \text{...(19.12)}$$

where $h_w$ = height of water from top of crest to bottom of trough in metres.

$V$ = wind velocity in km/hr.

$F$ = Fetch or straight length of water expanse in km.

The maximum pressure intensity due to wave action may by given by

$p_w = 2.4\, \gamma_w \,.\, h_w$ and acts at $\frac{h_w}{2}$ metres above the still water surface.

The pressure distribution may be assumed to be triangular, of height $\frac{5h_w}{3}$, as shown in Fig. 19.6.

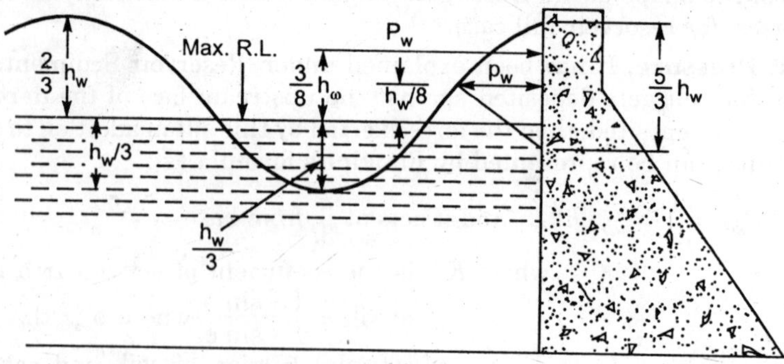


Fig. 19.6

Hence, the total force due to wave action ($P_w$)

$$= \frac{1}{2}\,(2.4\, \gamma_w \,.\, h_w) \,.\, \frac{5}{3} \,.\, h_w$$

or $$P_w = 2 \cdot \gamma_w \cdot h_w^2 = 2 \times 9.81\, h_w^2 \text{ kN/m}$$

$$= 19.62\, h_w^2 \text{ kN/m} \quad \text{...(19.14)}$$

This force acts at a distance $\frac{3}{8} h_w$ above the reservoir surface.

(6) **Ice Pressure.** The ice which may be formed on the water surface of the reservoir in cold countries, may sometimes melt and expand. The dam face has then to resist the thrust exerted by the expanding ice. This force acts linearly along the length of the dam and at the reservoir level. The magnitude of this force varies from 250 to 1500 kN/m$^2$ depending upon temperature variations. On an average, a value of 500 kN/m$^2$ may be allowed under ordinary conditions.

(7) **Weight of the Dam.** The weight of the dam body and its foundation is the major resisting force. In two dimensional analysis of a gravity dam, a unit length of the dam is considered. The cross-section can then be divided into rectangles and triangles. The weight of each along with their *c.gs.*, can be determined. The resultant of all these downward forces will represent the total weight of the dam acting at the *e.g.* of the dam.

**Combination of forces for Designs.** The design of a gravity dam should be checked for two cases, *i.e.* (*i*) when Reservoir is full ; and (*ii*) when Reservoir is empty.

(*i*) **Case I.** Reservoir full case :

When reservoir is full, the major forces acting are : weight of the dam, external water pressure, uplift pressure, and earthquake forces in serious seismic zones. The minor forces are : silt pressure, ice pressure and wave pressure. For the most conservative designs, and from purely theoretical point of view, one can say that a situation may arise when all the forces may act together. But such a situation will never arise and hence, all the forces are not generally taken together. U.S.B.R. has classified the 'normal load combinations' and 'extreme load combination, as given below :

(*a*) *Normal Load Combinations*

(*i*) Water pressure upto normal pool level, normal uplift, silt pressure and ice pressure. This class of loading is taken when ice force is serious.

(*ii*) Water pressure upto normal pool level, normal uplift, earthquake forces, and silt pressure.

(*iii*) Water pressure upto maximum reservoir level (maximum pool level), normal uplift, and silt pressure.

(*b*) *Extreme Load Combinations*

(*i*) Water pressure due to maximum pool level, extreme uplift pressure without any reduction due to drainage and silt pressure.

**Case II.** Reservoir empty case :

(*i*) Empty reservoir without earthquake forces to be computed for determining bending diagrams, etc. for reinforcement design, for grouting studies or other purposes.

(*ii*) Empty reservoir with a horizontal earthquake force produced towards the upstream has to be checked for non- development of tension at toe.

## 19.4. Modes of Failure and Criteria for Structural Stability of Gravity Dams

A gravity dam may fail in the following ways :

(1) By overturning (or rotation) about the toe.

(2) By crushing.

(3) By development of tension, causing ultimate failure by crushing.

(4) By shear failure called sliding.

The failure may occur at the foundation plane (*i.e.* at the base of the dam) or at any other plane at higher level.

(1) **Over-turning.** If the resultant of all the forces acting on a dam at any of its sections, passes outside the toe, the dam shall rotate and overturn about the toe. Practically, such a condition shall not arise, as the dam will fail much earlier by compression. The ratio of the righting moments about toe (anti clockwise) to the over turning moments about toe (clock-wise) is called the factor of safety against overturning. Its value, generally varies between 2 to 3.

(2) **Compression or crushing.** A dam may fail by the failure of its materials, *i.e.* the compressive stresses produced may exceed the allowable stresses, and the dam-material may get crushed. The vertical direct stress distribution at the base is given by the equation :

$$p = \text{Direct stress} + \text{Bending stress}.$$

$$\therefore \quad p_{\substack{max \\ min}} = \frac{\Sigma V}{B} \pm \frac{M}{I} y = \frac{\Sigma V}{B} \pm \frac{\Sigma V \cdot e}{B^2/6} = \frac{\Sigma V}{B}\left[1 \pm \frac{6e}{B}\right]$$

or

$$p_{\substack{max \\ min}} = \frac{\Sigma V}{B}\left[1 \pm \frac{6e}{B}\right] \quad \text{...(19.15)}$$

where $e$ = Eccentricity of the resultant force from the centre of the base.

$\Sigma V$ = Total vertical force.

$B$ = Base width.

**Note.** Resultant is nearer the toe and hence, maximum compressive stress is produced at the toe (Reservoir full case)

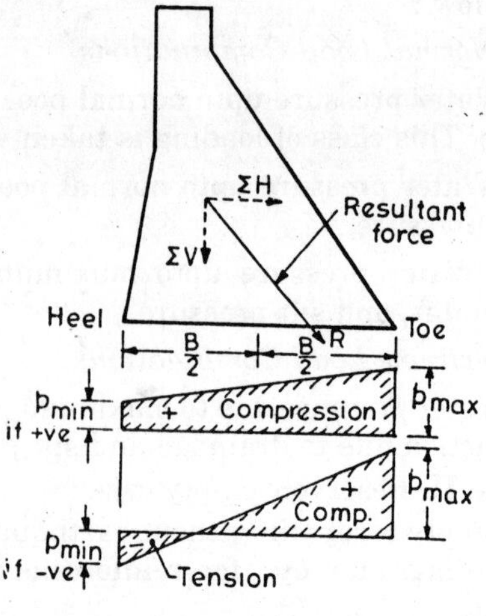


Fig. 19.7. (*a*) Vertical Stress Distribution for Reservoir Full case.

The maximum stress, *i.e.* $p_{max}$, will be produced on the end which is nearer to the resultant, as shown in Fig. 19.7 (*a*) and (*b*).

**Note.** The resultant is nearer the heel and hence, maximum compressive stress (+ve stress) is produced at the heel (Reservoir empty with horizontal earthquake wave moving away from reservoir-case).

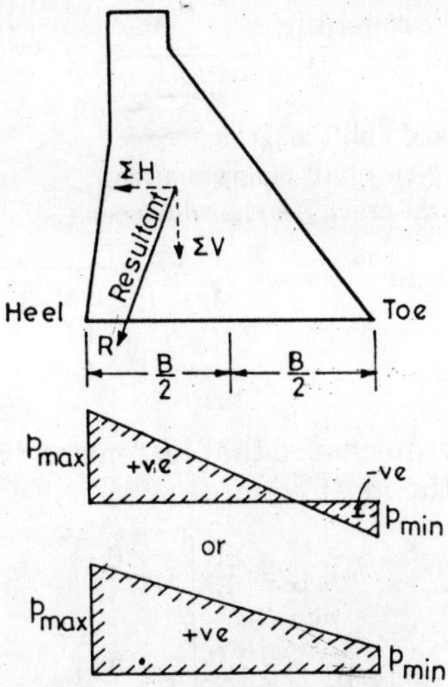


Fig. 19.7. (*b*) Vertical Stress Distribution for Reservoir Empty case.

If $p_{min}$ comes out to be negative, it means that tension shall be produced at the appropriate end.

If $p_{min}$ exceeds the allowable compressive stress of dam material [generally taken as 3000 kN/m$^2$ (30 kg/cm$^2$) for concrete]; the dam may crush and fail by crushing.

(3) **Tension.** Masonry and concrete gravity dams are usually designed in such a way that no tension is developed anywhere, because these materials cannot withstand sustained tensile stresses. If subjected to such stresses, these materials may finally crack. However, for achieving economy in designs of very high gravity dams, certain amount of tension may be permitted under severest loading condition. This may be permitted because of the fact that such worst loading conditions shall occur only momentarily for a little time and would neither last long nor occur frequently. The maximum permissible tensile stress for high concrete gravity dams, under worst leadings, may be taken as 500 kN/m$^2$ (5 kg/cm$^2$).

*Effect produced by tension cracks.* In a dam, when such a tension crack develops, say at the heel, crack width (or strictly speaking crack-area) looses contact with the bottom foundations, and thus, becomes ineffective.

Hence, the effective width $B$ (considering unit length) of the dam base will be reduced. This will increase $p_{max}$ at the toe.

Moreover, the uplift pressure diagram gets modified due to crack formation, as shown in Fig. 19.8, resulting in an increase in the uplift. Since the uplift increases and the net effective downward force reduces, the resultant will shift more towards the toe and thus further increasing the compressive stress at the toe and further lengthening the crack due to further tension development. The process continues ; the effective base width goes on reducing and compressive stress at the toe goes on increasing ; finally leading to the failure of the toe by direct compression. *Hence, a tension crack by itself does not fail the structure, but it leads to the failure of the structure by producing excessive compressive stresses.*

ABC = old uplift diagram
A′B′C′ = New uplift diagram after the crack AA′, has devloped.

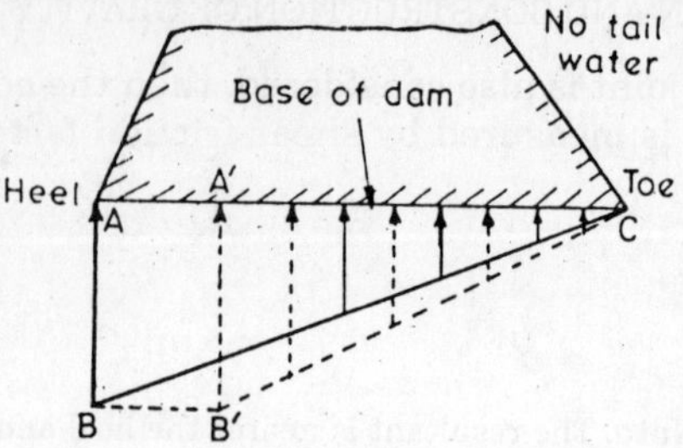

Fig. 19.8

In order to ensure that no tension is developed anywhere, we must ensure that $p_{min}$ is at the most equal to zero.

Since $$p_{\substack{max\\min}} = \frac{\Sigma V}{B}\left[1 \pm \frac{6e}{B}\right] \quad ...(19.15)$$

$$p_{min} = \frac{\Sigma V}{B}\left[1 - \frac{6e}{B}\right]$$

If $p_{min} = 0$, then $\frac{\Sigma V}{B}\left[1 - \frac{6e}{B}\right] = 0$

or $1 - \frac{6e}{B} = 0$ or $e = \frac{B}{6}$.

Hence, maximum value of eccentricity that can be permitted on either side of the centre is equal to $\left(\frac{B}{6}\right)$; which leads to the famous statement : *the resultant must lie within the middle third.*

(4) **Sliding.** Sliding (or shear failure) will occur when the net horizontal force above any plane in the dam or at the base of the dame exceeds the frictional resistance developed at that level.

The friction developed between two surfaces is equal to $\mu\Sigma V$., (Fig. 19.9) where $\Sigma V$ is the algebraic sum of all the vertical forces whether upward or downward, and $\mu$ is the coefficient of friction between the two surfaces. In order that no sliding takes place, the external horizontal forces ($\Sigma H$) must be less than the shear resistance $\mu \cdot \Sigma V$.

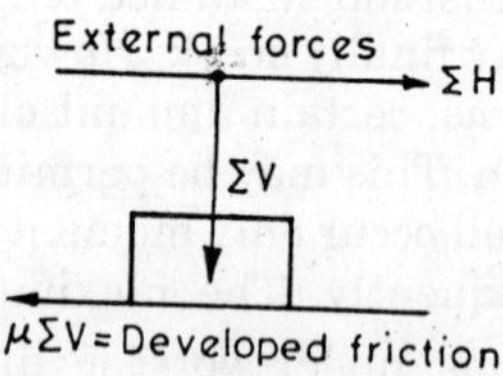

Fig. 19.9

or $\Sigma H < \mu \Sigma V$.

or $\frac{\mu \Sigma V}{\Sigma H} > 1$

$\frac{\mu \cdot \Sigma V}{\Sigma H}$ represents nothing but the factor of safety against sliding, which must be greater than unity.

$\therefore$ F.S.S. (Factor of safety against sliding) $= \frac{\mu \cdot \Sigma V}{\Sigma H}$.

In low dams, the safety against sliding should be checked only for friction, but in high dams, for economical precise designs, the shear strength of the joint, which is an additional shear resistance, must also be considered. If this shear resistance of the joint is also considered, then the equation for factor of safety against sliding which is measured by shear friction factor (S.F.F.) becomes

$$\text{S.F.F.} = \frac{\mu \Sigma V + B \cdot q}{\Sigma H} \qquad ...(19.16)$$

where $B$ = width of the dam at the joint,

$q$ = Average shear strength of the joint which varies from about 1400 kN/m$^2$ (14 kg/cm$^2$) for poor rocks to about 4000 kN/m$^2$ (40 kg/cm$^2$) for good rocks.

The value of $\mu$ generally varies from 0.65 to 0.75.

Attempts are always made to increase this shear strength ($q$) at the base and at other joints. For this purpose, foundation is stepped at the base, as shown in Fig. 19.10 and measures are taken to ensure a better bond between the dam base and the rock-foundation.

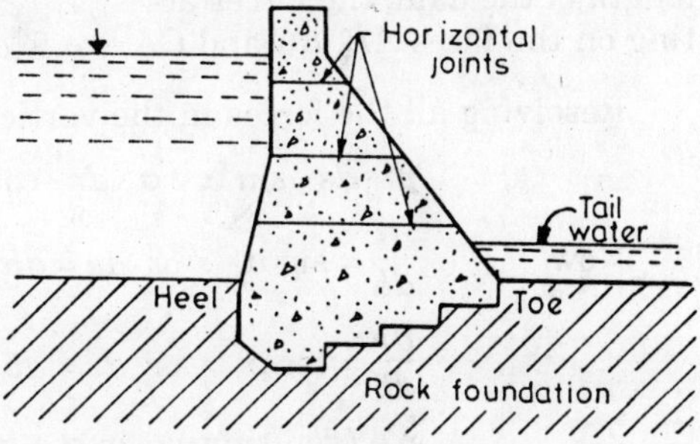

Fig. 19.10

During the construction of a dam, horizontal joints have to be left as shown in Fig. 19.10. The shear strength of these joints should be made as good as possible by ensuring better bond between the two surfaces. For this purpose, the lower surface must be thoroughly cleaned and a layer of neat cement or rich cement mortar should be spread before pouring the standard concrete mix for the upper layer. If these precautions of quality control are not adhered to in the filed, the assumption made in accounting for this shear strength in the design, will not be justified. That is why, for small dams, where quality control is less, this shear strength of the joint is not taken into account at all, while determining the shear friction factor or factor of safety against sliding.

**19.4.1. Principal and Shear Stresses.** The vertical stress intensity, $p_{max}$ or $p_{min}$ determined from the equation (19.15) is not the maximum direct stress produced anywhere in the dam. The maximum normal stress will, in fact, be the major principal stress that will be generated on the major principal plane. When the reservoir is full, the vertical direct stress [given by equation (19.15), and represented by $p_v$ in future] is maximum at the toe as the resultant is nearer to the toe. To study the principal stresses that will develop near the toe, let us consider a small element $ABC$ [See Fig. 19.11 ($a$) and ($b$)] near the toe of the dam. The element is so small that the stress intensities may be assumed to be uniform on its faces.

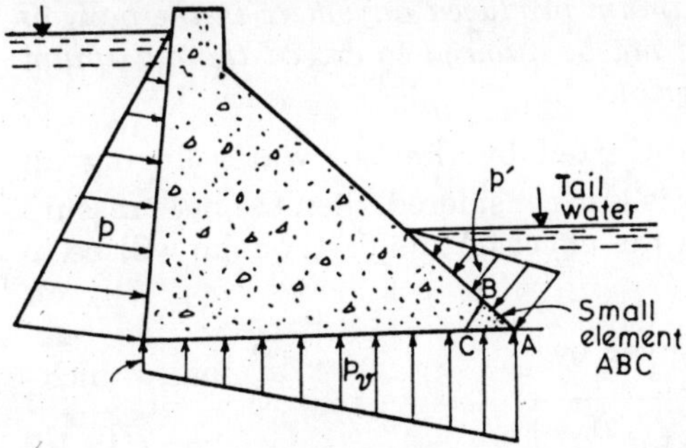

Fig. 19.11 ($a$)

Let the downstream face of the dam be inclined at an angle $\alpha$ to the vertical.

This face of the dam will act as a principal plane because the water pressure $p'$ acts at right angles to the face, and also there is no shear stress acting on this plane. Since the principal planes are at right angles to each other ; the plane $BC$ drawn at right angles to the face $AB$ will be the second principal plane. Let the stress acting on this plane be $\sigma$.

Let $ds$, $dr$ and $db$ be the lengths of $AB$, $BC$ and $CA$ respectively. $p'$ is the intensity of water pressure on face $AB$ and $p_v$ is the intensity of vertical pressure on face $AC$, and $\sigma$ is the intensity of normal stress (principal stress) on face $BC$. Considering unit length of the dam, the forces acting on the faces $AB$, $BC$ and $CA$ are $p'\,ds$, $\sigma\,dr$ and $p_v \cdot db$ respectively.

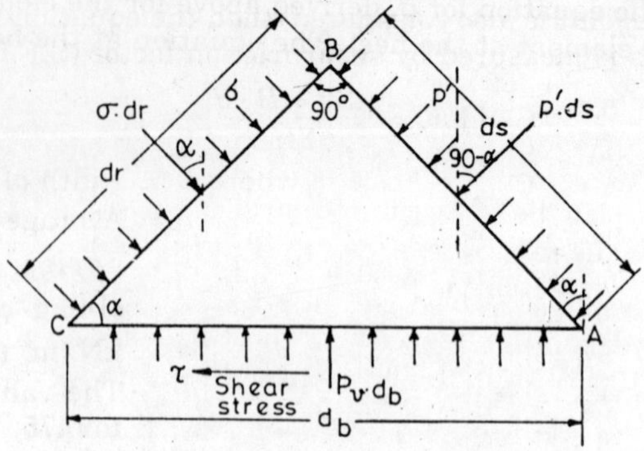


Fig. 19.11. (b) Enlarged view of small element $ABC$ of Fig. 19.11 (a).

Resolving all the forces in the vertical direction, we get

$$p' \cdot ds \cdot \sin\alpha + \sigma \cdot dr \cdot \cos\alpha = p_v \cdot db,$$

Now $$\frac{ds}{db} = \sin\alpha, \quad \text{or } ds = db \cdot \sin\alpha.$$

$$\frac{dr}{db} = \cos\alpha, \quad \text{or } dr = db \cdot \cos\alpha.$$

$\therefore$ $$p' \cdot (db \cdot \sin\alpha) \cdot \sin\alpha + \sigma \cdot (db \cdot \cos\alpha)\cos\alpha = p_v \cdot db$$

or $$p' \cdot \sin^2 a + \sigma \cdot \cos^2\alpha = p_v.$$

or $$\sigma = \frac{p_v - p' \cdot \sin^2\alpha}{\cos^2\alpha}$$

or $$\boxed{\sigma = p_v \cdot \sec^2\alpha - p'\tan^2\alpha} \quad ...(19.17)$$

For $\sigma$ to be maximum, $p'$ should be zero, *i.e.* when there is no tail water ; then in such a case

$$\sigma = p_v \cdot \sec^2\alpha \quad ...[19.17\ (a)]$$

Since $\sec^2\alpha$ is always more than 1, it follows, that $\sigma$ will be more than $p_v$. *This value of normal stress, which is the maximum produced anywhere in the body of the dam, must be calculated and should not be allowed to exceed the maximum allowable compressive stress of dam material.*

If the hydrodynamic pressure ($p_e'$) exerted by the tail water during an earthquake moving towards the reservoir is also considered, then the net pressure on the face $AB$ will be $(p' - p_e')$, because the effect of this earthquake will be to reduce the tail water pressure.

The principal stress ($\sigma$) can then be given by

or $$\boxed{\sigma_{at\ toe} = p_v \cdot \sec^2\alpha - (p' - p_e')\tan^2\alpha} \quad ...(19.18)$$

The equation for σ, derived above for the element at the toe is also applicable to the element at the heel. The equation at the heel is, therefore, given as :

$$\sigma_1 = \sigma_{at\ heel} = p_v \cdot \sec^2 \phi - (p + p_e) \tan^2 \phi \qquad ...(19.19)$$

where $\phi$ is the angle which the u/s face makes with vertical.

But at the heel, the pressure of water $p$ is always more than σ, and hence, $p$ will be the minor principal stress at the heel.

**Shear stress on the horizontal plane near the toe.** A shear stress τ will act on the face *CA* on which the vertical stress is acting. Resolving all the forces [Fig. 19.11 (*b*)] in the horizontal direction, we get

$$\sigma \cdot dr \sin \alpha - p' \cdot ds \cdot \cos \alpha = \tau_0 \cdot db$$

or $\sigma \cdot (db \cdot \cos \alpha) \sin \alpha - p' (db \cdot \sin \alpha) \cos \alpha = \tau_0 \cdot db$

or $\sigma \cdot \sin \alpha \cos \alpha - p' \sin \alpha \cos \alpha = \tau_0$

or $\tau_0 = (\sigma - p') \sin \alpha \cos \alpha$

Substituting the value of σ from equation (19.17), we get

$$\tau_0 = \left[\overline{p_v \sec^2 \alpha - p' \tan^2 \alpha} - p'\right] \sin \alpha \cos \alpha$$

$$\tau_0 = \left[p_v \sec^2 \alpha - p' (1 + \tan^2 \alpha)\right] \sin \alpha \cos \alpha = \left[(p_v - p') \sec^2 \alpha\right] \sin \alpha \cos \alpha$$

or $= [(p_v - p') \sec^2 \alpha \cdot \sin \alpha \cdot \cos \alpha]$

or $\boxed{\tau = (p_v - p') \tan \alpha}$ ...(19.20)

Neglecting tail water, shear stress is given by

$$\tau_0 = p_v \cdot \tan \alpha \qquad ...[19.20\ (a)]$$

If the effect of hydrodynamic pressure produced by an earthquake moving towards the reservoir, is also considered, the equation for shear stress on a horizontal plane near the toe becomes,

$$\tau_0 = \left[p_v - (p' - p_e')\right] \tan \alpha \qquad ...(19.21)$$

*Similarly*, shear stress at heel

$$= \tau_{0\ (heel)} = -\left[p_v - (p + p_e)\right] \tan \phi$$

–ve sign shows that the direction is reversed.

**Example 19.1.** *Fig. 19.12 (a) shows the section of a gravity dam (non-overflow portion) built of concrete.*

*Calculate (neglecting earthquake effects)*

(*i*) *The maximum vertical stresses at the heel and toe of the dam.*

(*ii*) *The major principal stress at the toe of the dam.*

(*iii*) *The intensity of shear stress on a horizontal plane near the toe.*

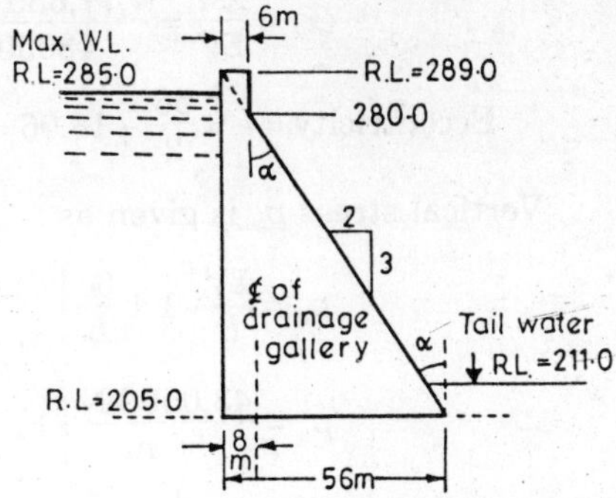


Fig. 19.12 (*a*)

*Assume weight of concrete = 23.5 kN/m³; and unit length of dam. Allowable stress in concrete may be taken 2500 kN/m²*

**Solution.** Assuming $\gamma_w = 9.81$ kN/m³ ; the various forces acting on the dam are drawn in Fig. 19.12 (*b*).

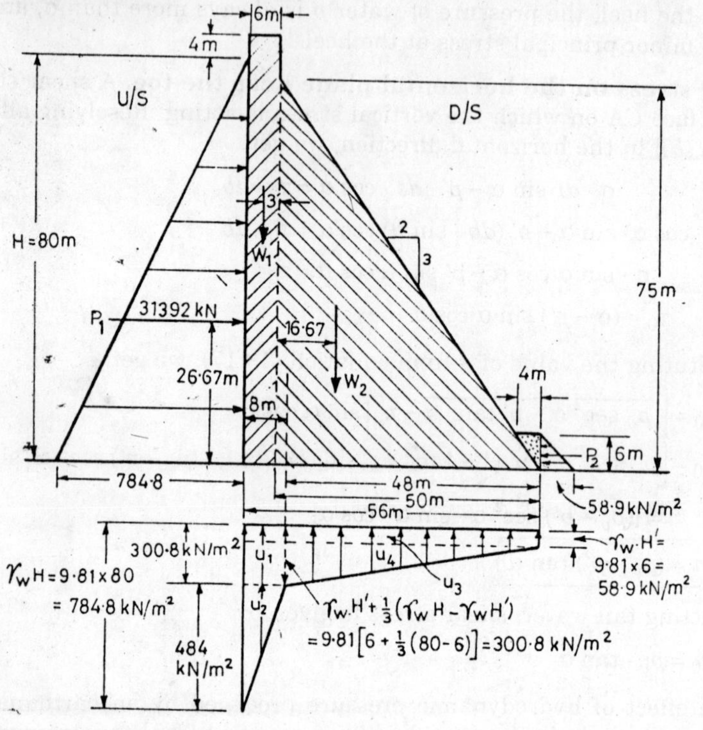


Fig. 19.12 (*b*)

Consider 1 m length of the dam.

The various forces and their moments about the toe are then calculated and tabulated in Table 19.1. From this table, we have

Distance of resultant from the toe

$$(\bar{x}) = \frac{\Sigma M}{\Sigma V} = \frac{7,77,639 \text{ kN} \cdot m}{43050 \text{ kN}} = 18.06 \text{ m}$$

$$\text{Eccentricity} = e = \frac{56}{2} - 18.06 = 28 - 18.06 = 9.94 \text{ m}$$

Vertical stress $p_v$ is given as :

$$p_v = \frac{\Sigma V}{B}\left[1 \pm \frac{6e}{B}\right]$$

$$\therefore \quad p_v = \frac{43,050 \text{ kN}}{56 \text{ m}}\left[1 \pm \frac{6 \times 9.94}{56}\right] = 768.8\ (1 \pm 1.065)$$

**Table 19.1**

| *Name of the force* | *Designation if given* | *Magnitude in kN* | *Lever arm in m* | *Moments about toe in kN.m* |
|---|---|---|---|---|
| **Vertical forces** | | | | |
| Downward weight of the dam | $W_1$ | (+) $84 \times 6 \times 1 \times 23.5 = 11{,}844$ | 53.0 | (+) 6,27,732 |
| | $W_2$ | (+) $\frac{1}{2} \times 50 \times 75 \times 1 \times 23.5 = 44{,}063$ | 33.33 | (+) 14,68,620 |
| Weight of water supported on d/s face | — | (+) $\frac{1}{2} \times 4 \times 6 \times 1 \times 9.81 = 118$ | 1.33 | (+) 157 |
| | | $\Sigma V_1 = 56{,}025$ | | $\Sigma M_1 =$ (+) 20,96,509 |
| Uplift pressures | $U_1$ | (–) $300.8 \times 8 \times 1 = 2406$ | 52.0 | (–) 1,25,112 |
| | $U_2$ | (–) $\frac{1}{2} \times 484 \times 8 \times 1 = 1936$ | 53.33 | (–) 1,03,247 |
| | $U_3$ | (–) $58.9 \times 48 \times 1 = 2827$ | 24.0 | (–) 67,848 |
| | $U_4$ | (–) $\frac{1}{2} \times 241.9 \times 48 \times 1 = 5806$ | 32.0 | (–) 1,85,792 |
| | | $\Sigma V_2 =$ (–) 12,975 | | $\Sigma M_2 =$ (–) 4,81,999 |
| | | $\Sigma V = 56{,}025 - 12{,}975 =$ **43050** | ,,, | |
| **Horizonal** *Water pressure* | | | ,,, | |
| On u/s face | $P$ | $\frac{1}{2} \times 784.8 \times 80 \times 1 = 31{,}392$ | 26.67 | (–) 8,37,225 |
| On d/s face | $P'$ | (–) $\frac{1}{2} \times 58.9 \times 6 =$ (–) 177 | 2.0 | (+) 354 |
| | | $\Sigma H$ (towards downstream) = 31,215 | | $\Sigma =$ (–) 8,36,871 |
| $\Sigma M =$ Net (+) moment = 20,96,509 – 4,81,999 – 8,36,871 = 7,77,639 kN-m | | | | |

∴ Max. vertical stress = $p_{max}$ at toe = 768.8 × 2.065 = **1587.6 kN/m²**

∴ Min. vertical stress = $p_{min}$ at heel = 768.8 × (–) 0.065 = (–) **49.97 kN/m²** **Ans.**

(*ii*) Major principal stress at toe (σ) is given by Eq. (19.17) as :

$$\sigma = p_{v\,(toe)} \sec^2 \alpha - p' \cdot \tan^2 \alpha$$

$$\text{here } p_{v\,(toe)} = 1587.6 \text{ kN/m}^2$$

$$p' = 58.9 \text{ kN/m}^2$$

$$\tan \alpha = 2/3$$

$$\sec \alpha^2 = 1 + \tan^2 \alpha = 1 + \frac{4}{9} = \frac{13}{9}$$

$$\therefore \quad \sigma = 1587.6 \times \frac{13}{9} = 58.9 \times \frac{4}{9}$$

$$= \mathbf{2267\ kN/m^2 < 2500\ kN/m^2\ (OK)} \quad \textbf{Ans.}$$

(*iii*) Intensity of shear stress on a horizontal plane near toe is given by equation (19.20)

$$\therefore \quad \tau_0 = \left[p_{v(toe)} - p'\right] \tan \alpha = 1587.6 - 58.9)\, \tfrac{2}{3} = \mathbf{1019.1\ kN/m^2}. \quad \textbf{Ans.}$$

## STABILITY ANALYSIS

The stability of a gravity dam can be approximately and easily analysed by two dimensional gravity method and can be precisely analysed by three dimensional methods such as slab analogy method, trial load twist method, or by experimental studies on models. Two dimensional gravity method is discussed below :

### 19.5. Gravity Method or Two Dimensional Stability Analysis

The preliminary analysis of all gravity dams can be made easily by isolating a typical cross-section of the dam of a unit width. This section is assumed to behave independently of the adjoining sections. In other words, the dam is considered to be made up of a number of cantilevers of unit width each, which act independently of each other. This assumption of independent functioning of each section, disregards the beam action in the dam as a whole.

If the vertical transverse joints of the dam are not grouted or keyed together, this assumption is nearly true. Hence, for wide *U*-shaped valleys, where transverse joints are not generally grouted, this assumption is nearly satisfied. But for narrow V- shaped valleys, where the transverse joints are generally keyed together and the entire length of the dam acts monolithically as a single body, this assumption may involve appreciable errors. In such cases, preliminary designs may be done by gravity method and precise final designs may be carried out by any of the available three dimensional methods.

The description of the three dimensional methods is beyond the scope of this book, and only the two dimensional analysis has been used for the design of gravity dams in this chapter.

**Assumptions.** The various assumptions made in the two dimensional designs of gravity dams are summarised below :

(*i*) The dam is considered to be composed of a number of cantilevers, each of which is 1 m thick and each of which acts independent of the other.

(*ii*) No loads are transferred to the abutments by beam action.

(*iii*) The foundation and the dam behave as a single unit ; the joint being perfect.

(*iv*) The materials in the foundation and body of the dam are isotropic and homogeneous.

(*v*) The stresses developed in the foundation and body of the dam are within elastic limits.

(*vi*) No movements of the foundations are caused due to transference of loads.

(*vii*) Small openings made in the body of the dam do not affect the general distribution of stresses and they only produce local effects as per St. Venant's principle.

**Procedure.** Two dimensional analysis can be carried out analytically or graphically.

(*a*) *Analytical Method.* The stability of the dam can be analysed in the following steps :

(*i*) Consider unit length of the dam.

(*ii*) Work out the magnitude and directions of all the vertical forces acting on the dam and their algebraic sum, *i.e.* $\Sigma V$.

(*iii*) Similarly, work out all the horizontal forces and their algebraic sum, *i.e.* $\Sigma H$.

(*iv*) Determine the lever arm of all these forces about the toe.

(*v*) Determine the moments of all these forces about the toe and find out the algebraic sum of all those moments, *i.e.* $\Sigma M$.

(*vi*) Find out the location of the resultant force by determining its distance from the toe, $\bar{x} = \dfrac{\Sigma M}{\Sigma V}$.

(*xii*) Find out the eccentricity ($e$) of the resultant ($R$) using $e = \dfrac{B}{2} - \bar{x}$. It must be less than $B/6$ in order to ensure that no tension is developed anywhere in the dam.

(*viii*) Determine the vertical stresses at the toe and heel using Eq. (19.15), *i.e.*

$$p_v = \frac{\Sigma V}{B}\left[1 \pm \frac{6e}{B}\right] \qquad \text{...(Eq. 19.15)}$$

*Sometimes stresses are found by ignoring uplift.*

(*ix*) Determine the maximum normal stresses, *i.e.* principal stresses at the toe and the heel using equations (19.18) to (19.20). They should not exceed the maximum allowable values. The crushing strength of concrete varies between 1500 to 3000 kN/m$^2$ depenting upon its grade M15 to M30.

(*x*) Determine the factor of safety against overturning as equal to $\dfrac{\Sigma \text{ Stabilising moment } (+)}{\Sigma \text{ Disturbing moment } (-)}$; +ve sign is used for anti-clockwise moments and –ve sign is used for clockwise moments.

(*xi*) Determine the factor of safety against sliding, using Sliding factor = $\dfrac{\mu \Sigma V}{\Sigma H}$.

Shear friction factor (S.F.F.) = $\dfrac{\mu \Sigma V + Bq}{\Sigma H}$.

Sliding factor must be greater than unity and S.F.F. must be greater than 3 to 5. The analysis should be carried out for reservoir full case as well as for reservoir empty case. The entire procedure has been illustrated in example 19.2.

(*b*) *Graphical method.* In the graphical method, the entire dam section is divided into a number of horizontal sections at some suitable intervals, particularly at the places where the slope changes, as shown in Fig. 19.13. For each section, the sum of the vertical forces ($\Sigma V$) and the sum of all the horizontal forces ($\Sigma H$) acting above that particular section, are worked out and the resultant force ($R$) is drawn, graphically. This is done for each section and a

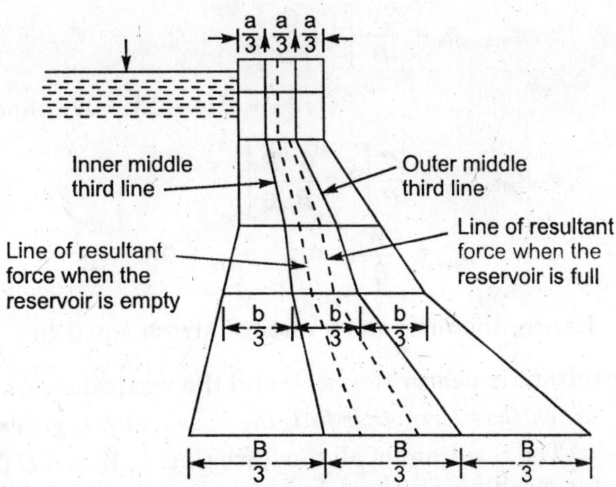


Fig. 19.13

line joining all the points where the individual resultants cut the individual sections, is drawn. This line represents the resultant force and should lie within the middle third, for no tension to develop. The procedure should be carried out for reservoir full case as well as for reservoir empty case. The resultant in both cases must show non-development of tension in the dam body.

## 19.6. Elementary Profile of a Gravity Dam

The elementary profile of a dam, subjected only to the external water pressure on the upstream side, will be a right-angled triangle, having zero width at the water level and a base width (*B*) at bottom *i.e.*, the point where the maximum hydrostatic water pressure acts. In other words, the shape of such a profile is similar to the shape of the hydrostatic pressure distribution (Fig. 19.14).

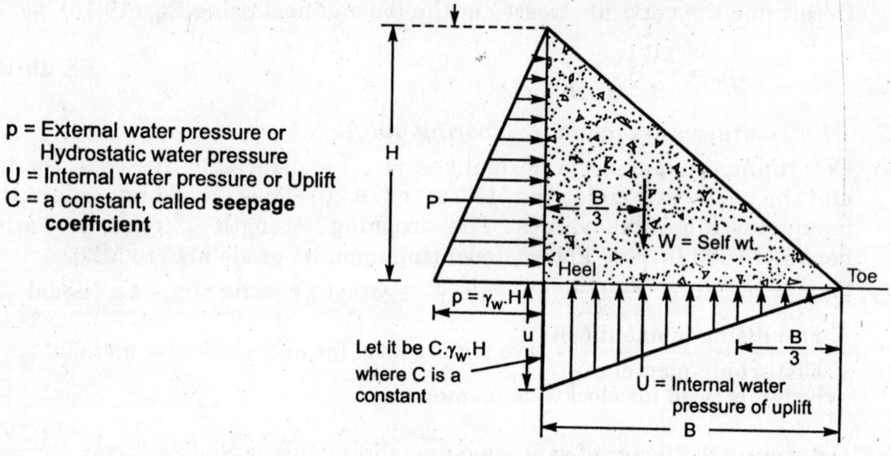


Fig. 19.14

When the reservoir is empty, the only single force acting on it is the self-weight (*W*) of the dam and it acts at a distance *B*/3 from the heel. This is the maximum possible innermost position of the resultant for no tension to develop. Hence, such a line of action of *W* is the most ideal, as it gives the maximum possible stabilising moment about the toe without causing tension at toe, when the reservoir is empty. *The vertical stress distribution at the base, when the reservoir is empty,* is given as :

$$p_{max/min} = \frac{\Sigma V}{B}\left[1 \pm \frac{6e}{B}\right]$$

Here $\Sigma V = W$, and $e = \frac{B}{6}$.

$$\therefore \quad p_{max/min} = \frac{W}{B}\left[1 \pm \frac{6.B}{B.6}\right]$$

or $$p_{max} = \frac{2W}{B} \quad \text{and} \quad p_{min} = 0.$$

Hence, the maximum vertical stress equal to $\frac{2W}{B}$ will act at the heel ($\because$ the resultant is nearer the heel) and the vertical stress at toe will be zero.

*When the reservoir is full, the base width is governed by* :

(*i*) The resultant of all the forces, *i.e. P, W* and *U* (Fig. 19.14) passes through the outer most middle third point (*i.e.* lower middle third point).

(*ii*) The dam is safe in sliding.

(*i*) For the 1st condition to be satisfied, we proceed as follows : Taking moments of all the forces (Fig. 19.14) about the lower middle third point (*i.e.* the point through which resultant is passing), we get

$$W\left(\frac{B}{3}\right) - U\left(\frac{B}{3}\right) - P\frac{H}{3} = R \times 0$$

or
$$(W - U)\,\frac{B}{3} - P\,\frac{H}{3} = 0$$

But
$$W = \frac{1}{2} \times B \times H \times 1 \times S_c \times \gamma_w$$

where $S_c$ = Sp. gravity of concrete, *i.e.* that of the material of the dam.

$\gamma_w$ = unit wt. of water = 9.81 kN/m$^3$

Let the uplift at the heel be $C \,.\, \gamma_w \,.\, H$, where $C$ is a constant, called the **seepage coefficient** which according to U.S.B.R. recommendation is taken equal to 1.0 in calculation and will be equal to zero when no uplift is considered.

$$\therefore \quad U = \left(\frac{1}{2} C \,.\, \gamma_w \,.\, H\right) B$$

and
$$P = \frac{1}{2}\,\gamma_w \,.\, H \,.\, H = \frac{\gamma_w H^2}{2}$$

$\therefore$ Equation $(W - U)\,\dfrac{B}{3} - P\dfrac{H}{3} = 0$, becomes

$$\left[\frac{1}{2} B \,.\, H \,.\, S_c \,.\, \gamma_w - \frac{1}{2} C \,.\, \gamma_w \,.\, H \,.\, B \,.\right]\frac{B}{3} - \frac{\gamma_w H^2}{2} \,.\, \frac{H}{3} = 0$$

or
$$\frac{B}{3} \times \frac{1}{2} B \,.\, H \,.\, \gamma_w \,.\, [S_c - C] = \frac{\gamma_w H^3}{6}$$

or
$$B^2 \,(S_c - C) = H^2$$

or
$$\boxed{B = \frac{H}{\sqrt{S_c - C}}} \qquad \text{...(19.22)}$$

Hence, if $B$ is taken equal to or greater than $\dfrac{H}{\sqrt{S_c - C}}$, no tension will be developed at the heel with full reservoir,

when $C = 1$

$$B = \frac{H}{\sqrt{S_c - 1}} \qquad \text{...[19.22 (a)]}$$

If uplift is not considered, $B = \dfrac{H}{\sqrt{S_c}}$ $(\because \; C = 0)$ ...[19.22 (b)]

(*ii*) For the II condition (*i.e.* dam is safe in sliding) to be satisfied ; the frictional resistance $\mu\Sigma V$ or $\mu\,(W - U)$ should be equal to or more than the horizontal forces $\Sigma H = P$.

or
$$\mu\,(W - U) \geq P$$

or
$$\mu\left(\frac{1}{2} BH \,.\, S_c \,.\, \gamma_w - \frac{1}{2} C \,.\, \gamma_w \,.\, H \,.\, B\right) \geq \frac{\gamma_w H^2}{2}$$

or $$\mu (S_c - C) \frac{1}{2} \cdot B \cdot H \cdot \gamma_w \geq \frac{\gamma_w H^2}{2}$$

or $$\mu (S_c - C) B \geq H$$

or $$B \geq \frac{H}{\mu (S_c - C)}$$

Under limiting condition

or $$\boxed{B = \frac{H}{\mu (S_c - C)}} \quad \text{...(19.23)}$$

If $C = 1$ $$B = \frac{H}{\mu (S_c - 1)} \quad \text{...[19.23(a)]}$$

If $C = 0$, *i.e.* no uplift is considered, then

$$B \geq \frac{H}{\mu S_c} \quad \text{...[19.23(b)]}$$

The value of $B$ chosen should be greater of the two values given by Equations (19.22) and (19.23).

Using $S_c = 2.4$ and $\mu = 0.7$ and $C = 0$, we get

$$B \text{ (by Equation 19.22)} = \frac{H}{\sqrt{2.4 - 0}} = \frac{H}{\sqrt{2.4}}$$

$$B \text{ (by Equation 19.23)} = \frac{H}{0.7 \times (2.4 - 0)} = \frac{H}{0.7 \times 2.4} = \frac{H}{1.68}$$

But $\frac{H}{1.68}$ is less than $\frac{H}{\sqrt{2.4}}$

∴ For all practical purposes, the base width may be taken as $\frac{H}{\sqrt{S_c}}$

*The vertical stress distribution when reservoir is full is given as :*

$$p_{max/min} = \frac{\Sigma V}{B}\left[1 \pm \frac{6e}{B}\right]$$

$$\text{where } \Sigma V = W - U$$
$$= \left(\frac{1}{2} B \cdot H \cdot 1 \cdot S_c \cdot \gamma_w - \frac{1}{2} C \cdot \gamma_w \cdot H \cdot B\right)$$
$$= \frac{1}{2} B \cdot \gamma_w \cdot H \cdot [S_c - C]$$
$$e = B/6$$

∴ $$p_{max/min} = \frac{\frac{1}{2} \cdot B \cdot \gamma_w \cdot H (S_c - C)}{B}\left[1 \pm \frac{6B}{6B}\right]$$

maximum stress will occur at toe, because the resultant is near the toe.

∴ $$p_{max} \text{ at toe} = \frac{1}{2} \gamma_w \cdot H (S_c - C)\, 2.0 = \gamma_w H (S_c - C)$$

$$\boxed{p_v \text{ at toe} = \gamma_w H (S_c - C)} \quad \text{...(19.24)}$$

$$P_{min} \text{ at heel} = 0.$$

The principal stress near the toe (σ) which is the maximum normal stress in the dam, is given by Equation (19.17)

$$\sigma = p_v \sec^2 \alpha - p' \tan^2 \alpha$$

when there is no tail water *i.e.*, $p' = 0$

$$\sigma = p_v \sec^2 \alpha$$

σ at toe, with full reservoir in elementary profile

$$= \gamma_w H (S_c - C) \sec^2 \alpha = \gamma_w H (S_c - C) [1 + \tan^2 \alpha]$$

$$= \gamma_w H (S_c - C) \left[1 + \frac{B^2}{H^2}\right]$$

$$\text{But } B = \frac{H}{\sqrt{S_c - C}} \text{ from Eq. (19.22)}$$

$$\text{or } \frac{B^2}{H^2} = \frac{1}{S_c - C}$$

$$\therefore \quad \sigma = \gamma_w H (S_c - C) \left[1 + \frac{1}{S_c - C}\right]$$

$$\text{or} \quad \boxed{\sigma = \gamma_w H (Sc - C + 1)} \quad \text{...(19.25)}$$

when $C = 1$, $S_c = 2.4$

$$\sigma = \gamma_w H (2.4 - 1 + 1) = \mathbf{2.4\, \gamma_w\, H.}$$

The shear stress $\tau_0$ at a horizontal plane near the toe is given by the equation (19.20) as :

$$\tau_0 = (p_v - p') \tan \alpha$$

$$\text{If} \quad p' = 0$$

$$\tau_0 = p_v \tan \alpha$$

$$\text{But } p_v = \gamma_w H (Sc - C) \quad \text{from Eq. (19.24)}$$

$$\therefore \quad \tau_0 = \gamma_w H (Sc - C) \tan \alpha$$

$$\text{or} \quad \tau_0 = \gamma_w \cdot H (S_c - C) \frac{B}{H}$$

$$= \gamma_w H (S_c - C) \frac{1}{\sqrt{S_c - C}}$$

$$\text{or} \quad \boxed{\tau_0 = \gamma_w H \sqrt{S_c - C}} \quad \text{...(19.26)}$$

When $C = 0$ and $S_c = 2.4$

$$\tau_0 = \gamma_w \cdot H \cdot \sqrt{2.4 - 1} = 1.18 \cdot \gamma_w \cdot H$$

## 19.7. High and Low Gravity Dams

The principal stress calculated for an elementary profile is given by Equation (19.25), *i.e.* $\sigma = \gamma_w H (S_c - C + 1)$. The value of principal stress calculated above varies only with $H$, as all other factors are fixed.

To avoid dam failure by crushing, the value of σ should be less than or at the most equal to the maximum allowable compressive stress of dam material. If $f$ represents the allowable stress of the dam material, then the maximum height

($H_{max}$) which can be obtained in an elementary profile, without exceeding the allowable compressive stresses of the dam material, is given as :

$$f = \gamma_w H (S_c - C + 1)$$

or
$$H = \frac{f}{\gamma_w (S_c - C + 1)}$$

The lowest value of $H$ will be obtained when $C = 0$, *i.e.* when uplift is neglected. Hence, for determining the limiting height and to be on a safer side, uplift is neglected.

$H_{max}$, *i.e.* maximum possible height is given as :

$$H_{max} = \frac{f}{\gamma_w (S_c + 1)} \quad \text{...(19.27)}$$

Hence, if the height of a dam having an elementary profile of a triangle, is more than that given by the Equation (19.27), the maximum compressive stress generated will exceed the allowable value. In order to keep it safe within limits, extra slopes on the upstream as well as on the downstream, below the limiting height will have to be given, as shown in Fig. 19.15.

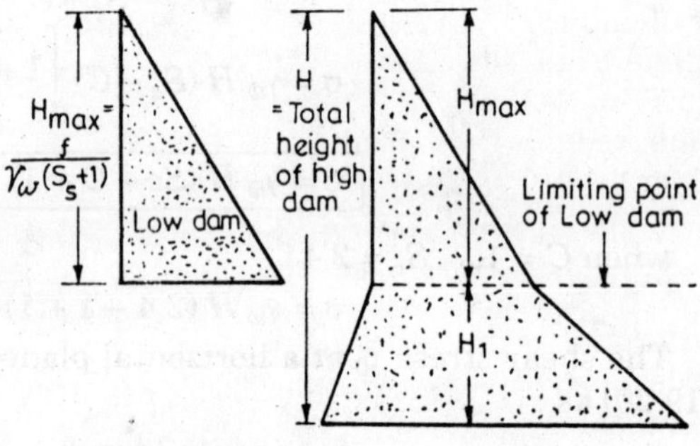


Low gravity dam High gravity dam

Fig. 19.15

This limiting height ($H_{max}$) given by Equation (19.27), draws a dividing line between a low gravity dam and a high gravity dram, which are purely technical terms to differentiate between them.

Hence, a low gravity dam is the one whose height is less than that given by Equation (19.27). If the height of the dam is more than this, it is known as a high gravity dam.

The limiting height of a low concrete gravity dam, constructed in concrete having strength equal to 3000 kN/m$^2$ is thus given :

$$H_{max} = \frac{f}{\gamma_w (S_c + 1)}$$

where $\gamma_w = 9.81$ kN/m$^3$

$S_c = 2.4$

$f = 3000$ kN/m$^2$

$$\therefore \quad H_{max} = \frac{f}{\gamma_w (S_c + 1)} = \frac{3000}{9.81 (2.4 + 1)} = \mathbf{90\ m}$$

## 19.8. Profile of a Dam from Practical Considerations

The elementary profile of a gravity dam, (*i.e.* a triangle with maximum water surface at apex) is only a theoretical profile. Certain changes will have to be made in this profile in order to cater to the practical needs. These needs are : (*i*) providing a straight top width, for a road construction over the top of the dam ; (*ii*) providing a free-board above the top water surface, so that water may not spill over the top of the dam due to wave action, etc.

The additions of these two provisions, will cause the resultant force to shift towards the heel. The resultant force, when the reservoir is empty, was earlier passing through the inner middle third point. This will, therefore, shift more towards the heel, crossing the inner middle third point and consequently, tension will be developed at the toe. In order to avoid the development of this tension, some masonry or concrete will have to be added to the upstream side, as shown in Fig. 19.16, which shows the typical section along with the possible dimensions that can be adopted for a low gravity dam section. It should, however, be checked for stability analysis.

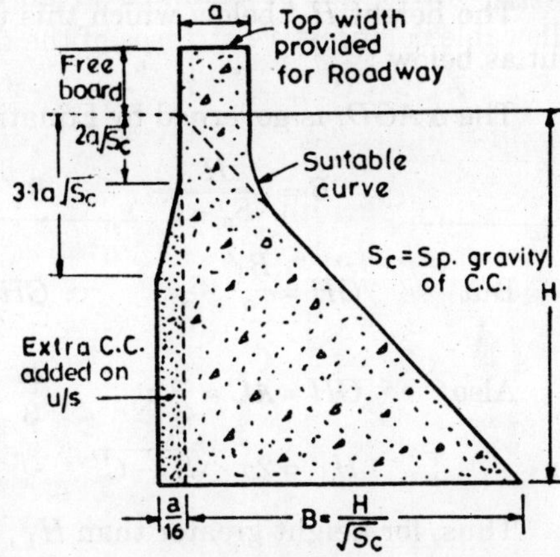


Fig. 19.16.
Typical section of a low gravity dam.

## 19.9. Design Considerations and Fixing the Section of a Dam

The free-board and top width for roadway should be selected as follows :

(*i*) **Freeboard.** The margin between the maximum reservoir level and top of the dam is known as freeboard. This must be provided in order to avoid the possibility of water spilling over the dam top due to wave action. This can also help as a safety for unforeseen floods, higher than the designed flood.

The freeboard is generally provided equal to $\frac{3}{2}h_w$. where $h_w$ is given by Equations (19.11) and (19.12). However, these days, a *free-board equal to 4 to 5% of the dam height is provided.*

(*ii*) **Top width.** The effects produced by the addition of top width at the apex of an elementary dam profile, and their remedies are explained below :

In Fig. 19.17, let *AEF* be the triangular profile of dam of height $H_1$. Let the element *ABQA* be added at the apex for providing a top width *a* for a road construction. Let $M_1$ and $M_2$ be the inner third and outer third points on base. Thus, $AM_1$ and $AM_2$ are the inner third and outer third lines. The weight of the element ($W_1$) will act through the *e.g.* of this triangle, *i.e.* along *C.M.* Let *CM* and $AM_1$ cross at *H*, and *CM* and $AM_2$ cross at *K*.

*Reservoir Empty Case.* We know that in the elementary profile, the resultant of the forces passes through the inner third point when the reservoir is empty. In such a case, the addition of $W_1$ above the plane *GHD* will shift this initial resultant towards the downstream side because $W_1$ lies downstream to the earlier resultant. Similarly, the addition of $\Delta W_1$ 'below the plane *GHD* will shift this initial resultant towards the upstream side, causing tension on the downstream. Hence, an upstream batter *GF* ' will have to be added below the plane *GHD* as shown in Fig. 19.17.

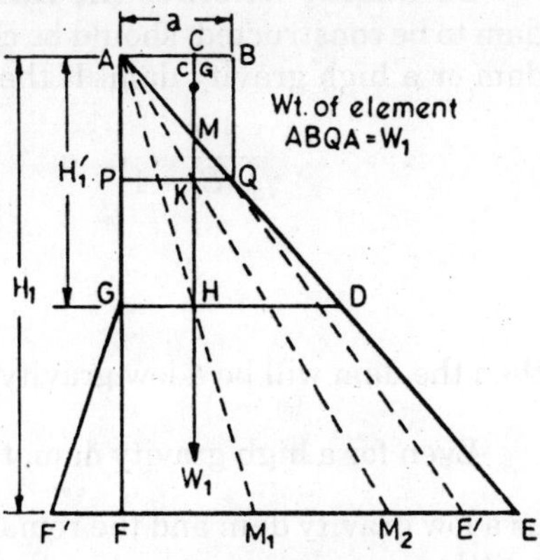


Fig. 19.17

The height $H_1'$ below which this upstream batter is required can be worked out as below :

The $\Delta AGD$, is governed by Equation (19.22), *i.e.*

$$B = \frac{H}{\sqrt{S_c - C}} \qquad \therefore\ B_1' = \frac{H_1'}{\sqrt{S_c - C}}$$

But $$GH = \frac{B_1'}{3} \qquad \therefore\ GH = \frac{B_1'}{3} = \frac{H_1'}{3\sqrt{S_c - C}}$$

Also $$GH = AC = \frac{2}{3}a \qquad \therefore\ \frac{2a}{3} = \frac{H_1'}{3 \cdot \sqrt{S_c - C}}$$

or $$H_1' = 2a \cdot \sqrt{S_c - C} \qquad ...(19.28)$$

Thus, for height greater than $H_1'$, u/s batter is necessary.

*Reservoir full case.* When the reservoir is full, the resultant of all the forces acting on the elementary profile passes through the outer third point. When $W_1$ is added to this initial resultant at any plane below the plane *PKQ*, final resultant will shift towards the upstream side because $W_1$ lies upstream of the initial resultant. In order to bring the resultant back to the outer third point from economy point of view, the slope of the d/s face may be flattened from *QE* to *QE'*.

Thus, an increases in top width, will increase the masonry in the added element and increase it on u/s face, but shall reduce it on d/s face. The most economical top width, without considering earthquake forces has been found by Creager to be equal to 14% of the dam height. Its useful value varies between 6 to 10 m and is generally taken approximately equal to $\sqrt{H}$, where $H$ is the height of max. water level above the bed.

## 19.10. Design of Gravity Dams

The section of gravity dam should be chosen in such a way that it is the most economical section and satisfies all the conditions and requirements of stability. Hence, after the section of the dam has been arrived at, the stability analysis for the dam must be carried out.

**To Decide whether the Dam is Low or High.** First of all, the height of the dam to be constructed, should be checked so as to ensure whether it is a low gravity dam or a high gravity dam. If the height of the dam is less than that given by

$$\frac{f}{\gamma_w (S_c + 1)}$$

(where $f$ is the permissible compressive stress of the dam material and $S_c$ is the Sp. gravity of the dam material) ;

then the dam will be a low gravity dam, otherwise it will be a high gravity dam.

Even for a high gravity dam, the upper height $H_1 = \dfrac{f}{\gamma_w (S_c + 1)}$ can be designed as a low gravity dam and the remaining lower portion can be designed as explained a little later.

**Design of Low Dam.** The economical section of low gravity dam of height $H_1$ after deciding the top width $a$ and freeboard, can be drawn as shown in Fig. 19.18.

The base width $B_1$ of the $\Delta AIJ$ can be chosen as given by Equation (19.22) as :

$$B_1 = \frac{H_1}{\sqrt{S_c - C}}$$

or by Equation (19.23), as : $B_1 = \dfrac{H_1}{\sqrt[\mu]{S_c - C}}$.

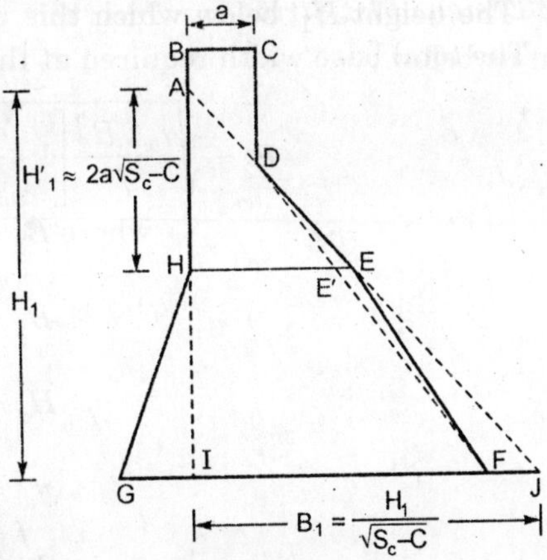


Fig. 19.18.

The upstream face can be kept vertical up to a height $H_1'$ to be determined by trial, and whose approximate 1st value may be chosen by Equation (19.28), as :

$$H_1' = 2a \cdot \sqrt{S_c - C}.$$

Below this height $H_1'$, the upstream face as well as the downstream face are sloped in such a manner that no tension is developed any-where in the dam, and the resultant forces remain as close to the outer third and inner third points as possible, for reservoir full and reservoir empty cases, respectively. This is accomplished by hit and trial method, and when it is so accomplished, all the stability requirements will be satisfied. The final shape of the dam will then be *ABCDEFGHA*.

The d/s face can be brought inward from the point *D* itself rather than from the point *E* (explained earlier). In such a case, the final dam shape will be *ABCDE′FGHA*. Sometimes, in practical conservative designs, the d/s slope is not at all brought inward and is kept the same as is obtained in an elementary profile. The final dam shape, in such a case, will be *ABCDEJGHA*.

**Note.** However, in most of the practical conservative designs, the low gravity dam is designed as per the provision of Fig. 19.16 and the d/s slope is not brought inward.

**Design of High Dam.** When the height of the dam exceeds $H_1$ given by $\dfrac{f}{\gamma_w(S_c - 1)}$; then its upper height equal to $H_1$ can be designed as low gravity dam as explained earlier, and its remaining height can be designed by dividing it into a number of suitable strips as shown in Fig. 19.19. The design of each strip can be carried out as per the formulas given below. The basis of these formula is that, the maximum normal stress (*i.e.* principal stress) should not exceed the allowable value (*f*), and at the same time, the section should be as economical as possible.

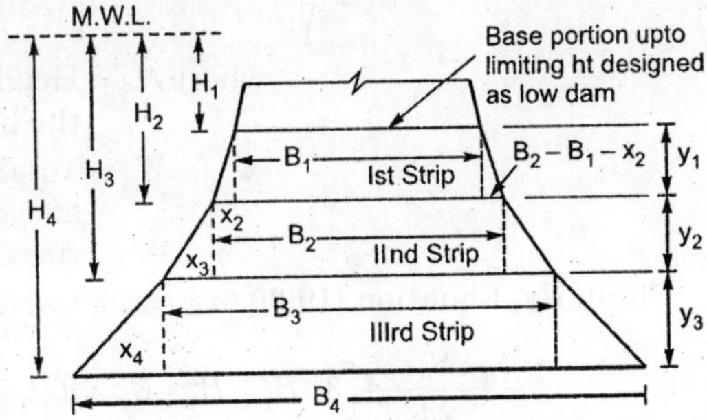


Fig. 19.19.

*Design of Ist Strip*

The total base width required at the bottom of the Ist strip ($B_2$) is given by

$$B_2 = \sqrt{\frac{\gamma_w \,.\, H_2^3}{f}\left[1 + \frac{\gamma_w^2 \,.\, H_2^4}{4 \,.\, W_2^2}\right]} \qquad \text{...[19.29}(a)]$$

where $B_1$ = Base width of low dam, *i.e.* the base width at top of Ist strip in m

$B_2$ = Base width required at the bottom of Ist strip in m.

$H_2$ = Height of dam portion from M.W.L. to the bottom of Ist strip in m.

$\gamma_w$ = Unit weight of water in kN/m³

$f$ = Allowable compressive stress of the dam material in kN/m³

$W_1$ = Total vertical weight of dam and water, above the top of Ist strip in kN/m run.

$W_2$ = Total vertical wt. of dam and water above the bottom of Ist strip in kN/m run.

The increase of base width required on the upstream side, at the bottom of Ist strip (say $X_2$), is given by the equation

$$\frac{\gamma_w \,.\, S_c \cdot y_1}{24}\left[3 \,.\, B_1^2 - B_2^2 + 6 \,.\, X_2(B_1 + B_2) + 2B_1B_2\right]$$

$$- \frac{\gamma_w \,.\, X_2}{12}\left[H_1 + H_2\right]\left[2 \,.\, B_2 - 3X_2\right] - W_1\left[\frac{B_2 - B_1}{3} - X_2\right] = 0 \qquad \text{...[19.30}(a)]$$

The increase in width required on u/s, *i.e.* $X_2$ can be determined from this equation. Total width $B_2$ is already known from Eq. 19.29($a$). Hence, the increase required on d/s can be calculated as ($B_2 - B_1 - X_2$).

*Design of IInd Strip*

Equation [19.29 ($a$)] for the Ist strip can be changed to Equation [19.29($b$)] for the IInd strip by changing suffixes as below

$$B_3 = \sqrt{\frac{\gamma_w \,.\, H_3^3}{f}\left[1 + \frac{\gamma_w^2 \,.\, H_3^4}{4\, W_3^2}\right]} \qquad \text{...[19.29 }(b)]$$

where $H_3$ = Height of the dam portion from M.W.L. to the bottom of the IInd strip in m.

$W_3$ = Total vertical wt. of the dam and water above the bottom of IInd strip in kN/m run.

Similarly, Equation [19.30 ($a$)] can be written as

$$\frac{\gamma_w \,.\, S_c \cdot y_2}{24}\left[3 \,.\, B_2^2 - B_3^2 + 6 \,.\, X_3(B_2 + B_3) + 2 \cdot B_2B_3\right]$$

$$- \frac{\gamma_w \,.\, X_3}{12}\left[H_2 + H_3\right]\left[2 \,.\, B_3 - 3X_3\right] - W_2\left[\frac{B_3 - B_2}{3} - X_3\right] = 0 \qquad \text{...[19.30 }(b)]$$

Total width $B_3$ required is known from Equation [19.29 (*b*)] and the increase required on u/s ($X_3$) is also known from equation [19.30 (*b*)]. The increase required on d/s can be worked out as equal to ($B_3 - B_2 - X_3$).

All the remaining strips are designed in this way till the base of the dam is reached.

**Example 19.2.** *Fig. 19.20 (a) shows the section of a gravity dam built of concrete. Examine the stability of this section at the base.*

*The earthquake forces may be taken as equivalent to 0.1 g for horizontal forces and 0.05 g for vertical forces. The uplift may be taken as equal to the hydrostatic pressure at the either ends and is considered to act over 60% of the area of the section.*

*A tail water depth of 6 m is assumed to be present when the reservoir is full and there is no tail water when the reservoir is empty.*

*Also indicate the values of various kinds of stresses that are developed at heel and toe. Assume the unit wt. of concrete as 24 kN/m³ ; and unit wt. of water = 10 kN/m³.*

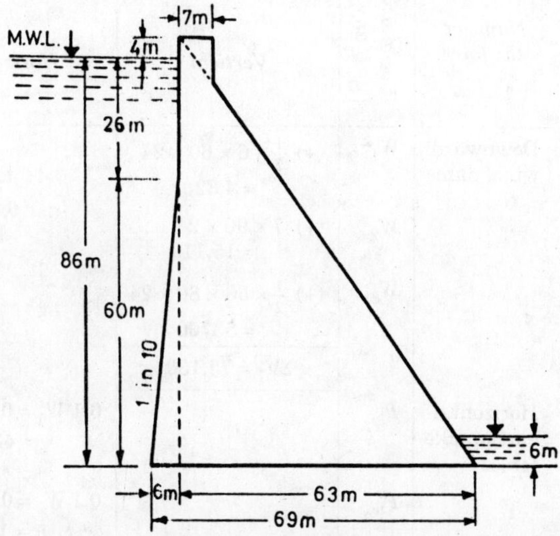


Fig. 19.20 (*a*)

**Solution.** The stability analysis shall be carried out for both the cases, *i.e.* (1) Reservoir Empty, and (2) Reservoir Full.

**Case (I) Reservoir Empty.** Consider 1 m length of the dam.

When the reservoir is empty, the various forces are worked out in Table 19.2 (*a*) with reference to Fig. 19.20 (*b*). Horizontal earthquake forces acting towards

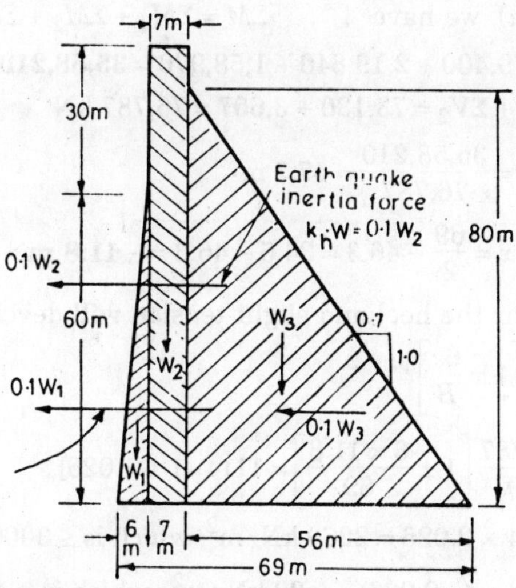


Fig. 19.20 (*b*). Reservoir empty case.

upstream are considered. Stability is examined for two sub-cases, *i.e.* (*a*) When vertical earthquake forces are additive to the weight of the dam ; (*b*). When vertical earthquake forces are subtractive to the dam weight.

**Table 19.2 (*a*)**

| *Name of the force* | *Designation if any* | *Magnitude of force in kN.* Vertical | Horizontal | *Lever arm m* | *Moments about the toe anti-clockwise (+ve) in kN.m.* |
|---|---|---|---|---|---|
| Downward wt. of dam | $W_1$ | (+) $\frac{1}{2} \times 6 \times 60 \times 24 = 4{,}320$ | | 65.0 | (+) 2,80,400 |
| | $W_2$ | (+) $7 \times 90 \times 24 = 15{,}110$ | | 59.5 | (+) 8,99,000 |
| | $W_3$ | (+) $\frac{1}{2} \times 56 \times 80 \times 24 = 53700$ | | 37.33 | (+) 20,00,000 |
| | | $\Sigma V_1 = 73{,}130$ | | | $\Sigma M_1 = (+)\ 31{,}79{,}400$ |
| Horizontal earthquake forces | $P_{w_1}$ | | $0.1\ W_1 = 0.1 \times 4320 = 432$ | 20.0 | (+) 8640 |
| | $P_{w_2}$ | | $0.1\ W_2 = 0.1 \times 1{,}511 = 1511$ | 45.0 | (+) 68000 |
| | $P_{W_3}$ | | $0.1\ W_3 = 0.1 \times 5{,}370 = 5370$ | 26.67 | (+) 1,43,200 |
| | | | $\Sigma H = 7313$ | | $\Sigma M_2 = 2{,}19{,}840$ |
| Vertical earthquake forces | | $\Sigma V_2 = 0.05 \times \Sigma V_1 = 0.05 \times 73130 = 3{,}657$ | | | $\Sigma M_2 = 0.05 \times \Sigma M_1 = 0.05 \times 31{,}79{,}400 = 1{,}58{,}970$ |

**Case (I).** (*a*) ***Reservoir empty and vertical earthquake forces are acting downward.***

From table 19.2 (*a*), we have $\Sigma M = \Sigma M_1 + \Sigma M_2 + \Sigma M_3$

$= 31{,}79{,}400 + 2{,}19{,}840 + 1{,}58{,}970 = \mathbf{35{,}58{,}210\ kN \cdot m}$

Also, $\Sigma V = \Sigma V_1 + \Sigma V_2 = 73{,}130 + 3{,}657 = 76{,}787$ kN

$$\bar{x} = \frac{\Sigma M}{\Sigma V} = \frac{35{,}58{,}210}{76{,}787} = 47.3 \text{ m}$$

$$e = \frac{B}{2} - \bar{x} = \frac{69}{2} - 46.3 = 34.5 - 46.3 = -\mathbf{11.8\ m} > \frac{B}{6}, \textit{ i.e. } 11.5 \text{ m.}$$

Resultant acts near the heel and slight tension will develop at toe.

$$p_{max/min} = \frac{\Sigma V}{B}\left[1 \pm \frac{6e}{B}\right]$$

$$\therefore\ p_{max/min} = \frac{76{,}787}{69}\left[1 \pm \frac{6 \times 11.8}{69}\right] = 1114\ [1 \pm 1.026]$$

$p_v$ at heel $= 1114 \times 2.026 = 2260$ kN/m$^2$; which is $\leq 3000$ (safe)

$p_v$ at toe $= 1114 \times (-0.026) = -29$ kN/m$^2$ ; which is $< 420$ (safe)

Average vertical stress

$$= \frac{\Sigma V}{B} = \frac{76787}{69} = 1114 \text{ kN/m}^2 \text{ ; which is} < 3000 \text{ (safe)}$$

Principal stress at toe,

$$\sigma = p_v \sec^2 \alpha \text{ ; } (\tan \alpha = 0.7)$$

$$= -29\,(1 + 0.49) = -29 \times 1.49 = -43 \text{ kN/m}^2 \text{ ; which is} < 420 \text{ (safe)}$$

Principle stress at heel

$$\sigma_1 = p_{v \cdot (heel)} \sec^2 \phi \quad \text{where } \tan \phi = 0.1$$

or $$\sec^2 \phi = 1 + \tan^2 \phi = 1 + 0.01 = 1.01.$$

or $$\sigma_1 = 2260 \times 1.01 = \mathbf{2280 \text{ kN/m}^2} \text{ ; which is} < 3000 \text{ (safe)}.$$

Shear stress at toe

$$\tau_{0(toe)} = p_{v(toe)} \tan \alpha$$

$$= -29 \times 0.7 = -20.3 \text{ kN/m}^2 \text{ ; which is} < 420 \text{ (safe)}$$

Shear stress at heel

$$\tau_{0(heel)} = p_{v \cdot (heel)} \tan \phi$$

$$= 2260 \times 0.1 = 226 \text{ kN/m}^2 \text{ ; which is} < 3000 \text{ (safe)}.$$

**Case I.** (*b*) *Reservoir empty and vertical earthquake forces are acting upward.*

Then $$\Sigma V = \Sigma V_1 - \Sigma V_3$$

$$= 73{,}130 - 3657 = 69473 \text{ kN}$$

$$\Sigma M = \Sigma M_1 + \Sigma M_2 - \Sigma M_3$$

$$= 31{,}79{,}400 + 2{,}19{,}840 - 1{,}58{,}970 = 32{,}40{,}270 \text{ kN} \cdot \text{m}.$$

$$\bar{x} = \frac{\Sigma M}{\Sigma V} = \frac{32{,}40{,}270}{69473} = \mathbf{46.7 \text{ m.}}$$

$$e = \frac{B}{2} - \bar{x} = 34.5 - 46.7 = (-)\,12.2 \text{ m} < \frac{B}{6}$$

**[– ve sign shows that resultant lies near the heel and, therefore, tension will develop at toe.]**

Average vertical stress

$$= \frac{\Sigma V}{B} = \frac{69{,}473}{69} = 1004 \text{ kN/m}^2$$

$$p_{max/min} = \frac{\Sigma V}{B}\left[1 \pm \frac{6e}{B}\right]$$

$$= \frac{69473}{69}\left[1 \pm \frac{6 \times 12.2}{69}\right] = 1004\,[1 \pm 1.06]$$

$$p_v \text{ at heel} = 1004 \times 2.06 = 2070 \text{ kN/m}^2 < 3000 \text{ (safe)}$$

$$p_v \text{ at toe} = (-)\,1004 \times 0.06 = -60.3 \text{ kN/m}^2 < 420 \text{ (safe)}$$

Principal stress at toe

$$= \sigma = p_{v(toe)} \sec^2 \alpha$$

$$= -60.3\,(1 + 0.49) = -60.3 \times 1.49 = 90 \text{ kN/m}^2$$

Shear stress at toe

$= \tau_0 = p_{v(toe)} \tan \alpha = -60.3 \times 0.7$

$= \mathbf{-42.21\ kN/m^2}$ ; which is < 420 (safe)

stresses at heel remain critical in this 1st case.

**Case II. When the reservoir is full**

Horizontal earthquake moving towards the reservoir causing upstream acceleration, and thus producing horizontal forces towards downstream is considered, as it is the worst case for this condition. Similarly, a vertical earthquake moving downward and thus, producing forces upward, *i.e.* subtractive to the weight of the dam is considered.

The uplift coefficient $C$ is taken as equal to 0.6, as given in the equation, and thus uplift pressure diagram as shown in Fig. 19.20 (*c*), is developed.

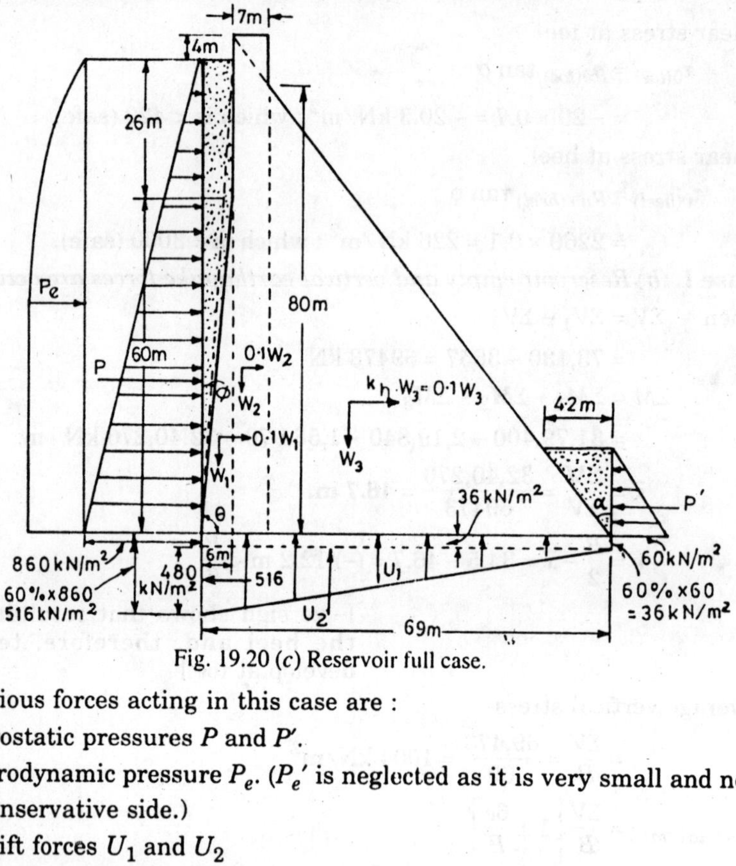


Fig. 19.20 (*c*) Reservoir full case.

The various forces acting in this case are :

(*i*) Hydrostatic pressures $P$ and $P'$.

(*ii*) Hydrodynamic pressure $P_e$. ($P_e'$ is neglected as it is very small and neglection is on conservative side.)

(*iii*) Uplift forces $U_1$ and $U_2$

(*iv*) Weight of the dam, $W_1$, $W_2$ and $W_3$.

(*v*) Horizontal inertial earthquake forces acting towards downstream, equal to 0.1 $W_1$, 0.1 $W_2$ and 0.1 $W_3$ at *c.gs.* of these weights $W_1$, $W_2$ and $W_3$ respectively.

(*vi*) A vertical force equal to 0.05 $W$ or (0.05 $\Sigma V_1$) acting upward.

**Calculation of $P_e$**

$P_e$ and the moment due to this hydrodynamic force is calculated, and then all the forces and their moments are tabulated in Table 19.2 (*b*).

**Calculation of $P_e$ from Zanger's formulas**

$$P_e = 0.726\, p_e\, H \quad \text{...(19.3)}$$

$$\text{where } p_e = C_m \cdot K_h \cdot \gamma_w \cdot H \quad \text{...(19.4)}$$

$$\text{and } \quad C_m = 0.735 \frac{\theta}{90^\circ}$$

Since the u/s inclined face is extended for more than half the depth, the overall slope up to the whole height may be taken.

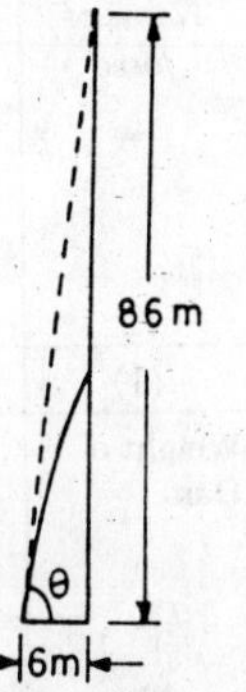

Fig. 19.20 (*d*)

$$\therefore \quad \tan\theta = \frac{86}{6} = 14.33$$

$$\theta = 81.9^\circ$$

$$\therefore \quad C_m = 0.735 \times \frac{81.9^\circ}{90^\circ} = 0.668.$$

$$p_e = 0.668 \times 0.1 \times 10 \times 86 = 57.5$$

$$P_e = 0.726 \times 57.5 \times 86 = \mathbf{3580\ kN.}$$

$$M_e = 0.412 \cdot P_e \cdot H = 0.412 \times 3580 \times 86 = \mathbf{1{,}26{,}500\ kN.m.}$$

**Case 2 (*a*) Reservoir full with all forces including uplift**

$$\Sigma M = [31{,}79{,}400 + 2{,}23{,}380 - 8{,}47{,}500 - 1{,}58{,}970 - 10{,}59{,}730 - 1{,}26{,}500 - 2{,}19{,}840]$$

$$= \mathbf{34{,}02{,}780 - 24{,}12{,}540 = 9{,}90{,}240\ kN/m.}$$

$$\Sigma V = 73130 + 3486 - 19030 - 3657 = 53929\ \text{kN}$$

$$\bar{x} = \frac{\Sigma M}{\Sigma V} = \frac{9{,}90{,}240}{53{,}929} = 18.36\ \text{m}$$

$$e = \frac{B}{2} - \bar{x} = 34.5 - 18.36 = 16.14 > \frac{B}{6}$$

The resultant is nearer the toe and tension is developed at the heel.

*Average vertical stress*

$$= \frac{\Sigma V}{B} = \frac{53929}{69} = 782\ \text{kN/m}^2.$$

$$p_{max/min} = \frac{\Sigma V}{B}\left[1 \pm \frac{6e}{B}\right]$$

$$= \frac{53929}{69}\left[1 \pm \frac{6 \times 18.32}{69}\right] = 782\ [1 \pm 1.595]$$

$p_v$ (at toe) = 782 × 2.595 = **2030 kN/m²** ; which is < 3000 (∴ Safe)

$p_v$ (at heel) = – 782 × 0.405

= – **316.7 kN/m²** ; which is < 420 kN/m² (∴ Safe)

Since the tensile stress developed is less than the safe allowable value, the dam is safe even when examined with seismic forces, under reservoir full condition.

Principal stress at toe

$$= \sigma = p_v \cdot \sec^2\alpha - p' \tan^2\alpha \quad \textit{i.e. } \text{Eq. (19.17)}$$

where $\tan\alpha = 0.7$, $p' = 60\ \text{kN/m}^2$ ; $p_v = 2030\ \text{kN/m}^2$

**Table 19.2 (*b*)**

| Name of force | Designation if any | Magnitude of force in kN: Vertical forces Downward = +ve Upward = −ve | Horizontal forces Towards Upstream = +ve Towards Downstream = −ve | Lever arm in m | Moments about toe in kN. Anticlock wise (+ve) and clockwise (− ve) in kN.m |
|---|---|---|---|---|---|
| (1) | (2) | (3) | (4) | (5) | (6) |
| Weight of Dam | $W_1$ | (+) $\frac{1}{2} \times 6 \times 60 \times 1 \times 24 = 4320$ | | 65.0 | (+) 2,80,400 |
| | $W_2$ | (+) $7 \times 90 \times 1 \times 24 = 15,110$ | | 59.5 | (+) 8,99,000 |
| | $W_3$ | (+) $\frac{1}{2} \times 56 \times 80 \times 1 \times 24 = 53,700$ | | 37.33 | (+) 20,00,000 |
| | | $\Sigma V_1 = (+)\ 73,130$ | | | $\Sigma M_1 = 31,79,400$ |
| Weight of water supported on u/s slope water on d/s slope. | — | (+) $26 \times 6 \times 1 \times 10 = 1560$ | | 66.0 | (+) 1,02,800 |
| | — | (+) $\frac{1}{2} \times 60 \times 6 \times 1 \times 10 = 1800$ | | 67.0 | (+) 1,10,400 |
| | — | (+) $\frac{1}{2} \times 6 \times 4.2 \times 1 \times 10 = 126$ | | 1.4 | (+) 180 |
| | | $\Sigma V_2 = (+)\ 3486$ | | | $\Sigma M_2 = (+)\ 2,23,380$ |
| Uplift forces | $U_1$ | (−) $69 \times 3.6 \times 10 = 2,480$ | | 34.5 | (−) 85,500 |
| | $U_2$ | (−) $\frac{1}{2} \times 69 \times 48 \times 10 = 16,550$ | | 46.0 | (−) 7,62,000 |
| | | $\Sigma V_3 = (-)\ 19,030$ | | | $\Sigma M_3 = (-)\ 8,47,500$ |
| Upward vertical earthquake forces 0.05 W | | $\Sigma V_4 = (-)\ 0.05 \cdot \Sigma V_1$ $= (-)\ 0.05 \times 73,130$ $= (-)\ 3,657$ | | | $= (-)\ 0.05 \cdot \Sigma M_1$ $= (-)\ 0.05 \times 31,79,400$ $\Sigma M_4 = (-)\ 1,58,970$ |
| Horizontal hydrostatic pressure | $P$ | | (−) $\frac{1}{2} \times 10 \times 86 \times 86 \times 1$ $= (-)\ 36,980$ | 28.67 | (−) 10,60,090 |
| | $P'$ | | (+) $\frac{1}{2} \times 10 \times 6 \times 6 \times 1$ $= (+)\ 180$ | 2.0 | (−) 360 |
| | | | $\Sigma H_1 = (-)\ 36,800$ | | $\Sigma M_5 = (-)\ 10,59,730$ |
| Horizontal hydro-dynamic pressure | $P_e$ | | Calculated separately earlier : $= (-)\ 3,580$ | | $\Sigma M_e = (-)\ 1,26,500$ (calculated separately earlier |
| | | | $\Sigma H_2 = (-)\ 3,580$ | | |
| Horizontal inertia forces due to earthquake | $P_{W_1}$ | | (−) $0.1\ W_1 = (-)\ 432$ | 20.0 | (−) 8,640 |
| | $P_{W_2}$ | | (−) $0.1\ W_2 = (-)\ 1,511$ | 45.0 | (−) 68,000 |
| | $P_{W_3}$ | | (−) $0.1\ W_3 = (-)\ 5,370$ | 26.67 | (−) 1,43,200 |
| | | | $\Sigma H_3 = (-)\ 7,313$ | | $\Sigma M_7 = (-)\ 2,19,840$ |

$$\Sigma H = \Sigma H_1 + \Sigma H_2 + \Sigma H_3 = (-)36,800 - 3580 - 7313 = (-)\ 47,693$$

$\Sigma H$ acts downstream (shown by −ve sign)

$$\sigma = 2030\,(1 + \tan^2 \alpha) - p' \tan^2 \alpha$$
$$= 2030\,(1 + 0.49) - 60 \times 0.49 = 2030 \times 1.49 - 29$$
$$= 3025 - 29 = \mathbf{2996\ kN/m^2 : which\ is < 3000} \quad \text{(just Safe)}$$

Principal stress at heel is

$$\sigma_1 = p_{v(heel)} \sec^2 \phi - (p + p_e) \tan^2 \phi \quad i.e. \text{ Eq. (19.19)}$$

where $\phi$ is the angle which the upstream face makes with the vertical

$$\tan \phi = 0.1$$

$$\therefore \quad \sigma_1 = -316.7\,[1 + (0.1)^2] - (860 + 57.5)\,(0.1)^2$$
$$= -316.7 \times 1.01 - 917.5 \times 0.01 = -319.9 - 9.2$$
$$= \mathbf{-329.1\ kN/m^2}\text{ ; which is } < 420 \text{ kN/m}^2 \quad \text{(Hence, safe)}$$

*Shear stress at toe*

$$\tau_{0(toe)} = (p_{v(toe)} - p') \tan \alpha = (2030 - 60)\ 0.7$$
$$= 1970 \times 0.7 = \mathbf{1379\ kN/m^2}.$$

*Shear Stress at heel*

$$\tau_{0(heel)} = -[p_{v(heel)} - (p + p_e)] \tan \phi$$
$$= -[-329.1 - (860 + 57.5)]\ 0.1$$
$$= -[-329.1 - 917.5]\ 0.1 = +1246.6 \times 0.1 = \mathbf{124.7\ kN/m^2}$$

*Factor of safety against overturning*

$$= \frac{\Sigma M\,(+)}{\Sigma M\,(-)} = \frac{34{,}02{,}780}{24{,}12{,}540} = 1.41 \text{ ; which is } < 1.5 \quad \text{(Hence, Unsafe)}$$

*Factor of safety against sliding*

$$= \frac{\mu \cdot \Sigma V}{\Sigma H}$$

where $\mu = 0.7$

$$\Sigma V = 53{,}929$$
$$\Sigma H = \Sigma H_1 + \Sigma H_2 + \Sigma H_3$$
$$= -36800 - 3580 - 7313 = -47{,}693 \text{ kN}$$

$$\textit{Sliding factor} = \frac{0.7 \times 53929}{47693} = 0.79, \text{ which is } < 1 \quad \text{(Hence, Unsafe)}$$

*Shear friction factor*

$$\text{S.F.F.} = \frac{\mu \cdot \Sigma V + B \cdot q}{\Sigma H}$$
$$= \frac{0.7 \times 53929 + 69 \times 1400}{47693}$$
$$= 2.81 \text{ ; which is less than } 3 \quad \text{(Hence, slightly unsafe)}$$

**Case 2 (*b*). Reservoir full, without uplift**

Sometimes, values of stresses at toe and heel are worked out when there is no uplift, as the vertical downward forces are maximum in this case. For this case, we shall calculate $\Sigma M$ and $\Sigma V$ by ignoring the corresponding values of $\Sigma V_3$ and $\Sigma M_3$ caused by uplift.

$$\therefore \quad \Sigma M = \Sigma M_1 + \Sigma M_2 + \Sigma M_4 + \Sigma M_5 + \Sigma M_6 + \Sigma M_7$$
$$= 31{,}79{,}400 + 2{,}23{,}380 - 1{,}58{,}970 - 10{,}59{,}730 - 1{,}26{,}500 - 2{,}19{,}840$$
$$= 34{,}02{,}780 - 15{,}65{,}040 = 18{,}37{,}740$$
$$\Sigma V = \Sigma V_1 + \Sigma V_2 + \Sigma V_4 = 73130 + 3486 - 3657 = 72{,}959 \text{ kN}$$
$$\bar{x} = \frac{\Sigma M}{\Sigma V} = \frac{18{,}37{,}740}{72{,}959} = 25.19 \text{ m}$$
$$e = \frac{B}{2} - \bar{x} = 34.5 - 25.9 = 9.31 \text{ m} > \frac{B}{6} \quad i.e. \quad \frac{69}{6} = 11.5 \text{ m}$$

Resultant is nearer the toe and no tension is developed any where.

$$p_{max/min} = \frac{\Sigma V}{B}\left[1 \pm \frac{6e}{B}\right]$$
$$= \frac{72{,}959}{69}\left[1 \pm \frac{6 \times 9.31}{69}\right] = 1057 \; [1 \pm 0.81]$$

$p_v$ at toe = 1057 × 1.81 = **1913 kN/m²** < 3000 (∴ Safe)

$p_v$ at heel = 1057 × 0.19 = **201 kN/m²** < 3000 (∴ Safe)

Principal stress at toe $= \sigma = p_v \cdot \sec^2 \alpha - p' \tan^2 \alpha$ ...(Eq. 19.17)

$$p' = 60, \tan \alpha = 0.7$$

$\therefore \sigma = 1913\,(1 + 0.49) - 60 \times 0.49 = 1913 \times 1.49 - 29 = 2821 \text{ kN/m}^2 < 3000$

(Hence, Unsafe)

*Principal stress at heel*

$$\sigma_1 = p_{v(heel)} \sec^2 \phi - (p + p_e) \tan^2 \phi \qquad \text{...(Eq. 19.19)}$$

where $\tan \phi = 0.1$

$$\therefore \quad \sigma_1 = 201(1 + 0.01) - (860 + 57.5) \times 0.01$$
$$= 203 - 9 = 194 \text{ kN/m}^2 < 420 \quad \text{(Safe)}$$

*Shear stress at toe*

$$\tau_0 = (p_v - p') \tan \alpha \qquad i.e. \text{ Eq. (19.20)}$$
$$= (1913 - 60)\, 0.7$$
$$= 1853 \times 0.7 = 1297 \text{ kN/m}^2 < 1400 \qquad (\therefore \text{ safe})$$

**Note.** Shear friction factor, etc. are not worked out here as they were more critical in the 1st case, *i.e.* in 'Reservoir full with uplift' case.

**Conclusion.** The dam is unsafe only in sliding and S.F.F., for which shear key etc. can be provided.

**Example 19.3.** *Examine the stability of the dam section given in the previous example, if there are no seismic forces acting on the dam. Also state the magnitude of maximum compressive stress and maximum shear stress that may develop under any conditions of loading in the dam and also state whether tension is developed anywhere or not.*

**Solution.** The figures calculated earlier in Table 19.2 (*a*) and (*b*) shall be used here.

**Case I. When the reservoir is empty**

$\Sigma V = \Sigma V_1$ from Table 10.2 (*a*) = 73130

$\Sigma M = \Sigma M_1$ from Table 19.2 (*b*) = 3179400

$$\therefore \quad \bar{x} = \frac{\Sigma M}{\Sigma V} = \frac{3179400}{73130} = 43.4 \text{ m}$$

$$e = \frac{B}{2} - \bar{x} = 34.5 - 43.4 = -8.9 \text{ m}$$

–ve sign means that the resultant is towards left side, *i.e.* nearer to the heel, and since $e < \frac{B}{6}$, no tension is developed

$$p_{\text{max/min}} = \frac{\Sigma V}{B}\left[1 \pm \frac{6e}{B}\right] = \frac{73130}{69}\left[1 \pm \frac{6 \times 8.9}{69}\right]$$

$$= 1060\ [1 \pm 0.774]$$

$$p_v \text{ at heel} = 1060(1 + 0.774) = 1060 \times 1.774 = \mathbf{1880\ kN/m^2}$$

$$p_v \text{ at toe} = 1060(1 - 0.774) = 1060 \times 0.226 = \mathbf{239\ kN/m^2}$$

Average vertical stress

$$= \frac{\Sigma V}{B} = \frac{73130}{69} = \mathbf{1060\ kN/m^2}$$

Principal stress at toe

$$\sigma = p_{v(toe)} \sec^2 \alpha$$

$$= 239(1 + 0.49) = 239 \times 1.49 = \mathbf{357\ kN/m^2}$$

Principal stress at heel,

$$\sigma = p_{v(heel)} \sec^2 \phi$$

where $\tan \phi = 0.1$

$$= 1880(1 + 0.01) = 1880 \times 1.01 = \mathbf{1896\ kN/m^2}$$

Shear stress at toe

$$\tau_0 = p_{v(toe)} \tan \alpha$$

$$= 239 \times 0.7 = 167.3 \text{ kN/m}^2$$

Shear stress at heel

$$\tau_{0(heel)} = p_{v(heel)} \tan \phi$$

$$= 1880 \times 0.1 = 188 \text{ kN/m}^2.$$

**Case II (*a*). When the reservoir is full with full uplift**

The values of forces, and moments etc. shall be used from the already calculated values in Table 19.2 (*b*).

$$\Sigma V = \Sigma V_1 + \Sigma V_2 + \Sigma V_3$$

(Seismic forces absent mean $\Sigma V_4$ and $\Sigma M_4$ are zero)

$$= 73130 + 3486 - 19030 = 76616 - 19030 = 57586 \text{ kN}$$

$$\Sigma M = \Sigma M_1 + \Sigma M_2 + \Sigma M_3 + \Sigma M_5$$

$$= 31{,}79{,}400 + 2{,}23{,}380 - 8{,}47{,}500 - 10{,}59{,}730$$

$$= 34{,}02{,}780 - 19{,}07{,}230 = 14{,}95{,}550 \text{ kN–m.}$$

$$\Sigma H = \Sigma H_1 = -\ 36800 \text{ kN}$$

$$\bar{x} = \frac{\Sigma M}{\Sigma V} = \frac{14{,}95{,}550}{57{,}586} = \mathbf{26\ m}$$

$$\frac{B}{2} - \bar{x} = 34.5 - 26 = \mathbf{8.5\ m}$$

The resultant is nearer the toe and $e$ is less than $B/6$, and hence, no tension is developed anywhere

$$\therefore \quad p_{max/min} = \frac{\Sigma V}{B}\left[1 \pm \frac{6e}{B}\right]$$

$$= \frac{57586}{69}\left[1 \pm \frac{6 \times 8.5}{69}\right] = 835[1 \pm 0.738]$$

$$p_v \text{ at toe} = 835(1 + 0.738) = 835 \times 1.738 = \mathbf{1451\ kN/m^2}$$

$$p_v \text{ at heel} = 835(1 - 0.738) = 835 \times 0.262 = \mathbf{219\ kN/m^2}$$

*Principal stress at toe*

$$\sigma = p_{v(toe)} \sec^2 \alpha - p' \tan^2 \alpha \qquad \text{...(19.17)}$$

$$= 1451(1 + 0.49) - 60 \times 0.49 = 1451 \times 1.49 - 29$$

$$= 2162 - 29 = \mathbf{2133\ kN/m^2}.$$

*Principal stress at heel*

$$\sigma_1 = p_{v(heel)} \sec^2 \phi - p \tan^2 \phi$$

where $\sec \phi = 0.1$

$$\sigma_1 = 219 \times (1 + 0.01) - (860) \times 0.01 = 221.2 - 8.6 = \mathbf{212.6\ kN/m^2}.$$

**Note.** This is the minor principal stress and major principal stress is equal to $p$, *i.e.* 860 kN/m$^2$.

*Shear stress at toe*

$$\tau_0 = \left[p_{v(toe)} - p'\right] \tan \alpha = (1451 - 60)\ 0.7 = 974 \text{ kN/m}^2.$$

*Shear stress at heel*

$$\tau_{0(heel)} = -(p_v - p) \tan \phi$$

$$= -(219 - 860)0.1 = 541 \times 0.1 = 54.1 \text{ kN/m}^2;$$

*Factor of safety against overturning*

$$= \frac{\Sigma M\,(+)}{\Sigma M\,(-)} = \frac{3402780}{1907230} = 1.78 > 1.5 \qquad (\therefore \text{ Safe})$$

*Factor of safety against sliding*

$$= \frac{\mu \Sigma V}{\Sigma H} = 0.7 \times \frac{57586}{36800} = 1.10 > 1 \qquad (\therefore \text{ Safe})$$

*Shear friction factor*

$$\text{S.F.F.} = \frac{\mu \Sigma V + B \cdot q}{\Sigma H} = \frac{0.7 \times 57586 + 69 \times 1400}{36800}$$

$$= 3.72 < 4 \text{ to } 5 \qquad (\therefore \text{ Unsafe})$$

**Case II** ($b$). **When the reservoir is full without uplift**

$$\Sigma V = \Sigma V_1 + \Sigma V_2 = 73130 + 3486 = 76616 \text{ kN}$$

$$\Sigma M = \Sigma M_1 + \Sigma M_2 + \Sigma M_5 = 3179400 + 223380 - 1059730$$

$$= 3402780 - 1059730 = 2343050 \text{ kN–m}$$

$$\Sigma H = \Sigma H_1 = (-)\ 36800 \text{ kN}$$

$$\bar{x} = \frac{\Sigma M}{\Sigma V} = \frac{2343050}{76616} = 30.6 \text{ m}$$

$$e = \frac{B}{2} - \bar{x} = 34.5 - 30.6 = \mathbf{3.9\ m}$$

Resultant is nearer the toe and $e < \frac{B}{6}$; hence, no tension is developed anywhere

$$p_{max/min} = \frac{\Sigma V}{B}\left[1 \pm \frac{6e}{B}\right] = \frac{76616}{69}\left[1 \pm \frac{6 \times 3.9}{69}\right]$$

$$= 1110\ [1 \pm 0.339]$$

$$p_v \text{ at toe} = 1110 \times 1.339 = \mathbf{1490\ kN/m^2}$$

$$p_v \text{ at heel} = 1110 \times 0.661 = \mathbf{734\ kN/m^2}$$

Average vertical stress

$$= \frac{\Sigma V}{B} = \frac{76619}{69} = \mathbf{1110\ kN/m^2}$$

*Principal stress at toe,*

$$\sigma = p_{v(toe)} \sec^2 \alpha - p' \tan^2 \alpha$$

$$= 1490\ (1 + 0.49) - 60 \times 0.49$$

$$= 2220 - 29 = \mathbf{2191\ kN/m^2}$$

*Principal stress at heel,*

$$\sigma_1 = p_{v(heel)} \sec^2 \phi - p \tan^2 \phi$$

$$= 734\ (1 + 0.01) - 860 \times 0.01$$

$$= 742 - 9 = \mathbf{733\ kN/m^2}$$

*Shear stress at toe*

$$\tau_0 = [p_{v(toe)} - p'] \tan \alpha$$

$$= (1490 - 60)\ 0.7 = 1430 \times 0.7 = 1001\ \text{kN/m}^2$$

*Shear stress at heel*

$$= -\ [p_{v(heel)} - p] \tan \phi$$

$$= -\ [734 - 860] \times 0.7 = 126 \times 0.7 = \mathbf{88.2\ kN/m^2}$$

**Conclusions.** We find that the dam is safe throughout except that the S.F.F. is equal to 3.72, while generally it should be between 4 to 5. The dam thus remains slightly unsafe in S.F.F. even when the seismic forces are not considered.

*The results of stability analysis are given below* :

| | | |
|---|---|---|
| The maximum shear stress developed in dam | = 1001 kN/m$^2$. | |
| Maximum compressive stress developed in dam | = 2191 kN/m$^2$ | |
| No tension is developed anywhere. | | **Ans.** |
| Factor of safety against sliding | = 1.10 | |
| S.F.F. | = 3.72 | |
| Factor of safety against overturning | = 1.78 | |

**Example 19.4.** *Design a concrete gravity dam for the following data :*

| | |
|---|---|
| *Maximum allowable compressive stress in concrete* | *= 3000 kN/m²*. |
| *Maximum reservoir level* | *= 200.0 m.* |
| *R.L. of bottom of dam* | *= 100.0 m* |
| *Specific gravity of concrete* | *= 2.4.* |
| *Unit wt. of water* | *= 10 kN/m³*. |

**Solution.** The practical profile of dam will have a freeboard of about say 3 to 4% of dam height. Use 3 m as freeboard. The freeboard cannot be calculated as the wave height, etc. are not given.

The R.L. of the top of the dam

$$= 200.0 + 3.0 = 203.0 \text{ m.}$$

The height of the low gravity dam

$$= H_1 = \frac{f}{\gamma_w (S_c + 1)} \quad \text{...(Eq. 19.26)}$$

In this question,

$$f = 3000 \text{ kN/m}^2 \ (i.e.\ 30 \text{ kg/cm}^2)$$

$$\gamma_w = 10 \text{ kN/m}^3$$

$$S_c = 2.4$$

$$\therefore \quad H_1 = \frac{3000}{10\,(2.4 + 1)} = 88.2 \text{ m}, \ < \text{Height of dam.}$$

Therefore, it is a high gravity dam.

Hence, the dam from RL 200.0 m to RL 200 – 88.2 = 111.8 m shall be designed as a low gravity dam, and the remaining bottom height of the dam from RL 111.8 m to 100.0 m shall be designed on the principles of high gravity dam.

**Design of low dam between RL 200.0 m to RL 111.8**

A section for this dam can be chosen as per the provisions of Fig. 19.16.

$$\text{Top width required} = a = \sqrt{\frac{H_1}{3.28}} = \sqrt{\frac{88.2}{3.28}} = \sqrt{26.9} = 5.18 \text{ m}$$

$$\text{Base width required} = \frac{H}{\sqrt{S}} = \frac{88.2}{\sqrt{2.4}} = 56.8 \text{ m.}$$

The upstream projection from the vertical face required

$$= \frac{a}{16} = \frac{5.18}{16} = 0.33 \text{ m (say)}$$

Total base width ($B_1$) provided

$$= 56.8 + 0.33 = 57.13 \text{ m.}$$

The u/s. batter starts at a depth of

$$2a\sqrt{S_c} = 2 \times 5.18 \times \sqrt{2.4} = 16.1 \text{ m.}$$

from below the M.W.L. and it ends at a depth of

$$3.1\,a\sqrt{S_c} = 3.1 \times 5.18\sqrt{2.4} = 24.9 \text{ m}$$

below the M.W.L. The section of this portion of dam is shown in Fig. 19.21 (*a*).

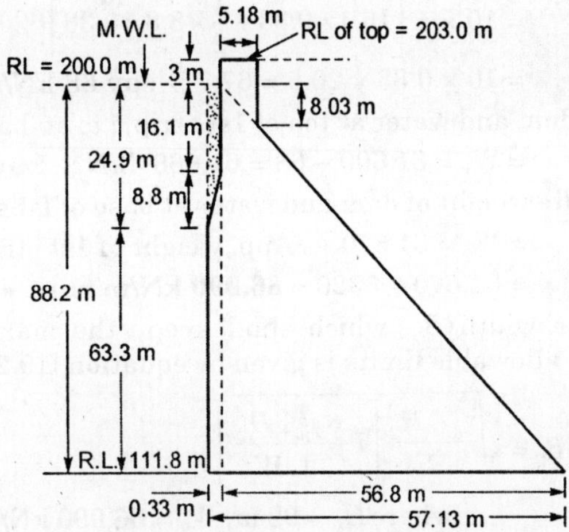


Fig. 19.21(*a*) Low dam portion from RL 200 m to RL 111.8 m.

**Design of the Dam from RL 111.8 m to RL 100 m**

Let us divide this length of 11.8 m into 3 strips. Let the depth of the Ist strip be 3.8 m and the depths of IInd and IIIrd strips be 4.0 m each.

**Design of the Ist strip**

***Consider 1 m length of the dam.***

*Weight of dam section up to RL 111.8 m. (i.e. weight of low dam)*

$$= 24 \times 1 \left[\frac{1}{2} \times 8.8 \times 0.33 + 63.3 \times 0.33 + \frac{1}{2} \times 56.8 \times 88.2 + 3 \times 5.18 + \frac{1}{2} \times 8.03 \times 5.18\right]$$

$$= 24\ [1.5 + 20.9 + 2{,}505 + 15.5 + 20.8]$$

$$= 24\ [2{,}563.7] = \mathbf{61{,}600\ kN/m\ run.}$$

Approximate width of bottom of Ist strip (say $B_2'$) is obtained by drawing a horizontal line at RL 108 m and by producing the already provided d/s face and u/s face of low dam, as shown in Fig. 19.21(*b*).

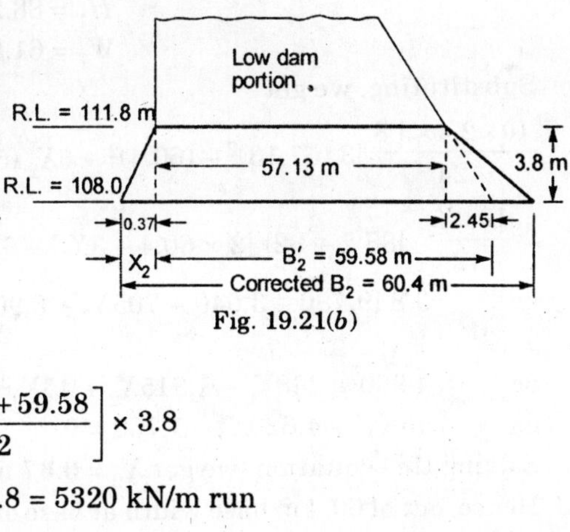


Fig. 19.21(*b*)

$$B_2' = 57.13 + \frac{56.8}{88.2} \times 3.8$$

$$= 57.13 + 2.45 = 59.58 \text{ m}$$

Approximate weight of 1st strip

$$= 24 \times 1 \left[\frac{57.13 + 59.58}{2}\right] \times 3.8$$

$$= 24 \times 58.35 \times 3.8 = 5320 \text{ kN/m run}$$

Weight of water resting on the u/s face

$$= 10 \times 1 \left[16.1 \times 0.33 + \frac{1}{2} \times 8.8 \times 0.33\right]$$

$= 10 \times 0.33 \times 20.5 = 67.7$; **say 68 kN/m run**

Total weight of dam and water at top of 1st strip, *i.e.* at base of small dam

$= W_1 = 61{,}600 + 68 = 61{,}668$ kN; **Say 61670 kN**

Total approximate weight of dam and water at base of Ist strip

$= W_2 = 61{,}670 +$ App. weight of 1st strip

$= 61{,}670 + 5320 =$ **66,990 kN/m run.**

The correct base width $B_2$, which shall keep the maximum compressive stresses within the allowable limits is given by equation [19.29 (*a*)], *i.e.*

$$B_2 = \sqrt{\frac{\gamma_w \cdot H_2^3}{f}\left[1 + \frac{\gamma_w^2 \cdot H_2^4}{4 \cdot W_2^2}\right]}$$

where $H_2 = 92$ m, $W_2 = 66{,}990$ kN/m run

$f = 3000$ kN/m², and $\gamma_w = 10$ kN/m³.

Substituting, we get

$$B_2 = \sqrt{\frac{10 \times (92)^3}{3000}\left[1 + \frac{10^2 \cdot (92)^4}{4 \cdot \times (66{,}990)^2}\right]}$$

$$= \sqrt{2596(1.4)} = \sqrt{3{,}634} = \mathbf{60.4\ m}$$

Now let us find out the projection $X_2$ of base $B_2$ on the u/s side. It is given by equation [19.30(*a*)], *i.e.*

$$\frac{\gamma_w \cdot S_s\, y_1}{24}\left[3B_1^2 - B_2^2 + 6X_2(B_1 + B_2) + 2B_1B_2\right]$$

$$-\frac{\gamma_w \cdot X_2}{12}[H_1 + H_2][2B_2 - 3X_2] - W_1\left[\frac{B_2 - B_1}{3} - X_2\right] = 0$$

where $y_1 = 3.8$ m

$B_1 = 57.13$ m, $B_2 = 60.4$ m

$H_1 = 88.2$ m, $H_2 = 92$ m

$W_1 = 61{,}670$ kN/m run

Substituting, we get

$$\frac{10 \times 2.4 \times 3.8}{24}[3\,(57.13)^2 - (60.4)^2 + 6X_2\,(57.13 + 60.4) + 2 \times 57.13 \times 60.4]$$

$$-\frac{10 \cdot X_2}{12}[88.2 + 92]\,[2 \times 60.4 - 3X_2] - 61{,}670\left[\frac{60.4 - 57.13}{3} - X_2\right] = 0.$$

or $\quad 3.8\,[9{,}780 - 3{,}640 + 705X_2 + 6{,}900] - 10X_2\,[1{,}815 - 45X_2]$

$- 61{,}670\,[1.09 - X_2] = 0$

or $\quad 4.960 + 268X_2 - 1{,}815X_2 + 45X_2^2 - 6{,}720 + 6{,}167X_2 = 0$

or $\quad 45X_2^2 + 4{,}620X_2 - 1{,}760 = 0$

Solving this equation, we get $\mathbf{X_2 = 0.37}$ **m (say).**

Hence, out of 60.4 m base width at base of 1st strip, 0.37 m shall be provided on u/s side as shown in Fig. 19.21(*b*). This will ensure that the resultant shall pass through the inner third point when the reservoir is empty and no tension will develop at toe.

**Design of IInd Strip.** It shall be designed on the same lines on which Ist strip has been designed.

Corrected weight of Ist strip

$$= 24 \times 1 \left[\frac{57.13 + 60.4}{2}\right] \cdot 3.8 = 1.2 \times 117.53 \times 3.8 = 5370 \text{ kN/m run.}$$

Weight of water resting on the u/s face of first strip

$$= 10 \cdot \left[88.2 \times 0.37 + \frac{1}{2} \times 3.8 \times 0.37\right] = 3.7\ [88.2 + 1.9] = 383 \text{ kN/m run.}$$

Total corrected weight at the base of 1st strip

$$W_2 = 61{,}670 + 5370 + 383 = 67{,}423 \text{ kN. } \mathbf{Say\ 67{,}420\ kN/m\ run.}$$

Depth of IInd strip = $y_2$ = 4.0 m.

Approximate base width $B_3'$ of the bottom of IInd strip

$$= 60.4 + \frac{56.8}{88.2} \times 4.0 + \frac{0.37}{3.8} \times 4.0$$

$$= 60.4 + 2.58 + 0.39 = 63.37 \text{ ; } \mathbf{say\ 63.4\ m.}$$

Approximate weight of IInd strip

$$= 24 \times 1 \left[\frac{60.4 + 63.4}{2}\right] \times 4 = 5940 \text{ kN/m run.}$$

Approximate weight of water resting on the u/s face of IInd strip, if the u/s slope given to Ist strip is extended in the IInd strip

$$= \left[\frac{92 + 96}{2} \times 0.39\right] 10 = \mathbf{370\ kN/m\ run.}$$

Hence, the total approximate weight of dam and water at base of IInd strip

$$= W_3 = 67{,}420 + 5{,}940 + 370 = \mathbf{73{,}730\ kN/m\ run.}$$

The corrected base width $B_3$ at bottom of IInd strip

$$B_3 = \sqrt{\frac{\gamma_w^2 \cdot H_3^3}{f}\left[1 + \frac{\gamma_w^2 H_3^4}{4W_3^2}\right]}$$

or
$$B_3 = \sqrt{\frac{10 \times (96)^3}{3000}\left[1 + \frac{10^2 (96)^4}{4 \times (73{,}730)^2}\right]}$$

$$= \sqrt{2{,}949(1 + 0.39)} = \sqrt{4{,}100} = \mathbf{64.0\ m.}$$

Now $X_3$ is obtained from the equation [19.30(*b*)], *i.e.*

$$\frac{\gamma_w \cdot S_s \cdot y_2}{24}\left[3B_2^2 - B_3^2 + 6X_3(B_2 + B_3) + 2B_2 B_3\right]$$

$$-\frac{\gamma_w \cdot X_3}{12}[H_2 + H_3][2B_3 - 3X_3] - W_2\left[\frac{B_3 - B_2}{3} - X_3\right] = 0$$

where $y_2 = 4.0$ m
$B_2 = 60.4$ m, $B_3 = 64.0$ m
$H_2 = 92$ m, $H_3 = 96$ m,
$W_2 = 67{,}420$ kN/m run, and $\gamma_w = 10$ kN/m$^3$

Substituting, we get

$$\frac{10 \times 2.4 \times 4.0}{24} [3\,(60.4)^2 - (64.0)^2 + 6 \cdot X_3\,(60.4 + 64.0) + 2 \times 60.4 \times 64.0]$$

$$- \frac{10 \cdot X_3}{12} [92 + 96]\,[2 \times 64.0 - 3X_3] - 67{,}420 \left[\frac{64.0 - 60.4}{3} - X_3\right] = 0.$$

or $\quad 4.0\,[10944 - 4096 + 747X_3 + 7731] - X_3\,[20053 - 470X_3]$

$$- 80{,}904 + 67420\,X_3.$$

or $\quad 58316 + 2988X_3 - 20053X_3 + 470X_3^2 - 80904 + 67420\,X_3 = 0$

or $\quad 124 - 36\,X_3 + X_3^2 - 172 + 143\,X_3$.

or $\quad X_3^2 + 107X_2 - 48 = 0$

Solving this equation, we get, $\boldsymbol{X_3 = 0.45}$ **m**

Hence, out of 64.2 m of base width of the base of IInd strip, 0.45 m shall be provided on u/s side as shown in Fig. 19.21(*c*).

**Design of last,** *i.e.* IIIrd strip

Corrected weight of IInd strip

$$= 24 \times 1 \left[\frac{60.4 + 64.0}{2}\right] \times 4.0 = 5980 \text{ kN/m run}$$

Weight of water resting on the u/s face of IInd strip

$$= 10 \left[92 \times 0.45 + \frac{1}{2} \times 4 \times 0.45\right] = 500 \text{ kN/m run}$$

Total corrected weight at the base of IInd strip

$$W_3 = 67{,}420 + 5980 + 500 = \mathbf{73{,}900 \text{ kN/m run}}$$

Depth of IIIrd strip = 4.0 m.

Approximate base width $B_4'$ at the bottom of IIIrd strip

$$= 64.0 + \frac{56.8}{88.2} \times 4.0 + 0.45$$
$$= 64.0 + 2.58 + 0.45 = 67.03 \text{ m.}$$

Approximate weight of III strip

$$= 24 \times 1 \left[\frac{64.0 + 67.03}{2}\right] \times 4.0 = 6280 \text{ kN/m run}$$

Approximate weight of water resting on the u/s face of IIIrd strip if the u/s slope given to IInd strip is extended in IIIrd strip

$$= 10 \left[\frac{96 + 100}{2} \times 0.45\right] = 441 \text{ kN/m run.}$$

Hence, the total approximate weight of dam and water at base of IIIrd strip

$$= W_4 = 73{,}900 + 6280 + 441 = \mathbf{80{,}621\ kN}$$

The corrected base width $B_4$ at the bottom of IIIrd strip is given by

$$B_2 = \sqrt{\frac{\gamma_w \cdot H_4^3}{f}\left[1 + \frac{\gamma_w^2 \cdot H_4^4}{4W_4^2}\right]}$$

$$\text{where} \quad f = 3000 \text{ kN/m}^2$$

$$H_4 = 100 \text{ m}$$

$$W_4 = 80{,}621 \text{ kN/m run.}$$

$$\therefore \quad B_4 = \sqrt{\frac{10 \times (100)^3}{3000}\left[1 + \frac{10^2 \times (100)^4}{4 \times (80621)^2}\right]}$$

$$= \sqrt{3{,}333[1 + 0.385]} = \sqrt{3{,}333(1.385)} = \sqrt{4{,}620} = 68.0 \text{ m}$$

Now $X_4$ is obtained from the equation

$$\frac{\gamma_w \cdot S_s \cdot y_3}{24}\left[3B_3^2 - B_4^2 + 6X_4(B_3 + B_4) - 2B_3B_4\right]$$

$$-\frac{\gamma_w \cdot X_4}{12}\left[H_3 + H_4\right][2B_4 - 3X_4] - W_3\left[\frac{B_4 - B_3}{3} - X_4\right] = 0$$

$$\text{where } y_3 = 4.0 \text{ m},$$

$$B_3 = 64.0 \text{ m}, \qquad B_4 = 68.0 \text{ m}$$

$$H_3 = 96 \text{ m}, \qquad H_4 = 100 \text{ m}$$

$$W_3 = 7{,}3900 \text{ kN/m run.}$$

Substituting, we get

$$\frac{10 \times 2.4 \times 4.0}{24}\,[3\,(64.0)^2 - (68.0)^2 + 6 \cdot X_4\,(64.0 + 68.4) + 2 \times 64.4 \times 68.0]$$

$$-\frac{10 \cdot X_4}{12}\,[96 + 100]\,[2 \times 68.0 - 3X_4] - 73{,}900\left[\frac{68.0 - 64.0}{3} - X_4\right] = 0.$$

or $\quad 4\,[12{,}288 - 4{,}624 + 792X_4 + 8.704] - 10X_4\,[2{,}220 - 49X_4]$

$$- 73{,}900\,[1.33 - X_4] = 0$$

or $\quad 65{,}470 + 3170X_4 - 22{,}200X_4 + 490X_4^2 - 98{,}530 + 73{,}900X_4 = 0$

or $\quad 6{,}547 + 317X_4 - 2{,}220X_4 + 49X_4^2 - 9{,}853 + 7{,}390X_4 = 0$

or $\quad 49X_4^2 + 5{,}487X_4 - 3{,}406 = 0$

or $\quad X_4^2 + 112X_4 - 69.5 = 0$

Solving, we get $\mathbf{X_4 = 0.62}$ **m.**

Hence, out of base width of 68.0 m, at the base of IIIrd strip, 0.62 m shall be provided on the u/s side, and the remaining shall be provided on d/s side as shown in Fig. 19.21(*c*).

The design of all the bottom strips is shown in Fig. 19.21 (*c*).

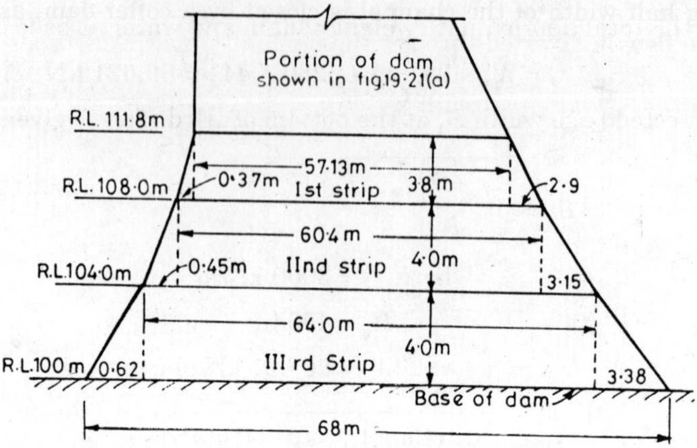


Fig. 19.21 (c)

## CONSTRUCTION OF GRAVITY DAMS

### 19.11. Diversion Problem in Dams Construction

Before the actual construction of a dam can start in a river channel, the water of the river channel must be temporarily diverted. It is advantageous to schedule the construction of the lower portion of the dam during normal periods of low flow so as to minimise the diversion problem. The diversion of river water can be accomplished in either of the following two ways :

(*i*) **Provision of a Diversion Tunnel.** If geological and topographical conditions are favourable, a diversion tunnel or a diversion open channel may be constructed to carry the entire flow around the dam site as shown in Fig. 19.22. The area in which construction work has to take place, is closed by coffer-dams. The diversion tunnel or channel will start from upstream of the upstream coffer-dam and will join the river again on the downstream of the downstream coffer-dam, as shown in Fig. 19.22.

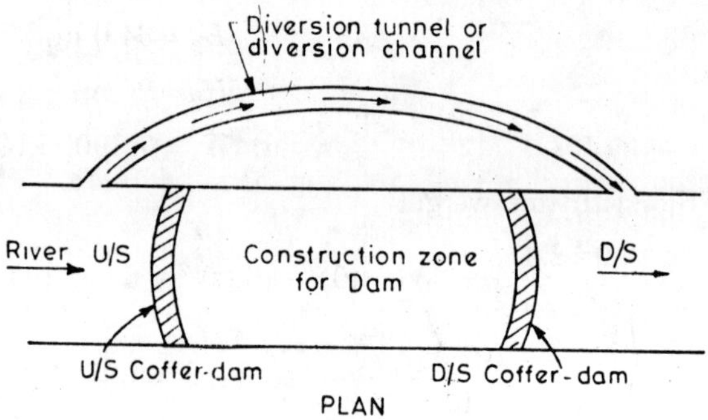


Fig. 19.22

(*ii*) **By constructing the dam in two stages.** The dam is sometimes constructed in two stages. In such a case, the flow is, first of all, diverted and confined to one side of the channel by constructing a semi-circular type of a coffer-dam as shown in Fig. 19.23. The construction work can be taken up in the water- free zone. When the work on the lower portion of the dam on half of its length in one side of the channel gets completed, the

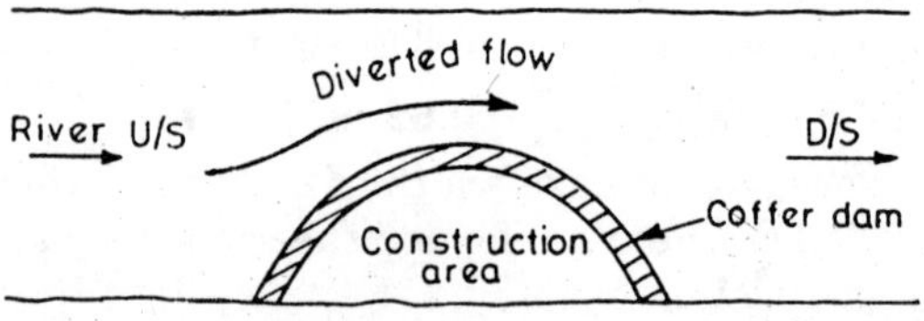


Fig. 19.23

remaining half width of the channel is closed by a coffer-dam, as shown in Fig. 19.24. The flow is diverted through the dam outlets or sometimes it may even be allowed to overtop the already constructed portion of the dam. The work will continue in the water-free zone.

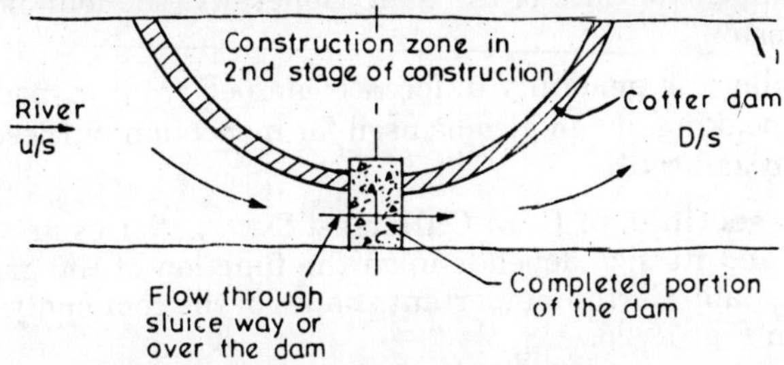


Fig. 19.24

## 19.12. Construction of Galleries in Gravity Dams

Galleries are the horizontal or sloping openings or passages left in the body of the dam. They may run longitudinally (*i.e.* parallel to dam axis) or transversely (*i.e.* normal to dam axis) and are provided at various elevations. All the galleries are interconnected by steeply sloping passages or by vertical shafts fitted with stairs or mechanical lifts. The size of a gallery will depend upon the size of the dam and the function of the gallery.

*Functions and Types of Galleries in Dams*

(1) **Foundation Gallery.** A gallery provided in a dam may serve one particular purpose or more than one purpose. For example, a gallery provided near the rock foundations, serves to drain off the water which percolates through the foundations. This gallery is called a *foundation gallery* or a *drainage gallery*. It runs longitudinally and is quite near to the upstream face of the dam. Its size usually varies from1.5 m × 2.2 m to 1.8 m × 2.4 m. Drain holes are drilled from the floor of this gallery after the foundation grouting has been completed. Seepage is collected through these drain holes. The size of the gallery should be sufficient to accommodate atleast a drilling machine. Besides draining off seepage water, it may be helpful for drilling and grouting of the foundations, when this can not be done from the surface of the dam.

(2) **Inspection Galleries.** The water which seeps through the body of the dam is collected by means of a system of galleries provided at various elevations (say at heights of 15 m or so) and interconnected by vertical shafts, etc. All these galleries, besides draining off seepage water, serve inspection purposes. They provide access to the interior of the dam and are, therefore, called *Inspection galleries*. However, galleries in dams are seldom provided for purely inspection purposes. They generally serve other purposes along with this purpose. Their main functions are summarised below :

(*i*) They intercept and drain off the water seeping through the dam body.

(*ii*) They provide access to dam interior for observing and controlling the behaviour of the dam.

(*iii*) They provide enough space for carrying pipes, etc. during artificial cooling of concrete, (see next article).

(*iv*) They provide access for grouting the contraction joints when this cannot be done from the face of the dam.

(v) They provide access to all the outlets and spillway gates, valves, etc. by housing their electrical and mechanical controls. All these gates, valves, etc. can hence, be easily controlled by men, from inside the dam itself.

(vi) They provide space for drilling and grouting of the foundations, then it cannot be done from the surface of the dam. Generally, the foundation gallery is used for this purpose.

Foundation gallery is generally differently named from inspection galleries, although strictly speaking, it can also be used for inspection purposes and may be called as inspection gallery.

**19.12.1. Cross-sections of Dam Galleries.** Dam galleries are formed as the concrete is placed and its size depends upon the function of the gallery and also upon the size of the dam. Certain important shapes of the commonly used galleries are shown below in Fig. 19.25.

The provision of a gallery in a dam body, changes the normal pattern of stresses in the body of the dam. Stress concentration may, therefore, occur at corners, and hence, in order to minimise this stress concentrations, the corners mast be rounded smoothly. Tension and compression zones may be worked out and proper reinforcements, etc. are provided to counteract them.

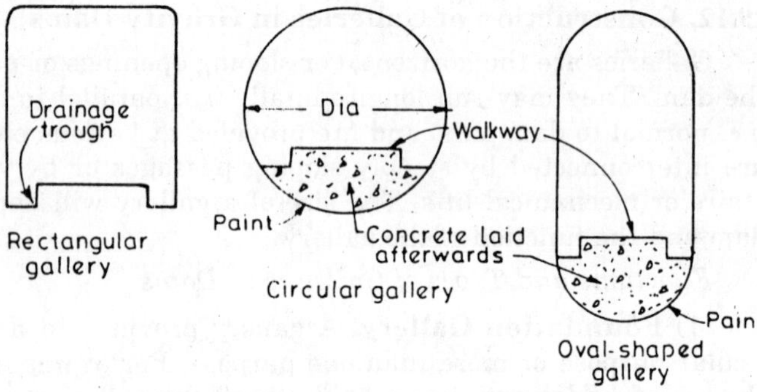


Fig. 19.25. Various Types of Dam Galleries.

## 19.13. Cracking of Concrete in Concrete Gravity Dams

When concrete sets, a tremendous amount of heat is liberated (due to heat of hydration of cement), which will raise the temperature inside the body of the dam. But the temperature outside the dam remains equal to the atmospheric temperature. Due to these temperature differences, temperature stresses get developed in the dam body. Besides, *due to shrinkage of concrete as it cools, shrinkage stresses get developed.* These temperature stresses and shrinkage stresses will cause the concrete to crack unless remedial measures are undertaken. Various measures, generally adopted in concrete gravity dams, to avoid this cracking are :

(i) Using minimum amount of cement in a given mix of specified strength. The quantity of cement can be decreased by better grading the aggregates.

(ii) 'Low lifts' should be used for concrete. When concrete is poured, it is poured up to a certain height in the first attempt. This height is called 'lift'. Generally, 1.5 m lift is used in modern dams. If this lift is reduced, more horizontal joints will get developed and also sufficient cooling time between two successive pours shall be obtained, thus reducing cracking.

(iii) By providing suitably spaced contraction joints, in addition to the normal construction joints.

(iv) Special low heat cements may be used.

(v) The materials which go into the concrete, may be cooled before mixing.

(*vi*) Further cooling is accomplished by circulating cold water through pipes embedded in concrete. This is quite an expensive measure and is adopted only for large gravity dams.

## 19.14. Joints in a Gravity Dam

The concreting of the dam is usually placed in blocks. The size of blocks will depend upon the size of dam and necessity of contraction joints required from the considerations of cracking of concrete. Their width is usually not more than 15 m or so, in large dams. Maximum height of a single pour of concrete (called 'lift') is usually about 1.5 m or so. The alternate blocks of the very first layer which is laid immediately over the rock foundation is taken as 0.75 m deep, as shown in Fig. 19.26. Dam sections are poured alternately so that each block is permitted to stand serveral days before another layer is poured on top of the first. Care must be taken to protect the individual dam sections from the drying action of air by ensuring full-fledged curing. After the shutterings are removed, the lateral surfaces of each section are painted with an asphaltic emulsion paint so as to prevent its adherence to the adjoining sections and thus to serve as construction joints to reduce cracking in concrete. The *Vertical joints* so developed are called **the transverse joints** and they run through the entire height and extend through the full width of the dam section. Transverse joints will thus divide the dam length into a number of vertical cantilevers each of which is independent of the other ; justifying the two dimensional analysis of gravity dams.

The *horizontal joints* called **longitudinal joints** are developed at each lift height and will extend through the entire width of the dam section. They shall run through the entire length of the dam but shall be staggered between transverse joints, as shown in Fig. 19.26.

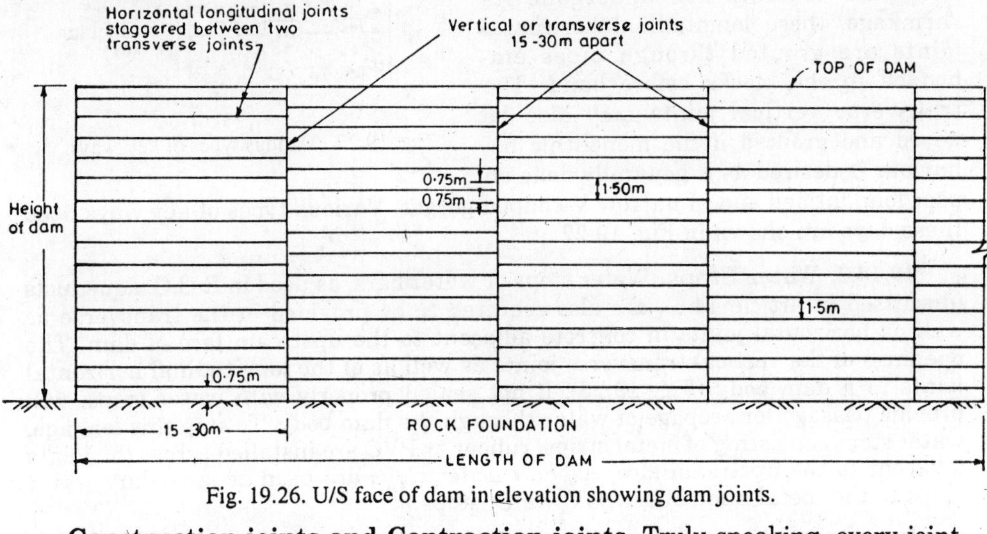


Fig. 19.26. U/S face of dam in elevation showing dam joints.

**Construction joints and Contraction joints.** Truly speaking, every joint whichsoever is left in the dam, shall be a construction joint and every *construction joint* will oppose contraction stresses, and hence will be a *contraction joint.* Therefore, there should be no difference between the two. But whether the joint left was needed from the considerations of practical difficulties in laying the concrete in a single stretch or it was left intentionally for making provisions for shrinkage and temperature stresses, sometimes defines the limits of these two terms. In other words, horizontal joints which were a must from considerations of lift, are sometimes called the construction joints ; while the joints which are mainly left for shrinkage of concrete are called the contraction joints. Although most of the times, these two terms are used to mean one and the same thing.

**19.14.1. Shear Keys or Key ways.** Where foundation conditions are such that undesirable differential settlements or displacements between adjacent concrete blocks may occur, shear keys are formed in transverse vertical joints between sections to carry the shear from one section to the adjacent one and make the dam act as a monolithic structure. These shear keys between the transverse joints are seldom provided on good foundations and in U-shaped valleys, as they will unjustify the two dimensional design of the dam.

Shear keys are however, always provided between the horizontal- longitudinal joints in order to transmit vertical shearing stresses across the section. After the concrete has undergone its shrinkage, these longitudinal horizontal joints are grouted through pipes embedded in the blocks before hand. The transverse vertical joints may also be keyed and grouted if the monolithic behaviour is desired as is generally done on poor foundations and in narrow V-shaped gorges. Various types of key ways used these days are shown in Fig. 19.27.

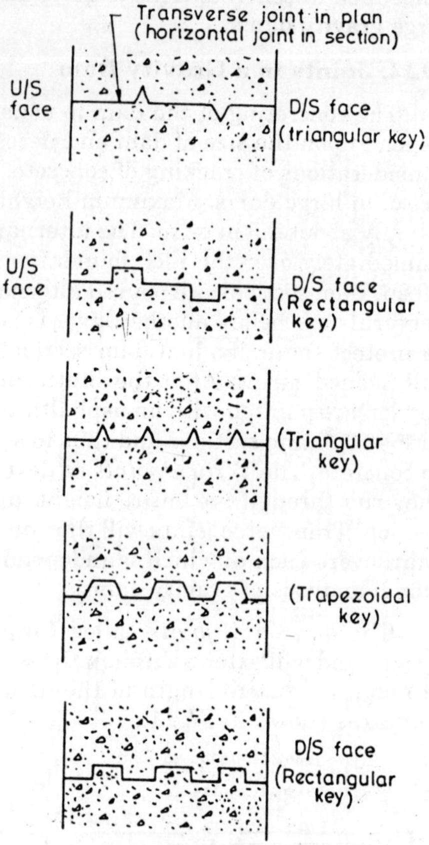


Fig. 19.27. Various types of key-ways.

**19.14.2. Water Stops.** Water stops or water bars, as used in R.C.C. aqueducts (discussed in article 14.6), are also required to be provided in the transverse as well as horizontal joints in concrete adjacent to the upstream face of dam. The openings of the vertical transverse joints as well as of the longitudinal horizontal joints in a dam body (Fig. 19.26), if not sealed properly with water stops, will provide passage for seepage of water through the dam body. To stop this leakage, water stops consisting of metal strips, rubber or PVC are installed across the joints adjacent to the upstream face. *Asphalt water stops* are used as secondary water stops to the metal and rubber/PVC water stops.

In **metal water stops**, copper or stainless steel strips of not less than 1.5 mm thickness and 300 mm width are used. Z-shape and M-shape metal stops, as shown in Fig. 19.28 (*a*) and (*b*), are generally used.

**Rubber water stops** are made from natural rubber confirming to the requirements given in table 19.3. The shape of such water stops may vary, depending upon the water head, etc. The usual minimum width of rubber water stops for dams is 300 mm.

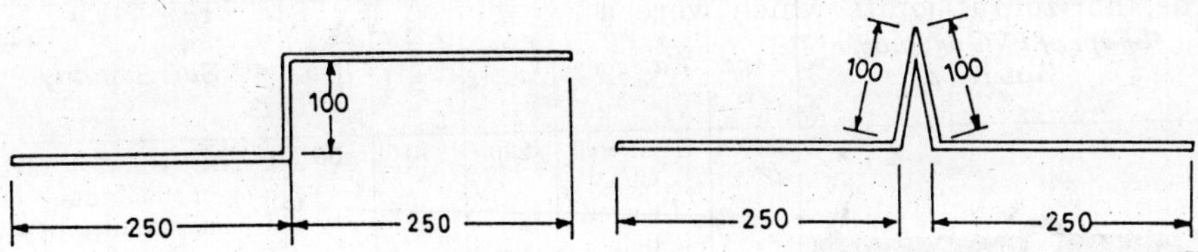


(*a*) Details of z-shaped metal water stop (*b*) Details of M-shaped metal water stop
Fig. 19.28. Various types of metal waterstops.

**PVC waterstops** should be fabricated from plastic compounds, the basin resin of which shall be polyvinyl chloride. The compound shall contain any additional resins, plasticizers, inhibitors or other material such that when the material is compounded, it shall meet the requirements given in table 19.3.

**Table 19.3. Performance Requirements of Rubber or PVC Water Stops**

| *S. No.* | *Characteristics* | *Unit* | *Values* |
|---|---|---|---|
| (*i*) | Tensile strength | $N/mm^2$ | 11.5, min |
| (*ii*) | Ultimate strength | % | 80, min |
| (*iii*) | Tear resistance | $N/mm^2$ | 4.9, min |
| (*iv*) | Stiffness in flexure | $N/mm^2$ | 2.46, min |
| (*v*) | Accelerated extraction | | |
| | (*a*) Tensile strength | $N/mm^2$ | 10.5, min |
| | (*b*) Ultimate alongation | % | 250, min |
| (*vi*) | Effect of alkali : 7 days | | |
| | (*a*) Weight increase | % | 0.10, max |
| | (*b*) Weight decrease | % | 0.10, max. |
| | (*c*) Hardness change | Point | ± 5 |
| (*vii*) | Effect of alkali : 28 days | | |
| | (*a*) Weight increase | % | 0.40, max |
| | (*b*) Weight decrease | % | 0.90, max |
| | (*c*) Dimension change | % | ± 1 |

Certain typical shapes in which rubber/PVC water stops, suitable for concrete gravity dams, may be made are shown in Fig. 19.29. The shape recommended by IS : 12200-1987 for dams is shown in Fig. 19.30.

| *Shape of PVC or Rubber Waterstop* | *Type* | *Category* | *Width* | *Thickness* | *Hydraulic Heads* | *Safe Stability* |
|---|---|---|---|---|---|---|
| (i)  | 230XA | Dumb Bell with Central Bulb | 230 | 10 | 50 | Used where large movement due to expansion/contraction is expected. Also for settlement joints and crumple sections |
| (ii)  | 230G | Serrated with Central bulb | 230 | 4-5 | 25 | For thicker concrete walls and with lower hydrostatic pressures for expansion joints. |
| (iii)  | 240H | Serrated with Central bulb | 240 | 8-10 | 50 | For thicker concrete wall and for higher hydrostatic pressure for expansion and construction joints. |
|  | 240RS | Ribbed and serrated | 240 | 5 | 25 | For construction joints |
| (v)  | 305F | Serrated with Central bulb | 305 | 8-12 | 60 | For very high hydrostatic pressure and where higher grading of concrete aggregates used for expansion joints in Dams etc. |
| (vi)  | 300N | Ribbed with Central bulb | 300 | 10-12 | 60 | Same as 305F but slightly thicker, specially designed for Narmada Project. |
| (vii)  | 230KD | Kicker | 230 | | 25 | Externally placed where re-inforcement in R.C.C. elements congested. Used for construction joints. |
| (viii)  | 230KE | Kicker with Central bulb | 230 | 5 | 25 | Externally placed in R.C.C. elements congested. Used for movement joints. |
| (ix)  | 320KZ | Kicker | 320 | 5 | 25 | Surface waterstops. Used for construction joints of long lengths. |

Fig. 19.29. Certain typical shapes in which rubber/PVC waterstops may be made to suit individual dam project.

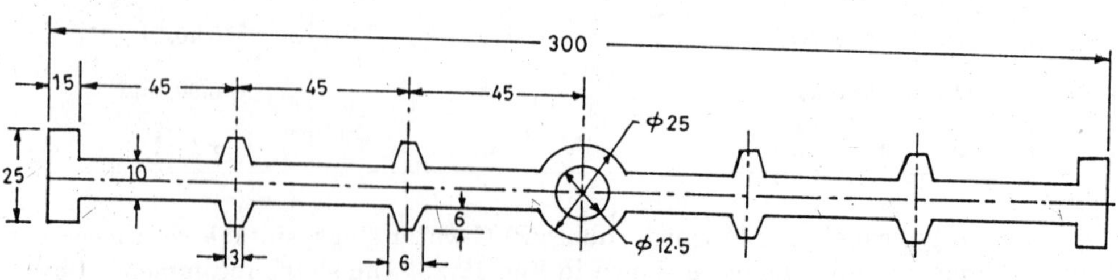


All Dimensions in mm

Fig. 19.30. Detailed dimensions of a rubber/PVC waterstop recommended by IS : 12200-1987.

Metal and rubber/PVC water stops are often used together to reduce leakage through the joints in a dam as shown in Fig. 19.31 to 19.34.

An **Asphalt water stop** is constructed by forming a well of square opening across the contraction joints and filling the opening with asphaltic compound. The well may be fitted in advance with a steam pipe or an electrical heat conductor for reliquefying the asphalt. The asphalt to be used should confirm to the specifications shown in table 19.4 below.

**Table 19.4. Recommended Specifications of Asphalt as per IS : 12200-1987 for Asphalt Joints**

| *S. No.* | *Property* | *Recommended Value* |
|---|---|---|
| 1. | Density | 1015–1065 kg/m$^3$ |
| 2. | Penetration at 25°C | 200–300 mm |
| 3. | Softening point (Ring and ball test) | 80–90°C |
| 4. | Brittleness test on 22 mm$^2$ specimen at 5°C, (energy absorbed) | 0.97 kgm |

The location, shape and dimensions of an asphalt water stop (square hole filled with asphalic compound) are shown within a typical sectional plan of a contraction joint in Fig. 19.31.

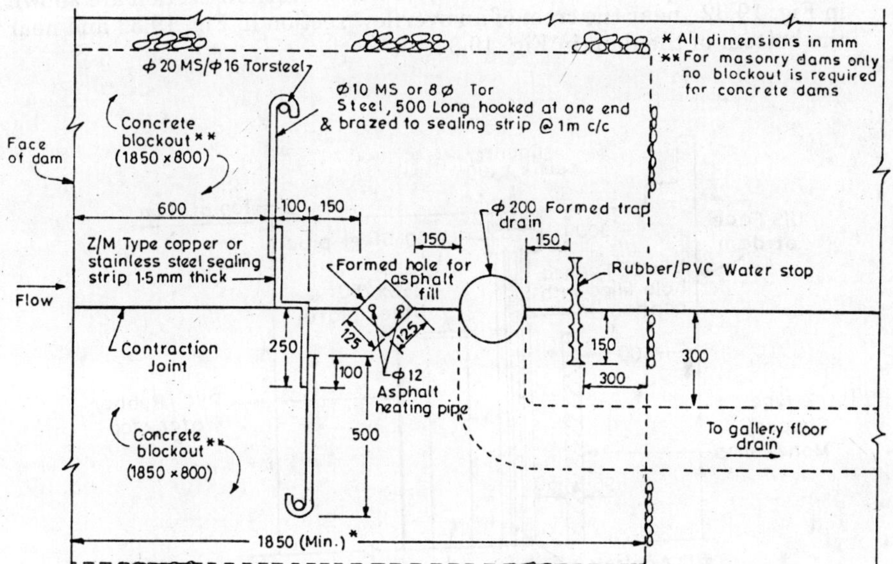


All Dimensions in mm

Fig. 19.31. Sectional plan at a contraction joint (square hole shows location of asphalt water stop)
Note : Dowel bars between concrete/masonry face of the blockout not shown.

#### 19.14.2.1. Installation of water stops

(1) The metal waterstops shall be erected with the help of anchor rods.

(2) In the case of masonry dams, the surface adjacent to the blockouts (shown by dotted lines in Fig. 19.31) shall be irregular and the joints in the masonry shall be raked out when mortar is green, with some stone protruding beyond dotted lines regularly in both directions. No such blockouts shall be provided in concrete dams where concreting on either side of the water seals is done along with the concreting of the rest of the concreting block.

(3) 25 mm dia dowel bars, 1500 mm long (500 mm in concrete and 1000 mm in masonry) 500 mm c/c in both directions shall be provided at the concrete/masonry interface of the blockout in case of a masonry dam to prevent shrinkage crack at the interface.

(4) The blockout may be concreted in lifts not more than 1.5 m. Minimum grade of concrete to be used in the blockout shall be M20.

(5) The blockout of one block may be concreted first and the joint face given a coat of coaltar black paint conforming to IS : 290-1961 and then only the blockout of the second block should be concreted, so as to have a clear contraction joint.

(6) Typical details of waterstop arrangement (at contraction joints between two monoliths of a dam) near the top of a non-overflow section are shown in Fig. 19.32 ; near the crest of an overflow section in Fig. 19.33 and near the bottom of the dam in Fig. 19.34.

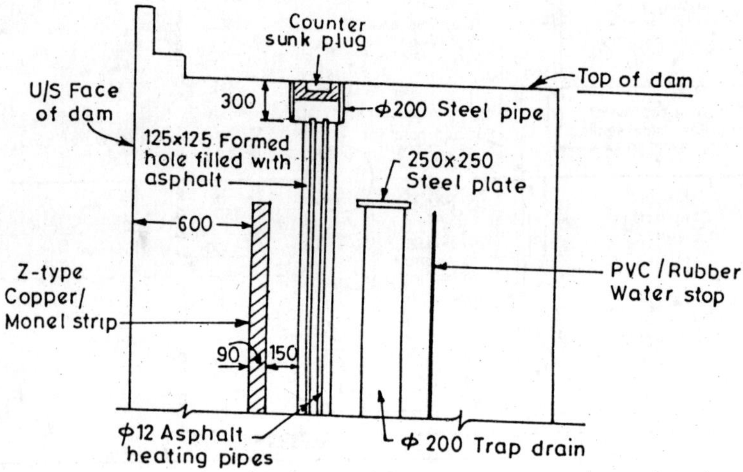


All Dimensions in mm

Fig. 19.32. Typical waterstop details near the top of non-overflow section of a dam.

Rubber/PVC waterstops shall be provided around galleries/adits at the contraction joints between two monoliths of a dam, as shown in Fig. 19.35.

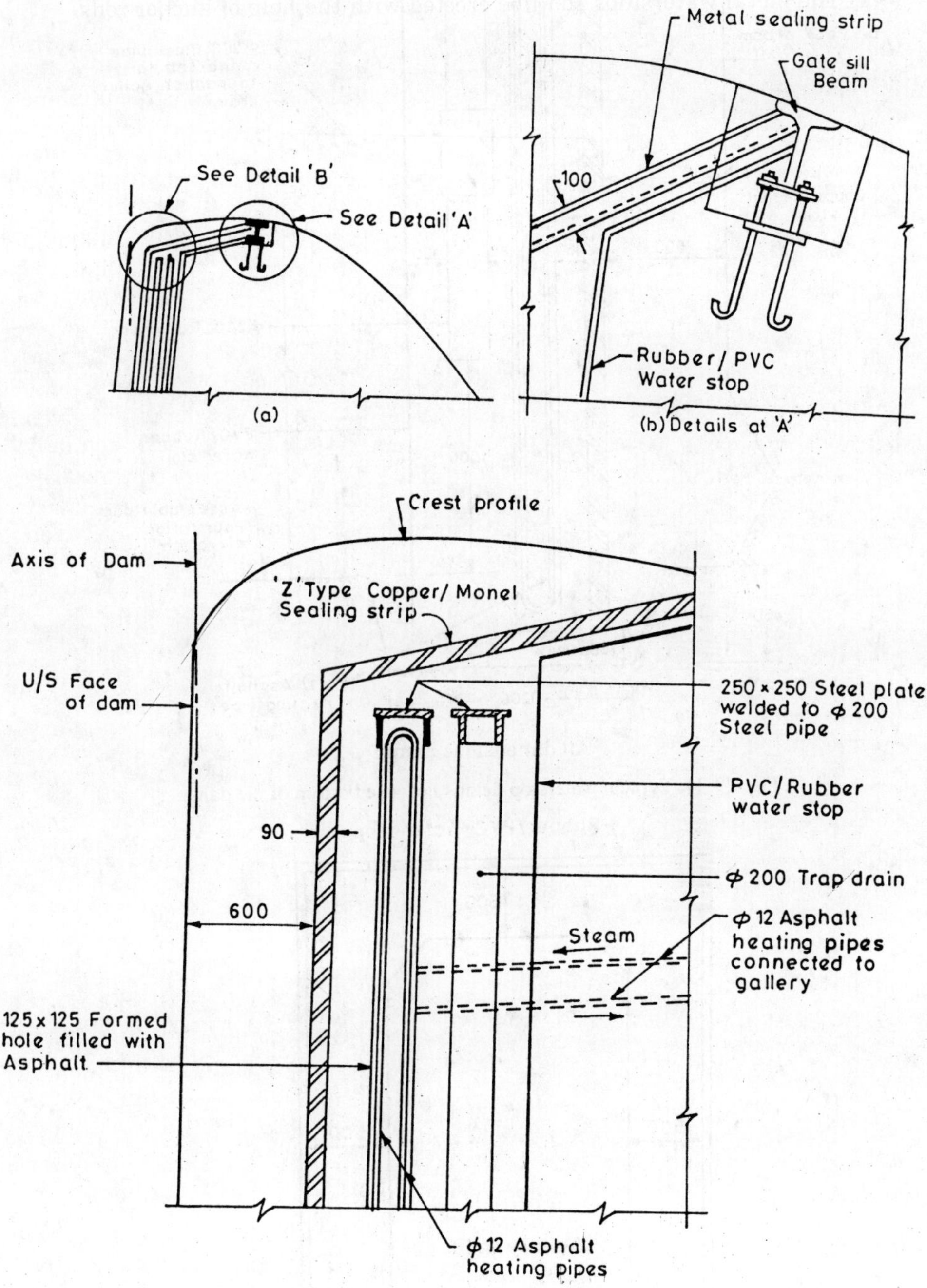


(c) Details at 'B'
All dimensions in mm

Fig. 19.33. Typical waterstop details near the crest of overflow section (spillway).

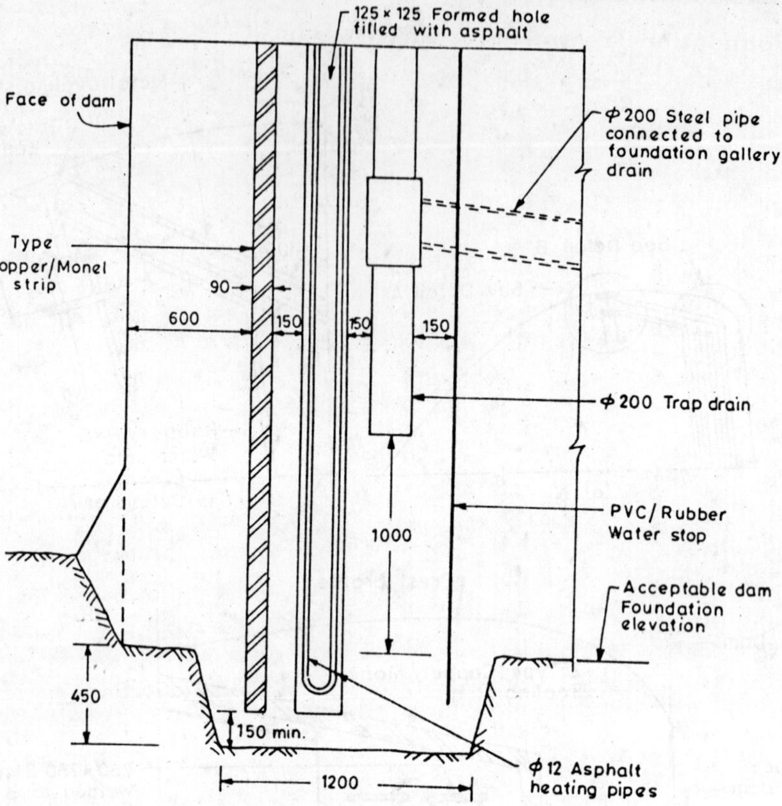


All dimensions in mm

Fig. 19.34. Typical waterstop details near the bottom of the dam.

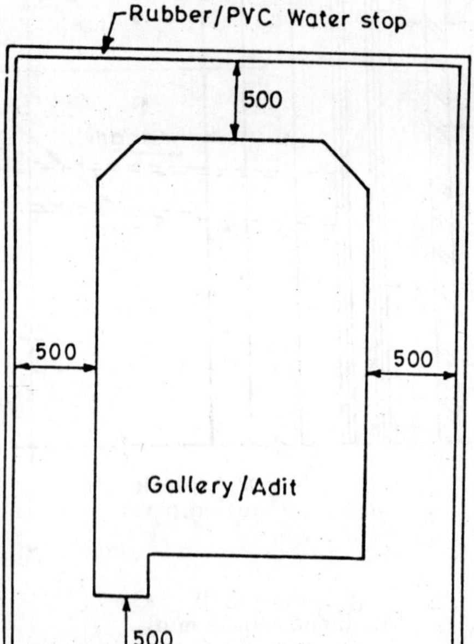


Fig. 19.35. Rubber/PVC waterstop around Dam Gallery/Adit at contraction joints.

## 19.15. Foundation Treatment for Gravity Dams

The material underlying the base of a dam, *i.e.* the foundations of the dam, must be strong enough and capable to withstand the foundation pressure exerted on it under various conditions of loading and in dry as well as wet condition. Most of the failures of the dams have occurred because of the failure of their underlying strata. A concrete gravity dam of California called St. Francis dam was about 62 m high and about 210 m long and failed soon after its completion. The cause of failure was found to be the presence of conglomerate in one abutment, which was weakened after exposure to moisture from the reservoir. Similarly, the failure of a 60 m high arch dam in France (called Malpasset dam) was attributed to the presence of a clay seam in the rock at one of the abutments. Austin Dam on the Colorado river in Texas failed in the year 1900, because large cavities had been dissolved in its limestone foundation.

All these examples have been quoted just to stress upon the readers and the designers, the importance of foundations and the need for their thorough investigation and remedial treatment, if any thing special is found adverse. Besides the special remedial measures in particular cases, the foundation treatment commonly adopted for all foundations can be divided into two steps :

(1) *Preparing the surface* ; and

(2) *Grouting the foundations.*

These treatments are briefly discussed below :

(1) **Preparing the Surface.** The surface preparation consists in removing the entire loose soil till a sound bed rock is exposed. The excavation should be carried out in such a way that the underlying rock is not damaged. The final surface obtained above is stepped, so as to increase the frictional resistance of the dam against sliding. The stepping of the foundation and provision of a shear key is shown in Fig. 19.36. The shear key may sometimes be provided in the centre but is generally provided at the heel.

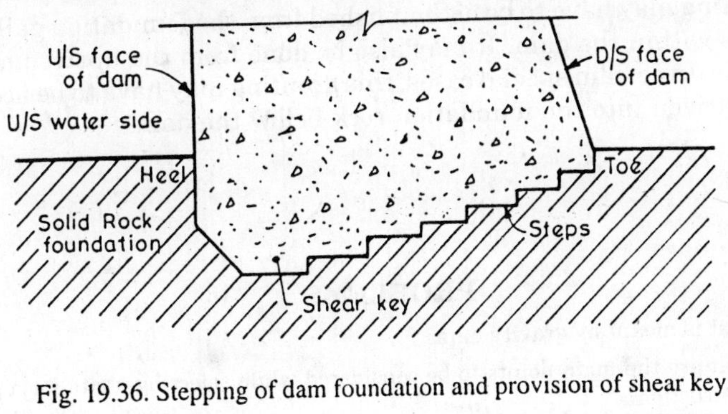


Fig. 19.36. Stepping of dam foundation and provision of shear key.

If faults, seams, or shattered rock zones are detected in the exploratory geological investigations, special steps and remedies must be taken to ensure their removal. They may have to be entirely excavated and back-filled with concrete grouting. The treatment will depend upon the specific needs.

The top foundation surface is thoroughly cleaned with wet sand blasting and washing before the concreting for dam section is started to be laid.

(2) **Grouting the Foundation.** The foundation grouting can be divided into :

(*a*) Consolidation grouting ; and

(*b*) Curtain grouting.

(*a*) *Consolidation grouting.* The entire foundation of the dam is consolidated by grouting. For this purpose, shallow holes (called *B* holes) are drilled through the foundation rock. The depths of these holes generally vary between 10 to 15 m. They are situated at about 5 to 20 m apart, in the general area of the heel of the dam. After the holes have been drilled, mixture of cement and water (called grout) is forced into the holes at low pressure of about 30 to 40 N/cm$^2$. This is accomplished, before any concreting for the dam section is laid. This low pressure grouting will result in a general consolidation of the foundations. These low pressure grout holes will later serve the purpose of a cut-off against leakage of high pressure grout, which is to be used after some concreting of the dam has taken place.

(*b*) *Curtain Grouting.* It helps in forming the principal barrier or a curtain against the seepage through the foundations, and thus reduce the uplift pressures. To accomplish this high pressure grouting, relatively deeper holes (called *A* holes) are drilled near the heel of the dam. The spacing of the holes may vary from 1.2 to 1.5 m. Holes are first of all, drilled and grouted at about 10 to 12 m apart, and then the intermediate holes are drilled and grouted. The depths of the holes vary from 30 to 40% of the total upstream water head for strong rock foundations, and may be as much as 70% of the water head for poor rocks. After the holes have been drilled, a mixture of cement and water (*i.e.* grout) is forced into the holes under high pressure. The grouting pressure may be kept as high as possible without lifting the foundation strata. Usually, the foundation pressure used in this high pressure grouting is equal to 2.5 $D$ N/cm$^2$, where $D$ is the depth of grouting in metres below the surface. This grouting is generally done in stages of depth equal to 15 m or so, and carried out only after some portion of the dam section has been laid.

This grouting may have to be accomplished from the foundation gallery or from other galleries within the dam. It may also be done from the upstream face of the dam, if possible. In certain special cases, this grouting may have to be accomplished from tunnels driven into the foundation rock below the dam.

## PROBLEMS

1. (*a*) What is meant by gravity dams ?

   (*b*) What are the main points to be considered while selecting a site for a gravity dam construction?

   (*c*) Explain briefly with neat sketches the different forces that may act on a gravity dam. Indicate their magnitudes, directions and locations.

2. (*a*) Explain in details the various forces causing instability in a gravity dam.

   (*b*) What is meant by the elementary profile of a gravity dam ; and how is it deduced ? What should be the maximum depth of elementary profile of a dam if the safe limit of stress on the masonry should not exceed 1500 kN per m$^2$ ?

   [**Ans.** 44 m, assuming unit weight of masonry = 24 kN/m$^3$]

**3.** Explain briefly how grout curtain and drainage affect uplift pressures in gravity dams.

Fig. 19.37 shows the section of a gravity dam (non-overflow portion) built of concrete. Calculate (neglecting earthquake effects):

(*i*) The maximum vertical stresses at the heel and toe of the dam.

(*ii*) The major principal stress at the toe of the dam.

(*iii*) The intensity of shear stress on a horizontal plane near the toe.

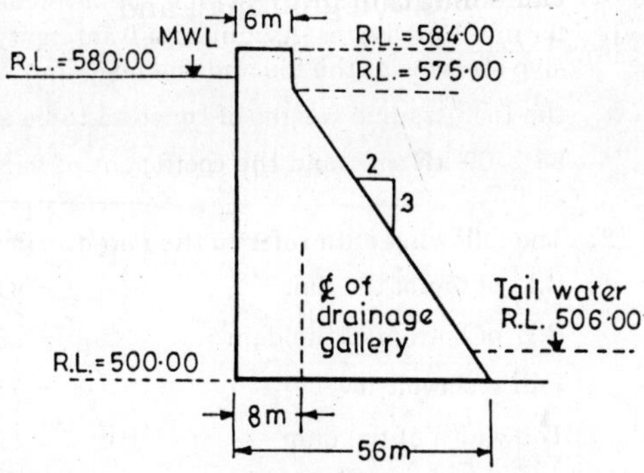


Fig. 19.37

Assume weight of concrete = 24 kN/m$^3$.

[**Hint :** Follow example 19.1]

**4.** (*a*) What do you mean by the elementary profile of a gravity dam ?

(*b*) What are the advantages and disadvantages of a gravity dam over the other types ?

(*c*) What are the methods adopted to reduce uplift in masonry dams ?

**5.** (*a*) What is meant by the term 'low dam' ? Determine the dimensions of the elementary profile of a low gravity dam.

(*b*) Explain the criteria that govern the design of a high gravity dam in different zones of its cross-section.

**6.** (*a*) What is meant by "concrete gravity dams" ? Draw a neat typical cross-section of such a dam. Name the highest dam of the world as well as that of India.

(*b*) What are the different ways by which a concrete gravity dam may fail, and how will you ensure its safety against each type of failure ?

**7.** (*a*) Differentiate between a 'low gravity dam' and a 'high gravity dam'.

(*b*) How does the practical profile of a low gravity dam differs from that of the theoretical one, and why ?

(*c*) Discuss step by step the analytical procedure that you will adopt for analysing the stability (two dimensional analysis) of gravity dams.

**8.** Write detailed notes on any two of the following :

(*i*) Forces acting on gravity dams.

(*ii*) Stability analysis of gravity dams.

(*iii*) Elementary profile of a gravity dam.

(*iv*) Design considerations and fixing the section of a gravity dam.

(*v*) Design criteria for the design of high gravity dams.

**9.** Write short notes on :

(*i*) Uplift force,

(*ii*) Drainage gallery.

(*iii*) Grout curtain,

(*iv*) Construction joint, and

(*v*) Earthquake forces on dams.

**10.** (*a*) Derive an expression for the limiting height of a low dam.

(*b*) Briefly explain the functions of the following :

(*i*) Drainage gallery

(*ii*) Construction joints in a dam, and

(*iii*) Ogee spillway.

[**Hint :** For (*iii*) Please refer chapter 21]

**11.** A concrete dam can be assumed to be trapezoidal in section having a top width of 2 m and bottom width of 10 m. Its height is 12 m and the upstream face has a batter of 1 : 10. Give an analysis of the stability of the dam for the base section for overturning and sliding in the full reservoir condition assuming no free-board allowance but allowing for uplift pressures. Assume uplift intensity factor as 100%. Also determine the compressive stresses at the toe and the heel, and major principal and shear stress developed at the toe. Assume weight of concrete to be 24 kN/m$^3$, unit shear strength of concrete to be 1400 kN/m$^3$, and the coefficient of friction between concrete and foundation soil to be 0.7.

**12.** The following data refer to the non-overflow section of a gravity dam :

| | |
|---|---|
| R.L. of top of the dam | = 315 m |
| R.L. of bottom of the dam | = 260 m |
| Full reservoir level | = 312 m |
| Top width of the dam | = 12 m. |

Upstream face is vertical. Downstream face is vertical upto R.L. 304 m ; and thereafter, the downstream face slopes at 0.7 (*H*) : 1 (*V*) up to base.

Drainage holes are located 8 m away from the upstream face.

Unit weight of masonry = 23 kN/m$^3$

Reduction of uplift at drainage hole = 50%

Coefficient of friction between masonry and foundation material = 0.8.

Determine (*i*) factor of safety against overturning ; (*ii*) factor of safety against sliding ; (*iii*) maximum pressure on foundation, and (*iv*) maximum principal stress in the masonry of the dam, at the base. Consider only the forces due to water thrust, uplift, earthquake (inertial forces due to weight of masonry only) and the self-weight.

**13.** (*a*) Discuss the evolution of the final profile of a gravity dam from its elementary triangular profile, and explain the main principles of its design.

(*b*) Write a brief note on the necessity and method of foundation treatment of dams.

**14.** (*a*) Explain how uplift considerations affect the design of gravity dams. What measures can be adopted to reduce the undesirable effects due to uplift in such cases ?

(*b*) Write short notes on :

(*i*) Dam galleries.

(*ii*) Cracking of concrete during the construction of concrete gravity dams, and remedial measures.

(*iii*) Provision of keyways in concrete gravity dams.

**15.** (*a*) Differentiate between rigid dams and non-rigid dams.

(*b*) How are pore pressures and uplift pressures controlled in rigid dams ?

(*c*) Describe briefly the investigations that are necessary, and the treatment which is commonly given to the foundations of gravity dams, so as to avoid their failures in general.

**16.** The cross-section of a low gravity dam is shown in Fig. 19.38.

Assuming the reservoir to be full, determine

(*i*) The normal stress ;

(*ii*) The principal stress ;

(*iii*) The shear friction factor at base.

Count full uplift as per U.S.B.R. recommendations. Neglect earthquake forces, wave pressure and silt pressure.

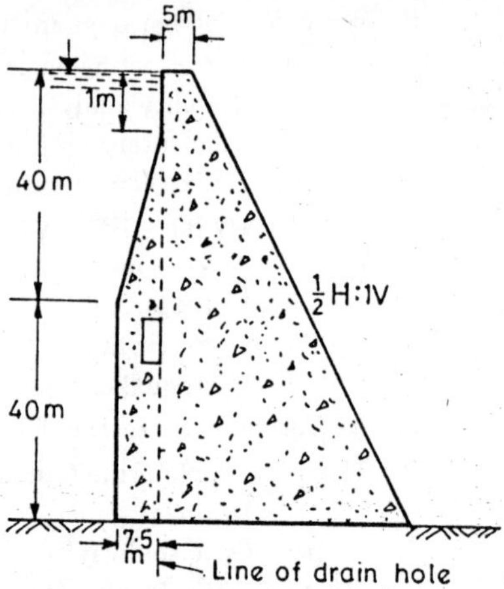


Fig. 19.38

**17.** Design the practical profile of a gravity dam made of stone masonry given the following data :

| | |
|---|---|
| R.L. of base of dam | = 198 m. |
| R.L. of HFL of reservoir | = 228 m. |
| Specific gravity of masonry | = 2.4. |
| Safe compressive stress in masonry | = 1200 kN/m$^2$. |

**18.** Fig. 19.39 shows the section of a concrete gravity dam. Check the stability of this dam section at the base. Assume any data not given and needed.

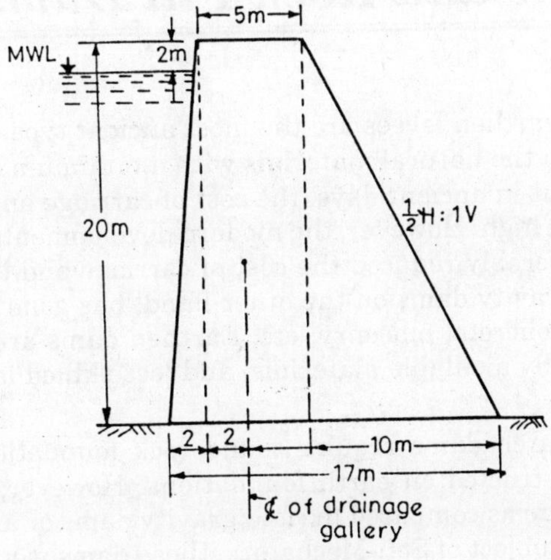


Fig. 19.39

# 20

# *Earthen Dams and Rock Fill Dams*

## 20.1. Introduction

Earthen dams and earthen levees are the most ancient type of embankments, as they can be built with the natural materials with a minimum of processing and primitive equipment. But in ancient days, the cost of carriage and dumping of the dam materials was quite high. However, the modern developments in earth moving equipments have considerably reduced the cost of carriage and laying of the dam materials. The cost of gravity dams on the other hand, has gone up because of an increase in the cost of concrete, masonry, etc. Earthen dams are still cheaper as they can utilise the locally available materials, and less skilled labour is required for them.

Gravity dams and arch dams require sound rock foundations, but earthen dams can be easily constructed on earth foundations. However, earth dams are more susceptible to failure as compared to rigid gravity dams or arch dams. Before the development of the subject of Soil-Mechanics, these dams were being designed and constructed on the basis of experience, as no rational basis for their design was available. This led to the failure of various such earthen embankments. However, in these days, these dams can be designed with a fair degree of theoretical accuracy, provided the properties of the soil placed in the dam, are properly controlled. This condition makes the design and construction of such dams, thoroughly interdependent. Continuous field observations of deformations and pore water pressures have to be made during the construction of such dams. Suitable modifications in the design, are then made during construction, depending upon these field observations.

## 20.2. Types of Earthen Dams

The earthen dam can be of the following three types :

1. *Homogeneous Embankment type*
2. *Zoned Embankment type*
3. *Diaphragm type.*

These three types of dams are described below :

(1) **Homogeneous Embankment Type.** The simplest type of an earthen embankment consists of a single material and is homogeneous throughout. Sometimes, a blanket of relatively impervious material may be placed on the upstream face. A purely homogeneous section is used, when only one type of

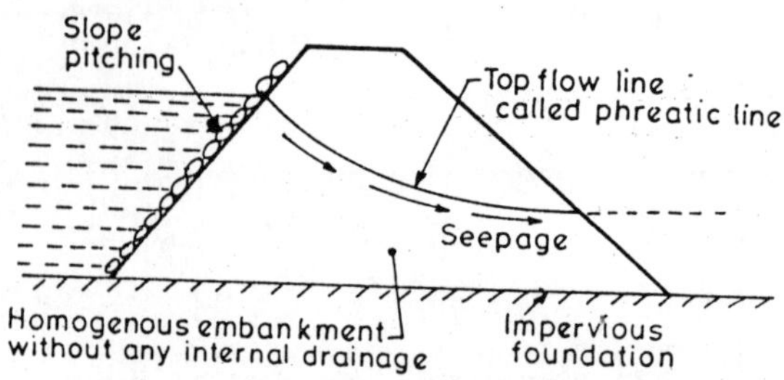

Fig. 20.1 (*a*). Homogeneous type embankment.

material is economically or locally available. Such a section is used for low to moderately high dams and for levees. Large dams are seldom designed as omogeneous embankments.

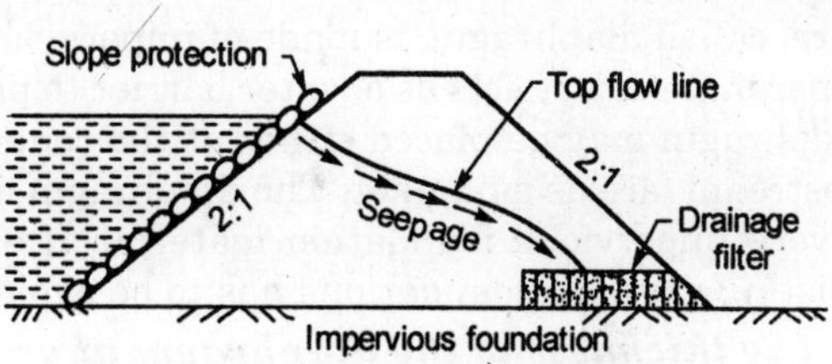

Fig. 20.1(*b*) Homogeneous embankment provided with a drainage filter.

A purely homogeneous section poses the problems of seepage, and huge sections are required to make it safe against piping, stability, etc. Due to this, a homogeneous section is generally added with an internal drainage system : such as a horizontal drainage filter [Fig. 20.1 (*b*)], rock toe, etc. The internal drainage system keeps the phreatic line (*i.e.* top seepage line) well within the body of the dam, and steeper slopes and thus, smaller sections can be used. the internal drainage is, therefore, always provided in almost all types of embankments.

(2) **Zoned Embankment Type.** Zoned embankments are usually provided with a central impervious core, covered by a comparatively pervious transition zone, which is finally surrounded by a much more pervious outer zone (Fig. 20.2).

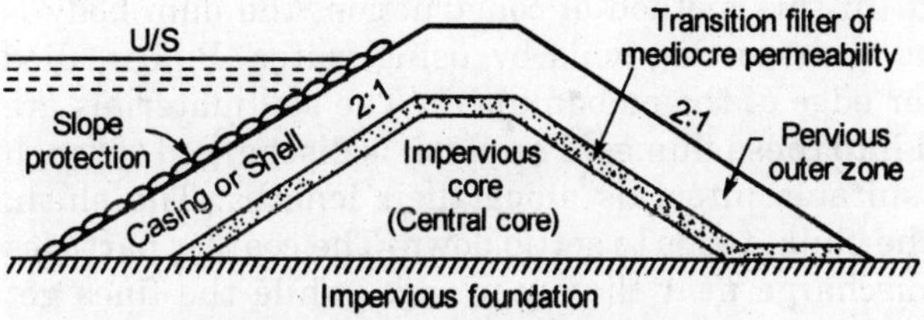

Fig. 20.2. Zoned type embankment.

The central core is supported and protected by shells. The upstream shell affords stability against sudden draw-down and the down-stream shell acts as a drain to control the line of seepage.

This type of embankments are widely constructed and the materials of the zones are selected depending upon their availabilities. ***Clay, inspite of it being highly impervious, may not make the best core, if it shrinks and swells too much. Due to this reason, clay is sometimes mixed with fine sand or fine gravel, so as to use it as the most suitable material for the central impervious core. Silts or silty clays may be used as the satisfactory central core materials.***

Freely draining *materials, such as coarse sands and gravels, are used in the outer shell*. Transition filters are provided between the inner zone and the outer zone, as shown in Fig. 20.2. This type of transition filters are always provided, whenever there is an abrupt change of permeability from one zone to the other.

The width of the central core is controlled by the availability of materials and design requirements, such as stability and seepage. If the variety of soils is available at a given location, there will be an ample range of materials available to choose for different sections of the embankment. In such situations, there will be no constraints on the height of the dam embankment, and the zoned type will invariably yield a more economical section. If the major part of the dam is composed of rock, then it is termed as a rockfill dam.

(3) **Diaphragm Type Embankments.** Diaphragm type embankments have a thin impervious core, which is surrounded by earth or rock fill. The impervious

core, called diaphragm, is made of impervious soils, concrete, steel, timber or any other material. It acts as a water barrier to prevent seepage through the dam. The diaphragm may be placed either at the centre as a central vertical core or at the upstream face as a blanket. The diaphragm must also be tied to the bed rock or to a very impervious foundation material, if excessive under-seepage through the existing previous foundations has to be avoided (Fig. 20.3).

***The thickness of the diaphragm at any level should be less than 3 m or lesser than the height of the embankment above the considered level.***

If the thickness of the diaphragm exceeds these limits, then the embankment will not be considered as of diaphragm type but of zoned embankment type.

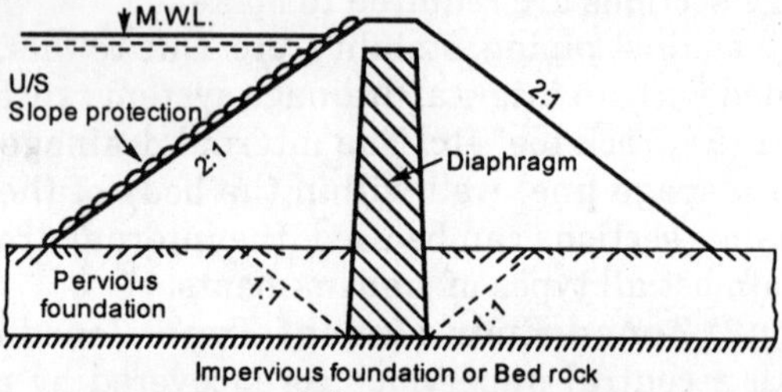

Fig. 20.3. Diaphragm type embankment.

## 20.3. Methods of Construction

There are two methods of constructing earthen dams ;

(1) *Hydraulic-fill Method* ; and

(2) *Rolled-fill Method.*

(1) **Hydraulic-fill Method.** In this method of construction, the dam body is constructed by excavating and transporting soils by using water. Pipes called flumes, are laid along the outer edge of the embankment. The soil materials are mixed with water and pumped into these flumes. The slush is discharged through the outlets in the flumes at suitable intervals along their lengths. The slush, flowing towards the centre of the bank, tends to settle down. The coarser particles get deposited soon after the discharge near the outer edge, while the fines get carried and settle at the centre, forming a zoned embankment having a relatively impervious central core.

Since the fill is saturated when placed, high pore pressures develop in the core material, and the stability of the dam must be checked for these pressures. This type of embankment is susceptible to settlement over long periods, because of slow drainage from the core.

Hydraulic-fill method is, therefore, seldom adopted these days. Rolled-fill method for constructing earthen dams is, therefore, generally and universally adopted in these modern days.

(2) **Rolled-fill Method.** The embankment is constructed by placing suitable soil materials in thin layers (15 to 30 cm) and compacting them with rollers. The soil is brought to the site from burrow pits and spread by bulldozers, etc. in layers. These layers are thoroughly compacted by rollers of designed weights. Ordinary road rollers can be used for low embankments (such as for levees or bunds) ; while power-operated rollers are to be used for dams. The moisture content of the soil fill must be properly controlled. The best compaction can be obtained at a moisture content somewhere near the optimum moisture content. (The optimum moisture content is the moisture required for obtaining optimum density in the fill). Compaction of coarse gravels cannot be properly done by rolling and is best done by vibrating equipment. Detail of rolling and compactiing different types of soils are available in "Soil Mechanics and Foundation Engineering" by the same author.

## 20.4. Shearing Strength of Soils

Before we describe the causes of failures of earthen dams and the criteria for their safe design ; we shall review a few important conceptions of soil-mechanics and their importance in the design of earth dams.

**Coulomb's Law.** The resistance offered by a soil mass against the shearing forces, is known as the shear strength of the soil. In order to ensure that the foundation soil and the soil of the body of the dam are stable, we must ensure that the shearing strength of this soil mass is sufficiently greater than the shear stresses developed within the soil mass by external forces. According to Coulomb's Law, the shear strength of any soil mass is provided by two factors, *i.e.*

(*i*) Cohesion between the soil grains.

(*ii*) Internal friction between the soil particles.

Thus, the unit shear strength ($\tau_f$) of a soil is given by

$$\tau_f = c + \sigma' \cdot \tan \phi \ (\text{kN/m}^2) \qquad ...(20.1)$$

where $c$ = Unit cohesion of soil in (kN/m$^2$)

$\sigma'$ = intergranular compressive stress, called the effective vertical stress (kN/m$^2$)

$\phi$ = Angle of internal friction of soil.

The constant $c$ and $\phi$ depend upon the type of soil and the moisture content in it. Sands and gravels are cohesionless soils and $c \approx 0$ for them ; while $\phi$ is approximatrely zero for saturated cohesive impervious clays.

## 20.5. Various Kinds of Densities and Their Relations

Every soil has pores or voids in it, which remain filled either with air or with water or with both, depending upon the void ratio ($e$) and the moisture content ($w$) of the soil. The void ratio, porosity, moisture content, etc. of a soil are defined as follows :

$$\text{Void ratio } (e) = \frac{\text{Volume of voids in a soil mass}}{\text{Volume of solids in the soil mass}} = \frac{V_v}{V_s}$$

$$\text{Porosity of soil } (n) = \frac{\text{Volume of voids}}{\text{Total volume of soil}} = \frac{V_v}{V}$$

$$= \frac{V_v}{V_s + V_v} = \frac{\dfrac{V_v}{V_s}}{1 + \dfrac{V_v}{V_s}} = \frac{e}{1+e}$$

or

$$e = \frac{1+e}{n}$$

$$\text{or Water content of soil } (w) = \frac{\text{Mass of moisture in soil mass}}{\text{Mass of solids in soil mass}}$$

$$= \frac{M_w}{M_s}$$

where $M_w$ = Mass of water in the soil mass.

$M_s$ = Mass of solids in the soil mass.

The unit weight of soil ($\gamma$) called bulk unit weight, is defined as the weight per unit volume of soil mass. If $W$ is the total weight of soil, then

$$\gamma = \frac{W}{V} = \frac{W_s + W_w}{V_s + V_v} \text{ (air does not have any weight)}$$

or
$$\gamma = \frac{W_s\left(1+\frac{W_w}{W_s}\right)}{V_s\left(1+\frac{V_v}{V_s}\right)} \qquad ...(A)$$

Now
$$\frac{W_s}{V_s} = \gamma_s = G \cdot \gamma_w$$

where $\gamma_s$ = unit weight of solids

$\gamma_w$ = unit weight of water, and

$G$ = specific gravity of solid soil particles $= \frac{\gamma_s}{\gamma_w}$

Substituting in (A), we get

$$\therefore \quad \gamma = \frac{W_s}{V_s}\left[\frac{1+\frac{W_w}{W_s}}{1+\frac{V_v}{V_s}}\right] = G \cdot \gamma_w \left[\frac{1+w}{1+e}\right]$$

Hence,
$$\gamma = G \cdot \gamma_w \cdot \left[\frac{1+w}{1+e}\right] \qquad ...(20.2)$$

If the soil mass is dry, $w = 0$

$$\therefore \quad \gamma_{dry} = \frac{G \cdot \gamma_w}{1+e} \qquad ...(20.3)$$

$$\therefore \quad \gamma = \gamma_{dry}(1+w)$$

*Degree of saturation (S) is defined as*

$$S = \frac{V_w}{V_v} = \frac{\text{Volume of water}}{\text{Volume of voids}}$$

Now
$$e = \frac{V_v}{V_s} = \frac{V_v}{V_s} \times \frac{V_w}{V_w} = \frac{V_v}{V_w} \times \frac{V_w}{V_v}$$

$$= \frac{1}{S}\left[\frac{\frac{M_w}{\gamma_w}}{\frac{M_s}{G \cdot \gamma_w}}\right] = \frac{1}{S} \cdot \frac{M_w \cdot S_s}{M_s} = \frac{1}{S} \cdot w \cdot G$$

$$\boxed{e = \frac{wG}{S}} \qquad ...(20.4)$$

For saturated soils $S = 1$,

$$\therefore \quad e = w \cdot G$$

$$\therefore \quad \boxed{\text{Moisture content, } w = \frac{e}{G} \text{ for saturated soil}} \qquad ...(20.5)$$

Now, if unit weight of saturated soil is designated as $\gamma_{sat}$, then

$$\gamma = \frac{G\gamma_w(1+w)}{(1+e)} \qquad \text{from Eq. (20.2)}$$

$$\gamma_{sat} = \frac{G \cdot \gamma_w \left(1 + \frac{e}{G}\right)}{1 + e} = \frac{G \cdot \gamma_w (G + e)}{G (1 + e)}$$

or

$$\boxed{\gamma_{sat} = \frac{\gamma_w (G + e)}{(1 + e)}} \quad ...(20.6)$$

Submerged unit weight of soil is defined as the unit weight of saturated soil minus the unit weight of water, and is designated as $\gamma_{sub}$.

$$\gamma_{sub} = \gamma_{sat} - \gamma_w = \frac{\gamma_w (G + e)}{1 + e} - \gamma_w$$

$$\gamma_{sub} = \gamma_w \left[\frac{G - 1}{1 + e}\right] \quad ...(20.7)$$

## 20.6. Pore-Water Pressure and its Significance in the Design of Earth Dams

We know that every soil has some voids or pores which are partly or fully filled with water. Let us consider a soil mass below the water-table *BB* in Fig. 20.4.

The soil below *BB* is fully saturated and all its pores between the solid particles of soil, are full of water. The soil below the water-table is, therefore, subjected to hydrostatic uplift. Hence, at any level (say *CC*), the total downward normal pressure ($\sigma$) exerted by the weight of the soil above this level, shall be supported partly by the intergranular pressure ($\sigma'$ or $\sigma_{eff}$) developed between dry particles of the soil, and partly by the hydrostatic pressure ($u$) due to the water present in the pores. Hence

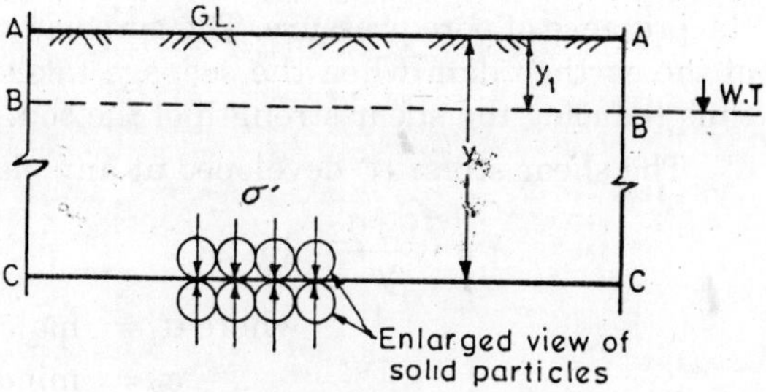


Fig. 20.4

$$\sigma = \sigma' + u \quad ...(20.8)$$

where, $\sigma$ is the total normal pressure on soil.

$\sigma'$ = Total effective pressure, *i.e.* the intergranular pressure, or the pressure which is transmitted from grain to grain of soil.

$u$ = Pore water pressure or *neutral pressure*. (It is the hydrostatic pressure due to presence of water in the soil pores.)

In Fig. 20.4, if the pressures are considered at a level say *CC* at a depth $y$ from the ground, we can easily write,

Total normal pressure ($\sigma$) at level *CC*

$$= \gamma_{dry}\, y_1 + \gamma_{sat}\, (y - y_1)$$

where $y_1$ is the depth of watertable below the ground level.

$$\therefore \quad \sigma = \gamma_{dry}\, y_1 + \gamma_{sat.}\, (y - y_1)$$

Pore water pressure $= u = \gamma_w (y - y_1)$

Hence, the net effective pressure ($\sigma'$) is given by

$$\sigma' = \sigma - u = \gamma_{dry}\, y_1 + \gamma_{sat}\,(y - y_1) - \gamma_w\,(y - y_1)$$
$$= \gamma_{dry}\, y_1 + \gamma_{sub}\,(y - y_1)$$

It, therefore, becomes evident that the effective normal stress is much less than the total normal stress, as a part of the total stress gets consumed by water as pore pressure. The effective stress is dependent on the submerged unit weight of soil.

*The water pressure or pore pressure acts equally in all directions. It does not press the soil grains against one another, and, therefore, does not lead to compression of the soil or an increase in its frictional resistance, that is why, it is called a 'neutral stress'.*

When the pore pressure is considered, the Coulomb's law will become

Unit Shear Strength of soil ($\tau_f$) $= c + \sigma' \cdot \tan\phi$

or
$$\boxed{\tau_f = c + (\sigma - u)\tan\phi} \quad \text{...(20.9)}$$

This clearly indicates that the shearing strength of a soil gets reduced due to the presence of pore pressure. The pore water pressure gets developed in the body of the earthen dam when the seepage takes place through the body of the dam, thus reducing the shear strength of the soil.

The shear stress ($\tau$) developed at any plane in an earth structure is given by

$$\tau = \frac{\sigma_1 - \sigma_2}{2} \cdot \sin 2\theta \quad \text{...(20.10)}$$

where $\sigma_1$ = major principal stress

$\sigma_2$ = minor principal stress

$\theta$ = is the angle between the plane considered for shear and the plane on which $\sigma_1$ acts.

It is apparent from the Equation (20.10), that the value of shearing stress remains unaltered, whether $\sigma_1$ and $\sigma_2$ are used or their effective components [$(\sigma_1 - u)$ and $(\sigma_2 - u)$] are used. But the shear strength of the soil gets reduced when effective component $(\sigma - u)$ is used in place of $\sigma$.

Hence, the stability of the dam against shear failure must be checked when the maximum pore water pressure is present

**Consolidation.** The development of pore pressure is important, even during the time of construction of an earth dam. When the fully or partly saturated soil is placed and rolled in the body of the dam, the entire applied external load is taken up by the water immediately, and transferred to the soil afterwards. The pore water thus gets compressed, and if it is unable to drain out freely (due to low permeability of the soil), the pore pressure rises. *This rise in pore pressure during compaction is known as* **'hydrostatic excess' pressure in pore water.** It further reduces the shearing strength of the soil and hence, the stability of the soil. As the excess water drains out, more and more consolidation will take place, as this pressure will be transferred to soil grains and the shear strength will tend to achieve its normal value. *Hence, the pore pressure temporarily reduces the shear strength of the soil during compaction* by preventing full compaction. But the shear strength gets recovered after the compaction is over, as the pore water is ultimately squeezed out.

In highly compressible soils, having low coefficient of permeability and moisture content above its optimum moisture content, this condition of 'hydrostatic excess' becomes very serious. Such soils are, therefore, more liable to fail during construction.

Similarly, appreciable consolidation of soil may takes place in fine grained compressible soils like clay, even after the construction is over, though sufficient compaction was done during construction.

Due to these reasons, pore pressure observations are often made during the construction period of an earth dam. If the hydrostatic excess of pore pressure rises to a dangerous amount, the construction may be stopped for some time till the excess water drains out and full natural compaction takes place. The construction may be restarted after this 'excess' is either fully dissipated or reduced to a safe value. The shear failure of the soil of the dam or its foundation is, therefore, very much connected with the development of pore pressure in the body of the dam and in the foundation, and must be properly checked and accounted for.

## 20.7. Causes of Failure of Earthen Dams

Earth dams are less rigid and hence more susceptible to failure. Every past failure of such a dam has contributed to an increase in the knowledge of the earth dam designers. Earthen dams may fail, like other engineering structures, due to improper designs, faulty constructions, lack of maintenance, etc. The various causes leading to the failure of earth dams can be grouped into the following three classes :

(1) Hydraulic failures

(2) Seepage failures

(3) Structural failures.

These causes are described below in details :

**20.7.1. Hydraulic failures.** About 40% of earth dam failures have been attributed to these causes. The failure under this category, may occur due to the following reasons:

(*a*) *By over topping.* The water may overtop the dam, if the design flood is under-estimated or if the spillway is of insufficient capacity or if the spillway gates are not properly operated. Sufficient freeboard should, therefore, be provided as an additional satety measure.

(*b*) *Erosion of upstream face.* The waves developed near the top water surface due to the winds, try to notch-out the soil from the upstream face and may even, sometimes, cause the slip of the upstream slope. Upstream stone pitching or riprap should, therefore, be provided to avoid such failures.

(*c*) *Cracking due to frost action.* Frost in the upper portion of the dam may cause heaving and cracking of the soil with dangerous seepage and consequent failure. An additional freeboard allowance upto a maximum of say 1.5 m should, therefore, be provided for dams in areas of low temperatures.

(*d*) *Erosion of downstream face by gully formation.* Heavy rains falling directly over the downstream face and the erosive action of the moving water, may lead to the formation of gullies on the downstream face, ultimately leading to the dam failure. This can be avoided by proper maintenance, filling the cuts from time to time especially during rainy season, by grassing the slopes and by providing proper berms at suitable heights (Fig. 20.5), so that the water has not to flow for considerable distances. The proper drainage arrangements are made for the removal of the rain water collected on the horizontal berms. Since the provision of berms ensures the collection and removal of water before it acquires high downward velocities, the consequent erosion caused by the moving water (run off) is considerably reduced.

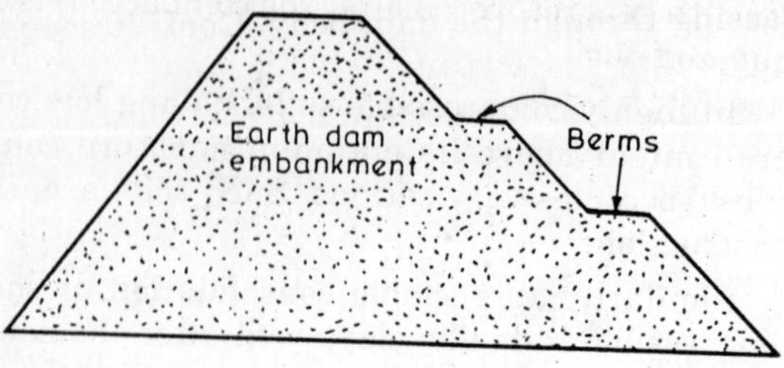

Fig. 20.5

(*e*) *Erosion of the d/s toe.* The d/s toe of the earth dam may get eroded due to two reasons, *i.e.* (*i*) the erosion due to cross currents that may come from the spillway buckets; and (*ii*) the erosion due to tail water. This erosion of the toe can be avoided by providing a downstream slope pitching or a riprap up to a height slightly above the normal tail water depth. Side walls of the spillway (called diaphragm walls) must be of sufficient height and length, as so to prevent the possibility of the cross flow towards the earthen embankment.

**20.7.2. Seepage Failures.** Controlled seepage or limited uniform seepage is inevitable in all earth dams, and ordinarily it does not produce any harm. However, uncontrolled or concentrated seepage through the dam body or through its foundation may lead to piping or sloughing and the subsequent failure of the dam. Piping is the progressive erosion and subsequent removal of the soil grains from within the body of the dam or the foundation of the dam. Sloughing is the progressive removal of soil from the wet downstream face. More than 1/3rd of the earth dams have failed because of these reasons.

(*a*) *Piping through foundations.* Sometimes, when highly permeable cavities or fissures or strata of coarse sand or gravel are present in the foundation of the dam, water may start seeping at a huge rate through them (Fig. 20.6). This concentrated flow at a high gradient, may erode the soil. This leads to increased flow of water and soil, ultimately resulting in a rush of water and soil, thereby creating hollows below the foundation. The dam may sink down into the hollow so formed, causing its failure.

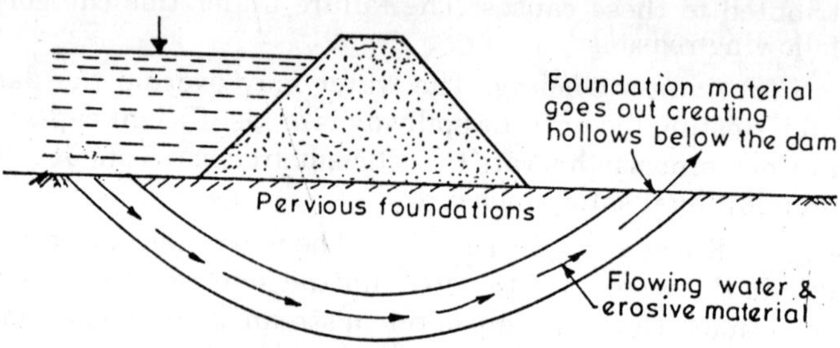

Fig. 20.6. Piping through the dam foundation.

(*b*) *Piping through the dam body.* When the concentrated flow channels get developed in the body of the dam, (Fig. 20.7) soil may be removed in the same manner as was explained in foundation piping, leading to the formation of hollows in the dam body, and subsequent subsidence of the dam. These flow channels may develop due to faulty construction, insufficient compaction, cracks developed in embankment due to foundation settlement, shrinkage cracks, animal burrows, etc. All these causes can be removed by better construction and better maintenance of the dam embankments.

Piping through the dam body, generally get developed near the pipe conduits passing through the dam body. Contact seepage along the outer side of conduits may either develop into piping, or seepage through leaks in the conduits may develop into piping. This can be avoided by thoroughly and properly compacting the soils near the outlet conduits and by preventing the possibilities of leakage through conduits, but preventing the formation of cracks in the conduits. These cracks in the conduits are caused by differential settlement and by overloading from the embankment. When these factors are controlled, automatically, the possibility of piping due to leakage through the conduits is reduced.

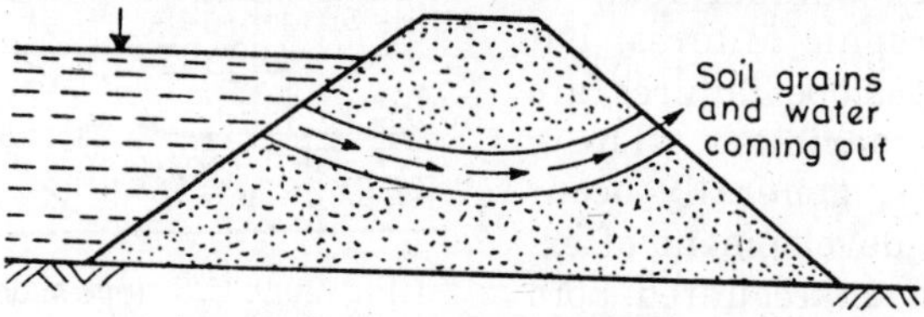

Fig. 20.7. Piping through the dam body.

(c) *Sloughing of D/S Toe.* The process behind the sloughing of the toe is somewhat similar to that of piping. The process of failure due to sloughing starts when the downstream toe becomes saturated and get eroded, producing a small slump or a miniature slide. The miniature slide leaves a relatively steep face which becomes saturated by the seepage from the reservoir and slumps again, forming a more unstable surface. The process continues till the remaining portion of the dam is too thin to withstand the horizontal water pressure, leading to the sudden failure of the dam.

**20.7.3. Structural failures.** About 25% of the dam failures have been attributed to structural failures. Structural failures are generally caused by shear failures, causing slides.

(a) *Foundation slide.* (*i.e.* overall stability of the dam). When the foundation of the earth dams are made of soft soils, such as fine silt, soft clay, etc., the entire dam may slide over the foundation. Sometimes, seams of fissured rocks, shales or soft clay, etc. may exits under the foundation, and the dam may slide over some of them, causing its failure. In this type of failure, the top of embankment gets cracked and subsides, the lower slope moves outward forming large mud waves near the heel, as shown in Fig. 20.8.

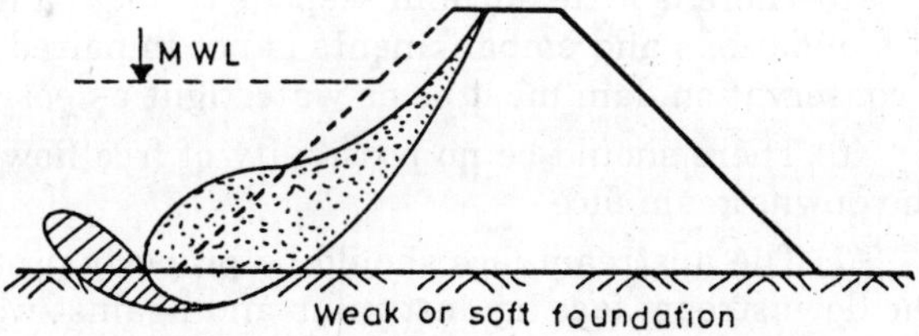

Fig. 20.8. Sliding due to soft or weak foundation.

Excessive pore water pressure in confined seams of sand and silt, artesian pressure in abutments, or hydrostatic excess developed due to consideration of clay seams embedded between sands or silts, etc. may reduce the shear strength of the soil, until it becomes incapable of resisting the induced shear stresses, leading to the failure of the dam foundation whithout warning. Loose sand foundations may fail by the liquefaction or flow slides.

(b) *Slide in Embankments.* When the embankment slopes are too steep for the strength of the soil, they may slide causing dam failure.

The most critical condition of the slide of the u/s slope is the sudden draw-down of the reservoir (Fig. 20.9); and the d/s slope is most likely to slide, when the reservoir is full (Fig. 20.10). The u/s slope failures seldom lead to catastrophic failures, but the d/s slope failures are very serious. These failures, generally occur due to development of excessive unaccounted pore pressures which may reduce the shearing strength of the soils as explained in the previous article. Many embankments may fail during the process of consolidation, at the time of construction or after the construction.

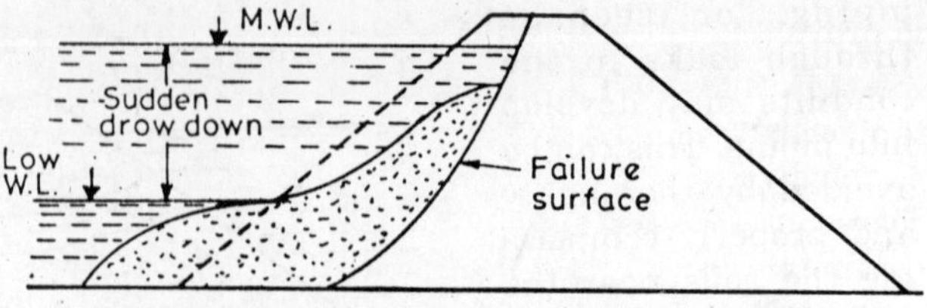


Fig. 20.9. U/S slope slide due to sudden draw-down.

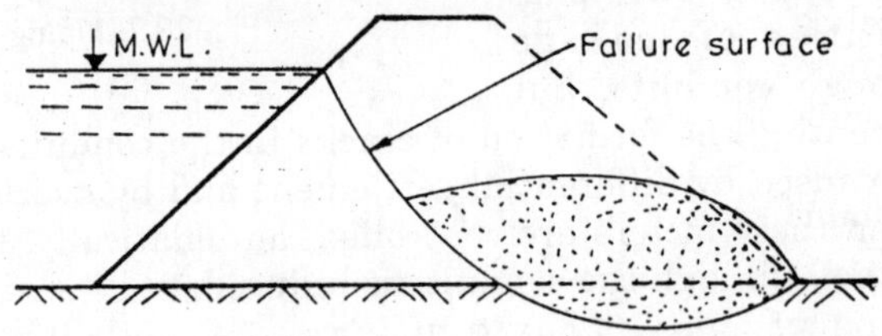


Fig. 20.10. D/S. slope slide during full reservoir condition.

## 20.8. Design Criteria for Earth Dams

(1) A fill of sufficiently low permeability should be developed out of the available materials, so as to best serve the intended purpose with minimum cost. Borrow pits should be as close to the dam site as possible, so as to reduce the leads.

(2) Sufficient spillway and outlets capacities should be provided so as to avoid the possibility of overtopping during design flood.

(3) Sufficient freeboard must be provided for wind set-up, wave action, frost action and earthquake motions.

(4) The seepage line (*i.e.* phreatic line) should remain well within the downstream face of the dam, so that no sloughing of the face occurs.

(5) There is little harm in seepage through a flood control dam. if the stability of foundations and embankments is not impaired, by piping, sloughing, etc. : but a conservation dam must be as watertight as possible.

(6) There should be no possibility of free flow of water from the upstream to the downstream face.

(7) The upstream face should be properly protected against wave action, and the downstream face against rains and against waves upto tail water. Provisions of horizontal berms at suitable intervals in the d/s face may be thought of, so as to reduce the erosion due to flow of rain water. Ripraps should be provided on the entire u/s slope and also on the d/s slope near the toe and up to slightly above the tail water so as to avoid erosion.

(8) The portion of the dam, downstream of the impervious core, should be properly drained by providing suitable horizontal filter drain, or toe drain, or chimney drain, etc.

(9) The upstream and downstream slopes should be so designed as to be stable under worst conditions of loading. These critical conditions occur for the u/s slope during sudden drawdown of the reservoir, and for the d/s slope during steady seepage under full reservoir.

(10) The u/s and d/s slope should be flat enough, as to provide sufficient base width at the foundation level, such that the maximum shear stress developed remains well below the corresponding maximum shear strength of the soil, so as to provide a suitable factor of safety.

(11) We know that the consolidation of the soil does not take place instantaneously when the compaction is done by external leadings. It takes place slowly as the excess pore water goes out and the load is transferred to the soil grains. In coarse gravels, the void openings are large enough so as to permit rapid escape of confined water and air, and full compaction may occur before the construction is over. But in fine grained impervious soils, the consolidation is slow. It, therefore, becomes necessary in such cases, as to provide an additional height of the fill. After consolidation, the the embankment will be of the desired height. Hence, a suitable allowance in the height of embankment (between 2 to 3% of dam height, determined by laboratory tests) must be made in fine grained soils so as to account for the consolidation that may take place upto years after construction. Dewatering the foundations may sometimes be used to accelerate the process of consolidation.

(12) Since the stability of the embankment and foundation is very critical during construction or even after construction (*i.e.* during the period of consolidation), due to development of excessive pore pressures and consequent reduction in shear strength of soil, the embankment slopes must remain safe under this critical condition also.

All the above criteria must be satisfied and accounted for, in order to obtain a safe design and construction of an earth dam.

## 20.9. Selecting a Suitable Preliminary Section for an Earth Dam

The preliminary design of an earth dam is done on the basis of existing dams of similar characteristics and the design is finalised by checking the adequacy of the selected section for the worst loading conditions. Empirical rules are frequently used in these designs.

A few recommendations, for selecting suitable values of top width, free board, u/s and d/s slopes, drainage arrangements, etc. are given below for preliminary designs :

**Freeboard.** Freeboard or minimum freeboard is the vertical distance between the maximum reservoir level and the top of the dam (*i.e.* the crown or crest of dam). The vertical distance between normal pool level or spillway crest and the top of the dam is termed as normal freeboard.

The minimum height of the freeboard for wave action is generally taken to be equal to 1.5 $h_w$, where $h_w$ is given by the equations (19.11) and (19.12). Most of the hydraulic failures of earth dams have occurred due to overtopping of dams. Hence, the freeboard must be sufficient enough, as to avoid any such possibility of overtopping. Values of freeboard, for various heights, recommended by U.S.B.R. are given in table 20.1.

**Table 20.1. U.S.B.R. Recommendations for Freeboard for Earth Dams**

| *Spillway Type* | *Height of Dam* | *Minimum freeboard Over MWL* |
|---|---|---|
| Uncontrolled (*i.e.* Free) Spillway | Any height | Between 2 m to 3 m |
| Controlled spillway | Height less than 60 m | 2.5 m above top of gates |
| Controlled spillway | Height more than 60 m | 3 m above top of gates |

An additional freeboard upto 1.5 m should be provided for dams situated in areas of low temperatures for frost action.

**Width.** The top width of large earthen dams should be sufficient to keep the seepage line well within the dam, when reservoir is full. It should also be sufficient to withstand earthquake shocks and wave action. For small dams, this top width is generally governed by minimum roadway width requirements.

The top width ($A$) of the earth dam can be selected as per the following recommendations :

$$A = \frac{H}{5} + 3 \text{ for very low dams} \quad ...(20.11)$$

$$A = 0.55\sqrt{H} + 0.2\,H \text{ for dams lower than 30 m} \quad ...(20.12)$$

$$A = 1.65\,(H + 1.5)^{1/3} \text{ for dams higher than 30 m} \quad ...(20.13)$$

where $H$ is the height of the dam.

**Upstream and Downstream slopes.** The side slopes depend upon various factors such as the type and nature of the dam, and foundation materials, height of dam, etc. etc. The recommended values of side slopes as given by Terzaghi are tabulated in Table 20.2.

**Table 20.2. Terzaghi's Side Slopes for Earth Dams**

| *Type of Material* | *U/S slope (H : V)* | *D/S slope (H : V)* |
|---|---|---|
| Homogeneous well graded | 2.5 : 1 | 2 : 1 |
| Homogeneous course silt | 3 : 1 | 2.5 : 1 |
| Homogeneous silty clay | | |
| (*i*) Height less than 15 m | 2.5 : 1 | 2 : 1 |
| (*ii*) Height more than 15 m | 3 : 1 | 2.5 : 1 |
| Sand or Sand and gravel with a central clay core | 3 : 1 | 2.5 : 1 |
| Sand or Sand and gravel with R.C. diaphragm | 2.5 : 1 | 2 : 1 |

The various dimensions of low earth dams for their preliminary sections, may sometimes be selected from the recommendations of Strange, as given in Table 20.3.

**Table 20.3. Preliminary Dimensions of Low Earth Dams (Strange's Recommendations)**

| *Height of dam in metres* | *Maximum freeboard in metres* | *Top width (A) in metres* | *U/S slope (H : V)* | *D/S slope (H : V)* |
|---|---|---|---|---|
| Up to 4.5 | 1.2 to 1.5 | 1.85 | 2 : 1 | 1.5 : 1 |
| 4.5 to 7.5 | 1.5 to 1.8 | 1.85 | 2.5 : 1 | 1.75 : 1 |
| 7.5 to 15 | 1.85 | 2.5 | 3 : 1 | 2 : 1 |
| 15 to 22.5 | 2.1 | 3.0 | 3 : 1 | 2 : 1 |

## SEEPAGE ANALYSIS

Seepage occurs through the body of all earthen dams and also through their pervious foundations. The amount of seepage has to be controlled in all conservation dams and the effects of seepage (*i.e.* position of phreatic line) has to be controlled for all dams, in order to avoid their failures.

The seepage through a pervious soil material, for two dimensional flow, is given by Laplacian equation

$$\frac{\partial^2 \phi}{\partial x^2} + \frac{\partial^2 \phi}{\partial y^2} = 0 \qquad \text{...(20.14)}$$

where $\phi = K \cdot h$ = Velocity potential
$K$ = Permeability of the soil
$h$ = Head causing flow.

The above equation is based on the following assumptions :

(*i*) Water is incompressible.

(*ii*) The soil is incompressible and porous. The size of the pore space do not change with time regardless of water pressure.

(*iii*) The quantity of water entering the soil in any given time is the same as the quantity flowing out of the soil.

(*iv*) Darcy's law is valid for the given soils.

(*v*) The hydraulic boundary conditions at the entry and exit are known.

A graphical solution of the above equation, (*i.e.* Eq. 20.11) suggests that the flow through the soil, following the above assumptions, can be represented by a *flow-net* ; which consists of two sets of curves, known as '*Equipotential lines*' (*i.e.* lines of equal energy) and '*stream lines*' (*i.e.* flow lines), mutually perpendicular to each other, as shown in Fig. 20.11.

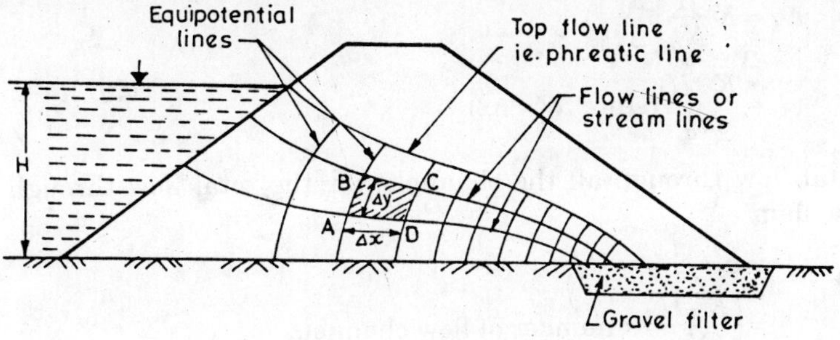


Fig. 20.11. Flow Net.

### 20.10. Seepage Discharge Through the Isotropic Soils

The amount of seepage can be easily computed from the flownet. Let us assume that the soil is isotropic, *i.e.* its permeability is constant in all directions, or $K_H = K_V$. (*i.e.* horizontal permeability is equal to the vertical permeability). The dam section is first of all plotted to a given scale (Same scale for horizontal and vertical directions). The flow net is drawn by free hand sketching the making suitable adjustments and corrections until the flow lines and equipotential lines intersect at right angles. It is convenient to draw only a limited number of flow lines and equipotential lines, such that the rate of flow between each pair of flow lines (called flow channels) is equal and the energy drop between any two successive potential lines is the same. The distance between the flow lines is made equal to the distance between the potential lines, thus forming a series of squares. Where the flow lines are curved, the squares formed will be distorted, but they will be more perfect as the number of lines is increased.

The seepage rate ($q$) can be computed from the flow net, using Darcy's Law. Applying the principle of continuity between each pair of flow lines, it is evident that the velocity must vary inversely with the spacing. Assuming the dam cross-section of Fig. 20.11 to have a unit width, we have :

The flow through the square $ABCD$ (called field) or through the flow channel containing this square

$$= \Delta q = K \cdot iA$$

$$= K\left(\frac{\Delta H}{\Delta x}\right)(\Delta y \times 1)$$

where $\Delta H$ is the energy drop between the two equipotential lines bounding the square $ABCD$ and $\Delta x$ and $\Delta y$ are defined in Fig. 20.11.

$$\therefore \quad \Delta q = K\left(\frac{\Delta H}{\Delta x}\right)\Delta y$$

But $$\Delta H = \frac{\text{Total drop, } i.e. \text{ total head causing flow}}{\text{Number of increments into which the total drop is equally divided}}$$

or $$\Delta H = \frac{H}{N_d}$$

where $N_d =$ Total number of drops in the complete flow-net.

$$\Delta q = K\frac{H}{N_d}\left(\frac{\Delta y}{\Delta x}\right)$$

$$= \frac{K \cdot H}{N_d} \text{ (since } \Delta y = \Delta x)$$

The total flow through all the channels, *i.e.* the total flow through the unit width of the dam

$$= q = \Sigma \cdot \Delta q$$

$$= \left(\frac{K \cdot H}{N_d}\right) \times \text{number of flow channels}$$

or $$\boxed{q = \frac{K \cdot H}{N_d} N_f} \qquad \text{...(20.15)}$$

This is the required expression, representing discharge passing through a flow net and is applicable only to isotropic soils (*i.e.* soils for which $K_H = K_V$).

## 20.11. Seepage Discharge for Non-isotropic Soils

If the permeability of the soil is different in the horizontal direction than that in the vertical direction ; the flow net is drawn in the same manner as was explained earlier for isotropic soils, with the only difference that the dam section shall be drawn to the same vertical scale *but to a transformed horizontal scale*. All horizontal dimensions shall be reduced by multiplying them by a factor equal to

$\sqrt{\frac{K_V}{K_H}}$. Flow net and squares will be drawn in the same manner, and number of flow channesl ($N_f$) and number of drops ($N_d$) shall be counted. The discharge can then be computed by the equation

$$q = \sqrt{K_H \cdot K_V}\ \frac{H \cdot N_f}{N_d} \qquad ...(20.16)$$

## 20.12. Line of Seepage or Phreatic Line in Earth Dams

Line of seepage or phreatic line or saturation line is defined as the line within the dam section below which there are positive hydrostatic pressures in the dam. The hydrostatic pressure on the phreatic line is equal to the atmospheric pressure and hence, equal to zero. Above the phreatic line, there is a zone of capillary saturation called capillary fringe, in which the hydrostatic pressures are negative. The appreciable flow through the dam body below the phreatic line, reduces the effective weight of this soil, and thus reduces the shear strength of the soil due to pore pressure. But on the other hand, the insignificant flow through the capillary fringe, leads to greater shear strength, because the capillary tension in water leads to increased intergranular pressure. The effects of the capillary fringe are thus on a slightly safer side and hence neglected.

It is, therefore, absolutely essential to determine the position of the phreatic line, as its position will enable us to determine the following things :

(*i*) It gives us a divide line between the dry (or moist) and submerged soil. The soil above the seepage line will be taken as dry and the soil below the seepage line shall be taken as submerged for computation of shear strength of soil.

(*ii*) It represents the top streamline and hence, helps us in drawing the flow net.

(*iii*) The seepage line determination, helps us to ensure that it does not cut the downstream face of the dam. This is extremely necessary for preventing softening or sloughing of the dam.

**20.12.1. Determination of Phreatic Line when the Dam section is Homogeneous and Provided with a Horizontal Filter.** It has been found by experiments that the seepage line is pushed down by the filter and it is very nearly parabolic except near its junction with the u/s face. Since the u/s face of the dam (*i.e.* *GB* in Fig. 20.12) becomes an equipotential line when fully covered with water (water shall rise up to the same piezometric head at every point of this line), the seepage line shall be perpendicular to this face near its junction point *B*. Let a base parabola with focus at *F* is drawn and produced so as to intersect the water surface at a point *A*. Cassagrande has shown that for dams with reasonably flat upstream slopes, $AB \approx 0.3\ HB$, where *H* is the projection of the point *G* (*i.e.* the end point of u/s slope) on the water surface. Knowing the point *A*, the base parabola *AIJC* can be drawn with its focus at *F*. It can then be corrected for the curve *BI* such that *BI* is perpendicular to *GB*, thus, *BIJC* will finally represent the seepage line.

*Equation of the base parabola.* The equation of the parabola can be determined from the basic property of the parabola *viz., the distance of any point P (x, y) on the parabola from its focus is the same as the distance of that point P (x, y) from a line called directrix.* Taking the focus (*F*) as the origin, the equation of the parabola can be written as

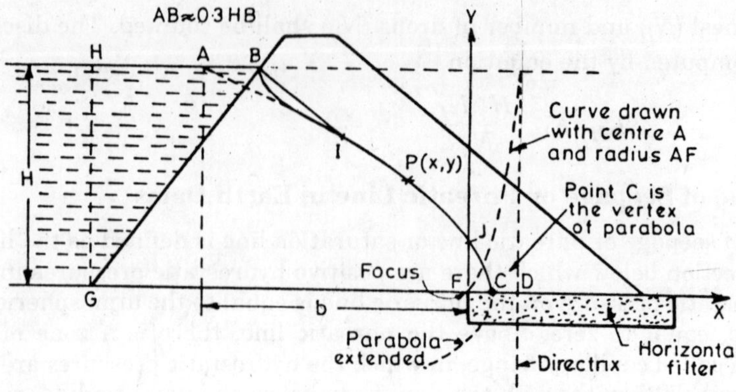


Point C is the vertex of the parabola
Fig. 20.12

$$\sqrt{x^2+y^2} = x + FD$$

where the vertical line through $D$ is the directrix. $ED$ is the distance of the focus from the directrix, called focal distance and is generally represented by $S$.

The equation of the parabola, then becomes

$$\sqrt{x^2+y^2} = x + S \quad \ldots(20.17)$$

If the horizontal distance between the already determined point $A$ and the focus ($F$) is taken as say $b$, then ($b$, $H$) represents the coordinates of the point $A$ on the parabola. Substituting in equation (20.17), we get

$$\sqrt{b^2+H^2} = b + S$$

or
$$S = \sqrt{b^2+H^2} - b \quad \ldots(20.18)$$

$S$ can be calculated from this equation.

The focal distance $S$ can also be measured by drawing an arc $FH$ with centre $A$ and radius equal to $AF$ so as to intersect the Horizontal water surface $AB$ produced at $H$. The vertical line $HD$ through $H$ will then represent the directrix. $FD$ will be equal to $S$. The centre point ($C$) of $FD$ will be the vertex of the parabola.

Also from equation (20.17) : when $x = 0$, $y_{(0)} = S$. Hence, the vertical ordinate $FJ$ at $F$ will be equal to $S$. Hence, $S$ is many a times, represented by $y_{(0)}$. Knowing the points $A$, $C$, $J$ and working out a few more points from the equation (20.17), the parabola can be easily drawn and corrected for the curve $BI$, so as to get the seepage line $BIJC$.

The discharge can also be calculated easily from the equation of the seepage line, without taking the trouble of drawing a flow net, as explained below :

Consider a unit width of the dam. Let $q$ be the seepage discharge per unit width of the dam. Then, according to Darcy's Law, $q = KiA$. When steady conditions have reached, the discharge crossing any vertical plane across the dam section will be the same. Hence, the values of $i$ and $A$ can be taken for any point on the seepage line

$$\therefore \qquad i = \frac{dy}{dx}$$

$$A = y \times 1 \text{ (i.e. saturated depth} \times \text{width)}$$

$$\therefore \qquad q = K\frac{dy}{dx} \cdot y \qquad ...(20.19)$$

But the equation of the parabola is

$$\sqrt{x^2 + y^2} = x + S \qquad ...(20.17)$$

or $$x^2 + y^2 = (x + S)^2$$

or $$y^2 = (x + S)^2 - x^2$$

or $$y = \sqrt{S^2 + 2xS}.$$

Equation (20.19) becomes

$$q = K\left[\frac{1}{2} \cdot (S^2 + 2xS)^{\frac{1}{2} - 1} \cdot 2S\right] \cdot \left[\sqrt{S^2 + 2xS}\right]$$

$$= K \cdot S \cdot (S^2 + 2xS)^{-\frac{1}{2}} \sqrt{S^2 + 2xS}$$

$$= K \cdot S \cdot \frac{\sqrt{S^2 + 2xS}}{\sqrt{S^2 + 2xS}} = K \cdot S.$$

or $$\boxed{q = KS} \qquad ...(20.20)$$

The coefficient of permeability $K$ and the focal distance ($S$) are known ; the discharge $q$ can be easily computed. This is an important equation. Strictly speaking, this equation is applicable only to dams with horizontal drainage but is used for other type of section also and gives quite close values of discharge.

**20.12.2. Determination of Phreatic Line when the Dam Section is Homogeneous (without Filter).** The phreatic line can be determined on the same principles as was done for dam with a filter case. The focus ($F$) of the parabola, in this case, will be the lowest point of the downstream slope as shown in Fig. 20.13. The base parabola $BIJC$ will cut the downstream slope at $J$ and extend beyond the dam toe up to the point $C$ (*i.e.* the vertex of the parabola).

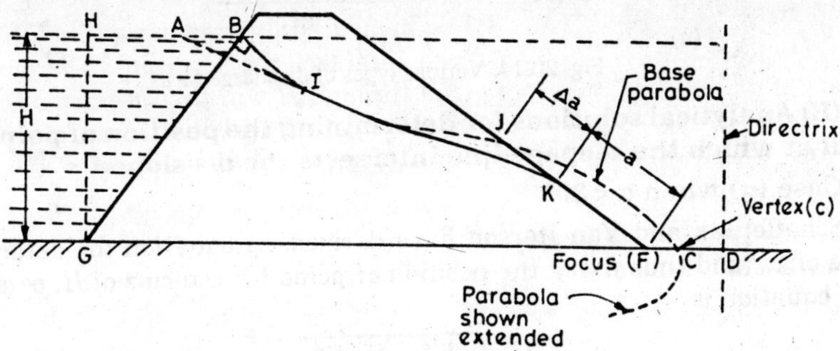


Fig. 20.13

The seepage line will, however, emerge out at $K$, meeting the downstream face tangentially there. *The portion KF is known as discharge face and always remains saturated.* The correction $JK$ (say $\Delta a$) by which the parabola is to be shifted downward can be determined as follows :

**(A) Graphical general solution.** Cassagrande has given a general solution to evaluate $\Delta a$ for various inclinations of discharge face. Let $a$ be the angle which the discharge face makes with the horizontal. The various values of $\frac{\Delta a}{a + \Delta a}$ have been given by Cassagrande, as shown in table 20.4.

**Table 20.4**

| $\alpha$ *in degrees* | $\frac{\Delta a}{a + \Delta a}$ | *Remarks* |
|---|---|---|
| 30° | 0.36 | *Note.* Intermediate values can be interpolated, or read out from a graph between $\alpha$ and $\frac{\Delta a}{a + \Delta a}$ plotted with the values given here. |
| 60° | 0.32 | |
| 90° | 0.26 | |
| 120° | 0.18 | |
| 135° | 0.14 | |
| 150° | 0.10 | |
| 180° | 0.0 | |

$(a + \Delta a)$ is the distance $FJ$ (*i.e.* the distance of the focus from the point where the parabola cuts the d/s face) and is known. $\Delta a$ can then be evaluated. $a$ and $\Delta a$ can be connected by a general equation.

$$\Delta a = (a + \Delta a)\left[\frac{180° - \alpha}{400°}\right] \qquad ...(20.21)$$

The value of $\alpha$ will be equal to 180° for a horizontal filter case and may be equal to or more than 90° in case a rock toe is provided at the downstream end, as shown in Fig. 20.14 (*a*). $\alpha$ will be less then 90° when no drainage is provided.

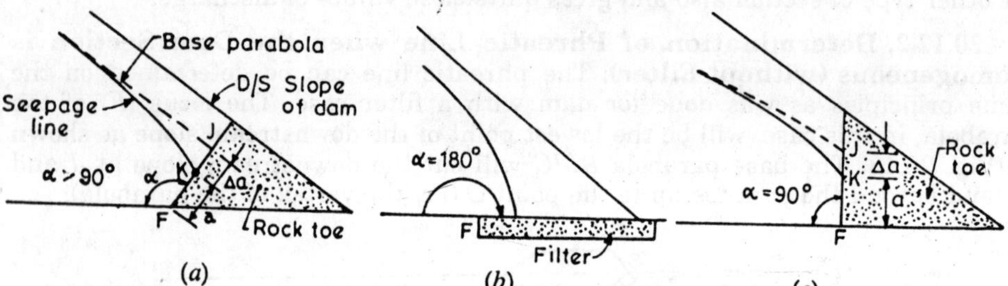


Fig. 20.14. Various types of discharge faces.

**(B) Analytical solutions for determining the position of point k, *i.e.* the point at which the seepage line intersects the d/s slope.**

**Case (*a*) when $\alpha$ < 30°**

Schaffernak and Van Iterson have derived an equation for determining the value of '$a$' (and thus fixing the position of point $K$) in terms of $H$, $b'$ and $\alpha$. Their final equation is

$$a = \frac{b'}{\cos\alpha} - \sqrt{\frac{b'^2}{\cos^2\alpha} - \frac{H^2}{\sin^2\alpha}} \qquad ...(20.22)$$

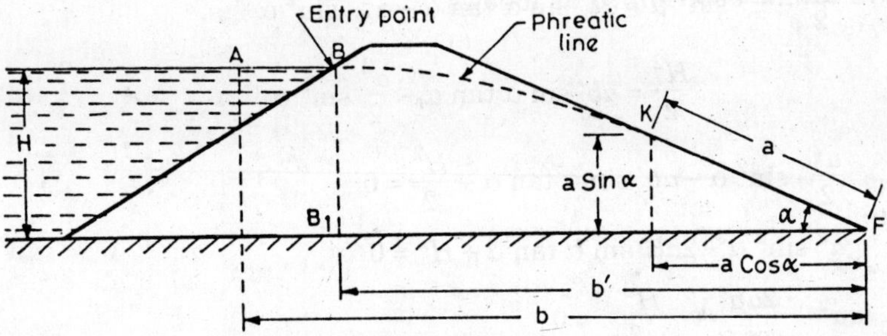


Fig. 20.15

The above equation has been obtained on the assumption that the hydraulic gradient ($i$) is equal to the slope of the phreatic line. This assumption is nearly true so long as the downstream slope is sufficiently flat (*i.e.* $\alpha < 30°$).

This equation can be easily derived on the basis of the above assumption as follows:

$$i = \text{Hydraulic gradient} = \frac{dy}{dx}$$

$$A = y \cdot 1$$

$$q = KiA = K \cdot \left(\frac{dy}{dx}\right) \cdot (y \cdot 1)$$

$$\text{But } \frac{dy}{dx} = \tan \alpha$$

$$\text{and } \quad y = a \sin \alpha$$

$$\therefore \quad q = K \cdot \tan \alpha \cdot a \cdot \sin \alpha$$

$$= K \cdot a \cdot \sin \alpha \cdot \tan \alpha$$

or $$K \cdot \frac{dy}{dx}(y) = K \cdot a \cdot \sin \alpha \tan a$$

or $$\frac{dy}{dx}(y) = a \sin \alpha \tan \alpha$$

or $$dy \cdot y = a \sin \alpha \tan \alpha \, dx$$

or $$y \cdot dy = a \sin \alpha \tan \alpha \, dx.$$

Integrating both sides between the limits

$$x = a \cos \alpha \quad \text{to} \quad x = b'$$

and $$y = a \sin \alpha \quad \text{to} \quad y = H,$$

we get $$\int_{y = a \sin \alpha}^{y = H} y \cdot dy = a \sin \alpha \cdot \tan \alpha \int_{x = a \cos \alpha}^{x = b'} dx$$

or $$\left| \frac{y^2}{2} \right|_{a \sin \alpha}^{H} = a \sin \alpha \tan \alpha \left| x \right|_{a \cos \alpha}^{b'}$$

or $$\frac{H^2 - a^2 \sin^2 \alpha}{2} = a \sin \alpha \tan \alpha \left[ b' - a \cos \alpha \right]$$

or $$\frac{H^2}{2} - \frac{a^2}{2} \sin^2 \alpha = ab' \sin \alpha \tan \alpha - a^2 \cdot \sin \alpha \cos \alpha \cdot \tan \alpha$$

or $$\frac{H^2}{2} - \frac{a^2}{2}\sin^2\alpha = ab'\sin\alpha\tan\alpha - a^2\cdot\sin^2\alpha$$

or $$\frac{H^2}{2} = ab'\sin\alpha\tan\alpha - \frac{a^2}{2}\sin^2\alpha$$

or $$\frac{a^2}{2}\cdot\sin^2\alpha - ab'\sin\alpha\tan\alpha + \frac{H^2}{2} = 0$$

or $$a^2\sin^2\alpha - 2ab'\sin\alpha\tan\alpha + H^2 = 0$$

or $$a^2 - \frac{2ab'}{\cos\alpha} + \frac{H^2}{\sin^2\alpha} = 0$$

or $$a = \frac{\frac{2b'}{\cos\alpha} \pm \sqrt{\left(\frac{2b'}{\cos\alpha}\right)^2 - \frac{4H^2}{\sin^2\alpha}}}{2}$$

Ignoring unfeasible +ve sign, we have

$$\boxed{a = \frac{b'}{\cos\alpha} - \sqrt{\frac{b^2}{\cos^2\alpha} - \frac{H^2}{\sin^2\alpha}}} \qquad ...(20.22)$$

This is the required equation.

**Case (*b*) When α lies between 30° and 60°**

Cassagrande has derived an equation for determining the value of '*a*' in terms of *b*, *H* and α. His final equation is

$$a = \sqrt{b^2 + H^2} - \sqrt{b^2 - H^2\cot^2\alpha} \qquad ...(20.23)$$

where *b* is defined in Fig. 20.16,

*H* is the head causing flow, and α is the angle which the d/s face makes with the horizontal (clockwise) as defined earlier.

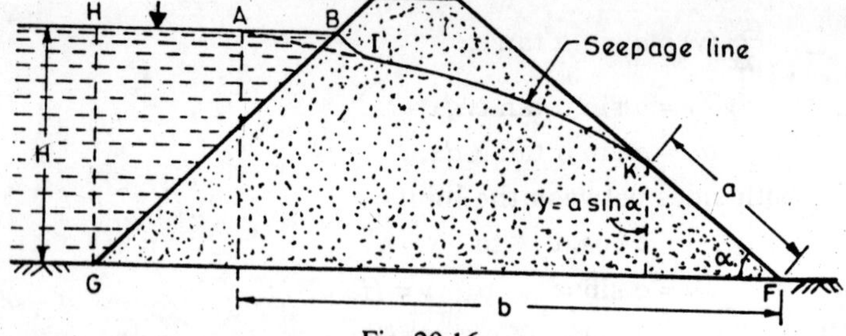


Fig. 20.16

Equation (20.23) gives satisfactory results for values of α less than 30°, but for steeper slopes, the deviations from the correct values become quite high and for such cases, Cassagrande has suggested the use of sin α in place of tan α.

In other words, the hydraulic gradient ($i$) is given by $\frac{dy}{ds}$ (*i.e.* sin α) of the seepage line and not by $\frac{dy}{dx}$ (*i.e.* tan α) as was taken in the previous case.

Therefore, $q = K\cdot i\cdot A$

or $$q = K \cdot \left(\frac{dy}{ds}\right) y \cdot 1 = K (\sin \alpha)(a \sin \alpha) \qquad \because \frac{dy}{ds} = \sin \alpha$$

or $$K \frac{dy}{ds} y = K \cdot a \cdot \sin^2 \alpha$$

or $$dy \cdot y = a \sin^2 \alpha \cdot ds.$$

integrating between the limits

$$y = a \sin \alpha, \qquad s = a$$
$$y = H \qquad s = S_0$$

where $S_0$ is the total length of parabola from the point $A$ to the point $F$.

or $$\int_{y = a \sin \alpha}^{y = H} y \cdot dy = a \cdot \sin^2 \alpha \int_{s = a}^{s = S_0} ds$$

or $$\left| \frac{y^2}{2} \right|_{a \sin \alpha}^{H} = a \sin^2 \alpha \cdot |s|_{s=a}^{s=S_0}$$

or $$\frac{H^2 - a^2 \sin^2 \alpha}{2} = a \sin^2 \alpha \, [S_0 - a]$$

or $$\frac{H^2}{2} - \frac{a^2}{2} \sin^2 \alpha = a \sin^2 \alpha \cdot S_0 - a^2 \sin^2 \alpha$$

or $$\frac{a^2}{2} \sin^2 \alpha - S_0 \sin^2 \alpha \cdot a + \frac{H^2}{2} = 0$$

or $$a^2 - 2S_0 \cdot a + \frac{H^2}{\sin^2 \alpha} = 0$$

or $$a = \frac{2 \cdot S_0 \pm \sqrt{4 \cdot S_0^2 - 4 \cdot \dfrac{H^2}{\sin^2 \alpha}}}{2}$$

Ignoring +ve sign, we get

$$a = S_0 - \sqrt{S_0{}^2 - \frac{H^2}{\sin^2 \alpha}} \qquad \text{...(20.24)}$$

The total length of the parabola $S_0$ can be approximately taken to be equal to $\sqrt{b^2 + H^2}$, then

$$S_0 = \sqrt{b^2 + H^2}$$

Substituting, we get

$$a = \sqrt{b^2 + H^2} - \sqrt{b^2 + H^2 - \frac{H^2}{\sin^2 \alpha}}$$

or $$a = \sqrt{b^2 + H^2} - \sqrt{b^2 - H^2 \left(\frac{1}{\sin^2 \alpha} - 1\right)}$$

$$\boxed{a = \sqrt{b^2 + H^2} - \sqrt{b^2 - H^2 \cot^2 \alpha}}$$ which is the required eqn. (20.23).

**Example 20.1.** *An earth dam made of a homogeneous material has the following data :*

| | |
|---|---|
| *Coefficient of permeability of dam material* | $= 5 \times 10^{-4}$ *cm/sec.* |
| *Level of top of dam* | = *200.0 m.* |
| *Level of deepest river bed* | = *178.0 m* |
| *H.F.L. of reservoir* | = *197.5 m.* |
| *Width of the top of dam* | = *4.5 m.* |
| *Upstream slope* | = *3 : 1.* |
| *Downstream slope* | = *2 : 1.* |

*Determine the phreatic line for this dam section and the discharge passing through the dam.*

**Solution.** Taking the focus ($F$) at the d/s toe of the dam as the origin, the equation of the base parabola is given by $\sqrt{x^2 + y^2} = x + S$

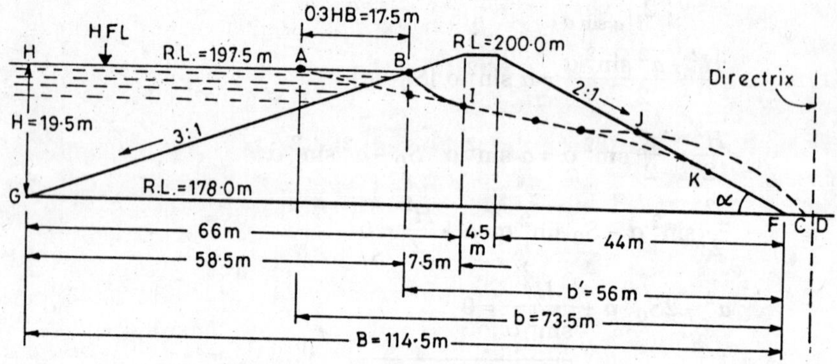


Fig. 20.17

where $S$ is the distance of the point $(x, y)$ from the directrix, called focal distance. Select the point $A$, in such a way that

$$AB \approx 0.3\ HB$$

or
$$AB \approx 0.3 \times 58.5 = 17.5 \text{ m.}$$

Plot the point $A$ at a distance 17.5 m from $B$ (Fig. 20.17). This is the start-point of the parabola. Now the coordinates of the point $A$ w.r.t. to $F$ as origin are (73.5 m, 19.5 m). Substituting these ordinates in the equation of the base parabola, we get

$$\sqrt{(73.5)^2 + (19.5)^2} = 73.5 + S.$$

or
$$\mathbf{S = 2.54\ m.}$$

The vertex ($C$) of the base parabola shall be situated at a distance equal to $\frac{S}{2}$ or 1.27 m from $F$, beyond the d/s toe of the dam, as shown in Fig. 20.17.

A few more coordinates of the base parabola at known distances ($x$) are worked out in table 20.4 using,

$$x^2 + y^2 = x^2 + S^2 + 2xS$$

or
$$y = \sqrt{S^2 + 2xS}$$

**Table 20.4**

| $x$ | $y^2 = 2xS + S^2 = 5.08x + 6.45$ | $y = \sqrt{S^2 + 2xS}$ |
|---|---|---|
| 0 | 6.45 | 2.54 |
| 10 | 57.25 | 7.68 |
| 20 | 108.05 | 10.39 |
| 30 | 158.85 | 12.61 |
| 40 | 209.65 | 14.48 |
| 44 | 229.97 | 15.16 |
| 48.5 | 252.83 | 15.92 |
| 56 | 290.93 | 17.04 |
| 60 | 311.25 | 17.64 |
| 70 | 362.05 | 19.02 |
| 73.5 | 380.00 | 19.50 |

The base parabola with all these ordinates, is then drawn.

Now, this parabola has to be corrected at entry and exit as explained earlier. At entry, the phreatic line is started from the point $B$ in such a way that it becomes at right angles to the u/s face $GB$ of the dam. A reverse curvature $BI$ is, therefore, given as shown in Fig. 20.17.

At exit, the point $K$ at which the phreatic line intersects the d/s face can be easily obtained by using the Eq. (20.21) as :

$$\Delta a = (a + \Delta a)\left(\frac{180 - \alpha}{400}\right)$$

where $\tan \alpha = \frac{1}{2}$; or $\alpha = 26° \cdot 54$

$(a + \Delta a) =$ Distance $FJ$, *i.e.* the distance of the focus from the point at which the base parabola intersects the d/s face, and is measured from Fig. 20.17 ≈ 25.6 m.

$$\therefore \quad \Delta a = 25.6\left[\frac{180 - 26.54}{400}\right] = 25.6 \times \frac{153.46}{400} = 9.84 \text{ m} \; ; \textbf{ say 9.8 m.}$$

$$a = 25.6 - 98 = 15.8 \text{ m.}$$

Knowing '$a$', the point $K$ is plotted and the phreatic line $BIK$ is completed.

*Discharge through the dam section* can be obtained from the equation (20.20) as :

$$q = K \cdot S, \qquad \text{where } K = 5 \times 10^{-4} \text{ cm/sec} = 5 \times 10^{-6} \text{ m/sec.}$$
$$S = 2.54 \text{ m}$$

$$\therefore \quad q = 5 \times 10^{-6} \times 2.54 \text{ m}^3\text{/m run/sec.}$$
$$= 12.7 \times 10^{-6} \text{ cumes/m length of dam}$$

Hence, $\quad q = 12.7 \times 10^{-6}$ cumecs/m length of dam. **Ans.**

**Note.** The value of '$a$' can also be determined from Eq. (20.22), if $(a + \Delta a)$ is not to be measured, as the dam section is not to be plotted to scale.

In that case $a = \dfrac{b'}{\cos a} - \sqrt{\left(\dfrac{b'}{\cos \alpha}\right)^2 - \left(\dfrac{H}{\sin \alpha}\right)^2}$

$$\text{where } b' = 56 \text{ m}$$
$$H = 19.5 \text{ m}$$
$$\sin 26.56° = 0.447$$
$$\cos 26.54° = 0.894.$$

$$\therefore \quad a = \frac{56}{0.894} - \sqrt{\left(\frac{56}{0.894}\right)^2 - \left(\frac{19.5}{0.447}\right)^2}$$
$$= 62.6 - \sqrt{3,920 - 1,900} = 62.6 - 44.9 = \mathbf{17.7\ m.}$$

Hence, $a = 17.7$ m which is slightly above the value obtained from Eq. (20.21) and is thus on a safer side.

**Example 20.2.** *A flow net is plotted for a homogeneous earthen dam of height 22 m and freeboard 2.0 m. The results obtained are,*

*Number of potential drops* = 10
*Number of flow channels* = 4.

*The dam has a horizontal filter of 30 m length at the downstream end and the coefficient of permeability of the dam material is* $5 \times 10^{-4}$ *cm/sec. Calculate the discharge per m run of the dam.*

**Solution.** The discharge through a dam section is approximately given by the Eq. (20.15) as :

$$q = K.H. \frac{N_f}{N_d}$$

$$\text{where } K = 5 \times 10^{-4} \text{ cm/sec} = 5 \times 10^{-6} \text{ m/sec.}$$
$$H = 22 - 2 = 20 \text{ m.}$$
$$N_f = 4$$
$$N_d = 10$$

$$\therefore \quad q = 5 \times 10^{-6} \times \frac{20 \times 4}{10}$$
$$= \mathbf{4.0 \times 10^{-6}} \textbf{ cumecs/m run of dam. Ans.}$$

**Example 20.3.** *For the dam section given in example 20.1, draw the seepage line if a horizontal filter of length equal to 25 m is provided inward from the downstream toe of the dam.*

**Solution.** The dam section is plotted from the dimensions given in example 20.1. The horizontal filter is also provided. Now, the left end of the filter will act as a focus and is designated as $F$ and is taken as origin.

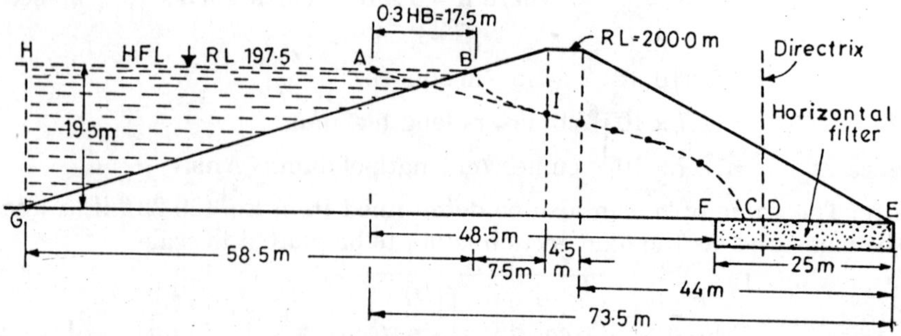


Fig. 20.18

The coordinates of any point $(x, y)$ on a parabola of equation

$$\sqrt{x^2 + y^2} = x + S$$

will now be for the point $A$.

The point $A$ is determined as in example 20.1. Its coordinates w.r. to $F$ as origin are (48.5 m, 19.5 m)

$$\therefore \quad \sqrt{(48.5)^2 + (19.5)^2} = 48.5 + S$$

or $$\sqrt{2,352 + 380} = 48.5 + S$$

or $$\mathbf{S = 3.77\ m.}$$

The vertex ($C$) of the parabola shall be situated at a distance equal to $\frac{S}{2}$, *i.e.* 1.83 m beyond the point $F$ and directrix shall be at a distance 3.77 m from $F$.

At point $F$, $\quad x = 0$

$\therefore \quad y = S = 3.77$ m.

A few more coordinates of the base parabola are worked out in table 20.5, using

$$x^2 + y^2 = x^2 + S^2 + 2xS$$

or $$y = \sqrt{2xS + S^2}$$

**Table 20.5**

| $x$ | $y^2 = (2xS + S^2) = 7.54x + 14.21$ | $y = \sqrt{2xS + S^2}$ |
|---|---|---|
| 0 | 14.21 | 3.77 |
| 10 | 89.61 | 9.47 |
| 15 | 127.31 | 11.27 |
| 19 | 157.47 | 12.54 |
| 23.5 | 191.40 | 13.85 |
| 31 | 247.95 | 15.75 |
| 40 | 315.81 | 17.78 |
| 48.5 | 380.00 | 19.5 |

The base parabola is drawn and correction at the entry point for the curve $BI$ is made by free hand sketching, in such a way that the phreatic line becomes $\perp$ to u/s face $GB$ of the dam. The exit point should also be corrected such that the phreatic line meets $\perp$ to base line $FCD$. The final phreatic line $BIC$ is thus drawn as shown in Fig. 20.18.

**20.12.3. Phreatic Line for a Zoned Section.** In case of a zoned section having a central impervious core, such as shown in Fig. 20.19 ; the effects of the outer zone can be neglected altogether.

The focus of the base parabola will, therefore, be located at the d/s toe of the core, as shown in Fig. 20.19. The phreatic line can then be drawn as usual with free hand corrections required at suitable places.

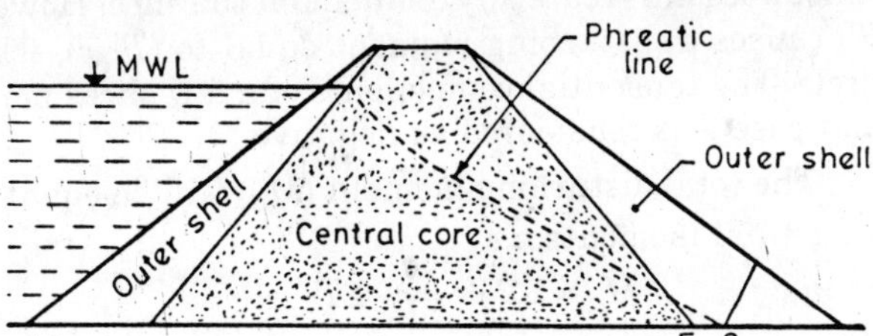


Fig. 20.19

## 20.13. Stability of Earthern Slopes

An earth embankment usually fails, because of the sliding of a large soil mass along a curved surface. It has been established by actual investigation of slides of railway embankments in Sweden, that the surface of slip is usually close to cylindrical, *i.e.* an arc of a circle in cross-section. The method which is described here and is generally used for examining the stability of slopes of an earthen embankment is called the *Swedish Slip Circle Method* or *The Slices Method*. The method thus assumes the condition of plane strain with failure along a cylindrical arc.

The location of the centre of the possible failure arc is assumed. The earth mass is divided into a number of vertical segments called slices. These verticals are usually equally spaced, though it is not necessary to do so. Depending upon the accuracy desired, six to twelve slices are generally sufficient.

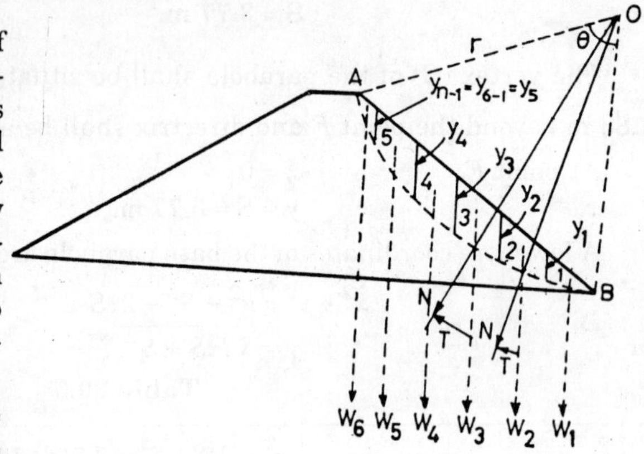


Fig. 20.20

Let $O$ be the centre and $r$ be the radius of the possible slip surface as shown in Fig. 20.20. Let the total arc $AB$ be divided into slices of equal width say $b$ metres each. The width of the last slice will be something different say let it be $m \cdot b$ metres.

Let these slices be numbered as 1, 2, 3, 4, ... and let the weight of these slices be $W_1, W_2, W_3, W_4$ ...

The forces between these slices are neglected and each slice is assumed to act independently as a vertical column of soil of unit thickness (⊥ to paper) and width $b$. The weight $W$ of each slice is assumed to act at its centre. The weight $W$ of each slice can be resolved into two components ; say a normal component ($N$) and a tangential component ($T$) such that

$$N = W \cos \alpha$$
$$T = W \sin \alpha$$

where $\alpha$ is the angle which the slope makes with the horizontal.

The normal component ($N$), will pass through the centre of rotation ($O$) and hence does not create any moment on the slice. However, the tangential component ($T$) causes a disturbing moment equal to $(T \cdot r)$, where $r$ is the radius of the slip circle. The tangential components of a few slices may create resisting moments; in that case $T$ is considered as negative.

The total disturbing moment ($M_d$) will be equal to the algebraic sum of all the tangential moments, *i.e.*

$$M_d = \Sigma T \cdot r = r \cdot \Sigma T$$

where $\Sigma T = (T_1 + T_2 + T_3 + ...)$

The resisting moment is supplied by the development of shearing resistance of the soil along the arcual surface $AB$. The magnitude of shear strength developed in each slice will depend upon the normal component ($N$) of that slice. Its magnitude will be

$$= c \cdot \Delta L + N \tan \phi$$

where $c$ is the unit cohesion, $\Delta L$ is the curved length of the slice, and $\phi$ is the angle of internal friction of soil.

This shear resistance is acting at a distance $r$ from $O$ and will provide a resisting moment

$$= r\,[c \cdot \Delta L + N \tan \phi]$$

The total resisting moment over the entire arc $AB$

$$M_r = r\,[\Sigma C \cdot \Delta L + \Sigma N \cdot \tan \phi]$$
$$= r\,[c \Sigma\, \Delta L + (\Sigma N) \tan \phi]$$
$$= r\,[c \cdot AB + (\Sigma N) \tan \phi]$$

where $\Sigma N$ is the sum of all the normal components.

$$= N_1 + N_2 + N_3 + \ldots$$

Length $AB$ of slip circle

$$= AB = \left[\frac{2\pi \cdot r}{360^\circ}\right] \times \theta$$

where $\theta$ is the angle in degrees, formed by the arc $AB$ at centre $O$.

Hence, the factor of safety against sliding

$$= \text{F.S.} = \frac{\text{Resisting moment}}{\text{Disturbing moment}} = \frac{M_r}{M_d}$$

or
$$= \frac{r\,[c \cdot AB + (\tan \phi)\, \Sigma N]}{r \cdot \Sigma T}$$

or
$$\text{F.S.} = \frac{[c \cdot AB + (\tan \phi)\, \Sigma N]}{(\Sigma T)} \qquad \ldots(20.24)$$

Eq. (20.24) can be worked out by working out $\Sigma T$ and $\Sigma N$ separately. This evaluation of $\Sigma N$ and $\Sigma T$ can be simplified as explained below :

If $y_1, y_2, y_3,\ldots$ are the vertical extreme ordinates (boundary ordinates) of the slices 1, 2, 3... respectively, then the respective weights can be written as

$$W_1 = \left(\frac{0 + y_1}{2}\right) \cdot b \cdot \gamma \cdot 1$$

where $\gamma$ is the unit weight of soil and unit width of the slice is considered.

$$W_2 = \left(\frac{y_1 + y_2}{2}\right) b \cdot \gamma$$

$$W_3 = \left(\frac{y_2 + y_3}{2}\right) b \cdot \gamma$$

$$W_4 = \left(\frac{y_3 + y_4}{2}\right) b \cdot \gamma$$

..............................

..............................

$$W_n = \left(\frac{y_{n-1} + 0}{2}\right) m \cdot b \cdot \gamma$$

where $n$ is the total number of slices.

$$\therefore \quad \Sigma W = W_1 + W_2 + W_3 + \ldots + W_n$$

$$= \left[ y_1 b + y_2 b + y_3 b + \ldots \frac{y_{n-1}}{2} b + \frac{y_{n-1}}{2} m \cdot b \right] \gamma$$

$$= \left[ y_1 + y_2 + y_3 + \ldots y_{n-1} \left( \frac{1+m}{2} \right) \right] \gamma \cdot b$$

Now $\quad \Sigma N = N_1 + N_2 + N_3 + \ldots$

$$= W_1 \cos \alpha + W_2 \cos \alpha + W_3 \cos \alpha + \ldots$$

$$\therefore \quad \Sigma N = \cos \alpha \, [W_1 + W_2 + W_3 + \ldots]$$

or $\quad \Sigma N = \cos \alpha \, (\Sigma W).$

Similarly $\Sigma T = \sin \alpha \, (\Sigma W)$ if all $T$'s are +ve.

$\Sigma W$ can be evaluated by adding all the vertical boundary ordinates of all the slices. (The last ordinate should be multiplied by $\frac{m+1}{2}$ before adding ) and then multiplying this summation (say $\Sigma y$) by $b \cdot \gamma$.

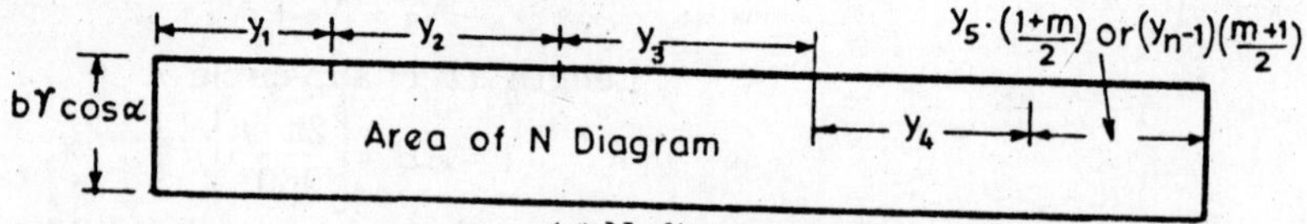


(*a*) N–diagram

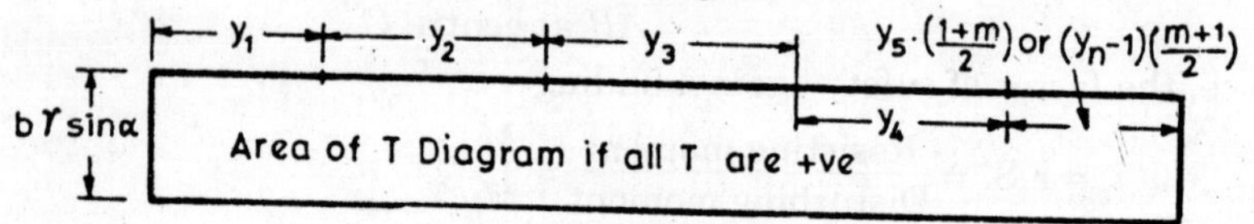


(*b*) T–diagram

Fig. 20.21

The area of $N$ diagram will represent $\Sigma N$ and that of $T$ diagram will represent $\Sigma T$.

As a general case, the value of $\Sigma N$ and $\Sigma T$ can be worked out in a tabular form as shown below in table 20.6.

**Table 20.6**

| *Slice number* | *Wt. of each slice* | $N = W \cos \alpha$ | $T = W \sin \alpha$ | $c \cdot \Delta L$ |
|---|---|---|---|---|
| 1 | $W_1$ | $N_1$ | $T_1$, | |
| 2 | $W_2$ | $N_2$ | $T_2$, | |
| 3 | $W_3$ | $N_3$ | $T_3$, | |
| ⋮ | | | | |
| $n$ | $W_n$ | $N_n$ | $T_n$, | |
| | | $\Sigma N$ | $\Sigma T$ | $\Sigma c \cdot \Delta L$ |

The F.S. is then calculated from equation

$$\text{F.S.} = \frac{\Sigma c \cdot \Delta L + (\Sigma N - \Sigma U) \tan \phi}{\Sigma T} \quad \ldots(20.25a)$$

**20.13.1. Location of the Centre of the Critical Slip Circle.** In order to find out the worst case, numerous slip circles should be assumed and factor of safety (F.S.) calculated for each circle, as explained earlier. The minimum factor of safety will be obtained for the critical slip circle. In order to reduce the number of trials, Fellenius has suggested a method of drawing a line ($PQ$), representing the locus of the critical slip circle.

The determination of this line $PQ$ for the d/s slope of an embankment is shown in Fig. 20.22($a$) and similarly, the determination of this line $PQ$ for u/s slope is shown in Fig. 20.22($b$). The point $Q$ is determined in such a way that its coordinates are $(4.5H, H)$ from the toe, as shown. The point $P$ is obtained with the help of directional angles $\alpha_1$ and $\alpha_2$ as shown. The value of $\alpha_1$ and $\alpha_2$ for various slopes, are tabulated in Table 20.7.

**Table 20.7**

| *Slope* | *Directional angles* | |
|---|---|---|
| | $\alpha_1$ *in degrees* | $\alpha_2$ *in degrees* |
| 1 : 1 | 27.5 | 37 |
| 2 : 1 | 25 | 35 |
| 3 : 1 | 25 | 35 |
| 4 : 1 | 25 | 35 |
| 5 : 1 | 25 | 35 |

After determining the locus of the critical slip circle, the critical slip circle can be drawn, keeping in view the following few points :

(*i*) Except for very small values of $\phi$, the critical arc passes through the toe of the slope.

(*ii*) If a hard stratum exists at a shallow depth under the dam, the critical arc cannot cross this stratum, but can only be tangential to it.

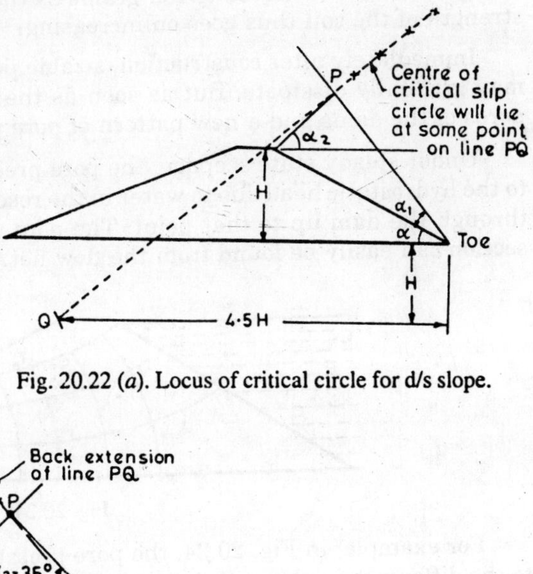


Fig. 20.22 ($a$). Locus of critical circle for d/s slope.

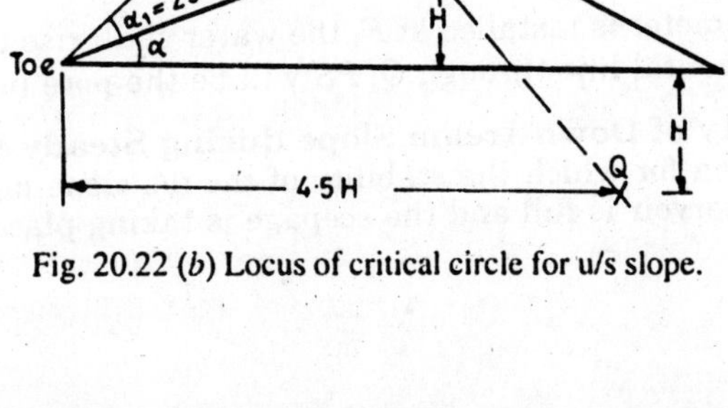


Fig. 20.22 ($b$) Locus of critical circle for u/s slope.

(*iii*) For very small values of $\phi$ (0 to 15°), the critical arc passes below the toe of the slope if the inclination of the slope is less than 53° (which is generally the case). The centre of the critical arc in such a case is likely to fall on a vertical line drawn through the centre of the slope, as shown in Fig. 20.23.

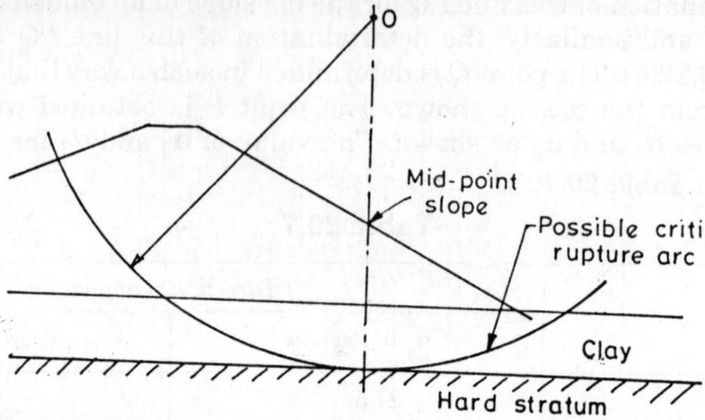


Fig. 20.23.

**20.13.2. Determination of Pore Pressures from Flow Net.** It was stated earlier that the soil is composed of voids which are filled with water and air. Whenever, any load is applied in the form of consolidation, it is taken up by water and gradually transferred to soil grains as the excess water drains out. The shear strength of the soil thus goes on increasing.

Immediately after construction, sizable pore pressures may be present, which may gradually dissipate. But as soon as the reservoir is filled, water enters the pores of the dams and a new pattern of pore pressure gets developed.

Under steady state seepage, the pore pressure at any point in a dam is equal to the hydrostatic head due to water in the reservoir minus the head loss in seepage through the dam up to that point. The pore pressure at any point within a dam section can easily be found from the flow net.

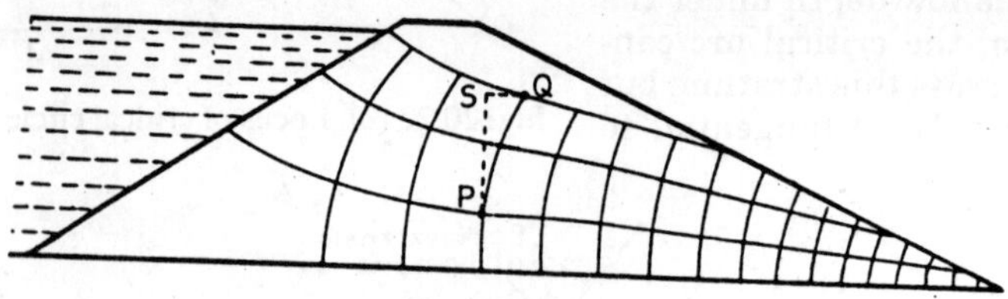


Fig. 20.24

For example, in Fig. 20.24, the pore pressure at any point say $P$ will be equal to the difference in elevation between $P$ and $Q$, where $Q$ is the point of intersection of the phreatic line with the equipotential line through the point $P$.

Hence, if a piezometer is installed at $F$, the water shall rise up to an elevation of $Q$. If $QS$ is a horizontal line through $Q$, $PS$ will be the pore pressure at $P$.

**20.13.3. Stability of Downstream Slope during Steady Seepage.** The most critical condition for which the stability of the d/s slope must be examined, occurs, when the reservoir is full and the seepage is taking place at full rate.

The seeping water below the phreatic line, exerts a pore pressure on the soil mass which lies below the phreatic line. Hence, if the slices of the critical arc, happen to include this submerged soil, [Fig. 20.25 (*a*)], the shear strength developed on those slices shall be correspondingly reduced. The net shear strength on such a slice shall be $= c\,\Delta L + (N - U)\tan\phi$, where $U$ is the pore pressure.

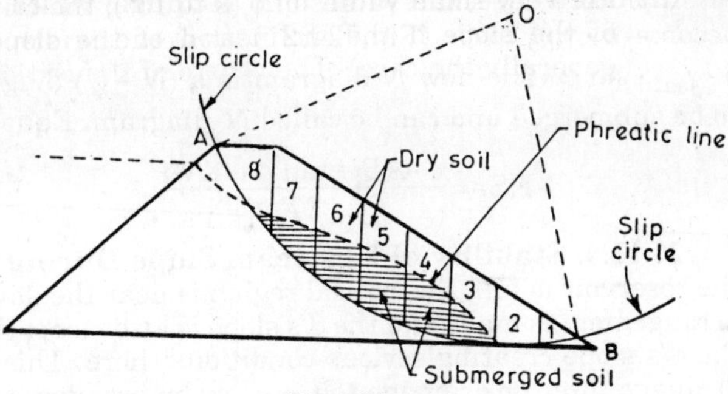


Fig. 20.25 (*a*)

The factor of safety (S.F.) for the entire slip circle is then given by the equation.

$$\text{F.S.} = \frac{c \cdot AB. + \tan\phi\,(\Sigma N - \Sigma U)}{\Sigma T} \qquad ...(20.25)$$

where $\Sigma U$ is the total pore pressure on the slip circle.

The pore pressure at a point is represented by the piezometric head at that point as explained earlier. The variation of the pore pressure along a failure arc is, therefore, obtained as explained below :

First of all, draw a flow net and thus determine the points of intersections of equipotential lines with the failure arc. At each point of intersection, measure the vertical ordinate from that intersection to the level at which that particular equipotential line cuts the phreatic line. The pore pressures represented by the vertical heights so obtained, are then plotted to a scale in a direction perpendicular to the sliding surface at the respective points of intersection.

The pore pressure distribution is thus shown in Fig. 20.25(*b*) (shaded area).

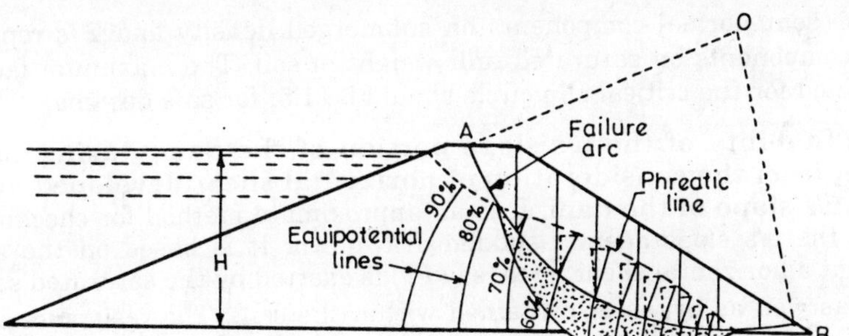


Fig. 20.25 (*b*)

The area of this diagram can be measured by a planimeter. The area of this diagram can also be calculated by ordinate method as was done for $N$ and $T$ cases taking the unit weight of water as 9.81 kN/m$^3$ ($\approx$ 10 kN/m$^3$). Knowing $\Sigma N$, $\Sigma U$ and $\Sigma T$, F.S. can be calculated easily by using equation 20.25.

**Approximate method.** In the absence of a flow net, the normal components, which are responsible for the shear strength of the soil, can be calculated on the basis of submerged unit weight of soil (*i.e.* $\gamma_{sub} = \gamma_{sat} - \gamma_w$). On the other hand, the values of tangential components *i.e.* $T$'s which are responsible for creating disturbing moments, should be calculated on the basis of saturated unit weight of soil.

In such a case, the width of $N$ rectangle which was taken as $b \cdot \gamma \cdot \cos \alpha$ will become $b \cdot \gamma_{sub} \cos \alpha$ (Fig. 20.21) and the width of $T$ rectangle will remain $b \cdot \gamma_{sat} \cdot \sin \alpha$. The new $N$ diagram is a $(N - U)$ diagram assuming the entire soil to be submerged and can be called $N'$ diagram. Equation (20.25) can be written as

$$\text{F.S.} = \frac{c \cdot AB + \tan \phi \cdot (\Sigma N')}{\Sigma T} \quad ...(20.26)$$

**20.13.4. Stability of Upstream Slope During Sudden Drawdown.** When the reservoir is full, the critical region is near the downstream face. If no drainage arrangement is made and the d/s slope is also steep, the phreatic line may intersect the d/s slope creating serious conditions there. This can be avoided by providing drainage filter or drainage toe, etc., or by broadening the base of the dam so that the head loss is great enough to bring the line of saturation beneath the d/s toe of the dam.

For the u/s slope, the critical condition can occur, when the reservoir is suddenly emptied. In such a case, the water level within the soil will remain as it was when the soil pores were full of water. The weight of this water within the soil, now tends to slide the u/s slope along a circular arc.

The tangential component of the saturated soil lying over the arc, will create a disturbing force ; while the normal component minus the pore pressure shall supply the shear strength of the soil. High pore pressures shall be developed in this case and although a true solution can be obtained from the flow net and pressure net, an approximate solution can be easily obtained, by considering the soil resting over the failure arc as saturated, for calculating $T$'s ; and as submerged for calculating $N$'s.

The factor of safety (F.S.) is finally obtained from the equation

$$\text{F.S.} = \frac{c \cdot AB + \tan \phi \, \Sigma N'}{\Sigma T}$$

$N$'s represent normal components on submerged density and $T$'s represent tangential components on saturated unit weight of soil. The maximum factor of safety obtained for the critical slip circle should be 1.5, for safe designs.

**20.13.5. Stability of the u/s slope portion of the dam, during sudden drawdown, from the consideration of horizontal shear developed at base under the u/s slope of the dam.** It is an approximate method for checking the stability of the u/s slope against sudden drawdown. It is based on the simple principle that a horizontal shear force (say $P_u$) is exerted by the saturated soil (*i.e.* by the soil as well as by water contained within the soil). The resistance to this force (say $R_u$) is provided by the shear resistance developed at the base of the soil mass, contained within the u/s triangular shoulder $GMN$ (Fig. 20.26).

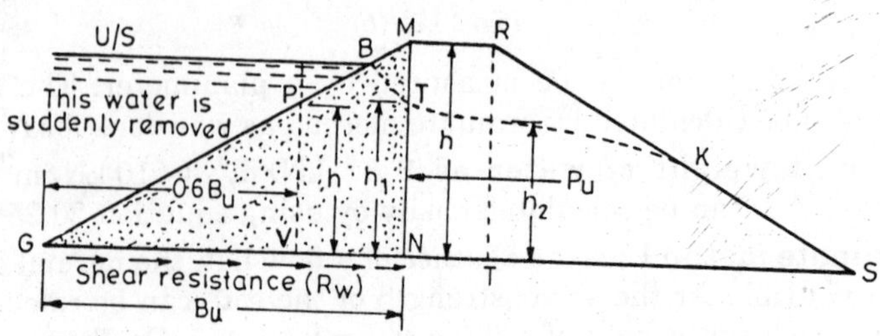


Fig. 20.26

Considering a unit length of the dam, the horizontal force $P_u$ is given by the equation

$$P_u = \left[\frac{\gamma_1 h^2}{2}\tan^2\left(45 - \frac{\phi}{2}\right) + \gamma_w \cdot \frac{{h_1}^2}{2}\right] \text{(say in kN)} \quad ...(20.27)$$

where $\gamma_1$ is the weighted density at the centre of triangular shoulder upstream and is given by

$$\gamma_1 = \frac{\gamma_{sub} \cdot h_1 + \gamma_{dry} \cdot (h - h_1)}{h} \quad ...(20.28)$$

Shear resistance ($R_u$) of u/s slope portion of the dam developed at base GN is given by

$$R_u = C + W \tan \phi \quad ...(20.29a)$$

where $W =$ the weight of the u/s triangular shoulder of dam

$C =$ The total cohesive force developed at base $GN$.

If $c$ is the unit cohesion of the dam soil, then $C = c \times (B_u \times 1)$ where $B_u$ = length $GN$.

The triangular profile of the u/s slope portion of dam has an area $GBTN$ as the submerged soil (*i.e.* the soil below the seepage line) and an area equal to $BMT$ as a dry area. The correct weight $W$ will be equal to ($\gamma_{sub} \times$ Area $GBTN + \gamma_{dry} \times$ Area $BMT$). These areas can be measured by a planimeter. If the measuring of the areas is to be avoided, the entire area may be taken as submerged. By so doing, the weight $W$ will be slightly reduced, and thus $W \tan \phi$ or $R_u$ or F.S. will be slightly reduced. Hence, the results obtained will be on a safer side.

In such a case,

$$W = [\text{Area of } \Delta\, GMN] \cdot \gamma_{sub} = \gamma_{sub} \cdot \left(\tfrac{1}{2} \cdot Bu \cdot h\right)$$

or

$$R_u = C + W \tan \phi$$

$$= c \cdot (B_u \times 1) + \left(\gamma_{sub} \cdot \tfrac{1}{2} \cdot B_u \cdot h\right) \tan \phi \quad ...(20.29)$$

Now $P_u$ and $R_u$ are known, the factor of safety against sliding can be easily calculated, using

$$\text{F.S.} = \frac{R_u}{P_u} \quad ...(20.30)$$

It should be more than 1.5.

The factor of safety calculated above is w.r. to average shear ($\tau_{av}$), which will be equal to

$$\tau_{av} = \left(\frac{P_u}{B_u \times 1}\right) \quad ...(20.31)$$

It has been found by photoelastic studies, that the maximum intensity of shear stress occurs at a distance 0.6 $B_u$ from the heel (*i.e.* 0.4 $B_u$ from the top shoulder) and is equal to 1.4 times the average shear intensity.

∴ Maximum shear stress induced

$$= \tau_{max} = 1.4 \left(\frac{P_u}{B_u}\right) \quad \text{...(20.32)}$$

which is developed at the point $V$ (Fig. 20.26) such that $GV = 0.6\ B_u$.

The unit shearing resistance developed at this point $V$ is give by

$$\tau_f = c + h' \gamma_{sub} \cdot \tan \phi$$

$$= c + 0.6\ h \cdot \gamma_{sub} \cdot \tan \phi \quad \text{...(20.33)}$$

F.S. at the point of the maximum shear

$$= \frac{\tau_f}{\tau_{max}} \quad \text{...(20.34)}$$

It should be more than 1.

**20.13.6. Stability of d/s slope under steady seepage from the considerations of horizontal shear at base under the d/s slope of the dam.** The stability of the d/s slope under steady seepage is generally tested with Swedish slip circle method. However, the F.S. against the horizontal shear forces can be evaluated on the same principles as was done for the d/s slope in the previous article.

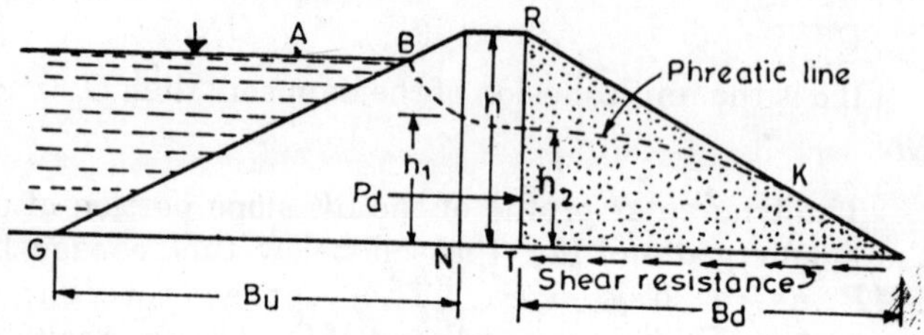


Fig. 20.27

With reference to Fig. 20.27.

The horizontal shear force $P_d$, in this case, is given by

$$P_d = \left[\frac{\gamma_2 \cdot h^2}{2} \tan^2\left(45 - \frac{\phi}{2}\right) + \frac{\gamma_w \cdot h_2{}^2}{2}\right] \quad \text{...(20.35)}$$

where $\gamma_2$ is the weighted density at the centre of the triangular shoulder downstream, given by

$$\gamma_2 = \frac{\gamma_{sub} \cdot h_2 + \gamma_{dry}\,(h - h_2)}{h} \quad \text{...(20.36)}$$

Shear resistance $R_d$ of d/s slope portion of dam, developed at base is given by

$$R_d = C + W \tan \phi,$$

where $W$ = The weight of the d/s slope portion of dam. (*i.e.* $\Delta\ RTS$)

$C$ = $c \times (B_d \times 1)$, where $c$ is the unit cohesion.

The triangular profile $RTS$ of the d/s slope portion of dam has an area say $A_1$ of dry soil above the seepage line and the area of submerged soil say $A_2$ below the seepage lines. These areas can be measured by a planimeter and then

$$W = [\gamma_{dry}\, A_1 + \gamma_{sub} \cdot A_2] \times 1$$

or

$$R_d = c \cdot B_d + \left[\gamma_{dry}\, A_1 + \gamma_{sub} \cdot A_2\right] \tan \phi \quad \text{...(20.37)}$$

If the measuring of the areas is to be avoided, the entire weight $W$ may be calculated on the basis of submerged soil, as it will be on a still safer side. In that case,

$$W = \gamma_{sub} \cdot (\tfrac{1}{2} \cdot B_d \cdot h)$$

Knowing $P_d$ and $R_d$, the factor of safety against shear can be easily determined as

$$\text{F.S.} = \frac{R_d}{P_d} \quad \text{...(20.38)}$$

The factor of safety at the point of maximum shear can also be determined in the same manner as was explained for the u/s slope portion.

**20.13.7. Stability of the foundation against Shear.** The generally available silt and clay foundations below the base of an earth dam, are sufficiently impervious, and there is generally no necessity of providing any treatment for under-seepage and piping for such foundations. But these foundations are weak in shear and must be investigated. In order to keep the shear stress developed at the foundations, within limits, the embankment-slopes may have to be flattened or berms on either side may be provided. If the available foundations are of plastic or unconsolidated clays, their shear strength will be very less and the matter should be seriously and thoroughly investigated.

The method given below, for determining the factor of safety against the foundation shear, is an approximate method and is based on the assumption that a soil has an equivalent liquid unit weight which would produce the same shear stress as the soil itself.

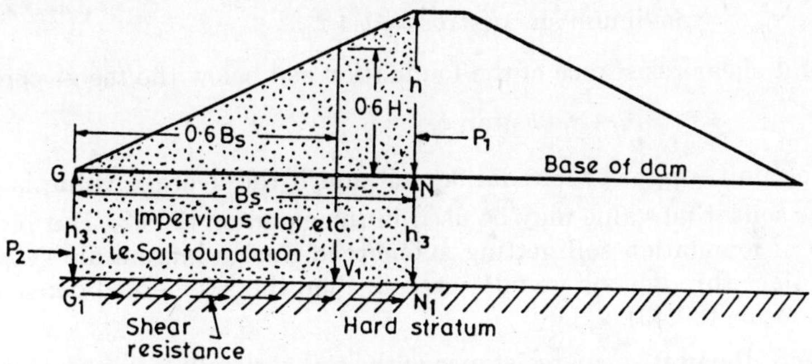


Fig. 20.28

The total horizontal shear force ($P$), under a slope of the dam is equal to the difference between the lateral thrust on a vertical through the top shoulder of the slope and a vertical through the toe of the slope.

or
$$P = (P_1 - P_2) = \frac{\gamma_{eq}\,(h + h_3)^2}{2} \tan^2\left(45 - \frac{\phi_1}{2}\right) - \frac{\gamma_{eq} \cdot h_3^{\,2}}{2} \times \tan^2\left(45 - \frac{\phi_1}{2}\right)$$

or
$$P = \gamma_{eq}\left[\frac{(h + h_3)^2 - h_3^{\,2}}{2}\right] \times \left[\tan^2\left(45° - \frac{\phi_1}{2}\right)\right] \text{ (kN)} \quad \text{...(20.39)}$$

where $h_3$ = the depth of foundation soil below the dam base, overlying the hard stratum below it.

$\gamma_{eq}$ = the equivalent unit weight of dry soil in foundation and dam. It is given by

$$\gamma_{eq} = \frac{\gamma_{dry} \text{ for dam material} \times h + \gamma_{dry} \text{ for foundation material} \times h_3}{h + h_3} \quad ...(20.40)$$

where $\phi_1$ = the equivalent angle of internal friction and is given by

$$\phi_1 = \tan^{-1}\left[\frac{c_f + \gamma_{eq}\,(h + h_3)\tan\phi_f}{\gamma_{eq}\cdot(h + h_3)}\right] \quad ...(20.41)$$

where $c_f$ and $\phi_f$ are the values of unit cohesion and angle of internal friction for the soil in the foundation.

The term, $\left[\gamma_{eq}\cdot\tan^2\left(45 - \frac{\phi_1}{2}\right)\right]$ is known as **equivalent liquid unit weight.**

Now the average shear stress at the base of the slope

$$(\tau_{av}) = \frac{P}{B_s} \quad ...(20.42)$$

where $B_s$ is the base width below the slope.

The value of $B_s$ will be equal to $B_u$ for u/s slope and $B_d$ for d/s slope. The minimum value will generate maximum stresses and hence, that particular slope should be considered which gives the minimum value of $B_s$, *i.e.* the slope which is less flat and is, therefore, the worst slope.

Maximum stress has been found by photoelastic studies to be 1.4 times the average stress and it occurs at a distance of 0.6 $B_s$ from the toe of the slope.

or $\tau_{max}$ = maximum shear stress = 1.4 $\tau_{av}$.

The unit shear resistance of the foundation soil below the toe at point $G_1$

$$= \tau_{f_1} = \left[c_f + \gamma_f\cdot h_3\tan\phi_f\right] \quad ...(20.43)$$

$\gamma_f$ is the unit weight of foundation soil and if the average value is given for impervious soils, that value may be used in the equation (20.43). But if there is a possibility of foundation soil getting submerged due to large scale seepage that may take place through the foundation soil, then the submerged density may be used in equation (20.43).

Similarly, the unit shear resistance of the soil vertically below the upper point of the considered slope (say at point $N_1$) is given by $\tau_{f_2}$.

$$\tau_{f_2} = c_f + \gamma_3\,(h + h_3)\tan\phi_f \quad ...(20.44)$$

where $\gamma_3$ = the equivalent unit weight of soil in the dam and foundation at the point $N_1$ and is given by

$$\gamma_3 = \frac{\gamma_f \times h_3 + \gamma_{dam} \times h}{h + h_3} \quad ...(20.45)$$

The values of $\gamma_f$ and $\gamma_{dam}$ may be taken as their dry densities or submerged densities depending upon the possibilities.

The average shear resistance

$$= \tau_f = \frac{\tau_{f_1} + \tau_{f_2}}{2} \quad ...(20.46)$$

Hence, overall factor of safety $= \dfrac{\tau_f}{\tau_{(av)}}$ ...(20.47)

This should be greater than 1.5.

The factor of safety at the point of maximum shear (*i.e.* the point $V_1$) must also be calculated as below :

The unit shear resistance at this point $V_1$

$$= \tau_{f(max)} = c_f + \gamma_4\,(h_3 + 0.6h)\tan\phi_f$$

where $\gamma_4$ is the equivalent weight of soil in dam and foundation and is given by

$$\gamma_4 = \frac{\gamma_f \times h_3 + \gamma_{dam} \times (0.6\,h)}{(h_3 + 0.6h)} \quad ...(20.48)$$

The dry or submerged unit weights may be used as explained earlier, in the above equation.

F.S. = Factor of safety at the point of maximum shear

$$= \frac{\tau_{f\,(max)}}{\tau_{max}} \quad ...(20.49)$$

This should be greater than unity.

**Example 20.4.** *An earthen dam made of homogeneous material has the following data :*

*Level of the top of the dam* = *200.00 m*
*Level of deepest river bed* = *178.0*
*H.F.L. of reservoir* = *197.5 m*
*Width of top of dam* = *4.5 m*
*Upstream slope* = *3 : 1*
*Downstream slope* = *2 : 1*
*Length of the horizontal filter from d/s toe, inwards* = *25 m*
*Cohesion of soil of dam* = *24 kN/ m*$^2$
*Cohesion of soil of foundation* = *54 kN/ m*$^2$
*Angle of internal friction of soil in the dam* = *25°*
*Angle of internal friction of soil in the foundation* = *12°*
*Dry weight of the soil in the dam* = *18 kN/m*$^3$
*Submerged weight of the soil in the dam* = *12 kN/m*$^3$
*Dry unit weight of the foundation soil* = *18.3 kN/m*$^3$
*Coefficient of permeability of soil in the dam* = $5 \times 10^{-6}$ *m/sec.*

*The foundation soil consists of 8 m thick layer of clay, having negligible coefficient of permeability. Check the stability of the dam and its foundations.*

**Solution.**

(1) **Overall stability of the dam section as a whole**

We will consider 1 m length of the dam. The section of the dam and the phreatic line is first of all drawn, as given in example 20.3 and shown in Fig. 20.29 (*a*). The dam section, etc. is generally drawn on a graph sheet so as to facilitate in measuring the areas above and below the seepage line, if planimeter is not available.

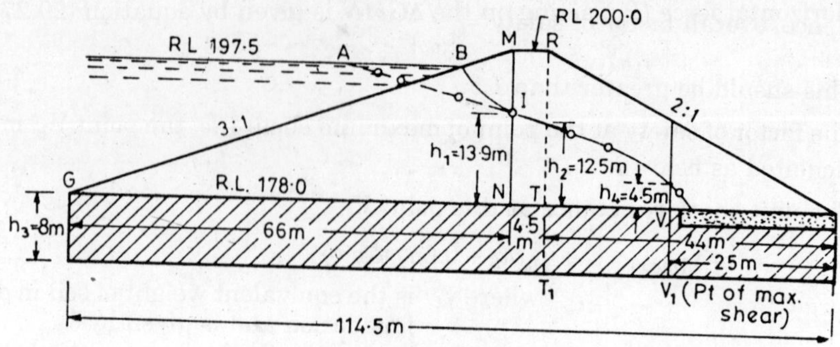


Fig. 20.29 (*a*)

The total area of dam section = $(114.5 + 4.5)\dfrac{22}{2}$ = 1,409 sq. m

The area above the seepage line is measured and is approximately found to be 380 $m^2$. (In the absence of a planimeter, graph can be used).

∴ Area below the seepage line = 1,409 – 380 = 1,029 sq. m

Now

Weight of the dry portion of the dam section

$= (380\ m^2 \times 1\ m \times 18\ kN/m^3 = 6830$ kN.

Weight of the submerged portion of the dam section

$= 1029\ m^2 \times 1\ m \times 12\ kN/m^3 = 12,350$ kN

Total weight of dam (called average weight)

= 6,830 + 12,350 = 19,180 kN

Shear resistance of the dam at the base

$= C + W \tan \phi$

where $C$ = Total cohesive strength of the soil at the base

$= c \times B \times 1 = (24 \times 114.5 \times 1)$ kN

$B$ = Total base width = 114.5 m

$W \tan \phi = 19,180 \tan 25°$

∴ Shear resistance at base,

$$R = 24 \times 114.5 \times 1 + 19180 \tan 25° = 11690 \text{ kN}$$

Horizontal force = Horizontal pressure of water.

$$= P = \tfrac{1}{2}\,\gamma_w\, h^2 = \tfrac{1}{2}\cdot 9.81\,(19.5)^2 = 1865 \text{ kN}$$

Factor of safety against failure due to horizontal shear at base

$$= \frac{11690}{1865} = 6.27 > 1.3 \ (\therefore \text{ Safe})$$

(2) **Stability of the u/s slope portion of dam (Under sudden drawdown)** *horizontal shear along the base under the u/s slope of dam*

Draw a vertical through the u/s extremity of the top width of dam [*i.e.* point *M*, Fig. 20.29 (*a*)] so as to cut the base of the dam at point *N*. This vertical *MN* cuts the seepage line at a point, the height of which is measured as $h_1$ = 13.6 m above the base of the dam.

Horizontal force ($P_u$) acting on the $\Delta GMN$ is given by equation (20.27) as :

$$P_u = \left[\frac{\gamma_1 h^2}{2} \tan^2\left(45 - \frac{\phi}{2}\right) + \gamma_w \cdot \frac{{h_1}^2}{2}\right]$$

where $\gamma_1$ = the weighted density at the centre of triangular shoulder upstream ($\Delta GMN$) and is given by equation (20.28) as :

$$\gamma_1 = \frac{\gamma_{sub} \cdot h_1 + \gamma_{dry}\,(h - h_1)}{h}$$

$$= \frac{12 \times 13.9 + 18\,(22.0 - 13.9)}{22.0}$$

$$= 14.7 \text{ kN/m}^3$$

$$\therefore \quad P_u = \frac{14.7 \times (22.0)^2}{2} \tan^2\left(45° - \frac{25°}{2}\right) + 9.81 \times \frac{(13.9)^2}{2} = 2391 \text{ kN}$$

Shear resistance $R_u$ of the u/s slope portion of dam developed at the base $GN$ is given by equation (20.29) as :

$$R_u = C + W \tan \phi$$

$= c\,(B_u \times 1) + (\gamma_{sub}\, \frac{1}{2} B_u\, h) \tan \phi$ ; neglecting the small dry soil area $BMI$, as it is very small and this neglection is on a safer side.

$$B_u = 66 \text{ m}$$

$$\therefore \quad R_u = 24 \times 66 + (12 \cdot \tfrac{1}{2} \cdot 66 \cdot 22.0) \tan 25°$$

$$= 1584 + 4062 = 5646 \text{ kN}$$

Factor of safety against horizontal shear along base under u/s slope

$$= \frac{R_u}{P_u} = \frac{5646}{2391} = 2.36 > 2.0 \qquad (\therefore \text{ safe})$$

Horizontal shear stress induced in the u/s slope portion of dam at base.

$$\tau_{av} = \frac{P_u}{B_u \times 1} = \frac{2391}{66} \text{ kN/m}^2 = 36.23 \text{ kN/m}^2$$

$$\tau_{max} = \text{Maximum shear}$$

$$= 1.4\, \tau_{av} = 1.4 \times 36.23 = 50.72 \text{ kN/m}^2$$

The maximum shear is developed at a point $0.6\, B_u$

$= 0.6 \times 66 = 39.6$ m away from point $G$

The unit shear resistance developed at this point

$$\tau_f = c + 0.6\, \gamma_{sub} \tan \phi$$

$$= 24 + 0.6 \times 22.0 \times 12 \tan 25° = 97.9 \text{ kN/m}^2$$

$$\therefore \quad \text{F.S.} = \frac{\tau_f}{\tau_{max}} = \frac{97.9}{50.72} = 1.93 > 1 \qquad (\therefore \text{ safe})$$

(3) **Stability of d/s portion of dam.** *Horizontal shear along base under the d/s slope of dam.*

Draw a vertical through the d/s extremity of the top width of dam (*i.e.* point $R$) to cut the base at point $T$ [Fig. 20.29 (*a*)]. Let this vertical cut the seepage line in a point, the height of which from the base is measured as $h_2 = 12.5$ *m*.

Horizontal force $P_d$ acting on the portion of downstream dam (*RTS*) during steady seepage is given by equation (20.35) as :

$$P_d = \left[\frac{\gamma_2 h^2}{2} \tan^2\left(45° - \frac{\phi}{2}\right) + \gamma_w \frac{h_2^2}{2}\right]$$

where $\gamma_2$ is the weighted density at the centre of the triangular shoulder *RTS* and given by equation (20.36) as :

$$\gamma_2 = \frac{\gamma_{sub}\, h_2 + \gamma_{dry}\,(h - h_2)}{h}$$

$$= \frac{12 \times 12.5 + 18 \times (22.0 - 12.5)}{22.0}$$

$$= 14.6 \text{ kN/m}^3$$

$$P_d = \frac{14.6\,(22)^2}{2} \tan^2\left(45° - \frac{25°}{2}\right) + 9.81 \cdot \frac{(12.5)^2}{2} = 2200 \text{ kN}$$

Shear resistance $R_d$ of the d/s slope portion of dam developed at base *TS* is given as :

$$R_d = C + W \tan\phi$$

The area $A_1$ of the dry soil within the $\Delta\, RTS$ above the seepage line ≈ 300 sq. m (from graph or planimeter).

The total area of the $\Delta\, RTS = \frac{1}{2} \times 44 \times 22 = 484 \text{ m}^2$

∴ Area of submerged soil

$$A_2 = 484 - 300 = 184 \text{ sq. m}$$

$$R_d = cB_d + \left[\gamma_{dry}\, A_1 + \gamma_{sub}\, A_2\right] \tan\phi$$

$$= 24 \times 44.0 + [18 \times 300 + 12 \times 184] \tan 25° = 4604 \text{ kN.}$$

F.S. against horizontal shear along base under d/s slope

$$= \frac{R_d}{P_d} = \frac{4604}{2200} = 2.09 > 2 \quad (\therefore \text{ Safe})$$

Average shear induced at base

$$= \frac{P_d}{B_d} = \frac{2200}{44} = 50 \text{ kN/m}^2$$

Maximum shear stress induced

$$\tau_{max} = 1.4 \times 50 = 70 \text{ kN/m}^2$$

The maximum shear stress is developed at a point 0.6 $B_d$

$$= 0.6 \times 44 = 26.4 \text{ m away from toe}$$

This unit shear resistance developed at this point

$$\tau_f = c + 0.6h\, \gamma_{sub} \tan\phi$$

(assuming the entire height as submerged as it will give safer results)

$$= 24 + 0.6 \times 22 \times 12 \tan 25° = 97.9 \text{ kN/m}^2$$

$$\therefore \quad \text{F.S.} = \frac{\tau_f}{\tau_{max}} = \frac{97.9}{70} = 1.40 > 1 \qquad (\therefore \text{ Safe})$$

(4) **Stability of the foundation soil**

Average compressive stress on foundation soil

$$= \frac{\text{Weight of dam}}{\text{Base area on which it acts}}$$

Since the compressive stress is maximum when the entire dam soil is dry, therefore, we will first calculate the dry weight of the dam.

Area of section of dam

= 1,409 sq. m (calculated earlier)

Dry weight of dam section

$$= 18 \times 1{,}409 = 25{,}362 \text{ kN}$$

Average compressive stress at base

$$= \frac{25362}{114.5} = 221.5 \text{ kN/m}^2$$

**Shear stress induced at base**

The total horizontal shear force ($P$) under the d/s slope of the dam (which is the worst case, *i.e.* the steepest slope) is given by equation (20.39) as :

$$P = \gamma_{eq}\left[\frac{(h+h_3)^2 - h_3^{\,2}}{2}\right]\left[\tan^2\left(45° - \frac{\phi_1}{2}\right)\right]$$

where $\gamma_{eq}$ is the equivalent weight of dry soil in foundation and dam

$$\gamma_{eq} = \frac{18h + 18.3\,h_3}{h + h_3}$$

[∵ Unit wt. of foundation soil of thickness $h_3 = 18.3$ kN/m$^3$]

where $h$ = 22 m

$h_3$ = 8 m.

$$\therefore \quad \gamma_{eq} = \frac{18 \times 22 + 18.3 \times 8}{22 + 8} = 18.1 \text{ kN/m}^3$$

$\phi_1$ is given by equation (20.41) as :

$$\gamma_{eq}\,(h + h_3) \tan \phi_1 = c_f + \gamma_{eq}\,(h + h_3) \tan \phi_f$$

or $\quad 18.1\,(22 + 8) \tan \phi_1 = 54 + 18.1\,(22 + 8) \tan 12°$

or $\quad \tan \phi_1 = 0.312$

or $\quad \phi_1 = 17.3°$

$$\therefore \quad P = 18.1\left[\frac{(22+8)^2 - (8)^2}{2}\right][\tan^2(45° - 8.65°)]$$

$$= \frac{18.1}{2}\,[900 - 64]\,[(0.737)^2] = 4100 \text{ kN}$$

Average shear stress induced at base of d/s slope

$$\tau_{av} = \frac{4100}{44} = 93.2 \text{ kN/m}^2$$

Maximum shear stress induced at $0.6 \times 44 = 26.4$ m away from the d/s toe inwards at point $V_1$ is given by

$$= \tau_{max} = 1.4 \times 93.2 = 130.4 \text{ kN/m}^2$$

**Shear resistance of the foundation soil below the d/s slope portion of dam**

Unit shear resistance $\tau_{f_1}$ below the toe at point $S_1$

$$= [c_f + \gamma_f \times h_3 \tan \phi_f]$$
$$= 54 + 18.3 \times 8 \times \tan 12°$$
$$\therefore \quad = 85.1 \text{ kN/m}^2$$

Unit shear resistance $\tau_{f_2}$ at point $T_1$

$$= c_f + \gamma_3 \, (h + h_3) \tan \phi_f$$

where
$$\gamma_3 = \frac{\gamma_{sub\ for\ dam} \times h_2 + \gamma_{dry\ for\ dam} \times (h - h_2) + \gamma_f h_3}{h + h_3}$$
$$= \frac{12 \times 12.5 + 18 \times 9.5 + 18.3 \times 8}{30} = 15.6 \text{ kN/m}^3$$

$$\therefore \quad \tau_{f_2} = 54 + 15.6 \, (22 + 8) \times \tan 12° = 153.5 \text{ kN/m}^2$$

The average unit shear resistance developed at foundation level in a length equal to $T_1S_1 = 44$ m, is given by

$$\tau_f = \frac{\tau_{f_1} + \tau_{f_2}}{2} = \frac{85.1 + 153.5}{2} = 119.3 \text{ kN/m}^2$$

Over all F.S. against shear

$$= \frac{\tau_f}{\tau_{av}} = \frac{119.3}{93.2} = 1.28 < 1.5 \qquad \textbf{(Hence, unsafe)}$$

The foundation soil is thus weaker to carry the load and hence the d/s slope will have to be flattened.

Shear resistance at the point of maximum shear, *i.e.* at point $V_1$ is given as :

$$(\tau_f)_{max} = c_f + (0.6h + h_3) \, \gamma_4 \tan \phi_f$$

$$\gamma_4 = \frac{12 \times 4.5 + 18 \, (0.6 \times 22 - 4.5) + 18.3 \times 8}{0.6 \times 22 + 8} = 16.8 \text{ kN/m}^3$$

$$(\tau_f)_{max} = 54 + [0.6 \times 22 + 8] \, 16.8 \tan 12° = 129.8 \text{ kN/m}^2$$

$$\text{F.S.} = \frac{[\tau_f]_{max}}{\tau_{max}} = \frac{129.8}{130.4} = 0.995 < 1.0 \qquad \text{(Hence, Unsafe)}$$

The foundation shear and F.S. can also be calculated below the u/s portion of dam soil, in the same manner as has been done for d/s slope portion, if required.

**Example 20.5.** *Check the stability of the d/s slope of the earthen dam section, given in example 20.4, on a possible slip circle. Determine the factor of safety available against such a slip. Also determine the net factor of safety that will be available, if by some how, the soil in the d/s triangular shoulder gets fully submerged. Also compare this net factor of safety with the factor of safety that can be obtained for similar conditions if the analysis was done by shear force determination at the base of the d/s slope.*

**Solution.** The dam section and phreatic line is first of all drawn as shown in Fig. 20.29 (*b*) and as given in previous example. The points *P* and *Q* are located for Fellenius construction such as explained earlier ; the line *PQ* represents the locus of the critical slip circle. A possible slip circular arc passing through the toe of the dam is then drawn with centre at $O_1$. The soil mass is then divided into slices of 5 metre width. It is found by chance that full 9 slices of 5 m width are

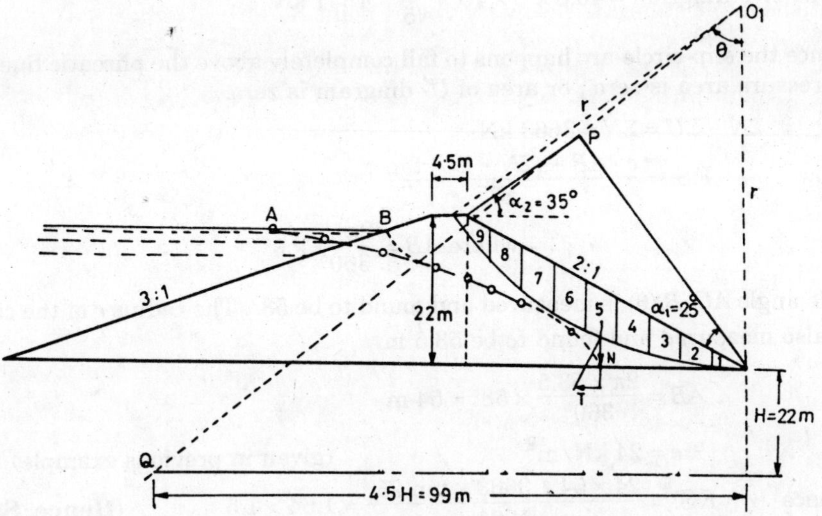


Fig. 20.29 (*b*)

**accommodated, making thereby the width of the last slice (which was *m.b*) equal to *b* or $m = 1$. By visual observations ; it is also found that the tangenital components (*T*'s) of the weights of all the slices (acting through their centres) happen to be +ve. This is because, the lines of action of all the weights are on one side (*i.e.* left side) of the centre $O_1$.**

**The end ordinates of all the slices say $y_1, y_2, y_3, ..., y_{n-1}$ or $y_8$ are measured and found to be 2.5 m, 4.5 m, 6 m, 7 m, 7.5 m, 6.0 m, and 4.5 m respectively.**

**The *N*-diagram is then drawn as shown in Fig. 20.30 (*a*). The dry density is used because the entire soil in the slices is dry.**

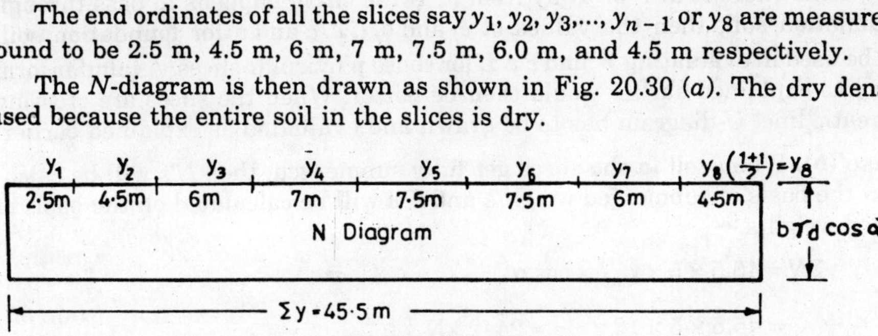


Fig. 20.30 (*a*). *N*-diagram.

Area of *N*-Diagram $= 45.5 \times 5 \times 18 \times \dfrac{2}{\sqrt{5}} = 3663$ kN

Similarly, *T*-Diagram is drawn as shown in Fig. 20.30 (*b*).

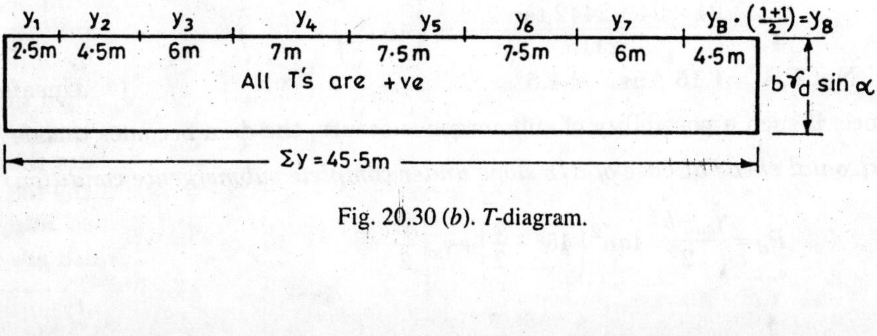


Fig. 20.30 (*b*). *T*-diagram.

$$\text{Area of } T\text{-diagram} = 45.5 \times 5 \times 18 \times \frac{1}{\sqrt{5}} = 1831 \text{ kN}$$

Since the slip-circle-arc happens to fall completely above the phreatic line, the pore pressure area is zero ; or area of $U$- diagram is zero.

$$\therefore \quad \Sigma N - \Sigma U = \Sigma N = 3663 \text{ kN}.$$

Now $$\text{F.S.} = \frac{c \cdot AB + (\Sigma N - \Sigma U) \tan \phi}{\Sigma T}$$

$$\text{where } AB = \frac{2\pi \cdot r}{360°} \times \theta°$$

The angle $AO_1B$ ($\theta$) is measured and found to be 58°. The radius $r$ of the curve $AB$ is also measured and found to be 53.5 m.

$$\therefore \quad AB = \frac{2\pi \times 53.5}{360°} \times 58° = 54 \text{ m}$$

$$c = 24 \text{ kN/m}^2 \qquad \text{(given in previous example)}$$

Hence $$\text{F.S.} = \frac{24 \times 54 + 3663 \times \tan 25°}{1831} = 1.64 > 1.5$$ **(Hence, Safe)**

The corresponding F.S. obtained in example 20.4 from shear analysis was 2.08. Hence, F.S. should be more than 1.5 in slip circle analysis and more than 2 in shear analysis for safe designs.

**Note.** The F.S. obtained above by slip circle method cannot be called as the critical F.S. or minimum F.S.; because the analysis has been carried out for only one circle. In fact, various other circles should be drawn by taking the centres on line $PQ$, somewhere near $P$ or away from it. If the circle happens to pass through the foundation soil, then the values of $c_f$ and $\phi_f$ (*i.e.* $c$ and $\phi$ for foundation soil) should be used in evaluating $W$ and $c\,\Delta L$ for those particular slices. A tabular form solution, as explained earlier would then be better. When the slices are crossing the phreatic line, $U$-diagram has to be drawn and evaluated as explained earlier.

**Case (*b*).** When soil in the slices get fully submerged, then $N$'s will be calculated on the basis of submerged weights and $T$ s will be calculated on the basis of saturated weights.

$$\therefore \quad \Sigma N = 45.5 \times 5 \times \gamma_{sub} \times \cos \alpha$$

$$= 45.5 \times 5 \times 12 \times \frac{2}{\sqrt{5}} = 2442 \text{ kN}$$

$$\Sigma T = 45.5 \times 5 \times \gamma_{sat} \sin \alpha$$

$$= 45.5 \times 5 \times 22 \times \sin 25° = 2115 \text{ kN}$$

Net $$\text{F.S.} = \frac{c \cdot AB + \Sigma N \tan \phi}{\Sigma T}$$

$$= \frac{24 \times 54 + 2442 \tan 25°}{2115} = 1.15$$

**Net F.S. = 1.15 Ans.** $< 1.5$ ($\therefore$ Unsafe)

Hence, if such a possibility of submergence occurs, the dam becomes unsafe.

*Horizontal shear at base of d/s slope under complete submergence condition.*

$$P_d = \frac{\gamma_{sub}\, h^2}{2} \tan^2\left(45° - \frac{\phi}{2}\right) + \gamma_w \frac{h^2}{2}$$

$$= \frac{12 \times (22)^2}{2} \tan^2 (45° - 12.5°) + \frac{9.81 \times (22)^2}{2} = 3553 \text{ kN}$$

$$R_d = c \cdot B_d + \gamma_{sub} \left(\tfrac{1}{2} \cdot B_d \cdot h\right) \tan \phi$$

$$= 24 \times 44.0 + 12 \times \tfrac{1}{2} \times 44 \times 22 \times \tan 25° = 3764 \text{ kN}$$

$$\text{F.S.} = \frac{3764}{3553} = 1.06 < 1.5 \qquad \textbf{(Hence, Unsafe)}$$

F.S. with slip circle analysis = 1.09
F.S. with horizontal shear analysis = 1.06 **Ans.**

**Conclusion.** The dam becomes unsafe as soon as the soil in the downstream shoulder gets submerged.

## SEEPAGE CONTROL IN EARTH DAMS

The water seeping through the body of the earthen dam or through the foundation of the earthen dam, may prove harmful to the stability of the dam by causing softening and sloughing of the slopes due to development of pore pressures. It may also cause piping either through the body or through the foundation, and thus resulting in the failure of the dam.

### 20.14. Seepage Control Through Embankments

Drainage filters called 'Drains' are generally provided in the form of (*a*) *rock toe* (*b*) *horizontal blanket* (*c*) *chimney drain*, etc. in order to control the seepage water. The provision of such filters reduces the pore pressure in the downstream portion of the dam and thus increases the stability of the dam, permitting steep slopes and thus affecting economy in construction. It also checks piping by migration of particles. These drains, consist of graded coarse material in which the seepage is collected and moved to a point where it can be safely discharged. In order to prevent movement of the fine material from the dam into the drain, the drain or filter material is graded from relatively fine on the periphery of the drain to coarse near the centre. A multi-layered filter, generally called **inverted filter** or *reverse filter* is provided as per the criteria suggested by Terzaghi for the design of such filters.

The various kinds of drains, which are commonly used are shown and described below :

**20.14.1. Rock Toe or Toe Filter** [Fig. 20.31 (*a*)]. The 'rock toe' consists of stones of size usually varying from 15 to 20 cm. A toe filter (graded in layers) is provided as a transition zone, between the homogeneous embankment fill and rock toe. Toe filter generally consists of three layers of fine sand, coarse sand, and gravel ;

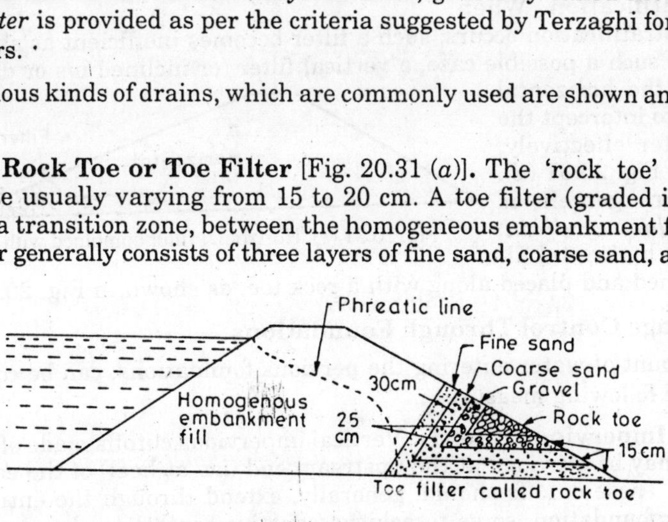


Fig. 20.31 (*a*). Rock Toe.

as per the filter criteria requirements. The height of the rock toe is generally kept between 25 to 35% of reservoir head. The top of the rock toe must be sufficiently higher than the tail water depth, so as to prevent the wave action of the tail water.

**20.14.2. Horizontal Blanket or Horizontal Filter.** [Fig. 20.31 (*b*) and (*c*)]. The horizontal filter extends from the toe (d/s end) of the dam, inwards, upto a distance varying from 25 to 100% of the distance of the toe from the centre line of the dam. Generally, a length equal to three times the height of the dam is sufficient. The blanket should be properly designed as per the filter criteria, and should be sufficiently pervious to drain off effectively.

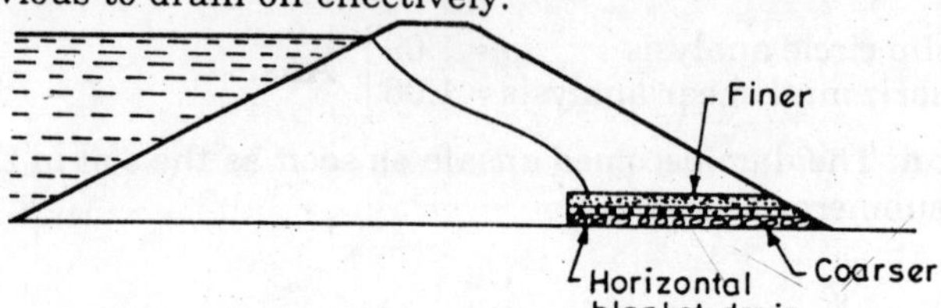


Fig. 20.31 (*b*). Horizontal Filter.

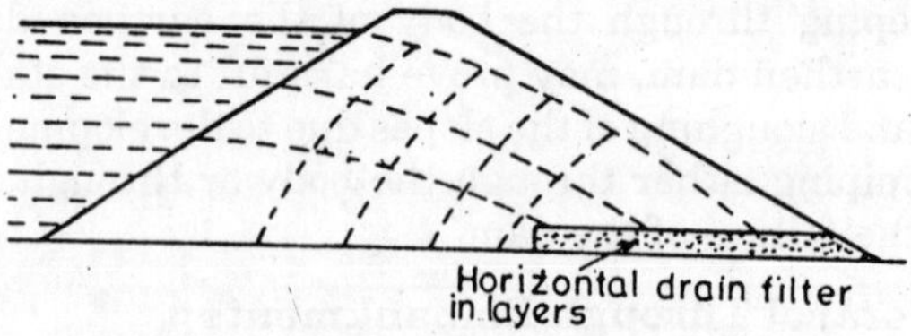


Fig. 20.31 (*c*). Inefficient 'Horizontal drain' in stratified embankments.

**20.14.3. Chimney Drain.** [Fig. 20.31 (*d*)]. The horizontal filter, not only helps in bringing the phreatic line down in the body of the dam but also provides drainage of the foundation and helps in rapid consolidation. But, the horizontal filter tries to make the soil more pervious in the horizontal direction and thus causes stratification. When large scale stratification occurs, such a filter becomes inefficient as shown in Fig. 20.31 (*c*). In such a possible case, a vertical filter (or inclined u/s or d/s) is placed along with the horizontal filter, so as to intercept the seeping water effectively, as shown in Fig. 20.31 (*d*). Such an arrangement is termed as *chimney drain.* Sometimes a horizontal filter is combined and placed along with a rock toe, as shown in Fig. 20.31 (*e*).

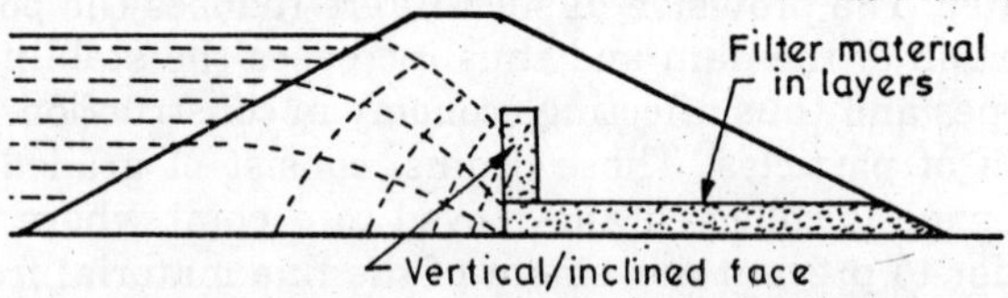


Fig. 20.31 (*d*). 'Chimney Drain' in Stratified Embankments.

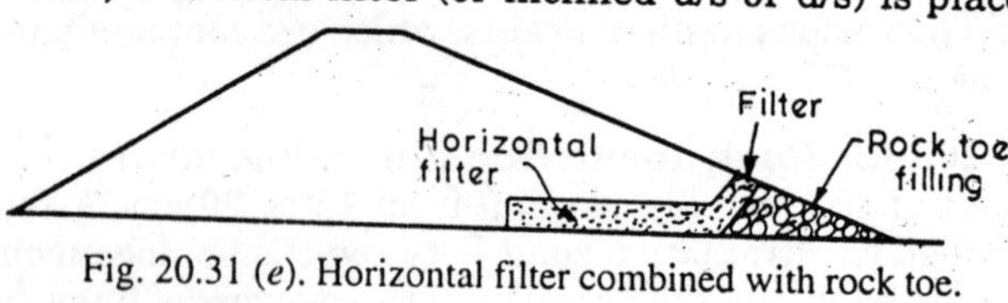


Fig. 20.31 (*e*). Horizontal filter combined with rock toe.

## 20.15. Seepage Control Through Foundations

The amount of water entering the pervious foundations, can be controlled by adopting the following measures :

**20.15.1. Impervious Cutoffs.** Vertical impervious cutoffs made of concrete or sheet piles may be provided at the upstream end (*i.e.* at heel) of the earthen dam (Fig. 20.32). These cutoffs should, generally, extend through the entire depth of the pervious foundation, so as to achieve effective control on the seeping water.

When the depth of the pervious foundation strata is very large, a cutoff, up to a lesser depth, called a *partial cutoff* may be provided. Such a cutoff reduces the seepage discharge by a smaller amount. So much so, that a 50% depth reduces the discharge by 25%, and 90% depth reduces the discharge by 65% or so.

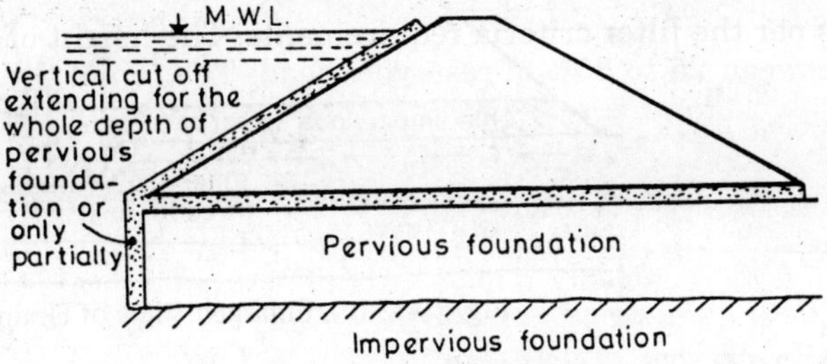


Fig. 20.32

**20.15.2. Relief Wells and Drain Trenches.** When large scale seepage takes place through the pervious foundation, overlain by a thin less pervious layer, there is a possibility that the water may boil up near the toe of the dam, as shown in Fig. 20.33 (*a*).

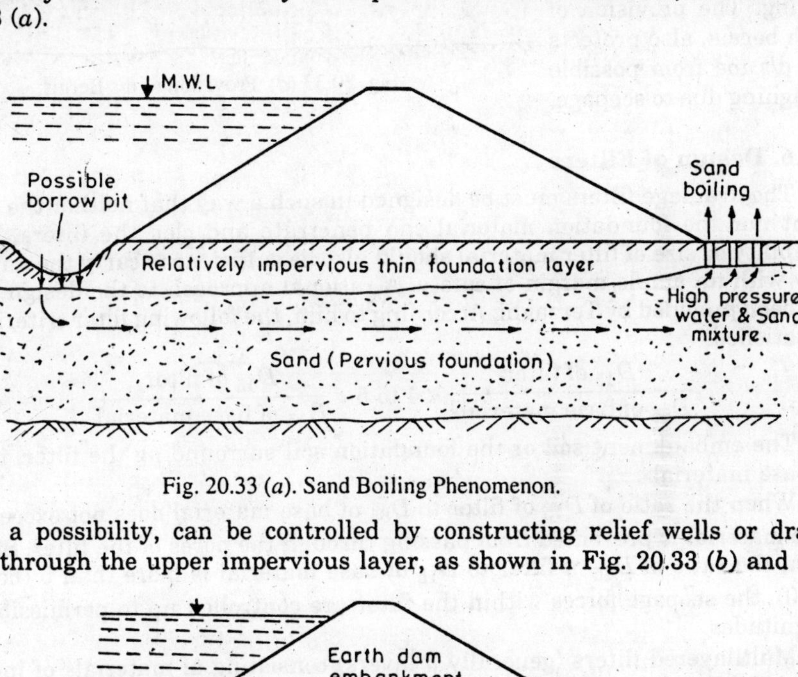


Fig. 20.33 (*a*). Sand Boiling Phenomenon.

Such a possibility, can be controlled by constructing relief wells or drain trenches through the upper impervious layer, as shown in Fig. 20.33 (*b*) and (*c*),

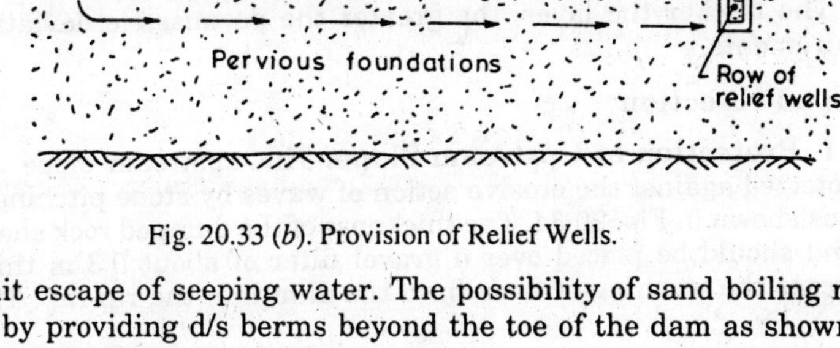


Fig. 20.33 (*b*). Provision of Relief Wells.

so as to permit escape of seeping water. The possibility of sand boiling may also be controlled by providing d/s berms beyond the toe of the dam as shown in Fig.

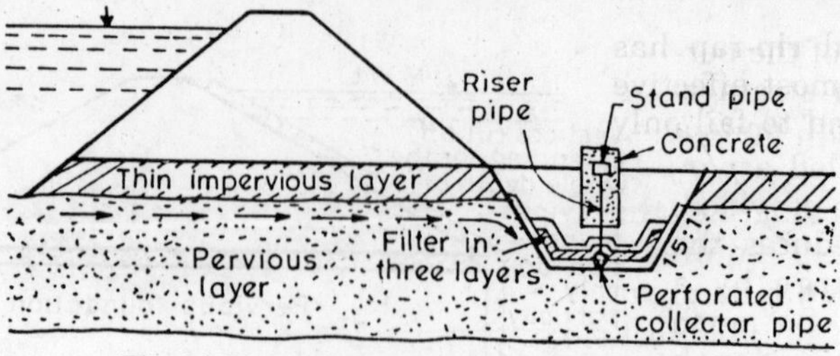


Fig. 20.33 (*c*). Enlarged View of Drain Trench.

20.33 (*d*). The weight of the overlying material, in such a case, is sufficient to resist the upward pressure and thus preventing the possibility of sand boiling. The provision of such berms, also protects the d/s toe from possible sloughing due to seepage.

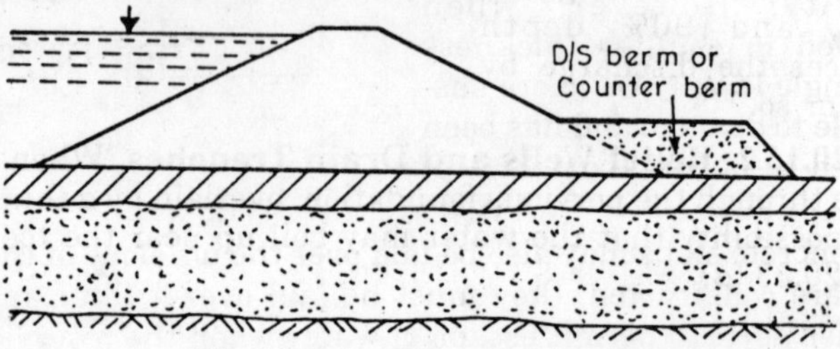


Fig. 20.33 (*d*). Provision of d/s Berms.

## 20.16. Design of Filters

The drainage filters must be designed in such a way that neither the embankment nor the foundation material can penetrate and clog the filters. The permeability or size of filter material should also be sufficient to carry the anticipated flow with an ample margin of safety. A rational approach to the design of filters has been provided by Terzaghi. According to him, the following filter criteria should be satisfied.

$$\frac{D_{15}\text{ of filter}}{D_{85}\text{ of base materials}} < 4 \text{ to } 5 < \frac{D_{15}\text{ of filter}}{D_{15}\text{ of base material}} \qquad ...(20.50)$$

The embankment soil or the foundation soil surrounding the filter, is known as base material.

When the ratio of $D_{15}$ of filter to $D_{85}$ of base material does not exceed 4 to 5, base material is prevented from passing through the pores of the filter. Similarly, when the ratio of $D_{15}$ of filter to $D_{15}$ of base material is more than 5 (between 5 to 40), the seepage forces within the filter are controlled up to permissible small magnitudes.

Multilayered filters (generally 3 layers) consisting of materials of increasing permeabilities from the bottom to top are, many a times, provided and are known as **inverted filters.** These filters are costly and should be avoided where possible. The minimum total thickness of filter is 1 m. However, if sufficient quantities of filter material are available at reasonable costs, thicker layers of filter may be provided. The thicker the layer, the greater the permissible deviation from the filter requirements.

## 20.17. Slope Protection

**20.17.1. Protection of Upstream Slope.** The upstream slope of the earth dam is protected against the erosive action of waves by stone pitching or by stone dumping, as shown in Fig. 20.34. The thickness of the dumped rock should be about I metre and should be placed over a gravel filter of about 0.3 m thickness. The filter prevents the washing of fines from the dam into the rip-rap. The provision of such a dumped rip-rap has been found to be most effective and has been found to fail only in 5% cases.

The stone pitching, *i.e.* the hand packed rip rap requires a lesser thickness and may prove more economical if suitable rock is available only in limited quantity. However, when provided in smaller thickness (*i.e.* single layer), it is more susceptible to damage and has been founded to fail in about 30% of cases.

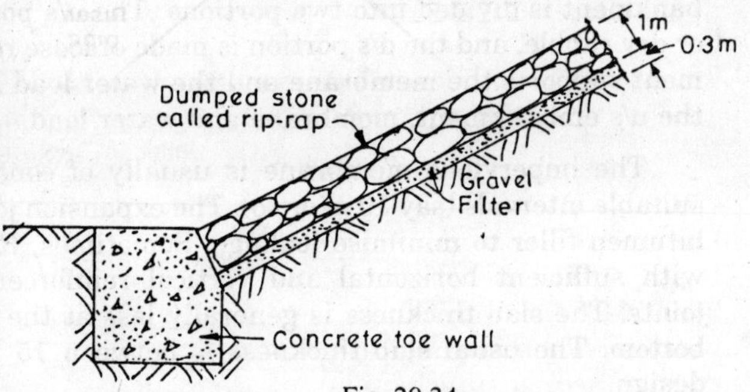


Fig. 20.34

Concrete slabs may also be laid over the u/s slope of the earth dam. When such slabs are constructed, they must be laid over a filter and weep holes should be provided so as to permit escape of water when the reservoir is drawn down. If the filter is not provided, the fines from the embankment may get washed away from the joints creating hollows beneath the slab and causing the consequent cracking and failure of the slab under its own weight. Concrete slab protections have been found to fail in about 36% cases, mainly because of non-providing of filter below them.

**20.17.2. Protection of Downstream Slope.** The downstream slope of the earthen dam is protected against the erosive action of waves upto and slightly above the water depth, in a similar manner as is explained above for u/s slope.

Moreover, the d/s slope should be protected against the erosive action of rain and its run-off by providing horizontal berms at suitable intervals say about 15 m or so (Fig. 20.5) so as to intercept the rain water and discharge it safely. Attempts should also be made so as to grass and plant the d/s slopes, soon after their construction.

## 20.18. Rockfill Dams

Rockfill dams have characteristics lying somewhere between the characteristics of gravity dams and those of earthen dams. In other words, they are less flexible than earthen dams and more flexible, then gravity dams. Their foundation requirements are not as strict and rigid as are required for gravity dams. But the foundation requirements are more rigid than those for earthen dams which can be constructed almost on any types of foundations. The steeper slopes are used in rockfill dams and hence, the base width is quite less. The smaller base width and the possibility of large scale seepage restricts the foundation requirements of such dams.

A typical section of a rockfill dams is shown in Fig. 20.35. It essentially consists of

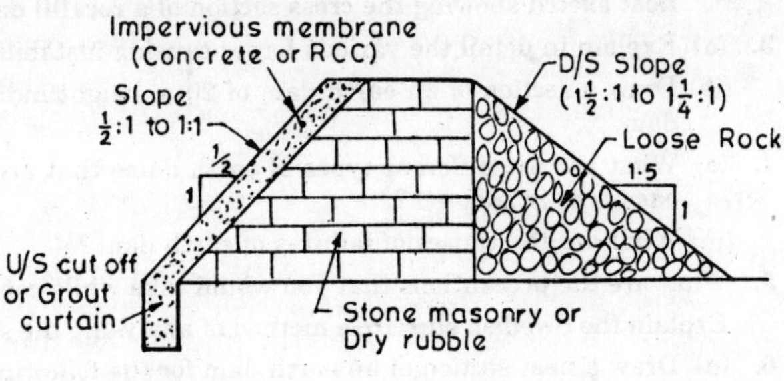


Fig. 20.35. Section of a rockfill dam.

an impervious membrane and embankment supporting the membrane. The embankment is divided into two portions. The u/s portion is made of stone masonry or dry rubble, and the d/s portion is made of loose rockfill. The u/s portion embankment supports the membrane and the water load : while the d/s portion supports the u/s embankment, membrane and water load.

The impervious membrane is usually of concrete, with expansion joints at suitable intervals (say 10 m or so). The expansion joints may be filled with suitable bitumen filler to minimise leakage. Sometimes, R.C.C. membranes are provided with sufficient horizontal and vertical reinforcements, without any expansion joints. The slab thickness is generally less at the top and is more near the dam bottom. The usual slab thickness is between 15 to 50 cm, depending upon the design.

The upstream face containing the membrance, is sloped at a slope varying from say $\frac{1}{2}$ : 1 ($\frac{1}{2}H$ : 1$V$) in low dams up to height of about 60 m or so, to 1 : 1 or $1\frac{1}{2}$ : 1 in high dams. The d/s slope of all the rockfill dams is kept at about 1.3 : 1, which represents the angle of repose of the rockfill.

Rockfill dams are generally cheaper than concrete dams and can be constructed rapidly if proper rock is available. The rock must be strong enough to withstand high intensity leadings even when wet. The size of loose rock may vary from small stones to 3 m or so. Rockfill dams are very useful in seismic regions, as they provide high resistance to seismic forces because of their flexible character. However, rockfill dams are liable to large settlements, which may lead to cracking of concrete membrane. Repairs for the membrane are, therefore, undertaken from time to time, as and when the necessity arises.

## PROBLEMS

1. (*a*) What are 'earthen dams' and under what circumstances are they preferred ?
   (*b*) Enumerate the different types of earthen dams, and draw neat sketches showing each type. **[Hint** ; See article 20.2]
   (*c*) Enumerate the two different methods which are adopted for constructing earthen dams. Which of these methods would you prefer and why ? **[Hint** : See article 20.3, preferring rolled fill method]
2. (*a*) What is meant by 'pore water pressure' : and what is its significance in the design of earthen dams?
   (*b*) What are 'rockfill dams' and what are their advantages over earthen dams. Draw a neat sketch showing the cross section of a rockfill dam.
3. (*a*) Explain in detail the various forces causing instability in a gravity dam.
   (*b*) Draw a section of an earth dam of 20 m height indicating the various parts of the dam.
4. (*a*) What are the different types of earth dams that are usually adopted. State where each type is adopted ?
   (*b*) What are the causes of failures of earth dam ?
5. What are the precautions that you would take while constructing an earth dam ?
   Explain the Swedish slip circle method of analysing the stability of an earth dam slopes.
6. (*a*) Draw a neat section of an earth dam for the following site :
   (*b*) Both silty clay and coarse sand are available at site. Hard stratum is available at about 5 m below the natural ground.

(c) Explain briefly how the stability of earthen slopes are checked by the slip circle method.

**7.** (a) Differentiate between 'Rigid dams' and 'Non-rigid dams' giving examples of each type.

(b) Explain with neat sketches how you would carry out the stability analysis of an earth dam.

**8.** (a) Differentiate between 'horizontal' and 'vertical' piping in earth dams. Suggest permanent measures to check vertical piping.

(b) Show with the help of simple outline sketches suitable broad designs of earth dams for different available materials and the governing geology of the site.

**Note.** At least four different designs should be given.

**9.** (a) Give a suitable design for a 50 m high dam for site where both clay silt and sand gravel are available in plenty and where foundation is pervious to a depth of 10 m. Assume suitable data. Give reasons favouring the suggested design.

**10.** (a) Briefly discuss the checks that are required to be made to investigate the stability of an earthen dam.

(b) An earthen dam has to be constructed to store a maximum depth of 12 m of water over river bed consisting of coarse sand and gravel up to a depth of 3 m below river bed followed thereafter by bard and sound rock. Clayee soil is available in plenty in the vicinity of the river. Draw and detail a suitable section of the dam at the river bed.

**11.** Explain how the following parameters affect design of an earth dam :

(i) optimum moisture content ;

(ii) $C$ and $\phi$ value of soil ; permeability of soil

(iii) sudden draw-down of the reservoir.

Illustrate with neat sketches the following parts of an earthen dam and state their functions briefly:

(i) Rock toe ;

(ii) horizontal drainage blanket ;

(iii) cut-off ;

(iv) Rip-rap.

**12.** Enumerate and explain by neat sketches the different ways by which the earthen dams may fail. Also suggest suitable precautions that should be undertaken to avoid each type of failure.

**13.** (a) Explain the meaning and importance of equipotential lines' and 'stream lines' in connection with seepage analysis of earthen dams.

(b) Derive an expression for calculating the seepage discharge through earthen dam bodies made of isotropic soils. What correction factor is required to be multiplied with this expression, if the dam soil is non-isotropic.

**14.** (a) Define and explain the term 'phreatic line' in earthen dams.

(b) How would you procced to determine the phreatic line through homogeneous earthen dams provided :

(i) with a horizontal filter ;

(ii) Without a horizontal filter.

**15.** (a) Explain and elaborate the importance of 'seepage' through earthen dams.

(b) What precautions and remedial measures would you undertake to control the 'seepage' through

(i) earthen dam body ;

(ii) through the dam foundation.

**16.** Write short notes on any five of the followings :

(i) Rock toe

(ii) Climney drain

(iii) Relief wells

(*iv*) Slope protection in earthen dams
(*v*) Rock fill dams
(*vi*) Consolidation of earthen dams
(*vii*) Pore pressure and its significance in relation to earthen dam construction
(*viii*) 'Seepage failures' of earthen dams
(*ix*) Rational design of drainage filters for earthen dams.

**17.** A section of a homogeneous earth dam is shown in Fig. 20.36. Calculate the seepage discharge per metre length, through the body of the dam. The coefficient of permeability of the dam material may be taken as $8 \times 10^{-5}$ m/sec. **[Ans.** $2.88 \times 10^{-1}$ cumecs/m]

**18.** An earthen dam made of homogeneous material has the following data :

**(A) Hydraulic data of dam**

Level of top of dam = 210 m.

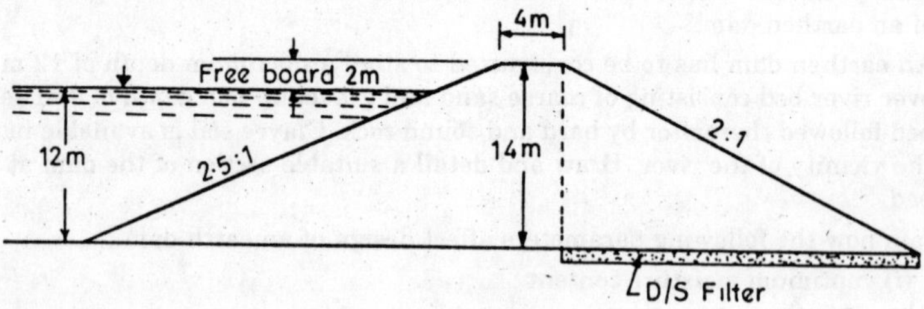


Fig. 20.36

Level of deepest river bed = 192 m.
HFL of reservoir = 208 m.
Width of top of dam = 10 m.
Upstream slope = 3 : 1
Downstream slope = 2.5 : 1
Length of horizontal filter from d/s toe inwards = 16 m.

**(B) Properties of the material of the dam**

Dry density = 18 kN/m$^3$
Saturation density = 21 kN/m$^3$
Average angle of friction = 30°
Average cohesion = 16 kN/m$^2$.

**(C) The foundation soil consists of 4 m thick layer of soil having the following properties :**

Average unit weight = 17 kN/m$^3$
Average cohesion = 54 kN/m$^2$.
Average angle of internal friction = 7°

Check the dam section for the following,:

(*i*) Sloughing of u/s slope during sudden drawdown.
(*ii*) Stability of foundation against shear.

[**Hint.** Follow example 20.4]

**19.** In order to determine the factor of safety of the d/s slope during steady seepage, the section of a homogeneous earth dam was drawn to scale of 1 cm = 10 m ; and the following results were obtained on a trial slip circle.

Area of N-diagram = 12.15 sq cm.

Area of T-diagram = 6.50 sq cm.

Area of U-diagram = 4.02 sq cm.

Length of arc = 11.60 cm.

The dam material has the following properties :

Effective angle of internal friction = 26°

Unit of cohesion = 19.62 $m^2$

Unit weight of soil = 19.62 $m^2$

Determine the factor of safety of the slope. **[Ans. F.S. 1.21]**

# 21

# *Spillways, Energy Dissipators, and Spillway Gates*

## 21.1. Introduction

A spillway is a structure constructed at a dam site, for *effectively disposing of the surplus water from upstream to downstream.* Just after the reservoir gets filled up, up to the normal pool level, water starts flowing over the top of the spillway crest (which is generally kept at normal pool level). Depending upon the inflow rate, water will start rising above the normal pool level, and at the same time, it will be let off over the spillway. The water can rise over the spillway crest, upto the maximum reservoir level, which can be estimated from the inflow flood hydrograph and the spillway characteristics, by the process of flood routing, explained earlier. Therefore, it is only the spillway, which will dispose of the surplus water and will not let the water rise above the maximum reservoir level. Had there been no such structure, over which the water would have overflown, the water level must have exceeded maximum reservoir level, and ultimately would have crossed the freeboard and thus overtopped the dam, causing the failure of the dam. *Hence, a spillway is essentially a safety valve for a dam.* It must be properly designed and must have adequate capacity to dispose of the entire surplus water at the time of the arrival of the worst design flood.

Many dams have failed (especially the earthen dams) because of the improperly designed or inadequate spillways.

## 21.2. Location of a Spillway

A spillway can be located either within the body of the dam, or at one end of it or entirely away from it, independently in a saddle. If a deep narrow gorge with steep banks, separated from a flank by a hillock with its level above the top of the dam (such as shown in Fig. 21.1), is available, the spillway can be best built independently of the dam.

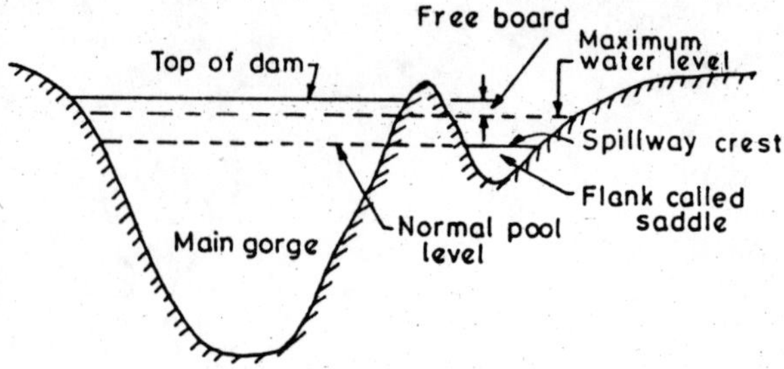


Fig. 21.1

Under such circumstances, a concrete or an earthen dam can be constructed across the main valley and a spillway can be constructed independently into the saddle. Sometimes, a concrete or a masonry dam along with its spillway can be constructed in the main valley, while the flank or flanks are closed by earthen dikes or embankments. The top level of such an embankment is kept at maximum reservoir level. The materials and designs of these embankments are such that they fail as soon as water overtops them. Hence, if by chance, either due to excessive flood above the design flood or due to failure of gates of main spillway, etc., the water rises above the maximum reservoir level, it shall overtop such embankment, which at once fails ; providing sufficient outlet for the disposal of excessive water. This type of a secondary safety arrangement is generally provided on large dams especially on earth and rockfill dams, and is *known as Subsidiary Spillway* or *Emergency Spillway* or *Breaching Section.*

The main spillway is constructed to dispose of the designed flood above the normal pool level and upto the maximum reservoir level. It is situated either within the dam, or at one end of it, or independently in a saddle away from the main dam. A separate independent spillway is generally preferred for earthen dams, although due to non-availability of sites, a concrete spillway is sometimes constructed within or at one of the ends of an earth dam. If the main spillway is situated in a flank, a secondary emergency spillway may be situated in another flank as shown in Fig. 21.2.

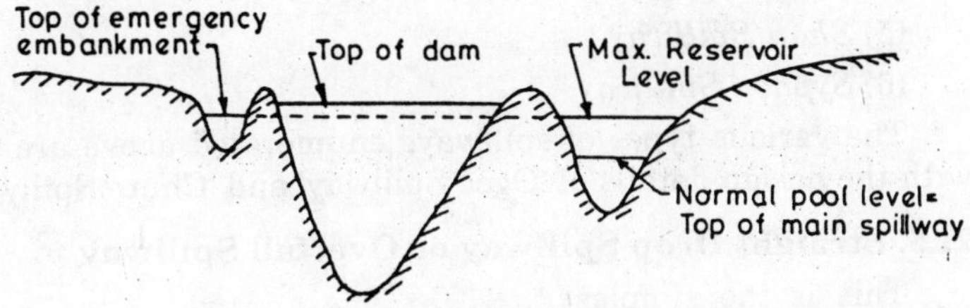


Fig. 21.2

## 21.3. Design Considerations for the Main Spillway

The main spillway, often called the spillway, is properly designed so as to dispose of the excess water without causing any damage to the dam, or to any of its appurtenant structures. The spillway structure should be structurally and hydraulically adequate and must not give way under worst and variable loading conditions.

The required discharging capacity of the spillway should be as closely estimated as possible. The underestimation will lead to overtopping of the main dam and its consequent damages ; while the over estimation will lead to unnecessarily costly constructions which shall never he utilised during the life of the dam, and hence, will remain a waste investment. However, on large dams, a conservative view is always preferred because the failure of a single dam due to inadequate capacity may result in the loss of numerous human lives to which no cost allocation can be made. Moreover, an emergency spillway or a breaching section is generally provided, the failure of which under necessary circumstances, may though cause serious erosion on the downstream, but shall protect the main dam from failure.

The water passing over the spillway and falling on the downstream side must not be allowed to erode the downstream soil, and hence, arrangements must be made for effectively dissipating the energy of the falling water.

## 21.4. Controlled and Uncontrolled Spillways

The flow of water over a spillway may be controlled by installing gates over the spillway crest. In such a case, the spillway is known as a controlled spillway. The out flow can be controlled in such spillways and hence, preferred in modern days. However, some spillways are left just by constructing their crest at normal pool level. As water will flow over such a spillway, depending upon the reservoir level and the corresponding head over the spillway, such uncontrolled spillways are guided only by the available water head, and hence, are called uncontrolled spillways. The comparative advantages and disadvantages of controlled and uncontrolled reservoirs have been discussed in chapter 18.

## VARIOUS TYPES OF SPILLWAYS

Depending upon the type of the structure constructed for disposing of the surplus water, the spillways can be of the following major types :

(1) *Straight Drop Spillway.*

(2) *Overflow Spillway* generally called *Ogee Spillway.*

(3) *Chute Spillway* often called *Trough Spillway* or Open channel Spillway.

(4) *Side Channel Spillway.*

(5) *Shaft Spillway.*

(6) *Syphon Spillway.*

The various types of spillways enumerated above are described below along with the design details of 'Ogee Spillway' and 'Chute Spillway'.

### 21.5. Straight Drop Spillway or Overfall Spillway

This is the simplest type of spillway and may be constructed on small bunds or on thin arch dams, etc. It is a low weir and simple vertical fall type structure, as shown in Fig. 21.3. The downstream face of the structure may be kept vertical or slightly inclined. The crest is sometimes extended in the form of an overhanging lip, which keeps small discharges away from the face of the overfall section. The water falls freely from the crest under the action of gravity. Since vacuum gets created in the underside portion of the falling jet, sufficient ventilation of the nappe is required in order to avoid pulsating and fluctuating effects of the jet. The design of such a spillway is done as that of a weir which was explained in the chapter on weirs. Sometimes, a secondary dam of low

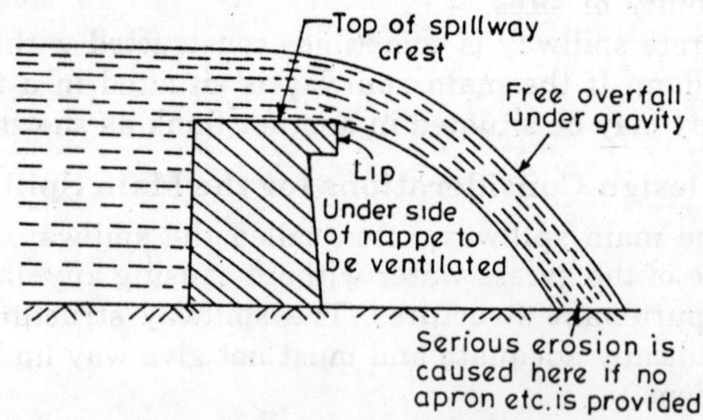


Fig. 21.3. (*a*) Straight drop spillway without d/s protection.

Fig. 21.3. (*b*) Straight drop spillway with d/s protection works.

height is constructed on the downstream side to create an artificial pool of water so as to dissipate the energy of the falling water.

## 21.6. Ogee Spillway or Overflow Spillway

Ogee spillway is an improvement upon the 'free overfall spillway, and is widely used with concrete, masonry, arch and buttress dams. Such a spillway can be easily used on valleys where the width of the river is sufficient to provide the required

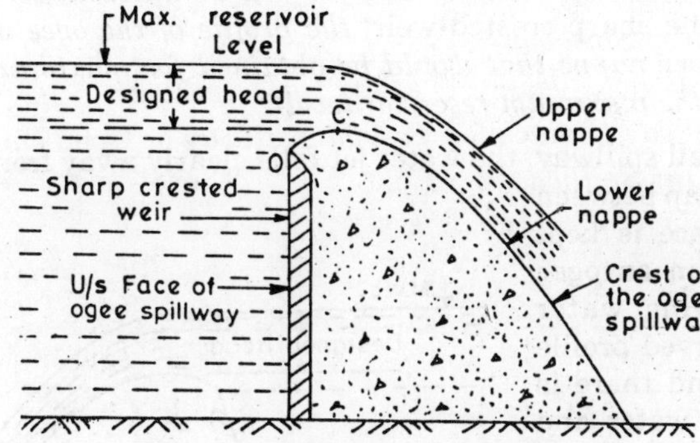


Fig. 21.4. (*a*) Section of an ogee spillway with vertical u/s face.

crest length and the river bed below can be protected from scour at moderate costs. *The profile of this spillway is made in accordance with the shape of the lower nappe of a free falling jet, over a duely ventilated sharp crested weir,* as shown in Fig. 21.4 (*a*).

Fig. 21.4. (*b*) Photoview of an Ogee Spillway of Amaravathi dam (earthern cum masonry dam) located on river Amaravathi (a tributary of river Cauvery) in Coimbatore District in Tamil Nadu state.

The shape of the lower nappe of freely falling jet over a sharp crested weir can be determined by the principle of projectile. It generally rises slightly (to point $C$) as it originates from the crest ($O$) of a sharp crested weir and then falls to make a parabolic form. Now, if the space between the sharp crested weir and the lower nappe is filled with concrete or masonry, the weir so formed will have a profile similar to an 'ogee' ($S$-shaped curve in section), and hence called an 'ogee weir' or an 'ogee spillway'. This lower nappe, will then become the crest of the spillway. Since the lower nappe of the free falling jet will be different for different heads over the crest of the sharp crested weir, *the profile of the ogee weir is generally confined to the lower nappe that would be obtained for maximum head over the spillway (i.e. upto the maximum reservoir level).*

In a free overfall spillway, the water jet falls clearly away from the face of the spillway, and the gap between the jet and the face is kept ventilated. While in an ogee spillway, the falling water glides over the curved profile of the spillway, and there is no space between water and crest of the spillway, under normal design conditions.

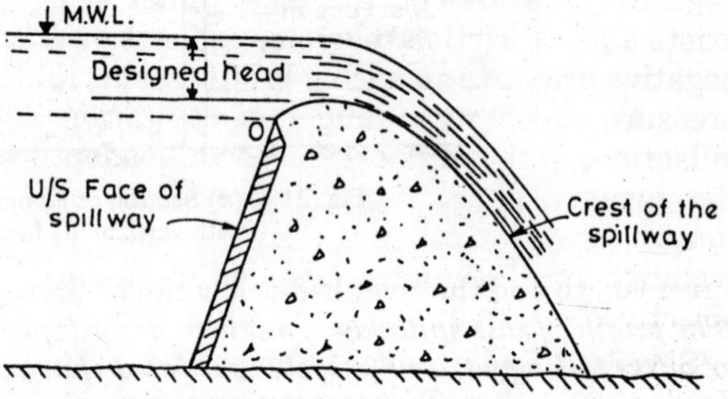

Fig. 21.5. Ogee spillway with inclined u/s face.

Normally, the upstream face of the spillway is kept vertical and the crest shape confirms to the lower nappe of a vertical sharp crested weir under maximum head. But if the upstream face of the spillway is kept sloping, the crest shape should also confirm to the lower nappe that would be obtained for an inclined sharp crested weir (Fig. 21.5).

**21.6.1. Cavitation.** The crest of the ogee spillway can be made to confirm only to one particular nappe that would be obtained at one particular head. This head is called the *designed head* and represented by, say, $H_d$. But in practice, the actual head of water on the spillway crest, called the *operating head,* may be less or more than the designed head. *If this operating head on the spillway is more than the designed head, the lower nappe of the falling jet may leave the ogee profile, thereby generating negative pressure at the point of separation.* The generation of vacuum or negative pressure (*i.e.* pressure below the atmospheric pressure) may lead to formation of bubbles or cavities in the water. These cavities or bubbles filled with air, vapour and other gases are formed in a liquid, whenever the absolute pressure (*i.e.* atmospheric pressure—vacuum pressure) of the liquid is close to its vapour pressure, so as to commence evaporation. Such a condition may arise when the head of water is more than the designed head and the consequent high velocity jet causes reduced pressures or negative pressures in the lower region of the water jet.

Such cavities, on moving downstream, may enter a region where the absolute pressure is much higher (*i.e.* more vacuum). This causes the vapour in the cavity to condense and return to liquid with a resulting implosion or collapse of the cavity. When the cavity collapses, extremely high pressures are generated. The continuous bombardment of these implosions will thus take place near the surface of the spillway, causing fatigue failure of its material. The small particles of concrete or masonry are thus broken away, causing formation of pits on its surface and giving the surface a spongy appearance. This damaging action of cavitation is called 'pitting'. The cavitation plus the vibrations from the alternate making and breaking of contact between the water and face of the spillway, may thus result in serious structural damages to the spillway crest. *Hence, it can be concluded that if the head of water over the spillway is more than the designed head, cavitation may occur. On the other hand, if the head of water over the spillway is less than the designed head, the falling jet would adhere to the crest of the ogee spillway, creating positive hydrostatic pressures and thereby reducing the discharge coefficient of the weir.*

**21.6.2. Designing the Crest of the Ogee Spillway.** The ogee spillways were being designed in the earlier periods, in accordance with the theoretical profile obtained for the lower nappe of a free falling jet. The profile was known as Bazin's profile. Theoretically, the adoption of such a profile, should cause no negative pressures on the crest under designed head. But in practice, there exists a lot of friction due to roughness on the surface of the spillway. Hence, negative pressure on such a profile seems inevitable. The presence of negative pressure causes the danger of cavitation and sometimes fluctuations and pulsations of the nappe. Hence, while adopting a profile for the spillway crest, the avoidance of negative pressures must be an objective along with consideration of other factors such as practicability, hydraulic efficiency, stability and economy. Depending upon research work based on these objectives, various modified profiles have been proposed these days.

Several standard ogee shapes have been developed by U.S. Army Corps of Engineers at their Waterways Experimental Station (WES). Such shapes are known as 'WES Standard Spillway Shapes'. The d/s profile can be represented by the equation

$$x^n = K \cdot H_d^{\,n-1} \cdot y \qquad ...(21.1)$$

where $(x, y)$ are the co-ordinates of the points on the crest profile with the origin at the highest point $C$ of the crest, called the apex.

$H_d$ is the *design head including the velocity head.*

$K$ and $n$ are constants depending upon the slope of the upstream face. The values of $K$ and $n$ for different values of upstream slope are tabulated in Table 21.1.

**Table 21.1**

| *Slope of the u/s face of the spillway* | $K$ | $n$ |
|---|---|---|
| Vertical, | 2.0 | 1.85 |
| $1:3\ (1H:3V)$ | 1.936 | 1.836 |
| $1:1\frac{1}{2}\ (1H:1\frac{1}{2}V)$ | 1.939 | 1.810 |

The d/s profile given by Eq. (21.1) is continued till it meets tangentially the straight sloping surface of the *d/s* face of the spillway, at the tangent point, determined by equating $\frac{dy}{dx} = \frac{V}{H}$.

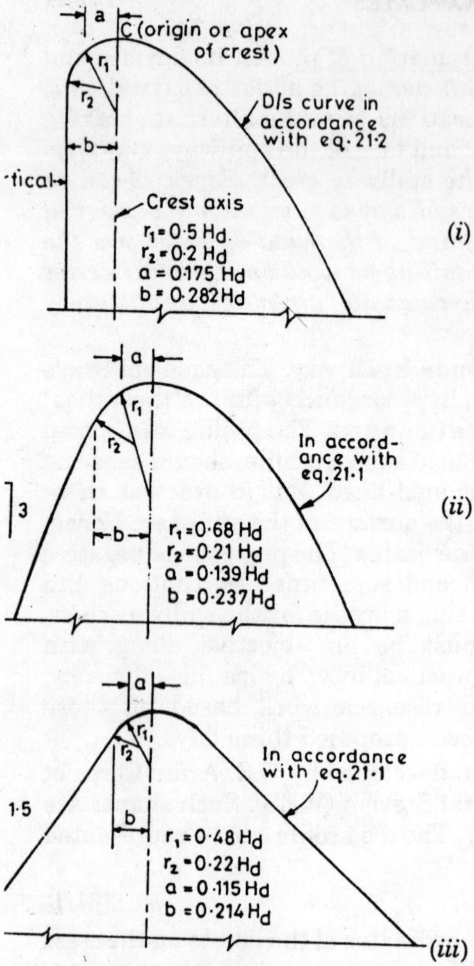


Fig. 21.6. WES profiles for Ogee Spillways for different u/s slopes.

Thus, for a spillway having a vertical u/s face, the d/s crest is given by the equation

$$x^{1.85} = 2 \cdot H_d^{0.85} \cdot y \qquad ...(21.2)$$

Different upstream curves were given by WES for different slopes, as shown in Fig. 21.6.

According to the latest studies of U.S. Army Corps, the u/s curve of the ogee spillway having a vertical u/s face, should have the following equation :

$$y = \frac{0.724\,(x + 0.27\,H_d)^{1.85}}{H_d^{0.85}} + 0.126\,H_d - 0.4315\,H_d^{0.375} \times (x + 0.27\,H_d)^{0.625} \qquad ...(21.3)$$

The u/s profile extends up to

$$x = -\,0.27\,H_d$$

Co-ordinates for the upper nappe for various WES shapes of ogee spillway are also available and can be utilised in the design of training walls and spillway bridges, etc.

The profile for an ogee spillway, having a vertical upstream face, can be determined on the basis of its WES profile, or Table 21.2 may be used for making the ogee profile.

**Table 21.2. Table for marking Ogee profile**

| $\frac{x'}{H_d}$ | $\frac{y}{H_d}$ | |
|---|---|---|
| | *Lower Nappe* | *Upper Nappe* |
| 0 | – 0.125 | 0.831 |
| 0.10 | – 0.033 | 0.807 |
| 0.25 | – 0.000 | 0.763 |
| 0.50 | – 0.034 | 0.668 |
| 0.75 | – 0.129 | 0.539 |
| 1.0 | – 0.283 | 0.373 |
| 1.5 | – 0.738 | – 0.088 |
| 2.0 | – 1.393 | – 0.743 |
| 3.0 | – 3.303 | – 2.653 |
| 4.0 | – 6.013 | – 5.363 |
| 5.0 | – 9.523 | – 8.873 |

$(x', y)$ are the co-ordinates of a point on the profile as shown in Fig. 21.7.

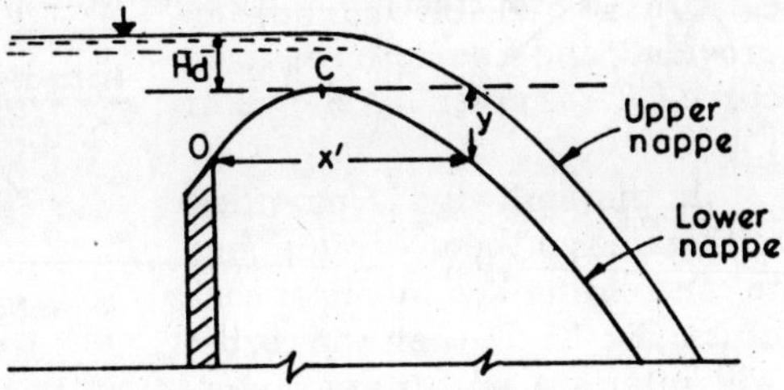


Fig. 21.7

The co-ordinates of the lower nappe determine the crest profile, while the plotting of the upper nappe is useful in determining the clearance for the spillway deck bridge and the top levels of training walls on the sides of the spillway.

After having plotted most of the profile of the ogee spillway, a smooth gradual reverse curvature is provided at the bottom of the downstream face, as shown in Fig. 21.8. The reverse curve turns the flow into the apron of a stilling basin or into the spillway discharge channel. A radius of about one-fourth of the spillway height is satisfactory for this reverse bottom curve.

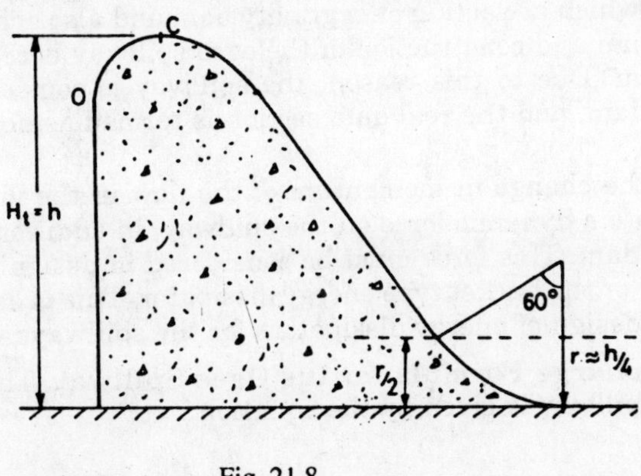


Fig. 21.8

**21.6.3. Practical Profile Consistent with the Profile of Gravity Dam.** When the profile for the crest of the ogee spillway is plotted over the triangular profile of the section of a gravity dam (non-overflow section), it is found that it goes beyond the downstream face of the dam, thus requiring thickening of the section for the spillway (Fig. 21.9). This extra concrete required, can be saved by shifting the curve of the nappe in a backward direction until this curve becomes tangential to the downstream face of the dam (Fig. 21.10). Hence, a saving can thus be affected, by providing a projecting corbel on the upstream face of the spillway section (Fig. 21.10). The construction of the spillway is thus carried out as if it was a non-overflow dam. Only the slight modifications are made after

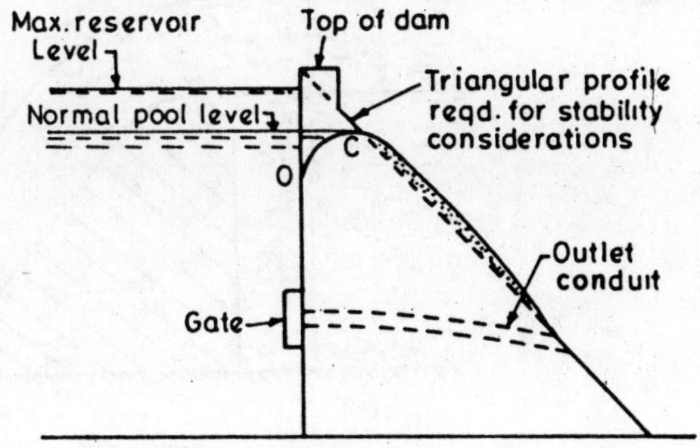


Fig. 21.9

reaching the required height (up to *O*) at which corbel is provided, and a smooth required curve *OCA* is given, as shown in Fig. 21.10.

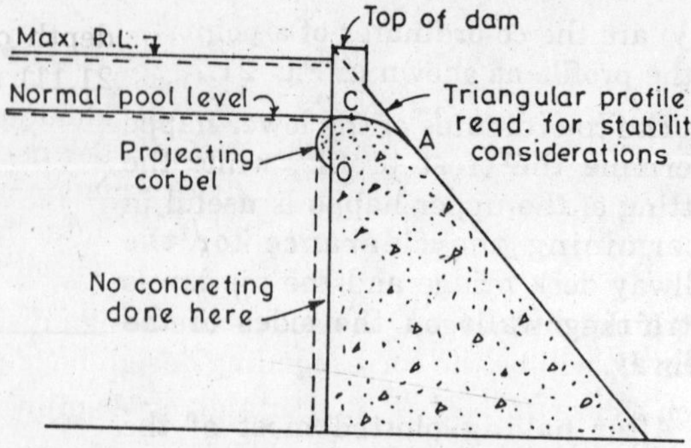


Fig. 21.10. Provision of Corbel.

If the spillway is provided with gates on the upstream face to control the flow in outlet conduits (Fig. 21.9), then the corbel will interfere with gate operation and hence the above concrete saving cannot be affected.

The structural design and stability requirements of ogee spillway are exactly the same as that of a gravity dam. The forces acting on a gravity dam also act on the ogee spillway and remain predominant. The pressure exerted on the crest of the spillway by the flowing water and the drag force caused by the fluid-friction are usually negligible as compared to other forces which are acting on a gravity dam and also act on an ogee spillway. Hence, the design and construction of the ogee spillway is consistent with that of the gravity dam. Due to this reason, the spillway is sometimes called overflow portion of the dam, and the real dam section is termed as non-overflow section of the dam.

However, the change in momentum of the flow in the vicinity of the reverse curve may create a dynamic force on the spillway, in addition to the forces acting on the gravity dam. This force must be considered in designing bottom of the d/s face of spillway or the bucket type energy dissipator. This is dealt separately while discussing the design of energy dissipators for the spillways.

**21.6.4. Discharge Formula for the Ogee Spillway.** The discharge passing over the ogee spillway is given by the equation :

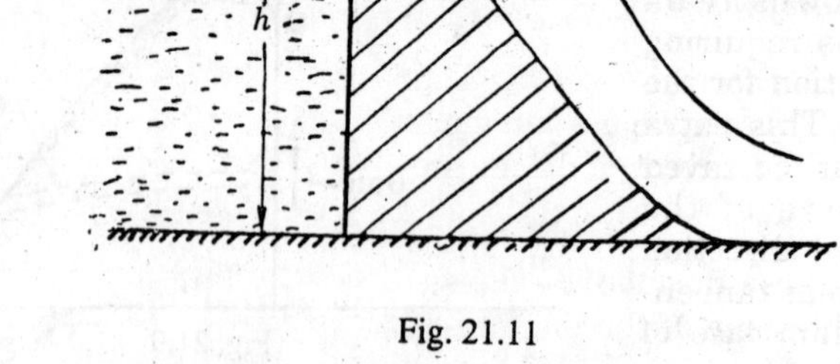


Fig. 21.11

$$Q = C \cdot L_e H_e^{3/2} \qquad \text{...(21.4)}$$

where $Q$ = Discharge

$L_e$ = Effective length of the spillway crest

$C$ = Coefficient of discharge which depends upon various factors such as relative depth of approach, [*i.e.* $d/H_d$ ratio (Fig. 21.11), relation of actual crest shape to the ideal nappe shape, slope of upstream face, downstream apron interference, and submergence, etc.

$H_e$ = Total head over the crest including the velocity head.

If the discharge $Q$ is used as the design discharge in Equation (21.4), then the term $H_e$ will be the corresponding design head ($H_d$) plus the velocity head ($H_a$). In such a case, $H_e = H_d + H_a$. For high ogee spillways, the velocity head is very small, and $H_e \approx H_d$.

**21.6.5. Variation of Coefficient of Discharge with Various Factors.** This is discussed below in details :

(*i*) *Depth of approach.* The coefficient of discharge firstly depends upon the depth of approach. In other words, it depends upon the height of the ogee weir ($h$) to the design head over the weir ($H_d$). If the height of the weir is more than 1.33 times the design head, the velocity of approach has been found to have a negligible effect upon discharge, and as such $H_d$ becomes equal to $H_e$ or $\frac{H_e}{H_d} = 1.0$. In such a case, the coefficient of discharge, say $C = C_d$, has been found to *be 2.2 in M.K.S. or S.I. Units* and 4.03 in *F.P.S. Units.*

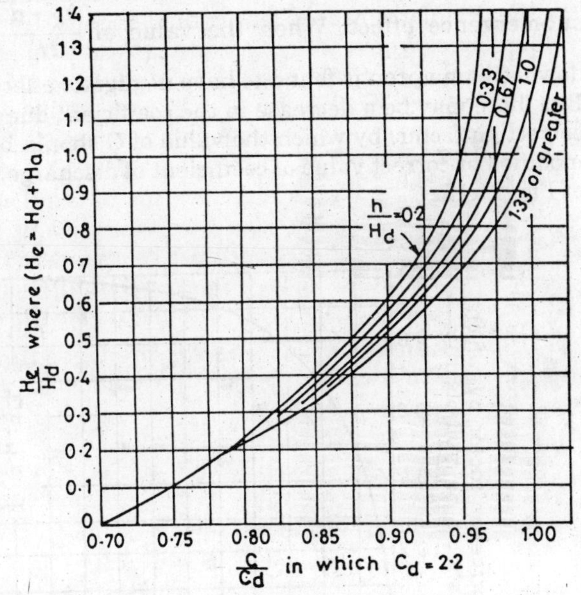


Fig. 21.12

However, in low spillways, with $\frac{h}{H_d} < 1.33$ ; the approach velocity is having an appreciable effect. The curves given in Fig. 21.12 can be used in such cases, to evaluate the coefficient of discharge $C$, using $C_d = 2.2$.

(*ii*) *U/s Slope.* The coefficient of discharge is also affected by the slope of the u/s face of the ogee weir. The values of $C_d$ and $C$ found up to now were for a vertical upstream face. If the u/s face is sloping, a correction factor by which the above values of $C$ should be multiplied can be obtained from the curves given in Fig. 21.13.

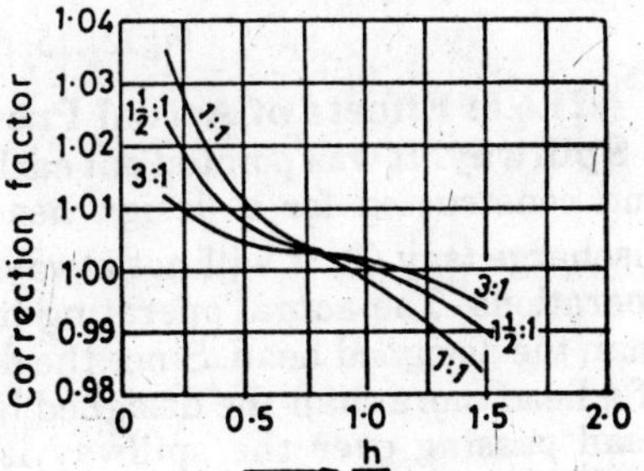


Fig. 21.13

*(iii) D/s apron interference and submergence effects.* The third important factor affecting the discharge coefficient of an ogee weir is the effect of downstream apron interference and downstream submergence.

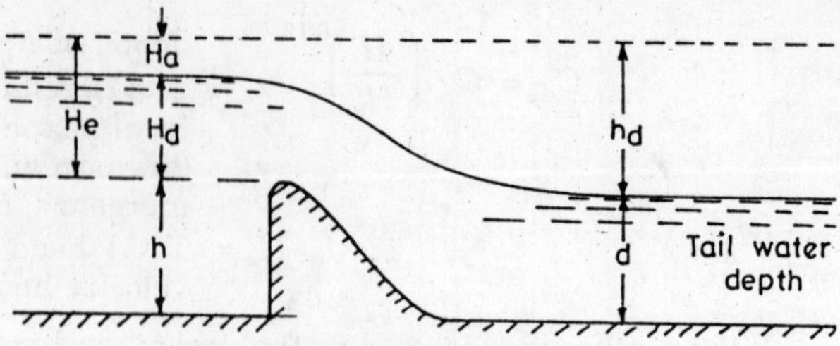


Fig. 21.14. Maximum tail water depth for a non-submerged weir.

When the tail water level is such that the top of the weir is covered by it, such that the weir cannot discharge freely ; the weir is then said to be a **submerged weir.**

Where the hydraulic jump occurs, the coefficient of discharge may decrease due to back pressure effect of the downstream apron and is independent of the submergence effect. When the value of $\frac{h_d + d}{H_e}$ (Fig. 21.15) exceeds 1.7, the downstream apron is found to have negligible effect on the coefficient of discharge. But there may be a decrease in the coefficient due to tail water submergence. The correction factor, by which the value of $C$ should be multiplied in order to get the modified or correct value of coefficient of discharge, can be obtained from the curve of Fig. 21.15.

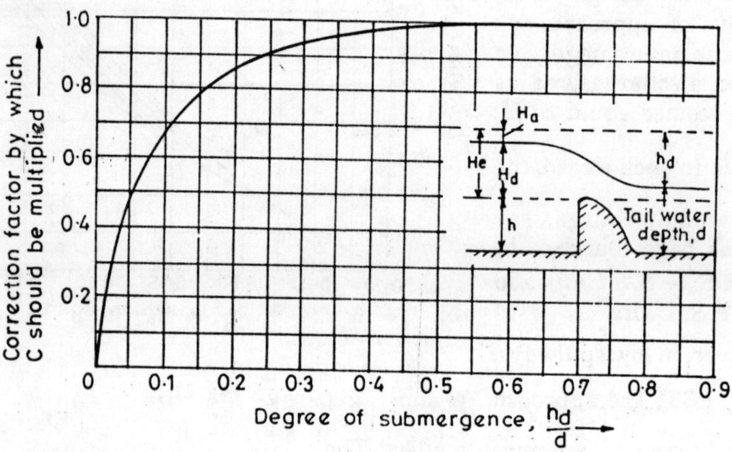


Fig. 21.15. Effect of submergence on $C$.

**21.6.6. Effects of Actual Prevailing Head on the Discharge Capacity of a Spillway.** It was pointed out earlier that when once a spillway has been designed and constructed for a design head ($H_d$), and for a corresponding coefficient of discharge (say $C$), it will not always find the same head over its crest in its actual operations. The actual operating head ($H$) i/c velocity head, may be less or more than the designed head. Since the design is done for maximum head, the possibility of a head more than the designed head is very meagre. When the actual operating head passing over the spillway is less than the designed head, the prevailing coefficient of discharge ($C_d$) tends to reduce, and is given by the equation

$$C_d = C \cdot \left(\frac{H}{H_e}\right)^{0.12} \qquad ...(21.5)$$

where $H_e$ is the designed head including velocity head.

Since an overflow spillway is sufficient in height (*i.e* $h > 1.33\ H_d$) ; the coefficient of discharge $C$ at designed head can be taken as 2.2. The prevailing coefficient of discharge at 50% head (*i/c* velocity head) will then be

$$C_d = 2.2 \times \left(\frac{0.5H_e}{H_e}\right)^{0.12} = 2.2 \times 0.92 = 2.02.$$

Similarly, for still lower heads, the coefficient of discharge goes on reducing and tends to become constant at about 1.7. (Because at very low heads, the velocity head becomes the governing factor, which tries to make $H$ a constant).

*The effective length ($L_e$) of ogee spillway.*

The effective length of the spillway crest ($L_e$) to be used in equation (21.4), is given by the equation

$$L_e = L - 2\,[K_p \cdot N + K_a]\,H_e \qquad ...(21.6)$$

where $L$ = the net clear length of the spillway crest

$K_p$ = Pier contraction coefficient

$K_a$ = Abutment contraction coefficient

$N$ = Number of piers

$H_e$ = Total design head on the crest including velocity head.

The values of $K_p$ and $K_a$ depend mainly upon the shape of the piers and that of the abutments. The greater is the divergence from streamlined flow, the greater is the contraction coefficient and lesser is the effective length of the crest. A 90° cut water nose pier is most efficient and has quite a low value of $K_p$, and is generally preferred. Values of $K_p$ and $K_a$ are given in Tables 21.3 and 21.4. Various shapes of piers are shown in Fig. 21.16.

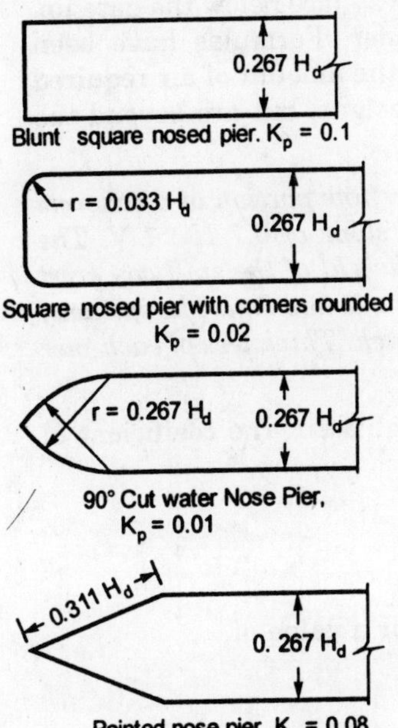


Fig. 2.16. Various shpaes of piers.

**Table 21.3.**

| *S. No.* | *Pier Shape* | *Contraction coefficient $K_p$* |
|---|---|---|
| 1. | Square nosed piers without any rounding | 0.1 |
| 2. | Square nosed piers with corners rounded on radius equal to 0.1 of pier thickness | 0.02 |
| 3. | Rounded nose piers and 90° cut water nosed piers | 0.01 |
| 4. | Pointed nose piers | 0.0 |

**Table 21.4**

| S. No. | *Shape of abutment* | *Contraction coefficient, $K_a$* |
|---|---|---|
| 1. | Square abutment with head wall at 90° to the direction of flow | 0.2 |
| 2. | Rounded abutment with head wall at 90° to the direction of flow | 0.1 |

**21.6.7. Aeration Arrangements in Gated Ogee Spillways.** The entire design of an overflow spillway has been done with the assumption that upper and lower nappe are subjected to full atmospheric pressure. But in practice, due to insufficient aeration, development of negative pressures takes place beneath the nappe due to removal of air by the falling jet. The development of negative pressures causes the danger of cavitation and induces fluctuation and pulsation effects on the jet, which may be very objectionable if the spillway is to be used as a discharge measuring device.

To control the development of negative pressures, aeration pipes 25 mm dia at say 3 m centre to centre may be provided along spillway face below the gate lip. These pipes can be connected to a bigger sized header. Formulas have been developed on the basis of model studies, for evaluating the amount of air required for aeration, and may have to be used in large dam designs, but are beyond the scope of this book.

**Example 21.1.** *Design a suitable section for the overflow portion of a concrete gravity dam having the downstream face sloping at a slope of 0.7 H : 1 V. The design discharge for the spillway is 8,000 cumecs. The height of the spillway crest is kept at RL 204.0 m. The average river bed level at the site is 100.0 m. The spillway length consists of 6 spans having a clear width of 10 m each. Thickness of each pier may be taken to be 2.5 m.*

**Solution.** Since the given spillway looks like a high weir, the coefficient of discharge may be assumed to be 2.2.

Now $\quad Q = C \cdot L_e H_e^{3/2}$

where $L_e = L - 2\,[N K_p + K_a]\, H_e$

Let us first work out the approximate value of $H_e$ for a value of

$$L_e \approx L = \text{clear waterway} = 6 \times 10 = 60 \text{ m.}$$

$\therefore \quad 8{,}000 = 2.2 \times 60\, H_e^{3/2}$

or $\quad H_e^{3/2} = \dfrac{8{,}000}{2.2 \times 60} = 60.6$

or $\quad H_e = (60.6)^{2/3} = 15.5 \text{ m.}$

The height of the spillway above the river bed (see Fig. 21.15)

$$= h = 204 - 100 = 104.0 \text{ m}$$

Since $\dfrac{h}{H_d}$, *i.e.* $\dfrac{104}{15.5} > 1.33$,

it is a high spillway, the effect of velocity head can, therefore, be neglected.

Since $\frac{h_d + d}{H_e} = \frac{H_e + h}{H_e} = \frac{15.5 + 104}{15.5} > 1.7;$

the discharge coefficient is not affected by tail water conditions, and the spillway remains a high spillway.

**U/s Slope.** The upstream face of the dam and spillway is proposed to be kept vertical. However, a batter of 1 : 10 will be provided from stability considerations in the lower part. This batter is small and will not have any effect on the coefficient of discharge.

**Effective length of spillway** ($L_e$) can now be worked out as

$$L_e = L - 2\,[N.K_p + K_a]\,H_e$$

Assuming that 90° cut water nose piers and rounded abutments shall be provided, we have

$$K_p = 0.01$$

and $K_a = 0.1$

No. of piers = $N = 5$.

Also assuming that the actual value of $H_e$ is slightly more than the approximate value worked out (*i.e.* 15.5 m), say, let it be 16.3 m, we have

$$\therefore \quad L_e = 60 - 2\,[5 \times 0.01 + 0.1] \times 16.3 = \mathbf{55.1\ m.}$$

Hence $Q = 2.2 \times 55.1 \times H_e^{3/2}$

or $8{,}000 = 2.2 \times 55.1 \times H_e^{3/2}$

or $H_e^{3/2} = \frac{8{,}000}{2.2 \times 55.1} \cong 66.0$

or $H_e = (66.0)^{2/3} = \mathbf{16.4\ m} \cong 16.3$ (assumed)

Hence, the assumed $H_e$ for calculating $L_e$ is all right. The crest profile will be designed for $H_d = 16.4$ m (neglecting velocity head).

**Note.** The velocity head ($H_a$) can also be calculated as follows :

$$\text{Velocity of approach} = V_a = \frac{8{,}000}{(60 + 5 \times 2.5)\,(104 + 16.4)}$$

$$= \frac{8{,}000}{72.5 \times 120.4} = 0.917 \text{ m/sec.}$$

$$H_a = \text{Velocity Head} = \frac{V_a^2}{2g} = \frac{(0.917)^2}{2 \times 9.81} = 0.043 \text{ m.}$$

This is very small and was, therefore, neglected.

*Downstream profile.* The W.E.S. d/s profile for a vertical u/s face is given by equation (21.2) as :

$$x^{1.85} = 2 \cdot H_d^{0.35} \cdot y$$

or $$y = \frac{x^{1.85}}{2(H_d)^{0.85}} = \frac{x^{1.85}}{2 \times (16.4)^{0.85}}$$

or $$y = \frac{x^{1.85}}{2 \times 10.8}$$

or $$y = \frac{x^{1.85}}{21.6} \qquad \text{...(21.7)}$$

**Before we determine the various co-ordinates of the d/s profile,we shall first determine the tangent point, upto which the above profile with extend.**

**The d/s slope of the dam is given to be 0.7 *H* : 1 *V*.**

**Hence,** $$\frac{dy}{dx} = \frac{1}{0.7}$$

**Differentiating the equation of the d/s profile w.r.to *x*, we get**

$$\frac{dy}{dx} = \frac{1.85x^{1.85-1}}{21.6} = \frac{1}{0.7}$$

or $$x^{0.85} = \frac{21.6}{1.85 \times 0.7} = 16.7$$

or $$x = 27.4 \text{ m}$$

$$\therefore \quad y = \frac{(27.4)^{1.85}}{21.6} = 21.15 \text{ m}$$

**The co-ordinates from $x = 0$ to $x = 27.4$ m, representing the *d/s* Ogee profile, are worked out in Table 21.5.**

**Table 21.5**

| *x (metres)* | $y = \frac{x^{1.85}}{21.6}$ *metres* |
|---|---|
| 1 | 0.05 |
| 2 | 0.17 |
| 3 | 0.35 |
| 4 | 0.60 |
| 5 | 0.91 |
| 6 | 1.27 |
| 7 | 1.69 |
| 8 | 2.17 |
| 9 | 2.70 |
| 10 | 3.28 |
| 12 | 4.59 |
| 14 | 6.11 |
| 16 | 7.82 |
| 18 | 9.72 |
| 20 | 11.82 |
| 22 | 14.09 |
| 24 | 16.56 |
| 26 | 19.20 |
| 27.4 | 21.15 |

*The u/s profile.* The u/s profile may be designed as per equation (21.3), as :

$$y = \frac{0.724\,(x + 0.27H_d)^{1.85}}{H_d^{0.85}} + 0.126\,H_d - 0.4315\,H_d^{0.375}\,(x + 0.27\,H_d)^{0.625}$$

Using $H_d$ = 16.4 m, we get

$$y = \frac{0.724\,[x + 0.27 \times 16.4]^{1.85}}{(16.4)^{0.85}} + 0.126\,(16.4)$$

$$- 0.4315\,(16.4)^{0.375}\,(x + 0.27 \times 16.4)^{0.625}$$

or $\quad y = 0.07\,(x + 4.44)^{1.85} + 2.07 - 1.234\,(x + 4.432)^{0.625}$ ...(21.8)

This curve should go upto $x = -\,0.27\,H_d$

or $\quad x = -\,0.27 \times 16.4 = -\,4.443$ m.

Various values of $x$ such as, $x = -\,0.5$, $x = -\,1.0$, $x = -\,2.0$, $x = -\,3.0$, $x = -\,4.0$, $x = -\,4.443$ are substituted in equation (21.8) and corresponding values of $y$ are worked out, as given below in Table 21.6.

**Table 21.6**

| *x in metres* | *y in metres* |
|---|---|
| – 0.5 | 0.020 |
| – 1.0 | 0.063 |
| – 2.0 | 0.27 |
| – 3.0 | 0.65 |
| – 4.0 | 1.34 |
| – 4.443 | 2.07 |

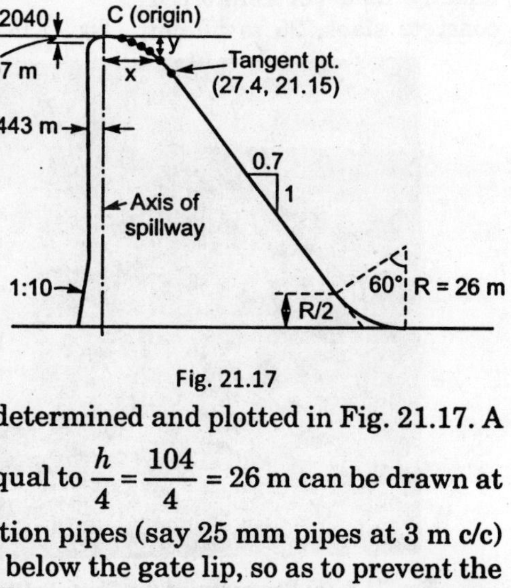


Fig. 21.17

The profile of the spillway has been determined and plotted in Fig. 21.17. A reverse curve at the toe with a radius equal to $\frac{h}{4} = \frac{104}{4}$ = 26 m can be drawn at angle 60°, as shown in Fig. 21.17. Aeration pipes (say 25 mm pipes at 3 m c/c) can be installed along the spillway face below the gate lip, so as to prevent the development of negative pressures. The energy dissipation arrangements have not been shown. They should be designed depending upon the position of the jump height curve and tail water curve, as explained afterwards. A sky jump bucket or an apron may be provided as per the prevailing conditions.

## 21.7. Chute Spillway or the Trough Spillway

An ogee spillway is mostly suitable for concrete gravity dams especially when the spillway is located within the dam body in the same valley. But for earthen and rockfill dams, a separate spillway is generally constructed in a flank or a saddle, away from the main valley, as explained earlier. Sometimes, even for gravity dams, a separate spillway is required because of the narrowness of the main valley. In all such circumstances, a separate spillway may have to be provided. The *Trough Spillway* or *Chute Spillway* is the simplest type of a spillway which can be easily provided independently and at low costs. It is lighter and adaptable to any type of foundations ; and hence provided easily on earth and rockfill dams. A chute spillway is sometimes known as a **waste weir**. If it is constructed in continuation to the dam at one end, it may be called a **flank weir**. If it is constructed in a natural saddle in a bank of the river separated from the main dam by a high ridge, it is called a **saddle weir.**

A chute spillway (Fig. 21.18) essentially consists of a steeply sloping open channel, placed along a dam abutment or through a flank or a saddle. It leads the water from the reservoir to the downstream channel below. The base for the channel is usually made of reinforced concrete slabs, 25 to 50 cm

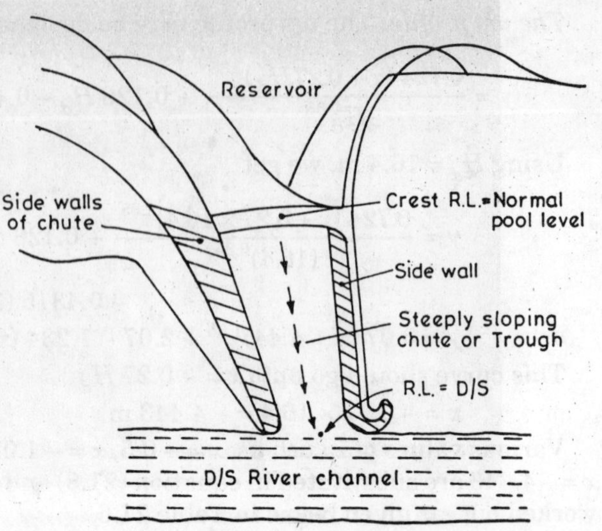


Fig. 21.18 (*a*). Simplified Line sketch of a Chute Spillway.

Fig. 21.8. (*b*) Photoview of the Chute Spillway of Anderson Ranch Dam, Idaho (USA).

thick. Light reinforcement of about 0.25% of the concrete area is provided in the top of the slabs in both directions. The chute is sometimes of constant width, but is usually narrowed for economy and then widened near the end to reduce the discharging velocity. Expansion joints are usually provided in the chute at intervals of about 10 metres in either direction. The expansion joints should be made watertight so as to avoid any under-seepage and its troublesome effects. Under-drains are also provided, so as to drain the water which may seep through the trough bottom and side walls. These under-drains may be in the form of perforated steel pipes, clay tiles or rock-filled trenches.

If the slope of the chute can confirm to available topography, the excavations shall be minimum. But the slope of the chute must be high enough, and should atleast be able to maintain super critical flow, to avoid unstable flow conditions.

**Fig. 21.18. (*c*) Photoview of two chutes provided at Bhakra Dam (India).**

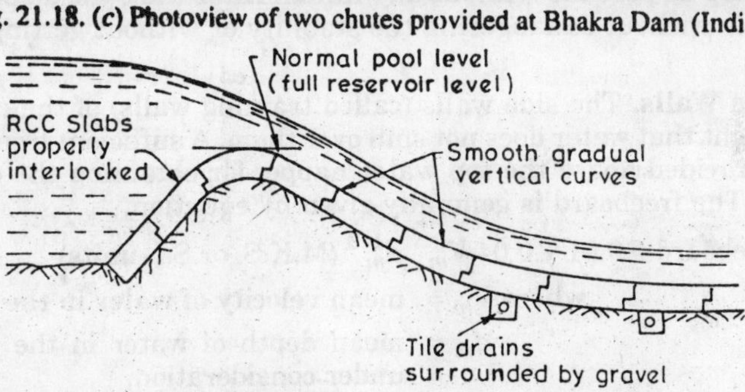


Fig. 21.18. (*d*) Section through a Chute Spillway.

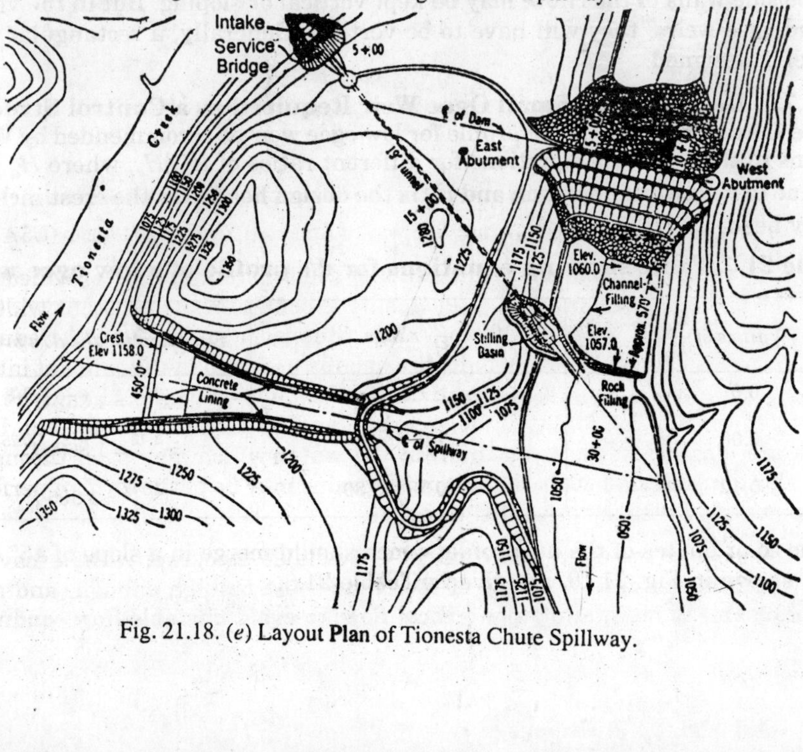


Fig. 21.18. (*e*) Layout **Plan** of Tionesta Chute Spillway.

When a vertical curve is provided at a point where the chute slope changes, it must be gradual and designed so as to avoid any separation of flow.

**21.7.1. Control Structure or a Low Ogee Weir.** Since the trough spillway is provided in a flank or a saddle, the height of spillway or ogee weir required to be constructed in that flank, will be small ; sometimes almost flat low weir shall be required depending on the natural levels of the bottom of the flank. If the flank bottom is at a level lower than the natural pool level, an ogee weir shall have to be constructed upto that level. If the flank bottom is at higher level than the normal pool level, excavations will have to be done upto that level. In such a case, the weir crest is normally left flat as it shall seldom be economical to excavate the rock just for the sake of constructing on 'ogee shape' for obtaining high coefficient of discharge.

**21.7.2. Chute Slope.** The water spilling over the control structure (*i.e.* ogee weir), then flows through the chute channel. The minimum slope of the chute is governed by the condition that supercritical flow must be maintained. The slope of the chute is kept just sufficient to meet this flow requirement from the crest for as long a distance as possible without any filling. After that, the slope is made as steep as possible without endangering the stability or without getting into heavy excavations.

**21.7.3. Side Walls.** The side walls (called training walls) of the chute should be of such a height that water does not spill over them. A sufficient freeboard must, therefore, be provided above the top water nappe, for obtaining the top levels of the side walls. The freeboard is generally given by equation,

$$\text{Freeboard} = 0.61 + 0.04\, V_m \cdot d_m^{1/3} \text{ (M.K.S. or S.I. units)} \qquad ...(21.9)$$

where $V_m$ = mean velocity of water in the chute reach

$d_m$ = mean depth of water in the chute reach under consideration.

The side walls of the chute may be kept vertical or sloping. But in the vicinity of gated ogee weirs, they will have to be vertical. Generally, a rectangular chute channel is designed.

**21.7.4. Design of the Small Ogee Weir Required as a Control Structure for the Chute Spillway.** The profile for low ogee weir as recommended by W.E.S. confirms to the following equations for different ratios of $H_a/H_e$, where $H_a$ is the head due to velocity of approach and $H_e$ is the design head over the crest including velocity head, *i.e.* $(H_e = H_d + H_a)$.

**Table 21.7. Table showing equations for d/s profile of a low ogee weir**

| *Values of* $\frac{H_a}{H_e}$ | $\frac{h}{H_e}$ *range* | *Equation for the d/s profile* |
|---|---|---|
| 0.0 | ≥ 1.0 | $x^{1.78} = 1.852\, H_e^{0.78} \cdot y$ |
| 0.08 | 1.0 – 0.58 | $x^{1.75} = 1.869\, H_e^{0.75} \cdot y$ |
| 0.12 | 0.58 – 0.30 | $x^{1.747} = 1.905\, H_e^{0.747} \cdot y$ |

The co-ordinates of the u/s profile, which should merge in a slope of 45° (*i.e.* 1 : 1), as shown in Fig. 21.19, are given in Table 21.8.

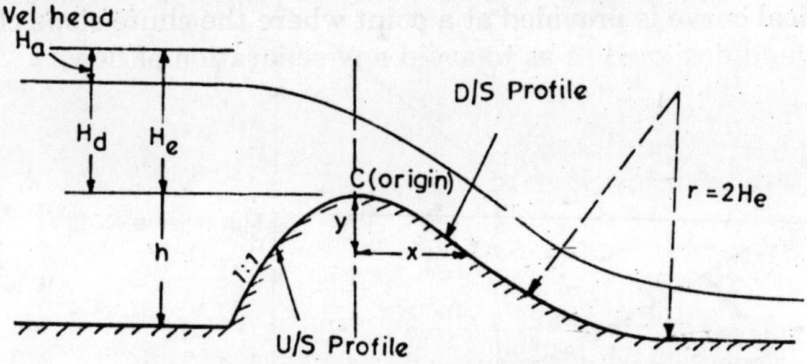


Fig. 21.19. Profile of the small ogee weir to be used as control structure for a chute spillway.

**Table 21.8. Co-ordinates of u/s profile for low ogee weir**

| $\frac{x}{H_e}$ | $\frac{y}{H_e}$ *for different values of* $\frac{H_a}{H_e}$ | | |
|---|---|---|---|
| | $\frac{H_a}{H_e} = 0.00$ | $\frac{H_a}{H_e} = 0.08$ | $\frac{H_a}{H_e} = 0.12$ |
| – 0.000 | 0.0000 | 0.0000 | 0.0000 |
| – 0.020 | 0.0004 | 0.0004 | 0.0004 |
| – 0.060 | 0.0036 | 0.0035 | 0.0035 |
| – 0.100 | 0.0103 | 0.0101 | 0.0099 |
| – 0.120 | 0.0150 | 0.0150 | 0.0147 |
| – 0.140 | 0.0207 | 0.0208 | 0.0199 |
| – 0.150 | 0.0239 | 0.0235 | 0.0231 |
| – 0.160 | 0.0275 | 0.0270 | 0.0265 |
| – 0.175 | 0.0333 | 0.0328 | 0.0325 |
| – 0.190 | 0.0399 | 0.0395 | 0.0390 |
| – 0.195 | 0.0424 | 0.0420 | — |
| – 0.200 | 0.0450 | — | — |

The reverse curve at the toe may have a radius equal to $2H_e$.

**21.7.5. Design of Vertical Curves of the Chute.** This is discussed below :

*Concave Curve.* Whenever the slope of chute changes from steeper to milder [Fig. 21.20 (*a*)], a concave vertical curve shall be provided. In no case, the radius of this curve should be less than 10 *d*, where *d* is the depth of water in metres.

*Convex curve.* Whenever the slope of the chute changes from milder to sleeper [Fig. 21.20 (*b*)], convex vertical curve shall have to be provided. The curvature should approximate to a parabolic shape given by the equation :

$$q = -x \tan \theta - \frac{x^2}{K\,[4\,(d + h_v)\cos^2 \theta]} \qquad \text{...(21.10)}$$

where $\theta$ is the slope angle of the floor upstream ;

$(d + h_v)$ is the specific energy of flow at the junction point ;

$K$ is constant which is $\geq 1.5$.

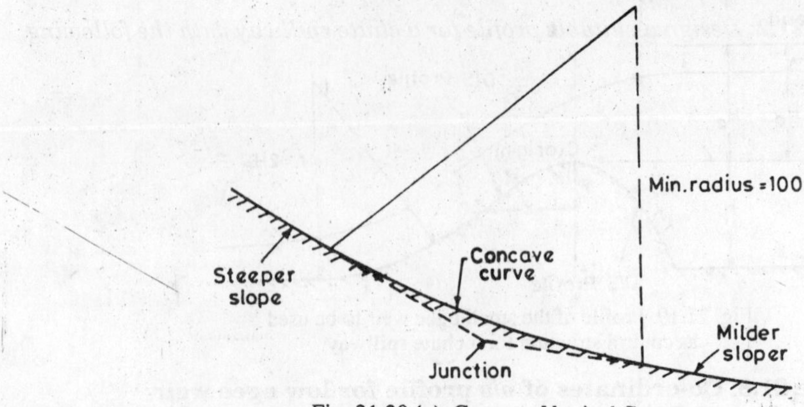


Fig. 21.20 (*a*). Concave Vertical Curve.

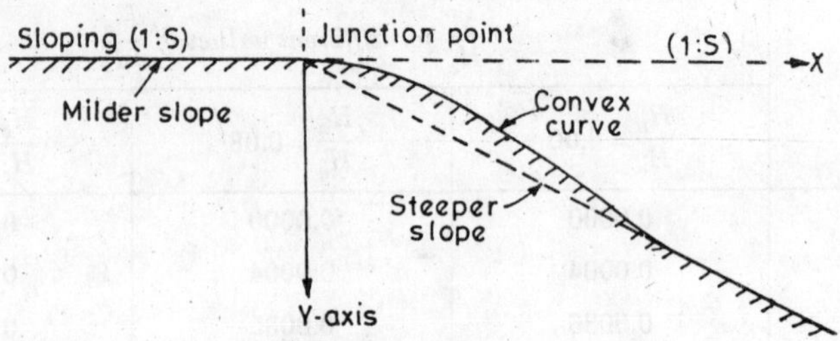


Fig. 21.20 (*b*). Convex Vertical Curve.

**21.7.6. Approach Channel of Chute Spillway.** An entrance channel called an approach channel, trapezoidal in shape with side slopes 1 : 1, may be constructed so as to lead the reservoir water up to the control structure (low ogee weir). If any curvature (in plan) is required, it is generally confined to the entrance channel, because the velocity of flow is low in this channel. The chute channel called the discharge channel or *discharge carrier* is generally kept straight in plan. If, however, any curvature is required to be provided, its floor should be super elevated to guide the high velocity flow around the bend, thus avoiding piling up of flow towards the outside of the chute. The friction head lost in the entrance channel upto the spillway crest can be calculated by Manning's formula, given as :

$$h_f = S_f \times L = \frac{n^2 \cdot V^2 \cdot L}{R^{4/3}} \quad \text{...(21.11)}$$

where $n$ = Manning's coefficient of roughness
Velocity in channel
$R$ = Hydraulic mean depth
$L$ = Length of channel
$S_f$ = Mean energy slope between two points.

The entire chute spillway can hence be divided into the following parts :

(*i*) Entrance channel
(*ii*) Control structure (Low Ogee weir)
(*iii*) Chute channel or Discharge carrier
(*iv*) Energy dissipation arrangements at the bottom in the form of a stilling basin.

The design principles for the various components have already been explained except those for energy dissipation. The energy dissipation arrangements required at the bottom of the spillway shall be explained a little later.

**Example 21.2.** *Design a suitable profile for a chute spillway with the following data :*

*Spillway crest level* = *200.0 m*

*Level of bottom of flank at which the low Ogee weir is to be constructed* = *192.0 m*

*Design discharge* = *5,000 cumecs.*

*Downstream tail water level corresponding to 5,000 cumecs = 103.0 m.*

*The spillway length consists of 5 spans of 10.0 m clear width each. The thickness of each spillway pier may be assumed to be 3 m. Assume any other data required.*

**Solution.**

**Design of approach channel :**

$$Q = C \cdot L_e H_e^{3/2}.$$

Assume the coefficient of discharge as 2.18, and taking $L_e$ as the clear width (approximately) = 10 × 5 = 50 m ;

and assuming $H_e \approx H$

we get $5{,}000 = 2.18\ (50)\ H^{3/2}$

or $H^{3/2} = \left(\dfrac{100}{2.18}\right) = 45.8 \text{ m.}$

or $H = (45.8)^{0.667} = 12.8 \text{ m.}$

Upstream water level = Crest level + $H$ = 200 + 12.8 = 212.8 m.

Bed level of river in flank = 192.0

∴ Water depth = 212.8 – 192.0 = **20.8 m.**

Assuming the trapezoidal approach channel with 1 : 1 side slopes,

The width of the channel = $B$

= Total length of spillway = 50 + 4 × 3 = 62 m.

Area of channel = $A = (B + y)\, y = (62 + 20.8)\ 20.8 = 1{,}720$ sq. m.

Velocity of approach = $V_a = \dfrac{5{,}000}{1{,}720} = 2.9 \text{ m/sec}$

Velocity Head $= H_a = \dfrac{V_a^2}{2g} = \dfrac{(2.9)^2}{2 \times 9.81} = \mathbf{0.43\ m.}$

The wetted perimeter $P$ of the channel

$= [B + 2\sqrt{2}\ (\text{water depth})]$

$= [62 + 2\sqrt{2} \times 20.8] = (62 + 58.8) = 120.8$ m.

$$R = \frac{A}{P} = \frac{1{,}720}{120.8} = \mathbf{14.2\ m.}$$

Assuming the length of the channel to be 160 m ; the head loss due to friction upto the spillway crest, is given by Manning's formula, as :

$$h_f = \frac{n^2 V^2 L}{R^{4/3}} \text{(assume } n = \text{Manning's rugosity Coefficient} = 0.019)$$

$$\therefore \quad h_f = \frac{(0.019)^2 \times (2.9)^2 \times 160}{(14.2)^{1.33}} = \mathbf{0.016\ m.}$$

Level of u/s *TEL*

= u/s water level + Velocity Head – Head lost up to spillway crest

= 212.8 + 0.43 – 0.016 = 213.214 ; say **213.21 m.**

Hence, head over the crest including velocity of approach

$= H_e = 213.21 - 200.0 = \mathbf{13.21\ m.}$

and $H_d = 13.21 - 0.43 = \mathbf{12.78\ m.}$

**Correct Coefficient of Discharge**

(*i*) *Correction due to height of weir,*

$$\frac{h}{H_d} = \frac{8}{12.78} = 0.63$$

$$\frac{H_e}{H_d} = \frac{13.21}{12.78} = 1.03$$

From Fig. 21.12 ; $\frac{C}{C_d} = 0.98.$

(*ii*) *Correction due to u / s slope of 45° (i.e. 1 : 1 slope)*

From Fig. 21.13 ; Correction factor = 1.008.

(*iii*) Assuming that the downstream apron elevation is maintained such that it does not affect $C_d$, we have,

The correct value of coefficient of discharge

$= C = 2.2 \times 0.98 \times 1.008 = 2.17$

Effective length $= L_e = 50 - 2\,[N \cdot K_p + K_a]\,H_e$

assuming $K_p = 0.01$ and $K_a = 0.1$

$$L_e = 50 - 2\,[4 \times 0.01 + 0.1] \times 13.21 = 50 - 3.7 = \mathbf{46.3\ m.}$$

The correct head $H_e$ will then be given by

$$5{,}000 = 2.17 \times 46.4 \times H_e^{3/2}$$

or $$H_e^{3/2} = \frac{5{,}000}{2.17 \times 46.3} = 49.7$$

or $$H_e = (49.7)^{0.667} = \mathbf{13.6\ m.}$$

Hence, corrected $H_d = 13.6 - 0.43 = 13.17$ m ; say 13.2 m.

Hence, the design shall be done for a design head equal to 13.2 m.

**Design of Crest Profile**

$$\frac{H_a}{H_e} = \frac{0.43}{13.2 + 0.43} = \frac{0.43}{13.63} = 0.0315$$

$$\frac{h}{H_e} = \frac{8}{13.6} = 0.58$$

Since $\frac{H_a}{H_e}$ lies between 0 and 0.08 and $\frac{h}{H_e} \approx 0.58$, we can use the equation $x^{1.75} = 1.869\, H_e^{0.75} \cdot y$ (from Table 21.7).

Hence, the d/s profile is given by the equation

$$x^{1.75} = 1.869\,(13.63)^{0.75} \cdot y$$

or $$x^{1.75} = 13.2\,y$$

or $$y = \frac{x^{1.75}}{13.2} \qquad \text{...(21.12)}$$

*Position of the d/s apron of spillway*

The apron or toe of spillway should be at such an elevation that it does not affect the coefficient of discharge.

or $$\frac{h_a + d}{H_e} \geq 1.7$$

or $$h_d + d \geq 1.7\ (13.63) = 23.2 \text{ m.}$$

Hence, maximum apron elevation

$$= \text{T.E.L.} - (h_d + d)$$

where T.E.L. = 212.63 × 0.016

= 213.614 m ; say **213.61 m**

∴ Max. apron elevation = 213.61 – 23.2 = 190.41 m.

Hence, provide the toe of the spillway (*i.e.* apron level) at *RL* **190.4 m.**

Discharge intensity downstream of spillway piers

$$= q = \frac{5{,}000}{50 + 12} = \frac{5{,}000}{62} = 80.7 \text{ m}^2/\text{sec.}$$

If $d$ is the width, then the velocity $= \dfrac{80.7}{d}$ m/s

$$\text{Specific energy} = d + \frac{V^2}{2g} = g + \frac{\left(\dfrac{80.7}{d}\right)^2}{2 \times 9.81}$$

But specific energy = 213.614 – 190.4 = 23.214

∴ $$d + \frac{\left(\dfrac{80.7}{d}\right)^2}{2 \times 9.81} = 23.214$$

or $$d + \frac{(80.7)^2}{19.62\, d^2} = 23.214$$

or $$d^3 + 331 = 23.21\, d^2$$

$$d^3 - 23.21\, d^2 + 331 = 0$$

Solving by hit and trial, we get

$$\boxed{d = 4.2 \text{ m}}$$

The d/s profile is designed between *RL* 200.0 m (crest level) and *RL* = 190.4 (apron level)

Hence, maximum ordinate $y$ = 200.0 – 190.4 = **9.6 m.**

Corresponding value of $x$ is obtained from eq. (21.12), as :

$$9.6 = \frac{x^{1.75}}{13.2}$$

or $$x^{1.75} = 126.7$$

or $$x = (126.7)^{0.57} = 15.08$$

The remaining co-ordinates of d/s profile between $x = 0$ and $x = 15.08$ m are worked out, as given in Table 21.9.

**Table 21.9. Coordinates of d/s Profile**

| *x in metres* | $y = \frac{x^{1.75}}{13.2}$ *in metres* |
|---|---|
| 0 | 0 |
| 0.5 | 0.023 |
| 1.0 | 0.077 |
| 2.0 | 0.25 |
| 4.0 | 0.86 |
| 6.0 | 1.74 |
| 8.0 | 2.88 |
| 10.0 | 4.03 |
| 12.0 | 5.84 |
| 14.0 | 7.9 |
| 15.08 | 9.6 |

The co-ordinates of the u/s profile are calculated in Table 21.10.

**Table 21.10. Co-ordinates of u/s Profile**
**$H_e$ = 13.61 m**

| $\frac{x}{H_e}$ | $\frac{y}{H_e}$ | $x$ | $y$ |
|---|---|---|---|
| – 0.000 | 0.0000 | 0 | 0 |
| – 0.020 | 0.0004 | – 0.272 | 0.0054 |
| – 0.060 | 0.0035 | – 0.817 | 0.0476 |
| – 0.100 | 0.0101 | – 1.361 | 0.1375 |
| – 0.120 | 0.0150 | – 1.633 | 0.2042 |
| – 0.140 | 0.0208 | – 1.905 | 0.2831 |
| – 0.150 | 0.0235 | – 2.042 | 0.3198 |
| – 0.160 | 0.0270 | – 2.178 | 0.3675 |
| – 0.175 | 0.0328 | – 2.382 | 0.4464 |
| – 0.190 | 0.0395 | – 2.586 | 0.5375 |
| – 0.200 | 0.0420 | – 2.722 | 0.5716 |

The u/s crest is joined at an angle of 45° to the bottom, as shown in Fig. 21.21.

**Design of the Chute or Discharge Carrier**

The critical depth $y_c = \sqrt[3]{\frac{q^2}{g}} = \sqrt[3]{\frac{(80.7)^2}{9.81}} = 8.77$ m.

The depth at the top of spillway ($d$) was calculated to be 4.2 m, which is less than $y_c$. Hence, the flow at the top of the spillway is supercritical.

The chute channel or the discharge carrier should now be given a milder slope for a little distance from toe, but in no case less than the critical slope, so that the flow remains super-critical.

Critical velocity = $V_c = \frac{q}{y_c} = \frac{80.7}{8.77} = 9.18$ m/sec.

But $V = \frac{1}{n} \cdot R^{2/3} \cdot S^{1/2}$

A rectangular channel with bottom width 62 m should be provided as the chute channel (discharge carrier) for the chute spillway.

At critical flow, $A = 62 \times 8.77$

$$P = 62 + 8.77 \times 2$$

$$\therefore \quad R = \frac{A}{P} = \frac{62 \times 8.77}{(62 + 8.77 \times 2)} = 6.82$$

Critical slope $S_c$ is, therefore, given as :

$$9.18 = \frac{1}{0.019} (6.82)^{\frac{2}{3}} \cdot S_c^{\,1/2}$$

or $$S_c^{\,1/2} = \frac{9.18 \times 0.019}{(6.82)^{2/3}} = \frac{9.18 \times 0.019}{3.61} = \frac{1}{20.7}$$

or $$S_c = \frac{1}{428}$$

A slope of say 1/250 is provided in 25 m distance from the top of the spillway.

Bed level at the end of $\frac{1}{250}$ slope $= 190.4 - \frac{25}{250} = 190.3$ m.

The reverse curve at toe may have a radius equal to

$$2H_e = 2 \times 13.63 = 27.22 \text{ m} \text{ ; say } 27.5 \text{ m}.$$

From this point onwards, *i.e.* $RL = 190.3$ m and upto $RL$ 103.0 m, the slope of the discharge carrier may be steepened depending upon the site contours. A slope such as 8 : 1 to 4 : 1 may be given in the initial stage and then a steeper slope of say 2 : 1 or 3 : 1 may be given. The entire design is now like that of a rectangular channel. Wherever, the slope is changed, transition curve (convex or concave) may be designed as explained earlier.

In the given question, since the site contours are not known, we shall assume that a slope of 6 : 1 is given for the first 21 m fall (*i.e.* up a $RL$ 169.3) and then a slope of say 2 : 1 is given for the rest of the reach. The depth, velocity, etc. at the end point of 1 in 250 slope may be taken to be the same, as they were at the toe of the spillway, because the small length of 25 m shall not produce much difference. With this assumption, the TEL at the starting point of new slope 6 : 1 ($RL = 190.3$) is equal to $213.21 - 0.1 = 213.11$ m.

The calculations of water depth, velocity, etc. can now be carried out for the entire reach ($RL = 190.3$ to $RL = 103.0$ m) of the discharge carrier by dividing the channel length into small reaches ; say of length 42 m each, as shown in Table 21.11. The depth ($d$) at the point is assumed in Column (6). Specific energy is calculated in Col. (9). The TEL at the end is calculated in Col. (10). The drop in energy line ($h_f$) in the interval is calculated in Col. (17). The drop ($h_f$) at the end of the interval is subtracted from the final calculated TEL at the beginning of interval in Col. (18). The Col. (18), is then compared with Col. (10). They should be almost equal. If difference is large, the assumed depth is changed till equivalence is obtained. The table is otherwise self explanatory.

**Table 21.11. Calculations of water depth on chute channel (*i.e.* Discharge Carrier)**

| S.N. | Distance from start of 6 : 1 slope | Length L | Drop in bed | Bed level | Depth (d) assumed) | Velocity $v = q/d = 80.7/d$ | Vel. head $= v^2/2g$ | Sp. energy $= d + v^2/2g$ | Bed level + Sp. energy = TEL calculated at the end | $A = Bd = 62d$ (Area) | $P = B + 2d = 62 + 2d$ wetted perimeter | $R = A/P$ | $R^{4/3}$ | $S_f = (n^2 . v^2)/R^{4/3}$ ; $n = 0.019$ | Average $S_f$ | $L . S_f = h_f$ | Actual TEL | Froude number $F = V/\sqrt{gd}$ |
|---|---|---|---|---|---|---|---|---|---|---|---|---|---|---|---|---|---|---|
| (1) | (2) | (3) | (4) | (5) | (6) | (7) | (8) | (9) | (10) | (11) | (12) | (13) | (14) | (15) | (16) | (17) | (18) | (19) |
| 1 | 0 | 0 | | 190.3 | 4.2 | 19.2 | 18.9 | 23.1 | 213.4 | 260 | 70.4 | 3.7 | 5.72 | 0.0232 | | | 213.53 | 2.97 |
| 2 | 42 | 42 | 7 | 183.3 | 3.7<br>3.7<br>3.62 | 21.4<br>22.4<br>22.3 | 24.2<br>25.7<br>25.3 | 27.9<br>29.3<br>28.92 | 211.2<br>212.6<br>212.22 | 229<br>223<br>224 | 69.4<br>69.2<br>69.24 | 3.31<br>3.23<br>3.24 | 4.94<br>4.78<br>4.8 | 0.0378<br>0.0378<br>0.0374 | 0.029<br>0.0305<br>0.0303 | 1.22<br>1.28<br>1.27 | 212.18<br>212.12<br>212.13 | <br><br>3.74 |
| 3 | 84 | 42 | 7 | 176.3 | 3.30<br>3.28 | 24.45<br>24.60 | 30.4<br>30.8 | 33.7<br>34.08 | 210.0<br>210.38 | 205<br>205 | 68.6<br>68.56 | 2.98<br>2.98 | 4.28<br>4.28 | 0.0504<br>0.0505 | 0.0439<br>0.0439 | 1.84<br>1.84 | 210.38<br>210.38 | <br>4.33 |
| 4 | 126 | 42 | 7 | 169.3 | 3.05 | 26.4 | 35.7 | 38.75 | 208.05 | 188 | 68.08 | 2.77 | 3.88 | 0.0652 | 0.0578 | 2.43 | 207.95 | 4.95 |
| **Start of Slope 2 : 1** | | | | | | | | | | | | | | | | | | |
| 5 | 168 | 45 | 21 | 148.3 | 2.5 | 32.3 | 53.2 | 55.7 | 204.0 | 155 | 67 | 2.32 | 3.07 | 0.1224 | 0.0938 | 3.94 | 204.11 | 6.46 |
| 6 | 210 | 42 | 21 | 127.3 | 2.21 | 36.5 | 67.9 | 70.11 | 197.41 | 136 | 66.4 | 2.05 | 2.56 | 0.1895 | 0.1559 | 6.54 | 197.46 | 7.84 |
| 7 | 258.6 | 48.6 | 24.3 | 103.0 | 1.98<br>2.015 | 40.7<br>40.0 | 84.7<br>81.85 | 86.68<br>83.865 | 189.98<br>186.865 | 122.6<br>125.0 | 66<br>66.04 | 1.86<br>1.895 | 2.29<br>2.35 | 0.259<br>0.245 | 0.2242<br>0.2172 | 10.9<br>10.54 | 186.51<br>186.87 | <br>9.00 |

**Design of Curve No. 1.** (At the junction of 250 : 1 and 6 : 1 slopes, a convex curve shall be provided).

The convex curve No. 1 can be designed as per Equation (21.10), as :

$$y = -x \tan\theta - \frac{x^2}{1.5\,[4\,(d + h_v)\cos^2\theta]}$$

$(d + h_v)$ is the sp. energy at the junction point

$$= 4.2 + \frac{\left(\frac{80.7}{4.2}\right)^2}{2 \times 9.81} = 23.21 \text{ m.}$$

$\tan\theta =$ slope of the angle of the floor u/s of the junction point $= \frac{1}{250}$

Since $\tan\theta$ is small, $\cos^2\theta$ will be approximately unity.

$$\therefore \quad y = -\frac{x}{250} - \frac{x}{6 \times 23.21} = \frac{-x}{250} - \frac{x^2}{139.3}$$

Differentiating w.r.t. to $x$, we get

$$\frac{dy}{dx} = \frac{1}{250} - \frac{x}{69.7}$$

The curve meets the downstream slope where

$$\frac{dy}{dx} = -\frac{1}{6}$$ (–ve sign shows that as $x$ increases, $y$ decreases).

Equating, we get $\frac{1}{250} + \frac{x}{69.7} = \frac{1}{6}$

or $$x = 69.7\left(\frac{1}{6} - \frac{1}{250}\right) = 69.7\left[\frac{244}{1,500}\right]$$

or **x = 11.3 m.**

Other co-ordinates of this curve for values of $x$ between $x = 0$ and $x = 11.3$ m, determined from the equation $y = \frac{x}{250} + \frac{x^2}{139.3}$ ($y$ is negative of course) are given in table 21.12.

**Table 21.12**

| *x in metres* | $y = \frac{x}{250} + \frac{x^2}{139.3}$ *in metres* |
|---|---|
| 0 | 0 |
| 1 | 0.011 |
| 3 | 0.077 |
| 5 | 0.20 |
| 7 | 0.371 |
| 9 | 0.619 |
| 10.0 | 0.78 |
| 11.3 | 0.97 |

**Design of curve No. II (convex).** *At the junction of 6 : 1 slope and 2 : 1 slope.*

From Table 21.11, the specific energy at this point is found to be

$$= d + h_v = 3.05 + 35.7 = 38.75 \text{ m}$$

$$\therefore \quad y = -\left\{x \tan\theta + \frac{x^2}{1.5[4(d + h_v)\cos^2\theta]}\right\}$$

$$\tan\theta = \frac{1}{6}\,;\ \cos\theta \approx 1$$

$$\therefore \quad y = -\left[\frac{x}{6} + \frac{x^2}{6 \times 38.75 \times 1}\right]$$

or
$$y = -\left[\frac{x}{6} + \frac{x^2}{232.5}\right]$$

$$\therefore \quad \frac{dy}{dx} = -\left[\frac{1}{6} + \frac{x}{116.25}\right]$$

It meets the downstream slope of 2 : 1 ; $\therefore\ \frac{dy}{dx} = -\frac{1}{2}$ (–ve sign shows that as $x$ increases, $y$ decreases.)

$$\therefore \quad \frac{1}{6} + \frac{x}{116.25} = \frac{1}{2}$$

or
$$\frac{x}{116.25} = \frac{1}{2} - \frac{1}{6} = \frac{1}{3}$$

or
$$x = \frac{116.25}{3} = 38.75 \text{ m}$$

are
$$\mathbf{x = 38.75 \text{ m.}}$$

It means that this convex curve shall become tangential to the slope of 2 : 1 after traversing a distance of 38.75 m. The co- ordinates of this curve can be found (between $x = 0$ and $x = 38.75$ m), using the equation $y = -\left[\frac{x}{6} + \frac{x^2}{232.5}\right]$ as shown in table 21.13.

**Table 21.13**

| *x in metres* | $y = \frac{x}{6} + \frac{x^2}{232.5}$ *in metres (y is downward)* |
|---|---|
| 0 | 0 |
| 1 | 0.17 |
| 5 | 0.95 |
| 10 | 2.10 |
| 15 | 3.47 |
| 20 | 5.05 |
| 25 | 6.85 |
| 30 | 8.87 |
| 35 | 11.10 |
| 38.75 | 12.93 |

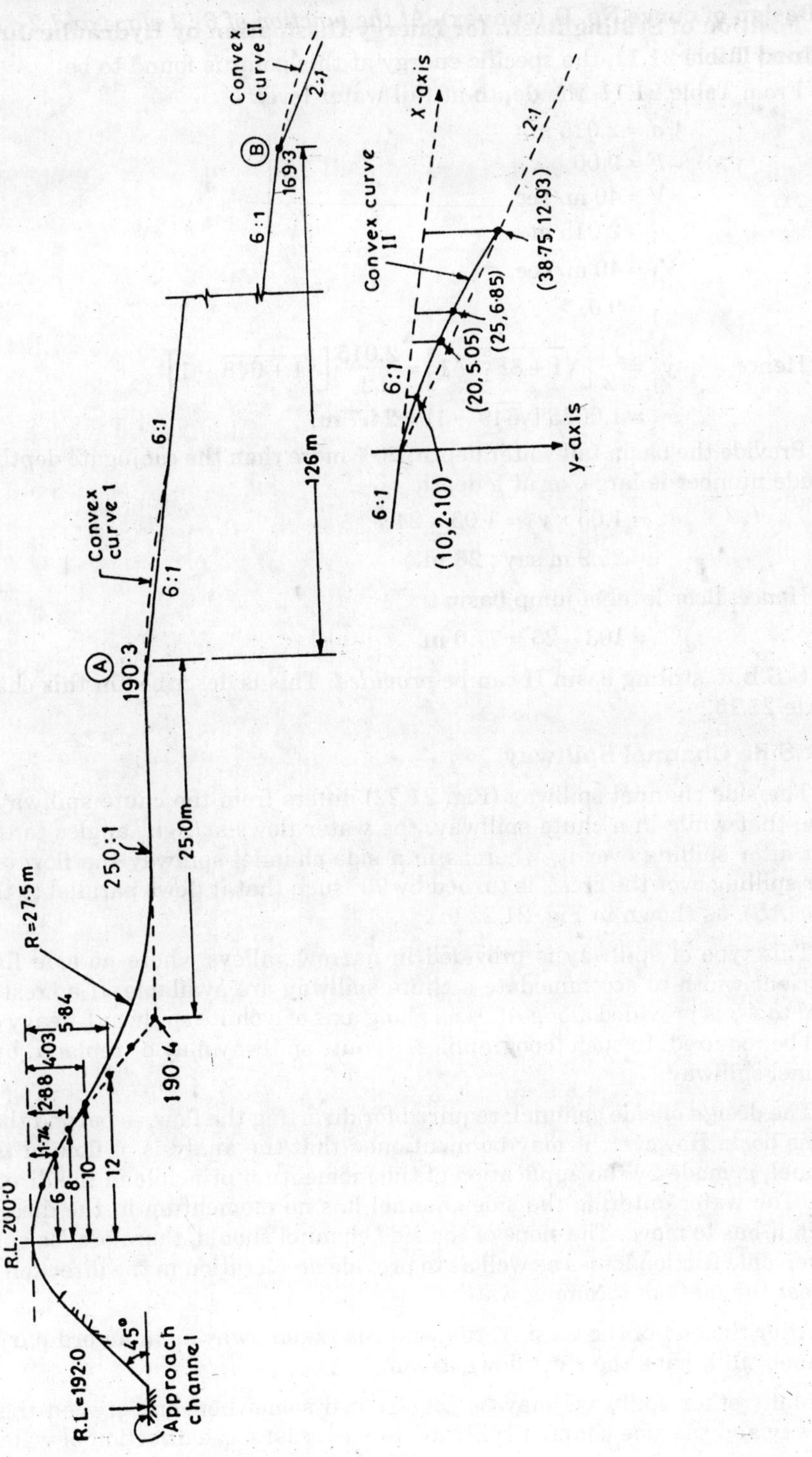


Fig. 21.21. Section of Chute Spillway.

**Position of Stilling Basin for Energy Dissipation by Hydraulic Jump (Explained later)***

From Table 21.11, the depth at tail water level

$$d = 2.015 \text{ m}$$
$$F = 9.00$$
$$V = 40 \text{ m/sec.}$$

$\therefore$

$$y_1 = 2.015 \text{ m}$$
$$V_1 = 40 \text{ m/sec.}$$
$$F_1 = 9.0.$$

Hence

$$y_2 = \frac{y_1}{2}\left[\sqrt{1+8F_1^2} - 1\right] = \frac{2.015}{2}\left[\sqrt{1+648} - 1\right]$$
$$= 1.0075\,[\sqrt{649} - 1] = \mathbf{24.7\ m.}$$

Provide the basin tail water depth, 50% more than the conjugate depth, as the Froude number is large, or at a depth

$$= 1.05 \times y_2 = 1.05 \times 24.7$$
$$= 25.9 \text{ m say ; } \mathbf{26\ m.}$$

Hence, floor level of jump basin

$$= 103 - 26 = \mathbf{77.0\ m.}$$

U.S.B.R. stilling basin II can be *provided.* This is described in this chapter in article 21.15.

## 21.8. Side Channel Spillway

The side channel spillway (Fig. 21.22) differs from the chute spillway in the sense that while in a chute spillway, the water flows at right angles to the weir crest after spilling over it, whereas in a side channel spillway the flow of water after spilling over the crest, is turned by 90° such that it flows parallel to the weir crest (*AB*), as shown in Fig. 21.22 (*a*).

This type of spillway is provided in narrow valleys where no side flanks of sufficient width to accommodate a chute spillway are available. If a crest length equal to *AB* is provided along *AC* (*i.e.*, along axis of a chute spillway), heavy cutting shall be required. In such topographies, a chute spillway may be replaced by a side channel spillway.

The design of side channel, required for diverting the flow, is beyond the scope of this book. However, it may be mentioned that the analysis of flow in the side channel, is made by the application of the momentum principle in the direction of flow. The water entering the side channel has no momentum in the direction in which it has to move. The slope of the side channel should, therefore, be sufficient to overcome friction losses as well as to provide acceleration in the direction of flow against the mass of incoming water.

After the end of the crest *A*, the water is taken away as in an ordinary chute channel, till it joins the river downstream.

Many other spillways may be constructed somewhere in between the chute spillway and the side channel spillway. In such cases, the direction of water after passing over the crest is changed somewhere between 0° and 90°.

---

* Readers are advised to go through this article, only after going through its theory as given in article 21.11.

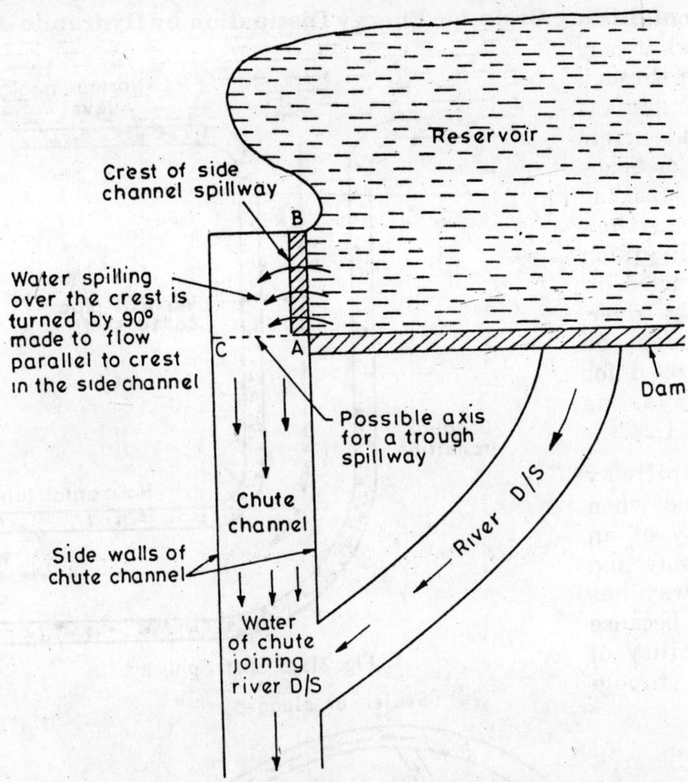


Fig. 21.22. (*a*) Simplified line sketch of a Side Channel Spillway.

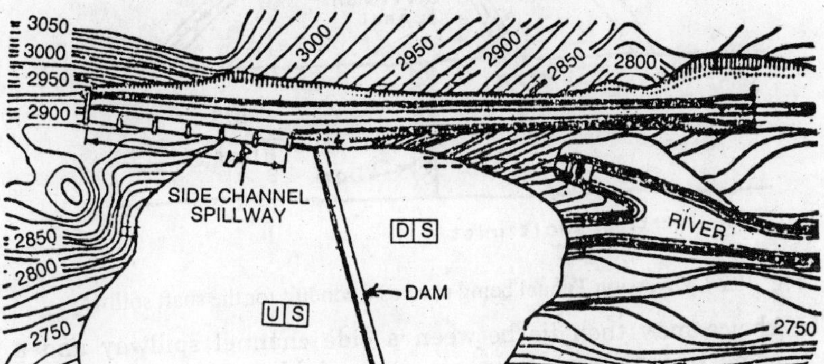


Fig. 21.22. (*b*) Layout plan of a Side Channel Spillway.

## 21.9. Shaft Spillway

In a shaft spillway (Fig. 21.23), the water from the reservoir enters into a vertical shaft which conveys this water into a horizontal tunnel which finally discharges the water into the river downstream. Sometimes, the vertical shaft may be excavated through some natural rocky island or rocky spur existing on the u/s of the river near the dam. Sometimes, artificial shafts may be constructed. For small heights, the shafts may be constructed entirely of metal or concrete, or clay tiles. But for larger heights, reinforced cement concrete may be used. For smaller heights, no special inlet design is necessary, but on large projects, a flared inlet called **morning glory** is often used.

The horizontal tunnel or the conduit may be taken either through the body of the dam (as may be done in concrete gravity dams) or below the foundations (as may be done in earthen dams). The diversion tunnels constructed for diversion of the river, may sometimes be planned and used for shaft spillways, as shown in Fig. 21.24.

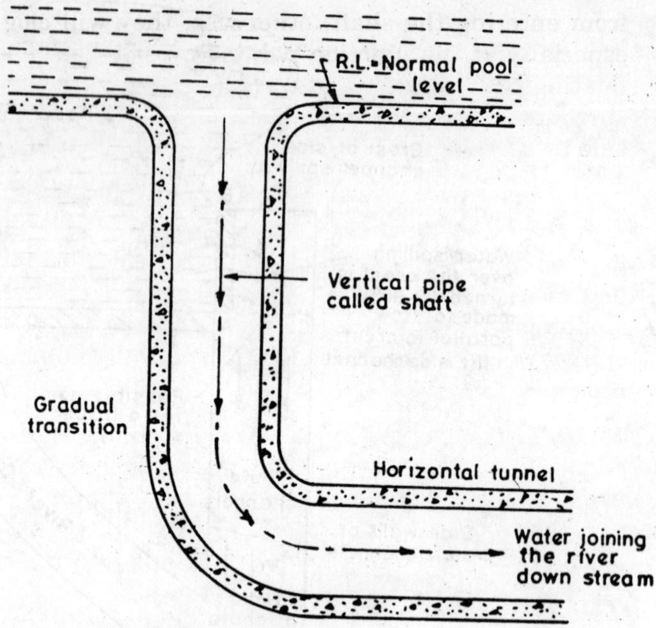


Fig. 21.23. Shaft Spillway.

A shaft spillway may be adopted when the possibility of an over-flow spillway and a trough spillway has been ruled out because of non-availability of space due to topography. The choice may then lie between a side channel spillway and a shaft spillway. If some suitable rock spur, etc. is available near the reservoir on the upstream, naturally, shaft spillway may become economical and consequently the first choice.

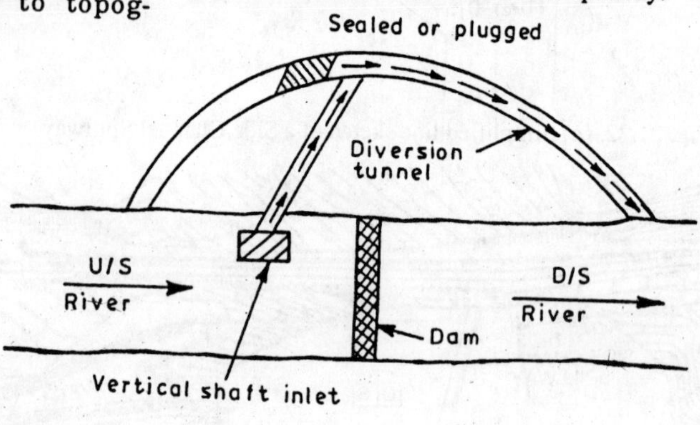


Fig. 21.24. Diversion Tunnel being used as a conduit for the shaft spillway.

The discharge through the shaft spillway does not increase at such a high rate as it increases in weir type spills. Hence, if the unestimated high floods occur, the shaft spillway may not prove as useful as a weir type spillway (such as overflow, chute or side channel) would have been. Hence, a shaft spillway design should be much more conservative than that of other weir type spillways.

A gradual transition must be provided between the vertical shaft and the horizontal conduit, in order to avoid danger of cavitation. Hydraulic analysis of shaft spillways is difficult and their functioning is generally tested in models. The entry of debris and other materials floating in the reservoir must be prevented from entering the shaft, otherwise, they will clog the shaft or the conduit, especially at the junction point. Properly designed or tested trash racks or floating booms may be provided for this purpose. Since, a shaft spillway is surrounded by water on all sides, it must be connected to the dam or the hill side by a bridge.

## SYPHON SPILLWAY

A siphon spillway essentially consists of a siphon pipe, one end of which is kept on the upstream side and is in contact with the reservoir, while the other end discharges water on the downstream side. Two typical installations of siphon pipes are shown in Figs. 21.25 and 21.26. Both these types of syphon spillways are the variations of a ***saddle syphon spillway*** or simply called a ***siphon spillway.***

### 21.10. Tilted Outlet Type of a Syphon Spillway

The siphon pipe in Fig. 21.25 has been installed within the body of the dam. When the valley is very narrow and no space is available for constructing a separate spillway, the siphon pipes can be installed within the dam body, as shown in Fig. 21.25. An air vent may be connected with the siphon pipe. The level of the air vent may be kept at normal pool level, while the entry point of the siphon pipe may be kept still lower, so as to prevent the entry of debris, etc. in the siphon. The outlet of the siphon may be submerged so as to prevent the entry of the air in the siphon from its d/s end.

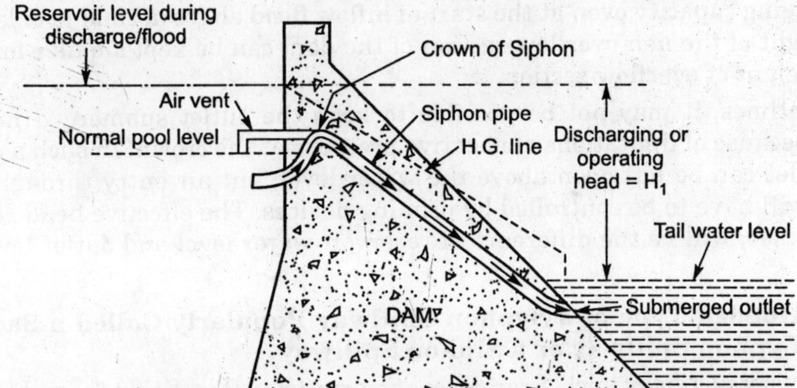


**Fig. 21.25.** Siphon pipe installed within the gravity dam.

When the water in the reservoir is upto or below the normal pool level, air enters the siphon through the vent and siphonic action cannot take place. When once the water level in the reservoir goes above the normal pool level, and if once the siphon is filled with water (*i.e.*, it is primed) ; the water will start flowing through the siphon by siphonic action. The outflow will continue till the water level in the reservoir falls back to normal pool level. As soon as it happens, the air will enter the siphon through the new exposed air vent, and the flow will stop.

**Discharge formula.** The velocity of flow at the outlet of the siphon can be obtained by equating the effective head $H_1$ (*i.e.*, the difference of water level in the reservoir and the tail water level, for submerged outlet) to the velocity head.

$$\therefore \qquad \frac{V^2}{2g} = H_1$$

$$\text{or} \qquad V = \sqrt{2gH_1}$$

$$\therefore \qquad Q = \text{Discharge} = C_d \cdot A\sqrt{2gH_1}.$$

Hence, the discharge through the siphon is given by the equation,

$$Q = C_d A \cdot \sqrt{2gH_1} \qquad \text{...(21.13)}$$

where $C_d$ is the co-efficient of discharge, the average value of which is about 0.65.

The above equation clearly shows that the discharge, through a siphon spillway is sufficiently independent of the water surface elevation of the reservoir. If the water surface in the reservoir rises, the discharge through the spillway is affected to a less extent because change in $H_1$ is small as compared to the corresponding change in head over an ogee type spillway. Hence, the discharge through the siphon is nearly always at capacity, when once the water level has risen above the normal pool level. This makes the siphon spillway particularly advantageous in disposing of sudden surges of water, such as may occur in canals and forebays when the outlet gates are closed rapidly. As the rise of water above the crest level is smaller (because of higher discharging capacity even at the start of inflow flood above normal pool level), the height of the non-overflow section of the dam can be kept smaller for the same height of overflow section.

Sometimes, it may not be possible to keep the outlet submerged in tail water because of limitations of negative pressure at the crown. In such a case, the outlet can be left open above the water level, but air entry through the outlet will have to be controlled by priming devices. The effective head $H_1$, in such a case, will be the difference of reservoir water level and outlet level of siphon.

## 21.11. Hooded Type of a Syphon Spillway Popularly Called a Saddle Syphon Spillway or a Siphon Spillway

The construction of such a ***saddle syphon spillway*** [Fig. 21.26 (*a*) and (*b*)] is more commonly adopted. In this case, a reinforced concrete hood is constructed over an ordinary overflow section of a gravity dam. The inlet or mouth of this hood is kept submerged in water, so as to prevent the entry of debris, ice, etc. A small depriming hood is kept above the main hood and both these hoods are connected by an air vent. The inlet of the deprimer hood is kept at normal pool level. The principle of functioning of both types of installations is essentially the same, except for the initial filling up of the siphon or ***'priming'***. At normal pool level (*i.e.* full reservoir level), the water stands up to the crest of the spillway. The lower limb or the leg of the main hood is generally kept submerged in the tail water, so that there is no air entry from the downstream end.

If now a flood enters the reservoir, the water level in the reservoir would start rising and a sheet of water would start flowing over the spillway crest.

Since the water level would then be above the inlet of the deprimer hood, the air entry at the inlet gets sealed. The air entry at the outlet is also sealed by tail water, etc. Hence, the water spilling over the crest, sucks all the remaining air from the hood within minutes. Siphonic action gets established after the air in the bend over the crest has been completely exhausted.

This action is known as ***Priming.*** The siphon runs full and water is discharged downstream under a greater effective (siphonic) head. During receding flood, when water level has gone down just to the reservoir level (Normal pool level), air enters through the mouth of the deprimer, dome, and the siphonic action is broken. This action is called **depriming** of the siphon and is achieved through the deprimer dome.

A photoview of this type of siphon spillway, called Sarala Sagar spillway, flanked by earth dam constructed on China Vagu stream (a tributary of Krishna river) is shown in Fig. 21.26 (*b*).

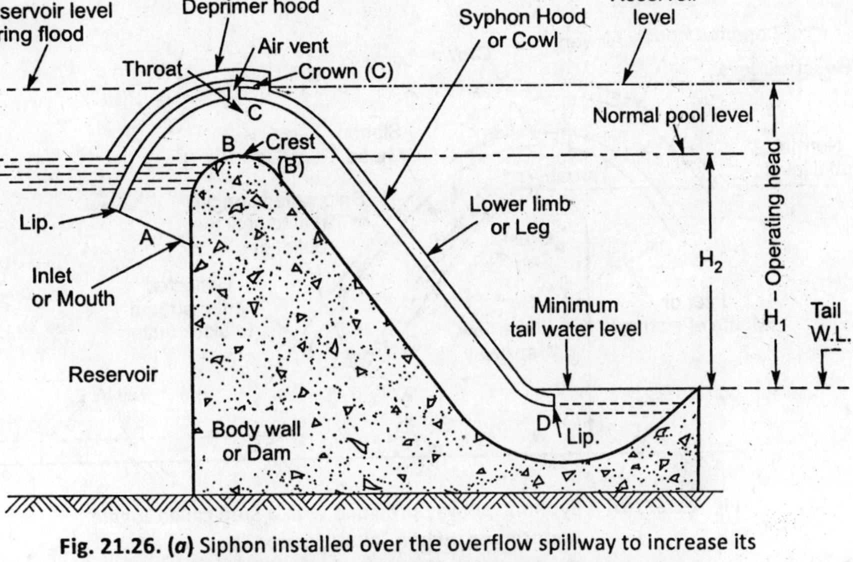


**Fig. 21.26. (*a*)** Siphon installed over the overflow spillway to increase its effectiveness and discharging capacity.

**Fig. 21.26 (*b*).** Photoview of a hooded type of siphon spillway (Sarla Sagar spillway across a tributary of Krishna river).

**21.11.1. Priming Devices for Saddle Syphons.** ***Priming*** is the action of the siphon spillway from the moment the water first spills over the crest to the instant when the whole siphon is filled with water. The loss of 3% of volume of air in the entire part of the siphon, may cause a rise 30 cm of level at the crown, which will be sufficient to start priming. With increase of flow, the capacity of the water to carry air with it increases, and in a short time, the entire air may be driven out, and priming may take place.

Priming is achieved by sealing the downstream outlet. When the outlet cannot be sealed by tail water, some other devices, called ***priming devices***, are used, which lead to an automatic priming of the siphon at a certain rise of water level above the crest. This rise of water level is called the ***priming depth***.

In Fig. 21.27 (*a*), the priming is accomplished by means of a step called ***Joggle*** or ***Tudal***, which deflects the sheet of water to strike against the lower end of the cover or hood, thus sealing the lower end or the outlet, from the atmosphere.

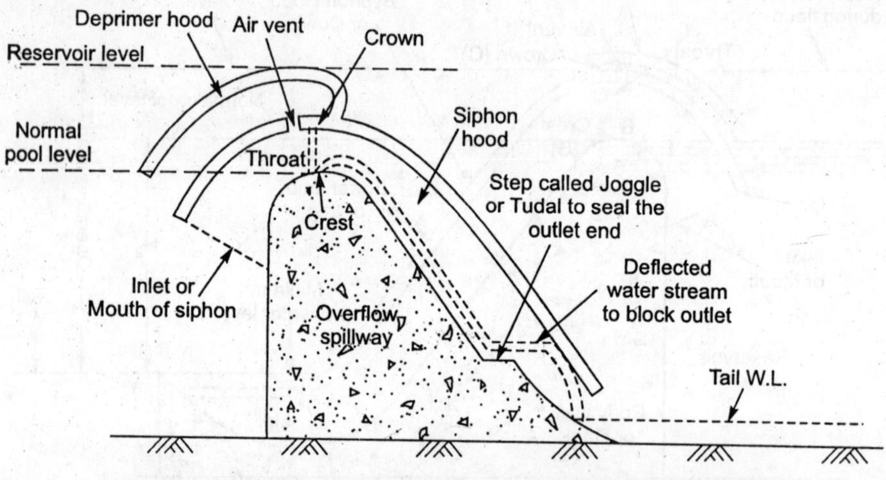


**Fig. 21.27. (*a*)** A Syphon spillway provided with a Step called Joggle for priming, when Tail WL is low.

Sometimes, an additional ***baby siphon*** may be installed as a priming device, as shown in Fig. 21.27 (*b*). Such a small auxiliary or baby siphon may be installed just below the crest. When the water level reaches the crest, the baby siphon will run full. The sheet of water issuing from it, is arranged to shoot across the lower limb of the main siphon, so as to seal it and prevent air entering from below. The priming thus takes place.

One installation was made at Maramsilli siphon (India) in 1921. The baby siphon is 0.3 m high and 2.4 m wide at throat, while the main siphon is 2.4 m high and 2.4 m wide. The priming is followed just in 15 to 30 seconds. Though priming is good, but the affective top water level for the consideration of storage is not the crest of the main siphon but the crest of the auxillary baby siphon. Hence, from the point of view of maximum discharge, the baby siphon may be looked upon as an obstruction.

As the flow over the main siphon increases, the jet through the baby siphon is depressed down, and care has to be taken to see that the combined jet does not strike below the tip of the lower leg of the hood. The baby siphon is expensive and difficult to construct.

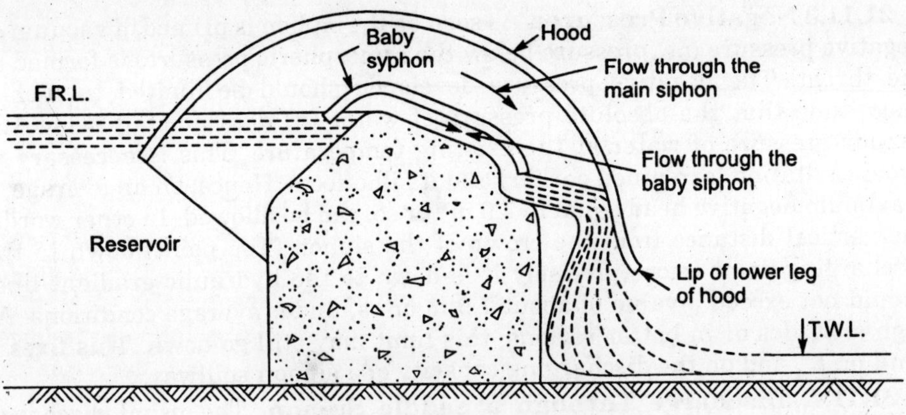


Fig. 21.27. (*b*) Maramsilli baby syphon.

Sometimes, a small ***obstruction weir*** or ***Priming weir*** may be constructed at the downstream end of the saddle siphon spillway, to ensure heading up of the tail water by creating an artificial pool, as shown in Fig. 21.27 (*c*). This raised tail water level will ensure to seal the outlet, to ensure proper priming.

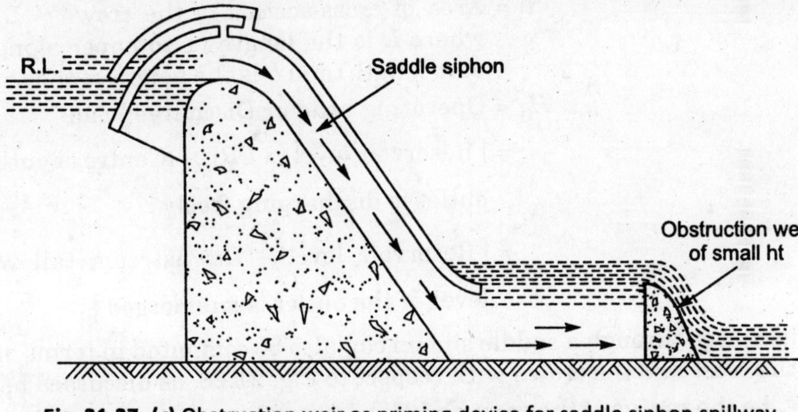


Fig. 21.27. (*c*) Obstruction weir as priming device for saddle siphon spillway.

**21.11.2 Depriming.** Depriming is the reverse process of priming. It is the action of the siphon from the time air starts entering the siphon through the ***air vent*** until the siphonic action is completely stopped. Depriming is provided by providing a **Deprimer**, which is a siphon breaker air vent provided in the main dome. It breaks the siphonic action. When the reservoir water surface is drawn below it.

If an air vent is not provided, siphonic action once initiated would continue till the reservoir level falls up to the level of inlet, which is provided much below the normal reservoir level.

The air vent consists of a vent pipe connected to the siphon with its open end in the plane of normal reservoir level. It can be provided either on the side or the top of the crown. The open end should be provided in a horizontal plane to facilitate instantaneous sealing and opening of the pipe. This prevents the development of unsteady flow, thus avoiding undesirable ***knocking of the siphon.***

**21.11.3 Negative Pressures.** As soon as the siphon is primed, a vacuum or negative pressure (*i.e.* pressure below the atmospheric pressure) is formed at the throat. The negative pressure developed should be limited to such a magnitude that the absolute pressure of water (σ) does not fall below the vapour pressure of water at the existing temperature. This is necessary to avoid cavitation (explained earlier) and its ill effects. Hence, on an average, a maximum negative head equal to 7.9 m* or so, can be allowed. In other words, the vertical distance from the crown of the siphon (Top point) down to the discharging point (more precisely it is down to the hydraulic gradient line) should not exceed a value of about 7.9 m or so, under average conditions. At high altit ıdes or in hotter regions, this limit may still go down. This fixes a limit on $H_1$ and on the discharging capacity of a siphon spillway.

**21.11.4. Discharge Through a Saddle Siphon.** The usual discharge formula for a siphon, as indicated earlier and given by Eq. (21.13), holds good for this type of saddle siphon also, as :

$$Q = C_d \cdot A \cdot \sqrt{2gH_1} \qquad \text{...(i.e. Eq. 21.13)}$$

where $C_d$ = coefficient of discharge, the average value of which is usually taken as 0.65.

$A$ = Area of cross-section of the crown = $L \times b$, where $L$ is the length (⊥ to paper along the dam length) and $b$ is the height of the throat.

$H_1$ = Operating head or Discharge head

= [Reservoir level (> FRL) – Centre of outlet, if outlet is discharging freely]

= [Reservoir level – Downstream tail water level, if the outlet is submerged]

The discharge through a saddle siphon can also be computed in terms of the radii of the crest and the crown with respect to Fig. 21.28, as discussed below.

Let $R_1$ be the radius of the crest and $R_2$ be the radius of the crown. Now, let us consider point $P$ on the vertical section at the throat, and let its radius be $R$. The $L$-section of the throat has been shown in Fig. 21.28 (*b*), which is rectangular, having length (or width) $L$ and height $b$.

---

*Absolute pressure = Atmospheric pressure – Negative pressure

Min. Absolute pressure = Atmospheric pressure – Max. Negative pressure

or Vapour pressure of water = Atmospheric pressure – Max. Negative pressure

∴ Limiting Max. Vacuum (*i.e.* Negative) Pressure

= Atmospheric pressure – Vapour pressure of water at the existing temperature (σ)

= 10.33 m – 2.43 m (*i.e.* app. value at 20°C from Appendix Table A-1)

= 7.9 m

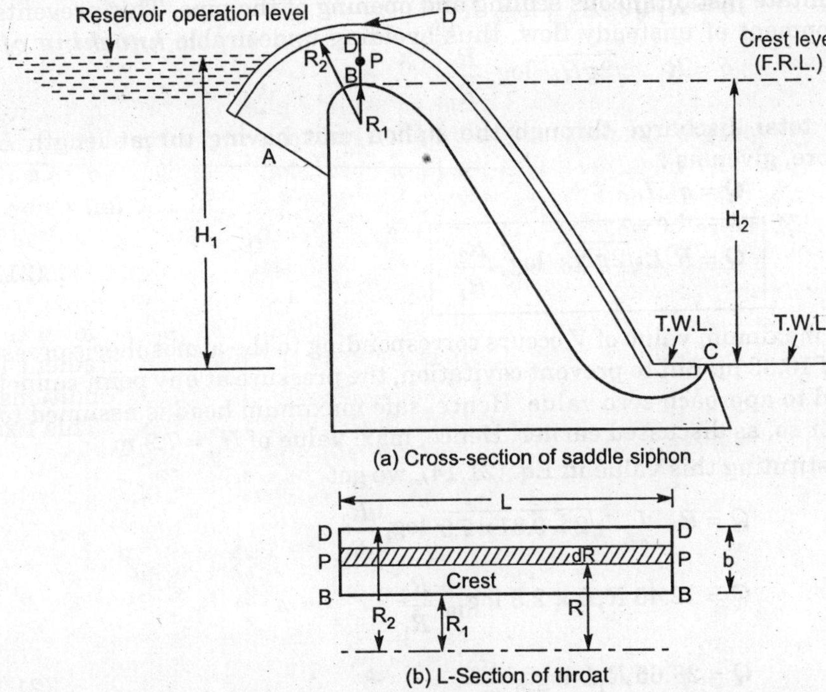


(a) Cross-section of saddle siphon

(b) L-Section of throat

**Fig. 21.28**

Let $V_1$, $V_2$ and $V$ be the velocities of flow at crest (Pt. $B$), crown (Pt. $D$) and point $P$, respectively. The flow through the throat is governed by free vortex flow, for which we have :

$$V \, . \, R = \text{constant}$$

$$\therefore \quad V_1 R_1 = V_2 R_2 = VR$$

***Let $H_0$ be the net operating head at the crest*** (Negative head, *i.e.*, head less than that of the atmospheric pressure). This head is calculated after making due allowances of all losses occurring in the siphon, *i.e.*, $H_0 = H_1$ – Losses through the siphon (say $H_L$).

Hence, $\quad V_1 = \sqrt{2gH_0}$

But $$V = \frac{V_1 \, . \, R_1}{R} = \sqrt{2gH_0} \times \frac{R_1}{R} = \frac{R_1}{R} \, . \, \sqrt{2gH_0}$$

Consider a thin rectangular strip of thickness $dR$ at point $P$, and let $\delta q$ be the discharge through this strip of unit length.

$$\therefore \quad \delta q = V \, . \, dR \, . \, 1 = \frac{R_1}{R} \, . \, \sqrt{2gH_0} \, . \, dR$$

The discharge through the throat per unit length is hence obtained by integrating the above equation between the limits of $R = R_1$ and $R = R_2$

$$\therefore \quad q = \int_{R_1}^{R_2} \frac{R_1}{R} \, . \, \sqrt{2gH_0} \, dR$$

or $$q = R_1 \sqrt{2gH_0} \int_{R_1}^{R_2} \frac{dR}{R}$$

or $$q = R_1 \sqrt{2gH_0} \mid \log_e R \mid_{R_1}^{R_2}$$

or $$q = R_1 \cdot \sqrt{2gH_0} \log_e \frac{R_2}{R_1}$$

The total discharge through the siphon unit having throat length $L$ is, therefore, given as :

$$Q = q \cdot L$$

or $$\boxed{Q \doteq R_1 L \sqrt{2gH_0} \log_e \frac{R_2}{R_1}} \quad \text{...(21.14)}$$

The maximum value of $V$ occurs corresponding to the atmospheric pressure head of 10.33 m. But to prevent cavitation, the pressure at any point cannot be allowed to approach zero value. Hence, safe maximum head is assumed to be 7.9 m or so, as discussed earlier. Hence, max. value of $H_0 \approx 7.9$ m

Substituting this value in Eq. (21.14), we get

$\therefore$ $$Q = R_1 \cdot L \cdot \sqrt{2 \times 9.81 \times 7.9} \log_e \frac{R_2}{R_1}$$

or $$Q = 12.45 \, R_1 L \times 2.3 \log_{10} \frac{R_2}{R_1}$$

or $$Q = 28.65 \, R_1 L \log_{10} \frac{R_2}{R_1} \quad \text{...(21.15)}$$

**21.11.5. Pressure Conditions in the Siphon Spillway.** With reference of Fig. 21.28, let

$H_1$ = Total operational head at any instant
= U/s W.L. – D/s W.L. (if the outlet end is submerged)

or = U/s W.L. – Level of centre of the outlet (if the outlet discharges freely)

$H_2$ = Difference in elevation levels of Throat (*i.e.* pt $B$) and outlet end (*i.e.* pt $C$)

$p_a$ = Atmospheric pressure

$H_L^*$ = Total head loss in the siphon*

$H_{L_1}$ = Head loss between the inlet and the throat.

$H_{L_2}$ = Head loss between the throat and the outlet end.

$\sigma$ = Absolute pressure at which air dissolved in water starts escaping and cavitation starts (average value of $\sigma$ may be taken as 2.43 m or so).

---

*The ***total head loss in the siphon*** ($H_L$) consists of the following component head losses :

(*i*) ***Inlet loss:*** Consisting of entry and velocity loss.

(*ii*) ***Upper leg loss*** or ***Throat loss:*** Consisting of head loss in change of velocity, friction and in conversion.

(*iii*) ***Lower leg loss:*** Consisting of friction loss.

(*iv*) ***Outlet portion loss:*** Consisting of friction and bend loss.

The maximum value of total head loss seldom exceeds 20% of the operating head $H_1$.

Now, let us apply Bernoulli's theorem between points $A$ and $C$ (*i.e.*, between inlet and outlet), taking the level of $TWL$ as datum, as :

$$p_a + H_1 = p_a + \frac{V_C^2}{2g} + H_L. \qquad ...(1)$$

where $H_L$ is the total head loss in siphon from $A$ to $C$

or $$\frac{V_C^2}{2g} = H_1 - H_L$$

or $$V_C = \sqrt{2g\,(H_1 - H_L)} \qquad ...(2)$$

If $A$ is the area of cross-section of the outlet bore, then the discharge through the spillway ($Q$) is :

$$Q = A \,.\, V_C = A \,.\, 2g\sqrt{H_1 - H_L} \qquad ...(3)$$

Equating this to Eq. (21.13), we get

$$A \,.\, \sqrt{2g(H_1 - H_L)} = C_d \,.\, A \,.\, \sqrt{2gH_1}$$

or $$2g(H_1 - H_L) = C_d^2 \,.\, (2gH_1)$$

or $$H_1 - H_L = C_d^2 \,.\, H_1$$

or $$H_L = (1 - C_d^2)\, H_1 \qquad ...(21.16)$$

Now, let $p_T$ = pressure head at the throat

$V_T$ = velocity of flow at the throat

Applying Bernoulli's theorem between the throat ($A$) and the exit ($C$), we get

$$p_T + \frac{V_T^2}{2g} + H_2 = p_a + \frac{V_C^2}{2g} + H_{L_2} \qquad ...(21.17)$$

($H_{L_2}$ is the head loss between throat and outlet)

Since the siphon is to run full bore (opening) and the area of section is constant between the throat and the outlet end, we have $V_T = V_C$.

$\therefore$ $$p_T + H_2 = p_a + H_{L_2}$$

or $$H_2 = (p_a - p_T) + H_{L_2} \qquad ...(21.18)$$

Since the pressure $p_T$ at the throat cannot be allowed to approach zero value, because the air dissolved in water begins to separate much earlier than the zero value, the value of $p_T$ may be assumed to be equal to $\sigma$ (max av. value = 2.43 m). Eq (5) then becomes

$\therefore$ $$H_2 = (p_a - \sigma) + H_{L_2} \qquad ...(21.19)$$

Substituting $p_a$ = atmospheric pressure = 10.33 m

$\sigma_{max}$ = 2.43 m, we get

$$H_2 \le (10.33 - 2.43) + H_{L_2}$$

or $$\boxed{H_2 \le 7.9 + H_{L_2}} \qquad ...(21.20)$$

This Eq. (21.20) limits the maximum difference in elevation between the throat (*i.e.*, crest pt. $A$) and the outlet end (pt. $C$). This limit can be slightly increased by reducing the area of the outlet end, in which case $V_C$ will be more than $V_T$ and Eq. (21.20) will get modified to

$$H_2 \le 7.9 + H_{L_2} + \left( \frac{V_c^2}{2g} - \frac{V_T^2}{2g} \right) \quad ...(21.20a)$$

The quantity $(p_a - p_T)$ or $(p_a - \sigma)$ in Eq. (21.18) or (21.18) above is known as the ***sub-atmospheric pressure head*** $(p_{a_T})$ at the crest of the throat. The maximum allowable sub-atmospheric pressure head is equal to the probable minimum atmospheric head at the place under consideration minus the minimum tolerable pressure head (σ).

Again, applying Bernoulli's theorem between the inlet end (point $A$) and the throat (point $B$) with crest (point $B$) as the datum, we get

$$p_a = p_T + \frac{V_T^2}{2g} + H_{L_1}$$

where $H_{L_1}$ = Head loss from inlet to throat

or $$\frac{V_T^2}{2g} = (p_a - p_T) - H_{L_1}$$

or $$\frac{V_T^2}{2g} = p_{aT} - H_{L_1} \quad ...(21.21)$$

or $$\frac{V_T^2}{2g} = (7.9 - H_{L_1}) \quad ...(21.21a)$$

**21.11.6. Classifications of Siphon Spillways:** Siphon spillways may be classified according to head, as :

(*i*) Medium;

(*ii*) High head; and

(*iii*) Low head siphons.

Their difference is clarified below :

(*i*) **Medium Head Syphon:** We have seen earlier in Eq. (21.20) that the difference in level between siphon crest and outlet end is given by:

$$H_2 \le 7.9 + H_{L_2} \quad ...(i.e. \text{ Eq. } 21.20)$$

This is true when the area of the syphon bore at the throat and at the exit end are equal. Hence, a medium head siphon spillway is the one in which the outlet bore area is the same as the bore area at the throat.

(*ii*) **High Head Siphon.** We have already explained that the limit of the value of $H_2$ can be increased if the velocity at the exit end $(V_C)$ is greater than the velocity at the throat $(V_T)$, the increase being equal to $\left[ \frac{V_C^2}{2g} - \frac{V_T^2}{2g} \right]$.

This can be achieved by reducing the outlet area by a nozzle or by tapering the lower part of the siphon bore or in some other manner by increasing the resistance to flow. *Such convergent siphons are known as high head siphons.*

(*iii*) **Low Head Siphons.** A low head siphon is the one in which the outlet bore is made divergent, with increased area. This makes the velocity $V_C$ lesser than $V_T$, thereby further decreasing the value of $H_2$ given by Eq. (21.20*a*).

**21.11.7. Qualities of a Good Siphon Spillway.** A good siphon spillway should have the following qualities :

(*i*) It should have quick priming.

(*ii*) *Priming depth* should be low, so that the siphon becomes operational even with a small rise of water level above the crest level (*i.e.*, FRL).

(*iii*) The priming and start of siphon should be automatic.

(*iv*) There should be no knocking at the times of priming or depriming (when water level goes just below the FRL). The flow through the siphon should be steady and continuous during its operation.

(*v*) The siphon should be self-cleansing, so that no silt or debris get deposited in the siphon bore.

(*vi*) When in operation, air should neither get any access to the siphon nor be entrained. The dissolved air in water should not get separated due to development of excess negative pressure. The flow should be smooth and free of vibrations.

(*vii*) The siphon should have high efficiency.

(*viii*) Its maintenance cost should be low, and all its parts should be capable of easy access and repairs.

(*ix*) The siphon should be simple in design and easy in construction.

**21.11.8. Advantages and Disadvantages of Siphon Spillways**

***Advantages and Uses***

(*i*) Its operation is automatic without any mechanical device.

(*ii*) In a Siphon spillways, discharge occurs at much larger head than in a normal spillway*. Such a spillway will, therefore, have a larger discharging capacity for the same rise in reservoir level.

Its ability to pass higher discharge at relatively low head, results in lower height of dam as well as less surrounding area to be acquired for reservoir submergence.

Siphon spillway can however be used where large capacities are not required (due to limitations caused by negative pressure considerations), where space is limited, and where fluctuations of reservoir level have to be maintained within close limits.

***Disadvantages and Limitations***

(*i*) Construction is expensive, and capacity is limited.

(*ii*) It causes sudden appearance of flood water downstream.

(*iii*) Large flood debris can sometimes block to outlet.

(*iv*) As the siphon is primed, the flow would result excessive vibrations in the dam body, which may cause expansion problems in the joints. Also the structure and foundation have to be strong enough to resist vibration stresses.

*A normal spillway flow is presented by the height of reservoir water over the spillway crest; whereas in a siphon spillway, the flow rate is governed by the difference in height of the inlet and outlet. However, there is a limit to this height from –ve pressure considerations.

(*v*) There is a possibility of cavitation for negative pressures, when the head difference between upstream and downstream levels is more than about 7.5 m or so.

(*vi*) Repair and maintenance of siphon spillways are difficult.

(*vii*) It has serious disadvantage due to occurrence of sudden surges and stoppages of outflow as a result of erratic siphonic action, thus causing severe fluctuations in the downstream river stage. A minor crack in the cover of the siphon would interference with the siphon. Therefore, a siphon spillway is usually constructed in batteries (*i.e.* units) so that the entire spillway is not affected even if cracks have developed in one or a few units.

(*viii*) Like other types of closed conduit spillways, a siphon spillway too is incapable of handing flows appreciably larger than the designed capacity. As such, siphon spillway, whenever used, is generally used as service spillway in conjunction with an auxiliary or emergency spillway.

## 21.12. Volute Siphon Spillway

Volute siphon spillway is a special type of siphon spillway designed in India by Mr. Ganesh Iyer. A sketch of this type of spillway is shown in Fig. 21.29. It consists of a vertical barrel or shaft, bent at the discharge end, and opened out in the form of a funnel at the top end. The top or lip of the funnel is kept at the full reservoir level (FRL), and a number of *volutes* (like blades of a centrifugal pump or turbine) are fixed in the funnel to induce a spiral motion to the water passing along them. A dome, supported on a number of pillars, is placed over the funnel. Over the main dome, is attached a *deprimer* dome. The entrance end of the deprimer hood or dome is kept just at the FRL.

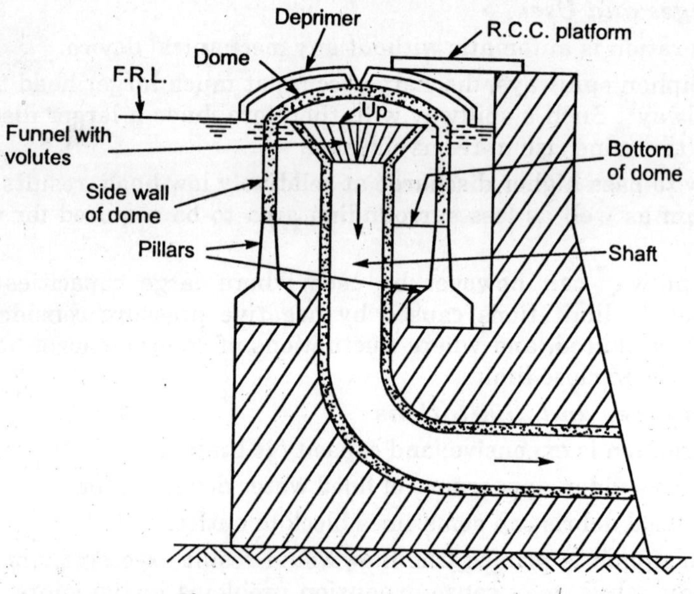


**Fig. 21.29.** Volute siphon spillway.

When the water level in the reservoir rises above FRL, it seals the entry of air through the deprimer hood, and sheet of water enters to the funnel form all sides. The spiral flow of water through the volutes causes suction, which sucks all the remaining air, and priming is, thus, achieved very quickly. The siphon shaft then runs full.

When the reservoir level falls to FRL, air enters through the deprimer dome, thereby breaking the siphonic action.

The discharge through the volute siphon is given as :

$$Q = C_d \,.\, A \,.\, \sqrt{2g \,.\, H_1} \quad \text{...(Eq. 21.13)}$$

where $C_d$ = coefficient of discharge

Also $$Q = A\sqrt{2g(H_1 - H_L)} \quad \text{...(21.22)}$$

where $H_L$ = Head loss through the siphon.

**Example 21.3.** *A saddle siphon spillway has the following data :*

*Full reservoir level = 485.0 m*

*Level of centre of siphon outlet = 479.6 m*

*High flood level = 485.9 m*

*High flood discharge = 570 cumec.*

*If the dimensions of the throat of the siphon are, width = 4.2 m and height = 1.9 m, determine the number of siphon units required to pass the flood safely. The siphon is to discharge freely in the air. Assume coefficient of discharge = 0.65.*

**Solution.** Maximum operating head = $H_1$ = *HFL* – Outlet level

= 485.9 – 479.6 = 6.3 m.

The discharge (*Q*) through the siphon (*Q*) is given as :

$$Q = C_d \,.\, A \,.\, \sqrt{2gH_1}$$

where *A* = Area of siphon bore

= 4.2 m × 1.9 m = 7.98 m$^2$

$C_d$ = 0.65 (given)

$$\therefore \quad Q = 0.65 \times 7.98\sqrt{2 \times 9.81 \times 6.3} = 57.67 \text{ m}^3\text{/s}$$

No. of units reqd. to pass 570 cumec discharge = $\dfrac{570}{57.67}$ = 9.88; **say 10 Ans.**

**Example 21.4.** *A saddle siphon spillway has a rectangular throat section 2.2 m wide and the radius of the crown and crest are 3 m and 1.6 m, respectively. With the normally adopted values of atmospheric pressure and the net operating head of 7.5 m at the crest, compute the discharge through the siphon.*

**Solution.** Use Eq. (21.14) as :

$$Q = R_1 L \,.\, \sqrt{2gH_0}\ \log_e \frac{R_2}{R_1}$$

Here $R_1$ = 1.6 m

$R_2$ = 3 m

$H_0$ = 7.5 m

$L$ = 2.2 m

$$\therefore \quad Q = 1.6 \times 2.2 \sqrt{2 \times 9.81 \times 7.5}\ \log_e \frac{3}{1.6}$$

$$= 42.7 \log_e \frac{3}{1.6} = \mathbf{26.84\ m^3/s.}\quad \textbf{Ans.}$$

**Example 21.5.** *A siphon spillway having a rectangular throat section 2.2 m wide and radius of the crown and crest as 3 m and 1.6 m respectively. The energy loses are 0.2, 0.2, 0.6 and 0.5 times the mean velocity head at the throat respectively in the inlet, upper leg, lower leg and outlet portion. The kinetic energy coefficient is 1.03. Local atmospheric pressure is 10 m of water. Determine :*

*(a) The discharge when the reservoir level is 0.3 m above the crest and the tail water level (outlet submerged) 6 m below it.*

*(b) The minimum pressure at the throat section. For submerged condition, outlet velocity is 50% of throat velocity.* (*Engineering Services, 2011*)

**Solution.** $L = 2.2$ m

$R_1 = 1.6$ m

$R_2 = 3$ m

$H_{L_1}$ = Head loss in the inlet, *i.e.* the head loss from entry to throat

$$= 0.2\ \alpha \frac{V_T^2}{2g}$$

where $\alpha$ = kinetic energy coefficient = 1.03 (given)

$p_a = 10$ m

(*i*) Applying Bernaulli's theorem between entry and throat, with crest as datum, we get

$$p_a = p_T + \alpha \,.\, \frac{V_T^2}{2g} + H_{L_1}$$

or
$$10 = p_T + \alpha \frac{V_T^2}{2g} + 0.2\ \alpha \frac{V_T^2}{2g}$$

or
$$p_T = \left[10 - \frac{1.2\ \alpha V_T^2}{2g}\right]$$

or
$$p_T = \left[10 - \frac{1.236\ V_T^2}{2g}\right] \qquad ...(i)$$

(*ii*) Also, applying Bernaulli's theorem between the throat and outlet, we get

$$p_T + \alpha \frac{V_T^2}{2g} + H_2 = p_a + \alpha \frac{V_C^2}{2g} + H_{L_2}$$

But $V_C = 0.5\ V_T$ (given)

$H_{L_2}$ = Head loss from throat to outlet

$$= (0.2 + 0.6 + 0.5)\alpha \,.\, \frac{V_T^2}{2g} = \frac{1.3\ \alpha\ V_T^2}{2g}$$

$$H_1 = 6 \text{ m}$$
$$H_2 = 6 - 0.3 = 5.7 \text{ m}$$

$$\therefore \quad p_T + \alpha \frac{V_T^2}{2g} + 5.7 = 10 + \alpha \cdot \frac{(0.5\,V_T)^2}{2g} + \frac{1.3\,\alpha \cdot V_T^2}{2g}$$

or $$p_T = (10 - 5.7) + \alpha \cdot \frac{V_T^2}{2g}\left[0.25 + 1.3 - 1\right]$$

or $$p_T = 4.3 + \frac{0.55\,\alpha V_T^2}{2g}$$

or $$p_T = 4.3 + \frac{0.55 \times 1.03\,V_T^2}{2g}$$

or $$p_T = 4.3 + \frac{0.567\,V_T^2}{2g} \quad ...(ii)$$

Equating (*i*) and (*ii*), we get

$$10 - \frac{1.236\,V_T^2}{2g} = 4.3 + \frac{0.567\,V_T^2}{2g}$$

or $$5.7 = 1.803\,\frac{V_T^2}{2g}$$

or $$V_T = \sqrt{\frac{5.7}{1.803} \times 2 \times 9.81} = 7.88 \text{ m/sec.}$$

$$\therefore \quad V_C = 0.5\,V_T = 0.5 \times 7.88 = 3.94 \text{ m/sec.}$$

$$Q = \text{Discharge} = V_C \times \text{Area of siphon bore}$$

or $$Q = V_C \times L \,.\, (R_2 - R_1)$$

or $$Q = 3.94 \text{ m/sec.} \times 2.2 \text{ m} \times (3 - 1.6) \text{ m}$$
$$= 3.94 \times 2.2 \times 1.4 = \mathbf{12.13 \text{ m}^3\text{/sec}} \quad \textbf{Ans.}$$

(*b*) From (*i*),

$$p_T = \left[10 - \frac{1.236\,V_T^2}{2g}\right]$$

$$\therefore \quad p_T = \left[10 - \frac{1.236 \times (7.88)^2}{2 \times 9.81}\right]$$

or $$p_T = 10 - 3.91 = \mathbf{6.09 \text{ m.}} \quad \textbf{Ans.}$$

## ENERGY DISSIPATORS

### 21.13. Energy Dissipation below Overflow Spillways

The water flowing over the spillway acquires a lot of kinetic energy by the time it reaches near the toe of the spillway (because of conversion of potential energy into kinetic energy). If arrangements are not made to dissipate this huge kinetic energy of water, and if the velocity of water is not reduced, large scale scour can take place on the downstream side near the toe of the dam and away from it. These arrangements are known as energy dissipation arrangements or ***energy dissipators.***

In general, the kinetic energy of this super-critical flow can be dissipated in two ways :

(*i*) By converting the super critical flow into sub-critical flow by *hydraulic jump.*

(*ii*) By directing the flow of water into air and then making it fall away from the toe of the structure. The energy is dissipated by the *aeration of jet and impact* of water on the river bed. Though some scour will take place, but it is too small or too far away from the dam to endanger it. Bucket type energy dissipators work on this principle.

**21.13.1. Hydraulic Jump Formation.** The phenomenon of hydraulic jump has already been explained in details in Chapter 10. It was mentioned therein, that a hydraulic jump can form in a horizontal rectangular channel, when the following relation is satisfied between the pre-jump depth ($y_1$) and post-jump depth ($y_2$), w.r.t. to Fig. 21.30.

$$y_2 = -\frac{y_1}{2} + \sqrt{\frac{y_1^2}{4} + \frac{2q^2}{g y_1}} \qquad \textit{i.e. } \text{Eq. (10.4)}$$

where $q$ is the discharge intensity.

For a given discharge intensity over a spillway, the depth $y_1$ is equal to $q/V_1$; and $V_1$ is determined by the drop $H_1$, being equal to $\sqrt{2gH_1}$.

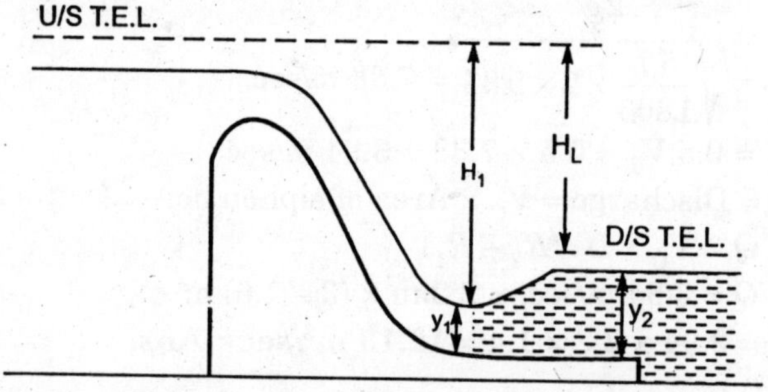


Fig. 21.30

Hence, for a given discharge intensity and given height of spillway, $y_1$ is fixed and thus $y_2$ (*i.e.* the depth required for the formation of hydraulic jump) is also fixed. But the availability of a depth equal to $y_2$ in the channel on the d/s cannot be guaranteed as it depends upon the tail water level, which depends upon the hydraulic dimensions and slope of the river channel below. The problem should, therefore, be analysed before any solution can be found. Hence, for different discharges, the tail water depth is found by actual gauge discharge observations and by hydraulic computations. The post jump depths ($y_2$) for all those discharges, are also computed from equation (10.4). If a graph is now plotted between $q$ and tail water depth, the curve obtained is known as the *Tail Water Curve (T.W.C.).* Similarly, if a curve is plotted on the same graph, between $q$ and $y_2$, the curve obtained is known as the *Jump Height Curve (J.H.C.)* or $y_2$ *curve.*

*Now there are five possibilities :*

(*a*) T.W.C. coinciding with $y_2$ curve at all discharges [Fig. 21.31 (*a*)].

(*b*) T.W.C. lying above the $y_2$ curve at all discharges [Fig. 21.31 (*b*)].

(*c*) T.W.C. lying below the $y_2$ curve at all discharges [Fig. 21.31 (*c*)].

(*d*) T.W.C. lying above the $y_2$ curve at smaller discharges and lying below the $y_2$ curve at larger discharges [Fig. 21.31 (*d*)].

(*e*) T.W.C. lying below the $y_2$ curve at smaller discharges and lying above the $y_2$ curve at larger discharges [Fig. 21.31 (*e*)].

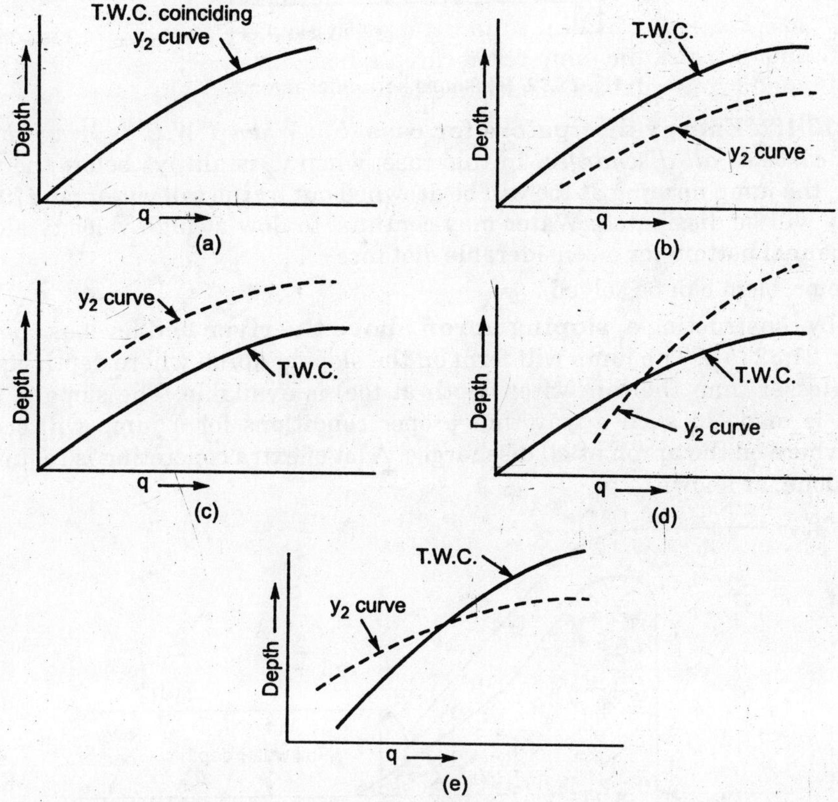


Fig. 21.31

Depending upon the relative positions of T.W.C. and $y_2$ curve, the energy dissipation arrangements can be provided below the spillway, as explained below for all these five cases.

**21.13.1.1. Energy dissipators for case** (*a*) : *When T.W.C. coincides with $y_2$ curve at all discharges.* This is the most ideal condition for jump formation. The hydraulic jump will form at the toe of the spillway at all discharges. In such a case, a simple concrete apron of length 5 $(y_2 - y_1)$ is generally sufficient to provide protection in the region of hydraulic jump, as shown in Fig. 21.32 (*a*).

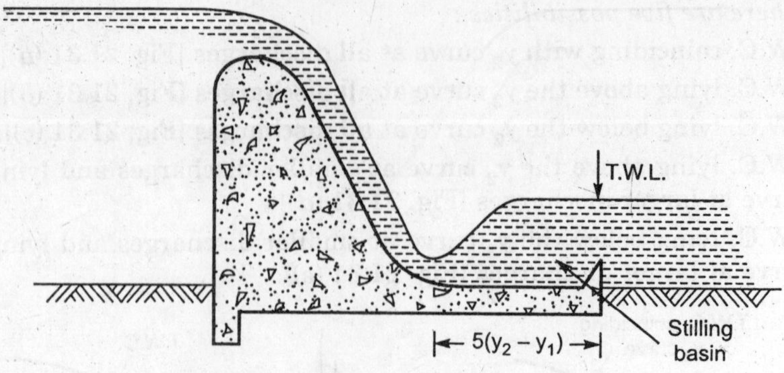


Fig. 21.32. (*a*) Simple horizontal apron.

**21.13.1.2. Energy dissipators for case** (*b*) : *When T.W.C. is lying above the* $y_2$ *curve at all discharges.* In this case, when $y_2$ is always below the tail water, the jump forming at toe will be drowned out by the tail water, and little energy will be dissipated. Water may continue to flow at high velocity along the channel bottom for a considerable distance.

The problem can be solved :

(*i*) by constructing a **sloping apron above the river bed** level as shown in Fig. 21.32 ($b_1$). The jump will form on the sloping apron where depth equal to $y_2$ (lesser than the tail water depth at toe) is available. The slope of the apron is made in such a way that proper conditions for a jump will occur somewhere on the apron at all discharges. A lot of extra concreting is required to be done, as shown.

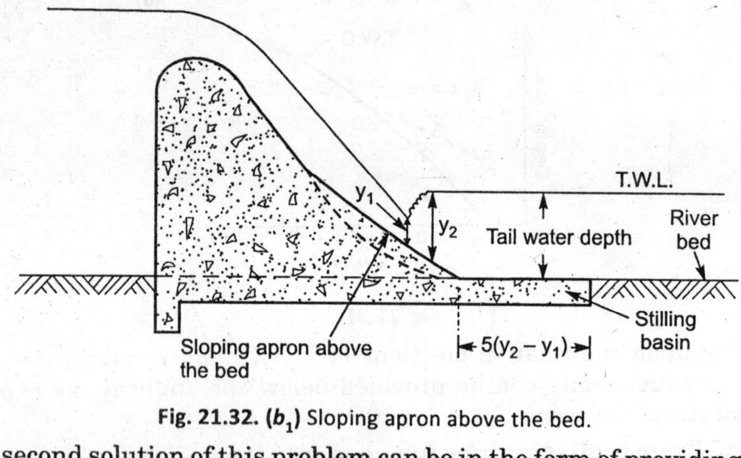


Fig. 21.32. ($b_1$) Sloping apron above the bed.

(*ii*) A second solution of this problem can be in the form of providing a **roller bucket** type of energy dissipator. It consists of an apron, which is upturned sharply at ends, as shown in Fig. 21.32 ($b_2$).

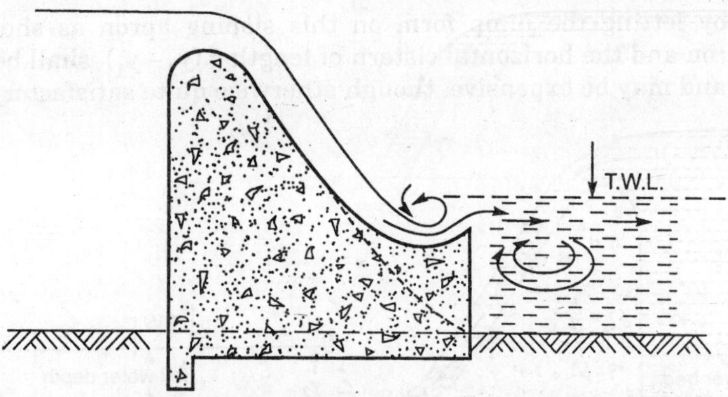


Fig. 21.32. ($b_2$) Roller bucket or submerged bucket.

Two main rollers are formed which dissipate the energy due to internal turbulence. In case of roller bucket, 45° bucket exit angle is most effective.

The roller which is formed downstream of the bucket, tends to move the scoured bed material towards the dam, thus, preventing serious scour at toe of the dam. Sometimes, the scoured material may enter the bucket under the action of u/s roller, and may cause severe abrasion. A ***dentated*** or ***slotted bucket lip*** may, therefore, have to be provided, so as to permit removal of material caught in the bucket.

**21.13.1.3. Energy dissipators for case** (*c*) : *When T.W.C. lies below the* $y_2$ *curve at all discharges.* (*i*) If the tail water is very low, the water may shoot up out of the above bucket, and fall harmlessly into the river at some, distance downstream of the bucket, as shown in Fig. 21.32 ($c_1$). *This bucket is then known as* **ski jump bucket** (also called *Flip bucket* or *Deflector bucket* or *Trajectory bucket*) and can be used for energy dissipation in case (*c*) : *i.e.* when the tail water depth is insufficient or low at all discharges. The ski jump bucket type of an energy dissipator requires sound and rocky river bed, because a part of the energy dissipation takes place by impact, although some of the energy is dissipated in air by diffusion and aeration.

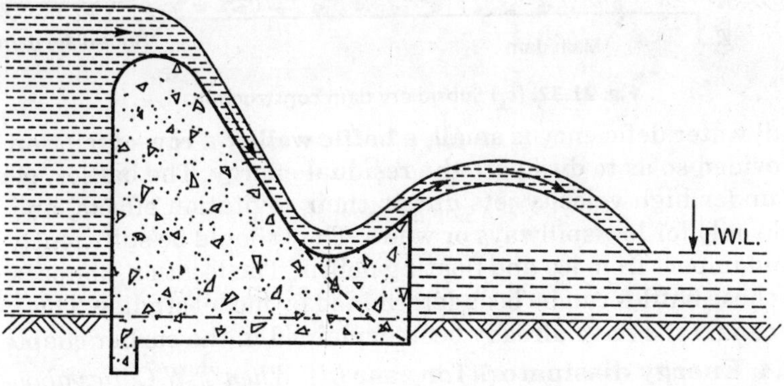


Fig. 21.32. ($c_1$) Ski jump bucket or Trajectory bucket or Deflector bucket or Flip bucket.

(*ii*) The second solution to the problem can be the *provision of a* **sloping apron** as *in case (b) but* **below the river bed,** as shown in Fig. 21.32 ($c_2$). The required depth $y_2$ which is greater than T.W. depth, can thus be made available by letting the jump form on this sloping apron as shown. This sloping apron and the horizontal cistern of length 5 $(y_2 - y_1)$ shall be entirely in cutting and may be expensive, though otherwise quite satisfactory.

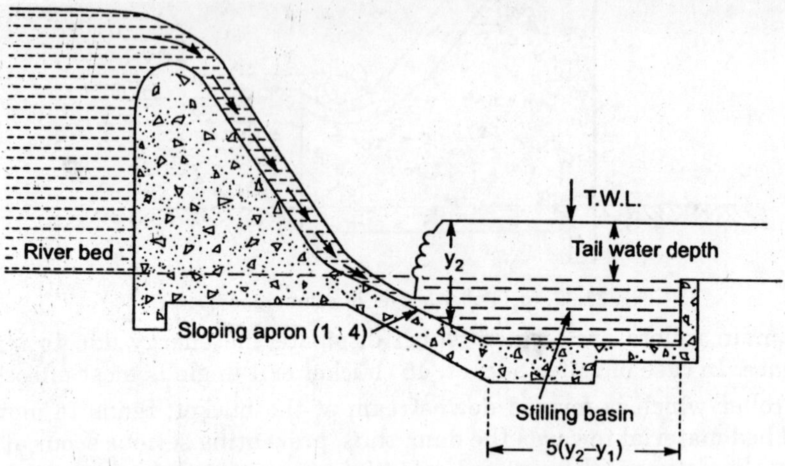


Fig. 21.32. $(c_2)$ Slopping apron below the bed.

*(iii)* The third solution to this problem may be the construction of a **subsidiary dam** below the main dam, so as to increase the tail water depth and cause a jump to form at the toe of the main dam, as shown in Fig. 21.32 $(c_3)$.

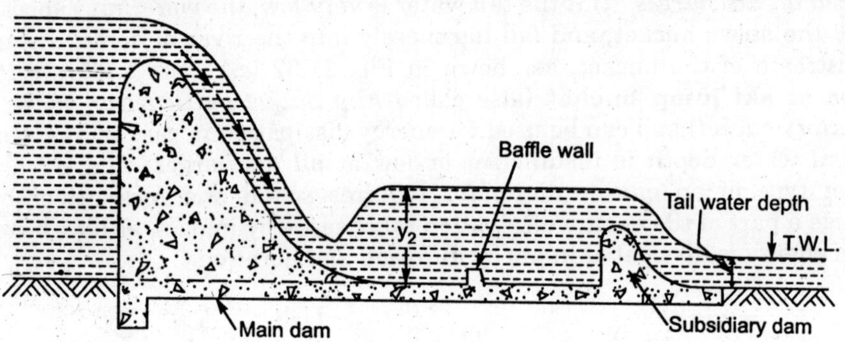


Fig. 21.32. $(c_3)$ Subsidiary dam construction.

If the tail water deficiency is small, a **baffle wall** or a **row of friction blocks** may be provided so as to dissipate the residual energy. The baffles, generally give way, under high velocity jets due to their cavitation effects, and hence, are suitable only for low spillways or weirs. They should be sufficiently strong to withstand impact from ice and floating debris. The location, shape, size and spacing of these baffles can be best determined by model studies. Their use for weirs and canal falls have already been explained in the earlier chapters.

**21.13.1.4. Energy dissipators for case** *(d)*. *When T.W.C. lies above the* $y_2$ *curve at low discharges and lies below the* $y_2$ *curve at high discharges.* In this case, at low discharges, the jump will be drowned and at high discharges, tail water depth is insufficient. The solution to the problem lies in providing a **sloping apron partly above and partly below the river bed** as shown in Fig. 21.32 *(d)*. The horizontal apron and end sill should also be provided.

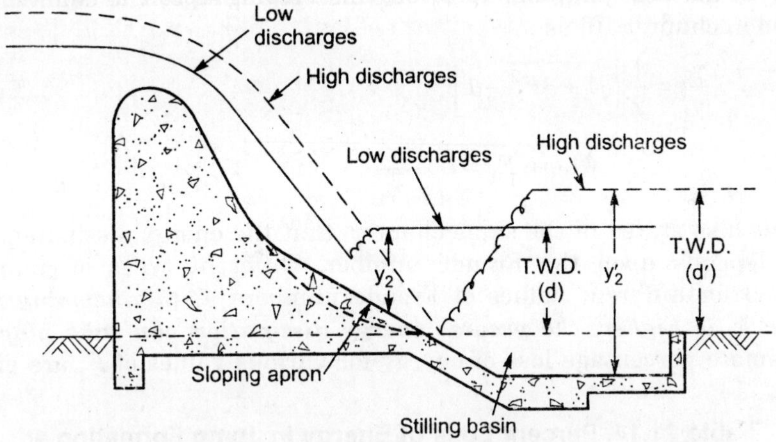


Fig. 21.32. (*d*) Sloping apron partly above and partly below the ground level.

At low discharges, the jump will form on the apron above the river bed, where the available depth is equal to the required depth and less than the T.W. depth. Similarly, at high discharges, the jump will form on the apron below the river bed, where the available depth is more than the T.W. depth and equal to the depth required for jump formation.

**21.13.1.5. Energy dissipators for case** (*e*). *When tail water depth is insufficient at low discharges and is greater at high discharges.* This case is just the reverse of case (*d*) and the same arrangement which was made in case (*d*) will serve the purpose. The only difference will be that at low discharges, the jump will form on the apron below the bed; and at high discharges, the jump will form on the apron at a point above the bed.

## 21.14. Energy Dissipation Below Other Types of Spillways

The five hydraulic conditions between tail water depth and post jump depth ($y_2$) discussed in the previous article, can occur below any type of spillway or even below the sluiceways or dam outlets, and erosion control can be obtained by the methods described. A chute or a shaft or a side channel spillway generally discharges water at a point far away from the dam. Hence, the protection is required only for the spillway, as the danger to the main dam is not there. Due to this reason, a *hydraulic stilling basin* is generally sufficient, and may be provided at the discharging point of the spillway. If sound rocks are available, a *Ski jump bucket* may be provided at low cost.

## 21.15. Energy Dissipation Below Sluiceways or Dam Outlets

Discharge through the sluiceways is generally small, and less protection is required below them. The water from the sluiceway is often allowed to fall directly into the stilling basin. Sometimes, the jet may be spread by a ***jet deflector*** at the outlet. Sometimes the energy dissipation arrangement of the main spillway may be utilised for sluiceways also.

## 21.16. Use of Hydraulic Jump as Energy Dissipator and Design of Stilling Basins

It was stated in chapter 10, that the hydraulic jump formation depends considerably upon the Froude number of the incoming flow ($F_1$). The pre-jump depth ($y_1$) and post-jump depth ($y_2$) are also governed by the equation (10.6) [derived in chapter 10] as :

$$\frac{y_2}{y_1} = \frac{1}{2}\left[\sqrt{1+8F_1^2} - 1\right] \quad \text{...(Eq. 10.6)}$$

$$\text{where } F_1 = \frac{V_1}{\sqrt{g \cdot y_1}}$$

It was also stated in the same chapter, that the energy dissipation in the jump, depends upon the Froude number. Different types of jump were described for different values of Froude numbers. ***If the incoming Froude number $F_1$ is higher, the greater energy dissipation can take place.*** The approximate percentage loss of energy for various values of $F_1$ are given in Table 21.14.

**Table 21.14. Percent Loss of Energy in Jump Formation at different values of Froude Nos.**

| $F_1$ | *% loss of energy* |
|---|---|
| 2.5 | 17 |
| 4.5 | 45 |
| 9.0 | 70 |
| 14.0 | 80 |
| 20.0 | 85 |

*However, the real problem in the design of stilling basins, is not the absolute dissipation of energy, but is the dissipation of this energy in as short a length as possible.* Hence, the effectiveness of a hydraulic jump has to be viewed in this light. The types of jumps for different ranges of Froude numbers, given in chapter 10, are again summarised below :

(*i*) When $F_1 < 2.5$, the jump is weak and energy loss is low. No blocks or other devices are provided in this range.

(*ii*) When $F_1$ lies between 2.5 to 4.5, as in case of weirs, barrages and canal falls, the jump is trouble-some and oscillating, which gives rise to heavy waves on the surface. ***Stilling basins*** with wave suppressor devices like ***chute blocks*** or ***friction blocks*** may be needed in this range. The length of the jump or the basin may be taken as 5 $(y_2 - y_1)$.

(*iii*) *When $F_1$ lies between 4.5 and 9.0 ; as is generally the case for dam spillways,* the jump performance is at its best. The jump is then called a steady jump. The length of the jump is almost constant and equal to $6y_2$ in this range. Hence for large spillways, where $y_2$ may be quite high, ***very long and expensive stilling basins may be required.*** Some auxiliary devices may also be introduced for further stabilising the flow and to reduce the length of the basin.

(*iv*) When $F_1 = 9.0$ and larger ; it is a strong jump. But the jump is likely to be rough and choppy. Hence, a ***bucket type of energy dissipator*** is preferred to a stilling basin of hydraulic jump type.

## 21.17. Standard Stilling Basins

Various types of stilling basins have been generalised for use on different types of works, by various agencies. The designs of these basins have been developed on the basis of long experience and on model studies, keeping in view the protection obtained consistent with the economy. These basins are not simple concrete aprons but are generally provided with **auxiliary devices** such as *chute blocks, sills, baffle walls,* etc. These devices can help in dissipating the energy of flow by offering resistance to flow and may stabilise the flow in a shorter length of the basin, thus affecting economy.

***In general, a stilling basin may be defined, as a structure in which the energy dissipating action is confined.*** If the phenomenon of hydraulic jump is basically used for dissipating this energy, it may be called a hydraulic jump type of stilling basin. The auxiliary devices may be used as additional measures for controlling the jump, etc.

Before we reproduce a few standard stilling basins, let us first describe, in brief, the effects produced by auxiliary devices.

**Chute Blocks.** Chute blocks are a kind of serrated device (*i.e.* row of small projections like teeth of saw) and provided at the entrance of the stilling basin. The incoming jet of water is furrowed and partly lifted from the floor, producing a shorter length of jump than what would have been without them. They also help in stabilising the flow and thus improve the jump performance (Fig. 21.33).

**Sills and Dentated Sills.** Sill (wall) or more preferably dentated sill is generally provided at the end of the stilling basin. The dentated sill diffuses the residual portion of high velocity jet reaching the end of the basin. They, therefore, help in dissipating residual energy and to reduce the length of the jump or the basin (Refer Fig. 21.33 and 21.45).

**Baffle Piers.** They are the blocks placed within the basin, across the basin floor. They help in breaking the flow and dissipate energy mostly by impact. These baffle piers, sometimes called **friction blocks** or ***basin blocks,*** are very useful in small structures, such as low spillways and weirs, etc. (Refer Fig. 21.33). They, however, give way due to cavitation, under the influence of high velocity jets, and hence are unsuitable for large works.

**21.17.1 ISI Standardised Stilling Basins.** The ISI (now BIS) has, vide their code IS: 4997-1968, laid down the criteria for the design of hydraulic jump type of stilling basins of rectangular cross-section with horizontal or sloping aprons, utilising auxiliary energy dissipation devices, such as chute blocks, basin or floor or friction blocks, and end sills, etc. *Stilling basin is a concrete structure, in which all or part of the energy dissipation action is confined, to avoid erosion of the downstream river bed.* In a stilling basin, the kinetic energy, first causes turbulence, and is ultimately lost as heat energy.

A ***hydraulic jump type stilling basin*** is a basin in which dissipation of energy is basically accomplished by hydraulic jump, which may be stabilised using *auxiliary devices* like chute blocks, friction, blocks and sill, etc. A hydraulic jump type of stilling basin may either have a *horizontal apron* or a *sloping apron*. Hence, the stilling basins may be divided into the following two basic types:

(A) *Hydraulic jump type stilling basins with horizontal apron; and*

(B) *Hydraulic jump type stilling basins with sloping apron.*

The second type of basins with sloping aprons *are required when TWC lies above the $y_2$ curve, i.e., the tail water or down stream water depth is more than $y_2$, causing the jump to be drowned out by the tail water, leading to much less energy, dissipation, and flow of water at high velocity along the river bottom for a considerable distance, causing harm and erosion to the river bed. Both these types of stiling basins are described below.*

**21.17.1.1. Hydraulic jump type stilling basins with horizontal apron.** Formation of hydraulic jump in stilling basins with horizontal apron is governed by the basic prejump and post-jump equation (10.6), i.e.

$$\frac{y_2}{y_1} = \frac{1}{2}\left[\sqrt{1+8F_1^2} - 1\right] \quad \text{...Eq. (10.6)}$$

where $F_1$ = Froude No. of the incoming flow = $\sqrt{\frac{V_1}{gy_1}}$

Such types of stilling basins (with horizontal apron) can be further sub-divided into the following two categories:

(I) *Stilling Basin type-I.* These are those stilling basins in which the Froud number of the incoming flow (*i.e.* $F_1$) is less than 4.5. This case is generally encountered on weirs, barrages and canal falls.

(II) *Stilling Basin Type II.* These are those stilling basins in which the Froud number of the incoming flow ($F_1$) is greater than 4.5. Such basins are a common feature of dam spillways.

We will now describe both these types of basins in details:

(1) **Indian Standard Stilling Basin Type I**. The BIS recommended stilling basin type I is shown in Fig. 21.33.

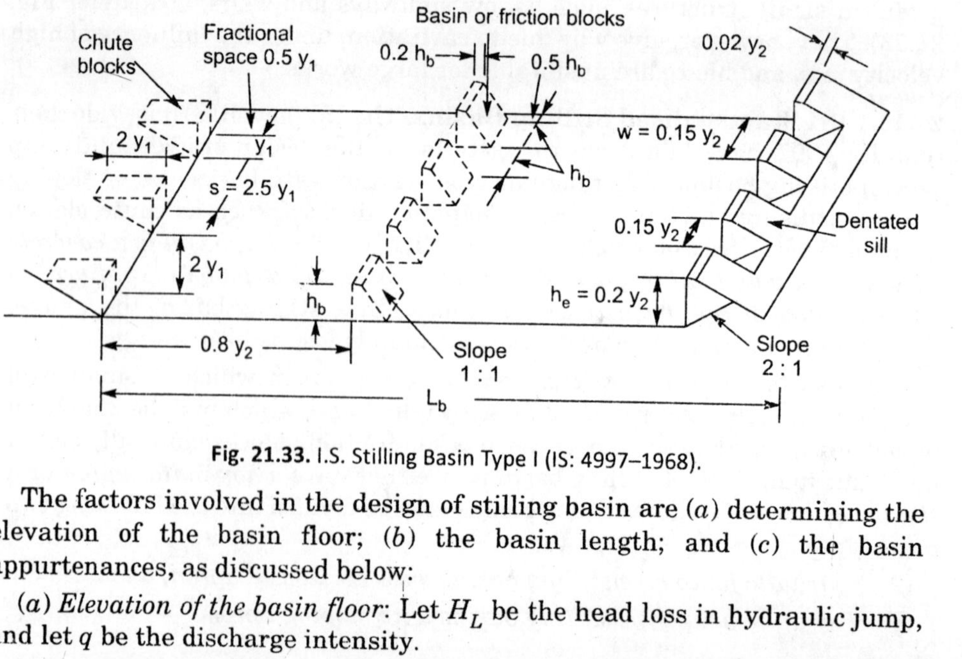


**Fig. 21.33.** I.S. Stilling Basin Type I (IS: 4997–1968).

The factors involved in the design of stilling basin are (*a*) determining the elevation of the basin floor; (*b*) the basin length; and (*c*) the basin appurtenances, as discussed below:

(*a*) *Elevation of the basin floor*: Let $H_L$ be the head loss in hydraulic jump, and let *q* be the discharge intensity.

To calculate $H_L$ from the known upstream and downstream total energy lines, the following procedure may be adopted.

**Step 1**. Where the basin is directly downstream from the crest or where the chute is not longer than the hydraulic head, $H_L$ may be assumed equal to the difference in the upstream and downstream total energy lines.

**Step 2.** Where the chute length exceeds the hydraulic head, $y_1$ should be first determined by taking $H_L$ equal to the difference between the upstream and downstream total energy lines. Then calculate friction losses in the chute by Manning's formula, and determine more accurate value of $H_L$, and then repeat the calculations to recalculate new values of $y_1$ and $H_L$. The computations may be repeated till the desired accuracy is reached.

**Step 3.** Knowing $H_L$ and $q$, calculate $y_c$, $y_1$ and $y_2$ from the following equations, or by using the Curve given in Fig. 21.34.

(*i*) $y_c = 3\sqrt{\dfrac{q^2}{g}}$ ...(Eq. 10.3)

(*ii*) $H_L = \dfrac{[y_2 - y_1]^3}{4y_1 \cdot y_2}$ ...(Eq. 10.12)

(*iii*) $y_2 = \dfrac{-y_1}{2} + \sqrt{\dfrac{y_1^2}{4} + \dfrac{2q^2}{g.y_1}}$ ...(Eq. 10.6)

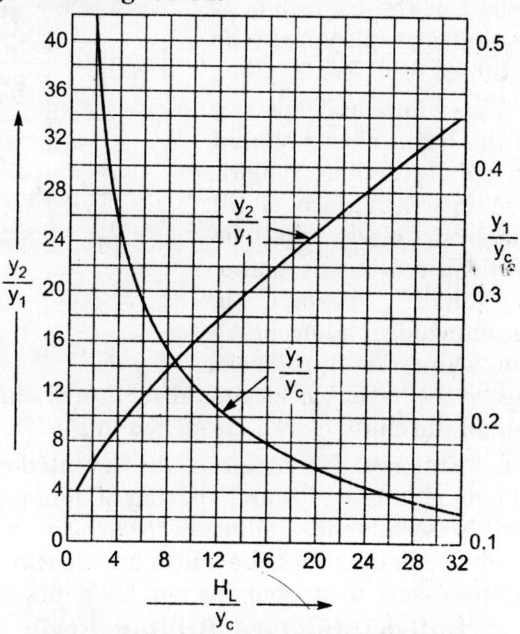


**Fig. 21.34.** Curve for determining Sequent depth (Horizontal apron).

Having computed $y_1$ and $y_2$, the *elevation of the basin* floor may be calculated by either deducting the specific energy at section 1–1 (*i.e.*, prejump section) from the total energy line at that section, or that at section 2–2 (post jump section) from the downstream total energy line.

(*b*) *Length and depth of the basin.* The length of the basin ($L_b$) may be determined from the curve given in Fig. 21.35.

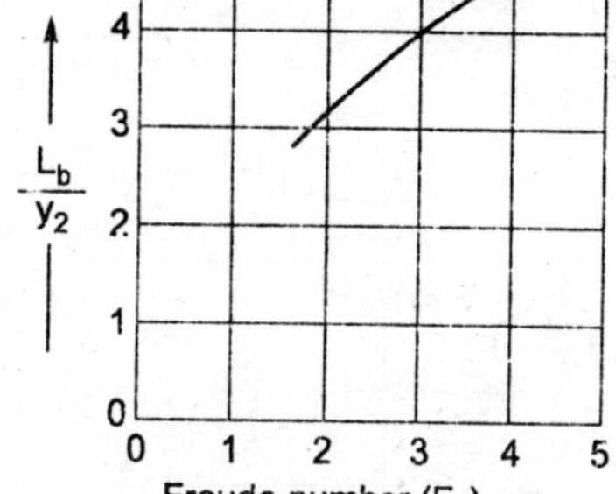


**Fig. 21.35.** Recommended length for Basin Type I

The basin will be provided with an *end sill* preferably a *dentated* one. In the boulder river reach, the sloping face of the end sill will be kept on the upstream side. Generally, the basin floor should not be raised above the level required from sequent depth consideration. If the raising of the floor becomes necessary due to site conditions, the same should not exceed 15% of $y_2$, and the basin in that case should be further supplemented by chute blocks and basin blocks. The basin blocks should not be used if the velocity of flow at the location of basin blocks exceed 15 m/sec, and in that case the floor of the basin should be kept at depth $y_2$ below the tail-water level. The tail-water depth should not generally exceed 10% of $y_2$.

(c) *Basin appurtenances*. The requirements for basin appurtenances, such as chute blocks, basin blocks, end sill are given in Fig. 21.33, and explained below:

(i) ***Chute blocks.*** The chute blocks should be kept at a height equal to $2y_1$ on the glacis slope. Their top length should also be equal to $2y_1$. The width of the chute blocks should be kept equal to $y_1$ and their spacing $2.5y_1$. A space equal to $0.5\ y_1$ should be left along each wall.

(ii) ***Basin blocks.*** The height of the basin blocks ($h_b$) in terms of $y_1$ may be obtained from Fig. 21.36.

The width and spacing of the basin blocks should be equal to their height. The upstream face of all the basin blocks shall be vertical and in one plane. A half space is recommended adjacent to the walls. The upstream face of the basin blocks should set at a distance of $0.8\ y_2$ from the downstream face of the chute blocks, as shown in Fig. 21.33.

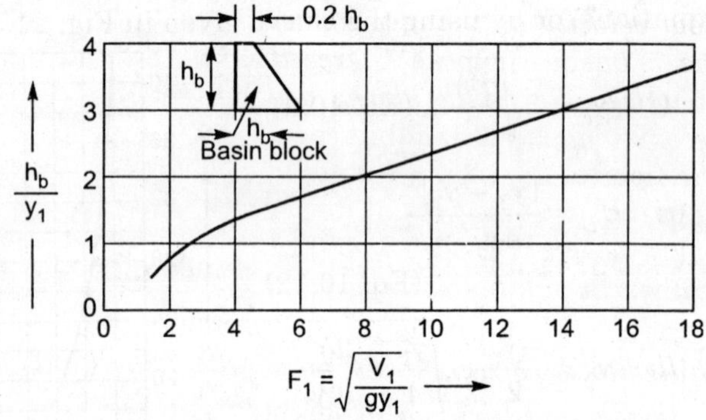


**Fig. 21.36**

(iii) ***End sill.*** The height of the dentated end sill is recommended as $0.2\ y_2$. The maximum width and spacing of dents shall be according to Fig. 21.33. A dent is recommended adjacent to each side wall. In the case of narrow basin, it is advisable to reduce the width and spacing but in the same proportion. It is not necessary to stagger the end sill dents with reference to chute blocks.

(2) **Indian Standard Stilling Basin Type II.** The IS recommended stilling basin type II is shown in Fig. 21.37.

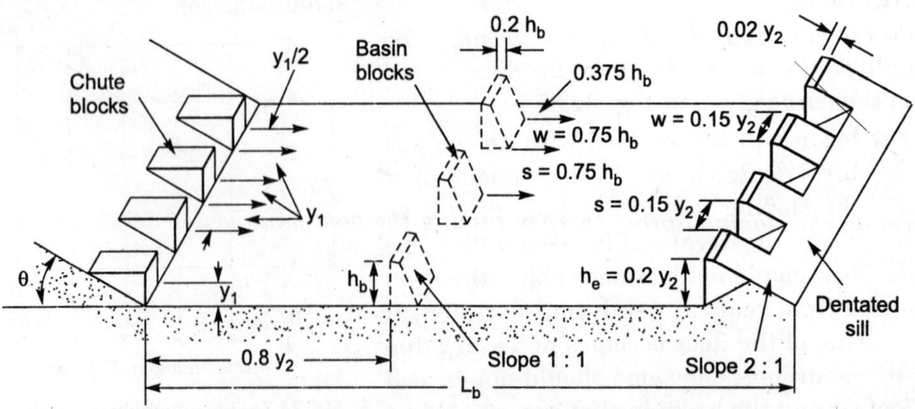


**Fig. 21.37.** IS stilling basin type II (IS: 4997:1968)

The various factors involved in the design of the type of basins, which are used ***when the Froude No. of the incoming flow ($F_1$) is more than 4.5*** (which is a common feature of dam spillways) are discussed below:

(*a*) *Basin length and depth*. The length of the basin ($L_b$) is determined from the curve given in Fig. 21.38.

The basin should be provided with chute blocks and dentated end sill, as shown in Fig. 21.37.

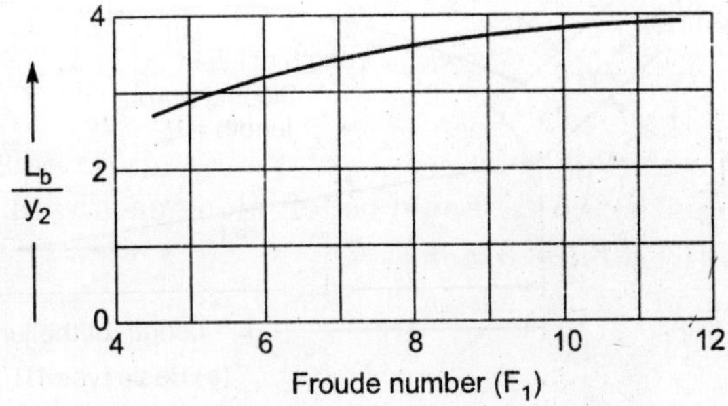


**Fig. 21.38.** Recommend length for IS stilling basin type II (IS: 4997 – 1968).

The maximum raising of the basin floor should not exceed 15% of $y_2$, and the basin in that case will be further supplemented by basin blocks. (Fig. 21.37). *However, when the flow velocity at the location of the basin blocks exceed 15 m/sec, no basin blocks are recommended*; and in that case, the floor of the basin should be kept at a depth equal to $y_2$ below the tail water level. The tail water depth should not generally exceed 10% of $y_2$.

(*b*) *Basin appurtenances*. Requirements for basin appurtenances, such as *chute blocks*, *basin blocks*, and *end sill* are given below.

(*i*) ***Chute blocks.*** The height, width and spacing of the chute blocks should be kept equal to $y_1$. The width and spacing may be varied to eliminate frictional blocks. A space equal to 0.5 $y_1$ is preferable along each wall.

(*ii*) ***Basin blocks.*** The height of the basin blocks in terms of $y_1$ can be obtained from Fig. 21.36. The width and spacing should be kept $\frac{3}{4}$th of the height. These should be placed at distance of 0.8 $y_2$ downstream from the chute blocks.

(*iii*) ***End sill.*** Same as for Basin type-I, described in article 21.17.1.1. (*c*) (*iii*)

**21.17.1.2. Hydraulic jump type stilling basins with sloping apron.** When the tail water is too high in comparison to sequent depth $y_2$, the jet left at the natural ground level would be drowned out by the higher tail water, heading to much less energy dissipation, and flow of water at high velocity along the river bottom for a considerable distance, causing harm and erosion of the river bed. *In such a case, a hydraulic jump type stilling basin with apron sloping from a higher level to the river bed level, shall preferably be constructed, as it would allow the formation of the hydraulic jump at a higher level on the sloping apron, thereby raising the post jump water depth level to somewhere near the tail water level.* This case was earlier explained in Fig. 21.32 ($b_1$).

Depending upon the tail water depth level in comparison to the $y_2$ depth level, IS: 4997–1968 has recommended the following two types of stilling basins:

(1) *Stilling basin type III.* [Fig. 21.39 (*a*)]. Such a standardised basin is recommended for the case when tail water curve is higher then $y_2$ curve at all discharges.

(2) *Stilling basin type IV.* [Fig. 21.39 (*b*)]. Such a basin is recommended for the case when tail water depth at maximum discharge exceeds $y_2$ depth considerably, but it is equal to or slightly higher than $y_2$ depth at lower discharges.

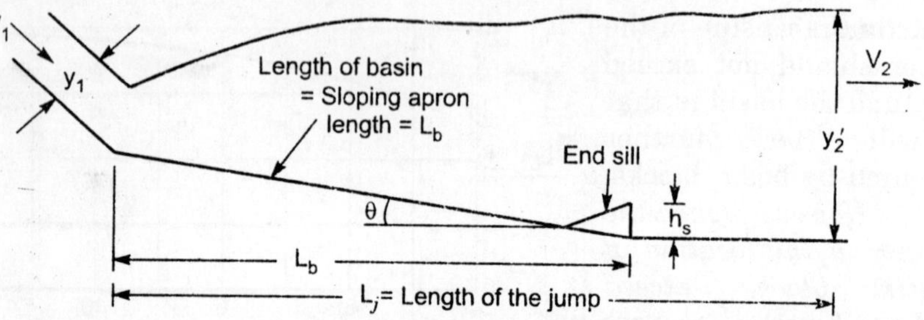


(a) Basin type III

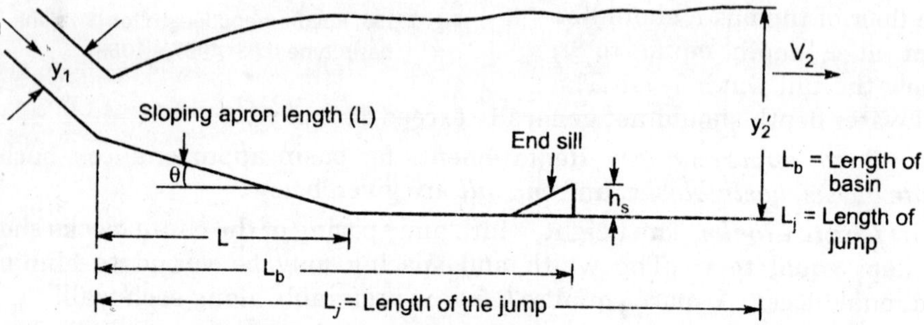


(b) Basin type IV

**Fig. 21.39.** I.S. Stilling Basin Type III and IV

**21.17.1.2.1 Conjugate depth on sloping apron.** Water depths at the beginning ($y_1$) and at the end of a hydraulic jump ($y'_2$) on a fully sloping apron (or glacis) is related by the following equation:

$$\frac{y'_2}{y_1} = \frac{1}{2\cos\theta}\left[\left[\frac{8F_1^2\cos^3\theta}{1-2K\tan\theta}+1\right]^{1/2}-1\right] \quad ...(21.23)$$

where $y'_2$ = conjugate depth sequent to $y_1$ for sloping apron, and may be determined from Fig. 21.40 (for basin III) or Fig. 21.41 (for basin IV).

$K$ = **Shape factor**. It is defined as a dimensionless parameter, which varies with the Froude number and the slope of the apron. This has been plotted against slope in Fig. 21.42 on the assumption that it is independent of Froude number.

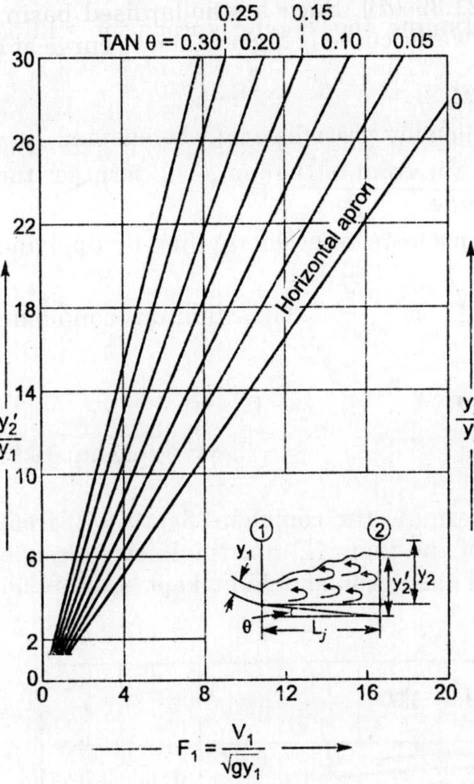


Fig. 21.40. Computation of $y'_2$ for basin III.

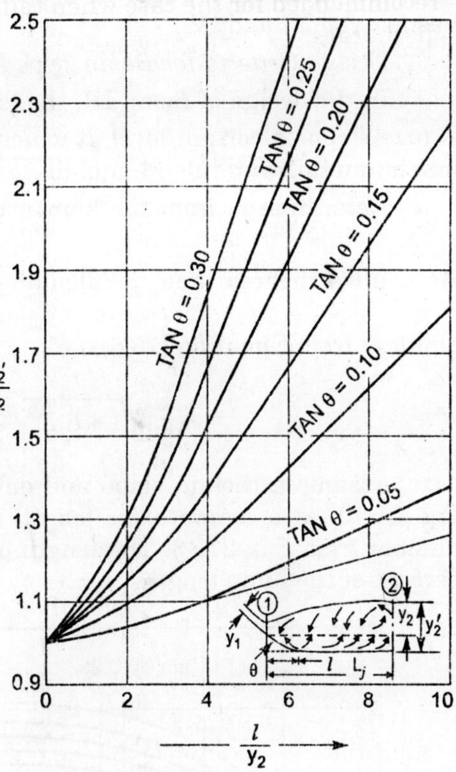


Fig. 21.41. Computation of $y'_2$ for basin IV

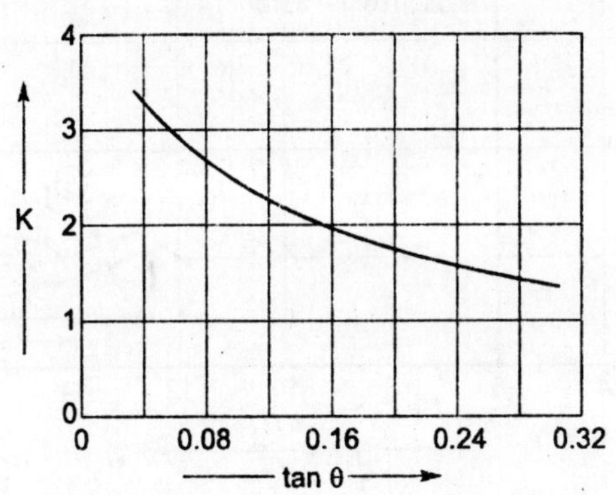


**Note:** Above curve is based on assumptions that $K$ is independent of $F_1$.

Fig. 21.42. Curve for determination of shape factor (K).

**21.17.1.2. *Design criteria for I.S. Basin III and Basin IV.*** It is not possible to standardise the design criteria for sloping aprons to the same extent as in the case of horizontal aprons. In this case, greater individual judgement is required. The slope and overall shape of the sloping apron shall be determined from (*i*) economic consideration; and (*ii*) the length being judged by the type and soundness of the downstream river bed. The following design criteria can however, serve as a guide in proportioning the sloping aprons for stilling basins type III and IV.

(1) *Design Criteria for Basin Type III.*

In the design of Basin III, the following procedure may be adopted

(*a*) Assume a certain level at which the front of the jump will form for the maximum tail-water depth and discharge.

(*b*) Determine $y_1$ from the known upstream total energy line by applying Bernoulli's theorem and to calculate $F_1$ as $\sqrt{\frac{V_1}{gy_1}}$. Then find out conjugate depth $y_2$ from Equation (10.6), *i.e.*,

$$\frac{y_2}{y_1} = \frac{1}{2}\left[\sqrt{1 + 8\,F_1^2} - 1\right] \qquad \text{...(Eq. 10.6)}$$

(*c*) Assume a certain slope and determine the conjugate depth ($y'_2$) from Fig. 21.40. Also compute the length of the jump ($L_j$) for the known Froude number from Fig. 21.43. The length of the apron should be kept equal to 60 percent of the jump length.

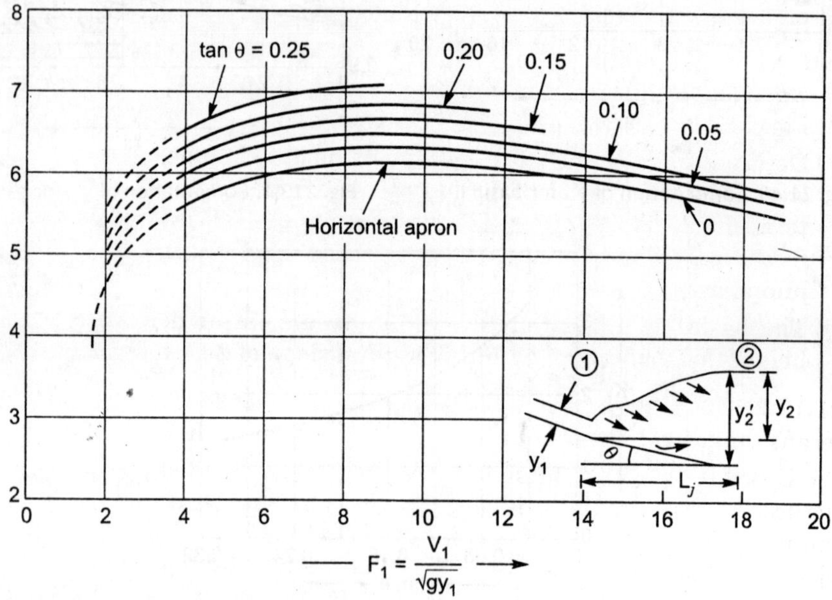


**Fig. 21.43.** Length of jump in terms of conjugate depth $y_2$ for Basin III and Basin IV

(*d*) Check whether the available tail-water depth at the end of the apron matches the conjugate depth $y'_2$. If not, change the slope or the level of upstream end of the apron or both. Several trials may be required before the slope and the location of the apron are compatible with the hydraulic requirements.

(*e*) The apron designed for maximum discharge may then be tested at lower discharges, say $\frac{1}{4}, \frac{1}{2}$ and $\frac{3}{4}$th of maximum. *If the tail-water depth is sufficient, or in excess of the conjugate depth for the intermediate discharges, the design is acceptable. If not a flatter slope at lower apron level should be tried or Basin IV* may be adopted.

(*f*) The basin should be supplemented by a solid or dentated end sill of height ($h_s$) equal to 0.05 to 0.2 $y_2$ with an upstream slope of 2 : 1 to 3 : 1.

(2) *Design Criteria for Basin Type IV.*

In the design of IS. Stilling Basin Type IV, the following procedure may be adopted.

(*a*) Determine the discharge at which the tail-water depth is most deficient.

(*b*) For the above discharge, determine the level and length of the apron on the basis of criteria given for horizontal apron.

(*c*) Assume a certain level at which the front of the jump will form for maximum tail water depth and discharge.

(*d*) Determine $y_1$ from the known upstream total energy line by applying Bernoulli's theorem and calculate $F_1$. Then find out conjugate depth $y_2$ from Eq. (10.6).

(*e*) Determine a suitable slope (by trail and error) so that the available tail-water depth matches the required conjugate depth $y'_2$ determined from Fig. 21.41.

(*f*) Determine the length of the jump for the above slope from Fig. 21.43. If the sum of the lengths of inclined portion and horizontal portion is equal to about 60 percent the jump length, the design is acceptable. If not, fresh trials may be done by changing the level of the upstream end of the jump formation.

(*g*) The basin should be supplemented by a solid or dentated end sill of height ($h_s$) 0.05 to 0.2 $y_2$ and upstream slope of 2 : 1 to 3 : 1.

**21.17.2. U.S.B.R. Basins.** U.S.B.R. has standardised stilling basins for different ranges of Froude numbers. The important of these basins are :

(1) **U.S.B.R. stilling basin II.** This is recommended for use on large structures, such as dam spillways, large canal structures, etc., when the incoming *Froude number* ($F_1$) *is more than 4.5.* The dimensions of the chute blocks, dentated sill, etc. are shown in Fig. 21.44. The length of the basin is related to the Froude number ($F_1$) as given in Table 21.15.

**Table 21.15**

| $F_1$ | *Length of the basin (L)* |
|---|---|
| 4 | 3.6 $y_2$ |
| 6 | 4 $y_2$ |
| 8 | 4.2 $y_2$ |
| 10 or more | 4.3 $y_2$ |

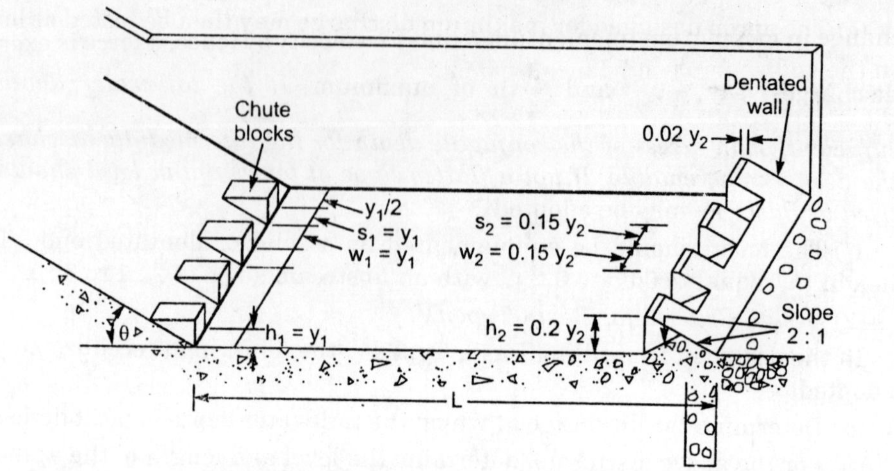


Fig. 21.44. U.S.B.R. Stilling basin II ($F_1 > 4.5$).

An economy in the length of the basin up to about 35% (4.3 $y_2$ in place of 6$y_2$) is thus obtained with auxiliary devices. The floor of the basin should be set at such a level as to provide 5% more water depth than $y_2$.

(2) **U.B.S.R. stilling basin IV.** This type of stilling basin is shown in Fig. 21.45. It is used for Froude number varying between 2.5 and 4.5, which generally occurs in canal weirs, canal falls, diversion dams, etc. This basin is applicable only to rectangular cross-sections. Since oscillating waves are generated in this range of Froude number, they are tried to be controlled at source by providing large chute blocks. The spacing between the blocks may be further reduced than what is shown in Fig. 21.45 (up to say 0.75 *W*) for better hydraulic performance. The floor of the basin may be set at such a level as to provide 5 to 10% more water depth than the theoretical $y_1$ . The length of the basin is generally kept equal to $5\,(y_2 - y_1)$.

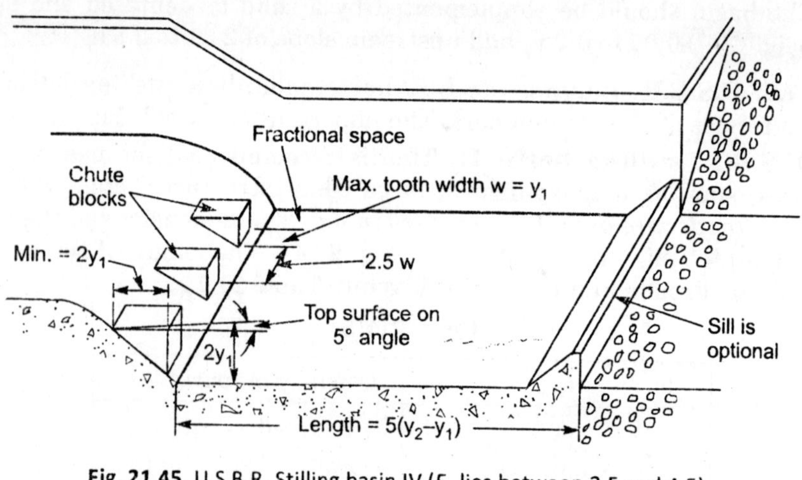


Fig. 21.45. U.S.B.R. Stilling basin IV ($F_1$ lies between 2.5 and 4.5).

## 21.18. Dynamic Force on Spillway Bucket and Spillway Bottom

When water flows over the curved bottom portion of an ogee spillway, there is a continuous change of velocity. Due to this change in velocity, there is a change in momentum (momentum = mass × velocity). Hence, a force is exerted on the spillway in accordance with Newton's second law of motion, *i.e.* rate of change of momentum is equal to this impressed force. This force is known as ***dynamic force.*** The spillway bottom where this curvature is provided is subjected to this force. The evaluation of this force is, therefore, necessary in the design of the bottom portion, or the design of the bucket, if it is provided as energy dissipator in an overflow spillway.

The magnitude of this force can be easily determined by using Newton's second law and by resolving all the forces acting on the curved element of water *AB* [Fig. 21.46 (*a*)] in horizontal and vertical directions. Mass of water when multiplied by net change in velocity in one direction will give the net component of this dynamic force in that direction. If *F* is this dynamic force, then

$$F_{(H)} = [V_{2(H)} - V_{1(H)}] \times \text{mass of water in the element}$$

$$F_{(V)} = [V_{2(V)} - V_{1(V)}] \times \text{mass of water in the element}$$

$$F = \sqrt{F_{(H)}^2 + F_{(V)}^2}$$

where $V_1$ and $V_2$ are the velocities at the entrance and exit of the curved element. $V_{1(H)}$ and $V_{2(H)}$ are their horizontal components, and $V_{1(V)}$ and $V_{2(V)}$ are their vertical components. An illustrative example has been solved below :

**Example 21.6.** *An overflow ogee spillway of height 15 m is discharging water with a head of 2 m over the crest. A reverse curvature of radius 4 m, subtending an angle of 60° at the centre, is provided at the spillway bottom as shown in Fig. 21.46 (a). Assuming the discharge coefficient for the spillway as 2.2, determine the magnitude of the dynamic force on the curve portion AB of the spillway.*

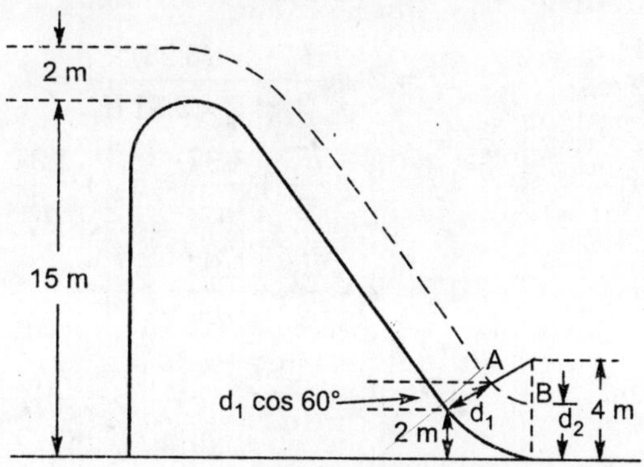


Fig. 21.46 (*a*)

**Solution.** The discharge over the spillway is given as :

$$Q = C \cdot L \cdot H^{3/2}$$

Discharge per unit length,

$$q = C \cdot H^{3/2}$$

$$= 2.2 \times (2)^{3/2} = 6.23 \text{ cumec/m}$$

Let $d_1$ and $d_2$ be the water depths, $V_1$ and $V_2$ be the flow velocities at *A* and *B* respectively.

Neglecting any loss of energy and the head due to velocity of approach and applying Bernoulli's theorem between u/s water surface and at sections *A* and *B*, respectively, we get

$$15 + 2 = 2 + d_1 \cos 60° + \frac{V_1^2}{2g} = d_2 + \frac{V_2^2}{2g}$$

But $V_1 = \frac{q}{d_1}$ ; and $V_2 = \frac{q}{d_2}$

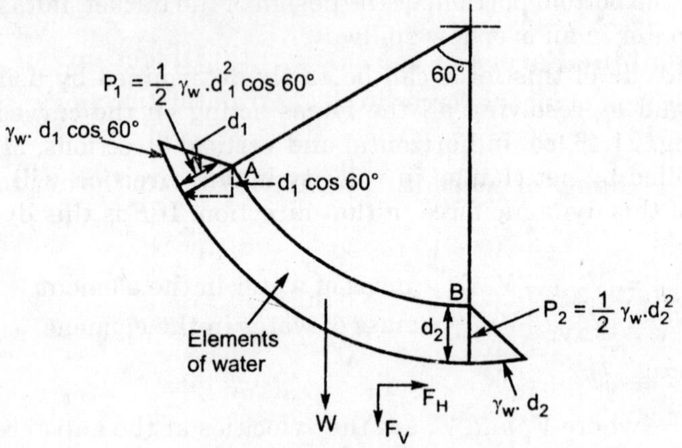


Fig. 21.46 (*b*)

$$\therefore \quad 17 = 2 + d_1 \cos 60° + \frac{q^2}{2gd_1^2} = d_2 + \frac{q^2}{2gd_2^2}$$

or $$17 = 2 + \frac{d_1}{2} + \frac{(6.23)^2}{2 \times 9.81\, d_1^2} = d_2 + \frac{(6.23)^2}{2 \times 9.81\, d_2^2}$$

or $$17 = 2 + \frac{d_1}{2} + \frac{1.97}{d_1^2} = d_2 + \frac{1.97}{d_2^2}$$

or $$17 = 2 + \frac{d_1}{2} + \frac{1.97}{d_1^2} \quad ...(i)$$

and $$17 = d_2 + \frac{1.97}{d_2^2} \quad ...(ii)$$

Eq. (*i*) yields :

$$15 = \frac{d_1}{2} + \frac{1.97}{d_1^2}$$

or $$30 d_1^2 = d_1^3 + 3.94$$

Solving by hit and trial, we get

$$\boxed{d_1 = 0.365 \text{ m}}$$

Similarly, solving equation (*ii*), we get

$$17 = d_2 + \frac{1.97}{d_2^2}$$

or $$17 d_2^2 = d_2^3 + 1.97$$

by hit and trial, we get $\boxed{d_2 = 0.344 \text{ m}}$

Now the hydrostatic forces $P_1$ and $P_2$ per unit length are given as :

$$P_1 = \tfrac{1}{2} \cdot \gamma_w \cdot d_1^2 \cos 60° = \tfrac{1}{2} \times 9.81 \times (0.365)^2 \times \tfrac{1}{2} = 0.327 \text{ kN/m run}$$

$$P_2 = \tfrac{1}{2} \cdot \gamma_w \cdot d_2^2 = \tfrac{1}{2} \times 9.81 \times (0.344)^2 = 0.58 \text{ kN/m run}$$

**Resolving all the Forces in the Horizontal Direction**

Algebraic sum of all the forces in the horizontal direction

$$= P_1 \cos 60° + F_H - P_2 \qquad ...(iii)$$

The rate of change of momentum per unit length in the horizontal direction

$$= \frac{\rho_w \cdot q}{g} [V_2 - V_1 \cos 60°]$$

$$= \frac{\rho_w \cdot q}{g} \left[ V_2 - \frac{V_1}{2} \right] \qquad ...(iv)$$

But $\quad V_1 = \dfrac{q}{d_1} = \dfrac{6.23}{0.365} = 17.06$ m/sec. ;

and $\quad V_2 = \dfrac{q}{d_2} = \dfrac{6.23}{0.344} = 18.12$ m/sec.

Equating (*iii*) and (*iv*), we get

$$P_1 \cos 60° + F_H - P_2 = \frac{\rho_w q}{g} \left[ V_2 - \frac{V_1}{2} \right]$$

or $$0.327 \times \tfrac{1}{2} + F_H - 0.58 = \frac{9.81 \times 6.23}{9.81} \left[ 18.12 - \frac{17.06}{2} \right] = 59.75$$

or $$F_H = \mathbf{60.16 \text{ kN}}$$

Absence of – ve sign indicates that the assumed direction of $F_H$ is correct.

**Resolving all the Forces in the Vertical Direction**

The weight $W$ of the water in the curved element $AB$

$$= \left[ \frac{2\pi \cdot R}{360°} \times \theta° \right] \cdot y_{mean} (\rho_w)$$

$$= \left[ \frac{2\pi \times 4}{360°} \times 60° \right] + \left[ \frac{0.365 + 0.344}{2} \right] \times 9.81 = 14.57 \text{ kN}$$

Resolving all the forces in the vertical direction, we get

$$W + P_1 \sin 60° - F_V = \frac{\gamma_w \cdot q}{g} [0 - V_1 \sin 60°]$$

or $$14.57 + 0.327 \times \frac{\sqrt{3}}{2} - F_V = \frac{9.81 \times 6.23}{9.81} [- 17.06 \sin 60°]$$

or $$14.85 - F_V = - 92.04$$

or $$F_V = \mathbf{106.89 \text{ kN}}$$

Absence of – ve sign indicates that the assumed direction of $F_V$ is correct and is vertically upward.

$$\text{Resultant force} = \sqrt{F_H^2 + F_V^2} = \sqrt{(60.16)^2 + (106.89)^2}$$
$$= \mathbf{122.64\ kN.}$$

Hence, the resultant dynamic force is equal to 122.64 kN and acts at an angle $= \tan^{-1}\left(\frac{106.89}{60.16}\right) = 60.6°$ with the horizontal in the upward direction, as shown in Fig. 21.46 (*c*).

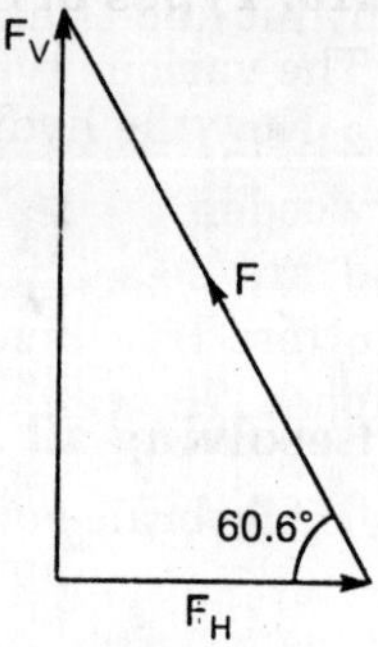

**Fig. 21.46 (*c*)**

## SPILLWAY CREST GATES

If a temporary barrier can be installed over the permanent raised crest of a spillway, additional water can be stored between the spillway crest and the top of the barrier during the fag-end of the rainy season. The small flows in excess of the barrier top level, may be permitted to pass over the barrier. If, however, large flood occurs, the barrier may be removed and full spillway capacity made available for the outflow.

Sometimes on large dams, regular gates may be installed over the permanent crest, so as to function like a movable additional crest. In such a case, the height of the permanent raised crest can be reduced and the balance provided by the movable crest (*i.e.* gate). If there is a permanent raised crest up to the gate top, the storage, of course would be equal to that of a gated crest ; but in times of serious floods, the rise in flood level would be much more as compared to what would have been in a gated crest. This is because, the gates would be opened during serious floods, so as to provide more head and hence larger discharge, and consequent lesser rise in flood levels. Hence, the top level of the non-overflow section and the value of land acquisition for the reservoir which has to be determined by the maximum rise of flood above the spillway crest, can be reduced by providing gated crest or controlled crests. In other words, the dam height can be reduced for the same useful storage, or more useful storage can be obtained for the same height, provided the dam spillway is controlled by gates, etc.

This saving in the dam height and land acquisition will be more, if more height of the gates is provided. This saving is, however, counter-balanced by the cost of the gates. The cost of the gates would be more if their height is increased. (The gate cost include the principal cost and OMR, *i.e.* operational, maintenance and repair costs.) An economic balance between these two factors must be worked out and the cheapest combination found before deciding the height of the permanent crest and the height of the temporary crest (*i.e.* gates). This is also governed by the limitations of maximum available gate heights.

Gates can be provided on all types of spillways except siphon spillways. In siphon spillways, the gates are not required, as the rise in flood level is already small compared to other types of spillways. The gates for earthen dams should be provided with caution, since the faulty operation or failure of their operation may lead to serious rise in flood levels, causing overtopping and failure of dam.

## 21.19. Types of Spillway Gates

The various types of spillway gates are described below in brief :

**21.19.1. Dropping Shutters or Permanent Flash Boards.** They consist of wooden panels usually 1.0 to 1.25 m high. They are hinged at the bottom and are supported against the water pressure by struts (Fig. 21.47). The shutters fall flat on the crest when the downstream supporting struts are tripped. Hence they are not suitable for curved crests.

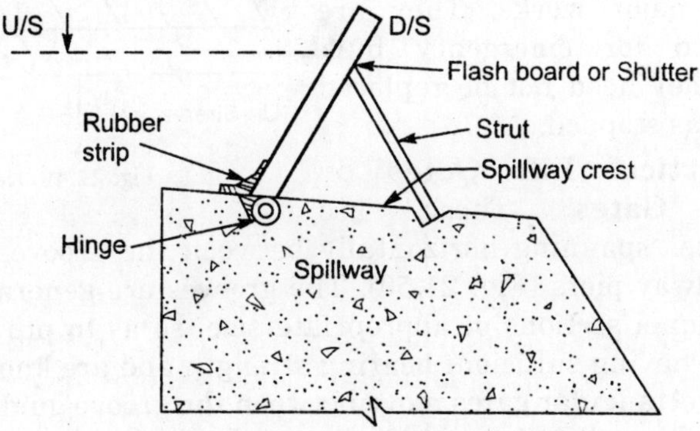


**Fig. 21.47.** Flash board or dropping shutter.

These shutters can be raised or lowered from an overhead cableway or a bridge. Various types of shutters which drop and hoist themselves automatically, have been designed these days. These automatic shutters work on the principle of counter weights acting against the water pressure. Automatic shutters do not function well when interferred by floating debris, ice, etc.

Sometimes, temporary flash boards, which shall fall as soon as overtopped by water, may be used for very minor works.

All kinds of flash-boards do have some disadvantages; and hence, used only on small spillways of minor importance.

**21.19.2. Stop Logs and Needles.** Stop logs consist of wooden beams or planks placed one upon the other and spanning in the grooves between the spillway piers (Fig. 21.48). They can be placed and removed either by hand or with hoisting mechanism. Considerable time may get wasted in removing

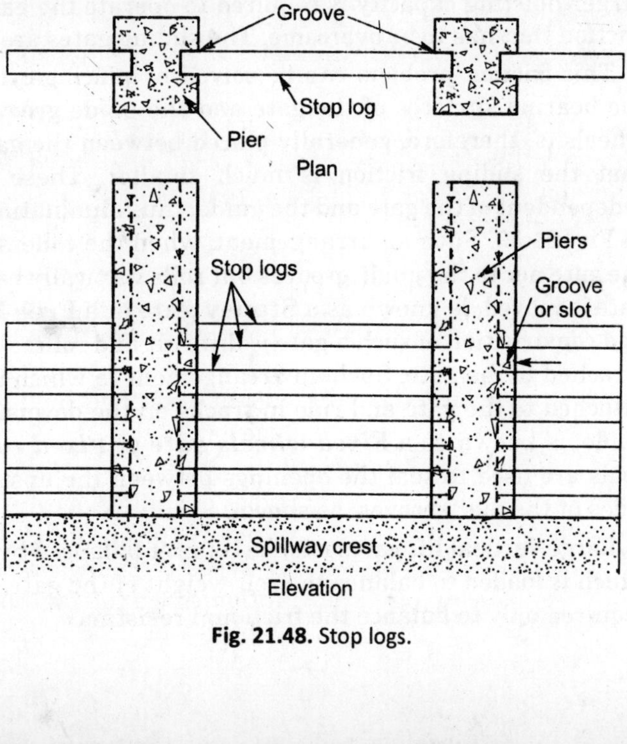


**Fig. 21.48.** Stop logs.

them, if they become jammed in the slots. Leakage between the logs is also a big problem. They are, therefore, used on very minor works.

***Needles.*** Needles are wooden logs kept side by side with their lower ends resting in a keyway on the spillway and upper ends supported by a bridge (Fig. 21.49). It is very difficult to handle these needles at the time of flow and hence they are not used on any major works. They are sometimes used for emergency bulk heads, where they need not be replaced until the flow has stopped.

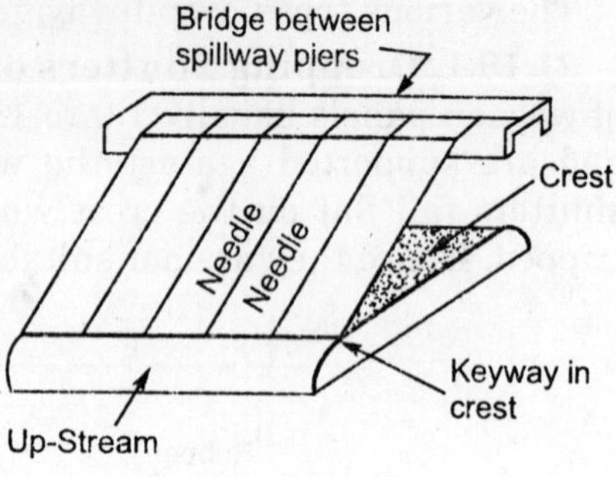


Fig. 21.49. Needles.

**21.19.3. Vertical Lift Gates or Rectangular Gates.** These are rectangular gates spanning horizontally between the grooves made in the supporting spillway piers (Fig. 21.50). The grooves are generally lined with rolled steel channel sections of appropriate size, so as to provide a smooth bearing surface having sufficient bearing strength and are known as ***groove guides.*** These rectangular gates move between the groove guides, and can be raised or lowered by a hoisting mechanism at the top.

The gates are often made of steel, although they may be made of concrete or wood. They are generally placed vertical, although they may be kept slightly inclined downstream.

Because of the hydrostatic force caused by the upstream water standing against the gate, large friction is developed between the gate and the downstream groove guides. Hence, if the gate is in direct contact with the guides, as is there in a sliding gate, large friction will be developed, and it will be very difficult to move the gate. Hence, in a ***sliding gate*** relatively larger hoisting capacity is required to operate the gate because of the sliding friction that has to be overcome. The sliding gates are, therefore, seldom used.

This friction problem can be solved by placing cylindrical rollers between the bearing surfaces of the gate and the guide grooves. A train of rollers or wheels is, therefore, generally placed between the gate and the d/s guide, so that the sliding friction is much smaller. These rollers may be placed independent of the gate and the guide, thus eliminating axle friction, as shown in Fig. 21.50. Such an arrangement, when the rollers are neither attached to the gate nor to the guide grooves but rolls vertically between the two when the gate is moved, is known as a **Stoney gate** or a **Free Roller gate**. The design and construction of such a gate is difficult and rollers are, therefore, generally attached to the gate. Such an arrangement in which the rollers or wheels are attached to the gate and ride in tracks on the downstream side of the groove guide, is known as a ***Fixed wheels gate*** or ***Fixed roller type gate.*** Rubber seals are used to seal the openings between the upstream leaf plate and the sides of the pier grooves, as shown.

Large vertical lift gates may be counter balanced by a counter weight beam, which is loaded to balance the self-weight of the gate. Hence, hoisting force is required only to balance the frictional resistance.

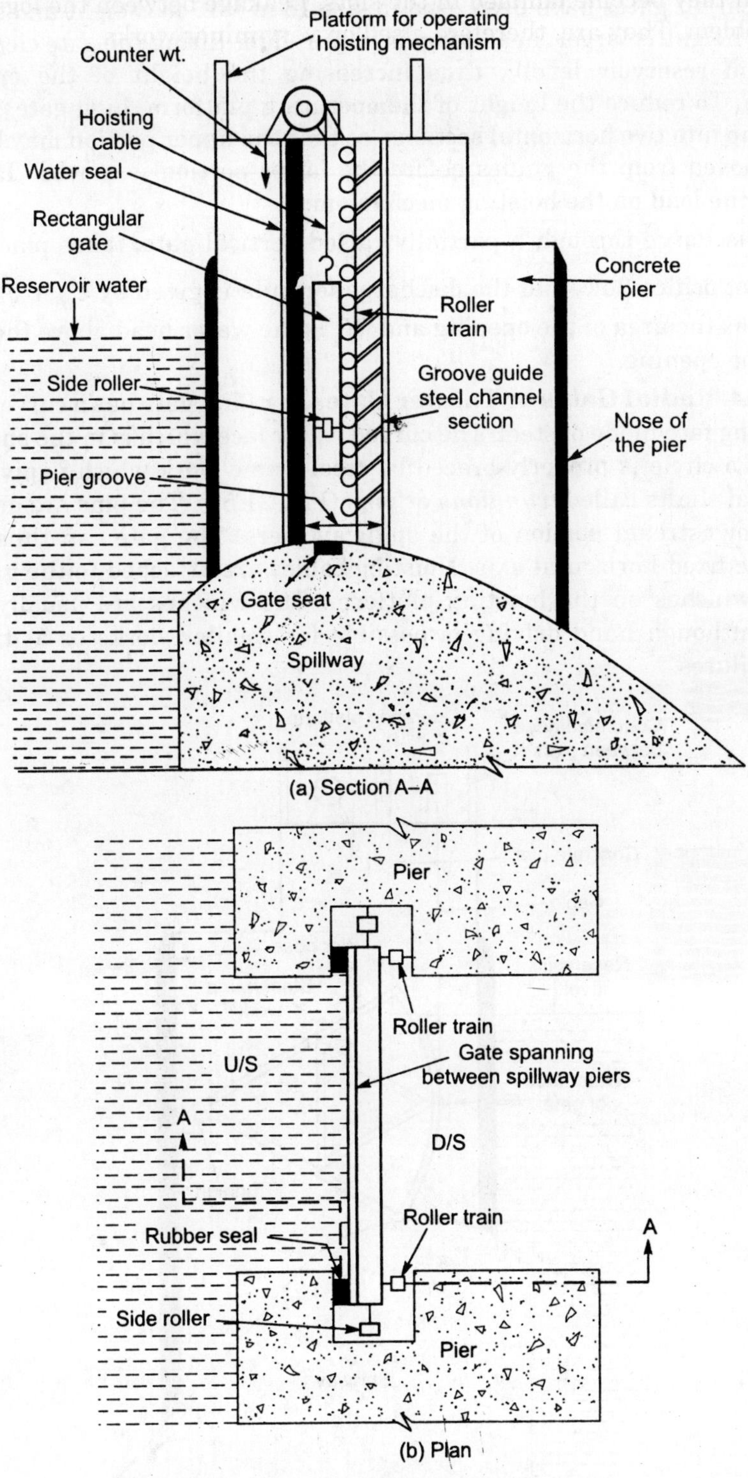


**Fig. 21.50.** 'Vertical stoney gate' or 'Free roller gate'.

Vertical lift gates have been used in size 15 m × 15 m (height and span). If the gate height is larger, head room is required for lifting the gate clear of the maximum reservoir level ; thus increasing the height of the operating platform. To reduce the height of the operating platform, high gates may be broken up into two horizontal sections, so that the upper portion may be lifted and removed from the guides before the lower portion is moved. This also reduces the load on the hoisting mechanism.

The discharge through a partially raised vertical gate, takes place by an undershot orifice flow, and the discharge formula is given by $C_d A \sqrt{2g \cdot H_1}$, where $A$ is the area of the opening and $H_1$ is the water head above the centre line of the opening.

**21.19.4. Radial Gates or Tainter Gates.** A radial gate has a curved water supporting face made of steel. The curved water face which is in the shape of a sector of a circle is properly braced by steel frame work which is pivoted on horizontal shafts called *trunnions* or *pins* (Fig. 21.51). The pins are anchored in the downstream portion of the spillway piers. The gate can thus rotate about the fixed horizontal axis. Hoisting cables are attached to the gate and lead to winches on the hoisting platform. The winches are usually motor driven, although hand driving is possible for smaller works or at times of power failures.

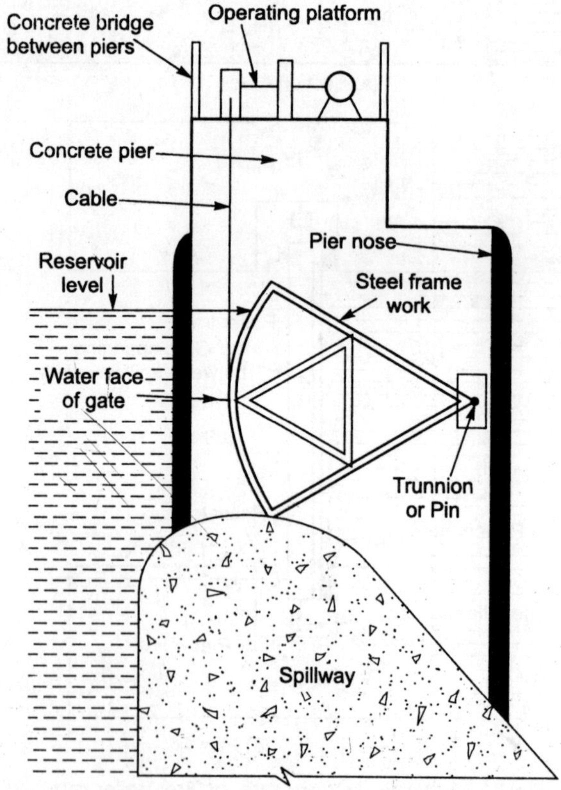


Fig. 21.51. Radial gate or Tainter gate.

The water-face segment is made concentric to the supporting pins so that the entire water thrust passes through the pins, thus creating no moment against the lifting of the gate. Hence, the lifting force is required only against the weight of the gate, the friction between the seals and the piers, and the frictional resistance at the pins. Counter weights, in order to counter balance the self weight, may also be used, which further reduces the lifting force. Moreover, the hoisting load is nearly constant for all gate openings. Hence, radial gates can be used with smaller lifting force for all heads, and hand operating hoisting mechanism may suffice for smaller works ; whereas in the vertical lift gates of the same size, power mechanism might be needed.

**21.19.5. Drum Gates.** Drum gates are useful for longer spans of the order of 40 m or so and medium heights say 10 m or so. The drum gate consists of a segment of a cylinder which may be raised above the spillway crest or may be lowered into the *recess* made into the top of spillway.

The U.S.B.R. drum gate (Fig. 21.52) is completely enclosed and is hinged at the upstream end. The buoyant forces due to head water pressure underneath the drum, aid in its lifting. In this type of dam gate, the drum is enclosed on all the three sides as well as on the ends, thus forming a water tight vessel. When the drum is lowered, it fits into the recess in such a way that the surface becomes coincident with the designed ogee shape of the crest.

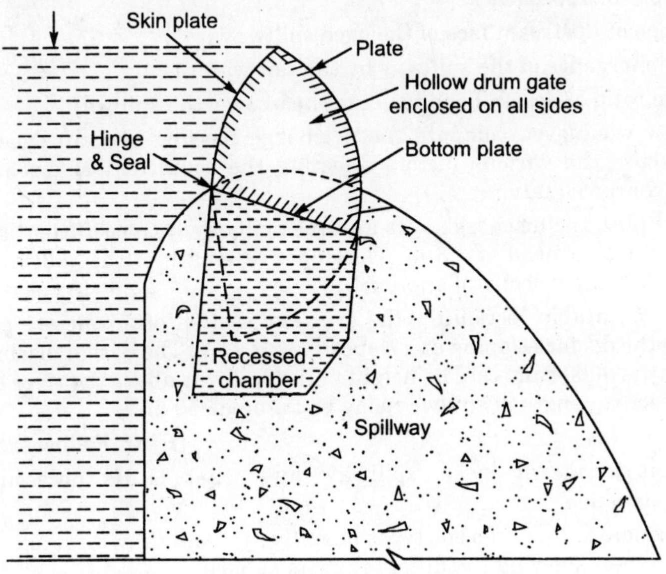


**Fig. 21.52.** Drum gate (U.S.B.R. type).

The other type of drum gate may have no bottom plate and shall be raised only by the buoyant action of water entering the recess, underneath the skin plate of drum.

The drum gates require large recess and hence, are not suitable for smaller spillways.

## PROBLEMS

1. (*a*) "A spillway is a safety valve in a dam". Discuss the statement.
   (*b*) Discuss the location of the 'main spillway' and 'subsidiary spillway' in gravity dams as well as in earthen dams.
   (*c*) Briefly describe an 'ogee spillway'.
2. (*a*) What are spillways and where are they provided ?
   (*b*) Enumerate the various types of spillways, and describe in details the most widely used type.
3. (*a*) What are the different kinds of spillways and how are they selected for individual conditions ? (*Madras University, 1975*)
   (*b*) Sketch an ogee profile and mark in it the different zones. (*Madras University, 1976*)
4. (*a*) What are spillways and what is their necessity ?
   (*b*) Enumerate the different type of spillways which are used in dam construction.
   (*c*) Discuss briefly the design principles that are involved in the design of an ogee spillway ; and a chute spillway.
5. (*a*) What is an ogee spillway and how is it designed ?
   (*b*) Discuss the necessity of the aeration arrangements that are required in gated ogee spillways.
6. Write down an equation for calculating the discharging capacity of an ogee spillway. How does 'coefficient of discharge' in this equation varies with the :
   (*i*) Depth of approach ;
   (*ii*) Slope of upstream face of the ogee spillway ;
   (*iii*) Submergence of the spillway by the tail water ;
   (*iv*) The ratio of design head to actual head over the spillway.
7. (*a*) How would you compute the discharge passing over an ogee spillway ? Discuss the various factors affecting the coefficient of discharge in the discharge equation.
   (*b*) Compute the discharge over an ogee spillway with coefficient of discharge $c = 2.3$ at a head of 3.8 m. The effective length of the spillway is 110 m. Neglect the velocity of approach. [**Ans.** 1872 cumecs]
8. Design a suitable section for the overflow section of a concrete gravity dam having the d/s face sloping at a slope of 0.7 H : 1 V. The design discharge for the spillway is 6000 cumecs. The height of the spillway above the river bed is 60 m. The effective length of spillway may be taken as 50 m. [**Hint.** Follow example 21.1]
9. (*a*) What is meant by a 'spillway' and what is its necessity in dam construction ?
   (*b*) Enumerate the different types of spillways, and draw neat sketches for all the types, showing the different parts of each.
10. (*a*) What is a siphon spillway ? Enumerate the two types of saddle siphon spillways, and describe with a neat sketch the component parts and functioning of one of these two types :
    (*b*) A siphon spillway has the following cross-section at its throat :
    Height of the throat = 1.5 m
    Width of the throat = 4 m.

At the design flow, the tail water elevation in 7 m below the summit of the siphon, and the head water elevation is 2 m above the summit, (*i*) Taking a coefficient of discharge as 0.6 ; determine the capacity of the syphon (*ii*) Determine the head that would be required on an ogee spillway 3.8 m long to discharge this flow, if coefficient of discharge is 2.25 ; (*iii*) What length of this ogee weir would be required to discharge the same flow with a head of 2.2 m on the crest.

[**Solution.** (*a*) Discharge through siphon spillway (*i.e.* capacity)

$$= 0.6\,(1.5 \times 4)\,\sqrt{2 \times 9.81 \times (7 + 2)} = 47.8 \text{ cumecs. } \textbf{Ans.}$$

(*b*) $47.8 = 2.25 \times 3.8\, H^{3/2}$ or $H = 3.15$ m. **Ans.**

(*c*) $4.78 = 2.25\, L\, (2.2)^{3/2}$ or $L = 6.5$ m. **Ans.**]

**11.** (*a*) Describe briefly the construction and functioning of a 'Hooded type' of siphon spillway. What is its discharge equation ?

(*b*) What is meant by priming ? Discuss the priming arrangements used in saddle siphon spillways.

**12.** (*a*) What is a 'chute spillway' ? Where is it preferred to ogee and other types of spillways ?

(*b*) Discuss briefly the component parts and their design for a chute spillway.

**13.** Discuss the design of the following components of a chute spillway :

(*i*) Approach channel

(*ii*) Control structure

(*iii*) Discharge carrier

(*iv*) Energy dissipation arrangement at the bottom.

**14.** Write short notes on :

(*i*) Volute siphon

(*ii*) Side channel spillway or Chute spillway or Shaft spillway

(*iii*) Ogee spillway

(*iv*) Priming devices for siphon spillways.

**15.** What is meant by an 'energy dissipator' ? Discuss the various methods used for energy dissipation below spillways.

**16.** Discuss briefly the various types of energy dissipators that are used for energy dissipation below overflow spillways, under different relative positions of T.W.C. and J.H.C.

**17.** (*a*) What is a 'spillway gate' and what are the merits and demerits of installing such gates ?

(*b*) Enumerate the important types of spillway gates. Describe with a neat sketch the construction and working of 'Tainter gate'.

**18.** (*a*) Discuss the merits and demerits of different types of spillway gates.

(*b*) Discuss with a neat sketch the construction, and functioning of a 'Stoney gate' as well as a 'Drum gate'.

**9.** Write detailed notes on the following :

(*i*) Energy dissipation below spillways

*or*

Spillway buckets and stilling basins.

(*ii*) Spillway gates.

# 22

# *Outlet Works Through Dams and River Intakes*

## 22.1. Sluiceways or Dam Outlets

The most of the water which is stored in a reservoir for irrigation, water supply or power generation purposes, is stored below the spillway crest level. The spillway is provided at normal pool level, such that the floods are discharged safely above the spillway. But, in order to draw water from the reservoir as and when needed, for irrigation, water supply, navigation or power purposes, it is absolutely necessary that outlet works are provided either through the body of the dam or adjacent to it through some hill side at one end of the dam. This water may be discharged into the downstream channel below the dam or may be transported at distances where required (to some power house, etc.) through pipes or canals. The opening, *i.e.* a pipe or a tunnel provided for this withdrawal of water is known as a *dam outlet or a sluiceway.*

The outlets of most of the dams consist of one or more sluiceways with their inlets at about minimum pool level. In most of the cases, a number of outlets are generally provided at different levels ; as a single large capacity outlet may be structurally unsatisfactory or difficult to construct. Moreover, by having more number of smaller capacity outlets, greater control on discharge can be obtained, which can be varied as and when desired. Hence, when wider fluctuations in the demand are anticipated, it is always, advisable to go in for more number of small capacity sluiceways, although it may prove to be somewhat costlier as compared to a few large capacity sluiceways.

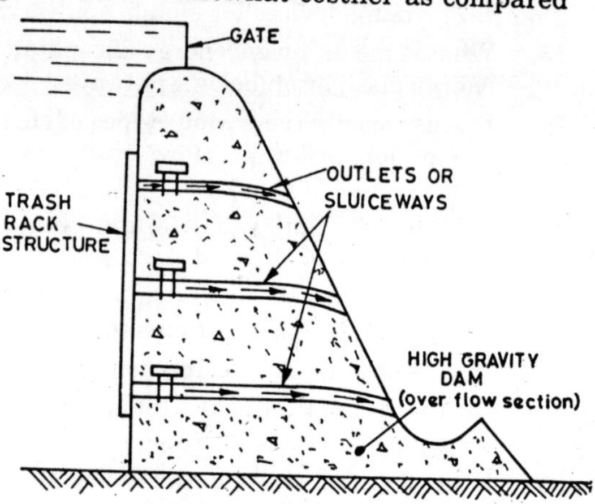


Fig. 22.1. Typical outlet arrangements through a concrete gravity dam.

A sluiceway is a pipe or a tunnel, circular or rectangular in section, that passes through the body of the dam or through some hillside at one end of the dam and discharges into the stream below. For masonry or concrete gravity dams, these sluiceways can easily pass through the body of the dam or the spillway (Fig. 22.1) ; but for earthen dams, it is preferred to place them outside the limits of the embankments. But, if no such adjacent hill site is available and there is no alternative left but to pass the sluiceway through the dam, *projecting collars* must be provided so as to reduce seepage along the outside of the conduit as shown in Fig. 22.2. The seepage is thus, reduced by increasing the length of the seepage path by at least 25 per cent.

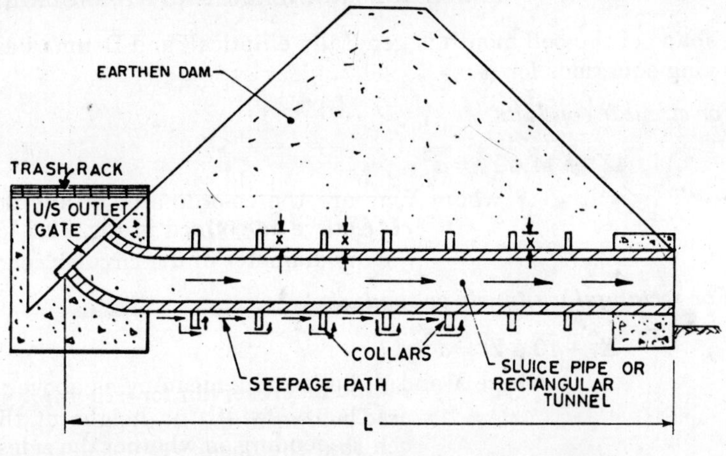


Fig. 22.2. Typical outlet arrangement through an earthen dam.

In Fig. 22.2, if $L$ is the total length of sluiceway from upstream to downstream, the length of the seepage path will be given as :

$$= L + (2x)N$$

where $x$ is the projection of each collar ; and
$N$ is the number of collars.

The increase in seepage path, *i.e.* $2 \cdot N \cdot x$ must generally be greater than $L/4$.

The interior of sluiceway tunnel or pipe, must be smooth, without any projections or cavities. The projections or cavities, etc. if at all present, may cause separation of flow from the boundary of the conduit, causing development of negative pressures and the consequent danger of cavitation*. The entrance of the sluiceway is also very important, since a square edged entrance [Fig. 22.3 (*a*)] is

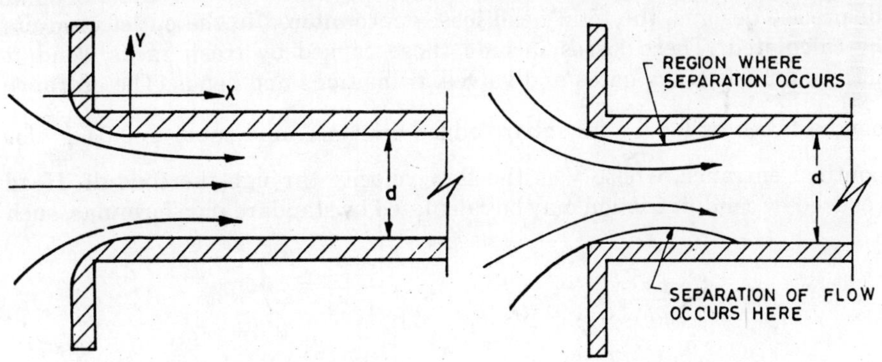


(*a*) Square edged entrance (*b*) Bell-mouthed entrance.

Fig. 22.3

likely to cause more separation of flow and more danger of consequent cavitation as compared to a bell mouthed or some other type of entrance. There have been several such cases where large scale damages have occurred due to cavitation in the vicinity of square-edged entrances. A bell-mouthed entrance [Fig. 22.3 (*b*)] is much better and superior to any other type, and the extra cost involved in shaping this entrance is usually justified except for small projects under low heads.

* Explained earlier in Chapter 21.

The shape of the bell mouth is generally elliptical, and Douma has suggested the following equations for it :

(*i*) *For circular conduits.*

$$4X^2 + 44.4\,Y^2 = d^2 \quad \text{...(22.1)}$$

where $X, Y$ are the co-ordinates of any point on the curve as defined in Fig. 22.2 (*b*) ; and
$d$ is diameter of the circular conduit.

(*ii*) *For rectangular tunnels or conduits*

$$X^2 + 10.4\,Y^2 = d^2 \quad \text{...(22.2)}$$

where $X$ and $Y$ have same meaning as above ; and
$d$ is the width or height of the conduit, depending on whether the sides or top and bottom curves are being designed.

## 22.2. Hydraulics of Outlet Works

The discharge passing through the dam outlet can be calculated easily, be using the equation :

$$Q = C_d \cdot A\,\sqrt{2gH} \quad \text{...[22.3(a)]}$$

where $Q$ = Discharge
$A$ = Area of outlet sluice
$H$ = Differential head causing flow, *i.e.* the difference of u/s and d/s water levels.
$C_d$ = Coefficient of discharge, whose value depends upon various factors such as the type of gate and trash rack provided, conduit friction, transitions etc.

For precise designs, the total head losses encountered in the outlet conduits may be calculated. These losses include those caused by trash racks, conduit entrance, conduit friction, gates and valves, transitions and bends. The entrance loss may be taken as $0.5\,\frac{V^2}{2g}$ for square edged entrance, and equal to $0.04\,\frac{V^2}{2g}$ for bell mouthed entrance, where $V$ is the flow velocity through the conduit. Head losses caused by conduit friction may be calculated by standard pipe formulas, such as :

$$h_L = \frac{f' \cdot LV^2}{2gd}$$

or

$$h_L = \frac{n^2 \cdot V^2 \cdot L}{R^{4/3}}$$

The head loss through the gate depends upon the type of gate and valve provided. A loss of about $0.2\,\frac{V^2}{2g}$ may be taken for fully open gate and butterfly valves. This head loss is taken as nil for ring follower gates. The head loss through the trash rack depends upon the velocity, and may be taken as shown in Table 22.1.

**Table 22.1. Head Loss through Trash Racks**

| *Velocity through trash rack in m / sec* | *Head loss in metres* |
|---|---|
| 0.15 | 0.006 |
| 0.30 | 0.03 |
| 0.45 | 0.09 |
| 0.62 | 0.15 |

The net effective head which is responsible for flow, should then be taken as $= H_{eff}$ = Differential head (H) – Head losses. The discharge can then be calculated easily, using

$$Q = A\sqrt{2g \cdot H_{eff}} \qquad ...[22.3\ (b)]$$

## 22.3. River Intakes

When water is withdrawn through a conduit from a river or a reservoir independently, and as such the entrance of the conduit is not an integral part of the dam or any other related structure, then an intake structure must be constructed at the entrance of the conduit. An intake structure may vary from a simple concrete block supporting the end of the conduit pipe to huge concrete towers, depending upon various factors such as, reservoir characteristics, capacity and discharge requirements, climatic conditions, etc. The basic function of an intake structure is to help in safely withdrawing water from the reservoir over a predetermined range of pool levels, and thus to protect the conduit from being damaged or clogged by ice, trash, debris, waves, etc.

Various types of intake structures are discussed below :

(1) **Simple Submerged Intakes.** A simple submerged intake consists of a simple concrete block, or a rock filled timber crib supporting the starting end of the withdrawal pipe, as shown in Figs. 22.4 (*a*) and (*b*).

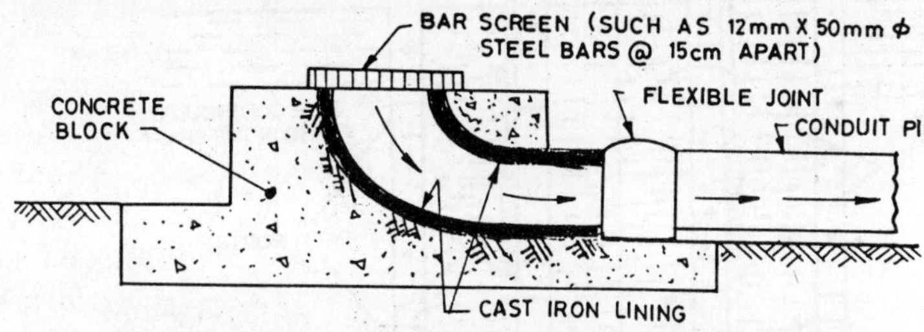


Fig. 22.4. (*a*) Simple concrete block submerged intake.

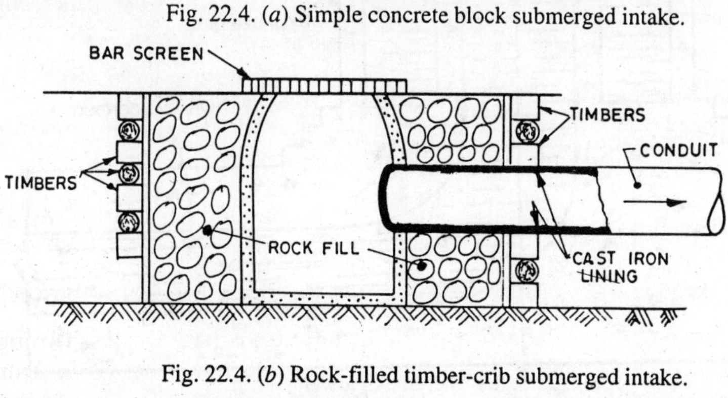


Fig. 22.4. (*b*) Rock-filled timber-crib submerged intake.

Such intake structures should be placed in the river or the reservoir at a place where they may not get buried under the sediment. These submerged intakes are very economical and donot obstruct navigation, and are, therefore, widely used on small works, and are particularly suitable as water-supply intakes from small rivers*. They are sometimes used as intakes to sluiceways of earthen dams with hydraulically operated gates for flow regulation. These intakes are not used on bigger projects as their main disadvantage is the fact that they are not easily accessible for repairing of their gates, etc.

(2) **Intake Towers.** Intake towers are generally used on large projects and where there are large fluctuations of water level. Openings at various levels, called *ports* are generally provided in these concrete towers, which may help in regulating the flow through the towers and permit some selection of the quality of water to be withdrawn. If the ports are submerged at all levels, then there is no problem of any clogging or damage by ice or debris, etc. However, the level of the lowest port should be high enough above the reservoir bed, so that the sediment is not drawn into them. There are two major types of intake towers, *viz.*

(*i*) Wet intake towers ; and

(*ii*) Dry intake towers.

(*i*) ***Wet intake towers.*** A typical section of a wet intake tower is shown in Fig. 22.5 (*a*). It consists of a concrete circular shell filled with water upto the reservoir

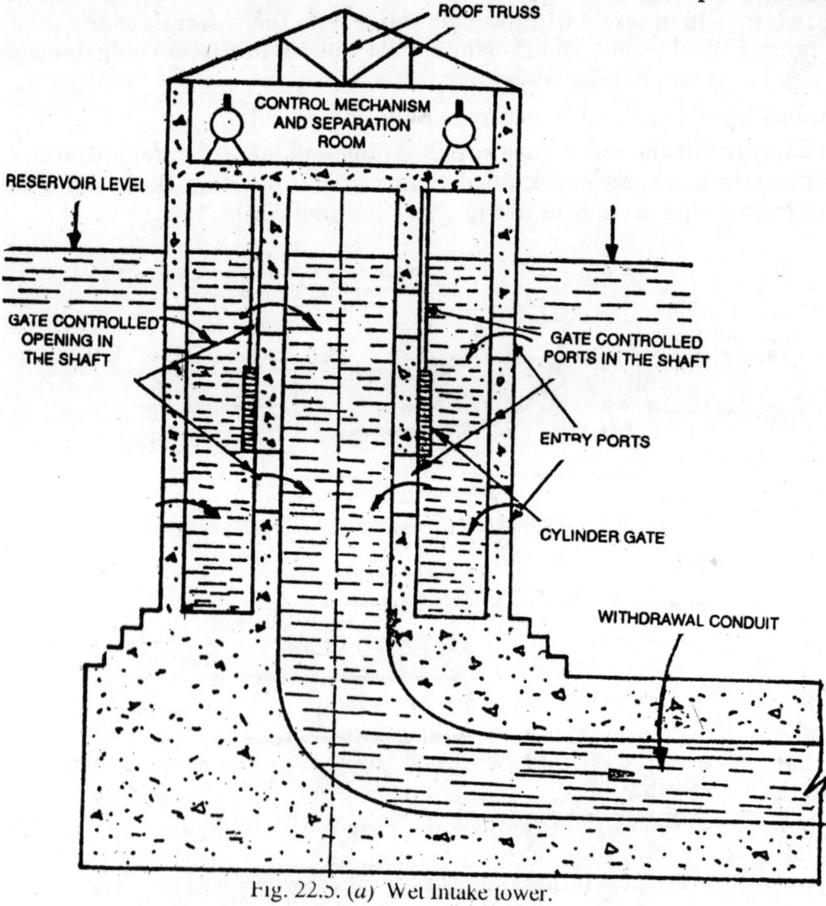


Fig. 22.5. (*a*) Wet Intake tower.

* More detailed description of river and canal intakes is given in "Water Supply Engineering" by the same author.

level, and has a vertical inside shaft which is connected to the withdrawal pipe. Openings are made into the inside shaft also. Gates are usually placed on the shaft, so as to control the flow of water into the shaft and the withdrawal conduit.

(*ii*) ***Dry Intake Towers.*** The essential difference between a wet intake tower and a dry intake tower is that, whereas in a wet intake tower, the water enters from the entry ports into the tower and then it enters into the conduit pipe through separate gate controlled openings ; in a dry intake tower, the water is directly drawn into the withdrawal conduit through the gated entry ports as shown in Fig. 22.5 (*b*). *A dry intake tower will, therefore, have no water inside* the tower if its

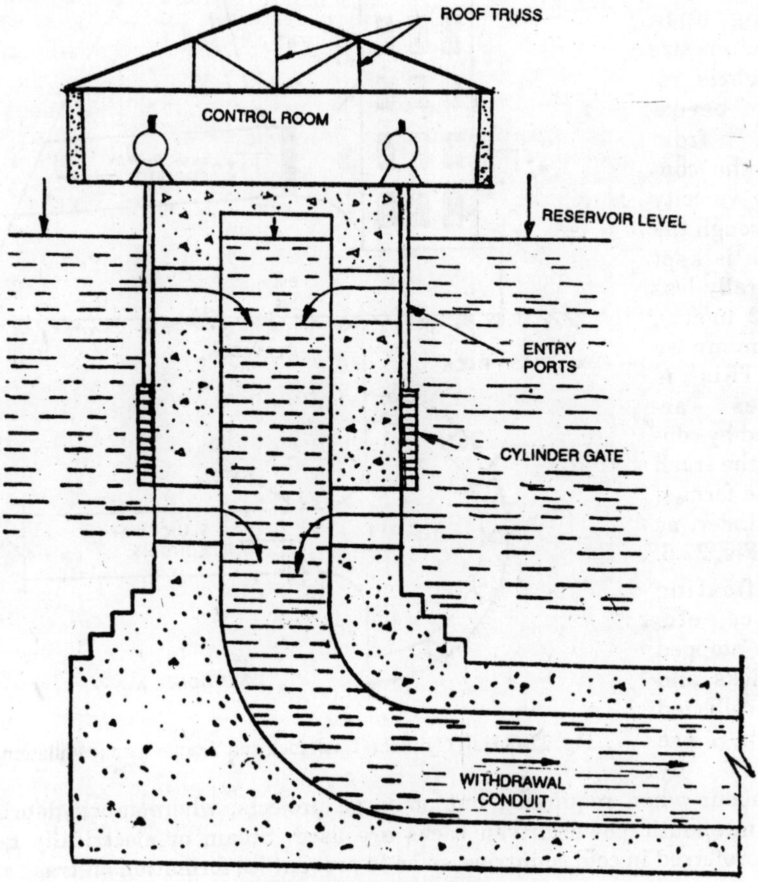


Fig. 22.5 (*b*). Dry Intake Tower.

gates are closed, whereas the wet intake tower will be *full of water even if its gates are closed*. When the entry ports are closed, a dry intake tower will be subjected to additional buoyant forces and hence, must be of heavier construction than the wet intake towers. However, the dry intake towers are useful and beneficial in the sense that water can be withdrawn from any selected level of the reservoir by opening the port at that level.

Intake towers are huge structures standing in the river, and hence should be located so as not to interfere with navigation, and must be properly designed so as to withstand the worst possible combination of various forces, such as hydrostatic pressures, wind and earthquake forces, and forces caused by waves, ice and debris, etc.

### 22.4. Trash Racks

The entrances to intakes and dam outlets are generally covered with trash racks so as to prevent the entry of debris, ice, etc. into the conduit. These racks are generally nothing but bar screens, made from steel bars spaced at 5 to 15 cm centre to centre (in both directions) depending upon the maximum size of the debris required to be excluded from entering the conduit. The velocity of flow through the trash rack is kept low (generally less than 0.62 m/sec), so as to minimise losses. This is sometimes accomplished by constructing the trash rack in the form of a half cylinder, as shown in Fig. 22.6.

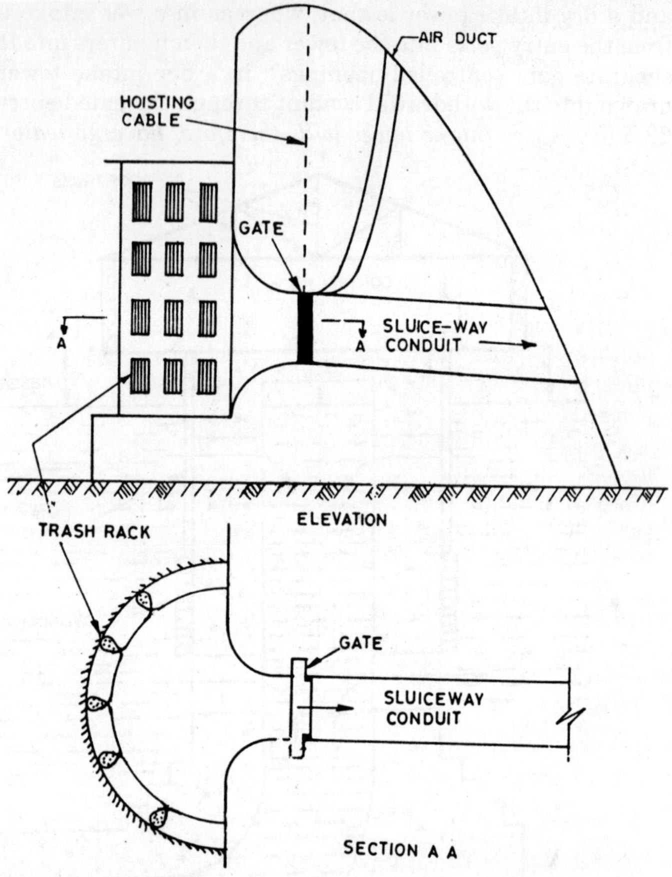


Fig. 22.6. Half Cylinder-Trash rack and Tractor gate installation.

The floating debris, ice, etc. which are stopped by the racks and thus get collected on them, are removed by manual labour when required ; but on large projects, where much debris are expected, automatic power-driven racks are used. Steam or electrically heated racks are preferred in cold countries, so as to prevent ice formation on trash racks.

## PROBLEMS

1. (*a*) What is meant by a 'Dam sluice' ? Why are such sluices necessary in dam construction ?

   (*b*) Show with neat sketches, the provision of sluiceways in

   (*i*) an earth dam.

   (*ii*) a concrete gravity dam.

2. Write short notes on :

   (*i*) Sluiceways in dams.

   (*ii*) River intakes.

   (*iii*) Trash Racks.

# 23

# *Pressure Conduits*

## 23.1. Definition, Etc.

A pressure conduit (such as a penstock) is a pipe which runs under pressure and, therefore, runs full. This type of conduits prove economical than canals or flumes, because they can generally follow shorter routes. Moreover, their biggest advantage is: that the water or any other fluid flowing through them is not exposed anywhere and hence, there are no chances or very less chances of its getting polluted. Hence, these pressure conduits are preferably used for city water supplies. Since the water wasted in percolation, evaporation, etc. is also saved, when water is carried through these conduits, they are preferably used when water is scarce. The flow of water through conduit pipes is generally turbulent, and hence, it will be considered so, while dealing with the hydraulics of flow through such pipes.

## 23.2. Hydraulics of Flow and Discharging Capacities of Pressure Conduits

The pressure conduits are designed in such a way that the available pressure head is just lost in overcoming the frictional resistance offered to the flow of water by the pipe. The velocities generated in these pipes should neither be too small to require a large size diameter pipe, nor should be so large as to cause excessive loss of pressure head.

The head loss by pipe friction may be found by using either of the following formulas :

(*i*) **Darcy-Weisbach Equation.** It states that

$$H_L = \frac{f' \cdot L \cdot V^2}{2gd}$$

where $H_L$ = Head loss in metres.
$L$ = Length of the pipe in metres.
$d$ = Diameter of the pipe in metres.
$V$ = Velocity of flow through the pipe in m/sec.
$f'$ = The friction factor, which depends upon Reynold number $\left(i.e.\ R_e = \frac{Vd}{v}\right)$ and relative roughness of the pipe.

The relative roughness $\left(\delta = \frac{2e}{d}\right)$ of a pipe depends on the absolute roughness (*e*) of the inside surface and the diameter of the pipe *d*. The approximate values of $f'$ are given by the following empirical relations :

$$f' = 0.04\left[1 + \frac{1}{35\, d}\right] \text{ for old pipes} \qquad ...(23.2)$$

and

$$f' = 0.02\left[1 + \frac{1}{35\, d}\right] \text{ for new pipes} \qquad ...(23.3)$$

The accurate value of $f'$ depends on $R_e$ and $\delta$, and may be given by formula, such as

$$f' = a + \frac{b}{R_e^m} \quad \text{...(23.4)}$$

where $a$, $b$ and $m$ are constants depending upon $\delta$.

For $R_e$ varying between 20,000 to 20,00,000, Schiller has given

$$f' = 0.005 + \frac{0.396}{R_e^{0.3}} \quad \text{...(23.5)}$$

and Nikuradse has given for values of $R_e$ varying between 20,000 to 32,40,000 as

$$f' = 0.0032 + \frac{0.221}{R_e^{0.237}} \quad \text{...(23.6)}$$

*(ii)* **Manning's formula.** Manning's formula is also applicable to turbulent flow in pressure conduits and yields good results, provided the roughness coefficient $n$ is accurately estimated. Head loss according to Manning's equation is given as :

$$H_L = \frac{n^2 \cdot V^2 \cdot L}{R^{4/3}} \quad \text{...(23.7)}$$

where $n$ = Manning's rugosity coefficient.
$L$ = Length of pipe in metres.
$V$ = Flow velocity through pipe in m/sec.
$R$ = Hydraulic mean depth of pipe

$$\left[ i.e., \; R = \frac{A}{P} = \frac{\frac{\pi d^2}{4}}{\pi d} = \frac{d}{4} \right] \text{ in metres}$$

*(iii)* **Hazen-William's formula.** This formula is widely used for pipe flows and states

$$V = 0.85 \, C_H \cdot R^{0.63} \, S^{0.54} \quad \text{...(23.8)}$$

where $C_H$ = Coefficient given by Table 23.1.
$R$ = Hydraulic mean depth of pipe, = $(d/4)$ in metres.
$S$ = Slope of the energy line.
$V$ = Flow velocity through the pipe in m/sec.

**Table 23.1. Values of $C_H$ for Hazen William's Formula**

| *Pipe material* | *Value of $C_H$* |
|---|---|
| Concrete (regardless of age) | 130 |
| Cast Iron, | |
| *New* | 130 |
| *5 years old* | 120 |
| *20 years old* | 100 |
| Welded steel (New) | 120 |
| Riveted steel (New) | 110 |
| Vitrified clay | 110 |
| Brick Sewers | 100 |
| Asbestos cement | 140 |

The carrying capacities of pipes decrease with time, because with the passage of time, their interior surfaces go on becoming more and more rough. Hence, it is customary to select higher values of $e$ and $n$ and lower values of $C_H$ for old pipes than are generally taken for new pipes.

## 23.3. Forces Acting on Pressure Conduits

Pressure pipes must be designed to withstand the following forces :

(1) *Internal pressure of water*

(2) *Water hammer pressures.*

(3) *Pressure due to external loads* (when buried under the ground).

(4) *Temperatures stresses* (when laid above the ground).

(5) *Longitudinal stresses* due to flow around bends or change in cross-section.

(6) *Flexural stresses* (when laid over supports at intervals or on bridges).

These forces are discussed below :

(1) **Internal Pressure of Water.** The pressure exerted on the walls of the pipe by the flowing water, in the form of Hoope's tension, is the internal pressure. The circumferential tensile stress produced is given as :

$$\sigma_1 = \frac{p_1 d}{2t} \text{ in kN/m}^2 \qquad \text{...(23.9)}$$

where $p_1$ = Internal static pressure in kN/m$^2$.
$d$ = Diameter of the pipe in metres.
$t$ = Thickness of the pipe shell in metres.
$\sigma_1$ = Circumferential tensile stress to be counteracted by providing Hoope's reinforcement.

(2) **Water Hammer Pressure.** When a liquid flowing in a pipe line is abruptly stopped by the closing of a valve, the velocity of the water column behind, is retarded, and its momentum is destroyed. This exerts a thrust on the valve and additional pressure on the pipe shell behind. The more rapid the closure of the valve, the more rapid is the change in momentum, and hence, greater is the additional pressure developed. The pressures so developed are known as *water-hammer pressures* and may be so high as to cause bursting of the pipe shell (due to increased circumferential tension) if not accounted for in the designs.

The maximum pressure developed in pipe lines due to water hammer is given by the formula

$$p_2 = \frac{14.762 \cdot V}{\sqrt{1 + \dfrac{K \cdot d}{t}}} \qquad \text{...(23.10)}$$

where $V$ = Velocity of water just before the closing of the valve in m/sec.
$d$ = Diameter of pipe in metres.
$t$ = Thickness of pipe shell in metres.
$K$ = Constant

$$= \frac{\text{Modulus of elasticity of pipe material}}{\text{Bulk modulus of elasticity of water}}$$

The value of $K$ for steel comes out to be 0.01, for cast iron = 0.02, and for cement concrete = 0.1.

The water hammer pressure can be calculated by using the above equation. But for design purposes, its value is generally taken as 840 $kN/m^2$ for small size pipes of 7.5 to 25 cm diameter, and is gradually reduced to 490 $kN/m^2$ for pipes above 1.2 m diameter, as given in Table 23.2.

**Table 23.2. Water Hammer Pressures**

| *Pipe diameter in cm* | *Water hammer Pressure in $kN/m^2$* |
|---|---|
| 7.5 to 25 | 840 |
| 30 to 40 | 770 |
| 50 | 630 |
| 60 | 600 |
| 75 | 560 |
| 90 | 490 |
| 105 to 135 | 490 |

The circumferential tensile stress caused by water hammer pressure may then be calculated easily by using

$$\sigma_2 = \frac{p_2 \cdot d}{2t}$$

(3) **Stress due to External Loads.** When large pipes are buried deep under the ground, the weight of the earth-fill may produce large stresses in the pipe material. The stress due to the external earth fill load is given by

$$f = 22.7 \frac{h \cdot d^2}{t} \quad ...(23.11)$$

where $h$ = depth of the earth-fill above the crown in metres.
$d$ = diameter of pipe in metres.
$f$ = stress produced in $kN/m^2$.

**Note.** In the above formula, it is assumed that the earth to the sides does not give any lateral support and weighs about 18.4 $kN/m^3$.

(4) **Temperatures Stresses.** When pipes are laid above the ground, they are exposed to the atmosphere and are, therefore, subjected to temperature changes. They expand during day time and contract at night. If this expansion or contraction is prevented due to fixation or friction over the supports, longitudinal stresses are produced in the pipe material. The amount of these stresses may be calculated by the formula :

$$f = E \cdot \alpha \cdot T \quad ...(23.12)$$

where $E$ = Modulus of elasticity of the pipe material.
$\alpha$ = Co-efficient of expansion of the pipe material.
$T$ = Change in temperature in °C.

Expansion joints at suitable intervals (say 20 m to 30 m or so) must be provided to counteract these stresses.

(5) **Stresses due to Flow Around Bends and Change in Cross-Section.** Whenever the velocity of a flow (either magnitude or direction) changes, there is a change in the momentum, and therefore, by Newton's Second Law, a force is exerted, which is proportional to the rate of change of momentum. The force required to bring this change in momentum comes from the pressure variation within the fluid and from forces transmitted to the fluid from the pipe walls.

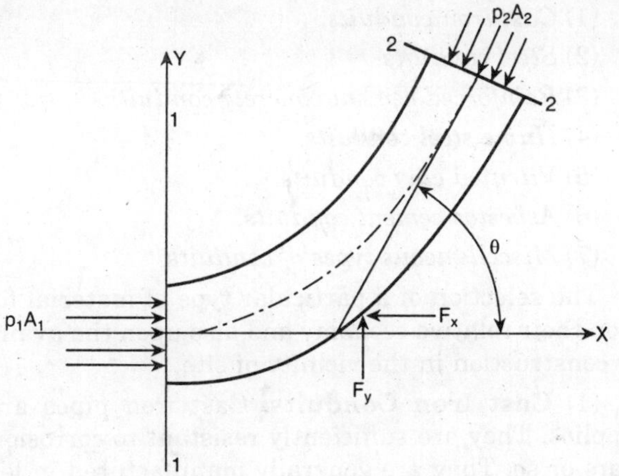

Fig. 23.1. Forces at a pipe bend.

The free-body diagram of various forces acting on the water contained in a horizontal pipe bend is shown in Fig. 23.1. Applying momentum equation, and resolving the forces in $x$ and $y$ direction, we get

$$p_1A_1 - F_x - p_2A_2 \cos\theta = \frac{\gamma_w Q}{g}\left[V_2 \cos\theta - V_1\right] \quad ...(23.13)$$

and

$$F_y - p_2A_2 \sin\theta = \frac{\gamma_w Q}{g}\left[V_2 \sin\theta\right] \quad ...(23.14)$$

where $p_1$ and $p_2$ are the pressures, $V_1$ and $V_2$ are the velocities at sections 1-1 and 2-2 respectively. The forces $F_x$ and $F_y$ are the forces which are transmitted from the pipe to the water. An equal and opposite force must, therefore, be developed in the form of stresses in the pipe wall. Similar forces will be developed when the cross-sectional area of the pipe is suddenly changed, as shown in Fig. 23.2.

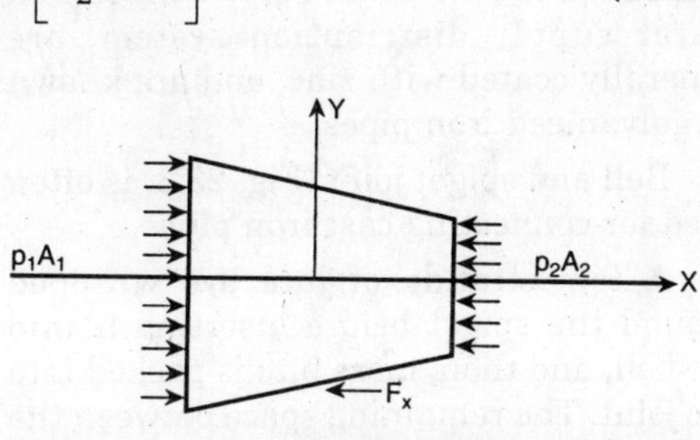

Fig. 23.2. Forces at a change in X-section of a pipe.

Due to these impressed external forces or stresses, the pipe line may be thrown out of alignment as and when such situations arise, unless held firmly by anchoring it in massive blocks of concrete or stone masonry.

(6) **Flexural Stresses.** Many a times, steel pipes are laid over concrete supports, built above the ground ; and sometimes the rain water, etc. may wash off the ground from below the pipes at intervals. Under all such circumstances, bending stresses get produced in the pipe, since the pipe then acts like a beam with loads resulting from the weight of the pipe, weight of water in the pipe and any other superimposed loads. The stresses caused by this beam action may be determined by usual methods of analysis applied to the beams. However, these stresses are generally negligible except for long spans or where there are huge superimposed loads.

## 23.4. Various Types of Pressure Conduits

Depending upon the construction material, the various types of pressure conduits are :

(1) *Cast iron conduits.*

(2) *Steel conduits.*

(3) *Reinforced cement concrete conduits.*

(4) *Hume steel conduits.*

(5) *Vitrified clay conduits.*

(6) *Asbestos cement conduits.*

(7) *Miscellaneous types of conduits.*

The selection of a particular type of material for the conduit depends mainly upon their relative economy and also upon the availability of material, labour, etc. for construction in the vicinity of site.

(1) **Cast Iron Conduits.** Cast iron pipes are widely used for city water supplies. They are sufficiently resistant to corrosion and may last as long as 100 years or so. They are generally manufactured in lengths of about 3.6 metres but may be manufactured in lengths up to 6 m or so on special orders. Cast iron pipes in different thicknesses, are generally available for withstanding different pressures upto a maximum of 2400 $kN/m^2$ (24.5 $kg/cm^2$).

In order to protect the pipes from corrosion, almost all the metal pipes are given some kind of protective coatings. Larger pipes are generally protected by bituminous linings or by cement mortar (1 : 2) linings. Smaller size pipes, such as those required for small service lanes in a water supply distribution system, are generally coated with zinc, and are known as galvanised iron pipes.

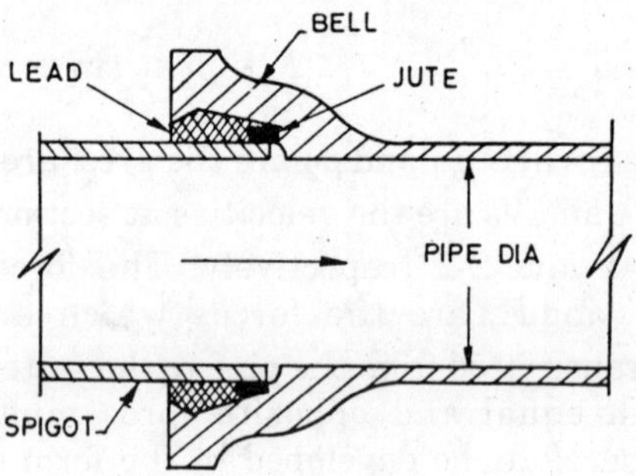

Fig. 23.3. Bell and spigot joint for cast iron pipes.

Bell and spigot joint (Fig. 23.3) is often used for connecting cast iron pipes.

A few strands of jute are wrapped around the spigot before inserting it into the bell, and then, more jute is packed into the joint. The remaining space between the bell and the spigot is finally filled with molten lead, which gets solidified and tightly chalked into the joint after cooling, and thus making a water-tight joint. The quantity of lead required varies from 3.5 to 4 kg for 15 cm dia pipe to about 45 to 50 kg for 1.2 metres dia pipe.

(2) **Steel conduits.** Steel pipes can be used for large diameters, say upto or above 6 metres or so. Steel plates of varying thicknesses for withstanding different pressures, are generally bent and welded or sometimes riveted. Welded pipes are smoother and stronger than riveted pipes, and are, therefore, generally used these days. Since steel is strong in tension, even large size diameter pipes can be made of thin shells. They are, therefore, lighter than cast iron pipes.

Galvanised steel pipes with circumferential corrugations are much more stronger than ordinary steel pipes. They are, therefore, used where large dia pipes of smaller wall thicknesses are required. Such pipes are usually manufactured in various sizes varying from 20 cm to 2 metres in diameter, and are lighter and can be more easily transported at distances.

Expansion joints are generally not required in steel pipes which are buried under the ground, because they are not subjected to large temperature changes. However, pipes which are exposed to the atmosphere may require expansion joints so as to minimise temperature changes.

Steel pipes get rusted quickly, which, reduces their life as well as carrying capacities. They are, therefore, protected on the inside as well as outside by protective coatings. Under ordinary conditions, steel pipes may last as long as 40 years or so. The various pipe lengths in case of steel pipes are connected by welded or riveted joints.

(3) **Reinforced Cement Concrete Conduits.** Ordinary cement concrete pipes are manufactured in small sizes (*i.e.* up to say 0.6 m diameter) ; while they are reinforced with steel for large dia pipes. They are easily available in sizes up to diameters say about 1.8 metres and may be got manufactured for larger diameters say up to about 4.5 metres, on special orders. These pipes may either be prepared at site by transporting various ingradients (*i.e.* cement, steel aggregates, water, etc.) or can be manufactured in factories and then transported to site. They are known as '*cast in situ*' pipes in the former case, and '*precast pipes*' in the latter case. Cast in situ pipes are useful when the site conditions are difficult, and where it may be difficult to carry the pipes. But since such pipes are cast at site, lesser supervision and check is possible, as compared to the case of precast pipes which are cast in the factories, and thus subjected to greater quality control and supervision.

R.C.C. pipes can be manufactured in three different ways, *viz.*

(*a*) pipes having bar and mesh reinforcement, and concrete poured by usual ordinary methods of concrete pouring, and tamped ;

(*b*) pipes made by rotating the mould or the form rapidly about the pipe axis. The mould contains concrete and fabricated reinforcement. The centrifugal force throws off the concrete which then spreads in a uniform layer over the internal surface of the mould and embed the reinforcement ; thus providing a high density watertight concrete surface. This type is known as *centrifugal type* ;

(*c*) the third type of pipes are made by lining thin cylindrical steel shells both internally and externally with rich cement concrete. These are stronger and more watertight than the first two. They are known as *cylinder type*.

R.C.C. pipes are generally made from 1 : 2 : 4 cement concrete with maximum size of aggregates as 6 mm. They are provided with circumferential reinforcement to carry hoop tension and a nominal longitudinal reinforncement equal to 0.25% of cross-sectional area of concrete. The thickness of R.C.C. pipes generally varies from 7.5 cm to 0.6 m for pipes of diameters* varying from 0.2 m to 2.75 m.

(4) **Hume Steel Conduits.** Hume steel pipes are R.C.C. pipes patented under this name. They consist of thin steel shells coated from inside with cement mortar by centrifugal process. The thickness of the inside coating varies from 1.2 cm to 3 cm depending upon the size of the pipe. They are also coated from outside so as to protect the steel shell from external weather or soil action. The thickness of external coating is 2.5 cm for pipes upto 1 metre in diameter and is 3.75 cm for pipes of larger diameters. The thickness of steel shell depends upon the size of the

* The standard mentioned diameter of pipes is always the internal diameter.

pipe and also upon the pressure to be borne by the pipe. Like all R.C.C. pipes, they are heavy and difficult to handle.

(5) **Vitrified Clay Conduits.** They are generally not used as pressure conduits, but are extensively used for carrying sewage and drainage at partial depths. These pipes are free from corrosion and provide a smooth hydraulically efficient surface. They are not used as pressure pipes, because clay is weak in tension, and formation of water-tight joints becomes difficult in them. Clay pipes are commonly made in lengths of about 0.6 m to 1.2 m or so.

(6) **Asbestos Conduits.** Asbestos, silica and cement are converted under pressure to a dense, homogeneous material possessing high strength, called *asbestos cement.* This material is used for casting these pipes. The asbestos fibre which is thoroughly mixed with cement serves as reinforcement. These pipes are generally available in different sizes, say from 10 cm to 90 cm in diameters and in about 4 metres lengths. These pipes are generally made in 4 different grades to withstand pressures of 350 $kN/m^2$ to 140 $kN/m^2$. These pipes are joined by means of special type of flexible joint called '*simplex joint*'. The assembly consists of a pipe sleeve and two rubber rings which are compressed between the pipe and the interior of the sleeve.

*Advantages.* The advantages of asbestos cement pipes are :

(*i*) They are light and hence easy to transport.
(*ii*) They can be easily assembled without skilled labour.
(*iii*) They are highly resistant to corrosion.
(*iv*) They are highly flexible and may permit as much as 12° deflection in laying them around curves.
(*v*) Expansion joints are not required as the coefficient of expansion is low and the joints are also flexible.
(*vi*) They are very smooth and thus provide a hydraulically efficient pipe. Their carrying capacities do not reduce with time.
(*vii*) They are very suitable to be used as small size distribution pipes.

*Disadvantages and limitations.* Their disadvantages are :

(*i*) They are costly.
(*ii*) These pipes do not have much strength and are brittle and soft. They are liable to get damaged by excavating tools or during transportation transits.
(*iii*) The rubber joint seals may deteriorate if exposed to gasoline or other petroleum products, and hence cannot be used for transporting petroleum products.

(7) **Miscellaneous Types of Conduits.** Various other materials which may be used for manufacturing pipe conduits are : copper, wrought iron, plastics, etc. *Copper pipes* are very costly although they are highly resistant to acidic as well as alkaline waters, they can be bent easily and do not sag due to heat. They are, therefore, very useful for carrying hot water in the interior of the buildings. Wrought iron pipes are lighter than cast iron pipes and can be more easily cut, threaded and worked. They are more costly but neat in appearance. They are generally manufactured in small sizes and are very useful for indoor works. However, they corrode quickly and are less durable. they are, therefore, generally protected by galvanising them with zinc coatings and they are then known as galvanised iron pipes.

*Plastic pipes* are lighter and free from corrosion. But they are of low strengths and less durable. Moreover, they can not withstand high temperatures exceeding 60°C or so. They may, however, be used for very minor works in house connections, etc. HDPE pipes are however, being increasingly used these days for various uses, such as in tubewell boarings, sewage mains, etc. etc. They are however, still not being made in large sizes suitable as penstock pipes.

## PROBLEMS

**1.** (*a*) What is meant by a pressure conduit ? Discuss briefly the forces acting on a pressure conduit.

(*b*) Write a brief note on the design capacities of pressure pipes.

**2.** What are the different construction materials which are used for manufacturing pressure pipes ? Discuss their comparative merits and demerits.

**3.** Write a note on Pressure conduits'.

# 24

# *Hydro-electric Power*

Faraday had shown that when a coil is rotated in a magnetic field, electricity is generated. Thus, in order to produce electrical energy, it is necessary that we should produce mechanical energy, which can be used to rotate the 'coil'. The mechanical energy is produced by running a prime mover (*known as turbine*) by the energy of fuels or flowing water. This mechanical power is converted into electrical power by electric generator which is directly coupled to the shaft of the turbine, and is thus run by the turbine. The electrical power, which is consequently obtained at the terminals of the generator, is then transmitted to the area where it is to be used for doing work.

The plant or machinery which is required to produce electricity (*i.e.* Prime mover + Electric generator) is collectively known as the *Power plant*. The building, in which the entire machinery along with other auxiliary units is installed, is known as the *Power house*.

### 24.1. Thermal and Hydropower

As stated earlier, the turbine blades can be made to run by the energy of fuels or flowing water. When fuel is used to produce steam for running the steam turbine, then the power generated is known as *Thermal power*. The fuel which is to be used for generating steam may either be (*i*) an ordinary fuel such as coal, fuel oil, gas, etc., or (*ii*) atomic fuel or nuclear fuel. Coal is simply burnt to produce steam from water and is the simplest and the oldest type of fuel. Diesel oil, etc. may also be used as fuels for producing steam. Atomic fuels such as uranium or thorium may also be used to produce steam. When conventional type of fuels such as coal, oil, etc. (called fossils) are used to produce steam for running the turbines, the power house is generally called an *Ordinary thermal power station* or *Thermal power station*. But when atomic fuel is used to produce steam, the power station, which is essentially a thermal power station, is called an *Atomic power station* or a *Nuclear power station*. In an ordinary thermal power station, steam is produced in a water boiler, while in the atomic power station, the boiler is replaced by a nuclear reactor and steam generator for raising steam. The electric power generated in both these cases is known as *Thermal power* and the scheme is called *Thermal power scheme*.

But, when the energy of the flowing water is used to run the turbines, then the electricity generated is called the *Hydroelectric power*. This scheme is known as *Hydel scheme*, and the power house is known as *Hydel power station* or *Hydroelectric power station*. In a hydel scheme, a certain quantity of water at a certain potential head, is essentially made to flow through the turbines. The head causing flow, runs the turbine blades, and thus producing electricity from the generator coupled to the turbine. In this chapter, we are concerned with hydel schemes only.

## 24.2. Classification of Hydel Plants

**24.2.1. Classification of Hydroplants on the Basis of Hydraulic Characteristics.** On the basis of this classification, the hydro plants may be divided into the following types:

(*i*) *Run-off river plants* ; (*ii*) *Storage plants*
(*iii*) *Pumped storage plants* ; (*iv*) *Tidal plants*

They are described below :

(*i*) **Run-off River Plants.** These plants are those which utilise the minimum flow in a river having no appreciable pondage on its upstream side. A weir or a barrage is sometimes constructed across a river simply to raise and maintain the water level at a pre-determined level within narrow limits of fluctuations, either solely for the power plant or for some other purpose where the power plant may be incidental. Such a scheme is essentially a low head scheme and may be suitable only on a perennial river having sufficient dry wheather flow of such a magnitude as to make the development worth while. [Pl. refer Fig. 24.2]

Run-off river plants generally have a very limited storage capacity, to supplement the normal flow. Such a small storage capacity, called **pondage,** is provided for meeting the hour to hour fluctuations of load or of streamflow over a day ; or occassionally, day to day fluctuations over a weekly cycle. When the available discharge at site is more than the demand (during off-peak hours), the excess water is temporarily stored in the pond on the upstream side of the barrage, which is then utilised during the peak hours.

The power stations constructed on diversion canals (Irrigation & Power Canals) called *Diversion canal plants*, can also be placed in this category.

The various examples of run-off river plants are : Ganguwal and Kotla power houses located on Nangal Hydel Channel, Mohammad Pur and Pathri power houses on Ganga Canal, and Sarda power house on Sarda Canal.

(*ii*) **Storage Plants.** A storage plant is essentially having an upstream storage reservoir of sufficient size, so as to permit sufficient carry-over storage from the monsoon season to the dry summer season, and thus to develop a firm flow substantially more than the minimum natural flow. In this scheme, a dam is constructed across the river, and the power house may be located at the foot of the dam such as in Bhakra, Hirakund, Rihand projects, etc. The power house may sometimes be located much away from the dam (on the down-stream side). In such a case, the power house is located at the end of tunnels which carry water from the reservoir. The tunnels are connected to the power house machines by means of pressure penstocks which may either be underground (as in Maithon and Koyna projects) or may be kept exposed (as in Kundah project).

When the power house is located near the dam, as is generally done in the low head installations ; it is known as *Concentrated fall hydroelectric development*. But when the water is carried to the power house at a considerable distance from the dam through a canal, tunnel, or penstock ; it is known as a *Divided fall development*.

(*iii*) **Pumped Storage Plants.** A pumped storage plant generates power during peak hours, but during the off-peak hours, water is pumped back from the tail water pool to the head water pool for future use. The pumps are run by some secondary power from some other plant in the system. The plant is thus primarily meant for assisting an existing thermal plant or some other hydel plant. A typical section for pumped storage plant is shown in Fig. 24.1.

During peak hours, the water flows from the reservoir to the turbine and electricity is generated. During off-peak hours, the excess power available from some other plant, is utilised for pumping water back from the tail pool to the head pool. This minor plant thus supplements the power of another major plant. In such a scheme, the same water is utilised again and again and no water is wasted.

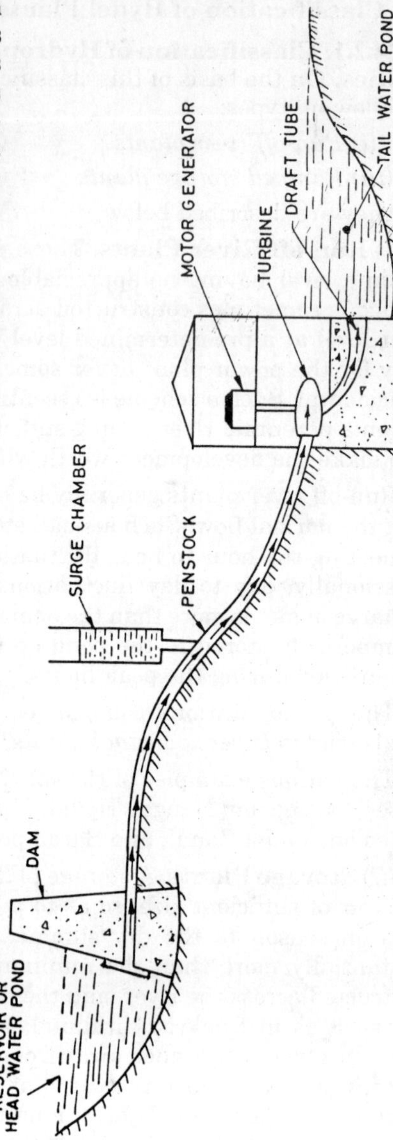


**Fig. 24.1.** Typical section *through a pumped storage plant.*

For heads varying between 15 to 90 m, reversible pump turbines have been devised, which can function both as turbine as well as a pump. Such reversible turbines can work at relatively high efficiencies and can help in reducing the cost of such a plant. Similarly, the same electrical machine can be used both as a generator as well as a motor by reversing the poles. The provision of such a scheme helps considerably in improving the load factor of the power system.

(*iv*) **Tidal Plants.** Tidal plants for generation of electric power are the recent and modern advancements, and essentially work on the principle that there is a rise in sea water during high tide period and a fall *during the low ebb period*. The water rises and falls twice a day ; each fall cycle occupying about 12 hours and 25 minutes. The advantage of this rise and fall of water is taken in a tidal plant. In other words, the tidal range, *i.e.* the difference between high and low tide levels is utilised to generate power. This is accomplished by constructing a basin separated from the ocean by a partition wall and installing turbines in openings through this wall.

Water passes from the ocean to the basin during high tides, and thus running the turbines and generating electric power. During low tide, the water from the basin runs back to ocean, which can also be utilised to generate electric power, provided, special turbines which can generate power for either direction of flow are installed. Such plants are useful at places where tidal range is high. *Rance power station* in France is an example of this type of power station. The tidal range at this place is of the order of 11 metres. This power house contains 9 units of 38,000 k. watts.

**24.2.2. Classification of Hydroplants on the Basis of Operating Head on Turbines.** On this basis, the plants may be divided into the following types :

(*i*) *Low head scheme* (head < 15 m).

(*ii*) *Medium head scheme* (head varies between 15 to 60 m).

(*iii*) *High head scheme* (head > 60 m).

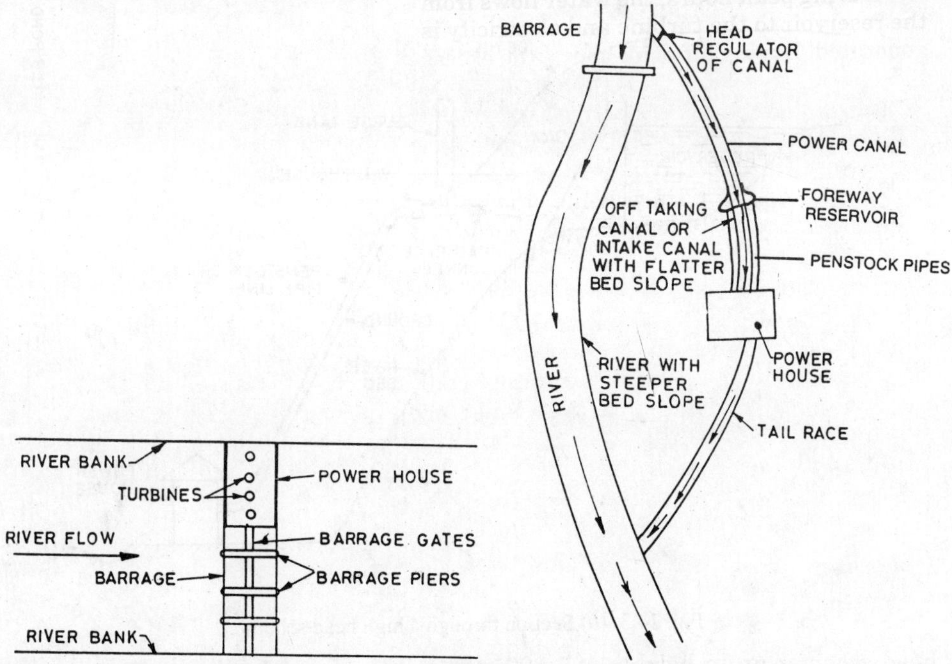


(*a*) Run of river plant (*b*) Diversion canal plant

Fig. 24.2. Low head schemes (in Plan)

In this scheme, a weir or a barrage is constructed to raise the water level, and the power house is constructed either in continuation with the barrage [Fig. 24.2 (*a*)] or at some distance downstream of the barrage, where water is taken to the power house through an intake canal, called the ***power canal*** or the *diversion canal* [Fig. 24.2 (*b*)].

(*ii*) **Medium Head Scheme.** A medium head scheme is one which uses water head varying between 15 to 60 metres or so. This scheme is thus essentially a dam reservoir scheme, although the dam height is mediocre. This scheme is having features somewhere between low head scheme and high head scheme.

(*iii*) **High Head Scheme.** A high head scheme is the one which uses water head of more than 60 m or so. A dam of sufficient height is, therefore, required to be constructed, so as to store water on the upstream side and to utilise this water throughout the year. High head schemes up to heights of 1,800 metres have been developed. The common examples of such a scheme are : Bhakra dam in (Punjab), Rihand dam in (U.P.), Hoover dam in (U.S.A.),

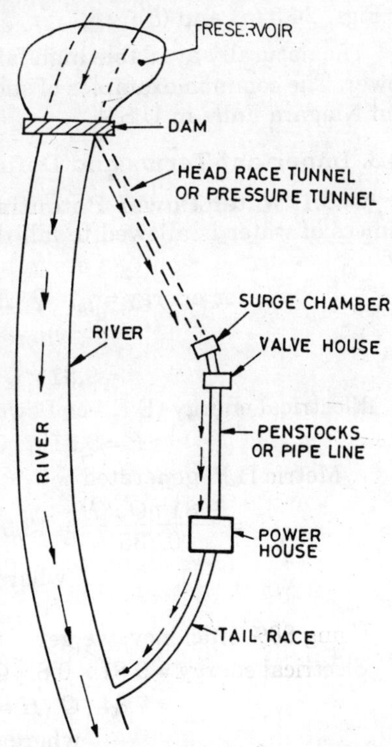


Fig. 24.3. (*a*) Plan of a high head scheme.

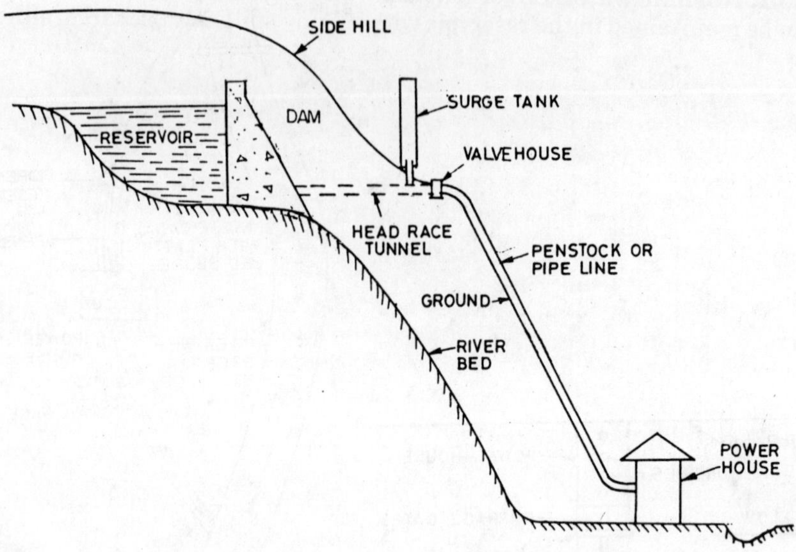


Fig. 24.3. (*b*) Section through a high head scheme.

head schemes up to heights of 1,800 metres have been developed. The common examples of such a scheme are : Bhakra dam in (Punjab), Rihand dam in (U.P.), Hoover dam in (U.S.A.), etc. A typical plan and section of such a scheme are shown in Figs. 24.3 (*a*) and (*b*).

The naturally available high falls can also be developed for generating electric power. The common examples of such power developments are : Jog Falls in India, and Niagara Falls in U.S.A.

## 24.3. Important Terms and Definitions Connected with Hydropower

**24.3.1. Water Power Potential.** The amount of power generated when $Q$ cumecs of water is allowed to fall through a head difference of $H$ metres is given by

$$\text{Water energy} = \gamma_w \cdot Q \cdot H \quad \text{...(24.1)}$$

where $\gamma_w$ is unit wt of water = 9.81 kN/m$^3$

$$= 9.81 \times Q \times H \text{ kN–m/sec. (i.e. k watts)}$$

Electrical energy (E.E.) or Power in k. watts, is therefore, given as :

$$P = 9.81\ \eta \cdot Q \cdot H \text{ (k watts)} \quad \text{...(24.2)}$$

∴ Metric H.P. generated

$$= \frac{9.81\ \eta Q \times H}{0.735} = 13.33\ \eta QH \quad (\because 1 \text{ m H.P.} = 0.735 \text{ kW}) \quad \text{...(24.2a)}$$

where $\eta$ is the overall efficiency of turbine, generator, etc.

Using 80% efficiency, we get

$$\text{Electrical energy} = 9.81 \times 0.8 \times Q \cdot H \text{ kW}$$

$$= 7.84 \cdot Q \cdot H \approx \mathbf{8\ QH\ kW} \quad \text{...(24.2b)}$$

where, $H$ is the design head in metres ; $Q$ is the design discharge in cumecs.

**24.3.2. Normal Water Level (N.W.L.).** The highest elevation of water level that can be maintained in the reservoir without any spillway discharge, either with a gated or a non-gated spillway, is known as Normal water level.

**24.3.3. Minimum Water Level (M.W.L.).** The elevation of water level which produces minimum net head on the power units (*i.e.* 65% of design head $H$) is known as minimum water level.

**24.3.4. Weighted Average Level (W.A.L.).** The level above and below which equal amounts of power are developed during an average year (*i.e.* 50% units between N.W.L. and W.A.L., and 50% units between W.A.L. and M.W.L.) is called weighted average level.

**24.3.5. Design Head.** It is the net head under which the turbine reaches peak efficiency at synchronous speed. Generally, the design head = W.A.L. – M.W.L. The difference of N.W.L. and M.W.L. is 125% of design head.

**24.3.6. Rated Head.** It is the head at which the turbine functioning at full gate opening will produce a power output, equal to that specified in the name plate of the turbine. This rated head should be equal to the design head of the turbine, so as to ensure maximum overall plant efficiency.

**24.3.7. Gross Head.** Gross head is the difference in the water level elevations at the point of diversion of water for the hydel scheme and the point of return of water back to the river.

**24.3.8. Operating Head.** Operating head is the simultaneous difference between the elevations of water surface in the *foreway* and the *tailrace*, after making due allowance for approach and exit velocity heads.

∴ Operating head = T.E.L. at forway entrance – T.E.L. at tailrace exit..(24.3)

**24.3.9. Net Head or Effective Head.** The effective head is the net head applied to the turbine, and is given by the difference of head at the point of entry and exit of turbine, and includes the respective velocity and pressure heads at both places.

**24.3.10. Installed Capacity of the Power House.** The total capacity in kilowatts or million killowatts of all the turbine-generator units installed in a power house, is called its Installed capacity.

**24.3.11. Dependable Capacity.** It is the load carrying capability of the power house with respect to the load characteristics during a specified time interval, and is decided by the power factor, capability and the load on the power house.

**24.3.12. Load Factor (L.F.).** Load factor is defined as the ratio of the average load over a certain period of time to the peak load during the same period. Depending upon the period chosen, we may have different load factors, such as daily, monthly or annual load factors.

$$\therefore \quad \text{Load factor} = \frac{\text{Average load over a certain period}}{\text{Peak load during that period}} \quad ...(24.4)$$

If the area under a load curve* is plotted, it would evidently represent the energy consumed in kilowatt hours (kWh). Thus, an annual load factor may also be defined as the ratio of the actual energy consumed during that year to the peak demand assumed to continue for one year.

∴ **Annual load factor**

$$= \frac{\text{Total yearly electrical units (kWh) produced}}{(\text{Max. power demand in kW}) \times 365 \times 24} \quad ...(24.5)$$

* A graph plotted between time in hrs (on $x$-axis) and load in kW (on $y$-axis) is called a **load curve.**

**Note.** The maximum load determines the capacity of the units, while the load factor gives an idea of the degree of utilisation of the capacity. For example, an annual load factor of 0.6 would indicate that the machines are producing 60% of their yearly rated capacity (maximum production capacity).

**24.3.13. Demand Factor.** A consumer is provided with the connected load of a certain rating (kilo-watts). He may or may not use the whole load at one time, as the may use only a part of it. The ratio of the maximum demand at any particular time to the connected load is known as the *demand factor*.

**24.3.14. Capacity Factor or Plant Factor.** It may be defined as the ratio of average output of the plant for a given period of time to the plant capacity, *i.e.*,

$$\text{Capacity factor} = \frac{\text{Average Load(over a given period of time)}}{\text{Plant capacity}} \quad ...(24.6\ a)$$

In other words, the capacity factor is the ratio of the energy actually produced by the plant in any given period to the energy it would be capable of producing at its full capacity during that period. For example, if a plant with a capacity of 10,000 kW were to produce 40,000 kWh when operating for 100 hours, then

$$\text{Capacity factor} = \frac{\frac{40,000}{100}}{10,000} = 0.4 \text{ or } 40\%$$

Hence, capacity factor

$$= \frac{\text{Energy actually produced in a given time}}{\text{Max. energy that can be produced by the plant during that given time}} \quad ...(24.6\ b)$$

The capacity factor and load factor would become identical, if the peak load is equal to the plant capacity ; *i.e.* when plant is allowed to be used up to its max. capacity, or there is no *reserve capacity*. However, load factor would be different, if the plant is not used to full capacity.

Hence, in the above example, if the peak load is equal to 8000 kW (as against the station capacity of 10,000 kW), then

$$\text{Load factor} = \frac{\frac{40,000}{100}}{8,000} = 50\% \text{ as against the capacity factor of } 40\%.$$

Also, the **reserve capacity** would then be

$$\text{Peak load} = \frac{\text{Plant capacity}}{1 + \text{R}} \quad ...(24.7)$$

where $R$ is Reserve capacity expressed as fraction of Plant capacity,

$$\therefore \quad 8000 = \frac{10,000}{1 + \text{R}}$$

or $$1 + \text{R} = \frac{10,000}{8,000}$$

or $$\text{R} = 0.25, \textit{ i.e. } 25\%.$$

(*i.e.* at 25% R.C. ; Peak load × 1.25 = Installed capacity

**24.3.15. Utilisation Factor or Plant Use Factor.** It is defined as :

$$\text{Utilisation factor (U.F.)} = \frac{\text{Water actually utilised for power production}}{\text{Water available in the river}} \quad ...(24.8)$$

It the water head is assumed to be constant, then the utilisation factor would be equivalent to :

$$\text{U.F.} = \frac{\text{Max. power utilised}}{\text{Max. power available}} \quad \text{...(24.9)}$$

The value of utilisation factor usually varies from 0.4 to 0.9 for a hydel plant, depending upon the plant capacity, load factor and storage.

**24.3.16. Firm Power.** The net amount of power which is continuously available from a plant without any break on firm or on guaranteed basis is known as *firm power*. This power should be available under the most adverse hydraulic conditions. The consumers can always be sure of getting this power.

**24.3.17. Secondary Power.** The excess power available over the firm power during the off peak hours or during monsoon, etc. is known as *secondary power*. There is no guarantee for secondary power, and it is supplied to the consumers on 'as and when available' basis.

**24.3.18. Power Factor.** It is defined as

$$\text{Power factor} = \frac{\text{Actual power in kilowatts (kW)}}{\text{Apparent power in kilo volt–amperes (KVA)}} \quad \text{...(24.10)}$$

The power factor can never be greater than unity. Its value depends upon the relationship between the inductance and resistance in the load. A load with very little inductance such as lighting bulbs, will have a power factor close to unity. The usual system load has a power factor varying between 0.8 to 0.9, but if various induction motors are installed in the load, the power factor may be as low as 0.5. Since the electrical machines are generally rated in KVA, the actual power developed depends much upon the power factor. This is the reason, why most of the Electric Supply Undertakings, do stress upon their consumers, to reduce their inductance and improve their power factors.

**Example 24.1.** *The load on a hydel plant varies from a minimum of 10,000 kW to a maximum of 35,000 kW. Two turbo-generators of capacities 22,000 kW each have been installed. Calculate :*

*(a) total installed capacity of the plant ;*

*(b) plant factor ;*

*(c) maximum demand ;*

*(d) load factor ;*

*(e) utilisation factor.*

**Solution.** (*a*) Since two generators, each of capacity 22000 kW are installed, we have

the *total installed capacity* $= 2 \times 22000$ kW = 44000 kW. **Ans.**

From eqn. (24.6*b*), we have

(*b*) *Plant factor* (*i.e.* capacity factor)

$$= \frac{\text{Energy actually produced in time } t}{\text{Max. energy that can be produced in time } t}$$

$$= \frac{\frac{(10000 + 35000)}{2} \times t}{44000 \times t} = \frac{22500}{44000} = 0.511, \textit{ i.e. } 51.1\%. \textbf{ Ans.}$$

(*c*) *Maximum demand* = 35000 kW (as given). **Ans.**

(*d*) *Load factor* $= \dfrac{\text{Average load over a certain period}}{\text{Peak load during that period}}$

$$= \frac{\dfrac{10000 + 35000}{2}\text{ kW}}{35000\text{ kW}} = \frac{22500}{35000} = 0.643 \textit{ i.e. } 64.3\%. \quad \textbf{Ans.}$$

(*e*) *Utilisation factor* $= \dfrac{\text{Max. power utilised}}{\text{Max. power available}}$

$$= \frac{35000\text{ kW}}{44000\text{ kW}} = 0.795, \textit{ i.e. } 79.5\%. \quad \textbf{Ans.}$$

**Example 24.2.** *A common load is shared by two hydel stations ; one being a base load station with 20 MW installed capacity, and the other being a stand-by station with 25 MW capacity. The yearly output of the stand-by station is $10 \times 10^6$ kWh and that of the base load plant as $110 \times 10^6$ kWh. The peak load taken by stand-by station is 12 MW and this station works for 2500 hours during the year. The base load station takes a peak of 18 MW. Find out :*

(*a*) *Annual load factors for both stations.*

(*b*) *Plant use factors for both stations.*

(*c*) *Capacity factors for both stations.*

**Solution.** We will work out all the required three factors for the base load plant first, and then for the other stand-by plant.

**For Base Load Plant.** For this base load station, data is given as :

Installed capacity $= 20\text{ MW} = 20 \times 10^3\text{ kW}$

Yearly output $= 110 \times 10^6\text{ kWh}$

Peak load taken $= 18\text{ MW} = 18 \times 10^3\text{ kW}.$

(*a*) Now, *Annual load factor*

$$= \frac{\text{Total energy (kWh) generated per year}}{\text{Max. power demand in kW} \times 365 \times 24}$$

$$= \frac{110 \times 10^6}{(18 \times 10^3) \times 365 \times 24} = \mathbf{0.698}, \textit{ i.e. } \mathbf{69.8\%}. \quad \textbf{Ans.}$$

(*b*) *Plant use factor* $= \dfrac{\text{Max. power utilised}}{\text{Max. power available}}$

$$= \frac{18000\text{ kW}}{20 \times 10^3\text{ kW}} = \mathbf{0.9}, \textit{ i.e. } \mathbf{90\%}. \quad \textbf{Ans.}$$

(*c*) *Capacity factor* $= \dfrac{\text{Av. energy utilised in a period (year)}}{\text{Max. energy that can be produced in that period}}$

$$= \frac{110 \times 10^6}{(20 \times 10^3) \times (24 \times 365)} = \mathbf{0.628}, \textit{ i.e. } \mathbf{62.8\%} \quad \textbf{Ans.}$$

**For stand by plant.** The data given for this plant is :

Installed capacity $= 25\text{ MW} = 25 \times 10^3\text{ kW}$

Yearly output in 2500 hrs $= 10 \times 10^6\text{ kWh}$

Peak load $= 12\text{ MW} = 12 \times 10^3\text{ kW}.$

No. of working hours during the year = 2500 hr.

(a) *Annual load factor*

$$= \frac{\text{Total kWh generated per year}}{\text{Max. power demand in kW} \times (24 \times 365)}$$

$$= \frac{10 \times 10^6 \text{ kWh}}{(12 \times 10^3 \text{ kW}) \times (24 \times 365 \text{ h})} = \mathbf{0.095}, \textit{i.e.}\ \mathbf{9.5\%}. \quad \textbf{Ans.}$$

(b) *Plant use factor* $= \dfrac{\text{Max. power utilised}}{\text{Max. available power}}$

$$= \frac{12 \times 10^3 \text{ kW}}{25 \times 10^3 \text{ kW}} = \mathbf{0.48}, \textit{i.e.}, \mathbf{48\%}. \quad \textbf{Ans.}$$

(c) *Capacity factor* $= \dfrac{\text{Av. output}}{\text{Station capacity}}$

$$= \frac{10 \times 10^6 \text{ kWh}}{(25 \times 10^3 \text{ kW}) \times 2500 \text{ h}} = \mathbf{0.16}, \textit{i.e.}\ \mathbf{16\%}. \quad \textbf{Ans.}$$

## 24.4. Principal Components of a Hydro-electric Scheme

A hydroelectric development scheme ordinarily includes a diversion structure, a conduit (penstock) to carry water to the turbines, turbines and governing mechanisms, generators, control and switching apparatus, housing for the equipment, transformers and transmission lines to the distribution centres. In addition to these major components, trash racks at the entrance to penstock, canal and penstock gates, a foreway, a surge tank, and other appurtenances may be required. A tailrace channel from the power house back to the river must be provided, if the power house is situated at such a place that the draft tubes cannot discharge water directly into the river.

No two power development schemes are exactly alike and each will have its own unique problems of design and construction. The choice of a particular type of a plant at a given site depends upon various factors, such as :

(*i*) general available topography of the area ;

(*ii*) available head ;

(*iii*) available flow ;

(*iv*) availability of other type of power stations in the vicinity :

(*v*) requirements of power for industries, etc.

(*vi*) Political influences of the area, etc.

The major components of a hydroelectric scheme are described below :

(1) **The Foreway.** A foreway is a storage basin or any other large body of water situated just infront of intake. Its main function is to temporarily store the water which is rejected by the plant due to reduced load during off-peak hours, and also to meet the instantaneous increased demand when the load is suddenly increased. The foreway, therefore, absorbs the short interval variations in water demand and corresponding fluctuations in power load.

In many cases, when canal leads the water to the turbines, the canal itself may be large enough to absorb these flow variations. When the canal is long, its section is sometimes enlarged at the end, so as to provide necessary temporary storage. In many other cases, when a reservoir is constructed by building a dam, and the water is taken to the power house directly from the reservoir through penstocks, the reservoir itself acts as a foreway.

(2) **Intake Structure.** The foreway is provided with some type of intake structure, to direct water into the penstocks. Various types of intakes are used depending upon the local conditions.

Intakes should be provided with **trash racks** so as to prevent the entry of debris into the penstocks, and thus to avoid the possible damage to *wicket gates* and turbine runners or to avoid the choking up of the nozzles of the impulse turbines. They are made of thin flat steel bars placed 10 to 30 cm apart. The permissible velocity of water entering the trash racks is 0.6 to 1.6 m/sec.

A floating boom is often placed in the storage reservoir so as to trap as much ice and floating debris as possible, and thus to avoid their entry into the canal. At places, where severe winters do occur, provision is made so as to minimise the ice trouble in the foreway. The trash racks may be electrically heated, so as to prevent clinging of ice with them. Sometimes, an air-bubbler system which agitates the water in the vicinity of the trash rack and brings warmer water to the surface, is also used. The problem of ice automatically gets reduced, if the foreway pond is large enough, and the intake is at considerable depth.

The foreway must be provided with a spillway or a wasteway, a bypass channel, etc., so that the water can be disposed of safely, if such a need arises at any time. Syphon spillways may often be used advantageously for this purpose.

Besides trash racks, other accessories of an intake structure are :

(*i*) *Rakes and trolley arrangements* required to clear trash racks.

(*ii*) *Ice removal equipment,* such as a *floating boom, electrical heating device* for trash racks, etc., if necessary.

(*iii*) *Penstock closing gates* with their hoisting mechanisms.

(3) **Penstocks.** Penstocks are the huge diameter pipes which carry water under pressure from the storage reservoir to the turbines. The structural design of a penstock is essentially similar to that of any other pressure pipe. But, because of the possibility of sudden load changes, design against water hammer* is essential. Short length penstocks are generally designed to take this extra pressure by providing heavier pipe wall. Slow-closing valves are then provided, so as to reduce these extra pressures. However, in case of long penstocks, a *surge tank* is generally provided, so as to absorb water hammer pressures and to provide water to meet sudden load increases.

Penstocks are generally made of steel or reinforced concrete, though wooden-stave pipes are also sometimes used. If the distance of the power house from the foreway is small, separate penstocks are generally used for each turbine. But long penstocks are generally branched at the lower end, so as to serve several turbines.

Penstock closing gates are usually provided at the entrance in the foreway. The gate can be closed to permit repair of penstock. An air vent and vent pipe connecting the top of the penstock with the open air, is provided downstream from the gate. Such a vent permits the entry of air into the penstock, as soon as the gates are closed and water drawn off through the turbine wheels. In the absence of such an arrangement, serious negative pressures will be exerted on the penstocks and may cause the penstocks to collapse. The water in the vent pipe should not get frozen, otherwise, entry of air will be prevented. Before opening the main head gate, the penstock, if empty, should be filled through a bypass, so that surge of water may not damage the turbines.

The penstocks should be located at such a level that sufficient water depth (called **water seal**) is provided above the penstock entrance in the foreway. If this

* Explained in previous chapter.

is not so, and too little water depth is available, vortices and whirlpools will tend to form, which may carry air into the penstocks and the turbine wheels, and thus lowering the turbine efficiency and causing undesirable pressure surges. The entrance of the penstocks should also be flared, so as to avoid any loss of head by contraction.

Sharp bends must be avoided in penstocks, because they cause loss of head and require special anchorages. The penstocks may either be buried under the ground or kept exposed. Exposed or uncovered penstocks are usually supported on cradles and carried across ravines on trestles. Uncovered penstocks are generally costlier because they require other appurtenances such as expansion joints, saddles, anchors, etc. ; but they are easily accessible for inspection and repairs. It is not advisable to partially bury the penstock pipes, as the region near the ground surface will be subjected to excessive deterioration due to local movements. In a few cases, the penstocks can be replaced by constructing a tunnel through a hill, if a considerable saving in length can be gained.

(4) **Surge Tank or Surge Chamber.** The simplest type of a surge chamber consists of a cylindrical chamber open to the atmosphere and connected to the penstocks, as close to the power house as possible (Fig. 24.4).

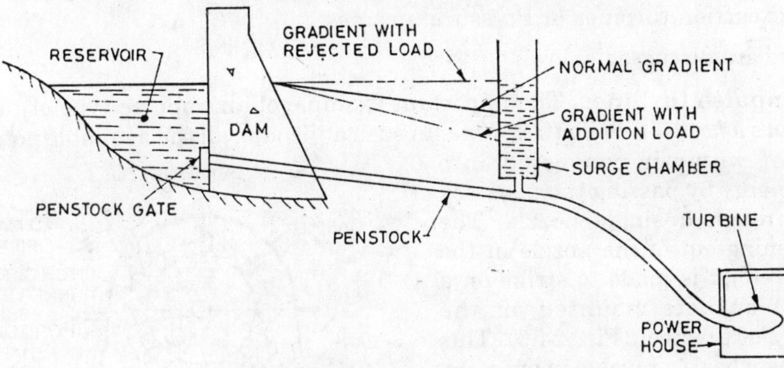


Fig. 24.4. Provisions of Surge chamber.

When the load is rejected by the power house turbines, the water level in the surge chamber rises, and decelerates the flow upstream of it. But when additional load comes, the immediate demand is met by drawing water from the surge chamber, which increases the flow gradient and thus accelerates the flow from the reservoir. When the load is steady, the water level in the surge chamber remains constant and at the level of normal hydraulic gradient line at that point, as shown in Fig. 24.4 A surge chamber, therefore, reduces the pressure fluctuations in the conduit pipe (*i.e.* penstocks) considerably, and thus prevents additional water hammer pressures from being exerted upon the walls of the conduit. The surge chamber should be high enough, to prevent any overflow of water, even at full load rejection.

Various types of surge chambers, such as (*i*) *Simple surge tank* (*ii*) *Throttled or Restricted orifice surge chamber,* (*iii*) *Differential surge chamber,* (*iv*) *Multiple surge chamber,* etc. are in use these days. The differential type of surge chamber is having a central riser pipe in addition to the main outer tank, as shown in Fig. 24.5.

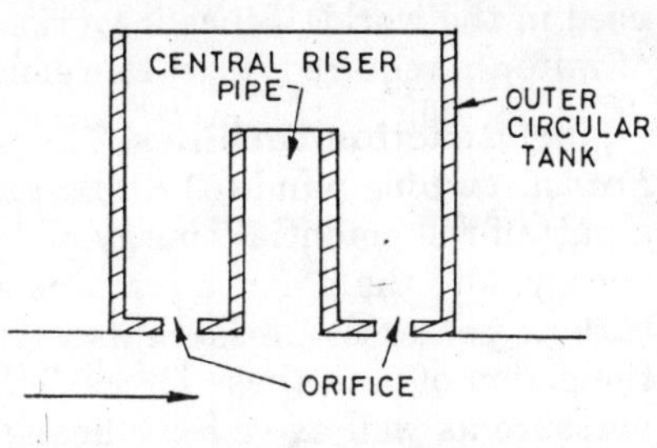


Fig. 24.5.
Differential type of surge chamber.

The main advantage of differential type of surge tank over a simple tank lies in the fact that retarding and accelerating heads are developed more promptly in the former case than in the latter case. Hence, for a given amount of stabilising effect, the capacity of a differential type of surge chamber may be less than that of a simple elementary type.

(5) **Hydraulic Turbines.** Turbines are machines which convert hydraulic energy into mechanical energy. The mechanical energy so developed by the turbine is then used to generate electric energy by directly coupling the shaft of the turbine with the generator.

In general, a turbine consists of a wheel (called runner) which is provided with specially designed blades or buckets. The water having large hydraulic energy is made to strike the runner, and thus causing it to rotate. This rotation of the turbine runner is passed on to the generator by coupling the generator and turbine together through the turbine shaft. This results in rotating the generator armature, and thus producing electrical power, called hydroelectric power.

Hydraulic turbines may be divided into two classes :

(*i*) Impulse turbines or Velocity turbines ; and

(*ii*) Reaction turbines or Pressure turbines.

These are discussed below :

(*i*) **Impulse turbines.** The important example of an impulse type of turbine is a *Pelton's wheel* or *a Pelton's turbine*. In such a turbine, all the available potential energy of water is converted into kinetic energy by passing the penstock water through a single nozzle. The water coming out of the nozzle in the form of free jet is made to strike on a series of buckets mounted on the periphery of a wheel (Fig. 24.6). This causes the wheel to revolve in open air, and water is in contact with only a part of the wheel at a time. The pressure of water is all along atmospheric, and thus there is no difference of pressure in the water at the inlet to the runner and at the outlet discharge to the tailrace. It also acts as a safeguard against accidents.

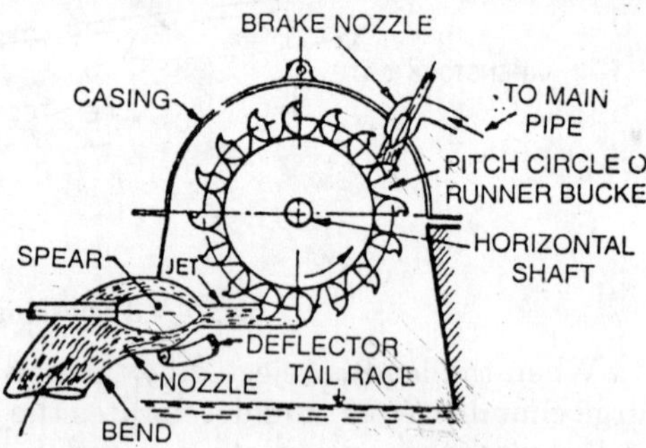

Fig. 24.6. Pelton's wheel (impulse turbine).

An impulse turbine is essentially a low speed turbine and is used for high heads of the order of 150 to 1,000 m (though heads of 1,700 to 1,800 m have been used in the world). Since it works under high heads, comparatively less quantity of water is required. It is, therefore used for high heads and low discharges.

(*ii*) **Reaction turbines.** The important examples of reaction turbines are : (*i*) *Francis turbine* ; and (*ii*) *Kaplan turbine.* A reaction turbine is one in which only a part of the potential energy of water is converted into velocity head (*i.e.* kinetic energy) and the balance remains as pressure head. Thus, the water entering the turbine possesses pressure as well as kinetic energy. The wheel is rotated under the action of both these forces. The water leaving the turbine also contains some pressure as well as velocity head. The pressure at the inlet is much higher than the pressure at the outlet. Since the entire flow takes place under pressure, a closed casing is absolutely necessary, so as to prevent any access of atmospheric air into the turbine. Since the water flows under pressure through such a turbine, the wheels of this turbine are submerged, and water enters all around the periphery of the wheel.

**Difference between Pelton's** (*i.e.* Impulse type) and **Francis** (*i.e.* Reaction type) **Turbines**

(1) In a Pelton's wheel, the total potential head is changed into kinetic head for affecting the motion of the runner ; while in a Francis wheel, only a part of it is converted.

(2) Water strikes only a few buckets at a time in Pelton's wheel ; while in Francis wheel, the water flows like that in a closed conduit. The runner is always full of water, and thus all the blades are simultaneously striken by water.

(3) In Pelton's wheel, the water falls freely to the atmosphere ; while in Francis wheel, the water is taken upto the tailrace by means of a closed draft tube, and thus, the whole passage of water is totally enclosed.

**Classification of turbines according to specific speed**

The specific speed ($N_s$) of a turbine is the speed of geometrically similar turbine that would develop one metric horse power under a head of one metre. Specific speed is given by the equation

$$N_s = \frac{N\sqrt{P_t}}{H^{5/4}} \quad \text{...(24.1)}$$

where $N$ = Normal working speed of the turbine in RPM

$P_t$ = The turbine output in metric horse power

$H$ = Net or effective head in metres.

Table 24.1 shows the specific speeds of various turbines.

**Table 24.1. Specific Speeds of Various Turbines**

| *S. No.* | *Type of turbine* | *Limit of head in metres* | *Specific Speed in Revolution per minute [i.e. R.P.M.]* |
|---|---|---|---|
| 1. | Pelton turbine | Over 150 (High heads) | 10 to 35 |
| 2. | Francis turbine | 25 to 450 (Medium heads) | 60 to 300 |
| 3. | Kaplan and Propeller turbines* | 1 to 70 (Low heads) | 300 to 1,000 |

**Choice of a particular type of turbine**

The selection of the type of a turbine primarily depends on head and load conditions. Different types of turbines can work satisfactorily only within a certain range of specific speeds (Table 24.1). At specific speeds greater than this usual range, there is a possibility of cavitation on the runner blades.

---

* Both Kaplan and Propeller turbines look alike and behave hydraulically in a similar fashion except that the Kaplan turbines have an additional arrangement of adjusting the inclination of the runner blades or vanes on the hub. This causes the Kaplan turbines to give higher efficiency even under part load conditions. *Kaplon turbines, cmpared to Propeller turbines, are therefore found to be more versatile, although costlier in their initial cost.*

The variability of load also dictates the choice of a particular type of turbine. Say for example, for a head of 150 to 450 m or so ; from table 24.1, both Pelton's as well as Francis turbines can be used. But here the variability of load will influence the choice of a particular type. For higher range of heads, Pelton's wheel is preferred for part load operation in comparison to Francis turbine, although the Pelton's turbine involves higher initial cost. The quality of water also sometimes dictates the type of turbine to be chosen. For example, in the head range of 150 to 450 m, the choice of a Francis turbine is automatically ruled out, if sand is present in the water, because its runner will not be able to withstand its erosive action.

(6) **The Power House.** A power house is a building consisting of a *substructure* to support the hydraulic and electrical equipment and a *superstructure* to house and protect this equipment. For most of the plants which are equipped with reaction turbines, the substructure usually consists of a concrete block extending from the foundation to the generator floor with waterways (*i.e.* the scroll casings and draft tubes) formed within it. They are cast integrally while pouring concrete. The elevation of the turbine with respect to tail water level should be determined by the necessity of avoiding cavitation.

The superstructure is a building which generally accommodates the generators and excitors on the ground floor, and the switch board and control room on the mezzanine floor. Vertical turbines are placed just below the floor level beneath the generators, while horizontal turbines are placed on the ground floor along side the generators. The generating units are almost always arranged in a straight line across the direction of flow. A travelling crane spanning the width of the power house, is generally provided in every power house, so as to facilitate the lifting of heavy machines.

(7) **The Draft Tube.** The draft tube is a conduit which connects the outlet of a reaction turbine runner to the tailrace. Water, as it emerges out of the runner, flows through this pipe of gradually increasing diameter and comes to the tailrace level. The advantages of having a draft tube instead of letting the water fall directly into the atmosphere are :

(*i*) Effective pressure head is increased by an amount

$$= H_s + \frac{V_2^2 - V_3^2}{2g}$$

as explained below :

Considering Fig. 24.7, we have

$$\frac{p_2}{\gamma_w} + Z + H_s + \frac{V_2^2}{2g} = \frac{p_a}{\gamma_w} + Z + \frac{V_2^2}{2g}$$

or $$\frac{p_2}{\gamma_w} = \frac{p_a}{\gamma_w} + \frac{V_3^2}{2g} - \frac{V_2^2}{2g} - H_s = \frac{P_a}{\gamma_w} - \left(H_s + \frac{V_2^2}{2g} - \frac{V_3^2}{2g}\right)$$

Hence, the negative pressure created at the outlet of the runner

$$= H_s + \frac{V_2^2 - V_3^2}{2g} ;$$

this should not exceed the vacuum pressure of water.

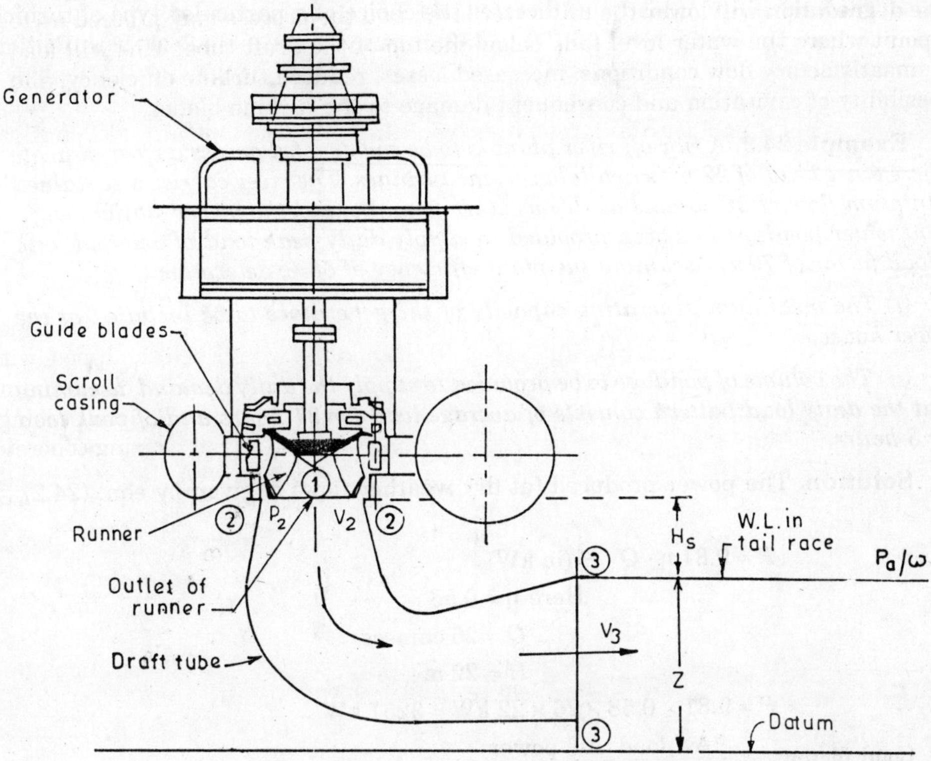


Fig. 24.7

So if by gradually increasing the diameter of this pipe, we are able to reduce the velocity $V_3 = 0$, then the net increase in head

$$= H_s + \frac{V_2^2}{2g}.$$

*(ii)* The draft tube permits the turbines to be installed at a higher level than the tailwater level, which facilitates the maintenance, etc. of the turbine.

*Outlet Gates.* The outlet of the draft tube should be provided with gates, so that the draft tube can be dewatered for repairs. If the span between tailrace piers is not too high, timber logs may be used for this purpose. However, for larger spans, steel stop logs or steel gates are commonly used. A hoisting mechanism is also required for lifting the gates or stop logs.

(8) **The Tailrace.** The channel into which the water is discharged after passing through the turbines is known as the *Tailrace*. If the power house is close to the stream, the outflow may be discharged directly into the stream. But when the stream is far off from the power house, one may have to construct a channel of considerable length between the power house and the stream. The tailrace must be designed properly and should not be neglected. In many low-head plants, more of the gross head may be utilised at very low cost if full attention is paid to the tailrace design.

Proper design and maintenance of the tailrace is necessary, so as to avoid excessive aggradation (*i.e.* silting) or degradation (*i.e.* scouring) of its bed. Aggradation will raise the tail water level and reduce the gross available head at the plant.

The degradation will lower the tail water level, and the process may proceed up to a point where the water level falls below the top of the draft tube. This will lead to unsatisfactory flow conditions, increased losses, reduced turbine efficiency, and possibility of cavitation and consequent damage to the turbine blades.

**Example 24.3.** *A run-off river plant is to be constructed across a river at a site where a net head of 22 m is available on the turbines. The river carries a sustained minimum flow of 26 cumecs as dry weather flow. Behind the power station, sufficient water pondage has been provided to supply daily peak load of demand with a load factor of 70%. Assuming the plant efficiency of 58%, determine :*

*(i) The maximum generating capacity of the generators to be installed at the power house.*

*(ii) The volume of pondage to be provided to supply the daily demand, assuming that the daily load pattern consists of average load for 21 hours and of peak load for 3 hours.*

**Solution.** The power produced (at dry weather flow) is given by eqn. (24.2), as

$$P = 9.81\,\eta \cdot Q \cdot H \text{ (in kW)}$$

Here $\eta = 0.58$

$Q = 26$ cumecs

$H = 22$ m

$$\therefore \quad P = 9.81 \times 0.58 \times 26 \times 22 \text{ kW} = 3251 \text{ kW}$$

$$\text{Load factor} = \frac{\text{Av. load (Av. power)}}{\text{Peak load (Peak power)}}$$

$$\therefore \quad 0.70 = \frac{3251}{\text{Peak load}}$$

$$\therefore \quad \text{Peak load} = \frac{3251}{0.7} = 4645 \text{ kW.}$$

(*i*) Assuming there is no reserve capacity, we have the *Maximum capacity of the generators to be installed* = 4645 kW. **Ans.**

(*ii*) Excess water from pondage is drawn in order to meet the excess demand (demand in excess of average) for 3 hours.

Excess power reqd. to be developed during 3 hours

= Peak demand – Av. demand

= 4645 – 3251 = 1394 kW

Excess discharge required for developing this excess power is given as

$$P = 9.81\,\eta \cdot Q \cdot H$$

$$\therefore \quad 1394 = 9.81 \times 0.58 \times Q \times 22$$

$$\therefore \quad Q = \frac{1394}{9.81 \times 0.58 \times 22} = 11.15 \text{ m}^3/\text{sec.}$$

This excess discharge is required for 3 hours each day.

∴ Required pondays per day

$= 11.15 \times 3 \times 60 \times 60 = 12.04 \times 10^4 \text{ m}^3$ **Ans.**

**Example 24.4.** *A run-off river plant with an installed capacity of 15,000 kW operates at 28% load factor when it serves as a peak load station :*

*(a) What should be the minimum discharge in the stream, so that it may serve as a base load station ? The plant efficiency may be assumed to be 80% when working under a head of 20 m.*

*(b) Also calculate the maximum load factor of the plant when the discharge in the stream is 35 cumecs.*

**Solution.** Installed capacity = 15,000 kW

$$\text{Load factor} = 28\%$$

$$\text{Load factor} = \frac{\text{Av. Load}}{\text{Peak Load}}$$

The installed capacity plant serving as peak load plant means that the Installed capacity = Peak load.

$$\therefore \quad \text{Peak load} = \text{Installed capacity} = 15000 \text{ kW}$$

$$\therefore \quad 0.28 = \frac{\text{Av. load}}{15000}$$

$$\therefore \quad \text{Av. load} = 15000 \times 0.28 = \mathbf{4200\ kW.}$$

(*a*) Now, when this plant has to serve as base load plant, it should develop average load, *i.e.* 4200 kW.

Hence, $Q$ reqd. to develop 4200 kW is given by :

$$P = 9.81\,\eta \cdot Q \cdot H$$

or

$$4200 = 9.81 \times 0.8 \times Q \times 20$$

$$\therefore \quad Q = \frac{4200}{9.81 \times 0.8 \times 20} = \mathbf{26.78\ cumecs.} \quad \textbf{Ans.}$$

(*b*) When discharge in the stream is 35 cumecs, then power developed

$$= 9.81\,\eta \cdot QH$$

$$= 9.81 \times 0.8 \times 35 \times 20 \text{ kW} = \mathbf{5488\ kW.}$$

$$\text{Load factor} = \frac{\text{Av. power}}{\text{Peak power}} = \frac{5488 \text{ kW}}{15000 \text{ kW}} = 0.366, \textit{ i.e. } 36.6\%.$$

Hence, maximum load factor at 35 cumecs flow in the stream = **36.6 %. Ans.**

**Example 24.5.** *A run-off river plant is installed on a river having a minimum flow of 15 $m^3$/sec. If the plant is used as a peak load plant operating only for 6 hours daily,* compute the firm capacity of the plant :*

*(a) without pondage ;*

*(b) with pondage but allowing 8% water to be lost in evaporation and other losses. Head at the plant is 16 m, and the plant efficiency may be assumed as 80%.*

**Solution.** (*a*) Power developed by 15 $m^3$/sec discharge is given by

$$P = 9.81\,\eta \cdot Q \cdot H$$

$$Q = 9.81 \times 0.8 \times 15 \times 16 = \mathbf{1882\ kW.}$$

This represents the firm capacity of the plant without pondage. **Ans.**

(*b*) With pondage, the total volume of water stored during 18 hours when the plant is not operating is

$$= 15 \times (18 \times 60 \times 60) \text{ m}^3 = 9.72 \times 10^5 \text{ m}^3$$

* Also called, *6 hours peaking plant.*

The loss of water due to evaporation, etc., per day

$$= 10\% \times 9.72 \times 10^5 \text{ m}^3 = 0.972 \times 10^5 \text{ m}^3$$

Net amount of available pondage per day

$$= (9.72 - 0.972)\, 10^5 \text{ m}^3 = 8.75 \times 10^5 \text{ m}^3$$

This daily pondage can supply a uniform $Q$, (reqd. for 6 hrs. only)

$$= \frac{8.75 \times 10^5}{6 \times 60 \times 60} \text{ m}^3/\text{sec} = 40.51 \text{ m}^3/\text{sec}.$$

∴ Total flow available for power generation (reqd. only for 6 hrs)

$$= (15 + 40.51) \text{ m}^3/\text{sec} = 55.51 \text{ m}^3/\text{sec}$$

∴ Power developed due to this discharge (*i.e.* firm power, because discharge is firm)

$$= 9.81 \cdot \eta \cdot Q \cdot H = 9.81 \times 0.8 \times 55.51 \times 16 = 6963 \text{ kW}.$$

Hence, the firm power developed by the plant (if pondage during rest hours is allowed) = **6963 kW. Ans.**

**Example 24.6.** *During a low water week, a river stream gets an average daily flow of 35 m³/sec with daily fluctuations requiring a pondage capacity of about 16% of the daily flow.*

*A hydroelectric plant is to be located on this river, which will operate for 5 days a week, 24 hours a day, but will supply power at varying rates, such that the daily load factor is of 60%, corresponding to which the pondage required is 0.2 times the meanflow to the turbines. On saturdays and sundays, all the flow is ponded for use on rest of the days.*

*If the effective head on the turbine, when the pond is full is to be 26 m, and the maximum allowable fluctuation in pond level is 1.2 m, determine :*

*(i) the surface area of the pond to satisfy all the operating conditions ;*

*(ii) the weekly output at the switchboards in kWh.*

*Assume the turbine efficiency as 82% and generator efficiency as 90%.*

**Solution.** $Q = 35 \text{ m}^3/\text{sec}.$

Now, Daily flow $= 35 \times 24 \times 60 \times 60 = 3.024 \times 10^6 \text{ m}^3$

Pondage reqd. for discharge fluctuations

$$= 16\% \text{ of daily flow} = \frac{16}{100} \times 3.024 \times 10^6 \text{ m}^3$$

$$= \mathbf{0.484 \times 10^6 \text{ m}^3} \quad ...(i)$$

The discharge at the turbines is caused by the daily flow of 35 cumecs plus the discharge caused by the water stored during the two weekly rest days, *i.e.* saturday and sunday.

Now, *total water stored in 2 rest days*

$$= 2 \times \text{Daily flow} = 2 \times 3.024 \times 10^6 \text{ m}^3$$

$$= \mathbf{6.048 \times 10^6 \text{ m}^3} \quad ...(ii)$$

This stored water is utilised during the 5 working days in the week. Hence, the discharge obtained per day from this storage

$$= \frac{6.048 \times 10^6}{5} \text{ m}^3 = 1.2096 \times 10^6 \text{ m}^3.$$

The streamflow caused by this storage

$$= \frac{1.2096 \times 10^6}{24 \times 60 \times 60} \text{ m}^3/\text{sec} = 14 \text{ m}^3/\text{sec}.$$

∴ Total average flow to the turbines

$$= 35 + 14 = \mathbf{49\ m^3/sec.}$$

Hence, pondage required for load fluctuations

$$= 0.2 \times \text{Mean daily flow to turbines}$$

$$= 0.2 \times 49 \times (24 \times 60 \times 60) = \mathbf{0.847 \times 10^6\ m^3} \quad ...(iii)$$

Hence, the total ponding required

$$= (i) + (ii) + (iii)$$

$$= [0.484 \times 10^6 + 6.048 \times 10^6 + 0.847 \times 10^6] \text{ m}^3 = 7.379 \times 10^6 \text{ m}^3.$$

(*i*) With maximum variation of 1.2 m in pond level, we have the min. surface area of pond reqd. to accommodate $7.379 \times 10^6$ m$^3$ of water

$$= \frac{7.379 \times 10^6}{1.2} \text{ m}^2 = 6.15 \times 10^6 \text{ m}^2$$

$$= \frac{6.15 \times 10^6}{10^4} \text{ hectares} = \mathbf{615\ hectares.} \quad \textbf{Ans.}$$

Since the allowable max. fluctuation in pond level is 1.2 m, and the head at full level is 26 m, we have the average head on turbines

$$= \frac{26 + (26 - 1.2)}{2} = \frac{26 + 24.8}{2} = \frac{50.8}{2} = 25.4 \text{ m.}$$

(*ii*) ∴ Max. Power generated = $9.81\, \eta \cdot Q \cdot H$.

where $\eta$ is the combined efficiency of turbines and generators

$$= 0.82 \times 0.90 = 0.738$$

$$Q = 49 \text{ m}^3/\text{sec}$$

$$H = 25.4 \text{ m}$$

$$\therefore \quad P_{max} = (9.81 \times 0.738 \times 49 \times 25.4) \text{ kW} = 9001.4 \text{ kW.}$$

With a daily load factor of 0.60, we have the average power developed = 0.6 × Max. power

$$= 0.6 \times 9001.4 = 5400.8 \text{ kW.}$$

∴ Weekly power developed at the switchboard

$$= (5400.8 \times 5 \text{ days}) \text{ kW day}$$

$$= (5400.8 \times 5 \times 24) \text{ kWh} = 6{,}48{,}100 \text{ kWh}$$

Hence, weekly output at switchboard

$$= \mathbf{6.481 \times 10^5\ kWh.} \quad \textbf{Ans.}$$

**Example 24.7.** *A run-of-stream station with an installed capacity of 15000 kW operates at 15% load factor when it serves as a peak load station. What should be the lowest discharge in the stream so that the station may serve as the base load station ? It is given that the plant efficiency is 75% when working under a head of 20 m.*

*(b) Also calculate the maximum load factor of the plant when the discharge in the stream rises to 20 cumecs.*

**Solution.** Installed capacity = 15000 kW

Load factor = 15%

Since the plant acts as peak load station, the installed capacity would be equal to the peak load.

Now, $\text{Load factor} = \dfrac{\text{Average Load}}{\text{Peak Load}}$

$\therefore \quad 0.15 = \dfrac{\text{Average Load}}{15000}$

$\therefore \quad$ Average load = 15000 × 0.15 = **2250 kW.**

In order that the plant can act as a base load station, it must supply 2250 kW average power. Using

$P = 9.81\,\eta \cdot Q \cdot H$, we have

$P = 2250$ kW

$\eta = 0.75$ (given)

$Q = ?$

$H = 20$ m.

$\therefore \quad 2250 = 9.81 \times 0.75 \times Q \times 20$

$\therefore \quad Q = \dfrac{2250}{9.81 \times 0.75 \times 20} = \mathbf{15.31.\ m^3/sec.}$

Hence, the stream must carry a minimum discharge of 15.31 cumecs, in order to make the plant work as base load station. **Ans.**

(*b*) When $Q = 20$ cumecs, then power developed by the plant is given as :

$P = 9.81\,\eta \cdot QH$

$= 9.81 \times 0.75 \times 20 \times 20 = 2940$ kW

$\text{Load factor} = \dfrac{\text{Av. Load}}{\text{Peak Load}} = \dfrac{2940 \text{ kW}}{15000 \text{ kW}} = 0.196$, *i.e.* **19.6%.** **Ans.**

**Example 24.8.** *The water turbines at a hydel storage plant produces 10,000 H.P.; when working under a net head of 30 m and with an overall efficiency of 80%. The inflow in the reservoir during a year is given below :*

**Table 24.1**

| *Month* | *Inflow in million cu. m (M. cu. m)* |
|---|---|
| Jan. | 90 |
| Feb. | 80 |
| March | 73 |
| April | 80 |
| May | 70 |
| June | 98 |
| July | 120 |
| Aug. | 80 |
| Sep. | 96 |
| Oct. | 105 |
| Nov. | 100 |
| Dec. | 75 |

*Find (a) the minimum reservoir capacity reqd. to satisfy the uniform demand of water ; (b) the total quantity of water wasted during the year. Assume the reservoir to be full at the beginning of November. Use analytical method.*

**Solution.** The H.P. produced by turbines = 10,000

Since 1 H.P. = 0.736 kW, we have the power produced by turbines in kW units

$$= 10{,}000 \times 0.736 = 7360 \text{ kW}.$$

Discharge reqd. to produce 7360 kW, $Q$, is given by

$$P = 9.81 \, \eta \, . \, Q.H.$$

where $\eta = 0.80$

$Q = ?$

$H = 30$ m

$$\therefore \quad 7360 = 9.81 \times 0.80 \times Q \times 30$$

or $$Q = \frac{7360}{9.81 \times 0.80 \times 30} = 31.29 \text{ m}^3\text{/sec.}$$

$\therefore$ Flow reqd. during each month

$$= 31.29 \times (30 \times 24 \times 60 \times 60) = 81 \times 10^6 \text{ m}^3 = \mathbf{81 \text{ M. cum.}}$$

It shows that 81 M. cum of flow is required to be compulsarily released in order to produce the requisite electric power. The inflow and this reqd. uniform outflow data is now analysed in Table 24.2 starting from Nov. onward (when reservoir is full) to workout the storage used, refill, net reservoir depletion, and the water wasted, as shown in this table. The table is otherwise self explanatory.

**Table 24.2**

| *Month* | *Inflow in M. cum* | *Outflow in M. cum* | *Draft used from storage in M. cum.* | *Storage refill in M. cum.* | *Reservoir depletion in M. cum* | *Water wasted M. cum* |
|---|---|---|---|---|---|---|
| Nov. | 100 | 81 | 0 | — | — | 19 |
| Dec. | 75 | 81 | 6 | — | 6 | — |
| Jan. | 90 | 81 | — | 6 | 0 | 3 |
| Feb. | 80 | 81 | 1 | — | 1 | — |
| March | 73 | 81 | 8 | — | 9 | — |
| April | 80 | 81 | 1 | — | 10 | — |
| May | 70 | 81 | 11 | — | 21 | — |
| June | 98 | 81 | — | 17 | 4 | — |
| July | 120 | 81 | — | 4 | 0 | 35 |
| Aug. | 80 | 81 | 1 | — | 1 | — |
| Sept. | 96 | 81 | — | 1 | — | 14 |
| Oct. | 105 | 81 | — | — | — | 24 |
| Σ | | | | | | 95 |

(*a*) From this table, it can be seen that the max. reservoir depletion is **21 M. cum**, which evidently represents the minimum reservoir capacity required to satisfy the uniform demand. **Ans.**

(*b*) Total quantity of water wasted = **95 M. cum. Ans.**

## 24.5. Comparison of Hydro-power with Thermal Power with Reference to Indian Conditions

Water is available in abundance in India, and most of our important cities and other industrial or populated areas are situated within a maximum distance of say 500 km. from the main concentrations of hydro-resources. Hydro-power can, therefore, be easily developed in India along with other utilisations of our water resources, such as irrigation, navigation, flood control, city water supply, etc. The hydropower has ultimately been found to be cheaper than the thermal power, although its exploitation entails large initial expenditure. It has been estimated, that on an average, the initial investment on a hydro project is about 25% more than that on a thermal project. But, the recurring cost on a hydro project is much less than that on a thermal project, and hence, the hydropower finally comes out to be more economical. Besides these economic considerations, hydropower has been mainly produced in India as a side product during the development of river valleys for irrigation and flood control.

Hydropower is much more important than thermal power in India, mainly because we do have almost inexhaustible water resources in our country, and the same is available in the vicinity of most of our needy areas. In sharp contrast to this, the coal, which is the only cheap and conventional source of thermal power, is confined only to a few regions, and has to be transported over long distances. Hence, in areas which are far away from coal bearing regions, hydropower is the only choice. Moreover, the amount of coal available in our country, is limited in quantity and, therefore, cannot be used infinitely without proper planning for the future. It can be stated that due to various inherent advantages which the hydropower possesses over thermal power, the use of hydropower has been found justifiable on overall economic considerations even in areas close to the collieries in Andhra Pradesh, Madhya Pradesh, Orissa, etc. But in areas, where hydropower has also been fully tapped, we have to think of nuclear or other thermal resources. *Pumped storage plants* may also be thought of in such areas, and their possibilities should also be investigated.

In addition to these advantages, the hydropower is considered to be environmental friendly as compared to thermal power, since thermal power generation causes tremendous air pollution, directly affecting the health of the public.

Besides all these advantages of hydropower over thermal power, there is one serious limitation of hydropower. The limitation is that the hydropower production is totally dependent upon rainfall. Though hydro-projects are designed on a very conservative basis of planning on minimum water that would be available for 90 out of 100 years from a catchment, still there may be some periods when reservoirs do not fill up. If this happens throughout the country, the power position will become difficult. This has happened in 1972 monsoon, when practically all the reservoirs were only half full, resulting in a sudden and drastic power shortage.

Taking such contingencies into account, it can finally be concluded that we must develop a combined system of power production. Hydropower should be exploited to as much extent as possible, and sufficient thermal power reserve should be kept to supplement the hydropower during the times of water crisis.

## 24.6. Hydropower Potentials of India

India has a vast potential for hydropower generation, particularly in the northern and north-eastern regions. As per an estimate made by the Central Electricity Authority, the hydropower potential or probable installed capacity of the country is assessed at **148.7 MkW (i.e. GW).**

This amounts to 148.7 × 0.6 ≈ ***84 MkW of power at 60% load factor****.

The hydropower capacity which can be installed on different river basins of India is shown in table 24.3.

**Table 24.3. Hydropotential of Different River Basins of India**

| *Basin* | *Probable Potential in MkW (GW) at 100% load factor, i.e. installed capacity* | *Installed capacity upto 31.3.2012 in MkW (GW)* |
|---|---|---|
| Indus basin | 33.832 | 11.08 |
| Ganga basin | 20.711 | 4.99 |
| Central Indian River Systems | 4.152 | 3.15 |
| Western flowing rivers of Southern India | 9.430 | 5.66 |
| Eastern flowing rivers of Southern India | 14.511 | 7.78 |
| Brahmaputra Basin | 66.665 | 1.85 |
| Total | **148.701** | **34.51** |

In addition, 56 No. of pumped storage project have also been identified with probable installed capacity of 94 MkW. In addition, hydropotential from small, mini and micro schemes has been estimated at 6.782 MkW from 1512 sites. Thus, in totality, India is endowed with hydro potential of about 250 MkW. However, exploitation has not been up to the desired level due to various constraints confronting the sector.

The development of hydropower in fact, began in India, sometimes in the year 1897, when a 300 kW power station was installed at Darjeeling. It was followed by a power station at Sivasamundram in Mysore State, in 1902. It was then followed by a 50 MW Tata Hydro Electric project of Bombay City in 1914. The process of installation of hydropower projects was slow before independence, but it gathered a lot of momentum after India became free. So much so, that the hydropower production which was hardly 0.8 GW before independence has gone to as high a value as 37.57 GW (MkW) by 31.03.2011, as shown in table 24.4 which gives the values of thermal, hydro and nuclear power capacities developed by the end of different plan periods, starting from the year 1971 to 2011.

Table 24.5 shows the values of identified and installed capacities of Hydropower in various states of India, as an 31/3/2005.

In India, the nuclear power generation is only about 3% of the total power generation, (including hydropower, thermal power, nuclear power, and power

---

*60% load factor means that the installed machines are run, on an average, to give output capacity equal to 60% of the station capacity. This is so, because machines will not always be run at the installed station capacity but will be run, on an average, at 60% of this capacity.

84 MkW is equivalant to annual electricity utilization of 84 × 365 × 24 × 0.6 (Power factor) = 441504 MkWh ≈ 441.5 BkWh or 441.5 TWh.

In India, the nuclear power generation is only about 2% of the total power generation, (including hydropower, thermal power, nuclear power, and power from non-conventional energy sources, like solar power, wind power, power from biomass and other wastes, etc.

**Table 24.4. Developed Power in India at 100% load factor (i.e. Installed capacity) (excluding the Renewable Power like Wind power, Solar power, Biomass power and power from small thydro projects)**

| *End of 5 yr. Plan* | *Power as on (Year end)* | *Installed capacity in MkW (GW)* | | | | *% of hydro power w.r.to total power* |
|---|---|---|---|---|---|---|
| | | *Thermal* | *Hydro* | *Nuclear* | *Total* | |
| End of 6th plan | 31.3.85 | 27.03 | 14.46 | 1.09 | 42.58 | 33.96 |
| End of 7th plan | 31.3.90 | 43.75 | 18.31 | 1.56 | 63.41 | 28.88 |
| End of 8th plan | 31.3.97 | 61.01 | 21.66 | 2.22 | 84.89 | 25.52 |
| End of 9th plan | 31.3.2002 | 74.43 | 26.27 | 2.72 | 103.42 | 25.40 |
| End of 10th plan | 31.3.2007 | 86.01 | 34.65 | 3.90 | 124.56 | 27.82 |
| End of 11th plan | 31.3.2012 | 131.60 | 38.99 | 4.78 | 175.37 | 22.23 |
| End of 12th plan (provisional) | 31.3.2017 | 215.21 | **39.69** | 5.78 | 265.12 | **16.65** |

**Table 24.5. Statewise Identified and Installed capacities of Hydropower in India as on 31.03.2017 (In terms of installed capacity - above 25 MW*)**

| *Region / State* | *Identified Capacity as per resessment study* | | *Capacity Under Operation* | | *Capacity Under Construction* | |
|---|---|---|---|---|---|---|
| | *Total (MW)* | *Above 25 MW (MW)* | *(MW)* | *%* | *(MW)* | *%* |
| *(1)* | *(2)* | *(3)* | *(4)* | *(5)* | *(6)* | *(7)* |
| **Northern** | | | | | | |
| Jammu & Kashmir | 14146 | 13543 | 3119.0 | 23.03 | 1265.5 | 9.34 |
| Himachal Pradesh | 18820 | 18540 | 9643.0 | 52.01 | 1997.0 | 10.77 |
| Punjab | 971 | 971 | 1096.3 | 100 | 206.0 | 21.22 |

*Hydro power plants having capacity of upto 100 kW are called **Micro hydro plants**. while those having capacity of 101 kW — 2000 kW are termed as **Mini hydro plants**; while those having capacity of 2001 kW to 25000 kW (*i.e.* 25 MW) are called **Small Hydro Plants**. All power plants of less than 25 MW capacities are included under Ministry of New and Renewable Energy.

| (1) | (2) | (3) | (4) | (5) | (6) | (7) |
|---|---|---|---|---|---|---|
| Haryana* | 64 | 64 | 0.0 | 0 | 0.0 | 0.00 |
| Rajasthan** | 496 | 483 | 411.0 | 85.09 | 0.0 | 0.00 |
| Uttarakhand | 18175 | 17998 | 3756.4 | 20.87 | 1430.0 | 7.95 |
| Uttar Pradesh | 723 | 664 | 501.6 | 75.54 | 0.0 | 0.00 |
| **Sub Total (NR)** | **53395** | **52263** | **18527.3** | **35.45** | **4898.5** | **9.37** |
| **WESTERN** | | | | | | |
| Madhya Pradesh | 2243 | 1970 | 2235.0 | 100 | 400.0 | 20.30 |
| Chhattisgarh | 2242 | 2202 | 120.0 | 5.45 | 0.0 | 0.00 |
| Gujarat# | 619 | 590 | 550.0 | 100 | 0.00 | 0.00 |
| Maharashtra | 3769 | 3314 | 2647.0 | 79.87 | 0.0 | 0.00 |
| Goa | 55 | 55 | 0.0 | 0.00 | 0.0 | 0.00 |
| **Sub total (WR)** | **8928** | **8131** | **5552.0** | **68.28** | **400.0** | **4.92** |
| **SOUTHERN** | | | | | | |
| Andhra Pradesh | 2366 | 2341 | 1610.0 | 68.77 | 960.0 | 41.01 |
| Telangana | 2058 | 2019 | 701.0 | 34.72 | 90.0 | 4.46 |
| Karnataka | 6602 | 6459 | 3657.4 | 56.62 | 0.0 | 0.00 |
| Kerala | 3514 | 3378 | 1881.5 | 55.70 | 100.0 | 2.96 |
| Tamilnadu | 1918 | 1693 | 1803.2 | 100 | 0.0 | 0.00 |
| **Sub Total (SR)** | **16458** | **15890** | **9653.1** | **60.75** | **1150.0** | **7.24** |
| **EASTERN** | | | | | | |
| Jharkhand | 753 | 582 | 170.0 | 29.21 | 0.0 | 0.00 |
| Bihar## | 70 | 40 | 0.0 | 0.00 | 0.0 | 0.00 |
| Odisha | 2999 | 2981 | 2142.3 | 71.86 | 0.0 | 0.00 |

*Eastern Yamuna Canal project (35 MW) has been developed in 2 stages each having Installed Capacity below 25 MW. Western Yamuna Canal project (64 MW) has been developed in 4 stages each having Installed Capacity below 25 MW

**Two schemes namely Mahi Bajaj Sagar I $ II were identified for I.C. of 97 MW has been developed with I.C of 140 MW. Gandhi Sagar (115 MW) scheme was identified in Rajasthan but has been developed in Madhya Pradesh with same capacity.

#Two schemes namely Ukai Dam and Sardar Sarovar were identified for an I.C. of 590 MW. However as per actual, the I.C. is 550 MW.

##Identified project namely East Gandak Canal has been developed with installed capacity below 25 MW

| *(1)* | *(2)* | *(3)* | *(4)* | *(5)* | *(6)* | *(7)* |
|---|---|---|---|---|---|---|
| West Bengal | 2841 | 2829 | 441.2 | 15.60 | 120.0 | 4.24 |
| Sikkim | 4286 | 4248 | 1965.0 | 46.26 | 1326.0 | 31.21 |
| **Sub Total (ER)** | **10940** | **10680** | **4718.5** | **44.18** | **1446.0** | **13.54** |
| **NORTH EASTERN** | | | | | | |
| Meghalaya | 2394 | 2298 | 282.0 | 12.27 | 40.0 | 1.74 |
| Tripura | 15 | 0 | 0.0 | 0.00 | 0.0 | 0.00 |
| Manipur | 1784 | 1761 | 105.0 | 5.96 | 0.0 | 0.00 |
| Assam | 680 | 650 | 375.0 | 57.69 | 0.0 | 0.00 |
| Nagaland | 1574 | 1452 | 75.0 | 5.17 | 0.0 | 0.00 |
| Arunachal Pd | 50328 | 50064 | 405.0 | 0.81 | 2854.0 | 5.70 |
| Mizoram | 2196 | 2131 | 0.0 | 0.00 | 60.0 | 2.82 |
| **Sub Total (NER)** | **58971** | **58356** | **1242.0** | **2.13** | **2954.0** | **5.06** |
| **All India** | **148701** | **145320** | **39692.8** | **27.31** | **10848.5** | **7.47** |
| Total power as installed capacity | **148.7** MkW | **145.32** MkW | **39.69** MkW | **27.31%** | **10.85** MkW | **7.47%** |
| Total power at 60% load factor | 89.22 MkW | 71.92 MkW | 23.81 MkW | 27.31% | 6.51 MkW | 7.47% |

**Note 1.** The data given in the table 24.5 does not include **pumped storage schemes (PSS).**

As a matter of fact, 9 PSS (4785.6 MW) are under operation, 2 PSS (1080 MW) are under construction, DPR of 1 PSS (500 MW) returned to project developer and 1 PSS (1000 MW) is Concurred by CEA

Moreover, these non-conventional renewable energy sources are increasingly being used in India to produce power. So much so that the power being generated from these renewable sources amounts to about 47 MkW (GW) as on 31.03.2017,*as against the power being produced from large hydroplants amounting to 39.69 MkW (GW) as on 31.03.2017. (Column 4 of Table 24.5)

---

*This includes power being generated from different renewable sources as:

Wind power = 29.15 MkW

Solar power = 9.56 MkW

Biomass power = 8.18 MkW

Waste to Energy (power) = 0.11 MkW

Total = 47.00 MkW (as on 31.03.2017)

In addition, 4.34 MkW power in being generated (as on 31.3.2017) from hydroplants of less than 25 MW 25000 kW (*i.e.* 25 MW), which were excluded in table 24.5, since such plants fall under the Ministry of New and Renewable Energy, and hence categorised under the power from Renewable Sources.

The nuclear power is highly cost intensive*, whereas thermal power involves huge fuel and *R* of *M* (Running of Maintenance) cost particularly for plants located far away from coal mines, or petro gas locations. The disposal of waste coal ash and air pollution caused by thermal power is also a costlier exercise, as compared to the present day cost being incurred for disposal of radioactive nuclear wastes, since such wastes can, through an expensive reprocessing, may lead to recovery of very costly plutonium, which may offset the cost of reprocessing and solve the problem of disposal of such nuclear radioactive wastes.

The Indian *Department of Atomic Energy* (DAE) is, therefore, planning to step up production of nuclear power as to boost its generation from 2% to 8% of the total electricity generation in the next 15 years. Whereas, the Deptt of Atomic Energy has calculated the nuclear power to be cheaper than the thermal power on an annual discounting of capital cost at about 8%, yet these calculations, based on various assumptions, have not been accepted by many, such as from Mr. Ramana, a fellow at *the centre for Interdisciplinary Studies in Environment and Development Bangalore,* and Mr. Amulya K.N. from the *International Energy Initiative.* ***Be that as it may, we need a mixed bag of various types of powers, since coal is an exhaustible source and cannot continue for ever.*** Hence, the aim of DAE to increase nuclear power production should not be opposed by saying that it is costlier than the thermal power or because of its high initial cost.

The latest emphasis of the present day NDA Govt. of India headed by Mr. Narender Modi, the Prime Minister, is however, to develop more and more **solar power,** the cost of which has come down drastically due to new policies of the Indian Govt.

Solar power capacity in India has, infact, gone up from mere 2.65 MkW to 9.56 MkW in just over two years. The unit tariff rate has also come down from ₹ 12 per unit to ₹ 2.62 per unit as in the recently auctioned Bhadla Solar Park in Rajasthan, compared to NTPC's average coal based power tariff of ₹ 3.20 p per unit. There are several reasons for this sharp drop in tariff in India. The cheaper solar panels with prices falling by 85 percent is one of the reasons. Another reason is the decision of the Indian Govt. (headed by **Mr. Narender Modi**) to provide security to the bidders that the solar power generated will be bought by the Govt. and payment ensured. This has been done by setting up **Energy Corporation of India** under the orbit of triportite agreement for payment of security against defaults by state distribution companies. Apart from ensuring purchase and payment, the Govt. has also taken care of the next big issue of land. Solar companies were finding land acquisition one of the biggest hurdles in setting up solar power plant. The Govt. has now allowed companies who specialise in land acquisition, and state Govts to allocate land which can be leased to parties interested in setting up solar unit. Moreover, most players are now entering into long term lease, even for solar panels, rather than purchasing them. These measures have brought down the price of solar power, drastically, leading to rise in creation and generation of solar power.

---

*The capital cost incurred on recent installation of two units of nuclear reactors of 220 MW each at the *Kaiga atomic power station ( Kaiga I & II)* was about ₹ 2900 crores with additional initial fuel loading cost of ₹ 1045 crores for heavy water and ₹ 84 crores for Uranium, as compared to ₹ 612 crores spent on installing *Raichur Thermal Power Station VII*, both plants being of similar size, cf recent vintage, and far away from coal supplies.

## PROBLEMS

1. (*a*) What is meant by Hydro-power ? Compare hydro-power with thermal power w.r. to Indian conditions.

   (*b*) Enumerate the different types of hydel plants, and describe the 'storage plant'.

2. (*a*) How do you classify a hydro-electric scheme on the basis of its operating head.

   (*b*) Define and differentiate between the following in connection with hydropower :

   (*i*) Firm and secondary power.

   (*ii*) Load factor, utilisation factor and plant factor.

   (*iii*) Installed and dependable capacity of a power house.

   (*iv*) Design head, rated head, gross head, operating head and effective head.

3. What are the principal components of a hydro-electric scheme ? Discuss the utility of each component.

4. (*a*) Write a brief note on the use and types of 'turbines' in a hydroelectric scheme.

   (*b*) What is a 'surge tank', and what are its types and uses ?

   (*c*) What is a 'draft tube', and what are its uses ?

5. Write short notes on :

   (*i*) Hydropower potential of India.

   (*ii*) Hydropower *vs.* thermal power.

   (*iii*) Surge tanks and their types.

   (*iv*) Hydraulic turbines and their types.

   (*v*) Trash racks.

# 25

# *Tank Irrigation*

## 25.1. Definition and General Introduction

Tank irrigation may be defined as the *storage irrigation scheme*, which utilises the water stored on the upstream side of a smaller* earth dam, called a *bund*. These earthen bund *reservoirs* are, thus, infact, called as *tanks*, specifically in South India, where such works are very common. This terminology is, therefore, limited to India only.

There is, thus, no technical difference between a 'reservoir' and a 'tank', except that a large-sized tank will be termed as a reservoir. Moreover, a reservoir will generally be formed by dams of any material (such as masonry dam, concrete dam, or earth dam, etc.) ; whereas, a tank is generally said to be formed by earth dams only (or strictly speaking by earthen bunds). These earthen bunds, spanning across the streams, are called ***tank bunds*** or ***tank banks***.

Most of the existing tanks of South India possess a *maximum* depth of 4.5 m, while a few are as deep as 7.5 to 9 m, and only a few exceptional ones exceed 11 m in depth. When the depth of the tank exceeds 12 m or so, the tank is generally referred to as a *reservoir*.

Like all earth dams, tank bunds are generally provided with *sluices* or *outlets* for discharging water from the tank for irrigation or other purposes. These *tank sluices* may be pipes or rectangular or arched openings passing near the base of the bund and through the body of the bund, and carrying the water into the downstream channel below the bund or transporting at distances where required, through pipes or canals. Sometimes, the ***supply sluices*** may not be passed through the body of the bund, and may be carried adjacent to it through some hill side at one end of the bund.

Similarly, as in the case of all dam reservoir projects, tanks are provided with arrangements for spilling away the excess surplus water that may enter into the tank, so as to avoid over-topping of the tank bund. These *surplus escape arrangements* may be in the form of a ***surplus escape weir***, provided in the body or at one end of the tank bund, or some other arrangement like a syphon spillway may be provided as is done in the case of earth dam projects. The surplus escape weir is a masonry weir (compared to an ogee spillway in an earth dam) with its top *i.e.* crest level at equal to *Full tank level* (F.T.L.). When tank is full upto FTL, and extra water comes in, it is discharged over the surplus escape weir. The length or capacity of this surplus escape weir will be so designed that water level in the tank does never exceed the *maximum water level* (M.W.L.). The top of the tank bund will be kept at a level, so as to provide a suitable free-board above this M.W.L.

Since the surplus escape weir is a masonry weir, it will have to be properly connected to the earthen bund by suitably designed bank connections.

---

*Generally less than 12 m or so in height.

## 25.2. Isolated Tanks and Tanks in Series

Most of the existing small sized tanks of south India form part of ***groups of tanks***, which are connected together in series, such that any tank either receives the surplus water of the upper tank(s), or sends its own surplus into some lower tank(s), or do both. However, when a tank neither receives water from an upper tank nor discharges its own surplus into a lower tank, it is called an ***isolated tank***. There do exist some isolated tanks also in South India.

Figure 25.1 shows a typical group of 15 tanks numbered 1 to 15, and an isolated tank A. It is evident that *considerable economy of water can be obtained from the system of grouping*, because the surplus water of the each tank and also the drainage of its wet cultivation are caught up by the next lower tank. Each tank of the group takes a share in the yield of the whole catchment above it, which can be classified as follows :

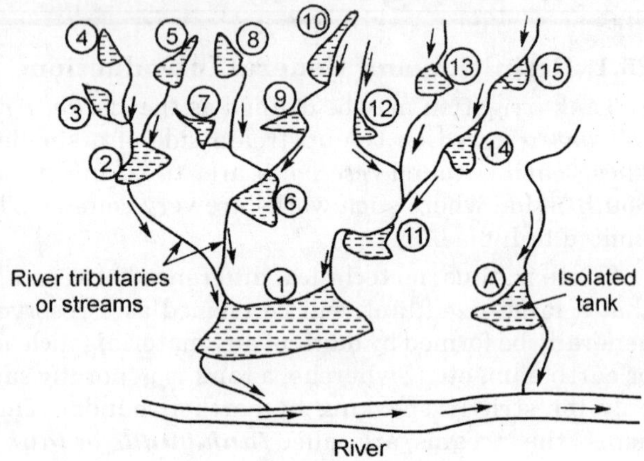


**Fig. 25.1.** Figure showing isolated tanks and tanks grouped in series.

(*i*) the ***free catchment***, which is the catchment area which only drains into the tank under consideration ; and

(*ii*) the ***combined catchment***, which is the area of the whole catchment above the tank under consideration.

The difference between the combined and the free catchment, thus, gives the area of catchment intercepted by upper tanks. The whole catchment of the highest tank, on each drainage, shall be its free catchment. Moreover, each tank will receive the whole run-off from its free catchment ; but from the remainder of its catchment it will receive only the balance run-off which remains after the upper tanks have been filled.

One of the disadvantage of grouping of tanks, however, is that, if a breach occurs in an upper tank, it exposes all the tanks in the series below, to the risk of similar failures. Hence, while making repairs or doing restoration works in such groups of tanks, it is necessary to consider each system of tanks collectively as a whole, and to carry out repairs from the upper end of each group. The extent of danger to lower tanks due to breach of an upper tank, however, depends upon their relative capacities. For example, if a small upper tank breaches and passes its storage into lower large-capacity tank, it may pose little danger even if the breach may completely empty the small tank. However, the danger will be imminent if the case is opposite, and the large sized upper tank breaches and passes its storage into the lower small sized tank.

The provision of suitable ***breaching sections**** in tanks forming part of a group is, therefore, considered as a very desirable precaution against their failure

---

***A 'breaching section' is a length kept lower and weaker than the remainder of the bound, so as to localise a breach in that length only.**

by breaching. This is because, a breach occurring in a tank bund at a well selected point (at the place of breaching section) will not empty the tank, and indeed frequently does little more than supplement its surplusing power to a reasonable degree to the greater security both of the tank itself and those below it in the group.

## 25.3. Capacity of Water Spread of a Tank

The *gross capacity of a tank* may be defined as the cubic content of water stored in the tank upto F.T.L. The *effective capacity of a tank*, will, however, be the cubic content of water stored between F.T.L. and the bottom or sill level of the lowest supply sluice.

These storage capacities can be computed easily by using the contour plan of the area of the water-spread ; the total capacity being the sum of the capacities between successive contours. The smaller the contour interval ($\Delta h$), the more accurate the capacity computation will be. This is because, if $A_1$ and $A_2$ represent the areas enclosed between two successive contours, then the cubic content between these contours is roughly taken as $\frac{A_1 + A_2}{2}(\Delta h)$.

The summation of all cubic contents between the successive contours will be the required storage capacity of the tank.

When the contour plan is not available, and only the area of the tank at F.T.L. is known, then the effective cubic content of the tank may be roughly computed as follows.

The area of tank at F.T.L. is multiplied by one-third of the depth from this level (F.T.L.) to the deep bed of the tank, or the level of the sill of the lowest sluice, whichever is higher of the two.

If the area ($A_1$) of the tank bed at the level of the sill of the lowest sluice, and the area ($A_2$) of the tank at F.T.L. are known even roughly, then the effective cubic content of the tank may be computed as $1/3\ [A_1 + A_2 + A_1A_2] \times h$, where $h$ is the height or difference in elevations between F.T.L. and sill level of the lowest sluice. This formula is based on the assumption that the water stored is in the shape of a frustum of a cone.

These principles on which the storage capacity of a reservoir is computed, were also discussed earlier under article 18.2.

## 25.4. Designing the Section of the Tank Bund

The tank bund, as you know, is nothing but a small-sized earth dam ; and hence, strictly speaking, its design and construction should be carried out in accordance with the principles applicable to earth dams (already discussed in details in chapter 20).

In accordance with these very principles, the tank bunds may be of three types; *i.e.*

Type A : *Homogeneous embankment type* (Fig. 20.1) ;

Type B : *Zoned embankment type* (Fig. 20.2) ; and

Type C : *Diaphragm type* (Fig. 20.3).

Most of the tank bunds of South India belong to type A, and they have been constructed with the soils excavated from pits in the immediate vicinity of each bund, and carried by head load to the bund.

As discussed in chapter 20, the large modern earth dams are designed after doing a lot of mathematical calculations for seepage and stability analysis, and keeping into consideration the latest advancements made in the field of soil mechanics. However, for small earthen bunds, generally no such calculations are carried out, and the sections are designed on the basis of practical considerations and experience derived from the success and failure of similar works, and of their working conditions.

The commonly adopted standards used for fixing the dimensions of tank bunds, in South India, are given in Table 25.1.

**Table 25.1. Common Dimensions of Tank Bunds**

| *S. No.* | *Depth of deep bed below F.T.L. i.e. maximum water depth in metres* | *Free board (in metres)* | *Width of top of bund (in metres)* |
|---|---|---|---|
| 1. | 1.5 to 3.0 | 0.9 | 1.2 |
| 2. | 3.0 to 4.5 | 1.2 | 1.5 |
| 3. | 4.5 to 6.0 | 1.5 | 1.8 |
| 4. | over 6.0 | 1.8 | 2.7 |

In favourable soils, such as red and white gravel, red and black loams, etc., the side slopes of the bund may be kept as ½ : 1 (½ *H* : 1 *V*) for smaller tanks with water depths not exceeding 2.5 m, and 2 : 1 for larger ones up to about 5 m in depth. In light sandy, or black cotton, or clayey soils, however, the slopes may be kept between 2 : 1 to 2½ : 1.

The upstream face of the tank bund is generally rivetted with stone apron or riprap (Fig. 25.2) so as to protect it against erosion, and if this is done, then the upstream slope is generally adopted is 1½ : 1 even upto 6 m depth. For inferior soils, or greater depths, however, the rivetted slope may be made flatter, say 2 : 1.

In this way, for average cases, a 1½ : 1 slope will generally be adopted for upstream face, and 2 : 1 slope for downstream face.

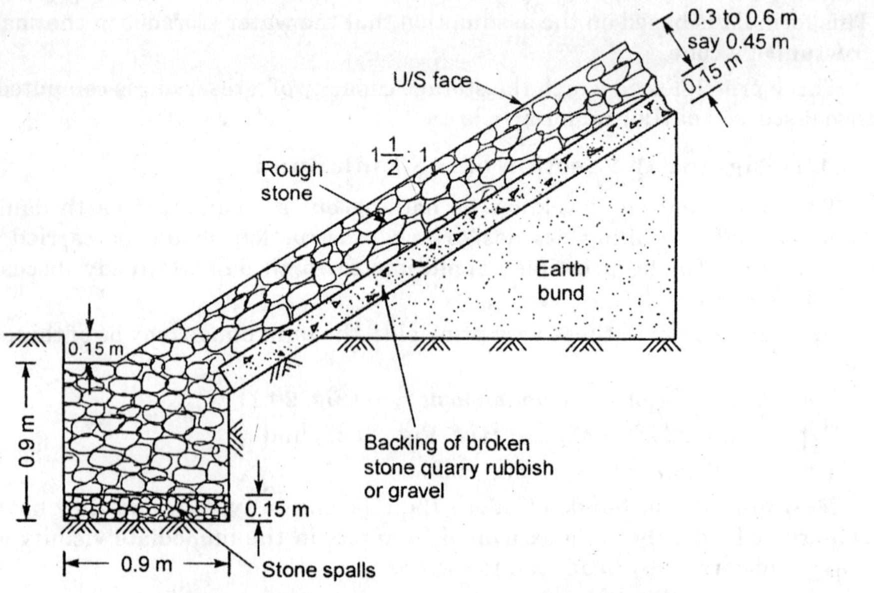


Fig. 25.2. Upstream revetment of tank bunds.

This practice is *contrary* to the standard recommendations adopted in various countries, where the upstream slope, even when revetted, is kept flatter than the downstream slope because of the soil being saturated. There are, however, thousands of tanks in Tamilnadu with water slopes of 1½ : 1, and failure by slipping of this slope is rare ; and hence, in normal cases, the local practice in this respect can be easily adopted. In very small tanks, and in case the upstream slope is heavily revetted, upstream faces have been given 1 : 1 or even steeper slopes in actual practice, but such steeper slopes are not recommended.

## 25.5. The Tank Weirs or Surplus Escape Weirs and Their Design Principles

**25.5.1. Definition.** As stated earlier, the excess surplus water is spilled from a tank, into the downstream channel, so as to avoid the rise of water in the tank above the M.W.L. In fact, the water will generally start spilling over the crest of this escape weir, as and when it rises above F.T.L. ; and the discharging capacity of this weir will be designed so as to pass the full maximum flood discharge (likely to enter the tank) with a depth over the weir equal to the difference between F.T.L. and M.W.L.

Although the effective storage capacity of a tank is limited by F.T.L., but the area submerged by the tank bund and revetment are all dependent on M.W.L. And hence, in order to restrict the dimensions of these, it is desirable to keep the difference between F.T.L. and M.W.L. to a smaller value. But, on the other hand, the smaller is this difference, the longer will be the surplus escape required in order to enable it to pass the given discharge. Hence, the difference ($H$) between F.T.L. and M.W.L. is fixed on a compromise basis, in each particular project, so as to obtain overall economy and efficiency. In small and medium sized tanks, the usual difference between F.T.L. and M.W.L. is kept from 0.3 to 0.6 metre, and it is rarely allowed to exceed 0.9 metre.

**25.5.2. Types of Tank Weirs and Their Cross-sections.** Tank escape weirs are similar to river-weirs (*i.e.* diversion weirs or anicuts), and hence, they may be classified into the following three general types :

*Type A* : *Masonry weirs with a vertical drop* ;

*Type B* : *Rockfill weirs with a sloping apron*; and

*Type C* : *Masonry weirs with a sloping masonary apron (glacis).*

These three types of river weirs have already been discussed in article 9.3.1.

Besides these three important types of river weirs, which are used as tank weirs also, the fourth type of a tank weir which is a combination of type A and type C, may also be used. In such a *D type weir,* a number of vertical steps are made (as in the case of a stepped fall) instead of providing a horizontal or sloping downstream apron. Such weirs of type D are called *weirs with stepped aprons.* The A and D type of weirs are most widely adopted, although every type of weir has been used in South Indian tanks.

The A type of weir *i.e.* weir with a masonry pucca horizontal floor, may sometimes by provided with depressed floors, so as to provide sufficient water-cushion for absorbing the impact of high water falls. Typical sections of the two sub-types of A type, and D type of weirs, and the conditions under which each is adopted are shown in Figs. 25.3 (*a*), 25.3 (*b*), and 25.4 respectively.

The general principles and considerations which guide the design of tank weir sections, are the same as in the case of river weirs founded on *rocks* or *loose*

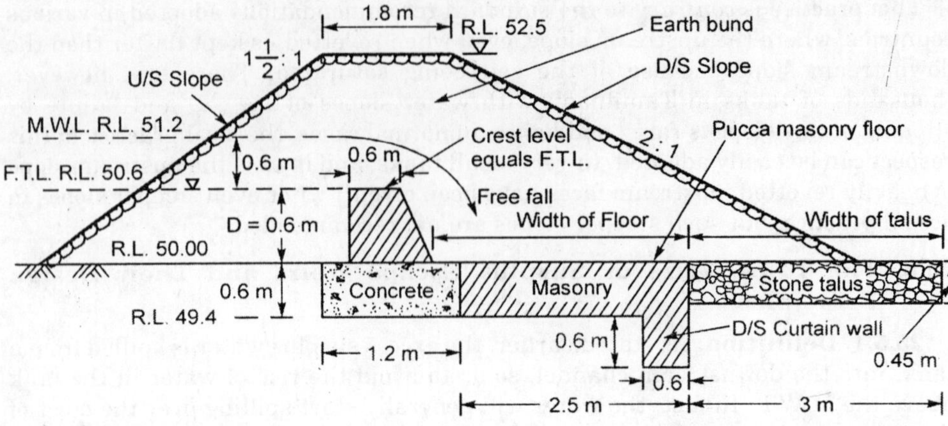


Fig. 25.3(*a*) A typical masonry tank weir with a vertical drop, with a horizontal undepressed floor, suitable for low drops of 0.6 to 0.9 metre or for even higher drop when the soil is a rock.

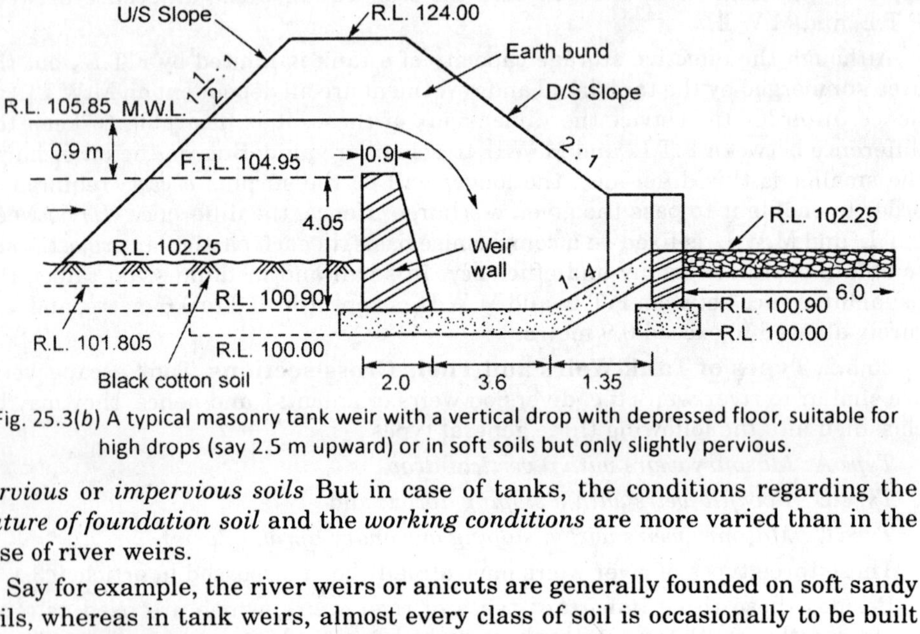


Fig. 25.3(*b*) A typical masonry tank weir with a vertical drop with depressed floor, suitable for high drops (say 2.5 m upward) or in soft soils but only slightly pervious.

*pervious* or *impervious soils* But in case of tanks, the conditions regarding the *nature of foundation soil* and the *working conditions* are more varied than in the case of river weirs.

Say for example, the river weirs or anicuts are generally founded on soft sandy soils, whereas in tank weirs, almost every class of soil is occasionally to be built on, except this rarely met foundation sand.

Similarly, the tail water conditions in anicuts differ from those in tank weirs. This is because, the tail water backing up during floods and affording some protection to the talus and to the soil at the downstream toe of the apron, are absent in the majority of tank weirs ; and, therefore, the tank weirs built on soft soil would, other things being equal, be more exposed to damage by scour at the downstream toe of the apron and retrogression of levels than river weirs.

On the other hand, the depth of water passing over the crests of tank weirs is generally far less than that over anicuts, and in the case of very many tanks, it is only, on an average, for a few days each year, that any flow at all passes over the weir, and while the duration of continuous heavy flow is frequently limited to a maximum of two to three days. Tank weirs are, thus, usually worked, far less heavily and far less continuously than anicuts, and this fact may be duly considered while designing them.

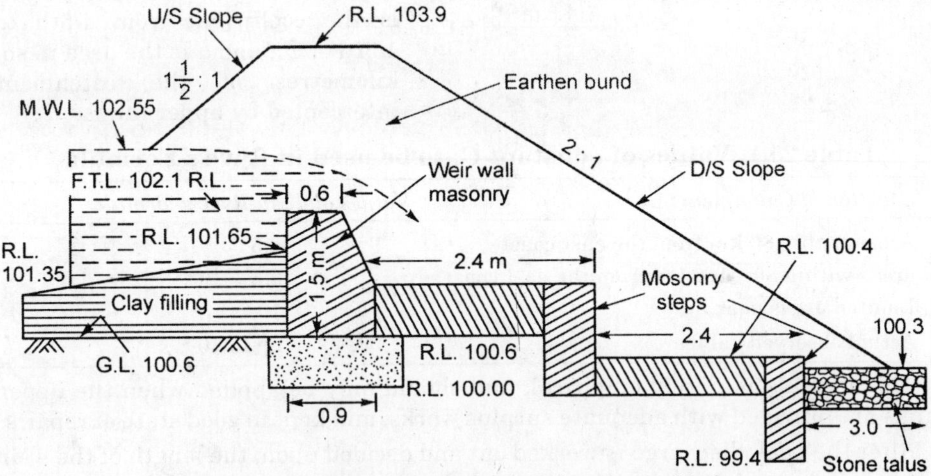


Fig. 25.4. A typical stepped-apron tank-weir, suitable for sites on soft soil when the drop exceeds 0.9 metre.

*Empirical Formulas for Determining Width of Floors of Tank Weirs.* The width of the horizontal floors of types $A$ and $D$ weirs, from the foot of the drop wall to the downstream edge of the floor should never be less than 2 $(D + H)$ where $D$ is the height of the drop wall, and $H$ is the maximum water head over the wall. In important works, this width may be increased to 3 $(D + H)$. The rough stone apron forming a *talus* below the last curtain wall may be of varying widths depending on the nature of the soil and the velocity and annual probable quantity and intensity of runoff ; it would generally vary from 2.5 $(D + H)$ to 5 $(D + H)$ according to varying conditions.

**25.5.3. Length of the Tank Weir.** In order to determine the length of the escape weir, it is, first of all, necessary to determine the *maximum flood discharge* that may enter into a tank, after it is filled up to full tank level. This peak discharge that may come from the free catchment of a tank can be fairly estimated by using the empirical formulas applicable to the given region, as discussed in article 7.17.2 of *Water Resources Engg. (Vol. I).* Generally, in South India, all tank weirs have been designed on the basis of Ryve's formula given by Equation:

$$Q_p = C_1 \,.\, A^{2/3} \qquad \ldots(25.1a)$$

where $C_1$ = Ryve's coefficient (Table 25.1)

$A$ = Catchment area in sq km

$Q_p$ = Peak flood discharge in cumecs

This formula is directly applicable for free catchments in all isolated tanks. However, for combined catchment of a tank forming a constituent unit of a group, the following modified formula is used for calculating peak discharge:

$$Q_p = C_1 \,.\, A^{2/3} - c_1 a^{2/3} \qquad \ldots(25.1b)$$

where $C_1$ is the coefficient in Ryve's formula (Refer Table 25.1)

$A$ is the area of the combined catchment in sq. kilometres

$c_1$ is the coefficient from 1/5th to 1/3rd of $C_1$; and $a$ is the area in sq. kilometres of the catchment intercepted by upper tanks.

**Table 25.1. Values of constant $C_1$ to be used in Ryve's Formula**

| *Location of Catchment* | *Values of $C_1$ in Ryve's formula* |
|---|---|
| Areas within 80 km from the east coast | 6.5 |
| areas within 80—160 km from the east coast | 8.5 |
| Limited areas near hills | 10.2 |
| Actual observed values | up to 40 |

This formula is purely empirical, and should only be applied when the upper tanks are provided with adequate surplus works, and kept in good state of repairs.

After the peak discharge is worked out and decided upon, the length of the weir can be worked out by using the appropriate discharge formula applicable to the weir type adopted in the project. In general, the discharge over a broad crested free weir and without any velocity of approach, is given as :

$$Q = C \cdot L \cdot H^{3/2} \qquad \text{...(25.2)}$$

where $L$ = the length of the weir ;

$H$ = the head of water over the weir (*i.e.* the difference between M.W.L. and F.T.L.), and

$C$ = a constant, the common values of which for three commonly used types of weirs in South India, are shown in Fig. 25.5.

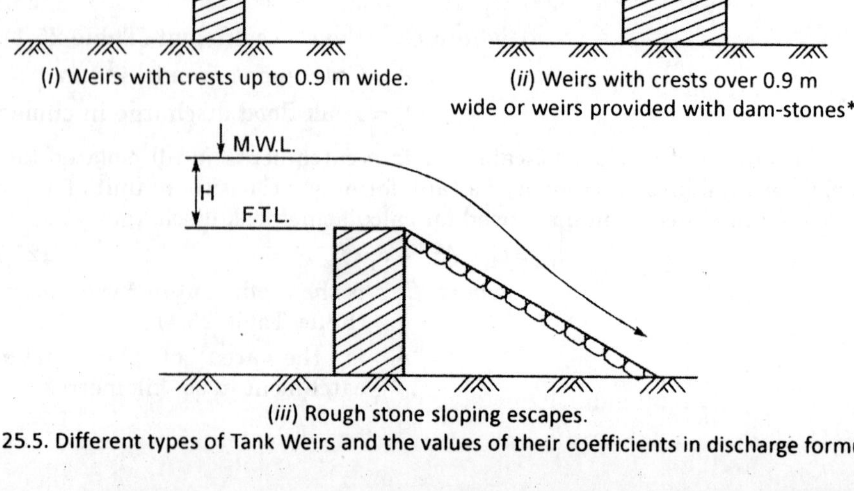


(*i*) Weirs with crests up to 0.9 m wide.

(*ii*) Weirs with crests over 0.9 m wide or weirs provided with dam-stones*

(*iii*) Rough stone sloping escapes.

Fig. 25.5. Different types of Tank Weirs and the values of their coefficients in discharge formulas.

*A common practice which has been resorted to for centuries in South India is to fix **damstones** in the crest wall of the weir. These stones are generally about 15 cm × 15 cm in section, and 0.75 to 0.9 m high, built into the masonry weir wall at about 0.45 to 0.6 m clear interval ; projecting above the weir crest to a height of 0.3 to 0.6 metre, their tops being generally at M.W.L.

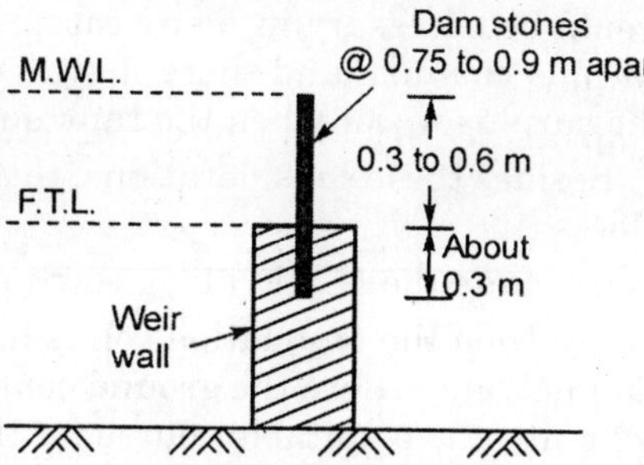

Fig. 25.6. Dam-stones.

The space between these stones are blocked up with clay and turf by cultivators, with the result that water can be held up in the tank up to M.W.L., before any surplusing takes place. When the tank receives more water, which just causes the water level in the tank to rise above M.W.L., the water over-tops the turf weir, which is rapidly washed away, and the surplus weir starts functioning with full head ($H$), without M.W.L. exceeding to any appreciable extent.

Whenever this occurs, all the water stored in the tank above the weir crest level, must however, be passed off before the turf dam can be rebuilt.

The dam stones cause obstruction to the discharge, and when they are installed, the length of the weir must be correspondingly increased, *i.e.* constant ($C$) in weir discharge formula reduced from 1.84 to 1.66.

Sometimes, instead of dam stones, shutters may be provided between F.T.L. and M.W.L., so that they can be lowered and raised as and when needed, thus serving the purpose of giving full pondage up to M.W.L. before the worst flood arrives for surplus disposal. They may, however, create operational problems, and necessitate special establishment.

If no dam stones or shutters are provided above F.T.L., and flood water starts spilling over the surplus work immediately after the tank water level exceeds F.T.L., then *strictly speaking*, the computed peak flood discharge should not be used to directly work out the length of the weir by the above simple and straight formulas. This is because of the fact that certain flood discharge will be absorbed between F.T.L. and M.W.L. This *flood absorption capacity* of a tank is, however, considered only in very big tanks or in large-sized tanks, and this is kept unaccounted as a further measure of safety in such works. A method of computing the allowance which may be made for the absorption capacity of big tanks has been worked out by Capt. Garret, and is given in his paper on the "General Theory of the Storage Capacity and Flood Regulation of Reservoirs" published in 1912 by the Government of India. The same may be referred to if needed along with its original tables given in the same paper.

**25.5.4. Choice of a Suitable Site for Locating a Tank Weir.** As pointed out earlier, a surplus weir can be located either in the body of the tank bund or at one end of it. When a saddle, disconnected with the tank bund is available, it must be preferred for locating the surplus weir, as compared to providing it within the bund body, but even in such a case, it will be necessary to locate it at one flank of the bund and connected with it.

If all the cultivation is on one side of the stream which is dammed by the bund, then it is preferable to locate the surplus weir on the other flank ; otherwise it would be necessary in many cases, to carry the distributary channels across the surplus channel, and there may also be danger of cultivation being damaged by the surplus water when the tank surpluses freely.

Besides these considerations, the most favourable sites for surplus weirs are at places ;

(*i*) where the N.S.L. along the line of the weir approximately equals F.T.L.

(*ii*) when the foundation soil is hard both on the weir site and along the runoff channel, and where the ground contours are such as to give clear approach to the weir site and a suitable course for the surplus water.

## 25.6. Tank Outlets or Tank Sluices and their Design Principles

As stated earlier, a *tank sluice* is an opening in the form of a culvert or a pipe running through or under the tank bund, and supplying water from the tank to the distributary channel below, to meet the irrigation or other water requirements, as and when needed. Suitable wing walls and other bank connections are also provided, as required at the head and tail end of the culvert.

In ordinary medium-sized sluices, ***masonry culverts*** with or without arch roofs are generally constructed (Refer Fig. 25.7). The size of the culvert (*i.e.* its cross-section) will depend upon the maximum quantity of water ($Q$)* it is required to convey, but in no case should be less than 0.6 metre wide and 0.75 metre high, so as to allow a man to enter it for examination and repairs or removal of obstructions. The size of the barrel should also be such as to limit the velocity through the sluice barrel to a maximum of 4.5 m/sec*, under the condition of plug hole being fully open and with the water at full tank level.

In case of very small sluices, earthen ware or cement or cast iron pipes may be used to take the place of masonry culverts. (Refer Fig. 25.8). Such sluices are called ***pipe sluices***. They are generally not adopted in tank bunds, where the depth below F.T.L. exceeds 2.5 metre or so. This is because, in such cases, the earthen ware pipes may get fractured, or leakage through their joints may take place, resulting in a breach, as the pipes can neither be examined nor repaired easily without cutting open the bund. Moreover, these pipes and especially the cast iron pipes are rarely found to be economical compared to masonry sluices.

Besides the above types of tank sluices, one more type of sluice *i.e.* ***a sluice with a tower head*** (Refer Fig. 25.9) is also sometimes provided. This form is sometimes found more economical than the types previously referred to, owing to the saving obtained by avoiding the heavy wing walls of the previous type. Moreover, here there is lesser danger of failure by cracking and bulging which frequently occurs in the wings of the other type. But, on the other hand, with such a sluice, additional length of the barrel is required. Tower-heads or wells are generally placed in the water slope of the bund, as thereby, the expanse of a bridge or causeway leading from the bank to the regulation platform is saved. But this has the disadvantage of making it impossible, while the tank has water, to have access to the portion of the barrel upstream of the well.

---

$$* \; V = \frac{Q}{A} = \frac{\text{Discharge}}{\text{Area of barrel}} = Cd \, . \sqrt{2g \, . \, h}$$

where $h$ is the height of the F.T.L. above the sill level of the sluice.

$Cd$ = 0.63 for free fall.

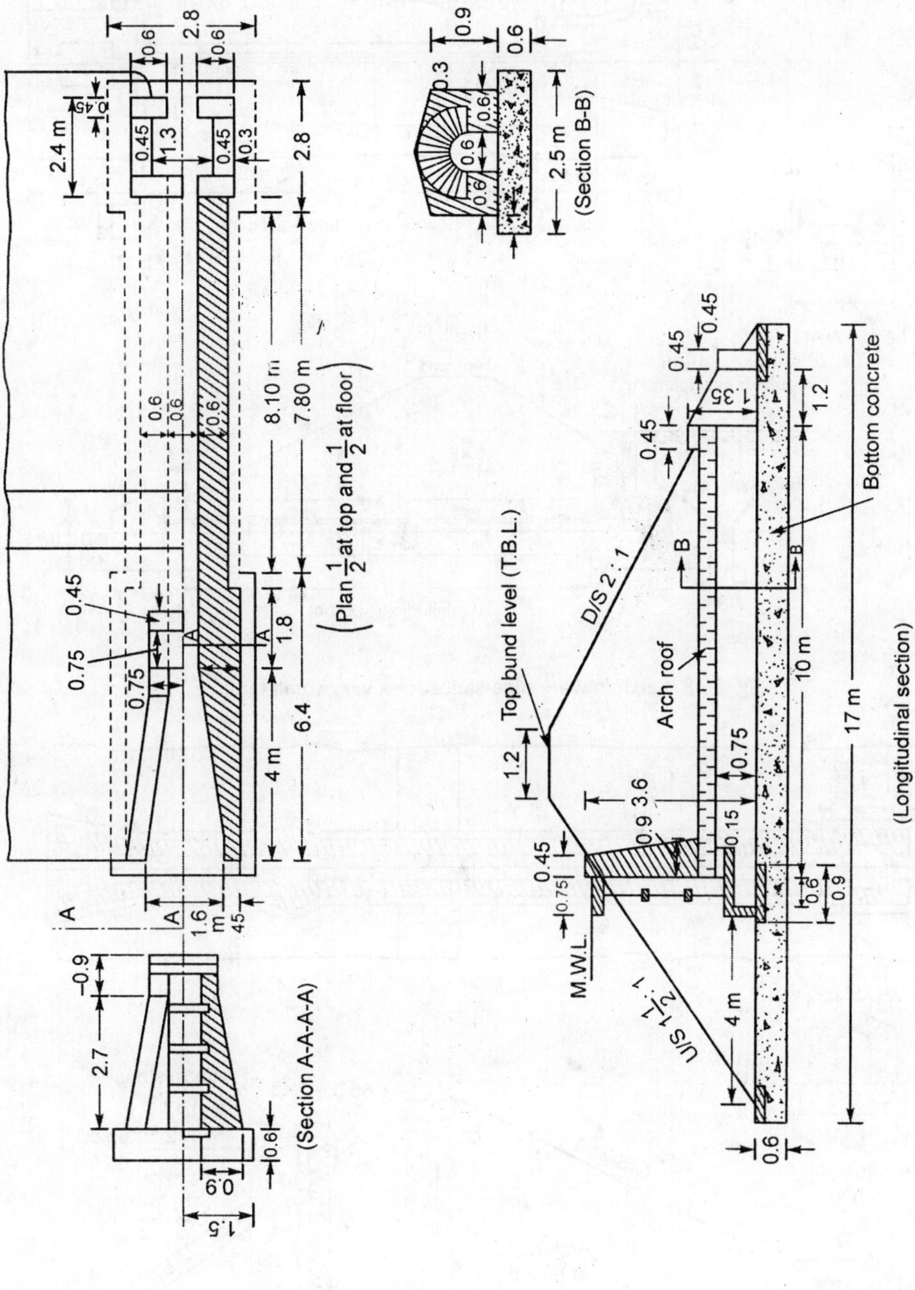


Fig. 25.7. Masonry supply sluice for a medium-sized tank.

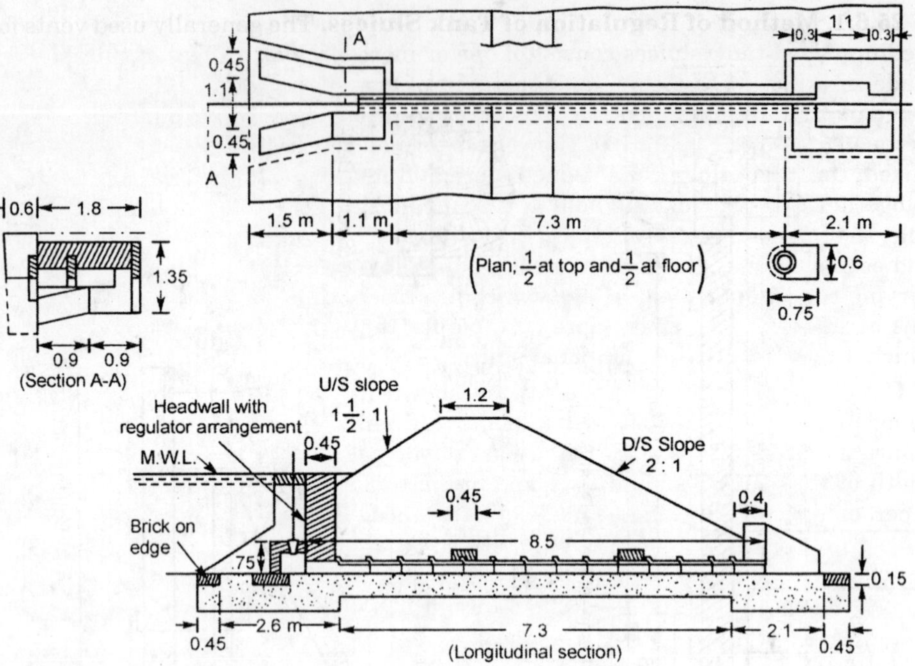


Fig. 25.8. Earthen ware pipe-sluice for a very small tank.

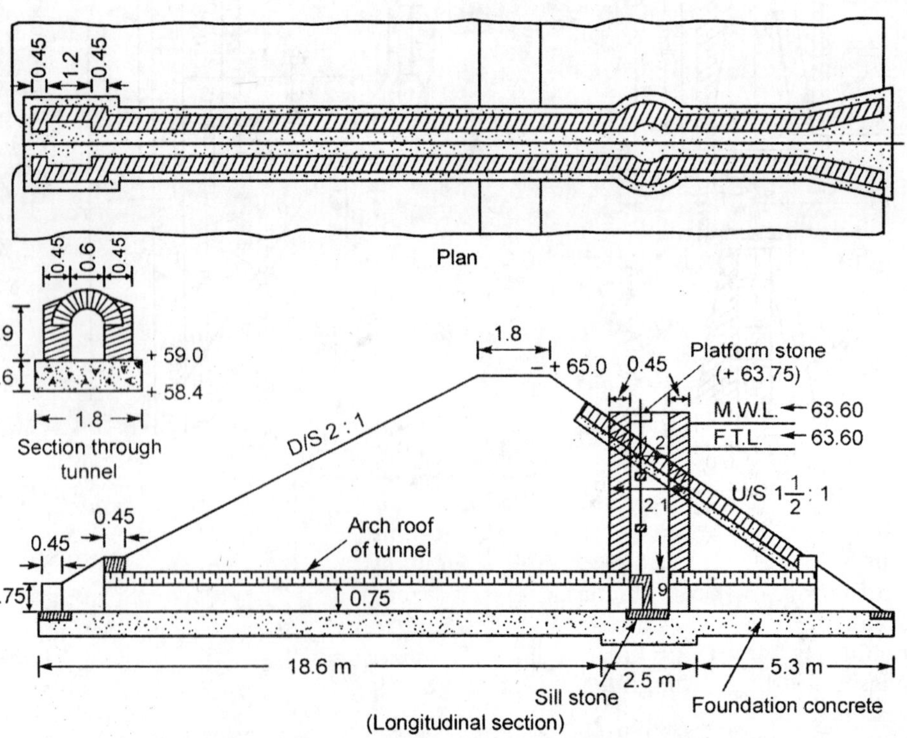


Fig. 25.9. Tank sluice with a tower-head.

**25.6.1. Method of Regulation of Tank Sluices.** The generally used vents for medium-sized tank sluices consist of one or more circular orifices of suitable size made in a horizontal stone, and the area of the opening is regulated by a long coned plug, the wider end of which is of the same diameter (allowing for clearance) as the vent (Refer Fig. 25.10). The vent is termed as the ***plug-hole.*** As the plug is raised, the hole or orifice is opened more and more, and when the plug is raised altogether clear of the plug hole by an amount not less than the diameter of the hole, the full discharge from the orifice will pass. This arrangement is both simple and economical, and affords an easily adjustable means of regulation to suit the varying conditions of water level which are inseparable from the reservoirs, and this arrangement is, thus, more suitable for tank sluices than the flat shutters which are generally used for canal sluices.

Details of plugs for tank-sluices are shown in Fig. 25.10, the standard sizes being 100, 150, 200, 250 and 300 mm dia, this being the size of the hole in the plug stone, the largest size of the cone part of the plug being about 6 mm less ; the width of the seating should be from 25 mm to 30 mm. The cones should have a taper of one to four, and be made of hard wood ; the smaller end of cone and the

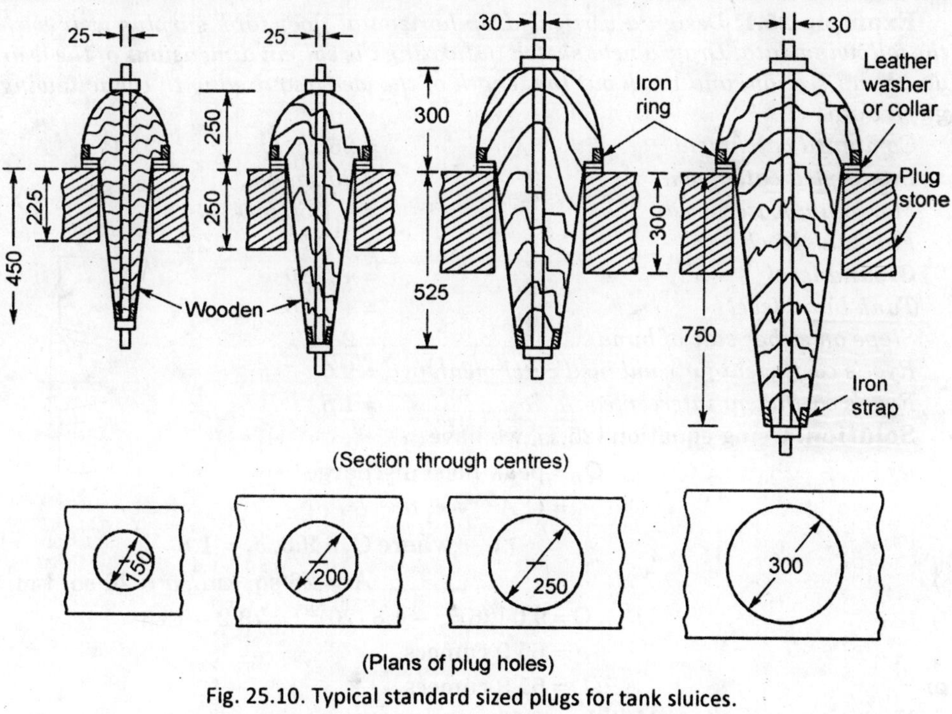


Fig. 25.10. Typical standard sized plugs for tank sluices.

greatest dia. of the plug should be cut cylindrical to take countersunk iron straps which should be shrunk on to prevent the plug from splitting. The lifting rod should be round iron, and secured by a split cotter pin at each end of the plug. When the rod is long, it should be passed through holes in guide beams fixed at intervals. The top end of the rod passes through a hole in the regulation platform; it should be flat iron with holes at about 75 mm interval through which an iron pin is passed to keep the plug suspended at whatever height it gives suitable elevation.

When the computed size of a tank sluice or plug hole is more than 250 to 300 mm in diameter, then it is generally advisable to provide two or more holes and plugs in the plug platform stone ; while for deep sluices, say over 7.5 m depth, two or more plug stones at different levels, each with separated plug holes, provide greater facility for easy regulation, and hence adopted.

The size of the orifices in the plug stone are generally calculated so as to pass the full supply with vents fully open and with 0.3 m of water standing over the plug stone platform. For regulation below this level, a rectangular vertical vent is provided which is closed by a shutter, which may be wooden but is generally a slab of stone. This vent is made of sufficient size to pass full supply with 0.15 m head. This shutter is brought into use only when the water in the tank is less than 1.2 m over the floor of the sluice; and at other times, the shutter is kept completely covering the orifice, and leakage past it should be stopped by filling silt against its upstream face. This vent will be regulated by men standing on the plug stone.

In case of large tanks, where the quantity of water to be released is great, and where adequate establishment is maintained, flat shutters working in grooves, and regulated by screw spears, may be installed.

For heads of over 9 m or so, special balanced valves or shutters moving on rollers are generally installed.

**Example 25.1.** *Design a vertical drop horizontal floor tank surplus weir with the following data. Draw a neat sketch indicating the salient dimensions of the weir and bund section, and work out the length of the weir required with the following given data:*

| | |
|---|---|
| *Combined catchment* | *= 26 sq. km.* |
| *Intercepted catchment* | *= 20 sq. km.* |
| *Maximum water level* | *= + 7.50 (m)* |
| *Full tank level* | *= + 6.70* |
| *Ground level* | *= + 5.80* |
| *Tank bund level* | *= + 1.80* |
| *Slope on either side of bund* | *= 2 : 1* |
| *Ryve's coefficient for combined catchment* | *= 9.0* |
| *Ryve's coefficient intercepted* | *= 1.8.* |

**Solution.** Using equation (25.1), we have

$$Q_P = \text{peak flood discharge}$$
$$= C_1 A^{2/3} - c_1 a^{2/3}$$

where $C_1 = 9.0$, $c_1 = 1.8$

$A = 26$ sq. km., $a = 20$ sq. km.

$$Q = 9.0\,(26)^{2/3} - 1.8\,(20)^{2/3} = 79.2 - 13.3$$
$$= 65.9 \text{ cumecs.}$$

or $$Q_p = 65.9 \text{ cumecs.}$$

Now, M.W.L. = 7.50 m

F.T.L. = 6.70 m

Ground level = 5.80 m

∴ Difference between M.W.L. and F.T.L. *i.e.* maximum head of water over surplus weir = $H$ = 7.50 – 6.70 = 0.8 m.

crest of weir is kept at F.T.L.

∴ Depth of drop *i.e.* height of weir crest

$$D = 6.7 - 5.8 = 0.9 \text{ m.}$$

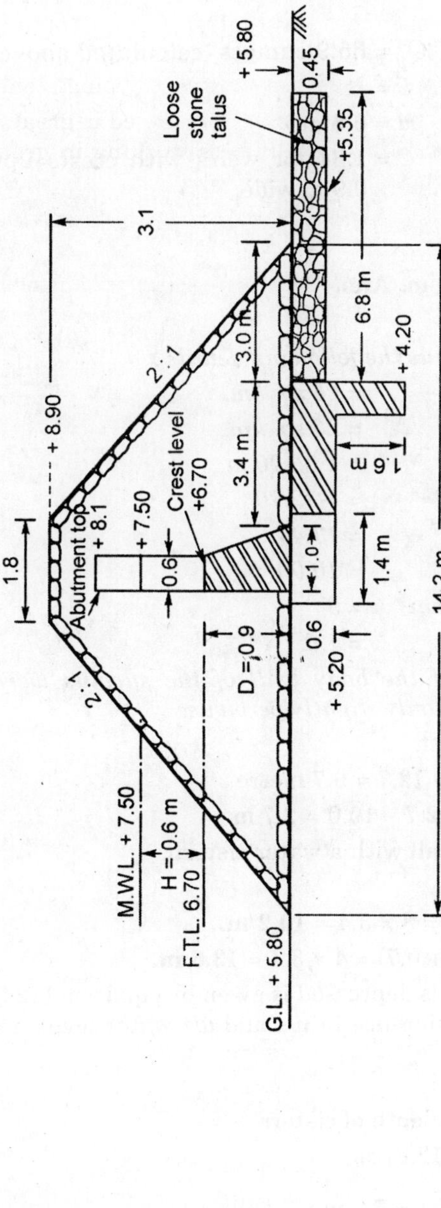


Fig. 25.11. Section of the designed tank surplus weir in example 25.1.

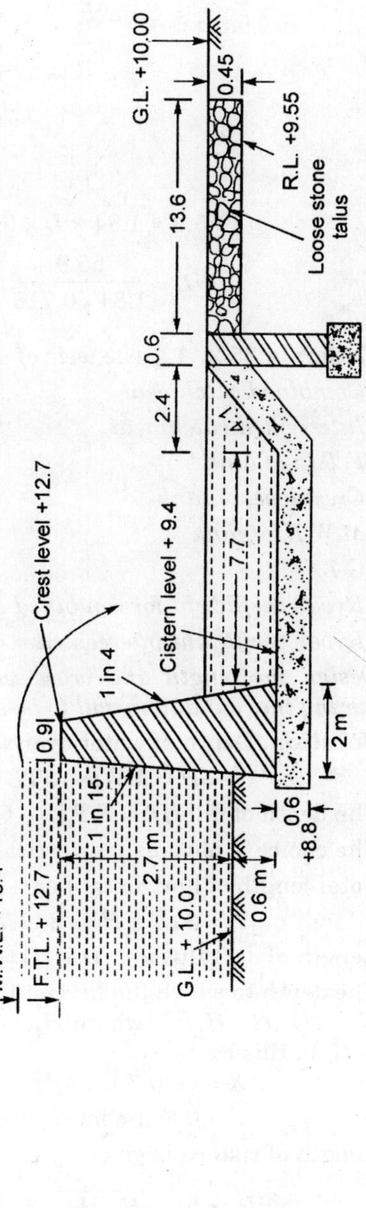


Fig. 25.12. Section of the designed tank weir in example 25.2.

Since the drop is small, we can adopt a masonry weir with a horizontal masonry floor (without depressed floor) *i.e.* type *A*, Fig. 25.3 (*a*), easily.

The widths of horizontal pucca floor and talus are calculated below :

The width of horizontal masonry floor = $2\,(D + H) = 2\,(0.9 + 0.8) = 3.4$ m.

The width of rough stone talus = $4\,(D + H) = 4\,(0.9 + 0.8) = 6.8$ m.

Use the top crest width equal to 0.6 m, and draw the sketch as shown in Fig. 25.11.

The length of the weir can be determined by using equation (25.2) as below :

$$Q_p = C \,.\, LH^{3/2}$$

where $Q_p$ = 65.9 cumecs (calculated above)
$L$ = ?
$H$ = 0.8 m.
$C$ = 1.84 for weirs with crests upto 0.9 m wide.

$$\therefore \quad 65.9 = 1.84 \times L \times (0.8)^{3/2}$$

or
$$L = \frac{65.9}{1.84 \times 0.716} \text{ m} = \mathbf{50\ m.\ Ans.}$$

**Example 25.2.** *A waste weir of a tank has the following details :*

| | |
|---|---|
| *Combined catchment* | *= 26 sq. km.* |
| *Intercepted catchment* | *= 20 sq. km.* |
| *F.T.L: of tank* | *= + 12.70 m.* |
| *Crest level of tank* | *= + 12.70* |
| *M.W.L. of tank* | *= 13.4* |
| *G.L.* | *= 10.00* |
| *Ryve's coefficient for combined catchment* | *= 9* |
| *Ryve's coefficient intercepted* | *= 1.5.* |

*Design the length and cross-section of the body wall of the surplus work, assuming the foundation soil to be soft but only slightly pervious.*

**Solution.** Depth of water over weir crest

$$= H = MWL - FTL = 13.4 - 12.7 = 0.7 \text{ metre}$$

The depth of fall ($D$) = F.T.L. – G.L. = 12.7 – 10.0 = 2.7 m.

The choice is obviously a vertical drop fall with a water cistern.

Total length of horizontal floor

$$= 3\,(D + H) = 3\,(2.7 + 0.7) = 3 \times 3.4 = \mathbf{10.2\ m.}$$

Length of d/s talus = $4\,(D + H) = 4\,(2.7 + 0.7) = 4 \times 3.4 =$ **13.6 m.**

The depth to which the horizontal floor is depressed is given by equation (12.9) as $X = 1/4\,(H \,.\, H_L)^{2/3}$, where $H_L$ is the difference in u/s and d/s water levels *i.e.* ($H + D$) in this case.

or
$$X = \tfrac{1}{4}\,(0.7 \times 3.4)^{2/3}$$
$$= 0.5 \text{ ; So let us use 0.6 m depth of cistern}$$

Length of cistern is given by equation (12.8) as

$$L_C = 5 \,.\, \sqrt{H \,.\, H_L} = 5 \,.\, \sqrt{0.7 \times 3.4} = 7.7 \text{ m}$$

The slope length of concrete floor in a slope of 4 : 1 will be = 4 × 0.6 = 2.4 m.

d/s cut-off wall thickness = 0.6 m

In this way, total length of floor provided

$$= 7.7 + 2.4 + 0.6 = 10.7 > 10.1 \text{ ; safe.}$$

Hence, provide the section of the waste weir, as shown in Fig. 25.12.

**Length of weir**

$$Q_p = C_1 A^{2/3} - c_1 a^{2/3}$$

where $C_1 = 9.0$

$c_1 = 1.5$

$A = 36$ sq. km.

$a = 20$ sq. km.

$$Q_p = 9\,(36)^{2/3} - 1.5\,(20)^{2/3}$$

$$= 9 \times 10.9 - 1.5 \times 7.4 = 98.1 - 11.1 = \textbf{87 cumecs}$$

Now $Q_p = CLH^{3/2}$

$$87 = 1.84 \times L\,.\,(0.7)^{3/2}$$

$$\therefore \quad L = \frac{87}{1.84 \times 0.948} = 49.8 \text{ m ; } \textbf{say 50 m}$$

Hence, use 50 m length of weir. **Ans.**

**Example 25.3.** *A tank has a catchment area of 120 sq. km. ; out of which 20 sq. km. make free catchment. Average annual rainfall on catchment is 75 cm. What should be the desirable gross storage for the tank ? If rice crop, with a duty of 640 hectares per cumec at the head of main canal and a crop period of 120 days, is to be irrigated from the water of this tank, find the area under the rice crop.*

*What should be the length of the waste weir to dispose of safely the high flood discharge, with a head of 0.6 m on the crest of the weir ? Use Ryve's formula, with coefficient as 7.5.*

*How will the live storage be affected if the tank gets filled 1½ times in a year.*

**Solution.** Average annual rainfall = 75 cm.

The rainfall of average bad year = 2/3rd of the average annual rainfall (say).

$$= \frac{2}{3} \times 75 = 50 \text{ cm.}$$

Run-off (as depth of water over the catchment) of this average bad year

$$= 20\% \text{ (say) of rainfall} = \frac{20}{100} \times 50 = 10 \text{ cm.}$$

The total catchment area of 120 sq. km. consists of independent catchment of 20 sq. km. and intercepted catchment of 100 sq. km.

Total volume of runoff in average bad year from free as well as intercepted catchment of the given tank

= full runoff from the free catchment + 1/7th of the runoff fr the intercepted catchment on the assumption that 6/7 of runoff from the intercepted catchment is impounded i upper tanks.

$$= \left[20(1000 \times 1000) \times \left(\frac{10}{1000}\right) + \frac{1}{7} \times 100(1000 \times 1000)\frac{10}{100}\right]$$

$$= \left[\frac{20 \times 10^6}{10} + \frac{1}{7} \times 100 \times 10^6 \times \frac{1}{10}\right] \text{m}^3$$

$= 10^6\ [2 + 1.43] = 3.43 \times 10^6\ \text{m}^3 = 3.43$ million cubic metre.

∴ Live storage = 1.1 × 3.43 = 3.77 M.m³

and, Gross storage = $\left(\frac{10}{9}\right)$ 3.77 = **4.19 M.m³. Ans.**

Water required at the head of the main canal to irrigate 640 hectares of rice crop

= 1 cumec for 120 days.
= 1 × (120 × 24 × 60 ×60) cu.m.
= 10.37 × 10⁶ cu.m. = 10.37 M.m³

Hence, the area of rice crop which can be irrigated from the full tank water

$$= \frac{\text{Effective storage of the tank}}{10.37} \times 640 \text{ hectares}$$

$$= \frac{0.9 \times \text{Live storage}}{10.37} \times 640 \text{ hectares}$$

(assuming the losses in the tank to be 10% of live storage).

$$= \frac{0.9 \times 3.77}{10.37} \times 640 = \textbf{209.4 hectares.} \quad \textbf{Ans.}$$

Now, the high flood discharge from the catchment into this tank (after the tank is full) is given by equation (25.2) as

$$Q_p = C_1 A^{2/3} - c_1 a^{2/3},$$

where $C_1 = 7.5$ (given)

$c_1 = \frac{1}{5} C_1$ (assumed) = 1.5

$A$ = Full combined catchment = 120 sq. km. (given)

$a$ = Intercepted catchment = 100 sq. km. (given)

∴ $Q_p = 7.5\,(120)^{2/3} - 1.5\,(100)^{2/3}$
$= 183 - 32.4 = 150.6$ cumecs.

Now, if $L$ is the length of the waste weir, then

$Q_p = C \,.\, L \,.\, H^{3/2} = 1.84 \,.\, L \,.\, H^{3/2}$

∴ $150.6 = 1.84 \times L \times (0.6)^{3/2}$

or $L$ = **176 m. Ans.**

If the tank fills 1½ times in a year, then the live storage

$$= \frac{3.77}{1.5} = \textbf{2.51 M.m}^3. \quad \textbf{Ans.}$$

## PROBLEMS

**1.** What is an irrigation tanks ? Differentiate between isolated tanks and group of tanks. How can your compute the storage capacity of an irrigation tank ?

**2.** (*a*) What are the causes of failure of tank bunds ?

[**Hint.** Refer article 20.7]

(*b*) Sketch a typical cross-section of a tank bund of type C. Under what circumstances such sections are chosen ?

(*c*) Explain briefly the function of a breaching section in a tank bund.

**3.** Design a tank surplus weir with the following data. Draw a neat sketch (not to scale) indicating the salient dimensions.

| | |
|---|---|
| Combined catchment | = 26 sq. km |
| Intercepted catchment | = 20 sq. km. |
| Full tank level | = + 6.70 (m) |
| Maximum water level | = + 7.50 m |
| Ground level | = + 5.80 m |
| Foundation level | = + 4.20 m |
| Tank bund level | = + 8.90 m |
| Top width of bund | = 1.8 m |
| Slope on either side of bund | = 2 : 1 |

Downstream level is + 4.80 in a distance of 8 m.

**4.** A tank surplus work has the following details :

| | |
|---|---|
| Combined catchment | = 26 sq. km. |
| Intercepted catchment | = 20 sq. km. |
| F.T.L. of tank | = 12.00 m. |
| Crest level of weir | = 12.00 m. |
| M.W.L. of tank | = 12.75 m. |
| G.L. | = 11.00 m. |
| Top of foundation concrete | = 9.60 m |
| Ryve's coefficient for combined catchment | = 9 |
| Ryve's coefficient intercepted | = 1.5 |

Design the length and cross-section of the body wall of the surplus work.

**5.** What is a waste weir, and where is it installed in a project of an irrigation tank ? Discuss its importance, and principles involved in designing such a work.

**6.** (*a*) What is a tank weir, and what are its types ? Discuss the conditions favourable to the selection of a particular type.

(*b*) How does the foundation and working conditions differ for a tank weir as against a river weir.

(*c*) Where should a tank weir be located in a tank irrigation project ?

**7.** (*a*) What are dam-stones and where are they installed ? What are their advantages and disadvantages ?

**8.** What is a surplus weir and its types in an irrigation tank project ? Draw net sketches of the sections of A type weirs with and without water cushion provision.

**9.** (*a*) What do you mean by a tank sluice, and where is it installed ?

(*b*) Describe the different types of tank sluices indicating the regulation arrangements.

(*c*) For the data given below, design and sketch a tank sluice :

| | |
|---|---|
| Discharge | = 0.25 cumec |
| Sill level of sluice | = + 20.00 m. |
| Top of bank | = + 56.00 m. |
| Assumed head of discharge | = 0.3 m. |
| Side slopes of banks in front and rear | = 1½ : 1 |

Assume any other data, if required.

**10.** Write short notes on :

(*i*) Free catchment and combined catchment

(*ii*) Isolated tanks and tanks in series :

(*iii*) Tank bunds

(*iv*) Tank sluices ; and

(*v*) Tank weirs.

# —26

# *Arch and Buttress Dams*

## ARCH DAMS

### 26.1. Definition and Types of Arch Dams

An arch dam may be defined as a solid wall, *curved in plan*, standing across the entire width of the river valley, in a single span*. This dam body is usually made of cement concrete, although rubble and stone masonry have also been used in the past.

This wall will structurally behave : partly as a *cantilever retaining wall standing up from its base, and partly, the load will be transferred to the two ends of the arch span by horizontal arch action*. The arch load will, thus, be transferred to the side walls of the canyon, which must be strong, stable and rocky.

The distribution of part of the load to the side walls of the canyon, reduces the load on the cantilever wall, thereby reducing its thickness, as compared to that in an ordinary gravity dam ; and that is the only benefit we derive from an arch dam in comparison to a gravity dam.

Evidently, the greater is the wall curvature (in plan), the greater will be the load that will be transferred to the sides of the canyon, and hence greater will be the economy in the dam thickness.

This economy in dam thickness can be further increased considerably by making the dam body not only curved in plan, but also curved in section. Such a non-vertical dam is known as **double curvature arch dam** or a **shell-arch dam**, because such dams are designed as ***shell-structures.*** Such three dimensional designs are quite complex, and are taught only at M-Tech level in Structural Engineering.

Since the design and construction of an arch dam is very complicate, requiring extraordinary skill for erecting shuttering in the field, it is generally preferred in practical life to construct gravity dams. And that is why, we find only one arch dam in our country**, as against several hundreds of gravity dams. This arch dam too, is not a simple arch dam, but a shell-arch dam.

Simple arch dams, which transfer a large part of their loading by cantilever action, may also be of different types, since their faces may be either vertical or curvilinear. Depending upon the shape consideration, simple arch dams can be divided into three types, *viz :*

(*i*) *Constant radius arch dams*

(*ii*) *Variable radius arch dams* ; and

(*iii*) *Constant angle arch dams.*

---

*When multiple or a number of arches are used, supported between intermediate piers, the dam is known as a *buttress dam.*

***Idukki dam,* across Pariyar river in Kerala Sate ; **Photoview shown in Fig. 26.1.**

Fig. 26.1. Photoview of Idukki Dam.

A concrete arch dam, constructed across Periyar river in Kerala State of India. The length of the dam at top is 366 m and maximum height above the lowest point of foundation is 169 m. The construction of the dam was completed during 1966-1974.

A *constant radius arch dam* is the simplest in design as well as construction, but uses the maximum concrete. A *constant angle arch dam* on the other hand, uses about 43% of the concrete used by a constant radius arch dam. The ***variable radius arch dam*** is an intermediate choice, using around 58% of the concrete used by constant radius arch dam.

The ***shell arch dams*** are much more economical than even the constant angle arch dams, as their sections can be quite thin. Say for example, the famous *Vajont dam* of Italy is only 22 m thick at its base, inspite of being 261.6 m in height. Similarly, the *ldukki dam* in India is only 45 m thick at its base, even with 170.7 m height. As compared to these thin sections, the famous *Hoover dam* of USA, which is a *constant radius arch dam (with upstream face vertical) is 201 m thick at its base, with only 222 m height. The difference between the three types of arch dams are explained below :*

**26.1.1. Constant Radius Arch Dams.** (Fig. 26.2). A constant radius arch dam is that, in which, the radii of the *outside** curved surface are equal at all elevations, from top to the bottom. The centres of all such circular arcs, called **extrodos**, will therefore, evidently lie on one vertical line. However, the **introdos** (*i.e.* inside** curved surface of the arch) has gradually decreasing radius from top to the bottom, so as provide increased concrete thickness towards the base for accounting the proportionally increasing hydrostatic water pressure of the reservoir. The dam body will, therefore, be triangular in cross-section with upstream face vertical, and a minimum thickness at the top.

Evidently, it is only the radii of the *introdos*, which decrease with depth ; while the centres of all such circular arcs continue to lie on the same vertical line, on which lie the centres of the *extrodos*. Hence, in such a dam, the centres of extrodos, introdos, as well as the centrelines of the horizontal arch rings. at various elevations, lie on a straight vertical line that passes through the centre of the horizontal arch ring at the crest. Such a dam is, therefore, sometimes called a **constant centre arch dam**, although strictly speaking, this centre is not at one point, but lies at different heights along one vertical line.

Evidently, the central angles of the arch rings of the introdos will vary at different elevations, due to the varying width of the river valley (see Fig. 26.2); the maximum being at the top of the dam, and the minimum at the bottom of the dam.

It has further been shown that the *best* or *most economical central angle* in an arch dam is the one whose value is equal to 133° – 34′*** But in a constant radius arch dam, such an angle value can be adopted only at one place, since the angle varies with height considerably, due to narrow V-shape of the valley. It is therefore considered prudent to keep the economical angle of 133°–34′ at about mid height. The angle at the top will, therefore, be more than this value; but due to topographical considerations, the angle at the top can not be fixed greater than about 150°. *Hence, the angle at the top should be such as to give the best average angle, but restricted to about 150°.*

---

*Water side or upstream side.

**Downstream side.

***This is derived in article 26.3.1.1, a little later.

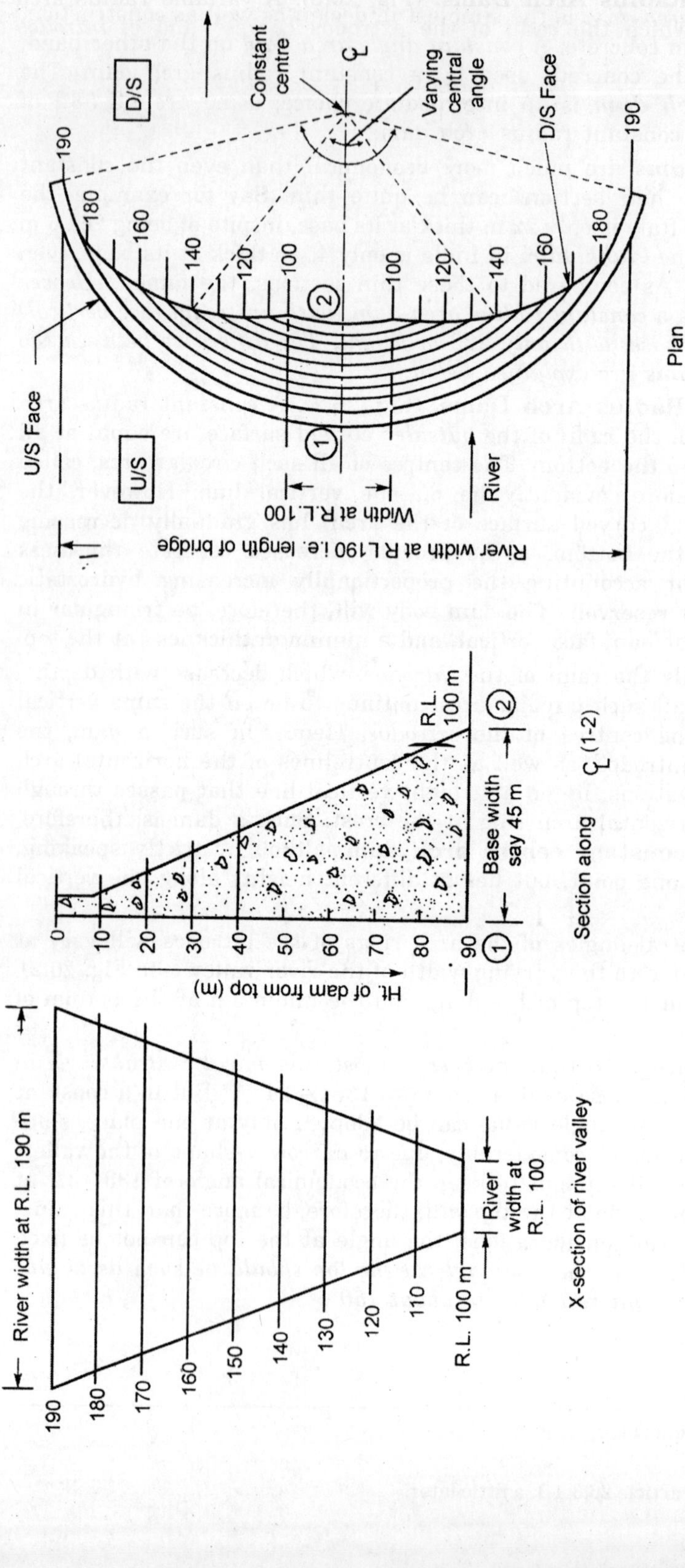


Fig. 26.2. Constant radius arch dam.

**26.1.2. Variable Radius Arch Dams.** (Fig. 26.3). A variable radius arch dam is the one in which the radii of the extrodos curves* and of *introdos curves*** vary at various elevations, being maximum at the top, and a certain minimum at its bottom. This makes the central angles as large as possible, so that the maximum arch efficiency may be obtained at all elevations.

In a typical design of such a dam, the downstream face of the dam at the central line (crown) is vertical ; while at all other locations, there is a batter on both the sides except at the abutments, where again, the upstream side becomes vertical. If overhangs are permitted, due to availability of stronger foundations, then the faces at the crown as well as abutments, may be provided with overhangs, affecting saving in the designed thicknesses.

Evidently, since in such an arch dam, the centres of the various arch rings at different elevations, do not lie on the same vertical line, it is also known as **variable centre arch dam**. Such dams are preferred for V-shaped valleys as compared to constant radius arch dams which may be preferred for comparatively wider U-shaped valleys.

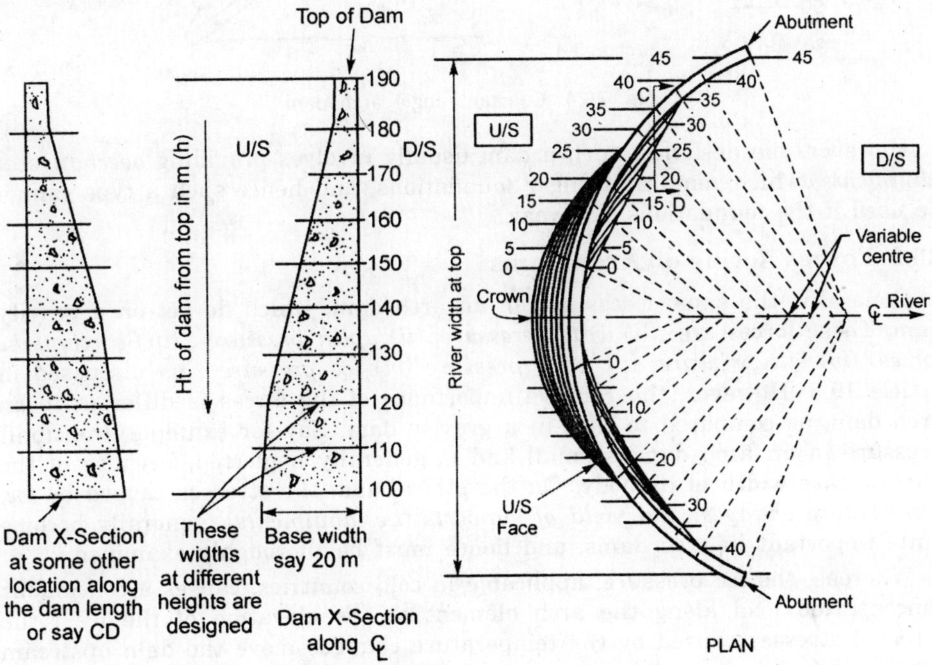


Fig. 26.3. Variable radius each dam.

**26.1.3. Constant Angle Arch Dams.** The constant angle arch dam is a special type of variable radius arch dam, in which the central angles of the horizontal arch rings are of the same magnitude at all elevations, as shown in Fig. 26.4. The design of such a dam can, thus, be made by adopting the best central angle of 133° – 34′ ; and hence such a dam proves to be the most economical, out of the three types of ordinary arch dams, as pointed out earlier also.

---

*arch rings corresponding to upsteam face.

** arch rings corresponding to downstream face.

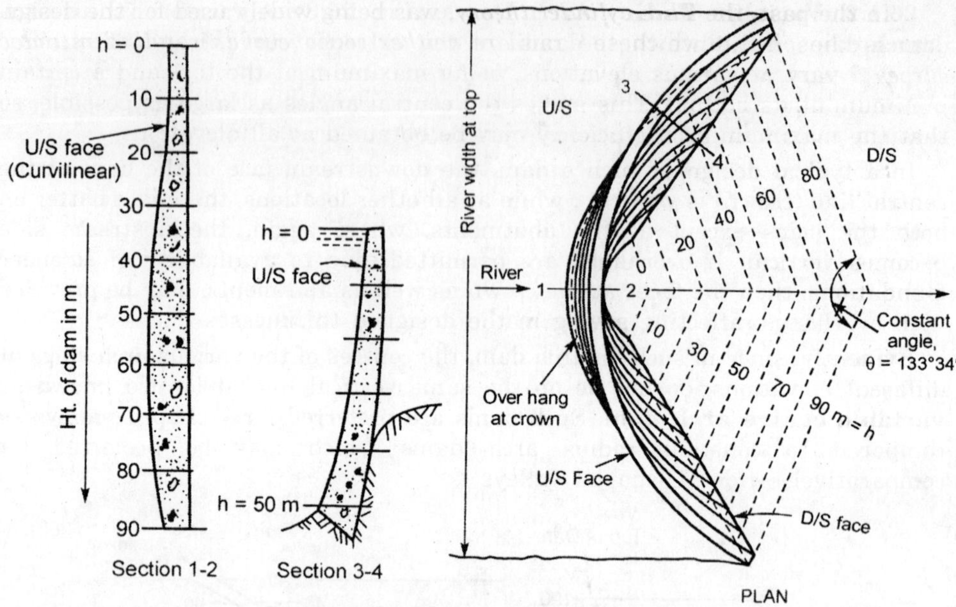


Fig. 26.4. Constant angle arch dam.

However, the design of such a dam usually involves providing *overhangs at abutments*, which require stronger foundations, and hence such a type cannot be used if the foundations are weak.

## 26.2. Forces Acting on Arch Dams

Generally, the same forces act on an arch dam, which do act on a gravity dam. These forces are : (*i*) *water pressure*; (*ii*) *uplift pressure* ; (*iii*) *earthquake forces*; (*iv*) *silt pressure*; (*v*) *wave pressure* (*vi*) *ice pressure* ; as discussed in article 19.3. However, the relative importance of the forces is different in an arch dam, as compared to that in a gravity dam. Say for example, the uplift pressure in an arch dam is small and is generally neglected, because of the narrow base width of its body. On the other hand, the stresses caused by *ice, temperature changes*, and *yield of supports (i.e. abutments)*, generally become quite important in arch dams, and hence must be thoroughly examined.

Whereas, the ice pressure, applicable in cold countries, causes a continuous concentrated load along the arch element at the elevation of the ice ; the internal stresses caused by the temperature changes move the dam upstream during the summer, and downstream during the winter. Hence, the **low water** *temperatures* become quite important in stress analysis, since these stresses act additive to the reservoir water pressure. Moreover, even the slight **yield of abutments** due to transfer of load by arch action, may also cause high internal stresses in the arch, and should therefore, be prudently accounted for.

## 26.3. Design of Arch Dams

Arch dams can be designed on the basis of any one of the following three methods:

(1) *Thin cylinder theory* ;

(2) *Theory of Elastic arches ; and*

(3) *The Trial load method.*

In the past, the *Thin cylinder theory*, was being widely used for the design of arch dams, but since these dams *are neither sections of thin cylinders nor are they free at the abutments*, as are the principal assumptions of the thin cylinder theory, this method gives erroneous results, and can only be relied upon for rough estimation of dimensions of the arches.

Similarly, in the *Elastic theory*, the dam is divided into horizontal arches, which are *assumed independent of each other*. This assumption is far from the actual position, since each ring is actually restricted in its movement by the ring below and above it. Hence, although the elastic theory takes into account the thickness of the arch rings and the restrictions at the abutments, it will not give precise results, because of the untrue assumption stated above.

The ***Trial load method***, which removes the above objections, is now-a-days, generally adopted, and leads to accurate estimation of the stresses in arches. In this method, the dam is divided into two systems of elements ; *viz.*

(*i*) *the horizontal arches* ; and

(*ii*) *the vertical cantilevers*, as shown in Fig. 26.5.

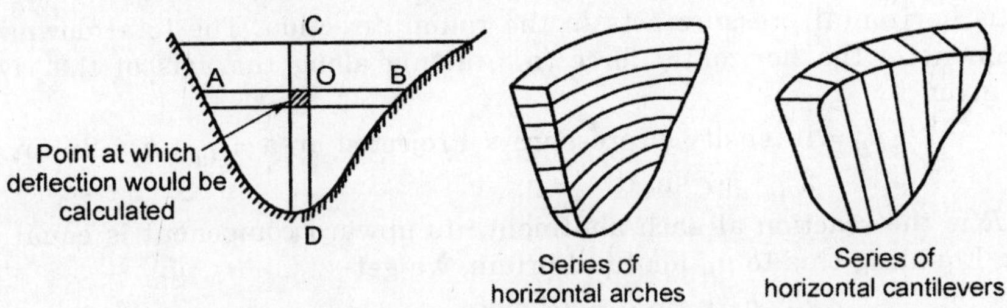


Fig. 26.5. Systems of elements for trial load method.

Each element is then supposed to take cartain percentage of the load due to pressure of water and silt, and the balance is assumed to be taken by the cross-elements. The actual distribution of the load between the **arch elements** and

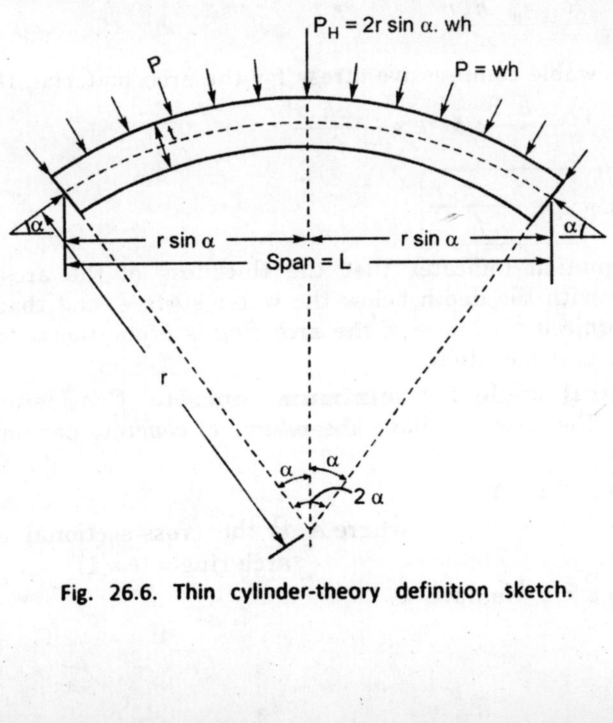


**Fig. 26.6. Thin cylinder-theory definition sketch.**

the **cantilever elements**, is then arrived at by several *trials*, due to which the method is called trial load method. In each trial, deflections for the arches are calculated at the points, where the cantilevers cross it. If the assumed distribution of load is correct, then the deflections calculated at these points from cantilever elements should, within reasonable limits, be the same. If they are not the same, then different distribution of loads is assumed, and the calculations are repeated, until the condition stipulated above is satisfied. The detailed solution of this method is complex, and is beyond the scope of this book.

**26.3.1. Thin Cylinder Theory.** In the thin cylinder theory, the stresses in an arch are assumed to be approximately the same as in a thin cylinder of equal outside radius.

Fig. 26.6 shows the action of a thin arch, having average radius $r$, angle subtended at the centre of the arch equal to $2\alpha$, and thickness $t$. Intensity of *hydrostatic pressure* $p$ at any depth $h$, is given by :

$$p = \gamma_w h \quad \text{...(i)}$$

where $\gamma_w$ = unit weight of water.

This horizontal pressure acts in the radial direction. The total downstream component of the horizontal force ($p_H$), acting along the axis of the river, is then given by

$$p_H = \text{Intensity of pressure} \times \text{Projected area} = \gamma_w \cdot h \times 2r \sin\alpha$$

$$= 2\gamma_w \, h \, r \sin\alpha \quad \text{...(ii)}$$

If $R$ is the reaction at each abutment, its upward component is equal to $2R \sin\alpha$. Equating this to $p_H$ for equilibrium, we get

$$2\gamma_w \cdot h \cdot r \sin\alpha = 2R \sin\alpha$$

or $$\gamma_w \cdot h\, r = R \quad \text{...(iii)}$$

This equation gives an expression for the abutment reaction, which is equal to the maximum compressive force induced in the arch.

If $\sigma$ is the compressive stress induced in the arch at the abutments, then

$$\sigma = \frac{R}{t} = \frac{\gamma_w \cdot h \cdot r}{t} \quad \text{...(iv)}$$

If $f_c$ is the allowable compressive stress for the arch material, then

$$f_c \gg \frac{\gamma_w \cdot h \cdot r}{t}; \text{ or } t \gg \frac{\gamma_w \cdot h \cdot r}{f_c}$$

Under Limits, $$\boxed{t = \frac{\gamma_w \cdot h \cdot r}{f_c}} \quad \text{...(26.1)}$$

The above equation indicates that the thickness of the arch ($t$) should increase linearly with the depth below the water surface, and that at a given elevation, the required thickness of the arch ring is proportional to the radius of the arch ($r$) at that elevation.

**26.3.1.1. Central angle for minimum concrete.** Considering the arch ring as shown in Fig. 26.6, we have the volume of concrete per unit height of the arch ring ($V$)

$$= r\,(2\alpha) \cdot A$$

where $A$ is the cross-sectional area of the arch ring = ($t \times 1$)

Since $A = t \times 1$, we have

$$V = r\,(2\alpha)\cdot t$$

Substituting the value of $t$ from Eq. (26.1), we get

$$V = r\cdot(2\alpha)\cdot\frac{\gamma_w\cdot h\cdot r}{f_c}$$

or $$V = \frac{\gamma_w\cdot h}{f_c}\cdot r^2\cdot(2\alpha) \quad ...(i)$$

Now, if $L$ is the span of the arch ring, then from Fig. 26.5, we observe that

$$\sin\alpha = \frac{L}{2r}$$

or $$r = \frac{L}{2\sin\alpha} \quad ...(ii)$$

Substituting this in Eq. ($i$) above, we get

$$V = \frac{\gamma_w\cdot h}{f_c}\left(\frac{L}{2\sin\alpha}\right)^2(2\alpha)$$

$$= \frac{\gamma_w\cdot h}{f_c}\cdot\frac{L^2}{2}\left[\frac{\alpha}{\sin^2\alpha}\right] \quad ...(iii)$$

where $\frac{\gamma_w\cdot h}{f_c}\cdot\frac{L^2}{2}$ is a constant = $K$ (say)

$\therefore$ $$V = K\cdot\left[\frac{\alpha}{\sin^2\alpha}\right]$$

or $$\frac{\partial V}{\partial\alpha} = K\left[\frac{\sin^2\alpha - 2\alpha\cdot\sin\alpha\cos\alpha}{\sin^4\alpha}\right]$$

For minimum volume of concrete ($V$)

$$\frac{\partial V}{\partial\alpha} = 0$$

$\therefore$ $$\sin^2\alpha = 2\alpha\cdot\sin\alpha\cdot\cos\alpha$$

or $$\sin\alpha = 2\alpha\cdot\cos\alpha$$

or $$\frac{\sin\alpha}{\cos\alpha} = 2\alpha$$

or $$\tan\alpha = 2\alpha$$

or $$2\alpha = \tan\alpha$$

$\therefore$ $$2\alpha = 133° - 34'$$

This shows that the most economical central angle is 133° – 34′, as was stated by us earlier.

**26.3.1.2. Limitations of thin cylinder theory.** The various limitations of this theory are :

(i) The arch sections are not thin cylinders. They are also not free at abutments, as assumed in this theory.

(ii) The theory does not consider shear and bending stresses in the arch.

(iii) The analysis is based on only the hydrostatic water pressure. *Temperature stresses and ice pressures, which are quite important in arch dams, get ignored in this theory.*

(iv) Stresses due to yielding of abutments and those due to rib shortening have not been accounted in this theory.

(v) Plastic flow of concrete and shrinkage in concrete have not been accounted for.

**26.3.2. The Elastic Theory for Design of Arch Dams.** As pointed out earlier, the archs are subjected to additional forces, such as :

(i) *temperature stresses*, due to temperature changes ;

(ii) *shrinkage stresses*, due to setting of concrete ;

(iii) stresses due to yielding of abutments ; and

(iv) stresses due to rib shortening.

The stresses at (i) and (ii) are quite evident, and need no explanation. It may however, be added that the effect of rise in temperature, as in summer season, produces opposite stresses, whereas, the stresses due to fall in temperature are additive, and hence, only need be accounted. The stress at (iii) above is also understood, since no abutment is rigid, and yields to an extent depending upon the nature of the rock. This yield increases the span of the arch, inducing additional stresses. The stress at (iv) above, can be explained as below :

In case of a cylinder, the pressure acts from all sides, and the entire ring is compressed so that the diameter of the cylinder decreases by a small amount, but *there* is no deformation. While in the case of an arch, *which is a part of a cylinder*, the ends being fixed, the distance between them (*i.e.* the span) does not change. This restriction of the change in span, causes additional stresses, known as *rib- shortening stresses. In a long thin arch, these stresses are small; but in the case of a thick arch of small angle, they are appreciable, and need to be accounted for.*

*All the above stresses are taken into consideration by the theory of elastic arches.*

**26.3.3. General formulas.** The formulas for elastic arches, usually given in the books on "Theory of Structures" or "Applied Mechanics" are reproduced here for the benefit of students, intending to solve arch dams by elastic theory.

With reference to Fig. 26.7.

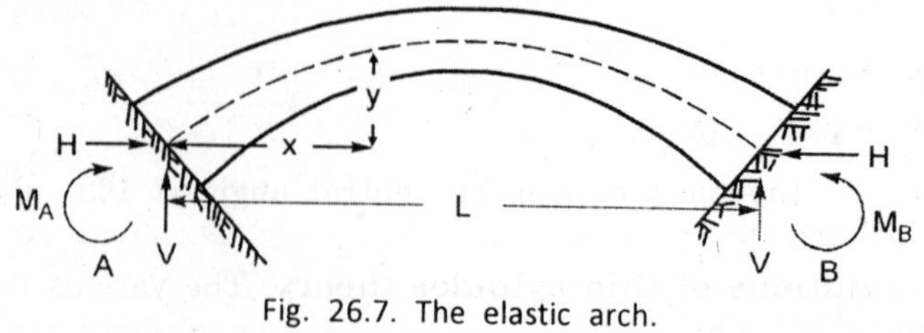


Fig. 26.7. The elastic arch.

$$M = \mu + M_A + (M_B - M_A)\frac{x}{L} + H \cdot y \qquad ...(26.2)$$

$$\int \frac{M}{E \cdot I} y ds = \int \frac{\mu \cdot y}{EI} \cdot ds + M_A \int \frac{y \cdot ds}{EI} + \frac{M_B - M_A}{L} \int \frac{xyds}{EI} + H \int \frac{y^2}{EI} ds = 0 \qquad ...(26.3)$$

$$\int \frac{Mx}{EI} ds = \int \frac{\mu \cdot y}{EI} \cdot ds + M_A \int \frac{x \cdot ds}{EI} + \frac{M_B - M_A}{L} \int \frac{x^2 \cdot ds}{EI} + H \int \frac{xy}{EI} ds = 0 \qquad ...(26.4)$$

and
$$\int \frac{M}{E \cdot I} ds = \int \frac{\mu}{EI} \cdot ds + M_A \int \frac{ds}{EI} + \frac{M_B - M_A}{L} \int \frac{x \cdot ds}{EI} + H \int \frac{yds}{EI} = 0 \qquad ...(26.5)$$

where $\mu$ = the bending moment on a straight horizontal freely supported beam

$M$ = the bending moment at any point on the arch

$M_A$ and $M_B$ = fixed end bending moments at both the ends of the arch

$H$ = horizontal thrust at the abutments.

For **symmetrical arch**, $M_A = M_B$; and the Eq. (26.4) vanishes ; while Eqs. (26.3) and (26.5) get simplified to :

$$\int \frac{M}{EI} . y . ds = \int \frac{\mu \cdot y}{EI} \cdot ds + M_A \int \frac{yds}{EI} + H \int \frac{y \cdot ds}{EI} = 0 \qquad ...[26.3(a)]$$

and
$$\int \frac{M}{EI} . ds = \int \frac{\mu}{EI} ds + M_A \int \frac{ds}{EI} + H \int \frac{y \cdot ds}{EI} = 0 \qquad ...[26.5(a)]$$

As the *contraction* due to temperature effect and *rib-shortening* due to thrust, is prevented by the abutments, these effects modify the above Eqs. (26.3), (26.4) and (26.5) as

$$\alpha \cdot TL - \int \frac{M}{EI} \cdot y \cdot ds = 0 \qquad ...[26.3(b)]$$

where $\alpha$ = Coefficient of thermal expansion

$T$ = Fall in temperature in °C

$$\int \frac{M}{EI} ds = 0 \qquad ...[26.4(b)]$$

$$\int \frac{M \cdot x}{EI} ds = 0 \qquad ...[26.5(b)]$$

and
$$M = M_A - V_x + H \cdot y \qquad ...(26.6)$$

in which the additional term $V$ stands for the vertical thrust of opposite nature at the two abutments, and is reckoned positive upwards.

For a symmetrical arch, $V = 0$ ; so that the Eq. (26.6) simplifies to

$$M = M_A + H \cdot y \qquad ...[26.6\ (a)]$$

and Eqs. [26.4(*b*)] and [26.5(*b*)] become

$$M_A \int \frac{y}{EI} \cdot ds + H \cdot \int \frac{y^2}{EI} ds - L \cdot \alpha T = 0 \qquad ...[26.4(c)]$$

and

$$M_A \int \frac{ds}{EI} + H \int \frac{y}{EI} ds = 0 \qquad ...[26.5(c)]$$

By solving these equations, the values of $M_A$, $H$ and $V$ are obtained.

*Rib-shortening* effect can be reckoned along with the other loads by adding its effect to the left side of Eq. (26.3).

## BUTTRESS DAMS

### 26.4. Definition and Types of Buttress Dams

An ordinary concrete gravity dam, as we know, is a solid body of mass concrete, somewhat triangular in section, running across the entire width of the river valley. Such a solid wall requires huge amount of concrete, which partly remains *unstressed to full extent*, thereby leading to wastage of concrete. The uplift pressure also acts on the entire width of the dam body, from its bottom, which further increases its size, without giving any additional benefit as a dam.

Efforts have, therefore, been made from time to time, to innovate methods for affecting economy in the use of concrete, by cutting down the concrete from those dam portions where it remains unstressed. Attempts have, therefore been made to provide *hollow gravity dams*. **Buttress dams**, are an improvement innovation over the hollow concrete gravity dams.

In such an innovated dam, therefore, *solid walls of specified thickness and section*, are constructed parallel to the flow, at some suitable intervals. These walls are called **buttresses.** *Inclined slabs* or *arch slabs* are then supported on upstream side on these buttresses, as shown in Fig. 26.8.

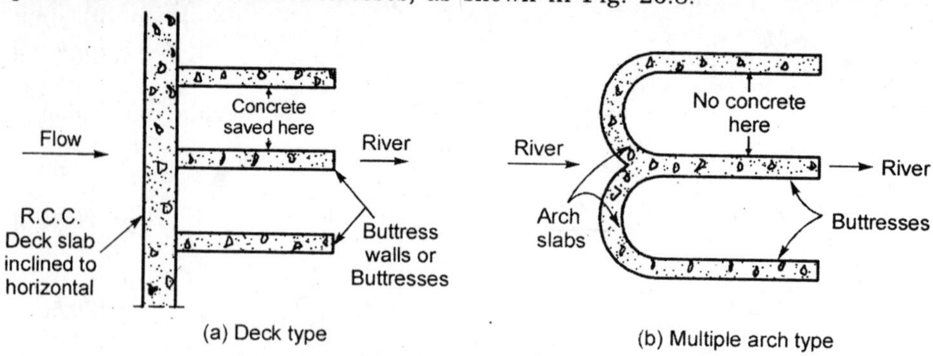


Fig. 26.8. Simplified line sketches of buttress dams.

Although, several types of buttress dams have been devised, but the most common types, are :

1. *Deck slab type ; and*
2. *Multiple arch type.*

Both these types of dams have been shown in their simplified line diagrams, in Figs. 26.8 (*a*) and (*b*) respectively.

*Slab type dams* are generally preferred for smaller heights, say from 20 to 50 m, or so. The highest such dam in the world is *Rodriguez dam* in Mexico, having a height of 73.2 m. Such dams are virtually referred to as buttress dams, or ***amberson dams*** (based upon the name of their inventor).

*The multiple arch dams*, on the other hand, are used for higher dam heights, say from 50 m onward, and the highest such dam of the world is *Manicougan-5 dam* in Canada, having a height of 210 m.

Detailed views of both these types of dams are shown in Figs. 26.9 and 26.12, respectively ; and are discussed below :

**26.4.1. Simply Supported Slab Type of Buttress Dams.** (Fig. 26.9). In such a type of dam, the R.C.C. deck slab is freely supported on the buttresses, which have corbels for the slab to rest on. The deck slab is inclined to horizontal by about 40° to 55°, so as to support the dead load of a portion of reservoir water, and thus, to provide the stabilising force, in addition to the 'self' weight of the dam, to avoid its sliding.

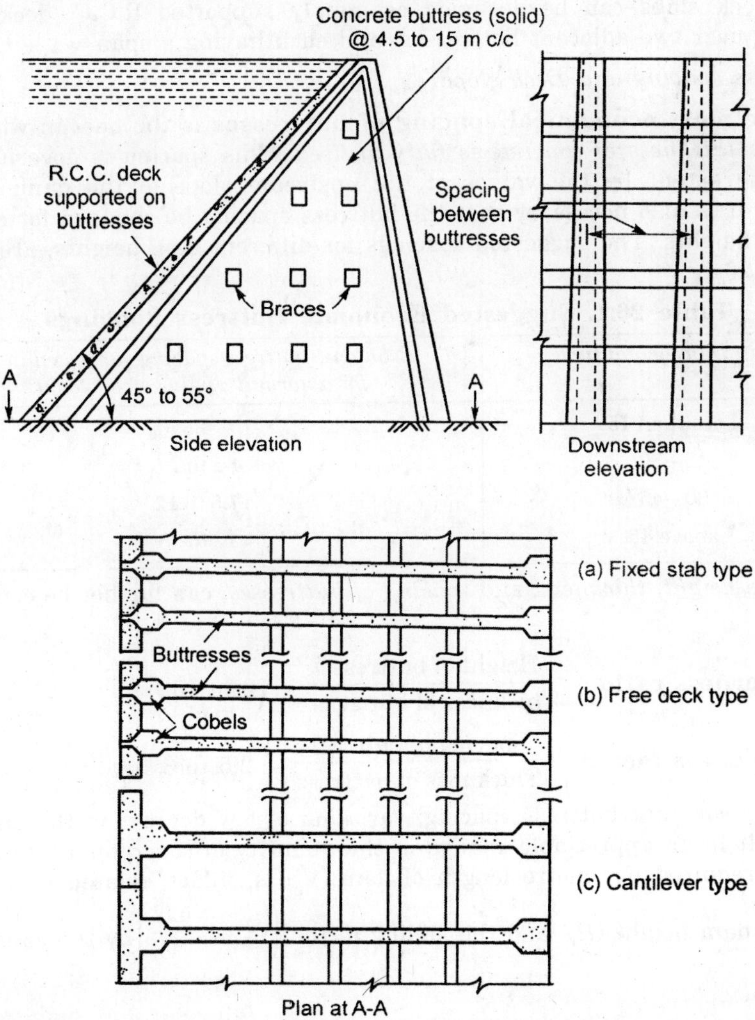


Fig. 26.9. Detailed views of a free deck (slab) type of a buttress dam.

**Design considerations.** The loading and safety criteria for buttress dams or buttresses, is the same as that for a gravity dam section, except that the provided *buttress thickness* '$t$' will take the load coming from the dam length $= x + t$ ; where $x$ is the clear spacing between the two consecutive buttresses. Hence, $t$ metre length of buttress section, will take the loads coming from $(x + t)$ metre length of dam, as against the unit metre length of gravity dam section taking load from the unit metre dam length. The resultant impact can be considered by increasing the unit weight of water by multiplying its actual value by a *surcharge factor* ($S$), defined as :

$$S = \frac{x+t}{t} \quad \text{...(26.7)}$$

Hence, the effective unit weight of water can be considered as $\gamma_w \left(\frac{x+t}{t}\right)$, and the section of the buttresses can be designed, exactly in the same manner, as a gravity dam section is designed, considering unit length and a continuous section.

The deck slabs can be designed as simply supported R.C.C. decks, each spanning over two adjacent buttresses, and each having a span $= x + t$.

*Buttress spacing and Deck slope*

(*i*) **The most economical spacing of buttresses** is the one, in which *the minimum thickness of concrete is fully utilised*. This spacing is governed to a very large extent by the values of the upstream slope of the dam ($\phi$). For economy, it is also necessary that the buttress spacing be changed for dams of different heights. The suggested spacings for different dam heights, are shown in Table 26.1.

**Table 26.1. Suggested Economic Buttress Spacings**

| *Mean Dam height in m* | *Economic buttress spacing suggested in m for a normal $\phi$ value of 40° to 50°* |
|---|---|
| Less than 15 | 4.5 |
| 15—30 | 4.5 to 7.5 |
| 30—45 | 7.5 to 12 |
| above 45 | 12 to 15 |

(*ii*) The *height, thickness*, and *spacing of buttresses*, can further be controlled by :

$$\textbf{Slenderness ratio} = \frac{\text{Height of buttress}}{\text{Thickness of buttress}} = 12 \text{ to } 15 \quad \text{...(26.8)}$$

$$\textbf{Massiveness factor} = \frac{\text{Spacing of buttress}}{\text{Thickness of buttress}} = 2.5 \text{ to } 3 \quad \text{...(26.9)}$$

(*iii*) For constant buttress spacing, variation of few degrees in the u/s slope may result in an appreciable change in the quantity of concrete. This concrete quantity required per metre length of dam ($V_c$) is, infact, related to *u/s* deck slope ($\phi$), dam height ($H$), and the sliding factor $\left(F_s \text{ i.e., } \frac{\Sigma H}{V}\right)$ by the equation:

$$V_c = 0.208\, H^2 \left[ \frac{1}{F_s} - \cot \phi \right] \qquad ...(26.10)$$

This equation clearly shows that for a fixed dam height ($H$) and a given value of $F_s$, the concrete quantity depends solely on $\phi$. Hence, for different values of $\phi$, at fixed values of $H$ and $F_s$, concrete quantities can be worked out. *Vertical lines*, like those shown in Fig. 26.10, are, thus, drawn for each value of the concrete quantity corresponding to each value of $\phi$.

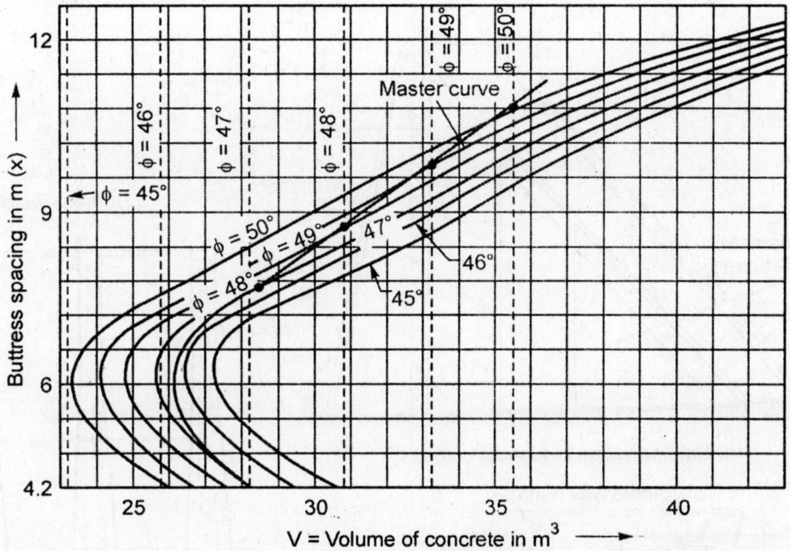


Fig. 26.10. Master curve for absolute economic value of buttress spacing.

Concrete quantities are also worked out for various angles of u/s slope ($\phi$) and different buttress spacings, for fixed decided dam height ($H$) ; and plotted to get a curve for each angle. A **master curve** is then drawn through the junctions of the curves and the vertical lines corresponding to each angle.

*The master curve* gives the **absolute economic value of buttress spacing** for different *u/s* slope ($\phi$) for the decided dam height.

Say for example, from the plotted curves of Fig. 26.10, we find that for a $\phi = 48°$, buttress spacing is about 8.8 m ; for a $\phi = 49°$, buttress spacing is about 10 m. *This reflects that by increasing the upstream slope, we can increase the buttress spacing.* This figure also reflects that lower values of $\phi$ (lesser than 47°) are not available on this particular master curve ; which shows that such lower values will not prove economical for this project. Hence, each individual project (or dam) has, thus, to be analysed for decided dam height, to determine the permissible economical values of $\phi$, and their corresponding values of buttress spacing, and thus finally to choose any one set of them.

**26.4.2. Multiple Arch Type of Buttres Dams.** As stated earlier, in this type of dams, arch slabs are constructed for *u/s* face of the dam, to be supported on buttresses, as shown in Figs. 26.11 and 26.12. Each buttress may be in the form of a single stiffened wall, as shown in Fig. 26.13($a$) ; or a double hollow wall, as shown in Fig. 26.13($b$).

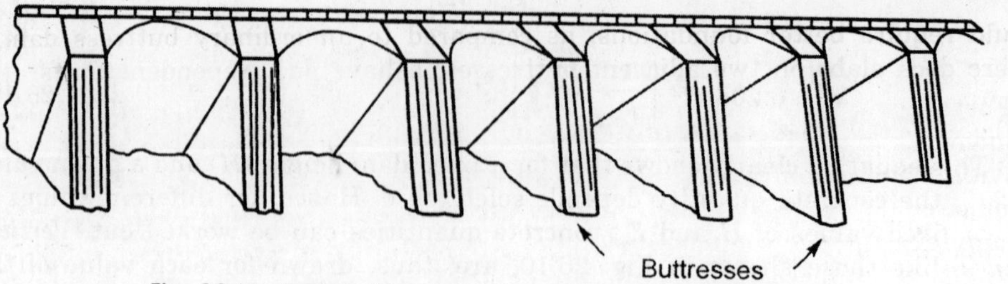


Fig. 26.11. Perspective view of a multi-arch buttress dam.

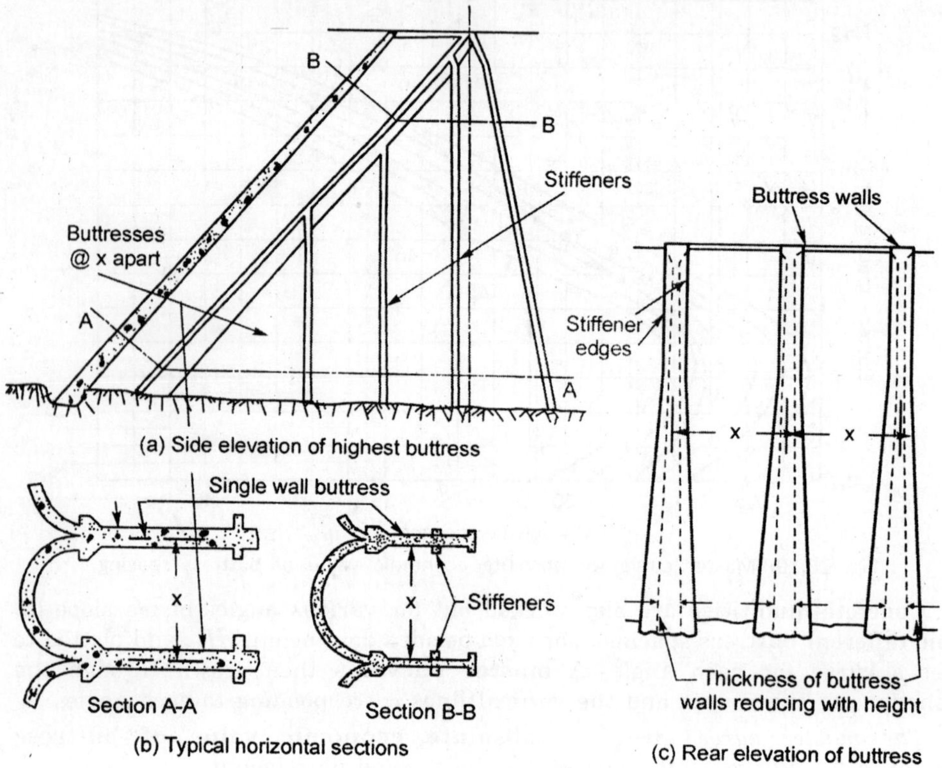


Fig. 26.12. Detailed views of a multiple arch dam.

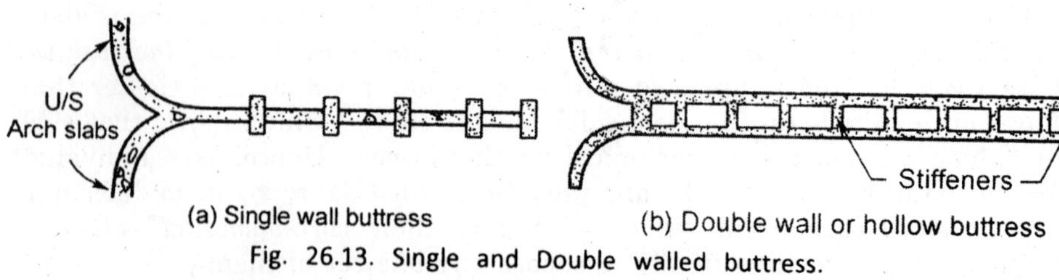


Fig. 26.13. Single and Double walled buttress.

***Meer Alam dam*** in India, built in about 1800 A.D., is the earliest recorded example of such a type of dam. As stated earlier, such dams are preferred for larger heights, as they prove more stable and flexible, as compared to the ordinary buttress dams (*i.e.* slab type).

The stability of any unit of a multiple arch dam depends on the adjacent units, so that any settlement in one would affect the others. Hence, such a dam would require better foundations, as compared to an ordinary buttress dam, where deck slabs on two adjacent buttresses, behave as independent units.

For short spans (10 to 15 m), circular arches of uniform thickness prove economical; whereas, for larger spans, arches of variable thickness are provided. If the central angle of each arch is kept between 180° to 150°, with semi circular or nearly semi circular arches, the horizontal thrust between adjoining arches gets eliminated.

For arched buttress dams, the range of buttress spacing generally adopted, is between 15 to 21 m, although as high a value as 42 m has been permitted in *Weber Greek dam* in California, USA ; a site plan of which is reflected in Fig. 26.14.

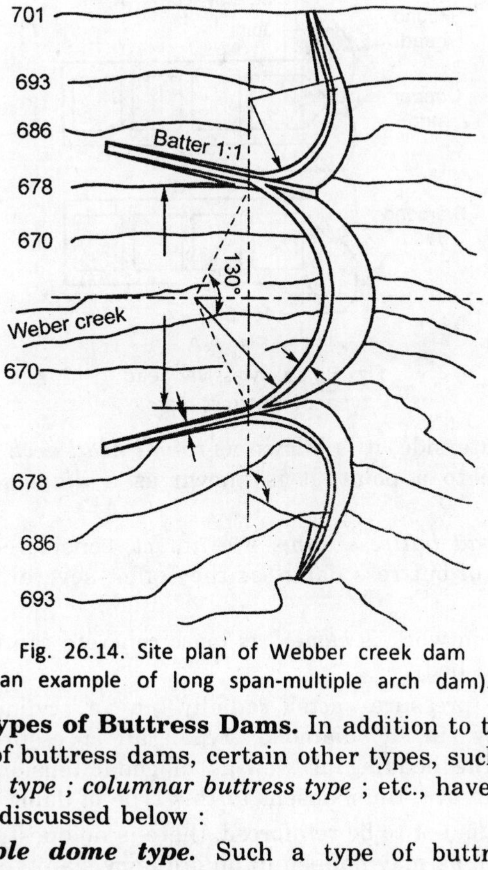


Fig. 26.14. Site plan of Webber creek dam (an example of long span-multiple arch dam).

**26.4.3. Other Types of Buttress Dams.** In addition to the above described two general types of buttress dams, certain other types, such as *multiple dome type ; massive head type ; columnar buttress type* ; etc., have also been adopted at some places, as discussed below :

**26.4.3.1. *Multiple dome type.*** Such a type of buttress dam is quite understandable, as it uses a number of *domes* instead of a number of arches as used in a multiple arch type of dam, and all other features remaining the same. Such a dam can help in increasing the buttress spacing, but requires sites, where stable foundations can be had at greater depths. The *Coolidge dam on San Carlos river in Arizona is an example of this type of a dam, which is 76.2 m high with buttresses @54.8 m apart.*

**26.4.3.2. *Massive head types.*** This type of buttress dam is the one, which does not use slabs or arches for upstream face, but the heads of the buttresses themselves are enlarged to meet each other, and thus, to form a continuous water supporting surface, as shown in Fig. 26.15. When the enlarged heads are

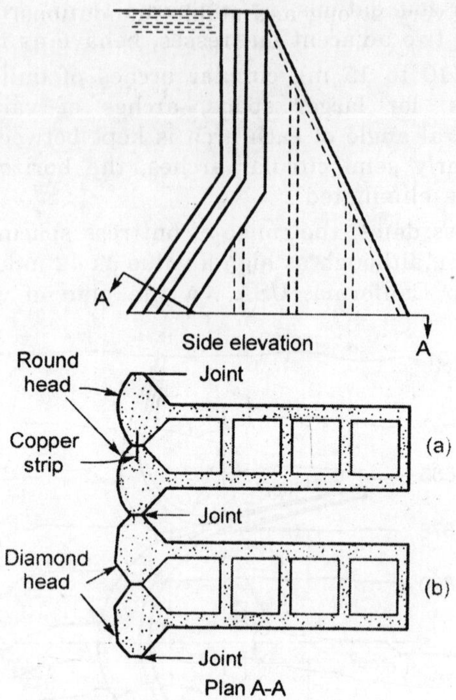


Fig. 26.15. Massive head type of buttress dam.

*cylindrical* on the outerside, it is known as *round head deck* (Fig. 26.15*a*) ; and when they converge to a point, it is known as a *diamond head deck* (Fig. 26.15*b*).

These *massive head buttress dams* are, infact, considered as improvement over all other types of buttress dams, as they offer, several **advantages**, such as :

(*i*) The construction work is easier, as mass concrete can be laid together in the entire dam body.

(*ii*) Since water pressure acts radially in a cylindrical type, and perpendicularly in a diamond type, all pressures are normally *compressive*. The bending as well as diagonal tension in the upstream part of the dam are, thus, absent in this type of dams.

(*iii*) Since the deck is not to be reinforced, there is no question of its failure by rusting of steel, as may happen in an ordinary slab type of buttress dam.

(*iv*) For smaller heights, they prove to give more economical buttress spacing, as compared to other two types of buttress dams.

(*v*) Such a dam body offers more resistance to sliding, because, it is considerably heavier, and has a greater sectional area along the horizontal planes.

**26.4.3.3. Columnar buttress type.** As shown in Fig. 26.16, this type is a modification of the ordinary slab type, as the deck here, is supported on columns. Very few dams have been built, since they offer the following *disadvantages*:

(*i*) they require very strong and stable foundations.

(*ii*) more skill is required in constructing the buttresses ; and even then, they cost practically as much as other conventional types.

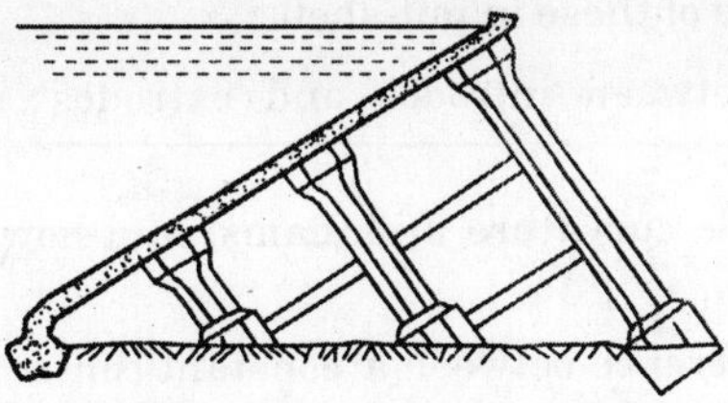

Fig. 26.16. Columnar buttress dam.

## PROBLEMS

**1.** (*a*) Define an arch dam and a double arch dam. What are the important types of arch dams, and what are their major differences ?

(*b*) What is meant by 'the best central angle of an arch dam', and what is its value ?

2. (*a*) Contrast the design criteria for a gravity dam and that for an arch dam by the cylinder theory. What are the relative advantages of constant radius, constant angle, and variable radius dams ?

(*b*) What are the considerations for the choice of a buttress dam in preference to other types ? When do we go in for the massive head type ?

(*c*) What is meant by the following :

(*i*) Economic buttress spacing

(*ii*) Best central angle of an arch dam.

**3.** (*a*) Distinguish between the constant radius and constant angle layouts of an arch dam. Obtain the value of the best central angle for the latter.

(*b*) Enumerate the different types of buttress dams ; and explain as to how a slab type of buttress dam differs in its design, as compared to a concrete gravity dam.

**4.** (*a*) State the condition under which you would recommend the construction of an arch dam ; and explain the various methods of layout of an arch dam.

(*b*) How do you ascertain the absolute economic value of buttress spacing for an ordinary buttress dam ?

**5.** State the conditions under which you would recommend the construction of an 'arch dam'. Draw neat sketches of various types of arch dams and state the circumstances under which you would recommend them.

**6.** (*a*) What is the 'master curve' as used in connection with buttress dams ? What is its use, and how is it drawn ?

(*b*) What are multiple arch dams, and how do they differ from ordinary arch dams and double arch dams ? Draw line sketches for all these types of dams.

**7.** (*a*) Massive head type of buttress dams are considered to be the best of all the available types of buttress dams. Why ?

(*b*) Enumerate the design methods or theories, that are used in the design of arch dams. Which of these theories or methods, is used to compute approximate result, and which one is used for obtaining precise results. Explain any one of these in nut-shell.

**8.** (*a*) Differentiate between 'introdos' and 'extrodos', as are applicable in arch dams.

(*b*) What are double curvature arch dams, and how do they differ from arch dams?

(*c*) What is the difference between a 'constant centre arch dam' and a 'constant angle arch dam'?

(*d*) What is meant by economic spacing of buttresses ; and how is it related to dam height and upstream slope of buttress dam ?

**9.** Write short notes on any four of the following :

(*i*) arch dams.

(*ii*) buttress dams.

(*iii*) methods of layout of an arch dam.

(*iv*) design principles involved in arch dams.

(*v*) master curve for economic buttress spacing.

(*vi*) best central angle for an arch dam.

(*vii*) design criteria for a slab type of buttress dam, and its comparison with the design of a gravity dam.

(*viii*) massive head type of buttress dams.

(*ix*) multiple arch dams.

(*x*) double curvature arch dams.

# —27

# *Chapterwise Multichoice Objective Questions*

**Q. 1. Irrigation is basically required in :**
(*a*) humid regions
(*b*) arid regions
(*c*) semi arid regions
(*d*) all of the above.

**Q. 2. Irrigation may not be needed in :**
(*a*) England
(*b*) Israel
(*c*) India
(*d*) none of the above.

**Q. 3. Pinpoint the correct statement :**
(*a*) irrigation helps in adopting mixed cropping
(*b*) 'mixed cropping' means sowing of a different crop after a particular crop has been grown
(*c*) over-irrigation may lead to saving in fertilizers
(*d*) irrigation helps in avoiding mixed cropping.

**Q. 4. 'Flood irrigation' method of irrigating fields, works best on :**
(*a*) level or gently rolling terrain
(*b*) steeply rolling terrain
(*c*) both (*a*) and (*b*)
(*d*) none of the above.

**Q. 5. In flood irrigation, the preferred method of applying irrigation water to the comparatively steeper rolling land, is :**
(*a*) check flooding
(*b*) boarder flooding
(*c*) wild flooding
(*d*) basin flooding.

**Q. 6. Fields enclosed by dikes, where rice is grown, are called :**
(*a*) paddies
(*b*) furrows
(*c*) basins
(*d*) none of the above.

**Q. 7. The method of growing crops on ridges, running on the sides of water ditches, is known as :**
(*a*) flood irrigation
(*b*) furrow irrigation
(*c*) check irrigation
(*d*) none of them.

**Q. 8. In a field under furrow irrigation, 'furrows' are referred to represent:**
(*a*) ridges on which crops are grown
(*b*) narrow ditches which carry irrigation water
(*c*) both (*a*) and (*b*)
(*d*) neither (*a*) nor (*b*).

**Q. 9. In a mildly water scarce area, the drip irrigation could be preferred for growing :**
(*a*) wheat
(*b*) fodder
(*c*) rice
(*d*) fruits and vegetables.

**Q. 10. The most suitable water for irrigation is :**
(*a*) C 1—S 1
(*b*) C 2—S 2
(*c*) C 4—S 1
(*d*) C 1—S 4.

**Q. 11. Addition of gypsum to the irrigation water is recommended to overcome difficulties posed by :**

(*a*) highly saline irrigation supplies
(*b*) irrigation supplies containing high quantities of sodium
(*c*) irrigation supplies containing heavy sediment
(*d*) all of the above.

**Q. 12. The electrical conductivity of medium saline water (C 2) at 25°C is of the order of :**

(*a*) 50 to 100 μΩ/cm (*b*) 100 to 250 μΩ/cm
(*c*) 250 to 750 μΩ/cm (*d*) 750 to 1500 μΩ/cm.

**Q. 13. The Sodium Absorption Ratio of an irrigation water is 8. This water will be called :**

(*a*) low sodium water (*b*) medium sodium water
(*c*) high sodium water (*d*) none of the above.

**Q. 14. High sodium content in irrigation water is :**

(*a*) generally good (*b*) generally bad
(*c*) generally good, but bad for a few crops
(*d*) generally bad, but good for a few crops.

**Q. 15. Salinity in irrigation water is measured by :**

(*a*) SAR value (*b*) Electrical-conductivity value
(*c*) pH-value (*d*) none of the above.

**Q. 16.** The time required to irrigate a strip of area 0.203 hectare by a stream discharge of 0.043 cumec, to provide an average depth of 6.35 cm to the field, is : (assume average rate of infiltration to be 5 cm/h)

(*a*) 2.75 hour (*b*) 1.35 hour
(*c*) 1.5 hour (*d*) 1.90 hour.

**Q. 17.** If the concentration of Na, Ca, and Mg in a water sample are 345, 60 and 18 mg/1, respectively, then the Sodium Absorption Ratio (SAR) of this water will be :

(*a*) 5 (*b*) 10.
(*c*) 39 (*d*) 55.

## CHAPTER 2 : Water Requirements of Crops

**Q. 18. With the increase in supplied irrigation water, the yield of crops:**

(*a*) increases continuously (*b*) decreases continuously
(*c*) increases up to a certain limit, and then becomes constant
(*d*) increases up to a certain limit, and then decreases.

**Q. 19. The most expected crops in a hot arid district of Rajasthan State in India, in the month of September, are :**

(*a*) Rice and Sugarcane (*b*) Bazra and Maize
(*c*) Wheat and Maize (*d*) Tobacco and Cotton.

**Q. 20. The maximum irrigation requirement of Rice crop is exhibited by its :**

(*a*) maximum delta value (*b*) maximum duty value
(*c*) minimum duty value (*d*) none of the above.

**Q. 21. The crop among the following, which is expected to have the maximum duty, is**

(*a*) Wheat (*b*) Rice
(*c*) Sugarcane (*d*) Cotton.

**Q. 22. Kor-Watering is the irrigation water supplied to a crop :**
(*a*) at the time of its sowing
(*b*) just before harvesting
(*c*) about three weeks after sowing
(*d*) about three weeks before harvesting.

**Q. 23. The duty of irrigation water for a given crop is maximum :**
(*a*) on the field (*b*) at the head of the main canal
(*c*) at the head of the water-course (*d*) none of them.

**Q. 24. The important Govt. reference made to the 'duty for a crop,' is usually related to its duty :**
(*a*) on the field (*b*) at the head of the main canal
(*c*) at the head of the water course (*d*) none of them.

**Q. 25. The duty at the end point of a canal minor, where the Govt. control usually ceases, is called :**
(*a*) duty on field (*b*) outlet duty
(*c*) flow duty (*d*) storage duty.

**Q. 26. The first important watering of crops is usually called :**
(*a*) paleo watering (*b*) kor-watering
(*c*) crop-watering (*d*) all of the above.

**Q. 27. The crop sequence, which cannot serve any useful purpose in 'Crop rotation', is :**
(*a*) Wheat—Juar—Gram (*b*) Rice—Gram—Rice
(*c*) Cotton—Wheat—Gram (*d*) Rice—Wheat—Cotton.

**Q. 28. The optimum depth of kor watering for rice is about :**
(*a*) 25 cm (*b*) 19 cm
(*c*) 13.5 cm (*d*) 9 cm.

**Q. 29. The kor period, within which a crop must receive its first major watering, will be :**
(*a*) less for humid climates (*b*) less for dry climates
(*c*) equal for all climates (*d*) none of the above.

**Q. 30. The crop, out of the following, which is not a cash crop, is :**
(*a*) Jute (*b*) Tea
(*c*) Rice (*d*) Sugarcane.

**Q. 31. In India, the cultivated area under Rabi season is generally $x$ times the area under Kharif season; where $x$ is :**
(*a*) 1 (*b*) 1/2
(*c*) 2 (*d*) none of them.

**Q. 32. The relationship between the duty D in ha/cumec, the water depth Δ in cm, and base period B in days, is given by :**
(*a*) $D = \frac{864B}{\Delta}$ (*b*) $D = \frac{8.64B}{\Delta}$
(*c*) $D = \frac{864\Delta}{B}$ (*d*) $D = \frac{8.64\Delta}{B}$

**Q. 33. Consumptive use of water for a crop represents :**
(*a*) The transpiration needs of the crop
(*b*) evaporation needs of the cropped area.
(*c*) evapotranspiration needs of the cropped area plus the minor quantity required in plant metabolism
(*d*) None of the above.

**Q. 34. The amount of irrigation water required to fulfil the evapo-transpiration and minor metabolic plant needs per unit of cropped area is called :**

(*a*) consumptive use
(*b*) consumptive irrigation requirement
(*c*) net irrigation requirement
(*d*) none of the above.

**Q. 35. When there is no percolation loss for leaching and other such soil needs, then the Net irrigation requirement (NIR) of a cropped area, will be equal to :**

(*a*) the consumptive use ($C_u$)
(*b*) $C_u - R_e$ : where $R_e$ is effective rainfall during cropping
(*c*) $\dfrac{C_u - R_e}{\eta_a}$ : where $\eta_a$ is water application efficiency
(*d*) none of the above.

**Q. 36. The Gross Irrigation Requirement (GIR) of water is equal to :**

(*a*) NIR $\div \eta_a$ (*b*) NIR $\div \eta_a \cdot \eta_c$
(*c*) NIR $\div \dfrac{\eta_a}{\eta_c}$ (*d*) none of above.

where $\eta_a$ = water application efficiency
$\eta_c$ = water conveyance efficiency

**Q. 37. The ratio of the water stored in the root zone of a crop, to the water actually delivered to the crop in the field, is known as;**

(*a*) water conveyance efficiency (*b*) water application efficiency
(*c*) water use efficiency (*d*) none of the above.

**Q. 38. The water which can be utilised by the crops from the soil is called :**

(*a*) field capacity water (*b*) capillary water
(*c*) hygroscopic water (*d*) none of the above.

**Q. 39. The moisture held by a well drained soil against gravity drainage, by the force of surface tension between the soil grains and water drops, is called:**

(*a*) field capacity water (*b*) hygroscopic water
(*c*) capillary water (*d*) water of adhesion.

**Q. 40. Permanent wilting point moisture content for a crop represents the :**

(*a*) hygroscopic water (*b*) capillary water
(*c*) field capacity water (*d*) none of the above.

**Q. 41. Available moisture for a crop is equal to :**

(*a*) field capacity moisture content—Wilting point moisture content
(*b*) field capacity moisture content—hygroscopic moisture content.
(*c*) both (*a*) and (*b*)
(*d*) none of the above.

**Q. 42. The optimum moisture content (m.c.) which is retained in the root zone of a soil, before applying irrigation water, is :**

(*a*) equal to : (the field capacity m.c.—wilting point m.c.)
(*b*) less than : (the field capacity m.c.—wilting point m.c.)
(*c*) more than : (the field capacity m.c.—wilting point m.c.)

(*d*) may be more or less than : (the field capacity m.c.—wilting point m.c.) depending upon the crop grown.

**Q. 43. Irrigation water is usually applied to the fields, when the available moisture in the root zone of the crop, gets depleted by :**

(*a*) 0—10% (*b*) 10—25%
(*c*) 50—80% (*d*) 100%.

**Q. 44. Frequency of irrigation is dependent upon the type of :**

(*a*) soil and crop (*b*) soil and climate
(*c*) soil, crop, and climate
(*d*) soil, crop, climate, and fertilizer.

**Q. 45.* A soil has a field capacity of 25%, permanent wilting point of 15%, and specific weight of 14.7 kN/m³. If the root zone depth of the grown crop is 90 cm, then its available moisture holding capacity is**

(*a*) 10 cm (*b*) 13.5 cm
(*c*) 16.67 cm (*d*) 20 cm.

**Q. 46. After how many days will you recommend supplying irrigation water to a crop having 0.8 m root zone depth, and grown in a soil having field capacity of 30%, permanent wilting = 15%, and density of soil = 1.5 gm/cc? The consumptive use for the crop is 5 mm/day, and only 60% of available moisture is permitted to be availed.**

(*a*) 9 days (*b*) 15 days
(*c*) 21 days (*d*) 40 days.

**Q. 47.* If the irrigation efficiency at a 10 hectare field is 80%, and the conveyance losses from the canal outlet is 10%, then the volume of water required at the canal outlet, for supplying 10 cm water depth in the field, will be equal to;**

(*a*) 10,390 kL (*b*) 7200 kL
(*c*) 13,900 kL (*d*) 1,39,000 kL.

**Q. 48. The Δ (in m) for a crop, having quantity duty D (in ha/M. cum), and base period B (in days) is given by :**

(*a*) $\Delta = \dfrac{8.64B}{D}$ (*b*) $\Delta = \dfrac{864B}{D}$ (*c*) $\Delta = \dfrac{100}{D}$ (*d*) $\Delta = \dfrac{10,000}{D}$

**Q. 49. The ratio of the water stored in the root zone during irrigation, to the water needed in the root zone prior to irrigation, is called :**

(*a*) efficiency of water use (*b*) efficiency of water storage
(*c*) efficiency of water application
(*d*) efficiency of water conveyance.

**Q. 50. If the irrigation water applied to a field penetrates uniformly throughout, then the water distribution efficiency is :**

(*a*) 1 (*b*) 0
(*c*) 0.5
(*d*) none of them, as more data is required to ascertain it.

**Q. 51. The efficiency of water application does not depend upon :**

(*a*) climatic conditions (*b*) type of the soil
(*c*) method of application
(*d*) geometry of the conveyance system.

---

*** Hints are given at the end of the chapter for solving starred objective questions.**

**Q. 52. The efficiency of water conveyance does not depend upon :**
(*a*) climatic conditions
(*b*) geometry of the conveyance system
(*c*) nature of the boundary of the conveyance system
(*d*) method of application of water.

**Q. 53. Which of the following is not correctly matched?**
(*a*) Rice—Kharif (*b*) Wheat—Rabi
(*c*) Barley—Kharif (*d*) Potatoe—Rabi.

## CHAPTER 3 : Canal Irrigation System

**Q. 54. If the intensity of irrigation for Kharif is 45% and that for Rabi is 60%; then the annual intensity of irrigation, is :**
(*a*) 60% (*b*) 100%
(*c*) 105% (*d*) none of them.

**Q. 55.* If in a certain irrigation project, and in a given year, 72% and 56% of the culturable command remained unirrigated in Kharif and Rabi seasons, respectively; then the intensity of irrigation for that year and for that project, would be :**
(*a*) 36% (*b*) 64%
(*c*) 72% (*d*) 128%.

**Q. 56. The gross irrigated area cannot exceed the cultivable commanded area :**
**(*a*) Correct** **(*b*) Incorrect.**

**Q. 57.* The area, on which crops are grown in a particular season, is called :**
(*a*) culturable commanded area (CCA) (*b*) gross sown area
(*c*) net sown area (*d*) none of the above.

**Q. 58.* The CCA for a particular State is 5 Mha; out of which, 4.5 Mha is being sown in Rabi season and 2.5 Mha in Kharif season. These areas are being irrigated to the extent of 90% and 80%, respectively. The annual intensity of irrigation for this State is :**
(*a*) 80.7% (*b*) 85%
(*c*) 100% (*d*) 121%.

**Q. 59. In plain areas, the irrigation canals are usually aligned along :**
(*a*) ridge lines (*b*) contour lines
(*c*) valley lines (*d*) across the contours.

**Q. 60. A ridge canal is also called a :**
(*a*) watershed canal (*b*) contour canal
(*c*) side slope canal (*d*) none of the above.

**Q. 61. The canal, which may frequently encounter cross-drainage works, will be a:**
(*a*) watershed canal (*b*) contour canal
(*c*) side slope canal (*d*) none of these.

**Q. 62. In a hilly district, where watershed line is very high as compared to the head-works, which type of alignment you would recommend for a proposed canal, if cross-drainage is to be avoided :**
(*a*) canal along the ridge line (*b*) canal along the contour line
(*c*) canal across the contour lines (*d*) none of them.

**Q. 63. The canal, which can irrigate only on one side, is a :**

(*a*) watershed canal (*b*) contour canal
(*c*) side slope canal (*d*) none of them.

**Q. 64. The capacity of an irrigation canal is usually controlled by :**

(*a*) average kharif demand (*b*) average rabi demand
(*c*) kor demand of rabi crops (*d*) kor demand of kharif crops.

**Q. 65. The optimum kor water depth for a kharif crop is 19 cm with an allowed kor water period of 3 weeks, the outlet discharge factor for this crop will be:**

(*a*) 955 ha/cumec (*b*) 782 ha/cumec
(*c*) 860 ha/cumec (*d*) none of them.

**Q. 66.* If the discharge required for different crops grown in a field is 0.4 cumec, and the capacity factor and time factor are 0.8 and 0.5, the design discharge for the distributary will be :**

(*a*) 0.80 cumec (*b*) 0.16 cumec
(*c*) 1 cumec (*d*) 1.24 cumecs.

**Q. 67.* An irrigation canal is to be designed to deliver 6.25 cumecs to meet the peak Rabi demand of a total of 5100 hectares of cropped area. The estimated canal losses are 25% of the head discharge. The duty on capacity of this canal will be :**

(*a*) 816 ha/cumec (*b*) 1020 ha/cumec
(*c*) 653 ha/cumec (*d*) 612 ha/cumec.

**Q. 68. 'Duty on capacity' is also called :**

(*a*) outlet duty (*b*) capacity factor
(*c*) full supply coefficient (*d*) quantity duty.

**Q. 69.* The 8000 hectares of gross commanded area of an irrigation project includes 20% of reserved forests, usar lands, roads, etc. The pastures and fallow lands are 10%. If the intensity of irrigation is 50%, the area to be irrigated is :**

(*a*) 4000 hectares (*b*) 3200 hectares
(*c*) 2800 hectares (*d*) none of them.

**Q. 70. The discharge carried by a minor distributary is usually less than:**

(*a*) 0.5 cumec (*b*) 1 cumec
(*c*) 2.5 cumecs (*d*) none of them.

**Q. 71. The discharge carried by a major distributary, is usually up to :**

(*a*) 5 cumecs (*b*) 10 cumecs
(*c*) 30 cumecs (*d*) 50 cumecs.

**Q. 72. Earthen irrigation canals, when aligned curvilinear in plan, must be on gentle curves. This adopted curve radius, should be :**

(*a*) more for canal carrying more discharge
(*b*) less for canal carrying more discharge
(*c*) equal for all discharges.

**Q. 73. The canal which is not supposed to do any irrigation is called :**

(*a*) main canal (*b*) water course
(*c*) major distributary (*d*) minor distributary.

**Q. 74. Unlined irrigation canals, when aligned on curvilinear routes in plan, will have to be pitched on :**

(*a*) both sides (*b*) concave side only
(*c*) convex side only (*b*) none of the above.

**Q. 75. The bed of an alluvial channel along the flow will always be :**
(*a*) flat (*b*) wavy
(*c*) duned and rippled (*d*) all of the above are possible.

**Q. 76. The bed form, which is not expected in an alluvial channel with sediment motion, is :**
(*a*) rippled (*b*) meandered
(*c*) duned (*d*) anti duned.

**Q. 77. The bed of an alluvial channel carrying silted water at a high velocity is expected to be :**
(*a*) rippled (*b*) duned
(*c*) flat (*d*) wavy.

**Q. 78. The anti-dunes develop on beds of alluvial streams, when Froude number is:**
(*a*) 0 (*b*) 0.5
(*c*) 1 (*d*) 1.2.

**Q. 79.* The critical shear stress τ, at which incipient motion of sediment takes place, is proportional to :**
(*a*) $\sqrt{d}$ (*b*) $d$
(*c*) $d^2$ where $d$ is grain size

**Q. 80.* In order to ensure that no scouring takes place in the bed of a channel of bed slope *S*, constructed in an incoherent alluvium of size 'd' cm, the flow velocity should be restricted to :**
(*a*) $4.85 \cdot d^{1/2} \cdot S^{-1/6}$ (*b*) $4.85 \cdot d^{-1/2} \cdot S^{1/6}$
(*c*) $0.48\, d^{1/2} \cdot S^{1/6}$ (*d*) $0.48 \cdot d^{1/2} \cdot S^{-1/6}$.

**Q. 81.*** Manning's rugosity coefficient is proportional to :
(*a*) $\sqrt{d}$ (*b*) $d$
(*c*) $d^{1/6}$ (*d*) $d^{2/3}$
(*e*) $d^{3/2}$; where $d$ is representative grain dia of bed surface.

**Q. 82. The finer sediment which does not originate from the stream bed, but originates from the catchment area, is best called as the :**
(*a*) bed load (*b*) suspended load
(*c*) wash load (*d*) none of these.

**Q. 83. The force exerted by the flowing water on the sediment particles to cause their motion, is called :**
(*a*) buoyant force (*b*) tractive force
(*c*) kinematic force (*d*) eddy force.

**Q. 84. The minimum size of sediment that may remain stable in an alluvial channel, carrying discharge intensity q, with hydraulic radius R, and bottom slope S, is :**
(*a*) $11\, RS$ (*b*) $1.35\, (q^2/g)^{1/2}$
(*c*) $\gamma\, RS$ (*d*) none of them.

**Q. 85.* The tractive force method of designing stable channels in non-cohesive soils, is applicable only when the channels :**
(*a*) carry sedimented water (*b*) carry clear water
(*c*) carry sedimented water, but sediment concentration does not exceed 500 ppm or so
(*d*) carry any type of water.

**Q. 86. In an earthen trapezoidal channel having side slopes of 1.5 H : 1 V, constructed in a non-cohesive soil, having $D_{75}$ particle size of 'd' m, and angle $\phi$ = 37°, no scouring will occur only when the hydraulic mean depth (R) of the channel is limited to :**

(*a*) $\dfrac{d}{33S_0}$ (*b*) $\dfrac{d}{11S_0}$

(*c*) $\dfrac{d}{21S_0}$ (*d*) none of them.

**Q. 87. If $\tau_c$ is the critical shear stress on a channel bed, then the critical shear on the side slopes (nH : IV) of this channel ($\tau_c'$) is given by :**

(*a*) $\tau_c' = \tau_c \cdot \sqrt{1 - \dfrac{\sin^2\theta}{\sin^2\phi}}$ (*b*) $\tau_c' = \tau_c \cdot \sqrt{1 + \dfrac{\sin^2\theta}{\sin^2\phi}}$

(*c*) $\tau_c' = \tau_c \cdot \sqrt{1 - \dfrac{\sin^2\phi}{\sin^2\theta}}$ (*d*) $\tau_c' = \tau_c \cdot \sqrt{1 + \dfrac{\sin^2\phi}{\sin^2\theta}}$

where $\phi$ = angle of repose of soil

$\theta = \tan^{-1}\left(\dfrac{1}{n}\right)$

**Q. 88. The Lacey's and Kennedy's empirical silt theories for designing irrigation channels in Indian non-cohesive alluvial soils, are meant to ensure :**

(*a*) no scouring in the channel

(*b*) no silting from out of the sedimented water entering the channel from the headworks

(*c*) both (*a*) and (*b*)

(*d*) neither (*a*) nor (*b*).

**Q. 89.* The critical velocity $V_0$ = 0.55 · m · $y^{0.64}$, as suggested by Kennedy for design of trapezoidal irrigation channels, is :**

(*a*) the maximum permissible velocity

(*b*) the minimum permissible velocity

(*c*) both (*a*) and (*b*)

(*d*) none of the above.

**Q. 90. Critical velocity Ratio for use in Kennedy's theory, is :**

(*a*) less than 1 (*b*) more than 1

(*c*) equal to 1 (*d*) all of the above.

**Q. 91. The Garret's diagrams are based on :**

(*a*) Lacey's theory (*b*) Khosla's theory

(*c*) Bligh's theory (*d*) Kennedy's theory.

**Q. 92. The critical velocity ratio was introduced in Kennedy's equation of critical velocity to take into account the effect of :**

(*a*) channel cross-section (*b*) climatic conditions

(*c*) silt grade (*d*) roughness of bed.

**Q. 93. The most important shape parameter in sediment analysis is :**

(*a*) sphericity (*b*) shape factor

(*c*) roundness (*d*) form factor.

**Q. 94. Lacey's silt factor is proportional to :**

(*a*) $\sqrt{d}$ (*b*) $d$

(*c*) $d^2$ ; where $d$ is the grain size.

**Q. 95. In mechanics of sediment transport, the parameter $\varepsilon = \frac{\eta}{\rho}$ (where $\eta$ is eddy viscosity and $\rho$ is density) is called :**

(*a*) kinematic viscosity (*b*) absolute viscosity

(*c*) eddy kinematic viscosity (*d*) none of the above.

**Q. 96. In sediment analysis, the parameter which indicates the amount of wear the grain particle has undergone during relative movement of silt and water, is**

(*a*) sphericity (*b*) shape factor

(*c*) roundness (*d*) form factor.

**Q. 97. The wetted perimeter P of a stable channel is proportional to :**

(*a*) $Q$ (*b*) $\sqrt{Q}$

(*c*) $Q^2$; where $Q$ is the discharge

**Q. 98. The units of Shield's Entrainment function are :**

(*a*) Newton/sq m (*b*) metre

(*c*) kg/m/sec (*d*) no dimensions.

**Q. 99. The cross-section of natural silt transporting channels tend to have the shape of :**

(*a*) trapezium (*b*) semi-ellipse

(*c*) semi-circle (*d*) none of them.

**Q. 100. In Lacey's regime theory, the flow velocity is proportional to :**

(*a*) $(Q f^2)^{1/2}$ (*b*) $\frac{Q}{f^2}$

(*c*) $(Qf^2)^{1/6}$.

**Q. 101. For all homogeneous fluids, the value of Von-Karman Universal Constant, is:**

(*a*) 0 (*b*) 1

(*c*) 0.4 (*d*) 0.8.

**Q. 102. The boundary shear stress, $\tau_0$ in an open channel is given by :**

(*a*) $\frac{\gamma h^2}{2}$ (*b*) $\gamma RS$

(*c*) $\frac{1}{n} R^{2/3} S^{1/2}$ where $h$ = depth

R = hydraulic radius

S = Slope

$\gamma$ = Specific weight of fluid

$n$ = Manning's roughness factor

**Q. 103. Lacey's regime theory is not applicable to a canal in :**

(*a*) true regime (*b*) initial regime

(*c*) final regime (*d*) none of them.

**Q. 104.*** When a channel is protected on the bed and sides with some protective material, and there is no possibility of change in its cross-section and longitudinal slope, then the channel is said to be in :

(*a*) initial regime (*b*) final regime

(*c*) permanent regime (*d*) true regime.

**Q. 105.*** Hydraulic depth is the ratio of :

(*a*) wetted area to wetted perimeter (*b*) wetted area to top width

(*c*) wetted area to bottom width
(*d*) wetted perimeter to top width.

**Q. 106. The unit of Chezy's coefficient C in Chezy's formula, is :**

(*a*) $\sqrt{m/s}$ (*b*) m/s
(*c*) m (*d*) Nil.

**Q. 107.** For a most economical trapezoidal channel section :

(*a*) hydraulic mean radius equals the depth of flow
(*b*) hydraulic mean radius equals half the depth of flow
(*c*) bottom width is twice the depth of flow
(*d*) bottom width is half the depth of flow.

**Q. 108.*The channels, whose boundaries are made of easily erodible loose soils, are called :**

(*a*) mobile boundary channels (*b*) rigid boundary channels
(*c*) non-alluvial channels (*d*) none of the above.

**Q. 109.*Lacey's scour depth for a stream, carrying a discharge of 3 cumecs per metre width and having a silt factor 1.2, is :**

(*a*) 1.32 m (*b*) 2.64 m
(*c*) 3.96 m (*d*) 4.32 m.

**Q. 110. Which one of the following is not the requirement of an ideal regime condition in Lacey's regime-theory ?**

(*a*) the discharge in the channel is constant
(*b*) the channel flows through the same soil grade, as that of the sediment entering the channel from the headworks
(*c*) the sediment grade and its amount entering the channel is constant.
(*d*) the silt grade should consist of clay sized particles.
(*e*) the flow should be uniform.

**Q. 111.*An irrigation canal proposed to be excavated in an incoherent alluvium of median size 0.4 mm, for carrying a discharge of 40 cumecs, should ideally have a bed slope of :**

(*a*) 1 in 6000 (*b*) 1 in 5200
(*c*) 1 in 2000 (*d*) 1 in 1265.

**Q. 112.*The silt factor 'f', as used by Lacey in his regime theory, was adjusted to be equal to 1 for the silt of Upper Bari-Doab alluvium. What was the median size of this alluvium ?**

(*a*) 0.28 mm (*b*) 0.32 mm
(*c*) 0.40 mm (*d*) 0.5 cm.

**Q. 113. Average sediment load in a regime canal of Lacey, is of the order of :**

(*a*) 50 ppm (*b*) 500 ppm
(*c*) 5000 ppm (*d*) 10,000 ppm.

**Q. 114.*The discharge carried by a regime canal of bed width 22 m and water depth 2.5 m, is about;**

(*a*) 22 cumecs (*b*) 34 cumecs
(*c*) 84 cumecs (*d*) none of them.

**Q. 115.*Two irrigation channels, A and B, are designed on Lacey's theory to carry the same discharge. The alluvium through which canal A has to pass, however, is coarser than that for canal B. In such a design, we expect :**

(*a*) channel *A* to be deeper (*b*) channel *B* to be deeper
(*c*) channel *A* to have larger wetted perimeter

(*d*) channel *B* to have larger cross-sectional area.

**Q. 116. The major drawback of Laceys regime theory, as used for the design of irrigation canals, is that :**

(*a*) it does not consider the quantum of sediment load, which is likely to flow into the canal

(*b*) It is applicable only to non-cohesive soils

(*c*) it is applicable only to alluvial areas of North India or Pakistan.

(*d*) none of the above.

**Q. 117. "An alluvial river scours and silts its bed automatically, depending upon the variation in the flow discharge". Which theory helps us to know this important conclusion?**

(*a*) Lacey's theory (*b*) Kennedy's theory

(*c*) Einstein's theory (*d*) none of them.

**Q. 118. Stable channels for a given bed-load transport, can be best designed on the basis of**

(*a*) Lacey's theory (*b*) Kennedy's theory

(*c*) Einstein's theory (*d*) all of the above.

**Q. 119. In an irrigation canal, berms are provided at :**

(*a*) NSL (*b*) FSL

(*c*) Bank level (*d*) none of the above.

**Q. 120. Counter berms are provided in an irrigation canal :**

(*a*) at the outer sides of the canal banks to allow movement of cattle

(*b*) at the inner sides of the canal banks at the ground level, to make bank line and bed line parallel

(*c*) at the outer sides of the canal banks to keep the saturation gradient line well within the bank section

(*d*) none of the above.

**Q. 121. Berms are provided in an irrigation canal at :**

(*a*) the outer edge of the canal banks to allow movement of cattle

(*b*) the inner sides of the canal bank at the ground level, to make inner bank line and bed line parallel

(*c*) the outer sides of the canal banks, to keep the phreatic line adequately covered by earth cover

(*d*) none of the above.

**Q. 122.*** An alluvial canal is excavated using side slopes of $n_1$ : 1(H : V) in excavation and $n_2$ : 1 in filling. If the depth of cutting is $d_1$, and top of the bank is at a height of $d_2$ above the bed, then the berm width required to have the bed line and bank line parallel to each other, despite fluctuations in the ground level, is :

(*a*) $(n_2 - n_1)\, d_2$ (*b*) $(n_2 - n_1)\, d_1$

(*c*) $(n_2 - n_1)\,(d_2 - d_1)$ (*d*) $n_2 \cdot d_2$.

**Q. 123.*** An alluvial irrigation canal is constructed with 1 : 1 (H : V) side slopes in cutting and 2 : 1 in filling, and the water depth upto FSL is expected to be *y*. A berm width equal to y is left initially, which with the passage of time, and when channel attains regime conditions, is likely to become :

(*a*) 1.1*y* (*b*) 1.25*y*

(*c*) 1.5*y* (*d*) 2*y*.

## CHAPTER 5 : Lining of Irrigation Canals and Economics of Lining

**Q. 124. A lined alluvial canal is best designed on the basis of :**

(*a*) Lacey's formula (*b*) Kennedy's formula

(*c*) Manning's formula (*d*) Continuity equation.

**Q. 125. Lining of irrigation channels :**

(*a*) increases water logging
(*b*) increases channel cross section
(*c*) increases command area
(*d*) increases chances of breaching.

**Q. 126.*An irrigation canal is lined along its 20 m perimeter @ Rs. 40 per m$^2$ of lining. The lining has resulted in net water saving @ 3 cumecs per M-m$^2$ of lined area, whereas each cumec of water increases the annual crop yield by Rs. 4 lakh. Assuming the life of lining as 40 years, and the saving in maintenance cost to be balanced by the interest on investment, the benefit cost ratio for the project is :**

(*a*) 1.2 (*b*) 1.0
(*c*) 0.83 (*d*) none of them.

**Q. 127. The transmission loss of valuable water in an ordinary alluvial irrigation canal in a North Indian State, in cumecs per million sq m of wetted area, is of the order of :**

(*a*) 0.5—0.6 (*b*) 2.5—3.5
(*c*) 25—35 (*d*) 250—350.

**Q. 128. A triangular lined canal section with corners rounded by a radius equal to the full supply depth of 4 m, is likely to have its hydraulic radius, as :**

(*a*) 4 m (*b*) 3 m
(*c*) 2 m
(*d*) cannot be ascertained, as side slopes are not given.

**Q. 129.***A standard triangular lined channel section with 1.5 H : 1 V side slopes has a capacity of 26 cumecs and full supply depth of 2.5 m. The lining material for the channel would be safe for withstanding mean velocity of:

(*a*) 1.8 m/s (*b*) 2.0 m/s
(*c*) 2.2 m/s (*d*) none of them.

**Q. 130. Safety ladders are provided in large irrigation canals, to :**

(*a*) enable the fish to pass from one place to another
(*b*) enable the cattle to cross the canal
(*c*) enable the swimmers to get out of the canal
(*d*) provide safe exit to avoid accidental drowning.

**Q. 131. The main advantage offered by Bentonite layer lining in an irrigation canal is that :**

(*a*) seepage loss is reduced
(*b*) cross-sectional area of the channel is reduced
(*c*) cost of land acquired is reduced
(*d*) earth work excavation is reduced.

**Q. 132. The free-board in lined canals is measured between :**

(*a*) FSL and top of lining (*b*) FSL and top of canal bank
(*c*) top of lining and top of canal bank (*d*) none of them.

**133. The minimum recommended free-board for lined canals carrying discharge of more than 10 cumecs is**

(*a*) 0.3 m (*b*) 0.6 m
(*c*) 0.75 m (*d*) 1.2 m.

Q. 134. Under-drainage arrangements in canals are necessarily required in :

(*a*) unlined canals (*b*) lined canals

(*c*) both (*a*) and (*b*)
(*d*) lined canals constructed on sandy soils alone.

**Q. 135. When elaborate under-drainage arrangements can not be provided, and the soil strata is sandy, the preferred material for canal lining, would be :**
(*a*) cement concrete lining (*b*) soil cement lining
(*c*) brick lining (*d*) shotcrete lining.

**Q. 136. Steel reinforcement in concrete lining is justified, mainly because :**
(*a*) it augments the structural strength of uncracked lining
(*b*) it enables the lining to take tensile loads
(*c*) it reduces the width of shrinkage cracks
(*d*) none of the above.

**Q. 137. Which among the following is not true about Asphaltic concrete lining?**
(*a*) it is fairly cheap
(*b*) it is flexible and readily confirms to subgrade
(*c*) it permits certain type of weed growth
(*d*) it decreases the rugosity coefficient to increase channel efficiency.

## CHAPTER 6 : Reclaimation of Water-logged and Saline Soils for Agricultural Purposes.

**Q. 138. A land is said to be water-logged, when :**
(*a*) the land is necessarily submerged under standing water
(*b*) there is a flowing water over the land
(*c*) the pH value of the soil becomes as high as 8.5
(*d*) the soil pores in the root zone get saturated with water, either by the actual watertable or by its capillary fringe.

**Q. 139. Water-logging of cropped land leads to reduced crop yields, due to :**
(*a*) ill-aeration of root zone, causing lack of oxygen to plants
(*b*) growth of water-loving plants interfering with the sown crop
(*c*) surrounding of the root zone by the resultant saline water, which extracts the good water from plant roots by osmosis
(*d*) all of the above.

**Q. 140.** Pick up the incorrect statement from the following :
(*a*) intensive irrigation should be avoided in areas susceptible to water-logging
(*b*) extensive irrigation should be adopted in areas susceptible to water logging
(*c*) lift irrigation can help alleviate water-logging susceptibilities.
(*d*) none of the above.

**Q. 141.** Which one of the followings does not contribute to water logging?
(*a*) inadequate drainage (*b*) seepage from unlined canals
(*c*) frequent flooding
(*d*) excessive tapping of ground water.

**Q. 142.** Which one of the followings, is not a remedial measure for water logging?
(*a*) good drainage for irrigated land
(*b*) conjunctive use of water in the basin

(*c*) lining of canals and water courses
(*d*) contour bunding.

**Q. 143. Pick up the incorrect statement :**
(*a*) moderately saline lands are represented by visible whiteish deposits of salts on the top and upper layers of soil
(*b*) $Na_2 CO_3$ is the most harmful salt for causing salinity
(*c*) Nacl is the least harmful salt for causing salinity
(*d*) a highly saline land is called an alkaline land, or a black alkali.
(*e*) none of the above.

**Q. 144. Point out the incorrect statement, out of the following :**
(*a*) salinity is caused by water-logging
(*b*) water logging is not caused by salinity
(*c*) salinity subsides, when once the water logging is removed
(*d*) none of the above.

**Q. 145. Alkaline soils are best reclaimed by :**
(*a*) leaching (*b*) addition of gypsum to soil
(*c*) providing good drainage
(*d*) addition of gypsum to soil and leaching.

**Q. 146. A recently reclaimed alkaline soil should preferably be sown with a salt resistant crop, like :**
(*a*) wheat (*b*) cotton
(*c*) barseem (*d*) any of the above.

**Q. 147.* A soil has an Exchangeable Sodium Percentage (ESP) of 16%, and its Electrical conductivity (EC) is 32 milli-mhos/cm. If the pH of this soil is 9.5, then the soil is classified as :**
(*a*) alkaline soil (*b*) saline soil
(*c*) saline alkali soil (*d*) none of the above.

**Q. 148. The soil becomes practically infertile when its pH value is about :**
(*a*) 0 (*b*) 7
(*c*) 11 (*d*) none of them.

**Q. 149. Which of the following crops can withstand the highest water table, which normally should be 0.7 to 2.5 m below the level of the cropped land?**
(*a*) wheat (*b*) lucerine
(*c*) cotton (*d*) rice.

**Q. 150. While growing crops in high water-table areas, open jointed drains, discharging into outlet drains, are sometimes laid below the cropped land to lower down the water-table and to remove the rain water effectively. These drains are called:**
(*a*) tile drains (*b*) french drains
(*c*) gravity outlets (*d*) surface drains.

**Q. 151. A tile drain is laid below a cropped land to remove excess irrigation water. The Drainage Coefficient of this drain, is usually expressed as;**
(*a*) cm of water depth removed from the drainage area per day
(*b*) cum of water removed per second
(*c*) percentage of applied water, which is intercepted by this drain
(*d*) none of the above.

**Q. 152. When a sub-surface open jointed drain is laid below a farm land to intercept excess irrigation water, then its capacity, depending upon soil type, should be of the order of :**

(*a*) 5—10% of applied water (*b*) 10—50% of applied water
(*c*) 50—80% of applied water (*d*) none of them.

**Q. 153. The method, which uses dead furrows on cropped farms for drainage of excess irrigation or rain water, is called :**
(*a*) surface inlet (*b*) tile drainage
(*c*) bedding (*d*) french drain.

## CHAPTER 8 : Rivers, their Behaviour, Control and Training

**Q. 154. Ghaggar river which appears and disappears in Rajasthan State of India, can be classified as a**
(*a*) meandering river (*b*) deltaic river
(*c*) virgin river (*d*) none of the above.

**Q. 155. The quantum of river discharge, which is large enough in magnitude as well as in frequency, so as to cause controlling effect on the river course and cross-section, is known as :**
(*a*) design discharge (*b*) median discharge
(*c*) maximum flood discharge (*d*) dominant discharge.

**Q. 156. The ratio of dominant discharge for a river to the peak discharge for that river, is of the order of :**
(*a*) 3/16 (*b*) 9/16
(*c*) 13/16 (*d*) 3/2.

**Q. 157. The 'meander length' for an alluvial river is :**
(*a*) the total channel length along its looped course
(*b*) the total channel length minus the direct straight length
(*c*) the axial length of one meander
(*d*) the looped length of one meander
(*e*) none of the above.

**Q. 158. The 'meander belt' for an alluvial river is :**
(*a*) the total river width between embankments
(*b*) the width between the outer edges of fully developed meander loop, measured perpendicular to river axis
(*c*) the same as meander width
(*d*) both (*b*) and (*c*).

**Q. 159. For a meandering alluvial river, the ratio of its 'channel length' to 'direct axial length' is always :**
(*a*) 1 (*b*) > 1
(*c*) < 1
(*d*) may be less or more than 1, depending upon the river.

**Q. 160. In a meandering river, the ratio of 'arcual channel length' to 'direct axial length', is called :**
(*a*) tortuosity (*b*) inverse of tortuosity
(*c*) cut off ratio (*d*) none of these.

**Q. 161. Tortuosity in a meandering river, is :**
(*a*) 1 (*b*) < 1
(*c*) > 1 (*d*) none of these.

**Q. 162. A river reach having tortuosity of 1.2, can be said to have :**
(*a*) 20% tortuosity of meander (*b*) 80% tortuosity of meander
(*c*) 120% tortuosity of meander (*d*) none of the above.

**Q. 163. In an alluvial meandering river in flood plains, the meandering length ($M_L$) is *x* times the meandering width ($M_B$), where *x* is roughly equal to :**

(a) 3 (b) 1/3
(c) 1 (d) 1/3 to 3.

**Q. 164. For a meandering alluvial river in flood plain, the meander length is about:**
(a) 6 W (b) 18 W
(c) 17 W (d) W.
where W is the normal river width

**Q. 165. For a meandering alluvial river in flood plain, the meander width is about:**
(a) 0.5 W (b) W
(c) 6 W (d) 17 W.
where W is the normal river width

**Q. 166.*An alluvial river has a dominant discharge of 1600 $m^3/s$ and a bed slope of 1 in 5000. The approximate value of meander belt for this river, would be:**
(a) 0.6 km (b) 6 km
(c) 60 km (d) none of these.

**Q. 167. An alluvial meandering river is likely to carry a peak discharge of 6000 cumecs. The extent of meandering belt in this river, estimated in km per km of meandering length, roughly equals.**
(a) 1/3 (b) 3
(c) 9 (d) 27.

**Q. 168. Meander ratio in an alluvial meandering river is given by :**
(a) $\frac{\text{meander length}}{\text{meander width}}$ (b) $\frac{\text{meander width}}{\text{meander length}}$
(c) $\frac{\text{meander width}}{\text{meander length}} \times 100$ (d) none of them.

**Q. 169.*The meander belt for an alluvial stream is estimated to be 3 km. The peak discharge in this river will roughly be of the order of :**
(a) 4000 cumecs (b) 700 cumecs
(c) 1200 cumecs (d) 1600 cumecs.

**Q. 170. In a meandering river reach, the deepest river portions will be available at:**
(a) the crossings
(b) the inner edges of meander loops
(c) the outer edges of meander loops
(d) none of them.

**Q. 171. Sinuosity of a meandering river is :**
(a) the same as tortuosity (b) inverse of tortuosity
(c) log of tortuosity (d) none of them.

**Q. 172. Pinpoint the incorrect statement :**
(a) guide-bunds are constructed to guide the river flow through a smaller width
(b) guide bunds are generally extended downstream of the hydraulic structure, in a length of about twice their upstream length
(c) guide banks prevent the out-flanking of the hydraulic structure by the changing river course
(d) aggrading rivers can be best controlled by soil conservation measures, including construction of check dams on their tributaries

(*e*) degrading rivers can be best controlled by building of cross-bars, weirs, etc., prior to other river training works.

**Q. 173. Aggrading rivers are :**

(*a*) silting rivers (*b*) scouring rivers
(*c*) rivers in regime (*d*) meandering rivers.

**Q. 174.** The river reach upstream of a newly built dam may behave, as :

(*a*) aggrading (*b*) degrading
(*c*) virgin (*d*) none of them.

**Q. 175. A barrage with 600 m length is designed across an alluvial river. The U/S and D/S lengths of the proposed guide banks of this barrage, will be about:**

(*a*) 750 m and 1500 m (*b*) 750 m and 750 m
(*c*) 750 m and 150 m (*d*) 750 m and 300 m

**Q. 176.*A pair of guide banks are to be constructed for an alluvial river at 400 m apart at a bridge site. The design discharge of the river is 5000 cumecs at 5 m water depth of HFL above the average bed level of the river. These guide banks need to be stone pitched on water side, with apron laid from the toe in the river bed. The average estimated depth and width of this apron, respectively, will be :**

(*a*) 2 m and 10 m (*b*) 1.5 m and 5 m
(*c*) 2 m and 20 m (*d*) 3 m and 20 m.

**Q. 177. The repelling groynes which are largely constructed projecting from river embankments, as anti-erosion works, are :**

(*a*) pointing upstream (*b*) pointing downstream
(*c*) perpendicular to the bank (*d*) none of these.

**Q. 178.*If a river in an alluvial plain has a dominant discharge of 1600 cumecs, the waterway for this bridge, in the absence of model testing, can be safely adopted, as equal to :**

(*a*) 190 m (*b*) 228 m
(*c*) 300 m (*d*) 480 m.

**Q. 179. The upstream angle of inclination of a repelling groyne with normal to the bank line, is of the order of :**

(*a*) 5 to 10° (*b*) 10 to 30°
(*c*) 30 to 50° (*d*) 70 to 90°

**Q. 180. Which one among the following is not the real cause of an alluvial river following a serpentine course?**

(*a*) variation of discharge
(*b*) availability of steeper land gradient than what is required for equilibrium.
(*c*) availability of inadequate land gradient than what is required for equilibrium
(*d*) none of these.

**Q. 181. An alluvial river increases its length by meandering due to;**

(*a*) variation of discharge (*b*) variation in land topography
(*c*) both (*a*) and (*b*) (*d*) none of the above.

**Q. 182. The factor which is not primarily responsible for meander in an alluvial river, is;**

(*a*) variation of discharge (*b*) excess silt load
(*c*) deficient silt load (*d*) none of them.

**Q. 183. The secondary factor. which is responsible for meandering in an alluvial river, is :**

(*a*) inadequate land gradient (*b*) deficient silt load

(*c*) constant discharge (*d*) none of them.

**Q. 184. Point out the incorrect statement in relation to the behaviour of a meandering alluvial river :**

(*a*) Primary cause for meandering is the silting action due to excess sediment at the available discharge and surface slope

(*b*) Secondary cause for meandering is the scouring action due to deficiency of sediment at the available discharge and surface slope

(*c*) Meandering tendency is more in an alluvial river during floods, because of high variation in discharge

(*d*) if there is no variation of discharge in an alluvial river, there will be no meandering

(*e*) none of them.

**Q. 185. Out of the following choices given below, choose impermeable spur(s) :**

(*a*) an earthen spur protected by stone apron

(*b*) an earthen spur, unprotected by stone apron

(*c*) a balli spur (*d*) a tree groyne

(*e*) (*a*) and (*b*) both (*f*) (*c*) and (*d*) both.

**Q. 186. Permeable spurs are best suitable for rivers, which :**

(*a*) carry heavy suspended load

(*b*) carry large bed load, but light suspended load

(*c*) need permanent protection to dikes

(*d*) need attracting the river current, for providing deeper channel

(*e*) flow in upper hilly reaches.

**Q. 187. Denehey's spur is :**

(*a*) a hockey-shaped earthen spur

(*b*) a *T*-shaped stone spur, as used in Australia

(*c*) a *T*-shaped earthen spur, as used in India

(*d*) a type of a balli spur, especially developed for Indian rivers.

**Q. 188. Which one of the following effects can not be attributed to have been caused by the construction of dikes along a river coarse?**

(*a*) faster travel of a flood wave

(*b*) higher flood levels along the river

(*c*) increased peak discharges all along downstream points

(*d*) decrease in surface slope of the river above the leveed reach

(*e*) increased bed levels in the river

(*f*) reduction in meander belt, causing steeper hydraulic gradients.

**Q. 189. Which one of the following effects produced by a Cut off in an alluvial river, is not an advantage to navigation :**

(*a*) shortened route and elimination of sharp bends

(*b*) shortened travel time, particularly at low and moderate river stages

(*c*) increased water depth at low river stages

(*d*) lowering of the flood stages and flood periods.

## CHAPTER 9 : Diversion Head Works

**Q. 190. Barrages constructed across alluvial rivers help in :**

(*a*) controlling floods (*b*) restoring river regime

(*c*) ensuring monsoon storage (*d*) all of them.

**Q. 191. Which one among the following is a correct choice in relation to a weir?**

(*a*) it is helpful in diverting excess water to a river from a canal
(*b*) it does not cause any heading up of water on its upstream side
(*c*) it stores water by raised counter-balanced gates
(*d*) it increases the chances of floods in the upstream areas
(*e*) none of them.

**Q. 192. If two canals are taken off from both the flanks of a river at the site of a diversion headwork, then the number of undersluices and divide walls, will respectively, be :**

(*a*) 1 and 1 (*b*) 1 and 2
(*c*) 2 and 1 (*d*) 2 and 2.

**Q. 193. Pinpoint the incorrect statement :**

(*a*) the old Okhla barrage near Delhi has recently been replaced by a new weir
(*b*) the cost of a barrage is usually higher than that of a weir
(*c*) a weir can pond up additional water on its upstream side, by using shutters on its top
(*d*) a barrage does not increase the susceptibility of flooding in upstream areas
(*e*) none of them.

**Q. 194. In a barrage project, a divide wall is provided to :**

(*a*) separate the lower crest 'undersluice side' from the higher crest 'weir side'
(*b*) separate the higher crest 'undersluice side' from the lower crest 'weir side'
(*c*) keep the cross-currents away from the barrage body
(*d*) serve none of the above purposes.

**Q. 195. A canal headworks has nothing to do with a :**

(*a*) weir (*b*) guide bank
(*c*) head regulator (*d*) safety ladder.

**Q. 196. In a diversion headworks project, the canal head regulator is usually aligned:**

(*a*) parallel to the barrage axis (*b*) perpendicular to divide wall
(*c*) parallel to the divide wall (*d*) none of them.

**Q. 197. The choice among the following, which does not control the discharging capacity to be provided for the undersluices, is :**

(*a*) that it should be at least double the canal discharge
(*b*) that it should be about 10 to 15% of maximum flood discharge
(*c*) that it should be able to pass winter freshets and low floods.
(*d*) that it should be equal to the dominant discharge of the river
(*e*) none of them.

**Q. 198.** A nadromous fish often move large distances in rivers in India :

(*a*) to upstream only (*b*) to downstream only
(*c*) to downstream in winter, and upstream in summer
(*d*) to upstream in winter, and downstream in summer.

**Q. 199. A fish ladder is provided in a canal project :**

(*a*) to catch the fish for commercial development
(*b*) to enable the fish to move freely in the river
(*c*) to serve the same purpose as a canal ladder.

(*d*) (*b*) and (*c*) both.

**Q. 200.** Head sluices are the gate controlled openings, in :
(*a*) the entire length of the barrage
(*b*) the under-sluice length of the barrage
(*c*) the regulator of the main off taking canal
(*d*) none of them.

**Q. 201. Silt excluders are constructed :**
(*a*) on the river bed downstream of the head regulator
(*b*) on the river bed upstream of the head regulator
(*c*) on the canal bed downstream of the canal head regulator
(*d*) none of these.

**Q. 202.** The silt exclusion device, constructed on the bed of the main canal, taking off from a headwork, is called :
(*a*) silt excluder (*b*) silt ejector
(*e*) both (*a*) and (*b*) (*d*) none of them.

**Q. 203. The tunnel openings provided infront of a canal head regulator at a Diversion headworks :**
(*a*) discharge sedimented water into the canal
(*b*) discharge sediment load into the under-sluices, from where it ejects out to the downstream river
(*c*) discharge clear water into the canal
(*d*) none of these.

**Q. 204. Canal headworks in the upper rocky reaches of the rivers, are uncommon, because :**
(*a*) more drops are required on the resulting canal system.
(*b*) costly head regulator is required
(*c*) more cross-drainage works are required on the resulting canal network
(*d*) all of the above.

## CHAPTER 10 : Hydraulic Jumps and its Usefulness in the Design of Irrigation Structures

**Q. 205. If $y_1$ and $y_2$ are water depths upstream and downstream of a hydraulic jump respectively, the loss of energy due to hydraulic jump is :**
(*a*) $(y_2 - y_1)^4/4y_1y_2$ (*b*) $(y_2 - y_1)^3/4y_1y_2$
(*c*) $(y_2 - y_1)^2/4y_1y_2$ (*d*) $(y_2 - y_1)/4y_1y_2$

**Q. 206. A hydraulic jump involves;**
(*a*) subcritical flow (*b*) super critical flow
(*c*) critical flow (*d*) all of the above.

**Q. 207. A hydraulic jump ensures :**
(*a*) change of subcritical flow to super critical flow
(*b*) change of super critical flow to subcritical flow
(*c*) change of subcritical flow to critical flow
(*d*) change of super critical flow to critical flow.

**Q. 208. The phenomenon of hydraulic jump leads to :**
(*a*) evolution of energy (*b*) dissipation of energy
(*c*) sometimes (*a*) and sometimes (*b*) (*d*) none of these.

**Q. 209. If a vertical line is drawn on the specific energy curve, corresponding to a particular value of specific energy, then the two depth values are obtained, which are known as :**

(*a*) initial and sequent depths (*b*) alternate depths
(*c*) conjugate depths (*d*) all of these.

**Q. 210. The sequent depth is always :**

(*a*) less than the alternate subcritical depth
(*b*) more than the alternate subcritical depth
(*c*) equal to the alternate subcritical depth
(*d*) none of the above.

**Q. 211. The formula connecting initial depth ($y_1$) with sequent depth ($y_2$) in a hydraulic jump, is given as :**

(*a*) $y_2 = \frac{-y_1}{2} + \sqrt{\frac{y_1^2}{4} + \frac{2q^2}{gy_1}}$ (*b*) $y_2 = \frac{y_1}{2} + \sqrt{\frac{y_1^2}{4} + \frac{2q^2}{gy_1}}$

(*c*) $y_2 = \frac{-y_1}{2} + \sqrt{\frac{y_1^2}{4} + \frac{q^2}{gy_1}}$ (*d*) $y_2 = \frac{y_1}{2} + \sqrt{\frac{y_1^2}{4} + \frac{q^2}{gy_1}}$

where $q$ is the discharge intensity in the channel at the section of the hydraulic jump.

**Q. 212.** The hydraulic jump that develops usually in barrages and canal head regulators, is of the type :

(*a*) strong jump (*b*) steady jump
(*c*) oscillating and weak jump (*d*) undular jump.

**Q. 213.** The formation of a hydraulic jump on a sloping glacis, as compared to that on a horizontal floor, is always :

(*a*) more definite & more efficient (*b*) more definite & less efficient
(*c*) less definite & more efficient (*d*) less definite & less efficient.

## CHAPTER 11 : Theory of Seepage and Design of Weirs and Barrages

**Q. 214. The safety of a hydraulic structure founded on pervious foundation can be ensured :**

(*a*) by providing sufficient length of its concrete floor
(*b*) by providing sufficient depth of its concrete floor
(*c*) by providing a downstream cutoff of some reasonable depth
(*d*) all of the above.

**Q. 215. Point out the correct meaning of 'piping', as applied to the design of barrages:**

(*a*) it refers to a network of pipes laid below the hydraulic structure to remove the seeping water
(*b*) it refers to the process of undermining of foundation and creation of hollows therein
(*c*) it refers to the uplift force caused by seeping water on the floor of the hydraulic structure
(*d*) none of the above.

**Q. 216. While designing the downstream floor thickness of a weir, the ordinate of the uplift pressure at a point is 2.8 m. If the relative density of concrete is taken to be 2.4, then the minimum thickness of the floor to be provided for resisting uplift pressure, without accounting safety factor, is :**

(*a*) 1.16 m (*b*) 2 m
(*c*) 0.8 m (*d*) none of them.

**Q. 217.*While designing the floor thickness of a weir in its upstream length, the average head causing uplift is 2.8 m. The provided floor thickness here, with suitable factor of safety, could best be :**

(*a*) 2 m (*b*) 2.5 m
(*c*) 0.8 m (*d*) 4.5 m.

**Q. 218. The critical exit gradient suggested in Khosla's theory of design of weirs and barrages, is :**

(*a*) less for more porous soils (*b*) more for more porous soils
(*c*) equal for all kinds of soils (*d*) none of them.

**Q. 219.*Khosla's safe exit gradient for design of weirs will be the lowest for the soil type :**

(*a*) fine sand (*b*) coarse sand
(*c*) shingle and gravels (*d*) none of them.

**Q. 220. Point out the incorrect statement in relation to the design of weirs and barrages :**

(*a*) Bligh's safe hydraulic gradient is the same as the Khosla's safe exit gradient
(*b*) the first streamline below a barrage section follows the bottom profile of the section
(*c*) equipotential lines are the lines joining the points of equal residual seepage head
(*d*) none of them.

**Q. 221. Which one of the following factors, is not to be considered in the design of major hydraulic structures founded on an alluvial North Indian River?**

(*a*) uplift caused by steady seepage
(*b*) piping caused by steady seepage
(*c*) uplift caused by high flood flow in the jump trough
(*d*) none of the above.

**Q. 222. Bligh's theory, as applied to the design of weirs and barrages on permeable foundations, account for :**

(*a*) hydrostatic forces only (*b*) hydrodynamic forces only
(*c*) both (*a*) and (*b*) (*d*) none of them.

**Q. 223. Retrogression is :**

(*a*) the back water effect of a weir
(*b*) the raising of the river bed upstream of the weir, during initial years of its construction
(*c*) the lowering of the river bed downstream of the weir, during initial years of its construction
(*d*) none of them.

**Q. 224. According to Khosla's theory of independent variables for seepage below a hydraulic structure, the exit gradient, in the absence of a downstream sheet pile, is :**

(*a*) 0 (*b*) 1
(*c*) $\infty$ (*d*) none of them.

**Q. 225. The value of Khosla's critical exit gradient for usually met alluvial sandy soils of our country, is about :**

(*a*) 0 (*b*) 1
(*c*) $\infty$ (*d*) 1/4 to 1/6.

**Q. 226. The value of Khosla's safe exit gradient for usually met alluvial river soils of our country is :**

(*a*) 0 (*b*) 1

(*c*) ∞ (*d*) 1/4 to 1/6.

**Q. 227. Critical exit gradient, as enunciated in Khosla's theory of design of weirs and barrages on pervious foundations, is :**

(*a*) the left out pressure in the seeping water at the downstream end point, where it emerges out on the river bed

(*b*) the rate of loss of pressure of the seeping water at the downstream emerging point, which is just enough to lift the soil grains at that point

(*c*) the actual pressure gradient of the seepage, at the down-stream emerging point

(*d*) none of the above.

**Q. 228. Safe exit gradient, as enunciated by Khosla's theory of design of weirs and barrages, is :**

(*a*) the actual seepage gradient available at the downstream exit point

(*b*) the seepage gradient at the downstream exit point, equalling the submerged weight of a unit volume of soil there

(*c*) about 1/4 to 1/6 times the gradient which is just able to lift the soil grain at the downstream exit point

(*d*) none of the above.

**Q. 229. The critical exit gradient, as applicable to barrages, is expressed by :**

(*a*) $w(1-n)(S_s - 1)$,

where $w$ = unit weight of water,
$n$ = porosity of soil
$S_s$ = specific gravity of soil grains

(*b*) $\frac{H}{d}\cdot\frac{1}{\pi\sqrt{\lambda}}$ where $H$, $d$, and $\lambda$ have their usual meaning

(*c*) $\frac{\text{Safe exit gradient}}{\text{Factor of safety}}$

(*d*) none of the above.

**Q. 230. The minimum thickness(t) of the downstream floor, as required in the design of weirs, can be expressed by the equation :**

(*a*) $\frac{h}{G+1}$ (*b*) $\frac{h}{G-1}$

(*c*) $\frac{h-t}{G-1}$ (*d*) $1.33\left(\frac{h}{G-1}\right)$

where $h$ = ordinate of hydraulic gradient line above the top of the floor
$G$ = Specific gravity of floor material.

**Q. 231.** The back water effect of a weir is best called :

(*a*) retrogression (*b*) afflux

(*c*) back water curve (*d*) none of them.

**Q. 232. A breast wall is usually provided :**

(*a*) in the weir section (*b*) in the under-sluice section

(*c*) in the main canal section (*d*) in the head regulator section.

**Q. 233. Just downstream of the pucca concrete floor of a barrage section, an inverted filter covered by c.c. blocks is laid in a length of about 1.5 to 2 D, where D is :**

(*a*) the Lacey's normal scour depth
(*b*) 1.5 times the Lacey's normal scour depth
(*c*) 1.5 times the Lacey's scour depth minus the downstream water depth
(*d*) none of the above.

**Q. 234. Just downstream of the concrete floor of a weir section, an inverted filter covered by c.c. blocks is laid. This filter should be such, as :**

(*a*) to let out the residual seepage
(*b*) not to allow the soil particles to escape
(*c*) not to get clogged in itself
(*d*) all of the above.

**Q. 235. The safety against any possible scour, on upstream or downstream side of the pucca floor of a hydraulic structure, is usually ensured by laying :**

(*a*) inverted filter (*b*) toewall
(*c*) rock toe (*d*) stone apron.

**Q. 236. For the head regulator, the most severe condition of uplift pressure on the floor occurs when :**

(*a*) the flow in the river is at flood level and canal is running at full supply depth
(*b*) the canal runs dry and the river flow is at high flood level
(*c*) the canal runs at full supply depth and the river flow is at pond level
(*d*) the canal runs dry and the river flow is at pond level.

## CHAPTER 12 : Canal Falls

**Q. 237. Canal drops are required to :**

(*a*) dissipate excess energy
(*b*) dissipate inadequate land slope
(*c*) dissipate excess land slope
(*d*) none of the above.

**Q. 238. A trapezoidal notch fall can maintain normal water depth in the upstream channel :**

(*a*) at any one given value of the design discharge
(*b*) at all the discharges
(*c*) at any two values of the design discharges
(*d*) at no discharge at all.

**Q. 239. The depth-discharge relationship of the upstream canal remains practically unaffected by the introduction of a fall of the type :**

(*a*) Ogee fall (*b*) Sarda type vertical fall
(*c*) Trapezoidal notch fall (*d*) none of the above.

**Q. 240. For low to moderate discharges of the order of 10-15 cumecs and fall height of 1 to 1.5 m, cheaper Sarda type canal falls can be recommended :**

(*a*) universally, without any if and buts
(*b*) if the canal runs with highly variable discharge
(*c*) when the canal fall is not to be used as a meter and discharge is fairly constant

(*d*) when the canal fall is not to be used as a meter, irrespective of variation in discharge.

**Q. 241.** The type of canal fall, which can be universally used without any if and buts for moderate discharges of the order of 40-60 cumecs and low fall heights of 1 to 1.5 m, is :

(*a*) vertical drop fall (*b*) ogee fall
(*c*) glacis fall (*d*) baffle fall.

**Q. 242.** The type of fall, which you may recommend for very high drops and very low discharges, is :

(*a*) Sarda type fall (*b*) Siphon well drop
(*c*) Straight glacis fall (*d*) Inglis fall.

**Q. 243. The canal fall, involving parabolic glacis, is called :**

(*a*) Straight glacis fall (*b*) Glacis fall
(*c*) Inglis fall (*d*) Montague fall.

**Q. 244. An Inglis fall, also called a Baffle fall, can be recommended for all discharges, provided :**

(*a*) the fall is more than 1.5 m (*b*) the fall is undrowned
(*c*) the fall is either flumed or unflumed (*d*) all of the above.

**Q. 245. The energy dissipation in a Sarda type canal drop is caused by :**

(*a*) hydraulic jump (*b*) friction blocks
(*c*) water pool (*d*) baffle wall.

**Q. 246. The best energy dissipation on the down-stream side of a canal drop, is caused in :**

(*a*) Sarda type fall (*b*) Glacis fall
(*c*) Ogee fall (*d*) Montague fall.

**Q. 247. A baffle wall and a water pool to dissipate energy, was introduced as an improvement, in a :**

(*a*) Glacis fall (*b*) Montague fall
(*c*) Inglis fall (*d*) none of the above.

**Q. 248. A meter fall of 1.8 m height is to be constructed on a canal carrying a discharge of 50 cumecs. The type of fall that you may recommend will be:**

(*a*) Glacis fall (*b*) Montague fall
(*c*) Inglis fall (*d*) none of the above.

**Q. 249. The energy dissipation in an Inglis fall is caused by :**

(*a*) a pool of water (*b*) a hydraulic jump
(*c*) neither (*a*) nor (*b*) (*d*) both (*a*) and (*b*).

**Q. 250. The length of the water cistern to be provided in a Sarda type fall, is :**

(*a*) $5\,(y_2 - y_1)$ (*b*) $5\sqrt{H.H_L}$

(*c*) $1/4(H.H_L)^{2/3}$ (*d*) $1.84 \cdot H^{3/2}$.

where $y_1$, $y_2$, $H$, $H_L$ have their usual meaning.

**Q. 251. The concrete glacis of a glacis fall, usually slopes H : V, as :**

(*a*) 2 : 1 (*b*) 1 : 2
(*c*) 1 : 5 (*d*) 5 : 1.

**Q. 252. In a glacis fall, the hydraulic jump forms on :**

(*a*) the horizontal floor (*b*) the sloping glacis
(*c*) either (*a*) or (*b*) (*d*) none of the above.

**Q. 253. Friction blocks on the downstream side of the pucca canal fall, can be used to dissipate the left out energy in the falling water, in case of :**

(*a*) a glacis fall
(*b*) a vertical drop fall
(*c*) a baffle fall
(*d*) (*a*) and (*b*) both
(*e*) (*a*), (*b*) and (*c*) all.

**Q. 254. A trapezoidal crest in a Sarda type canal fall is preferred and used in comparison to a rectangular crest, when :**

(*a*) the discharge is less than 14 cumecs
(*b*) the discharge is equal to or more than 14 cumecs
(*c*) the discharge is less than 30 cumecs
(*d*) the discharge is equal to or more than 30 cumecs.

**Q. 255. A cylinder fall, popularly known as syphon well drop, is suitable and economical for :**

(*a*) low discharges & low drops
(*b*) low discharges & high drops
(*c*) high discharges & low drops
(*d*) high discharges & high drops.

**Q. 256. In a baffle fall, the hydraulic jump forms, on :**

(*a*) the sloping glacis
(*b*) the horizontal baffle platform
(*c*) either (*a*) or (*b*)
(*d*) none of the above.

**Q. 257.** A baffle wall, as a sort of a low weir, is constructed in a baffle type canal fall at the end of the baffle platform and before the water cistern, inorder to head up water to its upstream side to :

(*a*) ensure formation of hydraulic jump to dissipate energy
(*b*) withstand actual impact of the high velocity jet to dissipate energy
(*c*) both (*a*) and (*b*)
(*d*) none of the above.

## CHAPTER-13 : Regulators Modules, and Miscellaneous Canal Structures

**Q. 258. The canal regulator, which is constructed at a diversion headworks, is called a :**

(*a*) cross regulator
(*b*) distributary head regulator
(*c*) canal module
(*d*) none of the above.

**Q. 259. The gated regulator, which is constructed in the parent canal near the site of an offtaking canal, is called a :**

(*a*) canal head regulator
(*b*) distributary head regulator
(*c*) cross regulator
(*d*) none of the above.

**Q. 260. A cross regulator helps in**

(*a*) increasing supply in the parent channel downstream
(*b*) increasing supply in the offtaking channel
(*c*) increasing water depth in the parent canal, upstream
(*d*) both (*b*) and (*c*)
(*e*) all (*a*), (*b*) and (*c*) above.

**Q. 261. Point out the choice among the following, which is not a function of a distributary head regulator :**

(*a*) it serves as a meter for measuring discharge in the offtaking canal
(*b*) it serves to control silt entry into the offtaking canal
(*c*) it helps in controlling and regulating supplies in the entire downstream canal network.
(*d*) it helps in controlling supplies in the offtaking canal.
(*e*) none of the above.

**Q. 262.** Khosla's theory of independent variables is used in the design of :
(*a*) weirs and barrages
(*b*) cross regulators and head egulators
(*c*) modules (*d*) both (*a*) and (*b*) above
(*e*) all (*a*), (*b*) and (*c*) above.

**Q. 263. The channel constructed to bypass the excess water entering a canal, is called a :**
(*a*) canal module (*b*) canal siphon
(*c*) canal escape (*d*) canal regulator.

**Q. 264.** The crest level of a distributary head regulator is usually kept :
(*a*) the same as the upstream bed level of the parent canal
(*b*) the same as the crest level of the cross-regulator at the site.
(*c*) 0.3 to 1.0 m higher than the upstream bed level of the parent canal
(*d*) none of the above.

**Q. 265.** The regulator constructed on a major distributary near its junction with a minor distributary, will be called a :
(*a*) distributary head regulator (*b*) cross-regulator
(*c*) regulating head (*d*) none of the above.

**Q. 266.** The arrangement made in a canal network, which acts as its safety valve, is:
(*a*) cattle crossing (*b*) canal ladder
(*c*) canal escape (*d*) canal module.

**Q. 267.** The arrangement provided in a canal to help the cattle safely cross the canal, is called a :
(*a*) canal ladder (*b*) canal crossing
(*c*) cattle crossing (*d*) all of the above.

**Q. 268. Point out the correct statement :**
(*a*) bed bars are constructed in canals at the toe of the canal lining, to prevent slippage of lining.
(*b*) cattle crossings, besides their usual function of helping the cattle to cross the canal, also provide safety to those who are swept away by currents.
(*c*) a canal escape helps in supplying irrigation water to the downstream water courses
(*d*) none of the above.

**Q. 269. Canal modules help in :**
(*a*) modulating and varying the canal discharge
(*b*) releasing desired quantity of water into the water courses
(*c*) releasing desired quantity of water into the minors
(*d*) all of the above.

**Q. 270. Canal outlets are also called :**
(*a*) canal escapes (*b*) canal modules
(*c*) canal offtakes (*d*) canal openings.

**Q. 271. The type of irrigation module, which makes equitable distribution of water more difficult, is a :**
(*a*) non-modular outlet (*b*) semi-module
(*c*) rigid module (*d*) none of them.

**Q. 272. A good irrigation module is the one, which :**
(*a*) draws heavy silt from the canal
(*b*) draws clear water from the canal
(*c*) draws fair share of silt from the canal

(*d*) none of the above.

**Q. 273. If the rate of change of discharge from an irrigation outlet is equal to the rate of change of discharge in the distributary, then the outlet is called :**

(*a*) flexible (*b*) proportional
(*c*) sensitive (*d*) none of these.

**Q. 274. The rate of change of discharge through an irrigation outlet becomes equal to the rate of change of water depth in the channel, when its :**

(*a*) flexibility is 1 (*b*) sensitivity is 1
(*c*) setting is 1 (*d*) sensitivity is zero.

**Q. 275. If the sensitivity of an irrigation module is 1/2, then 50% variation in canal water depth will cause *x*% variation in outlet discharge, where *x* is :**

(*a*) 100% (*b*) 50%
(*c*) 25% (*d*) 12.5%.

**Q. 276. A fully modular canal outlet has its :**

(*a*) sensitivity = 1, and flexibility = 1
(*b*) sensitivity = 1, and flexibility = 0
(*c*) sensitivity = 0, and flexibility = 1
(*d*) sensitivity = 0, and flexibility = 0.

**Q. 277. An irrigation outlet is said to be proportional, when its :**

(*a*) sensitivity = 1 (*b*) flexibility = 1
(*c*) setting = 1 (*d*) all of the above.

**Q. 278. If the percentage rise/fall in the outlet discharge of a canal distributary is more than the percentage rise/fall of discharge in the distributary, then the outlet is :**

(*a*) super sensitive (*b*) sub sensitive
(*c*) hyper proportional (*d*) sub proportional.

**Q. 279. An irrigation, outlet fixed at a height of 0.4 y above the canal DBL, where y is the full supply depth in the canal, will have its setting equal to :**

(*a*) 0.4 (*b*) 0.6
(*c*) 0.2 (*d*) none of these.

**Q. 280.** An irrigation outlet is said to be proportional, when its :

(*a*) setting = outlet index/channel index
(*b*) setting = channel index/outlet index
(*c*) setting = channel index × outlet index
(*d*) none of the above.

**Q. 281. For a wide trapezoidal channel, the channel index is :**

(*a*) 2/3 (*b*) 5/3
(*c*) 1/3 (*d*) none of them.

**Q. 282. A free pipe outlet is a :**

(*a*) rigid module (*b*) flexible module
(*c*) non-modular module (*d*) all of the above.

**Q. 283. The type of irrigation module, which is usually considered as the best in our country, is :**

(*a*) Adjustable orifice semi module (*b*) Punjab open flume outlet
(*c*) Khanna's rigid module (*d*) Kennedy's gauge module.

**Q. 284. The only irrigation semi module, through which the discharge is not proportional to $\sqrt{H}$, where H is the head causing flow through the module, is :**

(*a*) kennedy's gauge outlet (*b*) open flume outlet
(*c*) free pipe outlet
(*d*) adjustable orifice semi module.

## CHAPTER 14 : Cross Drainge Works

**Q. 285. The hydraulic structure which is constructed at the junction of a canal and a drain to dispose of drainage without disturbing canal supplies, is known as a;**

(*a*) cattle crossing (*b*) canal module
(*c*) canal regulator (*d*) canal crossing.

**Q. 286. The type of canal alignment, which involves maximum cross-drainage works, is a :**

(*a*) ridge canal (*b*) contour canal
(*c*) side slope canal (*d*) all of the above.

**Q. 287. The irrigation structure which is constructed at the crossing of a canal and a drain, to avoid the mixing of their respective discharges, is :**

(*a*) an aqueduct (*b*) a super passage
(*c*) a level crossing (*b*) (*a*) and (*b*) both
(*e*) (*a*), (*b*) and (*c*) all.

**Q. 288. When an irrigation canal is taken over a drainage channel, the crossing is called :**

(*a*) an aqueduct (*b*) a super passage
(*c*) a level crossing (*d*) none of them.

**Q. 289. An irrigation canal, freely flowing under a drainage channel, is specifically called a :**

(*a*) canal junction (*b*) canal crossing
(*c*) canal siphon (*d*) super passage.

**Q. 290. An irrigation canal flowing freely above a drainage, which in turn is flowing under pressure, is specifically, called a :**

(*a*) canal siphon (*b*) canal aqueduct
(*c*) siphon aqueduct (*d*) super passage.

**Q. 291. The crossing arrangement, preferably made at the junction of a huge canal and a river stream carrying short lived-high flood discharge at almost equal bed levels, is a :**

(*a*) super passage (*b*) aqueduct
(*c*) level crossing (*d*) canal siphon.

**Q. 292. The drainage water is sometimes allowed to join the canal water to augment canal supplies, through a hydraulic structure, called a :**

(*a*) canal outlet (*b*) canal inlet
(*c*) module (*d*) level crossing.

**Q. 293. The DBL of a canal and the HFL of a drain at their crossing point are, respectively, 216 m and 214 m. The adopted cross-drainage work here, will be a :**

(*a*) super passage (*b*) siphon
(*c*) aqueduct (*d*) siphon aqueduct.

**Q. 294. Pinpoint the correct statement in relation to the relative merits of the two cross-drainage works :**
(*a*) a super passage is preferred to an aqueduct
(*b*) a canal-syphon is preferred to a super-passage
(*c*) a super passage is preferred to a canal syphon
(*d*) a siphon aqueduct is inferior to a canal siphon, and is seldom used.

**Q. 295.** A cross-drainage work is called a siphon, when it carries the canal water:
(*a*) below the drainage under pressure
(*b*) below the drainage at atmospheric pressure
(*c*) above the drainage at atmospheric pressure
(*d*) none of the above.

**Q. 296. In a syphon aqueduct, the worst condition of uplift on the roof occurs, when:**
(*a*) the canal and drainage are flowing full
(*b*) the canal is flowing full and there is no drainage discharge
(*c*) the canal is empty and drainage is flowing full
(*d*) none of the above.

**Q. 297.* In an aqueduct provided with a pucca bottom floor, the uplift will occur :**
(*a*) on the roof slab (*b*) on the bottom floor
(*c*) on both the roof slab as well as the bottom floor
(*d*) no where, since the flow is free in the canal as well as in the drainage channel.

**Q. 298.* In a syphon aqueduct, the worst condition of uplift on the floor occurs, when:**
(*a*) the canal is full and the drainage empty, with watertable at drainage bed
(*b*) the canal is empty and the drainage full, with watertable at drainage bed
(*c*) the canal and drainage both full
(*d*) the canal is full and drainage empty, with watertable below the floor.

**Q. 299. Point out the incorrect statement :**
(*a*) aqueducts and super passages are usually not provided with pucca bottom floors
(*b*) aqueduct syphons and canal syphons are usually provided with pucca bottom floors
(*c*) in a level crossing, a cross regulator is provided on the canal below the crossing
(*d*) canals or drainage channels are usually flumed to affect economy at the sites of crossings, wherein the contraction transitions are not to be steeper than 22½° and the expansion transitions not to be steeper than 30°.
(*e*) none of the above.

**Q. 300. The head loss through a siphon barrel is usually given by Unwin's formula, by neglecting velocity of approach, as :**

$H_L = \left(1 + f_1 + f_2 \cdot \frac{L}{R}\right)\frac{V^2}{2g}$, **where $f_1$ and $f_2$ respectively represent the coefficient of head losses due to:**

(*a*) barrel friction and entry (*b*) entry and barrel friction

(*c*) barrel friction and exit (*d*) exit and barrel friction.

**Q. 301. A syphon aqueduct is constructed at a canal crossing site, where drainage HFL was 212.2 m, by allowing an afflux of 0.4 m at high flood discharge. At this site, the water level, downstream of the crossing at the same high flood, will be :**

(*a*) 212.2 m (*b*) 212.6 m

(*c*) 211.8 m

(*d*) none of them, as it cannot be predicted.

**Q. 302. The following data is available at the proposed site of a canal crossing :**

| ***Item*** | ***Drain*** | ***Canal*** |
|---|---|---|
| **B.L. (m)** | **252.2** | **248.0** |
| **FSL/HFL (m)** | **253.2** | **253.0** |
| **Discharge (*cumecs*)** | **2** | **400** |

**The most appropriate and economical cross-drainage work at the above site will be :**

(*a*) an aqueduct (*b*) a super passage

(*c*) a syphon aqueduct (*d*) a syphon.

## CHAPTER 15 : Construction of Culverts and Small Road Bridges Across Drains and Canals.

**Q. 303. Culverts are all those road bridges, whose spans are up to :**

(*a*) 6 m (*b*) 8 m

(*c*) 10 m (*d*) 12 m.

**Q. 304. A road bridge is proposed to be constructed across an alluvial stream. The design discharge at the bridge site is estimated to be 2500 m$^3$/s. It is decided to adopt a total effective waterway equal to 300 m for this bridge, as against the regime width of 237.5 m, due to afflux considerations. If the HFL at the site is at RL 214.0 m, then the RL of the open foundation can be suggested at :**

(*a*) 206.6 m (*b*) 207.2 m

(*c*) 208.5 m (*d*) 210.2 m.

**Q. 305. The relation between Scour depth (*R*) and Regime scour depth ($R_e$) for an alluvial river, having regime width (W) and contracted width (B), is given by :**

(*a*) $R = R_e \left(\frac{W}{B}\right)^{0.61}$ (*b*) $R = R_e \left(\frac{B}{W}\right)^{0.61}$

(*c*) $R = R_e \left(\frac{W}{B}\right)^{1/3}$ (*d*) $R = R_e \left(\frac{B}{W}\right)^{1/3}$

**Q. 306. The Lacey's scour depth *R*, given by equation $R = 0.473\left(\frac{Q}{f}\right)^{1/3}$, is applicable, when the river width equals :**

(*a*) actual river width (*b*) regime width

(*c*) any contracted width (*d*) all of the above.

**Q. 307. The Lacey's scour depth *R*, given by the equation $R = 1.35\left(\frac{q^2}{f}\right)^{1/3}$, is applicable, when the river width equals :**

(*a*) regime width (*b*) actual river width

(*c*) any contracted width (*d*) all of the above.

**Q. 308. Proposed RCC road bridge, having total height of its abutments or piers from top to bottom as H, can be proposed to have economical spans, each equal to :**

(*a*) $\frac{2}{3}$H (*b*) H

(*c*) 1.2 H (*d*) 1.5 H.

**Q. 309. A bridge culvert is to be constructed across a proposed lined canal. The afflux at this site for a contracted canal width, can be computed by :**

(*a*) Unwin's formula (*b*) Broad-crested weir formula

(*c*) Orifice formula (*d*) Molesworth formula.

**Q. 310. While going to Tirupati Tirumala Devasthanam from Chennai, highway road is found to be lowered at several places, over which natural drainage flow during rainy season. Such dips in the roads, are called :**

(*a*) freeways (*b*) road dips

(*c*) causeways (*d*) broadways.

## CHAPTER 16: River Navigation

**Q. 311. The minimum depth of water required for safe and economical navigation is:**

(*a*) 1 m (*b*) 2.0 m

(*c*) 2.7 m (*d*) 4.2 m

**Q. 312. The transit time taken by a barge tow, moving upstream in a navigable river, increases, if :**

(*a*) the flow velocity is more (*b*) the flow velocity is less.

**Q. 313. Inorder to improve the navigability in a river, the cutoffs are :**

(*a*) encouraged (*b*) discouraged.

**Q. 314. The belt conveyor forming an essential part of a ladder dredger, used for excavating river sand deposits, is used for :**

(*a*) excavation of sand

(*b*) transporting excavated sand to the surface, over small distances

(*c*) transporting excavated sand to the surface, over large distances

(*d*) none of these.

**Q. 315. The dredger which excavates sand deposits of a river, and discharges the excavated material with water over long distances, is called a :**

(*a*) ladder dredger (*b*) suction dredger

(*c*) dipper dredger (*d*) none of these.

**Q. 316. Lock and dam constructions for creation of a series of slack water pools along a river, for promoting its navigability, is adopted only when :**

(*a*) the river is perennial

(*b*) the river carries heavy sediment load

(*c*) the available water in the river is quite less

(*d*) none of the above.

**Q. 317. Timber is generally transported in our country through :**

(*a*) rail routes (*b*) road routes

(*c*) air routes (*d*) water routes.

**Q. 318. The oldest and cheapest mode of transporting cargo is by :**

(*a*) rail routes (*b*) road routes

(*c*) air routes (*d*) water routes.

**Q. 319. Ships or Steamers can ply to carry cargo, in Indian waterways, extending to about :**

(*a*) 500 km (*b*) 2500 km

(*c*) 10,000 km (*d*) 40,000 km.

## CHAPTER 17. Dams in General and a few Dams in Particular

**Q. 320. Aswan dam on Nile river in Egypt was perhaps the first modern dam of the world. Its construction was completed in the year :**

(*a*) 1664 (*b*) 1798

(*c*) 1902 (*d*) 1931.

**Q. 321. The only modern dam among the following, which has failed, is :**

(*a*) Koyna dam in India (*b*) Vega de Tera dam in spain

(*c*) Mauvoisin dam in Switzerland

(*d*) Trinity dam in California, USA.

**Q. 322. Earthen dams are :**

(*a*) rigid dams (*b*) non rigid dams

(*c*) overflow dams (*d*) diversion dams.

**Q. 323. When sand and gravel foundation strata is available at a proposed dam site of moderate height, the dam may be of the type:**

(*a*) earthen dam or rockfill dam (*b*) masonry gravity dam

(*c*) double arch dam (*d*) concrete gravity dam.

**Q. 324. The prominent modern dam, which has changed its name several times, is :**

(*a*) Nagarjuna sagar dam (*b*) Hoover dam

(*c*) Bhakra dam (*d*) Idukki dam.

**Q. 325. The highest concrete gravity dam of the world is Grand Dixence dam is Switzerland, whose height is :**

(*a*) 325 m (*b*) 317 m
(*c*) 285 m (*d*) 272 m.

**Q. 326. The only arch dam of our country is Idukki dam across Periyar river in Kerala State, whose height is :**
(*a*) 81 m (*b*) 143 m
(*c*) 149 m (*d*) 169 m.

**Q. 327. The highest concrete gravity dam of India, which is also the 2nd highest such dam of the world, is :**
(*a*) 285 m in height (*b*) 256 m in height
(*c*) 226 m in height (*d*) 206 m in height.

**Q. 328. Bhakra dam of our country is located in the State of :**
(*a*) Punjab (*b*) Himachal Pradesh
(*c*) Uttar Pradesh (*d*) Madhya Pradesh.

**Q. 329. The highest stone masonry gravity dam of India as well as of the world, is:**
(*a*) Ramganga dam (*b*) Nagarjuna Sagar dam
(*c*) Hirakud dam (*d*) Mettur dam.

## CHAPTER 18. Reservoirs and Planning for Dam Reservoirs

**Q. 330. A dam reservoir, not provided with gate controls on its spillway and other sluices, is called a :**
(*a*) detention basin (*b*) storage reservoir
(*c*) retarding basin (*d*) all of these.

**Q. 331. A dam reservoir, catering to flood control, irrigation, and water supply, although basically designed for irrigation alone, is a :**
(*a*) multipurpose reservoir (*b*) single purpose reservoir
(*c*) distribution reservoir (*d*) none of the above.

**Q. 332. A hydel power project has been envisaged to serve the water supply and irrigation needs of the area at its inception stage. The dam reservoir, so constructed, will be known as :**
(*a*) multipurpose reservoir (*b*) single purpose reservoir
(*c*) both (*a*) and (*b*) above (*d*) none of the above.

**Q. 333. The 'surcharge storage' in a dam reservoir is the volume of water stored between :**
(*a*) minimum and maximum reservoir levels
(*b*) minimum and normal reservoir levels
(*c*) normal and maximum reservoir levels
(*d*) none of the above.

**Q. 334. The 'useful storage' in a dam reservoir is the volume of water stored between:**
(*a*) minimum and maximum reservoir levels
(*b*) minimum and normal reservoir levels
(*c*) normal and maximum reservoir levels
(*d*) none of the above.

**Q. 335. The 'dead storage' in a dam reservoir, is the available volume for collection of silt and sediment, between :**

(*a*) bed level of the reservoir and minimum reservoir level
(*b*) bed level of the reservoir and the silt level in the reservoir
(*c*) bed level of the reservoir and the normal pool level
(*d*) none of the above.

**Q. 336. 'Bank storage' in a dam reservoir :**
(*a*) increases the computed reservoir capacity
(*b*) decreases the computed reservoir capacity
(*c*) sometimes increases and sometimes decreases the computed reservoir capacity
(*d*) has no effect on computed reservoir capacity.

**Q. 337. The water stored in a reservoir below the minimum pool level is called :**
(*a*) valley storage (*b*) bank storage
(*c*) surcharge storage (*d*) dead storage.

**Q. 338. For a flood control reservoir, the effective storage is equal to :**
(*a*) useful storage + valley storage
(*b*) useful storage + surcharge storage – valley storage
(*c*) useful storage + surcharge storage + valley storage
(*d*) useful storage – valley storage.

**Q. 339. The volume *V* of water stored in a reservoir in a depth y above a given datum, can be easily represented by an equation of the type:**
(*a*) $V = [\alpha + \beta y + \gamma y^2]$ where α, β, γ are constants
(*b*) $V = [\alpha y + \beta y^2 + \gamma y^3] + K$ *K* is reservoir capacity up to given datum
(*c*) neither (*a*) nor (*b*)
(*d*) both (*a*) and (*b*).

**Q. 340. Average yield of a storage reservoir is the arithmetic average of its :**
(*a*) firm yields over a long period
(*b*) secondary yields over a long period
(*c*) firm and secondary yields over a long period
(*d*) none of the above.

**Q. 341. If $A_1$ and $A_2$ are surveyed and known areas at a reservoir site at elevations of $E_1$ and $E_2$, respectively ; then the area, *A*, at any intermediate elevation E, is interpolated as :**

(*a*) $\sqrt{A} = \sqrt{A_1} + \frac{E - E_1}{E_2 - E_1}\left(\sqrt{A_2} - \sqrt{A_1}\right)$

(*b*) $\sqrt{A} = \sqrt{A_1} + \frac{E + E_1}{E_2 + E_1}\left(\sqrt{A_2} - \sqrt{A_1}\right)$

(*c*) $A = \sqrt{A_1} + \frac{E - E_1}{E_2 - E_1}\left[\sqrt{A_2} - \sqrt{A_1}\right]^2$

(*d*) none of the above.

**Q. 342. While planning a water supply reservoir, as compared to an irrigation reservoir, the design yield may be kept :**
(*a*) higher (*b*) lower
(*c*) equal
(*d*) lower or higher, as per designer's discretion.

**Q. 343. Yield of a reservoir represents :**
(*a*) the inflow into the reservoir (*b*) the capacity of the reservoir

(c) the outflow demand on the reservoir
(d) none of the above.

**Q. 344. The capacity of a storage reservoir can be decided by using :**
(a) the mass curve of inflow (b) the mass curve of outflow
(c) by both (a) and (b) together (d) none of the above.

**Q. 345. The maximum quantity of water that is estimated to remain available in a storage reservoir for supply, even during worst dry periods, is known as its :**
(a) firm yield (d) design yield
(c) reservoir yield (d) primary yield.

**Q. 346. Design yield of a storage reservoir is kept :**
(a) higher than its firm or safe yield
(b) lower than its firm or safe yield
(c) equal to its firm or safe yield
(d) higher or lower than the firm yield, depending upon the designer's intuition.

**Q. 347. The amount of water stored in a river channel without any artificial storage, is known as :**
(a) bank storage (b) river storage
(c) valley storage (d) dead storage.

**Q. 348. With the reduction in reservoir capacity over the passage of time, the trap efficiency**
(a) increases (b) decreases
(c) remains unaffected
(d) may increase or decrease, depending upon the reservoir characteristics.

**Q. 349. Trap efficiency of a storage reservoir is defined as :**

(a) $\dfrac{\text{Total annual sediment inflow}}{\text{Reservoir capacity}}$

(b) $\dfrac{\text{Total sediment deposited in a given period}}{\text{Total sediment inflow in that period}}$

(c) $\dfrac{\text{Total annual sediment deposited in the reservoir}}{\text{Dead storage capacity of the reservoir}}$

(d) none of the above.

**Q. 350. Capacity-inflow ratio for a storage reservoir is defined as :**

(a) $\dfrac{\text{Reservoir capacity}}{\text{average annual flood inflow}}$

(b) $\dfrac{\text{Reservoir capacity}}{\text{average annual sediment inflow}}$

(c) $\dfrac{\text{Dead storage capacity of the reservoir}}{\text{average annual sediment deposited}}$

(d) none of the above.

**Q. 351. With the increase in its capacity-inflow ratio, the trap efficiency of a reservoir:**
(a) increases (b) decreases
(c) remains unchanged
(d) may increase or decrease depending on the reservoir characteristics.

**Q. 352. The capacity-inflow ratio for a reservoir :**
(*a*) is a constant factor overtime
(*b*) increases with time
(*c*) decreases with time
(*d*) may increase or decrease with time.

**Q. 353. As the height of a proposed dam is increased, the cost per unit of storage:**
(*a*) increases (*b*) decreases
(*c*) initially increases and then decreases
(*d*) initially decreases and then increases.

**Q. 354. "Economical height of a Dam" is that height, for which the :**
(*a*) cost per unit of storage is minimum
(*b*) benefit cost ratio is maximum
(*c*) net benefits are maximum
(*d*) none of the above.

**Q. 355*. If 20% of the reservoir capacity is earmarked for dead storage in a storage reservoir of 30 M cum; and the average annual silt deposition in the reservoir is 0.1 M cum, then the useful life of the reservoir will start reducing after :**
(*a*) 60 years (*b*) 120 years
(*c*) 240 years (*d*) 300 years.

**Q. 356. The appropriate value of the estimated average life of a dam reservoir is :**
(*a*) 25 years (*b*) 50 years
(*c*) 75 years (*d*) none of these.

## CHAPTER 19 : Design and Construction of Gravity Dams

**Q. 357. The vertical component of the earthquake wave, which produces adverse effects on the stability of a dam, is, when it is acting in;**
(*a*) upward direction (*b*) downward direction
(*c*) both (*a*) and (*b*) (*d*) none of them.

**Q. 358. The horizontal component of an earthquake wave, producing instability in a dam is the one, which acts :**
(*a*) towards the reservoir (*b*) towards the dam
(*c*) both (*a*) and (*b*) (*d*) none of the above.

**Q. 359. The vertical downward earthquake acceleration, $\alpha_v$ = 0.1 g, acting on a gravity dam, will :**
(*a*) increase the resisting weight of the dam by 10%
(*b*) decrease the resisting weight of the dam by 10%
(*c*) increase the uplift by 10%
(*d*) none of the above.

**Q. 360. A gravity dam is subjected to hydro dynamic pressure, caused by:**
(*a*) the rising waters of the reservoir when a flood wave enters into it
(*b*) the rising waves in the reservoir due to high winds
(*c*) the increase in water pressure, momentarily caused by the horizontal earthquake, acting towards the reservoir.
(*d*) the increase in water pressure, momentarily caused by the horizontal earthquake, acting towards the dam.

**Q. 361. In a concrete gravity dam with a vertical upstream face, the stabilising force is provided by the :**

(*a*) weight of the dam
(*b*) the water supported against the upstream slope
(*c*) both (*a*) and (*b*)
(*d*) none of the above.

**Q. 362. In a concrete gravity dam, with a sloping upstream face, the resisting force is provided by the :**
(*a*) weight of the dam
(*b*) weight of the water supported on the upstream slope
(*c*) both (*a*) and (*b*)
(*d*) none of these.

**Q. 363. The factor among the following, which does not try to destabilise a masonry gravity dam, is :**
(*a*) water seeping below the foundation of the dam
(*b*) generation of waves by high winds
(*c*) deposition of silt in dead storage zone of the reservoir
(*d*) water standing against the downstream face of the dam
(*e*) water standing against the upstream face of the dam.

**Q. 364.*The horizontal destabilising force caused by the formation of waves in a storage reservoir having a fetch of 40 km, due to high wind of 160 km/h, is about :**
(*a*) 30 kN (*b*) 60 kN
(*c*) 90 kN (*d*) 130 kN.

**Q. 365.*According to IS: 6512—1984, the wind set up in a reservoir having maximum fetch of F km, and average reservoir depth along this fetch length as D m, as caused by the wind velocity of V km/h, is given by :**
(*a*) $\frac{V^2F}{62000\,D}$ (*b*) $\frac{V^2F}{D}$
(*c*) $\frac{V^2F}{6200D}$ (*d*) $\frac{V^2F}{620D}$

**Q. 366.*A reservoir, extending 20 km upstream and having a design wind velocity of 100 km/h, should be provided with a free-board of :**
(*a*) 1.2 m (*b*) 1.6 m
(*c*) 2.5 m (*d*) 3 m.

**Q. 367. Development of tensile stresses in a concrete or masonry gravity dam are usually not allowed, because it may lead to development of tension cracks, ultimately causing failure of the structure, by :**
(*a*) excessive seepage (*b*) excessive tensile stresses
(*c*) excessive compressive stresses (*d*) none of the above.

**Q. 368. The bottom portion of a concrete or a masonry gravity dam is usually stepped, in order to :**
(*a*) increase the overturning resistance of the dam
(*b*) increase the shear strength at the base of the dam
(*c*) decrease the shear stress at the base of the dam
(*d*) none of the above.

**Q. 369. The governing compressive stress in a concrete gravity dam, which should not be allowed to exceed the permissible value of about 3000 kN/m², while analysing full reservoir case, is :**

(*a*) the vertical maximum stress at the toe
(*b*) the major principal stress at the toe
(*c*) the shear stress at the toe
(*d*) none of the above.

**Q. 370. The two dimensional stability analysis of gravity dams proves better for U-shaped valleys than for *V*-shaped valleys, because the transverse joints in the dam body are generally :**
(*a*) not grouted in *U*-shaped valleys, but are keyed together in *V*-shaped valleys
(*b*) not grouted in *V*-shaped valleys, but are keyed together in *U*-shaped valleys.

**Q. 371. The small openings made in the huge body of a concrete gravity dam, such as sluices and inspection galleries, can be assumed to be causing only local effects without any appreciable effect on the distribution of stresses, as per the principle of :**
(*a*) Laplace (*b*) St. Venant
(*c*) Reynold (*d*) St. Francis.

**Q. 372. For usually values of permissible compressive stress and specific gravity of concrete, a high concrete gravity dam is the one, whose height exceeds :**
(*a*) 48 m (*b*) 70 m
(*c*) 88 m (*d*) none of these.

**Q. 373.* An elementary triangular concrete gravity dam, supporting 60 m height of reservoir water and full uplift, should have a minimum base width equal to:**
(*a*) 36 m (*b*) 39 m
(*c*) 51 m (*d*) 61 m.

**Q. 374. Inorder to economise on the provided section of a concrete gravity dam, attempts are made to reduce the uplift, by :**
(*a*) providing drainage gallery to collect seepage water
(*b*) constructing cut-off under upstream face
(*c*) pressure grouting in dam foundation
(*d*) constructing drainage channels between the dam and its foundation
(*e*) all of the above methods.

**Q. 375. Transverse joints in concrete gravity dams are the :**
(*a*) horizontal construction joints at each lift height
(*b*) vertical construction joints of full height and width
(*c*) diagonal construction joints for torsion
(*d*) none of the above.

**Q. 376. Leakage through the transverse joints in a gravity dam is prevented by :**
(*a*) shear keys (*b*) key ways
(*c*) water stops (*d*) none of these.

**Q. 377. Shear key is several times provided between the bottom of a masonry or concrete gravity dam and its foundation, to increase the frictional resistance of the dam against sliding. This key is usually provided :**
(*a*) near the toe (*b*) near the heel
(*c*) near the individual seams in the bed rock
(*d*) none of the above.

## CHAPTER 20 : Earthen Dams and Rock Fill Dams.

**Q. 378. Point out the correct statement with reference to earthen dams :**
(*a*) these dams are very costly as compared to other types
(*b*) they are less susceptible to failure as compared to rigid dams
(*c*) they can be constructed almost on every type of foundation
(*d*) highly skilled labour is generally not required.

**Q. 379. The most preferred type of an earthen dam section is the one, in which the:**
(*a*) entire embankment is made of one type of soil
(*b*) inner embankment is made of highly porous soil, surrounded by the outer shell of highly impervious soil, both separated by transition filter material of mediocre permeability
(*c*) inner embankment is made of highly impervious soil surrounded by the outer shell of highly pervious soil, both separated by transition filter material of mediocre permeability
(*d*) none of the above.

**Q. 380. When seepage takes place through the body of an earthen dam, it leads to:**
(*a*) development of pore pressures in the dam body
(*b*) reduction in the shear strength of the dam
(*c*) reduction in the developed shear stresses in the dam
(*d*) (*a*) and (*b*) both (*e*) (*a*), (*b*) and (*c*) all.

**Q. 381. The most preferred soil for the central impervious core of a zoned embankment type of an earthen dam, is**
(*a*) highly impervious clay (*b*) highly pervious gravel
(*c*) coarse sand (*d*) clay mixed with fine sand.

**Q. 382. Pure clayey soils are generally not preferred for the central impervious cores of zoned type of earthen dams, because :**
(*a*) clays are highly impervious (*b*) clays are highly pervious
(*c*) clays are susceptible to cracking (*d*) none of the above.

**Q. 383. During the construction of an earthen dam by hydraulic fill method, development of pore pressures become important in the :**
(*a*) central impervious core (*b*) pervious outer shell
(*c*) both (*a*) and (*b*) (*d*) none of the above.

**Q. 384. The process of laying and compacting earth in layers by power rollers under Optimum Moisture Conditions (OMC) for construction of earthen dams, is known as :**
(*a*) rolled fill method (*b*) hydraulic fill method
(*c*) OMC method (*d*) none of the above.

**Q. 385. Coordination between field and design engineers to ensure continuous field observations and modifications in design during construction, is more importantly required in case of :**
(*a*) concrete gravity dams (*b*) masonry gravity dams
(*c*) arch dams (*d*) earthen dams.

**Q. 386. When the water level standing against an earthen embankment, suddenly falls down, then there is an iminent risk of sliding failure, to the:**
(*a*) upstream slope (*b*) downstream slope
(*c*) both (*a*) and (*b*) (*d*) none of these.

**Q. 387. When the reservoir is full, the slope which is most likely to slide, is:**

(*a*) the upstream slope (*b*) the downstream slope
(*c*) both (*a*) and (*b*) (*d*) none of them.

**Q. 388. Provision of horizontal berms at suitable vertical intervals may be provided in the downstream face of an earthen dam, in order to :**

(*a*) allow the movement of cattle
(*b*) allow the inspection vehicles to move
(*c*) reduce the erosion caused by the flowing rain water
(*d*) none of the above.

**Q. 389. The only provision among the following, which can help control the seepage through the body of an earthen dam and, thus, to keep the phreatic line well within the dam width, is :**

(*a*) upstream impervious cut-off
(*b*) drain trench along the downstream toe
(*c*) relief wells
(*d*) chimney drain.

**Q. 390.* During the maintenance of an earthen dam, the apparent seepage through the foundation of the dam is best taken care of, by providing :**

(*a*) a chimney drain (*b*) a rock toe
(*c*) a drain trench along the downstream toe
(*d*) an upstream impervious cut-off.

**Q. 391. The base width of a rock fill dam, incomparison to that of an earthen dam, is :**

(*a*) much larger (*b*) much smaller
(*c*) sometimes larger sometimes smaller
(*d*) almost equal.

**Q. 392. On moderate foundations, and particularly in seismic areas, the type of dam which can preferably be considered for construction, is:**

(*a*) masonry gravity dam (*b*) earthen dam
(*c*) rock fill dam (*d*) arch dam.

## CHAPTER 21 : Spillways, Energy Dissipators, and Spillway Gates

**Q. 393. The 'safety valve' of a dam is its :**

(*a*) drainage gallery (*b*) inspection gallery
(*c*) spillway (*d*) outlet sluices.

**Q. 394. A ungated spillway starts functioning, as soon as the water level in the reservoir, crosses the :**

(*a*) maximum reservoir level (*b*) minimum reservoir level
(*c*) maximum conservation level (*d*) none of the above.

**Q. 395. The top line of an ogee spillway, which is kept at RL equal to the full reservoir level, is precisely called :**

(*a*) the crest (*b*) the apex of the crest
(*c*) the corbel (*d*) the chute.

**Q. 396. When the gated sluices are provided through the body of a dam spillway, then the ogee spillway structure shall normally be of :**

(*a*) corbel type (*b*) non-corbel type
(*c*) either (*a*) or (*b*) (*d*) none of these.

**Q. 397. The effective length of an ogee spillway will be different from its actual length, in case of :**

(*a*) gated spillway
(*b*) ungated spillway
(*c*) both gated as well as ungated spillway
(*d*) none of these.

**Q. 398. Which one of the following spillways is least suited to earthen dams ?**

(*a*) ogee spillway
(*b*) side channel spillway
(*c*) chute spillway
(*d*) shaft spillway.

**Q. 399. The spillway, which can be adopted with ease on gravity as well as earthen dams, is :**

(*a*) ogee spillway
(*b*) chute spillway
(*c*) both ogee as well as chute spillway
(*d*) none of these.

**Q. 400. The famous Bhakra dam of our country has been provided with:**

(*a*) trough spillway
(*b*) ogee spillway
(*c*) shaft spillway
(*d*) syphon spillway.

**Q. 401. Hydraulic jump is widely used for dissipation of energy in :**

(*a*) ogee spillways
(*b*) trough spillways
(*c*) side channel spillways
(*d*) all of these.

**Q. 402. The surplus reservoir water, after spilling over the spillway crest, flows on the chute :**

(*a*) parallel to the crest in a trough spillway
(*b*) parallel to the crest in a side channel spillway
(*c*) perpendicular to the crest in a side channel spillway
(*d*) obliquely to the crest in a chute spillway.

**Q. 403. When the crest of an ogee spillway is designed to be in accordance with the lower nappe of a free falling water jet over a duly ventilated sharp crested weir, then theoretically :**

(*a*) the pressure on the spillway crest will always be zero (*i.e.* atmospheric pressure)
(*b*) the pressure on the spillway crest will be zero at design head only
(*c*) the pressure on the spillway crest will always be negative
(*d*) the pressure on the spillway crest will always be positive.

**Q. 404. If the operating head on an ogee spillway is more than the design head, then:**

(*a*) the pressure on the crest will be zero
(*b*) the pressure on the crest will be negative, causing cavitation
(*c*) the pressure on the crest will be positive
(*d*) the discharge coefficient of the spillway will be reduced.

**Q. 405. In computing the spilling capacity of high ogee spillways, the velocity head is usually :**

(*a*) very small, and hence neglected
(*b*) very large, and hence cannot be neglected
(*c*) neither (*a*) nor (*b*).

**Q. 406.* In the functioning of an ogee spillway, the operating head :**

(*a*) frequently exceeds the design head
(*b*) rarely exceeds the design head
(*c*) never exceeds the design head
(*d*) has no connection with the design head.

**Q. 407. The discharge passing over an ogee spillway, per unit length of its appex line, is proportional to :**

(*a*) $H$ (*b*) $H^2$
(*c*) $H^{1/2}$ (*d*) $H^{3/2}$.

where $H$ is the head over the appex of its crest.

**Q. 408.*If a flood enters a dam reservoir at F.R.L., the efficiency of its ogee spillway will :**

(*a*) increase with the increasing head
(*b*) reduce with the increasing head
(*c*) remain constant with the increasing head
(*d*) vary with the increasing head.

**Q. 409. The W.E.S. downstream profile of an ogee spillway can be represented by an equation :**

$$\left(\frac{x}{H_d}\right)^n = K\left(\frac{y}{H_d}\right);$$

**where, the coefficients $K$ and $n$ for a vertical upstream spillway would, respectively, be :**

(*a*) 2.0 and 1.85 (*b*) 0.5 and 1.85
(*c*) 2.0 and 0.85 (*d*) 0.5 and 0.85

where $(x, y)$ are the coordinates on the crest profile with crest appex as origin.

**Q. 410. The portion of a chute spillway, which is known as its control structure, is :**

(*a*) low ogee weir (*b*) chute channel
(*c*) approach channel leading the water from the reservoir to the ogee weir
(*d*) stilling basin at its bottom.

**Q. 411. The spillway, which can be called as an 'overflow spillway', is essentially** :

(*a*) an ogee spillway (*b*) shaft spillway
(*c*) chute spillway (*d*) syphon spillway.

**Q. 412. The spillway, which may sometimes be called a 'waste weir', is :**

(*a*) an ogee spillway (*b*) a trough spillway
(*c*) a shaft spillway (*d*) all of the above.

**Q. 413. The spillways involving weir type spills, giving increased discharge with the increase in reservoir level, are :**

(*a*) ogee spillways (*b*) chute spillways
(*c*) side channel spillways (*d*) all of these.

**Q. 414. A shaft spillway is located :**

(*a*) inside the body of a gravity dam
(*b*) inside the upstream reservoir
(*c*) inside the downstream reservoir
(*d*) on side flanks of the main dam.

**Q. 415. The syphons installed within a gravity dam, to spill the surplus reservoir water, are known as :**

(*a*) hooded type syphon spillway
(*b*) tilted outlet type syphon spillway

(*c*) both (*a*) and (*b*) (*d*) none of these.

**Q. 416. The syphons installed over overflow dams, constitute what are known as :**

(*a*) tilted outlet type of syphon spillways
(*b*) hooded type of syphon spillways
(*c*) discharge carriers (*d*) baby syphons.

**Q. 417. Syphons installed over an overflow spillway will :**

(*a*) decrease the effectiveness of the spillway
(*b*) increase the effectiveness of the spillway
(*c*) not affect the effectiveness of the spillway
(*d*) all the above are possible.

**Q. 418. The only spillway among the following, through which the discharge does not increase as fast as it increases in all others, is:**

(*a*) chute spillway (*b*) side channel spillway
(*c*) ogee spillway (*d*) shaft spillway.

**Q. 419. The only spillway among the following, through which the discharge is almost at its capacity rate, even from the start of its functioning, is :**

(*a*) chute spillway (*b*) side channel spillway
(*c*) ogee spillway (*d*) syphon spillway.

**Q. 420.*An air vent is provided at F.R.L. to break the syphoning action at that level, in a :**

(*a*) hooded type of syphon spillway
(*b*) tilted outlet type of syphon spillway
(*c*) saddle syphon spillway
(*d*) all of the above.

**Q. 421. The energy dissipation at the toe of a spillway is affected basically by the use of hydraulic jump, in :**

(*a*) a roller bucket (*b*) a ski-jump bucket
(*c*) a sloping apron taken below the downstream river bed
(*d*) none of the above.

**Q. 422. The most ideal condition for energy dissipation in the design of spillways is the one, when :**

(*a*) the tail water rating curve lies above the jump rating curve at all discharges
(*b*) the tail water rating curve coincides with the jump rating curve at all discharges
(*c*) the tail water rating curve lies below the jump rating curve at all discharges
(*d*) the tail water rating curve either lies above or below the jump rating curve, depending upon the discharge.

**Q. 423. When the tail water depths in the river, downstream of a spillway are quite low, such that the tail water curve, at all discharges, lies below the post jump depth curve, then the energy dissipation can be affected best by :**

(*a*) a roller bucket (*b*) a ski-jump bucket
(*c*) either (*a*) or (*b*) (*d*) none of them.

**Q. 424. Energy dissipation arrangements made at the bottom of an overfall spillway, usually depend upon the relative positions of tail water curve (T.W.C.) and jump height curve (J.H.C.) and also**

**on Incoming Froude number. Such positions and suggested arrangements are given below in a haphazard manner, and you have to match them properly :**

| | |
|---|---|
| (*a*) T.W.C. coinciding with J.H.C. at all discharges, | (*i*) ski-jump bucket at higher incoming Froude number (> 9); and sloping apron below the d/s river bed at lower froude numbers (4.5 to 9) |
| (*b*) T.W.C. lying below the J.H.C. at all discharges | (*ii*) a single horizontal apron at toe of the spillway at d/s river bed |
| (*c*) T.W.C. lying above the J.H.C. at all discharges | (*iii*) a roller bucket at higher Froude numbers, and sloping apron above the d/s river bed at lower values of $F$ |
| (*d*) T.W.C. lying above the J.H.C. at low discharges, and below the J.H.C. at high discharges, and *vice versa* | (*iv*) a sloping apron partly below and partly above the d/s river bed. |

**Q. 425. A troublesome and oscillating hydraulic jump is usually formed in flows, involving the incoming Froude number in the range of 2.5-4.5, which normally is met in cases of :**

(*a*) weirs and barrages (*b*) overflow spillways of dams
(*c*) both (*a*) and (*b*) (*d*) none of them.

**Q. 426. A very steady and stable hydraulic jump is usually formed in flows, involving the approaching Froude number in the range of 4.5-9.0, which normally is met in cases of :**

(*a*) weirs and barrages (*b*) overflow spillways of dams
(*c*) both (*a*) and (*b*) (*d*) none of them.

**Q. 427. In case of very high overflow spillways, the approaching Froude number of the flow at the toe of the spillway may exceed 9.0, bringing the hydraulic jump to be in the category of a `strong jump'. In such a case, the energy dissipation arrangement preferably made at the toe of the spillway, may consists of :**

(*a*) bucket type dissipator (*b*) USBR stilling basin
(*c*) Indian standard stilling basin (*d*) none of the above.

**Q. 428. Standard USBR stilling basin II is useful for energy dissipation at the bottom of an overflow structure, if the approaching Froude number is :**

(*a*) less than 4.5 (*b*) more than 4.5.

**Q. 429. Standard stilling basin provided at the toe of a dam spillway for energy dissipation, is usually provided with auxiliary devices, like chute blocks and dentated sills, for the basic purpose of reducing the length of the stilling basin from about :**

(*a*) $6y_2$ to $4y_2$ (*b*) $4y_2$ to $2y_2$
(*c*) $6y_2$ to $2y_2$ (*d*) $3y_2$ to $y_2$.

where $y_2$ is the post jump depth

**Q. 430. The device which does not help in energy dissipation at the bottom of a hydraulic structure, over which water spills, is :**

(*a*) chute block (*b*) dentated sill
(*c*) morning glory (*d*) none of them.

**Q. 431. In a fixed roller of a spillway gate, the rollers are attached to the:**
(*a*) gate (*b*) groove guide
(*c*) either (*a*) or (*b*) (*d*) none of them.

**Q. 432. In vertical stoney spillway gate, the rollers are placed between the :**
(*a*) gate and the u/s groove guide (*b*) gate and the d/s groove guide
(*c*) either (*a*) or (*b*) (*d*) none of them.

**Q. 433. The spillway gate, which when lowered, can not be seen from a distance, is of the type :**
(*a*) sliding gate (*b*) roller gate
(*c*) tainter gate (*d*) USBR drum gate.

**Q. 434.* An ogee spillway of a concrete gravity dam having FRL of 325.0 m and MWL of 340.0 m, is provided with vertical gates between piers erected on the spillway. The effective length of the spillway is 60 m. The discharge through this spillway when gates are opened upto actual reservoir level of 331.0 m, will be :**
(*a*) 850 cumecs (*b*) 1700 cumecs
(*c*) 2800 cumecs (*d*) none of them.

## CHAPTER 22 : Outlet Works through Dams and River Intakes

**Q. 435.* A 2 m diameter sluiceway at RL 300.0 m, is provided through a concrete overflow dam section to release water to the downstream, where tail water level is not more than 275.0 m. The discharge through this outlet, when the water level in the reservoir is at FRL of 330.0 m, is of the order of :**
(*a*) 20 cumecs (*b*) 40 cumecs
(*c*) 50 cumecs (*d*) 80 cumecs.

**Q. 436.* A 4 m diameter tunnel has been constructed through an earthen dam, with a bell mouth entry. The water levels on U/s and D/s sides of the dam are at RL 226 m and 210 m, respectively. The discharge through this outlet will be of the order of :**
(*a*) 64 cumecs (*b*) 104 cumecs
(*c*) 170 cumecs (*d*) 216 cumecs.

**Q. 437. A trash rack is not required at the entrance of a :**
(*a*) syphon spillway (*b*) drum gate installation
(*c*) morning glory spillway (*d*) high head gate installation.

**Q. 438. The outlet provided in a dam body, to release water for downstream water demand, is known as :**
(*a*) spillway (*b*) sluiceway
(*c*) under-sluice (*d*) waterway.

**Q. 439. Bar screens, used to cover dam outlets to prevent entry of debris or ice into the sluiceway conduits, are called :**
(*a*) gate controlled ports (*b*) projecting collars
(*c*) trash racks (*d*) none of these.

**Q. 440. Which of the following statement is false ?**
(*a*) Water can be withdrawn from any selected level of reservoir using dry intake towers

(*b*) Intake towers have to be designed for worst combination of hydrostatic, wind, earthquake, and wave forces
(*c*) Dry intake towers are lighter in construction as compared to wet intake towers
(*d*) none of the above.

**Q. 441. A sluiceway conduit passing through an earthen dam has a length equal to *L*; and it has *n* projecting collars-each having projected length equal to *x*. The increased length of seepage path in this case is given by :**
(*a*) $L + 2nx$
(*b*) $L + nx$
(*c*) $2nx$
(*d*) none of them.

**Q. 442. Projecting collars are provided on sides of a rectangular tunnel of length (*L*), made through an earthen dam, to increase the seepage path. Their projection length (*X*) and numbers (*N*) are decided, as to provide increased seepage path equal to 2*NX*, where 2*NX* should generally be :**
(*a*) greater than $\frac{L}{4}$
(*b*) greater than $\frac{L}{2}$
(*c*) greater than *L*
(*d*) none of the above.

## CHAPTER 23 : Pressure Conduits

**Q. 443. The flow through Penstocks and Pressure conduits is, generally :**
(*a*) laminar
(*b*) turbulent
(*c*) both (*a*) and (*b*)
(*d*) none of these.

**Q. 444. The head lost in the flow of water through a penstock pipe of given length:**
(*a*) increases with the increase in flow velocity
(*b*) decreases with the increase in flow velocity
(*c*) decreases with the increase in the roughness of the pipe surface
(*d*) decreases with time.

**Q. 445. Hoop's reinforcement is provided in cement concrete pressure pipes, in order to counteract the;**
(*a*) water hammer pressure
(*b*) internal water pressure
(*c*) stresses caused by external backfills
(*d*) both (*a*) and (*b*)
(*e*) (*a*), (*b*) and (*c*) all.

**Q. 446. When a gate valve, installed in a pressure pipe, is suddenly closed, water hammer pressure is caused on the pipe shell, in the portion :**
(*a*) downstream of the valve
(*b*) upstream of the valve
(*c*) both (*a*) and (*b*)
(*d*) none of the above.

**Q. 447. Which of the following types of pressure conduits is preferably used for large heads?**
(*a*) P.C.C. pipes
(*b*) Cast iron pipes
(*c*) Prestressed concrete pipes
(*d*) Asbestos pipes.

## CHAPTER 24 : Hydro-electric Power

**Q. 448. A hydropower plant, developed at the site of a drop in an irrigation canal, is of the type :**
(*a*) run-off river plant
(*b*) storage plant
(*c*) pumped storage plant
(*d*) tidal plant.

**Q. 449. A nuclear power development scheme is essentially a :**
(*a*) thermal power scheme (*b*) hydropower scheme
(*c*) neither (*a*) nor (*b*) (*d*) both (*a*) as well as (*b*).

**Q. 450. Fossil fuels are used in :**
(*a*) a nuclear power plant (*b*) an ordinary thermal power plant
(*c*) a hydro power plant (*d*) none of them.

**Q. 451. Major hydropower generation in India is from :**
(*a*) run-off river plants (*b*) storage plants
(*c*) pumped storage plants (*d*) tidal plants.

**Q. 452. A run-off river plant for hydro power generation is essentially a :**
(*a*) high head scheme (*b*) low head scheme
(*c*) medium head scheme (*d*) any of these.

**Q. 453. A storage hydro plant essentially involves :**
(*a*) a barrage or a weir (*b*) a dam
(*c*) either (*a*) or (*b*) (*d*) neither (*a*) nor (*b*).

**Q. 454. The natural highwater falls may be used for generation of :**
(*a*) hydro-power (*b*) water-power
(*c*) both (*a*) and (*b*) (*d*) neither (*a*) nor (*b*).

**Q. 455. The power plant among the following, which is a storage plant, is:**
(*a*) Rance power station in France, located on sea
(*b*) Ganguwal power house, located on Nangal hydel channel in India
(*c*) Sarda power house, located on Sarda Canal in U.P. in India.
(*d*) none of these.

**Q. 456. The only statement, which is incorrect in regard to hydropower, is:**
(*a*) the system efficiency of a hydroplant is quite high
(*b*) the installation cost of a hydroplant is very high
(*c*) the running cost of a hydro power plant is very low
(*d*) the hydraulic turbines takes a lot of time in being put off and on.

**Q. 457. Identify the correct statement in regard to hydropower :**
(*a*) hydropower stations are generally labour oriented
(*b*) Gestation period for hydro-power plant is usually small
(*c*) the hydro generators give high efficiency over a wide range of load
(*d*) in a hydropower scheme, the firm power is usually high, as compared to total power.

**Q. 458. The minimum power, which a hydro-power plant can generate throughout the year, is called :**
(*a*) power plant capacity (*b*) power plant load
(*c*) firm power (*d*) water power.

**Q. 459. 400 cumecs of water is being released from a dam storage to meet the downstream demand, through the turbines of the connected hydroplant. The effective head of water acting on the turbines is 50 m. The efficiency of the hydroplant is 0.8. The electrical power generated from this plant, would be:**
(*a*) 1,56,800 MW (*b*) 156.8 M kW
(*c*) 156.8 MW (*d*) none of these.

**Q. 460. If the peak load for a power plant equals the plant capacity, then the ratio of the capacity factor to load factor will be :**
(*a*) 1 (*b*) 0
(*c*) $< 1$ (*d*) $>1$
(*e*) $-1$ (*f*) none of these.

**Q. 461.*If the peak load on a power plant having a capacity of 100 MW is 70 MW during a given week, and the energy produced is 58,80,000 kWh, the capacity factor for the plant for the week, will be :**

(*a*) 35% (*b*) 50%
(*c*) 70% (*d*) none of these.

**Q. 462.*The load factor for the data given in the Q. 461 will be :**

(*a*) 0.35 (*b*) 0.50
(*c*) 0.70 (*d*) none of these.

**Q. 463.*The utilisation factor for the data given in Q. 461, will be :**

(*a*) 0.35 (*b*) 0.50
(*c*) 0.70 (*d*) none of these.

**Q. 464. Pondage requirement in a hydropower plant includes the need :**

(*a*) to balance the varying demand
(*b*) to compensate for wastage and spillage
(*c*) to balance short time fluctuations in the flow
(*d*) all of these.

**Q. 465. A Pelton's turbine is a :**

(*a*) velocity turbine (*b*) reaction turbine
(*c*) pressure turbine (*d*) none of these.

**Q. 466. An impulse turbine, like a Pelton's wheel, is used for :**

(*a*) low heads & high discharges (*b*) high heads & low discharges
(*c*) low heads & low discharges
(*d*) high heads & high discharges.

**Q. 467. An impulse turbine, like a Pelton's turbine, is :**

(*a*) an open body exposed to atmosphere
(*b*) a closed body not exposed to atmosphere
(*c*) a partially open and partially closed body
(*d*) none of these.

**Q. 468. In a reaction turbine, like a Francis turbine or Kaplan turbine,**

(*a*) the potential head of water is totally changed into kinetic head
(*b*) the potential head of water is partially changed into kinetic head
(*c*) the potential head of water is not at all changed into kinetic head
(*d*) all of the above are possible.

**Q. 469. In a Francis turbine, the water**

(*a*) flows out openly
(*b*) flows out through a closed draft tube
(*c*) both (*a*) and (*b*), depending upon the type
(*d*) neither (*a*) nor (*b*).

**Q. 470.*You have to select turbines for a hydropower plant, having 150 m head of stored water behind the dam. The water in the reservoir is clear, and load on the power house is not likely to fluctuate much. Which type of turbines will you generally recommend?**

(*a*) Pelton's turbines (*b*) Francis turbines
(*c*) Kaplan turbines (*d*) Any of them.

**Q. 471. You have to select turbines for a hydropower plant, working on 150 m head. The water is sandy and load on the plant is highly variable. Which type of turbines will you generally recommend ?**

(*a*) Pelton's turbines (*b*) Francis turbines
(*c*) Kaplan turbines (*d*) Any of them will do.

**Q. 472. If two identical pumps, each capable to deliver a discharge *Q* against a head *H*, are connected :**
(*a*) in parallel, the resulting discharge is $2Q$ against a head $H$
(*b*) in parallel, the resulting discharge is $Q$ against a head of $2H$
(*c*) in series, the resulting discharge is $2Q$ against a head $H$
(*d*) in series, the resulting discharge is $Q$ against a head $2H$.

## CHAPTER 25 : Tank Irrigation

**Q. 473. The depth of water stored in an irrigation tank, rarely exceeds :**
(*a*) 4 m (*b*) 8 m
(*c*) 12 m (*d*) none of them.

**Q. 474. The storage created on the upstream side of a smaller stream by construction of low earthen bunds, is technically known as :**
(*a*) a reservoir (*b*) a tank
(*c*) a lake (*d*) none of these.

**Q. 475. The usual arrangement made in a tank-bund scheme to spill the surplus excess water, is known as a :**
(*a*) tank spillway (*b*) tank sluice
(*c*) tank weir (*d*) none of these.

**Q. 476. The provision of suitable breaching sections become more important in irrigation tanks, which are :**
(*a*) isolated (*b*) in groups
(*c*) both (*a*) and (*b*) (*d*) none of them.

**Q. 477. The cubic metres of water that can be stored in an irrigation tank between FTL (Full Tank Level) and Sill level of the lowest supply sluice, is known as:**
(*a*) the gross capacity (*b*) effective capacity
(*c*) the specific capacity (*d*) none of the above.

**Q. 478. If $A_1$ represents the area of a tank bed at bottom, and $A_2$ at FTL, then the storage capacity of this tank of height *H*, is given as :**
(*a*) $H\left[A_1 + A_2 + \sqrt{A_1A_2}\right]$ (*b*) $\frac{H}{2}\left[A_1 + A_2 + \sqrt{A_1A_2}\right]$
(*c*) $\frac{H}{3}\left[A_1 + A_2 + \sqrt{A_1A_2}\right]$ (*d*) none of them.

**Q. 479. The top level of a tank weir in a tank bund scheme, is kept at :**
(*a*) MTL (*b*) FTL
(*c*) either (*a*) or (*b*) (*d*) none of these.

**Q. 480. The capacity of an irrigation tank is sometimes increased by installing a temporary stone wall over the top of the tank weir. This fixture is known is :**
(*a*) saddle (*b*) dam stones
(*c*) breast wall (*d*) divide wall.

**Q. 481. The discharge through the sluice of a small irrigation tank is usually controlled by a :**

(*a*) dam stone (*b*) plug
(*c*) shutter gate (*d*) none of these.

**Q. 482. The mostly adopted earthen section for tank bunds is of the type:**
(*a*) homogeneous embankment type (*b*) zoned embankment type
(*c*) diaphragm type (*d*) none of the above.

**Q. 483.*The discharging capacity of a masonry waste weir of 50 m length and 1 m width provided in an earthen bund, for passing a flood with 0.8 m water depth over its crest, will be :**
(*a*) 12.25 cumecs (*b*) 25.4 cumecs
(*c*) 59.4 cumecs (*d*) 65.8 cumecs.

## CHAPTER 26 : Arch and Buttress Dams

**Q. 484. An arch dam, looks like a single arch, in :**
(*a*) plan (*b*) front elevation
(*c*) side elevation (*d*) none of them.

**Q. 485. An arch dam behaves as :**
(*a*) a cantilever vertical retaining wall, standing up from its base
(*b*) an arch, transferring loads at the two ends by horizontal arch action
(*c*) both (*a*) and (*b*) (*d*) none of the above.

**Q. 486. A non-vertical arch dam is known as a :**
(*a*) buttress dam (*b*) double curvature arch dam
(*c*) shell arch dam (*d*) both (*b*) and (*c*) above.

**Q. 487. Arch dams are suitable in locations where the canyon slope factor (length of arch at top/height of dam) is :**
(*a*) less than 8 (*b*) greater than 16

**Q. 488. The most economical type of arch dam in general is of :**
(*a*) constant radius type (*b*) variable radius type
(*c*) constant angle type (*d*) none of these.

**Q. 489. A constant angle arch dam, when compared to constant radius arch dam, utilises concrete quantity of about :**
(*a*) 43% (*b*) 130%
(*c*) 230% (*d*) none of them.

**Q. 490. Amongst the following types of dam sections, the thinnest and most economical section is expected for a :**
(*a*) constant angle arch dam (*b*) shell-arch dam
(*c*) constant radius arch dam (*d*) concrete gravity dam.

**Q. 491. Which of the following forces is the least important in the design of arch dams?**
(*a*) reservoir water force (*b*) uplift pressure
(*c*) temperature stresses (*d*) ice load
(*e*) yield stresses.

**Q. 492. A constant radius arch dam is also sometimes called a :**
(*a*) constant centre arch dam (*b*) constant angle arch dam
(*c*) variable angle arch dam (*d*) none of the above.

**Q. 493. The most economical central angle of the arch rings in an arch dam, is :**

(*a*) 123°—34′ (*b*) 133°—34′
(*c*) 143°—34′ (*d*) 153°—34′.

**Q. 494. The most economical central angle of the arch rings of an arch dam, can be adopted only at one place, preferably at mid height in an arch dam of the type :**

(*a*) constant angle arch dam (*b*) constant radius arch dam
(*c*) both (*a*) and (*b*) above (*d*) none of the above.

**Q. 495. In an arch dam, the 'extrodos curves' refer to the arch rings corresponding to the :**

(*a*) upstream face of the dam (*b*) downstream face of the dam
(*c*) either (*a*) or (*b*) (*d*) none of the above.

**Q. 496. A V-shaped valley with stronger foundations can suggest the choice of an arch dam of the type :**

(*a*) constant radius arch dam (*b*) variable radius arch dam
(*c*) constant angle arch dam (*d*) none of the above.

**Q. 497. The type of an arch dam, which generally requires 'overhangs at abutments', is of :**

(*a*) constant radius type (*b*) variable radius type
(*c*) constant angle type (*d*) none of the above.

**Q. 498. The most accurate method to design arch dams is based upon :**

(*a*) the thin cylinder theory (*b*) the theory of elastic arches
(*c*) the trial load method (*d*) all of the above.

**Q. 499. Which among the following is not an assumption of the 'Trial Load Analysis' method of design of arch dams ?**

(*a*) plane sections normal to the axis remain plane after flexure
(*b*) any horizontal arch ring is independent of the arch rings below and above
(*c*) modulus of elasticity of concrete and rock foundation is equal for compression and tension
(*d*) stresses are proportional to strains.

**Q. 500. The thin cylinder theory for designing arch dams, ignores :**

(*a*) temperature stresses (*b*) ice pressures
(*c*) yield stresses (*d*) all of the above.

**Q. 501. The only force considered in the design of arch dams by thin cylinder theory is :**

(*a*) water pressure (*b*) yield stresses
(*c*) shrinkage stresses (*d*) none of them.

**Q. 502. The temperature stresses, producing worst effects in the design of arch dams, are caused by :**

(*a*) rise of temperatures (*b*) fall of temperatures
(*c*) both (*a*) and (*b*) (*d*) none of the above.

**Q. 503. The stresses due to rib-shortening become quite important in the case of :**

(*a*) long thin arch dams
(*b*) thick arch dams of small angle

(*c*) both (*a*) and (*b*) above (*d*) none of the above.

**Q. 504. While designing a freely supported deck type of a buttress dam, the chosen spacing between buttresses is usually increased to achieve economy, from about 4.5 to 15 m or so, with the :**

(*a*) decrease in dam height from about 45 m to about 15 m

(*b*) increase in dam height from about 15 m to about 45 m.

**Q. 505. The load from the deck of a buttress dam is transmitted to the foundations, through the :**

(*a*) corbels (*b*) buttresses

(*c*) lateral braces (*d*) none of them.

**Q. 506. Which of the following components in a buttress dam are responsible for transmission of load from the sloping deck slab to the buttresses?**

(*a*) corbels (*b*) lateral braces

(*c*) counterforts (*d*) none of them.

**Q. 507. The most suitable foundation for buttress dams is :**

(*a*) pile foundation (*b*) well foundation

(*c*) mat foundation (*d*) none of them.

**Q. 508. The lateral braces are provided between buttresses, in a buttress dam, to :**

(*a*) transfer the load from the deck to the foundation

(*b*) transfer the load from the deck to the buttresses

(*c*) provide resistance against buckling of buttresses

(*d*) none of the above.

**Q. 509. No reinforcement is necessary in the :**

(*a*) fixed deck buttress dam (*b*) simple deck buttress dam

(*c*) cantilever deck buttress dam (*d*) massive head buttress dam.

**Q. 510. Which of the following dam is also known as Ambursen dam?**

(*a*) multiple arch buttress dam

(*b*) mushroom head buttress dam

(*c*) massive head buttress dam

(*d*) free deck buttress dam.

**Q. 511. Which of the following component of a buttress dam is also called a counterfort?**

(*a*) corbel (*b*) buttress

(*c*) lateral brace (*d*) sloping deck.

**Q. 512. Which of the following is incorrect about buttress dam?**

(*a*) it requires less construction material than a solid gravity dam of the same height

(*b*) it is less susceptible to deliberate damage or sabotage

(*c*) Ice pressure is not significant, because ice slides over the sloping deck

(*d*) the factor of safety is usually greater than for a gravity dam.

**Q. 513. Multiple dome buttress dam is a type of :**

(*a*) rigid buttress dam (*b*) deck slab buttress dam

(*c*) bulk head buttress dam (*d*) none of the above.

**Q. 514. In a simple deck slab buttress dam of a given height, the buttress spacing can be increased by :**

(*a*) increasing the upstream slope

(*b*) decreasing the upstream slope

(*c*) either (*a*) or (*b*)

(*d*) none of these.

**Q. 515. The buttresses of a free deck buttress dam are provided @ '*x*' m clear spacing, and buttress thickness is proposed to be '*t*' m. The buttress section can then be designed like the section of a concrete gravity dam, by considering the effective unit weight of water = surcharge factor (*S*) × actual unit weight of water ($\gamma_w$). The value of this surcharge factor is given as :**

(*a*) $S = \dfrac{x+t}{x}$ (*b*) $S = \dfrac{x-t}{x}$

(*c*) $S = \left(\dfrac{x+t}{1+x}\right)$ (*d*) none of these.

**Q. 516. Multiple arch dams are generally used for :**

(*a*) high heights (*b*) low heights

(*c*) medium heights (*d*) all of the above.

## ANSWERS OF THE OBJECTIVE QUESTIONS

| | | | | |
|---|---|---|---|---|
| 1. (*d*) | 2. (*a*) | 3. (*d*) | 4. (*a*) | 5. (*c*) |
| 6. (*a*) | 7. (*b*) | 8. (*b*) | 9. (*d*) | 10. (*a*) |
| 11. (*b*) | 12. (*c*) | 13. (*a*) | 14. (*b*) | 15. (*b*) |
| 16. (*b*) | 17. (*b*) | 18. (*d*) | 19. (*b*) | 20. (*c*) |
| 21. (*a*) | 22. (*c*) | 23. (*a*) | 24. (*c*) | 25. (*b*) |
| 26. (*b*) | 27. (*d*) | 28. (*b*) | 29. (*a*) | 30. (*c*) |
| 31. (*c*) | 32. (*a*) | 33. (*c*) | 34. (*b*) | 35. (*b*) |
| 36. (*b*) | 37. (*b*) | 38. (*b*) | 39. (*c*) | 40. (*a*) |
| 41. (*c*) | 42. (*b*) | 43. (*c*) | 44. (*c*) | 45. (*b*) |
| 46. (*c*) | 47. (*c*) | 48. (*c*) | 49. (*b*) | 50. (*a*) |
| 51. (*d*) | 52. (*d*) | 53. (*c*) | 54. (*c*) | 55. (*c*) |
| 56. (*b*) | 57. (*d*) | 58. (*d*) | 59. (*a*) | 60. (*a*) |
| 61. (*b*) | 62. (*c*) | 63. (*b*) | 64. (*d*) | 65. (*a*) |
| 66. (*c*) | 67. (*d*) | 68. (*c*) | 69. (*b*) | 70. (*c*) |
| 71. (*c*) | 72. (*a*) | 73. (*a*) | 74. (*b*) | 75. (*d*) |
| 76. (*b*) | 77. (*b*) | 78. (*d*) | 79. (*b*) | 80. (*d*) |
| 81. (*c*) | 82. (*c*) | 83. (*b*) | 84. (*a*) | 85. (*b*) |
| 86. (*c*) | 87. (*a*) | 88. (*c*) | 89. (*c*) | 90. (*d*) |
| 91. (*d*) | 92. (*c*) | 93. (*a*) | 94. (*a*) | 95. (*c*) |
| 96. (*c*) | 97. (*b*) | 98. (*d*) | 99. (*b*) | 100. (*c*) |
| 101. (*c*) | 102. (*b*) | 103. (*b*) | 104. (*c*) | 105. (*b*) |
| 106. (*a*) | 107. (*b*) | 108. (*a*) | 109. (*b*) | 110. (*d*) |
| 111. (*b*) | 112. (*b*) | 113. (*b*) | 114. (*b*) | 115. (*b*) |
| 116. (*a*) | 117. (*c*) | 118. (*c*) | 119. (*a*) | 120. (*c*) |

121. (b) 122. (b) 123. (c) 124. (c) 125. (c)
126. (a) 127. (b) 128. (c) 129. (c) 130. (d)
131. (a) 132. (a) 133. (c) 134. (b) 135. (c)
136. (c) 137. (d) 138. (d) 139. (d) 140. (d)
141. (d) 142. (d) 143. (e) 144. (c) 145. (d)
146. (c) 147. (a) 148. (c) 149. (d) 150. (a)
151. (a) 152. (b) 153. (c) 154. (c) 155. (d)
156. (b) 157. (c) 158. (d) 159. (b) 160. (a)
161. (c) 162. (a) 163. (b) 164. (a) 165. (d)
166. (b) 167. (b) 168. (b) 169. (b) 170. (c)
171. (a) 172. (b) 173. (a) 174. (a) 175. (c)
176. (a) 177. (a) 178. (c) 179. (b) 180. (d)
181. (c) 182. (c) 183. (b) 184. (d) 185. (e)
186. (a) 187. (c) 188. (e) 189. (d) 190. (b)
191. (d) 192. (d) 193. (a) 194. (a) 195. (d)
196. (c) 197. (d) 198. (c) 199. (b) 200. (c)
201. (b) 202. (b) 203. (b) 204. (d) 205. (b)
206. (d) 207. (b) 208. (b) 209. (b) 210. (a)
211. (a) 212. (c) 213. (b) 214. (d) 215. (b)
216. (b) 217. (c) 218. (a) 219. (a) 220. (a)
221. (d) 222. (a) 223. (c) 224. (c) 225. (b)
226. (d) 227. (b) 228. (c) 229. (a) 230. (b)
231. (b) 232. (d) 233. (c) 234. (d) 235. (d)
236. (b) 237. (c) 238. (c) 239. (c) 240. (c)
241. (c) 242. (b) 243. (d) 244. (d) 245. (c)
246. (d) 247. (a) 248. (c) 249. (d) 250. (b)
251. (a) 252. (b) 253. (e) 254. (b) 255. (b)
256. (b) 257. (c) 258. (d) 259. (c) 260. (d)
261. (c) 262. (d) 263. (c) 264. (c) 265. (b)
266. (c) 267. (c) 268. (d) 269. (b) 270. (b)
271. (a) 272. (c) 273. (b) 274. (b) 275. (c)
276. (d) 277. (b) 278. (c) 279. (b) 280. (a)
281. (b) 282. (b) 283. (a) 284. (b) 285. (d)
286. (b) 287. (d) 288. (a) 289. (d) 290. (c)
291. (c) 292. (b) 293. (c) 294. (c) 295. (a)
296. (c) 297. (b) 298. (a) 299. (e) 300. (b)
301. (a) 302. (d) 303. (b) 304. (a) 305. (a)
306. (b) 307. (d) 308. (d) 309. (d) 310. (c)
311. (c) 312. (b) 313. (a) 314. (b) 315. (b)
316. (c) 317. (d) 318. (d) 319. (b)
320. (c) 321. (b) 322. (b) 323. (a) 324. (b)
325. (c) 326. (d) 327. (c) 328. (b) 329. (b)
330. (c) 331. (b) 332. (a) 333. (c) 334. (b)
335. (a) 336. (a) 337. (d) 338. (b) 339. (b)
340. (c) 341. (a) 342. (b) 343. (c) 344. (c)
345. (a) 346. (b) 347. (c) 348. (b) 349. (b)
350. (a) 351. (a) 352. (c) 353. (d) 354. (a)
355. (a) 356. (c) 357. (b) 358. (a) 359. (b)

| | | | | |
|---|---|---|---|---|
| 360. (*c*) | 361. (*a*) | 362. (*c*) | 363. (*d*) | 364. (*d*) |
| 365. (*a*) | 366. (*c*) | 367. (*c*) | 368. (*b*) | 369. (*b*) |
| 370. (*a*) | 371. (*b*) | 372. (*c*) | 373. (*d*) | 374. (*e*) |
| 375. (*b*) | 376. (*c*) | 377. (*b*) | 378. (*c*) | 379. (*c*) |
| 380. (*d*) | 381. (*d*) | 382. (*c*) | 383. (*a*) | 384. (*a*) |
| 385. (*d*) | 386. (*a*) | 387. (*b*) | 388. (*c*) | 389. (*d*) |
| 390. (*c*) | 391. (*b*) | 392. (*c*) | 393. (*c*) | 394. (*c*) |
| 395. (*b*) | 396. (*b*) | 397. (*a*) | 398. (*a*) | 399. (*b*) |
| 400. (*a*) | 401. (*d*) | 402. (*b*) | 403. (*b*) | 404. (*b*) |
| 405. (*a*) | 406. (*b*) | 407. (*d*) | 408. (*a*) | 409. (*a*) |
| 410. (*a*) | 411. (*a*) | 412. (*b*) | 413. (*d*) | 414. (*b*) |
| 415. (*b*) | 416. (*b*) | 417. (*b*) | 418. (*d*) | 419. (*d*) |
| 420. (*d*) | 421. (*c*) | 422. (*b*) | 423. (*b*) | |
| 424. (*a*) (*ii*), (*b*) (*i*), (*c*) (*iii*), (*d*) (*iv*) | | | 425. (*a*) | 426. (*b*) |
| 427. (*a*) | 428. (*b*) | 429. (*a*) | 430. (*c*) | 431. (*a*) |
| 432. (*b*) | 433. (*d*) | 434. (*b*) | 435. (*c*) | 436. (*c*) |
| 437. (*b*) | 438. (*b*) | 439. (*c*) | 440. (*c*) | 441. (*c*) |
| 442. (*a*) | 443. (*b*) | 444. (*a*) | 445. (*d*) | 446. (*b*) |
| 447. (*c*) | 448. (*a*) | 449. (*a*) | 450. (*b*) | 451. (*b*) |
| 452. (*b*) | 453. (*b*) | 454. (*c*) | 455. (*d*) | 456. (*d*) |
| 457. (*c*) | 458. (*c*) | 459. (*c*) | 460. (*a*) | 461. (*a*) |
| 462. (*b*) | 463. (*c*) | 464. (*d*) | 465. (*a*) | 466. (*b*) |
| 467. (*a*) | 468. (*b*) | 469. (*b*) | 470. (*b*) | 471. (*a*) |
| 472. (*a*) | 473. (*c*) | 474. (*b*) | 475. (*c*) | 476. (*b*) |
| 477. (*b*) | 478. (*c*) | 479. (*b*) | 480. (*b*) | 481. (*b*) |
| 482. (*a*) | 483. (*c*) | 484. (*a*) | 485. (*b*) | 486. (*d*) |
| 487. (*a*) | 488. (*c*) | 489. (*a*) | 490. (*b*) | 491. (*b*) |
| 492. (*a*) | 493. (*b*) | 494. (*b*) | 495. (*a*) | 496. (*c*) |
| 497. (*c*) | 498. (*c*) | 499. (*b*) | 500. (*d*) | 501. (*a*) |
| 502. (*b*) | 503. (*b*) | 504. (*b*) | 505. (*b*) | 506. (*a*) |
| 507. (*c*) | 508. (*c*) | 509. (*d*) | 510. (*d*) | 511. (*b*) |
| 512. (*b*) | 513. (*a*) | 514. (*a*) | 515. (*b*) | 516. (*a*). |

## HINTS FOR SOLVING STARRED QUESTIONS

**Q. 16.** *Follow example 1.1, and compute :*
From eqn. (1.1)

$$T = 2.3 \cdot \frac{y}{f} \log_{10}\left(\frac{Q}{Q - f \cdot A}\right)$$

where $A = 0.203 \text{ ha} = 2030 \text{ m}^2$

$Q = 0.043 \text{ cumec} = 0.043 \times 60 \times 60 \text{ m}^3/\text{h}$

$= 154.8 \text{ m}^3/\text{h} = 5 \text{ cm/hr} = 0.05 \text{ m/h}$

$y = 6.35 \text{ cm} = 0.0635 \text{ m}$

$$T = 2.3 \times \frac{0.0635}{0.05} \log\left(\frac{154.8}{154.8 - 0.05 \times 2030}\right)$$ **1.35 h. Ans.**

**Q.17.** Combining wt. $= \dfrac{\text{Atomic wt}}{\text{Valency}}$

$\therefore$ Combining wt. of Na $= \dfrac{23}{1} = 23$

Combining wt. of Ca $= \dfrac{40}{2} = 20$

Combining wt of Mg $= \dfrac{24}{2} = 12$

$$Na^+ = \frac{\text{Na in mg/l}}{23} = \frac{345}{23} = 15 \text{ milli equivalent per litre}$$

$$Ca^{++} = \frac{60}{20} = 3 \text{ milli equivalent per litre}$$

$$Mg^{++} = \frac{18}{12} = 1.5 \text{ milli equivalent per litre}$$

From eqn. (1.4) :

$$SAR = \frac{Na^{++}}{\sqrt{\dfrac{Ca^{++} + Mg^{++}}{2}}} = \frac{15}{\sqrt{\dfrac{3+1.5}{2}}} = \frac{15}{1.5} = \mathbf{10.\ Ans.}$$

**Q.45.** Specific weight means unit weight = 14.7 kN/m$^3$.
Also, unit wt of water = 9.8 kN/m$^3$.
Available moisture holding capacity of root zone

$$= \frac{\gamma d}{\gamma_w} \text{ (F.C. m.c. – Wilting pt m.c.)}$$

$$= \left(\frac{14.7}{9.8}\right) \times (0.9 \text{ m}) [25\% - 15\%]$$

$$= 1.5 \times (0.9 \text{ m}) \times 10\% = 0.135 \text{ m} = \mathbf{13.5\ cm.\ Ans.}$$

**Q.47.** Water depth required at canal

$$= \frac{\text{Water depth required in the field}}{\eta_a \cdot \eta_c} = \frac{10 \text{ cm}}{0.8 \times 0.9} = 13.9 \text{ cm}$$

$\therefore$ Volume of water required for 10 hectare ($10 \times 10^4$ m$^2$) field

$$= \frac{13.9}{100} \text{ m} \times 10^5 \text{ m}^2 = 13{,}900 \text{ m}^3 = \mathbf{13{,}900\ kl.\ Ans.}$$

**Q. 48. Storage duty** or **Quantity duty** is the crop hectares, which can be fully grown/irrigated by 1 M. cum of water.
Thus, if $\Delta$(m) is the water depth required by the crop for its maturity, then $D$ hectares of crop will evidently require $(D \times 10^4)\ \Delta$ m$^3$ of water, which is 1 M. cum.

$$\therefore \quad D \times 10^4\ \Delta \text{ m}^3 = 1 \text{ M cum} = 10^6 \text{ m}^3$$

or $D \times \Delta = 100$ or $\Delta = \dfrac{100}{D}$

where $\Delta$ is in m, and $D$ is in ha/M. cum.

**Q. 55.** I.I. for Kharif = 100 – 72 = 28%
I.I. for Rabi = 100 – 56 = 44%
Total I.I. = 28 + 44 = **72%. Ans.**

**Q. 57.** Such an area may be called as a seasonal cropped area, or seasonal sown area, such as *Rabi area or Kharif area*. Net sown area is the total cropped area (which is only cropped once) in the entire year, *i.e.* in both the seasons.

**Q.58.** G.I. $A = 4.5 \times 0.9 + 2.5 \times 0.8 = 6.05$ M ha

A.I. $I = \frac{G.I.A.}{CCA} = \frac{6.05}{5} = 1.21$ ***i.e.* 121%. Ans.**

**Q.66.** Required discharge (mean) in the field = 0.4 cumec.

$$\text{Peak discharge required} = \frac{\text{Mean discharge}}{\text{Capacity factor}}$$

$$= \frac{0.4}{0.8} = 0.5 \text{ cumec} \qquad \left(\because C.F. = \frac{Q_{mean}}{Q_{peak}}\right)$$

$$\text{Design discharge for distributary} = \frac{\text{Required discharge for crops}}{\text{Time factor}}$$

(Since channel runs for lesser days than the crop days, the required channel discharge should be more) $= \frac{0.5}{0.5} =$ **1 cumec. Ans.**

**Q.67.** Discharge required at the head of canal $= \frac{6.25 \text{ cumecs}}{0.75} = 8.33$ cumecs

Area irrigated = 5100 hectares
Duty on capacity

$$= \frac{\text{Area to be irrigated during base period}}{\text{Design full supply discharge at its head during peak demand}}$$

$$= \frac{5100}{8.33} = \textbf{612 ha/cumecs. Ans.}$$

**Q.69.** G.C.A. = 8000 ha
CCA = GCA – Reserved forests, etc. = 8000 – 1600 = 6400 ha
Pastures and fallow lands are included in CCA, and hence not to be subtracted.
Area under irrigation
= CCA × In. of Irr. = 6400 × 0.5 = 3200 hectares. **Ans.**

**Q.79.** $\frac{\tau_0}{\gamma.d(S_s - 1)} = 0.056$

or $\tau_0 \propto d$. **Ans.**

**Q.80.** For non-scouring

$$d < 11\,RS$$

$$\text{or} \quad R_{max} = \frac{d}{11S} \qquad ...(i)$$

$$\therefore \quad V = \frac{1}{N} \cdot R^{2/3} \cdot S^{1/2}$$

$$\text{But} \quad N = \frac{1}{24}\, d^{1/6}$$

$$\therefore \quad V = \frac{24}{d^{1/6}} \cdot R^{2/3} \cdot S^{1/2}$$

$$V_{max} = \frac{24}{d^{1/6}} \cdot R_{\max}^{2/3} \cdot S^{1/2}$$

$$= \frac{24}{d^{1/6}} \cdot \left(\frac{d}{11S}\right)^{2/3} \cdot S^{1/2} = \frac{24}{(11)^{2/3}} \cdot d^{2/3-1/6} \cdot S^{1/2-2/3}$$

$$= 4.85\, d^{1/2} \cdot S^{-1/6}$$ where $d$ is in m

When $d$ is in cm *i.e.* $d_{cm}$, then

$$d = \frac{d_{cm}}{100}$$

$$\therefore \quad V_{max} = 4.85 \left(\frac{d_{cm}}{100}\right)^{1/2} \cdot S^{-1/6}$$

$$= 0.485\, d^{1/2} \cdot S^{-1/6}$$ where $d$ is in cm

or $V_{max} \approx 0.48\, d^{1/2} \cdot S^{-1/6}$. **Ans.**

**Q.81.** $N$ is given by Strickler's formula, as

$$N = \frac{1}{24} \cdot d^{1/6}$$ where $d$ is representative grain dia in m.

**Q.85.** Such channels are designed only for non-scouring. Hence, they must carry clear waters only. **Ans.**

**Q.86.** For 1.5 H : 1 V, $\tan\theta = \frac{y}{x} = \frac{1}{1.5}$

$\sin\theta = 0.554$

$$\frac{\tau_c'}{\tau_c} = \sqrt{1 - \frac{\sin^2\theta}{\sin^2\phi}} = \sqrt{1 - \left(\frac{0.554}{\sin 37°}\right)^2} = \sqrt{0.153} = 0.391.$$

For stability on banks

$0.75\, \gamma RS \leq 0.391\, \tau_c$

$\leq 0.391\, (\gamma RS)$

For stability on bed, $RS = \frac{d}{11}$

$0.75\, \gamma \cdot RS \leq 0.391 \cdot \gamma \cdot \frac{d}{11}$

or $RS \leq \frac{0.391}{11 \times 0.75} d$ or $RS \leq \frac{d}{21}$

or $R \leq \frac{d}{21S}$. **Ans.**

**Q.89.** If the actual velocity in the channel exceeds the optimum velocity ($V_0$), then scouring may occur ; and if the actual velocity is lesser than the optimum velocity ($V_0$), then silting may occur. This optimum velocity, thus, represents both maximum as well as minimum velocity.

**Q. 104 and Q. 108.** When a canal has a *fixed configuration* (*i.e.* its bed slope and cross-section do not change with time) due to its non-erodible surface, such as in a lined canal, then the channel is called in **permanent regime.** Besides no scouring, usually, there is no silting in such channels, because higher velocities are permitted, due to their inerodible surface. Hence, the silt amount leaving the channel is almost the same as the silt amount entering the channel. Such channels, whose boundaries are fixed and rigid, are called **rigid boundary canals,** and Lacey's theory is not applicable to

them. *The resistance offered to the flowing water in such a channel, is a function of the nature of the boundary surface alone.*
On the other hand, when the canals are constructed in erodible strata, their boundaries tend to change continuously with the flow of water in them, Such channels are evidently called **mobile boundary canals.** The material forming the boundaries of such channels are mostly loose soils, easily susceptible to erosion by flowing water. In such canals, higher velocities are usually not permissible, because this would lead to excessive scouring. Lower values of velocities will not be able to keep the entire silt load into suspension, which may lead to silting in them. Due to such scouring and silting of sediment, the sediment load transport at different sections along the channel, will usually be different. Silting and scouring may also result in modification of channel cross-section. *The resistance offered to the flow of water in such channels depends on two factors. viz :*
*(i) nature of the surface of boundaries, and*
*(ii) condition of the bed and banks of the channel.*
Such channels need be designed for non-silting and non-scouring velocity. Rivers and unlined alluvial canals are examples of mobile boundary channels.

**Q.109.** Scour depth $= 1.35 \left(\dfrac{q^2}{f}\right)^{1/3} = 1.35 \left(\dfrac{3^2}{1.2}\right)^{1/3} = 2.64$ m. **Ans.**

**Q.111.** From eqn. (4.14) :

$$f = 1.75\sqrt{d_{mm}} = 1.75\sqrt{0.4} = 1.107$$

From eqn. (4.19) :

$$\text{Regime bed slope} = \frac{f^{5/3}}{3340Q^{1/6}} = \frac{(1.107)^{5/3}}{3340(40)^{1/6}} = \frac{1}{5200}. \textbf{ Ans.}$$

**Q.112.**

$$1.0 = 1.75\sqrt{d_{mm}}$$

$$d_{mm} = \left(\frac{1}{1.75}\right)^2 = 0.326 \text{ mm} \approx 0.32 \text{ mm. } \textbf{Ans.}$$

**Q.114.** $P = b + \sqrt{5} \cdot y$ ; with ½ $H$ : 1 V slopes, which are generally attained in a regime canal, irrespective of $Q$ and $f$.

$$= 22 + \sqrt{5} \times 2.5 = 27.59 \text{ m}$$

$$P = 4.75\sqrt{Q} = 27.59$$

$$\therefore \quad Q = \left(\frac{27.59}{4.75}\right)^2 = 33.73 \text{ m}^3\text{/s} \approx 34 \text{ cumecs. } \textbf{Ans.}$$

**Q.115.** Given $\begin{bmatrix} Q_A = Q_B \\ f_A > f_B \end{bmatrix}$

Now since, $V = \left(\dfrac{Qf^2}{140}\right)^{1/6}$ and $P = 4.75\sqrt{Q}$

$V_A > V_B$ $\qquad P_A = P_B$

Since $Q$ is same, $A_A < A_B$

and since $R = \dfrac{A}{P}$

$R_A < R_B$ or $R_B > R_A$

*i.e.* **channel *B* has more hydraulic radius for the same wetted perimeter, and this happens only when the channel has more depth (deeper channels are more hydraulically efficient).**

$\therefore \quad y_B > y_A.$

Hence, choice (*b*) is the correct choice. **Ans.**

**Q.122.** By the geometry of Fig. 27.1, it can be seen that the points, *B*, *B*′, etc. will lie on the bank slope line, only when the berm equals $(n_2 - n_1)\, d_1$, where $d_1$ is the depth of cutting, which will vary according to ground level.

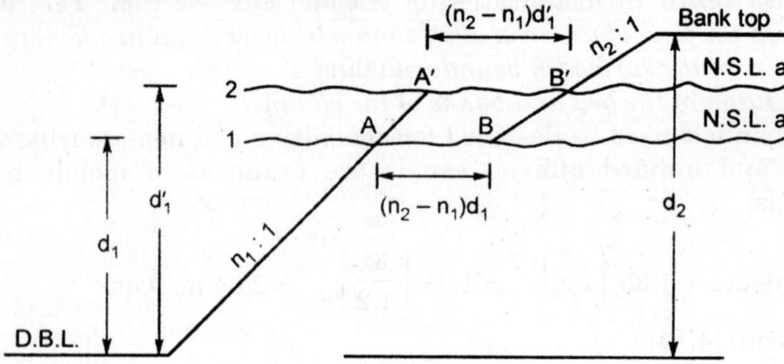


Fig. 27.1.

**Q.123.** It can be seen from Fig. 27.2 that when channel attains regime, ½ H : 1V side slopes will be formed, thereby increasing the initial berm width *AB* to *A*′*B*, where

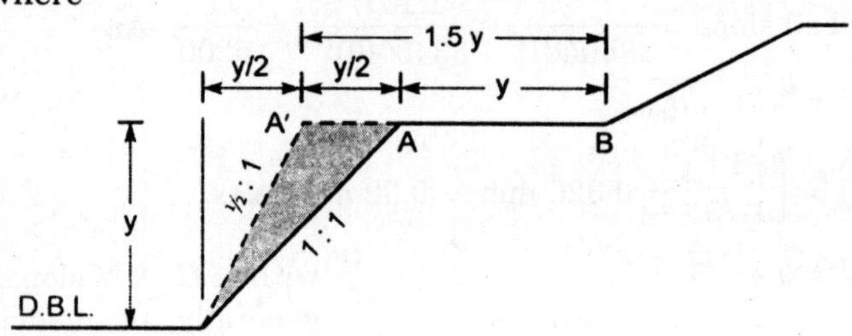


Fig. 27.2.

$A'B = A'A + AB = 0.5\, y + y = 1.5\, y.$ **Ans.**

**Q.126.** *Assume 1 km length*

$$\text{Annual benefit} = 1000 \text{ m} \times 20 \text{ m} \times \frac{3}{10^6} \text{ cumecs} \times \text{Rs. 4 lakh/yr cumecs}$$

$$= \text{Rs. } 24000 \text{ per yr}$$

$$\text{Annual cost} = \frac{100 \text{ m} \times 20 \text{ m}}{40 \text{ yr}} \times \text{Rs. } 40/\text{m}^2 = \text{Rs. } 20{,}000 \text{ per yr.}$$

$$\text{B.C. ratio} = \frac{24000}{20000} = 1.2. \ \textbf{Ans.}$$

**Q.129.** W.r.t. Fig. 27.3 ; $\tan\theta = \dfrac{1}{1.5}$

$\theta = 0.588$

$A = y^2 (\theta + \cot\theta)$

$= 2.5^2 (0.588 + 1.5) = 13.05 \text{ m}^2$

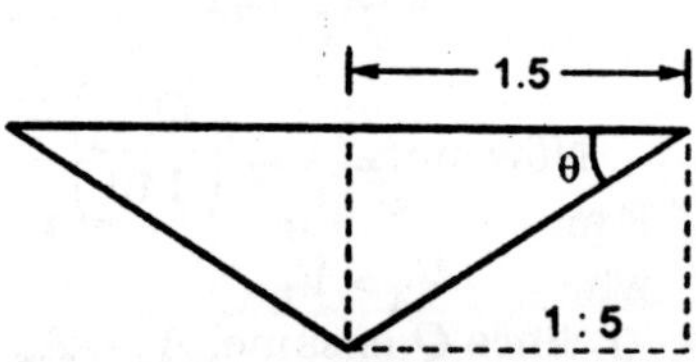


Fig. 27.3.

$$V = \frac{26}{13.00} \cong 2 \text{ m/s}. \textbf{ Ans.}$$

**Q.147.** The classification of saline and alkaline soils is as follows :

| *Sl. No.* | *Classification* | *Electrical Conductivity (EC) in micro-mho / cm* | *Exchangeable Sodium Percentage (ESP)* | *pH value* |
|---|---|---|---|---|
| 1. | **Saline soil** or *white alkali* | >4000 | < 15 | ≤ 8.5 |
| 2. | **Alkaline soil** or *Non-saline alkali* or *sodic soil* or *black alkali*, | < 4000 | > 15 | 8.5 to 10.0 |
| 3. | **Saline-alkali soil,** | > 4000 | > 15 | < 8.5 |

**Q.166.** $M_B \approx 150 \cdot \sqrt{Q_a} = 150\sqrt{1600} = 6000$ m *i.e.* 6 km. **Ans.**

**Q.169.** $M_B = 3 \text{ km} = 3000 \text{ m}$

Also, $M_B \approx 150 \cdot \sqrt{Q_{\text{dominant}}}$

$\therefore \quad 3000 = 150\sqrt{Q_{\text{dominant}}}$ or $Q_{\text{dominant}} = 400$ cumecs

$\therefore \; Q_{max} = \frac{16}{9} \times 400 = 711$ cumecs $\approx 700$ cumecs. **Ans.**

**Q.176.** Average thickness of apron

$= T = 1.9t = 1.9 \times 0.06\,(5000)^{1/3} = 1.95$ m ; say 2 m. **Ans.**

Average width of apron = 1.5 $D$

where $D = 1.5\,R - y$ (average),

where $R = 0.47\left(\frac{Q}{f}\right)^{1/3}$

or $R = 0.47\,(5000)^{1/3}$ (assume $f = 1$) = 8.04

$= 1.5 \times (1.5 \times 8.04 - 5)$

$= 10.6 \text{ m} \approx 10$ m (say). **Ans.**

**Q.178.** Waterway required

$= 1.2 \times 4.75 \cdot \sqrt{Q_{\max}}$

where 1.2 is safety factor of Lacey's regime width

$= 1.2 \times 4.75 \times \sqrt{1600 \times \frac{16}{9}}$ $\left(\text{since } Q_{\max} = \frac{16}{9} \cdot Q_{dom}\right)$

$= 304 \text{ m} \approx 300$ m. **Ans.**

**Q.216.** Only nominal thickness is required under upstream floor.

**Q.219.** ∵ Fine sand will get lifted most easily.

**Q.297.** Refer and study article 14.5.6.

**Q.298.** Refer and study article 14.5.6.

**Q.304.** $R = 1.35\left(\frac{q^2}{f}\right)^{1/3} = \left(\frac{2500}{300}\right)^{2/3}$ (assume $f = 1$)

$= 5.55$ m

Foundation may be taken down below *HFL* by 1.33*R*, or *R* + 1.2 m, whichever is more.

(*i*) 1.33 *R* = 7.38 m (*ii*) *R* + 1.2 m = 5.55 + 1.2 = 6.75 m < 7.38 m

Hence, foundation level = 214 – 7.38 = 206.62 m; say 206.6 m. **Ans.**

**Q.329.** Ramganga dam, although 128 m high, is earthen and rockfill dam. Nagarjuna Sagar dam is 125 m high, Hirakud dam is 59 m high, and Mettur dam is 70 m high ; they are all stone masonry gravity dams.

**Q.355.** Dead storage = 0.2 × 30 = 6 M. cum

Time during which 6 M. cum is filled by sediment

@0.1 M. cum/yr $= \frac{6}{0.1} = 60$ years. **Ans.**

**Q.364.** For $F > 32$ km, the wave *ht* is given by eqn. (19.12) as

$$h_w = 0.032 \cdot \sqrt{V \cdot F} \text{ m}$$
$$= 0.032 \sqrt{160 \times 40} = 2.56 \text{ m}$$

Force caused by waves $P_w$ is given by eqn. (19.14) as

$$P_w = 19.62\, h_w^{\,2} \text{ kN/m run of dam}$$
$$= 19.62 \times (2.56)^2 \text{ kN} = 128.6 \text{ kN}$$
$$\approx 130 \text{ kN}.$$ Hence, choice (*d*) is correct. **Ans.**

**Q.365.** The formula at choice (*a*) is correct. This may be remembered, as some numerical may be asked on this.

**Q.366.** From eqn. (19.11),

wave ht $(h_w) = 0.032 \sqrt{V \cdot F} + 0.763 - 0.271\, (F)^{1/4}$ for $F < 32$ km

$$\therefore \quad h_w = 0.032 \cdot \sqrt{100 \times 20} + 0.763 - 0.271\, (20)^{1/4}$$
$$= 1.431 + 0.763 - 0.573 = 1.62 \text{ m}$$

Free-board is generally provided equal to

$1.5\, h_w = 1.5 \times 1.62 = 2.45$ m, say 2.5 m. **Ans.**

**Q.373.** From eqn. (19.22) :

$$B = \frac{H}{\sqrt{S_s - c}} = \frac{H}{\sqrt{2.4 - 1}} = \frac{60}{\sqrt{1.4}}$$
$$= 50.7 \text{ m (with full uplift = 1)}$$

From eqn. (19.23) :

$$B = \frac{H}{\mu(S_s - c)} = \frac{60}{0.7(1.4)} = 61.22 \text{ m; say 61 m.}$$

Chosen $B$ should be greater of the two values, *i.e.* 61 m. **Ans.**

**Q.390.** Cut off can be provided only during construction. But if any seepage appears afterwards, a drain with inverted filter along toe line is provided to collect soil-free water, to check piping.

**Q.406.** Since an ogee spillway is designed for maximum head, up to *RL* of maximum reservoir level, there are almost no chances of operating head exceeding the design head. However, under rare emergency, when extraordinary flood is being passed near the end of monsoon season, it may become necessary to allow the reservoir level to even go above the Maximum Reservoir level, by encroaching upon the freeboard, in which case, the operating head may exceed the design head.

**Q.408.** The efficiency of spillway indirectly means its coefficient of discharge, which increases with the increasing head above full reservoir level (*FRL*) till it becomes about 2.2 at full design head (*i.e.* at MRL). Since the spillway starts working when water level just crosses *FRL*, the coefficient of discharge in early stages, is only about 1.7, *i.e.*, about 77% of its value at full design head. Hence, the spillway efficiency can be said to be increasing from 77% to 100%, as water level in the reservoir increases from F.R.L. to M.R.L.

**Q.420.** A saddle syphon is nothing but a syphon spillway, and includes both of its types. Air vent is the essential requirement of both types of syphon spillways.

**Q.434.** $Q = C_d \cdot A\sqrt{2gH_1}$

where $C_d$ = Coefficient of discharge through orifice

$H_1$ = Water head over the centre line of opening

$= RL331 - RL328 = 3$ m

$A$ = Area of opening

= 60 m × 6 m = 360 m²

$\therefore \quad Q = 0.62 \times 360\sqrt{2 \times 9.81 \times 3}$

= 1712 cumecs ≈ 1700 cumecs. **Ans.**

**Q.435.** $Q = C_d A \cdot \sqrt{2gH}$

$H$ = Water head over centre line of sluice (since it is a free overfall on downstream side)

$= RL330\ (FRL) - RL300$ m (sluice centre line level) = 30 m

$$A = \frac{\pi}{4} \times 2^2 = 3.14 \text{ m}^2$$

$C_d$ = Coefficient of discharge ≈ 0.62 for free overfall

$\therefore \quad Q = 0.62 \times 3.14\sqrt{2 \times 9.81 \times 30} = 47.23$ cumecs

say 50 cumecs (nearest appropriate choice given). **Ans.**

**Q.436.** $Q = C_d \cdot A \cdot \sqrt{2gH_L}$

$$= 0.8 \times \left(\frac{\pi}{4} \times 4^2\right)\sqrt{2 \times 9.81 \times (226 - 210)}$$

$= 0.8 \times 4\pi\sqrt{16.62 \times 16} = 164$ cumecs

≈ 170 cumecs (nearest appropriate choice). **Ans.**

To countercheck, let us see the absolute maximum discharge, which computes to be 204.8 cumecs at $C_d = 1$. In any case, even if all losses are neglected, $Q$ cannot exceed 204.8 cumecs, and hence the last choice is certainly incorrect, other choices at (*a*) and (*b*) are giving very low coefficient of discharge, such as $\frac{64}{204.8} = 0.3$ and $\frac{104}{204.8} = 0.5$, which are too low for submerged bell mouthed pipe flows.

**Q.461.** From eqn. [17.6(*a*)] :

$$\text{Capacity factor (C.F.)} = \frac{\text{Average load}}{\text{Plant capacity}}$$

$$\text{where, average load} = \frac{\text{Energy produced}}{\text{Time in h}} = \frac{58,80,000}{7 \times 24} \text{ kW}$$

$$= 35000 \text{ kW} = 35 \text{ MW}$$

$$\text{C.F.} = \frac{35 \text{ MW}}{100 \text{ MW}} = 0.35. \quad \textbf{Ans.}$$

**Q.462.** Load factor $= \dfrac{\text{Average load}}{\text{Peak load}} = \dfrac{35 \text{ MW}}{70 \text{ MW}} = 0.5.$ **Ans.**

**Q.463.** Utilisation Factor (*UF*)

$$= \frac{\text{Maximum power utilised}}{\text{Maximum power available}} = \frac{70 \text{ MW}}{100 \text{ MW}} = 0.7. \textbf{ Ans.}$$

**Q.470.** Francis turbines are cheaper than Pelton's turbines, and there is no special need for Pelton's turbines here, because part load (variable loads) are not likely to come and water is not sandy, which may certainly need Pelton's turbines. Hence, Francis turbines will suffice to our needs. **Ans.**

**Q.483.** $Q = C \cdot LH^{3/2} = 1.66 \times 50 \times (0.8)^{3/2}$

($\because$ When width of rectangular crest > 0.9 m, $C = 1.66$)

$= 59.4$ cumecs. **Ans.**

# —28

# *Objective Questions of the Engineering Services Exams.*

**Year 1999 (*Engineering Services*)**

**Q. 1.*** To generate 10,000 hp under a head of 81 m while working at a speed of 500 rpm, the turbine of choice would be :

(*a*) Pelton (*b*) Kaplan

(*c*) Bulb (*d*) Francis.

**Q. 2.** Consider the following statements :

1. Run of river plants can be located on any river
2. Runaway speed of a turbine is generally 180% of normal speed.
3. Underground power stations are suited to areas susceptible to land slides.
4. Higher the specific speed, higher will be the discharge and head.

Which of these statements are correct ?

(*a*) 1 and 2 (*b*) 2 and 3

(*c*) 1 and 4 (*d*) 2 and 4

**Q. 3.** Match List I and II and select the correct answer using the codes given below the lists

| *List I (Turbine)* | *List II (Specific speed)* |
|---|---|
| A Pelton | 1. 25 |
| B Propeller | 2. 75 |
| C Kaplan | 3. 500 |
| D Francis | 4. 800 |
| | 5. 900 |

*Codes*

| | *A* | *B* | *C* | *D* |
|---|---|---|---|---|
| (*a*) | 1 | 4 | 5 | 2 |
| (*b*) | 3 | 4 | 2 | 5 |
| (*c*) | 4 | 1 | 5 | 3 |
| (*d*) | 1 | 5 | 4 | 2 |

**Q. 4.** The spacing of tile drains to relieve water-logged land is directly proportional to the

(*a*) depth of drain below the ground surface

(*b*) depth of impervious strata from the drain

(*c*) depth of drain below the water level.

(*d*) coefficient of permeability of the soil to be drained.

**Q. 5.** In a flow-mass curve study, the demand line drawn from a ridge does not intersect the mass curve again. This implies that

(*a*) the reservoir is not full at the beginning
(*b*) the storage is not adequate.
(*c*) the demand cannot be met by the inflow, as the reservoir will not refill
(*d*) the reservoir is wasting water by spill.

**Q. 6.** The trap efficiency of a reservoir is a function of :
(*a*) inflow in the reservoir (*b*) ratio of inflow to storage capacity
(*c*) ratio of reservoir capacity to inflow.
(*d*) reservoir capacity.

**Q. 7.** In a linear reservoir, the
(*a*) volume varies linearly with elevation
(*b*) outflow rate varies linearly with storage
(*c*) storage varies linearly with time
(*d*) storage varies linearly with inflow rate

**Q. 8.** The best design of an arch dam is when :
(*a*) All horizontal water loads are transferred horizontally to the abutments.
(*b*) The dam is safe against sliding at various levels
(*c*) The load is divided between the arches and cantilevers, and the deflections at the conjugal points being equal.
(*d*) The deflections of the cantilevers are equal at different points.

**ANSWERS**

| | | | | | |
|---|---|---|---|---|---|
| 1. (*d*) | 2. (*b*) | 3. (*a*) | 4. (*d*) | 5. (*c*) | 6. (*c*) |
| 7. (*b*) | 8. (*c*). | | | | |

**Hints for Solving Starred Questions of the Year 1999**

**Q. 1.** From Eq. (24.1), we have

$$Ns = \frac{N.\sqrt{P_t}}{H^{5/4}}$$

$N$ = Normal speed of turbine in RPM = 500
$P_t$ = Turbine output in metric horse power = 10,000 hp.
$H$ = Effective head in m = 81 m

$$\therefore \quad Ns = \frac{500\sqrt{10,000}}{(81)^{5/4}} = 205.76$$

From Table 24.1, the turbine which is suitable for a head between 25–450 m and $N_s$ of 60 – 300 is Francis turbine; and hence the recommended turbine in this case is Francis type (Reaction type). Hence, choice (*d*) is correct. **Ans.**

**Year 2000 (*Engineering Services*)**

**Q. 1.** If a tangent drawn parallel to the demand line from a ridge point of a mass curve dose not intersect the mass curve again, it can be inferred that the :
(*a*) frequency of the flood entering into the reservoir is less.
(*b*) inflow into the reservoir cannot meet the demand.
(*c*) reservoir is overflowing resulting in wastage.
(*d*) reservoir can meet higher demand.

**Q. 2*.** In order to ensure that no scouring takes place in the bed of a channel of bed slope 'S' constructed on alluvial soil of particle size '*d*' cm, the flow velocity should be restricted to :

(*a*) $4.85d^{1/2} . S^{-1/6}$ (*b*) $4.85d^{-1/2} . S^{1/6}$

(*c*) $0.48d^{1/2} . S^{1/6}$ (*d*) $0.48d^{1/2} . S^{-1/6}$

**Q. 3.** Which one of the following sets is used to control the seepage through the foundations of an earth dam ?

(*a*) Chimney drain, upstream blanket and cut off trench

(*b*) Cut off sheet piles, upstream blanket and cut off trench.

(*c*) Upstream blanket, cut off sheet piles and chimney drains

(*d*) Relief wells, upstream blanket and chimney drain.

**Q. 4.** Consider the following statements :

1. In designing the hydraulic structure in permeable foundation by Khosla's theory, the slope correction is applicable to piles located at the ends of the sloped floor only in a structure with one or more sloped floors.
2. Khosla's theory can be correctly applied in alluvium of finite depth.
3. Length of flow has lesser effect on exit gradient than the depth of piles.
4. Intercepts between hydraulic gradient line and free water surface on the glacis and horizontal floor downstream is the unbalanced uplift in a structure.

Which of these statements related to the design of hydraulic structure are correct ?

(*a*) 1, 2 and 3 (*b*) 1, 3 and 4

(*c*) 2, 3 and 4 (*d*) 1, 2 and 4

**Q. 5.** Match *List-I* (Energy dissipation) with *List-II* (Water level and slope condition) and select the correct answer using the codes given below the lists :

| *List I* | *List II* |
|---|---|
| A. Roller bucket | 1. TWL (tail channel water level) is slightly above JWL (Jump height water level) and the slope of the channel is mild. |
| B. Ski-jump bucket | 2. TWL is considerably above JWL and the slope of the channel is mild. |
| C. Standing wave basin with depressed floor | 3. TWL is slightly below JWL and the slope of the channel is mild |
| D. Standing wave basin with raised floor | 4. TWL is considerably less than JWL and the slope of the channel is steep. |

*Codes*

| | *A* | *B* | *C* | *D* |
|---|---|---|---|---|
| (*a*) | 4 | 2 | 1 | 3 |
| (*b*) | 2 | 4 | 1 | 3 |
| (*c*) | 2 | 4 | 3 | 1 |
| (*d*) | 4 | 2 | 3 | 1 |

**Q. 6.** If the depth is 8.64 cm on a field over a base period of 10 days, then the duty is :

(*a*) 10 hectares per cum/s (*b*) 100 hectares per cum/s

(*c*) 864 hectares per cum/s (*d*) 1000 hectares per cum/s

## ANSWERS

1. (*b*) 2. (*a*) 3. (*b*) 4. (*b*) 5. (*c*) 6. (*d*)

## Hints for Solving Starred Questions of the year 2000

**Q. 2.** For designing a non scouring alluvial channel, we know from eqn. (4.13) that

$$d \leq 11\ RS$$

or $$R \leq \frac{d}{11S} \quad \text{or} \quad R_{max}\ \frac{d}{11S}$$

Also, $$V = \frac{1}{n} R^{2/3} . \sqrt{S}$$

or $$V_{max} = \frac{1}{n} R_{max}^{2/3} . \sqrt{S}$$

But $$n = \frac{1}{24} . d^{1/6} \quad \text{(Strickler's formula)}$$

∴ $$V_{max} = \frac{24}{d^{1/6}} . \left(\frac{d}{11S}\right)^{2/3} . \sqrt{S} = \frac{24}{(11)^{2/3}} . d^{2/3-1/6} . S^{1/2-2/3}$$

or $$V_{max} = 4.85 . d^{1/2} . S^{-1/6}$$

Hence, velocity in a non scouring channel should be restricted to above eqn., which makes choice (*a*) as the correct choice. **Ans.**

### Year 2001 (*Engineering Services*)

**Q. 1.** Match *List I* (Surface) with *List II* (Approximate range of Manning's $n$) and select the correct answer using the codes given below the lists :

| *List I* | *List II* |
|---|---|
| A. Gravel river bed with 50 mm diameter bed | 1. 0.02 to 0.022 |
| B. Badly maintained unlined irrigation channel with weed growth, | 2. 0.025 to 0.04 |
| C. Concrete lined channel | 3. 0.013 to 0.017 |
| D. Channel with brick lining. | 4. 0.04 to 0.08 |

*Codes :*

| | *A* | *B* | *C* | *D* |
|---|---|---|---|---|
| (*a*) | 4 | 3 | 2 | 1 |
| (*b*) | 4 | 2 | 3 | 1 |
| (*c*) | 1 | 3 | 2 | 4 |
| (*d*) | 1 | 2 | 3 | 4 |

**Q. 2.** If the life of a reservoir is determined by its capacity ($C$), volume of annual inflow into the reservoir ($I$) and concentration of sediment in the incoming flow ($C_s$). Life will be more if :

(*a*) $C$, $I$ and $C_s$ are high (b) $C$ and $I$ are high but $C_s$ is low

(*c*) $C$ is high but $I$ and $C_s$ are low (*d*) $C$, $I$ and $C_s$ are low.

**Q. 3.** In a curved reach of a meandering river, both deposition of sediments and erosion of bank occur. Which one of the following statements is true in this regard?

(*a*) Deposition of sediments occurs in the inner bank while the outer bank is subjected to erosion

(*b*) Deposition of sediments occur in the outer bank while the inner bank is subjected to erosion (*c*) In the direction of flow, the outer bank undergoes erosion first and the sediment is later deposited on the outer bank downstream

(*d*) In the direction of flow, the inner bank undergoes erosion first and the sediments are subsequently deposited on the inner bank downstream.

**Q. 4.** Gibb's module is a type of outlet which ensures :

(*a*) Constant discharge even if the water levels in the supply channel and water course fluctuate

(*b*) Variable discharge as per the need

(*c*) Constant discharge into the water course when the water levels in the supply channel vary

(*d*) Constant discharge for varying water levels in the water course for a given water level in the supply channel.

**Q. 5.** The Lacey's silt factor for a particular alluvium is 2.0. This alluvium would comprise :

(*a*) medium sand of size 0.5 mm (*b*) coarse sand of size 0.75 mm

(*c*) medium bajri of size 1.3 mm (*d*) coarse bajri of size 2.4 mm.

**Q. 6.** In a barrage on pervious foundation, sheet piles are provided both upstream and downstream of the barrage to reduce uplift pressure and to prevent piping. Which one of the following statements is true in this regard ?

(a) Compared to downstream sheet pile, the upstream sheet pile is more effective in reducing uplift and piping

(*b*) Compared to upstream sheet pile, the downstream sheet pile is more effective in reducing uplift and piping

(c) Downstream sheet pile is more effective in reducing uplift while the upstream sheet pile is more effective in reducing piping

(*d*) Upstream sheet pile is more effective in reducing uplift while the downstream sheet pile is more effective in reducing piping.

**Q. 7.** The following data pertains to a natural drain crossing an irrigation canal:

| *Item* | *Canal data* | *Drainage data* |
|---|---|---|
| Flow ($m^3/s$) | 5 | 500 |
| Bed level (m) | 120 | 116 |
| Depth of flow (m) | 0.8 | 10 |

Which one of the following types of cross-drainage should be recommended in this case ?

(*a*) Aqueduct (*b*) Syphon aqueduct

(*c*) Syphon (*d*) Super passage

**Q. 8.** Consider the following statements :

The downstream impervious floor of concrete for a barrage has ruptured. This can be due to :

1. insufficient length of upstream impervious floor
2. insufficient length of downstream impervious floor
3. insufficient depth of downstream pile
4. choking of inverted filter.

Which of these statements are correct ?

(*a*) 1 and 2 (*b*) 1 and 4

(*c*) 2, 3 and 4 (*d*) 1, 2, 3 and 4.

**Q. 9*.** Ten $m^3/s$ of water is diverted to a 32 hectare field for 4 hours. Soil proving after irrigation showed that 0.3 m of water has been stored in the root zone. Water application efficiency in this case would be :

(*a*) 96% (*b*) 66.67% (*c*) 48% (*d*) 24%.

**Q. 10.** Which one of the following statements is correct ?

In a river, silt excluder and silt ejector are constructed

(*a*) at a location after the head regulator and at the head of the canal, respectively

(*b*) at the head of the canal and at a location after the head regulator, respectively

(*c*) at the same location

(*d*) at specific locations depending upon diverse factors and their locations do not follow a set pattern.

## ANSWERS

| | | | | | |
|---|---|---|---|---|---|
| 1. (*b*) | 2. (*b*) | 3. (*a*) | 4. (*a*) | 5. (*c*) | 6. (*d*) |
| 7. (*b*) | 8. (*d*) | 9. (*b*) | 10. (*b*) | | |

## Hints for Solving Starred Questions of the year 2001

**Q. 9.** Water depth applied to fields

$$= \frac{10 \frac{m^3}{s} \times (4 \times 60 \times 60s)}{(32 \times 10^4 m^2)} = 0.45 \text{ m}$$

Water depth actually stored in root zone. = 0.3 m

$$\eta_a = \frac{0.3 \text{ m}}{0.45 \text{ m}} = \frac{2}{3} = \mathbf{66.67\%} \quad \textbf{Ans.}$$

## Year 2002 (*Engineering Services*)

**Q. 1.** The following items relate to hydrologic requirements in a hydropower system. Match List I with List II and select the correct answer using the codes given below the lists :

| | *List I* | *List II* |
|---|---|---|
| A | Stream flow in order of magnitude is plotted as ordinate and percent of time as abscissa | 1. Mass curve |
| B | Cumulative value of stream flow is plotted against time for whole period record | 2. Flow duration curve |
| C | Peat of available power against percent of time | 3. Hydrograph |
| D | Plot of flow in stream against time at specific interval | 4. Power duration curve |

*Codes :*

| | *A* | *B* | *C* | *D* |
|---|---|---|---|---|
| (*a*) | 2 | 3 | 4 | 1 |
| (*b*) | 4 | 1 | 2 | 3 |
| (*c*) | 2 | 1 | 4 | 3 |
| (*d*) | 4 | 3 | 2 | 1 |

**Q. 2.** An effective storage of a flood control reservoir is
(*a*) the amount of water which can be supplied from it in a particular interval of time
(*b*) the storage between the minimum and maximum reservoir levels under ordinary operating conditions
(*c*) the useful storage plus the surcharge storage less the valley storage
(*d*) the storage volume of flood water above the maximum reservoir level.

**Q. 3.** Consider the following :
Garret's diagram for the design of irrigation channel is based on
1. Kennedy's theory
2. Lacey's theory
3. Kutter's formula
4. Manning's formula.
Which one of these are correct ?
(*a*) 1 and 3 (*b*) 1 and 4 (*c*) 2 and 3 (*d*) 2 and 4.

**Q. 4.** The total number of dependent equations that form the Lacey's regime theory is
(*a*) 2 (*b*) 3 (*c*) 4 (*d*) 6.

**Q. 5.** Balanced depth of cutting of canal is
(*a*) half the total depth of a canal
(*b*) half of full supply depth
(*c*) the maximum cut that an excavator can take
(*d*) where volume of cutting is equal to volume of filling.

**Q. 6.** At a certain point in the floor of weir, the uplift pressure head due to seepage in 4.5 m. If the relative density of concrete is 2.5, the minimum thickness of floor required at this point to counteract the uplift pressure is
(*a*) 1 m (*b*) 2 m (*c*) 3 m (*d*) 4 m

**Q. 7.** A constant angle arch dam when compared to a constant radius arch dam utilizes concrete quantity of about
(*a*) 33% (*b*) 43% (*c*) 73% (*d*) 143%

**Q. 8.** Drowning ratio of canal outlet is the ratio of
(*a*) rate of change of discharge of outlet to that of distributary
(*b*) water depth above outlet crest to the full supply depth of the channel
(*c*) water depth above the crest on downstream to that on upstream of outlet
(*d*) rate of change of discharge of an outlet to the rate of change of water level of the channel.

**Q. 9*.** A canal was designed to supply the irrigation needs of 1000 ha of land growing rice of 140 days base period and having a delta of 130 cm. If the canal water is used to irrigate wheat of base period 119 days and having a delta of of 50 cm, the area that can be irrigated is ;
(*a*) 452 ha (*b*) 904 ha
(*c*) 1105 ha (*d*) 2210 ha

**Q. 10.** A canal fall is a control structure
(*a*) located at a place where the country slope is flatter than the canal bed slope
(*b*) located most economically where the depth of cutting is less than the balancing depth
(*c*) the location of which is independent of the command to be served
(*d*) designed to secure raising of water surface on its upstream.

**Q. 11.** Match List I (Equation) with List II (Applicability or principle of equation) and select the correct answer using the codes given below the lists :

| | *List I* | *List II* |
|---|---|---|
| A | Theim's equation | 1. is based on energy conservation principle |
| B | Dupuit's assumption | 2. is based on mass conservation principle |
| C | Bernoullis equation | 3. is applicable to steady flow towards a well in confined aquifer |
| D | Continuity equation | 4. is applicable to steady flow in an unconfined aquifer |

*Codes* :

| | A | B | C | D |
|---|---|---|---|---|
| (*a*) | 4 | 3 | 2 | 1 |
| (*b*) | 3 | 4 | 2 | 1 |
| (*c*) | 4 | 3 | 1 | 2 |
| (*d*) | 3 | 4 | 1 | 2 |

### ANSWERS

| | | | | | |
|---|---|---|---|---|---|
| 1. (*c*) | 2. (*c*) | 3. (*a*) | 4. (*d*) | 5. (*d*) | 6. (*c*) |
| 7. (*b*) | 8. (*c*) | 9. (*d*) | 10. (*b*) | 11. (*c*) | |

### Hints for Solving Sttared Questions of the Year 2002

**Q. 8.** Design discharge for canal

Rice Area to be irrigated = 1000 ha, $B$ for Rice = 140 days and $\Delta$ for rice = 130 cm

Duty ($D$) for rice

$$= \frac{864}{\Delta} = \frac{864 \times 140}{130} = 930.46 \text{ ha/cumec}$$

Discharge required for above duty

$$= \frac{1000}{930.46} \text{ cumec} = \mathbf{1.0747 \text{ cumec}}$$

When the above discharge of canal is used for irrigating wheat crop of

$$B = 119 \text{ days} \quad ; \quad \Delta = 50 \text{ cm.}$$

$$\text{Duty for wheat } (D) = \frac{864B}{\Delta} = \frac{864 \times 119}{50} = 2056.32 \text{ ha/cumec.}$$

The ha which can be irrigated by 1.0747 cumec.

$$= 2056.32 \times 1.0747 \text{ ha} = \mathbf{2210 \text{ ha}}$$

Hence, choice at (*d*) is correct. **Ans.**

### Year 2003 (*Engineering Services*)

**Q. 1.** The worst condition of uplift on the floor of a siphon aqueduct occurs when there is

(*a*) high flood flow in the drainage with canal dry
(*b*) full supply flow in the canal with drainage dry
(*c*) high flood flow in the drainage with canal running full
(*d*) water is at drainage bed and canal is dry.

**Q. 2.** A check dam is a

(*a*) flood control structure (*b*) soil conservation structure
(*c*) river training structure (*d*) water storage structure.

**Q. 3.** Leaching is a process

(*a*) by which alkali salts present in the soil are dissolved and drained away

(*b*) by which alkali salts in soil come up with water

(*c*) of draining excess water of irrigation

(*d*) which controls water-logging.

**Q. 4.** Directions

The following items consist of two statements, one labelled as the 'Assertion A' and the other as 'Reason (R)'. You are to examine these two statements carefully and select the answers to these items using the codes given below :

***Codes :***

(*a*) Both A and R are individually true and R is the correct explanation of A

(*b*) Both A and R are individually true but R is not the correct explaination of A

(*c*) A is true but R is false

(*d*) A is false but R is true.

**(i)** *Assertion* (*A*) : With tile lining of canals, permissible velocity of flow is lower than that with concrete lining.

*Reason* (*R*) : The surface of the lining becomes rough due to loss of surface material with high velocity.

**(ii)** *Assertion (A)* : Canal escape serves as a safety valve for a canal.

Reason (*R*) : Canal escape discharges the excess water in the parent canal due to sudden closure of outlets by farmers.

**Q. 5.** Which one of the following phenomena in a pipe flow is termed as water hammer ?

(*a*) The sudden rise of pressure in a long pipe due to sudden closure of valve

(*b*) the rise of a pressure in a pipe flow due to gradual closure of valve

(*c*) The rise of negative pressure

(*d*) The zero pressure in a pipe flow.

**Q. 6.** Flow duration curve is a plot of

(*a*) Flow against its time of occurrence in chronological order

(*b*) Flow in ascending order against percentage time in chronological order

(*c*) Flow that equalled or exceeded against percentage time

(*d*) Flow against duration of time for which it is sustained.

**Q. 7.*** Pondage in a hydropower station is defined as :

(*a*) impounding of considerable amount of excess water during seasons of surplus flow

(*b*) A regulating body of water in the form of relatively small amount of run-off to regulate flow variation in daily or weekly power requirements

(*c*) Excess run off to last for years

(*d*) Excess run off for a few hours only.

**Q. 8.** The specific speed of a turbine under a head of 150 m to develop 2000 HP while running at 300 rpm is

(*a*) 10—35 (*b*) 35—60 (*c*) 60—300 (*d*) 300—1000.

**Q. 9.** Which of the following pairs are correctly matched ?

1. Silt factor : Average size of silt particle.
2. Silt load : Volume of suspended sediments flowing with water in unit time.
3. Silt charge : Weight of silt per unit volume of water
4. Silt grade : Gradation between different silt particles

Select the correct answer using the codes given below :

**Codes**

(*a*) 1 and 4 (*b*) 3 and 4 (*c*) 2 and 3 (*d*) 1 and 2

**Q. 10.** Consider the following statements :

The function of a cut-off in an earth dam is to

1. Reduce uplift pressures on the dam
2. Prevent undermining of foundation
3. Reduce loss of stored water
4. Support the dam

Which of these statements are correct ?

(*a*) 1 and 3 (*b*) 2 and 4 (*c*) 2 and 3 (*d*) 3 and 4.

**Q. 13.** Match List I (Theory) with List II (Propounded By) and select the correct answer using the codes given below the Lists :

| List I (*Theory*) | List II (*Propounded By*) |
|---|---|
| A Exit gradient | 1. G. Lacey |
| B Alluvial canal | 2. L.K. Sherman |
| C Unit hydrograph | 3. A.N. Khosla |
| D Boundary layer | 4. C. Inglis |
| | 5. T.V. Karman |
| | 6. L. Prandtl |

**Codes :**

| | A | B | C | D |
|---|---|---|---|---|
| (*a*) | 1 | 3 | 2 | 6 |
| (*b*) | 6 | 2 | 3 | 5 |
| (*c*) | 3 | 1 | 2 | 6 |
| (*d*) | 3 | 1 | 4 | 2 |

**Q. 14.** The ideal condition for energy dissipation in the design of spillways is the one when the tail water rating curve

(*a*) Lies above jump curve at all discharges

(*b*) Coincides with the jump rating curve at all discharges

(*c*) Lies below jump rating curve at all discharges

(*d*) Lies either above or below the jump rating curve depending upon discharge.

## ANSWERS

| | | | | |
|---|---|---|---|---|
| 1. (*d*) | 2. (*b*) | 3. (*a*) | 4. (*i*)-(*a*), (*ii*)-(*a*) | 5. (*a*) |
| 6. (*c*) | 7. (*b*) | 8. (*a*) | 9. (*c*) | 10. (c) |
| 11. (*c*) | 12. (*b*) | | | |

## Hints for Solving Starred Quetions of the Year 2003

**Q. 9.** As pointed out in article 24.2.1, run-off river plants installed on main rivers or on diversion canals are usually provided with a small storage to meet the short term fluctuations of the water requirement of the plant. A balancing reservoir to store excess stream flow during periods of low consumption in the power house during off peak hours or days, and fot its release during excess load periods, is essentially required in all such low head power development schemes. Such a balancing reservoir of small capacity, called a ***pond***, is hence required near the power house.

In case of a power plant located on the river, the main weir itself provides on its upstream side the required pondage ; while in the case of a plant located on a diversion canal (power canal), the pond is required to be created at the end of the canal in the form of a **foreway reservoir**, from which the penstock pipes shall lead the water into the power house, as shown in Fig. 24.2 (*b*). No matter where the pond is located, its capacity has to be worked out considering the inflow and outflow patterns, just as in the case of storage reservoirs. As a matter of fact, the storage reservoirs provide storage of water for long term requirements, whereas, the *pondage* takes care of the hour to hour fluctuations over a day, or occassionally, day to day fluctuations over a weekly cycle. Provision of pondage ensures that the desired instantaneous rate of flow into the turbines under day to day changing conditions of power load as well as of inflow.

Short term fluctuations may arise due to the following reasons :

(*i*) The power consumption and hence the load on turbines may suddenly increase or decrease, needing in a few seconds either excess or lesser water, respectively. This change in water demand of the power house can be immediately met by the stored water available in the pond created for this purpose. This pond would provide extra water when needed, and retain excess flow when not needed in the plant.

(*ii*) The power load and, thus, the water demand of the power house, may be steady, but the supply of water may undergo a change. In natural streams, where some other power house is located on the upstream, the available flow may change depending upon the quantity of water released from the upstream power house.

Similarly, if the power house in question is located on a diversion channel, the inflow into the power house may get affected due to some breach(es) that may occur in the said diversion canal.

(*iii*) The water demand of turbines and the natural flow (supply) may change from time to time.

The pondage, in some cases, may also be provided for weekly demands. In most of the advanced countries, the free weekends result in decreased demand of power, reducing the water need of the given power house. The inflow during this period could be ponded for use during week days.

## Year 2004 (*Engineering Services*)

**Q. 1.** The delta for a crop having base period 120 days is 70 cm. What is the duty?

(*a*) 2480 hectare/cumec (*b*) 1481 hectare/cumec
(*c*) 148 hectare/cumec (*d*) 1.481 hectare/cumec.

**Q. 2.** For a culturable command area of 1000 hectare with intensity of irrigation of 50%, the duty on field for a certain crop is 2000 hectare/cumec. What is the discharge required at head of water course with 25% losses of water ?

(*a*) 3/16 cumec (*b*) 1/4 cumec
(*c*) 1/3 cumec (*d*) 1/2 cumec.

**Q. 3.** Which one of the following is the correct sequence in the increasing order of the Froude number of flow assumed by the bed form of an alluvial stream with movable bed material ?

(*a*) Ripple — Plane bed — Dune — Plane bed — Antidune
(*b*) Dune — Ripple — Plane bed — Antidune — Plane bed
(*c*) Plane bed — Ripple — Dune — Plane bed — Antidune
(*d*) Plane bed — Ripple — Antidune — Dune — Plane bed.

**Q. 4.** What is the moisture depth available for evapotranspiration in root zone of 1 m depth soil, if dry weight of soil is 1.5 gm/cc, field capacity is 30% and permanent wilting point is 10% ?

(*a*) 450 mm (*b*) 300 mm (*c*) 200 mm (*d*) 150 mm.

**Q. 5.** What is the regime scour depth for a channel in soil with silt factor of unity and carrying 8 $m^2/s$ of discharge intensity in accordance with Lacey's regime theory ?

(*a*) 3.6 m (*b*) 4 m (*c*) 5.4 m (*d*) 25.6 m.

**Q. 6.** Which one of the following equations represents the downstream curve of the "Ogee" spillway (where $x$ and $y$ are the co-ordinates of the crest profile measured from the apex of the crest and $H$ is the design head) ?

(*a*) $x^{1.85} = 2H^{0.85}\, y$ (*b*) $x = 2H^{1.85}\, y^{0.85}$

(*c*) $x^{0.85} = 2H^{1.85}\, y$ (*d*) $x = 2H^{0.85}\, y^{1.85}$.

**Q. 7.*** For a saddle siphon, the maximum operative head is 6.25 m. The width and height of the throat of the siphon are 4 m and 2 m respectively. The coefficient of discharge is 0.90. How many units are required to pass a flood of 300 cumec ? (Take $g$ = 10 $m/s^2$)

(*a*) One (*b*) Two (*c*) Three (*d*) Four.

**Q. 8.*** Which one of the following defines aridity index ($AI$) ?

(*a*) $AI = \dfrac{PET - AET}{PET} \times 100$ (*b*) $AI = \dfrac{PET}{AET} \times 100$

(*c*) $AI = AI = \dfrac{AET}{PET} \times 100$ (*d*) $AI = AI = \dfrac{AET - PET}{AET} \times 100$

## ANSWERS

1. (*b*) 2. (*c*) 3. (*c*) 4. (*b*) 5. (*c*) 6. (*a*)
7. (*d*) 8. (*a*).

### Hints for Solving Sterred Questions of the Year 2004

**Q. 7.** From Eq. (21.13), discharge through a saddle syphon is given as ;

$$Q = C_d \cdot A \cdot \sqrt{2gH_1}$$

where $C_d = 0.9$

$A$ = Area of flow width × Ht. of Throat = 4 m × 2 m = 8 $m^2$ (given)

$g$ = 10 $m/s^2$ (given)

$H_1$ = 6.25 m (given)

$$\therefore \quad Q = 0.9 \times 8\ \sqrt{2 \times 10 \times 6.25} = \mathbf{80.5\ m^3/s.}$$

Total discharge = 300 cumec (given)

$$\text{No. of units of saddle syphon} = \frac{300}{80.5} = 3.73\ ;\ \text{Say } 4.$$

Hence, choice (*d*) is correct. **Ans.**

**Q. 8. Aridity Index,** as its name implies, is an indicator of the dryness (aridness) of a soil/crop, due to which the plants do not get their full required quantity of water upto the field capacity. The max. available moisture [*i.e.*, Field capacity *mc* – Wilting point *mc*], thus, does not become available to the crops, which reduces their actual evapotranspiration (*AET*) in comparison to the potential (*i.e.* maximum) evapotranspiration (*PET*). The deficiency in evapotranspiration [*i.e.* (*PET* – *AET*)] when expressed as the ratio with *PET* is, hence, defined as the ***Aridity Index.*** Thus,

$$\text{Aridity Index} = AI = \frac{PET - AET}{PET} \ (\%)$$

Thus, if *PET* is 80 mm and *AET* is 20 mm, then, naturally, the crops will be facing $\frac{80 - 20}{80}$ = 75% dry conditions, since the aridity index is 75%.

Hence, choice (*a*) is correct. **Ans.**

## Year 2005 (*Engineering services*)

**Q. 1.** Which one of the following statements is correct ?

A forebay in a hydel system is provided at the junction of :

(*a*) the power channel and the tail race channel
(*b*) the tail race channel and the penstock
(*c*) the penstock and the turbine
(*d*) the power channel and the penstock.

**Q. 2.** The base width of a soil gravity dam is 25 m. The material of the dam has a specific gravity of 2.56 and the dam is designed as an elementary profile ignoring uplift. What is the approximate allowable height of the dam ?

(*a*) 64 m (*b*) 40 m (*c*) 164 m (*d*) 80 m

**Q. 3.** Match List I (*Relative Position of Canal and Drainage Channel*) with List II (*Type of Cross Drainage Work*) and select the correct answer using the code given below the Lists :

| *List I* | *List II* |
|---|---|
| A. Canal taken above the drainage channel at its grade | 1. Cansal siphon |
| B. Drainage channel taken above the canal at its bed slope | 2. Drainage siphon |
| C. Canal taken below the drainage channel | 3. Aqueduct |
| D. Drainage channel taken below the canal | 4. Super passage |

| | A | B | C | D |
|---|---|---|---|---|
| (*a*) | 2 | 4 | 1 | 3 |
| (*b*) | 3 | 1 | 4 | 2 |
| (*c*) | 2 | 1 | 4 | 3 |
| (*d*) | 3 | 4 | 1 | 2 |

**Q. 4.** Match List I with List II and select the correct answer using the code given below the Lists :

| *List I* | *List II* |
|---|---|
| A. Storage | 1. Small storage |
| B. Pondage | 2. Ratio of total inflow to the total number of days power plant operates in a week |
| C. Pondage factor | 3. Large reservoirs |
| D. Daily load factor | 4. (Average plant load) × (Peak load) |
| | 5. Ratio of average load to peak load |

| | A | B | C | D |
|---|---|---|---|---|
| (*a*) | 3 | 2 | 1 | 5 |
| (*b*) | 5 | 1 | 2 | 4 |
| (*c*) | 3 | 1 | 2 | 5 |
| (*d*) | 5 | 2 | 1 | 4 |

**Q. 5.** Match List I (*Factor*) with List II (*Ratio*) and select the correct answer using the code given below the Lists :

| *List I* | *List II* |
|---|---|
| A. Load factor | 1. $\dfrac{\text{Sum of individual maximum demands}}{\text{Simultaneous maximum demand}}$ |
| B. Capacity factor | 2. $\dfrac{\text{Maximum demands}}{\text{Station capacity}}$ |
| C. Diversity factor | 3. $\dfrac{\text{Average load}}{\text{Maximum load}}$ |
| D. Plant use factor | 4. $\dfrac{\text{Energy produced}}{\text{Installed capacity} \times \text{time in hours}}$ |

| | A | B | C | D |
|---|---|---|---|---|
| (*a*) | 1 | 2 | 3 | 4 |
| (*b*) | 3 | 4 | 1 | 2 |
| (*c*) | 1 | 4 | 3 | 2 |
| (*d*) | 3 | 2 | 1 | 4 |

**Q. 6.** The channel section can be designed on the basis of Lacey's Theory. The steps are mentioned below :

1. Finding out the perimeter
2. Finding out the velocity
3. Calculation of the silt factor
4. Finding out the area

What is the correct sequence of the steps ?

(*a*) 4-2-3-1 (*b*) 3-1-4-2 (*c*) 4-1-3-2 (*d*) 3-2-4-1

**Q. 7.** In Lacey's regime theory, the velocity of flow is proportional to :

(*a*) $Qf^2$ (*b*) $Q/f^2$ (*c*) $(Qf^2)^{1/6}$ (*d*) $(Q/f^2)^{1/6}$

**Q. 8.** Which of the following are the purposes of a groyne as a river training structure ?

1. It contracts a river channel to improve its depth.
2. It protects the river bank.
3. It does not allow silt to deposit in the vicinity.
4. It trains the flow along a certain course.

Select the correct answer using the code given below :

(*a*) 1 and 2 (*b*) 1, 2 and 3 (*c*) 1 and 3 (*d*) 2, 3 and 4

**Q. 9.** Denehy's groyne is a special type of groyne which is :

(*a*) Pointing upstream (*b*) Pointing downstream
(*c*) Hockey type (*d*) T-headed

**Q. 10.** In a laminar boundary layer, the velocity distribution can be assumed to be given, in usual notations, as

$$\frac{u}{v} = \frac{y}{\delta}$$

Which one of the following is the correct expression for the displacement thickness $\delta^*$ for this boundary layer ?

(*a*) $\delta^* = d$ (*b*) $\delta^* = \delta/2$ (*c*) $\delta^* = \delta/4$ (*d*) $\delta^* = \delta/6$

**Q. 11.** The Froude number of a hydraulic jump is 5.5. The jump can be classified as a/an :

(*a*) Undular jump (*b*) Oscillating jump
(*c*) Weak jump (*d*) Steady jump

## ANSWERS

| | | | | | |
|---|---|---|---|---|---|
| 1. (*d*) | 2. (*b*) | 3. (*a*) | 4. (*c*) | 5. (*b*) | 6. (*d*) |
| 7. (*c*) | 8. (*d*) | 9. (*d*) | 10. (*b*) | 11. (*b*) | |

## Year 2006 (*Engineering Services*)

**Q. 1.** Which one of the following statements is *not* correct ?
(*a*) Storage and pondage can be obtained from flow duration curve.
(*b*) Primary or firm power corresponds to maximum streamflow condition.
(*c*) Secondary power is occasionally called surplus power.
(*d*) Often, flash boards are put on dams to augment the pondage at low flow durations.

**Q. 2.** In the selection of turbine by specific speed or head, which one of the following statements is *not* correct ?
(*a*) For specific speed 10-35, Kaplan turbines
(*b*) For specific speed 60-300, Francis turbines
(*c*) For head 50-150 m, Francis turbines
(*d*) For head above 300, Pelton wheel

**Q. 3.** Penman's equation is based on
(*a*) Energy budgeting only
(*b*) Energy budgeting and water budgeting
(*c*) Energy budgeting and mass transfer
(*d*) Water budgeting and mass transfer

**Q. 4.** The basic equation of flood routing through a reservoir can be modified for discrete successive intervals $\Delta t$ by which one of the following ?

(*a*) $\left[\frac{I_1 + I_2}{2}\right]\Delta t + \left[S_1 + \frac{O_1\Delta t}{2}\right] = \left[S_2 + \frac{O_2\Delta t}{2}\right]$

(*b*) $\left[\frac{I_1 + I_2}{2}\right]\Delta t + \left[S_1 - \frac{O_1\Delta t}{2}\right] = \left[S_2 + \frac{O_2\Delta t}{2}\right]$

(*c*) $\left[\frac{I_1 + I_2}{2}\right]\Delta t + \left[S_1 + \frac{O_1\Delta t}{2}\right] = \left[S_2 - \frac{O_2\Delta t}{2}\right]$

(*d*) $\left[\frac{I_1 + I_2}{2}\right]\Delta t + \left[S_1 - \frac{O_1\Delta t}{2}\right] = \left[S_2 - \frac{O_2\Delta t}{2}\right]$

**Q. 5.** Which type of fall can be generally used for a moderate discharge of 40-60 cumecs and low fall heights of 1 to 1.5 m ?
(*a*) Vertical drop fall (*b*) Ogee fall
(*c*) Glacis fall (*d*) Baffle fall

**Q. 6.** If the sensitivity of an irrigation module is 0.5, then what per cent variation in outlet discharge will be caused by 50 percent variation in canal water depth ?

(*a*) 100% (*b*) 50%

(*c*) 25% (*d*) 12.5%

**Q. 7.** Consider the following terms relating to irrigation requirements :

1. Consumptive irrigation requirement
2. Net irrigation requirement
3. Field irrigation requirement
4. Gross irrigation requirement

For a given set up, which one of the following is the correct relation ?

(*a*) 1 > 2 > 3 > 4 (*b*) 1 < 2 < 3 < 4

(*c*) (1 = 2) > 3 > 4 (*d*) 1 < (2 = 3) < 4

**Q. 8.** A drainage coefficient

(*a*) Decides the choice of the method of the drainage.

(*b*) Decides the kind of crop that can be grown on the sand.

(*c*) Is the depth of water that can be removed from the drainage area in unit time.

(*d*) Is the flow of water from the soil into the tile laterals per unit time.

**Q. 9.** Match List-I with List-II and select the correct answer using the code given below the lists ($s$ = bed slope, $q$ = discharge intensity, $Q$ = Discharge) :

| *List-I* | *List-II* |
|---|---|
| A. Mean velocity in a Lacey regime channel | 1. $s^{1/2}$ |
| B. Mean velocity in a lined channel | 2. $s^{1/3}$ |
| C. Normal scour depth in an alluvial channel | 3. $q^{2/3}$ |
| D. Wetted perimeter on a Lacey regime channel | 4. $q^{1/2}$ |
| | 5. $Q^{1/2}$ |

***Code :***

| | A | B | C | D |
|---|---|---|---|---|
| (*a*) | 2 | 5 | 3 | 1 |
| (*b*) | 3 | 1 | 4 | 5 |
| (*c*) | 2 | 1 | 3 | 5 |
| (*d*) | 3 | 5 | 4 | 1 |

**Q. 10.** What type of cross drainage work is provided when the canal runs below the drain, with FSL of canal well below the bed of the drain ?

(*a*) Aqueduct (*b*) Super passage

(*c*) Level crossing (*d*) Siphon aqueduct

**ANSWERS**

| | | | | | |
|---|---|---|---|---|---|
| 1. (*b*) | 2. (*a*) | 3. (*c*) | 4. (*b*) | 5. (*c*) | 6. (*c*) |
| 7. (*b*) | 8. (*c*) | 9. (*c*) | 10. (*b*) | | |

## Year 2007 (*Engineering Services*)

**Q. 1.** The flood plain of a river carries a discharge of 2000 m³/s. What are the values of the meander length and dominant flow width ?

(*a*) 160 m, 48 m (*b*) 180 m, 42 m

(*c*) 200 m, 38 m (*d*) 220 m, 36 m

**Q. 2.** What are the recommended setting options of an adjustable proportional module worked with an open flume type outlet ?

1. 3/10 2. 9/10 3. 1/2 4. 3/5

Select the correct answer using the code given below :

(*a*) 1 and 2 (*b*) 1 and 3 (*c*) 2 and 3 (*d*) 2 and 4

**Q. 3.** Which one of the following is the purpose of providing the downstream sheet pile in a barrage ?

(*a*) To control failure due to piping by high value of exit gradient
(*b*) To control failure due to scour
(*c*) To stop failure due to sliding
(*d*) To stop failure due to uplift pressure

**Q. 4.** Consider the following statements associated with groynes :

1. Hydraulic behaviour of a system of groynes is influenced by the characteristics of particles that constitute the littoral drift.
2. Groyne is constructed approximately parallel to shore.

Which of the statements given above is/are correct ?

(*a*) 1 only (*b*) 2 only
(*c*) Both 1 and 2 (*d*) Neither 1 nor 2.

**Q. 5.** Which one of the following statements is *not* correct ?

A control section in an open channel is the site
(*a*) where the flow quantity can be controlled.
(*b*) at which flow is known to be critical.
(*c*) where the discharge can be measured.
(*d*) where the specific energy is determined.

**Q. 6.** A hydraulic jump occurs at the toe of a spillway. The depth before jump is 0.2 m. The sequent depth is 3.2 m. What is the energy dissipated in m (approximate) ?

(*a*) 27 (*b*) 10.5 (*c*) 15 (*d*) 42

**Q. 7*.** During a certain week a power plant turns out 8,400,000 kWhr and the peak load during the week is 100,000 kW. What is the load factor during the week?

(*a*) 40% (*b*) 45% (*c*) 50% (*d*) 60%

**Q. 8*.** A standard, ground-based evaporation pan, corresponding to Indian Standards, is installed at the banks of a surface reservoir. The water surface area on a particular day is 100 hectares. The recorded evaporation loss from the pan, on a certain day, is nearly 4.0 cm. What is the anticipated evaporation loss from the reservoir for that day ?

(*a*) (1.8 to 2) × $10^4$ $m^3$ per day (*b*) (2.5 to 2.75) × $10^4$ $m^3$ per day
(*c*) (3.0 to 3.25) × $10^4$ $m^3$ per day (*d*) (3.8 to 4.05) × $10^4$ $m^3$ per day

**Q. 9.** If the discharge required for different crops is 0.4 cumecs in the field and the capacity factor and time factors are 0.8 and 0.5 respectively, then what is the design discharge for the distributary at its head ?

(*a*) 0.80 Cumec (*b*) 0.16 Cumec
(*c*) 1.0 Cumec (*d*) 1.24 Cumecs

**Q. 10.** Consider the following statements :

An aqueduct is a cross drainage work in which

1. a canal is carried over the drainage channel.
2. a drainage channel is carried over the canal.
3. both drainage channel and canal are at the same level.

Which of the statements given above is/are correct ?

(*a*) 1 only (*b*) 1 and 2 only

(*c*) 2 and 3 only (*d*) 1, 2 and 3

**Q. 11*.** A soil sample has an exchangeable sodium percentage of 16%, its electrical conductivity is 32 milli-Mhos/cm and pH of 9.5. How is the soil classified?

(*a*) Saline soil (*b*) Saline-alkaline soil

(*c*) Alkaline soil (*d*) None of the above

**Q. 12*.** The sequent depth ratio of a hydraulic jump in a rectangular channel is 16.48. What is the Froude number (approximate) at the beginning of the jump ?

(*a*) 9.0 (*b*) 12.0 (*c*) 5.0 (*d*) 8.0

**Q. 13.** Groynes are adopted for river bank protection works. When it is placed inclined downstream in the direction of flow in the river, it is designated as which one of the following ?

(*a*) Repelling groyne

(*b*) Attracting groyne

(*c*) Neither repelling nor attracting groyne

(*d*) Fixed groyne

## ANSWERS

| | | | | | |
|---|---|---|---|---|---|
| 1. (*d*) | 2. (*d*) | 3. (*a*) | 4. (*d*) | 5. (*c*) | 6. (*b*) |
| 7. (*c*) | 8. (*c*). | 9. (*c*) | 10. (*a*) | 11. (*c*) | 12. (*b*) |
| 13. (*b*) | | | | | |

## Hints for Solving Starred Questions of the Year 2007

**Q. 7.** Load factor = $\dfrac{\text{Average load}}{\text{Peak load}}$

Average load in the given week = $\dfrac{\text{Total KWhr produced in the week}}{\text{No. of hr in a week of 7 days}}$

$$= \frac{84{,}00{,}000 \text{ kWhr}}{(7 \times 24) \text{ hr}} = 50{,}000 \text{ KW}$$

Peak load = 1,00,000 kW (given)

Load Factor $= \dfrac{50{,}000 \text{ kW}}{1{,}00{,}000 \text{ kW}} = \mathbf{0.5}$

Hence, choice (*c*) is correct **Ans.**

**Q. 8.** Pan coefficient for ISI Pan [article 7.34.2.4(*d*)] = 0.8

∴ Evoporation from Reservoir = Pan coefficient × Evaporation from Pan

= 0.8 × 4 cm = 3.2 cm

Area of reservoir = 100 hecteres = $100 \times 10^4$ m$^2$.

Vol. of water evaporded from reservoir = $(100 \times 10^4)$ m$^2$ × 3.2 cm = $3.2 \times 10^4$m$^3$

Hence, choice (*c*) is correct. **Ans.**

**Q. 11.** Pl see Table 1.3.

**Q 12.** From Eq. (1.7), we have

$$y_2 = \frac{y_1}{2}\left[-1 + \sqrt{1 + 8F_1^2}\right]$$

where $F_1$ is the incoming froude number *i.e.*, the Froude number at the beginning of the jump

$$\therefore \quad \frac{y_2}{y_1} = \frac{1}{2}\left[-1+\sqrt{1+8F_1^2}\right]$$

or $$16.48 = \frac{1}{2}\left[-1+\sqrt{1+8F_1^2}\right]$$

or $$\left[-1+\sqrt{1+8F_1^2}\right] = 32.96$$

$$\sqrt{1+8F_1^2} = 33.96$$

or $$1 + 8{F_1}^2 = 1153.28$$

or $$F_1 = 12.0$$

Hence, choice (*b*) is correct. **Ans.**

**Year 2008 (*Engineering Services*)**

**Q. 1.** Match List-I with List-II and select the correct answer using the code given below the lists :

| *List-I (Type of Power House)* | *List-II (Feature)* |
|---|---|
| A. Tidal Plant | 1. Utilizes reversible turbines and improves utilization factor |
| B. Storage Plant | 2. Utilizes current flow in stream and has pondage to improve load factor |
| C. Pumped Storage Plant | 3. Head on turbine varies with time cyclically |
| D. Run-of-river Plant | 4. Utilizes annually average flow with storage |

***Code :***

| | A | B | C | D |
|---|---|---|---|---|
| (*a*) | 3 | 4 | 1 | 2 |
| (*b*) | 3 | 2 | 1 | 4 |
| (*c*) | 1 | 2 | 3 | 4 |
| (*d*) | 1 | 4 | 3 | 2 |

**Q. 2.** The permissible tractive force in an erodible channel depends upon which of the following ?

1. Angle of repose of the material
2. Particle size
3. Sediment content of water
4. Wetted perimeter of channel

Select the correct answer using the code given below :

(*a*) 1, 2 and 4 (*b*) 1, 2 and 3

(*c*) 1 and 3 only (*d*) 2 and 4 only

**Q. 3.** Which one of the following is correct ?

Semi-module outlets are those outlets in which

(*a*) discharge gets affected by the change in water level of field channel

(*b*) discharge gets affected by the change in water level of the distributing channel but not with the change in water level of field channel

(*c*) discharge is independent of water levels in the distributing channel and the field channel

(*d*) none of these

## ANSWERS

1. (*a*) 2. (*a*) 3. (*b*)

## Year 2009 (*Engineering Services*)

**Q. 1.** Consider the following chemical emulsions :

1. Methyl alcohol
2. Cetyl alcohol
3. Stearyl alcohol
4. Kerosene

Which of the above chemical emulsions is/are used to minimize the loss of water through the process of evaporation ?

(*a*) 1 only (*b*) 1 and 4 (*c*) 2 and 4 (*d*) 2 and 3

**Q. 2.** Under which one of the following categories is the river Ganga classified in the reach through UP and Bihar ?

(*a*) Straight river (*b*) Meandering river
(*c*) Braided river (*d*) Deltaic river

**Q. 3.** Which of the following categories best describes the Hirakud reservoir ?

(*a*) Reservoir for irrigation and power
(*b*) Reservoir for flood control, power and irrigation
(*c*) Reservoir for irrigation and water supply
(*d*) Reservoir for recreation and fishery

**Q. 4.** During a particular stage of the growth of a crop, consumptive use of water is 2.2 mm/day. If the amount of water available in the soil is 25% of 80 mm depth of water, what is the frequency of irrigation ?

(*a*) 9 days (*b*) 13 days (*c*) 21 days (*d*) 25 days

**Q. 5.** In a canal irrigation project, 76% of the culturable command area (CCA) remained without water during Kharif season ; and 58% of CCA remained without water during Rabi season in a particular year. Rest of the areas got irrigated in each crop respectively. What is the intensity of irrigation for the project in that year ?

(*a*) 134% (*b*) 76% (*c*) 66% (*d*) 58%

**Q. 6.** What is the critical combination of vertical and horizontal earthquake accelerations to be considered for checking the stability of a gravity dam in reservoir full condition ?

(*a*) Vertically upward and horizontally downstream
(*b*) Vertically upward and horizontally upstream
(*c*) Vertically downward and horizontally upstream
(*d*) Vertically downward and horizontally downstream

**Q. 7*.** What is the height of wave which is likely to be generated by a wind of 80 km/hr in a reservoir having a fetch of 50 km

(*a*) 0.5 m (*b*) 1.0 m (*c*) 2.0 m (*d*) 3.0 m

**Q. 8.** Consider the following statements :

1. Giving equal weightages to horizontal and vertical creeps for design of weir foundations is one of the drawbacks of Kennedy's theory
2. Khosla's theory of design of foundations for weirs is based on potential theory.
3. Piping problem can be reduced by increasing the length of floor.
4. In Lane's weighted creep theory, horizontal creep is given less weightage compared to vertical creep.

Which of the above statements is/are correct ?

(*a*) 1 (*b*) 2, 3 and 4 (*c*) 2 and 4 only (*d*) 3 and 4 only

**Q. 9.** Consider the following statements related to undersluices provided in diversion weirs on permeable foundations :

1. They are fully gate-controlled and have crest at the same level as the weir crest when no silt excluders are provided.
2. They scour the silt deposited on the river bed in the pockets upstream of the canal head regulator.
3. It is not necessary to provide end pile line on the downstream end of the undersluice floor.
4. The discharge capacity of the undersluice is 10-15% of the maximum flood or two times the maximum discharge of the off taking canal or maximum winter discharge, whichever is the highest.

Which of the above statements is/are correct ?

(*a*) 1 (*b*) 2 and 4 only (*c*) 2, 3 and 4 (*d*) 3 and 4 only

## ANSWERS

| 1. (*d*) | 2. (*b*) | 3. (*b*) | 4. (*a*) | 5. (*c*) | 6. (*d*) |
|---|---|---|---|---|---|
| 7. (*c*) | 8. (*b*). | 9. (*b*) | | | |

## Hints for Solving Sterred Questions

**Q. 7.** Use Eq. (19.12) as :

$$h_w = 0.032\ \sqrt{v\,.\,F}$$

$$= 0.032\ \sqrt{80 \times 50} = \mathbf{2.02\ m}$$

Hence, the choice (*c*) is correct **Ans.**

## Year 2010 (*Engineering Services*)

**Q. 1.** Garrets' diagrams are used to :

(*a*) separate base flow from total runoff

(*b*) correct inconsistency in rainfall data

(*c*) determine reservoir capacity

(*d*) design channels.

**Directions :** *Each of the following given items consists of two statements one labelled as the 'Assertion A' and the other as 'Reason (R)'. You are to examine these two statements carefully and select the answers to these items using the codes given below :*

**Codes :**

(*a*) *Both A and R are individually true and R is the correct explanation of A.*

(*b*) *Both A and R are individually true but R is not the correct explanation of A.*

(*c*) *A is true but R is false.*

(*d*) *A is false but R is true.*

**Q. 2.** *Assertion (A) :* An economical channel section gives maximum discharge for a given cross-sectional area.

*Reason (R)* : An economical channel section has smooth surface for reduced friction.

**Q. 3.** *Assertion (A) :* In the border strip method of irrigation, the size of the strip depends on soil characteristics, slope of the land and discharge.

*Reason (R)* : Border strip method is a controlled type of subsurface irrigation method.

**Q. 4.** *Assertion (A)* : A seepage passing through the body of an earthen dam affects the weight of the dam.

*Reason (R)* : The specific weight of the submerged soil is not dependent on the porosity of soil.

**Q. 5.** *Assertion (A)* : Lining a canal is always beneficial and economical.

*Reason (R)* : The seepage losses are greatly reduced and extra water is available for irrigation.

**Q. 6*.** A homogeneous earth dam is 40 m high with 3 m as free-board and a 30 m horizontal filter at the downstream end. The flow net drawn in the dam section consists of 5 flow channels and 15 potential drops. If the dam is 500 m long and the permeability of the material of the dam is uniform $3 \times 10^{-5}$ m/s, the total discharge permeating through the body of the dam will be

(*a*) 0.0004 $m^3/s$ (*b*) 0.0036 $m^3/s$

(*c*) 0.0039 $m^3/s$ (*d*) 0.20 $m^3/s$.

**Q. 7*.** The discharge required for Rabi and Kharif crops are 0.4 $m^3/s$ and 0.3 $m^3/s$, respectively. The capacity and time factor are 0.8 and 0.5, respectively at each session. The design discharge of the distributary at its head is

(*a*) 0.8 $m^3/s$ (*b*) 0.16 $m^3/s$ (*c*) 1.0 $m^3/s$ (*d*) 1.24 $m^3/s$.

**Q. 8*.** The maximum height of a low gravity dam of elementary profile mode of concrete of relative density 2.5 and safe allowable stress of foundation material 3.87 MPa without considering uplift force is about ;

(*a*) 113 m (*b*) 217 m (*c*) 279 m (*d*) 325 m

**Q. 9*.** The wave height, in metres, generated on the surface of the reservoir, having a fetch length $F$ = 30 km, due to wind blowing on the surface of the reservoir at a velocity of 30 km/hr is :

(*a*) 0.26 m (*b*) 0.96 m (*c*) 0.52 m (*d*) 1.2 m.

**Q. 10.** A ski-jump bucket is generally used as an energy dissipator when the tail water

(*a*) is greater than 1.1 times the required conjugate depth for the formation of hydraulic jump ; and the river bed rock is 'good'

(*b*) depth is lesser than the depth required for the jump formation ; and the bed of the river channel is composed of 'sound' rock

(*c*) depth is equal to the depth required for the jump formation, and the river bed rock is 'good'.

(*d*) depth is 1.3 times the depth required for the jump formation and the river bed is composed of 'weak' rock.

**Q. 11.** A submerged pipe outlet is an example of :

(*a*) Semi modular outlet (*b*) Non-modular outlet

(*c*) Rigid module (*d*) Adjustable proportional module.

**Q. 12.** Heavy scour at the head and shank of guide banks can lead to undermining of the stone pitching and consequent failure of the guide bank. This situation is avoided by providing

(*a*) spurs (*b*) vertical cutoff

(*c*) marginal bunds (*d*) launching apron.

**Q. 13.** The two sides of a trapezoidal channel are lined with cement plaster, $n_w = 0.012$, while the earthen bed has $n_b = 0.025$, the area is divided into three parts as in the figure given below (Fig. 28.1) :

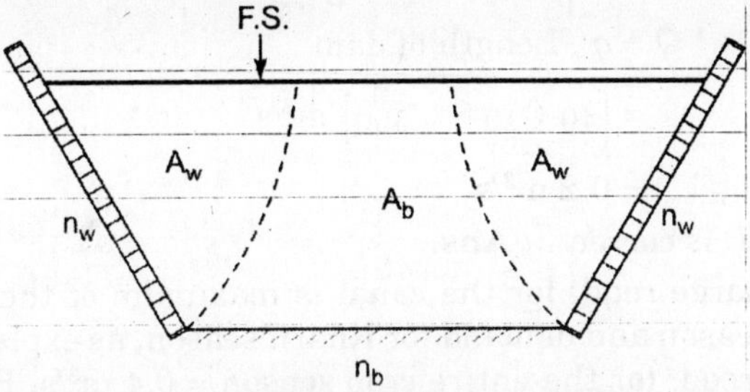


**Fig. 28.1**

Consider the following assumption to find the discharge through this section.

1. The average velocity is the same for all the 3 parts.
2. Discharge is the same in all the 3 parts.
3. Bed slope is the same for all the 3 parts.
4. Hydraulic mean radius is the same for all the 3 parts.

Which of the above assumptions are correct ?

(*a*) 1 and 3 (*b*) 1 and 4 (*c*) 2 and 3 (*d*) 2 and 4.

**Q. 14.** Consider the following statements :

The function of a launching apron at the end of downstream impervious apron of a weir is :

1. To protect the downstream pile from the scour holes progressing in the upstream direction
2. When the scour holes are formed, to provide for the stones of the falling apron to settle down in the holes and cover them.
3. To provide relief of the uplift pressure at the downstream end.
4. To provide extra length for proper formation of hydraulic jump.

Which of the above statements is/are correct ?

(*a*) 1, 2, 3 and 4 (*b*) 3 only (*c*) 2 only (*d*) 1 and 2 only.

**Q. 15*.** Consider the following statements :

1. High water training is undertaken to protect against damage due to floods.
2. Low water training is undertaken to provide sufficient depth for navigation
3. Mean water training is undertaken to provide efficient disposal of sediment.

Which of above statements is/are correct ?

(*a*) 1 and 2 only (*b*) 1, 2 and 3 (*c*) 2 and 3 only (*d*) 2 only.

## Answers

| | | | | |
|---|---|---|---|---|
| **1.** (*d*) | **2.** (*c*) | **3.** (*c*) | **4.** (*c*) | **5.** (*a*) |
| **6.** (*d*) | **7.** (*c*) | **8.** (*a*) | **9.** (*d*) | **10.** (*b*) |
| **11.** (*b*) | **12.** (*d*) | **13.** (*a*) | **14.** (*d*) | **15.** (*c*) |

## Hints for Solving Starred Questions

**Q. 6.** From Eq. (20.16), we have

$$\therefore \quad q = K\,H \,.\, \frac{N_f}{N_d}$$

$$\therefore \quad q = 3 \times 10^{-5} \frac{\text{m}}{\text{s}} \times 40 \text{ m} \times \frac{5}{15}$$

or $$q = 3 \times 10^{-5} \times 40 \times \frac{1}{3} \text{ m}^2/\text{s} = 40 \times 10^{-5} \text{ m}^2/\text{s}$$

$$Q = q \text{ . Length of dam}$$

$$= \left[40 \times 10^{-5} \times 500\right] \text{ m}^3/\text{s}$$

$$= 0.2 \text{ m}^3/\text{s}$$

Hence, choice (*d*) is correct. **Ans.**

**Q. 7.** The discharge reqd. for the canal is maximum of the two demands, *i.e.* demand for Rabi season and demand for Kharif season, as explained in article 3.7. Hence, discharge reqd. for the entire crop season = 0.4 $\text{m}^3/\text{s}$. But since the canal runs for fewer days than the crop period, the actual design discharge at outlet

$$= \frac{0.4 \text{ m}^3/\text{s}}{\text{Time factor}} = \frac{0.4}{0.5} = 0.8 \text{ m}^3/\text{s}.$$

Moreover, since the canal does not flow always to its designed capacity, the peak discharge reqd. in the canal to get mean discharge during entire crop period equal to 0.8 $\text{m}^3/\text{s}$, we have

$$\text{Peak discharge reqd.} \quad = \frac{0.8}{\text{Capacity factor}} = \frac{0.8}{0.8} = 1 \text{ m}^3/\text{s}.$$

Hence, choice (*c*) is correct. **Ans.**

**Q. 8.** The max. height of a triangular section low gravity dam is given by Eq. (19.27) as :

$$H_{max} = \frac{f}{\gamma_w (S_c + 1)} \quad \text{(when uplift is neglected)}$$

where $f$ = allowable stress in dam material
= 3.87 MPa = $3.87 \times 10^6$ Pa
= $3.87 \times 10^6$ N/m²
= $3.87 \times 10^3$ kN/m²

$\gamma_w$ = unit wt of water = 9.81 kN/m³

$S_c$ = sp. gravity of dam material = 2.5

$$\therefore \quad H_{max} = \frac{3.87 \times 10^3 \dfrac{\text{kN}}{\text{m}^2}}{9.81 \dfrac{\text{kN}}{\text{m}^3}[2.5 + 1]} = 112.7 \text{ m}$$

Hence, choice (*a*) is correct. **Ans.**

**Q. 9.** The wave height of the waves generated by blowing winds on the surface of a reservoir is given by Eq. (19.11) as :

$$h_w = 0.032\sqrt{V.F} + 0.763 - 0.271 (F)^{1/4} \text{ for } F < 32 \text{ km}$$

Here $F$ = 30 km

$V$ = wind velocity in km/hr = 30 km/hr (given)

$\therefore \quad h_w$ = ht. of wave from crest to trough

$\therefore \quad h_w = 0.032\sqrt{30 \times 30} + 0.763 - 0.271\,(30)^{1/4}$

$= 0.96 + 0.763 - 0.634 = \mathbf{1.09\ m.}$

Since the nearest given choice of 1.2 m is given at (*d*), it can be considered as the correct choice. **Ans.**

**Q. 15.** Pl. refer article 9.5.2.

### Year 2011 (*Engineering Services*)

**Q. 1.** Consider the following zones :

1. Saturation zone
2. Capillary zone
3. Intermediate zone
4. Soil water zone

Which of these do not relate to the zone of aeration in the soil profile ?

(*a*) 1 and 2 (*b*) 2 and 3 (*c*) 4 only (*d*) 1 only.

**Q. 2.** A constant centre (in plan view) arch dam is best suited for

(*a*) A V-shaped gorge (*b*) A U-shaped gorge

(*c*) Both U and V-shaped gorges (*d*) Multi-peaked gorges.

**Q. 3*.** An identified source of irrigation water has ion concentration of $Na^+$, $Ca^{++}$, and $Mg^+$ as 20, 10 and 8 milliequivalent per litre, respectively. The SAR of this water is approximately.

(*a*) 2.06 (*b*) 6.67 (*c*) 2.67 (*d*) zero.

**Q. 4*.** Two different channels, *M* and *N*, in two different sites are designed based on Lacey's theory, to carry same quantum of discharge. But the bed material of *M* is found to be finer than that of *N*.

(*a*) channel *M* will have steeper longitudinal slope

(*b*) channel *N* will have steeper longitudinal slope

(*c*) channels *M* and *N* can have same longitudinal slopes

(*d*) Silting is more in *M* than in *N*.

**Q. 5*.** In alluvial channels carrying clear water, the ratio of maximum tractive shear stress on the slides and that on the channel bed is approximately

(*a*) 0.5 (*b*) 1.76

(*c*) 0.76 (*d*) 1.5.

**Q. 6.** Match List-I with List-II and select the correct answer using the codes given below the lists :

| *List-I* | *List-II* |
|---|---|
| A Deflector | 1. Low-height wall across canal downstream of the fall |
| B Biff wall | 2. Vertical end wall with horizontal projection |
| C Cistern | 3. Short wall at downstream end of the fall |
| D Baffle wall | 4. Pond Level below the fall. |

| **Code :** | A | B | C | D |
|---|---|---|---|---|
| (*a*) | 1 | 2 | 4 | 3 |
| (*b*) | 3 | 2 | 4 | 1 |
| (*c*) | 1 | 4 | 2 | 3 |
| (*d*) | 3 | 4 | 2 | 1 |

**Q. 7.** Consider the following statements :

Mitra's hyperbolic transition design is based on the principle that

1. Flow depth in the canal, as well as the discharge is constant.
2. Width of the canal varies along with the discharge.
3. Rate of change of velocity per unit length of transition is constant throughout the length of the transition.

Which of these statements are correct ?

(*a*) 1, 2 and 3 (*b*) 1 and 2 only (*c*) 2 and 3 only (*d*) 1 and 3 only.

**Q. 8.** Which one of the groyne arrangements represent an attracting groyne' ?

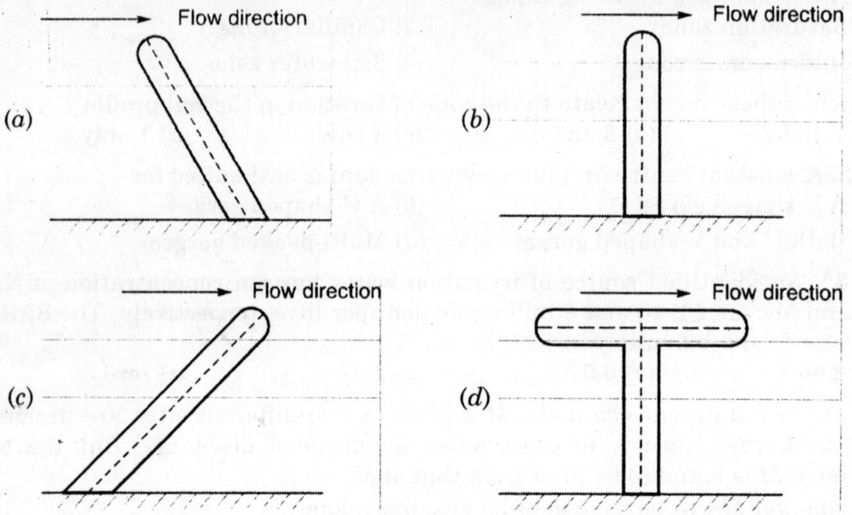


Fig. 28.2

**Q. 9.** Consider the following statements :

1. In a super passage, the drain runs over the canal
2. In a siphon, the drain runs below the canal
3. In a siphon aqueduct type II, the canal banks are made of RCC walls

Which of these statements is/are correct ?

(*a*) 1 only (*b*) 1 and 2 only (*c*) 2 and 3 only (*d*) 1 and 3 only.

**Q. 10.** A depth-discharge relationship of the canal section is maintained at a notch fall because the sill of the notches is :

(*a*) Level with downstream canal bed (*b*) Below the upstream canal bed

(*c*) Level with upstream canal bed (*d*) Above the upstream canal bed.

**Q. 11.** Flexibility of an outlet may be defined as the ratio of the rate of change of:

(*a*) Outlet discharge to the rate of change of water level of the parent channel

(*b*) Outlet discharge to the rate of change of the discharge of the parent channel

(*c*) Parent channel discharge to the rate of the change of the outlet discharge

(*d*) Parent channel water level to the rate of change of the outlet discharge.

**Q. 12.** Consider the following statements :

The general depth of scour calculated by Lacey's formula in a river represents the depth below the :

1. Maximum flood level in the river

2. Maximum flow water level in the river
3. Normal flow water level in the river
4. Existing river bed

Which of these statements is/are correct ?

(*a*) 1, 2, 3 and 4 (*b*) 1, 2, and 3 only
(*c*) 2, 3 and 4 only (*d*) 1 only.

**Directions.** *Each of the following items consists of two statements, one labelled as the 'Assertion (A)' and the other as 'Reason (R)'. You are to examine these two statements carefully and select the answers to these items using the codes given below :*

**Codes :**

(*a*) *Both A and R are individually true and R is the correct explanation of A*
(*b*) *Both A and R are individually true but R is not the correct explanation of A*
(*c*) *A is true but R is false*
(*d*) *A is false but R is true.*

**Q. 13.** *Assertion A* : In the case of water power plants, it is advisable to provide the surge tank as close to the turbine unit as possible.

*Reason (R)* : Purpose of surge tank is to provide the intended protection for the portion of the pen stock which lies on the upstream of it.

**Q. 14.** *Assertion A* : The duty of water decreases as the point of its measurement moves away from the field of application.

*Reason (R)* : Duty depends on soil characteristics.

**Q. 15*.** Best side slope for most economical trapezoidal section in open channel flow, wherein side slopes are defined by : $X$ horizontal to 1 vertical, is when $X$ equals

(*a*) 0.404 (*b*) 0.500 (*c*) 0.577 (*d*) 0.673.

**Answers**

| | | | | |
|---|---|---|---|---|
| **1.** (*d*) | **2.** (*a*) | **3.** (*b*) | **4.** (*b*) | **5.** (*c*) |
| **6.** (*a*) | **7.** (*d*) | **8.** (*c*) | **9.** (*d*) | **10.** (*d*) |
| **11.** (*b*) | **12.** (*b*) | **13.** (*a*) | **14.** (*c*) | **15.** (*c*) |

**Hints for Solving Starred Questions of the Year 2011**

**Q. 3.** From Eq. (1.4), we have

$$\text{SAR} = \frac{Na^+}{\sqrt{\dfrac{Ca^{++} + Mg^{++}}{2}}} = \frac{20}{\sqrt{\dfrac{10+8}{2}}} = \frac{20}{3} = 6.67.$$

Hence, choice (*b*) is correct. **Ans.**

**Q. 4.** Given that :

$d_{(M)} < d_{(N)}$ (where $d$ is dia of bed material, and $M$ and $N$ are channels)

$\therefore$ $f_{(M)} < f_{(N)}$ (where $f$ is silt factor $= 1.76\sqrt{d_{mm}}$)

But $S \propto f^{5/3}$ $\left(\text{given by Eq. 4.28 as } S = \dfrac{f^{5/3}}{3340\, Q^{1/6}}\right)$

$\therefore$ $S_{(M)} < S_{(N)}$ or $S_{(N)} > S_{(M)}$

Since slope for channel $N$ is more, channel $N$ will have more (*i.e.*, steeper slope). Hence, choice (*b*) is correct.

**Q. 5.** Max. Tractive shear stress on slides (*i.e.* slops) = $0.75\ \gamma_w . y . S$

And, Max Tractive shear stress on bed = $0.97\ \gamma_w . y . S$

$$\therefore \quad \frac{\text{Tractive shear on slope}}{\text{Tractive shear on bed}} = \frac{0.75\ \gamma_w \cdot y . S}{0.97\ \gamma_w \cdot y . S} = 0.76$$

Hence, choice (*c*) is correct. **Ans.**

**Q. 15.** Pl. refer Solved Example 4.6, wherein we have found that for a trapezoidal channel having side slopes $m : 1\ (H : V)$, the most economical section is obtained when $m = \frac{1}{\sqrt{3}}$. In this question, $m$ is represented by $X$. Hence $X = \frac{1}{\sqrt{3}}$ = **0.577.**

Hence choice (*c*) is correct. **Ans.**

## Year 2012 (*Engineering Services*)

**Q. 1.** The local scour depth in front of a semicircular shaped rectangular pier having width equal to W aligned parallel to the flow below the surrounding bed is

(*a*) 2.0 W (*b*) 1.5 W (*c*) 1.2 W (*d*) 1.0 W.

**Q. 2*.** A discharge of 72 m³/s is to be allowed through siphon spillways of 2 m width and 75 cm depth with working head of 8 m. The number of spillways to be provided will be (Take coefficient of discharge for the spillways = 0.64)

(*a*) 2 (*b*) 4 (*c*) 6 (*d*) 8.

**Directions :** *Each of the following items consists of two statements, one labelled as the 'Statement (I)' and the other as 'Statement (II)'. You are to examine these two statements carefully and select the answers to these items using the codes give below:*

**Codes :**

(*a*) Both Statement (I) and Statement (II) are individually true and statement (II) is the correct explanation of Statement (I).

(*b*) Both Statement (I) and Statement (II) are individually true but Statement (II) is *not* the correct explanation of Statement (I).

(*c*) Statement (I) is true but Statement (II) is false.

(*d*) Statement (I) is false but Statement (II) is true.

**Q. 3.** *Statement (I)* : In a flownet, each field must be a (curvilinear) square.

*Statement (II)* : Each flow–channel in a flownet has the same rate of flow.

**Q. 4.** *Statement (I)* : The best hydraulic section always has the minimum excavation.

*Statement (II)* : The best hydraulic section gives the minimum area for a given discharge.

### Answers

**1.** (*c*) **2.** (*c*) **3.** (*b*) **4.** (*a*)

### Hints for Solving Starred Questions of the Year 2012

**Q. 2.** Discharge through a saddle siphon is given as

$$Q = C_d \,.\, A\sqrt{2gH_1} \qquad \text{...(Eq. 21.13)}$$

$$\therefore \qquad 72\ \frac{\text{m}^3}{\text{s}} = 0.64\ .\ A \times \sqrt{2 \times 9.81 \times 8}$$

or $A = 8.98 \text{ m}^2$

But $A$ = (No. of siphon units, say $x$) × (Width × Height of Throat)

∴ $8.98 \text{ m}^2 = x \times 2 \text{ m} \times 0.75 \text{ m}$

or $$x = \frac{8.98}{2 \times 0.75} = \mathbf{6}$$

Hence choice (*c*) is correct. **Ans.**

## Year 2013 (*Engineering Services*)

**Q. 1.*** A rectangular channel carries a uniform flow with a Froude number of 2.83. The ratio of critical depth to normal depth of this flow is :

(*a*) 1.68 (*b*) 2.83

(*c*) 2.00 (*d*) 4.75

**Q. 2.*** A rice crop is to be irrigated in a field covering an area of 2400 ha, the duty and base period of rice are given as 860 ha/ cumec and 120 days respectively. The volume of water required in the field is nearly :

(*a*) 500 ha.m (*b*) 1400 ha.m

(*c*) 2000 ha.m (*d*) 2880 ha.m

**Q. 3.** In an irrigation project, in a certain year, 70% and 46% of the culturable command area in Kharif and Rabi, respectively, remained without water and rest of the area got irrigation water. The intensity of irrigation in that year for the project was :

(*a*) 116% (*b*) 84%

(*c*) 42% (*d*) 58%

**Q. 4.*** The gross command area for a distributary is 5000 hectares, 80% of which is culturable irrigable. The intensity of irrigation for Rabi is 50%; and for Kharif is 30%. The average duty at the head of the distributary is 2000 hectares/ cumec for Rabi season and 900 hectares/cumec for Kharif season. The discharge required at the head of the distributary from average demand considerations is :

(*a*) 1.00 cumec (*b*) 1.33 cumecs

(*c*) 2.33 cumecs (*d*) 3.33 cumecs

**Q. 5.** Consider the following statements :

1. A regime channel will have side slopes of value 0.5 H : 1 V.
2. Lacey's regime formula is applicable to regime channels with sediment concentration more than 5000 ppm.
3. For a Lacey regime channel, the Manning roughness coefficient is estimated by Strickler's formula.
4. The mean velocity in a Lacey channel is proportional to $R^{2/3}$ where $R$ = hydraulic radius.

Which of these statements are correct?

(*a*) 1 and 2 (*b*) 1 and 4

(*c*) 2 and 3 (*d*) 3 and 4

**Q. 6.*** The minimum size of gravel that will not move in the bed of a wide rectangular channel of depth 0.8 m and longitudinal slope 0.0041 is :

(*a*) 11 mm (*b*) 23 mm

(*c*) 36 mm (*d*) 47 mm

**Q. 7.** The flip bucket energy dissipator for a spillway is suitable where :

1. The tail water depth is low.
2. The rock on the downstream is fragile and is erodible.
3. The rock on the downstream is good and non-erodible.

(*a*) 1, 2 and 3 (*b*) 1 and 2 only
(*c*) 2 and 3 only (*d*) 1 and 3 only

**Q. 8.** Consider the following statements :

1. Groynes are constructed projecting from the bank into the river at the bank to be protected from flood.
2. Repelling groyne projects downstream into the river-flow from the point of its origin from the bank.
3. Attracting groyne projects upstream into the river from the point of its origin from the bank.
4. Perpendicular groyne projects normal to the river-flow from the point of its origin from the bank.

Which of these statements are correct?

(*a*) 1 and 4 (*b*) 1 and 2
(*c*) 2 and 3 (*d*) 3 and 4

**Directions :** *Each of the next given items consists of two statements, one labeled as the 'Statement (I)' and the other as 'Statement (II)'. You are to examine these two statements carefully and select the answers to these items using the codes given below :*

**Codes :**

(*a*) *Both Statement (I) and Statement (II) are individually true and Statement (II) is the correct explanation of Statement (I).*

(*b*) *Both Statement (I) and Statement (II) are individually true but Statement (II) is* **Not** *the correct explanation of Statement (I).*

(*c*) *Statement (I) is true but Statement (II) is false.*

(*d*) *Statement (I) is false but Statement (II) is true.*

**Q. 9.** *Statement (I)* : A rectangular bridge pier experience lateral force during flood flow.

*Statement (II)* : The bridge pier sheds vortices alternately from its either side into the flood flow.

**Q. 10.** *Statement (I)* : An economical channel section gives maximum discharge for a given cross-sectional area.

*Statement (II)* : An economical channel section has smooth surface with reduced friction.

**Q. 11.** *Statement (I)* : In the case of an open channel flow, for a hydraulically efficient section, the hydraulic radius is equal to 50% of depth of flow in the channel.

*Statement (II)* : Wetted perimeter is minimum for a hydraulically efficient section of an open channel.

## ANSWERS

| | | | | |
|---|---|---|---|---|
| **1.** (*c*) | **2.** (*d*) | **3.** (*b*) | **4.** (*b*) | **5.** (*b*) |
| **6.** (*d*) | **7.** (*d*) | **8.** (*a*) | **9.** (*c*) | **10.** (*c*) |
| **11.** (*a*) | | | | |

## Hints for Solving Starred Questions of the year 2013

**Q. 1.** $$F = \frac{V}{\sqrt{gy}} = \frac{q}{\sqrt{gy^3}}$$

$\therefore$ $$2.83 = \frac{q}{\sqrt{g\,y^3}} \qquad \text{or} \qquad q^2 = 8.(gy^3)$$

But $$y_c = \sqrt[3]{\frac{q^2}{g}} = \sqrt[3]{\frac{8gy^3}{g}} = 2y.$$

$\therefore$ $$\frac{y_c}{y} = 2.$$

Hence, choice (*c*) is correct. **Ans.**

**Q. 2.** Discharge reqd. $= \dfrac{2400 \text{ ha}}{\text{Duty}} = \dfrac{2400 \text{ ha}}{860 \text{ ha/cumec}} = \dfrac{2400}{860}$ m³/s = 27.9 m³/s

Volume of water reqd. $= 27.9 \dfrac{\text{m}^3}{\text{s}} \times$ Base period

$= 27.9 \dfrac{\text{m}^3}{\text{s}}$ (120 × 24 × 3600) = 2893 $10^4$ m³

= 2893 ha.m.

The nearest choice to this at (*d*) at 2880 ha.m is hence correct. **Ans.**

**Q. 4.** Discharge reqd. for Rabi $= \dfrac{\text{Area irrigated for Rabi}}{\text{Av. duty for Rabi}}$ = CCA × Intensity

$= \dfrac{(5000 \times 80\%) \text{ ha} \times 50\%}{2000 \text{ ha/cumec}} = 1$ m³/s ...(*i*)

Similarly, Discharge reqd. for Kharif

$= \dfrac{(5000 \times 80\%) \times 30\%}{900}$ m³/s

$= \dfrac{1200}{900} = 1.33$ m³/s ...(*ii*)

Discharge required is max. of two values, *i.e.* 1.33 m³/s. Hence choice (*b*) is correct. **Ans.**

**Q. 6.** Shear stress developed in a channel of given *R* and *S* :

$$\tau_0 = \gamma_w \,.\, R \,.\, S \qquad \text{...(Eq. 4.1)}$$

Here $\gamma$ = (9.81 × 1000) N/m³

$R = y$ for wide rectangular channel

= 0.8 m (given), $S$ = 0.041 (given)

$\therefore$ $$\tau_0 = (9.81 \times 1000)\,(0.8)\,(0.0041) \text{ N/m}^2$$

$= 32.18$ N/m² ...(*i*)

Also $\tau_c$ = Critical shear stress, *i.e.*, Minimum stress required to move a grain of size *d*(*mm*) in N/m²,

$= 0.687\ da\ (mm)$ ...(Eq. 4.45)

For the grain or gravel that will not move, the size $d$ in mm is hence given by :

$$\tau_0 = \tau_c$$

$$32.18 = 0.687\, d_a$$

or $$d_a = \frac{32.18}{0.687} = 47 \text{ mm.}$$

Hence, the nearest choice (*d*) is correct. **Ans.**

**Year 2014 (*Engineering Services*)**

**Q. 1.*** For an irrigated field having : Field capacity = 30%, Permanent wilting point = 10%, Permissible depletion of available moisture = 40%, Dry weight of soil = 14.70 kN/m$^3$, Unit weight of water = 9.8 kN/m$^3$ and Effective rainfall = 30 mm, what is the net irrigation requirement per meter depth of soil ?

(*a*) 300 mm (*b*) 150 mm

(*c*) 120 mm (*d*) 90 mm

**Q. 2.*** A lift irrigation scheme using a discharge of 96 m$^3$/hr is planned to raise a crop with an average $\Delta$ of 0.375 m. Intensity of irrigation is 60%. Assuming 3600 hours of working of the tubewell for a year, the culturable command area is

(*a*) 96 Ha (*b*) 48 Ha

(*c*) 154 Ha (*d*) 120 Ha

**Q. 3.*** A field measures 40 hectares. When 8 cumecs of water was supplied for 6 hours, 30 cm of water was stored in the root zone. The field application efficiency is nearly

(*a*) 70% (*b*) 80%

(*c*) 85% (*d*) 90%

**Q. 4.** In a Lacey regime channel :

1. The bed load is zero. 2. The suspended load is zero.
3. The bed slope is a function of the full supply discharge and the silt size.

Which of the above statements is/are correct?

(*a*) 1 and 2 (*b*) 3 only

(*c*) 2 and 3 (*d*) 2 only

**Q. 5.** A regime canal carries silt of median size 0.25 mm. The Lacey's silt factor of this silt is

(*a*) 0.66 (*b*) 0.88

(*c*) 0.44 (*d*) 0.22

**Q. 6.** A channel designed by Lacey's theory has a velocity of 0.88 m/sec. The silt factor is 1.1. Then hydraulic mean depth will be

(*a*) 1.95 m (*b*) 1.76 m

(*c*) 1.63 m (*d*) 1.50 m

**Q. 7.** At the base of a gravity dam section, the vertical stress at the toe was found to be 2.4 MPa. If the downstream face of the dam has a slope of 0.707 horizontal : 1 vertical, and if there is no tail water, the maximum principal stress at the toe of the dam is

(*a*) 1.7 MPa (*b*) 2.4 MPa

(*c*) 3.6 MPa (*d*) 4.8 MPa

**Q. 8.*** The top width and the bottom width at river thalweg level of a concrete gravity dam are 6 m and 24 m respectively. The vertical section of the dam at the thalweg alignment is 38 m above the thalweg. The upstream face is vertical and the depth of water on the upstream side is 36 m. The moment of hydrostatic force about the toe at that level is

(*a*) 77760 kN-m (*b*) 25292 kN-m
(*c*) 6480 kN-m (*d*) 51840 kN-m

**Q. 9.** Which of the following outlets are of semi-modular type ?

1. Khanna's module
2. Adjustable proportional module
3. Submerged pipe outlet
4. Open flume outlet
5. Kennedy's gauge outlet

(*a*) 2, 4 and 5 (*b*) 1, 2 and 4
(*c*) 2, 3 and 5 (*d*) 3, 4 and 5

**Q. 10.*** If the channel index at an irrigation outlet is 5/3, "setting" of an orifice type outlet in order to have proportionality is

(*a*) 0.90 (*b*) 0.67
(*c*) 0.30 (*d*) 0.15

***Directions:** Each of the next item(s) consists of two statements, one labelled as the 'Statement (I)' and the other as 'Statement (II)'. You are the examine these two statments corefully and select the answers to these items using the codes given below :*

**Codes :**

(*a*) Both Statement (I) and Statement (II) are individually true and Statement (II) is the correct explanation of Statement (I)
(*b*) Both Statement (I) and Statement (II) are individually true but Statement (II) is ***not*** the correct explanation of Statement (I)
(*c*) Statement (I) is true but Statement (II) is false
(*d*) Statement (I) is false but Statement (II) is true

**Q. 11.** *Statement (I)* : The sharp corners in the cross-section of a canal are rounded.

(*a*) *Statement (II)* : The corners may not become zones of stagnation.

## ANSWERS

| | | | | |
|---|---|---|---|---|
| **1.** (*d*) | **2.** (*c*) | **3.** (*a*) | **4.** (*b*) | **5.** (*b*) |
| **6.** (*b*) | **7.** (*b*) | **8.** (*a*) | **9.** (*a*) | **10.** (*c*) |
| **11.** (*a*) | | | | |

## Hints for Solving Starred Questions of the year 2014

**Q. 1.** Please follow Solved Example 2.17, 2.18 and Compute as under :
Max. storage capacity or available moisture

$$= \frac{\gamma_d \cdot d}{\gamma_w}\left[\frac{\text{Field capacity m.c.}}{100} - \frac{\text{Wilting pt. m.c.}}{100}\right]$$

$$= \frac{14.7 \text{ kN/m}^3 \times 1 \text{ m}}{9.81 \text{ kN/m}^3}[0.30 - 0.10] = 0.3 \text{ m}$$

Permissible depletion = 40%

∴ When available moisture depletes by 40% of 0.3 m *i.e.*, by 0.12 m, irrigation will be required. Hence, irrigation water will be supplied to fill up this depletion of 0.12 m.

∴ Gross irrigation required = 0.12 m = 12 cm = 120 mm.

Effective rainfall = 30 mm

∴ Net irrigation requirement = Gross irrigation requirement – Effective rainfall

= 120 mm – 30 mm = **90 mm**

Hence, choice (*d*) is correct. **Ans.**

**Q. 2.** Discharge of irrigation water from tubewell = 96 $m^3$/hr

Hr. of working tubewell = 3600 hr

∴ Total discharge/vol. of water obtained for irrigation from the tubewell

$$= 96 \text{ m}^3\text{/hr} \times 3600 \text{ hr}$$

$$= 3{,}45{,}600 \text{ m}^3 \qquad ...(i)$$

Depth of water required in $A$ $m^2$ of land area irrigated = 0.375 m

Vol. of water required by this $A$ $m^2$ of actually irrigated area

$$= 0.375 \text{ A m}^3 \qquad ...(ii)$$

Equating (*i*) and (*ii*), we get

$$0.375A = 3{,}45{,}600$$

$$\therefore \quad A = 9{,}21{,}600 \text{ m}^2$$

The culturable commanded area × Intensity of irrigation (%)

= Actual irrigated area

$$\therefore \quad \text{CCA} \times 0.60 = 9{,}21{,}600 \text{ m}^2$$

$$\therefore \quad \text{CCA} = \frac{9{,}21{,}600}{0.60} = 15{,}36{,}000 \text{ m}^2 = 153.6 \text{ ha; } i.e., \text{ 154 ha.}$$

Hence, choice (*c*) is correct. **Ans.**

**Q. 3.** Vol. of water given by a discharge of 8 $m^3$/sec (*i.e.*, cumecs) for 6 hours

$$= 8 \text{ m}^3\text{/s} \times (6 \times 3600) \text{ s} = 1{,}72{,}800 \text{ m}^3.$$

Area of field = 40 hectares = $40 \times 10^4$ $m^2$ = $4 \times 10^5$ $m^2$

Depth of water actually penetrating and storing in this field = 30 cm = 0.3 m.

Vol. of water actually stored = $(4 \times 10^5 \times 0.3)$ $m^3$ = $1.2 \times 10^5$ $m^3$ = 1,20,000 $m^3$

$$\text{Field application efficiency} = \frac{\text{Actual water stored in irrigation}}{\text{Water supplied to field}}$$

$$= \frac{1{,}20{,}000}{1{,}72{,}800} = 0.694$$

$$\approx 0.70$$

Hence, field application efficiency = 70%

Hence, choice (*a*) is correct. **Ans.**

**Q. 8.** ***Thalweg*** is the line connecting the lowest points of successive cross-sections along the course of a river valley. The given problem is simple, and can be solved as given below :

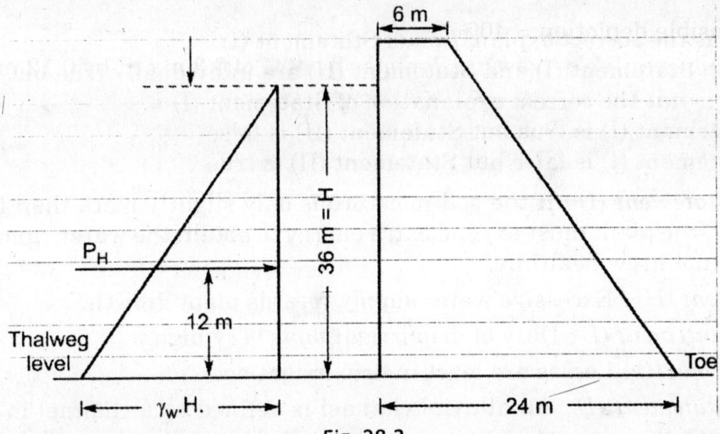


Fig. 28.3

Hydrostatic pressure (force)

$$P_H = \frac{1}{2} \cdot \gamma_w \cdot H^2 = \frac{1}{2} \times 10 \frac{\text{kN}}{\text{m}^3} \times (36 \text{ m} \times 36 \text{ m}) = 6480 \text{ kN/m run of dam}$$

Moment of $P_H$ about toe $= 6480 \text{ kN} \times 12 \text{ m} =$ **77760 kN m**

Hence, choice (*a*) is correct. **Ans.**

**Q. 10.** Setting $= \dfrac{H}{y} = \dfrac{\text{Outlet index } (m)}{\text{Channel index } (n)}$ (Pl. see page 453)

Since the discharge through an orifice type outlet is proportional to $\sqrt{H}$, the index $m = 1/2$.

Also, channel index = $n = 5/3$ (given)

$\therefore$ Setting $= \dfrac{m}{n} = \dfrac{1/2}{5/3} = 0.3$ Hence, choice (*c*) is correct. **Ans.**

## Year 2015 (*Engineering Services*)

**Q. 1.*** In a ski-jump bucket provided in an overflow spillway, the lip angle is 30°, and the actual velocity of flow entering the bucket is 30 m/s. The maximum vertical height attained by the trajectory of the jet, measured above the lip of the bucket, is nearly

(*a*) 45 m (*b*) 35 m (*c*) 22 m (*d*) 11 m

**Q. 2.** The discharge capacity required at the outlet to irrigate 3000 ha of sugarcane having a *kor* depth of 173 mm and a *kor* period of 30 days is

(*a*) 2.0 m³/s (*b*) 1.0 m³/s (*c*) 20 m³/s (*d*) 0.20 m³/s

**Q. 3.** By considering the channel index as $\frac{5}{3}$, the setting of an orifice type irrigation outlet to have proportionality, is

(*a*) 0.90 (*b*) 0.67 (*c*) 0.30 (*d*) 0.15

***Directions :** Each of the next given items consists of two statements, one labelled as the 'Statement (I)' and the other as 'Statement (II)'. Examine these two statements carefully and select the answers to these items using the codes given below :*

**Codes :**

(*a*) Both Statement (I) and Statement (II) are individually true and Statement

(II) is the correct explanation of Statement (I)

(*b*) Both Statement (I) and Statement (II) are individually true but Statement (II) is ***not*** the correct explanation of Statement (I)

(*c*) Statement (I) is true but Statement (II) is false

(*d*) Statement (I) is false but Statement (II) is true

**Q. 4.** *Statement (I) :* If the soil moisture is only slightly more than the wilting coefficient, the plant must expend extra energy to obtain the water ; and hence the plant will not grow healthily.

*Statement (II) :* Excessive water supply retards plant growth.

**Q. 5.** *Statement (I) :* Duty of drip irrigation is very high.

*Statement (II) :* Losses are least in drip irrigation.

**Q. 6.** *Statement (I) :* An alluvial channel is defined as a channel in watch the flow transports sediment of the same physical characteristics as the material in the wetted surface of the channel.

*Statement (II) :* This ensures that the channel cross-section and the channel slope do not change.

## ANSWERS

**1.** (*d*) **2.** (*a*) **3.** (*c*) **4.** (*b*) **5.** (*a*)
**6.** (*c*)

## Hints for Solving Starred Questions of the year 2015

**Q. 1.** A ski-jump bucket, shown in Fig. 21.32 ($c_1$) is again shown here in Fig. 28.4. Such an energy dissipator is suitable where the stream bed is composed of firm rock and the tail water depth is less than required for formation of hydraulic jump.

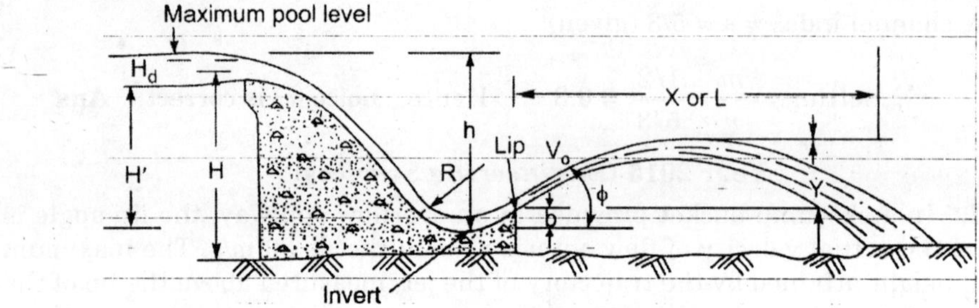


**Fig. 28.4.** Ski-jump bucket.

***Bucket Invert.*** The invert level of a ski-jump bucket is decided mainly from structural design point of view. If power houses are to be sited below the ski-jump bucket, then the invert should be fixed higher than the power house top. In certain cases, it happens that the tail water depth is lower than $y_2$, but the bucket invert is lower than tail water levels. In such cases, the bucket invert is decided is to provide a concrete cover of 1.5 to 3 m over the bed rock. The pure ski-jump or **flip**, as it is quite often called, is provided such that bucket lip is always higher than maximum tail water levels.

***Bucket radius (R).*** The following relation has been found to give a satisfactory value for the redius of the bucket :

(*i*) $$R = 0.6\sqrt{H'.H_d} \qquad ...(28.1)$$

where $H'$ = fall from the crest of the spillway to the bucket invert in m

$H_d$ = Head over the crest in m

(*ii*) Ven Te Chow's formula

$$R = 0.305 \times (10)^p \quad \text{...(28.2)}$$

where $$p = \frac{V + 6.4\, H_d + 4.88}{3.6\, H_d + 19.5} \quad \text{...(28.3)}$$

where $V$ = velocity at the toe of the spillway in m/sec

(*iii*) *R.S. Varshney's dimensionless formula*

$$\sqrt{F_1} = 0.09\frac{R}{y_1} + 1.96 \quad \text{...(28.4)}$$

where $F_1$ = Incoming Froud No.
$y_1$ = Initial depth
$R$ = Radius of the bucket

***Entrance and Exit Slopes.*** For entrance slope, the steepest spillway slope that should be used is $4V : 1H$. The **exit angle** or **lip angle** ($\phi$) is an important factor in determining the length of the trajectory. Theoratically, if friction, air retardance, etc are neglected, the following formula can be used to evaluate the horizontal length ($X$ or $L$) of the jet trajectory :

$$X \text{ or } L = \frac{V_0^2 \,.\, \sin 2\phi}{g} \quad \text{...(28.5)}$$

where $V_0$ = Actual velocity of flow entering the bucket
$\phi$ = Lip angle or Exit angle

The jet trajectory height ($Y$) is given by the relation :

$$Y = h \,.\, \sin^2 \phi \quad \text{...(28.6)}$$

$Y$ = Ht. of trajectory above the lip of bucket (Fig. 28.5)

where $h$ = height of max. water level above the bucket invert and can be $V_0 = \sqrt{2gh}$ , ...(28.7)

where $V_0$ is the actual velocity of flow entering the bucket

A judicious selection of exit angle ($\phi$) is necessary. An exit angle of 30° to 35° will make a good choice.

In the given question (*i.e.* Q. 1 of year 2015), we are given :

$$\phi = 30°,\ V_0 = 30 \text{ m/s},\ Y = ?$$

Using $V_0 = \sqrt{2gh}$ , we have ; 30 m/s = $\sqrt{2 \times 9.81 \,.\, h}$

or $h = 45.87$ m

Now $Y = ht$ of trajectory above the lip of bucket = $h \sin^2 \phi$ ...(Eq. 30.6)

∴ $Y = 45.87 \text{ m} \times (\sin 30°)^2$

$= 11.46$ m ; say 11 m

Hence, choice (*d*) is correct. **Ans.**

# —29

# *Conventional Questions of the Civil Services Exams.*

The Union Public Service Commission, Dholpur House, Shahjahan Road, New Delhi-110011, conducts every year two competitions, in which, you as civil engineering graduates (even with exam. results awaited) can participate for securing a direct class I Govt. appointment.

The first competition is named as **Combined Engineering Services (CES) Exam,** and the other competition is known as **Civil Services Exam.**, popularly called the *IAS competition.*

Press notifications for obtaining and submitting application forms are released by Secretary UPSC, through news papers including "Employment News" sometimes in Feb. every year. The exams are usually held in August.

On the basis of the first competition, you are likely to be appointed as Asstt. Executive Engineer or Asstt. Director or on an equivalent post in the civil engineering side of the various Government Departments like Railways, CPWD, P&T, Central Water Commission, etc. ; whereas, on the basis of the second competition, you are liable to become an IFS, IAS, IPS, or an officer on allied services, such as Asstt. Custom Collector or Asstt. Excise Collector, etc.

Most of the engineering degree students aspire to clear any one of these two competitions, for an excellent career full of promotional avenues, just after doing BE degree. The syllabuses for both these competitions, now a days, provide full opportunities to Civil Engineering degree holders, to make through these competitions.

In this chapter, we will take up the questions of the Civil Services Exams of the past 10 years pertaining to the subject of Water Resources Engg. (Vol. II) — Irrigation Engg. and Hydraulic Structures".

As far as the syllabus of the Civil Services Exam is concerned, the same has been described in "Water Resources Engg. (Vol. I)—Hydrology and Water Resources Engg." The students can refer that volume.

**Year 2009 (*Civil Services*)**

**Q.5**. (*b*) *Illustrate with the help of diagrams the basic three types of dam. Briefly mention the general design considerations to be followed for its safety.* (12)

(*c*) *Show with the help of a suitable diagram the component parts of an irrigation canal system and brielfy mention their respective functions.* (12)

**Q.6.** (*a*) *A trapezoidal irrigation canal with side slope of 1.5 H to 1 V is proposed to be lined with bricks to reduce seepage losses. It is required to carry a discharge of 29 m³/s of water. Find the wetted perimeter for minimum amount of lining and the required bed slope. The value of Manning's N is given as 0.015 and it is stipulated that the average velocity cannot exceed 1 m/s.* (20)

[**Solution.** $V_{max} = V = 1 \text{ m/s}$

$$Q = 29 \text{ m}^3/\text{s}$$

$$n = 0.015$$

$$\therefore \quad A = \frac{Q}{V} = \frac{29}{1} = 29 \text{ m}^2$$

For Trapezoidal lined section, the Area ($A$) and wetted perimeter ($P$) are given by Eq. (5.8) and (5.9) as

$$A = y\,[B + y\,\theta + y \cot \theta],$$

and

$$P = [B + 2y\,\theta + 2y \cot \theta]$$

Here

$$\tan \theta = \frac{V}{H} = \frac{1.0}{1.5} = 0.667$$

$$\therefore \quad \theta = 33°.69 = 33°.69 \times \frac{\pi}{180°} = 0.588$$

$$\therefore \quad A = y\,[B + y \times 0.588 + y \times 1.5]$$

or

$$29 = y\,[B + 2.088\, y]$$

or

$$B = \left[\frac{29}{y} - 2.088y\right]$$

$$P = [B + 2y \times 0.588 + 2y \times 1.5]$$

or

$$P = \left(\frac{29}{y} - 2.088y\right) + 1.176y + 3y$$

or

$$P = \frac{29}{y} + 2.088\, y$$

For $P$ to be minimum, we should have $\frac{dP}{dy} = 0$

$$\therefore \quad \frac{dP}{dy} = 29\,(-1y^{-2}) + 2.088 = 0$$

$$\therefore \quad \frac{29}{y^2} = 2.088$$

or

$$y = \mathbf{3.727\ m}$$

or

$$P_{\min} = \left[\frac{29}{3.727} + 2.088 \times 3.727\right] = \mathbf{15.56\ m.} \quad \textbf{Ans.}$$

Also

$$V = \frac{A}{n} R^{2/3} \cdot \sqrt{S} = \frac{1}{0.015} \times \left(\frac{29}{15.56}\right)^{2/3} \cdot \sqrt{S} = 1$$

or

$$S = \frac{1}{10194}. \quad \textbf{Ans.}\ ]$$

**Q.6** *(b) A crop is irrigated with reclaimed waste water whose salinity (ECw) measured by electrical conductivity is 1 mmho/cm. If the crop is irrigated to achieve a leaching fraction 0.20 i.e. (80% of water is lost through evaporation) determine the following :*

*(i) the salinity of the deep percolating water*

*(ii) the appropriate leaching fraction to maintain crop yield. The crop yield will suffer significant loss when TDS (total dissolved substance) of soil water exceeds 5000 mg/l.*

*Given : TDS (mg/l) = EC (mmho/cm) × 640.* (20)

[**Solution.** From Eq. (6.3), we have

LR = Leaching Requirement, *i.e.* Excess water above the evapotranspiration needs of a crop which flows down to the soil below:

$$= \frac{\text{EC of irrigation water } (EC(i))}{\text{EC of drained or leached water } [EC(d)]}$$

Here LR = 0.20

EC of irrigation water $= EC\ (i) = EC\ (w) = 1$ mili mho/cm

$$\therefore \quad 0.20 = \frac{1 \text{ m} \cdot \text{mho/cm}}{EC(d)}$$

or EC of drained water (deep paracolating water)

$$= \frac{1}{0.20} = \textbf{8 milli mho/cm.} \quad \textbf{Ans.}$$

*(ii) For maintaining crop yield*

TDS of soil water $\leq$ 5000 mg/L

But TDS (mg/L) = EC (m.mho/cm) × 640

$\therefore$ EC (*m* . mho/cm) × 640 $\leq$ 5000

or EC of soil water $\leq \dfrac{5000}{640} = 7.8125$

Hence, max EC of soil water or drained water for maintaining crop yield = 7.8125 m . mho/cm.

Now $$LR = \frac{EC(i)}{EC(d)} = \frac{1 \text{ m} \cdot \text{mho/cm}}{7.8125 \text{ m.mho/cm}} = 0.128 = 12.8\%$$

Hence a minimum of 12.8% of applied irrigation water must leach inside the soil to maintain crop yield to avoid crop loss **Ans.**]

**Q.8** (*b*) *Table below furnishes the kind of crop irrigated from the water reservoir by means of the canal system. The data about these crops are also furnished in the table*

| *Kind of crop* | *Crop period in days* | *Area under crops in ha* | *Duty of water in the field in ha / cumec* |
|---|---|---|---|
| *Cotton* | *240* | *3000* | *1430* |
| *Sugarcane* | *365* | *2000* | *860* |
| *Wheat* | *120* | *4000* | *2150* |
| *Rice* | *120* | *2000* | *860* |

*Assume 30% losses of water in the entire canal system and 15% losses in the reservoir. Find the live storage of the reservoir in ha.m.* (20)

[**Solution.** Pl. refer similar Solved Example 3.6, and compute as under :

Water reqd. for cotton $= \dfrac{\text{Area}}{\text{Duty}} = \dfrac{3000}{1430} = 2.098$ cumecs

Water reqd. for Sugarcane $= \dfrac{\text{Area}}{\text{Duty}} = \dfrac{2000}{860} = 2.326$ cumecs

Water reqd. for Wheat $= \dfrac{\text{Area}}{\text{Duty}} = \dfrac{4000}{2150} = 1.860$ cumecs

Water reqd. for Rice $= \dfrac{\text{Area}}{\text{Duty}} = \dfrac{2000}{860} = 2.326$ cumecs

Water vol. reqd. by cotton during the 240 days of crop period

$$= 2.098 \times (240 \times 24 \times 3600)^3$$

Water vol. reqd. by Sugarcane = $2.326 \times (365 \times 24 \times 3600)$ m$^3$

Water vol. reqd. by Wheat = $1.860 \times (120 \times 24 \times 3600)$ m$^3$

Water vol. reqd. by Rice $= 2.326 \times (120 \times 24 \times 3600)\ m^3$

Total water vol. reqd. by all four crops

$= [503.50 + 848.83 + 223.26 + 279.07]\ 86400\ m^3$

$= [1854.66 \times 86400]\ m^3$

$= \mathbf{160.24\ Mm^3}$

Counting losses of 30% and 15%, we get

Live storage of the reservoir

$$= \frac{160.24}{0.7 \times 0.85}$$

$= \mathbf{269.31\ Mm^3}$. **Ans.**

## Year 2010 (*Civil Services*)

**Q. 6.** (*a*) *Consider the following three mutually exclusive options of providing a cut-off under a weir with aprons on a permeable bed : (i) one only on the upstream end of the apron (ii) one only under the toe of the weir (iii) one only at the extreme downstream end of the apron. Sketch (illustrating all resultant effects) the possible flow net diagram in each of the three cases-showing four of the flowlines nearest to the weir-cum-cutoff construction.* (20)

**Q. 7.** (*a*) *A part of a solid gravity dam (sp. gr. of mass concrete – 2.64) is shown in its cross-section. Take uplift factor as 0.7. Draw a tabular statement for analysis of forces at horizontal section 26 m below water level. No ice, wind or seismic forces need to be considered.* (20)

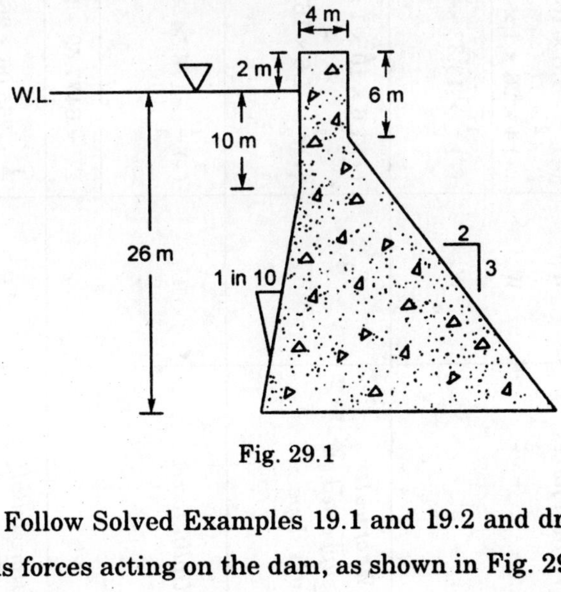


Fig. 29.1

**Solution.** Follow Solved Examples 19.1 and 19.2 and draw the diagram showing various forces acting on the dam, as shown in Fig. 29.2.

**Table 29.1 Solution Table for Q.7 (a) of Year 2010**

| *Name of the force* | *Designation* | *Magnitude in kN* | *Lever arm in m* | *Moments about the Toe in kN.m anticlockwise → +ve* |
|---|---|---|---|---|
| *Vertical forces* | $W_1$ | $(+)\ \frac{1}{2} \times 1.6 \times 16 \times 1 \times 25.9 = 331.52$ | 17.86 | 5,920.95 |
| | $W_2$ | $(+)\ 4 \times 26 \times 1 \times 25.9 = 2693.6$ | 15.33 | 41,292.89 |
| | $W_3$ | $(+)\ \frac{1}{2} \times 13.33 \times 20 \times 1 \times 25.9 = 3452.47$ | 8.89 | 30,692.46 |
| | | $\Sigma V_1 = 6477.59$ | | $\Sigma M_1 = 77906.3$ |
| *Wt of water supported on u/s slope* | $W_4$ | $(+)\ 1.6 \times 10 \times 1 \times 9.81 = 156.96$ | 18.13 | 2,845.68 |
| | $W_5$ | $(+)\ \frac{1}{2} \times 1.6 \times 16 \times 1 \times 9.81 = 125.57$ | 18.40 | 2,310.49 |
| | | $\Sigma V_2 = 282.53$ | | $\Sigma M_2 = 5156.17$ |
| *Uplift forces* | $U$ | $(-)\ \frac{1}{2} \times 178.54 \times 18.93 \times 1 = 1689.88$ | 11.95 | (–) 21326.28 |
| | | $\Sigma V_3 = (-)\ 1689.88$ | | $\Sigma M_3 = (-)\ 21326.28$ |
| | | $\Sigma V = 6477.59 + 282.53 - 1689.88 = 5070.24$ | | |
| *Horizontal hydrostatic water pressure (1) on U/S face* | $P$ | $\frac{1}{2} \times 255.06 \times 26 \times 1 = 3315.78$ | 8.67 | (–) 28747.81 |
| | | $\Sigma H$ (towards downstream) = 3315.78 | | $\Sigma M_4 = (-)\ 28747.81$ |
| | $\Sigma M$ = Net + ve Moment = 77906.3 + 5156.17 – 21326.28 – 28747.81 = **32983.38 kN** | | | |

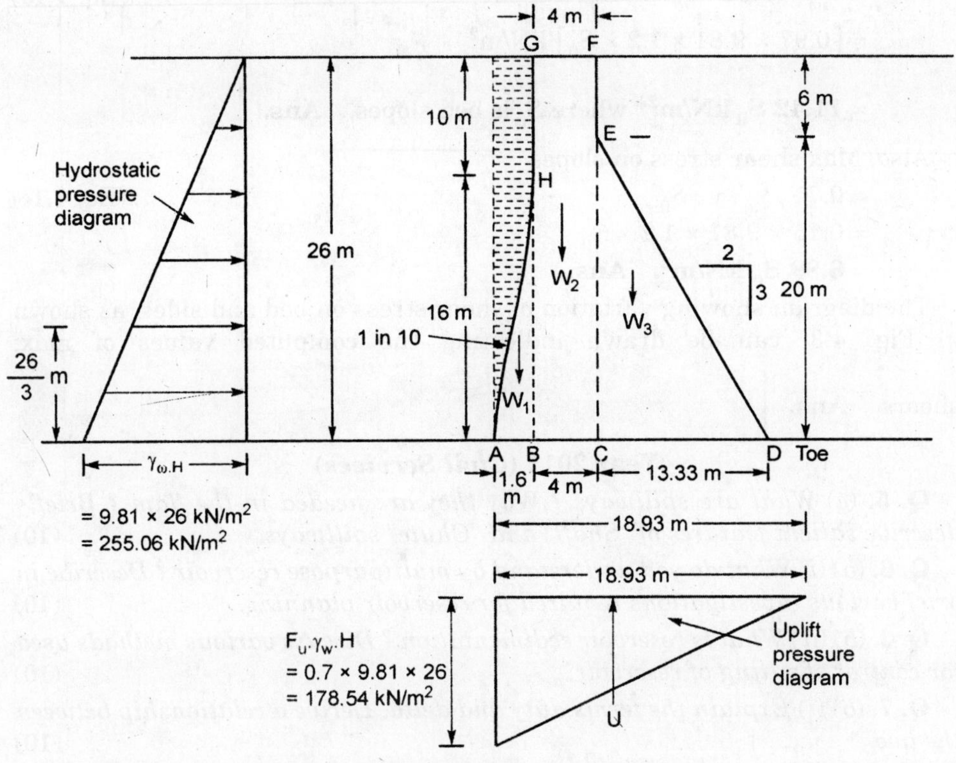


Fig. 29.2

Now, Unit wt. of concrete = 9.81 × 2.64 = 25.90 kN/m³

Compute and List the various forces per m run of dam in Table 29.1, which is the reqd. table. **Ans.**

**Q. 8.** (*a*) *Consider a trapezoidal rigid boundary channel of 3 m base width with side slopes 1 h to 1/2 v, with depth of flow being 1.2 m, and n = 0.012, with adequate free board. Sketch the shear stress distribution on the wetted perimeter. Explain the facts of zero and maximum, if both are so possible, shear stresses on each of the side slopes and on the bed-with reasons therefore.* (20)

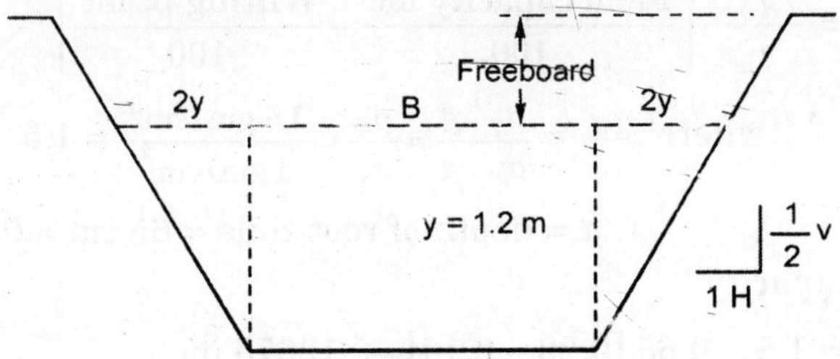


Fig. 29.3.

[ **Hint.** Refer article 4.4 and Fig. 4.3.

Max. shear stress on bed

$= \tau_{bm} = 0.97 \cdot \gamma_w \cdot y \cdot S_0$ ...(Eq. 4.1*b*)

$= [0.97 \times 9.81 \times 1.2 \times S_0]$ kN/m²

= **11.42 $S_0$ kN/m²*** where $S_0$ is bed slopes. **Ans.**

Also, Max shear stress on slopes

$= 0.75 \cdot \gamma_w \cdot y \cdot S_0$ ...(Eq. 4.1*c*)

$= 0.75 \times 9.81 \times 1.2 \times S_0$

= **8.89 $S_0$ kN/m³. Ans.**

The diagram showing variation of shear stress on bed and sides, as shown in Fig. 4.3, can be drawn indicating the computed values of max. shears. **Ans.**

### Year 2011 (*Civil Services*)

**Q. 5.** (*e*) *What are spillways ? Why they are needed in the dam ? Briefly describe salient features of "Shaft" and "Chute" spillways.* (10)

**Q. 6.** (*b*) (*i*) *What do you understand by multipurpose reservoir ? Describe in brief various investigations required for reservoir planning.* (10)

**Q. 6.** (*b*) (*ii*) *What is reservoir sedimentation ? Discuss various methods used for control of silting of reservoir.* (10)

**Q. 7.** (*b*) (*i*) *Explain the terms duty and delta. Derive a relationship between the two.* (10)

**Q. 7.** (*b*) (*ii*) *A soil has field capacity of 23% and wilting coefficient of 10%. If the root zone depth is 65 cm, determine the storage capacity of soil. Irrigation water is applied when moisture content falls to 10%. If water application efficiency is 70%, determine the water depth required to be applied in the field. The dry unit weight of soil is 1.5 gm/cm³.* (10)

**Solution.** Follow Solved Examples 2.16 and 2.17 and compute as under :

Maximum storage capacity of soil is the water stored between field capacity (23%) and wilting point (10%), and is given by equation :

Storage capacity or Available moisture

$$= \frac{\gamma_d \cdot d}{\gamma_w}\left[\frac{\text{Field capacity mc}}{100} - \frac{\text{Wilting pt mc}}{100}\right]$$

where $\dfrac{\gamma_d}{\gamma_w} = \dfrac{\rho_d \cdot g}{\rho_w \cdot g} = \dfrac{\rho_d}{\rho_w} = \dfrac{1.5 \text{ gm/cm}^3}{1 \text{ gm/cm}^3} = 1.5$

$d$ = depth of root zone = 65 cm = 0.65 m.

∴ Storage capacity

$= 1.5 \times 0.65\,[0.23 - 0.10] = 0.12675$ m

*The shear stress cannot be evaluated in absolute numerical figures, since neither the bed slope ($S_0$) is given nor it can be computed from Manning's formula, since discharge or velocity is not given.

= **12.675 cm. Ans.**

Since moisture content is allowed to vary between 23% and 10%, the deficiency created in this fall

$$= 1.5 \times 0.65\,[0.23 - 0.10] = 0.12675 \text{ m} = 12.675 \text{ cm}.$$

Hence, 12.675 cm is the net irrigation requirement.

∴ Field irrigation requirement

$$= \frac{\text{Net Irrigation requirement}}{\text{Water application efficiency}} = \frac{12.675 \text{ cm}}{0.70} = 18.107 \text{ cm}$$

Hence, the depth of irrigation water reqd. to be applied in the field

= **18.107 cm. Ans.**

**Q. 8.** (*b*) (*i*) *What do you understand by river training? What are the works that come under this category ?* (10)

**Q. 8.** (*b*) (*ii*) *Draw a sketch of a guide bank. Give the design features.* (10)

## Year 2012 (*Civil Services*)

**Q. 1.** (*d*) *During a particular stage of growth of a crop, consumptive use of water is 3.2 mm/day. Determine the interval between irrigations and depth of water to be applied when the amount of water available in the soil is (i) 35%, (ii) 75% and (iii) 0% of the maximum depth of available water in the root zone which is 100 mm. Assume irrigation efficiency to be 60%.* (12)

**Solution.** Pl. follow Similar Solved Example 2.24, and compute as under :

***(i)*** Readily available moisture in root zone

= 35% × 100 mm = **3.5 cm.**

Consumptive use

= 3.2 mm/day = 0.32 cm/day

∴ 0.32 cm of water is consumed by crop in = 1 day

3.5 cm of water will hence be consumed by crop in

$$= \frac{1}{0.32} \times 3.5 = 10.94 \approx 11 \text{ days.} \quad \textbf{Ans.}$$

Depth of water to be applied to the fields

$$= \frac{\text{Water required to be recouped}}{\text{Efficiency of irrigation application}} = \frac{3.5 \text{ cm}}{60\%} = \frac{3.5}{0.6} \text{ cm}$$

= **5.83 cm. Ans.**

Hence, 5.83 cm of irrigation water shall have to be applied to the fields at an interval of 11 days to recoup the moisture to original level. **Ans.**

***(ii)*** Readily available moisture in root zone

= 75% × 100 mm = **7.5 cm**

Consumptive use

= 3.2 mm/day = 0.32 cm/day

∴ 0.32 cm of water is consumed in = 1 day

7.5 cm of water will be consumed in

$$= \frac{1}{0.32} \times 7.5 \text{ days} = 23.43 \text{ day : say } \textbf{23 days.} \quad \textbf{Ans.}$$

Depth of water reqd.

$$= \frac{7.5 \text{ cm}}{60\ \%} = \frac{7.5}{0.6} \text{ cm} = \textbf{12.5 cm} \quad \textbf{Ans.}$$

*Hence, 12.5 cm of water depth is reqd. to be applied after 23 days to recoup the moisture to original level.*

**(*iii*)** Readily available moisture in root zone = 0% × 100 mm = 0.

Since there is no readily available moisture in root zone available for fulfilling consumptive use requirement, there is an imminent need for applying irrigation water.

Qty. of water reqd. to raise available water equal to its maximum value of 100 mm *i.e.*, 10 cm.

$$= \frac{10 \text{ cm}}{60\%} = \textbf{16.67 cm.}$$

Hence, 16.67 cm of irrigation water can be applied at an interval of

$$\frac{16.67 \text{ cm}}{0.32 \text{ cm/d}} = 52 \text{ days.} \quad \textbf{Ans}$$

**Q. 2.** (*b*) (*ii*) *State and explain factors which influence the silting of reservoirs. How this affects the useful life of reservoirs.? Also explain the terms 'live storage', 'dead storage' and 'effective storage' in this context.* (10)

**Q. 3.** (*c*) *What are the conditions based on which the effective length of the spillways is decided ? What are the various methods adopted in energy dissipation below a spillway ? Explain the most suitable energy dissipation arrangements that you would recommended for each of the following conditions.*

(*i*) *When Jump Height Curve (JHC) is always above Tail Water Curve (TWC).*

(*ii*) *When JHC and TWC nearly coincide, and*

(*iii*) *When JHC is always below TWC.* (20)

**Q. 4.** (*c*) (*i*) *What are the various causes of seepage and percolation losses from a canal ?* (5)

(*ii*) *Explain the term "Hydraulically Efficient Channel Section". Prove that the most efficient trapezoidal section is one half of a regular Hexagon.* (15)

[ **Hint:** A **regular hexagon*** will have all its six sides equal. Therefore, a trapezoidal channel will become a part of a regular hexagon if its two sides and bed width are equal.

---

*A **regular polygon** is a polygon that is equiangular (all angles are equal) and equilateral (all sides have the same length).

Now, for a most **hydraulically efficient channel,** the wetted perimeter (P) must be minimum for a given area (*A*). The expression for P in a tropezoidal channel involves two variables *i.e.*, *y* (depth) and *m* (when side slopes of the channel are *m* : 1 (H : V). In **Solved Example 4.6,** we have proved that for the **most hydraulically efficient trapezoidal channel,** $m = \frac{1}{\sqrt{3}}$.

Thus, for such a trapezoidal channel ABCD, shown in Fig. 29.4,

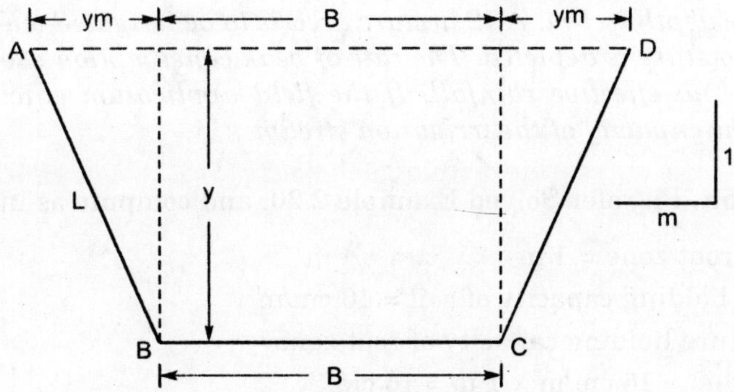


Fig. 29.4

We have the slant length *AB* = *CD* = *L*, where

$$(L) = \sqrt{y^2 + (ym)^2}$$

or $$L = \sqrt{1 + m^2} \cdot y \qquad ...(i)$$

Also from Eq. (1) in solved example 4.6, we have

$$B = 2\left[\sqrt{1 + m^2} - m\right] y \qquad ...(ii)$$

Substituting $m = \frac{1}{\sqrt{3}}$

as for most efficient channel, we have for the most efficient trapezoidal channel :

$$L = \sqrt{1 + \left(\frac{1}{\sqrt{3}}\right)^2}\, y = \sqrt{1 + \frac{1}{3}} \cdot y = \sqrt{\frac{4}{3}} \cdot y$$

or $$L = \frac{2}{\sqrt{3}} y \qquad ...(iii)$$

And *B* from (*ii*) is given as :

$$B = 2\left[\sqrt{1 + \frac{1}{3}} - \frac{1}{\sqrt{3}}\right] y = 2\left[\frac{2}{\sqrt{3}} - \frac{1}{\sqrt{3}}\right] y = \frac{2}{\sqrt{3}} y \qquad ...(iv)$$

Eq. (*iii*) and (*iv*) shows that

*L* = *B*. Hence, the three sides of such an efficient trepezoidal channel are equal, and hence the channel section is one half of a regular haxagon. **Ans.**

## Year 2013 (*Civil Services*)

**Q. 5.** (*c*) *Write briefly on the suitable river training measures to be adopted for the protection of upstream side of a barrage during flood time.* (10)

**Q. 6.** (*b*) *Compare the following in irrigation system :*

(*i*) *Water application efficiency vs. Water distribution efficiency.*

(*ii*) *Water storage efficiency vs. Consumption use efficiency.*

*The available moisture holding capacity of the soil of a field is 16 cm/m and the root zone depth is 1 m. A 12 hectare area is to be irrigated when 50% of the available moisture is depleted. The rate of peak consumption use is 4 mm/day and there is no effective rainfall. If the field application efficiency is 50%, determine the capacity of the irrigation stream.* (15)

[ **Solution.** Pl. refer Solved Example 2.20, and compute as under :

Depth of root zone = 1 m

Moisture holding capacity of soil = 16 cm/m

∴ Moisture holding capacity of root zone

= 16 cm/m × 1 m = 16 cm.

Since irrigation water is applied when available moisture is depleted to 50%, allowable depletion depth between two irrigations = 50% × 16 cm = 8 cm.

Now, Daily use of water = Consumptive use = 4 mm/day (given)

∴ Frequency of irrigation

$$= \frac{\text{Available moisture}}{\text{Moisture consumed per day}} = \frac{8 \text{ cm}}{4 \text{ mm/day}} = \frac{8 \text{ cm}}{0.4 \text{ cm/day}} = 20 \text{ days}$$

Net water depth reqd. to be applied while irrigating each time after 20 days each = 20 × 0.4 = 8 cm.

Now, field irrigation requirement

$$= \frac{\text{Net irrigation requirement}}{\text{Field application efficiency}} = \frac{8 \text{ cm}}{0.5} = 16 \text{ cm}$$

∴ Quantity of water reqd. in the field for irrigation of 12 ha plot

$$= 16 \text{ cm} \times 12 \text{ ha} = \frac{16}{100} \text{ m} \times 12 \times 10^4 \text{ m}^2 = \mathbf{19{,}200 \text{ m}^3}$$

Hence, 19,200 $m^3$ of irrigation water is required to irrigate 12 ha of field area, each time at 20 days interval. The required irrigation stream should be able to supply this much water.

To compute the discharge capacity of the irrigation channel, let us assume that the irrigation is carried over 8 hr period each time after 20 days. Then 19200 $m^3$ of water has to be supplied in 8 hr period. Thus, the canal discharge

$$= \frac{19200 \text{ m}^3}{(8 \times 60 \times 60) \text{ sec}} = \mathbf{0.67 \text{ cumec.}}$$

Hence, the irrigation stream should have a capacity of 0.67 cumec, and should supply irrigation water for 8 hours a day, each time after every 20 days. **Ans.** ]

**Q. 7.** (*a*) (*i*) *Explain the difference between contour canal and side slope canal.*

(*ii*) *What are the losses of water from canals ? State the various factors affecting their magnitude.*

(*iii*) *Propose with a sketch, a suitable cross drainage work at the location at a canal crossing a stream. Give reasons for the selection. The canal and stream data are as under :*

***Canal Data***

| | |
|---|---|
| *Full supply discharge* | *= 0.45 cumec* |
| *Full supply level* | *= R.L. 300.600* |
| *Canal bed level* | *= R.L. 300.000* |
| *Canal bed width* | *= 2 m* |

***Stream data***

| | |
|---|---|
| *High flood discharge* | *= 6 cumecs* |
| *High flood level* | *= R.L. 300.500* |
| *Stream bed level* | *= R.L. 299.400* |
| *Natural ground level* | *= R.L. 299.500* |

*State the characteristic features of the selected structure.* (20)

[**Hint.** Pl. refer Solved Example 14.5, and decide as under :

In this case, the canal bed level (300.0 m) is higher than the drainage bed level (299.4 m) by 0.6 m, and, therefore, we may think of taking the canal above the drainage. Further, the drainage FSL (300.5 m) is above the canal bed (300.0 m), we will have to syphon the drainage water, and thus a *syphon aqueduct* can be constructed at the site. However, since drainage discharge is much higher than the canal discharge, the syphoning of the drainage (wider drain) may not prove economical ; and on the other hand, the canal bed may be raised, so as to pass drainage water freely below the canal trough (as the difference in CBL and drainage HFL is only 0.5 m). In that case, the minimum canal level to be kept at the crossing will be equal to :

= Drainage HFL + Thickness of canal trough slab + Min. Free board

= 300.5 + 0.6 + 0.9 = 302.0 m

Therefore, the bed of the canal trough will have to be raised by an amount equal to 302.0 – 300.0 = 2.0 m at the site of the crossing.

Hence, hump shall be provided along the length of the trough with flat u/s slope and d/s slope to join the natural canal bed levels on both sides of the crossing. The crossing will then be designed as an **aqueduct.** The design procedure as used in Example 14.1 can be adopted to compute salient features like drainage waterway, canal waterway, canal trough, etc. However, since it is small sized canal with only 2 m bedwidth and drainage is also small having Lacey's perimeter ($P$) = $4.75\sqrt{Q}$ = $4.75\sqrt{6}$ = 11.6 m, no fluming may be required. **Ans.**]

**Q. 8.** (*b*) *State the circumstances under which the following types of energy dissipators will be suitable for spillways :*

(*i*) *Solid roller bucket*

(*ii*) *Ski-Jump* (10)

**Year 2014 (*Civil Services*)**

**Q. 5.** (*d*) *What do you understand by Capacity-Inflow Ratio in regard to Reservoir Sedimentation ? How life of a Reservoir is determined from C/I Ratio and Trap Efficiency.* (10)

**Q. 6.** (*c*) *Design an Alluvial Channel based on Lacey's theory from the following data :*

(*i*) *Discharge : 25 cumecs* (*ii*) *Lacey's silt factor : 1.1*

(*iii*) *Side slopes : 1/2 H to 1 V*

*Derive the Lacey's Waterway and Regime Scour Depth Formulae.* (15)

[**Hint:** Pl. follow Solved Example 4.10 ; and also refer to article 4.7.4.4(1) and 4.7.4.4(4), respectively for waterway and Regime Scour Depth, equations and their derivations from the fundamental equations of Lacey.]

**Q. 7.** (*c*) *What is Elementary profile for a gravity dam ? Discuss its shape and derive the expressions for its base width and stresses in it. Design the practical profile of a gravity Dam of stone masonry from the following data :*

(*i*) *R.L. of Base of Dam : 1250 m* (*ii*) *R.L. of H.F.L. : 1280 m*

(*iii*) *Specific gravity of masonry : 2.4*

(*iv*) *Safe compressive stress for dam material : 1200 kN/m²*. (20)

[**Hint:** Pl. refer article 19.6, and derive equation (19.22) as $B = \dfrac{H}{\sqrt{S_c - 1}}$ and equation (19.23*a*) as $B = \dfrac{H}{\mu(S_c - 1)}$. The value of $B$ chosen should be greater of these two values, which evidently is the first one, as discussed in the text.

Thus, the Base width ($B$) of elementary triangular profile of a dam $= \dfrac{H}{\sqrt{S_c - 1}}$ where $S_c$ is the sp. gravity of the dam material.

Also derive Eq. for max. stress at toe as derived in the text as Eqs. (19.24) as

$$p_v = \gamma_w \cdot H \cdot (S_c - 1) \text{ (when } C = 1\text{)}.$$

In the given numerical part of the question, we have

$$H = 1280 - 1250 = 30 \text{ m}$$

$$S_c = \text{Sp. gravity of dam material} = 2.4$$

Safe compressive stress for dam material = 1200 kN/m².

If we adopt a triangular dam profile with base width

$$B = \frac{H}{\sqrt{S_c - C}} = \frac{30 \text{ m}}{\sqrt{2.4 - 1}} = 25.4 \text{ m, (when full uplift is considered } (C = 1)$$

then $p_v$ at toe $= \gamma_w \cdot H \cdot (S_c - C)$

$$= 9.81 \frac{\text{kN}}{\text{m}^3} \times 30 \text{ m} (2.4 - 1) = 412 \text{ kN/m}^2 < 1200 \text{ kN/m}^2$$

($\because$ $C = 1$)

$\therefore$ Safe.

Hence, provide triangular section of dam with base width 25.4 m and ht. = 30 m. **Ans.**

**Q. 8.** (*c*) *The following data pertains to the healthy growth of a Crop :*

(*i*) *Field capacity of soil : 30%* (*ii*) *Permanent wilting percentage : 11%*

(*iii*) *Density of soil : 1300 kg/m³* (*iv*) *Effective depth of root zone : 700 mm*

(*v*) *Daily consumptive use of water for the given crop : 12 mm.*

*For healthy growth, moisture content must not fall below 25% of Water holding capacity between the field capacity and the permanent wilting point. Determine the watering interval in days.* (15)

**Solution.** Pl. follow Solved Example 2.16, and compute as under :

The available moisture = Field capacity – Permanent wilting

= 30% – 11% = 19%

For healthy growth, the balance moisture must not fall below 25% × 19% = 4.75%.

Hence, the readily available moisture = 19% – 4.75% = 14.25%

**Optimum moisture** = 30 – 14.25 = **15.75%.**

It means that the moisture will be refilled between 15.75% and 30% by irrigation.

Depth of water stored in root zone between these two limits

$$= \frac{\gamma_d \cdot d}{\gamma_w} \text{ [Field capacity m.c. – Optimum m.c.]}$$

$$= \frac{1300 \text{ kg/m}^3 \times 0.7 \text{ m}}{1000 \text{ kg/m}^3} \left[ 0.30 - 0.1575 \right] = 0.1297 \text{ m}$$

= 12.97 cm.

$\therefore$ Water available for evapotranspiration = 12.97 cm.

But 1.2 cm of water is consumed in 1 day

$\therefore$ 12.97 cm of water will be consumed in $\frac{1}{1.2} \times 12.97 = 10.8$ days

Hence, irrigation, water should be applied at 10 days interval. **Ans.**

**Q. 8.** (*c*) *Give the classification of energy dissipators. State the situations in which Hydraulic jump type and Bucket type dissipators are recommended based on relative positions of jump height curve and tail water rating curve.* (15)

## Year 2015 (*Civil Services*)

**Q. 5.** (*b*) *Define duty and delta. Derive a relationship between duty and delta. Using this relation, find delta for a crop with base period of 100 days. The duty of the crop is 432 ha / cumec.* (10)

[**Hint.** Derive Equation (2.2) as:

$$\Delta = \frac{864B}{D} \text{ cm} \qquad \text{where } \Delta \text{ is in cm}$$

$B$ = Base period in days
$D$ = Duty in ha/cumec

Here $B$ = 100 days ; $D$ = 432 ha/cumec

$$\therefore \quad \Delta = \frac{864 \times 100}{432} \text{ cm} = \mathbf{200\ cm.} \quad \textbf{Ans.}]$$

**Q. 5.** (*c*) *Explaining the objectives of river training, discuss the methods used for river training.*

**Q. 6.** (*c*) *The following data are collected to determine irrigation requirement. The water application efficiency is 65% and consumptive use coefficient in the growing season is 0.8. Find evapotranspiration and irrigation requirement.*

| *Month* | *Mean Monthly temp. (in °C)* | *Monthly % sunshine (in hr)* | *Effective rainfall (in cm)* |
|---|---|---|---|
| *November* | *18* | *7.20* | *2.6* |
| *December* | *15* | *7.15* | *2.8* |
| *January* | *13.5* | *7.30* | *3.5* |
| *February* | *14.5* | *7.10* | *2.0* |

(20)

[**Solution.** Pl. follow exactly similar Solved Example 2.9, and compute as: Blaney-Criddle equation for consumption use is given as:

$$C_u = k\Sigma f \qquad \text{where } f = \frac{p}{40}[1.8t + 32] \qquad \text{...(Eq. 2.9)}$$

Here $k = 0.8$ (given)

Values of $f$ are now computed in table 29.2.

**Table 29.2.**

| *Month* | $t$ *Mean Monthly temp. °C* | $p$ = *Monthly percent of annual sunshine hrs.* | *Re Effective rainfall in cm* | $f - \frac{p}{40}(1.8t + 32)$ |
|---|---|---|---|---|
| (*1*) | (*2*) | (*3*) | (*4*) | (*5*) |
| November | 18 | 7.2 | 2.6 | 11.6 |
| December | 15 | 7.15 | 2.8 | 10.5 |
| January | 13.5 | 7.30 | 3.5 | 10.3 |
| February | 14.5 | 7.10 | 2.0 | 10.3 |
| | | | Σ = 10.9 cm | Σ = 42.7 cm |

$C_u = k\Sigma f = 0.8 \times 42.7 =$ **34.20 cm**

Hence, evapotranspiration = **34.2 cm. Ans.**

Consumptive irrigation requirement (CIR) $= C_u - R_e$ ...(Eq. 2.6)

= 42.7 – 10.9 = **31.8 cm. Ans.**

Field Irrigation Requirement = FIR $= \frac{\text{CIR}}{\eta_a} = \frac{31.8}{0.65} =$ **48.9 cm. Ans.]**

**Q. 7.** *(a) Design and sketch the shape of an ogee spillway using the data given below:*

*Upstream head of water = 20 m*

*Slope of upstream face = 1 : 1.5 (H : V)*

*k = 1.939, n = 1.81* (15)

**[Solution.** Pl. refer solved Example 21.1, and compute as under:

Presumably, the given value of upstream head of water of 20 m is the value of head over the crest of the ogee spillway;

or $H_d = 20$ m

The *d/s* profile of the ogee spillway is now given by Eq. (21.1) as:

$$x^n = k \cdot H_d^{\,n-1} \cdot y$$ ...(Eq. 21.1)

where $n = 1.81$

$k = 1.939$

$H_d = 20$ m

$$\therefore \quad x^{1.81} = 1.939\,(20)^{0.81}\,y$$

or $21.95\,y = x^{1.81}$ ; or $y = \frac{x^{1.81}}{21.95}$

$$\frac{dy}{dx} = \frac{1.81\,x^{0.81}}{21.95}$$

Since the *d/s* slope value is not given, $\frac{dy}{dx}$ is not known, and hence the value of *x* and *y* for the tangent point cannot be determined. To determine the *d/s* profile, **we will have to assume this slope value,** as say 1*H* : 1.5*V* or 0.7*H* : 1*V* ; or $\frac{dy}{dx} = \frac{V}{H} = \frac{1}{0.7}$ (say) and then compute as in solved Example 21.1, the tangent point as $x = 27.4$ m, and $y = 21.15$ m. The ***d/s profile*** can be drawn as in Fig. 21.17. ***U/s profile*** of ogee spillway for the given value of *u/s* slope equal to 1H : $1\frac{1}{2}$ V, can be drawn as per Fig. 21.6.

The compound circular curve of *u/s* profile can be plotted with values of $r_1$, $r_2$, *a* and *b*, computed as follows:

[From Fig. 21.6 $r_1 = 0.48\,H_d = 0.48 \times 20 \text{ m} = 9.6$ m

$r_2 = 0.22\,H_d = 0.22 \times 20 \text{ m} = 4.4$ m

$a = 0.115\,H_d = 0.115 \times 20 \text{ m} = 2.3$ m

$b = 0.214\,H_d = 0.214 \times 20 \text{ m} = 4.28$ m.]

with these values, the *u/s* profile can be easily drawn as per Fig. [illegible] **Ans.]**

## Year 2016 (*Civil Services*)

**Q. 5.** (*c*) (*i*) *Explain the problems associated with expansive soils while laying the canal lining and remedial measures.* (6)

**Q. 5.** (*c*) (*ii*) *Discuss the purpose of safety ladders in lined canals along with details.* (4)

**Q. 6.** (*c*) (*ii*) *Under what conditions is the hydraulic jump type stilling basin type-1 proposed? Explain with figure.* (8)

**Q. 7:** (*b*) (*i*) *Explain how the water pressure is resisted by the following dams*

*1. Gravity dams*

*2. Earth dams.* (2)

**Q. 7.** (*b*) (*ii*) *How do you arrive the base width of an elementary profile of gravity dam?* (8)

**Q. 7.** (*b*) (*iii*) *Explain the reasons for cracking of concrete in concrete gravity dams and remedial measures.* (10)

**Q. 8.** (*a*) (*i*) *Discuss various maintenance problems posed by irrigation canals during their use and write the remedial measures.* (15)

**Q. 8.** (*c*) (*i*) *Differentiate between a weir and a barrage. How do you decide which structure among them is needed at a site of diversion headwork?* (5)

**Q. 8.** (*c*) (*ii*) *What are the causes of failure of weirs and their remedies?* (10)

## Year 2017 (*Civil Services*)

**Q. 5.** (*b*) (*i*) *Define permanent wilting point and field capacity. How is it useful in deciding the frequency of irrigation?* (5)

**Q. 7.** (*b*) (*ii*) *Determine the frequency of irrigation using the following data:*

*Consumptive use = 2.8 mm / day*

*Root zone depth = 80 mm*

*Irrigation efficiency = 65%*

*Available soil moisture = 50%*

*Also find the depth of water to be applied.* (5)

**[Solution.** Pl. refer similar solved example 2.24, and compute as under :

Readily available moisture (depth) in root zone.

= 50% of max depth of available water in root zone.

= 50% × 80 mm = 40 mm = 4 cm

Consumptive use = 2.8 mm/day = 0.28 cm/day

or 0.28 cm of water depth is consumed by crop in = 1 day

∴ 4 cm of water depth will be consumed by crop in $= \frac{1}{0.28} \times 4$

= 14.28 days ≈ 14 days.

Hence frequency of irrigation = **14 days. Ans.**

Also, the depth of water to be recouped through irrigation = 4 cm.

∴ Actual water depth required to be applied to the field

$$= \frac{\text{Water required to be recouped or absorbed}}{\text{Efficiency of irrigation application}}$$

$$= \frac{4 \text{ cm}}{0.65} = 6.15 \text{ cm}$$

Hence depth of water applied = **6.15 cm. Ans.]**

**Q. 6.** (*a*) (*ii*) *Explain step by step how reservoir capacity is decided using a mass inflow curve.* (5)

**Q. 6.** (*a*) (*ii*) *Distinguish between single-purpose reservoir and multi-purpose reservoir. Explain two important methods of allocation of water in a multi-purpose reservoir.* (5)

**Q. 6.** (*b*) (*i*). Determine the design discharge of a canal with the following data:

| *Crop* | *Base period (days)* | *Area (hectares)* | *Duty (hectares / cumec)* |
|---|---|---|---|
| Sugarcane | 320 | 850 | 580 |
| Overlap sugarcane in hot weather | 90 | 120 | 580 |
| Wheat (Rabi) | 120 | 600 | 1600 |
| Bajra (Monsoon) | 120 | 500 | 2000 |
| Vegetables (Hot weather) | 120 | 360 | 600 |

Time factor of the canal = 15/20

Capacity factor = 0.75 (5)

**[Solution.** Pl. follow solved example 3.6, and compute as under :

(*i*) Water required for sugarcane (whole year) = $\frac{\text{Area}}{\text{Duty}} = \frac{850 \text{ ha}}{580 \text{ ha/cumec}}$

= 1.466 cumec

(*ii*) Water required for overlap sugarcane (hot season) = $\frac{120}{580}$ = 0.207 cumec.

(*iii*) Water required for wheat (Rabi) = $\frac{600}{1600}$ = 0.374 cumec.

(*iv*) Water required for bajra (monsoon) = $\frac{500}{2000}$ = 0.40 cumec.

(*v*) Water required for vegetables (Hot season) = $\frac{360}{600}$ = 0.60 cumec.

∴ Total water required in each season is as follows :

(*a*) in hot season = 1.466 + 0.207 + 0.60 = 2.273 cumecs

(*b*) in monsoon season = 1.466 + 0.4 = 1.866 cumecs

(*c*) in rabi season = 1.466 + 0.374 = 1.84 cumecs

From the above data, it becomes clear that maximum water is required in hot season = **2.273 cumecs.**

Now, time factor = 15/20

As the canal runs for fever days than the crop period, the full supply discharge (based on average demand or duty) = $\frac{2.273}{15/20}$ = 3.03 cumecs.

Now, capacity factor = 0.75

Since capacity factor $= \frac{\text{Mean supply discharge in a canal}}{\text{Designed full capacity}}$

$$\therefore \quad 0.75 = \frac{3.03}{\text{Designed canal capacity}}$$

∴ Designed capacity of the canal is designed discharge

$$= \frac{3.03}{0.75} = 4.04 \text{ cumecs} \quad \textbf{Ans.}$$

**Q. 6.** (*b*) (*ii*) *Explain the modes of failure in gravity dams and earthen dam.* (5)

**Q. 6.** (*b*) (*iii*) *Write a note on reservoir sedimentation.* (5)

**Q. 6.** (*c*) (*i*) *Explain the function of a spillway. What are the advantages of an ogee spillway? Explain with a neet sketch.* (5)

**Q. 6.** (*c*) (*ii*) *Compute the discharge over an ogee spillway with the following data:*

*C = 2.4.*
*Head = 2 m*
*Length = 100 m*
*Crest is 8 m above the bottom* (5)

[**Solution.** From eq. (21.4), we have

$$Q = C \cdot L \cdot H_e^{3/2}$$

where $C$ = Coefficient of discharge = 2.4
$L$ = Length = 100 m
$H_d$ = 2 m
$h$ = 8 m

But $H_e = H_d$ + Vel head ($H_a$)

Neglecting vel head, we have

$$H_e = H_d = 2 \text{ m}$$

$$\therefore \quad Q = 2.4 \times 100\,(2)^{3/2} = \mathbf{678.82\ m^3/sec} \quad \textbf{Ans.}$$

**Q.6.** (*c*) (*iii*) *Write a note on stilling basin and river training works* (10)

**Q.8.** (*b*) (*i*) *Explain Lacey's regime theory. Define regime channel, initial regime and final regime.* (5)

**Q.8.** (*b*) (*ii*) *Explain the steps involved to design a channel using Lacey's theory.* (5)

**Q.8.** (*b*) (*iii*) *Compare Lavey's theory and Kennedy theory.* (5)

## Year 2018 (*Civil Services*)

**Q. 5.** (*b*) *What are the conditions which favour the selection of embankment dam?* (10)

**Q. 5.** (*c*) *What are the requirements for the success of an irrigation project.* (10)

**Q.7.** (*a*) *An earthen canal is to be designed to irrigate an area of 18000 ha of wheat during Rabi season. The duty at the head of the canal is determined to be 500 ha/m³/s. Design a trapezoidal canal with a side slope of 1.5 H : IV. Permissible velocity = 0.9 m/s; B/D ratio = 6, Manning n = 0.020* (15)

[**Solution.** Area of crop to be irrigated = 18000 ha

Duty at canal head = 500 ha/m³/s

$$\therefore \text{ Discharge required in canal} = \frac{\text{Area}}{\text{Duty}} = \frac{18000}{500} \text{ m}^3/\text{s} = 36 \text{ m}^3/\text{s}$$

Now, we have to design a trapezoidal canal with the following data :

$Q$ = Discharge = 36 m³/s

$V_{max}$ = Permissible velocity = 0.9 m/s

B/D ratio = 6

Side slopes = 1.5 H : 1V
Manning $n = 0.020$
Using $Q = V.A.$, we have

$$\text{Required flow area of canal} = A_{min} = \frac{Q}{V_{\max}} = \frac{36 \text{ m}^3/\text{s}}{0.9 \text{ m/s}} = 40 \text{ m}^2$$

But $A = \left[\frac{B + B + 3y}{2}\right] y$ (From Fig. 29.5.)

or $A = (B + 1.5\,y)\,y = 40 \text{ m}^2$.

or, $(B + 1.5y)y = 40$ ..(i)

Also $B/D = 6$; *i.e.*, $\frac{B}{y} = 6$

Substituting B = 6y in Eq. (i), we get

$(6y + 1.5y)\,y = 40$

or $7.5\,y^2 = 40 \text{ m}^2$

or $y = 2.30$ m

$\therefore$ $B = 6y = 6 \times 2.30$

$= 13.86$ say 14 m.

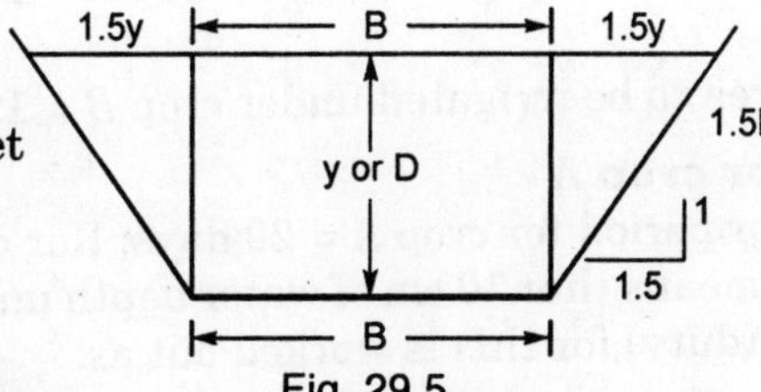

Fig. 29.5.

Hence, use an earthern canal with $B = 14$ m. and $y$ = water depth = 2.30 m with bed slope, yet to be determined.

The discharge capacity of this canal can be computed with Manning's formula, as:

$$Q = \frac{1}{n}.A.R.^{2/3}\sqrt{S}$$

where $n = 0.020$

$A = (B + 1.5y)\,y = (14 + 1.5 \times 2.30) \times 2.30$

$= 40.14 \text{ m}^2$

$P = B + 2\sqrt{(1.5y)^2 + y^2}$

or $P = 14 + 2\sqrt{(1.5 \times 2.30)^2 + (2.30)^2}$

or $P = 14 + 2\sqrt{11.90 + 5.29}$

or $P = 22.29$ m

$\therefore$ $R = \frac{A}{P} = \frac{40.34}{22.29} = 1.80$

$$36 \text{ m}^3/\text{s} = \frac{1}{0.020} \times (40.14)\,(1.80)^{2/3}.\sqrt{S}$$

$\therefore$ $36 \text{ m}^3/\text{s} = 2969.87\,\sqrt{S}$

or $\frac{1}{\sqrt{S}} = \frac{2969.87}{36}$ 82.50 or $\frac{1}{S} = 6805$

$\therefore$ Bad slope of canal $S = \frac{1}{6805}$

Hence use an earthern canal with bed width = 14 m

Water depth = 2.30 m, and bed slope = $\frac{1}{6805}$

Suitable free board of 0.75 m (Table 4.7) can be provided to get total depth of canal 2.30 + 0.75 = 3.05 m. Say 3.1 **Ans.]**

### Year 2019 (*Civil Services*)

**Q. 5.** (*c*) *A watercourse has a culturable commanded area of 1200 hectares. The intensity of irrigation for crop A is 40% and for crop B is 35%, both the crops being Rabi crops. Crop A has kor period of 20 days and crop B has kor period of 15 days. Calculate the discharge of the watercourse if the kor depth for crop A is 10 cm and for crop B, it is 16 cm.* *(10)*

**[Solution.** CCA = 1200 ha.

I.I. for crop $A$ = 40%

I.I. for crop $B$ = 35%

$$\text{Area to be irrigated under crop } A = 1200 \times \frac{40}{100} = 480 \text{ ha.}$$

$$\text{Area to be irrigated under crop } B = 1200 \times \frac{35}{100} = 420 \text{ ha.}$$

**For crop *A***

Kor period for crop $A$ = 20 days; Kor depth for crop $A$ = 10 cm

It means that 10 cm of water depth must be supplied in 20 days. The outlet factor (duty) for this is worked out as:

$$D = \frac{864B}{\Delta} = \frac{864 \times 20}{10} = 1728 \text{ ha/cumec}$$

where $D$ is duty in ha/cumec

$\Delta$ is in cm; and $B$ in days.

Now, discharge required to mature 480 hectares of land for an outlet factor of 1728 ha/cumec

$$= \frac{\text{Area in hactare}}{1728 \text{ ha/cumec}} = \frac{480}{1728} \text{ cumec}$$

$$= \mathbf{0.278 \text{ cumecs}} \quad \ldots(i)$$

**For crop *B***

Kor period for crop $B$ = 15 days ; Kor depth for crop $B$ = 16 cm

Outlet factor for this is worked out as

$$D = \frac{864B}{\Delta} = \frac{864 \times 15}{16} = 810 \text{ ha/cumec}$$

### Year 2020 (*Civil Services*)

**Q. 5.** (*c*) *A watercourse has culturable commanded area of 2600 hectares, out of which the intensities of irrigation for perennial sugarcane and rice crops are 20% and 40%, respectively. The duties for these crops at the head of watercourse are 750 hectares/cumec and 1800 hectares / cumec, respectively. Find the discharge required at the head of watercourse if the peak demand is 120% of the average requirement.* *(10)*

**[Solution.** Follow Exactly Similar Solved Example 3.3 and compute as under:

CCA = 2600 hactares

(*i*) Area to be irrigated under sugarcane

= 2600 ha × Int. of irrigation for sugarcane

= 2600 ha × 20% = **520 ha.**

(*ii*) Area to be irrigated under rice

= 2600 ha × Int. of irrigation for rice

Side slopes = 1.5 H : 1V

Manning $n = 0.020$

Using $Q = V.A.$, we have

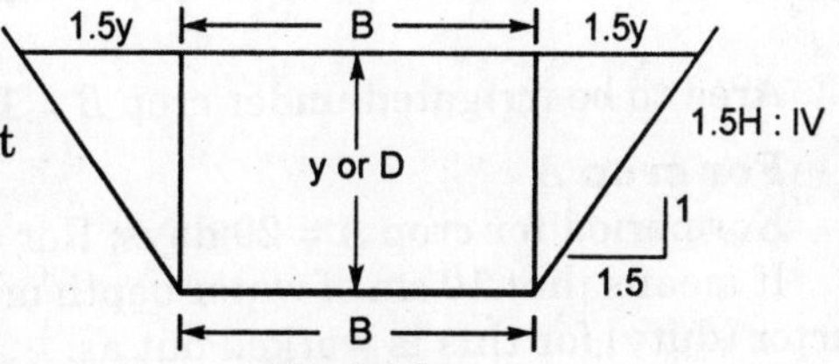

$= 2600 \text{ ha} \times 40\% = \mathbf{1040 \text{ ha.}}$

Average duty for sugarcane at the head of w.c.

$= 750 \text{ ha/cumec. (given)}$

$$\therefore \text{ Discharge required for surgarcane} = \frac{\text{Area}}{\text{Duty}} = \frac{520 \text{ ha.}}{750 \text{ ha/cumec}}$$

$$= \mathbf{0.69 \text{ cumec}} \qquad ...(i)$$

Also, Average duty for rice $= 1800$ ha/cumec

$$\therefore \text{ Discharge required for rice} = \frac{\text{Area}}{\text{Duty}} = \frac{1040 \text{ ha.}}{1800 \text{ ha/cumec}} = \mathbf{0.58 \text{ cumec}} \qquad ...(ii)$$

Total discharge required for sugarcane and rice

$= (i) + (ii) = 0.69 + 0.58 = \mathbf{1.27 \text{ cumecs}}$

The above value is an average demand. Since peak demand is 1.2 times the average demand, we have

Peak discharge required $= 1.2 \times 1.27 = \mathbf{1.52 \text{ cumecs.}}$

**Q. 6.** (*b*) *What are the main factors that affect the selection of type of dam?* (15)

**Q.7.**(*a*) *Design a trapezoidal alluvial channel using Kennedy's theory. The channel carries a discharge of 30 m³/s with critical velocity ratio 1.0 and Manning's roughness coefficient n equal to 0.0225. Assume that the bed slope is equal to 1 in 5000 and the trapesoidal channel has side slope of 1 H : 2V.* (15)

**[Solution.** Pl. refer Solved Example 4.7 and compute as under:

$Q = 30 \text{ m}^3/\text{s}$ ; CVR = m = 1.1

$n$ = Manning's roughness coefficient = 0.0225

$S = \dfrac{1}{5000}$ ; Side slopes = 1 H : 2V

Use Kennedy's equation as :

$$V_o = 0.55 \text{ m} \cdot y^{0.64} \qquad ...(\text{Eq. } 4.19)$$

where $y$ = water depth in the channel

$m = 1.1$ (given)

**Assume water depth (y = 1.8 m (say)**

$$\therefore \qquad V_o = 0.55 \times 1.1 \times (1.8)^{0.64} = 0.88 \text{ m/s}$$

$$A = \frac{Q}{V_o} = \frac{30 \text{ m}^3/\text{s}}{0.88 \text{ m/s}} = 34.1 \text{ m}^2$$

Now, for the channel section shown in Fig. 29.6, we have

$$A = \frac{B + (B + y)}{2} \cdot y$$

or

$$A = \left(B + \frac{y}{2}\right) y$$

or

$$A = \left(B + \frac{1.8}{2}\right) 1.8$$

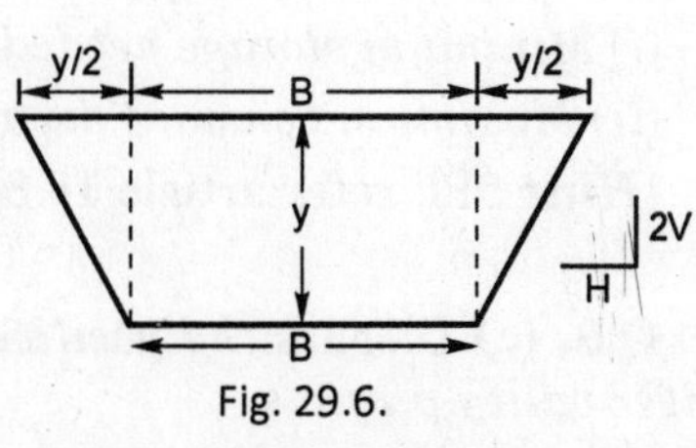

Fig. 29.6.

or $$34.1 = B + 0.9)\ 1.8$$

or $$B = 18.04 \text{ m}$$

Now $$P = B + 2 \cdot \sqrt{\left(\frac{y}{2}\right)^2 + y^2} = B + \sqrt{5} \cdot y$$

or $$P = 18.04 + \sqrt{5} \times 1.8 = 22.06$$

$\therefore$ $$R = \frac{A}{P} = \frac{34.1}{22.06} = 1.55 \text{ m}$$

Now, from Manning's formula, actual mean velocity ($V$) is given as:

$$V = \frac{1}{n} \cdot R^{2/3} \sqrt{S} \qquad \text{...(Eq. 4.21)}$$

$\therefore$ $$V = \frac{1}{0.0225} \times (1.55)^{2/3} \sqrt{\frac{1}{\sqrt{5000}}} = 0.84 \text{ m/s}$$

Since, $V_o$ of 0.88 m/s is more than $V$ of 0.84 m/s, we should reduce depth to reduce $V_o$. Let us now assume depth, $y$ = 1.4 m

$\therefore$ $$V_o = 0.55 \times 1.1 \times (1.4)^{0.64} = 0.75 \text{ m/s}$$

$$A = \frac{Q}{V_o} = \frac{30}{0.75} = 40 \text{ m}^2$$

$$A = \left(B + \frac{1.4}{2}\right) 1.4$$

or $$40 = (B + 0.7)\ 1.4$$

or $$B = 27.9 \text{ m}$$

$$P = B + \sqrt{5} \cdot y = 27.9 + \sqrt{5} \times 1.4 = 31 \text{ m}$$

$$R = \frac{A}{P} = \frac{40}{31} = 1.29$$

$$V = \frac{1}{0.0225} \times (1.29)^{2/3} \cdot \frac{1}{\sqrt{5000}} = 0.75 \text{ m/s}$$

Since $V_o$ tallies with $V$, the assumed depth $y$ of 1.4 m is OK.

Hence, use a channel with

$B$ = Bed width = 27.9 m, or say 28 m

$y$ = Water depth = 1.4 m

Side slopes = 2 $V$ : 1$H$ (or ½H : 1$V$)

**Q.8. (*b*)** *With the aid of neat sketches, describe how a flow-mass curve could be used for the determination of the following :*

(*i*) *Minimum storage needed to meet a constant demand*

(*ii*) *Maximum constant maintainable demand from a given storage.* (15)

[***Hint :*** Pl. refer article 18.8]

**Q.8. (*c*)** *Discuss the positive and negative impacts on environment due to river valley projects.* (15)

# —30

# *Conventional Questions of the Engineering Services Competition*

The **revised examination scheme** and syllabus for the Engg. Services Competition is given below :

**COMBINED ENGINEERING SERVICES EXAM.(Civil Engg.)**

**With effect from the year 2017, the examination scheme for this competition has been revised by the UPSC. This examination will now be conducted in the following three stages:**

| | | |
|---|---|---|
| **Stage I:** | *Preliminary Exam.* | 500 marks |
| **Stage II:** | *Main Exams.* | 600 marks |
| **Stage III:** | *Personality Test* | 200 marks |
| | Total | 1300 marks |

**These stages are explained below:**

**(*a*) Stage I:** Preliminary Stage I Exam under Stage I will consists of the following two papers:

| | | |
|---|---|---|
| **Paper-I:** | *General Studies and Engineering Aptitude Test* | *– 2 hrs – 200 marks* |
| **Paper-II:** | *Engineering Discipline Specific Paper* | *– 3 hrs – 300 marks* |
| | Subtotal | 500 marks |

**Note:** (1) Only those candidates qualifying at this stage will be permitted to appear for Stage-II examination.

**(2) Minimum qualifying marks for each paper shall be decided at the discretion of the UPSC.**

**(*b*) Stage II:** Main Stage II Exam will consists of the following two papers:

| | | |
|---|---|---|
| **Paper-I:** | *Engg. Discipline Specific Paper* | *– 3 hrs – 300 marks* |
| **Paper-II:** | *Engg. Discipline Specific Paper* | *– 3 hrs – 300 marks* |
| | Subtotal | 600 marks |

**Subtotal:** Stage-I + Stage-II = 1100 marks.

**Only those candidates qualifying at (Stage-I + Stage-II) shall be eligible for Stage-III *i.e.*, Personality Test (Interview) of 200 marks.**

**However, to determine final merit list, marks of Stage-I, + Stage-II, and of Stage-III, shall be added together, to draw merit list, out of total of 1300 marks.**

The revised syllabus for Civil Engineering Stream for both the stages is given below:

## BROAD CONTENTS OF THF GENERAL STUDIES AND ENGINEERING APTITUDE PAPER (Stage-I, Paper-I)–Common to all streams.

### General Studies and Engineering Aptitude

(Stage I–Paper I, Objective type, Common to all candidates, 2 hours duration, 200 Marks maximum)

***The questions from the following topics will be set in Paper-I of Stage-I***

1. Current issues of national and international importance relating to social, economic and industrial development.
2. Engineering Aptitude covering logical reasoning and analytical ability.
3. Engineering Mathematics and Numerical Analysis.
4. General Principles of Design, Drawing, Importance of Safety.
5. Standards and Quality practices in production, construction, maintenance and services.
6. Basics of Energy and Environment: Conservation, environmental pollution and degradation, climate change, environmental impact assessment.
7. Basics of Project Management.
8. Basics of Material Science and Engineering.
9. Information and Communication Technologies (ICT) based tools and their applications in Engineering such as networking, e-governance and technology based education.
10. Ethics and values in Engineering profession.

***Note: The paper in General Studies and Engineering Aptitude will include knowledge of relevant topics as may be expected from an engineering graduate, without special study.***

***Questions from all the 10 topics mentioned above shall be set. Marks for each topic may range from 5% to 15% of the total marks in the paper.***

**Syllabus for both the papers together for Stage-I objective type Paper-II, and separately for Stage-II conventional type Paper-I and Paper-II.**

***Civil Engineering Paper I (Revised w.e.f. the year 2017)***

**1. Building Materials:**

Stone, Lime, Glass, Plastics, Steel, FRP, Ceramics, Aluminium, Fly Ash, Basic Admixtures, Timber, Bricks and Aggregates: Classification, properties and selection criteria;

Cement: Types, Composition, Properties, Uses, Specifications and various tests; Lime and Cement Mortars and Concrete: Properties and various tests; Design of Concrete Mixes: Proportioning of aggregates and methods of mix design.

**2. Solid Mechanics:**

Elastic constants, stress, plane stress, strains, plain stress, Mohr's circle of stress and strain, elastic theories of failure, principle stresses, bending, shear and torsion.

**3. Structural Analysis:**

Basics of strength of materials, types of stresses and strain, bending moments and shear force, concept of bending and shear stresses; Analysis of determinate and indeterminate structures; Trusses, beams, plane frames; Rolling loads, influence lines, unit load method and other methods; Free and forced vibrations of single degree and multi degree freedom system; Suspended cables; Concepts and use of Computer Aided Design.

4. **Design of Steel Structures:**

Principles of Working Stress methods, Design of tension and compression members, Design of beams and beam column connections, built-up sections, girders, industrial roofs, Principles of Ultimate load design.

5. **Design of Concrete and Masonry Structures:**

Limit state design for bending, shear, axial compression and combined forces; Design of beams, Slabs, Lintels, Foundation, Retaining walls, Tanks, Staircases; Principles of pre-stressed concrete design including materials and methods; Earthquake resistant design of structures; Design of Masonry Structure.

6. **Construction Practice, Planning and Management**

Construction–Planning, Equipment, Site investigation and Management including Estimation with latest project management tools and network analysis for different types of works; Analysis of Rates of various types of works; Tendering Process and Contract Management, Quality Control, Productivity, Operation Cost, Land acquisition, Labour safety and welfare.

***Syllabus for Civil Engineering Paper-II (Revised wef year 2017)***

1. **Flow of Fluids, Hydraulic Machines and Hydro Power:**

**(*a*) Fluid Mechanics, Open Channel Flow, Pipe Flow:**

Fluid properties; Dimensional Analysis and Modeling; Fluid dynamics including flow kinematics and measurements; Flow net; Viscosity, Boundary layer and control, Drag, Lift, Principles in open channel flow, Flow controls. Hydraulic jump; Surges, Pipe networks.

**(*b*) Hydraulic Machines and Hydro Power**

Various pumps, Air vessels, Hydraulic turbines–types, classifications and performance parameters; Power house–classification and layout, storage, pondage, control of supply.

2. **Hydrology and Water Resources Engineering:**

Hydrological cycle, Ground water hydrology, Well hydrology and related data analysis; Streams and their gauging; River morphology; Flood, drought and their management; Capacity of Reservoirs.

Water Resources Engineering: Multipurpose uses of water, River basins and their potential; Irrigation systems, water demand assessment; Resources–storages and their yields; Water logging, canal and drainage design, Gravity dams, falls, weirs, energy dissipaters, barrage distribution works, cross drainage works and head-works and their design; Concept in canal design, construction and maintenance; River training, measurement and analysis of rainfall.

3. **Environmental Engineering:**

**(*a*) Water Supply Engineering:**

Sources, Estimation, quality standards and testing of water and their treatment; Rural, Institutional and industrial water supply; Physical, chemical and biological characteristics and sources of water, pollutants in water and its effects, Estimation of water demand; Drinking water standards, Water Treatment Plants, Water distribution networks.

**(*b*) Waste Water Engineering:**

Planning and design of domestic waste water, sewage collection and disposal; Plumbing Systems. Components and layout of sewerage system; Planning and design of Domestic Waste-water disposal system; Sludge management including treatment, disposal and re-use of treated effluents; Industrial waste waters and Effluent Treatment Plants including institutional and industrial sewage management.

**(*c*) Solid Waste Management:**

Sources and classification of solid wastes along with planning and design of its management system; Disposal system, Beneficial aspects of wastes and Utilization by Civil Engineers.

**(*d*) Air, Noise pollution and Ecology:**

Concepts and general methodology.

**4. Geo-technical Engineering and Foundation Engineering:**

**(*a*) Geo-technical Engineering:** Soil exploration–planning and methods, Properties of soil, classification, various tests and inter-relationships; Permeability and Seepage, Compressibility, consolidation and Shearing resistance, Earth pressure theories and stress distribution in soil; Properties and uses of geo-synthetics.

**(*b*) Foundation Engineering:** Types of foundation and selection criteria, bearing capacity, settlement analysis; design and testing of shallow and deep foundations; Slope stability analysis, Earthen embankments, Dams and Earth retaining structures: types, analysis and design, Principles of ground modifications.

**5. Surveying and Geology:**

**(*a*) Surveying:** Classification of surveys, various methodologies, instruments and analysis of measurement of distances, elevation and directions; Field astronomy, Global Positioning System; Map preparation, Photogrammetry; Remote sensing concepts; Survey Layout for culverts, canals, bridges, road/railway alignment and buildings, setting out of curves.

**(*b*) Geology:** Basic knowledge of Engineering geology and its application in projects.

**6. Transportation Engineering:**

**Highways:** Planning and construction methodology, Alignment and geometric design; Traffic Surveys and Controls; Principles of Flexible and Rigid pavements design.

**Tunneling:** Alignment, methods of construction, disposal of muck, drainage, lighting and ventilation.

**Railway Systems:** Terminology, planning, designs and maintenance practices; track modernization.

**Harbours:** Terminology, layouts and planning.

**Airports:** Layout, planning and design.

**Note:** *Evidently, the subject of Hydrology and Water Resources is covered in Paper-II of Main Exam (i.e., Stage-II) for conventional questions. Objective type questions from the subject will be covered in Paper-II of Stage-I (i.e., Preliminary Exam).*

In order to enable you to have a practice for solving questions of this important examination and for your facility and guidance, we are giving below the conventional questions that appeared during the last several years, since 1995, from *Irrigation Engg. portion*, in this competition. The objective type questions of previous years have been given in chapter 28.

## Certain Typical Questions from the Old Engineering Services Exams. from year 1996 to 2006.

**Q. 30.1.** (a) *Calculate the seepage through an earthen dam resting on an impervious foundation. The relevant data are given below:*

| | |
|---|---|
| *Height of the dam* | *= 60.0 m* |
| *Upstream slope* | *= 2.75 : 1 (H : V)* |
| *Downstream slope* | *= 2.50 : 1 (H : V)* |
| *Freboard* | *= 2.5 m* |
| *Crest width* | *= 8.0 m* |
| *Length of drainage blanket* | *= 120.0 m* |

*Coefficient of permeability of the embankment material in*

*X-direction* $= 8 \times 10^{-7}$ *m/s*

*Y-direction* $= 2 \times 10^{-7}$ *m/s* (*Engg. Services, 1996*)

[**Hint.** As a matter of fact such problems have generally been solved in "Soil Mechanics", and hence the students will have to refer and study article 7.23.1.1 in authors another book titled "Geotech Engineering – Soil Mechanics and Foundation Engineering", before attempting to solve this question.

According to the theory explained in that article, we have by Eqn. (7.67)

$q$ = discharge through dam = $K.S.$ ...(Eq. 7.67)

where $K = \sqrt{K_H + K_V}$ ...(Eq. 7.56)

$S$ = Focal distance of base parabola forming the top seepage line in the dam body.

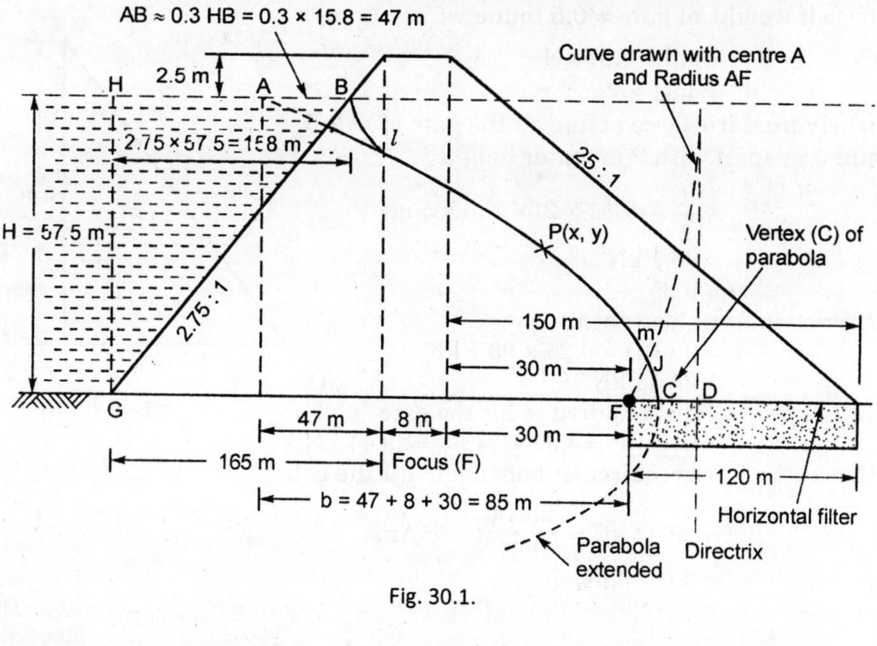


Fig. 30.1.

$S$ is further given by eq. (7.28) as

= Focal distance w.r. Fig. 7.59 in "Soil Mechanics"

$$= \sqrt{b^2 + H^2} - b \quad ...(\text{Eq. } 7.65)$$

where $H$ = max. water depth = 60 − 2.5 = 57.5 m

= 47 + 8 + 30 = 85 m (Please see Fig. 30.1.)

$$\therefore \quad S = \sqrt{85^2 + 57.5^2} - 85 = 17.65 \text{ m}$$

$$q = \sqrt{K_H . K_v} \cdot S$$

$$= \sqrt{8 \times 10^{-7} \times 2 \times 10^{-7}} \times 17.62 \text{ m}^3\text{/sec per m width of dam}$$

$= 4 \times 10^{-7} \times 17.62$ m³/s.m.

$= 70.48 \times 10^{-7}$ m³/sec. m = 0.610 m³/day.m

= 610 litres per day per m width of dam. **Ans.**

**Q. 30.2.** *A vertical lift gate 5 m × 2.5 m size weighing 0.5 Tonnes slides along guides (coefficient of friction 0.25) fitted on the side walls of an overflow spillway and its crest. What force will have to be exerted at the hoisting mechanism to lift the gate when the head of water over the crest is 2 m ?*

(*Engg. Services 1998*)

[**Solution.** In article 21.17.3, while describing the vertical lift gate, we have stated that when such a gate is sliding between groove guides not provided with rollers, force will be required to lift the gate at the hoisting mechanism for not only lifting the self wt. of the gate but also to overcome the friction that will be developed between the gate and the d/s groove guides due to the hydrostatic force caused by the water standing against the gate

(*i*) self weight of gate = 0.5 tonne wt.

= 0.5 × 9.81 kN

= 4.91 kN ...(*i*)

(*ii*) Hydrostatic force acting on the gate of ht 2.5 m and 5 m span with 2 m water height

$$P_H = \frac{1}{2} \times (9.81 \times 2) \times 2 \text{ m} \times 5 \text{ m}$$

= 98.1 kN

$\mu = 0.25$

∴ Friction force (downard)

$= \mu \cdot PH = 0.25 \times 98.1$ kN

= 24.53 kN ...(*ii*)

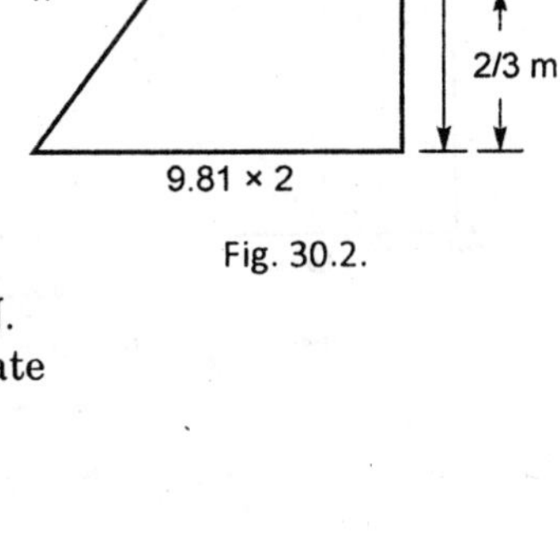

Fig. 30.2.

Total upward force required to lift the gate

= (*i*) + (*ii*) = 4.91 + 24.53 = 29.44 kN.

Hence, the force required at hoisting to lift the gate

$$= 29.44 \text{ kN} = \frac{29.44}{9.81} \text{t} \qquad \textbf{Ans.}$$

= **3 t** **Ans.**

**Q. 30.3.** *A river discharges 1000 cum / sec of water at high flood level of RL 103 m. A weir is constructed for flow diversion with a crest length of 255 m and total length of concrete floor as 40 m. The weir has to sustain the under-seepage at a maximum static head of 2.4 m. The silt factor and the safe exit gradient for the river bed material are 1.1 and 1 / 6, respectively. Determine the depth of cut off required at the downstream end of the concrete floor. Take the level of downstream concrete floor as RL 100.0 m. Check for exit gradient.*

(*Engg. Services, 1999*)

[**Solution.** High flood discharge = $Q$ = 1000 m$^3$/s.

Length of weir = $L$ = 255 m

$$q = \frac{Q}{L} = \frac{1000}{255} = 3.92 \text{ m}^2\text{/s}$$

$f$ = 1.1 (given)

$$R = 1.35\left(\frac{q^2}{f}\right)^{1/3} = 1.35\left(\frac{3.92^2}{1.1}\right)^{1/3} = 3.25 \text{ m}$$

Provide D/s cutoff up to 1.5 $R$ below d/s water level

1.5 $R$ = 1.5 × 3.25 = 4.88 m

D/s W.L. = ?

U/s WL = U/s HFL = 103.00 m (given)

$H$ = Max. static head causing seepage = 2.4 m (given)

∴ D/s W.L. = U/s HFL – $H$ = 103.0 – 2.4 = 100.6 m

Hence, R.L. of bottom of d/s cutoff = 100.6 – 4.88 = 95.72 m.

R.L. of D/s floor = 100.0 m (given)

∴ Depth of d/s cutoff = 100.0 – 95.72 = 4.28 m.

Hence $d$ = **4.28 m Ans.**

$$\text{Exit gradient } G_E = \frac{H}{d}\cdot\frac{1}{\pi\sqrt{\lambda}}$$

$$\text{where } \lambda = \frac{1+\sqrt{1+\alpha^2}}{2}$$

$$\text{where } \alpha = \frac{b}{d} = \frac{\text{Length of weir floor}}{\text{Depth of d/s cutoff}}$$

where, $b$ = 40 m (given)

$d$ = 4.28 m (calculated above)

$$\therefore \quad \alpha = \frac{40}{4.28} = 9.346$$

$$\therefore \quad \lambda = \frac{1+\sqrt{1+9.346^2}}{2} = 5.2$$

$$\therefore \quad G_E = \frac{2.4 \text{ m}}{4.28 \text{ m}}\,\frac{1}{\pi\,.\,\sqrt{5.2}} = \mathbf{0.078} \quad \textbf{Ans.}$$

$$= \frac{1}{12.77} < \frac{1}{6} \text{ (Safe exit gradient)}$$

Hence, the weir is safe from exit gradient considerations with bottom of d/s cut off at RL 95.72 m. **Ans.**

**Q. 30.4.** *An unlined irrigation canal has its bed and sides composed of cohesionless material having mean diameter 6 mm. Angle of repose of the material is 40°. The bed width of the canal is 5 m and the side slope 1.5 (H) : 1 V. Determine the maximum discharge that can be admitted into the canal without any sediment movement. Longitudinal slope of the canal is 1 in 5000, and Manning's n = 0.025.*

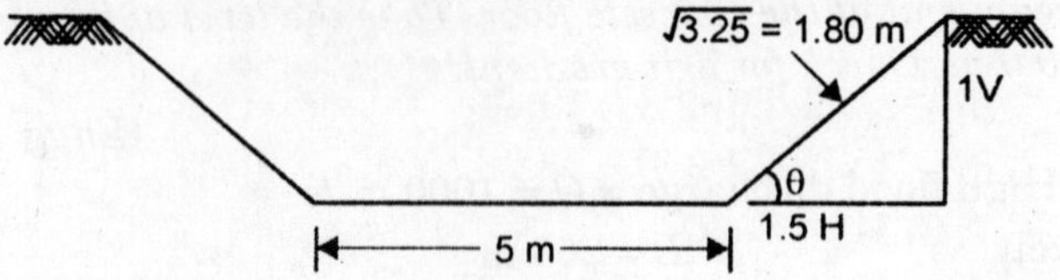


Fig. 30.3. *(Engg. Services, 2001)*

[**Solution.** Follow and refer Solved Example 4.4, and compute as under :

$$\phi = 40° \text{ (given)} \qquad \tan\theta = \frac{1}{1.5}$$

$$\sin\phi = \sin 40° = 0.6428 \qquad \sin\theta = \frac{1}{\sqrt{3.25}} = \mathbf{0.5547}$$

$$\frac{\tau_c'}{\tau_c} = \sqrt{1 - \frac{\sin^2\theta}{\sin^2\phi}} = \left[1 - \frac{(0.5547)^2}{(0.6428)^2}\right]^{1/2} = \sqrt{0.255} = 0.505.$$

∴ Min. shear stress required to dislodge the soil grain on side slope of canal is given by

$$\tau_c' = 0.505\ \tau_c$$

Hence, for stability, the shear stress actually going to be generated on slopes of the channel of given *R* & *S* must be less than or equal to 0.505 $\tau_c$ ; *i.e.*

$$\tau_0' \le 0.505\ \tau_c$$

The actual shear stress actually going to be generated on the side slopes of a channel of given *R* and *S* is given by Eqn. (4.17) as :

$$\tau_0' = 0.75\ \gamma_w \cdot R.S.$$

$$\therefore\ 0.75 \cdot \gamma_w \cdot R.S. \quad \le \quad 0.505 \cdot \tau_c$$

But $\tau_c = \gamma_w \cdot R.S. = \gamma_w \cdot \dfrac{d}{11}$ $\left(a\ R.S. = \dfrac{d}{11} \text{ from Eq. 4.13}\right)$

$$\therefore 0.75 \cdot \gamma_w \cdot R.S. \quad \le \quad 0.505\ \gamma_w \cdot \frac{d}{11}$$

or $RS \le \dfrac{0.505}{0.75 \times 11} d$ or $RS \le 0.0612\ d$

where *d* = mean particle dia = 6 mm = 0.006 m

*S* = 1 in 5000

or $RS \le 0.0612 \times 0.006 = 3.67 \times 10^{-4}$

or $R \times \dfrac{1}{5000} \le 3.67 \times 10^{-4}$ or $R \le 1.837$ m.

using 20% factor of safety, we have

max. $R = 1.837 \times 0.8 = 1.47$ m.

For 5 m wide trapezoidal channel with 1.5 (H) : 1 (V) side slopes, we have

$$R = \frac{A}{P} = \frac{(5 + 1.5y)y}{(5 + 2 \times 1.8y)} = \frac{(5 + 1.5y)y}{(5 + 3.6y)}$$

$$\therefore \quad 1.47 = \frac{(5 + 1.5y)y}{(5 + 3.6y)}$$

Solve by Hit and Trial, and find

$y \approx 2.3$ m.

$\therefore \quad A = (5 + 1.5 \times 2.3)\, 2.3 = 19.44$

$P = (5 + 2 \times 1.8 \times 2.3) = 13.28$

$R = \mathbf{1.46} \approx 1.47$ (ok)

$$Q_{max.} = \frac{1}{n} \cdot A \cdot R^{2/3} \cdot \sqrt{S}$$

$$= \frac{1}{0.025} \times 19.44 \times (1.46)^{2/3} \frac{1}{\sqrt{5000}} = \mathbf{14.15\ m^3/sec.} \quad \textbf{Ans.}$$

**Q. 30.5.** *Following particulars were recorded from a barrage :*

*(i) Maximum reservoir level* = *212 m*
*(ii) Pond level* = *211 m*
*(iii) Downstream high flood level in the river* = *210 m*
*(iv) Maximum design flood discharge* = *3500 $m^3/s$*
*(v) Crest level of the barrage* = *207 m*
*(vi) Crest level of the head regulator* = *208 m*
*(vii) Coefficient of discharge* = *2.10 $m^{1/2}$/sec for barrage*
= *1.50 $m^{1/2}$ sec for head regulator*
*(viii) River bed level* = *205 m*
*(ix) Design discharge of main canal* = *500 $m^3$/sec*

*Determine the number of gates required for the barrage and the head regulator if each gate has 10 m clear span. Neglect :*

*(1) end contractions due to piers and abutments ;*

*(2) velocity of approach.*

*If a stilling basin is provided downstream of the barrage for the energy dissipation, find the length and RL of the basin floor. Assume that the length of basin is 5 times the conjugate depth required for hydraulic jump. Neglect losses due to friction.* (*Engg. Services, 2001*)

**Solution.** (1) **Waterway for Barrage :**

**High flood discharge of 3500 m³/s has to pass through the gated openings provided at the barrage site, with the following data :**

**Max. water level at high flood U/S of barrage = 212.00 m**

**Crest level of barrage = 207.00 m**

**Opening available or Head causing flow above the crest of barrage**

**= 212.00 – 207.00 = 5.00 m** (**ignoring vel. head**)

**Now,** $Q = C \cdot L\, H^{3/2}$

**where $C$ = coeff. of discharge through the barrage**
**= 2.10 (given)**
**$L$ = Effective waterway of barrage (openings)**
**$H$ = 5.00 m**

$\therefore \quad 3500 = 2.10 \times L \times (5)^{3/2}$

or $\quad L = \dfrac{3500}{2.1 \times (5)^{3/2}} = 149.1 \text{ m} ; \text{ say } 150 \text{ m}$

Effective water way = Total water way (ignoring end contractions)

Hence, provide 15 bays, each of 10 m clear span.

$\therefore$ ***No. of gates required for barrage = 15 Ans.***

(2) **Water way for Canal Head Regulator:**

The discharge through canal regulator = 500 $m^3/s$

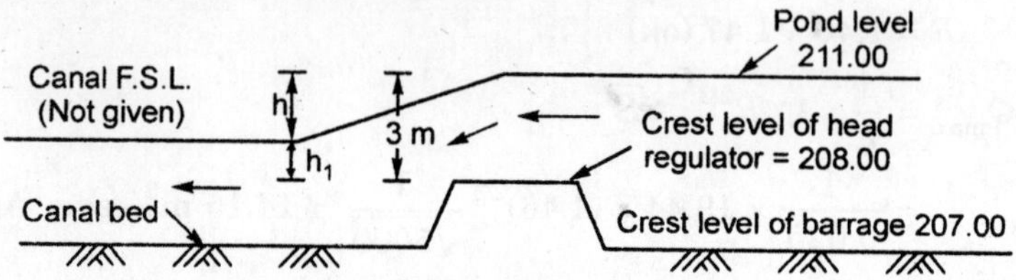


**Fig. 30.4.**

Coefficient of discharge for regulator opening = 1.5

Head over the crest of regulator at pond Level

$= 211.00 - 208.00 = 3.00 \text{ m}.$

$Q = C_1 \cdot L_1 \cdot H_1^{\,3/2}$

$\therefore \quad 500 = 1.5 \times L_1 \times (3)^{3/2}$

or $\quad L_1 = \dfrac{500}{1.5 \times (3)^{3/2}} = 64 \text{ m}.$

Provide 7 bays each of 10 m clear span, thereby providing 70 m water way.

No. of gates required for canal head regulator each of 10 m span = **7 Ans.**

(3) **Design of Stilling Basin:**

*U/S HFL* = 212.00

*D/S HFL* = 210.00

*U/S TEL* = 212.00 (ignoring velocity of approach)

*D/S TEL* = 210.00 (ignoring velocity of approach)

$H_L = 212.00 - 210.00 = $ **2 m**

***q*** = Discharge intensity through barrage bays

= 3500/150 = **23.34** $m^3/s/m$.

$E_{f_2}$ for $H_L = 2$ *m* and $q = 23.34$ cumecs, from Blench curves – (Plate 10.1) is found to be = **7.2** m

$\therefore$ Level at which jump will form

$= D/S\ TEL - E_{f_2} = 210.00 - 7.2$

= **202.8 m**

$y_2$ corresponding to $E_{f_2}$ of 7.2 m from plate 10.2 = 6.6 m.

Length of cistern = $5y_2 = 5 \times 6.6 = 33$ m.

Hence, provide 33 m long stilling basin at level say *202.6 m* (slightly lower than the calculated value of 202.8 m) as shown in Fig. 30.5.

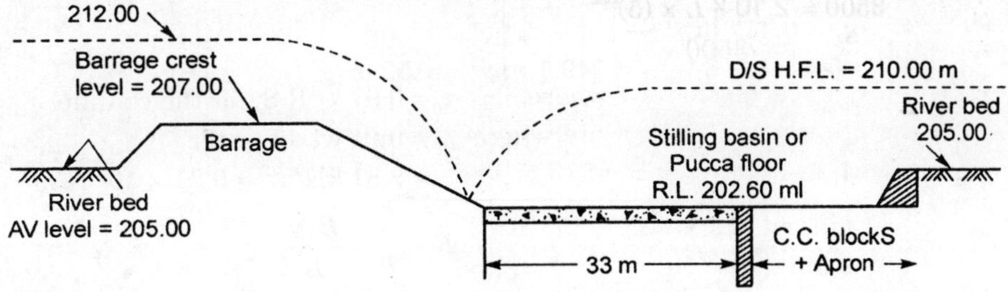


**Fig. 30.5.**

***Alternatively,*** *if Blench curves are not supplied in the Examination Hall,* then we can compute the value of $y_2$ as follows :

$$y_c = \sqrt[3]{\frac{q^2}{g}} \qquad \text{(from Eqn. 10.13)}$$

$$= \sqrt[3]{\frac{23.3^2}{9.81}} = \mathbf{3.81\ m}$$

Now let $\quad Y = y_2/ y_c \qquad$ (Eqn. 10.15)

and $\quad Z = H_L/y_c \qquad$ (Eqn. 10.16)

Compute value of $\quad Z = \dfrac{2\text{ m}}{3.81\text{ m}} = \mathbf{0.5248}$

Use Eqns. (10.19) and (10.20) to compute the value of $Y$ for known value of $Z$. Here $Z < 1$, hence Eqn. (10.19) is applicable

$\therefore \quad Y = 1 + 0.93556\ Z^{0.368} \quad (\text{for } Z < 1)$

or $\quad Y = 1 + 0.93556\ (0.5248)^{0.368} = \mathbf{1.738}$

$\therefore \quad Y = y_2/y_c = 1.738$

or $\quad y_2 = 1.738\ y_c = 1.738 \times 3.81\text{ m} = \mathbf{6.6\ m.}$

$5y_2$ = Length of cistern = 5 × 6.6 = **33 m. Ans.**

Also, $E_{f_2}$ can be calculated by Eq. (10.23) as :

$$\frac{E_{f_2}}{y_c} = Y + \frac{1}{2Y^2} = 1.738 + \frac{1}{2\,.\,(1.738)^2}$$

or $\quad E_{f_2} = 1.904 \times 3.81 = \mathbf{7.25\ m}$

Level at which jump will form

$= D/S\ TEL - E_{f_2} \qquad D/S\ TEL = D/S\ HFL$

$= D/S\ HFL - E_{f_2} \qquad$ (on neglecting vel. of approach)

$= 210.00 - 7.25 = 202.75$ m

Provide cistern at RL **202.6 m Ans.**

**Q. 30.6.** *A channel is 40 m wide, 2.6 m deep with slope of 1 in 5000. The representative bed material size $D_{65}$ is 0.64 mm, and mean diameter of the bed material is 0.38 mm. Mannings n = 0.021. Neglecting the side friction, find the bed load transported by the channel in tonnes per day by Meyer-Peter equation, considering B/D as large and $\tau_c$ = 0.075 d.* *(Engg. Services, 2003)*

**Solution.** Meyer-Peter's equation, giving bed load transport in N/m.s is given by Eq. (4.51) as :

$$g_b = 0.417\left[\tau_0\left(\frac{n'}{n}\right)^{3/2} - \tau_c\right]^{3/2}$$

where $\tau_0 = 0.97\ \gamma_w$ R.S. for the channel

where $\gamma_w$ = unit wt. of water

$= 9.81\ \text{kN/m}^3 = 9.81 \times 10^3\ \text{N/m}^3$

$$R = \frac{A}{P} = \frac{B.y}{B} = y$$

for wider channels having $B/y$ large = 2.6 m

$$S = \frac{1}{5000}$$

$\therefore$ $\tau_0 = 0.97 \times (9.81 \times 1000) \times 2.6 \times \dfrac{1}{5000} = \mathbf{4.948\ N/m^2}$

$n' = 1/24\ .\ d^{1/6}$

where $d$ is the representative size of the bed material in $m$, usually taken as **median size** ($d_{50}$) or $d_{65}$, which is given as equal to 0.64 mm.

$\therefore$ $d = 0.64 \times 10^{-3}$

$\therefore$ $n' = \dfrac{1}{24}(0.64 \times 10^{-3})^{1/6} = \mathbf{0.01223}$

***n* = 0.021 (given)**

$\tau_c$ is usually given as $\tau_c = 0.687\ .\ d_a$ when $\tau_c$ is in N/m², and $d_a$ is in mm. Here it is given that $\tau_c = 0.075d$. Eventually, the given eqn. is meant for $\tau_c$, when $\tau_c$ is expressed in kg/m². For S.I. units, the given eqn. will hence become

$$\tau_c = 0.075\ d_a \times 9.81\ \text{N/m}^2 = 0.736\ d_a\ (\text{N/m}^2)$$

$d_a$ is mean size of the bed particles, which is given as equal to 0.38 mm.

$\therefore$ $\tau_c = 0.736 \times 0.38\ \text{N/m}^2 = \mathbf{0.279\ N/m^2}$

Now, $$g_b = 0.417\left[4.948\left(\frac{0.01223}{0.021}\right)^{3/2} - 0.279\right]^{3/2}$$

$$= 0.417\ [2.199 - 0.279]^{3/2} = 0.417\ (1.92)^{1.5}$$

or $g_b = 1.109\ \text{N/m.s} = \dfrac{1.109}{9.81}$ kg/s/m

$\therefore$ Bed load transported by 40 m wide channel

$$= \frac{1.109}{9.81}\text{kg/s.m} \times 40\ \text{m} = 4.522\ \text{kg/sec} = \frac{4.522 \times 24 \times 3600}{1000}\ \text{t/day}$$

**= 390.69 tonnes/day. Ans.]**

**Q. 30.7.** *In a proposed reservoir, the areas containing within the contours are:*

| *Countor (in m) :* | *100* | *96* | *90* | *85* | *80* | *75* | *70* | *65* |
|---|---|---|---|---|---|---|---|---|
| *Area (in ha) :* | *32* | *26* | *24* | *18* | *15* | *13* | *7* | *2* |

*Using the method of end areas, calculate :*

*(i) capacity of the reservoir when it is full at 100 m level;*

*(ii) elevation of the water level when it is 60% full.*

*Ignore the volume below RL 65 m. Give your remarks about the method of 'End Areas'.* *(Engg. Services, 2006)*

**[Solution.** The ***Averaging End Areas Method,*** popularly called the ***End Areas Method*** is the simple Method of averaging of the areas. Use of this method involves averaging the end areas of successive ranges and multiplying by the distance between the ranges to obtain the intermediate volume. The total volume is computed by adding each intermediate volume for the entire reservoir height.

When areas at different contours are given, as here, the method is also known as ***Average contour area method.***

With this knowledge, you can easily solve this question by computing the mean areas between successive contours and multiplying those values with the contour interval between two successive contours, as shown in the solution table 30.1.

**Solution Table 30.1 for Question 30.7.**

| *Contour elevation m* | *Contour interval* | *Area at each contour in ha* | *Average Area* | *Volume enclosed between two consecutive contours (ha. m) (2) × (4)* | *Total Capacity Volume up to given elevation (computed from bottom to upward)* |
|---|---|---|---|---|---|
| *(1)* | *(2)* | *(3)* | *(4)* | *(5)* | *(6)* |
| 100 | | 32 | | | 596 |
| | 4 | | 29 | 116 | |
| 96 | | 26 | | | 480 |
| | 6 | | 25 | 150 | |
| 90 | | 24 | | | 330 |
| | 5 | | 21 | 105 | |
| 85 | | 18 | | | 225.0 |
| | 5 | | 16.5 | 82.5 | |
| 80 | | 15 | | | 142.5 |
| | 5 | | 14 | 70 | |
| 75 | | 13 | | | 72.5 |
| | 5 | | 10 | 50 | |
| 70 | | 7 | | | 22.5 |
| | 5 | | 4.5 | 22.5 | |
| 65 | | 2 | | | 0 (Given) |

(*i*) The capacity of the reservoir up to elevations between 65 m to 100 m are also computed by adding the partial volumes, treating the volume at 65 m elevation to be zero.

The capacity at RL 100 is worked out = **596 ham. Ans.**

(*ii*) The reservoir capacity (volume) to be filled for filling 60% of its full capacity = 60% × 596 ha.m = 355.8 ham. To determine the elevation corresponding to capacity of 355.8 ham, we can draw a curve (Fig. 30.6) between the computed values of capacity (col 6) to be on $x$-axis, and the corresponding values of elevation col (1) to be drawn on $y$-axis. From this curve (Fig. 30.6), the elevation corresponding to 355.8 ham is read out to be about **91 m. Ans.]**

**[Note.** In case graph sheet is not available to plot elevation capacity curve, then we can assume a linear relation of added capacity between RL 90 m (capacity = 330 ham) and RL 96 m (capacity = 480 ham) which shows 150 ham is added in 6 m elevation above RL 90 m. To add 355.8 – 330 = 25.8 ham, we need a rise of level

$$= \frac{6}{150} \times 25.8$$

$$= 1.032 \text{ m}$$

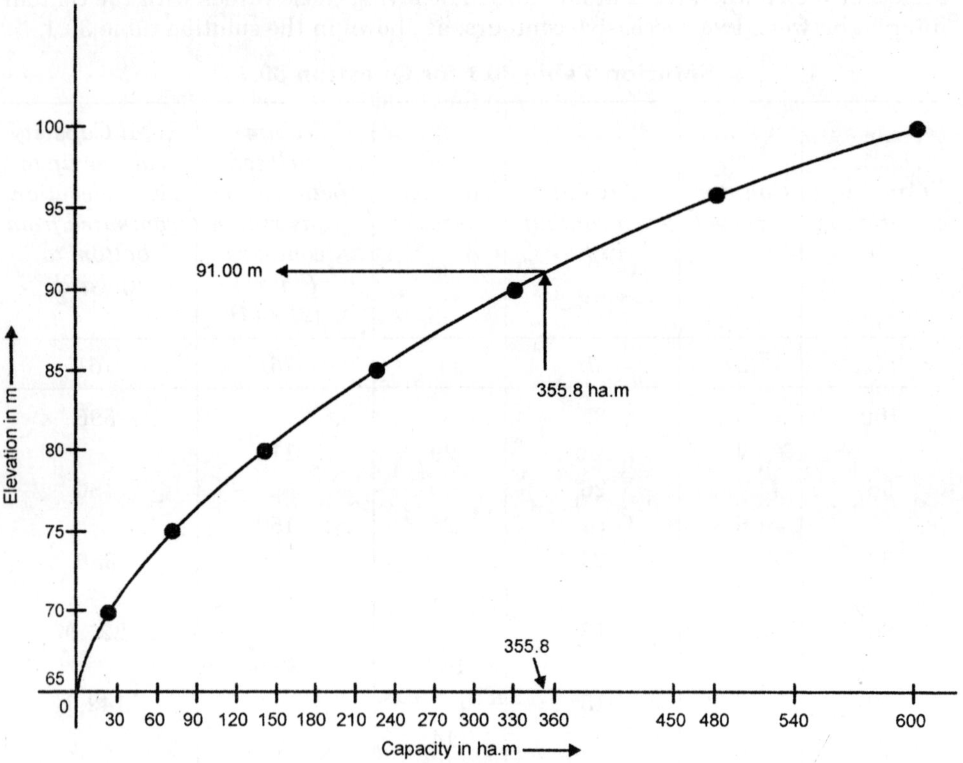


**Fig. 30.6**

Hence, elevation corresponding to 355.8 ham capacity

$$= 90 + 1.032$$

$$= \mathbf{91.032\ m}; \textit{ i.e. } \text{say } \mathbf{91\ m\ Ans.}$$

**Remarks on End Areas Method.** This method gives quite reliable results provided the contour interval is small and of the order of 0.6 m or so. However, the obtained results will not be very precise when contour interval is high as in the given question where it is 5 m. The capacity computation results obtained in such a case will only be approximate. **Ans.**

**Q. 30.8.** *A 5.0 m wide rectangular channel carries 15 $m^3/s$ of water with a velocity of 6 m/s. State whether hydraulic jump is a possibility. If yes, compute the height of the jump and power dissipated.* *(Engg. Services, 2010)*

**Solution.** $Q = 15 \text{ m}^3/\text{s}$

$V$ = Velocity = 6 m/s

$\therefore$ Area = $A = \dfrac{Q}{V} = \dfrac{15}{6} \text{ m}^2 = 2.5 \text{ m}^2$

But $A = B \cdot y_1 = 5 \cdot y_1$

$\therefore$ $5y_1 = 2.5 \text{ m}^2$

or $y_1 = 0.5$ m

Now, Froude Number, $F_1 = \dfrac{V_1}{\sqrt{g y_1}}$

or $F_1 = \dfrac{6}{\sqrt{9.81 \times 0.5}} = 2.71 > 1$

$\therefore$ Since $F_1 > 1$, it is a super critical flow, and hydraulic jump will form [Pl see Article 10.2].

Now, using Eq. (10.7), we have

$$y_2 = \frac{y_1}{2}\left[-1 + \sqrt{1 + 8F_1^2}\right]$$

$$\therefore \quad y_2 = \frac{0.5}{2}\left[-1 + \sqrt{1 + 8 \times (2.71)^2}\right] = 1.68 \text{ m}$$

Hence, ht. of jump = 1.68 m. **Ans.**

**Loss of Energy in jump formation** is given by $E_{f_1} - E_{f_2}$, which is also $H_L$, given by Eq. (10.12) as :

$$H_L = (E_{f_1} - E_{f_2}) = \frac{[y_2 - y_1]^3}{4 y_1 y_2} = \frac{(1.68 - 0.5)^3}{4 \times 1.68 \times 0.5} = 0.49 \text{ m}$$

Power dissipated in watt (*i.e.*, Nm/s or J/s)

$$= \gamma_w \cdot QH = (9.81 \times 1000)\, \frac{\text{N}}{\text{m}^3} \times 15\, \frac{\text{m}^3}{\text{s}} \times 0.49 \text{ m}$$

= 7210 Watts = **7.21 kW. Ans.**

**Q. 30.9.** *Show that the normal depth of flow in a triangular channel having side slopes zH : 1 V is given by*

$$y_n = 1.189 \left[\frac{Q \cdot n}{\sqrt{S_0}}\right]^{3/8} \cdot \left[\frac{z^2 + 1}{z^5}\right]^{1/8}$$

*(Engg. Services, 2010)*

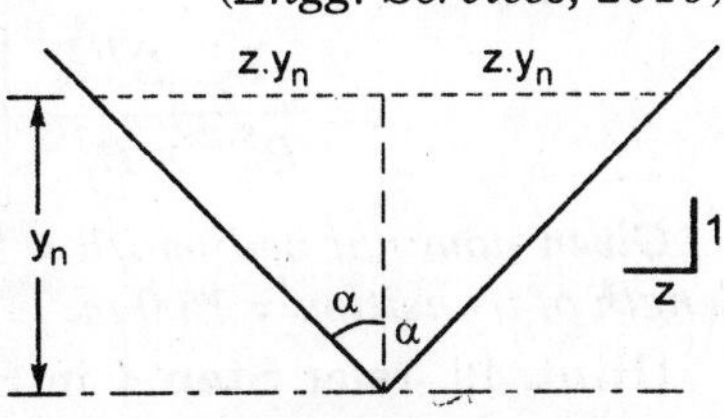


Fig. 30.7.

**Solution.** For a triangular channel shown in Fig. 30.7, we have

$$A = 2\left[\frac{1}{2} z \cdot y_n \cdot y_n\right] = z \cdot y_n^2 \quad \text{...(i)}$$

Also $$P = 2 \times \left[\sqrt{1+z^2} \,.\, y_n\right]$$

$$= 2y_n \,.\, \sqrt{1+z^2} \qquad ...(ii)$$

$$\therefore \quad R = \frac{A}{P} = \frac{z \,.\, y_n^2}{2 \,.\, y_n \,.\, \sqrt{1+z^2}} = \frac{z \,.\, y_n}{2 \,.\, \sqrt{1+z^2}} \qquad ...(iii)$$

Now, Discharge ($Q$) is given by Manning's formula as :

$$Q = \frac{1}{n} \,.\, A \,.\, R^{2/3} \,.\, \sqrt{S_0}$$

where $S_0$ = Bed slope
$n$ = Manning's coefficient

or $$Q = \frac{1}{n} \,.\, z \,.\, y_n^2 \left[\frac{z \,.\, y_n}{2\sqrt{1+z^2}}\right]^{2/3} \sqrt{S_0}$$

or $$Q \,.\, \left[2(1+z^2)^{1/2}\right]^{2/3} = \frac{1}{n} \,.\, z \,.\, y_n^2 \,.\, z^{2/3} \,.\, y_n^{2/3} \,.\, \sqrt{S_0}$$

or $$Q \,.\, (2)^{2/3}\,(1+z^2)^{1/3} = \frac{1}{n} \,.\, z^{5/3} \,.\, y_n^{8/3}\,\sqrt{S_0}$$

or $$y_n^{8/3} = \frac{n \,.\, Q(2)^{2/3} \,.\, (1+z^2)^{1/3}}{z^{5/3} \,.\, \sqrt{S_0}}$$

or $$y_n^{8/3} = \frac{Q \,.\, n}{\sqrt{S_0}}\left[\frac{1.587 \times (1+z^2)^{1/3}}{z^{5/3}}\right]$$

or $$y_n = \left[\frac{Q_n}{\sqrt{S_0}}\right]^{3/8} \cdot \frac{(1.587)^{3/8}\left(1+z^2\right)^{1/3 \times 3/8 = 1/8}}{z^{5/3 \times 3/8 = 5/8}}$$

or $$y_n = 1.189\left[\frac{Q_n}{\sqrt{S_0}}\right]^{3/8} \cdot \left[\frac{z^2+1}{z^5}\right]^{1/8}$$

which is the reqd. equation. **Ans.]**

**Q. 30.10.** *Design a transition using Mitra's hyperbolic transition given by*

$$B_x = \frac{B_c B_f \,.\, L_f}{L_f \,.\, B_c - (B_c - B_f)x}$$

*and compare the results using Chaturvedi's semi-cubical parabolic transition given by*

$$x = \frac{L_f \,.\, B_c^{3/2}}{B_c^{3/2} - B_f^{3/2}}\left[1 - \left(\frac{B_f}{B_x}\right)^{3/2}\right]$$

*Given nominal bed width = 25 m, width of flumed section = 10.0 m and total length of transition = 15.0 m.* *(Engg. Services, 2010)*

**[Hint.** Pl. refer **Step 4** in Solved Example 14.1, and let us first design the transition by using Mitras' equation, in which we have

$B_f$ = Bed width of flumed channel = 10 m

$B_c$ (or $B_n$) = Bed width of Normal channel = 25 m

$L_f$ = Length of flumed transition = 15 m

The Mitras' Equation then gives

$$B_x = \frac{25 \times 10 \times 15}{15 \times 25 - (25 - 10)x} = \frac{3750}{375 - 15x} \quad ...(i)$$

For various values of $x$ lying between 0 to 15 m, various values of $B_x$ are worked out as shown in Solution Table 30.2. The distance $x$ is measured from the flumed section, as shown in Fig. 30.9.

**Solution Table 30.2 (Based on Mitra's Equation)**

| $x$ in m | 0 | 3 | 6 | 9 | 12 | 15 |
|---|---|---|---|---|---|---|
| $B_x = \left[\frac{3750}{375 - 15x}\right]$ in m | 10 | 11.36 | 13.16 | 15.63 | 19.23 | 25 |

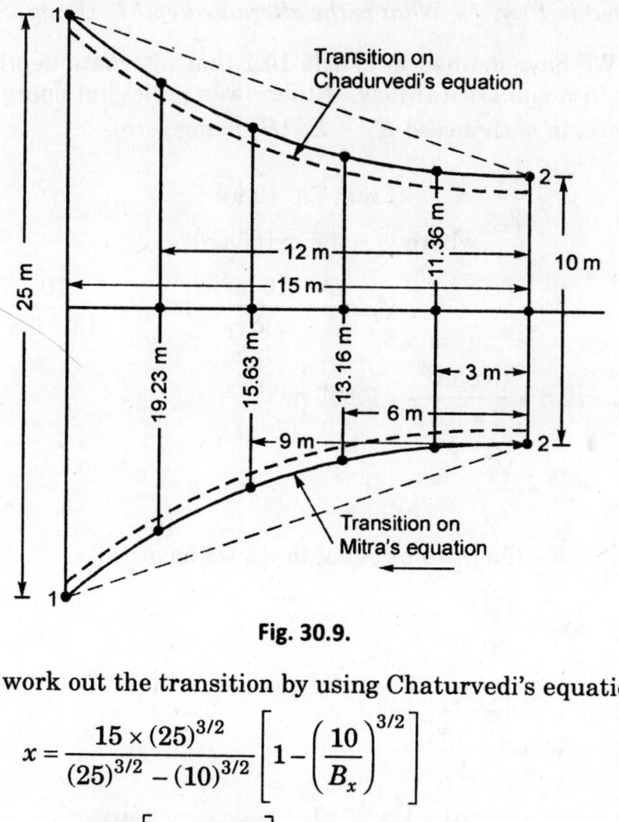


Fig. 30.9.

Now, let us work out the transition by using Chaturvedi's equation as :

$$x = \frac{15 \times (25)^{3/2}}{(25)^{3/2} - (10)^{3/2}} \left[1 - \left(\frac{10}{B_x}\right)^{3/2}\right]$$

or

$$x = \frac{1875}{93.38}\left[1 - \frac{31.62}{B_x^{3/2}}\right]$$

or $$x = 20.08\left[1 - \frac{31.62}{B_x^{3/2}}\right] \quad \text{or } [1 - 0.0498\,x] = \frac{31.62}{B_x^{3/2}}$$

or $$B_x = \left[\frac{31.62}{1 - 0.0498x}\right]^{2/3} \qquad ...(ii)$$

Let us now compute values of $B_x$ for values of $x$ as 3, 6, 9, 12 and 15 ($x$ measured from flumed section) in solution Table 30.3.

**Solution Table 30.3. (Based on Chaturvedi's Equation)**

| *x (in m)* | *0* | *3* | *6* | *9* | *12* | *15* |
|---|---|---|---|---|---|---|
| $B_x = \left[\frac{31.62}{1 - 0.0498x}\right]^{2/3}$ *in m* | *10* | *11.14* | *12.67* | *14.86* | *18.35* | *25* |

The transition can be plotted in Fig. 30.9 in dotted curve.

**Note.** It can be seen from table 30.2 and 30.3 that the Channel widths in transition length are not much different in use of the two equations. **Ans.]**

**Q. 30.11.** *A rectangular channel 15 m wide has a normal depth of 0.8 m. The discharge carried is 10 m³/s. What is the alternate depth?* ***(Engg. Services, 2010)***

**[Solution.** We have defined in article 10.1 that alternate depth is the depth $y'_2$ that occurs in a sub critical flow, if there was no loss of energy in the jump formation. Hence, in such a case $E_{f_1} = E_{f_2}$ ($H_L$ being zero).

Now $$E_{f_1} = y_1 + \frac{V_1^2}{2g} \qquad \text{(From Eq. 10.9)}$$

where $y_1 = 0.8$ m (given)

$$V_1 = \frac{Q}{A} = \frac{10\text{ m}^3/\text{s}}{by_1}; \text{ or } V_1 = \frac{10}{15 \times 0.8} = 0.833 \text{ m/s.}$$

∴ $$E_{f_1} = 0.8 + \frac{(0.833)^2}{2 \times 9.81} = 0.835 \text{ m}$$

But $$E_{f_2} = y_2 + \frac{V_2^2}{2g}$$

When $E_{f_2} = E_{f_1}$ (for $y_2$ to be equal to $y'_2$), we have

$$E_{f_1} = y'_2 + \frac{V_2^2}{2g}$$

or $$0.835 = y'_2 + \frac{\left(\frac{Q}{b \times y'_2}\right)^2}{2g}$$

or $$0.835 = y'_2 + \left(\frac{10}{15 \cdot y'_2}\right)^2 \times \frac{1}{2 \times 9.81}$$

or $$0.835 = y_2' + \frac{0.0227}{y_2'^2}$$

or $$0.835\, y_2'^2 = y_2'^3 + 0.0227$$

or $y_2'^3 - 0.835\, y_2'^2 - 0.227 = 0$

Solving by Hit and Trial, we get

$$y_2' = 0.865 \text{ m} \quad \textbf{Ans.}$$ ]

## Year Wise Questions from the Engineering Services Exams from the Year 2011 onwards.

### Year 2011 (*Engineering Services*)

**Q. 1.** (*a*) (*i*) *What is crop calender ? Illustrate the same.* (3)

[**Hint.** The ***crop calendar*** or the ***cropping calendar*** is a tool that provides timely information about the steps to be taken at different times of the crop period to successfully complete the harvesting of the crop from seeding to waterings, harvesting and storage. It, thus, contains information on planting, sowing, waterings, and harvesting times of the locally adopted crop(s) in specific agro-ecological zones.

We may, thus have different crop calendars for different crops for specific region(s) such as the crop calendar of soya bean crop, or the *crop calendar of rice* – starting from the fallow period and land preparation to crop establishment and maintenance to harvest and storage.

A typical monthwise crop calendar of crop waterings for important Indian crops in North India is shown in Fig. 30.10.

Two typical crop calendars for major food crops in India, and important crops of Italy, showing the timings (periods) of their sowing, growing and harvesting, are also shown in **Colour Fig. 30.1.**

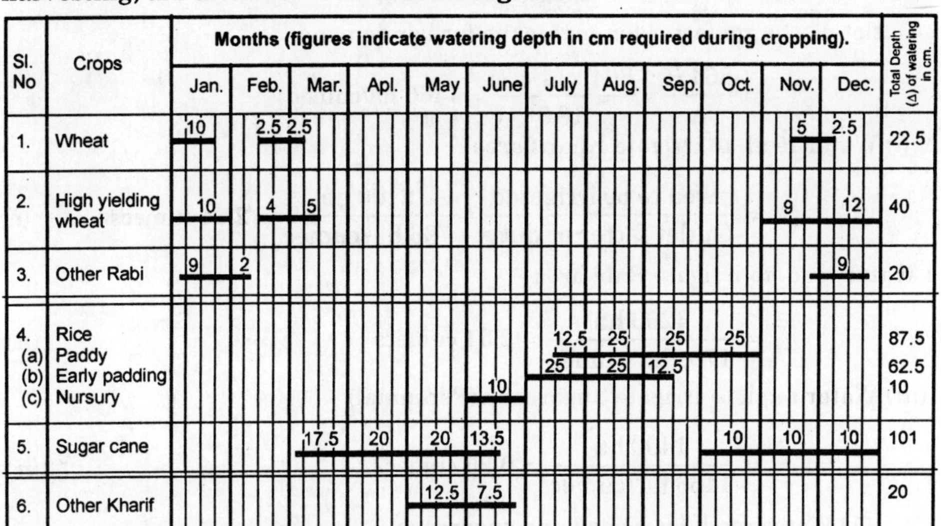


Fig. 30.10. Typical calendar of waterings for important Indian crops.

**Q.1.** *(a) (ii) Determine the discharge of a distributary at the tail end from the following data :*

*Gross Command Area (GCA) = 20000 ha*

*Cultivable Command Area (CCA) = 70%*

*Losses beyond the tail end = 1 $m^3 s^{-1}$*

| *Crop season* | *Crop* | *Irrigation intensity* | *Kor depth* | *Kor period* |
|---|---|---|---|---|
| *Kharif* | *Rice* | *15%* | *19 cm* | *2.5 weeks* |
| *Rabi* | *Wheat* | *30%* | *13.5 cm* | *4 weeks* |
| *Perennial* | *Sugarcane* | *10%* | *16.5 cm* | *4 weeks* |

(7)

**[Solution.** Pl. refer Solved Examples 3.1 and 3.2, and compute as under :

CCA = 70% × 20000 ha (GCA) = 14000 ha

Area to be irrigated for Kharif = CCA × Intensity of irrigation

= 14000 × 15% = 2100 ha

Similarly area to be irrigated for Rabi

= 14000 × 30% = 4200 ha

Also, area to be irrigated for Sugarcane (Perennial)

= 14000 × 10% = 1400 ha

The duty at the head of the water course *i.e.* outlet discharge factor is now computed as :

Outlet discharge factor for Kharif for kor demand

$$= \frac{864\,B}{\Delta} = \frac{864 \times (2.5 \times 7)}{19} \text{ ha/cumec} = 796 \text{ ha/cumec}$$

Outlet factor for Rabi for kor demand

$$= \frac{864\,B}{\Delta} = \frac{864 \times (4 \times 7)}{13.5} = 1792 \text{ ha/cumec}$$

Outlet discharge factor for sugarcane for kor demand

$$= \frac{864\,B}{\Delta} = \frac{864 \times (4 \times 7)}{16.5} = 1466 \text{ ha/cumec}$$

(*i*) Water reqd. to irrigate Kharif area

$$= \frac{\text{Area to be irrigated}}{\text{Outlet factor or duty}} = \frac{2100 \text{ ha}}{796 \text{ ha/cumec}} = 2.64 \text{ cumecs} \qquad ...(i)$$

Water reqd. to irrigate Rabi area

$$= \frac{4200 \text{ ha}}{1792 \text{ ha/cumec}} = 2.34 \text{ cumecs} \qquad ...(ii)$$

(*iii*) Water reqd. to irrigate Sugarcane (Perennial)

$$= \frac{1400 \text{ ha}}{1466 \text{ ha/cumec}} = 0.96 \text{ cumec} \qquad ...(iii)$$

∴ (1) Water reqd. for Kharif and sugarcane

= (*i*) + (*iii*)

= 2.64 + 0.96 = **3.60 cumecs** ...(1)

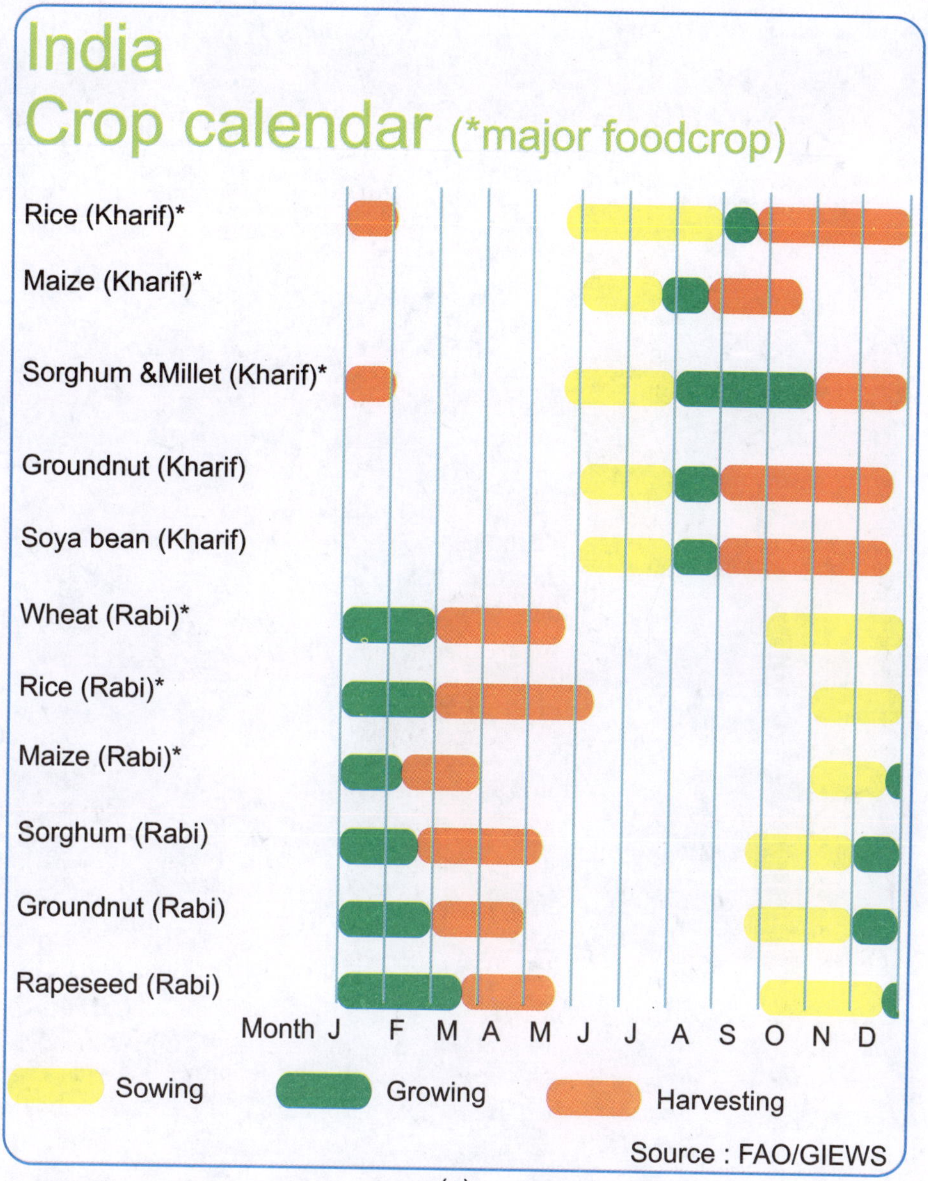


(*a*)

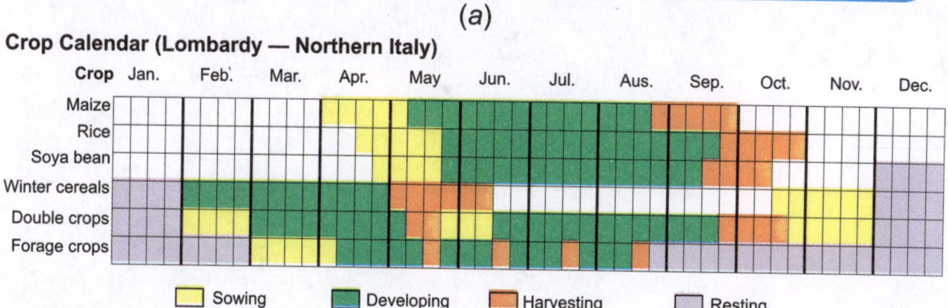


(*b*)

**Colour Fig. 30.1(*a*) & (*b*). Typical Crop Calendars**

(2) Water reqd. for Rabi and sugarcane

$= (ii) + (iii)$

$= 2.34 + 0.96 =$ **3.30 cumecs** ...(2)

The water reqd. is maximum of (1) and (2), and is **3.6 cumecs** at the head of the watercourse.

Since there is a loss of 1 $m^3$/s (or 1 cumec) beyond the tail end of distributory and upto water course, the water reqd. at tail end of distributary

$= 3.6 + 1.0 =$ **4.6 $m^3$/s. Ans.** ]

**Q. 3.** (*a*) (*ii*) *Define the following :*

(1) *Field irrigation requirement* (2) *Water application efficiency*

(3) *Water storage efficiency* (4) *Water distribution efficiency*

(5) *Consumptive use.* (5)

**Q. 4.** (*d*) *A hydraulic jump is formed in a 5 m wide outlet at a short distance downstream of control gate. If the flow depths are 10 m and 2 m in the u/s and d/s respectively of the sluice gate and the Q = 150 $m^3$ $s^{-1}$, determine*

(*i*) *Flow depth downstream of the jump*

(*ii*) *Thrust on the gate*

(*iii*) *Head losses in the jump.* **(10)**

[ **Solution.**

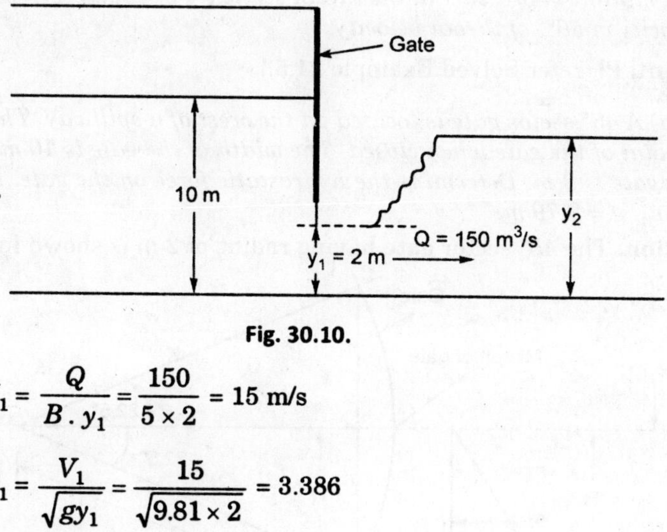


**Fig. 30.10.**

$$V_1 = \frac{Q}{B \cdot y_1} = \frac{150}{5 \times 2} = 15 \text{ m/s}$$

$$F_1 = \frac{V_1}{\sqrt{g y_1}} = \frac{15}{\sqrt{9.81 \times 2}} = 3.386$$

$$y_2 = \frac{y_1}{2}\left[\sqrt{1 + 8F_1^2} - 1\right] \quad \text{...(Eq. 10.7)}$$

$$= \frac{2}{2}\left[\sqrt{1 + 8 \times 3.386^2} - 1\right] = \mathbf{8.63\ m} \quad \textbf{Ans.} \quad \text{...}(i)$$

Now,
$$H_L = \frac{(y_2 - y_1)^3}{4 \cdot y_1 y_2} \quad \text{...(Eq. 10.12)}$$

$$= \frac{(8.63 - 2)^3}{4 \times 8.63 \times 2} = \mathbf{4.22\ m.} \quad \text{...}(iii)$$

∴ Head loss in the jump = **4.22 m Ans.**

Also, $V_2 = \dfrac{Q}{B \cdot y_2} = \dfrac{150}{5 \times 8.63} = 3.476$ m/s.

Thrust on the gate = $(V_1 - V_2)$ Mass of moving water

$$= (15 - 3.476)\,\frac{\text{m}}{\text{s}}\,[\rho_w \times Q] = 11.524\,\frac{\text{m}}{\text{s}}\left[1000\,\frac{\text{kg}}{\text{m}^3} \times 100\,\frac{\text{m}^3}{\text{s}}\right.$$

$$= (11.524 \times 1000 \times 100)\ \text{kg}\,\frac{\text{m}}{\text{sec}^2} = (11.524 \times 1000 \times 100)\ \text{N}$$

$$\left.= 11.524 \times 100\ \text{kN} = \mathbf{1152.4\ kN.\ Ans.}\right]$$

**Q. 5. (a)** (*i*) *What is priming ?* (1)

**Q. 5. (a)** (*ii*) *A siphon has a rectangular throat section 2.2 m wide and the radius of the crown and the crest are 3 m and 1.6 m, respectively. The energy losses are 0.2, 0.2, 0.6 and 0.5 times the mean velocity head at the throat respectively in the inlet, upper leg, lower leg and outlet portion. The kinetic energy coefficient is 1.03. Local atmospheric pressure is 10 m of water. Determine :*

(*a*) *The discharge when the reservoir level is 0.3 m above the crest and the tail water level (outlet submerged) 6 m below it.*

(*b*) *The minimum pressure at the throat section. For submerged condition, outlet velocity is 50% of throat velocity.* (9)

**[Hint.** Pl. refer Solved Example 21.5.]

**Q. 7.** (*a*) *A 45° sector gate is located on the crest of a spillway. The water is upto the mid-point of the gate when closed. The width of the gate is 10 m. The radius of the sector gate is 2 m. Determine the hydrostatic force on the gate. Mass density = 1000 kg/m³, g = 9.79 ms⁻².* (10)

**[Solution.** The 45° sector gate having radius of 2 m is shown in Fig. 30.11.

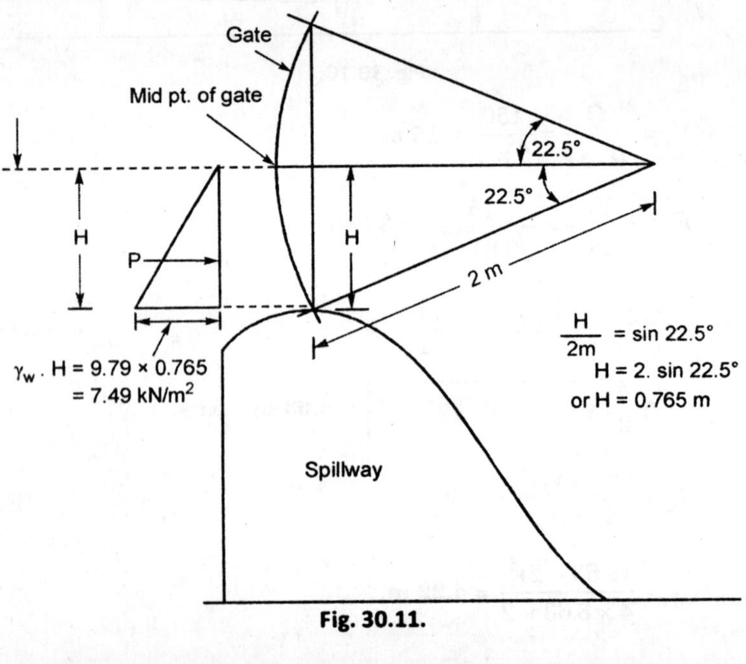


**Fig. 30.11.**

The hydrostatic force ($P$) exerted by water on the gate is also shown, and computed as :

$$P = \left[\frac{1}{2}(\gamma_w \,.\, H)\,.\, H\right] \times B \text{ (width of gate)}$$

where $\gamma_w = \rho \,.\, g = 1000 \dfrac{\text{kg}}{\text{m}^3} \times 9.79 \dfrac{\text{m}}{\text{s}^2}$

$$\left[\because\ g = 9.79 \text{ ms}^{-2} \text{(given)}\right]$$

$$= 9.79 \times 10^3 \left(\frac{\text{kgm}}{\text{s}^2}\right) \times \frac{1}{\text{m}^3}$$

$$= 9.79 \times 10^3 \text{ N/m}^3 = 9.79 \text{ kN/m}^3$$

$$H = 2 \sin 22.5° = 0.765 \text{ m}$$

$$B = 10 \text{ m}$$

or $$P = \left[\frac{1}{2} \times (9.79 \times 0.765) \times (0.765)\right] 10 \text{ kN}$$

or $P =$ **28.65 kN. Ans.** ]

## Year 2012 (*Engineering Services*)

**Q. 1.** (*b*) *What are the different types of cross drainage works that are required when a canal crosses a natural drainage? Draw a plan view of an aqueduct showing all details.* (8)

**Q. 2.** (*b*) *A solid gravity dam has 10 numbers of waste weir sluices, the width of each sluice being 4.0 m. The waste weir is designed for a high flood discharge of 100 m$^3$/s and the weir constant is given as 1.76. It is proposed to increase the line storage of the reservoir behind the dam by replacing the waste weir by a saddle siphon spillway having the crest at all units being at the same level. The discharge coefficient of this siphon spillway is 0.64 and the operated head for all the units is 7 m. Find the number of siphon units necessary to replace the waste weir and also find the extra line storage obtained by remodelling of the spillway. The HFL (high flood level) is the same in case of both types of spillway and the priming depth for the saddle spillway is 15 cm. The area of water spread from HFL to about 1.5 m below HFL may be taken as constant and equal to 4.5 million m$^2$.* (10)

[**Solution.** Total discharge through 10 Nos of waste weir sluices = 100 m$^3$/s

$\therefore$ Discharge through each waste weir sluice $= \dfrac{100}{10} = 10$ m$^3$/s

But discharge through a waste weir

$= CLH^{3/2}$

where $C = 1.76$ (given)

$L = 4$ m (given)

$\therefore$ $10 = 1.76 \times 4 \times H^{3/2}$

or $H =$ **1.26 m.**

$\therefore$ HFL = Normal Reservoir Level (NRL) + 1.26 m

Hence, water spread area between pool level (NRL) and HFL remains constant = 4.5 Mm$^2$ [$\because$ it is given to be the same between HFL and up to 1.5 m below it].

Now, Depth of priming = 15 cm = 0.15 m

Since, the siphon spillway will start removing water from the reservoir, only after the water level rises by 0.15 m (i.e. priming depth) over the reservoir level, the *increase in storage* in the reservoir

= 0.15 m × [Area of water spread between NRL and 0.15 m above it]

= 0.15 m × 4.5 $Mm^2$ = **0.675 $Mm^3$ Ans.**

Now, the discharge through one siphon unit

$$= C_d \cdot A \cdot H_1^{3/2}$$

where $C_d = 0.64$ (given)
$H_1$ = operating head = 7 m (given)
$A$ = Area of cross-section of bore of the hood at crown.

Now let each unit of siphon spillway is 4 m long* (*i.e.* equal to width of each sluice) and let the depth be 0.8 m*. Then $A = 0.8 \times 4 = 3.2\ m^2$

$\therefore \quad Q = 0.64 \times 3.2 \times (7)^{3/2} = 21.5\ m^3/s$

For passing 100 $m^3$/s of discharge, no. of units reqd. = **5. Ans.]**

**Q. 4.** (*a*) (*ii*) *An overflow spillway is 50 m high. At the design, the energy head is 2.5 m over the spillway. Find the sequent depth and energy loss in a hydraulic jump formed on a horizontal apron at the toe of the spillway. Neglect energy loss due to flow over the spillway face. Assume coefficient of discharge $C_d$ = 0.735 and the velocity at the toe, before the jump can be approximated as V = [2g (total head)]$^{1/2}$, where g is acceleration due to gravity.* (6)

[**Solution.** Discharge over the spillway is given as :

$$Q = C_d \cdot L_e \cdot H_e^{3/2} \quad \text{...(Eq. 21.4)}$$

$\therefore$ Discharge per m length of spillway ($q$) is given as :

$$q = C_d \cdot H_e^{3/2}$$

where $H_e$ = total head over the crest including vel head
= 2.5 m (given)
$C_d$ = 0.735 (given)

$\therefore \quad q = 0.735 \times (2.5)^{3/2}$

or $\quad q = 2.905\ m^3/s$ . per m length

Also $\quad V_1$ = velocity at Toe = $[2g \cdot (\text{Total head})]^{1/2}$ ...(given)

HereTotal head = 50 m + 2.5 m = 52.5 m

$$\therefore \quad V_1 = \sqrt{2 \times 9.81 \times 52.5} = 32.094 \text{ m/s}$$

$$\text{But} \quad y_1 = \frac{q}{V_1} = \frac{2.905}{32.094} = 0.0905 \text{ m}$$

$$\therefore \quad F_1 = \frac{V_1}{\sqrt{gy_1}} = \frac{32.094}{\sqrt{9.81 \times 0.0905}} = 34.059$$

Using Eq. (10.7) as : $y_2 = \frac{y_1}{2}\left[-1 + \sqrt{1 + 8F_1^2}\right]$, we get

---

***Since the length and height of bore of the hood is not given, it is necessary to assume the same to solve the question with the given data, which in the opinion of this author, is incomplete and insufficient.**

$$y_2 = \frac{0.0905}{2}\left[-1+\sqrt{1+8\times 34.059^2}\right] = 4.313 \text{ m}$$

Hence, sequent depth = $y_2$ = **4.313 m. Ans.**

Energy loss in the jump = $H_L = \dfrac{(y_2 - y_1)^3}{4.y_1.y_2}$ ...(Eq. 10.12)

$$= \frac{(4.313 - 0.0905)^3}{4\times 4.313\times 0.0905} = \textbf{48.22 m. Ans.]}$$

**Q. 5.** *(a) (ii) A triangular channel of apex angle 90° and a rectangular channel of the same material have the same bed slope. If the rectangular channel has the depth of flow equal to width of the channel and flow areas in both channels are the same, find the ratio of discharges in rectangular and triangular channels respectively. Use Manning's roughness equation for estimation of velocity.* (6)

[**Solution.** Let the triangular channel be designated as channel (1), and rectangular channel as channel (2). The geometry of channels are shown in Fig. 30.12.

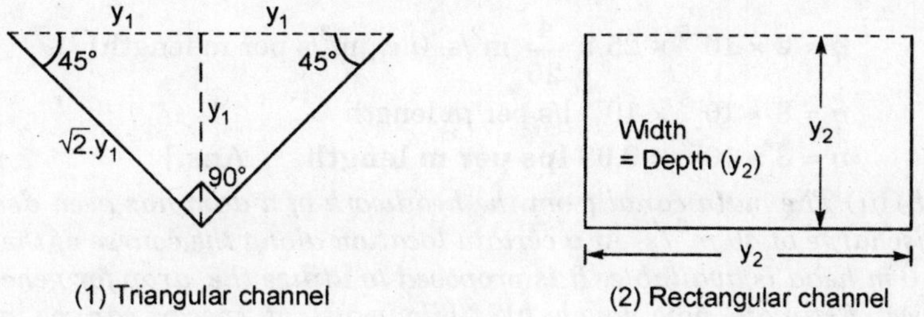


**Fig. 30.12.**

Area = $A_1 = 2\times\frac{1}{2}.y_1.y_1 = y_1^2$    Area = $A_2 = y_2 y_2 = y_2^2$

W.P. = $P_1 = 2.\sqrt{2}.y = 2\sqrt{2}\,y_1$    W.P. = $P_2 = 3y_2$

$R_1 = \dfrac{A_1}{P_1} = \dfrac{y_1^2}{2\sqrt{2}.y_1} = 0.354\,y_1$    $R_2 = \dfrac{A_2}{P_2} = \dfrac{y_2^2}{3y_2} = \dfrac{y_2}{3}$

But flow areas of both channels are equal.

$\therefore$ $A_1 = A_2$

or $y_1^2 = y_2^2$

or $y_1 = y_2 = (\text{say } y)$

$\therefore$ $R_1 = 0.354\,y_1 = 0.354\,y$

and $R_2 = \dfrac{y_2}{3} = 0.334\,y$

Discharge in Triangular channel (from Mannings formula) is given as

$$Q_1 = \frac{1}{n}.A_1R_1^{2/3}.\sqrt{S}$$

Similarly, discharge in the rectangular channel is given as

$$Q_2 = \frac{1}{n}.A_2.R_2^{2/3}\sqrt{S} \quad (\because\ n \text{ and } S \text{ are the same in both cases})$$

$$\therefore \quad \frac{Q_2}{Q_1} = \frac{\text{Discharge in rectangular channel}}{\text{Discharge in triangular channel}} = \frac{A_2R_2^{2/3}}{A_1R_1^{2/3}} = \left(\frac{R_2}{R_1}\right)^{2/3}$$

$(\because\ A_2 = A_1)$

$$\therefore \quad \frac{Q_2}{Q_1} = \left(\frac{0.334}{0.354}\right)^{2/3} = \mathbf{0.96.} \quad \mathbf{Ans.}]$$

**Q. 5.** (*b*) (*i*) *A homogenous earthen dam 42 m high is built to store water for irrigation water requirement with a freeboard of 2•m. A horizontal filter of 30 m length is provided at its downstream end. The coefficient of permeability of the material of the dam is given as 2 × $10^{-3}$ cm/sec. A flow net was constructed through the body of the dam and the following results were obtained :*

*No. of potential drops = 25; No. of flow channels = 4*

*Calculate the seepage flow per meter length of the dam in lps.* (6)

[**Solution:** The seepage discharge is given by Eq (20.15) as:

$$q = K.H. \frac{N_f}{N_d} \quad \text{...(Eq. 20.15)}$$

Here $K = 2 \times 10^{-3}$ cm/sec $= 2 \times 10^{-5}$ m/sec

$H = 42$ m; $N_f = 4$; $N_d = 25$

$\therefore \quad q = 2 \times 10^{-5} \times 25 \times \frac{4}{25}$ m²/s. (*i.e.*, m³/s per m length)

or $\quad q = 8 \times 10^{-5} \times 10^{-3}$ l/s per m length

or $\quad q = 8 \times 10^{-2} =$ **0.08 lps per m length. Ans.]**

**Q. 5.** (*b*) (*ii*) *The main canal from the headwork of a dam has been designed to carry a discharge of 40 m³ / s. At a certain location along the course of the canal, a drop of 4.0 m head is available. It is proposed to utilize the drop for generation of hydropower. Estimate how much kW (kilo watt) of energy can be generated assuming efficiency of machinery used as 75 percent.* (4)

[**Solution.** From Eq. (24.1),

Water energy generation

$= \gamma_w \cdot QH \qquad$ where $\gamma_w = 9.81$ kN/m³

$\therefore \quad$ W.E. $= 9.81\, QH = 9.81 \times 40 \times 4 = 1569.6$ kN m/sec $= 1569.6$ k.W

Electrical Energy or Power is given from Eq. (24.2) as

= Water Energy×Efficiency of Turbines, generator, etc.

= 1569.6 × 0.75 kW = **1177.2 kW. Ans.]**

**Q. 8.** (*b*) (*ii*) *Show with the help of a neat sketch the storage capacities alloted to various purposes in a multipurpose reservoir. Briefly explain what is meant by useful life of a reservoir and how it is estimated.* (6)

## Year 2013 (*Engineering Services*)

**Q. 1.** (*d*)

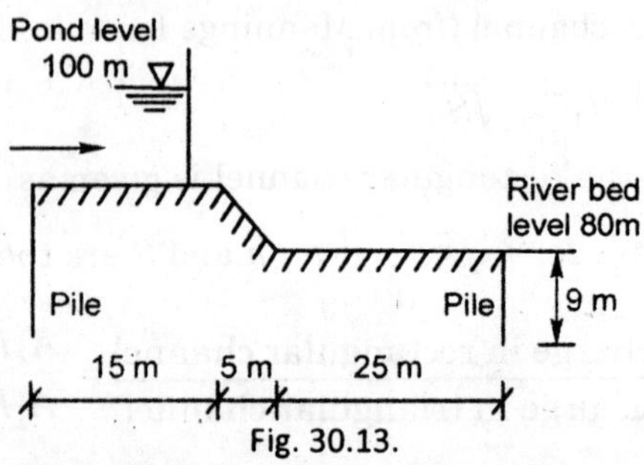


Fig. 30.13.

*Check the safety of a barrage shown above against the piping action. The safe exit gradient is* $\frac{1}{5}$. (4)

[**Solution.** $G_E$ = Exit gradient = $\frac{H}{d} \cdot \frac{1}{\pi\sqrt{\lambda}}$ ...(Eq. 11.19)

where $\lambda = \frac{1+\sqrt{1+\alpha^2}}{2}$ and $\alpha = \frac{b}{d}$

Substituting values, we get

$$\alpha = \frac{(15+5+25)\text{ m}}{9\text{ m}} = \frac{45}{9} = 5$$

$$\therefore \quad \lambda = \frac{1+\sqrt{1+5^2}}{2} = 3.0495$$

$$\therefore \quad G_E = \frac{20}{9} \frac{1}{\pi . \sqrt{3.0495}} = 0.405 > 0.2 \text{ (i.e. 1/5)}$$

Since actual exit gradient is more than the safe exit gradient, the barrage is not safe. **Ans.**

**Q. 4.** (*b*) *An overflow ogee spillway has six bays each of 8 m clear length and five rounded nose piers ($K_p$ = 0.01) each of 2.5 m thick. Both the abutments are sharp edged ($K_a$ = 0.2). The upstream face of the spillway is vertical. Compute the discharge that will pass over the spillway at the design head, $H_d$ = 3 m. The discharge coefficient for design head $C_o$ = 2.1.*

*Also compute the passing discharge over spillway for head over the crest H = 3.45 m. Neglect the approach velocity to the spillway. Discharge coefficients for head other than design head are given below :*

| $H/H_d$ | $C/C_o$ |
|---|---|
| 0.6 | 0.94 |
| 0.85 | 0.98 |
| 1.15 | 1.02 |
| 1.50 | 1.06 |

*C is discharge coefficient for head H.* (10)

[**Solution.** Please refer Solved Example 21.1 and compute as under :

Effective length of spillway crest ($L_e$) is given as :

$$L_e = L - 2\left[K_p.N + K_a\right] H_e \quad \text{...(Eq. 21.6)}$$

where $L$ = Net clear length of the spillway crest = 6 × 8 m = 48 m

$K_p$ = Pier contraction coefficient = 0.01

$N$ = Number of piers = 5

$K_a$ = Abutment contraction coefficient = 0.2

$H_e$ = Total design head on the crest including velocity head

= $H_d$ = 3 m (neglecting velocity head)

$\therefore$ $L_e = 48 - 2\,[0.01 \times 5 + 0.2]\,3 =$ **46.5 m**

Now, $Q = C \, . \, L_e \, . \, H_e^{\,3/2}$ ...(Eq. 21.4)

where $C$ = Coefficient of discharge = $C_0$ = 2.1 (given)

$\therefore$ Discharge $Q = 2.1 \times 46.5 \times (3)^{3/2}$

$=$ **507.4 m³/s Ans.**

When Head over the crest = 3.45 m instead of the design head of 3 m, then coefficient of discharge ($C$) is obtained from the given table for $\dfrac{H}{H_d} = \dfrac{3.45}{3} = 1.15$ as 1.02.

$\therefore$ $C$ for new head = $1.02 \times C_o = 1.02 \times 2.1 =$ **2.142**

$\therefore$ $Q = C\,L_e\,H_e^{\,3/2}$

or $Q = 2.142 \times 46.5 \times (3.45)^{3/2}$

or $Q =$ **638.27 m³/s. Ans.]**

**Q. 4.** (*c*) (*i*) *Derive the following Lacey's regime equations*

$$V = \left[\frac{Qf^2}{140}\right]^{\frac{1}{6}}$$

and $P = 4.75\sqrt{Q}$

*with the use of following basic equations of Lacey*

$$V = \sqrt{\frac{2}{5} fR}$$

$$Af^2 = 140\,V^5$$

*Here* $V$ = *velocity;*

$Q$ = *discharge;*

$f$ = *silt factor;*

$P$ = *wetted perimeter; and*

$R$ = *hydraulic mean radius.* (10)

[**Solution.** Given $V = \sqrt{\dfrac{2}{5} fR}$ ...(*i*)

$Af^2 = 140\,V^5$ ...(*ii*)

Using $V = \sqrt{\dfrac{2}{5} fR}$, we get $V^2 = \dfrac{2}{5} fR$

$\therefore$ $f = \dfrac{5}{2}\dfrac{V^2}{R}$

Also using (*ii*), we get

$Af^2 = 140\,V^5$ ...(*ii*)

$\therefore$ $140\,V^5 = Af^2$ where $f = \dfrac{5}{2}\dfrac{V^2}{R}$

$\therefore$ $140\,V^5 = A \, . \, \dfrac{25}{4} \, . \, \dfrac{V^4}{R^2}$

But $$R = \frac{A}{P}$$

$\therefore$ $$140\,V^5 = A\left(\frac{25}{4}\right)V^4 \cdot \frac{P^2}{A^2}$$

or $$140\,V = \frac{25}{4} \cdot \frac{P^2}{A}$$

or $$140\,V\,.\,A = \frac{25}{4}P^2$$

or $$140 \times \frac{4}{25}\,.\,Q = P^2 \qquad [\because\ V.A = Q]$$

or $$\frac{112}{5}\,.\,Q = P^2$$

or $$P = \sqrt{\frac{112}{5}Q} = \sqrt{22.4\,Q} = 4.75\sqrt{Q} \qquad \text{...}(iii)$$

Hence, $P = 4.75\sqrt{Q}$. **Ans.**

**Also** to derive equation $V = \left(\dfrac{Qf^2}{140}\right)^{1/6}$, we work out as follows from the given equation $(i)$

$$V = \sqrt{\frac{2}{5}\,.\,fR}$$

$\therefore$ $$V^2 = \frac{2}{5}\,.\,fR = \frac{2}{5}\,.\,f\left(\frac{A}{P}\right)$$

$\therefore$ $$V^2 = \frac{\frac{2}{5}\,.\,fA}{4.75\sqrt{Q}} \qquad \left[\because\ P = 4.75\sqrt{Q}\ \text{from } (iii) \text{ derived above}\right]$$

or $$V^2 = \frac{\frac{2}{5}\,.\,f\left(\frac{Q}{V}\right)}{4.75\sqrt{Q}} \qquad \left[\because\ Q = AV \ \text{ and } \ A = \frac{Q}{V}\right]$$

or $$V^3 = \frac{2\,f}{5 \times 4.75}\sqrt{Q}$$

Squaring, we get

$$V^6 = \frac{4}{25 \times 22.4}\,f^2\,.\,Q = \frac{Qf^2}{140}$$

$\therefore$ $$V = \left[\frac{Qf}{140}\right]^{1/6} \quad \textbf{Ans.} \qquad \text{...}(iv)$$

### Year 2014 (*Engineering Services*)

**Q. 3.** (*b*) *The intensity of irrigation is 20% for wheat and 10% for rice, where cultural commanded area of a distributary is 6000 ha. The Kor period for wheat is 3 weeks and for rice 2 weeks. Find the total outlet discharge required excluding losses in the channel. Assume depth for Kor watering as 9 cm and 25 cm for wheat and rice respectively.* (8)

**Solution.** Please follow Solved Examples 3.1 and 3.2, and compute as under:

Given CCA = 6000 ha

Intensity of irrigation for wheat (Rabi crop) = 20%

Intensity of irrigation for rice (Kharif crop) = 10%

Area to be irrigated for wheat = CCA × Intensity of irrigation

$$= 6000 \text{ ha} \times 20\% = 6000 \times \frac{20}{100} = 1200 \text{ ha}$$

Area to be irrigated for rice = 6000 ha × 10% = 600 ha

Kor period for wheat = 3 weeks, Kor watering depth for wheat = 9 cm

$$\therefore \quad \text{Outlet factor for wheat} = \frac{864\, B}{\Delta}$$

$$= \frac{864 \times (3 \times 7)}{9} = 2016 \text{ ha/cumec}$$

Similarly, outlet factor for rice (for which Kor period = 2 weeks and Kor depth = 25 cm)

$$= \frac{864 \times (2 \times 7)}{25} = 484 \text{ ha/cumec}$$

∴ Discharge required at the head of the distributary to irrigate wheat

$$= \frac{\text{Area of wheat to be irrigated}}{\text{Outlet factor for wheat}} = \frac{1200 \text{ ha}}{2016 \text{ ha/cumec}} = 0.60 \text{ cumec} \quad ...(i)$$

Similarly, discharge required at the head of the distributary to irrigate rice

$$= \frac{\text{Area of rice to be irrigated}}{\text{Outlet discharge for rice}} = \frac{600 \text{ ha}}{484 \text{ ha/cumec}} = 1.24 \text{ cumecs} \quad ...(ii)$$

The required discharge is max of (*i*) and (*ii*), *i.e.* 1.24 cumecs. **Ans.** ]

**Q. 5.** (*d*) (*ii*) *To estimate the seepage loss through a cofferdam foundation, the flow nets were constructed. The results of the flow net study gave $N_f$ = 7, $N_d$ = 17. The head of the water lost during seepage was 6 m. If K = 5 × $10^{-5}$ m/minute, compute the seepage loss per metre length of dam per day.* (4)

**Solution.** Please follow Solved Examples 20.2, and computes as under :

Seepage discharge is given by Eq. (20.15) as

$$q = \frac{K.H}{N_d}.N_f$$

where $q$ = Seepage per unit width = ?

$K$ = Permeability = 5 × $10^{-5}$ m/min

$H$ = Total head causing flow = 6 m

$N_f$ = No. of flow channels = 7

$N_d$ = No. of drops in the flow net = 17

Substituting values, we get

$$q = 5 \times 10^{-5} \frac{\text{m}}{\text{min}} \times 6 \text{ m} \times \frac{7}{17} = 12.35 \times 10^{-5} \text{ m}^2/\text{min}$$

$$= \frac{12.35}{60} \times 10^{-5} \text{ m}^2/\text{sec} = 2.06 \times 10^{-6} \text{ m}^2/\text{sec}$$

$$= 2.06 \times 10^{-6} \text{ m}^3 \text{ per sec per m run of dam}$$

$\therefore$ Discharge per day $= 2.06 \times 10^{-6} \times (24 \times 3600)$ m$^3$/m run of dam

$= 0.178$ m$^3$/m run = **178 litres/m run**

= 178 litres per m run of dam per day. **Ans.**

**Q. 6.** (*b*) (*ii*) *Obtain the diameter of the pipe outlet of a non-modular outlet:*

*Discharge through the outlet = 0.02 m*$^3$ *s*$^{-1}$

*Length of the outlet = 15 m*

*FSL of the distributary = 200.00 m*

*Available working head = 5 cm*

*Coefficient of discharge = 0.51* (4)

**Solution.** For all practical purposes, the discharge through the pipe outlet is approximately given by Eq. (13.22) as :

$$q = C_d \cdot A \cdot \sqrt{2g\, H_L}$$

where $q$ = Discharge through the outlet = 0.02 m³/s

$A$ = Area of pipe outlet $= \frac{\pi}{4} D^2 = ?$

$H_L$ = 5 cm = 0.05 m

$C_d$ = 0.51

$\therefore \quad 0.02 = 0.51 \times A \times \sqrt{2 \times 9.81 \times 0.05}$

or $\quad A = 0.0396 \text{ m}^2$

Dia of pipe $= D = \sqrt{0.0396 \times \frac{4}{\pi}} = 0.224 \text{ m} = 22.4 \text{ cm}$

Hence, **use 22.5 cm** dia pipe. **Ans.**

**Note :** Equation (13.21) for exact computation of velocity/discharge cannot be used, since the coefficient of friction for the pipe to be used is not given, although pipe length $l$ is given as 15 m, which has not been used in Eq. 13.22.

**Year 2015 (*Engineering Services*)**

**Q. 1.** (*c*) *Determine the maximum base width of the elementary profile of a gravity dam with the following data :*

| | |
|---|---|
| *Specific gravity of the dam material* | *= 2.4* |
| *Uplift intensity factor* | *= 1.0* |
| *Coefficient of static friction* | *= 0.80* |
| *Height of the dam* | *= 20.0 m* (4) |

**Solution.** The base width $B$ for the elementary profile of a gravity dam is given by Eq. (19.22) for "No tension to develop at base", and by Eq. (19.23) for the dam to be safe in sliding, both for reservoir full case. The value of $B$ for max. base width has to be greater of the two values.

From Eq. (19.22),

$$B = \frac{H}{\sqrt{S_c - C}} \quad \text{...(Eq. 19.22)}$$

and from Eq. (19.23),

$$B = \frac{H}{\mu(S_c - C)} \quad \text{...(Eq. 19.23)}$$

where $H = 20$ m, $\mu = 0.8$, $S_c = 2.4$, $C = 1.0$ (Given)

$$B \text{ from Eq. (19.22)} = \frac{20}{\sqrt{2.4 - 1}} = 16.89 \text{ m} \quad \text{...(i)}$$

$$B \text{ from Eq. (19.23)} = \frac{20}{0.8\,(2.4 - 1)} = 17.86 \text{ m} \quad \text{...(ii)}$$

The greater of the two values from (*i*) and (*ii*) = **17.86 m. Ans.**

**Q. 1.** (*d*) *There are four important theories with regard to formation of meanders— (i) disturbance theory, (ii) helicoidal flow theory, (iii) excess energy theory and (iv) instability theory. Friedkin and Werner subscribed to the idea of disturbance theory. In brief, explain their ideas.* (4)

**[Hint. Theories on Formation of Meanders in Rivers :** Several theories/ideas have been put forth by several investigators on the reasons for formation of meanders in rivers, since 1884 or so. The important ones among these include the following :

1. ***Earth's rotation theory,*** first suggested by Gilbert in 1884 and later by Quareishy in 1943.
2. ***Disturbance theory,*** first suggested by Griggs in 1906, and later by Tiffnay (1941), Hjulstrom (1945) and **Werner (1951).**
3. ***Helicoidal flow theory,*** first suggested by Pruschacincky and Liliavsky in 1954.
4. ***Excess energy theory,*** first suggested by Schoklitsch, Inglis and later by Joglekar in 1956, and then by Leopold and Wolman in 1960.

Out of the above theories, the most widely accepted theory is the **disturbance theory**, which was originally given by Griggs (1906) and later supported by ***Friedkin* (1945)** and ***Werner* (1951).** Friedkin (1945) subscribing to the idea of disturbance theory stated that meandering is primarily caused by *local bank erosion*, and consequent over-loading of silt in the river flow, leading to subsequent deposition of silt/sand on the downstream bank(s), as explained in article 8.4.3 in the text. Friedkin found that in alluvial streams, every phase of meandering represents a changing relationship among three closely related variables, which, though striving to reach a balance, seldom do so.

**These three variables are :** (*i*) the flow and hydraulic properties of the channel; (*ii*) the amount of silt/debris moving along the stream bed due to scour; and (*iii*) the rate of bank erosion.

In his laboratory experiments, Friedkin found four types of channels developing on alluvial rivers. In the first type, *with banks entirely resistant to erosion*, a deep narrow channel developed with extremely flat slopes. In the second type, *with slowly erodible banks*, a slowly meandering and relatively deep channel with fairly flat slopes developed. In the third type, *with easily erodible banks*, a wide shallow channel with fairly sleep slopes developed. In the fourth type *with extremely erodible banks*, a **braided stream pattern** was formed. Side slopes were extremely sleep. *There is an intermediate type of channel between the braided stream and the meandering.* Its wide and shallow channel is fairly straight and its flow is divided by only a few islands and bars.

The complexity of meandering prevents the formulation of set rules concerning it. It permits no more than a knowledge of the tendencies and counter-tendencies brought about by changes in each of the variables. Changes in one variable are opposed or limited by changes in another. They form a circle of dependency.

**Q. 2.** (*b*) *A guide bank is required for a bridge on a river. For a design flood discharge of 5000 m³/sec, compute the following :*

(*i*) *Lacey waterway*

(*ii*) *Total length between banks*

(*iii*) *u/s length of guide bund and d/s length of guide bund*

(*iv*) *Radius of u/s curved head and d/s curved head with angles.* (8)

[**Solution.** Pl. refer article 8.5.3(2) and Solved Example 8.1, and compute as under :

Lacey's clear effective waterway for guide bank is given by Lacey's Regime Perimeter ($P$), which is given as :

$$P = 4.75\sqrt{Q} \quad \text{...(Eq. 4.26)}$$

(*i*) ∴ $P = 4.75\sqrt{50{,}000}$ = **1062 m. Ans.**

(*ii*) Allowing 20% extra for piers etc., the net spacing between the two guide bunds ($L$) at the bridge site = $L = 1.2 \times 1062$ m = **1274. Ans.**

(*iii*) The length of the guide bund upstream of bridge (From Fig. 8.16)

= $1.5L$ (for $Q$ exceeding 20,000 cumecs)

∴ = $1.5 \times 1274$ = **1911 m. Ans.**

The length of the guide bund downstream of the bridge (From Fig. 8.16)

= $0.25\,L = 0.25 \times 1274$ = **31** (say). **Ans.**

(*iv*) Radius ($R$) of the curved head (upstream portion) may be kept from Fig. 8.16 = $0.45\,L$.

∴ $R = 0.45\,L = 0.45 \times 1274$ m = **574 m** with an angle of say **130°** (*i.e.*, between 120° to 145° as per values given in Fig. 8.16). **Ans.**

For the downstream curved portion of the guide bank, radius may be taken as equal to $R_1 = \dfrac{R}{2} = \dfrac{574}{2}$ = **287 m**, with sweep angle of 45° to 60°, say **60°. Ans.**]

These selected values are marked in Fig. 30.14.

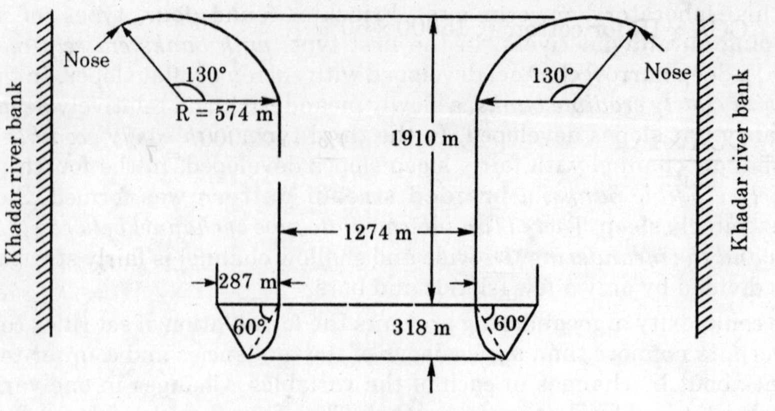


Fig. 30.14.

**Q. 3.** (*b*) *Three distributaries are used for irrigation. The details are given below. Find which one is less efficient :*

| *Discharge* | *15 $m^3$/sec* | *20 $m^3$/sec* | *25$m^3$/sec* |
|---|---|---|---|
| *CCA* | *15000 ha* | *25000 ha* | *30000 ha* |
| *Intensity of irrigation* | *60%* | *80%* | *60%* |
| *Base period* | *200 days (cotton)* | *120 days (wheat)* | *360 days (sugarcane)* |

(8)

**Solution. Let the three given distributaries, irrigating cotton, wheat and sugarcane, be respectively numbered as $D_1$, $D_2$ and $D_3$. Let us also assume that the Δ values for these three crops, be 50 cm, 40 cm and 120 cm, respectively for cotton, wheat and sugarcane (Table 2.1). Now, using equation (2.2), we have**

$$\Delta = \frac{864B}{D}$$ **where Δ is in cm, *B* (Base period) in days, and *D* is duty in hecteres/cumec**

or $$D = \frac{864B}{\Delta},$$

**where values for *B* are given in the question for all three crops.**

**The duties for the three crops, *i.e.*, cotton, wheat and sugarcane will, therefore, be :**

$$\text{Duty for cotton} = \frac{864 \times 200}{50} = 4 \times 864 \text{ ha/cumec}$$

$$\text{Duty for wheat} = \frac{864 \times 120}{40} = 3 \times 864 \text{ ha/cumec}$$

$$\text{Duty for sugarcane} = \frac{864 \times 360}{120} = 3 \times 864 \text{ ha/cumec.}$$

(*i*) *Outlet discharge required for cotton* $= \dfrac{\textbf{Area}}{\textbf{Duty}}$

$$= \frac{\text{CCA} \times \text{IR for cotton}}{\text{Duty for cotton}} = \frac{15000 \times 60\%}{4 \times 864} = \frac{9000\,\text{ha}}{4 \times 864\ \text{ha/cumec}} = 2.604\ \text{m}^3/\text{s}$$

*Outlet discharge required for wheat*

$$= \frac{\text{CCA} \times \text{IR for wheat}}{\text{Duty for wheat}} = \frac{25000 \times 80\%}{3 \times 864} = \frac{20000}{3 \times 864} = 7.71\ \text{m}^3/\text{s}$$

Also, outlet discharge required for sugarcane

$$= \frac{\text{CCA} \times \text{IR for sugarcane}}{\text{Duty for sugarcane}} = \frac{30000 \times 60\%}{3 \times 864} = \frac{18000}{3 \times 864} = 6.94\ \text{m}^3/\text{s}$$

Now,

(*i*) Water conveyance efficiency for distributary $D_1$

$$= \frac{\text{Water delivered to the cotton crop}}{\text{Water supplied at the canal at its head}} \times 100$$

$$= \frac{2.604}{15\,\text{m}^3/\text{s}} \times 100 = \mathbf{17.36\%} \qquad \text{...}(i)$$

(*ii*) Similarly, water conveyance efficiency for distributary $D_2$

$$= \frac{\text{Water delivered to wheat crop}}{\text{Water supplied the distributary at its head}}$$

$$= \frac{7.71\,\text{m}^3/\text{s}}{20\,\text{m}^3/\text{s}} \times 100 = \mathbf{38.55\%} \qquad \text{...}(ii)$$

(*iii*) Similarly, water conveyance efficiency for distributary $D_3$

$$= \frac{\text{Water delivered to sugarcane crop}}{\text{Water supplied in the distributary head}}$$

$$= \frac{6.94\,\text{m}^3/\text{s}}{25\,\text{m}^3/\text{s}} \times 100 = \mathbf{27.76\%} \qquad \text{...}(iii)$$

The lowest value from amongst the above calculated values (*i*), (*ii*) and (*iii*) is the value at (*i*). Hence, distributary No. $D_1$ (*i.e.*, first distributary) is the least efficient, while distributary No. $D_2$ (*i.e.*, second distributary) is the most efficient. **Ans.**]

**Q. 4.** (*a*) *A hydraulic jump occurs in a horizontal rectangular channel. Froude number before the jump is 12 and energy loss is 4 m. Estimate sequent depths, discharge intensity and Froude number after the jump.* (8)

[**Solution.** $F_1 = 12$ ; $H_L = 4$ m ; $q = ?$ ; $F_2 = ?$

$y_1$ and $y_2 = ?$

Using Eq. (10.7), we have

$$y_2 = \frac{y_1}{2}\left[-1 + \sqrt{1 + 8F_1^2}\right]$$

$$y_2 = \frac{y_1}{2}\left[-1 + \sqrt{1 + 8 \times 144}\right] = 16.48 y_1$$

Using Eq. (10.12), we have

$$H_L = \frac{[y_2 - y_1]^3}{4 \cdot y_1 y_2}$$

$$\therefore \quad 4 = \frac{(16.48y_1 - y_1)^3}{4 \cdot y_1 \cdot (16.48y_1)} = \frac{(15.48y_1)^3}{(16.48 \times 4)y_1^2}$$

or $4 = 56.27\, y_1$

or $y_1 = 0.071$ m. **Ans.**

$\therefore$ $y_2 = 16.48\, y_1 = 1648 \times 0.071$ m $=$ **1.171 m. Ans.**

Now, using Eq. (10.6), we have

$$y_2 = -\frac{y_1}{2} + \sqrt{\frac{y_1^2}{4} + \frac{2q^2}{gy_1}}$$

$$\therefore \quad 1.171 = (-)\frac{0.071}{2} + \sqrt{\frac{(0.071)^2}{4} + \frac{2q^2}{9.81 \times 0.071}}$$

$$1.2065 = \sqrt{0.020 + 2.87q^2}$$

Squaring both sides, we get

$$1.4556 = 0.020 + 2.87\, q^2$$

or $2.87\, q^2 = 1.4556 - 0.020 = 1.4356$

or $q =$ **0.707 m³/sec Ans.**

For computing $F_2$, we use Eq. (10.8) as :

$$y_1 = \frac{y_2}{2}\left[-1 + \sqrt{1 + 8F_2^2}\right]$$

$$\therefore \quad 0.071 = \frac{1.171}{2}\left[-1 + \sqrt{1 + 8F_2^2}\right]$$

or $\sqrt{1 + 8F_2^2} = 1.1213$

or $1 + 8F_2^2 = 1.257$

or $F_2 = \sqrt{\frac{1.257 - 1}{8}} =$ **0.179. Ans.**

**Q. 5.** (*b*) *Design a rigid boundary canal carrying 50 m³/sec. Take Manning's n = 0.014 and longitudinal slope of the canal as 12 per ten thousand. Channel is rectangular section.* (8)

**Solution.** Let us design a rectangular rigid boundary canal having base width = *b* and water depth *y*. The depth *y* may be assumed to be say 1.5 m or so,

$$A = b \cdot y$$

$$R = \frac{A}{P} = \frac{b \cdot y}{b + 2y}$$

Using Manning's formula :

$$Q = \frac{1}{n} A \cdot R^{2/3}, \sqrt{S}, \text{ we have}$$

$$50 \text{ m}^3/\text{s} = \frac{1}{0.014} (b \cdot y) \left(\frac{by}{b + 2y}\right)^{2/3} \cdot \sqrt{\frac{12}{10,000}}$$

As 1st approximation, let us assume that $b$ is much more than $y$, than $R = y$.

$$\therefore \quad 50 = \frac{(1.5)^{5/3}}{0.014} b \cdot \sqrt{\frac{12}{10000}}$$

$$\text{or} \quad b = \frac{50 \times 0.014}{(1.5)^{5/3}} \times \frac{1}{0.03464} = 10.28 \text{ m}$$

Say, let us use $b = 12$ m

Check for discharge

$$\therefore \quad R = \frac{by}{b + 2y} = \frac{12 \times 1.5}{12 + 2 \times 1.5} = 1.2$$

$$\therefore \quad Q = \frac{1}{0.014} (12 \times 1.5)(1.2)^{2/3} \cdot \sqrt{\frac{12}{10000}}$$

$$= 50.29 \text{ m}^3/\text{s} > 50 \text{ m}^3/\text{sec (OK)}$$

Hence, use a rigid boundary rectangular canal with bed width = 12 m and water depth = 1.5 m. **Ans.**]

**Q. 7.** (*b*) *Using Lacey's theory, design an irrigation channel carrying 30 m³/sec. Take silt factor as 1.0.* (8)

[**Hint.** Pl. follow exactly similar Solved Example 4.10.]

**Year 2016 (*Engineering Services*)**

**Q. 2.** (*b*) *What are hydraulic turbines ? Describe different types of hydraulic turbines with their characteristics and suitability of use. Differentiate between 'Pelton' and 'Francis turbines'.* (8)

[**Hint.** Pl. refer article 24.4 (5)]

**Q. 2.** (*d*) *Differentiate between 'low gravity' and 'high gravity' dams. Calculate the limiting height of low concrete gravity dam with the following data:*

*Compressive strength of concrete* = *2500 kN/m²*
*Unit weight of water* = *9.81 kN/m³*
*Specific gravity of the concrete* = *2.4.* (8)

[**Hint.** Pl. refer article 19.7.

The limitating height of low gravity dam is given by Eq. (19.27) as:

$$H_{\max} = \frac{f}{\gamma_w (S_c + 1)} \quad \text{...(Eq. 19.27)}$$

where $f$ = allowable stress of dam meterial (concrete) = 2500 kN/m²

$\gamma_w$ = unit wt. of water = 9.81 kN/m³

$S_c$ = sp. gravity of concrete = 2.4

$$\therefore \quad H_{max} = \frac{2500 \text{ kN/m}^2}{9.81 \frac{\text{kN}}{\text{m}^3}[2.4 + 1]} = \frac{2500}{9.81 \times 3.4} \text{ m}$$

= 74.95 m; say 75 m. **Ans.** ]

**Q. 3.** (*e*) *A hydraulic jump is formed in a horizontal rectangular channel. If post-jump flow depth = 5 m for a discharge per unit width = 5 m*$^3$*/s/m, then compute the pre-jump depth of flow and energy dissipated in the hydraulic jump.* (8)

[**Solution.** Here

$y_2 = 5$ m

$q = 5$ m$^3$/s/m

$y_1 = ?$

Energy dissipated = Loss of energy = $H_L$ = ?

Use Eq. (10.8) as:

$$y_1 = \frac{y_2}{2}\left[-1+\sqrt{1+8F_2^2}\right] \quad \text{...(Eq. 10.8)}$$

$$\text{where} \quad F_2 = \frac{V_2}{\sqrt{gy_2}}$$

$$\text{where} \quad V_2 = \frac{q}{y_2} \quad \text{...(Eq. 10.2)}$$

Substituting the given values, we get

$$V_2 = \frac{5 \text{ m}^3\text{/s/m}}{5 \text{ m}} = 1 \text{ m/s}$$

Also, $$F_2 = \frac{1}{\sqrt{9.81 \times 5}} = 0.143$$

$$\therefore \quad y_1 = \frac{5}{2}\left[-1+\sqrt{1+8\times 0.143^2}\right] = 0.196 \text{ m. } \textbf{Ans.}$$

Also, $H_L$ is given by Eq. (10.12) as:

$$H_L = \frac{[y_2 - y_1]^3}{4 \, . \, y_1 \, y_2} \quad \text{...(Eq. 10.12)}$$

$$\therefore \quad H_L = \frac{(5-0.196)^3}{4 \times 5 \times 0.196} \text{ m} = 28.28 \text{ m. } \textbf{Ans.}$$ ]

**Q. 6.** (*d*) *Using Lacey's theory, design an irrigation canal to carry a discharge of 30 m*$^3$*/s with silt factor = 1.05.* (8)

[**Hint.** Pl. follow exactly similar Solved Example 4.10. ]

**Q. 7.** (*d*) *An unlined channel having bed slope 1 in 4000, carries a certain discharge at flow depth = 5 m with hydraulic radius = 3.5 m. Determine the concentration of the suspended sediment load at 1.5 m above the channel bed if the concentration at 0.25 m above the bed is 400 ppm and the fall/settling velocity of the suspended particle is 0.3 m/s. Assume that Von Karman universal contant = 0.4.* (8)

[**Solution.** Pl. follow Solved Example 4.12, and compute as under :

Use Eq. (4.40) as:

$$\frac{c}{c_a} = \left[\frac{a(D-y)}{y(D-a)}\right]^{\frac{w_o}{KV^*}} \qquad \text{...(Eq. 4.40)}$$

where $S = \dfrac{1}{4000}$

$D = 5$ m

$R$ = Hydraulic radius = 3.5 m

$y$ = up distance from the bed of the channel, where sediment concentration ($c$) is desired = 1.5 m

$c_a$ = known sediment concentration = 400 ppm.

$a$ = known distance where sediment concentration is $c_a$ = 0.25 m

$w_o$ = fall velocity of sediment grain in still water = 0.3 m/s.

$K$ = Von-Karman universal constant = 0.4.

Now, $V^*$ = Shear friction velocity is given by Eq. (4.41) as:

$$V^* = \sqrt{\frac{\tau_0}{\rho_w}} = \sqrt{\frac{\gamma_w \,.\, R \,.\, S}{\rho_w}} = \sqrt{\frac{\rho_w \, g \, RS}{\rho_w}} = \sqrt{gRS}$$

where $R = 3.5$ m (given)

$$\therefore \qquad V^* = \sqrt{9.81 \times 3.5 \times \frac{1}{4000}} = 0.0926 \text{ m/s}$$

Now, substituting values in Eq. (4.40), we get

$$\frac{c}{400 \text{ ppm}} = \left[\frac{0.25(5-1.5)}{1.5(5-0.25)}\right]^{\frac{0.3}{0.4 \times 0.0926}}$$

or
$$\frac{c}{400} = \left[\frac{0.25 \times 3.5}{1.5 \times 4.75}\right]^{8.095} = (0.1228)^{8.095}$$

or
$$c = 1.69 \times 10^{-5} \text{ ppm. } \textbf{Ans.}$$

**[Author's Note: Fall velocity given as 0.3 m/s is too high. It should have been about 0.02 m/s.]**

**Q. 7.** (*e*) (*i*) *Define the following :*

1. *Potential evapotranspiration*
2. *Actual evapotranspiration*
3. *Field capacity*
4. *Permanent wilting point*

*How can evapotranspiration be measured for a given vegetation type ?* (8)

## Year 2017 (*Engineering Services*)

**Q. 1.** (*b*) *A trapezoidal channel with side slopes of 1 : 2 has to be designed to convey 8.0* $m^3/s$ *at a velocity of 1.0* $m/s$*, so that the amount of concrete lining for the bed and sides is the minimum.*

(*i*) *Calculate the area of lining required for one metre length of the channel.*

*(ii) If the roughless coefficient is n = 0.015, calculate the bed slop of the channel for uniform flow.* (12)

[**Solution.** Pl. refer Solved Example 4.5, and compute as under:

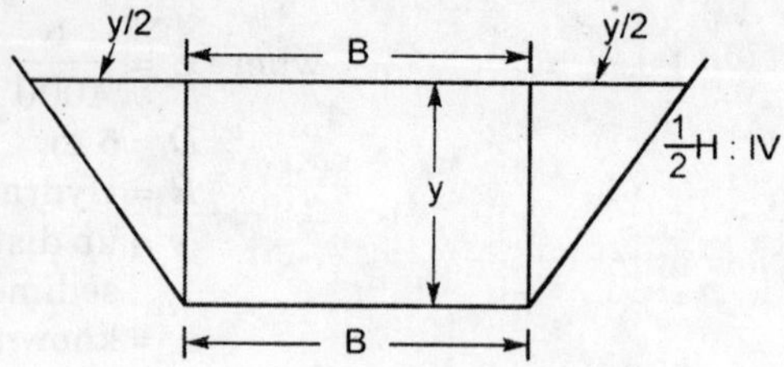

Fig. 30.15.

The given lined trapezoidal channel is to have side slopes of 1 : 2, which means $1H : 2V$; or $1/2\ H : 1\ V$, as shown in Fig. 30.15.

Hence, for such section, we have

Wetted perimeter

$$= P = B + 2\sqrt{\left(\frac{y}{2}\right)^2 + y^2} = B + 2 \cdot y\,(1.118)$$

$$P = B + 2.24\,y \quad ...(i)$$

Also $$A = \frac{[B + (B + y)]}{2}\,y$$

or $$A = (B + 0.5\,y)\,y \quad ...(ii)$$

or $$\frac{A}{y} = B + 0.5\,y$$

or $$B = \frac{A}{y} - 0.5\,y \quad ...(iii)$$

Substituting in (*i*), we get

$$P = \left[\frac{A}{y} - 0.5\,y\right] + 2.24\,y$$

or $$P = \frac{A}{y} + 1.74\,y \quad ...(iv)$$

For minimum perimeter, (with constant $A$), $\frac{dP}{dy} = 0$

or $$\frac{dP}{dy} = A\,(-1y^{-2}) + 1.74 \times 1$$

or $$\frac{dP}{dy} = -\frac{A}{y^2} + 1.74 = 0$$

or $$\frac{A}{y^2} = 1.74, \text{ or } A = 1.74\,y^2$$

Substituting in (*iv*), we get

$$P = \frac{1.74\,y^2}{y} + 1.74\,y = 3.48\,y$$

Also $$R = \frac{A}{P} = \frac{1.74y^2}{3.48\,y} = \frac{y}{2}.$$

***Thus, for a channel having $\frac{1}{2}$ H : 1 V side slopes, the most economical section will have***

$$A = 1.74y^2, \text{ and } R = \frac{y}{2}$$

Now in the given case

$$Q = 8 \text{ m}^3/\text{s}$$
$$V = 1 \text{ m/s}$$

Since $Q = A.V$, we have

$$8 \text{ m}^3/\text{s} = A.1 \text{ m/s}$$

or $\mathbf{A = 8\ m^2}$

But $A = 1.74\, y^2$

$\therefore$ $1.74\, y^2 = 8 \text{ m}^2$

or $y = \sqrt{\dfrac{8}{1.74}}$ m = 2.14 m

Also $R = \dfrac{y}{2} = \dfrac{2.14}{2} = 1.07$ m

Now from (*iv*), we have

$$P = \frac{A}{y} + 1.74\,y$$

or $P = \dfrac{8}{2.14} + 1.74 \times 2.14 = 7.46$ m

$\therefore$ Area of lining for 1 m length of channel

$= 7.46 \times 1 \text{ m} = 7.46\text{m}^2$ Ans.

(*ii*) Now, to determine bed slope ($S$), use Manning's Eq:

$$Q = \frac{1}{n}\,.\,A.\,R^{2/3}.\,\sqrt{S}$$

or $Q = \dfrac{1}{0.015}\, 8\,(1.07)^{2/3}.\,\sqrt{S}$

or $S = 2.06 \times 10^{-4}$

or $S = \dfrac{\mathbf{1}}{\mathbf{4864}}$ **Ans.**

**Q. 2.** (*b*) *(ii) A watercourse is designed to irrigate an area of 960 ha of rice. The transplantation of rice takes 15 days and during this, the total depth of water required is 40 cm. Find the duty of irrigation water on the field if there is an effective rainfall of 12 cm. Also, find the duty and discharge at the head of the watercourse assuming losses of water as 20 % in the watercourse.* (10)

**Solution.** Pl. follow similar Solved Example 3.5 and compute as under.

Total depth of water required for transplanting rice = 40 cm

Useful rainfall = 12 cm

∴ Extra water depth required to be supplemented by irrigation

$$= 40 - 12 = 28 \text{ cm}$$

Period in which this water is required = 15 days.

$$\therefore \quad \text{Duty of irrigation water} = \frac{864\,B}{\Delta} = \frac{864 \times 15}{28} = 462.86 \text{ hectares/cumec}$$

Assuming 20% losses in the water course (as given), we have

Duty at the head of the water course = 462.86 × 0.80 = 370.29 hectares /cumec. Ans.

Outlet discharge required

$$= \frac{\text{Area to be irrigated}}{\text{Duty}}$$

$$= \frac{960 \text{ hectares}}{370.29 \text{ hectares/cumec}} = \mathbf{2.59\ cumec.} \quad \textbf{Ans.}$$

**Q. 4.** (*a*) *A pipe of diameter d and length L leads from a reservoir. At the discharge end, a nozzle of diameter $d_n$ is fitted. At the inlet of pipe, head available is H. The nozzle is discharging into atmosphere. Prove that for maximum power transmitted by the nozzle $d_n = \sqrt[4]{\dfrac{d^5}{2fL}}$. Head loss in pipe is given by $h_f = \dfrac{fLV^2}{2gd}$ where f = Darcy's friction factor.* (20)

**Solution.** Power transmitted (*P*) by discharge (*Q*) at exit of nozzle, is given as:

$$P = \gamma_w \cdot Q\,[H_L]$$

where $H_L$ = Loss of head from inlet to outlet

or
$$P = \gamma_w \cdot Q\left[H - \frac{fLV^2}{2gd}\right]$$

where *V* is the vel. through the pipe of dia. *d* and length *L*

*H* = head at inlet pipe

or
$$P = \gamma_w \cdot \left[\frac{\pi}{4} \cdot d^2 \cdot V\right]\left[H - \frac{fLV^2}{2gd}\right]$$

or
$$P = \gamma_w \cdot \frac{\pi}{4} \cdot d^2\left[HV - \frac{fLV^3}{2gd}\right] \quad \ldots(i)$$

For maximum power, $\dfrac{dP}{dV} = 0$.

Differentiating (*i*), we get

$$\therefore \quad \frac{dP}{dV} = \gamma_w \cdot \frac{\pi}{4} \cdot d^2\left[H - \frac{3fLV^2}{2gd}\right] = 0$$

or $$\left[H - \frac{3 f LV^2}{2 gd}\right] = 0$$

or $$H = \frac{3 f LV^2}{2 gd} \quad \text{...(ii)}$$

Now, available head at the outlet of nozzle

$$= H - \frac{f LV^2}{2 gd} = \left[\frac{3 f LV^2}{2 gd} - \frac{f LV^2}{2 gd}\right] = \frac{2 f LV^2}{2 gd} \quad \text{...(iii)}$$

Also, if velocity through the nozzle is $V_n$, then we have the velocity head at the outlet of the nozzle $= \dfrac{V_n^2}{2g}$ ...(iv)

Equating (iii) and (iv), we get

$$\frac{2 f LV^2}{2 gd} = \frac{V_n^2}{2g} \quad \text{...(v)}$$

Also, $$\left(\frac{\pi}{4} . d^2\right) V = \left(\frac{\pi}{4} . d_n{}^2\right) . V_n$$

[$\because$ $A_1V_1 = A_2V_2$ at exit end between end point of pipe and nozzle].

or $$V_n = \left(\frac{d}{dn}\right)^2 . V \quad \text{...(vi)}$$

or $$\frac{V_n^2}{2g} = \frac{\left[\left(\frac{d}{dn}\right)^2 V\right]^2}{2g} = \left(\frac{d}{dn}\right)^4 . \frac{V^2}{2g}$$

Hence, from (iv), we have

$$\frac{2 f LV^2}{2 gd} = \left(\frac{d}{dn}\right)^4 . \frac{V^2}{2 g}$$

or $$\frac{2 fL}{d} = \frac{d^4}{dn^4}$$

or $$d_n{}^4 = \frac{d^5}{2 fL}$$

or $$d_n = \sqrt[4]{\frac{d^5}{2 fL}}$$

Hence, proved. **Ans.**

**Q. 4.** (b) *Explain the use of stilling basin as an energy dissipator and describe the various components of a stilling basin with their function with the help of sketches. List the four types of Indian standard basins.* (20)

**Q. 8.** *(iii) Briefly describe various geological considerations for the site selection of the dam.* (5)

## Year 2018 (*Engineering Services*)

**Q. 1.** *Design a most economical trapezoidal canal section for carrying discharge of 45 cumecs and side slope of 1½ : 1 at a velocity of 0.9 m / sec. Assume value of n = 0.022 (Manning's coefficient).* (12)

**[Solution.** The given channel section is shown in Fig. 30.16.

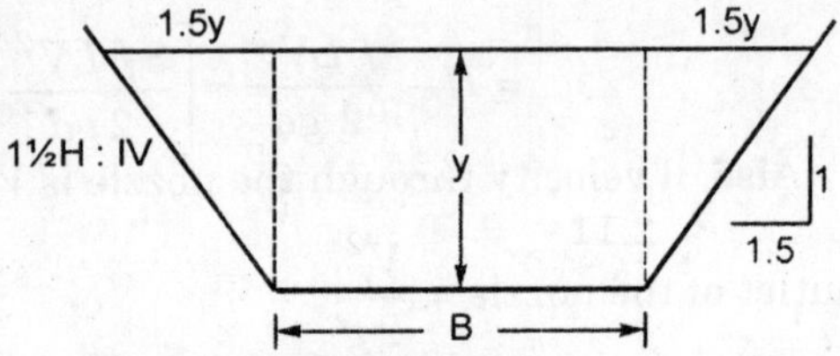


Fig. 30.16.

For such a channel section, we have

Wetted perimeter ($P$)

$$= B + 2 \cdot \sqrt{(1.5y)^2 + y^2}$$

or $$P = B + 2 \cdot \sqrt{3.25y^2}$$

or $$P = B + 3.61y \qquad ...(i)$$

Also $$A = \frac{B + (B + 3y)}{2} \cdot y$$

or $$A = (B + 1.5\,y)\,y$$

or $$\frac{A}{y} = B + 1.5y$$

or $$B = \frac{A}{y} - 1.5y \qquad ...(ii)$$

Substituting in (i) above, we get

$$P = \left[\frac{A}{y} - 1.5y\right] + 3.61y$$

or $$P = \frac{A}{y} + 2.11y \qquad ...(iii)$$

For *most economical section,* we should have *minimum perimeter* ($P$) (with constant A), and therefore, let us equate $\frac{dP}{dy} = 0$.

Also from *(iii)* $$\frac{dP}{dy} = A\,(-1y^{-2}) + 2.11 \times 1 = 0$$

$$\therefore -\frac{A}{y^2} + 2.11 = 0$$

or $$\frac{A}{y^2} = 2.11$$

or $$A = 2.11\,y^2 \qquad ...(iv)$$

Substituting in *(iii)*, we get

$$P = \frac{2.11y^2}{y} + 2.11\,y = 4.22\,y$$

Also, $R = \frac{A}{P} = \frac{2.11y^2}{4.22y} = \frac{y}{2}$

Now, in the given case, we have

$Q = 45 \text{ cumecs} = 45 \text{ m}^3/\text{s}$

$V = 0.9 \text{ m/sec}$

Since $Q = AV$, we have

$45 = A \cdot 0.9$

or $A = 50 \text{ m}^2$

But $A = 2.11\, y^2$

$\therefore \quad 2.11\, y^2 = 50$

or $y = \sqrt{\frac{50}{2.11}} \text{ m} = 4.87 \text{ m}$

Also, $R = \frac{y}{2} = \frac{4.87}{2} = 2.43 \text{ m}$

Also, $P = 4.22\, y = 4.22 \times 4.87 = 20.55 \text{ m}^3$

Now, to determine bed slope (S), we use Manning's formula for discharge, as:

$$Q = \frac{1}{n} . AR^{2/3} . S^{1/2}$$

where $n = 0.022$ (given)

$Q = 45 \text{ m}^3/\text{s}$ (given)

or $45 = \frac{1}{0.022} \times 50 \times (2.43)^{2/3} \cdot \sqrt{S}$

or $\sqrt{S} = 0.1095$

or $S = 1.199 \times 10^{-4}$

or $S = \frac{1}{8338}$ **Ans.**

Also, bed width $(B) = \frac{A}{y} - 1.5y = \frac{50}{4.87} - 1.5 \times 4.87 = 2.96$ m; say 3 m

Hence, use a trapezoidal channel with bed width = 3m.

Bed slope at $\frac{1}{8338}$, and water depth = 4.87 m **Ans.**

**Q. 4(*b*).** *Find the capacity of a reservoir for irrigating an area having G.C.A. = 40,000 ha. The cropping pattern consists of mainly the following crops.*

| *Crop* | *Period* | *Base period (days)* | *Outlet factor (ha/cumec)* | *Intensity of cropping %* | *Crop ratio* |
|---|---|---|---|---|---|
| *(1)* | *(2)* | *(3)* | *(4)* | *(5)* | *(6)* |
| *Sugarcane* | *Oct-Sept.* | *360* | *800* | *60* | *4* |
| *Cotton* | *May-Nov* | *180* | *1200* | *60* | *3* |
| *Wheat* | *Dec-April* | *120* | *1800* | *70* | *2* |
| *Gram* | *Dec-April* | *120* | *2000* | *70* | *3* |

*The area has 25% non-culturable area.*

*Assume*

*(i) Time factor = 7/10*

*(ii) Extra allowance for period of peak water use = 20%*

*(iii) Conveyance losses = 20%*

*(iv) Reservoir losses = 10%*

*Also calculate design discharge and capacity factor of the main canal.*

[**Hint.** There are inbuilt contradictions in the data given in this question.

(*i*) A close look at the values of crop periods shown in Col. (2) for all the four crops show that only Sugarcane is cropped over both the crop seasons. (*i.e.*, Kharif as well as Ravi seasons), during Oct-Sept. While cotton is grown over only in Kharif season (May-Nov.) and Wheat and Gram both are grown only during Ravi season (Dec.-April).

This means that the *crop ratios* (*i.e.*, area of a crop cropped in Kharif to the area of that crop cropped in Ravi) for Sugarcane, shown to be in Col. (6) may be OK, but it's values for wheat and gram given to be 2 and 3, respectively are not correct, as when the crop is grown only in Ravi, season its area cropped in Kharif is zero, and hence its crop ratio is also zero. Similarly, the given value of crop ratio for cotton as 3, is wrong.

(*ii*) *Another contradiction* : The total value of the intensity of cropping for all the four crops totals to 260%. Since intensity of cropping is the ratio of gross cropped area under all the crops to the total on net Cultivated/Cropped Area = CCA. Evidently, if all the growing crops are cropped twice and grown over both the crop periods, the intensity of cropping will be 200% as the grass cropped area will be twice the CCA. And when three out of four crops are being cropped only once. (in one crop season), the total intensity of cropping should be less then 200%, whereas given values totals to 260%.

Due to such inbuilt contradiction in given data, we are forced not to provide any solution to this question.]

## Year 2019 (*Engineering Services*)

**Q.1.(c)** *Draw the schematic diagram of a gravity dam and indicate the major forces acting on it. Draw the diagram of the up-lift force when (i) drain is not provided; and (ii) drain is provided.* (12)

**Q.2.(a)** *A trapezoidal chennel is to be designed to convey a discharge of 50 $m^3$/sec at a velocity of 2 m/sec. The bed width to depth ratio is 0.8. The side slopes are 1 H : 1 V. Calculate the bed width, depth of flow and bed slope of the channel. Assume Manning's coefficient, n = 0.02.*

[**Solution.** $Q = 50\ m^3/s$

$V = 2\ m/s$

Using $Q = V.A.$, we get

$50\ m^3/s = 2\ m/s \times A$

or $A = 25\ m^2$

From Fig. 30.17, we have

$$A = \frac{B+(B+2y)}{2}.y = (B+y)y$$

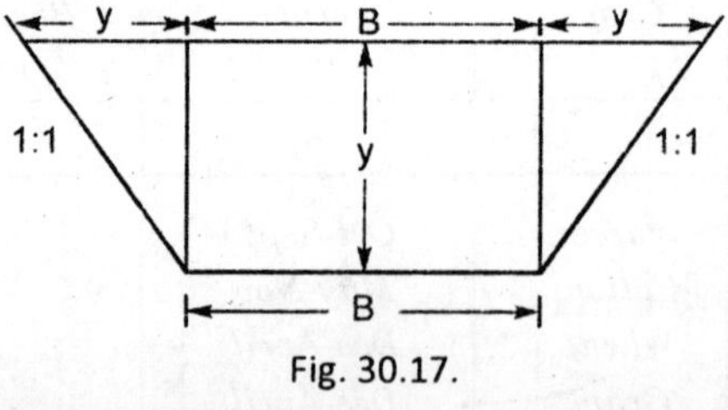


Fig. 30.17.

$\therefore \quad (B + y)y = 25 \text{ m}^2$

But $\quad \dfrac{B}{y} = 0.8$ (given)

$\therefore \quad B = 0.8\,y$

or $\quad (0.8y + y)\,y = 25$

or $\quad 1.8y^2 = 25$

or $\quad y = \mathbf{3.73\ m}$ ; and $B = 0.8 \times 3.73 = \mathbf{2.98\ m}$

Now $\quad P$ = wetted Perimeter = B + $2.\sqrt{2y^2}$

or $\quad P = 2.98 + 2\sqrt{2} \times 3.73 = 13.53$ m

But $\quad \boldsymbol{R} = \dfrac{A}{P} = \dfrac{25}{13.53} = 1.848$

From Manning formula, we have

$$V = \frac{1}{n} \cdot R^{2/3} \sqrt{S}$$

or $\quad 2 = \dfrac{1}{0.02} \times (1.848)^{2/3} \,.\, \sqrt{S}$

or $\quad S = \dfrac{1}{1417}$

Hence, $\quad B = 2.98$ m; $y = 3.73$ m; $S = 1$ in $1417$ — **Ans.**

**Q.4(*a*)** *Explain the term "initial regime" and "final regime", as explained in Lacey's theory of stable channels. Design a stable channel for carrying a discharge of 30 m³/s using Lacey's method, assuming a silt factor equal to 1.0.* (20)

[**Hint :** Pl. refer article 4.7.4.3, and Solved Example 4.10]

**Q.4(*b*)(*i*)** *Define field capacity, permanent wilting point and average moisture content. Explain how these will be useful in deciding the frequency of irrigation. (A schematic diagram showing less and more frequent irrigation is to be drawn for clarity.* (3+4+3=10)

**Q.4(*b*)(*ii*)** *In a hydraulic jump occurring in a horizontal channel, the Froude's number before the jump is 10.0 and energy loss is 3.2 m. Estimate sequent depths, discharge intensity and Froude's number after the jump.* (10)

[**Solution.** We know that incoming Froude No. $F_1$ is given as :

$$F_1 = \frac{V_1}{\sqrt{gy_1}} = \frac{\dfrac{q}{y_1}}{\sqrt{qy_1}} = \frac{q}{\sqrt{gy_1^3}} = 10.0 \text{ (given)} \qquad (i)$$

Also $\quad H_L$ = Energy loss = $\dfrac{(y_2 - y_1)^3}{4y_1 y_2} = 3.2$ m (given) ...(Eq. 10.1

From Eq. (*i*), $\dfrac{q^2}{gy_1^3} = 100$

or $\frac{q^2}{g} = 100\ y_1^3$

From Eq. (10.5), we have

$$y_1 \cdot y_2 (y_1 + y_2) = \frac{2q^2}{g} \quad \text{...(Eq. 10.5)}$$

or $y_1 \cdot y_2 (y_1 + y_2) = 2 \cdot 100\ y_1^3 = 200\ y_1^3$

Assume $\frac{y_2}{y_1} = x;$ or $y_2 = y_1 \cdot x$

$\therefore$ $y_1 \cdot (y_1 x)(y_1 + y_1 x) = 200\ y_1^3$

or $y_1^2 \cdot x (y_1)(1 + x) = 200\ y_1^3$

or $x(1 + x) = 200$

$\therefore$ $x = 13.65$ (from hit and trial)

or $y_2 = 13.65\ y_1$

From energy loss equation, *i.e.*, Eq. (10.12), we have

$$\frac{(y_2 - y_1)^3}{4 y_1 y_2} = 3.2 \text{ m}$$

Substituting $y_2 = 13.65\ y_1$, we get

or $\frac{(13.65 y_1 - y_1)^3}{4 \cdot y_1 \times (13.65 y_1)} = 3.2.$

or $\frac{(12.65 y_1)^3}{4 \times 13.65 \times y_1^2)} = 3.2$

or $y_1 = \mathbf{0.0863\ m}$ **Ans.**

$\therefore y_1 = 13.65 y_1 = 13.65 \times 0.0863 = 1.178$ m

Hence, sequent depths are **0.0863 m** and **1.178 m** **Ans.**

Also $\frac{q}{\sqrt{g \cdot y_1^3}} = 10.$

or $q = 10 \cdot \sqrt{g \cdot y_1^3} = 10\sqrt{9.81 \times (0.0863)^3}$

$= \mathbf{0.794\ m^3/sec}$ **Ans.**

Froude No. after the jump $= F_2 = \frac{q}{\sqrt{g \cdot y_2^3}} = \frac{0.794}{\sqrt{9.81 \times (1.178)^3}} = \mathbf{0.198}$ **Ans.**

## Year 2020 (*Engineering Services*)

**Q.1.(*d*)** *After how many days will you supply water to soil (clay loam) in order to ensure efficient irrigation of the given crop, if :*

*Field capacity of soil* = 27%
*Permanent wilting point* = 14%
*Density of soil* = 1.5 $g/cm^3$
*Effective depth of root zone* = 75 cm
*Daily consumptive use of water for the given crop* = 11 mm. (12)

[**Hint :** Follow Exactly Similar Solved Example 2.17 with only slight change in data values]

**Q.1.(*ii*)** *A channel has two sides vertical and semi-circular bottom of 2 m diameter. Calculate the discharge of water through the channel, when depth of flow is 2 m. Take C = 70 and slope of bed as 1 in 1000.* (6)

[**Solution.** Use Chezy's formula Eq. (4.22) as :

$$V = C \cdot \sqrt{R \cdot S}$$

or

$$Q = C \cdot A \cdot \sqrt{R \cdot S}$$

where C = 70 (given)

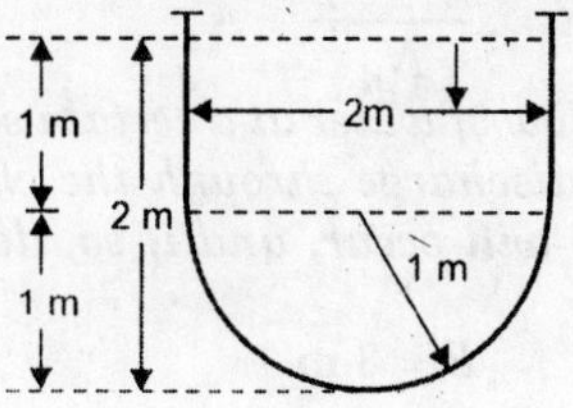


Fig. 30.18.

$$A = (2\text{ m} \times 1\text{ m}) + \frac{1}{2}\left[\frac{\pi}{4}(2)^2\right] = 3.57\text{ m}^2$$

and

$$P = (1\text{ m} + 1\text{ m}) + \frac{1}{2} \cdot \pi \times (2\text{ m}) = 5.14\text{ m}$$

$\therefore$

$$R = \frac{A}{P} = \frac{3.57}{5.14} = 0.69\text{ m}$$

$\therefore$

$$Q = 70 \times 3.57\sqrt{0.69 \times \frac{1}{1000}} = 6.59\text{ m}^3/\text{s} \quad \textbf{Ans.]}$$

**Q3.(*b*)** *What are the effects of water-lagging?* (5)

**Q4.(*c*) (*i*)** *What do you understand by galleries and shafts and why are they provided in gravity dams?* (12)

## Year 2021 (*Engineering Services*)

**Q2.(c)** *A homogeneous anisotropic embankment dam section is shown in the figure, the coefficient of permeability in the x and z directions being* $5 \times 10^{-8}$ *m/s and* $2 \times 10^{-8}$ *m/s respectively. Construct the flow net and determine the quantity of seepage through the dam. What is the pore water pressure at point 'P'?* *(20)*

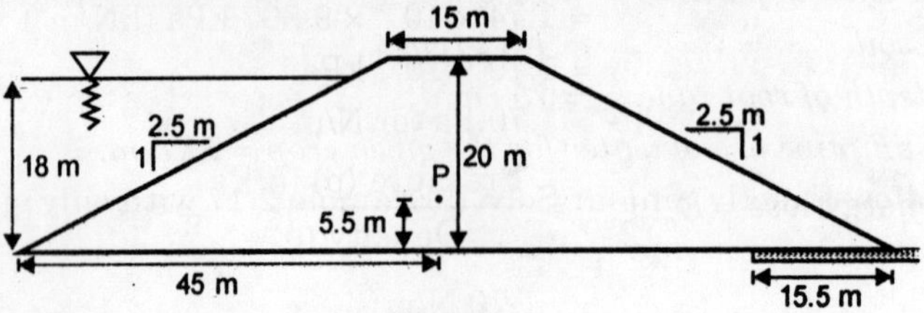


**Fig. 30.19.**

**[Hint:** The dam section may be plotted by multiplying the horizontal dimensions by $\sqrt{\frac{K_V}{K_H}} = \sqrt{\frac{2}{5}} = \sqrt{0.4}$ **= 0.632,** as stated in the text on page 732 under article 20.11, dam soil being a non-isotropic one. The flow net can then be drawn as explained in article 20.10. After counting flow channels ($N_f$) and number of drops ($N_d$), quantity of seepage can be computed using Eq. (20.16) as :

$$q = \sqrt{K_H \cdot K_V} \cdot \frac{H \cdot N_f}{N_d} \quad \text{...(Eq. 20.16)}$$

**Q4.(b)** (*i*) *The depth of flow of water at a certain section of a rectangular channel 3 m wide is 0.45 m. The discharge through the channel is 3.1* $m^3/s$*. Determine whether a hydraulic jump will occur, and if so, determine its height and loss of energy per kg of water.* *(10)*

**[Solution.**

$$B = 3 \text{ m}$$

$$y_1 = 0.45 \text{ m}$$

$$Q = 3.1 \text{ m}^3/\text{s}$$

$$q = \frac{Q}{B} = \frac{3.1}{3} = 1.03 \text{ m}^2/\text{s}$$

$$V_1 = \frac{q}{y_1} = \frac{1.03}{0.45} = 2.29 \text{ m/s}$$

Conjugate depth $y_2$ for Hydraulic Jump to form :

$$y_2 = \frac{-y_1}{2} + \sqrt{\frac{y_1^2}{4} + \frac{2q^2}{gy_1}} \quad \text{...(Eq. 10.6)}$$

or

$$y_2 = \frac{-0.45}{2} + \sqrt{\frac{(0.45)^2}{4} + \frac{2 \times (1.03)^2}{9.81 \times 0.45}}$$

$$= -0.225 + \sqrt{0.05063 + 0.481}$$

$$= -0.225 + 0.729 = \mathbf{0.504\ m.}$$

Thus depth will increase from 0.45 m to 0.504 m, being a very feable jump. Now $H_L$ = Head loss or Energy loss in H.J. formation is given as :

$$H_L = \frac{(y_2 - y_1)^3}{4 y_1 y_2} \quad \text{...(Eq. 10.12}$$

or $$H_L \text{ or } E_L = \frac{(0.504-0.45)^3}{4\times 0.45\times 0.504} = \mathbf{1.74 \times 10^{-4}\ m}$$

Thus energy loss in H.J. formation

$$= 1.74 \times 10^{-4} \text{ m head of water.}$$
$$= 1.74 \times 10^{-4} \times 9.807 \text{ kPa (kN/m}^2\text{)}$$
$$= 1.70 \times 10^{-3} \text{ kPa}$$
$$= 1.70 \text{ Pa (or N/m}^2\text{)}$$

$$\text{Pressure energy loss} = E_{pressure} = \frac{\text{Pressure (p) in Pa}}{\text{Density } (\rho)} = \text{Energy (J/kg)}$$

$$E_L \text{ in J/kg} = \frac{1.70 \text{ N/m}^2}{1000 \text{ kg/m}^3} = (1.7\times 10^{-3}) \text{ N.m/kg}$$

$$= \mathbf{1.7 \times 10^{-3}\ J/kg}$$ **Ans.]**

## Appendix Table A-I : Some Important Properties of Water

| *Temp.* °C | *Sp. gravity* | *Unit. wt. in kN/m³ i.e., Sp. gravity × 9.807* $(S_s \cdot \rho_w)$ | *Vapour pressure* *kN/m²* | *Viscosity* Dynamic or Absolute *kN-sec/m²* | *Viscosity* Kinematic $\left(v = \frac{\mu}{\rho}\right) m^2/sec$ |
|---|---|---|---|---|---|
| 0 | 0.99987 | 9.806 | 0.611 | $1.797 \times 10^{-6}$ | $1.78 \times 10^{-6}$ |
| 5 | 0.99999 | 9.807 | 0.872 | $1.517 \times 10^{-6}$ | $1.518 \times 10^{-6}$ |
| 10 | 0.99975 | 9.805 | 1.226 | $1.306 \times 10^{-6}$ | $1.310 \times 10^{-6}$ |
| 15 | 0.99907 | 9.800 | 1.703 | $1.126 \times 10^{-6}$ | $1.124 \times 10^{-6}$ |
| 20 | 0.99810 | 9.788 | 2.334 | $1.003 \times 10^{-6}$ | $1.009 \times 10^{-6}$ |
| 30 | 0.99574 | 9.765 | 4.609 | $0.802 \times 10^{-6}$ | $0.804 \times 10^{-6}$ |
| 40 | 0.99228 | 9.731 | 7.60 | $0.652 \times 10^{-6}$ | $0.654 \times 10^{-6}$ |
| 60 | 0.98338 | 9.644 | 19.34 | $0.470 \times 10^{-6}$ | $0.478 \times 10^{-6}$ |
| 80 | 0.97196 | 9.532 | 46.583 | $0.356 \times 10^{-6}$ | $0.366 \times 10^{-6}$ |
| 100 | 0.95865 | 9.401 | 101.50 | $0.282 \times 10^{-6}$ | $0.295 \times 10^{-6}$ |

## Appendix Table A-II : Other Important Properties of Water

*Standard Atmospheric Pressure*

$$= P_{atm} = 760 \text{ mm of Mercury}$$
$$= 1 \text{ bar (b)}$$
$$= 1000 \text{ m.b.}$$
$$= 101.325 \text{ kPa*}$$
$$= 10.33 \text{ m of water head**}$$

*Density of water* ($\rho$) *at 4°C*

in MKS units = 1000 kg/m³

in SI units = 1000 kg/m³

---

$$*P_{atm} = 760 \text{ mm of Hg} = \frac{760}{1000} (\text{m of Hg}) \times 13.595 \times 9.807 \text{ kN/m}^3$$

$\because$ Unit wt. of Hg $= S_s \cdot \rho_w = 13.595 \times 9.807$ kN/m³

$= 101.325$ kN/m² = **101.325 kPa**

$$**P_{atm} = \text{m of water height } (h_w) \times \text{unit wt. of water}$$
$$= h_w (m) \times 9.807 \text{ kN/m}^3 = 9.807\, h_w$$
$$101.325 = 9.807\, h_w$$
$$h_w = \mathbf{10.33 \text{ m.}}$$

*Unit weight of water (γ or w) at 4°C*

in MKS units = 1000 kgf/m$^3$, usually written as 1000 kg/m$^3$

in SI units = 9.807 kN/m$^3$

*Dynamic Viscosity of water at 4°C*

in MKS units = $1.598 \times 10^{-4}$ kgf.sec/m$^2$

in SI units = $1.567 \times 10^{-3}$ Ns/m$^2$

= $1.567 \times 10^{-2}$ Poise (P)

*Kinematic Viscosity of water (ν) at 4°C*

in MKS units = $1.567 \times 10^{-6}$ m$^2$/s

in SI units = $1.567 \times 10^{-6}$ m$^2$/s

= $1.567 \times 10^{-6}$ Stoke (St)

## Appendix Table A-III. Water Resources Potentials of River Basins of India

| Sl. No. | River Basin | Average Annual Surface Water Availability (50% dependability) | | Utilisable Surface Water (at 75% dependability) excluding ground water | Utilisable or Replenish-able Ground Water | Present Exploitation (storage) | |
|---|---|---|---|---|---|---|---|
| | | Water Reesources Potential b.cum (BCM) | Per Capita ($m^3$) | B.cum (BCM) | B.cum | Surface water | Ground water |
| (1) | (2) | (3) | (4) | (5) | (6) | (7) | (8) |
| | **(A) Major River Basins** | | | | | | |
| 1. | Indus | 73.30 | 1757 | 46.0 | 26.49 | 16.29 | 21.26 |
| 2. | Ganga-Brahmaputra-Meghna | | | | | | |
| | (a) Ganga | 525.02 | 1473 | 250.0 | 170.99 | 39.44 | 74.62 |
| | (b) Brahmaputra | 537.2 | 18417 | | | | |
| | (c) Barak | 48.4 (585.60) | 7646 | 24.0 | 26.55 | 2.33 | 4.74 |
| 3. | Godavari | 110.54 | 2026 | 76.3 | 46.65 | 25.12 | 15.71 |
| 4. | Krishna | 78.12 | 1312 | 58.0 | 26.41 | 41.80 | 11.91 |
| 5. | Cauvery | 21.36 | 666 | 19.0 | 12.30 | 8.60 | 7.62 |
| 6. | Brahmani-Baitarani | 28.48 | 2696 | 18.3 | 4.05 | 4.65 | 0.90 |
| 7. | Mahanadi | 66.88 | 2546 | 50.0 | 16.46 | 12.33 | 3.44 |
| 8. | Pennar | 6.31 | 648 | 6.9 | 4.93 | 2.65 | 2.27 |
| 9. | Mahi | 11.02 | 1057 | 3.1 | 7.19 | 4.72 | 3.53 |
| 10. | Sabarmati | 3.81 | 421 | 1.9 | — | 1.31 | |
| 11. | Narmada | 45.64 | 2855 | 34.5 | 10.93 | 7.23 | 3.64 |
| 12. | Tapi | 14.88 | 1091 | 14.5 | 8.27 | 9.41 | 4.30 |
| | **(B) Composite River Basins** | | | | | | |
| 1. | Subernarekha | 12.37 | 1392 | 6.8 | 1.82 | 0.67 | 0.42 |
| 2. | West Flowing rivers from Tapi to Tadri | 87.41 | 3194 | 11.9 | 8.27 | 11.27 | |
| 3. | West Flowing rivers from Tadri to Kanyakumari | 113.53 | 3539 | 24.3 | 8.52 | 10.24 | 6.51 |
| 4. | East Flowing rivers between Mahanadi and Pennar | 22.52 | 919 | 13.1 | 18.22 | 1.60 | 11.66 |
| 5. | East Flowing rivers between Pennár and Kanyakumari | 16.46 | 383 | 16.7 | 18.84 | 1.84 | 5.58 |
| 6. | West Flowing rivers of Kutch and Saurashtra including Luni | 15.10 | 631 | 15.0 | 11.23 | 4.73 | 6.59 |
| 7. | Area of inland drainage in Rajasthan desert | negl. | — | — | — | — | — |
| 8. | Minor rivers draining into Myanmar (Burma) and Bangladesh) | 31.0 | 14616 | — | — | 0.31 | — |
| | Total | 1869.35 | 2214 | 690.3 Say 690 | 431.42 Say 432 | **206-54*** | 184.70 |

* An additional live storage of 6.24 B.cum has been created through medium projects each having a capacity less than 10 Mcum (not included in above data) ; thus making a total storage of 212.78 B.cum ; say **213 B.cum**

## Appendix Table A-IV : Fresh Water Supplies and their Storages on Different Rivers in India

| Sl. No. | Name of the River Basin | Catchment area in Mha | Average annual flow in the river basins in B.cum 50% Dependability | Live Storage Created / Likely to be Created in B.cum Through Completed projects (upto 2004) | Through Ongoing projects | Through projects under consideration |
|---|---|---|---|---|---|---|
| 1 | 2 | | 4 | 5 | 6 | 7 |
| 1. | Indus (Up to Border) | 32.13 | 73.30 | 16.29 | 0.28 | 2.58 |
| 2. | (a) Ganga | 86.15 | 525.02 | 39.46 | 21.22 | 30.08 |
| | (b) Brahmaputra Barak and Others | 23.61 | 585.60 | 2.33 | 9.35 | 41.26 |
| 3. | Godavari | 31.28 | 110.54 | 25.12 | 6.21 | 5.84 |
| 4. | Krishna | 25.89 | 78.12 | 41.80 | 7.74 | 1.31 |
| 5. | Cauvery | 8.12 | 21.36 | 8.60 | 0.27 | 0.26 |
| 6. | Pennar | 5.52 | 6.31 | 2.65 | 2.17 | — |
| 7. | East Flowing Rivers from Mahandi to Godavari and Krishna to Pennar | 8.66 | 22.52 | 1.60 | 1.42 | 0.95 |
| 8. | East Flowing Rivers Between Pennar and Kanyakumari | 10.01 | 16.46 | 1.84 | 0.07 | — |
| 9. | Mahanadi | 14.16 | 66.88 | 12.33 | 1.87 | 10.09 |
| 10. | Brahamani and Baitarni | 5.18 | 28.48 | 4.65 | 0.88 | 8.72 |
| 11. | Subernarekha | — | 12.37 | 0.67 | 1.65 | 1.38 |
| 12. | Sabarmati | 2.17 | 3.81 | 1.31 | 0.66 | 0.10 |
| 13. | Mahi | 3.48 | 11.02 | 4.72 | 0.26 | 0.01 |
| 14. | West Flowing Rivers of Kutch, Saurashtra including Luni | 32.19 | 15.10 | 4.73 | 0.80 | 2.85 |
| 15. | Narmada | 9.88 | 45.64 | 7.23 | 16.38 | 0.47 |
| 16. | Tapi | 6.51 | 14.88 | 9.41 | 0.85 | 0.29 |
| 17. | West Flowing Rivers from Tapi to Tadri | 5.29 | 87.41 | 11.27 | 3.46 | 0.08 |
| 18. | West Flowing Rivers from Tadri to Kanyakumari | 5.62 | 113.53 | 10.24 | 1.32 | 1.45 |
| 19. | Area of Inland Drainage in Rajasthan desert | 6.0 | Negligible | — | — | — |
| 20. | Minor River Basins draining into Bangladesh and Burma | 3.63 | 31.00 | 0.31 | — | Negligible |
| | Total | | 1869.35 | 206.54* | 76.26 | 107.54 |

* Additional 6.24 B.cum storage is created through medium projects having live storage of less then 0.01 B.cum which have not been accounted here, thus creating a total storage of 212.78; say **213 B.cum**

## Appendix Table A-V : Statewise Position of Live Storages Created and Likely to be Created in B.cum (2004)

| S. No. | State | *Live storage Created in in B. cum through completed projects* | *Live storages to be created in B.cum through ongoing projects* | *Live storage to be created through projects under consideration* |
|---|---|---|---|---|
| 1. | Andhra Pradesh | 27.305 | 6.148 | 1.611 |
| 2. | Arunachal Pradesh | Nil | 0.241 | 37.934 |
| 3. | Assam | 0.012 | Nil | 0.726 |
| 4. | Bihar | 1.842 | 0.675 | 5.822 |
| 5. | Chatisgarh | 6.217 | 0.788 | 0.819 |
| 6. | Goa | 0.044 | 0.217 | Nil |
| 7. | Gujarat | 16.138 | 7.407 | 3.446 |
| 8. | Haryana | Nil | Nil | 0.258 |
| 9. | Himachal Pradesh | 13.917 | 0.189 | 0.986 |
| 10. | Jammu & Kashmir | Nil | 0.094 | 1.832 |
| 11. | Jharkhand | 2.472 | 6.878 | 0.476 |
| 12. | Karnataka | 33.631 | 1.413 | 0.066 |
| 13. | Kerala | 5.384 | 1.336 | 1.686 |
| 14. | Madhya Pradesh i/c Chhatisgarh | 17.156 | 16.776 | 7.335 |
| 15. | Maharashtra | 25.523 | 13.243 | 0.763 |
| 16. | Manipur | 0.397 | 8.450 | Nil |
| 17. | Meghalaya | 0.698 | Nil | 0.516 |
| 18. | Mizoram | Nil | 0.663 | 1.561 |
| 19. | Nagaland | 1.220 | Nil | 0.526 |
| 20. | Orissa | 17.225 | 1.998 | 21.100 |
| 21. | Punjab | 2.369 | Nil | Nil |
| 22. | Rajasthan | 8.285 | 1.426 | 1.807 |
| 23. | Sikkim | Nil | Nil | 0.001 |
| 24. | Tamil Nadu | 6.501 | 0.068 | 0.013 |
| 25. | Tripura | 0.312 | Nil | Nil |
| 26. | Uttranchal | 3.056 | 5.341 | 154.00 |
| 27. | Uttar Pradesh | 15.345 | 2.712 | 18.407 |
| 28. | West Bangal | 1.475 | 0.184 | Nil |
| 29. | Andaman & Nicobar island (U.T.) | Nil | Nil | Nil |
| 30. | Chandigarh (U.T.) | Nil | Nil | Nil |
| 31. | Dadar Nagar Haveli (U.T.) | Nil | Nil | Nil |
| 32. | Daman & Diu (U.T.) | Nil | Nil | Nil |
| 33. | N.C.T of Delhi | Nil | Nil | Nil |
| 34. | Lakshdeep (U.T.) | Nil | Nil | Nil |
| 35. | Pondicherry | 0.014 | Nil | Nil |
| | Total | 206.548 Say 206.54 B.cum* | 76.257 Say 76.26 B.cum | 107.545 Say 107.54 B.cum |

* An additional live storage capacity of 6.24 B.cum is estimated to have been created through medium projects, each having capacity of less than 0.010 B.cum (which are not included here, thus making total storage capacity of **212.78** ; say **213 B.cum**.

## Appendix Table A-VI : Levels/Capacities of Important Dam Reservoirs of India

| S. No. | Name of Reservoir | Location | Dead Storage level in m | Full reservoir level in m | Capacity at Dead Storage level in M.cum | Capacity at Full Reservoir level in M. cum |
|---|---|---|---|---|---|---|
| (1) | (2) | (3) | (4) | (5) | (6) | (7) |
| 1. | Nagarjunasagar | A.P. | 149.05 | 179.83 | 4719 | 11561 |
| 2. | Sriram Sagar | A.P. | 324.31 (MDDL) | 332.54 | 849 | 3454 |
| 3. | Srisailum | A.P. | 214.88 | 269.75 | 434 | 8722 |
| 4. | Ukai | Gujarat | 82.30 | 105.16 | 1411 | 8511 |
| 5. | Sabarmati | Gujarat | 170.90 | 189.58 | 54 | 908 |
| 6. | Kadana | Gujarat | 99.08 | 127.71 | 340 | 1542 |
| 7. | Krishnarajasagar | Karnataka | 18.29 | 38.04 | 125 | 1400 |
| 8. | Linganamakki | Karnataka | 527.30 | 554.43 | 295 | 4497 |
| 9. | Tungabhadra | Karnataka | 472. 44 | 49.74 | 2 | 3429 |
| 10. | Ghataprabha (Hidkal) | Karnataka | 629.11 | 662.95 | 60 | 1440 |
| 11. | Malaprabha | Karnataka | 619.35 | 633.83 | 96 | 1068 |
| 12. | Kabini | Karnataka | 674.57 | 694.33 | 100 | 553 |
| 13. | Bhadra | Karnataka | 30,48 | 56.70 | 241 | 26 |
| 14. | Idukki | Kerala | 681.23 | 732.43 | 292 | 997 |
| 15. | Gandhi Sagar | M.P. | 381.00 | 399.90 | 586 | 7413 |
| 16. | Tawa | M.P. | 334.24 | 355.40 | 263 | 2312 |
| 17. | Mahandadi | M.P. | 336.21 | 348.70 | 144 | 811 |
| 18. | Jayal wadi | Maharashtra | 452.17 | 463.91 | 352 | 2909 |
| 19. | Khadakvasla | Maharashtra | 574. 24 | 582.47 | 0 | 86 |
| 20. | Koyna | Maharashtra | 609.60 | 657.91 | 120 | 2797 |
| 21. | Hirakud | Orissa | 179.83 | 192.02 | 2318 | 8146 |
| 22. | Machkund | Orissa | 818.35 | 838.16 | 78 | 971 |
| 23. | Balimala | Orissa | 438.91 | 462.08 | 934 | 3610 |
| 24. | Salandi | Orissa | 50.30 | 80.30 | 8 | 566 |
| 25. | Gobind Sagar | H.P. | 445.62 | 515.11 | 95 | 9351 |
| 26. | Pong Dam | H.P. | 384.00 | 436.70 | 1288 | 8579 |
| 27. | Rana Pratap Sagar | Rajasthan | 342.90 | 352.91 | 1332 | 2905 |
| 28. | Lower Bhawani | Tamil Nadu | 248.41 | 280.22 | 0 | 929 |
| 29. | Mettur | Tamil Nadu | 204.22 | 240.79 | 62 | 2647 |
| 30. | Vaigai | Tamil Nadu | 257.56 | 279.20 | 1 | 193 |
| 31. | Parambikulam | Tamil Nadu | 534.31 | 556.26 | 125 | 505 |
| 32. | Adiyar | Tamil Nadu | 283.46 | 320.04 | 0 | 109 |
| 33. | Matatila | U.P. | 295.65 | 308.46 | 50 | 883 |
| 34. | Ramganga | U.P. | 323.00 | 365.30 | 395 | 2448 |
| 35. | Rihand | U.P. | 236.22 | 268.22 | 1640 | 10607 |
| 36. | Mayurkshi | West Bengal | 106.38 | 121.31 | 68 | 617 |
| 37. | Kangsabati | West Bengal | 120.43 | 134.14 | 138 | 52 |
| 38. | Maithon | DVC | 132.59 | 146.30* / 152.40 | 165 | 1275 |
| 39. | Panchet Hill | DVC | 119.48 | 120.47* / 427.94 | 170 | 1475 |
| 40. | Konar DVC | DVC | 410.57 |  | 61 | 336 |
| 41. | Tilaiya | DVC | 363.32 | 372.47 | 75 | 394 |

*In case of Maithon and Panchet reservoirs, F.R. levels are given without and with flood storage, both.

## Appendix Table A-VII : The Highest (Tallest) Dams of the World (above 220 m height)

| *S. No.* | *Name of the Dam* | *Height* | *Types of dam* | *Country where situated* | *River on which dam is constructed* | *Year in which completed* |
|---|---|---|---|---|---|---|
| 1. | **Jinping-I Dam*** | 305 m (1,001 ft) | Concrete Arch | China | Yalong | 2013 |
| 2. | Nurek Dam | 300 m (980 ft) | Embankment earth-fill | Tajikistan | Vakhsh | 1980 |
| 3. | Xiaowan Dam | 292 m (980 ft) | Concrete Arch | China | Lancang | 2010 |
| 4. | Xiluodu Dam | 285.5 m (937 ft) | Concrete Arch | China | Jinsha | 2013 |
| 5. | **Grande Dixence Dam*** | 285 m (935 ft) | **Concrete Gravity** | Switzerland | Dixence | **1964** |
| 6. | Enguri Dam | 271.5 m (891 ft) | Concrete Arch | Georgia | Enguri | 1987 |
| 7. | Yusufeli Dam | 270 m (890 ft) | Arch, double curvature | Turkey | Coruh | 2018 |
| 8. | Vajont Dam (disused) | 261.6 m (858 ft) | Concrete Arch | Italy | Vajont | 1959 |
| 9. | Nuozhadu Dam | 261.5 m (858 ft) | Embank-ment | China | Lancang | 2012 |
| 10. | Manuel Moreno Torres (chicoasen) Dam | 261 m (856 ft) | Embank-ment/earth-fill | Mexico | Grijalva | 1980 |
| 11. | **Tehri Dam** | 260.5 m (855 ft) | Embank-ment earth fill | **India** | Bhagirathi | **2006** |
| 12. | Mauvoisin Dam | 250 m (820 ft) | Concrete Arch | Switzerland | Begnes | 1957 |
| 13. | Laxiwa Dam | 250 m (820 ft) | Concrete Arch | China | Yellow | 2009 |
| 14. | Deriner Dam | 249 m (817 ft) | Concrete double-arch | Turkey | Coruh | 2012 |
| 15. | Alberto Lleras (Guavio) Dam | 243 m (797 ft) | Embankment rock-fill | Colombia | Guavio | 1989 |
| 16. | Mica Dam | 243 m (797 ft) | Embankment earth-fill | Canada | Columbia | 1973 |
| 17. | Gilgel Gibe III Dam | 243 m (797 ft) | Roller-compacted concrete gravity | Ethiopia | Omo | 2015 |
| 18. | Sayano Shushen-skaya Dam | 242 m (794 ft) | Concrete arch-gravity | Russia | Yenisei | 1985 |
| 19. | Ertan Dam | 240 m (790 ft) | Concrete arch-gravity | China | Yalong | 1999 |

*Grand Dixene dam on Dixence river is at present, the highest existing concrete gravity dam of the world; while Jiuping-1 dam in China on Yahlong river is the highest concrete dam in the world.

| S. No. | Name of the Dam | Height | Types of dam | Country where situated | River on which dam is constructed | Year in which completed |
|---|---|---|---|---|---|---|
| 20. | Changheba Dam | 240 m (790 ft) | Embankment concrete-face rock-fill | China | Dadu | 2016 |
| 21. | La Esmeralda Dam | 237 m (778 ft) | Embankment, rock-fill | Colombia | Bata | 1976 |
| 22. | Oroville Dam | 234.7 m (770 ft) | Embankment, earth-fill | USA | Feather | 1968 |
| 23. | El Cajon Dam | 234 m (768 ft) | Concrete double-arch | Honduras | Humuya | 1985 |
| 24. | Shuibuya Dam | 233 m (764 ft) | Embankment, concrete-face rock-fill | China | Qingjiang | 2008 |
| 25. | Chirkey Dam | 232.5 m (763 ft) | Concrete arch | Russia | Sulak | 1976 |
| 26. | Goupitan Dam | 232.5 m (763 ft) | Concrete double-arch | China | Wu Jiang | 2009 |
| 27. | Karun-4 Dam | 230 m (750 ft) | Concrete arch-gravity | Iran | Karun | 2010 |
| 28. | **Bhakra Dam*** | 226 m (741 ft) | Concrete gravity | India | Sutlej | 1963 |
| 29. | Luzzone Dam | 225 m (738 ft) | Concrete arch | Switzerland | Lago di Luzzone | 1963 |
| 30. | **Hoover Dam** | 221.46 m (726.6 ft) | Concrete arch-gravity | USA | Colorado | 1936 |
| 31. | Jianppinghe Dam | 221 m (725 ft) | Embankment, concrete-face rock-fill | China | Loushui | 2012 |
| 32. | Contra Dam | 220 m (720 ft) | Concrete arch | Switzerland | Verzasca | 1965 |
| 33. | La Yesca Dam | 220 m (720 ft) | Embankment concrete-face rock-fill | Mexico | Rio Grande de Santiago | 2012 |
| 34. | Mratinje Dam | 220 m (720 ft) | Concrete arch-gravity | Montenegro | Piva | 1976 |

*Bhakra dam is the highest concrete gravity dam of India, and is the second highest such dam in the world.

## Appendix Table A-VIII : Largest Dams* of the World

| *S. No.* | *Name of the Dam* | *Country where located* | *App. volume of fill in M.m³* | *Year completed* |
|---|---|---|---|---|
| 1. | Three Gorges Dam | China | 39,300 | 2012 |
| 2. | Syncrude Tailings Dam (Mildred MLSB) | Canada | 540 | 1995 |
| 3. | Chapeton Dam | Argentina | 296 | UC*** |
| 4. | Pati Dam | Argentina | 238 | UC |
| 5. | New Cornelia Tailings Dam | USA | 210 | 1973 |
| 6. | Tarbela Dam | Pakistan | 122 | 1976 |
| 7. | Kambaratinsk Dam | Kyrgyzstan | 112 | UC |
| 8. | Fort Peck Dam | Montana | 96 | 1940 |
| 9. | Lower Usuma Dam | Nigeria | 93 | 1990 |
| 10. | Cipasang Dam | Indonesia | 90 | 1990 |
| 11. | Ataürk Dam | Turkey | 85 | 1990 |
| 12. | Yacyreta-Apipe Dam | Paraguay/Argentina | 81 | 1998 |
| 13. | Guri (Raúl Leoni) Dam | Venezuela | 78 | 1986 |
| 14. | Rogun Dam | Tajikistan | 75 | 1985 |
| 15. | Oahe Dam | South Dakota | 70 | 1963 |
| 16. | Mangla Dam | Pakistan | 66 | 1967 |
| 17. | Gardiner Dam | Canada | 65 | 1968 |
| 18. | Afsluitdijk Dam | Netherlands | 63 | 1932 |
| 19. | Oroville Dam | California | 60 | 1968 |
| 20. | San Luis Dam | California | 59 | 1967 |
| 21. | Nurek Dam | Tajikistan | 58 | 1980 |
| 22. | **Nagarjuna Sagar Dam**** | India | 56 | 1976 |
| 23. | Garrison Dam | North Dakota | 51 | 1956 |
| 24. | Cochiti Dam | New Maxico | 48 | 1975 |
| 25. | Tabka (Thawra) Dam | Syria | 46 | 1976 |

* Largest by volume of the construction materials.

**Nagarjuna Sagar dam across Krishna river is the largest dam of India, and is the highest masonry dam of the world (124.7 m high). As against this, Hirakud dam across Mahanadi river is the longest earthen dam of the world (25.3 km long.)

*** UC = Under Construction

## Appendix Table A-IX : Crop Calendar of Sowing and Harvesting of Major Crops in Different States of India

| *States / Uts* | *Period* | *Kharif Paddy* | *Rabi Paddy* | *Summer Paddy* | *Kharif Bajra* | *Summer Bajra* | *Rabi Wheat* |
|---|---|---|---|---|---|---|---|
| 1 | 2 | 3 | 4 | 5 | 6 | 7 | 8 |
| Andhra Pradesh | Sowing | May-Jun | Nov-Dec | Mar-Apr | Jun(B)-Jul(M) | | |
| | Harvesting | Nov-Dec | May-Jun | July-Aug. | Aug(B)-Oct(B) | | |
| Assam | Sowing | Feb-Mar | Jun-July | Nov-Dec | | | Nov(B)-Dec(M) |
| | Harvesting | Jun-July | Nov-Dec | May-Jun | | | Mar(B)-Apr(E) |
| Bihar | Sowing | Jun-Sept. | Oct.-Nov | Feb-Mar | | | Nov(M)-Dec(E) |
| | Harvesting | | Apr-May | July-Aug. | | | Mar(M)-Apr(E) |
| Goa | Sowing | | | | | | |
| | Harvesting | | | | | | |
| Gujarat | Sowing | Jun-July | | | Jun(B)-July(E) | Feb(B)-Feb(E) | Oct(B)-Nov(E) |
| | Harvesting | Oct-Nov | | | Sept(B)-Nov(E) | May(B)-May(E) | Feb(B)-Mar(E) |
| Haryana | Sowing | Jun-July | | | Jun(M)-July(B) | | Oct(E)-Dec(B) |
| | Harvesting | Sept-Oct | | | Oct(M)-Nov(M) | | Apr(M)-Apr(E) |
| Himachal Pradesh | Sowing | May-Jun | | | | | Oct(B)-Dec(E) |
| | Harvesting | Oct | | | | | Apr(M)-Jun(E) |
| Jammu & Kashmir | Sowing | Apr-May | | | | | Oct(B)-Nov(E) |
| | Harvesting | Sept-Oct | | | | | May(B)-May(E) |
| Karnataka | Sowing | May-Jun | Sept-Oct | Jan-Feb | July(B)-Sept(E) | Jan(B)-Feb(E) | Oct(B)-Dec(E) |
| | Harvesting | Sept-Oct | Jan-Feb | May-Jun | Oct(B)-Nov(E) | Apr(B)-May(E) | Jan(B)-Feb(E) |
| Kerala | Sowing | Apr-May | Sept-Oct | Dec-Jan | | | |
| | Harvesting | Sept-Oct | Dec-Jan | Mar-Apr | | | |

B stands for beginning, M for middle and E for end.

| *States / Uts* | *Period* | *Kharif Paddy* | *Rabi Paddy* | *Summer Paddy* | *Kharif Bajra* | *Summer Bajra* | *Rabi Wheat* |
|---|---|---|---|---|---|---|---|
| 1 | 2 | 3 | 4 | 5 | 6 | 7 | 8 |
| Madhya Pradesh | Sowing | Jun-July | | | Jun(M)-July(E) | | Oct(M)-Dec(E) |
| | Harvesting | Oct-Nov | | | Sept(E)-Oct(E) | | Feb(B)-Mar(E) |
| Maharashtra | Sowing | Jun-July | | | Jun(B)-July(E) | | Oct(B)-Dec(E) |
| | Harvesting | Oct-Nov | | | Sept(B)-Oct(E) | | Feb(B)-Mar(E) |
| Orissa | Sowing | May-Jun | Jun-July | Dec-Jan | Jun-July | | Oct-Nov |
| | Harvesting | Sept-Oct | Nov-Dec | Apr-May | Sept-Oct | | Mar-Apr |
| Punjab | Sowing | May-Oct | | | | | Oct(B)-Nov(E) |
| | Harvesting | | | | | | Apr(B)-May(E) |
| Rajasthan | Sowing | Jun-July | | | Jun(B)-July(E) | | Nov(B)-Dec(E) |
| | Harvesting | Oct-Nov | | | Sept(B)-Oct(E) | | Mar(B)-May(E) |
| Uttar Pradesh | Sowing | Jun-July | Nov-Dec | | Jun(B)-July(E) | | Oct(B)-Jan(M) |
| | Harvesting | Oct-Nov | Apr-May | | Oct(B)-Nov(M) | | Apr(B)-Apr(M) |
| West Bengal | Sowing | | | | | | Nov(B)-Dec(E) |
| | Harvesting | | | | | | Mar(B)-Apr(E) |
| All India | Sowing | May-Aug | Dec-Jan | | Jun-July | Jan-Feb | Oct-Dec |
| | Harvesting | Sept-Jan | Apr-May | | Sept-Nov | Apr-May | Feb-Jun |

Source : Indian Council of Agricultural Research (Crop Science Division)

## Appendix Table A-X : Useful Conversions Between Different Units

### Length

1 metre = 3.281 ft ;
1 km = 0.622 mi ;
1 cm = 0.394 in ;
1 mm = 1000 μ (micron)

1 ft = 0.3048 m
1 mi = 1.6093 km
1 in = 2.54 cm
1 micron (μ) = $10^{-6}$ m

### Area

1 $m^2$ = 10.761 $ft^2$ ;
1 $km^2$ = $10^6$ $m^2$ = 100 ha ;
= 0.386 $mi^2$ = 247 acres

1 $ft^2$ = 0.0929 $m^2$
1 $mi^2$ = 2.59 $km^2$
= 251 ha
= 640 acre

1 ha = $10^4$ $m^2$
= 2.47 acres ;

1 acre = 4840 sq. yards
= 0.4047 ha

### Volume and Discharge

1 $m^3$ = 35.307 $ft^3$
1 ha.m = 8.13 acre ft ;

1 $ft^3$ = 0.02832 $m^3$
**1 acre ft** = 43560 $ft^3$
= 1233.6 $m^3$
= 0.123 ham
= $\frac{1}{2}$ **cusec day**

1 $m^3$ = 1000 litres (L)
1 L = 1.76 pints (imp)
= 0.22 gallon (imp)

**1 gallon** (*i.e.* imp gallon)
= 4.546 litres
1 U.S. gallon = 3.785 litres

1 cusec = 0.02832 cumec

1 cumec = 35.307 cusec ;
or 1 cusec = 28.32 L/s
or 1 cusec = 2.447 MLD (million litres per day)
or 1 cusec = 0.538 MGD (million gallon per day)
1 MGD = 1.86 cusec

### Force

1 kg wt = 9.807 N (Newton) ;
1 gm wt = 980.7 dynes
1 lb wt = 32.2 poundals
1 kg wt = 2.204 lb wt.

1 N = 0.102 kg wt

| 1 kg (wt) = 9.807 N = 2.204 lb (wt) = 9.807 × $10^5$ dynes |
|---|

*or*

| 1 N = 0.102 kg (wt) = 0.2247 lb (wt) = $10^5$ dynes |
|---|

1 poundal = 138.26 dynes = 14.102 gm (wt).

### Pressure

**Units :** N/$m^2$ = Pascal (Pa) ; kg/$m^2$ (*i.e.* kgf/$m^2$) : lb wt/$ft^2$

1 kg/$m^2$ = 9.807 N/$m^2$
= 9.807 Pa
1 kg/$cm^2$ = $10^4$ kg/$m^2$

1 Pa = 1 N/$m^2$ = 0.102 kg/$m^2$
1 kPa = 102 kg/$cm^2$
100 kPa = 10,200 kg/$m^2$
= 1.02 kg/$cm^2$

1 Atmosphere
= 760 mm of Hg
= 1 bar = 1000 mb
= 101.325 kPa = 1.033 kg/cm$^2$
= 10.33 m water head

| 1 kg/cm$^2$ = 98.07 kPa = 10 m of water | 100 kPa = 1.02 kg/cm$^2$ = 10.2 m of water |
|---|---|

1 lb/ft$^2$ = 47.88 kPa = 16.02 ft of water

100 kPa = 2.089 lb/ft$^2$ = 33.45 ft of water

1 psi = 1 lb/in$^2$ = 0.0703 kg/cm$^2$

1 lb/ft$^2$ = 4.882 kg/m$^2$

1 kg/cm$^2$ = 14.22 psi

1 kg/m$^2$ = 0.205 lb/ft$^2$

**Density (ρ)**

1 kg/m$^3$ = 0.0624 lb/cft

1 lb/cft = 16.02 kg/m$^3$

**Unit wt. (γ or w = ρ.g)**

**Units:** N/m$^3$, kg/m$^3$ (*i.e.*, kg-f/m$^3$); lb wt/ft$^3$

1 kg/m$^3$ (*i.e.* kg wt/m$^3$) = 9.807 N/m$^3$

1 N/m$^3$ = 0.102 kg/m$^3$

**Energy**

**Units:** Joule (N.m); kg (wt) m; lb (wt) m

| | | |
|---|---|---|
| | 1 Joule (j) | = 1 Nm |
| | 1 kg (wt) m | = 9.807 N.m |
| *i.e.* | 1 kg (wt) m | = 9.807 J; 1 J = 0.102 kg (wt) m |
| | 1 ft lb (wt) | = 0.138 kg (wt) m = 1.356 J |
| | 1 kg (wt) m | = 7.23 ft lb (wt) = 9.807 J |
| | 1 J | = 0.102 kg (wt) m = 0.737 ft lb (wt) |

**Power**

**Units:** J/s ; kgm/sec ; ft lb/sec

| | |
|---|---|
| 1 J/s | = 1 watt |
| 1 J/s | = 1 Nm/s = 1 watt = 0.102 kg (wt) m/sec = 0.737 ft lb (wt)/sec |
| 1 HP (fps) | = 550 ft lbs/sec = 746 J/s = 746 watts. |
| 1 HP (metric) | = 75 kg (wt) m/sec = 75 × 9.807 = 735 J/s = 735 watts |

**Dynamic Viscosity (μ or η)**

**Units :** N.s/m$^2$ ; kgf.sec/m$^2$ ; dyne sec/cm$^2$ (Poise)

1 N.s/m$^2$ = 1 Pa.s = 10 Poise = 0.102 kgf sec/m$^2$

1 dyne. sec/cm$^2$ (Poise) = 0.0102 kgf sec/m$^2$

**Kinematic Viscosity (ν)**

**Units :** m$^2$/s ; m$^2$/sec ; cm$^2$/sec (Stoke)

1 m$^2$/s = 10$^4$ cm$^2$/sec = 10$^4$ Stoke (S)

1 centi stoke (CS) = $\frac{1}{100} S$

# Bibliography


**General References :**

1. *Irrigation Principles and Practices* : *O. W. Isrelsen and V.E. Hansen* (John Wiley & Sons, New York)
2. *Fundamentals of Irrigation Engineering* : Dr. Bharat Singh (Nem Chand and Bros Roorkee U.P., India)
3. *Water Resources Engineering* : *R.K. Linsley and J.B. Faranzini* (McGraw Hill Book Co., New York)
4. *Theory and Design and Irrigation Structures* : *Varshney, Gupta & Gupta* (Nem Chand and Bros, Roorkee, U.P. India)

**Special References**

**Chapter 1**

5. *Design and Evaluation of Irrigation Methods* : *A.M. Michael, Shri Mohan and K.R. Swaminathan* (Water Technology Centre, IARI, New Delhi)
6. *Report of the Irrigation Commission 1972, Vol. I* : *Publication of* (Ministry of Irrigation and Agriculture, New Delhi)
7. *Proceedings of Work Shop on Sprinkler and Drip Irrigation Systems* : *Publication of* (Central Board of Irrigation and Power, New Delhi).

**Chapter 2**

8. *A Guide for Estimating Water Requirements of Crops* : *Publication of* (Water Management Division of Department of Agriculture, New Delhi)
9. *Symposium on "Optimum Requirements and Utilisation of water for Irrigated crops"* : *Publication No. 93* (Central Board of Irrigation and Power, New Delhi, 1963)
10. *Moisture Requirements in Agriculture* : *B.H. Roe* (Mc Graw Hill Book Co., New Delhi)
11. *Review of Work done on Water Requirements of Crops* : *Dastane Singh, Hukkeri and Vamdav* (Navbharat Prakashan, Poona)
12. *Water Resources of India and Their Utilisation in Agriculture* : *C. Dakshinamurthy, A.M. Michael* and *Sri Mohan*

**Chapter 3**

13. *Irrigation Engineering Vol. I (Canals and Barrages)* : *S. Leliavsky* (Chapman and Hall, London)
14. *Symposium on Irrigation Water Management* : *Publication No. 90* (Central Board of Irrigation and Power, New Delhi, 1967)
15. *The Fundamental Principles of Irrigation and Water Power* : *V.B. Priyani* (Charotar Book Stall, Anand, India 1967)

**Chapter 4**

16. *Hydraulics of Sediment Transport* : *Graf* (McGraw Hill Book Company, New York)

17. *Regime Flow in Incoherent Alluviums* : *Publication No. 20* by *Gerald Lacey (Central Board of Irrigation and Power, New Delhi, 1939)*

18. *"Symposium on Sediment Problems in Irrigation and Drainage Channels"* : *Publication No. 91* (Central Board of Irrigation and Power, New Delhi, 1965)

19. *Engineeering Hydraulics* : *Hunter Rouse* (John Wiley and Sons, NY)

20. *Irrigation Engineering Vol. IV (River and Canal Hydraulics)* : *S. Laliavsky* (Chapmann and Hall, London)

21. *Open Channel Flow* : *F.M. Handerson* (McMillian and Co.)

**Chapter 5**

22. *I.S. 3860-1966 : Specifications for Precast Cement Concrete Canal Linings* : *Publication of* (Bureau of Indian Standards, New Delhi)

23. *IS 3872-1966 : Code of Practice for Lining of Canals with Burnt Clay Tiles* : *Publication of* (Bureau of Indian Standards New Delhi)

24. *IS : 3873-1966 : Code of Practice for Laying insitu Cement Concrete Lining on Canals* : *Publication of* (Bureau of Indian Standards, New Delhi)

25. *IS 4515-1976 : Code of Practice for Boulder Lining for Canals* : *Publication of* (Bureau of Indian Standards, New Delhi)

26. *IS 4558-1968 Code of Practice for Under-Drainage of Lined Canals* : *Publication of* (Bureau of Indian Standards, Manak Bhavan, New Delhi)

27. IS 4701-1968 : *Code of Practice for Earthquakes on Canals* : *Publication of* (Bureau of Indian Standards, Manak Bhavan, New Delhi)

28. IS 4745-1968 : *Code of Practice for Design for Cross-section of Lined Canals* : *Publication of* (Bureau of Indian Standards, Manak Bhavan, New Delhi)

29. *Lower Cost Canal Linings* : *Publication of* (Bureau of Reclamation U.S. Deptt. Interior, Denver, Colorado, June 1948)

30. *Canal Lining in India* : *Kanwar Sain* (Proceedings of International Commission on Irrigation and Drainage, Third Congress San Francisco 1957, pp 145-175

31. *Investigation of Necessity for Drainage behind Canal Lining* : *A.S. Kalha and I.P. Kapila (Report* Central Design Office, Punjab Irrigation Branch, Simla, 1952)

**Chapter 6**

32. *Drainage Engineering* : *J.N. Luthin* (John Wiley and Sons, NY)

33. *Soil Conditions and Plant Growth* : *E.W. Russell* (Longman Green and Co., London)

34. *Symposium on "Water Logging — Causes and Measures for its Prevention, Vol. I and II* : *Publication No. 118* (Central Board of Irrigation and Power, New Delhi, 1972

## Chapter 7

35. *Report of the Committee on Pricing of Irrigation Water* : *Publication of* (Planning Commission, Govt. of India, New Delhi, 1992)

## Chapter 8

36. *Manual on River Behaviour Control and Training* : *Publication No. 60* (Central Board of Irrigation and Power, New Delhi, 1971)

37. Drainage and Flood Control Engineering (Chapter XI) : *G.W. Pickels* (McGraw Hill Book Company, NY, 1941)

## Chapter 9

38. *Irrigation Engineering, Vol. II (Canals and Barrages)* : *S. Leliavsky* (Chapmann and Hall, London)

39. *Irrigation Engineering, Vol. V (Weirs)* : *S. Leliavsky* (Chapman and Hall, London)

## Chapter 10

40. *Open Channel Hydraulics* : *Ven Te Chow* (McGraw Hill Book Co. International Students Edition)

41. The Standing Wave or Hydraulic Jump : *Publication No. 7 with supplement from Sh. R.M. Advani 1964* (Central Board of Irrigation and Power, New Delhi)

42. *Design of Weirs on Permeable Foundations* : *Publication No. 12 by A.N. Khosla, Dr. N.K. Bose, and Dr. E. Machenizie Taylor* (Central Board of Irrigation and Power, 1936 reprinted 1962, 1981)

## Chapter 12

43. *Irrigation Canal Falls* : *Publication No. 10 — Compiled by Mr. A.M.R. Montague* (Central Board of Irrigation and Power, 1935)

44. Draft Manual on Falls : *Publication of* (Central Water Commission, Delhi)

## Chapter 13

45. Irrigation (Vol. IV) : *K.B. Khushalani and Sh. Manohar Khushalani* (Oxford and IBH Publishing Co. Pvt, New Delhi)

## Chapter 14

46. *Design of Syphons* : *Publication No. 59 by Prof. N.S. Govinda Rao* (Central Board of Irrigation and Power, New Delhi, 1956)

47. *Irrigation Engineering, Vol. II (Syphons, Weirs and Locks)* : *S. Leliavsky* (Chapman and Hall, London)

## Chapter 15

48. *Design of small Bridges and Culverts* : *Goverdhan Lal* (CWC Publication)

49. *Guidelines for the Design of Small Bridges & Culverts* : Publication of (Indian Roads Congress, Jamnagar House, New Delhi-11)

50. *California Culvert Practice* : *Publication of* (State of California, Division of Highways, Publication Sacrmento California)

**Chapter 16**

51. *An introduction to Fluvial Hydraulics* : *S. Leliavsky* (Constable, London, 1955)

52. *Hand Book of Applied Hydraulics* : *Rich George* (McGraw Hill Book Co., New York)

**Chapter 17**

53. *Rivers in Harness (The Story of Dams)* : *Allan H. Culler* (A Division of Chilton Company, Publishers, New York)

54. *Silting of Reservoirs* : *A.N. Khosla* (Central Board of Irrigation and Power, New Delhi)

**Chapter 18**

55. *Investigation Manual for Storage Reservoirs* : *Publication No. P-58* (Central Board of Irrigation and Power, New Delhi)

56. *A Handbook of Applied Hydraulics* : *Publication of* (Central Board of Irrigation and Power, New Delhi)

57. *Manual on Capacity Survey of Reservoirs* : *Sh. B.N. Murthy*

**Chapter 19**

58. *Engineering for Dams (Vol. II) (Concrete Dams)* : *Creager, Justin and Hinds* (John Wiley and Sons, New York)

59. *Design of Small Dams* : *A Technical Publication of* (Govt. of U.S. Deptt. of Interior-Washington, & the Bureau of Reclaimation, Colorado, 1960, 1965)

60. *Treatise of Dams* : *USBR Manual* (The Bureau of Reclaimation Colorado, United States)

61. *Design and Construction of Dams* : *Edward Wegmann* (John Wiley and Sons, New York)

62. *IS-1893-1962 Recommendations for Earthquake Resistant Design of Structures* : *Publication of :* (Bureau of Indian Standards, New Delhi)

63. *Irrigation Engineering Vol. VI (Dams)* : *S. Lieliavsky* (Chapman and Hall, London)

**Chapter 20**

64. *Fundamentals of Soil Mechanics* : *D.W. Taylor* (Asia Publishing House, New Delhi)

65. *Geotech Engineering — Soil Mechanics and Foundation Engineering* : *Santosh Kumar Garg* (Khanna Publishers, New Delhi)

66. *Engineering for Dams, Vol. III (Earth, Rockfill, Steel and Timber Dams)* : *Creager, Justin and Hinds* (John Wiley and Sons, New York)

67. *Earth and Rock Dams* : Sheard, L. James and Others

**Chapter 21 and 22**

68. *Symposium on : "Spillway Capacity of Dams", and (ii) Various Methods of* Designing Spillway Capacity *Discharges"* : *Publication No. 65* (Central Board of Irrigation and Power, New Delhi

69. *Hydraulic Design of Spillways* : *Manual of* (U.S. Army Engineers EM 1102-2-1603)

70. *Dams and Control Works* : *USBR Publication* (Washington, DC, 1954)

71. *Hydraulic Energy Dissipators* : *I.A. Edwerd Elevatoroki* (Mc Graw Hill Book Co., New York)

72. *Symposium on Energy Dissipators* : *Publication No. 70* (Central Board of Irrigation & Power, New Delhi, 1955)

**Chapter 23**

73. *Corrosion and Corrosion Control* : *H.H. Uhlig* (John Wiley and Sons, New York)

74. *Fluid Mechanics with Engineering Applications* : *R.L. Daugherty and J.B. Franzini* (McGraw Hill Book Co., New York)

**Chapter 24**

75. *Hydropower Engineering* : *James Doland* (Ronald Publication, New York)

76. *Hydroelectric Engineering Practice, Vol. I to Vol. III* : *J. Gutherie Brown* (Blackie and Sons Ltd., London, 1958)

77. *Irrigation Engineering, Vol. III (Hydroelectric Engneering for Civil Engineers)* : *S. Leliavsky* (Chapman and Hall, London)

78. *Water Power Engineering* : *H.K. Barrows* (McGraw Hill Book Co., New York, 1943)

79. *Hydroelectric Hand Book* : *W.P. Creager and J.D. Justin,* (John Wiley and Sons, New York, 1963)

**Chapter 26**

80. *College of Engineering Manual on Irrigation* : *Cl. W.M. Ellis* (Govt. Press, Madras, 1950)

81. *Irrigation Planning for Intensive Cultivation* : *H.L. Sally* (Asia Publishing House, Bombay)

**Chapter 27**

82. *Irrigation (Vol. III)-Arch,* **Buttress, Earth and Rockfill** *Dams* : *K.B. Khushalani and Manohar* Khushalani, (Oxford and IBH Publishing Co. Pvt. Ltd., New Delhi)

# Index

## S

## Z